The Art of Electronics

Third Edition

At long last, here is the thoroughly revised and updated, and long-anticipated, third edition of the hugely successful *The Art of Electronics*. Widely accepted as the best single authoritative text and reference on electronic circuit design, both analog and digital, the first two editions were translated into eight languages, and sold more than a million copies worldwide. The art of electronics is explained by stressing the methods actually used by circuit designers – a combination of some basic laws, rules of thumb, and a nonmathematical treatment that encourages understanding why and how a circuit works.

Paul Horowitz is a Research Professor of Physics and of Electrical Engineering at Harvard University, where in 1974 he originated the Laboratory Electronics course from which emerged *The Art of Electronics*. In addition to his work in circuit design and electronic instrumentation, his research interests have included observational astrophysics, x-ray and particle microscopy, and optical interferometry. He is one of the pioneers of the search for intelligent life beyond Earth (SETI). He is the author of some 200 scientific articles and reports, has consulted widely for industry and government, and is the designer of numerous scientific and photographic instruments.

Winfield Hill is by inclination an electronics circuit-design guru. After dropping out of the Chemical Physics graduate program at Harvard University, and obtaining an E.E. degree, he began his engineering career at Harvard's Electronics Design Center. After 7 years of learning electronics at Harvard he founded Sea Data Corporation, where he spent 16 years designing instruments for Physical Oceanography. In 1988 he was recruited by Edwin Land to join the Rowland Institute for Science. The institute subsequently merged with Harvard University in 2003. As director of the institute's Electronics Engineering Lab he has designed some 500 scientific instruments. Recent interests include high-voltage RF (to 15 kV), high-current pulsed electronics (to 1200 A), low-noise amplifiers (to sub-nV and pA), and MOSFET pulse generators.

THE ART OF ELECTRONICS

Third Edition

Paul Horowitz HARVARD UNIVERSITY

Winfield Hill ROWLAND INSTITUTE AT HARVARD

CAMBRIDGE
UNIVERSITY PRESS

One Liberty Plaza, 20th Floor, New York, NY 10006, USA

Cambridge University Press is part of the University of Cambridge.

It furthers the University's mission by disseminating knowledge in the pursuit of education, learning, and research at the highest international levels of excellence.

www.cambridge.org
Information on this title: www.cambridge.org/9780521809269

© Cambridge University Press, 1980, 1989, 2015

First published 1980
Second edition 1989
Third edition 2015

14th printing 2019

Printed in the United Kingdom by TJ International Ltd. Padstow Cornwall

A catalog record for this publication is available from the British Library.

ISBN 978-0-521-80926-9 Hardback

Cambridge University Press has no responsibility for the persistence or accuracy of URLs for external or third-party Internet websites referred to in this publication and does not guarantee that any content on such websites is, or will remain, accurate or appropriate.

To Vida and Ava

In Memoriam: Jim Williams, 1948–2011

CONTENTS

ix

LIST OF TABLES

PREFACE TO THE THIRD EDITION

Moore's Law continues to assert itself, unabated, since the publication of the second edition a quarter century ago. In this new third (and final!) edition we have responded to this upheaval with major enhancements:

- an emphasis on devices and circuits for *A/D* and *D/A conversion* (Chapter 13), because embedded microcontrollers are everywhere
- illustration of specialized peripheral ICs for use with microcontrollers (Chapter 15)
- detailed discussions of logic family choices, and of interfacing logic signals to the real world (Chapters 10 and 12)
- greatly expanded treatment of important topics in the essential analog portion of instrument design:
  – precision circuit design (Chapter 5)
  – low-noise design (Chapter 8)
  – power switching (Chapters 3, 9, and 12)
  – power conversion (Chapter 9)

And we have added many entirely new topics, including:

- digital audio and video (including cable and satellite TV)
- transmission lines
- circuit simulation with SPICE
- transimpedance amplifiers
- depletion-mode MOSFETs
- protected MOSFETs
- high-side drivers
- quartz crystal properties and oscillators
- a full exploration of JFETs
- high-voltage regulators
- optoelectronics
- power logic registers
- delta–sigma converters
- precision multislope conversion
- memory technologies
- serial buses
- illustrative "Designs by the Masters"

In this new edition we have responded, also, to the reality that previous editions have been enthusiastically embraced by the community of practicing circuit designers, even though *The Art of Electronics* (now 35 years in print) originated as a course textbook. So we've continued the "how *we* do it" approach to circuit design; and we've expanded the depth of treatment, while (we hope) retaining the easy access and explanation of basics. At the same time we have split off some of the specifically course-related teaching and lab material into a separate *Learning the Art of Electronics* volume, a substantial expansion of the previous edition's companion *Student Manual for The Art of Electronics*.[1]

For the 14th printing, in addition to numerous corrections and improvements, a whole new Index of Parts has been added for the benefit of readers, and the Subject Index has been greatly expanded.

Digital oscilloscopes have made it easy to capture, annotate, and combine measured waveforms, a capability we have exploited by including some 90 'scope screenshots illustrating the behavior of working circuits. Along with those doses of reality, we have included (in tables and graphs) substantial quantities of highly useful measured data – such as transistor noise and gain characteristics (e_n, i_n, $r_{bb'}$; h_{fe}, g_m, g_{oss}), analog switch characteristics (R_{ON}, Q_{inj}, capacitance), op-amp input and output characteristics (e_n and i_n over frequency, input common-mode range, output swing, auto-zero recovery, distortion, available packages), and approximate prices (!) – the sort of data often buried or omitted in datasheets but which you need (and don't have the time to measure) when designing circuits.

We've worked diligently, over the 20 years it has taken to prepare this edition, to include important circuit design information, in the form of some 350 graphs, 50 photographs, and 87 tables (listing more than 1900 active components), the last enabling intelligent choice of circuit components by listing essential characteristics (both specified and measured) of available parts.

Because of the significant expansion of topics and depth of detail, we've had to leave behind some topics that were treated in the second edition,[2] notwithstanding the use of larger pages, more compact fonts, and most figures sized to fit in a single column. Some additional related mate-

---

[1] Both by Hayes, T. and Horowitz, P., Cambridge University Press, 1989 and 2016.

[2] Which, however, will continue to be available as an e-book.

rial that we had hoped to include in this volume (on real-world properties of components, and advanced topics in BJTs, FETs, op-amps, and power control) will instead be published in a forthcoming companion volume, *The Art of Electronics: The x-Chapters*. References in this volume to those x-chapter sections and figures are set in italics. A newly updated `artofelectronics.com` website will provide a home for a continuation of the previous edition's collections of *Circuit ideas* and *Bad circuits*; it is our hope that it will become a community, also, for a lively electronic circuit forum.

As always, we welcome corrections and suggestions (and, of course, fan mail), which can be sent to horowitz@physics.harvard.edu or to hill@rowland.harvard.edu.

With gratitude. Where to start, in thanking our invaluable colleagues? Surely topping the list is David Tranah, our indefatigable editor at the Cambridge University Press mother-ship, our linchpin, helpful LaTeXpert, wise advisor of all things bookish, and (would you believe?) *compositor*! This guy slogged through 1,905 pages of marked-up text, retrofitting the LaTeX source files with corrections from multiple personalities, then entering a few thousand index entries, and making it all work with its 1,500+ linked figures and tables. And then putting up with a couple of fussy authors. We are totally indebted to David. We owe him a pint of ale.

We are grateful to Jim Macarthur, circuit designer extraordinaire, for his careful reading of chapter drafts, and invariably helpful suggestions for improvement; we adopted every one. Our colleague Peter Lu taught us the delights of Adobe Illustrator, and appeared at a moment's notice when we went off the rails; the book's figures are testament to the quality of his tutoring. And our always-entertaining colleague Jason Gallicchio generously contributed his master Mathematica talents to reveal graphically the properties of delta–sigma conversion, nonlinear control, filter functions; he left his mark, also, in the microcontroller chapter, contributing both wisdom and code.

For their many helpful contributions we thank Bob Adams, Mike Burns, Steve Cerwin, Jesse Colman, Michael Covington, Doug Doskocil, Jon Hagen, Tom Hayes, Phil Hobbs, Peter Horowitz, George Kontopidis, Maggie McFee, Curtis Mead, Ali Mehmed, Angel Peterchev, Jim Phillips, Marco Sartore, Andrew Speck, Jim Thompson, Jim van Zee, GuYeon Wei, John Willison, Jonathan Wolff, John Woodgate, and Woody Yang. We thank also others whom (we're sure) we've here overlooked, with apologies for the omission. Additional contributors to the book's content (circuits, inspired web-based tools, unusual measurements, etc., from the likes of Uwe Beis, Tom Bruhns, and John Larkin) are referenced throughout the book in the relevant text.

Simon Capelin has kept us out of the doldrums with his unflagging encouragement and his apparent inability to scold us for missed deadlines (our contract called for delivery of the finished manuscript in December... of *1994!* We're only 20 years late). In the production chain we are indebted to our project manager Peggy Rote, our copy editor Vicki Danahy, and a cast of unnamed graphic artists who converted our pencil circuit sketches into beautiful vector graphics.

We remember fondly our late colleague and friend Jim Williams for wonderful insider stories of circuit failures and circuit conquests, and for his take-no-prisoners approach to precision circuit design. His no-bullshit attitude is a model for us all.

And finally, we are forever indebted to our loving, supportive, and ever-tolerant spouses Vida and Ava, who suffered through decades of abandonment as we obsessed over every detail of our second encore.

A note on the tools. Tables were assembled in Microsoft Excel, and graphical data was plotted with Igor Pro; both were then beautified with Adobe Illustrator, with text and annotations in the sans-serif Helvetica Neue LT typeface. Oscilloscope screenshots are from our trusty Tektronix TDS3044 and 3054 "lunchboxes," taken to finishing school in Illustrator, by way of Photoshop. The photographs in the book were taken primarily with two cameras: a Calumet Horseman 6×9 cm view camera with a 105 mm Schneider Symmar $f/5.6$ lens and Kodak Plus-X 120 roll film (developed in Microdol-X 1:3 at 75°F and digitized with a Mamiya multiformat scanner), and a Canon 5D with a Scheimpflug[3]-enabling 90 mm tilt-shift lens. The authors composed the manuscript in LaTeX, using the PCTeX software from Personal TeX, Incorporated. The text is set in the Times New Roman and Helvetica typefaces, the former dating from 1931,[4] the latter designed in 1957 by Max Miedinger.

Paul Horowitz
Winfield Hill
January 2015
Cambridge, Massachusetts

*   *   *   *   *

---

[3] What's *that*? Google it!

[4] Developed in response to a criticism of the antiquated typeface in *The Times* (London).

Legal Notice Addendum

In addition to the Legal Notice appended to the Preface to the Second Edition, we also make no representation regarding whether use of the examples, data, or other information in this volume might infringe others' intellectual property rights, including US and foreign patents. It is the reader's sole responsibility to ensure that he or she is not infringing any intellectual property rights, even for use which is considered to be experimental in nature. By using any of the examples, data, or other information in this volume, the reader has agreed to assume all liability for any damages arising from or relating to such use, regardless of whether such liability is based on intellectual property or any other cause of action, and regardless of whether the damages are direct, indirect, incidental, consequential, or any other type of damage. The authors and publisher disclaim any such liability.

PREFACE TO THE SECOND EDITION

Electronics, perhaps more than any other field of technology, has enjoyed an explosive development in the last four decades. Thus it was with some trepidation that we attempted, in 1980, to bring out a definitive volume teaching the art of the subject. By "art" we meant the kind of mastery that comes from an intimate familiarity with real circuits, actual devices, and the like, rather than the more abstract approach often favored in textbooks on electronics. Of course, in a rapidly evolving field, such a nuts-and-bolts approach has its hazards – most notably a frighteningly quick obsolescence.

The pace of electronics technology did not disappoint us! Hardly was the ink dry on the first edition before we felt foolish reading our words about "the classic [2Kbyte] 2716 EPROM...with a price tag of about \$25." They're so classic you can't even get them anymore, having been replaced by EPROMs 64 times as large, and costing less than half the price! Thus a major element of this revision responds to improved devices and methods – completely rewritten chapters on microcomputers and microprocessors (using the IBM PC and the 68008) and substantially revised chapters on digital electronics (including PLDs, and the new HC and AC logic families), on op-amps and precision design (reflecting the availability of excellent FET-input op-amps), and on construction techniques (including CAD/CAM). Every table has been revised, some substantially; for example, in Table 4.1 (operational amplifiers) only 65% of the original 120 entries survived, with 135 new op-amps added.

We have used this opportunity to respond to readers' suggestions and to our own experiences using and teaching from the first edition. Thus we have rewritten the chapter on FETs (it was too complicated) and repositioned it before the chapter on op-amps (which are increasingly of FET construction). We have added a new chapter on low-power and micropower design (both analog and digital), a field both important and neglected. Most of the remaining chapters have been extensively revised. We have added many new tables, including A/D and D/A converters, digital logic components, and low-power devices, and throughout the book we have expanded the number of figures. The book now contains 78 tables (available separately as *The Horowitz and Hill Component Selection Tables*) and over 1000 figures.

Throughout the revision we have strived to retain the feeling of informality and easy access that made the first edition so successful and popular, both as reference and text. We are aware of the difficulty students often experience when approaching electronics for the first time: the field is densely interwoven, and there is no path of learning that takes you, by logical steps, from neophyte to broadly competent designer. Thus we have added extensive cross-referencing throughout the text; in addition, we have expanded the separate *Laboratory Manual* into a *Student Manual* (*Student Manual for The Art of Electronics*, by Thomas C. Hayes and Paul Horowitz), complete with additional worked examples of circuit designs, explanatory material, reading assignments, laboratory exercises, and solutions to selected problems. By offering a student supplement, we have been able to keep this volume concise and rich with detail, as requested by our many readers who use the volume primarily as a reference work.

We hope this new edition responds to all our readers' needs – both students and practicing engineers. We welcome suggestions and corrections, which should be addressed directly to Paul Horowitz, Physics Department, Harvard University, Cambridge, MA 02138.

In preparing this new edition, we are appreciative of the help we received from Mike Aronson and Brian Matthews (AOX, Inc.), John Greene (University of Cape Town), Jeremy Avigad and Tom Hayes (Harvard University), Peter Horowitz (EVI, Inc.), Don Stern, and Owen Walker. We thank Jim Mobley for his excellent copyediting, Sophia Prybylski and David Tranah of Cambridge University Press for their encouragement and professional dedication, and the never-sleeping typesetters at Rosenlaui Publishing Services, Inc. for their masterful composition in TEX.

Finally, in the spirit of modern jurisprudence, we remind you to read the legal notice here appended.

Paul Horowitz
Winfield Hill
March 1989

Legal notice

In this book we have attempted to teach the techniques of electronic design, using circuit examples and data that we believe to be accurate. However, the examples, data, and other information are intended solely as teaching aids and should not be used in any particular application without independent testing and verification by the person making the application. Independent testing and verification are especially important in any application in which incorrect functioning could result in personal injury or damage to property.

For these reasons, we make no warranties, express or implied, that the examples, data, or other information in this volume are free of error, that they are consistent with industry standards, or that they will meet the requirements for any particular application. **THE AUTHORS AND PUBLISHER EXPRESSLY DISCLAIM THE IMPLIED WARRANTIES OF MERCHANTABILITY AND OF FITNESS FOR ANY PARTICULAR PURPOSE,** even if the authors have been advised of a particular purpose, and even if a particular purpose is indicated in the book. The authors and publisher also disclaim all liability for direct, indirect, incidental, or consequential damages that result from any use of the examples, data, or other information in this book.

PREFACE TO THE FIRST EDITION

This volume is intended as an electronic circuit design textbook and reference book; it begins at a level suitable for those with no previous exposure to electronics and carries the reader through to a reasonable degree of proficiency in electronic circuit design. We have used a straightforward approach to the essential ideas of circuit design, coupled with an in-depth selection of topics. We have attempted to combine the pragmatic approach of the practicing physicist with the quantitative approach of the engineer, who wants a thoroughly evaluated circuit design.

This book evolved from a set of notes written to accompany a one-semester course in laboratory electronics at Harvard. That course has a varied enrollment – undergraduates picking up skills for their eventual work in science or industry, graduate students with a field of research clearly in mind, and advanced graduate students and post-doctoral researchers who suddenly find themselves hampered by their inability to "do electronics."

It soon became clear that existing textbooks were inadequate for such a course. Although there are excellent treatments of each electronics specialty, written for the planned sequence of a four-year engineering curriculum or for the practicing engineer, those books that attempt to address the whole field of electronics seem to suffer from excessive detail (the handbook syndrome), from oversimplification (the cookbook syndrome), or from poor balance of material. Much of the favorite pedagogy of beginning textbooks is quite unnecessary and, in fact, is not used by practicing engineers, while useful circuitry and methods of analysis in daily use by circuit designers lie hidden in application notes, engineering journals, and hard-to-get data books. In other words, there is a tendency among textbook writers to represent the theory, rather than the art, of electronics.

We collaborated in writing this book with the specific intention of combining the discipline of a circuit design engineer with the perspective of a practicing experimental physicist and teacher of electronics. Thus, the treatment in this book reflects our philosophy that electronics, as currently practiced, is basically a simple art, a combination of some basic laws, rules of thumb, and a large bag of tricks. For these reasons we have omitted entirely the usual discussions of solid-state physics, the h-parameter model of transistors, and complicated network theory, and reduced to a bare minimum the mention of load lines and the s-plane. The treatment is largely nonmathematical, with strong encouragement of circuit brainstorming and mental (or, at most, back-of-the-envelope) calculation of circuit values and performance.

In addition to the subjects usually treated in electronics books, we have included the following:

- an easy-to-use transistor model;
- extensive discussion of useful subcircuits, such as current sources and current mirrors;
- single-supply op-amp design;
- easy-to-understand discussions of topics on which practical design information is often difficult to find: op-amp frequency compensation, low-noise circuits, phase-locked loops, and precision linear design;
- simplified design of active filters, with tables and graphs;
- a section on noise, shielding, and grounding;
- a unique graphical method for streamlined low-noise amplifier analysis;
- a chapter on voltage references and regulators, including constant current supplies;
- a discussion of monostable multivibrators and their idiosyncrasies;
- a collection of digital logic pathology, and what to do about it;
- an extensive discussion of interfacing to logic, with emphasis on the new NMOS and PMOS LSI;
- a detailed discussion of A/D and D/A conversion techniques;
- a section on digital noise generation;
- a discussion of minicomputers and interfacing to data buses, with an introduction to assembly language;
- a chapter on microprocessors, with actual design examples and discussion – how to design them into instruments, and how to make them do what you want;
- a chapter on construction techniques: prototyping, printed circuit boards, instrument design;

- a simplified way to evaluate high-speed switching circuits;
- a chapter on scientific measurement and data processing: what you can measure and how accurately, and what to do with the data;
- bandwidth narrowing methods made clear: signal averaging, multichannel scaling, lock-in amplifiers, and pulse-height analysis;
- amusing collections of "bad circuits," and collections of "circuit ideas";
- useful appendixes on how to draw schematic diagrams, IC generic types, *LC* filter design, resistor values, oscilloscopes, mathematics review, and others;
- tables of diodes, transistors, FETs, op-amps, comparators, regulators, voltage references, microprocessors, and other devices, generally listing the characteristics of both the most popular and the best types.

Throughout we have adopted a philosophy of naming names, often comparing the characteristics of competing devices for use in any circuit, and the advantages of alternative circuit configurations. Example circuits are drawn with real device types, not black boxes. The overall intent is to bring the reader to the point of understanding clearly the choices one makes in designing a circuit – how to choose circuit configurations, device types, and parts values. The use of largely nonmathematical circuit design techniques does not result in circuits that cut corners or compromise performance or reliability. On the contrary, such techniques enhance one's understanding of the real choices and compromises faced in engineering a circuit and represent the best approach to good circuit design.

This book can be used for a full-year electronic circuit design course at the college level, with only a minimum mathematical prerequisite; namely, some acquaintance with trigonometric and exponential functions, and preferably a bit of differential calculus. (A short review of complex numbers and derivatives is included as an appendix.) If the less essential sections are omitted, it can serve as the text for a one-semester course (as it does at Harvard).

A separately available laboratory manual, *Laboratory Manual for the Art of Electronics* (Horowitz and Robinson, 1981), contains twenty-three lab exercises, together with reading and problem assignments keyed to the text.

To assist the reader in navigation, we have designated with open boxes in the margin those sections within each chapter that we feel can be safely passed over in an abbreviated reading. For a one-semester course it would probably be wise to omit, in addition, the materials of Chapter 5 (first half), 7, 12, 13, 14, and possibly 15, as explained in the introductory paragraphs of those chapters.

We would like to thank our colleagues for their thoughtful comments and assistance in the preparation of the manuscript, particularly Mike Aronson, Howard Berg, Dennis Crouse, Carol Davis, David Griesinger, John Hagen, Tom Hayes, Peter Horowitz, Bob Kline, Costas Papaliolios, Jay Sage, and Bill Vetterling. We are indebted to Eric Hieber and Jim Mobley, and to Rhona Johnson and Ken Werner of Cambridge University Press, for their imaginative and highly professional work.

Paul Horowitz
Winfield Hill
April 1980

FOUNDATIONS

1.1 Introduction

The field of electronics is one of the great success stories of the 20th century. From the crude spark-gap transmitters and "cat's-whisker" detectors at its beginning, the first half-century brought an era of vacuum-tube electronics that developed considerable sophistication and found ready application in areas such as communications, navigation, instrumentation, control, and computation. The latter half-century brought "solid-state" electronics – first as discrete transistors, then as magnificent arrays of them within "integrated circuits" (ICs) – in a flood of stunning advances that shows no sign of abating. Compact and inexpensive consumer products now routinely contain many millions of transistors in VLSI (very large-scale integration) chips, combined with elegant optoelectronics (displays, lasers, and so on); they can process sounds, images, and data, and (for example) permit wireless networking and shirt-pocket access to the pooled capabilities of the Internet. Perhaps as noteworthy is the pleasant trend toward increased performance per dollar.[1] The cost of an electronic microcircuit routinely decreases to a fraction of its initial cost as the manufacturing process is perfected (see Figure 10.87 for an example). In fact, it is often the case that the panel controls and cabinet hardware of an instrument cost more than the electronics inside.

On reading of these exciting new developments in electronics, you may get the impression that you should be able to construct powerful, elegant, yet inexpensive, little gadgets to do almost any conceivable task – all you need to know is how all these miracle devices work. If you've had that feeling, this book is for you. In it we have attempted to convey the excitement and know-how of the subject of electronics.

In this chapter we begin the study of the laws, rules of thumb, and tricks that constitute the art of electronics as we see it. It is necessary to begin at the beginning – with talk of voltage, current, power, and the components that make up

electronic circuits. Because you can't touch, see, smell, or hear electricity, there will be a certain amount of abstraction (particularly in the first chapter), as well as some dependence on such visualizing instruments as oscilloscopes and voltmeters. In many ways the first chapter is also the most mathematical, in spite of our efforts to keep mathematics to a minimum in order to foster a good intuitive understanding of circuit design and behavior.

In this new edition we've included some intuition-aiding approximations that our students have found helpful. And, by introducing one or two "active" components ahead of their time, we're able to jump directly into some applications that are usually impossible in a traditional textbook "passive electronics" chapter; this will keep things interesting, and even exciting.

Once we have considered the foundations of electronics, we will quickly get into the active circuits (amplifiers, oscillators, logic circuits, etc.) that make electronics the exciting field it is. The reader with some background in electronics may wish to skip over this chapter, since it assumes no prior knowledge of electronics. Further generalizations at this time would be pointless, so let's just dive right in.

1.2 Voltage, current, and resistance

1.2.1 Voltage and current

There are two quantities that we like to keep track of in electronic circuits: voltage and current. These are usually changing with time; otherwise nothing interesting is happening.

Voltage (symbol V or sometimes E). Officially, the voltage between two points is the cost in energy (work done) required to move a unit of positive charge from the more negative point (lower potential) to the more positive point (higher potential). Equivalently, it is the energy released when a unit charge moves "downhill" from the higher potential to the lower.[2] Voltage is also called

---

[1] A mid-century computer (the IBM 650) cost $300,000, weighed 2.7 tons, and contained 126 lamps on its control panel; in an amusing reversal, a contemporary energy-efficient lamp contains a computer of greater capability *within its base*, and costs about $10.

[2] These are the *definitions*, but hardly the way circuit designers think of voltage. With time, you'll develop a good intuitive sense of what voltage really is, in an electronic circuit. Roughly (*very* roughly) speaking, voltages are what you apply to cause currents to flow.

potential difference or *electromotive force* (EMF). The unit of measure is the *volt*, with voltages usually expressed in volts (V), kilovolts ($1\,kV = 10^3\,V$), millivolts ($1\,mV = 10^{-3}\,V$), or microvolts ($1\,\mu V = 10^{-6}\,V$) (see the box on prefixes). A joule (J) of work is done in moving a coulomb (C) of charge through a potential difference of 1 V. (The coulomb is the unit of electric charge, and it equals the charge of approximately 6×10^{18} electrons.) For reasons that will become clear later, the opportunities to talk about nanovolts ($1\,nV = 10^{-9}\,V$) and megavolts ($1\,MV = 10^6\,V$) are rare.

Current (symbol I). Current is the rate of flow of electric charge past a point. The unit of measure is the ampere, or amp, with currents usually expressed in amperes (A), milliamperes ($1\,mA = 10^{-3}\,A$), microamperes ($1\,\mu A = 10^{-6}\,A$), nanoamperes ($1\,nA = 10^{-9}\,A$), or occasionally picoamperes ($1\,pA = 10^{-12}\,A$). A current of 1 amp equals a flow of 1 coulomb of charge per second. By convention, current in a circuit is considered to flow from a more positive point to a more negative point, even though the actual electron flow is in the opposite direction.

Important: from these definitions you can see that currents flow *through* things, and voltages are applied (or appear) *across* things. So you've got to say it right: always refer to the voltage *between* two points or *across* two points in a circuit. Always refer to current *through* a device or connection in a circuit.

To say something like "the voltage through a resistor ..." is nonsense. However, we do frequently speak of the voltage *at a point* in a circuit. This is always understood to mean the voltage between that point and "ground," a common point in the circuit that everyone seems to know about. Soon you will, too.

We *generate* voltages by doing work on charges in devices such as batteries (conversion of electrochemical energy), generators (conversion of mechanical energy by magnetic forces), solar cells (photovoltaic conversion of the energy of photons), etc. We *get* currents by placing voltages across things.

At this point you may well wonder how to "see" voltages and currents. The single most useful electronic instrument is the oscilloscope, which allows you to look at voltages (or occasionally currents) in a circuit as a function of time.[3] We will deal with oscilloscopes, and also voltmeters, when we discuss signals shortly; for a preview see Appendix O, and the multimeter box later in this chapter.

---

[3] It has been said that engineers in other disciplines are envious of electrical engineers, because we have such a splendid visualization tool.

In real circuits we connect things together with wires (metallic conductors), each of which has the same voltage on it everywhere (with respect to ground, say).[4] We mention this now so that you will realize that an actual circuit doesn't have to look like its schematic diagram, because wires can be rearranged.

Here are some simple rules about voltage and current:

1. The sum of the currents into a point in a circuit equals the sum of the currents out (conservation of charge). This is sometimes called Kirchhoff's current law (KCL). Engineers like to refer to such a point as a *node*. It follows that, for a series circuit (a bunch of two-terminal things all connected end-to-end), the current is the same everywhere.

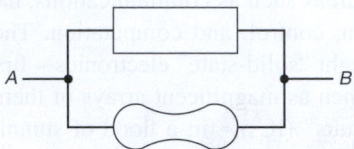

Figure 1.1. Parallel connection.

2. Things hooked in parallel (Figure 1.1) have the same voltage across them. Restated, the sum of the "voltage drops" from A to B via one path through a circuit equals the sum by any other route, and is simply the voltage between A and B. Another way to say it is that the sum of the voltage drops around any closed circuit is zero. This is Kirchhoff's voltage law (KVL).

3. The power (energy per unit time) consumed by a circuit device is

$$P = VI \qquad (1.1)$$

This is simply (energy/charge) × (charge/time). For V in volts and I in amps, P comes out in watts. A watt is a joule per second ($1\,W = 1\,J/s$). So, for example, the current flowing through a 60 W lightbulb running on 120 V is 0.5 A.

Power goes into heat (usually), or sometimes mechanical work (motors), radiated energy (lamps, transmitters), or stored energy (batteries, capacitors, inductors). Managing the heat load in a complicated system (e.g., a large computer, in which many kilowatts of electrical energy are converted to heat, with the energetically insignificant by-product of a few pages of computational results) can be a crucial part of the system design.

---

[4] In the domain of high frequencies or low impedances, that isn't strictly true, and we will have more to say about this later, and in Chapter *1x*. For now, it's a good approximation.

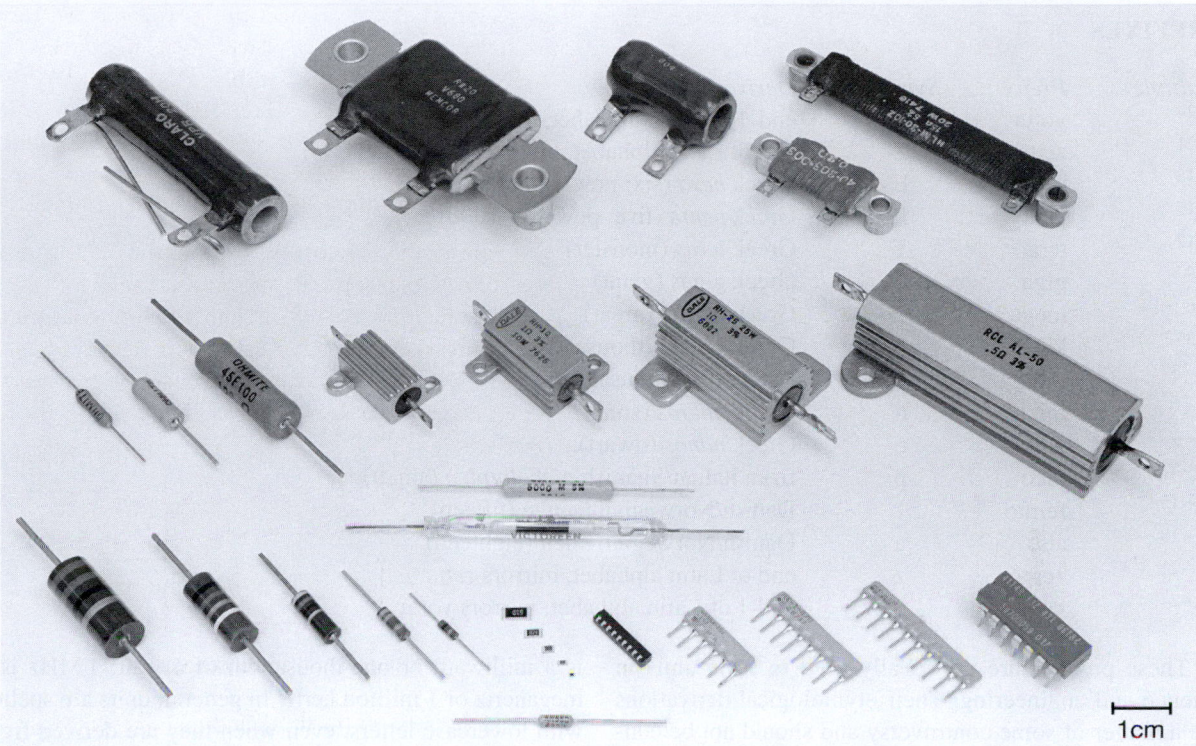

Figure 1.2. A selection of common resistor types. Top row, left to right (wirewound ceramic power resistors): 20W vitreous enamel with leads, 20W with mounting studs, 30W vitreous enamel, 5W and 20W with mounting studs. Middle row (wirewound power resistors): 1W, 3W, and 5W axial ceramic; 5W, 10W, 25W, and 50W conduction-cooled ("Dale-type"). Bottom row: 2W, 1W, $\frac{1}{2}$W, $\frac{1}{4}$W, and $\frac{1}{8}$W carbon composition; surface-mount thick-film (2010, 1206, 0805, 0603, and 0402 sizes); surface-mount resistor array; 6-, 8-, and 10-pin single in-line package arrays; dual in-line package array. The resistor at bottom is the ubiquitous RN55D $\frac{1}{4}$W, 1% metal-film type; and the pair of resistors above are Victoreen high-resistance types (glass, 2 GΩ; ceramic, 5 GΩ).

Soon, when we deal with periodically varying voltages and currents, we will have to generalize the simple equation $P = VI$ to deal with *average* power, but it's correct as a statement of *instantaneous* power just as it stands.

Incidentally, don't call current "amperage"; that's strictly bush league.[5] The same caution will apply to the term "ohmage"[6] when we get to resistance in the next section.

1.2.2 Relationship between voltage and current: resistors

This is a long and interesting story. It is the heart of electronics. Crudely speaking, the name of the game is to make and use gadgets that have interesting and useful *I*-versus-*V* characteristics. Resistors (*I* simply proportional to *V*),

capacitors (*I* proportional to rate of change of *V*), diodes (*I* flows in only one direction), thermistors (temperature-dependent resistor), photoresistors (light-dependent resistor), strain gauges (strain-dependent resistor), etc., are examples. Perhaps more interesting still are *three-terminal* devices, such as transistors, in which the current that can flow between a pair of terminals is controlled by the voltage applied to a third terminal. We will gradually get into some of these exotic devices; for now, we will start with the most mundane (and most widely used) circuit element, the resistor (Figure 1.3).

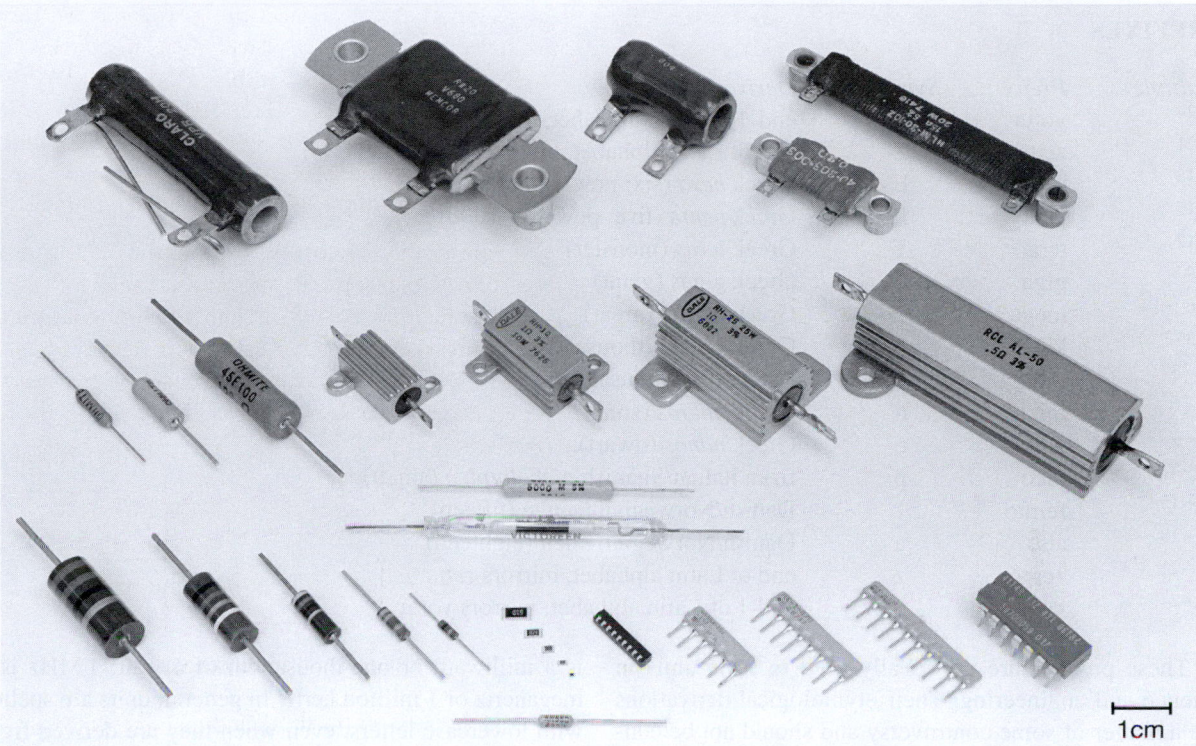

Figure 1.3. Resistor.

A. Resistance and resistors

It is an interesting fact that the current through a metallic conductor (or other partially conducting material) is proportional to the voltage across it. (In the case of wire

---

[5] Unless you're a power engineer working with giant 13 kV transformers and the like – those guys are allowed to say amperage.

[6] ...also, Dude, "ohmage" is not the preferred nomenclature: *resistance*, please.

PREFIXES

Multiple	Prefix	Symbol	Derivation
10^{24}	yotta	Y	end-1 of Latin alphabet, hint of Greek *iota*
10^{21}	zetta	Z	end of Latin alphabet, hint of Greek *zeta*
10^{18}	exa	E	Greek *hexa* (six: power of 1000)
10^{15}	peta	P	Greek *penta* (five: power of 1000)
10^{12}	tera	T	Greek *teras* (monster)
10^{9}	giga	G	Greek *gigas* (giant)
10^{6}	mega	M	Greek *megas* (great)
10^{3}	kilo	k	Greek *khilioi* (thousand)
10^{-3}	milli	m	Latin *mille* (thousand)
10^{-6}	micro	μ	Greek *mikros* (small)
10^{-9}	nano	n	Greek *nanos* (dwarf)
10^{-12}	pico	p	from Italian/Spanish *piccolo/pico* (small)
10^{-15}	femto	f	Danish/Norwegian *femten* (fifteen)
10^{-18}	atto	a	Danish/Norwegian *atten* (eighteen)
10^{-21}	zepto	z	end of Latin alphabet, mirrors *zetta*
10^{-24}	yocto	y	end-1 of Latin alphabet, mirrors *yotta*

These prefixes are universally used to scale units in science and engineering. Their etymological derivations are a matter of some controversy and should not be considered historically reliable. When abbreviating a unit with a prefix, the symbol for the unit follows the prefix without space. Be careful about uppercase and lowercase letters (especially m and M) in both prefix and unit: 1 mW is a milliwatt, or one-thousandth of a watt; 1 MHz is a megahertz or 1 million hertz. In general, units are spelled with lowercase letters, even when they are derived from proper names. The unit name is not capitalized when it is spelled out and used with a prefix, only when abbreviated. Thus: hertz and kilohertz, but Hz and kHz; watt, milliwatt, and megawatt, but W, mW, and MW.

conductors used in circuits, we usually choose a thick-enough gauge of wire so that these "voltage drops" will be negligible.) This is by no means a universal law for all objects. For instance, the current through a neon bulb is a highly nonlinear function of the applied voltage (it is zero up to a critical voltage, at which point it rises dramatically). The same goes for a variety of interesting special devices – diodes, transistors, lightbulbs, etc. (If you are interested in understanding why metallic conductors behave this way, read §§4.4–4.5 in Purcell and Morin's splendid text *Electricity and Magnetism*).

A resistor is made out of some conducting stuff (carbon, or a thin metal or carbon film, or wire of poor conductivity), with a wire or contacts at each end. It is characterized by its resistance:

$$R = V/I; \qquad (1.2)$$

R is in ohms for V in volts and I in amps. This is known as Ohm's law. Typical resistors of the most frequently used type (metal-oxide film, metal film, or carbon film) come in values from 1 ohm ($1\,\Omega$) to about 10 megohms ($10\,M\Omega$). Resistors are also characterized by how much power they can safely dissipate (the most commonly used ones are rated at 1/4 or 1/8 W), their physical size,[7] and by other parameters such as tolerance (accuracy), temperature coefficient, noise, voltage coefficient (the extent to which R depends on applied V), stability with time, inductance, etc. See the box on resistors, Chapter *1x*, and Appendix C for further details. Figure 1.2 shows a collection of resistors, with most of the available morphologies represented.

Roughly speaking, resistors are used to convert a

---

[7] The sizes of *chip resistors* and other components intended for surface mounting are specified by a four-digit size code, in which each pair of digits specifies a dimension in units of $0.010''$ (0.25 mm). For example, an 0805 size resistor is $2\,mm \times 1.25\,mm$, or $80\,mils \times 50\,mils$ (1 mil is $0.001''$); the height must be separately specified. To add confusion to this simple scheme, the four-digit size code may instead be *metric* (sometimes without saying so!), in units of 0.1 mm: thus an "0805" (English) is also a "2012" (metric).

RESISTORS

Resistors are truly ubiquitous. There are almost as many types as there are applications. Resistors are used in amplifiers as loads for active devices, in bias networks, and as feedback elements. In combination with capacitors they establish time constants and act as filters. They are used to set operating currents and signal levels. Resistors are used in power circuits to reduce voltages by dissipating power, to measure currents, and to discharge capacitors after power is removed. They are used in precision circuits to establish currents, to provide accurate voltage ratios, and to set precise gain values. In logic circuits they act as bus and line terminators and as "pullup" and "pull-down" resistors. In high-voltage circuits they are used to measure voltages and to equalize leakage currents among diodes or capacitors connected in series. In radiofrequency (RF) circuits they set the bandwidth of resonant circuits, and they are even used as coil forms for inductors.

Resistors are available with resistances from $0.0002\,\Omega$ through $10^{12}\,\Omega$, standard power ratings from 1/8 watt through 250 watts, and accuracies from 0.005% through 20%. Resistors can be made from metal films, metal-oxide films, or carbon films; from carbon-composition or ceramic-composition moldings; from metal foil or metal wire wound on a form; or from semiconductor elements similar to field-effect transistors (FETs). The most commonly used resistor type is formed from a carbon, metal, or oxide film, and comes in two widely used "packages": the cylindrical *axial-lead* type (typified by the generic RN55D 1% 1/4 W metal-film resistor),[8] and the much smaller *surface-mount* "chip resistor." These common types come in 5%, 2%, and 1% tolerances, in a standard set of values ranging from $1\,\Omega$ to $10\,M\Omega$. The 1% types have 96 values per decade, whereas the 2% and 5% types have 48 and 24 values per decade (see Appendix C). Figure 1.2 illustrates most of the common resistor packages.

Resistors are so easy to use and well behaved that they're often taken for granted. They're not perfect, though, and you should be aware of some of their limitations so that you won't be surprised someday. The principal defects are variations in resistance with temperature, voltage, time, and humidity. Other defects relate to inductance (which may be serious at high frequencies), the development of thermal hot spots in power applications, or electrical noise generation in low-noise amplifiers. We treat these in the advanced Chapter *1x*.

voltage to a current, and vice versa. This may sound awfully trite, but you will soon see what we mean.

B. Resistors in series and parallel

From the definition of R, some simple results follow:

1. The resistance of two resistors in series (Figure 1.4) is

$$R = R_1 + R_2. \tag{1.3}$$

By putting resistors in series, you always get a *larger* resistor.

2. The resistance of two resistors in parallel (Figure 1.5) is

$$R = \frac{R_1 R_2}{R_1 + R_2} \quad \text{or} \quad R = \frac{1}{\dfrac{1}{R_1} + \dfrac{1}{R_2}}. \tag{1.4}$$

By putting resistors in parallel, you always get a *smaller* resistor. Resistance is measured in ohms (Ω), but in practice we frequently omit the Ω symbol when referring to resistors that are more than $1000\,\Omega$ ($1\,k\Omega$). Thus, a $4.7\,k\Omega$ resistor is often referred to as a 4.7k resistor, and a $1\,M\Omega$

Figure 1.4. Resistors in series.

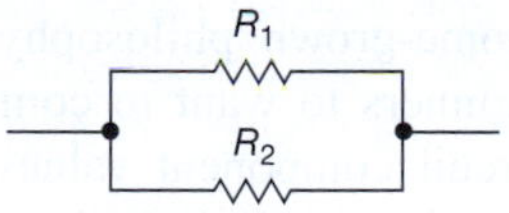

Figure 1.5. Resistors in parallel.

resistor as a 1M resistor (or 1 meg).[9] If these preliminaries bore you, please have patience – we'll soon get to numerous amusing applications.

Exercise 1.1. You have a 5k resistor and a 10k resistor. What is their combined resistance (a) in series and (b) in parallel?

Exercise 1.2. If you place a 1 ohm resistor across a 12 volt car battery, how much power will it dissipate?

Exercise 1.3. Prove the formulas for series and parallel resistors.

---

[8] Conservatively rated at 1/8 watt in its RN55 military grade ("MIL-spec"), but rated at 1/4 watt in its CMF-55 industrial grade.

[9] A popular "international" alternative notation replaces the decimal point with the unit multiplier, thus 4k7 or 1M0. A $2.2\,\Omega$ resistor becomes 2R2. There is an analogous scheme for capacitors and inductors.

Exercise 1.4. Show that several resistors in parallel have resistance

$$R = \cfrac{1}{\cfrac{1}{R_1} + \cfrac{1}{R_2} + \cfrac{1}{R_3} + \cdots} \qquad (1.5)$$

Beginners tend to get carried away with complicated algebra in designing or trying to understand electronics. Now is the time to begin learning intuition and shortcuts. Here are a couple of good tricks:

Shortcut #1 A large resistor in series (parallel) with a small resistor has the resistance of the larger (smaller) one, roughly. So you can "trim" the value of a resistor up or down by connecting a second resistor in series or parallel: to trim *up*, choose an available resistor value below the target value, then add a (much smaller) series resistor to make up the difference; to trim *down*, choose an available resistor value above the target value, then connect a (much larger) resistor in parallel. For the latter you can approximate with proportions – to lower the value of a resistor by 1%, say, put a resistor 100 times as large in parallel.[10]

Shortcut #2 Suppose you want the resistance of 5k in parallel with 10k. If you think of the 5k as two 10k's in parallel, then the whole circuit is like three 10k's in parallel. Because the resistance of *n* equal resistors in parallel is 1/*n*th the resistance of the individual resistors, the answer in this case is 10k/3, or 3.33k. This trick is handy because it allows you to analyze circuits quickly in your head, without distractions. We want to encourage mental designing, or at least "back-of-the-envelope" designing, for idea brainstorming.

Some more home-grown philosophy: there is a tendency among beginners to want to compute resistor values and other circuit component values to many significant places, particularly with calculators and computers that readily oblige. There are two reasons you should try to avoid falling into this habit: (a) the components themselves are of finite precision (resistors typically have tolerances of ±5% or ±1%; for capacitors it's typically ±10% or ±5%; and the parameters that characterize transistors, say, frequently are known only to a factor of 2); (b) one mark of a good circuit design is insensitivity of the finished circuit to precise values of the components (there are exceptions, of course). You'll also learn circuit intuition more quickly if you get into the habit of doing approximate calculations in your head, rather than watching meaningless numbers pop up on a calculator display. We believe strongly that reliance on formulas and equations early in your electronic circuit

education is a fine way to prevent you from understanding what's really going on.

In trying to develop intuition about resistance, some people find it helpful to think about *conductance*, $G = 1/R$. The current through a device of conductance G bridging a voltage V is then given by $I = GV$ (Ohm's law). A small resistance is a large conductance, with correspondingly large current under the influence of an applied voltage. Viewed in this light, the formula for parallel resistors is obvious: when several resistors or conducting paths are connected across the same voltage, the total current is the sum of the individual currents. Therefore the net conductance is simply the sum of the individual conductances, $G = G_1 + G_2 + G_3 + \cdots$, which is the same as the formula for parallel resistors derived earlier.

Engineers are fond of defining reciprocal units, and they have designated as the unit of conductance the siemens ($S = 1/\Omega$), also known as the mho (that's ohm spelled backward, given the symbol ℧). Although the concept of conductance is helpful in developing intuition, it is not used widely;[11] most people prefer to talk about resistance instead.

C. Power in resistors

The power dissipated by a resistor (or any other device) is $P = IV$. Using Ohm's law, you can get the equivalent forms $P = I^2 R$ and $P = V^2/R$.

Exercise 1.5. Show that it is not possible to exceed the power rating of a 1/4 watt resistor of resistance greater than 1k, no matter how you connect it, in a circuit operating from a 15 volt battery.

Exercise 1.6. Optional exercise: New York City requires about 10^{10} watts of electrical power, at 115 volts[12] (this is plausible: 10 million people averaging 1 kilowatt each). A heavy power cable might be an inch in diameter. Let's calculate what will happen if we try to supply the power through a cable 1 foot in diameter made of pure copper. Its resistance is $0.05\,\mu\Omega$ (5×10^{-8} ohms) per foot. Calculate (a) the power lost per foot from "$I^2 R$ losses," (b) the length of cable over which you will lose all 10^{10} watts, and (c) how hot the cable will get, if you know the physics involved ($\sigma = 6 \times 10^{-12}\,\text{W}/\text{K}^4\,\text{cm}^2$). If you have done your computations correctly, the result should seem preposterous. What is the solution to this puzzle?

---

[10] With an error, in this case, of just 0.01%.

[11] Although the elegant *Millman's theorem* has its admirers: it says that the output voltage from a set of resistors (call them R_i) that are driven from a set of corresponding input voltages (V_i) and connected together at the output is $V_{\text{out}} = (\sum V_i G_i)/\sum G_i$, where the G_i are the conductances ($G_i = 1/R_i$).

[12] Although the "official" line voltage is 120 V ±5%, you'll sometimes see 110 V, 115 V, or 117 V. This loose language is OK (and we use it in this book), because (a) the median voltage at the wall plug is 3 to 5 volts lower, when powering stuff; and (b) the *minimum* wall-plug voltage is 110 V. See ANSI standard C84.1.

D. Input and output

Nearly all electronic circuits accept some sort of applied *input* (usually a voltage) and produce some sort of corresponding *output* (which again is often a voltage). For example, an audio amplifier might produce a (varying) output voltage that is 100 times as large as a (similarly varying) input voltage. When describing such an amplifier, we imagine measuring the output voltage for a given applied input voltage. Engineers speak of the *transfer function* **H**, the ratio of (measured) output divided by (applied) input; for the audio amplifier above, **H** is simply a constant (**H** = 100). We'll get to amplifiers soon enough, in the next chapter. However, with only resistors we can already look at a very important circuit fragment, the *voltage divider* (which you might call a "de-amplifier").

1.2.3 Voltage dividers

We now come to the subject of the voltage divider, one of the most widespread electronic circuit fragments. Show us any real-life circuit and we'll show you half a dozen voltage dividers. To put it very simply, a voltage divider is a circuit that, given a certain voltage input, produces a predictable fraction of the input voltage as the output voltage. The simplest voltage divider is shown in Figure 1.6.

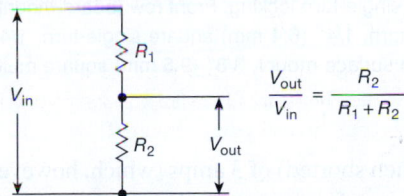

$$\frac{V_{\text{out}}}{V_{\text{in}}} = \frac{R_2}{R_1 + R_2}$$

Figure 1.6. Voltage divider. An applied voltage V_{in} results in a (smaller) output voltage V_{out}.

An important word of explanation: when engineers draw a circuit like this, it's generally assumed that the V_{in} on the left is a voltage that you are applying to the circuit, and that the V_{out} on the right is the resulting output voltage (produced by the circuit) that you are measuring (or at least are interested in). You are supposed to know all this (a) because of the convention that signals generally flow from left to right, (b) from the suggestive names ("in," "out") of the signals, and (c) from familiarity with circuits like this. This may be confusing at first, but with time it becomes easy.

What is V_{out}? Well, the current (same everywhere, assuming no "load" on the output; i.e., nothing connected across the output) is

$$I = \frac{V_{\text{in}}}{R_1 + R_2}.$$

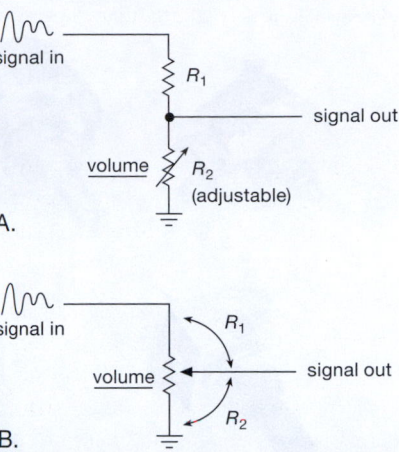

Figure 1.7. An adjustable voltage divider can be made from a fixed and variable resistor, or from a potentiometer. In some contemporary circuits you'll find instead a long series chain of equal-value resistors, with an arrangement of electronic switches that lets you choose any one of the junctions as the output; this sounds much more complicated – but it has the advantage that you can adjust the voltage ratio *electrically* (rather than mechanically).

(We've used the definition of resistance and the series law.) Then, for R_2,

$$V_{\text{out}} = IR_2 = \frac{R_2}{R_1 + R_2} V_{\text{in}}. \tag{1.6}$$

Note that the output voltage is always less than (or equal to) the input voltage; that's why it's called a divider. You could get amplification (more output than input) if one of the resistances were negative. This isn't as crazy as it sounds; it is possible to make devices with negative "incremental" resistances (e.g., the component known as a *tunnel diode*) or even true negative resistances (e.g., the negative-impedance converter that we will talk about later in the book, §6.2.4B). However, these applications are rather specialized and need not concern you now.

Voltage dividers are often used in circuits to generate a particular voltage from a larger fixed (or varying) voltage. For instance, if V_{in} is a varying voltage and R_2 is an adjustable resistor (Figure 1.7A), you have a "volume control"; more simply, the combination R_1R_2 can be made from a single variable resistor, or *potentiometer* (Figure 1.7B). This and similar applications are common, and potentiometers come in a variety of styles, some of which are shown in Figure 1.8.

The humble voltage divider is even more useful, though, as a way of *thinking* about a circuit: the input voltage and upper resistance might represent the output of an amplifier, say, and the lower resistance might represent the input of

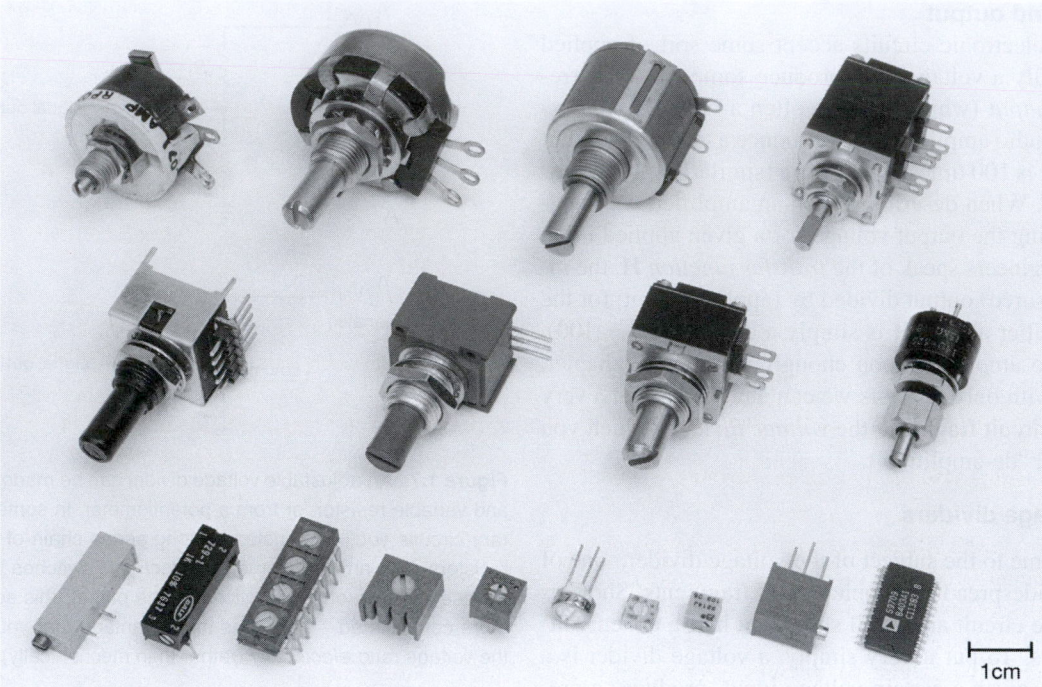

Figure 1.8. Most of the common potentiometer styles are shown here. Top row, left to right (panel mount): power wirewound, "type AB" 2W carbon composition, 10-turn wirewound/plastic hybrid, ganged dual pot. Middle row (panel mount): optical encoder (continuous rotation, 128 cycles per turn), single-turn cermet, single-turn carbon, screw-adjust single-turn locking. Front row (board-mount trimmers): multiturn side-adjust (two styles), quad single-turn, 3/8″ (9.5 mm) square single-turn, 1/4″ (6.4 mm) square single-turn, 1/4″ (6.4 mm) round single-turn, 4 mm square single-turn surface mount, 4 mm square multiturn surface mount, 3/8″ (9.5 mm) square multiturn, quad nonvolatile 256-step integrated pot (E^2POT) in 24-pin small-outline IC.

the following stage. In this case the voltage-divider equation tells you how much signal gets to the input of that last stage. This will all become clearer after you know about a remarkable fact (Thévenin's theorem) that will be discussed later. First, though, a short aside on voltage sources and current sources.

1.2.4 Voltage sources and current sources

A perfect *voltage source* is a two-terminal "black box" that maintains a fixed voltage drop across its terminals, regardless of load resistance. This means, for instance, that it must supply a current $I = V/R$ when a resistance R is attached to its terminals. A real voltage source can supply only a finite maximum current, and in addition it generally behaves like a perfect voltage source with a small resistance in series. Obviously, the smaller this series resistance, the better. For example, a standard 9 volt alkaline battery behaves approximately like a perfect 9 volt voltage source in series with a $3\,\Omega$ resistor, and it can provide a maximum

current (when shorted) of 3 amps (which, however, will kill the battery in a few minutes). A voltage source "likes" an open-circuit load and "hates" a short-circuit load, for obvious reasons. (The meaning of "open circuit" and "short circuit" sometimes confuse the beginner: an open circuit has nothing connected to it, whereas a short circuit is a piece of wire bridging the output.) The symbols used to indicate a voltage source are shown in Figure 1.9.

A perfect *current source* is a two-terminal black box that maintains a constant current through the external circuit, regardless of load resistance or applied voltage. To do this it must be capable of supplying any necessary voltage across its terminals. Real current sources (a much-neglected subject in most textbooks) have a limit to the voltage they can provide (called the *output-voltage compliance,* or just *compliance*), and in addition they do not provide absolutely constant output current. A current source "likes" a short-circuit load and "hates" an open-circuit load. The symbols used to indicate a current source are shown in Figure 1.10.

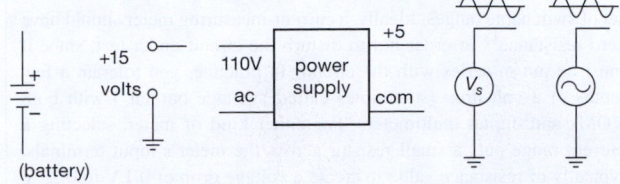

Figure 1.9. Voltage sources can be either steady (dc) or varying (ac).

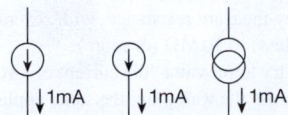

Figure 1.10. Current-source symbols.

A battery is a real-life approximation to a voltage source (there is no analog for a current source). A standard D-size flashlight cell, for instance, has a terminal voltage of 1.5 V, an equivalent series resistance of about $0.25\,\Omega$, and a total energy capacity of about 10,000 watt–seconds (its characteristics gradually deteriorate with use; at the end of its life, the voltage may be about 1.0 V, with an internal series resistance of several ohms). It is easy to construct voltage sources with far better characteristics, as you will learn when we come to the subject of feedback; this is a major topic of Chapter 9. Except in the important class of devices intended for portability, the use of batteries in electronic devices is rare.

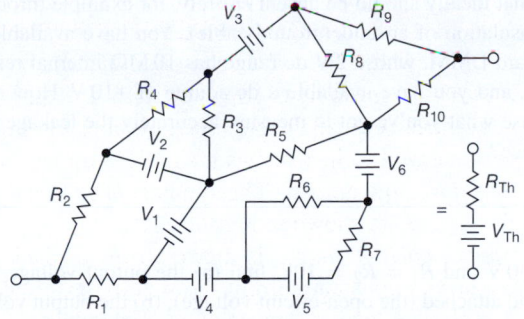

Figure 1.11. The Thévenin equivalent circuit.

1.2.5 Thévenin equivalent circuit

Thévenin's theorem states[12] that any two-terminal network of resistors and voltage sources is equivalent to a single

---

[12] We provide a proof, for those who are interested, in Appendix D.

resistor R in series with a single voltage source V. This is remarkable. Any mess of batteries and resistors can be mimicked with one battery and one resistor (Figure 1.11). (Incidentally, there's another theorem, Norton's theorem, that says you can do the same thing with a current source in parallel with a resistor.)

How do you figure out the Thévenin equivalent R_{Th} and V_{Th} for a given circuit? Easy! V_{Th} is the open-circuit voltage of the Thévenin equivalent circuit; so if the two circuits behave identically, it must also be the open-circuit voltage of the given circuit (which you get by calculation, if you know what the circuit is, or by measurement, if you don't). Then you find R_{Th} by noting that the short-circuit current of the equivalent circuit is V_{Th}/R_{Th}. In other words,

$$V_{Th} = V \text{ (open circuit)},$$
$$R_{Th} = \frac{V \text{ (open circuit)}}{I \text{ (short circuit)}}. \qquad (1.7)$$

Let's apply this method to the voltage divider, which must have a Thévenin equivalent:

1. The open-circuit voltage is

$$V = V_{in}\frac{R_2}{R_1+R_2}.$$

2. The short-circuit current is

$$V_{in}/R_1.$$

So the Thévenin equivalent circuit is a voltage source,

$$V_{Th} = V_{in}\frac{R_2}{R_1+R_2}, \qquad (1.8)$$

in series with a resistor,

$$R_{Th} = \frac{R_1R_2}{R_1+R_2}. \qquad (1.9)$$

(It is not a coincidence that this happens to be the parallel resistance of R_1 and R_2. The reason will become clear later.)

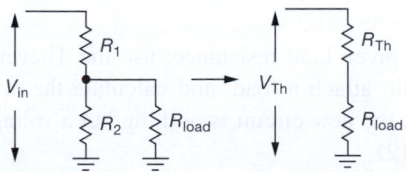

Figure 1.12. Thévenin equivalent of a voltage divider.

From this example it is easy to see that a voltage divider is not a very good battery, in the sense that its output voltage drops severely when a load is attached. As an example, consider Exercise 1.10. You now know everything you need to know to calculate exactly how much the output will

MULTIMETERS

There are numerous instruments that let you measure voltages and currents in a circuit. The oscilloscope is the most versatile; it lets you "see" voltages versus time at one or more points in a circuit. Logic probes and logic analyzers are special-purpose instruments for troubleshooting digital circuits. The simple multimeter provides a good way to measure voltage, current, and resistance, often with good precision; however, it responds slowly, and thus it cannot replace the oscilloscope where changing voltages are of interest. Multimeters are of two varieties: those that indicate measurements on a conventional scale with a moving pointer, and those that use a digital display.

The traditional (and now largely obsolete) VOM (volt-ohm-milliammeter) multimeter uses a meter movement that measures current (typically $50\,\mu A$ full scale). (See a less-design-oriented electronics book for pretty pictures of the innards of meter movements; for our purposes, it suffices to say that it uses coils and magnets.) To measure *voltage*, the VOM puts a resistor in series with the basic movement. For instance, one kind of VOM will generate a 1 V (full-scale) range by putting a 20k resistor in series with the standard $50\,\mu A$ movement; higher voltage ranges use correspondingly larger resistors. Such a VOM is specified as $20,000\,\Omega/V$, meaning that it looks like a resistor whose value is 20k multiplied by the full-scale voltage of the particular range selected. Full scale on any voltage range is 1/20,000 amps, or $50\,\mu A$. It should be clear that one of these voltmeters disturbs a circuit less on a higher range, since it looks like a higher resistance (think of the voltmeter as the lower leg of a voltage divider, with the Thévenin resistance of the circuit you are measuring as the upper resistor). Ideally, a voltmeter should have infinite input resistance.

Most contemporary multimeters use electronic amplification and have an input resistance of $10\,M\Omega$ to $1000\,M\Omega$ when measuring voltage; they display their results digitally, and are known collectively as digital multimeters (DMMs). A word of caution: sometimes the input resistance of these meters is very high on the most sensitive ranges, dropping to a lower resistance for the higher ranges. For instance, you might typically have an input resistance of $10^9\,\Omega$ on the 0.2 V and 2 V ranges, and $10^7\,\Omega$ on all higher ranges. Read the specifications carefully! However, for most circuit measurements these high input resistances will produce negligible loading effects. In any case, it is easy to calculate how serious the effect is by using the voltage-divider equation. Typically, multimeters provide voltage ranges from a volt (or less) to a kilovolt (or more), full scale.

A multimeter usually includes *current*-measuring capability, with a set of switchable ranges. Ideally, a current-measuring meter should have zero resistance[13] in order not to disturb the circuit under test, since it must be put in series with the circuit. In practice, you tolerate a few tenths of a volt drop (sometimes called "voltage burden") with both VOMs and digital multimeters. For either kind of meter, selecting a current range puts a small resistor across the meter's input terminals, typically of resistance value to create a voltage drop of 0.1 V to 0.25 V for the chosen full-scale current; the voltage drop is then converted to a corresponding current indication.[14] Typically, multimeters provide current ranges from $50\,\mu A$ (or less) to an amp (or more), full scale.

Multimeters also have one or more batteries in them to power the resistance measurement. By supplying a small current and measuring the voltage drop, they measure resistance, with several ranges to cover values from $1\,\Omega$ (or less) to $10\,M\Omega$ (or more).

Important: don't try to measure "the current of a voltage source," by sticking the meter across the wall plug; the same applies for ohms. This is a leading cause of blown-out meters.

Exercise 1.7. What will a $20,000\,\Omega/V$ meter read, on its 1 V scale, when attached to a 1 V source with an internal resistance of 10k? What will it read when attached to a 10k–10k voltage divider driven by a "stiff" (zero source resistance) 1 V source?

Exercise 1.8. A $50\,\mu A$ meter movement has an internal resistance of 5k. What shunt resistance is needed to convert it to a 0–1 A meter? What series resistance will convert it to a 0–10 V meter?

Exercise 1.9. The very high internal resistance of *digital* multimeters, in their voltage-measuring ranges, can be used to measure extremely low *currents* (even though the DMM may not offer a low current range explicitly). Suppose, for example, you want to measure the small current that flows through a $1000\,M\Omega$ "leakage" resistance (that term is used to describe a small current that ideally should be absent entirely, for example through the insulation of an underground cable). You have available a standard DMM, whose 2 V dc range has $10\,M\Omega$ internal resistance, and you have available a dc source of +10 V. How can you use what you've got to measure accurately the leakage resistance?

---

drop for a given load resistance: use the Thévenin equivalent circuit, attach a load, and calculate the new output, noting that the new circuit is nothing but a voltage divider (Figure 1.12).

Exercise 1.10. For the circuit shown in Figure 1.12, with

$V_{in}=30\,V$ and $R_1 = R_2 = 10k$, find (a) the output voltage with no load attached (the open-circuit voltage); (b) the output voltage with a 10k load (treat as a voltage divider, with R_2 and R_{load} combined into a single resistor); (c) the Thévenin equivalent circuit; (d) the same as in part (b), but using the Thévenin equivalent circuit [again, you wind up with a voltage divider; the answer should agree with the result in part (b)]; (e) the power dissipated in each of the resistors.

A. Equivalent source resistance and circuit loading

As we have just seen, a voltage divider powered from some fixed voltage is equivalent to some smaller voltage source

---

[13] This is the opposite of an ideal voltage-measuring meter, which should present infinite resistance across its input terminals.

[14] A special class of current meters known as *electrometers* operate with very small voltage burdens (as little at 0.1 mV) by using the technique of feedback, something we'll learn about in Chapters 2 and 4.

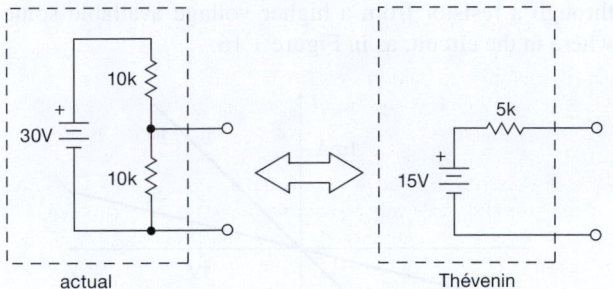

Figure 1.13. Voltage divider example.

in series with a resistor. For example, the output terminals of a 10k–10k voltage divider driven by a perfect 30 volt battery are precisely equivalent to a perfect 15 volt battery in series with a 5k resistor (Figure 1.13). Attaching a load resistor causes the voltage divider's output to drop, owing to the finite *source resistance* (Thévenin equivalent resistance of the voltage divider output, viewed as a source of voltage). This is often undesirable. One solution to the problem of making a stiff voltage source ("stiff" is used in this context to describe something that doesn't bend under load) might be to use much smaller resistors in a voltage divider. Occasionally this brute-force approach is useful. However, it is usually best to construct a voltage source, or power supply, as it's commonly called, using active components like transistors or operational amplifiers, which we will treat in Chapters 2–4. In this way you can easily make a voltage source with internal (Thévenin equivalent) resistance as small as milliohms (thousandths of an ohm), without the large currents and dissipation of power characteristic of a low-resistance voltage divider delivering the same performance. In addition, with an active power supply it is easy to make the output voltage adjustable. These topics are treated extensively in Chapter 9.

The concept of equivalent internal resistance applies to all sorts of sources, not just batteries and voltage dividers. Signal sources (e.g., oscillators, amplifiers, and sensing devices) all have an equivalent internal resistance. Attaching a load whose resistance is less than or even comparable to the internal resistance will reduce the output considerably. This undesirable reduction of the open-circuit voltage (or signal) by the load is called "circuit loading." Therefore you should strive to make $R_{\text{load}} \gg R_{\text{internal}}$, because a high-resistance load has little attenuating effect on the source (Figure 1.14).[15] We will see numerous circuit

examples in the chapters ahead. This high-resistance condition ideally characterizes measuring instruments such as voltmeters and oscilloscopes.

A word on language: you frequently hear things like "the resistance looking into the voltage divider" or "the output sees a load of so-and-so many ohms," as if circuits had eyes. It's OK (in fact, it's a rather good way of keeping straight which resistance you're talking about) to say what part of the circuit is doing the "looking."[16]

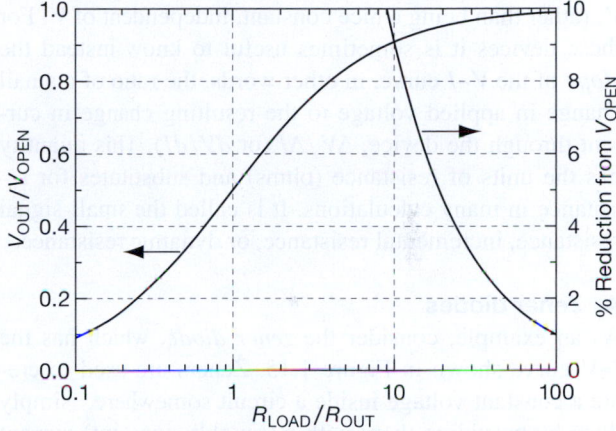

Figure 1.14. To minimize the attenuation of a signal source below its open-circuit voltage, keep the load resistance large compared with the output resistance.

B. Power transfer

Here is an interesting problem: what load resistance will result in maximum power being transferred to the load for a given source resistance? (The terms *source resistance, internal resistance,* and *Thévenin equivalent resistance* all mean the same thing.) It is easy to see that either $R_{\text{load}}=0$ or $R_{\text{load}}=\infty$ results in zero power transferred, because $R_{\text{load}}=0$ means that $V_{\text{load}}=0$ and $I_{\text{load}}=V_{\text{source}}/R_{\text{source}}$, so that $P_{\text{load}}=V_{\text{load}}I_{\text{load}}=0$. But $R_{\text{load}}=\infty$ means that $V_{\text{load}}=V_{\text{source}}$ and $I_{\text{load}}=0$, so that again $P_{\text{load}}=0$. There has to be a maximum in between.

Exercise 1.11. Show that $R_{\text{load}} = R_{\text{source}}$ maximizes the power in the load for a given source resistance. Note: skip this exercise if you don't know calculus, and take it on faith that the answer is true.

---

[15] There are two important exceptions to this general principle: (1) a current source has a high (ideally infinite) internal resistance and should drive a load of relatively low load resistance; (2) when dealing with ra-

dio frequencies and transmission lines, you must "match impedances" (i.e., set $R_{\text{load}}=R_{\text{internal}}$) in order to prevent reflection and loss of power. See Appendix H on transmission lines.

[16] The urge to anthropomorphize runs deep in the engineering and scientific community, despite warnings like "don't anthropomorphize computers ... they don't like it."

Lest this example leave the wrong impression, we would like to emphasize again that circuits are ordinarily designed so that the load resistance is much greater than the source resistance of the signal that drives the load.

1.2.6 Small-signal resistance

We often deal with electronic devices for which I is not proportional to V; in such cases there's not much point in talking about resistance, since the ratio V/I will depend on V, rather than being a nice constant, independent of V. For these devices it is sometimes useful to know instead the *slope* of the V–I curve, in other words, the ratio of a small change in applied voltage to the resulting change in current through the device, $\Delta V/\Delta I$ (or dV/dI). This quantity has the units of resistance (ohms) and substitutes for resistance in many calculations. It is called the small-signal resistance, incremental resistance, or dynamic resistance.

A. Zener diodes

As an example, consider the *zener diode*, which has the I–V curve shown in Figure 1.15. Zeners are used to create a constant voltage inside a circuit somewhere, simply done by providing them with a (roughly constant) current derived from a higher voltage within the circuit.[17] For example, the zener diode in Figure 1.15 will convert an applied current in the range shown to a corresponding (but fractionally narrower) range of voltages. It is important to know how the resulting zener voltage will change with applied current; this is a measure of its "regulation" against changes in the driving current provided to it. Included in the specifications of a zener will be its dynamic resistance, given at a certain current. For example, a zener might have a dynamic resistance of $10\,\Omega$ at $10\,\mathrm{mA}$, at its specified zener voltage of $5\,\mathrm{V}$. Using the definition of dynamic resistance, we find that a 10% change in applied current will therefore result in a change in voltage of

$$\Delta V = R_{\mathrm{dyn}}\Delta I = 10 \times 0.1 \times 0.01 = 10\,\mathrm{mV}$$

or

$$\Delta V/V = 0.002 = 0.2\%,$$

thus demonstrating good voltage-regulating ability. In this sort of application you frequently get the zener current

---

[17] Zeners belong to the more general class of *diodes* and *rectifiers*, important devices that we'll see later in the chapter (§1.6), and indeed throughout the book. The ideal diode (or rectifier) acts as a perfect conductor for current flow in one direction, and a perfect insulator for current flow in the reverse direction; it is a "one-way valve" for current.

through a resistor from a higher voltage available somewhere in the circuit, as in Figure 1.16.

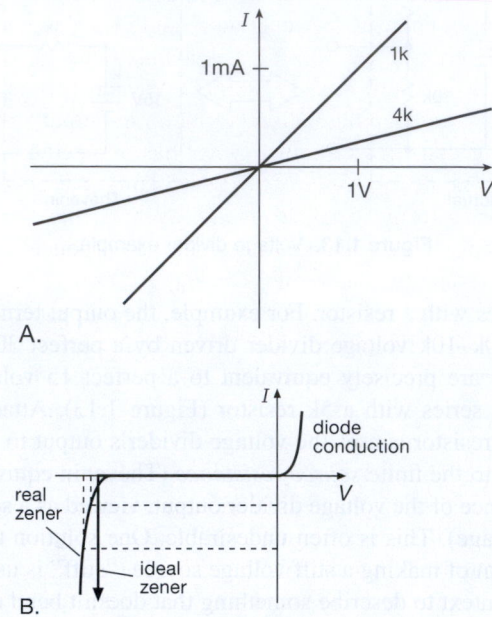

Figure 1.15. *I*–*V* curves: A. Resistor (linear). B. Zener diode (nonlinear).

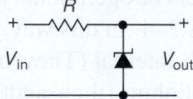

Figure 1.16. Zener regulator.

Then,

$$I = \frac{V_{\mathrm{in}} - V_{\mathrm{out}}}{R}$$

and

$$\Delta I = \frac{\Delta V_{\mathrm{in}} - \Delta V_{\mathrm{out}}}{R},$$

so

$$\Delta V_{\mathrm{out}} = R_{\mathrm{dyn}}\Delta I = \frac{R_{\mathrm{dyn}}}{R}(\Delta V_{\mathrm{in}} - \Delta V_{\mathrm{out}})$$

and finally

$$\Delta V_{\mathrm{out}} = \frac{R_{\mathrm{dyn}}}{R + R_{\mathrm{dyn}}}\Delta V_{\mathrm{in}}.$$

Aha – the voltage-divider equation, again! Thus, for *changes* in voltage, the circuit behaves like a voltage divider, with the zener replaced by a resistor equal to its dynamic resistance at the operating current. This is the utility of incremental resistance. For instance, suppose in the preceding circuit we have an input voltage ranging between 15 and 20 V, and we use a 1N4733 (5.1 V, 1W zener diode) in order to generate a stable 5.1 V power supply. We choose $R = 300\,\Omega$, for a maximum zener current of 50 mA: $(20\,V - 5.1\,V)/300\,\Omega$. We can now estimate the output-voltage regulation (variation in output voltage), knowing that this particular zener has a specified maximum dynamic resistance of $7.0\,\Omega$ at 50 mA. The zener current varies from 50 mA to 33 mA over the input-voltage range; this 17 mA change in current then produces a voltage change at the output of $\Delta V = R_{\mathrm{dyn}}\Delta I$, or 0.12 V.

It's a useful fact, when dealing with zener diodes, that the dynamic resistance of a zener diode varies roughly in inverse proportion to current. It's worth knowing, also, that there are ICs designed to substitute for zener diodes; these "two-terminal voltage references" have superior performance – much lower dynamic resistance (less than $1\,\Omega$, even at currents as small as 0.1 mA; that's a thousand times better than the zener we just used), and excellent temperature stability (better than 0.01%/°C). We will see more of zeners and voltage references in §§2.2.4 and 9.10.

In real life, a zener will provide better regulation if driven by a current source, which has, by definition, $R_{\mathrm{incr}} = \infty$ (the same current, regardless of voltage). But current sources are more complex, and therefore in practice we often resort to the humble resistor. When thinking about zeners, it's worth remembering that low-voltage units (e.g., 3.3 V) behave rather poorly, in terms of constancy of voltage versus current (Figure 1.17); if you think you need a low voltage zener, use a two-terminal reference instead (§9.10).

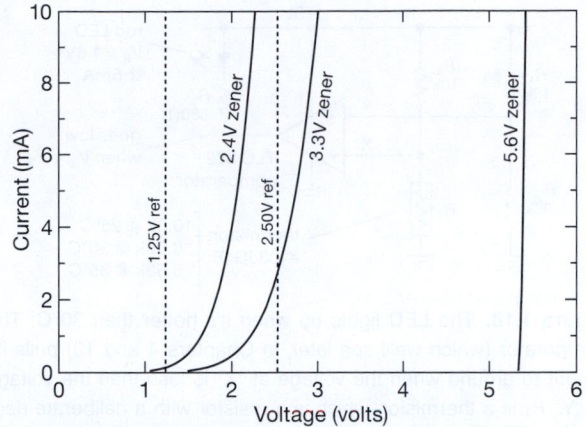

Figure 1.17. Low-voltage zeners are pretty dismal, as seen in these measured *I* vs. *V* curves (for three members of the 1N5221–67 series), particularly in contrast to the excellent measured performance of a pair of "IC voltage references" (LM385Z-1.2 and LM385Z-2.5, see §9.10 and Table 9.7). However, zener diodes in the neighborhood of 6 V (such as the 5.6 V 1N5232B or 6.2 V 1N5234B) exhibit admirably steep curves, and are useful parts.

1.2.7 An example: "It's too hot!"

Some people like to turn the thermostat way up, annoying other people who like their houses cool. Here's a little gadget (Figure 1.18) that lets folks of the latter persuasion know when to complain – it lights up a red light-emitting diode (LED) indicator when the room is warmer than 30°C (86°F). It also shows how to use the humble voltage divider (and even humbler Ohm's law), and how to deal with an LED, which behaves like a zener diode (and is sometimes used as such).

The triangular symbol is a *comparator*, a handy device (discussed in §12.3) that switches its output according to the relative voltages at its two input terminals. The temperature sensing device is R_4, which decreases in resistance by about 4%/°C, and which is 10kΩ at 25°C. So we've made

it the lower leg of a voltage divider ($R_3 R_4$), whose output is compared with the temperature-insensitive divider $R_1 R_2$. When it's hotter than 30°C, point "X" is at a lower voltage than point "Y," so the comparator pulls its output to ground.

At the output there's an LED, which behaves electrically like a 1.6 V zener diode; and when current is flowing, it lights up. Its lower terminal is then at $5\,V - 1.6\,V$, or +3.4 V. So we've added a series resistor, sized to allow 5 mA when the comparator output is at ground: $R_5 = 3.4\,V/5\,mA$, or 680 Ω.

If you wanted to, you could make the setpoint adjustable by replacing R_2 with a 5k pot in series with a 5k fixed resistor. We'll see later that it's also a good idea to add some *hysteresis*, to encourage the comparator to be decisive. Note that this circuit is insensitive to the exact power-supply voltage because it compares *ratios*. Ratiometric techniques are good; we'll see them again later.

1.3 Signals

A later section in this chapter will deal with capacitors, devices whose properties depend on the way the voltages and currents in a circuit are *changing*. Our analysis of dc circuits so far (Ohm's law, Thévenin equivalent circuits, etc.) still holds, even if the voltages and currents are changing in time. But for a proper understanding of alternating-current (ac) circuits, it is useful to have in mind certain common

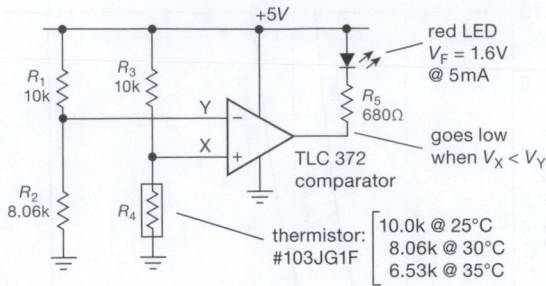

Figure 1.18. The LED lights up when it's hotter than 30°C. The comparator (which we'll see later, in Chapters 4 and 12) pulls its output to ground when the voltage at "X" is less than the voltage at "Y." R_4 is a thermistor, which is a resistor with a deliberate negative temperature coefficient; that is, its resistance decreases with increasing temperature – by about 4%/°C.

types of *signals*, voltages that change in time in a particular way.

1.3.1 Sinusoidal signals

Sinusoidal signals are the most popular signals around; they're what you get out of the wall plug. If someone says something like "take a $10\,\mu$V signal at 1 MHz," they mean a sinewave. Mathematically, what you have is a voltage described by

$$V = A \sin 2\pi f t \qquad (1.10)$$

where A is called the amplitude and f is the frequency in hertz (cycles per second). A sinewave looks like the wave shown in Figure 1.19. Sometimes it is important to know the value of the signal at some arbitrary time $t = 0$, in which case you may see a *phase* ϕ in the expression:

$$V = A \sin(2\pi f t + \phi).$$

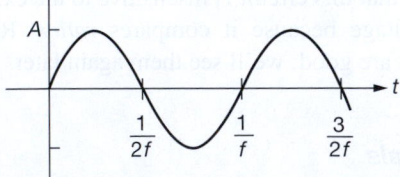

Figure 1.19. Sinewave of amplitude A and frequency f.

The other variation on this simple theme is the use of *angular frequency*, which looks like this:

$$V = A \sin \omega t.$$

Here ω is the angular frequency, measured in radians per

second. Just remember the important relation $\omega = 2\pi f$ and you won't go wrong.

The great merit of sinewaves (and the cause of their perennial popularity) is the fact that they are the solutions to certain linear differential equations that happen to describe many phenomena in nature as well as the properties of linear circuits. A linear circuit has the property that its output, when driven by the sum of two input signals, equals the sum of its individual outputs when driven by each input signal in turn; i.e., if $\mathcal{O}(A)$ represents the output when driven by signal A, then a circuit is linear if $\mathcal{O}(A + B) = \mathcal{O}(A) + \mathcal{O}(B)$. A linear circuit driven by a sinewave always responds with a sinewave, although in general the phase and amplitude are changed. No other periodic signal can make this statement. It is standard practice, in fact, to describe the behavior of a circuit by its *frequency response*, by which we mean the way the circuit alters the amplitude of an applied sinewave as a function of frequency. A stereo amplifier, for instance, should be characterized by a "flat" frequency response over the range 20 Hz to 20 kHz, at least.

The sinewave frequencies we usually deal with range from a few hertz to a few tens of megahertz. Lower frequencies, down to 0.0001 Hz or lower, can be generated with carefully built circuits, if needed. Higher frequencies, up to say 2000 MHz (2 GHz) and above, can be generated, but they require special transmission-line techniques. Above that, you're dealing with microwaves, for which conventional wired circuits with lumped-circuit elements become impractical, and exotic waveguides or "striplines" are used instead.

1.3.2 Signal amplitudes and decibels

In addition to its amplitude, there are several other ways to characterize the magnitude of a sinewave or any other signal. You sometimes see it specified by *peak-to-peak amplitude* (pp amplitude), which is just what you would guess, namely, twice the amplitude. The other method is to give the *root-mean-square amplitude* (rms amplitude), which is $V_{\text{rms}} = (1/\sqrt{2})A = 0.707A$ (this is for sinewaves only; the ratio of pp to rms will be different for other waveforms). Odd as it may seem, this is the usual method, because rms voltage is what's used to compute power. The nominal voltage across the terminals of a wall socket (in the United States) is 120 volts rms, 60 Hz. The *amplitude* is 170 volts (339 volts pp).[18]

---

[18] Occasionally you'll encounter devices (e.g., mechanical moving-pointer meters) that respond to the *average* magnitude of an ac signal.

A. Decibels

How do you compare the relative amplitudes of two signals? You could say, for instance, that signal X is twice as large as signal Y. That's fine, and useful for many purposes. But because we often deal with ratios as large as a million, it is better to use a logarithmic measure, and for this we present the decibel (it's one-tenth as large as something called a bel, which no one ever uses). By definition, the ratio of two signals, in decibels (dB), is

$$dB = 10 \log_{10} \frac{P_2}{P_1}, \qquad (1.11)$$

where P_1 and P_2 represent the *power* in the two signals. We are often dealing with signal *amplitudes*, however, in which case we can express the ratio of two signals having the same waveform as

$$dB = 20 \log_{10} \frac{A_2}{A_1}, \qquad (1.12)$$

where A_1 and A_2 are the two signal amplitudes. So, for instance, one signal of twice the amplitude of another is $+6\,$dB relative to it, since $\log_{10} 2 = 0.3010$. A signal 10 times as large is $+20\,$dB; a signal one-tenth as large is $-20\,$dB.

Although decibels are ordinarily used to specify the ratio of two signals, they are sometimes used as an absolute measure of amplitude. What is happening is that you are assuming some reference signal level and expressing any other level in decibels relative to it. There are several standard levels (which are unstated, but understood) that are used in this way; the most common references are (a) 0 dBV (1 V rms); (b) 0 dBm (the voltage corresponding to 1 mW into some assumed load impedance, which for radiofrequencies is usually $50\,\Omega$, but for audio is often $600\,\Omega$; the corresponding 0 dBm amplitudes, when loaded by those impedances, are then 0.22 V rms and 0.78 V rms); and (c) the small noise voltage generated by a resistor at room temperature (this surprising fact is discussed in §8.1.1). In addition to these, there are reference amplitudes used for measurements in other fields of engineering and science. For instance, in acoustics, 0 dB SPL (sound pressure level) is a wave whose rms pressure is $20\,\mu$Pa (that's 2×10^{-10} atm); in audio communications, levels can be stated in dBrnC (relative noise reference weighted in frequency by "curve C"). When stating amplitudes this way, it is best to be specific about the 0 dB reference amplitude;

say something like "an amplitude of 27 decibels relative to 1 V rms," or abbreviate "27 dB re 1 V rms," or define a term like "dBV."[19]

Exercise 1.12. Determine the voltage and power ratios for a pair of signals with the following decibel ratios: (a) 3 dB, (b) 6 dB, (c) 10 dB, (d) 20 dB.

Exercise 1.13. We might call this amusing exercise "Desert Island dBs": in the table below we've started entering some values for power ratios corresponding to the first dozen integral dBs, using the results for parts (a) and (c) of the last exercise. Your job is to complete the table, without recourse to a calculator. A possibly helpful hint: starting at 10 dB, go down the table in steps of 3 dB, then up in a step of 10 dB, then down again. Finally, get rid of yucky numbers like 3.125 (and its near relatives) by noticing that it's charmingly close to π.

dB	ratio (P/P_0)
0	1
1	
2	
3	2
4	
5	
6	4
7	
8	
9	8
10	10
11	

1.3.3 Other signals

A. Ramp

The ramp is a signal that looks like the one shown in Figure 1.20A. It is simply a voltage rising (or falling) at a constant rate. That can't go on forever, of course, even in science fiction movies. It is sometimes approximated by a finite ramp (Figure 1.20B) or by a periodic ramp (known as a *sawtooth*, Figure 1.20C).

B. Triangle

The triangle wave is a close cousin of the ramp; it is simply a symmetrical ramp (Figure 1.21).

C. Noise

Signals of interest are often mixed with *noise*; this is a catch-all phrase that usually applies to random noise of thermal origin. Noise voltages can be specified by their

---

For a sinewave the relationship is $V_{avg} = V_{rms}/1.11$. However, such meters are usually calibrated so that they indicate the rms sinewave amplitude. For signals other than sinewaves their indication is in error; be sure to use a "true rms" meter if you want the right answer.

---

[19] One of the authors, when asked by his nontechnical spouse how much we spent on that big plasma screen, replied "36 dB\$."

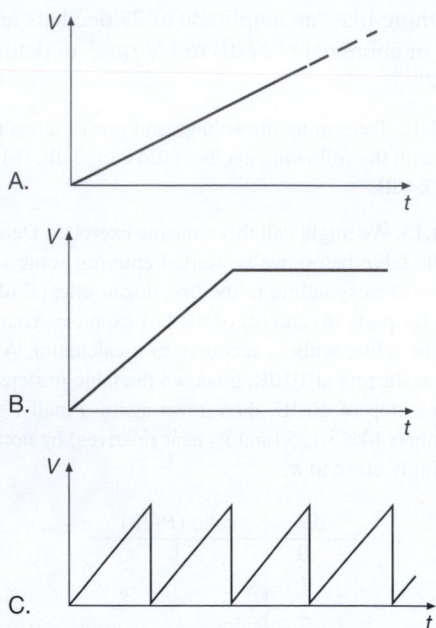

Figure 1.20. A: Voltage-ramp waveform. B: Ramp with limit. C: Sawtooth wave.

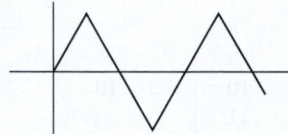

Figure 1.21. Triangle wave.

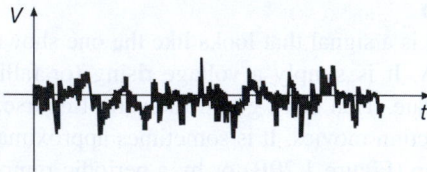

Figure 1.22. Noise.

frequency spectrum (power per hertz) or by their amplitude distribution. One of the most common kind of noise is *band-limited white Gaussian noise*, which means a signal with equal power per hertz in some band of frequencies and that exhibits a Gaussian (bell-shaped) distribution of amplitudes when many instantaneous measurements of its amplitude are made. This kind of noise is generated by a resistor (Johnson noise or Nyquist noise), and it plagues sensitive measurements of all kinds. On an oscilloscope it appears as shown in Figure 1.22. We will discuss noise and low-noise techniques in considerable detail in Chapter 8.

D. Square wave

A square wave is a signal that varies in time as shown in Figure 1.23. Like the sinewave, it is characterized by amplitude and frequency (and perhaps phase). A linear circuit driven by a square wave rarely responds with a square wave. For a square wave, the peak amplitude and the rms amplitude are the same.

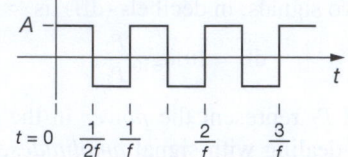

Figure 1.23. Square wave.

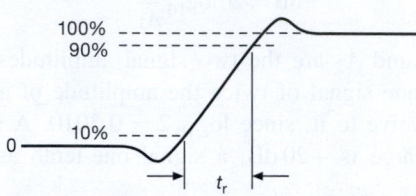

Figure 1.24. Rise time of a step waveform.

The edges of a square wave are not perfectly square; in typical electronic circuits the *rise time* t_r ranges from a few nanoseconds to a few microseconds. Figure 1.24 shows the sort of thing usually seen. The rise time is conventionally defined as the time required for the signal to go from 10% to 90% of its total transition.

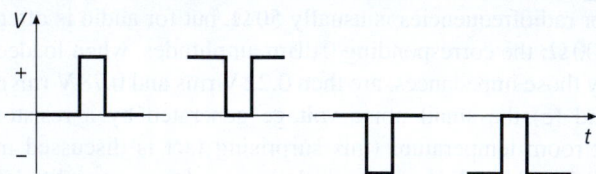

Figure 1.25. Positive- and negative-going pulses of both polarities.

E. Pulses

A pulse is a signal that looks like the objects shown in Figure 1.25. It is defined by amplitude and pulse width. You can generate a train of periodic (equally spaced) pulses, in which case you can talk about the frequency, or pulse repetition rate, and the "duty cycle," the ratio of pulse width to repetition period (duty cycle ranges from zero to 100%). Pulses can have positive or negative polarity; in addition, they can be "positive-going" or "negative-going." For instance, the second pulse in Figure 1.25 is a negative-going pulse of positive polarity.

F. Steps and spikes

Steps and spikes are signals that are talked about a lot but are not so often used. They provide a nice way of describing what happens in a circuit. If you could draw them, they would look something like the example in Figure 1.26. The step function is part of a square wave; the spike is simply a jump of vanishingly short duration.

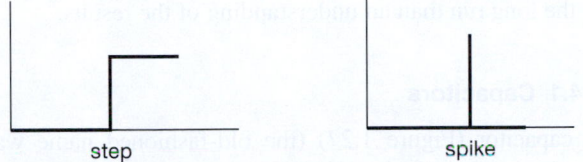

    step   spike

Figure 1.26. Steps and spikes.

1.3.4 Logic levels

Pulses and square waves are used extensively in digital electronics, in which predefined voltage levels represent one of two possible states present at any point in the circuit. These states are called simply HIGH and LOW, and correspond to the 1 (true) and 0 (false) states of Boolean logic (the algebra that describes such two-state systems).

Precise voltages are not necessary in digital electronics. You need only to distinguish which of the two possible states is present. Each digital logic family therefore specifies legal HIGH and LOW states. For example, the "74LVC" digital logic family runs from a single +3.3 V supply, with output levels that are typically 0 V (LOW) and 3.3 V (HIGH), and an input decision threshold of 1.5 V. However, actual outputs can be as much as 0.4 V away from ground or from +3.3 V without malfunction. We'll have much more to say about logic levels in Chapters 10 through 12.

1.3.5 Signal sources

Often the source of a signal is some part of the circuit you are working on. But for test purposes a flexible signal source is invaluable. They come in three flavors: signal generators, pulse generators, and function generators.

A. Signal generators

Signal generators are sinewave oscillators, usually equipped to give a wide range of frequency coverage, with provision for precise control of amplitude (using a resistive divider network called an *attenuator*). Some units let you *modulate* (i.e., vary in time) the output amplitude ("AM" for "amplitude modulated") or frequency ("FM" for "frequency modulated"). A variation on this theme is the *sweep generator*, a signal generator that can sweep its output frequency repeatedly over some range. These are handy for testing circuits whose properties vary with frequency in a particular way, e.g., "tuned circuits" or filters. Nowadays these devices, as well as most test instruments, are available in configurations that allow you to program the frequency, amplitude, etc., from a computer or other digital instrument.

For many signal generators the signal source is a *frequency synthesizer*, a device that generates sinewaves whose frequencies can be set precisely. The frequency is set digitally, often to eight significant figures or more, and is internally synthesized from a precise standard (a stand-alone quartz-crystal oscillator or rubidium frequency standard, or a GPS-derived oscillator) by digital methods we will discuss later (§13.13.6). Typical of synthesizers is the programmable SG384 from Stanford Research Systems, with a frequency range of 1 μHz to 4 GHz, an amplitude range of −110 dBm to +16.5 dBm (0.7 μV to 1.5 V, rms), and various modulation modes such as AM, FM, and ΦM; it costs about $4,600. You can get synthesized sweep generators, and you can get synthesizers that produce other waveforms (see *Function Generators*, below). If your requirement is for no-nonsense accurate frequency generation, you can't beat a synthesizer.

B. Pulse generators

Pulse generators make only pulses, but what pulses! Pulse width, repetition rate, amplitude, polarity, rise time, etc., may all be adjustable. The fastest ones go up to gigahertz pulse rates. In addition, many units allow you to generate pulse pairs, with settable spacing and repetition rate, or even programmable patterns (they are sometimes called pattern generators). Most contemporary pulse generators are provided with logic-level outputs for easy connection to digital circuitry. As with signal generators, these come in the programmable variety.

C. Function generators

In many ways function generators are the most flexible signal sources of all. You can make sine, triangle, and square waves over an enormous frequency range (0.01 Hz to 30 MHz is typical), with control of amplitude and dc offset (a constant-dc voltage added to the signal). Many of them have provision for frequency sweeping, often in several modes (linear or logarithmic frequency variation versus time). They are available with pulse outputs (although not with the flexibility you get with a pulse generator), and some of them have provision for modulation.

Traditional function generators used analog circuitry, but contemporary models generally are synthesized digital function generators, exhibiting all the flexibility of a function generator along with the stability and accuracy of a frequency synthesizer. In addition, they let you program an "arbitrary" waveform, specifying the amplitude at a set of equally spaced points. An example is the Tektronix AFG3102, with a lower frequency limit of 1 *micro*hertz, which can make sine and square waves to 100 MHz, pulses and "noise" to 50 MHz, and arbitrary waveforms (up to 128k points) to 50 MHz. It has modulation (five kinds), sweep (linear and log), and burst modes (1 to 10^6 cycles), and everything is programmable, including frequency, pulse width and rise times, modulation, and amplitude (20 mV to 10 Vpp); it even includes some bizarre built-in waveforms such as $\sin(x)/x$, exponential rise and fall, Gaussian, and Lorentzian. It has two independent outputs and costs about $5k. For general use, if you can have only one signal source, the function generator is for you.

1.4 Capacitors and ac circuits

Once we enter the world of *changing* voltages and currents, or "signals," we encounter two very interesting circuit elements that are useless in purely dc circuits: capacitors and inductors. As you will see, these humble devices, combined with resistors, complete the triad of passive linear circuit elements that form the basis of nearly all circuitry.[20] Capacitors, in particular, are essential in nearly every circuit application. They are used for waveform generation, filtering, and blocking and bypass applications. They are used in integrators and differentiators. In combination with inductors, they make possible sharp filters for separating desired signals from background. You will see some of these applications as we continue with this chapter, and there will be numerous interesting examples in later chapters.

Let's proceed, then, to look at capacitors in detail. Por-

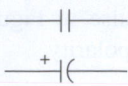

Figure 1.27. Capacitors. The curved electrode indicates the negative terminal of a polarized capacitor, or the "outer foil" of a wrapped-film capacitor.

tions of the treatment that follows are necessarily mathematical in nature; the reader with little mathematical preparation may find the math review in Appendix A helpful. In any case, an understanding of the details is less important in the long run than an understanding of the results.

1.4.1 Capacitors

A capacitor (Figure 1.27) (the old-fashioned name was *condenser*) is a device that has two wires sticking out of it and has the property

$$Q = CV. \tag{1.13}$$

Its basic form is a pair of closely-spaced metal plates, separated by some insulating material, as in the rolled-up "axial-film capacitor" of Figure 1.28. A capacitor of C farads with V volts across its terminals has Q coulombs of stored charge on one plate and $-Q$ on the other. The capacitance is proportional to the area and inversely proportional to the spacing. For the simple parallel-plate capacitor, with separation d and plate area A (and with the spacing d much less than the dimensions of the plates), the capacitance C is given by

$$C = 8.85 \times 10^{-14} \, \varepsilon A / d \text{ F}, \tag{1.14}$$

where ε is the dielectric constant of the insulator, and the dimensions are measured in centimeters. It takes a lot of area, and tiny spacing, to make the sort of capacitances commonly used in circuits.[21] For example, a pair of 1 cm² plates separated by 1 mm is a capacitor of slightly less than 10^{-12} F (a picofarad); you'd need 100,000 of them just to create the 0.1 μF capacitor of Figure 1.28 (which is nothing special; we routinely use capacitors with many microfarads of capacitance). Ordinarily you don't need to calculate capacitances, because you buy a capacitor as an electronic component.

To a first approximation, capacitors are devices that might be considered simply frequency-dependent resistors.

---

[20] Readers of the scientific journal *Nature* (London) were greeted, in 2008, with an article titled "The missing memristor found" (D. B. Strukov et al., **453**, 80, 2008), purporting to have found a heretofore missing "fourth fundamental [passive circuit] element." We are skeptical. However the controversy is ultimately resolved, it should be noted that the memristor is a *non*linear element; there are only three *linear* passive 2-terminal circuit elements.

[21] And it doesn't hurt to have a high dielectric constant, as well: air has ε=1, but plastic films have ε=2.1 (polypropylene) or 3.1 (polyester). And certain ceramics are popular among capacitor makers: ε=45 ("C0G" type) or 3000 ("X7R" type).

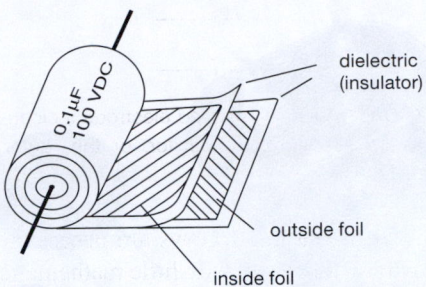

Figure 1.28. You get a lot of area by rolling up a pair of metallized plastic films. And it's great fun unrolling one of these axial-lead Mylar capacitors (ditto for the old-style golf balls with their lengthy wound-up rubber band).

They allow you to make frequency-dependent voltage dividers, for instance. For some applications (bypass, coupling) this is almost all you need to know, but for other applications (filtering, energy storage, resonant circuits) a deeper understanding is needed. For example, ideal capacitors cannot dissipate power, even though current can flow through them, because the voltage and current are 90° out of phase.

Before launching into the details of capacitors in the following dozen pages (including some necessary mathematics that describes their behavior in time and in frequency), we wish to emphasize those first two applications – bypass and coupling – because they are the most common uses of capacitors, and they are easy to understand at the simplest level. We'll see these in detail later (§§1.7.1C and 1.7.16A), but no need to wait – it's easy, and intuitive. Because a capacitor looks like an open circuit at dc, it lets you couple a varying signal while blocking its average dc level. This is a *blocking* capacitor (also called a *coupling* capacitor), as in Figure 1.93. Likewise, because a capacitor looks like a short circuit at high frequencies, it suppresses ("bypasses") signals where you don't want them, for example on the dc voltages that power your circuits, as in Figure 8.80A (where capacitors are suppressing signals on the +5 V and −5 V dc supply voltages, and also on the base terminal of transistor Q_2).[22] Demographically, these two applications account for the vast majority of capacitors that are wired into the world's circuits.

Taking the derivative of the defining equation 1.13, you get

$$I = C\frac{dV}{dt}. \tag{1.15}$$

So a capacitor is more complicated than a resistor: the current is not simply proportional to the voltage, but rather to the rate of change of voltage. If you change the voltage across a farad by 1 volt per second, you are supplying an amp. Conversely, if you supply an amp, its voltage changes by 1 volt per second. A farad is an enormous capacitance, and you usually deal in microfarads (μF), nanofarads (nF), or picofarads (pF).[23] For instance, if you supply a current of 1 mA to 1μF, the voltage will rise at 1000 volts per second. A 10 ms pulse of this current will increase the voltage across the capacitor by 10 volts (Figure 1.29).

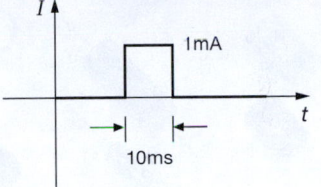

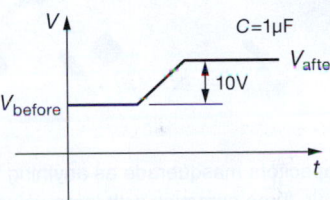

Figure 1.29. The voltage across a capacitor changes when a current flows through it.

When you charge up a capacitor, you're supplying energy. The capacitor doesn't get hot; instead, it stores the energy in its internal electric fields. It's an easy exercise to discover for yourself that the amount of stored energy in a charged capacitor is just

$$U_C = \frac{1}{2}CV^2, \tag{1.16}$$

where U_C is in joules for C in farads and V in volts. This is an important result; we'll see it often.

Exercise 1.14. Take the energy challenge: imagine charging up a capacitor of capacitance C, from 0 V to some final voltage V_f. If you do it right, the result won't depend on how you get there,

---

[22] Ironically, these essential bypass capacitors are so taken for granted that they are usually omitted from schematic diagrams (a practice we follow in this book). Don't make the mistake of omitting them also from your actual circuits!

[23] To make matters confusing to the uninitiated, the units are often omitted on capacitor values specified in schematic diagrams. You have to figure it out from the context.

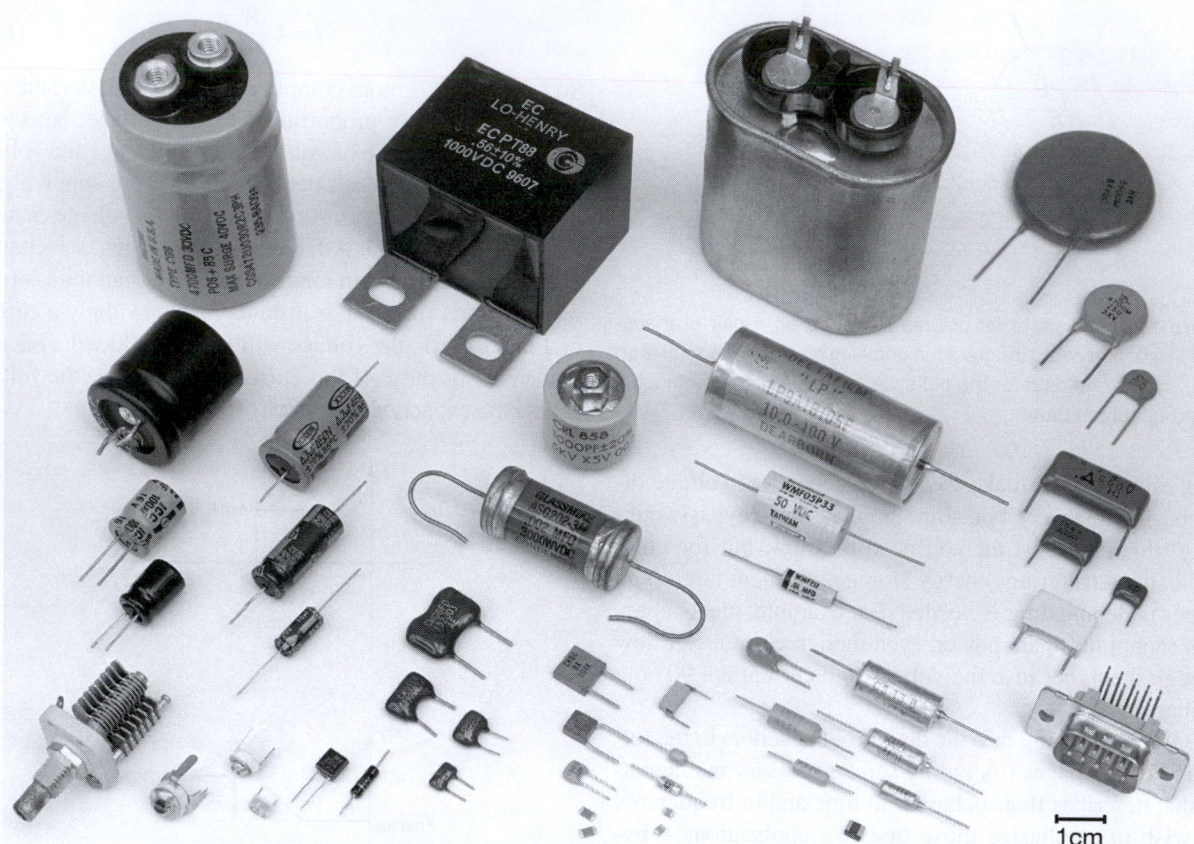

1cm

Figure 1.30. Capacitors masquerade as anything they like! Here is a representative collection. In the lower left are small-value variable capacitors (one air, three ceramic), with large-value polarized aluminum electrolytics above them (the three on the left have *radial leads*, the three on the right have *axial leads*, and the specimen with screw terminals at top is often called a *computer electrolytic*). Next in line across the top is a low-inductance film capacitor (note the wide strap terminals), then an oil-filled paper capacitor, and last, a set of disc ceramic capacitors running down the right. The four rectangular objects below are film capacitors (polyester, polycarbonate, or polypropylene). The D-subminiature connector seems misplaced – but it is a *filtered* connector, with a 1000 pF capacitor from each pin to the shell. To its left is a group of seven polarized tantalum electrolytics (five with axial leads, one radial, and one surface-mount). The three capacitors above them are axial-film capacitors. The ten capacitors at bottom center are all ceramic types (four with radial leads, two axial, and four surface-mount *chip capacitors*); above them are high-voltage capacitors – an axial-glass capacitor, and a ceramic *transmitting capacitor* with screw terminals. Finally, below them and to the left are four mica capacitors and a pair of diode-like objects known as *varactors*, which are voltage-variable capacitors made from a diode junction.

so you don't need to assume constant current charging (though you're welcome to do so). At any instant the rate of flow of energy into the capacitor is VI (joules/s); so you need to integrate $dU = VIdt$ from start to finish. Take it from there.

Capacitors come in an amazing variety of shapes and sizes (Figure 1.30 shows examples of most of them); with time, you will come to recognize their more common incarnations. For the smallest capacitances you may see examples of the basic parallel-plate (or cylindrical piston) construction. For greater capacitance, you need more area and

closer spacing; the usual approach is to plate some conductor onto a thin insulating material (the dielectric), for instance, aluminized plastic film rolled up into a small cylindrical configuration. Other popular types are thin ceramic wafers (ceramic chip capacitors), metal foils with oxide insulators (electrolytic capacitors), and metallized mica. Each of these types has unique properties; for a brief rundown, see the section on capacitors in Chapter *1x*. In general, ceramic and polyester types are used for most noncritical circuit applications; capacitors with polycarbonate, polystyrene, polypropylene, Teflon, or glass dielectric are used in demanding applications; tantalum capacitors are used where greater capacitance is needed; and aluminum electrolytics are used for power-supply filtering.

A. Capacitors in parallel and series

The capacitance of several capacitors in parallel is the sum of their individual capacitances. This is easy to see: put voltage V across the parallel combination; then

$$C_{total}V = Q_{total} = Q_1 + Q_2 + Q_3 + \cdots$$
$$= C_1V + C_2V + C_3V + \cdots$$
$$= (C_1 + C_2 + C_3 + \cdots)V$$

or

$$C_{total} = C_1 + C_2 + C_3 + \cdots. \tag{1.17}$$

For capacitors in series, the formula is like that for resistors in parallel:

$$C_{total} = \cfrac{1}{\cfrac{1}{C_1} + \cfrac{1}{C_2} + \cfrac{1}{C_3} + \cdots} \tag{1.18}$$

or (two capacitors only)

$$C_{total} = \frac{C_1 C_2}{C_1 + C_2}. $$

Exercise 1.15. Derive the formula for the capacitance of two capacitors in series. *Hint*: because there is no external connection to the point where the two capacitors are connected together, they must have equal stored charges.

The current that flows in a capacitor during charging ($I = C\,dV/dt$) has some unusual features. Unlike resistive current, it's not proportional to voltage, but rather to the rate of change (the "time derivative") of voltage. Furthermore, unlike the situation in a resistor, the power ($V \times I$) associated with capacitive current is not turned into heat, but is stored as energy in the capacitor's internal electric field. You get all that energy back when you discharge the capacitor. We'll see another way to look at these curious properties when we talk about *reactance*, beginning in §1.7.

1.4.2 *RC* circuits: *V* and *I* versus time

When dealing with ac circuits (or, in general, any circuits that have changing voltages and currents), there are two possible approaches. You can talk about V and I versus time, or you can talk about amplitude versus signal frequency. Both approaches have their merits, and you find yourself switching back and forth according to which description is most convenient in each situation. We begin our

study of ac circuits in the *time domain*. Starting with §1.7, we will tackle the *frequency domain*.

What are some of the features of circuits with capacitors? To answer this question, let's begin with the simple *RC* circuit (Figure 1.31). Application of the capacitor rules gives

$$C\frac{dV}{dt} = I = -\frac{V}{R}. \tag{1.19}$$

This is a differential equation, and its solution is

$$V = Ae^{-t/RC}. \tag{1.20}$$

So a charged capacitor placed across a resistor will discharge as in Figure 1.32. Intuition serves well here: the current that flows is (from the resistor equation) proportional to the remaining voltage; but the slope of the discharge is (from the capacitor equation) proportional to that current. So the discharge curve has to be a function whose derivative is proportional to its value, i.e., an exponential.

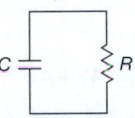

Figure 1.31. The simplest *RC* circuit.

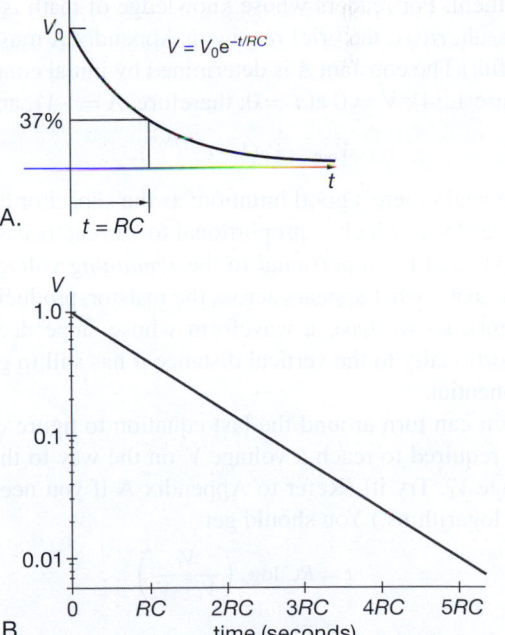

A.

B.

Figure 1.32. *RC* discharge waveform, plotted with (A) linear and (B) logarithmic voltage axes.

A. Time constant

The product RC is called the *time constant* of the circuit. For R in ohms and C in farads, the product RC is in seconds. A microfarad across 1.0k has a time constant of 1 ms; if the capacitor is initially charged to 1.0 V, the initial current is 1.0 mA.

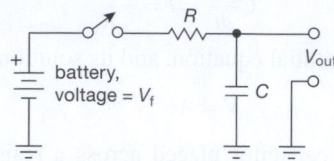

Figure 1.33. *RC* charging circuit.

Figure 1.33 shows a slightly different circuit. At time $t = 0$, someone connects the battery. The equation for the circuit is then

$$I = C\frac{dV}{dt} = \frac{V_f - V_{out}}{R},$$

with the solution

$$V_{out} = V_f + A e^{-t/RC}.$$

(Please don't worry if you can't follow the mathematics. What we are doing is getting some important results, which you should remember. Later we will use the results often, with no further need for the mathematics used to derive them. For readers whose knowledge of math is somewhat, uh, *rusty*, the brief review in Appendix A may prove helpful.) The constant A is determined by initial conditions (Figure 1.34): $V = 0$ at $t = 0$; therefore, $A = -V_f$, and

$$V_{out} = V_f(1 - e^{-t/RC}). \tag{1.21}$$

Once again there's good intuition: as the capacitor charges up, the slope (which is proportional to current, because it's a capacitor) is proportional to the *remaining* voltage (because that's what appears across the resistor, producing the current); so we have a waveform whose slope decreases proportionally to the vertical distance it has still to go – an exponential.

You can turn around the last equation to figure out the time required to reach a voltage V on the way to the final voltage V_f. Try it! (Refer to Appendix A if you need help with logarithms.) You should get

$$t = RC \log_e\left(\frac{V_f}{V_f - V}\right) \tag{1.22}$$

B. Decay to equilibrium

Eventually (when $t \gg RC$), V reaches V_f. (Presenting the "5RC rule of thumb": a capacitor charges or decays to

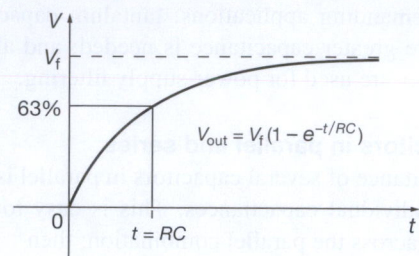

Figure 1.34. *RC* charging waveform.

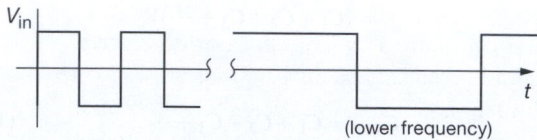

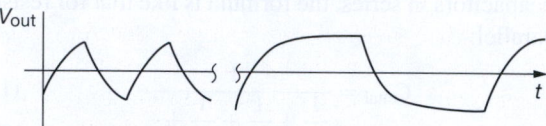

Figure 1.35. Output (lower waveforms) across a capacitor, when driven by square waves through a resistor.

within 1% of its final value in five time constants.) If we then change the battery voltage to some other value (say, 0 V), V will decay toward that new value with an exponential $e^{-t/RC}$. For example, replacing the battery's step input from 0 to $+V_f$ with a square-wave input $V_{in}(t)$ would produce the output shown in Figure 1.35.

Exercise 1.16. Show that the rise time (the time required for going from 10% to 90% of its final value) of this signal is 2.2RC.

You might ask the obvious next question: what about $V(t)$ for arbitrary $V_{in}(t)$? The solution involves an inhomogeneous differential equation and can be solved by standard methods (which are, however, beyond the scope of this book). You would find

$$V(t) = \frac{1}{RC}\int_{-\infty}^{t} V_{in}(\tau)e^{-(t-\tau)/RC}d\tau.$$

That is, the RC circuit averages past history at the input with a weighting factor of

$$e^{-\Delta t/RC}.$$

In practice, you seldom ask this question. Instead, you deal in the *frequency domain*, in which you ask how much of each frequency component present in the input gets through. We will get to this important topic soon (§1.7). Before we do, though, there are a few other interesting circuits we can analyze simply with this time-domain approach.

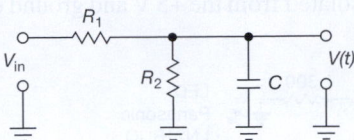

Figure 1.36. Looks complicated, but it's not! (Thévenin to the rescue.)

C. Simplification by Thévenin equivalents

We could go ahead and analyze more complicated circuits by similar methods, writing down the differential equations and trying to find solutions. For most purposes it simply isn't worth it. This is as complicated an *RC* circuit as we will need. Many other circuits can be reduced to it; take, for example, the circuit in Figure 1.36. By just using the Thévenin equivalent of the voltage divider formed by R_1 and R_2, you can find the output $V(t)$ produced by a step input for V_{in}.

Exercise 1.17. In the circuit shown in Figure 1.36, $R_1 = R_2 = 10$k, and $C = 0.1\,\mu$F. Find $V(t)$ and sketch it.

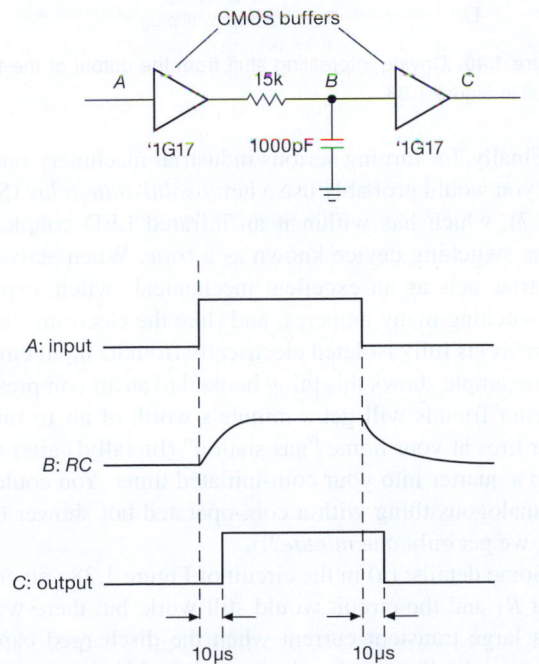

Figure 1.37. Producing a delayed digital waveform with the help of an *RC* and a pair of LVC-family logic buffers (tiny parts with a huge part number: SN74LVC1G17DCKR!).

D. A circuit example: time-delay circuit

Let's take a short detour to try out these theoretical ideas on a couple of real circuits. Textbooks usually avoid such pragmatism, especially in early chapters, but we think it's fun to apply electronics to practical applications. We'll need to introduce a few "black-box" components to get the job done, but you'll learn about them in detail later, so don't worry.

We have already mentioned logic levels, the voltages that digital circuits live on. Figure 1.37 shows an application of capacitors to produce a delayed pulse. The triangular symbols are "CMOS[24] buffers." They give a HIGH output if the input is HIGH (more than one-half the dc power-supply voltage used to power them), and vice versa. The first buffer provides a replica of the input signal, but with low source resistance, to prevent input loading by the *RC* (recall our earlier discussion of circuit loading in §1.2.5A). The *RC* output has the characteristic decays and causes the output buffer to switch 10 μs after the input transitions (an *RC* reaches 50% output after a time $t = 0.7RC$). In an actual application you would have to consider the effect of the buffer input threshold deviating from one-half the supply voltage, which would alter the delay and change the output pulse width. Such a circuit is sometimes used to delay a pulse so that something else can happen first. In designing circuits you try not too often to rely on tricks like this, but they're occasionally handy.

E. Another circuit example: "One Minute of Power"

Figure 1.38 shows another example of what can be done with simple *RC* timing circuits. The triangular symbol is a *comparator*, something we'll treat in detail later, in Chapters 4 and 10; all you need to know, for now, is that (a) it is an IC (containing a bunch of resistors and transistors), (b) it is powered from some positive dc voltage that you connect to the pin labeled "V_+," and (c) it drives its output (the wire sticking out to the right) either to V_+ or to ground, depending on whether the input labeled "+" is more or less positive than the input labeled "−," respectively. (These are called the *non-inverting* and the *inverting* inputs, respectively.) It doesn't draw any current from its inputs, but it happily drives loads that require up to 20 mA or so. And a comparator is decisive: its output is either "HIGH" (at V_+) or "LOW" (ground).

Here's how the circuit works: the voltage divider R_3R_4 holds the (−) input at 37% of the supply voltage, in this case about +1.8 V; let's call that the "reference voltage."

---

[24] Complementary metal-oxide semiconductor, the dominant form of digital logic, as we'll see from Chapter 10 onward.

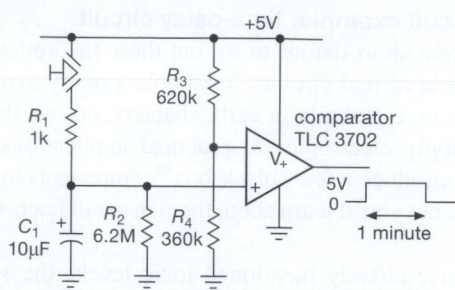

Figure 1.38. *RC* timing circuit: one push → one minute!.

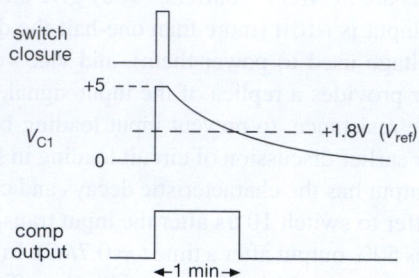

Figure 1.39. Producing a delayed digital waveform for the circuit of Figure 1.38. The voltage V_{C1} has a rise time of $R_1 C_1 \approx 10$ ms.

So if the circuit has been sitting there for a while, C_1 is fully discharged, and the comparator's output is at ground. When you push the START button momentarily, C_1 charges quickly (10 ms time constant) to +5 V, which makes the comparator's output switch to +5 V; see Figure 1.39. After the button is released, the capacitor discharges exponentially toward ground, with a time constant of $\tau = R_2 C_1$, which we've set to be 1 minute. At that time its voltage crosses the reference voltage, so the comparator's output switches rapidly back to ground. (Note that we've conveniently chosen the reference voltage to be a fraction $1/e$ of V_+, so it takes exactly one time constant τ for that to happen. For R_2 we used the closest standard value to $6\,M\Omega$; see Appendix C.) The bottom line is that the output spends 1 minute at +5 V, after the button is pushed.

We'll add a few details shortly, but first let's use the output to do some interesting things, which are shown in Figures 1.40A–D. You can make a self-stopping flashlight key fob by connecting its output to an LED; you need to put a resistor in series, to set the current (we'll say much more about this later). If you prefer to make some noise, you could connect a *piezo beeper* to beep continuously (or intermittently) for a minute (this might be an end-of-cycle signal for a clothes dryer). Another possibility is to attach a small electromechanical *relay*, which is just an electrically operated mechanical switch, to provide a pair of contacts

that can activate pretty much any load you care to switch on and off. The use of a relay has the important property that the load – the circuit being switched by the relay – is electrically isolated from the +5 V and ground of the timing circuit itself.

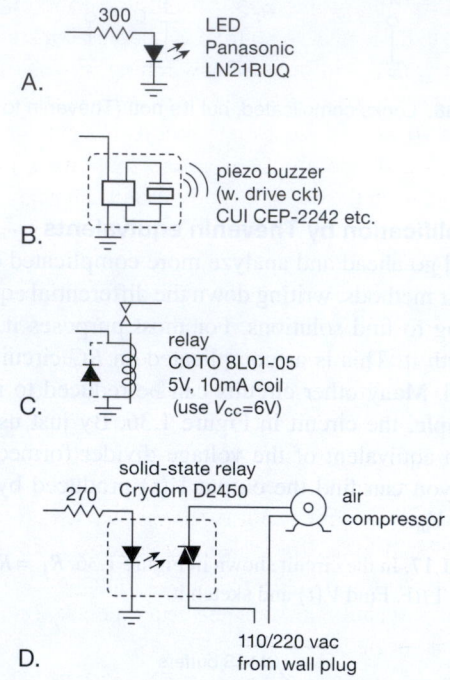

Figure 1.40. Driving interesting stuff from the output of the timer circuit in Figure 1.38.

Finally, for turning serious industrial machinery on and off, you would probably use a hefty *solid-state relay* (SSR, §12.7), which has within it an infrared LED coupled to an ac switching device known as a *triac*. When activated, the triac acts as an excellent mechanical switch, capable of switching many amperes, and (like the electromechanical relay) is fully isolated electrically from its input circuit. The example shows this thing hooked to an air compressor, so your friends will get a minute's worth of air to inflate their tires at your home "gas station" (literally!) after they drop a quarter into your coin-initiated timer. You could do an analogous thing with a coin-operated hot shower (but, hey, we get only *one minute*?!).

Some details: (a) in the circuit of Figure 1.38 you could omit R_1 and the circuit would still work, but there would be a large transient current when the discharged capacitor was initially connected across the +5 V supply (recall $I = C\, dV/dt$: here you would be trying to produce 5 V of "dV" in roughly 0 s of "dt"). By adding a series resistor R_1 you limit the peak current to a modest 5 mA while charging the capacitor fast enough ($>99\%$ in $5RC$ time constants, or 0.05 s). (b) The comparator output would likely bounce around a bit (see Figure 4.31), just as the (+) input crosses the reference voltage in its leisurely exponential promenade toward ground, owing to unavoidable bits of electrical noise. To fix this problem you usually see the circuit arranged so that some of the output is coupled back to the input in a way that reinforces the switching (this is officially called *hysteresis*, or *positive feedback*; we'll see it in Chapters 4 and 10). (c) In electronic circuits it's always a good idea to *bypass* the dc supply by connecting one or more capacitors between the dc "rail" and ground. The capacitance is noncritical – values of $0.1\,\mu\mathrm{F}$ to $10\,\mu\mathrm{F}$ are commonly used; see §1.7.16A.

Our simple examples above all involved turning some load on and off. But there are other uses for an electronic *logic* signal, like the output of the comparator, that is in one of two possible binary states, called HIGH and LOW (in this case $+5\,\mathrm{V}$ and ground), 1 and 0, or TRUE and FALSE. For example, such a signal can enable or disable the operation of some other circuit. Imagine that the opening of a car door triggers our 1 min HIGH output, which then enables a keypad to accept a security code so you can start the car. After a minute, if you haven't managed to type the magic code, it shuts off, thus ensuring a certain minimum of operator sobriety.

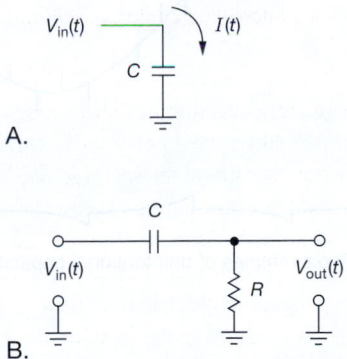

A.

B.

Figure 1.41. Differentiators. A. Perfect (except it has no output terminal). B. Approximate (but at least it has an output!).

1.4.3 Differentiators

You can make a simple circuit that differentiates an input signal; that is, $V_{\mathrm{out}} \propto dV_{\mathrm{in}}/dt$. Let's take it in two steps.

1. First look at the (impractical) circuit in Figure 1.41A: The input voltage $V_{\mathrm{in}}(t)$ produces a current through the capacitor of $I_{\mathrm{cap}} = C\,dV_{\mathrm{in}}/dt$. That's just what we want – if we

could somehow use the current through C as our "output"! But we can't.[25]

2. So we add a small resistor from the low side of the capacitor to ground, to act as a "current-sensing" resistor (Figure 1.41B). The good news is that we now have an output proportional to the current through the capacitor. The bad news is that the circuit is no longer a perfect mathematical differentiator. That's because the voltage across C (whose derivative produces the current we are sensing with R) is no longer equal to V_{in}; it now equals the difference between V_{in} and V_{out}. Here's how it goes: the voltage across C is $V_{\mathrm{in}} - V_{\mathrm{out}}$, so

$$I = C\frac{d}{dt}(V_{\mathrm{in}} - V_{\mathrm{out}}) = \frac{V_{\mathrm{out}}}{R}.$$

If we choose R and C small enough so that $dV_{\mathrm{out}}/dt \ll dV_{\mathrm{in}}/dt$, then

$$C\frac{dV_{\mathrm{in}}}{dt} \approx \frac{V_{\mathrm{out}}}{R}$$

or

$$V_{\mathrm{out}}(t) \approx RC\frac{d}{dt}V_{\mathrm{in}}(t).$$

That is, we get an output proportional to the rate of change of the input waveform.

To keep $dV_{\mathrm{out}}/dt \ll dV_{\mathrm{in}}/dt$, we make the product RC small, taking care not to "load" the input by making R too small (at the transition the change in voltage across the capacitor is zero, so R is the load seen by the input). We will have a better criterion for this when we look at things in the frequency domain (§1.7.10). If you drive this circuit with a square wave, the output will be as shown in Figure 1.42.

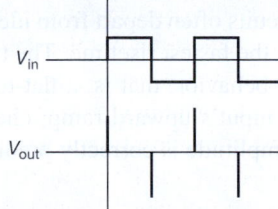

Figure 1.42. Output waveform (bottom) from differentiator driven by a square wave.

Differentiators are handy for detecting *leading edges* and *trailing edges* in pulse signals, and in digital circuitry you sometimes see things like those depicted in Figure 1.43. The RC differentiator generates spikes at the transitions of the input signal, and the output buffer converts

---

[25] Devotees of the cinema will be reminded of Dr. Strangelove's outburst: "The whole point of a doomsday machine is lost ... if you keep it *secret!*"

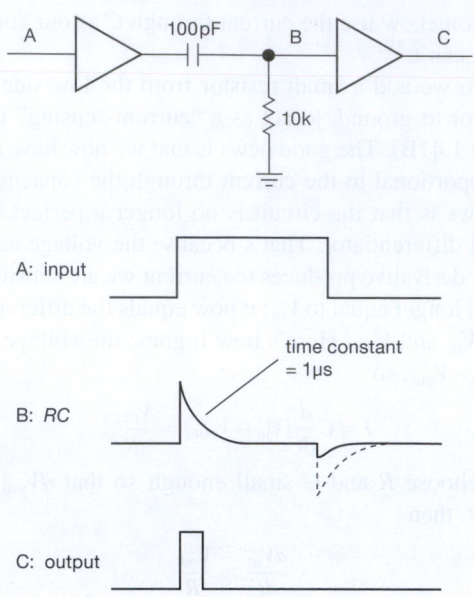

Figure 1.43. Leading-edge detector.

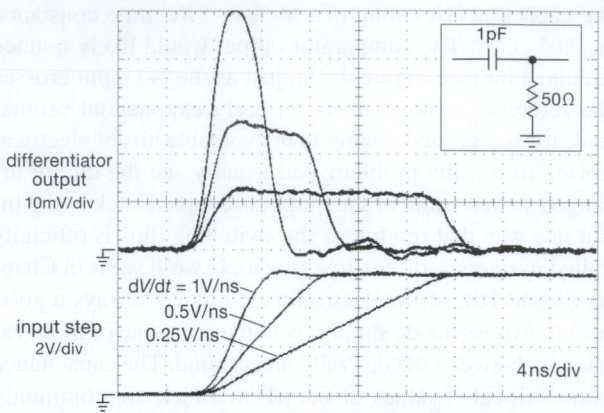

Figure 1.44. Three fast step waveforms, differentiated by the *RC* network shown. For the fastest waveform (10^9 volts per second!), imperfections in the components and measuring instruments cause deviation from the ideal.

the spikes to short square-topped pulses. In practice, the negative spike will be small because of a diode (a handy device discussed in §1.6) built into the buffer.

To inject some real-world realism here, we hooked up and made some measurements on a differentiator that we configured for high-speed signals. For this we used $C=1$ pF and $R=50\,\Omega$ (the latter is the world-wide standard for high-speed circuits, see Appendix H), we drove it with a 5 V step of settable slew rate (i.e., dV/dt). Figure 1.44 shows both input and output waveforms, for three choices of dV_{in}/dt. At these speeds (note the horizontal scale: 4 *nano*seconds per division!) circuits often depart from ideal performance, as can be seen in the fastest risetime. The two slower steps show reasonable behavior; that is, a flat-top output waveform during the input's upward ramp; check for yourself that the output amplitude is correctly predicted by the formula.

A. Unintentional capacitive coupling

Differentiators sometimes crop up unexpectedly, in situations where they're not welcome. You may see signals like those shown in Figure 1.45. The first case is caused by a square wave somewhere in the circuit coupling capacitively to the signal line you're looking at; that might indicate a missing resistor termination on your signal line. If not, you must either reduce the source resistance of the signal line or find a way to reduce capacitive coupling from the offending square wave. The second case is typical of what you might see when you look at a square wave, but have a

broken connection somewhere, usually at the scope probe. The very small capacitance of the broken connection combines with the scope input resistance to form a differentiator. Knowing that you've got a differentiated "something" can help you find the trouble and eliminate it.

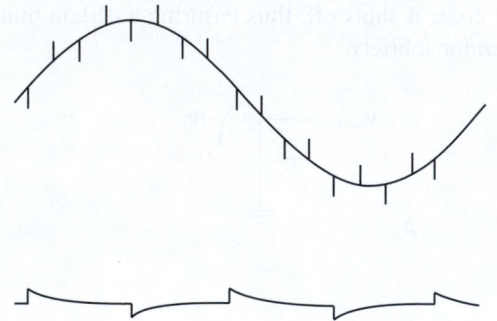

Figure 1.45. Two examples of unintentional capacitive coupling.

1.4.4 Integrators

If *RC* circuits can take derivatives, why not integrals? As before, let's take it in two steps.

1. Imagine that we have an input signal that is a time-varying *current* versus time, $I_{in}(t)$ (Figure 1.46A).[26] That input current is precisely the current through the capacitor, so $I_{in}(t)=C\,dV(t)/dt$, and therefore $V(t)=\frac{1}{C}\int I_{in}(t)\,dt$.

---

[26] We're used to thinking of signals as time-varying *voltages*; but we'll see how we can convert such signals to proportional time-varying *currents*, by using "voltage-to-current converters" (with the fancier name "transconductance amplifiers").

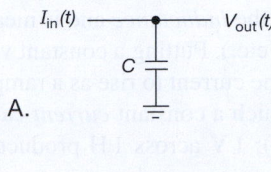

A.

B.

Figure 1.46. Integrator. A. Perfect (but requires a *current* input signal). B. Approximate (see text).

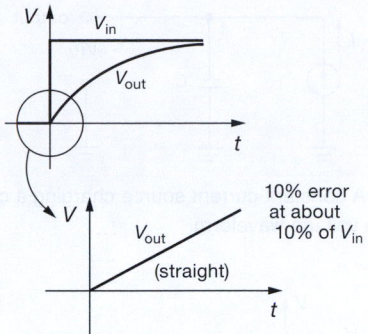

Figure 1.47. Integrator approximation is good when $V_{out} \ll V_{in}$.

That's just what we wanted! Thus a simple capacitor, with one side grounded, is an integrator, *if* we have an input signal in the form of a current $I_{in}(t)$. Most of the time we don't, though.

2. So we connect a resistor in series with the more usual input *voltage* signal $V_{in}(t)$, to convert it to a current (Figure 1.46B). The good news is that it works, sort of. The bad news is that the circuit is no longer a perfect integrator. That's because the current through C (whose integral produces the output voltage) is no longer proportional to V_{in}; it is now proportional to the difference between V_{in} and V. Here's how it goes: the voltage across R is $V_{in}-V$, so

$$I = C\frac{dV}{dt} = \frac{V_{in}-V}{R}.$$

If we manage to keep $V \ll V_{in}$, by keeping the product RC large,[27] then

$$C\frac{dV}{dt} \approx \frac{V_{in}}{R}$$

or

$$V(t) = \frac{1}{RC}\int^t V_{in}(t)\,dt + \text{constant.}$$

That is, we get an output proportional to the integral over time of the input waveform. You can see how the approximation works for a square-wave input: $V(t)$ is then the exponential charging curve we saw earlier (Figure 1.47). The first part of the exponential is a ramp, the integral of a constant; as we increase the time constant RC, we pick off a smaller part of the exponential, i.e., a better approximation to a perfect ramp.

Note that the condition $V \ll V_{in}$ is the same as saying that I is proportional to V_{in}, which was our first integrator

circuit. A large voltage across a large resistance approximates a current source and, in fact, is frequently used as one.

Later, when we get to operational amplifiers and feedback, we will be able to build integrators without the restriction $V_{out} \ll V_{in}$. They will work over large frequency and voltage ranges with negligible error.

The integrator is used extensively in analog computation. It is a useful subcircuit that finds application in control systems, feedback, analog–digital conversion, and waveform generation.

A. Ramp generators

At this point it is easy to understand how a ramp generator works. This nice circuit is extremely useful, for example in timing circuits, waveform and function generators, analog oscilloscope sweep circuits, and analog-digital conversion circuitry. The circuit uses a constant current to charge a capacitor (Figure 1.48). From the capacitor equation $I = C(dV/dt)$, you get $V(t) = (I/C)t$. The output waveform is as shown in Figure 1.49. The ramp stops when the current source "runs out of voltage," i.e., reaches the limit of its compliance. On the same figure is shown the curve for a simple RC, with the resistor tied to a voltage source equal to the compliance of the current source, and with R chosen so that the current at zero output voltage is the same as that of the current source. (Real current sources generally have output compliances limited by the power-supply voltages used in making them, so the comparison is realistic.) In the next chapter, which deals with transistors, we will design some current sources, with some refinements to follow in the chapters on operational amplifiers (op-amps) and FETs. Exciting things to look forward to!

---

[27] Just as with the differentiator, we'll have another way of framing this criterion in §1.7.10.

Exercise 1.18. A current of 1 mA charges a 1 μF capacitor. How long does it take the ramp to reach 10 volts?

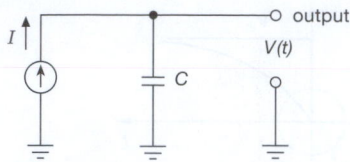

Figure 1.48. A constant-current source charging a capacitor generates a ramp voltage waveform.

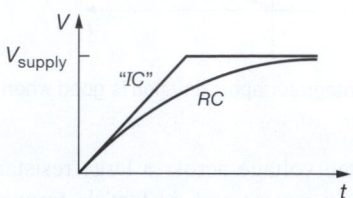

Figure 1.49. Constant-current charging (with finite compliance) versus *RC* charging.

1.4.5 Not quite perfect...

Real capacitors (the kind you can see, and touch, and pay money for) generally behave according to theory; but they have some additional "features" that can cause problems in some demanding applications. For example, all capacitors exhibit *some series resistance* (which may be a function of frequency), and *some series inductance* (see the next section), along with some frequency-dependent parallel resistance. Then there's a "memory" effect (known as *dielectric absorption*), which is rarely discussed in polite society: if you charge a capacitor up to some voltage V_0 and hold it there for a while, and then discharge it to $0\,V$, then when you remove the short across its terminals it will tend to drift back a bit toward V_0.

Don't worry about this stuff, for now. We'll treat in detail these effects (and other oddities of real-world components) in the advanced topics Chapter *1x*.

1.5 Inductors and transformers

1.5.1 Inductors

If you understand capacitors, you shouldn't have great trouble with inductors (Figure 1.50). They're closely related to capacitors: the rate of current change in an inductor is proportional to the voltage applied across it (for a capacitor it's the other way around – the rate of *voltage* change is proportional to the *current* through it). The defining equation for an inductor is

$$V = L\frac{dI}{dt}, \tag{1.23}$$

where L is called the *inductance* and is measured in henrys (or mH, μH, nH, etc.). Putting a constant voltage across an inductor causes the current to rise as a ramp (compare with a capacitor, in which a constant *current* causes the *voltage* to rise as a ramp); 1 V across 1 H produces a current that increases at 1 amp per second.

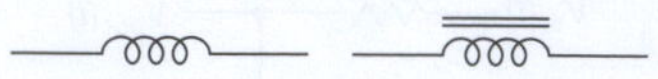

Figure 1.50. Inductors. The parallel-bar symbol represents a *core* of magnetic material.

Just as with capacitors, the energy invested in ramping up the current in an inductor is stored internally, here in the form of magnetic fields. And the analogous formula is

$$U_{\mathrm{L}} = \frac{1}{2}LI^2, \tag{1.24}$$

where U_{L} is in joules (watt seconds) for L in henrys and I in amperes. As with capacitors, this is an important result, one which lies at the core of switching power conversion (exemplified by those little black "wall-warts" that provide power to all manner of consumer electronic gadgets). We'll see lots more of this in Chapter 9.

The symbol for an inductor looks like a coil of wire; that's because, in its simplest form, that's all it is. Its somewhat peculiar behavior comes about because inductors are magnetic devices, in which two things are going on: current flowing through the coil creates a magnetic field aligned along the coil's axis; and then changes in that field produce a voltage (sometimes called "back EMF") in a way that tries to cancel out those changes (an effect known as Lenz's law). The inductance L of a coil is simply the ratio of magnetic flux passing through the coil divided by the current through the coil that produces that flux (multiplied by an overall constant). Inductance depends on the coil geometry (e.g., diameter and length) and the properties of any magnetic material (the "core") that may be used to confine the magnetic field. That's all you need to understand why the inductance of a coil of given geometry is proportional to the square of the number of turns.

Exercise 1.19. Explain why $L \propto n^2$ for an inductor consisting of a coil of n turns of wire, maintaining fixed diameter and length as n is varied.

We'll get into some more detail in the Chapter *1x*. But it's worth displaying a semi-empirical formula for the approximate inductance L of a coil of diameter d and length l, in which the n^2 dependence is on display:

$$L \approx K \frac{d^2 n^2}{18d + 40l} \quad \mu\mathrm{H},$$

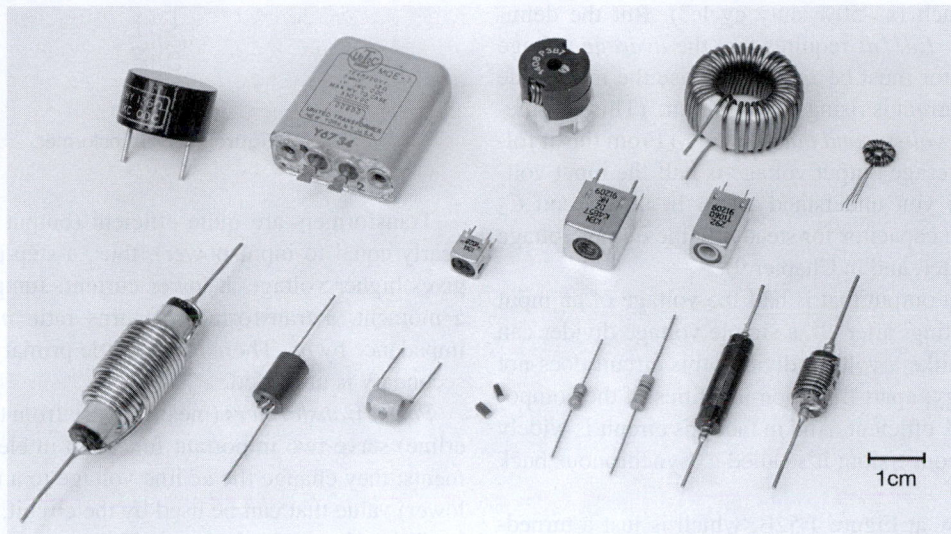

Figure 1.51. Inductors. Top row, left to right: encapsulated toroid, hermetically-sealed toroid, board-mount pot core, bare toroid (two sizes). Middle row: slug-tuned ferrite-core inductors (three sizes). Bottom row: high-current ferrite-core choke, ferrite-bead choke, dipped radial-lead ferrite-core inductor, surface-mount ferrite chokes, molded axial-lead ferrite-core chokes (two styles), lacquered ferrite-core inductors (two styles).

where $K = 1.0$ or 0.395 for dimensions in inches or centimeters, respectively. This is known as Wheeler's formula and is accurate to 1% as long as $l > 0.4d$.

As with capacitive current, inductive current is not simply proportional to voltage (as in a resistor). Furthermore, unlike the situation in a resistor, the power associated with inductive current ($V \times I$) is not turned into heat, but is stored as energy in the inductor's magnetic field (recall that for a capacitor the power associated with capacitive current is likewise not dissipated as heat, but is stored as energy in the capacitor's electric field). You get all that energy back when you interrupt the inductor's current (with a capacitor you get all the energy back when you discharge the voltage to zero).

The basic inductor is a coil, which may be just a loop with one or more turns of wire; or it may be a coil with some length, known as a solenoid. Variations include coils wound on various core materials, the most popular being iron (or iron alloys, laminations, or powder) and ferrite (a gray, nonconductive, brittle magnetic material). These are all ploys to multiply the inductance of a given coil by the "permeability" of the core material. The core may be in the shape of a rod, a toroid (doughnut), or even more bizarre shapes, such as a "pot core" (which has to be seen to be understood; the best description we can come up with is a doughnut mold split horizontally in half, if doughnuts were made in molds). See Figure 1.51 for some typical geometries.

Inductors find heavy use in radiofrequency (RF) circuits, serving as RF "chokes" and as parts of tuned circuits (§1.7.14). A pair of closely coupled inductors forms the interesting object known as a transformer. We will talk briefly about them shortly.

An inductor is, in a real sense, the opposite of a capacitor.[28] You will see how that works out later in the chapter when we deal with the important subject of *impedance*.

A. A look ahead: some magic with inductors

Just to give a taste of some of the tricks that you can do with inductors, take a look at Figure 1.52. Although we'll understand these circuits a lot better when we go at them in Chapter 9, it's possible to see what's going on with what we know already. In Figure 1.52A the left-hand side of inductor L is alternately switched between a dc input voltage V_{in} and ground, at some rapid rate, spending equal times

---

[28] In practice, however, capacitors are much more widely used in electronic circuits. That is because practical inductors depart significantly from ideal performance – by having winding resistance, core losses, and self-capacitance – whereas practical capacitors are nearly perfect (more on this in Chapter 1x). Inductors are indispensable, however, in *switching power converters*, as well as in tuned *LC* circuits for RF applications.

connected to each (a "50% duty cycle"). But the defining equation $V = L\,dI/dt$ requires that the *average* voltage across an inductor must be zero, otherwise the magnitude of its average current is rising without limit. (This is sometimes called the *volt-second balance* rule.) From this it follows that the average output voltage is half the input voltage (make sure you understand why). In this circuit C_2 acts as a storage capacitor for steadying the output voltage (more on this later, and in Chapter 9).

Producing an output that is half the voltage of an input is not very exciting; after all, a simple voltage divider can do that. But, unlike a voltage divider, this circuit does not waste any energy; apart from non-idealities of the components, it is 100% efficient. And in fact this circuit is widely used in power conversion; it's called a "synchronous buck converter."

But look now at Figure 1.52B, which is just a turned-around version of Figure 1.52A. This time, volt-second balance requires that the output voltage be *twice* the input voltage. You can't do *that* with a voltage divider! Once again, the output capacitor (C_1 this time) serves to hold the output voltage steady by storing charge. This configuration is called a "synchronous boost converter."

These and other switching converters are discussed extensively in Chapter 9, where Table 9.5 lists some fifty representative types.

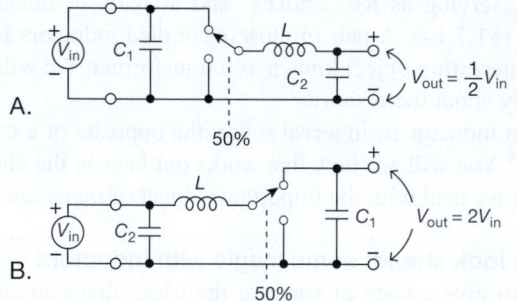

Figure 1.52. Inductors let you do neat tricks, such as *increasing* a dc input voltage.

1.5.2 Transformers

A transformer is a device consisting of two closely coupled coils (called primary and secondary). An ac voltage applied to the primary appears across the secondary, with a voltage multiplication proportional to the turns ratio of the transformer, and with a current multiplication inversely proportional to the turns ratio. Power is conserved. Figure 1.53 shows the circuit symbol for a laminated-core transformer (the kind used for 60 Hz ac power conversion).

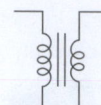

Figure 1.53. Transformer.

Transformers are quite efficient (output power is very nearly equal to input power); thus, a step-up transformer gives higher voltage at lower current. Jumping ahead for a moment, a transformer of turns ratio n increases the impedance by n^2. There is very little primary current if the secondary is unloaded.

Power transformers (meant for use from the 115 V powerline) serve two important functions in electronic instruments: they change the ac line voltage to a useful (usually lower) value that can be used by the circuit, and they "isolate" the electronic device from actual connection to the powerline, because the windings of a transformer are electrically insulated from each other. They come in an enormous variety of secondary voltages and currents: outputs as low as 1 volt or so up to several thousand volts, current ratings from a few milliamps to hundreds of amps. Typical transformers for use in electronic instruments might have secondary voltages from 10 to 50 volts, with current ratings of 0.1 to 5 amps or so. A related class of transformers is used in electronic power conversion, in which plenty of power is flowing, but typically as pulse or square waveforms, and at much higher frequencies (50 kHz to 1 MHz is typical).

Transformers for signals at audio frequencies and radio frequencies are also available. At radio frequencies you sometimes use tuned transformers if only a narrow range of frequencies is present. There is also an interesting class of transmission-line transformers. In general, transformers for use at high frequencies must use special core materials or construction to minimize core losses, whereas low-frequency transformers (e.g., ac powerline transformers) are burdened instead by large and heavy cores. The two kinds of transformers are in general not interchangeable.

A. Problems, problems...

This simple "first-look" description ignores interesting – and important – issues. For example, there are inductances associated with the transformer, as suggested by its circuit symbol: an effective parallel inductance (called the *magnetizing inductance*) and an effective series inductance (called the *leakage inductance*). Magnetizing inductance causes a primary current even with no secondary load; more significantly, it means that you cannot make a "dc transformer." And leakage inductance causes a voltage drop that depends on load current, as well as bedeviling circuits that have fast pulses or edges. Other departures from ideal performance include winding resistance, core losses, capacitance, and magnetic coupling to the outside world. Unlike capacitors (which behave nearly ideally in most circuit applications), the deficiencies of inductors have significant effects in real-world circuit applications. We'll deal with these in Chapter *1x* and Chapter 9.

1.6 Diodes and diode circuits

We are not done with capacitors and inductors! We have dealt with them in the *time domain* (*RC* circuits, exponential charge and discharge, differentiators and integrators, and so on), but we have not yet tackled their behavior in the *frequency domain*.

We will get to that soon enough. But this is a good time to take a break from *"RLC"* and put our knowledge to use with some clever and useful circuits. We begin by introducing a new device, the *diode*. It's our first example of a nonlinear device, and you can do nifty things with it.

1.6.1 Diodes

The circuit elements we've discussed so far (resistors, capacitors, and inductors) are all *linear*, meaning that a doubling of the applied signal (a voltage, say) produces a doubling of the response (a current, say). This is true even for the reactive devices (capacitors and inductors). These components are also *passive*, as opposed to *active* devices, the latter exemplified by transistors, which are semiconductor devices that control the flow of power. And they are all two-terminal devices, which is self-explanatory.

Figure 1.54. Diode.

The diode (Figure 1.54) is an important and useful two-terminal passive *nonlinear* device. It has the *V–I* curve shown in Figure 1.55. (In keeping with the general philosophy of this book, we will not attempt to describe the solid-state physics that makes such devices possible.)

The diode's arrow (the anode terminal) points in the direction of forward current flow. For example, if the diode is in a circuit in which a current of 10 mA is flowing from anode to cathode, then (from the graph) the anode is approximately 0.6 V more positive than the cathode; this is called the "forward voltage drop." The reverse current,

which is measured in the nanoamp range for a general-purpose diode (note the hugely different scales in the graph for forward and reverse current), is almost never of any consequence until you reach the reverse breakdown voltage (also called the peak inverse voltage, PIV), typically 75 volts for a general-purpose diode like the 1N4148. (Normally you never subject a diode to voltages large enough to cause reverse breakdown; the exception is the zener diode we mentioned earlier.) Frequently, also, the forward voltage drop of about 0.5 to 0.8 V is of little concern, and the diode can be treated as a good approximation to an ideal one-way conductor. There are other important characteristics that distinguish the thousands of diode types available, e.g., maximum forward current, capacitance, leakage current, and reverse recovery time; Table 1.1 includes a few popular diodes, to give a sense of the capabilities of these little devices.

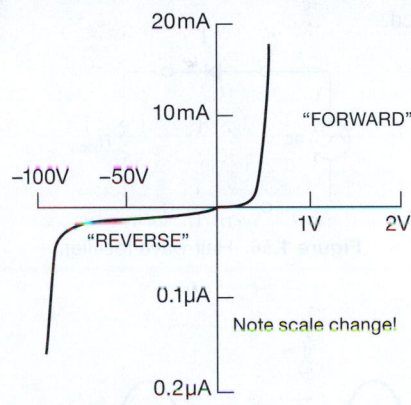

Figure 1.55. Diode *V–I* curve.

Before jumping into some circuits with diodes, we should point out two things: (a) a diode doesn't have a resistance (it doesn't obey Ohm's law). (b) If you put some diodes in a circuit, it won't have a Thévenin equivalent.

1.6.2 Rectification

A rectifier changes ac to dc; this is one of the simplest and most important applications of diodes (which are sometimes called rectifiers). The simplest circuit is shown in Figure 1.56. The "ac" symbol represents a source of ac voltage; in electronic circuits it is usually provided by a transformer, powered from the ac powerline. For a sinewave input that is much larger than the forward drop (about 0.6 V for silicon diodes, the usual type), the output will look like that in Figure 1.57. If you think of the diode as a one-way conductor, you won't have any trouble

Table 1.1 Representative Diodes

Part #	V_R (max) (V)	I_R (typ, 25°C) (A @ V)		V_F @ I_F (mV)	(mA)	Capacitance (pF @ V_R)		SMT[a] p/n	Comments
Silicon									
PAD5	45	0.25pA	20V	800	1	0.5pF	5V	SSTPAD5	metal + glass can
1N4148	75	10nA	20V	750	10	0.9pF	0V	1N4148W	jellybean sig diode
1N4007	1000	50nA	800V	800	250	12pF	10V	DL4007	1N4004 lower V
1N5406	600	<10μA	600V	1.0V	10A	18pF	10V	none	heat through leads
Schottky[b]									
1N6263	60	7nA	20V	400	1	0.6pF	10V	1N6263W	see also 1N5711
1N5819	40	10μA	32V	400	1000	150pF	1V	1N5819HW	jellybean
1N5822	40	40μA	32V	480	3000	450pF	1V	none	power Schottky
MBRP40045	45	500μA	40V	540	400A	3500pF	10V	you jest!	Moby dual Schottky

Notes: (a) SMT, surface-mount technology. (b) Schottky diodes have lower forward voltage and zero reverse-recovery time, but more capacitance.

understanding how the circuit works. This circuit is called a *half-wave rectifier*, because only half of the input waveform is used.

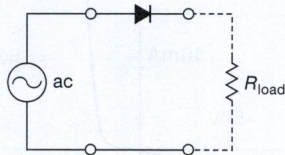

Figure 1.56. Half-wave rectifier.

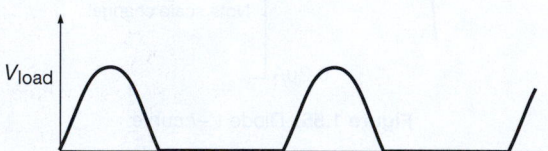

Figure 1.57. Half-wave output voltage (unfiltered).

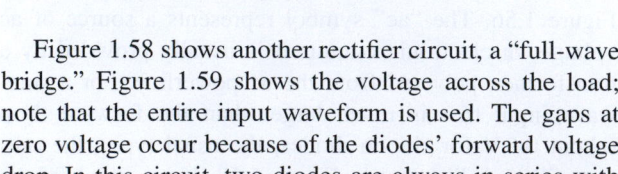

Figure 1.58. Full-wave bridge rectifier.

Figure 1.58 shows another rectifier circuit, a "full-wave bridge." Figure 1.59 shows the voltage across the load; note that the entire input waveform is used. The gaps at zero voltage occur because of the diodes' forward voltage drop. In this circuit, two diodes are always in series with the input; when you design low-voltage power supplies, for

which the diode drop becomes significant, you have to remember that.[29]

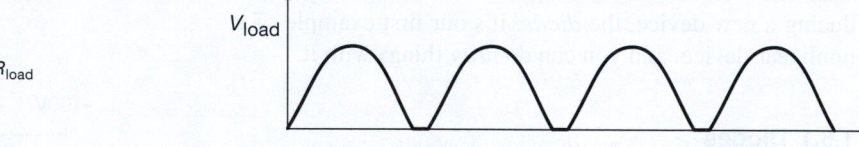

Figure 1.59. Full-wave output voltage (unfiltered).

1.6.3 Power-supply filtering

The preceding rectified waveforms aren't good for much as they stand. They're "dc" only in the sense that they don't change polarity. But they still have a lot of "ripple" (periodic variations in voltage about the steady value) that has to be smoothed out in order to generate genuine dc. This we do by attaching a relatively large value capacitor (Figure 1.60); it charges up to the peak output voltage during the diode conduction, and its stored charge ($Q = CV$) provides the output current in between charging cycles. Note that the diodes prevent the capacitor from discharging back through the ac source. In this application you should think of the capacitor as an energy storage device, with stored energy $U = \frac{1}{2}CV^2$ (recall §1.4.1; for C in farads and V in volts, U comes out in joules, or equivalently, watt seconds).

The capacitor value is chosen so that

$$R_{\text{load}}C \gg 1/f,$$

---

[29] The diode drop can be eliminated with *active switching* (or *synchronous switching*, a technique in which the diodes are replaced by transistor switches, actuated in synchronism with the input ac waveform (see §9.5.3B).

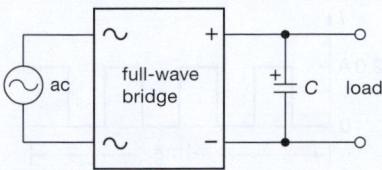

Figure 1.60. Full-wave bridge with output storage ("filter") capacitor.

(where f is the ripple frequency, here 120 Hz) in order to ensure small ripple by making the time constant for discharge much longer than the time between recharging. We make this vague statement clearer now.

A. Calculation of ripple voltage

It is easy to calculate the approximate ripple voltage, particularly if it is small compared with the dc (see Figure 1.61). The load causes the capacitor to discharge somewhat between cycles (or half-cycles, for full-wave rectification). If you assume that the load current stays constant (it will, for small ripple), you have

$$\Delta V = \frac{I}{C}\Delta t \qquad \left(\text{from } I = C\frac{dV}{dt}\right). \qquad (1.25)$$

Just use $1/f$ (or $1/2f$ for full-wave rectification) for Δt (this estimate is a bit on the safe side, because the capacitor begins charging again in less than a half-cycle). You get[30]

$$\Delta V = \frac{I_{\text{load}}}{fC} \qquad \text{(half-wave)},$$

$$\Delta V = \frac{I_{\text{load}}}{2fC} \qquad \text{(full-wave)}.$$

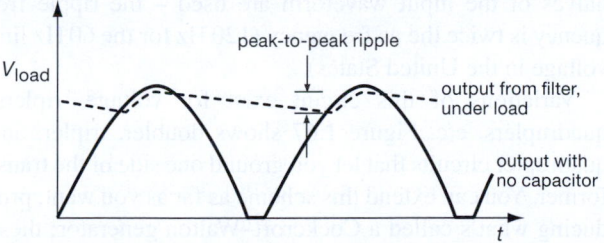

Figure 1.61. Power-supply ripple calculation.

If you wanted to do the calculation without any approximation, you would use the exact exponential discharge formula. You would be misguided in insisting on that kind

---

[30] While teaching electronics we've noticed that students love to memorize these equations! An informal poll of the authors showed that two out of two engineers don't memorize them. Please don't waste brain cells that way – instead, learn how to derive them.

of accuracy, though, for two reasons. (a) The discharge is an exponential only if the load is a resistance; many loads are not. In fact, the most common load, a *voltage regulator*, looks like a constant-current load. (b) Power supplies are built with capacitors with typical tolerances of 20% or more. Realizing the manufacturing spread, you design conservatively, allowing for the worst-case combination of component values.

In this case, viewing the initial part of the discharge as a ramp is in fact quite accurate, especially if the ripple is small, and in any case it errs in the direction of conservative design – it overestimates the ripple.

Exercise 1.20. Design a full-wave bridge rectifier circuit to deliver 10 V dc with less than 0.1 V (pp) ripple into a load drawing up to 10 mA. Choose the appropriate ac input voltage, assuming 0.6 V diode drops. Be sure to use the correct ripple frequency in your calculation.

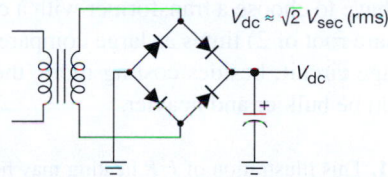

Figure 1.62. Bridge rectifier circuit. The polarity marking and curved electrode indicate a polarized capacitor, which must not be allowed to charge with the opposite polarity.

1.6.4 Rectifier configurations for power supplies

A. Full-wave bridge

A dc power supply with the bridge circuit we just discussed looks as shown in Figure 1.62. In practice, you generally buy the bridge as a prepackaged module. The smallest ones come with maximum current ratings of 1 A average, with a selection of rated minimum breakdown voltages going from 100 V to 600 V, or even 1000 V. Giant bridge rectifiers are available with current ratings of 25 A or more.

B. Center-tapped full-wave rectifier

The circuit in Figure 1.63 is called a center-tapped full-wave rectifier. The output voltage is half what you get if you use a bridge rectifier. It is not the most efficient circuit in terms of transformer design, because each half of the secondary is used only half the time. To develop some intuition on this subtle point, consider two different configurations that produce the same rectified dc output voltage: (a) the circuit of Figure 1.63, and (b) the same transformer, this time with its secondary cut at the center tap and rewired with the two halves in parallel, the resultant combined secondary winding connected to a full-wave bridge. Now, to deliver the same output power, each half winding in (a), during its conduction cycle, must supply the same current as the parallel pair in (b). But the power dissipated in the winding resistances goes like I^2R, so the power lost to heating in the transformer secondary windings reduced by a factor of 2 for the bridge configuration (b).

Here's another way to see the problem: imagine we use the same transformer as in (a), but for our comparison circuit we replace the pair of diodes with a bridge, as in Figure 1.62, and we leave the center tap unconnected. Now, to deliver the same output *power*, the current through the winding during that time is twice what it would be for a true full-wave circuit. To expand on this subtle point: heating in the windings, calculated from Ohm's law, is I^2R, so you have four times the heating for half the time, or twice the average heating of an equivalent full-wave bridge circuit. You would have to choose a transformer with a current rating 1.4 (square root of 2) times as large compared with the (better) bridge circuit; besides costing more, the resulting supply would be bulkier and heavier.

Exercise 1.21. This illustration of I^2R heating may help you understand the disadvantage of the center-tapped rectifier circuit. What fuse rating (minimum) is required for passing the current waveform shown in Figure 1.64, which has 1 amp average current? *Hint*: a fuse "blows out" by melting a metallic link (I^2R heating), for steady currents larger than its rating. Assume for this problem that the thermal time constant of the fusible link is much longer than the time scale of the square wave, i.e., that the fuse responds to the value of I^2 averaged over many cycles.

C. Split supply

A popular variation of the center-tapped full-wave circuit is shown in Figure 1.65. It gives you split supplies (equal plus and minus voltages), which many circuits need. It is an efficient circuit, because both halves of the input waveform are used in each winding section.

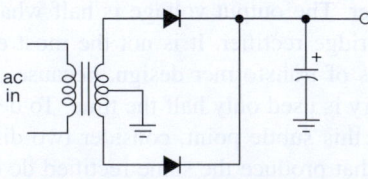

Figure 1.63. Full-wave rectifier using center-tapped transformer.

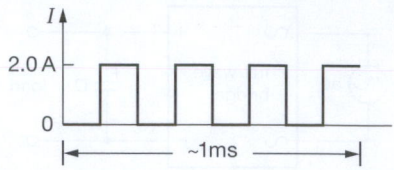

Figure 1.64. Illustrating greater I^2R heating with discontinuous current flow.

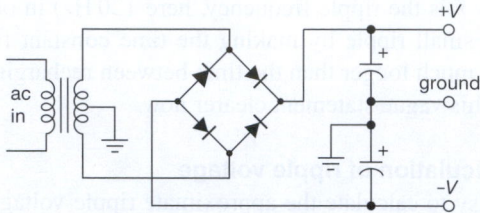

Figure 1.65. Dual-polarity (split) supply.

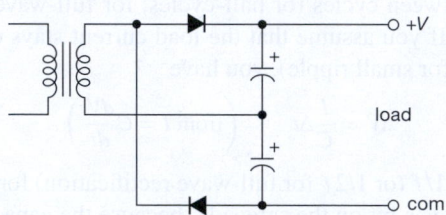

Figure 1.66. Voltage doubler.

D. Voltage multipliers

The circuit shown in Figure 1.66 is called a voltage doubler. Think of it as two half-wave rectifier circuits in series. It is officially a full-wave rectifier circuit because both halves of the input waveform are used – the ripple frequency is twice the ac frequency (120 Hz for the 60 Hz line voltage in the United States).

Variations of this circuit exist for voltage triplers, quadruplers, etc. Figure 1.67 shows doubler, tripler, and quadrupler circuits that let you ground one side of the transformer. You can extend this scheme as far as you want, producing what's called a Cockcroft–Walton generator; these are used in arcane applications (such as particle accelerators) and in everyday applications (such as image intensifiers, air ionizers, laser copiers, and even bug zappers) that require a high dc voltage but hardly any current.

1.6.5 Regulators

By choosing capacitors that are sufficiently large, you can reduce the ripple voltage to any desired level. This brute-force approach has three disadvantages.

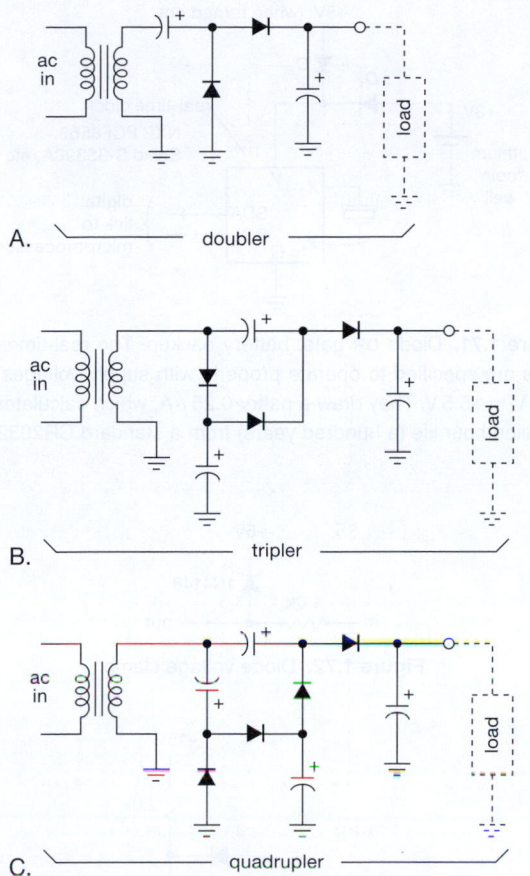

A.   doubler

B.   tripler

C.   quadrupler

Figure 1.67. Voltage multipliers; these configurations don't require a floating voltage source.

- The required capacitors may be prohibitively bulky and expensive.
- The very short interval of current flow during each cycle[31] (only very near the top of the sinusoidal waveform) produces more I^2R heating.
- Even with the ripple reduced to negligible levels, you still have variations of output voltage that are due to other causes, e.g., the dc output voltage will be roughly proportional to the ac input voltage, giving rise to fluctuations caused by input line voltage variations. In addition, changes in load current will still cause the output voltage to change because of the finite internal resistances of the transformer, diode, etc. In other words, the Thévenin equivalent circuit of the dc power supply has $R > 0$.

A better approach to power-supply design is to use enough capacitance to reduce ripple to low levels (perhaps

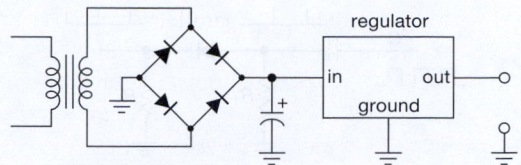

Figure 1.68. Regulated dc power supply.

10% of the dc voltage), then use an active *feedback circuit* to eliminate the remaining ripple. Such a feedback circuit "looks at" the output, making changes in a controllable series resistor (a transistor) as necessary to keep the output voltage constant (Figure 1.68). This is known as a "linear regulated dc power supply."[32]

These voltage regulators are used almost universally as power supplies for electronic circuits. Nowadays complete voltage regulators are available as inexpensive ICs (priced under $1). A power supply built with a voltage regulator can be made easily adjustable and self-protecting (against short circuits, overheating, etc.), with excellent properties as a voltage source (e.g., internal resistance measured in milliohms). We will deal with regulated dc power supplies in Chapter 9.

1.6.6 Circuit applications of diodes

A. Signal rectifier
There are other occasions when you use a diode to make a waveform of one polarity only. If the input waveform isn't a sinewave, you usually don't think of it as a rectification in the sense of a power supply. For instance, you might want a train of pulses corresponding to the rising edge of a square wave. The easiest way is to rectify the differentiated wave (Figure 1.69). Always keep in mind the 0.6 V (approximately) forward drop of the diode. This circuit, for instance, gives no output for square waves smaller than 0.6 V pp. If this is a problem, there are various tricks to circumvent this limitation. One possibility is to use *hot carrier diodes* (Schottky diodes), with a forward drop of about 0.25 V.

A possible *circuit solution* to this problem of finite diode drop is shown in Figure 1.70. Here D_1 compensates D_2's forward drop by providing 0.6 V of *bias* to hold D_2 at the threshold of conduction. Using a diode (D_1) to provide the bias (rather than, say, a voltage divider) has several

---

[31] Called the *conduction angle*.

[32] A popular variant is the regulated *switching* power converter. Although its operation is quite different in detail, it uses the same feedback principle to maintain a constant output voltage. See Chapter 9 for much more on both techniques.

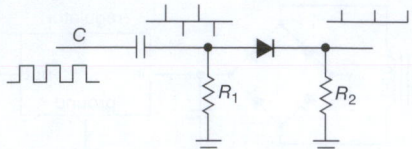

Figure 1.69. Signal rectifier applied to differentiator output.

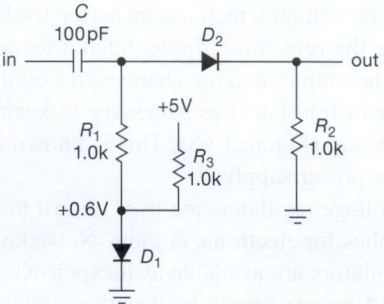

Figure 1.70. Compensating the forward voltage drop of a diode signal rectifier.

advantages: (a) there is nothing to adjust, (b) the compensation will be nearly perfect, and (c) changes of the forward drop (e.g., with changing temperature) will be compensated properly. Later we will see other instances of matched-pair compensation of forward drops in diodes, transistors, and FETs. It is a simple and powerful trick.

B. Diode gates

Another application of diodes, which we will recognize later under the general heading of *logic*, is to pass the higher of two voltages without affecting the lower. A good example is *battery backup*, a method of keeping something running (e.g, the "real-time clock" chip in a computer, which keeps a running count of date and time) that must continue running even when the device is switched off. Figure 1.71 shows a circuit that does the job. The battery does nothing until the +5 V power is switched off; then it takes over without interruption.

C. Diode clamps

Sometimes it is desirable to limit the range of a signal (i.e., prevent it from exceeding certain voltage limits) somewhere in a circuit. The circuit shown in Figure 1.72 will accomplish this. The diode prevents the output from exceeding about +5.6 V, with no effect on voltages less than that (including negative voltages); the only limitation is that the input must not go so negative that the reverse breakdown voltage of the diode is exceeded (e.g., −75 V for a 1N4148). The series resistor limits the diode current during

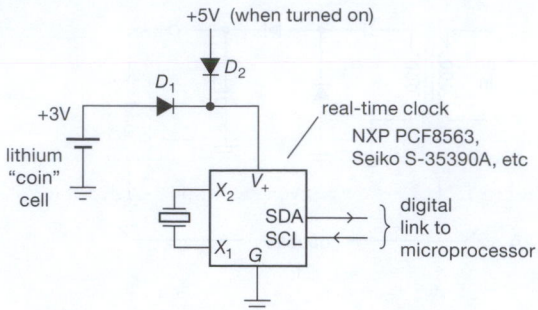

Figure 1.71. Diode OR gate: battery backup. The real-time clock chips are specified to operate properly with supply voltages from +1.8 V to +5.5 V. They draw a paltry 0.25 µA, which calculates to a 1-million-hour life (a hundred years) from a standard CR2032 coin cell!

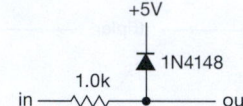

Figure 1.72. Diode voltage clamp.

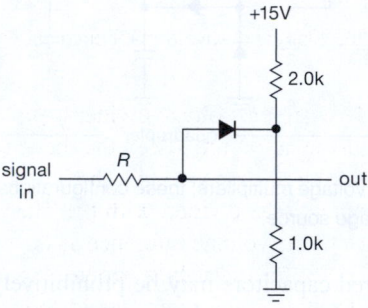

Figure 1.73. Voltage divider providing clamping voltage.

clamping action; however, a side effect is that it adds 1 kΩ of series resistance (in the Thévenin sense) to the signal, so its value is a compromise between maintaining a desirable low source (Thévenin) resistance and a desirable low clamping current. Diode clamps are standard equipment on all inputs in contemporary CMOS digital logic. Without them, the delicate input circuits are easily destroyed by static electricity discharges during handling.

Exercise 1.22. Design a symmetrical clamp, i.e., one that confines a signal to the range −5.6 to +5.6 V.

A voltage divider can provide the reference voltage for a clamp (Figure 1.73). In this case you must ensure that the resistance looking into the voltage divider (R_{vd}) is small compared with R because what you have looks as shown

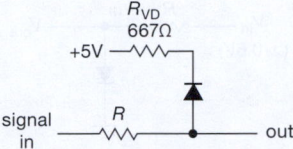

Figure 1.74. Clamping to voltage divider: equivalent circuit.

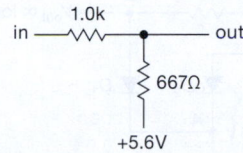

Figure 1.75. Poor clamping: voltage divider not stiff enough.

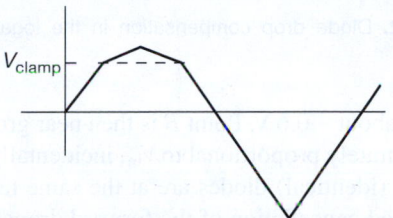

Figure 1.76. Clamping waveform for circuit of Figure 1.73.

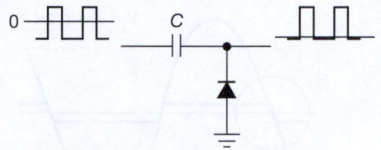

Figure 1.77. dc restoration.

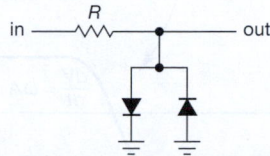

Figure 1.78. Diode limiter.

in Figure 1.74 when the voltage divider is replaced with its Thévenin equivalent circuit. When the diode conducts (input voltage exceeds clamp voltage), the output is really just the output of a voltage divider, with the Thévenin equivalent resistance of the voltage reference as the lower resistor (Figure 1.75). So, for the values shown, the output of the clamp for a triangle-wave input would look as shown in Figure 1.76. The problem is that the voltage divider doesn't provide a stiff reference, in the language of electronics. A stiff voltage source is one that doesn't bend easily, i.e., it has low internal (Thévenin) resistance.

In practice, the problem of finite impedance of the voltage-divider reference can be easily solved by use of a transistor or an op-amp. This is usually a better solution than using very small resistor values, because it doesn't consume large currents, yet it provides a voltage reference with a Thévenin resistance of a few ohms or less. Furthermore, there are other ways to construct a clamp, using an op-amp as part of the clamp circuit. You will see these methods in Chapter 4.

Alternatively, a simple way to stiffen the clamp circuit of Figure 1.73, for time-varying signals only, is to add a so-called *bypass capacitor* across the lower (1 kΩ) resistor. To understand this fully we need to know about capacitors in

the frequency domain, a subject we'll take up shortly. For now we'll simply say that you can put a capacitor across the 1k resistor, and its stored charge acts to maintain that point at constant voltage. For example, a 15 μF capacitor to ground would make the divider look as if it had a Thévenin resistance of less than 10 Ω for frequencies above 1 kHz. (You could similarly add a bypass capacitor across D_1 in Figure 1.70.) As we'll learn, the effectiveness of this trick decreases at low frequencies, and it does nothing at dc.

One interesting clamp application is "dc restoration" of a signal that has been ac coupled (capacitively coupled). Figure 1.77 shows the idea. This is particularly important for circuits whose inputs look like diodes (e.g., a transistor with grounded emitter, as we'll see in the next chapter); otherwise an ac-coupled signal will just fade away, as the coupling capacitor charges up to the signal's peak voltage.

D. Limiter

One last clamp circuit is shown in Figure 1.78. This circuit limits the output "swing" (again, a common electronics term) to one diode drop in either polarity, roughly ±0.6 V. That might seem awfully small, but if the next stage is an amplifier with large voltage amplification, its input will always be near 0 V; otherwise the output is in "saturation" (e.g., if the next stage has a gain of 1000 and operates from ±15 V supplies, its input must stay in the range ±15 mV in order for its output not to saturate). Figure 1.79 shows what a limiter does to oversize sinewaves and spikes. This clamp circuit is often used as input protection for a high-gain amplifier.

E. Diodes as nonlinear elements

To a good approximation the forward current through a diode is proportional to an exponential function of the

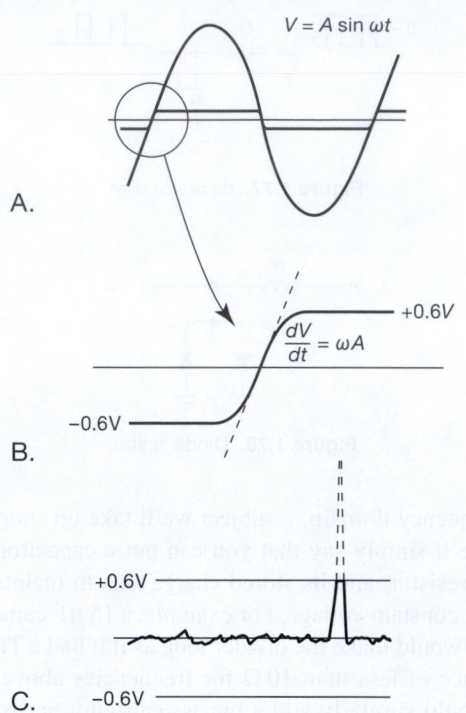

A.

B.

$$\frac{dV}{dt} = \omega A$$

C.

Figure 1.79. A. Limiting large-amplitude sinewaves; B. details; and C. spikes.

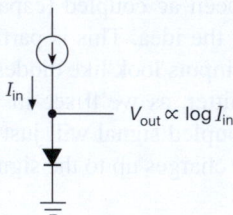

Figure 1.80. Exploiting the diode's nonlinear V–I curve: logarithmic converter.

voltage across it at a given temperature (for a discussion of the exact law, see §2.3.1). So you can use a diode to generate an output voltage proportional to the logarithm of a current (Figure 1.80). Because V hovers in the region of 0.6 V, with only small voltage changes that reflect input current variations, you can generate the input current with a resistor if the input voltage is much larger than a diode drop (Figure 1.81).

In practice, you may want an output voltage that isn't offset by the 0.6 V diode drop. In addition, it would be nice to have a circuit that is insensitive to changes in temperature (a silicon diode's voltage drop decreases approximately 2 mV/°C). The method of diode drop compensation is helpful here (Figure 1.82). R_1 makes D_2 conduct, holding

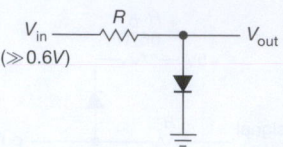

Figure 1.81. Approximate log converter.

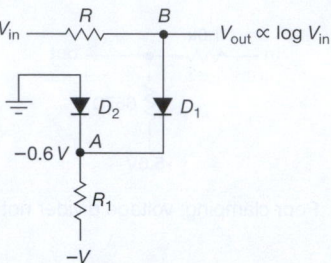

Figure 1.82. Diode drop compensation in the logarithmic converter.

point A at about -0.6 V. Point B is then near ground (making I_{in} accurately proportional to V_{in}, incidentally). As long as the two (identical) diodes are at the same temperature, there is good cancellation of the forward drops, except, of course, for the difference owing to input current through D_1, which produces the desired output. In this circuit, R_1 should be chosen so that the current through D_2 is significantly larger than the maximum input current in order to keep D_2 in conduction.

In Chapter *2x* we will examine better ways of constructing logarithmic converter circuits, along with careful methods of temperature compensation. With such methods it is possible to construct logarithmic converters accurate to a few percent over six decades or more of input current. A better understanding of diode and transistor characteristics, along with an understanding of op-amps, is necessary first. This section is meant to serve only as an introduction for things to come.

1.6.7 Inductive loads and diode protection

What happens if you open a switch that is providing current to an inductor? Because inductors have the property

$$V = L\,dI/dt,$$

it is not possible to turn off the current suddenly, because that would imply an infinite voltage across the inductor's terminals. What happens instead is that the voltage across the inductor rises abruptly and keeps rising until it forces current to flow. Electronic devices controlling inductive

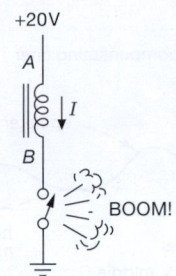

Figure 1.83. Inductive "kick."

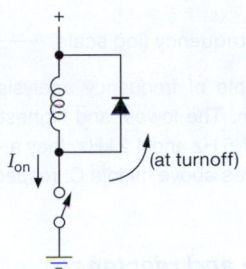

Figure 1.84. Blocking inductive kick.

loads can be easily damaged, especially the component that "breaks down" in order to satisfy the inductor's craving for continuity of current. Consider the circuit in Figure 1.83. The switch is initially closed, and current is flowing through the inductor (which might be a relay, as described later). When the switch is opened, the inductor tries to keep current flowing from A to B, as it had been. In other words, it tries to make current flow out of B, which it does by forcing B to a high positive voltage (relative to A). In a case like this, in which there's no connection to terminal B, it may go 1000 V positive before the switch contact "blows over." This shortens the life of the switch and also generates impulsive interference that may affect other circuits nearby. If the switch happens to be a transistor, it would be an understatement to say that its life is shortened; its life is *ended*.

The best solution usually is to put a diode across the inductor, as in Figure 1.84. When the switch is on, the diode is back-biased (from the dc drop across the inductor's winding resistance). At turn-off the diode goes into conduction, putting the switch terminal a diode drop above the positive supply voltage. The diode must be able to handle the initial diode current, which equals the steady current that had been flowing through the inductor; something like a 1N4004 is fine for nearly all cases.

The only disadvantage of this simple protection circuit is that it lengthens the decay of current through the inductor, because the rate of change of inductor current

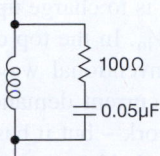

Figure 1.85. *RC* "snubber" for suppressing inductive kick.

is proportional to the voltage across it. For applications in which the current must decay quickly (high-speed actuators or relays, camera shutters, magnet coils, etc.), it may be better to put a resistor across the inductor, choosing its value so that $V_{supply} + IR$ is less than the maximum allowed voltage across the switch. For the fastest decay with a given maximum voltage, a zener with series diode (or other voltage-clamping device) can be used instead, giving a linear-like ramp-down of current rather than an exponential decay (see discussion in Chapter *1x*).

For inductors driven from ac (transformers, ac relays), the diode protection just described will not work, because the diode will conduct on alternate half-cycles when the switch is closed. In that case a good solution is an *RC* "snubber" network (Figure 1.85). The values shown are typical for small inductive loads driven from the ac powerline. Such a snubber should be included in all instruments that run from the ac powerline, because the power transformer is inductive.[33]

An alternative to the *RC* snubber is the use of a bidirectional zener-like voltage-clamping element. Among these the most common are the bidirectional "TVS" (transient voltage suppressor) zener and the metal-oxide varistor ("MOV"); the latter is an inexpensive device that looks something like a disc ceramic capacitor and behaves electrically like a bidirectional zener diode. Both classes are designed for transient voltage protection, are variously available at voltage ratings from 10 to 1000 volts, and can handle transient currents up to thousands of amperes (see Chapter *9x*). Including a transient suppressor (with appropriate fusing) across the ac powerline terminals makes good sense in a piece of electronic equipment, not only to prevent inductive spike interference to other nearby instruments but also to prevent occasional large powerline spikes from damaging the instrument itself.

1.6.8 Interlude: inductors as friends

Lest we leave the impression that inductance and inductors are things only to be feared, let's look at the circuit in

---

[33] As explained in §9.5.1, you should choose a capacitor rated for "across-the-line" service.

Figure 1.86. The goal is to charge up the capacitor from a source of dc voltage V_{in}. In the top circuit (Figure 1.86A) we've done it the conventional way, with a series resistor to limit the peak current demanded from the voltage source. OK, it does work – but it has a drawback that can be serious, namely that half the energy is lost as heat in the resistor. By contrast, in the circuit with the inductor (Figure 1.86B) no energy is lost (assuming ideal components); and, as a bonus, the capacitor gets charged to twice the input voltage. The output-voltage waveform is a sinusoidal half-cycle at the resonant frequency $f = 1/2\pi\sqrt{LC}$, a topic we'll see soon (§1.7.14).[34],[35]

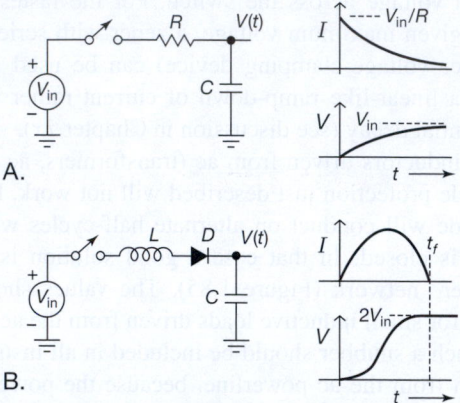

A.

B.

Figure 1.86. Resonant charging is lossless (with ideal components) compared with the 50% efficiency of resistive charging. Charging is complete after t_f, equal to a half-cycle of the resonant frequency. The series diode terminates the cycle, which would otherwise continue to oscillate between 0 and $2V_{in}$.

---

[34] A mechanical analogy may be helpful here. Imagine dropping packages onto a conveyor belt that is moving at speed v; the packages are accelerated to that speed by friction, with 50% efficiency, finally reaching the belt speed v, at which speed they ride into the sunset. That's resistive charging. Now we try something completely different, namely, we rig up a conveyor belt with little catchers attached by springs to the belt; and alongside it we have a second belt, running at twice the speed ($2v$). Now when we drop a package onto the first conveyor it compresses a spring, then rebounds at $2v$; and it makes a soft landing onto the second conveyor. No energy is lost (ideal springs), and the package rides off into the sunset at $2v$. That's reactive charging.

[35] Resonant charging is used for the high-voltage supply in flashlamps and stroboscopes, with the advantages of (a) full charge between flashes (spaced no closer than t_f), and (b) no current immediately after discharge (see waveforms), thus permitting the flashlamp to "quench" after each flash.

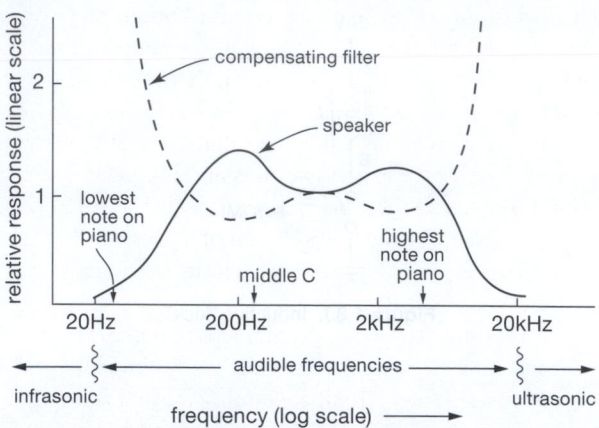

Figure 1.87. Example of frequency analysis: "boom box" loudspeaker equalization. The lowest and highest piano notes, called A0 and C8, are at 27.5 Hz and 4.2 kHz; they are four octaves below A440 and four octaves above middle C, respectively.

1.7 Impedance and reactance

Warning: this section is somewhat mathematical; you may wish to skip over the mathematics, but be sure to pay attention to the results and graphs.

Circuits with capacitors and inductors are more complicated than the resistive circuits we talked about earlier, in that their behavior depends on frequency: a "voltage divider" containing a capacitor or inductor will have a frequency-dependent division ratio. In addition, circuits containing these components (known collectively as *reactive* components) "corrupt" input waveforms such as square waves, as we saw earlier.

However, both capacitors and inductors are *linear* devices, meaning that the amplitude of the output waveform, whatever its shape, increases exactly in proportion to the input waveform's amplitude. This linearity has many consequences, the most important of which is probably the following: *the output of a linear circuit, driven with a sinewave at some frequency f, is itself a sinewave at the same frequency (with, at most, changed amplitude and phase)*.

Because of this remarkable property of circuits containing resistors, capacitors, and inductors (and, later, linear amplifiers), it is particularly convenient to analyze any such circuit by asking how the output voltage (amplitude and phase) depends on the input voltage *for sinewave input at a single frequency*, even though this may not be the intended use. A graph of the resulting *frequency response*, in which the ratio of output to input is plotted for each sinewave frequency, is useful for thinking about many kinds of waveforms. As an example, a certain "boom-box" loud-speaker might have the frequency response shown in Figure 1.87, in which the "output" in this case is of course sound pressure, not voltage. It is desirable for a speaker to have a "flat" response, meaning that the graph of sound pressure versus frequency is constant over the band of audible frequencies. In this case the speaker's deficiencies can be corrected by the introduction of a passive filter with the inverse response (as shown) within the amplifiers of the radio.

As we will see, it is possible to generalize Ohm's law, replacing the word "resistance" with "impedance," in order to describe any circuit containing these linear passive devices (resistors, capacitors, and inductors). You could think of the subject of impedance (generalized resistance) as Ohm's law for circuits that include capacitors and inductors.

Some terminology: impedance ($\mathbf{Z}$) is the "generalized resistance"; inductors and capacitors, for which the voltage and current are always 90° out of phase, are *reactive*; they have *reactance* (X). Resistors, with voltage and current always in phase, are *resistive*; they have *resistance* (R). In general, in a circuit that combines resistive and reactive components, the voltage and current at some place will have some in-between phase relationship, described by a complex impedance: impedance = resistance + reactance, or $\mathbf{Z} = R + jX$ (more about this later).[36] However, you'll see statements like "the impedance of the capacitor at this frequency is . . ." The reason you don't have to use the word "reactance" in such a case is that impedance covers everything. In fact, you frequently use the word "impedance" even when you know it's a resistance you're talking about; you say "the source impedance" or "the output impedance" when you mean the Thévenin equivalent resistance of some source. The same holds for "input impedance."

In all that follows, we will be talking about circuits driven by sinewaves at a single frequency. Analysis of circuits driven by complicated waveforms is more elaborate, involving the methods we used earlier (differential equations) or decomposition of the waveform into sinewaves (Fourier analysis). Fortunately, these methods are seldom necessary.

1.7.1 Frequency analysis of reactive circuits

Let's start by looking at a capacitor driven by a sinewave voltage source $V(t) = V_0 \sin \omega t$ (Figure 1.88). The current

---

[36] But, in a nutshell, the magnitude of $\mathbf{Z}$ gives the ratio of amplitudes of voltage to current, and the polar angle of $\mathbf{Z}$ gives the phase angle between current and voltage.

is

$$I(t) = C\frac{dV}{dt} = C\omega V_0 \cos \omega t,$$

i.e., a current of amplitude $\omega C V_0$, with its phase leading the input voltage by 90°. If we consider amplitudes only, and disregard phases, the current is

$$I = \frac{V}{1/\omega C}.$$

(Recall that $\omega = 2\pi f$.) It behaves like a frequency-dependent resistance $R = 1/\omega C$, but in addition the current is 90° out of phase with the voltage (Figure 1.89).

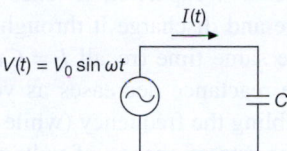

Figure 1.88. A sinusoidal ac voltage drives a capacitor.

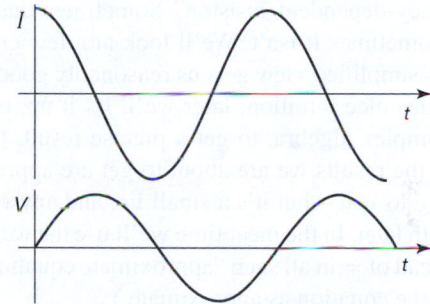

Figure 1.89. Current in a capacitor leads the sinusoidal voltage by 90°.

For example, a $1\,\mu\text{F}$ capacitor put across the 115 volts (rms) 60 Hz powerline draws a current of rms amplitude:

$$I = \frac{115}{1/(2\pi \times 60 \times 10^{-6})} = 43.4\,\text{mA (rms)}.$$

Soon enough we will complicate matters by explicitly worrying about *phase shifts* and the like – and that will get us into some complex algebra that terrifies beginners (often) and mathophobes (always). Before we do that, though, this is a good time to develop intuition about the frequency-dependent behavior of some basic and important circuits that use capacitors, ignoring for the time being the troublesome fact that, when driven with a sinusoidal signal, currents and voltages in a capacitor are not in phase.

As we just saw, the ratio of *magnitudes* of voltage to

current, in a capacitor driven at a frequency ω, is just

$$\frac{|V|}{|I|} = \frac{1}{\omega C},$$

which we can think of as a sort of "resistance" – the magnitude of the current is proportional to the magnitude of the applied voltage. The official name for this quantity is *reactance*, with the symbol X, thus X_C for the reactance of a capacitor.[37] So, for a capacitor,

$$X_C = \frac{1}{\omega C}. \tag{1.26}$$

This means that a larger capacitance has a smaller reactance. And this makes sense, because, for example, if you double the value of a capacitor, it takes twice as much current to charge and discharge it through the same voltage swing in the same time (recall $I = C\,dV/dt$). For the same reason the reactance decreases as you increase the frequency – doubling the frequency (while holding V constant) doubles the rate of change of voltage and therefore requires that you double the current, thus half the reactance.

So, roughly speaking, we can think of a capacitor as a "frequency-dependent resistor." Sometimes that's good enough, sometimes it isn't. We'll look at a few circuits in which this simplified view gets us reasonably good results, and provides nice intuition; later we'll fix it up, using the correct complex algebra, to get a precise result. (Keep in mind that the results we are about to get are approximate; we're *lying* to you – but it's a small lie, and anyway we'll tell the truth later. In the meantime we'll use the weird symbol $\asymp$ instead of $=$ in all such "approximate equations," and we'll flag the equation as approximate.)

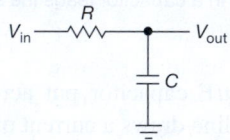

Figure 1.90. Lowpass filter.

A. *RC* lowpass filter (approximate)

The circuit in Figure 1.90 is called a *lowpass filter*, because it passes low frequencies and blocks high frequencies. If you think of it as a frequency-dependent voltage divider, this makes sense: the lower leg of the divider (the capacitor) has a decreasing reactance with increasing frequency, so the ratio of V_{out}/V_{in} decreases accordingly:

---

[37] Later we'll see *inductors*, which also have a 90° phase shift (though of the opposite sign), and so are likewise characterized by a reactance X_L.

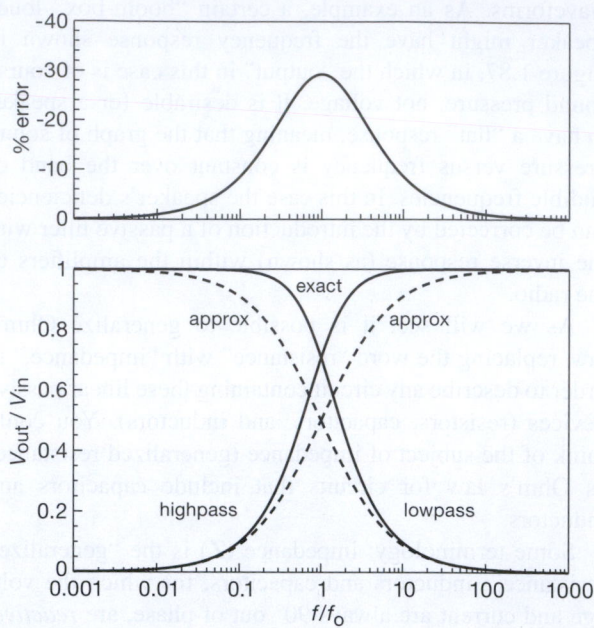

Figure 1.91. Frequency response of single-section *RC* filters, showing the results both of a simple approximation that ignores phase (dashed curve), and the exact result (solid curve). The fractional error (i.e., dashed/solid) is plotted above.

$$\frac{V_{out}}{V_{in}} \asymp \frac{X_C}{R + X_C} = \frac{1/\omega C}{R + 1/\omega C} = \frac{1}{1 + \omega RC} \quad \text{(approximate!)} \tag{1.27}$$

We've plotted that ratio in Figure 1.91 (and also that of its cousin, the *highpass filter*), along with their exact results that we'll understand soon enough in §1.7.8.

You can see that the circuit passes low frequencies fully (because at low frequencies the capacitor's reactance is very high, so it's like a divider with a smaller resistor atop a larger one) and that it blocks high frequencies. In particular, the crossover from "pass" to "block" (often called the *breakpoint*) occurs at a frequency ω_0 at which the capacitor's reactance $(1/\omega_0 C)$ is equal to the resistance R: $\omega_0 = 1/RC$. At frequencies well beyond the crossover (where the product $\omega RC \gg 1$), the output decreases inversely with increasing frequency; that makes sense because the reactance of the capacitor, already much smaller than R, continues dropping as $1/\omega$. It's worth noting that, even with our "ignoration of phase shifts," the equation (and graph) for the ratio of voltages is quite accurate at both low and high frequencies and is only modestly in error around the crossover frequency, where the correct ratio is $V_{out}/V_{in} = 1/\sqrt{2} \approx 0.7$, rather than the 0.5 that we got.[38]

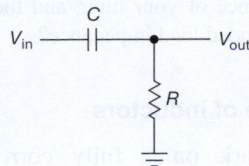

Figure 1.92. Highpass filter.

B. *RC* highpass filter (approximate)

You get the reverse behavior (pass high frequencies, block low) by interchanging R and C, as in Figure 1.92. Treating it as a frequency-dependent voltage divider, and ignoring phase shifts once again, we get (see Figure 1.91)

$$\frac{V_{out}}{V_{in}} \simeq \frac{R}{R + X_C} = \frac{R}{R + 1/\omega C} = \frac{\omega RC}{1 + \omega RC} \quad (approximate!)$$

(1.28)

High frequencies (above the same crossover frequency as before, $\omega \gg \omega_0 = 1/RC$) pass through (because the capacitor's reactance is much smaller than R), whereas frequencies well below the crossover are blocked (the capacitor's reactance is much larger than R). As before, the equation and graph are accurate at both ends, and only modestly in error at the crossover, where the correct ratio is, again, $V_{out}/V_{in} = 1/\sqrt{2}$.

C. Blocking capacitor

Sometimes you want to let some band of signal frequencies pass through a circuit, but you want to block any steady dc voltage that may be present (we'll see how this can happen when we learn about amplifiers in the next chapter). You can do the job with an *RC* highpass filter if you choose the crossover frequency correctly: a highpass filter always blocks dc, so what you do is choose component values so that the crossover frequency is *below* all frequencies of interest. This is one of the more frequent uses of a capacitor and is known as a dc *blocking capacitor*.

For instance, every stereo audio amplifier has all its inputs capacitively coupled, because it doesn't know what dc level its input signals might be riding on. In such a coupling application you always pick R and C so that all frequencies of interest (in this case, 20 Hz to 20 kHz) are passed

---

[38] Of course, it fails to predict anything about phase shifts in this circuit. As we'll see later, the output signal's phase lags the input by 90° at high frequencies, going smoothly from 0° at low frequencies, with a 45° lag at ω_0 (see Figure 1.104 in §1.7.9).

without loss (attenuation). That determines the product RC: $RC > 1/\omega_{min}$, where for this case you might choose $f_{min} \approx 5\,\text{Hz}$, and so $RC = 1/\omega_{min} = 1/2\pi f_{min} \approx 30\,\text{ms}$.

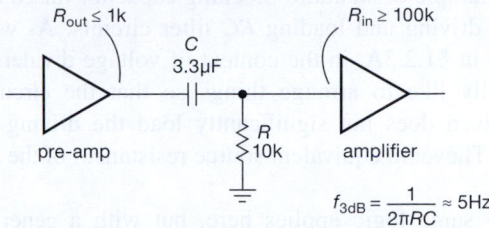

Figure 1.93. "Blocking capacitor": a highpass filter for which all signal frequencies of interest are in the passband.

You've got the product, but you still have to choose individual values for R and C. You do this by noticing that the input signal sees a load equal to R at signal frequencies (where C's reactance is small – it's just a piece of wire there), so you choose R to be a reasonable load, i.e., not so small that it's hard to drive, and not so large that the circuit is prone to signal pickup from other circuits in the box. In the audio business it's common to see a value of $10\,\text{k}\Omega$, so we might choose that value, for which the corresponding C is $3.3\,\mu\text{F}$ (Figure 1.93). The circuit connected to the output should have an input resistance much greater than $10\,\text{k}\Omega$, to avoid loading effects on the filter's output; and the driving circuit should be able to drive a $10\,\text{k}\Omega$ load without significant attenuation (loss of signal amplitude) to prevent circuit loading effects by the filter on the signal source. It's worth noting that our approximate model, ignoring phase shifts, is perfectly adequate for blocking capacitor design; that is because the signal band is fully in the passband, where the effects of phase shifts are negligible.

In this section we've been thinking in the frequency domain (sinewaves of frequency f). But it's useful to think in the time domain, where, for example, you might use a blocking capacitor to couple pulses, or square waves. In such situations you encounter waveform *distortion*, in the form of "droop" and overshoot (rather than the simple amplitude attenuation and phase shift you get with sinusoidal waves). Thinking in the time domain, the criterion you use to avoid waveform distortion in a pulse of duration T is that the time constant $\tau = RC \gg T$. The resulting droop is approximately T/τ (followed by a comparable overshoot at the next transition).

You often need to know the reactance of a capacitor at a given frequency (e.g., for design of filters). Figure 1.100 in §1.7.8 provides a very useful graph covering large ranges of capacitance and frequency, giving the value of $X_C = 1/2\pi fC$.

D. Driving and loading *RC* filters

This example of an audio-blocking capacitor raised the issue of driving and loading *RC* filter circuits. As we discussed in §1.2.5A, in the context of voltage dividers, you generally like to arrange things so that the circuit being driven does not significantly load the driving resistance (Thévenin equivalent source resistance) of the signal source.

The same logic applies here, but with a generalized kind of resistance that includes the reactance of capacitors (and inductors), known as *impedance*. So a signal source's impedance should generally be small compared with the impedance of the thing being driven.[39] We'll have a precise way of talking about impedance shortly, but it's correct to say that, apart from phase shifts, the impedance of a capacitor is equal to its reactance.

What we want to know, then, are the input and output impedances of the two simple *RC* filters (lowpass and highpass). This sounds complicated, because there are four impedances, and they all vary with frequency. However, if you ask the question the right way, the answer is simple, and the same in all cases!

First, assume that in each case the right thing is being done to the other end of the filter: when we ask the input impedance, we assume the output drives a high impedance (compared with its own); and when we ask the output impedance, we assume the input is driven by a signal source of low internal (Thévenin) impedance. Second, we dispose of the variation of impedances with frequency by asking only for the *worst-case* value; that is, we care what only the *maximum* output impedance of a filter circuit may be (because that is the worst for driving a load), and we care about only the *minimum* input impedance (because that is the hardest to drive).

Now the answer is astonishingly simple: in all cases the worst-case impedance is just *R*.

Exercise 1.23. Show that the preceding statement is correct.

So, for example, if you want to hang an *RC* lowpass filter onto the output of an amplifier whose output resistance is $100\,\Omega$, start with $R = 1k$, then choose C for the breakpoint you want. Be sure that whatever loads the output has an input impedance of at least 10k. You can't go wrong.

Exercise 1.24. Design a two-stage "bandpass" *RC* filter, in which

the first stage is highpass with a breakpoint of 100 Hz, and the second stage is lowpass with a breakpoint of 10 kHz. Assume the input signal source has an impedance of $100\,\Omega$. What is the worst-case output impedance of your filter, and therefore what is the minimum recommended load impedance?

1.7.2 Reactance of inductors

Before we embark on a fully correct treatment of impedance, replete with complex exponentials and the like, let's use our approximation tricks to figure out the reactance of an inductor.

It goes as before: we imagine an inductor L driven by a sinusoidal voltage source of angular frequency ω such that a current $I(t) = I_0 \sin \omega t$ is flowing.[40] Then the voltage across the inductor is

$$V(t) = L\frac{dI(t)}{dt} = L\omega I_0 \cos \omega t.$$

And so the ratio of *magnitudes* of voltage to current – the resistance-like quantity called *reactance* – is just

$$\frac{|V|}{|I|} = \frac{L\omega I_0}{I_0} = \omega L.$$

So. for an inductor,

$$X_\mathrm{L} = \omega L. \tag{1.29}$$

Inductors, like capacitors, have a frequency-dependent reactance; however, here the reactance *increases* with increasing frequency (the opposite of capacitors, where it *decreases* with increasing frequency). So, in the simplest view, a series inductor can be used to pass dc and low frequencies (where its reactance is small) while blocking high frequencies (where its reactance is high). You often see inductors used this way, particularly in circuits that operate at radio frequencies; in that application they're sometimes called *chokes*.

1.7.3 Voltages and currents as complex numbers

At this point it is necessary to get into some complex algebra; you may wish to skip over the math in some of the following sections, taking note of the results as we derive them. A knowledge of the detailed mathematics is not necessary for understanding the remainder of the book. Very little mathematics will be used in later chapters. The section ahead is easily the most difficult for the reader with little mathematical preparation. *Don't be discouraged!*

As we've just seen, there can be phase shifts between

---

[39] With two important exceptions, namely, transmission lines and current sources.

[40] We take the easy path here by specifying the *current*, rather than the voltage; we are rewarded with a simple derivative (rather than a simple integral!).

the voltage and current in an ac circuit being driven by a sinewave at some frequency. Nevertheless, as long as the circuit contains only *linear* elements (resistors, capacitors, inductors), the magnitudes of the currents everywhere in the circuit are still proportional to the magnitude of the driving voltage, so we might hope to find some generalization of voltage, current, and resistance in order to rescue Ohm's law. Evidently a single number won't suffice to specify the current, say, at some point in the circuit, because we must somehow have information about both the magnitude and phase shift.

Although we can imagine specifying the magnitudes and phase shifts of voltages and currents at any point in the circuit by writing them out explicitly, e.g., $V(t) = 23.7\sin(377t + 0.38)$, it turns out that we can meet our requirements more simply by using the algebra of complex numbers to *represent* voltages and currents. Then we can simply add or subtract the complex number representations, rather than laboriously having to add or subtract the actual sinusoidal functions of time themselves. Because the actual voltages and currents are real quantities that vary with time, we must develop a rule for converting from actual quantities to their representations, and vice versa. Recalling once again that we are talking about a single sinewave frequency, ω, we agree to use the following rules.

1. Voltages and currents are *represented* by the complex quantities $\mathbf{V}$ and $\mathbf{I}$. The voltage $V_0\cos(\omega t + \phi)$ is to be represented by the complex number $V_0 e^{j\phi}$. Recall that $e^{j\phi} = \cos\phi + j\sin\phi$, where $j = \sqrt{-1}$.

2. We obtain *actual* voltages and currents by multiplying their complex number representations by $e^{j\omega t}$ and then taking the real part: $V(t) = \mathcal{R}e(\mathbf{V}e^{j\omega t})$, $I(t) = \mathcal{R}e(\mathbf{I}e^{j\omega t})$.

In other words,

circuit voltage versus time	complex number representation
$V_0\cos(\omega t + \phi)$	$V_0 e^{j\phi} = a + jb$

multiply by
$e^{j\omega t}$ and
take real part

(In electronics, the symbol j is used instead of i in the exponential in order to avoid confusion with the symbol i meaning small-signal current.) Thus, in the general case, the actual voltages and currents are given by

$$V(t) = \mathcal{R}e(\mathbf{V}e^{j\omega t})$$
$$= \mathcal{R}e(\mathbf{V})\cos\omega t - \mathcal{I}m(\mathbf{V})\sin\omega t$$

$$I(t) = \mathcal{R}e(\mathbf{I}e^{j\omega t})$$
$$= \mathcal{R}e(\mathbf{I})\cos\omega t - \mathcal{I}m(\mathbf{I})\sin\omega t.$$

For example, a voltage whose complex representation is

$$\mathbf{V} = 5j$$

corresponds to a (real) voltage versus time of

$$V(t) = \mathcal{R}e[5j\cos\omega t + 5j(j)\sin\omega t]$$
$$= -5\sin\omega t \text{ volts.}$$

1.7.4 Reactance of capacitors and inductors

With this convention we can apply complex Ohm's law correctly to circuits containing capacitors and inductors, just as for resistors, once we know the reactance of a capacitor or inductor. Let's find out what these are. We begin with a simple (co)sinusoidal voltage $V_0\cos\omega t$ applied across a capacitor:

$$V(t) = \mathcal{R}e(V_0 e^{j\omega t}).$$

Then, using $I = C(dV(t)/dt)$, we obtain

$$I(t) = -V_0 C\omega\sin\omega t = \mathcal{R}e\left(\frac{V_0 e^{j\omega t}}{-j/\omega C}\right),$$

$$= \mathcal{R}e\left(\frac{V_0 e^{j\omega t}}{\mathbf{Z}_C}\right)$$

i.e., for a capacitor

$$\mathbf{Z}_C = -j/\omega C \quad (= -jX_C);$$

$\mathbf{Z}_C$ is the impedance of a capacitor at frequency ω; it is equal in magnitude to the reactance $X_C = 1/\omega C$ that we found earlier, but with a factor of $-j$ that accounts for the 90° leading phase shift of current versus voltage. As an example, a 1 μF capacitor has an impedance of $-2653j\,\Omega$ at 60 Hz, and $-0.16j\,\Omega$ at 1 MHz. The corresponding reactances are $2653\,\Omega$ and $0.16\,\Omega$.[41] Its reactance (and also its impedance) at dc is infinite.

If we did a similar analysis for an inductor, we would find

$$\mathbf{Z}_L = j\omega L \quad (= jX_L).$$

A circuit containing only capacitors and inductors always has a purely imaginary impedance, meaning that the voltage and current are always 90° out of phase – it is purely reactive. When the circuit contains resistors, there is also

---

[41] Note the convention that the reactance X_C is a real number (the 90° phase shift is implicit in the term "reactance"), but the corresponding impedance is purely imaginary: $\mathbf{Z} = R - jX$.

a real part to the impedance. The term "reactance" in that case means the imaginary part only.

1.7.5 Ohm's law generalized

With these conventions for representing voltages and currents, Ohm's law takes a simple form. It reads simply

$$\mathbf{I} = \mathbf{V}/\mathbf{Z},$$
$$\mathbf{V} = \mathbf{IZ},$$

where the voltage represented by $\mathbf{V}$ is applied across a circuit of impedance $\mathbf{Z}$, giving a current represented by $\mathbf{I}$. The complex impedance of devices in series or parallel obeys the same rules as resistance:

$$\mathbf{Z} = \mathbf{Z_1} + \mathbf{Z_2} + \mathbf{Z_3} + \cdots \quad \text{(series)}, \quad (1.30)$$

$$\mathbf{Z} = \cfrac{1}{\cfrac{1}{\mathbf{Z_1}} + \cfrac{1}{\mathbf{Z_2}} + \cfrac{1}{\mathbf{Z_3}} + \cdots} \quad \text{(parallel)}. \quad (1.31)$$

Finally, for completeness we summarize here the formulas for the impedance of resistors, capacitors, and inductors:

$$\mathbf{Z_R} = R \qquad \text{(resistor)},$$
$$\mathbf{Z_C} = -j/\omega C = 1/j\omega C \quad \text{(capacitor)}, \quad (1.32)$$
$$\mathbf{Z_L} = j\omega L \qquad \text{(inductor)}.$$

With these rules we can analyze many ac circuits by the same general methods we used in handling dc circuits, i.e., application of the series and parallel formulas and Ohm's law. Our results for circuits such as voltage dividers will look nearly the same as before. For multiply-connected networks we may have to use Kirchhoff's laws, just as with dc circuits, in this case using the complex representations for V and I: the sum of the (complex) voltage drops around a closed loop is zero, and the sum of the (complex) currents into a point is zero. The latter rule implies, as with dc circuits, that the (complex) current in a series circuit is the same everywhere.

Exercise 1.25. Use the preceding rules for the impedance of devices in parallel and in series to derive the formulas (1.17) and (1.18) for the capacitance of two capacitors (a) in parallel and (b) in series. *Hint:* in each case, let the individual capacitors have capacitances C_1 and C_2. Write down the impedance of the parallel or series combination; then equate it to the impedance of a capacitor with capacitance C. Then find C.

Let's try out these techniques on the simplest circuit imaginable, an ac voltage applied across a capacitor, which we looked at earlier, in §1.7.1. Then, after a brief look at power in reactive circuits (to finish laying the groundwork),

we'll analyze (correctly, this time) the simple but extremely important and useful *RC* lowpass and highpass filter circuits.

Imagine putting a $1\,\mu\text{F}$ capacitor across a 115 volts (rms) 60 Hz powerline. What current flows? Using complex Ohm's law, we have

$$\mathbf{Z} = -j/\omega C.$$

Therefore, the current is given by

$$\mathbf{I} = \mathbf{V}/\mathbf{Z}.$$

The phase of the voltage is arbitrary, so let us choose $\mathbf{V} = A$, i.e., $V(t) = A\cos\omega t$, where the amplitude $A = 115\sqrt{2} \approx 163$ volts. Then

$$\mathbf{I} = j\omega C A \approx -0.061\sin\omega t.$$

The resulting current has an amplitude of 61 mA (43 mA rms) and leads the voltage by 90°. This agrees with our previous calculation. More simply, we could have noticed that the impedance of the capacitor is negative imaginary, so whatever the absolute phase of V, the phase of I_{cap} must lead by 90°. And in general the phase angle between current and voltage, for any two-terminal *RLC* circuit, is equal to the angle of the (complex) impedance of that circuit.

Note that if we wanted to know just the magnitude of the current, and didn't care what the relative phase was, we could have avoided doing any complex algebra: if

$$\mathbf{A} = \mathbf{B}/\mathbf{C},$$

then

$$A = B/C,$$

where A, B, and C are the magnitudes of the respective complex numbers; this holds for multiplication, also (see Exercise 1.26). Thus, in this case,

$$I = V/Z = \omega C V.$$

This trick, which we used earlier (because we didn't know any better), is often useful.

Surprisingly, there is no power dissipated by the capacitor in this example. Such activity won't increase your electric bill; you'll see why in the next section. Then we will go on to look at circuits containing resistors and capacitors with our complex Ohm's law.

Exercise 1.26. Show that, if $\mathbf{A} = \mathbf{BC}$, then $A = BC$, where A, B, and C are magnitudes. *Hint:* represent each complex number in polar form, i.e., $\mathbf{A} = Ae^{j\theta}$.

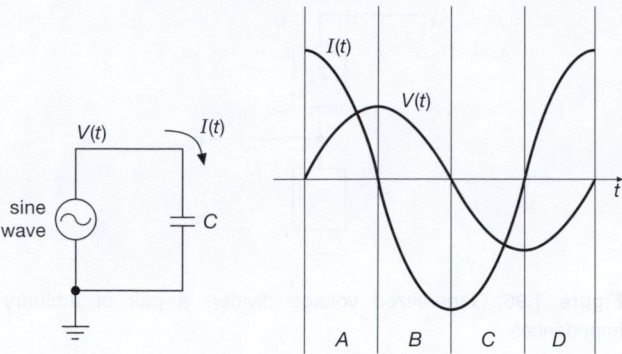

Figure 1.94. The power delivered to a capacitor is zero over a full sinusoidal cycle, owing to the 90° phase shift between voltage and current.

1.7.6 Power in reactive circuits

The instantaneous power delivered to any circuit element is always given by the product $P = VI$. However, in reactive circuits where V and I are not simply proportional, you can't just multiply their amplitudes together. Funny things can happen; for instance, the sign of the product can reverse over one cycle of the ac signal. Figure 1.94 shows an example. During time intervals A and C, power is being delivered to the capacitor (albeit at a variable rate), causing it to charge up; its stored energy is increasing (power is the rate of change of energy). During intervals B and D, the power delivered to the capacitor is negative; it is discharging. The average power over a whole cycle for this example is in fact exactly zero, a statement that is always true for any purely reactive circuit element (inductors, capacitors, or any combination thereof). If you know your trigonometric integrals, the next exercise will show you how to prove this.

Exercise 1.27. Optional exercise: prove that a circuit whose current is 90° out of phase with the driving voltage consumes no power, averaged over an entire cycle.

How do we find the average power consumed by an arbitrary circuit? In general, we can imagine adding up little pieces of VI product, then dividing by the elapsed time. In other words,

$$P = \frac{1}{T}\int_0^T V(t)I(t)\,dt, \qquad (1.33)$$

where T is the time for one complete cycle. Luckily, that's almost never necessary. Instead, it is easy to show that the average power is given by

$$P = \mathcal{R}e(\mathbf{VI}^*) = \mathcal{R}e(\mathbf{V}^*\mathbf{I}), \qquad (1.34)$$

where $\mathbf{V}$ and $\mathbf{I}$ are complex rms amplitudes (and an asterisk means *complex conjugate* – see the math review, Appendix A, if this is unfamiliar).

Let's take an example. Consider the preceding circuit, with a 1 volt (rms) sinewave driving a capacitor. We'll do everything with rms amplitudes, for simplicity. We have

$$\mathbf{V} = 1,$$

$$\mathbf{I} = \frac{\mathbf{V}}{-j/\omega C} = j\omega C,$$

$$P = \mathcal{R}e(\mathbf{VI}^*) = \mathcal{R}e(-j\omega C) = 0.$$

That is, the average power is zero, as stated earlier.

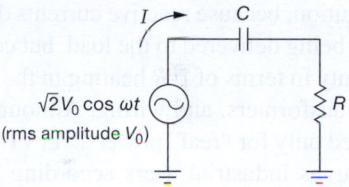

Figure 1.95. Power and power factor in a series *RC* circuit.

As another example, consider the circuit shown in Figure 1.95. Our calculations go like this:

$$\mathbf{Z} = R - \frac{j}{\omega C},$$

$$\mathbf{V} = V_0,$$

$$\mathbf{I} = \frac{\mathbf{V}}{\mathbf{Z}} = \frac{V_0}{R - (j/\omega C)} = \frac{V_0\,[R + (j/\omega C)]}{R^2 + (1/\omega^2 C^2)},$$

$$P = \mathcal{R}e(\mathbf{VI}^*) = \frac{V_0^2 R}{R^2 + (1/\omega^2 C^2)}.$$

(In the third line we multiplied numerator and denominator by the complex conjugate of the denominator in order to make the denominator real.) The calculated power[42] is less than the product of the magnitudes of $\mathbf{V}$ and $\mathbf{I}$. In fact, their ratio is called the *power factor*:

$$|\mathbf{V}|\,|\mathbf{I}| = \frac{V_0^2}{[R^2 + (1/\omega^2 C^2)]^{1/2}},$$

$$\text{power factor} = \frac{\text{power}}{|\mathbf{V}|\,|\mathbf{I}|}$$

$$= \frac{R}{[R^2 + (1/\omega^2 C^2)]^{1/2}}$$

---

[42] It's always a good idea to check limiting values: here we see that $P \to V^2/R$ for large C; and for small C the magnitude of the current $|I| \to V_0/X_C$, or $V_0\omega C$, thus $P \to I^2R = V_0^2\omega^2 C^2 R$, in agreement at both limits.

in this case. The power factor is the cosine of the phase angle between the voltage and the current, and it ranges from 0 (purely reactive circuit) to 1 (purely resistive). A power factor of less than 1 indicates some component of reactive current.[43] It's worth noting that the power factor goes to unity, and the dissipated power goes to V^2/R, in the limit of large capacitance (or of high frequency), where the reactance of the capacitor becomes much less than R.

Exercise 1.28. Show that all the average power delivered to the preceding circuit winds up in the resistor. Do this by computing the value of V_R^2/R. What is that power, in watts, for a series circuit of a 1 μF capacitor and a 1.0k resistor placed across the 115 volt (rms), 60 Hz powerline?

Power factor is a serious matter in large-scale electrical power distribution, because reactive currents don't result in useful power being delivered to the load, but cost the power company plenty in terms of I^2R heating in the resistance of generators, transformers, and wiring. Although residential users are billed only for "real" power [$\mathcal{R}e(\mathbf{VI}^*)$], the power company charges industrial users according to the power factor. This explains the capacitor yards that you see behind large factories, built to cancel the inductive reactance of industrial machinery (i.e., motors).

Exercise 1.29. Show that adding a series capacitor of value $C = 1/\omega^2L$ makes the power factor equal 1.0 in a series RL circuit. Now do the same thing, but with the word "series" changed to "parallel."

1.7.7 Voltage dividers generalized

Our original voltage divider (Figure 1.6) consisted of a pair of resistors in series to ground, input at the top and output at the junction. The generalization of that simple resistive divider is a similar circuit in which either or both resistors are replaced with a capacitor or inductor (or a more complicated network made from R, L, and C), as in Figure 1.96. In general, the division ratio V_{out}/V_{in} of such a divider is not constant, but depends on frequency (as we have already seen, in our approximate treatment of the lowpass and highpass filters in §1.7.1). The analysis is straightforward:

$$\mathbf{I} = \frac{\mathbf{V}_{in}}{\mathbf{Z}_{total}},$$

$$\mathbf{Z}_{total} = \mathbf{Z}_1 + \mathbf{Z}_2$$

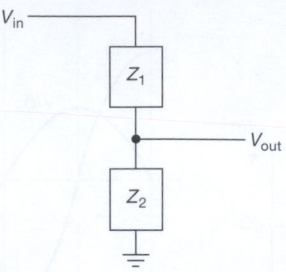

Figure 1.96. Generalized voltage divider: a pair of arbitrary impedances.

$$\mathbf{V}_{out} = \mathbf{I}\mathbf{Z}_2 = \mathbf{V}_{in}\frac{\mathbf{Z}_2}{\mathbf{Z}_1 + \mathbf{Z}_2}.$$

Rather than worrying about this result in general, let's look at some simple (but very important) examples, beginning with the RC highpass and lowpass filters we approximated earlier.

1.7.8 RC highpass filters

We've seen that by combining resistors with capacitors it is possible to make frequency-dependent voltage dividers, owing to the frequency dependence of a capacitor's impedance $\mathbf{Z}_C = -j/\omega C$. Such circuits can have the desirable property of passing signal frequencies of interest while rejecting undesired signal frequencies. In this subsection and the next we revisit the simple lowpass and highpass RC filters, correcting the approximate analysis of §1.7.1; though simple, these circuits are important and widely used. Chapter 6 and Appendix E describe filters of greater sophistication.

Referring back to the classic RC highpass filter (Figure 1.92), we see that the complex Ohm's law (or the complex voltage-divider equation) gives

$$\mathbf{V}_{out} = \mathbf{V}_{in}\frac{R}{R - j/\omega C} = \mathbf{V}_{in}\frac{R(R + j/\omega C)}{R^2 + (1/\omega^2C^2)}.$$

(For the last step, multiply top and bottom by the complex conjugate of the denominator.) Most often we don't care about the phase of V_{out}, just its amplitude:

$$V_{out} = (\mathbf{V}_{out}\mathbf{V}_{out}^*)^{1/2}$$
$$= \frac{R}{[R^2 + (1/\omega^2C^2)]^{1/2}}V_{in}.$$

Note the analogy to a resistive divider, where

$$V_{out} = \frac{R_2}{R_1 + R_2}V_{in}.$$

Here the impedance of the series RC combination

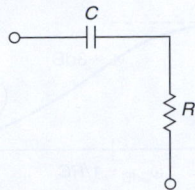

Figure 1.97. Input impedance of unloaded highpass filter.

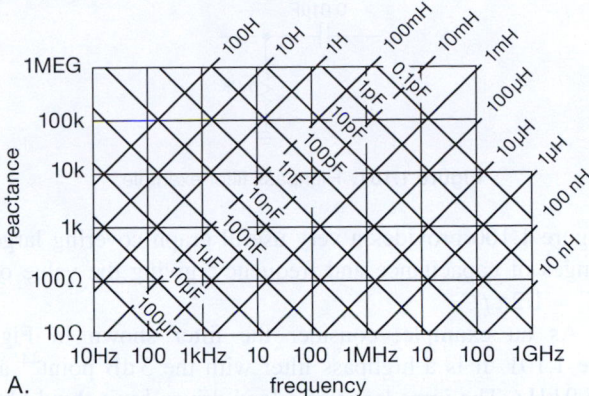

A.

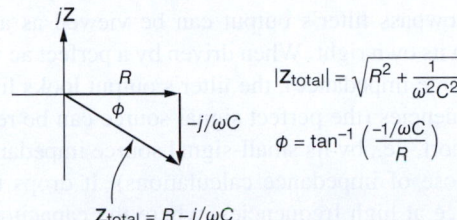

$$|Z_{\text{total}}| = \sqrt{R^2 + \frac{1}{\omega^2 C^2}}$$

$$\phi = \tan^{-1}\left(\frac{-1/\omega C}{R}\right)$$

$$Z_{\text{total}} = R - j/\omega C$$

Figure 1.98. Impedance of series RC.

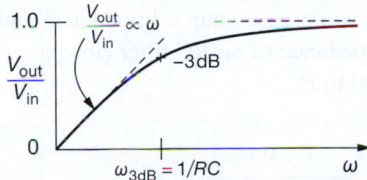

Figure 1.99. Frequency response of highpass filter. The corresponding phase shift goes smoothly from $+90°$ (at $\omega = 0$), through $+45°$ (at $\omega_{3\text{dB}}$), to $0°$ (at $\omega = \infty$), analogous to the lowpass filter's phase shift (Figure 1.104).

(Figure 1.97) is as shown in Figure 1.98. So the "response" of this circuit, ignoring phase shifts by taking magnitudes of the complex amplitudes, is given by

$$V_{\text{out}} = \frac{R}{[R^2 + (1/\omega^2 C^2)]^{1/2}} V_{\text{in}}$$

$$= \frac{2\pi f R C}{[1 + (2\pi f R C)^2]^{1/2}} V_{\text{in}} \qquad (1.35)$$

and looks as shown in Figure 1.99 (and earlier in Figure 1.91).

Note that we could have gotten this result immediately by taking the ratio of the *magnitudes* of impedances, as in Exercise 1.26 and the example immediately preceding it; the numerator is the magnitude of the impedance of the lower leg of the divider (R), and the denominator is the magnitude of the impedance of the series combination of R and C.

As we noted earlier, the output is approximately equal to the input at high frequencies (how high? $\omega \gtrsim 1/RC$) and goes to zero at low frequencies. The highpass filter is

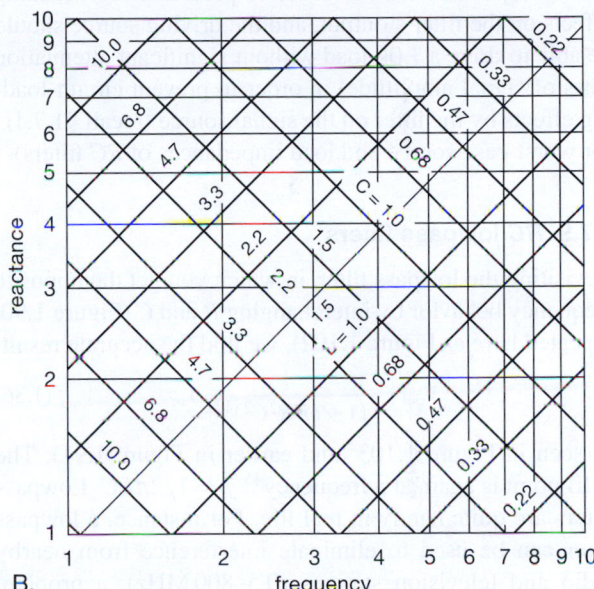

B.

Figure 1.100. A: Reactance of inductors and capacitors versus frequency; all decades are identical, except for scale. B: A single decade from part A expanded, with standard 20% component values (EIA "E6") shown.

very common; for instance, the input to the oscilloscope can be switched to "ac coupling." That's just an RC highpass filter with the bend at about 10 Hz (you would use ac coupling if you wanted to look at a small signal riding on a large dc voltage). Engineers like to refer to the $-3\,\text{dB}$ "breakpoint" of a filter (or of any circuit that behaves like a filter). In the case of the simple RC high-pass filter, the $-3\,\text{dB}$ breakpoint is given by

$$f_{3\text{dB}} = 1/2\pi RC.$$

You often need to know the impedance of a capacitor at a given frequency (e.g., for the design of filters).

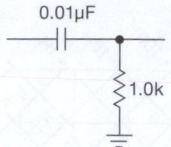

Figure 1.101. Highpass filter example.

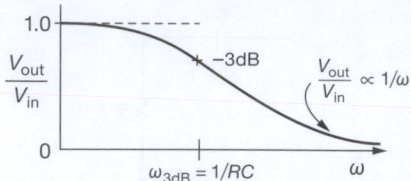

Figure 1.103. Frequency response of lowpass filter.

Figure 1.100 provides a very useful graph covering large ranges of capacitance and frequency, giving the value of $|\mathbf{Z}| = 1/2\pi fC$.

As an example, consider the filter shown in Figure 1.101. It is a highpass filter with the 3 dB point[44] at 15.9 kHz. The impedance of a load driven by it should be much larger than 1.0k in order to prevent circuit loading effects on the filter's output, and the driving source should be able to drive a 1.0k load without significant attenuation (loss of signal amplitude) in order to prevent circuit loading effects by the filter on the signal source (recall §1.7.1D for worst-case source and load impedances of *RC* filters).

1.7.9 *RC* lowpass filters

Revisiting the lowpass filter, in which you get the opposite frequency behavior by interchanging R and C (Figure 1.90, repeated here as Figure 1.102), we find the accurate result

$$V_{out} = \frac{1}{(1 + \omega^2 R^2 C^2)^{1/2}} V_{in} \qquad (1.36)$$

as seen in Figure 1.103 (and earlier in Figure 1.91). The 3 dB point is again at a frequency[45] $f = 1/2\pi RC$. Lowpass filters are quite handy in real life. For instance, a lowpass filter can be used to eliminate interference from nearby radio and television stations (0.5–800 MHz), a problem that plagues audio amplifiers and other sensitive electronic equipment.

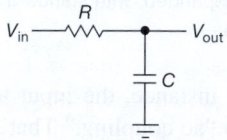

Figure 1.102. Lowpass filter.

Exercise 1.30. Show that the preceding expression for the response of an *RC* lowpass filter is correct.

---

[44] One often omits the minus sign when referring to the −3 dB point.

[45] As mentioned in §1.7.1A, we often like to define the breakpoint frequency $\omega_0 = 1/RC$, and work with frequency ratios ω/ω_0. Then a useful form for the denominator in eq'n 1.36 is $\sqrt{1 + (\omega/\omega_0)^2}$. The same applies to eq'n 1.35, where the numerator becomes ω/ω_0.

The lowpass filter's output can be viewed as a signal source in its own right. When driven by a perfect ac voltage (zero source impedance), the filter's output looks like R at low frequencies (the perfect signal source can be replaced with a short, i.e., by its small-signal source impedance, for the purpose of impedance calculations). It drops to zero impedance at high frequencies, where the capacitor dominates the output impedance. The signal driving the filter sees a load of R plus the load resistance at low frequencies, dropping to just R at high frequencies. As we remarked in §1.7.1D, the worst-case source impedance and the worst-case load impedance of an *RC* filter (lowpass or highpass) are both equal to R.

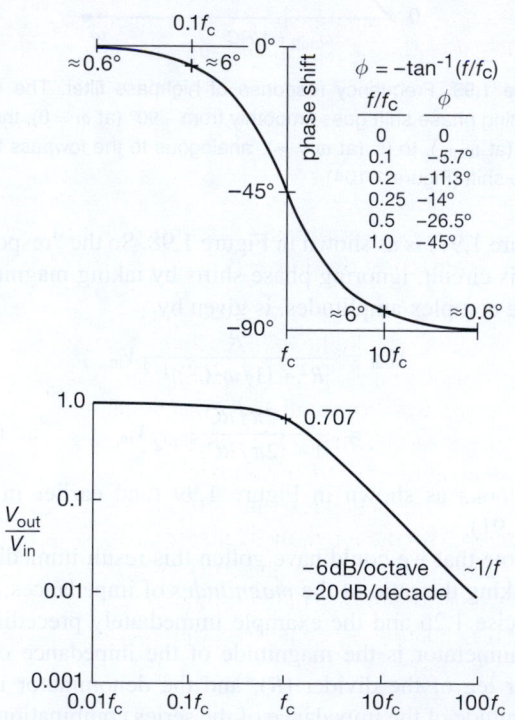

Figure 1.104. Frequency response (phase and amplitude) of lowpass filter plotted on logarithmic axes. Note that the phase shift is −45° at the −3 dB point and is within 6° of its asymptotic value for a decade of frequency change.

In Figure 1.104, we've plotted the same lowpass filter response with logarithmic axes, which is a more common way that it's done. You can think of the vertical axis as decibels, and the horizontal axis as octaves (or decades). On such a plot, equal distances correspond to equal ratios. We've also plotted the phase shift, using a linear vertical axis (degrees) and the same logarithmic frequency axis. This sort of plot is good for seeing the detailed response even when it is greatly attenuated (as at right); we'll see a number of such plots in Chapter 6, when we treat active filters. Note that the filter curve plotted here becomes a straight line at large attenuations, with a slope of -20 dB/decade (engineers prefer to say "-6 dB/octave"). Note also that the phase shift goes smoothly from $0°$ (at frequencies well below the breakpoint) to $-90°$ (well above it), with a value of $-45°$ at the -3 dB point. A rule of thumb for single-section RC filters is that the phase shift is $\approx 6°$ from its asymptotic value at $0.1f_{3\,\mathrm{dB}}$ and at $10f_{3\,\mathrm{dB}}$.

Exercise 1.31. Prove the last assertion.

An interesting question is the following: is it possible to make a filter with some arbitrary specified amplitude response and some other arbitrary specified phase response? Surprisingly, the answer is no: the demands of causality (i.e., that response must follow cause, not precede it) force a relationship between phase and amplitude response of realizable analog filters (known officially as the Kramers–Kronig relation).

1.7.10 *RC* differentiators and integrators in the frequency domain

The RC differentiator that we saw in §1.4.3 is exactly the same circuit as the highpass filter in this section. In fact, it can be considered as either, depending on whether you're thinking of waveforms in the time domain or response in the frequency domain. We can restate the earlier time-domain condition for its proper operation ($dV_{\mathrm{out}} \ll dV_{\mathrm{in}}$) in terms of the frequency response: for the output to be small compared with the input, the signal frequency (or frequencies) must be well below the 3 dB point. This is easy to check: suppose we have the input signal $V_{\mathrm{in}} = \sin \omega t$. Then, using the equation we obtained earlier for the differentiator output, we have

$$V_{\mathrm{out}} = RC\frac{d}{dt}\sin \omega t = \omega RC \cos \omega t,$$

and so $dV_{\mathrm{out}} \ll dV_{\mathrm{in}}$ if $\omega RC \ll 1$, i.e., $RC \ll 1/\omega$. If the input signal contains a range of frequencies, this must hold for the highest frequencies present in the input.

The RC integrator (§1.4.4) is the same circuit as the lowpass filter; by similar reasoning, the criterion for a good integrator is that the lowest signal frequencies must be well above the 3 dB point.

1.7.11 Inductors versus capacitors

Instead of capacitors, inductors can be combined with resistors to make lowpass (or highpass) filters. In practice, however, you rarely see RL lowpass or highpass filters. The reason is that inductors tend to be more bulky and expensive and perform less well (i.e., they depart further from the ideal) than capacitors (see Chapter *1x*). If you have a choice, use a capacitor. One important exception to this general statement is the use of ferrite beads and chokes in high-frequency circuits. You just string a few beads here and there in the circuit; they make the wire interconnections slightly inductive, raising the impedance at very high frequencies and preventing oscillations, without the added series resistance you would get with an RC filter. An *RF choke* is an inductor, usually a few turns of wire wound on a ferrite core, used for the same purpose in RF circuits. Note, however, that inductors are *essential* components in (a) *LC* tuned circuits (§1.7.14), and (b) switch-mode power converters (§9.6.4).

1.7.12 Phasor diagrams

There's a nice graphical method that can be helpful when we are trying to understand reactive circuits. Let's take an example, namely, the fact that an RC filter attenuates 3 dB at a frequency $f = 1/2\pi RC$, which we derived in §1.7.8. This is true for both highpass and lowpass filters. It is easy to get a bit confused here, because at that frequency the reactance of the capacitor equals the resistance of the resistor; so you might at first expect 6 dB attenuation (a factor of 1/2 in voltage). That is what you would get, for example, if you were to replace the capacitor with a resistor of the same impedance magnitude. The confusion arises because the capacitor is reactive, but the matter is clarified by a phasor diagram (Figure 1.105). The axes are the real (resistive) and imaginary (reactive) components of the impedance. In a series circuit like this, the axes also represent the (complex) voltage, because the current is the same everywhere. So for this circuit (think of it as an RC voltage divider) the input voltage (applied across the series RC pair) is proportional to the length of the hypotenuse, and the output voltage (across R only) is proportional to the length of the R leg of the triangle. The diagram represents the situation at the frequency where the capacitor's reactance equals R, i.e., $f = 1/2\pi RC$, and shows that the ratio of output voltage to input voltage is $1/\sqrt{2}$, i.e., -3 dB.

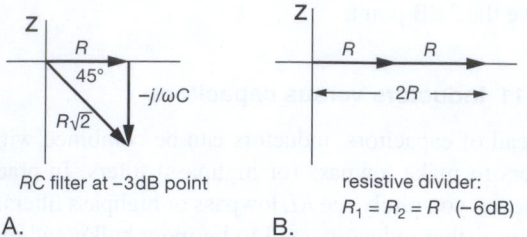

A.   *RC* filter at -3dB point

B.   resistive divider:
$R_1 = R_2 = R$ (-6 dB)

Figure 1.105. Phasor diagram for lowpass filter at 3 dB point.

The angle between the vectors gives the phase shift from input to output. At the 3 dB point, for instance, the output amplitude equals the input amplitude divided by the square root of 2, and it leads by $45°$ in phase. This graphical method makes it easy to read off amplitude and phase relationships in *RLC* circuits. You can use it, for example, to get the response of the highpass filter that we previously derived algebraically.

Exercise 1.32. Use a phasor diagram to derive the response of an *RC* high-pass filter: $V_{out} = V_{in}R/\sqrt{R^2 + (1/\omega^2 C^2)}$.

Exercise 1.33. At what frequency does an *RC* lowpass filter attenuate by 6 dB (output voltage equal to half the input voltage)? What is the phase shift at that frequency?

Exercise 1.34. Use a phasor diagram to obtain the lowpass filter response previously derived algebraically.

In the next chapter (§2.2.8) we'll see a nice example of phasor diagrams in connection with a constant-amplitude phase-shifting circuit.

1.7.13 "Poles" and decibels per octave

Look again at the response of the *RC* lowpass filter (Figures 1.103 and 1.104). Far to the right of the "knee" the output amplitude is dropping proportional to $1/f$. In one octave (as in music, one octave is twice the frequency) the output amplitude will drop to half, or -6 dB; so a simple *RC* filter has a 6 dB/octave falloff. You can make filters with several *RC* sections; then you get 12 dB/octave (two *RC* sections), 18 dB/octave (three sections), and so on. This is the usual way of describing how a filter behaves beyond the cutoff. Another popular way is to say a "three-pole filter," for instance, meaning a filter with three *RC* sections (or one that behaves like one). (The word "pole" derives from a method of analysis that is beyond the scope of this book and that involves complex

transfer functions in the complex frequency plane, known by engineers as the "*s*-plane." This is discussed in the advanced volume, in Chapter *1x*.)

A caution on multistage filters: you can't simply cascade several identical filter sections in order to get a frequency response that is the concatenation of the individual responses. The reason is that each stage will load the previous one significantly (since they're identical), changing the overall response. Remember that the response function we derived for the simple *RC* filters was based on a zero-impedance driving source and an infinite-impedance load. One solution is to make each successive filter section have much higher impedance than the preceding one. A better solution involves active circuits like transistor or operational amplifier (op-amp) interstage "buffers," or active filters. These subjects will be treated in Chapters 2–4, 6, and 13.

1.7.14 Resonant circuits

When capacitors are combined with inductors or are used in special circuits called active filters, it is possible to make circuits that have very sharp frequency characteristics (e.g., a large peak in the response at a particular frequency), compared with the gradual characteristics of the *RC* filters we've seen so far. These circuits find applications in various audio and RF devices. Let's now take a quick look at *LC* circuits (there will be more on them, and active filters, in Chapter 6 and in Appendix E).

A. Parallel and series *LC* circuits

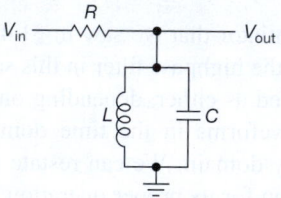

Figure 1.106. *LC* resonant circuit: bandpass filter.

First, consider the circuit shown in Figure 1.106. The impedance of the *LC* combination at frequency f is just

$$\frac{1}{\mathbf{Z}_{LC}} = \frac{1}{\mathbf{Z}_L} + \frac{1}{\mathbf{Z}_C} = \frac{1}{j\omega L} - \frac{\omega C}{j}$$
$$= j\left(\omega C - \frac{1}{\omega L}\right),$$

i.e.,

$$\mathbf{Z}_{LC} = \frac{j}{(1/\omega L) - \omega C}.$$

In combination with R it forms a voltage divider. Because of the opposite behaviors of inductors and capacitors, the impedance of the parallel LC goes to infinity at the *resonant frequency*

$$f_0 = 1/2\pi\sqrt{LC} \qquad (1.37)$$

(i.e., $\omega_0 = 1/\sqrt{LC}$), giving a peak in the response there. The overall response is as shown in Figure 1.107.

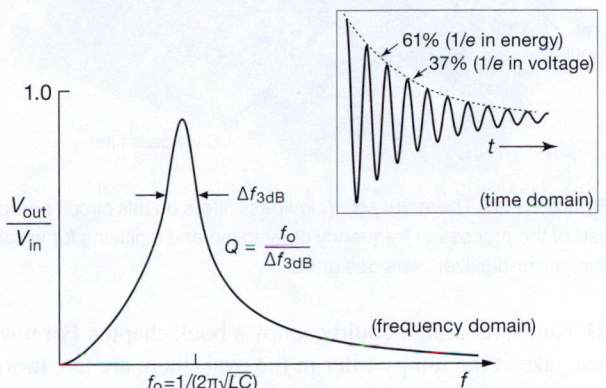

Figure 1.107. Frequency response of parallel LC "tank" circuit. The inset shows the time-domain behavior: a damped oscillation ("ringing") waveform following an input voltage step or pulse.

In practice, losses in the inductor and capacitor limit the sharpness of the peak, but with good design these losses can be made very small. Conversely, a Q-spoiling resistor is sometimes added intentionally to reduce the sharpness of the resonant peak. This circuit is known simply as a parallel LC resonant circuit (or "tuned circuit," or "tank") and is used extensively in RF circuits to select a particular frequency for amplification (the L or C can be variable, so you can tune the resonant frequency). The higher the driving impedance, the sharper the peak; it is not uncommon to drive them with something approaching a current source, as you will see later. The *quality factor Q* is a measure of the sharpness of the peak. It equals the resonant frequency divided by the width at the -3 dB points. For a parallel RLC circuit, $Q = \omega_0 RC$.[46]

Another variety of LC circuit is the series LC (Figure 1.108). By writing down the impedance formulas involved, and assuming that both the capacitor and inductor are ideal, i.e., that they have no resistive losses,[47] you can convince yourself that the impedance of the LC goes to zero

---

[46] Or, equivalently, $Q = R/X_C = R/X_L$, where $X_L = X_C$ are the reactances at ω_0.

[47] We'll see in Chapter *1x* that real components depart from the ideal, often expressed in terms of an effective series resistance, ESR.

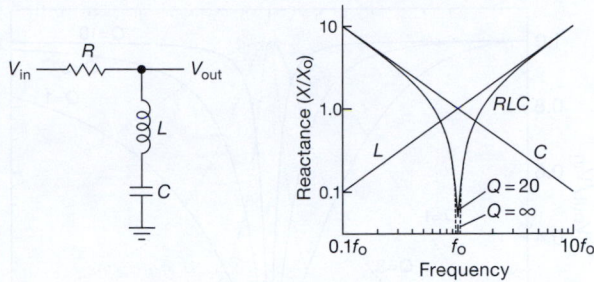

Figure 1.108. LC notch filter ("trap"). The inductive and capacitive reactances behave as shown, but the opposite sign of their complex impedances causes the series impedance to plummet. For ideal components the reactance of the series LC goes completely to zero at resonance; for real-world components the minimum is non-zero, and usually dominated by the inductor.

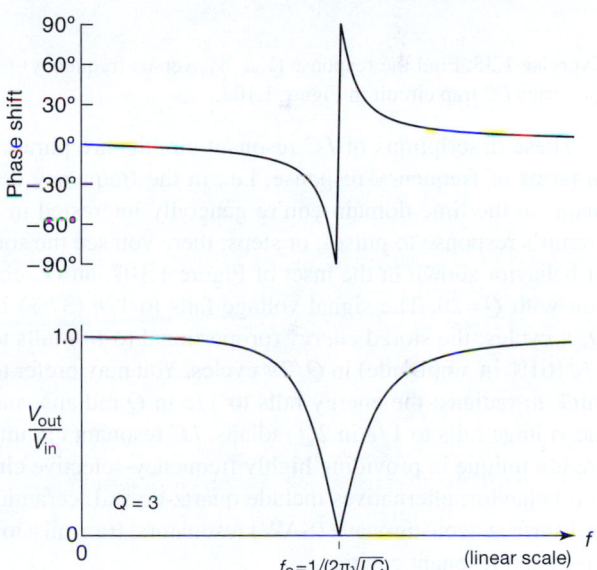

Figure 1.109. Frequency and phase response of the series LC trap. The phase changes abruptly at resonance, an effect seen in other resonator types (see for example Figure 7.36).

at resonance ($f_0 = 1/2\pi\sqrt{LC}$). Such a circuit is a "trap" for signals at or near the resonant frequency, shorting them to ground. Again, this circuit finds application mainly in RF circuits. Figure 1.109 shows what the response looks like. The Q of a series RLC circuit is $Q = \omega_0 L/R$.[48] To see the impact of increasing Q, look at the accurate plots of tank and notch response in Figure 1.110.

---

[48] Or, equivalently, $Q = X_L/R = X_C/R$, where $X_L = X_C$ are the reactances at ω_0.

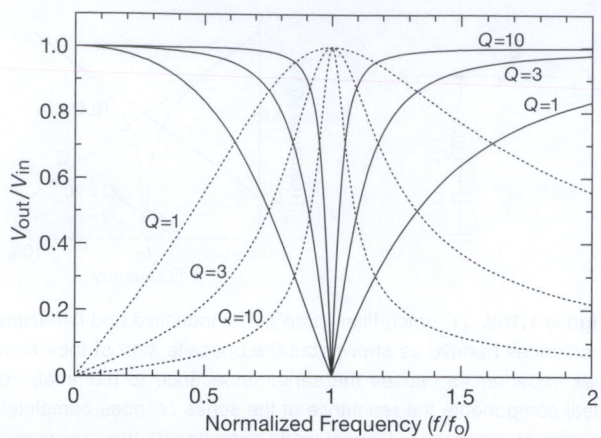

Figure 1.110. Response of *LC* tank (dotted curves) and trap (solid curves) for several values of quality factor, *Q*.

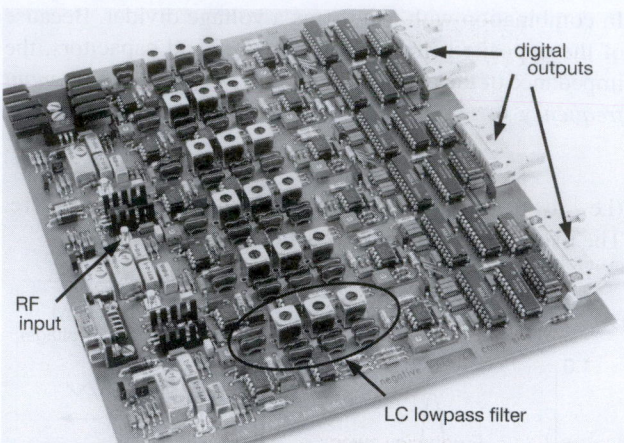

Figure 1.111. There are six *LC* lowpass filters on this circuit board, part of the process of frequency conversion and digitizing for which this "mixer-digitizer" was designed.

Exercise 1.35. Find the response (V_{out}/V_{in} versus frequency) for the series *LC* trap circuit in Figure 1.108.

These descriptions of *LC* resonant circuits are phrased in terms of frequency response, i.e., in the frequency domain. In the time domain you're generally interested in a circuit's response to pulses, or steps; there you see the sort of behavior shown in the inset of Figure 1.107, an *LC* circuit with $Q=20$. The signal voltage falls to $1/e$ (37%) in Q/π cycles; the stored *energy* (proportional to v^2) falls to $1/e$ (61% in amplitude) in $Q/2\pi$ cycles. You may prefer to think in radians: the energy falls to $1/e$ in Q radians, and the voltage falls to $1/e$ in $2Q$ radians. *LC* resonant circuits are not unique in providing highly frequency-selective circuit behavior; alternatives include quartz-crystal, ceramic, and surface acoustic-wave (SAW) resonators; transmission lines; and resonant cavities.

1.7.15 *LC* filters

By combining inductors with capacitors you can produce filters (lowpass, highpass, bandpass) with far sharper behavior in frequency response than you can with a filter made from a simple *RC*, or from any number of cascaded *RC* sections. We'll see more of this, and the related topic of active filters, in Chapter 6. But it's worth admiring now how well this works, to appreciate the virtue of the humble inductor (an often-maligned circuit component).

As an example, look at Figure 1.111, a photograph of a "mixer-digitizer" circuit board that we built for a project some years back (specifically, a radio receiver with 250 million simultaneous channels). There's lots of stuff on the board, which has to frequency-shift and digitize three

RF bands; its design could occupy a book chapter. For now just gaze at the lumpy filter in the oval (there are five more on the board), comprising three inductors (the square metal cans) and four capacitors (the pairs of shiny oblongs). It's a lowpass filter, designed to cut off at 1.0 MHz; it prevents 'aliases' in the digitized output, a subject we'll visit in Chapter 13.

How well does it work? Figure 1.112 shows a "frequency sweep," in which a sinewave input goes from 0 Hz to 2 MHz as the trace goes from left to right across the screen. The sausage shapes are the "envelope" of the sinewave output, here comparing the *LC* filter with an *RC* lowpass filter with the same 1 MHz cutoff (1 kΩ and 160 pF). The *LC* wins, hands down. The *RC* is pathetic by comparison. It's not even good English to call 1 MHz its "cutoff": it hardly cuts anything off.

1.7.16 Other capacitor applications

In addition to their uses in filters, resonant circuits, differentiators, and integrators, capacitors are needed for several other important applications. We treat these in detail later in the book, mentioning them here only as a preview.

A. Bypassing

The impedance of a capacitor decreases with increasing frequency. This is the basis of another important application: *bypassing*. There are places in circuits where you want to allow a dc voltage, but you don't want signals present. Placing a capacitor across that circuit element (usually a resistor) will help to kill any signals there. You choose the (noncritical) capacitor value so that its

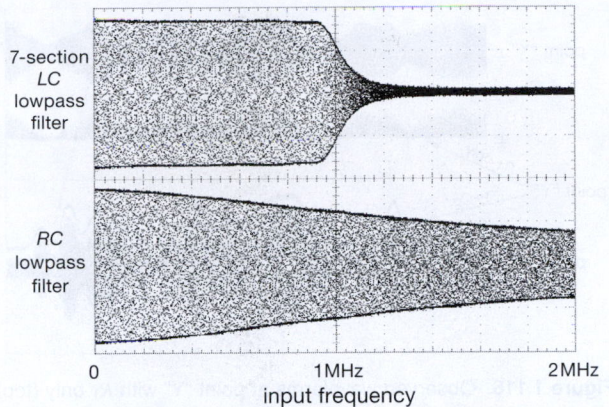

Figure 1.112. Frequency sweep of the *LC* lowpass filter shown in Figure 1.111 compared with an *RC* lowpass filter with the same 1 MHz cutoff frequency. The dark outline is the amplitude envelope of the fast swept sinewave, which achieves a sandpaper appearance in this digital 'scope capture.

impedance at signal frequencies is small compared with what it is bypassing. You will see much more of this in later chapters.

B. Power-supply filtering

We saw this application in §1.6.3, to filter the ripple from rectifier circuits. Although circuit designers often call them *filter* capacitors, this is really a form of bypassing, or energy storage, with large-value capacitors; we prefer the term *storage* capacitor. And these capacitors really are large – they're the big shiny round things you see inside most electronic instruments. We'll get into dc power-supply design in detail in Chapter 9.

C. Timing and waveform generation

As we've seen, a capacitor supplied with a constant current charges up with a ramp waveform. This is the basis of ramp and sawtooth generators, used in analog function generators, oscilloscope sweep circuits, analog–digital converters, and timing circuits. *RC* circuits are also used for timing, and they form the basis of delay circuits (monostable multivibrators). These timing and waveform applications are important in many areas of electronics and will be covered in Chapters 3, 6, 10, and 11.

1.7.17 Thévenin's theorem generalized

When capacitors and inductors are included, Thévenin's theorem must be restated: any two-terminal network of resistors, capacitors, inductors, and signal sources is equivalent to a single complex impedance in series with a single signal source. As before, you find the (complex) impedance

and the signal source (waveform, amplitude, and phase) from the open-circuit output voltage and the short-circuit output current.

1.8 Putting it all together – an AM radio

In our circuit course we tie together the topics of this chapter by hooking up a simple AM radio. The signal that's transmitted is a sinewave at the station's frequency in the AM band (520–1720 kHz), with its amplitude varied ("modulated") according to the audio waveform (Figure 1.113). In other words, an audio waveform described by some function $f(t)$ would be transmitted as a RF signal $[A + f(t)]\sin 2\pi f_c t$; here f_c is the station's "carrier" frequency, and the constant A is added to the audio waveform so that the coefficient $[A + f(t)]$ is never negative.

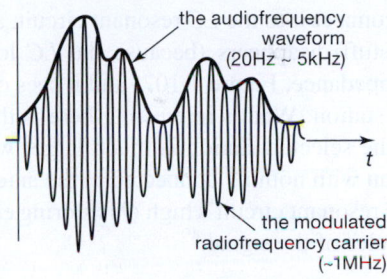

Figure 1.113. An AM signal consists of an RF carrier ($\sim$1 MHz) whose amplitude is varied by the audio-frequency signal (speech or music; audible frequencies up to $\sim$5 kHz). The audio waveform is dc offset so that the envelope does not cross zero.

At the receiver end (that's *us!*) the task is to select this station (among many) and somehow extract the modulating *envelope*, which is the desired audio signal. Figure 1.114 shows the simplest AM radio; it is the "crystal set" of yesteryear. It's really quite straightforward: the parallel *LC* resonant circuit is tuned to the station's frequency by the variable capacitor C_1 (§1.7.14); the diode D is a half-wave rectifier (§1.6.2), which (if ideal) would pass only the positive half-cycles of the modulated carrier; and R_1 provides a light load, so that the rectified output follows the half-cycles back down to zero. We're almost done. We add small capacitor C_2 to prevent the output from following the fast half-cycles of carrier (it's a storage capacitor, §1.7.16B), choosing the time constant $R_1 C_2$ to be long compared with a carrier period ($\sim$1 μs), but short compared with the period of the highest audio frequency ($\sim$200 μs).

Figures 1.115 and 1.116 show what you see when you probe around with a 'scope. The *bare* antenna shows plenty of low-frequency pickup (mostly 60 Hz ac powerline), and a tiny bit of signal from all the AM stations at once. But

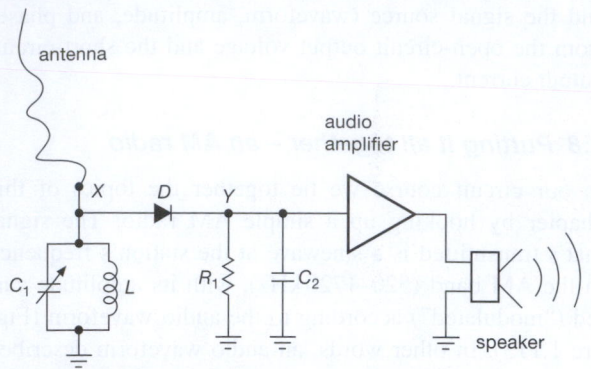

Figure 1.114. The simplest AM receiver. Variable capacitor C_1 tunes the desired station, diode D picks off the positive envelope (smoothed by R_1C_2), and the resulting weak audio signal is amplified to drive the loudspeaker, loudly.

when you connect it to the LC resonant circuit, all the low-frequency stuff disappears (because the LC looks like a very low impedance, Figure 1.107) and it sees only the selected AM station. What's interesting here is that the amplitude of the selected station is much larger with the LC attached than with nothing connected to the antenna: that's because the resonant circuit's high Q is storing energy from multiple cycles of the signal.[49]

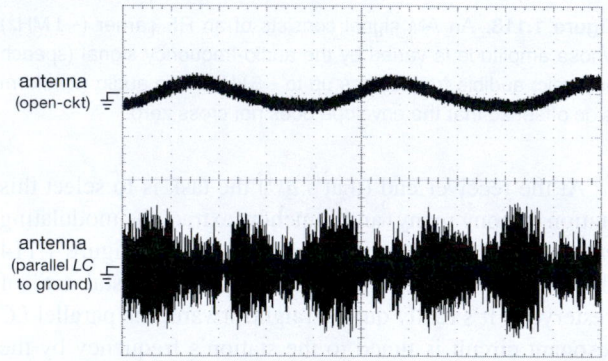

Figure 1.115. Observed waveforms at point "X" from the bare antenna (top) and with the LC connected. Note that the low-frequency junk disappears and that the radio signal gets *larger*. These are single-shot traces, in which the ∼1 MHz radiofrequency carrier appears as a solid filled area. Vertical: 1 V/div; horizontal: 4 ms/div.

The audio amplifier is fun, too, but we're not ready for it. We'll see how to make one of those in Chapter 2 (with discrete transistors), and again in Chapter 4 (with operational amplifiers, the Lego™ block of analog design).

---

[49] There are more complicated ways of framing this, but you don't really want to know just yet…

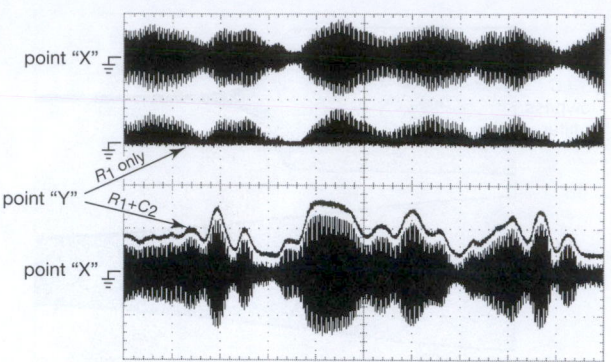

Figure 1.116. Observed waveforms at point "Y" with R_1 only (top) and with smoothing capacitor C_2 included (bottom). The upper pair is a single-shot capture (with the ∼1 MHz carrier appearing as solid black), and the lower pair is a separate single-shot capture, in which we have offset the rectified wave for clarity. Vertical: 1 V/div; horizontal: 1 ms/div.

And one amusing final note: in our class, we like to show the effect of probing point "X" with a length of BNC (bayonet Neill–Concelman) cable going to a 'scope input (that's how we start out, in the first week). When we do that, the cable's capacitance (about 30 pF/ft) adds to C_1, lowering the resonant frequency and so tuning to a different station. If we choose right, it changes *languages* (from English to Spanish)! The students howl with laughter – a language-translating electronic component. Then we use an ordinary 'scope probe, with its ∼10 pF of capacitance: no change of station, nor of language.

1.9 Other passive components

In the following subsections we would like to introduce briefly an assortment of miscellaneous but essential components. If you are experienced in electronic construction, you may wish to proceed to the next chapter.

1.9.1 Electromechanical devices: switches

These mundane but important devices seem to wind up in most electronic equipment. It is worth spending a few paragraphs on the subject (and there's more in Chapter *1x*). Figures 1.117 and 1.118 show some common switch types.

A. Toggle switches

The simple toggle switch is available in various configurations, depending on the number of poles; Figure 1.119 shows the usual ones (SPST indicates a single-pole single-throw switch, SPDT indicates a single-pole double-throw

Figure 1.117. Switch Smorgasbord. The nine switches at right are momentary-contact ("pushbutton") switches, including both panel-mounting and PCB-mounting types (PCB, printed-circuit board). To their left are additional types, including lever-actuated and multipole styles. Above them are a pair of panel-mounting binary-coded *thumbwheel* switches, to the left of which is a matrix-encoded hexadecimal keypad. The switches at center foreground are toggle switches, in both panel-mounting and PCB-mounting varieties; several actuator styles are shown, including a locking variety (fourth from front) that must be pulled before it will switch. The rotary switches in the left column illustrate binary-coded types (the three in front and the larger square one), and the traditional multipole–multiposition configurable wafer switches.

Figure 1.118. Board-mounted "DIP switches." Left group, front to back and left to right (all are SPST): single station side-action toggle; three-station side-action, two-station rocker, and single-station slide; eight-station slide (low-profile) and six-station rocker; eight-station slide and rocker. Middle group (all are hexadecimal coded): six-pin low-profile, six-pin with top or side adjust; 16-pin with true and complement coding. Right group: 2 mm×2 mm surface-mount header block with movable jumper ("shunt"), $0.1''\times0.1''$ (2.54 mm×2.54 mm) through-hole header block with shunts; 18-pin SPDT (common actuator); eight-pin dual SPDT slide and rocker; 16-pin quad SPDT slide (two examples).

switch, and DPDT indicates a double-pole double-throw switch). Toggle switches are also available with "center OFF" positions and with up to four poles switched simultaneously. Toggle switches are always "break before make," e.g., the moving contact never connects to both terminals in an SPDT switch.

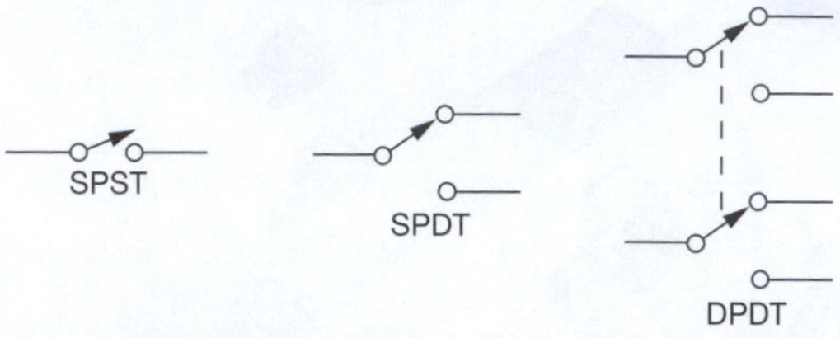

Figure 1.119. Fundamental switch types.

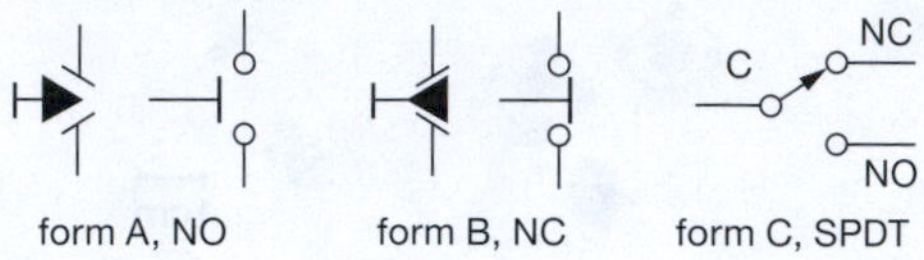

Figure 1.120. Momentary-contact (pushbutton) switches.

B. Pushbutton switches

Pushbutton switches are useful for momentary-contact applications; they are drawn schematically as shown in Figure 1.120 (NO and NC mean normally open and normally closed). For SPDT momentary-contact switches, the terminals must be labeled NO and NC, whereas for SPST types the symbol is self-explanatory. Momentary-contact switches are always "break before make." In the electrical (as opposed to electronic) industry, the terms form A, form B, and form C are used to mean SPST (NO), SPST (NC), and SPDT, respectively.

C. Rotary switches

Rotary switches are available with many poles and many positions, often as kits with individual wafers and shaft hardware. Both *shorting* (make-before-break) and *non-shorting* (break-before-make) types are available, and they can be mixed on the same switch. In many applications the shorting type is useful to prevent an open circuit between switch positions, because circuits can go amok with unconnected inputs. Nonshorting types are necessary if the separate lines being switched to one common line must not ever be connected to each other.

Sometimes you don't really want all those poles, you just want to know how many clicks (detents) the shaft has been turned. For that a common form of rotary switch en-

codes its position as a 4-bit binary quantity, thereby saving lots of wires (only five are needed: the four bits, and a common line). An alternative is the use of a *rotary encoder*, an electromechanical panel-mounting device that creates a sequence of N pulse pairs for each full rotation of the knob. These come in two flavors (internally using either mechanical contacts or electro-optical methods), and typically provide from 16 to 200 pulse pairs per revolution. The optical varieties cost more, but they last forever.

D. PC-mounting switches

It's common to see little arrays of switches on printed-circuit (PC) boards, like the ones shown in Figure 1.118. They're often called *DIP switches*, referring to the integrated circuit dual in-line package that they borrow, though contemporary practice increasingly uses the more compact *surface-mount technology* (SMT) package. As the photograph illustrates, you can get coded rotary switches; and because these are used for set-and-forget internal settings, you can substitute a multipin *header* block, with little slide-on "shunts" to make the connections.

E. Other switch types

In addition to these basic switch types, there are available various exotic switches such as Hall-effect switches, reed switches, proximity switches, etc. All switches carry maximum current and voltage ratings; a small toggle switch might be rated at 150 volts and 5 amps. Operation with inductive loads drastically reduces switch life because of arcing during turn-off. It's always OK to operate a switch *below* its maximum ratings, with one notable exception: since many switches rely on substantial current flow to clean away contact oxides, it's important to use a switch that is designed for "dry switching" when switching low-level signals;[50] otherwise you'll get noisy and intermittent operation (see Chapter *1x*).

F. Switch examples

As an example of what can be done with simple switches, let's consider the following problem: suppose you want to sound a warning buzzer if the driver of a car is seated and one of the car doors is open. Both doors and the driver's seat have switches, all normally open. Figure 1.121 shows a circuit that does what you want. If one OR the other door is open (switch closed) AND the seat switch is closed, the buzzer sounds. The words OR and AND are used in a logic sense here, and we will see this example again in

---

[50] These use gold contact plating.

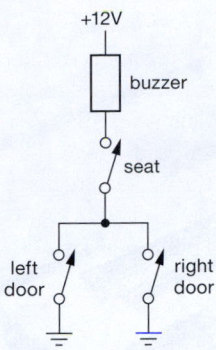

Figure 1.121. Switch circuit example: open door warning.

Chapters 2, 3, and 10 when we talk about transistors and digital logic.

Figure 1.122 shows a classic switch circuit used to turn a ceiling lamp on or off from a switch at either of two entrances to a room.

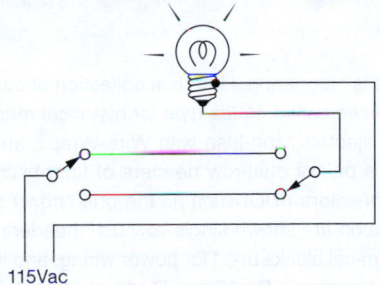

Figure 1.122. Electrician's "three-way" switch wiring.

Exercise 1.36. Although few electronic circuit designers know how, every *electrician* can wire up a light fixture so that any of *N* switches can turn it on or off. See if you can figure out this generalization of Figure 1.122. It requires two SPDT switches and *N*−2 DPDT switches.

1.9.2 Electromechanical devices: relays

Relays are electrically controlled switches. In the traditional electromechanical relay, a coil pulls in an armature (to close the contacts) when sufficient coil current flows. Many varieties are available, including "latching" and "stepping" relays.[51] Relays are available with dc or ac

---

[51] In an amusing historical footnote, the stepping relay used for a century as the cornerstone of telephone exchanges (the "Strowger selector") was invented by a Topeka undertaker, Almon Strowger, evidently because he suspected that telephone calls intended for his business were being routed (by the switchboard operators in his town) to a funeral home competitor.

excitation, and coil voltages from 3 volts up to 115 volts (ac or dc) are common. "Mercury-wetted" and "reed" relays are intended for high-speed (∼1 ms) applications, and giant relays intended to switch thousands of amps are used by power companies.

The *solid-state relay* (SSR) – consisting of a semiconductor electronic switch that is turned on by a LED – provides better performance and reliability than mechanical relays, though at greater cost. SSRs operate rapidly, without contact "bounce," and usually provide for smart switching of ac power (they turn on at the moment of zero voltage, and they turn off at the moment of zero current). Much more on these useful devices in Chapter 12.

As we'll learn, electrically controlled switching of signals within a circuit can be accomplished with transistor switches, without having to use relays of any sort (Chapters 2 and 3). The primary uses of relays are in remote switching and high-voltage (or high-current) switching, where it is important to have complete electrical isolation between the control signal and the circuit being switched.

1.9.3 Connectors

Bringing signals in and out of an instrument, routing signal and dc power around between the various parts of an instrument, providing flexibility by permitting circuit boards and larger modules of the instrument to be unplugged (and replaced) – these are the functions of the *connector*, an essential ingredient (and usually the most unreliable part) of any piece of electronic equipment. Connectors come in a bewildering variety of sizes and shapes.[52] Figures 1.123, 1.124, and 1.125 give some idea of the variety.

A. Single-wire connectors

The simplest kind of connector is the simple pin jack or banana jack used on multimeters, power supplies, etc. It is handy and inexpensive, but not as useful as the shielded-cable or multiwire connectors you often need. The humble binding post is another form of single-wire connector, notable for the clumsiness it inspires in those who try to use it.

B. Shielded-cable connectors

To prevent capacitive pickup, and for other reasons we'll go into in Appendix H, it is usually desirable to pipe signals around from one instrument to another in shielded coaxial cable. The most popular connector is the BNC type that

---

[52] A search for "connector" on the DigiKey website returns 116 categories, with approximately 43,000 individual varieties in stock.

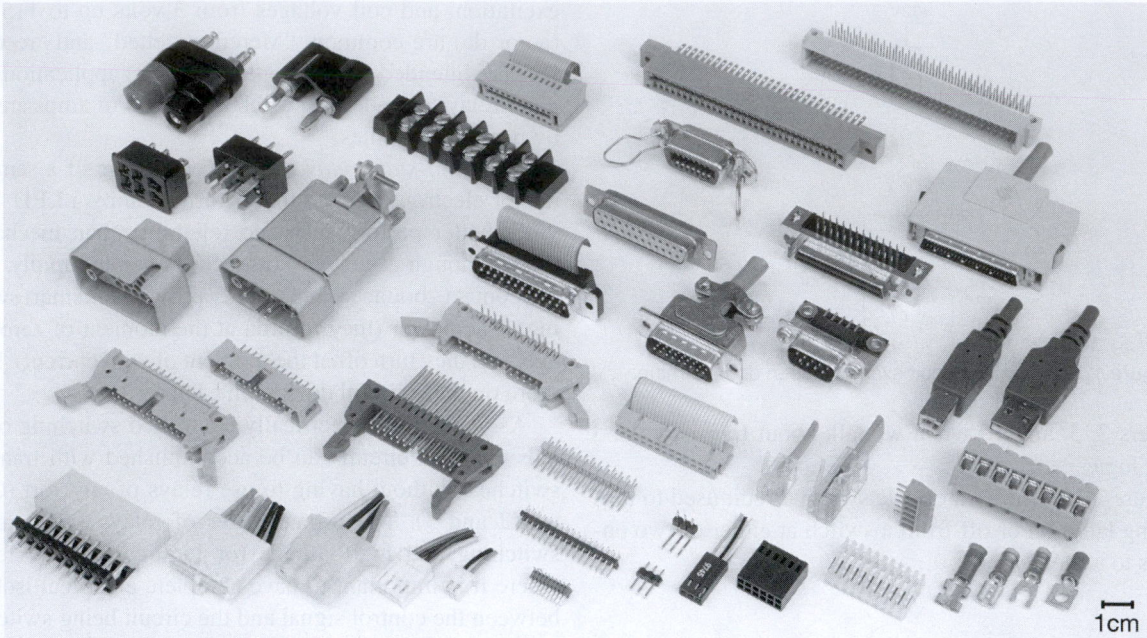

1cm

Figure 1.123. Rectangular connectors. The variety of available multipin connectors is staggering. Here is a collection of common specimens: the five connectors at lower left are multipin nylon power connectors (sometimes called *Molex-type* for historical reasons). Above them are four dual-row box headers (0.1″ spacing, shown with and without latch ejectors, and also with Wire-Wrap® and right-angle tails), and to their right an open ("unshrouded") 0.1″ dual-row header, along with a pair of dual-row headers of finer pitch (2 mm and 1.27 mm). These dual-row male connectors mate with *insulation displacement* connectors (IDC) such as the one shown attached to a short length of ribbon cable (just above the unshrouded header). Just below the ribbon are shown single-row 0.1″ headers, with mating shells (AMP MODU) that accept individual wire leads. At bottom right are several terminal blocks used for power wiring, and four "Faston"-type crimpable spade lugs. Above them are USB connectors, and to their left are the common RJ-45 and RJ-11 modular telephone/data connectors. The popular and reliable D-subminiature connectors are at center, including (right to left) a pair of 50-pin micro-D (cable plug, PCB socket), the 9-pin D-sub, 26-pin high-density, and a pair of 25-pin D-subs (one IDC). Above them are (right to left) a 96-pin VME backplane connector, a 62-pin card-edge connector with solder tails, a "Centronics-type" connector with latching bail, and a card-edge connector with ribbon IDC. At top left is a miscellany – a mating pair of "GR-type" dual banana connectors, a mating pair of Cinch-type connectors, a mating pair of shrouded Winchester-type connectors with locking jackscrews, and (to their right) a screw-terminal barrier block. Not shown here are the *really* tiny connectors used in small portable electronics (smartphones, cameras, etc); you can see a fine example in Figure 1.131.

adorns most instrument front panels. It connects with a quarter-turn twist and completes both the shield (ground) circuit and inner conductor (signal) circuit simultaneously. Like all connectors used to mate a cable to an instrument, it comes in both panel-mounting and cable-terminating varieties.

Among the other connectors for use with coaxial cable are the TNC ("threaded Neill–Concelman," a close cousin of the BNC, but with threaded outer shell), the high-performance but bulky type N, the miniature SMA and SMB, the subminiature LEMO and SMC, and the high-voltage MHV and SHV. The so-called phono jack used in audio equipment is a nice lesson in bad design, because the

inner (signal) conductor mates *before* the shield (ground) when you plug it in; furthermore, the design of the connector is such that both shield and center conductor tend to make poor contact. You've undoubtedly *heard* the results! Not to be outdone, the television industry has responded with its own bad standard, the type-F coax "connector," which uses the unsupported inner wire of the coax as the pin of the male plug, and a shoddy arrangement to mate the shield.[53]

We hereby induct these losers into the Electronic

---

[53] Advocates of each would probably reply "This is our most *modestly* priced receptacle."

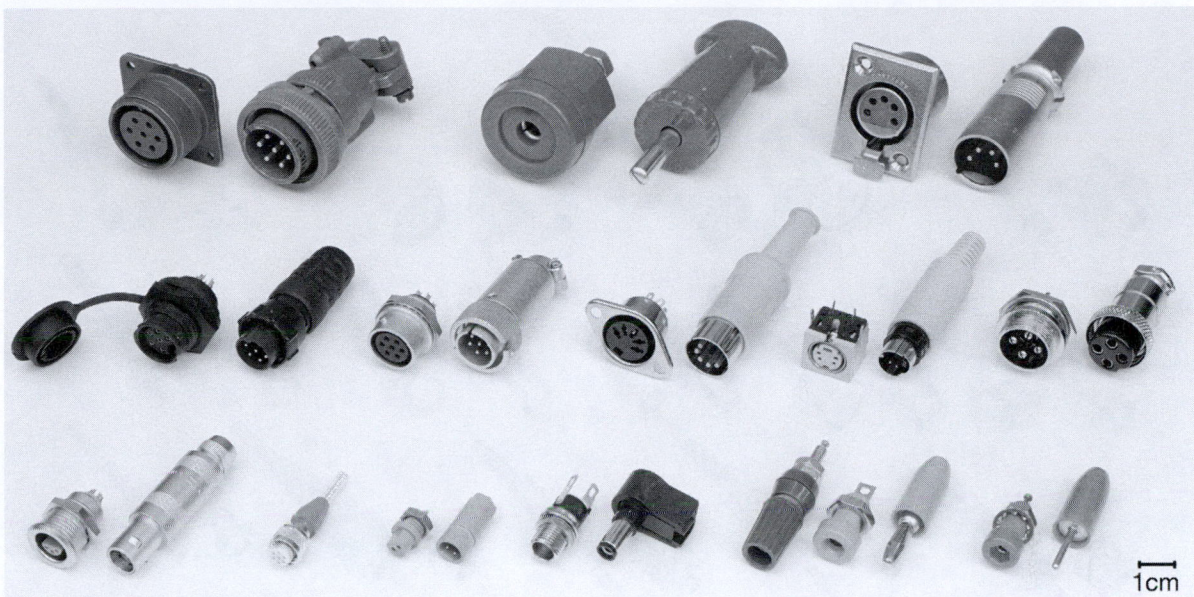

Figure 1.124. Circular connectors. A selection of multipin and other "non-RF" connectors; the panel-mounting receptacle is shown to the left of each cable-mounted plug. Top row, left to right: "MS"-type (MIL-C-5015) rugged connector (available in hundreds of configurations), high-current (50 A) "Supericon," multipin locking XLR. Middle row: weatherproof (Switchcraft EN3), 12 mm video (Hirose RM), circular DIN, circular mini-DIN, 4-pin microphone connector. Bottom row: locking 6-pin (Lemo), microminiature 7-pin shielded (Microtech EP-7S), miniature 2-pin shrouded (Litton SM), 2.5 mm power, banana, pin jack.

Components Hall of Infamy, some charter members of which are shown in Figure 1.126.

C. Multipin connectors

Very frequently electronic instruments demand multiwire cables and connectors. There are literally dozens of different kinds. The simplest example is a three-wire "IEC" powerline cord connector. Among the more popular are the excellent type-D subminiature, the Winchester MRA series, the venerable MS type, and the flat ribbon-cable mass-termination connectors. These and others are shown in Figure 1.123.

Beware of connectors that can't tolerate being dropped on the floor (the miniature hexagon connectors are classic) or that don't provide a secure locking mechanism (e.g., the Jones 300 series).

D. Card-edge connectors

The most common method used to make connection to printed-circuit cards is the card-edge connector, which mates to a row of gold-plated contacts at the edge of the card; common examples are the motherboard connectors that accept plug-in computer memory modules. Card-edge connectors may have from 15 to 100 or more connections,

and they come with different lug styles according to the method of connection. You can solder them to a "motherboard" or "backplane," which is itself just another PCB containing the interconnecting wiring between the individual circuit cards. Alternatively, you may want to use edge connectors with standard solder-lug terminations, particularly in a system with only a few cards. A more reliable (though more costly) solution is the use of "two-part" PCB connectors, in which one part (soldered onto the board) mates with the other part (on a backplane, etc); an example is the widely used VME (VersaModule Eurocard) connector (upper right-hand corner of Figure 1.123).

1.9.4 Indicators

A. Meters

To read out the value of some voltage or current, you have a choice between the time-honored moving-pointer type of meter and digital-readout meters. The latter are more expensive and more accurate. Both types are available in a variety of voltage and current ranges. There are, in addition, exotic panel meters that read out such things as VUs (volume units, an audio dB scale), expanded-scale ac volts (e.g., 105 to 130 V), temperature (from a thermocouple),

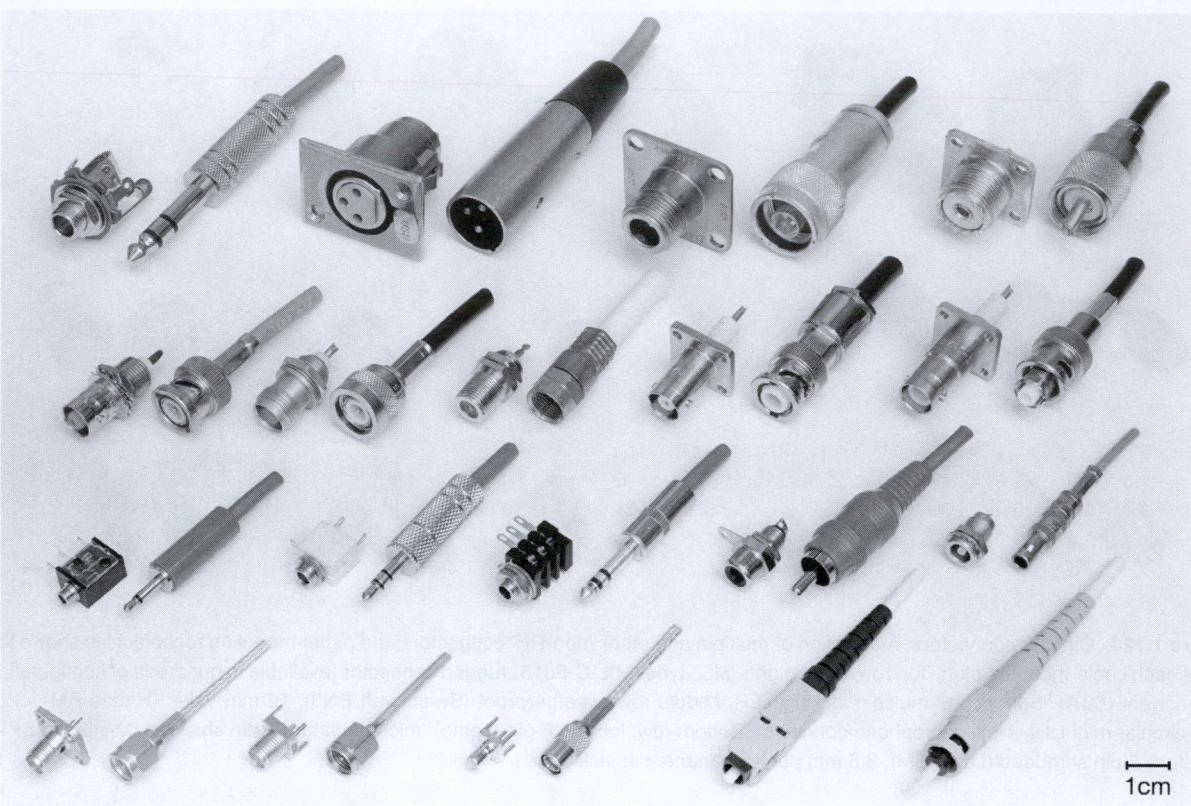

Figure 1.125. RF and shielded connectors. The panel-mounting receptacle is shown to the left of each cable-mounted plug. Top row, left to right: stereo phone jack, audio "XLR" type; N and UHF (RF connectors). Second row down: BNC, TNC, type F; MHV and SHV (high voltage). Third row down: 2.5 mm (3/32″) audio, 3.5 mm stereo, improved 3.5 mm stereo, phono ("RCA type"), LEMO coaxial. Bottom row: SMA (panel jack, flexible coax plug), SMA (board-mount jack, rigid coax plug), SMB; SC and ST (optical fiber).

percentage motor load, frequency, etc. Digital panel meters often provide the option of logic-level outputs, in addition to the visible display, for internal use by the instrument.

As a substitute for a dedicated meter (whether analog or digital), you increasingly see an LCD (liquid-crystal display) or LED panel with a meter-like pattern. This is flexible and efficient: with a graphic LCD display module (§12.5.3) you can offer the user a choice of "meters," according to the quantity being displayed, all under the control of an embedded controller (a built-in microprocessor; see Chapter 15).

B. Lamps, LEDs, and displays

Flashing lights, screens full of numbers and letters, eerie sounds – these are the stuff of science fiction movies, and except for the last, they form the subject of lamps and displays (see §12.5.3). Small incandescent lamps used to be standard for front-panel indicators, but they have been re-

placed with LEDs. The latter behave electrically like ordinary diodes, but with a forward voltage drop in the range of 1.5 to 2 volts (for red, orange, and some green LEDs; 3.6 V for blue[54] and high-brightness green; see Figure 2.8). When current flows in the forward direction, they light up. Typically, 2 mA to 10 mA produces adequate brightness. LEDs are cheaper than incandescent lamps, they last pretty much forever, and they come in four standard colors as well as "white" (which is usually a blue LED with a yellow fluorescent coating). They come in convenient panel-mounting packages; some even provide built-in current limiting.[55]

LEDs can also be used for digital displays, for example

---

[54] The invention of the gallium nitride blue LED was the breakthrough product of a lone and unappreciated employee of Nichia Chemical Industries, Shuji Nakamura.

[55] And of course, for both residential and commercial area lighting, LEDs have now largely relegated to the dustbin of history the century-old hot-filament incandescent lamp.

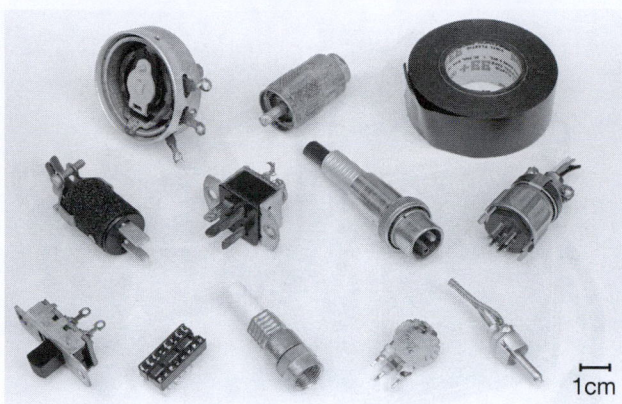

Figure 1.126. Components to avoid. We advise against using components like these, if you have a choice (see text if you need convincing!). Top row, left to right: low-value wirewound pot, type UHF connector, electrical tape ("just say no!"). Middle row: "cinch-type" connectors, microphone connector, hexagon connectors. Bottom row: slide switch, cheap IC socket (not "screw-machined"), type-F connector, open-element trimmer pot, phono connector.

as 7-segment numeric displays or (for displaying letters as well as numbers – "alphanumeric") 16-segment displays or dot-matrix displays. However, if more than a few digits or characters need to be displayed, LCDs are generally preferred. These come in line-oriented arrays (e.g., 16 characters by 1 line, up to 40 characters by 4 lines), with a simple interface that permits sequential or addressable entry of alphanumeric characters and additional symbols. They are inexpensive, low power, and visible even in sunlight. Back-lighted versions work well even in subdued light, but are not low power. Much more on these (and other) *optoelectronic* devices in §12.5.

1.9.5 Variable components

A. Variable resistors

Variable resistors (also called volume controls, potentiometers, pots, or trimmers) are useful as panel controls or internal adjustments in circuits. A classic panel type is the 2-watt-type AB potentiometer; it uses the same basic material as the fixed carbon-composition resistor, with a rotatable "wiper" contact. Other panel types are available with ceramic or plastic resistance elements, with improved characteristics. Multiturn types (3, 5, or 10 turns) are available, with counting dials, for improved resolution and linearity. "Ganged" pots (several independent sections on one shaft) are also manufactured, although in limited variety, for applications that demand them. Figure 1.8 shows a representative selection of pots and trimmers.

For use inside an instrument, rather than on the front panel, *trimmer pots* come in single-turn and multiturn styles, most intended for printed-circuit mounting. These are handy for calibration adjustments of the "set-and-forget" type. Good advice: resist the temptation to use lots of trimmers in your circuits. Use good design instead.

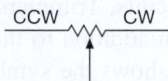

Figure 1.127. Potentiometer (three-terminal variable resistor).

The symbol for a variable resistor, or pot, is shown in Figure 1.127. Sometimes the symbols CW and CCW are used to indicate the clockwise and counterclockwise ends.

An all-electronic version of a potentiometer can be made with an array of electronic (transistor) switches that select a tap in a long chain of fixed resistors. As awkward as that may sound, it is a perfectly workable scheme when implemented as an IC. For example, Analog Devices, Maxim/Dallas Semiconductor, and Xicor make a series of "digital potentiometers" with up to 1024 steps; they come as single or dual units, and some of them are "nonvolatile," meaning that they remember their last setting even if power has been turned off. These find application in consumer electronics (televisions, stereos) where you want to adjust the volume from your infrared remote control, rather than by turning a knob; see §3.4.3E.

One important point about variable resistors: don't attempt to use a potentiometer as a substitute for a precise resistor value somewhere within a circuit. This is tempting, because you can trim the resistance to the value you want. The trouble is that potentiometers are not as stable as good (1%) resistors, and in addition they may not have good resolution (i.e., they can't be set to a precise value). If you must have a precise and settable resistor value somewhere, use a combination of a 1% (or better) precision resistor and a potentiometer, with the fixed resistor contributing most of the resistance. For example, if you need a 23.4k resistor, use a 22.6k 1% fixed resistor (a standard value) in series with a 2k trimmer pot. Another possibility is to use a series combination of several precision resistors, selecting the last (and smallest) resistor to give the desired series resistance.

As we'll see later (§3.2.7), it is possible to use FETs as voltage-controlled variable resistors in some applications. Another possibility is an "optophotoresistor" (§12.7). Transistors can be used as variable-gain amplifiers, again controlled by a voltage. Keep an open mind when design brainstorming.

Figure 1.128. Variable capacitor.

B. Variable capacitors

Variable capacitors are primarily confined to the smaller capacitance values (up to about 1000 pF) and are commonly used in RF circuits. Trimmers are available for in-circuit adjustments, in addition to the panel type for user tuning. Figure 1.128 shows the symbol for a variable capacitor.

Diodes operated with applied reverse voltage can be used as voltage-variable capacitors; in this application they're called *varactors*, or sometimes *varicaps* or *epicaps*. They're very important in RF applications, especially phase-locked loops, automatic frequency control (AFC), modulators, and parametric amplifiers.

C. Variable inductors

Variable inductors are usually made by arranging to move a piece of core material in a fixed coil. In this form they're available with inductances ranging from microhenrys to henrys, typically with a 2:1 tuning range for any given inductor. Also available are rotary inductors (coreless coils with a rolling contact).[56]

D. Variable transformers

Variable transformers are handy devices, especially the ones operated from the 115 volt ac line. They're usually configured as "autotransformers," which means that they have only one winding, with a sliding contact. They're also commonly called Variacs (the name given to them by General Radio), and they are made by Technipower, Superior Electric, and others. Figure 1.129 shows a classic unit from General Radio. Typically they provide 0 to 135 volts ac output when operated from 115 volts, and they come in current ratings from 1 amp to 20 amps or more. They're good for testing instruments that seem to be affected by powerline variations, and in any case to verify worst-case performance. *Important Warning*: don't forget that the output is not electrically isolated from the powerline, as it would be with a transformer!

---

[56] An interesting form of variable inductor of yesteryear was the *variometer*, a rotatable coil positioned within a fixed outer coil and connected in series with it. As the inner coil was rotated, the total inductance went from maximum (four times the inductance of either coil alone) all the way down to zero. These things were *consumer* items, listed for example in the 1925 Sears Roebuck catalog.

Figure 1.129. A powerline variable transformer ("Variac") lets you adjust the ac input voltage to something you are testing. Here a 5 A unit is shown, both clothed and undressed.

1.10 A parting shot: confusing markings and itty-bitty components

In our electronics course,[57] and indeed in day-to-day electronics on the bench, we encounter a wonderful confusion of component markings. Capacitors in particular are just, well, perverse: they rarely bother specifying *units* (even though they span 12 orders of magnitude, picofarads to farads), and for ceramic SMT varieties they dispense with any markings whatsoever! Even worse, they are still caught up in the transition from printing the value as an integer (e.g., "470" meaning 470 pF) versus using exponent notation (e.g., "470" meaning 47×10^0, i.e., 47 pF). Figure 1.130 shows exactly that case! Another trap for the unwary (and sometimes the wary, as well) is the date-code *gotcha*: the 4-digit code (yydd) can masquerade as a part number, as in the four examples in the photo. And, as components become smaller and smaller, there's precious little room for all but the briefest of markings; so, following the pharmaceutical industry, manufacturers invent a short

---

[57] Physics 123 ("Laboratory Electronics") at Harvard University: "Half course (fall term; repeated spring term). A lab-intensive introduction to electronic circuit design. Develops circuit intuition and debugging skills through daily hands-on lab exercises, each preceded by class discussion, with minimal use of mathematics and physics. Moves quickly from passive circuits, to discrete transistors, then concentrates on operational amplifiers, used to make a variety of circuits including integrators, oscillators, regulators, and filters. The digital half of the course treats analog–digital interfacing, emphasizing the use of microcontrollers and programmable logic devices (PLDs)." See http://learningtheartofelectronics.com/wp-content/uploads/2016/07/p123.pdf.

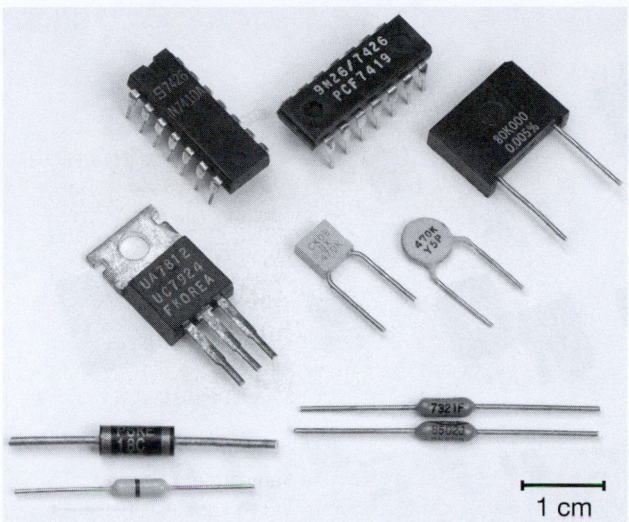

Figure 1.130. Confusion Central! The three ICs are each marked with both a part number (e.g., UA7812) and a "date code" (e.g., UC7924, signifying the 24th week of 1979). Unfortunately, both are perfectly valid part numbers (a +12 V or a −24 V regulator). The resistor pair (actually two views of identically marked resistors) suffers from the same problem: it could be 7.32 KΩ±1%, or it could be 85.0 kΩ±5% (it's the former, but who would know?). The pair of ceramic capacitors are both marked 470K (470,000 of something?), but, surprise, the "K" means 10% tolerance; and, bigger surprise, the square cap is 47 pF, the round one is 470 pF. And what is one to make of a black box labeled 80K000 (pronounced "eighty-koooh"), or a diode with two cathodes (and no anode?), or a resistor with a single black band in the center?

alphanumeric code for each component. And that's all you get. For example, National's LMV981 op-amp comes in several 6-pin packages: the SOT23 is marked "A78A," the smaller SC70 says "A77," and the really tiny microSMD blurts out a single letter "A" (or "H" if it's free of lead). Not much to go on.

1.10.1 Surface-mount technology: the joy and the pain

While we're complaining, let's whine just a bit about the difficulty of prototyping circuits with tiny surface-mount components. From an *electrical* point of view they are excellent: low inductance, and compact. But they are nearly impossible to wire up in prototype breadboard fashion, in the way that was easy with "through-hole" (or "leaded" – pronounced lee′-ded) components, such as resistors with axial leads (a wire sticking out each end), or integrated circuits in DIP (dual in-line) cases. Figure 1.131 gives

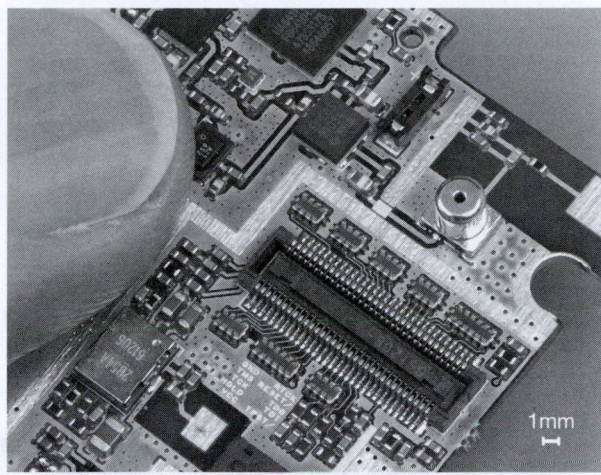

Figure 1.131. We're "all thumbs" when working with surface-mount technology (SMT). This is a corner of a cellphone circuit board, showing small ceramic resistors and capacitors, integrated circuits with ball-grid connecting dots on their undersides, and the Lilliputian connectors for the antenna and display panel. See also Figure 4.84.

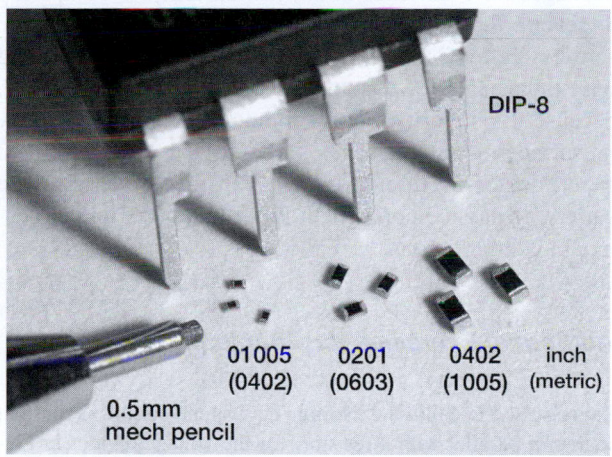

Figure 1.132. How small can these things get?! The "01005"-size SMT (0.016″×0.008″, or 0.4mm×0.2mm) represents the industry's greatest insult to the experimenter.

a sense of the scale of these little components, and Figure 1.132 displays the true horror of the tiniest of these – the "01005"-size chip components (0402 metric) that measure 200μm×400μm: not much thicker than a human hair, and indistinguishable from dust!

Sometimes you can use little adapter carriers (from companies like Bellin Dynamic Systems, Capital Advanced Technologies, or Aries) to convert an SMT integrated circuit to a fake DIP. But the densest surface-mount

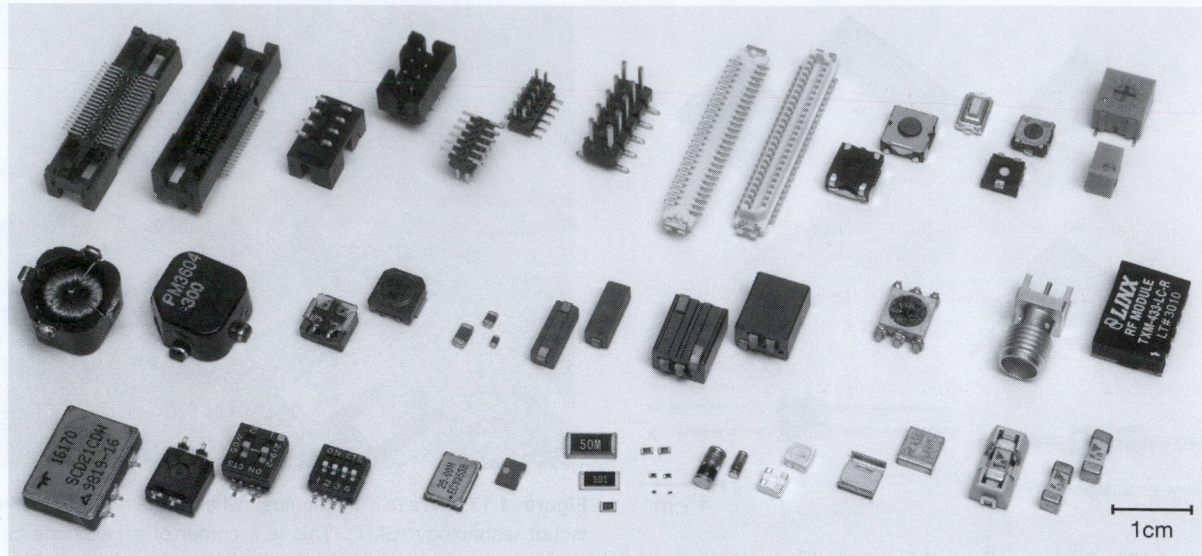

Figure 1.133. A taste of the world of passive components in surface-mount packages: connectors, switches, trimmer pots, inductors, resistors, capacitors, crystals, fuses.... If you can name it, you can probably get it in SMT.

packages have no leads at all, just an array of bumps (up to several thousand!) on the underside; and these require serious "reflow" equipment before you can do anything with them. Sadly, we cannot ignore this disturbing trend, because the majority of new components are offered only in surface-mount packages. Woe to the lone basement experimenter–inventor! Figure 1.133 give a sense of the variety of passive component types that come in surface-mount configurations.

Additional Exercises for Chapter 1

Exercise 1.37. Find the Norton equivalent circuit (a current source in parallel with a resistor) for the voltage divider in Figure 1.134. Show that the Norton equivalent gives the same output voltage as the actual circuit when loaded by a 5k resistor.

Exercise 1.38. Find the Thévenin equivalent for the circuit shown in Figure 1.135. Is it the same as the Thévenin equivalent for Exercise 1.37?

Exercise 1.39. Design a "rumble filter" for audio. It should pass frequencies greater than 20 Hz (set the $-3\,dB$ point at 10 Hz). Assume zero source impedance (perfect voltage source) and 10k (minimum) load impedance (that's important so that you can choose R and C such that the load doesn't affect the filter operation significantly).

Exercise 1.40. Design a "scratch filter" for audio signals (3 dB

down at 10 kHz). Use the same source and load impedances as in Exercise 1.39.

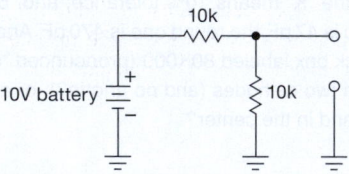

Figure 1.134. Example for Norton equivalent circuit.

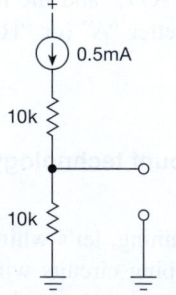

Figure 1.135. Example for Thévenin equivalent circuit.

Exercise 1.41. How would you make a filter with R's and C's to give the response shown in Figure 1.136?

Exercise 1.42. Design a bandpass RC filter (as in Figure 1.137);

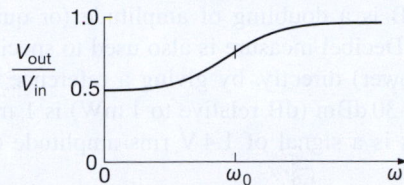

Figure 1.136. High-emphasis filter response.

f_1 and f_2 are the 3 dB points. Choose impedances so that the first stage isn't much affected by the loading of the second stage.

Exercise 1.43. Sketch the output for the circuit shown in Figure 1.138.

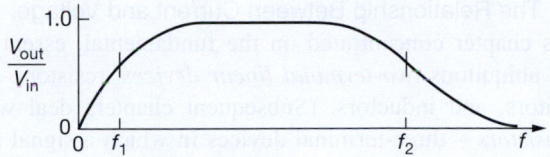

Figure 1.137. Bandpass filter response.

Exercise 1.44. Design an oscilloscope "$\times 10$ probe" to use with a scope whose input impedance is $1\,\mathrm{M\Omega}$ in parallel with $20\,\mathrm{pF}$ by figuring out what goes inside the probe handle in Figure 1.139. Assume that the probe cable adds an additional $100\,\mathrm{pF}$ and that the probe components are placed at the tip end (rather than at the scope end) of the cable. The resultant network should have

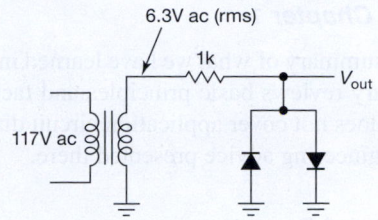

Figure 1.138. Circuit for Exercise 1.43.

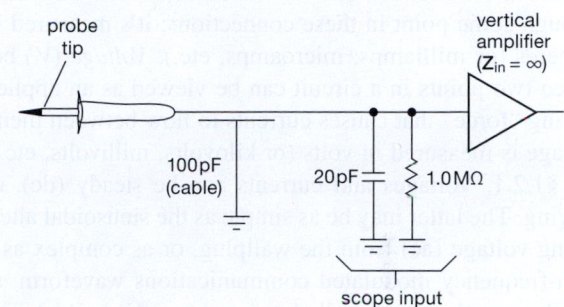

Figure 1.139. Oscilloscope $\times 10$ probe.

20 dB ($\times 10$ voltage division ratio) attenuation at all frequencies, including dc. The reason for using a $\times 10$ probe is to increase the load impedance seen by the circuit under test, which reduces loading effects. What input impedance (R in parallel with C) does your $\times 10$ probe present to the circuit under test when used with the scope?

Review of Chapter 1

An A-to-H summary of what we have learned in Chapter 1. This summary reviews basic principles and facts in Chapter 1, but it does not cover application circuit diagrams and practical engineering advice presented there.

¶A. Voltage and Current.

Electronic circuits consist of components connected together with wires. *Current* (I) is the rate of flow of charge through some point in these connections; it's measured in amperes (or milliamps, microamps, etc.). *Voltage* (V) between two points in a circuit can be viewed as an applied driving "force" that causes currents to flow between them; voltage is measured in volts (or kilovolts, millivolts, etc.); see §1.2.1. Voltages and currents can be steady (dc), or varying. The latter may be as simple as the sinusoidal alternating voltage (ac) from the wallplug, or as complex as a high-frequency modulated communications waveform, in which case it's usually called a *signal* (see ¶B below). The algebraic sum of currents at a point in a circuit (a *node*) is zero (Kirchhoff's current law, KCL, a consequence of conservation of charge), and the sum of voltage drops going around a closed loop in a circuit is zero (Kirchhoff's voltage law, KVL, a consequence of the conservative nature of the electrostatic field).

¶B. Signal Types and Amplitude.

See §1.3. In digital electronics we deal with *pulses*, which are signals that bounce around between two voltages (e.g., +5 V and ground); in the analog world it's *sinewaves* that win the popularity contest. In either case, a periodic signal is characterized by its frequency f (units of Hz, MHz, etc.) or, equivalently, period T (units of ms, μs, etc.). For sinewaves it's often more convenient to use *angular* frequency (radians/s), given by $\omega = 2\pi f$.

Digital amplitudes are specified simply by the HIGH and LOW voltage levels. With sinewaves the situation is more complicated: the amplitude of a signal $V(t) = V_0 \sin \omega t$ can be given as (a) *peak* amplitude (or just "amplitude") V_0, (b) *root-mean-square* (rms) amplitude $V_{rms} = V_0/\sqrt{2}$, or (c) peak-to-peak amplitude $V_{pp} = 2V_0$. If unstated, a sinewave amplitude is usually understood to be V_{rms}. A signal of rms amplitude V_{rms} delivers power $P = V_{rms}^2/R_{load}$ to a resistive load (regardless of the signal's waveform), which accounts for the popularity of rms amplitude measure.

Ratios of signal amplitude (or power) are commonly expressed in *decibels* (dB), defined as dB $= 10\log_{10}(P_2/P_1)$ or $20\log_{10}(V_2/V_1)$; see §1.3.2. An amplitude ratio of 10 (or power ratio of 100) is 20 dB; 3 dB is a doubling of power; 6 dB is a doubling of amplitude (or quadrupling of power). Decibel measure is also used to specify amplitude (or power) directly, by giving a reference level: for example, -30 dBm (dB relative to 1 mW) is 1 microwatt; +3 dBVrms is a signal of 1.4 V rms amplitude (2 Vpeak, 4 Vpp).

Other important waveforms are square waves, triangle waves, ramps, noise, and a host of *modulation* schemes by which a simple "carrier" wave is varied in order to convey information; some examples are AM and FM for analog communication, and PPM (pulse-position modulation) or QAM (quadrature-amplitude modulation) for digital communication.

¶C. The Relationship Between Current and Voltage.

This chapter concentrated on the fundamental, essential, and ubiquitous *two-terminal linear devices*: resistors, capacitors, and inductors. (Subsequent chapters deal with *transistors* – three-terminal devices in which a signal applied to one terminal controls the current flow through the other pair – and their many interesting applications. These include amplification, filtering, power conversion, switching, and the like.) The simplest linear device is the *resistor*, for which $I = V/R$ (Ohm's Law, see §1.2.2A). The term "linear" means that the response (e.g., current) to a combined sum of inputs (i.e., voltages) is equal to the sum of the responses that each input would produce: $I(V_1 + V_2) = I(V_1) + I(V_2)$.

¶D. Resistors, Capacitors, and Inductors.

The resistor is clearly linear. But it is not the only linear two-terminal component, because linearity does not require $I \propto V$. The other two linear components are *capacitors* (§1.4.1) and *inductors* (§1.5.1), for which there is a time-dependent relationship between voltage and current: $I = C\,dV/dt$ and $V = L\,dI/dt$, respectively. These are the *time domain* descriptions. Thinking instead in the *frequency domain*, these components are described by their *impedances*, the ratio of voltage to current (as a function of frequency) when driven with a sinewave (§1.7). A linear device, when driven with a sinusoid, responds with a sinusoid of the same frequency, but with changed amplitude and phase. Impedances are therefore complex, with the real part representing the amplitude of the response that is in-phase, and the imaginary part representing the amplitude of the response that is in quadrature (90° out of phase). Alternatively, in the polar representation of complex impedance ($Z = |Z|e^{i\theta}$), the magnitude $|Z|$ is the ratio of magnitudes ($|Z| = |V|/|I|$) and the quantity θ is the phase shift between V and I. The impedances of the three linear 2-terminal components are $Z_R=R$, $Z_C=-j/\omega C$, and $Z_L=j\omega L$, where (as always) $\omega=2\pi f$; see §1.7.5. Sinewave current through a resistor is in phase with voltage, whereas for a capacitor it leads by 90°, and for an inductor it lags by 90°.

¶E. Series and Parallel.

The impedance of components connected in series is the sum of their impedances; thus $R_{series}=R_1+R_2+\cdots$, $L_{series}=L_1+L_2+\cdots$, and $1/C_{series}=1/C_1+1/C_2+\cdots$. When connected in parallel, on the other hand, it's the *admittances* (inverse of impedance) that add. Thus the formula for capacitors in parallel looks like the formula for resistors in series, $C_{parallel}=C_1+C_2+\cdots$; and vice versa for resistors and inductors, thus $1/R_{parallel}=1/R_1+1/R_2+\cdots$. For a pair of resistors in parallel this reduces to $R_{parallel}=(R_1R_2)/(R_1+R_2)$. For example, two resistors of value R have resistance $R/2$ when connected in parallel, or resistance $2R$ in series.

The power dissipated in a resistor R is $P=I^2R=V^2/R$. There is no dissipation in an ideal capacitor or inductor, because the voltage and current are 90° out of phase. See §1.7.6.

¶F. Basic Circuits with R, L, and C.

Resistors are everywhere. They can be used to set an operating current, as for example when powering an LED or biasing a zener diode (Figure 1.16); in such applications the current is simply $I=(V_{supply}-V_{load})/R$. In other applications (e.g., as a transistor's load resistor in an amplifier, Figure 3.29) it is the *current* that is known, and a resistor is used to convert it to a voltage. An important circuit fragment is the *voltage divider* (§1.2.3), whose unloaded output voltage (across R_2) is $V_{out}=V_{in}R_2/(R_1+R_2)$.

If one of the resistors in a voltage divider is replaced with a capacitor, you get a simple *filter*: lowpass if the lower leg is a capacitor, highpass if the upper leg is a capacitor (§§1.7.1 and 1.7.7). In either case the -3 dB transition frequency is at $f_{3dB}=1/2\pi RC$. The ultimate rolloff rate of such a "single-pole" lowpass filter is -6 dB/octave, or -20 dB/decade; i.e., the signal amplitude falls as $1/f$ well beyond f_{3dB}. More complex filters can be created by combining inductors with capacitors, see Chapter 6. A capacitor in parallel with an inductor forms a *resonant circuit*; its impedance (for ideal components) goes to infinity at the resonant frequency $f=1/(2\pi\sqrt{LC})$. The impedance of a *series LC* goes to zero at that same resonant frequency. See §1.7.14.

Other important capacitor applications in this chapter (§1.7.16) include (a) *bypassing*, in which a capacitor's low impedance at signal frequencies suppresses unwanted signals, e.g., on a dc supply rail; (b) *blocking* (§1.7.1C), in which a highpass filter blocks dc, but passes all frequencies of interest (i.e., the breakpoint is chosen below all signal frequencies); (c) *timing* (§1.4.2D), in which an *RC* circuit (or a constant current into a capacitor) generates a sloping waveform used to create an oscillation or a timing interval; and (d) *energy storage* (§1.7.16B), in which a capacitor's stored charge $Q=CV$ smooths out the ripples in a dc power supply.

In later chapters we'll see some additional applications of capacitors: (e) *peak detection* and *sample-and-hold* (§§4.5.1 and 4.5.2), which capture the voltage peak or transient value of a waveform, and (f) the *integrator* (§4.2.6), which performs a mathematical integration of an input signal.

¶G. Loading; Thévenin Equivalent Circuit.

Connecting a load (e.g., a resistor) to the output of a circuit (a "signal source") causes the unloaded output voltage to drop; the amount of such *loading* depends on the load resistance, and the signal source's ability to drive it. The latter is usually expressed as the *equivalent source impedance* (or *Thévenin impedance*) of the signal. That is, the signal source is modeled as a perfect voltage source V_{sig} in series with a resistor R_{sig}. The output of the resistive voltage divider driven from an input voltage V_{in}, for example, is modeled as a voltage source $V_{sig}=V_{in}R_2/(R_1+R_2)$ in series with a resistance $R_{sig}=R_1R_2/(R_1+R_2)$ (which is just $R_1\|R_2$). So the output of a 1kΩ–1kΩ voltage divider driven by a 10 V battery looks like 5 V in series with 500 Ω.

Any combination of voltage sources, current sources, and resistors can be modeled perfectly by a single voltage source in series with a single resistor (its "Thévenin equivalent circuit"), or by a single current source in parallel with a single resistor (its "Norton equivalent circuit"); see Appendix D. The Thévenin equivalent source and resistance values are found from the open-circuit voltage and short-circuit current as $V_{Th}=V_{oc}$, $R_{Th}=V_{oc}/I_{sc}$; and for the Norton equivalent they are $I_N=I_{sc}$, $R_N=V_{oc}/I_{sc}$.

Because a load impedance forms a voltage divider with the signal's source impedance, it's usually desirable for the latter to be small compared with any anticipated load impedance (§1.2.5A). However, there are two exceptions: (a) a *current source* has a high source impedance (ideally infinite), and should drive a load of much lower impedance; and (b) signals of *high frequency* (or fast risetime), traveling through a length of cable, suffer reflections unless the load impedance equals the so-called "characteristic impedance" Z_0 of the cable (commonly 50 Ω), see Appendix H.

¶H. The Diode, a Nonlinear Component.

There are important two-terminal devices that are not linear, notably the *diode* (or *rectifier*), see §1.6. The ideal diode conducts in one direction only; it is a "one-way valve." The onset of conduction in real diodes is roughly at 0.5 V in the "forward" direction, and there is some small leakage current in the "reverse" direction, see Figure 1.55. Useful diode circuits include power-supply *rectification* (conversion of ac to dc, §1.6.2), signal rectification (§1.6.6A), *clamping* (signal limiting, §1.6.6C), and *gating* (§1.6.6B). Diodes are commonly used to prevent polarity

reversal, as in Figure 1.84; and their exponential current versus applied voltage can be used to fashion circuits with logarithmic response (§1.6.6E).

Diodes specify a maximum safe reverse voltage, beyond which avalanche breakdown (an abrupt rise of current) occurs. You don't go there! But you can (and should) with a *zener diode* (§1.2.6A), for which a reverse breakdown voltage (in steps, going from about 3.3 V to 100 V or more) is specified. Zeners are used to establish a voltage within a circuit (Figure 1.16), or to limit a signal's swing.

BIPOLAR TRANSISTORS

2.1 Introduction

The transistor is our most important example of an "active" component, a device that can amplify, producing an output signal with more power in it than the input signal. The additional power comes from an external source of power (the power supply, to be exact). Note that *voltage* amplification isn't what matters, since, for example, a step-up transformer, a "passive" component just like a resistor or capacitor, has voltage gain but no power gain.[1] Devices with power gain are distinguishable by their ability to make oscillators, by feeding some output signal back into the input.

It is interesting to note that the property of power amplification seemed very important to the inventors of the transistor. Almost the first thing they did to convince themselves that they had really invented something was to power a loudspeaker from a transistor, observing that the output signal sounded louder than the input signal.

The transistor is the essential ingredient of every electronic circuit, from the simplest amplifier or oscillator to the most elaborate digital computer. Integrated circuits (ICs), which have largely replaced circuits constructed from discrete transistors, are themselves merely arrays of transistors and other components built from a single chip of semiconductor material.

A good understanding of transistors is very important, even if most of your circuits are made from ICs, because you need to understand the input and output properties of the IC in order to connect it to the rest of your circuit and to the outside world. In addition, the transistor is the single most powerful resource for interfacing, whether between ICs and other circuitry or between one subcircuit and another. Finally, there are frequent (some might say too frequent) situations in which the right IC just doesn't exist, and you have to rely on discrete transistor circuitry to do the job. As you will see, transistors have an excitement all their own. Learning how they work can be great fun.

There are two major species of transistors: in this chapter we will learn about bipolar junction transistors (BJTs), which historically came first with their Nobel Prize-winning invention in 1947 at Bell Laboratories. The next chapter deals with "field-effect" transistors (FETs), the now-dominant species in digital electronics. To give the coarsest comparison, BJTs excel in accuracy and low noise, whereas FETs excel in low power, high impedance, and high-current switching; there is, of course, much more to this complex subject.

Our treatment of bipolar transistors is going to be quite different from that of many other books. It is common practice to use the *h*-parameter model and equivalent circuit. In our opinion that is unnecessarily complicated and unintuitive. Not only does circuit behavior tend to be revealed to you as something that drops out of elaborate equations, rather than deriving from a clear understanding in your own mind as to how the circuit functions; you also have the tendency to lose sight of which parameters of transistor behavior you can count on and, more important, which ones can vary over large ranges.

In this chapter we will instead build up a very simple introductory transistor model and immediately work out some circuits with it. Its limitations will soon become apparent; then we will expand the model to include the respected Ebers–Moll conventions. With the Ebers–Moll equations and a simple three-terminal model, you will have a good understanding of transistors; you won't need to do a lot of calculations, and your designs will be first rate. In particular, they will be largely independent of the poorly controlled transistor parameters such as current gain.

Some important engineering notation should be mentioned. Voltage at a transistor terminal (relative to ground) is indicated by a single subscript (C, B, or E): V_C is the collector voltage, for instance. Voltage between two terminals is indicated by a double subscript: V_{BE} is the base-to-emitter voltage drop, for instance. If the same letter is repeated, that means a power-supply voltage: V_{CC} is the (positive) power-supply voltage associated with the collector,

---

[1] It is even possible to achieve modest voltage gain in a circuit comprising only resistors and capacitors. To explore this idea, surprising even to seasoned engineers, look at Appendix J on SPICE.

and V_{EE} is the (negative) supply voltage associated with the emitter.[2]

Why transistor circuits are difficult

For those learning electronics for the first time, this chapter will be difficult. Here's why: all the circuits in the last chapter dealt with *two-terminal devices*, whether linear (resistors, capacitors, inductors) or nonlinear (diodes). So there was only one voltage (the voltage between the terminals) and only one current (the current flowing through the device) to think about. Transistors, by contrast, are *three-terminal devices*, which means there are two voltages and two currents to juggle.[3]

2.1.1 First transistor model: current amplifier

Let's begin. A bipolar transistor is a three-terminal device (Figure 2.1), in which a small current applied to the base controls a much larger current flowing between the collector and emitter. It is available in two flavors (*npn* and *pnp*), with properties that meet the following rules for *npn* transistors (for *pnp* simply reverse all polarities):

1. **Polarity** The collector must be more positive than the emitter.
2. **Junctions** The base–emitter and base–collector circuits behave like diodes (Figure 2.2) in which a small current applied to the base controls a much larger current flowing between the collector and emitter. Normally the base–emitter diode is conducting, whereas the base–collector diode is reverse-biased, i.e., the applied voltage is in the opposite direction to easy current flow.
3. **Maximum ratings** Any given transistor has maximum values of I_C, I_B, and V_{CE} that cannot be exceeded without costing the exceeder the price of a new transistor (for typical values, see the listing in Table 2.1 on page 74, Table 2.2 on page 106, and Table 8.1 on pages 501–502 There are also other limits, such as power dissipation ($I_C V_{CE}$), temperature, and V_{BE}, that you must keep in mind.
4. **Current amplifier** When rules 1–3 are obeyed, I_C is roughly proportional to I_B and can be written as

$$I_C = h_{FE}I_B = \beta I_B, \tag{2.1}$$

---

[2] In practice, circuit designers use V_{CC} to designate the positive supply and V_{EE} the negative supply, even though logically they should be interchanged for *pnp* transistors (where all polarities are reversed).

[3] You might think that there would be three voltages and three currents; but it's slightly less complicated than that, because there are only two independent voltages and two independent currents, thanks to Kirchhoff's voltage and current laws.

where β, the current gain (sometimes called[4] h_{FE}), is typically about 100. Both I_B and I_C flow to the emitter. Note: the collector current is not due to forward conduction of the base–collector diode; that diode is reverse-biased. Just think of it as "transistor action."

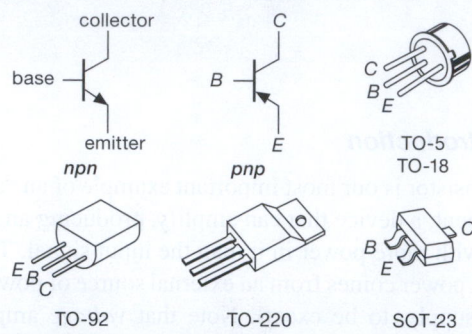

Figure 2.1. Transistor symbols and small transistor package drawings (not to scale). A selection of common transistor packages are shown in Figure 2.3.

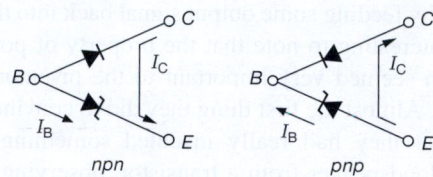

Figure 2.2. An ohmmeter's view of a transistor's terminals.

Rule 4 gives the transistor its usefulness: a small current flowing into the base controls a much larger current flowing into the collector.

An important warning: the current gain β is not a "good" transistor parameter; for instance, its value can vary from 50 to 250 for different specimens of a given transistor type. It also depends on the collector current, collector-to-emitter voltage, and temperature. *A circuit that depends on a particular value for beta is a bad circuit.*

Note particularly the effect of rule 2. This means you can't go sticking an arbitrary voltage across the base–emitter terminals, because an enormous current will flow if the base is more positive than the emitter by more than about 0.6 to 0.8 V (forward diode drop). This rule also implies that an operating transistor has $V_B \approx V_E + 0.6\,\mathrm{V}$ ($V_B = V_E + V_{BE}$). Again, polarities are normally given for *npn* transistors; reverse them for *pnp*.

Let us emphasize again that you should not try to think of the collector current as diode conduction. It isn't,

---

[4] As the "*h*-parameter" transistor model has fallen out of popularity, you tend often to see β (instead of h_{FE}) as the symbol for current gain.

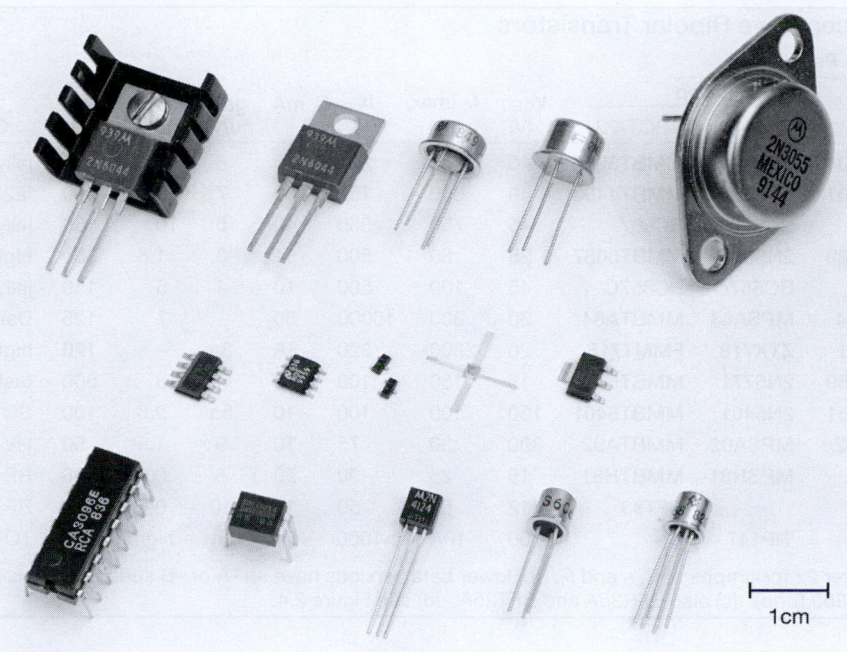

Figure 2.3. Most of the common packages are shown here, for which we give the traditional designations. Top row (power), left to right: TO-220 (with and without heatsink), TO-39, TO-5, TO-3. Middle row (surface mount): SM-8 (dual), SO-8 (dual), SOT-23, ceramic SOE, SOT-223. Bottom row: DIP-16 (quad), DIP-4, TO-92, TO-18, TO-18 (dual).

because the collector–base diode normally has voltages applied across it in the reverse direction. Furthermore, collector current varies very little with collector voltage (it behaves like a not-too-great current source), unlike forward diode conduction, in which the current rises very rapidly with applied voltage.

Table 2.1 on the following page includes a selection of commonly used bipolar transistors, with the corresponding curves of current gain[5] in Figure 2.4, and a selection of transistors intended for power applications is listed in Table 2.2 on page 106. A more complete listing can be found in Table 8.1 on pages 501–502 and Figure 8.39 in Chapter 8.

2.2 Some basic transistor circuits

2.2.1 Transistor switch

Look at the circuit in Figure 2.5. This application, in which a small control current enables a much larger current to

flow in another circuit, is called a transistor switch. From the preceding rules it is easy to understand. When the mechanical switch is open, there is no base current. So, from rule 4, there is no collector current. The lamp is off.

When the switch is closed, the base rises to 0.6 V (base–emitter diode is in forward conduction). The drop across the base resistor is 9.4 V, so the base current is 9.4 mA. Blind application of Rule 4 gives $I_C = 940$ mA (for a typical beta of 100). That is wrong. Why? Because rule 4 holds only if Rule 1 is obeyed: at a collector current of 100 mA the lamp has 10 V across it. To get a higher current you would have to pull the collector below ground. A transistor can't do this, and the result is what's called *saturation* – the collector goes as close to ground as it can (typical saturation voltages are about 0.05–0.2 V, see Chapter *2x*.) and stays there. In this case, the lamp goes on, with its rated 10 V across it.

Overdriving the base (we used 9.4 mA when 1.0 mA would have barely sufficed) makes the circuit conservative; in this particular case it is a good idea, since a lamp draws more current when cold (the resistance of a lamp when cold is 5 to 10 times lower than its resistance at operating current). Also, transistor beta drops at low collector-to-base voltages, so some extra base current is necessary to bring

---

[5] In addition to listing typical betas (h_{FE}) and maximum allowed collector-to-emitter voltages (V_{CEO}), Table 2.1 includes the cutoff frequency (f_T, at which the beta has decreased to 1) and the feedback capacitance (C_{cb}). These are important when dealing with fast signals or high frequencies; we'll see them in §2.4.5 and Chapter *2x*.

Table 2.1   Representative Bipolar Transistors

npn		pnp		V_{CEO}	I_C (max)	h_{FE} @ mA		gain	C_{cb}^a	f_T^a	
TO-92	SOT-23	TO-92	SOT-23	(V)	(mA)	(typ)		curve[d]	(pF)	(MHz)	Comments
2N3904	MMBT3904	2N3906	MMBT3906	40	150	200	10	6	2.5	300	jellybean
2N4401	MMBT4401	2N4403	MMBT4403	40	500	150	150	7	7	300	'2222 and '2907 dies
BC337	BC817	BC327	BC807	45	750	350	40	5	10	150	jellybean
2N5089	MMBT5089	2N5087	MMBT5087	25	50	500	1	3	1.8	350	high beta
BC547C	BC847C	BC557C	BC857C	45	100	500	10	4	5	150	jellybean[b]
MPSA14	MMBTA14	MPSA64	MMBTA64	30	300	10000	50	-	7	125	Darlington
ZTX618	FMMT618	ZTX718	FMMT718	20	2500	320	3A	3a	-	120	high I_C, small pkg
PN2369	MMBT2369	2N5771	MMBT5771	15	150	100	10	10	3	500	fast switch, gold doped
2N5551	MMBT5551	2N5401	MMBT5401	150	100	100	10	5a	2.5	100	SOT-223 available
MPSA42	MMBTA42	MPSA92	MMBTA92	300	30	75	10	9	1.5	50	HV small signal
MPS5179	BFS17	MPSH81	MMBTH81	15	25	90	20	8	0.9	900	RF amplifier
—	BFR93[c]	—	BFT93[c]	12	50	50	15	10	0.5	4000	RF amp
TIP142	—	TIP147	—	100	10A	>1000	5A	-	high	low	TO-220, Darlington

Notes: (a) see Chapter *2x* for graphs of C_{cb} and f_T. (b) lower beta versions have an -A or -B suffix; low-noise versions are BC850 (*npn*) and BC860 (*pnp*). (c) also BFR25A and BFT25A. (d) see Figure 2.4.

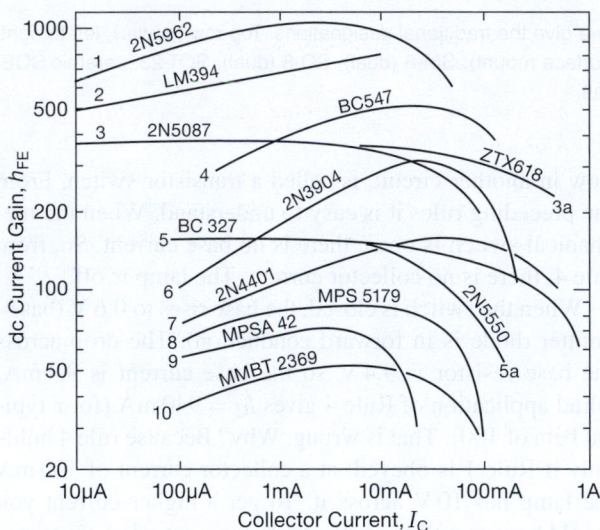

Figure 2.4. Curves of typical transistor current gain, β, for a selection of transistors from Table 2.1. These curves are taken from manufacturers' literature. You can expect production spreads of +100%, −50% from the "typical" values graphed. See also Figure 8.39 for measured beta plots for 44 types of "low-noise" transistors.

a transistor into full saturation. Incidentally, in a real circuit you would probably put a resistor from base to ground (perhaps 10k in this case) to make sure the base is at ground with the switch open. It wouldn't affect the ON operation, because it would sink only 0.06 mA from the base circuit.

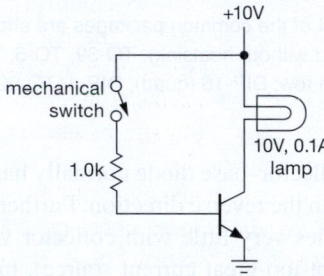

Figure 2.5. Transistor switch example.

There are certain cautions to be observed when designing transistor switches:

1. Choose the base resistor conservatively to get plenty of excess base current, especially when driving lamps, because of the reduced beta at low V_{CE}. This is also a good idea for high-speed switching, because of capacitive effects and reduced beta at very high frequencies (many megahertz).[6]

2. If the load swings below ground for some reason (e.g., it is driven from ac, or it is inductive), use a diode in series with the collector (or a diode in the reverse direction to ground) to prevent collector–base conduction on negative swings.

3. For inductive loads, protect the transistor with a diode

---

[6] A small "speed-up" capacitor – typically just a few picofarads – is often connected across the base resistor to improve high-speed performance.

across the load, as shown in Figure 2.6.[7] Without the diode the inductor will swing the collector to a large positive voltage when the switch is opened, most likely exceeding the collector–emitter breakdown voltage, as the inductor tries to maintain its "on" current from V_{CC} to the collector (see the discussion of inductors in §1.6.7).

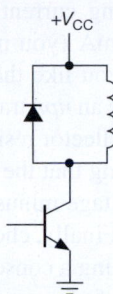

Figure 2.6. Always use a suppression diode when switching an inductive load.

You might ask why we are bothering with a transistor, and all its complexity, when we could just use that mechanical switch alone to control the lamp or other load. There are several good reasons: (a) a transistor switch can be driven *electrically* from some other circuit, for example a computer output bit; (b) transistor switches enable you to switch very rapidly, typically in a small fraction of a microsecond; (c) you can switch many different circuits with a single control signal; (d) mechanical switches suffer from wear, and their contacts "bounce" when the switch is activated, often making and breaking the circuit a few dozen times in the first few milliseconds after activation; and (e) with transistor switches you can take advantage of remote *cold switching*, in which only dc control voltages snake around through cables to reach front-panel switches, rather than the electronically inferior approach of having the signals themselves traveling through cables and switches (if you run lots of signals through cables, you're likely to get capacitive pickup as well as some signal degradation).

A. "Transistor man"
The cartoon in Figure 2.7 may help you understand some limits of transistor behavior. The little man's perpetual task in life is to try to keep $I_C = \beta I_B$; however, he is only allowed to turn the knob on the variable resistor. Thus he can go from a short circuit (saturation) to an open circuit (transistor in the OFF state), or anything in between, but he isn't allowed to use batteries, current sources, etc.

---

[7] Or, for faster turn-off, with a resistor, an *RC* network, or zener clamp; see §1.6.7.

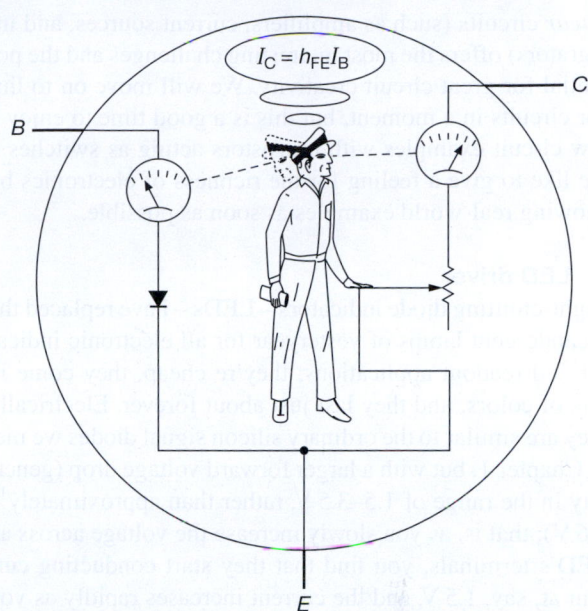

Figure 2.7. "Transistor man" observes the base current, and adjusts the output rheostat in an attempt to maintain the output current β times larger; h_{FE} and β are used interchangeably.

One warning is in order here: don't think that the collector of a transistor looks like a resistor. It doesn't. Rather, it looks approximately like a poor-quality constant-current sink (the value of current depending on the signal applied to the base), primarily because of this little man's efforts.

Another thing to keep in mind is that, at any given time, a transistor may be (a) cut off (no collector current), (b) in the active region (some collector current, and collector voltage more than a few tenths of a volt above the emitter), or (c) in saturation (collector within a few tenths of a volt of the emitter). See the discussion of transistor saturation in Chapter *2x* for more details.

2.2.2 Switching circuit examples
The transistor switch is an example of a *nonlinear* circuit: the output is not proportional to the input;[8] instead it goes to one of two possible states (cut off, or saturated). Such two-state circuits are extremely common[9] and form the basis of digital electronics. But to the authors the subject of

---

[8] A mathematician would define linearity by saying that the response to the sum of two inputs is the sum of the individual responses; this necessarily implies proportionality.

[9] If you took a census, asking the transistors of the world what they are doing, at least 95% would tell you they are switches.

linear circuits (such as amplifiers, current sources, and integrators) offers the most interesting challenges and the potential for great circuit creativity. We will move on to linear circuits in a moment, but this is a good time to enjoy a few circuit examples with transistors acting as switches – we like to give a feeling for the richness of electronics by showing real-world examples as soon as possible.

A. LED driver

Light-emitting diode indicators – LEDs – have replaced the incandescent lamps of yesteryear for all electronic indicator and readout applications; they're cheap, they come in lots of colors, and they last just about forever. Electrically they are similar to the ordinary silicon signal diodes we met in Chapter 1, but with a larger forward voltage drop (generally in the range of 1.5–3.5 V, rather than approximately[10] 0.6 V); that is, as you slowly increase the voltage across an LED's terminals, you find that they start conducting current at, say, 1.5 V, and the current increases rapidly as you apply somewhat more voltage (Figure 2.8). They light up, too! Typical "high-efficiency" indicator LEDs look pretty good at a few milliamps, and they'll knock your eye out at 10–20 mA.

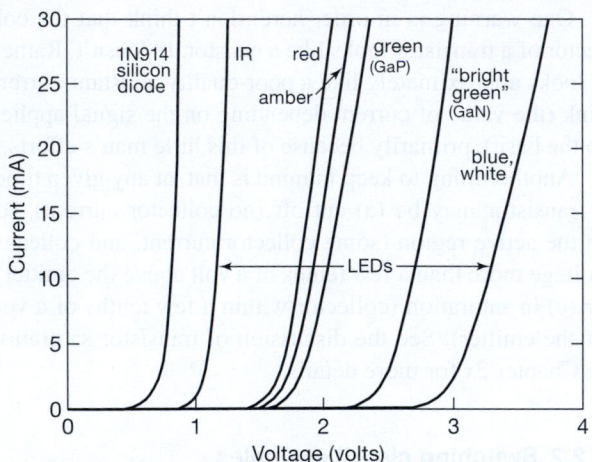

Figure 2.8. Like silicon diodes, LEDs have rapidly increasing current versus applied voltage, but with larger forward voltage drops.

We'll show a variety of techniques for driving LEDs in Chapter 12; but we can drive them already, with what we know. The first thing to realize is that we can't just switch a voltage across them, as in Figure 2.5, because of their steep *I* versus *V* behavior; for example, applying 5 V across an

LED is guaranteed to blow it out. We need instead to treat it gently, coaxing it to draw the right current.

Let's assume that we want the LED to light in response to a digital signal line when it goes to a HIGH value of +3.3 V (from its normal resting voltage near ground). Let's assume also that the digital line can provide up to 1 mA of current, if needed. The procedure goes like this: first, choose an LED operating current that will provide adequate brightness, say 5 mA (you might want to try a few samples, to make sure you like the color, brightness, and viewing angle). Then use an *npn* transistor as a switch (Figure 2.9), choosing the collector resistor to provide the chosen LED current, realizing that the voltage drop across the resistor is the supply voltage minus the LED forward drop at its operating current. Finally, choose the base resistor to ensure saturation, assuming a conservatively low transistor beta ($\beta \geq 25$ is pretty safe for a typical small-signal transistor like the popular 2N3904).

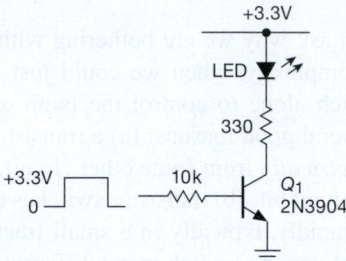

Figure 2.9. Driving an LED from a "logic-level" input signal, using an *npn* saturated switch and series current-limiting resistor.

Note that the transistor is acting as a saturated switch, with the collector resistor setting the operating current. As we'll see shortly, you can devise circuits that provide an accurate *current* output, largely independent of what the load does. Such a "current source" can also be used to drive LEDs. But our circuit is simple, and effective. There are other variations: we'll see in the next chapter that a MOSFET-type[11] transistor is often a better choice. And in Chapters 10–12 we'll see ways to drive LEDs and other optoelectronic devices directly from digital integrated circuits, without external discrete transistors.

Exercise 2.1. What is the LED current, approximately, in the circuit of Figure 2.9? What minimum beta is required for Q_1?

B. Variations on a theme

For these switch examples, one side of the load is connected to a positive supply voltage, and the other side is

---

[10] The larger drop is due to the use of different semiconductor materials such as GaAsP, GaAlAs, and GaN, with their larger bandgaps.

[11] metal-oxide semiconductor field-effect transistor.

switched to ground by the *npn* transistor switch. What if you want instead to ground one side of the load and switch the "high side" to a positive voltage?

It's easy enough – but you've got to use the other polarity of transistor (*pnp*), with its emitter at the positive rail, and its collector tied to the load's high side, as in Figure 2.10A. The transistor is cut off when the base is held at the emitter voltage (here +15 V), and switched into saturation by bringing the base toward the collector (i.e., toward ground). When the input is brought to ground, there's about 4 mA of base current through the 3.3 kΩ base resistor, sufficient for switching loads up to about 200 mA ($\beta > 50$).

An awkwardness of this circuit is the need to hold the input at +15 V to turn off the switch; it would be much better to use a lower control voltage, for example, +3 V and ground, commonly available in digital logic that we'll be seeing in Chapters 10–15. Figure 2.10B shows how to do that: *npn* switch Q_2 accepts the "logic-level" input of 0 V or +3 V, pulling its collector load to ground accordingly. When Q_2 is cut off, R_3 holds Q_3 off; when Q_2 is saturated (by a +3 V input), R_2 sinks base current from Q_3 to bring it into saturation.

The "divider" formed by $R_2 R_3$ may be confusing: R_3's job is to keep Q_3 off when Q_2 is off; and when Q_2 pulls its collector low, most of its collector current comes from Q_3's base (because only ∼0.6 mA of the 4.4 mA collector current comes from R_3 – make sure you understand why). That is, R_3 does not have much effect on Q_3's saturation. Another way to say it is that the divider would sit at about +11.6 V (rather than +14.4 V), were it not for Q_3's base–emitter diode, which consequently gets most of Q_2's collector current. In any case, the value of R_3 is not critical and could be made larger; the tradeoff is slower turn-off of Q_3, owing to capacitive effects.[12]

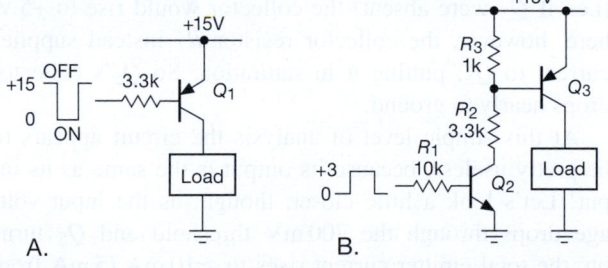

Figure 2.10. Switching the high side of a load returned to ground.

---

[12] But don't make it too small: Q_3 would not switch at all if R_3 were reduced to 100 Ω (why?). We were surprised to see this basic error in an instrument, the rest of which displayed circuit design of the highest sophistication.

C. Pulse generator – I

By including a simple *RC*, you can make a circuit that gives a pulse output from a step input; the time constant $\tau = RC$ determines the pulse width. Figure 2.11 shows one way. Q_2 is normally held in saturation by R_3, so its output is close to ground; note that R_3 is chosen small enough to ensure Q_2's saturation. With the circuit's input at ground, Q_1 is cut off, with its collector at +5 V. The capacitor C_1 is therefore charged, with +5 V on its left terminal and approximately +0.6 V on its right terminal; i.e., it has about 4.4 V across it. The circuit is waiting for something to happen.

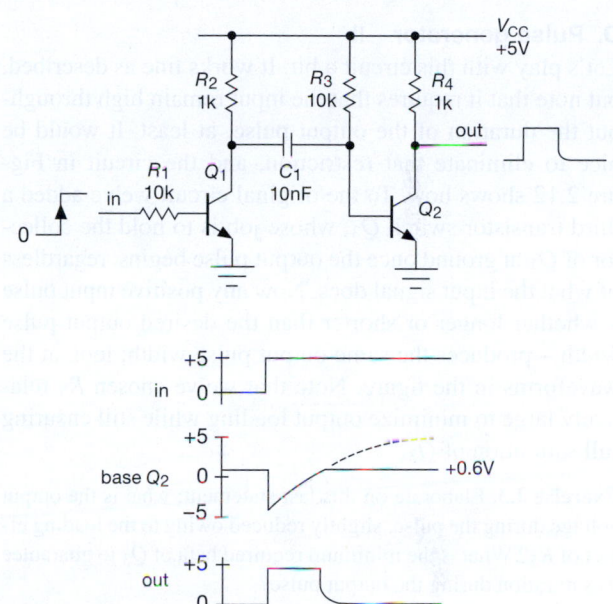

Figure 2.11. Generating a short pulse from a step input waveform.

A +5 V positive input step brings Q_1 into saturation (note the values of R_1 and R_2), forcing its collector to ground; because of the voltage across C_1, this brings the base of Q_2 momentarily negative, to about −4.4 V.[13] Q_2 is then cutoff, no current flows through R_4, and so its output jumps to +5 V; this is the beginning of the output pulse. Now for the *RC*: C_1 can't hold Q_2's base below ground forever, because current is flowing down through R_3, trying to pull it up. So the right-hand side of the capacitor charges toward +5 V, with a time constant $\tau = R_3 C_1$, here equal to 100 μs. The output pulse width is set by this time constant

---

[13] A caution here: this circuit should not be run from a supply voltage greater than +7 V, because the negative pulse can drive Q_2's base into reverse breakdown. This is a common oversight, even among experienced circuit designers.

and is proportional to τ. To figure out the pulse width accurately you have to look in detail at the circuit operation. In this case it's easy enough to see that the output transistor Q_2 will turn on again, terminating the output pulse, when the rising voltage on the base of transistor Q_2 reaches the ≈ 0.6 V V_{BE} drop required for turn-on. Try this problem to test your understanding.

Exercise 2.2. Show that the output pulse width for the circuit of Figure 2.11 is approximately $T_{pulse}=0.76R_3C_1=76\mu s$. A good starting point is to notice that C_1 is charging exponentially from -4.4 V toward +5 V, with the time constant as above.

D. Pulse generator – II

Let's play with this circuit a bit. It works fine as described, but note that it requires that the input remain high throughout the duration of the output pulse, at least. It would be nice to eliminate that restriction, and the circuit in Figure 2.12 shows how. To the original circuit we've added a third transistor switch Q_3, whose job is to hold the collector of Q_1 at ground once the output pulse begins, regardless of what the input signal does. Now any positive input pulse – whether longer or shorter than the desired output pulse width – produces the same output pulse width; look at the waveforms in the figure. Note that we've chosen R_5 relatively large to minimize output loading while still ensuring full saturation of Q_3.

Exercise 2.3. Elaborate on this last statement: what is the output voltage during the pulse, slightly reduced owing to the loading effect of R_5? What is the minimum required beta of Q_3 to guarantee its saturation during the output pulse?

E. Pulse generator – III

For our final act, let's fix a deficiency of these circuits, namely a tendency for the output pulse to turn off somewhat slowly. That happens because Q_2's base voltage, with its leisurely $100\,\mu s$ RC time constant, rises smoothly (and relatively slowly) through the turn-on voltage threshold of ≈ 0.6 V. Note, by the way, that this problem does not occur at the turn-*on* of the output pulse, because at that transition Q_2's base voltage drops abruptly down to approximately -4.4 V, owing to the sharp input step waveform, which is further sharpened by the switching action of Q_1.

The cure here is to add at the output a clever circuit known as a *Schmitt trigger*, shown in its transistor implementation[14] in Figure 2.13A. It works like this: imagine a time within the positive output pulse of the previous circuits, so the input to this new Schmitt circuit is high (near

---

[14] We'll see other ways of making a Schmitt trigger, using op-amps or comparators, in Chapter 4.

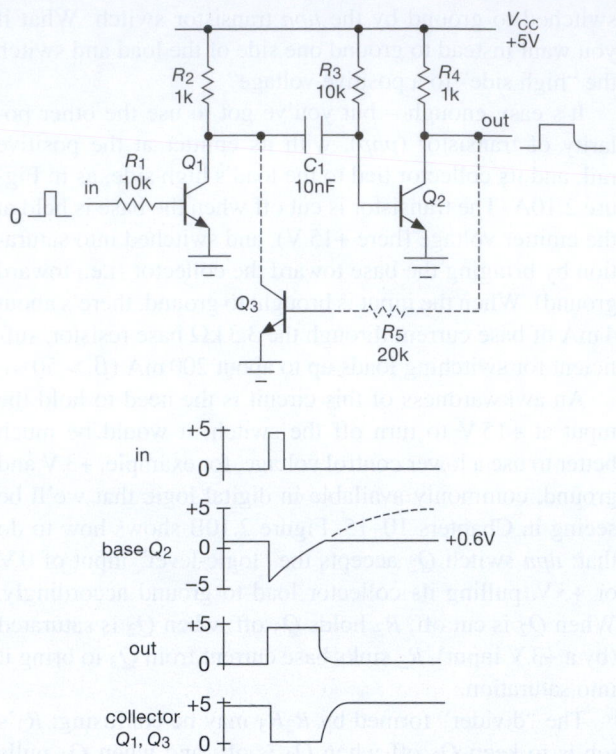

Figure 2.12. Generating a short pulse from a step or pulse input.

+5 V). That holds Q_4 in saturation, and so Q_5 is cut off, with the output at +5 V. The emitter current of Q_4 is about 5 mA, so the emitter voltage is approximately +100 mV; the base is a V_{BE} higher, approximately +700 mV.

Now imagine the trailing edge of the input pulse waveform, whose voltage smoothly drops toward ground. As it drops below 700 mV, Q_4 begins to turn off, so its collector voltage rises. If this were a simple transistor switch (i.e., if Q_5 were absent) the collector would rise to +5 V; here, however, the collector resistor R_7 instead supplies current to Q_5, putting it in saturation. So Q_5's collector drops nearly to ground.

At this simple level of analysis the circuit appears to be pretty useless, because its output is the same as its input! Let's look a little closer, though: as the input voltage drops through the 700 mV threshold and Q_5 turns on, the total emitter current rises to ≈ 10 mA (5 mA from Q_5's collector current, and another ≈ 5 mA from its base current, both of which flow out the emitter). The drop across the emitter resistor is now 200 mV, which means that the input threshold has increased to about +800 mV. So the input voltage, which had just dropped below 700 mV, now finds itself well below the new threshold, causing the

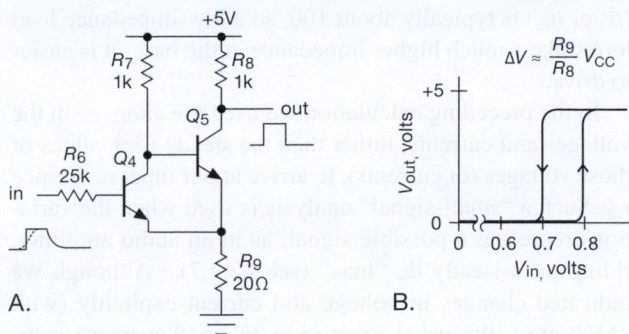

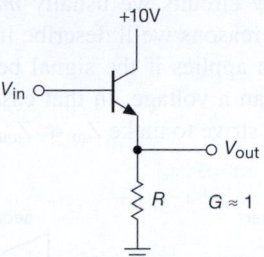

Figure 2.13. A "Schmitt trigger" produces an output with abrupt transitions, regardless of the speed of the input waveform.

output to switch abruptly. This "regenerative" action is how the Schmitt trigger turns a slowly moving waveform into an abrupt transition.

A similar action occurs as the input rises through this higher threshold; see Figure 2.13B, which illustrates how the output voltage changes as the input voltage passes through the two thresholds, an effect known as *hysteresis*. The Schmitt trigger produces rapid output transitions as the input passes through either threshold. We'll see Schmitt triggers again in Chapters 4 and 10.

There are many enjoyable applications of transistor switches, including "signal" applications like this (combined with more complex digital logic circuits), as well as "power switching" circuits in which transistors operating at high currents, high voltages, or both, are used to control hefty loads, perform power conversion, and so on. Transistor switches can also be used as substitutes for mechanical switches when we are dealing with continuous ("linear" or "analog") waveforms. We'll see examples of these in the next chapter, when we deal with FETs, which are ideally suited to such switching tasks, and again in Chapter 12, where we deal with the control of signals and external loads from logic-level signals.

We now move on to consider the first of several *linear* transistor circuits.

2.2.3 Emitter follower

Figure 2.14 shows an example of an *emitter follower*. It is called that because the output terminal is the emitter, which follows the input (the base), less one diode drop:

$$V_E \approx V_B - 0.6 \text{ volts.}$$

The output is a replica of the input, but 0.6 to 0.7 V less positive. For this circuit, V_{in} must stay at +0.6 V or more, or else the output will sit at ground. By returning the emit-

ter resistor to a negative supply voltage, you can permit negative voltage swings as well. Note that there is no collector resistor in an emitter follower.

Figure 2.14. Emitter follower.

At first glance this circuit may appear quite thoroughly useless, until you realize that the input impedance is much larger than the output impedance, as will be demonstrated shortly. This means that the circuit requires less power from the signal source to drive a given load than would be the case if the signal source were to drive the load directly. Or a signal of some internal impedance (in the Thévenin sense) can now drive a load of comparable or even lower impedance without loss of amplitude (from the usual voltage-divider effect). In other words, an emitter follower has current gain, even though it has no voltage gain. It has *power* gain. Voltage gain isn't everything!

A. Impedances of sources and loads
This last point is very important and is worth some more discussion before we calculate in detail the beneficial effects of emitter followers. In electronic circuits, you're always hooking the output of something to the input of something else, as suggested in Figure 2.15. The signal source might be the output of an amplifier stage (with Thévenin equivalent series impedance Z_{out}), driving the next stage or perhaps a load (of some input impedance Z_{in}). In general, the loading effect of the following stage causes a reduction of signal, as we discussed earlier in §1.2.5A. For this reason it is usually best to keep $Z_{out} \ll Z_{in}$ (a factor of 10 is a comfortable rule of thumb).

In some situations it is OK to forgo this general goal of making the source stiff compared with the load. In particular, if the load is always connected (e.g., within a circuit) and if it presents a known and constant Z_{in}, it is not too serious if it "loads" the source. However, it is always nicer if signal levels don't change when a load is connected. Also, if Z_{in} varies with signal level, then having a stiff source

($Z_{out} \ll Z_{in}$) ensures linearity, where otherwise the level-dependent voltage divider would cause distortion.[15]

Finally, as we remarked in §1.2.5A, there are two situations in which $Z_{out} \ll Z_{in}$ is actually the wrong thing to do: in radiofrequency circuits we usually *match* impedances ($Z_{out} = Z_{in}$), for reasons we'll describe in Appendix H. A second exception applies if the signal being coupled is a *current* rather than a voltage. In that case the situation is reversed, and we strive to make $Z_{in} \ll Z_{out}$ ($Z_{out} = \infty$, for a current source).

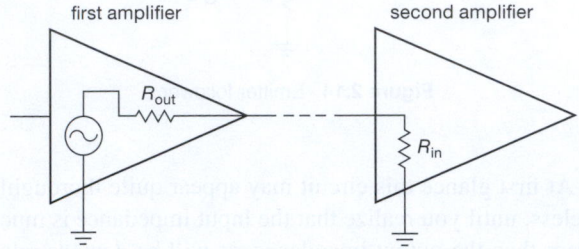

Figure 2.15. Illustrating circuit "loading" as a voltage divider.

B. Input and output impedances of emitter followers

As we've stated, the emitter follower is useful for changing impedances of signals or loads. To put it starkly, that's really the whole point of an emitter follower.

Let's calculate the input and output impedances of the emitter follower. In the preceding circuit we consider R to be the load (in practice it sometimes *is* the load; otherwise the load is in parallel with R, but with R dominating the parallel resistance anyway). Make a voltage change ΔV_B at the base; the corresponding change at the emitter is $\Delta V_E = \Delta V_B$. Then the change in emitter current is

$$\Delta I_E = \Delta V_B / R,$$

so

$$\Delta I_B = \frac{1}{\beta + 1}\Delta I_E = \frac{\Delta V_B}{R(\beta + 1)}$$

(using $I_E = I_C + I_B$). The input resistance is $\Delta V_B / \Delta I_B$. Therefore

$$r_{in} = (\beta + 1)R. \tag{2.2}$$

The transistor small-signal (or "incremental") current gain

---

[15] We use the boldface symbol **Z** when the complex nature of impedance is important. In common usage the term "impedance" can refer loosely to the *magnitude* of impedance, or even to a purely real impedance (e.g., transmission-line impedance); for such instances we use the ordinary math-italic symbol Z.

(β, or h_{fe}) is typically about 100, so a low-impedance load looks like a much higher impedance at the base; it is easier to drive.

In the preceding calculation we used the *changes* in the voltages and currents, rather than the steady (dc) values of those voltages (or currents), to arrive at our input resistance r_{in}. Such a "small-signal" analysis is used when the variations represent a possible signal, as in an audio amplifier, riding on a steady dc "bias" (see §2.2.7). Although we indicated changes in voltage and current explicitly (with "ΔV," etc.), the usual practice is to use lowercase symbols for small-signal variations (thus $\Delta V \leftrightarrow v$); with this convention the above equation for ΔI_E, for example, would read $i_E = v_B / R$.

The distinction between dc current gain (h_{FE}) and small-signal current gain (h_{fe}) isn't always made clear, and the term beta is used for both. That's alright, since $h_{fe} \approx h_{FE}$ (except at very high frequencies), and you never assume you know them accurately, anyway.

Although we used resistances in the preceding derivation, we could generalize to complex impedances by allowing ΔV_B, ΔI_B, etc., to become complex numbers. We would find that the same transformation rule applies for impedances:

$$\mathbf{Z}_{in} = (\beta + 1)\mathbf{Z}_{load}. \tag{2.3}$$

We could do a similar calculation to find that the output impedance $\mathbf{Z}_{out}$ of an emitter follower (the impedance looking into the emitter) driven from a source of internal impedance $\mathbf{Z}_{source}$ is given by

$$\mathbf{Z}_{out} = \frac{\mathbf{Z}_{source}}{\beta + 1}. \tag{2.4}$$

Strictly speaking, the output impedance of the circuit should also include the parallel resistance of R, but in practice $\mathbf{Z}_{out}$ (the impedance looking into the emitter) dominates.

Exercise 2.4. Show that the preceding relationship is correct. *Hint*: hold the source voltage fixed and find the change in output current for a given forced change in output voltage. Remember that the source voltage is connected to the base through a series resistor.

Because of these nice properties, emitter followers find application in many situations, e.g., making low-impedance signal sources within a circuit (or at outputs), making stiff voltage references from higher-impedance references (formed from voltage dividers, say), and generally isolating signal sources from the loading effects of subsequent stages.

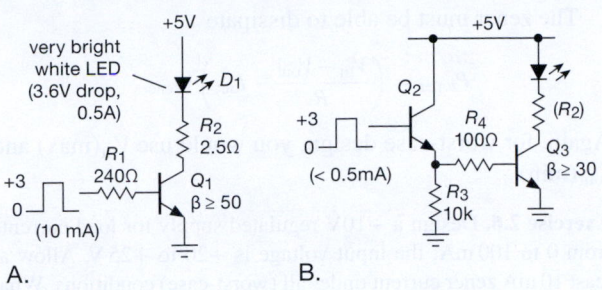

A.   B.

Figure 2.16. Putting an emitter follower in front of a switch makes it easy for a low-current control signal to switch a high-current load.

Exercise 2.5. Use a follower with the base driven from a voltage divider to provide a stiff source of +5 volts from an available regulated +15 V supply. Load current (max) = 25 mA. Choose your resistor values so that the output voltage doesn't drop more than 5% under full load.

C. Follower drives switch

Figure 2.16 shows a nice example of an emitter follower rescuing an awkward circuit. We're trying to switch a really bright white LED (the kind you use for "area lighting"), which drops about 3.6 V at its desired 500 mA of forward current. And we've got a 0–3 V digital logic signal available to control the switch. The first circuit uses a single *npn* saturated switch, with a base resistor sized to produce 10 mA of base current, and a 2.5 Ω current-limiting resistor in series with the LED.

This circuit is OK, sort of. But it draws an uncomfortably large current from the control input; and it requires Q_1 to have plenty of current gain at the full load current of 0.5 A. In the second circuit (Figure 2.16B) an emitter follower has come to the rescue, greatly reducing the input current (because of its current gain), and at the same time relaxing the minimum beta requirement of the switch (Q_3). To be fair, we should point out that a low-threshold MOSFET provides an even simpler solution here; we'll tell you how, in Chapters 3 and 12.

D. Important points about followers

Current flow in one direction only. Notice (§2.1.1, rule 4) that in an emitter follower the *npn* transistor can only *source* (as opposed to *sink*) current. For instance, in the loaded circuit shown in Figure 2.17 the output can swing to within a transistor saturation voltage drop of V_{CC} (about +9.9 V), but it cannot go more negative than −5 volts. That is because on the extreme negative swing, the transistor can do no better than to turn off completely, which it does at −4.4 volts input (−5 V out-

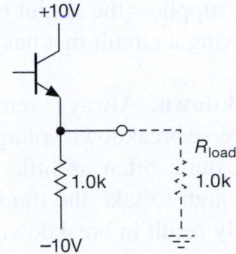

Figure 2.17. An *npn* emitter follower can source plenty of current through the transistor, but can sink limited current only through its emitter resistor.

put, set by the divider formed by the load and emitter resistors). Further negative swing at the input results in back-biasing of the base–emitter junction, but no further change in output. The output, for a 10 volt amplitude sinewave input, looks as shown in Figure 2.18.

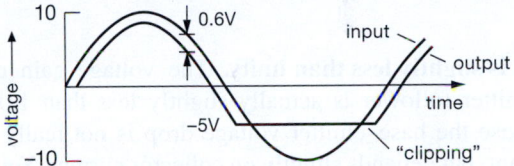

Figure 2.18. Illustrating the asymmetrical current drive capability of the *npn* emitter follower.

Another way to view the problem is to say that the emitter follower has a low value of *small-signal* output impedance, whereas its large-signal output impedance is much higher (as large as R_E). The output impedance changes over from its small-signal value to its large-signal value at the point where the transistor goes out of the active region (in this case at an output voltage of −5 V). To put this point another way, a low value of small-signal output impedance doesn't necessarily mean that the circuit can generate large signal swings into a low resistance load. A low small-signal output impedance doesn't imply a large output current capability.

Possible solutions to this problem involve either decreasing the value of the emitter resistor (with greater power dissipation in resistor and transistor), using a *pnp* transistor (if all signals are negative only), or using a "push–pull" configuration, in which two complementary transistors (one *npn*, one *pnp*) are used (§2.4.1). This sort of problem can also come up when the load that an emitter follower is driving contains voltage or current sources of its own, and thus can force a current in the "wrong" direction. This happens most often with

regulated power supplies (the output is usually an emitter follower) driving a circuit that has other power supplies.

Base-emitter breakdown. Always remember that the
base–emitter reverse breakdown voltage for silicon transistors is small, quite often as little as 6 volts. Input
swings large enough to take the transistor out of conduction can easily result in breakdown (causing permanent degradation of current gain β) unless a protective
diode is added (Figure 2.19).

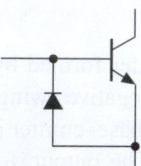

Figure 2.19. A diode prevents base–emitter reverse voltage breakdown.

Gain is slightly less than unity. The voltage gain of an
emitter follower is actually slightly less than 1.0, because the base–emitter voltage drop is not really constant, but depends slightly on collector current. You will
see how to handle that later in the chapter, when we have
the Ebers–Moll equation.

2.2.4 Emitter followers as voltage regulators

The simplest regulated supply of voltage is simply a zener
(Figure 2.20). Some current must flow through the zener,
so you choose

$$\frac{V_{in}(min) - V_{out}}{R} > I_{out}(max).$$

Because V_{in} isn't regulated, you use the lowest value of V_{in}
that might occur. Designing for satisfactory operation under the worst combination (here minimum V_{in} and maximum I_{out}) is known as "worst-case" design. In practice,
you would also worry about component tolerances, line-
voltage limits, etc., designing to accommodate the worst
possible combination that would ever occur.

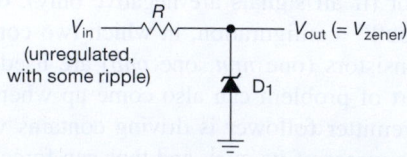

Figure 2.20. Simple zener voltage regulator.

The zener must be able to dissipate

$$P_{zener} = \left(\frac{V_{in} - V_{out}}{R} - I_{out}\right) V_{zener}.$$

Again, for worst-case design, you would use $V_{in}(max)$ and
$I_{out}(min)$.

Exercise 2.6. Design a $+10\dot{V}$ regulated supply for load currents
from 0 to 100 mA; the input voltage is $+20$ to $+25$ V. Allow at
least 10 mA zener current under all (worst-case) conditions. What
power rating must the zener have?

This simple zener-regulated supply is sometimes used
for noncritical circuits or circuits using little supply current. However, it has limited usefulness, for several reasons:

- V_{out} isn't adjustable or settable to a precise value.
- Zener diodes give only moderate ripple rejection and regulation against changes of input or load, owing to their
finite dynamic impedance.
- For widely varying load currents a high-power zener is
often necessary to handle the dissipation at low load current.[16]

By using an emitter follower to isolate the zener, you
get the improved circuit shown in Figure 2.21. Now the situation is much better. Zener current can be made relatively
independent of load current, since the transistor base current is small, and far lower zener power dissipation is possible (reduced by as much as a factor of β). The collector
resistor R_C can be added to protect the transistor from momentary output short circuits by limiting the current, even
though it is not essential to the emitter follower function.
Choose R_C so that the voltage drop across it is less than the
drop across R for the highest normal load current (i.e., so
that the transistor does not saturate at maximum load).

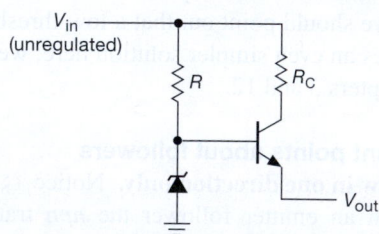

Figure 2.21. Zener regulator with follower, for increased output
current. R_C protects the transistor by limiting maximum output current.

---

[16] This is a property shared by all *shunt regulators*, of which the zener is
the simplest example.

Exercise 2.7. Design a $+10\,\mathrm{V}$ supply with the same specifications as in Exercise 2.6. Use a zener and emitter follower. Calculate worst-case dissipation in transistor and zener. What is the percentage change in zener current from the no-load condition to full load? Compare with your previous circuit.

A nice variation of this circuit aims to eliminate the effect of ripple current (through R) on the zener voltage by supplying the zener current from a current source, which is the subject of §2.2.6. An alternative method uses a lowpass filter in the zener bias circuit (Figure 2.22). R is chosen such that the series pair provides sufficient zener current. Then C is chosen large enough so that $RC \gg 1/f_{\mathrm{ripple}}$.[17]

Later you will see better voltage regulators, ones in which you can vary the output easily and continuously by using feedback. They are also better voltage sources, with output impedances measured in milliohms, temperature coefficients of a few parts per million per degree centigrade, and other desirable features.

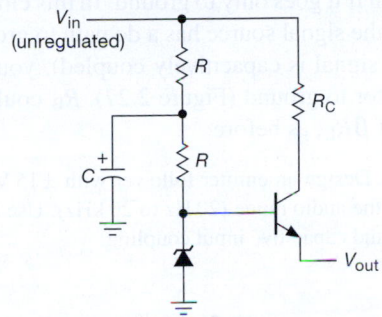

Figure 2.22. Reducing ripple in the zener regulator.

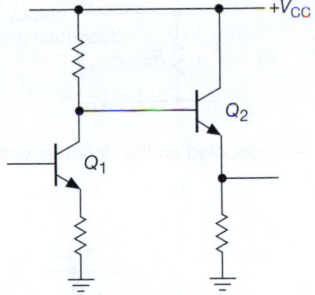

Figure 2.23. Biasing an emitter follower from a previous stage.

2.2.5 Emitter follower biasing

When an emitter follower is driven from a preceding stage in a circuit, it is usually OK to connect its base directly to the previous stage's output, as shown in Figure 2.23.

---

[17] In a variation of this circuit, the upper resistor is replaced with a diode.

Because the signal on Q_1's collector is always within the range of the power supplies, Q_2's base will be between V_{CC} and ground, and therefore Q_2 is in the active region (neither cut off nor saturated), with its base–emitter diode in conduction and its collector at least a few tenths of a volt more positive than its emitter. Sometimes, though, the input to a follower may not be so conveniently situated with respect to the supply voltages. A typical example is a capacitively coupled (or ac-coupled) signal from some external source (e.g., an audio signal input to a stereo amplifier). In that case the signal's average voltage is zero, and direct coupling to an emitter follower will give an output like that in Figure 2.24.

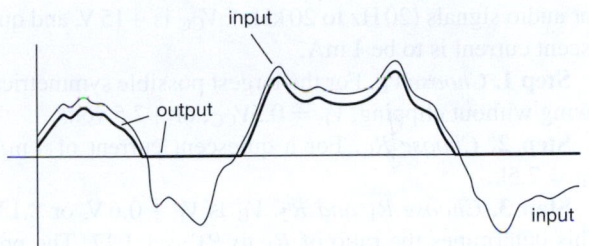

Figure 2.24. A transistor amplifier powered from a single positive supply cannot generate negative voltage swings at the transistor output terminal.

It is necessary to *bias* the follower (in fact, any transistor amplifier) so that collector current flows during the entire signal swing. In this case a voltage divider is the simplest way (Figure 2.25). R_1 and R_2 are chosen to put the base halfway between ground and V_{CC} when there is no input signal, i.e., R_1 and R_2 are approximately equal. The process of selecting the operating voltages in a circuit, in the absence of applied signals, is known as setting the *quiescent point*. In this case, as in most cases, the quiescent point is chosen to allow maximum symmetrical signal swing of the output waveform without *clipping* (flattening of the top or bottom of the waveform). What values should R_1 and R_2 have? Applying our general principle (§1.2.5A, §2.2.3A), we make the impedance of the dc bias source (the impedance looking into the voltage divider) small compared with the load it drives (the dc impedance looking into the base of the follower). In this case,

$$R_1 \| R_2 \ll \beta R_{\mathrm{E}}.$$

This is approximately equivalent to saying that the current flowing in the voltage divider should be large compared with the current drawn by the base.

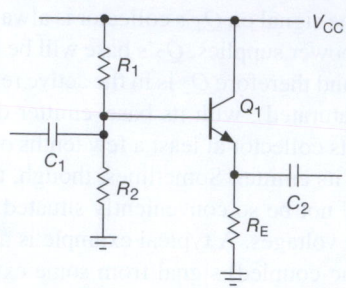

Figure 2.25. An ac-coupled emitter follower. Note base bias voltage divider.

A. Emitter follower design example

As an actual design example, let's make an emitter follower for audio signals (20 Hz to 20 kHz). V_{CC} is +15 V, and quiescent current is to be 1 mA.

Step 1. *Choose* V_E. For the largest possible symmetrical swing without clipping, $V_E = 0.5V_{CC}$, or +7.5 volts.

Step 2. *Choose* R_E. For a quiescent current of 1 mA, $R_E = 7.5k$.

Step 3. *Choose* R_1 *and* R_2. V_B is $V_E + 0.6$ V, or 8.1 V. This determines the ratio of R_1 to R_2 as 1:1.17. The preceding loading criterion requires that the parallel resistance of R_1 and R_2 be about 75k or less (one-tenth of $7.5k \times \beta$). Suitable standard values are $R_1 = 130k$, $R_2 = 150k$.

Step 4. *Choose* C_1. The capacitor C_1 forms a highpass filter with the impedance it sees as a load, namely the impedance looking into the base in parallel with the impedance looking into the base voltage divider. If we assume that the load this circuit will drive is large compared with the emitter resistor, then the impedance looking into the base is βR_E, about 750k. The divider looks like 70k. So the capacitor sees a load of about 63k, and it should have a value of at least 0.15 μF so that the 3 dB point will be below the lowest frequency of interest, 20 Hz.

Step 5. *Choose* C_2. The capacitor C_2 forms a highpass filter in combination with the load impedance, which is unknown. However, it is safe to assume that the load impedance won't be smaller than R_E, which gives a value for C_2 of at least 1.0 μF to put the 3 dB point below 20 Hz. Because there are now two cascaded highpass filter sections, the capacitor values should be increased somewhat to prevent excessive attenuation (reduction of signal amplitude, in this case 6 dB) at the lowest frequency of interest. $C_1 = 0.47 \,\mu$F and $C_2 = 3.3 \,\mu$F might be good choices.[18]

---

[18] These values may seem curiously "unround." But they are chosen from the widely available EIA "E6" decade values (see Appendix C); and in fact "round-number" values of 0.5 μF and 3.0 μF are harder to find.

From our simple transistor model, the output impedance at the emitter is just $Z_{out} = R_E \| [(Z_{in} \| R_1 \| R_2)/\beta]$, where Z_{in} is the (Thévenin) output resistance of the signal that drives this circuit. So, taking $\beta \approx 100$, a signal source with 10 kΩ output resistance would result in an output impedance (at the emitter) of about 87 Ω. As we'll see later in the chapter (§2.3), there's an effect (the intrinsic emitter impedance, r_e) that adds an additional resistance of $0.025/I_E$ effectively in series with the emitter; so the output impedance here (with 10 kΩ source) would be about 110 Ω.

B. Followers with split supplies

Because signals often are "near ground," it is convenient to use symmetrical positive and negative supplies. This simplifies biasing and eliminates coupling capacitors (Figure 2.26).

Warning: you must always provide a dc path for base bias current, even if it goes only to ground. In this circuit it is assumed that the signal source has a dc path to ground. If not (e.g., if the signal is capacitively coupled), you must provide a resistor to ground (Figure 2.27). R_B could be about one-tenth of βR_E, as before.

Exercise 2.8. Design an emitter follower with ±15 V supplies to operate over the audio range (20 Hz to 20 kHz). Use 5 mA quiescent current and capacitive input coupling.

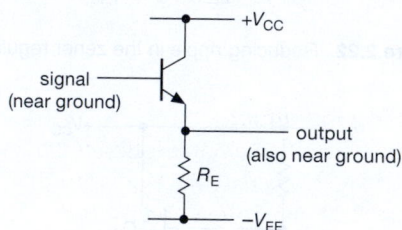

Figure 2.26. A dc-coupled emitter follower with split supply.

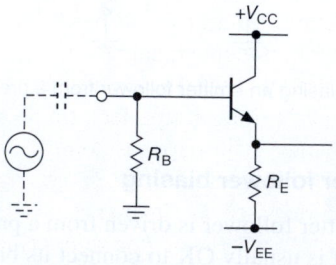

Figure 2.27. Always provide a dc bias path.

C. Bad biasing

You sometimes see sadness-inducing circuits like the disaster shown in Figure 2.28. The designer chose R_B by assuming a particular value for beta (100), estimating the base current, and then hoping for a 7 V drop across R_B. This is a bad design; beta is not a good parameter and will vary considerably. By using voltage biasing with a stiff voltage divider, as in the detailed example presented earlier, the quiescent point is insensitive to variations in transistor beta. For instance, in the previous design example the emitter voltage will increase by only 0.35 V (5%) for a transistor with $\beta = 200$ instead of the nominal $\beta = 100$. And, as with this emitter follower example, it is just as easy to fall into this trap and design bad transistor circuits in the other transistor configurations (notably the common-emitter amplifier, which we will treat later in this chapter).

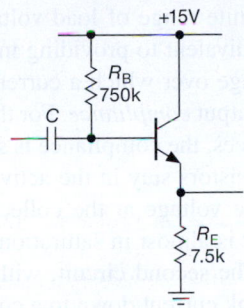

Figure 2.28. Don't do this!

D. Cancelling the offset – I

Wouldn't it be nice if an emitter follower did not cause an offset of the output signal by the $V_{BE} \approx 0.6$ V base–emitter drop? Figure 2.29 shows how to cancel the dc offset, by cascading a *pnp* follower (which has a positive V_{BE} offset) with an *npn* follower (which has a comparable negative V_{BE} offset). Here we've configured the circuit with ±10 V symmetrical split supplies; and we've used equal-value emitter resistors so that the two transistors have a comparable quiescent current for an input signal near 0 V.

This is a nice trick, useful to know about and often helpful. But the cancellation isn't perfect, for reasons we'll see later in the chapter (V_{BE} depends somewhat on collector current, and on transistor size, §2.3), and again in Chapter 5. But, as we'll see in Chapter 4, it *is* in fact rather easy to make a follower, using *operational amplifiers*, with nearly perfect zero offset (10 μV or less); and as a bonus you get input impedances in the gigaohms (or more), input currents in the nanoamps (or less), and output impedances

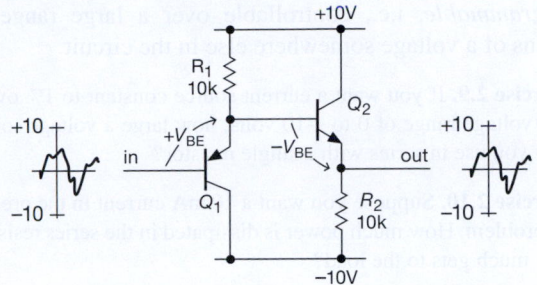

Figure 2.29. Cascading a *pnp* and an *npn* follower produces approximate cancellation of the V_{BE} offsets.

measured in fractions of an ohm. Take a look ahead at Chapter 4.

2.2.6 Current source

Current sources, although often neglected, are as important and as useful as voltage sources. They often provide an excellent way to bias transistors, and they are unequaled as "active loads" for super-gain amplifier stages and as emitter sources for differential amplifiers. Integrators, sawtooth generators, and ramp generators need current sources. They provide wide-voltage-range pullups within amplifier and regulator circuits. And, finally, there are applications in the outside world that require constant current sources, e.g., electrophoresis or electrochemistry.

A. Resistor plus voltage source

The simplest approximation to a current source is shown in Figure 2.30. As long as $R_{load} \ll R$ (in other words, $V_{load} \ll V$), the current is nearly constant and is approximately

$$I \approx V/R.$$

The load doesn't have to be resistive. A capacitor will charge at a constant rate, as long as $V_{cap} \ll V$; this is just the first part of the exponential charging curve of an RC.

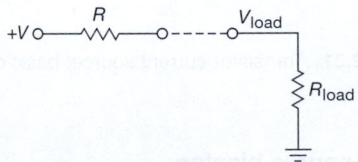

Figure 2.30. Current-source approximation.

There are several drawbacks to a simple resistor current source. To make a good approximation to a current source, you must use large voltages, with lots of power dissipation in the resistor. In addition, the current isn't easily *programmable*, i.e., controllable over a large range by means of a voltage somewhere else in the circuit.

Exercise 2.9. If you want a current source constant to 1% over a load voltage range of 0 to +10 volts, how large a voltage source must you use in series with a single resistor?

Exercise 2.10. Suppose you want a 10 mA current in the preceding problem. How much power is dissipated in the series resistor? How much gets to the load?

B. Transistor current source

Happily, it is possible to make a very good current source with a transistor (Figure 2.31). It works like this: applying V_B to the base, with $V_B > 0.6$ V, ensures that the emitter is always conducting:

$$V_E = V_B - 0.6 \text{ volts}.$$

So

$$I_E = V_E / R_E = (V_B - 0.6 \text{ volts}) / R_E.$$

But, since $I_E \approx I_C$ for large beta,

$$I_C \approx (V_B - 0.6 \text{ volts}) / R_E, \qquad (2.5)$$

independent of V_C, as long as the transistor is not saturated ($V_C \gtrsim V_E + 0.2$ volts).

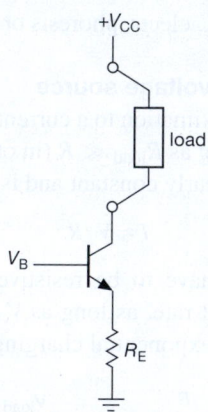

Figure 2.31. Transistor current source: basic concept.

C. Current-source biasing

The base voltage can be provided in a number of ways. A voltage divider is OK, as long as it is stiff enough. As before, the criterion is that its impedance should be much less than the dc impedance looking into the base (βR_E). Or you can use a zener diode (or a two-terminal IC reference like the LM385), biased from V_{CC}, or even a few forward-

biased diodes[19] in series from base to the corresponding emitter supply. Figure 2.32 shows some examples. In the last example (Figure 2.32C), a *pnp* transistor *sources* current to a load returned to ground. The other examples (using *npn* transistors) should properly be called current *sinks*, but the usual practice is to refer to them all loosely as "current sources."[20] In the first circuit, the voltage-divider impedance of ~1.3k is stiff compared with the impedance looking into the base of about 100k (for $\beta = 100$), so any changes in beta with collector voltage will not much affect the output current by causing the base voltage to change. In the other two circuits the biasing resistors are chosen to provide several milliamps to bring the diodes into conduction.

D. Compliance

A current source can provide constant current to the load only over some finite range of load voltage. To do otherwise would be equivalent to providing infinite power. The output voltage range over which a current source behaves well is called its output *compliance*. For the preceding transistor current sources, the compliance is set by the requirement that the transistors stay in the active region. Thus in the first circuit the voltage at the collector can go down until the transistor is almost in saturation, perhaps +1.1 V at the collector. The second circuit, with its higher emitter voltage, can sink current down to a collector voltage of about +5.1 V.

In all cases the collector voltage can range from a value near saturation all the way up to the supply voltage. For example, the last circuit can source current to the load for any voltage between zero and about +8.6 V across the load. In fact, the load might even contain batteries or power supplies of its own, which could carry the collector beyond the supply voltage (Figure 2.32A,B) or below ground (Figure 2.32C). That's OK, but you must watch out for transistor breakdown (V_{CE} must not exceed BV_{CEO}, the specified collector–emitter breakdown voltage) and also for excessive power dissipation (set by $I_C V_{CE}$). As you will see in §§3.5.1B, 3.6.4C, and 9.4.2, there is an additional safe-operating-area constraint on power transistors.

Exercise 2.11. You have +5 and +15 V regulated supplies available in a circuit. Design a 5 mA *npn* current sink using the +5 V to bias the base. What is the output compliance?

---

[19] A red LED, with its forward voltage drop of ≈1.6 V, is a convenient substitute for a string of three diodes.

[20] "Sink" and "source" simply refer to the direction of current flow: if a circuit *supplies* (positive) current to a point, it is a *source*, and vice versa.

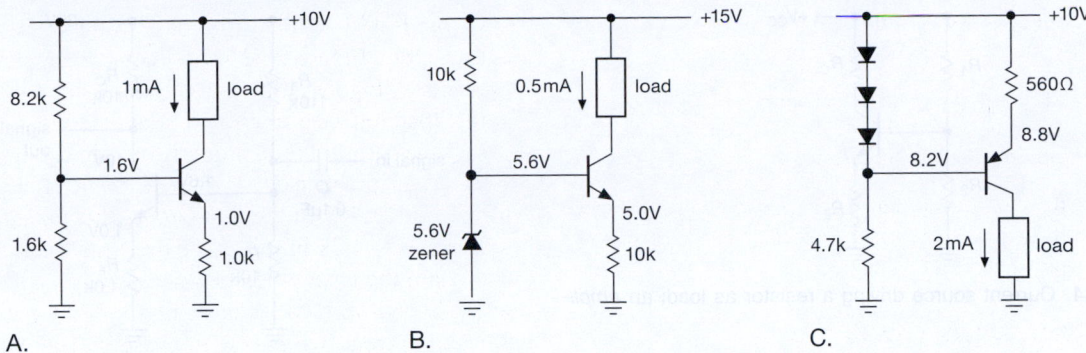

Figure 2.32. Transistor current-source circuits, illustrating three methods of base biasing; *npn* transistors *sink* current, whereas *pnp* transistors *source* current. The circuit in C illustrates a load returned to ground. See also Figure 3.26.

A current source doesn't have to have a fixed voltage at the base. By varying V_B you get a voltage-programmable current source. The input signal swing v_{in} (recall that lowercase symbols mean *variations*) must stay small enough so that the emitter voltage never drops to zero, if the output current is to reflect input-voltage variations smoothly. The result will be a current source with variations in output current proportional to the variations in input voltage, $i_{out} = v_{in}/R_E$. This is the basis of the amplifier we'll see next (§2.2.7).

E. Cancelling the offset – II
It's a minor drawback of these current source circuits that you have to apply a base voltage that is offset by $V_{BE} \approx 0.6$ V from the voltage that you want to appear across the emitter resistor; and it is of course the latter that sets the output current. It's the same offset issue as with an emitter follower; and you can use the same trick (§2.2.5D) to bring about approximate cancellation of the offset in situations in which that is a problem.

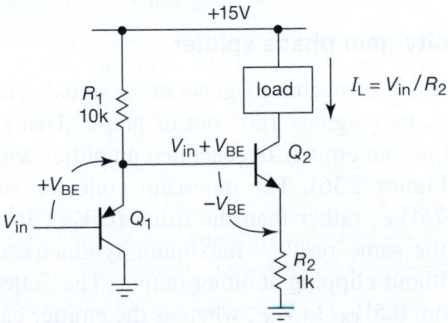

Figure 2.33. Compensating the V_{BE} drop in a current source.

Look at Figure 2.33. It has our standard current-source output stage Q_2, with the current set by the voltage across the emitter resistor: $I_L = V_E/R_2$. So Q_2's base needs to be

a V_{BE} higher (the offset), but that's just what the *pnp* input follower does anyway. So, voilà, the voltage at Q_2's emitter winds up being approximately equal to V_{in} that you apply; and so the output current is simply $I_L = V_{in}/R_2$, with no ifs, ands, buts, or V_{BE} offsets. Cute!

We hasten to point out, though, that this is not a particularly accurate cancellation, because the two transistors will in general have different collector currents, and therefore somewhat different base–emitter drops (§2.3). But it's a first-order hack, and a lot better than nothing. And, once again, the magic of operational amplifiers (Chapter 4) will provide a way to make current sources in which the output current is accurately programmed by an input voltage, without that pesky V_{BE} offset.

F. Deficiencies of current sources
These transistor current-source circuits perform well, particularly when compared with a simple resistor biased from a fixed voltage (Figure 2.30). When you look closely, though, you find that they do depart from the ideal at some level of scrutiny – that is, the load current does show some (relatively small) variation with voltage. Another way to say the same thing is that the current source has a finite ($R_{Th} < \infty$) Thévenin equivalent resistance.

We discuss the causes of these deficiencies, and some very clever circuit fixes, later in the chapter, and also in Chapter 2x.

2.2.7 Common-emitter amplifier
Consider a current source with a resistor as load (Figure 2.34). The collector voltage is

$$V_C = V_{CC} - I_C R_C$$

We could capacitively couple a signal to the base to cause the collector voltage to vary. Consider the example in

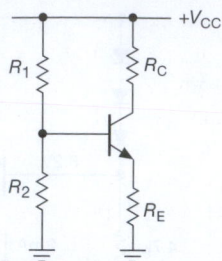

Figure 2.34. Current source driving a resistor as load: an *amplifier*!

Figure 2.35. Blocking capacitor C is chosen so that all frequencies of interest are passed by the highpass filter it forms in combination with the parallel resistance of the base biasing resistors[21]; that is,

$$C \geq \frac{1}{2\pi f(R_1 \| R_2)}.$$

The quiescent collector current is 1.0 mA because of the applied base bias and the 1.0k emitter resistor. That current puts the collector at +10 volts (+20 V, minus 1.0 mA through 10k). Now imagine an applied wiggle in base voltage v_B. The emitter follows with $v_E = v_B$, which causes a wiggle in emitter current

$$i_E = v_E/R_E = v_B/R_E$$

and nearly the same change in collector current (β is large). So the initial wiggle in base voltage finally causes a collector voltage wiggle

$$v_C = -i_C R_C = -v_B(R_C/R_E)$$

Aha! It's a *voltage amplifier*, with a voltage amplification (or "gain") given by

$$\text{gain} = v_{\text{out}}/v_{\text{in}} = -R_C/R_E \qquad (2.6)$$

In this case the gain is −10,000/1000, or −10. The minus sign means that a positive wiggle at the input gets turned into a negative wiggle (10 times as large) at the output. This is called a *common-emitter amplifier* with emitter degeneration.

A. Input and output impedances of the common-emitter amplifier

We can easily determine the input and output impedances of the amplifier. The input signal sees, in parallel, 110k, 10k, and the impedance looking into the base. The latter is

---

[21] The impedance looking into the base itself will usually be much larger because of the way the base resistors are chosen, and it can generally be ignored.

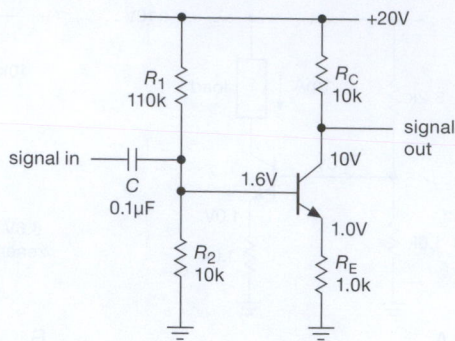

Figure 2.35. An ac common-emitter amplifier with emitter degeneration. Note that the output terminal is the collector rather than the emitter.

about 100k (β times R_E), so the input impedance (dominated by the 10k) is about 8k. The input coupling capacitor thus forms a highpass filter, with the 3 dB point at 200 Hz. The signal driving the amplifier sees 0.1 μF in series with 8k, which to signals of normal frequencies (well above the 3 dB point) just looks like 8k.

The output impedance is 10k in parallel with the impedance looking into the collector. What is that? Well, remember that if you snip off the collector resistor, you're simply looking into a current source. The collector impedance is very large (measured in megohms), and so the output impedance is just the value of the collector resistor, 10k. It is worth remembering that the impedance looking into a transistor's collector is high, whereas the impedance looking into the emitter is low (as in the emitter follower). Although the output impedance of a common-emitter amplifier will be dominated by the collector load resistor, the output impedance of an emitter follower will not be dominated by the emitter load resistor, but rather by the impedance looking into the emitter.

2.2.8 Unity-gain phase splitter

Sometimes it is useful to generate a signal and its inverse, i.e., two signals 180° out of phase. That's easy to do – just use an emitter-degenerated amplifier with a gain of −1 (Figure 2.36). The quiescent collector voltage is set to $0.75V_{\text{CC}}$, rather than the usual $0.5V_{\text{CC}}$, in order to achieve the same result – maximum symmetrical output swing without clipping at either output. The collector can swing from $0.5V_{\text{CC}}$ to V_{CC}, whereas the emitter can swing from ground to $0.5V_{\text{CC}}$.

Note that the phase-splitter outputs must be loaded with equal (or very high) impedances at the two outputs to maintain gain symmetry.

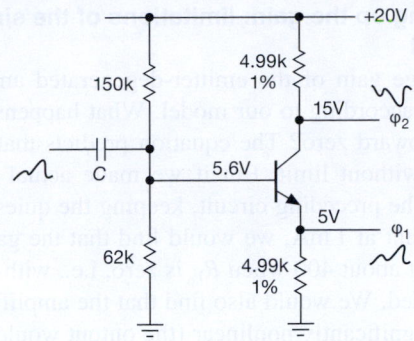

Figure 2.36. Unity-gain phase splitter.

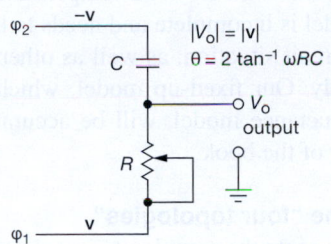

Figure 2.37. Constant-amplitude phase shifter.

A. Phase shifter

A nice use of the phase splitter is shown in Figure 2.37. This circuit gives (for a sinewave input) an output sinewave of adjustable phase (from zero to 180°) and with constant amplitude. It can be best understood with a phasor diagram of voltages (§1.7.12); with the input signal represented by a unit vector along the real axis, the signals look as shown in Figure 2.38.

Signal vectors v_R and v_C must be at right angles, and they must add to form a vector of constant length along the real axis. There is a theorem from geometry that says that the locus of such points is a circle. So the resultant vector (the output voltage) always has unit length, i.e., the same amplitude as the input, and its phase can vary from nearly zero to nearly 180° relative to the input wave as R is varied from nearly zero to a value much larger than X_C at the operating frequency. However, note that the phase shift depends on the frequency of the input signal for a given setting of the potentiometer R. It is worth noting that a simple RC highpass (or lowpass) network could also be used as an adjustable phase shifter. However, its output amplitude would vary over an enormous range as the phase shift was adjusted.

An additional concern here is the ability of the phase-splitter circuit to drive the RC phase shifter as a load. Ideally, the load should present an impedance that is large

compared with the collector and emitter resistors. As a result, this circuit is of limited utility where a wide range of phase shifts is required. You will see improved phase-splitter techniques in Chapter 4, where we use op-amps as impedance buffers, and in Chapter 7, where a cascade of several phase-shifter sections generates a set of "quadrature" signals that extends the phase-shifting range to a full 0° to 360°.

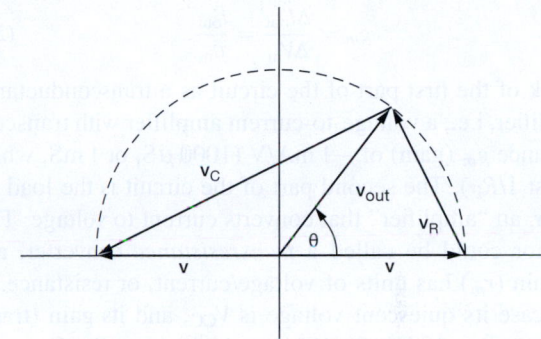

Figure 2.38. Phasor diagram for phase shifter, for which $\theta = 2\arctan(\omega RC)$.

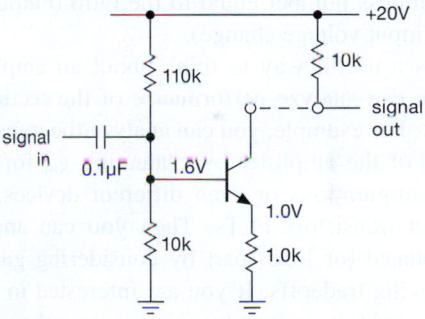

Figure 2.39. The common-emitter amplifier is a transconductance stage driving a (resistive) load.

2.2.9 Transconductance

In the preceding section we figured out the operation of the emitter-degenerated amplifier by (a) imagining an applied base voltage swing and seeing that the emitter voltage had the same swing, then (b) calculating the emitter current swing; then, ignoring the small base current contribution, we got the collector current swing and thus (c) the collector voltage swing. The voltage gain was then simply the ratio of collector (output) voltage swing to base (input) voltage swing.

There's another way to think about this kind of amplifier. Imagine breaking it apart, as in Figure 2.39. The first part is a voltage-controlled current source, with quiescent current of 1.0 mA and gain of −1 mA/V. Gain means the ratio of output to input; in this case the gain has units of current/voltage, or 1/resistance. The inverse of resistance is called *conductance*.[22] An amplifier whose gain has units of conductance is called a *transconductance* amplifier; the ratio of changes $\Delta I_{out}/\Delta V_{in}$ (usually written with small-signal changes indicated in lowercase: i_{out}/v_{in}) is called the transconductance, g_m:

$$g_m = \frac{\Delta I_{out}}{\Delta V_{in}} = \frac{i_{out}}{v_{in}}. \tag{2.7}$$

Think of the first part of the circuit as a transconductance amplifier, i.e., a voltage-to-current amplifier with transconductance g_m (gain) of −1 mA/V (1000 μS, or 1 mS, which is just $1/R_E$). The second part of the circuit is the load resistor, an "amplifier" that converts current to voltage. This resistor could be called a *transresistance* converter, and its gain (r_m) has units of voltage/current, or resistance. In this case its quiescent voltage is V_{CC}, and its gain (transresistance) is 10 V/mA (10 kΩ), which is just R_C. Connecting the two parts together gives you a voltage amplifier. You get the overall gain by multiplying the two gains. In this case the voltage gain $G_V = g_m R_C = -R_C/R_E$, or −10, a unitless number equal to the ratio (output voltage change)/(input voltage change).

This is a useful way to think about an amplifier, because you can analyze performance of the sections independently. For example, you can analyze the transconductance part of the amplifier by evaluating g_m for different circuit configurations or even different devices, such as field-effect transistors FETs. Then you can analyze the transresistance (or load) part by considering gain versus voltage swing tradeoffs. If you are interested in the overall voltage gain, it is given by $G_V = g_m r_m$, where r_m is the transresistance of the load. Ultimately the substitution of an active load (current source), with its extremely high transresistance, can yield single-stage voltage gains of 10,000 or more. The *cascode* configuration, which we will discuss later, is another example easily understood with this approach.

In Chapter 4, which deals with operational amplifiers, you will see further examples of amplifiers with voltages or currents as inputs or outputs: voltage amplifiers (voltage to voltage), current amplifiers (current to current), and transresistance amplifiers (current to voltage).

---

[22] The inverse of reactance is *susceptance* (and the inverse of impedance is *admittance*), and has a special unit, the *siemens* ("S," not to be confused with lowercase "s," which means seconds), which used to be called the *mho* (ohm spelled backward, symbol "℧").

A. Turning up the gain: limitations of the simple model

The voltage gain of the emitter-degenerated amplifier is $-R_C/R_E$, according to our model. What happens as R_E is reduced toward zero? The equation predicts that the gain will rise without limit. But if we made actual measurements of the preceding circuit, keeping the quiescent current constant at 1 mA, we would find that the gain would level off at about 400 when R_E is zero, i.e., with the emitter grounded. We would also find that the amplifier would become significantly nonlinear (the output would not be a faithful replica of the input), the input impedance would become small and nonlinear, and the biasing would become critical and unstable with temperature. Clearly our transistor model is incomplete and needs to be modified to handle this circuit situation, as well as others we will talk about presently. Our fixed-up model, which we will call the transconductance model, will be accurate enough for the remainder of the book.

B. Recap: the "four topologies"

Before jumping into the complexity just ahead, let's remind ourselves of the four transistor circuits we've seen, namely the switch, emitter follower, current source, and common-emitter amplifier. We've drawn these very schematically in Figure 2.40, omitting details like biasing, and even the polarity of transistor (i.e., *npn* or *pnp*). For completeness we've included also a fifth circuit, the *common-base amplifier*, which we'll meet soon enough (§2.4.5B).

2.3 Ebers–Moll model applied to basic transistor circuits

We've enjoyed seeing some nice feats that can be accomplished with the simplest BJT model – switch, follower, current source, amplifier – but we've run up against some serious limitations (hey, would you believe, *infinite* gain?!). Now it's time to go a level deeper, to address these limitations. The material that follows will suffice for our purposes. And – good news – for many BJT applications the simple model you've already seen is completely adequate.

2.3.1 Improved transistor model: transconductance amplifier

The important change is in rule 4 (§2.1.1), where we said earlier that $I_C = \beta I_B$. We thought of the transistor as a current amplifier whose input circuit behaved like a diode. That's roughly correct, and for some applications it's good enough. But to understand differential amplifiers,

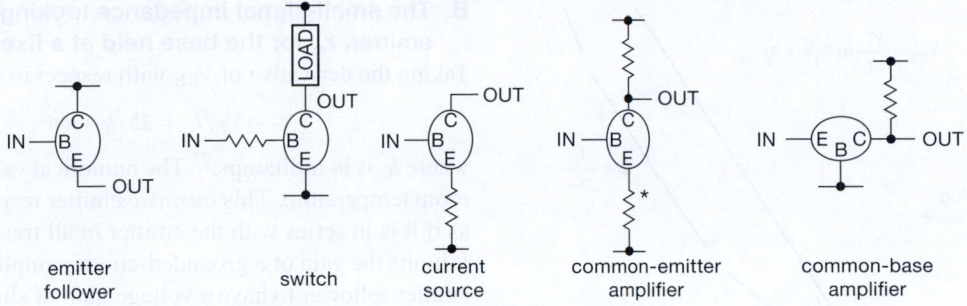

Figure 2.40. Five basic transistor circuits. Fixed voltages (power supplies or ground) are indicated by connections to horizontal line segments. For the switch, the load may be a resistor, to produce a full-swing voltage output; for the common-emitter amplifier, the emitter resistor may be bypassed or omitted altogether.

logarithmic converters, temperature compensation, and other important applications, you must think of the transistor as a *transconductance* device – collector current is determined by base-to-emitter *voltage*.

Here's the modified rule 4.

4. Transconductance amplifier When rules 1–3 (§2.1.1) are obeyed, I_C is related to V_{BE} by[23]

$$I_C = I_S(T)\left(e^{V_{BE}/V_T} - 1\right), \quad (2.8)$$

or, equivalently,

$$V_{BE} = \frac{kT}{q}\log_e\left(\frac{I_C}{I_S(T)} + 1\right), \quad (2.9)$$

where

$$V_T = kT/q = 25.3\,\text{mV} \quad (2.10)$$

at room temperature (68°F, 20°C), q is the electron charge (1.60×10^{-19} coulombs), k is Boltzmann's constant (1.38×10^{-23} joules/K, sometimes written k_B), T is the absolute temperature in degrees Kelvin (K=°C+273.16), and $I_S(T)$ is the *saturation current* of the particular transistor (which depends strongly on temperature, T, as we'll see shortly). Then the base current, which also depends on V_{BE}, can be approximated by

$$I_B = I_C/\beta,$$

where the "constant" β is typically in the range 20 to 1000, but depends on transistor type, I_C, V_{CE}, and temperature. $I_S(T)$ approximates the reverse leakage current (roughly 10^{-15} A for a small-signal transistor like the 2N3904). In the active region $I_C \gg I_S$, and therefore

the -1 term can be neglected in comparison with the exponential:

$$I_C \approx I_S(T)e^{V_{BE}/V_T}. \quad (2.11)$$

The equation for I_C is known as the Ebers–Moll equation.[24] It also describes approximately the current versus voltage for a diode, if V_T is multiplied by a correction factor m between 1 and 2. For transistors it is important to realize that the collector current is accurately determined by the base–emitter voltage, rather than by the base current (the base current is then roughly determined by β), and that this exponential law is accurate over an enormous range of currents, typically from nanoamps to milliamps. Figure 2.41 makes the point graphically.[25] If you measure the base current at various collector currents, you will get a graph of β versus I_C like that in Figure 2.42. Transistor beta versus collector current is discussed further in Chapter 2x.

Although the Ebers–Moll equation tells us that the base–emitter voltage "programs" the collector current, this property is not easy to use in practice (biasing a transistor by applying a base voltage) because of the large temperature coefficient of base–emitter voltage. You will see later how the Ebers–Moll equation provides insight and solutions to this problem.

2.3.2 Consequences of the Ebers–Moll model: rules of thumb for transistor design

From the Ebers–Moll equation (2.8) we get these simple (but handy) "ratio rules" for collector current: $I_{C2}/I_{C1} = \exp(\Delta V_{BE}/V_T)$ and $\Delta V_{BE} = V_T\log_e(I_{C2}/I_{C1})$. We

[23] We indicate the important temperature dependence of I_S by explicitly showing it in functional form – "$I_S(T)$".

[24] J. J. Ebers & J. L. Moll, "Large-signal behavior of junction transistors," *Proc. IRE* **42**, 1761 (1954).

[25] This is sometimes called a Gummel plot.

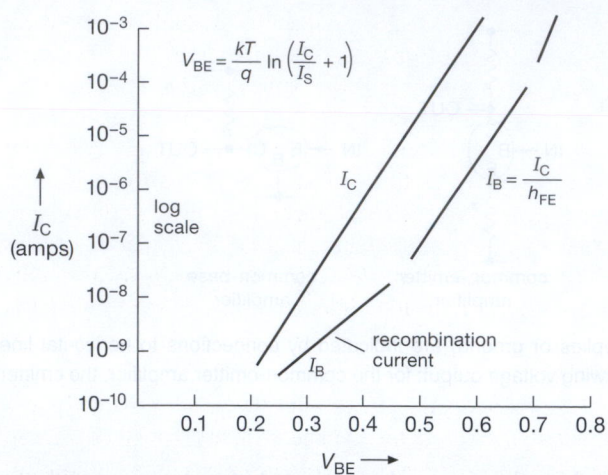

Figure 2.41. Transistor base and collector currents as functions of base-to-emitter voltage V_{BE}.

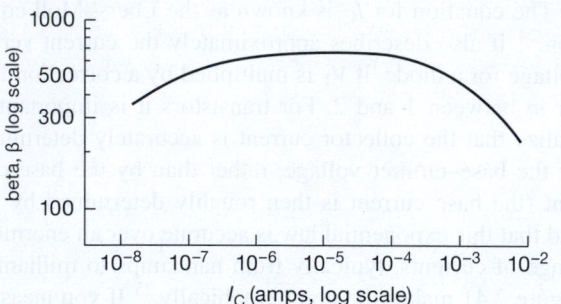

Figure 2.42. Typical transistor current gain (β) versus collector current.

also get the following important quantities we will be using often in circuit design.

A. The steepness of the diode curve.

How much do we need to increase V_{BE} to increase I_C by a factor of 10? From the Ebers–Moll equation, that's just $V_T \log_e 10$, or 58.2 mV at room temperature. We like to remember this as *base–emitter voltage increases approximately 60 mV per decade of collector current.* (Two other formulations: collector current doubles for each 18 mV increase in base–emitter voltage; collector current increases 4% per millivolt increase in base–emitter voltage.) Equivalently, $I_C = I_{C0}e^{\Delta V/25}$, where ΔV is in millivolts.[26]

B. The small-signal impedance looking into the emitter, r_e, for the base held at a fixed voltage.

Taking the derivative of V_{BE} with respect to I_C, you get

$$r_e = V_T/I_C = 25/I_C \text{ ohms,} \qquad (2.12)$$

where I_C is in milliamps.[27] The numerical value $25/I_C$ is for room temperature. This *intrinsic* emitter resistance, r_e, acts as if it is in series with the emitter in all transistor circuits. It limits the gain of a grounded-emitter amplifier, causes an emitter follower to have a voltage gain of slightly less than unity, and prevents the output impedance of an emitter follower from reaching zero. Note that the transconductance[28] of a grounded emitter amplifier is just

$$g_m = I_C/V_T = 1/r_e \ (=40I_C \text{ at room temp).} \qquad (2.13)$$

C. The temperature dependence of V_{BE}.

A glance at the Ebers–Moll equation suggests that V_{BE} (at constant I_C) has a positive temperature coefficient because of the multiplying factor of T in V_T. However, the strong temperature dependence of $I_S(T)$ more than compensates for that term, such that V_{BE} (at constant I_C) *decreases* about 2.1 mV/°C. It is roughly proportional to $1/T_{abs}$, where T_{abs} is the absolute temperature. Sometimes it's useful to cast this instead in terms of the temperature dependence of I_C (at constant V_{BE}): I_C *increases* about 9%/°C; it doubles for an 8°C rise.

There is one additional quantity we will need on occasion, although it is not derivable from the Ebers–Moll equation. It is known as the Early effect,[29] and it sets important limits on current-source and amplifier performance.

D. Early effect.

V_{BE} (at constant I_C) varies slightly with changing V_{CE}. This effect is caused by the variation of effective base width as V_{CE} changes, and it is given, approximately, by

$$\Delta V_{BE} = -\eta \Delta V_{CE}, \qquad (2.14)$$

where $\eta \approx 10^{-4}$–10^{-5}. (As an example, the *npn* 2N5088 has $\eta = 1.3 \times 10^{-4}$, thus a 1.3 mV change of V_{BE} to maintain constant collector current when V_{CE} changes by 10 V.)

---

[26] The "25" in this and the following discussion is more precisely 25.3 mV, the value of $k_B T/q$ at room temperature. It's proportional to absolute temperature – engineers like to say "PTAT," pronounced *pee'-tat*. This has interesting (and useful) consequences, for example the opportunity to make a "silicon thermometer." We'll see more of this in Chapter 2x, and again in Chapter 9.

[27] We like to remember the fact that $r_e = 25\,\Omega$ at a collector current of 1 mA. Then we just scale inversely for other currents: thus $r_e = 2.5\,\Omega$ at $I_C = 10$ mA, etc.

[28] At the next level of sophistication we'll see that, since the quantity r_e is proportional to absolute temperature, a grounded emitter amplifier whose collector current is PTAT has transconductance (and gain) independent of temperature. More in Chapter 2x.

[29] J. M. Early, "Effects of space-charge layer widening in junction transistors," *Proc. IRE* **40**, 1401 (1952). James Early died in 2004.

This is often described instead as a linear increase of collector current with increasing collector voltage when V_{BE} is held constant; you see it expressed as

$$I_C = I_{C0}\left(1 + \frac{V_{CE}}{V_A}\right), \qquad (2.15)$$

where V_A (typically 50–500 V) is known as the Early voltage.[30] This is shown graphically in Figure 2.59 in §2.3.7A. A low Early voltage indicates a low collector output resistance; *pnp* transistors tend to have low V_A, see measured values in Table 8.1. We treat the Early effect in more detail in Chapter 2x.[31]

These are the essential quantities we need. With them we will be able to handle most problems of transistor circuit design, and we will have little need to refer to the Ebers–Moll equation itself.[32]

2.3.3 The emitter follower revisited

Before looking again at the common-emitter amplifier with the benefit of our new transistor model, let's take a quick look at the humble emitter follower. The Ebers–Moll model predicts that an emitter follower should have nonzero output impedance, even when driven by a voltage source, because of finite r_e (item B in the above list). The same effect also produces a voltage gain slightly less than unity, because r_e forms a voltage divider with the load resistor.

These effects are easy to calculate. With fixed base voltage, the impedance looking back into the emitter is just $R_{out} = dV_{BE}/dI_E$; but $I_E \approx I_C$, so $R_{out} \approx r_e$, the intrinsic emitter resistance [recall $r_e = 25/I_C(\text{mA})$]. For example, in Figure 2.43A, the load sees a driving impedance of $r_e = 25\,\Omega$, because $I_C = 1$ mA. (This is paralleled by the emitter resistor R_E, if used; but in practice R_E will always be much larger than r_e.) Figure 2.43B shows a more typical situation, with finite source resistance R_S (for simplicity we've omitted the obligatory biasing components –

base divider and blocking capacitor – which are shown in Figure 2.43C). In this case the emitter follower's output impedance is just r_e in series with $R_s/(\beta + 1)$ (again paralleled by an unimportant R_E, if present). For example, if $R_s = 1$k and $I_C = 1$ mA, $R_{out} = 35\,\Omega$ (assuming $\beta = 100$). It is easy to show that the intrinsic emitter r_e also figures into an emitter follower's *input* impedance, just as if it were in series with the load (actually, parallel combination of load resistor and emitter resistor). In other words, for the emitter follower circuit the effect of the Ebers–Moll model is simply to add a series emitter resistance r_e to our earlier results.[33]

The voltage gain of an emitter follower is slightly less than unity, owing to the voltage divider produced by r_e and the load. It is simple to calculate, because the output is at the junction of r_e and R_{load}: $G_V = v_{out}/v_{in} = R_L/(r_e + R_L)$. Thus, for example, a follower running at 1 mA quiescent current, with 1k load, has a voltage gain of 0.976. Engineers sometimes like to write the gain in terms of the transconductance, to put it in a form that holds for FETs also (see §3.2.3A); in that case (using $g_m = 1/r_e$) you get $G_V = R_L g_m/(1 + R_L g_m)$.

2.3.4 The common-emitter amplifier revisited

Previously we got wrong answers for the voltage gain of the common-emitter amplifier with emitter resistor (sometimes called emitter degeneration) when we set the emitter resistor equal to zero; recall that our wrong answer was $G_V = -R_C/R_E = \infty$!

The problem is that the transistor has $25/I_C(\text{mA})$ ohms of built-in (intrinsic) emitter resistance r_e that must be added to the actual external emitter resistor. This resistance is significant only when small emitter resistors (or none at all) are used.[34] So, for instance, the amplifier we considered previously will have a voltage gain of $-10\text{k}/r_e$, or -400, when the external emitter resistor is zero. The input impedance is not zero, as we would have predicted earlier (βR_E); it is approximately βr_e, or in this case (1 mA quiescent current) about 2.5k.[35]

---

[30] The connection between Early voltage and η is $\eta = V_T/(V_A + V_{CE}) \approx V_T/V_A$; see Chapter 2x.

[31] Previewing some of the results there, the Early effect (a) determines a transistor's collector output resistance $r_o = V_A/I_C$; (b) sets a limit on single-stage voltage gain; and (c) limits the output resistance of a current source. Other things being equal, *pnp* transistors tend to have low Early voltages, as do transistors with high beta; high-voltage transistors usually have high Early voltages, along with low beta. These trends can be seen in the measured Early voltages listed in Table 8.1.

[32] The computer circuit-analysis program SPICE includes accurate transistor simulation with the Ebers–Moll formulas and Gummel–Poon charge models. It's a lot of fun to "wire up" circuits on your computer screen and set them running with SPICE. For more detail see the application of SPICE to BJT amplifier distortion in Chapter 2x.

[33] There's more, if you look deeper: at high frequencies (above f_T/β) the effective current gain drops inversely with frequency; so you get a linearly rising output impedance from an emitter follower that is driven with low R_s. That is, it looks like an inductance, and a capacitive load can cause ringing or even oscillation; these effects are treated in Chapter 2x.

[34] Or, equivalently, when the emitter resistor is bypassed with a capacitor whose impedance at signal frequencies is comparable with, or less than, r_e.

[35] These estimates of gain and input impedance are reasonably good, as long as we stay away from operation at very high frequencies or

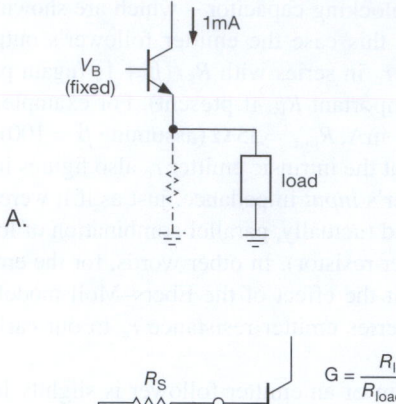

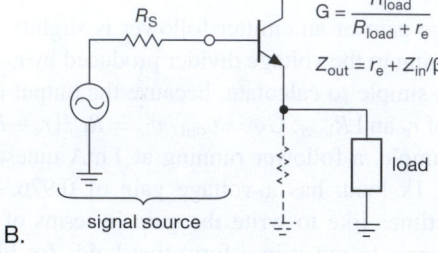

$$G = \frac{R_{load}}{R_{load} + r_e}$$

$$Z_{out} = r_e + Z_{in}/\beta$$

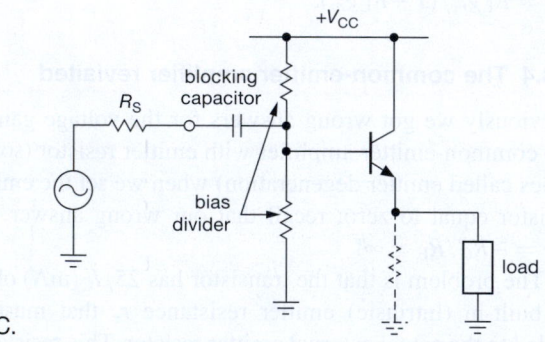

Figure 2.43. Output impedance of emitter followers (see text).

The terms "grounded emitter" and "common emitter" are sometimes used interchangeably, and they can be confusing. We will use the phrase "grounded-emitter amplifier" to mean a common-emitter amplifier with $R_E = 0$ (or equivalent bypassing). A common-emitter amplifier stage may have an emitter resistor; what matters is that the emitter circuit is common to the input circuit and the output circuit.

---

from circuits in which the collector load resistor is replaced with a current source "active load" ($R_C \to \infty$). The ultimate voltage gain of a grounded-emitter amplifier, in the latter situation, is limited by the Early effect; this is discussed in more detail Chapter *2x*.

A. Shortcomings of the single-stage grounded emitter amplifier

The extra voltage gain you get by using $R_E = 0$ comes at the expense of other properties of the amplifier. In fact, the grounded-emitter amplifier, in spite of its popularity in textbooks, should be avoided except in circuits with overall negative feedback. In order to see why, consider Figure 2.44.

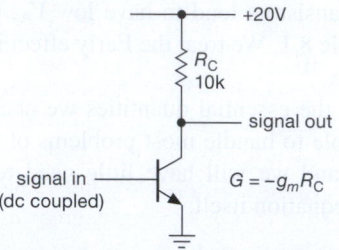

Figure 2.44. Common-emitter amplifier without emitter degeneration.

1. Nonlinearity. The voltage gain is $G = -g_m R_C = -R_C/r_e = -R_C I_C(\text{mA})/25$, so for a quiescent current of 1 mA, the gain is -400. But I_C varies as the output signal varies. For this example, the gain will vary from -800 ($V_{out} = 0$, $I_C = 2$ mA) down to zero ($V_{out} = V_{CC}$, $I_C = 0$). For a triangle-wave input, the output will look as in Figure 2.45. The amplifier has lots of distortion, or poor linearity. The grounded-emitter amplifier without feedback is useful only for small-signal swings about the quiescent point. By contrast, the emitter-degenerated amplifier has a gain almost entirely independent of collector current, as long as $R_E \gg r_e$, and can be used for undistorted amplification even with large-signal swings.

It's easy to estimate the distortion, both with and without an external emitter resistor. With a *grounded* emitter, the incremental (small-signal) gain is $G_V = -R_C/r_e = -I_C R_C/V_T = -V_{drop}/V_T$, where V_{drop} is the instantaneous voltage drop across the collector resistor. Because the gain is proportional to the drop across the collector resistor, the nonlinearity (fractional change of gain with swing) equals the ratio of instantaneous swing to average quiescent drop across the collector resistor: $\Delta G/G \approx \Delta V_{out}/V_{drop}$, where V_{drop} is the average, or quiescent, voltage drop across the collector resistor R_C. Because this represents the extreme variation of gain (i.e., at the peaks of the swing), the overall waveform "distortion" (usually stated as the amplitude of the residual waveform after subtraction of the strictly linear component) will be smaller by roughly a factor of 3. Note that the distortion depends on only the ratio of swing to quiescent drop, and not directly on the operating current, etc.

As an example, in a grounded emitter amplifier powered from +10 V, biased to half the supply (i.e., $V_{\text{drop}} = 5$ V), we measured a distortion of 0.7% at 0.1 V output sinewave amplitude and 6.6% at 1 V amplitude; these values are in good agreement with the predicted values. Compare this with the situation with an added external emitter resistor R_E, in which the voltage gain becomes $G_V = -R_C/(r_e + R_E) = -I_C R_C/(V_T + I_C R_E)$. Only the first term in the denominator contributes distortion, so the distortion is reduced by the ratio of r_e to the total effective emitter resistance: the nonlinearity becomes $\Delta G/G \approx (\Delta V_{\text{out}}/V_{\text{drop}})[r_e/(r_e + R_E)] = (\Delta V_{\text{out}}/V_{\text{drop}})[V_T/(V_T + I_E R_E)]$; the second term is the factor by which the distortion is reduced. When we added an emitter resistor, chosen to drop 0.25 V at the quiescent current – which by this estimate should reduce the nonlinearity by a factor of 10 – the measured distortion of the previous amplifier dropped to 0.08% and 0.74% for 0.1 V and 1 V output amplitudes, respectively. Once again, these measurement agree well with our prediction.

Exercise 2.12. Calculate the predicted distortion for these two amplifiers at the two output levels that were measured.

As we remarked, the nonlinearity of a common-emitter amplifier, when driven by a triangle wave, takes the form of the asymmetric "barn roof" distortion sketched in Figure 2.45.[36] For comparison we took a real-life 'scope (oscilloscope) trace of a grounded emitter amplifier (Figure 2.46); we used a 2N3904 with a 5k collector resistor to a +10 V supply, biased (carefully!) to half the supply. With a ruler we estimated the incremental gain at output voltages of +5 V (halfway to V_+) and at +7.5 V, as shown, where the collector current is 1 mA and 0.5 mA, respectively. The gain values are in pretty good agreement with the predictions ($G = R_C/r_e = I_C(\text{mA})R_C/25\,\Omega$) of $G = -200$ and $G = -100$, respectively. By comparison, Figure 2.47 shows what happened when we added a 225 Ω emitter resistor: the gain is reduced by a factor of 10 at the quiescent point ($G = R_C/(R_E + r_e) \approx R_C/250\,\Omega$), but with much improved linearity (because changes in r_e contribute little to the overall resistance in the denominator, which is now dominated by the fixed 225 Ω external emitter resistor).

For sinusoidal input, the output contains all harmonics of the fundamental wave. Later in the chapter we'll see how to make differential amplifiers with a pair of transistors; for these the residual distortion is symmetric, and contains only the odd harmonics. And in Chapter 2x we'll see some very clever methods for cancelling distortion in differential

---

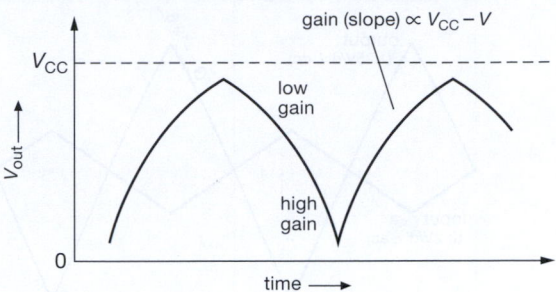

Figure 2.45. Nonlinear output waveform from grounded-emitter amplifier.

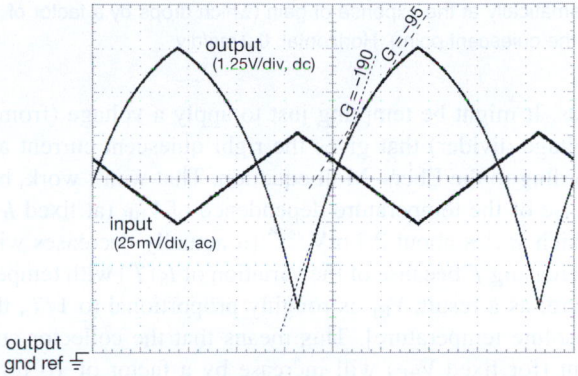

Figure 2.46. *Real life!* The grounded-emitter amplifier of Figure 2.44, with $R_C = 5$k, $V_+ = +10$ V, and a 1 kHz triangle wave input. Top and bottom of the screen are +10 V and ground for the dc-coupled output trace (note sensitive scale for the ac-coupled input signal). Gain estimates (tangent lines) are at V_{out} values of $0.5V_+$ and $0.75V_+$. Horizontal: 0.2 ms/div.

amplifiers, along with the use of SPICE simulation software for rapid analysis and circuit iteration. Finally, to set things in perspective, we should add that any amplifier's residual distortion can be reduced dramatically by use of *negative feedback*. We'll introduce feedback later in this chapter (§2.5), after you've gained familiarity with common transistor circuits. Feedback will finally take center stage when we get to *operational amplifiers* in Chapter 4.

2. Input impedance. The input impedance is roughly $Z_{\text{in}} = \beta r_e = 25\beta/I_C(\text{mA})$ ohms. Once again, I_C varies over the signal swing, giving a varying input impedance. Unless the signal source driving the base has low impedance, you will wind up with nonlinearity because of the nonlinear (variable) voltage divider formed from the signal source and the amplifier's input impedance. By contrast, the input impedance of an emitter-degenerated amplifier is nearly constant, and high.

3. Biasing. The grounded emitter amplifier is difficult to

---

[36] Because the gain (i.e., the slope of V_{out} versus V_{in}) is proportional to the distance from the V_{CC} line, the shape of the curve is in fact an exponential.

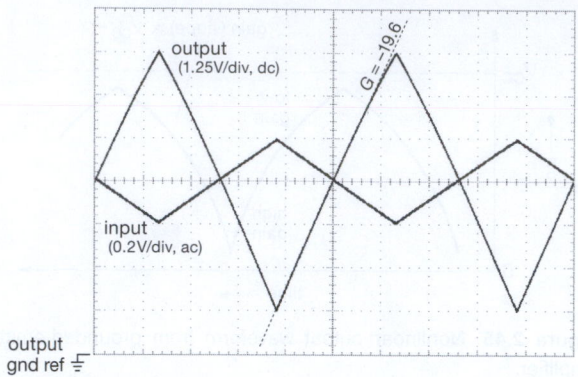

Figure 2.47. Adding a 225 Ω emitter resistor improves the linearity dramatically at the expense of gain (which drops by a factor of 10 at the quiescent point). Horizontal: 0.2 ms/div.

bias. It might be tempting just to apply a voltage (from a voltage divider) that gives the right quiescent current according to the Ebers–Moll equation. That won't work, because of the temperature dependence of V_{BE} (at fixed I_C), which varies about 2.1 mV/°C [it actually decreases with increasing T because of the variation of $I_S(T)$ with temperature; as a result, V_{BE} is roughly proportional to $1/T$, the absolute temperature]. This means that the collector current (for fixed V_{BE}) will increase by a factor of 10 for a 30°C rise in temperature (which corresponds to a 60 mV change in V_{BE}), or about 9%/°C. Such unstable biasing is useless, because even rather small changes in temperature will cause the amplifier to saturate. For example, a grounded emitter stage biased with the collector at half the supply voltage will go into saturation if the temperature rises by 8°C.

Exercise 2.13. Verify that an 8°C rise in ambient temperature will cause a base-voltage-biased grounded emitter stage to saturate, assuming that it was initially biased for $V_C = 0.5V_{CC}$.

Some solutions to the biasing problem are discussed in the following sections. By contrast, the emitter-degenerated amplifier achieves stable biasing by applying a voltage to the base, most of which appears across the emitter resistor, thus determining the quiescent current.

B. Emitter resistor as feedback

Adding an external series resistor to the intrinsic emitter resistance r_e (emitter degeneration) improves many properties of the common-emitter amplifier, but at the expense of gain. You will see the same thing happening in Chapters 4 and 5, when we discuss *negative feedback*, an important technique for improving amplifier characteristics by feeding back some of the output signal to reduce the effective input signal. The similarity here is no coincidence –

the emitter-degenerated amplifier itself uses a form of negative feedback. Think of the transistor as a transconductance device, determining collector current (and therefore output voltage) according to the voltage applied between the base and emitter; but the input to the amplifier is the voltage from base to *ground*. So the voltage from base to emitter is the input voltage, *minus a sample of the output* (namely $I_E R_E$). That's negative feedback, and that's why emitter degeneration improves most properties of the amplifier (here improved linearity and stability and increased input impedance.[37]) Later in the chapter, in §2.5, we'll make these statements quantitative when we first look at feedback. And there are great things to look forward to, with the full flowering of feedback in Chapters 4 and 5!

2.3.5 Biasing the common-emitter amplifier

If you must have the highest possible gain (or if the amplifier stage is inside a feedback loop), it is possible to arrange successful biasing of a common-emitter amplifier. There are three solutions that can be applied, separately or in combination: bypassed emitter resistor, matched biasing transistor, and dc feedback.

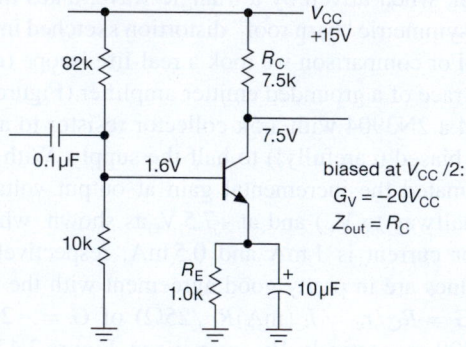

Figure 2.48. A bypassed emitter resistor can be used to improve the bias stability of a grounded-emitter amplifier.

A. Bypassed emitter resistor

You can use a bypassed emitter resistor, biasing as for the degenerated amplifier, as shown in Figure 2.48. In this case R_E has been chosen about $0.1R_C$, for ease of biasing; if R_E is too small, the emitter voltage will be much smaller than the base–emitter drop, leading to temperature instability of the quiescent point as V_{BE} varies with temperature. The emitter bypass capacitor is chosen to make its impedance small compared with r_e (not R_E – why?) at the

---

[37] And, as we'll learn, the output impedance would be reduced – a desirable feature in a voltage amplifier – if the feedback were taken directly from the collector.

lowest frequency of interest. In this case its impedance is 25 Ω at 650 Hz. At signal frequencies the input coupling capacitor sees an impedance of 10k in parallel with the base impedance, in this case $\beta \times 25\,\Omega$, or roughly 2.5k. At dc, the impedance looking into the base is much larger (β times the emitter resistor, or about 100k).

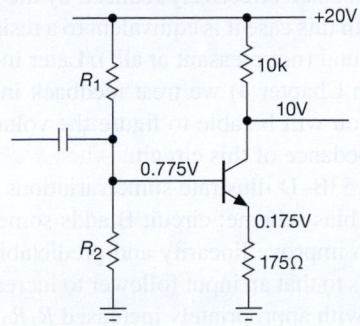

Figure 2.49. Gain-of-50 stage presents bias stability problem.

A variation on this circuit consists of using two emitter resistors in series, one of them bypassed. For instance, suppose you want an amplifier with a voltage gain of 50, quiescent current of 1 mA, and V_{CC} of +20 volts, for signals from 20 Hz to 20 kHz. If you try to use the emitter-degenerated circuit, you will have the circuit shown in Figure 2.49. The collector resistor is chosen to put the quiescent collector voltage at $0.5V_{CC}$. Then the emitter resistor is chosen for the required gain, including the effects of the r_e of $25/I_C(\text{mA})$. The problem is that the emitter voltage of only 0.175 V will vary significantly as the ~0.6 V of base–emitter drop varies with temperature ($-2.1\,\text{mV/}^\circ\text{C}$, approximately), since the base is held at constant voltage by R_1 and R_2; for instance, you can verify that an increase of 20°C will cause the collector current to increase by nearly 25%.

Exercise 2.14. Show that this statement is correct.

The solution here is to add some bypassed emitter resistance for stable biasing, with no change in gain at signal frequencies (Figure 2.50). As before, the collector resistor is chosen to put the collector at 10 volts ($0.5V_{CC}$). Then the unbypassed emitter resistor is chosen to give a gain of 50, including the intrinsic emitter resistance $r_e = 25/I_C(\text{mA})$. Enough bypassed emitter resistance is added to make stable biasing possible (one-tenth of the collector resistance is a good guideline). The base voltage is chosen to give 1 mA of emitter current, with impedance about one-tenth the dc impedance looking into the base (in this case about 100k). The emitter bypass capacitor is chosen to have low impedance compared with $180+25\,\Omega$ at the lowest signal frequencies. Finally, the input coupling ca-

pacitor is chosen to have low impedance compared with the *signal-frequency* input impedance of the amplifier, which is equal to the voltage-divider impedance in parallel with $\beta \times (180+25)\,\Omega$ (the 820 Ω is bypassed and looks like a short at signal frequencies).

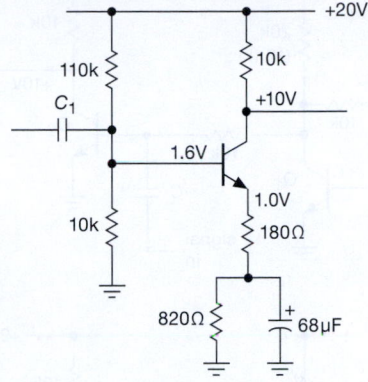

Figure 2.50. A common-emitter amplifier combining bias stability, linearity, and large voltage gain.

An alternative circuit splits the signal and dc paths (Figure 2.51). This lets you vary the gain (by changing the 180 Ω resistor) without bias change.

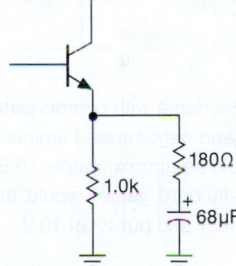

Figure 2.51. Equivalent emitter circuit for Figure 2.50.

B. Matched biasing transistor
You can use a matched transistor to generate the correct base voltage for the required collector current; this ensures automatic temperature compensation (Figure 2.52).[38] Q_1's collector is drawing 1 mA, since it is guaranteed to be near ground (about one V_{BE} drop above ground, to be exact); if Q_1 and Q_2 are a matched pair (available as a single device, with the two transistors on one piece of silicon), then Q_2 will also be biased to draw 1 mA, putting its collector at

[38] R. Widlar, "Some circuit design techniques for linear integrated circuits," *IEEE Trans. Circuit Theory* **CT-12**, 586 (1965). See also US Patent 3,364,434.

+10 volts and allowing a full ± 10 V symmetrical swing on its collector. Changes in temperature are of no importance as long as both transistors are at the same temperature. This is a good reason for using a "monolithic" dual transistor.

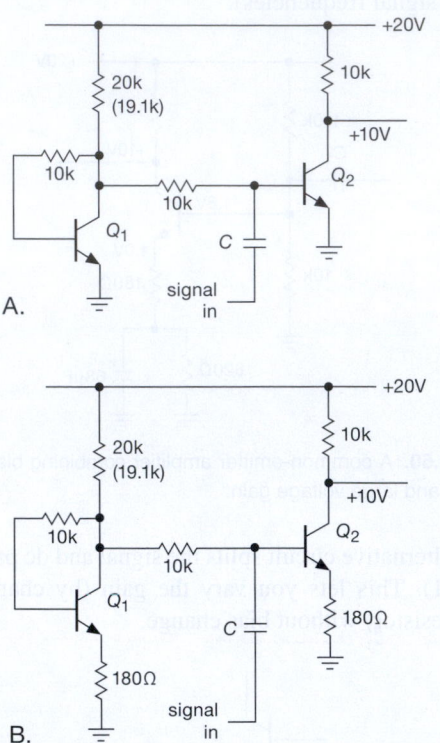

Figure 2.52. Biasing scheme with compensated V_{BE} drop, for both grounded emitter (A) and degenerated emitter (B) stages. With the values shown, V_C would be approximately 10.5 V; reducing the 20k resistor to 19.1k (a standard value) would take into account the effects of V_{BE} and finite β and put V_C at 10 V.

C. Feedback at dc

You can use dc feedback to stabilize the quiescent point. Figure 2.53A shows one method. By taking the bias voltage from the collector, rather than from V_{CC}, you get some measure of bias stability. The base sits one diode drop above ground – and because its bias comes from a 10:1 divider, the collector must be at 11 diode drops above ground, or about 7 volts. Any tendency for the transistor to saturate (e.g., if it happens to have unusually high beta) is stabilized, since the dropping collector voltage will reduce the base bias. This scheme is acceptable if great stability is not required. The quiescent point is liable to drift a volt or so as the ambient (surrounding) temperature changes, because the base–emitter voltage has a significant temperature coefficient (Ebers–Moll, again). Better stability is possible if

several stages of amplification are included within the feedback loop. You will see examples later in connection with feedback.

A better understanding of feedback is really necessary to understand this circuit. For instance, feedback acts to reduce the input and output impedances. The input signal sees R_1's resistance effectively reduced by the voltage gain of the stage. In this case it is equivalent to a resistor of about $200 \,\Omega$ to ground (not pleasant at all!). Later in this chapter (and again in Chapter 4) we treat feedback in enough detail so that you will be able to figure the voltage gain and terminal impedance of this circuit.

Figures 2.53B–D illustrate some variations on the basic dc-feedback bias scheme: circuit B adds some emitter degeneration to improve linearity and predictability of gain; circuit C adds to that an input follower to increase the input impedance (with appropriately increased $R_1 R_2$ divider values and changed ratio to accommodate the additional V_{BE} drop); and circuit D combines the methods of Figure 2.51 with circuit B to achieve greater bias stability.

Note that the base bias resistor values in these circuits could be increased to raise the input impedance, but you should then take into account the non-negligible base current. Suitable values might be $R_1 = 220\text{k}$ and $R_2 = 33\text{k}$. An alternative approach might be to bypass the feedback resistance in order to eliminate feedback (and therefore lowered input impedance) at signal frequencies (Figure 2.54).[39]

D. Comments on biasing and gain

One important point about grounded emitter amplifier stages: you might think that the voltage gain can be raised by increasing the quiescent current, since the intrinsic emitter resistance r_e drops with rising current. Although r_e does decrease with increasing collector current, the smaller collector resistor you need to obtain the same quiescent collector voltage just cancels the advantage. In fact, you can show that the small-signal voltage gain of a grounded-emitter amplifier biased to $0.5V_{CC}$ is given by $G = 20V_{CC}$ (in volts), independent of quiescent current.

Exercise 2.15. Show that the preceding statement is true.

If you need more voltage gain in one stage, one approach is to use a current source as an *active load*. Because its impedance is very high, single-stage voltage gains of

---

[39] But *caution*: the cascaded *RC* sections (33k into $10 \,\mu\text{F}$, 33k into the input capacitor) can cause peaking or instability, unless care is taken (for example by avoiding similar *RC* products).

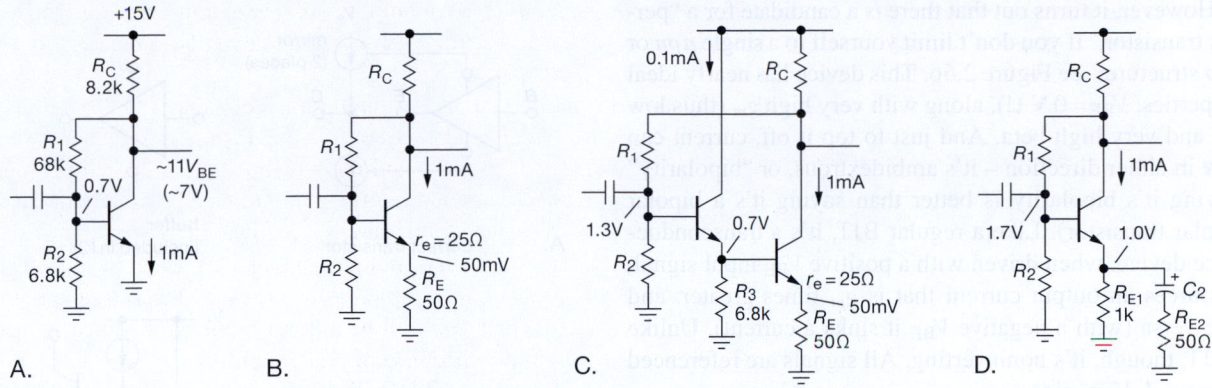

Figure 2.53. Bias stability is improved by feedback.

1000 or more are possible.[40] Such an arrangement cannot be used with the biasing schemes we have discussed, but must be part of an overall dc feedback loop, a subject we will discuss in Chapter 4. You should be sure such an amplifier looks into a high-impedance load; otherwise the gain obtained by high collector load impedance will be lost. Something like an emitter follower, an FET, or an op-amp presents a good load.

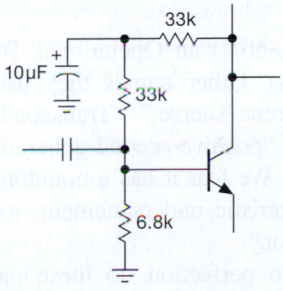

Figure 2.54. Eliminating impedance-lowering feedback at signal frequencies.

In RF amplifiers intended for use only over a narrow frequency range, it is common to use a parallel LC circuit as a collector load. In that case a very high voltage gain is possible since the LC circuit has high impedance (like a current source) at the signal frequency, with low impedance at dc. Because the LC is "tuned," out-of-band interfering signals (and distortion) are effectively rejected. Additional bonuses are the possibility of peak-to-peak (pp) output swings of $2V_{CC}$, and the use of transformer coupling from the inductor.

---

[40] Ultimately limited by the transistor's finite collector output resistance (a consequence of the Early effect); see the Early effect discussion in Chapter 2x.

Exercise 2.16. Design a tuned common-emitter amplifier stage to operate at 100 kHz. Use a bypassed emitter resistor, and set the quiescent current at 1.0 mA. Assume $V_{CC} = +15$ V and $L = 1.0$ mH, and put a 6.2k resistor across the LC to set $Q = 10$ (to get a 10% bandpass; see §1.7.14). Use capacitive input coupling.

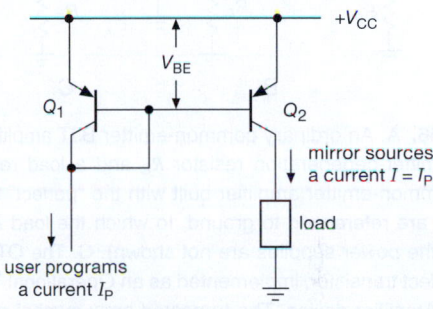

Figure 2.55. Classic bipolar-transistor matched-pair current mirror. Note the common convention of referring to the positive supply as V_{CC}, even when *pnp* transistors are used.

2.3.6 An aside: the perfect transistor

Looking at BJT transistor properties like the non-zero (and temperature-dependent) V_{BE}, the finite (and current-dependent) emitter impedance r_e and transconductance g_m, the collector current that varies with collector voltage (Early effect) etc., one is tempted to ask which transistor is better? Is there a "best" transistor, or perhaps even a *perfect* transistor? If you go through our transistor tables, e.g., Tables 2.1 and 2.2, and especially Table 8.1 for small-signal transistors, you'll see there is no best transistor candidate. That's because all physical bipolar transistors are subject to the same device physics, and their parameters tend to scale with die size and current, etc.

However, it turns out that there *is* a candidate for a "perfect transistor," if you don't limit yourself to a single *npn* or *pnp* structure; see Figure 2.56. This device has nearly ideal properties: $V_{BE}=0$ V (!), along with very high g_m (thus low r_e), and very high beta. And just to top it off, current can flow in either direction – it's ambidextrous, or "bipolarity" (saying it's bipolarity is better than saying it's a bipolar bipolar transistor). Like a regular BJT, it's a transconductance device: when driven with a positive V_{BE} input signal, it sources an output current that is g_m times greater, and vice versa (with a negative V_{BE} it sinks a current). Unlike a BJT, though, it's noninverting. All signals are referenced to ground. Very nice.

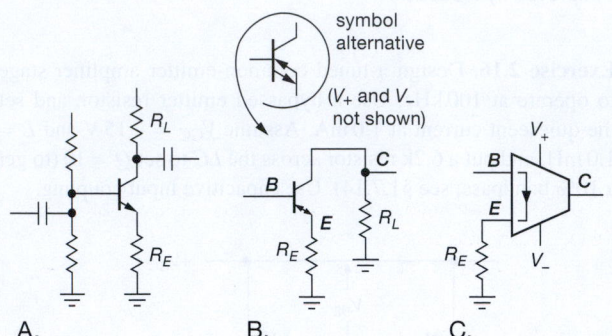

Figure 2.56. A. An ordinary common-emitter BJT amplifier stage, with an emitter-degeneration resistor R_E and a load resistor R_L. B. In a common-emitter amplifier built with the "perfect" transistor, all signals are referenced to ground, to which the load R_L is also returned (the power supplies are not shown). C. The OTA symbol for the perfect transistor, implemented as an Operational Transconductance Amplifier device. The truncated apex symbol means the device has a current output.

How does the perfect transistor work? Figure 2.57 shows a four-transistor circuit known as a *diamond transistor stage*. This circuit is a variation of the cascaded *pnp-npn* emitter follower in Figure 2.29: a complementary *npn-pnp* input follower is wired in parallel, and biased with current sources; the emitter outputs ($2V_{BE}$ apart) drive a matched push–pull output follower, which therefore runs at the same quiescent current. The common node is the effective emitter, E. Finally, a pair of current mirrors brings the two individual collector currents to a common output, the effective collector, C, where the output current is zero if the input voltage (between terminals B and E) is zero. As with an ordinary BJT, any current into (or out of) the emitter has to appear at the collector. The part does require two power supply connections. We'll have more to say about this interesting component in Chapters *2x* and *4x*.

Texas Instruments calls their perfect transistor (its part

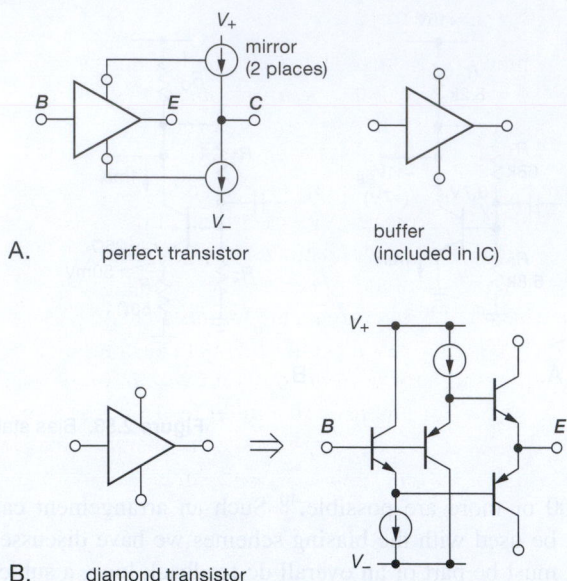

Figure 2.57. A. The OPA860 perfect transistor includes a diamond transistor (the triangle) and a pair of current mirrors. A second diamond transistor acts as an output buffer. B. The diamond transistor consists of a complementary pair of matched offset-cancelled emitter followers.

number is OPA860[41]) an Operational Transconductance Amplifier (OTA). Other names they use are "Voltage-Controlled Current source," "Transconductor," "Macro Transistor," and "positive second-generation current conveyor" (CCII+). We fear it has a branding identity crisis, so, with characteristic understatement, we're calling it a "perfect transistor."

How close to perfection do these parts come? The OPA860 and OPA861 perfect transistors have these specs: $V_{os}=3$ mV typ (12 mV max), $g_m=95$ mS, $r_e=10.5\,\Omega$, $Z_{out}=54$k$\Omega\|2$ pF, $Z_{in}=455$k$\Omega\|2$ pF, $I_{out(max)}=\pm15$ mA. Its maximum gain is 5100. Hardly perfect, but, hey, not half bad. You can create many nice circuits with these puppies (e.g., active filter, wideband current summing circuit, or integrator for nanosecond-scale pulses); see the OPA860 datasheet for details.

---

[41] TI's OP861 version omits the output buffer and is available in a small SOT-23 package. That's one of our favorite surface-mount package styles, available for many of the other transistors mentioned in our tables. Knowledgeable readers will recognize the circuitry from inside a current-feedback, or CFB op-amp. Some of these devices (for example the AD844) make the internal node available.

2.3.7 Current mirrors

The technique of matched base–emitter biasing can be used to make what is called a *current mirror*, an interesting current-source circuit that simply reverses the sign of a "programming" current. (Figure 2.55). You program the mirror by sinking a current from Q_1's collector. That causes a V_{BE} for Q_1 appropriate to that current at the circuit temperature and for that transistor type. Q_2, matched to Q_1,[42] is thereby programmed to source the same current to the load. The small base currents are unimportant.[43]

One nice feature of this circuit is voltage compliance of the output transistor current source to within a few tenths of a volt of V_{CC}, as there is no emitter resistor drop to contend with. Also, in many applications it is handy to be able to program a current with a current. An easy way to generate the control current I_P is with a resistor (Figure 2.58). Because the bases are a diode drop below V_{CC}, the 14.4k resistor produces a control current, and therefore an output current, of 1 mA. Current mirrors can be used in transistor circuits whenever a current source is needed. They're very popular in integrated circuits, where (a) matched transistors abound and (b) the designer tries to make circuits that will work over a large range of supply voltages. There are even resistorless IC op-amps in which the operating current of the whole amplifier is set by one external resistor, with all the quiescent currents of the individual amplifier stages inside being determined by current mirrors.

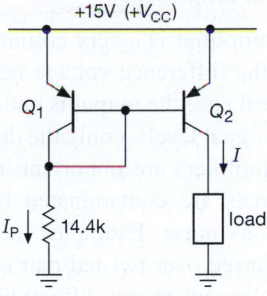

Figure 2.58. Programming current-mirror current.

A. Current-mirror limitations due to the Early effect

One problem with the simple current mirror is that the output current varies a bit with changes in output voltage, i.e.,

the output impedance is not infinite. This is because of the slight variation of V_{BE} with collector voltage at a given current in Q_2 (which is due to the Early effect); said a different way, the curve of collector current versus collector–emitter voltage at a fixed base-emitter voltage is not flat (Figure 2.59). In practice, the current might vary 25% or so over the output compliance range – much poorer performance than that of the current source with an emitter resistor discussed earlier.

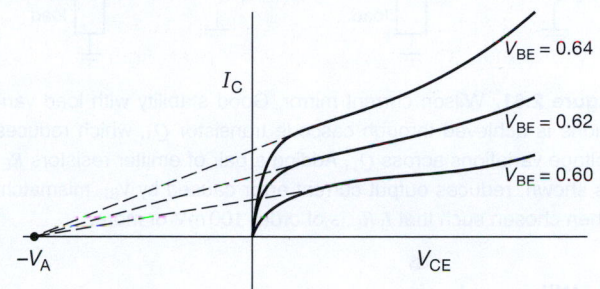

Figure 2.59. Early effect: collector current varies with V_{CE}. (Interestingly, you get a very similar curve, with comparable V_A, if you apply instead a family of constant base *currents*.)

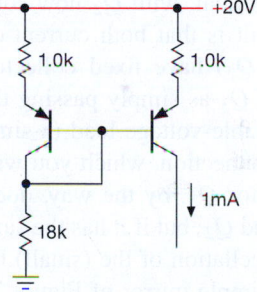

Figure 2.60. Improved current mirror with emitter resistors.

One solution, if a better current source is needed (it often isn't), is the circuit shown in Figure 2.60. The emitter resistors are chosen to have at least a few tenths of a volt drop; this makes the circuit a far better current source, since the small variations of V_{BE} with V_{CE} are now negligible in determining the output current. Again, matched transistors should be used. Note that this circuit loses its effectiveness if operation over a wide range of programming current is intended (figure out why).[44]

---

[42] A monolithic dual transistor is ideal; Table 8.1b on page 502 lists most available matched transistors. Some, such as the DMMT3904 and 3906, are matched to 1 mV and are quite affordable, $0.36 in small quantities.

[43] This circuit is often called the Widlar current mirror; see the reference on page 97 and US Patent 3,320,439.

[44] Current sources and current mirrors are discussed in more detail in Chapter *2x*.

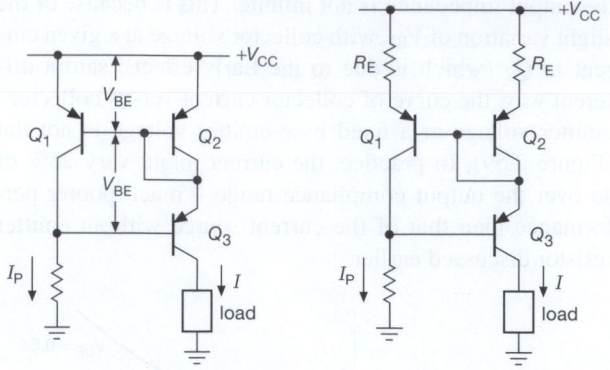

Figure 2.61. Wilson current mirror. Good stability with load variations is achieved through cascode transistor Q_3, which reduces voltage variations across Q_1. Adding a pair of emitter resistors R_E, as shown, reduces output current error caused by V_{BE} mismatch, when chosen such that $I_P R_E$ is of order 100 mV or more.

B. Wilson mirror

Another current mirror with improved consistency of current is shown in the clever circuit of Figure 2.61. Q_1 and Q_2 are in the usual mirror configuration, but Q_3 now keeps Q_1's collector fixed at two diode drops below V_{CC}. That circumvents the Early effect in Q_1, whose collector is now the programming terminal, with Q_2 now sourcing the output current. The result is that both current-determining transistors (Q_1 and Q_2) have fixed collector–emitter drops; you can think of Q_3 as simply passing the output current through to a variable-voltage load (a similar trick is used in the cascode connection, which you will see later in the chapter). Transistor Q_3, by the way, does not have to be matched to Q_1 and Q_2; but if it has the same beta, then you get an exact cancellation of the (small) base current error that afflicts the simple mirror of Figure 2.55 (or the beta-enhanced mirror in Chapter 2x).

Exercise 2.17. Show that this statement is true.

There are additional nice tricks you can do with current mirrors, such as generating multiple independent outputs, or an output that is a fixed multiple of the programming current. One trick (invented by the legendary Widlar) is to unbalance the R_E's in Figure 2.61; as a rough estimate, the output current ratio is approximately the ratio of resistor values (because the base-emitter drops are roughly equal). But to get it right you need to take into account the difference of V_{BE}'s (because the transistors are running at different currents), for which the graph in Figure 2.62 is helpful. This graph is also useful for estimating the current unbalance in a current mirror built with discrete (i.e.,

not matched) transistors. We treat current mirrors further in Chapter 2x (§§2x.3 and 2x.11).

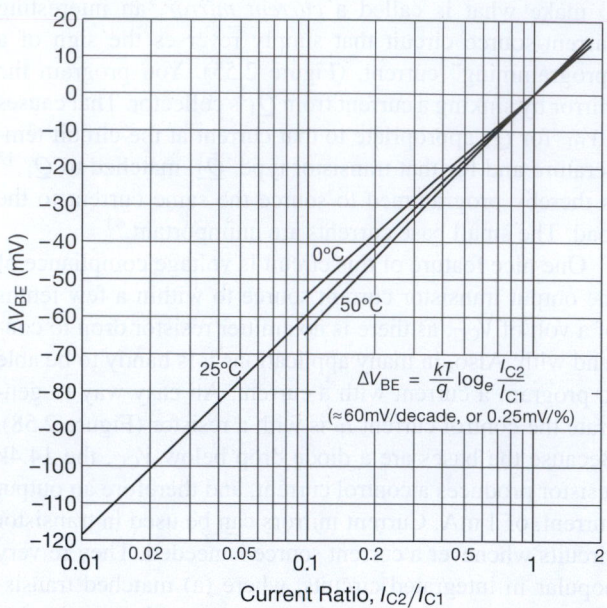

Figure 2.62. Collector current ratios for matched transistors, as determined by the difference in applied base-emitter voltages. See Table 8.1b for low-noise matched BJTs.

2.3.8 Differential amplifiers

The differential amplifier is a very common configuration used to amplify the difference voltage between two input signals. In the ideal case the output is entirely independent of the individual signal levels – only the difference matters.

Differential amplifiers are important in applications in which weak signals are contaminated by "pickup" and other miscellaneous noise. Examples include digital and RF signals transferred over twisted-pair cables, audio signals (the term "balanced" means differential, usually $600\,\Omega$ impedance, in the audio business), local-area-network signals (such as 100BASE-TX and 1000BASE-T Ethernet), electrocardiogram voltages, magnetic disk head amplifiers, and numerous other applications. A differential amplifier at the receiving end restores the original signal if the interfering "common-mode" signals (see below) are not too large. Differential amplifiers are universally used in operational amplifiers, an essential building block that is the subject of Chapter 4. They're very important in dc amplifier design (amplifiers that amplify clear down to dc, i.e., have no coupling capacitors) because their symmetrical design is inherently compensated against thermal drifts.

Some nomenclature: when both inputs change levels together, that's a *common-mode* input change. A differential change is called *normal mode*, or sometimes *differential mode*. A good differential amplifier has a high *common-mode rejection ratio* (CMRR), the ratio of response for a normal-mode signal to the response for a common-mode signal of the same amplitude. CMRR is usually specified in decibels. The common-mode input range is the voltage level over which the inputs may vary. The differential amplifier is sometimes called a "long-tailed pair."

Figure 2.63 shows the basic circuit. The output is taken off one collector with respect to ground; that is called a *single-ended output* and is the most common configuration. You can think of this amplifier as a device that amplifies a difference signal and converts it to a single-ended signal so that ordinary subcircuits (followers, current sources, etc.) can make use of the output. (If, instead, a differential output is desired, it is taken between the collectors.)

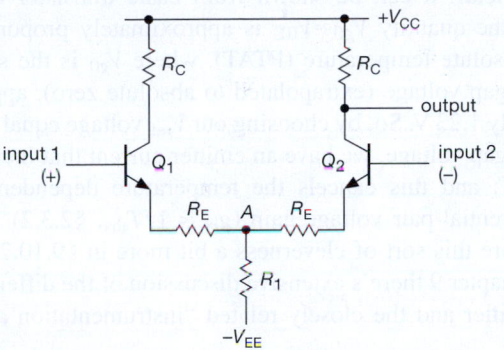

Figure 2.63. Classic transistor differential amplifier.

What is the gain? That's easy enough to calculate: imagine a symmetrical input signal wiggle, in which input 1 rises by v_{in} (a small-signal variation) and input 2 drops by the same amount. As long as both transistors stay in the active region, point A remains fixed. You then determine the gain as with the single transistor amplifier, remembering that the input change is actually twice the wiggle on either base: $G_{diff} = R_C/2(r_e + R_E)$. Typically R_E is small, $100\,\Omega$ or less, or it may be omitted entirely. Differential voltage gains of a few hundred are possible.

You can determine the common-mode gain by putting identical signals v_{in} on both inputs. If you think about it correctly[45] (remembering that R_1 carries both emitter currents), you'll find $G_{CM} = -R_C/(2R_1 + R_E)$. Here we've ignored

nored the small r_e, because R_1 is typically large, at least a few thousand ohms. We really could have ignored R_E as well. The CMRR is thus roughly $R_1/(r_e + R_E)$. Let's look at a typical example (Figure 2.64) to get some familiarity with differential amplifiers.

Collector resistor R_C is chosen for a quiescent current of $100\,\mu A$. As usual, we put the collector at $0.5V_{CC}$ for large dynamic range. Q_1's collector resistor can be omitted, since no output is taken there.[46] R_1 is chosen to give total emitter current of $200\,\mu A$, split equally between the two sides when the (differential) input is zero. From the formulas just derived, this amplifier has a differential gain of 10 and a common-mode gain of 0.55. Omitting the 1.0k resistors raises the differential gain to 50, but drops the (differential) input impedance from about 250k to about 50k (you can substitute Darlington transistors[47] in the input stage to raise the impedance into the megohm range, if necessary).

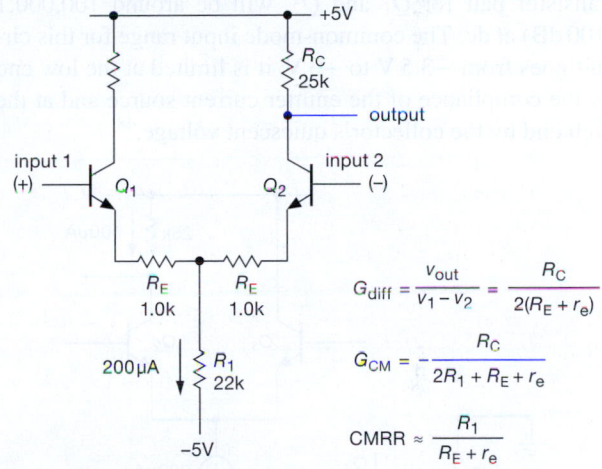

Figure 2.64. Calculating differential amplifier performance.

$$G_{diff} = \frac{v_{out}}{v_1 - v_2} = \frac{R_C}{2(R_E + r_e)}$$

$$G_{CM} = -\frac{R_C}{2R_1 + R_E + r_e}$$

$$CMRR \approx \frac{R_1}{R_E + r_e}$$

Remember that the maximum gain of a single-ended grounded emitter amplifier biased to $0.5V_{CC}$ is $20V_{CC}$ (in volts). In the case of a differential amplifier the maximum differential gain ($R_E = 0$) is half that figure, or (for arbitrary quiescent point) 20 times the voltage (in volts) across the collector resistor. The corresponding maximum CMRR (again with $R_E = 0$) is equal to 20 times the voltage (in volts) across R_1. As with the single-ended common-emitter amplifier, the emitter resistors R_E reduce distortion, at the

---

[45] *Hint:* replace R_1 with a parallel pair, each of resistance $2R_1$; then notice that you can cut the wire connecting them together at point A (because no current flows); take it from there.

[46] *Can* be omitted, but at the expense of accurately balanced base-emitter drops: you get better balance if you retain both collector resistors (avoiding Early effect); but you suppress Miller effect (§2.4.5) at input 1 if you omit Q_1's collector resistor.

[47] See §2.4.2.

expense of gain. See the extensive discussion of BJT amplifier distortion in Chapter *2x*.

Exercise 2.18. Verify that these expressions are correct. Then design a differential amplifier to run from ± 5 V supply rails, with $G_{\text{diff}} = 25$ and $R_{\text{out}} = 10$k. As usual, put the collector's quiescent point at half of V_{CC}.

A. Biasing with a current source

The common-mode gain of the differential amplifier can be reduced enormously by the substitution of a current source for R_1. Then R_1 effectively becomes very large and the common-mode gain is nearly zero. If you prefer, just imagine a common-mode input swing; the emitter current source maintains a constant total emitter current, shared equally by the two collector circuits, by symmetry. The output is therefore unchanged. Figure 2.65 shows an example. The CMRR of this circuit, using an LM394 monolithic transistor pair for Q_1 and Q_2, will be around 100,000:1 (100 dB) at dc. The common-mode input range for this circuit goes from -3.5 V to $+3$ V; it is limited at the low end by the compliance of the emitter current source and at the high end by the collector's quiescent voltage.[48]

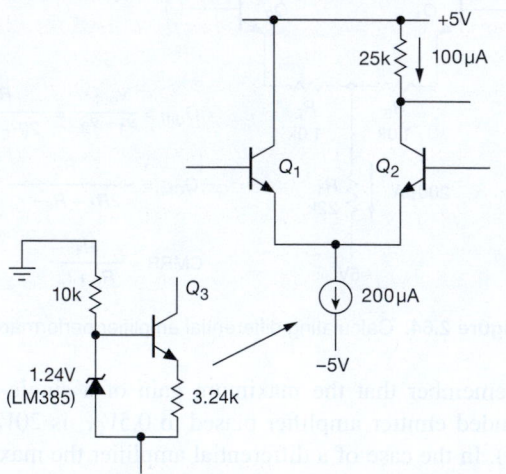

Figure 2.65. Improving CMRR of the differential amplifier with a current source.

Be sure to remember that this amplifier, like all transistor amplifiers, must have a dc bias path to the bases. If the input is capacitively coupled, for instance, you must have

base resistors to ground. An additional caution for differential amplifiers, particularly those without inter-emitter resistors: bipolar transistors can tolerate only 6 volts of base–emitter reverse bias before breakdown; thus, applying a differential input voltage larger than this will destroy the input stage (if there is no inter-emitter resistor). An inter-emitter resistor limits the breakdown current and prevents destruction, but the transistors may be degraded nonetheless (in beta, noise, etc.). In either case the input impedance drops drastically during reverse conduction.

An interesting aside: the emitter current sink shown in Figure 2.65 has some variation with temperature, because V_{BE} decreases with increasing temperature (to the tune of approximately -2.1 mV/°C, §2.3.2), causing the current to increase. More explicitly, if we call the 1.24 V zener-like reference "V_{ref}," then the drop across the emitter resistor equals $V_{\text{ref}} - V_{\text{BE}}$; the current is proportional, thus increasing with temperature. As it happens, this is in fact *beneficial*: it can be shown from basic transistor theory that the quantity $V_{g0} - V_{\text{BE}}$ is approximately proportional to absolute temperature (PTAT), where V_{g0} is the silicon bandgap voltage (extrapolated to absolute zero), approximately 1.23 V. So, by choosing our V_{ref} voltage equal to the bandgap voltage, we have an emitter current that increases PTAT; and this cancels the temperature dependence of differential-pair voltage gain ($g_m \propto 1/T_{\text{abs}}$, §2.3.2). We'll explore this sort of cleverness a bit more in §9.10.2. And in Chapter 9 there's extensive discussion of the differential amplifier and the closely related "instrumentation amplifier."

B. Use in single-ended dc amplifiers

A differential amplifier makes an excellent dc amplifier, even for single-ended inputs. You just ground one of the inputs and connect the signal to the other (Figure 2.66). You might think that the "unused" transistor could be eliminated. Not so! The differential configuration is inherently compensated for temperature drifts, and even when one input is at ground that transistor is still doing something: a temperature change causes both V_{BE}'s to change the same amount, with no change in balance or output. That is, changes in V_{BE} are not amplified by G_{diff} (only by G_{CM}, which can be made essentially zero). Furthermore, the cancellation of V_{BE}'s means that there are no 0.6 V drops at the input to worry about. The quality of a dc amplifier constructed this way is limited only by mismatching of input V_{BE}'s or their temperature coefficients. Commercial monolithic transistor pairs and commercial differential amplifier ICs are available with extremely good matching (e.g., the MAT12 *npn* monolithic matched pair has a typical drift of

---

[48] You can make good current sinks also with JFETs (see the discussion in §3.2.2C), but BJTs are better for this task in many ways. See for example Figure 3.26, where we show four configurations of BJT current sinks that improve upon the JFET alternative.

V_{BE} between the two transistors of 0.15 μV/°C). See Table 8.1b on page 502 for a listing of matched BJTs.

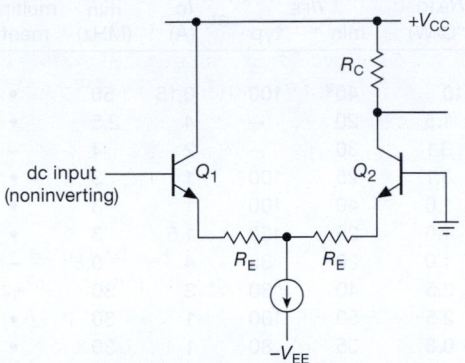

Figure 2.66. A differential amplifier can be used as a precision single-ended dc amplifier.

Either input could have been grounded in the preceding circuit example. The choice depends on whether or not the amplifier is supposed to invert the signal. (The configuration shown is preferable at high frequencies, however, because of the *Miller effect*; see §2.4.5.) The connection shown is noninverting, and so the inverting input has been grounded. This terminology carries over to op-amps, which are versatile high-gain differential amplifiers.

C. Current-mirror active load

As with the simple grounded-emitter amplifier, it is sometimes desirable to have a single-stage differential amplifier with very high gain. An elegant solution is a current-mirror active load (Figure 2.67). Q_1Q_2 is the differential pair with emitter current source. Q_3 and Q_4, a current mirror, form the collector load. The high effective collector load impedance provided by the mirror yields voltage gains of 5000 or more, assuming no load at the amplifier's output.[49] Such an amplifier is very common as the input stage in a larger circuit, and is usually used only within a feedback loop, or as a comparator (discussed in the next section). Be sure to keep the load impedance of such an amplifier very high, or the gain will drop enormously.

D. Differential amplifiers as phase splitters

The collectors of a symmetrical differential amplifier generate equal signal swings of opposite phase. By taking outputs from both collectors, you've got a phase splitter. Of course, you could also use a differential amplifier with both

---

[49] The dc gain is limited primarily by the Early effect; see §2.3.2 and the discussion in Chapter 2x.

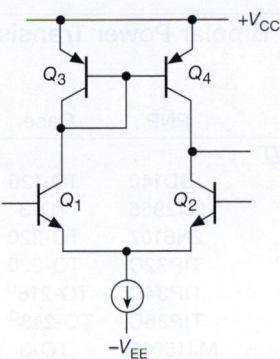

Figure 2.67. Differential amplifier with active current mirror load.

differential inputs and differential outputs. This differential output signal could then be used to drive an additional differential amplifier stage, with greatly improved overall common-mode rejection.

E. Differential amplifiers as comparators

Because of its high gain and stable characteristics, the differential amplifier is the main building block of the *comparator* (which we saw in §1.4.2E), a circuit that tells which of two inputs is larger. They are used for all sorts of applications: switching on lights and heaters, generating square waves from triangles, detecting when a level in a circuit exceeds some particular threshold, class-D amplifiers and pulse-code modulation, switching power supplies, etc. The basic idea is to connect a differential amplifier so that it turns a transistor switch on or off, depending on the relative levels of the input signals. The linear region of amplification is ignored, with one or the other of the two input transistors cut off at any time. A typical hookup is illustrated in §2.6.2 by a temperature-controlling circuit that uses a resistive temperature sensor (thermistor).

2.4 Some amplifier building blocks

We've now seen most of the basic – and important – transistor circuit configurations: switch, follower, current source (and mirror), and common-emitter amplifier (both single-ended and differential). For the remainder of the chapter we look at some circuit elaborations and their consequences: push-pull, Darlington and Sziklai, bootstrapping, Miller effect, and the cascode configuration. We'll finish with an introduction to the wonderful (and essential) technique of *negative feedback*. Chapter 2x deals with follow-on transistor circuits and techniques at a greater level of sophistication.

Table 2.2 Bipolar Power Transistors[a]

NPN	PNP	Case	V_{CEO} max (V)	I_C max[b] (A)	P_{diss} max[b,h] (W)	$R_{\theta JC}$[c] (°C/W)	h_{FE} min	h_{FE} typ	at I_C (A)	f_T min (MHz)	multiple manf?
standard BJT											
BD139	BD140	TO-126	80	1.5	12.5	10	40[e]	100	0.15	50	•
2N3055	2N2955	TO-3	60	15	115	1.5	20	--	4	2.5	•
2N6292	2N6107	TO-220	70	7	40	3.1	30	--	2	4	•
TIP31C	TIP32C	TO-220	100	3	40	3.1	25	100	1	3	–
TIP33C	TIP34C	TO-218[d]	100	10	80	1.6	40	100	1	3	•
TIP35C	TIP36C	TO-218[d]	100	25	125	1.0	25	150	1.5	3	•
MJ15015	MJ15016	TO-3	120	15	180	1.0	20	35	4	0.8[g]	–
MJE15030	MJE15031	TO-220	150	8	50	2.5	40	80	3	30	•,z
MJE15032	MJE15033	TO-220	250	8	50	2.5	50	100	1	30	•
2SC5200	2SA1943	TO-264	230	17	150	0.8	55	80	1	30	•
2SC5242[k]	2SA1962[k]	TO-3P	250	s	s	s	s	s	s	s	•
MJE340	MJE350	TO-126	300	0.5	20	6	30	--	0.05	--	•
TIP47	MJE5730	TO-220	250	1	40	3.1	30	--	0.3	10	•
TIP50[u]	MJE5731A[u]	TO-220	400	s	s	s	s	--	s	s	•
MJE13007	MJE5852	TO-220	400[f]	8	80	1.6	8[g]	20[g]	2	14[t]	–
Darlington											
MJD112	MJD117	DPak	100	2	20	6.3	1000	2000	2	25	•
TIP122	TIP127	TO-220	100	5	65	1.9	1000	--	3	--	•
TIP142	TIP147	TO-218	100	10	125	1.0	1000	--	5	--	•
MJ11015	MJ11016	TO-3	120	30	200	0.9	1000	--	20	4	•
MJ11032	MJ11033	TO-3	120	50	300	0.6	1000	--	25	--	•
MJH11019	MJH11020	TO-218	200[v]	15	150	0.8	400	--	10	3	•

Notes: (a) sorted more or less by voltage, current and families; see also additional tables in *Chapter 2x*. (b) with case at 25C. (c) P_{diss}(reality) $=(T_{J[your-max-value]} - T\text{amb}) / (R_{\theta JC} + R_{\theta CS} + R_{\theta SA})$; this is a much lower number than the "spec," especially if you're careful with T_J max, say 100°C. (d) similar to TO-247. (e) higher gain grades are available. (f) much higher V_{CES} "blocking" capability (compared with Vceo), e.g. 700V for MJE13007. (g) higher for the PNP device. (h) P_{diss}(max) $= (150°C-25°C) / R_{\theta JC}$; this is a classic datasheet specsmanship value. (k) larger pkg version of above. (s) same as above. (t) typical. (u) higher voltage version of above. (v) there are also 150V and 250V versions. (z) if these are hard to get, try the '028 and '029 versions (120V rather than 150V).

2.4.1 Push–pull output stages

As we mentioned earlier in the chapter, an *npn* emitter follower cannot sink current and a *pnp* follower cannot source current. The result is that a single-ended follower operating between split supplies can drive a ground-returned load only if a high quiescent current is used.[50] The quiescent current must be at least as large as the maximum output current during peaks of the waveform, resulting in high quiescent power dissipation. For example, Figure 2.68 shows a follower circuit to drive an 8 Ω loudspeaker load with up to 10 watts of audio.

An explanation of the driver stage: the *pnp* follower Q_1 is included to reduce drive requirements and to cancel Q_2's V_{BE} offset (0 V input gives approximately 0 V output).

---

[50] An amplifier in which current flows in the output transistor over the full waveform swing is sometimes called a "class-A" amplifier.

Q_1 could, of course, be omitted for simplicity. The hefty current source in Q_1's emitter load is used to ensure that there is sufficient base drive to Q_2 at the top of the signal swing. A resistor as emitter load would be inferior because it would have to be a rather low value (50 Ω or less) in order to guarantee at least 50 mA of base drive to Q_2 at the peak of the swing, when load current would be maximum and the drop across the resistor would be minimum; the resultant quiescent current in Q_1 would be excessive.

The output of this example circuit can swing to nearly ±15 volts (peak) in both directions, giving the desired output power (9 V rms across 8 Ω). However, the output transistor dissipates 55 watts with no signal (hence the heatsink symbol), and the emitter resistor dissipates another 110 watts. Quiescent power dissipation many times greater than the maximum output power is characteristic of this kind of class-A circuit (transistor always in conduction);

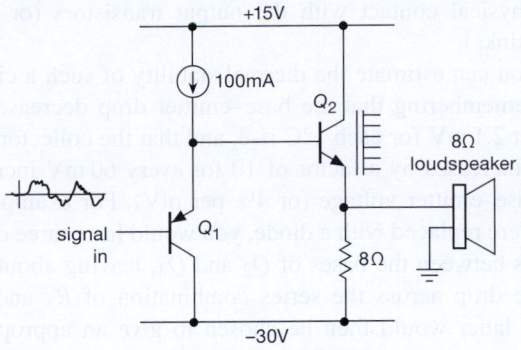

Figure 2.68. A 10 W loudspeaker amplifier, built with a single-ended emitter follower, dissipates 165 W of quiescent power!

this obviously leaves a lot to be desired in applications in which any significant amount of power is involved.

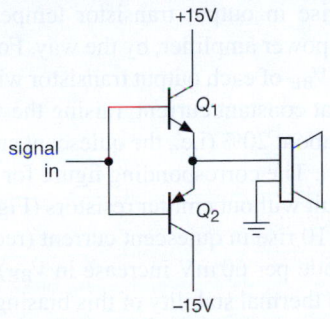

Figure 2.69. Push–pull emitter follower.

Figure 2.69 shows a *push–pull* follower doing the same job. Q_1 conducts on positive swings, Q_2 on negative swings. With zero input voltage, there is no collector current and no power dissipation. At 10 watts of output power there is less than 10 watts of dissipation in each transistor.[51]

A. Crossover distortion in push-pull stages

There is a problem with the preceding circuit as drawn. The output trails the input by a V_{BE} drop; on positive swings the output is about 0.6 V less positive than the input, and the reverse for negative swings. For an input sine wave, the output would look as shown in Figure 2.70. In the language of the audio business, this is called *crossover distortion*. The best cure (feedback offers another method, although by itself it is not entirely satisfactory; see §4.3.1E) is to bias the push–pull stage into slight conduction, as in Figure 2.71.

The bias resistors R bring the diodes into forward conduction, holding Q_1's base a diode drop above the input

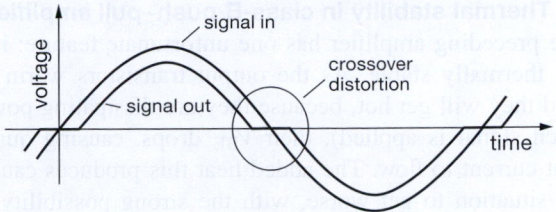

Figure 2.70. Crossover distortion in the push–pull follower.

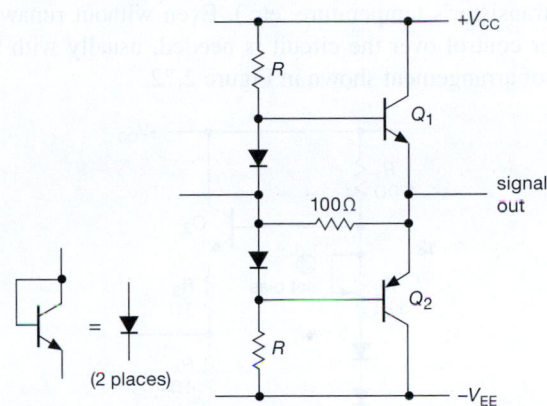

Figure 2.71. Biasing the push–pull follower to eliminate crossover distortion.

signal and Q_2's base a diode drop below the input signal. Now, as the input signal crosses through zero, conduction passes from Q_2 to Q_1; one of the output transistors is always on. The value R of the base resistors is chosen to provide enough base current for the output transistors at the peak output swing. For instance, with ± 20 V supplies and an 8 Ω load running up to 10 watts sinewave power, the peak base voltage is about 13.5 volts and the peak load current is about 1.6 amps. Assuming a transistor beta of 50 (power transistors generally have lower current gain than small-signal transistors), the 32 mA of necessary base current will require base resistors of about 220 Ω (6.5 V from V_{CC} to base at peak swing).

In this circuit we've added a resistor from input to output (this could have been done in Figure 2.69 as well). This serves to eliminate the "dead zone" as conduction passes from one transistor to the other (particularly in the first circuit), which is desirable especially when this circuit is included within a larger feedback circuit. However, it does not substitute for the better procedure of linearizing by biasing, as in Figure 2.71, to achieve transistor conduction over the full output waveform. We have more to say about this in Chapter 2x.

---

[51] An amplifier like this, with half-cycle conduction in each of the output transistors, is sometimes called a "class-B" amplifier.

B. Thermal stability in class-B push–pull amplifiers

The preceding amplifier has one unfortunate feature: it is not thermally stable. As the output transistors warm up (and they will get hot, because they are dissipating power when signal is applied), their V_{BE} drops, causing quiescent current to flow. The added heat this produces causes the situation to get worse, with the strong possibility of what is called *thermal runaway* (whether it runs away or not depends on a number of factors, including how large a "heatsink" is used, how well the diode's temperature tracks the transistor's temperature, etc.). Even without runaway, better control over the circuit is needed, usually with the sort of arrangement shown in Figure 2.72.

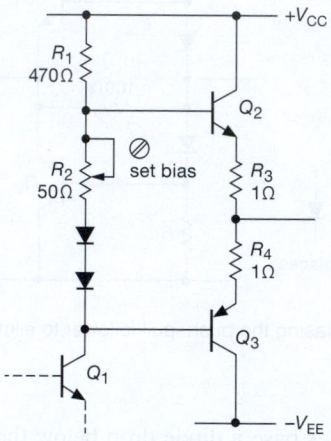

Figure 2.72. Adding (small) emitter resistors improve thermal stability in the push–pull follower.

For variety, the input is shown coming from the collector of the previous stage; R_1 now serves the dual purpose of being Q_1's collector resistor, and also providing current to bias the diodes and bias-setting resistor in the push–pull base circuit. Here R_3 and R_4, typically a few ohms or less, provide a "cushion" for the critical quiescent current biasing: the voltage between the bases of the output transistors must now be a bit greater than two diode drops, and you provide the extra with adjustable biasing resistor R_2 (often replaced with a third series diode, or, better, with the more elegant biasing circuit of Figure 2.78 on page 111). With a few tenths of a volt across R_3 and R_4, the temperature variation of V_{BE} doesn't cause the current to rise very rapidly (the larger the drop across R_3 and R_4, the less sensitive it is), and the circuit will be stable. Stability is improved by mounting the diodes[52]

---

[52] Or, better, diode-connected transistors: connect base and collector together as "anode" with emitter as "cathode."

in physical contact with the output transistors (or their heatsinks).

You can estimate the thermal stability of such a circuit by remembering that the base–emitter drop decreases by about 2.1 mV for each 1°C rise, and that the collector current increases by a factor of 10 for every 60 mV increase in base–emitter voltage (or 4% per mV). For example, if R_2 were replaced with a diode, you would have three diode drops between the bases of Q_2 and Q_3, leaving about one diode drop across the series combination of R_3 and R_4. (The latter would then be chosen to give an appropriate quiescent current, perhaps 100 mA for an audio power amplifier.) The worst case for thermal stability occurs if the biasing diodes are not thermally coupled to the output transistors.

Let us assume the worst and calculate the increase in output-stage quiescent current corresponding to a 30°C temperature rise in output transistor temperature. That's not a lot for a power amplifier, by the way. For that temperature rise, the V_{BE} of each output transistor will decrease by about 63 mV at constant current, raising the voltage across R_3 and R_4 by about 20% (i.e., the quiescent current will rise by about 20%). The corresponding figure for the preceding amplifier circuit without emitter resistors (Figure 2.71) will be a factor of 10 rise in quiescent current (recall that I_C increases a decade per 60 mV increase in V_{BE}), i.e., 1000%. The improved thermal stability of this biasing arrangement (even without having the diodes thermally coupled to the output transistors) is evident. And you'll do significantly better when the diodes (or diode-connected transistors, or, best of all, V_{BE}-referenced biasing as shown in Figure 2.78) ride along on the heatsink.

This circuit has the additional advantage that, by adjusting the quiescent current, you have some control over the amount of residual crossover distortion. A push–pull amplifier biased in this way to obtain substantial quiescent current at the crossover point is sometimes referred to as a "class-AB" amplifier, meaning that both transistors conduct simultaneously during a significant portion of the cycle. In practice, you choose a quiescent current that is a good compromise between low distortion and excessive quiescent dissipation. Feedback, introduced later in this chapter (and exploited shamelessly, and with joy, in Chapter 4), is almost always used to reduce distortion still further.

We will see a further evolution of this circuit in §2.4.2, where we supplement it with the intriguingly named techniques of V_{BE}-referenced biasing, collector bootstrapping, and β-boosting complementary Darlington output stage.

C. "Class-D" amplifiers

An interesting solution to this whole business of power dissipation (and distortion) in class-AB linear power amplifiers is to abandon the idea of a linear stage entirely and use instead a *switching* scheme: imagine that the push–pull follower transistors Q_2 and Q_3 in Figure 2.72 are replaced with a pair of transistor *switches*, with one ON and the other OFF at any time, so that the output is switched completely to $+V_{CC}$ or to $-V_{CC}$ at any instant. Imagine also that these switches are operated at a high frequency (say at least 10 times the highest audio frequency) and that their relative timing is controlled (by techniques we'll see later, in Chapters 10–13) such that the *average* output voltage is equal to the desired analog output. Finally, we add an *LC* lowpass filter to kill the high switching signal, leaving the desired (lower-frequency) analog output intact.

This is a *Class-D*, or *switching* amplifier. It has the advantage of very high efficiency, because the switching transistors are either off (no current) or in saturation (near-zero voltage); that is, the power dissipated in the switching transistors (the product $V_{CE} \times I_C$) is always small. There's also no worry about thermal runaway. The downsides are the problems of emission of high-frequency noise, switching feedthrough to the output, and the difficulty of achieving excellent linearity.

Class-D amplifiers are nearly universal in inexpensive audio equipment, and they are increasingly finding their way into high-end audio equipment. Figure 2.73 shows measured waveforms of an inexpensive (and tiny!) class-D amplifier IC driving a 5 Ω load with a sinewave at the high end of the audio range (20 kHz). This particular IC uses a 250 kHz switching frequency, and can drive 20 watts each into a pair of stereo speakers; pretty much everything you need (except for the output *LC* filters) is on the chip, which costs about $3 in small quantities. Pretty neat.

2.4.2 Darlington connection

If you hook two transistors together as in Figure 2.74, the result – called a *Darlington connection*[53] (or *Darlington pair*) – behaves like a single transistor with beta equal to the product of the two transistor betas.[54] This can be very handy where high currents are involved (e.g., voltage regulators or power amplifier output stages), or for input stages

---

[53] Sidney Darlington, US Patent 2,663,806: "Semiconductor Signal Translating Device." Darlington wanted the patent to cover any number of transistors in one package, but the lawyers at Bell Laboratories overruled him, thus forgoing a patent that would have covered every IC.

[54] At the operating current of each transistor, of course.

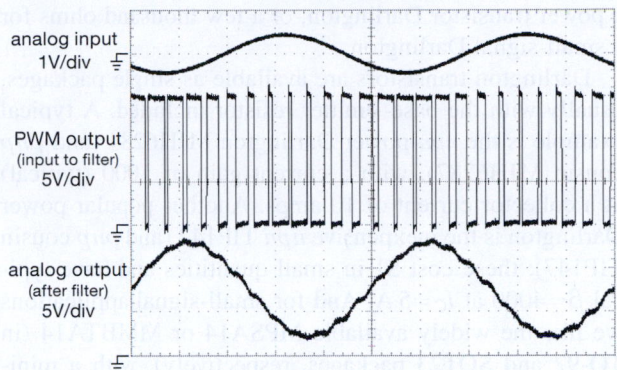

Figure 2.73. Class-D amplifier waveforms: a 20 kHz input sinewave controls the "duty cycle" (fraction of time the output is HIGH) of a push–pull switched output. These waveforms are from a TPA3123 stereo amplifier chip running from +15 V, and show the prefiltered PWM (pulse-width modulated) output, and the final smoothed output after the *LC* lowpass output filter. Horizontal: 10 μs/div.

of amplifiers where very high input impedance is necessary.

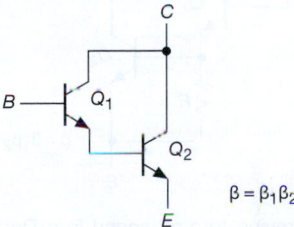

$$\beta = \beta_1 \beta_2$$

Figure 2.74. Darlington transistor configuration.

For a Darlington transistor the base-emitter drop is twice normal and the saturation voltage is at least one diode drop (since Q_1's emitter must be a diode drop above Q_2's emitter). Also, the combination tends to act like a rather slow transistor because Q_1 cannot turn off Q_2 quickly. This problem is usually taken care of by including a resistor from base to emitter of Q_2 (Figure 2.75). Resistor R also prevents leakage current through Q_1 from biasing Q_2 into conduction;[55] its value is chosen so that Q_1's leakage current (nanoamps for small-signal transistors, as much as hundreds of microamps for power transistors) produces less than a diode drop across R, and so that R doesn't sink a large proportion of Q_2's base current when it has a diode drop across it. Typically R might be a few hundred ohms in

---

[55] And, by stabilizing Q_1's collector current, it improves the predictability of the Darlington's total V_{BE}.

a power transistor Darlington, or a few thousand ohms for a small-signal Darlington.

Darlington transistors are available as single packages, usually with the base–emitter resistor included. A typical example is the *npn* power Darlington MJH6284 (and *pnp* cousin MJH6287), with a current gain of 1000 (typical) at a collector current of 10 amps. Another popular power Darlington is the inexpensive *npn* TIP142 (and *pnp* cousin TIP147): these cost $1 in small quantities and have typical $\beta=4000$ at $I_C=5$ A. And for small-signal applications we like the widely available MPSA14 or MMBTA14 (in TO-92 and SOT23 packages, respectively), with a minimum beta of 10,000 at 10 mA and 20,000 at 100 mA. These 30 volt parts have no internal base–emitter resistor (so you can use them at very low currents); they cost less than $0.10 in small quantities. Figure 2.76 shows beta versus collector current for these parts; note the pleasantly high values of beta, but with substantial dependence both on temperature and on collector current.

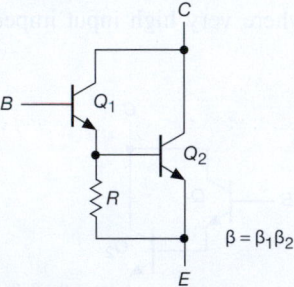

Figure 2.75. Improving turn-off speed in a Darlington pair. (The beta formula is valid as long as R does not rob significant base current from Q_2.)

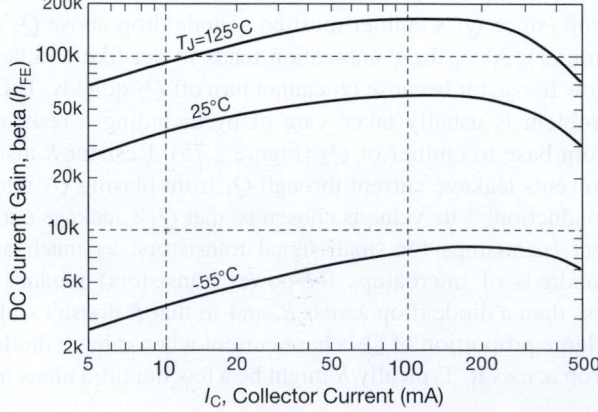

Figure 2.76. Typical beta versus collector current for the popular MPSA14 *npn* Darlington (adapted from the datasheet).

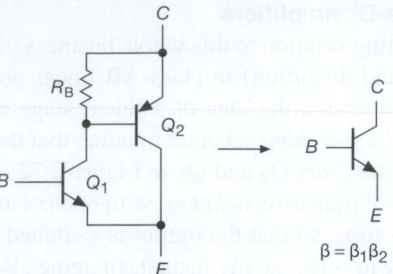

Figure 2.77. Sziklai connection ("complementary Darlington").

A. Sziklai connection

A similar beta-boosting configuration is the Sziklai connection,[56] sometimes referred to as a complementary Darlington (Figure 2.77). This combination behaves like an *npn* transistor, again with large beta. It has only a single base–emitter drop, but (like the Darlington) it also cannot saturate to less than a diode drop. A resistor from base to emitter of Q_2 is advisable, for the same reasons as with the Darlington (leakage current; speed; stability of V_{BE}). This connection is common in push–pull power output stages in which the designer may wish to use one polarity of high-current output transistor only. However, even when used as complementary polarity pairs, it is generally to be preferred over the Darlington for amplifiers and other linear applications; that is because it has the advantage of a single V_{BE} drop (versus two), and that voltage drop is stabilized by the base–emitter resistor of the output transistor. For example, if R_B is chosen such that its current (with a nominal V_{BE} drop across it) is 25% of the output transistor's base current at peak output, then the driver transistor sees a collector current that ranges over only a factor of 5; so its V_{BE} (which is the Sziklai's V_{BE}) varies only 40 mV ($V_T \ln 5$) over the full output current swing. The Sziklai configuration is discussed in more detail in Chapter 2x (see §2x.10); and you'll find nice examples of circuits that rely on the

---

[56] George C. Sziklai, "Symmetrical properties of transistors and their applications," *Proc. IRE* **41**, 717–24 (1953), and US patents 2,762,870 and 2,791,644. His new complementary configuration is buried as Figure 8, where he remarks that "The complementary symmetry of transistors finds an interesting application when it is applied to the cascading of push-pull amplifier stages." The circuit evidently was devised by Sziklai, Lohman, and Herzog, for a transistorized TV demonstration at RCA; the common wisdom was that transistors weren't good enough for the task. In early ICs, where only poor *pnp* transistors were available, an additional *npn* was added, in Sziklai fashion, to boost the current capability of the *pnp*; the combination was called a "composite lateral *pnp*."

Sziklai's unique properties in that chapter's section "BJT amplifier distortion: a SPICE exploration."

Figure 2.78 shows a nice example of a push–pull Sziklai output stage. This has an important advantage compared with the Darlington alternative, namely that the biasing of the Q_3Q_5 pair into class-AB conduction (to minimize crossover distortion) has just two base–emitter drops, rather than four; and, more importantly, Q_3 and Q_5 are running cool compared with the output transistors (Q_4 and Q_6), so they can be relied upon to have a stable base–emitter drop. This allows higher quiescent currents than with the conventional Darlington, where you have to leave a larger safety margin; bottom line, lower distortion.[57]

In this circuit Q_2 functions as an "adjustable V_{BE} multiplier" for biasing, here settable from 1.7 to 3 V_{BE}'s; it is bypassed at signal frequencies. Another circuit trick is the "bootstrapping" of Q_1's collector resistor by C_1 (see §2.4.3), raising its effective resistance at signal frequencies and increasing the amplifier's loop gain to produce lower distortion.

B. Superbeta transistor

The Darlington connection and its near relatives should not be confused with the so-called superbeta transistor, a device with very high current gain achieved through the manufacturing process. A typical superbeta transistor is the 2N5962, with a guaranteed minimum current gain of 450 at collector currents from $10\,\mu\text{A}$ to $10\,\text{mA}$ (see, for example, Table 8.1a on page 501). Superbeta matched pairs are available for use in low-level amplifiers that require matched characteristics, for example the differential amplifier of §2.3.8. Legendary examples are the LM394 and MAT-01 series; these provide high-gain *npn* transistor pairs whose V_{BE}'s are matched to a fraction of a millivolt (as little as $50\,\mu\text{V}$ in the best versions) and whose betas are matched to about 1%. The MAT-03 is a *pnp* matched pair (see Table 8.1b on page 502). Some commercial op-amps use superbeta differential input stages to achieve input (i.e., base bias) currents as low as 50 *pico*amps this way; examples are the LT1008 and LT1012.

2.4.3 Bootstrapping

When biasing an emitter follower, for instance, you choose the base voltage-divider resistors so that the divider presents a stiff voltage source to the base, i.e., their par-

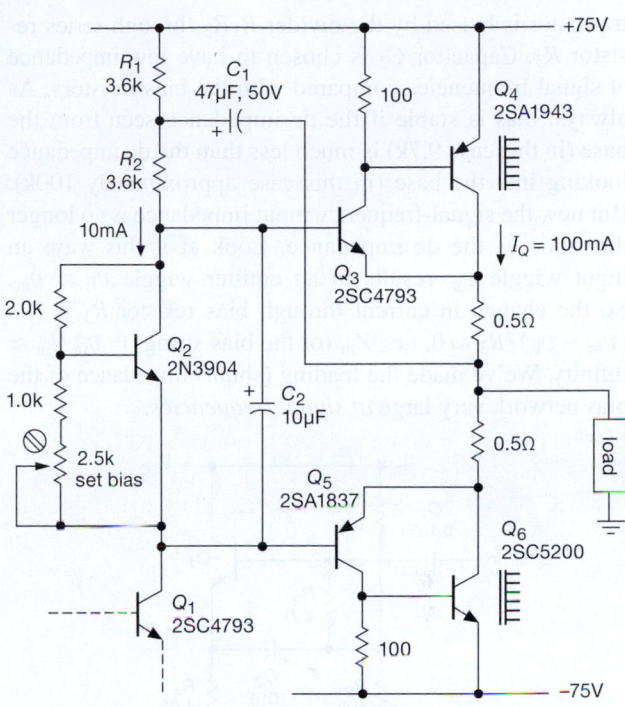

Figure 2.78. Push–pull power stage with Sziklai-pair output transistors, capable of output swings to $\pm70\,\text{V}$ and output currents of $\pm2\,\text{A}$ peak.

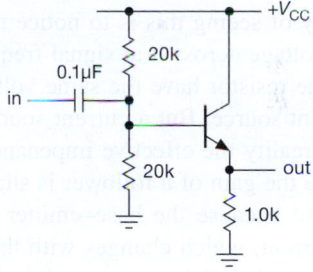

Figure 2.79. Bias network lowers input impedance.

allel impedance is much less than the impedance looking into the base. For this reason the resulting circuit has an input impedance dominated by the voltage divider – the driving signal sees a much lower impedance than would otherwise be necessary. Figure 2.79 shows an example. The input resistance of about 9.1k is mostly due to the voltage-divider impedance of 10k. It is always desirable to keep input impedances high, and anyway it's a shame to load the input with the divider, which, after all, is only there to bias the transistor.

"Bootstrapping" is the colorful name given to a technique that circumvents this problem (Figure 2.80). The

---

[57] To handle higher power, a common practice is to connect in parallel several identical Q_3Q_4 stages (each with its $0.5\,\Omega$ emitter resistor), and similarly for Q_5Q_6. See §2.4.4.

transistor is biased by the divider R_1R_2 through series resistor R_3. Capacitor C_2 is chosen to have low impedance at signal frequencies compared with the bias resistors. As always, bias is stable if the dc impedance seen from the base (in this case 9.7k) is much less than the dc impedance looking into the base (in this case approximately 100k). But now the signal-frequency input impedance is no longer the same as the dc impedance. Look at it this way: an input wiggle v_{in} results in an emitter wiggle $v_E \approx v_{in}$. So the change in current through bias resistor R_3 is $i = (v_{in} - v_E)/R_3 \approx 0$, i.e., Z_{in} (of the bias string) $= v_{in}/i_{in} \approx$ infinity. We've made the loading (shunt) impedance of the bias network very large *at signal frequencies*.

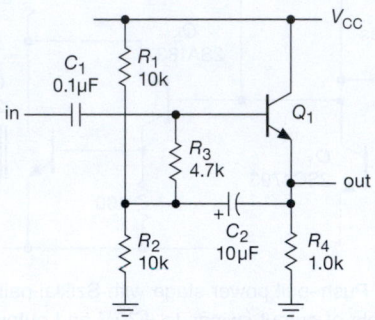

Figure 2.80. Raising the input impedance of an emitter follower at signal frequencies by bootstrapping the base bias divider.

Another way of seeing this is to notice that R_3 always has the same voltage across it at signal frequencies (since both ends of the resistor have the same voltage changes), i.e., it's a current source. But a current source has infinite impedance. In reality the effective impedance is less than infinity because the gain of a follower is slightly less than unity. That is so because the base–emitter drop depends on collector current, which changes with the signal level. You could have predicted the same result from the voltage-dividing effect of the impedance looking into the emitter [$r_e = 25/I_C(mA)$ ohms] combined with the emitter resistor. If the follower has voltage gain A (slightly less than unity), the effective value of R_3 at signal frequencies is

$$R_3/(1-A).$$

The voltage gain of a follower can be written $A = R_L/(R_L + r_e)$, where R_L is the total load seen at the emitter (here $R_1 \parallel R_2 \parallel R_4$), so the effective value of bias resistor R_3 at signal frequencies can be written as $R_3 \rightarrow R_3(1 + R_L/r_e)$. In practice the value of R_3 is effectively increased by a hundred or so, and the input impedance is then dominated by the transistor's base impedance. The emitter-degenerated amplifier can be bootstrapped in the same way,

since the signal on the emitter follows the base. The bias divider circuit is driven by the low-impedance emitter output at signal frequencies, which is what isolates the input signal from this usual task, and makes possible the beneficial increase of input impedance.

A. Bootstrapping collector load resistors

The bootstrap principle can be used to increase the effective value of a transistor's collector load resistor, if that stage drives a follower. That can increase the voltage gain of the stage substantially – recall that $G_V = -g_m R_C$, with $g_m = 1/(R_E + r_e)$. This technique is used in Figure 2.78, where we bootstrapped Q_1's collector load resistor (R_2), forming an approximate current-source load. This serves two useful functions: (a) it raises the voltage gain of Q_1, and (b) it provides base drive current to $Q_3 Q_4$ that does not drop off toward the top of the swing (as would a resistive load, just when you need it most).

2.4.4 Current sharing in paralleled BJTs

It's not unusual in power electronics design to find that the power transistor you've chosen is not able to handle the required power dissipation, and needs to share the job with additional transistors. This is a fine idea, but you need a way to ensure that each transistor handles an equal portion of the power dissipation. In §9.13.5 we illustrate the use of transistors *in series*. This can simplify the problem, because we know they'll all be running at the same current. But it's often more attractive to divide up the current by connecting the transistors in parallel, as in Figure 2.81A.

There are two problems with this approach. First, we know the bipolar transistor is a transconductance device, with its collector current determined in a precise way by its base-to-emitter voltage V_{BE}, as given by the Ebers–Moll equations 2.8 and 2.9. As we saw in §2.3.2, the temperature coefficient of V_{BE} (at constant collector current) is about -2.1 mV/°C; or, equivalently, I_C *increases* with temperature for a fixed V_{BE}.[58] This is unfortunate, because if the junction of one of the transistors becomes hotter than the rest, it takes more of the total current, thereby heating up even more. It's in danger of the dreaded thermal runaway.

The second problem is that transistors of the same part number are not identical. They come off the shelf with

---

[58] This result comes directly from $\partial I_C/\partial T = -g_m \partial V_{BE}/\partial T$, which, after substituting $g_m = I_C/V_T$ tells us that the fractional change of collector current is just $(\partial I_C/\partial T)/I_C = -(\partial V_{BE}/\partial T)/V_T$. So the collector current increases fractionally by about 2.1mV/25mV (or 8.4%) per °C – a rather large amount!

differing values of V_{BE} for a given I_C. This is true even for parts made at the same time on the same fabrication line, and from the same silicon wafer. To see how large a variation you are likely to get, we measured 100 adjacent ZTX851 transistors on a reel, with an observed spread of about 17 mV, shown in Figure 8.44. This really represents a "best case," because you cannot be certain that a batch of incoming transistors derive from a single lot, much less a single wafer. When you first build something, the V_{BE}'s of "identical" transistors may be within 20–50 mV of each other, but that matching is lost when one of them has to be replaced someday. It's always safer to assume a possible 100 mV or so spread of base-emitter voltages. Recalling that $\Delta V_{BE}=60$ mV corresponds to a factor of ten current ratio, it's clear that you cannot get away with a direct parallel connection like Figure 2.81A.

The usual solution to this problem is the use of small resistors in the emitters, as shown in Figure 2.81B. These are called *emitter-ballasting* resistors, and their value is chosen to drop at least a few tenths of a volt at the higher end of the anticipated operating current range. That voltage drop must be adequate to swamp the V_{BE} spread of the individual transistors, and is ordinarily chosen somewhere in the range of 300–500 mV.

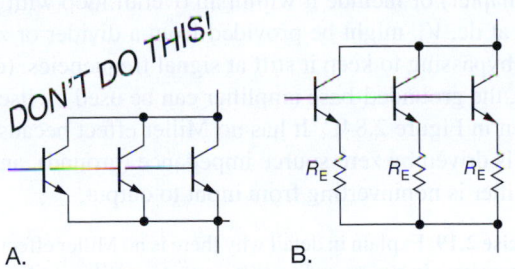

Figure 2.81. To equalize the currents of parallel transistors, use emitter ballasting resistors R_E, as in circuit B.

At high currents the resistors may suffer from an inconveniently-high power dissipation, so you may want to use the current-sharing trick shown in Figure 2.82. Here the current-sensing transistors Q_4–Q_6 adjust the base drive to the "paralleled" power transistors Q_1–Q_3 to maintain equal emitter currents (you can think of Q_4–Q_6 as a high-gain differential amplifier with three inputs). This "active ballast" technique works well with power Darlington BJTs, and it works particularly well with MOSFETs (see Figure 3.117), thanks to their negligible input (gate) current, thus making

MOSFETs a good choice for circuits with lots of power dissipation.[59]

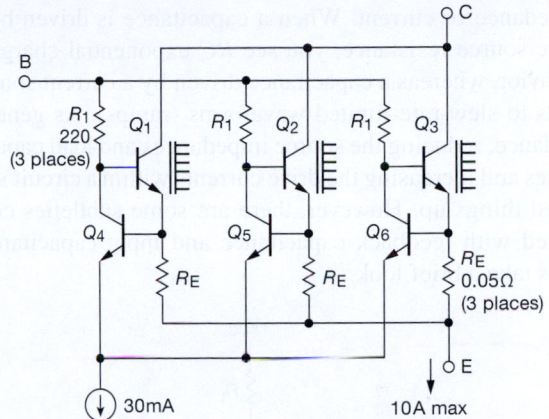

Figure 2.82. Active ballasting of parallel transistors Q_1–Q_3 via feedback from current sensing transistors Q_4–Q_6 lets you configure a parallel power transistors with very low drops across the emitter resistors.

2.4.5 Capacitance and Miller effect

In our discussion so far we have used what amounts to a dc, or low-frequency, model of the transistor. Our simple current amplifier model and the more sophisticated Ebers–Moll transconductance model both deal with voltages, currents, and resistances seen at the various terminals. With these models alone we have managed to go quite far, and in fact these simple models contain nearly everything you will ever need to know to design transistor circuits. However, one important aspect that has serious impact on high-speed and high-frequency circuits has been neglected: the existence of capacitance in the external circuit and in the transistor junctions themselves. Indeed, at high frequencies the effects of capacitance often dominate circuit behavior; at 100 MHz a typical junction capacitance of 5 pF has an impedance of just 320 Ω!

In this brief subsection we introduce the problem, illustrate some of its circuit incarnations, and suggest some methods of circumventing its effects. It would be a mistake to leave this chapter without realizing the nature of this problem. In the course of this brief discussion we will encounter the infamous *Miller effect*, and the use of configurations such as the cascode to overcome it.

---

[59] Another nice feature of MOSFETs is their lack of second breakdown, thus a wider safe-operating area; see §3.6.4C.

A. Junction and circuit capacitance

Capacitance limits the speed at which the voltages within a circuit can swing ("slew rate"), owing to finite driving impedance or current. When a capacitance is driven by a finite source resistance, you see RC exponential charging behavior, whereas a capacitance driven by a current source leads to slew-rate-limited waveforms (ramps). As general guidance, reducing the source impedances and load capacitances and increasing the drive currents within a circuit will speed things up. However, there are some subtleties connected with feedback capacitance and input capacitance. Let's take a brief look.

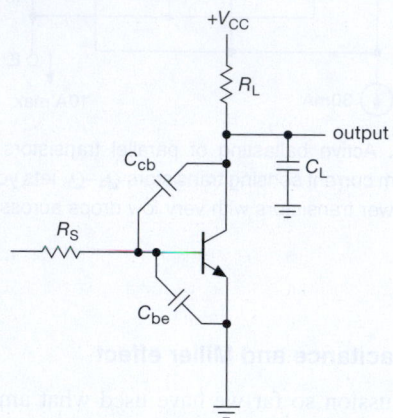

Figure 2.83. Junction and load capacitances in a transistor amplifier.

The circuit in Figure 2.83 illustrates most of the problems of junction capacitance. The output capacitance forms a time constant with the output resistance R_L (R_L includes both the collector and load resistances, and C_L includes both junction and load capacitances), giving a rolloff starting at some frequency $f = 1/2\pi R_L C_L$.

The same is true for the input capacitance, C_{be}, in combination with the source impedance R_S. Of greater significance, at high frequencies the input capacitance robs base current, effectively decreasing the transistor's beta. In fact, transistor datasheets specify a cutoff frequency, f_T, at which the beta has decreased to unity – not much of an amplifier anymore! We discuss this further in Chapter *2x*.

B. Miller effect

The feedback impedance C_{cb} is another matter. The amplifier has some overall voltage gain G_V, so a small voltage wiggle at the input results in a wiggle G_V times larger (and inverted) at the collector. This means that the signal source sees a current through C_{cb} that is $G_V + 1$ times as large as if C_{cb} were connected from base to ground; i.e., for the pur-

pose of input rolloff frequency calculations, the feedback capacitance behaves like a capacitor of value $C_{cb}(G_V + 1)$ from input to ground. This effective increase of C_{cb} is known as the Miller effect. It often dominates the rolloff characteristics of amplifiers, because a typical feedback capacitance of 4 pF can look like several hundred picofarads to ground.

There are several methods available for beating the Miller effect: (a) you can decrease the source impedance driving a grounded-emitter stage by using an emitter follower. Figure 2.84 shows three other possibilities; (b) the differential amplifier circuit with no collector resistor in Q_1 (Figure 2.84A) has no Miller effect; you can think of it as an emitter follower driving a grounded-base amplifier (see below); (c) the famous cascode configuration (Figure 2.84B) elegantly defeats the Miller effect. Here Q_1 is a grounded-emitter amplifier with R_L as its collector resistor: Q_2 is interposed in the collector path to prevent Q_1's collector from swinging (thereby eliminating the Miller effect) while passing the collector current through to the load resistor unchanged. The input labeled V_+ is a fixed-bias voltage, usually set a few volts above Q_1's emitter voltage to pin Q_1's collector and keep it in the active region. This circuit fragment is incomplete, because biasing is not shown; you could either include a bypassed emitter resistor and base divider for biasing Q_1 (as we did earlier in the chapter) or include it within an overall loop with feedback at dc. V_+ might be provided from a divider or zener, with bypassing to keep it stiff at signal frequencies. (d) Finally, the grounded-base amplifier can be used by itself, as shown in Figure 2.84C. It has no Miller effect because the base is driven by zero source impedance (ground), and the amplifier is noninverting from input to output.

Exercise 2.19. Explain in detail why there is no Miller effect in either transistor in the preceding differential amplifier and cascode circuits.

Capacitive effects can be somewhat more complicated than this brief introduction might indicate. In particular: (a) the rolloffs that are due to feedback and output capacitances are not entirely independent; in the terminology of the trade there is *pole splitting*; (b) the transistor's input capacitance still has an effect, even with a stiff input signal source. In particular, current that flows through C_{be} is not amplified by the transistor. This base current "robbing" by the input capacitance causes the transistor's small-signal current gain h_{fe} to drop at high frequencies, eventually reaching unity at a frequency known as f_T. (c) To complicate matters, the junction capacitances depend on voltage: a dominant portion of C_{be} changes proportionally with

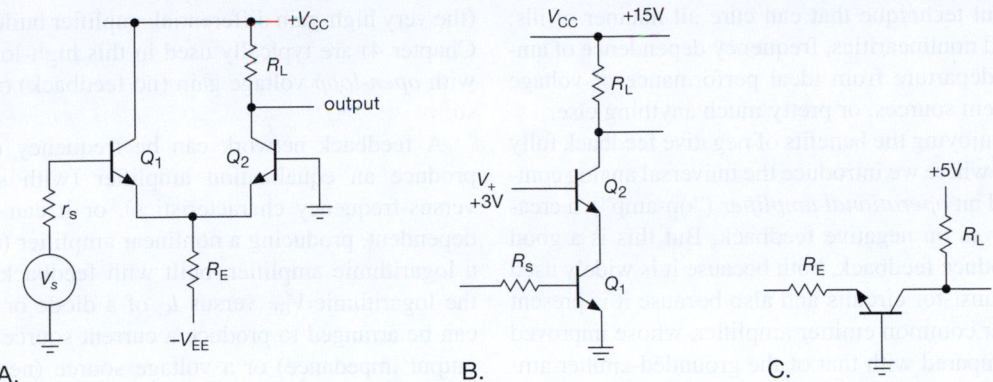

Figure 2.84. Three circuit configurations that avoid the Miller effect. A. Differential amplifier with inverting input grounded. B. Cascode connection. C. Grounded base amplifier.

operating current, so f_T is given instead.[60] (d) When a transistor is operated as a switch, effects associated with charge stored in the base region of a saturated transistor cause an additional loss of speed.

The Miller effect looms large in high-speed and wideband circuits, and we'll be seeing it again and again in subsequent chapters.

2.4.6 Field-effect transistors

In this chapter we have dealt exclusively with BJTs, characterized by the Ebers–Moll equation. BJTs were the original transistors, and they are widely used in analog circuit design. However, it would be a mistake to continue without a few words of explanation about the other kind of transistor, the FET, which we will take up in detail in Chapter 3.

The FET behaves in many ways like an ordinary bipolar transistor. It is a three-terminal amplifying device, available in both polarities, with a terminal (the *gate*) that controls the current flow between the other two terminals (*source* and *drain*). It has a unique property, though: the gate draws no dc current, except for leakage. This means that extremely high input impedances are possible, limited only by capacitance and leakage effects. With FETs you don't have to worry about providing substantial base current, as was necessary with the BJT circuit design of this chapter. Input currents measured in *pico*amperes are commonplace. Yet the FET is a rugged and capable device, with voltage and current ratings comparable to those of bipolar transistors.

Most of the available devices fabricated with BJTs

(matched pairs, differential and operational amplifiers, comparators, high-current switches and amplifiers, and RF amplifiers) are also available with FET construction, often with superior performance. Furthermore, digital logic, microprocessors, memory, and all manner of complex and wonderful large-scale digital chips are built almost exclusively with FETs. Finally, the area of micropower design is dominated by FET circuits. It is not an exaggeration to say that, demographically, almost all transistors are FETs.[61]

FETs are so important in electronic design that we devote the next chapter to them before treating operational amplifiers and feedback in Chapter 4. We urge the reader to be patient with us as we lay the groundwork in these first three difficult chapters; that patience will be rewarded many times over in the succeeding chapters, as we explore the enjoyable topics of circuit design with operational amplifiers and digital integrated circuits.

2.5 Negative feedback

We've hinted earlier in the chapter that feedback offers a cure to some vexing problems: biasing the grounded-emitter amplifier (§2.3.4 and 2.3.5), biasing the differential amplifier with current-mirror active load (§2.3.8C), and minimizing crossover distortion in push–pull followers (§2.4.1A). It's even better than that – *negative feedback*

---

[60] See values of f_T versus collector current for 25 transistors, plotted and tabulated in Chapter 2x's section titled "BJT Bandwidth and f_T."

[61] Lest this outpouring of enthusiasm leave the wrong impression, we hasten to point out that BJTs are alive and well, largely because they are unbeatable when it comes to characteristics like accuracy and noise (the subjects of Chapters 5 and 8). They excel also in transconductance (i.e., gain). Those muscular power FETs suffer from rather high input capacitance; and, as discrete parts, you cannot get small-signal MOSFETs, only *power* MOSFETs.

is a wonderful technique that can cure all manner of ills: distortion and nonlinearities, frequency dependence of amplifier gain, departure from ideal performance of voltage sources, current sources, or pretty much anything else.

We'll be enjoying the benefits of negative feedback fully in Chapter 4, where we introduce the universal analog component called an *operational amplifier* ("op-amp"), a creature that thrives on negative feedback. But this is a good place to introduce feedback, both because it is widely used in discrete transistor circuits and also because it is present already in our common emitter amplifier, whose improved linearity (compared with that of the grounded-emitter amplifier) is due to negative feedback.

2.5.1 Introduction to feedback

Feedback has become such a well-known concept that the word has entered the general vocabulary. In control systems, feedback consists of comparing the actual output of the system with the desired output and making a correction accordingly. The "system" can be almost anything: for instance, the process of driving a car down the road, in which the output (the position and velocity of the car) is sensed by the driver, who compares it with expectations and makes corrections to the input (steering wheel, throttle, brake). In amplifier circuits the output should be a multiple of the input, so in a feedback amplifier the input is compared with an attenuated version of the output.

As used in amplifiers, negative feedback is implemented simply by coupling the output back in such a way as to cancel some of the input. You might think that this would only have the effect of reducing the amplifier's gain and would be a pretty stupid thing to do. Harold S. Black, who attempted to patent negative feedback in 1928, was greeted with the same response. In his words, "Our patent application was treated in the same manner as one for a perpetual-motion machine."[62] True, it does lower the gain, but in exchange it also improves other characteristics, most notably freedom from distortion and nonlinearity, flatness of response (or conformity to some desired frequency response), and predictability. In fact, as more negative feedback is used, the resultant amplifier characteristics become less dependent on the characteristics of the open-loop (no-feedback) amplifier and finally depend on the properties only of the feedback network itself. Operational amplifiers

(the very high-gain differential amplifier building blocks of Chapter 4) are typically used in this high-loop-gain limit, with *open-loop* voltage gain (no feedback) of a million or so.

A feedback network can be frequency dependent, to produce an equalization amplifier (with specific gain-versus-frequency characteristics), or it can be amplitude dependent, producing a nonlinear amplifier (an example is a logarithmic amplifier, built with feedback that exploits the logarithmic V_{BE} versus I_C of a diode or transistor). It can be arranged to produce a current source (near-infinite output impedance) or a voltage source (near-zero output impedance), and it can be connected to generate very high or very low input impedance. Speaking in general terms, the property that is sampled to produce feedback is the property that is improved. Thus, if you feed back a signal proportional to the output current, you will generate a good current source.[63]

Let's look at how feedback works, and how it affects what an amplifier does. We will find simple expressions for the input impedance, output impedance, and gain of an amplifier with negative feedback.

2.5.2 Gain equation

Look at Figure 2.85. To get started we've drawn the familiar common-emitter amplifier with emitter degeneration. Thinking of the transistor in the Ebers–Moll sense, the small-signal voltage from base to emitter (ΔV_{BE}) programs the collector current. But ΔV_{BE} is less than the input voltage V_{in}, because of the drop across R_E. If the output is unloaded, it's easy to get the equation in the figure. In other words, the common-emitter amplifier with emitter degeneration is a grounded-emitter amplifier with negative feedback, as we hinted earlier.

This circuit has some subtleties, which we'd like to sidestep for now by looking instead at the more straightforward configuration shown in Figure 2.85B. Here we've drawn a differential amplifier (with differential gain A), with a fraction of its output signal subtracted from the circuit input v_{in}. That fraction, of course, is given simply by

---

[62] See the fascinating article in *IEEE Spectrum*, December 1977. His patent for negative feedback (No. 2,102,671, modestly titled "Wave translation system") was granted in 1937, nine years after his initial filing.

[63] Feedback can also be *positive*; that's how you make an oscillator, for instance. As much fun as that may sound, it simply isn't as important as negative feedback. More often it's a nuisance, because a negative-feedback circuit may have large enough phase shifts at some high frequency to produce positive feedback and oscillations. It is surprisingly easy to have this happen, and the prevention of unwanted oscillations is the object of what is called *compensation*, a subject we treat briefly at the end of Chapter 4.

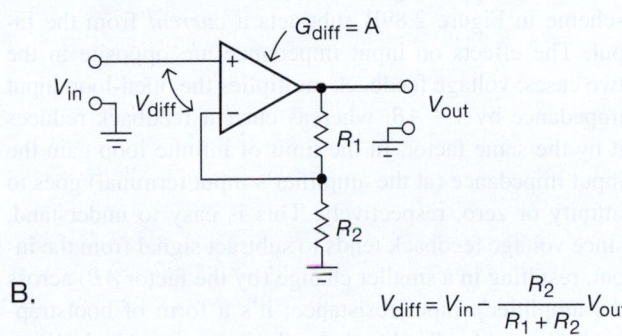

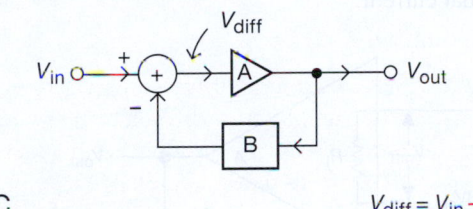

A.

$$\Delta V_{BE} = \Delta V_{in} - \frac{R_E}{R_C}\Delta V_{out}$$

$G_{diff} = A$

B.

$$V_{diff} = V_{in} - \frac{R_2}{R_1 + R_2}V_{out}$$

C.

$$V_{diff} = V_{in} - BV_{out}$$

Figure 2.85. Negative feedback subtracts a fraction of the output from the input: A. Common-emitter amplifier. B. Differential amplifier configured as a noninverting voltage amplifier. C. Conventional block diagram.

the voltage divider equation, as shown. This is a very common configuration, widely used with op-amps (Chapter 4), and known simply as a "noninverting amplifier."

When talking about negative feedback, it's conventional to draw a diagram like Figure 2.85C, in which the feedback fraction is simply labeled B. This is useful because it allows more generality than a voltage divider (feedback can include frequency-dependent components like capacitors, and nonlinear components like diodes), and it keeps the equations simple. For a voltage divider, of course, B would simply be equal to $R_2/(R_1+R_2)$.

Let's figure out the gain. The amplifier has open-loop voltage gain A, and the feedback network subtracts a fraction B of the output voltage from the input. (Later we will generalize things so that inputs and outputs can be currents

or voltages.) The input to the gain block is then $V_{in} - BV_{out}$. But the output is just the input times A:

$$A(V_{in} - BV_{out}) = V_{out}.$$

In other words,

$$V_{out} = \frac{A}{1+AB}V_{in},$$

and so the closed-loop voltage gain, V_{out}/V_{in}, is just

$$G = \frac{A}{1+AB}. \qquad (2.16)$$

Some terminology: the standard designations for these quantities are as follows: G = closed-loop gain, A = open-loop gain, AB = loop gain, $1+AB$ = return difference, or desensitivity. The feedback network is sometimes called the beta network (no relation to transistor beta, h_{fe}).[64]

2.5.3 Effects of feedback on amplifier circuits

Let's look at the important effects of feedback. The most significant are predictability of gain (and reduction of distortion), changed input impedance, and changed output impedance.

A. Predictability of gain

The voltage gain is $G = A/(1+AB)$. In the limit of infinite[65] open-loop gain A, $G = 1/B$. For finite gain A, feedback acts to reduce the effects of variations of A (with frequency, temperature, amplitude, etc.). For instance, suppose A depends on frequency as in Figure 2.86. This will surely satisfy anyone's definition of a poor amplifier (the gain varies over a factor of 10 with frequency). Now imagine we introduce feedback, with $B = 0.1$ (a simple voltage divider will do). The closed-loop voltage gain now varies from 1000/[1 +(1000×0.1)], or 9.90, to 10,000/[1 + (10,000×0.1)], or 9.99, a variation of just 1% over the same range of frequency! To put it in audio terms, the original amplifier is flat to ±10 dB, whereas the feedback amplifier is flat to ±0.04 dB. We can now recover the original gain of 1000 with nearly this linearity simply by cascading three such stages.

It was for just this reason (namely, the need for extremely flat-response telephone repeater amplifiers) that

---

[64] We'll see later that amplifiers used with feedback commonly have significant lagging phase shifts from input to output. So the open-loop voltage gain A should properly be represented as a complex number. We'll treat this in §2.5.4; for now we'll adopt the simplification that the amplifier's output voltage is proportional to its input voltage.

[65] Which is not a bad approximation for an op-amp, whose typical open-loop gain is in the neighborhood of $A_{OL} \approx 10^6$.

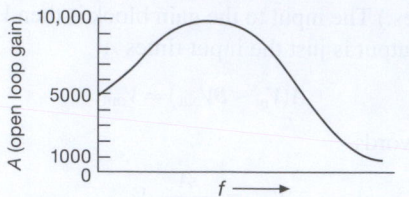

Figure 2.86. Amplifier with open-loop gain A that varies widely with frequency f.

negative feedback in electronics was invented. As the inventor, Harold Black, described it in his first open publication on the invention [*Elec. Eng.*, **53**, 114, (1934)], "by building an amplifier whose gain is made deliberately, say 40 decibels higher than necessary (10,000-fold excess on energy basis) and then feeding the output back to the input in such a way as to throw away the excess gain, it has been found possible to effect extraordinary improvement in constancy of amplification and freedom from nonlinearity." Black's patent is spectacular, with dozens of elegant figures; we reproduce one of them here (Figure 2.87), which makes the point eloquently.

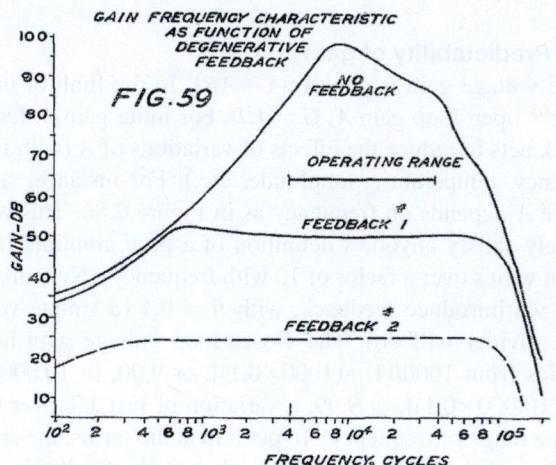

Figure 2.87. Harold Black explains it in his historic 1937 patent, with the unassuming title "Wave translation system."

It is easy to show, by taking the partial derivative of G with respect to A (i.e., $\partial G/\partial A$), that relative variations in the open-loop gain are reduced by the desensitivity:

$$\frac{\Delta G}{G} = \frac{1}{1+AB}\frac{\Delta A}{A}. \qquad (2.17)$$

Thus, for good performance the loop gain AB should be much larger than 1. That's equivalent to saying that the open-loop gain should be much larger than the closed-loop gain.

A very important consequence of this is that nonlinearities, which are simply gain variations that depend on signal level, are reduced in exactly the same way.

B. Input impedance
Feedback can be arranged to subtract a voltage or a current from the input (these are sometimes called *series feedback* and *shunt feedback*, respectively). The noninverting amplifier configuration we've been considering, for instance, subtracts a sample of the output voltage from the differential *voltage* appearing at the input, whereas the feedback scheme in Figure 2.89B subtracts a *current* from the input. The effects on input impedance are opposite in the two cases: voltage feedback multiplies the open-loop input impedance by $1+AB$, whereas current feedback reduces it by the same factor. In the limit of infinite loop gain the input impedance (at the amplifier's input terminal) goes to infinity or zero, respectively. This is easy to understand, since voltage feedback tends to subtract signal from the input, resulting in a smaller change (by the factor AB) across the amplifier's input resistance; it's a form of bootstrapping. Current feedback reduces the input signal by bucking it with an equal current.

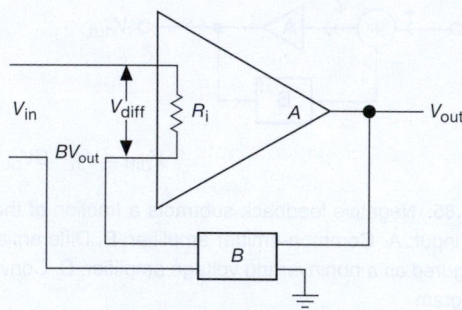

Figure 2.88. Series-feedback input impedance.

Series (voltage) feedback
Let's see explicitly how the effective input impedance is changed by feedback. We illustrate the case of voltage feedback only, since the derivations are similar for the two cases. We begin with a differential amplifier model with (finite) input resistance as shown in Figure 2.88. An input V_{in} is reduced by BV_{out}, putting a voltage $V_{\text{diff}} = V_{\text{in}} - BV_{\text{out}}$ across the inputs of the amplifier. The input current is therefore

$$I_{\text{in}} = \frac{V_{\text{in}} - BV_{\text{out}}}{R_i} = \frac{V_{\text{in}}\left(1 - B\dfrac{A}{1+AB}\right)}{R_i} = \frac{V_{\text{in}}}{(1+AB)R_i},$$

giving an effective input impedance

$$Z_{in} = V_{in}/I_{in} = (1+AB)R_i.$$

In other words, the input impedance is boosted by a factor of the loop gain plus one. If you were to use the circuit of Figure 2.85B to close the feedback loop around a differential amplifier whose native input impedance is 100 kΩ and whose differential gain is 10^4, choosing the resistor ratio (99:1) for a target gain of 100 (in the limit of infinite amplifier gain), the input impedance seen by the signal source would be approximately 10 MΩ, and the closed-loop gain would be 99.[66]

Shunt (current) feedback

Look at Figure 2.89A. The impedance seen looking into the input of a voltage amplifier with current feedback is reduced by the feedback current, which opposes voltage changes at the input.[67] By considering the current change produced by a voltage change at the input, you find that the input signal secs a parallel combination of (a) the amplifier's native input impedance R_i and (b) the feedback resistor R_f divided by $1+A$. That is,

$$Z_{in} = R_i \| \frac{R_f}{1+A}$$

(see if you can prove this). In cases of very high loop gain (e.g., an op-amp), the input impedance is reduced to a fraction of an ohm, which might seem bad. But in fact this configuration is used to convert an input current into an output voltage (a "transresistance amplifier"), for which a low input impedance is a good characteristic. We'll see examples in Chapters 4 and *4x*.

By the addition of an input resistor (Figure 2.89B) the circuit becomes an "inverting amplifier," with input resistance as shown. You can think of this (particularly in the high-loop-gain limit) as a resistor feeding a current-to-voltage amplifier. In that limit R_{in} approximately equals R_1 (and the closed-loop gain approximately equals $-R_2/R_1$).

It is a straightforward exercise to derive an expression for the closed-loop voltage gain of the inverting amplifier with finite loop gain. The answer is

$$G = -A(1-B)/(1+AB)$$

where B is defined as before, $B = R_1/(R_1+R_2)$. In the limit of large open-loop gain A, $G = 1-1/B$ (i.e., $G = -R_2/R_1$).

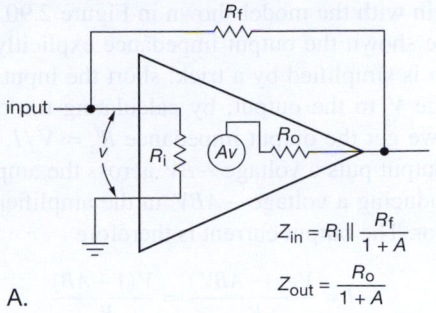

A.

$$Z_{in} = R_i \| \frac{R_f}{1+A}$$

$$Z_{out} = \frac{R_o}{1+A}$$

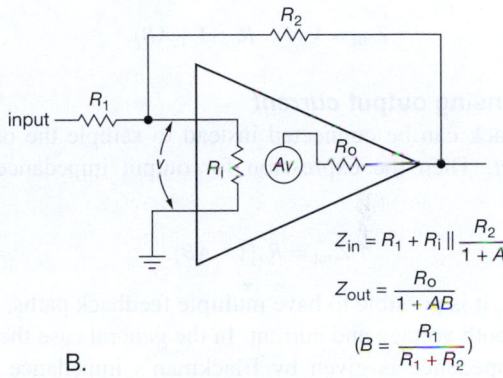

B.

$$Z_{in} = R_1 + R_i \| \frac{R_2}{1+A}$$

$$Z_{out} = \frac{R_o}{1+AB}$$

$$\left(B = \frac{R_1}{R_1+R_2}\right)$$

Figure 2.89. Input and output impedances for (A) transresistance amplifier and (B) inverting amplifier.

Exercise 2.20. Derive the foregoing expressions for input impedance and gain of the inverting amplifier.

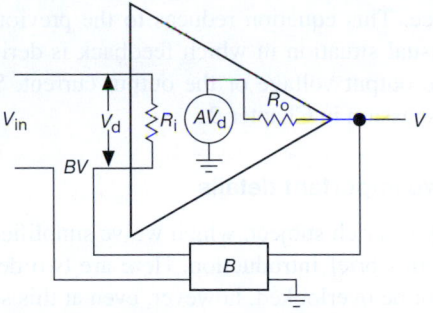

Figure 2.90. Output impedance.

C. Output impedance

Again, feedback can extract a sample of the output voltage or the output current. In the first case the open-loop output impedance will be reduced by the factor $1+AB$, whereas in the second case it will be increased by the same factor. We illustrate this effect for the case of voltage sampling.

---

[66] Of course, knowing that the open-loop gain is approximately 10^4, you might bump the resistor ratio up to 100:1 to compensate. With an op-amp there's no need: with a typical open-loop gain of $\sim 10^6$, the closed-loop gain would be $G_{CL} = 99.99$.

[67] As in the circuit of Figure 2.53 in §2.3.5C.

We begin with the model shown in Figure 2.90. This time we have shown the output impedance explicitly. The calculation is simplified by a trick: short the input and apply a voltage V to the output; by calculating the output current I, we get the output impedance $R_o' = V/I$. Voltage V at the output puts a voltage $-BV$ across the amplifier's input, producing a voltage $-ABV$ in the amplifier's internal generator. The output current is therefore

$$I = \frac{V - (-ABV)}{R_o} = \frac{V(1 + AB)}{R_o}$$

giving an effective output impedance[68] of

$$Z_{out} = V/I = R_o/(1 + AB).$$

D. Sensing output *current*

Feedback can be connected instead to sample the output *current*. Then the expression for output impedance becomes

$$Z_{out} = R_o(1 + AB).$$

In fact, it is possible to have multiple feedback paths, sampling both voltage and current. In the general case the output impedance is given by Blackman's impedance relation[69]

$$Z_{out} = R_o \frac{1 + (AB)_{SC}}{1 + (AB)_{OC}},$$

where $(AB)_{SC}$ is the loop gain with the output shorted to ground and $(AB)_{OC}$ is the loop gain with no load attached. Thus feedback can be used to generate a desired output impedance. This equation reduces to the previous results for the usual situation in which feedback is derived from either the output voltage or the output current. See additional discussion in Chapter *2x*.

2.5.4 Two important details

Feedback is a rich subject, which we've simplified shamelessly in this brief introduction. Here are two details that should not be overlooked, however, even at this somewhat superficial level of understanding.

A. Loading by the feedback network

In feedback computations, you usually assume that the beta network doesn't load the amplifier's output. If it does, that must be taken into account in computing the open-loop gain. Likewise, if the connection of the beta network at the amplifier's input affects the open-loop gain (feedback removed, but network still connected), you must use the modified open-loop gain. Finally, the preceding expressions assume that the beta network is unidirectional, i.e., it does not couple any signal from the input to the output.

B. Phase shifts, stability, and "compensation"

The open-loop amplifier gain A is central in the expressions we've found for closed-loop gain and the corresponding input and output impedances. By default one might reasonably assume that A is a real number – that is, that the output is in phase with the input. In real life things are more complex,[70] because of the effects of circuit capacitances (and Miller effect, §2.4.5), and also the limited bandwidth (f_T) of the active components themselves. The result is that the open-loop amplifier will exhibit lagging phase shifts that increase with frequency. This has several consequences for the closed-loop amplifier.

Stability

If the open-loop amplifier's lagging phase shift reaches 180°, then negative feedback becomes *positive* feedback, with the possibility of oscillation. This is not what you want! (The actual criterion for oscillation is that the phase shift be 180° at a frequency at which the loop gain AB equals 1.) This is a serious concern, particularly in amplifiers with plenty of gain (such as op-amps). The problem is only exacerbated if the feedback network contributes additional lagging phase shift (as it often will). The subject of *frequency compensation* in feedback amplifiers deals directly with this essential issue; you can read about it in §4.9.

Gain and phase shift

The expressions we found for closed-loop gain and for the input and output impedances contain the open-loop gain A. For example, the voltage amplifier with series feedback (Figures 2.85B&C, 2.88, and 2.90) has closed-loop gain $G_{CL} = A/(1 + AB)$, where $A = G_{OL}$, the amplifier's open-loop gain. Let's imagine that the open-loop gain A is 100, and that we've chosen $B = 0.1$ for a target closed-loop gain of $G_{CL} \approx 10$. Now, if the open-loop amplifier had no phase shifts, then $G_{CL} \approx 9.09$, also without phase shift. If instead the amplifier has a 90° lagging phase shift, then A is pure

---

[68] If the open-loop gain A is real (i.e. no phase shift), then the output impedance Z_{out} will be real (i.e., resistive: R_{out}). As we'll see in Chapter 4, however, A can be (and often is) complex, representing a lagging phase shift. For op-amps the phase shift is 90° over most of the amplifier's bandwidth. The result is an *inductive* closed-loop output impedance. See for example Figure 4.53 in Chapter 4.

[69] R. B. Blackman, "Effect of feedback on impedance," *Bell. Sys. Tech. J.* **22**, 269 (1943).

[70] That's a pun, get it?

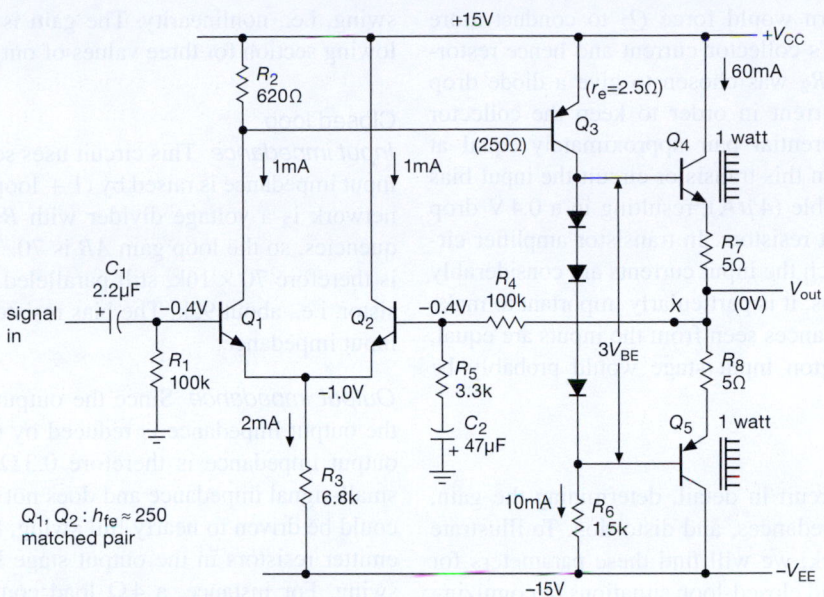

Figure 2.91. Transistor power amplifier with negative feedback.

imaginary ($A = -100j$), and the closed-loop gain becomes $G_{CL} = 9.90 - 0.99j$. That's a magnitude $|G_{CL}| = 9.95$, with a lagging phase shift of approximately 6°. In other words, the effect of a pretty significant (halfway to oscillation!) open-loop phase shift turns out, in fact, to be favorable: the closed-loop gain is only 0.5% less than the target, compared with 9% for the case of the same amplifier without phase shift. The price you pay is some residual phase shift and, of course, an approach to instability.

As artificial as this example may seem, it in fact reflects a reality of op-amps, which usually have an ~90° lagging phase shift over almost their entire bandwidth (typically from ~10 Hz to 1 MHz or more). Because of their much higher open-loop gain, the amplifier with feedback exhibits very little phase shift, and an accurate gain set almost entirely by the feedback network. Much more on this in Chapter 4, and in §4.9.

Exercise 2.21. Verify that the above expressions for G_{CL} are correct.

2.5.5 Two examples of transistor amplifiers with feedback

Let's look at two transistor amplifier designs to see how the performance is affected by negative feedback. There's a bit of complexity in this analysis ... don't be discouraged![71]

---

[71] Those fearful of discouragement may wish to skip over this section in a first reading.

Figure 2.91 shows a complete transistor amplifier with negative feedback. Let's see how it goes.

A. Circuit description

It may look complicated, but it is extremely straightforward in design and is relatively easy to analyze. Q_1 and Q_2 form a differential pair, with common-emitter amplifier Q_3 amplifying its output. R_6 is Q_3's collector load resistor, and push–pull pair Q_4 and Q_5 form the output emitter follower. The output voltage is sampled by the feedback network consisting of voltage divider R_4 and R_5, with C_2 included to reduce the gain to unity at dc for stable biasing. R_3 sets the quiescent current in the differential pair, and since overall feedback guarantees that the quiescent output voltage is at ground, Q_3's quiescent current is easily seen to be 10 mA (V_{EE} across R_6, approximately). As we discussed earlier (§2.4.1B), the diodes bias the push–pull pair into conduction, leaving one diode drop across the series pair R_7 and R_8, i.e., 60 mA quiescent current. That's class-AB operation, good for minimizing crossover distortion, at the cost of 1 watt standby dissipation in each output transistor.

From the point of view of our earlier circuits, the only unusual feature is Q_1's quiescent collector voltage, one diode drop below V_{CC}. That is where it must sit in order to hold Q_3 in conduction, and the feedback path ensures that it will. (For instance, if Q_1 were to pull its collector closer to ground, Q_3 would conduct heavily, raising the output voltage, which in turn would force Q_2 to conduct more heavily, reducing Q_1's collector current and hence restoring the status quo.) R_2 was chosen to give a diode drop at Q_1's quiescent current in order to keep the collector currents in the differential pair approximately equal at the quiescent point. In this transistor circuit the input bias current is not negligible ($4\,\mu A$), resulting in a 0.4 V drop across the 100k input resistors. In transistor amplifier circuits like this, in which the input currents are considerably larger than in op-amps, it is particularly important to make sure that the dc resistances seen from the inputs are equal, as shown (a Darlington input stage would probably be better here).

B. Analysis

Let's analyze this circuit in detail, determining the gain, input and output impedances, and distortion. To illustrate the utility of feedback, we will find these parameters for both the open-loop and closed-loop situations (recognizing that biasing would be hopeless in the open-loop case). To get a feeling for the linearizing effect of the feedback, the gain will be calculated at +10 volts and −10 volts output, as well as at the quiescent point (0 V).

Open loop

Input impedance We cut the feedback at point X and ground the right-hand side of R_4. The input signal sees 100k in parallel with the impedance looking into the base. The latter is h_{fe} times twice the intrinsic emitter resistance plus the impedance seen at Q_2's emitter caused by the finite impedance of the feedback network at Q_2's base. For $h_{fe} \approx 250$, $Z_{in} \approx 250 \times [(2 \times 25) + (3.3k/250)]$; i.e., $Z_{in} \approx 16k$.

Output impedance Since the impedance looking back into Q_3's collector is high, the output transistors are driven by a 1.5k source (R_6). The output impedance is about 15Ω ($\beta \approx 100$) plus the 5Ω emitter resistance, or 20Ω. The intrinsic emitter resistance of 0.4Ω is negligible.

Gain The differential input stage sees a load of R_2 paralleled by Q_3's base resistance. Since Q_3 is running 10 mA quiescent current, its intrinsic emitter resistance is 2.5Ω, giving a base impedance of about 250Ω (again, $\beta \approx 100$). The differential pair thus has a gain of

$$\frac{250\|620}{2 \times 25\Omega} \quad \text{or} \quad 3.5.$$

The second stage, Q_3, has a voltage gain of 1.5k Ω/2.5Ω, or 600. The overall voltage gain at the quiescent point is 3.5×600, or 2100. Since Q_3's gain depends on its collector current, there is substantial change of gain with signal

swing, i.e., nonlinearity. The gain is tabulated in the following section for three values of output voltage.

Closed loop

Input impedance This circuit uses series feedback, so the input impedance is raised by (1 + loop gain). The feedback network is a voltage divider with $B=1/30$ at signal frequencies, so the loop gain AB is 70. The input impedance is therefore $70 \times 16k$, still paralleled by the 100k bias resistor, i.e., about 92k. The bias resistor now dominates the input impedance.

Output impedance Since the output *voltage* is sampled, the output impedance is reduced by (1 + loop gain). The output impedance is therefore 0.3Ω. Note that this is a small-signal impedance and does not mean that a 1Ω load could be driven to nearly full swing, for instance. The 5Ω emitter resistors in the output stage limit the large signal swing. For instance, a 4Ω load could be driven only to 10 Vpp, approximately.

Gain The gain is $A/(1+AB)$. At the quiescent point that equals 30.84, using the exact value for B. To illustrate the gain stability achieved with negative feedback, the overall voltage gain of the circuit with and without feedback is tabulated at three values of output level at the end of this paragraph. It should be obvious that negative feedback has brought about considerable improvement in the amplifier's characteristics, although in fairness it should be pointed out that the amplifier could have been designed for better open-loop performance, e.g., by using a current source for Q_3's collector load and degenerating its emitter, by using a current source for the differential-pair emitter circuit, etc. Even so, feedback would still make a large improvement.

	Open loop			Closed loop		
V_{out}	−10	0	+10	−10	0	+10
Z_{in}	16k	16k	16k	92k	92k	92k
Z_{out}	20Ω	20Ω	20Ω	0.3Ω	0.3Ω	0.3Ω
Gain	1360	2100	2400	30.60	30.84	30.90

C. Series-feedback pair

Figure 2.92 shows another transistor amplifier with feedback. Thinking of Q_1 as an amplifier of its base-emitter voltage drop (thinking in the Ebers–Moll sense), the feedback samples the output voltage and subtracts a fraction of it from the input signal. This circuit is a bit tricky because Q_2's collector resistor doubles as the feedback network. Applying the techniques we used earlier, one can show that G(open loop) ≈ 200, loop gain ≈ 20, Z_{out}(open

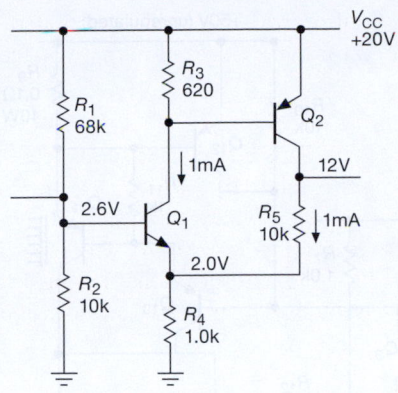

Figure 2.92. Series-feedback pair.

this circuit to ensure stability (i.e., to prevent oscillation), particularly if the output were capacitively bypassed (as it should be), for reasons we will see later in connection with feedback loop stability (§4.9).

We'll see much more of this subject in Chapter 9.

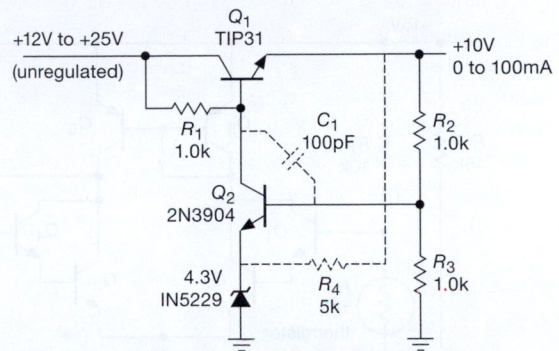

Figure 2.93. Feedback voltage regulator.

loop) $\approx 10k$, Z_{out}(closed loop) $\approx 500\,\Omega$, and G(closed loop) ≈ 9.5.

Exercise 2.22. Go for it!

2.6 Some typical transistor circuits

To illustrate some of the ideas of this chapter, let's look at a few examples of circuits with transistors. The range of circuits we can cover at this point is necessarily limited, because real-world circuits usually incorporate op-amps (the subject of Chapter 4) and other useful ICs – but we'll see plenty of transistors used alongside ICs in those later chapters.

2.6.1 Regulated power supply

Figure 2.93 shows a very common configuration. R_1 normally holds Q_1 on; when the output reaches 10 volts, Q_2 goes into conduction (base at 5 V), preventing further rise of output voltage by shunting base current from Q_1's base. The supply can be made adjustable by replacing R_2 and R_3 with a potentiometer. In this *voltage regulator* (or "regulated dc supply") circuit, negative feedback acts to stabilize the output voltage: Q_2 "looks at" the output and does something about it if the output isn't at the right voltage.

A few details: (a) Adding a biasing resistor R_4 ensures a relatively constant zener current, so that the zener voltage does not change significantly with load current. It is tempting to provide that bias current from the input, but it is far better to use the regulated output. A warning is in order: whenever you use an output voltage to make something happen within a circuit, make sure that the circuit will start up correctly; here, however, there is no problem (why not?). (b) The capacitor C_1 would probably be needed in

2.6.2 Temperature controller

The schematic diagram in Figure 2.94 shows a temperature controller based on a *thermistor* sensing element, a device that changes resistance with temperature. Differential Darlington Q_1–Q_4 compares the voltage of the adjustable reference divider R_4–R_6 with the divider formed from the thermistor and R_2. (By comparing *ratios* from the same supply, the comparison becomes insensitive to supply variations; this particular configuration is called a Wheatstone bridge.) Current mirror Q_5Q_6 provides an active load to raise the gain, and mirror Q_7Q_8 provides emitter current. Q_9 compares the differential amplifier output with a fixed voltage, saturating Darlington $Q_{10}Q_{11}$ (which supplies power to the heater) if the thermistor is too cold. R_9 is a current-sensing resistor that turns on protection transistor Q_{12} if the output current exceeds about 6 amps; that steals base drive from $Q_{10}Q_{11}$, preventing damage. And R_{12} adds a small amount of positive feedback, to cause the heater to snap on and off abruptly; this is the same trick (a "Schmitt trigger") as in Figure 2.13.

2.6.3 Simple logic with transistors and diodes

Figure 2.95 shows a circuit that performs a task we illustrated in §1.9.1F: sounding a buzzer if either car door is open and the driver is seated. In this circuit the transistors all operate as switches (either OFF or saturated). Diodes D_1 and D_2 form what is called an OR gate, turning off Q_1 if either door is open (switch closed). However, the collector of

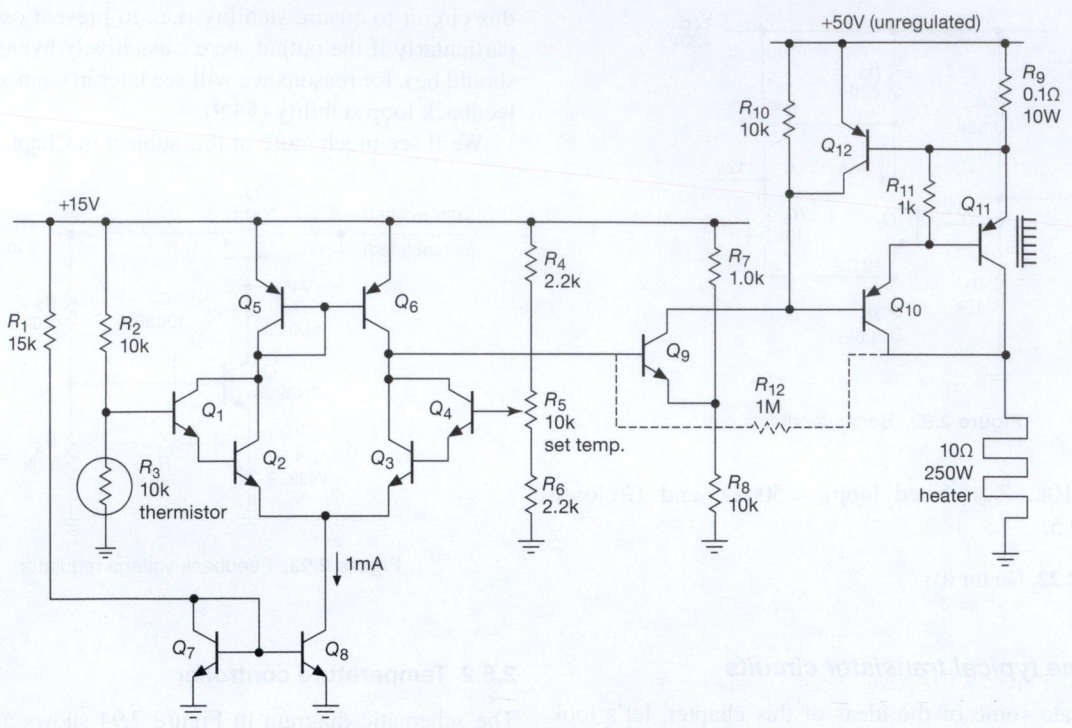

Figure 2.94. Temperature controller for 250 W heater.

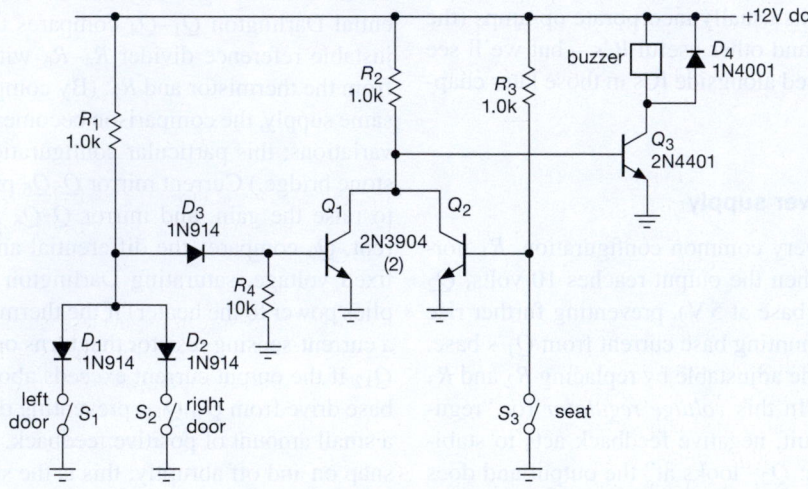

Figure 2.95. Both diodes and transistors are used to make digital logic "gates" in this seat-belt buzzer circuit.

Q_1 stays near ground, preventing the buzzer from sounding unless switch S_3 is also closed (driver seated); in that case R_2 turns on Q_3, putting 12 volts across the buzzer. D_3 provides a diode drop so that Q_1 is OFF with S_1 or S_2 closed, and D_4 protects Q_3 from the buzzer's inductive turn-off transient. In Chapters 10–15 we discuss logic circuitry in detail.

Additional Exercises for Chapter 2

Exercise 2.23. Design a transistor switch circuit that allows you to switch two loads to ground by means of saturated *npn* transistors. Closing switch A should cause both loads to be powered, whereas closing switch B should power only one load. *Hint:* use diodes.

Exercise 2.24. Consider the current source in Figure 2.96. (a) What is I_{load}? What is the output compliance? Assume V_{BE} is 0.6 V. (b) If β varies from 50 to 100 for collector voltages within the output compliance range, how much will the output current vary? (There are two effects here.) (c) If V_{BE} varies according to $\Delta V_{BE} = -0.0001 \, \Delta V_{CE}$ (Early effect), how much will the load current vary over the compliance range? (d) What is the temperature coefficient of output current assuming that β does not vary with temperature? What is the temperature coefficient of output current assuming that β increases from its nominal value of 100 by 0.4%/°C?

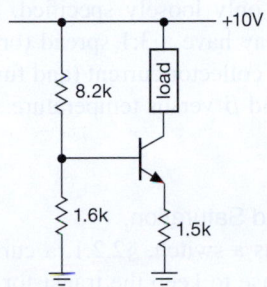

Figure 2.96. Current source exercise.

Exercise 2.25. Design a common-emitter *npn* amplifier with a voltage gain of 15, V_{CC} of +15 V, and I_C of 0.5 mA. Bias the collector at $0.5V_{CC}$, and put the low-frequency 3 dB point at 100 Hz.

Exercise 2.26. Bootstrap the circuit in the preceding problem to raise the input impedance. Choose the rolloff of the bootstrap appropriately.

Exercise 2.27. Design a dc-coupled differential amplifier with a voltage gain of 50 (to a single-ended output) for input signals near ground, supply voltages of ±15 volts, and quiescent currents of 0.1 mA in each transistor. Use a current source in the emitter and an emitter follower output stage.

Exercise 2.28. In this problem you will ultimately design an amplifier whose gain is controlled by an externally applied voltage (in Chapter 3 you will see how to do the same thing with FETs). (a) Begin by designing a long-tailed pair differential amplifier with emitter current source and no emitter resistors (undegenerated). Use ±15 V supplies. Set I_C (each transistor) at 100 μA, and use $R_C = 10$k. Calculate the voltage gain from a single-ended input (other input grounded) to a single-ended output. (b) Now modify the circuit so that an externally applied voltage controls the emitter current source. Give an approximate formula for the gain as a function of controlling voltage. (In a real circuit you might arrange a second set of voltage-controlled current sources

to cancel the quiescent-point shift that gain changes produce in this circuit, or a differential-input second stage could be added to your circuit.)

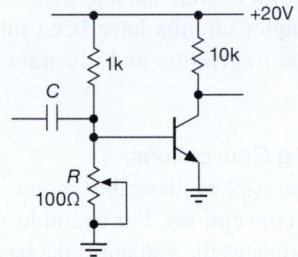

Figure 2.97. Bad biasing.

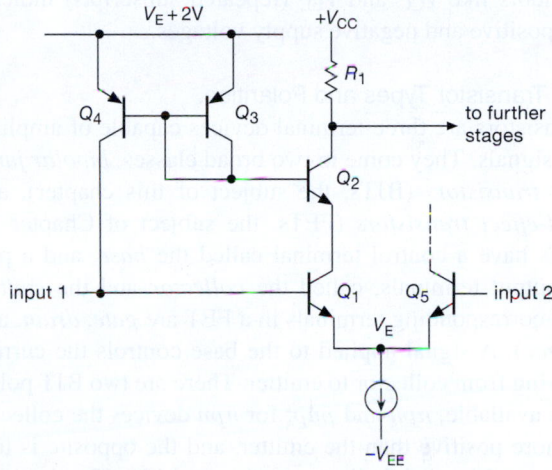

Figure 2.98. Base-current cancellation scheme used in precision operational amplifiers. Bias-current cancellation is discussed in detail in Chapter 4x.

Exercise 2.29. Disregarding the lessons of this chapter, a disgruntled student builds the amplifier shown in Figure 2.97. He adjusts R until the quiescent point is $0.5V_{CC}$. (a) What is Z_{in} (at high frequencies where $Z_C \approx 0$)? (b) What is the small-signal voltage gain? (c) What rise in ambient temperature (roughly) will cause the transistor to saturate?

Exercise 2.30. Several commercially available precision op-amps (e.g., the venerable OP-07) use the circuit in Figure 2.98 to cancel input bias current (only half of the symmetrical-input differential amplifier is shown in detail; the other half works the same way). Explain how the circuit works. *Note*: Q_1 and Q_2 are a beta-matched pair. *Hint*: it's all done with mirrors.

Review of Chapter 2

An A-to-W review of Chapter 2. This review doesn't follow the exact topic order in the chapter: here we first cover transistor theory, then circle back to discuss some applications. In the chapter circuits have been interspersed with theory to provide motivation and illustrate how to use the theory.

¶A. Pin-Labeling Conventions.

The introduction (§2.1) describes some transistor and circuit-labeling conventions. For example, V_B (with a single subscript) indicates the voltage at the base terminal, and similarly I_B indicates current flowing into the base terminal. V_{BE} (two subscripts) indicates base-to-emitter voltage. Symbols like V_{CC} and V_{EE} (repeated subscripts) indicate the positive and negative supply voltages.

¶B. Transistor Types and Polarities.

Transistors are three-terminal devices capable of amplifying signals. They come in two broad classes, *bipolar junction transistors* (BJTs, the subject of this chapter), and *field-effect transistors* (FETs, the subject of Chapter 3). BJTs have a control terminal called the *base*, and a pair of output terminals, called the *collector* and the *emitter* (the corresponding terminals in a FET are *gate*, *drain*, and *source*). A signal applied to the base controls the current flowing from collector to emitter. There are two BJT polarities available, *npn* and *pnp*; for *npn* devices the collector is more positive than the emitter, and the opposite is true for *pnp*. Figure 2.2 illustrates this and identifies intrinsic diodes that are part of the transistor structure, see ¶D and ¶F below. The figure also illustrates that the collector current and the (much smaller) base current combine to form the emitter current.

Operating modes Transistors can operate as *switches* – turned ON or OFF – or they can be used as *linear* devices, for example as amplifiers, with an output current proportional to an input signal. Put another way, a transistor can be in one of three states: *cutoff* (non-zero V_{CE} but zero I_C), *saturated* (non-zero I_C but near-zero V_{CE}), or in the *linear region* (non-zero V_{CE} and I_C). If you prefer prose (and using "voltage" as shorthand for collector-to-emitter voltage V_{CE}, and "current" as shorthand for collector current I_C), the cutoff state has voltage but no current, the saturated state has current but near-zero voltage, and the linear region has both voltage and current.

¶C. Transistor Man and Current Gain.

In the simplest analysis, §2.1.1, the transistor is simply a current amplifier, with a *current gain* called *beta* (symbol β, or sometimes h_{FE}). A current into the base causes

a current β times larger to flow from collector to emitter, $I_C = \beta I_B$, if the external circuit allows it. When currents are flowing, the base-emitter diode is conducting, so the base is ~0.65 V more positive (for *npn*) than the emitter. The transistor doesn't *create* the collector current out of thin air; it simply throttles current from an available supply voltage. This important point is emphasized by our "transistor man" creation (Figure 2.7), a little homunculus whose job is to continuously examine the base current and attempt to adjust the collector's current to be a factor of β (or h_{FE}) times larger. For a typical BJT the beta might be around 150, but beta is only loosely specified, and a particular transistor type may have a 3:1 spread (or more) in specified beta at some collector current (and further 3:1 spreads of β versus I_C and β versus temperature, see for example Figure 2.76).

¶D. Switches and Saturation.

When operated as a switch, §2.2.1, a current must be injected into the base to keep the transistor "ON." This current must be substantially more than $I_B = I_C/\beta$. In practice a value of 1/10th of the maximum expected collector current is common, but you could use less, depending on the manufacturer's recommendations. Under this condition the transistor is in *saturation*, with 25–200 mV across the terminals. At such low collector-to-emitter voltages the base-to-collector diode in Figure 2.2 is conducting, and it robs some of the base-current drive. This creates an equilibrium at the saturation voltage. We'll return in ¶K to look at some circuit examples. See also the discussion of transistor saturation in Chapter *2x*.

¶E. The BJT is a Transconductance Device.

As we point out in §2.1.1, "A circuit that depends on a particular value for beta is a bad circuit." That's because β can vary by factors of 2 to 3 from the manufacturer's nominal datasheet value. A more reliable design approach is to use other highly-predictable BJT parameters that take into account that it is a *transconductance* device. In keeping with the definition of transconductance (an output current proportional to an input voltage), a BJT's collector current, I_C, is controlled by its base-to-emitter voltage, V_{BE}, see §2.3. (We can then rely on $I_B = I_C/\beta$ to estimate the base current, the other way around from the simple approach in ¶C.) The transconductance view of BJTs is helpful in many circumstances (estimating gain, distortion, tempco), and it is essential in understanding and designing circuits such as differential amplifiers and current mirrors. However, in many situations you can circumvent the beta-uncertainty problem with circuit design tricks such as dc feedback or emitter degeneration, without explicitly invoking Ebers–Moll

(¶F). Note also that, just as it would be a bad idea to bias a BJT by applying a base current calculated from I_C/β (from an assumed β), it would be even worse to attempt to bias a BJT by applying a calculated V_{BE} (from an assumed I_s, see ¶F); more on this in ¶Q, below. We might paraphrase this by saying "a circuit that depends on a particular value for I_s, or for operation at a precise ambient temperature, is a bad circuit."

¶F. Ebers–Moll.

Figure 2.41 shows a typical *Gummel plot*, with V_{BE} dictating I_C, and thus an approximate I_B. Equations (2.8) and (2.9) show the exponential (or logarithmic) nature of this relationship. A simple form of the equation, $I_C=I_s\exp(V_{BE}/V_T)$ and its inverse, $V_{BE}=V_T\log_e(I_C/I_s)$, where the constant $V_T=25$ mV at 25°C, reveals that collector current is determined by V_{BE} and a parameter I_s, the latter related to the transistor die size and its current density. I_s is a very small current, typically some 10^{11} times smaller than I_C. The Ebers–Moll formula accurately holds for the entire range of silicon BJT types, for example those listed in Table 8.1. The integrated-circuit (IC) industry relies on Ebers–Moll for the design of their highly-successful BJT linear circuits.

¶G. Collector Current versus Base Voltage: Rules of Thumb.

See §2.3.2. It's useful to remember a few rules of thumb, which we can derive from Ebers–Moll: I_C increases by a factor of ten for a ≈60 mV increase in V_{BE}; it doubles for an ≈18 mV V_{BE} increase, and it increases 4% for a 1 mV V_{BE} increase.

¶H. Small Signals, Transconductance and r_e.

See §2.3.2B. It's convenient to assume operation at fixed I_C, and look for the effect of small changes ("small signals"). First, thinking about the rules of thumb above, we can calculate (eq'n 2.13,) the transconductance, $g_m=\partial I_C/\partial V_{BE}=I_C/V_T$. This evaluates to $g_m=40$ mS at 1 mA, with g_m proportional to current. To put it another way, we can assign an effective internal resistance r_e in series with the emitter, $r_e=1/g_m=V_T/I_C$, see eq'n 2.12. (The small r indicates *small signal*.) A useful fact to memorize: r_e is about 25 Ω at a collector current of 1 mA, and it scales inversely with current.

¶I. Dependence on Temperature.

See §2.3.2C. In ¶F we said $V_T=25$ mV at 25°C, which suggests it's not exactly a constant, but changes with temperature. Because $V_T=kT/q$ (§2.3.1), you might guess that V_{BE} is proportional to absolute temperature, thus a temper-ature coefficient of about +2mV/°C (because $V_{BE}\approx600$ mV at $T=300$K). But the scaling parameter I_s has a large opposite tempco, producing an overall tempco of about -2.1 mV/°C. Memorize this fact also! Because V_T is proportional to absolute temperature, the tempco of transconductance at fixed collector current is inversely proportional to absolute temperature (recall $g_m=I_C/V_T$), and thus drops by about 0.34%/°C at 25°C.

¶J. Early Effect.

See §2.3.2D. In our simple understanding so far, base voltages (or currents) "program" a BJT's collector current, independent of collector voltage. But in reality I_C increases slightly with increasing V_{CE}. This is called the *Early effect*, see eq'n 2.14 and Figure 2.59, which can be characterized by an *Early voltage* V_A, a parameter independent of operating current; see eq'n 2.15. If the Early voltage is low (a common drawback of *pnp* transistors) the effect can be quite large. For example, a *pnp* 2N5087 with $V_A=55$ V has $\eta=4\times10^{-4}$, and would experience a 4 mV shift of V_{BE} with a 10 V change of V_{CE}; if instead the base voltage were held constant, a 10 V increase of collector voltage would cause a 17% increase of collector current. We hasten to point out there are circuit configurations, such as *degeneration*, or the *cascode*, that alleviate the Early effect. For more detail see the discussion in Chapter 2*x*.

Circuit Examples

With this summary of basic BJT theory, we now circle back and review some circuit examples from Chapter 2. One way to review the circuits is to flip through the chapter looking at the pictures (and reading the captions), and refer to the associated text wherever you are uncertain of the underlying principles.

¶K. Transistor Switches.

BJT switches are discussed in §2.2.1, and circuit examples appear in Figures 2.9 (driving an LED), 2.10 (high-side switching, including level shifting), and 2.16 (with an emitter-follower driver). Simply put, you arrange to drive a current into the base to put the transistor into solid saturation for the anticipated collector load current (i.e., $I_B\gg I_C/\beta$), bringing its collector within tens of millivolts of the emitter. More like this appears in Chapter 12 (Logic Interfacing). Looking forward, the use of *MOSFET* switches often provide a superior switching solution (§§3.4.4 and 3.5); their control terminal (the gate) conveniently requires *no* static gate current, though you may have to provide significant transient currents to charge its gate capacitance during rapid switching.

¶L. Transistor Pulsers.

Basic timer and pulse generator circuits are shown in Figures 2.11 (pulse from a step) and 2.12 (pulse from a pulse). These are simple, but not terribly accurate or stable; better to use a dedicated timer or pulse generator IC, see §7.2.

¶M. Schmitt Trigger.

A *Schmitt trigger* is a threshold level-detecting circuit (Figure 2.13) with hysteresis to prevent multiple transitions when noisy input signals go though the threshold(s). Although you can make a Schmitt trigger circuit with discrete transistors, good design practice favors the use of dedicated *comparator* ICs, see §§4.3.2 and 12.3.

¶N. Emitter Follower.

The emitter follower is a linear amplifier with an ideal voltage gain of unity, see §2.2.3. The beta of the transistor increases the follower's input impedance and reduces its output impedance, see §2.2.3B and eq'n 2.2. There's more detail in §2.3.3 and Figure 2.43, where the effect of the intrinsic emitter resistance r_e is taken into account. In simplified form $R_{out} = r_e + R_s/\beta$, where R_s is the signal source resistance seen at the base. The dc output voltage is offset from the dc input by V_{BE}, about 0.6 V to 0.7 V, unless a cancelling circuit is used, see §2.2.3D and Figure 2.29. Emitter followers are also used as voltage regulators, see §2.2.4 and Figures 2.21 and 2.22. A precision alternative is the *op-amp follower*, see §4.2.3 in Chapter 4.

¶O. Current Source (or Current Sink).

In contrast to the familiar *voltage source* (which delivers a constant voltage regardless of load current, think of a battery), a *current source* delivers a constant current regardless of the load's voltage drop, see §2.2.6 and Figure 2.31; there's no everyday "battery equivalent." Transconductance devices like BJTs, with their relatively constant collector output currents, are natural candidates for making current sources. For the simplest current source, the base is biased with a voltage, say V_b, with respect to a reference point (often ground), and the emitter is connected through a resistor to the same reference. For an *npn* transistor with ground reference the output (sinking) current will be $I_C = (V_b - V_{BE})/R_E$, see Figure 2.32. For better stability and predictability the V_{BE} term can be cancelled, see Figure 2.33. The operating voltage range of a current source is called its *compliance range*, set on the low end by collector saturation, and on the high end by the transistor's breakdown voltage or by power-dissipation issues. Current sources are frequently created using current-mirror circuits, see ¶P below. Precise and stable current sources can

be made with op-amps (§4.2.5); there are also dedicated current-source integrated circuits (§9.3.14).

¶P. Current Mirrors.

A current mirror (§2.3.7) is a three-terminal current-source circuit that generates an output current proportional to an input "programming" current. In a typical configuration (Figures 2.55 and 2.58) the mirror attaches to a dc rail (or to ground), reflecting the programming current, the latter perhaps set by a resistor. The circuit often omits any emitter resistors, thus achieving compliance to within a fraction of a volt of the rail. Ordinarily you wouldn't attempt to apply exactly the right V_{BE} to generate a prescribed I_C (à la Ebers–Moll); but that's exactly what you're doing here. The trick is that one transistor (Q_1) of the matched pair inverts Ebers–Moll, creating from the programming current I_P exactly the right V_{BE} to re-create the same current in the output transistor Q_2. Cute!

These circuits assume matched transistors, such as you would find inside an IC (recall from ¶G that even a 1 mV difference of V_{BE} produces a 4% change of current). Figure 2.62 graphs base-emitter voltage difference versus collector current ratio, $\Delta V_{BE} = V_T \log_e(I_{C2}/I_{C1})$. You can exploit this effect to generate a "ratio mirror," as discussed in Chapter *2x*.

As nice as it looks, the basic current mirror of Figure 2.55 suffers from Early-effect change of output current when the output voltage changes. The effect is particularly serious with *pnp* transistors: in the example of a 2N5087 in ¶J above, the 4 mV change of V_{BE} (for a 10 V output change) would cause a 17% current error. One solution (Figure 2.60) is to add emitter degeneration resistors, at the expense both of compliance near the reference rail and of dynamic range. A more elegant solution is the Wilson mirror (Figure 2.61), which defeats Early effect by exploiting the ever-useful *cascode* configuration (Figure 2.84B). Cascode transistor Q_3 passes output transistor Q_2's collector current to the load, while Q_2 operates with a fixed V_{CE} of one diode drop (its own V_{BE}). The Wilson mirror's ingenious configuration also cancels base-current errors (an ordinary mirror with BJTs having $\beta = 100$ has a current error of 2%). Degeneration resistors can be added, as shown in circuit B, for additional suppression of Early effect, but they would be omitted in a "pure Wilson mirror." Linear ICs are full of Wilson mirrors. See Chapter *2x* for a discussion of *bipolarity* current mirrors.

¶Q. Common-Emitter Amplifiers.

See §§2.2.7 and 2.3.4, and Figures 2.35, 2.48 and 2.50. The simplest form of BJT amplifier has a grounded emitter, a load resistor R_L from the collector to a supply V_+, and a dc bias plus a small signal voltage applied to the base. The gain is $G_V = -R_L/r_e$. If the base bias is carefully set so that the collector current pulls the collector halfway to ground, then $I_C = V_s/2R_L$, $r_e = V_T/I_C = 2R_L V_T/V_s$, and so the voltage gain (recall $V_T \approx 25$ mV) is $G_V = -20V_s$, where V_s is in units of volts. For $V_s = 20$ V, for example, the voltage gain is -400.

That's a lot of gain! Unless the signals are small, however, there's a serious problem: the gain is inverse in r_e, thus proportional to I_C. But the latter changes as the output voltage swings up and down, producing first-order changes in gain, with resulting severe distortion (Figure 2.46). This can be alleviated (at the expense of gain) by adding *emitter degeneration* in the form of an emitter resistor R_E. The gain is then $G_V = -R_L/(R_E + r_e)$, with greatly reduced effect of varying r_e; see Figure 2.47, where emitter degeneration was added to reduce the gain by a factor of ten ($R_E = 9r_e$). This is also a form of negative feedback, see §2.3.4B and ¶W below. You can think of this circuit as a classic current source (¶O) driving a resistor as load; the voltage gain is the current source's transconductance multiplied by the load resistance, $G_V = g_m R_L$, where $g_m = -1/r_e$.

We've sidestepped the important issue of setting the base bias voltage to produce the desired quiescent collector current. But we don't know the appropriate voltage V_{BE}, and a small change has a big effect, see ¶G above (e.g., a 60 mV uncertainty in V_{BE}, which is about what you might encounter from different batches of a given transistor, produces a $10\times$ error in I_C!). There are many circuit solutions, see §2.3.5, but the simplest involves adding emitter degeneration at dc, bypassed as necessary to produce higher gain at signal frequencies (Figure 2.50 and 2.51). Another approach is to use a matching transistor to set the bias, analogous to the current mirror (Figure 2.52); this method is inherent in the widely-used *differential amplifier* (Figure 2.65). A third approach is to exploit feedback to set the bias (Figures 2.53 and 2.54), a method that figures centrally in op-amp circuits (Chapter 4).

¶R. Differential Amplifiers.

The differential amplifier (§2.3.8) is a symmetrical configuration of two matched transistors, used to amplify the difference of two input signals. It may include emitter degeneration (Figure 2.64), but need not (Figure 2.65). For best performance the emitter pulldown resistor is replaced by a current source, and (for highest gain) the resistive collector load is replaced by a current mirror (Figure 2.67). Differential amplifiers should reject strongly any common-mode input signal, achieving a good common-mode rejection ratio (CMRR, the ratio G_{diff}/G_{CM}). Differential amplifiers can be used to amplify single-ended input signals (ground

the other input), where the inherent cancellation of V_{BE} offsets allows accurate dc performance (§2.3.8B). Ordinarily you use only one output from a differential amplifier; that is, it is used to to convert a balanced input to a single-ended output. But you can use both outputs (a "fully-differential amplifier," §5.17) to drive a balanced load, or to create a pair of signals $180°$ out of phase (a *phase splitter*). See also the sections on the emitter-input differential amplifier and on BJT amplifier distortion in Chapter *2x*, and §5.13–§5.16 (precision differential and instrumentation amplifiers).

¶S. Comparators.

A differential amplifier with lots of gain G_{diff} is driven into differential saturation with a small differential input (§2.3.8E). For example, just a few millivolts of input difference is adequate to saturate the output if $G_{diff} = 1000$ (easily accomplished with a current-mirror collector load). When operated in this way, the differential amplifier is a voltage *comparator*, a circuit used widely to sense thresholds or compare signal levels; it's the basis of analog-to-digital conversion, and figures importantly in Chapter 12 (see §12.3 and Tables 12.1 and 12.2).

¶T. Push–Pull Amplifiers.

A single transistor conducts in one direction only (e.g., an *npn* transistor can only sink current from its collector, and source current from its emitter). That makes it awkward to drive a heavy load with alternating polarity (e.g., a loudspeaker, servomotor, etc.), although it can be done, wastefully, with a single-ended stage ("class-A") with high quiescent current, see Figure 2.68. The push-pull configuration uses a pair of transistors connected to opposite supply rails (§2.4.1), an arrangement that can supply large output currents of either polarity with little or no quiescent current. Figure 2.69 shows a push-pull follower with complementary polarities, and with zero quiescent current ("class-B"); this produces some crossover distortion, which can be eliminated by biasing the pair into quiescent conduction ("class-AB," Figure 2.71),. The output transistors can be beta-boosting configurations like the Darlington or Sziklai (¶U), see for example Figure 2.78. The push-pull configuration is widely used in logic circuits (see Figure 10.25), gate driver ICs (see Figure 3.97), and in combination with op-amps to deliver greater output currents (see Figure 4.26).

¶U. The Darlington and Sziklai Connections.

These simple combinations of two transistors create a 3-terminal equivalent transistor with $\beta = \beta_1\beta_2$. The Darlington (Figures 2.74 and 2.75) cascades two transistors of the same polarity and has a base-emitter drop of $2V_{BE}$; the Sziklai (Figure 2.77) pairs opposite polarities, and has a single base-emitter drop (which is only weakly dependent on output current, thanks to R_B). For either configuration a resistor R_B should be connected across the output transistor's base-emitter terminals. For more about this subject see see the discussion in Chapter *2x*.

¶V. Miller Effect.

Like all electronic components, transistors have inter-terminal capacitances, designated (by terminal pairs) C_{be}, C_{ce}, and C_{cb}.[72] While C_{be} and C_{ce} slow the input and output waveforms by creating lowpass filters with the source and load resistances, the effect of the feedback capacitance C_{cb} is more insidious: it creates an additional input capacitance to ground equal to C_{cb} multiplied by the stage's inverting voltage gain, thus its effective input capacitance becomes $C_{eff}=(G_V+1)C_{cb}$. This is the infamous *Miller effect* (§2.4.5B), whose impact can be devastating in high-speed and wideband amplifiers. Some circuit solutions include

---

[72] These have many aliases (a common set uses initials for "in" and "out" instead of "base" and "collector," thus C_{ie}, C_{oe}, and C_{ob}, respectively), see the section on BJT bandwidth in Chapter *2x*.

the grounded-base amplifier, the differential amplifier, and the cascode configuration (see the discussion of cascode in Chapter *2x*).

¶W. Negative Feedback.

If there were a Nobel prize for grand-concepts-in-circuit-design, it would surely go to Harold Black for his elegant elucidation of *negative feedback*. In its simplest form, it consists of subtracting, from the input signal, a fraction B of an amplifier's output signal V_{out} (Figure 2.85). If the amplifier's open-loop gain is A, then the closed-loop gain becomes (eq'n 2.16) $G_{cl}=A/(1+AB)$. The quantity AB, which generally is large compared with unity, is called the *loop gain*, and it (more precisely the quantity $1+AB$) is the multiplier by which negative feedback improves the amplifier's performance: improved linearity and constancy of gain, and (in this *series feedback* circuit configuration) raised input impedance and lowered output impedance; see §2.5.3.

Feedback is the essence of linear design, and it is woven deeply into the DNA of op-amp circuits (the subject of Chapter 4), and power circuits (Chapter 9). With negative feedback you can make amplifiers with 0.0001% distortion, voltage sources with 0.001 Ω output impedance, and many other wonders too magnificent here to relate. Stay tuned. Better yet, *read on!*

FIELD-EFFECT TRANSISTORS

3.1 Introduction

Field-effect transistors (FETs) are different from the bipolar transistors[1] that we talked about in the last chapter. Broadly speaking, however, they are similar devices, which we might call *charge-control devices*: in both cases (Figure 3.1) we have a three-terminal device in which the conduction between two electrodes depends on the availability of charge carriers, which is controlled by a voltage applied to a third *control electrode*.

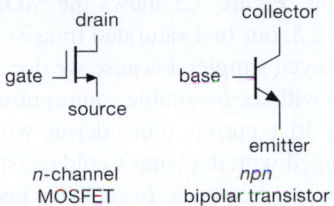

Figure 3.1. The *n*-channel MOSFET and its *npn* transistor analog.

Here's how they differ: in a bipolar transistor the collector–base junction is back-biased, so no current normally flows. Forward-biasing the base–emitter junction by ≈ 0.6 V overcomes its diode "contact potential barrier," causing electrons to enter the base region, where they are strongly attracted to the collector. Although some base current results, most of these "minority carriers" are captured by the collector. This results in a collector current, controlled by a (much smaller) base current. The collector current is proportional to the rate of injection of minority carriers into the base region, which is an exponential function of V_{BE} (the Ebers–Moll equation). You can think of a bipolar transistor as a current amplifier (with roughly constant current gain β) or as a transconductance device (Ebers–Moll: collector current programmed by base-emitter *voltage*).

In an FET, as the name suggests, conduction in a *channel* is controlled by an *electric field*, produced by a voltage applied to the *gate* electrode. There are no forward-biased junctions, so the gate draws no current. This is per-

haps the most important advantage of the FET. As with BJTs, there are two polarities, *n-channel* FETs (conduction by electrons) and *p-channel* FETs (conduction by holes). These two polarities are analogous to the familiar *npn* and *pnp* bipolar transistors, respectively. In addition, however, FETs tend to be confusing at first because they can be made with two different kinds of gates (thus JFETs and MOSFETs) and with two different kinds of channel doping (leading to *enhancement* and *depletion* modes). We'll sort out these possibilities shortly.

First, though, some motivation and perspective. The FET's nonexistent gate current is its most important characteristic. The resulting high input impedance (which can be greater than $10^{14}\ \Omega$) is essential in many applications, and in any case it makes circuit design simple and fun. For applications like analog switches and amplifiers of ultra-high input impedance, FETs have no equal. They can be easily used by themselves or combined with bipolar transistors to make integrated circuits. In the next chapter we'll see how successful that process has been in making nearly perfect (and wonderfully easy to use) *operational amplifiers*, and in Chapters 10–14 we'll see how digital electronics has been revolutionized by MOSFET integrated circuits. Because many FETs using very low current can be constructed in a small area, they are especially useful for very large-scale integration (VLSI) digital circuits such as microprocessors, memory, and "application-specific" chips of the sort used in cellphones, televisions, and the like. At the other end of the spectrum, robust high-current MOSFETs (50 amps or more) have replaced bipolar transistors in many applications, often providing simpler circuits with improved performance.

3.1.1 FET characteristics

Beginners sometimes become catatonic when directly confronted with the confusing variety of FET types. That variety arises from the combined choices of polarity (*n-channel* or *p-channel*), form of gate insulation [semiconductor *junction* (JFET) or oxide *insulator* (MOSFET)], and channel doping (*enhancement* or *depletion* mode). Of the

---

[1] Often called BJTs, for "bipolar junction transistors," to distinguish them from FETs.

131

eight resulting possibilities, six *could* be made, and five actually are. Four of those five are of major importance.

It will aid understanding (and sanity), however, if we begin with one type only, just as we did with the *npn* bipolar transistor. Once comfortable with FETs, we'll have little trouble with their family tree.

A. FET V–I curves

Let's look first at the *n*-channel enhancement-mode MOSFET, which is analogous to the *npn* bipolar transistor (Figure 3.2). In normal operation the drain (~collector) is more positive than the source (~emitter). No current flows from drain to source unless the gate (~base) is brought positive with respect to the source. Once the gate is thus "forward-biased," there will be drain current, all of which flows to the source. Figure 3.2 shows how the drain current I_D varies with drain-source voltage V_{DS} for a few values of controlling gate-source voltage V_{GS}. For comparison, the corresponding "family" of curves of I_C versus V_{BE} for an ordinary *npn* bipolar transistor is shown. Evidently there are a lot of similarities between *n*-channel MOSFETs and *npn* bipolar transistors.

Like the *npn* transistor, the FET has a high incremental drain impedance, giving roughly constant current for V_{DS} greater than a volt or two. By an unfortunate choice of language, this is called the "saturation" region of the FET (a better term is "current saturation") and corresponds to the "active" region of the bipolar transistor. Analogous to the bipolar transistor, a larger gate-to-source bias produces a larger drain current. And, analogous to bipolar transistors, FETs are not perfect transconductance devices (constant drain current for constant gate-source voltage): just as the ideal Ebers–Moll transconductance characteristic of bipolar transistors is degraded by the Early effect (§2.3.2D and §2x.8), there's an analogous departure from the ideal transconductance behavior for FETs, characterized by a finite drain output resistance r_o (more usually called $1/g_{os}$, see §3.3.2 and §3x.4).

So far, the FET looks just like the *npn* transistor. Let's look closer, though. For one thing, over the normal range of currents the saturation drain current increases rather modestly with increasing gate voltage (V_{GS}). In fact, it is approximately proportional to $(V_{GS} - V_{th})^2$, where V_{th} is the gate threshold voltage at which drain current begins ($V_{th} \approx 1.63V$ for the FET in Figure 3.2); compare this mild quadratic law with the steep exponential transistor law, as given to us by Ebers and Moll. Second, there is *zero* dc gate current, so you mustn't think of the FET as a device with current gain (which would be infinite). Instead, think of the FET as a transconductance device, with gate-source

voltage programming the drain current, as we did with the bipolar transistor in the Ebers–Moll treatment. Recall that the transconductance g_m is simply the ratio i_d/v_{gs} (using the convention of lowercase letters to indicate "small-signal" changes in a parameter; e.g., $i_d/v_{gs} = \delta I_D/\delta V_{GS}$). Third, the gate of a MOSFET is truly insulated from the drain-source channel; thus, unlike the situation for bipolar transistors (or JFETs, as we'll see), you can bring it positive (or negative) at least 10 V or more without worrying about diode conduction. Finally, the FET differs from the bipolar transistor in the so-called *linear* (low-voltage) region of the graph, where it behaves rather accurately like a resistor, *even for negative* V_{DS}; this turns out to be quite useful because the equivalent drain-source resistance is, as you might guess, programmed by the gate-source voltage.

B. Two examples

FETs will have more surprises in store for us. But before getting into more details, let's look at two simple switching applications. Figure 3.3 shows the MOSFET equivalent of Figure 2.5, our first saturated transistor switch. The FET circuit is even simpler, because we don't have to concern ourselves with the inevitable compromise of providing adequate base drive current (considering worst-case minimum β combined with the lamp's cold resistance) without squandering excessive power. Instead, we just apply a full-swing dc voltage drive to the cooperative high-impedance gate. As long as the switched-on FET behaves like a resistance that is small compared with the load, it will bring its drain close to ground; typical power MOSFETs have $R_{ON} < 0.1\,\Omega$, which is fine for this job.

We demonstrate this circuit in our electronics course, but we put a resistor in series with the gate. The students are surprised when they discover its resistance – $10\,\mathrm{M}\Omega$ – which implies a "beta" of at least 100,000. They are even more surprised when they notice that the light stays on when the gate is then open-circuited: the gate voltage is held on the gate's capacitance, and will stay that way for the rest of the hour's lecture.[2] That implies that the gate current is well below a *pico*ampere!

Figure 3.4 shows an "analog switch"[3] application, which cannot be done at all with bipolar transistors. The idea here is to switch the conduction of a FET from open-circuit (gate reverse-biased) to short-circuit (gate forward-biased), thus blocking or passing the analog signal (we'll see plenty of reasons to do this sort of thing later). In this

---

[2] The gate capacitance "remembers" whatever voltage was last applied. So you can have it stay on, stay off, or even stay at half-brightness, with no noticeable change even with the gate floating.

[3] Also called a "linear switch."

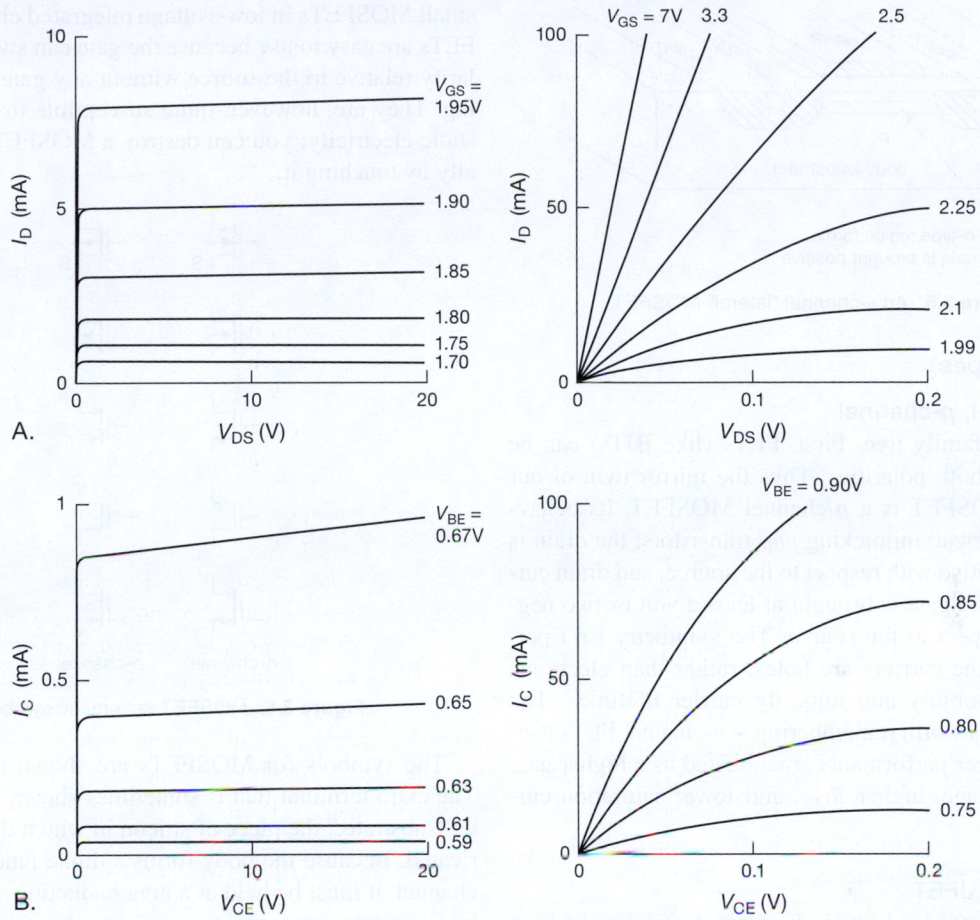

Figure 3.2. Measured MOSFET/transistor characteristic curves: A. VN0106 (similar to the popular 2N7000) *n*-channel MOSFET: I_D versus V_{DS} for various values of V_{GS}. B. 2N3904 *npn* bipolar transistor: I_C versus V_{CE} for various values of V_{BE}.

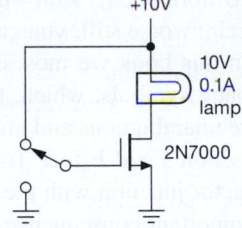

Figure 3.3. MOSFET power switch.

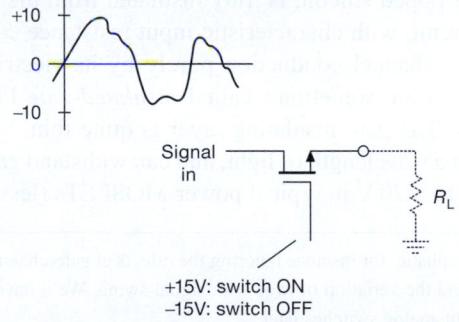

Figure 3.4. MOSFET analog (signal) switch.

case we just arrange for the gate to be driven more negative than any input signal swing (switch *open*), or a few volts more positive than any input signal swing (switch *closed*). Bipolar transistors aren't suited to this application, because the base draws current and forms diodes with the emitter and collector, producing awkward clamping action. The MOSFET is delightfully simple by comparison, need-

ing only a voltage swing into the (essentially open-circuit) gate.[4]

---

[4] It's only fair to mention that our treatment of this circuit has been some-

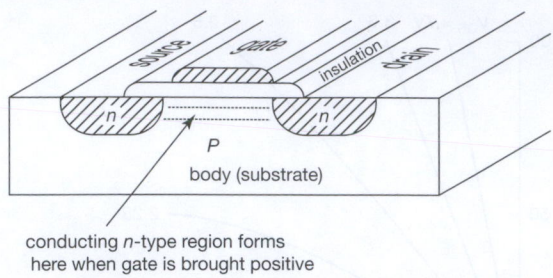

conducting *n*-type region forms
here when gate is brought positive

Figure 3.5. An *n*-channel "lateral" MOSFET.

3.1.2 FET types

A. *n*-channel, *p*-channel

Now for the family tree. First, FETs (like BJTs) can be
fabricated in both polarities. Thus the mirror twin of our
n-channel MOSFET is a *p*-channel MOSFET. Its behavior is symmetrical, mimicking *pnp* transistors: the drain is
normally negative with respect to the source, and drain current flows if the gate is brought at least a volt or two negative with respect to the source. The symmetry isn't perfect because the carriers are holes, rather than electrons,
with lower mobility and minority carrier lifetime.[5] The
consequence is worth remembering – *p*-channel FETs usually have poorer performance, manifested as a higher gate
threshold voltage, higher R_{ON}, and lower saturation current.[6]

B. MOSFET, JFET

In a MOSFET ("Metal-Oxide-Semiconductor Field-Effect
Transistor") the gate region is separated from the conducting channel by a thin layer of SiO_2 (glass) grown onto
the channel (Figure 3.5). The gate, which may be either
metal or doped silicon, is truly insulated from the source–
drain circuit, with characteristic input resistance $>10^{14}\,\Omega$.
It affects channel conduction purely by its electric field.
MOSFETs are sometimes called *insulated-gate* FETs, or
IGFETs. The gate insulating layer is quite thin, typically
less than a wavelength of light, and can withstand gate voltages up to $\pm20\,V$ in typical power MOSFETs (less for the

---

what simplistic, for instance ignoring the effects of gate-channel capacitance and the variation of R_{ON} with signal swing. We'll have more to
say about analog switches later.

[5] These are semiconductor parameters of importance in transistor performance.

[6] In the case of so-called "complementary pairs" (an *n*-channel and a *p*-
channel part with similar voltage and current ratings), the *p*-channel
part is usually built with a larger area in order to match the performance
of the *n*-channel part. You can see the evidence in the datasheet in the
form of greater capacitance for the *p*-channel part.

small MOSFETs in low-voltage integrated circuits). MOS-
FETs are easy to use because the gate can swing either polarity relative to the source without any gate current flowing. They are, however, quite susceptible to damage from
static electricity; you can destroy a MOSFET device literally by touching it.

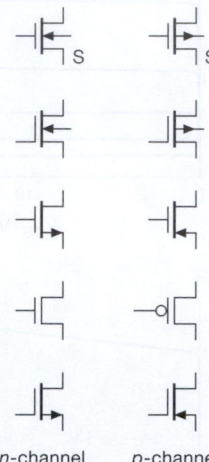

n-channel   *p*-channel

Figure 3.6. MOSFET schematic symbols.

The symbols for MOSFETs are shown in Figure 3.6.
The extra terminal that is sometimes shown is the "body,"
or "substrate," the piece of silicon in which the FET is fabricated. Because the body forms a diode junction with the
channel, it must be held at a nonconducting voltage. It can
be tied to the source or to a point in the circuit more negative (positive) than the source for *n*-channel (*p*-channel)
MOSFETs. It is common to see the body terminal omitted; furthermore, engineers often use the symbol with the
symmetrical gate. Unfortunately, with what's left you can't
tell source from drain; worse still, you can't tell *n*-channel
from *p*-channel! In this book we most often use the bottom pair of schematic symbols, which, though somewhat
unconventional, are unambiguous and uncluttered.[7]

In a JFET (Junction Field-Effect Transistor) the gate
forms a semiconductor junction with the underlying channel. This has the important consequence that *a JFET gate
should not be forward biased with respect to the channel,
to prevent gate current*. For example, diode conduction
will occur as the gate of an *n*-channel JFET approaches
$+0.6\,V$ with respect to the more negative end of the channel (which is usually the source). The gate is therefore operated reverse-biased with respect to the channel, and no

---

[7] In current practice, logic designers like to use the second pair up from
the bottom, while power MOSFET users prefer the second pair down
from the top.

current (except diode leakage) flows in the gate circuit. The circuit symbols for JFETs are shown in Figure 3.7. Once again, we favor the symbol with offset gate to identify the source (though JFETs and small integrated MOSFETs are symmetrical, power MOSFETs are quite asymmetrical, with very different capacitances and breakdown voltages).

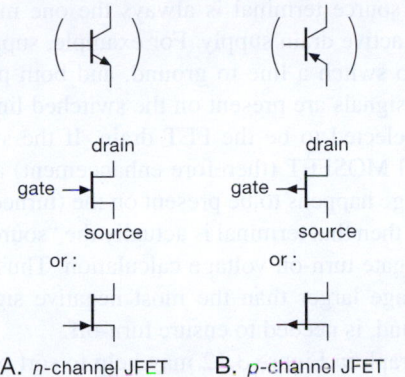

Figure 3.7. JFET schematic symbols: A. *n*-channel JFET. B. *p*-channel JFET.

C. Enhancement, depletion

The *n*-channel MOSFETs with which we began the chapter were nonconducting, with zero (or negative) gate bias, and were driven into conduction by bringing the gate positive with respect to the source. This kind of FET is known as *enhancement mode*. The other possibility is to manufacture the *n*-channel FET with the channel semiconductor "doped" so that there is plenty of channel conduction even with zero gate bias, and the gate must be reverse-biased by a few volts to cut off the drain current. Such a FET is known as *depletion mode*. MOSFETs can be made in either variety, because the gate, being insulated from the channel, can swing either polarity. But JFETs, with their gate-channel diode, permit only reverse gate bias, and therefore are made only in depletion mode.

A graph of drain current versus gate-source voltage, at a fixed value of drain voltage, may help clarify this distinction (Figures 3.8 and 3.9). The enhancement-mode device draws no drain current until the gate is brought positive (these are *n*-channel FETs) with respect to the source, whereas the depletion-mode device is operating at nearly its maximum value of drain current when the gate is at the same voltage as the source. In some sense the two categories are artificial, because the two curves are identical except for a shift along the V_{GS} axis. In fact, it is possible to manufacture "in-between" MOSFETs. Nevertheless,

the distinction is an important one when it comes to circuit design.

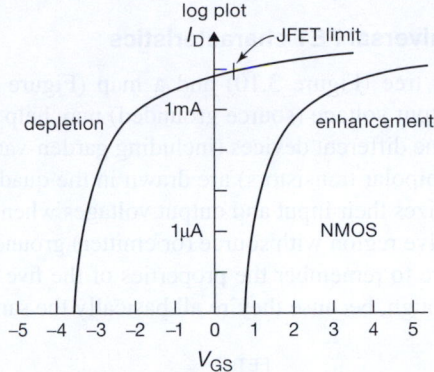

Figure 3.8. Transfer characteristics (I_D versus V_{GS}) for a JFET (depletion-mode) and a MOSFET (enhancement-mode) transistor. See also the measured curves in Figure 3.19.

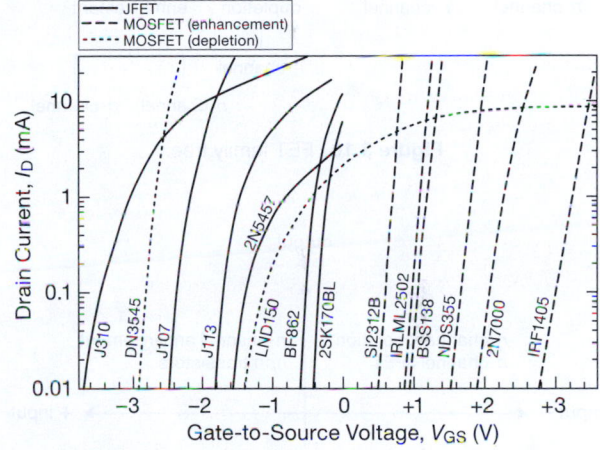

Figure 3.9. Lending some authenticity to Figure 3.8's notional sketch: measured I_D versus V_{GS} for a selection of *n*-channel FETs.

Note that JFETs are always depletion-mode devices and that the gate cannot be brought more than about 0.5 V more positive (for *n*-channel) than the source, since the gate-channel diode will conduct. MOSFETs *can* be either enhancement or depletion, but in practice the dominant species is enhancement, with a sprinkling of depletion-mode MOSFETs.[8] Most of the time, then, you need worry only about (a) depletion-mode JFETs and

---

[8] In the form of *n*-channel GaAs FETs, "dual-gate" cascodes for radiofrequency applications, and a selection of high-voltage depletion-mode power MOSFETs (such as the Supertex lateral LND150 or vertical DN3435, as well as offerings by six other manufacturers).

(b) enhancement-mode MOSFETs. Each come in the two polarities, *n*-channel and *p*-channel.

3.1.3 Universal FET characteristics

A family tree (Figure 3.10) and a map (Figure 3.11) of input–output voltage (source grounded) may help simplify things. The different devices (including garden-variety *npn* and *pnp* bipolar transistors) are drawn in the quadrant that characterizes their input and output voltages when they are in the active region with source (or emitter) grounded. You don't have to remember the properties of the five kinds of FETs, though, because they're all basically the same.

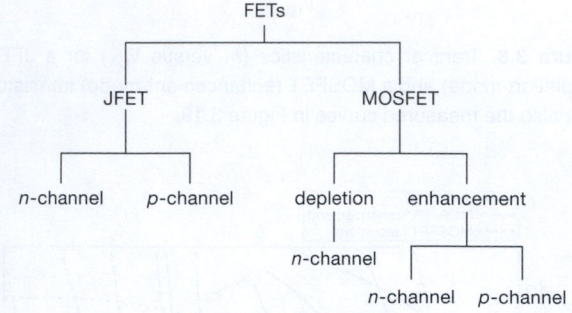

Figure 3.10. FET family tree.

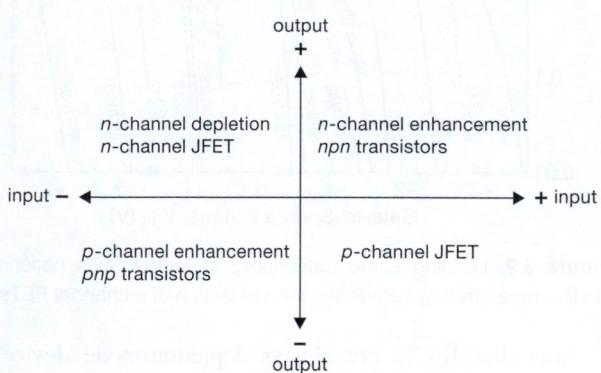

Figure 3.11. Transistor "polarity map."

First, with the source grounded, a FET is turned on (brought into conduction) by bringing the gate voltage "toward" the active drain supply voltage. This is true for all five types of FETs, as well as the bipolar transistors. For example, an *n*-channel JFET (which is necessarily depletion mode) uses a positive drain supply, as do all *n*-type devices. Thus a positive-going gate voltage tends to turn on the JFET. The subtlety for depletion-mode devices is that the gate must be (negatively) back-biased for zero drain

current, whereas for enhancement-mode devices zero gate voltage is sufficient to give zero drain current.

Second, because of the near symmetry of source and drain, either terminal can act as the effective source (exception: not true for power MOSFETs, in which the body is internally connected to the source). When thinking of FET action, and for purposes of calculation, remember that the effective source terminal is always the one most "away" from the active drain supply. For example, suppose a FET is used to switch a line to ground, and both positive and negative signals are present on the switched line, which is usually selected to be the FET drain. If the switch is an *n*-channel MOSFET (therefore enhancement) and a negative voltage happens to be present on the (turned-off) drain terminal, then that terminal is actually the "source" for purposes of gate turn-on voltage calculation. Thus a negative gate voltage larger than the most negative signal, rather than ground, is needed to ensure turn-off.

The graph in Figure 3.12 may help to sort out all these confusing ideas. Again, the difference between enhancement and depletion is merely a question of displacement along the V_{GS} axis, i.e., whether there is a lot of drain current or no drain current at all when the gate is at the same potential as the source. The *n*-channel and *p*-channel FETs are complementary in the same way as *npn* and *pnp* bipolar transistors.

In Figure 3.12 we have used standard symbols for the important FET parameters of saturation current and cutoff voltage. For JFETs the value of drain current with the gate shorted to source is specified on the datasheets as I_{DSS} and is nearly the maximum drain current possible. (I_{DSS} means current from <u>d</u>rain to <u>s</u>ource with the gate <u>s</u>horted to the source. Throughout the chapter you will see this notation, in which the first two subscripted letters designate the pair of terminals and the third specifies the condition.) For enhancement-mode MOSFETs the analogous specification is $I_{D(ON)}$, given at some forward gate voltage ("I_{DSS}" would be zero for any enhancement-mode device).

For JFETs the gate-source voltage at which the drain current is brought essentially to   zero[9] is called the "gate-source cutoff voltage," $V_{GS(OFF)}$, or (sometimes) the "pinch-off voltage," V_P, and is typically in the range of $-1\,V$ to $-5\,V$ (positive for *p*-channel, of course). The analogous quantity is not normally specified for enhancement-mode MOSFETs;[10] instead, datasheets specify the

---

[9] Usually chosen to be 10 nA; a pinchoff-voltage test circuit is described in §4.3.4.

[10] We will use the symbol V_{th} to designate the analogous idealized "gate-source cutoff voltage" for MOSFETs, which we will need in some

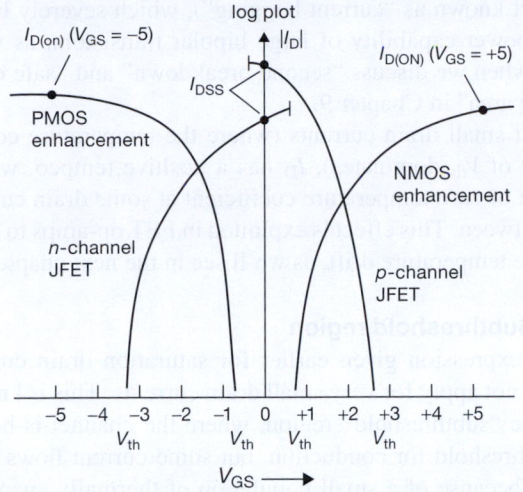

Figure 3.12. Important gate voltages and drain currents.

enhancement-mode MOSFET.[11] We remarked that FETs behave like pretty good transconductance devices over most of the graph (i.e., I_D nearly constant for a given V_{GS}), except at small V_{DS}, where they approximate a resistance (i.e., I_D proportional to V_{DS}). In both cases the applied gate-source voltage controls the behavior, which can be well described by the FET analog of the Ebers–Moll equation. Let's look now at these two regions a bit more closely; we'll revisit this important subject in greater detail in §3.3 and again in the advanced-topics Chapter *3x*.

Figure 3.13 shows the situation schematically. In both regions the drain current depends on $V_{GS} - V_{th}$, the amount by which the applied gate-source voltage exceeds the threshold (or pinch-off) voltage. The linear region, in which drain current is approximately proportional to V_{DS}, extends up to a voltage $V_{DS(sat)}$, after which the drain current is approximately constant. The slope in the linear region, I_D/V_{DS}, is proportional to the gate bias, $V_{GS} - V_{th}$. Furthermore, the drain voltage at which the curves enter the "saturation region," $V_{DS(sat)}$, is approximately $V_{GS} - V_{th}$, making the saturation drain current, $I_{D(sat)}$, proportional to $(V_{GS} - V_{th})^2$, the quadratic law we mentioned earlier. For reference, here are the universal FET drain-current formulas:

$$I_D = 2\kappa[(V_{GS} - V_{th})V_{DS} - V_{DS}^2/2] \quad \text{(linear region)} \quad (3.1)$$

$$I_D = \kappa(V_{GS} - V_{th})^2 \quad \text{(saturation region)} \quad (3.2)$$

"gate-source threshold voltage," $V_{GS(th)}$, at which the onset of drain current has reached a small but arbitrary threshold value, typically 0.25 mA. $V_{GS(th)}$ is typically in the range of 0.5–5 V, in the "forward" direction, of course.

With FETs it is easy to get confused about polarities. For example, *n*-channel devices, which usually have the drain positive with respect to the source, can have positive or negative gate voltage and positive (enhancement) or negative (depletion) threshold voltages. To make matters worse, the drain can be (and often is) operated negative with respect to the source. Of course, all these statements go in reverse for *p*-channel devices. In order to minimize confusion, we will always assume that we are talking about *n*-channel devices unless explicitly stated otherwise. Likewise, because MOSFETs are nearly always enhancement mode, and JFETs are always depletion mode, we'll omit those designations from now on.

3.1.4 FET drain characteristics

In Figure 3.2 we showed a family of curves of I_D versus V_{DS} that we measured for a VN0106, an *n*-channel

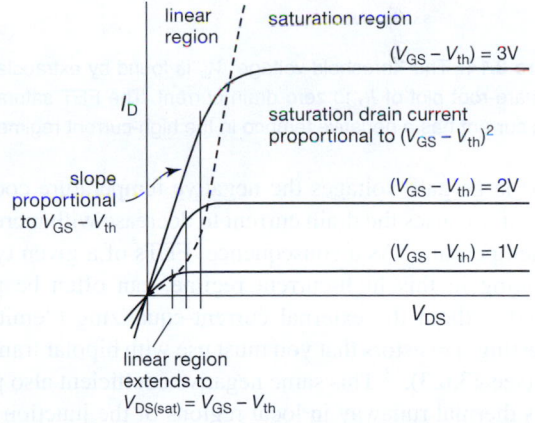

Figure 3.13. Linear and saturation regions of FET operation.

If we call $V_{GS} - V_{th}$ (the amount by which the gate-source voltage exceeds the threshold) the "gate drive," the

discussion that follows. In the electronics literature the symbol V_T is used for this quantity, called the "threshold voltage"; but we prefer to avoid the same symbol that is used for the "thermal voltage" V_T in the Ebers–Moll equation, where $V_T = kT/q \approx 25$ mV. And don't confuse V_{th} with $V_{GS(th)}$: V_{th} is obtained from extrapolating a $\sqrt{I_D}$ versus V_{GS} plot; it's not found in datasheets, but it's quite useful. By contrast, $V_{GS(th)}$ is not terribly useful, but it's the quantity you find in datasheets.

[11] The VN0106 is not widely available. It is similar to the very popular 2N7000 or BS170 (in the TO-92 package) and to the 2N7002, BSS138, or MMBF170 (in the SMT packages).

important results are that (a) the resistance in the linear region is inversely proportional to the gate drive, (b) the linear region extends to a drain-source voltage approximately equal to the gate drive, and (c) the saturation drain current is proportional to the square of the gate drive. These equations assume that the body is connected to the source. Note that the "linear region" is not really linear because of the V_{DS}^2 term; we'll show a clever circuit fix later.

The scale factor κ depends on particulars such as the geometry of the FET, oxide capacitance, and carrier mobility.[12] It has a temperature dependence $\kappa \propto T^{-3/2}$, which alone would cause I_D to decrease with increasing temperature. However, V_{th} also depends slightly on temperature (2–5 mV/°C); the combined effect produces the curve of drain current versus temperature, as shown in Figure 3.14.

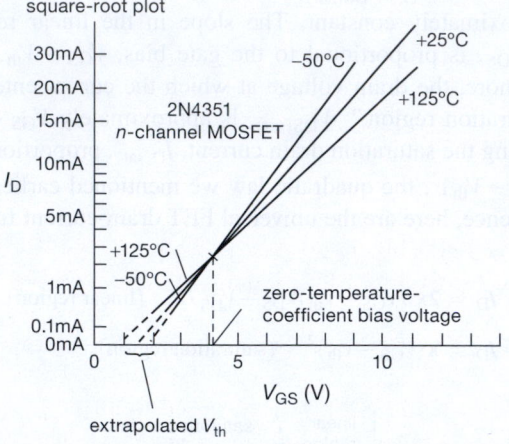

Figure 3.14. The "threshold voltage" V_{th} is found by extrapolating a square-root plot of I_D to zero drain current. The FET saturation drain current has a negative tempco in the high-current regime.

At large gate voltages the negative temperature coefficient of κ causes the drain current to decrease with increasing temperature. As a consequence, FETs of a given type, operating in this high-current regime, can often be paralleled without the external current-equalizing ("emitter-ballasting") resistors that you must use with bipolar transistors (see §3.6.3).[13] This same negative coefficient also prevents thermal runaway in local regions of the junction (an

---

[12] You'll commonly see the symbol k used here. We prefer κ, to avoid confusion with the Boltzmann constant k that figures in the Ebers–Moll equation for bipolar transistor behavior. The SPICE model for JFETs calls this parameter β (and for V_{th} it uses the parameter "VTO").

[13] Some cautions apply, most notably with ordinary ("vertical") power MOSFETs in linear applications, where they are operated at drain currents well below the region of negative temperature coefficient – see §3.5.1B and §3.6.3. In such applications (e.g., audio power amplifiers)

---

effect known as "current hogging"), which severely limits the power capability of large bipolar transistors, as we'll see when we discuss "second breakdown" and "safe operating area" in Chapter 9.

At small drain currents (where the temperature coefficient of V_{th} dominates), I_D has a positive tempco, with a point of zero temperature coefficient at some drain current in between. This effect is exploited in FET op-amps to minimize temperature drift, as we'll see in the next chapter.

A. Subthreshold region

Our expression given earlier for saturation drain current does not apply for very small drain currents. This is known as the "subthreshold" region, where the channel is below the threshold for conduction, but some current flows anyway because of a small population of thermally energetic electrons. If you've studied physics or chemistry, you probably know in your bones that the resulting drain current is exponential (with some scale factor) in the difference voltage $V_{GS} - V_{th}$.

We measured some MOSFETs over nine decades of drain current (1 nA to 1 A) and plotted the result as a graph of I_D versus V_{GS} (Figure 3.15). The region from 1 nA to 1 mA is quite precisely exponential; above this subthreshold region the curves enter the normal "quadratic" region. For the *n*-channel MOSFET (Supertex type VN01, similar to the ever-popular 2N7000) we checked out a sample of 20 transistors (from four different manufacturing runs spread over two years), plotting the extreme range to give you an idea of the variability (see next section). Note the somewhat poorer characteristics (V_{th}, $I_{D(ON)}$) of the "complementary" VP01 (similar to the popular BS250).

JFETs exhibit similar behavior, as illustrated in the measured data of Figure 3.16 (though V_{GS} is necessarily limited to reverse-bias voltage polarity, or at most to a forward-bias less than a diode drop). The quadratic region, where $I_D \propto (V_{GS} - V_{th})^2$, is seen most clearly by plotting the *square root* of drain current versus gate voltage; see Figure 3.14 and Figure 3.51 later in the chapter.

3.1.5 Manufacturing spread of FET characteristics

Before we look at some circuits, let's take a look at the range of FET parameters (such as I_{DSS} and $V_{GS(th)}$), as well as their manufacturing "spread" among devices of the same nominal type, in order to get a better idea of the FET. Unfortunately, many of the characteristics of FETs show a much greater process spread than the corresponding

---

the alternative "lateral" MOSFET is popular, owing to its stabilizing negative coefficient.

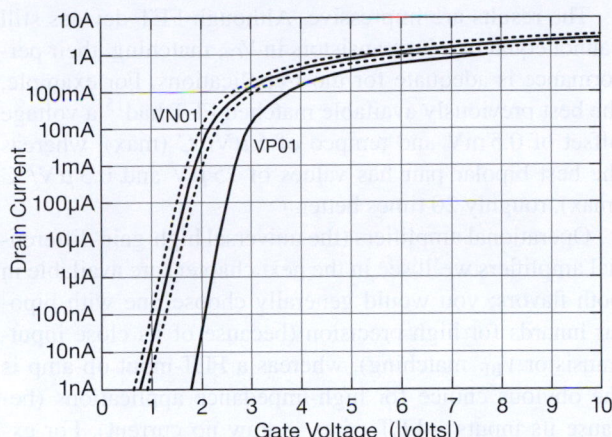

Figure 3.15. Measured MOSFET saturation drain current versus gate-source voltage. For the VN01 the dotted curves are the extreme specimens, and the solid curve is the median, from a group of 20 MOSFETs.

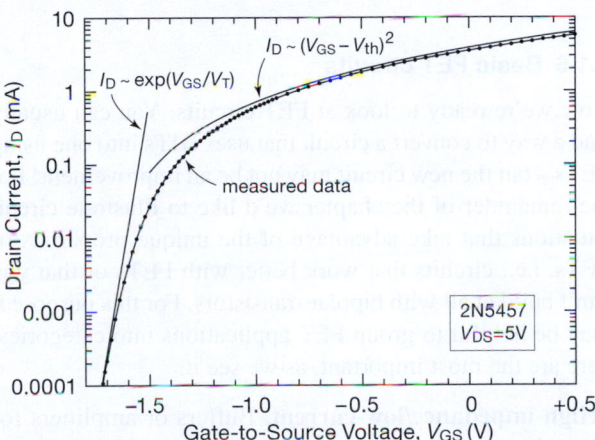

Figure 3.16. Five decades of measured drain current versus gate-to-source voltage for the *n*-channel 2N5457 JFET. In the subthreshold region the drain current is exponential, like a BJT, with nearly the same scale factor V_T (kT/q, or 25.3 mV at room temperature); at higher currents it becomes quadratic (the calculated curve has been offset by +10% for clarity).

characteristics of bipolar transistors, a fact that the circuit designer must keep in mind. For example, the 2N7000 (a typical *n*-channel MOSFET) has a specified $V_{GS(th)}$ of 0.8–3 V ($I_D = 1$ mA), compared with the analogous V_{BE} spread of 0.63–0.83 V (also at $I_C = 1$ mA) for a small *npn* bipolar transistor. Here's what you can expect:

FET Characteristics: Manufacturing Spread

Characteristic	Available Range	Spread
I_{DSS}, $I_{D(ON)}$	1 mA to 500 A	×5
$R_{DS(ON)}$	0.001 Ω to 10k	×5
g_m @ 1 mA	500–3000 μS	×5
V_P (JFETs)	0.5–10 V	5 V
$V_{GS(th)}$ (MOSFETs)	0.5–5 V	2 V
$BV_{DS(OFF)}$	6–1000 V	
$BV_{GS(OFF)}$	6–125 V	

$R_{DS(ON)}$ is the drain-source resistance (linear region, i.e., small V_{DS}) when the FET is conducting fully, e.g., with the gate grounded in the case of JFETs or with a large applied gate-source voltage (usually specified as 10 V) for MOSFETs. I_{DSS} and $I_{D(ON)}$ are the saturation-region (large V_{DS}) drain currents under the same turned-on gate drive conditions. V_P is the pinch-off voltage (JFETs), $V_{GS(th)}$ is the turn-on gate threshold voltage (MOSFETs), and the BV's are breakdown voltages. As you can see, a JFET with a grounded source may be a good current source, but you can't predict very well what the current will be. Likewise, the V_{GS} needed to produce some value of drain current can vary considerably, in contrast to the predictable (≈ 0.6 V) V_{BE} of bipolar transistors. Figure 3.17 illustrates this latter point graphically: we measured the V_{GS} values at a drain current of 1 mA for a hundred pieces each (hey, they're pretty cheap: about $0.10 each) of three popular JFET types (the 2N5457–59 series, graded by their I_{DSS}). The spread of gate-source voltages, within each type, is about 1 V. For comparison, look at the analogous plot for BJTs in Figure 8.44; there the spread is just 10–20 mV.

A. Matching of characteristics

As you can see, FETs are inferior to bipolar transistors in V_{GS} predictability, i.e., they have a large spread in the V_{GS} required to produce a given I_D. Devices with a large process spread will, in general, have a larger offset (voltage unbalance) when used as differential pairs. For instance, typical run-of-the-mill bipolar transistors might show a spread in V_{BE} of 25 mV or so, at some collector current, for a selection of off-the-shelf transistors. The comparable "official" figure (as specified on datasheets) for MOSFETs is more like 1 V to 2 V![14] Because FETs have some very desirable characteristics otherwise, it is worthwhile putting in some

---

[14] In practice, we've found considerably better matching within a single batch of MOSFETs, sometimes as tightly matched as 50 mV or so. On the other hand, a more typical spread within one batch is several hundred millivolts, as illustrated later in Figure 3.41. If matching is

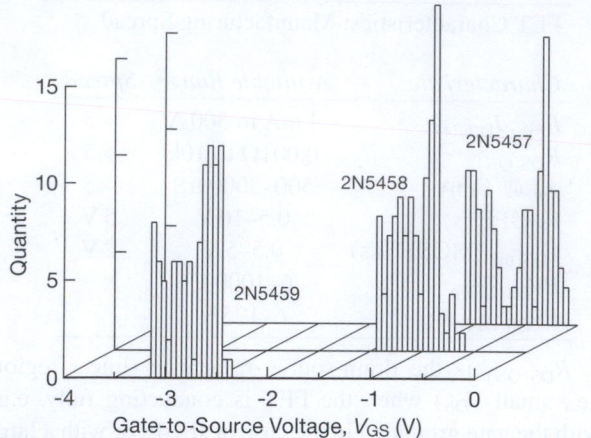

Figure 3.17. We wore our fingers to the bone (a "digital" measurement?) collecting V_{GS} values (at $V_{DS}=5\,V$ and $I_D=1\,mA$) for 300 JFETs in the popular 2N5457–59 series. Compare with the analogous histograms in Figure 8.44.

extra effort to reduce these offsets in specially manufactured matched pairs. IC designers use techniques like interdigitation (two devices sharing the same general piece of IC real estate) and thermal-gradient cancellation schemes to improve performance (Figure 3.18).

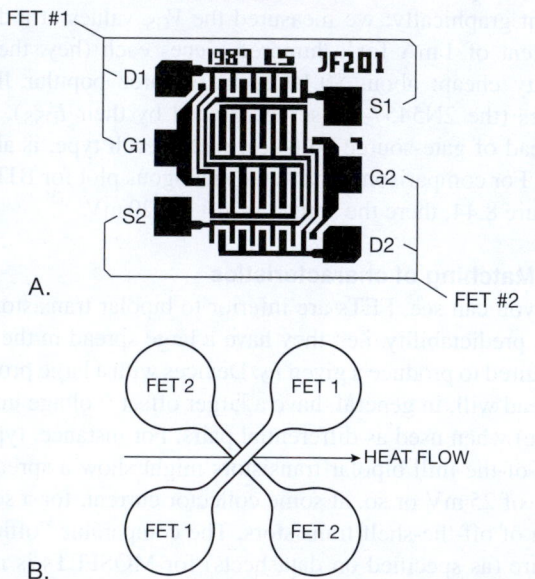

Figure 3.18. Techniques for transistor matching: A. Interdigitation (Courtesy of Linear Integrated Systems.) B. Temperature gradient cancellation.

important in some application (when several transistors are used in parallel, for example), you should measure the actual parts.

The results are impressive. Although FET devices still cannot equal bipolar transistors in V_{GS} matching, their performance is adequate for most applications. For example, the best previously available matched FET had[15] a voltage offset of 0.5 mV and tempco of 5 μV/°C (max), whereas the best bipolar pair has values of 25 μV and 0.3 μV/°C (max), roughly 20 times better.

Operational amplifiers (the universal high-gain differential amplifiers we'll see in the next chapter) are available in both flavors; you would generally choose one with bipolar innards for high precision (because of its close input-transistor V_{BE} matching), whereas a FET-input op-amp is the obvious choice for high-impedance applications (because its inputs – FET gates – draw no current). For example, the inexpensive JFET-input LF411 and LF412 that we will use as our all-around op-amp in the next chapter has a typical input (leakage) current of 50 pA and costs $0.60; the popular MOSFET-input TLC272 costs about the same and has a typical input (leakage) current of only 1 pA! Compare this with a common bipolar op-amp, the LM324, with typical input (bias) current of 45,000 pA (45 nA).[16]

3.1.6 Basic FET circuits

Now we're ready to look at FET circuits. You can usually find a way to convert a circuit that uses BJTs into one using FETs – but the new circuit may not be an improvement! For the remainder of the chapter we'd like to illustrate circuit situations that take advantage of the unique properties of FETs, i.e., circuits that work better with FETs or that you can't build at all with bipolar transistors. For this purpose it may be helpful to group FET applications into categories; here are the most important, as we see it.

High-impedance/low-current. Buffers or amplifiers for applications in which the base current and finite input

---

[15] Sadly, these parts are no longer available. But the art of transistor matching is alive and well in the innards of op-amps, for which the best JFET specimen has an offset and tempco of 0.1 mV and 1 μV/°C, respectively, whereas the best BJT specimen has 0.01 mV and 0.1 μV/°C, i.e., 10 times better.

[16] BJT enthusiasts would cry "foul!," and point out that you can use superbeta BJTs, combined with bias-current cancellation schemes, to bring the input current down to 25 pA; they would further point out that FET-input current (which is leakage) rises dramatically with temperature, whereas BJT-input current (which is honest bias current) is stable or even tends to decrease slightly (see Figure 3.48). FET enthusiasts would prevail, though, with the rejoinder that MOSFET-input amplifiers like the dual LMC6042 have typical input currents of 2 *femto*amps (that's 0.000002 nA!).

impedance of BJTs limit performance. Although you *can* build such circuits with discrete FETs, current practice favors using integrated circuits built with FETs. Some of these use FETs as a high-impedance front-end for an otherwise bipolar design, whereas others use FETs throughout. When available FET ICs do not provide adequate performance, a hybrid approach (discrete JFET front-end, assisted by an op-amp) can push the performance envelope.

Analog switches. MOSFETs are excellent voltage-controlled analog switches, as we hinted in §3.1.1B. We'll look briefly at this subject. Once again, you should generally use dedicated "analog switch" ICs, rather than building discrete circuits.

Digital logic. MOSFETs dominate microprocessors, memory, special-purpose VLSI, and most high-performance digital logic. They are used exclusively in micropower logic and low-power portable devices. Here, too, MOSFETs make their appearance in integrated circuits. We'll see why FETs are preferable to BJTs.

Power switching. Power MOSFETs are usually preferable to ordinary bipolar power transistors for switching loads, as we suggested in our first circuit of the chapter. For this application you use *discrete* power FETs.

Variable resistors; current sources. In the "linear" region of the drain curves, FETs behave like voltage-controlled resistors; in the "saturation" region they are voltage-controlled current sources. You can exploit this intrinsic behavior of FETs in your circuits.

Generalized replacement for bipolar transistors. You can use FETs in oscillators, amplifiers, voltage regulators, and radiofrequency circuits (to name a few), where bipolar transistors are also normally used. FETs aren't *guaranteed* to make a better circuit – sometimes they will, sometimes they won't. You should keep them in mind as an alternative.

Now let's look at these subjects. We'll adopt a slightly different order, for clarity.

3.2 FET linear circuits

*A **note** to the reader:* This section and the next (§§3.2 and 3.3) deal primarily with *JFETs*, which are well suited to linear applications such as current sources, followers, and amplifiers. If you need a low-noise amplifier with extremely high input impedance, the JFET is your friend (and maybe your *only* friend). Readers wishing to move directly to MOSFETs, starting with FET switches, may wish to

Table 3.1 JFET Mini-table[a] (see also JFET Table 3.7)

Part #	I_D Curve	Idss (mA)	$V_{GS(off)}$ min (V)	max (V)	measured at 1mA V_{GS} (V)	g_m (mS)	G_{max}[b] (V/V)	C_{rss} typ (pF)	R_{ON} typ (Ω)
2N5484	A	1–5	–0.3	–3	–0.73	2.3	180	1	-
2N5485	B	4–10	–0.4	–4	–1.7	2.1	110	1	-
2N5486	C	8–20	–2	–6	–2.4	2.1	50	1	-
2N5457	D	1–5	–0.5	–6	–0.81	2.0	200	1.5	-
2N5458	E	2–9	–1	–7	–2.3	2.3	170	1.5	-
2N5459	F	4–16	–2	–8	–2.8	2.0	100	1.5	-
BF862	G	10–25	–0.3	–1.2	–0.40	12	250	1.9	-
J309	H	12–30	–1	–4	–1.6	4.2	300	2	50
J310	J	24–60	–2	–6.5	–3.0	4.3	100	2	50
J113	K	2–	–0.5	–3	–1.5	5.7	140	3	50
J112	L	5–	–1	–5	–3.3	5	100	3	30
PN4393	M	5–30	–0.5	–3	–0.83	6.2	100	3.5	100
PN4392	N	25–75	–2	–5	–2.6	5.4	130	3.5	60
LSK170B	P	6–12	–0.2	–2	–0.09	11	160	5	-
J110	Q	10–	–0.5	–4	–1.2	6.1	220	8	18
J107	R	100–	–0.5	–4.5	–2.6	8.2	340	35	8
J105	-	500–	–4.5	–10	–8.7	6.4	60	35	3
IF3601	S	30–	–0.04	–3	–0.24	27	1400	300	-

Notes: (a) sorted by family C_{rss}, and within each family by increasing I_{DSS}. (b) $G_{max}=g_m/g_{os}$, the maximum grounded-source voltage gain into a current source as drain load; G_{max} is proportional to V_{DS} (tabulated values are at $V_{DS}=5V$), and for most JFETs G_{max} is relatively constant over varying I_D.

skip over these JFET materials,[17] and proceed directly to §3.4 on page 171, where we launch into the MOSFET-dominated subjects of signal switching, digital logic, and power switching.

3.2.1 Some representative JFETs: a brief tour

Table 3.1 lists a small selection of representative *n*-channel JFETs.[18] Let's take a look at what you get.

This selection includes only *n*-channel JFETs, the dominant polarity. Complements with similar characteristics are sometimes available, for example the *p*-channel 2N5460–62 for the *n*-channel 2N5457–59; see Table 3.7 on page 217 for additional examples.

Many JFETs come in families of three or four parts, graded by I_{DSS} and $V_{GS(off)}$, which alleviates somewhat

---

[17] You'll want to study this material, though, if you want to understand MOSFET linear amplifiers, because we address topics like the significance of an FET's transconductance and output conductance, and their variation with drain voltage and current.

[18] The expanded Table 3.7 on page 217 includes many more JFETs; later in this chapter there are analogous tables of MOSFETs (Tables 3.4a and 3.4b, pages 188–191, Table 3.5, page 206, and Table 3.6, page 210).

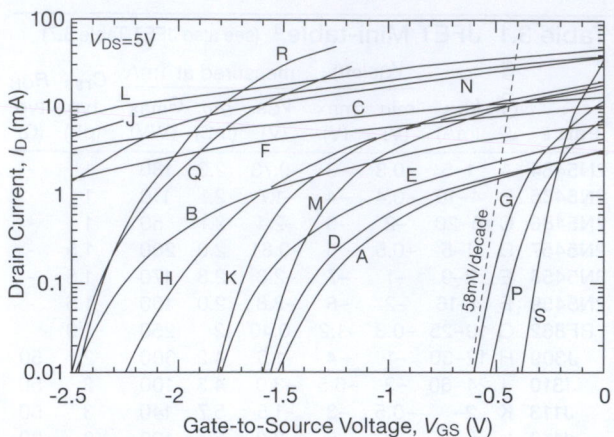

Figure 3.19. Measured drain current versus gate-source voltage for the JFETs in Table 3.1 on the page before.

the annoying circuit design problems created by the wide spread of those parameters. But even those graded families may present a spread of as much as 5:1 (or more). Note also that JFETs intended for switching applications (the ones specifying R_{on}) may specify only a *minimum* value of I_{DSS}: what can you say, for example, about the likely value for a J110 (specified as $I_{DSS} = 10$ mA, minimum)? Answer: not much – our sample measured 122 mA!

In many applications (amplifiers, followers) you want lots of transconductance gain, g_m. JFET datasheets usually specify g_m at the part's I_{DSS}, but that's not terribly useful if you don't know what I_{DSS} is. Moreover, the listed g_m at I_{DSS} is afflicted with the usual specification spread, typically 5:1 or so. Unlike BJTs, for which the transconductance is predictably given by $g_m = 1/r_e = I_C/V_T$ (where $V_T = kT/q \approx 25.3$ mV), the transconductance of different JFET types can vary by an order of magnitude, even when each is operated at the same drain current. In Table 3.1 on the page before we've listed measured values of g_m, all at a standard current of 1 mA.[19] At these currents their transconductance is much less than that of a BJT (where $g_m = 40$ mS at 1 mA), though they compete well at very low currents (the subthreshold region). This behavior can be seen in the different slopes of the measured I_D versus V_{GS} curves of Figure 3.19.

The column labeled G_{max} lists the voltage gain when used as a grounded-source amplifier with current-source load; in that case the effective load resistance is related to a quantity called g_{os}, the output conductance seen looking into the drain with the gate voltage held constant (analo-

gous to the Early effect in BJTs; more on this in Chapter *3x*). Here, too, there is a wide spread among JFET types.

A parameter that is important in low-level amplification is a JFET's input noise voltage, not listed here but treated in detail in Chapter 8. The standout happens to be the IF3601 (an amazing $e_n = 0.3$ nV/$\sqrt{Hz}$), but your pact with the devil is the high 300 pF capacitance of the large-area junction.[20]

There's much more to say about the inhabitants of the JFET zoo, as we'll see in connection with Table 3.7 on page 217. Chapter 8 discusses JFETs in connection with noise (§§8.6 and 8.6.5), with a table of relevant parts (Table 8.2 on page 516).

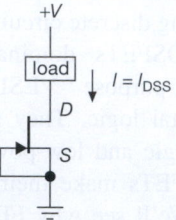

Figure 3.20. An *n*-channel JFET current sink.

3.2.2 JFET current sources

JFETs are used as current sources within integrated circuits (particularly op-amps), and also sometimes in discrete designs. The simplest JFET current source is shown in Figure 3.20; we chose a JFET, rather than a MOSFET, because it needs no gate bias (it's depletion mode). From a graph of FET drain characteristics (Figure 3.21) you can see that the current will be reasonably constant for V_{DS} larger than a couple of volts. However, because of I_{DSS} spread, the current is unpredictable. For example, the MMBF5484 (a typical *n*-channel JFET) has a specified I_{DSS} of 1–5 mA. Still, the circuit is attractive because of the simplicity of a two-terminal constant-current device. If that appeals to you, you're in luck. You can buy "current-regulator diodes" that are nothing more than JFETs with gate tied to source, sorted according to current. They're the current analog of a zener (voltage-regulator) diode. Here are the characteristics of the 1N5283–1N5314 series:[21]

---

[19] In the normal "quadratic" region of drain current, transconductance varies approximately as $\sqrt{I_D}$, see §3.3.3.

[20] See Table 8.2 for the IF3601 and IF3602 (dual). Runners-up in the low-noise competition are the LSK170B and the BF862, with considerably lower capacitances.

[21] Said to be available from several manufacturers. Alternatives include the MS5283, MV5283, and MX5283 series from Microsemi; the SST502–SST511 and CR160–CR470 series from Vishay; and the J500–J511, J553–J557, and U553–U557 series from InterFET. Alternative sources: Central Semiconductor and Linear Integrated Systems.

Characteristic	Value
Currents available	0.22–4.7 mA
Tolerance	±10%
Temperature coefficient	±0.4%/°C
Voltage range	1–2.5 V min, 100 V max
Current regulation	5% typical
Impedance	1M typ (for 1 mA device)

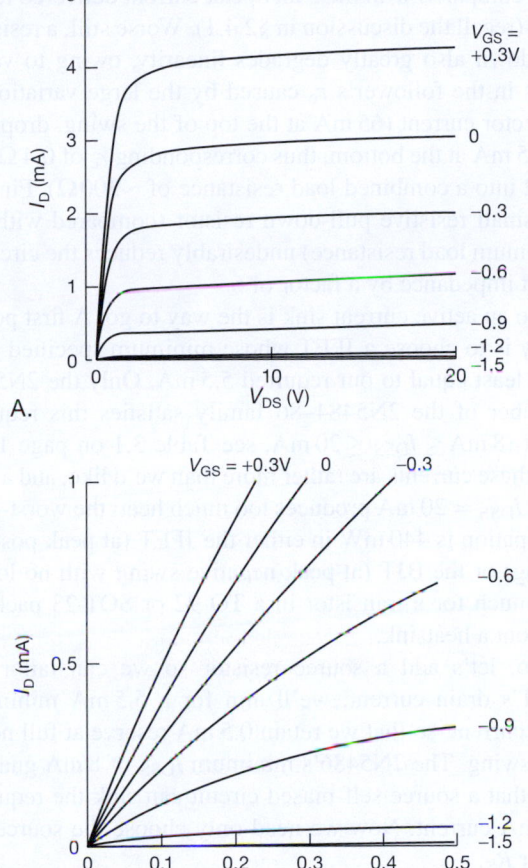

Figure 3.21. Measured JFET characteristic curves. 2N5484 *n*-channel JFET: I_D versus V_{DS} for various values of V_{GS}. See also Figure 3.47.

We measured I versus V for a 1N5294 (rated at 0.75 mA), applying 1 ms voltage pulses at 100 ms intervals to prevent heating. Figure 3.22A shows good constancy of current up to the breakdown voltage ($\sim$145 V for this particular specimen). You can see also the effect of heating when voltage is applied continuously in a dc measurement, caused by the negative temperature coefficient of drain current. Figure 3.22B shows that the device reaches full cur-

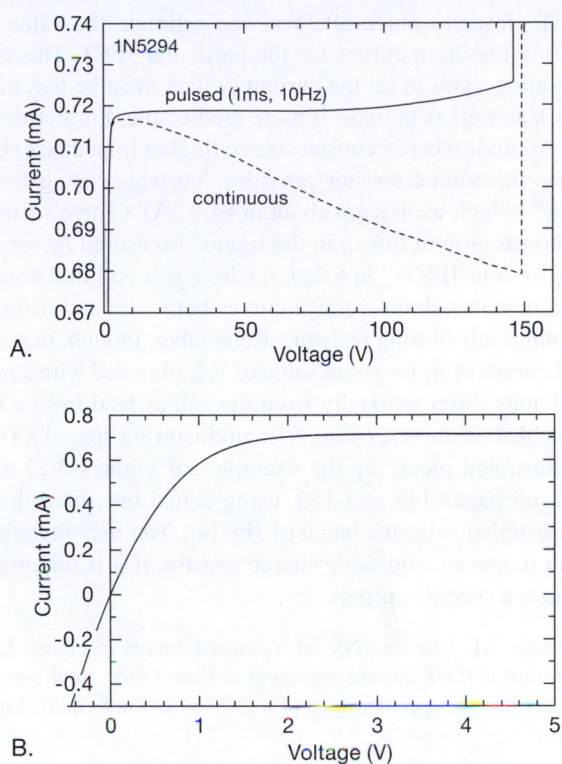

Figure 3.22. 1N5294 current regulator diode.

rent with somewhat less than 1.5 V across it (here both pulsed and dc curves are plotted, demonstrating negligible thermal effects with less than 0.4 mW dissipation). We'll show how to use these devices to make a cute triangle-wave generator in §7.1.3E. And we'll have much more to say about current sources in §4.2.5 and §9.3.14.

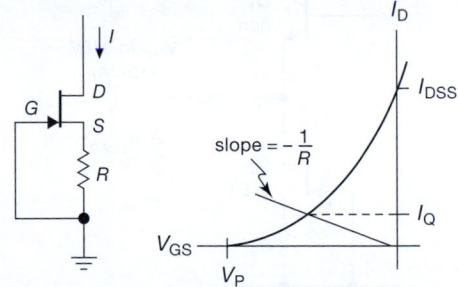

Figure 3.23. JFET current sink ($I=V_{GS}/R$) for $I_D < I_{DSS}$.

A. Source self-biasing

A variation of the previous circuit (Figure 3.23) gives you an adjustable current source. The self-biasing resistor R back-biases the gate by $I_D R$, reducing I_D and bringing the

JFET closer to pinch-off. You can estimate the value of R from the drain curves for the particular JFET. This circuit allows you to set the current (which must be less than I_{DSS}), as well as to make it more predictable. Furthermore, the circuit is a better current source (higher impedance) because the source resistor provides "current-sensing feedback" (which we'll learn about in §4.2.5A). (There's a nice demonstration of this in in the figure "Measured I_D versus V_{DS} for four JFETs" in §3x.4.3, where you will find drain–current versus drain–voltage curves both with and without a source self-biasing resistor.) Remember, though, that actual curves of I_D for some value of V_{GS} obtained with a real FET may differ markedly from the values read from a set of published curves, owing to manufacturing spread. (This is illustrated nicely by the examples of Figures 3.25 and 3.41 on pages 145 and 156, using actual measured drain characteristics from a batch of JFETs.) You may therefore want to use an adjustable source resistor, if it is important to have a specific current.

Exercise 3.1. Use the 2N5484 measured curves in Figure 3.21 to design a JFET current source to deliver 1 mA. Now ponder the fact that the specified I_{DSS} of a 2N5484 is 1 mA (min), 5 mA (max).

B. Example: emitter follower pull-down

Let's look at an example to explore further this problem of unpredictability of JFET zero-bias drain current, I_{DSS} (or, equivalently, the difficulty of predicting the gate-source bias needed to produce a desired drain current).

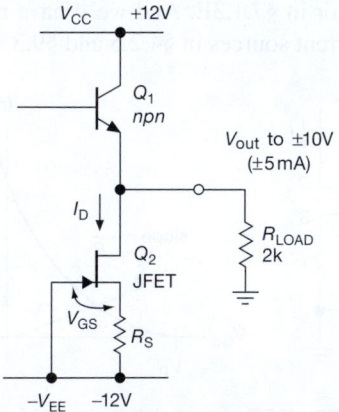

Figure 3.24. Design example: *npn* emitter follower with JFET current sink.

Figure 3.24 shows a BJT emitter follower, running between ±12 V split supplies, with a current-sinking JFET pull-down to the negative rail. We specify that the circuit

must be able to deliver a full swing of ±10 V into a 2 kΩ load (that's ±5 mA load current). You might at first think of using a simple pull-down resistor R_E to the −12 V rail. But the output swing requirement makes things difficult, because you would need to keep R_E less than 400 Ω (we might choose 365 Ω, a standard 1% value) to get a full negative swing; and that low resistance would produce a relatively high quiescent current (at 0 V output) of 33 mA (thus ∼400 mW quiescent dissipation both in Q_1 and in R_E), compared with the 5 mA peak current delivered to the load (recall the discussion in §2.4.1). Worse still, a resistive pulldown also greatly degrades linearity, owing to variations in the follower's r_e caused by the large variation of collector current (65 mA at the top of the swing, dropping to 0.5 mA at the bottom, thus corresponding r_e of 0.4 Ω and 50 Ω into a combined load resistance of ∼300 Ω). Finally, the small resistive pull-down resistor (compared with the minimum load resistance) undesirably reduces the circuit's input impedance by a factor of 6.

So an active current sink is the way to go. A first possibility is to choose a JFET whose minimum specified I_{DSS} is at least equal to our required 5.5 mA. Only the 2N5486 member of the 2N5484–86 family satisfies this requirement (8 mA ≤ I_{DSS} ≤ 20 mA, see Table 3.1 on page 141). But these currents are rather more than we'd like, and a part with I_{DSS} = 20 mA produces too much heat: the worst-case dissipation is 440 mW in either the JFET (at peak positive swing) or the BJT (at peak negative swing with no load), too much for a transistor in a TO-92 or SOT-23 package without a heatsink.

So, let's add a source resistor so we can tailor the JFET's drain current; we'll aim for a 5.5 mA minimum sink current, so that we retain 0.5 mA reserve at full negative swing. The 2N5486's minimum I_{DSS} of 8 mA guarantees that a source self-biased circuit can sink the required 5.5 mA current. Now we need only choose the source resistor R_S.

The problem is that datasheet curves of I_D versus V_{GS} (called "transfer characteristics"), when provided at all, do not show the full range of possibilities; instead they show curves typical of parts with two or three selected values of I_{DSS} within the allowable range. And sometimes all you get are tabulated limits for I_{DSS} and for $V_{GS(off)}$.[22] But you can measure some JFETs to get a sense of things. We did this, and Figure 3.25 shows measured I_D versus V_{GS} curves for seven 2N5486s from different manufacturers

---

[22] It *is* possible to extrapolate from published (or measured) curves by estimating k and V_{th} for the simple quadratic law $I_D = k(V_{GS} - V_{th})^2$. See the discussion in Chapter *3x*.

and batches.[23] Assuming this represents the full range of variability (it doesn't, quite, as seen from the minimum I_{DSS} of 9.2 mA) then we can swing a load line up from the origin until the lowest intersection is above $I_D = 5.5$ mA. That's an R_S of 140 Ω (shown), for which the range of drain current is 5.7 mA (minimum) to 9.5 mA (maximum).

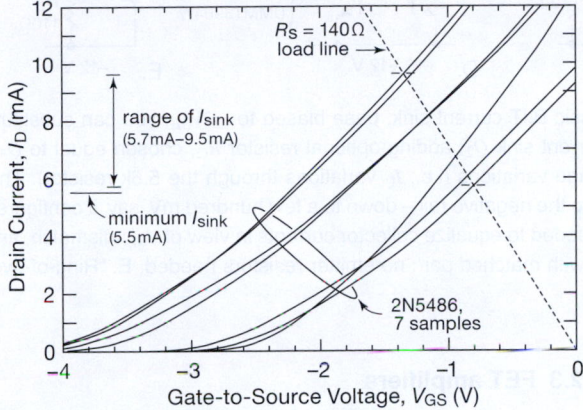

Figure 3.25. Choosing a source resistor R_S to bias a JFET current sink to produce $I_{sink} \geq 5.5$ mA.

The good news is that the circuit will work; the bad news is that the range of sink currents is nearly 2:1 (taking into account the possibility of production parts whose curves span a somewhat wider range than seen in these seven parts). But the good news, again, is that even for a JFET at the high end of the range (thus $I_{sink} \approx 10$ mA), the follower's worst-case dissipation is limited to 220 mW (at peak negative swing, no load), and the JFET's worst-case dissipation is likewise limited to 220 mW (at peak positive swing). This is well within allowable dissipation for a TO-92 transistor (350 mW at 25°C ambient).

C. Current sinks for JFET amplifiers
Stepping back a bit, one might ask whether a JFET current sink, with its 2:1 spread of quiescent current, was a good choice. True, it works. But you can do better with a simple BJT current sink, five versions of which are shown in Figure 3.26. These use more parts, but sink a predictable current. And if you *really* care about minimizing parts count, then you can always use the alternative of a JFET selected for a narrow range of I_{DSS}, with no self-biasing resistor,

i.e., Figure 3.24 with $R_S = 0$ (the 2N5485 specifies I_{DSS} as 4 mA–10 mA; you might select parts from 5.5–8 mA).[24]

This example illustrates the down side of the loose drain current (and corresponding gate voltage) specifications characteristic of all JFETs. As attractive as it may seem to drop in a JFET when you need a current source, it's problematic. But JFETs come into their own when you need an amplifier with high input impedance and low noise – although the loosey-goosey specs are still challenging, the results are worth the bother. We'll see examples presently.

D. Imperfect current source
A JFET current source, even if built with a source resistor, shows some variation of output current with output voltage; i.e., it has finite output impedance, rather than the desirable infinite Z_{out}.[25] The measured curves of Figure 3.21, for example, suggest that, over a drain voltage range of 5–20 V, a 2N5484 shows a drain current variation of 5% when operated with gate tied to source (i.e., I_{DSS}). This might drop to 2% or so if you use a source resistor. An elegant solution is the use of a cascode transistor to suppress drain voltage variations in the current-setting transistor. This can be used both for BJT current sources (it's shown in §2x.3) and for JFET current sources, as shown in Figure 3.27. The idea (as with BJTs) is to use a second JFET to hold constant the drain-source voltage of the current source. Q_1 is an ordinary JFET current source, shown in this case with a source resistor. Q_2 is a JFET of larger I_{DSS}, connected "in series" with the current source. It passes Q_1's (constant) drain current through to the load, while holding Q_1's drain at a fixed voltage – namely the gate-source voltage that makes Q_2 operate at the same current as Q_1. Thus Q_2 shields Q_1 from voltage swings at its output; since Q_1 doesn't see drain voltage variations, it just sits there and provides constant current. If you look back at the Wilson mirror (Figures 2.61, 3.26D), you'll see that it uses this same voltage-clamping idea.

You may recognize this JFET circuit as the "cascode," which is normally used to circumvent Miller effect (§2.4.5). A JFET cascode is simpler than a BJT cascode,

---

[23] Making a mere cameo appearance here, compared with their full performance in Figures 3.55 and 3.56 and associated discussion.

[24] An expensive alternative is to use a pre-sorted two-terminal "current regulator diode" like those in the footnote 3.2.2 on page 142. These appear to be a vanishing species, and the range of currents is quite limited. (A double complaint, reminiscent of the dialog "The food there is so bad." "Yes, and such small portions.")

[25] This is important also for JFET amplifiers (§§3.2.3A and 3.3.2). For further details look at the discussion in Chapter 3x.

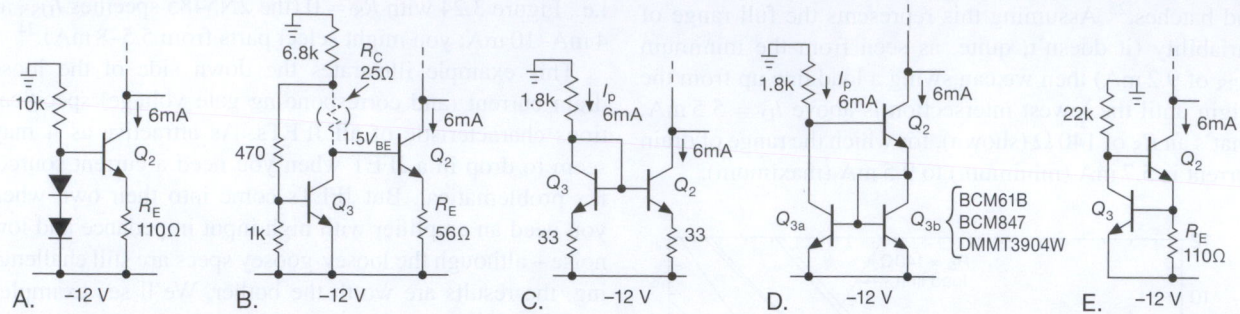

Figure 3.26. Alternatives to the JFET pulldown of Figure 3.24. A. Classic BJT current sink, base biased to $\sim 2V_{\mathrm{BE}}$; you can substitute a red LED for the diode pair. B. Q_3 creates a "$1.5V_{\mathrm{BE}}$" base bias for current sink Q_2; adding optional resistor R_C, chosen equal to Q_3's r_e, cleverly compensates for the latter's change in V_{BE} with supply-voltage variations (i.e., I_C variations through the 5.6k resistor). This configuration is useful if the current-sink output must operate very close to the negative rail – down to a few hundred mV, say, if configured as a "$1.25V_{\mathrm{BE}}$" bias. C. Current mirror with ≈ 200 mV emitter ballasting (needed to equalize collector currents in view of V_{BE} mismatch, and to suppress the Early effect output-current variations). D. Wilson mirror with matched pair; no emitter resistors needed. E. "Ring-of-two" current source. See Figure 2.32 for other current source circuits.

however, because you don't need a bias voltage for the gate of the upper FET: because it's depletion mode, you can simply connect the upper gate to the lower source terminal (compare with Figure 2.84); Q_2's gate-to-source voltage at the operating current (set by Q_1 with its R_S) then sets Q_1's drain-to-source operating voltage: $V_{\mathrm{DS}1} = -V_{\mathrm{GS}2}$. A nice additional benefit is that the resultant circuit is a *two-terminal* current source.

It is important to realize that a good bipolar transistor current source will give far better predictability and stability than a JFET current source. Furthermore, the op-amp-assisted current sources we'll see in the next chapter are better still. For example, a FET current source might vary 5% over a typical temperature range and load-voltage variation, even after being set to the desired current by trimming the source resistor, whereas an op-amp/transistor (or op-amp/FET) current source is predictable and stable to better than 0.5% without great effort.

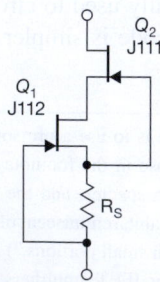

Figure 3.27. Cascode JFET current sink.

3.2.3 FET amplifiers

Source followers and common-source FET amplifiers are analogous to the emitter followers and common-emitter amplifiers made with bipolar transistors that we talked about in the previous chapter. However, the absence of dc gate current makes it possible to realize very high input impedances. Such amplifiers are essential when dealing with the high-impedance signal sources encountered in measurement and instrumentation. For some specialized applications you may want to build followers or amplifiers with discrete FETs; most of the time, however, you can take advantage of FET-input op-amps. In either case it's worth knowing how they work.

With JFETs it is convenient to use the same self-biasing scheme as with JFET current sources (§3.2.2), with a single gate-biasing resistor to ground (Figure 3.28); MOSFETs require a divider from the drain supply, or split supplies, just as we used with BJTs. The gate-biasing resistors can be quite large (a megohm or more), because the gate leakage current is measured in picoamps to nanoamps.

A. Transconductance

The absence of gate current makes *transconductance* (the ratio of output current to input voltage: $g_m = i_{\mathrm{out}}/v_{\mathrm{in}}$) the natural gain parameter for FETs. This is in contrast to bipolar transistors in the last chapter, where we at first flirted with the idea of current gain, or beta ($i_{\mathrm{out}}/i_{\mathrm{in}}$), then introduced the transconductance-oriented Ebers–Moll model: it's useful to think of BJTs either way, depending on the application.

FET transconductance can be estimated from the

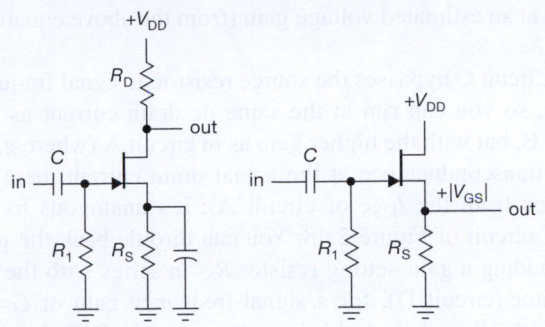

Figure 3.28. Common-source amplifier and source follower. For both configurations the source voltage is above ground, because of the source current flowing through R_S, with a quiescent point $V_S = V_{GS} = R_S I_D(V_{GS})$.

characteristic curves, either by looking at the increase in I_D from one gate-voltage curve to the next on the family of curves (Figures 3.2 or 3.21), or, more simply, from the slope of the I_D versus V_{GS} transfer characteristics curve (Figures 3.15 or 3.51). The transconductance depends on drain current (we'll see how, shortly) and is, of course,

$$g_m(I_D) = i_d/v_{gs}.$$

(Remember that lowercase letters indicate quantities that are small-signal variations.) From this we get the voltage gain,

$$G_{voltage} = v_d/v_{gs} = -R_D i_d/v_{gs}$$

i.e.,

$$G = -g_m R_D. \tag{3.3}$$

just the same as the bipolar-transistor result in §2.2.9, with load resistor R_C replaced with R_D. Typically, small-signal FETs have transconductances in the neighborhood of 10 millisiemens[26] (mS) at a few milliamps.[27] Because g_m depends on drain current, there will be some variation of gain (nonlinearity) over the waveform as the drain current varies, just as we have with grounded emitter amplifiers (where $g_m = 1/r_e$, proportional to I_C).

In the following discussion we'll be using the concept of FET gate drive, $V_{GS} - V_{th}$. Recall that V_{th} is the extrapolated gate threshold voltage we discussed in §§3.1.3 and 3.1.4.

The variation of g_m with drain current is easy to calculate and highly useful when designing JFET followers and amplifiers. For operation above subthreshold ($I_D >$

$I_{DSS}/25$, say), we've seen that the drain current is quadratic in the gate drive

$$I_D = \kappa(V_{GS} - V_{th})^2, \tag{3.4}$$

from which the transconductance ($g_m = i_d/v_{gs} = \partial I_D/\partial V_{GS}$) is seen to be

$$g_m = 2\kappa(V_{GS} - V_{th}) = 2\sqrt{\kappa I_D}. \tag{3.5}$$

In other words, in the "quadratic region" of drain current, g_m is proportional to gate drive, increasing approximately linearly from pinchoff to its specified value at I_{DSS}; alternatively, you can say that it is proportional to the square root of drain current.[28] This is a helpful rule, particularly because the datasheets specify g_m only at its maximum value, at I_{DSS}; we'll use it shortly.[29]

As an example, if (as is often the case) you're operating a JFET in its quadratic region and you want to estimate the transconductance at some drain current I_D, then if you know g_m at some other drain current I_{D0} (which may be I_{DSS}) you can exploit the square root dependence on drain current in eq'n 3.5 to find, simply

$$g_m/g_{m0} = (I_D/I_{D0})^{\frac{1}{2}}. \tag{3.6}$$

FETs in general have considerably lower transconductance than bipolar transistors,[30] which makes them less impressive as amplifiers and followers; we treat this in more detail in §3x.2. However, their outstanding characteristic of extremely low input (gate) current, often of order a picoampere or less, makes it worthwhile to develop circuit solutions that circumvent the problems of low gain (e.g., current source as drain load), or that enhance their effective transconductance ("transconductance enhancer").

At this point it's helpful to see some JFET amplifier examples.

---

[26] Formerly millimhos, or m℧.

[27] This is substantially less than that of a BJT at the same current; the latter has $g_m = 40$ mS at 1 mA, and 200 mS at 5 mA, for example. There's further discussion in §3.3.3 and §3x.2.

[28] Be careful about signs: in these equations V_{th} and V_{GS} are negative (for n-channel JFETs), but V_{th} is more negative, thus a positive value for g_m. As long as you respect signs, these expressions work for n-channel or p-channel and for enhancement or depletion modes. Note that the value κ is not given on datasheets, but can be determined empirically for a given part type and manufacturer. Generally speaking, within a given batch or type of JFET you'll find variations in V_{th}, with κ being relatively constant. Thus a measurement of I_{DSS} and V_{th} allows you to calculate κ from eq'n 3.4, under the assumption that the quadratic region of drain current extends all the way to I_{DSS} (it usually does).

[29] See §3x.2 for further discussion of transconductance versus drain current.

[30] Except in the low drain-current ("subthreshold") region; see Figure 3.54 and analogous figures in §3x.2.

B. JFET amplifier configurations

Figure 3.29 shows the basic configurations for a JFET common-source amplifier stage. In circuit A the JFET is running at its I_{DSS}, with R_D sized small enough so that the drain is at least a volt or two above ground for the maximum specified I_{DSS}. (This is often an annoying constraint, given the loose ratio of specified $I_{DSS(max)}/I_{DSS(min)}$ – commonly 5:1 for most JFETs, see Tables 3.1 on page 141, 8.2 on page 516, and 3.7 on page 217; presently we'll see ways to handle this awkward situation.) The resistor across the input can be very large – 100 MΩ or more – with an input blocking capacitor (for an ac-coupled amplifier); or it can be omitted altogether for a dc-coupled signal near ground. For this circuit the ideal voltage gain is $G=g_m R_D$, where g_m is the transconductance at the operating drain current; it is analogous to the BJT grounded-emitter amplifier of Figure 2.44.[31]

To illustrate actual component values and performance we've chosen the exemplary BF862, because of its high transconductance (45 mS typical at I_{DSS}) and tight I_{DSS} spec (10–25 mA); it happens also to be a low-noise part, as we'll see in Chapter 8. The drain resistor R_D is sized to maintain a minimum 2.5 V across Q_1 (for specified $I_{DSS(max)}$); the typical voltage gain is then $G=-g_m R_D \approx -13$ (inverting).

By adding a source resistor R_S circuit B lets you run at a drain current less than I_{DSS}, in the manner of Figures 3.23 and 3.25. But the source degeneration reduces the gain, to $G=-R_D/(R_S+1/g_m)$. This is analogous to the degenerated BJT common-emitter amplifier of Figure 2.49 (but with simpler self-biasing because the gate-source junction is back-biased), with $1/g_m$ replacing r_e (you can think of $1/g_m$ as an "intrinsic source impedance" of the JFET, analogous to the intrinsic emitter resistance r_e of the BJT).[32] Illustrating with the same BF862, we aim for a drain current of 2 mA by choosing a source self-bias resistor $R_S=200\,\Omega$, estimating that ~0.4 V of gate back-bias is about right.[33] Estimating that $g_m \approx 20$ mS at this drain current,[34] we ar-

rive at an estimated voltage gain (from the above equation) of $G \approx -8$.

Circuit C bypasses the source resistor at signal frequencies, so you can run at the same dc drain current as circuit B, but with the higher gain as in circuit A (where g_m is the transconductance at the actual drain current, here reduced from the I_{DSS} of circuit A); it's analogous to the BJT circuit of Figure 2.48. You can throttle back the gain by adding a gain-setting resistor $R_{S'}$ in series with the capacitor (circuit D), for a signal-frequency gain of $G=-R_D/(R_S\|R_{S'}+1/g_m)$; this is analogous to the BJT circuit of Figure 2.50. Or you can step on the throttle by adding a second stage of voltage gain, as in circuit E, with the second stage common-emitter amplifier multiplying the first-stage gain of any of the previous single-stage circuits by a factor $R_C I_C/V_T$ (i.e., R_C/r_e), where as always $V_T=kT/q \approx 25$ mV. This approximation assumes Q_2 is driven by a voltage source, i.e., $R_D \ll \beta r_e$; most of the gain of the combination comes from the BJT, with its high transconductance. The similar-looking circuit F creates a three-terminal hybrid creature, with the BJT's g_m contributing to achieve a high effective overall transconductance; in this configuration the BJT is a "transconductance enhancer." This configuration is closely analogous to the BJT complementary Darlington (Sziklai) of Figure 2.77, and is treated in more detail in §3x.2.

C. Adding a cascode

The last four circuits show how to implement a drain-clamping cascode to the common-source stage. This elegant configuration is usually featured as a way to circumvent the Miller effect (the effective multiplication of drain-gate capacitance by the stage's voltage gain); that's the way it was presented in Figure 2.84, where it made its first appearance. Here it indeed accomplishes that (it's a "Miller killer"), which is helpful in keeping the input impedance high. But it's better than that: (a) it also lets you keep the drain-to-source voltage low (avoiding the precipitous rise in "impact-ionization" gate current, §3.2.8); and (b) by clamping the drain-source voltage it circumvents the "g_{os} effect" (finite output impedance r_o, caused by I_D dependence on V_{DS}); the voltage gain is then not degraded, and is simply $G=g_m R_D$. This latter benefit is reminiscent of the use of a cascode transistor in the Wilson current mirror (Figure 2.61); it's an "Early killer." There's more detail in Chapter 3x (§3x.4), including experimental results from four JFET exemplars whose g_m and g_{os} were measured; then their gain predictions were compared with measured amplifier gains (both with and without cascode).

In circuit G the BJT's base bias sets the JFET's

---

[31] However, owing to the finite output impedance of the JFET (called r_o, or $1/g_{os}$), the drain load resistor effectively sees a parallel resistance of r_o, so the gain is reduced to $G=g_m(R_D\|r_o)$; this has a negligible effect for the component values here. This is analogous to the Early effect in BJTs, and becomes important for large values of R_D, or (especially) when R_D is replaced by a current source. Lots of discussion in §3x.4.

[32] This time we're ignoring the JFET's finite output impedance; see §3x.4 if you're curious why and for much more detail (if it's detail you really want... be careful what you wish for).

[33] Our confidence is bolstered, a bit, by having measured a sample's I_D versus V_{GS}.

[34] We measured this, too.

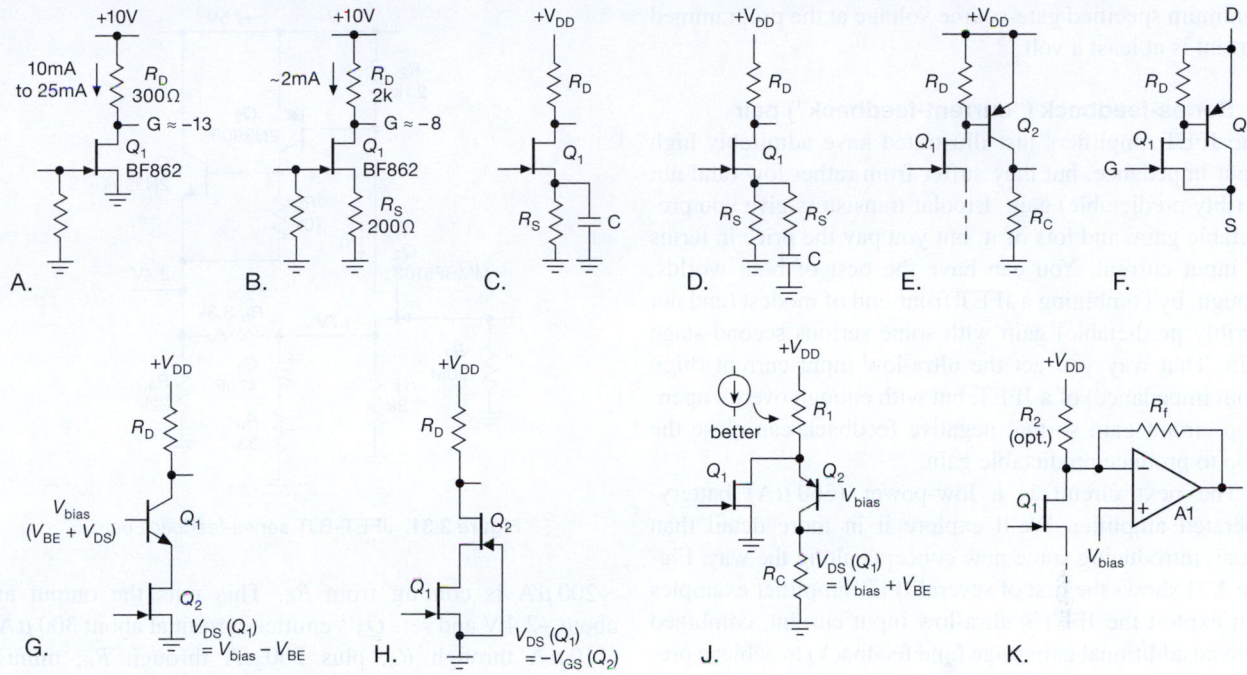

Figure 3.29. Simple single-supply JFET common-source amplifiers.

drain-to-source operating voltage. It's simpler to use a second JFET (Q_2 in circuit H), which must be chosen to have a larger V_{GS} back-bias than Q_1 at the same drain current, though the generally loose gate voltage specifications make this an uncertain proposition. Circuit J is an "inverted cascode," in which Q_1's drain current variations divert current from Q_2; it's a helpful circuit to know about when you find yourself bumping into the positive rail with a conventional cascode. Finally, in circuit K an op-amp transimpedance stage (current-to-voltage; see §4.3.1) substitutes for the drain-clamping cascode transistor Q_2: feedback through R_f maintains its inverting input (the "−" input) at the bias voltage while producing an output voltage $V_o = -I_D R_f + V_{bias}$; the optional resistor R_2 lets you add an offset to reposition the quiescent output voltage according to your whim.

The circuits in Figure 3.29 operate from a single positive supply voltage; they're simple, but, given the characteristically loose I_{DSS} and V_{GS} specifications of JFETs, they suffer from significant uncertainty of operating current. If you have available a negative supply voltage as well, there are several ways to rig things up to ensure predictable biasing. Look at Figure 3.30A, where the operating current of the *n*-channel JFET is set by the source pulldown resistor, $I_D = -(V_- + V_{GS})/R_S$, or approximately V_-/R_S for negative supply voltages large compared with

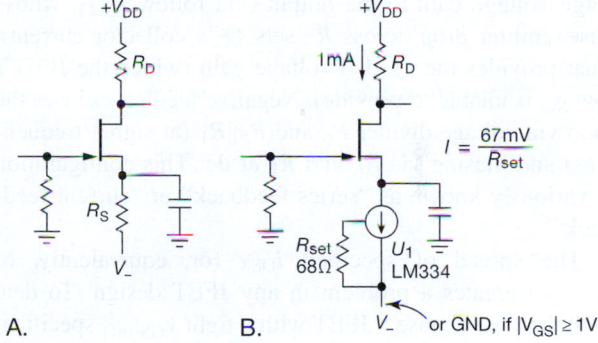

Figure 3.30. A negative supply rail allows predictable source-pulldown biasing of JFET common-source amplifiers.

the JFET's gate-source voltage. As in Figure 3.29C, the bypass capacitor lets signal frequencies partake of the full JFET gain, i.e., $G_V = -g_m R_D$, where g_m is the transconductance at the operating current. A more elegant solution is the use of a current-sink pulldown, as in Figure 3.30B. The LM334 is an inexpensive ($\sim$\$0.50) resistor-programmable current source (see §9.3.14B), here configured for 1 mA ($I \approx 0.067/R_{set}$). With this circuit there's no uncertainty about the operating current (it does not depend on V_{GS}); better still, the LM334 operates down to 1 V drop, so you can operate with a single positive supply if the JFET's minimum specified gate-source voltage at the programmed current is at least a volt.[35]

D. Series-feedback ("current-feedback") pair

The JFET amplifiers just illustrated have admirably high input impedance, but they suffer from rather low (and not terribly predictable) gain. Bipolar transistors give you predictable gain, and lots of it; but you pay the price in terms of input current. You can have the best of both worlds, though, by combining a JFET front-end of modest (and not terribly predictable) gain with some serious second-stage gain. That way you get the ultra-low input current (high input impedance) of a JFET, but with enough overall open-loop circuit gain so that negative feedback can close the loop to produce predictable gain.

The next circuit is a low-power ($660\,\mu\text{A}$) battery-operated amplifier. We'll explore it in more detail than usual, introducing some new concepts along the way. Figure 3.31 shows the first of several JFET amplifier examples that exploit the JFET's ultra-low input current, combined with an additional gain stage (and feedback) to achieve predictable and stable voltage gain. It is similar to the bipolar series-feedback pair illustrated in Figure 2.92: Q_1 is a common-source amplifier, with BJT Q_2 providing second-stage voltage gain to the output (via follower Q_3, whose base–emitter drop across R_3 sets Q_2's collector current). That provides the needed voltage gain (which the JFET's low g_m is unable to provide). Negative feedback closes the loop via voltage divider R_6 and $R_5\|R_1$ (at signal frequencies) and biasing via R_6 and R_1 at dc. This configuration is variously known as "series feedback" or "current feedback."

The spread of specified I_{DSS} (or, equivalently, of $V_{GS(off)}$) creates a problem in any JFET design. To deal with that, we choose a JFET with a tight $V_{GS(off)}$ specification (-1.2 to $-2.7\,\text{V}$), and we run it at a drain current well below I_{DSS} (10 mA minimum) so that the gate-source voltage is close to $V_{GS(off)}$. The feedback path sets the signal gain. With some careful thought (and some juggling and iteration), the same feedback path can be made to establish the (dc-coupled) bias condition.

Here's how it goes. The gate is at ground; we start by assuming a source voltage of approximately 1.7 V, and choose R_1 for $500\,\mu\text{A}$. Of that, about $300\,\mu\text{A}$ comes from the JFET's drain current (Q_2's V_{BE} across R_2); thus

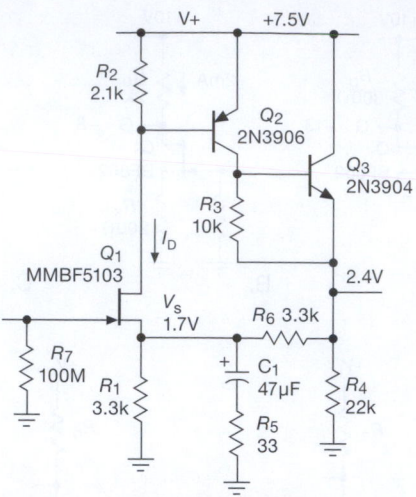

Figure 3.31. JFET–BJT series-feedback pair.

$\sim\!200\,\mu\text{A}$ is coming from R_6. This puts the output at about $+2.4$ V and sets Q_3's emitter current at about $300\,\mu\text{A}$ ($110\,\mu\text{A}$ through R_4, plus $200\,\mu\text{A}$ through R_6, minus $60\,\mu\text{A}$ from R_3).

That's the self-consistent situation, under the assumed V_{GS} of -1.7 V. For a different V_{GS}, the dc output voltage would change accordingly; it could range from $+1.3$ to $+4$ V over the extreme specified range of gate threshold voltage. This would degrade the maximum possible output swing, but that would usually be OK for an amplifier handling small signals (if not, R_1 could be selected for different V_{GS} ranges of JFET parts).[36]

The nominal gain at signal frequencies is approximately 100, set by R_5 (blocked by C_1): $G=1+R_6/(R_5\|R_1)$. The low-frequency -3 dB point is at 100 Hz (where the reactance of C_1 equals R_5). The high-frequency -3 dB point is not so easily calculated, but a SPICE model puts it at approximately 800 kHz (it measured 720 kHz in our breadboard, where there are some added parasitic capacitances). The latter is due primarily to the RC rolloff of Q_1's output signal impedance of 2.1 kΩ (i.e., R_2) driving Q_2's input capacitance of $\sim\!4$ pF, with the latter greatly magnified by the Miller effect.

For large signal driving impedances R_{sig} the amplifier's

---

[35] The effective capacitance of an LM334 current source/sink is 10 pF, small enough to ignore for most purposes. We calculated this from the slew-rate plot in the datasheet. TI's application note LB-41 has additional useful information about the LM334.

[36] A better way to ensure proper biasing is to replace R_1 with a 0.5 mA current sink. A JFET comes to mind (we've got JFETs on the mind!), but, given their unpredictable dc characteristics, a much better choice would be one of the BJT current sinks illustrated in Figure 3.26. Another way of handling this thorny problem is to use a slow feedback loop to stabilize I_D to a desired value less than the specified minimum I_{DSS}.

bandwidth is reduced,[37] owing to an input capacitance of ~5 pF; see Figure 3.32. This is due primarily to the JFET's drain-to-gate capacitance (plus wiring capacitance), given that the source terminal is bootstrapped by feedback. There are numerous tricks to deal with this effect (if more bandwidth is desired), including a cascode in the JFET's drain (which can be bootstrapped to further suppress its input capacitance) and a cascode in the Q_2 BJT gain stage. Some of these techniques are discussed in Chapter *3x*.

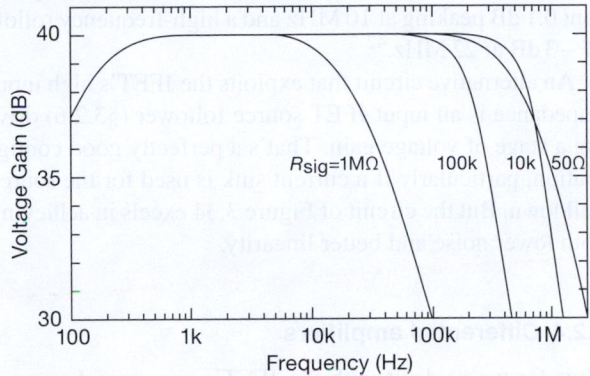

Figure 3.32. Measured gain versus frequency for the amplifier of Figure 3.31. The f_{3dB} with $R_{sig}=1\,M\Omega$ shows that $C_{in}=7\,pF$.

Design equations and design hints
Collecting it together in one place:

$$G = 1 + \frac{R_6}{R_5 \| R_1} \approx 1 + \frac{R_6}{R_5} \quad \text{(ac gain)},$$

$$G_{OL} = g_{m1}R_2\,g_{m2}R_3\,g_{m3}(R_4\|R_6), \quad \text{(open-loop gain}[38])$$

$$I_D = \frac{V_{BE2}}{R_2} \approx \frac{0.7}{R_2} \quad \text{(JFET bias)},$$

$$I_{C2} = \frac{V_{BE3}}{R_3} \approx \frac{0.65}{R_3} \quad (Q_2 \text{ bias}),$$

$$V_{out} = V_S\left(1 - \frac{R_6}{R_1}\right) + \frac{R_6}{R_2}V_{BE2} + \frac{R_6}{R_1}|V_{EE}| \quad \text{(output bias)}.$$

---

[37] Though paradoxically for R_{sig} values of a few kΩ it is *extended* somewhat, owing to some response "peaking."

[38] This expression overestimates the open-loop gain by neglecting the gain-limiting Early effect in Q_2, the stage where most of the circuit's overall gain resides. The measured 2N3906 Early voltage $V_A \approx 25\,V$ (§*2x.8*) implies a maximum voltage gain of the Q_2Q_3 stage of ~1000 (compared with its ideal $G \approx 2500$), thus an overall open-loop gain of ~5000. This is ample for the modest ×100 closed-loop gain.

For single-supply operation the last term is zero. Basically, R_2 sets I_D, and the ratio R_6/R_1 sets V_{out}. For single-supply operation ($V_{EE}=0$) use a small value for R_6 if the JFET has substantial V_{GS} at its operating current and larger values for R_6 for lower V_{GS} parts. The latter is tricky, because the "leverage" of R_6/R_1 can push V_{out} all over the map. Choose R_4 to help set I_{C3} after you've dealt with V_{out}. It may be necessary to select R_1 to go with batches of parts with similar V_{GS}. A negative V_{EE} supply helps with biasing and also permits output swings both sides of ground.

Another way to handle the uncertainty in V_{out} is to apply a positive bias to the gate, as in Figure 3.33A. This adds a positive offset of V_B at the source terminal (whose voltage is now $V_S = V_B - V_{GS}$, where V_{GS} is negative for an *n*-channel JFET), making V_{GS} less important, and thus a smaller fractional uncertainty in V_S. While you're at it, you can easily bootstrap the gate bias divider,[39] as in Figure 3.33B, to raise the input impedance.

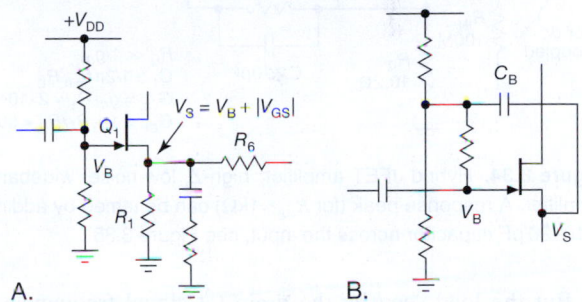

Figure 3.33. A. A positive bias at the gate of Q_1 in Figure 3.31 improves the predictability of V_{out}. B. Add a bootstrap to raise R_{in}.

E. Simple "hybrid" JFET amplifier

With the assistance of an op-amp (the magnificent centerpiece of Chapter 4) you can do wonders. Put simply, an op-amp is a "very high-gain difference amplifier in a bottle," intended to be fodder for feedback as the universal core of pretty much any analog circuit. It's *pure engine*: a turbocharged Harley unicycle with dual intakes. This example and the next show a couple of ways to use an op-amp's properties in support of a JFET amplifier. Look first at Figure 3.34. Here we've chosen the excellent 2SK170B (with LSK170B as a second source) for the front-end: it has lots of transconductance (about 25 mS at its I_{DSS} of 6–12 mA), along with very low noise voltage (~1 nV/√Hz). We run it at zero gate voltage; and we deal with the 2:1

---

[39] You can, of course, bootstrap the gate resistor even when the gate is biased at ground, via R_7 in Figure 3.31.

spread of specified I_{DSS} (a tighter specification than provided for most JFETs) by choosing the drain load resistor R_D small enough to avoid dc saturation even at I_{DSS} (max). The actual drain voltage is unimportant because we use ac coupling to the second stage (via C_1). Ignoring for the moment the second stage (and setting $R_g = 0$), the front-end voltage gain would be $G = g_m R_D$, or roughly $G \approx 25$, with perhaps $\pm 25\%$ uncertainty from JFET manufacturing process variations.

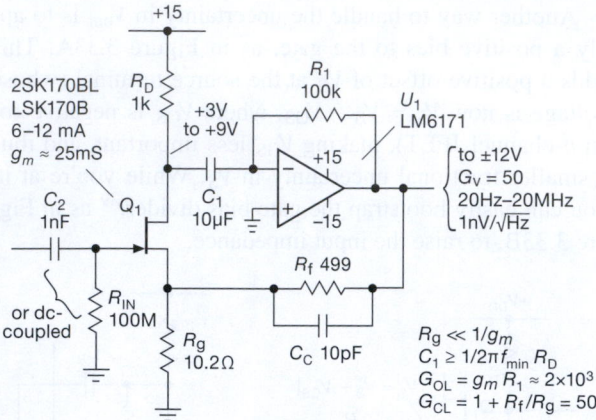

Figure 3.34. Hybrid JFET amplifier: high-Z, low-noise, wideband amplifier. A response peak (for $R_{sig} \sim 1k\Omega$) can be tamed by adding a 10–20 pF capacitor across the input, see Figure 3.35.

But the load seen by the drain (at signal frequencies) is in fact the low-impedance input of the second stage, which is an op-amp configured here as a current-to-voltage ("transresistance") converter (see §4.3.1C). Its "gain" (ratio of output voltage to input current, thus units of resistance) is just R_1, making the overall open-loop gain $G = g_m R_1$ (again assuming no feedback, and $R_g = 0$). So, for the circuit values shown, the open-loop gain $G_{OL} \approx 2500$.

Now we close the loop via R_f, subtracting from the input a fraction $R_g/(R_g + R_f)$, for an ideal closed-loop gain $G_{CL} = 1 + R_f/R_g = 50$. The loop gain (ratio of open-loop to closed-loop gains) is about 50, adequate for good linearity and predictability of gain (see §2.5.3). Note the low resistance of R_g: it should be small enough so that the open-loop gain is not much reduced (thus $R_g < 1/g_m$); and it should also be small enough so that its Johnson noise contribution is insignificant (§8.1). For a JFET transconductance of 25 mS, the first constraint limits R_g to somewhat less than 40 Ω; and for this JFET's noise voltage of $\sim 1\,\mathrm{nV}/\sqrt{\mathrm{Hz}}$ the second constraint limits R_g to somewhat less than 25 Ω. Thus we choose the low $\sim 10\,\Omega$ value. With the values shown, the open-loop gain is reduced by $\sim 20\%$, the in-

put noise voltage is increased by $\sim 8\%$, and the feedback network loads the op-amp to ± 20 mA at full swing.[40]

The op-amp here was chosen for its wide bandwidth (it falls to unity gain at ~ 100 MHz), so the circuit's closed-loop gain rolls off at about 20 MHz,[41] as seen in the measured data of Figure 3.35. A bonus is the impressive output drive capability: up to 100 mA, and full ± 10 V swing to nearly 10 MHz. The small compensation capacitor C_c enhances stability: with no compensation we measured 5 dB of peaking at 16 MHz; adding C_c resulted in an insignificant 0.1 dB peaking at 10 MHz and a high-frequency rolloff of -3 dB at 22 MHz.[42]

An alternative circuit that exploits the JFET's high input impedance is an input JFET source follower (§3.2.6) driving a stage of voltage gain. That's a perfectly good configuration, particularly if a current sink is used for the source pulldown. But the circuit of Figure 3.34 excels in achieving both lower noise and better linearity.

3.2.4 Differential amplifiers

Thus far we've dealt with the JFET's uncertain I_D versus V_{GS} relationship by restricting ourselves to ac-coupled amplifier designs. But we can do better: a JFET matched pair lets us make dc-coupled amplifiers of respectable performance. And of course their very low input current means that these circuits can serve as high-input-impedance front-end stages for bipolar differential amplifiers, as well as for the important op-amps and comparators we'll meet in the next chapter. As we mentioned earlier, the substantial V_{GS} offsets of FETs will generally result in larger input voltage offsets and offset drifts than with a comparable amplifier

---

[40] If you're not happy with that, you could double or triple R_f and R_g, at the expense of slightly higher amplifier noise. See §8.6 for much more on low-noise JFET design.

[41] When driven with a low signal impedance. With a higher impedance signal source, the rolloff is dominated by the circuit's input capacitance: Q_1 has a drain-to-gate capacitance C_{rss} of 6 pF, thus the observed -3 dB bandwidth of ~ 400 kHz with a 100 kΩ source. Happily, the larger gate-to-source capacitance ($C_{iss} \approx 30$ pF) is bootstrapped into insignificance by feedback. "And what of the evil Miller effect?" you wonder. That's suppressed here, because the op-amp holds its $(-)$ input fixed (a "virtual ground") via feedback through R_1; see §4.3.1C.

[42] In any feedback circuit there's the possibility of some "peaking" in the response at some frequency (or, in the worst case, a full-blown oscillation). This circuit exhibits modest peaking for input impedances in the range of a few kΩ, as seen in Figure 3.35. Its amplitude can be tamed by adding ~ 10–20 pF of input shunt capacitance, with some consequent reduction of bandwidth.

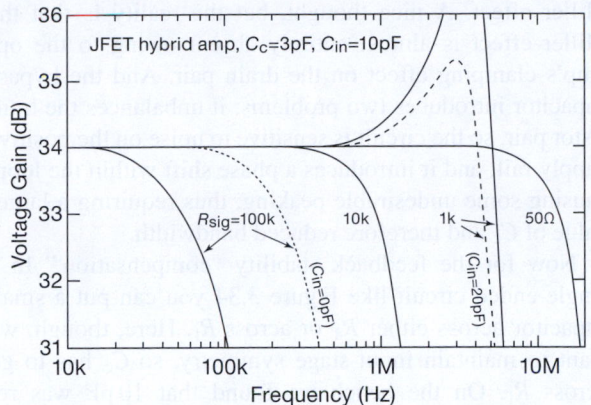

Figure 3.35. Measured gain versus frequency for the amplifier of Figure 3.34. The solid curves are with a 10 pF shunt capacitor across the input, a compromise value for good performance over a wide range of input signal impedances (omit the capacitor for high source impedance, increase it for the worst-case $R_{sig} \approx 1 k\Omega$).

constructed entirely with bipolar transistors, but of course the input impedance will be raised enormously.

Figure 3.36 shows the simplest configurations, analogous to the simple BJT differential amplifiers of Figures 2.63 and 2.67. The differential gain of the classic long-tailed pair in Figure 3.36A (defined as $\Delta V_{out}/\Delta V_{in}$, with $R_1 = R_2$ and differential output as shown) is just $G = g_m R_1$; the common-mode rejection is greatly improved with a current sink substituting for the source resistor R_S. A drawback of this circuit is the uncertainty in gain (owing to the uncertainty in transconductance); and the gain is modest, owing to the limited transconductance of JFETs in general. You can circumvent the gain limitation by replacing the drain load resistor(s) with a current mirror, as in Figure 3.36B. This circuit fragment is not bias stable, however: it must be accompanied by a following stage that is configured to provide dc feedback.

These circuits suffer also from the Miller effect (§2.4.5), whose multiplying action on the feedback capacitance C_{rss} acts to increase the effective input capacitance, thus (in combination with the signal's source impedance) reducing the bandwidth. The current mirror in circuit (B) clamps Q_1's drain, but the Miller effect is present still at Q_2's input.[43] As with BJT amplifier stages, an effective method of eliminating the Miller effect is the use of a cascode transistor (either JFET or BJT) in the drain(s), a desirable configuration discussed later and in Chapter 3x in the section

---

[43] Unless that output drives a transimpedance stage that clamps its voltage, as in Figure 3.31 or 3.34.

"Bandwidth of the Cascode.". By clamping the drain voltage, the cascode also eliminates a reduction in gain seen in simple circuits like that of Figure 3.36A: that's because the JFET's drain-current depends somewhat on drain voltage (the upward slope in plots of drain-current versus drain-source voltage, which you can think of as a finite output impedance), which can reduce the ideal $g_m R_1$ gain by as much as 25%.

For this and other reasons the cascode is highly recommended, even when bandwidth is not an issue. Curious about those "other reasons"? Consider this: the gate current in JFETs, normally down in the picoamperes, rises precipitously with drain-to-source voltage – see Figure 3.49 on page 164 where picoamps become *micro*amps! A cascode lets you clamp the drain to a low operating voltage, suppressing this effect. This is nicely illustrated in the figure toward the end of §3x.4 in Chapter 3x.

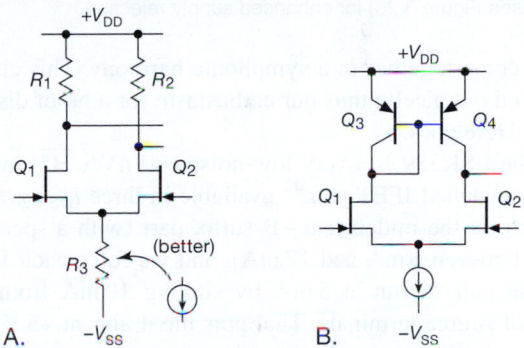

Figure 3.36. Simplest JFET differential amplifiers.

A. Example: a dc-coupled hybrid JFET amplifier

In the earlier hybrid amplifier design example (Figure 3.34) we dodged the issue of unpredictable JFET drain current by settling for an ac-coupled amplifier. That's fine for something like an audio or RF amplifier; but sometimes you'd like response all the way down to dc.

You can achieve that by exploiting a matched-pair JFET differential input stage in a fully dc-coupled arrangement, as in Figure 3.37. The overall configuration is a common-source differential amplifier with the input signal connected to one side. Its differential voltage output drives a wideband op-amp, whose output (divided by the gain-setting voltage divider) provides negative feedback to the other terminal of the input pair. As with the previous circuit, the closed-loop gain is $G_{CL} = 1 + R_f/R_g = 50$, with a small loop compensation capacitor C_c chosen for best response without peaking.

The devil's in the details, which, if you're lucky,

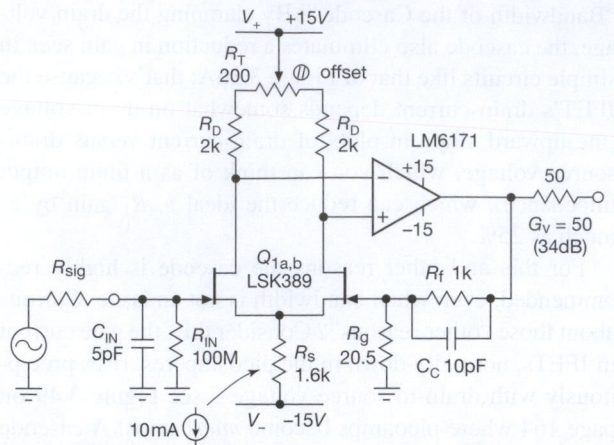

Figure 3.37. An op-amp closes the loop around a JFET matched pair to create a dc-coupled wideband low-noise amplifier with high input impedance. Replace source bias resistor R_S with a current sink (see Figure 3.26) for enhanced supply rejection.

may come together in a symphonic harmony. This circuit worked out nicely; thus our enthusiasm for a bit of discussion. Here goes.

The LSK389 is a very low-noise ($\sim 1\,\mathrm{nV}/\sqrt{\mathrm{Hz}}$) monolithic matched JFET pair,[44] available in three I_{DSS} grades. We chose the midcurrent –B suffix part (with a specified I_{DSS} between 6 mA and 12 mA), and we force each JFET of the pair to run at 5 mA by sinking 10 mA from the pair of source terminals. That puts the drains at +5 V (the op-amp enforces equality), with an open-loop gain (from single-ended input to differential output) of approximately $G = g_m(R_D + 0.5R_T) \approx 40$. To this the op-amp contributes its substantial open-loop gain (90 dB, i.e., ×30,000), in a scary-looking configuration whose stability might seem seriously to be in doubt. But not to worry – the ÷50 divider in the return loop safely limits the loop gain,[45] with any tendency toward instability easily tamed with appropriate choice of C_c.

According to the datasheet, the input pair Q_{1ab} has "Tight Matching." But that's on the scale of JFETs, not BJTs – here that "matching" is a whopping ±20 mV maximum (about 100× the matching of a good BJT pair), which the circuit's $G = 50$ would amplify to an output offset of 1 V! Hence the offset trim R_T, with enough range to balance the worst-case input offset.

Our initial design included a bypass capacitor from the drain of the input transistor to ground, to suppress the

Miller effect. A nice thought, but the reality is that the Miller effect is almost entirely absent owing to the op-amp's clamping effect on the drain pair. And the bypass capacitor introduces two problems: it unbalances the transistor pair, so the circuit is sensitive to noise on the positive supply rail; and it introduces a phase shift within the loop, causing some undesirable peaking, thus requiring a larger value of C_c and therefore reduced bandwidth.

Now for the feedback stability "compensation." In a single-ended circuit like Figure 3.34 you can put a small capacitor across either R_1 or across R_f. Here, though, we want to maintain input-stage symmetry, so C_c has to go across R_f. On the bench we found that 10 pF was required to eliminate peaking in the frequency response, when tested with a low-level sinewave input; Figure 3.38 shows the data.

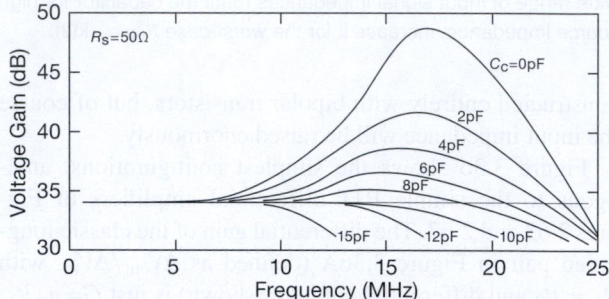

Figure 3.38. Selecting the compensation capacitor C_c for the amplifier of Figure 3.37. A value of 8 or 10 pF works well.

A final bit of compensation is the shunt capacitor C_{in} across the input to suppress some modest peaking that is seen for signal sources of $\sim 1\,\mathrm{k\Omega}$ impedance (R_{sig}). Here 5 pF did the job nicely (though it increased the circuit's input capacitance to ~ 20 pF). Figure 3.39 shows the measured gain versus frequency for the finished circuit (with the component values shown in Figure 3.37), for nine values of R_{sig} (spanning four decades of resistance).

As with the previous circuit, the LM6171 op-amp provides a full ±10 V output swing to nearly 10 MHz. The 50Ω resistor in series with the output ensures stability into capacitive loads; it also provides "back termination" into 50Ω coax cable (see the Appendix H on Transmission Lines). This op-amp is not particularly quiet ($e_n \sim 12\,\mathrm{nV}/\sqrt{\mathrm{Hz}}$), but that's good enough: the circuit's input-stage gain of ~ 40 reduces the op-amp's noise contribution to $\sim 0.3\,\mathrm{nV}/\sqrt{\mathrm{Hz}}$ when referred to the input. The amplifier's overall noise is $\sim 2\,\mathrm{nV}/\sqrt{\mathrm{Hz}}$.[46]

---

[44] A replacement for Toshiba's legendary (and discontinued) 2SK389.

[45] To approximately the same value as that of the op-amp alone, were it connected as a stable unity-gain follower.

[46] If it's *quiet* you want, you can substitute the large-die IF3602 dual

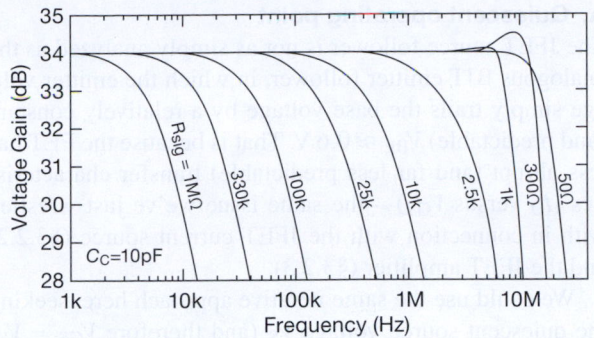

Figure 3.39. Measured gain versus frequency for the amplifier of Figure 3.37, for a range of values of signal source impedance R_{sig}.

This is a pretty good amplifier! With a bit of refinement (primarily substituting a low-noise current sink for R_{S}), two such amplifiers, combined in a so-called "instrumentation amplifier" (INA) configuration, will outperform any available integrated INA in terms of both noise and speed; see Figure 8.49 in §8.6.3.

B. Comparison with JFET-input op-amps

We can hear it already: "Yeah, yeah... you guys like to show off your clever circuits, with lots of discrete components. But nowadays that kind of circuit art is obsolete, because you can get all that performance, and more, in readily available integrated circuits – op-amps, in particular."

But, *can* you? Table 3.2 lists the currently available op-amps with JFET inputs that have a chance of competing. How does their performance compare with that of our dc-coupled hybrid amplifier of Figure 3.37? Let's see...

Output swing Only the first three op-amps can swing over the full ±15 V range; OK, but...

Bandwidth ...the fastest of those "high-voltage" op-amps has 80 MHz GBW, thus for $G=50$ a bandwidth of less than 2 MHz; and the two fastest op-amps, which can match our amplifier's bandwidth, can swing only ±4 V or so. By contrast, our amplifier's GBW is 4 GHz (40 times the op-amp's $f_{\text{T}}=100$ MHz), and thus excess gain (e.g., 400 at 10 MHz) with lower resulting distortion.

Noise Our amplifier's noise of $\sim 2\,\text{nV}/\sqrt{\text{Hz}}$ is 6 dB quieter than the best of Table 3.2's offerings.

Cost $5–10 for the op-amp solution, roughly the same for the better-performing hybrid amplifier ($2.50 for the LM6171, $3.25 for the LSK389 dual JFET).

JFET. That will reduce the input noise to $\sim 0.7\,\text{nV}/\sqrt{\text{Hz}}$, but with a greatly increased input capacitance (about 300 pF!). (And thermal noise from the feedback network will degrade this unless R_{g} is decreased to $\sim 5\,\Omega$; see §8.1 and Figure 8.80A.)

Table 3.2 Selected Fast JFET-input Op-amps[a]

	Supply		I_{bias}	e_{n}		Slew	
	Voltage range	I_{Q} typ	25°C typ	1kHz typ	GBW typ	Rate typ	Cost qty 25
Part #	(V)	(mA)	(pA)	(nV/$\sqrt{\text{Hz}}$)	(MHz)	(V/µs)	($US)
OPA604A	9–50	5	50	10	20	25	2.93
OPA827A	8–40	5	15	4	22	28	9.00
ADA4637	9–36	7	1	6	80[d]	170	10.12
OPA656	9–13	14	2	7[b]	230	290	5.59
OPA657	9–13	14	2	7	1600[d]	700	10.01
ADA4817	5–10.6	19	2	4[c]	1050	870	4.93

Notes: (a) candidates for wideband low-noise amplifier. (b) low $e_{\text{n}}C$ noise: C_{in}=2.8pF. (c) lowest $e_{\text{n}}C$ noise: C_{in}=1.5pF. (d) decomp, G_{CL} >7.

The contest so far? The hybrid amplifier is winning, in terms of the combined performance metric of bandwidth, output swing, and noise voltage. But we're not done yet...

Offset voltage The op-amps win, here, with out-of-the-box V_{os} values of 2 mV (max) for the three fastest parts; the hybrid amp requires manual trimming of its 20 mV worst-case untrimmed offset, if better is needed.

Parts count A win, again, for the op-amps.

Input capacitance Just 1.5 pF for the ADA4817, versus 10 pF or more for the hybrid (the price we paid for 2× lower noise).

Input current 20 pA (max) for the ADA4817 (but that's a low-voltage part), versus 200 pA for the hybrid (*unfair!* – that's specified at a large negative bias, $V_{\text{GS}} = -30\,\text{V}$)

The verdict? A split decision: the JFET op-amp solution is simple and can deliver plenty of speed (or plenty of swing, but not both), along with untrimmed accuracy and very low-input capacitance (thus low "$e_{\text{n}}C$" noise; see Chapter 8). The hybrid approach delivers speed and swing and lowest noise voltage; but it requires manual trim, it's more complicated, and it has more input capacitance. Note also that an op-amp is a more flexible building block in general, providing, for example, a wide common-mode input-voltage range that our hybrid circuit does not have; but that is not needed here, because the input is always close to ground (owing to the circuit's gain of 50).

3.2.5 Oscillators

In general, FETs have characteristics that make them useful substitutes for bipolar transistors in almost any circuit that can benefit from their uniquely high input impedance and low bias current. A particular instance is the use of a JFET amplifier stage to implement a high-stability LC or crystal oscillator; we'll show examples in §7.1.5D.

3.2.6 Source followers

Because of the relatively low transconductance of FETs, it's often better to use a FET "source follower" (analogous to an emitter follower) as an input buffer to a conventional BJT amplifier, rather than trying to make a common-source FET amplifier directly. You still get the high input impedance and zero dc input current of the FET, and the BJT's large transconductance lets you achieve high single-stage gain. Furthermore, discrete FETs (i.e., those that are not part of an integrated circuit) tend to have higher inter-electrode capacitance than BJTs, leading to greater Miller effect (§2.4.5B) in common-source amplifiers; the source-follower configuration, like the emitter follower, has no Miller effect.

FET followers, with their high input impedance, are commonly used as input stages in oscilloscopes as well as in other measuring instruments. There are many applications in which the signal-source impedance is intrinsically high, e.g., capacitor microphones, pH probes, charged-particle detectors, or microelectrode signals in biology and medicine. In these cases, a FET input stage (whether discrete or part of an integrated circuit) is a good solution. Within circuits there are situations in which the following stage must draw little or no current. Common examples are analog "sample-and-hold" and "peak detector" circuits, in which the level is stored on a capacitor and will "droop" if the next amplifier draws significant input current. In all these applications the advantage of negligible input current of a FET more than compensates for its low transconductance, making source followers (or even common-source amplifiers) attractive alternatives to the bipolar emitter follower.

Figure 3.40 shows the simplest source follower, which ideally should produce an accurate replica of the input waveform while drawing essentially zero input current. Let's figure out important things like its quiescent operating point, its exact voltage gain, its output impedance, and the voltage offset from input to output.

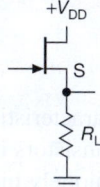

Figure 3.40. *N*-channel JFET source follower. Unlike the *npn* BJT emitter follower (in which the output trails the input by $V_{BE} \approx 0.6$ V), the output is here more *positive* than the input.

A. Quiescent operating point

The JFET source follower is not as simply analyzed as the analogous BJT emitter follower, in which the emitter voltage simply trails the base voltage by a relatively constant (and predictable) $V_{BE} \approx 0.6$ V. That is because the FET has less abrupt (and far less predictable) transfer characteristics (I_D versus V_{GS}) – the same issue we've just wrestled with in connection with the JFET current source (§3.2.2) and the JFET amplifier (§3.2.3).

We could use the same iterative approach here, seeking the quiescent source voltage V_S (and therefore $V_{GS} = V_S$) that produces a source current I_S (and therefore $I_D = I_S$) consistent with that V_S. And we could do this by interpolating between curves like those in Figure 3.21A (a family of I_D versus V_{DS} for several V_{GS}'s), or by sliding up and down a transfer characteristic curve like that in Figure 3.41 (a single curve of I_D versus V_{GS}, at some fixed V_{DS}), until we find the point at which $I_D R_L = -V_{GS}$.

But there is an elegant graphical method, used extensively during the days of vacuum tubes, that lets you find the operating point immediately: the method of "load lines."

B. Load lines

To find the operating point for the source follower in Figure 3.40, we simply notice that the load resistor R_L imposes its rules on allowable V_{GS} versus I_S, namely, Ohm's law: $I_S R_L = -V_{GS}$. We can plot this constraint on the same graph as the transfer curve of Figure 3.41, as a straight line with slope $-1/R_L$; note that it goes "backward,"

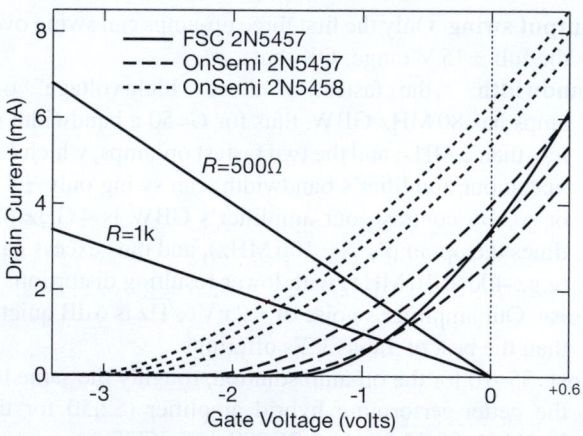

Figure 3.41. Measured transfer curves for a set of 2N5457 and 2N5458 *n*-channel JFETs at $V_{DS}{=}10$ V. These measurements extend beyond I_{DSS}, with V_{GS} taken 0.6 V into positive territory. The OnSemi curves show the parts with the lowest, middle, and highest I_{DSS} from a batch of ten each.

because $V_S = -V_{GS}$. The operating point has to be consistent with this constraint and simultaneously with the transfer characteristic of the JFET. In other words, the operating point is the intersection of the two curves. In this case, with $R_L = 1k$, the quiescent point is at $V_S = +1.6$ V (and, from the lowest 2N5458 curve, $I_D = 1.6$ mA).

Lest one be tempted to fall too quickly in love with this technique, we hasten to point out that the characteristic curves for a particular type of JFET exhibit a large spread. For the 2N5458 illustrated in Figure 3.41, for example, the specification allows I_{DSS} to be anywhere between 2 and 9 mA (and the pinchoff voltage $V_{GS(off)}$ can range from -1.0 to -7.0 V). In practice, it is rare to find devices at the extremes, and there tends to be good consistency in a single manufacturing batch (as indicated by the date code stamped on the parts); for example, by measuring a batch of 10 2N5458's (Figure 3.41) we determined that the quiescent point in this circuit would range from 1.52 to 1.74 V.

C. Output amplitude and voltage gain

We can figure out the output amplitude, as we did for the emitter follower in Section 2.3.3, using the transconductance. We have

$$v_s = R_L i_d$$

since i_g is negligible; but

$$i_d = g_m v_{gs} = g_m(v_g - v_s),$$

so

$$v_s = \left[\frac{R_L g_m}{(1 + R_L g_m)} \right] v_g.$$

That is, the gain is

$$G = \frac{1}{1 + \dfrac{1}{g_m R_L}}. \tag{3.7}$$

For $R_L \gg 1/g_m$ it is a good follower ($v_s \approx v_g$), with gain approaching, but always less than, unity. We are not near that limit in this example, in which the measured value $g_m = 1.9$ mS implies a voltage gain of $G_V = 0.66$ into the 1 kΩ load, far from the ideal of unity gain. Furthermore, the variation of transconductance over the signal swing results in undesirable nonlinearity. One solution is to use a JFET with higher transconductance, or (better) add a BJT transconductance enhancer (Figure 3.29F and §3x.2). But in situations in which the external load impedance is high, an elegant solution is to use a current sink as an active load, as we'll see presently (§3.2.6F).

D. Input impedance

Our hope that JFET source followers have infinite input impedance is largely fulfilled, but they do have some gate leakage current (see §3.2.8) and input capacitance (see Table 3.7 on page 217). Gate leakage can become a problem at drain-gate voltage greater than about 5 V (Figure 3.49), so be sure to check the JFET's datasheet and, if necessary, consider adding a cascode to limit V_{DG}.

A follower's frequency response when driven by signals of high source impedance is limited by input capacitance, $f_{3dB} = 1/2\pi C_{in}$, where $C_{in} = C_{iss} + C_{rss} + C_{stray}$. The gate-source capacitance C_{iss} is generally about two to five times higher than the gate-drain capacitance C_{rss}, but fortunately it's bootstrapped by the follower action and is effectively reduced to $(1 - G_V)C_{iss}$. If you follow our advice (below) so that G_V is nearly 1.0, only the JFET's C_{rss} remains to limit the bandwidth. But it's possible to bootstrap the drain and reduce the effect of C_{rss} by a factor of 5. This leaves C_{stray} as "the last man standing" to limit the bandwidth – but you may be able to knock him down too by "guarding" most of the input wiring capacitance (i.e., using the follower's output signal to drive the cable's shield; see the discussion of signal guarding in §5.15.3).

E. Output impedance

The preceding equation for v_s is precisely what you would predict if the source follower's output impedance were equal to $1/g_m$ (try the calculation, assuming a source voltage of v_g in series with $1/g_m$ driving a load of R_L). This is exactly analogous to the emitter-follower situation, in which the output impedance was $r_e = 25/I_C$, or $1/g_m$. It can be easily shown explicitly that a source follower has output impedance $1/g_m$ by figuring the source current for a signal applied to the output with the gate grounded (Figure 3.42). The drain current is

$$i_d = g_m v_{gs} = g_m v,$$

so

$$r_{out} = v/i_d = 1/g_m, \tag{3.8}$$

typically a few hundred ohms at currents of a few milliamps.[47]

In general, FET source followers aren't nearly as stiff as emitter followers. The exception is at very low currents,

---

[47] In practice, a more convenient way to measure the follower output impedance is to inject a signal *current* and measure the resulting source voltage, as in Figure 3.42B. Get the current from a signal generator, with a series resistor R_{sig} much larger than r_{out}, taking care to keep v_{out} small, say ~50 mV; then the equation in the figure gives you r_{out}.

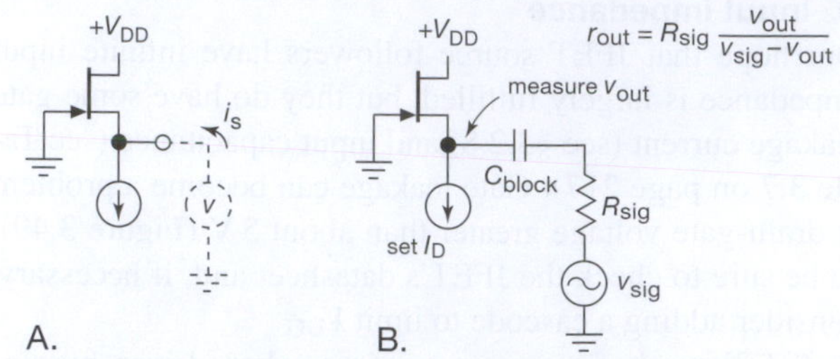

Figure 3.42. Calculating source-follower output impedance.

in the subthreshold region, where the transconductance of some JFETs approaches that of a BJT operated at the same current; see Figure 3.54 on page 168.

In this example, with $g_m = 1.9$ mS, the impedance looking back into the JFET's source is $r_{out} = 525\,\Omega$, which combines with the parallel 1k source load resistor to produce an output impedance of $345\,\Omega$, quite a bit higher than the analogous value of $r_e = 16\,\Omega$ for a BJT operating at the same 1.6 mA.

We were able to calculate the voltage gain and output impedance reasonably accurately in this example because we took the trouble to measure the I_D versus V_{GS} characteristic curves. It's worth pointing out, however, that the manufacturer's datasheet for the 2N5458 gives us little help here: it gives no characteristic curves for the 2N5458, only for the lower-current part (2N5457); and for the 2N5458 it specifies g_m only at I_{DSS}, where it gives a range of 1.5 mS to 5.5 mS. From these limits, along with the I_{DSS} and $V_{GS(OFF)}$ limits above, we would not be able to form a good estimate of the operating transconductance, because the operating point with a fixed value of source load resistor is ill-determined. We could do better by assuming that we adjust R_S to make $I_D = 1.6$ mA, say; then, using the fact that $g_m \propto \sqrt{I_D}$, the specified limits of I_{DSS} and of g_m (at I_{DSS}) guarantee that g_m lies in the range 0.6 mS to 4.9 mS.[48] Our measured value of g_m falls nicely within this range, being rather close to the geometric mean of these limits.

There are two drawbacks to this circuit.

1. The relatively high output impedance means that the output swing may be significantly less than the input swing, even with high load impedance, because R_L alone forms a divider with the source's output impedance. Furthermore, because the drain current is changing over the signal waveform, g_m and therefore the output impedance will

---

[48] In fact, one can narrow that estimate somewhat, because g_m and I_{DSS} are correlated: a JFET sample with unusually high g_m will lie at the high end of the I_{DSS} distribution as well.

vary, producing some nonlinearity (distortion) at the output. The situation is improved if FETs of high transconductance are used, of course, but a combination FET–BJT follower (or FET–BJT "g_m enhancer," Figure 3.29F) is often a better solution.

2. Because the V_{GS} needed to produce a certain operating current is a poorly controlled parameter in FET manufacture, a source follower has an unpredictable dc offset, a serious drawback for dc-coupled circuits.

(An additional problem is caused by the fact that an FET's drain current depends to some degree on drain-to-source voltage. You might call this the "g_{os} effect," which also acts to reduce the gain from the ideal $G = 1$. It is discussed later in §§3.3.2 and in the Chapter *3x* section "Bandwidth of the Source Follower with a Capacitive Load.")

Perhaps this is a good place to pause and realize that many of the circuits we've been considering would be easier to implement, and would work better, if we had access to a negative supply voltage. But often that's not the case; so, in the spirit of real-world circuit design constraints (and as a useful learning exercise) we're slogging through the extra difficulties posed by single-supply JFET follower design. But if you do have a negative supply available, by all means use it!

F. Active load

The addition of a few components improves the source follower enormously. Work with us here, as we take it in stages (Figure 3.43).

First we replace the load resistor (called R_S in Figure 3.43A) with a (pull-down) current sink (circuit B). (You can think of this as the previous case, with infinite R_S.) The constant source current makes V_{GS} approximately constant, thus reducing nonlinearities. A nice trick (circuit B′) has a BJT follower doing double duty, both providing low output impedance while sinking a (roughly) constant current of V_{BE}/R_B.

We still have the problem of unpredictable (and therefore nonzero) offset voltage (from input to output) of V_{GS} (or $V_{GS} + V_{BE}$, for circuit B′). Of course, we could simply adjust I_{sink} to the particular value of I_{DSS} for the given FET (in the first circuit) or adjust R_B (in the second). This is a poor solution for two reasons: (a) it requires individual adjustment for each FET; and (b) even so, I_D may vary by a factor of two over the normal operating temperature range for a given V_{GS}.

A better circuit uses a matched FET pair to achieve zero offset (circuit C). Q_1 and Q_2 are a matched pair, on a single chip of silicon, for example the excellent LSK389 (see

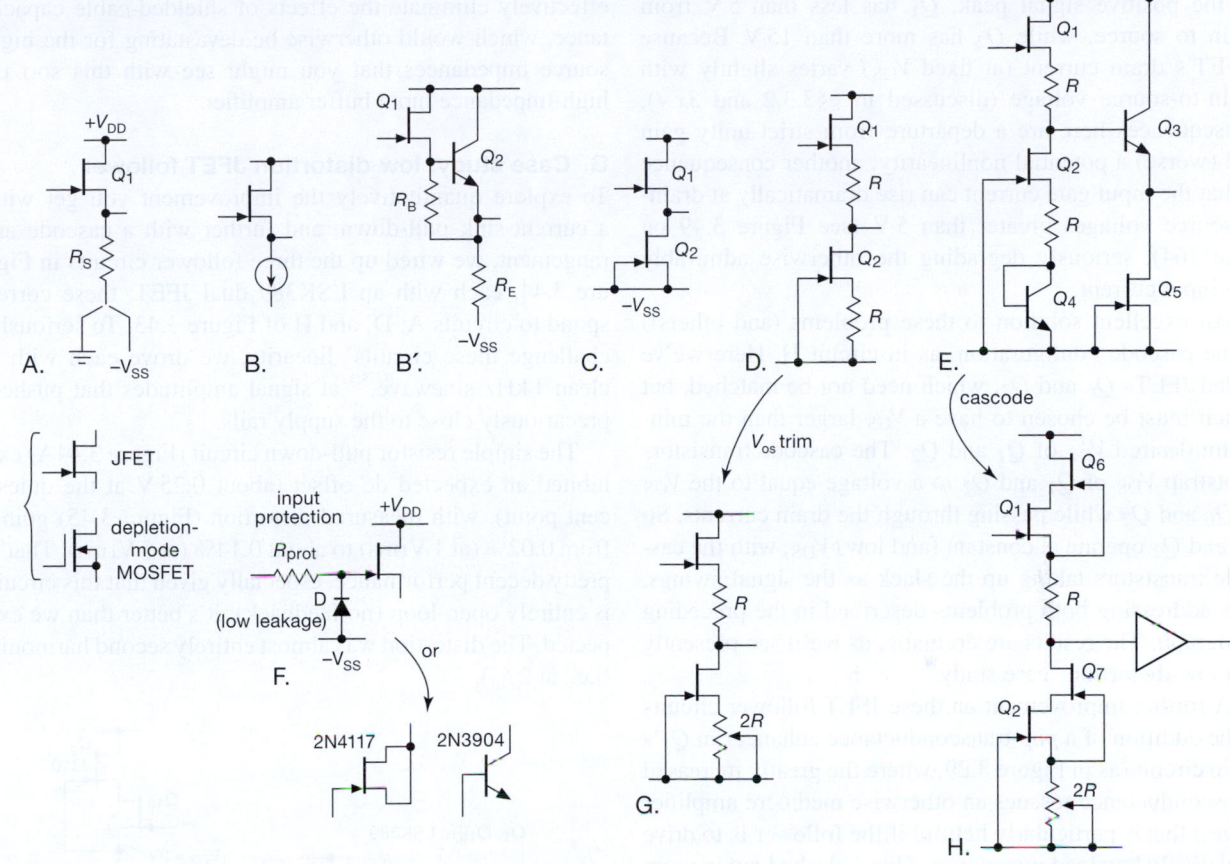

Figure 3.43. JFET unity-gain source followers – from simplest to best.

Table 3.7 on page 217), with Q_2 sinking a current of I_{DSS}, i.e., its drain-current corresponding to $V_{GS}=0$. But the JFETs are matched, so $V_{GS}=0$ for both transistors: voilà, Q_1 is a follower with zero offset. Because Q_2 tracks Q_1 in temperature, the offset remains near zero, independent of temperature.

You usually see the preceding circuit with source resistors added (circuit D). A little thought should convince you that the upper resistor R is necessary and that equal-value resistors guarantee that $V_{out} = V_{in}$ if Q_1 and Q_2 are matched. This circuit modification gives better I_D predictability, it allows you to set the drain current to some value less than I_{DSS}, and the source degeneration gives improved linearity. The variation of circuit G lets you trim the (already small) residual offset voltage caused by imperfect matching of the $Q_1 Q_2$ pair; the LSK389, for example, specifies a worst-case mismatch (at 1 mA drain current) of $\Delta V_{GS} = 20$ mV.[49]

Circuit E adds a BJT output follower (Q_3), with a JFET

current sink pull-down (Q_5). Transistor Q_4 adds a compensating V_{BE} in Q_2's source to maintain approximately zero dc offset from input to output.

Circuits A–D all share a problem, namely, the drain-to-source voltage across Q_1 varies directly with input signal. This can cause several undesirable effects. For example, imagine that circuit C is run between ±10 V supply rails and that the input signal swings between +5 and −5 V.

the two JFETs, but in circuit D there is a V_{DS} mismatch that depends on the input signal voltage relative to the supply rails. To estimate the resulting offset voltage of the follower you need to know the JFETs' output conductance (g_{os}), which causes an input–output offset in the follower that is proportional to the V_{DS} mismatch. That parameter is not specified on this JFET's datasheet, but from our measurements (see Table 3.7 on page 217) we know that $g_{os} \approx 100\,\mu S$, which causes a follower offset of $\Delta V = \Delta V_{DS}/G_{max}$; here that amounts to ≈ 60 mV for a 10V difference in V_{DS}, quite a bit larger than the 20 mV maximum untrimmed offset of the JFET pair (when the V_{DS}'s are balanced). The cure? Circuit H, a cascode in each JFET to hold each V_{DS} constant. Thunderous applause, yet again, for the remarkable cascode.

---

[49] But a tricky "gotcha": the offset specification assumes equal V_{DS} for

At the positive signal peak, Q_1 has less than 5 V from drain to source, while Q_2 has more than 15 V. Because a FET's drain current (at fixed V_{GS}) varies slightly with drain-to-source voltage (discussed in §§3.3.2 and *3x.4*), consequences here are a departure from strict unity gain and (worse) a potential nonlinearity; another consequence is that the input gate current can rise dramatically at drain-to-source voltages greater than 5 V (see Figure 3.49 on page 164), seriously degrading the otherwise admirably low input current.

An excellent solution to these problems (and others!) is the cascode configuration, as in circuit H. Here we've added JFETs Q_6 and Q_7, which need not be matched, but which must be chosen to have a V_{GS} larger than the minimum desired V_{DS} of Q_1 and Q_2. The cascode transistors bootstrap V_{DS} of Q_1 and Q_2 to a voltage equal to the V_{GS} of Q_6 and Q_7 while passing through the drain currents. So Q_1 and Q_2 operate at constant (and low) V_{DS}, with the cascode transistors taking up the slack as the signal swings, thus addressing both problems described in the preceding paragraph. The results are dramatic, as we'll see presently in a low-distortion "case study."

A further improvement on these JFET follower circuits is the addition of a *pnp* transconductance enhancer in Q_1's drain circuit (as in Figure 3.29, where the greatly increased transconductance rescues an otherwise mediocre amplifier stage); this is particularly helpful if the follower is to drive a relatively low load impedance. This is fleshed out in more detail in §*3x.2*.

JFETs can handle plenty of forward gate current, but they are easily damaged by reverse breakdown. When that possibility exists, it's a good idea to add gate protection, as in circuit F. The series resistor R_{prot} limits current through the clamp diode D (which should be a low-leakage part like the 1N3595, if low input current is important). You can use the base–collector junction of an ordinary BJT, or the gate-channel diode of a JFET; see the plot of measured diode reverse leakage currents in §*1x.7*. But there's a compromise here: a large value of R_{prot} safely limits the clamp current, but it introduces excessive Johnson (thermal) noise, a serious issue in low-noise applications. The use of a depletion-mode MOSFET current limiter solves this problem elegantly; see §5.15.4 for details.

Note that the JFETs in these examples can be replaced with depletion-mode MOSFETs, which are available with voltage ratings to 1000 V; in that case it's necessary to protect the gate against both forward and reverse overvoltages greater than ±20 V.

In a further variation of these circuits, you can use the output signal to drive an inner "guard" shield in order to effectively eliminate the effects of shielded-cable capacitance, which would otherwise be devastating for the high source impedances that you might see with this sort of high-impedance input buffer amplifier.

G. Case study: low-distortion JFET follower

To explore quantitatively the improvement you get with a current-sink pull-down, and further with a cascode arrangement, we wired up the three follower circuits in Figure 3.44, each with an LSK389 dual JFET; these correspond to circuits A, D, and H of Figure 3.43. To seriously challenge these circuits' linearity, we drove each with a clean 1 kHz sinewave,[50] at signal amplitudes that pushed precariously close to the supply rails.

The simple resistor pull-down circuit (Figure 3.44A) exhibited an expected dc offset (about 0.25 V at the quiescent point), with measured distortion (Figure 3.45) going from 0.02% (at 1 Vrms) to about 0.14% (at 5 Vrms). That's pretty decent performance, especially given that this circuit is entirely open-loop (no feedback); it's better than we expected. The distortion was almost entirely second harmonic (i.e., at $2f_{in}$).

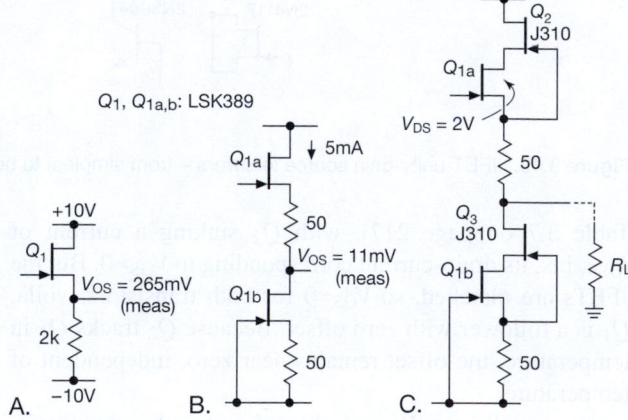

Figure 3.44. Three candidates for the JFET follower low-distortion medal-of-honor.

Adding an LSK389 matched JFET current sink with source degeneration (Figure 3.44B) made a nice improvement: a dc offset of about 10 mV and a measured distortion reduced by a factor of ten (20 dB), with the distortion now almost entirely third harmonic ($3f_{in}$). We're in serious audiophile territory here. Finally, adding a cascode (the

---

[50] From an SRS DS360 "Ultra-low distortion function generator": distortion less than 0.0003%. We measured the output distortion with a ShibaSoku 725B distortion analyzer.

J310's V_{GS} is much larger than that of Q_1, so that the latter runs at $V_{DS} \approx 2\,\text{V}$) improves the linearity by another 20 dB, bumping into the measurement floor of our modest apparatus.[51] The low drain-to-source voltage across Q_{1a} imposed by the cascode also ensures a low input gate current.[52]

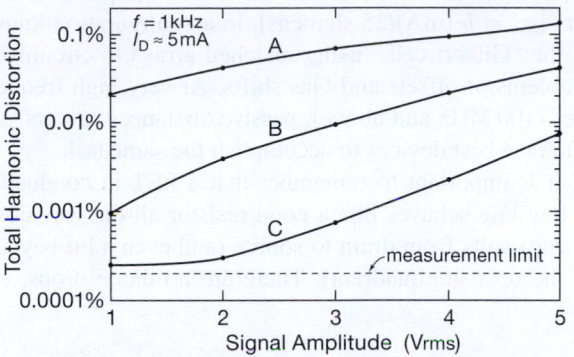

Figure 3.45. Measured distortion versus signal amplitude for the JFET followers of Figure 3.44, with $R_L = 1\,\text{M}$.

3.2.7 FETs as variable resistors

Figure 3.21 showed the region of JFET characteristic curves (drain current versus V_{DS} for a small family of V_{GS} voltages), both in the normal ("saturated") regime and in the "linear" region of small V_{DS}. We showed the equivalent pair of graphs for a MOSFET at the beginning of the chapter (Figure 3.2). The I_D-versus-V_{DS} curves are approximately straight lines for V_{DS} smaller than $V_{GS} - V_{th}$, and they extend in both directions through zero, i.e., the device can be used as a voltage-controlled resistor for small signals of either polarity. From our equation for I_D versus V_{GS} in the linear region (§3.1.4, eq'n 3.1) we easily find the ratio (I_D/V_{DS}) to be

$$\frac{1}{r_{DS}} = 2\kappa \left[(V_{GS} - V_{th}) - \frac{V_{DS}}{2} \right]. \qquad (3.9)$$

The last term represents a nonlinearity, i.e., a departure from resistive behavior (resistance shouldn't depend on signal voltage). However, for drain voltages substantially less than the amount by which the gate is above threshold ($V_{DS} \to 0$), the last term becomes unimportant, and the FET behaves approximately like a resistance:

---

[51] Oscillator courtesy of eBay; distortion analyzer courtesy of the MIT Flea Market.

[52] These impressively low distortions were measured into a high impedance. If you want to drive a substantial load, you may need to add a "g_m enhancer" to Q_{1a}; see §§*3x.2* and *3x.4* in Chapter *3x*.

$$r_{DS} \approx 1/[2\kappa(V_{GS} - V_{th})]. \qquad (3.10)$$

Because the device-dependent parameter κ isn't a quantity you are likely to know, it's more useful to write r_{DS} as

$$r_{DS} \approx r_{G0}(V_{G0} - V_{th})/(V_G - V_{th}), \qquad (3.11)$$

where the resistance r_{DS} at any gate voltage V_G is written in terms of the (known) resistance r_{G0} at some gate voltage V_{G0}.

Exercise 3.2. Derive the preceding "scaling" law.

From either formula you can see that the conductance ($=1/r_{DS}$) is proportional to the amount by which the gate voltage exceeds threshold. Another useful fact is that $r_{DS} = 1/g_m$, i.e., the channel resistance in the *linear* region is the inverse of the transconductance in the *saturated* region. This is a handy thing to know, because either g_m or r_{DS} is a parameter nearly always specified on FET data sheets.

Exercise 3.3. Show that $r_{DS} = 1/g_m$ by finding the transconductance from the saturation drain-current formula in §3.1.4.

Typically, the values of resistance you can produce with FETs vary from a few tens of ohms (as low as $0.001\,\Omega$ for power MOSFETs) all the way up to an open circuit. A typical application might be an automatic-gain-control (AGC) circuit in which the gain of an amplifier is adjusted (by means of feedback) to keep the output within the linear range. In such an AGC circuit you must be careful to put the variable-resistance FET at a place in the circuit where the signal swing is small, preferably less than 200 mV or so.

The range of V_{DS} over which the FET behaves like a good resistor depends on the particular FET and is roughly proportional to the amount by which the gate voltage exceeds threshold. Typically you might see nonlinearities of about 2% for $V_{DS} < 0.1(V_{GS} - V_{th})$, and perhaps 10% nonlinearity for $V_{DS} \approx 0.25(V_{GS} - V_{th})$. Matched FETs make it easy to design a ganged variable resistor to control several signals at once. You can also find some JFETs specifically intended for use as variable resistors (e.g., the InterFET 2N4338–41 series, and VCR series), with nominal ON-resistances specified at some V_{GS} (usually 0V).

A. Linearizing trick

It is possible to improve the linearity and, simultaneously, the range of V_{DS} over which a FET behaves like a resistor, by a simple compensation scheme. Look at the expression 3.9 for $1/r_{DS}$; you can see that the linearity will be nearly perfect if you can add to the gate voltage a voltage equal to one half the drain-source voltage. Figure 3.46 shows two circuits to do exactly that.

In the first, the JFET forms the lower half of a resistive voltage divider, thus forming a voltage-controlled attenuator (or "volume control"). R_1 and R_2 improve the linearity by adding a voltage of $0.5V_{DS}$ to V_{GS}, as just discussed. The JFETs shown have an ON-resistance (gate grounded) of $60\,\Omega$(max), giving the circuit an attenuation range of 0 to 40 dB. In the second circuit, the JFET's controllable resistance forms the lower leg of the gain-setting feedback divider in an op-amp noninverting voltage amplifier (the voltage gain is $G=[10k/R_{FET}]+1$).

The linearization of r_{DS} with a resistive gate-divider circuit, as above, is remarkably effective. In Figure 3.47 we've compared actual measured curves of I_D versus V_{DS} in the linear (low-V_{DS}) region for FETs, both with and without the linearizing circuit. The linearizing circuit is essential for low-distortion applications with signal swings of more than a few millivolts. We used it for amplitude control in the oscillator circuit of Figure 7.22, where the JFET was combined with a series resistor to create a low-distortion gain trimmer; the measured distortion of just 0.0002%.

When considering FETs for an application requiring a gain control, e.g., an AGC or "modulator" (in which the amplitude of a high-frequency signal is varied at an audio rate, say), it is worthwhile to look also at "analog-

multiplier" ICs. These are high-accuracy devices with good dynamic range that are normally used to form the product of two voltages. One of the voltages can be a dc control signal, setting the multiplication factor of the device for the other input signal, i.e., the gain. Analog multipliers exploit the g_m-versus-I_C characteristic of bipolar transistors [$g_m = I_C(\text{mA})/25$ siemens], in a configuration known as the "Gilbert cell," using matched arrays to circumvent problems of offsets and bias shifts. At very high frequencies (100 MHz and above), passive "balanced mixers" are often the best devices to accomplish the same task.

It is important to remember that a FET in conduction at low V_{DS} behaves like a good resistor all the way down to zero volts from drain to source (and even a bit beyond, in the opposite quadrant). There are no diode drops, sat-

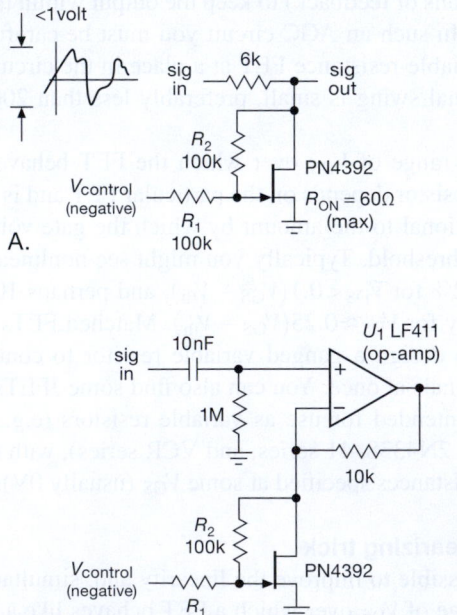

Figure 3.46. Linearizing the JFET variable resistor.

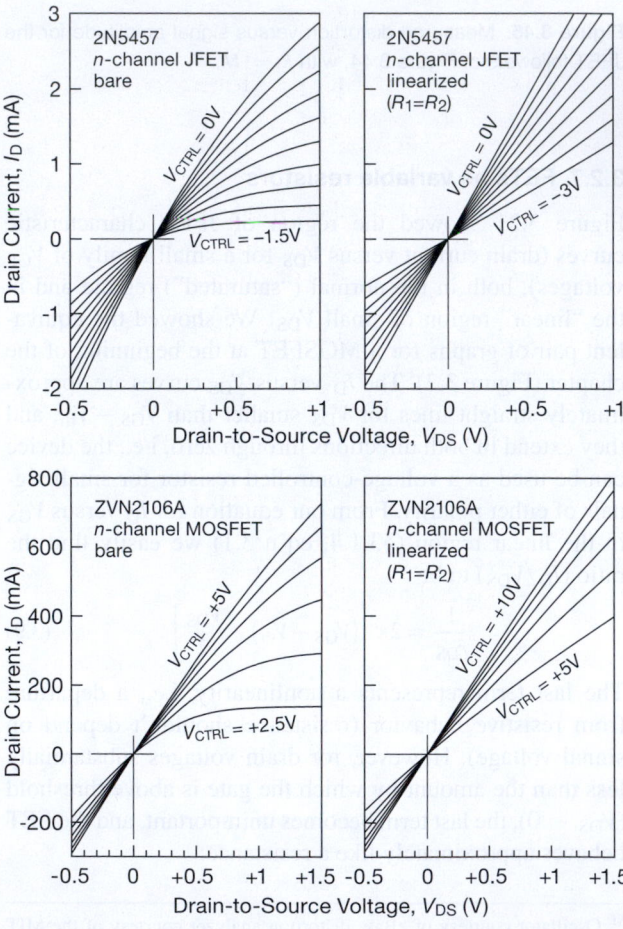

Figure 3.47. Measured curves of I_D versus V_{DS} for both (top) a JFET and (bottom) a MOSFET, showing the linearizing effect of a resistor pair (as in Figure 3.46). Note the relatively high current scale for the MOSFET.

uration voltages, or the like to worry about. We will see op-amps and digital logic families (CMOS) that take advantage of this nice property, giving outputs that saturate cleanly to the power supplies.

3.2.8 FET gate current

We said at the outset that FETs in general, and MOS-FETs in particular, have essentially zero steady-state gate current. This is perhaps the most important property of FETs, and it was exploited in the high-impedance amplifiers and followers in the previous sections. It will prove essential, too, in applications to follow – most notably analog switches and digital logic.

Of course, at some level of scrutiny we might expect to see some gate current. It's important to know about gate current, because a naive zero-current model is guaranteed to get you in trouble sooner or later. In fact, finite gate current arises from several mechanisms. (a) Even in MOSFETs the silicon dioxide gate insulation is not perfect, leading to leakage currents in the picoampere range. (b) In JFETs the gate "insulation" is really a back-biased diode junction, with the same impurity and junction leakage-current mechanisms as ordinary diodes. (c) Furthermore, JFETs (n-channel in particular) suffer from an additional effect known as "impact-ionization" gate current, which can reach astounding levels. (d) Finally – and most important for high-speed circuits – both JFETs and MOSFETs have *dynamic* gate current, caused by ac signals driving the gate capacitance; this can cause Miller effect, just as with bipolar transistors.[53] We'll deal with this important topic later, in §§3.5 and 3.5.4.

In most cases gate input currents are negligible in comparison with BJT base currents. However, there are situations in which a FET may actually have *higher* input current. Let's look at the numbers.

A. Gate leakage

The low-frequency input impedance of a FET amplifier (or follower) is limited by gate leakage. JFET datasheets usually specify a breakdown voltage, BV_{GSS}, defined as the voltage from gate to *channel* (source and drain connected together) at which the gate current reaches $1\,\mu$A. For smaller applied gate-channel voltages, the gate leakage current, I_{GSS}, again measured with the source and drain connected together, is considerably smaller, dropping

quickly to the picoampere range for gate-drain voltages well below breakdown. With MOSFETs you must never allow the gate insulation to break down; instead, gate leakage is specified as some maximum leakage current at a specified gate-channel voltage. Integrated circuit amplifiers with FETs (e.g., FET op-amps) use the misleading term "input bias current," I_B, to specify input leakage current; it's usually in the picoampere range.

The good news is that these leakage currents are in the picoampere range at room temperature. The bad news is that they increase rapidly (in fact, exponentially) with temperature, roughly doubling every 10°C. By contrast, BJT base current isn't leakage, it's bias current, and in fact tends to *decrease* slightly with increasing temperature. The comparison is shown graphically in Figure 3.48, a plot of input current versus temperature for several IC amplifiers (op-amps). The FET-input op-amps have the lowest input currents at room temperature (and below), but their input current rises rapidly with temperature, crossing over the curves for amplifiers with carefully designed BJT input stages like the LM10 and LT1012. These BJT op-amps, along with "premium" low-input-current JFET op-amps like the OPA111 and OPA627, are fairly expensive. However, we also included everyday "jellybean" op-amps like the bipolar LM358 and JFET LF411/2 in the figure to give

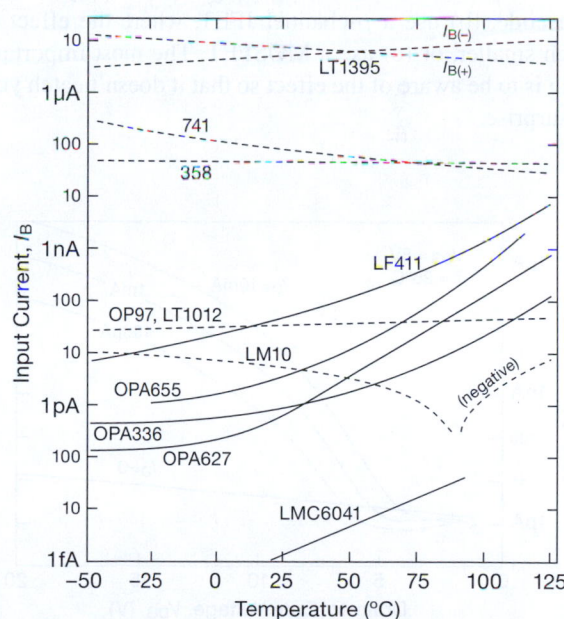

Figure 3.48. The input current of a FET amplifier is gate leakage, which doubles every 10°C. The FET-input amplifiers in this plot (solid lines) are easily spotted by their characteristic upward slope.

---

[53] In extreme cases, for example in high-voltage power switching, it can require *amperes* of gate drive current to switch a large MOSFET on the time scale of nanoseconds. This is not a trivial effect!

an idea of input currents you can expect with inexpensive (less than a dollar) op-amps.

B. JFET impact-ionization current

In addition to conventional gate leakage effects, *n*-channel JFETs suffer from rather large gate leakage currents when operated with substantial V_{DS} and I_D (the gate leakage specified on datasheets is measured under the unrealistic conditions that both $V_{DS}=0$ and $I_D=0$!). Figure 3.49 shows what happens. The gate leakage current remains near the I_{GSS} value until you reach a critical drain-gate voltage, at which point it rises precipitously. This extra "impact-ionization" current is proportional to drain current, and it rises exponentially with voltage and temperature. The onset of this current occurs at drain-gate voltages of about 25% of BV_{GSS}, and it can reach gate currents of a microamp or more. Obviously a "high-impedance buffer" with a microamp of input current is worthless. That's what you would get if you used a BF862 as a follower, running 1 mA of drain current from a 20 volt supply.

This extra gate leakage current afflicts primarily *n*-channel JFETs, and it occurs at higher values of drain-gate voltage. Some cures are to (a) operate at low drain-gate voltage, either with a low-voltage drain supply or with a cascode, (b) use a *p*-channel JFET, where the effect is much smaller, or (c) use a MOSFET. The most important thing is to be aware of the effect so that it doesn't catch you by surprise.

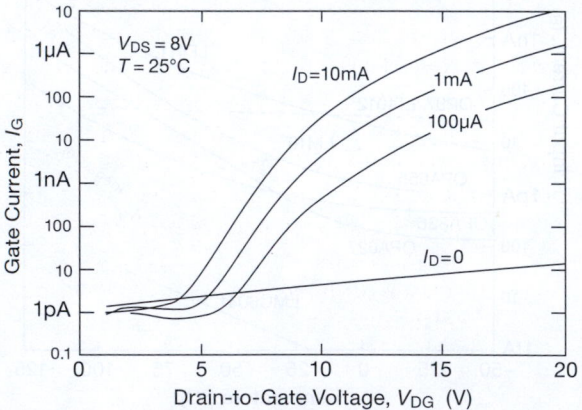

Figure 3.49. JFET gate leakage increases disastrously at higher drain-gate voltages and is proportional to drain current, as seen in the datasheet curves for the otherwise excellent BF862 *n*-channel JFET.

C. Dynamic gate current

Gate leakage is a dc effect. Whatever is driving the gate must also supply an *ac* current, because of gate capacitance. Consider a common-source amplifier. Just as with bipolar transistors, you can have the simple effect of input capacitance to ground (called C_{iss}), and you can have the capacitance-multiplying Miller effect (which acts on the feedback capacitance C_{rss}). There are two reasons why capacitive effects are more serious in FETs than in bipolar transistors. First, you use FETs (rather than BJTs) because you want very low input current; thus the capacitive currents loom relatively larger for the same capacitance. Second, FETs often have considerably larger capacitance than equivalent bipolar transistors.

To appreciate the effect of capacitance, consider a FET amplifier intended for a signal source of 100k source impedance. At dc there's no problem, because the picoampere currents produce only microvolt drops across the signal source's internal impedance. But at 1 MHz, say, an input capacitance of 5 pF presents a shunt impedance of about 30k, seriously attenuating the signal. In fact, *any* amplifier is in trouble with a high-impedance signal at high frequencies, and the usual solution is to operate at low impedance ($50\,\Omega$ is typical) or use tuned *LC* circuits to resonate away the parasitic capacitance. The point to understand is that the FET amplifier doesn't look like a $10^{12}\,\Omega$ load at signal frequencies.

As another example, imagine switching a 5 amp high-voltage load with a power MOSFET (there aren't any high-power *JFETs*), in the style of Figure 3.50. One might naively assume that the gate could be driven from a digital logic output with low current-sourcing capability, for example the so-called 4000-series CMOS logic, which can supply an output current of the order of 1 mA with a swing from ground to $+10$ V. In fact, such a circuit would be a disaster, because 1 mA of gate drive into the 200 pF average feedback capacitance of the IRF740 would stretch the output switching speed to a leisurely 50 μs.[54]

Even worse, the dynamic gate currents ($I_{gate}=C\,dV_D/dt$) would force currents back into the logic device's output, possibly destroying it by means of a perverse effect known as "SCR latchup" (more of which in Chapters 10 and 11);

---

[54] Our model is unacceptably crude, because feedback capacitance varies rapidly with drain voltage (see §3.5.4A). It's OK to use a constant value for feedback capacitance C_{rss} for small-signal calculations; but for a switching application like this, you need to go to the datasheet for values of *gate charge*, which take into account the nonlinear behavior of the capacitances. In this example the datasheet specifies $Q_G \approx 40$ nC of gate charge, producing a switching time of $t = Q_G/i = 40\,\mu$s at a (dynamic) drive current i of 1 mA.

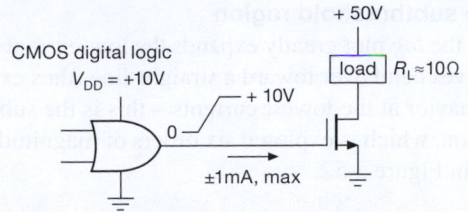

Figure 3.50. Dynamic gate current example: driving a fast-switching load.

for this and other reasons a series resistor (not shown in the figure) is usually added between the driving device and the MOSFET's gate. Bipolar power transistors have somewhat lower capacitances and therefore somewhat lower dynamic input currents (but still in the same ballpark); but when you design a circuit to drive a 5 amp power BJT, you're *expecting* to provide a few hundred milliamps or so of base drive (via a Darlington or whatever), whereas with a FET you tend to take for granted low input current. In this example – in which you would have to supply a couple of *amperes* of gate drive current to bring about the 25 ns switching speed that the MOSFET is capable of – the ultra-high-impedance FET has lost some of its luster.

Exercise 3.4. Estimate the switching times for the circuit of Figure 3.50, with 1 amp of gate drive current, assuming either (a) an average feedback capacitance of 200 pF, or (more accurately) (b) a required gate charge of 40 nC.

3.3 A closer look at JFETs

In §3.1.4 we introduced the landscape of FET operating regions: for drain voltages of a volt or more (to get beyond the "linear" resistive region) there is the conventional operating region in which the saturation drain current[55] I_D is proportional to $(V_{GS} - V_{th})^2$ and (at much lower drain currents) the subthreshold region in which I_D is exponential in V_{GS}.

That's the simple picture. Because JFETs are the devices of choice for accurate or low-noise (or both) circuits with high input impedance, it's worth looking more closely at their idiosyncrasies, ideally with measurements of actual devices.

We've undertaken an exhaustive (and exhausting!) review of most of the available JFETs, collecting sample batches of each, often from multiple manufacturers. Table 3.7 on page 217 includes most of them, with some measured parameters (I_{DSS}, and g_{os}, g_m, and V_{GS} at a useful drain current) alongside the datasheet specifications.[56] See also Table 8.2 on page 516, which lists a nice selection of low-noise JFETs.

Advanced JFET topics are treated in Chapter *3x*; here we discuss a few essential subjects – JFET operating regions (including the often neglected *subthreshold* region), JFET transconductance, and JFET capacitance.

3.3.1 Drain current versus gate voltage

A persistent issue with JFET circuit design is *parameter spread*. This is nicely illustrated in Figures 3.51 and 3.52, where we've plotted measured I_D versus V_{GS} for six samples of the 2N5457 *n*-channel JFET (three from each of two manufacturers[57]), and three samples of the related 2N5458 (from the same 2N5457–59 family). In each case we chose the parts with the highest, lowest, and median values of measured I_{DSS} in a batch of ten parts. For these measurements we ventured into positive gate voltages (up nearly to a diode drop, the onset of gate conduction), well beyond the usual zero-bias limit; nothing terrible happens, but in general this is a practice that should be avoided.

Let's look more closely at the data for these nine parts to understand different aspects of their performance and the impact on JFET circuit designs.

A. The quadratic region

The *linear* plot is what you usually see in datasheets and in textbooks. That scaling reveals well the quadratic behavior of drain current at a significant fraction of the zero-bias drain current (I_{DSS}), where JFETs are most often operated [eq'n 3.4]. You can see also the variation of I_{DSS} among samples (the specified range for the half-dozen 2N5457 samples is indicated by a vertical bar), as well as the somewhat steeper slope for the Fairchild 2N5457 samples. The slope is, of course, just the transconductance gain, $g_m = dI_D/dV_{GS}$, increasing linearly with V_{GS} [eq'n 3.5] in this regime of where the drain current is quadratic in $V_{GS} - V_{th}$.

---

[55] The term "saturation" can be confusing: for FETs it is used to denote *current* saturation, the region of drain voltages more than a volt or so where the drain current is approximately constant. By contrast, for BJTs the term "saturation" denotes *voltage* saturation (an ON switch), in which the collector voltage is close to zero. It never hurts to add the qualifying adjective (though people seldom do).

[56] See also Table 8.2 for noise parameters. These JFET tables represent several months' work – of comparing datasheet specifications, of checking distributors' stock, and of lab measurements. We found it revealing and rewarding, and we hope it is of use to the reader.

[57] We used dashed lines for the Fairchild parts, to untangle the crowd.

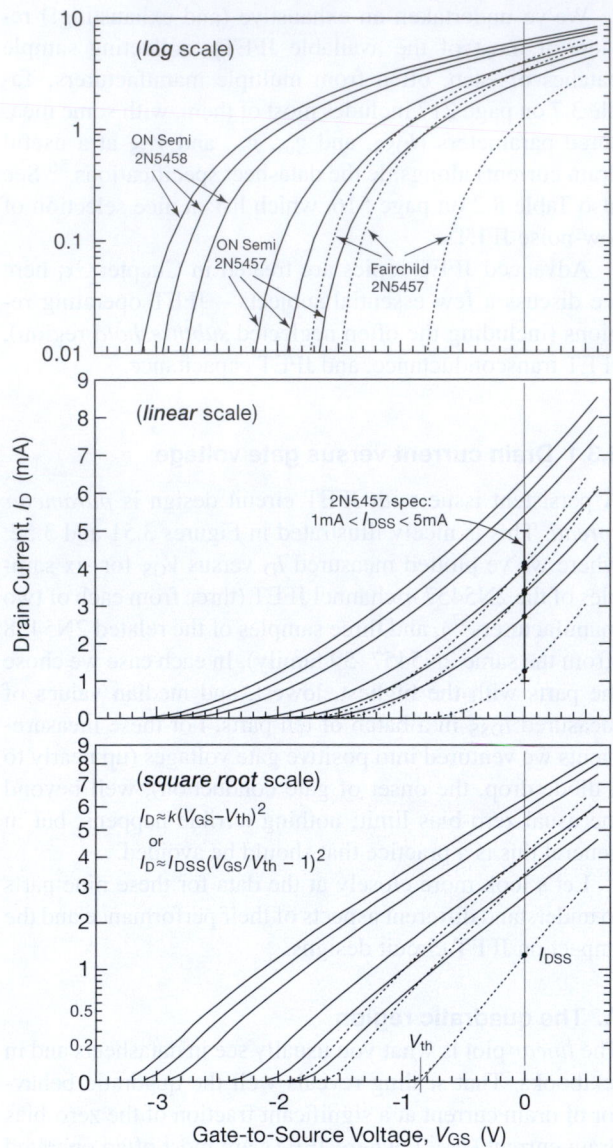

Figure 3.51. Drain current versus gate voltage for nine parts from the 2N5457–9 n-channel JFET family, operating with $V_{DS}=5\,V$. The same data is plotted on linear, log, and square-root axes. The ON Semi and Fairchild parts both meet the I_{DSS} spec (middle panel) despite their substantially different curves. Note that the measurements extend beyond I_{DSS}, to $V_{GS}+0.6\,V$.

Finding V_{th}: square-root plot

Look next at the *square-root* plot, which gives the quadratic region a chance to stretch out its limbs. The extrapolation to zero drain current defines the threshold voltage V_{th}. At that voltage the current is not *zero* – it's just down near the top of the subthreshold region.

B. The subthreshold region

Finally, the *log* plot greatly expands the low-current region. The curves bend over toward a straight-line (thus exponential) behavior at the lowest currents – this is the subthreshold region, which is explored six orders of magnitude more deeply in Figure 3.52.

C. The deep subthreshold region

From the extended curves in Figure 3.52 we can see that JFETs continue to do their stuff down to *pico*amp drain currents, conforming accurately to an exponential drain-current law (analogous to the BJT's Ebers–Moll), which can be written as

$$I_D = I_0 \exp(V_{GS}/nV_T) \tag{3.12}$$

with the same $V_T=kT/q\approx25\,mV$ as in Ebers–Moll (but with an added fudge factor n). The measured data in Figure 3.52 corresponds to a near-unity value for n ($n=1.05$). In other words, at very low drain currents, a JFET has nearly the same transconductance as a BJT operating at that same collector current (as confirmed in the measured transconductance data of Figure 3.54 on page 168).[58] Note that the Fairchild parts (in a gesture of egalitarian spirit) are no longer anomalous – unlike their behavior in the quadratic region (where they were the transconductance champs), at low currents they exhibit the same transconductance (slope) as the other parts.

3.3.2 Drain current versus drain-source voltage: output conductance

Drain current (at constant V_{GS}) *does* have some dependence upon drain-to-source voltage, in contrast to the idealized picture that you usually see (and to the promulgation of which we are guilty parties: see §3.1.4 and Figure 3.13). You can think of this effect as analogous to Early effect in BJTs (§2x.8), or, equivalently, as a finite output impedance r_o (or, more commonly, finite output *conductance* $g_{os}=1/r_o$) seen at the drain terminal when the gate-source voltage is held constant. This effect limits the maximum gain of a grounded-source amplifier configured with

---

[58] But not much bandwidth! For instance, at $I_D=10\,pA$ a 2N5457 would have a paltry gain–bandwidth product f_T of just 140 Hz ($f_T=g_m/2\pi C_{iss}$; see "Bandwidth of the Cascode, BJT vs FET" in Chapter 3x). It's nice to know that JFETs (like BJTs) work well at very low currents, a useful characteristic for micropower and nanopower applications. But don't forget that device capacitances loom large at low currents, so current-starved designs are slower than circuits running at normal currents.

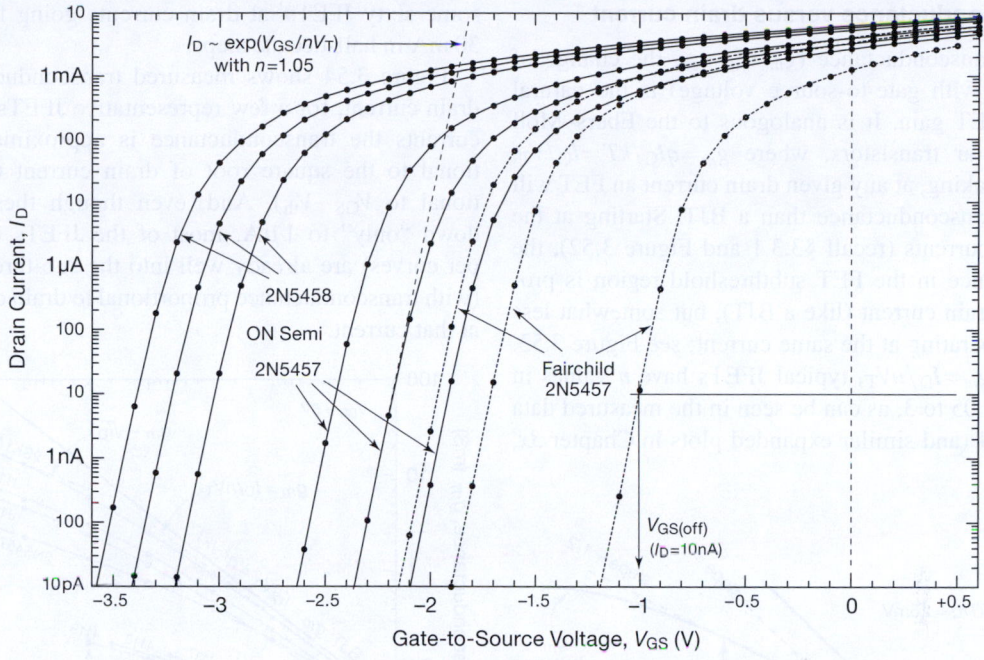

Figure 3.52. Log plot of measured drain currents (dots) versus gate voltage for the same JFETs as in Figure 3.51. JFETs work fine at 10 pA, far below the arbitrary 10 nA drain current that conventionally defines the gate-source cutoff voltage $V_{\mathrm{GS(off)}}$.

a current source as drain load (that's the $G_{\max}$ parameter in Table 3.1 on page 141), and you need to take it into account if your gain approaches $G_{\max}$. It also further degrades the performance of a source follower (already degraded by low $g_m R_{\mathrm{L}}$; see §3.2.6) if $G_{\max}$ is comparable to or less than $g_m R_{\mathrm{L}}$. These consequences are discussed in detail in Chapter 3x (§3x.4); but they are important enough to be worthy of mention here.

A. Degraded gain and linearity in the common-source amplifier

This "g_{os}" effect" limits the maximum gain of the common-source amplifier, by effectively putting a resistance $r_o = 1/g_{\mathrm{os}}$ across the drain load impedance. For a simple resistive drain load R_{D} that reduces the gain from the ideal $G = g_m R_{\mathrm{D}}$ to $G = g_m (R_{\mathrm{D}} \parallel r_o)$, or

$$G = g_m R_{\mathrm{D}} \frac{1}{1 + g_{\mathrm{os}} R_{\mathrm{D}}} \qquad (3.13)$$

A further unwelcome consequence is some nonlinearity, owing to the dependence of g_{os} upon drain voltage. See §3x.4 if you want to know more about this.

B. Gain error in the source follower

The "g_{os}" effect" also acts to reduce the gain of the source follower from the ideal $G = 1$. This is most noticeable

with relatively light loads, where $R_{\mathrm{L}} \gg 1/g_m$ (and therefore where you would expect the voltage gain to be very close to unity). Taking this effect into account, the voltage gain of the simple JFET follower of Figure 3.40 becomes

$$G = \frac{1}{1 + \dfrac{1}{g_m R_{\mathrm{L}}} + \dfrac{1}{G_{\max}}} \qquad (3.14)$$

where $G_{\max}$ is the transconductance times the drain resistance ($r_o = 1/g_{\mathrm{os}}$), so $G_{\max} = g_m/g_{\mathrm{os}}$ at the operating voltage and current. (We defined $G_{\max}$ because it's roughly constant with current[59] and therefore more useful than the current-dependent r_o and g_{os}). Thus, for example, you need both $g_m R_{\mathrm{L}}$ and $G_{\max}$ greater than 100 for $<1\%$ gain error – or you need to use some tricks, such as an active load, a cascode connection, or a "g_m enhancer." Looking back to the tutorial source follower progression in Figure 3.43, that is the reason for the use of an active load and a cascode connection in circuit H. Table 3.1 lists measured values of $G_{\max}$ for common JFETs. For further enlightenment read the exposé in §3x.4.

---

[59] Looking more closely, both g_m and g_{os} depend on drain current, approximately proportional to $\sqrt{I_{\mathrm{D}}}$ (§§3.3.3, 3x.2); their ratio $G_{\max}$ is relatively flat with drain current, but approximately proportional to drain-source voltage (§§3.3.4, 3x.4).

3.3.3 Transconductance versus drain current

Recall that transconductance ($g_m \equiv i_d/v_{gs}$, the change of drain current with gate-to-source voltage) is the natural measure of FET gain. It is analogous to the Ebers–Moll view of bipolar transistors, where $g_m = qI_C/kT = I_C/V_T$. Generally speaking, at any given drain current an FET will have lower transconductance than a BJT. Starting at the lowest drain currents (recall §3.3.1 and Figure 3.52), the transconductance in the FET subthreshold region is proportional to drain current (like a BJT), but somewhat less than a BJT operating at the same current; see Figure 3.53. If you write $g_m = I_D/nV_T$, typical JFETs have n values in the range of 1.05 to 3, as can be seen in the measured data of Figure 3.54 (and similar expanded plots in Chapter *3x*, see §*3x.2*).

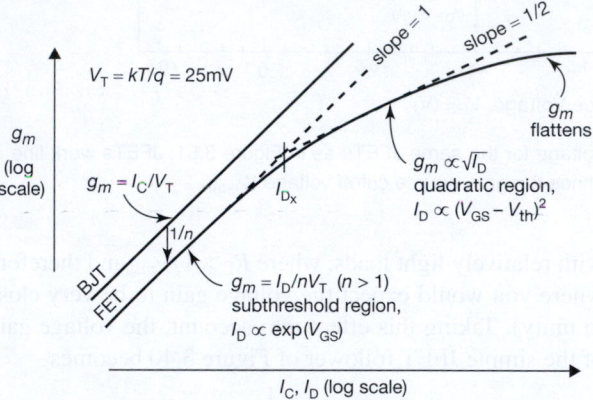

Figure 3.53. Transconductance versus drain current: in the subthreshold region an FET's transconductance is proportional to current (like a BJT); at normal currents where $I_D \propto (V_{GS}-V_{th})^2$ it goes like $\sqrt{I_D}$; then at still higher operating currents it flattens toward a maximum.

At higher currents, in the FET's "quadratic region" of drain current where $I_D \propto (V_{GS}-V_{th})^2$, the transconductance becomes proportional to the square root of drain current, falling away below the BJT's increasing ($\propto I_C$) transconductance. At still higher currents the transconductance flattens further. For JFETs the story ends here; but for MOSFETs (where you can bring the gate voltage high enough to reach constant maximum drain current), the transconductance heads back down, for example with the LND150 whose I_D versus V_{GS} curve is plotted in Figure 3.9.

It's not hard to measure transconductance directly – see §*3x.3* for circuit details of one simple method (where the secret sauce is a cascode stage to clamp the drain voltage). We used that circuit to measure the transconductance of some sixty JFETs, at drain currents going from $1\,\mu$A to $30\,$mA in half-decade steps.

Figure 3.54 shows measured transconductance versus drain current, for a few representative JFETs.[60] At higher currents the transconductance is approximately proportional to the square root of drain current (thus proportional to $V_{GS}-V_{th}$). And, even though these curves go down "only" to $1\,\mu$A, most of the JFETs (the four upper curves) are already well into the sub-threshold region (with transconductance proportional to drain current) down at that current.

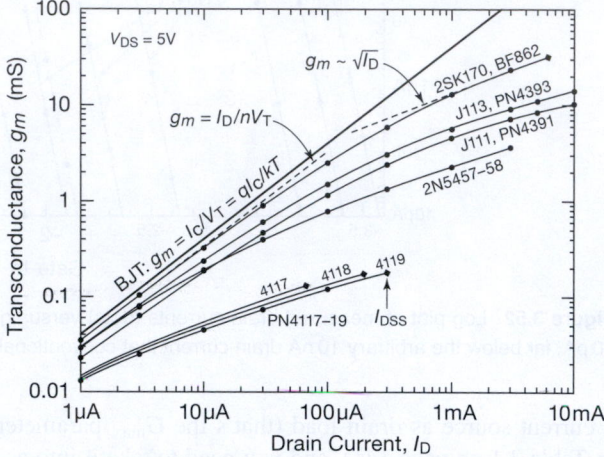

Figure 3.54. Measured low-frequency transconductance for some representative *n*-channel JFETs. See also the analogous plot in §3x.2.

So, what's going on with the PN4117–19? Well, these are very small JFETs (you can tell by their very low I_{DSS} values), so even at a drain current of $1\,\mu$A the current *density* is high enough to bring them over into their quadratic region. Put another way, in the preceding discussion we should have been referring to "current density" (rather than simple drain current) as the measure of whether a given JFET is in its subthreshold region or its quadratic region.

Note that this holds also for MOSFETs: a large power MOSFET (with drain current ratings in the hundreds of amps) will be running in its subthreshold region if used in a linear circuit application.[61] As we'll see in §3.6.3, this has an important effect on thermal stability when multiple MOSFETs are connected in parallel in power applications, because in the subthreshold region MOSFETs have a *positive* thermal coefficient of drain current with temperature;

---

[60] See the rich thicket of measured curves in the analogous plot in §3x.2.
[61] SPICE-lovers beware: the simulation models for power MOSFETs are nearly useless in the subthreshold region.

this can cause serious mischief when not fully appreciated by the circuit designer.

A. Transconductance within a JFET family

From Figure 3.54 and the preceding discussion it would appear that you cannot really predict a JFET's transconductance at any given operating current; that graph shows a 50:1 variation of g_m at a given current. Worse still, the data of Figure 3.51 suggests that you cannot even predict with any reasonable certainty the operating current of a given JFET part type for a given gate bias (or vice-versa).

As it turns out, the situation is not so bleak. Within a family (or extended family) of similar JFETs the transconductance depends primarily (and predictably) upon drain current, even though the corresponding gate voltage may be all over the map. Take a look at Figure 3.55, which shows measured drain current curves for a varied selection of seven 2N5486 JFETs (manufactured over a period of

35 years[62]), along with a sample of lower-current siblings (2N5484 and 2N5485) and a related cousin (SST4416, also featured for RF applications, and with similar low capacitance and comparable range of I_{DSS}).

These measured curves exhibit a wide variation of behavior for this selection of parts (there's a 3:1 spread of $V_{GS(off)}$, and a 5:1 spread of I_{DSS}), suggesting a similar unpredictability of transconductance gain.[63] But that suggestion would be misleading. When you measure their transconductances, you get the curves of Figure 3.56: within this deliberately varied selection of ten related parts there's at most a $\pm20\%$ peak variation of transconductance at a given drain current.

The moral: within a JFET family of *similar part types*, the drain current (at whatever gate voltage is needed to get that current) is reasonably predictive of the transconductance. And the practical consequence is that a JFET amplifier circuit will have reasonably predictable gain if it is arranged with bias feedback to set the drain current to a desired value (which should be chosen no greater than the specified minimum I_{DSS}, unless you are willing to sort your incoming parts).

But a piece of advice: for better gain predictability in JFET amplifiers it's a good idea (a) to use some source degeneration (§*3x.4*), preferably in combination with a

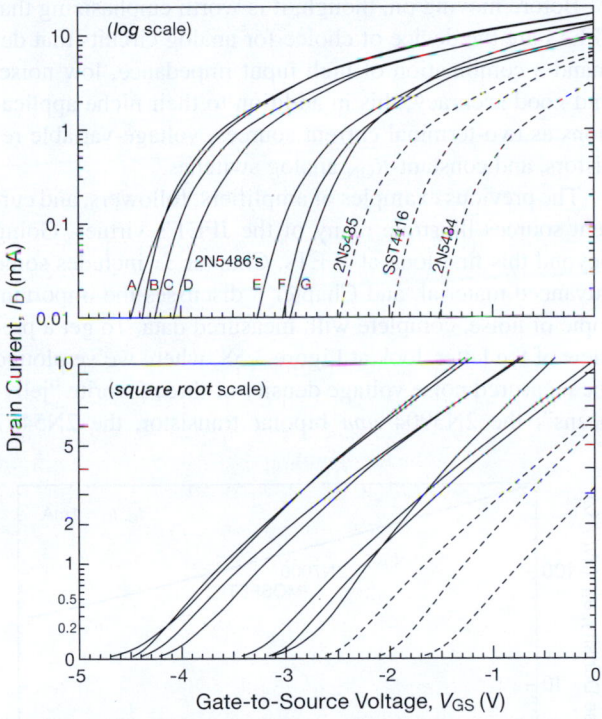

Figure 3.55. Measured drain current versus gate-source voltage for a collection of seven 2N5486 JFETs of varying vintages and manufacturers (along with three smaller JFET types, indicated with dashed lines), plotted on both log and square-root axes, illustrating the wide spread of threshold voltages V_{th} (the straight-line extrapolation of the latter's curves). Compare the large spread seen here with Figure 3.56, where the same set of JFETs exhibit only minor variations in transconductance at any given drain current.

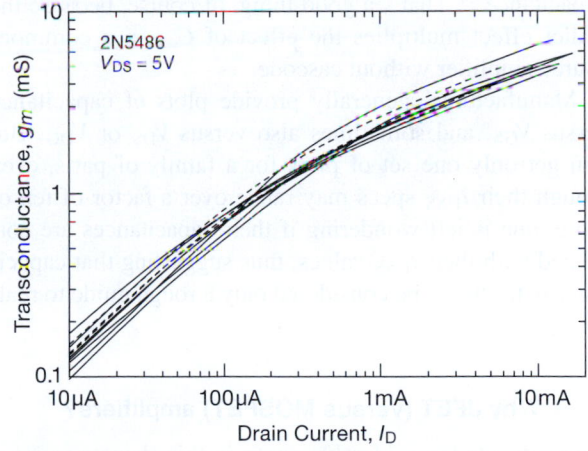

Figure 3.56. Measured transconductance versus drain current for the same set of JFETs as in Figure 3.55.

---

[62] A, Intersil, date code 7328 (28th week of 1973); B, Central Semiconductor, contemporary; C and E, Fairchild, date code BF44; D, Vishay SST5486, contemporary; F, Vishay, date code 0536; G, Motorola, 1990 vintage.

[63] There are some helpful correlations, though: the gate threshold voltage V_{th} is predictive of the zero-bias drain current I_{DSS}.

transconductance-boosting enhancer circuit, or (b) to use overall feedback with enough minimum loop gain (with worst-case g_m) to ensure the gain accuracy you need (as in Figures 3.31, 3.34, and 3.37).

3.3.4 Transconductance versus drain voltage

The transconductance of a FET at a given drain current is relatively independent of drain *voltage*, except for drain-to-source voltages below a volt or two. This contrasts with its strong dependence on drain *current*. See the discussion in Chapter *3x* (§*3x.2*, with measured data).

3.3.5 JFET capacitance

Just as with bipolar transistors (and, later, MOSFETs), the capacitance seen between the terminals of a JFET depends on the (reverse) bias; this is often called a "nonlinear capacitance" and decreases markedly with increasing reverse-bias. Figure 3.57 shows datasheet plots of feedback and input capacitances for two common *n*-channel JFETs. These values – just a few picofarads – are typical of small-signal JFETs and quite a bit smaller than you see in power MOS-FETs (compare with Figure 3.100 on page 197). JFETs are usually symmetrical, but owing to the larger reverse bias, the gate-drain capacitance is smaller than the gate-source capacitance.[64] That's a good thing, of course, because the Miller effect multiplies the effect of C_{rss} in a common-source amplifier without cascode.

Manufacturers generally provide plots of capacitance versus V_{GS}, and sometimes also versus V_{DS} or V_{DG}. But you get only one set of plots for a family of parts, even though their I_{DSS} specs may range over a factor of ten or more; one is left wondering if their capacitances are correlated with their I_{DSS} values, thus suggesting that capacitance data should be considered only a rough guide to reality.

3.3.6 Why JFET (versus MOSFET) amplifiers?

We've devoted considerable space in this chapter to *junction* FETs (JFETs), a topic that often receives minimal discussion in standard references, in which the emphasis is on insulated-gate FETs (i.e., *MOS*FETs). We'll soon shift the focus toward the latter, for good reasons: small integrated MOSFETs are dominant in (a) low-voltage and low-power analog circuits (op-amps, portable electronics, RF circuits,

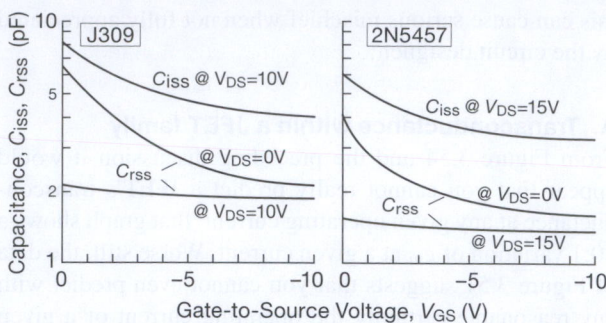

Figure 3.57. Input and feedback capacitance of JFETs.

and the like); (b) analog switching; (c) logic circuits, microprocessors, and memory; and, in the form of discrete packaged power transistors, (d) power switching and linear power applications. These are all hot topics in contemporary electronics, and MOSFETs are the dominant species of transistor on the planet, by a huge margin.

Before moving on, though, it is worth emphasizing that JFETs are the device of choice for analog circuits that demand a combination of high input impedance, low noise, and good accuracy; this in addition to their niche applications as two-terminal current sources, voltage-variable resistors, and constant-R_{ON} analog switches.

The previous examples of amplifiers, followers, and current sources illustrate many of the JFET's virtues. Going beyond this first look at JFETs, Chapter *3x* includes some advanced material, and Chapter 8 discusses the important topic of noise, complete with measured data. To get a preview of the latter, look at Figure 3.58, where we've plotted the measured noise voltage density of three favorite "jellybeans": the 2N3904 *npn* bipolar transistor, the 2N5457

---

[64] By "gate" and "source" we mean the pins your circuit uses for those functions, not the actual labels on the pins.

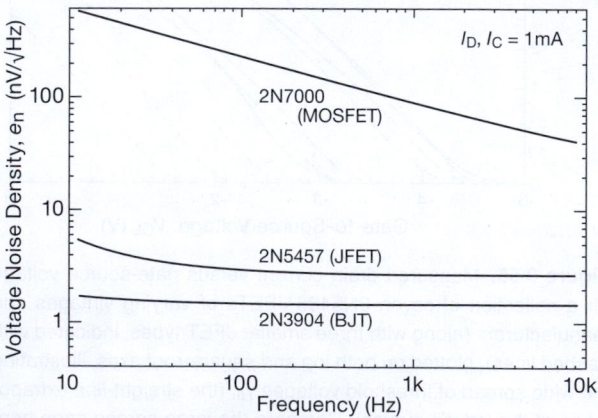

Figure 3.58. Noise voltage of three popular jellybeans, illustrating the poor low-frequency noise properties of MOSFETs.

JFET and the 2N7000 MOSFET. These parts make no pretense of premium membership; they cost a pittance,[65] and they're not aimed at low-noise applications. But they illustrate an important trend: MOSFETs are inherently noisy at low frequencies, by as much as 40 dB relative to their BJT and JFET cousins.[66] Don't use them for low-level audio applications – but power MOSFETS can be used for the muscular output stage of an audio amplifier.

3.4 FET switches

The two examples of FET circuits that we gave at the beginning of the chapter were both *switches*: a logic-switching application and a linear signal-switching circuit. These are among the most important FET applications, and they exploit the FET's unique characteristics: high gate impedance and bipolarity resistive conduction clear down to zero volts. In practice you usually use MOSFET integrated circuits (rather than discrete transistors[67]) in all digital logic and linear-switch applications, and it is only in power switching applications that you ordinarily resort to discrete FETs. Even so, it is essential (and fun!) to understand the workings of these chips; otherwise you're almost guaranteed to fall prey to some mysterious circuit pathology.

3.4.1 FET analog switches

A common use of FETs, particularly MOSFETs, is as analog switches. Their combination of low ON-resistance (all the way to zero volts), extremely high OFF resistance, low leakage currents, and low capacitance makes them ideal as voltage-controlled switch elements for analog signals. An ideal analog, or linear, switch behaves like a perfect mechanical switch: in the ON state it passes a signal through to a load without attenuation or nonlinearity; in the OFF state it is an open circuit. It should have negligible capacitance to ground and negligible coupling from the signal applied to the control input.

Let's look at an example (Figure 3.59). Q_1 is an *n*-channel enhancement-mode MOSFET, and it is nonconducting when the gate is grounded or negative. In that state the drain–source resistance (R_{off}) is typically more

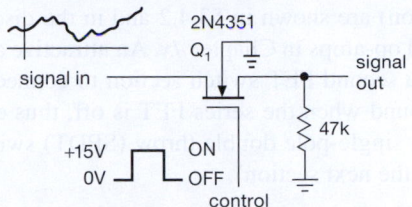

Figure 3.59. nMOS analog switch, with body terminal and diode shown.

than 10,000 M, and no signal gets through (though at high frequencies there will be some coupling through drain-source *capacitance*; more on this later). Bringing the gate to +15 V puts the drain-source channel into conduction, typically 20–200 Ω in FETs intended for use as analog switches. The gate signal level is not at all critical, as long as it is sufficiently more positive than the largest signal (to maintain R_{ON} low), and it could be provided from digital logic circuitry (perhaps using a FET or BJT to generate a full-supply swing) or even from an op-amp running from a +15 V supply. Swinging the gate negative (as from a bipolarity op-amp output) doesn't hurt, and in fact has the added advantage of allowing the switching of analog signals of either polarity, as will be described later. Note that the FET switch is a bidirectional device; signals can go either way through it. Ordinary mechanical switches work that way, too, so it should be easy to understand.

The circuit as shown will work for positive signals up to about 10 V; for larger signals the gate drive is insufficient to hold the FET in conduction (R_{ON} begins to rise), and negative signals would cause the FET to turn on with the gate grounded (it would also forward-bias the channel–body junction). If you want to switch signals that are of both polarities (e.g., signals in the range -10 V to $+10$ V), you can use the same circuit, but with the gate driven from -15 V (OFF) to $+15$ V (ON); the body terminal should then be tied to -15 V.

With any FET switch it is important to provide a load resistance in the range of 1k to 100k in order to reduce capacitive feedthrough of the input signal that would otherwise occur during the OFF state. The value of the load resistance is a compromise: low values reduce feedthrough, but they begin to attenuate the input signal because of the voltage divider formed by R_{ON} and the load. Because R_{ON} varies over the input signal swing (from changing V_{GS}), this attenuation also produces some undesirable nonlinearity. Excessively low load resistance appears at the switch input, of course, loading the signal source as well. Several possible solutions to this problem (multiple-stage switches, R_{ON}

---

[65] Roughly $0.02, $0.05, and $0.04, respectively, in 1000-piece quantities.

[66] According to John Willison, this phenomenon may be associated with intermittent trapping and release of charges on the insulated gate.

[67] It's hardly your choice – you are *forced* to, because discrete small-signal MOSFETs are a vanishing breed (perhaps qualifying for protected species status).

cancellation) are shown in §3.4.2 and in the discussion of rail-to-rail op-amps in Chapter *4x*. An attractive alternative is to use a second FET switch section to connect the output to ground when the series FET is off, thus effectively forming a single-pole double throw (SPDT) switch (more on this in the next section).

A. CMOS linear switches

Frequently it is necessary to switch signals that may go nearly to the supply voltages. In that case the simple *n*-channel switch circuit just described won't work, because the gate is not forward-biased at the peak of the signal swing. The solution is to use paralleled complementary MOSFET (CMOS) switches (Figure 3.60). The triangular symbol is a digital inverter, which we'll discuss shortly; it inverts a HIGH input to a LOW output, and vice versa. When the control input is high, Q_1 is held ON for signals from ground to within a few volts of $+V_{DD}$ (where R_{ON} starts increasing dramatically). Q_2 is likewise held ON (by its grounded gate) for signals from $+V_{DD}$ to within a few volts of ground (where its R_{ON} increases dramatically). Thus signals anywhere between $+V_{DD}$ and ground are passed through with low series resistance (Figure 3.61). Bringing the control signal to ground turns off both FETs, providing an open circuit. The result is an analog switch for signals all the way from ground to $+V_{DD}$. This is the basic construction of the classic 4066 CMOS "transmission gate." It is bidirectional, like the switches described earlier: either terminal can be the input.

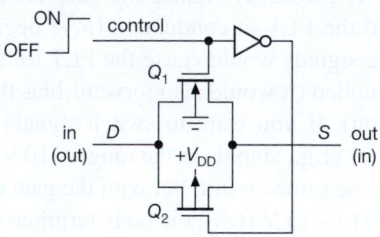

Figure 3.60. CMOS analog switch.

There is a variety of integrated circuit CMOS analog switches available, with different operating voltage ranges and with various switch configurations (e.g., several independent sections with several poles each). Going back to the prehistoric, there's the classic CD4066 "analog transmission gate," which belongs to the original 4000 series of digital CMOS logic and that acts as an analog switch for signals between ground and a single positive supply.[68]

---

[68] In that role it's happy to switch digital signals as well, thus its family membership.

More commonly, though, you'll choose a dedicated analog switch IC, for example a member of the industry-standard DG211 family. These parts (and their many variations; see Table 3.3 on page 176) are particularly convenient to use: they accept logic-level (0V = LOW, >2.4V = HIGH) control signals, they will handle analog signals to ±15 V (compared with only ±7.5V for the 4000 series), they come in a variety of configurations, and they have relatively low ON resistance (25 Ω or less for some members of these families, and down to a fraction of an ohm for lower voltage switches). Analog Devices, Intersil, Maxim, Vishay–Siliconix, and other manufacturers offer a nice range of analog switches.

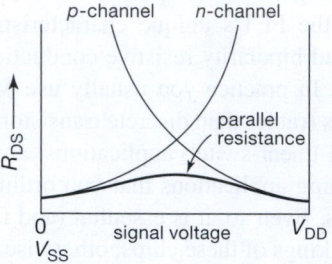

Figure 3.61. ON-resistance of the CMOS analog switch.

B. JFET analog switches

Although most available analog switches are built with a pair of paralleled MOSFETs of complementary polarities – the CMOS architecture just described – it is possible to build analog switches with JFETs, and there are some advantages.

The basic circuit (Figure 3.62) uses a single *n*-channel JFET, Q_1, as the analog switch. Its conduction is controlled by a transistor switch, Q_2, which pulls the gate down to a large negative voltage (-15 V, say) to cut off conduction in the JFET (switch OFF). Shutting off Q_2 lets the gate float to the source voltage, putting the (depletion-mode) JFET into full conduction (switch ON). The gate resistor R_1 is made deliberately large so that the output signal line does not see significant loading in the OFF state; its value is a compromise, because a larger resistance incurs a longer turn-on delay. With a low-impedance signal source (e.g., the output of an op-amp) it may be preferable to put the resistor on the input side (i.e., feed the signal in from the right).

Because there is only a single *n*-channel JFET, this switch cannot accept input signals all the way to the negative supply: signal voltages that come closer to the negative

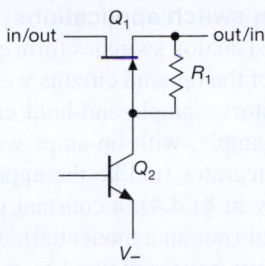

Figure 3.62. JFET analog switch.

supply than the FET's $V_{GS(off)}$ will bring it back into conduction.[69] There is no such limitation at the positive end.

A nice feature of this JFET analog switch is the constancy of R_{ON} with signal level: because the gate stays at the source voltage, there is *no* variation of R_{ON} with signal voltage; the JFET does not even know that the signal is varying! This nice characteristic is on display in Figure 3.63, which compares a plot of R_{ON} versus V_{sig} for a JFET analog switch (SW06) with that of the DG211 CMOS switch.[70]

In practice it's inconvenient to have to drive the switch control with a signal near the negative supply rail (and dou-

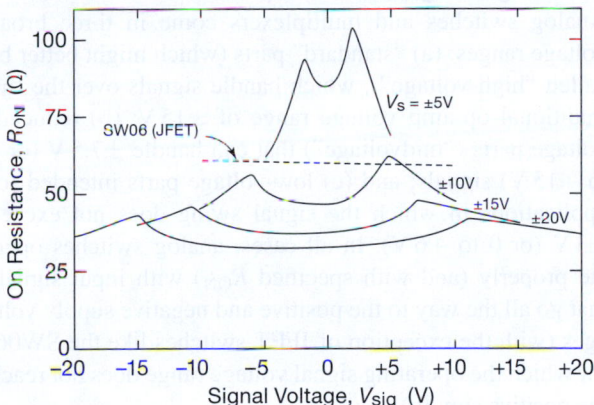

Figure 3.63. Contrasted with a CMOS analog switch like the DG211 (solid curves, shown for different supply voltages), the JFET analog switch maintains admirably flat R_{ON} over signal level.

---

[69] In fact, you've got to stay a bit further away from the actual $V_{GS(off)}$: recall (§3.1.3) that it's defined as the gate–source voltage that results in a small (but nonzero) drain current, usually $I_D = 10\,nA$.

[70] The slight variation of R_{ON} is caused by substrate effects: the SW06 is an integrated circuit, built on a silicon substrate; so the JFET and associated components have some inkling of the absolute signal level. If even that small variation is unacceptable, you can rig up a discrete implementation (as in Figure 3.62); you'll see such circuitry used in some of Agilent's accurate digital multimeters.

---

bly so for CMOS switches). Instead, you would probably use a level-shifting circuit so a logic-level input that goes between 0 V and +3 V, say, would activate the switch. Figure 3.64 shows a simple way to do that, using comparators with "open-collector" outputs (§12.3) to drive the gates of discrete JFET switches. For *integrated* analog switches, this sort of control signal circuitry is ordinarily built in. The SW06 JFET analog switch (and the nearly extinct DG180–189 family from Vishay–Siliconix) includes such drivers, along with some other elegant tricks.[71]

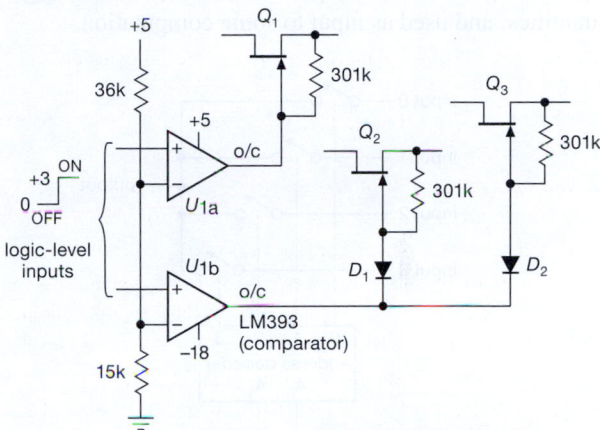

Figure 3.64. A comparator with open collector outputs, powered from +5 V and −18 V, converts a 0-to-3 V logic-level input into a JFET gate drive that swings to −18 V. This method is used in some of Agilent's digital multimeters (see "Designs by the Masters," 13.8.6) to accommodate analog signals over a full ±12 V range. The added diodes allow a control signal to drive more than one JFET switch.

JFET analog switches are inherently more rugged than their CMOS cousins, for which performance-degrading protection circuits are required for robustness against overvoltage faults. They do suffer from high charge injection, however (see §3.4.2E). Despite their nice features, integrated JFET switches and multiplexers (see next subsection) are almost extinct, with fine examples like the SW-01, SW-7510, and MUX-08 series from Precision Monolithics (now Analog Devices) gone forever (but happily the SW06 lives on!).

C. Multiplexers

A nice application of FET analog switches is the "multiplexer" (or MUX), a circuit that allows you to select any of several inputs, as specified by a digital control signal.

---

[71] Such as an internal MOSFET switch that disconnects the gate resistor (R_1 in Figure 3.62) when the switch is OFF, to eliminate circuit loading.

The analog signal present on the selected input will be passed through to the (single) output. Figure 3.65 shows the basic scheme. Each of the switches SW0 through SW3 is a CMOS analog switch. The "select logic" decodes the address and *enables* (jargon for "turns on") the addressed switch only, disabling the remaining switches. Such a multiplexer is usually used in conjunction with digital circuitry that generates the appropriate addresses (lots more in Chapters 10 and 11). A typical situation might involve a data-acquisition instrument in which a number of analog input voltages must be sampled in turn, converted to digital quantities, and used as input to some computation.

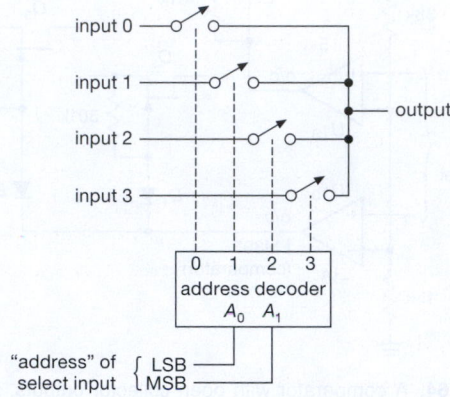

Figure 3.65. Analog multiplexer.

Because analog switches are bidirectional, an analog multiplexer such as this is also a "demultiplexer": a signal can be fed into the "output" and will appear on the selected "input." When we discuss digital circuitry in Chapters 10 and 11, you will see that an analog multiplexer such as this can also be used as a "digital multiplexer–demultiplexer," because logic levels are, after all, nothing but voltages that happen to be interpreted as binary 1s and 0s.

Typical of analog multiplexers are the industry-standard DG408–09 and DG508–09 series (and their many improved versions), 8- or 16-input MUX circuits that accept logic-level address inputs and operate with analog voltages up to ±15 V. The 4051–4053 devices in the CMOS digital family are analog multiplexers–demultiplexers with up to 8 inputs, but with 15 Vpp maximum signal levels; they have a V_{EE} pin (and internal level shifting) so that you can use them with bipolarity analog signals and unipolarity (logic-level) control signals. We are especially fond of the low-voltage '4053 series, with three SPDT switches. Our interest is evidently shared by others, with a large number of interesting parts available – see Table 13.7 (4053-style SPDT Switches) on page 917.

D. Other analog switch applications

Voltage-controlled analog switches form essential building blocks for some of the op-amp circuits we'll see in the next chapter – integrators, sample-and-hold circuits, and peak detectors. For example, with op-amps we will be able to build a "true" integrator (unlike the approximation to an integrator we saw in §1.4.4): a constant input generates a linear ramp output (not an exponential), etc. With such an integrator you must have a method to reset the output; a FET switch across the integrating capacitor does the trick. We won't try to describe these applications here. Because op-amps form essential parts of the circuits, they fit naturally into the next chapter. Great things to look forward to!

3.4.2 Limitations of FET switches

Analog switches aren't perfect – they have nonzero resistance when ON, and nonzero leakage when OFF, as well as capacitive feedthrough and charge injection during changes in switch state. You can see some of the variety in Table 3.3 on page 176). Let's take a look at these limitations.

A. Voltage range and latchup

Analog switches and multiplexers come in three broad voltage ranges: (a) "standard" parts (which might better be called "high-voltage"), which handle signals over the full traditional op-amp voltage range of ±15 V; (b) reduced-voltage parts ("midvoltage") that can handle ±7.5 V (or 0 to $+15$ V) signals; and (c) low-voltage parts intended for applications in which the signal swing does not exceed ±3 V (or 0 to $+6$ V). In all cases, analog switches operate properly (and with specified R_{ON}) with input signals that go all the way to the positive and negative supply voltages (with the exception of JFET switches like the SW06, for which the operating signal voltage range does not reach the positive supply).

However, input signals *beyond* the supply rails are another story. All CMOS integrated circuits have some form of input protection circuit, because otherwise the gate insulation is easily destroyed (see §3.5.4H on handling precautions). The usual protection network is shown in Figure 3.66. Although it may use distributed diodes, the network is equivalent to clamping diodes to V_{SS} and to V_{DD}, combined with resistive current limiting. If you drive the inputs (or outputs) more than a diode drop beyond the supply voltages, the diode clamps go into conduction, making the inputs (or outputs) look like a low impedance to the respective supplies. Worse still, the chip can be driven into "SCR latchup," a terrifying (and destructive) condition we'll describe in more detail in §10.8. For now, all you need to know about it is that you don't want it! SCR latchup is triggered by input currents (through the protection network) of roughly 20 mA or more. Thus you must be careful not to drive the analog inputs as much as a diode drop beyond the rails.[72] This means, for instance, that for most parts you must be sure the power supply voltages are applied before any signals that have significant drive-current capability; alternatively, you can use series diodes in the supply lines, so that input signals applied before the dc power do not produce input current.

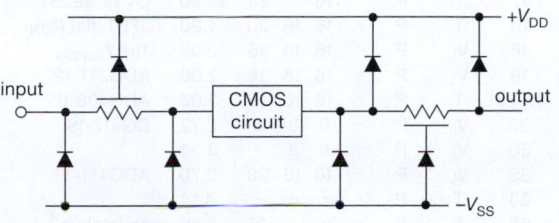

Figure 3.66. CMOS input–output protection networks. The series resistor at the output is often omitted.

The trouble with diode–resistor protection networks is that they compromise switch performance by increasing R_{ON}, shunt capacitance, and leakage. A different approach makes use of "dielectric isolation" to eliminate SCR latchup without the serious performance compromises inherent in traditional protection networks. Both methods result in a "protected" (or *fault-protected*) analog switch, in which you may safely overdrive the inputs without damage. Note, however, that the output does not follow the input beyond the rails.[73]

For example, the MAX4508 multiplexer adds fault protection to the standard DG508A 8-input analog multiplexer, making it tolerant to ±30 V input swings; it has

an $R_{ON(typ)}$ of 300 Ω. The AD7510DI-series of "Protected Analog Switches" from Analog Devices uses dielectric isolation to achieve input signal fault protection to ±25 V beyond the power supplies, while maintaining a respectable 75 Ω $R_{ON(typ)}$ in the normal operating signal range. Watch out, though – fault protection is the exception in the analog switch arena, and most analog switch ICs are not forgiving!

Maxim offers a nice outboard solution that you can put in front of an unprotected switch (or any other analog component), in the form of its multichannel "signal-line protector" ICs (the three- and eight-channel MAX4506–07).[74] These accept input signal swings to ±36V (whether powered or unpowered), are free of latchup regardless of power sequencing, and they pass through signals that are properly within the supply limits while clamping their outputs to the supply rails (which can be ±8 V to ±18 V split supplies, or a +9 to +36 V single supply) when there's an input beyond the rails. They even have the grace to open-circuit the input when thus overdriven – see Figure 3.67. The price you pay (above their literal $4 and $6 cost, respectively) is an ON-resistance in the range of 50–100 Ω (depending as usual on total supply voltage) and an input capacitance of 20 pF (thus a rolloff at ∼100 MHz).

B. ON-resistance
CMOS switches operated from a relatively high supply voltage (±15 V, say) will have low R_{ON} over the entire signal swing, because one or the other of the transmission FETs will have a forward gate bias of at least half the

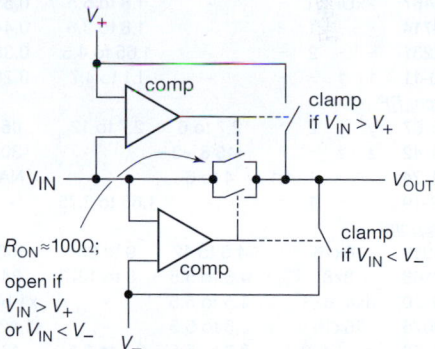

Figure 3.67. Maxim's MAX4506–7 fault-protected "Signal-Line Protectors" prevent signal swings beyond the rails, both clamping the output and disconnecting the input when overdriven. The series analog switches are normally ON, but both are switched OFF if V_{IN} goes beyond either rail, controlled by logic shown in the more complete MAX4508 datasheet.

---

[72] This prohibition goes for *digital* CMOS ICs as well as the analog switches we have been discussing.

[73] With a few exotic exceptions, for example the MAX14778 "Dual ±25V Above- and Below-the-Rails 4:1 Analog Multiplexer." This puppy, which runs on a single +3 to +5 V supply, is not only fault protected to ±25 V, it operates properly with signal voltages over that full range! How do they do it?! Turns out they have included an on-chip "charge-pump" voltage converter (§3.4.3D) to power the guts. Even more remarkable, this device has a very low ON-resistance (1.5 Ω) that is remarkably flat with signal voltage (0.003 Ω) over the full ±25 V range. And it can handle up to 300 mA of signal current. This device is intended for applications in which external large-swing signals need to be switched by a circuit with only a low-voltage single supply. Sadly for us experimenters, it comes only in the tiny TQFN (Thin Quad Flat-pack, No leads) package, requiring a reflow oven for soldering onto a circuit board.

[74] Analog Devices offers a similar single-channel part, the ADG465, in a convenient SOT23-6 package.

Table 3.3 Analog Switches

Part #[a]	SPST (NO)	SPST (NC)	SPDT[u]	MUX[v]	split (±V)	single (+V)	Ron typ @ Vsupply (Ω)	Vsupply (V)	Q_{inj} typ (pC)	Cap typ, ON (pF)	Logic[d]	Control[e]	SOT-23	DIP	SOIC	other	Price qty 25 ($US)	Comments
high voltage																		
MAX4800-02	-	-	-	8:1	40 to 100	c	22	±100	600	36	V_L	S	-	-	-	28	16.18	*really HV!*[1]
MAX326-27	4	4	-	-	5 to 18	10 to 30	1500	±15	2	6	T	P	-	16	16	16	6.76	low leakage[2]
MAX4508-09	-	-	-	84	4.5 to 20	9 to 36	300	±15	2	28, 22	T	P	-	16	16	-	6.78	OV to ±30V
MAX354-55	-	-	-	84	4.5 to 18	4.5 to 36	285	±15	80	28, 14	T	P	-	16	16	-	8.43	OV to ±25V
DG508-09	-	-	-	84	5 to 20	10 to 36	180	±15	2	18, 11	T	P	-	16	16	16	2.56	low leakage[3]
ADG1211-13	w	w	-	-	5 to 15	10 to ?	120	±15	0.3	2.6	T	P	-	-	-	16	4.63	low C, Q_{inj}
ADG1221-23	x	x	-	-	5 to 16.5	5 to 16.5	120	±15	0.1	3	T	P	-	-	-	10	3.04	low flat Q_{inj}
AD7510-12DI	4	4	2	-	5 to 15	-	75	±15	30	17	T	P	-	16	-	20	13.90	OV rails±25V
SW06	4	-	-	-	12 to 18	-	60	±15	-	15	T	P	-	16	16	20	4.30	JFET, flat R_{ON}
DG441-42	4	4	-	-	4.5 to 22	5 to 24	50	±15	2	16	V_L	P	-	16	16	16	2.05	1μA I_{supply}
DG211-12	4	4	-	-	4.5 to 22	5 to 22	45	±15	1	16	V_L	P	-	16	16	16	2.08	ADG211-12
DG408-09	-	-	-	84	5 to 20	5 to 30	40	±15	20	37, 25	T	P	-	16	16	16	3.06	ADG408-09
ADG417-19	1	1	1	-	5 to 20	5 to 20	25	±15	3	30	V_L	P	-	8	8	8	2.72	DG417-19
MAX317-19	1	1	1	-	4.5 to 20	10 to 30	20	±15	3	30	V_L	P	-	8	8	-	3.34	
DG411-13	w	w	-	-	4.5 to 20	10 to 30	17	±15	5	35	V_L	P	-	16	16	20	2.70	ADG411-13
DG447-48	1	1	-	-	4.5 to 20	7 to 36	13	±15	10	30	T	P	-	-	6	-	1.14	
ADG5412-13	y	y	-	-	9 to 22	9 to 40	10	±15	240	60	T	P	-	-	-	16	5.66	no latchup[4]
DG467-68	1	1	-	-	4.5 to 20	7 to 36	5	±15	21	76	T	P	-	-	8	-	0.75	ADG467-68
mid-voltage																		
DG4051-53	-	-	-	842	2.5 to 5	2.7 to 12	66	±5	0.25	3.4	V+	P	-	-	16	16	1.35	Table 13.7
74HC4051-53	-	-	-	842	2.5 to 5	2 to 10.5	40	±5		5, 25	V+	P	-	16	16	16	0.41	Table 13.7
MAX4541-44	x	x	1	-	-	2.7 to 12	30	+5	1	13, 20	T	P	8	8	8	8	1.33	
ISL5120-23	x	x	1	-	-	2.7 to 12	19	+5	3	28	T	P	8,6	-	8	-	1.71	
ISL43210	-	-	1	-	-	2.7 to 12	19	+5	3	28	T	P	6[s]	-	-	-	1.33	
ADG619-20[k]	-	-	1	-	2.7 to 5.5	2.7 to 5.5	7	+5	6	95	T	P	8	-	-	8	2.56	
low-voltage																		
ADG708-09	-	-	-	84	2.5	1.8 to 5.5	3	+5	3	96, 48	T	P	-	-	16	-	3.01	
ADG719	-	-	1	-	-	1.8 to 5.5	2.5	+5	-	27	T	P	6	-	-	-	1.76	SC70='749
MAX4624-25[k]	-	-	1	-	-	1.8 to 5.5	0.65	+5	65	100	T	P	6	-	-	-	2.10	
ADG884	-	-	2	-	-	1.8 to 5.5	0.3	+5	125	300	T	P	-	-	-	10	2.42	(5)
ISL84467	2xDPDT		-		-	1.8 to 5.5	0.55	+5	126	102	V+	P	-	-	16	16	1.19	80mΩ match
ISL84714	-	-	2	-	-	1.8 to 3.6	0.44	+3	20	100	V+	P	6	-	-	-	1.24	5mΩ match
NLAS52231	-	-	2	-	-	1.65 to 4.5	0.38	+3	53	85	V+	P	-	-	-	8	0.90	(6)
ISL43L110-11	1	1	-	-	-	1.1 to 4.7	0.25	+3	72	160	V+	P	5[s]	-	-	-	0.65	lowest R_{ON}
T-switch, RF																		
MAX4565-67	y	y	2	-	2.7 to 6	2.7 to 12	46	±5	25	6	T	P	-	16,20	16,20	-	4.74	video[7]
DG540-42	z	z	-	-	+15 & -3	-	30	nom	25	14	T	P	-	16	16	20	5.00	
AD8170, 74	-	-	1	4:1	4 to 6	-	NA	-	NA	1.1	T	P	-	8,14	-	-	5.11[n]	MUX+amp[8]
ADG918-19	-	-	1	-	-	1.65 to 2.75	-	-	-	1.6	V+	P	-	-	-	8	2.52	RF[9]
crosspoint																		
AD75019	16x16				4.5 to 12	9 to 25	150	±12	-	10 m	T	S	-	-	-	44	26.20	20 MHz
ADG2188	8x8				4.5 to 5.5	8 to 12	34	±5	3	9.5	V_L	I	-	-	-	32	9.37	200 MHz
MAX4359-60	4x4, 8x8				4.5 to 5.5	-	x1 buf	-	NA	8	T	P, ser	-	40	24	44	9.68[n]	35 MHz
MAX9675	16x16				2.5 to 5.5	-	x1,x2 buf	-	(g)	5	V_L	S	-	-	-	100	24.14	110 MHz
MAX4550, 70	dual 4x2				2.7 to 5.5	2.7 to 5.5	43	±5	7	11[o]	V+	S or I	-	2	-	28	6.39	audio[10]

Notes: (a) listed within categories by decreasing R_{ON}; all are CMOS except SW06; parts in **bold italic** are widely used "jellybeans." (b) numerals represent quantity of each switch type in a single package; letters refer to explanatory footnotes for successively numbered parts. (c) V_{neg} at least −15V, V_{pos} at least +40V, total no more than 200V. (d) T=TTL thresholds; V_L=external logic threshold supply; V+="CMOS" threshold, depends on positive analog supply voltage. (e) P=parallel logic inputs; I=I²C serial; S=SPI serial. (g) 50mV glitch. (h) 0.1dB to 14MHz, −95dB xtalk at 20KHz & R_L=10k, THD+N=0.014% (R_L=1k, f=1kHz). (k) second p/n is make-before-break. (m) min or max. (n) higher p/n is ≈50% more. (o) switch OFF. (s) SC-70. (u) SPDT are break-before-make unless noted otherwise. (v) 84=8:1 & dual 4:1; 842=8:1, dual 4:1, & triple 2:1. all dual 4:1 have single 2-bit address. (w) 4xNC, 4xNO, 2 each. (x) 2xNO, 2xNC, 1 each. (y) 4xNO, 2 each. (z) 2xNO, 2 each.

Comments: (1) Supertex HV2203. (2) 1pA typ. (3) 3pA typ; I_S=10μA typ. MAX308-09 is similar. (4) 8kV HBM ESD. (5) 0.4Ω@V_S=+3V; 400mA; −3dB/50Ω@18MHz. (6) low R_{ON}, e.g. speaker switch. (7) −3dB@350MHz, −90dB xtalk@10MHz. (8) 250 MHz; ext gain set. (9) −3dB at 4GHz, -30dB xtalk at 4GHz. (10) clickless.

supply voltage. However, when operated with lower supply voltages, the switch's R_{ON} value will rise, the maximum occurring when the signal is about halfway between the supply and ground (or halfway between the supplies, for dual-supply voltages). Figure 3.68 shows why. As V_{DD} is reduced, the FETs begin to have significantly higher ON-resistance (especially near $V_{GS} = V_{DD}/2$), because enhancement-mode FETs can have a $V_{GS(th)}$ of at least a few volts, and a gate-source voltage of as much as 5 to 10 volts may be required to achieve low R_{ON}. Not only will the parallel resistances of the two FETs rise for signal voltages between the supply voltage and ground, but also the peak resistance (at half V_{DD}) will rise as V_{DD} is reduced, and for sufficiently low V_{DD} the switch will become an open circuit for signals near $V_{DD}/2$.

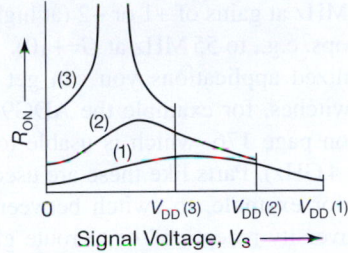

Figure 3.68. CMOS analog switch R_{ON} increases at low supply voltage.

There are various tricks used by the designers of analog switch ICs to keep R_{ON} low and approximately constant (for low distortion) over the signal swing. For example, the original 4016 analog switch used the simple circuit of Figure 3.60, producing R_{ON} curves that look like those in Figure 3.69. In the improved 4066 switch, the designers added a few extra FETs so that the n-channel body voltage follows the signal voltage, producing the R_{ON} curves of Figure 3.70. The "volcano" shape, with its depressed central

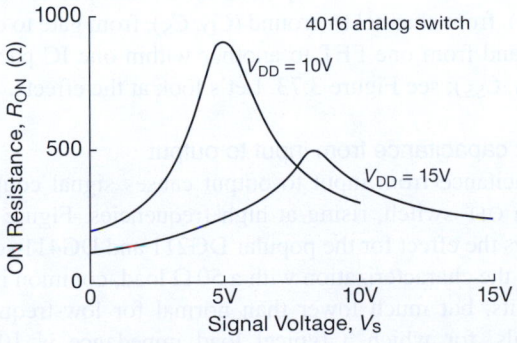

Figure 3.69. ON-resistance for 4016 CMOS switch.

R_{ON}, replaces the "Everest" shape of the 4016. Improved switches, like the industry standard DG408–09, intended for serious analog applications, succeed even better, with low and flat R_{ON} curves that deviate less than 10% or so over the signal voltage range. This is often achieved at the expense of increased "charge transfer" (see the later section on *glitches*).

Looking through manufacturers' selection tables for analog switches, you'll find standard-voltage units with R_{ON} as low as several ohms and flatness to a few tenths of an ohm; low-voltage switches can be found with R_{ON} as low as $0.25\,\Omega$, and flatness of $0.03\,\Omega$. This static performance comes at a real cost though, namely, high capacitance and high charge injection (see discussion below, and Table 3.3 on the preceding page). If your application requires low distortion into moderate load impedances, the better approach may be to choose a switch with excellent "ON-resistance flatness" spec ($R_{FLAT(ON)}$), and accept a higher overall R_{ON} with its lower capacitance.

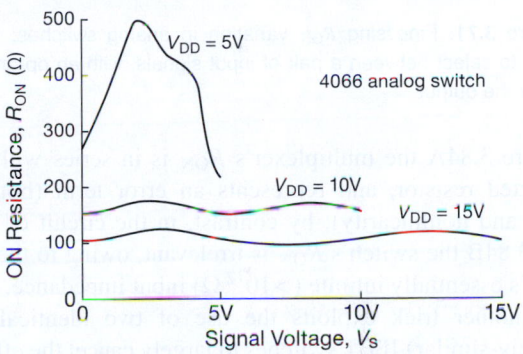

Figure 3.70. ON-resistance for the improved 4066 CMOS switch; note change of scale from previous figure.

Keep in mind, too, that in some cases you can finesse the problem altogether with a different choice of circuit configuration, as in Figure 3.71, which shows three approaches to a circuit that selects one of two input signals. Circuit A's gain is $R_2/(R_1 + R_{ON})$, so a variation of R_{ON} with signal amplitude produces changes of gain, and thus nonlinearity. Circuit B is better, because the switch output is held at ground by feedback around the op-amp; but the ON-resistance still reduces the gain somewhat, degrading the circuit's precision. Circuit C is blissfully unaware of R_{ON}, owing to the op-amp's very high input impedance; it's the most linear and precise of all.

This lesson can be applied to other circuit configurations as well. As an example, take a look at Figure 3.84 on page 183, where an analog multiplexer is used to select an amplifier's overall voltage gain. In the circuit of

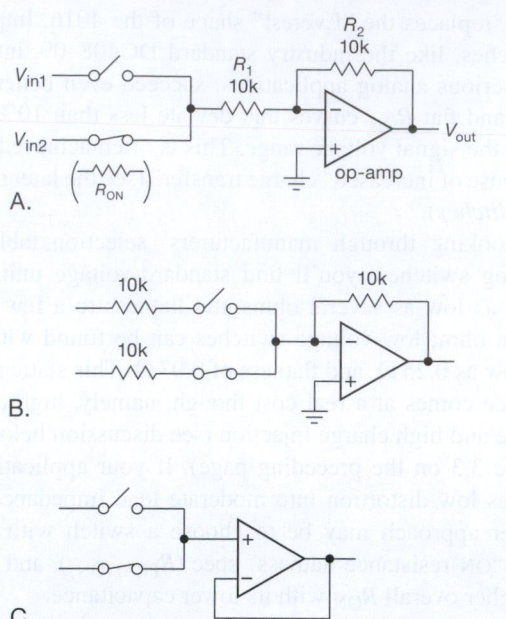

Figure 3.71. Finessing R_{ON} variation in analog switches: three ways to select between a pair of input signals, with an op-amp to buffer the output.

Figure 3.84A the multiplexer's R_{ON} is in series with the selected resistor, and represents an error term (both in gain and nonlinearity); by contrast, in the circuit of Figure 3.84B the switch's R_{ON} is irrelevant, owing to the op-amp's essentially infinite ($>10^{12}\,\Omega$) input impedance.

Another trick exploits the use of two identical (or closely-similar) JFET switches to largely cancel the effects of R_{ON}. See Chapter *4x*'s section "JFET linear switch with R_{ON} compensation" for an explanation and illustration of this elegant technique.

C. Speed

High-voltage FET analog switches have ON-resistances R_{ON} generally in the range of 20 to 200 Ω.[75] In combination with substrate and stray capacitances, this resistance forms a lowpass filter that limits operating speeds to frequencies of the order of 10 MHz or less (Figure 3.72). FETs with lower R_{ON} tend to have larger capacitances (up to 50 pF or more), so no gain in speed results (unless the designer has made other design tradeoffs). Much of the rolloff is due to protection components – current-limiting series resistance and capacitance of shunt diodes.

[75] As we remarked, you can get switches with lower R_{ON}, as low as 0.25 Ω, at the expense of some combination of increased capacitance, increased charge injection, and reduced operating voltage range.

However, low-voltage analog switches do better in terms of bandwidth (as is usually the case with smaller-geometry semiconductors): a contemporary ± 2.5 V analog switch like the popular ADG719 has 2.5 Ω of ON-resistance, 27 pF of capacitance, and 400 MHz bandwidth. There is also a class of analog switches and multiplexers targeted specifically at video and RF applications. These include both passive ("unbuffered") MUX/switches, and MUX/switches combined with an amplifier ("active" or "buffered"). Active MUX/switches operate on +5 V or ± 5 V supplies and have fixed voltage gains of either $\times 1$ or $\times 2$ (the latter are intended for driving a 50 Ω or 75 Ω transmission line through a series matching resistor, which attenuates the output by a factor of 2); in some cases you set the gain with an external resistor pair. An example of the latter is the AD8174 4-input multiplexer, with a bandwidth of 270 MHz at gains of +1 or +2 (at higher gains the bandwidth drops, e.g., to 55 MHz at $G=+10$).

For specialized applications you can get some really fast analog switches, for example the ADG918–19 listed in Table 3.3 on page 176, which is usable to 2 GHz (it's down 3 dB at 4 GHz). Parts like these are used in wireless applications, for example, to switch between two signal sources in "diversity reception" or to route gigahertz signals through a choice of filter paths. To reduce crosstalk, these wideband switches usually employ a T-switch topology (see Figure 3.77 in the next subsection).

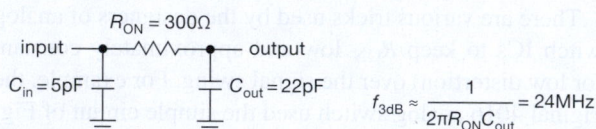

Figure 3.72. The parasitic RC's of a CMOS switch limit the analog signal bandwidth.

D. Capacitance

FET switches exhibit capacitance from input to output (C_{DS}), from channel to ground (C_D, C_S), from gate to channel, and from one FET to another within one IC package (C_{DD}, C_{SS}); see Figure 3.73. Let's look at the effects.

C_{DS}: capacitance from input to output

Capacitance from input to output causes signal coupling in an OFF switch, rising at high frequencies. Figure 3.74 shows the effect for the popular DG211 and DG411 series. Note the characterization with a 50 Ω load, common in RF circuits, but much lower than normal for low-frequency signals, for which a typical load impedance is 10k or more. Even with a 50 Ω load, the feedthrough becomes

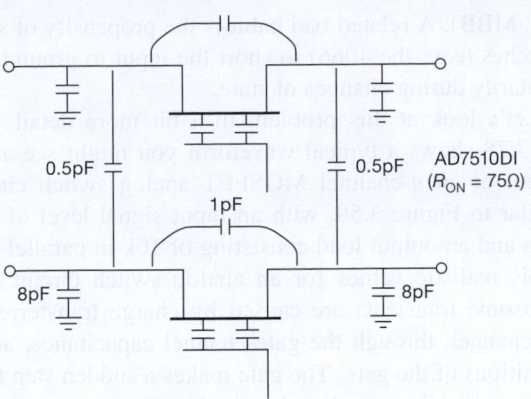

Figure 3.73. Capacitances between isolated sections of the AD7510 quad analog switch cause signal crosstalk.

significant at high frequencies (1 pF has a reactance of 5k at 30 MHz, giving −40 dB of feedthrough). And, of course, there is significant attenuation (and nonlinearity) driving a 50 Ω load, because for these parts R_{ON} is typically 45 Ω and 17 Ω, respectively. With a 10k load the feedthrough situation is much worse, of course.

Exercise 3.5. Calculate the feedthrough into 10k at 1 MHz, assuming $C_{DS} = 1$ pF.

In most low-frequency applications capacitive feedthrough is not a problem. If it is, the best solution is to use a pair of cascaded switches (Figure 3.75) or, better still, a combination of series and shunt switches, enabled alternately (Figure 3.76). The series cascade doubles the attenuation (in decibels) at the expense of additional R_{ON}, whereas the series-shunt circuit (effectively an SPDT configuration) reduces feedthrough by dropping the effective load resistance to R_{ON} when the series switch

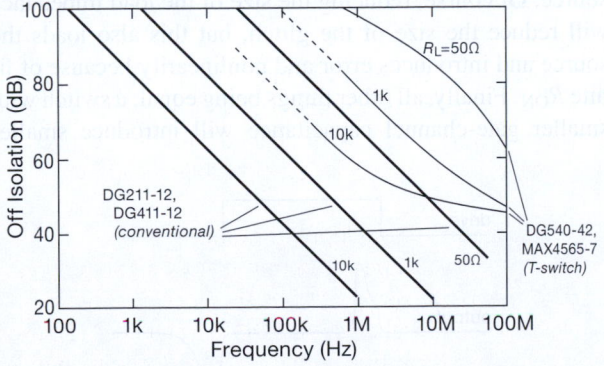

Figure 3.74. High-frequency feedthrough in analog switches. There is less feedthrough with a low load resistance and less still with a "T-switch" configuration.

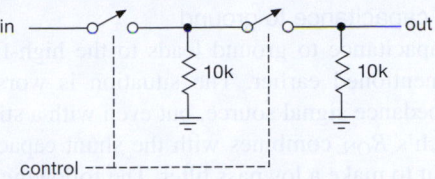

Figure 3.75. Cascaded analog switches for reduced feedthrough.

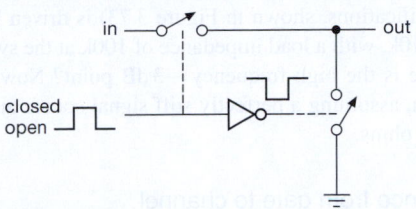

Figure 3.76. SPDT analog switch configuration for reduced feedthrough.

is off. Some commercial analog switches are built with a T-network of three switches (Figure 3.77) to achieve low feedthrough for signals going in either direction; from the outside you can't even tell that they've used this trick, except by noticing the excellent isolation specifications as in Figure 3.74 (unless, of course, they brag about it on the datasheet).

Exercise 3.6. Recalculate switch feedthrough into 10k at 1 MHz, assuming $C_{DS} = 1$ pF and $R_{ON} = 50\Omega$, for the configuration of Figure 3.76.

Most CMOS SPDT switches have controlled break-before-make (BBM) characteristics so that the signal sources are not momentarily connected during switching. In some cases, however, you need the reverse, i.e., make-before-break (MBB), for example in a gain-selecting feedback circuit like Figure 3.84B. To deal with this, some CMOS switches are available in both flavors, for example the ADG619 and ADG620 (BBM and MBB, respectively, as noted in Table 3.3).

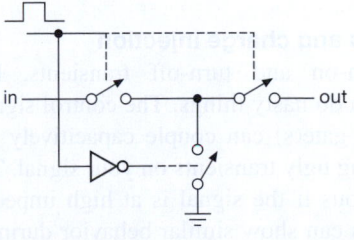

Figure 3.77. A T-switch further reduces high-frequency feedthrough.

C_D, C_S: capacitance to ground

Shunt capacitance to ground leads to the high-frequency rolloff mentioned earlier. The situation is worst with a high-impedance signal source, but even with a stiff source the switch's R_{ON} combines with the shunt capacitance at the output to make a lowpass filter. The following problem shows how it goes.

Exercise 3.7. An AD7510 (here chosen for its complete capacitance specifications, shown in Figure 3.73) is driven by a signal source of 10k, with a load impedance of 100k at the switch's output. Where is the high-frequency −3 dB point? Now repeat the calculation, assuming a perfectly stiff signal source and a switch R_{ON} of 75 ohms.

Capacitance from gate to channel

Capacitance from the controlling gate to the channel causes a different effect, namely, the coupling of nasty little transients into your signal when the switch is turned on or off. This subject is worth some serious discussion, so we'll defer it to the next section on glitches.

C_{DD}, C_{SS}: capacitance between switches

If you package several switches on a single piece of silicon the size of a grain of rice, it shouldn't surprise you if there is some coupling between channels ("crosstalk"). The culprit, of course, is cross-channel capacitance. The effect increases with frequency and with signal impedance in the channel to which the signal is coupled. Here's a chance to work it out for yourself.

Exercise 3.8. Calculate the coupling, in decibels, between a pair of channels with $C_{DD} = C_{SS} = 0.5$ pF (Figure 3.73) for the source and load impedances of the previous exercise. Assume that the interfering signal is 1 MHz. In each case calculate the coupling for (a) OFF switch to OFF switch, (b) OFF switch to ON switch, (c) ON switch to OFF switch, and (d) ON switch to ON switch.

It should be obvious from this example why most broadband RF circuits use low signal impedances, usually 50 Ω. If crosstalk is a serious problem, don't put more than one signal on one chip.

E. Glitches and charge injection

During turn-on and turn-off transients, FET analog switches can do nasty things. The control signal being applied to the gate(s) can couple capacitively to the channel(s), putting ugly transients on your signal. The situation is most serious if the signal is at high impedance levels. Multiplexers can show similar behavior during transitions of the input address as well as a momentary connection between inputs if the turn-off delay exceeds the turn-on delay

(i.e., MBB). A related bad habit is the propensity of some switches (e.g., the 4066) to short the input to ground momentarily during changes of state.

Let's look at this problem in a bit more detail. Figure 3.78 shows a typical waveform you might see at the output of an *n*-channel MOSFET analog switch circuit, similar to Figure 3.59, with an input signal level of zero volts and an output load consisting of 10k in parallel with 20 pF, realistic values for an analog switch circuit. The handsome transients are caused by charge transferred to the channel, through the gate-channel capacitance, at the transitions of the gate. The gate makes a sudden step from one supply voltage to the other, in this case between ±15 V supplies, transferring a slug of charge

$$Q = C_{GC}[V_G(\text{finish}) - V_G(\text{start})].$$

Here, C_{GC} is the gate-channel capacitance, typically around 5 pF. Note that the amount of charge transferred to the channel ("charge injection") depends only on the total voltage change at the gate, not on its rise time. Slowing down the gate signal gives rise to a smaller-amplitude glitch of longer duration, with the same total area under its graph. Lowpass filtering of the switch's output signal has the same effect. Such measures may help if the peak amplitude of the glitch must be kept small, but in general they are ineffective in eliminating gate feedthrough. In some cases the gate-channel capacitance may be predictable enough for you to cancel the spikes by coupling an inverted version of the gate signal through a small adjustable capacitor.

The gate-channel capacitance is distributed over the length of the channel, which means that some of the charge is coupled back to the switch's input. As a result, the size of the output glitch depends on the signal-source impedance and is smallest when the switch is driven by a voltage source. Of course, reducing the size of the load impedance will reduce the size of the glitch, but this also loads the source and introduces error and nonlinearity because of finite R_{ON}. Finally, all other things being equal, a switch with smaller gate-channel capacitance will introduce smaller

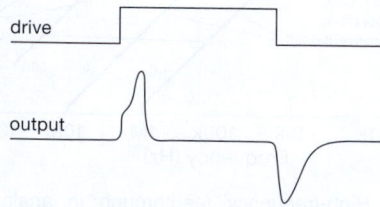

Figure 3.78. Charge-transfer glitches, on a greatly expanded scale.

switching transients, although you pay a price in the form of increased R_{ON}.

Figure 3.79 shows an interesting comparison of gate-induced charge transfers for a collection of analog switches, including JFETs. For the CMOS switches the internal gate signals are making a full rail-to-rail swing (e.g., $\Delta V = 30\,\text{V}$ for switches running from $\pm 15\,\text{V}$); for the JFET switch the gate swings from $-15\,\text{V}$ to the signal voltage. The JFET switch shows a strong dependence of glitch size on signal, because the gate swing is proportional to the level of the signal above $-15\,\text{V}$. Well-balanced CMOS switches have relatively low feedthrough because the charge contributions of the complementary MOSFETs tend to cancel out (one gate is rising while the other is falling). Just to give scale to these figures, it should be pointed out that 30 pC corresponds to a 3 mV step across a $0.01\,\mu\text{F}$ capacitor. That's a rather large filter capacitor, and you can see that this is a real problem, since a 3 mV glitch is pretty large when dealing with low-level analog signals.

In Figure 3.80 we've plotted, on an expanded scale, the charge injection scene for a selection of analog switches that exhibit particularly low charge injections. Switches optimized for low charge injection will usually brag about it on the datasheet's headline. For example, the Analog

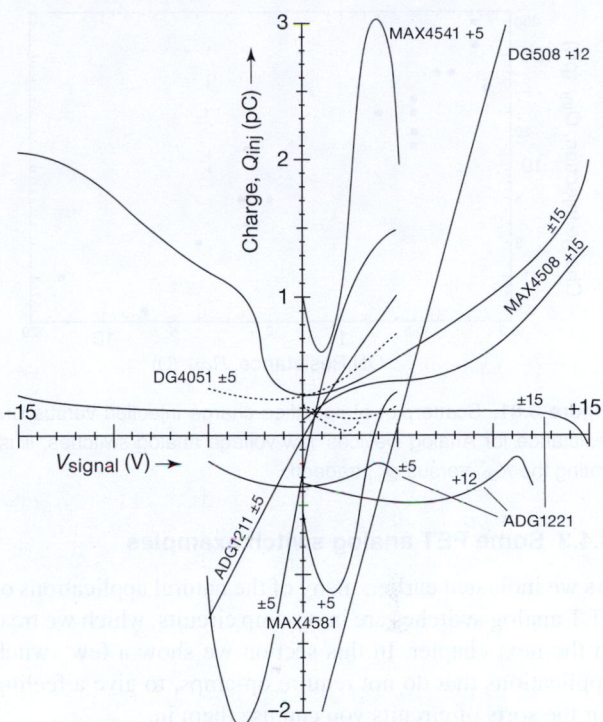

Figure 3.80. Need an analog switch with low charge injection? Here are some candidates, plotted on a greatly expanded scale. The three dotted curves are for the DG4051 with $\pm 5\text{V}$, $+5\text{V}$, and $+3\text{V}$ supplies. Check the datasheet for analogous plots for the DG4053 triple SPDT switch.

Devices ADG1221-series datasheet shouts, in bold letters, "**Low Capacitance, Low Charge Injection, $\pm 15\,\text{V}/+12\,\text{V}$ *i*CMOS® Dual SPST Switches**"; quite a mouthful, but quite a switch!

As might be expected, switches with lower ON-resistance generally exhibit greater charge injection. Figure 3.81 shows this trend, in a scatterplot of datasheet values of Q_{inj} versus R_{ON} for the low-voltage CMOS analog switches currently offered by Analog Devices.

F. Other switch limitations

Some additional characteristics of analog switches that may or may not be important in any given application are switching time, settling time, BBM delay, channel leakage current (both ON and OFF), device quiescent current, input current during overvoltage, R_{ON} matching among multiple channels, and temperature coefficient of R_{ON}. We'll show unusual restraint by ending the discussion at this point, leaving the reader to look into these details when the circuit application demands it.

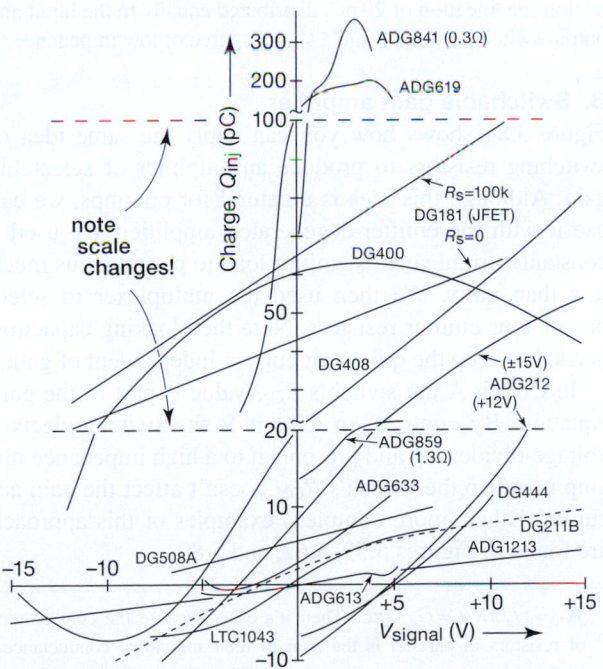

Figure 3.79. Charge transfer for various FET linear switches as a function of signal voltage, taken from the respective datasheets.

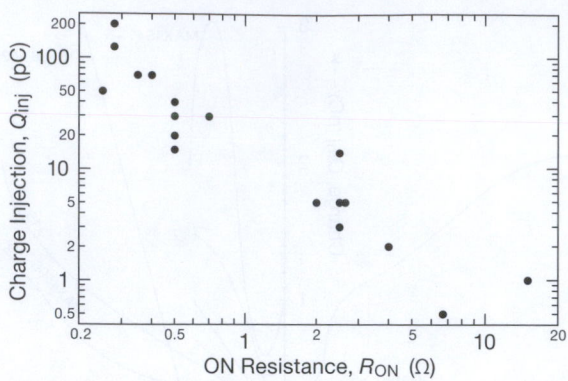

Figure 3.81. Scatterplot of specified charge injection versus ON-resistance for Analog Devices' low-voltage analog switches, illustrating the R_{ON} versus Q_{inj} tradeoff.

3.4.3 Some FET analog switch examples

As we indicated earlier, many of the natural applications of FET analog switches are in op-amp circuits, which we treat in the next chapter. In this section we show a few switch applications that do not require op-amps, to give a feeling for the sorts of circuits you can use them in.

A. Switchable *RC* lowpass filter

Figure 3.82 shows how you could make a simple *RC* low-pass filter with selectable 3 dB points. We've used a multiplexer to select one of four preset resistors, using a 2-bit (digital) address. We chose to put the switch at the input, rather than after the resistors, because there is less charge injection at a point of lower signal impedance. Another possibility, of course, is to use FET switches to select the *capacitor*. To generate a very wide range of time constants you might have to do that, but the switch's finite R_{ON} would limit attenuation at high frequencies, to a maximum of R_{ON}/R_{series}. We've also indicated a unity-gain buffer, following the filter, since the output impedance is high. You'll see how to make "perfect" followers (precise gain, high Z_{in}, low Z_{out}, and no V_{BE} offsets, etc.) in the next chapter. Of course, if the amplifier that follows the filter has high input impedance, you don't need the buffer.

Figure 3.83 shows a simple variation in which we've used four independent switches, rather than a 4-input multiplexer. With the resistors scaled as shown, you can generate 16 equally spaced 3 dB frequencies by turning on binary combinations of the switches.[76]

---

[76] An easy way to see that the 3 dB frequencies are integral multiples of the lowest setting is to rewrite f_{3dB} in terms of the *conductance* of the parallel resistance R_p of the selected resistors:

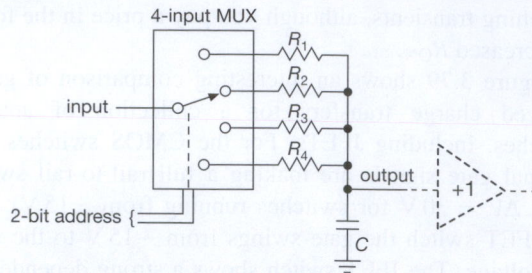

Figure 3.82. Analog-MUX selectable *RC* lowpass filter.

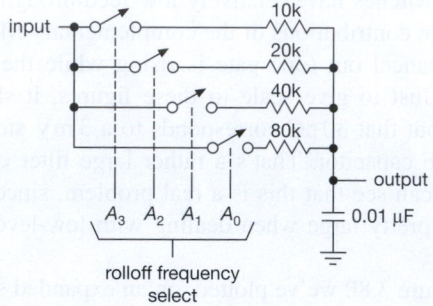

Figure 3.83. *RC* lowpass filter with choice of 15 equally spaced cutoff frequencies.

Exercise 3.9. What are the 3 dB frequencies for this circuit? Estimate the gain-switching glitch amplitude, assuming a charge injection specification of 20 pC, distributed equally to the input and output switch terminals, and a signal source of low impedance.

B. Switchable gain amplifier

Figure 3.84 shows how you can apply the same idea of switching resistors to produce an amplifier of selectable gain. Although this idea is a natural for op-amps, we can use it with the emitter-degenerated amplifier. We used a constant-current sink as emitter load to permit gains much less than unity. We then used the multiplexer to select one of four emitter resistors. Note the blocking capacitor, needed to keep the quiescent current independent of gain.

In Circuit A the switch's R_{ON} value is part of the gain equation. By contrast, in Circuit B the switch selects a voltage-divider tap and presents it to a high impedance op-amp input, so the switch's R_{ON} doesn't affect the gain accuracy. Other (more complex) examples of this approach are found in Figures 5.59, 5.62, and 5.80.

$f_{3dB}=1/2\pi R_p C=G_p/2\pi C$. Then it's easy, because the conductance of resistors in parallel is the sum of their individual conductances. So for this circuit $f_{3dB}=nG_{80k}/2\pi C=199n$ Hz, where $G_{80k}=12.5\mu S$, $C=10$ nF, and n is the integer [1..15] represented by the selected switches A_n.

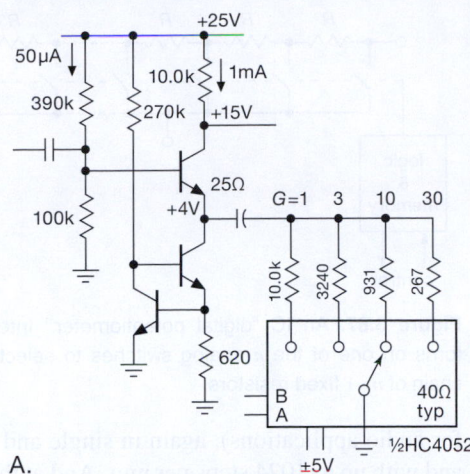

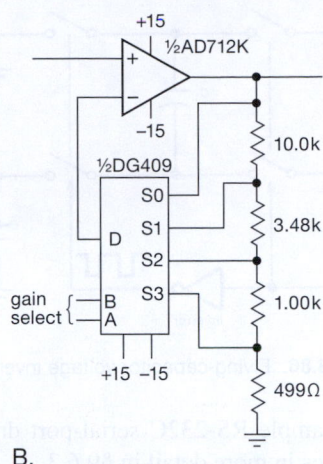

Figure 3.84. A. An analog multiplexer selects appropriate emitter degeneration resistors to achieve decade-switchable gain. B. A similar technique, but with the versatile "op-amp" building block (the hero of Chapter 4).

C. Sample-and-hold

Figure 3.85 shows how to make a "sample-and-hold" (S/H) circuit, which comes in handy when you want to convert an analog signal to a stream of digital quantities ("analog-to-digital conversion") – you've got to hold each analog level steady while you figure out how big it is. The circuit is simple: a unity-gain input buffer generates a low-impedance copy of the input signal, forcing it across a small capacitor. To hold the analog level at any moment, you simply open the switch. The high input impedance of the second buffer (which should have FET input transistors to keep input current near zero) prevents loading of the capacitor, so it holds its voltage until the FET switch is again closed.

Exercise 3.10. The input buffer must supply current to keep the capacitor following a varying signal. Calculate the buffer's peak output current when the circuit is driven by an input sinewave of 1 V amplitude at 10 kHz.

You can do considerably better by closing a feedback loop around the S/H circuit; take a look at §4.5.2. Better still, buy a complete integrated circuit S/H (e.g., the AD783 has an internal hold capacitor, settles to 0.01% in 0.25μs, and droops less than 0.02μV/μs) – let someone else do the hard work!

D. Flying-capacitor voltage converter

Here's a nice way (Figure 3.86) to generate a needed negative power-supply voltage in a circuit that is powered by a single positive supply. The pair of FET switches on the left connects C_1 across the positive supply, charging it to V_{in}, while the switches on the right are kept open.[77] Then the input switches are opened, and the switches on the right are closed, connecting charged C_1 across the output, transferring some of its charge onto C_2. The switches are diabolically arranged so that C_1 gets turned upside down, generating a *negative* output! This particular circuit, often referred to as a *charge-pump dc–dc converter*, originated as the Intersil 7660 voltage converter chip, and is widely available in improved variants, including voltage-doubling versions and regulated versions. You find them, also, as built-in portions of larger integrated circuits that require dual supply

---

[77] The device labeled "inverter" turns a logic HIGH voltage into a logic LOW voltage, and vice versa. We'll show you how to make one in the next section (and we'll really get you up to speed on these in Chapters 10–14!).

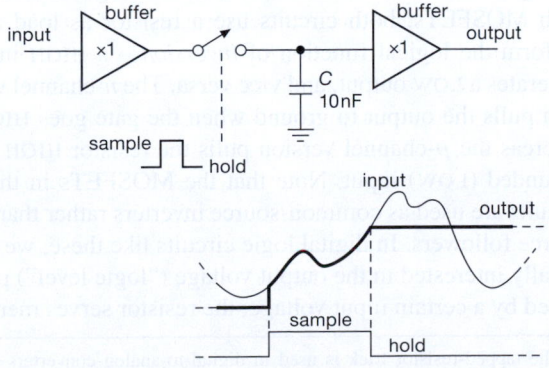

Figure 3.85. Sample-and-hold.

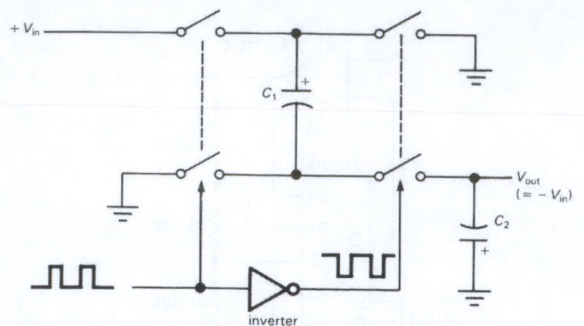

Figure 3.86. Flying-capacitor voltage inverter.

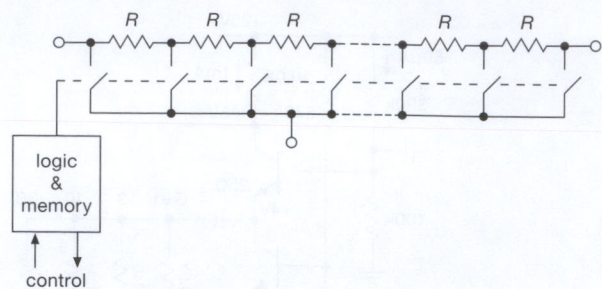

Figure 3.87. An IC "digital potentiometer." Internal digital logic turns on one of the n analog switches to select a tap along the chain of $n-1$ fixed resistors.

voltages, for example RS-232C serial-port drivers. We'll visit these devices in more detail in §9.6.3.

E. Digital potentiometer

It's nice to be able to turn a pot electrically – for example, to adjust the volume control of a television set by the remote control "clicker." This kind of application is common, and the semiconductor industry has responded with a variety of electrically settable pots, known variously as *EEPOT*, E^2POT, or just plain *digital pot*. A digital pot consists of a long resistor chain, with an array of FET switches that connects the selected tap to the output pin (Figure 3.87); the tap is selected by a digital input (Chapters 10 and onward).[78] Digital pots come in single, dual, and multiple units; many have "nonvolatile" memory, to retain the position of the pot after power has been turned off. Some have nonlinear taps, for example for audio volume controls, for which it's best to have equal step sizes in decibels (that is, each step produces the same *fractional* increase in voltage-divider ratio). Note that, whatever the configuration, the switch's R_{ON} appears as a series resistance at the output ("wiper") pin.

As an example, the Analog Devices website lists some 50 digital potentiometers, with from 32 to 1024 steps (256-step models seem to be the most popular), and with one to six channels (singles and duals seem most popular); they use a serial data connection (only two or three pins are needed, regardless of the length of the control data), and they average about $1 (1000-piece quantity). The selection from Maxim/Dallas includes linear and log taper units (the term *taper* predates digital pots and refers to the resistance versus shaft rotation characteristic of a pot; the log taper is

for audio applications), again in single and multiple units, and with up to 1024 steps per unit. And at the bottom of Intersil's selection table, after all the usual digital pots, you'll even find digital *capacitors*![79]

3.4.4 MOSFET logic switches

The *other* kinds of FET switch applications are *logic* and *power switching* circuits. The distinction is simple: in analog signal switching you use a FET as a series switch, passing or blocking a signal that has some range of analog voltage. The analog signal is usually a low-level signal at insignificant power levels. In logic switching, on the other hand, MOSFET switches open and close to generate full swings between the power supply voltages. The "signals" here are really digital, rather than analog – they swing between the power supply voltages, representing the two states HIGH and LOW. In-between voltages are not useful or desirable; in fact, they're not even *legal!* Finally, "power switching" refers to turning on or off the power to a load such as a lamp, relay coil, or motor winding; in these applications, both voltages and currents tend to be large. We'll take logic switching first.

Figure 3.88 shows the simplest kind of logic switching with MOSFETs: both circuits use a resistor as load and perform the logical function of *inversion* – a HIGH input generates a LOW output, and vice versa. The n-channel version pulls the output to ground when the gate goes HIGH, whereas the p-channel version pulls the resistor HIGH for grounded (LOW) input. Note that the MOSFETs in these circuits are used as common-source inverters rather than as source followers. In digital logic circuits like these, we are usually interested in the output voltage ("logic level") produced by a certain input voltage; the resistor serves merely

---

[78] There are two varieties here: one uses a serial digital data protocol, so that the desired tap position can be sent, as a number, from a controlling microprocessor; the other kind has UP and DOWN pins, with internal memory to hold the current tap position.

[79] The tapped-resistor trick is used in digital-to-analog converters (see §13.2.1), and many ADCs use digital capacitors (see §13.7.

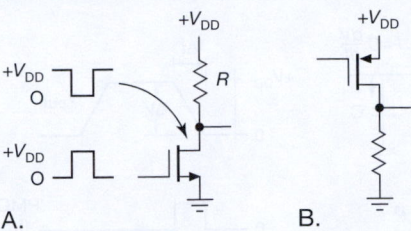

Figure 3.88. nMOS and pMOS logic inverters, with resistive "pullups".

as a passive drain load, to make the output swing to the drain supply when the FET is off. If, on the other hand, we replace the resistor with a lightbulb, relay, print-head hammer, or some other hefty load, we've got a power-switching application (Figure 3.3). Although we're using the same "inverter" circuit, in the power-switching application we're interested instead in turning the load on and off.

A. CMOS inverter

The nMOS and pMOS inverters of the preceding circuits have the disadvantage of drawing current in the ON state and having relatively high output impedance in the OFF state. You can reduce the output impedance (by reducing R), but only at the expense of increased dissipation, and vice versa. Except for current sources, of course, it's never a good idea to have high output impedance. Even if the intended load is high impedance (another MOSFET gate, for example), you are inviting capacitive noise pickup problems, and you will suffer reduced switching speeds for the ON-to-OFF ("trailing") edge (because of stray loading capacitance). In this case, for example, the nMOS inverter with a compromise value of drain resistor, say 10k, would produce the waveform shown in Figure 3.89.

The situation is reminiscent of the single-ended emitter follower in §2.4.1, in which quiescent power dissipation and power delivered to the load were involved in a similar compromise. The solution there – the push–pull configuration – is particularly well suited to MOSFET switching. Look at Figure 3.90, which you might think of as a push-pull switch: input at ground cuts off the bottom transistor and turns on the top transistor, pulling the output HIGH. A HIGH input ($+V_{DD}$) does the reverse, pulling the output to ground. It's an inverter with low output impedance in *both* states, and no quiescent current whatsoever. It's called a CMOS (complementary MOS) inverter, and it is the basic structure of all digital CMOS logic, the logic family that has become universal in large and very-large scale integrated circuits (LSI, VLSI), and has largely replaced earlier logic families (with names like transistor–transistor logic,

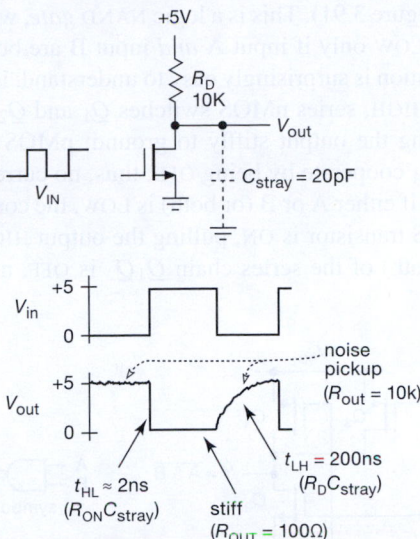

Figure 3.89. High off-impedance in the nMOS inverter causes long rise times and susceptibility to capacitively-coupled noise.

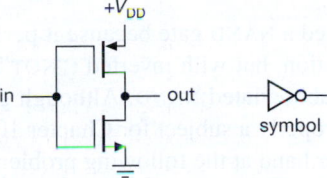

Figure 3.90. CMOS logic inverter and circuit symbol.

TTL) based on bipolar transistors. Note that the CMOS inverter is two complementary MOSFET switches *in series*, alternately enabled, whereas the CMOS analog switch (treated earlier in the chapter) is two complementary MOSFET switches *in parallel*, enabled simultaneously.

Exercise 3.11. The complementary MOS transistors in the CMOS inverter are both operating as common-source inverters, whereas the complementary bipolar transistors in the push-pull circuits of §2.4.1 (e.g., Figure 2.69) are (noninverting) emitter followers. Try drawing a "complementary BJT inverter" analogous to the CMOS inverter. Why won't it work?

B. CMOS gates

We'll be seeing much more of digital CMOS in the chapters on digital logic and microprocessors (Chapters 10–14). For now, it should be evident that CMOS is a low-power logic family (with *zero* quiescent power) with high-impedance inputs, and with stiff outputs that swing the full supply range. Before leaving the subject, however, we can't resist the temptation to show you one additional CMOS circuit (Figure 3.91). This is a logic NAND *gate*, whose output goes LOW only if input A *and* input B are both HIGH. The operation is surprisingly easy to understand: if A and B are both HIGH, series nMOS switches Q_1 and Q_2 are both ON, pulling the output stiffly to ground; pMOS switches Q_3 and Q_4 cooperate by being OFF; thus, no current flows. However, if either A or B (or both) is LOW, the corresponding pMOS transistor is ON, pulling the output HIGH; since one (or both) of the series chain Q_1Q_2 is OFF, no current flows.

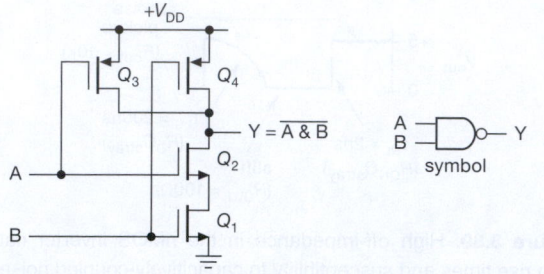

Figure 3.91. CMOS NAND gate and circuit symbol.

This is called a NAND gate because it performs the logical AND function, but with inverted ("NOT") output – it's a NOT–AND, abbreviated NAND. Although gates and their variants are properly a subject for Chapter 10, you will enjoy trying your hand at the following problems.

Exercise 3.12. Draw a CMOS AND gate. *Hint*: AND = NOT NAND.

Exercise 3.13. Draw a NOR gate: the output is LOW if either A *or* B (or both) is HIGH.

Exercise 3.14. You guessed it – draw a CMOS OR gate.

Exercise 3.15. Draw a 3-input CMOS NAND gate.

The CMOS digital logic we'll be seeing later is constructed from combinations of these basic gates. The combination of very low-power dissipation and stiff rail-to-rail output swing makes CMOS logic the family of choice for most digital circuits, accounting for its popularity. Furthermore, for micropower circuits (such as wristwatches and small battery-powered instruments), it's the only game in town.

Lest we leave the wrong impression, however, it's worth noting that CMOS logic is not *zero* power. There are two mechanisms of current drain:

(a) During transitions, a CMOS output must supply a transient current $I = CdV/dt$ to charge any capacitance it sees (Figure 3.92). You get load capacitance both from wiring ("stray" capacitance) and from the input capacitance of ad-

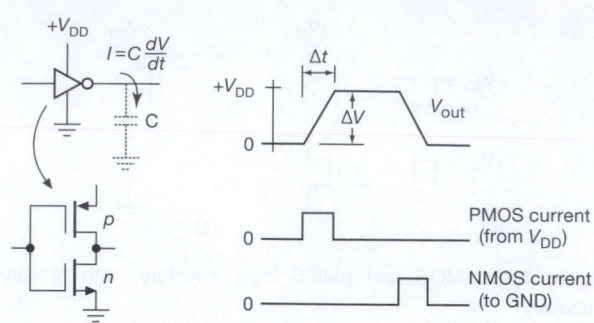

Figure 3.92. Capacitive charging current. The average supply current is proportional to switching rate, and equals CVf.

ditional logic that you are driving. In fact, because a complicated CMOS chip contains many internal gates, each driving some on-chip internal capacitance, there is some current drain in any CMOS circuit that is making transitions, even if the chip is not driving any external load. Not surprisingly, this "dynamic" current drain is proportional to the rate at which transitions take place.

(b) The second mechanism of CMOS current drain is shown in Figure 3.93: as the input jumps between the supply voltage and ground, there is a region where both MOSFETs are conducting, resulting in large current spikes from V_{DD} to ground. This is sometimes called "class-A current," "shoot-through," or "power-supply crowbarring." You will see some consequences of this in Chapters 10–12. As long as we're dumping on CMOS, we should mention that an additional disadvantage of CMOS (and, in fact, of

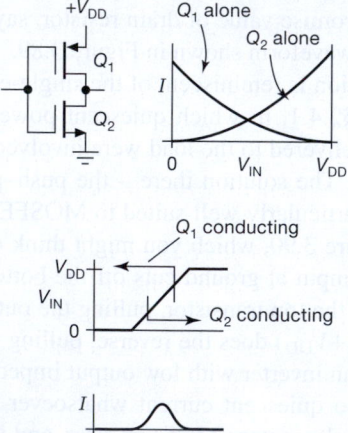

Figure 3.93. When the input gate voltage of a CMOS inverter is intermediate between V_{DD} and ground, both MOSFETs are partially conducting, causing "class-A conduction," also known as "shoot-through" current.

all MOSFETs) is their vulnerability to damage from static electricity. We'll have more to say about this in §3.5.4H.

3.5 Power MOSFETs

MOSFETs work well as saturated switches, as we suggested with our simple circuit in §3.1.1B. Power MOSFETs are now available from many manufacturers, making the advantages of MOSFETs (high input impedance, easy paralleling, absence of "second breakdown") applicable to power circuits. Generally speaking, power MOSFETs are easier to use than conventional bipolar power transistors. However, there are some subtle effects to consider, and cavalier substitution of MOSFETs in switching applications can lead to prompt disaster. We've visited the scenes of such disasters and hope to avert their repetition. Read on for our handy guided tour.

FETs were feeble low-current devices, barely able to run more than a few tens of milliamps, until the late 1970s, when the Japanese introduced "vertical-groove" MOS transistors. Power MOSFETs are now made by all the manufacturers of discrete semiconductors (e.g., Diodes-Inc, Fairchild, Intersil, IR, ON Semiconductor, Siliconix, Supertex, TI, Vishay, and Zetex, along with European companies like Amperex, Ferranti, Infineon, NXP, and ST, and many of the Japanese companies such as Renesas and Toshiba); they are called, variously, VMOS, TMOS, vertical DMOS, and HEXFET. Even in conventional transistor power packages such as TO-220, TO-247, and D-PAK they can handle surprisingly high voltages (up to 1500 V or more), and peak currents over 1000 amps (continuous currents of 200 A), with R_{ON} below 0.001 Ω. Small-power MOSFETs sell for much less than a dollar, and they're available in all the usual transistor packages. You can also get arrays (multiple MOSFETs) in standard multipin IC packages such as the traditional dual in-line package (DIP) and the smaller surface-mount varieties such as SOT-23, SOIC, and TSOP. Ironically, it is now discrete *low-level* MOSFETs that are hard to find, there being no shortage of power MOSFETs; Table 3.4a on page 188 lists a selection of small *n*-channel MOSFETs up to 250 V, Table 3.4b on pages 189–191 lists other *n*-channel MOSFET sizes and voltages, and Table 3.6 on page 210 has a nice selection of depletion-mode power MOSFETs. There are additional types listed in the MOSFET tables in Chapter *3x*.

3.5.1 High impedance, thermal stability

Two important advantages of the power MOSFET compared with the bipolar power transistor, are (a) its extremely high input impedance (essentially infinite at dc),

and (b) its inherent thermal stability. As simple as these might seem, there's more to say, and some important cautions.

A. Input impedance

First, the "infinite" input impedance holds only at dc because of substantial input capacitance, which can run to 1000–10,000 pF in typical power MOSFETs. In addition, for *switching* applications you have also to worry about feedback capacitance, i.e., drain-to-gate capacitance (also called *reverse transfer capacitance*, C_{rss}), because the Miller effect (§2.4.5) boosts the effective value by the voltage gain. In §3.5.4 we'll discuss this further and display some waveforms showing how the Miller effect fights your efforts to bring about rapid switching. Jumping to the bottom line, you may have to supply several *amperes* of gate drive current to switch power loads in the tens of nanoseconds that MOSFETs can inherently achieve – hardly the characteristics of an infinite input-impedance device!

B. Thermal stability

Second, there are two mechanisms affecting thermal stability in MOSFETs, namely an increase in R_{ON} with increasing temperature, and, *at the higher end of the transistor's drain current only*, a decrease in drain current (at constant V_{GS}) with increasing temperature; see Figure 3.14 and Figures 3.115, and 3.116 in §3.6.3. This latter effect is very important in power circuits and is worth understanding: the large junction area of a power transistor (whether BJT or FET) can be thought of as a large number of small junctions in parallel (Figure 3.94), all with the same applied voltages. In the case of a bipolar power transistor, the positive temperature coefficient of collector current at fixed V_{BE} (approximately $+9\%/°C$, see §2.3) means that a local hot spot in the junction will have a higher current density, thus producing additional heating. At sufficiently high V_{CE} and I_C, this "current hogging" can cause local thermal runaway. As a result, bipolar power transistors are limited to a "safe operating area" (on a plot of collector current versus collector voltage) smaller than that allowed by transistor power dissipation alone. The important point here is that the *negative* temperature coefficient of MOS drain current, when operating at relatively high currents, prevents these junction hot spots entirely. MOSFETs also have no second breakdown, and their safe operating area (SOA) is limited only by power dissipation (see Figure 3.95, where we've compared the SOAs of an *npn* and an nMOS power transistor of the same I_{max}, V_{max}, and P_{diss}). This is one reason MOSFETs are favored in linear power applications such as audio power amplifiers.

Table 3.4a MOSFETs — Small *n*-channel (to 250V), and *p*-channel (to 100V)

small n-channel to 250V

nMOS type	Pkg[p]	SMT[x]	V_{DSS} (V)	P_D[c] (W)	I_D[y] (A)	R_{DS}[r]@V_{GS} (mΩ)	(V)	Q_G[s] (nC[t])	C_{iss} (pF[t])	Cost[q] ($US)
ZVN4424	TO-92	•	240	0.7	0.3	4.3Ω	2.5	8	110	0.85
BSP89	SOT-223	•	240	1.5	0.4	2.8Ω	10	–	100	0.48
ZVNL120	TO-92	•	200	0.7	0.2	6Ω	3	2	55	0.53
BS107A	TO-92	–	200	0.4	0.2	5Ω	10	–	60	0.31
FQT4N20L	SOT-223	•	200	2.2	0.7	1.0Ω	4.5	4	240	0.34
FQT7N10L	SOT-223	•	100	2	1.2	300	5	4.6	220	0.37
ZXMN10A08E	SOT-23	•	100	1.1	0.6	200	10	7.8	500	0.57
ZXMN10A08G	SOT-223	•	100	2	1.5	200	10	7.7	405	0.48
VN2222LL	TO-92	–	60	0.4	0.1	7.5Ω	10	–	<60	0.36
VN10KN3	TO-92	–	60	0.7	0.2	6.6Ω	5	1.1	48	**0.18**
RHU002N06T	SOT-323	•	60	0.2	0.1	2.8Ω	4	1	15	0.21
2N7000	TO-92	–	60	0.4	0.2	2.5Ω	5	1	20	**0.17**
2N7002	SOT-23	•	60	0.2	0.1	2.5Ω	4.5	0.9	20	**0.16**
2N7002W	SOT-323	•	60	0.28	0.1	2.5Ω	4.5	0.7	25	**0.15**
NDS7002A	SOT-23	•	60	0.36	0.20	1.3Ω	4.5	0.8	80	**0.27**
Si1330EDL	SOT-323	•	60	0.18	0.15	1.4Ω	4.5	0.4	–	0.38
BSS138	SOT-23	•	50	0.36	0.25	1.0Ω	4.5	0.95	27	**0.15**
ZVN2106A	TO-92	•	60	0.7	0.3	800	10	1.5	75	0.49
ZVN4306A	TO-92	•	60	0.85	1.0	320	5	3.5	350	1.11
NDT3055	SOT-223	•	60	3	1.7	84	10	9	250	0.34
PHT8N06LT	SOT-223	•	55	8.3	5[z]	10	5	11.2	500	0.75
IRF7470	SO-8	•	40	1.7	10	7	4.5	29	3400	0.76
2SK3018	SOT-323	•	30	0.2	0.05	5Ω	4	–	13	0.20
FDV303N	SOT-23	•	25	0.35	0.4	330	2.7	1.1	50	**0.23**
IRLML2030	SOT-23	•	30	1.3	0.9	123	4.5	1	110	0.25
NDS355AN	SOT-23	•	30	0.5	1.0	105	4.5	3.5	195	**0.29**
FDN337N	SOT-23	•	30	0.5	1.3[z]	70	2.5	4.2	300	**0.16**
FDT439N	SOT-223	•	30	1.3	4	55	4.5	10.7	500	0.60
NTR4170N	SOT-23	•	30	0.8	2	50	4.5	4.8	430	0.12
PMV40UN	SOT-23	•	30	1.9	2[u]	45	2.5	5.5	445	0.39
NDT451AN	SOT-223	•	30	3	4	42	4.5	11	720	0.72
IRLML0030	SOT-23	•	30	1.3	2[u]	33	4.5	2.6	380	0.18
NTLJS4114N	WDFN	•	30	1.9	2.5[v]	26	2.5	5	650	0.30
NTMS4800N	SO-8	•	30	0.75	5	20	4.5	7.7	940	0.22
IRF7807Z	SO-8	•	30	2.5	10	14.5	4.5	7.2	770	0.61
FDS6680A	SO-8	•	30	1.0	7	10	4.5	16	1600	0.68
FDS8817NZ	SO-8	•	30	2.5	10	7	4.5	17	1800	0.75
NDS331N	SOT-23	•	20	0.5	0.8	150	2.7	2.2	160	**0.30**
FDG327NZ	SC70-6	•	20	0.42	1.2	90	1.8	2.1	410	0.40
IRLML2502	SOT-23	•	20	1.25	1.25	50	2.5	3.8	740	**0.30**
Si2312CDS	SOT-23	•	20	0.8	2[u]	35	1.8	3.8	870	0.28
Si2312BDS	SOT-23	•	20	0.8	2[v]	30	2.5	3.8	770	**0.40**
IRF6201	SO-8	•	20	2.5	15	2.1	2.5	130	8600	1.06

p-channel to 100V

pMOS type	Pkg[p]	SMT[x]	V_{DSS} (V)	P_D[c] (W)	I_D[y] (A)	R_{DS}[r]@V_{GS} (mΩ)	(V)	Q_G[s] (nC[t])	C_{iss} (pF[t])	Cost[q] ($US)
small										
FQT5P10	SOT-223	•	100	2	0.5	820	10	6.3	190	0.38
VP0106N3	TO-92	–	60	0.7	0.2	8Ω	5	0.5	45	0.55
BS250P	TO-92	•	45	0.7	0.2	9Ω	10	–	60	**0.61**
ZVP2106A	TO-92	•	60	0.7	0.25	3Ω	10	1.8	100	0.61
BSS84	SOT-23	•	50	0.3	0.13	3Ω	5	1	25	**0.26**
NDS0605	SOT-23	•	60	0.4	0.18	1.3Ω	4.5	0.8	79	**0.27**
FDV304P	SOT-23	•	25	0.4	0.3	1.2Ω	2.7	0.75	63	**0.27**
FDN358P	SOT-23	•	30	0.5	1.5	161	4.5	4	182	0.34
ZXMP4A16G	SOT-223	•	40	2	3	83	4.5	14	1000	0.93
IRF7205	SO-8	–	30	2.5	3	60	10	27	870	**0.37**
NTR4171P	SOT-23	•	30	0.5	1.5	60	4.5	16	720	**0.13**
DMP4050	SO-8	•	40	1.6	3	55	4.5	6.9	670	0.54
Si4435DDY	SO-8	•	30	2.5	6	28	4.5	15	1350	**0.53**
IRF7424	SO-8	•	30	2.5	7	20	4.5	75	4000	0.73
Si4463DY	SO-8	•	30	3	7[z]	13	2.5	28	5800	1.03
LP0701N3	TO-92	–	16.5	1	0.4	1.7Ω	3	1.6	120	0.82
ZXM61P02F	SOT-23	•	20	0.6	0.5	550	2.7	1.8	150	0.29
NDS332P	SOT-23	•	20	0.5	0.7	350	2.7	2.4	195	**0.30**
IRLML6402	SOT-23	•	20	1.3	2.2	80	2.5	5	630	**0.33**
SI3443DV	TSOP-6	•	20	2	3.4	70	2.7	5.5	610	0.42
Si2315BDS	SOT-23	•	20	0.57	2.1	71	1.8	8	715	**0.43**
FDS6575	SO-8	•	20	1.5	6[z]	11	2.5	53	4950	1.19
CSD25401	SON	•	20	2.8	8[z]	14	2.5	5.5	1100	1.96
IRLML6401	SOT-23	•	12	1.3	2	125	1.8	6	830	0.19
IRF7702	TSSOP-8	•	12	1.5	6	15	2.5	30	3500	0.90
IRF7420	SO-8	•	12	2.5	8	15	2.5	24	3500	0.66
IRF7410G	SO-8	•	12	2.5	12	8	2.5	55	8700	**0.97**
IRF7210	SO-8	•	12	2.5	10	7	2.5	115	17200	**1.03**
large										
IRF9540	TO-220	–	100	150	19	120	10	40	1400	**2.20**
IRF9540N	TO-220	–	*	140	20	110	*	60	1300	**1.28**
IRFP9140	TO-247	–	100	180	19	120	10	38	1400	3.07
IRFP9140N	TO-247	–	*	140	18	111	*	60	1300	**1.42**
IRF5210	TO-220	–	100	200	25	50	10	115	2700	1.93
IXTR90P10P	TO-247	–	100	190	55	20	10	120	5800	8.10
IXTK170P10	TO-264	–	100	890	130	10	10	240	12600	14.20
IXTN170P10	SOT-227	–	*	*	170	*	*	*	*	19.37
SUD08P06	DPak	•	60	25	2[z]	158	4.5	7	450	0.75
NTD2955G	DPak	•	60	55	2[z]	155	10	14	500	0.45
NTP2955	TO-220	–	*	62	11	*	*	*	*	**0.69**
MJE2955[b]	*TO-220*	–	*60*	*75*	*–*	*80*[b]	*–*	*–*	*–*	*0.64*
FQB11P06	D²Pak	–	60	53	2[z]	140	10	13	420	0.61
FQP27P06	TO-220	–	60	120	19	55	10	33	1100	0.72
FQB27P06	D²Pak	–	*	*	5[z]	*	*	*	*	0.92
IRF4905	TO-220	–	55	200	50	16	10	120	3400	**2.04**
STB80PF55	D²Pak	–	55	2.4	7	16	10	190	5500	2.53
STP80PF55	TO-220	–	*	300	55	*	*	*	*	1.78
SUM55P06	D²Pak	•	60	125	7[z]	15	10	76	3500	2.75
IRF9Z34	TO-220	–	55	68	15	10	10	23	620	**0.97**
SUP90P06	TO-220	–	60	250	90	9	4.5	90	9200	3.06
SUP75P05	TO-220	–	55	250	75	8	10	140	8500	5.42
IRFP064V	TO-247	–	60	250	80	5.5	10	175	6800	1.89
IRF9204	TO-220	–	40	143	35	20	4.5	150	7700	1.47
MTP50P03HDL	TO-220	–	30	125	30	20	5	74	3500	3.37
FDD6637	DPak	•	35	57	7[z]	14	4.5	25	2400	0.71
SUP75P03	TO-220	–	30	187	75	5.5	10	140	9000	1.91
IPB80P03P4L-04	D²Pak	•	30	137	16	4.7	4.5	60	8700	1.02

Notes: (•) same as row above. (b) BJT for comparison. (c) P_{diss} for T_{case}=25°C. (m) max. (p) the I²-PAK (TO-262) is a "sawed-off TO-220" stand-up part with 0.1" pitch (three leads plus tab), while the D²-PAK (TO-263) is an SMT version (2 leads plus tab); the I-PAK (TO-251) is a smaller version of the I²-PAK (i.e., stand-up, 3 leads plus tab, 0.09" pitch), with its corresponding D-PAK (TO-252) SMT version (2 leads plus tab). (q) qty 100; inexpensive "jellybean" parts are indicated in **boldface**. (r) R_{DS} typ for T_J=25°C, multiply by 1.5 if hot; BJT value is for I_B=0.3A. (s) total gate charge to V_{GS}; switching loss = $Q_G V_{GS} f$. (t) typical. (u) with 6 cm² PCB copper. (v) with 0.4 cm² PCB copper. (x) surface-mount part, or SMT available if through-hole indicated (e.g., TO-92 or TO-220). (y) guideline conservative estimate, saturated switch at V_{GS}, T_{case}=70°C. (z) with 2–5 cm² PCB copper; add heatsink for higher current.

This table shows selected representative MOSFETs. The left nMOS column lists TO-92 and small surface-mount parts. The right column lists all pMOS parts up to 100 V. The lists are sorted by decreasing switch spec, $R_{DS(ON)}$. Ignore parts with inadequate V_{DSS}, and evaluate parts with a good switch I_D spec margins. Study candidate datasheets for viability. Amplifiers and linear regulators rely on the P_D spec. But $R_{\Theta JC}$=125°C/P_D, and the junction temp will be $T_J=T_A+P_D(R_{\Theta JC}+R_{\Theta JA})$, where the latter term is your heatsink thermal path. Both R_Θ terms vary widely for different packages, and the P_D spec is useful only in that context. You may find that high-voltage parts have lower $R_{\Theta JC}$.

Table 3.4b *n*-channel Power MOSFETs, 55V to 4500V[a] (page 1 of 3)

Part #[k]	Package[p]	Manufacturer[n]	Surface mount	V_{DSS} 25°C (V)	P_{diss}[c] (W)	I_D (V_{GS}=10V) pulse (A)	25°C[b] (A)	70°C[z] (A)	$R_{\theta JC}$ (°C/W)	$R_{DS(on)}$[e] typ[r] (mΩ)	max (mΩ)	at V_{DS} (V)	Gate zener	Superjunction[k]	Charge Q_G[v,s] typ (nC)[t]	Q_{GD}[s] typ (nC)[t]	Capacitance[s,s2] (V_{DS}=25V[k]) C_{iss} typ (pF)	C_{oss} typ (pF)	C_{rss} typ (pF)	Cost[q] US $	year of intro[y]
2.5 to 4.5kV																					
IXTT02N450	TO-268	Ix	-	4500	113	0.6	0.2	*0.14*	1.1	480Ω	625Ω	10	-	-	10.6	5.5	246	19	5.8	17.05	2013
IXTT1N450HV	TO-268HV	Ix	-	4500	520	3	1	*0.8*	0.24	72Ω	85Ω	10	-	-	40	20	1730	78	28	28.67	2013
IXTL2N450	i5-Pak	Ix	-	4500	220	8	2	*1.3*	0.56	16Ω	20Ω	10	-	-	180	83	6860	267	105	88.80	2013
IXTH02N250	TO-247	Ix	-	2500	83	0.6	0.2	*0.14*	1.5	385Ω	450Ω	10	-	-	7.4	5.3	116	8	3	9.52	2013
1500V																					
2SK1317	TO-3P	R	-	1500	100	7	2.5	*1.5*	1.25	9000	12000	10	-	-	-	-	990	125	60	5.46	2004
STP3N150	TO-220	ST	-	1500	140	10	2.5	1.6	0.89	6000	9000	10	-	-	29.3	17	939	102	13.2	4.64	2008
STP4N150	TO-220	ST	-	1500	160	12	4	2.5	0.78	5000	7000	10	-	-	30	9	1300	120	12	4.15	2003
IXTH6N150	TO-247	Ix	-	1500	540	24	6	4	0.23	*2905*	3500	10	-	-	67	36	2230	170	64	6.07	2010
1200V																					
IXTY02N120	D-Pak	Ix	•	1200	33	0.6	0.2	0.2	3.80	60Ω	75Ω	10	-	-	4.7	3.2	104	8.6	1.9	2.92	2009
IXTP1N120P	TO-220	Ix	-	1200	63	1.8	1	*0.7*	2	16Ω	20Ω	10	-	-	17.6	10.6	550	25	5.4	2.46	2007
IXTP3N120	TO-220	Ix	-	1200	200	12	3	*2.5*	0.62	*3735*	4500	10	-	-	42	21	1100	110	40	4.65	2003
STP6N120K3	TO-220	ST	-	1200	150	20	6	3.8	0.83	1950	2400	10	•	-	39	23.5	1050	90	3	4.28	2010
IXFH16N120P	TO-247	Ix	-	1200	680	35	16	8	0.19	850	950	10	-	-	120	47	6900	390	48	11.02	2007
IXFX26N120P	TO-264	Ix	-	1200	960	60	26	*19*	0.13	*380*	460	10	-	-	225	96	16000	735	58	19.28	2007
IXFN32N120	SOT-227	Ix	-	1200	780	128	32	*20*	0.16	*290*	350	10	-	-	400	188	15900	1000	260	27.15	2001
CMF20120D[x]	TO-247	Cr	-	1200	215	90	42	24	0.44	80	100	20	-	-	91	43	1915	120	13	32.05	2011
800-1000V																					
IXTY01N100	D-Pak	Ix	-	1000	25	0.4	0.1	*0.1*	5	60Ω	80Ω	10	-	-	6.9	3	54	6.9	2	0.93	2004
IXTA05N100	D-Pak	Ix	•	1000	40	3	0.75	*0.5*	3.1	15Ω	17Ω	10	-	-	7.8	4.1	260	22	8	1.68	2006
IXTP1N100	TO-220	Ix	-	1000	54	6	1.5	0.9	2.3	8300	11000	10	-	-	14.5	7.5	400	37	13	0.93	2002
FQD2N100	D-Pak	F	-	1000	50	6.4	1.6	1	2.5	7100	9000	10	-	-	12	6.5	400	40	5	0.79	2001
STD2NK100Z	D-Pak	ST	•	1000	70	7.4	1.85	1.16	1.8	6250	8500	10	•	-	16	9	499	53	9	2.19	2006
STP2NK100Z	TO-220	ST	•	1000	70	7.4	1.85	1.16	1.8	6250	8500	10	•	-	16	9	499	53	9	2.16	2006
IRFBG20	TO-220	IR	-	1000	54	5.6	1.4	0.86	2.3	8000	11000	10	-	-	27	15	500	52	17	1.02	1993
IRFBG30	TO-220	IR	-	1000	125	12	3.1	2	1	4000	5000	10	-	-	50	30	980	140	50	1.31	1993
STP5NK100Z	TO-220	ST	-	1000	125	14	3.5	2.2.	1	2700	3700	10	•	-	42	22	1154	106	21	2.52	2005
FQA8N100	TO-3P	F	-	1000	225	32	8	5	0.56	1200	1450	10	-	-	53	23	2475	195	16	2.28	2005
IXTX24N100	TO-247	Ix	-	1000	568	96	24	16	0.22	333	400	10	-	-	267	142	8700	785	315	15.53	2009
FQP9N90C	TO-220	F	-	900	205	32	8	2.8	1.85	1120	1400	10	-	-	45	18	2100	175	14	1.73	2002
FQD1N80	DPak	F	•	800	45	4	1	0.63	2.78	15000	20000	10	-	-	5.5	3.3	150	20	2.7	0.22	1999
STQ1NK80	TO-92	ST	-	800	3	5	0.25	0.16	40	13000	16000	10	•	-	7.7	4.5	160	26	7	0.35	2004
STN1NK80	SOT-223	ST	•	800	2.5	5	*	*	50	*	*	*	•	-	*	*	*	*	*	0.75	2004
STD1NK80	DPak	ST	•	800	45	5	*	*	2.78	*	*	*	•	-	*	*	*	*	*	0.64	2004
SPP02N80C3	TO-220	Inf	-	800	42	6	2	1.2	3	2400	2700	10	-	-	12	6	290	130	6	1.04	2005
FQP7N80C	TO-220	F	-	800	167	26	6.6	4.2	0.75	1570	1900	10	-	-	27	11			10	1.25	2000
SPP11N80C3	TO-220	Inf	-	800	156	33	11	7.1	0.8	390	450	10	-	-	50	25	1600	800	40	2.72	2003

Notes: (a) sorted by voltage groups, then by descending $R_{DS(on)}$; parts with rated below 55V are not included; many MOSFETs have an alphanumeric part-number scheme, e.g., 10N60, where 10 is the continuous current capability, N means *n*-channel, and the last number is V_{DSS}/10, so 60 means 600 volts; for these parts a prefix letter usually indicates the package type. (∗) same as row above. (b) the maximum continuous current with T_C = 25°C (manifestly impossible), assuming T_J = 150°C; in some cases the current is limited by the package (bonding wires). (c) P_{diss} for T_{case}=25°C from datasheet. (e) at T_J = 25°C. (k) parts in *italics* have super-junction technology, specs are at V_{DS} = 50 or 100V. (m) max. (n) Cr = Cree, F = Fairchild, Inf = Infineon, IR = International Rectifier, Ix = IXYS, N = NXP, R = Renesas, ST = STMicroelectronics, Su = Supertex, To = Toshiba, V = Vishay. (p) package types (check datasheet to see which pkgs the mfgr offers for that part); plastic power types with three leads are (largest first): TO-264 with 0.215" lead spacing (to match TO-3 center hole and insulated side mounting notches), TO-247 (same spacing, optional center hole), TO-3P (same spacing and center hole, both to match TO-3), TO-220 (smaller with 0.2" tab with hole, 0.1" lead spacing, very popular); the I2-PAK (TO-262) is a "sawed-off TO-220" stand-up part with 0.1" pitch (three leads plus tab), while the D2-PAK (TO-263) is an SMT version (2 leads plus tab); the I-PAK (TO-251) is a smaller version of the I2-PAK (i.e., stand-up, 3 leads plus tab, 0.09" pitch), with its corresponding D-PAK (TO-252) SMT version (2 leads plus tab); SOT-223 three leads 0.09" spacing with short tab; power SO-8P is similar to SOIC-8 with a metal "power pad"; LFPAK=SOT669 has SO-8 footprint replacing 4 pins with drain tab; the SOT-227 package measures 1x1.5 in, has four #4 captive-nut spade-lug lead connections, and a super-useful insulated metal heat-sink plate. (q) qty 100. (r) R_{DS} typ for T_J = 25°C; if hot, scale by 1.5 to 2x for low-voltage parts, or by 2.2 to 3.5x for high-voltage parts, see Chapter 3x. (s) total gate charge to VGS; gate switching loss = $Q_G V_{GS} f$. (s2) drain capacitive switching $I_{oss} = C_{oss} V_{DS}^2 f$. (t) typical. (v) at the V_{GS} at which $R_{DS(on)}$ is specified. (w) newer -N version costs less, but has higher $R_{\theta JC}$. (x) this is a silicon-carbide (SiC) rather than silicon MOSFET; these have lower capacitance but require higher gate voltages. (y) est'd year-of-introduction; see "A 30-year MOSFET Saga" in chapter 3x. (z) guideline conservative estimate, saturated switch at V_{GS}, T_C = 70°C.

Usage comments: • when scanning the list for low C_{oss}, or maximum power, or low R_{on}, etc., you may find a better part for your spec, with much higher V_{DSS} than required -- especially true for high-power capability, a linear servo or pass transistor. • 500V parts have been superceded by 600V parts by many manufacturers. • linear apps often need power dissipation capability, this is often better with older large-die parts; newer designs use narrow V-grooves to achieve low $R_{DS(on)}$, and therefore require much less die area. There are good candidates in the table.

Table 3.4b n-channel Power MOSFETs, 55V to 4500V[a] (page 2 of 3)

Part #[k]	Package[p]	Manuf.[n]	Surf. mount	V$_{DSS}$ 25°C (V)	P$_{diss}$[c] (W)	I$_D$ pulse (A)	I$_D$ 25°C[b] (A)	I$_D$ 70°C[z] (A)	R$_{\theta JC}$ (°C/W)	R$_{DS(on)}$[e] typ[r] (mΩ)	R$_{DS(on)}$ max (mΩ)	at V$_{DS}$ (V)	Gate zener[k]	Super-junction[k]	Q$_G$[v,s] typ (nC)[t]	Q$_{GD}$[s] typ (nC)[t]	C$_{iss}$ typ (pF)	C$_{oss}$ typ (pF)	C$_{rss}$ typ (pF)	Cost[q] US $	year of intro[y]
600V																					
VN2460N3, N8	TO-92	Su	•	600	1	0.5	0.16	0.12	125	25000	25000	5	-	-	5.5	4	120	10	5	0.84	2000
FQN1N60C	TO-92	F	-	600	3	1.2	0.3	0.18	50	9300	11500	10	-	-	4.8	2.7	130	19	3.5	0.31	2003
FQD1N60C	DPak	F	•	600	28	4	1	1.6	4.5	*	*	*	-	-	*	*	*	*	*	0.47	2003
SPD01N60c3	DPak	IR	•	600	11	1.6	0.8	0.5	11	5600	6000	10	-	-	2.2	0.9	100	40	2.5	0.60	2003
IXTP2N60P	TO-220	Ix	-	600	55	4	2	1.4	2.25	4200	5500	10	-	-	7	2.1	240	28	3.5	0.79	2005
FQP2N60C	TO-220	F	-	600	54	8	2	1.35	2.32	3600	4700	10	-	-	8.5	4.1	180	20	4.3	0.60	2002
FQP3N60C	TO-220	F	-	600	75	12	3	1.8	1.67	2800	3400	10	-	-	10.5	4.5	435	45	5	0.70	2005
FQP5N60C	TO-220	F	-	600	100	18	4.5	2.6	1.25	2000	2500	10	-	-	15	6.6	515	55	6.5	0.75	2003
FDP5N60NZ	TO-220	F	-	600	100	18	4.5	2.7	1.25	1650	2000	10	•	-	10	4	450	50	5	0.82	2011
SPD03N60	DPak	Inf	•	600	38	9.6	3.2	2	3.3	1280	1400	10	-	-	13	6	400	150	5	0.84	2002
FDP7N60NZ	TO-220	F	-	600	147	26	6.5	3.9	0.85	1050	1250	10	-	-	13	5.6	550	70	7	1.10	2010
STP7NM60	TO-220	ST	-	600	45	20	5	3	2.8	840	900	10	-	•	14	7.7	363	25	1.1	1.31	2008
FQB7N60	D2Pak	F	•	600	142	30	7.4	4.7	0.88	800	1000	10	-	-	29	14.5	1100	135	16	1.14	2000
FDP10N60NZ	TO-220	F	-	600	185	40	10	6	0.68	640	750	10	-	•	23	8	1110	130	10	1.49	2010
STP8N65M5	TO-220	ST	-	650	70	28	7	4.4	1.8	560	600	10	-	•	15	6	690	18	2	4.48	2009
FCP7N60N	TO-220	F	-	600	64	20	6.8	4.3	1.95	460	520	10	-	•	18	6	719	30	2.1	1.60	2009
STP11N65M5	TO-220	ST	-	650	85	36	9	5.6	1.5	430	480	10	-	•	17	8.5	644	18	2.5	1.67	2012
FCP9N60N	TO-220	F	-	600	83	27	9	5.7	1.5	330	385	10	-	•	22	7.1	930	35	2	2.37	2009
TK10E60W	TO-220	To	-	600	100	39	9.7		1.25	327	380	10	-	•	20	9.5	700	20	2.3	2.04	2012
FQA19N60	TO-3P	F	-	600	300	74	18.5	11.7	0.42	300	380	10	-	-	70	33	2800	350	35	2.71	2000
SPP15N60CFD	TO-220	Inf	-	600	156	33	13	8.4	0.80	280	330	10	-	•	63	38	1820	520	21	2.87	2006
STP15NM60ND	TO-220	ST	-	600	125	56	14	9	1	270	299	10	-	•	40	22	1250	65	5	3.42	2007
IPP60R280C6	TO-220	Inf	-	600	104	40	14	8.7	1.2	250	280	10	-	•	43	22	950	60		1.95	2009
SPP15N60C3	TO-220	Inf	-	600	156	45	15	9.4	0.8	250	280	10	-	•	63	29	1660	540	40	2.33	2002
FCP13N60N	TO-220	F	-	600	116	39	13	8.2	1.07	244	258	10	-	•	30	9.5	1325	50	3	2.41	2009
SiHP15N60E	TO-220	V	-	600	180	39	15	9.6	0.7	230	280	10	-	•	38	17	1350	70	5	1.77	2011
STP16N65M5	TO-220	ST	-	650	90	48	12	7.3	1.4	230	279	10	-	•	31	13	1250	30	3	2.96	2011
IPP60R250CP	TO-220	Inf	-	600	104	40	12	8	1.2	220	250	10	-	•	26	9	1200	54		2.56	2010
STP22NM60N	TO-220	ST	-	600	125	64	16	10	1	200	220	10	-	•	44	25	1330	84	4.6	3.82	2009
IPP60R199CP	TO-220	Inf	-	600	139	51	16	10	0.9	180	199	10	-	•	32	11	1520	72	2	2.88	2006
IRFP27N60K	TO-247	IR	-	600	500	110	27	18	0.29	180	220	10	-	-	105	50	4660	460	41	5.53	2002
TK16E60W	TO-220	To	-	600	130	63	16.8	11	0.96	160	190	10	-	•	38	16	1350	35	4	2.86	2012
STP21N65M5	TO-220	ST	-	650	125	68	17	10.7	1	150	179	10	-	•	50	23	1950	46	3	2.33	2010
SiHP24N65E	TO-220	V	-	700	250	70	24	16	0.5	120	145	10	-	•	81	37	2740	122	4	2.45	2009
IXFH50N60P3	TO-247	Ix	-	600	1040	125	50	35	0.12	118	145	10	-	-	94	23	6300	630	2.5	4.95	2011
IPP60R099CP	TO-220	Inf	-	600	255	93	31	19	0.5	90	99	10	-	•	60	20	2800	130	100	4.68	2005
SPW47N60C3	TO-247	Inf	-	600	415	141	47	30	0.3	60	70	10	-	•	252	121	9800	2200	145	5.00	2004
FCH47N60N	TO-247	F	-	600	368	141	47	30	0.34	52	62	10	-	•	115	34	5037	200	2.5	9.97	2010
IXKN75N60C	SOT-227	Ix	-	600	568	-	75	60	0.22	30	36	10	-	•	500	220	15000	6000	300	33.33	2003
FCH76N60N	TO-247	F	-	600	543	228	76	48	0.23	28	36	10	-	•	218	66	9310	370	3.3	18.80	2010
500V																					
IRF820	TO-220	IR	-	500	50	8	2.5	1.6	2.5	2500	3000	10	-	-	16	8.2	360	92	37	0.43	1980
IRF830	TO-220	IR	-	500	74	18	4.5	2.9	1.7	1200	1500	10	-	-	26	12	610	160	68	0.59	1980
IRF840	TO-220	IR	-	500	125	32	8	5.1	1	800	850	10	-	-	52	22	1300	310	120	0.70	1982
FDP5N50NZ	TO-220	F	-	500	78	18	4.5	2.7	4.1	1380	1500	10	•	-	9	4	330	50	4	0.95	2010
STD5NK50Z	DPak	ST	•	500	70	17.6	4.4	2.7	1.78	1220	1500	10	•	-	20	10	535	75	17	1.06	2002
FDP7N50	TO-220	F	-	500	89	28	7	4.2	1.4	760	900	10	-	-	12.8	5.8	720	95	9	1.07	2006
FQP9N50	TO-220	F	-	500	147	36	9	5.7	0.85	580	730	10	-	-	28	12.5	1100	160	20	0.72	2002
IRFP450	TO-247	IR	-	500	190	56	14	9	0.65	350	400	10	-	-	110	55	2600	720	340	2.12	1993
IRFP460	TO-247	IR	-	500	280	80	20	13	0.45	230	270	10	-	-	150	70	4200	870	350	2.58	1993
STW20NK50Z	TO-247	ST	-	500	190	68	20	12.6	0.66	230	270	10	•	-	85	42	2600	328	72	3.15	2002
FDA28N50	TO-3P	F	-	500	310	112	28	17	0.4	122	155	10	-	-	80	32	3866	576	42	3.81	2007
FCP22N60N	TO-220	F	-	500	205	66	22	13.8	0.61	140	165	10	-	•	45	14.5	1950	76	3	4.67	2007
FDH44N50	TO-247	F	-	500	750	176	44	32	0.2	110	120	10	-	-	90	31	5335	645	40	6.80	1998
IXFN80N50P	SOT-227	Ix	-	500	700	200	80	45	0.18	54	65	10	-	-	195	64	12700	1280	120	18.77	2004
IXFB100N50P	TO-264	Ix	-	500	1250	250	100	75	0.10	44	49	10	-	-	240	78	20000	1700	140	16.18	2005

Table 3.4b n-channel Power MOSFETs, 55V to 4500V[a] (page 3 of 3)

Part #[k]	Package[p]	Manuf[n]	SM	V_{DSS} 25°C (V)	P_{diss}[c] (W)	I_D pulse (A)	I_D 25°C[b] (A)	I_D 70°C[z] (A)	$R_{\theta JC}$ (°C/W)	$R_{DS(on)}$ typ[r] (mΩ)	$R_{DS(on)}$ max (mΩ)	at V_{DS} (V)	Gate zener	Super-junction[k]	$Q_G^{v,s}$ typ (nC)[t]	Q_{GD}^s typ (nC)[t]	C_{iss} typ (pF)	C_{oss} typ (pF)	C_{rss} typ (pF)	Cost[q] US$	year of intro[y]
200–400V																					
IRF710	TO-220	IR	-	400	36	6	2	1.2	3.5	3100	3600	10	-	-	5.7	2.2	170	34	6.3	0.38	1980
IRF720	TO-220	IR	-	400	50	13	3.3	2.1	2.5	1300	1800	10	-	-	15	7	410	120	47	0.28	1980
IRF730	TO-220	IR	-	400	74	22	5.5	3.5	1.7	740	1000	10	-	-	24	13	700	170	64	0.53	1980
IRF740	TO-220	IR	-	400	125	40	10	6.3	1.0	435	550	10	-	-	43	21	1400	330	120	0.67	1981
STP7NK40Z	TO-220	ST	-	400	70	22	5.4	3.4	1.78	850	1000	10	-	-	19	10	535	82	18	1.13	2002
STP11NK40Z	TO-220	ST	-	400	110	36	9	5.67	1.14	490	550	10	-	-	32	18.5	930	140	30	1.45	2003
FQP17N40	TO-220	F	-	400	170	64	16	10.1	0.74	210		10	-	-	45	21.7	1800	270	30	1.45	2006
IRFP244	TO-220	V	-	250	150	60	15	9.7	0.83	180	280	10	-	-	63.0	39	1400	320	73	2.99	1997
FQP16N25	TO-220	F	-	250	250	64	16	10	0.88	180	230	10	-	-	27	15	920	190	23	1.36	2000
FQA30N40	TO-3P	F	-	400	290	120	30	19	0.43	107	140	10	-	-	90	46	3400	580	60	3.53	1999
FDP33N25	TO-220	F	-	250	235	132	33	20	0.53	77	94	10	-	-	36.8	17	1640	30	39	1.23	2006
FDP2710	TO-220	F	-	250	403		50	31	0.48	36	42	10	-	-	78	18	5470	426	97	2.92	2007
FDA69N25	TO-3P	F	-	250	480	276	69	44	0.28	34	41	10	-	-	77	37	3570	780	84	2.50	2006
IXTK120N25P	TO-264	Ix	-	250	700	300	120	80	0.18	19	24	10	-	-	185	80	8000	1300	220	8.16	2004
IRF610	TO-220	IR	-	200	36	10	3.3	2.1	3.5	1250	1500	10	-	-	6.3	3.2	140	53	15	0.28	1980
IRL620	TO-220	IR	-	200	50	21	5.2	3.3	2.5	630	800	5	-	-	8.2	5.5	360	91	27	1.03	1993
IRF620	TO-220	IR	-	200	50	18	5.2	3.3	2.5	550	800	10	-	-	10.6	5	260	100	30	0.38	1980
IRL630	TO-220	IR	-	200	74	36	9	5.7	1.7	290	400	5	-	-	24	24	1100	220	70	1.08	1989
IRF630	TO-220	IR	-	200	74	36	9	5.7	1.7	220	400	10	-	-	27	14	800	240	76	0.46	1980
IRL640	TO-220	IR	-	200	125	68	17	11	1.0	125	180	5	-	-	42	24	1800	400	120	0.78	1992
IRF640	TO-220	IR	-	200	125	72	18	11	1.0	130	180	10	-	-	45	24	1300	430	130	0.59	1981
PSMN102-200Y	LFPak	N	•	200	113	65	21.5	13.6	1.1	86	102	10	-	-	31	10	1568	170	55	0.92	2008
IRFP260N	TO-247	IR	-	200	300	200	50	35	0.50	35	40	10	-	-	234	110	4057	603	161	2.15	2009
IRFP4668	TO-247	IR	-	200	520	520	130	92	0.29	8	9.7	10	-	-	161	52	10720	810	160	4.88	2008
55–100V																					
IRF510	TO-220	IR	-	100	43	20	5.6	4	3.5	410	540	10	-	-	5.2	2.2	180	81	15	0.35	1980
IRF520	TO-220	IR	-	100	60	37	9.2	6.5	2.5	200	270	10	-	-	10.3	3.9	360	150	34	0.38	1980
IRF530	TO-220	IR	-	100	88	56	14	10	1.7	100	160	10	-	-	16.2	7	670	250	60	0.44	1980
IRF540	TO-220	IR	-	100	130	110	28	20	1.0	50	77	10	-	-	47	17	1700	560	120	0.60	1980
FQP33N10	TO-220	F	-	100	127	132	33	23	1.18	40	52	10	-	-	38	18	1150	320	62	0.89	1995
PSMN039-100YS	LFPak	N	•	100	74	112	28	20	1	31	40	10	-	-	23	8	1847	86	64	0.57	2010
FQP44N10	TO-220	F	-	100	146	174	43	31	1	30	39	10	-	-	48	24	1400	425	85	1.02	2000
SUP85N10	TO-220	V	-	100	250	240	85	60	0.6	10	12	5	-	-	105	23	6550	665	265	4.88	2000
HUF75652G3	TO-247	F	-	100	515	1200	75 h	75	0.29	6.7	8	10	-	-	393	74	7585	2345	630	4.87	1998
IRFB4110	TO-220	IR	-	100	370	670	180	120	0.4	3.7	4.5	10	-	-	150	43	9620	670	250	2.84	2005
IRFZ14	TO-220	IR	-	60	43	40	10	7.2	3.5	135	200	10	-	-	9.7	4.7	300	160	29	0.44	1986
IRFZ24	TO-220	IR	-	60	60	68	12	17	2.5	68	100	10	-	-	19	8	640	360	79	0.53	1986
IRFZ34	TO-220	IR	-	60	88	120	30	21	1.7	42	50	10	-	-	30	15	1200	600	100	0.65	1986
IRFZ44[w]	TO-220	IR	-	60	150	200	50	36	1.0	24	28	10	-	-	42	17	1900	920	170	0.77	1986
IRLZ44N	TO-220	IR	-	55	110	160	47	33	1.4	20	25	5	-	-	28	17	1700	400	150	0.47	1992
NDP6060L	TO-220	F	-	60	100	144	48	24	1.5	20	28	5	-	-	43	21	1630	460	150	1.63	1995
IRL3705N	TO-220	IR	-	55	170	310	89	63	0.9	11	12	5	-	-	95	49	3600	870	320	1.30	2004
IRFP054N	TO-247	IR	-	60	170	290	81	57	0.9	10	12	10	-	-	130	53	2900	880	330	1.36	1996
IRL2505	TO-220	IR	-	55	200	360	104	74	0.75	9	10	5	-	-	130	67	5000	1100	390	1.65	1996
STP80NF55-08	TO-220	ST	-	55	300	320	80 h	80	0.5	6.5	8	10	-	-	112	40	3740	830	265	2.57	2007
IRF3205Z	TO-220	F	-	55	170	440	110	78	0.9	4.9	6.5	10	-	-	76	30	3450	550	310	1.26	2001
IRF1405	TO-220	IR	-	55	330	680	169	118	0.45	4.6	5.3	10	-	-	170	62	5480	1210	280	1.76	2001
FDP025N06	TO-220	F	-	60	395	1060	265	120	0.38	1.9	2.5	10	-	-	174	50	11190	1610	750	3.41	2006
FDP020N06B	TO-220	F	-	60	333	1252	313	221	0.45	1.65	2.0	10	-	-	87	34	16100	3640	127	3.79	2011

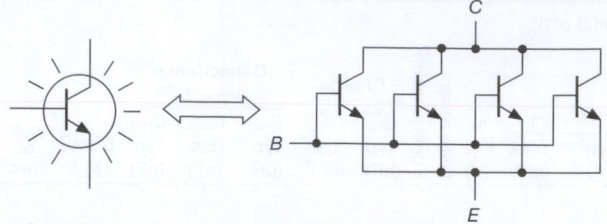

Figure 3.94. A large-junction-area transistor can be thought of as many paralleled small-area transistors.

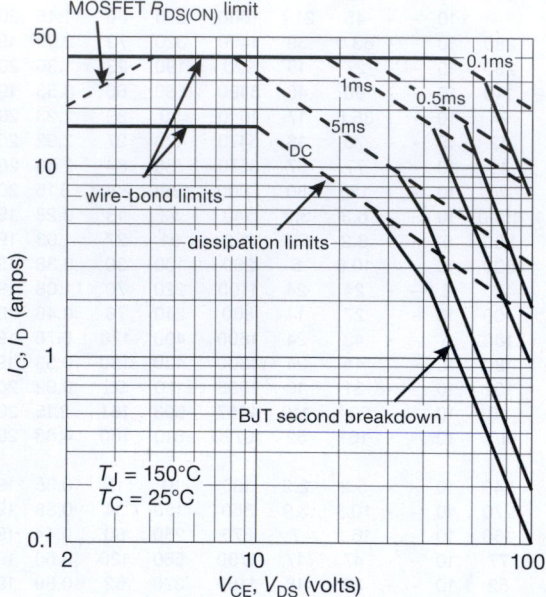

Figure 3.95. Power MOSFETs do not suffer from second breakdown: comparing safe operating areas (SOAs) of a 160 W BJT (MJH6284) and MOSFET (RFP40N10).

This negative temperature coefficient ("tempco") of I_D (at fixed V_{GS}) has produced some bad advice in the amplifier community, in particular the statement that it's always OK to connect a set of power MOSFETs in parallel without the current-equalizing "emitter-ballasting" resistors that are necessary with bipolar transistors.[80] You *could* do that, if the MOSFETs were to be operated in the high-current regime where you get the stabilizing negative tempco. But in practice you usually can't operate up there anyway, because of power-dissipation limitations that we'll see in §9.4.1A. And at the lower currents where the tempco is positive and destabilizing, one of a set of paralleled

MOSFETs will tend to hog the current and suffer excessive power dissipation, often leading to early failure. The solution is to use a small source-ballasting resistor in each of the paralleled MOSFETs (which should be of the same type, and from the same manufacturer), chosen to drop about a volt at the operating current.

By contrast, you *can* parallel power MOSFETs in *switching* applications. That's because the MOSFET is here operated in the ohmic region of low V_{DS} (characterized by approximately constant resistance R_{ON}, as opposed to the higher-voltage "current-saturation" region where the transistor is characterized by approximately constant I_D): it is the positive tempco of R_{ON} that stabilizes the current sharing in paralleled power MOSFETs. No ballasting resistors are needed, or even desirable. We'll say more about this in §3.6.3.

3.5.2 Power MOSFET switching parameters

Most power MOSFETs are enhancement type, available in both *n*- and *p*-channel polarities. Relevant parameters are the breakdown voltage V_{DSS} (ranging from 12 V to 4.5 kV for *n*-channel, and to 500 V for *p*-channel); the channel on-resistance $R_{DS(on)}$ (as low as 0.8 mΩ); the current- and power-handling ability (as much as 1000 A and 1000 W); and the gate capacitances C_{rss} and C_{iss} (as much as 2000 pF and 20,000 pF, respectively).

We need to rain on this parade! These impressively high current and power ratings are usually specified at 25°C case temperature, allowing the junction temperature to rise to 175°C (while the impressively low R_{on} is specified at 25°C *junction* temperature!). Unless you're resident at the South Pole, these are completely unrealistic conditions during continuous high-power switching.[81] See the discussion in §3.5.4D.

3.5.3 Power switching from logic levels

You often want to control a power MOSFET from the output of digital logic. Although there are logic families that generate swings of 10 V or more (the "legacy" 4000-series CMOS), the most common logic families (known generically as CMOS) use supply voltages of +5, +3.3, or +2.5 V, and generate output levels close to that voltage or to ground (HIGH and LOW, respectively).[82] Figure 3.96

---

[80] Because of their *positive* tempco of I_C at constant V_{BE}, see §§2.3 and 2x.2.

[81] In their defense, datasheets do provide derating coefficients, but the derated current and power somehow are neglected when the front-page banner is typeset.

[82] The family known as TTL operates from +5 V, but its HIGH output can be as low as +2.4 V, a characteristic shared by some other +5 V parts.

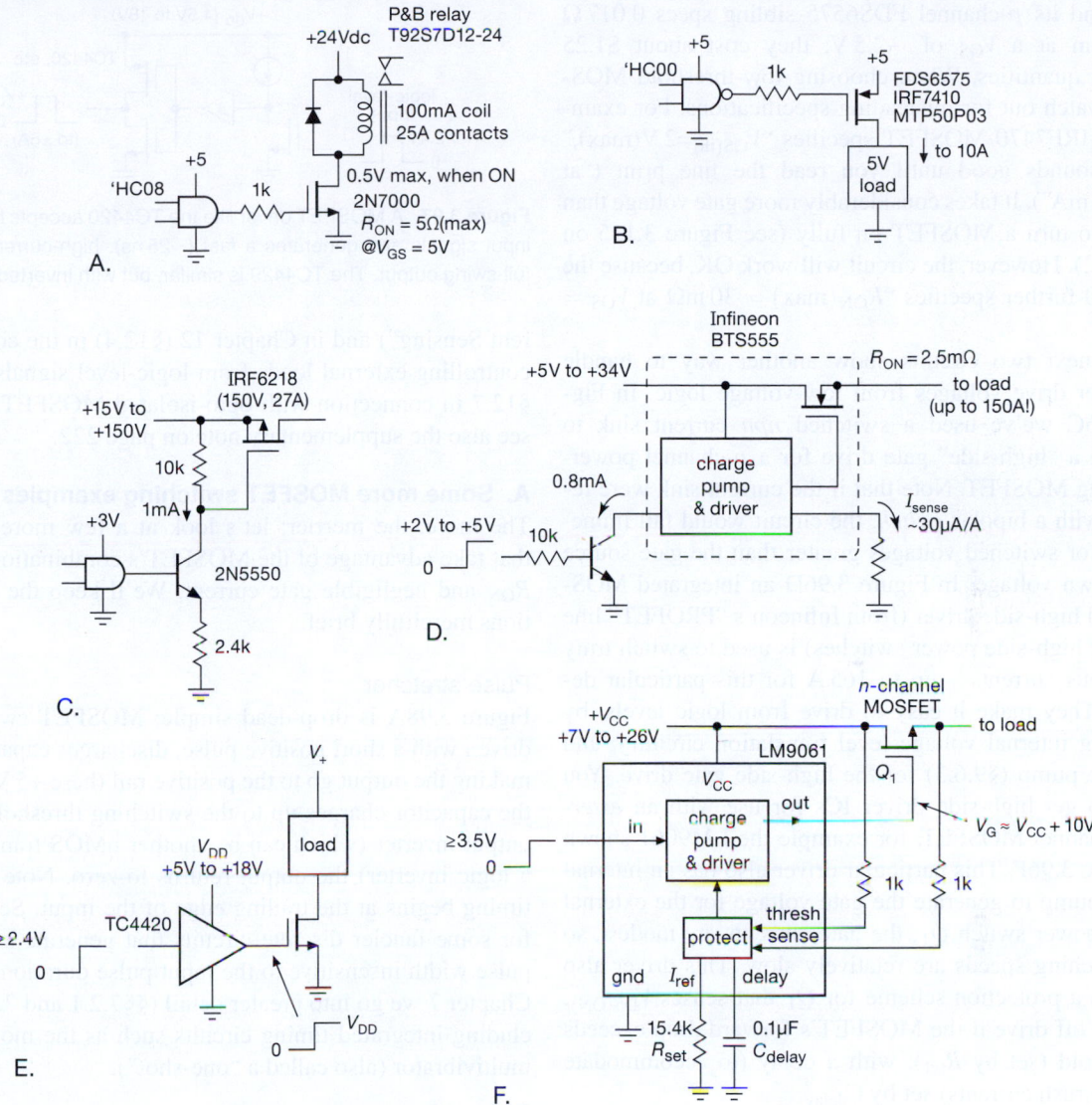

Figure 3.96. MOSFETs can switch power loads when driven from digital logic levels. See also Figure 3.106 on page 204.

shows how to switch loads from these logic families. In the first circuit (Figure 3.96A), the +5 V gate drive will fully turn on a garden-variety MOSFET, so we chose the 2N7000, an inexpensive transistor ($0.04 in quantity!) that specifies $R_{ON} < 5\,\Omega$ at $V_{GS} = 4.5$ V. The diode protects against inductive spike (§1.6.7). The series gate resistor, though not essential, is a good idea, because MOSFET drain-gate capacitance can couple the load's inductive transients back to the delicate CMOS logic (more on this soon).

For variety, in the second circuit (Figure 3.96B) we've used the *p*-channel MTP50P03HDL, driving a load returned to ground. In a commonly used technique called

power switching, the "load" could be additional circuitry, powered up electrically on command. The '50P03 specifies a maximum R_{ON} of 0.025 Ω at $V_{GS}=-5$ V and can handle a 50 amp load current; for lower R_{ON} you could select the IRF7410 (0.007 Ω, 16 A, $1.50); see Table 3.4a on page 188.

Lower voltage logic is increasingly popular in digital circuits. The switching configurations of Figures 3.96A & B can be used for lower voltages, but be sure to use MOSFETs with specified "logic thresholds." For example the 20 V 16 A *n*-channel FDS6574A from Fairchild specifies a maximum R_{ON} of 0.009 Ω at a paltry V_{GS} of

1.8 V, and its *p*-channel FDS6575 sibling specs 0.017 Ω maximum at a V_{GS} of −2.5 V; they cost about $1.25 in small quantities. When choosing low-threshold MOSFETs, watch out for misleading specifications. For example, the IRF7470 MOSFET specifies "$V_{GS(th)}$=2 V(max)," which sounds good until you read the fine print ("at I_D=0.25 mA"). It takes considerably more gate voltage than $V_{GS(th)}$ to turn a MOSFET on fully (see Figure 3.115 on page 212). However, the circuit will work OK, because the IRF7470 further specifies "$R_{ON}(max) = 30$ mΩ at $V_{GS} = 2.8$ V."

The next two circuits show another way to handle the lower drive voltages from low-voltage logic. In Figure 3.96C we've used a switched *npn* current sink to generate a "high-side" gate drive for a *p*-channel power-switching MOSFET. Note that if the current sink were replaced with a bipolar *switch*, the circuit would fail immediately for switched voltages greater than the gate-source breakdown voltage. In Figure 3.96D an integrated MOSFET and high-side driver (from Infineon's "PROFET" line of smart high-side power switches) is used to switch truly prodigious currents – up to 165 A for this particular device.[83] They make it easy to drive from logic levels, by including internal voltage level translation circuitry, and a charge pump (§9.6.3) for the high-side gate drive. You can also get high-side driver ICs for use with an *external n*-channel MOSFET, for example the LM9061 shown in Figure 3.96F. This particular driver also has an internal charge pump to generate the gate voltage for the external nMOS power switch Q_1; the gate currents are modest, so the switching speeds are relatively slow. This driver also includes a protection scheme for Q_1 that senses $V_{DS(ON)}$, shutting off drive if the MOSFET's forward drop exceeds a threshold (set by R_{set}), with a delay (to accommodate higher inrush currents) set by C_{delay}.

Finally, Figure 3.96E shows how you can sidestep this issue entirely and ensure healthy gate drive voltage *and* current, by using a "MOSFET gate driver" chip like the TC4420. It accepts logic-level input (threshold guaranteed less than +2.4 V), and produces a muscular full-swing output with internal push–pull MOSFETs of its own (Figure 3.97). It can source or sink several amperes of gate current, ensuring fast switching with the large capacitive loads that power MOSFETs present (see §3.5.4B). The trade-off here is cost (about $1) and complexity. Table 3.8 on page 218 lists a selection of nice gate-driver ICs. We'll see these again in detail in Chapter *9x* ("High-side Cur-

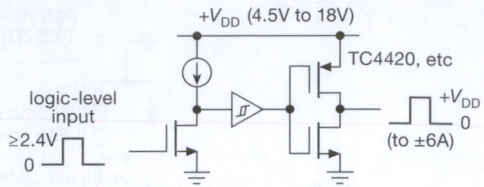

Figure 3.97. A MOSFET driver like the TC4420 accepts logic-level input signals and generates a fast (∼25 ns), high-current (±6 A), full-swing output. The TC4429 is similar, but with inverted output.

rent Sensing") and in Chapter 12 (§12.4) in the context of controlling external loads from logic-level signals, and in §12.7 in connection with opto-isolated MOSFET drivers; see also the supplementary note on page 222.

A. Some more MOSFET switching examples

The more, the merrier: let's look at a few more circuits that take advantage of the MOSFET's combination of low R_{ON} and negligible gate current. We'll keep the descriptions mercifully brief.

Pulse stretcher

Figure 3.98A is drop-dead simple: MOSFET switch Q_1, driven with a short positive pulse, discharges capacitor C_1, making the output go to the positive rail (here +5 V); when the capacitor charges up to the switching threshold of the output inverter (which can be another nMOS transistor or a logic inverter) the output returns to zero. Note that the timing begins at the trailing edge of the input. See §2.2.2 for some fancier discrete circuits that generate an output pulse width insensitive to the input pulse duration. And in Chapter 7 we go into greater detail (§§7.2.1 and 7.2.2), including integrated timing circuits such as the monostable multivibrator (also called a "one-shot").

Relay driver

An electromechanical relay (more detail in Chapter *1x*) switches its contacts in response to an energizing current in the coil. The latter has some rated voltage that is guaranteed to switch the contacts and hold them in the energized position. For example, the relays specified in Figure 3.98B have a coil rating of +5Vdc, at which they draw 185 mA (i.e., a coil resistance of 27 Ω).[84] In some sense the rated voltage is a compromise: enough to operate the relay reliably, but without excessive current. But you can cheat a bit and get faster closure if you overdrive the coil momentarily,

---

[83] One of our acquaintances uses the powerful BT555 switch in their "Jaws of Life"-type rescue device.

[84] There's more: it specifies a "must operate" coil voltage of 3.75 V and a "must release" voltage of 0.5 V. Lots of specs, too, about the switched contacts: configuration, voltage and current ratings, endurance, and so on.

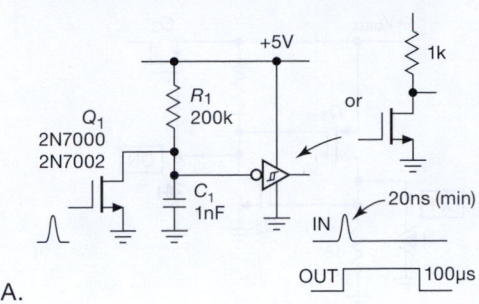

A.

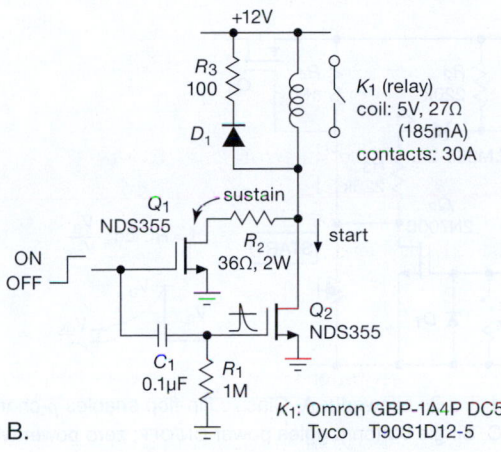

B.

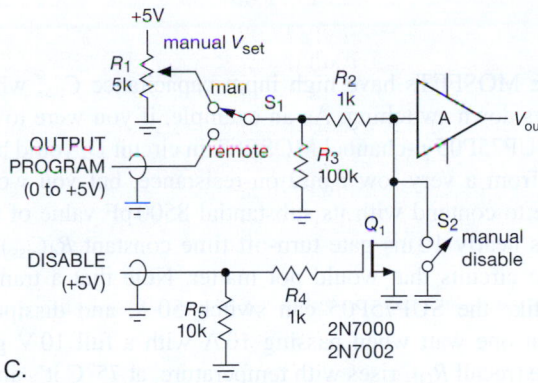

C.

Figure 3.98. Useful – and simple – MOSFET applications: A. pulse stretcher; B. relay driver, with initial pulsed overdrive; C. programmable voltage supply with disable control.

Control of a programmable supply

It's nice to control things remotely, with a brainy computer in charge. You might assemble (or buy) a voltage source that accepts a low-level analog input, in the manner of Figure 3.98C, in which the "A" symbol represents a dc amplifier that outputs a voltage $V_{out}=AV_{in}$, perhaps capable of substantial output current as well. But it's always good to provide a way to disable external control, so things don't go nuts when the computer crashes or is booting up (or gets taken over by an evildoer). The figure shows a simple way to implement a manual DISABLE control (which duplicates the external DISABLE input); might as well have a manual voltage mode as well, as shown.

On/Off battery control

It's convenient to power battery-operated instruments with a 9 volt battery: easy to get; provides plenty of voltage headroom; can be used to create a split supply, see §4.6.1B; but you've got to conserve the battery's vital juices, which run out after about 500 mAh.

Figure 3.99 shows some ways to implement power switching with MOSFETs. Circuit A is the classic flip-flop (two buttons: SET, and RESET; it's called an "SR flip-flop"). The OFF button shuts off Q_2, whose drain goes HIGH, holding Q_1 on and simultaneously holding pass transistor Q_3 off; you can easily make yourself believe that the ON button does the opposite (and you would be right). Two buttons are OK (though we can do better, stay tuned), but this circuit has the disadvantage of drawing current in either state. You could minimize the standby current by using a 10 MΩ resistor for R_1, say; then it draws 1 μA when off, which calculates out to a 50-year battery lifetime – far longer than its ∼5 year shelf life.

But there's a better way. Look at Circuit B, where the complementary pair flip-flop Q_1Q_2 draws *no* current (other than nanoamp-scale leakage) in the OFF state. The next step is Circuit C (also zero power when OFF), which achieves the minimum button count, in which a single pushbutton acts as an ON/OFF "toggle." This circuit is a bit tricky, because you have to juggle several time constants appropriately.[85] But the basic concept is simple, and elegant: charge a capacitor from the inverted output of the flip-flop's control input, then momentarily connect the charged capacitor to the control input to make it toggle.

---

[85] The capacitor's charging time to the new state $\tau_C=R_1C_1$ should be ∼100 ms to allow for switch bounce; R_3C_1 should be much shorter, and the discharge time constant of output transistor's gate capacitance $(R_3+R_4)C_g$ should be faster still. Here we chose 100 ms, 2 ms, and ∼0.4 ms, respectively.

as shown in the figure. Here Q_2 applies 12 V for an initial ∼0.1 s, after which Q_1 sustains the by-now firmly closed contacts with the rated 5 V across the coil. Diode D_1 provides a conduction path for the inductive current at release, with series resistor R_3 allowing ∼20 V during current decay for faster release (see the discussion in Chapter *1x* to to understand why this is a good thing).

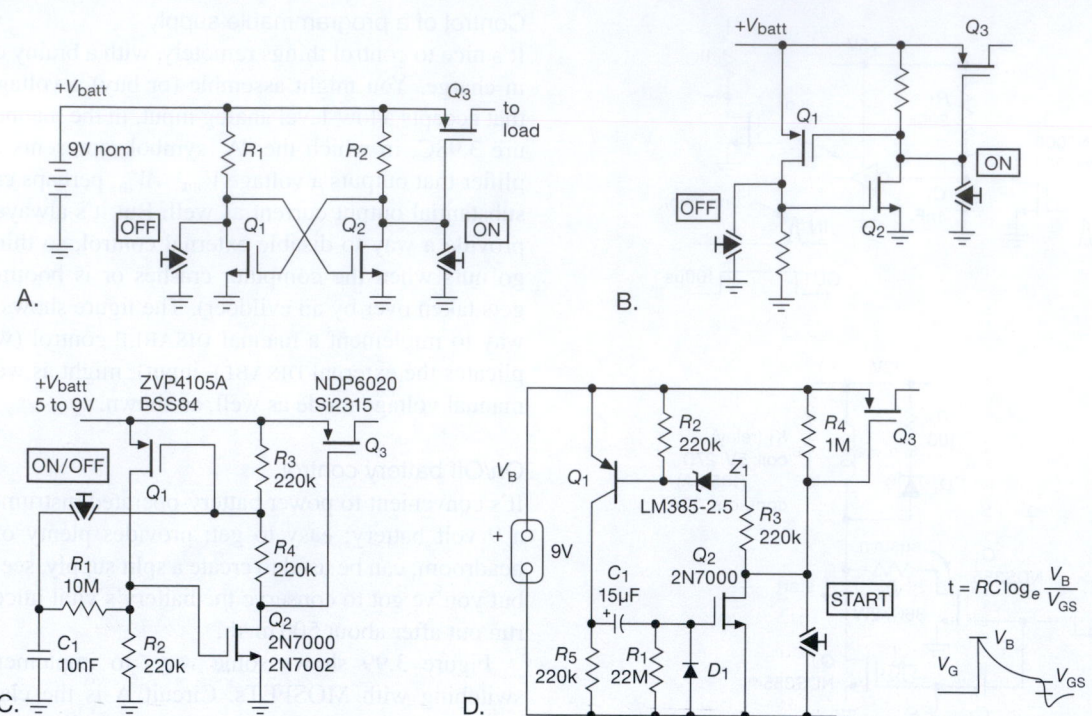

Figure 3.99. ON/OFF battery power control with MOSFETs (no fancy integrated circuits allowed!): A. Classic flip-flop enables p-channel series switch; separate ON and OFF buttons. B. Ditto, but zero power when off. C. Single button toggles power ON/OFF; zero power when off. For each MOSFET two choices are listed: the upper is a TO-92 through-hole type, the lower a SOT23 surface-mount type. D. "Five minutes of power" (approximately); zero power when off. The p-channel MOSFET pass transistors can be as large as you need, see Table 3.4a.

Finally, for the forgetful we suggest something like Circuit D, a one-button control that turns off automatically (to zero power) after roughly five minutes. Here Q_1Q_2 form a complementary flip-flop, held in the ON state by ac coupling through C_1; once ON the latter discharges through R_1, with a time constant of 330 s. This is an approximation to the actual shutoff time interval, which is set in detail by the ratio of Q_2's gate switching voltage to the actual battery voltage.[86] There's a bit of complexity here, in the form of the zener Z_1, which was forced upon us by the need to keep Q_3 fully ON during the critical interval when the flip-flop is deciding to switch OFF. We chose a BJT (rather than a MOSFET) for Q_1 because of its well-defined turn-on voltage; even so, there are the usual headaches here, caused by the uncertainty in gate threshold voltages of Q_2 and Q_3.

In all four of the circuits in Figure 3.99 you can use as large a p-channel pass transistor as needed; see Table 3.4a on page 188 for suggestions. But remember that large MOSFETs have high input capacitance C_{iss}, which slows down switching. As an example, if you were to use an SUP75P05 p-channel MOSFET in circuit D, you'd benefit from a very low 8 mΩ on-resistance, but you would have to contend with its substantial 8500 pF value of C_{iss} (thus nearly 10 ms gate turn-off time constant R_4C_{iss}). In these circuits that would not matter. Note that a transistor like the SUP75P05 can switch 50 A, and dissipates about one watt when passing 10 A with a full 10 V gate drive (recall R_{ON} rises with temperature, at 75°C it's about 10 mΩ); at most a small 2 W clip-on heatsink is needed (see §9.4.1).[87]

3.5.4 Power switching cautions

As nice as MOSFETs are, designing circuits with them is not entirely simple, owing to numerous details that can bedevil you. We simply summarize some of the important issues here and delve further into power switching in Chapter *3x* and Chapter 9.

---

[86] A 9 V alkaline battery starts life at about 9.4 V, and reaches old age at 6 V (1V/cell), or very old age at 5.4 V (0.9V/cell).

[87] If you want to work backward from your available heatsinking capability, use $I = \sqrt{P/R_{ON}}$.

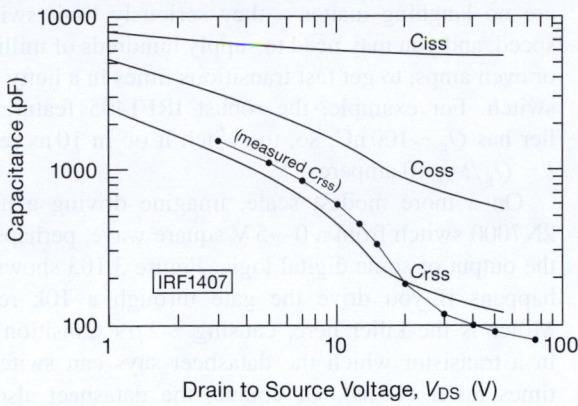

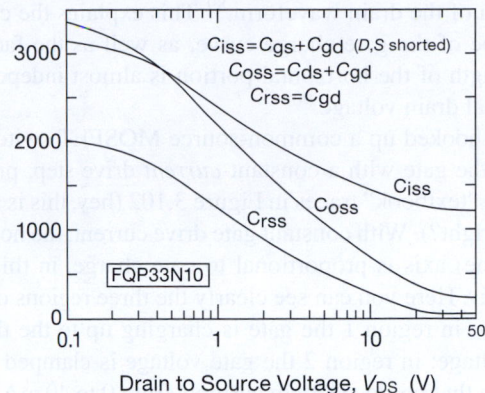

Figure 3.100. Interelectrode capacitances in two power MOSFETs, from graphical data in their respective datasheets. The feedback capacitance C_{rss}, though smaller than the input capacitance C_{iss}, is effectively multiplied by the Miller effect and typically dominates in switching applications.

A. MOSFET gate capacitance

Power MOSFETs have essentially infinite input *resistance*, but they have plenty of input capacitance and also feedback capacitance, such that fast switching may require literally amperes of gate drive current.[88] Although you may not care about speed in many applications, you still have to worry, because low gate drive current causes dramatically higher power dissipation (from the $VI\Delta t$ product during extended switching transitions); it may also permit oscillations during the slow transition. The various interelectrode capacitances are *nonlinear* and increase with decreasing voltage, as shown in Figure 3.100. The capacitance from gate to ground (called C_{iss}) requires an input current of $i=C_{iss}\,dV_{GS}/dt$, and the (smaller) feedback capacitance (called C_{rss}) produces an input current $i=C_{rss}\,dV_{DG}/dt$. The latter usually dominates in a common-source switch, because ΔV_{DG} is usually much larger than the ΔV_{GS} gate drive, effectively multiplying the feedback capacitance by the voltage gain (Miller effect). A nice way to look at this is in terms of gate *charge*, next.

B. Gate charge

In a common-source switch, charging of the gate-source and gate-drain capacitances require an input gate driving current whenever the gate voltage is changing. Additionally, during drain voltage transitions the Miller effect contributes additional gate current. These effects are often plotted as a graph of "gate charge versus gate-source voltage," as in Figure 3.101.

The initial slope is the charging of C_{iss}. The horizontal portion begins at the turn-on voltage, where the rapidly

Figure 3.101. Gate charge versus V_{GS}. The newer small-geometry ("shrunk-die") IRF520N has a lower threshold voltage, but comparable gate charge. Note in all cases the larger capacitance (reduced slope of V_{GS} versus Q_g) to the right of the "Miller shelf," the result of larger interelectrode capacitances at low V_{DS} (Figure 3.100).

falling drain forces the gate driver to supply additional charge to C_{rss} (Miller effect). If the feedback capacitance were independent of voltage, the length of the horizontal portion would be proportional to initial drain voltage, after which the curve would continue upward at the original slope. In fact, the "nonlinear" feedback capacitance C_{rss} rises rapidly at low voltage (Figure 3.100), which means that most of the Miller effect occurs during the low-voltage

---

[88] Newer-generation MOSFETs generally exhibit somewhat lower capacitances, but their smaller size allows less power dissipation, so for high-

power applications you may be forced to a larger part, thus giving up the capacitance advantage.

portion of the drain waveform.[89] This explains the change in slope of the gate charge curve, as well as the fact that the length of the horizontal portion is almost independent of initial drain voltage.[90]

We hooked up a common-source MOSFET switch and drove the gate with a constant-*current* drive step, producing the "textbook" traces in Figure 3.102 (hey, this is a textbook, right?). With constant gate drive current, the horizontal (time) axis is proportional to gate charge, in this case 3 nC/div. Here you can see clearly the three regions of gate activity: in region 1 the gate is charging up to the threshold voltage; in region 2 the gate voltage is clamped at the voltage that produces drain currents from 0 to 40 mA (40 V positive rail, 1k load resistor); after the drain is brought to ground the gate resumes its upward voltage ramp, but with reduced slope (owing to increased input capacitance at zero drain voltage).

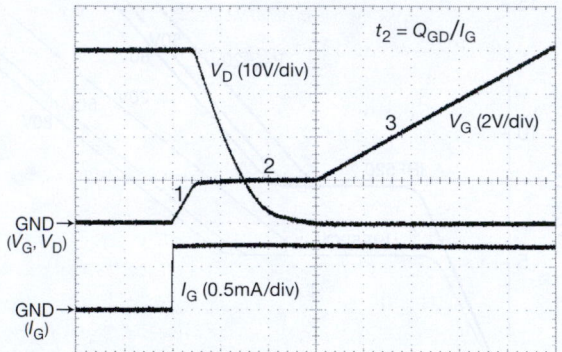

Figure 3.102. Gate charge. Waveforms of an IRLZ34N *n*-channel MOSFET, wired as a common-source switch (1k load to +40 V), with 0.75 mA gate drive. The horizontal scale of 4 μs/div therefore corresponds to 3 nC/div of gate charge.

Notice also that the drain voltage trace is curved, caused by the increasing drain-gate capacitance as it heads toward ground: with constant applied input I_G, the increasing C_{rss} mandates decreasing dV_D/dt (to keep the product, i.e., the feedback current, equal to the input current).

The Miller effect and gate charge in MOSFET switches

are no laughing matter – they seriously limit switching speed, and you may need to supply hundreds of milliamps, or even amps, to get fast transitions times in a hefty power switch. For example, the robust IRF1405 featured earlier has $Q_g \sim 100$ nC; so, to switch it on in 10 ns requires $I = Q_g/t = 10$ amperes![91]

On a more modest scale, imagine driving a humble 2N7000 switch from a $0 \rightarrow 5$ V square wave, perhaps from the output of some digital logic. Figure 3.103 shows what happens if you drive the gate through a 10k resistor. Miller is the killer here, causing ~ 2 μs transition times in a transistor which the datasheet says can switch 200 times faster (10 ns). Of course, the datasheet also says $R_{GEN} = 25\,\Omega$... and they mean it!

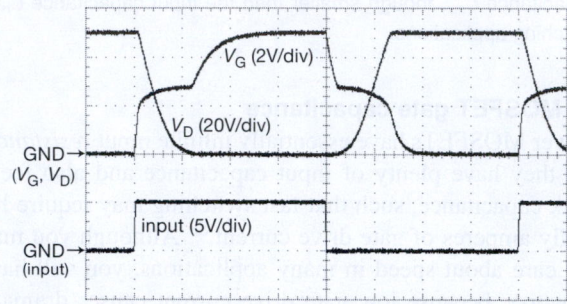

Figure 3.103. A 2N7000 MOSFET switch (1k load to +50 V), whose gate is driven by a 5 V (logic-level) voltage step through a 10k series resistor. The Miller effect stretches the switching time to ~ 2 μs. Horizontal scale: 2 μs/div.

There's plenty more to say about gate charge in MOSFETs: dependence on load current, the shape of the "Miller plateau," variations among MOSFET types, and measurement techniques. This is discussed in detail in the advanced material in Chapter *3x*.

C. MOSFET drain capacitances

In addition to the gate-to-ground capacitance C_{iss}, MOSFETs also have a gate-drain feedback capacitance C_{dg} (usually called C_{rss}), and an output capacitance (called C_{oss}) that is the combined capacitances from drain to gate C_{dg} and drain to source C_{ds}. As just seen, the effect of the feedback capacitance C_{rss} is evident in the gate-charge waveforms in Figure 3.102. The output capacitance is important, too: it is the capacitance that must be charged and discharged each

---

[89] This effect can be rather abrupt in power MOSFETs, as seen in the C_{rss} versus V_{DS} plot for the IRF1407 (Figure 3.100). In fact, our measured data shows even steeper behavior than that of the datasheet's plot. As explained in Chapter *3x*, this is due to the effective formation of a cascode within the MOSFET, wherein a depletion-mode JFET acts to clamp the drain of the active MOSFET, isolating the latter's gate and thus greatly reducing the feedback capacitance.

[90] The height (V_{GS}) of the horizontal portion depends modestly on drain current; see the figures in §*3x.12*.

[91] Quite often it is the output *transition* time that you care about; that is, the time spent just in region 2 (apart from the delay time in region 1, or the gate overcharging time in region 3); for this reason MOSFET datasheets separately specify Q_{gd}, the gate-to-drain "Miller" charge. For the IRF1405, for example, $Q_{gd} = 62$ nC, thus requiring a gate input current of 6.2 A to bring about a 10 ns transition time.

switching cycle, which, if not reactively recycled, consumes power $P = C_{oss}fV_{DD}^2$, which can become significant at high switching frequencies. See §§9.7.2B for details.

D. Current and power ratings

MOSFET datasheets specify a maximum continuous drain current, but this is done assuming an unrealistic 25°C case temperature. It is calculated from $I_{D(max)}^2 R_{DS(ON)} = P_{max}$, substituting a maximum power (see §9.4) $P_{max} R_{\Theta JC} = \Delta T_{JC} = 150°C$, where they have assumed $T_{J(max)} = 175°C$ (thus a 150°C ΔT_{JC}), and they use the value of $R_{DS(ON)}$ (max) at 175°C from an R_{DS} tempco plot (e.g., see Figure 3.116). That is, $I_{D(max)} = \sqrt{\Delta T_{JC} / R_{\Theta JC} R_{ON}}$. Some datasheets list power and drain current at a more realistic 75°C or 100°C case temperature. This is better, but you don't really want to run your MOSFET junction at 175°C, so we recommend using a still lower maximum I_D rating for continuous current, and corresponding dissipated power.

E. Body diode

With rare exceptions[92] power MOSFETs have the body connected to the source terminal. Because the body forms a diode with the channel, this means that there is an effective diode between drain and source (Figure 3.104) (some manufacturers even draw the diode explicitly in their MOSFET symbol so that you won't forget). This means that you cannot use power MOSFETs bidirectionally, or at least not with more than a diode drop of reverse drain-source voltage. For example, you couldn't use a power MOSFET to zero an integrator whose output swings both sides of ground, and you couldn't use a power MOSFET as an analog switch for bipolarity signals. This problem does not occur with *integrated circuit* MOSFETs (analog switches, for example), where the body is connected to the most negative power-supply terminal.

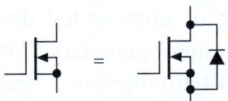

Figure 3.104. Power MOSFETs connect body to source, forming a drain-source diode.

The MOSFET's body diode exhibits the same reverse-recovery effect as ordinary discrete diodes. If biased into forward conduction, it will require some duration of reversed current flow to remove the stored charge, ending with a sharp "snap-off." This can cause curious misbehavior, analogous to the rectifier snap-off transients discussed

in the colorfully-named section "Transformer + rectifier + capacitor = giant spikes!" in Chapter 9x.[93]

F. Gate-source breakdown

Another trap for the unwary is the fact that gate-source breakdown voltages (±20 V is a common figure) are lower than drain-source breakdown voltages (which range from 20 V to more than 1000 V). This doesn't matter if you're driving the gate from the small swings of digital logic, but you get into trouble immediately if you think you can use the drain swings of one MOSFET to drive the gate of another.

G. Gate protection

As we discuss next, all MOSFET devices are extremely susceptible to gate oxide breakdown, caused by electrostatic discharge. Unlike JFETs or other junction devices, in which the junction avalanche current can safely discharge the overvoltage, MOSFETs are damaged irreversibly by a single instance of gate breakdown. For this reason it is a good idea to use a series gate resistor of 1k or so (assuming that speed is not an issue), particularly when the gate signal comes from another circuit board. This greatly reduces the chances of damage; it also prevents circuit loading if the gate is damaged, because the most common symptom of a damaged MOSFET is substantial dc gate current.[94] You can get additional protection by using a pair of clamp diodes (to $V+$ and to ground), or a single clamp zener to ground, downstream of the gate resistor (which can then be of much lower resistance, or omitted altogether); but note that a zener clamp adds some input capacitance.[95] It's also a good idea to avoid floating (unconnected) MOSFET

---

[92] Such as the 2N4351 and the SD210 series of lateral MOSFETs. These are available from Linear Systems, Fremont, CA.

[93] This can be a serious problem in certain types of switching circuits, where inductive currents continue to flow after termination of switch conduction. It can be addressed by adding an external diode between drain and source, which will be in parallel with the MOSFET's intrinsic drain-source diode. You can use a Schottky diode for voltages below roughly 60 V, but at higher voltages even Schottky diodes have too much voltage drop and thus fail to take over from the FET's intrinsic diode. To deal with this there are power MOSFETs available that include special soft-recovery diodes in their design. These diodes have low reverse-recovery-charge Q_{rr}, and hence faster recovery time t_{rr}, so they don't get stuffed with as much charge from this persistent inductive current. They may also have a slower snapoff, further reducing the spike energy.

[94] A MOSFET with damaged gate may exhibit drain conduction when it should be in a non-conducting state: leakage current from drain to (damaged) gate brings the drain down to a voltage that produces the V_{GS} corresponding to the drain current.

[95] Power MOSFETs used to incorporate internal zener protection, but now it's rare: the zener itself became a dominant failure mechanism! MOSFETs with internal gate zener diodes are marked in the "Gate zener" column of Table 3.4b on page 189.

gates, which are susceptible to damage when floating (there is then no *circuit* path for static discharge, which otherwise provides a measure of safety). This can happen unexpectedly if the gate is driven from another circuit board. A good practice is to connect a pull-down resistor (say 100k to 1M) from gate to source of any MOSFETs whose gates are driven from an off-card signal source. This also ensures that the MOSFET is in the off state when disconnected or unpowered.

H. MOSFET handling precautions

The MOSFET gate is insulated by a layer of glass (SiO_2) a hundred nanometers thick (less than a wavelength of light). As a result, it has very high resistance, and no resistive or junction-like path that can discharge static electricity as it is building up. In a classic situation you have a MOSFET (or MOSFET integrated circuit) in your hand. You walk over to your circuit, stick the device into its socket, and turn on the power, only to discover that the FET is dead. You killed it! You should have grabbed onto the circuit board with your other hand before inserting the device. This would have discharged your static voltage, which in winter can reach thousands of volts.[96] MOS devices don't take kindly to "carpet shock," which is officially called *electrostatic discharge* (ESD). For purposes of static electricity, you can be approximated by the "human body model" (HBM), which is 100 pF in series with 1.5k;[97] in winter your capacitance may charge to 10 kV or more with a bit of shuffling about on a fluffy rug, and even a simple arm motion with shirt or sweater can generate a few kilovolts. Here are some scary-looking numbers:

Typical Electrostatic Voltages[a]

| | Electrostatic Voltage | |
| | 10%–20% humidity (V) | 65%–90% humidity (V) |
Action		
Walk on carpet	35,000	1,500
Walk on vinyl floor	12,000	250
Work at bench	6,000	100
Handle vinyl envelope	7,000	600
Pick up poly bag	20,000	1,200
Shift position on foam chair	18,000	1,500

(a) adapted from Motorola Power MOSFET Data Book.

---

[96] "Smokey, my friend, you are entering a world of pain."

[97] A bit simplistic, though. The HBM, charged to 2.5 kV, peaks at 1.7 A, with a time constant of 150 ns. There are other models, for example the "machine" model (several cycles of 12 kHz, up to 6 A), or the "charge device model" (CDM), which recognizes that a portion of a charged object with less series resistance can discharge directly into the circuit with 6 A pulses 2 ns wide. See also §12.1.5.

Although any semiconductor device can be clobbered by a healthy spark, MOS devices are particularly susceptible because the energy stored in the gate-channel capacitance, when it has been brought up to breakdown voltage, is sufficient to blow a hole through the delicate gate oxide insulation. (If the spark comes from your finger, your additional 100 pF only adds to the injury.) Figure 3.105 (from a series of ESD tests on a power MOSFET[98]) shows the sort of mess this can make. Calling this "gate breakdown" gives the wrong idea; the colorful term "gate *rupture*" might be closer to the mark!

The electronics industry takes ESD very seriously. It is probably the leading cause of nonfunctional semiconductors in instruments fresh off the assembly line. Books are published on the subject, and you can take courses on it.[99] MOS devices, as well as other susceptible semiconductors[100] should be shipped in conductive foam or bags, and you have to be careful about voltages on soldering irons, etc., during fabrication. It is best to ground soldering irons, table tops, etc., and use conductive wrist straps. In addition, you can get "antistatic" carpets, upholstery, and even clothing (e.g., antistatic smocks containing 2% stainless steel fiber). A good antistatic workstation includes humidity control, air ionizers (to make the air slightly conductive, which keeps things from charging up), and educated workers. In spite of all this, failure rates increase dramatically in winter.

Once a semiconductor device is safely soldered into its circuit, the chances for damage are greatly reduced. In addition, most small-geometry MOS devices (e.g., CMOS digital integrated circuits, but not power MOSFETs) have protection diodes in the input gate circuits. Although the internal protection networks of resistors and clamping diodes (or sometimes zeners) compromise performance somewhat, it is often worthwhile to choose those devices because of the greatly reduced risk of damage by static electricity. In the case of unprotected devices, for example power MOSFETs, small-geometry (low current) devices tend to be the most troublesome, because their low input capacitance is easily brought to high voltage when it comes in contact with a charged 100 pF human. Our personal experience with the small-geometry VN13 MOSFET

---

[98] The MOSFET is an MTM6N60, for which $C_{iss} = 1100$ pF. This forms a capacitive divider with the HBM's 100 pF, attenuating the 1 kV 150 ns spike to about 80 V. But that's still way over the part's 20 V maximum V_{GS} rating.

[99] Of course, we academics love to give courses on just about anything.

[100] Which includes just about everything: small-geometry RF bipolar transistors are very delicate; and you can zap a plain ol' BJT if you hit it hard enough.

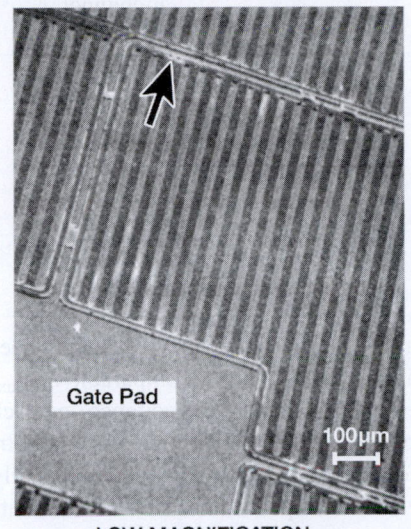

LOW MAGNIFICATION

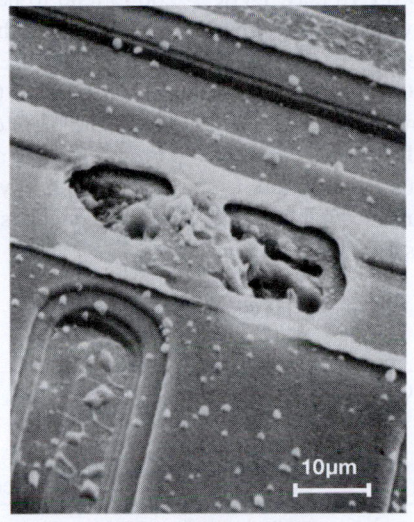

HIGH MAGNIFICATION

Figure 3.105. Scanning electron micrograph of a 6 amp MOSFET destroyed by 1 kV charge on "human body equivalent" (1.5k in series with 100 pF) applied to its gate. (Courtesy of Motorola, Inc.)

was been so dismal in this regard that we stopped using it in production instruments.

It is hard to overstate the problem of gate damage caused by breakdown in MOSFETs. Chip designers realized the seriousness of the problem, and routinely rate the "ESD tolerance" of their devices. Typically MOS ICs survive 2 kV, applied with the HBM of a charged 100 pF capacitor in series with a 1.5k resistor, and the datasheet will say so. Devices that might be exposed to external impulses (e.g., interface and line drivers) are sometimes rated to 15 kV (for example, Maxim RS-422/485 and RS-232 interface chips with an "E" suffix, and many of the analogous parts from other manufacturers).

See also the discussion of input protection in Chapter 12 (§12.1.5).

I. MOSFETs in parallel

Sometimes you need to use several power transistors in parallel, either to handle greater currents or to be able to dissipate more power, or both. As we discussed earlier, bipolar transistors, owing to their +9%/°C tempco of collector current with temperature, need emitter-ballasting resistors to ensure that the current is distributed equally among the participating transistors. For MOSFETs, as we mentioned in §3.5.1, the situation is different: sometimes you can connect them in parallel without any resistors (e.g., as saturated switches), and sometimes you can't (as linear power devices[101]). There's also the related topic of thermal run-

away. These are important topics, and deserve the more extended discussion in §3.6.3.

3.5.5 MOSFETs versus BJTs as high-current switches

Power MOSFETs are attractive alternatives to conventional power BJTs most of the time. They're comparable in price, simpler to drive, and they don't suffer from second breakdown and consequently reduced safe-operating-area (SOA) constraints (Figure 3.95).

Keep in mind that, for small values of drain voltage, an ON MOSFET behaves like a small resistance (R_{ON}), rather than exhibiting the finite saturation voltage ($V_{CE(sat)}$) of its bipolar transistor cousin. This can be an advantage, because the "saturation voltage" goes clear to zero for small drain currents. There is a general perception that MOSFETs don't saturate as well at high currents, but our research shows this to be largely false. In the following table we've chosen comparable pairs (*npn* versus *n*-channel MOSFET), for which we've looked up the specified $V_{CE(sat)}$ or $R_{DS(ON)}$. The *low-current* MOSFET is comparable to its "small-signal" *npn* cousin, but in the range of 6–10 A and 0–100 V, the MOSFET does better. Note particularly that very large base currents are needed to bring the bipolar power transistor into good saturation – 10% or more of the collector current (thus as much as 1 A!) – compared with the (zero-current) 10 volt bias at which MOSFETs are usually specified. Note also that

---

[101] Exception: "lateral" power MOSFETs, such as the 2SK1058.

high-voltage MOSFETs (say, $BV_{DS} > 200V$) tend to have larger $R_{DS(ON)}$, with larger temperature coefficients, than the lower-voltage units; here IGBTs excel over MOSFETs above 300 to 400 volts. We've listed capacitances in the table, because power MOSFETs traditionally had more capacitance than BJTs of the same rated current. In some applications (particularly if switching speed is important) you might want to consider the product of capacitance and saturation voltage as a figure of merit.

parameter	*n*-channel FQP9N25	*p*-channel FQP9P25
V_{max}	250 V	250 V
I_{max}	9.4 A	9.4 A
$R_{ON}(max)$	0.42 Ω	0.62 Ω
$C_{rss}(typ)$	15 pF	27 pF
$C_{iss}(typ)$	540 pF	910 pF
$Q_g(typ)$	15.5 nC	29 nC
$T_{JC}(max)$	1.39 °C/W	1.04 °C/W
Price (qty 1k)	$0.74	$0.97

Note that the *p*-channel device, having been fabricated with a larger area to achieve comparable $I_{D(max)}$, winds up with inferior (i.e., larger) capacitance, gate charge, R_{ON}, and pricing. It is also slower and has lower transconductance, according to the datasheet. Paradoxically, the *p*-channel device has improved thermal conductivity (see §9.4.1A), presumably resulting from the larger required chip size.

3.5.6 Some power MOSFET circuit examples

Enough theory! Let's look at a few circuit examples with power MOSFETs.

A. Some basic power switches

Figure 3.106 shows six ways to use a MOSFET to control the dc power to some subcircuit that you want to turn on and off. If you have a battery-operated instrument that needs to make some measurements occasionally, you might use Circuit A to switch the power-hungry microprocessor off except during those intermittent measurements. Here we've used a pMOS switch, turned on by a 1.5 V logic swing to ground; the particular part shown is specified for low gate voltage, in particular $R_{ON}=17$ mΩ (max) at $V_{GS}=-1.5$ V. The "1.5 V logic" is micropower CMOS digital circuitry, kept running even when the microprocessor is shut off (remember, CMOS logic has zero static dissipation).

An important point: you have to worry about proper switch operation at lower voltages, if the "1.5 V supply" is in fact an alkaline battery, with an end-of-life voltage of ~ 1.0 V. In that case you may be better off using a *pnp* transistor – see the discussion in "Low-voltage switching: MOS versus BJT" in Chapter *3x*.

In the second circuit (B), we're switching dc power to a load that needs $+12$ V at considerable current; maybe it's a radio transmitter, or whatever. Because we have only a 3.3 V logic swing available, we've used a small *npn* current sink to generate an 8 V negative-going swing (relative to $+12$ V) to drive the pMOS gate. Note the high-value

---

BJT-MOSFET-IGBT Comparison[a]

class	part#	V_{sat} 25°C (V)	V_{sat} 125°C (V)	C_r[b] (pF)	price
60 V,	2N4401[N]	0.75	0.8	8	$0.06
0.5 A	2N7000[V]	0.6	0.95	25	$0.09
60 V,	TIP42A[N]	1.5	1.7	50	$0.63
6 A	IRFZ34E[V]	0.25	0.43	50	$1.03
100 V,	TIP142[D]	3.0	3.8	low	$1.11
10 A	IRF540N[V]	0.44	1.0	40	$0.98
400 V,	2N6547[N]	1.5	2.5	125	$2.89
10 A	FQA30N40[V]	1.4	3.2	60	$3.85
600 V,	STGP10NC60[I]	1.75	1.65	12	$0.86
10 A					

(a) $I_B=I_C/10$, $V_{GS}=10$ V, except $I_B=I_C/250$ for Darlington.
(b) C_{ob} or C_{rss}.
(D) Darlington. (I) IGBT. (N) *npn* BJT. (V) vertical nMOS.

Remember that power MOSFETs can be used as BJT substitutes for linear power circuits, for example audio amplifiers and voltage regulators (we'll treat the latter in Chapter 9). Power MOSFETs are also available as *p*-channel devices, although there tends to be a much greater variety available among the (better-performing) *n*-channel devices. The available *p*-channel MOSFETs go only to 500 V (or occasionally 600 V), and generally cost more for comparable performance in some parameters ($V_{DS(max)}$ and $I_{D(max)}$, say), with reduced performance in other parameters (capacitance, R_{ON}). Here, for example, are specifications for a pair of complementary MOSFETs from Fairchild, matched in voltage and current ratings, and packaged in the same TO-220 power package.

collector resistor, perfectly adequate here because the pMOS gate draws no dc current (even a beefy 10 A brute), and we don't need high switching speed in an application like this.

The third circuit (C) is an elaboration of circuit B, with short-circuit current limiting courtesy of the *pnp* transistor. That's always a good idea in power-supply design – for example, it's impressively easy to slip with the oscilloscope probe. In this case, the current limiting also prevents momentary short-circuiting of the +12 V supply by the initially uncharged bypass capacitor. See if you can figure out how the current limiting circuit works.

Exercise 3.16. How does the current-limiting circuit work? How much load current does it allow?

An interesting detail: in Circuits B and C we could have hooked up the driver transistor as a *switch* (instead of a current source), omitting the emitter resistor and adding a current-limiting base resistor of 100k or so. But that circuit would create problems if you attempted to operate the circuit from a higher supply voltage, owing to the limited gate breakdown voltages of MOSFETs (± 20 V or less). It would also defeat the current-limit scheme of Circuit C. You could fix those problems by adding a resistor directly in series with the collector, tailoring its value for correct gate drive; but the current-source scheme we've used solves these problems automatically, and it can be used to switch 24 or 48 V with no component changes.

Exercise 3.17. You have a dc source using full-wave rectified 120 Vac. Design a 155 to 175 V version of Figure 3.106C to pulse 0.5 A into a flash string consisting of 38 white LEDs in series. Explain your choice for R_1 and R_2 and the ratio R_2/R_1. Select Q_1 and Q_2 and evaluate their power dissipation. Use Table 2.1 on page 74, along with the MOSFET tables in this chapter. *Extra credit*: evaluate Q_2's worst-case heating with a 10 ms maximum flash length (hint: use the datasheet's plot of "Transient Thermal Impedance").

There is still a problem (uh, an "issue"?) with Circuit C, namely the large power dissipation in pass transistor Q_2 under fault conditions such as a shorted output. The brute-force approach (which we've adopted more often than we care to admit) is to use a husky MOSFET with enough heatsinking to handle $P = V_{IN}I_{lim}$; that works OK for modest voltages and currents. Better to add foldback current limiting, as in Figure 12.45C on page 823. But ideally we'd want something like a pass transistor with internal thermal limiting.[102] That's one benefit of devices like that in Circuit E.

A popular alternative, at least for low voltage switching, is the use of a low R_{ON} analog switch (recall Table 3.3 on page 176), as in Circuit D. The switch listed there operates with power-supply voltages from 1.1 V to 4.5 V, with a worst-case R_{ON} that is plenty good enough to power loads up to 100 mA or so. It may seem strange to be using an analog switch, designed with complementary *n*-channel and *p*-channel MOSFETs for good signal properties over the full rail-to-rail range, as a simple positive-voltage power switch; but these things are inexpensive, and they take care of the logic interfacing and other details for you, so why not?

In Circuit E we've shown the interesting alternative of an *n*-channel MOSFET switch, for which you need to generate a gate drive that is more positive than the input supply voltage, preferably by a healthy 10 volts or so. You can get "high-side driver" ICs for this job, in varieties that cope with speed and voltage tradeoffs (for example the LM9061 in Figure 3.96; see also §§3.5.3, 12.4.2, 12.4.4, and Table 12.5 on page 826). Here we've gone a step further, using a high-side driver that includes the power MOSFET as well. It gets its gate drive signal with an internal oscillator and charge-pump converter (of the sort we saw in §3.4.3D). This particular device is intended for low-voltage operation, and includes internal current limiting and overtemperature protection.

Why bother with all this, when a *p*-channel MOSFET is easier to drive? Although the use of an nMOS switch with high-side drive does add complexity, it benefits from the better characteristics and much wider variety of *n*-channel MOSFETs; it is generally the preferred scheme.

Finally, Circuit F shows how to switch a negative supply rail to a load; it's analogous to Circuit B, but with an *n*-channel switch and a grounded-base *pnp* transistor to convert a positive logic level into a sourcing current that creates a 10 V gate swing across R_2. You can (and probably should) add current limiting in the manner of Circuit C.

B. Floating power switches

Sometimes you need to switch a voltage (and its load) that's "floating" far from ground. For example, you might want to test the pulsed power capability of a resistor while sensing the current at the low side; or you might want to make millisecond-scale pulsed measurements of a transistor to circumvent heating effects; or you might want a general-purpose floating two-terminal switch that can handle ac or dc. In such situations you can't use the basic ground-referenced schemes of Figure 3.106. Figure 3.107 shows two straightforward approaches, both using an optoisolator (§12.7) to convey the switching command from its ground-referenced home to the floating switch circuit.

---

[102] If instead we were switching the *low* side (with an *n*MOS switch), we could use a protected MOSFET; see Table 12.4 on page 825.

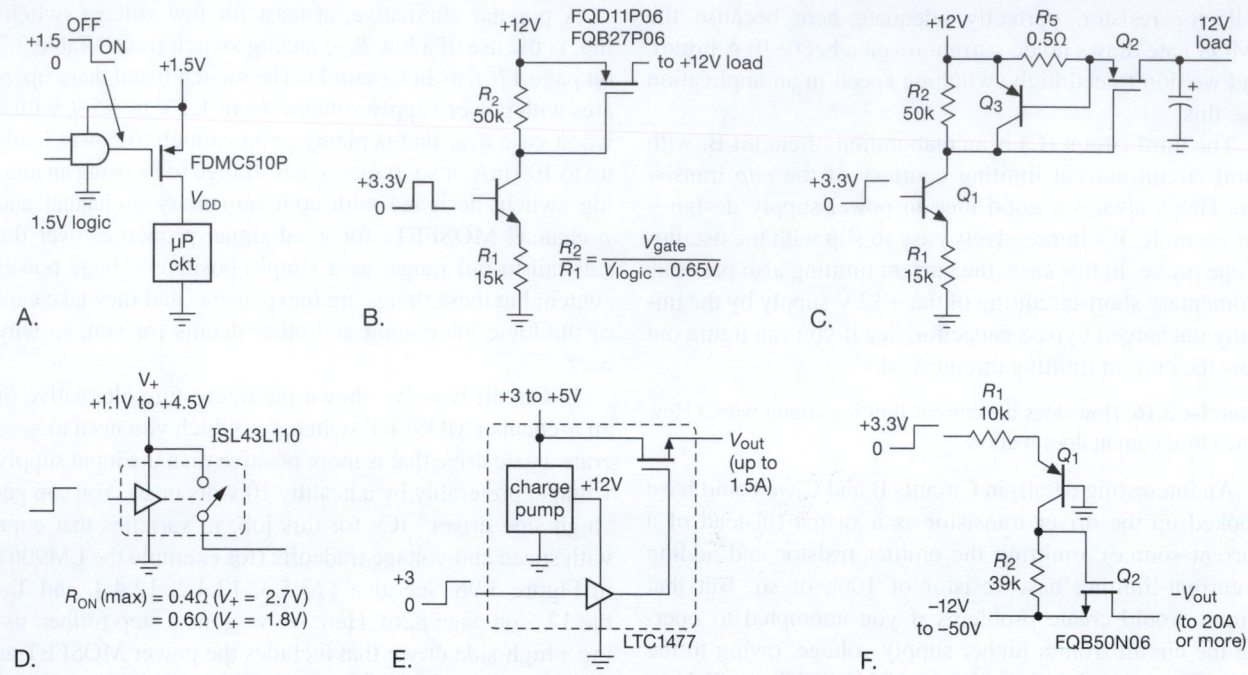

Figure 3.106. dc power switching with MOSFETs.

In Circuit A the gates of a pair of series-connected *n*-channel power MOSFETs are driven by a push–pull BJT follower that receives its base drive signal from a self-generating ("photovoltaic") optocoupler, U_2. The latter uses a series-connected photovoltaic stack to generate a floating ∼8 V signal in response to a 10 mA LED input drive current (see Figure 12.91, and discussion in §1x.7), with some internal circuitry to enhance the turn-off time. The gate driver pair Q_1Q_2 could be omitted at the expense of greater switching time (see below). These drivers reduce the effective load capacitance of the MOSFETs by a factor of beta, so that the resulting switching times (with typical power MOSFETs for Q_3 and Q_4) are limited by the optocoupler's intrinsic speed, of order 200 μs.

Of course the gate drivers Q_1 and Q_2 need a floating voltage source, here provided by a second inexpensive optovoltaic generator U_1, which need not be fast (assuming the circuit is not operated at a rapid switching rate) since it serves only to keep C_1 charged to ∼8 V. You can substitute a floating 9 V battery for U_1: it can provide lots more current than U_1's feeble ∼20 μA output, but of course you have to replace it from time to time (an alkaline 9 V "1604"-style battery is good for about 500 mAh, and has a shelf life of 5 years or so). This circuit can switch either polarity – when ON, the series MOSFETs sum to $2R_{ON}$ (the body diodes conduct only during ON–OFF transitions, or

at very high currents). Note that this circuit is an "unprotected" switch – there's no provision for current or power limiting of the output transistors.

Circuit B addresses this vulnerability and leverages the benefits of the BTS555 integrated protected switch. Here we've taken the simple approach of a floating 9 V battery to supply its internal circuitry's operating power (15 μA typical when off, 1 mA on). This thing is protected against pretty much anything bad that you can throw at it. Its switching speed is comparable to that of Circuit A (typically 300 μs on, 100 μs off), and it's good for lots of current (100 A or more); but it's limited to 34 V across its switch terminals. See §12.4.4 for more details, and Table 12.5 on page 826 for additional component suggestions.

Returning to Circuit A in Figure 3.107, what sort of performance can you get with readily available *n*-channel MOSFETs? Here's a selection of candidates (see Table 3.5 on page 206), culled from the many thousands of possibilities,[103] spanning the full range of voltages.[104]

Several trends are clearly evident:

---

[103] A search today for *n*-channel discrete MOSFETs finds 20,330 types at Digi-Key, 11,662 at Mouser, and 4,607 at Newark. A bit of an overestimate, because different package options are listed separately – but you get the point.

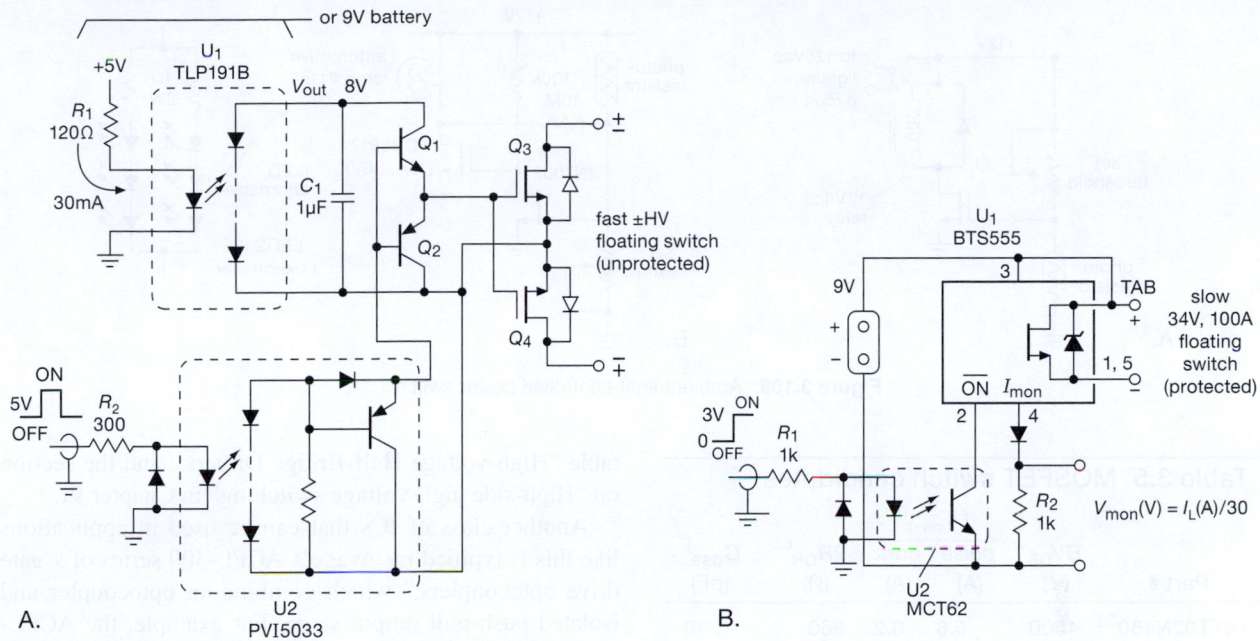

Figure 3.107. Floating MOSFET power switches: A. bipolarity, unprotected; B. unipolarity, limited voltage, protected.

(a) There's a severe tradeoff of R_{ON} versus voltage rating – for the listed types the ON-resistance spans a $\sim$100,000:1 range over the $\sim$100:1 voltage range.

(b) You also pay a price, literally, for very high-voltage parts; for example, the listed 4.5 kV part costs \$22.

(c) Higher current parts have greater output capacitance (which is what you see across the switch terminals when OFF), even here where we've carefully selected parts to minimize the $R_{ON}C_{oss}$ figure-of-merit tradeoff. They also have larger input capacitance and gate charge, which are relevant to switching speed.

(d) And, important data is *missing*! You need to go to the datasheets for important information such as thermal resistance, pulse current and pulse energy specifications, gate charge, and the like. The data presented here are at most advisory, and you need to use detailed specifications in the circuit context to predict actual performance. For example, the "maximum pulsed current" specs generally apply to pulse lengths somewhat shorter than this circuit can produce; and the R_{ON} spec assumes 10 V of gate drive, also somewhat greater than we have here.

Let's finish this example by estimating the switching speed of the circuit of Figure 3.107A. Imagine we want 600 V capability and choose the middle-of-the-road FCP22N60N, a MOSFET that delivers a good combination

of ON-resistance and capacitance at modest cost (about \$5, qty 100). For switching speed the relevant parameter here is gate charge ($Q_{GS}+Q_{GD}$), approximately 25 nC according to tabulated and graphical data. That must be supplied by the isolating driver U_2, boosted by the current gain β of Q_1 and Q_2. From U_2's datasheet we can estimate the output sourcing current (from its "Typical Response Time" plot) as approximately $3\mu A$. If for the moment we imagine that Q_1 and Q_2 were omitted, with U_2 driving the MOSFET gates directly, the turn-on time would be $t \approx Q_{gate}/I_{U2}$, or 8.3 ms. Now magically restore the BJT drivers, and the estimated switching time drops by a factor of beta; for typical $\beta \sim 200$ it becomes $\sim 40\,\mu s$.[105]

Not so fast! Look again at the datasheet for U_2, you'll find that the turn-on time bottoms out at about $100\,\mu s$; likewise, its intrinsic turn-off time is about $350\,\mu s$, even with small load capacitance. Those numbers dominate the performance of Circuit A, for nearly every MOSFET listed, assuming of course that the BJT driver Q_1Q_2 is included. If you can tolerate slower switching, you can simplify things by omitting the drivers and their floating power source.

If you need *faster* switching, there are many integrated high-side driver chips that can do the job, for example the series of "high-voltage gate-driver ICs" from Interna-

---

[104] See the MOSFET tables in this chapter and in Chapter *3x* for additional data on these and other power MOSFETs.

[105] If you look at the MOSFET drivers in Table 3.8 on page 218, you'll see that the ZXGD3002–04 are simply a pair of very-high-gain *npn* and *pnp* BJTs, in SOT23-6 packages, perfect for Q_1 and Q_2.

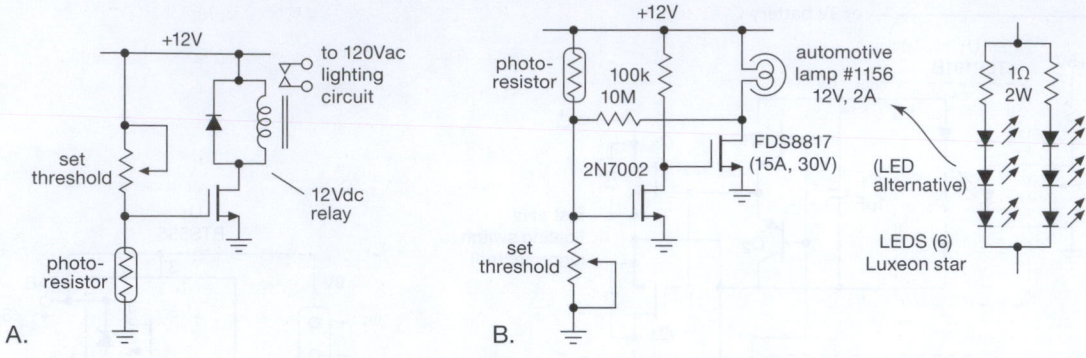

Figure 3.108. Ambient-light-controlled power switch.

Table 3.5 MOSFET switch candidates[a]

Part #	BV_{DS} (V)	$I_{D(max)}$[b] pulse (A)	$I_{D(max)}$[b] cont (A)	$2R_{ON}$[c] (Ω)	C_{oss}[t] (pF)
IXTT02N450$^{\$\$}$	4500	*0.6*	**0.2**	960	19
IXTH02N250$^{\$}$	2500	*0.6*	**0.2**	770	9
STW4N150	1500	*12*	**2**	10	120
IXTP3N120	1200	*12*	**3**	6.5	100
IXFH16N120P$^{\$}$	1200	*35*	**10**	1.7	390
IRFBG20	1000	*5.6*	**1**	16	52
IRFBG30	1000	*12*	**2**	8	140
IXFH12N100$^{\$}$	1000	*48*	**5**	2	320
IPP60R520CP	650	*17*	**4**	1	32
FCP22N60N	600	*66*	**12**	0.28	76
FCH47N60N$^{\$}$	600	*140*	**30**	0.1	200
IRF640N	200	*72*	**12**	0.24	190
FQP50N06L	60	*210*	**25**	0.08[d]	450
IRLB3034	40	*1400*	**125**	0.003[e]	2000
FDP8860	30	*1800*	**100**	0.004[d]	1700

Notes: (a) all are in TO-220 or TO-247 pkgs. (b) *italics* designate maximum pulsed drain current, for pulse width specified in the part's datasheet (e.g., 80μs); **boldface** designates maximum continuous drain current at T_J=70°C. (c) at V_{GS}=10V, unless marked otherwise. (d) at V_{GS}=5V. (e) at V_{GS}=4.5V. (t) typ. ($) not inexpensive. ($$) expensive.

tional Rectifier. These use internal high-voltage transistors to send the control signals up to the high side, with maximum voltage ratings most commonly of 600 V. These typically have switching times in the range of 100 ns to 1 μs. They are intended for cyclic applications such as pulse-width-modulated bridge drivers and use high-side charge pumps to develop the over-the-rail gate drive voltage; but you can adapt them for pulsed applications by substituting a flying 9 V battery, as we've done here. See Chapter *3x*'s

table "High-voltage Half-Bridge Drivers," and the section on "High-side high-voltage switching" in Chapter *9x*.

Another class of ICs that can be used in applications like this is typified by Avago's ACPL-300 series of "gate drive optocouplers," which combine an optocoupler and isolated push-pull output stage. For example, the ACPL-W343's output stage can source or sink 3 A (minimum), with 40 ns rise and fall times (into a load of 25 nF in series with 10 Ω), and isolation good to 2 kV. You have to supply 15–30 V isolated dc for the output stage,[106] in the manner of Figure 3.107B, with the usual bypass capacitor (for peak output currents); the quiescent current is 2 mA, good for 200 hours of operation if you use a pair of 9 V batteries. See §12.7.3 and Figure 12.87 for additional circuit discussion and suggestions.

C. Some unusual switching examples
Light-at-night
Figure 3.108A shows a simple MOSFET switching example, one that takes advantage of the high gate impedance. You might want to turn on exterior lighting automatically at sunset. The photoresistor has low resistance in sunlight, high resistance in darkness. You make it part of a resistive divider, driving the gate directly (no dc loading!). The light goes on when the gate voltage reaches the value that produces enough drain current to close the relay. Sharp-eyed readers may have noticed that this circuit is not particularly precise or stable; that's OK, because the photoresistor undergoes an enormous change in resistance (from 10k to 10M, say) when it gets dark. Note that the MOSFET may have to dissipate some power during the time the gate bias is inching up, since we're operating in the linear region;

---

[106] A bit high for MOSFET gate driving (these are targeted at IGBTs); but you can get parts with lower minimum output supply voltage, for example the HCPL-3180 or the PS9506 from Renesas (both can operate with 10 V min).

but it's switching only a relay, not the power load, so this is of little concern. The circuit's lack of a precise and stable threshold means that the light may turn on a few minutes early or late – again, no big deal. But an additional worry is the behavior of the relay, which is not happy with a marginal coil-driving voltage (which holds the contacts closed with less than the rated mechanical force, thus potentially shortening the relay's life; see the discussion of relays in Chapter *9x*).

These problems are remedied in Figure 3.108B, where a pair of cascaded MOSFETs delivers much higher gain, augmented by some positive feedback through the 10M resistor; the latter adds hysteresis, which causes the circuit to snap on regeneratively as it reaches threshold.

Hefty piezo driver

Figure 3.109 shows a real power MOSFET job: a 200 watt amplifier to drive a piezoelectric underwater transducer at 200 kHz. We've used a pair of hefty nMOS transistors, driven alternately to create ac drive in the (high-frequency) transformer primary. The series inductor in the secondary resonates with the transducer's capacitance to step up the voltage across the piezo to several kilovolts. The TC4425A is a handy "3A Dual High-Speed Power MOSFET Driver" (like the TC4420 in Figure 3.97), which takes a logic-level input (0 V=LOW, $\geq$2.4 V=HIGH), and generates a full-swing (0 to +V_{DD}) output pair, one inverted and the other noninverted; see Table 3.8 on page 218. It's needed to overcome capacitive loading, since the MOSFETs must be turned on fully in a fraction of a microsecond. The diodes shunted across the series gate resistors cause a rapid turn-off to prevent undesirable conduction overlap of the power transistors.

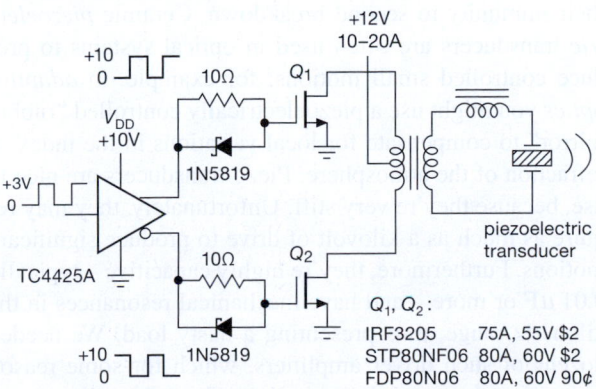

Figure 3.109. MOSFET piezo power driver.

In the figure: +12V 10–20A; +10 0; 10Ω Q_1; V_{DD} +10V; 1N5819; +3V 0; 10Ω Q_2; TC4425A; +10 0 1N5819; piezoelectric transducer; Q_1, Q_2: IRF3205 75A, 55V $2 / STP80NF06 80A, 60V $2 / FDP80N06 80A, 60V 90¢

3.5.7 IGBTs and other power semiconductors

The contemporary power MOSFET is a versatile transistor for both power switching applications (e.g., dc power control, or dc–dc switching converters), and for linear power applications (such as audio amplifiers). But there are some drawbacks and some useful alternatives.

A. Insulated-gate bipolar transistor (IGBT)

The IGBT is an interesting MOSFET-bipolar hybrid, most simply described as an integrated complementary-Darlington-like (Sziklai) connection of an input MOSFET with a power bipolar transistor (Figure 3.110). So it has the input characteristics of a MOSFET (zero dc gate current), combined with the output characteristics of a power bipolar transistor; note, however, that it cannot saturate to less than V_{BE}. Unlike MOSFETs, IGBTs do not have an intrinsic reverse diode, so inductive ringing, etc., can easily exceed the reverse voltage rating (e.g., 20 V). Many IGBTs include an internal "anti-parallel" diode to protect against this problem.[107]

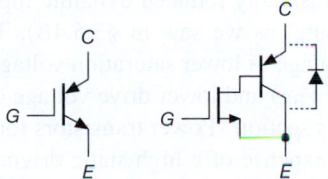

Figure 3.110. IGBT symbol and simplified equivalent circuit showing the optional "anti-parallel" diode.

Nearly all available IGBTs come in the nMOS-*pnp* polarity only, and thus behave as *n*-type devices.[108] They are generally high-voltage and high-power devices, available in discrete transistor power packages like the TO-220, TO-247, and in surface-mount packages like the D^{2}PAK and SMD-220, with ratings to 1200 V and 100 A. For higher currents you can get them in larger rectangular power "modules," with higher voltage ratings and with current ratings to 1000 A or more.

IGBTs excel in the arena of high voltage switching, because high-voltage MOSFETs suffer from greatly increased R_{ON}: an approximate rule-of-thumb for MOSFETs is that the R_{ON} increases as the square of the voltage

---

[107] Some parts are available with and without the added diode, indicated for example by a -D suffix on the part number.

[108] Currently the only *p*-type IGBTs we know of are the Toshiba GT20D200 series.

rating.[109] For example, let us compare two power products from International Rectifier (along with a BJT of comparable ratings):

Type		MOSFET	IGBT	BJT
Type		IRFPG50	IRG4PH50S	TT2202
V_{max}		1000 V	1200 V	1500 V
I_{max}	dc	6.1 A	57 A	10 A
	pulse	24 A	114 A	25 A
R_{ON} (typ)	25°C	1.5 Ω	–	–
	150°C	4 Ω	–	–
V_{ON}	25°C	23 V	1.2 V	1 V (@8 A)
(typ, 15 A)	150°C	60 V	1.2 V	1 V (@8 A)

These are comparably priced (about $5) and packaged (TO-247), have similar input characteristics (2.8 nF and 3.6 nF input capacitance), and the resulting saturation voltages V_{ON} when switching 15 A are shown for the same full input drive of $V_{in} = +15$ V. The IGBT is the clear winner in this high-voltage and high-current regime.[110] And, when compared with a power BJT, it shares the MOSFET advantage of high static input impedance (though still exhibiting the drastically reduced dynamic input impedance during switching, as we saw in §3.5.4B). The BJT does have the advantage of lower saturation voltage (the IGBT's V_{ON} is at least V_{BE}) and lower drive voltage (see the figure in Chapter *3x*'s section "Power transistors for linear amplifiers"), at the expense of a high static driving current; the latter drawback is exacerbated at high currents, where BJT beta drops rapidly. A saturated BJT also suffers from slow recovery owing to stored charge in the base region.

With the very high voltages and currents encountered with IGBTs, it is mandatory to include fault protection in the circuit design: an IGBT that may be required to switch up to a 50 A load from a 1000 V supply will be destroyed in milliseconds if the load becomes short-circuited, owing to the 50 kW (!) power dissipation. The usual method is to shut off the drive if V_{CE} has not dropped to just a few volts after 5 μs or so of input drive (see Figure 12.87B). We'll revisit these three power transistor technologies in Chapter *3x*.

B. Thyristors

For the utmost in *really* high-power switching (we're talking kiloamperes and kilovolts), the preferred devices are

the *thyristor* family, which include the unidirectional "silicon controlled rectifiers" (SCRs) and the bidirectional "triacs." These three-terminal devices behave somewhat differently from the transistors we've seen (BJTs, FETs, and IGBTs): once triggered into conduction by a small control current (a few milliamps) into their control electrode (the *gate*), they remain ON until external events bring the controlled current (from *anode* to *cathode*) to zero. They are used universally in house-current lamp dimmers, where they are switched on for a fraction of each half-cycle of ac line voltage, thus varying the *conduction angle*.

Thyristors are available in ratings from 1 A to many thousands of amperes, and voltage ratings from 50 V to many kilovolts. They come in small transistor packages, the usual transistor power packages, larger modules, and really scary "hockey puck" packages that are capable of switching megawatts. These are hefty devices; you can hurt yourself just by dropping one on your foot.

3.6 MOSFETs in linear applications

Although we dealt extensively with linear applications in this chapter's treatment of *JFETs*, our discussion of MOSFETs has concentrated almost entirely on switching applications. Lest we leave the wrong impression, we address in this section some applications of discrete power MOSFETs to linear applications, particularly those that benefit from their unique properties. See also additional applications to linear voltage regulators in Chapter 9 (e.g., Figures 9.17, 9.20, 9.104, 9.110, and 9.113), and advanced material in §*3x.5*.

3.6.1 High-voltage piezo amplifier

A nice application of MOSFETs as linear amplifiers takes advantage of available types with high voltage ratings, and their immunity to second breakdown. Ceramic *piezoelectric* transducers are often used in optical systems to produce controlled small motions; for example, in *adaptive optics* you might use a piezoelectrically controlled "rubber mirror" to compensate for local variations in the index of refraction of the atmosphere. Piezo transducers are nice to use, because they're very stiff. Unfortunately, they may require as much as a kilovolt of drive to produce significant motions. Furthermore, they're highly capacitive – typically 0.01 μF or more – and have mechanical resonances in the kilohertz range, thus presenting a nasty load. We needed dozens of such driver amplifiers, which for some reason cost a few thousand dollars apiece if you buy them commercially.

---

[109] You find exponents from 1.6 to 2.5 in the literature; the lower end of this range is likely to be more accurate; see our plots in Chapter *3x*.

[110] Where it excels also in maintaining high transconductance, compared with the MOSFET. The advantage goes to the IGBT, starting around 200 V. See also the comparison table in §3.5.5.

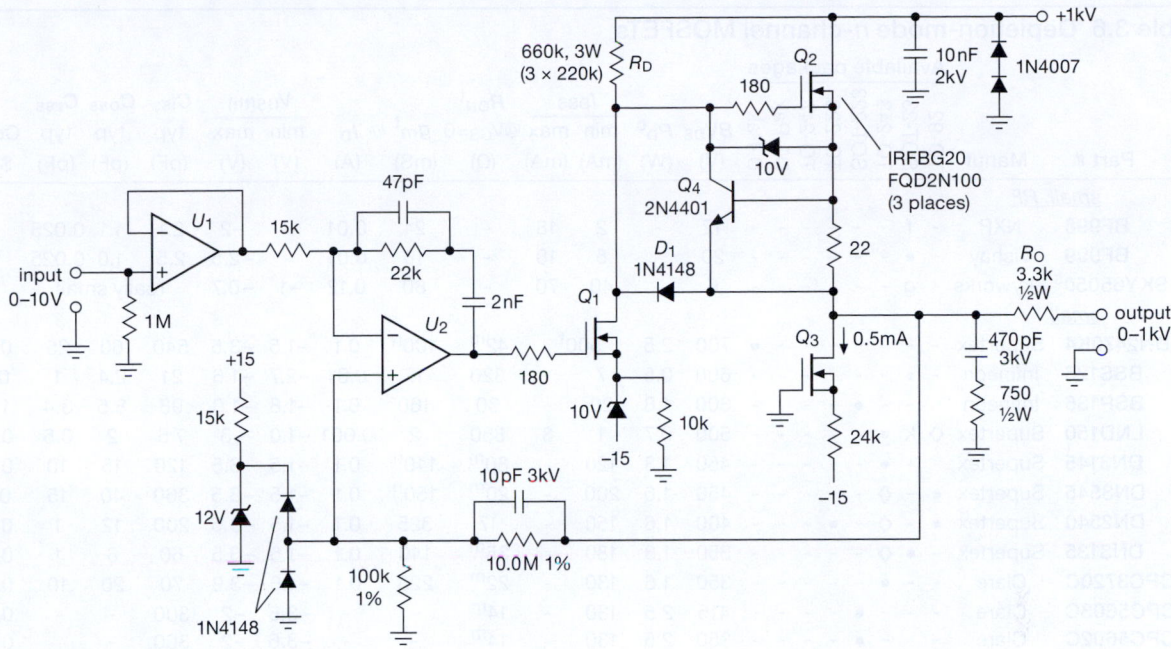

Figure 3.111. 1 kV low-power piezo driver with totem-pole output stage. A similar design is used for the high-voltage regulated dc supply shown in Figure 9.110.

We solved our problem with the circuit shown in Figure 3.111. The IRFBG20 is an inexpensive (~$2) MOSFET, good for 1 kV and 1.4 A; the similar FQD2N100 (1 kV, 1.6 A) costs about $0.85. The first transistor is a common-source inverting amplifier, driving a source follower with active current-sink load. The *npn* transistor is a current limiter and can be a low-voltage unit, since it floats on the output. One subtle feature of the circuit is the fact that it's actually push–pull, even though it looks single-ended: you need plenty of current (20 mA) to push 10,000 pF around at 2 V/μs; the output transistor can *source* current, but the pulldown resistor can't sink enough (look back to §2.4.1, where we motivated push–pull with the same problem). In this circuit the driver transistor is the pull-down, with conduction through the diode D_1![111] The rest of the circuit involves feedback with op-amps, a forbidden subject until the next chapter; in this case the magic of feedback makes the overall circuit linear (100 V of output per volt of input), whereas without it the output voltage would depend on the (nonlinear) I_D-versus-V_{GS} characteristic of the input transistor. A nice improvement to this circuit consists of replacing the 660kΩ 3 W pullup resistor (whose current drops at high output voltages, e.g.,

to 0.15 mA at 900 V) with a depletion-mode MOSFET current source set to, say, 0.25 mA. See the discussion below (§3.6.2C, and also Figure 3.23, Table 3.6, and §§9.3.14C and *3x.6*).

For a detailed analysis of a precision high-voltage amplifier with bipolarity output capability, see the section with that name in Chapter *4x*. For an analysis of relevant issues, such as MOSFET transconductance at low currents, see §3x.5.2, and for the response of a MOSFET source follower into a capacitive load see §3x.8.

Exercise 3.18. Modify this circuit so the high-voltage output can be turned on and off, under control of an input signal (0 V for off, +3V for on).

3.6.2 Some depletion-mode circuits

Depletion-mode MOSFETs are the neglected siblings of the far more popular enhancement-mode MOSFETs. They can do some nice tricks, though, that are worth knowing about. And they are available in high voltage (to 1 kV) and high current (to 6 A) varieties. Table 3.6 on the next page lists nearly all available parts of this somewhat rarefied species. Here are some applications that exploit their property of conduction at zero gate voltage.

---

[111] This is called a "totem-pole" output stage, and became popular in the early 1970s in bipolar TTL logic, see Figure 10.25A.

Table 3.6 Depletion-mode *n*-channel MOSFETs

Part #	Manuf	TO-92	SOT-23	TO-243	SOT-223	TO-220	TO-247	D2-Pak	D-Pak	BV_{DS} (V)	$P_D{}^c$ (W)	I_{DSS} min (mA)	I_{DSS} max (mA)	$R_{ON}{}^t$ @V_{GS}=0 (Ω)	gm^t (mS)	@ I_D (A)	$V_{GS(th)}$ min (V)	$V_{GS(th)}$ max (V)	C_{iss} typ (pF)	C_{oss} typ (pF)	C_{rss} typ (pF)	Cost^y $US
small, RF																						
BF998	NXP	-	f	-	-	-	-	-	-	12	-	2	18	-	24	0.01	-	-2	2.1	1.1	0.025	-
BF999	Vishay	-	•	-	-	-	-	-	-	20	-	5	18	-	16	0.01	-	-2.5	2.5	1.0	0.025	-
SKY65050^s	Skyworks	-	g	-	-	-	-	-	-	6	-	40	70	-	80	0.12	-1	-0.7	really small			-
small																						
DN2470K4	Supertex	-	-	-	-	-	-	-	•	700	2.5	500^t		42^m	100^n	0.1	-1.5	-3.5	540	60	25	0.81
BSS126	Infineon	-	•	-	-	-	-	-	-	600	0.5	7	-	320	17	0.01	-2.7	-1.6	21	2.4	1	0.13
BSP135	Infineon	-	-	-	•	-	-	-	-	600	1.8	20	-	30	160	0.1	-1.8	-1.0	98	8.5	3.4	1.38
LND150	Supertex	◊	k	•	-	-	-	-	-	500	0.7	1	3	850	2	0.001	-1.0	-3	7.5	2	0.5	0.58
DN3145	Supertex	-	-	•	-	-	-	-	-	450	1.3	120	-	60^m	140^n	0.1	-1.5	-3.5	120	15	10	0.68
DN3545	Supertex	•	-	◊	-	-	-	-	-	450	1.6	200	-	20^m	150^n	0.1	-1.5	-3.5	360	40	15	0.74
DN2540	Supertex	•	-	◊	-	•	-	-	-	400	1.6	150	-	17	325	0.1	-1.5	-3.5	200	12	1	0.81
DN3135	Supertex	-	•	◊	-	-	-	-	-	350	1.3	180	-	35^m	140	0.1	-1.5	-3.5	60	6	1	0.62
CPC3720C	Clare	-	-	•	-	-	-	-	-	350	1.6	130	-	22^m	225	0.1	-1.6	-3.9	70	20	10	0.37
CPC5603C	Clare	-	-	-	•	-	-	-	-	415	2.5	130	-	14^m	-	-	-3.6	-2	300	-	-	0.69
CPC5602C	Clare	-	-	-	•	-	-	-	-	350	2.5	130	-	14^m	-	-	-3.6	-2	300	-	-	0.60
DN3535	Supertex	-	-	•	-	-	-	-	-	350	1.6	200	-	10^m	200^n	0.1	-1.5	-3.5	360	40	10	0.68
BSS139	Infineon	-	•	-	-	-	-	-	-	250	0.4	30	-	12.5	130	0.08	-1.4	-	60	6.7	2.6	0.57
BSP129	Infineon	-	-	-	•	-	-	-	-	240	1.8	50	-	6.5	360	0.28	-2.1	-1.0	82	12	6	0.47
DN3525N8	Supertex	-	-	•	-	-	-	-	-	250	1.6	300	-	6^m	225^n	0.15	-1.5	-3.5	270	20	5	0.66
CPC3703	Clare	-	-	•	-	-	-	-	-	250	1.6	300	-	4^m	225^n	0.1	-1.6	-3.9	327	51	27	0.57
BSP149	Infineon	-	-	-	•	-	-	-	-	200	1.8	140	-	1.7	800	0.48	-1.8	-1.0	326	41	17	0.80
BSS159	Infineon	-	•	-	-	-	-	-	-	50	0.4	70	-	4	160	0.16	-2.5	0.0	70	15	6	0.31
large					see note x																
IXTx01N100D^x	IXYS	-	-	-	-	P	-	-	Y	1000	25	100^t	-	90	150	0.1	-2.5	-	120	25	5	0.75
IXTx08N100D2^x	IXYS	-	-	-	-	P	-	-	Y	1000	60	100^t	-	21^m	560	0.4	-2	-4	325	24	6.5	0.69
IXTx1R6N100D2^x	IXYS	-	-	-	-	P	-	-	Y	1000	100	100^t	-	21^m	1100	0.8	-2.5	-4.5	645	43	11	1.66
IXTx3N100D2^x	IXYS	-	-	-	-	P	-	A	-	1000	125	3000	-	5.5^m	4200	3	-2.5	-4.5	1020	68	17	2.11
IXTx6N100D2^x	IXYS	-	-	-	-	P	H	A	-	1000	300	6000	-	5.5^m	4200	3	-2.5	-4.5	2650	167	41	4.66
IXTx02N50D2^x	IXYS	-	-	-	-	P	-	-	Y	500	25	250^t	-	20	150	0.2	-2	-5	120	25	5	1.05
IXTx08N50D2^x	IXYS	-	-	-	-	P	-	A	-	500	60	800	-	4.6^m	570	0.4	-2	-4	312	35	11	1.62
IXTx1R6N50D2^x	IXYS	-	-	-	-	P	-	-	Y	500	100	1600^t	-	2.3	1750	0.8	-2	-4	645	65	17	1.66
IXTx3N50D2^x	IXYS	-	-	-	-	P	-	A	-	500	125	1600^t	-	1.5	2100	1.5	-2	-4	1070	102	24	2.13
IXTx6N50D2^x	IXYS	-	-	-	-	P	H	A	-	500	300	6000	-	0.5^m	4500	3	-2	-4	2800	255	64	4.66
IXTH20N50D^{x,e}	IXYS	-	-	-	-	-	H	-	-	500	400	1500^{t,e}	-	0.5^m	7500	10	-1.5	-3.5	2500	400	100	8.61

Notes: (c) P_D at T_C=25°C, for package marked ◊. (e) the IXTH20N50D delivers most of its current capability in the enhancement-mode region. (f) has two gates, 4-lead SOT-143 package. (g) SC-70 package. (k) unusual pinout; also try LND250K1. (m) maximum. (n) minimum. (s) SKY part's full name: SKY65050-372LF. (t) typical. (x) substitute the letter listed in the package column for the "x" in the part number; for example, IXTP01N100D is a TO-220 package. (y) quantity 100.

A. Input protection

Low-level circuits (such as sensitive amplifiers) don't like to be driven beyond their power supply rails. One simple protection scheme uses a series resistor at the input, with a pair of downstream clamp diodes to the amplifier's supply rails. That's OK for small overdrive; but it's unsatisfac-tory if the input may go to a few hundred volts (think *powerline!*), because the large resistor value (~100k or so, to limit the fault current and dissipated power) compromises the signal bandwidth and noise. Figure 3.112 shows how to use a pair of depletion-mode MOSFETs (instead of a large resistor) as the series element. The particular part shown is small (SOT23, SOT89, or TO-92), inexpensive ($0.60), and able to withstand momentary inputs to ± 500 V. The pair looks like a series resistance of ~ 1.7k (twice $R_{DS(ON)}$) until the input goes beyond the amplifier's rails, whereupon it limits the current through the clamp diodes to ~ 2 mA. See some subtleties in "A Riff on Robust Input Protection" (§5.15.5).

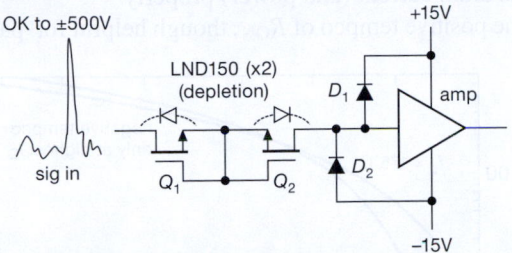

Figure 3.112. Protecting a low-level input from outrageous over-voltage "faults." Under normal conditions the series-connected depletion-mode MOSFETs Q_1 and Q_2 (with their intrinsic body diodes shown) conduct with an effective series resistance of $R_{ON} \sim 1$ kΩ each. An input signal beyond the amplifier's ± 15 V rails is clamped by diodes D_1 or D_2, with Q_1 and Q_2 safely limiting the clamp current to their $I_{DSS} \sim 2$ mA. See also Figure 5.81.

B. HV capacitor discharge

Human contact with a circuit at a few hundred volts can be, well, a *shocking* experience. That's why it's considered good manners to arrange things so that storage capacitors charged to such voltages are discharged promptly after power is removed. Capacitors, after all, have pretty good memory – they can stay charged for hours, or even years (that's how bits are stored in "flash memory," see §14.4.5).

The traditional approach is to put a "bleeder" resistor across the storage capacitor, sized to discharge it in ~ 10 s or so. Good enough. But it is not really satisfactory when you have a large-value capacitor, for example one used to store the energy for a short-duration high-voltage pulse generator. Figure 3.113 shows such an application, with a 100 μF storage capacitor charged to +400 V by a low-power dc–dc converter (say 10 W), the latter powered from a low-voltage dc supply that also powers the other pulse-generator circuitry.

What you'd like is a bleeder resistor that is connected only when the external power is removed. Here depletion-mode MOSFET Q_1 is held in the nonconducting state when the supply is powered ($V_{GS} = -9$ V), but is sent into conduction ($V_{GS} \approx 0.6V$) when the +12 V is absent. It's rated at 500 V, 3 A I_{DSS}(min), and costs about $2. (You can get depletion-mode MOSFETs up to 1 kV.) We don't need 3 A (which would discharge 100 μF in just 13 ms); but we *do* need a MOSFET large enough to absorb the stored energy,

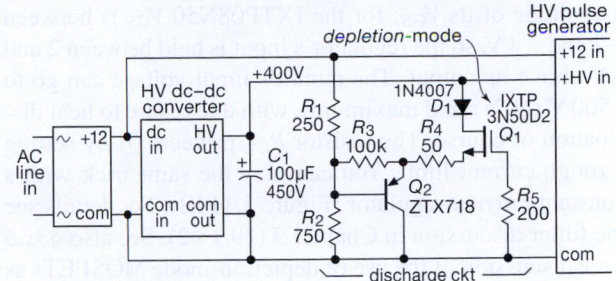

Figure 3.113. Depletion-mode MOSFET Q_1 discharges the 100 μF high voltage capacitor C_1 when power is removed; when powered it is inactive.

here 8 joules – this part can absorb a 25–50 J pulse without exceeding $T_{J(max)}$ (see Chapter 9x). Follower Q_2 boosts the discharge current, which otherwise would be just a few milliamps (set by bleeder R_2).

C. Current source

Depletion-mode power MOSFETs make excellent 2-terminal current sources, capable of high voltages (to 1000 V for some parts, see Table 3.6) and many watts of power dissipation. They extend the basic idea, seen earlier with JFETs (§3.2.2, Figures 3.20 and 3.23), to higher voltage and power levels. Because these applications are associated with power, we defer the discussion to Chapter 9 (§9.3.14C), where you can see that the circuits are the same as with JFETs (Figure 9.36), and you can delight in curves of measured current versus voltage (Figures 9.40 and 9.41).[112] Such a depletion-mode MOSFET current source is ideal for an application like the high-voltage piezo driver we just saw (§3.6.1), where it can replace the primitive 660k power-resistor pullup and thereby supply approximately constant driver-stage drain current over the signal swing.

D. Extending regulator V_{IN}

Sometimes you need to extend the permissible range of dc input voltage to some low-voltage device. Figure 3.114A shows an example: a linear voltage regulator (§9.3) that provides +3.3 V (for example) from a higher dc input. Such regulators have limited maximum input voltage range – perhaps +20 to +30 V (if made with BJTs), or as little as +6 V (if made with CMOS). Here the n-channel depletion-mode MOSFET Q_1 is wired as a follower, providing at the regulator's input a voltage that is greater than V_{OUT} by the

---

[112] There's plenty more detail fleshed out in Chapter 3x (§3x.6), where we show tricks for raising the output impedance (i.e., constancy of current) with the ever-wonderful cascode, raising operating voltage with a series stack, and reducing power dissipation.

magnitude of its V_{GS}; for the IXTP08N50 V_{GS} is between −2 and −4 V, so the regulator's input is held between 2 and 4 V above its output. The circuit's input voltage can go to +500 V (Q_1's rated maximum), with due regard to heat dissipation of course. The resistor R_{CL} protects Q_1 by setting a rough current limit. You can play the same trick with a constant-*current* regulator (Figure 3.114B). For details see the fuller discussion in Chapter 9 (§9.13.2). See also §3x.6 for a discussion of the use of depletion-mode MOSFETs as current sources, especially at high voltages.

3.6.3 Paralleling MOSFETs

You sometimes hear the statement that power MOSFETs can be paralleled directly (without ballasting resistors in the source leads), because their negative temperature coefficient of I_D at fixed V_{GS} guarantees automatic redistribution of drain currents in a paralleled array. Furthermore, the story goes, the same property prevents thermal runaway.

A. As *switches* – Yes!

Power MOSFETs exhibit negative tempco of I_D – but only at high drain currents (or, more accurately, at relatively

large values of V_{GS}), as seen in Figure 3.115. For *switching* applications, in which you are operating at essentially zero V_{DS} (limited by R_{ON}), the large gate drive puts the device into the region of negative I_D tempco, so you *can* (and should) simply tie multiple MOSFETs in parallel, with no ballasting resistors.[113] Here R_{ON} increases with increasing temperature (Figure 3.116), and the parallel connection shares drain current (and power) properly.

The positive tempco of R_{ON}, though helpful for parallel

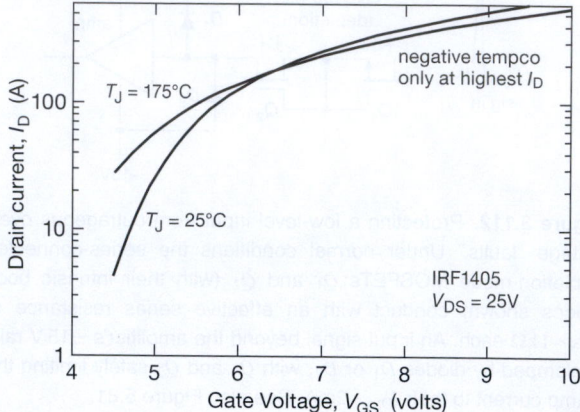

Figure 3.115. Transfer characteristics (I_D versus V_{GS}) for the IRF1405 *n*-channel power MOSFET. Note that the temperature coefficient is *positive* except at the highest drain currents (>175 A); for *linear* applications you would rarely exceed 10 A of drain current.

operation of MOSFET switches, creates a new problem, namely the possibility of *thermal runaway*. See §3.6.4.

B. In linear power circuits – No!

Here the situation is more complicated: in most *linear* applications (e.g., audio power amplifiers, in which there is substantial voltage V_{DS} across the transistor) you are operating in the positive tempco region of relatively low drain currents – because otherwise the power dissipation ($I_D V_{DS}$) would be much greater than allowable by thermal considerations (i.e., excessive junction temperature; see §9.4.1A). For example, the transistor of Figure 3.115 is limited to 200 watt dissipation at a case temperature of 75°C; so, in a circuit with 25 V across it, the average drain current is limited to 8 A, in which I_D has a large positive tempco. So, for practical linear applications – in which you operate

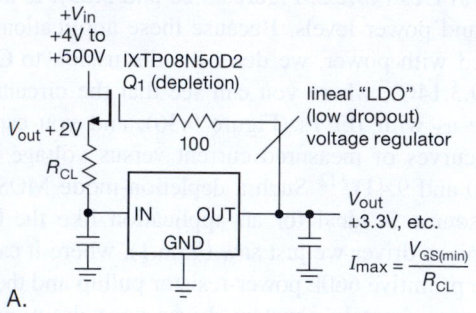

A.

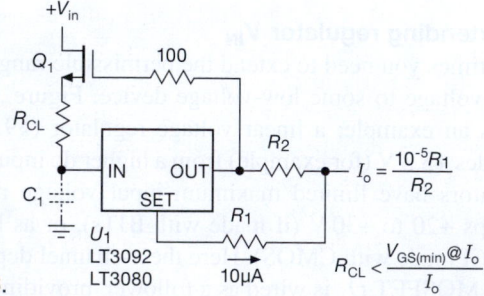

B.

Figure 3.114. A. A high voltage depletion-mode MOSFET extends the input voltage range of a series voltage regulator. B. An analogous circuit for a current source. For good performance at high frequencies C_1 should be small, or even eliminated altogether. See Figure 9.104, where the *i*s are dotted, *t*s are crossed, and all is explained.

---

[113] However, each FET should have its own series gate resistor to prevent oscillation during the switching transitions; these are typically in the range of a few ohms to a few tens of ohms, and they should ordinarily be used as well for single switching MOSFETs. Ferrite beads on the gate or drain leads can also be helpful to tame oscillations.

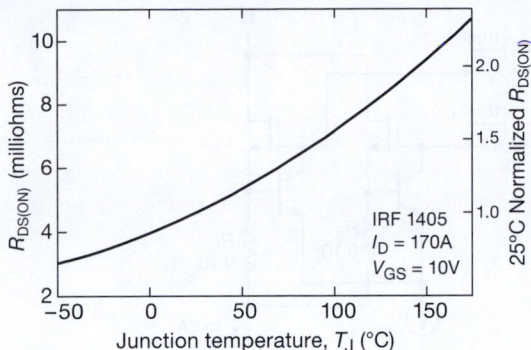

Figure 3.116. ON-resistance increases with increasing temperature: R_{ON} versus temperature for the IRF1405 *n*-channel power MOSFET.

with substantial V_{DS} – unequal current sharing of paralleled MOSFETs is in fact exacerbated. And, because you're using multiple transistors exactly because a single one won't handle the power, the circuit is in serious trouble; a single transistor will likely hog an excessive share of the current, putting its dissipation well over the limit set by thermal resistance and heatsinking.

Source-ballast resistors

The solution is to add small ballasting resistors in the individual source leads, chosen roughly so that the voltage drop across them is at least comparable to the scatter in gate-source operating voltages (Figure 3.117A). We've found that a few tenths of a volt drop is frequently adequate for MOSFETs of a given type, from a single manufacturing batch or from transistors selected for matched V_{GS};[114] however, the datasheet specs would suggest (conservatively) larger drops – a volt or two at full operating current. Unless you are willing to worry about matched transistors (both during initial construction, and later replacement), you should take the conservative approach to produce a robust design, with source-ballasting resistors sized to drop a volt or two at the currents where power dissipation becomes important.

This example illustrates a frequent designer's quandary, namely a choice between a conservative circuit that meets the strict worst-case design criterion, and is therefore *guaranteed* to work, and a better-performing circuit design that doesn't meet worst-case specifications, but is overwhelmingly likely to function without problems. There are times when you will find yourself choosing the latter, ignoring the little voice whispering into your ear.

Active feedback

This current-matching problem exemplifies a typical circuit tradeoff of robustness versus performance: a conservatively large ballast drop produces increased R_{ON} and power dissipation. As is often the case, a clever circuit can recover the lost benefits.

Figure 3.117B shows a nice solution, another in our "Designs from the Masters" series; just a snippet this time, but a valuable snippet. The small current-sense resistors in the MOSFET source leads provide active feedback through a primitive differential amplifier. Compared with a conservative source-ballasting circuit (Figure 3.117A), in which the source resistors are chosen to provide a 2 V drop (at a nominal 1 A per transistor operating current), the active circuit uses much smaller $0.1\,\Omega$ sense resistors, providing 100 mV of drop, which is applied to the *npn* differential pair to adjust the gate voltages as needed to equalize the source currents. This circuit requires larger gate drive voltage, which is seldom a problem; in exchange, it minimizes the voltage drop and impedance in the MOSFET's high-current path. This scheme is well suited to relatively slow circuits, for example the series pass element in a linear power supply. Note that this arrangement is easily expanded to any number of MOSFETs.[115]

There's a pleasant exception to this general characteristic of positive I_D tempco in power MOSFETs: *lateral* devices (as opposed to the *vertical* structure of nearly all power MOSFETs) exhibit negative tempco beginning at very low gate voltage (and very low I_D); see Figure 3.118. Lateral power MOSFETs do not attain the high breakdown voltage and low R_{ON} ratings of vertical power MOSFETs, but they are favored in linear power applications such as audio amplifiers for their linearity and thermal stability. A popular choice is the 2SK1058 (*n*-channel) and 2SJ162 (*p*-channel) complementary pair from Renesas (Hitachi), limited to 160 V and 7 A; their R_{ON} is an unimpressive $\sim$1 Ω. This is not of great concern in the linear amplifier context, where they do not operate near voltage-saturation; but it's high enough that you often see several of them used in parallel. See the section "Power transistors for linear amplifiers" in Chapter *3x* for further discussion.

The positive tempco of I_D in power MOSFETs creates an additional problem, namely the possibility of *thermal runaway.*

---

[114] In the example in §3.6.3A you might put four IRF1405s in parallel, with $0.1\,\Omega$ 10 W resistors in each source lead to handle a total current of 25A.

[115] We found this cute circuit trick used in some HP (subsequently Agilent, now Keysight) E3610-series linear power supplies. It's much simpler than using individual op-amps to bias each transistor, as some MOSFET manufacturers suggest. A different way to benefit from the higher power-dissipation capability of multiple transistors is to wire them in *series*, see for example Figure 9.111. A series connection guarantees equal current distribution.

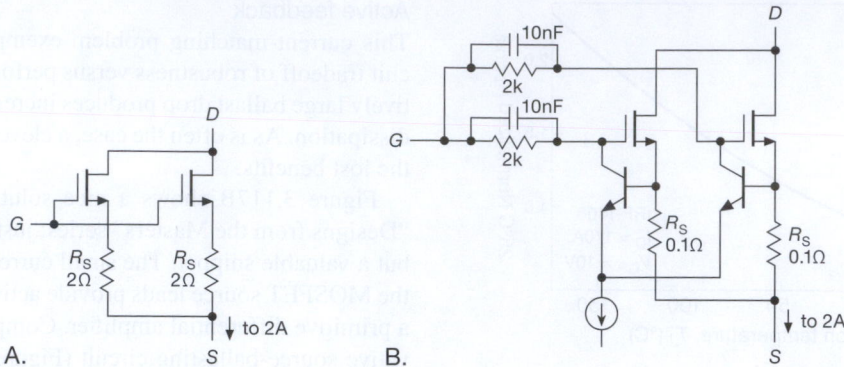

Figure 3.117. Paralleling power MOSFETs: A. with source ballasting resistors; B. with sense resistors and active feedback.

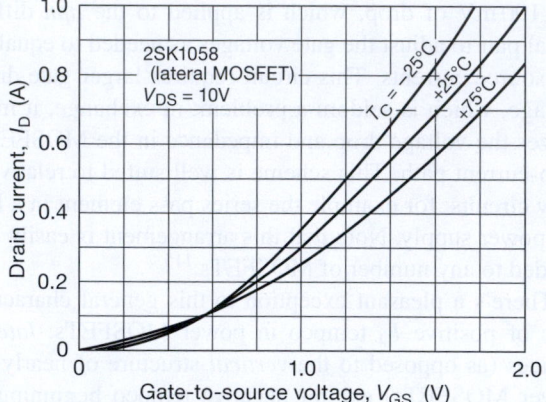

Figure 3.118. Transfer characteristics (I_D versus V_{GS}) for the 2SK1058 *lateral* n-channel power MOSFET, popular for use in high fidelity audio power amplifiers. Here the temperature coefficient is *negative* over most of the operating region.

3.6.4 Thermal runaway

Up to now, we've avoided the R-word, because "thermal runaway" is quite independent of whether transistors are used in parallel; it refers particularly to circuit configurations in which the power dissipation produces a rise in temperature that in turn raises the power that must be dissipated. Two important examples are the push–pull linear amplifier and the saturated power switch.

A. Push–pull power amplifier

In the class-AB push–pull power amplifier, commonly used in audio output stages, the push–pull pair is biased with substantial quiescent current (typically ~100 mA) to preserve linearity during waveform crossover. The quiescent current varies with temperature because both I_D (with MOSFETs) and I_C (with bipolar transistors) have positive

temperature coefficients at constant drive voltage. Depending on the circuit configuration and the degree of heatsinking, the output transistors may or may not reach a stable temperature; if they don't, you've got thermal runaway (independent of whether or not you've paralleled multiple transistors).

We saw this earlier in §2.4.1B, where we introduced the push–pull audio power amplifier built with complementary *bipolar* transistors. Because bipolar transistors have a positive tempco of collector current at fixed V_{BE},[116] the usual approach is to bias the bases apart with a voltage source that tracks the tempco of the output stage V_{BE}'s – typically by using diodes or transistor base–emitter junctions, thermally coupled to the output stage heatsink – often in conjunction with small emitter resistors in the output stage (Figure 3.119B).

Power MOSFETs in linear push–pull amplifiers present the same problem, because they are operated in the region of positive tempco of I_D (§3.6.3B). You can use the same trick (bias generator with tracking negative tempco, perhaps in combination with small-value output-stage source resistors; see the section "Power transistors for linear amplifiers" in Chapter *3x*). However, the problem is nicely finessed by using lateral power MOSFETs, whose negative tempco of I_D (Figure 3.118) beginning at $I_D \approx 100$ mA guarantees no thermal runaway. The usual approach is to bias the output stage gates apart with a (settable) constant dc voltage, as shown in Figure 3.119, bypassed at signal frequencies.[117] The bias is typically set for a quiescent current I_Q close to the zero-tempco crossover (100 mA for

---

[116] Or, alternatively, a negative tempco of V_{BE} for constant I_C.

[117] The figure shows bare-bones circuitry. In practice the bipolar transistors would be configured as Darlington or Sziklai pairs, and the single-ended driver stage might be replaced with a symmetrical pair of drivers, driven from the differential input stage. For a 150 W amplifier you would probably use paralleled pairs of transistors to stay within

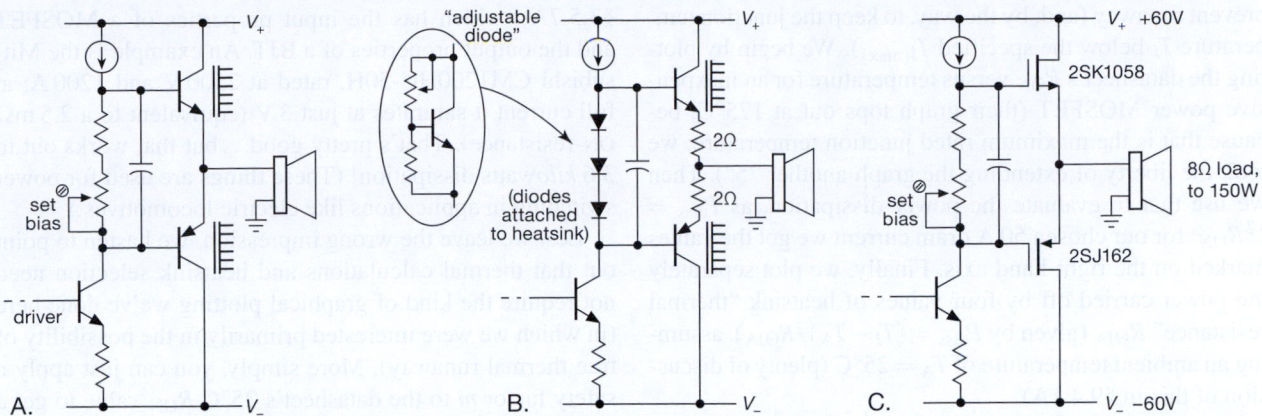

Figure 3.119. Thermal stability in push–pull power amplifiers – simplified output stage configurations. A. Fixed V_{BE} biasing promotes runaway, owing to positive I_C tempco in bipolar output stage. B. Tracking thermally-coupled bias generator tames runaway. C. Stable quiescent current in lateral MOSFETs biased at fixed V_{GS}; no thermal compensation is needed.

the 2SK1058/2SJ162 complementary pair), ensuring that I_Q remains relatively constant as the amplifier warms up.[118]

B. Saturated switch

It's widely believed that MOSFETs are immune to thermal runaway when used for power *switching*. The thought process goes like this: "These puppies have really low R_{ON} when driven to full conduction, so they hardly need any heatsinking; besides, if they do heat up (while carrying some large but bounded current), the thing will stabilize at some elevated temperature, because the power carried off by the heatsink increases roughly proportional to the rise above ambient temperature, and eventually catches up to the power being dissipated; plus, hey, these things are *tough!*"

Nice thoughts. But the reality can be different. That's because R_{ON} isn't constant, but increases with temperature (Figure 3.116); so the switch dissipates more power as it heats up, and, if the heatsink is too small, the heat it carries off may never catch up – in which case the process runs away!

To lend some perspective: you don't need to have actual thermal *runaway* to cause overheating and destruction – an undersized heatsink[119] will do the job just fine by al-

lowing the junction temperature to soar above $T_{J(max)}$. And, as we'll see shortly, the better approach is to reduce power dissipation by reducing R_{ON}, rather than by piling on larger heatsinks. With this cautionary comment, let's see how actual thermal runaway could occur in an ill-advised design.

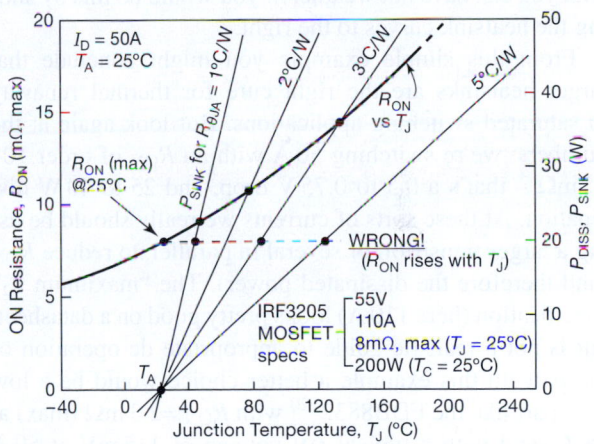

Figure 3.120. Thermal runaway in a MOSFET switch. The curved line plots maximum ON-resistance and corresponding power dissipation at 50 A for an IRF3205 *n*-channel power MOSFET. The straight lines plot power removed by three choices of heatsink thermal resistance $R_{\Theta JA}$. Thermal runaway occurs with the smallest heatsink, where there is no graphical intersection.

Figure 3.120 shows an easy graphical way to see what's happening and to figure out how much heatsink you need to

---

allowable junction temperature; no ballasting resistors would be needed for the MOSFET version.

[118] Because lateral MOSFETs may be hard to get, a regular power MOSFET can be used in place of the BJT "V_{BE} diode" (as in Figure 3.119B) to bias a complementary pair of ordinary power MOSFETs. This approach prevents thermal runaway because the tempco of a MOSFET is higher at low currents than at high currents, see Figure 3.115.

[119] Or none at all! The impressive $I_{D(max)}$ ratings on the datasheet could

tempt you to omit the heatsink entirely, even in a power-switching circuit operating at substantial drain currents.

prevent runaway (and, by the way, to keep the junction temperature T_J below the specified $T_{J(max)}$). We begin by plotting the datasheet's R_{ON} versus temperature for an inexpensive power MOSFET (their graph tops out at 175°C, because that is the maximum rated junction temperature; we took the liberty of extending the graph another 75°). Then we use that to evaluate the power dissipation, as $P_{diss} = I^2 R_{ON}$; for our chosen 50 A drain current we got the values marked on the right-hand axis. Finally, we plot separately the power carried off by four values of heatsink "thermal resistance" $R_{\Theta JA}$ (given by $P_{diss} = (T_J - T_A)/R_{\Theta JA}$), assuming an ambient temperature of $T_A = 25$°C (plenty of discussion of this in §9.4.1A).

The heatsinks carry off an amount of power proportional to the temperature rise above ambient, as plotted; the transistor generates power according to its graph. The intersection (if any!) is the equilibrium temperature, which in this case is about 45°C or 75°C, for the two larger heatsinks. But the smallest heatsink has no intersection – it cannot carry off as much heat as the transistor generates, at any temperature: *thermal runaway!* In real life you should assume that the ambient temperature will be higher (equipment is put into racks, or stacked with other equipment; and, you can have hot weather!): you would do this by sliding the heatsink curves to the right.

From this simple example you might conclude that larger heatsinks are the right cure for thermal runaway in saturated switching applications. But look again at the numbers: we're switching 50 A with an R_{ON} of order 10–15 mΩ – that's a 0.5 to 0.75 V drop, and 25 to 40 W dissipation. At these sorts of currents we really should be using a larger transistor, or several in parallel, to reduce R_{ON} (and therefore the dissipated power). The "maximum I_D" specification (here 110 A) looks pretty good on a datasheet, but is not a realistic guide to appropriate dc operation of the part. In this example a better choice would be a low R_{ON} part like the FDB8832[120] with $R_{ON}=2.3$ mΩ (max) at 25°C, and with a typical ON voltage of 115 mV at 50 A and power dissipation of 5.8 W.[121] This is a 30 V part (high-voltage MOSFETs have higher R_{ON}); if you wanted to switch somewhat higher voltages with low R_{ON} and P_D, your choices are to use a high-power MOSFET module[122] or (less expensive) several conventional MOSFETs in parallel. For voltages above 400 V or so the transistor of choice is the IGBT (insulated-gate bipolar transistor; see

§3.5.7A), which has the input properties of a MOSFET and the output properties of a BJT. An example is the Mitsubishi CM1200HC-50H, rated at 2500 V and 1200 A: at full current it saturates at just 3 V (equivalent to a 2.5 mΩ ON-resistance). That's pretty good… but that works out to 3.6 *kilo*watts dissipation! (These things are used for power switching in applications like electric locomotives.)

Lest we leave the wrong impression, we hasten to point out that thermal calculations and heatsink selection need not require the kind of graphical plotting we've done here (in which we were interested primarily in the possibility of true thermal runaway). More simply, you can just apply a safety factor m to the datasheet's 25°C R_{ON} value to get a reasonable estimate of R_{ON} at the maximum junction temperature (150°C); from that you get

$$T_J \approx T_A + I_D^2 \cdot m R_{ON(25C)} \cdot R_{\Theta JA}. \qquad (3.15)$$

The multiplier m varies somewhat with the voltage rating of the MOSFET; based on data from many datasheets (see the graph and discussion in the section "MOSFET ON-resistance versus temperature" in Chapter *3x*), it ranges from roughly $m \approx 1.5$ (for low-voltage MOSFETs) to roughly $m \approx 2.5$ (for high-voltage MOSFETs). As a practical rule-of-thumb, you'll be safe if you use $m=2$ for MOSFETs rated to 100 V and $m=2.5$ for those of higher voltages (at least to 1kV).

C. Second breakdown and safe operating area

It's worth emphasizing a related thermal effect ("second breakdown") that was discussed earlier in §3.5.1B: power transistors fail (usually[123]) if operated beyond their maximum voltage, their maximum current, or their maximum junction temperature (the latter dependent on power dissipation, pulse duration, heatsink thermal resistance, and ambient temperature; see §9.4.2). The boundaries define the *safe operating area*, or SOA, for example as shown in Figure 3.95. Bipolar transistors suffer from an additional failure mode known as *second breakdown*, an incompletely understood instability characterized by local heating, reduction of the breakdown voltage, and, often, destruction of the junction. It is second breakdown that imposes the additional constraint on the bipolar SOAs in Figure 3.95. Happily, MOSFETs are less likely to suffer from second breakdown, which contributes to their popularity in power circuits.[124] Note that for both kinds of transistors the maximum current and power limits are higher for short pulses.

---

[120] From the same manufacturer, Fairchild Semiconductor.

[121] Rising to 3.6 mΩ, 180 mV, and 9 W (maximum) at $T_J=150$°C.

[122] These come in husky "SOT-227" packages, with screw terminals on the top, an isolated metal base, and with names like "ISOTOP" and "miniBLOC."

[123] Or, perhaps more accurately, they are not guaranteed *not* to fail!

[124] Some newer fine-geometry types *are* susceptible, however; see IR App Note AN-1155.

Table 3.7 Junction Field-Effect Transistors (JFETs)[a]

Part #[b]	N or P	Jelly?	TO-92: 2N,PN	SOT23: MMBF	SOT23: PMBF	MMBF_LT	VDSS max (V)	IDSS min (mA)	IDSS max (mA)	IDSS meas (mA)	RON max (Ω)	VGS(off)[c] min (V)	VGS(off)[c] max (V)	VGS@ID meas (V)	VGS@ID (mA)	gm min (mS)	gm max (mS)	gm @ID (mA)	gm meas (mS)	gm @ID (mA)	Gmax[e]	Ciss typ (pF)	Crss typ (pF)
PN4117	N	•	A	C	-	-	40	0.03	0.09	0.07	-	-0.6	-1.8	-0.33	0.03	0.07	0.21	z	0.09	0.03	420[r]	1.2	0.3
'4118	N	-	A	C	-	-	40	0.08	0.24	0.20	-	-1	-3	-1.33	0.1	0.08	0.25	z	0.13	0.1	260[r]	1.2	0.3
'4119	N	-	A	C	-	-	40	0.20	0.60	0.30	-	-2	-6	0.0	0.3	0.10	0.33	z	0.18	0.3	140[r]	1.2	0.3
BFT46	N	-	-	C	-	-	25	0.20	1.5	0.63	-	-	-1.2	-0.16	0.3	1	-	z	1.7	0.3	190[s]	3.5	0.8
BF511	N	-	-	D	-	-	20	0.7	3	4.2	-	-	-1.5	-0.75	1	4	-	z	2.7	1	120	-	0.3
2N5457	N	•	A	C	-	-	25	1	5	3.5	-	-0.5	-6	-0.81	1	1	5	z	2.3	1	220	4.5	1.5
'5458	N	-	A	C	-	-	25	2	9	4.1	-	-1	-7	-0.97	1	1.5	5.5	z	2.2	1	190	4.5	1.5
'5459	N	-	A	C	-	-	25	4	16	9.9	-	-2	-8	-1.82	3	2	6	z	2.9	3	100	4.5	1.5
2N5460	P	•	A	C	-	-	25	-1	-5	3.4	0.75		6	+0.97	1	1	4	z	1.9	1	260	4.5	1.2
'5461	P	•	A	C	-	-	25	-2	-9	2.7	-	1	7.5	+0.67	1	1.5	5	z	2	1	210	4.5	1.2
'5462	P	•	A	C	-	-	25	-4	-16	5.9	-	1.8	9	+4.15	3	2	6	z	2.5	3	30	4.5	1.2
MMBF4416	N	•	A	C	*C*	C	30	5	15	5.9	-	-	-6	-0.19	5	4.5	7.5	z	3.9	5	70	4	0.8
2N5484	N	-	A	C	-	C	25	1	5	3.3	-	-0.3	-3	-0.73	1	3	6	z	2.3	1	230	10	2.2
'5485	N	-	A	C	-	-	25	4	10	6.6	-	-0.5	-4	-1.65	1	3.5	7	z	2.1	1	150	10	2.2
'5486	N	-	A	C	-	-	25	8	20	14	-	-2	-6	-2.61	1	4	8	z	2.1	1	75	10	2.2
2SK170BL	N	-	B	-	-	-	40	6	12	6.1	-	-0.2	-1.5	-0.04	5	22[t]		z	29	5	470	30	6
LSK170B	N	-	B	C	-	-	40	6	12	7.6	-	-0.2	-2	-0.17	3	10[t]		1	20	3	160	20	5
LSK170C	N	-	B	C	-	-	40	10	20	13	-	-0.2	-2	-0.26	5	10[t]		1	24	5	90	20	5
BF861B	N	-	-	*C*	-	-	25	6	15	8	-	-0.5	-1.5	-0.47	1	16	25	z	16	5	150	7.5	
BF545C	N	-	-	*C*	-	-	30	12	25	19	-	-3.2	-7.8	-1.80	5	3.0	6.5	z	3.7	5	30	1.7	0.8
BF862	N	-	-	*C*	-	-	20	10	25	12	-	-0.3	-1.2	-0.21	5	35	45[t]	z	26	5	270	10	1.9
PF5103	N	-	A	C	-	-	40	10	40	19	30	-1.2	-2.7	-1.00	5	7.5		2	10	5	160	16	6
PN4391	N	•	A	C	C	C	40	50	150	115	30	-4	-10	-7.15	5	12[t]		5	8.8	5	30	12	3.5
'4392	N	•	A	C	C	C	40	25	75	38	60	-2	-5	-1.67	5	16[t]		10	10	5	130	12	3.5
'4393	N	•	A	C	C	C	40	5	30	16	100	-0.5	-3	-1.25	1	13[t]		10	6.2	1	150	12	3.5
J105	N	-	A	*C*	-	-	25	500	-	-	3	-4.5	-10	-8.39	5	40[t]		5	37	10	60	160[m]	35[m]
J106	N	-	A	*C*	-	-	25	200	-	-	6	-2	-6	-2.42	5	53[t]		5	43	10	230	160[m]	35[m]
J107	N	-	A	*C*	-	-	25	100	-	-	8	-0.5	-4.5	-1.93	5	75[t]		5	48	10	340	160[m]	35[m]
J108	N	•	A	C	C	-	25	80	-	325	8	-3	-10	-5.83	5	37[t]		5	31	10	60	85	15
J109	N	•	A	C	C	-	25	40	-	201	12	-2	-6	-2.85	5	26[t]		5	32	10	160	85	15
J110	N	•	A	C	C	-	25	10	-	122	18	-0.5	-4	-1.80	5	20[t]		5	34	10	220	85	15
J111	N	-	A	C	C	-	35	20	-	115	30	-3	-10	-7.6	5	-	-	-	8.4	5	30	28	5
J112	N	•	A	C	C	-	35	5	-	47	50	-1	-5	-2.8	5	6.7[t]		1	9.5	5	100	28	5
J113	N	-	A	C	C	-	35	2	-	21	100	-0.5	-3	-1.0	5	8[t]		1	11	5	100	28	5
J174	P	-	B	C	C	-	30	-20	-135	26	85	-5	-10	+2.08	5	4.5	-	5	-	-	15	13	6
J175	P	•	B	C	C	C	30	-7	-60	13	125	-3	-6	+1.58	1	-	-	-	-	-	30	13	6
J176	P	-	B	C	C	C	30	-2	-25	6.1	250	-1	-4	+0.86	1	6.3	-	5	-	-	40	13	6
J177	P	-	B	C	C	C	30	-1.5	-20	4.2	300	-0.8	-2.5	+0.62	1	-	-	-	-	-	50	13	6
J308	N	•	A	C	C	-	25	12	60	35	-	-1	-6.5	-	-	8	-	10	12	5	120	4	2
J309	N	-	A	*C*	*C*	C	25	12	30	23	-	-1	-4	-1.2	5	10	20	10	11	5	300	4	2
J310	N	-	A	*C*	*C*	C	25	24	60	39	-	-2	-6.5	-2.4	5	8	18	10	8.9	5	100	4	2
dual JFETs																							
LS840-42	N		F				60	0.5	5	3.3	-	-1	-4.5	-0.85	1	0.5	1	0.2	2.1	1	180	4	1.2
'843-5	N		F				60	1.5	15	-	-	-1	-3.5	-	-	1	1.5[t]	0.5	-	-	-	8[m]	3[m]
LSK389A	N		F, J				40	2.6	6.5	-	-	-0.15	-2	-	-	8	20[t]	3	-	-	-	25	5.5
'389B	N		F, J				40	6	12	12	-	-0.15	-2	-0.24	5	8	20[t]	3	23	5	170	25	5.5
'389C	N		F, J				40	10	20	-	-	-0.15	-2	-	-	8	20[t]	3	-	-	-	25	5.5
LS5912	N		F, J, K[p]				25	7	40	18	-	-1	-5	-1.75	5	4	10	5	5.7	5	70	5	1.2
IFN146	N		F[v]				40	-	30	6	-	-0.3	-1.2	-0.19	1	30	40[t]	z	25	5	660	75[m]	15[m]

(a) listed generally by increasing IDSS, but also by part number within a family (e.g., J105–J113); see also Table 8.2 for noise parameters.
(b) for families of related parts, **boldface** designates the family matriarch. (c) usually specified at ID=1nA or 10nA, though sometimes at 10µA or even 200µA (e.g., for the J105–J113 "switches"); it doesn't much matter, given the wide range of specified VGS(off). (d) see the accompanying pinout figure; all JFETs appear to be symmetric (source and drain are interchangeable), but *italic* designates a datasheet pinout in which the S and D terminals are interchanged. (e) Gmax=gm/gos, the maximum grounded-source voltage gain into a current source as drain load; listed values measured at ID=1mA and VDS=5V, unless noted otherwise. Gmax is proportional to VDS, and for most JFETs Gmax is relatively constant over varying ID. Use tabulated Gmax to find gos=gm/Gmax. (m) maximum. (p) several PDIP-8 pkgs available. (r) at ID=30µA. (s) at ID=300µA. (t) typical. (v) variant of "F" pinout: G and D terminals interchanged. (z) at IDSS.

Table 3.8 Low-side MOSFET Gate Drivers[a]

Part #	Mfg[d]	# channels	V_{min} (V)	V_{max} (V)	I_{pk} (A)	$t_d+0.5t_r$ (ns, typ)	C_{load} (nF)	logic thresh[p]	source below gnd?	current limit	UVLO?	enable?	inv?	non-inv?	output rail-to-rail?	TO220, Dpak	DIP	SOIC, MSOP	SOT23	smaller	Comments
TC4426-28	MC+	2	4.5	18	1.5	55	1	T	-	-	-	-	n	n	•	-	•	•	-	•	G,H
TC4423-25	MC+	2	4.5	18	3	70	1.8	T	-	-	-	-	n	n	•	-	•	•	-	•	G,H
TC4420,29	MC+	2	4.5	18	6	80	2.5	T	-	-	-	-	•	•	•	-	•	•	-	•	G
TC4421-22	MC+	1	4.5	18	9	85	10	T	-	-	-	-	•	•	•	•	•	•	-	•	G,J
FAN3111	F	1	4.5	18	1	20	0.5	C	-	-	c	c	c	c	•	-	-	•	•	-	
FAN3100C,T	F	1	4.5	18	2	20	1	C,T	-	•	c	c	c	c	•	-	-	•	•	-	A
FAN3180	F	1	5	18	2	30	1	T	-	-	-	•	•	•	•	-	-	•	•	-	B
FAN3216-17	F	2	4.5	18	2	25	2.2	T	-	-	-	•	•	•	•	-	-	•	•	-	D
FAN3226-29C,T	F	2	4.5	18	2	25	1	C,T	-	-	c	c	c	c	•	-	-	•	•	-	C,E
FAN3213-14	F	2	4.5	18	4	20	2.2	T	-	-	-	•	•	•	•	-	-	•	•	-	C
FAN3223-25C,T	F	2	4.5	18	4	25	2.2	C,T	-	-	c	c	c	c	•	-	-	•	•	-	E
FAN3121-22	F	1	4.5	18	9	21	10	T	-	-	-	•	•	•	•	-	-	•	•	-	
IRS44273L	IR	1	12	20	1.5	50	1	T	-	-	-	-	•	•	•	-	-	•	•	-	
IR25600	IR	2	6	20	1.5	75	1	T	-	-	-	-	•	•	•	-	-	•	•	-	
MAX17600-05	MA	2	4	14	4	15	1	C5,T	-	-	-	•	n	n	•	-	-	•	-	•	H
MAX5054-57	MA	2	4	15	4	38	5	C,T	-	-	-	c	c	c	•	-	-	•	-	•	
MAX5048A,B	MA	1	4	12.6	7.6[h]	18	1	C,T	-	-	•	c	c	c	•	-	-	•	-	•	
UCC37323-25[k]	TI	2	4.5	15	4	47	1.8	T	-	-	-	•	•	•	•	-	•	•	-	•	
UCC27517	TI	1	4.7	20	4	17	1.8	T	-	-	•	c	c	c	•	-	-	•	•	•	
UCC27516-19	TI	1	4.7	20	4	17,21	1.8	T,C	-	-	•	•	•	•	•	-	-	•	•	•	
UCC27523-26	TI	2	4.7	20	5	17	1.8	T	-	-	-	•	•	•	•	-	-	•	-	•	E,H
UCC37321-22[k]	TI	1	4	15	9	50	10	T	-	-	-	•	•	•	•	-	•	•	-	•	
MIC44F18-20	MI	1	4.5	13.2	6	24	1	T	-	-	-	•	•	•	•	-	-	•	•	•	
ADP3623-25	A	2	4.5	18	4	28	2.2	T	-	-	-	•	•	•	•	-	-	•	-	•	H,P
LM5110	TI	2	3.5	14	5[f]	38	2	T	•	-	-	-	n	n	•	-	-	•	-	•	H,L
LM5112	TI	2	3.5	14	7[g]	38	2	T	•	-	-	-	n	n	•	-	-	•	-	•	H,L,M
LM5114	TI	1	4	12.6	7.6[h]	16	1	C	-	-	•	c	c	c	•	-	-	•	-	•	
ISL89367	IN	2	4.5	16	6	45	10	F	-	-	•	n	o	o	•	-	-	•	-	•	N
ISL89160-62	IN	2	4.5	16	6	45	10	C5,T	-	-	-	•	•	•	•	-	-	•	-	•	O
MC34151	O	2	6.5	18	1.5	50	1	T	-	-	-	-	•	•	•	-	•	•	-	•	
IR2121	IR	1	12	18	2[e]	200	3.3	T	-	•	-	-	•	•	•	-	•	•	-	-	F
UC3708	TI	2	5	35	3	37	1	T	-	-	-	-	•	•	•	-	•	•	-	-	
IXDD602	IX	1	4.5	35	2	50	1	C5	-	-	-	-	•	•	•	-	-	•	-	•	H,R
IXDD604	IX	1	4.5	35	4	40	1	C5	-	-	-	-	•	•	•	-	-	•	-	•	H,R
IXDD609	IX	1	4.5	35	9	60	10	C5	-	-	-	-	•	•	•	-	•	•	-	•	R
IXDD614	IX	1	4.5	35	14	70	15	C5	-	-	-	-	•	•	•	•	•	•	-	-	R
IXDD630	IX	1	10	35	30	65	5.6	C5	-	-	-	-	•	•	•	•	•	-	-	-	K,R
ZXGD3002-04	D	1	-	20,40	9,5	11	1	-	-	-	-	-	•	-	-	-	-	-	•	-	M,S

Notes: (a) sorted by family, within family sorted by Iout; except for ZXGD3000-series, all devices swing rail-to-rail, or nearly so. (b) into Cload at V_S=12V. (c) input gate with inv and non-inv inputs. (d) A=Analog Devices; D=Diodes,Inc; F=Fairchild; IN=Intersil; IR=International Rectifier; IX=Ixys/Clare; L=LTC; MA=Maxim; MC=Microchip; MI=Micrel; O=OnSemiconductor; S=STMicroelectronics; TI=Texas Instruments. (e) 1A source, 2A sink. (f) 3A source, 5A sink. (g) 3A source, 7A sink. (h) 1.3A source, 7.6A sink. (k) 37xxx for 0 to 70°C, 27xxx for -40°C to 105°C. (n) see part-specific comments. (o) XOR input sets optional invert. (p) C=CMOS; C5=5V CMOS; F=flexible, set by V_{ref-} and V_{ref+} input pins; T=TTL.

Comments: (A) suffix specifies logic threshold. (B) includes 3.3V LDO output. (C) 2ns td channel match. (D) 1ns td channel match. (E) dual inv+en, dual non-inv+en, dual inputs. (F) source-resistor current-sense input terminal, suitable for driving an IGBT. (G) industry std, many mfgrs. (H) dual inv, dual non-inv, or one each. (J) for 8-pin pkgs, n- and p-ch drains on separate pins. (K) t_r, t_f = 50ns into 68nF. (L) output swing to neg rail, can be 5V below logic gnd. (M) n- and p-ch drains on separate pins. (N) resistor-programmed edge-delay timers; 2-input AND signal inputs. (O) ISL89163-65 same, but include enable inputs; ISL89166-68 same, but include resistor-programmed edge-delay timer inputs. (P) overtemp protection and output. (R) full p/n is IXDx6..., where x = N, I, D and F for non-inv, inv, dual non-inv+en, or one of each. (S) series is one each high-current high-gain npn and pnp transistor emitter-followers for pullup and down.

Review of Chapter 3

An A-to-Z summary of what we have learned in Chapter 3. This summary reviews basic principles and facts in Chapter 3, but it does not cover application circuit diagrams and practical engineering advice presented there.

¶A. FETs.

In Chapter 3 we explored the world of Field-Effect Transistors, or FETs. FETs have a conducting channel with terminals named *Drain* and *Source*. Conduction in the channel is controlled by an electric field created by a third *Gate* electrode (§3.1). As with bipolar transistors (BJTs), FETs are transconductance devices (see ¶G below), which means the drain *current* (assuming sufficient drain-to-source voltage) is controlled by the gate *voltage*.

¶B. *n*-channel and *p*-channel.

Like BJTs with their *npn* and *pnp* types, FETs come in both *n*- and *p*-channel polarities (§3.1.2). In either case the channel conductance increases if the gate voltage is taken toward the drain voltage. For example, for an *n*-channel FET with a positive drain voltage, the channel can be turned on with a sufficient positive-going voltage, and cutoff with a sufficient negative-going voltage. That's not to say the *n*-channel device requires positive and negative voltages to turn on and off. A threshold voltage V_{th} can be defined where the FET is just slightly turned on, and the channel responds to gate voltages above and below V_{th} for control.

¶C. Enhancement and Depletion Modes.

See Figure 3.8. Enhancement-mode devices have a high enough threshold voltage V_{th} that they are nonconducting (i.e., off) when their gate voltage is at $V_{GS}=0$ V. To bring such a FET into conduction, the gate of is brought positive (if *n*-channel) or negative (if *p*-channel). Depletion-mode devices, by contrast, have their threshold voltage well into the "off" direction, thus they are conducting (i.e., on) with their gate-voltage at $V_{GS}=0$ V. Thus for example you must apply a considerable negative gate voltage V_{GS} to turn off an *n*-channel depletion-mode FET. See Figure 3.9 where drain current versus gate voltage is shown for a selection of *n*-channel devices. FETs can be fabricated with the transfer curve shifted left or right (more about this in ¶H below). Figures 3.10 and 3.11 show convenient maps of the FET types.

¶D. MOSFETs and JFETs.

In *metal-oxide* FETs (MOSFETs) the gate electrode is fully insulated from the channel, and can be taken positive or

negative, typically up to ±20 V. In junction FETs (JFETs) the semiconductor gate contacts the channel and acts as a diode junction, so it is insulated only in the reverse direction. Therefore JFETs are necessarily depletion mode devices; one cannot make an enhancement-mode JFET. Figures 3.6 and 3.7 show FET symbols.

¶E. FET Characteristics, Gate and Drain.

See Figure 3.13. A FET's channel conductance and current is controlled primarily by its gate voltage, but it's also affected by the drain voltage V_{DS}. At very low drain voltages the channel acts like a resistor, whose value is controlled by the gate (§3.1.2 and §3.2.7); this is called the *linear* region. At higher drain voltages the drain current levels off, being controlled by the gate voltage and only weakly dependent upon drain voltage; this is called the *saturated* region. In the saturated region the FET drain acts like a current source (or sink), and the device is characterized by its transconductance g_m (see ¶G below). MOSFETs are often used as switches. In this mode of operation a large gate voltage (e.g., 10 V) is applied to make the channel resistance low enough to approximate a closed switch. More on FET switches in sections ¶¶O–Q below.

¶F. Square-law.

Over a large region of gate voltages greater than V_{th}, and for drain voltages above a volt or so (i.e., in the saturated region), a FET's drain current behaves like a square-law device; that is, its drain current is proportional to the square of the excess gate-drive voltage $(V_{GS}-V_{th})^2$, see Figure 3.14 and eq'n 3.2. This is sometimes called the *quadratic* region. The threshold voltage V_{th} is generally determined with an extrapolated $\sqrt{I_D}$ plot, as the figure shows. For V_{GS} below threshold the FET is in the subthreshold region; see ¶I below.

¶G. Transconductance and Amplifiers.

Transconductance g_m is the change in output drain current caused by a change in gate voltage: $g_m=i_D/v_{GS}$ (the lowercase i and v signify small signals). Common-source FET amplifiers (§3.2.3, Figures 3.28 and 3.29) have voltage gain $G=-g_m R_D$, where R_D is the drain load resistance. In contrast to BJTs (where $g_m \propto I_C$), the transconductance of FETs rises only as $\sqrt{I_D}$ in the important quadratic region; see Figures 3.53 and 3.54. As a consequence FET amplifiers with resistive drain loads have lower gain when designed to operate at higher current, because R_D is generally chosen inversely proportional to drain current. The FET's internal output resistance also acts as a load resistance, thus limiting gain ("G_{max}") even with an ideal current-source drain load; see §3.3.2 eq'n 3.13, and Table 3.1.

When used as a *follower*, an FET has an output impedance $r_{out}=1/g_m$, see ¶K below.

¶H. Biasing JFET Amplifiers.

JFETs are well suited for making signal amplifiers (by contrast there are few viable small discrete *MOS*FETs), and they work especially well in low-noise amplifiers. But there's one very painful issue analog designers face: the uncertain value of the gate operating voltage for any given part. Scanning the min and max columns for $V_{GS(off)}$ in the JFET Table 3.1 on page 141, we see values for a particular JFET that range from -1 V to -7 V, or -0.4 V to -4 V. The latter is a 10:1 ratio! Figure 3.17 shows V_{GS} histograms for 300 parts, 100 each for three different JFET types in a family. Here we see gate voltage spreads of about 1 V, which you might rely upon if you buy a batch of parts from one manufacturer and measure them. But, *caution*: Figures 3.51 and 3.52 show how the same part type may vary when purchased from different manufacturers. To deal with the uncertainty, special biasing schemes are often required in FET amplifier circuits. Figures 3.25 and 3.41 show examples of the load-line concept for analyzing amplifier biasing.

¶I. Subthreshold Region.

The simple FET formula of eq'n 3.2 predicts zero drain current when the gate voltage reaches threshold ($V_{GS}=V_{th}$). In reality the drain current is not zero, and transitions smoothly to a subthreshold region (see Figure 3.16) where the FET looks more like a BJT, with its exponential Ebers-Moll characteristic (§2.3.1). In this region (where I_D rises exponentially with V_{GS}) we're glad to see a higher $g_m \propto I_D$; but sadly the FET proportionality constant is usually $2\times$ to $5\times$ smaller than for BJTs, see Figure 3.53.

¶J. Self-biased Amplifiers.

Depletion-mode MOSFETs (and all JFETs) operate with a reverse voltage on their gates, which allows them to be self-biased (§3.2.6A). The source terminal is "higher" than the gate terminal, so a source resistor connected between them sets the drain current to $I_D=V_{GS}/R$. This is also a convenient way to make a 2-terminal current source, but the tolerance will be poor due to the wide variability in V_{GS}, see ¶H. Alternately the V_{GS} voltage available at the source pin may be used to operate a current-setting IC like the LM334.

¶K. Source Followers.

Source followers (§3.2.6), Figure 3.40 have a nominal gain of 1, analogous to the BJT emitter follower. Because of their lower g_m, however, they have considerably higher output resistance, $r_{out}=1/g_m$, so the ideal unity gain is reduced by load resistance, see eq'n 3.7.

¶L. FETs as Variable Resistors.

At low drain voltages ($V_{DS}\ll V_{GS}$) FETs act as variable resistors programmed by the gate voltage. Because the slope varies with V_{DS}, however, the resistance is somewhat nonlinear. But there's a simple trick to linearize this resistance, by exploiting the quadratic behavior of FETs, see Figures 3.46 and 3.47.

¶M. FET Gate Current.

The gate of a JFET forms a diode junction with the channel; it's normally reverse-biased, with some non-zero dc leakage current (§3.2.8). This current roughly doubles for every $10°C$ temperature increase; furthermore it increases dramatically at high drain currents and drain voltages due to impact ionization, see Figure 3.49. MOSFET gates do not suffer from either of these leakage-current-increasing effects. In contrast to the generally negligible dc gate leakage, the input capacitance C_{iss} of FETs (which can be quite high, many hundreds of pF for large power MOSFETs) often presents a substantial ac load. Use a gate-driver chip (Table 3.8) to provide the high transient currents needed for rapid switching.

¶N. JFET Switches.

JFETs can be used as analog-signal switches, as in the *n*-channel switch of Figure 3.62. The switch is OFF when the gate is taken at least V_{th} below than the most negative input signal. To turn the switch ON the gate voltage must be allowed to equal the source. JFETs are symmetrical, so e.g., for an n-channel part, the "source" would be the most negative pin. Large-die JFETs work well as power switches up to 100 mA; Table 3.1 lists parts with R_{ON} as low as 3 Ω.

¶O. CMOS Switches.

CMOS signal switches are made with a parallel pair of complementary *n*- and *p*-channel MOSFETs. This reduces R_{ON} as shown in Figure 3.61, and beneficially causes cancellation of most of the injected charge transfer (§3.4.2E), see Figure 3.79. The injected charge scales roughly inversely proportional to R_{ON} (Figure 3.81), so there's a tradeoff between desirably low on-resistance and desirably low self-capacitance. As an example, Table 3.3 lists a switch with an impressive $R_{ON}=0.3\,\Omega$ – but it's burdened with a whopping 300 pF of self-capacitance. A T-switch configuration can be used to reduce the signal feedthrough at high-frequencies, see Figure 3.77.

¶P. CMOS Logic Gates.

See Figure 3.90. A series pair of complementary (*n*- and *p*-channel) small-geometry MOSFETs between the positive rail and ground forms the simplest logic inverter (Figure 3.90); more switches can be arranged to make CMOS logic gates (e.g., Figure 3.91, §3.4.4), with the attractive property of nearly zero static power, except when switching. CMOS logic is covered extensively in Chapters 10 and 12, and is the basis for all contemporary digital processors.

¶Q. MOSFET Power Switches.

Most power MOSFETs (§3.5) are enhancement type, available in both *n*- and *p*-channel polarities. They are very popular for use as high-current high-voltage power switches. A few relevant parameters are the breakdown voltage V_{DSS} (ranging from 20 V to 1.5 kV for *n*-channel, and to 500 V for *p*-channel); the channel on-resistance $R_{DS(on)}$ (as low as 2 mΩ); the power-handling ability (as high as 1000 W with the case held unrealistically at 25°C); and the gate capacitance C_{iss} (as high as 10,000 pF), which must be charged and discharged during MOSFET switching, see ¶S below. Table 3.4a lists representative small-package *n*-channel parts rated to +250 V and *p*-channel parts of all sizes to −100 V; Table 3.4b extends the *n*-channel selection to higher voltage and current; more complete tables are found in Chapter 3x.

¶R. Maximum Current.

MOSFET datasheets list a maximum continuous rated current, specified however at an unrealistic 25°C case temperature. This is calculated from $I_{D(max)}^2 R_{DS(ON)} = P_{max}$, substituting a maximum power $P_{max} R_{\Theta JC} = \Delta T_{JC} = 150°C$ (see §9.4), where they have assumed $T_{J(max)} = 175°C$ (thus a 150°C ΔT_{JC}), and they use the value of $R_{DS(ON)}$ (max) at 175°C from an R_{DS} tempco plot (e.g., see Figure 3.116). That is, $I_{D(max)} = \sqrt{\Delta T_{JC}/R_{\Theta JC} R_{ON}}$. Some datasheets show the calculation for a more realistic 75°C or 100°C case temperature. Even so, you don't really want to run your MOSFET junction at 175°C, so we recommend using a lower maximum continuous I_D and corresponding P_{diss}.

¶S. Gate Charge.

The capacitances in power MOSFETs that slow down switching are most easily analyzed with gate-charge plots, like Figure 3.101. First consider turn-ON: as current flows

into the gate capacitance $C_{iss} + C_{rss}$ (dominated by C_{iss}) the gate voltage rises. There is a switching delay, because the FET's drain remains off until the gate voltage is high enough for the FET to sink the drain current. Then the drain voltage starts to fall, as seen in Figures 3.102 and 3.103. The falling drain creates a reverse gate current $I = C_{rss} dV_D/dt$ that prevents further increase in the gate voltage. Put another way, the falling slew-rate $dV_D/dt = I_G/C_{rss}$ is set by the gate current available to charge the feedback (Miller) capacitance C_{rss}. When V_{DS} reaches zero the gate resumes charging, now at a slower rate because the C_{rss} contribution to total gate capacitance is larger at $V_{DS} = 0$, see Figure 3.100). The MOSFET does not reach its intended low value of $R_{DS(ON)}$ until the gate attains its full drive voltage. Turn-off proceeds similarly. MOSFET datasheets include values for C_{iss} and C_{rss}, but the latter is typically at $V_{DS} = 25$ V, so you need to go to the datasheet plots of capacitances versus drain voltage.

¶T. MOSFET Gate Damage.

MOSFET gates typically have ±20 V to ±30 V maximum ratings (but as little as ±10 V for low-threshold types), beyond which the very thin metal-oxide gate-channel insulator can be permanently damaged, see Figure 3.105. Be sure to discharge static charge before installation of discrete MOSFETs and MOS ICs.

¶U. FET versus BJT for Power Switching.

See §3.5.4H; see also ¶Z below.

¶V. MOSFET Switch Polarity.

Both *n*- and *p*-channel polarities of MOSFETs can be used to switch a voltage, see Figure 3.106 where most of the circuits show a conventional approach with a *p*-channel FET switching a positive voltage. But circuit E shows an *n*-channel FET doing the same task, with an additional voltage source powering the gate (the better-performing *n*-channel FET is preferred if it can be easily used, see §3.1.2). Figure 3.107 illustrates the use of photodiodes to power the high-side gates, to make "floating" switches.

¶W. Power MOSFET Amplifiers.

Unlike bipolar power transistors, power MOSFETs have a wide safe-operating area (SOA) and do not suffer from second breakdown (see Figure 3.95), which is due to a localized thermal-runaway heating problem. Figure 3.119 shows typical class-AB biasing techniques necessary for use in linear power amplifiers.

¶X. Depletion-mode Power MOSFETs.

Although most power MOSFETs are enhancement-mode types, *n*-channel depletion-mode types are available; §3.5.6D shows some applications. See also Table 3.6 on page 210.

¶Y. Paralleling Power MOSFETs.

When used as switches, yes, but when used in power amplifiers, no, at least not without high-value source-ballast resistors! Figure 3.117B shows an elegant active-feedback workaround for use with regulator pass elements.

¶Z. IGBTs.

IGBTs are an alternative to power MOSFETs, see §3.5.7 where we show a comparison between power MOSFETs, IGBTs and BJTs. They're primarily useful at voltages above 300 V and switching rates below 100 kHz, though there are some nice IGBTs for use at RF, for example the IRGB4045, good for 150 W or more at 20 MHz.

Supplement: MOSFET gate drivers

In this chapter we introduced MOSFET gate drivers (see for example Figure 3.96), and we listed a good selection in Table 3.8 on page 218. These drivers are powered from a single positive rail (typically 4.5–18 V), accept logic-level input, and produce full-swing outputs with substantial peak current capability (up to 10 A for some parts) from their internal push-pull MOSFET output stage. They make cameo appearances later, e.g., in Figures 12.42B, 12.44H, and 15.10. Those figures illustrate some circuit basics, but there is more to know. Here we gather together some useful circuit tricks.[125]

Figure 3.121A shows the general case. Bypass capacitor C_1 is essential, given the substantial peak drive currents during MOSFET turn-on (e.g., driving a gate capacitance[126] C_{iss} of 1000 pF through 10 V in 10 ns requires one ampere); note the return path of both the driver and of C_1 to the MOSFET's source terminal. C_1 is usually 0.1 µF, which drops 0.1 V while charging the gate. Diode D_1 protects the driver from latchup, which can occur with very rapid drain slew-rates in combination with some source–terminal inductance; something like a 1N5819 (1 A Schottky) is good. The series gate resistors limit the transient gate current,

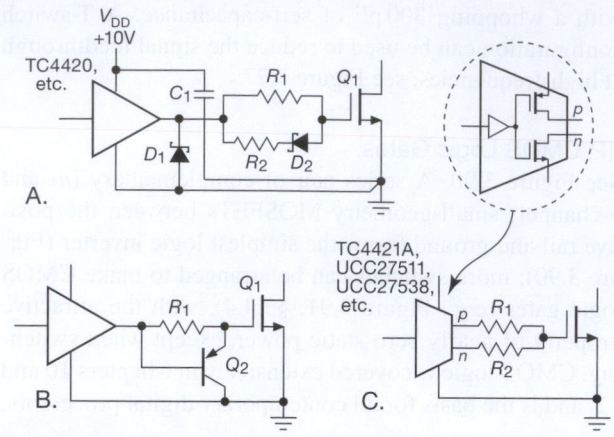

Figure 3.121. MOSFET gate driver configurations. A. Full configuration: capacitor C_1 supplies the transient sourcing current during turn-on, while D_1 protects the driver from dV/dt latchup; the gate resistors reduce turn-on and turn-off slew rates (R_2 is often omitted). B. Q_2 boosts turn-off gate drive. C. Driver with separate outputs eliminates need for D_2.

slowing the switching time; this is useful to reduce system transients when you don't need the fastest switching speeds. Typically R_2 is smaller than R_1 (and is often zero), for faster turn-off; typical resistor values are in the range of 5 Ω–100 Ω.

The circuits in Figures 3.121B and C show some variations: Transistor Q_2 in circuit B boosts gate sinking current for dV/dt protection, for example in bridge configurations where the upper MOSFET's slew rate can source enough current ($C_{rss} dV/dt$) to turn back on the lower MOSFET, thus creating rail-to-rail shoot-through; and circuit C shows how drivers with separate sourcing and sinking outputs[127] let you tailor source and sink gate resistances without an extra diode.

---

[125] This short section has been brought over from Chapter 3x.

[126] Better stated in terms of total gate charge, i.e., $Q_g = 10$ nC. See §3.5.4B and Figures 3.101 and 3.102.

[127] For example, TC4421/22, TC4431/32, TC4451/52, LM5114, MAX5048, MAX15070, ZXGD3002/04, and UCC27531/38.

OPERATIONAL AMPLIFIERS

CHAPTER 4

4.1 Introduction to op-amps – the "perfect component"

In the previous three chapters we learned about circuit design with "discrete components," both active and passive. Our basic building blocks were transistors, both bipolar (BJT) and field-effect (FET), along with the resistors, capacitors, and other components that are needed to set bias, couple and block signals, create load impedances, and so on.

With those tools we have gone quite far. We've learned how to design simple power supplies, signal amplifiers and followers, current sources, dc and differential amplifiers, analog switches, power drivers and regulators, and even some rudimentary digital logic.

But we've also learned to struggle with imperfections. Voltage amplifiers suffer from nonlinearity (a grounded-emitter amplifier with a 1 mV input signal has ~1% distortion), which you can trade off against voltage gain (by adding emitter degeneration); differential amplifiers have input unbalance, typically tens of millivolts (with bipolar transistors), ten times more with discrete junction-FETs (JFETs); in bipolar design you have to worry about input current (often substantial), and the ever-present V_{BE} and its variation with temperature; in FET design you trade absence of input current for unpredictability of V_{GS}; and so on.

We've seen hints that things can be better, in particular the remarkable linearizing effects of negative feedback (§2.5.3), and its ability to make overall *circuit* performance less dependent on *component* imperfections. It is negative feedback that gives the emitter-degenerated amplifier its linearity advantage over the grounded-emitter amplifier (at the cost of voltage gain). And in the high-loop-gain limit, negative feedback promises circuit performance largely independent of transistor imperfections.

Promised, but not yet delivered: the high-gain amplifier blocks we need to get high loop gain in a feedback arrangement still involve substantial design efforts – the hallmark of complex circuits implemented with discrete (as opposed to integrated) components.

With this chapter we enter the promised land! The op-amp is, essentially, a "perfect part": a complete integrated amplifier gain block, best thought of as a dc-coupled differential amplifier with single-ended output, and with extraordinarily high gain. It also excels in precise input symmetry and nearly zero input current. Op-amps are designed as "gain engines" for negative feedback, with such high gain that the circuit performance is set almost entirely by the feedback circuitry. Op-amps are small and inexpensive, and they should be the starting point for nearly every analog circuit you design. In most op-amp circuit designs we're in the regime where they are essentially perfect: with them we will learn to build nearly perfect amplifiers, current sources, integrators, filters, regulators, current-to-voltage converters, and a host of other modules.

Op-amps are our first example of *integrated circuits* – many individual circuit elements, such as transistors and resistors, fabricated and interconnected on a single "chip" of silicon.[1] Figure 4.1 shows some IC op-amp packaging schemes.

4.1.1 Feedback and op-amps

We first met negative feedback in Chapter 2, where we saw that the process of coupling the output back, in such a way as to cancel some of the input signal, improved characteristics such as linearity, flatness of response, and predictability. As we saw quantitatively, the more negative feedback that is used, the less the resultant amplifier characteristics depend on the characteristics of the open-loop (no-feedback) amplifier, ultimately depending only on the properties of the feedback network itself. Operational amplifiers are typically used in this *high-loop-gain* limit, with *open-loop* voltage gain (no feedback) of a million or so.

A feedback network can be frequency-dependent, to produce an equalization amplifier (for example the

---

[1] The first operational amplifiers were made with vacuum tubes, followed by implementations with discrete transistors. See §4x.1 for a description (with photo and schematics) of a once-popular vacuum-tube op-amp, the Philbrick K2-W.

Figure 4.1. Op-amps (and other linear ICs) come in a bewildering variety of "packages," most of which are represented in this photograph. Top row, left to right: 14-pin plastic dual in-line package (DIP), 8-pin plastic DIP ("mini-DIP"). Middle row: 14-pin thin-shrink small-outline package (TSSOP), 8-pin small-outline package (SO-8), 8-pin TSSOP ("μMAX"). Bottom row: 5-pin small-outline transistor package (SOT23), 6-ball chip-scale package (CSP – top and bottom views), 5-pin SC-70. The 14-pin packages hold quad op-amps (i.e., four independent op-amps), the 8-pin packages hold duals, and the rest are singles. (TSSOP and smaller packages courtesy of Travis Eichhorn, Maxim Semiconductor.)

treble and bass "tone control" stage of amplification that you find in most audio systems); or it can be amplitude-dependent, producing a nonlinear amplifier (a popular example is a logarithmic amplifier, built with feedback that exploits the logarithmic V_{BE} versus I_C of a diode or transistor). It can be arranged to produce a current source (near-infinite output impedance) or a voltage source (near-zero output impedance), and it can be connected to generate very high or very low input impedance. Speaking in general terms, the property that is sampled to produce feedback is the property that is improved. Thus, if you feed back a signal proportional to the output current, you will generate a good current source.

As we remarked in §2.5.1, feedback can be arranged intentionally to be *positive*, for example to make an oscillator, or, as we'll see later, to make a Schmitt trigger circuit. That's the *good* kind of positive feedback. The bad kind occurs, uninvited (and unwelcome), when a negative-feedback circuit is burdened with sufficient accumulated phase shifts at some frequency to produce overall positive feedback, and oscillations. This can occur for a variety of reasons. We'll discuss this important subject, and see how to prevent unwanted oscillations by *frequency compensation*, the topic of §4.9 at the end of the chapter.

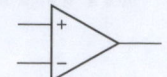

Figure 4.2. Op-amp symbol.

Having made these general comments, we now look at a few feedback examples with op-amps.

4.1.2 Operational amplifiers

The operational amplifier is a very high-gain dc-coupled differential amplifier with a single-ended output. You can think of the classic long-tailed pair (§2.3.8) with its two inputs and single output as a prototype, although real op-amps have much higher gain (typically 10^5 to 10^6) and lower output impedance, and they allow the output to swing through most or all of the supply range (you often use a split supply, for example ± 5 V). Operational amplifiers are available in literally thousands of types, with the universal symbol shown in Figure 4.2, where the $(+)$ and $(-)$ inputs do as expected: the output goes positive when the noninverting input $(+)$ goes more positive than the inverting input $(-)$, and vice versa. The $(+)$ and $(-)$ symbols don't mean that you have to keep one positive with respect to the other, or anything like that; they just tell you the relative phase of the output (which is important to keep negative feedback *negative*). Using the words "noninverting" and "inverting," rather than "plus" and "minus" helps avoid confusion. Power-supply connections are frequently not displayed, and there is no ground terminal. Operational amplifiers have enormous voltage gain, and they are *never* (well, hardly ever) used without feedback. Think of an op-amp as fodder for feedback. The open-loop gain is so high that, for any reasonable closed-loop gain, the characteristics depend on only the feedback network. Of course, at some level of scrutiny this generalization must fail. We will start with a naïve view of op-amp behavior and fill in some of the finer points later, when we need to.

There are literally thousands of different op-amps available, offering various performance tradeoffs that we will explain later (look ahead to Tables 4.2a,b, 5.5, or 8.3 if you want to see a small sample of what's available). A very good all-around performer is the popular LF411 ("411" for short), originally introduced by National Semiconductor. Like many op-amps, it is a wee beastie packaged in the so-called mini-DIP (dual in-line package) or SOIC (small-outline IC), and it looks as shown in Figure 4.3. It is inexpensive (less than $1) and easy to use; it comes in an improved grade (LF411A) and also in a version containing

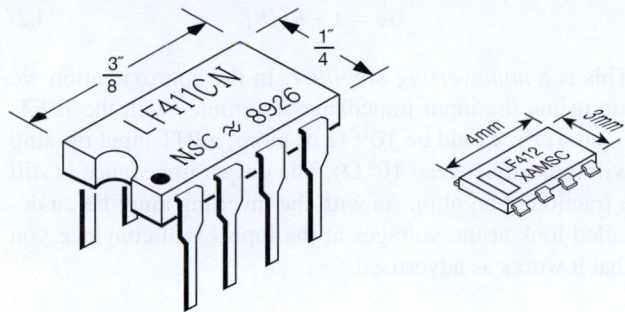

Figure 4.3. Mini-DIP and SOIC packages.

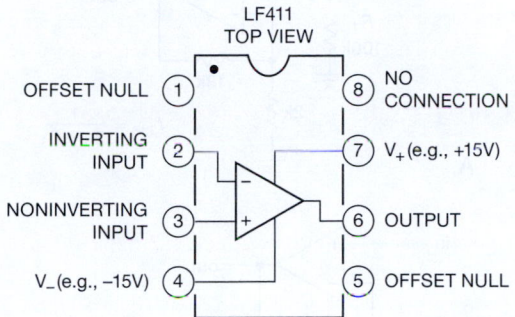

Figure 4.4. Pin connections for LF411 op-amp in 8-pin DIP.

two independent op-amps (LF412, called a "dual" op-amp). We will adopt the LF411/LF412 throughout this chapter as our "standard" op-amp, and we recommend it (or perhaps the versatile LMC6482) as a good starting point for your circuit designs.

Inside the 411 is a piece of silicon containing 24 transistors (21 BJTs, 3 FETs), 11 resistors, and 1 capacitor. (You can look ahead to Figure 4.43 on page 243 to see a simplified circuit diagram of its innards.) The pin connections are shown in Figure 4.4. The dot in the upper-left-hand corner, or notch at the end of the package, identifies the end from which to begin counting the pin numbers. As with most electronic packages, you count pins counterclockwise, viewing from the top. The "offset null" terminals (also known as "balance" or "trim") have to do with correcting (externally) the small asymmetries that are unavoidable when making the op-amp. More about this later in the chapter.

4.1.3 The golden rules

Here are the simple rules for working out op-amp behavior with external negative feedback. They're good enough for almost everything you'll ever do.

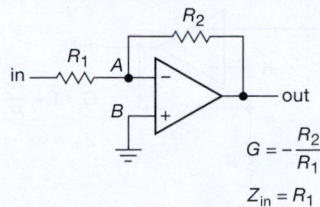

$$G = -\frac{R_2}{R_1}$$

$$Z_{in} = R_1$$

Figure 4.5. Inverting amplifier.

First, the op-amp voltage gain is so high that a fraction of a millivolt between the input terminals will swing the output over its full range, so we ignore that small voltage and state golden rule I.

I. The output attempts to do whatever is necessary to make the voltage difference between the inputs zero.

Second, op-amps draw very little input current (about 50 pA for the inexpensive JFET-input LF411, and often less than a picoamp for MOSFET-input types); we round this off, stating golden rule II.

II. The inputs draw no current.

One important note of explanation: golden rule I doesn't mean that the op-amp actually changes the voltage at its *inputs*. It can't do that. (How could it, and be consistent with golden rule II?) What it does is "look" at its input terminals and swing its output terminal around so that the external feedback-network brings the input differential to zero (if possible).

These two rules get you quite far. We illustrate with some basic and important op-amp circuits, and these will prompt a few cautions listed in §4.2.7.

4.2 Basic op-amp circuits

4.2.1 Inverting amplifier

Let's begin with the circuit shown in Figure 4.5. The analysis is simple, if you remember your golden rules.

1. Point B is at ground, so rule I implies that point A is also.
2. This means that (a) the voltage across R_2 is V_{out} and (b) the voltage across R_1 is V_{in}.
3. So, using rule II, we have $V_{out}/R_2 = -V_{in}/R_1$.

In other words, the voltage gain ($G_V \equiv V_{out}/V_{in}$) is

$$G_V = -R_2/R_1 \qquad (4.1)$$

Later you will see that it's sometimes better not to ground B directly, but through a resistor – but don't worry about that now.

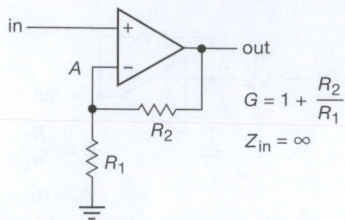

Figure 4.6. Noninverting amplifier.

Our analysis seems almost too easy! In some ways it obscures what is actually happening. To understand how feedback works, just imagine some input level, say +1 volt. For concreteness, imagine that R_1 is 10k and R_2 is 100k. Now, suppose the output decides to be uncooperative, and sits at zero volts. What happens? R_1 and R_2 form a voltage divider, holding the inverting input at +0.91 volts. The op-amp sees an enormous input unbalance, forcing the output to go negative. This action continues until the output is at the required −10.0 volts, at which point both op-amp inputs are at the same voltage, namely ground. Similarly, any tendency for the output to go more negative than −10.0 volts will pull the inverting input below ground, forcing the output voltage to rise.

What is the input impedance? Simple. Point A is always at zero volts (it's called a *virtual ground*). So $Z_{in} = R_1$. At this point you don't yet know how to figure the output impedance; for this circuit, it's a fraction of an ohm.

Note that this analysis is true even for dc – it's a dc amplifier. So if you have a signal source that has a dc offset from ground (collector of a previous stage, for instance), you may want to use a coupling capacitor (sometimes called a blocking capacitor, since it blocks dc but couples the signal). For reasons you will see later (having to do with departures of op-amp behavior from the ideal), it is usually a good idea to use a blocking capacitor if you're interested only in ac signals anyway.

This circuit is known as an *inverting amplifier*. Its one undesirable feature is the low input impedance, particularly for amplifiers with large (closed-loop) voltage gain, where R_1 tends to be rather small. That is remedied in the next circuit (Figure 4.6).

4.2.2 Noninverting amplifier

Consider Figure 4.6. Again, the analysis is simplicity itself:

$$V_A = V_{in}.$$

But V_A comes from a voltage divider: $V_A = V_{out}R_1/(R_1 + R_2)$. Set $V_A = V_{in}$, and you get a voltage gain of

$$G_V = 1 + R_2/R_1. \tag{4.2}$$

This is a *noninverting amplifier*. In the approximation we are using, the input impedance is infinite (with the JFET-input 411 it would be $10^{12}\,\Omega$ or more; a BJT-input op-amp will typically exceed $10^8\,\Omega$). The output impedance is still a fraction of an ohm. As with the inverting amplifier, a detailed look at the voltages at the inputs will convince you that it works as advertised.

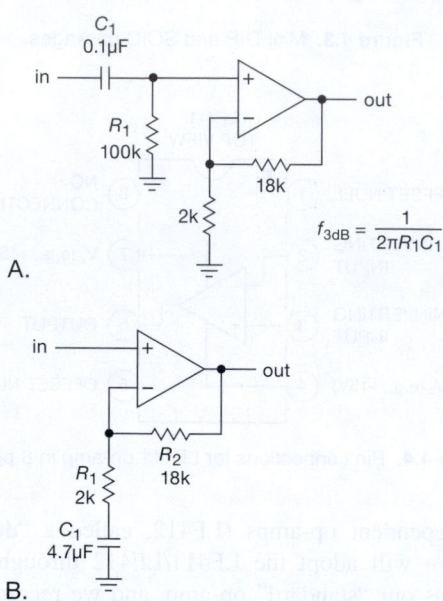

Figure 4.7. Amplifiers for ac signals: A. ac-coupled noninverting amplifier, B. blocking capacitor rolls off the gain to unity at dc.

A. An ac amplifier

The basic noninverting amplifier, like the inverting amplifier earlier, is a dc amplifier. If the signal source is ac-coupled, you must provide a return to ground for the (very small) input current, as in Figure 4.7A. The component values shown give a voltage gain of 10 and a low-frequency 3 dB point of 16 Hz.

If only ac signals are being amplified, it is often a good idea to "roll off" the gain to unity at dc, especially if the amplifier has large voltage gain, to reduce the effects of finite "input offset voltage" (§4.4.1A). The circuit in Figure 4.7B has a low-frequency 3 dB point of 17 Hz, the frequency at which the impedance of the capacitor C_1 equals R_1, or 2.0k. Note the large capacitor value required. For noninverting amplifiers with high gain, the capacitor in this ac amplifier configuration may be undesirably large. In that

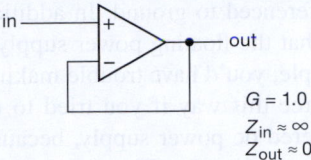

Figure 4.8. Op-amp follower.

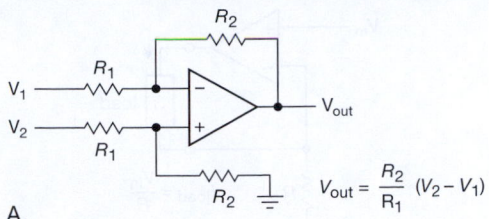

A.

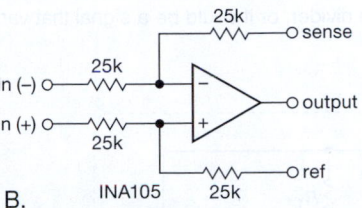

B.

Figure 4.9. Classic difference amplifier: A. Op-amp with matched resistor ratios. B. Integrated version, with uncommitted "sense" and "reference" pins. In the best grade (INA105A) the resistor ratio is matched to better than 0.01%, with a temperature coefficient better than 5 ppm/°C.

case it may be preferable to omit the capacitor and trim the offset voltage to zero, as we will discuss later. An alternative is to raise R_1 and R_2, perhaps using a T network for the latter (Figure 4.66 on page 259).

In spite of its desirable high input impedance, the noninverting amplifier configuration is not necessarily to be preferred over the inverting amplifier configuration in all circumstances. As we will see later, the inverting amplifier puts less demand on the op-amp, and therefore gives somewhat better performance. In addition, its virtual ground provides a handy way to combine several signals without interaction. Finally, if the circuit in question is driven from the (stiff) output of another op-amp, it makes no difference whether the input impedance is 10k (say) or infinity, because the previous stage has no trouble driving it in either case.

4.2.3 Follower

Figure 4.8 shows the op-amp version of an emitter follower. It is simply a noninverting amplifier with R_1 infinite and R_2 zero (gain $= 1$). An amplifier of unity gain is sometimes called a *buffer* because of its isolating properties (high input impedance, low output impedance).

4.2.4 Difference amplifier

The circuit in Figure 4.9A is a *difference amplifier* (sometimes called a *differential amplifier*) with gain R_2/R_1. This circuit requires precise resistor matching to achieve high common-mode rejection ratios (CMRR). You may be lucky and find a batch of 100k 0.01% resistors at an electronics flea market or surplus outlet; otherwise you can buy precision resistor *arrays*, with close matching of ratios and temperature coefficients.[2] All your difference amplifiers

will have unity gain, but that's easily remedied with further (single-ended) stages of gain. If you can't find good resistors (or even if you can!), you should know that you can buy this circuit as a conveniently packaged difference amplifier, with well-matched resistors; examples are the INA105 or AMP03 ($G = 1$), INA106 ($G=10$ or 0.1), and INA117 or AD629 ($G = 1$ with input dividers; input signals to ±200 V) from TI/Burr-Brown and Analog Devices (many more are listed in Table 5.7 on page 353). The unity-gain INA105 configuration is shown in Figure 4.9B, with its uncommitted "sense" and "reference" pins. You get the classic difference amplifier by connecting *sense* to the output and *ref* to ground. But the additional flexibility lets you make all sorts of nifty circuits, such as a precision unity-gain inverter, noninverting gain-of-2 amplifier, and noninverting gain-of-0.5 amplifier. We treat difference amplifiers in greater detail in §5.14.

Exercise 4.1. Show how to make these three circuits with an INA105.

There are, in addition, more sophisticated differential amplifier configurations, known officially as "instrumentation amplifiers"; they are discussed in detail in §§5.15 and 5.16, along with a listing in Table 5.8 on page 363.

---

[2] For example, the BI Technologies type 664 thin-film quad (four resistors of the same value) in an 8-lead surface-mount IC package (SOIC); these come in accuracies to 0.1%, ratio tracking to 0.05%, and tracking temperature coefficients to ±5 ppm/°C. They are inexpensive (about $2 for the best grade), and available from Mouser Electronics, among others. Companies like Vishay have offerings with astonishingly good performance: their best resistor arrays specify worst-case ratio tracking to 0.001%, and tracking temperature coefficient (tempco) to ±0.1 ppm/°C.

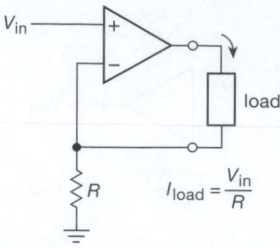

Figure 4.10. Basic op-amp current source (floating load). V_{in} might come from a voltage divider, or it could be a signal that varies with time.

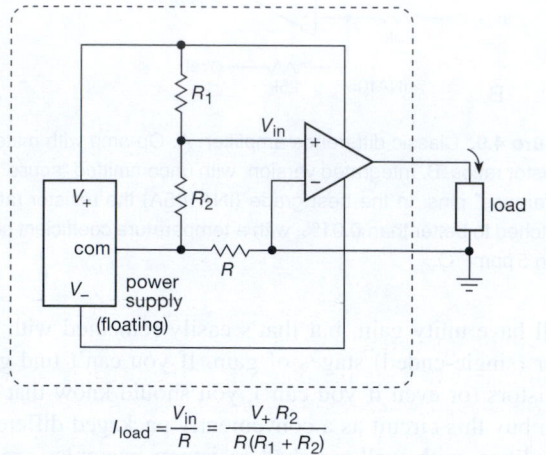

$$I_{load} = \frac{V_{in}}{R} = \frac{V_+ R_2}{R(R_1 + R_2)}$$

Figure 4.11. Current source with grounded load and floating power supply.

4.2.5 Current sources

The circuit in Figure 4.10 approximates an ideal current source, without the V_{BE} offset of a transistor current source. Negative feedback results in V_{in} at the inverting input, producing a current $I = V_{in}/R$ through the load. The major disadvantage of this circuit is the "floating" load (neither side grounded). You couldn't generate a usable sawtooth wave with respect to ground with this current source, for example. One solution is to float the whole circuit (power supplies and all) so that you can ground one side of the load (Figure 4.11). The circuit in the box is the previous current source, with its power supplies shown explicitly. R_1 and R_2 form a voltage divider to set the current. If this circuit seems confusing, it may help to remind yourself that "ground" is a relative concept. Any one point in a circuit could be called ground. This circuit is useful for generating currents into a load that is returned to ground, but it has the disadvantage that the control input is now floating, so you cannot program the output current with an in-

put voltage referenced to ground. In addition, you've got to make sure that the floating power supply is truly floating – for example, you'd have trouble making a microamp dc current source this way if you tried to use a standard wall-plug-powered dc power supply, because capacitance between windings in its transformer would introduce reactive currents, at the 60 Hz line frequency, that might well exceed the desired microamp output current; one possible solution would be to use batteries. Some other approaches to this problem are presented in Chapter 9 (§9.3.14) in the discussion of constant-current power supplies.[3]

A. Current sources for loads returned to ground

With an op-amp and external transistor it is possible to make a simple high-quality current source for a load returned to ground; a little additional circuitry makes it possible to use a programming input referenced to ground (Figure 4.12). In the first circuit, feedback forces a voltage $V_{CC} - V_{in}$ across R, giving an emitter current (and therefore an output current) $I_E = (V_{CC} - V_{in})/R$. There are no V_{BE} offsets, or their variations with temperature, with I_C, with V_{CE}, etc., to worry about. The current source is imperfect (ignoring op-amp errors: I_B, V_{OS}) only insofar as the small base current may vary somewhat with V_{CE} (assuming the op-amp draws no input current), not too high a price to pay for the convenience of a grounded load; a Darlington for Q_1 would reduce this error considerably. This error comes about, of course, because the op-amp stabilizes the *emitter* current, whereas the load sees the *collector* current. A variation of this circuit, using a MOSFET instead of a bipolar transistor, avoids this problem altogether, since FETs draw no dc gate current (but large power MOSFETs have plenty of input capacitance, which can cause problems; see the comment at the end of this subsection).

With this circuit the output current is proportional to the voltage drop below V_{CC} applied to the op-amp's noninverting input; in other words, the programming voltage is referenced to V_{CC}, which is fine if V_{in} is a fixed voltage generated by a voltage divider, but an awkward situation if an external input is to be used. This is remedied in the second circuit, in which a similar current source with an *npn* transistor is used to convert an input voltage (referenced to ground) to a V_{CC}-referenced input to the final current

---

[3] Another limitation of op-amp current-source circuits is their degraded performance at higher frequencies: an op-amp's output is inherently low impedance, (typically a push–pull follower, with $R_{out} \sim 100\,\Omega$, look ahead to Figure 4.43), so a current-source circuit must rely on feedback (which declines with increasing frequency) to raise the op-amp's output impedance. See further discussion in §§4.2.5B, and 4.4.4.

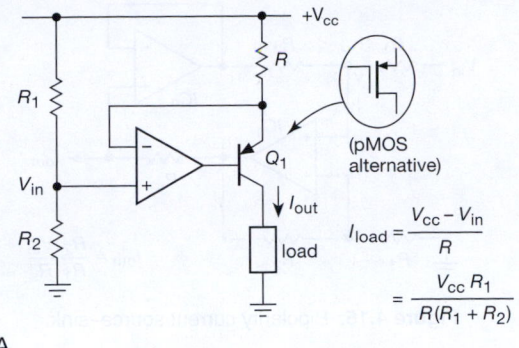

$$I_{\text{load}} = \frac{V_{\text{cc}} - V_{\text{in}}}{R}$$

$$= \frac{V_{\text{cc}} R_1}{R(R_1 + R_2)}$$

A.

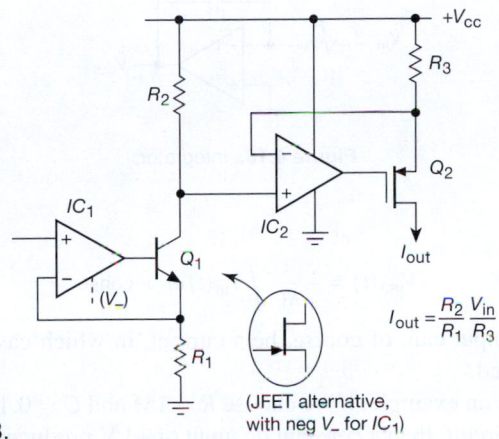

$$I_{\text{out}} = \frac{R_2}{R_1} \frac{V_{\text{in}}}{R_3}$$

B.

Figure 4.12. Current sources for grounded loads that don't require a floating power supply. The op-amps may need to have rail-to-rail input and output capability (RRIO); see text.

source; for the latter we've used a *p*-channel MOSFET for variety (and to eliminate the small base-current error you get with bipolar transistors). Op-amps and transistors are inexpensive. Don't hesitate to use a few extra components to improve performance or convenience in circuit design.

One important note about these circuits: at low output currents the voltage across the emitter (or source) resistors may be quite small, which means that the op-amps must be able to operate with their inputs near or at the positive supply voltage. For example, in the circuit of Figure 4.12B IC_2 needs to operate with its inputs close to the positive supply rail. Don't assume that a given op-amp will do this, without explicit permission from the datasheet! The LF411's datasheet waffles a bit on this, but grudgingly admits that it will work, albeit with degraded performance, with the inputs at the positive rail. (It will not, however, work down to the negative rail; but with IC_1 powered from split supply voltages there's no problem there.) By contrast, op-amps like the LMC7101 or LMC6482 *guarantee* proper

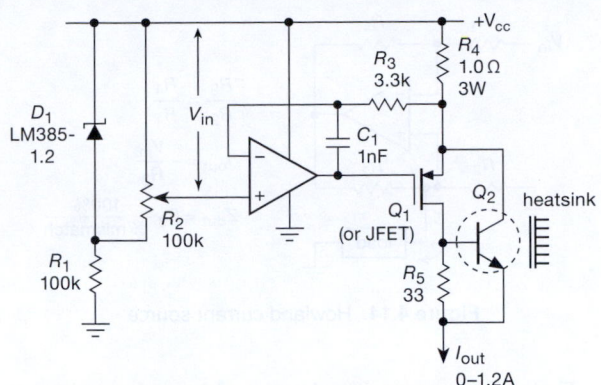

Figure 4.13. FET–bipolar current source suitable for high currents.

operation all the way to (and a bit beyond) the positive rail (see the "Swing to supplies?" column in Table 4.2a on page 271). Alternatively, the op-amp could be powered from a separate V_+ voltage higher than V_{CC}.

Exercise 4.2. What is the output current in the last circuit for a given input voltage V_{in}? (Did we get it right in the figure?)

Figure 4.13 shows an interesting variation on the op-amp–transistor current source. Although you can get plenty of current with a simple power MOSFET, the high inter-electrode capacitances of high-current FETs may cause problems. When a relatively low-current MOSFET[4] is combined with a high-current *npn* power transistor, this circuit has the advantage of zero base current error (which you get with FETs) along with much smaller input capacitance. In this circuit, which is analogous to the "complementary Darlington" (or Sziklai circuit; see §2.4.2A), bipolar transistor Q_2 kicks in when the output current exceeds about 20 mA.

Lest we leave the wrong impression, we emphasize that the simpler MOSFET-only circuit (in the manner of Figure 4.12B) is a preferable configuration, given the major drawback of power BJTs, namely their susceptibility to "second breakdown" and consequent limit on safe operating area (as we saw in §3.5.1B, see particularly Figure 3.95). Big power MOSFETs have large input capacitance, so in such a circuit you should use a network like Figure 4.13's $R_3 C_1$ to prevent oscillation.

B. Howland current source
Figure 4.14 shows a nice "textbook" current source. If the resistors are chosen so that $R_3/R_2 = R_4/R_1$, then it can be shown that $I_{\text{load}} = -V_{\text{in}}/R_2$.

Exercise 4.3. Show that the preceding result is correct.

---

[4] Such as a BS250P or BSS84, see Table 3.4a on page 188.

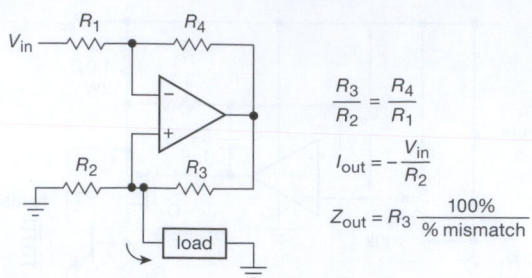

Figure 4.14. Howland current source.

$$\frac{R_3}{R_2} = \frac{R_4}{R_1}$$

$$I_{out} = -\frac{V_{in}}{R_2}$$

$$Z_{out} = R_3 \frac{100\%}{\% \text{ mismatch}}$$

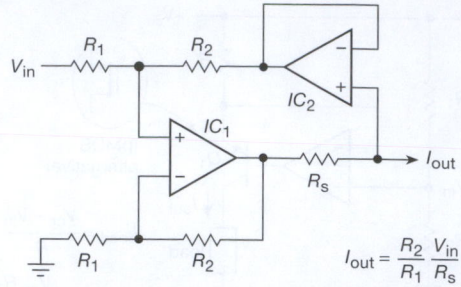

Figure 4.15. Bipolarity current source–sink.

$$I_{out} = \frac{R_2}{R_1} \frac{V_{in}}{R_s}$$

This sounds great, but there's a hitch: the resistor ratios must be matched exactly; otherwise it isn't a perfect current source. Even so, its performance is limited by the op-amp's common-mode rejection ratio (CMRR, §2.3.8). For large output currents the resistors must be small, and the compliance is limited. Also, at high frequencies (where the loop gain is low, as we'll learn shortly) the output impedance can drop from the desired value of infinity to as little as a few hundred ohms (the op-amp's open-loop output impedance). These drawbacks limit the applicability of this clever circuit.

You can convert this circuit into a noninverting current source by grounding R_1 (where V_{in} is shown) and applying the control input voltage V_{in} instead to R_2.

Figure 4.15 is a nice improvement on the Howland circuit, because the output current is sourced through a sense resistor R_s whose value you can choose independently of the matched resistor array (with resistor pairs R_1 and R_2). The best way to understand this circuit is to think of IC_1 as a difference amplifier whose output *sense* and *reference* connections sample the drop across R_s (i.e., the current); the latter is buffered by follower IC_2 so there is no current error.

For this configuration you can exploit the internal precision matched resistors in an integrated difference amplifier: use something like an INA106 for R_1, R_2, and IC_1, wired "backwards" (for G=0.1) to reduce the drop across the sense resistor. See §5.14 and Table 5.7 on page 353.

4.2.6 Integrators

Op-amps allow you to make nearly perfect integrators, without the restriction that $V_{out} \ll V_{in}$. Figure 4.16 shows how it's done. Input current V_{in}/R flows through C. Because the inverting input is a virtual ground, the output voltage is given by

$$V_{in}/R = -C(dV_{out}/dt)$$

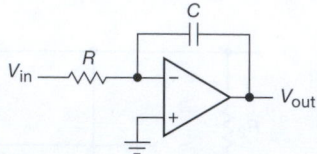

Figure 4.16. Integrator.

or

$$V_{out}(t) = -\frac{1}{RC} \int V_{in}(t)\, dt + \text{const.} \qquad (4.3)$$

The input can, of course, be a current, in which case R is omitted.

As an example, if we choose $R = 1M$ and $C = 0.1\,\mu\text{F}$ in this circuit, then a constant dc input of $+1$ V produces $1\,\mu\text{A}$ of current into the summing junction, hence an output voltage that is ramping downward at $dV_{out}/dt = -V_{in}/RC = -10$ V/s. To say it algebraically, for a constant V_{in} or constant I_{in},

$$\Delta V_{out} = -\frac{V_{in}}{RC}\Delta t = -\frac{I_{in}}{C}\Delta t.$$

We rigged up the integrator of Figure 4.16, with $R = 1\,\text{M}\,\Omega$ and $C = 1$ nF, and drove it with the simple test waveform shown in Figure 4.17. Without having taken a math class, the thing knows calculus!

Sharp-eyed readers may have noticed that this circuit doesn't have any feedback at dc, and so there's no way for it to have a stable quiescent point: for *any* nonzero input voltage V_{in}, the output is going *somewhere!* As we'll see shortly, even with V_{in} exactly at zero volts, the output tends to wander off, owing to op-amp imperfections (non-zero input current, and "offset voltage"). These latter problems can be minimized by careful choice of op-amp and circuit values; but even so you usually have to provide some way to reset the integrator. Figure 4.18 shows how this is commonly done, either with a reset switch (both discrete JFET and integrated CMOS analog switch examples are shown)

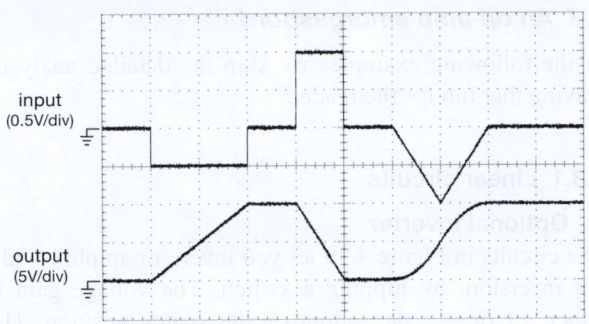

Figure 4.17. Integrator waveforms. The output can go anywhere it wants to, unlike our simple *RC* "integrator" of §1.4.4. Horizontal: 10 ms/div.

or with a large-value feedback resistor across the integrating capacitor. Closing a reset switch[5] (Figures 4.18A,B) zeroes the integrator by rapidly discharging the capacitor, while allowing perfect integration when open. The use of a feedback resistor (Figure 4.18D) produces stable biasing by restoring feedback at dc (where the circuit behaves like a high-gain inverting amplifier), but the effect is to roll off the integrator action at very low frequencies, $f < 1/R_f C$. An additional series analog switch at the input (Figure 4.18C) lets you control the intervals during which the integrator is active; when that switch is open the integrator output is frozen at its last value.

You don't have to worry about zeroing the integrator, of course, if it's part of a larger circuit that does the right thing. We'll see a beautiful example shortly (§4.3.3), namely an elegant triangle-wave generator, in which an untamed integrator is just what you want.

This first look at the op-amp integrator assumes that the op-amp is perfect, in particular that (a) the inputs draw no current, and (b) the amplifier is balanced with both inputs at precisely the same voltage. When our op-amp honeymoon is over we'll see that real op-amps do have some input current (called "bias current," I_B), and that they exhibit some voltage imbalance (called "offset voltage," V_{OS}). These imperfections are not large – bias currents of picoamps are routine, as are offset voltages of less than a millivolt – but they can cause problems with circuits like integrators, in which the effect of a small error grows with time. We'll deal with these essential topics later in the chapter (§4.4), after you're comfortable with the basics.

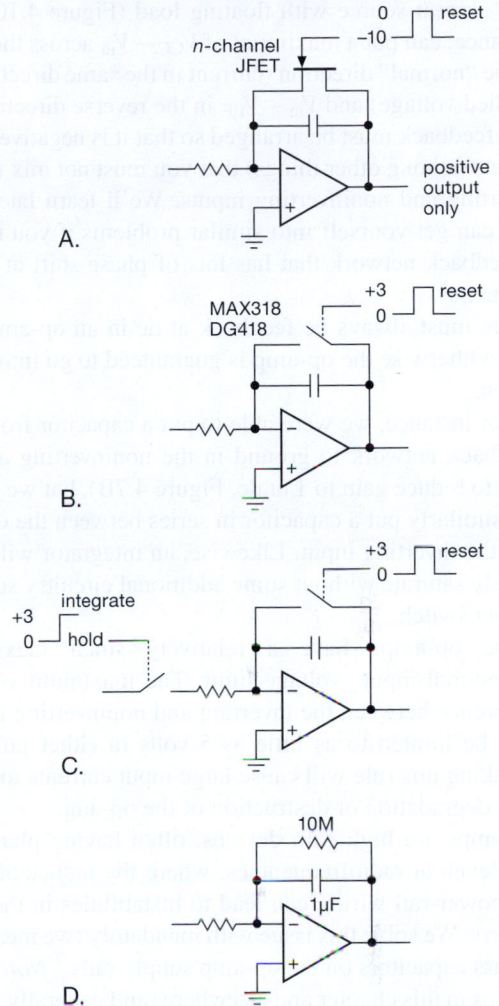

Figure 4.18. Op-amp integrators with reset switches.

4.2.7 Basic cautions for op-amp circuits

- In all op-amp circuits, golden rules I and II (§4.1.3) are obeyed only if the op-amp is in the active region, i.e., inputs and outputs are not saturated at one of the supply voltages.

  For instance, overdriving one of the amplifier configurations will cause output clipping at output swings near V_{CC} or V_{EE}. During clipping, the inputs will no longer be maintained at the same voltage. The op-amp output cannot swing beyond the supply voltages (typically it can swing only to within 2 V of the supplies, though certain op-amps are designed to swing all the way to one supply or the other, or to both; the latter are known as "rail-to-rail output" op-amps). Likewise, the output compliance of an op-amp current source is set by the same limitation.

---

[5] Refer back to §3.4 for a detailed discussion of FET switches.

The current source with floating load (Figure 4.10), for instance, can put a maximum of $V_{CC} - V_{in}$ across the load in the "normal" direction (current in the same direction as applied voltage) and $V_{in} - V_{EE}$ in the reverse direction.[6]

- The feedback must be arranged so that it is negative. This means (among other things) that you must not mix up the inverting and noninverting inputs. We'll learn later that you can get yourself into similar problems if you rig up a feedback network that has lots of phase shift at some frequency.

- There must always be feedback at dc in an op-amp circuit. Otherwise the op-amp is guaranteed to go into saturation.

  For instance, we were able to put a capacitor from the feedback network to ground in the noninverting amplifier (to reduce gain to 1 at dc, Figure 4.7B), but we could not similarly put a capacitor in series between the output and the inverting input. Likewise, an integrator will ultimately saturate without some additional circuitry such as a reset switch.

- Some op-amps have a relatively small maximum differential-input voltage limit. The maximum voltage difference between the inverting and noninverting inputs may be limited to as little as 5 volts in either polarity. Breaking this rule will cause large input currents to flow, with degradation or destruction of the op-amp.

- Op-amps are high-gain devices, often having plenty of gain even at radiofrequencies, where the inductances in the power-rail wiring can lead to instabilities in the amplifiers. We solve this issue with mandatory (we mean it!) bypass capacitors on the op-amp supply rails.[7] *Note*: The figures in this chapter and elsewhere (and generally in the real world) do not show bypass capacitors, for simplicity. You have been warned.

We take up some more issues of this type in §4.4, and again in Chapter 5 in connection with precision circuit design.

---

[6] The load could be rather strange, e.g., it might contain batteries, requiring the reverse sense of voltage to get a forward current; the same thing might happen with an inductive load driven by changing currents.

[7] When we were young we were taught that each op-amp needed its own set of bypass capacitors. But with experience we've come to realize that one pair of capacitors can work to stabilize nearby op-amps. Furthermore, local wiring inductance with multiple sets of bypass capacitors can lead to resonances, which allow one op-amp to interfere with another. For example, if $L=25$ nH and $C=0.01\,\mu$F, then $f_{LC}=10$ MHz, and $X_{LC}=1.6\,\Omega$. The impedance peak at resonance will be Q times higher. You can solve this problem by adding an additional parallel lossy bypass capacitor, such as a small electrolytic. Its equivalent series resistance, of order $0.5\,\Omega$ or more, acts to damp the resonant Q.

4.3 An op-amp smorgasbord

In the following examples we skip the detailed analysis, leaving that fun for the reader.

4.3.1 Linear circuits

A. Optional inverter

The circuits in Figure 4.19 let you invert, or amplify without inversion, by flipping a switch. The voltage gain is either $+1$ or -1, depending on the switch position. The "switches" can be CMOS analog switches[8], which let you control the sense of inversion with a (digital) signal. The clever variation of Figure 4.20 lets you vary the gain continuously from follower to inverter. And when the pot R_1 is at mid-position, the circuit does nothing at all!

Exercise 4.4. Show that the circuits in Figure 4.19 work as advertised.

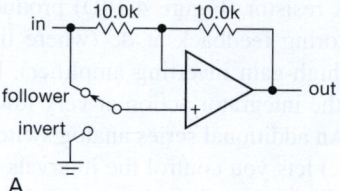

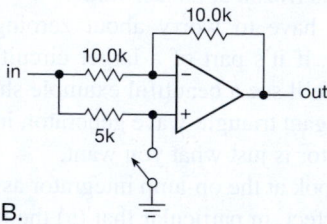

Figure 4.19. Optional inverters; $G = \pm 1.0$

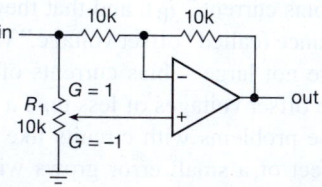

Figure 4.20. Follower-to-inverter: continuously adjustable gain from $G = +1$ to $G = -1$.

---

[8] For example the ADG419 or MAX319 ±20 V SPDT switches in convenient 8-pin packages, see §3.4 and Table 3.3 on page 176.

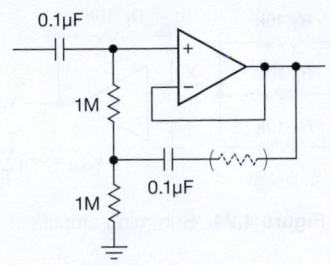

Figure 4.21. Op-amp follower with bootstrap.

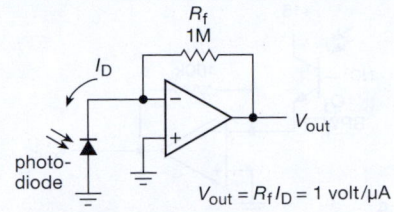

Figure 4.22. Photodiode amplifier.

B. Follower with bootstrap

As with transistor amplifiers, the bias path can compromise the high input impedance you would otherwise get with an op-amp, particularly with ac-coupled inputs, for which a resistor to ground is mandatory. If that is a problem, the bootstrap circuit shown in Figure 4.21 is a possible solution. As in the transistor bootstrap circuit (§2.4.3), the $0.1\,\mu F$ capacitor makes the upper 1M resistor look like a high-impedance current source to input signals. The low-frequency rolloff for this circuit will begin at about $10\,Hz$, dropping at 12 dB per octave for frequencies somewhat below this.[9] This circuit may exhibit some frequency peaking, analogous to the Sallen-and-Key circuit of §4.3.6; this can be tamed by adding a resistor of 1–10k in series with the feedback capacitor.

The very low input current (and therefore high input impedance) of FET-input op-amps generally make bootstrapping unnecessary; you can use 10 M or larger resistors for the input bias path in ac-coupled amplifiers.

C. Ideal current-to-voltage converter

Remember that the humble resistor is the simplest *I*-to-*V* converter. However, it has the disadvantage of presenting a nonzero impedance to the source of input current; this can be fatal if the device providing the input current has very little compliance or does not produce a constant current as the output voltage changes. A good example is a *photovoltaic cell*, a diode junction that has been optimized as a light detector. Even the garden-variety signal diodes you use in circuits have a small photovoltaic effect (there are amusing stories of bizarre circuit behavior finally traced to this effect). Figure 4.22 shows the good way to convert current to voltage while holding the input strictly at ground. The inverting input is a virtual ground; this is fortunate, because a photovoltaic diode can generate only a few tenths

of a volt. This particular circuit has an output of 1 volt per microamp of input current. (With BJT-input op-amps you sometimes see a resistor connected between the noninverting input and ground; its function will be explained shortly in connection with op-amp shortcomings.)

Of course, this *transresistance* configuration can be used equally well for devices that source their current from some positive excitation voltage, such as V_{CC}. Photodiodes and phototransistors (both devices that source current from a positive supply when exposed to light) are often used this way (Figure 4.23). The photodiode has lower photocurrent, but excels in linearity and speed; very fast photodiodes can operate at *giga*hertz speeds. By contrast, the phototransistor has a considerably higher photocurrent (owing to transistor beta, which boosts the native collector-to-base photocurrent), with poorer linearity and speed. You can even get photo-Darlingtons, which extend this trend.

In real-world applications it is usually necessary to include a small capacitor across the feedback resistor, to ensure stability (i.e., prevent oscillation or ringing). This is because the capacitance of the detector, in combination with the feedback resistor, forms a lowpass filter; the resulting lagging phase shift at high frequencies, combined with the op-amp's own lagging phase shift (see §4.9.3), can add up to $180°$, thus producing overall *positive* feedback, and thus oscillation. We treat this interesting problem in some detail in Chapter *4x* ("Transresistance amplifiers"); be sure to read that section carefully if you are building amplifiers for photodiodes. (And analogous stability problems occur, for similar reasons, when you drive capacitive loads with op-amps; see §4.6.1B.)

Exercise 4.5. Use a 411 and a 1 mA (full scale) meter to construct a "perfect" current meter (i.e., one with zero input impedance) with 5 mA full scale. Design the circuit so that the meter will never be driven more than $\pm 150\%$ full scale. Assume that the 411 output can swing to ± 13 volts (± 15 V supplies) and that the meter has $500\,\Omega$ internal resistance.

---

[9] You might be tempted to reduce the input coupling capacitor since its load has been bootstrapped to high impedance. However, this can generate a peak in the frequency response, in the manner of an active filter (see §6.3).

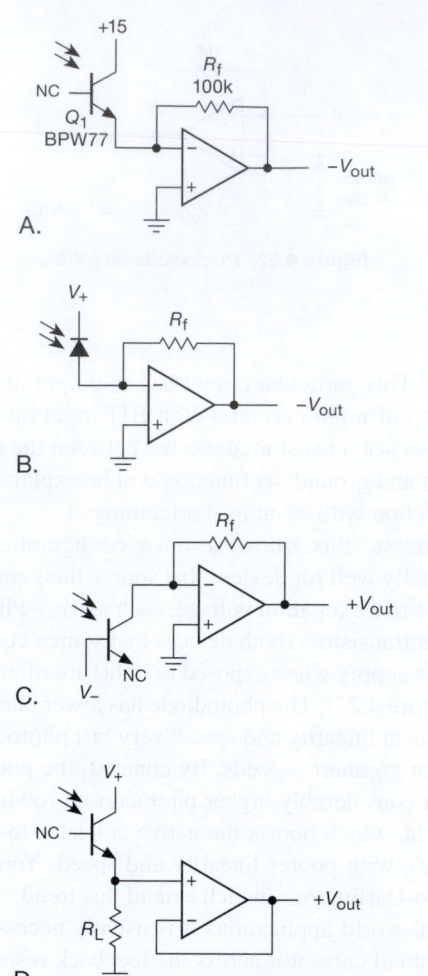

Figure 4.23. Photodiode amplifiers with reverse bias: A. Photo-transistor; note base terminal is not used. B. Photodiode. C. Phototransistor used as photodiode; for variety we show it current *sinking*. D. Phototransistor with load resistor driving voltage follower.

D. Summing amplifier

The circuit shown in Figure 4.24 is just a variation of the inverting amplifier. Point X is a virtual ground, so the input current is $V_1/R_1 + V_2/R_2 + V_3/R_3$. With equal resistor values you get $V_{out} = -(V_1 + V_2 + V_3)$. Note that the inputs can be positive or negative. Also, the input resistors need not be equal; if they're unequal, you get a weighted sum. For instance, you could have four inputs, each of which is +1 volt or zero, representing binary values 1, 2, 4, and 8. By using input resistors of 10k, 5k, 2.5k, and 1.25k, you will get a negative output in volts equal to the binary count input. This scheme can be easily expanded to several dig-

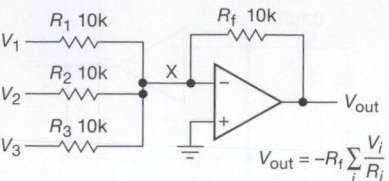

Figure 4.24. Summing amplifier.

its. It is the basis of digital-to-analog conversion, although a different input circuit (an $R–2R$ ladder) is usually used.

Exercise 4.6. Show how to make a two-digit digital-to-analog converter (DAC) by appropriately scaling the input resistors in a summing amplifier. The digital input represents two digits, each consisting of four lines that represent the values 1, 2, 4, and 8 for the respective digits. An input line is either at +1 volt or at ground, i.e., the eight input lines represent 1, 2, 4, 8, 10, 20, 40, and 80. With ±15 V supplies, the op-amp's outputs generally cannot swing beyond ±13 volts; you will have to settle for an output in volts equal to one-tenth the value of the input number.

E. Power booster

For high output current, a power transistor follower can be hung on an op-amp output (Figure 4.25). In this case a noninverting amplifier has been drawn, though a follower can be added to any op-amp configuration. Notice that feedback is taken from the emitter; thus feedback enforces the desired output voltage in spite of the V_{BE} drop. This circuit has the usual problem that the follower output can only *source* current. As with transistor circuits, the remedy is a push–pull booster (Figure 4.26). We'll see later that the limited speed with which the op-amp can move its output (slew rate) seriously limits the speed of this booster in the crossover region, creating distortion. For slow-speed applications you don't need to bias the push–pull pair into quiescent conduction, because feedback will take care of most of the crossover distortion. Complete power booster ICs are available, e.g. the LT1010 and BUF633/4. These

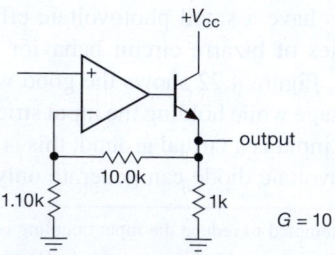

Figure 4.25. Single-ended emitter follower boosts op-amp output current (sourcing only).

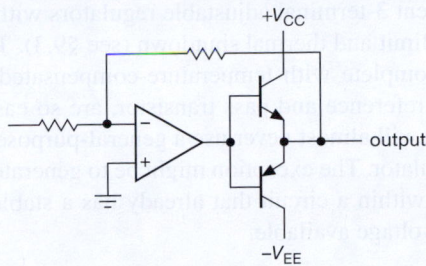

Figure 4.26. Push–pull follower boosts op-amp output current, both sourcing and sinking. You commonly see a small resistor (~100 Ω) connected between the bases and emitters to reduce crossover nonlinearity by maintaining feedback throughout the signal swing. See Figure 2.71 for improved output-stage biasing.

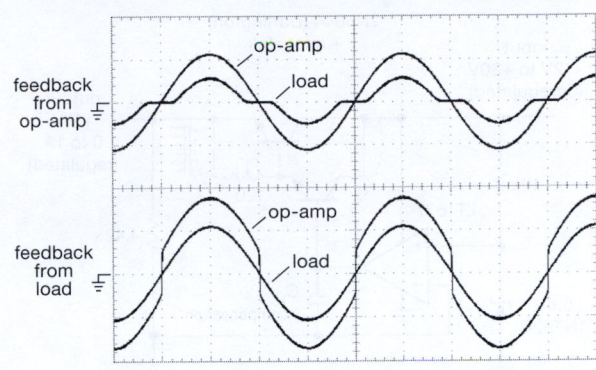

Figure 4.27. Feedback cures crossover distortion in the push–pull follower. Vertical: 1 V/div; horizontal: 2 ms/div.

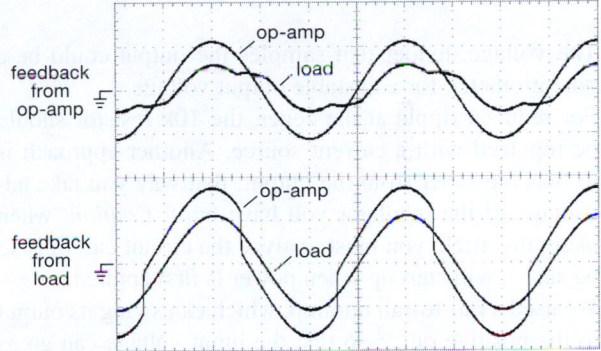

Figure 4.28. Same as Figure 4.27, but loaded with a loudspeaker of 6 Ω nominal impedance.

are unity-gain push-pull amplifiers capable of 200 mA of output current, and operation to 20–100 MHz (see §5.8.4, and also the discussion (and table) of unity-gain buffers in Chapter *4x.*); they are carefully biased for low open-loop crossover distortion, and include on-chip protection (current limit, and often thermal shutdown as well). As long as you ensure that the op-amp driving them has significantly less bandwidth, you can include them inside the feedback loop without any worries.[10]

Feedback and the push-pull booster

The push–pull booster circuit illustrates nicely the linearizing effect of negative feedback. We hooked up an LF411 op-amp as a noninverting unity-gain follower, driving a BJT push–pull output stage, and we loaded the output with a 10 Ω resistor to ground. Figure 4.27 shows the output signals at the op-amp and at the load, with an input sinewave of 1 V amplitude at 125 Hz. For the upper pair of traces we (foolishly) took the feedback from the op-amp's output, which produced a fine replica of the input signal; but the load sees severe crossover distortion (from the $2V_{BE}$ dead zone). With the feedback coming from the push–pull output (where the load is connected) we get what we want, as seen in the lower pair of traces. The op-amp cleverly creates an exaggerated waveform to drive the push–pull follower, with precisely the right shape to compensate for the crossover.

Figure 4.28 shows what these waveforms look like when we try driving an actual loudspeaker, a load that is more complicated than a resistor (because it's both a "motor" and a "generator," it exhibits resonances and other nasty properties; it's also got a reactive crossover network, and

an inductive coil to propel the cone). Once again, the magic of feedback does the job, this time with an op-amp output that is charmingly unsymmetrical.[11]

F. Power supply

An op-amp can provide the gain for a feedback voltage regulator (Figure 4.29). The op-amp compares a sample of the output with the zener reference, changing the drive to the Darlington "pass transistor" as needed. This circuit supplies a stable 10 volt output ("regulated"), at up to 1 amp load current. Some notes about this circuit:

---

[10] But beware a common error: a working circuit is upgraded by substituting a faster op-amp, whereupon the "improved" circuit oscillates!

[11] We should note, in fairness, that the fine performance seen here is at a rather low frequency (we chose it close to the speaker's bass resonance, to illustrate how clever feedback can be). But the situation degrades at high frequencies, owing to finite slew rate and falling loop gain (topics we'll see in §4.4). It's far better to eliminate most crossover distortion in the push–pull stage itself, by proper "class-AB" biasing (see Figure 2.71 in §2.4.1A), or by using an external unity-gain buffer (see Figure 4.87, and §5.8.4); then using feedback to suppress any residual distortion.

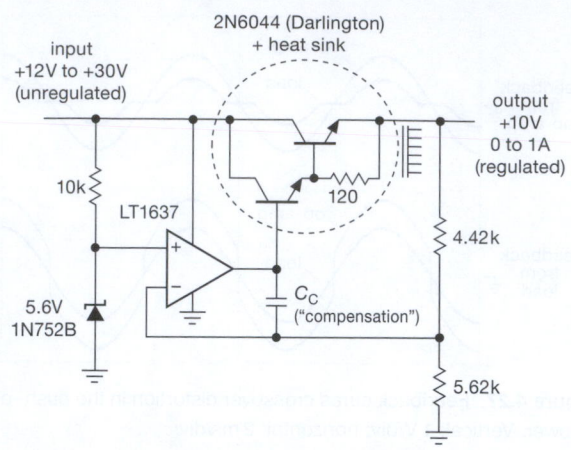

input
+12V to +30V
(unregulated)

2N6044 (Darlington)
+ heat sink

output
+10V
0 to 1A
(regulated)

10k

LT1637

5.6V
1N752B

120

C_C
("compensation")

4.42k

5.62k

Figure 4.29. Voltage regulator.

- The voltage divider that samples the output could be a potentiometer, for adjustable output voltage.
- For reduced ripple at the zener, the 10k resistor should be replaced with a current source. Another approach is to bias the zener from the output; that way you take advantage of the regulator you have built. *Caution*: when using this trick, you must analyze the circuit carefully to be sure it will start up when power is first applied.
- We used a rail-to-rail op-amp, which can swing its output to the positive rail,[12] so that the input voltage can go as low as +12 V without putting the Darlington pass transistor into saturation. With a 411, by contrast, you would have to allow another 1.5–2 V of margin, because the op-amp's output cannot get closer than that to the positive supply rail.
- The circuit as drawn could be damaged by a temporary short circuit across the output, because the op-amp would attempt to drive the Darlington pair into heavy conduction. Regulated power supplies should always have circuitry to limit "fault" current (see §9.1.1C for more details).
- Without the "compensation capacitor" C_C the circuit would likely oscillate when the dc output is bypassed (as it would be when powering a circuit) because of the additional lagging phase shift. Capacitor C_C ensures stability into a capacitive load, a subject we'll visit in §§4.6.1B, 4.6.2, and 9.1.1C.
- Integrated circuit voltage regulators are available in tremendous variety, from the time-honored 723 to the

---

[12] Our suggested LT1637 is a 44-volt "over-the-top" op-amp that exhibits strikingly higher input-bias currents when its input is near the positive rail (as much as $I_B = 20\mu A$, about 100 times its normal bias current). The LT1677, with $I_B = 0.2\mu A$, might be a better choice.

convenient 3-terminal adjustable regulators with internal current limit and thermal shutdown (see §9.3). These devices, complete with temperature-compensated internal voltage reference and pass transistor, are so easy to use that you will almost never use a general-purpose op-amp as a regulator. The exception might be to generate a stable voltage within a circuit that already has a stable power-supply voltage available.

In Chapter 9 we discuss voltage regulators and power supplies in detail, including special ICs intended for use as voltage regulators.

4.3.2 Nonlinear circuits

A. Comparator – an introduction

It is quite common to want to know which of two signals is larger, or to know when a given input signal exceeds a predetermined voltage. For instance, the usual method of generating triangle waves is to supply positive or negative currents into a capacitor, reversing the polarity of the current when the amplitude reaches a preset peak value. Another example is a digital voltmeter. In order to convert a voltage to a number, the unknown voltage is applied to one input of a comparator, with a linear ramp (capacitor + current source) applied to the other. A digital counter counts cycles of an oscillator during the time that the ramp is less than the unknown voltage and displays the result when equality of amplitudes is reached. The resultant count is proportional to the input voltage. This is called single-slope integration; in most sophisticated instruments a dual-slope integration is used (Chapter 13).

The simplest form of comparator is a high-gain differential amplifier, made either with transistors or with an op-amp (Figure 4.30). In this circuit there's no feedback – the op-amp goes into positive or negative saturation according to the difference of the input voltages. Because of the enormous voltage gain of op-amps (typically 10^5–10^6), the inputs will have to be equal to within a fraction of a millivolt in order for the output not to be saturated. Although an ordinary op-amp can be used as a comparator (and frequently is), there are special ICs intended for use as comparators. They let you set the output voltage levels independently of the voltages used to power the comparator (e.g., you can have output levels of 0 V and +5 V from a comparator powered from ±15 V); and they are generally much faster, because they are not trying to be op-amps, i.e., linear amplifiers intended for use with negative feedback. We'll talk about them in detail in Chapter 12 (§§12.1.7 and 12.3, and Table 12.2).).

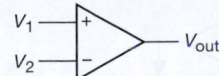

Figure 4.30. Comparator: an op-amp without feedback.

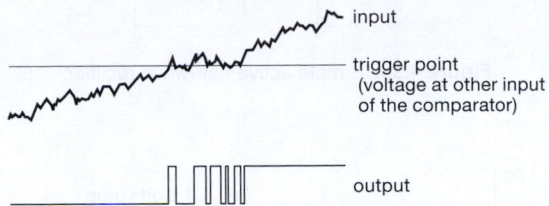

Figure 4.31. Comparator without hysteresis produces multiple transitions from noisy input signal.

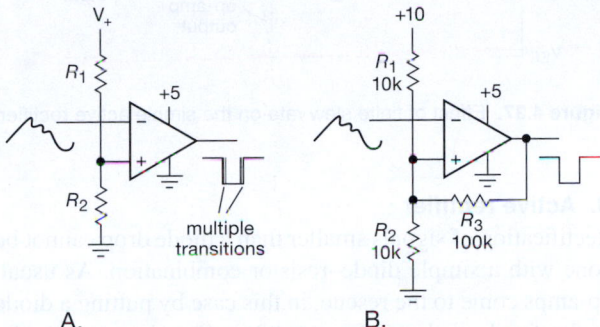

Figure 4.32. Positive feedback prevents multiple comparator transitions. A. Comparator without feedback. B. Schmitt trigger configuration uses positive feedback to prevent multiple output transitions. Special comparator ICs are generally preferable, and are drawn with the same symbol.

B. Schmitt trigger

The simple comparator circuit in Figure 4.30 has two disadvantages. For a very slowly varying input, the output swing can be rather slow. Worse still, if the input is noisy, the output may make several transitions as the input passes through the trigger point (Figure 4.31). Both these problems can be remedied by use of *positive* feedback (Figure 4.32). The effect of R_3 is to make the circuit have two thresholds, depending on the output state. In the example shown, the threshold when the output is at ground (input high) is 4.76 volts, whereas the threshold with the output at +5 volts is 5.0 volts. A noisy input is less likely to produce multiple triggering (Figure 4.33). Furthermore, the positive feedback ensures a rapid output transition, regardless of the speed of the input waveform. (A small "speed-up" capacitor of 10–100 pF is often connected across R_3 to enhance switching speed still further.) This configuration is known

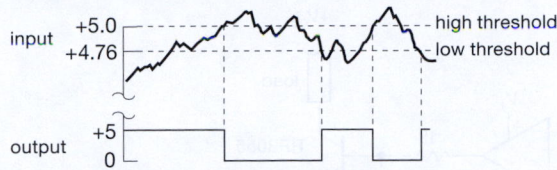

Figure 4.33. Hysteresis tames noise-prone comparator.

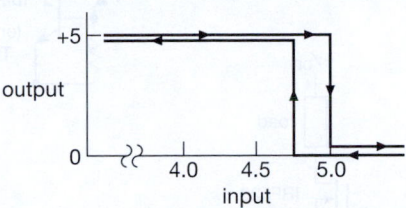

Figure 4.34. Output versus input ("transfer function") for Schmitt trigger.

as a Schmitt trigger, a function that we saw earlier in a discrete transistor implementation (Figure 2.13).

The output depends both on the input voltage and on its recent history, an effect called *hysteresis*. This can be illustrated with a diagram of output versus input, as in Figure 4.34. The design procedure is easy for Schmitt triggers that have a small amount of hysteresis. Use the circuit of Figure 4.32B. First choose a resistive divider (R_1, R_2) to put the threshold at approximately the right voltage; if you want the threshold near ground, just use a single resistor from noninverting input to ground. Next, choose the (positive) feedback resistor R_3 to produce the required hysteresis, noting that the hysteresis equals the output swing, attenuated by a resistive divider formed by R_3 and $R_1\|R_2$. Finally, if you are using a comparator with "open-collector" output, you must add an output pullup resistor small enough to ensure a nearly full supply swing, taking account of the loading by R_3 (read about comparator outputs in §12.3, and see Table 12.2). For the case in which you want thresholds symmetrical about ground, connect an offsetting resistor of appropriate value from the noninverting input to the negative supply. You may wish to scale all resistor values to keep the output current and impedance levels within a reasonable range.

C. Power-switching driver

The output of a comparator or Schmitt trigger switches abruptly between high and low voltages; it's not a continuous (or "linear") signal. You might want to use its output to turn a substantial load on or off. Examples might be a relay, laser, or motor.

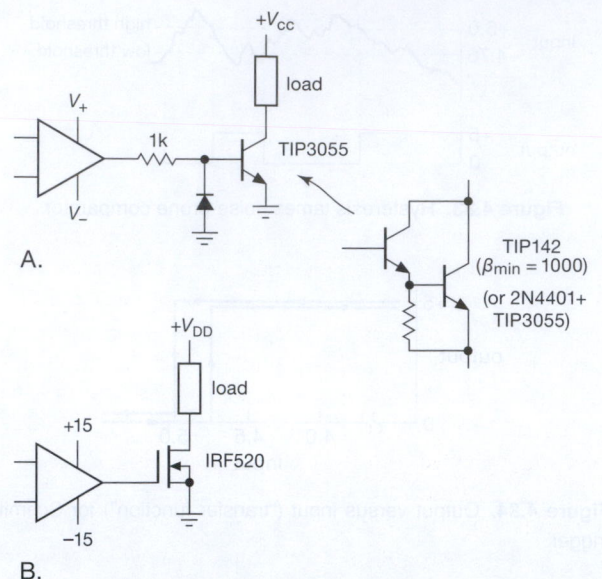

A.

B.

Figure 4.35. Power switching with an op-amp; A. With bipolar *npn*; note base current limit and reverse protection, B. With power MOSFET; note simplified drive circuit.

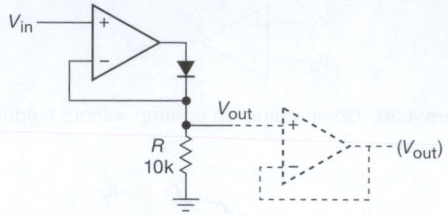

Figure 4.36. Simple active half-wave rectifier.

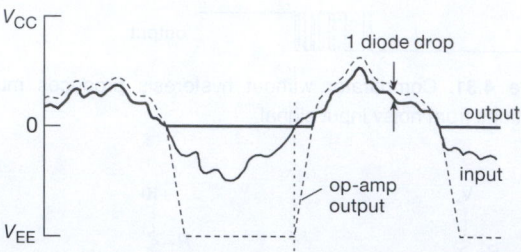

Figure 4.37. Effect of finite slew rate on the simple active rectifier.

For loads that are either on or off, a switching transistor can be driven from a comparator or op-amp. Figure 4.35A shows how. Note the diode to prevent reverse base–emitter breakdown (op-amps powered from dual supply rails easily swing more than the −6 V base–emitter breakdown voltage); it would be omitted if the op-amp's negative supply were no more than −5 V. The TIP3055 is a jellybean classic power transistor for noncritical high-current applications, though you'll find plenty of variety of available types with improved maximum voltage, current, power dissipation, and speed (see the listing in Table 2.2 on page 106). A Darlington can be used if currents greater than about 1 amp need to be driven.

In general, however, you're better off using an *n*-channel power MOSFET, in which case you can dispense with the resistor and diode altogether (Figure 4.35B). The IRF520[13] is a near-classic – but the variety of readily available power MOSFETs is overwhelming (see Table 3.4); in general you trade off high breakdown voltage against low ON-resistance.

When switching external loads, don't forget to include a reverse diode if the load is inductive (§1.6.7).

---

[13] Along with its higher-current cousins, the IRF530 and IRF540 and the higher-voltage relatives (IRF620–640 and IRF720–740) that fill out the orderly family tree; see "A 30-year MOSFET saga" (§*3x.11*).

D. Active rectifier

Rectification of signals smaller than a diode drop cannot be done with a simple diode–resistor combination. As usual, op-amps come to the rescue, in this case by putting a diode in the feedback loop (Figure 4.36). For V_{in} positive, the diode provides negative feedback; the circuit's output follows the input, coupled by the diode, but without a V_{BE} drop. For V_{in} negative, the op-amp goes into negative saturation and V_{out} is at ground. R could be chosen smaller for lower output impedance, with the tradeoff of higher op-amp output current. A better solution is to use an op-amp follower at the output, as shown, to produce very low output impedance regardless of the resistor value.

There is a problem with this circuit that becomes serious with high-speed signals. Because an op-amp cannot swing its output infinitely fast, the recovery from negative saturation (as the input waveform passes through zero from below) takes some time, during which the output is incorrect. It looks something like the curve shown in Figure 4.37. The output (heavy trace) is an accurate rectified version of the input (light trace), except for a short time interval after the input rises through zero volts. During that interval the op-amp output is racing up from saturation near −V_{EE}, so the circuit's output is still at ground. A general-purpose op-amp like the 411 has a *slew rate* (maximum rate at which the output can change) of 15 V/μs; recovery from negative saturation therefore takes about 1 μs (when operating from ±15 V supplies), which may introduce significant output

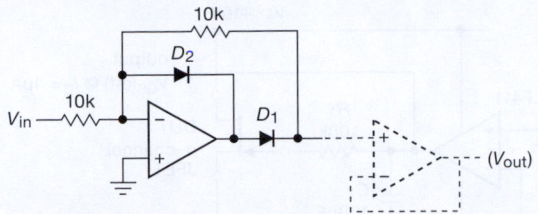

Figure 4.38. Improved active half-wave rectifier.

error for fast signals. A circuit modification improves the situation considerably (Figure 4.38).

D_1 makes the circuit a unity-gain inverter for negative input signals. D_2 clamps the op-amp's output at one diode drop below ground for positive inputs, and since D_1 is then back-biased, V_{out} sits at ground. The improvement comes because the op-amp's output swings only two diode drops as the input signal passes through zero. Because the op-amp output has to slew only about 1.2 volts instead of V_{EE} volts, the "glitch" at zero crossings is reduced more than 10-fold. This rectifier is inverting, incidentally. If you require a noninverted output, attach a unity-gain inverter to the output.

The performance of these circuits is improved if you choose an op-amp with a high slew rate. Slew rate also influences the performance of the other op-amp applications we've discussed, for instance the simple voltage amplifier circuits. Shortly we'll take a closer look at the ways in which real op-amps depart from the ideal – input current, offset voltage, bandwidth and slew rate, and so on – because you need to know about those limitations if you want to design good circuits. With that knowledge we'll also look at some active *full-wave* rectifier circuits to complement these half-wave rectifiers.[14] First, though, we'd like to demonstrate some of the fun of designing with op-amps by showing a few real-world circuit examples.

4.3.3 Op-amp application: triangle-wave oscillator

These op-amp circuit fragments that we've been exploring – amplifiers, integrators, Schmitt triggers, etc. – are interesting enough; but the real excitement in circuit design comes when you creatively put pieces together to make a complete "something." A nice example that we can handle now is a triangle-wave *oscillator*. Unlike any other circuits so far, this one has no input signal; instead it creates an output signal, in this case a symmetrical triangle wave of

---

[14] And in Chapter *4x* we'll see additional non-linear circuit applications of op-amps, for example a logarithmic amplifier and a "chaotic" Lorenz-attractor circuit.

1 volt amplitude. As a by-product you also get a square wave, for free. (We'll see many more examples of oscillators in Chapter 7).

The idea is first to use an integrator (with a constant-dc input voltage) to generate a ramp; we need to turn the ramp around when it reaches its ±1 V limits, so we let the integrator output (the ramp) drive a Schmitt trigger, with thresholds at ±1 V. The output of the Schmitt, then, is what ought to determine the direction of the ramp. Aha! Just use its output (which switches between the supply rail voltages) as the input to the integrator.

Figure 4.39 shows a circuit implementation. It's easiest to start with IC_2, which is wired as a *noninverting* Schmitt trigger (it looks like an inverting amplifier, but it's not – note that feedback goes to the non-inverting input), for a reason we'll see soon. This configuration is used less frequently than the conventional inverting circuit of Figure 4.32B, because of its lower input impedance (and substantial input current reversal at threshold). Importantly, the LMC6482 has rail-to-rail output swing, so with ±5 V power supplies its thresholds are at ±1 V, set by a 5:1 ratio of R_3 to R_2.

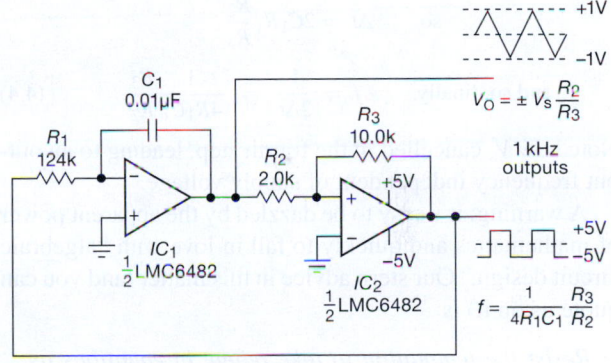

Figure 4.39. Triangle-wave oscillator.

The Schmitt's ±5 V output is the input to the integrator IC_1. We chose C_1 to be a convenient value of 0.01 μF, then calculated R_1 to ramp through 2 V in a half period (0.5 ms), using $5\,V/R_1 = I_{in} = C_1[dV/dt]_{ramp}$. The calculated resistor value of 125 kΩ (in the figure we show the nearest standard 1% "E96" resistor value; see Appendix C) came out reasonable, given real-world op-amp characteristics, as we'll learn later in the chapter. If it hadn't, we would have changed C_1; this is typically how you get to your final circuit component values.

Exercise 4.7. Confirm that value of R_1 is correct, and that the Schmitt trigger thresholds are at ±1 V.

Now the reason for connecting IC_2 as a *noninverting* Schmitt trigger becomes clear: if IC_2's output is at -5 V, say, then the triangle wave is ramping upward toward the Schmitt's $+1$ V threshold, at which point the Schmitt's output will switch to $+5$ V, reversing the cycle. If we had instead used the more conventional inverting Schmitt configuration, the oscillator would not oscillate; in that case it would "latch up" at one limit, as you can verify by walking through one cycle of operation.

The expressions for output frequency and amplitude are shown in the figure. It's interesting to note that the frequency is independent of supply voltage; but if you alter the resistor ratio R_2/R_3 to change the output amplitude, you will also change the frequency. Sometimes it is good to develop algebraic expressions for circuit operation, to see such dependencies. Here's how it goes in this case:

$$\frac{dV}{dt} = \frac{I}{C} = \frac{V_S/R_1}{C_1},$$

$$\text{so} \quad \Delta t = C_1\frac{R_1}{V_S}\Delta V,$$

$$\text{but} \quad \Delta V = 2\frac{R_2}{R_3}V_S,$$

$$\text{so} \quad \Delta t = 2C_1R_1\frac{R_2}{R_3},$$

$$\text{and so, finally,} \quad f = \frac{1}{2\Delta t} = \frac{1}{4R_1C_1}\frac{R_3}{R_2}. \quad (4.4)$$

Note how V_s cancelled in the fourth step, leading to an output frequency independent of supply voltage.

A warning: it's easy to be dazzled by the apparent power of mathematics and quickly to fall in love with "algebraic circuit design." Our stern advice in this matter (and you can quote us on it) is:

Resist the temptation to take refuge in equations as a substitute for understanding how a circuit really works.

4.3.4 Op-amp application: pinch-off voltage tester

Here's another nice application of op-amps: suppose you want to measure a batch of JFETs in order to put them into groups that are matched in pinch-off voltage $V_{GS}(\text{off})$ (sometimes called V_P, see §3.1.3). This is useful because the large spread of specified V_P sometimes makes it difficult to design a good amplifier.[15] We'll assume that you want to find the gate-source back-bias that results in a drain

---

[15] You can as well use this same circuit to match the threshold voltage, V_{GSth}, of a set of MOSFETs.

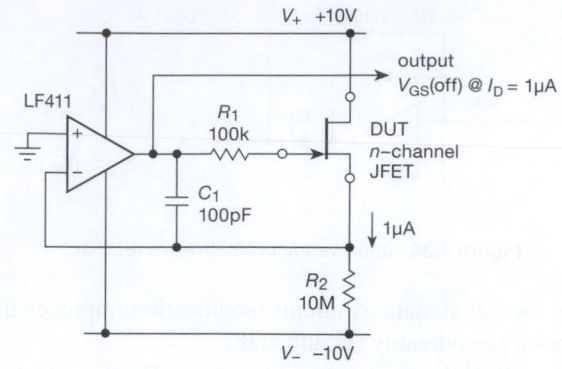

Figure 4.40. Simple pinch-off voltage tester.

current of $1\,\mu\text{A}$ with the drain at $+10$ V and the source grounded.

If you didn't know about op-amps, you could imagine (a) grounding the source, (b) hooking up a sensitive current meter from the drain to a $+10$ V supply, and then (c) adjusting the gate voltage with a variable negative supply to a value that produces $1\,\mu\text{A}$ of measured drain current.

Figure 4.40 shows a better way. The device under test (you'll often see the acronym DUT) has its drain tied to $+10$ V; but the source lead, instead of being grounded, is tied to the inverting input (virtual ground) of an op-amp whose noninverting input is grounded. The op-amp controls the gate voltage, thus holding the source at ground. Because the source is pulled down to -10 V through a 10M resistor, the source current (and therefore the drain current) is $1\,\mu\text{A}$. The op-amp's output is the same as the gate voltage, so the output of this circuit is the pinch-off voltage you wanted to know.

A few details:

- We chose power supply voltages of ±10 V for the op-amp to make the rest of the circuit simpler, since we wanted to measure V_P with $+10$ V on the drain. That's OK, because most op-amps work well over a range of supply voltages (in fact, the trend is toward lower operating voltages, driven by the market for battery-powered consumer devices). But if you have only ±15 V available, you would have to generate $+10$ V within your circuit, either with a voltage divider, a zener, or a 3-terminal voltage regulator (see Chapter 9).

- We put a 100k resistor (R_1) as protection in series with the gate to prevent any significant gate current from flowing during plug-in transients, etc. This can introduce a lagging phase shift around the loop at high frequencies (as can the rather large pull-down resistor R_2), so we added a small feedback capacitor C_1 to maintain stability.

We talk about this business of stability toward the end of the chapter, in §4.9.

- For this circuit to work properly, it's important that the op-amp's inverting input not load the source terminal, for example by drawing anything approaching a microamp of current. As we'll learn shortly, this is not always the case. For this example our general-purpose 411 op-amp, with its JFET input transistors, is fine (with input currents in the picoamperes); but an op-amp that uses bipolar transistors for its input stage would generally have input currents in the 10's to 100's of nanoamps, and should be avoided for a low-current application like this.
- The drain current at which pinch-off voltage is specified is not always $1\,\mu$A. You'll see $V_{GS(off)}$ specified at values of drain current ranging from 1 nA to tens of microamps, depending on the size of the JFET, and the whim of the manufacturer. (In an informal survey of datasheets we found 1 nA to be the most popular, followed by $1\,\mu$A, 10 nA, and 0.5 nA, with five other values used occasionally.) It would be easy to modify the circuit to accommodate higher test currents; but to go to 10 nA, say, you would need a 1 GΩ resistor for R_2! In that case a better solution is to return the pull-down resistor to a lower voltage, say -0.1 V, which you could generate with a voltage divider from the -10 V negative supply. You'd have to worry again about op-amp input currents with such a small test current.

Exercise 4.8. Show how to make the pinch-off tester operate from ±15 V supplies, with the measurement still made at $V_D = +10$ V; assume that the largest resistor value available is 10MΩ.

Exercise 4.9. Modify the pinch-off tester circuit of Figure 4.40 so that you can measure V_{GS} at three values of drain current, namely $1\,\mu$A, $10\,\mu$A, and $100\,\mu$A, by setting a 3-position switch. Assume that the largest resistor value you can conveniently get is 10 MΩ.

Exercise 4.10. Now change the circuit so that it measures $V_{GS(off)}$ at $I_D{=}1$ nA. Assume you can get 100 MΩ 5% resistors.

4.3.5 Programmable pulse-width generator

When triggered by a short input pulse, the circuit in Figure 4.41 generates an output pulse[16] whose width is set by 10-turn pot R_1. Here's how it works.

IC$_1$, IC$_2$, and Q_1 form a current source that charges timing capacitor C, as we'll detail below. IC$_3$ is a versatile timer IC, whose many exploits we will enjoy in Chapter 7. It holds C discharged (through a saturated MOSFET

switch whose drain drives the DIS pin to ground) and simultaneously holds the output at ground, until it receives a negative-going trigger pulse at its TRIG input pin; at that point it releases DIS and switches its output to V_+, in this case +5 V.

The current source now charges C with a positive-going ramp, according to $I = C\,dV/dt$. This continues until the capacitor voltage, which also drives IC$_3$'s TH input, reaches a voltage equal to 2/3 of the supply voltage, $V_{TH} = \frac{2}{3}V_+$; at this point IC$_3$ abruptly pulls DIS back to ground, simultaneously switching its output to ground. This completes the cycle.

The current source is an elegant circuit. We want to source a current into the capacitor, with compliance from ground to at least +3.3 V (2/3 of +5 V), with linear control by a pot that returns to ground. For reasons we'll see presently, we want the programmed current proportional to the supply voltage V_+. In this circuit Q_1 is the current source, with IC$_2$ controlling its base to hold its emitter at +5 V. IC$_1$ is an inverting amplifier referenced to +5 V; it pivots its output to a voltage that exceeds +5 V by an amount proportional to the current flowing through R_1 and R_2. That excess voltage appears across R, generating the output current. You'll know you understand how it works by doing the following problem.

Exercise 4.11. Calculate the current sourced by Q_1 by calculating the output voltage of IC$_1$ as a function of R_X (the sum of R_1 and R_2), R_3, and V_+. Now use it to calculate the output pulse width, knowing that IC$_3$ switches when the voltage at TH reaches $\frac{2}{3}V_+$.

This circuit is an illustration of the use of *ratiometric* techniques: for a given setting of R_1, both the capacitor charging current I and the timer IC threshold voltage V_{TH} individually depend on supply voltage V_+; but their variation is such that the final pulse width T does not depend on V_+. That is why the current source was designed with $I \propto V_+$. The use of ratiometric techniques is an elegant way to design circuits with excellent performance, often without requiring precise control of power-supply voltages.

4.3.6 Active lowpass filter

The simple *RC* filters we saw back in Chapter 1 have a soft rolloff; that is, their response versus frequency does not progress sharply from a passband to a stopband. Perhaps surprisingly, this behavior cannot be remedied by simply cascading multiple stages, as we'll see in detail in Chapter 6 (and particularly in connection with *active filters*, §6.3). Much better filter performance can be achieved if

---

[16] More about pulsers, for those interested, in §§7.1.4B and 7.2.

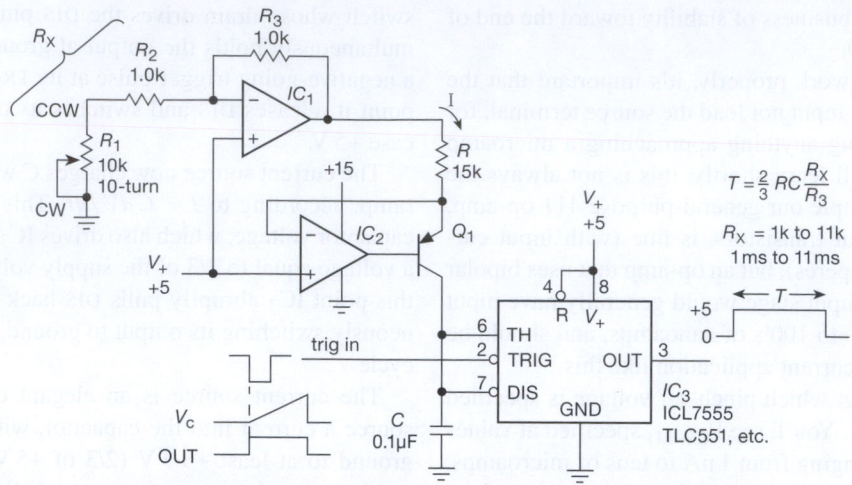

$$T = \frac{2}{3} RC \frac{R_X}{R_3}$$

R_X = 1k to 11k
1ms to 11ms

Figure 4.41. Pulse generator with programmable width.

you include both inductors and capacitors, or, equivalently, if you "activate" the filter design by using op-amps.

Figure 4.42 shows an example of a simple and even partly intuitive filter. This configuration is known as a Sallen-and-Key filter, after its inventors. The unity-gain amplifier can be an op-amp connected as a follower, or a unity-gain *buffer* IC, or just an emitter follower. This particular filter is a second-order lowpass filter. Note that it would be simply a pair of cascaded passive *RC* lowpass filters, except for the fact that the bottom of the first capacitor is bootstrapped by the output. It is easy to see that at high frequencies (well beyond $f = 1/2\pi RC$) it falls off just like a cascaded *RC*, i.e., at −12 dB/octave, because the output is essentially zero (and therefore the first capacitor's lower end is effectively grounded). As we lower the frequency and approach the passband, however, the bootstrap action tends to reduce the attenuation, thus giving a sharper "knee" to the curve of response versus frequency. We've plotted the response versus frequency, with three "tunings" of the *R* and *C* values.[17]

Of course, such hand-waving cannot substitute for honest analysis, which luckily has been done for a prodigious variety of nice filters. And contemporary general-purpose SPICE-based analog simulation tools, or special filter analysis software, let you design and view filter response curves with relative ease.

[17] Butterworth and two Chebyshevs (0.1 dB and 0.5 dB passband ripple), going from flattest to peakiest response; for the Butterworth, for example, the component values are C_1=10 nF, C_2=2 nF, R_1=12.7k, and R_2=100k. Active filters are discussed in detail in Chapter 6.

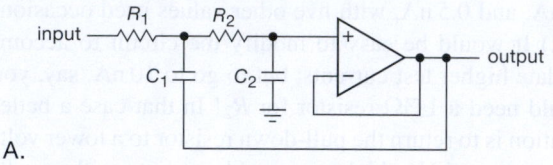

A.

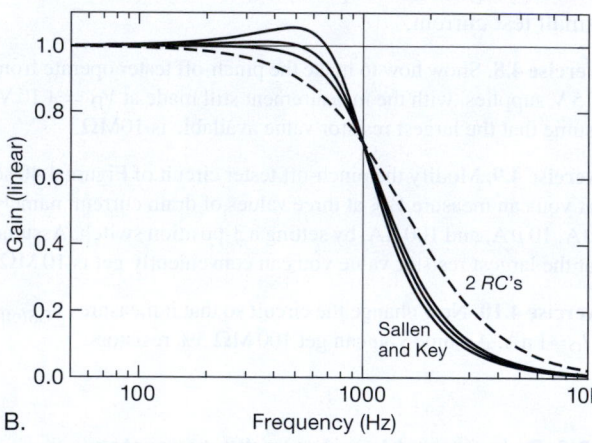

B.

Figure 4.42. Sallen-and-Key active lowpass filter: A. Schematic; B. frequency response, compared with a cascade of two passive *RC* sections.

4.4 A detailed look at op-amp behavior

We've hinted that op-amps aren't perfect, and that the performance of circuits such as active rectifiers and Schmitt triggers is limited by op-amp speed, or "slew rate." For those applications a high-speed op-amp is often required.

But slew rate is just one of a half-dozen important parameters of op-amps, which include input offset voltage, input bias current, input common-mode range, noise, bandwidth, output swing, and supply voltage and current. To state the situation fairly, op-amps are remarkable devices, with near-ideal performance for most applications you are likely to encounter. To put it quantitatively, think of the difficulty of designing, with discrete transistors and other components, a high-gain dc differential amplifier that has an input current less than a picoamp, an offset from perfect balance less than a millivolt, a bandwidth of several megahertz, and that operates with its inputs anywhere between the two supply voltages. You can get such an op-amp for a dollar; it comes in a tiny package measuring $1.5\,\mathrm{mm} \times 3\,\mathrm{mm}$, and it draws less than a milliamp.

But op-amps *do* have performance limitations – that's why there are literally thousands of available types – and in general you're faced with a tradeoff: you can get much lower bias current (for example), at the expense of offset voltage. A good understanding of op-amp limitations and their influence on circuit design and performance will help you choose your op-amps wisely and design with them effectively.

To motivate the subject, imagine that you've been asked to design a dc amplifier, so that small voltages ($0-10\,\mathrm{mV}$) can be seen on a handsome analog meter scale. And it should have at least $10\mathrm{M}\Omega$ input resistance, and be accurate to 1% or so. No problem, you say...I'll just use the non-inverting amplifier configuration (to get high input resistance), with lots of gain ($\times 1000$, say, so $10\,\mathrm{mV}$ is amplified to $10\,\mathrm{V}$). Speed is not an issue, so you don't worry about slew rate. With supreme confidence you draw up the circuit (with an LF411 op-amp), your technician builds it, and ... your boss fires you! The thing was a disaster: it read 20% of full-scale with no input attached, and it drifted like crazy when carried outside. It does work OK – as a *paperweight*.[18]

To get started, look at Figure 4.43, a simplified schematic of the LF411. Its circuit is relatively straightforward, in terms of the kinds of transistor circuits we discussed in the last two chapters. It has a JFET differential input stage, with current-mirror active load, buffered with an *npn* follower (to prevent loading of the high-gain input stage) driving a grounded-emitter *npn* stage (with current-source active load). This drives the push–pull emitter follower output stage (Q_7Q_8), with current-limiting circuitry

(R_5Q_9 and R_6Q_{10}) to protect against output short-circuit.[19] The curious feedback capacitor C_C ensures stability; we'll learn about it later. This circuit displays the internal circuitry characteristic of the typical op-amp, and from it we can see how and why op-amp performance departs from ideal.

Exercise 4.12. Explain how the current-limiting circuitry in Figure 4.43 works. What is the maximum output current?

Exercise 4.13. Explain the function of the two diodes in the output stage.

Let's look at these problems, what the consequences are for circuit design, and what to do about it.

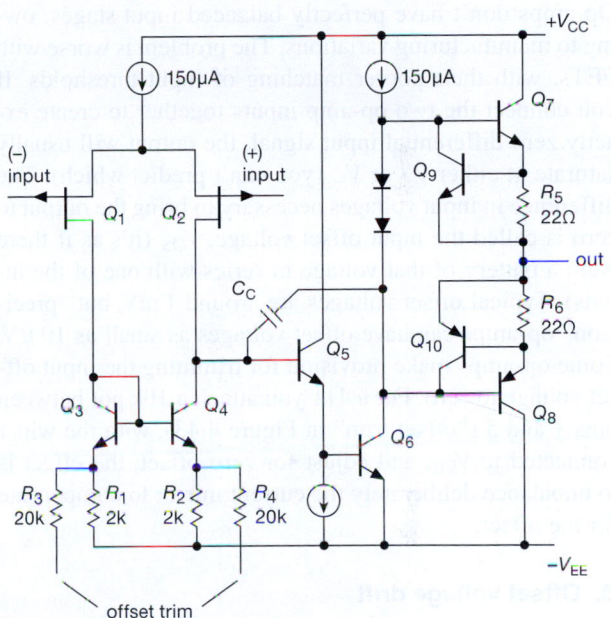

Figure 4.43. Simplified schematic of the LF411 op-amp.

4.4.1 Departure from ideal op-amp performance

The ideal op-amp has these characteristics:

- Input current = 0 (input impedance = ∞).
- $V_{\mathrm{out}} = 0$ when both inputs are at precisely the same voltage (zero "offset voltage").
- Output impedance (open loop) = 0.
- Voltage gain = ∞.
- Common-mode voltage gain = 0.

---

[18] We'll revisit this example in §4.4.3, and again in more detail in Chapter 5.

[19] The LF411's *detailed* schematic reveals a more elaborate negative current-limit configuration; check it out on the datasheet, to see if you can understand how it works.

- Output can change instantaneously (infinite slew rate).
- Absence of added "noise."

All of these characteristics should be independent of temperature and supply voltage changes.

In the following paragraphs we describe how real op-amps depart from these ideals. As you struggle through the fact-filled sections, you may want to refer to Table 4.1 to maintain perspective. Tables 4.2a,b, 5.5, and 8.3 may be helpful also, for seeing some actual numbers. And we'll revisit these in more detail in Chapter 5 (§§5.7 and 5.8) in connection with the design of precision circuits.

A. Input offset voltage

Op-amps don't have perfectly balanced input stages, owing to manufacturing variations. The problem is worse with FETs, with their poorer matching of input thresholds. If you connect the two op-amp inputs together to create exactly zero differential input signal, the output will usually saturate at either V_+ or V_- (you can't predict which). The difference in input voltages necessary to bring the output to zero is called the input offset voltage, V_{OS} (it's as if there were a battery of that voltage in series with one of the inputs). Typical offset voltages are around 1 mV, but "precision" op-amps can have offset voltages as small as 10 μV. Some op-amps make provision for trimming the input offset voltage to zero. For a 411 you attach a 10k pot between pins 1 and 5 ("offset trim" in Figure 4.43), with the wiper connected to V_{EE}, and adjust for zero offset; the effect is to unbalance deliberately the current mirror to compensate for the offset.

B. Offset voltage drift

Of greater importance for precision applications is the *drift* of the input offset voltage with temperature and time, since any initial offset could be manually trimmed to zero. A 411 has a typical offset voltage of 0.8 mV (2 mV maximum), with a tempco of $\Delta V_{OS}/\Delta T = 7\,\mu V/°C$ and unspecified coefficient of offset drift with time. The OP177A, a precision op-amp, is laser-trimmed for a maximum offset of 10 *microvolts, with a temperature coefficient of 0.1 μV/°C (max) and long-term drift of 0.2 μV/month (typical) – roughly a hundred times better in both offset and tempco.

C. Input current

The input terminals sink (or source, depending on the op-amp type) a small current called the input bias current, I_B, which is defined as half the sum of the input currents with the inputs tied together (the two input currents are approximately equal and are simply the base or gate currents of the input transistors). For the JFET-input 411 the bias current is typically 50 pA (200 pA max) at room temperature (but as much as 4 nA at 70°C), while a typical BJT-input op-amp like the OP27 has a bias current of 15 nA, varying little with temperature. As a rough guide, BJT-input op-amps have bias currents in the tens of nanoamps, whereas JFET-input op-amps have input currents in the tens of picoamps (i.e., 1000 times lower), and MOSFET-input op-amps have input currents of typically a picoamp or less. Generally speaking, you can ignore input current with FET op-amps, but not with bipolar-input op-amps.[20]

The significance of input bias current is that it causes a voltage drop across the resistors of the feedback network, bias network, or source impedance. How small a resistor this restricts you to depends on the dc gain of your circuit and how much output variation you can tolerate. For example, an LF412's maximum input current of 200 pA means that you can tolerate resistances (seen from the input terminals) up to ∼5 MΩ before you have to worry about it at the 1 mV level.

We will see more about how this works later. If your circuit is an integrator, bias current produces a slow ramp even when there is no external input current to the integrator.

Op-amps are available with input bias currents down to a nanoamp or less for bipolar-transistor-input circuit types, or down to a fraction of a picoamp ($10^{-6}\,\mu A$) for MOSFET-input circuit types. The very lowest bias currents are typified by the BJT-input LT1012, with a typical input current of 25 pA, the JFET-input OPA129, with an input current of 0.03 pA, and the MOSFET LMC6041, with an input current of 0.002 pA. At the other end, very fast BJT op-amps like the THS4011/21 (∼300 MHz) have input currents of 3 μA. In general, BJT op-amps intended for high-speed operation have higher bias currents.

D. Input offset current

Input offset current is a fancy name for the difference of the input currents between the two inputs. Unlike input bias current, the offset current, I_{OS}, is a result of manufacturing variations, since an op-amp's symmetrical input circuit would otherwise result in identical bias currents at the two inputs. The significance is that, even when it is driven by identical source impedances, the op-amp will see unequal voltage drops and hence a difference voltage between its inputs. You will see shortly how this influences design.

Typically, the offset current is somewhere between one-

---

[20] There's a nice trick, called *bias-current cancellation*, exploited in some BJT op-amps to achieve input currents as low as 10s of picoamps. Look back to Figure 2.98; this is discussed further in Chapter *4x*.

Table 4.1 Op-amp Parameters[a]

Parameter	bipolar (BJT) jellybean	bipolar (BJT) premium	JFET-input jellybean	JFET-input premium	CMOS jellybean	CMOS premium	Units
V_{OS} (max)	3	0.025	2	0.1	2	0.1	mV
TCV_{OS} (max)	5	0.1	20	1	10	3	µV/°C
I_B (typ)	50nA	25pA	50pA	40fA	1pA	2fA	@ 25°C
e_n (typ)	10	1	20	3	30	7	nV/√Hz @ 1kHz
f_T (typ)	2	2000	5	400	2	10	MHz
SR (typ)	2	4000	15	300	5	10	V/µs
V_S (min)[b]	5	1.5	10	5	2	1	V
V_S (max)[b]	36	44	36	36	15	15	V

Notes: (a) typical and "best" values of important op-amp performance parameters. (b) total supply: $V_+ - V_-$.

Typical and "best" values of important op-amp performance parameters. In this chart we list values for run-of-the-mill ("jellybean") parts, and for the best op-amp you can get *for each individual parameter*. That is, you cannot get a single op-amp that has the combination of excellent performance shown in any of the "premium" columns. In this chart you can clearly see that bipolar op-amps excel in precision, stability, speed, wide supply voltage range, and noise, at the expense of bias current; JFET-input types are intermediate, with CMOS op-amps displaying the lowest bias current.

half and one-tenth the bias current. For the 411, $I_{offset}=$ 25 pA, typical. However, for bias-compensated op-amps (like the OPA177), the specified offset current and bias current are comparable, for reasons we'll see in the advanced Chapter 5.

E. Input impedance

Input impedance refers to the small-signal[21] differential input resistance (impedance looking into one input, with the other input grounded), which is usually much less than the common-mode resistance (a typical input stage looks like a long-tailed pair with current source). For the FET-input 411 it is about $10^{12}\,\Omega$, whereas for BJT-input op-amps like the LT1013 it is about 300 MΩ. Because of the input bootstrapping effect of negative feedback (it attempts to keep both inputs at the same voltage, thus eliminating most of the differential-input signal), Z_{in} in practice is raised to very high values and usually is not as important a parameter as input bias current.

F. Common-mode input range

The inputs to an op-amp must stay within a certain voltage range, typically less than the full supply range, for proper operation. If the inputs go beyond this range, the gain of the op-amp may change drastically, even reversing sign! For a 411 operating from ±15 volt supplies, the guaranteed common-mode input range is ±11 volts minimum.

[21] *Not* V_{in}/I_{bias}!

However, the manufacturer claims that the 411 will operate with common-mode inputs all the way to the positive supply, though performance may be degraded. Bringing either input down to the negative supply voltage causes the amplifier to go berserk, with symptoms like phase reversal[22] and output saturation to the positive supply. From the circuit in Figure 4.43 you can see why the LF411 cannot possibly operate with input voltages to the negative rail, because that would put the source terminals of the input JFET pair below the negative rail, taking them out of the active region. This is discussed further in Chapter *4x*, along with some good war stories.

There are many op-amps available with common-mode input ranges down to the negative supply, e.g., the bipolar LT1013 and the CMOS TLC2272 and LMC6082; these are often referred to as "single-supply op-amps" or "ground-sensing op-amps" (see §4.6.3). There are also some op-amps whose common-mode input range includes the positive supply, e.g., the JFET LF356. With the trend toward lower supply voltages for battery-powered equipment, op-amp designers have come up with varieties that accommodate input signals over the full range between supply voltages; these are called rail-to-rail, because supply voltages

[22] The popular and inexpensive ($0.07 in quantity) LM358 and LM324 single-supply op-amps suffer from input phase reversal for inputs more than 400 mV below the negative rail. Improved replacements like the LT1013 and LT1014 fix this problem (and also an output crossover-distortion problem).

are often called supply *rails*.[23] Examples are the CMOS LMC6482 and TLV2400 series, and the bipolar LM6132, LT1630, and LT6220 series. These have the additional nice feature of being able to swing their outputs all the way to the rails (see the subsection on output swing below). These would seem to be ideal op-amps; however, as we discuss in §§5.7, 5.9, and 5.10, rail-to-rail op-amps typically make compromises that affect other characteristics, notably off-set voltage, output impedance, and supply current. There are, in addition, a few (*very* few) op-amps that operate properly for input voltages *above* the positive rail (for example, the "over-the-top" LT1637, listed in Table 4.2a on page 271).

In addition to the *operating* common-mode range, there are maximum allowable input voltages beyond which damage will result. For the 411 they are ±15 volts (but not to exceed the negative supply voltage, if it is less).

G. Differential input range

Some bipolar op-amps allow only a limited voltage between the inputs, sometimes as small as ±0.5 volt, although most are more forgiving, permitting differential inputs nearly as large as the supply voltages. Exceeding the specified maximum can degrade or destroy the op-amp.

H. Output swing versus load resistance

The LF411, typical of many op-amps, cannot swing its output closer than a volt or two from either supply rail, even when lightly loaded ($R_L > 5k$, say). That's because the output stage is a push–pull emitter follower, so even a full rail-to-rail drive to its bases would leave the output a diode drop short of both rails; the drive circuitry has its own difficulties getting close to the rails as well, and the current-limit sense resistors R_5 and R_6 impose an additional voltage drop, which accounts for the shortfall.

For low values of load resistance, the internal current-limit circuit will set the maximum swing. For example, the 411 can swing its output to within about 2 volts of V_{CC} and V_{EE} into load resistances greater than about 1k. Load resistances significantly less than that will permit only a small swing. This is frequently shown on datasheets as a graph of peak-to-peak output voltage swing V_{om} as a function of

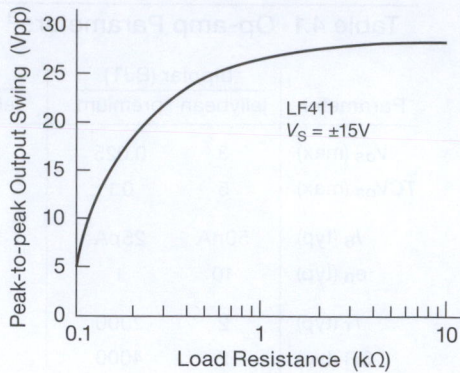

Figure 4.44. Maximum peak-to-peak output swing versus load (LF411).

load resistance, or sometimes just a few values for typical load resistances. Figure 4.44 shows the datasheet's graph for the LF411. Many op-amps have asymmetrical output drive capability, with the ability to sink more current than they can source (or vice versa). For that reason you often see maximum output swing plotted, versus load current, as separate curves for output sourcing and sinking current into a load. Figure 4.45 shows such graphs for the LF411.

Some op-amps can produce output swings all the way down to the negative supply (e.g., the bipolar LT1013 and CMOS TLC2272), a particularly useful feature for circuits operated from a single positive supply, because output swings all the way to ground are then possible. Finally, op-amps with CMOS transistor outputs in a common-source amplifier configuration[24] (e.g., the LMC6xxx series) can swing all the way to both rails. For such op-amps a much more useful graph plots how close the output can get to each power-supply rail as a function of load current (both sourcing and sinking). An example is shown in Figure 4.46 for the CMOS rail-to-rail LMC6041. Note the effective use of log–log axes, so you can read off accurately the fact that this op-amp can swing to within 1 mV of the rails when supplying $10\,\mu A$ of output current, and that its output resistance is approximately $80\,\Omega$ (sinking) and $100\,\Omega$ (sourcing). You can find bipolar op-amps that share this property, without the limited supply voltage range of the CMOS op-amps (usually ±8 V max), for example, the LM6132/42/52 family and the LT1636/7.

I. Output Impedance

Output impedance R_o means the op-amp's intrinsic output impedance *without feedback* (see Figure 2.90). For the 411 it is about $40\,\Omega$, but with some low-power op-amps it can

---

[23] The term "Rail-to-Rail®" is apparently a registered trademark of Nippon Motorola Ltd, though we believe it has been in common use in electronics for decades. This may turn out to be an unwise proprietary claim from their point of view, just as the trademarking of "TRI-STATE®" by National Semiconductor simply drove the industry to adopt the nonproprietary term "3-state" in written references (and, in most cases, to stick with "tristate" in spoken conversation).

[24] Or bipolar-transistor outputs in a common-emitter configuration.

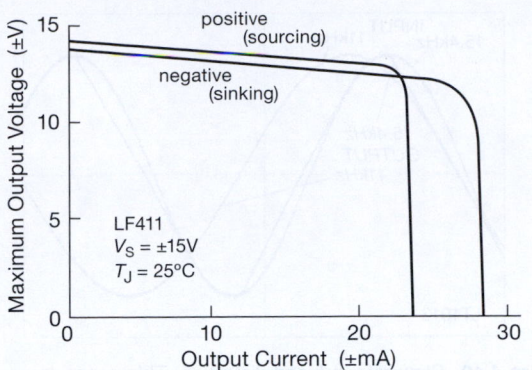

Figure 4.45. Maximum output voltage (both sourcing and sinking) versus load current (LF411). The maximum output current capability decreases by ~25% at $T_j = 125°C$.

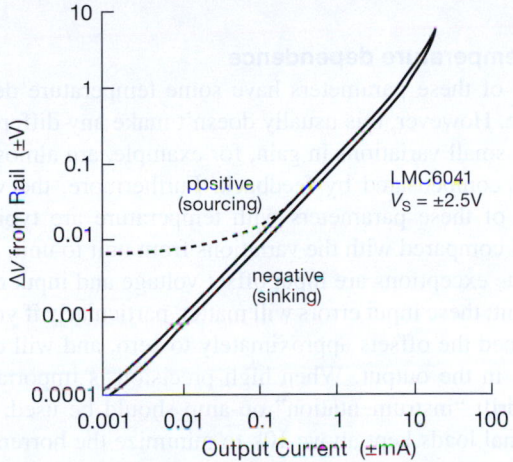

Figure 4.46. Maximum swing (as ΔV from the respective rails) versus load current for a CMOS rail-to-rail output op-amp. The solid curves are *measured* values; you can't always trust datasheets – in this case the datasheet's *sourcing* curve (dashed curve) is evidently in error.

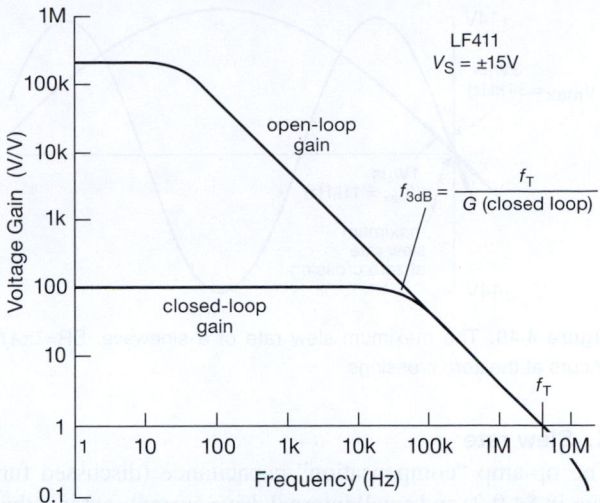

Figure 4.47. LF411 gain versus frequency ("Bode plot").

be as high as several thousand ohms, a characteristic shared by some op-amps with rail-to-rail outputs. Feedback lowers the output impedance into insignificance (or raises it, for a current source), by a factor of the loop gain AB (see §2.5.3C); so what usually matters more is the maximum output current, with typical values of ±20 mA or so (but much higher for the rarified group of "high current" op-amps, see Table 4.2b on page 272).

J. Voltage gain, bandwidth, and phase shift
Typically the voltage gain A_{vo} (sometimes called A_{VOL}, A_V, G_V, or G_{VOL}) at dc is 100,000 to 1,000,000 (often specified

in decibels, thus 100 dB to 120 dB), dropping to unity gain at a frequency (called f_T, or sometimes gain-bandwidth product, GBW), most often in the range of 0.1 MHz to 10 MHz. This is usually given as a graph of open-loop voltage gain as a function of frequency, on which the f_T value is clearly seen; see, for example, Figure 4.47, which shows the curve for our favorite LF411.

For *internally compensated* op-amps this graph is simply a 6 dB/octave rolloff beginning at some fairly low frequency (for the 411 it begins at about 10 Hz), an intentional characteristic necessary for stability, as we'll see in §4.9. This rolloff (the same as a simple *RC* lowpass filter) results in a constant 90° lagging phase shift from input to output (open-loop) at all frequencies above the beginning of the rolloff, increasing to 120° to 160° as the open-loop gain approaches unity. Because a 180° phase shift at a frequency where the voltage gain equals 1 will result in positive feedback (oscillations), the term "phase margin" is used to specify the difference between the phase shift at f_T and 180°.

There's a price to pay for greater bandwidth f_T, namely, higher transistor operating currents, and therefore higher op-amp supply currents. You can get op-amps with supply currents of less than 1 μA, but they have f_T's down around 10 kHz! In addition to high *supply* currents, very fast op-amps can have relatively high input bias currents, often more than a microamp, owing to their bipolar input stages operating at high collector current. Don't use fast op-amps if you don't need them – in addition to the drawbacks just mentioned, their high gain at high frequency makes it easier for your circuit to oscillate.

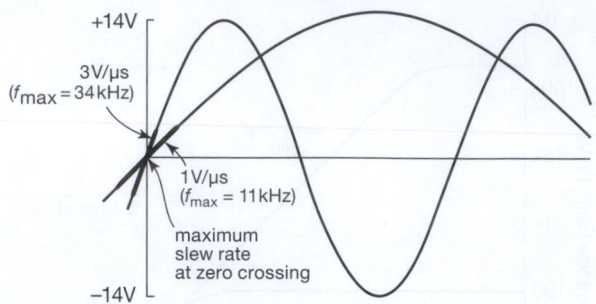

Figure 4.48. The maximum slew rate of a sinewave, SR=$2\pi A f$, occurs at the zero crossings.

K. Slew rate

The op-amp "compensation" capacitance (discussed further in §4.9.2) and small internal drive currents act together to limit the rate at which the output can change, even when a large input unbalance occurs. This limiting speed is usually specified as *slew rate* or slewing rate (SR). For the 411 it is 15 V/μs; low-power op-amps typically have slew rates less than 1 V/μs, whereas a high-speed op-amp might slew at hundreds of volts per microsecond. The slew rate limits the amplitude of an undistorted sine-wave output swing above some critical frequency (the frequency at which the full supply swing requires the maximum slew rate of the op-amp), thus the "output voltage swing as a function of frequency" graph (seen in datasheets; see for example Figure 4.54). A sine wave of frequency f hertz and amplitude A volts requires a minimum SR of $2\pi A f$ volts per second, with the peak slewing occurring at the zero crossings (Figure 4.48). Figure 4.49 shows a 'scope trace illustrating real-world "slew-rate distortion."

For externally compensated op-amps, the slew rate depends on the compensation network used. In general, it will be lowest for "unity-gain compensation," increasing to perhaps 30 times faster for $\times 100$ gain compensation. This is discussed further in §4.9.2B and in Chapter *4x*.[25] As with gain–bandwidth product f_T, higher SR op-amps run at higher supply currents.

An important note: slew rate is ordinarily specified for a unity-gain configuration (i.e., a follower) with a full-swing step input. So there is a large differential drive at the op-amp's input, which really gets the currents flowing in there. The slew rate will be considerably less for a small input, say 10 mV.

---

[25] Where we show, among other things, the fact that slew rate in conventional BJT op-amps is limited by bandwidth: $S=0.32 f_T$. Happily, this can be circumvented, with a bit of *un*conventional design.

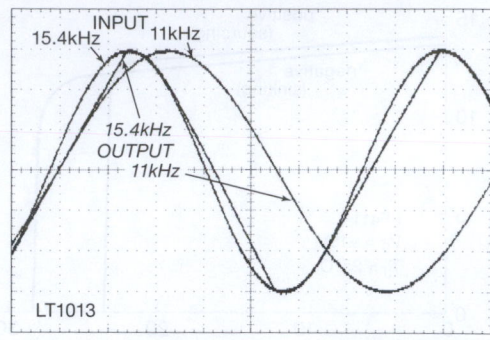

Figure 4.49. Slew-rate-induced distortion. This scope trace of an LT1013 op-amp follower, for which the datasheet specifies a typical SR of 0.4 V/μs, shows the input and output waveforms for a sinewave whose peak SR is 0.6 V/μs ($A = 6.0$ V, f=15.4 kHz); also shown is a slower sinewave, which overlays its (identical) output ($A = 6.0$ V, f=11 kHz: SR = 0.4 V/μs). Scales: 2 V/div, 10 μs/div.

L. Temperature dependence

Most of these parameters have some temperature dependence. However, this usually doesn't make any difference, since small variations in gain, for example, are almost entirely compensated by feedback. Furthermore, the variations of these parameters with temperature are typically small compared with the variations from unit to unit.

The exceptions are input offset voltage and input offset current; these input errors will matter, particularly if you've trimmed the offsets approximately to zero, and will cause drifts in the output. When high precision is important, a low-drift "instrumentation" op-amp should be used, with external loads kept above 10k to minimize the horrendous effects on input-stage performance caused by temperature gradients. We will have much more to say about this subject in Chapter 5.

M. Supply voltage and current

Traditionally, most op-amps were designed for ± 15 V power supplies, with a smattering of "single-supply" op-amps that operated on single supplies (i.e., $+V$ and ground), typically from $+5$ V to $+15$ V. The traditional split-supply op-amps were somewhat flexible; for example, the third-generation LF411 accepts supplies from ± 5 V to ± 18 V. Most of these early op-amps ran at supply currents of a few milliamps.

There has been an important trend to lower-current and especially lower-voltage operation to accommodate battery-powered equipment. So, for example, it is now common to see op-amps that operate with total supply voltages (the span from V_+ to V_-) of 5 V, or even 3 V, and run on supply currents of 10 μA to 100 μA. These are usually

gain: $G_v = -\dfrac{R_2}{R_1}$

dc error: $\Delta V_{out} = (1 + \dfrac{R_2}{R_1}) V_{os} + I_B R_2$

Figure 4.50. Inverting amplifier.

built with 100% CMOS circuitry, but there are some bipolar designs as well. These are usually rail-to-rail output stages – obviously such op-amps cannot afford the luxury of the "no-closer-than-2-volts-from-either-rail" mantra!

When considering these op-amps, watch out for unusually low maximum supply voltage restrictions. Many such op-amps are limited to as little as 10 V total supply (i.e., ± 5 V), and an increasing number are limited to 5 volts or less. Also, note that an op-amp with microamp *quiescent* current will necessarily draw plenty of current if you ask it to supply that amount of current to an attached load; output current doesn't come out of thin air.

N. Miscellany: CMRR, PSRR, e_n, i_n

For completeness, we should mention here that op-amps are also limited in common-mode rejection ratio (CMRR) and power-supply rejection ratio (PSRR), i.e., their incomplete rejection of common-mode input variations and power-supply fluctuations. This becomes more important at high frequencies, where the loop gain is decreasing and where the compensation capacitor C_C couples negative-rail fluctuations into the signal chain.

In addition, op-amps are not noiseless – they introduce both voltage noise (e_n) and current noise (i_n) at their input. These become significant limitations primarily in connection with precision circuits and low-noise amplifiers, and they will be treated in Chapters 5 and 8.

4.4.2 Effects of op-amp limitations on circuit behavior

Let's go back and look at the inverting amplifier with these limitations in mind. We'll see how they affect performance, and we'll learn how to design effectively in spite of them. With the understanding we'll get from this example, you should be able to handle other op-amp circuits. Figure 4.50 shows the circuit again.

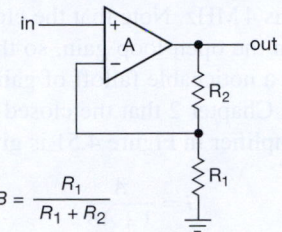

$B = \dfrac{R_1}{R_1 + R_2}$

Figure 4.51. Op-amp noninverting amplifier with finite open-loop gain.

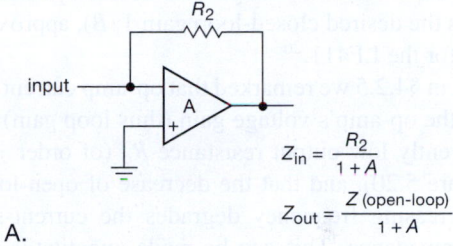

$Z_{in} = \dfrac{R_2}{1 + A}$

$Z_{out} = \dfrac{Z \,(\text{open-loop})}{1 + A}$

A.

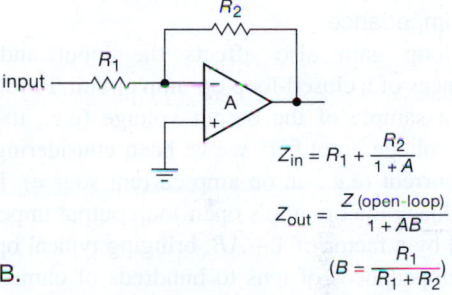

$Z_{in} = R_1 + \dfrac{R_2}{1 + A}$

$Z_{out} = \dfrac{Z \,(\text{open-loop})}{1 + AB}$

$(B = \dfrac{R_1}{R_1 + R_2})$

B.

Figure 4.52. Input and output impedances: A. transresistance amplifier, B. inverting voltage amplifier.

A. Open-loop gain

Finite open-loop gain affects bandwidth, input and output impedances, and linearity. We saw this earlier, in the context of discrete transistor amplifiers, when we introduced negative feedback in Chapter 2 (§2.5.3). That material forms an essential background to what follows here; be sure to review it if you are foggy on this stuff.

Bandwidth

Because of finite open-loop gain, the voltage gain of the amplifier with feedback (closed-loop gain) will begin dropping at a frequency where the open-loop gain approaches R_2/R_1 (Figure 4.47). For garden-variety op-amps like the 411, this means that you're dealing with a relatively low-frequency amplifier; the open-loop gain is down to 100 at

40 kHz, and f_T is 4 MHz. Note that the closed-loop gain is always less than the open-loop gain, so the overall amplifier will exhibit a noticeable falloff of gain at a fraction of f_T. Recall from Chapter 2 that the closed-loop gain of the noninverting amplifier in Figure 4.51 is given by

$$G = \frac{A}{1+AB},$$

where B is the fraction of the output fed back, in this case $B = R_1/(R_1+R_2)$. The output will therefore be down 3 dB at the frequency where the magnitude of the loop gain AB is unity (i.e., where the magnitude of the open-loop gain A equals the desired closed-loop gain $1/B$), approximately 40 kHz for the LF411.[26]

Back in §4.2.5 we remarked that op-amp current sources rely on the op-amp's voltage gain (thus loop gain) to raise its inherently low output resistance R_o (of order ~100 Ω, see Figure 5.20), and that the decrease of open-loop gain with increasing frequency degrades the current-source's output impedance. This can be made quantitative: Z_{out} at increasing frequencies is of the form $R_o \cdot f_T/f$.

Output impedance

Finite loop gain also affects the input and output impedances of a closed-loop op-amp circuit. Feedback can extract a sample of the output voltage (e.g., the noninverting voltage amplifiers we've been considering) or the output current (e.g., an op-amp current source). For voltage feedback the op-amp's open-loop output impedance is lowered by a factor of $1+AB$, bringing typical open-loop output impedances of tens to hundreds of ohms down to milliohms (for large loop gain), but rising back up to open-loop values as the loop gain falls to unity at higher frequencies.

This linear rise in closed-loop output impedance is nicely illustrated in Figure 4.53, adapted from the LT1055 datasheet. You can see how greater loop gain (feedback configured for lower closed-loop gain) produces correspondingly lower output impedance; and you can see the linear rise up to the op-amp's native R_{out} (sometimes designated r_o), here about 60 Ω. Note also that an impedance that rises linearly with frequency is like an inductor. And, in fact, that's just what the output looks like for signals in this frequency range. This can have important conse-

quences, for example, creating a series LC resonant circuit when the op-amp's load is capacitive.

The effect of the lowered loop gain (at high frequencies) is to degrade the beneficial effects of negative feedback. So a *voltage* amplifier suffers from increased output impedance, as we've seen. And the reverse is true for an amplifier with feedback that senses output *current*: here feedback normally acts to *raise* the native output impedance by a factor of loop gain (that's *good*: you want high output impedance in a current source), which then drops back to its open-loop values as the loop gain falls. Some op-amps (most notably those with rail-to-rail outputs) use an output stage with intrinsically high output impedance; for these op-amps a high loop gain is essential to achieve low output impedance.

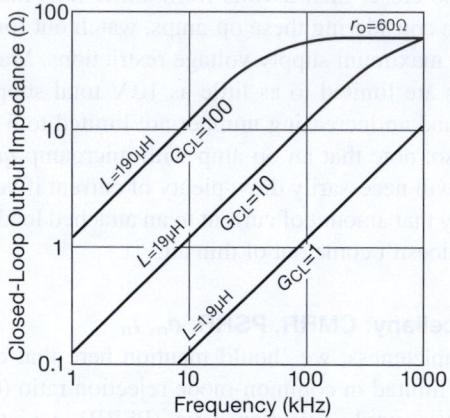

Figure 4.53. An op-amp's closed-loop output impedance rises approximately linearly with frequency over a large portion of its bandwidth, thus behaving like an inductance $L_{out} \approx r_0 G_{CL}/2\pi f_T$. After the loop gain drops to unity, Z_{out} looks like the op-amp's open-loop output resistance r_0. These curves were adapted from the LT1055 datasheet.

Input impedance

The input impedance of a noninverting amplifier is raised by a factor of $1+AB$ from its open-loop value, a matter usually of little consequence because of the high native input impedances of op-amps.

The *inverting* amplifier circuit is different from the noninverting circuit and has to be analyzed separately. It's best to think of it as a combination of an input resistor driving a shunt feedback stage (Figure 4.52). The shunt stage alone has its input at the "summing junction" (the inverting input of the amplifier), where the currents from feedback and input signals are combined (this amplifier connection is really a "transresistance" configuration; it converts a current

---

[26] The open-loop gain A has a lagging 90° phase shift over most of the op-amp's bandwidth, as can be seen from a Bode plot like Figure 4.47, i.e., you can approximate the open-loop gain, then, by $A(f)=j \cdot f_T/f$. That's why the closed-loop gain is down 3 dB, and not 6 dB, when the loop gain AB has unit magnitude.

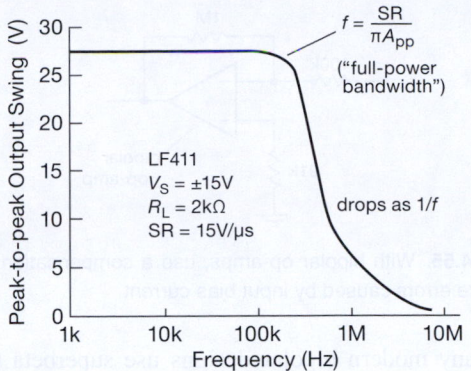

Figure 4.54. Peak-to-peak output swing versus frequency (LF411).

input to a voltage output). Feedback reduces the impedance looking into the summing junction, R_2, by a factor of $1+A$ (see if you can prove this). In cases of very high loop gain, the input impedance is reduced to a fraction of an ohm, a good characteristic for a current-input amplifier.

The classic op-amp inverting amplifier connection is a combination of a shunt feedback transresistance amplifier and a series input resistor, as in the figure. As a result, the input impedance equals the sum of R_1 and the impedance looking into the summing junction. For high loop gain, R_{in} approximately equals R_1.

It is a straightforward exercise to derive an expression for the closed-loop voltage gain of the inverting amplifier with finite loop gain. The answer is

$$G = -A(1-B)/(1+AB), \qquad (4.5)$$

where B is defined as before, $B = R_1/(R_1 + R_2)$. In the limit of large open-loop gain A, $G = -1/B + 1$ (i.e., $G = -R_2/R_1$).

Exercise 4.14. Derive the foregoing expressions for input impedance and gain of the inverting amplifier.

Linearity

In the limit of infinite loop gain, a feedback circuit's behavior depends on only the feedback network; the native nonlinearities of the op-amp (e.g., voltage dependence of gain, crossover distortion, and so on) are compensated by feedback. These defects reappear as loop gain is reduced, for example at higher frequencies. It is for this reason that you have to choose your op-amps with care, for example if you want to design low-distortion audio amplifier circuits. Op-amps intended for this sort of application have carefully designed output stages, and often they specify distor-

tion as a function of frequency and gain. An example is the excellent AD797, which specifies a maximum distortion of 0.0003% at 20 kHz and 3 V (rms) output.

B. Slew rate
Because of limited slew rate, the maximum undistorted sinewave output swing drops above a certain frequency. Figure 4.54 shows the curve for a 411, with its 15 V/μs slew rate. For slew rate S, the output amplitude is limited to $A(pp) \leq S/\pi f$ for a sinewave of frequency f, thus explaining the $1/f$ dropoff of the curve. The flat portion of the curve reflects the power-supply limits of output voltage swing. An easy formula to remember is[27]

$$S_{min} = \omega A = 2\pi f A \qquad (4.6)$$

where S_{min} is the minimum required SR for a sinewave of amplitude A (that's half the peak-to-peak amplitude: $A_{PP} = 2A$) and angular frequency ω; recall that $\omega = 2\pi f$. As an aside, the slew-rate limitation of op-amps can be usefully exploited to filter sharp noise spikes from a desired signal, with a technique known as *nonlinear lowpass filtering*: if the slew rate is deliberately limited, the fast spikes can be dramatically reduced with little distortion of the underlying signal.

C. Output current
Because of limited output-current capability, an op-amp's output swing is reduced for small load resistances, as we saw in Figure 4.44. For precision applications it is a good idea to avoid large output currents in order to prevent on-chip thermal gradients produced by excessive power dissipation in the output stage.

D. Offset voltage
Because of input offset voltage, a zero input produces an output[28] of $V_{out} = G_{dc}V_{OS} = (1 + R_2/R_1)V_{OS}$. For an inverting amplifier with a voltage gain of 100 built with a 411, the output could be as large as ± 0.2 volt when the input

---

[27] Readers comfortable with calculus will recognize this simply as the magnitude of the time derivative of a sinusoid, which brings out one factor of ω.

[28] Note that the relevant gain is the *noninverting* gain; that is because the V_{OS} error acts not at the *circuit's* input, but at the *op-amp's* input terminals. So the effect is as if the error V_{OS} were applied to the noninverting terminal of the amplifier.

is grounded ($V_{OS} = 2$ mV max). Solutions: (a) If you don't need gain at dc, use a capacitor to drop the gain to unity at dc, as in Figure 4.7B. In this case you could do that by capacitively coupling the input signal. (b) Trim the voltage offset to zero with the manufacturer's recommended trimming network. (c) Use an op-amp with smaller V_{OS}. (d) Trim the voltage offset to zero with an external trimming network, as for example in §4.8.3 (see Figure 4.91).

E. Input bias current

Even with a perfectly trimmed op-amp (i.e., $V_{OS} = 0$), our inverting amplifier circuit will produce a nonzero output voltage when its input terminal is connected to ground. That is because the finite input bias current, I_B, produces a voltage drop across the resistors, which is then amplified by the circuit's voltage gain. In this circuit the inverting input sees a driving impedance of $R_1 \| R_2$, so the bias current produces a voltage $V_{in} = I_B(R_1 \| R_2)$, which is then amplified by the gain at dc, $1 + R_2/R_1$ (see footnote 28); the result is an output error voltage of $V_{out} = I_B R_2$.

With FET-input op-amps the effect is usually negligible, but the substantial input current of bipolar op-amps (and also *current-feedback* op-amps; see Chapter *4x*) can cause real problems. For example, consider an inverting amplifier with $R_1 = 10$k and $R_2 = 1$M; these are reasonable values for an audiofrequency inverting stage, where we might like to keep Z_{in} at least 10k. If we chose the low-noise bipolar NE5534 ($I_B = 2\,\mu$A, max), the output (for grounded input) could be as large as $100 \times 2\,\mu$A$\times 9.9$k, or 1.98 volt ($G_{dc}I_B R_{unbalance}$), which is unacceptable. By comparison, for our jellybean LF411 (JFET-input) op-amp, the corresponding worst-case output (for grounded input) is 0.2 mV; for most applications this is negligible, and in any case is dwarfed by the V_{OS}-produced output error (200 mV, worst-case untrimmed, for the LF411).

There are several solutions to the problem of bias-current errors. If you must use an op-amp with large bias current, it is a good idea to ensure that both inputs see the same dc driving resistance, as in Figure 4.55. In this case, 91k is chosen as the parallel resistance of 100k and 1M. In addition, it is best to keep the resistance of the feedback network small enough so that bias current doesn't produce large offsets; typical values for the resistance seen from the op-amp inputs are 1k to 100k or so. A third cure involves reducing the gain to unity at dc, as in Figure 4.7B.

In most cases, though, the simplest solution is to use op-amps with negligible input current. Op-amps with JFET- or MOSFET-input stages generally have input currents in the picoamp range (watch out for its rapid rise versus temperature, though, roughly doubling every 10°C),

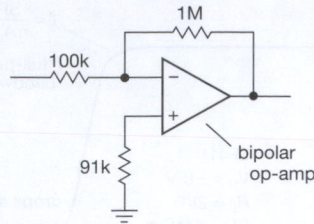

Figure 4.55. With bipolar op-amps, use a compensation resistor to reduce errors caused by input bias current.

and many modern bipolar designs use superbeta transistors or bias-cancellation schemes to achieve bias currents nearly as low and *decreasing* slightly with temperature. With these op-amps, you can have the advantages of bipolar op-amps (precision, low noise) without the annoying problems caused by input current. For example, the precision low-noise bipolar OP177 has $I_B < 2$ nA, and the bias-compensated bipolar LT1012 has $I_B = \pm 25$ pA (typ). Among inexpensive FET op-amps, the JFET LF411 has $I_B = 50$ pA (typ), and the MOSFET TLC270 series, priced under a dollar, has $I_B = 1$ pA (typ).

F. Input offset current

As we just described, it is usually best to design circuits so that circuit impedances, combined with op-amp bias current, produce negligible errors. However, occasionally it may be necessary to use an op-amp with high bias current, or to deal with signals of extraordinarily high Thévenin impedances. Examples of high-bias-current amplifiers are current-feedback op-amps (e.g., the AD844), low-noise (e_n) op-amps (e.g., the AD797), and wideband op-amps (e.g., the LM7171), each with input currents of several *microamps.*

In these cases the best you can do is to balance the dc driving resistances seen by the op-amp at its input terminals. There will still be some error at the output ($G_{dc}I_{offset}R_{source}$) that is due to unavoidable asymmetry in the op-amp input currents. In general, I_{offset} is smaller than I_{bias} by a factor of 2 to 20 (with bipolar op-amps generally showing better matching than FET op-amps).

G. Limitations imply tradeoffs

In the preceding paragraphs we discussed the effects of op-amp limitations, taking the example of the simple inverting voltage amplifier circuit. Thus, for example, op-amp input current caused a *voltage* error at the output. In a different op-amp application you may get a different effect; for example, in an op-amp integrator circuit, a finite input current produces an output *ramp* (rather than a constant) with zero applied input. As you become familiar with op-amp circuits you will be able to predict the effects of op-amp limitations in a given circuit and therefore choose which op-amp to use in a given application. In general, there is no "best" op-amp (even when price is no object): for example, op-amps with the very lowest input currents (MOSFET types) generally have larger voltage offsets and greater noise, and vice versa. Good circuit designers choose their components with the right tradeoffs to optimize performance, without going overboard on unnecessary "gold-plated" parts.

To help put this discussion of "op-amp realities" into perspective, you may want to look again at Table 4.1 on page 245, where we've summarized the kinds of performance you can expect from op-amps that might be described as average, or "jellybean" (for example, the LF412 is a jellybean JFET op-amp), and from those that are among the best available ("premium") *for each given parameter*. Sadly, you can't get an op-amp that combines all the characteristics in a "premium" column; *engineering is the art of compromise.*

The limitations of op-amp performance we have talked about will have an influence on component values in nearly all circuits. For instance, the feedback resistors must be large enough so that they don't load the output significantly, but they must not be so large that input bias current produces sizable offsets. High impedances in the feedback network cause both loading effects, and destabilizing phase shifts, from stray capacitances; they also increase susceptibility to capacitive pickup of interfering signals. These tradeoffs typically dictate resistor values of 2k to 100k with general-purpose op-amps.

Similar sorts of tradeoffs are involved in almost all electronic design, including the simplest circuits constructed with transistors. For example, the choice of quiescent current in a transistor amplifier is limited at the high end by device dissipation, increased input current, excessive supply current, and reduced current gain, whereas the lower limit of operating current is limited by leakage current, reduced current gain, and reduced speed (from stray capacitance in combination with the high resistance values). For these reasons you typically wind up with collector currents in the range of a few tens of microamps to a few tens of milliamps (higher for power circuits, sometimes a bit lower in "micropower" applications), as mentioned in Chapter 2.

In later chapters we look more carefully at some of these problems in order to convey a good understanding of the tradeoffs involved.

Exercise 4.15. Draw a dc-coupled inverting amplifier with a gain of 100 and $Z_{in} = 10k$. Include compensation for input bias current and show offset voltage trimming network (10k pot between pins 1 and 5, wiper tied to V_-). Now add circuitry so that $Z_{in} \geq 10^8 \, \Omega$.

4.4.3 Example: sensitive millivoltmeter

To put flesh on these bones, let's look at a very simple design example – a dc amplifier with lots of gain, high input impedance, and (for variety in today's too-digital world) an *analog* zero-center panel meter readout. We'll aim for ±10 mV full-scale sensitivity, and 10 megohms input impedance.

Figure 4.56 shows the initial design, where we assume we've got ±5 V supplies available (more on this later), and we use a noninverting gain of 100 to produce a ±1 V op-amp output at fullscale. That drives a 100–0–100 µA zero-center meter movement, which we decorate with a relabeled scale that reads "−10mV···0···10 mV."

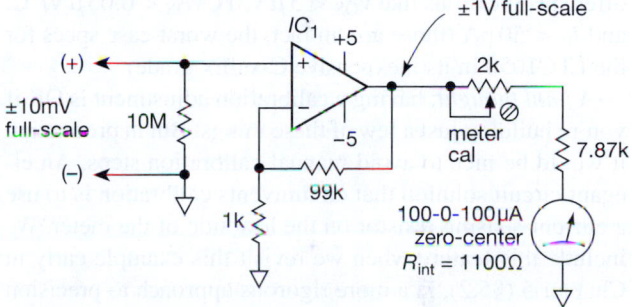

Figure 4.56. Sensitive millivoltmeter with analog readout; see text for choice of IC_1.

It looks simple, and it is. But it won't work well if we're not careful. Imagine we choose our default LF411 for IC_1, figuring that the low bias current of a JFET-input op-amp is just what we need. We short the input leads together and find to our horror that the meter reads way off center, as much as ±2 mV. That's because the 411 has V_{OS}=2 mV (max). Ideally we'd like the thing to read zero with the input leads shorted or open, where "zero" might realistically mean no more than 1% of the full-scale reading.

OK, we add an offset trimmer and adjust it until the output reads zero with shorted input. We leave it on the bench, go to lunch, then come back and find it now reads −0.2 mV with shorted input. That's because it sat in the sun, warmed up by 10°C, and therefore drifted by 200 µV (the LF411 has a temperature coefficient of offset voltage TCV_{OS}=20 µV/°C, max). Well, we won't use it in the sun!

So, we wait for it to cool down, and note with satisfaction that it reads zero again.

Now we go to test it on a voltage divider, but we find that when we remove the short (test leads open circuited) the meter reads +2 mV! This time the problem is bias current, specified as 200 pA (max) at room temperature. That's not much current, but it develops 2000 μV across the 10M input resistor, which the meter dutifully reports as an input.

We could solve the V_{OS} problem with a precision bipolar op-amp, but we'd be in worse trouble with I_B. We need an op-amp with $V_{OS} < 100\,\mu$V and $I_B < 10$ pA. The solution is a precision FET-input op-amp, for example, the OPA336 (125 μV untrimmed, and 10 pA), which is just about good enough without trimming, and certainly fine if we're willing to trim the initial offset.

A better solution here is to use a "chopper" op-amp like the LTC1050C or AD8638 (see Table 5.6). These are sometimes called "zero-drift," or "auto-zeroing" amplifiers. We'll learn about them shortly, and in more detail in the next chapter; for now all you need to know is that they offer specifications like $V_{OS} < 5\,\mu$V, $TCV_{OS} < 0.05\,\mu$V/°C, and $I_B < 50$ pA (those are, in fact, the worst-case specs for the LTC1050 in its inexpensive C-suffix grade).

A final thought: having a calibration adjustment is OK if you're building just a few of these things. But in production it would be nice to avoid manual calibration steps. An elegant circuit solution that circumvents calibration is to use a current-sensing resistor on the low side of the meter. We include this feature when we revisit this example early in Chapter 5 (§5.2), in a more rigorous approach to precision design.

4.4.4 Bandwidth and the op-amp current source

Back in §4.2.5 we remarked that op-amp current sources rely on the op-amp's voltage gain (thus loop gain) to raise its inherently low output resistance R_o (of order ~100 Ω, see Figure 5.20), and that the decrease of open-loop gain with increasing frequency degrades the current-source's output impedance. Put another way, the op-amp current source is a peculiar circuit, because an op-amp virtue (inherently low output impedance, i.e., a voltage source) becomes a vice, which must be punished with the cudgel of plentiful loop gain. This can be made quantitative: because of finite bandwidth f_T, the output impedance of an op-amp current source at increasing frequencies is of the form $R_o \cdot f_T / f$, dropping ultimately to the op-amp's native output resistance R_o at the unity-gain frequency f_T.

Likewise, finite slew rate affects the current-source's output impedance, making it look like a shunt capacitance.

Here's how to think about it: an ideal current source with a *real* capacitive load slews at a rate $S = dV/dt = I/C$; so a current source that is afflicted with a maximum slew rate S looks like an ideal current source burdened with an effective shunt capacitance $C_{eff} = I_{out}/S$. For example, a 10 mA current source made with an op-amp of slew rate 1 V/μs has an effective capacitive load of 10 nF; this is rather large, compared even with a large MOSFET.

4.5 A detailed look at selected op-amp circuits

The performance of the next few circuits is affected significantly by the limitations of op-amps; we will go into a bit more detail in their description.

4.5.1 Active peak detector

There are numerous applications in which it is necessary to determine the peak value of some input waveform. The simplest method is a diode and capacitor (Figure 4.57). The highest point of the input waveform charges up C, which holds that value while the diode is back-biased.

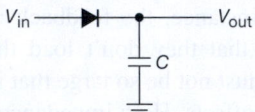

Figure 4.57. Passive peak detector.

This method has some serious problems. The input impedance is variable and is very low during peaks of the input waveform. Also, the diode drop makes the circuit insensitive to peaks less than about 0.6 volt, and inaccurate (by one diode drop) for larger peak voltages. Furthermore, since the diode drop depends on temperature and current, the circuit's inaccuracies depend on the ambient temperature and on the rate of change of output; recall that $I = C(dV/dt)$. An input emitter follower would improve the first problem only.

Figure 4.58A shows a better circuit, which exhibits the benefits of feedback. By taking feedback from the voltage at the capacitor, the diode drop doesn't cause any problems. The sort of output waveform you might get is shown in Figure 4.59.

Op-amp limitations affect this circuit in three ways.

(a) A finite op-amp slew rate causes a problem, even with relatively slow input waveforms. To understand this, note that the op-amp's output goes into negative saturation when the input is less positive than the output

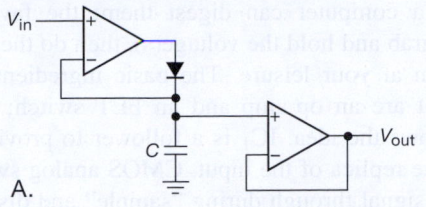

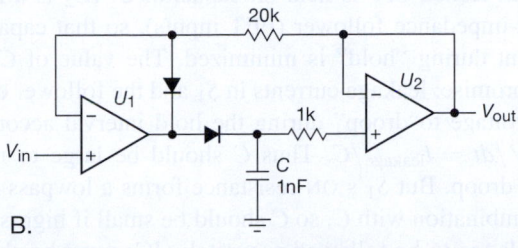

Figure 4.58. A. Op-amp peak detector (more accurately, a "peak tracker"). B. Improved peak tracker responds to short peaks, because the input op-amp does not have to slew from negative saturation.

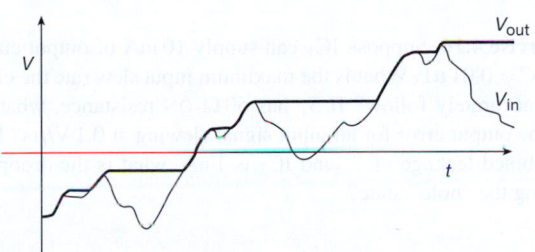

Figure 4.59. Peak detector output waveform.

(try sketching the op-amp voltage on the graph; don't forget about diode forward drop). So the op-amp's output has to race back up to the output voltage (plus a diode drop) when the input waveform next exceeds the output. At slew rate S, this takes roughly $(V_{out} - V_-)/S$, where V_- is the negative supply voltage. Improved circuit 4.58B solves this problem.

(b) Input bias current causes a slow discharge (or charge, depending on the sign of the bias current) of the capacitor. This is sometimes called "droop," and it is best avoided by using op-amps with very low bias current. For the same reason, the diode must be a low-leakage type (e.g., the FJH1100, with less than 1 pA reverse current at 20 V, an "FET diode" such as the PAD5, or a diode-connected JFET such as the 2N4417; see the diode discussion in Chapter *1x*), and the follow-

ing stage must also present high impedance (ideally it should also be an FET or FET-input op-amp).

(c) The maximum op-amp output current limits the rate of change of voltage across the capacitor, i.e., the rate at which the output can follow a rising input. Thus the choice of capacitor value is a compromise between low droop and high output slew rate.

For instance, a $1 \, \mu F$ capacitor used in this circuit with the common LM358 (which would be a poor choice because of its high bias current) would droop at $dV/dt = I_B/C = 0.04 \, V/s$ (using the "typical" value $I_B = 40 \, nA$; the worst-case value of $I_B = 500 \, nA$ produces a droop of $0.5 \, V/s$) and would follow input changes only up to $dV/dt = I_{output}/C = 0.02 \, V/\mu s$. This maximum follow rate is much less than the op-amp's slew rate of $0.5 \, V/\mu s$, being limited by the maximum output current of 20 mA driving $1 \, \mu F$. By decreasing C you could achieve greater output slewing rate, at the expense of greater droop. A more realistic choice of components would be the popular TLC2272 MOSFET-input op-amp as driver and output follower (1 pA typical bias current) and a value of $C = 0.01 \, \mu F$. With this combination you would get a droop of only $0.0001 \, V/s$ and an overall circuit slew rate of $2 \, V/\mu s$. For even better performance, use a MOSFET op-amp like the LMC660 or LMC6041, with a typical input current of 2 *femto*amps. Capacitor leakage (or diode leakage or both) may then limit performance even if unusually good capacitors are used, e.g., polystyrene or polypropylene (see Chapter *1x*).[29]

A. Resetting a peak detector

In practice it is usually desirable to reset the output of a peak detector in some way. One possibility is to put a resistor across the peak-hold capacitor so that the circuit's output decays with a time constant RC. In this way it holds only the most recent peak values. A better method is to put a transistor switch across C; a short pulse to the base then zeros the output. An FET switch is often used instead. For example, in Figure 4.58 you could connect an n-channel MOSFET such as a 2N7000 across C; bringing the gate

---

[29] There's more to capacitor "leakage" than one might at first suspect: an effect known as *dielectric absorption* ("DA") can cause serious mischief in circuits that rely on ideal capacitor performance. It manifests itself rather clearly in the following simple experiment: charge a tantalum capacitor up to 10 volts or so, let it sit there for a while, then rapidly discharge it by momentarily putting a $100 \, \Omega$ resistor across it. Remove the resistor and watch the capacitor's voltage on a high-impedance voltmeter. You will be amazed to see the capacitor *charge back up*, reaching perhaps a volt or so after a few seconds! This unhelpful effect is treated in more detail in Chapter *1x* and §5.6.2.

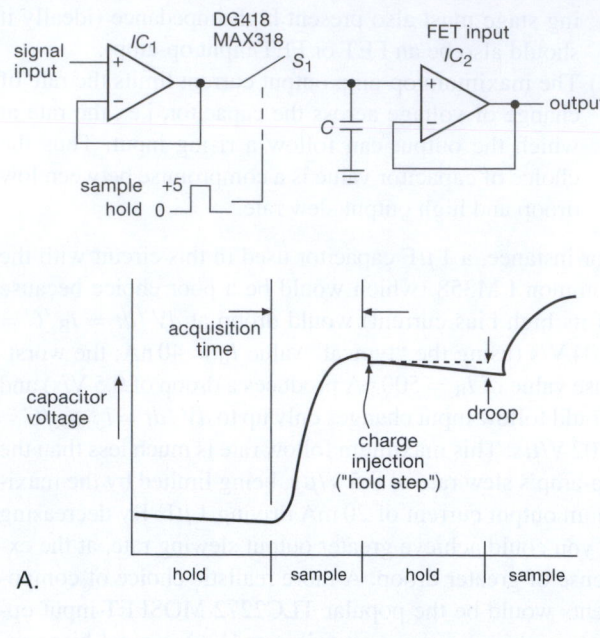

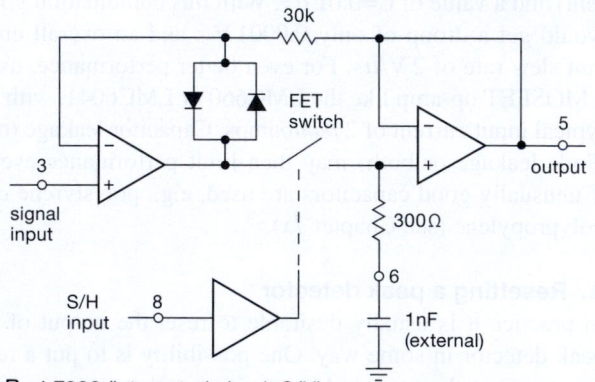

B. LF398 (integrated circuit S/H)

Figure 4.60. Sample-and-hold: A. Standard configuration, with exaggerated waveform; B. LF398 single-chip S/H.

momentarily positive then zeros the capacitor voltage. An integrated CMOS analog switch (such as the MAX318, with a small series resistor to limit the current) can be used instead of a discrete nMOS (*n*-type metal-oxide semiconductor) transistor.

4.5.2 Sample-and-hold

Closely related to the peak detector is the "sample-and-hold" (S/H) circuit (sometimes called "follow-and-hold"). These are especially popular in digital systems in which you want to convert one or more analog voltages to num-

bers so that a computer can digest them: the favorite method is to grab and hold the voltage(s), then do the digital conversion at your leisure. The basic ingredients of an S/H circuit are an op-amp and an FET switch; Figure 4.60A shows the idea. IC_1 is a follower to provide a low-impedance replica of the input. CMOS analog switch S_1 passes the signal through during "sample" and disconnects it during "hold." Whatever signal was present when S_1 was turned OFF is held on capacitor C. IC_2 is a high-input-impedance follower (FET inputs), so that capacitor current during "hold" is minimized. The value of C is a compromise: leakage currents in S_1 and the follower cause C's voltage to "droop" during the hold interval according to $dV/dt = I_{\text{leakage}}/C$. Thus C should be large to minimize droop. But S_1's ON-resistance forms a lowpass filter in combination with C, so C should be small if high-speed signals are to be followed accurately. IC_1 must be able to supply C's charging current ($I = C dV/dt$) and must have sufficient slew rate to follow the input signal. In practice, the slew rate of the whole circuit will usually be limited by IC_1's output current and S_1's ON-resistance.

Exercise 4.16. Suppose IC_1 can supply 10 mA of output current and $C = 0.01\,\mu\text{F}$. What is the maximum input slew rate the circuit can accurately follow? If S_1 has 50 Ω ON resistance, what will be the output error for an input signal slewing at 0.1 V/μs? If the combined leakage of S_1 and IC_2 is 1 nA, what is the droop rate during the "hold" state?

For both the S/H circuit and the peak detector, an op-amp drives a capacitive load. When designing such circuits, make sure you choose an op-amp that is stable at unity gain when loaded by the capacitor C. Some op-amps, (e.g., the LT1457, a member of Linear Technology's "CLOAD^{TM} stable" op-amps) are specifically designed to drive large (0.01 μF) capacitive loads directly. Some other tricks you can use are discussed in §4.6.1B and in Chapter *4x*.

You don't have to design S/H circuits from scratch, because there are nice monolithic ICs that contain all the parts you need. National's LF398 is a popular part, containing the FET switch and two op-amps in an inexpensive (\$1.25) 8-pin package. Figure 4.60B shows how to use it. Note how feedback closes the loop around *both* op-amps. There are plenty of fancy S/H chips available, if you need better performance than the LF398 offers. For example, the AD783 from Analog Devices includes an internal capacitor and guarantees a maximum acquisition time of 0.4 μs for 0.01% accuracy following a 5 volt step.

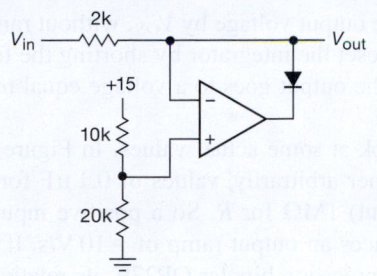

Figure 4.61. Active clamp.

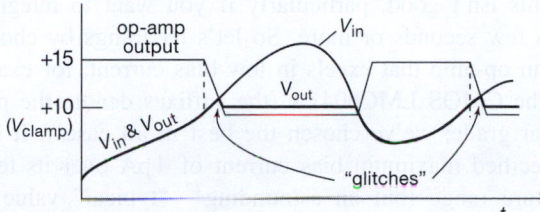

Figure 4.62. Finite slew-rate causes clamp output "glitches".

4.5.3 Active clamp

Figure 4.61 shows a circuit that is an active version of the clamp function we discussed in Chapter 1. For the values shown, $V_{in} < +10$ volts puts the op-amp output at positive saturation, and $V_{out} = V_{in}$. When V_{in} exceeds $+10$ volts the diode closes the feedback loop, clamping the output at 10 volts. In this circuit, op-amp slew-rate limitations allow small glitches as the input reaches the clamp voltage from below (Figure 4.62).

Exercise 4.17. The active clamp in Figure 4.61 suffers from a slew-rate speed limitation similar to that of the peak tracker of Figure 4.58A. Figure out an improvement to the clamp circuit, analogous to the trick used in Figure 4.58B.

4.5.4 Absolute-value circuit

The circuit shown in Figure 4.63 gives a positive output equal to the magnitude of the input signal; it is a full-wave rectifier. As usual, the use of op-amps and feedback eliminates the diode drops of a passive full-wave rectifier.

You can imagine situations in which you want an output proportional to the *logarithm* of the absolute value. A simple circuit change might be to substitute a diode (or transistor with base tied to collector) for the feedback resistor of the second op-amp, exploiting the Ebers–Moll relation between diode voltage and summing-junction current. As we will see in Chapter *4x*, this is the basis of the *logarithmic amplifier*; the circuit needs a few

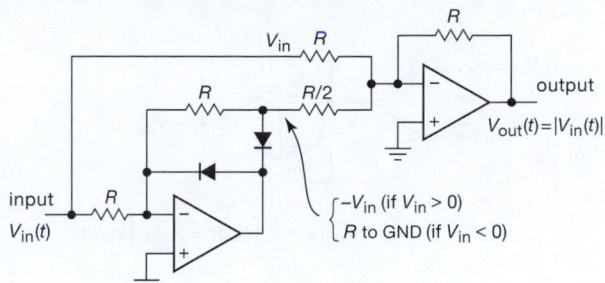

Figure 4.63. Active full-wave rectifier.

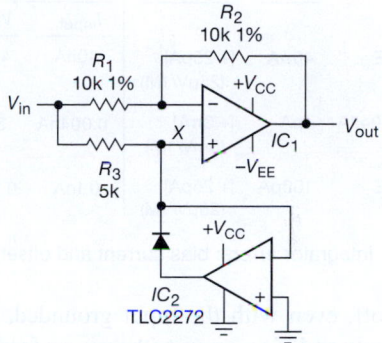

Figure 4.64. Another full-wave rectifier; note that ground is the negative supply voltage for IC_2.

additional components, however, to compensate for the temperature coefficient of V_{BE}.

Exercise 4.18. Figure out how the circuit in Figure 4.63 works. *Hint*: apply first a positive input voltage and see what happens; then do negative.

Figure 4.64 shows another absolute-value circuit. It is readily understandable as a simple combination of an optional inverter (IC_1) and an active clamp (IC_2). For negative input levels the clamp holds point X at ground, making IC_1 a unity-gain inverter; for positive input levels, the clamp is out of the circuit, with its output at negative saturation, making IC_1 a follower. Thus the output is equal to the absolute value of the input voltage. By running IC_2 from a single positive supply, you avoid problems of slew-rate limitations in the clamp, since its output moves over only one diode drop. Note that no great accuracy is required of R_3.

4.5.5 A closer look at the integrator

We introduced the op-amp integrator in §4.2.6, before dealing with input bias current and offset voltage. One problem with that circuit (Figure 4.16) is that the output tends

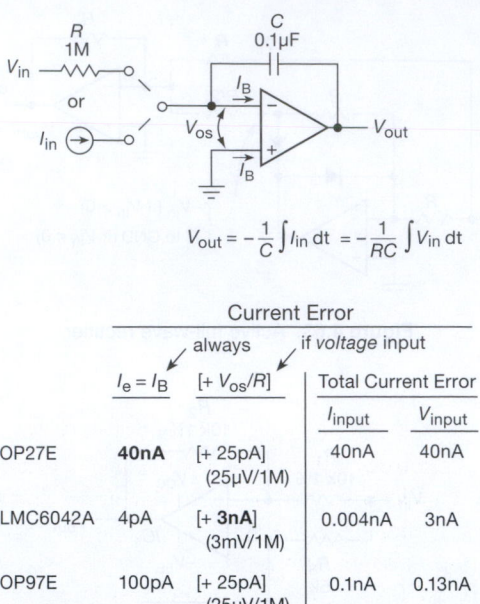

$$V_{out} = -\frac{1}{C}\int I_{in}\,dt = -\frac{1}{RC}\int V_{in}\,dt$$

	Current Error			
		always	if *voltage* input	
	$I_e = I_B$	$[+\,V_{os}/R]$	Total Current Error	
			I_{input}	V_{input}
OP27E	**40nA**	[+ 25pA]	40nA	40nA
		(25μV/1M)		
LMC6042A	4pA	[+ **3nA**]	0.004nA	3nA
		(3mV/1M)		
OP97E	100pA	[+ 25pA]	0.1nA	0.13nA
		(25μV/1M)		

Figure 4.65. Integrator errors: bias current and offset voltage.

to wander off, even with the input grounded, because of op-amp offsets and bias current (there's no feedback at dc, which violates the third item in §4.2.7). This problem can be minimized by using a FET op-amp for low input current and offset, trimming the op-amp input offset voltage, and using large R and C values. Furthermore, in applications in which the integrator is zeroed periodically by closing a switch placed across the capacitor (Figures 4.18A–C), only the drift over short time scales matters.

It's worth looking at this in a bit more detail. Look at the integrator in Figure 4.65, shown with a choice of voltage input V_{in} (which, in the absence of op-amp errors, produces a current into the summing junction of $I = V_{in}/R$), or a current input I_{in} (in which case you omit the input resistor R). The ideal integrator produces an output

$$V_{out}(t) = -\frac{1}{C}\int I_{in}(t)\,dt = -\frac{1}{RC}\int V_{in}(t)\,dt.$$

It's easy enough to figure out the effect of the op-amp's input errors I_B and V_{OS}. Let's first take the case of an integrator circuit with *current* input.[30] The op-amp's bias current I_B adds (or subtracts) from the true input current I_{in}; in the absence of any external input current the integrator's output would ramp at a rate $dV_{out}/dt = I_B/C$. The effect of op-amp input offset voltage, on the other hand, is simply

to offset the output voltage by V_{OS}, without ramping;[31] so, when you reset the integrator by shorting the feedback capacitor C, the output goes to a voltage equal to V_{OS} rather than zero.

Let's look at some actual values. In Figure 4.65 we've chosen, rather arbitrarily, values of 0.1 μF for C and (for voltage input) 1MΩ for R. So a positive input current of 1 μA produces an output ramp of -10 V/s. If we were to choose the precision bipolar OP27E, its relatively high input current of ± 40 nA (max) would cause an output ramp of as much as $dV_{out}/dt = I_B/C = \pm 0.4$ V/s.

This isn't good, particularly if you want to integrate for a few seconds or more. So let's fix things by choosing an op-amp that excels in low bias current, for example the CMOS LMC6041A (the suffixes denote the particular grade; we've chosen the best in all cases). It has a specified maximum bias current of 4 pA over its temperature range (but an astounding[32] "typical" value of 2 fA, or 2×10^{-15} A). Now the worst-case output ramp, in the absence of any input signal current, is reduced to $dV_{out}/dt = I_B/C = \pm 40$ μV/s. The "typical" ramping rate is 2000 times smaller, if the specs can be believed; that's a mere 0.02 μV/s.

At this point, the lesson seems to be that the best op-amp for any integrator is the one with the smallest bias current I_B. But, alas, life is more complicated. In particular, if the integrator is wired for *voltage* input, with a series input resistor R, then the op-amp's offset voltage V_{OS} now produces a ramp when the input to the circuit is held at ground. Imagine the input is grounded ($V_{in} = 0$), and think about it this way: the op-amp strives to align its inputs with a voltage V_{OS} between them; that small voltage then produces a current $I = V_{OS}/R$ through the input resistor. That current has to come through the feedback capacitor, that is, the output must ramp to produce the current needed to satisfy the op-amp's twisted belief that its inputs should differ by V_{OS}. Another way to say it is that the current acts just like an input current of $I = -V_{OS}/R$.

Now the choice of op-amp is not so clear! Look at Figure 4.65 again. The CMOS op-amp with its very low bias current has a pretty large offset voltage, $V_{OS} = 3$ mV (max). So in this circuit it can produce an equivalent input current of 3 nA (3 mV across 1 MΩ); that's nearly a thousand times

---

[30] Examples of signals that are naturally in the form of a current include those from a photodiode, a PMT, or an ion detector, or from dielectric, semiconductor or nanomaterial measurements.

[31] If the input signal is not a true current source, but rather comes from a voltage V_{in} in series with a resistor R_{in}, then the op-amp's V_{os} causes an additional small error current of V_{os}/R_{in}.

[32] Particularly when considering that its input is protected by diode clamps to the supply rails. How did these magicians accomplish this? (The same way you shake hands with a gorilla – *very carefully!*).

larger than the worst-case contribution of its bias current, and it's getting into the same ballpark as the input current of the OP27E bipolar op-amp we considered first.

If minimum drift is needed with these particular circuit values, the solution is to choose an op-amp with the best compromise of low bias current and low offset voltage; to be precise, it should have the minimum value of the total worst-case error current $I_E = I_B + V_{OS}/R$. A good choice would be the bipolar OP97E, a precision (low-offset) op-amp with internal bias-cancellation circuitry. It sports maximum values of $I_B = 0.1$ nA and $V_{OS} = 25\,\mu V$; the corresponding worst-case current error is $I_E = 0.125$ nA, which is 25 times better than that of the LMC6041A and 320 times better than that of the OP27.

Note that the relative contribution of V_{OS} and I_B to integrator error is scaled by the value of R. So you can simply choose a larger resistor value if you have an op-amp with excellent I_B but only modest V_{OS}.

If the residual drift of an integrator circuit is still too large for a given application, or if long-term accuracy is unimportant, one solution is to put a large resistor R_2 across C to provide dc feedback for stable biasing, as shown in Figure 4.18D. The effect is to roll off the integrator action at very low frequencies, $f < 1/R_2C$. The feedback resistor may become rather large in this sort of application. Figure 4.66 shows a trick for producing the effect of a large feedback resistor using smaller values in any op-amp circuit. In this case – an inverting amplifier circuit – the feedback network behaves like a single $10\,M\Omega$ resistor, thus producing a voltage gain of -100. This technique has the advantage of using resistors of convenient values without the problems of stray capacitance, etc., that occur with very large resistor values. Note that this "T-network" trick may increase the effective input offset voltage if used in a transresistance configuration (§4.3.1C). For example, the circuit of Figure 4.66, driven from a high-impedance source (e.g., the current from a photodiode, with the input resistor omitted), has an output offset of 100 times V_{OS}, whereas the same circuit with a $10\,M\Omega$ feedback resistor has an output equal to V_{OS} (assuming the offset that is due to input current is negligible).

4.5.6 A circuit cure for FET leakage

Sometimes a circuit technique is so elegant and fascinating that we feel compelled to tell others about it. That's the case for the circuit in this section, brought over and updated from our 2nd edition. Read it and delight in its cleverness; but then check our remarks in the concluding paragraph.

In the integrator with a FET reset switch (Figure 4.18),

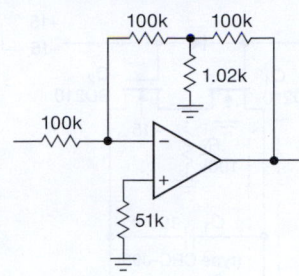

Figure 4.66. "T-network" simulates large-value resistor (here $10M\Omega$).

drain-source leakage sources a small current into the summing junction even when the FET is OFF. With an ultra-low-input-current op-amp and low-leakage capacitor, this can be the dominant error in the integrator. For example, the excellent LMC6001A CMOS-input "electrometer" op-amp has a maximum input current of 0.025 pA, and a high-quality $0.1\,\mu F$ metallized Teflon or polystyrene capacitor specifies leakage resistance as 10^7 *megohms*, minimum. Thus the integrator, exclusive of reset circuit, keeps stray currents at the summing junction below 1 pA (for a worst-case 10 V full-scale output), corresponding to an output dV/dt of less than 0.01 mV/s. Compare this with the leakage contribution of a MOSFET such as the SD210 (enhancement mode), which specifies a maximum leakage current of 10 nA at $V_{DS}=10$ V and $V_{GS}=-5$ V! In other words, the reset FET can contribute up to 10,000 times as much leakage as everything else combined.

Figure 4.67 shows a clever circuit solution. Although both n-channel MOSFETs are switched together, Q_1 is switched with gate voltages of 0 and $+15$ volts so that gate leakage (as well as drain-source leakage) is entirely eliminated during the OFF state (zero gate voltage). In the ON state the capacitor is discharged as before, but with twice R_{ON}. In the OFF state, Q_2's small leakage passes to ground through R_2 with negligible drop. There is no leakage current at the summing junction because Q_1's source, drain, and substrate are all at the same voltage. (Sharp-eyed readers may have noticed that the virtual ground at the op-amp's inverting input is imperfect to the extent of its offset voltage V_{OS}.[33] This can be trimmed to eliminate completely any leakage current from Q_1.)

The ultimate limit to capacitor "droop" in this circuit, once FET switch leakage has been eliminated, is set by the op-amp's input current and by the capacitor's self-discharge. The capacitor shown has a specified[34] leakage

---

[33] "There isn't a *literal* connection, Dude."

[34] In careful multiyear measurements of leakage in some polyester and

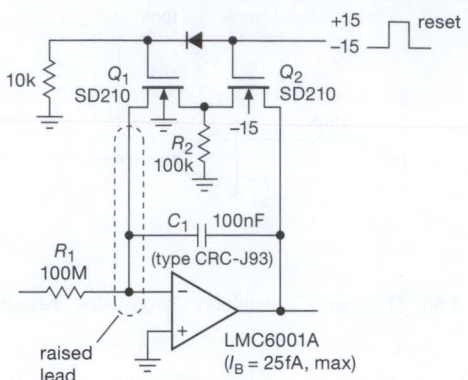

Figure 4.67. MOSFET leakage defeated by clever circuit.

resistance of 10^7 MΩ, i.e., 10^{13} Ω. The resulting leakage currents, of order 10^{-16} A (following reset), are entirely negligible compared with op-amp bias currents. For the op-amp shown, the specified bias current is 25 fA maximum (10 fA typ) at 25°C; that bias current produces a maximum droop of 0.25 μV/s. There are no op-amps with lower specified maximum bias current currently available; but you can find op-amps whose "typical" bias current is lower – for example, the LMC6041, an inexpensive op-amp whose datasheet proclaims I_B =2 fA (typ) at 25°C (no maximum is given at room temperature; I_B =4 pA max over the full temperature range). What is one to make of an op-amp whose typical input current is 2000 times smaller than the guaranteed maximum? Just that the manufacturer *knows* that it's very good, but it's too painful to test in production. You'd do well to use these inexpensive units in such a circuit if you're willing to screen the op-amps yourself; otherwise pay the bounty for a unit that has a guaranteed limit (but note that the LMC6001A, such a unit, has a typical I_B that is five times higher than that of the less expensive LMC6041).

When designing circuits where low input current is needed, watch out for temperature effects: all FET op-amps (both JFET and CMOS types) exhibit dramatic increases in input current with rising temperature, typically doubling each 10°C; the LMC6001A's guaranteed maximum bias current jumps from 25 fA at 25°C to 2000 fA at 85°C. At high temperatures, the input (leakage) currents

polypropylene capacitors at low voltages, Tom Bruins of Agilent found actual leakage resistances to be some 10,000 times larger than specified. Using 160 V capacitors at 10 V, he measured time constants (in seconds, sometimes called "megohm-microfarad products") of 2×10^7 seconds (polyester) and $>10^9$ seconds (polypropylene). In fact, what is commonly called "leakage" in such capacitors appears mostly to be dielectric absorption currents; see §5.6.2 and Chapter *1x*.

of FET op-amps may often be higher than the input (bias) currents of low-I_B bipolar types; that is because leakage currents rise exponentially with temperature, whereas transistor bias currents remain roughly constant (or decrease slightly). Look back at Figure 3.48 for a good illustration; see also Figures 5.6 and 5.38.

It may be difficult to find discrete low-capacitance MOSFETs with substrate pins; currently the SD210 family (with SST-prefix SMT versions) is available from Linear Systems (Fremont, CA). The two-switch T-configuration is sound, though it may be challenging to find suitable switch components without substrate-diode conduction, etc. These MOSFETs work well, but if they become "unobtanium" we suggest you modify the circuit to use JFETs, in the manner of Figure 5.5.

4.5.7 Differentiators

Differentiators are similar to integrators, but with R and C reversed (Figure 4.68). Because the inverting input is at ground, the rate of change of input voltage produces a current $I = C(dV_{in}/dt)$ and hence an output voltage

$$V_{out} = -RC\frac{dV_{in}}{dt}. \tag{4.7}$$

Differentiators are bias stable, but they generally have problems with noise and instabilities at high frequencies because of the op-amp's high gain and internal phase shifts. For this reason it is necessary to roll off the differentiator action at some maximum frequency. The usual method is shown in Figure 4.69. The choice of the rolloff components R_1 and C_2 depends on the noise level of the signal and the bandwidth of the op-amp, with larger values providing greater stability and less noise, at the expense of differentiator bandwidth. A minimum recommended value for R_1 is given by $R_1 = 0.5\sqrt{R_2/C_1 f_T}$; C_2 can be added for further noise reduction, with a starting value of $C_2 \approx C_1 R_1/R_2$. At high frequencies ($f \gg 1/2\pi R_1 C_1$) this circuit becomes an integrator because of R_1 and C_2. We'll explain what's going on here in more detail in §4.9.3.

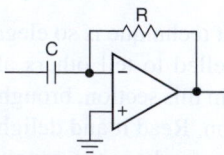

Figure 4.68. Op-amp differentiator (noisy, probably unstable!).

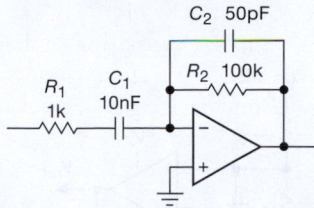

Figure 4.69. Adding R_1 and C_2 stabilizes the basic op-amp differentiator (consisting of C_1, R_2, and the op-amp); they also reduce high-frequency noise.

4.6 Op-amp operation with a single power supply

Op-amps don't *require* ±15 volt regulated supplies. They can be operated from split supplies of lower voltages[35] or from unsymmetrical supply voltages (e.g, +12 and −3), as long as the total supply voltage ($V_+ − V_−$) is within specifications (see Table 4.1 on page 245 for generic values, and Tables 4.2a,b on pages 271–272 for specific parts). Unregulated supply voltages are often adequate because of the high "power-supply rejection ratio" you get from negative feedback (for the 411 it's 90 dB typ). But there are many occasions when it would be nice to operate an op-amp from a single supply, say +9 volts. This can be done with ordinary op-amps by generating a "reference" voltage above ground, if you are careful about minimum supply voltages, output-swing limitations, and maximum common-mode input range.

In many cases, however, you can simplify these circuits by taking advantage of a class of op-amps designed for single supply operation. With characteristic directness, engineers call these "single-supply op-amps." Their common feature is that both their input common-mode range and their output swing extends to the negative supply rail (i.e., ground, when run from a single positive supply). A subclass of these can swing their outputs to *both* supplies ("rail-to-rail outputs"), and some of those even permit input swings to both rails ("rail-to-rail I/O"). Keep in mind, though, that operation with symmetrical split supplies should be considered the normal op-amp technique for most applications.

4.6.1 Biasing single-supply ac amplifiers

For a general-purpose op-amp like the 411, the inputs and output can typically swing to within about 1.5 volts of either supply. With $V_−$ connected to ground, you can't have either of the inputs or the output at ground; that is, it won't work properly if you drive the inputs to ground, and it simply cannot swing its output to ground.

Thus one reason why the circuit in Figure 4.70 won't work is that the ac-coupled low-level signal from the microphone is centered on ground, where the op-amp will not work. But even if the op-amp's input common-mode range included the negative rail (ground, here), we'd still be in trouble, because in this circuit the amplified output would also be centered on ground (so that it would have to swing both above and below ground). It is important to understand that this problem with the output cannot be solved in this manner – an op-amp simply cannot swing beyond its supply rails. Even an op-amp with rail-to-rail inputs *and* outputs would not work.

Exercise 4.19. Draw a sketch of the output waveform from the circuit of Figure 4.70, when driven with a 10 mV input sinewave, assuming that the op-amp is of the special class with rail-to-rail inputs and outputs.

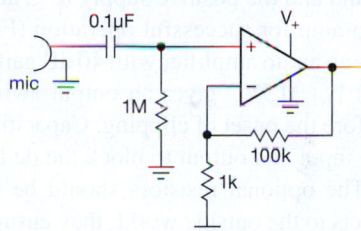

Figure 4.70. Defective single-supply microphone amplifier.

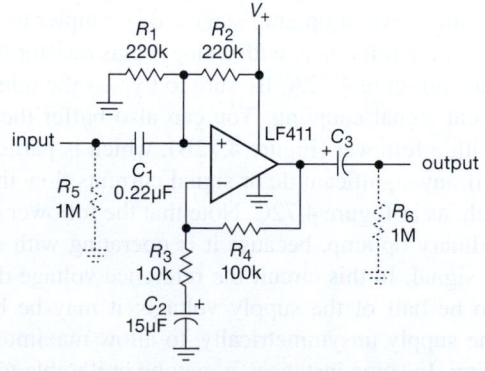

Figure 4.71. A reference voltage at $\frac{1}{2}V_+$ (created by divider R_1R_2) allows single-supply operation with an ordinary op-amp.

---

[35] In a world of lower and lower operating voltages, op-amps that can work at ±15 V are now called "high-voltage op-amps."

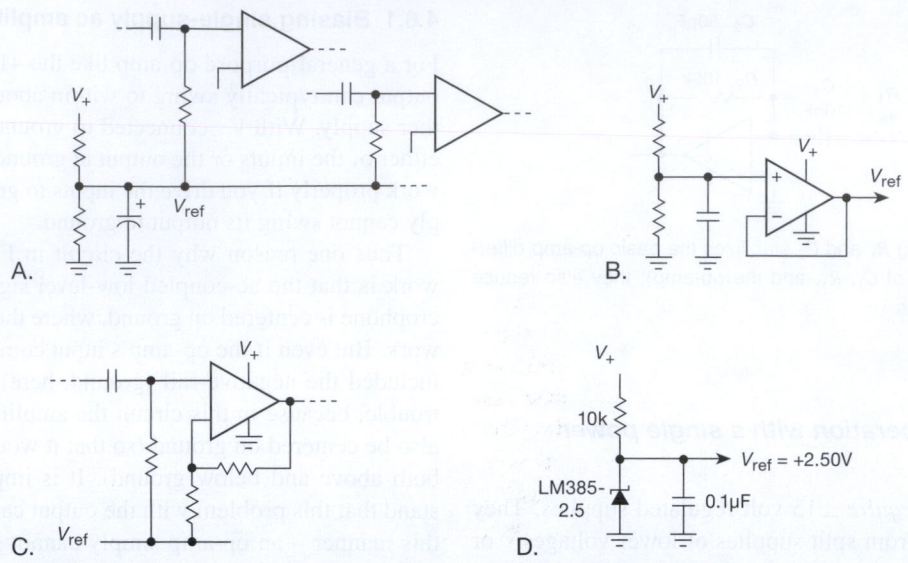

Figure 4.72. Biasing schemes for single-supply operation A. Common reference (also known, confusingly, as a "virtual ground") for multiple stages; note bypass capacitor. B. Follower generates low-impedance reference. C. The reference can serve as the return path for feedback, with significant signal currents. D. Zener-type fixed voltage reference.

A. Reference voltage

One solution is to generate a *reference voltage* somewhere between ground and the positive supply (e.g. at half of V_+) to bias the op-amp for successful operation (Figure 4.71). This circuit is an audio amplifier with 40 dB gain. Choosing $V_+=12$ V and $V_{ref}=0.5V_+$ gives an output swing of about 9 volts pp before the onset of clipping. Capacitive coupling is used at the input and output to block the dc level, which equals V_{ref}. The optional resistors should be used if this circuit connects to the outside world; they ensure that there is no dc voltage at the input and output, which prevents loud clicks and pops when external stuff is connected.

The reference voltage can be generated at the op-amp input with a simple resistive divider, as shown. If the circuit requires several op-amp stages, it is simpler to generate a common reference, with a single bias resistor to each stage, as in Figure 4.72A. Be sure to bypass the reference, to prevent signal coupling. You can also buffer the reference with a follower (Figure 4.72B), which is particularly useful if any significant dc or signal currents flow through that path, as in Figure 4.72C. Note that the follower can be any ordinary op-amp, because it is operating with a mid-supply signal. In this circuit the reference voltage doesn't have to be half of the supply voltage; it may be best to split the supply unsymmetrically, to allow maximum signal swing. In some instances it may be preferable to put it at a fixed voltage from one rail, using an IC zener-like fixed

reference, as in Figure 4.72D; that rail is then a regulated supply with respect to the common reference.

Contemporary circuit design is moving to lower supply voltages, often in the form of a single positive supply. For operation with a single +5 V supply, for example, a conventional op-amp like the 411 simply will not do: Not only can its outputs not swing typically closer than 1.5 V to the supply rails; in fact, it is not even specified for operation from a total supply voltage of less than 10 V. So for such circuits you should use op-amps designed for low-voltage operation. These are often called "single-supply" op-amps and come in several forms, some of which include specified operation of both inputs and outputs down to the negative rail; others feature output swings to both *rails*, of which a subset permits both inputs and outputs to go to both rails. We'll deal with these shortly, in §4.6.3.

B. Supply splitter

The circuit in Figure 4.72C suggests a different approach to operation with a battery. Instead of piping a line called V_{ref} around as a signal common, with the negative battery terminal called ground, why not ground the "reference" output, effectively splitting the single supply into a positive–negative pair? This is a common technique in battery operated equipment, and is shown in Figure 4.73. The battery voltage is split by the resistive divider, which feeds a follower to generate a low-impedance common voltage. To

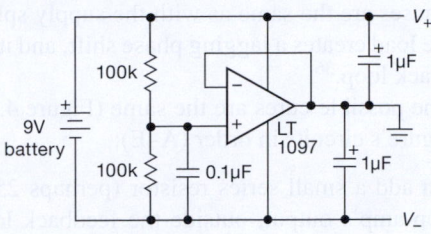

Figure 4.73. Op-amp split-supply generator. A follower generates a low-impedance mid-battery output voltage, which becomes circuit ground.

the outside world that common voltage is "ground," with both ends of the battery floating.

The output should be bypassed, as always, to maintain low-impedance supply rails, relative to ground, at signal frequencies. That is necessary because ground is usually the common return for filters, biasing networks, loads, etc. Look at almost any normal split-supply circuit and you'll find dc and signal currents flowing into and out of ground.

This raises an interesting problem, which we discuss in detail in Chapter *4x* and in §9.1.1C, namely that the op-amp's output resistance, in combination with the by-pass capacitor, creates a lagging phase shift at high frequencies that can cause the feedback loop to go into oscillation. Some op-amps are designed to circumvent this problem, for example the LT1097 shown in the figure (whose datasheet states that it is stable with any capacitive load). Even so, this circuit exhibits a peak in its output impedance versus frequency (Figure 4.74), and a related effect, namely, a ringing transient with that same characteristic frequency (Figure 4.75); you can think of these effects as the not-quite-banished ghost of an oscillation. As the figures show, a small series damping resistor at the op-amp's output (Figure 4.76A) effectively stops this resonant behavior, at the expense of an increase in dc output impedance.

If increased output impedance is undesirable (which often it is not), another approach is to take "slow" feedback downstream of the damping resistor (which preserves accurate dc performance, i.e., low dc output impedance), with a parallel "fast" feedback path from the upstream side (Figure 4.76B) to prevent ringing. You can see the result in Figure 4.75, where we used R_1=2.7Ω, R_2=10k, and C=2.7 nF: the initial transient looks just like that with a 2.7 Ω damping resistor, but then returns to the correct dc level because dc feedback is taken from the point of load. A third possibility is to "overcompensate" the op-amp, for which the LT1097 provides hospitality in the form of a convenient "overcomp" pin; adding a capacitor to ground from this

pin increases the phase margin by shifting the dominant pole downward in frequency (§4.9).

There's a nice integrated solution from Texas Instruments, the TLE2425 and TLE2426 "rail-splitter" ICs. These come in a convenient 3-terminal TO-92 (small transistor) package, draw less than 0.2 mA quiescent current, are stable into any capacitive load greater than 0.33 μF, and can source or sink an unbalanced current of 20 mA (Figure 4.77). The TLE2426 splits the rails 50% with an internal resistive divider, whereas the TLE2425 uses an internal voltage reference to put the output common 2.50 V above the negative rail.

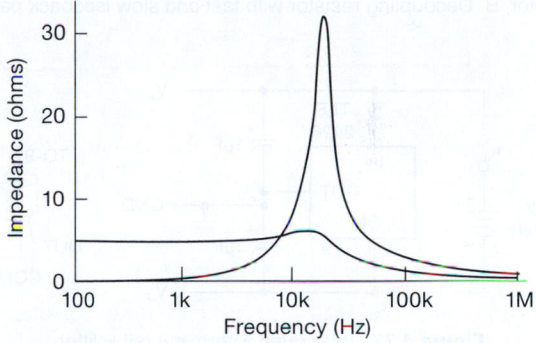

Figure 4.74. One effect of the capacitive load is a bump in the output impedance, which is greatly reduced with a 5 Ω damping resistor; see text.

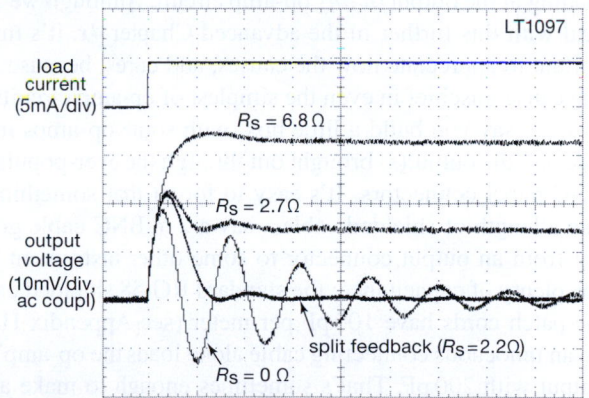

Figure 4.75. Measured output voltage transient of the rail-splitter circuit of Figure 4.76A caused by a 4.5 mA load current step, with several values of series damping resistor. The latter eliminates ringing, at the expense of dc output impedance. An alternative is the "split-feedback" scheme of Figure 4.76B. Scales: 5 mA/div and 10 mV/div; 40 μs/div.

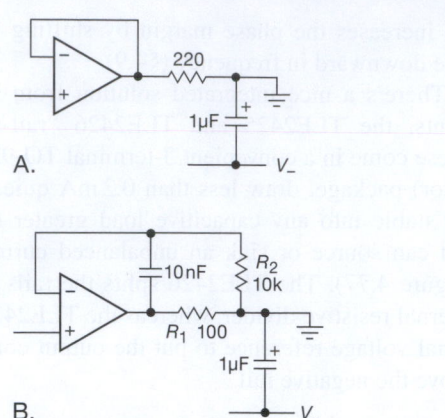

A.

B.

Figure 4.76. Stabilizing the split-supply generator: A. Decoupling resistor, B. Decoupling resistor with fast and slow feedback paths.

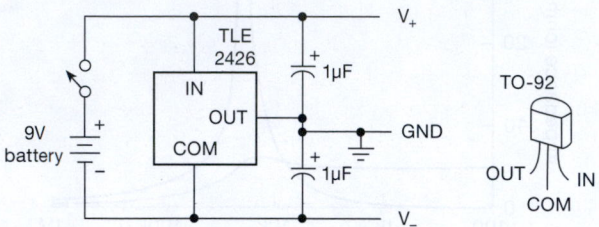

Figure 4.77. Integrated 3-terminal rail splitter.

4.6.2 Capacitive loads

This particular example of a supply-splitter illuminates a more general problem, namely, the effect of capacitive loading at the output of *any* op-amp circuit. Although we'll deal with this further in the advanced Chapter *4x*, it's important to appreciate now the causes, and cures, because it can cause mischief in even the simplest of op-amp circuits.

Let's say you build a little box, with some op-amps inside and the output(s) brought out through the ever-popular BNC panel connectors. It's easy to forget that something like a length of shielded cable – say a 2 m BNC cable going from an output connector to some other instrument – has plenty of capacitance: the standard RG-58 shielded cable patch cords have 100 pF per meter (see Appendix H). So an innocuous connecting cable alone loads the op-amp's output with 200 pF. That's sometimes enough to make an op-amp follower oscillate (we rig up just such a demonstration in our circuit design class, where an LF411 follower screams loudly when asked to drive 8 ft of cable). And even if it doesn't break into oscillation, it will likely exhibit response peaks at high frequencies, evident as overshoot and ringing.

The causes are the same as with the supply splitter: the capacitive load creates a lagging phase shift, and it's within the feedback loop.[36]

And the possible cures are the same (Figure 4.78); taking the figure's circuits in order (A–E):

- You can add a small series resistor (perhaps 25–100 Ω) at the op-amp's output, outside the feedback loop. (It's quite common to see a 50 Ω output resistor, which forms a matched source to "50 Ω cable"; see Appendix H.) This is OK, and easy; but it does mean that feedback does not act on the actual output signal, which may be significant with nasty loads, or at high frequencies, etc.

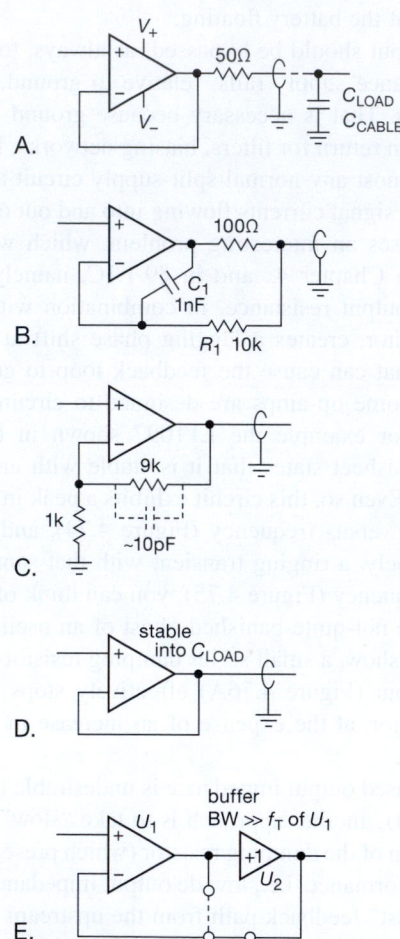

Figure 4.78. Driving capacitive loads.

---

[36] You can also think of this as an effect of the op-amp's inductive-like output impedance (see Figure 4.53) combining with the capacitance to form a resonance, with all its phase shifts, again within the feedback loop.

- You can split the feedback loop, as shown, so that feedback comes directly from the op-amp's output at high frequencies, where instability lurks. And at lower frequencies the feedback accurately controls the signal seen by the load. This is not really a compromise, because those high frequencies are exactly where the thing would oscillate anyway if you were to allow feedback from the load.
- You can reduce the loop gain, for example by increasing the closed-loop gain, to regain stability.
- You can seek an op-amp that guarantees stability into the range of load capacitances you expect. Many op-amps provide good data in the form of plots of "Stability versus Capacitive Load." Figure 4.79 shows an example, taken from the datasheet for the LMC6482.
- You can add a unity-gain buffer, with its low native output impedance, either within or outside the feedback loop. If you add it inside the loop, you need to worry about phase shifts introduced by the buffer; it should have significantly higher f_T than the op-amp, and it's often a good idea to include a 50–100 Ω series resistor at the buffer's input (not shown). You may need to rolloff the op-amp's response with a small capacitor, as in Figure 4.87 on page 274.

4.6.3 "Single-supply" op-amps

As we just remarked, some op-amps are designed specifically to allow inputs and outputs to go to the negative rail. These are called "single-supply" (or "ground-sensing") op-amps, the idea being that their negative rail is actually tied to ground. The input range actually extends slightly below ground, typically to −0.3 V. In some cases the outputs can swing also to the positive rail ("rail-to-rail output"), and a subset of these permits input swings to (and slightly beyond) both rails ("rail-to-rail input"). Linear Technology has introduced an exotic new twist – op-amps that permit input swings well beyond the positive rail (they call them "Over-The-Top™" amplifiers).

These amplifiers can simplify single-supply circuits because you don't need a midsupply reference, rail splitter, etc. But you have to remember that the output cannot go *below* ground – so you can't build an audio amplifier like Figure 4.70, whose output would need to swing both sides of ground. Before looking at the characteristics of these op-amps in more detail, let's look at a design example.

A. Example: single-supply photometer

Figure 4.80 shows a typical example of a circuit for which single-supply operation is convenient. We discussed a similar circuit earlier under the heading of current-to-voltage converters (and we will go further in Chapter *4x*). Because a photocell circuit might well be used in a portable light-measuring instrument, and because the output is known to be positive only, this is a good candidate for a battery-operated single-supply circuit. R_1 sets the full-scale output at 5 volts for an input photocurrent of 0.5 μA. The small feedback capacitor is added to ensure stability, as we'll explain in §4.9.3. No offset voltage trim is needed in this circuit, since the worst-case untrimmed offset of 10 mV corresponds to a negligible 0.2% of full-scale meter indication. The TLC27L1 is an inexpensive micropower (10 μA supply current) CMOS op-amp with input and output swings to the negative rail. Its low input current (0.6 pA, typ, at room temperature[37]) makes it good for low-current applications like this. If you choose a bipolar op-amp for this kind of low signal-current circuit, better performance at

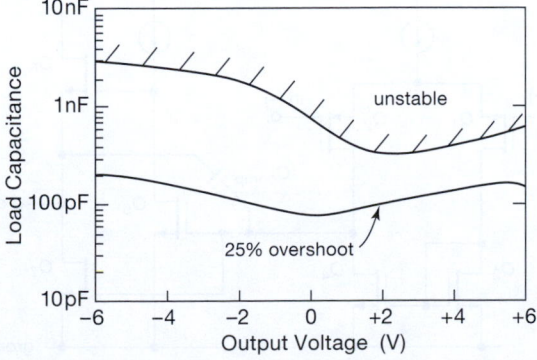

Figure 4.79. Stability versus capacitive load for a LMC6482 op-amp follower with R_{load}=2k and ±7.5 V supplies.

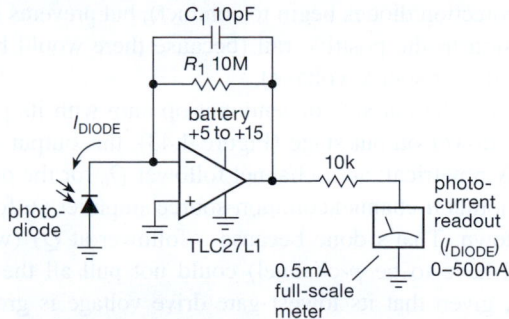

Figure 4.80. Single-supply photometer.

---

[37] Usually taken as 25°C on datasheets. This is a bit warmer than typical office or lab space (77°F, in the King's units), but you can rationalize that choice by saying that it allows for some heating inside the electronics enclosure.

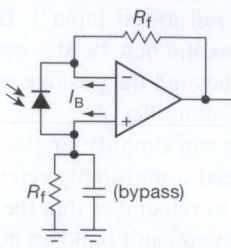

Figure 4.81. Photodiode amplifier with simple bias current cancellation.

low light levels results if the photodiode is connected as in the circuit shown in Figure 4.81.

It's worth noting that the "current budget" of this circuit is dominated by the output current that drives the meter, which can go as high as $500\,\mu A$. It's easy to overlook a point like that, blithely assuming that the battery need provide only the op-amp's $10\,\mu A$ quiescent current. At $10\,\mu A$ a standard 9 V battery lasts 40,000 hours (5 years), whereas with continuous operation at $500\,\mu A$ it would last a month.

B. Single-supply op-amp innards

It's helpful to look at the circuitry of a typical single-supply op-amp, both to understand how these types achieve operation to one or both rails, and also to appreciate some of the subtleties and pitfalls of designing them into your circuits. Figure 4.82 is a simplified schematic of the very popular TLC270 series of CMOS single-supply op-amps. The input stage is a p-channel MOSFET differential amplifier with current-mirror active load. The use of enhancement-mode p-channel input transistors lets the inputs go to the negative rail (and a bit beyond, until the omnipresent input protection diodes begin to conduct), but prevents input operation to the positive rail (because there would be no forward gate-source voltage).

Unlike the classic conventional op-amp with its push–pull follower output stage (Figure 4.43), this output stage is unsymmetrical: an n-channel follower Q_6 for the pullup and another n-channel common-source amplifier Q_7 for the pull-down. That's done because a follower at Q_7 (which would have to be p-channel) could not pull all the way down, given that its lowest gate drive voltage is ground. This unsymmetrical output requires the common-source driver Q_5 for Q_6's gate, with matching threshold voltages for Q_5 and Q_7 to set the output-stage quiescent current. The feedback capacitor C_{comp} is for frequency compensation (see §4.9.2). This output stage can saturate at ground, with an impedance of Q_7's R_{ON}; but it can't reach V_+, because Q_6 is an n-channel MOSFET follower.

Exercise 4.20. What sets the source voltage of Q_1 and Q_2 when the inputs are approximately at ground? And what determines the high end of the input range? Why is the latter always below V_+?

Exercise 4.21. What sets the maximum positive voltage to which Q_6 can pull the output, assuming the op-amp is lightly loaded?

This same output-stage structure – follower pullup with amplifier pull-down – can be built with bipolar transistors; an example is the popular LT1013/LT1014 single-supply dual–quad op-amps, improved variants of the classic LM358/LM324 op-amps. A note of caution: don't make the mistake of assuming that you can make *any* op-amp's output work down to the negative rail simply by providing an external current sink. In most cases the circuitry driving the output stage does not permit that. Look for explicit permission in the datasheet!

One way to achieve rail-to-rail outputs – i.e., operation to *both* supply rails – is to replace the n-channel follower pullup Q_6 in Figure 4.82 with a p-channel common-source amplifier; then each transistor can saturate to its respective rail. This requires some driver circuitry changes, of course. An analogous circuit can be built with bipolar transistors – common emitter pnp pullup and npn pull-down. Contemporary examples include the CMOS TLC2270, LMC6000, and MAX406 series, and the bipolar LM6132, LT1881, and MAX4120 series. As we'll see in Chapter *4x*, there are other ways to make single-supply and rail-to-rail outputs. These amplifiers differ in important ways, and you must watch out for misleading statements about output swing to the negative rail (ground).

These output stages are pretty straightforward, and not surprising. But they don't generalize to the *input* stage. How, indeed, can you possibly achieve rail-to-rail input capability? To complete the picture without going into any

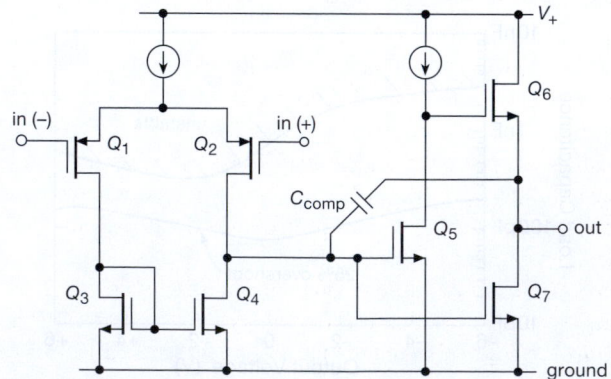

Figure 4.82. Simplified schematic of the TLC271-series single-supply op-amp.

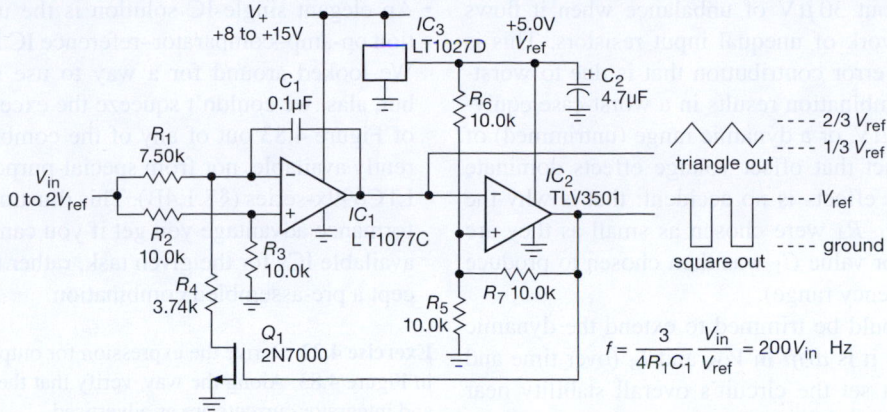

Figure 4.83. Precision voltage-controlled waveform generator.

detail, the trick is to design an amplifier with two independent input stages, one *p*-channel (or *pnp*) and the other *n*-channel (or *npn*). We talk about them, and some other nifty tricks (such as putting on-chip voltage generators to create bias supplies beyond the rails, in combination with a conventional op-amp like Figure 4.43), in Chapter *4x*. Single-supply op-amps are indispensable in battery-operated equipment.

4.6.4 Example: voltage-controlled oscillator

Figure 4.83 shows a clever circuit, borrowed from the application notes of several manufacturers. IC_1 is an integrator, rigged up so that the capacitor current ($V_{in}/15k$) changes sign, but not magnitude, when Q_1 conducts. IC_2 is connected as a Schmitt trigger, with thresholds at one-third and two-thirds of V_{ref}. The *n*-channel MOSFET Q_1 is here used as a switch, pulling the bottom side of R_4 to ground when IC_2's output is HIGH and leaving it open-circuited when the output is LOW.

A nice feature of this circuit is its operation from a single positive supply. The TLV3501 is a CMOS comparator with rail-to-rail output swing, which means that the output of the Schmitt goes all the way from V_{ref} to ground; this ensures that the thresholds of the Schmitt don't drift, as they would with an op-amp of conventional output-stage design, with its ill-defined limits of output swing. In this case the result is stable frequency and amplitude of the triangle wave. Note that the frequency depends on only the *ratio* V_{in}/V_{ref}; this means that if V_{in} is generated from V_{ref} by a resistive divider (made from some sort of resistive transducer, say), the output frequency won't vary with V_{ref}, only with changes in resistance. This is another example

of ratiometric techniques; circuit designers like to use this trick to minimize dependence on power-supply voltages.

Some additional points.

- Both the frequency conversion coefficient and the output swing amplitude are set by the reference voltage that powers IC_2 (V_{ref}), in this case a precise and stable +5.00 V provided by the 3-terminal voltage reference IC_3. This voltage could be left unregulated if the control voltage is arranged to be proportional to it, as described above. The output amplitude would, however, still be dependent on that supply rail. The solution just shown is preferable.
- The integrator op-amp, IC_1, is a "precision" op-amp, with a maximum offset voltage of $60\,\mu V$. It was chosen to provide accurately proportional frequency down near zero volts input. You can think of this instead in terms of *dynamic range* of the frequency control: input offset voltage in the integrator op-amp produces an error in frequency equivalent to a value of V_{in} of twice that offset voltage (because of the divider R_2R_3); to say it another way, at an input voltage $V_{in} \approx 2V_{OS}$, the output frequency will be in error by 100% (it could be as large as twice the programmed frequency and as low as zero). So the ratio of maximum to minimum frequency is roughly equal to V_{ref}/V_{OS}. The LT1077C shown in the figure provides a dynamic range of nearly 100,000:1 (the ratio $V_{ref}/V_{OS} = 5\,V/60\,\mu V$).
- The integrator op-amp must operate down to zero volts input; i.e., it must be a "single-supply" (or "ground-sensing") op-amp, of which the LT1077C is an example.
- Input current I_B of the integrator op-amp also causes an error, most serious when the control voltage V_{in} is near zero volts. The LT1077C has well-matched inputs with $I_B(max)=11\,nA$, which causes a worst-case error

equivalent to about $30\,\mu$V of unbalance when it flows through the network of unequal input resistors. This is smaller than the error contribution that is due to worst-case V_{OS}; the combination results in a worst-case equivalent error of $90\,\mu$V, or a dynamic range (untrimmed) of 50,000:1. The fact that offset voltage effects dominate over bias current effects is no accident: that is why the resistor values R_1–R_4 were chosen as small as they are (and the capacitor value C_1 was then chosen to produce the desired frequency range).

• The LT1077C could be trimmed to extend the dynamic range; ultimately it is *drift* in V_{OS} and I_B (over time and temperature) that set the circuit's overall stability near zero frequency.

• The TLV3501 is an unusually fast (4.5 ns) comparator with rail-to-rail output swing. However, its supply voltage is limited to +5.5 V maximum. If you wanted to run that portion of the circuit at a higher voltage, you could substitute a fast rail-to-rail op-amp like the CA3130. The latter part has been around a long time and is nearing extinction;[38] but it excels in speed for a low-power op-amp because it is *uncompensated* (see §4.9.2B). It would not be suitable for the input op-amp, however, because it is not stable as an integrator, for reasons we will see shortly. It also has large input offset voltage.

• Another possibility is to replace the Schmitt trigger circuit with a CMOS 555-type timer IC, for example, an ICL7555. These have stable input thresholds at one-third and two-thirds of the supply rail, and rail-to-rail fast output swing.

• Switch alternatives: IC switches like the SD210 or 74HC4066 (the latter belongs to the 74HC family of digital logic) could replace the discrete MOSFET Q_1; their lower capacitance would improve operation at high frequencies.

• Another possibility, if power consumption matters more than maximum frequency or dynamic range, is to use low-power CMOS rail-to-rail op-amps for both ICs, for example, a TLC2252 dual op-amp ($35\,\mu$A per channel). In this case scale up the resistor values, particularly in the input stage, because CMOS op-amps have negligible input current for this application.

• If the use of a dual op-amp in a single package seems particularly appealing, then a good overall choice is the bipolar LM6132, with rail-to-rail inputs and outputs and a slew rate of 14 V/μs; in the same family you can get faster op-amps (LM6142, LM6152), at the cost of higher input and supply currents.

---

[38] See "Here Yesterday, Gone Today" on page 273.

• An elegant single-IC solution is the use of a combination op-amp–comparator–reference IC like the MAX951. We looked around for a way to use such a chip here, but, alas, we couldn't squeeze the excellent performance of Figure 4.83 out of any of the combination chips currently available, nor from special-purpose timers like the LTC699x-series (§7.1.4B). This illustrates the circuit performance advantage you get if you can combine the best available ICs for the given task, rather than having to accept a pre-assembled combination.

Exercise 4.22. Derive the expression for output frequency shown in Figure 4.83. Along the way, verify that the Schmitt thresholds and integrator currents are as advertised.

4.6.5 VCO implementation: through-hole versus surface-mount

Traditionally, electronic components were made with wire leads sticking out the ends (e.g., "axial-lead" resistors and capacitors), or rows of pins sticking down (e.g., ICs with DIP – "dual in-line" – packaging). Contemporary practice has shifted strongly toward "surface-mount" components, in which the connections are made directly to contacts on a ceramic or plastic package. See, for example, the photographs of resistors in Chapter 1 (Figure 1.2), of op-amps earlier in this chapter (Figure 4.1), or of small logic (Figure 10.23) in Chapter 10.

The *good news* is that surface-mount technology (SMT) lets you make smaller gadgets; and it is better *electrically* as well, because of reduced inductances in the smaller packages.

The *bad news* is that SMT makes it difficult to wire up a circuit on the spur of the moment on a prototype "breadboard" (of either the solder-in or plug-in style), an exercise that is fast and easy with through-hole components. The problem is compounded by the fact that many new high-performance components (e.g., op-amps) are available *only* in surface-mount packages.

In a nutshell, your choices boil down to (a) sticking with through-hole components (if you can get the parts you need) and enjoying the easy prototyping and ability to build a one-off gadget quickly; (b) going with the flow, and using mostly SMT components, laying out a printed circuit board for each circuit you want to build; or (c) trying to retain the best of both worlds by prototyping with through-hole components, where available, and using SMT adapters (or "carriers") for the SMT components that you cannot get in through-hole packages. The latter are tiny circuit boards on which you solder an SMT component, whose leads connect to a row of pins, producing a faux through-hole component. We've struggled with this whole business and have concluded that this last option, though attractive in principle, is fast fading away, because of the decreasing availability of through-hole components.

To give a glimpse of the tradeoffs, we laid out the voltage-controlled oscillator (VCO) circuit of Figure 4.83 on printed-circuit boards, exploring the alternatives of (a) through-hole components, (b) relatively large SMT components, and (c) small SMT components. Figure 4.84 shows them at actual size; we've shown only the component outlines and "pads" (metal foil patterns that make the connections to the components). For the through-hole board we used standard DIP op-amp and comparator and axial-lead 1/4-watt resistors; for the large SMT we used SOIC-8 op-amp and comparator and 0805 SMT resistors; and for the small SMT we used SOT-23 op-amp and comparator and the smaller 0603 resistors.[39] The latter is 4.5 times smaller than the through-hole board. And there is no penalty in performance; in fact, smaller components generally deliver somewhat better performance owing to smaller parasitic inductances.

4.6.6 Zero-crossing detector

This example illustrates the use of a single-supply *comparator*, a close kin to the single-supply op-amp. Like the latter, it will operate with input signals all the way to the lower supply rail, which often is ground. The circuit, shown in Figure 4.85, generates an output square wave for use with 5 V "TTL" logic (0 to +5 V range) from an input wave of any amplitude up to 150 volts rms. The LM393 is a comparator IC (like the TLV3501 we used in the last example), specialized for this sort of application; it cannot be used as an amplifier, in the manner of an op-amp, because its internal phase shifts are not tailored ("compensated," see §4.9) to permit feedback without oscillation. It also has an "open-collector" output, which you must pull up externally to a supply rail, as shown. Its internal circuit is shown in Figure 4.86; note the overall similarity to an op-amp (Figure 4.43), with the important omission of the compensation capacitor C_C, and the lack of a "pullup" transistor at the output. We treat comparators in more detail in §12.3.

---

[39] The 4-digit designation gives the length and width, in units of 0.01″ (0.25 mm); so, for example, an 0603 resistor is 0.06″×0.03″ (1.5 mm×0.75 mm). We think that's pretty small, but the industry hasn't stopped there: standard sizes include 0402 and 0201 (the latter a mere dust mote, 0.5 mm×0.25 mm!).

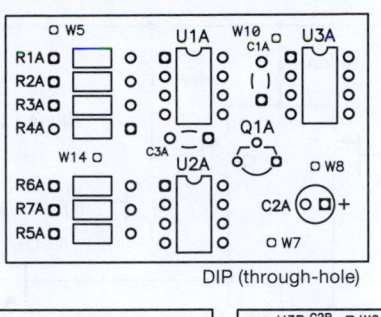

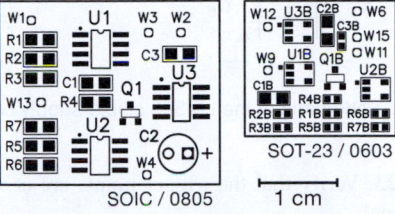

Figure 4.84. Printed-circuit layouts for the VCO of Figure 4.83. The use of small surface-mount components reduces the board area to 22% of the analogous board using through-hole components. Additional benefits are a greater selection of available parts, and better electrical performance.

Resistor R_1, combined with D_1 and D_2, limits the input swing to −0.6 volt to +5.6 volts, approximately; its power rating is set by the maximum rms input voltage. Resistive divider R_2R_3 is necessary to limit negative swing to less than 0.3 volt, the limit for a 393 comparator. R_5 and R_6 provide hysteresis for this Schmitt trigger circuit, with R_4 setting the trigger points symmetrically about ground. The input impedance is nearly constant because of the large R_1 value relative to the other resistors in the input attenuator. A 393 is used because its inputs can go all the way to ground, making single-supply operation simple.

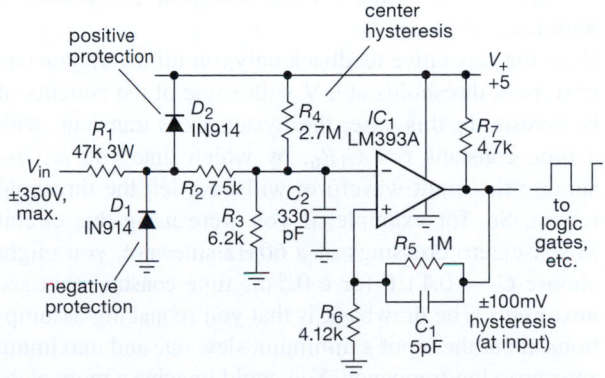

Figure 4.85. Zero-crossing level detector with input protection.

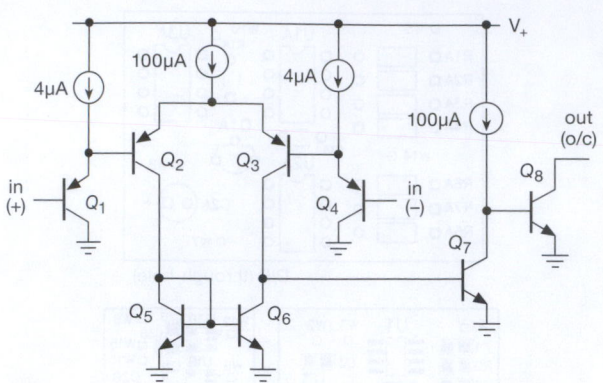

Figure 4.86. Schematic of the LM393 single-supply comparator.

Exercise 4.23. Verify that the trigger points are at ± 100 mV at the input signal.

Some additional points.

- The vintage LM393 severely limits the allowable swing below ground, because the output will switch polarity if the input goes below -0.3 V, a pathology tactfully called *phase reversal* in the datasheet. That is prevented here by diode D_1 and divider $R_2 R_3$; alternatively the low side of D_1 could be biased a diode drop above ground, as in Figure 5.81. Resistor R_3 could be omitted if a modern comparator like the LT1671 were used; the latter also has internal active pullup to $+5$ V, so you would omit pull-up resistor R_7 as well.
- We intentionally set the Schmitt thresholds symmetrically around ground, but that may not be the best choice. For example, you might want output transitions accurately synchronized with the exact zero crossings of the input waveform. Omitting R_4 would set the negative-going input threshold exactly at 0 V; alternatively, you could set the positive-going threshold to 0 V with a properly-chosen value for R_4 (test your understanding with Exercise 4.24).
- By using capacitive feedback only (omitting R_5), you can have both thresholds at 0 V with some of the benefits of hysteresis. In this case, the hysteresis is transient, with a time constant $\tau = C_1 R_6$, by which time you are assuming the input waveform will have left the threshold region. So, for example, if you were using this circuit to sense zero crossings of a 60 Hz sinewave, you might choose $C_1 = 0.1\,\mu$F for a 0.5 ms time constant (but see next item). The drawback is that you're making assumptions about the input's minimum slew rate and maximum zero-crossing frequency. You could imagine a more elaborate scheme, with additional comparators, such that the input threshold is restored to 0 V after the input wave-

form passes a second higher threshold. This design challenge would yield a precise zero-crossing circuit (for both waveform slopes) with no restrictions on input speed, etc.
- Be careful if you decide to increase the value of the speed-up capacitor C_1 – this capacitor causes a negative-going transient at the inverting input of the comparator, and if the capacitor is much larger than a few picofarads you may cause phase reversal at the comparator's output (a pathology of many comparators, including the LM393). In that case it's best to use a modern comparator whose datasheet specifically brags that it is free of phase reversal; an example is the MAX989.

Exercise 4.24. What value of R_4 in Figure 4.85 puts the positive-going input threshold at 0 V?

Exercise 4.25. Try designing a hysteretic circuit, with several comparators, such that *both* thresholds are precisely at 0 V, under the assumption that the input waveform always travels a minimum of 50 mV beyond ground before coming back.

4.6.7 An op-amp table

We've collected in Table 4.2a on the facing page a representative selection of useful op-amps, including many of our favorites. You can get an idea of the price and performance of parts that are in wide use. Better yet, use this table as a starting point in your next design! More comprehensive op-amp tables are located in the chapter on precision design (Table 5.4, High-speed op-amps; Table 5.5, Precision op-amps; Table 5.6, Auto-zero op-amps), and in the chapter on noise (Table 8.3, Low-noise op-amps).

4.7 *Other amplifiers and op-amp types*

In this first op-amp chapter we've met the "standard" split-supply op-amp, implemented variously with bipolar transistors, JFETs and MOSFETs. We've also seen examples of single-supply op-amps, some with rail-to-rail outputs (and even rail-to-rail inputs).

There are other choices, some of which we'll look at in Chapters *4x* and 5. It's worth listing them here, because one or more of them may be the best solution to a design problem that looks initially like it needs an op-amp.

Current-feedback op-amps

These look a lot like ordinary ("voltage-feedback") op-amps, but differ by having a low-impedance inverting input terminal that is a current summing junction. They excel in wideband circuits with moderate to high voltage gain; see the discussion in Chapter *4x*.

Table 4.2a Representative Operational Amplifiers (see also Tables 5.2–5.6 and 8.3)

Part #[a]	#/pkg	Package DIP	SOIC-8	SOT-23	cost qty 25 ($US)	Total Supply min (V)	max (V)	I_Q typ[b] (mA)	f_T typ (MHz)	SR typ (V/µs)	V_{os} max (µV)	I_{bias} typ (nA)	e_n typ (nV/√Hz)	IN +	IN −	OUT +	OUT −	Comments
LM358, 324	2,4	●	●		0.16	3	32	1	1	0.5	7000	45	40		●		●	single-supply jellybean
LT1013, 1014	2,4	●	●		1.30	4	44	0.7	0.8	0.4	40	12	22		●		●	precision single-supply
LT1077A	1	●	●		4.11	2.2	44	0.05	0.23	0.08	40	7	27		●		●	low-power bipolar, also OP193
LMC6482A, 84A	2,4	●	●	●	1.73	3	16	1.3	1.5	1.3	750	20fA	37	●	●	●	●	CMOS jellybean, LMC7101 SOT-23
TLC2272A, 74	2,4	●	●		1.57	4	16	2.2	2.2	3.6	950	0.001	9		●	●	●	CMOS
LMC6442A	2		●		2.00	2.2	16	0.002	0.01	0.004	3000	5fA	170	●	●	●	●	micro-power!
LMC6041, 42, 44	1,2,4	●	●		1.48	4.5	16	0.02	0.08	0.015	3000	2fA	83		●	●	●	for low power, LMC6061 has I_Q=20µA
LMC6081A, 82, 84	1,2,4	●	●		2.72	4	16	0.45	1.3	1.5	350	10fA	22		●	●	●	pico-power, operates to V_{CC}+5V
TLV2401, 02	1,2		●	●	1.42	2.5	16	0.0009	0.005	0.002	1200	0.1	500	●	●	●	●	similar to LMC6482
LMC7101A	1			●	0.93	2.7	16	0.5	1	1	3000	0.001	37	●	●	●	●	
LF411, 412C	1,2	●	●		0.72	7	36	4.5	3	13	2000	0.05	18					JFET, TI, dual cheaper than single
LF347B	4	●	●		0.58	7	36	8	3	13	5000	0.05	18					low cost JFET, 15¢ per op-amp
LT1057A, 1058	2,4	●	●		6.30	7	40	3.4	5	24	450	0.05	13					improved LF412, also see AD712
OPA727, 2727	1,2		●		2.58	4	13	4.3	20	30	150	0.085	6		●	●	●	e-trim CMOS
OPA376, 2376	1,2		●	●	2.03	2.2	7	0.76	5.5	2	25	0.2pA	7.5		●	●	●	e-trim CMOS
TLC272C, 274C	2,4	●	●		0.69	3	16	1.4	2.2	5.3	10mV	0.1pA	27		●		●	consider TLV27x family
OPA129B	1	●			10.15	10	36	1.2	1	2.5	2000	30fA	17					electrometer, mass spec, pH probe
LT1012A	1	●	●		5.11	2.4	40	0.37	0.4	0.2	25	0.025	14		●			low I_B bipolar
LTC1050C	1	●	●		4.30	6	18	1.1	2.5	4	5	0.01	1.6µV[c]		●			chopper
LT1637	1	●	●	●	2.32	2.7	44	0.19	1	0.35	350	20	27	+	●	●	●	"over-the-top": V_{in} to V_{EE} + 44V
LT1097	1	●	●		2.33	2.6	40	0.35	0.7	0.2	50	0.04	14					CLOAD stable, comp pin
OPA177	1	●	●		2.30	6	44	1.5	0.6	0.3	60	0.5	7					improved OP-07
OPA277, 2277, 4227	1,2,4	●	●		3.60	4	36	0.8	1	0.8	20	0.5	8					improved OP-07, see also OPA227
LM6132A, 34	2,4	●	●		2.92	2.7	35	0.8	11	14	2000	110	27	●	●	●	●	early RRIO
AD797A	1	●	●		8.36	10	36	8.2	80	20	80	250	0.9					low distortion, low noise
ADA4000-1, 2, 4	1,2,4	●	●	●	1.46	8	36	1.3	5	20	1700	0.005	16					JFET
LT6220, 21, 22	1,2,4		●	●	2.61	2.5	12.6	0.9	60	20	200	15[h]	10	●	●	●	●	
OPA627A	1	●	●		24.50	9	36	7	16	55	250	0.002	5.6					low-noise JFET
OPA657	1		●	●	12.90	8	13	14	1600	700	1800	0.002	4.8					fast JFET
OPA454	1		●		4.88	10	120	3.2	2.5	13	4000	0.002	35					high voltage, also OPA445 miniDIP
THS4011, 12	1,2	●	●		5.60	9	33	7.8	290	310	6000	2000	7.5					fast VFB[e]; THS4021/22 decomp, G>10
LM7171A	1	●	●		3.60	8	36	6.5	200	4100	4000	2700	14					CFB[e], 100mA output current
EL5165	1		●	●	2.32	5	12.6	5	1400	6000	5000	8500	1.7					CFB[e]
AD8011	1		●	●	3.98	3	12.6	5	570[d]	3500	5000	5000	2					low-power two-stage CFB

Notes: (a) *Italicized* part numbers have corresponding number of op-amps per package. (b) quiescent current per package. for the **boldface** part number (that with the least number of op-amps; e.g.,1mA for the LM358). (c) peak-to-peak noise voltage. (d) GBW for G_V=10. (e) VFB=voltage feedback, CFB=current feedback. (h) rises to 250nA at the negative rail.

Table 4.2b Monolithic Power and High-Voltage Op-Amps[a]

Type	Mfg	Total supply min (V)	Total supply max (V)	I_Q typ (mA)	Diff'l input[b] max (V)	FET	Ext comp	f_T typ (MHz)	Slew rate typ (V/µs)	I_{out}(max) typ (A)	P_{diss} (50°C) max (W)	Therm lim	Prog. curr lim	shutdown	Package	Cost qty 25 ($US)	Comments
low power																	
LME49726	TI	2.5	6	0.7	full	•	-	6.8	3.7	0.35	1[u]	-	-	-	MSOP	1.29	A
OPA567	TI	2.7	7.5	3.4	full	•	-	1.2	1.2	2.2	12.5[n]	•	•	•	QFN	5.53	B
OPA569	TI	2.7	7.5	3.4	full	•	-	1.2	1.2	2.2	25[n]	•	•	•	SO-20	7.41	B,C
AD8010	Analog	10	12.6	16	1.2	-	-	230	800	0.2	1.3[u]	-	-	-	SO-16	6.69	D
LM6171	TI	5	36	2.5	10	-	-	100	3000	0.12	0.7	-	-	-	DIP, SO	4.27	E
LTC2057HV	LTC	4.8	65	0.8	full	•	-	1.5	0.45	0.02	low	-	-	-	SO-8	3.32	F
ADA4700	Analog	10	100	1.7	full	-	-	3.5	20	0.03	2.5	•	-	-	SO-8	6.00	
OPA445	TI	20	100	4.2	80	•	-	2	10	0.015	0.6	-	-	-	DIP, SO-8	10.07	G
OPA454	TI	10	100	3.2	full[g]	•	•	2.5	13	0.12	7.5[n]	•	-	•	SO-8	6.09	
LTC6090	LTC	9.5	140	2.8	full	•	-	12	21	0.05	15[n]	•	-	•	SO,TTSOP	4.87	H
medium power																	
L2720W	ST	4	28	10	full	-	-	1.2	2	1	5	•	-	-	SO-16	1.02	I
ISL1532A	Intersil	10	30	3.5[o]	full	-	-	50	400	1	1	-	-	•	SSOP-20	1.43[r]	J
THS3120	TI	9	33	7	4	-	-	130	900	0.47	15[n]	-	-	•	MSOP-8	5.57	K
LT1794	LTC	10	36	26	full	-	-	200	600	0.72	25[n]	•	-	•	SO-20	8.09	J
LT1206	LTC	10	36	12[o]	full	-	•	60	900	0.5	15[p]	•	-	•	DIP,TO-220	5.88	K,L
LT1210	LTC	8	36	35	full	-	•	35	900	2	15[p]	•	-	•	TO-220-7	8.75	K,M
L272	FSC	4	40	8	full	-	-	0.35	1	1	5	•	-	-	DIP, SO-16	2.08	N
PA75	Apex	5	40	8	full	-	-	1.4	1.4	2.5	19	-	-	-	TO-220-7	28.88	O
TDA7256	ST	10	50	80	full	-	-	9	10	3	35	•	-	•	TO-220-11	3.42	P
LM1875	TI	16	60	70	full	-	-	5.5	8	4	25	•	-	-	TO-220-5	2.75	P
OPA552	TI	8	60	7	full	•	-	12	24[d]	0.2	25[p]	•	-	-	DIP, DDPak	5.70	Q
LM675	TI	20	60	18	full	-	-	5.5	8	3	40	•	-	-	TO-220-5	4.82	R
OPA547	TI	8	60	10	full	•	-	1	6	0.5	25	•	-	•	TO-220-7	9.57	S
OPA548	TI	8	60	17	full	-	-	1	10	3	30	•	-	•	TO-220-11	13.22	S
OPA549	TI	8	60	26	full	-	-	0.9	9	8	53	•	-	•	TO-220-11	20.65	S
OPA453	TI	20	80	4.5	full	•	-	7.5	23[d]	0.05	25	•	-	-	TO-220-7	5.50	T
OPA541	TI	20	80	20	full	-	-	2	10	10	90	-	•	-	TO-3, SIP-11	19.28	
LM3886	TI	18	84[s]	50	60	-	-	8	19	11.5	75	•	-	•	TO-220-11	5.94	P
TDA7293	ST	24	120	30	30	-	-	-	15	6.5	75	•	-	•	TO-220-15	5.49	P
higher voltage																	
PA340	Apex	20	350	2.2	16	•	•	10	32[k]	0.06	16	•	-	-	DDPak-7	21.45	U
PA90	Apex	30	400	10	20	•	•	100	300[e]	0.2	18	-	•	-	SIP-12 +tab	188.00[v]	U
PA15	Apex	100	450	2.0	25	•	•	5.8	20[d]	0.2	18[n]	-	-	-	SIP-10	185.00[v]	U
PA98	Apex	30	450	21	25	•	•	100	1000[e]	0.2	18	-	•	-	SIP-12 +tab	272.00[v]	U
PA97	Apex	100	900	0.6	20	•	•	1	8[e]	0.01	3[n]	-	-	-	SIP-12	176.00[v]	V

Notes: (a) within categories, sorted by maximum voltage, then output current; the Apex parts are hybrid, and neither PCB nor instrument-box types are listed. (b) not to exceed total supply voltage. (c) P_{diss} with case at T_C = 50C, based on $R_{\theta JC}$. (d) when comp for $G>10$. (e) when comp for $G>100$. (g) internal JFETs limit current to 4mA. (h) see notes. (k) for C_C=4.7pF, $G\geq10$. (n) provided you can get the heat out of the package! (o) adjustable. (p) power package. (r) qty 1k. (s) 94V w/o signal. (u) to ambient. (v) unit qty; see distributor prices (and your banker) for larger qty.

Comments: (A) dual, RRO. (B) RRIO. (C) with current monitor. (D) video. (E) VFB with CFB. (F) auto-zero, 4µV. (G) has V_{os} trim; also in TO-99. (H) dual='6091. (I) update of L272. (J) dual, ADSL driver. (K) current-feedback, CFB. (L) can drive 10nF capacitive loads. (M) 1.1A min. (N) Fairchild's version. (O) amp+buffer. (P) audio amplifier. (Q) slower OPA551 for G=1. (R) classic workhorse. (S) current-limit adjustment with resistor or external current. (T) slower OPA452 for G=1. (U) MOSFET output. (V) "inexpensive."

"Zero-drift" op-amps

These unusual op-amps, which include auto-zero and chopper-stabilized amplifiers, are tailored for precision (low-V_{OS}) applications. They use internal MOS switches to measure and correct for input offset error. These are the only amplifiers with values of untrimmed V_{OS} down to $5\,\mu V$ or less. See Table 5.6 on page 335.

High-voltage, high-power op-amps

You can get op-amps with maximum output currents of 25 amps or more, or with power supply voltages to 1 kV or

"Here Yesterday, Gone Today"

In its untiring quest for better and fancier chips, the semiconductor industry can sometimes cause you great pain. It might go something like this: you've designed and prototyped a wonderful new gadget; debugging is complete, and you're ready to go into production. When you try to order the parts, you discover that a crucial IC has been discontinued by the manufacturer! An even worse nightmare goes like this: customers have been complaining about late delivery on some instrument that you've been manufacturing for many years. When you go to the assembly area to find out what's wrong, you discover that a whole production run of boards is built, except for one IC that "hasn't come in yet." You then ask purchasing why they haven't expedited the order; turns out they have, just haven't received it. Then you learn from the distributor that the part was discontinued six months ago and that none is available!

Why does this happen, and what do you do about it? We've generally found four reasons that ICs are discontinued.

1. *Obsolescence:* Much better parts come along, and it doesn't make much sense to keep making the old ones. This has been particularly true with digital memory chips (e.g., small static RAMs and EPROMs, which are superseded by denser and faster versions each year), though linear ICs have not entirely escaped the purge. In these cases there is often a pin-compatible improved version that you can plug into the old socket.

2. *Not selling enough:* Perfectly good ICs sometimes disappear. If you are persistent enough, you may get an explanation from the manufacturer – "there wasn't enough demand," or some such story. You might characterize this as a case of "discontinued for the convenience of the manufacturer." We've been particularly inconvenienced by Harris's discontinuation of their splendid

HA4925 – a fine chip, the fastest quad comparator, now gone, with no replacement anything like it. In our first edition we reported that Harris also discontinued the HA2705 – another great chip, the world's fastest low-power op-amp, gone without a trace! Since that time, Maxim came out with the MAX402, similarly a fast low-power op-amp. Lots of us used it; then – whammo – can't get it! Sometimes a good chip is discontinued when the wafer fabrication line changes over to a larger wafer size (e.g., from the original 3″ diameter wafer to a 5″ or 6″ wafer).

3. *Lost schematics:* You might not believe it, but sometimes the semiconductor house loses track of the schematic diagram of some chip and can't make any more! This apparently happened with the Solid State Systems SSS-4404 CMOS 8-stage divider chip.

4. *"Upgraded" production line:* Sometimes a manufacturer will replace older test equipment (which may have been working just fine) with the latest and greatest new stuff. Problem is, the programs to run the new testers aren't finished yet. So, the wafer line *could* be making lots of chips ... but there's no way to test them. This scenario appears to have played out in the case of the magnificent OPA627, one of our all-time favorites (there was nearly a year in which you couldn't get these puppies, but, thankfully, it's back in production).

5. *Manufacturer out of business:* This also happened to the SSS-4404! If you're stuck with a board and no available IC, you've got several choices. You can redesign the board (and perhaps the circuit) to use something that is available. This is probably best if you're going into production with a new design or if you are running a large production of an existing board. A cheap and dirty solution is to make a little "daughterboard" that plugs into the empty IC socket and includes whatever it takes to emulate the nonexistent chip. Although this latter solution isn't terribly elegant, it gets the job done.

more! These are specialized (and expensive) devices, extremely useful for applications such as piezo drivers, servo drivers, and so on. See Table 4.2b on the facing page for some favorites.

Micropower op-amps

At the other end of the spectrum, you can get op-amps with quiescent currents as low as a microamp or less. These things aren't blazingly fast – the LMC6442, with

$I_Q = 10\,\mu A$ per amplifier, has an f_T of 10 *kilo*hertz, and a slew rate[40] of 0.004 V/μs – but they do let you run a portable instrument just about forever on a single battery.

Instrumentation amplifiers

These are integrated differential amplifiers with settable voltage gain. They contain several op-amps internally and

---

[40] The manufacturer would *never* use "V/μs" on a datasheet of such a sluggish op-amp – look for V/*milli*second instead.

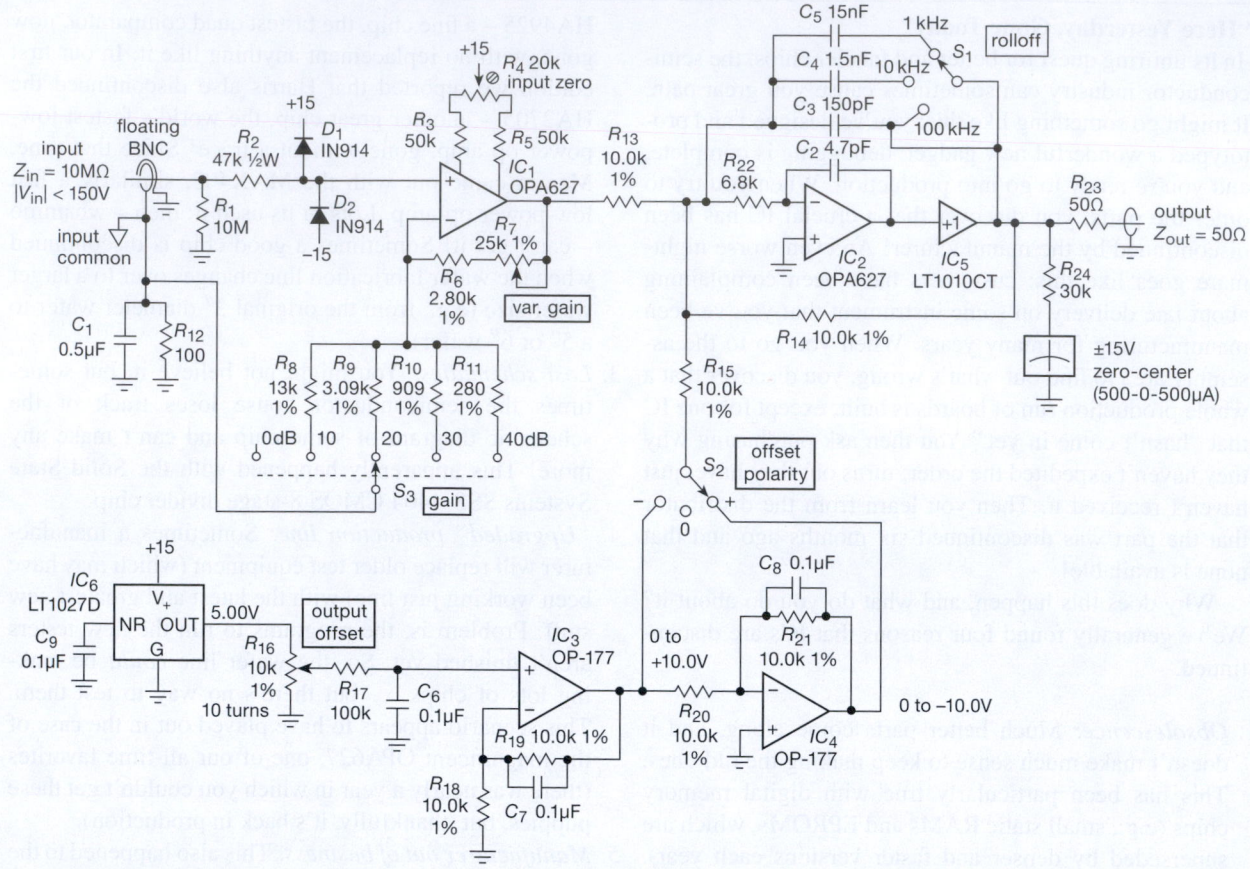

Figure 4.87. Laboratory dc amplifier with output offset. Op-amp power supply connections and bypass capacitors are not shown explicitly, a common practice in circuit schematics.

excel in stability and common-mode rejection. Instrumentation amplifiers are discussed in §5.15.

Video and radiofrequency amplifiers

Specialized amplifiers for use with video signals, or with communications signals at frequencies from 10 MHz to 10 GHz, are widely available as fixed-gain amplifier modules. At these frequencies you generally don't use op-amps.

Dedicated amplifier variants

Microphone preamps, speaker amplifiers, stepping motor drivers, and the like are available as customized ICs with superior characteristics and ease of use.

4.8 Some typical op-amp circuits

4.8.1 General-purpose lab amplifier

Figure 4.87 shows a dc-coupled "decade amplifier" with settable gain, bandwidth, and wide-range dc output offset.

IC_1 is a low-noise JFET-input op-amp with noninverting gain from unity (0 dB) to ×100 (40 dB) in accurately calibrated 10 dB steps; a vernier is provided for variable gain. IC_2 is an inverting amplifier; it allows offsetting the output over a range of ±10 volts, accurately calibrated by the 10-turn pot R_{16} by injecting current into the summing junction. C_3–C_5 set the high-frequency rolloff, because it is often a nuisance to have excessive bandwidth (and noise). IC_5 is a power booster for driving low-impedance loads or cables; it can provide ±150 mA output current.

Some interesting details: a 10 MΩ input resistor is small enough, since the bias current of the OPA627 is 10 pA (maximum, at room temperature), thus producing a 0.1 mV error with open input. R_2, in combination with clamp diodes D_1 and D_2, limits the input voltage at the op-amp to the range $V_- - 0.6$ V to $V_+ + 0.6$ V. With the protection components shown, the input can go to ±150 volts without damage. The JFET-input OPA627 was chosen for its combination of low input current ($I_B = 1$ pA, typ), modest precision (V_{OS}=100 μV, max), low noise (e_n=5 nV/$\sqrt{Hz}$, typ), and wide bandwidth (f_T=16 MHz, typ); the latter is needed to preserve some loop gain at the high-frequency end of the instrument (100 kHz) when running at full gain (40 dB).

The output stage is an inverter with a unity-gain power buffer inside the feedback loop. The vintage LT1010 has plenty of slew rate, bandwidth, and muscle, with less than 10 Ω open-loop output impedance (which of course is lowered by feedback; see §2.5.3C). Both it and the OPA627 have enough slew rate (75 V/μs and 55 V/μs, respectively) to generate a full $\pm$15 V output swing at the full 100 kHz bandwidth of the instrument. A power buffer like this is good for isolating capacitive loads from the op-amp (more on this in Chapter *4x*; see also §§4.6.1B and 4.6.2); furthermore, *it* takes the heat when driving a hefty load, which keeps IC$_2$ cool, an important consideration with precision (low-V_{OS}) op-amps. It takes lots of drive compared with an op-amp – up to 0.5 *milli*amp – but that's no problem when you're driving it with an op-amp.

The offset circuit consists of a precision LT1027 3-terminal voltage reference IC. We'll learn more about these in Chapter 9; they generate a highly stable voltage output when powered from a noncritical dc rail that is at least 2 volts higher than their specified output voltage. This particular part comes in several grades, the best of which (LT1027A) has a maximum error of 1 mV, and is guaranteed to drift less than 2 ppm/°C; for this application we'd save some money by choosing the inexpensive "D" grade (5.0 V$\pm$2.5 mV; 5 ppm/°C). The OP177 is a highly stable precision op-amp (V_{OS} < 10 μV, TCV$_{OS}$ < 0.1 μV/°C in its best grade) that provides a stable offsetting voltage. Capacitor C_6 bypasses noise on the reference voltage, and C_7 and C_8 reduce amplifier noise by limiting the bandwidth of the amplifiers. For a dc application like this you don't need, and don't want, lots of bandwidth. We'll talk in detail about this sort of precision design in Chapter 5.

Some additional points.

- In a circuit like this the input protection network can limit the ultimate bandwidth, because R_2 forms a lowpass filter in combination with the combined input capacitance of IC$_1$, diode capacitance, and associated wiring capacitance. In this case the total capacitance is approximately 12 pF, which puts the 3 dB point at 300 kHz, well above the 100 kHz high-frequency limit of the instrument. To use a similar protection circuit in a wideband amplifier, you could reduce the value of R_2, put a small capacitor (47 pF, say) across it, or both. You could also use clamp diodes with lower capacitance, for example a 1N3595 or a PAD5 (see Chapter *1x*).

- A really useful general-purpose laboratory amplifier should have true *differential* inputs. This is best done with an instrumentation amplifier, rather than an op-amp; see §5.15. Here we've compromised with a "pseudo-differential" configuration, in which the input common terminal (which is the return path for feedback), floating from circuit ground with a 100 Ω resistor, is allowed to accommodate a small amount of signal from the input source. A better arrangement, though still not symmetrically differential, is shown in Figure 4.88, where a difference amplifier (IC$_7$) uses the floating input common as a reference. Note the use of a chassis-isolated BNC panel connector.

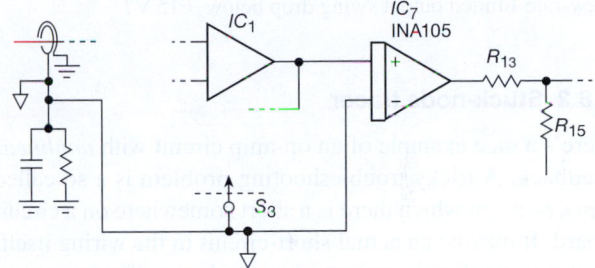

Figure 4.88. Difference amp cancels error from signal on input common.

- In many situations it is preferable to introduce the dc offset at the input rather than at the output. Then you can change gain, without adjusting the offset, to zoom in on a portion of the input signal. This requires a much larger range of offset voltage, and other circuit changes as well. We'll see an example in Chapter 5.

- Watch out for op-amps that exhibit phase reversal when their inputs go more than 0.3 V below V_-; in such cases a restrictive input clamp must be used to prevent negative swings below that limit. This is a common defect of many op-amps, which the excellent OPA627 does not share.

- Contemporary instrumentation usually provides for remote operation, with digital control from a computer. This circuit, however, uses mechanical controls for gain, bandwidth, and offset. You could replace the mechanical switches with analog switches, and use a DAC to generate the offset, to adapt this instrument to digital control.

- The rolloff capacitors C_3–C_5 close the loop around the output amplifier pair (IC$_2$+IC$_5$) at high frequencies, which is beneficial in terms of reducing noise. But it also promotes instability, owing to the combined phase shifts of the two amplifiers. This arrangement is still OK, though, as long as the bandwidth of buffer IC$_5$ is much greater than that of amplifier IC$_2$.

But that is not the case here: the OPA627 op-amp has a unity gain bandwidth of f_T=16 MHz, at which it specifies a 75° phase margin. But the LT1010 buffer adds about 50° of additional lagging phase shift, pushing the amplifier close to instability (see §4.9 for an explanation of phase margin and stability). The solution here is to use a small feedback capacitor around the op-amp (4.7 pF, C_2), which closes the high-frequency feedback path directly. This rolls off its gain to unity at about 1 MHz, at which frequency the buffer contributes less than 5° additional lagging phase shift.

Exercise 4.26. Check that the gain is as advertised. How does the variable-offset circuitry work? At what frequency would the slew-rate-limited output swing drop below ±15 V?

4.8.2 Stuck-node tracer

Here's a nice example of an op-amp circuit with *nonlinear* feedback. A tricky troubleshooting problem is a so-called *stuck node*, in which there is a short somewhere on a circuit board. It may be an actual short-circuit in the wiring itself, or it may be that the output of some device (for example a digital logic gate, see Chapter 10) is held in a fixed state. It's hard to find, because anywhere you look on that line, you measure zero volts to ground.

A technique that does work, however, is to use a sensitive voltmeter to measure voltage drops *along* the stuck trace. A typical signal trace on a printed-circuit board might be 0.010″ wide and 0.0013″ thick (1 ounce per square foot), which has a resistance along the trace of 53 mΩ/in. So if there's a device holding the line to ground somewhere and you inject a diagnostic current of 10 mA dc somewhere else, there will be a voltage drop of 530 μV per inch in the direction of the stuck node.

Let's design a stuck-node tracer. It should be battery powered so that it can float anywhere on the powered circuit under test. It should be sensitive enough to indicate a drop of as little as ±100 μV on its zero-center meter, with larger meter deflections for larger drops. Ideally it should have a nonlinear scale, so that even for voltage drops of tens of millivolts the meter will not go off scale. And with some care it should be possible to design a circuit that draws so little battery current that we can omit the on/off switch: 9 V batteries or AA-size cells give nearly their full shelf life of several years at continuous drain currents of less than 20 μA (they have capacities of about 500 mAh and 2500 mAh, respectively).

With a floating supply provided by batteries, the simplest circuit is a high gain inverting amplifier driving a zero-center meter (Figure 4.89). Because the input and output are both intrinsically bipolarity, it's probably best to use a pair of AA cells, running the op-amp from ±1.5 volt unregulated supplies. The back-to-back Schottky diodes reduce the gain gracefully at large output swings and prevent pegging; Figure 4.90 plots the resulting meter deflection versus V_{in}.

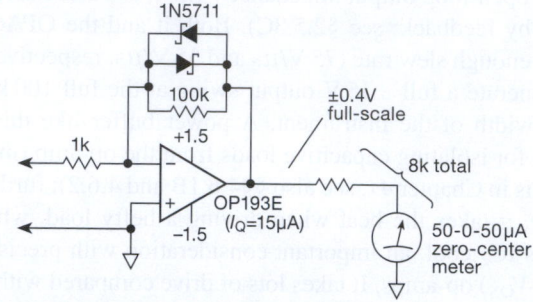

Figure 4.89. Stuck-node tracer: high-gain floating dc amplifier with nonlinear feedback.

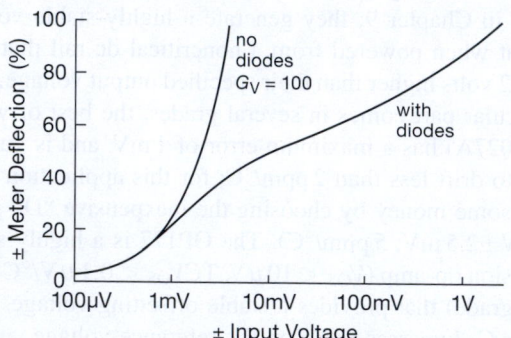

Figure 4.90. Stuck-node tracer achieves large dynamic range through nonlinear feedback.

The major difficulty in this design is in achieving an input offset of less than 100 μV while maintaining micropower current drain, all with supply voltages of just ±1.5 volts. The OP193 is specified to operate down to 2 V total supply voltage, and its output stage swings to the negative rail and to within a volt of the positive rail. In its best grade ("E" suffix) its offset voltage is 75 μV, maximum. Its quiescent current of just 15 μA ensures that the batteries will last their full shelf life, since that current would provide continuous operation for over 150,000 hours from a 2500 mAh battery.

Some additional points.

• One subtle problem with this circuit is that an alkaline battery at the end of its life is down to about 1.0 V terminal voltage; so you would have insufficient headroom to provide full-scale positive output voltage (+0.5 V), given the all-*npn* output stage. A solution is to use a higher battery voltage (e.g., 3 V lithium cells, or multiple alkaline 1.5 V AA cells). But operation from a single pair of AA cells is an elegance worth preserving. In this case you would do better to use an op-amp with true rail-to-rail output, for example the CMOS OPA336. The latter has a quiescent current of $20\,\mu$A, operates down to 2.3 V total supply voltage, and has an untrimmed offset voltage of $125\,\mu$V. Its input voltage range goes to the negative rail and to within 1 volt of the positive rail; the latter is fine here, because we have chosen an inverting amplifier configuration with both inputs at 0 V.

- We rather artificially constrained the circuit design by choosing a zero-center analog meter and then insisting on using just a pair of AA alkaline cells. In real life you'd probably be happier with an *audio* output, with the pitch increasing with input voltage drop; then you could keep your eye on the circuit as you probe around. For this job you'd probably use a simple current-controlled oscillator, built with an op-amp relaxation oscillator or a 555-type timer IC (Chapter 7); for a noncritical application you don't need the linearity and stability of the VCO we designed in Figure 4.83.

- Don't forget the "rail-splitter" techniques we discussed in §4.6.1B; you can always use those tricks to create a split plus and minus rail, for example, from a single 9 V battery. With ±4.5 V rails, you have a much wider range of op-amps to choose from. We were forced to choose from a rather small selection that run on 2 V total supply, draw only tens of microamps supply current, and have "precision" low input offset voltage. Once you have 5 V total supply available (a 9 V battery is down to 6 V at the end of its life), there are literally hundreds of available op-amps, dozens of which run on micropower current drain and have precision low offsets. See, for example, Table 5.5 on page 320.

4.8.3 Load-current-sensing circuit

Figure 4.91 shows a hefty (10 kW!) power supply driving a 100 amp load; the illustrated circuit provides a voltage output proportional to load current, for use with a current regulator, metering circuit, or whatever. The output current is sensed with a *current shunt*, a calibrated manganin-metal 4-terminal power resistor R_S, of resistance $0.0005\,\Omega$, whose "Kelvin connection" of four leads ensures that the sensed voltage does not depend on a low-resistance bond to the sensing terminals (as would be the case if you tried

to do the same thing with a conventional 2-terminal resistor). The voltage drop goes from 0 to 50 mV, with probable common-mode offset caused by the effects of resistance in the ground lead (note that the power supply is connected to chassis ground at the output). For that reason the op-amp is wired as a differential amplifier, with a gain of 200. Voltage offset is trimmed externally with R_8, as the venerable LM358A doesn't have internal trimming circuitry. A zener reference with a few percent stability is adequate for trimming, because the trimming is itself a small correction (you hope!). The supply voltage, V_+, could be unregulated, since the power-supply rejection of the op-amp is more than adequate, 85 dB (typ) in this case.

Some additional points.

- Chassis ground and circuit ground would be connected together, somewhere. But there could easily be a volt or so separating circuit ground from the sensing point along the high-current negative return, because of the very large currents flowing. For that reason we connected the negative supply lead of the op-amp to the more negative end of the current-shunt output. This ensures that the common-mode voltage appearing at the op-amp input never goes below its negative rail; it is a "single-supply" op-amp, with operating common-mode range to its negative rail.

- Low offset voltage is important in this application; for example, to achieve 1% accuracy in a current measurement made at 10% of full-scale load current (i.e., a 10 A load, producing a sense voltage of 5 mV) requires an offset voltage no greater than $50\,\mu$V! We chose the vintage LM358A for our initial design, because it costs only 20 cents. But its poor untrimmed offset (3 mV, max) necessitates external manual trimming; and its lack of external trim terminals forced us to use lots of components. The need for manual trimming might not seem important if you're just building one of these for your lab; but in production it's an extra step, requiring a test setup and procedure, as well as additional parts inventory, etc.

- So, you might choose instead the LT1006, a single op-amp that lets you trim externally with a single 10k pot. However, its improved performance (V_{OS}=80 μV, max, untrimmed) in the least expensive grade – 40 times better than the LM358A – means that you hardly need to trim at all. Carrying this idea further, you could choose instead the LT1077A, a single-supply op-amp with $40\,\mu$V maximum untrimmed offset; it too can be trimmed externally.

- For the utmost in accuracy you should use a chopper-stabilized ("zero-drift") op-amp, for example the LTC1050C. It has $5\,\mu$V maximum offset voltage in the cheapest grade (combined with sub-nanoamp input

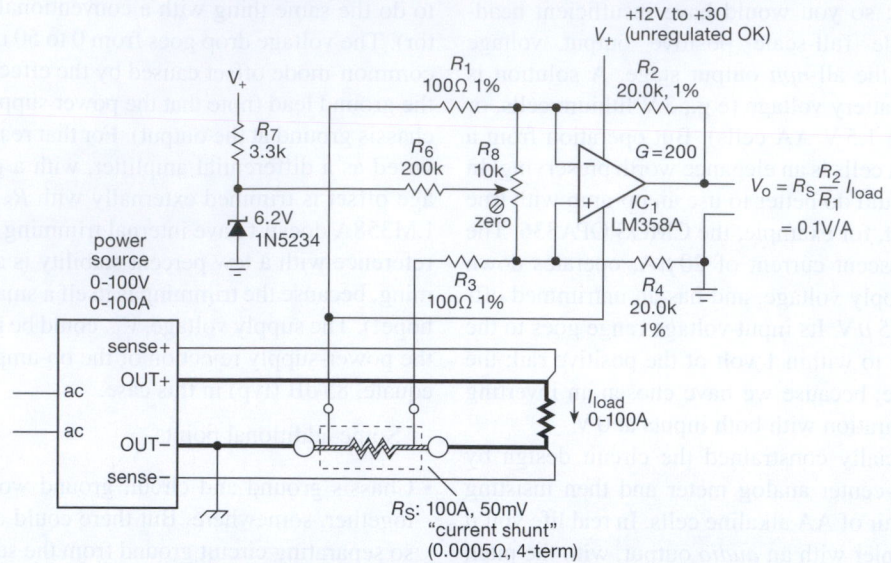

Figure 4.91. High-power current-sensing amplifier.

bias current, which doesn't matter here). This op-amp includes on-chip capacitors for its chopper, and operates from a single-supply (with input common-mode range to the negative rail), just like the LM358. Its $5\,\mu V$ offset voltage corresponds to 1% accuracy at 1% of full scale; that's a dynamic range of 10,000:1, not bad for a simple circuit. See Table 5.6 on page 335 for auto-zero op-amp choices.

- Finally, an interesting design alternative is to do *high-side* current sensing. That is, the shunt is connected instead to the OUT+ power terminal. This has the advantage of keeping all the circuit grounds (power supply, and load) connected together. We'll see how to do this in the advanced Chapter *9x*.

4.8.4 Integrating suntan monitor

We nerds don't ordinarily go to the beach. But when we do, we like to rely on some electronics to tell us when to turn over. What we want to monitor, of course, is the *integrated* dose of tan-producing (UV-rich) sunlight.

There are many ways to accomplish this; in fact, we'll revisit this task when we turn to mixed-signal (analog + digital) electronics (in Chapter 13), and again when we're looking for nifty things to do with microcontrollers (in Chapter 15). Here we want to show how an op-amp integrator can be used to build a suntan monitor circuit.

The idea is to integrate (accumulate) the photocurrent from a sensor whose output is proportional to the intensity

of tanning sunlight. We'll imagine that we have a photodiode, optically filtered to pass only the UV rays of interest, with an output short-circuit current of $\sim 1\,nA$ (nominal) in full sunlight; we'll assume that the photocurrent might range down to a tenth of this value, or so, in hazy sun.

A. First try: direct integration

The circuit in Figure 4.92 is a reasonable first try. It uses a single-supply CMOS micropower op-amp ($10\,\mu A$ per amplifier), powered from a 9 V battery, to integrate the (negative) photocurrent. A nanoamp produces a positive-going ramp of 0.5 mV/s at the op-amp's output, which we connect to a Schmitt trigger comparator with settable positive threshold. The LM385-2.5Z micropower two-terminal (zener-type) voltage reference then gives us a range of 0 to 1.5 hours ($\sim 5000s$) full sunlight equivalent (let's call it "FSE"), at which point the comparator output goes to ground, driving the piezo alarm. The latter draws 15 mA, a substantial battery load, but it is *very* loud, so even a dozing nerd will quickly enough shut the thing off (via the "reset" button). This circuit draws about $50\,\mu A$ when integrating, good for about 8000 hours of operation (a 9 V battery has a capacity of 500 mAh at low drain). 8000 hours is about a year, so that's a lot of tanning; the battery will die of old age first.

Exercise 4.27. The LM385 requires a minimum of $10\,\mu A$ of current for proper operation. What does the circuit provide, at the end of battery life (6 V)?

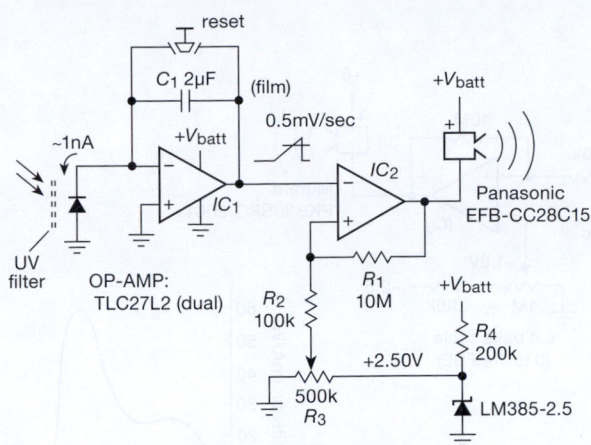

Figure 4.92. Integrating suntan monitor, first try.

Exercise 4.28. How much hysteresis does Schmitt trigger IC_2 provide? How will that affect the operation?

B. Second try: two-step conversion

One problem with the last circuit is that the unfiltered photodiode current is at least a few microamps, in direct sunlight. Trying to cut down the light by a factor of a thousand is risky, because you get light leaks, etc., that cause large errors.

The circuit in Figure 4.93 fixes that, by first converting the photocurrent (however large its magnitude) to a voltage, then integrating that in a second stage where we can choose an input resistor to generate a current in the nanoamp range. Now, however, we've got to use split supplies. That's because whichever polarity we choose for the transresistance (current to voltage) amplifier's output (by connecting the photodiode appropriately), the subsequent integrator's output will be the opposite polarity; integrators invert. In our circuit we used a 2.5 V reference to split the 9 V battery; most of the current in the circuit is between the positive and negative rails, so the reference needs less than $20\,\mu A$ of bias.[41] In this circuit we've shown a two-pole power switch, wired so the integrating capacitor is held reset until power is turned on.

The integrator output triggers a Schmitt comparator, as before, driving the mighty-lunged piezo screamer. Note that its large drive current is rail-to-rail; it does not pass through our ground reference. The circuit's operating cur-

---

[41] Alternatively, we could have used a TLE2425 3-terminal "rail splitter" (§4.6.1B), which, however, would consume $170\,\mu A$. Although that would dominate the power budget, the thing would still run for 2000 hours (about 3 months) of continuous operation.

rent is about $60\,\mu A$, good for nearly a year of continuous operation.

A final note: the LMC6044 is a quad, rail-to-rail output, micropower op-amp ($10\,\mu A$/amplifier). So if a stiff ground reference were needed, the unused op-amp section could be configured as in Figure 4.73, with the stabilizing trick of Figure 4.76A.

C. The "Mark-III" suntan integrator

It's always fun to see how elegantly you can shrink down circuit complexity. In this case there's a nice trick you can use to eliminate the two-stage integration, namely, a "current divider." Figure 4.94 shows how it's done: the photocurrent drives a pair of resistors, bridging the same voltage (because the inverting input is a virtual ground); the current divides proportional to the relative conductance, in this case in the ratio of 1000:1 if pot R_2 is turned to minimum resistance. That means that a photocurrent of $1\,\mu A$ would inject a current of 1 nA into the integrator. If you prefer, you can think of the circuit as a resistive load (R_1 in series with R_2, which easily dominates R_3), which develops a voltage $V_{in} = I_{diode}(R_1 + R_2)$; that voltage is the input to the integrator, via R_3. Because the voltage developed by the photocurrent can range up to nearly a volt, it's necessary to back-bias the detector diode, in this case with a forward-biased diode D_2, which generates a -0.4 V rail.

The integrator's positive-going output ramp drives Schmitt comparator A_2, with fixed comparison voltage provided by reference D_1. Its output drives the by-now usual piezo alarm.

Now for the elegance: it turns out you can get, packaged in a single small IC, a combination op-amp, comparator, and voltage reference. The MAX951 shown is just one of several such offerings, and it fills the bill here. It is because of the internal connection of D_1 and IC_2's inverting input that we were forced to put the suntan control at the input, rather than at the comparator.

A few additional comments.

- The accuracy of the current divider depends on the accuracy of the virtual ground. The op-amp shown has a maximum offset voltage of 3 mV, so at 10% full sunlight and with the control set to minimum resistance (maximum bake cycle), the error is about 30% (10 mV signal, 3 mV offset). In other words, the circuit elegance involves a compromise in performance, relative to the more straightforward (some might say heavy-handed) approach in Figure 4.93, where the error is about 3% at minimum sunlight.

- Diode D_2 will be forward biased by the IC's quiescent

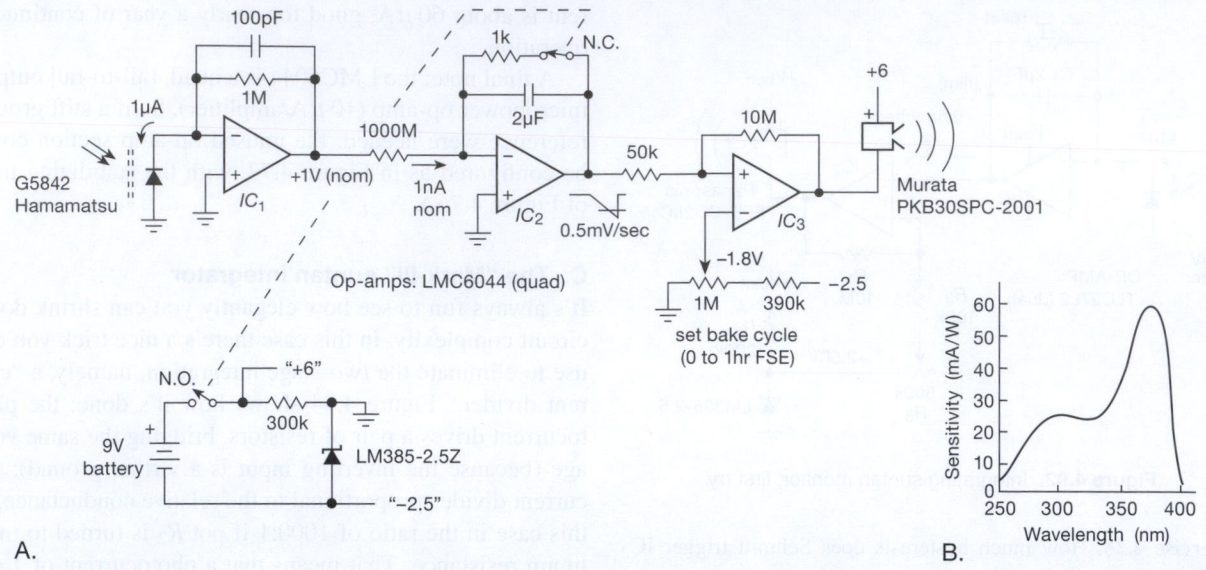

Figure 4.93. Integrating suntan monitor, second try; FSE is "full sunlight equivalent." A. Schematic. B. Spectral response of the Hamamatsu G5842 photodiode, whose short-circuit photocurrent in sunlight is about 1 μA.

current of $7\,\mu A$ as long as the photocurrent is less than this value. Thus the biasing resistor R_6 can be omitted unless a maximum photocurrent of more than about $5\,\mu A$ is anticipated.

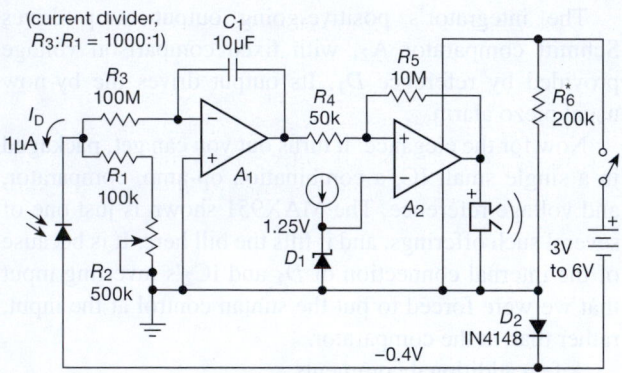

Figure 4.94. Integrating suntan monitor, third try. A_1, A_2, and D_1 all live inside a MAX951 multifunction chip. A_2 is a comparator.

• The MAX951 has a specified operating voltage range of 2.7 V to 7 V. The low-voltage operation is a pleasant bonus; but for this particular IC it also means that we cannot run directly from a 9 V battery, unless we use a voltage regulator (see Chapter 9) to reduce the supply voltage to 7 V or less. This illustrates an important lesson, namely, that you have to watch out for low maximum supply-voltage ratings when using ICs intended for low-voltage operation. It also illustrates the trend of IC

manufacturers toward lower supply voltages for their new product designs.

4.9 Feedback amplifier frequency compensation

We first met feedback in Chapter 2 (§2.5), where we saw its beneficial effects on stability and predictability of amplifier gain, and the reduction of an amplifier's inherent non-linearities. We saw, also, how it affects input and output impedances of amplifiers: for example, by sensing output voltage, and using series feedback at the input, the input impedance is raised and the output impedance is lowered, both by a factor of the loop gain. All is not rosy, though: the combination of gain with feedback creates the possibility of oscillation. Here, in the context of op-amps, we continue the treatment of negative feedback, looking at the important subject of *frequency compensation* – the business of preventing oscillation in amplifiers with negative feedback. The material in §2.5 is a necessary background for the sections that follow.

Let's begin by looking at a graph of open-loop voltage gain versus frequency for several op-amps: you'll typically see something like the curves in Figure 4.95. From a superficial look at such a *Bode plot* (a log–log plot of open-loop gain and phase versus frequency) you might conclude that the OP27 is an inferior op-amp, since its open-loop gain drops off so rapidly with increasing frequency. In fact, that rolloff is built into the op-amp intentionally and is

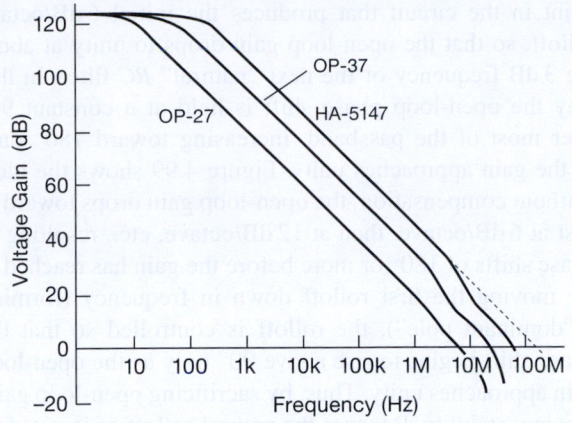

Figure 4.95. Open-loop gain versus frequency for three similar op-amps.

recognizable as the same −6 dB/octave curve characteristic of an *RC* lowpass filter. The OP37, by comparison, is identical to the OP27 except that it is *decompensated* (and similarly for the discontinued[42] HA-5147). Op-amps are most often internally compensated, with decompensated and uncompensated varieties sometimes available. Let's take a look at this business of frequency compensation.

4.9.1 Gain and phase shift versus frequency

An op-amp (or, in general, any multistage amplifier) will begin to roll off at some frequency because of the lowpass filters formed by signals of finite source impedance driving capacitive loads within the amplifier stages. For instance, it is common to have an input stage consisting of a differential amplifier, perhaps with current-mirror load (see the LF411 schematic in Figure 4.43), driving a common-emitter second stage. For now, imagine that the capacitor labeled C_C in that circuit is removed. The high output impedance of the input stage Q_2, in combination with the combined capacitance seen at its output, forms a lowpass filter whose 3 dB point might fall somewhere in the range of 100 Hz to 10 kHz.

The decreasing reactance of this capacitance with increasing frequency gives rise to the characteristic 6 dB/octave rolloff: at sufficiently high frequencies (which may be below 1 kHz), the capacitive loading dominates the collector load impedance, resulting in a voltage gain $G_V = g_m X_C$, i.e., the gain drops off as $1/f$. It also produces a 90° lagging phase shift at the output relative to the input signal. (You can think of this as the tail of an *RC* low-

pass filter characteristic, where R represents the equivalent source impedance driving the capacitive load. However, it is not necessary to have any actual resistors in the circuit.)

In a multistage amplifier there will be additional rolloffs at higher frequencies, caused by lowpass filter characteristics in the other amplifier stages, and the overall open-loop gain will look something like that shown in Figure 4.96. The open-loop gain begins dropping at 6 dB/octave at some low frequency f_1, because of capacitive loading of the first-stage output. It continues dropping off with that slope until an internal *RC* of another stage rears its ugly head at frequency f_2, beyond which the rolloff goes at 12 dB/octave, and so on.

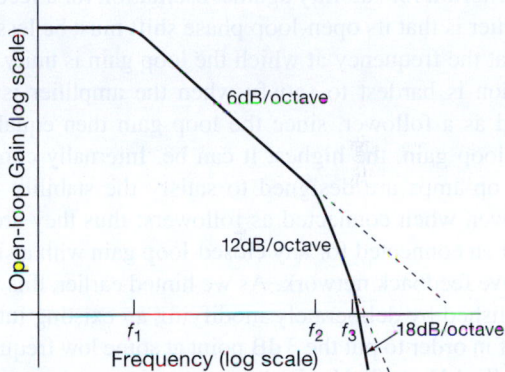

Figure 4.96. Multistage amplifier: gain versus frequency.

What is the significance of all this? Remember that an *RC* lowpass filter has a phase shift that looks as shown in Figure 4.97. Each lowpass filter within the amplifier has a similar phase-shift characteristic, so the overall phase

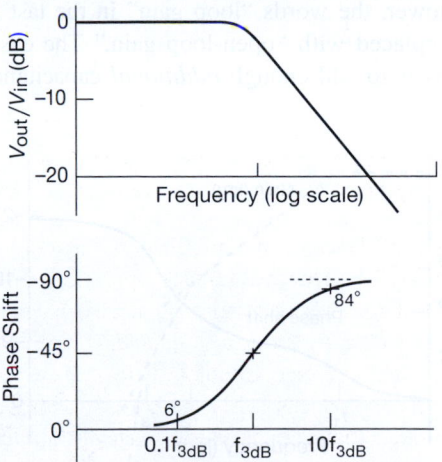

Figure 4.97. Bode plot: gain and phase versus frequency.

shift of the hypothetical amplifier will be as shown in Figure 4.98.

Now here's the problem: if you were to connect this amplifier as an op-amp follower, for instance, it would oscillate. That's because the open-loop phase shift reaches 180° at some frequency at which the gain is still greater than 1 (negative feedback becomes positive feedback at that frequency). That's all you need to generate an oscillation, as any signal whatsoever at that frequency builds up each time around the feedback loop, just like a public address system with the gain turned up too far.

A. Stability criterion

The criterion for stability against oscillation for a feedback amplifier is that its open-loop phase shift must be less than 180° at the frequency at which the loop gain is unity. This criterion is hardest to satisfy when the amplifier is connected as a follower, since the loop gain then equals the open-loop gain, the highest it can be. Internally compensated op-amps are designed to satisfy the stability criterion even when connected as followers; thus they are stable when connected for any closed-loop gain with a simple resistive feedback network. As we hinted earlier, this is accomplished by deliberately modifying an existing internal rolloff in order to put the 3 dB point at some low frequency, typically 1 Hz to 20 Hz. Let's see how that works.

4.9.2 Amplifier compensation methods

A. Dominant-pole compensation

The goal is to keep the open-loop phase shift much less than 180° at all frequencies for which the loop gain is greater than 1. Assuming that the op-amp may be used as a follower, the words "loop gain" in the last sentence can be replaced with "open-loop gain." The easiest way to do this is to add enough *additional* capacitance at the

point in the circuit that produces the initial 6 dB/octave rolloff, so that the open-loop gain drops to unity at about the 3 dB frequency of the next "natural" *RC* filter. In this way the open-loop phase shift is held at a constant 90° over most of the passband, increasing toward 180° only as the gain approaches unity. Figure 4.99 shows the idea. Without compensation, the open-loop gain drops toward 1, first at 6 dB/octave, then at 12 dB/octave, etc., resulting in phase shifts of 180° or more before the gain has reached 1. By moving the first rolloff down in frequency (forming a "dominant pole"), the rolloff is controlled so that the phase shift begins to rise above 90° only as the open-loop gain approaches unity. Thus, by sacrificing open-loop gain, you buy stability. Because the natural rolloff of lowest frequency is usually caused by the Miller effect in the stage driven by the input differential amplifier, the usual method of dominant-pole compensation consists simply of adding additional feedback capacitance around the second-stage transistor, so the combined voltage gain of the two stages is $g_m X_C$ or $g_m/2\pi f C_{comp}$ over the compensated region of the amplifier's frequency response (Figure 4.100). In practice, Darlington-connected transistors would probably be used for both stages.

By putting the dominant-pole unity-gain crossing at the 3 dB point of the next rolloff, you get a phase margin of about 45° in the worst case (follower), since a single *RC* filter has a 45° lagging phase shift at its 3 dB frequency, i.e., the phase margin equals $180° - (90° + 45°)$, with the 90° coming from the dominant pole.

An additional advantage of using a Miller-effect pole for

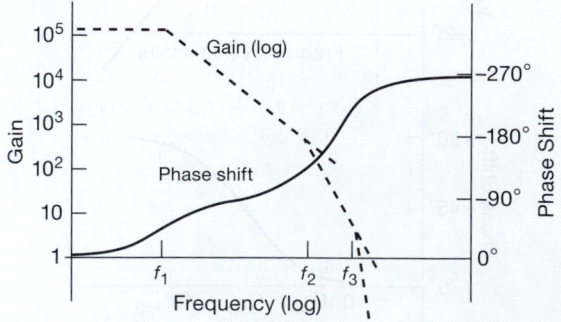

Figure 4.98. Gain and phase in a multistage amplifier.

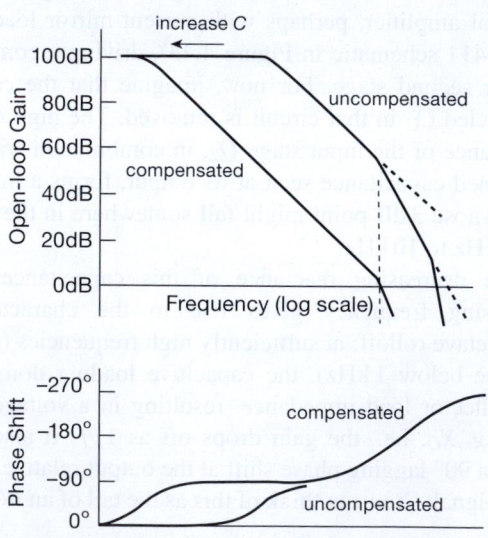

Figure 4.99. "Dominant-pole" compensation.

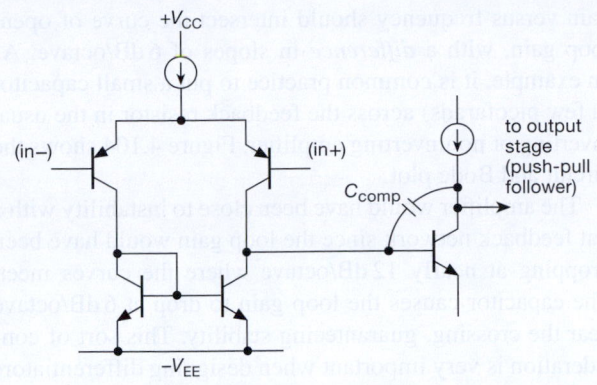

Figure 4.100. Classic op-amp input stage with compensation.

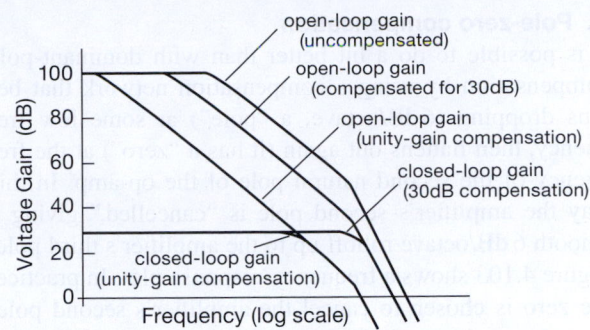

Figure 4.102. Stability is easier to achieve with larger closed-loop gain.

compensation is that the compensation is inherently insensitive to changes in voltage gain with temperature or manufacturing spread of gain: higher gain causes the feedback capacitance to look larger, moving the pole downward in frequency in exactly the right way to keep the unity-gain crossing frequency unchanged. In fact, the actual 3 dB frequency of the compensation pole is quite irrelevant; what matters is the point at which it intersects the unity-gain axis (Figure 4.101).

B. Decompensated and uncompensated op-amps

If an op-amp is used in a circuit with closed-loop gain greater than unity (i.e., not a follower), it is not necessary to put the pole (the term for the "corner frequency" of a lowpass filter, see Chapter *1x*) at such a low frequency as the stability criterion is relaxed because of the lower loop gain. Figure 4.102 shows the situation graphically.

For a closed-loop gain of 30 dB, the loop gain (which is the ratio of the open-loop gain to the closed-loop gain) is less than for a follower, so the dominant pole can be placed at a higher frequency. It is chosen so that the open-loop gain reaches 30 dB (rather than 0 dB) at the fre-

quency of the next natural pole of the op-amp. As the graph shows, this means that the open-loop gain is higher over most of the frequency range, and the resultant amplifier will work at higher frequencies. Some op-amps are available in "decompensated" (a better word might be "undercompensated") versions, which are internally compensated for closed-loop gains greater than some minimum ($A_V > 5$ in the case of the OP37); these specify a minimum closed-loop gain, and require no external capacitor. Another example is the THS4021/2, a decompensated version ($G_V \geq 10$) of the unity-gain stable THS4011/2. These are really speedy op-amps, with an f_T of 300 MHz (for the "slow" THS4011/2), and greater than 1 GHz for the THS4021/2. For the decompensated versions the manufacturer (TI) supplies recommended external capacitance values (sometimes in combination with a resistor; see below) for a selection of minimum closed-loop gains.[43] Decompensated or uncompensated op-amps are worth using if you need the added bandwidth and your circuit operates at high gain; see Chapter *4x* for further discussion.

Some intuition: it may at first seem paradoxical that an op-amp circuit configured for a *low*-gain circuit is more prone to oscillation than one configured for *high*-gain. But it makes sense: the better stability of an op-amp connected for a closed-loop gain of $G_{CL}=100$ (40 dB), say, comes about because the feedback network (resistive divider) attenuates signals by a factor of 100. So it's harder to sustain an oscillation going around the loop, compared with a follower (in which the feedback has unity gain).

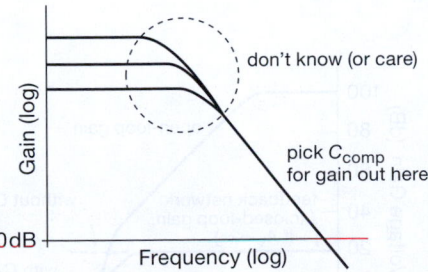

Figure 4.101. The compensation capacitor is chosen to set the open-loop unity-gain frequency; the *low*-frequency gain is unimportant.

---

[43] In some cases external compensation components are required for any plausible closed-loop gain; these are properly called "uncompensated" op-amps.

C. Pole-zero compensation

It is possible to do a bit better than with dominant-pole compensation by using a compensation network that begins dropping (6 dB/octave, a "pole") at some low frequency, then flattens out again (it has a "zero") at the frequency of the second natural pole of the op-amp. In this way the amplifier's second pole is "cancelled," giving a smooth 6 dB/octave rolloff up to the amplifier's third pole. Figure 4.103 shows a frequency-response plot. In practice, the zero is chosen to cancel the amplifier's second pole; then the position of the first pole is adjusted so that the overall response reaches unity gain at the frequency of the amplifier's third pole. A good set of datasheets for an op-amp with external compensation will often give suggested component values (an R and a C) for pole-zero compensation, as well as the usual capacitor values for dominant-pole compensation. Moving the dominant pole downward in frequency actually causes the second pole of the amplifier to move upward somewhat in frequency, an effect known as "pole splitting." The frequency of the cancelling zero is then chosen accordingly.

4.9.3 Frequency response of the feedback network

In all of the discussion thus far we have assumed that the feedback network has a flat frequency response; this is usually the case, with the standard resistive voltage divider as a feedback network. However, there are occasions when some sort of equalization amplifier is desired (integrators and differentiators are in this category) or when the frequency response of the feedback network is modified to improve amplifier stability. In such cases it is important to remember that the Bode plot of *loop* gain versus frequency is what matters, rather than the curve of open-loop gain. To make a long story short, the curve of ideal closed-loop

gain versus frequency should intersect the curve of open-loop gain, with a *difference* in slopes of 6 dB/octave. As an example, it is common practice to put a small capacitor (a few picofarads) across the feedback resistor in the usual inverting or noninverting amplifier. Figure 4.104 shows the circuit and Bode plot.

The amplifier would have been close to instability with a flat feedback network since the loop gain would have been dropping at nearly 12 dB/octave where the curves meet. The capacitor causes the loop gain to drop at 6 dB/octave near the crossing, guaranteeing stability. This sort of consideration is very important when designing differentiators because an ideal differentiator has a closed-loop gain that *rises* at 6 dB/octave; it is necessary to roll off the differentiator action at some moderate frequency, preferably going over to a 6 dB/octave rolloff at high frequencies. Integrators, by comparison, are very friendly in this respect, owing to their 6 dB/octave closed-loop rolloff. It takes real talent to make a low-frequency integrator oscillate!

Exercise 4.29. Show on a Bode plot that the value of stabilizing resistor R_1 in Figure 4.69 stops the differentiator action (i.e., flattens the curve of closed-loop gain) before the crossing point of open-loop and closed-loop gains. Explain our value of minimum recommended resistance R_1.

A. What to do

In summary, you are generally faced with the choice of internally compensated or uncompensated op-amps. It is simplest to use the compensated variety, and that's the usual choice. You might begin by considering the

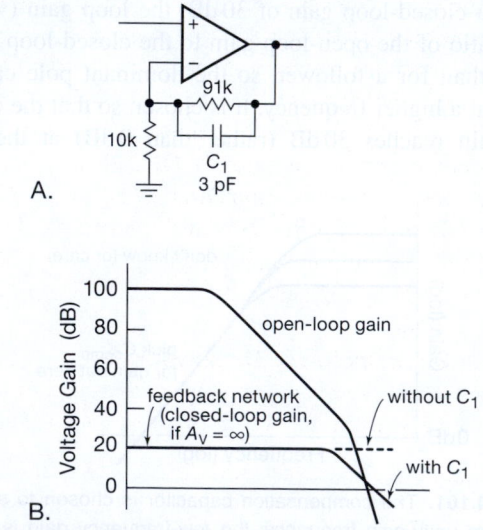

A.

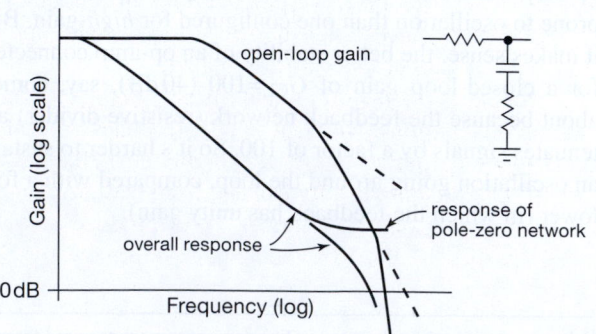

Figure 4.103. Cancelling the amplifier's second pole in "pole-zero" compensation.

Figure 4.104. A small feedback capacitor enhances stability.

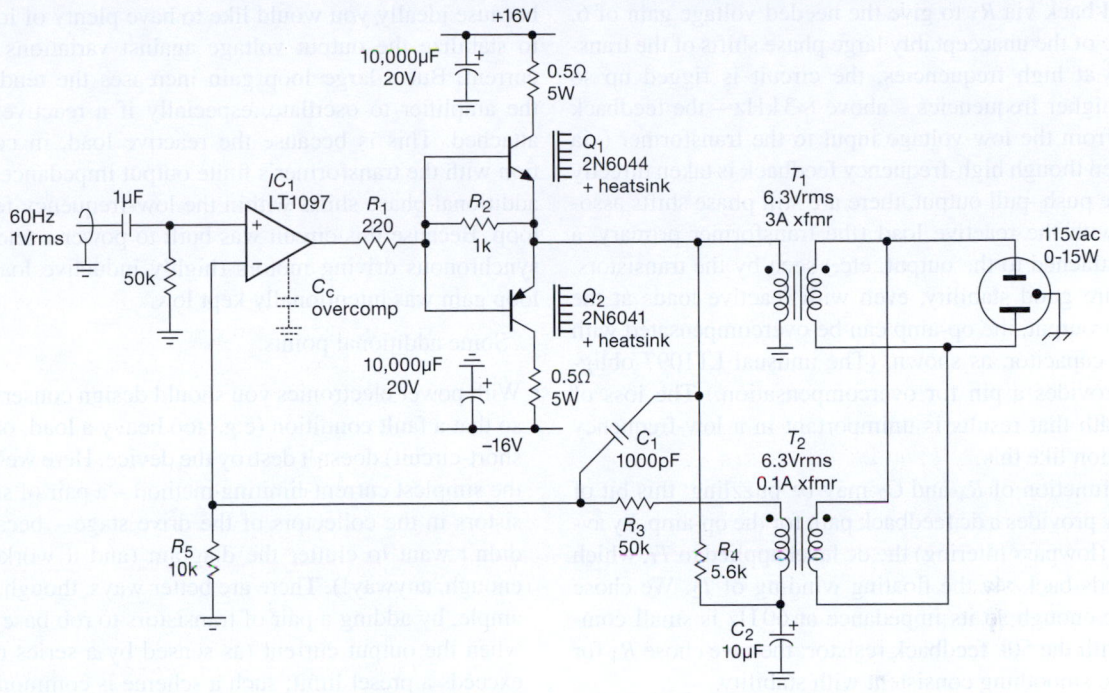

Figure 4.105. Output amplifier for 60 Hz power source. The push–pull output transistors Q_1 and Q_2 are power Darlingtons in a plastic power package.

conventional LF411 (JFET, ± 5 V to ± 15 V supply) or an improved version (the LT1057), or the rail-to-rail input and output LMC6482 (CMOS, $+3$ V to $+15$ V supply), or perhaps the accurate and quiet LT1012, all internally compensated for unity gain. If you need greater bandwidth or slew rate, look for a faster compensated op-amp (see Table 4.2a on 271 for some choices). If it turns out that nothing is suitable and the closed-loop gain is greater than unity (as it usually is), you can use a decompensated (or uncompensated) op-amp, perhaps with an external capacitor as specified by the manufacturer for the gain you are using. Using our previous example, the popular OP27 low-noise precision op-amp (unity-gain-compensated) has $f_T = 8$ MHz and a slew rate of 2.8 V/μs; it is available as the decompensated OP37 (minimum gain of 5), with $f_T = 63$ MHz and a slew rate of 17 V/μs.[44]

B. Example: precision 60 Hz power source
Uncompensated op-amps, or op-amps with a compensation pin, also give you the flexibility of *over*compensating, a

simple solution to the problem of additional phase shifts introduced by other stuff in the feedback loop. Figure 4.105 shows an example. This is a low-frequency amplifier designed to generate a precise and stable 115 volt ac power output from a variable frequency 60 Hz low-level sinewave input.[45] The op-amp is wired as an ac-coupled noninverting amplifier, with its output driving a Darlington push–pull emitter-follower output stage Q_1Q_2, which in turn drives the low-voltage winding of a small power transformer, T_1, whose windings are in the ratio of 6.3 V:115 V. In this way we generate 115 V ac output without high-voltage op-amps or transistors. Of course, we pay the price in proportionally higher drive current; here the transistors need to supply about 3 A (rms) to produce a 15 W output.

To generate low distortion and a stable output voltage under load variations, we want to take feedback from the actual 115 V output sinewave. It is highly desirable, however, to keep the output fully isolated from circuit ground. So we use a second transformer T_2 to produce a low-voltage replica of the 115 V output waveform, which is

---

[44] And, before it was discontinued, the similar "more-decompensated" HA-5147 (minimum gain of 10), with $f_T = 120$ MHz and a slew rate of 35 V/μs.

[45] The original design was used to drive an astronomical telescope at sidereal (star-tracking) rate. Interesting trivia: contrary to popular belief, the Earth turns on its axis once every 23 hours, 56 minutes, and 4.1 seconds; figure out why it isn't 24:00:00!

then fed back via R_3 to give the needed voltage gain of 6. Because of the unacceptably large phase shifts of the transformers at high frequencies, the circuit is rigged up so that at higher frequencies – above ∼3 kHz – the feedback comes from the low-voltage input to the transformer (via C_1). Even though high-frequency feedback is taken directly from the push–pull output, there are still phase shifts associated with the reactive load (the transformer primary, a motor attached to the output, etc.) seen by the transistors. To ensure good stability, even with reactive loads at the 115 volt output, the op-amp can be overcompensated with a small capacitor, as shown. (The unusual LT1097 obligingly provides a pin for overcompensation.) The loss of bandwidth that results is unimportant in a low-frequency application like this.

The function of R_4 and C_2 may be puzzling: this bit of circuitry provides a dc feedback path for the op-amp, by averaging (lowpass filtering) the dc level applied to T_1, which then feeds back via the floating winding of T_2. We chose C_2 large enough so its impedance at 60 Hz is small compared with the 50k feedback resistor; then we chose R_4 for adequate smoothing consistent with stability.

The performance of this amplifier is quite satisfying. Figure 4.106 shows the output regulation, i.e., the change of rms output amplitude versus load. For comparison we show the comparable curve when feedback is taken exclusively from the driving winding of T_1, from which you can see that the desired feedback path improves output amplitude regulation, under load variations from zero to full power, from a mediocre 10% to just 0.2%. The output sinewave is very clean, with measured distortion well below 1% under all load conditions, including driving a synchronous motor (which represents a reactive load).

An application such as this represents a compromise, because ideally you would like to have plenty of loop gain to stabilize the output voltage against variations in load current. But a large loop gain increases the tendency of the amplifier to oscillate, especially if a reactive load is attached. This is because the reactive load, in combination with the transformer's finite output impedance, causes additional phase shifts within the low-frequency feedback loop. Because this circuit was built to power a telescope's synchronous driving motors (highly inductive loads), the loop gain was intentionally kept low.

Some additional points.

• With power electronics you should design conservatively so that a fault condition (e.g., too heavy a load, or even a short-circuit) doesn't destroy the device. Here we've used the simplest current-limiting method – a pair of small resistors in the collectors of the drive stage – because we didn't want to clutter the diagram (and it worked well enough, anyway!). There are better ways, though, for example, by adding a pair of transistors to rob base current when the output current (as sensed by a series resistor) exceeds a preset limit; such a scheme is commonly used within the integrated circuitry of op-amps themselves – see Figure 4.43. As we'll explain in §9.13.3, there are still better protective circuits. The problem with simple current-limiting protection is that a short-circuit load would cause the transistors to experience the limit current with the full supply voltage across them; the resulting power dissipation is far greater than the maximum under ordinary operation, which requires conservative heatsinking and component selection. *Foldback* current limiting would be better, though a bit more complicated.

• A push–pull follower with the bases tied together has a crossover region in which the feedback loop is effectively broken (see §2.4.1A). With Darlington transistors the crossover region is four V_{BE}'s, about 2.5 V. The resistor R_2 in Figure 4.105 ensures that there is always some linear coupling from the op-amp to T_1, to prevent the feedback loop from rattling around under light load. Better still would be diode biasing, in the manner of Figures 2.71 or 2.72; see also the discussion of push-pull follower output stages in Chapter 2x.

• There is an elegant way to use a normal ±15 V op-amp to generate larger voltage swings, by replacing the emitter followers in Figure 4.105 with a "pseudo-Darlington" configuration with modest noninverting gain (also known as a "series feedback pair," see §2.5.5C), say a factor of 5. Then you can run the power output stage from a ±75 V supply while powering the op-amp from conventional ±15 V.

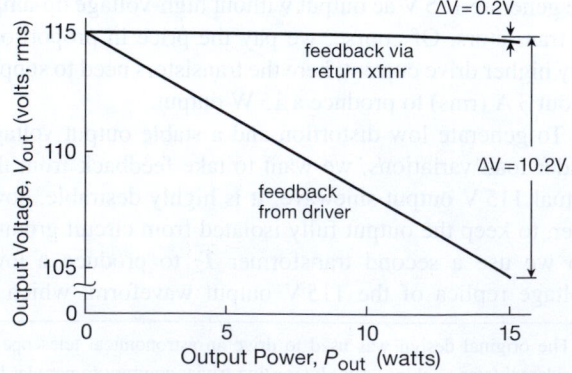

Figure 4.106. Measured output voltage versus load for 60 Hz power source.

C. Motorboating

In ac-coupled feedback amplifiers, stability problems can also crop up at very low frequencies, because of the accumulated *leading* phase shifts caused by several capacitively coupled stages. Each blocking capacitor, in combination with the input resistance (from bias strings and the like), causes a leading phase shift that equals 45° at the low-frequency 3 dB point and approaches 90° at lower frequencies. If there is enough loop gain, the system can go into a low-frequency oscillation picturesquely known as "motorboating." With the widespread use of dc-coupled amplifiers, motorboating is almost extinct. However, old-timers can tell you some good stories about it.

Additional Exercises for Chapter 4

Exercise 4.30. Design a "sensitive voltmeter" to have $Z_{in} = 1\,M\Omega$ and full-scale sensitivities of 10 mV to 10 V in four ranges. Use a 1 mA meter movement and an op-amp. Trim voltage offsets if necessary, and calculate what the meter will read with input open, assuming (a) $I_B = 25\,pA$ (typical for a 411) and (b) $I_B = 80\,nA$ (typical for a 741). Use some form of meter protection (e.g., keep its current less than 200% of full scale), and protect the amplifier inputs from voltages outside the supply voltages. What do you conclude about the suitability of the 741 for low-level high-impedance measurements?

Exercise 4.31. Design an audio amplifier, using an OP27 op-amp (low noise, good for audio), with the following characteristics: gain = 20 dB, $Z_{in} = 10k$, −3 dB point = 20 Hz. Use the non-inverting configuration, and roll off the gain at low frequencies in such a way as to reduce the effects of input offset voltage. Use proper design to minimize the effects of input bias current on output offset. Assume that the signal source is capacitively coupled.

Exercise 4.32. Design a unity-gain phase splitter (see §2.2.8 in Chapter 2) using 411s. Strive for high input impedance and low output impedances. The circuit should be dc-coupled. At roughly what maximum frequency can you obtain full swing (27 V pp, with ±15 V supplies), owing to slew-rate limitations?

Exercise 4.33. El Cheapo brand loudspeakers are found to have a treble boost, beginning at 2 kHz (+3 dB point) and rising 6 dB/octave. Design a simple *RC* filter, buffered with AD611 op-amps (another good audio chip) as necessary, to be placed be-

tween preamp and amplifier to compensate for this rise. Assume that the preamp has $Z_{out} = 50k$ and that the amplifier has $Z_{in} = 10k$, approximately.

Exercise 4.34. A 741 is used as a simple comparator, with one input grounded; i.e., it is a zero-crossing detector. A 1 volt amplitude sine wave is fed into the other input (frequency=1 kHz). What voltage(s) will the input be when the output passes through zero volts? Assume that the slew rate is 0.5 V/μs and that the op-amp's saturated output is ±13 V.

Exercise 4.35. The circuit in Figure 4.107 is an example of a "negative-impedance converter." (a) What is its input impedance? (b) If the op-amp's output range goes from V_+ to V_-, what range of input voltages will this circuit accommodate without saturation?

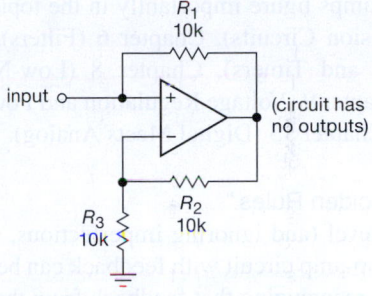

Figure 4.107. Negative-impedance converter.

Exercise 4.36. Consider the circuit in the preceding problem as the 2-terminal black box (Figure 4.108). Show how to make a dc amplifier with a gain of +10. Why can't you make a dc amplifier with a gain of −10? (Hint: the circuit is susceptible to a latchup condition for a certain range of source resistances. What is that range? Can you think of a remedy?)

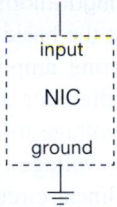

Figure 4.108. Negative-impedance connector as a 2-terminal device

Review of Chapter 4

An A-to-O summary of what we have learned in Chapter 4. This summary reviews basic principles and facts in Chapter 4, but it does not cover application circuit diagrams and practical engineering advice presented there.

¶A. The Ideal Op-amp.

In Chapter 4 we explored the world of Operational Amplifiers ("Op-amps"), universal building blocks of analog circuits. A good op-amp approaches the ideal of an infinite-gain wideband noiseless dc-coupled difference amplifier with zero input current and zero offset voltage. Op-amps are intended for use in circuits with negative feedback, where the feedback network determines the circuit's behavior. Op-amps figure importantly in the topics of Chapter 5 (Precision Circuits), Chapter 6 (Filters), Chapter 7 (Oscillators and Timers), Chapter 8 (Low-Noise Techniques), Chapter 9 (Voltage Regulation and Power Conversion), and Chapter 13 (Digital Meets Analog).

¶B. The "Golden Rules."

At a basic level (and ignoring imperfections, see ¶¶K–M below), an op-amp circuit with feedback can be simply understood by recognizing that feedback from the output operates to (I) make the voltage difference between the inputs zero; and, at the same level of ignoration, (II) the inputs draw no current. These rules are quite helpful, and for dc (or low-frequency circuits) they are in error only by typical offset voltages of a millivolt or less (rule I), and by typical input currents of order a picoamp for FET types or tens of nanoamps for BJT types (rule II).

¶C. Basic Op-amp Configurations.

In §4.2 and §4.3 we met the basic linear circuits (detailed in ¶¶D–F below): inverting amplifier, non-inverting amplifier (and follower), difference amplifier, current source (transconductance, i.e., voltage-to-current), transresistance amplifier (i.e., current-to-voltage), and integrator. We saw also two important *non*-linear circuits: the Schmitt trigger, and the active rectifier. And in §4.5 we saw additional circuit building blocks: peak detector, sample-and-hold, active clamp, active full-wave rectifier (absolute-value circuit), and differentiator.

¶D. Voltage Amplifiers.

The *inverting amplifier* (Figure 4.5) combines input current V_{in}/R_1 and feedback current V_{out}/R_2 into a summing junction; it has voltage gain $G_V = -R_2/R_1$ and input impedance R_1. In the *noninverting amplifier* (Figure 4.6) a fraction of the output is fed back to the inverting input; it has voltage gain $G_V = 1 + R_2/R_1$ and near-infinite input impedance. For the *follower* (Figure 4.8) the feedback gain is unity, i.e., the resistive divider is replaced by a connection from output to inverting input. The *difference amplifier* (Figure 4.9) uses a pair of matched resistive dividers to generate an output $V_{out} = (R_2/R_1)\Delta V_{in}$; its input impedance is $R_1 + R_2$, and its common-mode rejection depends directly on the accuracy of the resistor matching (e.g., $\sim$60 dB with $\pm$0.1% resistor tolerance). Difference amplifiers are treated in greater detail in §5.14. A pair of input followers can be used to achieve high input impedance, but a better 3-op-amp configuration is the *instrumentation amplifier*, see §5.15.

¶E. Integrator and Differentiator.

The *integrator* (Figure 4.16) looks like an inverting amplifier in which the feedback resistor is replaced by a capacitor; thus the input current V_{in}/R_1 and feedback current $C\,dV_{out}/dt$ are combined at the summing junction. Ignoring the imperfections in ¶K below, the integrator is "perfect," thus any non-zero average dc input voltage will cause the output to grow and eventually saturate. The integrator can be reset with a transistor switch across the feedback capacitor (Figure 4.18); alternatively you can use a large shunt resistor to limit the dc gain, but this defeats the integrator operation at low frequencies ($f \lesssim 1/R_f C$). The integrator's input impedance is R_1.

The op-amp *differentiator* (Figure 4.68) is a similar configuration, but with R and C interchanged. Without additional components (Figure 4.69) this configuration is unstable (see ¶O, below).

¶F. Transresistance and Transconductance Amplifiers.

By omitting the input resistor, an inverting voltage amplifier becomes a *transresistance amplifier*[46], i.e., a current-to-voltage converter (Figure 4.22). Its gain is $V_{out}/I_{in} = -R_f$, and (ignoring imperfections) the impedance at its input (which drives the summing junction) is zero. Capacitance at the input creates issues of stability, bandwidth, and noise; see §8.11 and the discussion in Chapter *4x*. Transresistance amplifiers are widely used in photodiode applications.

A *transconductance amplifier* (Figures 4.10–4.15) converts a voltage input to a current output; it is a voltage-controlled current source. The simplest form uses an op-amp and one resistor (Figure 4.10), but works only with a floating load. The Howland circuit and its variations (Figures 4.14 and 4.15) drive a load returned to ground, but

---

[46] Or *transimpedance* amplifier.

their accuracy depends on resistor matching. Circuits with an external transistor (Figures 4.12 and 4.13) drive loads returned to ground, do not require resistor matching, and, in contrast to the other circuits, benefit from the intrinsically high output impedance of the transistor. In Chapter *4x* we describe a nice variation on the transistor-assisted current source that achieves both high speed and bipolarity output (i.e., sinking and sourcing)

¶G. Nonlinear Circuits: Peak Detector, S/H, Clamp, Rectifier.

Because of their high gain, op-amps provide accuracy to nonlinear functions that can be performed with passive components alone; in these circuits one or more diodes select the regions in which feedback acts. The *peak detector* (Figure 4.58) captures and holds the highest (or lowest) voltage since the last reset; the *sample-and-hold* (S/H) circuit (Figure 4.60) responds to an input pulse by capturing and holding the value of an input signal voltage; the *active clamp* (Figure 4.61) bounds a signal to a maximum (or minimum) voltage; the *active rectifier* creates accurate half-wave (Figures 4.36 and 4.38) or full-wave (Figures 4.63 and 4.64) outputs. In practice the performance of these circuits is limited by the finite slew rate and output current of real op-amps (see ¶M, below).

¶H. Positive Feedback: Comparator, Schmitt Trigger, and Oscillator.

If the feedback path is removed, an op-amp acts as a *comparator*, with the output responding (by saturating near the corresponding supply rail) to a reversal of differential input voltage of a millivolt or less (Figure 4.32A). Adding some positive feedback (Figure 4.32B) creates a *Schmitt trigger*, which both speeds up the response and also suppresses noise-induced multiple transitions. Op-amps are optimized for use with negative feedback in linear applications (notably by a deliberate internal −6 dB/octave rolloff "compensation," see ¶O below), so special comparator ICs (lacking compensation) are preferred, see §12.3 and Tables 12.1 and 12.6. A combination of positive feedback (Schmitt trigger) and negative feedback (with an integrator) creates an *oscillator* (Figure 4.39), a subject treated in detail in Chapter 7.

¶I. Single-Supply and Rail-to-rail Op-amps.

For some op-amps both the input common-mode range and the output swing extend all the way down to the negative rail, making them particularly suited for operation with a single positive supply. Rail-to-rail op-amps allow input swings to both supply rails, or output swings to both rails,

or both; see Table 4.2a. The latter are especially useful in circuits with low supply voltages.

¶J. Some Cautions.

In linear op-amp circuits, the Golden Rules (see ¶B, above) will be obeyed only if (a) feedback is negative and (b) the op-amp stays in the active region (i.e., not saturated). There must be feedback at dc, or the op-amp will saturate. Power supplies should be bypassed. Stability is degraded with capacitive loads, and by lagging phase shifts in the feedback path (e.g., by capacitance at the inverting terminal). And, most important, real op-amps have a host of limitations (¶¶K–N, below) that bound attainable circuit performance.

¶K. Departures from Ideal Behavior.

In the real world op-amps are not perfect. There is no "best" op-amp, thus one must trade off a range of parameters: input imperfections (offset voltage, drift, and noise; input current and noise; differential and common-mode range), output limitations (slew rate, output current, output impedance, output swing), amplifier characteristics (gain, phase shift, bandwidth, CMRR and PSRR), operating characteristics (supply voltage and current), and other considerations (package, cost, availability). See §4.4, Tables 4.1, 4.2a, and 4.2b, the more extensive tables in Chapters 5, and 8, and ¶¶L–N below.

¶L. Input Limitations.

The *input offset voltage* (V_{os}), ranging from about $25\,\mu\text{V}$ ("precision" op-amp) to 5 mV, is the voltage unbalance at the input terminals. It's an important parameter for precision circuits, and circuits with high closed-loop dc gain; the error seen at the output is $G_{CL}V_{os}$). Some op-amps provide pins for external trimming of offset voltage (e.g., see Figure 4.43).

The *offset voltage drift*, or *tempco* (TCV$_{os}$, or $\Delta V_{os}/\Delta T$), is the temperature coefficient of offset voltage; it ranges from about $0.1\,\mu\text{V}/°\text{C}$ ("precision" op-amp) to $10\,\mu\text{V}/°\text{C}$. Even if you're lucky and have an op-amp with low V_{os} (or you've trimmed it to zero), TCV$_{os}$ represents the growth of offset with changing temperature.

The *input noise voltage density* (e_n) represents a noisy voltage source in series with the input terminals. It ranges from about $1\,\text{nV}/\sqrt{\text{Hz}}$ (low-noise bipolar op-amp) to $100\,\text{nV}/\sqrt{\text{Hz}}$ or more (micropower op-amps). Noise voltage is important in audio and precision applications.

The *input bias current* (I_B) is the (non-zero) dc current at the input terminals. It ranges from a low of about 5 fA (CMOS low-bias op-amps, and "electrometer" op-amps) to

50 nA (typical[47] BJT op-amps) to a high of $10\,\mu\text{A}$ (wideband BJT-input op-amps). Bias current flowing through the circuit's dc source resistance causes a dc voltage offset; it also creates a current error in integrators and transresistance amplifiers.

The *input noise current* (i_n) is the equivalent noise current added at the input. For most op-amps[48] it is simply the shot noise of the bias current ($i_n = \sqrt{2qI_B}$); it ranges from about $0.1\,\text{fA}/\sqrt{\text{Hz}}$ (CMOS low-bias op-amps, "electrometer" op-amps) to $1\,\text{pA}/\sqrt{\text{Hz}}$ (wideband BJT op-amps). Input noise current flowing through the circuit's ac source impedance creates a noise voltage, which can dominate over e_n. The ratio $r_n = e_n/i_n$ is the op-amp's *noise resistance*; for signal source impedances greater than r_n the current noise dominates.

Op-amps function properly when both inputs are within the *input common-mode voltage range* (V_{CM}), which may extend to the negative rail ("single-supply" op-amps), or to both rails ("rail-to-rail" op-amps). Beware: many op-amps have a more restricted *input differential voltage range*, sometimes as little as just a few volts.

¶M. Output Limitations.

The *slew rate* (SR) is the op-amp's dV_{out}/dt with an applied differential voltage at the input. It is set by internal drive currents charging the compensation capacitor, and ranges from about $0.1\,\text{V}/\mu\text{s}$ (micropower op-amps) to $10\,\text{V}/\mu\text{s}$ (general purpose op-amps) to $5000\,\text{V}/\mu\text{s}$ (high-speed op-amps). Slew rate is important in high-speed applications generally, and in large-swing applications such as A/D and D/A converters, S/H and peak detectors, and active rectifiers. It limits the large-signal output frequency: a sinewave of amplitude A and frequency f requires a slew rate of $\text{SR} = 2\pi A f$; see Figure 4.54.

Op-amps are small devices, with *output current* deliberately limited to prevent overheating; see for example Figure 4.43, where R_5Q_9 and R_6Q_{10} limit the output sourcing and sinking currents to $I_{lim} = V_{BE}/R \approx 25\,\text{mA}$, illustrated in Figure 4.45. If you need more output current, there are a few high-current op-amps available; you can also add an external unity-gain power buffer like the LT1010 (I_{out} to $\pm 150\,\text{mA}$), or a discrete push-pull follower.

The open-loop *output impedance* of an op-amp is generally in the neighborhood of $100\,\Omega$, which is reduced by the loop gain to fractions of an ohm at low frequencies. Because an op-amp's open-loop gain G_{OL} falls as $1/f$ over

most of its bandwidth (see ¶O below), however, the circuit's *closed-loop* output impedance rises approximately proportional to frequency; it looks inductive (Figure 4.53).

In general the *output swing* for an op-amp like Figure 4.43 extends only to within a volt or so from either rail. Many CMOS and other low-voltage op-amps, however, specify unloaded rail-to-rail output swings, see Figure 4.46.

Op-amps can be grouped into several *supply voltage* ranges: "low-voltage" op-amps have a maximum total supply voltage (i.e., $V_+ - V_-$) around 6 V, and generally operate down to 2 V; "high-voltage" op-amps allow total supply voltages to 36 V, and generally operate down to 5–10 V. In between there is a sparse class of what might be called "mid-voltage" op-amps, with total supply voltages in the neighborhood of 10–15 V. See Table 5.5. There are also op-amps that are truly high-voltage (to hundreds of volts), see Table 4.2b.

¶N. Gain, Phase Shift, and Bandwidth.

Op-amps have large dc *open-loop gain* $G_{OL(dc)}$, typically in the range of 10^5–10^7 (the latter being typical of "precision" op-amps, see Chapter 5). To ensure stability (see ¶O, below) the op-amp's open-loop gain falls as $1/f$, reaching unity at a frequency f_T (see Figure 4.47). This limits the closed-loop *bandwidth* to $\text{BW}_{CL} \approx f_T/G_{CL}$. Over most of the operating frequency range the op-amp's open-loop *phase shift* is $-90°$, eliminated in the closed-loop response by feedback.

¶O. Feedback Stability, "Frequency Compensation," and Bode Plots

Finally, negative feedback can become *positive* feedback, promoting instability and oscillations, if the accumulated phase shift reaches $180°$ at a frequency at which the loop gain is ≥ 1. This topic is foreshadowed in §4.6.2 in connection with capacitive loads, and it is discussed in detail in §4.9. The basic technique is *dominant-pole compensation*, in which a deliberate $-6\,\text{dB/octave}$ (i.e., $\propto 1/f$) rolloff is introduced within the op-amp in order to bring the gain down to unity at a frequency lower than that at which additional unintended phase shifts rear their ugly heads (Figure 4.99). Most op-amps include such compensation internally, such that they are stable at all closed-loop gains (the unity-gain follower configuration is most prone to instability, because there is no attenuation in the feedback path). "Decompensated" op-amps are less aggressively compensated, and are stable for closed-loop gains greater than some minimum (often specified as $G > 2$, 5, or 10;

---

[47] The input current of "bias-compensated" BJT op-amps is typically around 50 pA.

[48] But not "bias-compensated" BJT op-amps, see §8.9.

Figure 4.95). Compensated op-amps exhibit an open-loop lagging phase shift of 90° over most of their frequency range (beginning as low as 10 Hz or less). Thus an external feedback network that adds another 90° of lagging phase shift at a frequency where the loop gain is unity will cause oscillation.

A favorite tool is the *Bode Plot*, a graph of gain (log) and phase (linear) versus frequency (log); see Figure 4.97. The *stability criterion* is that the difference of slopes between the open-loop gain curve and the ideal closed-loop gain curve, at their intersection, should ideally be 6 dB/octave, but in no case as much as 12 dB/octave.

PRECISION CIRCUITS

In the preceding chapters we dealt with many aspects of analog circuit design, including the circuit properties of passive devices, transistors, FETs, and op-amps, the subject of feedback, and numerous applications of these devices and circuit methods. In all our discussions, however, we have not yet addressed the question of the best that can be done, for example, in minimizing amplifier errors (nonlinearities, drifts, etc.) and in amplifying weak signals with minimum degradation by amplifier "noise." In many applications these are the most important issues, and they form an important part of the art of electronics. Therefore, in this chapter, we look at methods of precision circuit design (deferring the issue of noise in amplifiers to Chapter 8).

Chapter overview

This is a *big* chapter – and an important one. It deals with a range of topics, which need not be read in order. As guidance, we offer this outline: we start with a careful examination of errors in circuits made with operational amplifiers, and explore the use of an error budget. We explore issues of unspecified parameters and "typical" versus "worst-case" component errors, and discuss ways to deal with them. Along the way we deal with some neglected topics such as diode leakage at the sub-picoamp level, "memory effect" in capacitors, distortion and gain nonlinearity, and an elegant way to remedy *phase* error in amplifiers. We discuss op-amp distortion in detail, with comparative graphs and test circuits.

Next we discuss the dark side of rail-to-rail op-amps: their open-loop output impedance, and input common-mode crossover errors. We provide detailed selection tables for precision, chopper and high-speed op-amps, and comparative graphs charting their noise, bias current, and distortion. We show how to interpret the multitude of op-amp parameters, and we discuss the tradeoffs you'll have to make.

For those working in the low microvolt and nanovolt territory, we show the devastating effects of $1/f$ noise, and how auto-zero (AZ) op-amps solve this problem; but there's a tradeoff – the current noise of these devices that is often overlooked. As an interlude we look in some de-

tail at the cleverness of the front-end stage of an exemplary precision digital multimeter.

Then we advance to difference and instrumentation amplifiers – these go to the head of the class both in terms of digging out a difference signal in the presence of common-mode input, and in terms of gain accuracy and stability. We show their internal designs and how they're used, with extensive tables and graphs comparing popular parts. Finally we take up fully differential amplifiers – these have differential inputs *and* outputs, and an output "common-mode control" input pin. Once again, we organize tables, internal circuit diagrams, and guidance for their use with high-performance ADCs.

For readers looking for the basics, this chapter can be skipped over in a first reading. Its material is not essential for an understanding of later chapters.

5.1 Precision op-amp design techniques

In the field of measurement and control there is often a need for circuits of high precision. Control circuits should be accurate, stable with time and temperature, and predictable. The usefulness of measuring instruments likewise depends on their accuracy and stability. In almost all electronic subspecialties we always have the desire to do things more accurately – you might call it the joy of perfection. Even if you don't always actually *need* the highest precision, you can still delight in a full understanding of what's going on.

5.1.1 Precision versus dynamic range

It is easy to get confused between the concepts of *precision* and *dynamic range*, especially because some of the same techniques are used to achieve both. Perhaps the difference can best be clarified by some examples: a 5-digit multimeter has high precision; voltage measurements are accurate to 0.01% or better. Such a device also has wide dynamic range; it can measure millivolts and volts on the same scale. A precision decade amplifier (one with selectable gains of 1, 10, and 100, say) and a precision voltage reference may have plenty of precision, but not necessarily much dynamic range. An example of a device with wide dynamic range but only moderate accuracy might be a 6-decade logarithmic amplifier (log-amp) built with carefully trimmed op-amps but with components of only 5% accuracy; even with accurate components a log-amp might have limited accuracy because of lack of log conformity (at the extremes of current) of the transistor junction used for the conversion, or because of temperature-induced drifts.

Another example of a wide-dynamic-range instrument (greater than 10,000:1 range of input currents) with only moderate accuracy requirement (1%) is the coulomb meter described in §9.26 of the previous edition of this book. It was originally designed to keep track of the total charge put through an electrochemical cell, a quantity that needs to be known only to approximately 5% but that may be the cumulative result of a current that varies over a wide range. It is a general characteristic of wide-dynamic-range design that input offsets must be carefully trimmed in order to maintain good proportionality for signal levels near zero; this is also necessary in precision design, but, in addition, precise components, stable references, and careful attention to every possible source of error must be used to keep the sum total of all errors within the so-called error budget.

5.1.2 Error budget

A few words on *error budgets*. There is a tendency for the beginner to fall into the trap of thinking that a few strategically placed precision components will result in a device with precision performance. On rare occasions this will be true. But even a circuit peppered with 0.01% resistors and expensive op-amps won't perform to expectations if somewhere in the circuit there is an input offset current multiplied by a source resistance that gives a voltage error of 10 mV, for example. With almost any circuit there will be errors arising all over the place, and it is essential to tally them up, if for no other reason than to locate problem areas where better devices or a circuit change might be needed. Such an error budget results in rational design, in many cases revealing where an inexpensive component will suffice, and eventually permitting a careful estimate of performance.

To add some spice to the subject, we note that there is some controversy in the engineering community surrounding this business of error budgets. One camp (which we might characterize as strict constructionists) insists that you allow for the true worst case, or you are guilty of violating good engineering practice. For example, if there are

18 gain-determining resistors of $\pm1\%$ tolerance in a circuit, then the guaranteed performance must be specified as $\pm18\%$. The response of the other camp (which we might characterize as pragmatists) is "balderdash – it's overly constraining to allow an extremely unlikely possibility to limit the performance of a circuit design; and one can deal with such eventualities with component test procedures, finished circuit performance testing, and so on." We'll revisit this controversy (and take sides in the debate) after running through an introductory example.

5.2 An example: the millivoltmeter, revisited

To motivate the discussion of precision circuits, let's revisit a circuit from the previous chapter. There we flirted, briefly, with issues of precision in §4.4.3, mostly to illustrate the effects of input offset voltage V_{OS} and input bias current I_B in a low-level dc application (a 0–10 mV millivoltmeter with 10 MΩ input resistance[1]). Back then, with wide-eyed naivety, we were astonished to see that our trusty LF411 op-amp was wholly inadequate to the task; it had way too much offset, and too much input current as well. We found a solution in the form of either a precision low-bias op-amp (an OPA336) or a chopper (also known as "auto-zeroing") amplifier (an LTC1050).

As we'll soon see, our celebration of that "solution" was premature: we pronounced victory with an op-amp whose I_B alone caused the maximum allowable zero-input error of 1%. A careful design must take into account the cumulative effect of multiple sources of error.

5.2.1 The challenge: 10 mV, 1%, 10 MΩ, 1.8 V single supply

To make the problem more interesting, let's further constrain the specifications. This time we'll ask that the 0–10 mV meter operate from a single +3 V battery (either a lithium cell or a pair of alkaline AAA cells); that forces us to worry about "single-supply" operation, in which the op-amp must work down to zero volts at both input and output. Furthermore, it must work down to the end-of-life voltage of alkaline cells, which you see stated variously as 1.0 V/cell, or 0.9 V/cell; that means operation down to +1.8 V total supply voltage. And, as before, let's require an input resistance of 10 MΩ and insist that it indicates 0 mV ($\pm1\%$ of full scale) when the input is either shorted

---

[1] Note that it can be used as a sensitive *current* meter: with its 10M input resistance and 1% accuracy, it can measure currents down to 10 pA ($1\% \times 10\,\text{mV}/10\,\text{M}\Omega = 10\,\text{pA}$).

or open. Note that this "zero-error" specification is different from a full-scale accuracy ("scale error") specification: we might be happy with ±5% full-scale accuracy, but we'd be most unhappy with a meter that reads 5% of full scale (here 0.5 mV) when there's nothing connected to it.

Following the suggestion from last chapter's design, let's use current-sensing feedback, so the design is independent of the internal resistance of the analog meter. Figure 5.1 shows the circuit.

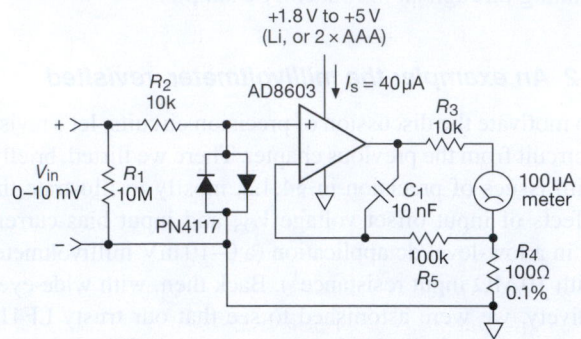

Figure 5.1. Accurate millivoltmeter powered with single lithium cell. The input protection clamp uses low-leakage diode-connected PN4117 JFETs.

5.2.2 The solution: precision RRIO current source

We use a precise current-sensing resistor R_4, in this case a 0.1% 100 Ω resistor. Sounds exotic, but in fact these things are commonplace: the ever-helpful DigiKey website shows over 100,000 in stock, from five different suppliers, at prices down to $0.20 (in quantities of 10). Note the routing of the input common ("−" terminal) connection directly to the low side of the sense resistor, a precaution that becomes increasingly important with small-value sense resistors, in which the wiring resistance of the ground return may add significant error.[2] Because the meter likely presents an inductive load (it's a moving coil in a magnetic field, which is both inductive in its own right and reactive through its motor-like electromechanical properties), we took the precaution of dividing the feedback path in the usual way (through R_5 at low frequencies, C_1 at high frequencies; see for example Figure 4.76). The 10k output resistor R_3 limits meter current for off-scale inputs.

The more challenging parts of this design are the input

protection network (which we blithely ignored in Chapter 4's example), and, most critically, the choice of op-amp. First, the protection network: the requirements seem easy – clamp to a nondestructive op-amp input voltage (during input overvoltages), and draw less than ∼10 pA leakage current at the full-scale input voltage (10 mV), in both forward and reverse directions. (That amount of diode current would reduce the input resistance by 1%.) As it turns out, the datasheets don't ordinarily tell you how much current a diode draws at very low voltages. But if you go measure it, you'll be surprised at what you find (Figure 5.2). Everyone's favorite jellybean signal diode (1N914, 1N4148) is rather leaky, looking roughly like a 10 MΩ resistor at low voltages.[3] There are some specialized low-leakage diodes like the (somewhat hard to get) PAD-1 or PAD-5 that do much better; but you can do as well by using a diode-connected low-leakage JFET like the n-channel PN4117 (i.e., tie the source and drain together to form the cathode, and use the gate as the anode), or you can just use the diode terminal pairs from an ordinary npn transistor.[4] In this circuit the upstream 10k resistor R_2 limits the clamp current, while having no effect on the circuit accuracy.

And now for the op-amp. This was the stumbling block in Chapter 4, and it has gotten only more difficult here, with the single low-voltage supply. We can separate the errors into a "zero" error and an overall scale factor error. The latter is the easy part: the *circuit's* gain is accurately determined, so we merely require an accurate meter movement (if we don't want any trim adjustments; or we could reduce the sense resistor and add a resistor to produce a trimmable gain greater than unity). It's the "zero" requirement that is the tough part, because of the high sensitivity plus the high mandated input resistance. We require a worst-case combined effect of input offset and bias current of 100 μV and 10 pA *individually*. That is, each alone would cause a zero error of 0.1 mV ($V_{err}=V_{OS}+I_BR_1$), so each must be smaller so that the worst-case combination meets specifications.

We looked at some promising contemporary op-amp offerings, which we've listed (along with the usual suspects) in Table 5.1 on page 296. The inexpensive jellybeans in the

---

[2] Small-value sense resistors used for accurate measurement of high currents are available as *4-wire* resistors. This arrangement is known as a *Kelvin connection*, and the sense resistor is sometimes called a *shunt*.

[3] For all these diodes, the straight-line portion at low voltages represents a resistance in parallel with a non-conducting diode; so the low-leakage 1N3595 looks like 10,000 MΩ for $V \lesssim 10$ mV. These currents can be estimated from other diode parameters, namely reverse-bias leakage current at low reverse voltage or forward current at specified forward voltage. More on this in Chapter 1x.

[4] Or our friend John Larkin's favorite, the collector-base junction of the BFT25, a low-cost 5 V npn microwave transistor. Its leakage is less than 10 fA when reverse biased, and <40 fA when forward biased up to 50 mV.

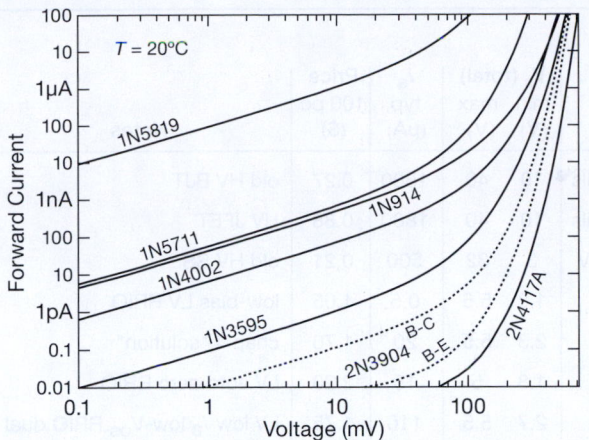

Figure 5.2. Diode datasheets are often skimpy with data like this, showing the low end of the current versus forward applied voltage. See Chapter *1x* for lots more detail.

first three rows are hopeless for this job, with both offset voltage and bias current far worse than required (listed in the bottom row). And they all fail to operate at +1.8 V; and the first two fail the "single-supply" requirement. So much for el-cheapo op-amps.

The LPV521 is characteristic of the new breed of "low-voltage, low-bias, low-power" op-amps. It does those things just fine, but, in common with many CMOS op-amps, its V_{OS} specs are only so-so. It's possible to do considerably better. For example, the CMOS precision OPA336 was our hero earlier; but look – its worse-case I_B consumes our entire error budget, and it would require V_{OS} trimming anyway, which is not so easily done in a single-supply setup (unlike our friend the LF411, this op-amp has no trim pins). We brushed off such niceties earlier; but this time let's accept full responsibility for producing products that meet specs. That means adopting some serious worst-case discipline.

What choices are left? Most designers would reach for an auto-zero (chopper) amplifier for a precision application that does not require bandwidth and that is tolerant of noise. The best candidate we found is the ADA4051, with excellent V_{OS} specs, but five times too much I_B, if you believe the "maximum" specs. (Competing auto-zero amplifiers had greater bias current, greater minimum supply voltage, or both.) The last two op-amps are contemporary CMOS op-amps that qualify as "precision," owing to careful design and production-line offset trimming. Both meet the error budget goals (but see below). We chose the AD8603 because it meets the 1.8 V supply spec, and as a bonus it operates at 35% the supply current of the LTC6078

(a dual op-amp, with no single version we could find).[5] As listed in Table 5.5, the AD8603 runs at $40\,\mu A$ quiescent current, and has 1 pA (max) input current and $50\,\mu V$ (max) offset voltage at 25°C.

Are we done? Not quite. The specs in the table are for operation at 25°C. It can get plenty warmer, with impressive increases in bias current for CMOS devices (for which the "bias" current is leakage current). Manufacturers will ordinarily provide worst-case (and sometimes typical) values at the high end of the operating temperature range (e.g., 85°C for "industrial" temperature range devices). For our chosen AD8603 they specify $I_B(\text{max}) = 50\,\text{pA}$ at 85°C. No worst-case data for intermediate temperatures are given; but that value is consistent with a doubling every 10°C, so we can be confident that the circuit will meet specs[6] up to 50°C. It's worth noting that manufacturers are sometimes rather lazy in setting worst-case leakage specs, as for example with the LPV521 in the table, whose ratio of "maximum" to "typical" I_B is 100:1. One industry guru attributes that to an unwillingness to test production parts to a tighter spec.

Exercise 5.1. We designed a ±10 mV meter in Chapter 4 (Figure 4.56), but our design in Figure 5.1 is unipolar (i.e., 0 to +10 mV). Your boss has asked you to modify the design for ±10 mV capability, using a ±100 μA zero-center meter and retaining the feature of operating with either a single lithium cell (or pair of AAA alkaline cells). *Hint*: you may wish to look at the AD8607 dual member of the AD8603 op-amp family. *Extra credit*: after finishing, you decide to impress your boss by enhancing the design so it will work with a 0–100 μA meter.

5.3 The lessons: error budget, unspecified parameters

From this rather simple first example we've learned some important basic principles: (a) first, you've got to identify and quantify the sources of error within a circuit in order to create an error budget; and (b) strict worst-case design requires that all components (passive and active) be operated

---

[5] Five other parts we considered (see Tables 5.5, pages 320–321, and 5.6, page 335): the bipolar LT1077A, lower I_s and V_{OS} but too high V_s (min); the bipolar LT6003, I_s only 1 μA, but too much offset voltage (a 5% error); the CMOS LMP2232A, poor 1.5% offset voltage; the MAX9617 auto-zero op-amp, with 0.1% (versus the AD8603's 0.5%) offset voltage, but whoa, its input *current* through 10 MΩ amounts to a 1.4% offset; and the ISL28133 auto-zero, but its 3% offset (from input current) again reveals the input current weakness of auto-zero amplifiers.

[6] The datasheet does provide a plot of "typical" I_B versus temperature, confirming its exponential behavior of a doubling every 10°C.

Table 5.1 Millivoltmeter Candidate Op-amps

Part #	Input	V_{os} typ (μV)	V_{os} max (μV)	I_{bias} typ (pA)	I_{bias} max (pA)	V_{cm} neg lim (V)	V_{out}	V_s (total) min (V)	V_s (total) max (V)	I_s typ (μA)	Price 100 pc ($)	Notes
uA741	BJT	2000	6000	80k	500k	2	1.5V from rails	10	40	1500	0.27	old HV BJT
LF411	JFET	800	2000	50	200	3	1.5V from rails	10	40	1800	0.88	HV JFET
LM358A	BJT	2000	3000	45k	100k	0	0 to $V_+ - 1.2$V	3	32	500	0.21	old HV SS
LPV521	CMOS	100	1000	0.01	1	−0.1	R-R	1.8	5.5	0.5	1.05	low-bias LV RRIO
OPA336	CMOS	60	125	1	10	−0.2	R-R	2.3	5.5	20	1.70	chap 4 "solution"
ADA4051	CMOS	2	17	5	50	0	R-R	1.8	5	15	2.20	LV auto-zero RRIO
LTC6078	CMOS	7	25	0.2	1	0	R-R	2.7	5.5	110	1.75	LV low-I_B low-V_{os} RRIO dual
AD8603	CMOS	12	50	0.2	1	−0.3	R-R	1.8	5	40	1.40	LV low-I_B low-V_{os} RRIO
Ckt Limit		-	100	-	10	0	*0 to anything*	1.8	>3.6			*must budget contributions*

within their datasheet specifications, and that the effects of their guaranteed worst-case errors be added (as unsigned magnitudes) to determine the overall circuit's performance.

So far, so good; and in this example the op-amp choice allowed us to meet (and exceed) our target zero-error specifications (1% of full scale, with input open or shorted), even under guaranteed worst-case values of I_B and V_{OS}.

A. Unspecified parameters: a pragmatic approach

But, looking a little closer, we've also seen in this design example that an unspecified parameter (forward diode current at low voltages, 0–10 mV) figured into the overall error.[7] What should one do under such circumstances?

The authors belong to the camp of "pragmatists" on this matter: first, you may have to be creative in reading the datasheet (as we did in the case of op-amp input current), particularly when the manufacturer's worst-case numbers represent a statement that "I don't want to test this parameter, so I'll put a conservative guess in the datasheet"; this is particularly relevant with parameters like leakage current, where the limits of automated test equipment (ATE) and testing time constraints encourage conservative worst-case datasheet specifications. Second, you may have to do some testing[8] of poorly specified (or unspecified) parameters (as we did with forward diode current). It may be sufficient to establish that the circuit effects of the unspecified parameter are completely insignificant (as here, when the current through the clamp diodes was less than 0.01 pA, or three

orders of magnitude below budget); or, if it's a closer call, you may have to set up a testing regime of incoming components to ensure that you meet specs. And third, you may have to deal with a situation in which there are many components contributing to the overall error budget by simply validating the performance of the overall circuit, subassembly, or complete instrument at final test.

This approach may appear cavalier. But the fact is that there are many situations in which you simply cannot meet challenging specifications while staying within the published worst-case specifications (or lack of specifications). Two examples help make this point: one of the authors designed and manufactured a line of battery-powered oceanographic instruments, intended for long-duration (from weeks to as long as a year) submerged observations and data logging. A typical instrument might have 200 or more 4000B-series CMOS ICs. The datasheet lists the 25°C quiescent current[9] as "0.04 μA (typ), 10 μA (max)." Great. So 200 of these puppies probably draw a total of 8 μA, but they *could* draw (in a wildly improbable scenario) as much as 2 mA. A year's worth of operation would require 70 mAh (using typical values), but, under strict worst-case rules, we would have to allow for 17.5 Ah (amp-hours). Here's the rub: the substantial battery pack for these volume-constrained deep submergeable pressure housings provided only 5 Ah of capacity (with some safety margin of derating). And 80% of the battery capacity was budgeted for the sensors and recorders. So strict worst-case

---

[7] As did the op-amp input current at modestly elevated temperatures.

[8] If you buy in large quantities the manufacturer may be willing to do these tests.

[9] Amusingly, the specs are the same for simple parts such as gates, or complex parts such as counters or arithmetic logic units.

design would require quadrupling the battery pack (and expanding the pressure case), or, alternatively, removing a substantial volume of the instrumentation payload. The solution was (and still is) obvious: build the subcircuits, and test them for conforming quiescent current. They invariably worked just fine, and the testing served mostly to identify modules in which there was a defective component, usually caused by improper handling of the sensitive CMOS components.

A second example is a commercial instrument, namely a sensitive electrometer from Keithley. These things will measure currents down to *femto*amps (10^{-15}A), which requires a front-end stage of extraordinarily low bias current. They accomplish this with a JFET follower matched pair as input stage to a conventional precision op-amp, in a current-to-voltage configuration (the input is a summing junction, at zero volts). And to keep the gate current low, they operate the JFETs at a very low drain voltage of just +0.55 V, with the source terminal sitting just a fraction of a volt below the drain. Now, nowhere in the JFET datasheet will you find anything telling you what happens at such low voltages; and they won't tell you what the gate leakage is likely to be. You can throw up your hands and say that such an instrument cannot be made. Or you can do what Keithley did, which is to find a good source of JFETs and qualify them with in-house testing so that you can get on with the job.

Both examples illustrate that there are situations in which you simply cannot meet your design requirements while staying within the manufacturer's published worst-case specs. Having said that, we note that there are some engineers who will not deviate from strict worst-case specified component parameters in their circuit designs. They don't want to use special parts, and they won't touch such stuff with a 10-foot (3m) pole. We invite you to choose what you would do.

5.4 Another example: precision amplifier with null offset

Having warmed up with the millivoltmeter, let's tackle a more complex design, one in which there are multiple error challenges. We describe the design choices and errors of this particular circuit within the framework of precision design in general, thus rendering painless what could otherwise become a tedious exercise.

We designed a precision amplifier (Figure 5.3) that lets you "freeze" the value of the input signal, amplifying any subsequent changes from that level by gains of exactly 1, 10, or 100. This might come in particularly handy in an experiment in which you wish to measure a small change in some quantity (e.g., light transmission or RF absorption) as some condition of the experiment is varied. It is ordinarily difficult to get accurate measurements of small changes in a large dc signal, owing to drifts and instabilities in the amplifier. In such a situation a circuit of extreme precision and stability is required.

Here we show the example of a *strain gauge* sensor, which consists of a strain-sensitive resistive bridge whose elements change resistance (slightly!) in response to mechanical strain. A common resistance value is $350\,\Omega$; and the sensitivity is such that, when biased with +5 V, the differential output voltage across the bridge changes by $\pm 10\,$mV in response to the rated full-scale mechanical strain.[10] This small differential voltage sits on a dc level of +2.5 V, so you've got to begin with a good differential amplifier.

An important note at the outset: digital techniques offer an attractive alternative to the purely analog circuitry used here. A skilled designer would likely make use of precision analog/digital conversion techniques, perhaps in a hybrid implementation (in which a stable DAC is used to create the nulling signal within an analog circuit like ours), or perhaps in an all-digital scheme that relies on the intrinsic precision of a high resolution ADC alone.[11] Regardless, our all-analog example offers a smorgasbord of important lessons in precision design. But the reader can confidently look forward to exciting revelations in chapters to come.

5.4.1 Circuit description

The front-end begins with an *instrumentation amplifier* U_1, a configuration of three op-amps that we'll talk about later (§5.15); these are differential-input amplifiers that excel in achieving high common-mode rejection, and allow gain selection with a single resistor (one or more are often provided internally). Here we've selected one with a good combination of low input current, offset drift, and noise, for reasons we'll explain later. Its gain of $\times 100$ is followed with a noninverting $\times 10$ gain stage (U_2), for an overall gain of $\times 1000$; that produces a full-scale output of ± 10 V as input to the nulling circuitry (U_3–U_5). If the input signal were single ended (e.g., from a thermocouple, photosensor, microwave absorption detector, or whatever), you would omit U_1, bringing in the signal at point "X," and adjusting the gain of U_2 accordingly.

---

10 The strain-gauge sensor sensitivity is "2mV-per-volt"; that's pretty low. There are semiconductor strain sensors with higher sensitivity, but they may not be as stable.

11 There's an example of the latter in §13.9.11C.

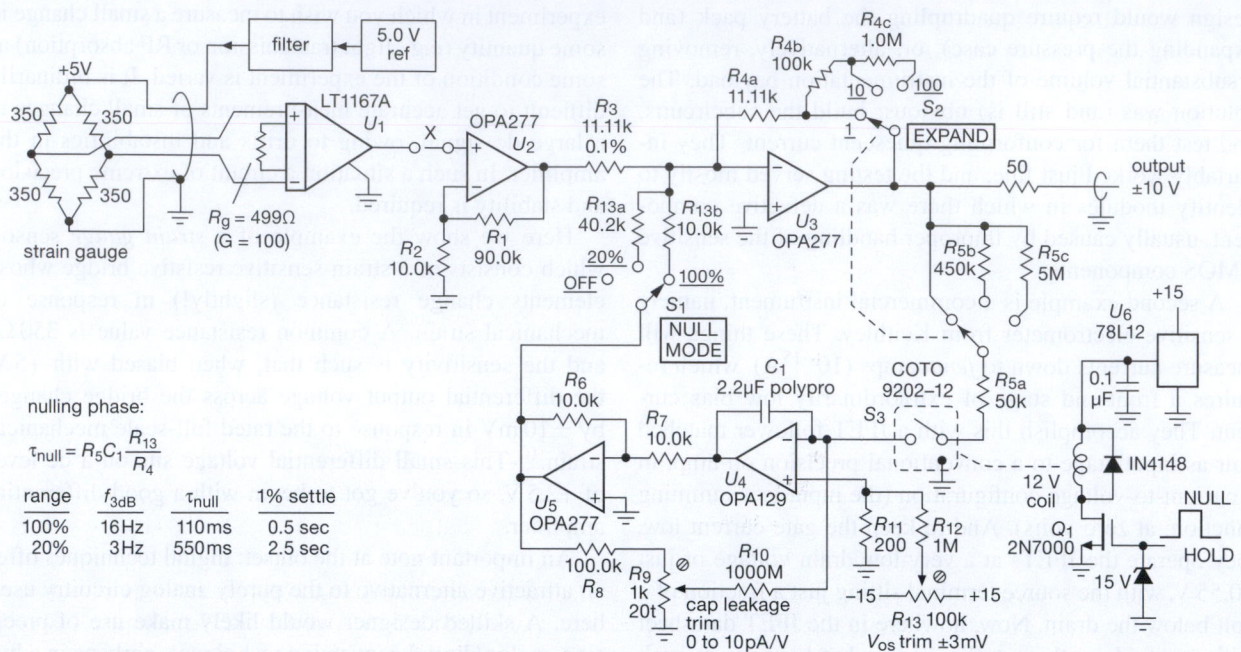

Figure 5.3. Autonulling dc laboratory amplifier. Gain-setting resistors are 0.1% tolerance.

The nulling circuitry works as follows: the amplifier stage U_3 is configured in the inverting configuration, permitting a dc offset according to a current added to the summing junction. Nulling takes place when relay (switch) S_3 is activated by a logic HIGH to coil driver Q_1's gate. Then U_4 charges the analog "memory" capacitor (C_1) as necessary to maintain zero output. No attempt is made to follow rapidly changing signals, because in the sort of application for which this was designed the signals are essentially dc, and some averaging is a desirable feature. When the switch is opened, the voltage on the capacitor remains stable, resulting in an output signal from U_3 proportional to the wanderings of the input thereafter. The gain about the input null level can be increased in decade steps (switch S_2) to expand changes; the gain of the nulling integrator is switched accordingly to keep the feedback bandwidth constant. Switch S_1 selects the full-scale nulling range (100%, 20%, or none).

There are a few additional features that we should describe before going on to explain in detail the principles of precision design as applied here: (a) U_5, in addition to providing the needed inversion of the nulling level, participates in a first-order leakage-current compensation scheme: the tendency of C_1 to discharge slowly through its own leakage ($\geq 100{,}000\,\mathrm{M\Omega}$, corresponding to a time constant of ≥ 3 days) is compensated by a small charging cur-

rent through R_{10} proportional to the voltage across C_1; and (b) integrator U_4 was selected for very low input current I_B (to minimize droop during "hold"), for which the tradeoff is relatively poor offset voltage V_{OS}; so we added an external offset trim (R_{11}–R_{13}). This is not terribly critical, in any case, because an offset here merely causes a nonzero null of the same magnitude.

5.5 A precision-design error budget

For each category of circuit error and design strategy, we will devote a few paragraphs to a general discussion, followed by illustrations from the preceding circuit. Circuit errors can be divided into the categories of (a) errors in the external network components, (b) op-amp (or amplifier) errors associated with the input circuitry, and (c) op-amp errors associated with the output circuitry. Examples of the three are resistor tolerances, input offset voltage, and errors that are due to finite slew rate, respectively.

Let's start by setting out our error budget. It is based on a desire to keep input drift (from temperature and power-supply variations) down to the $10\,\mu\mathrm{V}$ level, and nulled drift (primarily from capacitor "droop," along with temperature and supply variations) below $1\,\mu\mathrm{V/min}$ (referred to the input, or RTI). As with any budget, the individual items are arrived at by a process of tradeoffs, based on what can be done with available technology. In a sense the budget represents the end result of the design, rather than the starting point. However, it will aid our discussion to have it now.

It's important to understand that the items in such a budget come from several sources: (a) parameters that are specified in the datasheet; (b) estimates of poorly specified (or unspecified) parameters; and (c) parameters that you may not even realize are important.[12] We might paraphrase these, respectively, as the *knowns*, the *known unknowns*, and the *unknown unknowns*.

5.5.1 Error budget

These are all in the form of worst-case voltage errors (at 25°C) *referred to the instrument input*.

1. ×100 Difference amplifier (U_1: LT1167A)

Offset voltage	40 μV
Noise voltage (0.1–10 Hz)	0.28 μVpp (typ – no "max" spec)
Temperature	0.3 μV/°C
Power supply	28 nV/100 mV change
Input offset current ×R_s	0.11 μV/350 Ω of R_s

2. ×10 Gain amplifier (U_2: OPA277)

Offset voltage	0.5 μV
Temperature	10 nV/°C
Time	2 nV/month (typ – no "max" spec)
Power supply	1 nV/100 mV change
Bias current	0.3 μV
Load-current heating	5 nV at full scale (5 mW, 0.1°C/mW)

3. Output amplifier (U_3: OPA277)

Offset voltage	50 nV
Temperature	1 nV/°C
Time	0.2 nV/month (typ – no "max" spec)
Power supply	0.1 nV/100 mV change
Bias current	30 nV
Load-current heating	5 nV at full scale (1 kΩ load)

4. Hold amplifier (U_4: OPA129)

U_4 offset tempco	10 nV/°C
Power supply	10 nV/100 mV change
Capacitor droop	0.4 μV/min
(see current error budget)	
Charge transfer	1.1 nV

Current errors through C_1 (needed for the preceding voltage error budget) are as follows:

Capacitor leakage	
Maximum (uncompensated)	(100 pA)
Typical (compensated)	10 pA
U_4 input current	0.25 pA
U_4's nulled V_{OS}/R_{10}	0.1 pA
Relay S_3 OFF leakage	10 pA (1 pA typ)
Printed-circuit-board leakage	5.0pA

Not bad, though you *could* complain about the 40 μV input offset – but we'd reply that a few tens of microvolts of *static* offset is of no concern in a nulling instrument, it's only the *drift* (with time and temperature) that matters. The various items in the budget will make sense as we discuss the choices faced in this particular design. We organize these by the categories of circuit errors listed earlier: network components, amplifier input errors, and amplifier output errors.

We'll address these errors quantitatively, in the context of Figure 5.3, beginning in §5.7.6, after looking at error sources a bit more generally in the next section.

5.6 Component errors

The degrees of precision of reference voltages, current sources, amplifier gains, etc., all depend on the accuracy and stability of the resistors used in the external networks. Even where precision is not involved directly, component accuracy can have significant effects, e.g., in the common-mode rejection of a differential amplifier made from an op-amp (see §§4.2.4 and 5.14), where the ratios of two pairs of resistors must be accurately matched. The accuracy and linearity of integrators and ramp generators depend on the properties of the capacitors used, as do the performances of filters, tuned circuits, etc. As we will see presently, there are places where component accuracy is crucial, and there are other places where the particular component value hardly matters at all.

Components are generally specified with an initial accuracy, as well as the changes in value with time (stability) and temperature. In addition, there are specifications of voltage coefficient (nonlinearity) and bizarre effects such as "memory" and dielectric absorption (for capacitors). Complete specifications also include the effects

---

[12] We discovered an example of the latter while measuring femtoamp leakage currents in a nice shielded test enclosure: after the box had been opened to change anything inside, it took quite a while for the measurements to settle down. Turns out that the process of moving things caused some rearrangement of surface charge on the Teflon-insulated wires, with a long relaxation time. Pease talks about this in his article "What's All This Teflon Stuff, Anyhow?" – see the footnote references in §5.10.7. We've experienced a similar bizarre manifestation with analog panel meters, where a swipe of the hand across the glass face can cause the needle to move way upscale...and *stay* there!

of temperature cycling and soldering, shock and vibration, short-term overloads, and moisture, with well-defined conditions of measurement. In general, components of greater initial accuracy will have their other specifications correspondingly better, in order to provide an overall stability comparable to the initial accuracy. However, the overall error that is due to all other effects combined can exceed the initial accuracy specification. Beware!

As an example, RN55C 1% tolerance metal-film resistors have the following specifications: temperature coefficient (tempco), $50 \, \text{ppm/}^\circ\text{C}$ over the range -55°C to $+175^\circ\text{C}$; soldering, temperature, and load cycling, 0.25%; shock and vibration, 0.1%; moisture, 0.5%. By way of comparison, the legacy 5% carbon-composition resistors (Allen–Bradley type CB) have these specifications: tempco, 3.3% over the range $25–85^\circ\text{C}$; soldering and load cycling, +4%, −6%; shock and vibration, ±2%; moisture, +6%. From these specs it should be obvious why you can't just select (using an accurate digital ohmmeter) carbon resistors that happen to be within 1% of their marked value for use in a precise circuit, but are obliged to use 1% resistors (or better) designed for long-term stability as well as initial accuracy. For the utmost in precision it is necessary to use ultra-precise resistors or resistor arrays, such as Susumu's RG-series of SMT (surface-mount technology) resistors (tolerance to 0.02%, tempco to $5 \, \text{ppm/}^\circ\text{C}$), Vishay's MPM-series metal-film networks (absolute tolerance to 0.05%, matched to 0.01%; absolute tempco to $25 \, \text{ppm/}^\circ\text{C}$, matching tempco to $2 \, \text{ppm/}^\circ\text{C}$), or their even better "Bulk Metal Foil" types (absolute tolerance to 0.005%, matched to 0.001%; absolute tempco to $0.2 \, \text{ppm/}^\circ\text{C}$, matching tempco to $0.1 \, \text{ppm/}^\circ\text{C}$).

5.6.1 Gain-setting resistors

In the preceding circuit (Figure 5.3), 0.1% resistors are used in the gain-setting network, R_1–R_4, for accurately predictable gain. As we'll see shortly, the value of R_3 is a compromise, with small values reducing offset current error in U_3 but increasing heating and thermal offsets in U_2. Note that 1% resistors are used in the offset attenuator network, R_5–R_{13}; here absolute accuracy is irrelevant, and the stability of 1% metal-film resistors is altogether adequate.

5.6.2 The holding capacitor

A. Leakage
The largest error term in this circuit, as the error budget shows, is capacitor leakage in the holding capacitor, C_1. Capacitors intended for low-leakage applications give a leakage specification, sometimes as a leakage resistance,

sometimes as a time constant (megohm-microfarads). In this circuit C_1 must have a value of at least a few microfarads in order to keep the charging rate from other current error terms small (see budget). In that range of capacitance, film capacitors (polystyrene, polypropylene, and polyester) have the lowest leakage. Polypropylene capacitors (from manufacturers such as Epcos, Kemet, Panasonic, Vishay, and Wima, generally with voltage ratings of 200–600 V) often have dc leakage specified in units of megohm–microfarads, with values in the range of 10,000–100,000 MΩ μF; thus for a capacitance of $2.2 \, \mu\text{F}$ this amounts to a parallel equivalent leakage resistance of at least 4.5–45 GΩ.

Even so, and adopting a plausible value of, say, 100 GΩ, that's equivalent to a leakage current of 100 pA at full output (10 V), corresponding to droop rates of roughly 3 mV/min at the output, the largest error term by far. For that reason we added the leakage-cancellation scheme described earlier. It is fair to assume that the effective leakage can be reduced to 10% of the capacitor's worst-case leakage specification (in practice, we can probably do much better). No great stability is required in the cancellation circuit, given the modest demands made of it. As we will see later when we discuss voltage offsets, R_{10} is kept intentionally large so that input voltage offsets in U_4 aren't converted to a significant current error.

B. Dielectric absorption
We're not done with the capacitor, yet. An important effect, quite apart from resistive leakage, is capacitor "memory," officially known as *dielectric absorption*.[13] This is the tendency of capacitors to return, to some extent, to a previous state of charge, as shown in the measured data of Figure 5.4 (each capacitor was held at +10 V for a day or more, then discharged to 0 V for 10 s, then open-circuited and observed while it did its thing); see also the discussion in §§*1x.3*, 4.5.1, 4.5.6, and 13.8.4.

5.6.3 Nulling switch

In the previous edition of this book, the analogous circuit (Figure 7.1, page 393) used MOSFETs (instead of relay S_3) to activate the nulling circuit. That choice provided plenty of education, because we had to worry about (a) MOSFET channel leakage, devastatingly large at about ∼1 nA, and (b) gate charge injection, of order 100 pC in that circuit. The solutions there were (a) the use of a series-connected

---

[13] It is not entirely clear that what is called "leakage" in high-quality capacitors is in fact distinct from dielectric absorption; see the footnote in §4.5.6.

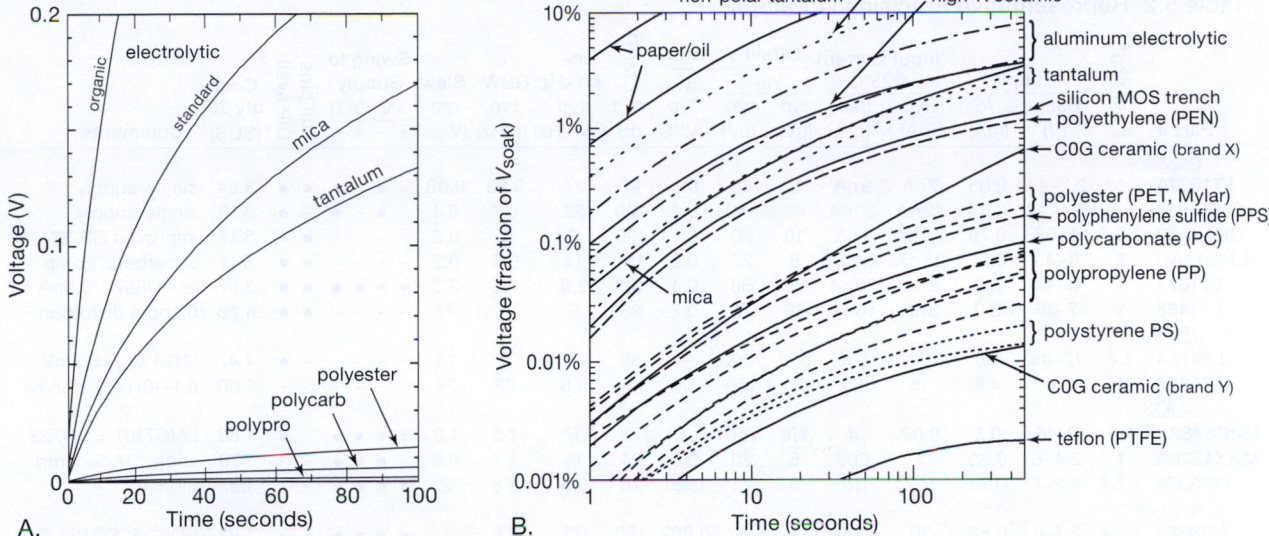

Figure 5.4. Capacitors exhibit *memory effect* (dielectric absorption), a tendency to return to a previous state of charge. This is highly unhelpful in applications (such as analog sample-and-hold) in which a capacitor is used to retain an analog voltage. A. linear plot, showing the basic effect; B. log–log plot, revealing four decades of dirty laundry. Teflon is the uncontested winner; but it's hard to find, so the plastic film types (PS and PP) are generally your best choice. Ceramic C0G can be excellent, but beware brand variations.

MOSFET pair, such that the downstream MOSFET had all four terminals (source, drain, gate, and substrate) ordinarily at zero volts, and (b) a sufficiently large holding capacitor such that the error was negligible, along with the observation that charge transfer was not of great concern because it resulted in a small offset of the auto-zero.

This time we have taken a more pragmatic (but less educational) approach, by using instead a small signal *relay*. The Coto 9202-12 is a small (4 mm×6 mm × 18 mm) shielded relay, energized by 12 Vdc at 18 mA, with a specified OFF resistance of $10^{12} \, \Omega$ minimum ($10^{13} \, \Omega$ typical). The worst-case $R_{\rm off}$ value corresponds to a droop rate of 0.3 mV/min, but ten times less for "typical" $R_{\rm off}$. Relays isolate better than transistor switches (higher $R_{\rm off}$ and lower $C_{\rm off}$, here < 1 pF), and they have better ON performance as well (lower $R_{\rm on}$ than a low-capacitance analog switch, here $<0.15 \, \Omega$).

Of course, there *is* capacitance between the coil and the contacts, and thus an opportunity for the same sort of charge transfer as with a MOSFET switch (where the full-swing transitions at the gate couple capacitively to the drain and source). As we remarked in Chapter 3 (§3.4.2E), the total charge transferred is independent of the transition time and depends only on the total control-voltage swing and the coupling capacitance: $\Delta Q = C_{\rm coup}\Delta V_{\rm control}$. In this circuit, charge transfer results in a simple voltage error of the auto-zero, because the charge is converted to a volt-

age in the holding capacitor C_1. It's easy to estimate the error: the Coto relay specifies a coil-to-contact capacitance of 0.2 pF (for our grounded-shield configuration), and thus a corresponding charge transfer of $\Delta Q = 2.4$ pC when the 12 V coil is energized.[14] That produces a voltage step of $\Delta V_C = \Delta Q/C_1 = 1.1 \, \mu$V across the 2.2 μF capacitor C_1. This is completely within our error budget; in fact, we've likely overestimated the effect, because our calculation assumed that the entire coil undertook a 12 V step, whereas the average step is half that value.

For readers who harbor a deep-seated dislike of mechanical relays, we show in Figure 5.5 a switch implementation with series-connected JFETs. During HOLD the JFET gates are back-biased to -5 V, by the somewhat tortured level-shifting circuit Q_1–Q_4. The reader is invited to estimate the magnitude of droop (use the datasheet's $I_{\rm D(off)}=0.1$ pA) and of charge injection (use the datasheet's $C_{\rm rss}=0.3$ pF) for this circuit.

5.7 Amplifier input errors

The deviations of op-amp input characteristics from the ideal that we discussed in Chapter 4 (finite values of input impedance and input current, voltage offset, common-mode rejection ratio, and power-supply rejection ratio, and

---

[14] Assuming care is taken in the wiring layout to maintain the low 0.2 pF capacitance of the HOLD signal.

Table 5.2 Representative Precision Op-amps

Part #	# per pkg[a]	Supply[p] Range (V)	I_Q (mA)	Input Current @25°C typ (pA)	max (pA)	Offset Voltage V_{os} typ (μV)	V_{os} max (μV)	ΔV_{os} typ (μV/°C)	CMRR min (dB)	e_n @1kHz typ (nV/√Hz)	GBW typ (MHz)	Slew typ (V/μs)	Swing to Supply IN +	IN −	OUT +	OUT −	null pins	DIP avail	Cost qty 25 ($US)	Comments
bipolar																				
LT1077A	1	2.2–44	0.05	7nA	9nA	9	40	0.4	97	27	0.23	0.08	-	●	-	●	●	●	3.84	single-supply
LT1013	2,4	3.4–44	0.35	12nA	20nA	40	150	0.4	100	22	0.7	0.4	-	●	-	●	-	●	3.13	single-supply
OPA277P	1,2,4	4–36	0.79	0.5nA	1nA	10	20	0.1	130	8	1	0.8	-	-	-	-	●	●	3.17	improved OP-27
LT1012AC	1	8–40	0.37	25pA	0.1nA	8	25	0.2	114	14	0.5	0.2	-	-	-	-	●	●	5.11	superbeta, comp
LT1677	1	3–44	2.8	2nA	20nA	20	60	0.4	109	3.2	7.2	2.5	●	●	●	●	●	●	3.07	or AD8675, 0.5nA
LT1468	1	7–36	3.9	3nA	10nA	30	75	0.7	96	5	90	23	-	-	-	-	●	●	4.26	0.7ppm distortion
JFET																				
LF412A[b]	1,2	12–44	1.8	50	200	500	1000	-	80	25	4	15	-	-	-	-	-	●	4.47	'411A V_{os} <0.5mV
OPA827	1	8–40	4.8	15	50	75	150	1.5	104	3.8	22	28	-	-	-	-	-	-	9.00	0.1–10Hz: 0.25μVpp
CMOS																				
LMC6482A[b]	2,4	3–16	0.5	0.02	4	110	750	1	70	37	1.5	1.3	●	●	●	●	-	●	1.88	LMC7101 = SOT23
MAX4236A	1	2.4–6	0.35	1	500	5	20	0.6	84	14	1.7	0.3	-	●	●	●	-	-	1.78	shdn, '37 decomp
OPA376	1,2,4	2.2–7	0.76	0.2	10	5	25	0.26	76	7.5	5.5	2	●	●	●	●	-	-	1.32	etrim
auto-zero																				
AD8628	1,2,4	2.7–6	0.85	30	100	1	5	0.002	120	22	2.5	1	●	●	●	●	-	-	1.92	SOIC-8, SOT23-5

Notes: (a) **boldface** indicated number in a package for the part # listed. (b) not precision, listed for comparison. (p) I_Q, typical, per amplifier.

their drifts with time and temperature) generally constitute serious obstacles to precision circuit design, and they force trade-offs in circuit configuration, component selection, and the choice of a particular op-amp. The point is best made with examples, as we will do shortly. Note that these errors, or their analogs, exist for amplifiers of discrete design as well.

While reading through the following discussion, you may find it helpful to refer to Tables 5.2 and 5.3 where we list ten favorite precision op-amps (plus two inexpensive and not-precision comparees). Those listings add quantitative flesh to a somewhat abstract set of bony teachings.

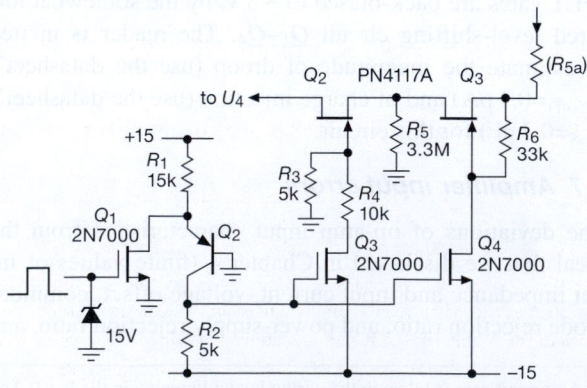

Figure 5.5. Electronic switch replacement of mechanical relay S_3 in Figure 5.3. A lot of work, and no improvement in performance.

5.7.1 Input impedance

Let's discuss briefly the error terms just listed. The effect of finite input impedance is to form voltage dividers in combination with the source impedance driving the amplifier, reducing the gain from the calculated value. Most often this isn't a problem, because the input impedance is bootstrapped by feedback, raising its value enormously. As an example, the OPA277P precision op-amp (with BJT-input, not FET-input stage) has a typical "differential-mode input impedance," of $100\,M\Omega$. In a circuit with plenty of loop gain, feedback raises the input impedance to the datasheet's "common-mode input impedance" $250{,}000\,M\Omega$. If that's not good enough, FET-input op-amps have astronomical values of R_{in}, for example $10^{13}\,\Omega$ (differential) and $10^{15}\,\Omega$ (common-mode) for the OPA129 that's used in this circuit.

5.7.2 Input bias current

More serious is the input bias current. Here we're talking about currents measured in nanoamps, and this already produces voltage errors of microvolts for source impedances as small as $1\,k\Omega$. Again, FET op-amps come to the rescue, but with generally increased voltage offsets as part of the bargain. Bipolar superbeta op-amps such as the LT1012 can also have surprisingly low input currents. As an example, compare the OPA277 precision bipolar op-amp with the LT1012 (bipolar, optimized for low bias current), the OPA124 (JFET, precision and low bias), the OPA129 (ultra-low-bias JFET), and the LMC6001 (CMOS, lowest-bias op-amp); these are some of the best you can get at the

Table 5.3 Nine Low-input-current Op-amps

Part #	Supply		Input current @25°C		V_{OS} max	TCV_{OS}	
	V_{total} (V)	I_Q (μA)	typ (pA)	max (pA)	(μV)	typ (μV/°C)	max (μV/°C)
bipolar							
OPA277P	10–36	790	500	1000	20	0.1	0.15
superbeta							
LT1012AC	8–40	370	25	100	25	0.2	0.6
AD706	4–36	750	50	200	100	0.2	1.5
JFET							
OPA124PB	10–36	2500	0.35	1	250	1	2
OPA129B	10–36	1200	0.03	0.1	2000	3	10
MOSFET							
MAX9945	4.8–40	400	0.05	–	5000	2	–
CMOS, low-voltage							
LMP7721	1.8–6	1300	0.003	0.02	150	1.5	4
LMC6001A	5–16	450	0.01	0.025	350	2.5	10
ADA4530-1	4.5–16	900	<0.001	0.02	50	0.13	0.5

time of writing, and we've chosen the best grade of each one (Table 5.3; and see also Table 5.5 on pages 320–321 for greater detail, along with a wider selection of precision op-amps with low offset voltage; as well as relevant tables in Chapter *4x*).

Well-designed FET amplifiers have extremely low bias currents, but with much larger offset voltages, compared with the precision OPA277. Because the offset voltage can always be trimmed, what matters more is the drift with temperature. In this case the FET amplifiers are 4 to 20 times worse. The op-amp with the lowest input current uses MOSFETs for the input stage. MOSFET op-amps are popular because of the proliferation of inexpensive units like TI's TLC270 series, as well as the ultra-low-bias-current devices like National's LMC6000-series parts. However, in contrast to JFETs or bipolar transistors, MOSFETs can have very large drifts of offset voltage with time, an effect that is discussed below. So the improvement in current errors you buy with a FET op-amp can be wiped out by the larger voltage error terms. With any circuit in which bias current can contribute significant error, it is often wise to ensure that both op-amp input terminals see the same dc source resistance (see, e.g., Figure 4.55); then the op-amp's *offset current* becomes the relevant specification. But be aware that a number of precision op-amps use a "bias-compensation" scheme to cancel (approximately) the input current in order to make that error term smaller (look back at Exercise 2.24 on page 125 to see how it's done). With op-amps of this type you generally don't gain anything by matching the dc resistances seen by the two inputs, as the residual bias current and the offset current are comparable, in a bias-compensated op-amp.

A. Variation with temperature

One additional point to keep in mind when using FET-input op-amps is that the input "bias" current is actually gate

leakage current, and it rises dramatically with increasing temperature: it roughly doubles for every 10°C increase in chip temperature, as seen in Figure 5.6. Because FET op-amps often run warm (our jellybean LF412, for example, dissipates 100 mW when run from ±15 V supplies), the actual input current may be considerably higher than the 25°C figures you see on the datasheet.[15] By contrast, the input current of a BJT-input op-amp is actual base current, relatively constant with temperature. So a FET-input op-amp with impressive input-current specs on paper may not give such an improvement over a good superbeta bipolar unit. As the graph shows, for example, the LT1057 JFET-input op-amp with its ~3 pA input current (at 25°C) will have an input current of about 100 pA at 75°C chip temperature, which is higher than the input current of the superbeta LT1012 at the same temperature. And our jellybean LF412 JFET op-amp has an input current that is comparable to that of the LT1012 at 25°C and many times higher at elevated temperatures.

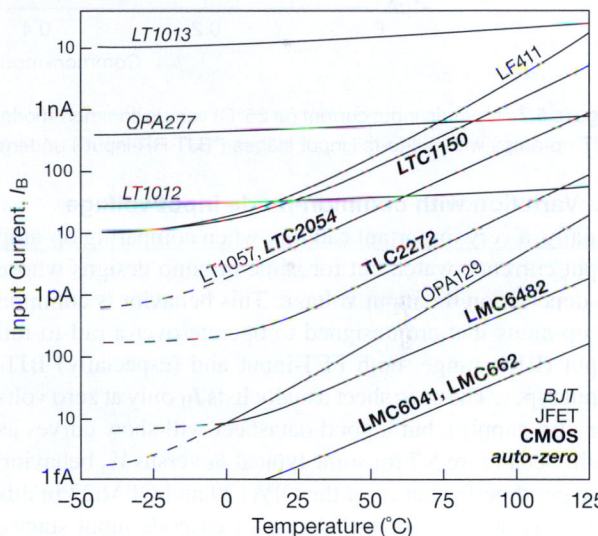

Figure 5.6. Op-amp input current versus temperature, plotted from datasheet values. See also Figures 5.38 and 3.48. JFET-input op-amps are indicated in plain ("roman") type, BJT op-amps are *italic*, CMOS op-amps are **bold**, and auto-zero op-amps are ***bold slant***.

---

[15] Making this quantitative, the LF412's maximum quiescent current is 6.5 mA, thus 195 mW dissipation when run from ±15 V supplies. In a DIP-8 package that produces a 22°C temperature rise (the thermal resistance $R_{\Theta JA}$=115°C/W), with a consequent quadrupling of the specified I_B=200 pA (max). If the op-amp were driving a load, you'd have yet more dissipation. To put this in perspective, though, note that the driving impedance seen at the op-amp's input would have to be greater than 1 MΩ in order for this current-induced error to exceed the 1 mV (typ) input offset-voltage error.

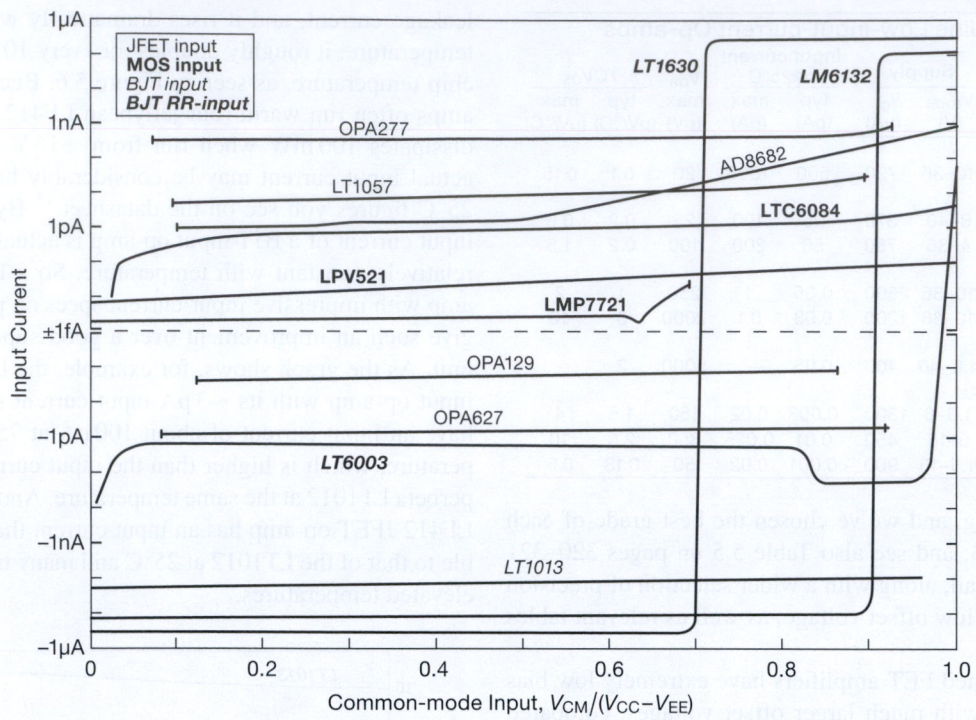

Figure 5.7. Op-amp input current (at 25°C) versus common-mode input voltage, plotted over their operating range from datasheet values; BJT op-amps with rail-to-rail input stages ("BJT RR-input") undergo an abrupt reversal of input-current polarity.

B. Variation with common-mode input voltage

Finally, a very important caution: when comparing op-amp input currents, watch out for some op-amp designs whose I_B depends on the input voltage. This behavior is common in op-amps that are designed to operate over a rail-to-rail input (RRI) range, both FET-input and (especially) BJT-input types. The spec sheet usually lists I_B only at zero volts (or mid-supply), but a good datasheet will show curves as well. See Figure 5.7 for some typical I_B versus V_{in} behavior. The good performance of the OPA129 and OPA627 in this regard is due in part to their use of cascode input stages. The LMP7721 stands out not only for its 20 fA maximum input current, but for its prize-winning trajectory on this graph.

5.7.3 Voltage offset

Voltage offsets at the amplifier input are obvious sources of error. Op-amps differ widely in this parameter, ranging from "precision" op-amps offering worst-case V_{OS} values generally in the 10s of microvolts to ordinary jellybean op-amps like the LF412 with V_{OS} values of 2–5 mV. At the time of writing,[16] the champion (by a slim margin) in the

(nonchopper, see below) world of low offsets is the bipolar OPA277P ($\pm 20\,\mu$V, max), which, astonishingly, is equaled by the *CMOS* MAX4236A (though the latter's drift is $12\times$ worse, as one might expect).

Although many good single op-amps (but not duals or quads) have offset-adjustment terminals, it is still wise to choose an amplifier with inherently low initial offset V_{OS} max, for several reasons. First, op-amps designed for low initial offset tend to have correspondingly low offset drift with temperature and with time. Second, a sufficiently precise op-amp eliminates the need for external trimming components (a trimmer takes up space, needs to be adjusted initially, and may change with time). Third, offset voltage drift and common-mode rejection are degraded by the unbalance caused by an offset-adjustment trimmer.

Figure 5.8 illustrates how a trimmed offset has larger drifts with temperature. It shows also how the offset adjustment is spread over the trimmer pot rotation, with best

---

[16] In the previous edition of this book we awarded that honor to the MAX400M, with its specified worst-case V_{OS} of $10\,\mu$V. With confi-

dence we added that "we expect to see further incremental improvements in this area." That confidence was evidently misplaced: the Maxim website now says of the MAX400 "This product was manufactured for Maxim by an outside wafer foundry using a process that is no longer available. It is not recommended for new designs. The datasheet remains available for existing users." *Sic transit...*

resolution near the center, especially for large values of trimmer resistance. Finally, you'll generally find that the recommended external trimming network provides far too much range, making it nearly impossible to trim V_{OS} down to a few microvolts; even if you succeed, the adjustment is so critical it won't stay trimmed for long. Another way to think about it is to realize that the manufacturer of a precision op-amp has *already* trimmed the offset voltage, in a custom test jig using "laser-zapping" techniques; you may be unable to do any better yourself. Our advice is (a) to use precision op-amps for precision circuits, and (b) if you must trim them further, arrange a narrow-range trim circuit similar that shown in Figure 5.3, with values adjusted to produce a full-scale range of $\pm 50\,\mu$V, linear in trimmer rotation (e.g., $R_{11}=33\,\Omega$ and $R_{12}=10$M). Figure 5.9 shows how to arrange narrow-range external offset trims for both inverting and noninverting amplifier configurations.

Because voltage offsets can be trimmed to zero, what ultimately matters is the drift of offset voltage with time, temperature, and power-supply voltage. Designers of precision op-amps work hard to minimize these errors. You get the best performance from bipolar-input (as opposed to FET) op-amps in this regard, but input-current effects may then dominate the error budget. As shown in Table 5.2 on page 302 the best op-amps keep drifts below 1 μV/°C; the OPA277P typifies the best drift specification (for a non-chopper op-amp): $\Delta V_{OS}=0.2\,\mu$V/°C, max.

Another factor to keep in mind is the drift caused by self-heating of the op-amp when it drives a low-impedance load. It is often necessary to keep the load impedance above 10k to prevent large errors from this effect. As usual, that may compromise the next stage's error budget from the effects of bias current! We'll see just such a problem in

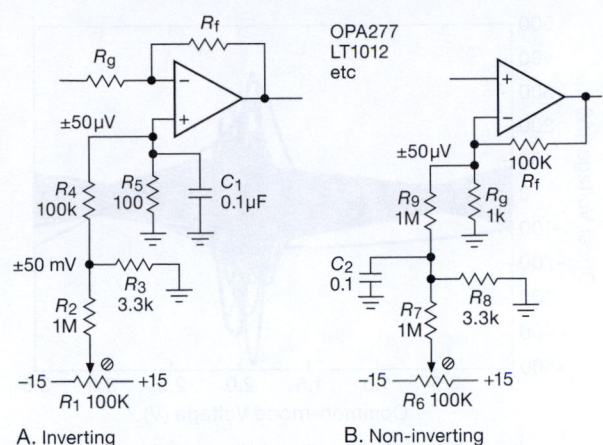

Figure 5.9. Narrow-range external trimming networks for precision op-amps.

this design example. For applications in which drifts of a few microvolts are important, the related effects of thermal gradients (from nearby heat-producing components) and thermal emf's (from voltages across junctions of dissimilar metals) become important. This will come up again when we discuss the ultraprecise *chopper-stabilized* amplifier in §5.11.

An important caution: when datasheets specify the particular measurement conditions for a parameter like V_{OS}, they mean it! A sobering example is shown in Figure 5.10, a plot of V_{OS} versus V_{CM} for the AD8615 op-amp, whose datasheet declares (on the front page) "Low offset voltage: 65 μV max," but whose tabular data reveal that the measurement conditions are "$V_{CM} = 0.5$ V and 3.0 V."

5.7.4 Common-mode rejection

Insufficient common-mode rejection ratio (CMRR) degrades circuit precision by effectively introducing a voltage offset as a function of dc level at the input. This effect is usually negligible, because it is equivalent to a small gain change, and in any case it can be overcome by choice of configuration: an inverting amplifier is insensitive to op-amp CMRR, in contrast with a noninverting amplifier. However, in "instrumentation amplifier" applications you are looking at a small differential signal riding on a large dc offset, and a high CMRR is essential. In such cases you have to be careful about circuit configurations and, in addition, you must choose an op-amp with a high-CMRR specification. Once again, a superior op-amp like the OPA277 can solve your problems, with a CMRR (min)

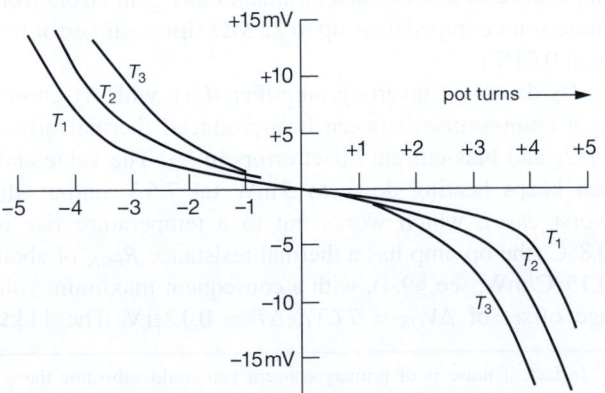

Figure 5.8. Typical op-amp offset versus offset-adjustment potentiometer rotation for several temperatures.

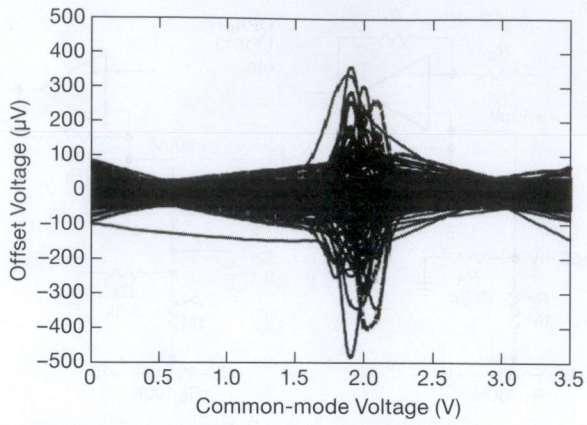

Figure 5.10. This op-amp specifies a maximum offset voltage of $\pm 60\,\mu$V. But it also specifies the following conditions: V_s=3.5 V, and V_{CM}=0.5 V or 3.0 V. Moral: don't ignore the footnotes!

at dc of 130 dB, compared with our jellybean LF411's meager specification of 70 dB. We discuss high-gain differential and instrumentation amplifiers later in the chapter, beginning at §5.13.

5.7.5 Power-supply rejection

Changes in power-supply voltage cause small op-amp errors. As with most op-amp specifications, the power-supply rejection ratio (PSRR) is referred to a signal at the *input*. For example, the OPA277 has a specified PSRR of 126 dB at dc, meaning that a 1 volt change in one of the power-supply voltages causes a change at the output equivalent to a change in the differential input signal of 0.5 μV.

The PSRR drops with increasing frequency, approximately tracking the behavior of the open-loop gain, and a graph documenting this scurrilous behavior is often given on the datasheet. For example, the PSRR (relative to the negative rail) of our favorite OPA277 begins dropping at 1 Hz and is down to 95 dB (typ) at 60 Hz and 50 dB at 10 kHz. This rarely presents a problem, because power-supply noise is also decreasing at higher frequencies if you have used good bypassing. However, 120 Hz ripple could present a problem if an unregulated supply is used.

It is worth noting that the PSRR will not, in general, be the same for the positive and negative supplies. Thus the use of dual-tracking regulators doesn't necessarily bring any benefits. Note also that PSRR is often specified for $G = 1$, and can be considerably worse at higher gains; in fact, op-amps have been found that exhibit *gain* (!) from one rail to the output at moderate gain settings.

5.7.6 Nulling amplifier: input errors

Now we're ready to embark on a detailed discussion of the most serious error issues in the amplifier of Figure 5.3. The circuit begins with an optional precision instrumentation amplifier front-end U_1 (more in §5.15), here chosen for its stable and accurate differential gain of 100×, low input current, and adequately low noise (9 nV/$\sqrt{\text{Hz}}$ typ at 10 Hz). Its worst-case offset voltage and tempco ($\pm 40\,\mu$V, 0.3 μV/°C) specs are a factor of two worse than a precision op-amp like the OPA277 (in the best grade), but its 120 dB (min) CMRR as a difference amplifier, combined with 0.08% worst-case gain accuracy, 50 ppm/°C (max) gain tempco, and low-voltage noise, make it a good front-end for a low-level bridge application like this. Although not important with the low source impedance in this example, its input current is satisfyingly low for a BJT-input amplifier, at just 0.35 nA max.[17]

For single-ended inputs U_1 is omitted, and the signal is brought in at point "×" (add a 470 Ω series resistor, with a pair of low-leakage clamp diodes – see Figure 5.2 – to the rails, for overdrive protection). The OPA277's accuracy and stability rule here, though it is tempting to consider substituting a FET-input part; but the $\gtrsim$10× poorer V_{OS} tempco specification more than offsets the advantage of low input current, except with sources of very high impedance. The OPA277's 1 nA (max) bias current gives an error of 1 μV/1 kΩ source impedance, whereas the best-in-class JFET OPA627B (at \$35 apiece!), though providing negligible current error with its 5 pA (max) input current, would exhibit voltage offset drifts as large as 3 μV/4°C (4°C is considered a typical laboratory ambient temperature range of variation). In this circuit one could happily add an offset trim to U_2, preferably in the manner of Figure 5.9. As mentioned earlier, feedback bootstraps the input impedance to 250 GΩ and eliminates any gain errors from finite source impedance, up to 25 MΩ (for a gain error less than 0.01%).

U_2 drives an inverting amplifier (U_3), with R_3 chosen as a compromise between heat-produced thermal offsets in U_2 and bias-current offset errors in U_3. The value chosen keeps heating down to 5 mW (at 7.5 V output, the worst case), which works out to a temperature rise of 0.8°C (the op-amp has a thermal resistance $R_{\Theta JA}$ of about 0.15°C/mW, see §9.4), with a consequent maximum voltage offset of $\Delta V_{OS} = TCV_{OS}\Delta T = 0.12\,\mu$V. The 11 k$\Omega$

---

[17] In fact, if noise is of primary concern you could substitute the ×4 quieter INA103 instrumentation amplifier at the front-end, paying the price in input offset current: a whopping 1 μA (thus $\pm 350\,\mu$V) of static offset voltage created by the 350 Ω differential source resistance here.

source impedance seen by U_3 results in an error due to bias-current offset, but, with U_3 inside a feedback loop with U_4 and U_5 trimming the overall offset to zero, all that matters is the drift in the current error term. The OPA277 provides a graph of typical bias-current change with temperature (not often specified by manufacturers), from which the error result of $0.2\,\mu\text{V}/4°\text{C}$ in the error budget is calculated. Reducing the value of R_3 would improve this term, at the expense of the heating term in U_2.

The dc input impedance of U_3 comes closer to presenting a problem. To estimate the error, we compare U_3's differential input impedance of $100\,\text{M}\Omega$ with the worst-case (i.e., with gain set for $\times 100$) impedance seen driving its input. The latter is just the feedback resistance (1M) divided by the loop gain $G_{\text{OL}}/G_{\text{CL}}$, thus $10\,\Omega$. So the worst-case loading effect is 1 part in 10^7, three orders of magnitude less than 0.01% error. This is one of the toughest examples we could think of, and even so the op amp input impedance presents no problem, thus demonstrating that, in general, you can ignore the effects of op-amp input impedances.

Drifts in offset voltage in both U_2 and U_3 over time, temperature, and power-supply variations affect the final error equally and are tabulated in the budget. It is worth pointing out that they are all automatically cancelled at each "zeroing" cycle, and only short-term drifts matter anyway. These errors are all in the microvolt range, thanks to a good choice of op-amp. U_4 has larger drifts, but it must be a FET type to keep capacitor current small, as already explained. Note that errors in U_4's output are amplified by the gain setting of U_3; thus they are specified as *input* errors in the budget.

Note the general philosophy of design that emerges from this example: you work at the problem areas, choosing configurations and components as necessary to reduce errors to acceptable values. Tradeoffs and compromises are involved, with some choices depending on external factors (e.g., the use of a FET-input op-amp for U_2 would be preferable for source impedances greater than about $10\,\text{k}\Omega$).

5.8 Amplifier output errors

As we discussed in Chapter 4, op-amps have some serious limitations associated with the output stage. Limited slew rate, output crossover distortion (§2.4.1A), and finite open-loop output impedance can all cause trouble, and they can cause precision circuits to display astoundingly large errors if not taken into account.

5.8.1 Slew rate: general considerations

As we mentioned in §4.4.1K, an op-amp can swing its output voltage only at some maximum rate. This effect originates in the frequency-compensation circuitry of the op-amp, as we will soon explain in a bit more detail. One consequence of a finite slew rate is to limit the output swing at high frequencies to a maximum of $V\text{pp} = S/\pi f$, as shown in §4.4.2 and plotted here in Figure 5.11.

A second consequence is best explained with the help of a graph of slew rate versus differential-input signal (Figure 5.12). The point to be made here is that a circuit that demands a substantial slew rate must operate with a substantial voltage error across the op-amp's input terminals. This can be disastrous for a circuit that pretends to be highly precise: the feedback loop is in error, more so as the output slews more rapidly, thus producing a distorted output waveform. (Look ahead at the measured distortion plots in Figure 5.19 on page 311 to see this effect, for example in the LT1013 for which $S=0.8\,\text{V}/\mu\text{s}$.)

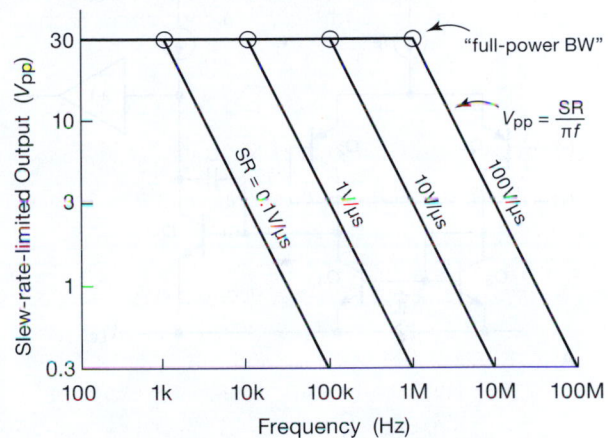

Figure 5.11. Maximum output swing versus frequency.

Let's look at the innards of an op-amp in order to get some understanding of the origin of slew rate (see §4x.9 for a more extensive discussion). The vast majority of op-amps can be summarized with the notional "Widlar circuit" shown in Figure 5.13. A differential input stage,[18] loaded with a current mirror, drives a stage of large voltage gain with a compensation capacitor from output to input. The output stage is a unity-gain push–pull follower. The compensation capacitor C is chosen to bring the open-loop gain of the amplifier down to unity before the phase shifts

---

[18] We've simplified it slightly: the input stage of Widlar's original LM101 used a *pnp* differential pair, but it was configured as a common-base amplifier driven by an *npn* follower pair.

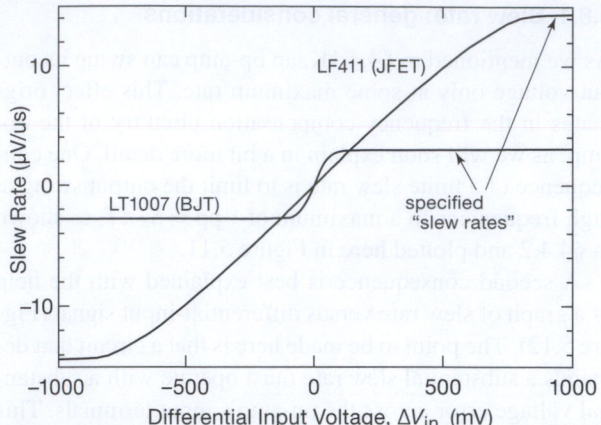

Figure 5.12. A substantial differential-input voltage is required to produce the full op-amp slew rate, as shown in these measured data. For BJT-input op-amps it takes ~60 mV to reach full slew rate; for JFETs and MOSFETs it's more like a volt.

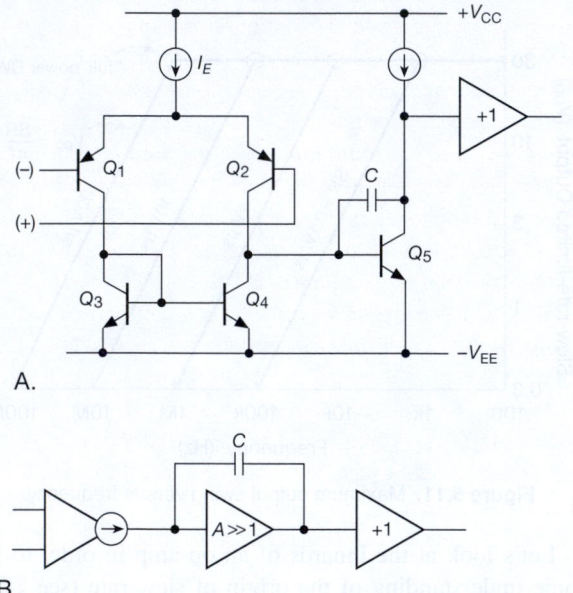

Figure 5.13. Typical op-amp internal compensation scheme.

caused by the other amplifier stages have become significant. That is, C is chosen to put f_T, the unity-gain bandwidth, near the frequency of the next amplifier rolloff pole, as described in §4.9. The input stage has very high output impedance, and it looks like a current source to the next stage.

The op-amp is slew-rate limited when the input signal drives one of the differential-stage transistors nearly to cutoff, driving the second stage with the total emitter current

I_E of the differential pair. For a BJT input stage this occurs with a differential input voltage of about 60 mV, at which point the ratio of currents in the differential stage is 10:1. At this point Q_5 is slewing its collector as rapidly as possible, with all of I_E going into charging C. The transistor Q_5 and C thus form an integrator, with a slew-rate-limited ramp as output. It's not hard to derive an expression for the slew rate, knowing how bipolar transistors work – see the discussion in §4x.9. The bottom line is that the classic BJT-input op-amp circuit of Figure 5.13 has a slew rate S given by $S \approx 0.3 f_T$.

To get a higher slew rate, then, you can choose an op-amp with greater bandwidth f_T; if you are operating at closed-loop gains greater than unity, you can use a decompensated op-amp (with its higher f_T value). But there are ways (as explained in §4x.9) to beat the limit $S \approx 0.3 f_T$ (which assumed a unity-gain-compensated op-amp with a BJT differential input configured for maximum gain, i.e., with $R_E=0$). Namely: (a) use an op-amp with reduced input-stage transconductance (either a FET-input op-amp or a BJT-input op-amp with emitter degeneration); (b) use an op-amp with a different input-stage circuit, specifically designed for enhanced slew rate – examples are the "cross-coupled transconductance reduction" technique (used in the TLE2142 family; see the cross-coupled input stage circuit shown in §4x.9), and the Butler "wide dynamic range transconductance stage" (used for example in the OP275 and OP285; see the Butler input stage circuit in §4x.9); (c) use a current-feedback (CFB) op-amp, or a CFB variant (with a buffered inverting input) that mimics an ordinary voltage-feedback (VFB) op-amp.

These tricks work. If we define an enhancement factor m (i.e., $S=0.3m f_T$), the LF411 (with JFET input) plotted in Figure 5.12 has $m = 12$, compared with the bipolar LT1007 ($m=1.0$); the TLE2141 (with cross-coupled BJT input stage) has $m = 25$, and the OP275/285 (with Butler input stages) have $m = 8$; the LT1210 (a CFB op-amp) has $m = 55$ with the recommended feedback resistor; and the LT1351 (a CFB in VFB's clothing) has $m = 220$.

For a deeper look at slew rate, turn to the extended discussion in Chapter 4x (§4x.9).

5.8.2 Bandwidth and settling time

Slew rate measures how rapidly the output voltage can change. The op-amp slew-rate specification usually assumes a large differential input voltage (60 mV or more), which (in spite of its potential for creating output distortion) is not unreasonable, given that an op-amp whose output isn't where it's supposed to be will have its input driven hard by feedback, assuming a reasonable amount of loop gain. Of perhaps equal importance in high-speed precision applications is the time required for the output to get where it's going following an input change. This *settling-time* specification (the time required to get within the specified accuracy of the final value and stay there; see Figure 5.14) is always given for devices such as digital-to-analog converters, where precision is the name of the game, but it is not normally specified for op-amps.

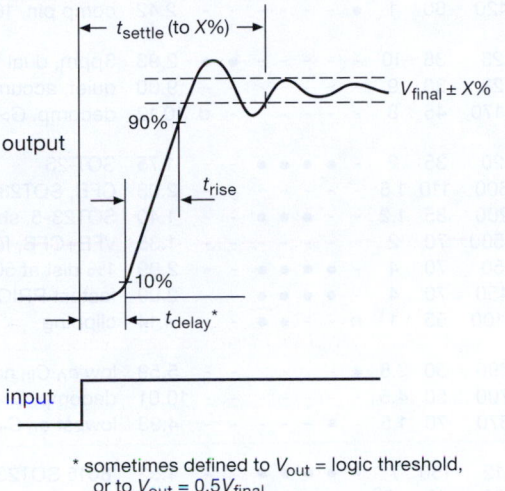

Figure 5.14. Settling time defined.

We can estimate op-amp settling time by considering first a different problem, namely, what would happen to a perfect voltage step somewhere in a circuit if it were followed by a simple *RC* lowpass filter (Figure 5.15). It is a simple exercise to show that the filtered waveform has the settling times shown. This is a useful result, because you often limit bandwidth with a filter to reduce noise (more on that later in the chapter). To extend this simple result to an op-amp, just remember that a compensated op-amp has a 6 dB/octave rolloff over most of its frequency range, just like a lowpass filter. When connected for closed-loop gain G_{CL}, its "bandwidth" (the frequency at which the loop gain drops to unity) is approximately given by

$$f_{3dB} = f_T/G_{CL}.$$

As a general result, a system of bandwidth B has response time $\tau \approx 1/(2\pi B)$; thus the equivalent "time constant" of the op-amp is

$$\tau \approx G_{CL}/2\pi f_T.$$

The settling time is then roughly 5–10τ.

Let's try our prediction on a real case. The TLE2141

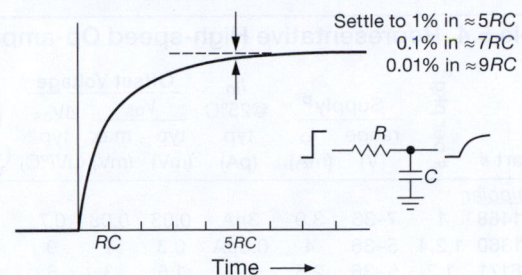

Figure 5.15. Settling time of an *RC* lowpass filter.

from TI is a precision fast-settling op-amp, with an f_T of 5.9 MHz. Our simple formula then estimates the inverting-configuration (i.e., $G=2$) response time to be 54 ns, thus a settling time of 378 ns (7τ) to 0.1%. This is in good agreement with the datasheet's value of 340 ns.

There are several points worth making: (a) our simple model gives us only a lower bound for the actual settling time in a real circuit; you should always check the slew-rate-limited rise time, which may dominate. (b) Even if slew rate is not a problem, the settling time may be much longer than our idealized "single-pole" model, depending on the op-amp's compensation and phase margin. (c) The op-amp will settle more quickly if the frequency compensation scheme used gives a plot of open-loop phase shift versus frequency that is a nice straight line on a log–log graph (as in Figure 5.17); op-amps with wiggles in the phase-shift graph are more likely to exhibit overshoot and ringing, as in the upper waveform shown in Figure 5.14. (d) A fast settling time to 1%, say, doesn't necessarily guarantee a fast settling time to 0.01%, since there may be a long tail (Figure 5.16). (e) There's no substitute for an actual settling-time specification from the manufacturer.

Table 5.4 lists a selection of high-speed op-amps suitable for applications that demand high f_T, high slew rate, fast settling time, and reasonably low offset voltage.

5.8.3 Crossover distortion and output impedance

Some op-amps (for example the classic single-supply LM324/358) use a simple push–pull follower output stage, without biasing the bases two diode drops apart, as we discussed in §2.4.1. This leads to "class-B" distortion near zero output, because the driver stage has to slew the bases through $2V_{BE}$ as the output current passes through zero (Figure 5.18). This crossover distortion can be substantial, particularly at higher frequencies where the loop gain is reduced; see the measured data in Figure 5.19. It is greatly reduced in op-amp designs that bias the output

Table 5.4 Representative High-speed Op-amps[x]

Part #	# per pkg[a]	Supply[p] range (V)	I_Q (mA)	I_{in} @25°C typ (pA)	V_{os} typ (mV)	V_{os} max (mV)	ΔV_{os} typ (μV/°C)	e_n typ[r] (nV/√Hz)	GBW typ (MHz)	Slew typ (V/μs)	I_{out} typ (mA)	C_{in} (pF)	dist. graph	Swing IN +	IN −	OUT +	OUT −	null pins	DIP avail	cost qty 25 ($US)	Comments
bipolar																					
LT1468	1	7–36	3.9	3nA	0.03	0.08	0.7	5	90	23	22	4	−	−	−	−	−	•	•	4.26	0.7 ppm dist
LT1360	1,2,4	5–36	4	0.3μA	0.3	1	9	9	50	800	34	3	−	−	−	−	−	•	•	2.75	C-Load™
LM6171	1,2	5–36	2.5	1μA	1.5	3	6	12	100	3600	90	-	•	−	−	−	−	−	•	2.57	VFB+CFB
AD844	1	9–36	6.5	0.2μA	0.05	0.3	1	9	330[g]	2000	60	2	−	−	−	−	−	•	•	5.23	CFB, comp pin
AD8021[b]	1	4.5–26	7	7.5μA	0.4	1	0.5	2.1	925	420	60	1	•	−	−	−	−	−	−	2.42	comp pin, 16-bits
JFET																					
OPA604A	1,2	9–50	5.3	50	1	5	8	10	20	25	36	10	−	−	−	−	−	•	•	2.93	3ppm, dual '2604
OPA827A	1	8–40	4.8	15	0.08	0.15	1.5	3.8	22	28	30	9	−	−	−	−	−	−	•	9.00	quiet, accurate
ADA4637	1	9–36	7.0	1	0.12	0.3	1	6.1	80	170	45	8	−	−	−	−	−	−	d	10.12	decomp, G>7
low-voltage bipolar																					
LT6220	1,2,4	2.2–13	0.9	15nA	0.07	0.35	1.5	10	60	20	35	2	−	•	•	•	•	−	−	1.75	SOT-23
LMH6723	1,2,4	4.5–13	1	2μA	1	3	-	4.3	370	600	110	1.5	−	−	•	•	•	−	−	2.03	CFB, SOT23-5
ADA4851	1,2,4	3–12.6	2.5	2.2μA	0.6	3.4	4	10	125	200	85	1.2	−	−	•	•	•	−	−	1.40	SOT23-5, shdn pin
LT1818	1,2	4–12.6	9	2μA	0.2	1.5	10	6	400	2500	70	2	−	−	−	−	−	−	−	1.35	VFB+CFB, fast
LT6200	1,2	3–12.6	16.5	10μA	0.2	1.2	8	0.95	165	50	70	4	−	•	•	•	•	−	−	2.99	1% dist at 50MHz
LT6200-10	1	3–12.6	16.5	10μA	0.2	1.2	8	0.95	1600	450	70	4	−	•	•	•	•	−	−	2.99	fastest RRIO
OPA698[e]	1	5–13	16	3μA	2	5	15[m]	5.6	450	1100	55	1	n	−	−	•	•	−	−	4.14	clipping
low-voltage JFET																					
OPA656	1	9–13	14	2	0.25	1.8	2	7	230	290	50	2.8	•	−	−	−	−	−	−	5.59	low $e_n{\cdot}C_{in}$ noise
OPA657	1	9–13	14	2	0.25	1.8	2	7	1600	700	50	4.5	−	−	−	−	−	−	−	10.01	decomp, G>7
ADA4817	1,2	5–10.6	19	2	0.4	2	7	4	1050	870	70	1.5	−	•	•	−	−	−	−	4.93	lowest $e_n{\cdot}C_{in}$
CMOS																					
AD8616	2	2.7–6	1.7	0.2	0.02	0.06	1.5	7	24	12	150	7	−	•	•	•	•	−	−	1.52	'8615 SOT23-5
LMP7717	1,2	1.8–6	1.15	0.05	0.01	0.15	1	6.2	88	28	15	15[c]	−	−	•	•	•	−	−	2.18	decomp, G>10
OPA350	1,2,4	2.5–7	5.2	0.5	0.15	0.5	4	7	38	22	40	6.5	−	•	•	•	•	−	•	1.67	6ppm

Notes: (a) **boldface** indicates number in a package for the part number listed. (b) for G<20 use ext C_C chosen to set f_{3dB}=200MHz. (c) 15mA sinking, 47mA sourcing. (d) for DIP-8 see OPA637. (e) OPA699 decomp. (g) at G=10. (m) max. (n) distortion plot for OPA699. (p) I_Q, typical, per amplifier. (r) at 1kHz. (x) see also the fast op-amp table in *Chapter 4x*.

push–pull pair into slight conduction ("class-AB"), for example the LT1013, which is an improved version of the LM324. The right choice of op-amp can have enormous impact on the performance of low-distortion audio amplifiers. Perhaps this problem has contributed to what the audiophiles refer to as "transistor sound." Some modern op-amps, particularly those intended for audio applications, are designed to produce extremely low crossover distortion. Examples are the LT1028, the AD797, and the excellent "LME49000" series from NSC, e.g., the LME49710. The latter, for example, has less than 0.0001% distortion over the full audio band of 20–20 kHz. (That's the claim, anyway; we may be overly gullible!) These amplifiers all have very low noise voltage, as well; the LT1028, for example, vies for the title of world noise-voltage champion, with $e_n = 1.7\,\text{nV}/\sqrt{\text{Hz}}$ (max) at 10 Hz. See the expanded plots of op-amp distortion in Figures 5.43 and 5.44, where various op-amps compete for the title of king of low distortion. High-voltage op-amps have an advantage below

10 kHz, whereas the low-voltage types have an advantage above 200 kHz.

The open-loop output impedance of a typical split-supply op-amp is highest when the output is near ground, because the output transistors are operating at their lowest current into the (ground-returned) load. The output impedance also rises at high frequency as the transistor gain drops off, and it may rise slightly at very low frequencies because of thermal feedback on the chip.

It is easy to neglect the effects of finite open-loop output impedance, thinking that feedback will cure everything. But when you consider that some op-amps have open-loop output impedances of a few hundred ohms, it becomes clear that the effects may not be negligible, especially at low to moderate loop gains. Figures 5.20 and 5.21 shows some typical graphs of op-amp output impedance, both with and without feedback.

The finite output impedance contributes also to instability when driving capacitive loads, as we discussed in

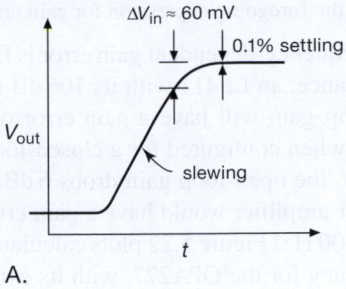

A.

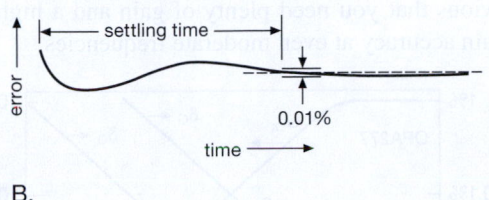

B.

Figure 5.16. A. The slewing decreases when input error approaches 60 mV. B. Settling to high precision can be surprisingly lengthy.

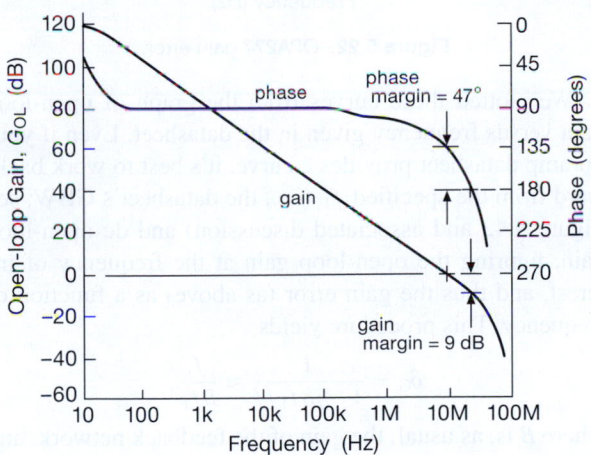

Figure 5.17. OP-42 gain and phase versus frequency.

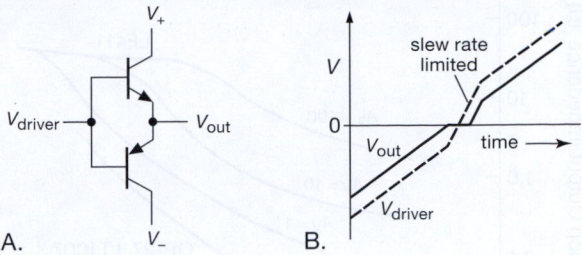

Figure 5.18. Crossover distortion in class-B push–pull output stage.

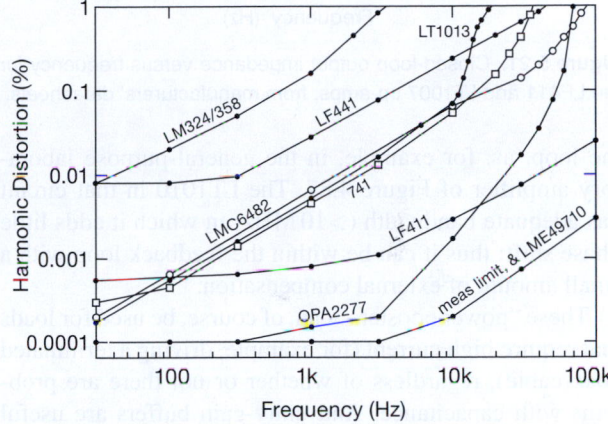

Figure 5.19. Measured harmonic distortion versus frequency for several popular op-amps (1 Vrms output, unloaded). See also Figures 5.43 and 5.44.

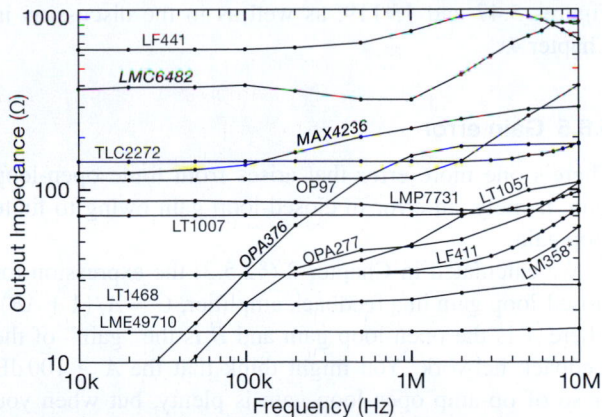

Figure 5.20. Measured open-loop output impedance versus frequency for a selection of op-amps. Parts shown in **bold** have CMOS output circuits. * With output pulldown resistor.

5.8.4 Unity-gain power buffers

If the technique of split feedback paths is unacceptable, one solution is to add a unity-gain high-current buffer inside

§4.6.2, owing to the additional lagging phase shift within the feedback loop that is created by R_{out} in combination with C_{load}. Several common solutions were shown in Figure 4.78, including a split feedback path or the inclusion of a unity-gain buffer within the loop. The latter is worth a mention here.

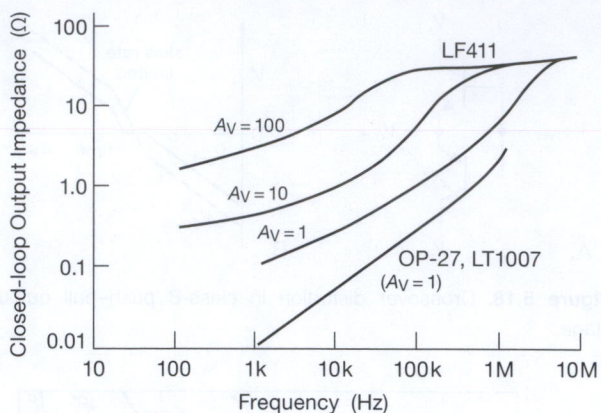

Figure 5.21. Closed-loop output impedance versus frequency for the LF411 and LT1007 op-amps, from manufacturers' datasheets.

the loop, as, for example, in the general-purpose laboratory amplifier of Figure 4.87. The LT1010 in that circuit has adequate bandwidth ($>10\,\mathrm{MHz}$) in which it adds little phase shift; thus it can be within the feedback loop with a small amount of external compensation.

These "power boosters" can, of course, be used for loads that require high current (for example, driving a terminated coax cable), regardless of whether or not there are problems with capacitance. And unity-gain buffers are useful even with loads of only moderate current, in the context of precision circuit design, because they prevent thermal drifts by keeping the heat out of the low-offset amplifier. You can see a couple of examples of power boosters in Figures 5.47 and 13.119, as well as in the discussion in Chapter *4x*.

5.8.5 Gain error

There's one more error that arises from finite open-loop gain, namely, an error in closed-loop gain owing to finite loop gain.

We calculated in Chapter 2 (§2.5.2) the expression for closed-loop gain in a feedback amplifier, $G = A/(1+AB)$, where A is the open-loop gain and B is the "gain" of the feedback network. You might think that the $A \geq 100\,\mathrm{dB}$ or so of op-amp open-loop gain is plenty, but when you try to construct extremely precise circuits you are in for a surprise. From the preceding gain equation it is easy to show that the "gain error," defined as

$$\delta_G = \text{gain error} \equiv \frac{G_{\text{ideal}} - G_{\text{actual}}}{G_{\text{ideal}}},$$

is just equal to $1/(1+AB)$ and ranges from 0 for $A = \infty$ to 1 (100%) for $A = 0$.

Exercise 5.2. Derive the foregoing expression for gain error.

The resulting frequency-dependent gain error is far from negligible. For instance, an LF411 with its 106 dB of low-frequency open-loop gain will have a gain error of 0.5% at low frequencies when configured for a closed-loop gain of 1000. Worse yet, the open-loop gain drops 6 dB/octave above 20 Hz, so our amplifier would have a gain error of a whopping 10% at 500 Hz! Figure 5.22 plots calculated gain error versus frequency for the OPA277, with its extraordinary 140 dB of low-frequency open-loop gain, when configured for closed-loop gains of 100 and 1000. It should be obvious that you need plenty of gain and a high f_T to maintain accuracy at even moderate frequencies.

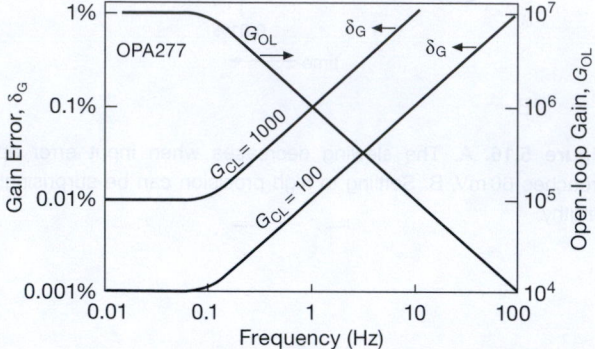

Figure 5.22. OPA277 gain error.

We plotted these curves from the graph of open-loop gain versus frequency given in the datasheet. Even if your op-amp datasheet provides a curve, it's best to work backward from the specified f_T (i.e., the datasheet's GBW; see Figure 5.42 and associated discussion) and dc open-loop gain, figuring the open-loop gain at the frequency of interest, and thus the gain error (as above) as a function of frequency. This procedure yields

$$\delta_G = \frac{1}{1 - jBf_T/f} \approx \frac{f}{Bf_T},$$

where B is, as usual, the gain of the feedback network, and the approximation is valid for the useful case $Bf_T/f \gg 1$. Of course, in some applications, such as filters, B may also depend on frequency.

Exercise 5.3. Derive the foregoing result for $\delta_G(f)$.

5.8.6 Gain nonlinearity

Op-amps have lots of open-loop gain at low frequencies, and the excess (G_{OL}/G_{CL}) is the loop-gain feedback mechanism that contributes to accuracy and the reduction of the op-amp's intrinsic nonlinearities, as discussed first in §2.5.3. Ideally, then, we want lots of open-loop gain in a precision circuit. And that's why auto-zero amplifiers (§5.11) and precision op-amps are built with high open-loop gains, for example ∼160 dB for the auto-zero LMP2021, and ∼150 dB for the precision LT1007.

For *accuracy*, then, we want lots of loop gain. For purposes of *linearity*, however, it's OK to have less loop gain – what matters more is the intrinsic linearity of the op-amp, combined with an open-loop gain characteristic that changes linearly (if at all) with output swing. The intrinsic linearity is strongly influenced by the output-stage design, particularly when the amplifier is driving a load: crossover distortion is always bad, as is an output stage that is asymmetric in its source/sink capabilities (like the LM358, with a Darlington *npn* pullup and single *pnp* pull-down). And a poor layout within the chip can create nonlinearities from the thermal offsets produced by local heating when driving a load.

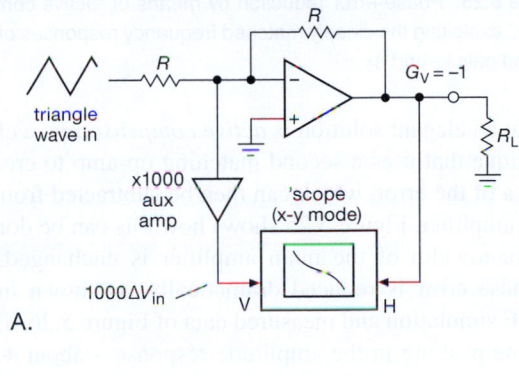

A.

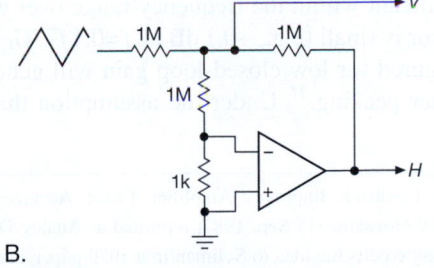

B.

Figure 5.23. Low-frequency gain nonlinearity test circuit. A. Notional: an auxiliary amplifier makes visible the μV-scale differential input voltage versus output swing. B. Circuit used by Pease for measurements in AN-1485 (see footnote on the current page).

In a nice set of measurements, Bob Pease[19] explored the low-frequency gain nonlinearity of a selection of op-amp

types (sadly, none from other manufacturers), operating as unity-gain inverters with a full-swing output; he made measurements both when the op-amps were unloaded, and when they were driving a 1 kΩ load. The basic scheme is shown in Figure 5.23A, where a 'scope looks at the amplifier input error versus output swing. For Pease's actual measurements he used the subtle variant in Figure 5.23B, in which the op-amp amplifies its error by ×1000, delivering the bad news directly.

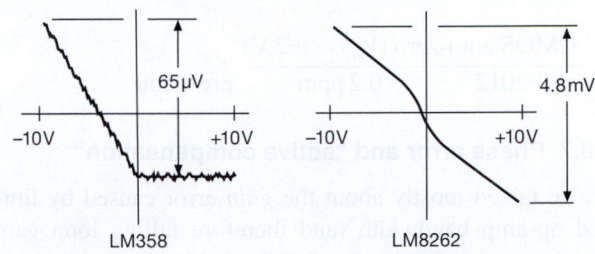

Figure 5.24. Gain nonlinearity traces for two op-amps with output-stage deficiencies. In these *x–y* displays the vertical axis shows the (small) differential input signal required to produce the (full-swing) output signal indicated on the horizontal axis. To estimate gain error, divide the vertical deviation from a best-fit straight line by the full-swing output.

The sorts of things you see (with a loaded op-amp) are shown in Figure 5.24, where we've sketched Pease's traces for the aforementioned LM358 (afflicted with asymmetric source/sink) and the LM8262 (fast, but afflicted with some crossover distortion). An exemplary op-amp like the very low distortion LM4562 presents an ideal nearly-horizontal straight line. The LF411 (single) and LF412 (dual), our JFET jellybeans, show an interesting contrast: according to Pease, the LF411 chip layout is sub-optimal (in terms of gain and thermal effects), with great effort rewarded by better results in the dual LF412.

Here are some of his summary results for op-amps driving a somewhat lighter load (4 kΩ). In general, the measured gain nonlinearity, when *unloaded*, was far smaller than these listed values. Keep in mind that these measurements were made at very low frequencies (generally just a few hertz), where the loop gain is maximum.

---

[19] National Semiconductor App Note AN-1485: *The Effect of Heavy Loads on the Accuracy and Linearity of Operational Amplifier Circuits*

(or, *"What's All this Output Impedance Stuff, Anyhow?"*). Gain nonlinearity data can be found in some datasheets, for example the AD620 instrumentation amplifier.

HV BJT ($V_{sig}=\pm10\,V$)		
LM8262	12 ppm	xover dist.
LM358	1 ppm	asym. output stage
LF411	1.4 ppm	poor layout – thermal
LF412	0.3 ppm	better layout
LM4562	0.025 ppm	pro-audio, $G_{OL}=10^7$

CMOS RRO ($V_{sig}=\pm4\,V$)		
LMC6482	1.1 ppm	jellybean
LMC6062	0.2 ppm	precision

CMOS auto-zero ($V_{sig}=\pm2\,V$)		
LMP2012	0.2 ppm	precision

5.8.7 Phase error and "active compensation"

We've talked mostly about the *gain* error caused by limited op-amp bandwidth (and therefore falling loop gain with increasing frequency). But limited loop gain also produces *phase* error, which can be important in applications such as video, interferometry, and so on. And the effect is not at all negligible – recall (§1.7.9) that a single *RC*-like rolloff creates a phase shift of ~6° at a frequency of $f_C/10$, and ~0.6° at $f_C/100$; the latter is two full decades below the −3 dB breakpoint. If we model an op-amp's rolloff of open-loop gain similarly (a "single-pole" rolloff), we can expect comparable phase shifts.

In this approximation, the resultant phase shift for an op-amp voltage amplifier is thus given by

$$\phi = \tan^{-1}\left(\frac{f}{f_C}\right) \approx \frac{f}{f_C}\ \text{(radians)},$$

where the −3 dB breakpoint f_C is the frequency at which the loop gain has fallen to unity: $f_C = f_T/G_{CL}$. Here G_{CL} is the closed-loop gain (as set by the feedback network), and f_T is the gain–bandwidth product (GBW) of the op-amp (for a single-pole rolloff that's the same as the frequency at which the open-loop gain is unity; but for typical op-amps, with more complicated rolloffs, you want to use the GBW figure). Multiply by 57.3 ($180/\pi$) to get the answer in degrees. The approximate result (the last expression) is reasonably accurate for small-to-moderate phase shifts, up to 0.5 radian, say.

There are several ways to address this problem. The simplest is to use an amplifier of greater bandwidth. If you don't want to (or cannot) do that, another possibility is to introduce an *RC* network in the feedback path to cancel the phase error (in *s*-plane language, you are introducing a zero to cancel a pole). This can be effective, but requires "tuning" of the compensation network to match the fre-

quency response of the particular op-amp specimen itself; and because the op-amp's characteristics change with temperature, the network must do likewise. A third possibility is to cascade two stages, each configured for lower gain (and therefore smaller phase error).

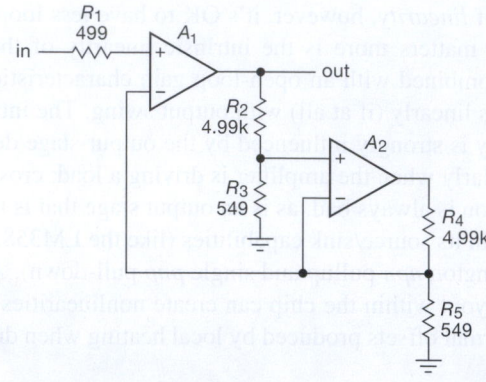

Figure 5.25. Phase-error reduction by means of "active compensation," exploiting the closely matched frequency responses of dual op-amp pair A_1 and A_2.

But an elegant solution is *active compensation*, a clever technique that uses a second matching op-amp to create a replica of the error, which can then be subtracted from the main amplifier. Figure 5.25 shows how this can be done.[20] The bandwidth of the main amplifier is unchanged, but its phase error is reduced dramatically, as shown in the SPICE simulation and measured data of Figure 5.26. There is some peaking in the amplitude response – about +3 dB at the frequency at which the phase error is 45° – but generally insignificant within the frequency range over which the phase error is small (e.g., +0.1 dB at $f=0.1f_T/G_{CL}$). A circuit configured for low closed-loop gain will generally exhibit greater peaking.[21] Under the assumption that the

---

[20] See "Active Feedback Improves Amplifier Phase Accuracy," by J. Wong, *EDN Magazine*, 17 Sept 1987; reprinted as Analog Devices AN-107. Wong credits the idea to Soliman in a 1979 paper, and Soliman credits the idea to Brackett and Sedra in a 1976 paper. But Wong's paper is the most useful reference for understanding the configuration of Figure 5.25.

[21] In SPICE simulations we found that the peaking increased to ~7 dB for the LF412 model configured for $G=2$; this can be tamed by adding a compensation capacitor C_c across feedback resistor R_2. Choosing C_c to match the op-amp's f_T (i.e., $C_c = 1/2\pi f_T R_2$) reduced the peaking to 4 dB, at the expense of tripling the (pretty small) phase error. In his article, James Wong warns that the technique may result in an unstable amplifier for low gains, below $G=5$ for example. He shows also how the technique can be further improved if A_2 is made from two amplifiers.

amplifiers are matched, it can be shown that this technique produces a phase shift given approximately by

$$\phi \approx \left(\frac{f}{f_C}\right)^3 \text{ (radians)},$$

again accurate for small-to-moderate phase shifts ($\lesssim 30°$, say, i.e., the small-angle approximation).

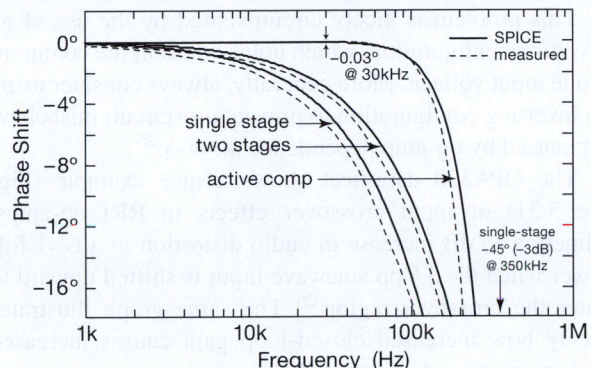

Figure 5.26. SPICE simulation and measured data of the phase shift versus frequency for the circuit of Figure 5.25, implemented with an LF412 dual JFET op-amp. For comparison the analogous data are plotted for both a single $G{=}10$ stage and for a cascade of two stages, each with $G{=}\sqrt{10}$. The measured part's f_C was 295 kHz, somewhat lower than the SPICE model's 350 kHz.

Real op-amps are not perfectly matched. To see how a mismatch in f_T affects the phase compensation, we ran a SPICE simulation with an f_T mismatch of $\pm 10\%$ (Figure 5.27). Evidently our test-bench part, chosen quite at random (dip fingers into parts bin, grasp first part touched, extract, measure), has a considerably better f_T matching, as suggested by Wong: "Monolithically matched dual or quad op-amps can provide the frequency-matching characteristics (to within 1% to 2%) necessary for the success of the active-feedback approach."[22]

It's interesting to compare predicted phase shifts for several scenarios mentioned at the outset: (a) a single amplifier of given bandwidth (call it f_{T0}, 3 MHz for the LF412), configured for a closed-loop gain $G = 10$; (b) two cascaded stages, each with $G = \sqrt{10}$; (c) the active compensation method of Figure 5.25; and (d) a single amplifier of greater bandwidth ($10f_{T0}$, say). Here are the calculated results:

---

[22] We went back to the bench and measured a handful of LF412 dual op-amps. Among different specimens the f_T values ranged over $\pm 20\%$, but within any single part the f_T's of its two op-amps matched typically to 0.1%, with one outlier showing a 1.5% mismatch.

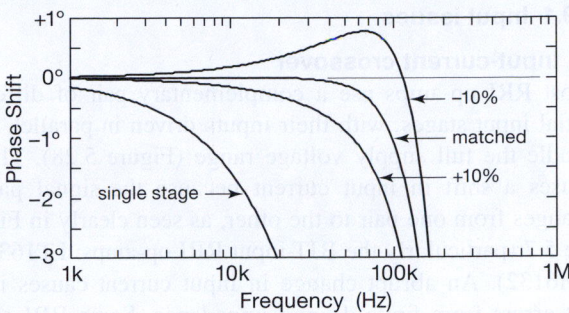

Figure 5.27. Active compensation of phase error requires matched op-amp bandwidths, as seen in this SPICE simulation for which the f_T of the compensation op-amp A_2 has been varied $\pm 10\%$ relative to that of the signal path op-amp A_1.

	Single, $G{=}10$	2-stage, each $G{=}\sqrt{10}$	Active, comp	Single, $f_T{=}10f_{T0}$
$0.001f_{T0}$	$-0.57°$	$-0.36°$	$-0.00006°$	$-0.006°$
$0.003f_{T0}$	$-1.7°$	$-1.1°$	$-0.0015°$	$-0.17°$
$0.01f_{T0}$	$-5.7°$	$-3.6°$	$-0.06°$	$-0.57°$
$0.03f_{T0}$	$-17.2°$	$-10.9°$	$-1.5°$	$-1.7°$
$0.1f_{T0}$	$-45°$	$-36°$	$-45°$	$-5.7°$

It's clear that the remarkable (and underutilized) technique of active compensation represents an efficient use of resources. The noninverting $G = 2$ case looks especially useful, e.g., to drive backterminated $75\,\Omega$ video cables.[23]

5.9 RRIO op-amps: the good, the bad, and the ugly

In Chapter 4 (§§4.4.1, 4.4.2, and 4.6.3) we introduced rail-to-rail op-amps, including (a) op-amps that operate properly with common-mode inputs over the full supply voltage range (RRI), (b) op-amps that can swing their outputs over the full supply range (RRO), and (c) op-amps that can do both (RRIO). With lower supply voltages increasingly in vogue, you see many new op-amps with these desirable capabilities.

Desirable, but to be used with caution. These benefits come at a cost, which we'll discuss here in the context of precision design (with further discussion in Chapter 4x). In circuits that strive for accuracy there are some hidden compromises in the designs of these op-amps about which the datasheet may be, uh, understated (or completely silent). Here are the important ones.

---

[23] Or sometimes called "double-terminated", as in Figure 12.110. We suggest trying your amplifier of choice, taking care to terminate the second op-amp with $150\,\Omega$.

5.9.1 Input issues

A. Input-current crossover

Most RRI op-amps use a complementary pair of differential input stages, with their inputs driven in parallel, to handle the full supply voltage range (Figure 5.28). This causes a shift in input current because the signal path changes from one pair to the other, as seen clearly in Figure 5.7 (particularly the BJT-input RRI op-amps: LT1630, LM6132). An abrupt change in input current causes input errors from finite driving impedance. Some RRI op-amps avoid this problem by using an on-chip charge pump to generate a supply voltage beyond the rail, so a single input amplifier allows rail-to-rail inputs. Examples are the OPA360-series,[24] the AD8505 and ADA4505, the MAX4162-series, and the MAX4126-series. Except for the BJT-input MAX4126, these all use MOS inputs.

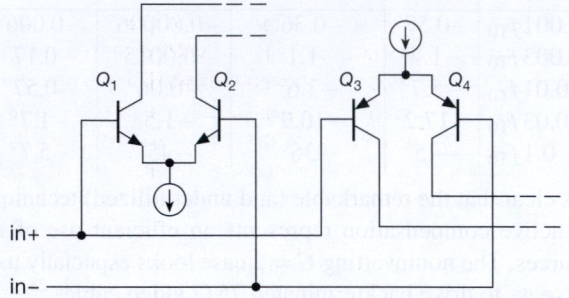

Figure 5.28. A typical rail-to-rail input circuit consists of a pair of complementary differential amplifiers, with downstream circuity to select the active pair's output.

In situations where you need RRO but don't need full rail-to-rail *input* (a voltage amplifier with $G>2$, say), be sure to consider an RRO op-amp with input extending to the negative rail only (sometimes called "ground sensing"). Note also that, by using an op-amp in an inverting circuit configuration, you avoid this problem completely (but you probably would not choose an RRI op-amp for such a configuration anyway).

B. Input offset-voltage crossover

The dual input stages of RRI op-amps cause similar mischief in terms of their input offset voltage V_{OS}, as seen in Figure 5.29. The abrupt change can occur close to either end of the supply range, as seen in the LMP7701 and LMP7731 op-amps from the same manufacturer. These curves were adapted from their respective datasheets,

---

[24] Playfully named "Zer∅-Crossover" amplifiers, or ZCOs.

which typically display a figure showing a tangle of overlapping curves measured on multiple op-amp samples (if they're willing to show any data at all about this seamy topic). Here you can see by comparison the uncomplicated (and downright *boring*) behavior of an RRI op-amp with an on-chip charge pump powering a single input amplifier. This variation of V_{OS} with V_{CM} is not only undesirable, it is also unpredictable, as you can see in Figure 5.30.

This problem is nicely circumvented by the use of an inverting configuration, which holds constant the common-mode input voltage. More generally, always consider using an inverting configuration to prevent *any* circuit misbehavior caused by op-amp dependence on V_{CM}.[25]

The OPA350 datasheet shows a nice example (Figure 5.31) of input crossover effects in RRI op-amps, namely a 17 dB increase in audio distortion in a $G=1$ follower when the 3 Vpp sinewave input is shifted upward to enter the crossover region.[26] The same graph illustrates nicely how increased closed-loop gain causes increased distortion owing to decreased loop gain.

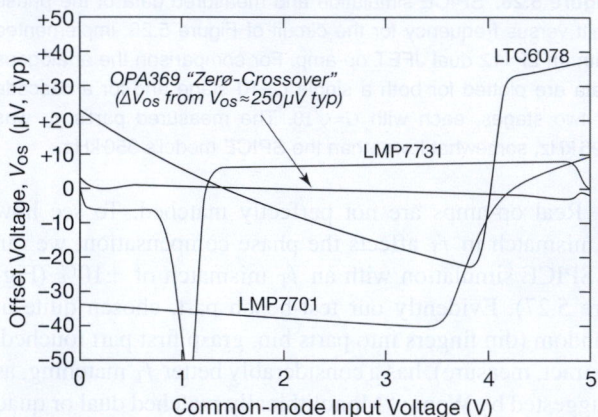

Figure 5.29. Op-amps with rail-to-rail inputs usually exhibit a shift of V_{OS} as the input voltage passes control from one input pair to the other. The OPA369 circumvents this by using a single input pair, powered beyond the rail by an on-chip charge pump.

5.9.2 Output issues

A. Output impedance

The output stage of a conventional (not RRO) op-amp is ordinarily a complementary push–pull follower (or some

---

[25] As Jim Williams liked to say, "Use an inverting configuration, unless you can't."

[26] See also Bonnie Baker's article (in the *Baker's Best* series) "Where did all that racket come from?" in *EDN Magazine*, 23 April 2009, available at edn.com.

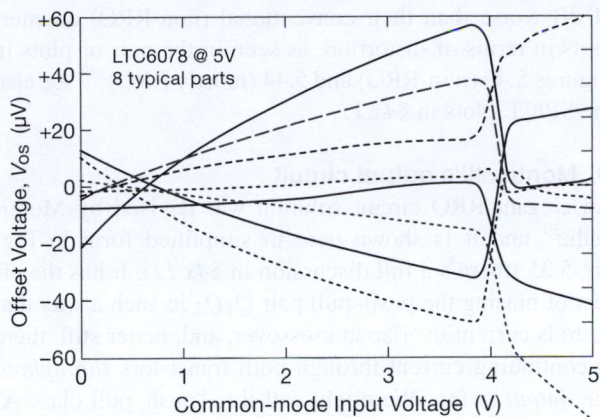

Figure 5.30. The shift of offset voltage in an RRI op-amp can be unpredictable (even as to the *sign* of the effect!), as seen in these data, adapted from the unusually forthcoming manufacturer's datasheet.

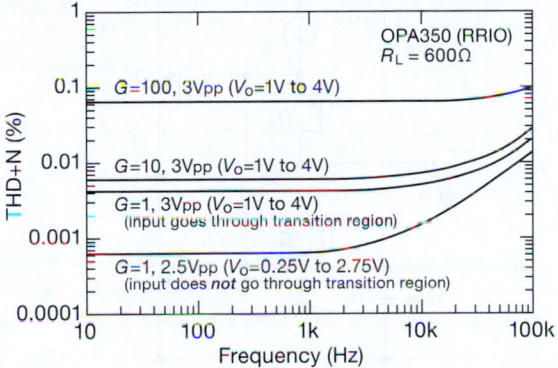

Figure 5.31. Distortion versus frequency for the OPA350 RRIO op-amp. The two lowest curves show the dramatic increase in distortion when the input signal enters the input crossover region. Increasing closed-loop gain causes further distortion because of reduced loop gain.

variation thereupon), biased with some conduction overlap to prevent crossover distortion at mid-supply (see §5.8.3). By contrast, the output complementary pair in an RRO op-amp is configured as a push–pull common-source *amplifier*; see Figure 5.32. That's necessary for the output to reach the rails (absent a second set of beyond-the-rails supply voltages). But it creates problems, owing to its inherently high output impedance.

The high $Z_{\rm out}$ means that the output-stage gain (and therefore the loop gain) depends on the value of load resistance; and a capacitive load creates large phase shifts, compromising the loop stability (see, for example, Figure 4.79). These problems are addressed in part by use of

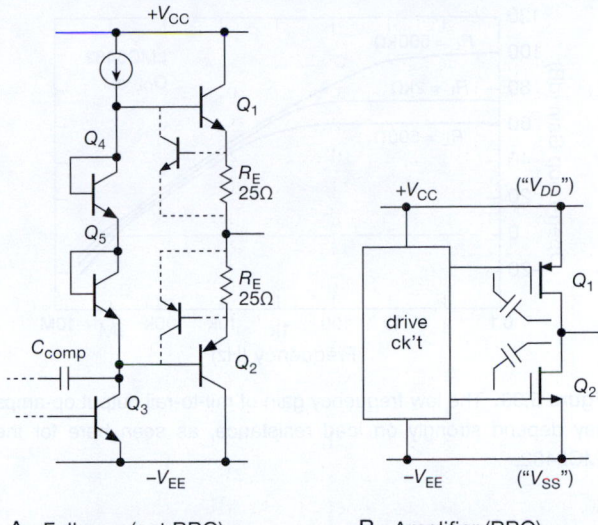

A. Follower (not RRO)   **B. Amplifier (RRO)**

Figure 5.32. The classic (not rail-to-rail) op-amp output stage is a push–pull unity-gain follower with inherently low output impedance, biased (via Q_4Q_5) to suppress crossover distortion; it has straightforward biasing and current limiting. By contrast, a rail-to-rail output stage (usually implemented in CMOS) is a push–pull common-source amplifier ($G>1$) with inherently high output impedance; it requires considerable trickery in its biasing and current limiting.

internal feedback around the output stage (the capacitors in Figure 5.32B), so that the gain and output impedances are reasonably well controlled except at low frequencies – see for example Figures 5.33 and 5.34.[27]

B. Saturation at the rails
Some "rail-to-rail output" op-amps (in particular, those with a BJT output stage) don't quite make it the last few millivolts; that's because the output transistor's saturation voltage is not zero. (This is not usually a problem with MOSFET outputs, which look like an $R_{\rm on}$ to one rail or the other when driven full range.) Usually this doesn't matter, because what you care about most is getting full use of a limited supply voltage (when operating with low-voltage supplies). But it does matter, for example, if you've got a single-supply setup in which the op-amp is driving an ADC whose conversion range goes clear down to ground.

In such a case be sure to check the specifications. Some

---

[27] It's unusual to see plots (or even tabulated values) of *open-loop* output impedance on datasheets; and in cases where a graph is shown, it rarely extends to very low frequencies. It is likely that other op-amps, including some with conventional (follower) output stages, also exhibit a rise in open-loop output impedance at very low frequencies. This is rarely of concern, though, owing to the very high loop gain down there.

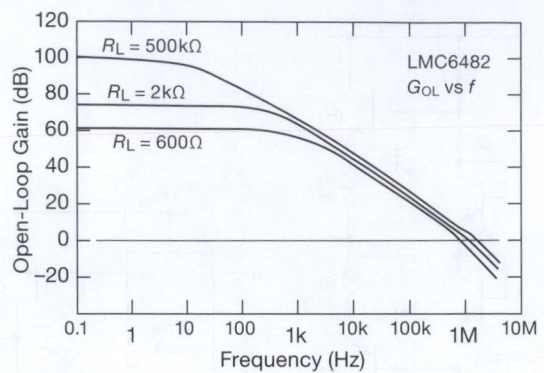

Figure 5.33. The low-frequency gain of rail-to-rail output op-amps may depend strongly on load resistance, as seen here for the LMC6482.

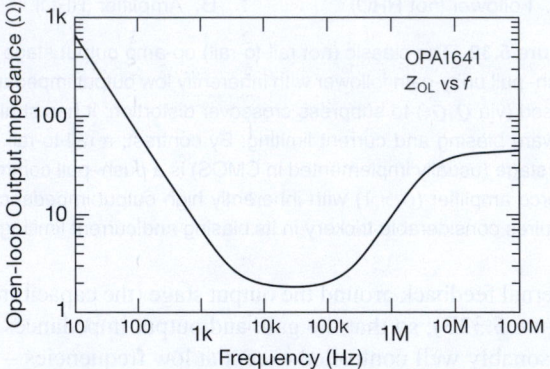

Figure 5.34. For some RRO op-amps the open-loop output impedance rises markedly at low frequencies, owing to internal capacitive negative feedback around the output stage that becomes ineffective at low frequencies. But, not to worry, there's lots of loop gain at low frequencies in typical op-amp applications.

RRO op-amps will warn you that the output will not reach the negative rail (e.g., 10 mV for the bipolar LT6003); others will instruct you to add an external pull-down resistor or current sink (e.g., the bipolar LT1077, which saturates to 3 mV with no pull-down, and 0.1 mV with a 5kΩ pull-down). Op-amps with clean MOSFET saturation will tell you not to worry – the unloaded output will go all the way to ground (e.g., $\lesssim$0.1 mV for the CMOS AD8616 or AD8691).

C. Distortion

The rail-to-rail output stage (Figure 5.32B) presents real challenges to the chip designer when it comes to quiescent biasing and reduction of crossover distortion. Despite heroic efforts, these amplifiers generally perform some 20–

40 dB worse than their conventional (non-RRO) counterparts in terms of distortion, as seen in the pair of plots in Figures 5.43 (non-RRO) and 5.44 (mostly RRO);[28] see also the SPICE plots in §4x.11.

D. Monticelli's output circuit

An elegant RRO circuit solution was devised by Monticelli,[29] and it is shown here in simplified form in Figure 5.35 (there's a full discussion in §4x.11). It has the effect of biasing the push–pull pair Q_1Q_2 in such a way that there is current overlap at crossover, and, better still, there is continuing current through both transistors *throughout the output swing*. We might call this "push–pull class-A" mode (though it seems to have been named already: "class-AA"). It is used, for example, in the CMOS OPA365 and in the BJT OPA1641. And it works – these parts have −114 dB and −126 dB harmonic distortion, respectively.

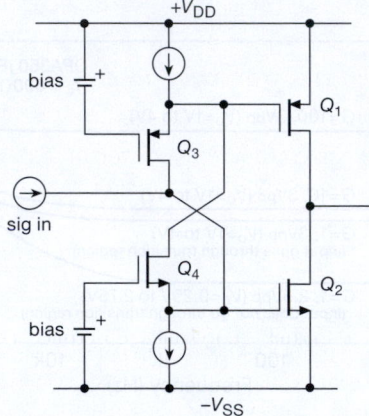

Figure 5.35. The Monticelli rail-to-rail output circuit.

Here's a capsule description of the Monticelli circuit's operation: first, think of Q_3 and Q_4 each as unity-gain current amplifiers whose source terminal is the "summing

---

[28] In fairness, we note that some of the poorer "distortion" results (which are actually THD+N – distortion plus noise) may be due to the lower supply voltages of the RRO op-amps, necessitating lower signal levels, thus causing noise to loom larger.

[29] See his patent US4570128, and his IEEE *JSSC* paper (SC-21, #6, 1986), in which he says "The output stage (Figure 8) must solve a level shifting problem that has plagued rail-to-rail designs for some time. Elaborate solutions have been proposed that combine multiple embedded feedback loops that are in effect op amps within op amps. To succeed as a general-purpose quad, a simpler solution had to be found." Although originally developed at NSC, this circuit (or close variations) is popular with op-amp designers at Analog Devices and at TI (even before it swallowed NSC).

junction" (because the gate is held at fixed voltage). Now imagine an increasing input signal current, which reduces the net current sunk at Q_4's source. This reduces its V_{GS}, which increases Q_2's V_{GS}, thereby increasing the output pull-down current. Meanwhile, the reduced drain current in Q_4 causes less of Q_3's source current to be diverted, thus increasing the V_{GS} of Q_3; that causes a reduction in Q_1's V_{GS}, and therefore a lower output pullup current. The overall quiescent current is set by the dc bias applied to Q_3 and Q_4. So it's a nicely balanced circuit, with a single-ended current input and a push–pull current output.

This is one cool circuit! In §4x.11 there's a more complete description, including SPICE simulations of a BJT implementation, and comparison with a conventional (not rail-to-rail) class-AB push–pull emitter follower. This inherently symmetrical circuit also works well with differential current drives to the drains of both Q_3 and Q_4, a configuration you'll often see.

5.10 Choosing a precision op-amp

If there's no such thing as a perfect operational amplifier, then that's especially true for precision op-amps. Although sufficient perfection may be achieved in a few parameters, the design tradeoffs required for achieving this invariably degrade other parameters. For example, if we need a very quiet medium-frequency op-amp, a world-class quiet IC, we won't be able to enjoy world-class low-input-bias currents.[30] That's because the amplifier will use bipolar input transistors, which will have to be operated at fairly high collector currents, and you know what that means for the base currents (e.g., look at the LT1028). Another example: if we want micropower operating current, we won't be able to enjoy world-class fast settling time, because we won't be able to have a high f_T and fast slew rates; that takes power, and lots of it.

In this section we take an in-depth look at the process of choosing a precision op-amp that is right for the job at hand, linked closely to a broad selection of exemplary parts in Tables 5.5 (pages 320–321) and 5.6 (on page 335). If you've got a circuit design that you've been struggling with, this section should hit the spot. The nitty-gritty level of detail that follows is essential to the careful design that distinguishes an excellent circuit from a compromise-ridden also-ran. For the casual reader, on the other hand,

the level of detail in the following treatment may be, well, "not superficial enough."[31]

As we begin our tour of precision op-amp parameters and their significance, we invite you to bury yourself in the data. With your circuit design goals in mind, start with an important op-amp parameter and look for the best choices. After zeroing in on a winning value, you can examine other parameters for that op-amp: do some of the other parameters for our winning op-amp now look like poor choices? Maybe your op-amp isn't a winner after all. Or perhaps you've got to return to your design goals and adjust them in accordance with reality, and repeat the process. Remember always that "engineering is the art of compromise."

5.10.1 "Seven precision op-amps"

Seven is a nice number, and, in preparation for the extended discussion of the very practical issue of choosing a precision op-amp, we provide in Table 5.5 (pages 320–321) a comparison of the important specifications for an updated listing of seven of our favorite precision op-amps. The problem is, we just couldn't restrict ourselves to a mere seven – it's closer to seven *dozen*! Spend some time with it (and check off your own seven faves!) – it will give you a good feeling for the trade-offs you face in high-performance design with op-amps. Note particularly the trade-offs of offset voltage (and drift) versus input current for the best bipolar and JFET op-amps. You also get the lowest noise voltage from bipolar op-amps, trending downward with increasing bias current; we'll see why that happens later in Chapter 8 when we discuss noise. The awards for low-input *current*, however, always go to the FET op-amps, again for reasons that will become clear later. In general, choose FET op-amps for low-input current and current noise; choose bipolar op-amps for low-input voltage offset, drift, and voltage noise.

Among FET-input op-amps, those using JFETs dominate the scene, particularly where precision combined with low noise is needed (but not *all* JFET op-amps: note that our jellybean favorites, the LF411/412, are not precise enough to qualify for membership in the table). That dominance is being challenged, though, by some low-voltage CMOS parts like the factory-trimmed MAX4236A and OPA376, and by parts like the TLC4501A that use tricks such as auto-zeroing at power-up.[32]

---

[30] In §8.6.3 we show a discrete op-amp circuit where both of these goals are achieved.

[31] A phrase lifted from a student's reply in an end-of-course questionnaire: "This course was not superficial enough for me."

[32] There was traditionally a problem peculiar to MOSFETs, which has been largely solved through process improvements. MOS transistors

Table 5.5 "Seven" precision op-amps (page 1: high voltage)

Part #	# per pkg	Supply Range (V)	Supply I_q (mA)	Input Current @25°C typ (pA)	Input Current @25°C max (pA)	V_{os} typ (µV)	V_{os} max (µV)	ΔV_{os} typ (µV/°C)	ΔV_{os} max (µV/°C)	CMRR min (dB)	V_{npp} dc (µV)	e_n 1kHz (nV/√Hz)	i_n 1kHz (fA/√Hz)	GBW typ (MHz)	Slew typ (V/µs)	Settle (µs)	C_{in} pF	Cost qty 25 ($US)	Comments
HV bipolar																			
LT1077A	1	2.2-44	0.05	7 nA	9 nA	9	40	0.4	1.6	97	0.5	27	65	0.23	0.08	-	low	3.84	single-supply
LT1490A	2,4	2.5-44	0.04	1 nA	8 nA	110	500	2	4	84	1	50	15	0.2	0.07	-	4.6	3.25	over-the-top
AD8622A	2,4	4-36	0.22	45 nA	200 nA	10	125	0.5	1.2	125	0.2	11	150	0.56	0.48	-	5.5	5.33	single-supply
LT1013[f]	2,4	3.4-44	0.35	12 nA	20 nA	40	150	0.4	2	100	0.55	22	70	0.7	0.4	16	low	3.13	single-supply
OPA277	1,2,4	4-36	0.79	0.5 nA	1 nA	10	20	0.1	0.15	130	0.22	8	200	1	0.8	16	low	3.17	improved OP-27
TLE2141A	1,2,4	4-44	3.5	0.7 µA	1.5 µA	175	500	1.7	-	85	0.5	10.5	1900	5.9	45	0.4	low	1.15	cross-coupled slew
LT1677	1	3-44	2.8	2 nA[e]	20 nA[e]	20	60	0.4	2	109	0.09	3.2	1200	7.2	2.5	5	low	3.07	an "RRIO LT1007"
AD8675	1,2	9-36	2.5	0.5 nA	2 nA	10	75	0.2	0.6	114	0.1	2.8	300	10	2.5		4.2	2.22	0.6ppm dist, dual=76
OPA2209	1,2,4	4.5-40	2.2	1 nA	4.5 nA	35	150	1	3	120	0.13	2.2	500	18	6.4	2.1	-	4.04	0.25ppm dist, SOT23
LT1007[g]	1	4-44	2.7	10 nA	35 nA	10	25	0.2	0.6	117	0.06	2.5	400	8	2.5	-	4	2.48	'37 decomp 60MHz
ADA4004	1,2,4	9-36	2.2	40 nA	90 nA	40	125	0.7	1	110	0.15	1.8	3500	12	2.7	-	4	4.20	family, SOT23
LT1468	1	7-36	3.9	3 nA	10 nA	30	75	0.8	2.2	96	0.3	5	600	90	23	0.8	4	4.26	0.7ppm dist
AD8597	1,2	9-36	4.8	40 nA	210 nA	10	120	0.2	0.8	120	0.08	1.1	4300	10	16	2.0	12	3.71	1ppm dist
LT1028A[h]	1	8-44	7.4	25 nA	90 nA	10	40	0.2	0.8	108	0.04	0.85	4700	75	15	-	5	6.48	lower I_B than AD697
HV superbeta																			
LT6010A[n]	1,2,4	2.7-40	0.14	20	110	10	35	0.2	0.8	107	0.4	14	100	0.35	0.11	45	4	2.22	replace LT1012, w/RRO
LT1012AC[m]	1	2.4-40	0.37	25	100	8	25	0.2	0.6	114	0.5	14	(20)	0.5	0.2	10	low	5.11	has overcomp pin
OP97E	1,2,4	4.5-40	0.40	30	100	10	25	0.2	0.6	114	0.5	14	(20)	0.9	0.2	0.9	low	5.52	dual='297, quad=497
AD706	2,4	4-36	0.8	50	200	30	100	0.3	1.5	106	0.5	15	50	0.8	1.5	1.0	-	4.05	AD704 quad
LT1884A	2,4	3.5-40	0.85	150	400	25	50	0.3	0.8	114	0.4	9.5	50	2	0.9	10	2	5.23	LT1882 slower, lower I_B
HV JFET																			
AD795	1	8-36	1.3	1	2	100	500	3	10	90	1	9	0.6	1.6	1	11	2.2	7.97	replace OPA111
OPA124PB	1	10-36	2.5	0.35	1	100	250	1	2	100	1.6	8	0.5	1.5	1.6	10	3	6.40	substrate pin
OPA140	1,2,4	4.5-40	1.8	0.5	10	30	120	0.35	1	126	0.25	5.1	0.8	11	20	0.9	10	3.75	0.5ppm dist
AD711C	1,2	4.5-40	2.5	15	25	100	250	2	5	86	0.25	16	10	4	20	1.0	5.5	2.19	AD712 low-cost dual
LT1055C[x]	1	20-40	2.8	10	50	120	700	3	12	85	0.6	15	1.8	4.5	12	1.8	4	2.52	1057 dual, 1058 quad
ADA4000	1,2,4	5-36	2	5	40	200	1700	2	-	80	1	16	10	4.5	20	-	5.5	1.46	jellybean, AD711 subst
OPA192	1,2,4	4.5-40	1	5	20	5	25	0.2	0.5	120	1.3	5.5	1.5	10	20	0.9	0.6	3.87	CMOS, e-Trim™
OPA134	1,2,4	5-36	4	5	100	500	2000	2	-	86	-	8	3	8	20	1		1.60	0.8ppm dist, jellybean
OPA1641	1,2,4	4.5-40	1.8	2	20	1000	3500	-	-	120	-	5.1	0.8	20	11		8	2.76	0.5ppm dist, family
AD8620A	1,2	10-27	2.5	2	10	85	250	0.5	1	90	1.8	6	5	25	60	0.6	15	11.86	caution, ±12V max
OPA827	1	8-40	4.8	15	50	75	150	1.5	-	104	0.25	3.8	2.2	22	28	0.55	9	9.00	cheaper than 627
OPA627B	1	10-36	7	5	50	40	100	0.4	0.8	106	0.6	4.5	1.6	16	55	0.55	7	30.00	ADA4627B 2nd-source
OPA637B	1	10-36	7	1	5	40	100	0.4	0.8	106	0.6	4.5	1.6	80	135	0.45	7	30.00	ADA4637B 2nd-source
AD549KH	1	10-36	0.6	75 fA	0.1	150	250	2	5	90	4	35	0.5	3	3	5	1	28.00	-L=60fA, TO-99 pkg
OPA129B	1	10-36	1.2	30 fA[z]	0.1	500	2000	3	10	80	4	17	0.1	2	2.5	-	2	12.00	lowest I_B for HV part
HV chopper																			
LTC1150	1	4.8-32	0.8[s]	10	100	0.5	10	0.01	0.05	110	1.8	high	(1.8)	2.5	3	-	n	6.00	only HV AZ w/int caps
ADA4638	1	4.5-33	0.85	45	90	6	4.5	-	0.08	130	1.2	66	100	1.5	1.5	4	9		5kHz noise spike
OPA2188	1,2,4	4-40	0.42	160	850	6	25	0.03	0.09	120	0.25	8.8	750	2	0.8	27	9.5		new

Table 5.5 "Seven" precision op-amps (page 2: low voltage)

Part #	# per pkg[a]	Supply Range (V)	I_Q (mA)[p]	Iin typ (pA)[t]	Iin max @25°C (pA)	V_{os} typ (µV)	V_{os} max (µV)	ΔV_{os} typ (µV/°C)	ΔV_{os} max (µV/°C)	CMRR min (dB)	V_{npp} dc (µV)[b]	e_n 1kHz (nV/√Hz)	i_n 1kHz (fA/√Hz)[k]	GBW typ (MHz)	Slew typ (V/µs)	Settle (µs)[d]	dist graph	C_{in} (pF)	IN+	IN−	OUT+	OUT−	null pins	comp pin	shdn pin	DIP avail	Cost qty 25 ($US)	Comments
LV bipolar																												
LT6003	1,2,4	1.5-18	0.001	40	140	175	500	2	5	73	3	325	12	0.003	0.001	slow		6	•	•	•	•					1.60	SOT-23, '6004 dual
EL8176	1	2.4-6	0.055	0.5nA	2nA	25	100	0.7	-	90	1.5	28	160	0.4	0.13	-		-	•	•	•	•			•		2.30	SOT-23, charge pump
LMP7731	1,2	1.8-6	2.2	1.5nA	30nA	6	500	0.5	5.5	101	0.08	2.9	1100	22	2.4	0.3		2		•	•	•					1.98	SOT-23, 3nV/√Hz
LT6220	1,2,4	2.2-13	0.9	15nA	150nA	70	350	1.5	5	85	0.5	10	800	60	20	0.3		2	•	•	•	•					1.75	SOT-23
LT6230	1,2,4	3-12.6	3.3	5µA	10µA	100	500	0.5	3	96	0.18	1.1	2400	215	70	0.05		7			•	•		•			2.72	-10=decomp, 1.3GHz
LV JFET																												
OPA656	1	9-13	14	2	20	250	1800	2	12	80	-	7	1.3	230	290	0.02	•	2.8			•	•					5.59	'657 for 1.6GHz G>5
LV CMOS																												
LMP2232A	1,2,4	1.8-6	0.01	0.02	1	10	150	0.3	0.75	81	2.3	60	0.1[c]	0.13	0.06	slow		-	•	•	•	•					3.29	'31 single, '34 quad
TSV611A	1,2	1.8-6	0.01	1	10	-	800	2	-	61	-	105	-	0.12	0.04	-		-	•	•	•	•					0.82	'612 dual
AD8603	1,2,4	1.8-6	0.04	0.2	1	12	50	1	4.5	85	2.3	25	50	0.4	0.1	23		-	•	•	•	•					1.36	SOT-23, 1pA
LTC6078	2,4	2.7-6	0.055	0.2	1	7	25	0.2	0.7	95	1	18	0.25[c]	0.75	0.05	24		23	•	•	•	•					3.54	V_{os} degrades near V+
LTC6081	2	2.7-6	0.33	0.2	1	-	70	0.2	0.8	93	1.3	13	0.5	3.6	1	6		18	•	•	•	•					4.75	degrade $V_{cm}>V_{cc}-1.5$V
MAX4236A	1	2.4-6	0.35	1	500	5	20	0.6	-	84	0.2	14	0.6	1.7	0.3	1		7		•	•	•					1.78	not AZ, '4237 decomp
LMC6482A	2,4	3-16	0.5	0.02	4	110[v]	750	1	-	70	-	37	30	1.5	1.3	2		7.5	•	•	•	•				•	1.88	'7101=SOT-23
OPA376	1,2,4	2.2-7	0.76	0.2	10	5	25	0.26	1	76	0.8	7.5	0.25[c]	5.5	2	2		3	•	•	•	•					1.32	CMOS etrim™
OPA364	1,2	1.8-5.5	0.85	1	10	-	500	3	-	74	10	17	0.6	7	5	1.5		13	•	•	•	•			•		2.18	chg-pump, 20ppm dist
TLC4501A	1,2,4	4-7	1	1	60	10	40	1	-	90	1.5	12	0.6	4.7	2.5	2.2		3		•	•	•				•	3.06	0.3s self-cal at pwr-on
OPA743	1,2,4	3.5-13	1.1	0.05	10	1500	7000	8	4.5	66	11	30	2.5	7	10	15		8	•	•	•	•				•	1.58	jellybean
LMP7715	1,2	1.8-6	1.15	0.05	1	10	150	1	4	85	-	5.8	10	17	9.5	-		4	•	•	•	•			•		2.05	decomp, G>10
LMP7717	1,2	1.8-6	1.15	0.05	1	10	150	1	4	85	-	6.2	10	88	28	1		15	•	•	•	•					2.18	0.6ppm dist
AD8692	1,2,4	2.7-6	0.85	0.2	1	400	2000	0.3	6	68	1.6	8	50	10	5	1		15	•	•	•	•					1.36	jellybean, '15 SOT-23
AD8616	1,2,4	2.7-6	1.7	0.2	1	23	60[f]	1.5	7	80	2.4	7	50	24	12	0.5		5	•	•	•	•					1.52	jellybean, '15 SOT-23
OPA350	1,2,4	2.5-7	5.2	0.5	10	150	500	4	7	74	4	5.8	4	38	22	0.5		7	•	•	•	•			•		1.67	0.6ppm dist
OPA380	1,2	2.7-7	7.5	3	50	4	25	0.03	0.1	100	3	22	10	90	80	2	•	6.5	•	•	•	•					5.39	transimp, w/auto-zero
LMC6001A	1	5-16	0.45	10fA	25fA	-	350	2.5	10	83	-	22	0.13	1.3	1.5	-		3								•	12.19	
LMP7721	1	1.8-6	1.3	3fA[y]	20fA	26	150	1.5	4	83	-	7	10	17	10	-	•	-	•	•	•	•					11.89	very low e_n for 20fA!
LV chopper																												
MAX9617	1	1.6-6	0.06	10	140	0.8	10	0.01	0.12	116	0.42	42	100	1.5	0.7	-		-	•	•	•	•					1.60	charge pump
AD8638	1,2	5-16	1	1.5	40	3	9	0.01	0.06	118	1.2	60	noisy	1.35	2.5	3		4	•	•	•	•			•		3.31	auto-zero; SOT23
LTC2050H	1	2.7-11	0.8	7	50	0.5	3	-	0.03	120	1.5	3	2	3	2	-		1.7	•	•	•	•			•	•	2.19	replaces LTC1050
AD8551	1,2,4	2.7-6	0.7	10	50	1	5	0.01	0.04	120	1	42	(2)	1.5	0.4	50	•	-	•	•	•	•					2.34	
OPA735	1,2	2.7-13	0.6	100	200	1	5	0.01	0.05	115	2.5	135	40	1.8	1.5	-		10	•	•	•	•					3.38	auto-zero, "chopper"

Notes: (a) boldface indicates number in a package for the part number listed. (b) 0.01Hz–10Hz or 0.1Hz–10Hz. (c) calculated. (d) usually to 0.01%. (e) for $V_{EE}+1.4$V $< V_{CM} < V_{CC}-0.7$V. (f) LTC suggests LT1490/1. (g) LTC suggests LT1677. (h) at 1kHz or 10kHz (i.e., above the 1/f corner), except 10Hz for chopper op-amps. (k) at 1kHz or 10kHz. LTC suggests LT6200/30. (m) LT1097 cheaper. (n) dual & quad have degraded I_B and V_{os}; '6013=decomp. (o) values in parenthesized (*thin italics*) should not be relied upon; measured values often are as much as 5x–100x larger; see discussion in Chapter 8. (p) per amplifier. (s) can be reduced to 200µA. (t) typical. (v) V_{os} is insensitive to V_{CM}. (x) the (hard-to-get) -A version has V_{os}=50µV typ, 150µV max. (y) special pinout for guard. (z) special pinout for guard+substrate pin.

Finally, the so-called *chopper-stabilized* (here and in Table 5.6 on page 335) amplifiers form the most important exception to the generalization that FET op-amps, particularly MOSFET types, suffer from larger initial offsets and much larger drifts of V_{OS} with temperature and time than do bipolar-transistor op-amps. In fact, these devices (known also as *auto-zero* or *zero-drift* amplifiers) are the amplifiers with the *smallest* offset voltage and drift, typically in the $\pm 1\,\mu\text{V}$ and $\pm 0.05\,\mu\text{V}/^\circ\text{C}$ range. They use MOSFET analog switches and amplifiers to sense, and correct, the residual offset error of an ordinary op-amp (which itself is often built with MOSFETs, on the same chip). This is not without compromise, however: chopper-stabilized amplifiers have some unpleasant characteristics that make them unsuited for many applications, as we'll see in §5.11.

5.10.2 Number per package

The first column in Table 5.5 gives the choices available for the number of devices per package (the number in **boldface** shows which choice matches the part number). We generally list single op-amp parts, even though in practice the dual op-amps are more useful and popular (in some cases distributors don't even stock the single types). Special features such as pins for external offset-nulling, compensation, and shutdown are available only for the single op-amp package types and are indicated in the right-hand columns. Generally the specs are identical for the different dies and appear on the same datasheet, but not always.

5.10.3 Supply voltage, signal range

It's likely your first consideration will be supply-voltage range and signal levels. High-voltage parts (able to operate from ± 15 V, i.e., 30 V total) are listed first in the table,

with parts ordered more or less by I_Q, the quiescent supply current, in each category. Battery-powered applications benefit from low supply currents, but some low-drift applications do as well, because op-amp self-heating temperature effects will be less. Some parts offer a version with a power-shutdown (SHDN) pin. For example, the LT6010 current drops from $135\,\mu\text{A}$ to $12\,\mu\text{A}$ in shutdown (but a *gotcha*: the shutdown pin itself takes another $15\,\mu\text{A}$) and it takes $25\,\mu\text{s}$ to turn ON or OFF. Other parts do better, e.g., the OPA364 draws $0.9\,\mu\text{A}$ when off.

Circuits operating with high-voltage supplies benefit from using high signal levels, such as ± 10 V full scale. An offset voltage, say $V_{OS}=40\,\mu\text{V}$, is a smaller fraction of 20 Vpp than it is of 0 to 4 V. With the exception of chopper op-amps, you don't get any offset-voltage improvement for low-voltage parts.

Low-voltage parts finish the table (most are 5.5 V maximum total supply, but some permit 11 V or higher, suitable for ± 5 V operation), but it's important to realize that many "high-voltage" parts are designed and specified to work well at low voltages, even under 3 V. Some work well with ± 3 V to ± 5 V supplies and should not be rejected simply because they can also work at higher voltages. But be warned, you need to examine the common-mode input range and output-swing range. For example, although a 44 V op-amp like the LT1490 works with 3 V supplies and allows rail-rail inputs and outputs, another fine 44 V LTC op-amp, the low-noise LT1007 (which works down to 4 V) is limited to inputs and outputs no closer than 2 V from the rails – nearly useless when running from ± 2 V. Clearly it's not meant to be a low-voltage part. Your quick guide to these issues is the "swing to supply" columns; the LT1490 has all four checks whereas the LT1007 has no checks.

5.10.4 Single-supply operation

If you're running with low supply voltages, you may want to use a single-supply power arrangement. Op-amps capable of single-supply use have at a minimum the ability to operate their inputs and outputs to the negative rail (i.e., ground). Many permit operation with outputs also to the positive rail, and claim rail-to-rail outputs on the front page of the datasheet. But be warned, there's usually a performance degradation when outputs are near the supply rails. Some op-amps offer zero-volt or below-the-rail operation if you add a pulldown resistor.[33]

Seven high-voltage op-amps on our list offer single-

---

are susceptible to a unique debilitating effect that neither FETs nor bipolar transistors have. It turns out that sodium-ion impurity migration and/or phosphorus polarization effects in the gate insulating layer can cause offset voltage drifts under closed-loop conditions, in extreme cases as much as 0.5 mV over a period of years. The effect is increased for elevated temperatures and for a large applied differential-input signal, with some datasheets showing a typical 5 mV change of V_{OS} over 3000 hours of operation at 125°C with 2 V across the input. This sodium-ion disease can be alleviated by introducing phosphorus into the gate region. Texas Instruments, for example, uses a phosphorus-doped polysilicon gate in its "LinCMOS" series of op-amps (TLC270-series) and comparators (TLC339 and TLC370-series). These popular inexpensive parts come in a variety of packages and speed/power selections and maintain respectable offset voltages with time (50 μV eventual offset drift per volt of differential input).

---

[33] Many op-amps can do this, but without saying so. That's because their pullup transistor and drivers work all the way down to the negative

supply operation; some, like the LT1013, excel at it. Two offer full rail-to-rail input-output, or RRIO, operation. The fast-slewing TLE2141 is especially interesting (fast settling, but high bias current), as is the low-noise LT1677 (lower bias current, but slow slewing and settling). All but two of the low-voltage parts offer single-supply operation.

There are some precision op-amps with low supply currents, down to 10–60 μA, although this will severely limit your choices for other parameters. There's even a respectable 0.85 μA (and 1.8 V) op-amp, the LT6003. Some op-amp types, such as the JFETs, don't offer any low-power parts. Nonetheless you might choose one for its low noise and low bias current.

5.10.5 Offset voltage

Perhaps the single parameter most often associated with precision amplifiers is input-voltage error. To measure small offset voltages, use the op-amp's gain to magnify the effect, as shown in Figure 5.36. Offset voltage was our required parameter to gain entry into the table; few op-amps with maximum offset voltages above 250 μV made the grade. There are plenty of parts with <10 μV *typical* offset voltages, but "typical" isn't a reliable spec when you're in the business of manufacturing precision instruments. Bipolar input stages have an advantage over JFET and CMOS in the table, but they suffer from higher input bias currents. The "superbeta" parts are a pleasant exception, especially at high temperatures (see Figure 5.38), but none of these parts has inputs that operate to either supply rail.

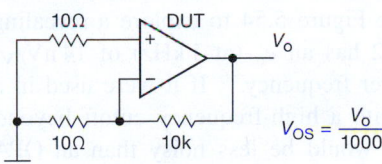

Figure 5.36. Offset-voltage test circuit. The ×1000 voltage gain makes submillivolt offsets in the device under test (DUT) easily measurable. Input *current* effects are negligible, owing to the small (10 Ω) resistances seen at the op-amp's inputs. Add a 200 Ω resistor at the output if you want to drive a cable.

Many low-offset parts have disappeared since the second edition of our book, especially in the JFET category. They've become too expensive for the market and have lost the competition with low-voltage chopper and auto-zero op-amps. The latter have a few token representatives in this

table, but they have their own selection table (Table 5.6 on page 335) that you should examine if they look good for your design. They have their own problem issues, such as current noise, which we discuss in §5.11.

Offset voltage variation with common-mode input voltage is a serious issue for some parts, especially for RRIO op-amps (see Figure 5.29), and is not addressed on the table. It's always important to follow up an initial choice from the table with a careful examination of the datasheet. For example, the OPA364 and MAX9617 use internal charge pumps to power their input stages, eliminating this problem completely.

Offset voltage drift with temperature is an important parameter when stability of measurements matters. This parameter is not production tested. The maximum drift spec may not be very reliable, and some manufacturers have stopped providing such a spec.

Offset voltage drift with time is a parameter that used to appear in precision op-amp datasheets, with values of order 300–400 nV/month; a few high-performance parts like the LT1007 claimed 200 nV/month, and chopper-stabilized op-amps generally claim drifts of 50 nV/month.[34] This is somewhat uncharted territory, and some people claim that drift slows with time, or perhaps it is more akin to a random walk, in either case suggesting that a drift specification should perhaps have units of nV/$\sqrt{\text{month}}$.

5.10.6 Voltage noise

Voltage noise is the in-band variation of op-amp input–offset voltage that's indistinguishable from signal. It's useful to view it as a "noise spectral density" function $e_n(f)$, which tells you the rms noise voltage in a 1 Hz bandwidth (see §8.2.1) centered at frequency f. Figure 5.37 shows an idealized plot of input voltage-noise density as a function of frequency for some of the op-amps in the table. For most op-amps $e_n(f)$ is essentially flat for frequencies above its "$1/f$ corner frequency," with e_n increasing below the corner frequency, approximately as $1/\sqrt{f}$. (Auto-zero, or "chopper-stabilized," op-amps are not shown. They behave differently, because low-frequency "noise" is removed by the auto-zero process, so their e_n is flat at low frequencies. We'll discuss them shortly; see §5.11.)

The voltage–noise column in Table 5.5 (pages 320–321) shows e_n at its commonly specified frequency of 1 kHz,

---

[34] Although it's possible to do considerably better, for example the measured 6 nV/month reported by Bob Pease for the LMP2011. See Table 8.3 and Figures 8.60, 8.61, 8.110, and 8.110. The IF3602 is a large-geometry dual JFET available from InterFet, shown for comparison.

---

rail, without sourcing current to the output. Some require a minimum pull-down current, e.g., 0.5 mA for the OPA364.

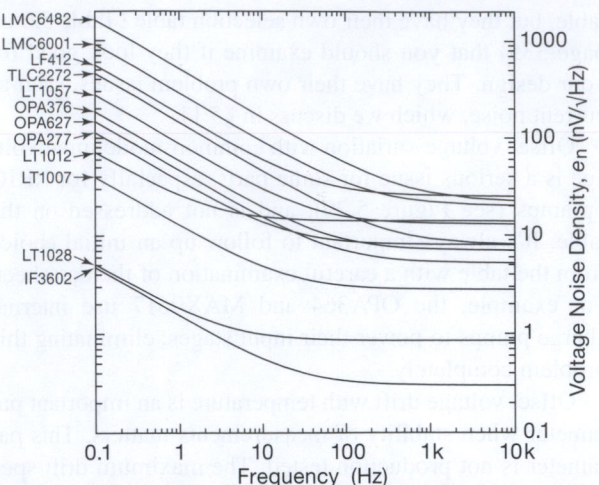

Figure 5.37. Voltage noise density e_n for a selection of representative op-amps, showing the increase of noise power below the $1/f$ corner frequency. Some op-amps that have good specs at 1 kHz don't look so good at 0.1 Hz. Figure 5.54 shows the resulting voltage noise v_n when such a noise density is integrated over frequency.

comfortably in the flat region above the $1/f$ corner of most op-amps; e_n varies from $0.85\,\mathrm{nV}/\sqrt{\mathrm{Hz}}$ for the high-current LT1028 to $325\,\mathrm{nV}/\sqrt{\mathrm{Hz}}$ for the current-starved LT6003.

We discuss noise in greater detail below, and in Chapter 8, but let's start with the simple relationship relating voltage-noise *density* e_n (given in units of $\mathrm{nV}/\sqrt{\mathrm{Hz}}$) to the value of *integrated* total voltage noise V_n (in units of nV, or μV; and either rms or peak-to-peak) over some frequency passband. In the flat (i.e., white noise) region of e_n at frequencies above the $1/f$ corner, the integrated noise voltage is simply $V_n = e_n\sqrt{\mathrm{BW}}$.

As we'll see, for circuits with bandwidths of 1–10 kHz or more, the noise voltage V_n is dominated by the noise density at high frequencies. Datasheets for most op-amps give an e_n value at 1 kHz, but some also specify it at 10 kHz, 100 kHz, or even 1 MHz; and they usually provide graphical plots of e_n versus frequency. Because the integrated noise voltage is dominated by e_n at the high-frequency end of the operating range, be sure to use that value (or a frequency-weighted eyeball average) in the simple formula above. Using a high-frequency e_n value is particularly important for transimpedance amplifiers, which suffer from "$e_n\omega C$" current noise ($i_n = e_n 2\pi f C_{in}$) at high frequencies.

A. "1/f" noise

We discuss $1/f$ noise in greater detail in Chapter 8 (where we show how to determine the $1/f$ noise corner, etc., in §8.13.4, but here we address the practical question "What effect do the scary-looking rising noise-density curves in Figure 5.37 have on my circuit noise?" The noise density e_n is indeed higher at low frequencies, but that density gets multiplied by a smaller frequency span. Put another way, the total noise *voltage* (as contrasted with the noise *density* e_n) contributed by an op-amp depends both on its e_n and on the circuit's bandwidth. More precisely, the mean square noise voltage is the integral of e_n^2 over the bandpass:

$$v_n^2 = \int_{f_a}^{f_b} e_n^2(f)\,df,$$

where $e_n(f)$ is the noise spectral density (often plotted in datasheets), and the bandpass (or observation band) extends from f_a to f_b. We then get the rms noise voltage by taking the square root of v_n^2.

We perform the integrations and show the devastating effects of $1/f$ noise in the integrated-noise plots later, in Figure 5.54 (in the context of auto-zero op-amps). More bandwidth means more noise, and all of the curves rise as $\sqrt{f}$ at the high end. The op-amps are ranked in order by their high-frequency e_n values; it's interesting to compare their positions in Figure 5.37 with the rankings in Figure 5.54. At the low frequency end, the noise voltage of conventional op-amps levels off, because their rising $1/f$ noise density makes up for the reduced bandwidth,[35] whereas the noise voltage of the auto-zero amplifiers continues its downward trend.

Let's use Figure 5.54 to explore a revealing example. The LT1012 has an e_n (at 1 kHz) of $14\,\mathrm{nV}/\sqrt{\mathrm{Hz}}$, and a 2.5 Hz corner frequency.[36] If it were used in a precision amplifier with a high-frequency cutoff beyond 1 Hz, for example, it would be less noisy than an OPA277, even though the latter has a lower noise density at 1 kHz of $8\,\mathrm{nV}/\sqrt{\mathrm{Hz}}$, because the latter has a 20 Hz noise corner (indicated by black dots). But the OPA277 wins back some respect when we see that it's dramatically quieter than a competing $9\,\mathrm{nV}/\sqrt{\mathrm{Hz}}$ part, the TLC2272, which suffers from a much higher 330 Hz noise corner.[37]

---

[35] Well, not exactly: if the noise power density were truly to continue rising as $1/f$, the integral would diverge at zero frequency (dc). For Figure 5.54 we set the low frequency limit to be 0.01 Hz.

[36] How to determine the corner frequency? See the discussion in §8.13.4.

[37] The LT1012 and OPA277 are BJT parts, and the TLC2272 is a CMOS op-amp. The '2272 boasts a miniscule 60 pA max bias current, far better than the '277's 1 nA, but not much better than the '1012's amazing 100 pA.

The integrated-noise plots are revealing, but it's helpful to have a single number in the table to evaluate our op-amps. The "$V_n$pp" parameter in the table is the peak-to-peak voltage noise over a 0.1–10 Hz band. This shows an op-amp's "dc noise" as seen in the flat part of the curves in Figure 5.54. The values range up to 11 μVpp (but parts like the LMC6482 avoid the competition by not listing any spec at all). The LT1028 is the winner with 35 nVpp, but the LMP7731 is a notable part at 80 nVpp, with its RRIO capabilities and its SOT-23 package. The ADA4075, with its 1 mV offset, didn't make the precision table, but its 60 nVpp noise level is attractive.

Precision op-amps that suffer from $1/f$ noise (i.e., all except auto-zero types) have a $V_n$pp spec whose lower frequency limit is usually 0.1 Hz. If you need a lower starting frequency (such as 0.01 Hz, used in the plots), multiply the low-frequency noise-voltage value listed by the square root of the number of additional low-end decades you want (that's an interesting $1/f$ noise factoid). As long as the $1/f$ corner is multiple decades higher, you can ignore the white-noise contribution to the spectrum.

The $V_n$pp parameter is your primary clue about an op-amp's long-term drift performance.

5.10.7 Bias current

The available input bias currents range from femtoamps to microamps (nine orders of magnitude!). In some applications this is the parameter that rules out entire classes of op-amp choices. Parts featuring very low typical input currents often have unimpressive *maximum* specs; this is due to the difficulty and expense of automated testing at currents below 10 pA or so. For example, the $1.88 low-cost LMC6482A CMOS op-amp has a 20 fA "typical" spec, but the datasheet shows a 4 pA maximum value – that's 200× worse.[38] However if you're willing to pay over $10, you can get an LMP7721 with a 20 fA maximum spec.

As we've said earlier (and will say again), the "bias" current of JFET and CMOS op-amps is a *leakage* current, and it increases exponentially with temperature; see Figure 5.38. That's the bad news. The *good* news (as we've also said before) is that there are some op-amps (like the LT1012 and AD706) that have low JFET-like input currents, but which have BJT inputs and therefore enjoy better

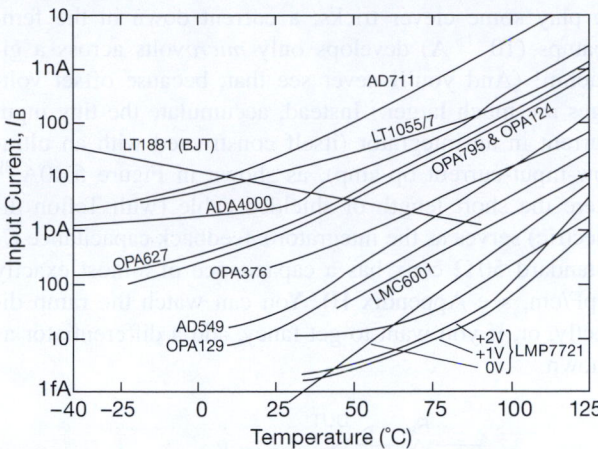

Figure 5.38. Input current versus temperature for a representative set of op-amps from Table 5.5 on pages 320–321, taken from manufacturers' datasheets. See also Figures 5.6 and 3.48.

high-temperature performance and improved offset voltages and drift, see Figure 5.6.

Low-input-current op-amps generally have higher offset voltage and offset voltage drift, and they're usually more noisy. The OPA627 and ADA4627 JFET op-amps are exceptions, and they've served us well, but they're expensive. Happily the new JFET OPA827 has lower noise and offers some price relief. The AD743, not in the table because of its 1 mV offset voltage, sports a 2.9 nV/$\sqrt{\text{Hz}}$ spec. Looking at CMOS parts (which are low voltage), we find that the LMP7715 is the best low-noise contender, at 5.8 nV/$\sqrt{\text{Hz}}$, but a low-cost part like the AD8616, with 1 pA bias and 7 nV/$\sqrt{\text{Hz}}$ noise, may offer a good compromise.

High-speed op-amps often have high input bias currents, typically 200 nA to as much as 20 μA. They also tend to have high offset voltages, above 0.5 mV, so most parts in this category didn't even make it into the precision table, but appear in the high-speed op-amp category in Table 5.4 on page 310. TI's fast OPA656 and '657 low-voltage JFET op-amps, with 2 pA typical bias current, 290 V/μs slew rate, and 20 ns settling time, demanded and were granted residence in both tables. The OPA380 is a 90 MHz op-amp meant for fast transimpedance applications, featuring 50 pA and 25 μV maximum offset. This part uses an auto-zero circuit to achieve 25 μV offset, but avoids excess current noise with an isolating filter (Figure 5.41).

A. Measuring bias current

To measure input currents (or offset currents) down to the nanoamp level or so, you can use the simple circuit in Figure 5.39. For *really* small currents, though, you've got

---

[38] Note 13 on the LMC6482 datasheet says "Guaranteed limits are dictated by tester limitations and not device performance. Actual performance is reflected in the typical value." That's illuminating, but not entirely helpful for the designer of a mass-production instrument.

to play some clever tricks: a current down in the femtoamps (10^{-15} A) develops only *micro*volts across a gigaohm! (And you'll never see that, because offset voltages are much larger.) Instead, accumulate the tiny input current in an integrator (itself constructed with an ultra-low-input-current op-amp), as shown in Figure 5.40A.[39] Here the short length of shielded cable (with Teflon dielectric) serves as the integrator's feedback capacitance C_1 (standard $50\,\Omega$ coax has a capacitance of almost exactly 1 pF/cm, see Appendix H). You can watch the ramp directly, or, if you want to get fancy, add a differentiator as shown.

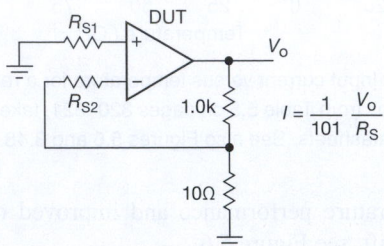

Figure 5.39. Input current (set one of the R_s's to zero) and input offset current ($R_{s1} = R_{s2}$) test circuit. Use R_s values large enough so that the voltage developed across them is at least some 10s of millivolts, so that errors due to offset voltage can be ignored. Add a $200\,\Omega$ resistor at the output if you want to drive a cable.

A somewhat simpler method, which has worked well for us, is to wire the op-amp as a follower, with a small capacitor from the (+) input to ground (Figure 5.40B); the op-amp's input current then generates an input ramp, faithfully reproduced at the output. At first we struggled with memory effects in mica and film capacitors, but we finally settled on an air (variable) capacitor, of the sort that were used to tune AM radios in the good ol' days. With the capacitor set to 365 pF we got an output ramp of 0.20 mV/s from an LMC6482, and thus an input current of 73 fA. You can "reset" this circuit by collapsing the supply rails. Be sure to put the whole business in a metal box: these open inputs are really sensitive beasts!

---

[39] Based on nice techniques worked out by Paul Grohe and Bob Pease at National Semiconductor. See Paul Rako's article "Measuring Nanoamperes" (*EDN*, 26 April 2007), and two riffs by Pease from his series in Electronic Design: "What's All This Teflon Stuff, Anyhow?" (14 Feb 1991) and "What's All This Femtoampere Stuff, Anyhow" (2 Sept 1993). Nicely readable versions currently available at http://electronicdesign.com/test-amp-measurement/whats-all-teflon-stuff-anyhow and .../whats-all-femtoampere-stuff-anyhow.

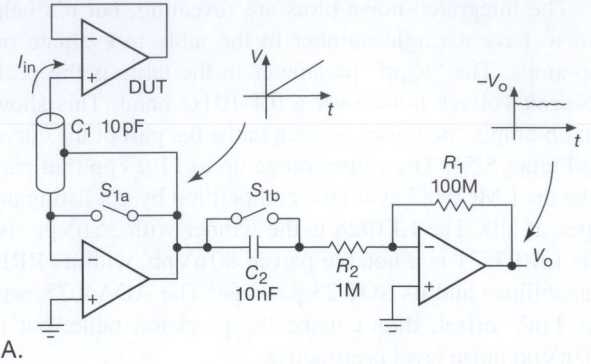

A.

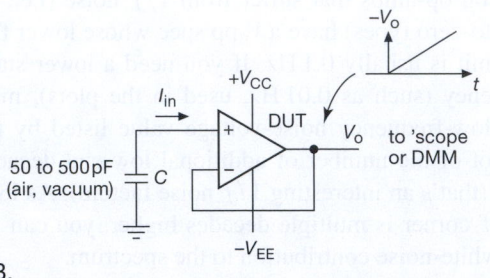

B.

Figure 5.40. Integrate the input current to measure in the picoamp range (and below). A. With a separate MOS-input op-amp integrator whose input current is in the low femtoamps (e.g., an LMP7721, I_B=3 fA typ, 20 fA max). Use electromechanical relays (not MOS switches) for S_1, for example the COTO 9202 series shown in Figure 5.3. B. More simply, let the device's input current charge a small capacitor, and observe the ramp at the G=1 output.

5.10.8 Current noise

The op-amp's input current noise density i_n flows through the source impedance seen at the amplifier's input terminals, contributing an equivalent noise voltage density $i_n Z_s$; this is often negligible compared with the amplifier's e_n. We can define a "noise impedance" for the op-amp, $Z_n \equiv e_n/i_n$, so that we can safely ignore current noise when the source impedance $Z_s \ll Z_n$.

Typical i_n values range from 0.1 fA/$\sqrt{\text{Hz}}$ to 50 fA/$\sqrt{\text{Hz}}$ for CMOS and JFET op-amps, and up to 5 pA/$\sqrt{\text{Hz}}$ for low-noise BJT-input op-amps that operate at relatively high input currents. The superbeta BJT LT1012 (with its low input current) does considerably better, at 20 fA/$\sqrt{\text{Hz}}$. But note that the high-i_n LT1028 is the *voltage*-noise winner, for which we pay the penalty of high current noise. Its noise impedance $Z_n = 850\,\Omega$, which means that unusually low circuit resistances must be used, say $300\,\Omega$ or less, to obtain the full benefit from its low voltage noise. By contrast, the BJT LT1013 single-supply op-amp has $Z_n = 315\,\text{k}\Omega$, a comfortably high value.

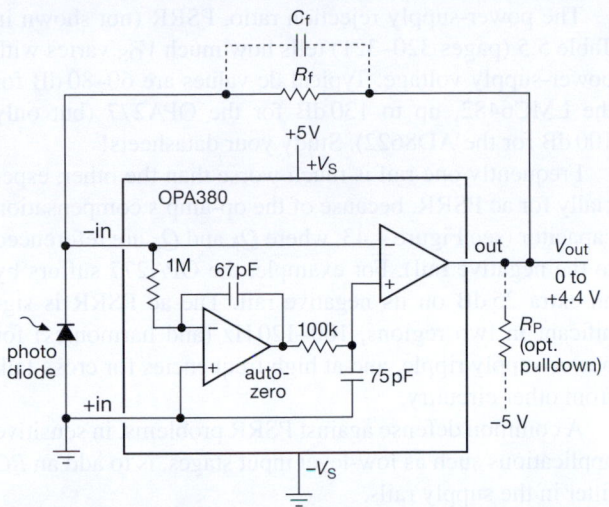

Figure 5.41. The OPA380 achieves 4 μV typical voltage offset using an auto-zero feature, yet it has only 10 fA/$\sqrt{\text{Hz}}$ current noise at 10 kHz. It is ideal for transimpedance applications such as the photodiode preamp shown here.

The manufacturer's spec gives the current-noise density at a high frequency like 1 kHz or 10 kHz, chosen to be well above the current-noise $1/f$ corner frequency. The current-noise $1/f$ corner typically comes at much higher frequencies than the voltage-noise $1/f$ corner. For example, the OPA277 has a 20 Hz voltage-noise corner, but a 200 Hz current-noise corner; for an LT1007 it's 2 Hz and 120 Hz. The differing $1/f$ corner frequencies mean that we'll get different Z_n values at low frequencies. Returning to the LT1028, for example, its relatively higher current noise at low frequencies lowers Z_n to 212 Ω at 10 Hz. This requires that we push down further our circuit resistances for optimum performance, to 100 Ω or less.

At high frequencies the current noise may consist largely of the fundamental (and unavoidable) *shot noise*, the statistical fluctuations of electron flow (eq'n 8.6). For an input bias (or leakage) current I_B that lower bound is $i_n = \sqrt{2qI_B}$; for a bias current of 10 pA that evaluates to $i_n=1.8$ fA/$\sqrt{\text{Hz}}$ (from which you can conveniently scale up or down by the square root of I_B). The typical and maximum bias-current specs vary widely, as we saw; evidently many manufacturers simply list the calculated shot-noise value corresponding to the typical bias-current spec. For example, the LT1013 has I_B=12 nA typ, from which we can calculate a current-noise density of 62 fA/$\sqrt{\text{Hz}}$; the manufacturer's spec is 70 fA/$\sqrt{\text{Hz}}$.

An important exception comes in the case of BJT-input op-amps with bias-current cancellation circuits: that greatly reduces the dc input current, but not the current noise. For example, the LT1007, with its quiet e_n=2.5 nV/$\sqrt{\text{Hz}}$, has a 10 nA bias-current spec, from which we calculate a shot-noise current of 56 fA/$\sqrt{\text{Hz}}$, but the manufacturer's spec is i_n=400 fA/$\sqrt{\text{Hz}}$. That's seven times too high! What's going on? To achieve low-noise voltage they run the input transistors at high collector currents. That creates a high base current, which would be the op-amp's input current if they hadn't used the old base-current-cancellation trick (§4x.10). So the dc bias current is small, but the current noise is large. The ultralow-e_n LT1028 also uses bias-current cancellation, keeping its bias current down to 25 nA, but with a current noise 10 times higher than the calculated shot-noise value. And for the LT6010, whose input current is brought down to just 20 *pico*amps (the kind of low currents you see in FET-input op-amps), the current noise is 40 times larger than calculated shot noise.

In other words, it's important to realize that the input *noise* current of a bias-cancelled op-amp will be considerably larger than you would expect if you were to calculate the shot noise arising from the net (i.e., cancelled) input bias current. Rather, you need to calculate the shot noise from the uncancelled base currents (and then apply a factor of $\sqrt{2}$ to account for the additional noise in the cancellation current). For example, using the LT6010's value of $I_B = \pm 20$ pA (typ), you'd incorrectly estimate a shot noise current of $i_n \approx 2.5$ fA/$\sqrt{\text{Hz}}$, whereas the datasheet lists a typical value (at 1 kHz, well above the $1/f$ corner) of 100 fA/$\sqrt{\text{Hz}}$; similarly for the LT1028 (which specifies 1000 fA/$\sqrt{\text{Hz}}$, versus the 90 fA/$\sqrt{\text{Hz}}$ you would incorrectly estimate from the net I_B). Table 5.5 does not tell you if a BJT amplifier employs bias-current cancellation, but there's a convenient bias-cancel column in the low-noise BJT op-amp Table 8.3a on page 522. Op-amps with bias-current cancellation generally do not have rail-to-rail input stages.[40]

A caution: some datasheets list greatly optimistic values for i_n, evidently making exactly this error. For example, the bias-cancelled LT1012's datasheet shows a typical i_n leveling off to 6 fA/$\sqrt{\text{Hz}}$ (beyond the $1/f$ corner), which is what you'd calculate from the specified net (i.e., cancelled) input current of ± 100 pA max, whereas you would expect a value about 10 times larger (assuming the uncancelled

---

40 The LT1677 listed in Table 8.3a is an exception. Its datasheet has a graph labelled "Input Bias Current Over the Common Mode Range," showing that the bottom 1.4 V and top 0.7 V of the common-mode range suffer from high bias currents. The op-amp's offset voltage is also degraded in these regions. But, hey, we warned you about that in §5.9.1!

base current is about $100\times$ larger). We were skeptical of the datasheet's claimed i_n, so we measured it (along with others that appeared to be similarly in error), and found[41] $i_n \approx 55\,\text{fA}/\sqrt{\text{Hz}}$. This error shares some of the characteristics of an *epidemic*, having infected also the datasheets of auto-zero op-amps. For example, the exemplary AD8628A (listed in Table 5.6 on page 335) specifies an input noise-current density of $5\,\text{fA}/\sqrt{\text{Hz}}$; imagine our surprise when we measured a value 12 times larger. Not to be outdone, the MCP6V06 auto-zero op-amp's specification of $0.6\,\text{fA}/\sqrt{\text{Hz}}$ is rather at odds with its measured $170\,\text{fA}/\sqrt{\text{Hz}}$. See the discussion in §§5.11.

It's important to note that, in the case of chopper and auto-zero op-amps, the input current noise spec is usually given at a low 10 Hz frequency, because that's below the region of very high switch charge-injection current noise (see §3.4.2E). If you take the trouble to measure an auto-zero's input noise, you'll see something like the plots shown in Figure 5.52. This is an unfortunate situation, with insufficient guidance (and perhaps deliberate obfuscation) from manufacturers. We discuss this further in §5.11.

5.10.9 CMRR and PSRR

The common-mode rejection ratio, CMRR, tells how much the input offset voltage V_{OS} varies with common-mode input voltage. The problem, of course, is that such a change in V_{OS} masquerades as a change in the input signal voltage.

The CMRR values vary from 70 dB (min) for our favorite LMC6482 (an inexpensive CMOS dual op-amp), up to 130 dB for the precision OPA277. Degradation of CMRR at high frequencies often matters, and there's usually a plot in the op-amp's datasheet (check out a few for yourself, for example these two; we show CMRR plots for other types of op-amp in Figures 5.73 and 5.82). For example, the LMC6482 typical CMRR starts falling upward of 1 kHz, and it's down to 80 dB at 10 kHz. It's interesting that both the OPA277 and the AD8622 (another expensive high performer at dc) degrade to about 80 dB by 10 kHz, joining our jellybean CMOS friend. Other parts do better, such as the LT1007 (114 dB typ at 10 kHz). And to repeat a warning we made elsewhere: the CMRR spec often applies to only a limited common-mode range; read the datasheet carefully.

Note this universal cure: a time-honored way to avoid CMRR troubles is to use an inverting configuration.

---

[41] Datasheets for the closely similar OP-97 and LT1097 make the same error, evidently corrected in that of the later LT6010 (the recommended successor to the LT1012).

The power-supply rejection ratio, PSRR (not shown in Table 5.5 (pages 320–321) tells how much V_{OS} varies with power-supply voltage. Typical dc values are 60–80 dB for the LMC6482, up to 130 dB for the OPA277 (but only 100 dB for the AD8622). Study your datasheets!

Frequently one rail is much worse than the other, especially for ac PSRR, because of the op-amp's compensation capacitor (see Figure 4.43, where Q_5 and Q_6 are referenced to the negative rail). For example, the OPA277 suffers by an extra 25 dB on its negative rail. The ac PSRR is significant in two regions, 100–120 Hz (and harmonics) for power-supply ripple, and at high frequencies for cross-talk from other circuitry.

A common defense against PSRR problems, in sensitive applications such as low-level input stages, is to add an *RC* filter in the supply rails.

5.10.10 GBW, f_T, slew rate and "m," and settling time

It's tempting to think one can never have too much GBW (gain–bandwidth product, or f_T, its original name that we favor, see Figure 5.42). After all, a higher GBW means greater loop gain, and higher loop gain means lower error (gain, phase, distortion). What's more, with higher f_T we're well on our way to having a faster slew rate, via the formula $S = 0.32 m f_T$ as discussed in detail in §4x.9.

Furthermore, a faster slew rate means a greater full-power bandwidth (FPBW): a sinewave $V(t) = A\sin\omega t$ has a peak slew rate $S = \omega A$, so FPBW $= S/\pi V_{pp}$. Finally, because the first step on the way to waveform settling is the slewing delay $t = \Delta V/S$, a higher f_T is an important step (and often the primary determinant), toward a faster settling-time spec. The data in Tables 5.4 (page 310), and 5.5 (pages 320–321) let you explore the essential question "what price higher bandwidth?"

A. An aside: GBW and f_T
First, a short riff on "GBW" and "f_T." Figure 5.42 shows a plot of open-loop gain versus frequency for a THS4021 wideband op-amp. This is a "decompensated" op-amp, stable for closed-loop gains ≥ 10, with a textbook Bode plot. The term GBW properly describes the product of open-loop gain times frequency in the region in which the gain is dropping at 6 dB/octave (i.e., $G_{OL} \propto 1/f$). Its extrapolation crosses the $G_{OL} = 0$ dB axis at a frequency equal to GBW. At that frequency, however, the gain is less than unity, owing to the effect of additional higher-frequency poles in the amplifier. Strictly speaking, the symbol f_T is used for the (lower) frequency at which $G_{OL} = 1$.

But we *like* the simpler variable f_T, and so do plenty of other folks; so it's loosely used in place of GBW. Perhaps this is excusable, given that f_T is pretty close in value to GBW for op-amps that are compensated for stability at unity gain (this describes most op-amps). In any case, unless stated otherwise, we will use f_T to mean GBW.

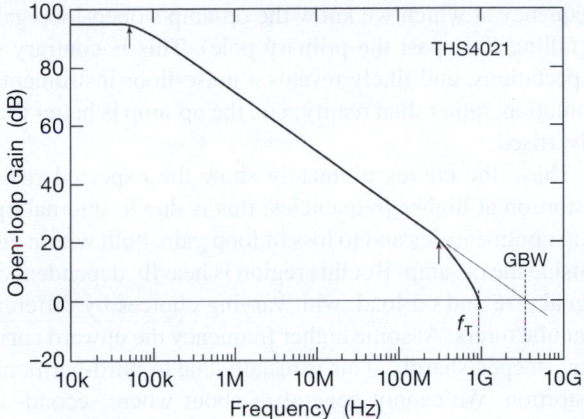

Figure 5.42. An op-amp's gain–bandwidth product (GBW) is the frequency at which the extrapolation of the open-loop gain curve crosses the unity-gain axis. It's often loosely called "f_T," though the latter is properly the frequency at which the closed-loop gain is unity. The arrows indicate the dominant and second poles. The data here are taken from the THS4021 datasheet, which also shows the phase shift reaching 180° by 400 MHz.

B. Milking stools and barstools

If offset voltage is one leg of the precision op-amp stool,[42] bandwidth and speed certainly comprise another. Many of the fast op-amps we wanted to include in the precision Table 5.5 (pages 320–321) were passed over because they had too much offset voltage – so some of them have gotten their own Table 5.4 (page 310). For example, in that table we have the LT6200, with 165 MHz GBW, 0.95 nV/$\sqrt{\text{Hz}}$, and 1.0 mV of offset. It's a conventional voltage-feedback (VFB) op-amp offering 50 V/μs slewing and 140 ns settling time, so whatever is wrong with it? Only 50 V/μs, with $m = 1$? A limited 10 V (±5 V) supply with a high 16.5 mA current? And a *very* high 40 μA (!) max bias current? The point here is that there's a price to pay for high f_T, and maybe this part is not so attractive after all. But at least the LT6200 has less than 1 nV/$\sqrt{\text{Hz}}$ of noise,

1% distortion at 50 MHz, and it's even RRIO to boot.[43] And there's the LT6200-10 variant, with 1.6 GHz of GBW. Nice!

We did include some admirable high-speed op-amps in Table 5.5, (pages 320–321), overlooking in one case an, uh, *underwhelming* V_{OS}. One such favorite of ours is the OPA656, a member of a small family of highly useful op-amps offered by the Burr–Brown division at TI: it combines 230 MHz GBW, 290 V/μs slew rate, and 20 ns settling time. With its 2 pA JFET inputs having less than 3 pF of input capacitance, we readily forgive its 1.8 mV offset voltage, and we're happy enough with its 7 nV/$\sqrt{\text{Hz}}$ of input voltage noise. It is excellent for a transimpedance amplifier of the sort used with photodiodes – see §§4x.3 and 4x.9. The OPA656 even has a distortion plot in Figure 5.44 (see next subsection), where we see it has less than 0.1% distortion to 10 MHz and beyond. And it has a 1.6 GHz cousin, the OPA657. When we need higher operating voltage, and crave lower e_n, we turn to the 4.5 nV/$\sqrt{\text{Hz}}$ OPA637; it has somewhat greater capacitance (7 pF), and less bandwidth (80 MHz). And there's the 90 MHz OPA380 shown above. Very fine products from BB/TI.

On the subject of low distortion, Linear Technology offers the LT1468 (90 MHz, 75 μV, ±15 V supply), which claims 0.7 ppm distortion with a 10 V signal; and with its 0.8 μs settling time, it's a good candidate to feed hungry ADCs. Not to be outdone, National Semiconductor offers the LMP7717, a CMOS op-amp with an 88 MHz f_T, yet drawing just 1 mA, working down to a 1.8 V total supply (!), and offering 1 pA I_B, 6 nV/$\sqrt{\text{Hz}}$ e_n, and 150 μV V_{OS} input specs, along with rail-to-rail outputs. Parts like these suggest that maybe we can have our speed + precision cake and eat it too.

5.10.11 Distortion

Although much of the precision analog-design field concerns dc and low frequencies, there are applications that require accuracy at higher speeds: audio and video, communications, scientific measurement, and so on. With falling op-amp loop gain, input errors are rising, output impedance is rising, and slew-rate limitations may come into play. We need a way to evaluate an op-amp's performance at mid to high frequencies. Some manufacturers help by providing harmonic distortion curves in their datasheets. If

---

[42] Three-legged, or four? We've found that city folks don't know why it is that milking stools have three legs, whereas barstools have four. Those with rural upbringing can tell you, in a flash.

[43] Yeah, OK, but read carefully: on page 10 of the datasheet you'll see that there are ~1 mV shifts of offset voltage when the input is within 1.5 V of the rail! A powerful incentive to use inverting mode!

it's painful going through hundreds of datasheets looking for tabulated specs to compare, it's doubly painful leafing through their back pages looking for distortion curves.

Here we provide some "value added" (hey, this book isn't cheap!) by compiling distortion plots from the datasheets of fifty selected high-performance op-amps: Figure 5.43 (for high-voltage op-amps) and Figure 5.44 for low-voltage and RR-output op-amps (including some HV types). The op-amps listed in Tables 5.4 (page 310) and 5.5 (pages 320–321) have a check in the "dist graph" column if they appear in one of these graphs. We've also measured the distortion and made plots for some popular older op-amps that don't have datasheet plots; see Figure 5.19.

Burr–Brown/TI's OPA134 and OPA627, along with National Semi's LME49990 and other LME49700-series op-amps, are the winners in the high-voltage category. LTC's LT1468 stands out as well. Analog Devices' AD8021 stands out at high frequencies, and they often recommend it for driving ADCs. TI's THS3061, which has an impressive 7000 V/μs slew rate, looks pretty good above 100 kHz, and as a bonus it can deliver 145 mA into 50 Ω. The OPA1632 and LME49724 are fully differential amplifiers, see §5.17.

The low-voltage and RR op-amp table comprises mostly parts running from ± 5V supplies, or lower. Most of these have rail-to-rail outputs, demonstrating that RRO op-amps can compete in the precision arena. Some low-voltage op-amps are at a disadvantage relative to their HV cousins, because they have to use abnormally low signal levels. Consider the graph's OPA1641 JFET winner, at 0.5 ppm. It's a high-voltage part with RR outputs, and it gets tested with 8.5 Vpp signals, a luxury not available to the low-voltage RRO parts. The OPA376 is the low-voltage winner in this category, at 3 ppm, while being tested with 2.8 Vpp. It's interesting that both of these op-amps use the Monticelli output stage (see Figure 5.35 and §4x.11).

A. Distortion: some caveats

Some caution is advised here: it's tempting to look at the distortion plots and think that you know how the op-amps compare. But some of the distortion measurements need to be taken with more than a grain of salt, and a few caveats are in order. First, there are no de facto standards for op-amp distortion, and manufacturers have chosen different operating conditions.[44] Some use THD, others THD+N (total harmonic distortion plus noise), and still others may

concentrate on specific distortion products, e.g., the 2nd or 3rd harmonic. These hidden choices affect an op-amp's standing on the charts.

Second, the distortion plots sometimes reveal artifacts of the measurement process – for example, the curves in Figures 5.43 and 5.44 begin with a flat distortion profile starting at dc, which, however, often continues well past the frequency at which we know the op-amp's open-loop gain is falling (i.e., past the primary pole). This is contrary to expectations, and likely reveals a noise-floor instrumental limitation, rather than reality; i.e., the op-amp is better than advertised.

Third, the curves ultimately show the expected rising distortion at higher frequencies; this is due to internal op-amp nonlinearities and to loss of loop gain, both within and outside the op-amp. But this region is heavily dependent on signal size and on load, with varying choices by different manufacturers. At some higher frequency the upward curve may steepen sharply. This is usually due to third-harmonic distortion. We cannot generalize about where second- or third-harmonic distortion will dominate for any given op-amp, but it seems that second-order distortion is the most common culprit. This is surprising, given an effort in many op-amps to fully balance the design.

Fourth, when you're in the <10 ppm territory all kinds of strange things can bite you. As wise guru Jim Williams once remarked, "If you think you've measured something to 1 ppm, you're probably wrong." Figure 5.45 illustrates an often-overlooked issue. Here we have a precision OPA1641 op-amp that's capable of distortion performance below 1 ppm at 1 kHz, and in noninverting mode the op-amp has just over 20 ppm of distortion at 100 kHz. It's a JFET op-amp with 8 pF of input capacitance (reasonably low, especially considering the op-amp's low 5 nV/$\sqrt{\text{Hz}}$ noise spec). In an admirable display of candor, the datasheet warns us that "The *n*-channel JFETs in the FET input stage exhibit a varying input capacitance with applied common-mode input voltage," and it provides plots of increasing distortion arising from an input source resistance driving the op-amp's dynamically-changing capacitance. For example, with $R_s = 600\,\Omega$ the 100 kHz distortion increases dramatically, to 100 ppm. They suggest carefully matched input impedances to reduce this type of distortion (an effect that is not confined to this particular op-amp: be warned!). Better yet, use the inverting configuration.

Finally, a look at the test circuit used by many manufacturers to make sub-100 ppm distortion measurements (Figure 5.46). The trick is to reduce op-amp's loop gain by a factor of 100, thus increasing the distortion by the same factor; the reported distortion is then gotten by dividing the

---

[44] Manufacturers use different voltage levels (2 Vpp, 3 Vrms, 10 V peak, and 20 Vpp), different loads (100 Ω, 600 Ω, 2k, 10k and open circuit), different common-mode voltages, different analyzer filters, and even different gains for their measurements.

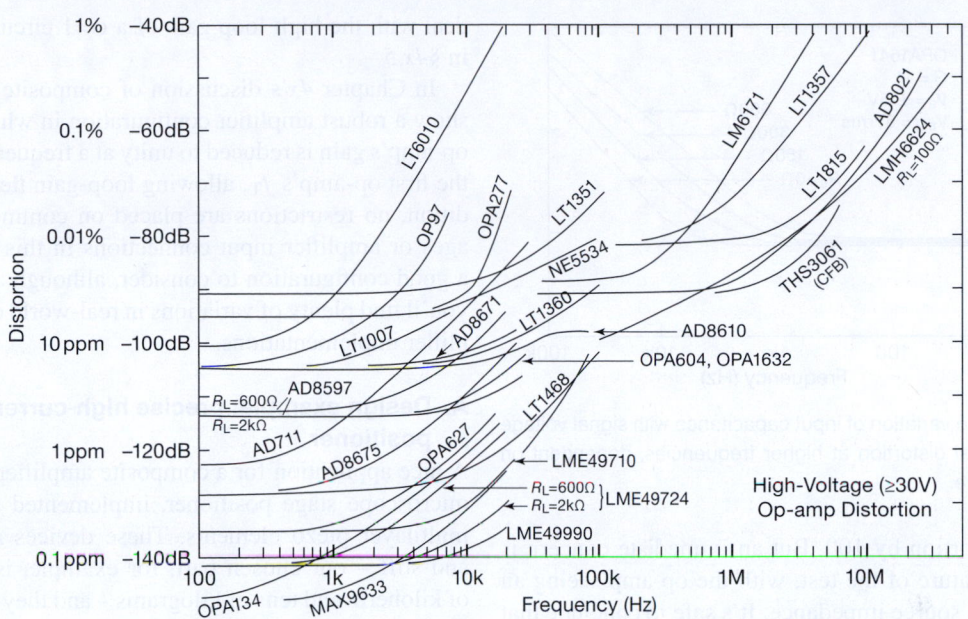

Figure 5.43. Harmonic distortion versus frequency for a selection of "high-voltage" (≥30 V total supply) op-amps, from manufacturers' datasheets.

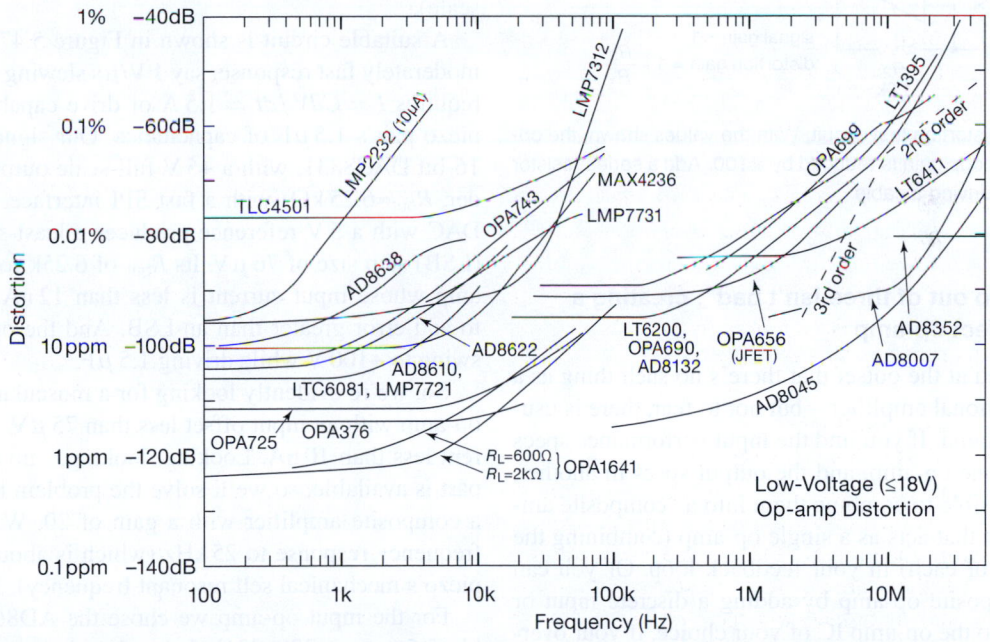

Figure 5.44. Harmonic distortion versus frequency for a selection of "low-voltage" (≤18 V total supply) op-amps, from manufacturers' datasheets. Most of these have rail-to-rail output stages. See also Figure 5.19.

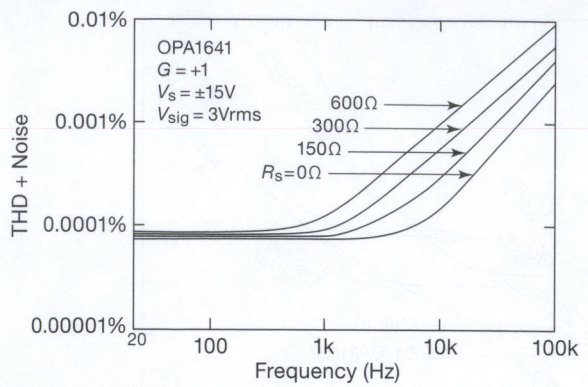

Figure 5.45. The variation of input capacitance with signal voltage causes additional distortion at higher frequencies, dependent on source resistance.

measured distortion by 100. But an immediate concern is the artificial nature of the test, with the op-amp seeing an artificially low source impedance. It's safe to conclude that we all can likely benefit from further work in this area.[45]

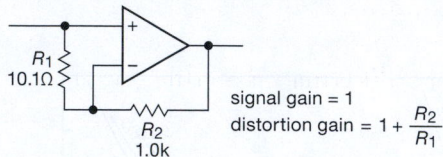

signal gain = 1
distortion gain = $1 + \dfrac{R_2}{R_1}$

Figure 5.46. Distortion test circuit. With the values shown, the op-amp's effective loop gain is reduced by ×100. Add a series resistor at the output if driving a cable.

5.10.12 "Two out of three isn't bad": creating a perfect op-amp

We recognized at the outset that there's no such thing as a perfect operational amplifier – but not to fear, there is usually a workaround. If you find the input performance specs you need in one op-amp, and the output specs in another, it may be possible to combine them into a "composite amplifier" circuit that acts as a single op-amp (combining the best features of each) in your feedback loop. Or you can create a composite op-amp by adding a discrete input or output stage to the op-amp IC of your choice. If your overall feedback circuit has very high gain, such as a G=10,000 amplifier, you may not need to worry about compensation (e.g., see Figure 5.61). However, it's not terribly difficult to

deal with the high loop gain of a G=1 circuit, as we show in §*4x.5*.

In Chapter *4x*'s discussion of composite amplifiers we show a robust amplifier configuration in which the second op-amp's gain is reduced to unity at a frequency well above the first op-amp's f_T, allowing loop-gain flexibility. In addition, no restrictions are placed on common-mode voltages or amplifier input connections in this approach. It's a good configuration to consider, although as often as not you'll find plenty of variations in real-world composite amplifier implementations.

A. Design example: precise high-current piezo positioner

A nice application for a composite amplifier is a precision microscope stage positioner, implemented with a pair of multilayer piezo elements. These devices are both swift and stiff – our chosen part, for example, is good to tens of kilohertz and tens of kilograms – and they provide stable and accurate positioning (at the nanometer scale) over their limited motion (here 6 μm). On the down side, they present a difficult load, being highly capacitive (here 0.75 μF) and requiring relatively high drive voltage (here 100 V full-scale).

A suitable circuit is shown in Figure 5.47. We'd like a moderately fast response, say 1 V/μs slewing speed, which requires $I = CdV/dt = 1.5$ A of drive capability into the piezo pair's 1.5 μF of capacitance. Our signal source is a 16-bit DAC8831, with a +5 V full-scale output (R-$2R$ ladder, R_{out}=6.25 kΩ) with a fast SPI interface. Running the DAC with a 5 V reference produces a least-significant bit (LSB) step size of 76 μV. Its R_{out} of 6.25k requires an op-amp whose input current is less than 12 nA in order not to add error greater than an LSB. And the op-amp has to swing to +100 V while driving 1.5 μF.

So, we're evidently looking for a muscular 150 V, 1.5A op-amp with an input offset less than 75 μV, and bias current less than 10 nA. Looking, looking... no joy! No such part is available, so we'll solve the problem by rigging up a composite amplifier with a gain of 20. We'll specify a frequency response to 25 kHz (which is about 20% of the piezo's mechanical self resonant frequency).

For the input op-amp we chose the AD8675 from Table 5.5 (pages 320–321). It has low input errors (75 μV and 2 nA max), and enough output swing to drive a high-voltage $G = 20$ stage. For our output amplifier we chose the Apex PB51, a power driver capable of 300 V and 1.5 A (but subject to a safe-operating-area constraint, for example limited to 130 V drop when driving 2 A for 100 ms). Its maximum input errors are 1.75 V (!) and 70 μA (!) – that's

---

[45] OK, we plead guilty-as-charged to that conclusion favored by all academics – "needs more study (and a grant proposal is in the mail)."

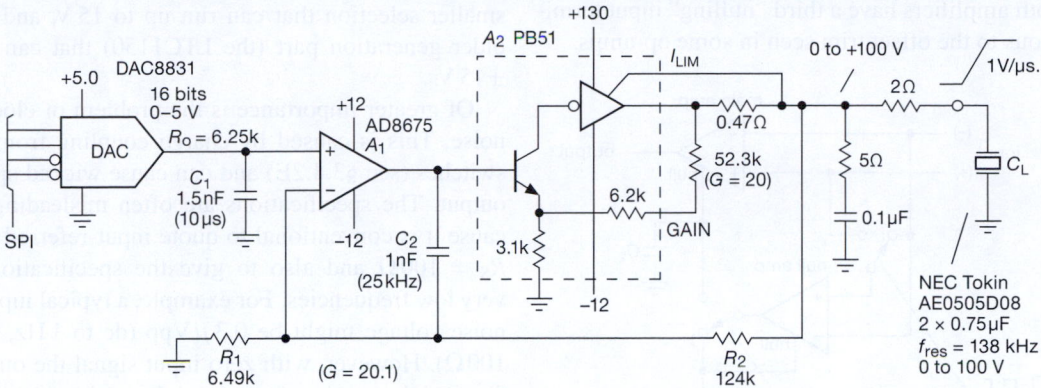

Figure 5.47. Precision composite amplifier for driving a $1.5\,\mu F$ piezo positioner: output to 100 V and 1.5 A, with $75\,\mu V$ maximum offset, 2 nA maximum input current, and response to 25 kHz. The specified piezo actuator moves $6\,\mu m$ per 100 V.

why they call it a *driver* rather than an op-amp! Its gain is set with an external resistor, here $52.3\,k\Omega$ (for $G = 20$) to match the desired overall gain.

The single feedback path $R_1 R_2$ sets $G = 20$ for the composite amplifier. A general scheme for compensating composite amplifiers is suggested in §4x.5. Here we use a different scheme, with C_2 isolating the input op-amp A_1 from the output amplifier A_2 at frequencies above 25 kHz. That way we don't have to worry about A_2's response at high frequencies, where it's struggling with its capacitive load. This is really a "custom" configuration, forced on us by the, uh, *challenging* load; what is has in common with other composite amplifier configurations is the single overall feedback path that determines its gain over most of its operating regime.

To get an understanding of circuit operation, imagine a 2 V input step from the DAC. This causes a 2 V step at the output of A_1 (which acts like a follower at high speeds, owing to C_2), which is approximately the right signal to tell A_2 to drive current into C_L and move its output by 40 volts (that's why we chose $G=20$ for the output stage). As amplifier A_2 approaches this goal, op-amp A_1 takes charge and presents correction signals to cause the output to hone in on the precise value.

A few additional circuit details: the DAC's response is slowed by C_1 to a $10\,\mu s$ time constant – no point in jolting the amplifier, given its limited bandwidth. The series *RC* across the output promotes stability, both by reducing A_2's open-loop gain at high frequencies and by providing lossy damping; it's widely used in audio power amplifiers. Note that substantial bypass capacitors (as much as $100\,\mu F$, not shown) are needed with such large signal currents.

Further examples of the composite amplifier technique

can be found in §4x.5 and in Figures 5.58, 5.59, 5.61, 8.49, 8.50, 8.78, 8.80, and 13.48.

And other examples of techniques that exploit the "two-out-of-three" concept of adding external amplifier blocks are (a) discrete JFET front-end for a BJT op-amp (e.g., Figure 5.58), (b) output unity-gain buffer (§5.8.4), and (c) bootstrapped power supplies to extend voltage range, or to improve CMRR (e.g., Figure 5.79).[46]

5.11 Auto-zeroing (chopper-stabilized) amplifiers

Even the best of precision low-offset op-amps cannot match the stunning V_{OS} performance of the so-called "chopper-stabilized" or "auto-zero" (also called "zero-drift") op-amps. Ironically, these interesting amplifiers are built with CMOS, otherwise famous for its mediocrity when it comes to offset voltage or drift. The trick here is to put a second *nulling* op-amp on the chip, along with some MOS analog switches and offset-error storage capacitors. One of several possible configurations is shown in Figure 5.48. The main op-amp functions as a conventional (imperfect) amplifier. The nulling op-amp's job is to monitor the input offset of the main amplifier, adjusting a slow correction signal as needed in an attempt to bring the input offset exactly to zero. Because the nulling amplifier has an offset error of its own, there is an alternating cycle of operation in which the nulling amplifier corrects its own offset

---

[46] If you're considering piezo positioners for a precision application, be aware that they exhibit some nonlinearity and hysteresis when driven from a voltage source. These issues are said to be ameliorated when the drive signal is quantified by *charge* instead of voltage. See Chapter *3x* for a precision current-drive circuit that circumvents this problem and makes fast linear piezo steps.

voltage. Both amplifiers have a third "nulling" input terminal, analogous to the offset trim seen in some op-amps.

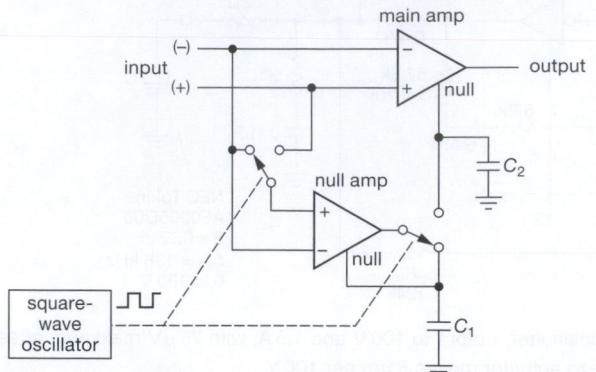

Figure 5.48. The original ICL7650 and ICL7652 auto-zero ("chopper-stabilized") op-amps. Capacitors C_1 and C_2 are external.

The auto-zeroing cycle goes like this. (a) Disconnect the nulling amplifier from the input, short its inputs together, and connect its output to C_1, the holding capacitor for its correction signal; the nulling amplifier now has zero offset. (b) Now connect the nulling amplifier across the input and connect its output to C_2, the holding capacitor for the main amplifier's correction signal; the main amplifier now has zero offset (assuming that the nulling amplifier has not drifted). The MOS analog switches are controlled by an on-board oscillator, typically running in the range of 1–50 kHz.

5.11.1 Auto-zero op-amp properties

Auto-zero op-amps do best what they are optimized for, namely delivering V_{OS} values (and tempcos) 5–50 times better than the best precision bipolar op-amp (see Table 5.6 on the next page). What's more, they do this while delivering full op-amp speed and bandwidth.[47] They also have extraordinarily high open-loop gain at low frequencies (typically 130–150 dB, a consequence of their "composite amplifier architecture"); and, happily, they are inexpensive, particularly when compared with conventional precision op-amps.

That's the good news. The bad news is that auto-zero amplifiers have a number of diseases that you must watch out for. Being CMOS devices, most of them have a severely limited supply voltage – often 6 V total supply, with a

smaller selection that can run up to 15 V, and just one[48] older generation part (the LTC1150) that can operate at ±15 V.

Of greater importance is the problem of clock-induced noise. This is caused by charge coupling from the MOS switches (see §3.4.2E) and can cause wicked spikes at the output. The specifications are often misleading here, because it is conventional to quote input-referred noise with $R_S = 100\,\Omega$ and also to give the specification for only very low frequencies. For example, a typical input-referred noise voltage might be 0.3 μVpp (dc to 1 Hz, with $R_S = 100\,\Omega$). However, with zero input signal the output waveform might consist of a train of 5 μs-wide 10 mV spikes of alternating polarity!

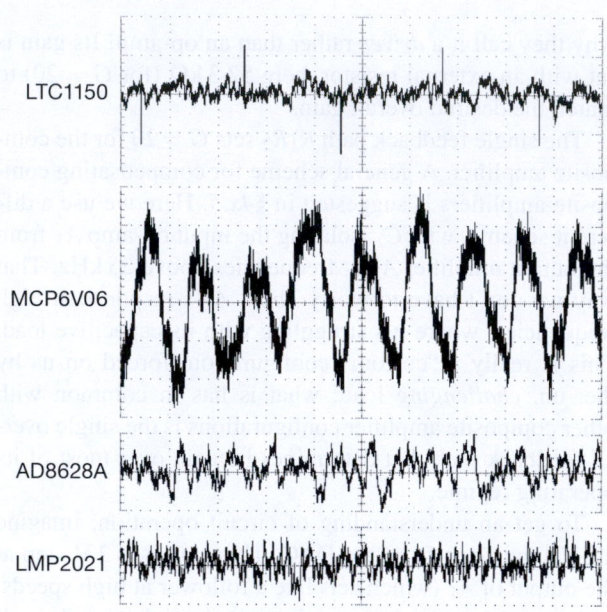

Figure 5.49. Output waveforms from four auto-zero op-amps, configured for $G = 100$, with the input connected to ground through a 100 Ω resistor. Vertical: 2 mV/div; horizontal: 100 μs/div.

The internal switching also causes spikes of *input* current, which means that input signals of high source impedance R_S will exhibit larger input-referred spikes. Figures 5.49 and 5.50 show this behavior, measured with R_S of 100 Ω and 1 MΩ, in several auto-zero op-amps configured

---

[47] Unlike an earlier generation of synchronous amplifiers that were also called "chopper amplifiers," but that had bandwidth limited to a fraction of the chopping clock frequency.

[48] Some old high-voltage types that require external (correction-signal) capacitors may still be available.

Table 5.6 Chopper and Auto-zero Op-amps

Part #	# per pkg	Supply (V)	I_Q typ (µA)	I_{bias}[a] @25°C typ (pA)	I_{bias} @25°C max (pA)	Voff @25°C typ (µV)	Voff @25°C max (µV)	Tempco typ (nV/°C)	Tempco max (nV/°C)	CMRR min (dB)	V_{npp} dc[d] (µV)	Noise[†,u] e_n 1kHz (nV/√Hz)	Noise i_n 10Hz (fA/√Hz)	HF Noise (>kHz)	GBW typ (MHz)	Slew (V/µs)	t_{settle} typ (µs)	$t_{recovery}$ typ (µs)	Swing IN −	IN +	OUT +	OUT −	Cost qty 25 ($US)	Comments
ADA4051-1	1,2	1.8-6	15	20	70	2	15	20	100	110	2	95	100	40	0.13	0.04	110	120	●	●	●	●	1.86	auto-zero, -2=dual
OPA333	1,2	1.8-7	17	70	200	2	10	20	50	106	1.1	55	100	>20	0.35	0.16	45	80	●	●	●	●	2.57	auto-cal, dual=2333
ISL28133	1,2	1.65-6.5	18	30	300	2	8	20	75	118	1.1	65	79	8	0.4	0.1	35	?	●	●	●	●	1.69	dual=28233
OPA330	1,2	1.8-7	21	200	500	8	50	20	250	100	1.1	55	100	>20	0.35	0.16	45	80	●	●	●	●	2.56	auto-cal, dual=2330
MAX9617	1,2	1.6-6	59	10	140	0.8	10	5	120	122	0.42	42	100	50	1.5	0.7	?	?	●	●	●	●	1.60	RRIO, charge pump
OPA378	1,2	2.2-6	125	150	550	20	50	100	250	100	0.4	20	200	15	0.9	0.4	9	4	●	●	●	c	2.16	auto-cal, '2378=dual
LTC2054	1,2	2.7-6	140	1	150	0.5	3	20	30	120	1.6	85	-	-	0.5	0.5	-	4000	●	●	-	-	2.25	dual=2055
AD8538	1,2	2.7-6	150	15	25	5	13	30	100	115	1.2	52[h]	-	1,15	0.43	0.4	5	50	●	●	●	-	1.80	self-calibrating, duals
MCP6V06	1,2	1.8-6.5	200	1	-	-	3	-	50	140	1.7	82	*(0.6)*	1	1.3	0.5	300	100	●	●	●	-	1.42	low-cost, dual = 6V07
LTC1049	1	5-16	200	15	50	2	10	20	-	110	3	100	*(2)*	10	0.8	0.8	-	6000	●	-	-	-	2.85	miniDIP pkg avail
OPA335	1,2	1.8-7	285	70	200	1	5	20	50	110	1.4	55	*(2)*	10	2	1.6	6	50	●	●	●	●	2.56	auto-cal, dual=2335
MAX4238	1	2.7-6	600	1	-	0.1	2	10	-	120	1.2	30	-	?	2	0.35	1000	3300	●	●	●	●	1.62	'4239=decomp, G=5
OPA734	1,2	2.7-13	600	100	200	1	5	10	50	115	2.5	135	40	17	2	1	50	8	●	●	●	●	2.56	auto-cal, dual=2734
AD8551	1,2,4	2.7-6	700	10	50	1	5	5	40	120	1	42	*(2)*	3.5	1.5	0.4	50	50	●	●	●	●	2.34	auto-zero
AD8572	1,2,4	2.7-6	700	10	50	1	5	5	40	120	1.3	51	*(2)*	2.3	1.5	0.4	-	50	●	●	●	●	3.34	auto-zero
LTC2050HV	1	2.7-11	750	25	75	0.5	3	-	30	115	1.5	-	-	?	3	2	-	2000	●	●	-	-	3.20	OK for ±5V supplies
AD8628	1,2,4	2.7-6	850	30	100	1	5	2	20	120	0.5	22	*(5)*	15	2.5	1	-	50	●	●	●	●	1.92	self-cal, dual, quads
LMP2011	1,2,4	2.5-5.8	930	3	-	0.12	25	15	-	95	0.85	35[h]	-	25	3	4	-	50	●	●	-	-	2.55	copper leadframe
TLC4501A	1,2	4-6	1000	1	60	10	40	1000	-	90	1.5	12	*0.6*	none	4.7	2.5	2	0	-	●	-	-	2.88	self-zero at powerup
AD8638	1,2	5-16	1000	1.5	40	3	9	10,40	60	118	1.2	60	-	8	1.35	2.5	3	50	●	●	●	-	2.29	self-calibrating, duals
LTC1050	1	4.75-16	1000	10	30	0.5	5	10	50	114	1.6	11	*(1.8)*	2.5	2.5	4	-	3000	●	●	●	-	2.85	favorite
LMP2021	1,2	2.4-6	1100	25	100	0.4	20	49	20	105	0.2	14	350	30	5	2.5	1	50	●	●	●	-	3.38	EMI-rej, '2022=dual
MAX4236A[b]	1	2.4-6	350	1	500	5	20	600	2000	84	0.2	7	0.6	none	1.7	0.3	-	0	●	●	●	●	1.78	comp low Vos CMOS
AD8616[b]	1,2,4	2.7-6	1.7	0.2	1	23	60[f]	1500	4000	80	2.4	2	2	none	24	12	0.5	0	●	●	●	●	1.52	comp CMOS
LT1028[b]	1	8-44	7400	25nA	95nA	10	40	200	800	108	0.04	0.9	4700	none	75	15	-	0	-	-	-	-	6.31	comp low-noise BJT
high-voltage																								
LTC1150	1	4.8-32	800	10	60	0.5	10	10	50	110	1.8	-	-	0.5	1.8	1.5	-	20ms	-	●	n	n	5.84	OK for ±15V supplies
LT1012A[b]	1	2.4-40	380	25	100	8	25	200	600	114	0.5	14	20	none	0.8	0.2	-	0	-	-	-	-	5.11	comp low I_B bipolar
LT1007A[b]	1	7-44	3000	10nA	35nA	10	25	200	1000	117	0.06	2.5	1500	none	8	1.7	-	0	-	-	-	-	5.83	comp low e_n bipolar

Notes: (a) check datasheets for plots of I_{bias} vs common-mode input voltage. (b) conventional (not auto-zero) precision op-amps, for comparison. (c) crossover region. (d) 0.01Hz to 10Hz. (f) at V_{CM}=0.5V and 3.0V (with V_S=5.5V), but as much as 400µV near V_{CM} near ±2.5V. (g) 1nV/°C at V_S=2.5V. (h) 150nV/√Hz hump above 2kHz. (n) near to rails w/o load. (t) typical. (u) current noise values indicated with *light italics* should not be relied upon; measured values appear to be greater by factors of 100 or more, see discussion in Chapter 8.

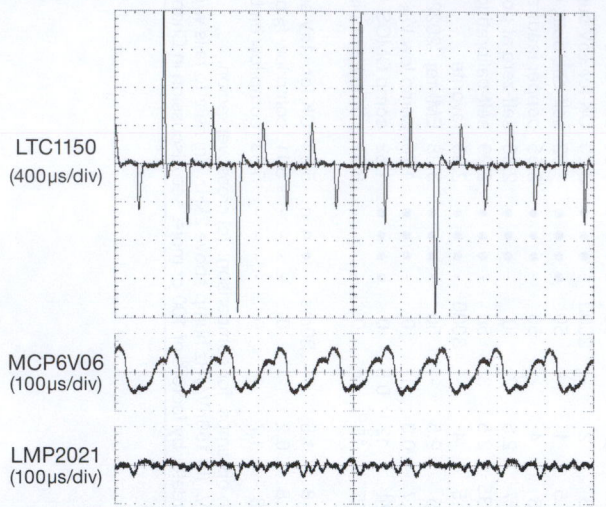

Figure 5.50. Output waveforms from three auto-zero op-amps, configured for $G = 100$, with the input connected to ground through a 1 MΩ resistor. Vertical: 100 mV/div.

for a voltage gain of 100.[49] There is considerable variation among these parts, with the conventional auto-zero configuration (Figure 5.48, used in the LTC1150 and MCP6V06) exhibiting greater clock feedthrough compared with that of alternative designs (as in the AD8628A and LMP2021) that are intended to reduce these undesirable effects.[50]

The datasheets *do* reveal this unseemly behavior, indirectly, in plots of voltage noise versus frequency.[51] Figure 5.51 shows a pair of such plots for two auto-zero products from Analog Devices: the AD8551 has a ~4 kHz fixed-frequency oscillator, whereas their AD8571 has a deliberately variable (spread-spectrum) oscillator to eliminate sharp spectral lines (which can create undesirable intermodulation with nearby signal frequencies). Note, by

---

[49] The waveforms in the latter show 8 nApp current spikes for the LTC1150, 1 nApp noise for the MCP6V06 (despite its impressive spec: $0.6\,\mathrm{fA}/\sqrt{\mathrm{Hz}}$ at 10 Hz), and 0.2 nApp "rumble" for the LMP2021 (which sports a $0.35\,\mathrm{pA}/\sqrt{\mathrm{Hz}}$ current-noise spec).

[50] From the AD8628A datasheet: "The AD8628/AD8629/AD8630 family uses both auto-zeroing and chopping in a patented ping-pong arrangement to obtain lower low frequency noise together with lower energy at the chopping and auto-zeroing frequencies, maximizing the signal-to-noise ratio for the majority of applications without the need for additional filtering. The relatively high clock frequency of 15 kHz simplifies filter requirements for a wide, useful noise-free bandwidth."

[51] Watch out, though, for claimed values of noise *current* – the low values listed in many datasheets are completely incorrect, sometimes by factors of ×10 to ×100, evidently having been calculated a priori as the shot noise corresponding to the dc input current; see the discussion in §5.10.8 and in §8.9.1F.

---

the way, that these plots specify an input signal of *zero* source impedance.

It's always instructive to make some actual measure-

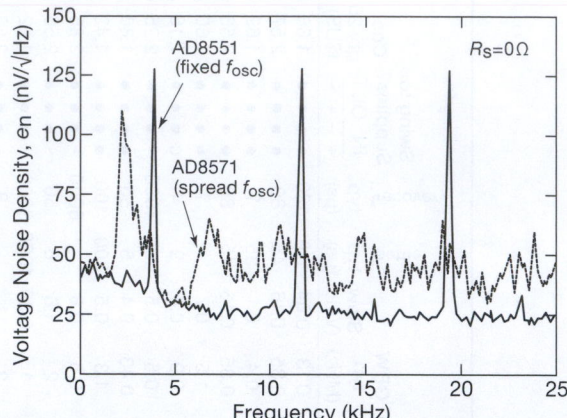

Figure 5.51. Spectra of noise voltage, adapted from their datasheets, for a pair of auto-zero op-amps. The AD8571 varies its oscillator frequency in order to suppress sharp spectral features.

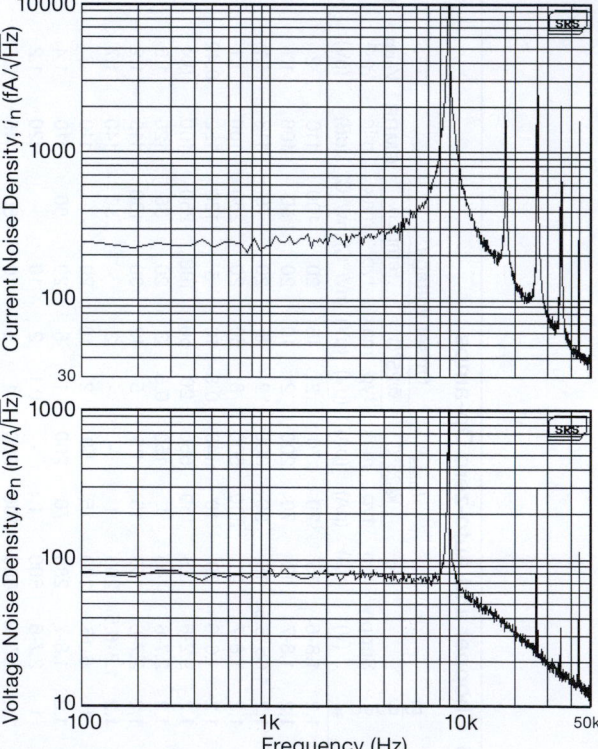

Figure 5.52. Measured voltage-noise density (bottom) and current-noise density (top) for an MCP6V06 auto-zero amplifier. Switch-induced clock noise at 9 kHz (and harmonics) is prominent.

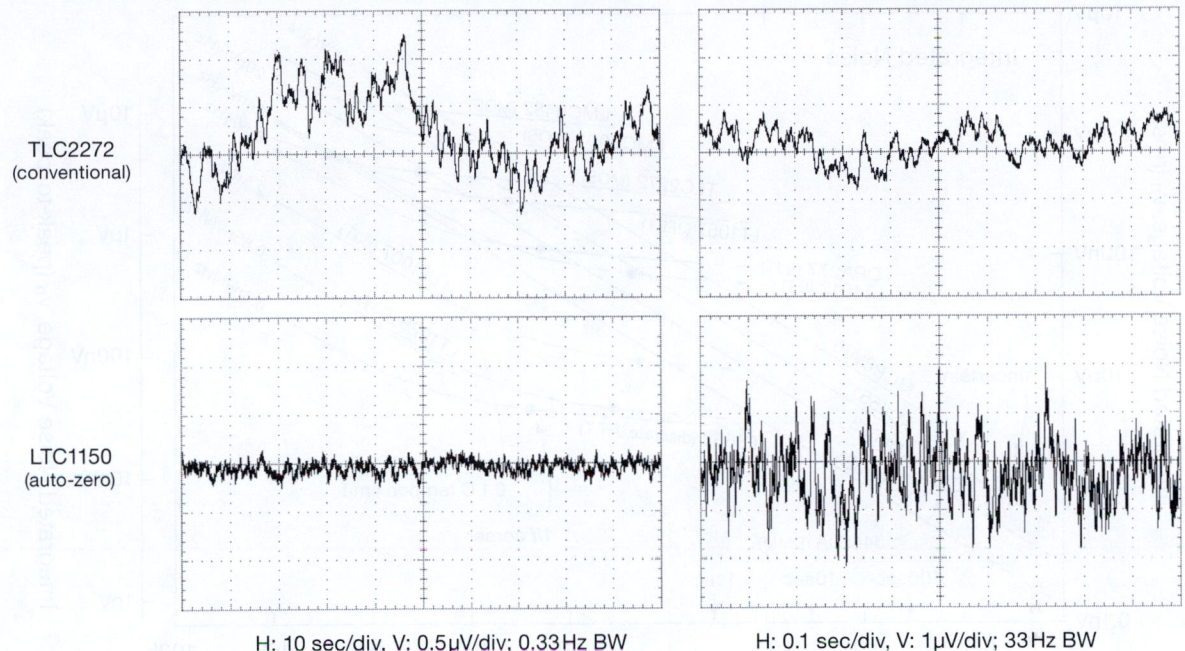

TLC2272
(conventional)

LTC1150
(auto-zero)

H: 10 sec/div, V: 0.5 μV/div; 0.33 Hz BW   H: 0.1 sec/div, V: 1 μV/div; 33 Hz BW

Figure 5.53. At very low frequencies a chopper-stabilized op-amp has lower noise than a conventional op-amp, but with 100× greater bandwidth it has more noise, as seen in these measured traces. See also Figure 5.54

ments yourself, if for no other reason than for "fact checking" the manufacturer. We ran some spectral noise plots for a half-dozen auto-zero amplifiers, with a particular interest in chopper-induced narrowband noise at the clock frequency and its harmonics. For these measurements we took data with $R_s = 0$ (to reveal the input voltage noise e_n), and then with $R_s = 1 M\Omega$ (to reveal input current noise i_n). Figure 5.52 shows the results, for one specimen from our collection of auto-zero amplifiers. The low-frequency measured e_n agrees well with the datasheet's value of $82 \, nV/\sqrt{Hz}$, but, as noted above, the measured current-noise density i_n is far greater than the specified value of $0.6 \, fA/\sqrt{Hz}$ – a factor of ×400 in this case.

For low-frequency applications you can (and should) RC-filter the output to a bandwidth of a few hundred hertz, which will suppress output spikes. This spiky input-current noise is also of no importance in applications with low input impedances, in integrating applications (e.g., integrating ADCs; see §13.8.3), or in applications in which the output is intrinsically slow (e.g., a thermocouple circuit with a meter at the output). In fact, if you want only very slow output response, and therefore lowpass-filter the output to extremely low frequencies (below 1 Hz), a chopper amplifier will actually have *less* noise than a conventional low-noise op-amp; see Figures 5.53 and 5.54.

Another way to put it is that auto-zero amplifiers have lots of *wideband* voltage noise (~$50 \, nV/\sqrt{Hz}$ at 1 kHz, compared with just a few $nV/\sqrt{Hz}$ for a good low-noise op-amp), but their noise density holds constant at very low frequencies, as contrasted with the ~$1/f$ ("flicker-noise") divergence of conventional op-amps (and everything else; see Chapter 8). For example, a conventional low-noise BJT op-amp like the LT1007 has $e_n = 2.5 \, nV/\sqrt{Hz}$ (typ) at 1 kHz, but its noise power density rises as $1/f$ below its "corner frequency" of 2 Hz, thus $e_n \sim 100 \, nV/\sqrt{Hz}$ at 0.001 Hz. Compare that with an auto-zero like the AD8551, with roughly flat $e_n = 42 \, nV/\sqrt{Hz}$: the latter will have far smaller fluctuations on time scales of minutes. In fact, the AD8551's datasheet even specifies a peak-to-peak noise voltage from "0 Hz to 1 Hz" of $0.32 \, \mu V$ (typ); no conventional op-amp would dare to project its drift out to infinite time!

A final problem with auto-zero amplifiers is their unfortunate overload recovery. What happens is this: the auto-zeroing circuit, in attempting to bring the input difference voltage to zero, implicitly assumes there is overall feedback operating. If the amplifier's output saturates (or if there is no external circuit to provide feedback), there will be a large differential input voltage, which the nulling amplifier sees as an input offset error; it therefore blindly

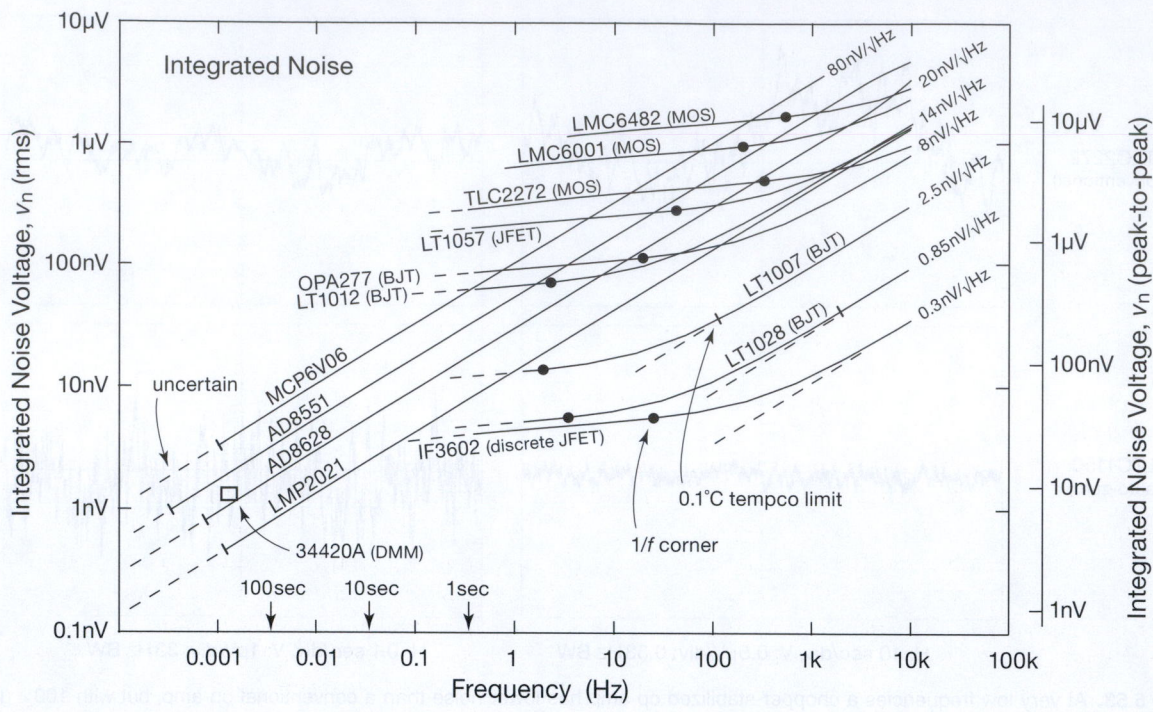

Figure 5.54. Integrated RMS voltage noise versus amplifier bandwidth. The "zero-drift" aspect of auto-zero amplifiers causes their low-frequency integrated noise voltage v_n to fall at low frequencies, proportional to the square root of the lowpass bandwidth. In contrast, the rising noise density $e_n \propto 1/\sqrt{f}$ of conventional op-amps causes a plateau in the integrated noise voltage v_n below the $1/f$ corner frequency, as seen in these computed curves. (We chose a lower limit of 0.01 Hz, when integrating the latter because the unbounded integral is divergent. This plot lets you see the region where the auto-zero op-amp wins or loses, compared with the op-amp of your choice. You can draw in estimated plots for other parts if you know e_n and the $1/f$ corner frequency. See §8.13.4.) See Figure 5.37 for the spectral noise-density plots behind these curves, and see Figure 8.63 for graphs of three dozen more op-amp types.

generates a large correction voltage that charges up the correction capacitors to a large voltage before the nulling amplifier itself finally saturates. Recovery is slow – t_r can extend up to several milliseconds. One "cure" is to sense when the output is approaching saturation, and clamp the input to prevent it. You can prevent saturation in chopper amplifiers (and in ordinary op-amps, as well) by bridging the feedback network with a bidirectional zener (two zeners in series), which clamps the output at the zener voltage, rather than letting it limit at the supply rail; this works best in the inverting configuration.

Alternatively, you can do an end run around this problem by choosing a part with fast recovery time, for example the OPA378 or OPA734 (with $t_r = 4\,\mu s$ and $8\,\mu s$, respectively).

5.11.2 When to use auto-zero op-amps

• Slow but accurate measurements from transducers: weigh scales, thermocouples, current shunts

• Accurate in-circuit dc conditioning, for example creating precise sets of voltages from a voltage reference

• "Normal-bandwidth" applications that want low-voltage and low-I_B CMOS, can tolerate broadband noise, require low offset voltages ($\ll 1\,mV$), and don't want to pay the cost premium of precision CMOS op-amps.

5.11.3 Selecting an auto-zero op-amp

Table 5.6 on page 335 lists a nice selection of currently available auto-zero op-amps, plus a few conventional op-amps for comparison. This is a good place to go first when you need an auto-zero amplifier. It's also a good place to learn about some of the common properties, and some of the quirks, of these amplifiers. Here are some comments, to get you started.

Supply voltage All but one of the parts are low voltage, 5.5–6 V max, and many will operate down to 2 V or less. Five can operate from ±5 V supplies. The supply currents range from $15\,\mu$A to 1.1 mA. The parts are listed approximately by supply current.

Input current Auto-zero amplifiers are built with CMOS, so the input currents are typically in the picoamps. We might expect the input currents to be in the single picoamp territory like other CMOS op-amps. Although there are a few parts for which this is true (e.g., MAX4238, MCP6V06 and LTC2054), most have considerably higher currents, up to 0.5 nA max, no doubt because of input-switch charge coupling. Even most conventional JFET op-amps do better except at high temperatures, where auto-zeros are usually much better. For example, I_B of the auto-zero LMP2021 typically stays below 75 pA at $125°$C (for any common-mode voltage), compared with the (conventional) JFET OPA124 (0.5 nA) and LF412 (10 nA). Auto-zero input currents are not as low as the best conventional CMOS parts (with their femtoamp currents), but considerably better than conventional precision BJT-input parts like the LT1028 or LT1007 on the table.[52]

Offset voltage This is where auto-zero amplifiers really shine, with maximum offset voltages ranging from 0.1–$5\,\mu$V typical (and $2\,\mu$V–$25\,\mu$V maximum – in the precision design business you've often got to pay serious attention to "maximum" specifications). A few conventional (not auto-zero) parts can approach this figure ($20\,\mu$V for the CMOS MAX4236A, $25\,\mu$V for the BJT LT1012A and LT1007), but they cannot begin to match the excellent tempco of the auto-zero parts (generally in the range of 5–20 nV/$°$C), gained, of course, by continual offset correction. Conventional op-amps also suffer from the devastating effects of $1/f$ noise, which sets performance floors in the 10–100 nV region; see Figure 5.54 and associated discussion.

The auto-zero amplifiers have typical offset-voltage tempco drifts from 4–100 nV/$°$C. The maximum specs range up to 250 nV/$°$C (and beyond? many parts don't list a maximum spec). The AD8628 and LMP2021 are the winners in this category. But these parts consume about 1 mA and thus can be expected to have more self-

heating of the die than say the $60\,\mu$A MAX9617 with its 5 nV/$°$C spec. No manufacturer can afford to perform temperature tests on production parts, so these specs should be taken with a grain of salt.

Is this performance believable? At the level of nV/$°$C you have to worry seriously about thermocouple effects in external connections, and even within the chip's lead frame itself: typical thermal EMFs are of order 5–$40\,\mu$V/$°$C – that's $1000\times$ (or more) the specified tempco of these auto-zero op-amps!

Voltage noise Auto-zero amplifiers exhibit higher broadband noise than conventional op-amps, owing to their CMOS inputs and associated switching elements. The voltage noise density e_n at 1 kHz (the usual benchmark) is of order 50–100 nV/$\sqrt{\text{Hz}}$, bested by many conventional CMOS parts, and all BJT parts. But, unlike conventional op-amps, the noise density does not rise at low frequencies, so the low-frequency integrated noise voltage (which you can think of as fluctuations, or drift) is better than even the best low-noise op-amps (as seen in Figures 5.53 and 5.54). Speaking approximately, the integrated-noise voltage falls as the $1/\sqrt{t}$ (or proportional to the square root of lowpass frequency).

In addition to e_n, a useful parameter in the table is the 0.1–10 Hz peak-peak voltage noise (v_n) specification. The AD8628, MAX9617, and OPA378[53] do very well with $0.5\,\mu$V to $0.4\,\mu$Vpp noise specs, but the LMP2021 is the clear winner with an e_n of 11 nV/$\sqrt{\text{Hz}}$ and a v_n of $0.26\,\mu$Vpp. This part is available in a convenient SOT23 package.

A note of caution, here: the very low-frequency extrapolation (i.e., long-term drift) must eventually become dominated by other drift sources (e.g., diffusion of impurities); see Bob Pease's column, "What's All This Long-Term Stability Stuff, Anyhow?" (published in *Electronic Design*, 20 July 2010).

Noise current The noise current density i_n must be at least the shot noise value (given by $i_n = \sqrt{2qI_B}$, with $q = 1.6\times10^{-19}$ C; thus 1.8 fA/$\sqrt{\text{Hz}}$ for a bias current of $I_B=10$ pA) corresponding to the input current I_B, generally in the range of a few fA/$\sqrt{\text{Hz}}$. In fact, for most of these parts it is much larger – by as much as factors of 10–100.

The current noise is $125\times$ higher than shot noise for

---

[52] Some parts warn that for high source impedances the bias current may change dramatically as a function of input *capacitance*! For example, the input current of an LMP2021 with $R_s = 1$ GΩ varies from -25 to $+25$ pA for an input shunt capacitance C_s ranging from 2 to 500 pF. Note that such input currents create large offsets with such high source resistances: 25 pA into 1 GΩ is 25 mV. A graph in the datasheet shows that the input current I_B goes through zero for $C_s =22$ pF. Other manufacturer's parts show similar effects. In a transimpedance amplifier with high R_F, using a large feedback capacitor C_F can dramatically reduce the bias-current error.

[53] A favorite of Phil Hobbs, who gushes "The OPA378 is a really beautiful zero-drift chopper-CAZ mutation that doesn't exhibit switching noise, and has constant 35 nV [per root hertz] noise down to DC. I've used it in an etalon-locked diode laser for downhole applications, and it's a thing of great charm."

the LMP2021, which was the winner of the voltage-noise competition. Parts that claim to do well in this regard (i.e., with input noise current approximately equal to the calculated shot noise[54]) include the AD8572, AD8551, and LTC1050. The MCP6V06 is the winner of the current-noise competition, with $0.6\,fA/\sqrt{Hz}$. This spec predicts $2\,\mu V$ from current noise through a $1\,G\Omega$ resistor in a 10 Hz bandwidth, about equal to the part's $1.7\,\mu V$ voltage noise v_n. The TLC4501A does equally well because it does its auto-zero only once, at powerup. But we know that the TLC4501A, unlike the MCP6V06, will do well at higher frequencies and with wider bandwidths because it has no busy auto-zero oscillator and switches. But it will do poorly at long time scales because of $1/f$ noise and multiple sources of drift.

A caution, loyal reader, as you make your choices – seven otherwise-attractive parts in the table, like the AD8538 and LMP2011, don't have any current-noise specs or plots. You may need to go to the bench to get your answer.

Slew rate and settling time For the listed parts the slew rates range from 0.04 to $2.5\,V/\mu s$, and the gain–bandwidths range from 0.13 to 4.7 MHz. The faster parts are meant to compete for use in ordinary op-amp sockets. For these parts, the settling time t_s is dominated by slew rate. But there are anomalies, for example the MCP6V06[55] and MAX4238, whose settling times are one or two orders of magnitude longer than the competition. This may be related to recovery time – the parts with recovery times in the milliseconds have very long settling times (the MAX4238), or they aren't willing to say (five other parts).

Input voltage range Most auto-zero op-amps do not support input voltages to the positive rail (though they all are rail-to-rail *output*). The MCP6V06, OPA333, ISL28133, MAX9617, and most of the Analog Devices parts are notable for full rail-to-rail input operation, without V_{OS} or CMRR degradation. The MAX9617 achieves this using an internal above-the-rail charge-pump power supply. Note also that the V_{OS} specifications may be conditioned on a restricted range of input voltage – most of the way to V_+ for some, others only partway. *Be sure to read the fine print in the specification!* For example, the OPA335 datasheet says "$(V_-) - 0.1V < V_{cm} < (V_+) - 1.5V$" next to its 130 dB

CMRR spec, and "$V_{cm} = V_s/2$" next to its $1\,\mu V$ offset-voltage spec.

Packages A few of the (older) Linear Technology types are available in DIP-8 packages for easy breadboarding. Otherwise you can use a SOIC-to-DIP or SOT23-to-DIP adapter (check out offerings from Aries or Bellin Dynamic Systems).

5.11.4 Auto-zero miscellany

A. ac-coupled "chopper amp"

When considering auto-zeroing chopper amplifiers, be sure you don't confuse this technique with another "chopper" technique, namely the traditional low-bandwidth chopper amplifier in which a small dc signal is converted to ac ("chopped") at a known frequency, amplified in ac-coupled amplifiers, then finally demodulated by multiplying with the same waveform used to chop the signal initially (Figure 5.55). This scheme is quite different from the full-bandwidth auto-zeroing technique we've been considering, in that it rolls off at signal frequencies approaching the clock frequency, typically just a few hundred hertz. You sometimes see it used in chart recorders and other low-frequency instrumentation.

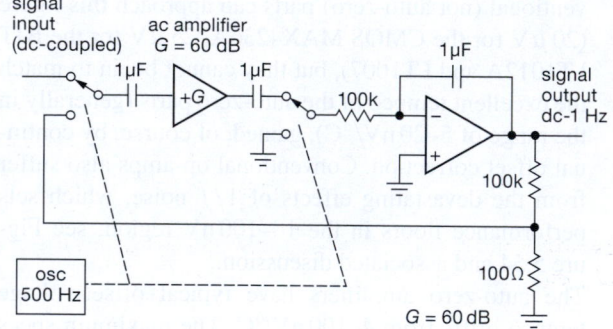

Figure 5.55. An ac-coupled chopper amplifier.

B. Thermal offsets

When you build dc amplifiers with submicrovolt offset voltages, you should be aware of *thermal offsets*, which are little thermally driven batteries produced by the junction of dissimilar metals. You get a Seebeck-effect "thermal EMF" when you have a pair of such junctions at different temperatures. In practice you usually have joints between wires with different plating; a thermal gradient, or even a little draft, can easily produce thermal voltages of a few microvolts. Even similar wires from different manufacturers can produce thermal EMFs of $0.2\,\mu V/°C$, 10 to a 100 times the

---

[54] At 10 Hz, who knows about higher frequencies?!

[55] The curious part number reminds us old timers of a favorite vacuum tube of yesteryear.

typical drift spec of the auto-zero amplifiers in Table 5.6 on page 335! The best approach is to strive for symmetrical wiring and component layouts, and then avoid drafts and gradients.

Here, for approximate guidance when worrying about parasitic thermocouples, Peltier shifts, and the like, are some thermoelectric-pair voltages (from Agilent AN-1389-1):

Copper-to-	Approx μV/°C
Copper	<0.3
Cd–Sn solder	0.2
Sn–Pb solder	5
Gold	0.5
Silver	0.5
Brass	3
Be–Cu	5
Aluminum	5
Kovar	42
Silicon	500
Copper oxide	1000

C. Power-up self-calibrating op-amps

Texas Instruments has an interesting approach to circumventing clocking noise in auto-zero amplifiers, namely to do it only once! Their TLC4501 family of "Self-calibrating (Self-Cal™) precision CMOS rail-to-rail output operational amplifiers" comes to life at power-up, performing an auto-zero and holding the correcting offset in an on-chip DAC. The good news is that you get no chopping noise and pretty good offset specs ($10\,\mu$V typ, $40\,\mu$V max), at least compared with those of typical CMOS op-amps. The less good news, as you might expect, is that the drift of offset with temperature is unspectacular (you might say "worst in class"), at ±1000 nV/°C typ, compared with $\sim$20 nV/°C for true auto-zero op-amps (see Table 5.6 on page 335).

D. The non-chopper competition

You can't beat the $\sim$1 μV (typ) offset specs of these auto-zero and chopper amplifiers, but you can do pretty well with the very best factory-trimmed precision op-amps. Bipolar op-amps like the LT1007 or LT1012 do quite well, at $10\,\mu$V (typ), and the remarkable CMOS MAX4236A has an offset spec of $5\,\mu$V typ (and $20\,\mu$V max). Note, however, that these factory-trimmed op-amps cannot come close to the drift specs of a true auto-zero: ±200 nV/°C (typ) for the bipolar types, and ±600 nV/°C for the CMOS, compared with $\sim$20 nV/°C for true auto-zero.

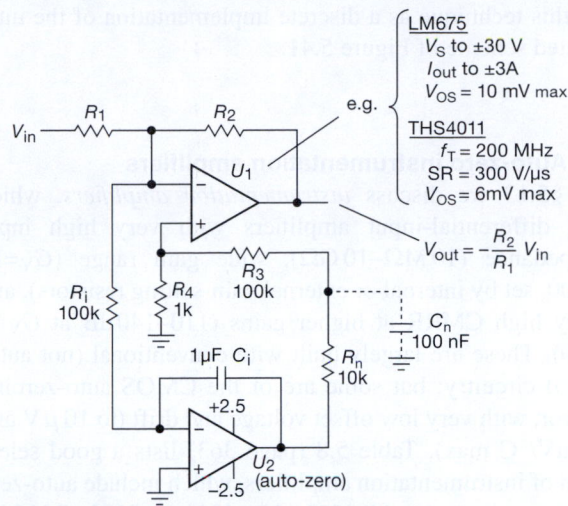

Figure 5.56. External auto-zeroing. But watch out for U_1's high bias current.

E. External auto-zero

You can use an auto-zero op-amp as an external offset trim for a conventional op-amp. This can be handy when you need a high-voltage, high-power, or high-speed op-amp whose input offset is too large. Figure 5.56 shows the scheme, which works most naturally with the inverting configuration, as shown.

The (low-voltage) auto-zero op-amp U_2 is configured as an integrator, looking at the error voltage at the inverting input of the conventional op-amp (U_1). The integrator's output is attenuated by $\times100$ to trim the voltage at the noninverting accordingly. With the values shown, the integrator's output responds according to $dV/dt = -\Delta V/R_iC_i$; thus a 10 μV error produces a $-100\,\mu$V/s integrator output and a $-1\,\mu$V/s correction at the noninverting terminal of U_1. A long time constant is both desirable (op-amp offsets drift only very slowly), and necessary (to prevent loop oscillation). Here the correction range is set by divider R_3R_4, so that a ±1 V integrator output produces a ±10 mV correction. The LM675 is a nice high-power op-amp (3 A output current, supply voltage to ±30 V, with sophisticated on-chip safe-operating-area and thermal protection), but with a maximum offset voltage of ±10 mV. The auto-zero reduces that by a factor of 1000. Similarly, the THS4011 is a fast op-amp ($f_T = 200$ MHz, SR = 300 V/μs) with a maximum offset voltage of ±6 mV. An additional noise filter R_nC_n at the output of the auto-zero, as shown, may be needed to suppress switching noise in the (slow) correction loop when this technique is used with small signals and low-noise parts like the THS4011 (7.5 nV/$\sqrt{\text{Hz}}$). You can think of this technique as a discrete implementation of the integrated scheme of Figure 5.41.

F. Auto-zero instrumentation amplifiers

In §5.13 we discuss *instrumentation amplifiers*, which are differential-input amplifiers with very high input impedance ($10\,M\Omega$–$10\,G\Omega$), wide gain range (G_V=1–1000, set by internal or external gain-setting resistors), and very high CMRR at higher gains (110–140 dB at G_V = 100). These are largely built with conventional (not auto-zero) circuitry; but some are of the CMOS auto-zeroing flavor, with very low offset voltage and drift (to $10\,\mu V$ and $20\,nV/°C$ max). Table 5.8 (page 363) lists a good selection of instrumentation amplifiers, which include auto-zero types such as the AD8553, AD8230, AD8293, INA333, LTC2053, and MAX4209. Some *conventional* instrumentation amplifiers that compete in this low-drift arena are the LTC1167/8 ($40\,\mu V$, $50\,nV/°C$ max) and the AD8221 ($25\,\mu V$, $300\,nV/°C$ max).

G. Do it yourself

If you like to get down into the guts of circuits, take a look at the LTC1043 "Precision Instrumentation Switched Capacitor Building Block." It lets you make your own high-CMRR differential amplifier. That's just one of its many tricks, which include switched-capacitor filters, oscillators, modulators, lock-in amplifiers, sample-and-hold, frequency-to-voltage conversion, and "flying capacitor" voltage inversion, multiplication, and division. The datasheet makes great bedtime reading.

5.12 Designs by the masters: Agilent's accurate DMMs

This is another in the series of featured "Designs by the Masters," in which we take a close look at some exemplary circuit designs. Think of these as *master classes* in circuit design. You can learn a lot by cracking open a well-designed instrument. A fine example is provided by the excellent digital multimeters from Agilent, specifically their 34401A (6.5-digit) and 34420A (7.5-digit) benchtop meters. In Chapter 13 (§13.8.6) we discuss the precision "multislope ADC" technique they use. Here, in the context of precision *analog* design, we look in some detail at the clever front ends they've designed, from schematics help-fully provided in their service manuals.[56] Let's see how the real pros do it!

5.12.1 It's *impossible!*

At first glance, the task is impossible. Here's why.

Accuracy We need accuracy and linearity at the parts-per-million level, in a meter whose full-scale ranges go down to a fraction of a volt (100 mV for the 34401A, 1 mV for the 34420A). That's far down in the nanovolt range.

Low noise Accuracy is useless if the instrumental noise makes successive measurements bounce around by many LSBs. So we need input voltage noise levels down at the nanovolt level for the most sensitive ranges.

High input impedance A voltmeter should have high input impedance to minimize circuit loading. So, for measurements at the ppm level, you'd like R_{in} to be something like a million times as large as typical circuit impedances. That puts you in the gigaohm range, with input currents down in the picoamps.

Hence the quandary: gigaohms and picoamps means FETs. Conventional FET op-amps won't deliver this performance, though, owing to relatively large offset voltage, drift, and voltage noise. AZ op-amps (see Table 5.6 on page 335) are considerably more accurate, but they suffer from plentiful current noise. And discrete JFETs (those with large area can have very low voltage noise, less than $1\,nV/\sqrt{Hz}$) with their uncertain I_D versus V_{GS} characteristics (§3.1.5) would seem to be hopeless at the fractional microvolt level. End of discussion.

5.12.2 Wrong – it *is* possible!

But it *can* be done. The trick is to realize that a digital instrument (with its on-board microprocessor brain) can calibrate away offsets (with a "zero" measurement) and scale errors (with a "full-scale" measurement), so that what matters is not the presence of offsets *per se*, but their stability (drift) during the measurement time.[57] That allows the use

---

[56] In the good ol' days, manufacturers proudly published their circuits. Nowadays it's far less common – for example, circuit diagrams are absent from the service manual for the Agilent 34410A (successor to the 34401A). Happily, some manufacturers (e.g., Stanford Research Systems) continue to display their circuit ingenuity, with full schematics and parts lists.

[57] A further trick is to average many such calibrate-measure cycles (up to 2 minutes' worth, in the 34420A) to beat down the scatter.

of discrete dual JFETs, with their unbeatable combination of low e_n and low I_B, in a JFET-enhanced op-amp configuration. That way you can get the high loop gain you need for linearity, particularly in the sensitive ranges where the front-end gain is 1000 or 10,000.

That's not the end of the story. You need precise resistor networks with low voltage coefficient, a circuit configuration that maintains its accuracy over a large common-mode input range (to ±10 V), and of course a voltage reference whose stability determines the instrument's overall accuracy.

5.12.3 Block diagram: a simple plan

These instruments leverage the power of "embedded control" (an on-board microcontroller) to deliver great performance from an architecture of great simplicity. The basic scheme (Figure 5.57) is simplicity itself: it consists of a single amplifier, configured in the familiar noninverting op-amp connection, with floating ground referenced to the (−) input jack. The microcontroller is the boss, here: its code implements the high-accuracy ADC (§13.8.6), and takes care of the multiple on-the-fly calibrations that are needed to wring part-per-million (or better) performance out of a collection of inexpensive parts. Let's dive into the guts of these two DMMs to see how it all works.

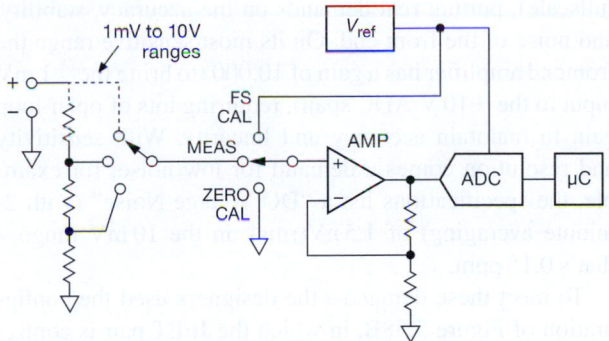

Figure 5.57. The Agilent DMMs: extreme simplicity... at the block diagram level.

5.12.4 The 34401A 6.5-digit front end

The 34401A made its debut in 1991, surprising the T&M (test and measurement) world with astonishingly good performance (resolution to 6.5 digits, measurements to 1000/s, accuracy to 20 ppm) at an affordable price (~$1k). The input amplifier (preceded by protection circuitry, and atten-

uators for the 100 V and 1000 V ranges[58]) provides gains of ×100 (100 mV range), ×10 (1 V range), and ×1 (10 V range) with $R_{in} > 10\,G\Omega$; the input attenuator kicks in for the 100 V and 1000 V ranges, for which $R_{in} = 10\,M\Omega$.

The basic structure is a low-noise precision op-amp (an OP-27), driven by a JFET source-follower pair, as shown in Figure 5.58A. (The configuration in Figure 5.58B, where a JFET common-source differential amplifier replaces the follower, is used in the 34420A to provide the additional loop gain and lower voltage noise needed for its more sensitive 1 mV and 10 mV full-scale ranges). The BJT-input op-amp provides lots (120 dB) of stable (0.2 μV/°C) low-noise (3 nV/√Hz) gain, but at a price: an unacceptable input current of ±15 nA, with correspondingly high input current noise (1.7 pA/√Hz). The JFET follower cures problems of input current and noise, at the expense of offset stability (40 μV/°C!) and significant added voltage noise (10 nV/√Hz). The tradeoff sounds bad – but it's good enough for this instrument. (It's *not* good enough for the more accurate and sensitive 34420A, as we'll see presently.)

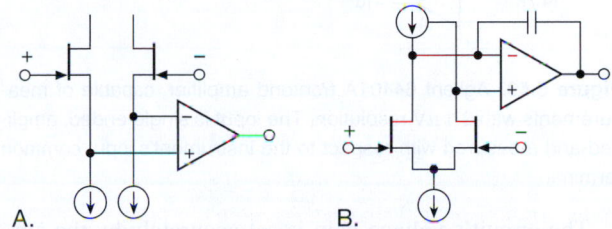

Figure 5.58. Basic JFET-enhanced op-amp configurations for the Agilent DMMs: A. source follower, used in the 34401A; B. common-source differential amplifier, used in the 34420A.

The full circuit is shown in Figure 5.59. Note first the bootstrapped drain supply for the JFET pair: Q_2 maintains a constant drain-source voltage across Q_1 (equal to Q_2's V_{GS} at the operating current, the latter held constant by the complicated-looking current sinks on Q_1's source terminals). This is essential, because the JFET pair Q_1 is anything but precise (would you believe, $V_{OS}(max)$=40 mV?!), and so the variation of this mediocre offset voltage with varying input signal (thus varying V_{DS}) would surely torpedo any expectation of accuracy. But by bootstrapping the drains to follow the sources, the transistors don't even know that there's any input signal variation; they don't know, so they can't mess things up. Furthermore, the low operating voltage (1–2 V) keeps the gate leakage small

---

[58] All have 20% "overrange," e.g., ±12 V on the "10 V" range.

and unchanging with input voltage variations over the full ± 15 V input signal range. Clever![59]

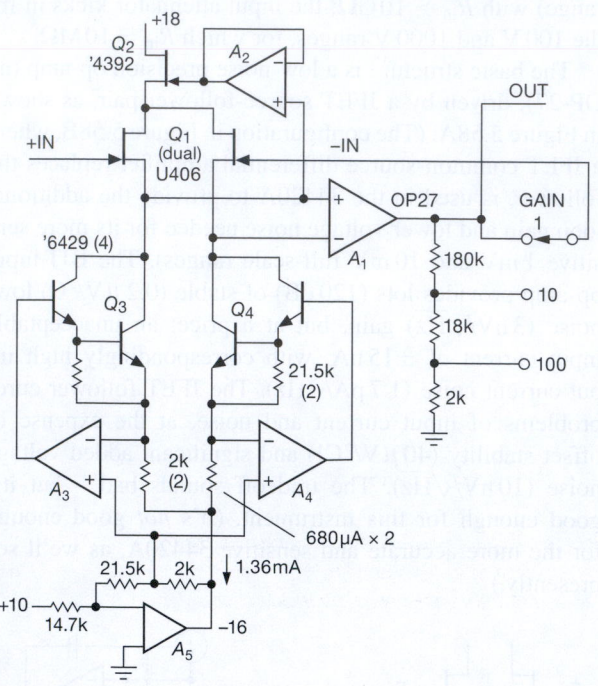

Figure 5.59. Agilent 34401A frontend amplifier, capable of measurements with 0.1 μV resolution. The input is single-ended, amplified and measured with respect to the instrument's input common terminal.

The circuit's voltage gain is set accurately by the analog switch and matched resistor network, implemented in a custom gain-switching IC.

The source pull-down circuitry is a current sink pair, based on the stable +10 V reference that is used also for the downstream ADC (see §13.8.6). It's easier to understand in the redrawn form of Figure 5.60, in which only one of the current sink pair is shown and the Darlington is replaced with a single *npn* transistor. The left-hand op-amp generates a voltage across R_2 of $V_{REF}R_2/R_1$; hence the sink current shown in the figure. In its DMM, Agilent uses a matched network for R_2 and the pair of R_3's (one for each source pull-down). The extra resistor R_4 offsets the emitter voltage downward, to $V_E = -V_{REF}R_4/R_1$, to provide the needed compliance for input signals that range over ± 15 V (± 12 V operating range, plus an additional 3 V to accommodate ripple and noise). If you plug in the resistor values

---

[59] It's necessary that Q_1's V_{GS} at 0.7 mA be less than Q_2's V_{GS} at 1.4 mA, because the difference is Q_1's V_{DS} operating voltage. It's likely that Agilent has an incoming batch inspection to ensure that this condition is met.

---

from Figure 5.59, you'll find that the compliance extends down to -14 V (emitter is at -14.6 V), and that the individual source pull-down currents are 680 μA. The designers used Darlington transistors to keep the base current error small (roughly $I_c/4500$, assuming a transistor beta of 200).

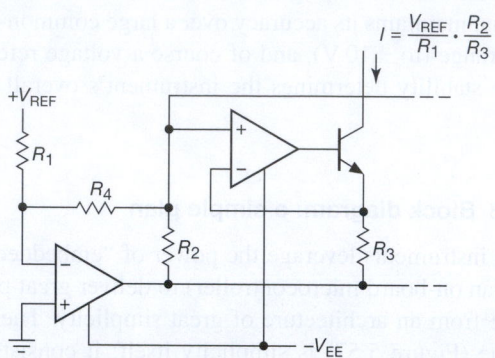

$$I = \frac{V_{REF}}{R_1} \cdot \frac{R_2}{R_3}$$

Figure 5.60. Agilent 34401A reference-based current sink.

5.12.5 The 34420A 7.5-digit frontend

With the 6.5-digit 34401A as a warmup, let's look at its wiser sibling, the 34420A 7.5-digit DMM. It boasts both improved resolution and greater sensitivity (1 mV fullscale), putting real demands on the accuracy, stability, and noise of the front end. On its most sensitive range the frontend amplifier has a gain of 10,000 (to bring the ± 1 mV input to the ± 10 V ADC span), requiring lots of open-loop gain to maintain accuracy and linearity. With sensitivity and resolution comes a demand for low noise; for example, the specifications list a "DC Voltage Noise" (with 2-minute averaging) of 1.5 nV(rms) on the 10 mV range – that's 0.15 ppm.

To meet these demands, the designers used the configuration of Figure 5.58B, in which the JFET pair is configured as a common-source differential amplifier for greater loop gain and reduced noise. The full amplifier circuit is shown in Figure 5.61.

Once again the JFETs are operated at constant current (2 mA each), with bootstrapped drains (held $V_Z - 2V_{BE} - 1$ V above the source, i.e., $V_{DS} \approx 2.5$ V). They chose JFETs with much larger geometry for greatly reduced noise voltage (an impressively low $e_n = 0.4$ nV/$\sqrt{Hz}$ at 10 Hz). These are *monster* JFETs: $I_{DSS} = 50$ mA min, 1000 mA max (how's that for a parameter spread?!), with an input capacitance of order 500 pF, and an unenviable offset voltage spec of ± 100 mV (with no tempco

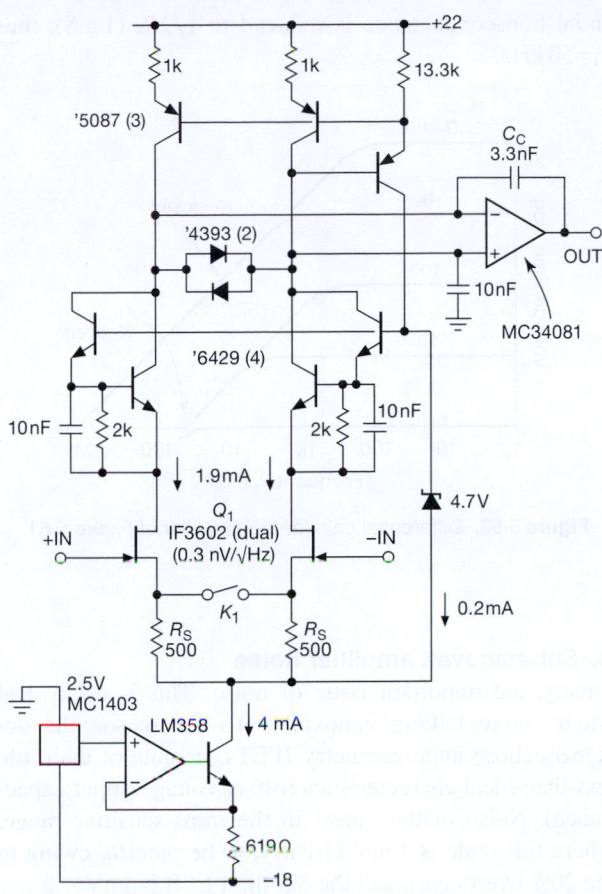

Figure 5.61. Agilent 34420A frontend gain block, used for measurements of 0.1 nV resolution with $G=10{,}000$ (gain-switching feedback shown in Figure 5.62).

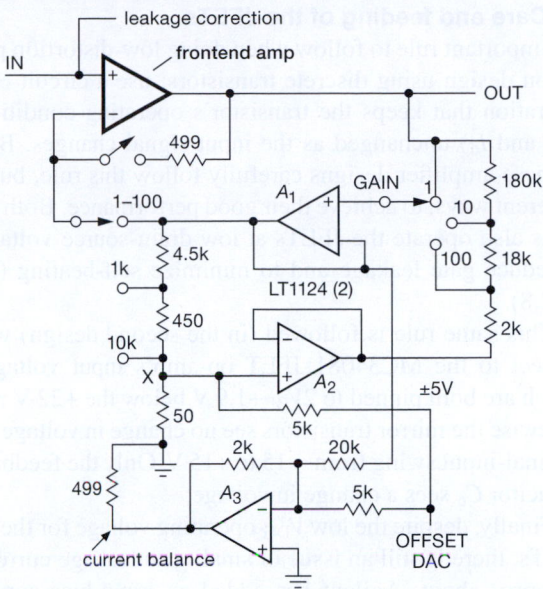

Figure 5.62. Agilent 34420A range-switching feedback circuit, wrapped around the "amplifier" of Figure 5.61. See §13.3.3 and Figure 13.15 for the leakage-correction circuit.

specified). This latter parameter would not appear to bode well for nanovolt measurements! (As we'll see presently, there's provision for continually trimming the measured offset.) These things are brutes, but they sure are quiet. We'll return to this circuit shortly, to deal with issues of gain, bandwidth, and noise. First, though, a look at the overall gain-setting loop.

A. Two-stage gain-setting loop

What's shown in Figure 5.61 is the bare amplifier, which sits in the feedback and trimming circuit of Figure 5.62. The gain selection is made with two stages of precision attenuators of high stability (these are custom modules), isolated by precision op-amp A_1 whose offset can be read and cancelled as part of the measurement cycle (by comparing the amplifier's output with the two possible $G=1$ configurations). The offset of A_2 can also be measured, from the

amplifier's output when the offset DAC is set to zero and the gain to ×100.

The "offset DAC" is used to dispatch the big elephant in the room (the JFET pair's offset, which can range to ~50 mV). This it does by generating a voltage output (±5 V range), which offsets the node marked "x" over a range of ±50 mV. Follower A_2 replicates that offset at the bottom of the right-hand gain-setting divider, establishing an effective ground-reference point for the first-stage divider. Here's a subtle point: with a circuit of this precision (in the 1 mV range, the LSB is just 1 nV) you have to worry about stuff you normally ignore – for example the effect of voltage drops through ground path resistance. Here the offset DAC may sink or source up to 1 mA (5 V across 5 kΩ), which would push "ground" around by an unacceptable 100 nV if its path had, say, 0.1 mΩ of resistance. That's why the designers added A_3, whose output pushes a balancing current of opposite sign into the same ground node; if that current is matched to 1%, the error is reduced to 1 nV.

This raises a related point: when talking nanovolt stability, don't we have to worry about A_1's drift? Quite so – but the effect of splitting the gain-setting attenuator into two sections is to reduce that drift by the attenuation of the left-hand divider, i.e., ×100 in the most sensitive range.

B. Care and feeding of the JFETs

An important rule to follow when doing low-distortion precision design using discrete transistors: use a circuit configuration that keeps the transistor's operating conditions (V_{ds} and I_d) unchanged as the input signal changes. Both of these amplifier designs carefully follow this rule, but in different ways, to achieve their good performance. Both designs also operate the JFETs at low drain-source voltages to reduce gate leakage and to minimize self-heating (see §3.2.8).

This same rule is followed (in the second design) with respect to the MC34081 JFET op-amp's input voltages, which are both pinned to $2V_{BE}+1.9$ V below the +22 V rail. Likewise the mirror transistors see no change in voltage for a signal-input swing from -15 to $+15$ V. Only the feedback capacitor C_c sees a change in voltage.

Finally, despite the low V_{DG} operating voltage for the Q_1 JFETs, there is still an issue of small gate leakage currents to worry about. Agilent has added an input-bias-current correction circuit, with an 8-bit DAC, to solve this problem. We discuss how this interesting circuit works in §13.3.3.

C. Amplifier gain: ×1 to ×10,000, stable to 0.1 ppm

Turning back to the amplifier (Figure 5.61), we can understand some nice subtleties. A closed-loop gain of ×10,000 is needed in the 1 mV range, for which lots of open-loop gain is needed. The JFET differential amplifier provides gain ahead of the op-amp; though the gain is not easily calculated at dc (it depends on the impedance of the current-mirror drain load), we can estimate its gain–bandwidth product f_T by noting that compensation capacitor C_c makes the differential gain roll off according to $G = g_m X_C/2$, where g_m is the transconductance of each JFET at the operating current (relay K_1 is closed for the 1 mV and 10 mV ranges, removing the 500 Ω source degeneration resistors). f_T is the frequency at which the gain of the composite amplifier has fallen to unity, i.e., $f_T = g_m/4\pi C_c$. To estimate g_m, we note that these JFETs are operating well down in the subthreshold region (their I_{DSS} is typically 300 mA), where FETs behave more like BJTs (I_D exponential in V_{GS}; see Figure 3.15), with their transconductance proportional to drain current and with g_m only somewhat less than a BJT operating at the same current. For the IF3602 JFET running at $I_D=2$ mA, then, we can estimate $g_m\approx60$ mS (a BJT would have $g_m=40I_c$ mS), thus $f_T=1.5$ MHz.

Running this backward, we find that the open-loop gain is about 10^6 at 1 Hz, as seen in Figure 5.63. Source degeneration is enabled for the low-gain ranges, to maintain stability. The rolloff is easy to calculate, because the differ-

ential transconductance is reduced to $1/2R_s$ (1 mS), thus $f_T=50$ kHz.

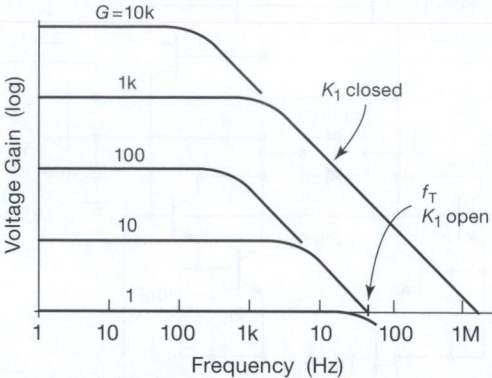

Figure 5.63. Differential gain for the amplifier of Figure 5.61.

D. Sub-nanovolt amplifier noise

Finally, the important issue of noise. This is a big deal when you're talking nanovolts; it's the reason the designers chose huge-geometry JFETs, in spite of their, uh, less-than-ideal characteristics (offset voltage, input capacitance). Noise matters most in the most sensitive range, where full scale is 1 mV (1.2 mV, to be precise, owing to the 20% overrange), and the 6.5-digit LSB is 1 nV.

There are several noise sources here. The JFETs contribute about 1 nVrms in a 3 Hz band around 1 Hz ($e_n=0.4$ nV/$\sqrt{\text{Hz}}$ each, multiplied by 1.4 for uncorrelated noise). To explore further the measurement fluctuations, take a look at Figure 5.54, where we show the voltage noise of various op-amps, of both conventional and chopper-stabilized varieties. The traces show the integrated rms noise voltage up to a cutoff frequency (x-axis), including the effect of the component's $1/f$ noise. The IF3602 is the lowest-noise part on the graph. If we assume an integration time of 100 PLC (powerline cycles) or 1.67 s, to achieve 7.5-digit performance, that interval corresponds to a 0.6 Hz cutoff frequency, and about 3 nV rms of noise. If we average 64 such measurements over a two-minute period, we could hope for the rms fluctuations to be reduced by 8×, to about 0.4 nV. Agilent claims 1.3 nV in their datasheet, evidently allowing for some nonrandom variations and other errors.

Noise in the current sink is of lesser importance because the differential stage cancels it to a high degree; that's a good thing, because this design uses a noisy voltage reference! (the MC1403 is an early bandgap design, with unspecified noise voltage).[60] The 34420A's digitizing capability drops from 7.5 to 6.5 digits when operating faster than 20 PLC, or 1.5 readings/sec, and further drops to 5.5 digits above 25 rd/s, and 4.5 digits above 250 rd/s, so rising high-frequency amplifier noise wouldn't be noticed.

E. Going beyond the specifications
When pushing the limits of the possible, you often find that the job cannot be done while respecting worst-case component specifications. Here, for example, the critical JFET transistor pair has a worst-case specified gate leakage current of 500 pA (at 25°C), whereas the instrument specifies a maximum input current of 50 pA. What to do?

If you are a major manufacturer, you can often persuade the supplier to screen parts to a tighter specification. In any case, you can do the job yourself. Be aware, though, that there's generally no guarantee of process continuity, and the availability of better-than-specified parts; worse still, the special parts you need may be discontinued altogether! One possibility, if you're willing to hazard a guess as to an instrument's long-term popularity, is to buy a lifetime supply of a critical part.

5.13 Difference, differential, and instrumentation amplifiers: introduction
These terms describe a class of dc-coupled amplifiers that accept a differential signal-input pair (call them V_{in+} and V_{in-}), and output either a single-ended signal or a differential-output pair that is accurately proportional to the difference: $V_{out} = G_V \Delta V_{in} = G_V(V_{in+} - V_{in-})$. Their shared claim to fame is high common-mode rejection, combined with excellent accuracy and stability of voltage gain. Here are their distinctive features, as commonly understood among circuit designers.

Difference amplifier differential in, single ended out; op-amp plus two matched resistor pairs (Figure 4.9, §4.2.4, and Figure 5.65); CMRR 90–100 dB; accurate but low gain (G_V=0.1–10); input impedance 25–100k, intended

to be driven by a low impedance; inputs typically can go beyond rails.

Instrumentation amplifier differential in, single ended out; very high input impedance (10 MΩ–10 GΩ), wide gain range (G_V=1–1000), and very high CMRR at higher gains (110–140 dB at G_V=100); §5.15, e.g., Figure 5.77.

Differential amplifier differential or single ended in, differential out; most are low voltage, fast settling, and wideband; ideal for twisted-pair cable drivers and fast differential-input ADCs; §5.17, e.g., Figure 5.95.

An obvious application is the recovery of a signal that is inherently differential, but that rides on some common-mode level or that is afflicted by common-mode interference. Figure 5.64 shows an example of each.

The first example is the strain gauge we saw earlier (§5.4), a bridge arrangement of resistors that converts strain (elongation) of the material to which it is attached into resistance changes; the net result is a small change in differential-output voltage when powered by a fixed dc-bias voltage. The resistors all have approximately the same resistance, typically 350 Ω, but they are subjected to differing strains. The full-scale sensitivity is typically ±2 mV per volt, so that the full-scale output is ±10 mV for 5 V dc excitation. This small differential-output voltage (proportional to strain) rides on a +2.5 V dc level. The differential-input amplifier must have extremely good CMRR in order to amplify the millivolt differential signals while rejecting the ~2.5 V common-mode signal and its variations. For example, suppose you want a maximum error of 0.1% of full scale. That's ±0.01 mV, riding on 2500 mV, which amounts to a CMRR of 250,000:1, or 108 dB. This overestimates the needed CMRR: in practice you would perform a "zero calibration," so that the CMRR need be adequate only to reject *variations* in the +5 V bridge bias; something like 60 dB would suffice here.[61]

The second example (Figure 5.64B) comes from the world of professional audio, where you encounter some pretty impressive challenges. In a concert recording situation, for example, you may have microphones hanging from a high ceiling, with connecting cables 100 m or more in length. The peak signal levels may be around a volt, dropping to a millivolt during quiet portions of the music. But you've got to keep powerline pickup and

---

[60] For single-ended amplifiers we would want the current-source noise i_n to be less than $e_n(amp)g_m$. We can use the expression $e_n(ref)/e_n(amp) = g_m R_S$ to determine the allowable voltage noise in the current-source reference. For this circuit that ratio is 37, thus only 11 nV/$\sqrt{Hz}$ for a noise contribution comparable to that of the 0.3 nV/$\sqrt{Hz}$ JFET. The MC1403 reference is about 20× worse than this. Evidently the Agilent engineers are relying on the matched noise currents in the two JFETs to cancel to better than 5%, enforced by the 1% current-mirror resistors. At frequencies above about 10 Hz, however, the 10 nF capacitor defeats this cancellation.

[61] We'll see the strain gauge again, in connection with analog-to-digital converters, in §13.9.11C (Figure 13.67). A similar biased bridge arrangement is used in the platinum resistance temperature detector (RTD), which is the sensor used in the microcontroller-based thermal controller in §15.6.

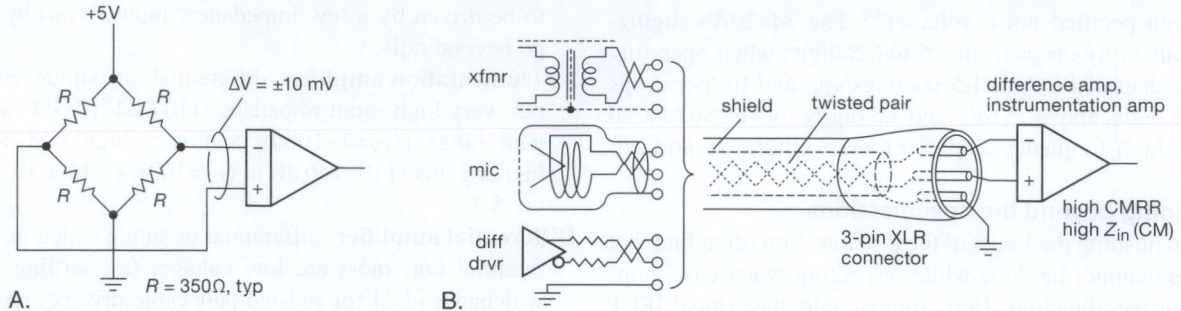

Figure 5.64. Inherently differential signals, for which good common-mode rejection is required. (a) Strain gauge. (b) Audio balanced line pair.

other annoyances (e.g., switching noise from lighting dimmers) another 40 dB below that – the human ear is distressingly sensitive to extraneous sounds. Add up the dBs – we need 100,000:1 suppression of pickup ($<10\,\mu$V)! Seems impossible; but recording engineers have been doing this successfully for decades by the simple expedient of transporting audio signals on a balanced differential pair (with a standard signal impedance of $150\,\Omega$ or $600\,\Omega$). For this they use a well-shielded twisted-pair cable, terminated in the legendary 3-pin XLR latching connector (which can take plenty of abuse – it's got a tough metal shroud, good strain relief, and so on). And to keep the signal fully balanced they use either a high-quality audio transformer or a well-designed differential-output driver (at the transmit end), and another transformer or a well-designed differential-input amplifier (at the receive end).

Lest we leave a misleading impression, we hasten to note that differential-input amplifiers are helpful also in situations in which the signals themselves are not inherently differential. Two common examples are accurate low-side current sensing (Figure 5.68a), and the use of differential-input amplifiers when sending signals between instruments (Figure 5.67). In the latter, the flexibility provided by a difference amplifier's output REF pin enables us to avoid ground loops while transporting a signal between a pair of instruments whose local grounds are not identical.

The tricks involved in making good instrumentation amplifiers and, more generally, high-gain differential amplifiers are similar to the precision techniques discussed earlier. Bias current, offsets, and CMRR errors are all important. Let's begin by discussing the design of difference amplifiers for less critical applications, working up to the most demanding instrumentation requirements and their circuit solutions.

5.14 Difference amplifier

Let's look first at the difference amplifier: its basic operation, some applications, a closer look at its performance parameters, and finally some clever circuit variations.

5.14.1 Basic circuit operation

The classic difference amplifier (Figure 5.65) consists of an op-amp with matched resistor pairs R_f and R_i, for which the differential gain is

$$G_{\text{diff}} \equiv \frac{V_{\text{out}}}{V_{\text{in}+} - V_{\text{in}-}} = R_f / R_i.$$

Assuming an ideal op-amp, the common-mode rejection

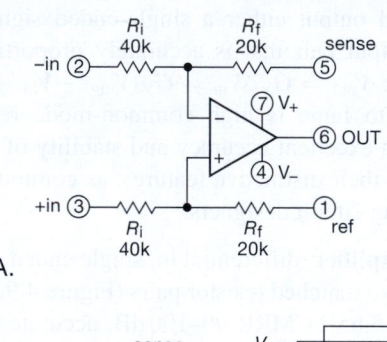

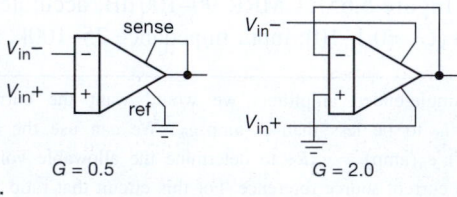

Figure 5.65. Classic difference amplifier, with the resistor values used in the AD8278/9 (G=0.5 or 2).

is limited by the matching of the R_f/R_i ratio in the two paths; with discrete resistors of 1% tolerance, for example,

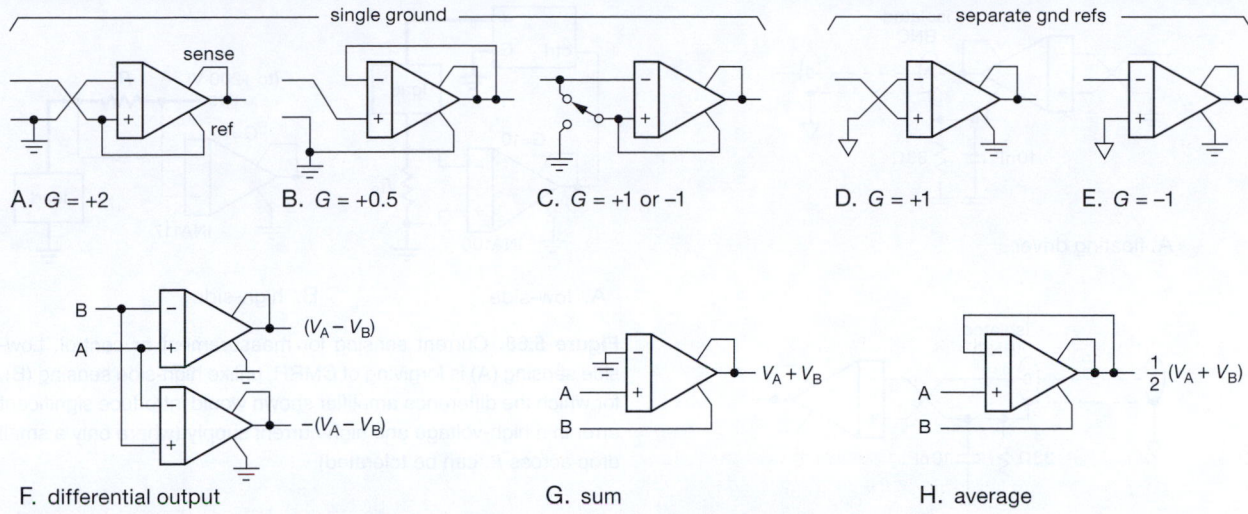

Figure 5.66. The tricks that a unity-gain difference amplifier can play. Note the separate ground symbols for (D) and (E).

you could expect a CMRR of ~40 dB at low frequencies (where the op-amp itself has a higher CMRR and where effects of capacitive imbalance are negligible). That's adequate for situations in which only a modest CMRR is needed, for example low-side current sensing. For better performance you could trim one of the resistors or use resistors of tighter tolerance (for example the commonly available 0.1% types, typically $0.10–$0.20 in 100-pc quantity, or the inexpensive Susumu RG-series, with 0.05% tolerance and 10 ppm/°C tempco, about $1); or, better, use matched resistor pairs (e.g., the not-inexpensive Vishay MPM-series, with ratio tolerances down to 0.01% and ratio tempcos to 2 ppm/°C or the LT5400 matched quad, with similar ratio tolerance and with ratio tempco of 1 ppm/°C max).

But don't get carried away... because there are plentiful offerings of complete integrated difference amplifiers of excellent performance, costing a lot less than you'll spend rolling your own. We've listed many of these in Table 5.7 on page 353. In the "normal" configuration the SENSE line is connected to the output and the REF line is connected to circuit ground. But you can run it in reverse, as shown in Figure 5.65B. To give a sense of the performance you get with these integrated difference amplifiers, the worst-case specifications for the AD8278B in the figure are gain accuracy of ±0.02% (for G=0.5 or G=2); gain tempco of 1 ppm/°C; offset voltage and tempco of 100 μV and 1 μV/°C; and a CMRR of 80 dB. It costs about $3.

5.14.2 Some applications

A. Single-ended input

It's perfectly OK to use a difference amplifier with a single-ended input, for example, to get a precise and stable gain. Figure 5.66 shows some simple configurations. Note that differential nature of its inputs is not "wasted" when a difference amplifier is used as in (D) and (E), because its independent REF pin accommodates small differences in ground potential at input and output. Put another way, the output voltage, relative to *its* ground, is precisely ±1.0 (in this case) times the input voltage, relative to *its* ground.

B. Ground-loop isolation

This property is just what you need for the ground-loop isolating application of Figure 5.67. In the first circuit, the driver side permits its output reference to float to the potential of the (non-floating) receive side. The small-value resistor and bypass capacitor allow a small voltage difference when forced by the receive end. Both sides are happy. In the second circuit we've allowed the receive side's input common to float, as forced by the (non-floating) driver. These circuits resolve minor ground-loop problems in single-ended cable connections; but they are no substitute for the fully isolated and/or balanced approach that's needed in demanding applications like professional audio or video.

C. Current sensing

Figure 5.68 shows difference amplifiers used for low-side and high-side current sensing, perhaps as part of a constant-

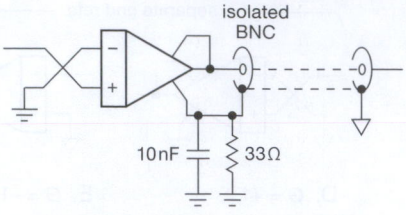

A. floating driver

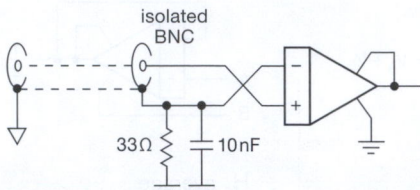

B. floating receiver

Figure 5.67. Exploiting the REF pin to prevent powerline-frequency ground loops in connected instruments.

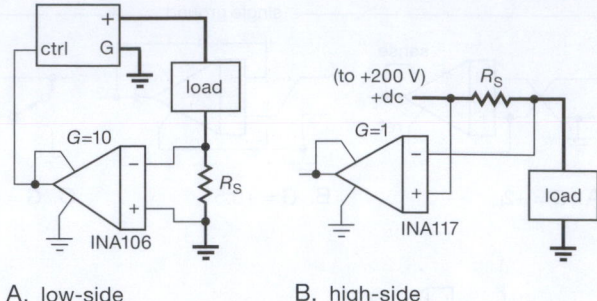

A. low-side   B. high-side

Figure 5.68. Current sensing for measurement or control. Low-side sensing (A) is forgiving of CMRR, unlike high-side sensing (B), for which the difference amplifier shown would introduce significant error in a high-voltage and high-current supply (where only a small drop across R_s can be tolerated).

current control, or simply for precise load-current monitoring. At first glance it may seem unnecessary to use a difference amplifier for the low-side sensing configuration, because the sense resistor returns to circuit ground. But imagine that we're dealing with lots of power, say up to 10 A load current. We'd use a low-value precision sense resistor, perhaps $0.01\,\Omega$, to keep its power dissipation below 1 W. Even though one side of it is connected to ground, it would be unwise to use a single-ended amplifier, because a connection resistance of just a milliohm would contribute 10% error! A differential-input amplifier is the solution, connected as shown to a 4-wire "Kelvin-connected" sense resistor. Note that the difference amplifier need not have particularly good CMRR, because the low side isn't moving far from ground.

The same is not true of the high-side configuration (Figure 5.68B), where the common-mode voltage is far greater than the differential voltage. Here we've specified a unity-gain difference amplifier intended for high-voltage applications, which permits common-mode input voltages to $\pm200\,$V (its internal circuit uses a 20:1 resistive divider at the front end, see §5.14.3). Let's look at the numbers: the dc voltage can range from 0 to 200 V; so the specified minimum CMRR of 86 dB (1:20,000) would interpret that as equivalent to a differential input variation of 10 mV. Is that bad? You bet! To maintain 1% accuracy of sensed current we'd need to size R_s to drop 1 V at full load current. That's a lot of "voltage burden," and it's a whopping 10 W

of dissipated power, if this were a powerful 10 A dc supply as in the other example. So this scheme is adequate for a low-current 200 V supply; but there are better ways to do high-side sensing, for example by floating the amplifier and relaying the output down to earth as a current or (via an opto-isolator) as a digitized quantity.

D. Current sources

Figure 5.69 shows how to rig up a difference amplifier so that the (differential) input signal controls the voltage drop across a series sense resistor R_s; in other words, it's a current source. You're welcome to use a single-ended control input, if you want. The output current can be of either polarity, and these circuits don't know, or care, whether the load returns to ground or to some other potential.

The op-amp follower U_2 should be chosen for an input bias current that is small compared with the minimum load current, and an offset voltage that is small compared with the drop across R_s at minimum load current. You might begin by choosing R_s for ~1 V drop at $I_L(\text{max})$, or smaller for high $I_L(\text{max})$ or for low supply voltage, then see if $R_sI_L(\text{min})$ is at least $100\,\mu$V or so. For a large dynamic range, say $I_L(\text{max})/I_L(\text{min})\geq10^4$, you'll need low-offset amplifiers for both U_1 and U_2; candidates are listed in Tables 5.2 (page 302), 5.3 (page 303), 5.5 (pages 320–321) and 5.6 (page 335).

The circuit in Figure 5.69A is good for low currents. For load currents greater than ~5 mA use a power buffer (Figure 5.69B); that can be a wideband (for stability) unity-gain IC buffer like the LT1010, or (if you need only one polarity of output current) a MOSFET or BJT follower. A favorite trick is to use a 3-terminal adjustable regulator (like an LM317) as a "power buffer," thus taking advantage of its internal thermal and overcurrent protection. To use it this way, you drive the ADJ pin, and the OUT pin "follows" 1.25 V higher (should this be called a voltage *leader*?!). As always, feedback takes care of the offset.

We like using $G = 10$ difference amplifiers, such as INA106 or INA143, connected "backwards" for $G = 0.1$, because the sense voltage is then one-tenth of the programming voltage and thus doesn't eat up so much of the output compliance range. In common with other current sources whose output comes from an op-amp, these configurations become more like *voltage* sources at high frequencies, where op-amp compensation and slew-rate effects dominate (see §§4.2.5 and 4.4.4).[62] Better performance at high frequencies can be gotten with an active current source based on an instrumentation amplifier, configured with an inherently high impedance output terminal, as shown in Figure 5.87 in §5.16.9.

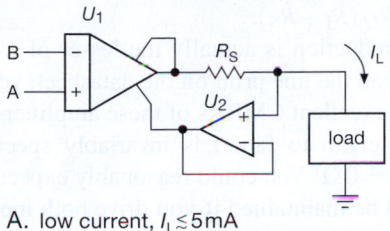

A. low current, $I_L \lesssim 5$ mA

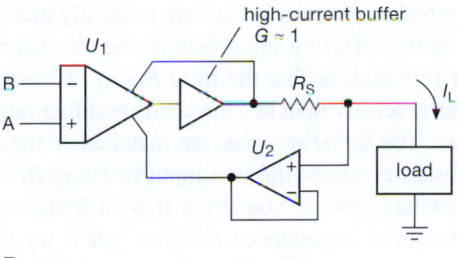

B. "any" current, R_S: mΩ to GΩ

Figure 5.69. Precision current sources: $I_L = G_{diff}(V_A - V_B)/R_S$. The (bipolarity) output current in (A) is limited to U_1's I_{out} (max). Adding a unity-gain power buffer (an IC wideband buffer, or a transistor follower) in (B) allows large output currents (follower U_2 can be omitted if R_S is less than 0.2 Ω or so). These circuits don't know, or care, where the load returns.

E. High-level line driver

Professional audio lives and breathes *differential* analog signaling, in the form of balanced lines terminated (usually) in a nominal 600 Ω bridging resistance. And the levels are substantial: pro-audio equipment adopts a "0 dB" standard of 1.23 Vrms,[63] and you'll generally see additional specified headroom of 16 dB to 20 dB without clipping. So a +20 dB level is 12.3 Vrms, or a differential amplitude of 17.4 V (34.8 Vpp). That requires some serious attention to line drivers, which should not compromise the low noise and distortion qualities of the program material.

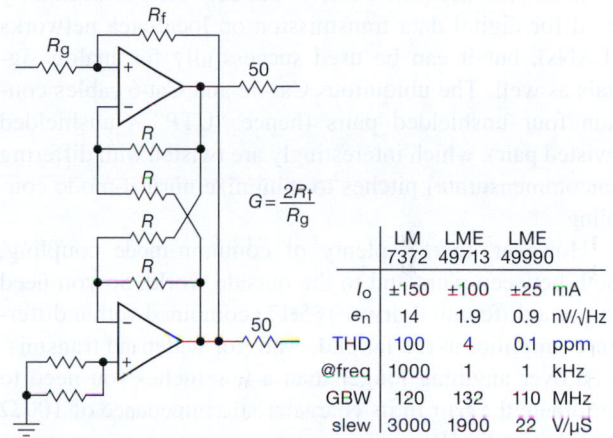

Figure 5.70. Differential high-level line driver for professional audio.

Figure 5.70 shows a nice $G = 2$ circuit for the job, based around a pair of unity-gain difference amplifiers. These are implemented here with wideband op-amps of substantial output-current capability, allowing for an overall gain (differential-output voltage divided by single-ended input voltage) other than $\times 2$. The particular listed op-amps all run from supply voltages to ± 18 V, producing output swings (on each line of the pair) to ± 15 V or so (significantly greater than the ± 9 V corresponding to a 20 dB pro-audio headroom). Three op-amp choices are shown, of comparable bandwidth ($\sim$100 MHz) but aimed at different applications. The LM7372 is well specified out to 10 MHz and intended for video and broadcast applications, whereas the LME49xxx parts are optimized for audio bandwidths, with pretty impressive specifications out to 20 kHz. It's not often (more like *never!*) that you see amplifiers with total harmonic distortion (THD) specifications of 0.1 ppm

---

[62] You can define an effective current-source output *capacitance* $C_{eff} = I_{out}/S$ (where S is the output slew rate) as a way to characterize this shortcoming.

[63] The base unit for audio level is "0 dBu"(sometime written "0 dBv"), an rms voltage corresponding to 1 mW into a 600 Ω load; that works out to 0.775 Vrms. Pro-audio's 0 dB level is +4 dBu, hence 1.23 Vrms; home audio is considerably less muscular, at -10 dBu, or 0.25 Vrms.

($-140\,\text{dB}$), here combined with very low voltage noise ($1.4\,\text{nV}/\sqrt{\text{Hz}}$ at 10 Hz).[64]

Some interesting alternatives are the DRV134 and the LME49724, which integrate a fully differential output circuit, capable of comparable performance, into one IC. The latter are examples of *fully-differential amplifiers*, with differential inputs as well; one input can be grounded to work with single-ended signal sources. We'll see these later, in §5.17.

F. Wideband analog over twisted pair

Twisted-pair network cable ("Cat-5e," etc.) is ordinarily used for digital data transmission on local area networks (LANs), but it can be used successfully for analog signals as well. The ubiquitous Cat-5e and Cat-6 cables contain four unshielded pairs (hence "UTP" – unshielded twisted pair), which interestingly are twisted with differing (incommensurate) pitches to minimize normal-mode coupling.

However, there's plenty of common-mode coupling, both between pairs and to the outside world. So you need to use a differential driver (§5.17) combined with a difference amplifier at the far end. And, for wideband transmission over anything longer than a few inches you need to terminate the pair in its characteristic impedance of $100\,\Omega$ (see Appendix H).

Figure 5.71 shows how to use a difference amplifier as an analog "line receiver" for such signals (we'll show the driving end later, in §5.17.2). This circuit comes from the AD8130's datasheet, illustrating the use of some "peaking" to compensate for the signal attenuation at high frequencies in the rather long (300m!) cable. The amplifier has $G = 3$ at low frequencies, with $R_4 C_1$ kicking in around 1.6 MHz (where C_1's reactance equals R_3) to boost the sagging response. The result of this simple "equalizer" is to produce an overall response flat to $\pm 1\,\text{dB}$ from dc to 9 MHz. The $\times 3$ low-frequency gain is needed (a) to compensate for the signal that is lost because of the resistive divider consisting of R_1 and the cable pair's $50.5\,\Omega$ round-trip resistance (a factor of 1.5), and (b) to double the output signal so it can drive a "back-terminated" video cable (see Appendix H).

5.14.3 Performance parameters

In the preceding sections we glossed over some important issues: input impedance (both differential and common

mode) and its effect on gain when driven with finite source impedance or its effect on CMRR when driven from an unbalanced source impedance; common-mode input range; and amplifier bandwidth and its effect on CMRR. It's time to peel the onion.

A. Input impedance

From Figure 5.65 it might seem that the input impedance of a difference amplifier is R_i (or maybe some multiple of that), and that all is well if the signal driving it has a source (Thévenin) impedance R_S that is much less, perhaps our usual seat-of-the-pants criterion of $R_S \lesssim 10 R_i$.

Not so! These amplifiers revel in their precise gain (the selections in Table 5.7 on the next page have worst-case gain errors of 0.1% or better), and that gain accuracy is compromised unless R_S is smaller than the amplifier's R_I by at least that factor. That's because the signal's source impedance is effectively in series with R_I, reducing the gain by the factor $R_I/(R_I + R_S)$.

That gain reduction is actually the lesser of two problems. If you read the fine print on the datasheet, you'll discover that the excellent CMRRs of these amplifiers (surely their greatest claim to fame) is invariably specified for (drumroll) $R_S = 0\,\Omega$! You could reasonably expect that the CMRR would be maintained if you drive both inputs with equal source impedances, because that should maintain the resistor ratio match. You could expect that, but you would be disappointed: the CMRR degrades rapidly with rising source impedances, even if they are precisely matched.

Why is that? During manufacture the internal resistors are laser trimmed, so that the ratio R_F/R_I of the upper resistor pair precisely matches the corresponding ratio of the lower pair. The *ratios* are what are matched, at the expense of the absolute values; the two input resistors R_I may differ somewhat.[65] So, if you drive it with a signal pair of matched source impedances R_S, you get a mismatch in the feedback ratios, hence degraded CMRR. Bottom line: drive these puppies from an op-amp output, or from a signal source of very low R_S (e.g., a low-value current-sense resistor).

That's not always possible, of course. What can be done to raise the input impedance? The first thought might be simply to raise all the resistor values by some large factor. That has several drawbacks, the most serious of which are (a) the Johnson noise contribution

---

[64] You've got to keep source impedances quite low in order not to degrade such a low e_n; even a $100\,\Omega$ resistor has an open-circuit voltage noise of $1.3\,\text{nV}/\sqrt{\text{Hz}}$. See Chapter 8.

[65] And the overall resistor scaling is typically good to only $\pm 20\%$ of the nominal value on the datasheet: absolute resistance value is sacrificed on the altar of resistor ratio matching.

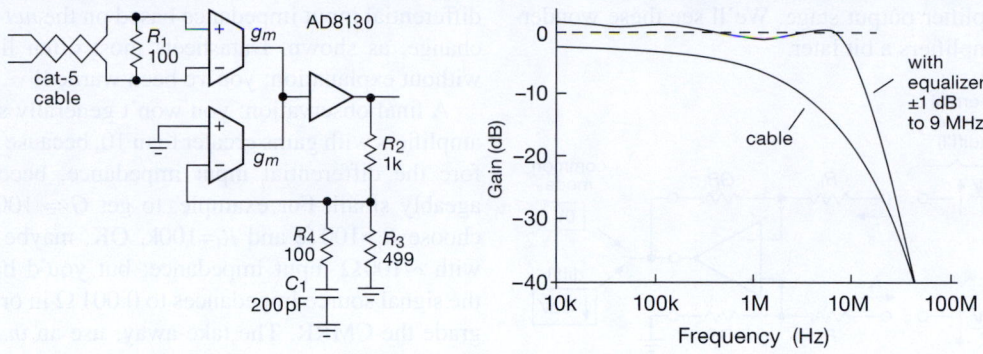

Figure 5.71. Differential line receiver for wideband analog over twisted pair. The "treble boost" provided by C_1R_4 compensates for the high-frequency rolloff of a 300m length of Cat-5 cable, as shown. See also Figure 5.101.

Table 5.7 Selected Difference Amplifiers

Part #	Config	Gain	ΔG max (%)	V_{CM}^a ±max (Vpp)	Offset Voltage typ (µV)	Offset Voltage max (µV)	CMRR typ (dB)	CMRR min (dB)	Curve	Diff'l Z_{in} (kΩ)	$V_{n(pp)}$ dc^b (µV)	e_n^r (nV/√Hz)	BW −3dB (MHz)	Settle 0.01% (µs)	Filter?	Supply Range (V)	I_S (mA)	Cost ($US)	DIP?	Comments
INA105K	A	1.0	0.025	20	50	500	100	72	1	50	2.4	60	1	5	-	10–36	1.5	8.88	●	legacy
AMP03G	A	1.0	0.008	20	25	750	95	75	1	50	2	20^c	3	1	-	10–36	2.5	5.86	●	
INA134	A	1.0	0.075	26	75	250	90	74	4	50	7^d	52	3.1	3	-	8–36	2.4	2.36	●	audio, < 5ppm dist
INA154	A	1.0	0.1	25	75	1500	90	74	4	50	2.6	52	3.1	3	-	8–36	2.4	2.36	-	low cost
INA132P	A	1.0	0.075	28	75	250	90	76	5	80	1.6	65^c	0.3	88	-	2.7–36	0.16	4.62	●	
AD8271B	A^m	0.5,2	0.02	>18	300	600	92	80	10	20	1.5	38	15	0.55	-	5–36	2.6	3.50	-	G_{diff} = 0.5,1,2
AD8273	A	0.5,2	0.05	>18	200	1400	86	77	7	24	7^d	52	20	0.75	-	5–36	2.5	3,13	-	dual audio, 6ppm dist
THAT1206	F	0.50	0.5	26	-	10mV	90	70	-	48	-	28^n	34	-	-	24–40	4.7	4.75	●	Z_{CM}=10M, <6ppm dist
AD8278B	A	0.5,2	0.02	>18	50	100	100	80	1	80	1.4	47	1	9	-	2–36	0.2	2.72	-	G_{diff} = 0.5,2; dual '79
INA106	A	10	0.025	11	50	200	100	86	1	20	1	30	5	10	-	10–36	1.5	11.00	●	0.10, legacy
INA143U	A	10	0.1	15	100	500	96	86	4^h	20	1	30	0.15	9	-	4.5–36	1.0	3.36	-	0.10, dual=INA2143
LT1991A	A^m	1,4,10	0.06	27	15	50	100	77	8^h	90	0.25	46	0.11^f	48	f	2.7–40	0.10	2.50	-	3,9,12, includes filter
LT1996A	A^m	9–117	0.07	27^q	15	50	100	80	-	33^q	0.25	18	0.04	85	f	2.7–40	0.10	5.72	-	9,27,81, includes filter
LT1995	A^m	1–7	0.2	31^q	600	4000	87	75	-	4^q	-	14	32	0.1	-	2.7–40	7	3.78	-	1,2,4,6
INA146	B	0.1^o	0.1	100	1000	5000	80	70	11	200	10	550	0.55	80	●	4.5–36	0.57	4.25	●	high-voltage inputs
INA117P	C	1.0	0.02	200	120	1000	94	86	8	800	25	550	0.2	10	-	10–36	1.5	5.54	●	high-voltage inputs
AD629B	C	1.0	0.03	270	100	500	96	86	6	800	15	550	0.5	15	-	5–36	1.2	7.21	●	high-voltage inputs
INA149	C	1.0	0.02	275	350	1100	100	90	1	800	20	550	0.5	7	-	4–40	0.8	6.00	-	high-voltage inputs
AD8479	C	1.0	0.01	600	500	1000	96	90	u	2000	30	1600	0.13	11	-	5–36	0.55	9.66	●	highest CM voltage
AD628a	C	ext	0.1	120	-	1500	-	75	-	220	15	300	0.6	40	●	4.5–36	1.2	3.36	-	HV inputs, filter+gain
low-voltage																				
AD8275B	D	0.2	0.024	27	150	500	96	80	2	108	1^{c,e}	40^e	15	0.45	-	3.3–15	1.9	4.23	-	5, with V_{ref}/2 offset
AD830	E^g	1–10	0.6	24	1500	3000	100	90	12	370	-	27	85	0.025	p	8–36	14	4.86	-	G=1+R_2/R_1, V_{in} < 2V
AD8129	E^g	10–100	0.6	20	200	800	105	92	12	1000	-	4.6	200	0.02	p	4.5–25	10	2.90	-	G=1+R_2/R_1, V_{in} < 2V
AD8130	E^g	1–20	0.6	20	400	1800	105	90	12	6000	-	12.3	250	0.02	p	4.5–25	11	2.90	-	G=1+R_2/R_1, V_{in} < 2V
EL5172	E^g	1–20	1.5	20	7mV	25mV	95	75	-	300	-	26	250	0.01	p	4.7–12	5.6	1.27	-	G=1+R_2/R_1, V_{in} < 4V
INA152EA	A	1.0	0.1	18	250	1500	94	80	3	40	2.4	87	0.8	25	-	2.7–20	0.5	3.60	-	0.5,1,2, RRIO
MAX4198	A	1.0	0.1	0,5	30	500	90	74	9^h	50	7.8	58	0.175	34	-	2.7–7	0.05	2.38	-	

Notes: (a) maximum common-mode voltage. (b) 0.01–10Hz or 0.1–10Hz. (c) at 100Hz. (d) 20 to 20kHz. (e) RTO (referred to output). (f) includes 4pF filter caps. (g) circuit E from Fig. 5.89. (h) CMRR curve flattens at BW frequency. (k) RTI (referred to input) unless noted. (m) multiple resistors. (n) noise = −107dBu, 20kHz BW. (o) includes extra opamp filter stage. (p) for G>1. (q) depends on gain. (r) at 1kHz. (t) typical. (u) between 5 & 6.

$(e_n$=0.13 $R^{\frac{1}{2}}$ nV/√Hz, thus a devastating 130 nV/√Hz for resistor values of ~1 MΩ; see Chapter 8), and (b) the bandwidth penalty caused by distributed stray capacitances. The second thought would be to put a pair of precision op-

amp followers at the inputs. That's OK, but there's a better way still, in circuit configurations such as the "three-op-amp instrumentation amplifier," in which a front-end differential stage of high CMRR and high impedance drives a difference amplifier output stage. We'll see these wonderfully useful amplifiers a bit later.

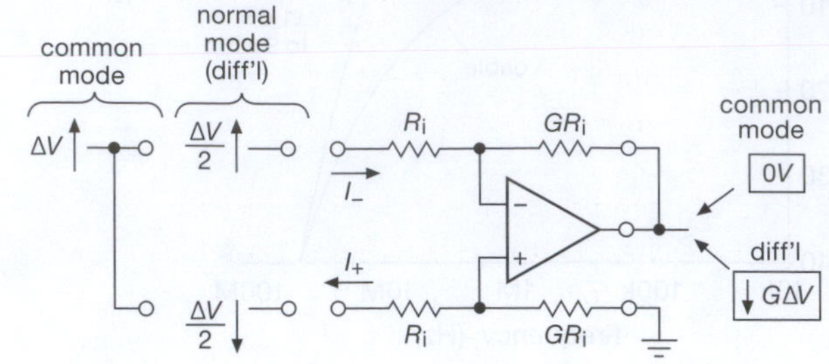

common mode: Z_{in} (each) $= (G+1)R_i$

Z_{in} (combined) $= \frac{1}{2}(G+1)R_i$

differential mode: $Z_{in}(-) \equiv \frac{\Delta V}{\Delta I} = 2\frac{G+1}{2G+1}R_i$

$Z_{in}(+) \equiv \frac{\Delta V_-}{\Delta I_+} = \frac{G+1}{2}R_i$

$"Z_{in}" \equiv \frac{\Delta V}{\Delta(I_- - I_+)} = \frac{G+1}{G}R_i$

Figure 5.72. The several input impedances of the difference amplifier.

Circling back around to the initial question, what exactly is the "input impedance"? There are several answers – look at Figure 5.72. By *common-mode* input impedance we mean the incremental impedance seen at either input[66] when both are driven together. With common-mode drive the two impedances are equal (apart from minor mismatching, as described above), because the output is fixed.

The *differential*-input impedance is a longer story. Taken individually (by grounding the other input), the two inputs exhibit different R_{in}: the inverting input connects to a virtual ground by way of R_i, so its $R_{in} = R_i$, whereas the noninverting input sees $R_{in} = R_i + R_f$. For a difference amplifier with $G=10$, for example, these differ by a factor of eleven. This is a useful result, particularly if you plan to use a difference amplifier in a single-ended configuration like Figures 5.66D and E. A purist could argue, though, that we've done it wrong: a single-ended input change is really a combination of a purely differential input plus a common-mode offset of half that value. To satisfy such a person we've calculated the expressions in Figure 5.72, based upon a "pure" (symmetrical) differential input. Even when defined this way the input impedances seen at the two inputs are different. This leads us to a final definition of

differential input impedance based on the *net* input-current change, as shown. Datasheets most often list this value, without explanation: you've been warned!

A final observation: you won't generally see difference amplifiers with gains greater than 10, because R_i, and therefore the differential input impedance, becomes unmanageably small. For example, to get $G = 1000$ you might choose R_i=100 Ω and R_f=100k. OK, maybe you can live with ~100 Ω input impedance; but you'd have to match the signal source impedances to 0.001 Ω in order not to degrade the CMRR. The take-away: use an *instrumentation amplifier* (§5.15), not a difference amplifier, for high-gain differential-input applications.

B. Common-mode input range

The voltage divider formed by R_i and R_f permits the standard difference amplifier to accept input signals *beyond* the supply rails: protection diodes are at the internal op-amp inputs, so the input signals could in principle swing as far as $\pm V_S(G+1)/G$ with supply voltages of $\pm V_S$ (figure out why). For example, the AD8278 has $R_i = 40$k and $R_f = 20$k, so it can be connected for $G = 0.5$ ("normal") or $G = 2$ (reversed). It specifies the common-mode input range for both gains: $-3(V_S+0.1)$ to $+3(V_S-1.5)$ for $G = 0.5$ (that is, about ±40 V with ±15 V supplies, which can be very useful indeed!), and $-1.5(V_S+0.1)$ to $+1.5(V_S-1.5)$ for $G = 2$ (i.e., ±20 V with ±15 V supplies). But check the specs – not all difference amplifiers let you go that far.

A few of the difference amplifiers in Table 5.7 on the preceding page) do better still; for example, the INA117 has a common-mode range of ±200 V while maintaining a differential gain of unity. This it does by using a pair of 20:1 voltage dividers at the input to bring the ±200 V signal within the op-amp's common-mode range of ±10 V (the circuit is shown in Figure 5.75C). The price it pays is degraded offset and noise: typical values of $120\,\mu$V and $550\,$nV/$\sqrt{Hz}$, compared with $25\,\mu$V and $20\,$nV/$\sqrt{Hz}$ for the conventional AMP03.[67]

An important point: when using difference amplifiers with large input voltage ranges, beware of the large equivalent input errors created by imperfect common-mode rejection. For example, the AD629B specifies a typical dc CMRR of 96 dB, but you've got to consider the worst-case (minimum) value of 86 dB. With that CMRR, a 200 V

---

[66] Some manufacturers specify half that value, i.e., the impedance with both terminals tied together; the datasheet usually tells you which they mean.

[67] There's another way to boost V_{CM} without such compromise, by using a second op-amp to cancel the common-mode signal; see Figure 7.27 in the previous edition of this book. We are unaware of any commercial difference amplifiers that use this trick.

common-mode input has an input-referred differential error of 10 mV, completely swamping the 0.5 mV specified maximum offset voltage. Put another way, the error that is due to finite CMRR is larger than the specified V_{OS} for input $|V_{CM}| > 10$ V. And the situation is worse still at signal frequencies: you might imagine using such a difference amplifier for powerline current monitoring. At 60 Hz the CMRR of the INA117 (similar to that of the AD629B at dc) degrades to 66 dB (min). So the 160 V peak powerline signal produces a huge 80 mV input error. And ideally you'd want to monitor currents on powerlines at higher voltages still, perhaps to 400 V. There are better ways to do this; take a look at §13.11.1 if you're curious.

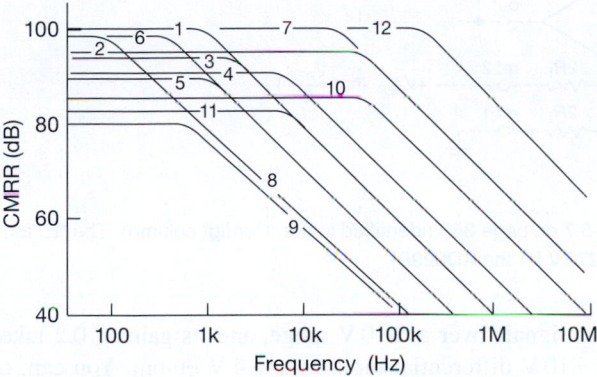

Figure 5.73. Common-mode rejection ratio versus frequency for the difference amplifiers in Table 5.7 on page 353 (identified in the "Curve" column).

C. Bandwidth

Difference amplifiers are built with op-amps, frequency compensated for stability with the by-now familiar $1/f$ open-loop gain rolloff. As with any op-amp circuit, loop gain is responsible for good behavior and the loss of loop gain at higher frequencies not only limits the bandwidth of a difference amplifier (and its linearity, constancy of gain, low output impedance, etc.), it also degrades the all-important CMRR. Figure 5.73 shows this behavior, for the difference amplifiers listed in Table 5.7 (page 353). Not surprisingly, amplifiers with greater closed-loop bandwidth (therefore with op-amps of greater f_T) maintain high CMRR to higher frequencies.

Note that some amplifiers perform well near dc but poorly at high frequencies, for example classic parts like the INA105, AD829 and LT1991, whereas others not so well rated at dc may look better at high frequencies, such as the AD8271 and the MAX4198. A few are outstanding performers from dc to high frequencies, according to the

plots, such as the AD8273[68] and our favorite AD8275. The transconductance-balancing AD8129 does very well, but is limited to small inputs and high gains. This information is not shown in the datasheet specifications; it's necessary to look for performance plots to make the comparisons.

Common-mode rejection at higher frequencies is degraded also by the effects of stray inductance and by asymmetrical capacitive loading. It is necessary to balance the circuit capacitances to achieve good CMRR at high frequencies. This may require careful mirror-image placement of components. Even when so symmetrized, the decreasing input capacitive shunt reactance at higher frequencies creates an increasing sensitivity to any unbalance in signal source impedance.[69]

5.14.4 Circuit variations

A. Filter node

The difference amplifiers in Table 5.7 (page 353) include one (the INA146) with a node brought out for noise (low-pass) filtering of the difference stage ($G = 0.1$) by a capacitor to ground (Figure 5.75B). It includes a second single-ended stage, with gain set by a pair of external resistors; so you can have overall gains from 0.1 to 100. The low first-stage gain gives you a large ±100 V common-mode range, though at the expense of noise and offset.

B. Offset trim

Integrated difference amplifiers are factory trimmed to pretty good precision, with typical values in the 25–100 μV range (but with worst-case offsets an order of magnitude larger). As with any op-amp circuit, you can rig up an external trim, as in Figure 5.74A. Here R_2R_3 divides the trimmer's ±15 V range to ±1 mV at the REF pin; R_1 balances the 10 Ω added resistance to ground, to preserve the CMRR. The amplifier's gain is slightly reduced by the ratio $R_f/(R_f + R_2)$; for a typical R_f of 25k, that's 0.04%, in the same range as the amplifier's specified gain accuracy. If this bothers you, use a smaller R_2.

---

[68] Although the datasheet *graphs* for the AD8273 show a best-in-class CMRR of 100 dB below 40 kHz, the *tabulated* typical CMRR is shown as only 86 dB. We like the plot better, but the designer may be stuck with the worst-case spec of 77 dB. Some users may want to perform incoming inspections to settle the issue.

[69] The professional audio people are keenly aware of these effects, and they don't mince words. As stated pithily by Whitlock and Floru in an Audio Engineering Society paper, "**Noise rejection in a balanced system has absolutely NOTHING to do with signal symmetry** (equal and opposite signal voltage swings). It is the balance of common-mode **impedances** that defines a balanced system!" [emphasis in the original].

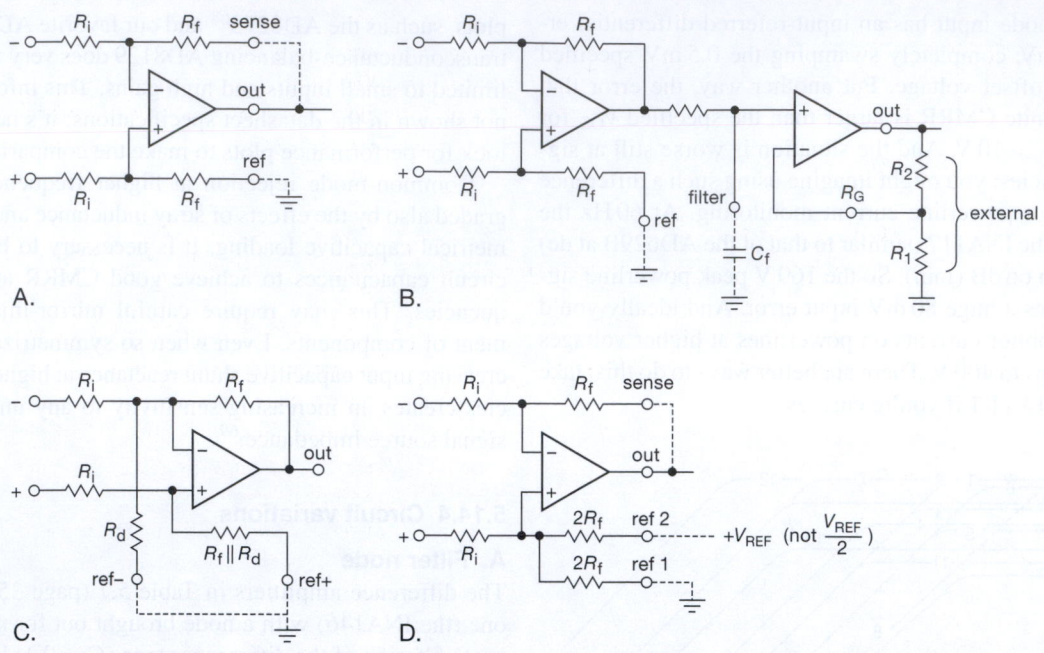

Figure 5.75. Circuit configurations for the difference amplifiers in Table 5.7 on page 353 (identified in the "Config" column). The "E" form is shown in Figure 5.89. The "C" form is used for high voltages (e.g., ± 270 V for the AD629B).

C. CMRR trim

Likewise, you can trim out residual CMRR (caused by slight mismatch of resistor ratios R_f/R_i in the two paths) with the circuit of Figure 5.74B. It's important to limit the trim range, to permit an accurate and stable trim to something considerably better than the off-the-shelf 80 dB (worst-case) CMRR specification. You can't get any old value of trimmer, and it's best not to use values less than $100\,\Omega$ (even if you can find them) if you care about stability. Here we've chosen standard resistor values and a $100\,\Omega$ trimmer to produce a resistance range of 20–$30\,\Omega$ from the REF terminal to ground, providing a $\pm 5\,\Omega$ symmetrical variation around R_1. For the 25k resistor values of this unity-gain difference amplifier (rather typical; see Table 5.7 on page 353), this corresponds to a trim range adequate to null an initial CMRR of 75 dB. You can, of course, add an offset null to this circuit, as indicated.

D. Single-supply offset

One of the difference amplifiers in Table 5.7 (page 353) helpfully splits the reference feedback resistor into a parallel pair (Figure 5.75D), so it's easy to offset the output voltage range. For example, you could run the amplifier on a single +5 V supply, with REF2 driven from a clean reference of that same voltage. With no difference signal the output will be +2.5 V. The amplifier can accommodate in-

put signals over a ± 10 V range, and its gain of 0.2 takes a ± 10 V differential input to a 0–4 V output. You can, of course, use a lower reference voltage. It's often convenient to use $V_{\text{ref}} = 4.096$ V when driving an ADC; this makes the step size come out in round numbers, e.g., 1 mV/step for a 12-bit conversion.

5.15 Instrumentation amplifier

The difference amplifiers of the previous section are inexpensive, and fine for many applications; and they have the nice feature of accepting inputs beyond the rails. But they have limited gain (≤ 10) and CMRR ($\lesssim 85$ dB min), their resistors make them somewhat noisy (20 nV/$\sqrt{\text{Hz}}$ to 50 nV/$\sqrt{\text{Hz}}$), and their relatively low input resistance ($\sim$10k to $\sim$100k) limits their utility to situations where the driving signals are of low impedance (op-amp outputs, low-Z balanced lines, low-R sense resistors).

If you need lots of gain, or a high input impedance, or superior CMRR, you need something different. It's called an *instrumentation amplifier*. These impressive devices have input impedances upward of $10^9\,\Omega$, gains from unity to 1000 or so, low voltage noise (down to $\sim$1 nV/$\sqrt{\text{Hz}}$), and worst-case CMRRs of 100–120 dB (look ahead to Table 5.8 on page 363).

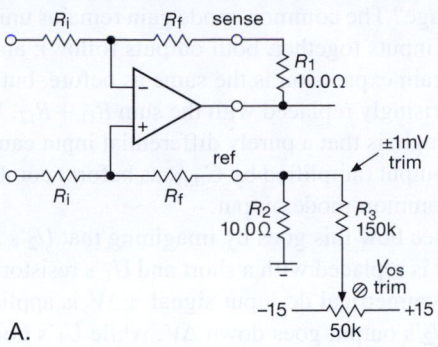

A.

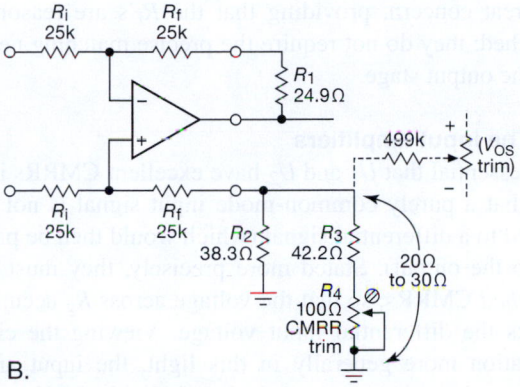

B.

Figure 5.74. Trimming the offset and CMRR of a difference amplifier.

5.15.1 A first (but naive) guess

High input impedance – that's easy, just add op-amp followers to the difference amplifier (Figure 5.76); and then the resistors R_i and R_f can be smaller, reducing their Johnson noise contribution.

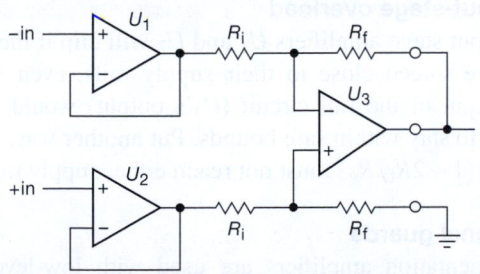

Figure 5.76. A first stab at improving the difference amplifier.

Indeed, this circuit has the enormous input impedance we expect from an op-amp follower, so there is no longer a problem from any reasonable source impedance.[70] But it

---

[70] At least at dc. At higher frequencies it again becomes important to

does not improve the CMRR, which is still limited by the resistor ratio matching of R_f/R_i: it's really hard to do better than 100,000:1 with on-chip laser trimming (both initial trim and stability with time and temperature). In fact, this circuit degrades the CMRR somewhat, with two more amplifiers in the signal path.

5.15.2 Classic three-op-amp instrumentation amplifier

The circuit in Figure 5.77 is much better. It is the standard "three-op-amp instrumentation amplifier," one of several configurations that provide the desirable combination of high CMRR, high R_{in}, low e_n, and plenty of gain when you need it. The input stage is a clever configuration of two op-amps that provides high differential gain and unity common-mode gain without requiring close resistor matching. Its differential output represents a signal with substantial reduction in the comparative common-mode signal (when configured for $G_{diff} \gg 1$), and it is used to drive a conventional differential amplifier circuit. The latter is usually arranged for unity gain and is used to generate a single-ended output while removing the common-mode signal.

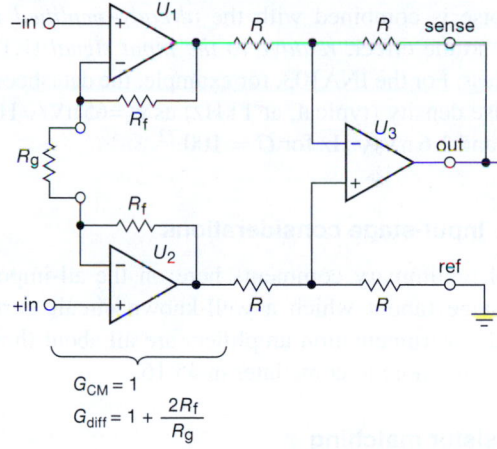

$$G_{CM} = 1$$
$$G_{diff} = 1 + \frac{2R_f}{R_g}$$

Figure 5.77. The classic three-op-amp instrumentation amplifier.

It's worth looking more closely at this circuit. We've hinted that it can deliver very high CMRR and very low e_n. But that is true *only when configured for large differential*

---

have matched source impedances relative to the common-mode signal, because the input capacitance of the circuit forms a voltage divider in combination with the source resistance. "High frequencies" may even mean 60 Hz and its harmonics, because common-mode ac powerline pickup is a common nuisance.

gain. To see why, imagine we configure it for $G_{diff} = 1$, by omitting the gain-setting resistor R_g. Then we've just got the previous circuit (Figure 5.76), i.e., a buffered unity-gain difference amplifier. It has the same limitations of CMRR (set by resistor matching) and noise (from U_3's resistors).

Now imagine we set $G_{diff} = 100$. We would do this by choosing R_g so that $1 + 2R_f/R_g = 100$, i.e., $R_g = 2R_f/(G-1)$. For the INA103, for example, $R_f = 3\,k\Omega$, so we'd use $R_g = 60.6\,\Omega$. The INA103 conveniently includes a resistor of that value,[71] so for $G_{diff} = 100$ you need only tie a pair of pins together. Let's look again at the CMRR and noise scene.

First, the CMRR: the front end has a differential gain of 100 and a common-mode gain of unity. In other words, it passes on to the difference amplifier stage a signal that has received 40 dB of CMRR blessings. Another 80 dB of CMRR in the output stage, and we've got the promised 120 dB CMRR. These numbers are typical of available instrumentation amplifiers, as listed in Table 5.8 (page 363) and graphed in Figure 5.82. For the INA103, for example, the datasheet lists CMRR=86 dB/72 dB (typ/min) for $G = 1$, and 125 dB/100 dB (typ/min) for $G = 100$.

Second, the noise voltage: the output stage still contributes Johnson noise from its resistor array, along with noise inherent in its amplifier. That cannot be helped. But that noise is combined with the *already-amplified* input signal, so the effect, *relative to the input signal* (RTI), is 100× less. For the INA103, for example, the datasheet lists the noise density (typical, at 1 kHz) as $e_n = 65\,nV/\sqrt{Hz}$ for $G = 1$ and $1.6\,nV/\sqrt{Hz}$ for $G = 100$.[72]

5.15.3 Input-stage considerations

Several preliminary comments here on the all-important input stage (about which a well-known circuit guru remarked "instrumentation amplifiers are all about their inputs"), with more to come later in §5.16.

A. Resistor matching
The circuit looks handsome with its symmetrical matched R_f's, but that requirement did not intrude in the discussion above. What is the effect of mismatched feedback resistors

in the first stage? The common-mode gain remains unity (if you tie both inputs together, both outputs follow); and the differential gain expression is the same as before, but with $2R_f$ not surprisingly replaced with the sum $R_{f1} + R_{f2}$. What changes, though, is that a purely differential input causes a differential output (amplified by G_{diff}, as before) combined with some common-mode output.

You can see how this goes by imagining that U_2's feedback resistor is replaced with a short and U_1's resistor with $2R_f$, and a symmetrical dc input signal $\pm\Delta V$ is applied to the inputs: U_2's output goes down ΔV, while U_1's goes up $(1 + 4R_f/R_g)\Delta V$. That's the correct differential output, but with a common-mode offset of $(2R_f/R_g)\Delta V$. This is not of great concern, providing that the R_f's are reasonably matched; they do not require the precise matching needed for the output stage.

B. The input amplifiers
It is essential that U_1 and U_2 have excellent CMRRs in order that a purely common-mode input signal is not converted to a differential signal (which would then be passed on to the output). Stated more precisely, they must have *matched* CMRRs, so that the voltage across R_g accurately tracks the differential input voltage. Viewing the circuit operation more generally in this light, the input amplifiers need not have extremely low individual offset voltages – what matters is that their offset voltages are accurately *matched* and remain so with changes in common-mode voltage. This gives rise to several circuit variants in which the "op-amps" U_1 and U_2 are configurations with well-matched base–emitter drops between each input and the corresponding R_g pin; see for example Figure 5.88C below.

C. Input-stage overload
The input stage amplifiers U_1 and U_2 will clip if their outputs are forced close to their supply rails, even though the output of the full circuit (U_3's output) would be expected to stay within safe bounds. Put another way, $V_{CM} \pm 0.5V_{diff}(1 + 2R_f/R_g)$ must not reach either supply rail.

D. Signal guards
Instrumentation amplifiers are used with low-level signals, often conveyed by shielded cables to minimize noise. This adds input capacitance, thereby limiting the bandwidth (particularly with signals of moderate to high source impedance). Of perhaps greater importance, it degrades the CMRR at signal frequencies: the cable's shunt capacitance forms a voltage divider with the signal's source impedance, separately for each input; so if you have a

---

[71] More precisely, it includes an on-chip resistor, ratio matched to R_f, to produce an overall guaranteed gain of 100.0±0.25%.

[72] The datasheet separates out the front-end and second-stage contributions, $1\,nV/\sqrt{Hz}$ and $65\,nV/\sqrt{Hz}$, respectively. From these you can calculate the input-referred noise $e_n(RTI) = \{e_n(in)^2 + [e_n(out)/G]^2 + 4kTR_g\}^{1/2}$. The last term is the square of the Johnson noise voltage $e_n = 0.13\sqrt{R_g}\,nV/\sqrt{Hz}$.

signal pair with unbalanced source impedances (a common situation), common-mode signal variations will create some differential signal input.[73] Finally, leakage currents become significant with signals of very high (MΩ to GΩ) source impedance. A nice technique for greatly reducing both the effective cable capacitance and any leakage current is to drive the shield actively with a "guard" voltage (Figure 5.78).

If there is a common shield surrounding the signal pair, the idea is to drive it with a buffered replica of the common-mode signal, as in circuit (A); a small series resistor is generally a good idea, for stability. A grounded outer shield can be used to eliminate any noise coupling to the guard, if that becomes necessary. This circuit requires access to the first-stage outputs, which you rarely get in an integrated instrumentation amplifier. Some amplifiers oblige by including this circuitry internally and providing a "data-guard" output pin, as in (B). If not, you can derive a common-mode signal yourself, as in (C).

Common-mode guarding reduces greatly the capacitive loading of the signal pair and therefore improves the CMRR (by minimizing the conversion of common-mode to normal-mode signals). But it does not reduce the effects of cable capacitance (and leakage) on *normal-mode* (differential) signals themselves. To accomplish that you need to shield the signals individually, driving each shield with a replica of the signal it shields, as in circuit (D). This is the familiar "bootstrap," acting here to reduce both the capacitance and leakage current seen by each signal. It thus minimizes high-frequency rolloff, and dc error in signals of high source impedance.[74] And, as with common-mode guarding, it also minimizes the degradation of CMRR by effectively eliminating the cables' capacitances.

Some instrumentation amplifiers provide these individual guard outputs, for example the INA116 shown, which is evidently intended for very low current measurements (it boasts a 3 *femto*amp typical input current). If not, you can roll your own, as in circuit (E), exploiting the fact that the R_g nodes follow the inputs. A caution, though: there is no guarantee that the signals at the R_g pins are not offset from the input signals, as for example in the configurations we'll see later in Figures 5.88C and 5.89F. With such an offset the bootstrap would be effective in minimizing capacitance, but much less so for leakage currents.

---

[73] Put another way, the cable's capacitance degrades the CMRR by creating differential phase shifts between the two signals, owing to their unbalanced source impedances.

[74] Instruments for measuring very low currents – "electrometers" and "source measure units" – include guard outputs and (usually) special BNC-like "triax" connectors for use with triaxial shielded cables.

E. Bootstrapped power supply

Figure 5.79 shows a trick that is analogous to signal guarding (you could call it "power-supply guarding"), occasionally helpful if you need to enhance further the CMRR of an instrumentation amplifier. U_3 buffers the common-mode signal level, driving the common terminal of a small floating split supply for U_1 and U_2. This bootstrapping scheme effectively eliminates the input common-mode signal from U_1 and U_2, because they see no swing (due to common-mode signals) at their inputs relative to their power supplies. U_3 and U_4 need not be bootstrapped if this is a discrete implementation. This scheme can do wonders for the CMRR, at least at dc. At increasing frequencies you have the usual problems of presenting matched impedances to the input capacitances.

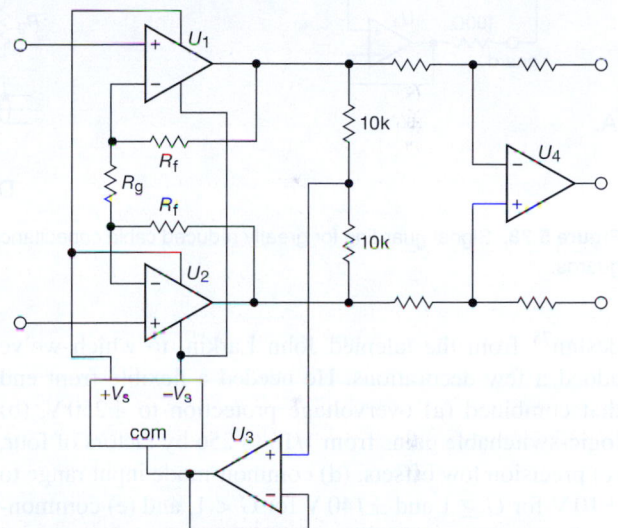

Figure 5.79. Instrumentation amplifier with bootstrapped input power supply for high CMRR.

5.15.4 A "roll-your-own" instrumentation amplifier

Integrated instrumentation amplifiers are excellent performers, and you usually can save a lot of work (and expense, and PCB real estate) by taking advantage of the broad selection of available parts, for which Table 5.8 (page 363) is a good starting point. But sometimes you need additional capability, for example a wider range of gains, or precise trimming of offsets and CMRR, or protection against outrageous abuse visited upon the input terminals.

Figure 5.80 is an example (another in the series of Designs by the Masters), based upon a commercialized

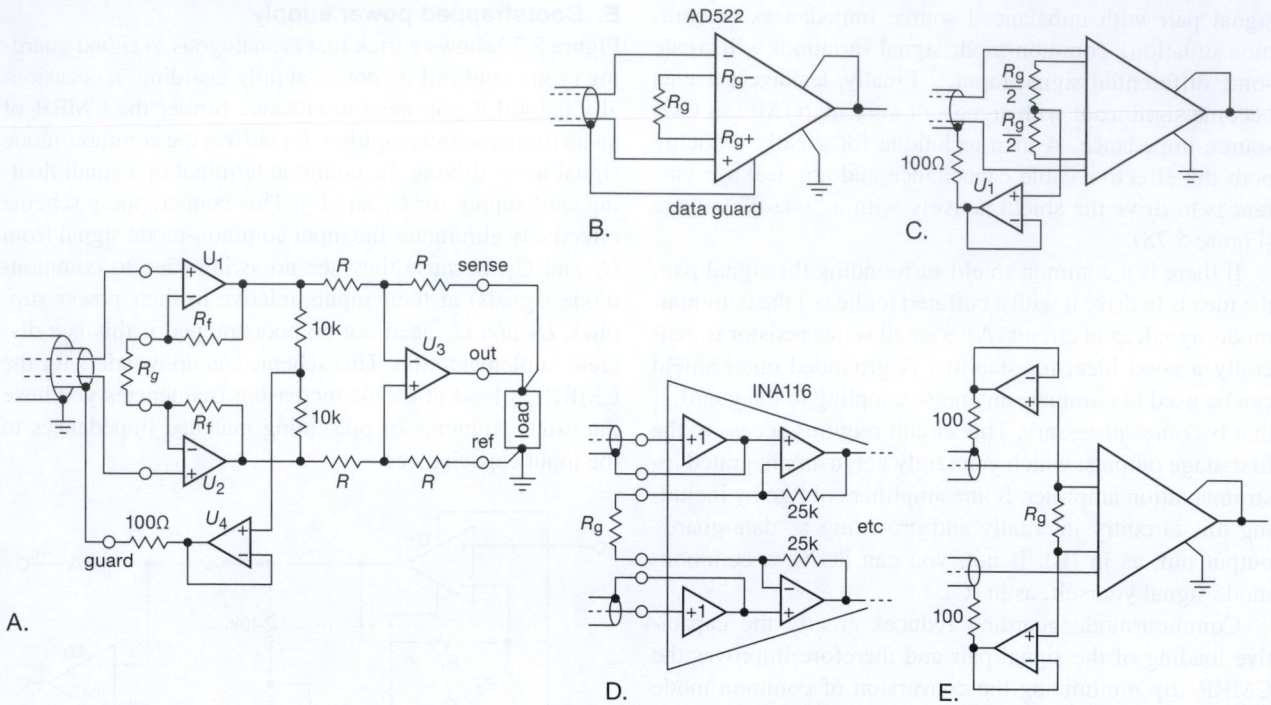

Figure 5.78. Signal guarding for greatly reduced cable capacitance. (A)–(C) are common-mode guards; (D) and (E) are individual signal guards.

design[75] from the talented John Larkin, to which we've added a few decorations. He needed a flexible front end that combined (a) overvoltage protection to ±250 V, (b) logic-switchable gains from 1/16 to 256 by factors of four, (c) precision low offsets, (d) common-mode input range to ±10 V for $G \geq 1$ and ±140 V for $G < 1$, and (e) common-mode rejection of 120 dB at high gain.

The overall structure is the familiar three-op-amp configuration, with U_1 and its symmetrical twin (not shown) as the differential front end, which drives the unity-gain difference amplifier output stage U_3. The gain is set by analog switches U_2 and its twin, which select a tap on the resistor string R_6–R_{10}. For example, when the "×64" position is selected, $R_g = 201.1\,\Omega$ and $R_f = 6.411\,k\Omega$, thus a gain $G_{\text{diff}} = 1 + 2R_f/R_g = 64.8$.[76]

There's plenty more going on, here. Let's take it from

left to right. The lossy chip inductor L_1 (these are often rated by their impedance, mostly resistive, at 100 MHz) combined with capacitor C_1 suppresses high-frequency interference that is outside the amplifier's bandwidth but able to cause nonlinear mischief in op-amp input stages. C_1's value is not critical, but its tight tolerance keeps the input impedance balanced in order not to compromise high-frequency CMRR. Relays K_1 and K_2 are used for frequent in-system offset and gain calibrations, essential for establishing and maintaining gain accuracies better than 0.1% (with a circuit using 1% resistor values) and zero offsets of 10 μV or better.[77] Relay K_3 switches in a ×16 gain attenuation for input signals beyond ±10 V. The input impedance is then set by R_1+R_2, with R_3 balance trim for good CMRR. R_1's value is a compromise between high R_{in} and bandwidth: here the 33.2 k, loaded by a downstream capacitance of ~10 pF, puts the 3 dB rolloff at 500 kHz, about right for the product's 200 kHz specified bandwidth. But a higher input impedance would be nice, perhaps a nice round-number 1 MΩ (preferably with R_1 as a series pair

---

[75] Highland Technology's V490 VME Multi-Range Digitizer.

[76] The gain-setting resistors are chosen from standard "E96" 1% resistor values, so the actual gains differ from round-number values (with ±1% tolerance they would never be perfect, anyway). This front end would form part of a data-acquisition system, with overall gain and offset data held in software, from a calibration procedure carried out by relays K_1 and K_2.

[77] Larkin sings the praises of Fujitsu FTR-B3GA4.5Z DPDT relays, with their sub-picofarad capacitances.

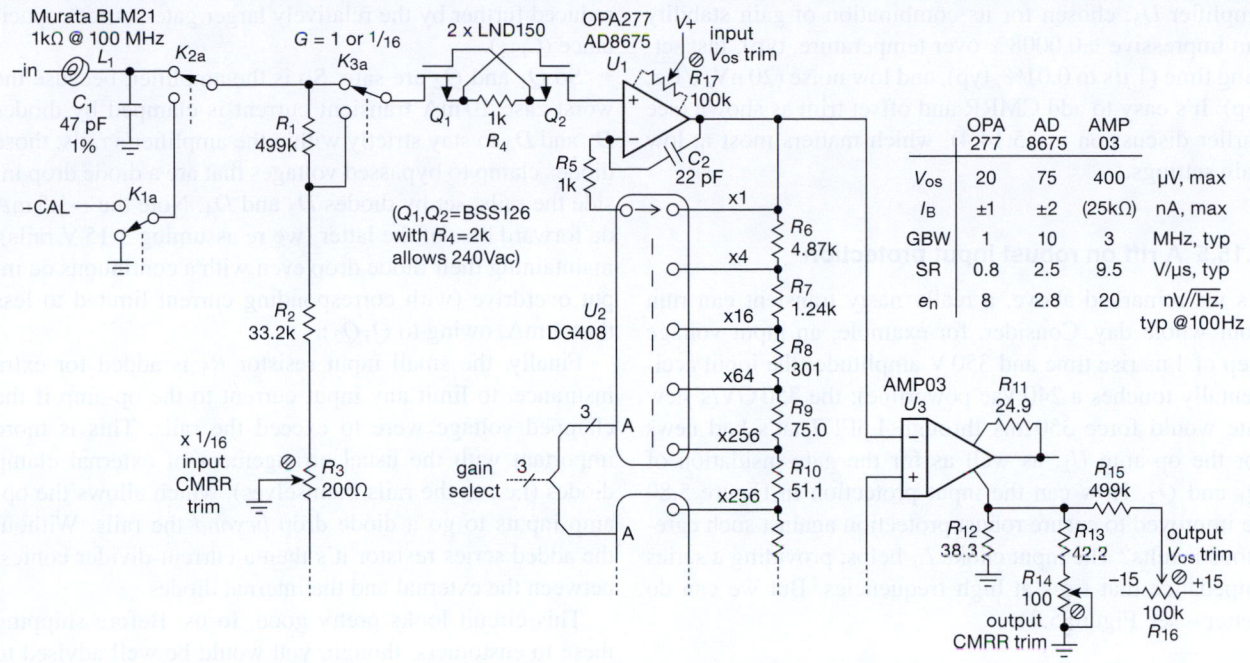

Figure 5.80. A "discrete" instrumentation amplifier design that combines precision, high CMRR, large common-mode voltage range, digital gain selection, and protection against ±250 V inputs. The symmetrical path to the noninverting input of U_3 is omitted to save paper.

of 464k, to accommodate input overvoltages without damage).

The curious Q_1Q_2 twins are a nice touch: these are *depletion-mode* high-voltage MOSFETs (see Table 3.6) from Supertex or Infineon,[78] used here as a ∼0.5 mA bidirectional current limiter to protect the op-amps (whose internal clamp diodes are unfazed by a few milliamps of input current; the AD8675, for example, specifies a maximum I_{in} of ±5 mA).[79] Here R_4 serves to reduce the saturation current from its maximum I_{DSS} of 3 mA (thus 750 mW dissipation at 250 V input, too much for a little SOT23 transistor), the tradeoff being an additional 1 kΩ of noise-adding input resistance in series with the unavoidable ∼1.7k on-resistance of the series pair of zero-biased MOSFETs. This circuit provides a reasonable level of overdrive protection; but an abrupt input step to +500 V, say, could well couple a large enough current transient (through the

capacitance of Q_1 and Q_2) to damage the amplifier, or produce gate breakdown in the MOSFETs themselves.

The op-amps are precision bipolar types with input-current cancellation; note the tradeoff, for the two choices, of precision versus speed and bias current. These are laser trimmed for low offset voltage and are probably good enough without human intervention; but they include offset-trimming terminals to which you can attach a trimpot as shown. That sounds like a good idea – but be cautious, because the external trimming range you get when you do that is generally too much! For the AD8675, for example, the trimmer adjusts ±3500 μV (nearly 50 times the maximum untrimmed offset). So it may be a touchy (and unstable) adjustment to improve upon the already-trimmed V_{OS}.

The analog multiplexer selects a tap on the gain-setting string, which makes the gain insensitive to the on-resistance of the switches (here ∼100 Ω). That's the right way to do it; the *wrong* way would be to use a separate feedback resistor to each switch position, with R_{10} connected between the common terminals of the multiplexers. The small feedback capacitor C_2 ensures stability, given the lagging phase shift within the loop contributed by the ∼40 pF of switch capacitance to ground.

Finally, we arrive at our destination via the difference

---

[78] They're rated at 500 V, and come in three small package styles (TO-92, SOT23, and SOT89 with tab). The alternative BSS126 from Infineon is rated at 600 V, and costs less ($0.15 in small quantities). It comes only in the SOT-23 package, whereas the LND150 is available in three package styles, including a 1.5 W TO-243 small power package.

[79] Be sure to check that the op-amp does not suffer from phase reversal (see, e.g., §4.6.6), if you care about the output during input overdrive. A robust cure is to use a pair of input clamp diodes, as in Figure 5.81.

amplifier U_3, chosen for its combination of gain stability (an impressive $\pm 0.0008\%$ over temperature, typ), fast settling time (1 μs to 0.01%, typ), and low noise (20 nV/$\sqrt{\text{Hz}}$, typ). It's easy to add CMRR and offset trim as shown (see earlier discussion in §5.14.4), which matters most at low gain settings.

5.15.5 A riff on robust input protection

As we remarked above, a really nasty transient can ruin your whole day. Consider, for example, an input voltage step of 1 ns rise time and 350 V amplitude (the input accidentally touches a 240 Vac powerline): the 350 GV/s slew rate would force 350 mA through 1 pF! That's bad news for the op-amp U_1, as well as for the gate insulation of Q_1 and Q_2. How can the input protection in Figure 5.80 be improved to ensure robust protection against such egregious insults? The input choke L_1 helps, providing a series impedance that rises at high frequencies. But we can do better – see Figure 5.81.

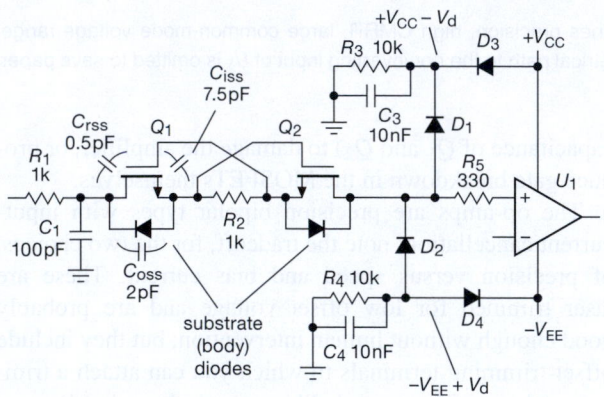

Figure 5.81. Bulletproofing the amplifier frontend with slew-rate control, diode clamping, and input-current limiting.

The first step is to limit the incoming slew rate, with R_1C_1: a 500 V step causes a maximum $dV/dt = 5$ V/ns slew rate across C_1,[80] with no significant degradation of bandwidth or noise. Looking at it most simply, that maximum slew rate can produce a transient current through the drain-source capacitance (C_{OSS}) of at most $I = C_{\text{OSS}} dV/dt = 10$ mA, thus a drop across R_2 of at most 10 V. That is well below the gate-source rating of ± 20 V; and it's a conservative estimate, because the actual gate-source voltage is

---

[80] And a momentary 500 V, 250 W surge in R_1, which should be a bulk-composition type, or several SMT resistors in series, to handle both the voltage and the energy transient; see Chapter *1x*.

reduced further by the relatively larger gate-source capacitance (C_{ISS}).

So Q_1 and Q_2 are safe. So is the amplifier, because the worst-case 10 mA transient current is clamped by diodes D_1 and D_2 to stay strictly within the amplifier's rails: those diodes clamp to bypassed voltages that are a diode drop inside the rails, set by diodes D_3 and D_4. Note the $\sim$1.5 mA dc forward bias of the latter (we're assuming ± 15 V rails), maintaining their diode drop even with a continuous dc input overdrive (with corresponding current limited to less than 1 mA, owing to Q_1Q_2).

Finally, the small input resistor R_5 is added for extra insurance, to limit any input current to the op-amp if the clamped voltage were to exceed the rails. This is more important with the usual arrangement of external clamp diodes (i.e., to the rails themselves), which allows the op-amp inputs to go a diode drop *beyond* the rails. Without the added series resistor it's then a current-divider contest between the external and the internal diodes.

This circuit looks pretty good, to us. Before shipping these to customers, though, you would be well advised to put it on the bench and beat it up, brutally. Circuits can surprise you.

5.16 Instrumentation amplifier miscellany

There's much to love about these amplifiers, whose nice variety is evident in the collection of Table 5.8 (with additional "programmable gain" types in Table 5.9 on pages 370–371). When you need some real performance, those tables are worth serious study! Here we collect some important advice on their care and feeding; refer often to Table 5.8 for enrichment as you read the following sections.[81]

5.16.1 Input current and noise

Instrumentation amplifiers (INAs) must condition the input signal without disturbing it; thus high input impedance Z_{in}, low input current I_{in}, and a low level of current noise i_n. There's the usual tradeoff – the lower voltage offset and voltage noise of BJTs versus the lower input current and current noise of FETs. Some bipolar-input INAs (for example the LT1167/8) use the trick of bias-current cancellation to achieve sub-nanoamp input currents. Conversely, the auto-zero INAs do best in terms of offset voltage, but they pay a price in current noise and chopping artifacts.

For some amplifiers the input current noise is close to

---

[81] See also related material in the second edition, pp. 422-428.

Table 5.8 Selected Instrumentation Amplifiers

Part #	Circuit[q]	Supply range (V)	I_s typ (mA)	I_B max[a] (nA)	I_{os} max[a] (nA)	V_{os} max (μV)	ΔV_{os} typ (μV/C)	ΔG max (%)[g]	Z_{in} (Ω)	CMRR G=100 typ (dB)[j]	CMRR G=100 min (dB)	CMRR Unbal?	CMRR G=1 min (dB)	CMRR Curve[z]	$V_{n(pp)}$[c] (μV)	e_n[x] (nV/√Hz)	i_n[x] (fA/√Hz)	R_n[x] (kΩ)	BW[d] −3dB (kHz)	Settle[d] 0.01% (μs)	Slew (V/μs)	Filter?	DIP avail?	Needs Rg?	IN +	IN −	OUT +	OUT −	Gain, comments
high-voltage, BJT																													
INA103	A	18–50	9	12μA	1000	250	1.0	0.07	60M	125	100	0	72	2	–	1.0	2000	0.5	800	3.5	15	–	–	int	–	–	–	–	1, 100, adjustable
INA163	A	9–36	10	12μA	1000	250	1.0	0.1	60M	116	100	0	80	3	–	1.0	800	1.3	800	3.5	15	–	●	●	–	–	–	–	G = 1 + 6k/Rg
INA217	A	9–36	10	12μA	1000	250	1.0	0.2	60M	116	100	0	80	3	–	1.3	800	1.6	800	3.5	15	–	●	●	–	–	–	–	G = 1 + 5k/Rg
THAT1510	C	10–40	6	14μA	1400	250	–	6	30M	120	105	0	45	3	–	1.0	2300	0.43	7000	–	28	p	●	●	–	–	–	–	G = 1 + 10k/Rg
INA166	A2	9–36	10	12μA	1000	250	2.5	1	60M	120	–	–	–	4	–	1.3	800	1.6	450	3.5	15	p	●	●	–	–	–	–	G = 2000 + 60k/Rg
AD8428B	C	8–36	6.7	50	10	25	0.3	0.05	6M	153	140	0	–	27	0.05	1.3	1500	0.9	3500	0.75	50	p	●	●	–	–	–	–	G = 2000
AD627B	C2	2.2–36	0.06	10	1	150	0.1	0.25	20G	96	83	1k	–	16	0.56	38	50	760	3	290	0.05	–	●	●	–	–	–	●	G = 5 + 200k/Rg
INA128	A	4.7–36	0.70	5	5	50	0.2	0.5	10G	125	120	1k	80	1	0.2	8	300	27	200	9	4	–	●	●	–	–	–	–	G = 1 + 50k/Rg
INA126	B	2.7–36	0.18	25	2	250	0.5	0.1	1G	94	83	0	–	15	0.7	35	60	590	9	160	0.4	p	●	●	–	–	–	●	G = 5 + 80k/Rg
LT1789-10	C	2.2–36	0.12	40	0.2	750	0.2	0.27	1.6G	113	98	1k	–	14	1.1	48	62	775	12	190	0.06	–	●	●	–	–	n	●	G = 1 + 200k/Rg
AD8221B	C	4.6–36	0.90	0.4	0.4	25	0.3[d]	0.15	100G	–	110	1k	80	17	0.25	8	40	200	100	10	2	–	●	●	–	–	–	●	G = 1 + 49.4k/Rg
LT1167A	C	4.6–40	0.9	0.35[e]	0.35[e]	40	0.05	0.08	1T	125	120	1k	90	5	0.28	8	60	130	120	14	1.2	–	●	●	–	–	–	●	G = 1 + 49.4k/Rg
LT1168A	C	4.6–40	0.42	0.25[e]	0.3[e]	40	0.05	0.5[t]	1T	135	120	1k	90	5	0.28	10	30	330	13	30	0.5	–	●	●	–	–	–	●	G = 1 + 49.4k/Rg
AD620B	C	4.6–36	0.9	1	0.5	50	0.3	0.15	10G	130	120	1k	80	5	0.28	9	100	90	120	15	1.2	–	●	●	–	–	–	–	inexpensive
AD8227	C	2.2–36	0.35	27	1.5	200	0.2	0.3	0.8G	–	105	1k	–	6	0.5	24	100	240	50	35	0.8	–	●	●	–	–	–	●	G = 5 + 80k/Rg
JFET																													
INA110B	A	12–36	3	0.1	0.05	250	2	0.1	5T	116	106	1k	80	7	1	15	1.8	8.3M	470	5	17	–	●	int	–	–	–	–	1, 10, 100, 200, 500
AD8220B	C	4.5–36	0.75	0.01	6pA	125	5[d]	0.2	10T	90	84	1k	80	8	0.8	14	1	14M	120	8	2	–	–	●	–	–	–	●	G = 1 + 50k/Rg
MOSFET																													
INA116	A	9–36	1	25fA[v]	25fA[v]	2000	20	0.5	10T[i]	94	86	1k	80	5a	2	28	0.1	300M	70	400	0.8	–	●	●	–	–	–	●	guard, G = 1 + 50k/Rg
low-voltage, BJT																													
AD8129A	E	4.5–25	10	2000	400	800	2	0.6	6M	110	90	0	–	18	–	4.5	1000	4.5	270M[s]	0.02°	930	p	–	–	–	–	–	–	250MHz, G≥10
AD8223B	C	3–24	0.65	25	2	100	1	0.3	2G	110	90	1k	78	20	0.6	30	70	430	50	18	0.2	–	●	●	–	–	–	●	low-cost, G=5(1+R2/R1)
MAX4194-7	A2	2.7–7.5	0.09	20	3	225	0.5	0.5	1G	115	95	1k	78	10a	0.6	8.7	–	–	1.5	40	–	–	●	●	–	n	–	–	G = 1, 10, 100, var
AD623B	A	5–12	0.38	25	2	100	0.1	0.35	2G	110	105	1k	77	13	1.5	35	1500	23	10	20	0.3	p	●	●	–	–	–	●	G = 1 + 100k/Rg
ISL28270	E[k]	2.4–5.5	0.12	2	1	150	0.7	0.5[t]	3M	110	90	1k	90	12	3.5	60	370	160	240	–	0.5	–	–	●,–	–	–	●	G ≥ 100, dual
INA337	F2[k]	2.7–5.5	2.4	2	2	100	0.1	0.2	10G	120	106	1k	106	10	0.8	33	150	220	10	160	filter	p	●	–	–	–	●	n	G = 2 R2/R1
CMOS																													
AD8236	B	1.8–5.5	0.04	10pA	5pA	3500	2.5	0.2	440G	110	100	0	–	25	4	76	15	5M	0.8	1000	0.01	–	–	●	–	–	●	●	G = 5 + 420k/Rg
INA321V	B2	2.7–7.5	0.04	10pA	10pA	500	7	0.1	10T	94	90	0	80	9	20	100	3	33M	20	12	0.4	p	–	–	n	n	●	●	G = 5 (1 + R2/R1)
ISL28272	E[k]	2.4–5.5	0.12	30pA	30pA	500	0.7	0.2[t]	1G	100	80	0	106	13	6	78	0.2	390M	100	–	0.5	–	–	–	–	–	●	●	dual, G ≥ 100[u]
auto-zero																													
AD8230	D	8–16	2.7	1	0.3	10	0.05	0.04	cap[w]	120	110	1k	110	19	3	240	–	–	2.5	slow	2	p	●	●	–	–	●	●	2kHz chopper noise
LTC2053	D	2.5–5.5	0.95	10	3	10	0.05	ext	cap[w]	120	105	1k	100	20	2.5[c]	170[b]	–	–	5	4000	0.2	p	●	●	–	–	●	●	3kHz chopper noise
MAX4209	E	2.9–5.5	1.4	0.001[t]	0.001[t]	20	0.05	0.25	2G	135	106	1k	106	21	2.5	170	–	–	10	120	0.08	p	●	●	n	n	●	●	G=100, AZ, '08 var G
INA333	A	1.8–7	0.05	0.2	0.2	25	0.1	0.25	100G	115	100	1k	80	22	1.0	50	100	500	3.5	400	0.2	●	●	●	n	n	●	●	G = 1 + 100k/Rg
AD8553	F	1.8–6	1.1	1	2	20	0.02	0.3	50M	140	120	0	–	23	0.7	30[h]	–	–	1	2400	slow	●	–	●	–	–	r	r	use 1kHz filter
AD8293	F	1.8–6	1.1	2	4	50	0.02	1	50M	140	110	na	na	24	0.7[c]	35	–	–	0.5	2400	filter	●	–	int	–	–	r	r	G = 80, 160 AZ-amp

Notes: (a) at 25°C. (b) at 100Hz. (c) 0.01–10Hz or 0.1–10Hz. (d) at G=100 max, not 1000. (e) if available. (f) max. (g) gain error, at G=100. (h) plus high-freq AZ noise. (i) plus high-freq AZ noise. (j) has input guard pins. (k) has charge pump to enhance CM performance. (m) plus filter stage and output clamp. (n) to within 50mV–200mV of rail. (o) to 0.1%. (p) can filter if gain greater than minimum. (q) see Figures 5.88 & 5.89. (r) V_{out} to V_{EE}, but V_{ref} 0.8V min. (s) G=10. (t) typical. (u) ISL28271 for G≥10; both are decomp. (v) typical bias, offset = 1fA, 3fA. (w) warning: high bias current. (x) at 1kHz. (y) dual=INA2321. (z) see Figure 5.82.

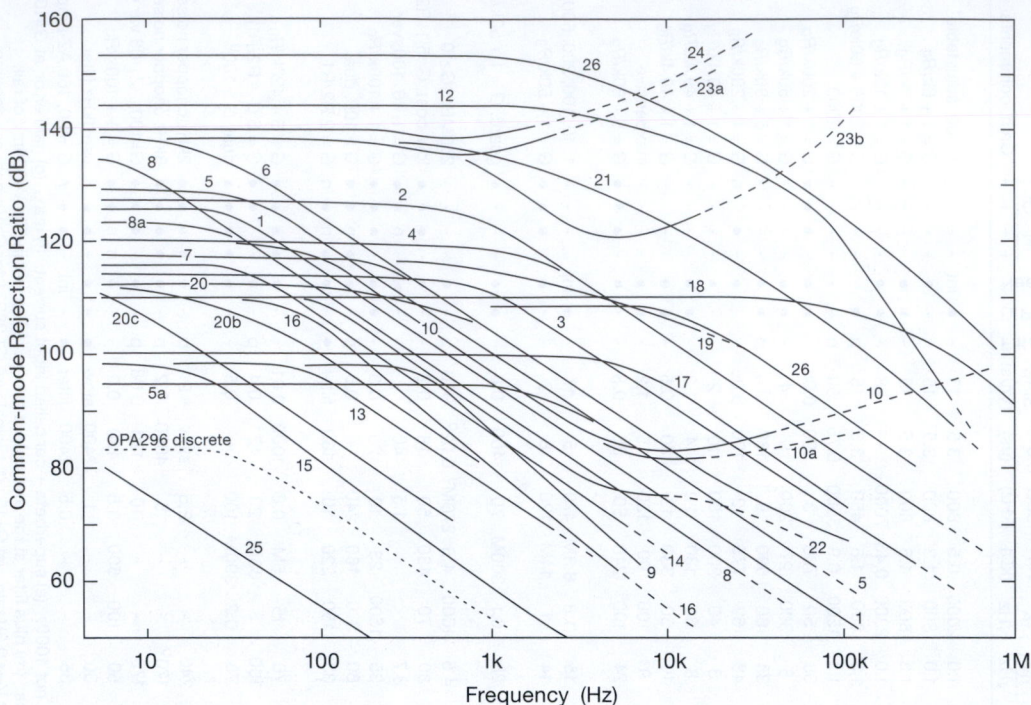

Figure 5.82. Common-mode rejection ratio versus frequency, for the instrumentation amplifiers listed in Tables 5.8 and 5.9. The "OP296 discrete" curve is measured data of a two-op-amp discrete implementation (Figure 5.88, circuit B) shown in the AD627 instrumentation amplifier (curve 16) datasheet as testimonial to the superiority of integrated amplifiers. Curves 23a and 23b show the effect of filter choices (1 kHz and 10 kHz, respectively) for the same amplifier (the AD8553). The dashed line extensions indicate a region well past the amplifier's cutoff frequency.

the shot-noise limit (as in §5.10.8), but this is greatly exceeded for auto-zero amplifiers, and BJT-input amplifiers that use bias-current cancellation.

5.16.2 Common-mode rejection

Instrumentation amplifiers must often deal with small difference signals riding on much larger common-mode voltages, requiring a high CMRR.

To get a sense of the problem, consider an INA used with a strain-gauge bridge powered from 5 V: the INA's common-mode signal input is 2.5 V, with a typical full-scale output of 10 mV (i.e., 2 mV/V). So a signal that is 0.1% of full-scale is just 10 μV. That's -108 dB with respect to the 2.5 V common-mode signal! Just seven of the INAs in Table 5.8 meet this spec for their minimum CMRR (though all but five meet it for their *typical* CMRR). An amplifier that fails to meet the CMRR spec will simply exhibit an output error larger than 0.1% of full scale, entirely down to insufficient CMRR. Keep this in perspective, though: most of the INAs in the table have more than

10 μV of input offset voltage, anyway. So we can say for many applications the listed INAs do reasonably well (and some exceptionally well), at least at dc.

A. CMRR versus frequency
A much more severe test of an INA's CMRR is its ability to reject common-mode signals at high frequencies. The plots in Figure 5.82 show this degradation, beginning somewhere in the range of 100 Hz to 5 kHz. In contrast to the (dc) strain-gauge application, imagine measuring the voltage across low-value current-sense resistors monitoring the windings of a three-phase motor. If the ac drive frequency is low enough (say 50 to 60 Hz), the INA may be up to the task. But if the motor windings are driven from a pulse-width-modulated (PWM) controller, with say 40 kHz pulses, high-frequency CMRR degradation may render an INA unusable for the task. So Figure 5.82's curves may be useful not only for choosing the best INA for the job, but in fact to determine whether an INA may be used at all.

Some of the curves of CMRR versus frequency level off after falling at the usual 6 dB/octave. This happens because the INA includes provision for bandwidth filtering at an internal node (with an external capacitor) that attenuates the common-mode feedthrough by the same 6 dB/octave, thus cancelling out the input-stage CMRR degradation. For example, we see this for the AD8293 and AD8553 (curves 24 and 23a) with 1 kHz filtering, and for the latter and the INA337 (curves 23b and 10) with 10 kHz filtering. This is seen also for some micropower parts, with their very limited bandwidth, for example the MAX4194 (curve 10a), which runs at 90 μA and has a 1.5 kHz bandwidth. That's where its response falls with $G = 100$. You can use this principle to advantage by adding a post filter in applications where fast response isn't necessary, for example the three-phase motor current monitor.

B. Case study: increasing CMRR with upstream gain

Here's a nice example from our research lab, where we needed to control the current through a magnet coil with a high degree of stability, to bring about a "Bose–Einstein condensate" of cold atoms, and (in a first experiment of its kind) slow the speed of light to that of a bicycle.[82] The currents ranged up to 875 A (!), and we wanted something close to ∼10 ppm stability in the controlled current, which was sensed with a low-side 4-wire current shunt of 100 $\mu\Omega$ resistance (thus 87.5 mV full-scale signal[83]). With such high currents flowing, we had to deal with as much as a volt or so of common-mode signal, upon which the difference signal (87.5 mV maximum, but often much less) was riding. To control that current to 10 ppm thus required some 140 dB of CMRR, and of course very low input offset drift. It also required a low input noise voltage, ideally less than 0.1 μVpp integrated low-frequency noise in order to achieve 10 ppm.

The combination of very high CMRR and very low noise and offset is a tall order for any instrumentation amplifier. The solution here is to put some low-noise and low-drift gain upstream of the INA, as shown in Figure 5.83. A precision low-noise op-amp like the LT1028A (typical drift and low-frequency noise of 0.1 μV/°C and 35 nVpp) is wired as a $G = -50$ composite amplifier (§4x.5), within which conventional op-amp A_2 is configured with $G = 5$. The compensation capacitor C_C limits the bandwidth (and therefore the noise) to about 10 kHz. The precision op-amp

is run at lower voltage to reduce thermal effects (offset-voltage drift, thermal-gradient errors), with its ±8 V rails regulated down (with floating 3-terminal linear regulators) from the ground-referenced ±15 V that powers A_2 and A_3.

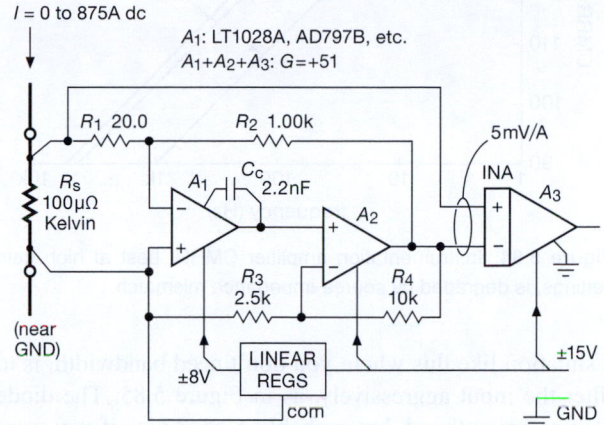

Figure 5.83. Current sensing with low drift and low noise. Op-amps A_1 and A_2 form a composite amplifier, with a precision low-noise first stage and a noncritical second stage. The output signal provides feedback to the high-current power supply.

5.16.3 Source impedance and CMRR

Instrumentation amplifiers excel in high input impedance, but that does not confer automatic immunity from the effects of mismatched signal source impedances (which so severely degrade CMRR in *difference* amplifiers, with their relatively low input impedance; see §5.14.3A). Most datasheets are bashful about displaying their dirty laundry in this regard, so we must applaud Analog Devices' candor, as shown in Figure 5.84. Note the greater effect at higher gain settings (where there's more to lose, because the CMRR is so good), and at higher frequencies (where the amplifier's input impedance is falling, owing to capacitance).

5.16.4 EMI and input protection

Whether you "roll your own" or use an integrated instrumentation amplifier, you've got to think about protection both from overloads and from electromagnetic interference (EMI). An example from real life: a colleague uses temperature-monitoring thermocouples in his lab, connected with the usual unshielded wire pairs to an INA front end running at high gain. All was well until a particular switching power supply was energized, at which point the thing went haywire.

The problem, of course, was common-mode EMI, coupled onto the long unshielded cable. And the solution, in

---

[82] L. Vestergaard Hau, S. E. Harris, Z. Dutton, and C. H. Behroozi, "Light speed reduction to 17 metres per second in an ultracold atomic gas," *Nature* **397** 594–598 (1999).

[83] A smallish voltage, but already an uncomfortably large power dissipation (∼75 W), requiring a temperature stabilized oil bath.

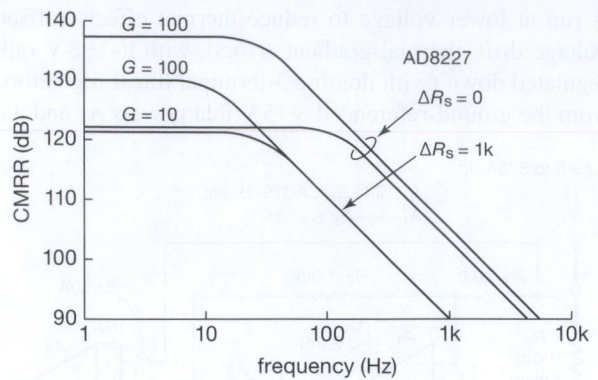

Figure 5.84. Instrumentation amplifier CMRR, best at high gain settings, is degraded by source-impedance mismatch.

a situation like this where you don't need bandwidth, is to filter the input aggressively, as in Figure 5.85. The diode clamps are optional, but probably a good idea if you want the amplifier to survive an unusual input event. In situations where you *do* need bandwidths that include interfering signals, you're unlikely to get away with unshielded cable; use shielded wiring, and pay attention to grounding paths.

It is hard to overstate the seriousness of EMI/RFI: RF signals leaking into your inputs cause rectification inside BJT op-amps, and thus dc offsets. Cables or PCB traces can exhibit narrowband (high-Q) resonances, enhancing these effects. If you notice changes of offset voltage when you grip a cable, or touch a circuit node with a pencil, or just wave your hands around, you're probably seeing RF coupling (the other possibility is a circuit oscillation). Lossy ferrite beads are excellent for attenuating RF, as well as for reducing the Q of unwanted wiring resonances; but they're not a panacea, and often you need to resort to additional filtering.

5.16.5 Offset and CMRR trimming

Instrumentation amplifiers that provide both SENSE and REF pins (e.g., the INA103) can be externally trimmed, if needed, for both offset voltage and CMRR, as shown previously for the difference amplifier (Figure 5.74). More often you're given only the REF pin, though. That's enough to trim the offset voltage, but note that any offset applied to the REF pin must have a source impedance of no more than $\sim 10^{-6}R_f$ (i.e., a few *milli*ohms) in order not to compromise the amplifier's 100 dB+ CMRR. This is best done with a precision op-amp, as in Figure 5.86.

Some INAs helpfully include output-stage offset trim terminals, to be used with an external trimpot. A few of

these (e.g., the INA110 and others) even provide separate trim pairs for the front-end and output stages.

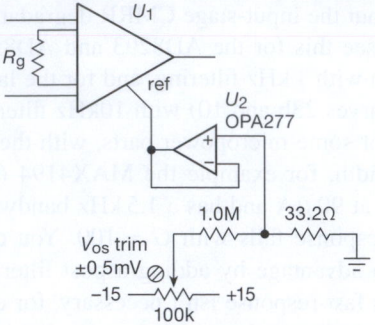

Figure 5.86. Trimming the offset of an INA that doesn't provide a SENSE pin, offset pins, or a buffered REF pin.

5.16.6 Sensing at the load

As with the difference amplifier, the REF and SENSE pins can be connected directly at the load, as in Figure 5.78(a), to eliminate errors from wiring resistance and unrelated ground currents.

5.16.7 Input bias path

Instrumentation amplifiers earn honest bragging rights from their very high input impedance, but, just as with opamps, you've got to provide a dc return path. If you don't, the amplifier will saturate. This occurs naturally in circuits like the strain gauge of Figure 5.64A, but not in something like a thermocouple (Figure 5.85). For the latter you can use a resistor from one of the inputs to ground (or to mid-supply, for a single-supply amplifier), or you can use a bias resistor from each input to ground to preserve symmetry.

5.16.8 Output voltage range

If an INA is operated at low gain and with its common-mode input near the rails (but legally within the specified operating range), the internal amplifier may saturate, causing the INA's output to go to an incorrect voltage. For example, look at the "Maximum Output Voltage vs. Common-Mode Input" graph in the datasheet for the AD623 (this amplifier is like type A in Figure 5.88, except with *pnp*-input emitter followers). If the common-mode input is at 0 V with $G=10$ (which is legitimate), the maximum output capability is only 1.0 V! *You have been warned!*

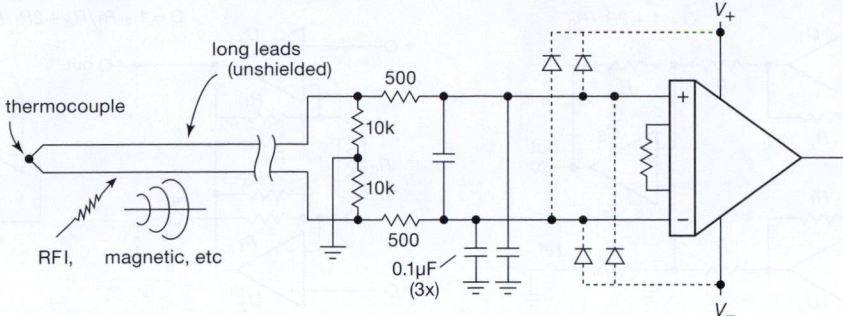

Figure 5.85. Electromagnetic interference couples nicely into long unshielded cables, which you can think of as *antennas*. Use lowpass filtering at high-gain inputs, with optional diode clamping to the rails. Note the 10 kΩ resistor pair, to set the dc level of the floating differential input.

5.16.9 Application example: current source

The excellent CMRR of instrumentation amplifiers, combined with very low input capacitance (typically ∼2 pF), lets you design an active current source in which the current sense resistor rides on the high side of the output, with a "flying" INA converting its voltage drop to a ground-referenced output. Such a circuit is shown in Figure 5.87A, exploiting the >80 dB CMRR of the AD8221B out to 50 kHz, where the stabilizing compensation network R_1C_C rolls off. The MOSFET, configured as a common-source transconductance stage, has inherently high output impedance, which makes it possible to maintain good current-source behavior at higher frequencies. This improves upon a current-source circuit like that in Figure 5.69 (§5.14.2D), in which the performance degrades with frequency owing to the op-amp's compensation and limited slew rate. In this circuit the source-degeneration resistor R_3 acts to reduce the MOSFET transconductance, to enhance stability (some breadboard testing, with anticipated load impedances, would not be inadvisable). See Table 3.4 for selected *p*-channel MOSFETs.

For this circuit the op-amp output must be able to swing to the positive rail (the LT1490 is RRIO), and the AD8221B needs a few volts of negative supply voltage (−3 V is adequate) because its common-mode input voltage does not extend to the negative rail.[84]

The AD8221B's low offset voltage (25 µV max) provides a large dynamic range, corresponding to an output error of just 0.25 mA out of the full-scale range of 5 A (20,000:1).

---

[84] The AD8227 variant allows V_{CM} to the negative rail, so you could run it single supply; but you pay a price in larger V_{OS} and I_B, and its CMRR degrades at a lower frequency.

For lower output currents you can substitute a smaller bipolar transistor, as in Figure 5.87B; its lower capacitance permits greater loop bandwidth (we've configured it here as a follower). The AD8221B's low input

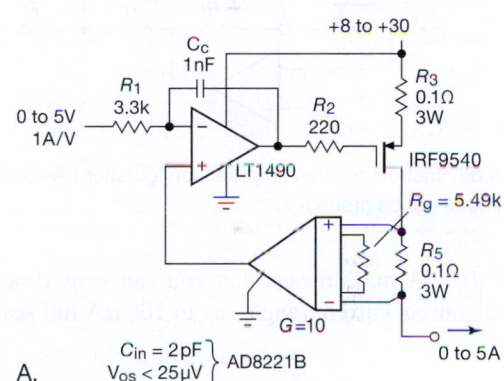

A.

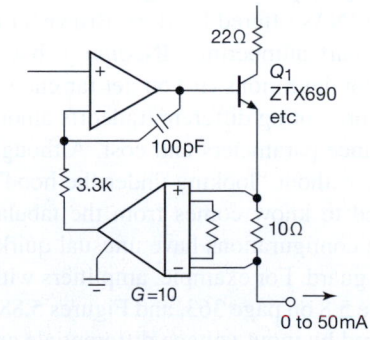

B.

Figure 5.87. Precision current source with flying instrumentation amplifier. A. Power MOSFET for 5 A full-scale current, B. a small high-gain (β∼500 over 0.1–50 mA) BJT provides greater bandwidth.

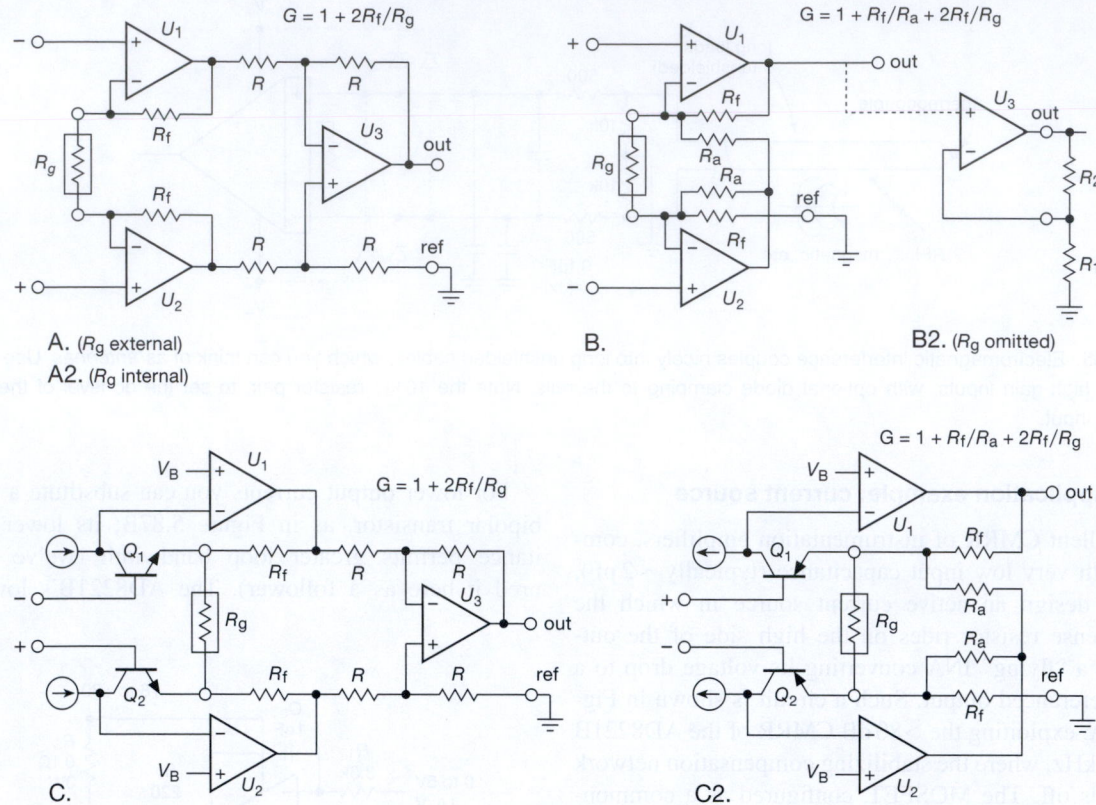

Figure 5.88. Instrumentation amplifier configurations A–C, as listed in Table 5.8 (page 363). For these and other differencing circuits the "ref" pin need not be grounded.

current (0.4 nA max) means that you can scale down the full-scale output current range, say to 100 μA full scale.

5.16.10 Other configurations

The classic 3-op-amp circuit of Figure 5.77 is widely used, notably in the INAs offered by Burr–Brown/TI (recognizable by their part numbering "INA*nnn*"); but you'll see other circuit configurations (if you get far enough into the datasheets) representing different tradeoffs among the various performance parameters and cost. Although you can get along fine without "looking under the hood"(most of what you need to know comes from the tabulated data), some of these configurations have unusual quirks that can catch you off-guard. For example, amplifiers with configuration E (Table 5.8 on page 363, and Figures 5.88 and 5.89) can be damaged by input voltage differentials greater than ± 0.5 V (!),[85] and the amplifiers with configuration F do not

operate with the REF tied to ground (even though intended for low-voltage single-supply operation). Quite apart from these kinds of worries, curiosity impels a brief look at these circuits.

There's a general principle in play in nearly all of these circuits (D and E excepted): (a) the voltage across a gain-setting resistor R_g is precisely the same as the input differ-ence voltage, creating a current $I_g = \Delta V_{in}/R_g$; and (b) that current is used to generate an accurately proportional out-put voltage $V_{out} \propto I_g$. Classic configuration A puts this in clear view: the input op-amps (or equivalent – they need not *be* fully featured op-amps) enforce a matched ΔV_{in} across R_g, with the resultant current flowing through the two R_f's; thus a differential output $\Delta V_{out} = (\Delta V_{in}/R_g)(R_g + 2R_f)$. The

---

[85] Some of these have back-to-back clamp diodes across the inputs (true also of some op-amps and comparators), for which the damage is

caused by excessive input current; others tolerate greater differential inputs (as much as a few volts), though most of the E-configuration parts are considerably more restrictive than those with the other circuit topologies. What matters more, from the user's point of view, is the maximum differential input without performance degradation.

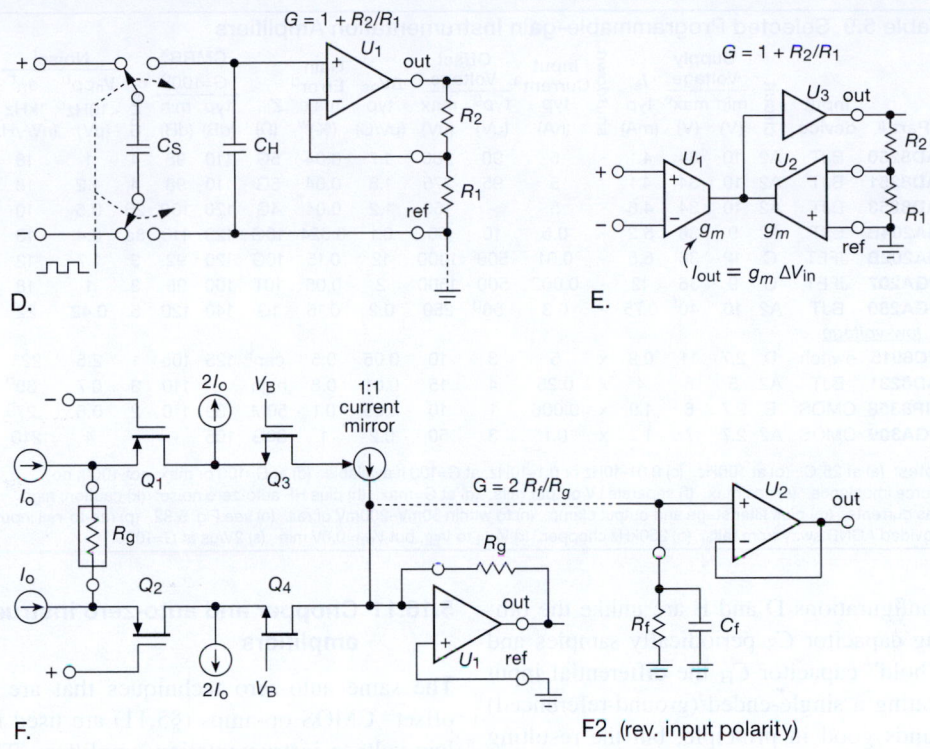

Figure 5.89. Instrumentation amplifier configurations D–F, as listed in Table 5.8 on page 363.

unity-gain difference amplifier converts this to a single-ended output, with gain $G = 1 + 2R_f/R_g$.

Configuration C works similarly, but here the matched emitter followers $Q_1 Q_2$ create the ΔV_{in} replica across R_g, with the op-amps serving to ensure equal emitter currents (and thus no contribution to the differential output).[86] In this circuit and following, V_B is a reference "bias" voltage.

One nice thing about the C and C2 configurations is that a small RFI-defeating capacitor can be placed across base-to-emitter of the input transistors, because they're exposed pins. Keep those capacitors small – 100 pF or less – so the amplifier's bandwidth and stability are not degraded.[87]

The clever configuration B is different: it is more economical, requiring only two op-amps and fewer trimmed

resistors, but its performance suffers, with poorer CMRR (particularly at higher frequencies). (The reader also suffers, trying to figure out how this tangle of a circuit works.) Configuration C2 is the discrete differential-pair analog of B (just as C was for A), with similarly underwhelming specifications.

Configuration F continues the theme of replicating ΔV_{in} across R_g, with the resultant unbalanced current at U_1's summing junction being converted to a single-ended output. In this circuit Q_3 and Q_4 form a "folded cascode," pinning the drains of Q_1 and Q_2 while passing their currents through (offset by twice the quiescent current, thus sinking). This circuit requires accurate matching of the current sources and of the current mirror (stated more precisely, it requires constancy of the current sources and mirror over common-mode variations). Evidently that can be accomplished with good design (and the assistance of circuit tricks like the cascode), given the impressive 140 dB (typ) CMRR specification.[88]

---

[86] To achieve low input currents with the C configuration, LTC uses superbeta BJTs with base-current cancellation in some of the listed parts ($I_B \approx 50$ pA); Analog Devices does even better using JFETs, but with greater offset and noise. Some of the TI/Burr–Brown INAs listed as A types may in fact use the C configuration; their datasheets are silent on the circuit details.

[87] BJT-input amplifiers are prone to RFI upset, because their inputs are forward-biased base–emitter (diode) junctions. And RFI is a real problem in these low-level circuits with inputs from remote sensors. Better to use a JFET-input amplifier if you are bedeviled by RFI.

[88] Because they contain only a few MOSFETs, current sources, and current mirrors, devices of configuration F can be quite inexpensive. For example, the AD8293 (an AZ with fixed $G=80$ or 160) sells for only $0.97 (qty 100).

Table 5.9 Selected Programmable-gain Instrumentation Amplifiers

Part #	Input device	circuit	Supply Voltage min (V)	Supply Voltage max[e] (V)	I_s typ (mA)	shutdown	Input Current[a] typ (nA)	Offset Voltage typ[g] (µV)	Offset Voltage max (µV)	ΔV_{os} typ (µV/C)	Gain Error[d] typ (%)[w]	Gain Error[d] max	Z_{in} (Ω)	CMRR[x] G=100[d] typ (dB)	CMRR[x] G=100[d] min (dB)	curve	Noise[t] V_{npp}[t] <10Hz[c] (µV)	Noise[t] e_n[t] 1kHz (nV/√Hz)
AD8250	BJT	A2	10	34	4.1	-	5	90	260	1.7	0.04	5G	110	98	4	1	18
AD8251	BJT	A2	10	34	4.1	-	5	95	275	1.8	0.04	5G	110	98	4	1.2	18
AD8253	BJT	A2	10	34	4.6	-	5	-	160	1.2	0.04	4G	120	100	26	0.5	10
PGA204B	BJT	A2	9	36	5.2	-	0.5	10	50	0.1	0.024	10G	123	110	8a	0.4	13
PGA202B	JFET	C	12	36	6.5	-	0.01	500	1000	12	0.15	10G	120	92	3	1.7	12
PGA207	JFET	C	9	36	12	-	0.002	500	1500	2	0.05	10T	100	95	3	1	18
PGA280	BJT	A2	10	40[f]	0.75	-	0.3	50[q]	250	0.2	0.15	1G	140	120	5	0.42	22
low-voltage																	
LTC6915	switch	D	2.7	11	0.9	x	5	3	10	0.05	0.5	cap[k]	125	105	1	2.5	225
AD8231	BJT	A2	3	6	4	x	0.25	4	15	0.01	0.8	high	-	110	3	0.7	39[h]
LMP8358	CMOS	E	2.7	6	1.9	x	0.006	1	10	0.05	0.1	50M	139	110	2	0.6	27[h]
PGA309	CMOS	A2	2.7	7	1.2	x	0.1	3	50	0.2	1	30G	105	-	-	4	210

Notes: (a) at 25°C. (b) at 100Hz. (c) 0.01–10Hz or 0.1–10Hz, at G=100 if available. (d) at G=100 or max, not 1000; no source imbalance. (e) abs max. (f) separate LV output rails. (g) at G=max. (h) plus HF auto-zero noise. (k) caution: high bias currents. (m) plus filter stage and output clamp. (n) to within 50mV–200mV of rail. (o) see Fig. 5.82. (p) rail-to-rail input, provided AGND away from rails. (q) 250kHz chopper. (r) V_{out} to V_{EE}, but V_{ref}=0.8V min. (s) 2V/µs at G=1000.

Parts using configurations D and E are unlike the others. In D a flying capacitor C_S periodically samples and conveys to the "hold" capacitor C_H the differential input voltage, thus creating a single-ended (ground-referenced) replica. This sounds good in principle; but the resulting noise is high, and the slow commutating rate (3–6 kHz, for the two exemplars in Table 5.8 on page 363) limits the bandwidth and extends the settling time. This technique is also vulnerable to aliasing at half the chopping frequency. However, these parts are inexpensive, and they may be well suited for some dc applications.

Finally, in E the output currents from a pair of differential-input transconductance amplifiers are combined and forced to equality: one amplifier sees the input signal pair, and the other sees a divided fraction of the output, producing a single-ended output voltage, as shown. This low-cost configuration (no laser-trimmed resistor pairs, etc.) is limited to small differential input signals (thus high gain), and generally exhibits relatively poor gain accuracy.[89] Interestingly, the fastest amplifier in Table 5.8, by far, uses this configuration.

5.16.11 Chopper and auto-zero instrumentation amplifiers

The same auto-zero techniques that are used in "zero-offset" CMOS op-amps (§5.11) are used in some CMOS low-voltage instrumentation amplifiers. These are recognizable by their very low offset-voltage specifications, down in the tens of microvolts, where non-AZ CMOSs do not venture (see Table 5.8 on page 363). These amplifiers attain excellent CMRR also, but they pay the price in broadband noise, switching-frequency noise,[90] and (sometimes) input bias and noise currents. The amplifiers are useful particularly in low-frequency applications, for example as input stages for integrating ADCs (see Figures 13.67), or combined with lowpass filtering.

5.16.12 Programmable gain instrumentation amplifiers

You set the voltage gain of a simple op-amp circuit with external resistors. In contrast, *difference* amplifiers (Table 5.7 on page 353) ordinarily are configured with fixed gain, set by a precision internal matched resistor network. And the gain of *instrumentation* amplifiers (Table 5.8 on page 363) is ordinarily set by a single gain-setting resistor R_g. Note, though, that some of the amplifiers in Table 5.8 include several internal R_g gain-setting resistors, allowing accurate gain selection with only an external jumper wire. Taking this one step further, you can get *programmable-*

---

[89] Some E types (e.g., the AD8130, a variant of the AD8129 in Table 5.8) are specified and characterized only for G=1. These are especially useful as differential line receivers, etc., but they are generally limited in swing, typically in the range of 3 to 4 Vpp (the AD8237, with its flying switched capacitors, is an exception). See also §12.10 for a discussion of differential signaling in the *digital* context.

[90] Which may not be evident from the datasheets, which sometimes omit spectral noise plots.

Table 5.9 Selected Programmable-gain Instrumentation Amplifiers (cont'd)

Part #	Gain choices	BW[d] −3dB (MHz)	Slew Rate (V/µs)	Settle[d] 0.01% (µs)	Swing to Supplies? IN +	IN −	OUT[v] +	OUT −	mux?	filter?	out sense	Interface	Comments	Cost ($ US)
AD8250	1,2,5,10	3	25	0.65	-	-	-	-	-	-	-	pins	fast settle to 10ppm	7.78
AD8251	1,2,4,8	2.5	25	0.68	-	-	-	-	-	-	-	pins	fast settle to 10ppm	7.78
AD8253	1,10,100,1000	0.55	20 s	1.5	-	-	-	-	-	-	-	pins	fast settle to 10ppm	7.68
PGA204B	1,10,100,1000	0.01	0.7	100	-	-	-	-	-	-	●	pins	PGA205 for 1,2,4,8	17.09
PGA202B	1,10,100,1000	1	20	2	-	-	-	-	-	●	●	pins	PGA203 for 1,2,4,8	14.34
PGA207	1,2,5,10	0.6	25	3.5	-	-	-	-	-	-	●	pins	PGA206 for 1,2,4,8	18.36
PGA280	1/8–128, by x2	0.05	2	40	-	-	f	f	2	-	-	SPI	output supply 2.7–5V	6.46
low-voltage														
LTC6915	1,2,4...4096	slow	0.2	5ms	x	x	x	x	-	●	●	SPI,pins	chopper, high-gain	4.58
AD8231	1,2,4...128	0.07	1.1	slow	n	n	x	x	-	-	-	pins	auto-zero	3.08
LMP8358	10–1k, by 1-2-5	0.68	6.5	4	-	x	x	x	-	●	-	SPI,pins	auto-zero, prog filter	5.27
PGA309	4–128, 2.7–1152	0.4	0.02	slow	-	n	n	x	●	-	-	SPI	sensor interface[u]	7.05

(t) typical. (u) 148-page user manual; includes ADC for sensor span, offset tempco cal, etc.; parameters stored in external SOT23 EEPROM. (v) caution: the RR output parts often don't allow the REF pin to be close to V_{EE}; be sure to check the datasheet! (w) part with large gain error assume you'll perform gain calibration. (x) the CMRR spec is typically at 60Hz; check the plots if you care about performance at higher frequencies.

gain amplifiers (PGAs) in which that selection is made by a digital input code (either applied as a parallel logic-level code to a set of pins, or as a multibit serial code through a serial port like SPI or I²C; see Chapters 14 and 15). These are, in essence, integrated versions of the discrete digitally programmable instrumentation amplifier of Figure 5.80.

Some examples of stand-alone instrumentation PGAs are the PGA204/5, LMP8358, and PGA280, with more listed in Table 5.9. The PGA202/3 (JFET) and PGA204/5 (BJT) are traditional "high-voltage" (to ±18 V) parts that accept a 2-bit parallel code (on two pins) to select gains of 1, 10, 100, or 1000 (PGA202/4) or 1, 2, 4, or 8 (PGA203/5). The more recent LMP8358 is a low-voltage single-supply part (2.7–5.5 V) with auto-zero and gains from 10 to 1000 in a 1–2–5 sequence (i.e., 10, 20, 50, 100, 200, 500, 1000), and with ambidextrous programming – the three pins can be used either as a 3-bit gain-setting parallel port or a 3-wire SPI serial port that programs both the gain and some additional parameters such as input polarity reversal, fault detection, and bandwidth. It is fast (8 MHz) and accurate ($V_{OS} = 10\,\mu$V, max).

Finally, the very elegant PGA280 addresses the need for a PGA whose input signals can range over ±10 V or more, but with a separately powered output stage matched to contemporary low-voltage single-supply ADCs and microcontrollers. The inputs can range over ±15.5 V (with ±18 V supplies), with the output supplied from the same +2.7 V

to +5 V supply that powers the ADC or µC.[91] This solves the messy problem of protecting the inputs of a low-voltage part when it is driven from an IC running from higher voltages. There's a REF-like pin that sets the output mid-span voltage; the output is actually *differential*, with a pair of complementary outputs, but you can ignore one and treat it as single-ended (with a slight loss of accuracy).

This amplifier has excellent performance: it programs through a digital serial port, with selectable gains from 1/8 to 128 by factors of two. It combines low offset voltage (auto-zeroing: ±15 µV max, at $G = 128$), high input impedance (>1 GΩ typ), low gain drift (±3 ppm/°C max), and excellent CMRR (gain-dependent: 130–140 dB typ). Among its other tricks are an on-chip 2-input multiplexer (two differential-input pairs), an uncommitted byte-wide bidirectional digital port, and various signal conditioning and fault-detection options.

Figure 5.90 shows the sort of application for which this part would be ideal: our colleagues have developed an experimental robotic grasping hand[92] that is driven by a single torque motor through a set of passive linkages and couplings. For control you'd like to know the applied

---

[91] Or you can power the output from a split supply, staying within that total supply range.

[92] See, for example, Dollar and Howe, "The Highly Adaptive SDM Hand: Design and Performance Evaluation", *International Journal of Robotics Research* 29, (5), 585–597 (2010), available at the web page of The Harvard BioRobotics Laboratory: biorobotics.harvard.edu.

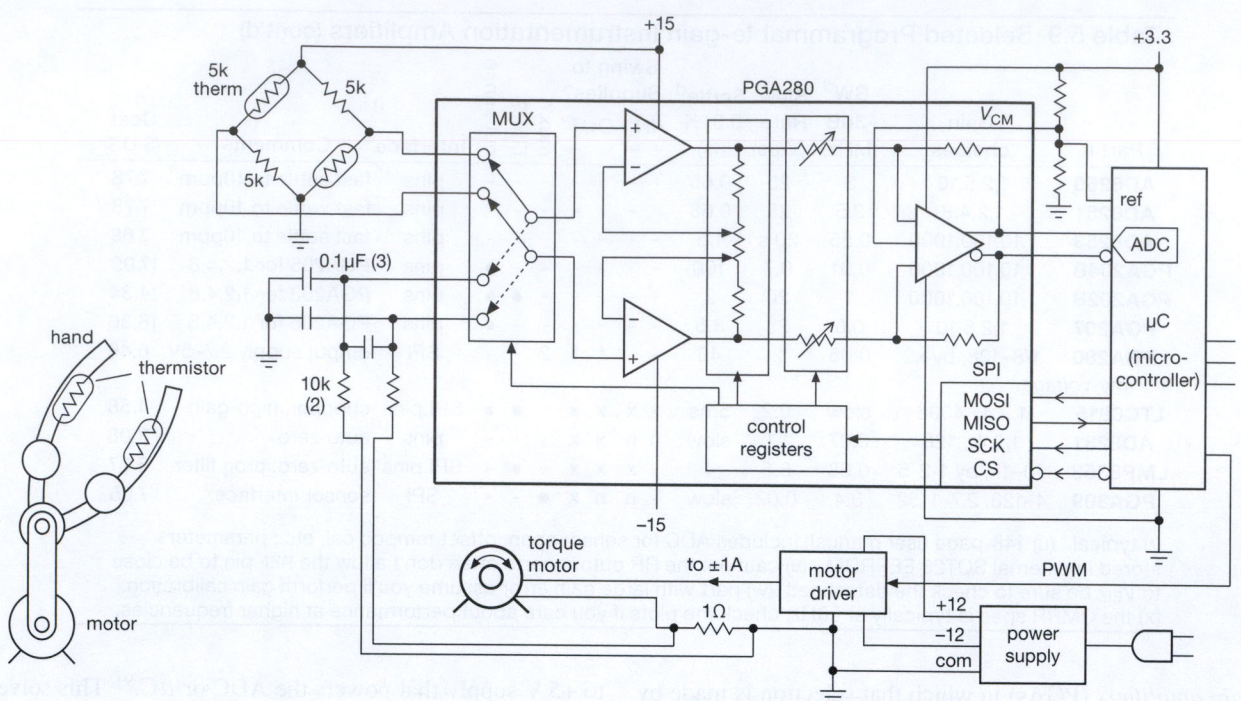

Figure 5.90. Two-channel level-shifting PGA application: reading out the applied torque and thermal response of a grasping robotic hand. This chopper-stabilized (250 kHz) instrumentation PGA works from ±15 V rails, but delivers its output to a 3.3 V ADC – *that's* being helpful! Recommended.

torque (from the motor current), and something about the contact pressure (if any) with the object being grasped. The PGA280's wide input compliance and range of gains, combined with 2-channel differential inputs, make this an easy task: we imagine a pair of thermistors, self-heated a few degrees above ambient, that sense the temperature at the grasping points, and a low-side sense resistor in the motor return line. The low-voltage output stage connects to the on-chip ADC of a microcontroller, which controls the channel switching and gain of the PGA. We've rigged the thermistors so that first contact makes an upward step in differential voltage, followed by a second step in the same direction when the other "finger" makes contact. (The step size provides further information about the object's material.) Note the differential "4-wire" connection at the motor-current-sense resistor, to eliminate errors from ground connection resistance; we've filtered that signal, because the use of pulse-width modulation tends to create high-frequency noise.

When looking around for a good instrumentation amplifier, be sure to consider specialized PGA variants that are intended to serve as front ends for low-level sensors and the like, for example the PGA309[93] or PGA2310.[94]

PGAs are quite popular as portions of more complex ICs, for example within ADCs (Figure 13.67 again, and Figures 13.70 and 13.71) and microcontrollers (Figure 15.10).

5.16.13 Generating a differential output

Both instrumentation amplifiers and difference amplifiers are used to convert a differential input signal to a single-ended output. That's fine, if that's what you want (and it usually is). There are some situations, though, in which you need a differential output signal, for example when

---

[93] Whose 148-page User's Guide describes it thus: "The PGA309 is a smart programmable analog signal conditioner designed for resistive bridge sensor applications. It is a complete signal conditioner with bridge excitation, initial span and offset adjustment, temperature adjustment of span and offset, internal/external temperature measurement capability, output over-scale and under-scale limiting, fault detection, and digital calibration."

[94] "The PGA2310 is a high-performance, stereo audio volume control designed for professional and high-end consumer audio systems."

driving several varieties of analog-to-digital converter (ADC, a major subject of Chapter 13). Most simply, this can be done by adding a unity-gain inverter to the single-ended output, as in Figure 5.91A.[95] This works, for sure, but the gain accuracy will be degraded unless the resistor pair is matched to at least the precision and stability of the driving amplifier. This pitfall is circumvented in circuit B, where the unity-gain inverter forces the SENSE output pin to a symmetrical voltage relative to ground (or some other reference voltage). With this circuit the gain accuracy is retained; the effect of any resistor mismatch is simply to offset the symmetry of the outputs about ground (or reference voltage), which is generally of less importance because of the differential-input nature of the device being driven.

Both circuits, however, share the drawback of introducing a time delay (or phase shift), owing to the finite bandwidth of the inverting stage. One solution is to use a pair of matched amplifiers, as in Figure 5.66F. But a better way, particularly when plenty of bandwidth and rapid settling is needed (as with fast ADCs), is to use a *differential amplifier* (or *fully differential amplifier*, to emphasize the distinction), a term that has come to mean an amplifier with differential inputs *and* outputs; see §5.17. The PGA280 of the previous section is such an amplifier (though its designers officially call it an instrumentation amplifier).

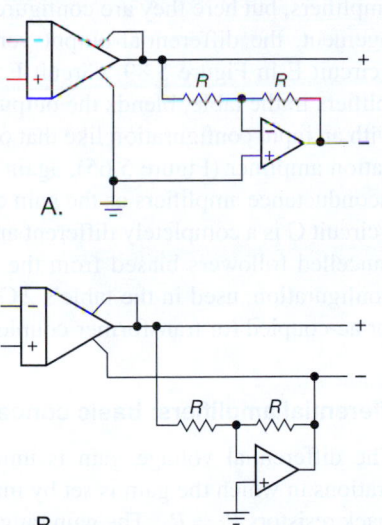

A.

B.

Figure 5.91. Generating a differential output from an instrumentation amplifier or difference amplifier. Method B maintains gain accuracy.

---

[95] See also the parts listed under "single-ended to differential" in Table 5.10 (page 375).

5.17 Fully differential amplifiers

The term "fully differential amplifier" (or sometimes "differential in/out amplifier," or just "differential amplifier") is used to describe an amplifier with differential input *and* differential output, along with an additional input pin ("V_{OCM}") that sets the common-mode voltage of the output pair. We favor the "fully" term, to distinguish clearly from difference amplifiers and instrumentation amplifiers, both of which have single-ended outputs.

For some important applications you need to create a balanced differential *output*, from either a differential or a single-ended input signal. This is often the case with ADCs that have complementary inputs; see Figure 13.65 (a "charge-redistribution" ADC), Figure 5.102 (pipeline flash), Figure 13.28 (flash), Figure 13.37 (SAR), and Figure 13.68 (delta–sigma). For that application the important performance parameters are likely to be settling time, gain accuracy and stability, the ability to set the common-mode output voltage, and the ability to drive a rail-to-rail swing into a low-voltage ADC.

Other applications where differential signals are widely used include analog signaling over twisted pairs (e.g., through existing Cat-5-style network cable); telecom applications such as ADSL and HDSL links; oscilloscope input stages; and RF communications subcircuits such as IF and baseband blocks.

You can, of course, create a differential-output signal pair by using single-ended amplifiers (op-amps, difference amplifiers, and instrumentation amplifiers), as illustrated in Figures 5.66F, 5.70, 5.91, and 13.37. But you do better, particularly in terms of speed and noise, with an integrated differential amplifier, which also lets you set the common-mode output voltage (i.e., the midpoint of the output swing); this capability is particularly useful when driving differential-input ADCs powered from a single supply, because of their fussy insistence on common-mode input voltages.

Table 5.10 on page 375 includes a good selection of currently available differential amplifiers, keyed to the circuit diagrams in Figures 5.94–5.98. These illustrate the unflagging creativity of the human species, and some nice circuit tricks as well.

Let's take a quick tour. Circuit A, intended for single-ended inputs, is simply a differential amplifier "kit," with an input op-amp whose gain you set with external resistors. You can wire it as a noninverting amplifier (therefore high input impedance), or you can configure it as an inverting amplifier (for example, to handle a large input swing by setting the gain to less than unity). The noninverting input of A_2, conveniently of high impedance, lets you set the output common-mode output voltage. The LT6350 is a low-noise, low-distortion amplifier of this configuration, with the added feature of rail-to-rail outputs.[96] Figure 5.92 shows how you would use it to drive an ADC, in this case the LTC2393 that cooperates by providing a mid-scale dc reference output (V_{CM}).[97] The amplifier runs from the same +5 V and ground supply voltages, which eliminates the frequent worry about driving the ADC's input clamp diodes into conduction. The input lowpass filter to the ADC $(R_1 R_2 C_1)$ serves two functions: (a) it is an anti-alias filter, limiting the input bandwidth to ∼150 kHz; and (b) it provides the recommended shunt input capacitance to suppress the effects of the ADC's internal switching transients (which afflict many ADCs, including "charge redistribution SAR" converters, and delta–sigma converters – see Chapter 13).

Circuit B is a symmetrical balanced configuration optimized for pro-audio: well-balanced high-level drive (>15 Vrms) with low distortion into balanced pairs, and stability into the load capacitances you get with long cables (to 10,000 pF or more). Figure 5.93 shows a typical application, in this case generating the high-level low-distortion balanced output needed for pro-audio cable driving, first discussed in §5.14.2E. Of note here is the very high *common-mode* output impedance, which preserves signal balance by allowing the receive end to override the driver's default (symmetrical about ground at the driver end, set weakly by the 10k resistors). In fact, it's even OK to ground one side at the receive end, in order to generate a single-ended signal there. We'll have more to say about this in §5.17.1.

Circuit C is the configuration of Figure 5.66F, with high-impedance (buffered) common-mode voltage-setting inputs (V_{OCM}). The table entries with this configuration are low-noise wideband amplifiers, good for video and communications applications.

Circuit D is a popular configuration, sometimes with internal gain-setting resistors, sometimes external. The dif-

---

[96] The ADA4922-1 is a faster, 0.05% version, with a fixed unity-gain input stage.

[97] The ADC has an internal reference of reasonable accuracy (0.5%), but it lets you attach an external reference of better performance (e.g., the LT1790–4.096, with accuracy and drift of ±0.05% and 10 ppm/°C, max). Sounds good – but it would not accomplish much here, because the gain accuracy of the system is limited by that of the LT6350 amplifier (±0.6%, max). You could use the ADA4922-1 instead.

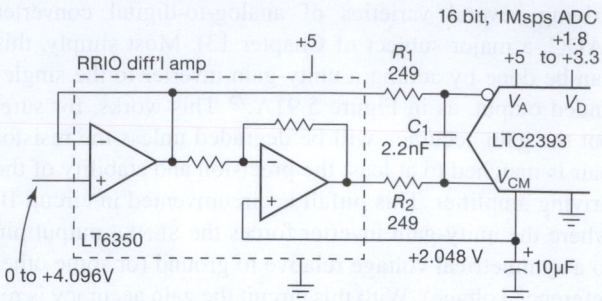

Figure 5.92. ADC driver circuit using a configuration A differential amplifier. The ADC provides a mid-span V_{CM} output reference, here used to set the amplifier's common-mode level. See also the better-performing AD4922.

ferential amplifier within the feedback loop consists of a symmetrical pair of transconductance amplifiers (voltage-to-current) generating a voltage across a resistive load, with voltage followers to generate the low-impedance output pair. The V_{OCM} input lets you assert the common-mode output voltage, which otherwise defaults to mid-supply (in which case it's good to attach a bypass capacitor). The V_{OCM} input bandwidth typically is comparable to that of the amplifier.

Circuit E continues the theme of differential transconductance amplifiers, but here they are configured in a feedback arrangement, the differential-output version of the analogous circuit E in Figure 5.89. Circuit F, used in the fastest amplifiers in the table, blends the output configuration of D with an input configuration like that of the classic instrumentation amplifier (Figure 5.65), again with differential transconductance amplifiers as the gain element.

Finally, circuit G is a completely different animal, a pair of offset-cancelled followers biased from the V_{OCM} input pin. This configuration, used in the table's 2 GHz entry, is intended for ac-coupled (or transformer-coupled) inputs.

5.17.1 Differential amplifiers: basic concepts

A. Gain The differential voltage gain is unity for most configurations in which the gain is set by matched pairs of feedback resistors $R_f = R_g$. The gain range column in Table 5.10 (page 375) identifies parts with fixed gains, minimum gains, or a set of gain selections. In some cases the exact gain is affected by source-impedance and termination-matching issues; see §5.17.4 and the formulas in Figure 5.104.

Table 5.10 Selected Differential Amplifiers

Part #	Circuit	Gain Range	Gain Set by	ΔG (%)	Bandwidth −3dB (MHz)	Bandwidth 0.1dB (MHz)	Speed Slew Rate (V/µs)	Speed Settle 0.05% (ns)	V_{out} diff'l max (Vpp[b])	Offset Voltage typ (mV)	Offset Voltage max (mV)	Input bias (µA)	Z_{in}[h] diff (Ω)	Noise e_n, typ @1MHz (nV/√Hz)	Supply Range (V)	I_s typ (mA)	RRO	Cost qty 25 ($US)	Comments
high voltage																			
PGA280	q	1/8–128	PGA	0.15	1.5	-	1	40µs	10	0.05	0.25	0.3nA	1G	22	10-40	0.8	●	6.46	INA, SPI, low V_{out}
THAT1606	D3	2.0	fixed	2	10	-	15	-	67	4	15	-	5k	25	8-40	4.9	-	3.91	audio[t], 7ppm dist
AD8270	C	0.5,1,2	fixed	0.08	15	-	30	700	55	0.2	0.75	0.5	20k	26[d]	5-36	2.3	-	3.56	precision resistors
LME49724	D	1.0 up	R_f/R_g	-	50	-	18	200	52	0.2	1	0.06	R_g[h]	2.1	5-36	10	-	3.53	audio, 0.3ppm dist
OPA1632	D	1–10	R_f/R_g	-	180	40	50	200	52	0.5	3	2	R_g[h]	1.3	5-33	14	-	4.39	pro audio
low voltage																			
LMP7312	D3	0.1-2	PGA	0.04	0.53	-	1.4	-	RR	-	0.1	c	160k	7.5	±5	2	y	6.43	2-ch-input, SPI
LTC1992	D	1–10	R_f/R_g	-	3.2	-	1.5	-	RR	0.25	2.5	2pA	30k	45[d]	2.7-12	0.7	w	2.80	R_f = 30k to 50k
LTC1992-x	D3	1–10	p/n	0.3	3.2	-	1.5	-	RR	0.25	2.5	2pA	30k	45[d]	2.7-12	0.7	w	5.23	-x for G=1,2,5,10
AD8137	D	1.0 up	R_f/R_g	-	76	10	450	100	RR	0.7	2.6	0.5	R_g[h]	8[d]	2.7-12	2.6	●	1.98	R_f=1k-10k, cheap
THS4521	D	1–10	R_f/R_g	-	145	20	490	13[k]	RR	0.2	2	0.65	R_g[h]	4.6[d]	2.5-5.5	1.1	w	2.81	R_f=1k; '4522 dual
EL5170	E3	2.0	2.0	1	100	12	1100	20	6.0	6	25	6	300k	28[d]	5-12	7.4	-	1.44	cheap, '5370 triple
LTC6605-14	D3	1, 4, 5	1,4,5	0.3	14[e]	f	f	f	RR	0.25	1	12	400/G	2.6	2.7-5.5	16	w	9.81	dual, 16 bit ready
LT6600-x	D2	1–8	402/R_g	7	10[g]	f	f	f	4.8	5	25	30	R_g[h]	10[d]	2.7-11	36	-	3.90	4th-order filters
LTC6601	D3	1–7	1 to 7	3	6-27[i]	f	f	f	RR	0.25	1.5	12	400/G	2.2	2.7-5.5	16	w	5.58	pin-select filter
AD8138	D	1.0 up	R_f/R_g	-	320	30	1150	16	15.6	1	2.5	3.5	R_g[h]	5	3-10	20	-	4.12	R_f=500Ω
LMH6551	D	1.0 up	R_f/R_g	-	370	50	2400	18	15.6	0.5	4	4	R_g[h]	6[d]	3.3-10	13	-	3.50	R_f=365Ω
EL5173	E3	2.0	2.0	0.5	450	60	900	10	6.0	3	30	11	150k	25[d]	4.8-12	12	-	2.80	EL5373 triple
EL5177	E	1.0 up	R_f,R_g	1.5	550	120	1100	10	6.0	1.4	25	14	150k	21	4.8-12	12	-	1.85	G=1+R_f/R_g, cheap
AD8139	D	1.0 up	R_f/R_g	-	410	45	800	45	RR	0.15	0.5	2.3	R_g[h]	2.2	4.5-12	25	-	6.00	R_f=200Ω
AD8132	D	1.0 up	R_f/R_g	-	350	90	1200	15	14.4	1	3.5	3	R_g[h]	8[d]	2.7-11	12	-	3.07	R_f=348Ω
LT6402-20	C	10	fixed	10	300	30	400	8[k]	7.0	1	6.5	c	100	1.9	4.0-5.5	30	-	3.90	opt 75MHz filter
LTC6404-4	D	4 up	R_f/R_g	-	600	450	1200	13	RR	0.5	2	23	R_g[h]	1.5	2.7-5.5	30	w	4.91	R_f=402; ver -1,-2[j]
THS4520	D	1–10	R_f/R_g	-	620	30	570	7	RR	0.25	2.5	6.5	R_g[h]	2	3-6	14	●	4.07	R_f=499Ω
PGA870	s	0.3-10	PGA	4	650	100	2900	5	4.8	5[z]	30[z]	c	150	6	4.8-6	143	-	9.70	0.5dB atten, pins
ADA4932-1	D	1.0 up	R_f/R_g	-	560	300	2800	9	15.0	0.5	2.2	2.5	R_g[h]	3.6[d]	3-10	10	-	4.72	R_f=499Ω; -2=dual
LT1993-10	C	10	fixed	12	700	50	1100	4	7	1	6.5	c	100	1.7	5	100	-	4.20	-2,-4 for G=2,4
ADA4950-1	D	1, 2, 3	pins	1.7	750	210	2900	9	15	0.2	2.5	c	500/G	5.5[d]	3-11	9.5	-	4.78	-2=dual
LMH6553	D	1.0 up	R_f/R_g	-	900	50	2300	10	15.4	-	-	50	R_g[h]	1.2	5-12	29	-	6.43	CFB, 274Ω; clamp
ADA4938-1	D	1–10	R_f/R_g	-	1000	150	4700	6.5	15.2	1	4	13	R_g[h]	2.6	4.5-11	37	-	6.06	R_f=200Ω
LMH6552	D	1.0 up	R_f/R_g	-	1500	450	3800	10	15.4	-	-	80	R_g[h]	1.1	5-12	23	-	8.14	CFB, 357Ω
THS4513	D	1.0 up	R_f/R_g	-	1600	700	5100	16	5.6	1	4	8	R_g[h]	2.2	3-6	35	-	10.34	R_f=348Ω
LMH6555	D3	4.84	4.84	3	1200	180	1300	2.2[k]	1.0	15	50	c	78	4[d]	3.3	120	-	10.55	assumes R_S=50Ω
THS4511	D	1–10	R_f/R_g	-	1600	620[r]	4900	3.3[k]	5.2	1	4	8	R_g[h]	2	3.5-5.5	39	v	9.13	R_f=349Ω
ADA4937-1	D	1–5	R_f/R_g	-	1900	200	6000	7	6	0.5	2.5	21	R_g[h]	2.2[d]	3-10	40	-	6.06	R_f=200Ω; -2=dual
THS4508	D	2–10	R_f/R_g	-	2000	400	6600	2[k]	5.2	1	4	8	R_g[h]	2.3	3.8-5.5	39	v	9.13	R_f=349Ω
THS4509	D	2–10	R_f/R_g	-	2000	300	6600	10	5.6	1	4	8	R_g[h]	1.9	3-6	38	-	9.13	R_f=349Ω
LTC6416	G	0.98	1.0	2	2000	300	3400	1.8[k]	4.2	0.5	5	5	12k	1.8	2.7-3.9	42	-	8.27	emitter follower
THS770006	D	2	fixed	3	2400	350	3100	2.2	4.9	0.5	5	c	100	1.7	5	100	-	10.45	Z_{out}=27Ω at 1GHz
AD8352	F	1–8	R_g	12	2500	190	8000	2[k]	6	-	6	5	3k	2.7	3.5-5	37	-	5.65	specs for G=10dB
ADL5561	D3	2,4,6	pins	0.5	2900	600	9800	2[k]	4.3	0.25	-	c	800/G	1.7	3-3.6	40	-	5.89	good for IF-strip
ADL5562	D3	2,4,6	pins	0.7	3300	270	9800	2[k]	4.3	0.25	-	c	800/G	1.6	3-3.6	80	-	5.88	good for IF-strip
ADA4960-1	F	1–10	R_f/R_g	6	5000	300	8700	1.4	3.5	-	36	20	10k	5.4	5	60	-	11.47	specs for G=6dB
single-ended to differential[m]																			
DRV134	B	2.0	fixed	2	1.5	-	15	2500	60	1	10	-	10k	35	9-40	5.2	-	4.21	audio[t], 5ppm dist
ADA4922-1	A2	2.0	fixed	0.05	38	-	260	580	43	0.18	0.55	3.5	11M	6	9-26	5.4	n	5.81	18-bit ADC driver
low voltage																			
LT6350	A	2.0 up	opamp	0.6	33	-	48	240	RRIO	0.1	0.7	1.2	4M	4.1[d]	2.7-12	4.8	●	3.42	precision
ADA4941-1	A	2.0 up	opamp	1	30	-	22	300	RR	0.2	0.8	3	24M	5.1[d]	2.7-12	2.5	w	4.45	odd f_{3dB} specs

Notes: (a) fixed gain, programmable gain (PGA), or gain set by input resistor R_g. (b) RR output means V_{out} diff = 2 × V_{supply} max. (c) included in V_{os} spec. (d) includes feedback resistor noise. (e) matched 2nd-order antialias filters; 7, 10MHz avail. (f) set by filter. (g) 6600-x specifies 2.5 to 20MHz 4th-order filters. (h) nominally Z_{in} = R_g, where G = R_f/R_g; but for "D" circuit types it's greater than R_g, see later section on Differential Amplifier Input Impedance. (i) filter, 6 to 27MHz strappable. (j) -4 version compensated for G<4. (k) settle to 1%. (m) most differential-to-differential amplifiers can convert single-ended input to differential output. (n) near. (o) instrumentation amplifier figures. (q) inst-amp input, differential-amp output. (r) G=2. (s) R-2R ladder input, differential-amp output. (t) high common-mode Z_{out}, like an isolation transformer. (v) inputs to −V_{EE}. (w) RRO and inputs to −V_{EE} or within 0.2V of −V_{EE}. (y) RRO and input beyond rails, to ±15V. (z) RTO.

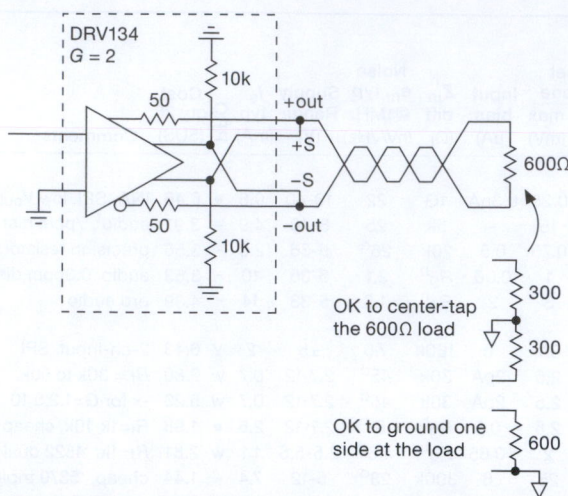

Figure 5.93. Balanced audio driver with a high common-mode output impedance, so that the receiver (load end) sets the common-mode voltage. $V_{CM(out)}$ defaults to 0 V if the load is left floating.

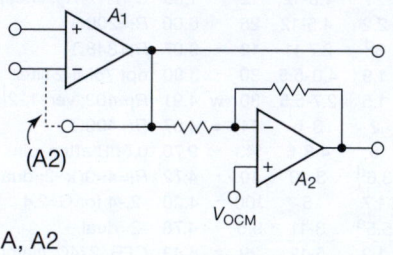

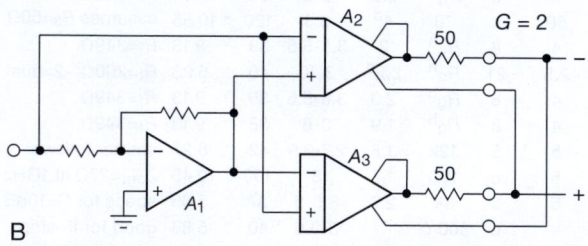

Figure 5.94. Differential-output amplifier configurations A and B, as listed in Table 5.10 on page 375. The uncommitted input op-amp can be configured as a noninverting amplifier (or follower), or an inverting amplifier.

B. Input impedance The input impedance of amplifiers with configuration D is equal to R_g, making them unsuitable for high gains because the input impedance becomes unmanageably low: the signal source is heavily loaded, the effective R_g is increased by the source impedance R_S, and the CMRR is degraded by source-

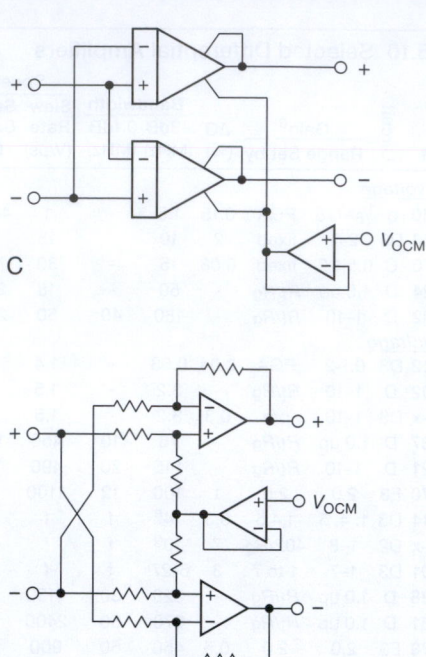

Figure 5.95. Fully differential amplifier configuration C, as listed in Table 5.10 on page 375. The symmetry is evident in the redrawn version, C', where the gain is $G = 2R_f/R_g$.

impedance imbalance. The exact value of R_g (and therefore the input impedance) will be affected by source termination and loading considerations; see §5.17.4 and the formulas in Figure 5.104.

C. Single-ended input Most fully differential amplifiers work fine with single-ended inputs, i.e., with the "−" input grounded. But you may wish to use $G = 2$ or higher to achieve full peak-to-peak drive into a differential ADC.[98]

D. Common-mode rejection With differential pairs at both input and output, there are *two* measures of common-mode rejection: *differential* V_{out} versus common-mode V_{in}, which is usually quite good (e.g., 80 dB up to 1 MHz); and *common-mode* V_{out} versus common-mode V_{in}, which can be significantly poorer (e.g., 50 dB up to 1 MHz, degrading above that). But the latter is not terribly worrisome if the receiving device (e.g., an ADC) has good common-mode rejection itself. Resistor matching is important for configurations

---

[98] This configuration is also available with a voltage gain less than unity. In a burst of cuteness, Analog Devices calls their AD8475 (with gains of 0.4 and 0.8) a *funnel amplifier* (get it?). With it you can reduce a 20 Vpp signal to a differential pair of 4 Vpp signals, for input to a low-voltage ADC.

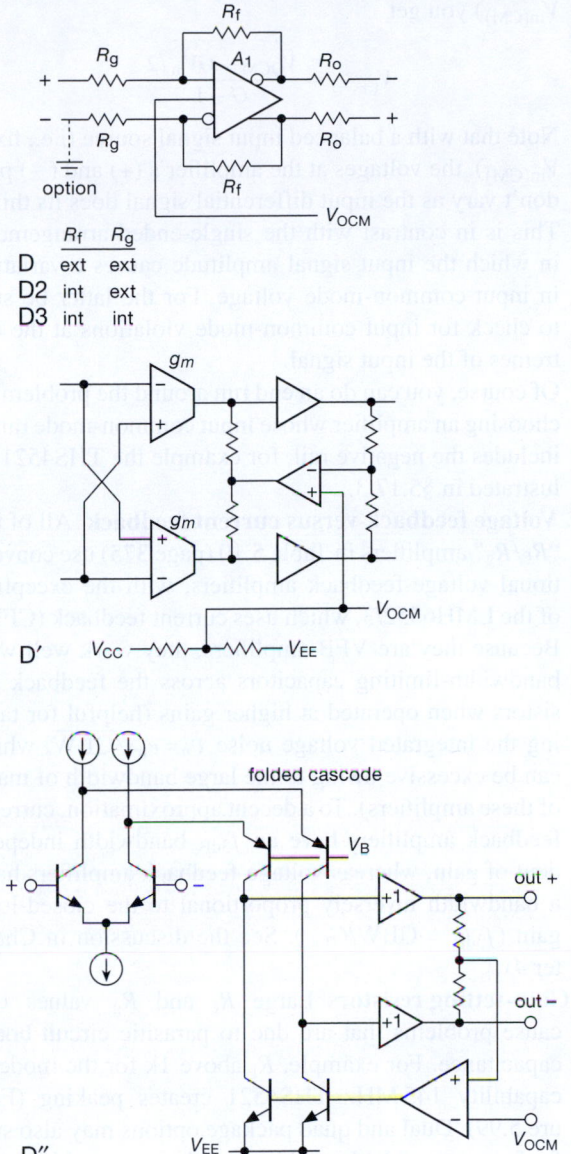

Figure 5.96. Fully differential amplifier configuration D, as listed in Table 5.10 on page 375; the gain is $G=R_f/R_g$. A typical configuration for output amplifier A_1 is shown in D′, and TI's version for their THS45xx series is D″; the THS4508/11/21, for example, use polarity complements (*pnp* input pair, etc.), permitting operation down to $V_{in}=V_{EE}-0.2$ V.

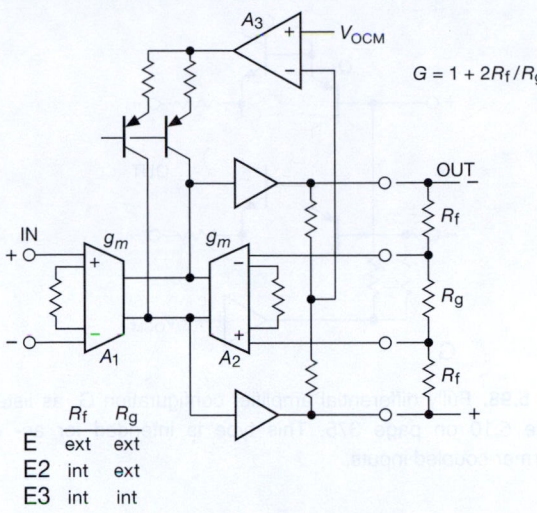

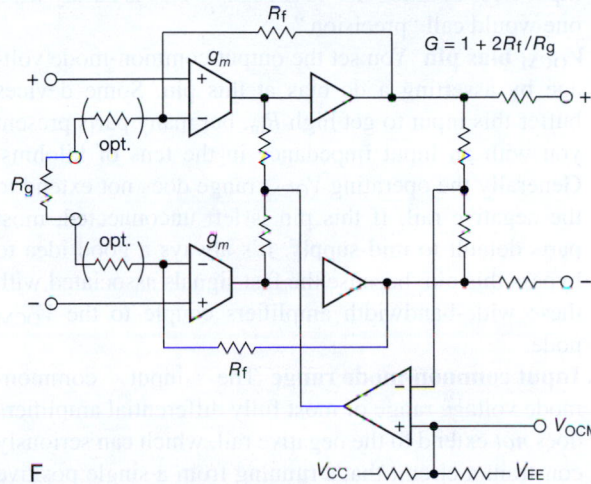

Figure 5.97. Fully differential amplifier configurations E and F, as listed in Table 5.10 on page 375.

with external R_f and R_g gain-setting resistors (configuration D); see §5.17.5 for more details.

E. Single-ended output The datasheets of some differential amplifiers describe operation with a single-ended output.[99] When it's operated in single-ended output mode, however, you care about the *output* offset voltage ΔV_{OCM} (i.e., the output error with respect to the V_{OCM} reference), which translates back to an input-referred error of $\Delta V_{OCM}/G$. For the LMP7312 the output offset is ±20 mV, far larger than the maximum input offset of ±0.1 mV. This is a low-gain amplifier ($G=0.1$ to 2), so this output offset looks like a corresponding worst-case

[99] For example, the datasheet for the "Precision SPI-Programmable AFE with Differential/Single-Ended Input/Output" LMP7312 says that "the output can be configured in both single-ended and differential modes with the output common mode voltage set by the user."

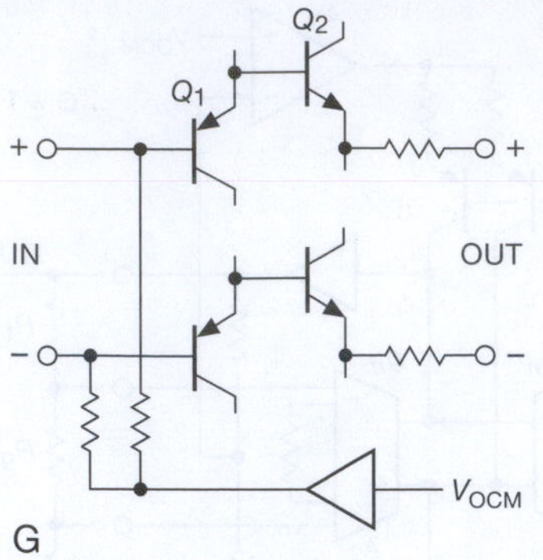

Figure 5.98. Fully differential amplifier configuration G, as listed in Table 5.10 on page 375. This type is intended for ac- or transformer-coupled inputs.

input error of ± 200 mV to ± 10 mV! This is hardly what one would call "precision."

F. V_{OCM} **bias pin** You set the output common-mode voltage by asserting a dc bias at this pin. Some devices buffer this input to get high R_{in}, but many parts present you with an input impedance in the tens of kilohms. Generally the operating V_{OCM} range does not extend to the negative rail. If this pin is left unconnected, most parts default to mid-supply. It's always a good idea to bypass this pin, because the fast signals associated with these wide-bandwidth amplifiers couple to the V_{OCM} node.

G. Input common-mode range The input common-mode voltage range of most fully differential amplifiers does *not* extend to the negative rail, which can seriously constrain a circuit that's running from a single positive supply. This isn't as bad as it sounds, though: the outputs, sitting up there around a positive common-mode output voltage (set by the dc you apply to the V_{OCM} input pin), bring the input terminals up by the voltage dividers formed by R_f and R_g. This effect is largest when operating at low gains; when operating at higher gains it's best to check that the input common-mode range is not violated. Assuming that there's plenty of loop gain (i.e., that $G_{OL} \gg G$) the (equal) voltage at the amplifier's inverting and noninverting pins is

$$V_{(+,-)} = \frac{V_{OCM} + GV_{in(CM)}}{G+1},$$

where the differential gain $G = R_f/R_g$, and $V_{in(CM)}$ is the common-mode voltage of the (differential) input signal source. If the input is single ended (with the other differential input grounded) then (substituting $V_{in}/2$ for

$V_{in(CM)}$) you get

$$V_{(+,-)} = \frac{V_{OCM} + GV_{in}/2}{G+1}.$$

Note that with a balanced input signal source (i.e., fixed $V_{in(CM)}$), the voltages at the amplifier's $(+)$ and $(-)$ pins don't vary as the input differential signal does its thing. This is in contrast with the single-ended arrangement, in which the input signal amplitude causes a variation in input common-mode voltage. For the latter be sure to check for input common-mode violations at the extremes of the input signal.

Of course, you can do an end run around the problem by choosing an amplifier whose input common-mode range includes the negative rail, for example the THS4521 illustrated in §5.17.3.

H. Voltage feedback versus current feedback All of the "R_f/R_g" amplifiers in Table 5.10 (page 375) use conventional voltage-feedback amplifiers, with the exception of the LMH6552/3, which uses current feedback (CFB). Because they are VFB amplifiers, they work well with bandwidth-limiting capacitors across the feedback resistors when operated at higher gains (helpful for taming the integrated voltage noise $v_n = e_n\sqrt{GBW}$, which can be excessive owing to the large bandwidth of many of these amplifiers). To a decent approximation, current-feedback amplifiers have an f_{3dB} bandwidth independent of gain, whereas voltage-feedback amplifiers have a bandwidth inversely proportional to the closed-loop gain ($f_{3dB} = GBW/G_{CL}$). See the discussion in Chapter *4x*.

I. Gain-setting resistors Large R_f and R_g values can cause problems that are due to parasitic circuit-board capacitance. For example, R_f above 1k for the modest-capability 145 MHz THS4521 creates peaking (Figure 5.99). Dual and quad package options may also suffer from unavoidable peaking problems caused by lead-frame issues, so it's generally better to choose fixed-gain types in multiple-amplifier packages. Large values of R_f and R_g also create (a) loss of speed, (b) increased input offset error from the relatively large bias currents characteristic of fast bipolar amplifiers, and (c) increased input-referred voltage noise, both from resistor Johnson noise and from the noise voltage developed across R_f by the amplifier's input noise current.

To put some flesh on these latter bones, consider the THS4521 again, with our (somewhat extreme) straw-man $R_f = R_g = 100$k. From Figure 5.99 you can see the $\times 10$ reduced bandwidth and peaking. This peaking (which occurs with VFB amplifiers but not CFB types)

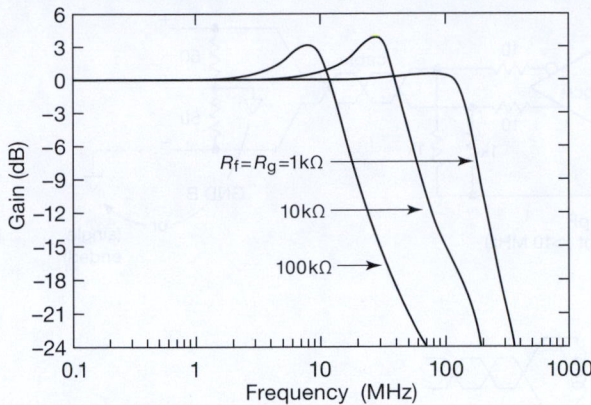

Figure 5.99. Large gain-setting resistor values create peaking in the frequency response, as shown in this datasheet plot for the THS4521 configured for unity gain ($R_f = R_g$).

can be tamed by putting a small capacitor across each feedback resistor, but you'll lose a bit more bandwidth in the process.[100] Turning this vice into a virtue, we note that you may wish to add feedback capacitors anyway, to reduce the bandwidth intentionally.

As for offset voltage, this part has an input bias current of 650 nA (typ), which would create a drop of 65 mV across 100k. But the bias currents are reasonably well matched, with an offset current spec of $\Delta I_B = \pm 50$ nA (typ), thus creating an input offset of 5 mV. That's much better, but it does seriously degrade the amplifier's typical $V_{OS} = \pm 0.2$ mV (± 2 mV max). You need to keep R_f and R_g less than 10k to preserve the amplifier's accuracy.

Finally, noise. There are two contributions, the resistors' Johnson noise ($e_n = \sqrt{4kTR} = 0.13\sqrt{R}$ nV/$\sqrt{\text{Hz}}$), and the noise voltage developed by the amplifier's current noise ($e_n = i_n R_f$). For $R_f = 100$k the Johnson noise voltage is 40 nV/$\sqrt{\text{Hz}}$, and the noise produced by the amplifier's $i_n = 0.6$ pA/$\sqrt{\text{Hz}}$ is 65 nV/$\sqrt{\text{Hz}}$. These disastrously degrade the amplifier's typical 4.6 nV/$\sqrt{\text{Hz}}$ (taking the usual root-sum-of-squares, the total added noise voltage is 76 nV/$\sqrt{\text{Hz}}$). The table below summarizes these figures, and also those corresponding to $R_f R_g = 10$k and $R_f = R_g = 1$k.[101]

Bottom line: compared with the nominal 1k value, the use of 100k gain-setting resistors reduces your bandwidth $\times 10$, increases the typical offset voltage $\times 25$, and increases the typical input-referred noise voltage $\times 16$. You wouldn't want to do this. But you could reasonably use something like 2.49k, 4.99k, or perhaps 10k, buying increased input impedance at the expense of modest degradation of bandwidth, noise, and accuracy.

R_f, R_g	-3 dB Bandwidth (MHz)	Offset voltage* (mV, typ)	$\sqrt{4kTR}$ (nV/$\sqrt{\text{Hz}}$)	$i_n R_f$ (nV/$\sqrt{\text{Hz}}$)	Total* (nV/$\sqrt{\text{Hz}}$)
			Input-referred noise		
1k	150	± 0.2	4	0.7	4.6
10k	45	± 0.5	13	6.5	15
100k	15	± 5	40	65	76

* Includes V_{OS} and e_n of the amplifier.

J. Common-mode output impedance The voltage asserted onto the V_{OCM} pin sets the common-mode output voltage. Put another way, differential amplifiers have a low common-mode output impedance. That's usually what you want; after all, that's the reason there *is* a V_{OCM} pin. But this can create difficulties if the output is sent to a remote load that needs to establish its preferred common-mode level. That is the case in balanced audio (or video), sent substantial distances over balanced twisted-pair cable.

Take a look at Figure 5.100B. By driving the V_{OCM} pin with the average output voltage, you create an amplifier that cooperates by letting the load lead the dance. In fact, the load can even unbalance the signal intentionally (by grounding one side), in which case the other output swings symmetrically around ground with the full desired differential output voltage.[102] There are a few differential amplifiers that are designed specifically for this kind of application, with an internal configuration that creates a high common-mode output impedance. We saw one example (the DRV134, similar to the SSM2142) in Figure 5.93. Another excellent one is the THAT1606, from the curiously named THAT Corporation.[103]

The traditional solution has been to use an isolating transformer, which also can do the job of converting between single-ended and balanced signals (this is known as a "balun," for *bal*anced–*un*balanced) as shown. But

---

[100] You can look at this another way: the manufacturer's recommended gain-setting resistor values are chosen such that a small amount of peaking is exploited to extend the amplifier's natural bandwidth.

[101] Many parts (the D and E configurations) let you add feedback capacitors to reduce the bandwidth. With some parts this may introduce instability at low gains, but with others it may improve the stability, especially when larger resistor values are used.

[102] Omit the small bypass capacitor shown, if this mode of operation is anticipated.

[103] A member of the hard-to-Google corporate name club, which includes *AND Displays*, and *ON Semiconductor*. (Try it: you'll get more than ten billion Google hits for "AND" or "ON.")

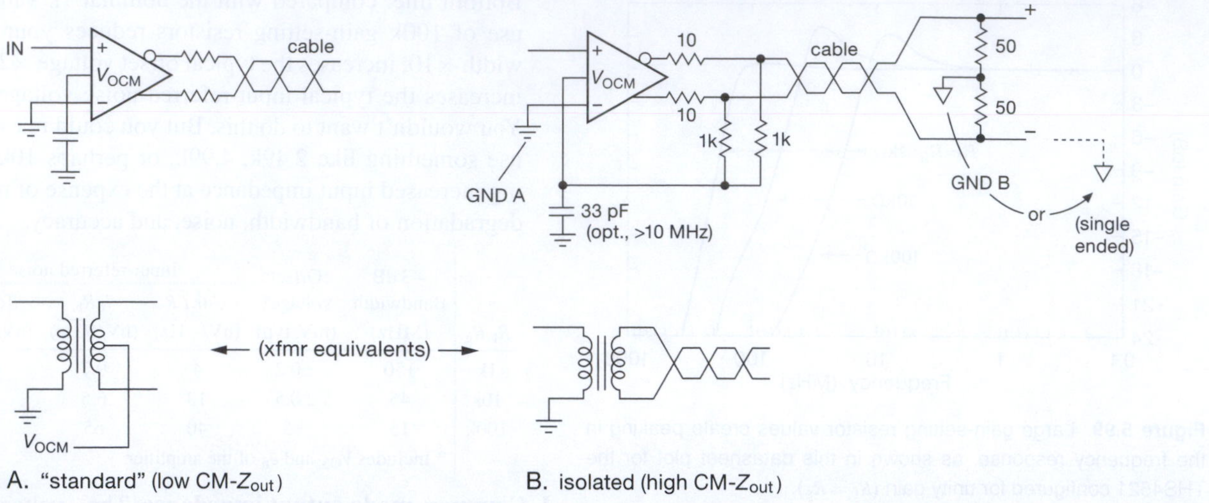

A. "standard" (low CM-Z_{out})   B. isolated (high CM-Z_{out})

Figure 5.100. The common-mode output voltage of a differential amplifier is set by the V_{OCM} pin (usually defaulting to mid-supply if undriven), producing a low common-mode output impedance, as in A. But for balanced audio or video applications you want the (distant) *load* to be the boss. You can trick a differential amplifier, as in B, simulating the very high CM-Z_{out} of an isolating transformer or balun.

transformers are bulky, limited in bandwidth and linearity, and not inexpensive. High CM-Z_{out} differential amplifiers can be an attractive alternative.

5.17.2 Differential amplifier application example: wideband analog link

We conclude the discussion of differential amplifiers with several application examples: a wideband analog link over twisted-pair cable, and a riff on driving differential-input ADCs.

In §5.14.2F we illustrated the use of a difference amplifier as the receiving end of an analog link over differential twisted-pair cable. In that circuit R_4C_1 creates a rising response at high frequencies ("equalization"), to compensate for the increasing cable attenuation. To complete the link, of course, we need a differential driver.

Figure 5.101 shows the whole enchilada, here implemented with Intersil's EL5170/72 differential line driver/receiver pair. They make triple units (EL5370/72) also, with application to color video. Bandwidths to tens of megahertz are easily achievable over tens of meters of Cat-5 cable, with modest equalization at the receiver end. Coaxial cable is considerably better, and two $50\,\Omega$ coax lines can replace the $100\,\Omega$ differential pair. As always, the balanced signal combines with excellent receiver CMRR (here 95 dB typ) to provide high rejection of interference.

5.17.3 Differential-input ADCs

Many analog-to-digital converters require differential signal inputs. This is almost universally true of high-speed converters (e.g., pipelined flash converters), as well as the varieties known as "charge-redistribution SAR" and "delta-sigma" (ADCs, in all their glory, are the major subject of Chapter 13). And in many cases the input is hardly benign – the internal switched capacitors introduce charge transients at the input terminals, mandating some external shunt capacitance. An additional annoyance is the requirement that the driver must be able to swing the inputs over the full conversion range (which may include ground), but without driving them beyond the ADC's power rails (with the risk of damage from input clamp conduction and possible SCR latchup).

A. First iteration: single-supply ADC driver with V_{OCM} offset

Figure 5.102 shows two iterations of an input stage for fast single-supply ADCs with differential inputs. Our first design was based around the AD9225 12-bit 25 Msps pipelined flash converter, which runs from a single +5 V analog supply, and has a separate digital supply pin for interfacing to microcontrollers running from +3 V to +5 V. Its input span is programmable, either 0–2 V or 0–4 V, and it provides a mid-span dc output (+1 V or +2 V) that can be used to set the differential amplifier's common-mode output (via the V_{OCM} pin).

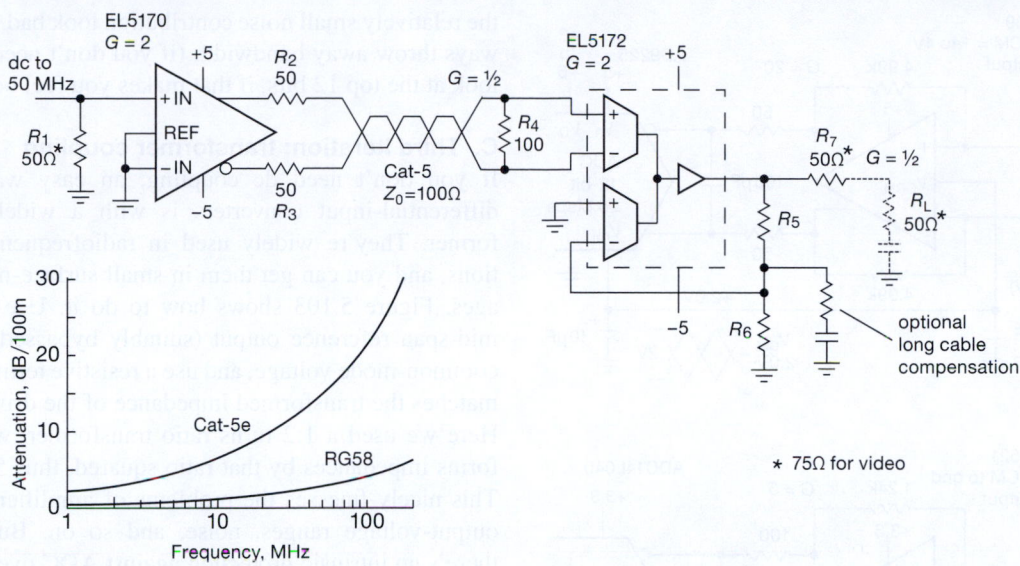

Figure 5.101. Wideband analog link over Cat-5 network cable. The EL5370/72 package three similarly performing drivers–receivers in a single IC, convenient for sending analog video (RGB, S-Video, or YPbPr component video) over a single cable (which has four twisted pairs). See also Figure 5.71.

By running the differential amplifier from the same +5 V and ground, we are assured that the ADC's inputs cannot be driven beyond the rails. We chose the AD8139 differential amplifier for its low noise (2.2 nV/$\sqrt{\text{Hz}}$), ample bandwidth ($\sim$15 MHz at $G = 20$), and ability to swing its output rail-to-rail (to drive the ADC over its full-input span). We used the recommended pair of series resistors to suppress amplifier ringing from charge transients at the ADC input; and we added a shunt capacitor to both reduce these transients and also provide a second stage of anti-alias filtering.[104]

That's all good news. The bad news is that this amplifier, in common with most differential amplifiers, does not include ground in its input common-mode operating range: you've got to stay a volt away from either rail. So you cannot simply tie one input to ground and drive the other with a small signal around ground.[105] The amplifier *does* let you run from split supplies (e.g., ±5 V), which solves the input signal-level problem; but then you have to worry about

power-supply sequencing and the risk of driving negative current into the ADC's clamp diodes.

B. Second iteration: single-supply ADC driver with $V_{\text{in(CM)}}$ to ground

What to do? Find a single-supply amplifier that operates with inputs to the negative rail! That's what we did in the second circuit, whose THS4521's common-mode input range includes ground ("NRI" – negative rail input), and in fact guarantees proper operation with inputs to -0.1 V.[106] It also has the needed rail-to-rail output, but it is somewhat noisier and slower than the AD8139 (4.6 nV/$\sqrt{\text{Hz}}$, and a gain of only ×5, to maintain 18 MHz bandwidth).

We teamed it with the ADC14L040, a more accurate and faster ADC (14-bit, 40 Msps) that runs on a single +3.3 V supply and uses less power (235 mW versus 335 mW). The ADC's span is ±0.5 V, centered on an allowable midspan voltage of +0.5 V to +2.0 V. We could have used the ADC's reference-derived +1.5 V output to drive the amplifier's V_{OCM} pin, as before; but when that pin is not driven the amplifier defaults to mid-supply

---

[104] Repeating some important advice: when designing with high-speed ICs, it's particularly important to pay attention to special instructions in the datasheet; the AD9255, for example, devotes nearly a full page to a discussion of input R's and C's.

[105] Except when operating at low gain: here that would require $G \leq 1$, so that the AD8139's input terminals are brought up to 1 V or more by the $R_f R_g$ divider. See the discussion in §5.17.1.

[106] Other differential amplifiers whose feedback inputs can operate at or near ground are indicated with **w** or **v** in Table 5.10 (page 375), and include the LTC1992, LTC6605, LTC6601, LTC6404, THS4508, and THS4511. These parts span the bandwidth range from 3 MHz to 2000 MHz.

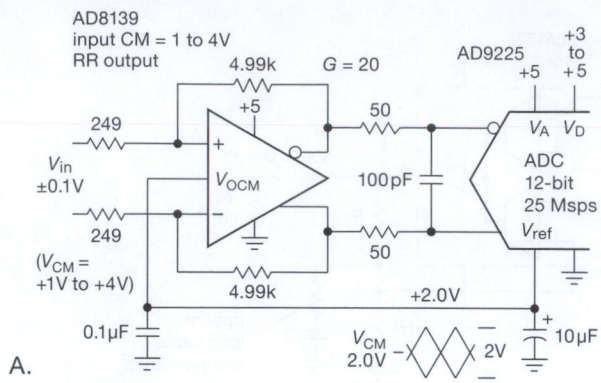

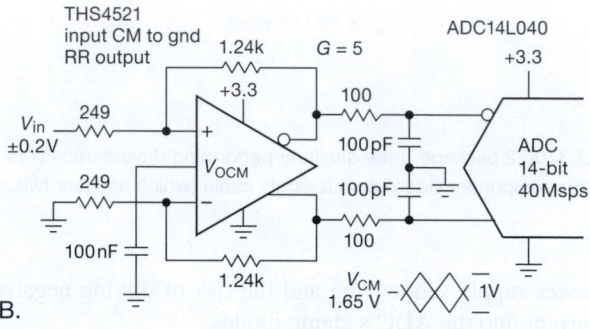

Figure 5.102. The care and feeding of single-supply differential-input ADCs. A. The AD9225 provides a midscale V_{ref} output to set the amplifier's common-mode output; but the AD8139's inputs cannot go closer than 1 V from either rail. B. The THS4521 is unusual in allowing input voltages to the negative rail (here ground).

(+1.65 V), which is fine. As before we added the recommended decoupling filter.

Given the ADC's higher resolution, it's worth checking to see how the noise voltage contributed by the amplifier and resistors compares with the converter's step size. Taking into account the input gain, the step size is $400\,\text{mV}/2^{14}$, or $25\,\mu\text{V}$. The amplifier's noise density ($4.6\,\text{nV}/\sqrt{\text{Hz}}$) combined with the (uncorrelated) resistor noise ($2.7\,\text{nV}/\sqrt{\text{Hz}}$) is about $5.3\,\text{nV}/\sqrt{\text{Hz}}$, or about $18\,\mu\text{Vrms}$ in the $\sim\!12\,\text{MHz}$ effective bandwidth of the amplifier and RC filter. In other words, the noise voltage is comparable to the converter's LSB step size. This is OK, though it would be nice to have it somewhat lower.[107] Perhaps a way to think about it is that the circuit's virtues of speed and resolution have made

the relatively small noise contribution look bad. You can always throw away bandwidth (if you don't need it), or just look at the top 12 bits, if that makes you feel better.

C. Third Iteration: transformer coupling

If you don't need dc coupling, an easy way to drive differential-input converters is with a wideband transformer. They're widely used in radiofrequency applications, and you can get them in small surface-mount packages. Figure 5.103 shows how to do it. Use the ADC's mid-span reference output (suitably bypassed) to set the common-mode voltage, and use a resistive termination that matches the transformed impedance of the driving source. Here we used a 1:2 turns ratio transformer, which transforms impedances by that ratio squared, thus $50\,\Omega\!:\!200\,\Omega$. This nicely finesses the problems of amplifier input- and output-voltage ranges, noise, and so on. But note that there's no intrinsic protection against ADC overdrive.

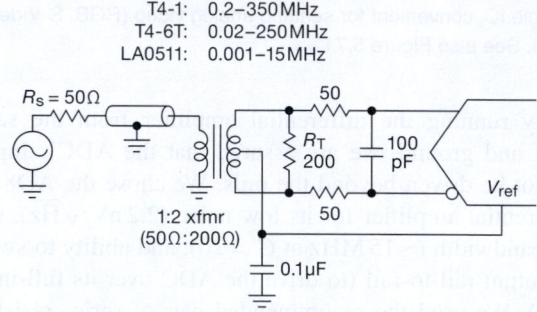

Figure 5.103. Wideband transformers can drive differential-input ADCs. They have excellent CMRR, and are available with frequency spans of 10,000:1.

5.17.4 Impedance matching

Differential amplifiers are commonly used for wide-bandwidth applications, in which the (single-ended) input must be properly terminated to match the signal's source impedance (usually $50\,\Omega$). This is particularly important when the signal arrives through a length of transmission line, in order to prevent reflections (see Appendix H). This is not difficult, as long as you keep your wits about you.[108]

Figure 5.104 shows the situation when differential amplifiers of the D type (Figure 5.96) are used. The extra resistor R_T is chosen so that the signal source sees an input impedance equal to R_S (that is, $R_T \parallel R_{in} = R_S$). Note

---

[107] But sometimes a little bit of noise can be a *good* thing, as it can improve ADC linearity and dynamic range by way of "dithering." See, for example, *The Art of Digital Audio* by John Watkinson (3rd ed., 2001); or Vanderkooy and Lipshitz "Dither in digital audio," *J. Audio Eng. Soc.*, **35**, (12), 966–975, (1987).

[108] And perhaps read the helpful Analog Devices MT-076 "Differential Driver Analysis."

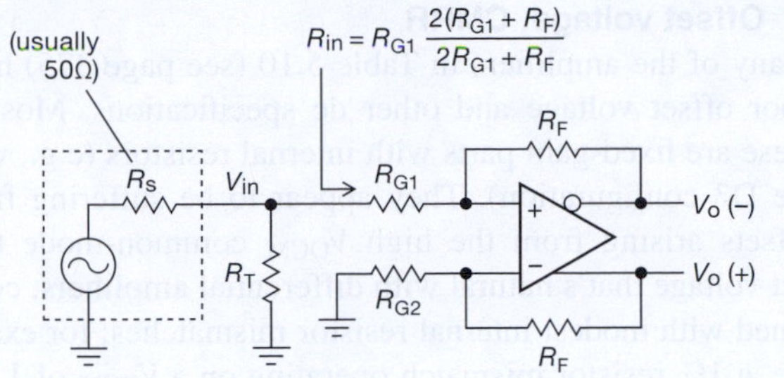

for given R_S, G, and R_{G1},

$R_T \parallel R_{in} = R_S$ (terminate)

$R_{G2} = R_{G1} + R_S \| R_T$ (balance)

$R_F = G\,\dfrac{R_{G2}\,(R_S + R_T)}{2R_T}$ (gain)

$$\left(G \equiv \frac{V_o(+) - V_o(-)}{V_{in}} = \frac{2R_T}{R_T + R_S}\,\frac{R_F}{R_{G2}} \right)$$

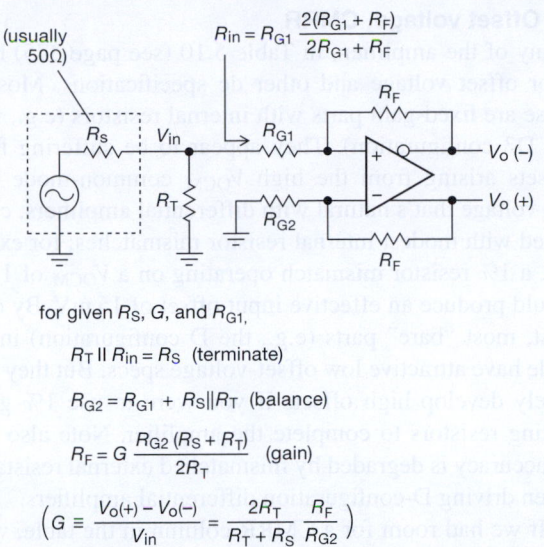

Figure 5.104. Terminated single-ended to differential amplifier: design equations.

especially that the amplifier's noninverting input is not a virtual ground, so R_{in} is somewhat larger than R_{G1} alone, according to the equation in the figure. The differential amplifier has the usual equal-value feedback resistors R_F, but the gain-setting resistors R_G are unequal, to take account of the finite impedance at the drive point (marked V_{in}). That is, R_{G2} must be larger by the parallel resistance $R_S \parallel R_T$. Finally, the feedback resistors R_F must be adjusted upward to bring the gain back to the desired value.

Note that the gain is defined in terms of V_{in}, i.e., with respect to the *loaded* input signal amplitude (*not* the open-circuit source amplitude). This makes good sense, because signal amplitudes (from signal generators, etc.) are normally specified as their properly terminated amplitudes.

As an example, for a 50 Ω source, $G = 2$, and $R_{G1} = 200\,\Omega$, you would find (choosing nearest 1% standard resistor values) $R_T = 60.4\,\Omega$, $R_{G2} = 226\,\Omega$, and $R_F = 412\,\Omega$.

Unlike the situation at high frequencies, it is not necessary to terminate a signal source when operating at low frequencies (e.g., audio). In that case R_T is omitted and R_{G2} is simply $R_{G1} + R_S$. The gain, defined now in terms of the *open-circuit* source signal amplitude, is just $G = R_F/(R_{G2})$.

5.17.5 Differential amplifier selection criteria

All differential amplifiers are not created equal. There are plenty of subtleties lurking, associated with tradeoffs of bandwidth, accuracy, output drive capability, supply voltage, and the like. Here's a summary collection of such considerations, tied closely to the differential amplifiers listed in Table 5.10 on page 375.

A. Supply voltage, RR output capability

The first group in Table 5.10 lists the high-voltage differential amplifiers, with ±12 V to ±15 V supplies (though some can work down to ±5 V). These are usually used with bipolar (split) supplies, but most have the V_{OCM} common-mode output capability to drive single-supply ADCs. This common-mode capability distinguishes the parts on this table from other types. Most parts have internal rail-splitting divider resistors (which require a bypass capacitor) to establish the common-mode output voltage, but this can be overridden by a dc mid-span output provided from the ADC (see Figure 13.28 in §13.6.2). Be sure to check both datasheets – sometimes an op-amp will be required (in the manner of Figure 5.86).

Keep in mind that "V_{out} diff'l max (Vpp)" means $(V_{a+} - V_{b-}) + (V_{b+} - V_{a-})$; that is, twice the peak-to-peak output swing of any one output.

These parts have high differential-output capability, >50 Vpp (each output going ±12.5 V), and higher still for ±18 V supplies, thus well suited for line-driving applications. The differential THAT1606, OPA1632, and LME49724, and the single-ended-input DRV134 are all intended for pro-audio (see §13.9.11D). As described before, the differential types can also be used with single-ended inputs. For lowest distortion, all four of these parts should be driven with a low-impedance source, such as an op-amp output.

Next in Table 5.10 (page 375) are the low supply-voltage amplifiers. Most high-frequency, low supply-voltage differential amplifiers are limited to a maximum of ±5 V supplies, or even less. Many cannot be used with total supply voltages greater than 5 V, or even 3.3 V in some cases. Some can run from a single supply as low as +2.7 V or +3.3 V, others need at least +5 V.

Many of the low- to mid-frequency low-voltage parts have rail-to-rail (RRO) outputs, convenient for single-supply ADCs, which don't allow signals beyond their supply rails: simply power the amplifier from the same supply rail as the ADC. But note especially that high-frequency RRO types may suffer degraded high-frequency capability when used near their supply rails. For example the LTC6404, with a featured 600 MHz bandwidth, reveals even at 10 MHz distortion that climbs dramatically when the output approaches within 400 mV of the rails.

An alternative to the use of low-voltage RR outputs to protect ADC inputs is to use an amplifier with output-voltage clamping. That's a nice feature of the LMH6553.

This part is also a CFB amplifier, good for wide bandwidth at high gains, but bad for noise (see the next subsection).

B. Common-mode input range, and the negative rail

Most of the parts have summing junctions that must be operated at least a volt or more above the negative supply rail (the low-power THS4521 is an exception). However, as explained in §§5.17.1G and 5.17.3), that does not necessarily prevent the input *signals* from going down to ground, most particularly when the amplifier is operated in a fully differential configuration at low gains ($G = 1$ or $G = 2$, for example).

Nine of the parts (marked **w** or **v** in Table 5.10's "RRO" column – see page 375) allow use of their summing junctions down to $-V_{EE}$. The datasheets will declare something like "Common-Mode Input Range Includes the Negative Rail" or "NRI" on the front page. In most cases performance is not degraded under this condition, in contrast to the loss of performance when RRO-capable amplifiers are run to the extremes of output swing.[109] This is especially useful when fully differential amplifiers are used as single-ended to differential converters, with the $(-)$ input grounded, as in Figure 5.102B. But be careful – if either summing-junction input is taken further than the specified -0.2 V below $-V_{EE}$, polarity reversal at the output may occur, similar to the situation with the legacy (but still wildly popular) LM324/358 single-supply op-amps.[110]

C. Low Z_{in}

Most of these amplifiers present rather low input impedances to their signal sources, especially when configured for high gains, because the specified R_f gain-setting resistor values are low, and Z_{in} is roughly R_f/G (exceptions: LTC6416, and the EL5170 family). Most parts with higher input impedances are noisier, mostly because of resistor Johnson noise (exception: the AD8352, which uses the F rather than the D configuration).

Signal impedance matching is often a concern, especially at high frequencies, say 30–100 MHz and above, even with short PCB traces. An amplifier's low Z_{in} complicates the problem of matching the impedance of the signal source, and it also affects the amplifier's gain. See §5.17.4 for helpful formulas.

---

[109] Some parts (for example the THS4008 and THS4511) even specify "$V_{S-} = 0$" and "input referenced to ground" as the operating condition for their specifications.

[110] For the latter you can substitute the improved LT1013/1014, which eschew that nasty habit and provide better performance overall; but no such solution is available for differential amplifiers.

D. Offset voltage, CMRR

Many of the amplifiers in Table 5.10 (see page 375) have poor offset voltage and other dc specifications. Most of these are fixed-gain parts with internal resistors (e.g., with the D3 configuration). They appear to be suffering from offsets arising from the high V_{OCM} common-mode output voltage that's natural with differential amplifiers, combined with modest internal resistor mismatches; for example, a 1% resistor mismatch operating on a V_{OCM} of 1.5 V would produce an effective input offset of 15 mV. By contrast, most "bare" parts (e.g., the D configuration) in the table have attractive low offset-voltage specs. But they will surely develop high offsets if you were to use 1% gain-setting resistors to complete the amplifier. Note also that dc accuracy is degraded by mismatched external resistance when driving D-configuration differential amplifiers.

If we had room for a CMRR column in the table, we'd see a similar bare- versus fixed-gain dichotomy, and for the same reason. Taking two examples from their respective datasheets, the bare ADA4932 has a 100 dB typical CMRR, compared with 64 dB for the similar fixed-gain ADA4950. Likewise for the bare LTC1992 (90 dB) and the fixed-gain LTC1992-10 (60 dB).

In many fully differential applications CMRR isn't important. But if it matters in your design, use 0.1% resistors, or matched resistor arrays. And be sure to take special care with circuit-board wiring capacitances, which really matter at high frequencies: for example, to achieve -80 dB matching at 1 MHz with 500 Ω gain-setting resistors, you've got to match capacitances to a difficult 0.03 pF! And no good deed goes unpunished – the CMRR will degrade by 20 dB for each decade increase in frequency.

A better solution would be to use an E- or F-configuration part. Intersil's EL5170-series parts have good CMRR, e.g., 80 dB at 1 MHz (but poor 25 mV offset), and Analog Device's AD8352 boasts 60 dB at 100 MHz and 6 mV offset. Both are much less affected by unbalanced input resistances than D-configuration types.

E. Fixed gain, external resistor-settable gain

One good reason to select a fixed-gain amplifier is that some of them have better gain accuracy than can be easily and inexpensively achieved with discrete resistors – for example, attractive worst-case gain errors of $\pm 0.04\%$ for the NSC LMP3712, $\pm 0.15\%$ for the TI (Burr–Brown) PGA280 (both programmable-gain parts), and $\pm 0.08\%$ for the Analog Devices pin-programmed AD8270.

The simplicity of fixed-gain types may seem appealing, but some external-resistor types have their attractive aspects, e.g., TI's THS4520 and Analog's ADA4932 draw much lower supply currents than their competitors. The THS4520 can be used to make $G=10$ amplifiers with 120 MHz bandwidth.

Fixed-gain types are easier to use at high frequencies, because they avoid painful wiring and pin-capacitance problems. But most have poor absolute gain accuracy, ΔG in Table 5.10 (exception: the ADL5561). Most don't allow bandwidth-limiting filter capacitors to be added, and most restrict you to low gain values (exceptions: the PGA870 and LT1993-10, but note their high power consumption).

F. VFB, CFB, f_{3dB}, GBW, and filters

Few of the fixed-gain amplifiers offer special filtering capability. Three LTC offerings are an exception, especially the LT6600-x, with five fixed 4th-order filter frequencies (alliteration!) available from 2.5 MHz to 20 MHz.

All of the adjustable-gain types employ voltage-feedback-mode op-amps, with two exceptions (the LMH6552 and LMH6553). For gains $G \geq 4$ they follow the GBW rule, namely $f_{-3dB} =$ GBW/G. But be aware that the "Bandwidth" value in the table is generally considerably higher (1.5× or more) than the part's GBW, because it is determined at unity gain, where the amplifier benefits from response peaking that extends its −3 dB rolloff frequency. (The manufacturer may want to show off its best-looking values, but the peaking for $G = 1$ may be so severe with some parts that you may not want to use them that way.) You may have to study datasheet response plots, etc., to determine the actual true (and very useful) GBW value. Because these are VFB op-amp types, they are stable with bandwidth-limiting filter capacitors added across the feedback resistors. You can increase the R_f value (which increases the input impedance) and add a small parallel C_f capacitor to bring peaking under control or a larger one to provide a bandwidth filter for your signal.[111]

G. Response peaking, GBW, and 0.1 dB bandwidth

Gain peaking is a primary killer of good "bandwidth-to-0.1 dB" ratings. The 0.1 dB bandwidth numbers may be

much improved for higher gains, where low-gain peaking is eliminated. Take, for example, the ALD5561, which boasts a −3 dB bandwidth of 2900 MHz at its minimum gain ($G = 2$), but its −0.1 dB bandwidth is a disappointing 200 MHz (i.e., just 7% of its −3 dB bandwidth). However, at its maximum gain ($G = 6$), where the −3 dB is reduced somewhat (to 1800 MHz), its −0.1 dB bandwidth *improves* to 600 MHz (i.e., 33% of its −3 dB bandwidth). This behavior is displayed nicely in the datasheet's graphs (see Figure 5.105). It is possible that the settling time is also improved (from a lack of ringing), although this isn't specified.

Note that some of the parts in the table are available in dual configurations (noted in the Comments column in Table 5.10), which can be helpful in providing matched time-delay responses, important for many applications.

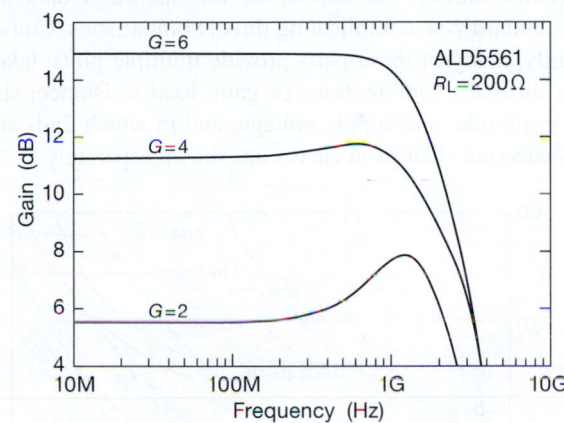

Figure 5.105. Amplifier peaking at low gain settings produces an extended "−3 dB bandwidth," at the expense of flatness of response.

H. Slew rate, settling time, large-signal bandwidth

Just as with some high-speed op-amps, the datasheets indicate a much wider bandwidth for small ($\sim$100 mV) signals than for large ($\sim$2 V) signals. This is a slew-rate issue: amplifier output-swing capabilities are reduced as their slew-rate limits are approached. For example, Analog Devices' low-power ADA4932 has a 2800 V/μs specified slew rate, which implies that 1 V amplitude sinewave outputs are possible to $f = S/2\pi A = 445$ MHz. Indeed the part's datasheet shows a −3 dB response to 560 MHz (or even 1 GHz with smaller R_f) for 100 mV output, but only to 360 MHz for 2 Vpp. Higher slew-rate parts are available,

---

[111] Taking the example of the LMH6553 and LMH6552 CFB amplifiers, these are specified with $R_f=274\,\Omega$ and $357\,\Omega$, at which the respective bandwidths are 900 and 1500 MHz, and the slew rates are 2300 and 3800 V/μs. These specs are for $G=1$, but with CFB op-amps you can dramatically increase their gain without losing too much bandwidth. For example the 1500 MHz LMH6552 claims to have 800 MHz of bandwidth still at $G=4$. For higher gain you may prefer not to reduce R_i much, but rather to increase R_f. With CFB amplifiers a primary effect of increasing R_f is to reduce the slew rate proportionally; but, hey, you had plenty to begin with! Increasing R_f with CFB does cause an increase in noise.

even to $10\,\mathrm{kV/\mu s}$ (the ALD5561), implying a 2 Vpp capability to 1.5 GHz.[112]

I. Distortion

Two of the high-voltage parts (the OPA1632 and LME49724), often used for professional music applications, have their distortion performance plotted in Figure 5.43. We would hope that a fully differential circuit, with differential input signals, might have lower distortion than competing single-ended circuits, at least for the symmetrical 2nd-harmonic. And indeed the differential LME49724 does very well, in the $-140\,\mathrm{dB}$ territory. However, the single-ended LME49990 and OPA134 op-amps do better on the graph.

Figure 5.106 plots distortion (from manufacturers' datasheets) for a few of the differential amplifiers in Table 5.10 (page 375) for frequencies to 100 MHz. As we cautioned earlier, the conditions for distortion data are not standardized, complicating direct comparisons. Consequently, many of these parts provide multiple plots, taken with different combinations of gain, load resistance, signal amplitude, and supply voltage, and in which 2nd- and 3rd-harmonic distortion curves are shown separately.

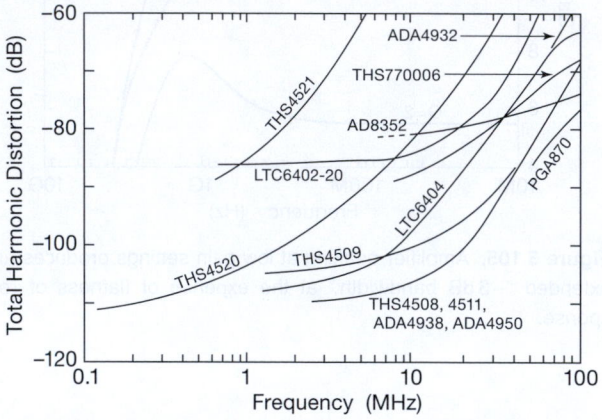

Figure 5.106. Total harmonic distortion (THD) versus frequency, taken from the respective datasheets, for a selection of the differential amplifiers in Table 5.10 on page 375.

Be careful when evaluating distortion (a parameter that we've not listed in the table). For example, the ADA4932 has 560 MHz (or 360 MHz) bandwidths, as discussed earlier, but the luster of its front-page claim to being a low-distortion amplifier ($-90\,\mathrm{dB}$ at 20 MHz) is tarnished some-

what when you discover that it deteriorates by a factor of $10\times$ by 50 MHz, well below its 360 MHz bandwidth.

As we saw with the op-amp distortion plots in Figures 5.43 and 5.44, there's a strong correlation between speed (high GBW, fast slew rate) and low distortion at high frequencies. This is especially clear in Figure 5.106 above 1 MHz. For example, the 145 MHz THS4521 that we've often mentioned performs poorly even in the under-5 MHz-region, compared with its TI relative, the 1.6 GHz THS4511 (note that both have NRI front ends, i.e., operating common-mode input range to the negative rail).[113] Four of the best parts in this class maintain better than $-100\,\mathrm{dB}$ distortion out to 20 MHz, compared with just 7 MHz for the best-in-class AD8045 op-amp in Figure 5.44. In other words, at high frequencies (say above 10 MHz) fully differential amplifiers excel in low distortion compared with single-ended op-amps.

As we will see in Chapter 13, 16-bit ADCs are available with sampling rates to 250 Msps (e.g., the AD9467, see Table 13.4), justifying a need for better than 0.01% linearity ($-80\,\mathrm{dB}$ distortion) at frequencies approaching 100 MHz. Happily, manufacturers of differential amplifiers are rising to this challenge.[114]

J. Noise, high 1/f noise corners

We conclude with a few comments about noise. We've tabulated amplifier *voltage* noise, but not *current* noise. But we do list the input bias current I_B, which is roughly predictive of current noise, which must equal or exceed the shot-noise contribution of $i_n = \sqrt{2qI_B}$. Note that CFB amplifiers, with their exceptionally high input currents, have much greater input current noise, generally as much as $10\times$ greater than VFB amplifiers.

Looking at Table 5.10 (page 375), we see many amplifiers with noise densities in the range of $25-45\,\mathrm{nV}/\sqrt{\mathrm{Hz}}$. Assuming no increase above say 10 MHz (check the datasheet plots) this corresponds to a wideband input-voltage noise of $V_n = e_n\sqrt{\mathrm{BW}}$, which evaluates to 175–$700\,\mathrm{\mu Vrms}$ for bandwidths of 50–250 MHz. This is quite a bit larger than the $30\,\mathrm{\mu V}$ LSB step size of a 16-bit ADC with a 2 Vpp full-scale differential input. While some

---

[112] The AD8351 (not in our table, but similar to the AD8352), offers 3 GHz bandwidth with 13 V/ns slew rate and 2 Vpp capability to nearly 2 GHz.

[113] In fairness, the THS4521 gets by on a miserly 1.1 mA of supply current and has RR outputs, whereas the THS4511 takes 39 mA and lacks RROs.

[114] But are they rising to the challenge posed by NSC's ADC12D1800, a 12-bit 3600 Msps converter? Or the even greater challenge posed by the even-speedier ADCs that surely will be available by the time this book's ink has dried? Such ADCs will likely have to be driven with transformers.

dither is a good thing, it's clear that even at $G = 1$ these amplifiers are too noisy for some applications.

There are many other amplifiers in the table with e_n specs down in the 1.1–5 nV/$\sqrt{\text{Hz}}$ range. But these are the bare amplifier summing-junction noise specs, without the necessary feedback resistors taken into account. Amplifier gains of 5 or 10 are generally assumed, not the least to overcome the amplifier's own output-stage noise. Many of the amplifiers specify R_f values of 350–500 Ω. For $G = 1$ the input resistor R_g would have the same value, and its Johnson noise of 2.4–2.8 nV/$\sqrt{\text{Hz}}$ would dominate the intrinsic noise of the quieter amplifiers. However, for $G = 10$ the 35–50 Ω resistor noise is under 1 nV/$\sqrt{\text{Hz}}$, which would not badly degrade the completed amplifier's noise.

Finally, many amplifiers have good looking noise specs, but we must warn you to examine the noise-versus-frequency curves on their datasheets; many have very high $1/f$ noise corners. This is especially true for current noise, with some $1/f$ corners way up at 1 MHz or beyond. An example of a troublesome spec might be the THS4508, with its *pnp* input transistors (for operation to GND) and 4.7 pA/$\sqrt{\text{Hz}}$ of current noise at 1 MHz. This creates 1.6 nV/$\sqrt{\text{Hz}}$ across a 349 Ω resistor, which is OK compared with the part's $e_n = 2.3$ nV/$\sqrt{\text{Hz}}$. But if you were to use 1k resistors, the corresponding current-induced noise voltage would be 4.7 nV/$\sqrt{\text{Hz}}$, dominating the amplifier's e_n. Depending on the application, this might be of concern. Or it might not.

Review of Chapter 5

An A-to-M summary of what we have learned in Chapter 5. This summary reviews basic principles and facts in Chapter 5, but it does not cover application circuit diagrams and practical engineering advice presented there.

¶A. Precision and Dynamic Range.

A *precision* circuit is one that exhibits (through careful design and choice of components) both initial *accuracy*, and also *stability* (i.e., maintenance of accuracy over time and temperature). A precision circuit may (but need not) exhibit wide *dynamic range* (the ratio of signal amplitudes over which it operates); see §5.1.

¶B. Error Budget.

When designing a precision circuit, you need to keep track of numerous error contributions (from voltage offsets, current-induced offsets, component tolerances, and the like); this is best tallied in an *error budget*, which promotes design discipline, and which assists in helping you spot the dominant error sources; see the example in §5.5.

¶C. Strict versus Pragmatic.

Strict worst-case design mandates that all components be operated within their datasheet specifications, and that the effects of all their worst-case errors be added (as unsigned magnitudes) to determine the circuit's performance. The benefit of such conservatism is a circuit guaranteed to perform within specifications (assuming proper design); the downside is that it may be impossible to achieve design goals with available and/or affordable components, taking their worst-case specifications in a worst-case arithmetical combination (and noting that some critical performance parameters may be unspecified, or show only "typical" values). We favor a pragmatic approach (§5.3), taking with a grain of salt some of the published worst-case parameters (for example, input leakage current, where testing limits encourage highly conservative worst-case specs), or adopting reasonable estimates of unspecified parameters. It may be sufficient to establish that the circuit effects of the unspecified parameter are completely insignificant; or, if it's a closer call, you may have to set up a testing regime of incoming components to ensure that you meet specs. If have a situation in which there are many components contributing to an overfull error budget, you may have to validate the performance of the overall circuit, subassembly, or complete instrument at final test.

¶D. Component Errors – Resistors.

Taking a simple example, resistor accuracy (including the effects of tempco) sets a limit on the accuracy of an amplifier's gain. But it's not that simple, because amplifier gain usually depends on a resistance *ratio*; so the situation is greatly improved if you use a resistor pair with accurate ratio and matched tempco, and it's the *ratio* mismatch and the tempco of the *ratio* that are limiting. Likewise, the CMRR limit of a difference amplifier depends on resistor-pair matching, $\mathrm{CMRR(dB)} \approx 20\log(100/p)$, where p is the ratio mismatch in percent. To give scale to these statements, typical metal-film resistors come in accuracy ranges of 0.05%–1%, with a typical tempco spec of 25–100 ppm/°C; resistor arrays intended for high precision have accuracies in the range of 0.01%–0.05%, tempcos down to 1 ppm/°C or better, and matching accuracies and tempcos down to 0.01%–0.001% and 1–0.1 ppm/°C, respectively. Real resistors depart from the ideal also in *linearity*, i.e., they exhibit some resistance change with applied voltage. See §5.6.1; see also §*1x.2* for further discussion of resistors and their vicissitudes (e.g., long-term drift, and parasitic inductance and capacitance).

¶E. Component Errors – Capacitors and Switches.

Capacitors have several amusing traits that affect the accuracy of integrators and of sample-and-hold circuits; these include resistive leakage (thus exponential decay), and more seriously *dielectric absorption* (memory effect, see §5.6.2, §*1x.3*, and the dielectric-absorption plots in that section.) These application circuits include analog switches (for integrator reset, and sample-to-hold switching), which introduce errors through leakage and charge injection (§§3.4.2E and 5.6.3).

¶F. Amplifier Input Errors.

This is where most of your troubles are located. The primary qualification for membership in the category of *precision op-amps* is a small offset voltage V_{os}, and a correspondingly low tempco TCV_{OS} (sometimes called ΔV_{OS}). The offset voltage operates at the input, so the RTI (referred to the input) error is simply V_{OS}; at the output of an amplifier of voltage gain G_V the V_{OS}-induced error is $\Delta V_{out} = G_V V_{OS}$. In an integrator circuit with input resistor R_{in}, the input offset voltage is equivalent to an input current error of $\Delta I_{in} = V_{OS}/R_{in}$. To give an idea of scale, a typical precision op-amp has an offset voltage of 10–50 μV, and a tempco of 0.1–0.5 μV/°C. *Auto-zero* op-amps (see ¶I) improve on these figures by about a factor of ten; see Table 5.5 on pages 320–321 and Table 5.6 on 335. See also §5.10.5.

But there's more. Input bias current I_B flowing through

the source resistance R_S seen at the amplifier's input produces an RTI voltage error of $\Delta V_{in} = I_B R_S$. For precision *bipolar* op-amps, whose bias currents are of order 10 nA, this becomes serious for source resistances greater than a few kilohms (where $I_B R_S$ amounts to a few tens of microvolts, in the same ballpark as a precision op-amp's V_{OS}). High-R_S situations thus mandate low-bias op-amps, usually those with JFET or MOSFET inputs, or (for moderate source resistance) a bipolar precision op-amp with bias-cancellation (where I_B is of order 1 nA). *A warning*: the very low bias current of FET-input op-amps rises dramatically at higher temperatures (see Figure 5.6), where its input current may even exceed that of a bias-cancelled bipolar op-amp. See also §5.10.7.

Looking deeper, additional error sources at op-amp inputs include I_B variation with common-mode input voltage V_{CM} (§5.7.2, Figure 5.7), V_{OS} variation with V_{CM} (i.e., CMRR, §5.7.3, §5.10.9, and Figures 5.10, 5.29, and 5.30), PSRR, and input noise (e_n, i_n, §§5.10.6 and 5.10.8).

¶G. Amplifier Output Errors.
While much of precision analog design concerns dc and low frequencies, some applications require accuracy at higher speeds: audio and video, communications, scientific measurement, and so on. With falling op-amp loop gain, input errors are rising, output impedance is rising, and slew-rate limitations may come into play, along with reduced suppression of output-stage crossover distortion, gain nonlinearity, and phase error. And overshoot, ringing, and settling time are critical in dynamic applications.

These can be made quantitative. The gain error $\delta_G \equiv (G_{ideal} - G_{actual})/G_{ideal} = 1/(1 + AB)$, where B is the feedback fraction around open-loop gain A; see §5.8.5. The closed-loop bandwidth is $f_{3dB} = f_T/G_{CL}$, corresponding to a time constant of approximately $\tau \approx G_{CL}/2\pi f_T$, which (if well compensated, with good phase margin) produces a settling time of order 5–10τ; see §§5.8.2 and 5.10.10. An op-amp's *actual* settling time can be considerably longer, and there's no substitute for real data; see Table 5.5 on pages 320–321. A well-compensated op-amp exhibits a closed-loop phase error of $\phi \approx f/f_C$, where $f_C = f_T/G_{CL}$ is the frequency at which the closed-loop gain has fallen to unity.[115] Op-amp distortion depends strongly on the output-stage circuit; see the extensive plots in Figures 5.43 and 5.44.

---

[115] Throughout the book we use f_T as shorthand for the proper term *gain–bandwidth product* (GBP, GBW, or GBWP), which is the extrapolated unity-gain crossover frequency.

¶H. Rail-to-rail Op-amps.
It's tempting to choose op-amps with rail-to-rail common-mode input range (RRI), rail-to-rail output (RRO), or both (RRIO), especially for operation at low supply voltages.

RRI op-amps. But there are drawbacks associated with the duplicate complementary input stages that can seriously compromise precision, notably an abrupt change of both I_B and V_{OS} at the input crossover voltage (§5.9.1). Some RRIO op-amps (e.g., the OPA360-series) circumvent this problem by using an on-chip charge pump. If you don't need full rail-to-rail input, choose an op-amp whose input common-mode range extends only to one rail (usually the negative rail).

RRO op-amps. Op-amps that feature rail-to-rail output have their own issues, stemming from the use of a common-source (or common-emitter) output-stage topology, rather than the conventional source follower (or emitter follower). These include increased distortion, and higher output impedance (thus gain and phase shift are more susceptible to load impedance). However, many RRO op-amps mitigate this latter problem by using internal capacitive feedback to lower the open-loop output impedance at high frequencies, as seen for example in Figure 5.34.

¶I. Auto-zero Op-amps.
This subclass of op-amps includes on-chip offset-nulling circuitry that operates cyclically (with an on-chip oscillator) to trim the input offset voltage (§5.11). As the entries in Table 5.6 on page 335 demonstrate, this produces typical V_{OS} values around a microvolt, with tempcos around 10 nV/°C, roughly an order of magnitude better than the best conventional precision op-amps. Notable, too, is the absence of a $1/f$ low-frequency rising noise voltage (see for example Figures 5.53 and 5.54).

That's the good news. The bad news is that the switching action produces noise-spectra peaks at the switching frequency and its harmonics (Figure 5.51), superposed on a low-frequency noise voltage floor that is already considerably higher than that of conventional low-noise op-amps (compare Table 5.6 on page 335 with Table 5.5 on pages 320–321). The input switching circuitry also results in relatively high input noise *current*, compared with low-noise precision JFET-input op-amps.

Auto-zero op-amps are built with CMOS, and (except for some legacy parts) are generally limited to low supply voltages ($\leq$7 V total supply). A shining exception is the recent LTC2057HV (4.75 V to 60 V total supply!). Another caution: because of their internal voltage-storage correction capacitors, auto-zero op-amps can exhibit slow recovery from saturation, as long as several milliseconds.

¶J. Difference, Differential, and Instrumentation Amplifiers: Taxonomy.

These share the property of accurate and stable differential gain, with high common-mode rejection: $V_{out} = G_V \Delta V_{in} = G_V (V_{in+} - V_{in-})$. In common usage, the terms are distinguished as follows.

Difference amplifier. Differential-in, single-ended out; op-amp plus two matched resistor pairs (Figure 5.65, §5.14); CMRR 90–100 dB; accurate but low gain ($G_V = 0.1$–10); input impedance 25–100k, intended to be driven by a low impedance; inputs typically can go beyond rails.

Instrumentation amplifier. Differential-in, single-ended out; very high input impedance (10MΩ to 10GΩ), wide gain range ($G_V = 1$–1000), and very high CMRR at higher gains (110–140 dB at $G_V = 100$); see §5.15 and Figure 5.77.

Fully differential amplifier. Differential or single-ended in, differential-out; most are low voltage, fast-settling, and wideband; ideal for twisted-pair cable drivers and fast differential-input ADCs; see §5.17.

¶K. Difference Amplifiers.

The classic *difference amplifier*, typified by the original INA105 (see Table 5.7 on page 353, §5.14), consists of an op-amp plus a pair of matched resistor dividers, with SENSE and REF inputs (at the bottom of the divider strings) in addition to the V_{in+} and V_{in-} signal inputs. With REF grounded and SENSE tied to the output, the gain is $G_{diff} \equiv V_{out}/\Delta V_{in} = R_f/R_i$, where R_f and R_i are the feedback and input resistors, respectively. Figure 5.66 shows circuit variations that exploit alternative connections of the SENSE and REF pins.

Difference amplifiers are simple, and good enough for many applications. Some permit common-mode inputs well beyond the rails (e.g., ±200 V for the INA117). But difference amplifiers suffer from relatively low input impedance (tens of kΩ), limited gain (typically in the range of $G = 1$–10), relatively high noise voltage, unimpressive CMRR (typically $\lesssim 90$ dB), and degradation of both gain and CMRR when driven with signals of non-zero source impedance.

¶L. Instrumentation Amplifiers.

These drawbacks are nicely remedied in the *instrumentation amplifier* configuration of three op-amps (Figure 5.77, §5.15). The input impedance is high (10 MΩ–1 TΩ), and the gain (which can be as large as ×1000) is set with a single external resistor (or pin-selectable choice of internal resistors). When configured for high gain, most instrumentation amplifiers deliver a typical CMRR around 120 dB, and an input noise voltage e_n less than $10 \, nV/\sqrt{Hz}$; they do not permit common-mode outputs beyond the rails. Instrumentation amplifiers are available in BJT, JFET, MOSFET, auto-zero, and programmable-gain types; see Tables 5.8 on page 363 and 5.9 on pages 370–371.

¶M. Fully Differential Amplifiers.

Unlike difference amplifiers and instrumentation amplifiers, fully differential amplifiers (§5.17) generate a balanced *differential output* centered on a settable common-mode voltage. This is useful for driving fast ADCs with complementary inputs, balanced transmission lines, and RF communication circuits. Befitting these applications, they tend to be wideband (up to 1 GHz or more), high slew-rate (1000 V/μs or more), fast settling (a few nanoseconds), and quiet ($5 \, nV/\sqrt{Hz}$ or less). There are many internal circuit topologies (labeled **A–G** in §5.17); see Table 5.10 for a representative selection.

FILTERS

6.1 Introduction

With only the techniques of transistors and op-amps it is possible to delve into a number of interesting areas of linear (as contrasted with digital) circuitry. We believe that it is important to spend some time doing this now, in order to strengthen understanding of some of these difficult concepts (transistor behavior, feedback, op-amp limitations, etc.) before introducing more new devices and techniques and getting into the large area of digital electronics. Therefore, in this chapter we treat the topic of filters, and particularly *active filters*. The latter use resistors and capacitors, in combination with amplifiers (usually op-amps), to produce filters with well-defined frequency response. As we'll see, these filters (along with the classic *LC* passive filters that they can emulate) can be much sharper than the simple *RC* filters we saw in Chapter 1.

The three following chapters will continue with additional topics in analog electronics: Chapter 7 (Oscillators and timers), Chapter 8 (Low-noise techniques), and Chapter 9 (Voltage regulation and power conversion). Then, following two chapters on digital logic, we revisit analog electronics, happily harmonized with the intervening digital teachings, in Chapter 12 (Logic interfacing), Chapter 13 (Digital meets analog), and Chapter 15 (Microcontrollers).

6.2 Passive filters

In Chapter 1 we began a discussion of filters made from resistors and capacitors. Those simple *RC* filters produced gentle highpass or lowpass gain characteristics, with a 6 dB/octave falloff well beyond the −3 dB point. By cascading highpass and lowpass filters, we showed how to obtain bandpass filters, again with gentle 6 dB/octave "skirts." Such filters are sufficient for many purposes, especially if the signal being rejected by the filter is far removed in frequency from the desired signal passband. Some examples are bypassing of radiofrequency signals in audio circuits, "blocking" capacitors for elimination of dc levels, and separation of modulation from a communications "carrier."

6.2.1 Frequency response with *RC* filters

Often, however, filters with flatter passbands and steeper skirts are needed. This happens whenever signals must be filtered from other interfering signals nearby in frequency. The obvious next question is whether or not (by cascading a number of identical lowpass filters, say) we can generate an approximation to the ideal "brick-wall" lowpass frequency response, as in Figure 6.1.

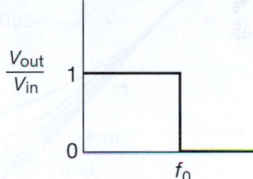

Figure 6.1. Ideal brick-wall lowpass filter.

We know already that simple cascading won't work, because each section's input impedance will seriously load the previous section, degrading the response. But with buffers between each section (or by arranging to have each section of much higher impedance than the one preceding it), it would seem possible. Nonetheless, the answer is no. Cascaded *RC* filters do produce a steep *ultimate* falloff, but the "knee" of the curve of response versus frequency is not sharpened. We might restate this as "many soft knees do not a hard knee make." To make the point graphically, we plotted some graphs of gain response (i.e., $V_{\text{out}}/V_{\text{in}}$) versus frequency for lowpass filters constructed from 1, 2, 4, 8, 16, and 32 identical *RC* sections, perfectly buffered (Figure 6.2).

The first graph shows the effect of cascading several *RC* sections, each with its 3 dB point at unit frequency. As more sections are added, the overall 3 dB point is pushed downward in frequency, as you could easily have predicted.[1] To compare filter characteristics fairly, the rolloff

---

[1] This downward shift in rolloff frequency is sometimes called the "shrinkage factor"; for a cascade of n identical and buffered *RC* lowpass sections, the 3 dB frequency is given by $f_{3\text{dB}}(n)/f_{3\text{dB}}(1) = \sqrt{2^{1/n} - 1}$.

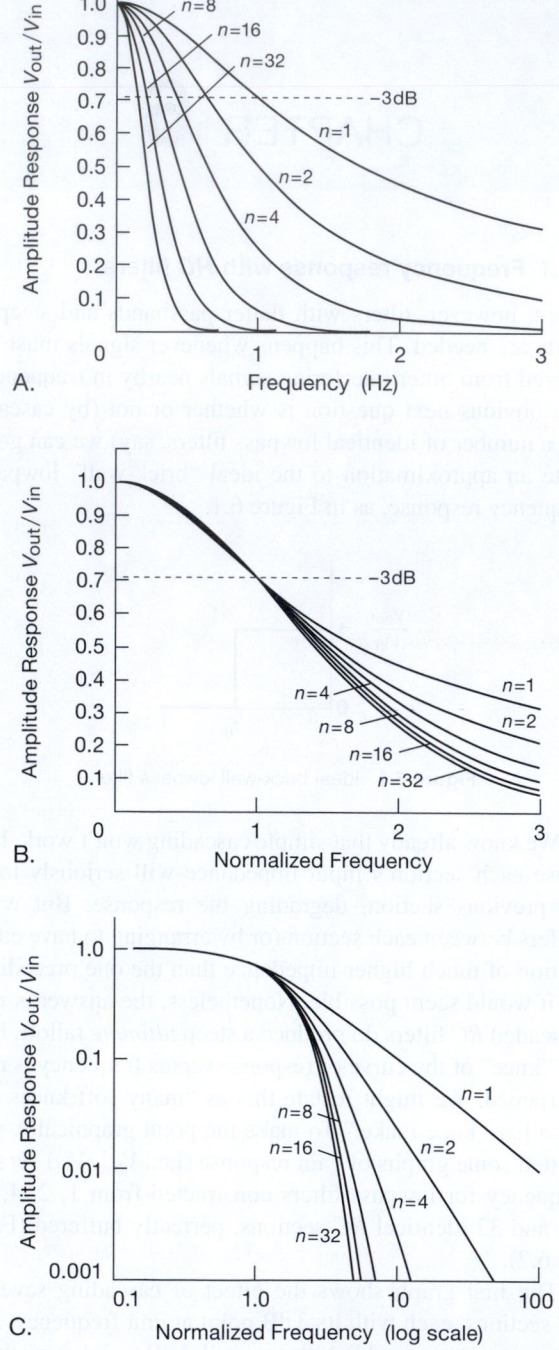

Figure 6.2. Frequency responses of multisection *RC* filters. Graphs A and B are linear plots, whereas C is logarithmic. The filter responses in B and C have been normalized (or scaled) for 3 dB attenuation at unit frequency.

frequencies of the individual sections should be adjusted so that the overall 3 dB point is always at the same frequency. For this reason, the other graphs in Figure 6.2 are all "normalized" in frequency, meaning that the −3 dB point (or breakpoint, however defined) is at a frequency of 1 radian per second (or at 1 Hz). To determine the response of a filter whose breakpoint is set at some other frequency, simply multiply the values on the frequency axis by the actual breakpoint frequency f_c. In general, we will also stick with the log–log graph of frequency response when talking about filters because it tells the most about the frequency response. It lets you see the approach to the ultimate rolloff slope, and it permits you to read off accurate values of attenuation. In this case (cascaded *RC* sections), the normalized graphs in Figures 6.2B and 6.2C demonstrate the soft knee characteristic of passive *RC* filters.

It's interesting to look also at the *phase shift* of an *RC* lowpass cascade, again adjusted to put the overall 3 dB points at unit frequency; these are plotted in Figure 6.3. The lagging phase shift reaches $90° \times n$ asymptotically, for *n* cascaded sections, as you might expect (recall the smooth transition from 0° to 90° lagging phase shift of a single *RC* section, Figure 1.104). Perhaps non-intuitively, however, the phase shift at the 3 dB point grows progressively with larger cascades. Phase-shift characteristics are important, as we'll see presently, because they determine the filter's in-band waveform distortion.

A. Degradation of ultimate attenuation: non-ideal capacitors

Unlike ideal capacitors, *real* capacitors exhibit some extra "parasitic" elements – most prominently an effective series resistance (ESR) and an effective series inductance (ESL). So at very high frequencies (where the capacitor's ESR becomes comparable to the capacitive reactance $1/\omega C$) a real *RC* filter stops rolling off. We modeled this by using SPICE (see Appendix J) for the cascaded multisection *RC* filter, see Figure 6.4. For this comparison we assumed that you want to do some *RC* filtering of a dc rail that supplies a low-level stage, to suppress higher frequency switching noise, coupled signals, and the like. So we gave ourselves a "budget" of 100 Ω total series resistance (consistent with a load current of a few milliamps); and we limited ourselves to 20 μF total capacitance (to maintain reasonable physical size). Then we ran simulations of three filters: a single 100 Ω, 20 μF *RC* stage; a 2-section filter with 50 Ω and 10 μF in each section; and a 4-section filter with 25 Ω and 5 μF in each section. We ran plots of the response of these three filters, first with perfect capacitors (no ESR), then with realistic ESR values, taken from capacitor datasheets (e.g., $1\,\Omega$ for a $5\,\mu$F electrolytic capacitor rated at 100 V).

You can see the effect of the series resistance, namely a loss of ultimate attenuation at high frequencies, where the impedance of the capacitors asymptotes to the ESR value, rather than continuing to fall as $1/f$. Nevertheless, it is clear that spreading the total capacitance into several filter sections makes good sense.

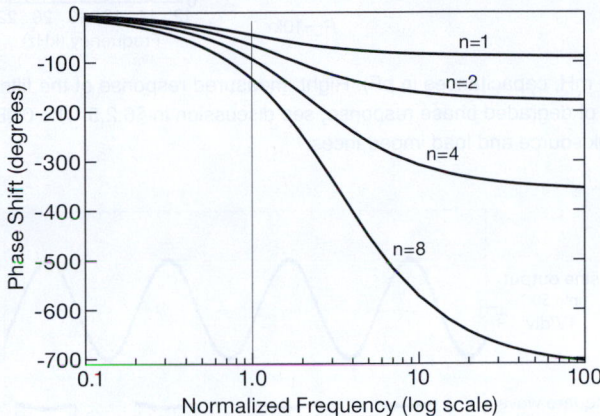

Figure 6.3. Phase shift versus frequency for the multisection *RC* lowpass filters of Figure 6.2C.

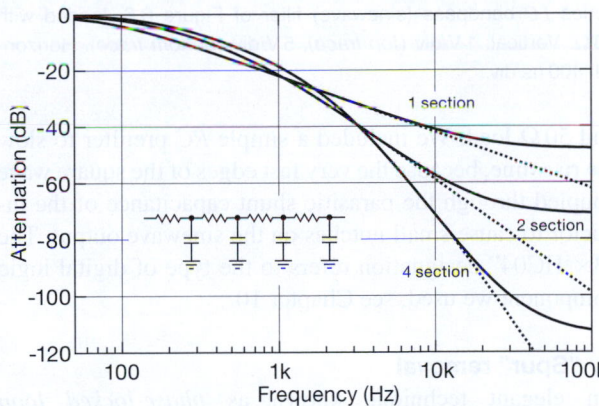

Figure 6.4. Real capacitors include some irreducible series resistance, which limits the ultimate attenuation of *RC* filters. This SPICE simulation compares ideal (dotted curves) and real (solid curves) cascaded *RC* lowpass filters.

6.2.2 Ideal performance with *LC* filters

As we pointed out in Chapter 1, filters made with inductors and capacitors can have very sharp responses (§1.7.14). We discussed the parallel *LC* resonant circuit as an example, as

well as the series *LC* trap. And we showed a dramatic comparison of an *RC* and an *LC* lowpass filter, each with the same 1 MHz cutoff frequency (Figure 1.112). By including inductors in the design, it is possible to create filters with any desired flatness of passband combined with sharpness of transition and steepness of falloff outside the band. Figure 6.5 shows an example of a telephone filter and its stunning bandpass characteristics.[2]

Obviously the inclusion of inductors into the design brings about some magic that cannot be performed without them. In the terminology of network analysis, that magic consists of the use of "off-axis poles" (see Chapter *1x*). Even so, the complexity of the filter increases according to the required flatness of passband and steepness of falloff outside the band, accounting for the large number of components used in the preceding filter. The transient response and phase-shift characteristics are also generally degraded as the amplitude response is improved to approach the ideal brick-wall characteristic.

6.2.3 Several simple examples

The impressive Orchard and Sheahan filter of Figure 6.5 is a frighteningly complex design, showing what can be done with sophisticated classic *LC* filter synthesis.[3] But you don't have to be a filter wizard to make "good enough" filters[4] that solve most of the problems you are likely to

---

[2] Not shown is its less-than-stunning in-band-phase characteristics: 495° phase lag at 16.5 kHz, rising nonlinearly (*writhing* might be a better term) to 1270° at 19.5 kHz. Fortunately, phase has little effect upon audio intelligibility.

[3] Based on Figures 11 and 12 from Orchard, H. J., and Sheahan, D. F., "Inductorless Bandpass Filters," *IEEE Journal of Solid-State Circuits*, Vol. SC-5, No. 3 (1970), where the designers illustrated an active-filter implementation of this passive-element design. The essence of their paper was that you could implement such an *LC* filter with better performance and smaller size by replacing the real inductors with gyrators (§6.2.4C). In their implementation the inductors were implemented with inductorless "Riordan gyrators," with each π-inductor (set of three inductors, including the floating inductor) requiring one quad op-amp, nine resistors and two capacitors. The authors state that inductor *Q*'s greater than 1000 are practical, with the available op-amp gain–bandwidth product being the primary limitation. Their 1970-technology gyrator implementation (occupying one cubic inch) was far superior to what was possible with conventional inductors. Interested readers may wish to read R. H. S. Riordan, "Simulated inductors using differential amplifiers," *Electronic. Letters* **3**, pp. 50–51 (Feb. 1967).

[4] *Better is the enemy of good enough.* (Proverb attributed, variously, to the Soviet Admiral Sergei Gorshkov, to Carl von Clausewitz, and to Voltaire.)

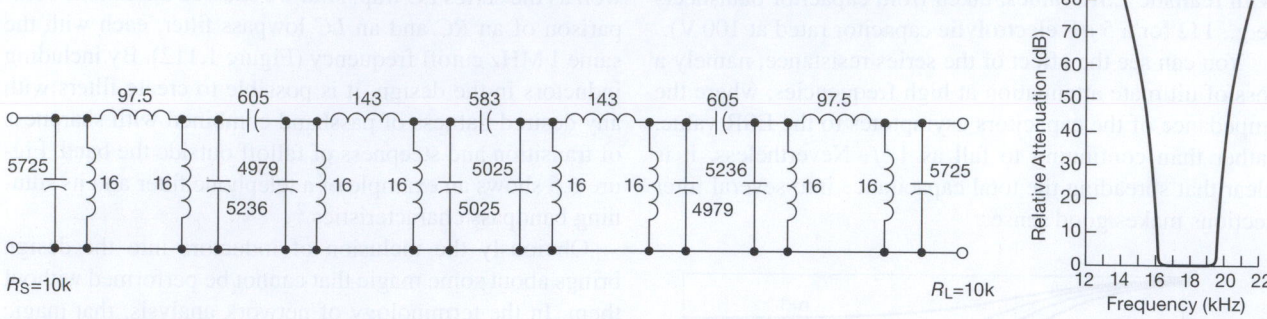

Figure 6.5. Left: An unusually good *LC* bandpass filter (inductances in mH, capacitances in pF). Right: measured response of the filter circuit. The admirably sharp frequency response comes at the expense of degraded phase response; see discussion in §6.2.5. The 0 dB value in the response curve corresponds to ~9 dB of loss, assuming 10k source and load impedances.

encounter. Here we show three simple filters that we used in recent designs at our radiotelescope observatory.

A. Sinewave from digital square wave

With digital electronics it is very easy to make and manipulate pulses or square waves of precise frequency. But at our observatory we wanted sine waves, not square waves. Figure 6.6 shows a simple way to produce a sinewave output from a fixed-frequency square wave, namely the use of a tuned series *LC*. It behaves like a very low impedance at resonance[5] ($f_0 = 1/2\pi\sqrt{LC}$), rising on either side (asymptotically as $1/f$ at low frequencies, and as f at high frequencies).

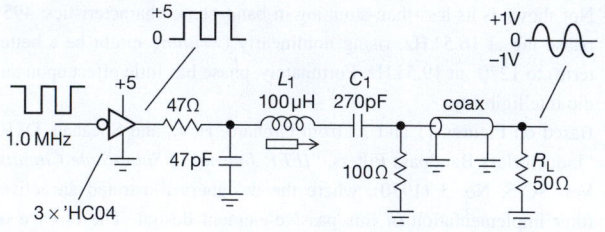

Figure 6.6. A series *LC* bandpass filter converts a square wave into a sinewave suitable for driving a 50 Ω load.

Here we chose the product *LC* for resonance at 1.0 MHz, and the value of *L* such that its impedance at 3 MHz (the next frequency component of a 1 MHz square wave, which has only odd harmonics) is large compared with the 50 Ω load impedance. For $L_1 = 100\,\mu H$, the reactance at 3 MHz is $X_L = 2\pi f L \approx 2\,k\Omega$.

Figure 6.7 shows the measured performance. The slight bowing of the square wave is due to loading by the filter

---

[5] Where it would have zero impedance were it not for losses in the inductor and capacitor.

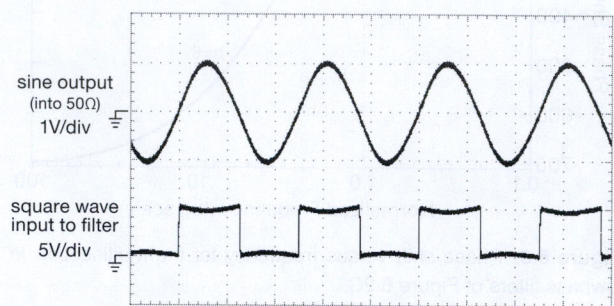

Figure 6.7. Input (*lower trace*) and output (*upper trace*) of the series *LC* bandpass (sinewave) filter of Figure 6.6, loaded with 50 Ω. Vertical: 1 V/div (*top trace*), 5 V/div (*bottom trace*). Horizontal: 400 ns/div.

and 50 Ω load. We included a simple *RC* prefilter to slow the rise time, because the very fast edges of the square wave coupled through the parasitic shunt capacitance of the inductor to cause small notches on the sinewave output. The "3×'HC04" designation refers to the type of digital logic component we used; see Chapter 10.

B. "Spur" removal

An elegant technique known as *phase-locked loop (PLL) frequency synthesis*, discussed later in Chapter 13 (§§13.13.6A and 13.13.6B), permits you to generate a desired precise frequency of your choosing, beginning with a standard "reference" frequency, for example 10.0 MHz. Figure 6.8 shows a portion of a 78.0 MHz PLL synthesizer we built, in block diagram form. The basic idea is to use a voltage-tunable oscillator (VCO), and compare an integer subdivision of its desired output frequency with a different subdivision of the reference frequency such that those frequencies will agree when the output frequency is correct. A frequency error produces a correction signal to steer the

VCO toward the correct oscillation frequency. Here we divided the 10 MHz reference by 50 (producing 200 kHz), to be compared with the VCO's output after division by 390; those divided frequencies will agree when the VCO is oscillating at 78.0 MHz.

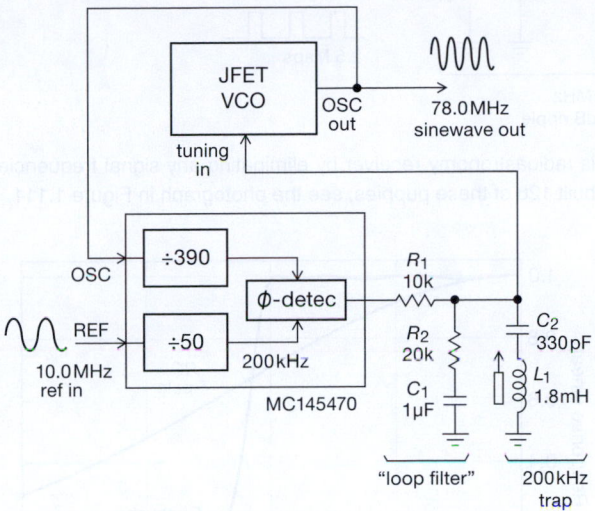

Figure 6.8. A series *LC* "trap" suppresses spurs at the 200 kHz reference frequency in this phase-locked loop (PLL) oscillator.

We engineered a simple but quite good JFET oscillator (shown later, in Figure 7.29), with its output energy almost entirely at its central frequency. It was so "clean" that the dominant undesired component of its output was a bit of residual energy at $78.0 \text{ MHz} \pm 0.2 \text{ MHz}$, caused by the 200 kHz internal comparison frequency. The simple cure here was to put a series *LC* trap, tuned to 200 kHz, across the analog tuning voltage, as shown. The three other components ($R_1 R_2 C_1$) form the classic PLL *loop filter*, as we'll see in §13.13.

C. Anti-alias lowpass filter

An analog waveform can be *digitized*, by sampling its amplitude periodically and converting each sample to a numeric quantity. We'll see later (Chapter 13, e.g., Figure 13.60) that the process can introduce artifacts, both from the finite accuracy with which the amplitudes are quantized, and from the finite rate at which those samples are taken. These artifacts can be suppressed, to any required degree, by adequate choice of quantization depth (amplitude accuracy) and rate (sampling frequency).

The important fact, for this filter example, is that the signal being digitized must not contain signals whose frequency exceeds half the sampling rate f_S; this is known as

the *Nyquist criterion*.[6] The usual way this is accomplished is by passing the predigitized signals through a lowpass "anti-aliasing" filter, whose cutoff ensures thorough attenuation of signals above the Nyquist frequency $f_S/2$. This usually requires a sharp filter cutoff, because otherwise you would have to go to a much higher sampling rate to escape signals that pass through the soft cutoff; furthermore you want a filter that is flat throughout the desired signal passband.

In this radiotelescope receiver example (Figure 6.9) we use a *mixer* (a device that multiplies two signals together to produce its output) to convert a 2 MHz band of signal frequencies centered on 78 MHz (the "IF" band) to a band centered on dc (known as "baseband"). A mixer can do this sort of frequency shifting, because the product of two sinusoids is a pair of waves at the sum and difference frequencies: $\cos(\omega_1 t)\cos(\omega_2 t) = \frac{1}{2}\{\cos(\omega_1 - \omega_2)t + \cos(\omega_1 + \omega_2)t\}$. Here the signal band drives one input to the mixer, and a fixed oscillator at 78 MHz (the "local oscillator," or LO) drives the other input. The difference frequency at the output of the mixer is the baseband,[7] in which the band from dc to 1 MHz contains the signals we want to digitize in this example.[8]

Here we amplified the baseband, then passed it through a serious anti-aliasing filter, specifically a "7-section *LC* Chebyshev lowpass filter with 1.0 MHz cutoff frequency and 0.1 dB peak-to-peak ripple."[9] We designed the filter with a weird input and output impedance (378 Ω) to take advantage of standard value tunable inductors. The filter removes signal components above 1 MHz, and this filtered baseband is then amplified (again) and digitized (by the device labeled ADC – analog-to-digital-converter) at a sampling rate of 2.5 Msps (megasamples/s). The corresponding Nyquist frequency of 1.25 MHz is well into the stopband of the very sharp lowpass filter; in fact, the calculated and measured performance are in close agreement, demonstrating that the input signals are reduced by 20 dB at that

---

[6] Violating this rule produces *aliasing*, the creation, in the digitized output, of nonexistent in-band frequency components; see §13.5.1B.

[7] The sum frequency, centered on 156 MHz, is discarded in subsequent filtering.

[8] So we can Fourier transform them to get a radio spectrum. More precisely, the baseband contains frequencies from "−1 MHz" to +1 MHz, which a single mixer folds into a single dc–1 MHz band; but we recover the unfolded baseband by using a *pair* of mixers, driven with sine and cosine LO signals. The pair of filtered baseband signals, commonly called *I* and *Q* (for *in-phase* and *quadrature*), are individually digitized to create the complex input time series to the (complex) Discrete Fourier Transform.

[9] This is the filter we used for the linear swept response of Figure 1.112.

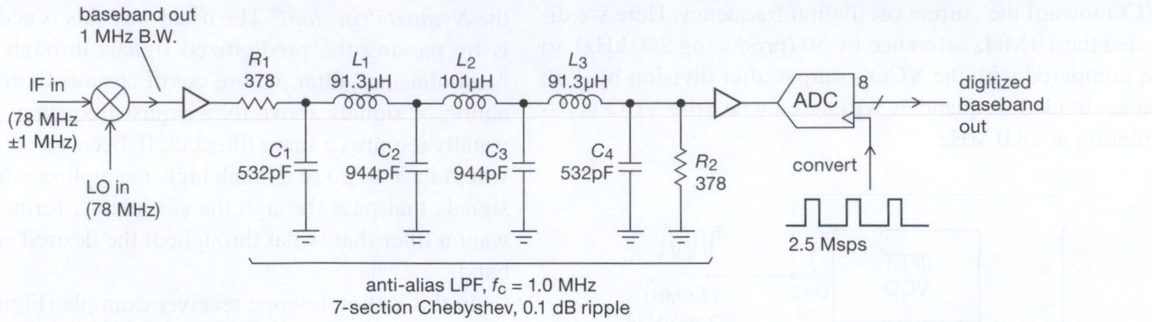

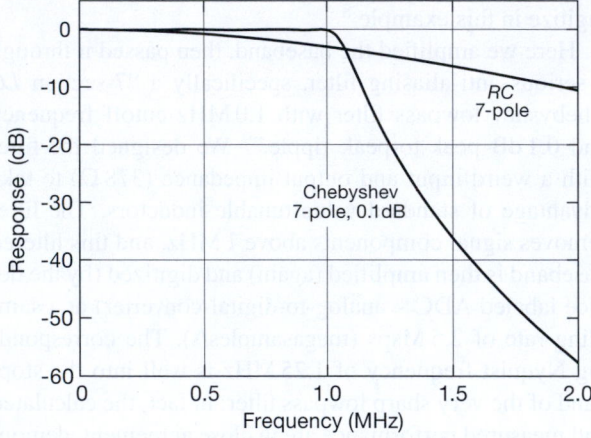

Figure 6.9. A sharp 7-section *LC* lowpass filter prevents aliasing in this radioastronomy receiver by eliminating any signal frequencies above the Nyquist frequency (1.25 MHz, or half the sampling rate). We built 126 of these puppies; see the photograph in Figure 1.111.

frequency and that the worst-case aliased signals (at 1.5 MHz) are reduced by an additional 16 dB. This is stunning performance for a filter that is easily designed and constructed, especially when compared with an *RC* filter of similar component count, where the attenuation at $1.25f_c$ is just 1.6 dB relative to that at f_c; Figures 6.10 and 6.11 make the comparison graphically.

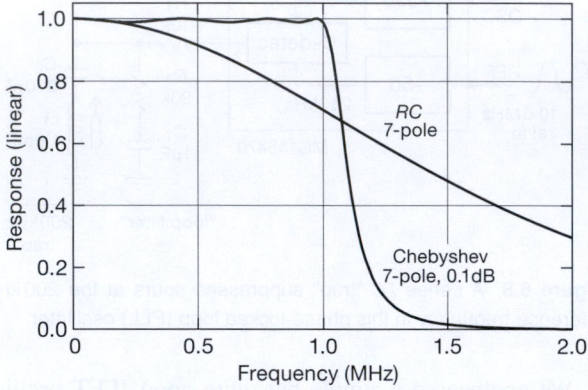

Figure 6.11. The same filter pair of Figure 6.10, here plotted on a linear scale. The Chebyshev's passband "ripple" (of +0 dB/−0.1 dB, or ±0.6% in amplitude) is more easily seen, but the details of stopband attenuation are lost.

Frequently you'll want an anti-alias filter between the amplifier and the ADC input. For example, if the sampling frequency is 25 MHz, you may want a steep rolloff input filter starting at 10 MHz. Texas Instruments has a nice application note describing how to convert a single-ended filter to differential form (SLWA053B: *Design of differential filters for high-speed signal chains*, available at www.ti.com).

Figure 6.10. The abrupt cutoff of the 7-section *LC* filter of Figure 6.9 compared with the soft rolloff of a 7-section *RC* filter with the same 1 MHz cutoff.

D. Passive differential filter

Most high-frequency ADCs have differential inputs, see §13.6.2, and many require a low input-signal source impedance, terminated in many cases by a differential capacitor. We discussed low-impedance high-frequency differential-output amplifiers in §5.17, where, for example, Figure 5.102 shows a differential lowpass filter consisting of two 50 Ω resistors and a 100 pF capacitor, as specified for the AD9225 25 MSps ADC (see also Figure 13.28).

6.2.4 Enter active filters: an overview

The synthesis of filters from passive components (*R*, *L*, and *C*) is a highly developed subject, with a rich literature of traditional handbooks (e.g., the authoritative work by Zverev; see Appendix N), now supplemented by elegant software tools that make such designs a routine task. However, inductors as circuit elements frequently leave much to be desired. They are often bulky and expensive, and they depart from the ideal by being "lossy," i.e., by having significant series resistance, as well as other "pathologies" such as nonlinearity, distributed winding capacitance, and susceptibility to magnetic pickup of interference. Furthermore, the inductances needed for low-frequency filters may dictate unmanageably large components. Finally, classic filters made with L's and C's are not electrically tunable.

What is needed is a way to make inductorless filters with the characteristics of ideal RLC filters. Ideally we might hope for tunability, either by an analog tuning voltage or a varying pulse frequency.

By using op-amps as part of the filter design, we can synthesize any RLC filter characteristic without using inductors. Such inductorless filters are known as *active filters* because of the inclusion of an active element (the amplifier). We'll see another class of active filter – the *switched capacitor* filter – that adds MOSFET switches to produce, in effect, a frequency-tunable resistor. These deliver performance similar to that of the standard active filter (sometimes called a "continuous-time" filter), but with the added feature of precise tuning of its characteristic frequency breakpoints (with an externally applied clocking frequency) over a wide range. (This tunability comes at a price, though, namely some switching noise and a reduced dynamic range; see §6.3.6.)

Active filters can be used to make lowpass, highpass, bandpass, and band-reject filters, with a choice of filter types according to the important features of the response, e.g., maximal flatness of passband, steepness of skirts, or uniformity of time delay versus frequency (more on this shortly). In addition, "allpass filters" with flat amplitude response but tailored phase versus frequency can be made (they're also known as "delay equalizers"), as well as the opposite – a filter with constant phase shift but tailored amplitude response.

A. Negative-impedance converter, gyrator, and generalized impedance converter

Three interesting circuit elements that should be mentioned in any overview are the negative-impedance converter (NIC), the gyrator, and the generalized impedance converter[10] (GIC). These devices can mimic the properties of inductors while using only resistors and capacitors in addition to op-amps.

Once you can do that, you can build inductorless filters with the ideal properties of any RLC filter, thus providing at least one way to make active filters.

---

[10] Also known as a generalized immittance converter.

B. Negative-impedance converter

The NIC converts an impedance to its *negative*, whereas the gyrator converts an impedance to its *inverse*. The following exercises will help you discover for yourself how that works out.

Exercise 6.1. Show that the circuit in Figure 6.12 is a negative-impedance converter, in particular that $Z_{in} = -Z$. *Hint*: apply some input voltage V and compute the input current I. Then take the ratio to find $Z_{in} = V/I$.

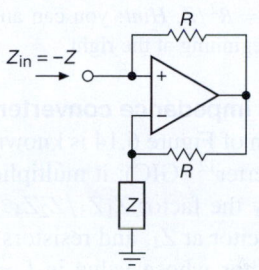

Figure 6.12. Negative-impedance converter.

The NIC therefore converts a capacitor to a "backward" inductor:

$$Z_C = 1/j\omega C \rightarrow Z_{in} = j/\omega C, \qquad (6.1)$$

i.e., it is inductive in the sense of generating a current that lags the applied voltage, but its impedance has the wrong frequency dependence (it goes down, instead of up, with increasing frequency).

C. Gyrator

The gyrator, on the other hand, converts a capacitor to a true inductor:

$$Z_C = 1/j\omega C \rightarrow Z_{in} = j\omega CR^2, \qquad (6.2)$$

i.e., an inductor with inductance $L = CR^2$.

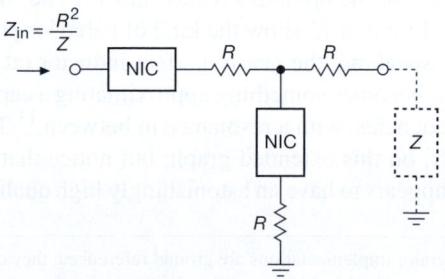

Figure 6.13. Gyrator implemented with NICs.

The existence of the gyrator makes it intuitively reasonable that inductorless filters can be built to mimic any filter using inductors by simply replacing each inductor with a gyrated capacitor.[11] The use of gyrators in just that manner is perfectly OK; and in fact the telephone filter illustrated previously, though designed as a classic *LC* filter, was implemented with gyrators (in a configuration known as a *Riordan* gyrator, which looks different from Figure 6.13). In addition to simple gyrator substitution into pre-existing *RLC* designs, it is possible to synthesize many other filter configurations.

Exercise 6.2. Show that the circuit in Figure 6.13 is a gyrator, in particular that $Z_{\text{in}} = R^2/Z$. *Hint:* you can analyze it as a set of voltage dividers, beginning at the right.

D. Generalized impedance converter

The configuration of Figure 6.14 is known as a generalized impedance converter[12] (GIC); it multiplies the impedance attached at Z_5 by the factor $Z_1 Z_3 / Z_2 Z_4$. So, for example, if you put a capacitor at Z_4, and resistors everywhere else, you get an inductor whose value is $L = (R_1 R_3 R_5 / R_2)C$; that is, it becomes a gyrator. But you can do more amusing things with a GIC: for example, if you put capacitors at Z_3 and Z_5, you wind up with a frequency-dependent negative resistor (FDNR). Filter implementations with GIC-implemented FDNRs have been popular in the audio design field, where it is claimed that they have superior noise and distortion characteristics compared with something like a Sallen-and-Key filter (next section). The field of inductorless filter design is both lively and rich with detail, with new designs appearing in the journals every month.

Performance limits

As with any op-amp circuit, the performance of gyrators and GICs at high frequencies depends on the op-amp's bandwidth (and other characteristics). So a GIC configured as an inductor (capacitor at Z_4, resistors elsewhere) will stop looking like an inductor at frequencies greater than a few percent of the op-amp's bandwidth f_T. The simulation results in Figure 6.15 show the kind of behavior you'll see. Roughly speaking, the nearly perfect inductor (at low frequencies) becomes something approximating a capacitor at high frequencies, with a resonance in between.[13] This may look ugly, on this extended graph; but notice that the "inductor" appears to have an astonishingly high quality factor

---

[11] Most gyrator implementations are ground referenced; they can replace an inductor that is returned to ground, but not a floating inductor.

[12] Or, equivalently, a generalized immittance converter.

[13] The peaking can be eliminated by adding a resistor in series with the gyrator's capacitor, roughly equal to its reactance at the peaking frequency.

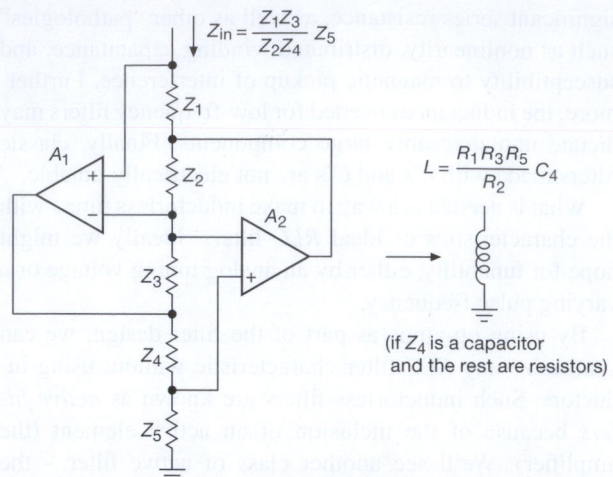

Figure 6.14. Generalized impedance converter. If Z_4 is a capacitor, the circuit behaves like an inductor, with value as shown. From A. Antoniou, *IEE Proc.*, **116**, 1838–1850 (1969).

$Z_{\text{in}} = \dfrac{Z_1 Z_3}{Z_2 Z_4} Z_5$

$L = \dfrac{R_1 R_3 R_5}{R_2} C_4$

(if Z_4 is a capacitor and the rest are resistors)

Q of about 2×10^5 at 1 kHz if you assume that the 4.6 mΩ impedance floor on the graph properly represents the inductor's loss (i.e., equivalent series resistance, ESR). (In reality there are other losses, so realizable Q values are in the range of 1000… still pretty darn good for an inductor that's a fraction of a henry !). And, for the highest-bandwidth op-amp (f_T=50 MHz) the capacitance is just 2.3 pF; you could never make a 160 mH inductor with such a tiny "winding capacitance," nor with such a high self-resonant frequency.

Gyrators are used in real-world filters: in an App Note, Texas Instruments suggests using multiple GIC stages to make anti-alias filters.[14] And Stanford Research Systems uses four GIC stages acting as an $R+LC$ ladder to make an 8-zero 9-pole elliptical lowpass filter for their SR830 DSP-based lock-in amplifier, "so that all frequency components greater than half the sampling frequency are attenuated by at least 96 dB." The A/D samples at 256 kHz and the filter passes signals from dc to 102 kHz; they've allowed themselves a 25% frequency margin to get the attenuation down to 96 dB.[15] The full schematic of the filter is included in the

---

[14] Application Note AB-026A, by Rick Downs (TI document sbaa001, 1991).

[15] According to SRS, "The architecture of the filter is based on a singly terminated passive *LC* ladder filter. *L*'s are simulated with active gyrators formed by op-amp pairs. Passive *LC* ladder filters have the special characteristic of being very tolerant of variations in component values. Because no section of the ladder is completely isolated from the other, a change in value of any single component affects the entire ladder. The design of the *LC* ladder however, is such that the characteristics of the rest of the ladder will shift to account for the change in such a

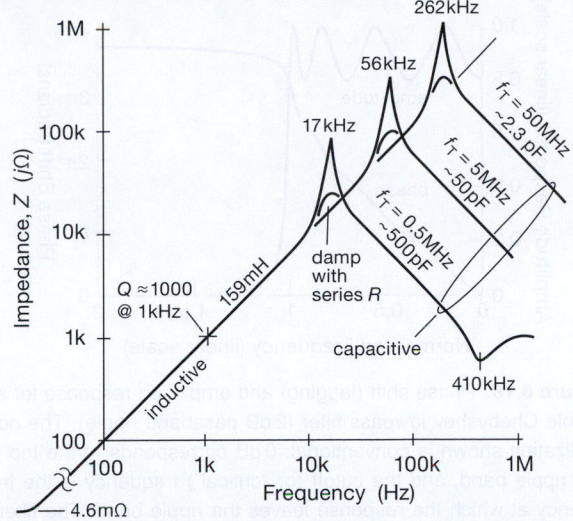

Figure 6.15. Finite op-amp bandwidth degrades the ideal GIC inductor, which becomes capacitive at frequencies beginning at a small fraction of f_T, as seen in these plots derived from SPICE simulations. When compared with a *physical* inductor, with its winding capacitance and self-resonant frequency, the GIC inductor's analogous capacitance and self-resonant frequency (which depend on the op-amp's bandwidth) can be significantly better, as suggested in these plots (which, however, are predicated on the use of an ideal capacitor).

instrument's wonderfully informative manual – a hallmark of all **SRS** products.

E. Sallen-and-Key filter

Figure 6.16 shows an example of a simple and even partly intuitive filter topology, an example of which we saw earlier in §4.3.6. These are known as Sallen-and-Key filters, after the inventors.[16] The unity-gain amplifier can be an op-amp connected as a follower, or just an emitter follower or source follower. The particular filters shown are 2-pole lowpass and highpass filters. Taking the example of the lowpass filter (Figure 6.16A), note that it would be simply two cascaded *RC* lowpass filter sections, except for the fact that the bottom of the first capacitor is bootstrapped by the output. It is easy to see that at very high frequencies it falls off just like a cascaded *RC*, because the output is essentially

---

way as to minimize its effect on the ladder. Not only does this loosen the requirement for extremely high accuracy resistors and capacitors, but it also makes the filter extremely stable despite wide temperature variations. As such, the anti-aliasing filter used in the SR830 does not ever require calibration to meet its specifications."

[16] R. P. Sallen and E. L. Key, "A practical method of designing RC active filters," *IRE Trans. Circuit Theory*, **2** (1), 74–85 (1955).

zero. As the output rises at decreasing frequency, however, the bootstrap action tends to reduce the attenuation, giving a sharper knee. Of course, such hand-waving cannot substitute for honest analysis, which luckily has already been done for a prodigious variety of nice filters. We will come back to active filter circuits in §6.3, after a short introduction to filter performance parameters and filter types.

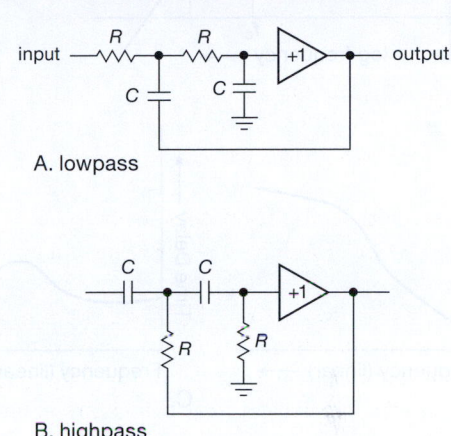

Figure 6.16. Sallen-and-Key lowpass and highpass active filters. The ultimate performance of these simple-looking filters is affected by the non-zero output impedance of the follower, see Figure 6.36.

6.2.5 Key filter performance criteria

There are some standard terms that keep appearing when we talk about filters and try to specify their performance. It is worth getting it all straight at the beginning.

A. Frequency domain

The most obvious characteristic of a filter is its gain versus frequency, typified by the sort of lowpass characteristic shown in Figure 6.17.

The *passband* is the region of frequencies that are relatively unattenuated by the filter. Most often the passband is considered to extend to the -3 dB point, but with certain filters (most notably the "equiripple" types) the end of the passband may be defined somewhat differently. Within the passband the response may show variations or *ripples*, defining a *ripple band*, as shown. The *cutoff frequency*, f_c, is the end of the passband. The response of the filter then drops off through a *transition region* (also colorfully known as the *skirt* of the filter's response) to a *stopband*, the region of significant attenuation. The stopband may be defined by some minimum attenuation, e.g., 40 dB.

Along with the gain response, the other parameter of importance in the frequency domain is the *phase shift* of the

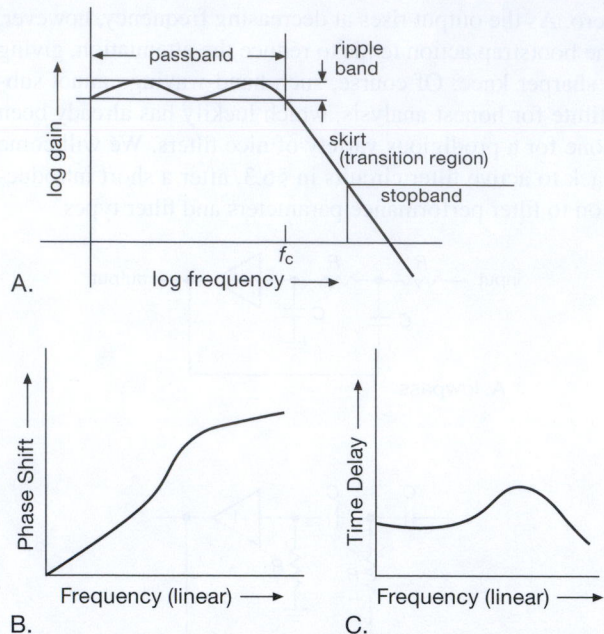

Figure 6.17. Filter characteristics versus frequency.

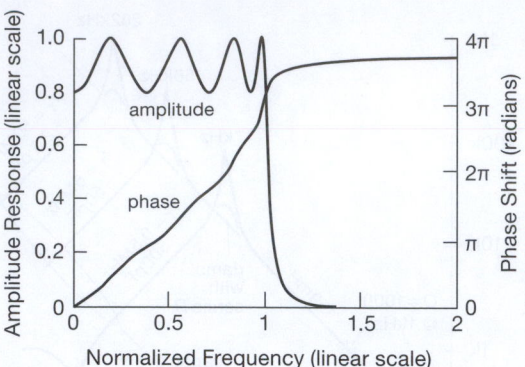

Figure 6.18. Phase shift (lagging) and amplitude response for an 8-pole Chebyshev lowpass filter (2 dB passband ripple). The normalization shown is conventional: 0 dB corresponds to the top of the ripple band, and the cutoff (or "critical") frequency is the frequency at which the response leaves the ripple band. The filter's actual dc gain is unity (0 dB); for even-order filters (like this one) the ripple rises from dc, whereas for odd-order filters the ripple drops from dc.

output signal relative to the input signal. In other words, we are interested in the *complex* response of the filter, which usually goes by the name of $\mathbf{H}(\mathbf{s})$, where $\mathbf{s} = j\omega$, and $\mathbf{H}$ and $\mathbf{s}$ are complex. Phase is important because a signal entirely within the passband of a filter will emerge with its waveform distorted if the time delay of different frequencies in going through the filter is not constant. Constant time delay corresponds to a phase shift increasing linearly with frequency ($\Delta t = -d\phi/d\omega = -\frac{1}{2\pi}d\phi/df$); hence the term *linear-phase filter* applied to a filter that is ideal in this respect. Figure 6.18 shows a typical graph of amplitude response and phase shift for a lowpass filter that is definitely not a linear-phase filter. Graphs of phase shift versus frequency are best plotted on a linear frequency axis.

B. Time domain

As with any ac circuit, filters can be described in terms of their *time-domain* properties: rise time, overshoot, ringing, and settling time. This is of particular importance where steps or pulses may be present. Figure 6.19 shows a typical lowpass-filter step response. Here, *rise time* is, as usual, the time required to go from 10% to 90% of the final value. Of greater interest is the *settling time*, which is the time required to get within some specified amount of the final value *and stay there*. The *delay time* is the time duration from the input step to the output reaching

50% of its final value.[17] *Overshoot* and *ringing* are self-explanatory terms for some undesirable properties of filters. The phase-shifting characteristics of filters imply a corresponding time delay, which you sometimes see plotted (or tabulated) as *group delay* versus frequency.[18]

6.2.6 Filter types

Suppose you want a lowpass filter with flat passband and sharp transition to the stopband. The ultimate rate of falloff, well into the stopband, will always be 6*n* dB/octave, where *n* is the number of "poles." You need one capacitor (or inductor) for each pole, so the required ultimate rate of falloff of filter response determines, roughly, the complexity of the filter.

Now, assume that you have decided to use a 6-pole lowpass filter. You are guaranteed an ultimate rolloff of 36 dB/octave at high frequencies. It turns out that the filter design can now be optimized for maximum flatness of passband response, at the expense of a slow transition from passband to stopband. Alternatively, by allowing some

---

[17] Sometimes t_d is defined to 10% (rather than 50%) output.

[18] That term comes from wave analysis in dispersive materials, where one distinguishes *phase velocity* and *group velocity*. The latter refers to the speed with which a group of frequencies, together making up a characteristic wave shape, moves through the medium. The group delay is the analogous quantity, expressed as a time delay T_g, for a signal passing through a filter. The connection between phase shift and group delay is $T_g = -d\phi/d\omega = -\frac{1}{2\pi}d\phi/df$.

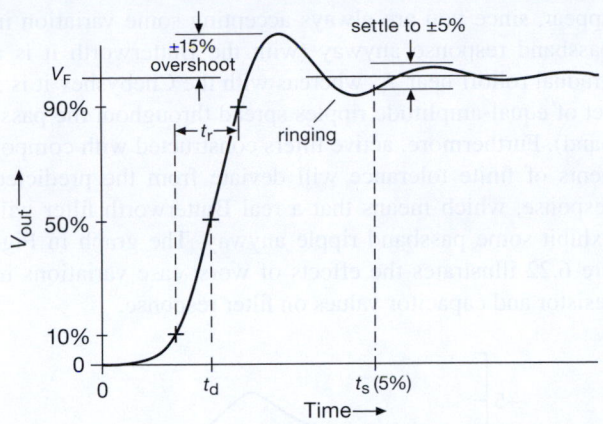

Figure 6.19. Time-domain filter characteristics. A simple *RC* low-pass filter, for example, would have no overshoot or ringing, and would be characterized by a rise time of $t_r = 2.2RC$ ($\approx 0.35/f_{3dB}$), a delay time of $t_d = 0.69RC$, and a settling time (to 1%) of $t_s = 4.6RC$.

ripple in the passband characteristic, the transition from pass-band to stopband can be steepened considerably. A third criterion that may be important is the ability of the filter to pass signals within the passband without distortion of their waveforms caused by phase shifts. You may also care about rise time, overshoot, and settling time. Generally speaking, you've got to make tradeoffs among these characteristics – a filter with a sharp cutoff will exhibit poor time-domain properties such as ringing and phase shifts.

There are filter designs available to optimize each of these characteristics, or combinations of them. In fact, rational filter selection will not be carried out as just described; rather, it normally begins with a set of requirements on passband flatness, attenuation at some frequency outside the passband, and whatever else matters. You will then choose the best design for the job, using the number of poles necessary to meet the requirements.[19] In the next few sections we introduce the three popular classics – the Butterworth filter (maximally flat passband), the Chebyshev filter (steepest transition from passband to stopband), and the Bessel filter (maximally flat time delay). Each of these filter responses can be produced with a variety of different filter circuits, some of which we discuss later. They are

all available in lowpass, highpass, bandpass, and band-stop (notch) versions.[20]

A. Butterworth and Chebyshev filters

The Butterworth filter produces the flattest passband response, at the expense of steepness in the transition region from passband to stopband. As you will see later, it has only mediocre phase and transient characteristics. The amplitude response is given by

$$\frac{V_{out}}{V_{in}} = \frac{1}{[1+(f/f_c)^{2n}]^{\frac{1}{2}}}, \qquad (6.3)$$

where *n* is the order of the filter (number of poles). Increasing the number of poles flattens the passband response and steepens the stopband falloff, as shown in Figure 6.20.

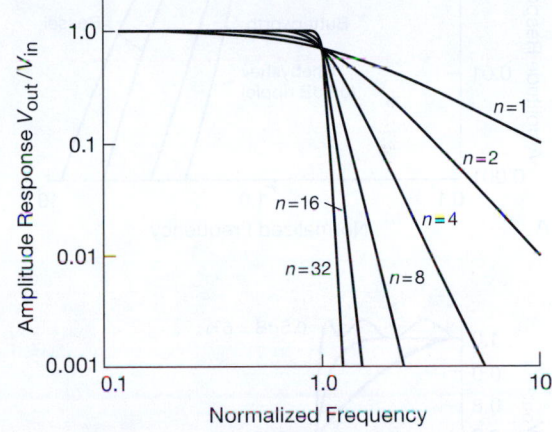

Figure 6.20. Normalized lowpass Butterworth filter response curves. Note the improved attenuation characteristics for the higher-order filters.

The Butterworth filter trades off everything else for maximum flatness of response. It starts out extremely flat at zero frequency and bends over near the cutoff frequency f_c (which is usually the −3 dB point).

In most applications, all that really matters is that the wiggles in the passband response be kept less than some amount, say 1 dB. The Chebyshev filter responds to this reality by allowing some ripples throughout the passband, with greatly improved sharpness of the knee (compared with the "maximally flat" Butterworth, for example). A Chebyshev filter is specified in terms of its number of poles

---

[19] Traditionally you used a good filter-design handbook, with its tables and graphs. Now the job is far easier, thanks to elegant filter-design software that guides you in the choice of filter and then finishes the job with a complete circuit implementation. Such programs fall under the general rubric of CAD, for computer-aided design. Both Linear Technology and Texas Instruments offer free programs on their websites; they are called FilterCAD™ and FilterPro,™ respectively; see also the helpful LTC documents AN38 and DN245.

[20] Filters can be made, also, to perform *equalization* (a specified amplitude and/or phase versus frequency profile that is none of these simple filter types). Among these, the *phase equalizer* (or *delay equalizer*) is noteworthy, in having a specified phase response combined with a flat frequency response; it is also called an *all-pass* filter.

and passband ripple. By allowing greater passband ripple, you get a sharper knee. The amplitude is given by

$$\frac{V_{\text{out}}}{V_{\text{in}}} = \frac{1}{[1+\varepsilon^2 C_n^2(f/f_c)]^{\frac{1}{2}}} , \qquad (6.4)$$

where C_n is the Chebyshev polynomial of the first kind of degree n, and ε is a constant that sets the passband ripple. Like the Butterworth (but in even greater degree), the Chebyshev has phase and transient characteristics that are far from ideal.

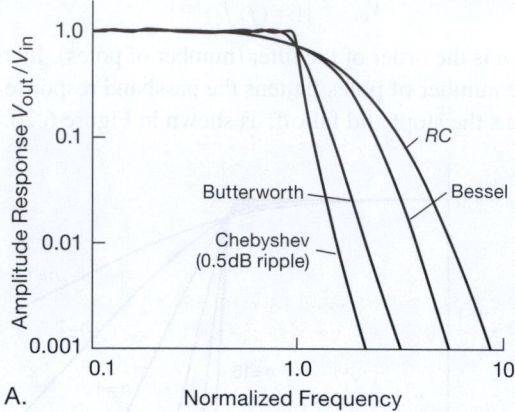

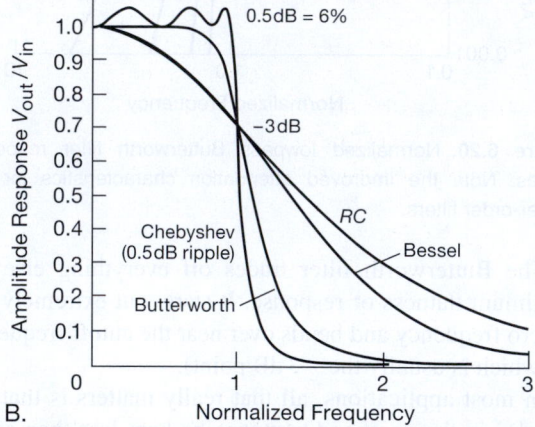

Figure 6.21. Comparison of some common 6-pole lowpass filters. The same filters are plotted on both linear and logarithmic scales. The actual gains of the filters are shown, rather than the "top-adjusted" 0 dB convention.

Figure 6.21 presents graphs comparing the responses of Chebyshev and Butterworth 6-pole lowpass filters. As you can see, they're both tremendous improvements over a 6-pole *RC* filter.

As a practical reality, the Butterworth, with its "maximally flat" passband, may not be as attractive as it might

appear, since you are always accepting some variation in passband response anyway (with the Butterworth it is a gradual rolloff near f_c, whereas with the Chebyshev it is a set of equal-amplitude ripples spread throughout the passband). Furthermore, active filters constructed with components of finite tolerance will deviate from the predicted response, which means that a real Butterworth filter will exhibit some passband ripple anyway. The graph in Figure 6.22 illustrates the effects of worst-case variations in resistor and capacitor values on filter response.

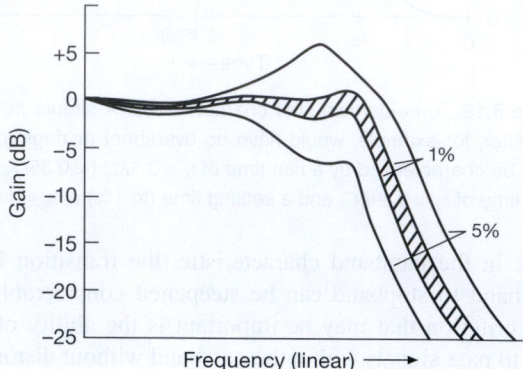

Figure 6.22. The effect of component tolerance on active-filter performance.

Viewed in this light, the Chebyshev is a very rational filter design. It manages to improve the situation in the transition region by spreading equal-size ripples[21] throughout the passband, the number of ripples increasing with the order of the filter. Even with rather small ripples (as little as 0.1 dB) the Chebyshev filter offers considerably improved sharpness of the knee compared with the Butterworth. To make the improvement quantitative, suppose that you need a filter with flatness to 0.1 dB within the passband and 20 dB attenuation at a frequency 25% beyond the top of the passband. By actual calculation, that will require a 19-pole Butterworth, but only an 8-pole Chebyshev.

The idea of accepting some passband ripple in exchange for improved steepness in the transition region, as in the equiripple Chebyshev filter, is carried to its logical limit in the so-called elliptic (or Cauer) filter by trading ripple in both passband *and* stopband for an even steeper transition region than that of the Chebyshev filter.[22] Such a filter does the job, if you are satisfied with an amplitude characteristic that reaches and maintains some minimum attenuation throughout the stopband (rather than continuing to fall

---

[21] It is sometimes called an equiripple filter.
[22] Or fewer filter sections to achieve a given steepness.

off with a $6n$ dB/octave ultimate slope). The payback is a simpler filter, with better phase and amplitude characteristics (see below). With computer-aided design techniques, the design of elliptic filters is as straightforward as for the classic Butterworth and Chebyshev filters.

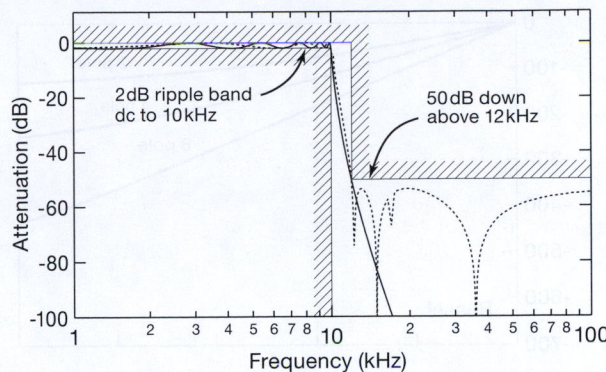

Figure 6.24. Lowpass filter example: a 6-pole elliptic filter with both passband and stopband ripple (dashed curve) meets the performance specifications shown here, whereas you would need an 11-pole Chebyshev (which has ripple only in its passband) or a 32-pole Butterworth ("maximally flat" – no ripple in passband or stopband) to do as well.

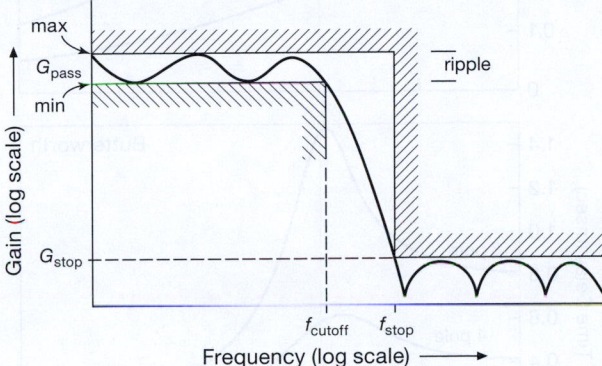

Figure 6.23. Specifying filter frequency-response parameters.

Figure 6.23 shows how you specify filter frequency response graphically. In this case (a lowpass filter) you indicate the allowable range of filter gain (i.e., the ripple) in the passband, the minimum frequency at which the response leaves the passband, the maximum frequency at which the response enters the stopband, and the minimum attenuation in the stopband. As an example, Figure 6.24 compares the responses for Chebyshev and elliptic lowpass filter implementations to meet a specified performance, here requiring an 11-pole Chebyshev or a 6-pole elliptic (to meet the same specifications with a Butterworth would require a 32-pole implementation!). The simpler elliptic filter has the better phase characteristics, but its response does not continue to fall off monotonically with frequency once it has reached the specified stopband attenuation.

B. Bessel filter

As we've suggested, the amplitude versus frequency response of a filter does not tell the whole story. A filter characterized by a flat amplitude response may exhibit rapidly changing phase shifts, which produce unequal time delays for signals within its passband. The result is that a signal in the passband will suffer distortion of its waveform. In situations where the shape of the waveform is paramount, a linear-phase filter (or constant-time-delay filter) is desirable. A filter whose phase shift varies linearly with frequency is equivalent to a constant time delay for signals within the passband; i.e., the waveform is not distorted. The

Bessel filter (also called the Thomson filter)[23] has maximally flat time delay within its passband in analogy with the Butterworth, which has maximally flat amplitude response.

To see the kind of improvement in time-domain performance you get with the Bessel filter, look at Figure 6.25 for a comparison of phase shift and time delay versus frequency for the Bessel filter compared with two classic filters that exhibit more abrupt frequency characteristics (Butterworth and Chebyshev). The poor time-delay performance of the Butterworth (and to a greater extent of the Chebyshev) gives rise to effects such as waveform distortion and overshoot when driven with pulse signals – see Figure 6.26. On the other hand, the price you pay for the Bessel's constancy of time delay is an amplitude response with even less steepness than that of the Butterworth or Chebyshev in the transition region between passband and stopband. An important point: adding sections to a Bessel filter (i.e., making it of higher order) does not significantly increase the steepness of transition into the stopband; it does, however, improve the phase linearity (constancy of time delay), as well as increasing the ultimate rate of falloff, reaching the usual $6n$ dB/octave asymptotic limit (look ahead to Figure 6.30).

---

[23] That's the legendary German mathematician Friedrich Bessel (1784–1846) who, though not a practicing circuit designer, developed the mathematics. The label Bessel–Thomson recognizes Thomson's application to filters: Thomson, W. E., "Delay Networks having Maximally Flat Frequency Characteristics," *Proceedings of the Institution of Electrical Engineers*, Part III, **96** 44, pp. 487–490 (1949).

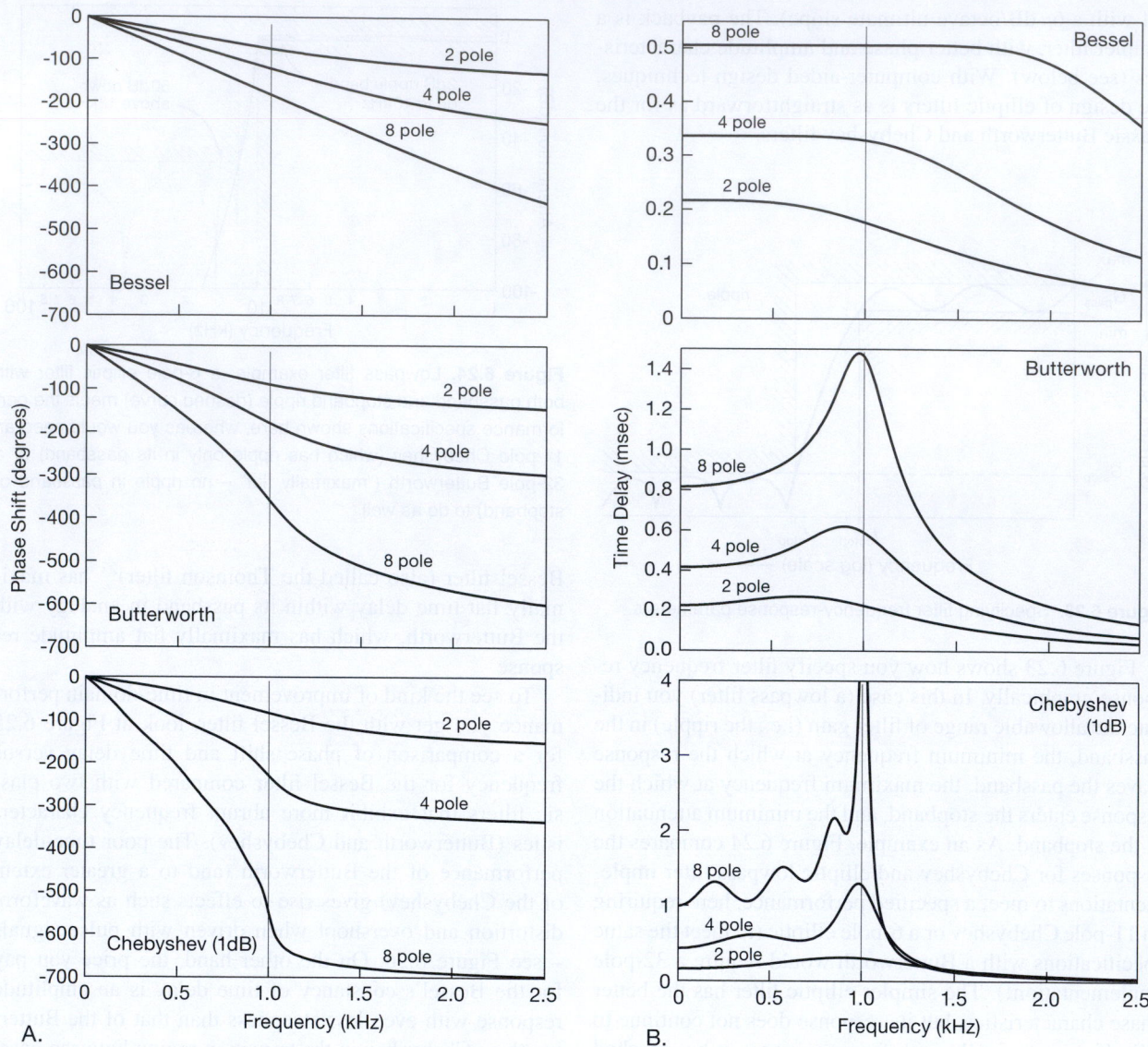

Figure 6.25. A. Phase shift *vs.* frequency for three lowpass filter types, each configured with a cutoff frequency of 1 kHz (vertical line). B. Time delay *vs.* frequency for the adjacent filters; note change of vertical-scale units and linear frequency axis. If you like normalized units, use f/f_c for the horizontal axes, and t_d/T for the time delay.

There are numerous filter designs that attempt to improve on the Bessel's good time-domain performance by compromising some of the constancy of time delay for improved rise time and amplitude-versus-frequency characteristics. The Gaussian filter has phase characteristics nearly as good as those of the Bessel, with improved step response. In another class there are interesting filters that allow uniform ripples in the passband time delay (in analogy with the Chebyshev's ripples in its amplitude response)

and yield approximately constant time delays even for signals well into the stopband; these are sometimes called simply "linear phase" filters, characterized with a parameter that sets the phase ripple (e.g., 0.5°) within the passband. Another approach to the problem of making filters with uniform time delays is to use allpass filters (also known as delay equalizers). These have constant amplitude response with frequency, with a phase shift that can be tailored to individual requirements. Thus they can be used to improve the time-delay constancy of any filter, including Butterworth and Chebyshev types.

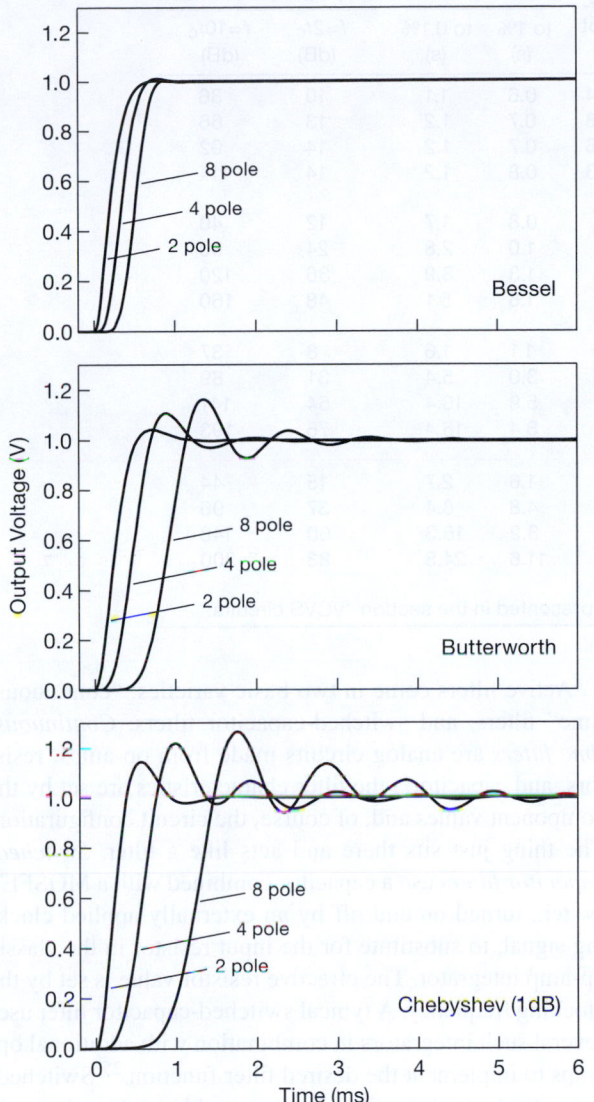

Figure 6.26. Response to a 1 V step input at $t = 0$, for the three 1 kHz lowpass filters of the previous figures.

C. Filter comparison

In spite of the preceding comments about the Bessel filter's frequency response, it still has vastly superior properties in the time domain compared with the Butterworth and Chebyshev. The Chebyshev, with its highly desirable amplitude-versus-frequency characteristics, actually has the poorest time-domain performance of the three. The Butterworth is in between in both frequency- and time-

domain properties. Table 6.1 on the next page and Figures 6.26 and 6.27 give more information about time-domain performance for these three kinds of filters to complement the frequency-domain graphs presented earlier. They make it clear that the Bessel is a desirable filter when performance in the time domain is important.

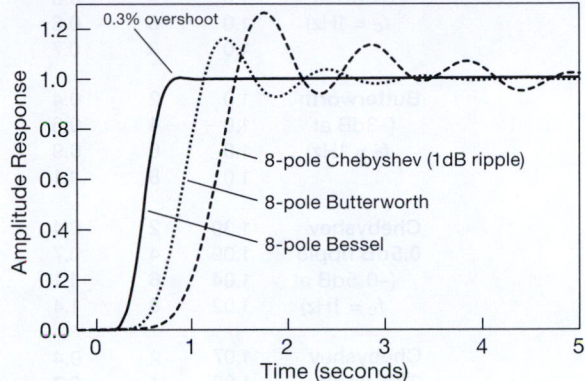

Figure 6.27. Step-response comparison for 8-pole lowpass filters normalized to 1 Hz cutoff frequency.

6.2.7 Filter implementation

We'll see in the next section how to implement these classic filters with R's, C's, and op-amps. These are called *active filters*, and have the advantage of requiring no inductors. This is good, because inductors tend to be bulky, imperfect, and not inexpensive.

However, for use at frequencies above roughly 100 kHz, it is often preferable to build passive filters like the anti-alias lowpass filter example we showed in Figure 6.9. You have two choices: you can build your own, or you can buy what you need. To make your own, you can use any of numerous design tables (we give a set in Appendix E) or filter-design software (see §6.3.8) to calculate the L and C values for the particular filter you want. If you are making only a few, you may wish to use slug-tuned inductors (adjusted with an inductance meter or bridge), and either 1% capacitors or hand-trimmed pairs of paralleled capacitors, to get the precision you need.

Alternatively, you can just throw money at the problem: there are dozens of manufacturers of standard and custom filters, and they are happy to build anything you want. At the low end of the frequency spectrum (say below 100 MHz) they will use lumped elements (L's and C's); above that you'll get coaxial or cavity filters. If the filter you want is a standard unit (e.g., in the Mini-Circuits Labs catalog), it will be inexpensive and generally

Table 6.1 Time-domain Performance Comparison for Lowpass Filters[a]

Type	f_{3dB} (Hz)	Poles	Step risetime (0 – 90%) (s)	Over-shoot (%)	Settling time to 1% (s)	Settling time to 0.1% (s)	Stopband attenuation $f=2f_C$ (dB)	Stopband attenuation $f=10f_C$ (dB)
Bessel	1.0	2	0.4	0.4	0.6	1.1	10	36
(–3dB at	1.0	4	0.5	0.8	0.7	1.2	13	66
f_C = 1Hz)	1.0	6	0.6	0.6	0.7	1.2	14	92
	1.0	8	0.7	0.3	0.8	1.2	14	114
Butterworth	1.0	2	0.4	4	0.8	1.7	12	40
(–3dB at	1.0	4	0.6	11	1.0	2.8	24	80
f_C = 1Hz)	1.0	6	0.9	14	1.3	3.9	36	120
	1.0	8	1.1	16	1.6	5.1	48	160
Chebyshev	1.39	2	0.4	11	1.1	1.6	8	37
0.5dB ripple	1.09	4	0.7	18	3.0	5.4	31	89
(–0.5dB at	1.04	6	1.1	21	5.9	10.4	54	141
f_C = 1Hz)	1.02	8	1.4	23	8.4	16.4	76	193
Chebyshev	1.07	2	0.4	21	1.6	2.7	15	44
2dB ripple	1.02	4	0.7	28	4.8	8.4	37	96
(–2dB at	1.01	6	1.1	32	8.2	16.3	60	148
f_C = 1Hz)	1.01	8	1.4	34	11.6	24.8	83	200

Notes: (a) a design procedure for these filters is presented in the section "VCVS circuits."

delivered from stock. Otherwise you will pay at least a hundred dollars, and wait at least a few weeks. Some manufacturers we've used are Lark Engineering, Mini-Circuits Laboratories, Trilithic (Cir-Q-Tel), and TTE.

6.3 Active-filter circuits

A lot of ingenuity has been used in inventing clever active circuits, each of which can be used to generate response functions such as the Butterworth, Chebyshev, etc. You might wonder why the world needs more than one active-filter circuit. The reason is that various circuit realizations excel in one or another desirable property, so there is no all-around best circuit.

Active filters can be built using discrete op-amps as the active elements.[24] In that case you must provide the resistors and capacitors that set the filter characteristics. These passive components must generally be accurate and stable, particularly in filters with sharp frequency characteristics. An attractive alternative is to take advantage of the rich variety of IC active filters, in which most of the hard work has been done, including the on-chip integration of matched passive components.

Active filters come in two basic varieties: "continuous-time" filters, and switched-capacitor filters. *Continuous-time filters* are analog circuits made from op-amps, resistors, and capacitors; the filter characteristics are set by the component values and, of course, the circuit configuration. The thing just sits there and acts like a filter. *Switched-capacitor filters* use a capacitor combined with a MOSFET switch, turned on and off by an externally applied clocking signal, to substitute for the input resistor in the classic op-amp integrator. The effective resistor value is set by the clocking frequency. A typical switched-capacitor filter uses several such integrators in combination with additional op-amps to implement the desired filter function.[25] Switched-capacitor filters have the advantages of being simply tuned over a wide range (by the applied clocking frequency), of maintaining stable characteristics, and of being particularly easy to fabricate as ICs. However, they are generally noisier (i.e., with smaller dynamic range), have higher distortion, and can introduce switching artifacts such as aliasing and clock feedthrough.

Some of the features to look for in active filters are (a) small numbers of parts, both active and passive, (b) ease

---

[24] Or even discrete transistor followers, as in the simple Sallen-and-Key filter.

[25] The resulting circuit configuration is often identical to that of some continuous-time active filter, for example the type known as "state-variable" or "biquad."

of adjustability, (c) small spread of parts values, especially the capacitor values, (d) undemanding use of the op-amp, especially requirements on slew rate, bandwidth, and output impedance, (e) the ability to make high-Q filters, (f) electrical tunability, and (g) sensitivity of filter characteristics to component values and op-amp gain (in particular, the gain–bandwidth product, f_T). In many ways, the last feature is one of the most important. A filter that requires parts of high precision is difficult to adjust, and it will drift as the components age; in addition, there is the nuisance that it requires components of good initial accuracy. The VCVS circuit probably owes most of its popularity to its simplicity and its low parts count, but it suffers from high sensitivity to component variations. By comparison, recent interest in more complicated filter realizations is motivated by the benefits of insensitivity of filter properties to small component variability.

In this section we present several circuits for lowpass, highpass, and bandpass active filters. We begin with the popular VCVS, or controlled-source type, then show the state-variable designs available as ICs from several manufacturers, and finally mention the twin-T sharp rejection filter.

Most of the new active-filter ICs being introduced are of the switched-capacitor type, owing to their ease of use, small size, low cost, excellent stability, and (in some cases) complete absence of required external components. We conclude the chapter with a discussion of them.

6.3.1 VCVS circuits

The voltage-controlled voltage-source (VCVS) filter, also known simply as a controlled-source filter, was devised by Sallen and Key (and introduced in simplified form in §6.2.4E). It's a variation of the simpler unity-gain circuit shown earlier (Figure 6.16), in which the unity-gain follower is replaced with a noninverting amplifier of gain greater than unity. Figure 6.28 shows the circuits for lowpass, highpass, and bandpass realizations. The resistors at the outputs of the op-amps create a noninverting voltage amplifier of voltage gain K, with the remaining R's and C's contributing the frequency response properties for the filter. These are 2-pole filters, and they can be Butterworth, Bessel, etc., by suitable choice of component values, as we show later. Any number of VCVS 2-pole sections may be cascaded to generate higher-order filters. When that is done, the individual filter sections are, in general, not identical. In fact, each section represents a quadratic polynomial factor of the nth-order polynomial describing the overall filter.

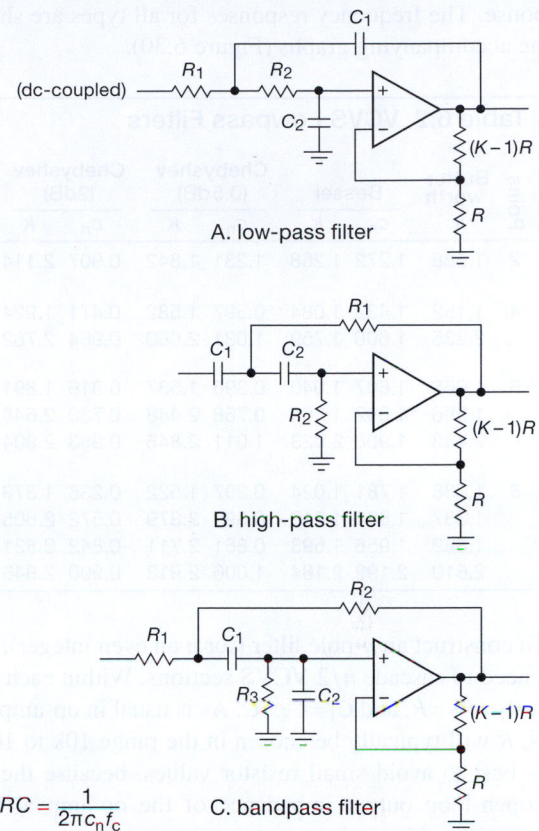

$$RC = \frac{1}{2\pi c_n f_c}$$

Figure 6.28. VCVS active-filter circuits.

There are design equations and tables in most standard filter handbooks for all the standard filter responses, usually including separate tables for each of a number of ripple amplitudes for Chebyshev filters. In the next section we present an easy-to-use design table for VCVS filters of Butterworth, Bessel, and Chebyshev responses (0.5 dB and 2 dB passband ripple for Chebyshev filters) for use as lowpass or highpass filters. Bandpass and band-reject filters can be made from combinations of these.

6.3.2 VCVS filter design using our simplified table

To use Table 6.2 to make a lowpass or a highpass filter, begin by deciding which filter response you need. As we mentioned earlier, the Butterworth may be attractive if maximum flatness of passband is desired, the Chebyshev gives the fastest rolloff from passband to stopband (at the expense of some ripple in the passband), and the Bessel provides the best phase characteristics, i.e., constant signal delay in the passband, with correspondingly good step response. The frequency responses for all types are shown in the accompanying graphs (Figure 6.30).

Table 6.2 VCVS Lowpass Filters

Poles	Butterworth K	Bessel c_n	Bessel K	Chebyshev (0.5 dB) c_n	Chebyshev (0.5 dB) K	Chebyshev (2 dB) c_n	Chebyshev (2 dB) K
2	1.586	1.272	1.268	1.231	1.842	0.907	2.114
4	1.152	1.432	1.084	0.597	1.582	0.471	1.924
	2.235	1.606	1.759	1.031	2.660	0.964	2.782
6	1.068	1.607	1.040	0.396	1.537	0.316	1.891
	1.586	1.692	1.364	0.768	2.448	0.730	2.648
	2.483	1.908	2.023	1.011	2.846	0.983	2.904
8	1.038	1.781	1.024	0.297	1.522	0.238	1.879
	1.337	1.835	1.213	0.599	2.379	0.572	2.605
	1.889	1.956	1.593	0.861	2.711	0.842	2.821
	2.610	2.192	2.184	1.006	2.913	0.990	2.946

To construct an n-pole filter (for n an even integer), you will need to cascade $n/2$ VCVS sections. Within each section, $R_1 = R_2 = R$, and $C_1 = C_2 = C$. As is usual in op-amp circuits, R will typically be chosen in the range 10k to 100k. (It is best to avoid small resistor values, because the rising open-loop output impedance of the op-amp at high frequencies adds to the resistor values and upsets calculations.) Then all you need to do is set the gain, K, of each stage according to the table entries. For an n-pole filter there are $n/2$ entries, one for each section.

A. Butterworth lowpass filters
If the filter is a Butterworth, all sections have the same values of R and C, given simply by $RC = 1/2\pi f_c$, where f_c is the desired -3 dB frequency of the entire filter. To make a 6-pole lowpass Butterworth filter, for example, you cascade three of the lowpass sections shown previously, with gains of 1.07, 1.59, and 2.48 (preferably in that order, to avoid dynamic range problems), and with identical R's and C's to set the 3 dB point.

B. Bessel and Chebyshev lowpass filters
To make a Bessel or Chebyshev filter with the VCVS, the situation is only slightly more complicated. Again we cascade several 2-pole VCVS filters, with prescribed gains for each section. Within each section we again use $R_1 = R_2 = R$ and $C_1 = C_2 = C$. However, unlike the situation with the Butterworth, the RC products for the different sections are different and must be scaled by the normalizing factor c_n (given for each section in Table 6.2 on this page) according

to $RC = 1/2\pi c_n f_c$. Here f_c is again the -3 dB point for the Bessel filter, whereas for the Chebyshev filter it defines the end of the passband, i.e., it is the frequency at which the amplitude response falls out of the ripple band on its way into the stopband. For example, the response of a Chebyshev lowpass filter with 0.5 dB ripple and $f_c = 100$ Hz will be flat within $+0$ dB to -0.5 dB from dc to 100 Hz, with 0.5 dB attenuation at 100 Hz and a rapid falloff for frequencies greater than 100 Hz. Values are given for Chebyshev filters with 0.5 dB and 2.0 dB passband ripple; the latter have a somewhat steeper transition into the stopband (Figure 6.30).

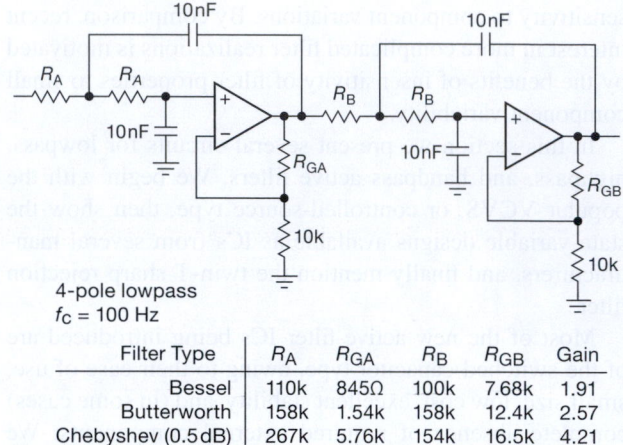

Filter Type	R_A	R_{GA}	R_B	R_{GB}	Gain
Bessel	110k	845Ω	100k	7.68k	1.91
Butterworth	158k	1.54k	158k	12.4k	2.57
Chebyshev (0.5dB)	267k	5.76k	154k	16.5k	4.21

Figure 6.29. VCVS lowpass filter example. Resistor values shown are the nearest standard 1% values (known as "E96").

An example
As an illustration, Figure 6.29 shows a VCVS implementation of a 4-pole lowpass filter with $f_c = 100$ Hz; resistor values for three filter characteristics are listed, calculated as just described. We used a similar filter (6-pole Butterworth, $f_c = 90$ Hz) to create a precision 50–70 Hz sinewave from a digital square wave that was referenced to a crystal oscillator; the output was amplified and used to drive an astronomical telescope.[26]

C. Highpass filters
To make a highpass filter, use the highpass configuration shown previously, i.e., with the R's and C's interchanged. For Butterworth filters, everything else remains unchanged (use the same values for R, C, and K). For the Bessel and Chebyshev filters, the K values remain the same, but the

---

[26] *The Art of Electronics*, 2nd edition, pp. 249 and 549.

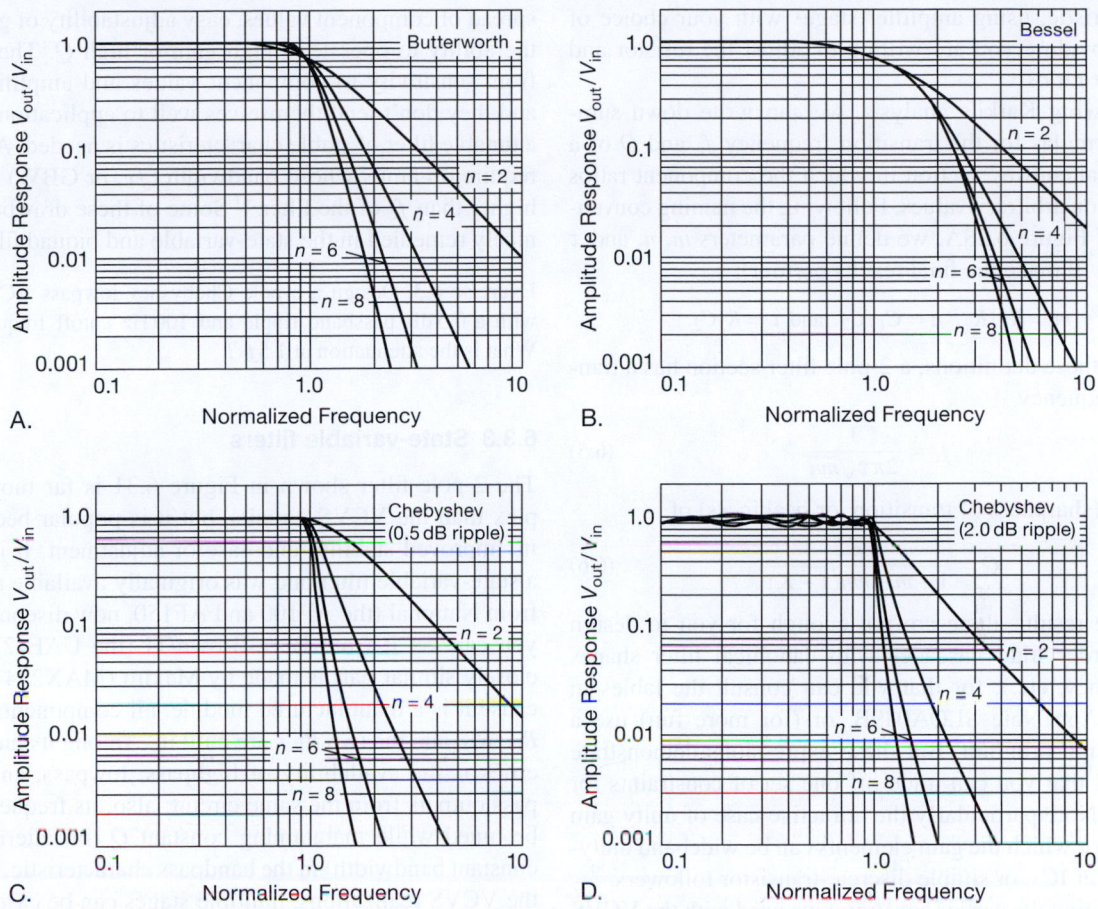

Figure 6.30. Normalized frequency response graphs for the 2-, 4-, 6-, and 8-pole filters in Table 6.2. The Butterworth and Bessel filters are normalized to 3 dB attenuation at unit frequency, whereas the Chebyshev filters are normalized to 0.5 dB and 2 dB attenuations. As explained earlier, the top of the ripple band in the Chebyshev plots has been set to unity.

normalizing factors c_n must be inverted, i.e., for each section the new c_n equals $1/(c_n$ listed in Table 6.2 on the preceding page).

A bandpass filter can be made by cascading overlapping lowpass and highpass filters. A band-reject filter can be made by summing the outputs of nonoverlapping lowpass and highpass filters. However, such cascaded filters won't work well for high-Q filters (extremely sharp bandpass filters) because there is great sensitivity to the component values in the individual (uncoupled) filter sections. In such cases a high-Q single-stage bandpass circuit (e.g., the VCVS bandpass circuit illustrated previously, or the state-variable and biquad filters in the next section) should be used instead. Even a single-stage 2-pole filter can produce a response with an extremely sharp peak. Information on such filter design is available in the standard references.

D. Generalizing the Sallen-and-Key filter

A design simplification in these Sallen-and-Key (or VCVS) filter circuits was the use of identical resistor and capacitor values within each 2-pole filter stage; but with that simplification came a set of oddball amplifier gains, as seen in the "Gain" column in Figure 6.29.

Often you want to set the filter's gain, for example to prevent saturation, or so that you can change the filter characteristics (by a change of component value) without altering the gain. When you constrain the gain, however, you have to relax the component ratio constraint. You can learn all about this in a pair of nice Application Notes by James Karki from TI.[27] The bottom line is that (just as with the preceding VCVS circuits) you can create any filter

---

[27] "Analysis of the Sallen-and-Key Architecture," SLOA024B (2002) and "Active Low-Pass Filter Design," SLOA049B (2002).

characteristic, using amplifier stages with your choice of gain, providing you are willing to adjust the resistor and capacitor ratios.

Following Karki's analysis, we can write down summary formulas for the transition frequency f_c and Q of a 2-pole Sallen–Key section in which the component ratios can take on arbitrary values. Following the naming convention[28] of Figure 6.28A, we define parameters m, n, and τ (which will make the final results prettier):

$$m = R_1/R_2, \quad n = C_1/C_2, \text{ and } \tau = R_2 C_2.$$

With these definitions, a 2-pole filter section has a transition frequency

$$f_c = \frac{1}{2\pi\tau\sqrt{mn}} \tag{6.5}$$

and a Q (sharpness of transition, or peakiness) of

$$Q = \frac{\sqrt{mn}}{1 + m + mn(1 - K)}. \tag{6.6}$$

These results alone are not enough for you to design higher-order filter cascades with canonical filter shapes (Chebyshev, etc.); for that you can consult the tables in Karki's App Note SLOA049B, or (for more fun) use a filter-design program. But these expressions demonstrate the point that you can trade off one set of constraints for another. Note particularly the attractive case of unity gain ($K=1$), for which the gain elements can be wideband unity-gain buffer ICs, or simple discrete-transistor followers.[29]

Revisiting the earlier constraint we used with the VCVS table (i.e., $R_1 = R_2 = R$, $C_1 = C_2 = C$), these formulas reduce to the simple forms

$$f_c = \frac{1}{2\pi RC}, \quad Q = \frac{1}{3 - K}, \tag{6.7}$$

for which the circuit goes unstable ($Q \to \infty$) when $K=3$. Note that such a circuit, with the gain K further constrained to unity (i.e., a follower, as we illustrated in Figure 6.16 to introduce the idea of an active filter) produces a pretty anemic filter, with a Q of just one half.

E. Summary

VCVS filters minimize the number of components needed (two poles per op-amp) and offer the additional advantages of noninverting gain, low output impedance, small spread of component values, easy adjustability of gain, and the ability to operate at high gain or high Q. They suffer from sensitivity to component values and amplifier gain, and they don't lend themselves well to applications where a tunable filter of stable characteristics is needed. And they require op-amps whose bandwidth (f_T, or GBW) is much higher than f_c of the filter.[30] Some of these drawbacks are nicely remedied in the state-variable and biquad filters.

Exercise 6.3. Design a 6-pole Chebyshev lowpass VCVS filter with a 0.5 dB passband ripple and 100 Hz cutoff frequency f_c. What is the attenuation at $1.5 f_c$?

6.3.3 State-variable filters

The 2-pole filter shown in Figure 6.31 is far more complex than the VCVS circuits, but it is popular because of its improved stability and ease of adjustment. It is called a state-variable filter and was originally available as an IC from National (the AF100 and AF150, now discontinued); you can get it from Burr–Brown/TI (the UAF42), and a closely similar part is made by Maxim (MAX274–5). Because it is a manufactured module, all components except R_G, R_Q, and the two R_F's are built in. Among its nice properties is the availability of highpass, lowpass, and bandpass outputs from the same circuit; also, its frequency can be tuned while maintaining constant Q (or, alternatively, constant bandwidth) in the bandpass characteristic. As with the VCVS realizations, multiple stages can be cascaded to generate higher-order filters. The frequency can be made adjustable with a dual pot for the R_F pair. But, given the inverse ($1/R$) frequency tuning, you may prefer a linear scheme like that shown in Figure 6.34, where you could use either a dual pot or a dual multiplying DAC (see §6.3.3C).

Extensive design formulas and tables are provided by the manufacturers for the use of these convenient ICs. They show how to choose the external resistor values to make Butterworth, Bessel, and Chebyshev filters for a wide range of filter orders, with lowpass, highpass, bandpass, or band-reject responses. Among the nice features of these hybrid ICs is integration of the capacitors into the module,[31] so that only external resistors need be added.

---

[28] A caution: we have used the labels ("refdes") for the R's and C's as defined in the original Sallen and Key paper, but in many references (including Karki's) the labels for C_1 and C_2 are interchanged.

[29] You can turn these equations around to find m in terms of n, given a target value for Q: define a quantity $\alpha = (n/2Q^2) - 1$; then $m = \alpha + \sqrt{\alpha^2 - 1}$.

[30] This is stated, by various authorities, as requiring something like $f_T \geq 50 Q^2 f_c$, where the Q of an op-amp section is given in terms of its gain K (as listed in the table) by $Q = 1/(3 - K)$. You may see the bandwidth criterion (which is approximate in any case) stated alternatively as $f_T \geq 50 f_c$, or $f_T \geq 50 K f_c$; they're all in the same ballpark.

[31] And, of course, no inductors are needed in this (or any other) active-filter implementation.

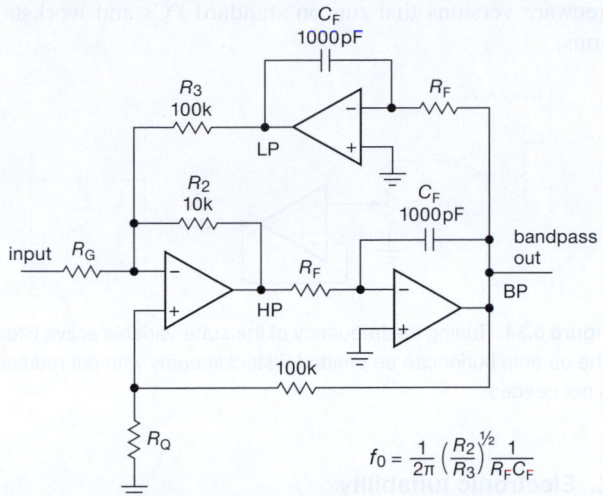

$$f_0 = \frac{1}{2\pi}\left(\frac{R_2}{R_3}\right)^{1/2}\frac{1}{R_F C_F}$$

Figure 6.31. State-variable active filter.

A. Bandpass filters

The state-variable circuit, in spite of its large number of components, is a good choice for sharp (high-Q) bandpass filters. It has low component sensitivities, does not make great demands on op-amp bandwidth, and is easy to tune. For example, in the circuit of Figure 6.31 used as a bandpass filter, the two resistors R_F set the center frequency, and R_Q and R_G together determine the Q and band-center gain:

$$R_F = 5.03 \times 10^7 / f_0 \quad \text{ohms,} \tag{6.8}$$

$$R_Q = 10^5 / (3.48Q + G - 1) \quad \text{ohms,} \tag{6.9}$$

$$R_G = 3.16 \times 10^4 Q / G \quad \text{ohms.} \tag{6.10}$$

So you could make a tunable-frequency, constant-Q filter by using a 2-section variable resistor (pot) for R_F. Alternatively, you could make R_Q adjustable, producing a fixed-frequency, variable-Q (and, unfortunately, variable-gain) filter.

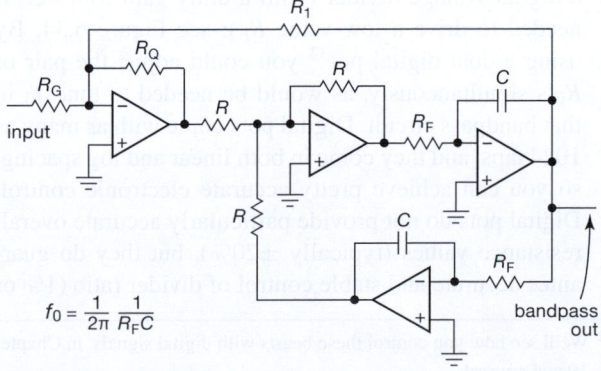

$$f_0 = \frac{1}{2\pi}\frac{1}{R_F C}$$

Figure 6.32. A filter with independently settable gain and Q.

Exercise 6.4. Calculate resistor values in Figure 6.32 to make a bandpass filter with $f_0 = 1\,\text{kHz}$, $Q = 50$, and $G = 10$.

Figure 6.32 shows a useful variant of the state-variable bandpass filter. The bad news is that it uses four op-amps; the good news is that you can adjust the bandwidth (i.e., Q) without affecting the midband gain. In fact, both Q and gain are set with a single resistor each. The Q, gain, and center frequency are completely independent and are given by these simple equations:

$$f_0 = 1/2\pi R_F C, \tag{6.11}$$

$$Q = R_1/R_Q, \tag{6.12}$$

$$G = R_1/R_G, \tag{6.13}$$

$$R \approx 10\text{k} \quad (\text{noncritical, matched}). \tag{6.14}$$

Biquad filter

A close relative of the state-variable filter is the so-called biquad filter, shown in Figure 6.33. This circuit also uses three op-amps and can be constructed from the state-variable ICs mentioned earlier. It has the interesting property that you can tune its frequency (via R_F) while maintaining constant *bandwidth* (rather than constant Q). Here are the design equations:

$$f_0 = 1/2\pi R_F C, \tag{6.15}$$

$$\text{BW} = 1/2\pi R_B C, \tag{6.16}$$

$$G = R_B/R_G. \tag{6.17}$$

The Q is given by f_0/BW and equals R_B/R_F. As the center frequency is varied (via R_F), the Q varies proportionately, keeping the bandwidth f_0/Q constant.

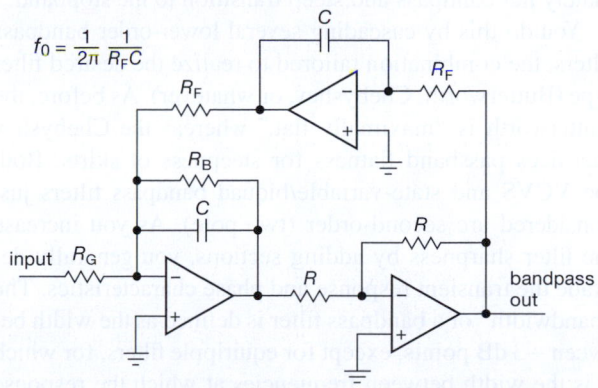

$$f_0 = \frac{1}{2\pi}\frac{1}{R_F C}$$

Figure 6.33. Biquad active filter.

When you design a biquad filter from scratch (rather than with an active-filter IC that already contains most of the parts), the general procedure goes something like this.

(1) Choose an op-amp whose bandwidth f_T is at least 10 to 20 times Gf_0.
(2) Pick a round-number capacitor value in the vicinity of $C = 10/f_0 \, \mu F$, where f_0 is in Hz.
(3) Use the desired center frequency to calculate the corresponding R_F from eq'n 6.15.
(4) Use the desired bandwidth to calculate R_B from eq'n 6.16.
(5) Use the desired band-center gain to calculate R_G from eq'n 6.17.

You may have to adjust the capacitor value if the resistor values become awkwardly large or small. For instance, in a high-Q filter you may need to increase C somewhat to keep R_B from becoming too large (or you can use the T-network trick described in §4.5.5). Note that R_F, R_B, and R_G each act as op-amp loads, and so they ought not to become less than, say, 5k. When juggling component values, you may find it easier to satisfy requirement (1) by decreasing integrator gain (increase R_F) and simultaneously increasing the inverter-stage gain (increase the 10k feedback resistor).

As an example, suppose we want to make a filter with the same characteristics as in Exercise 6.4 on page 411. We would begin by provisionally choosing $C = 0.01\mu F$. Then we find $R_F = 15.9k$ ($f_0 = 1$ kHz) and $R_B = 796k$ ($Q = 50$; BW=20 Hz). Finally, $R_G = 79.6k$ ($G = 10$).

Exercise 6.5. Design a biquad bandpass filter with $f_0 = 60$ Hz, BW=1 Hz, and $G = 100$.

B. Higher-order bandpass filters

As with our earlier lowpass and highpass filters, it is possible to build higher-order bandpass filters with approximately flat bandpass and steep transition to the stopband.

You do this by cascading several lower-order bandpass filters, the combination tailored to realize the desired filter type (Butterworth, Chebyshev, or whatever). As before, the Butterworth is "maximally flat," whereas the Chebyshev sacrifices passband flatness for steepness of skirts. Both the VCVS and state-variable/biquad bandpass filters just considered are second-order (two pole). As you increase the filter sharpness by adding sections, you generally degrade the transient response and phase characteristics. The "bandwidth" of a bandpass filter is defined as the width between -3 dB points, except for equiripple filters, for which it is the width between frequencies at which the response falls out of the passband ripple channel.

You can find tables and design procedures for constructing complex filters in standard books on active filters, or in the datasheets for active-filter ICs. There are also some very nice filter-design programs, including shareware and

freeware versions that run on standard PCs and workstations.

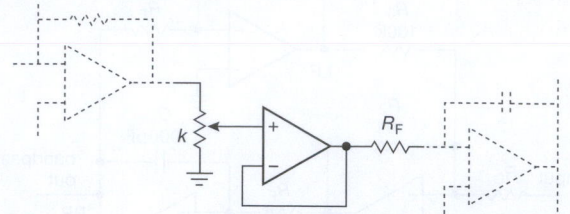

Figure 6.34. Tuning the frequency of the state-variable active filter. The op-amp buffer can be omitted if strict linearity with pot rotation is not needed.

C. Electronic tunability

Sometimes you want electrical tunability (or switchability) so you can change filter characteristics under control of a signal (rather than having to turn the shaft on a variable resistor). An example might be an anti-alias lowpass filter that precedes a digitizer, in which the digitizing rate f_{samp} can be varied over some range. In that case the filter's f_c must be set to follow the Nyquist frequency, $f_c \approx f_{samp}/2$ (see §§6.2.3C, 6.3.7A, and 13.5.1B). In active-filter circuits like the VCVS you can do this, to a limited extent, by using analog switches to select among a small set of fixed resistors, each of which substitutes for one of the resistors in the filter. But the state variable filters provide a particularly convenient way to accomplish both switchability and continuous tuning, in one of several ways.

Digital potentiometer As we discussed in §3.4.3E, you can get convenient ICs that contain a long string of matched resistors, with MOSFET switches to select the voltage-divider tap (by means of digital control[32]). So you can effectively alter the value of a programming resistor (e.g., R_F in Figure 6.31) by preceding it with such a digital voltage divider (with a unity-gain follower, if needed to drive a low-value R_F); see Figure 6.34. By using a dual digital pot[33] you could adjust the pair of R_F's simultaneously, as would be needed to tune f_0 in that bandpass circuit. Digital pots come with as many as 1024 taps, and they come in both linear and log spacing, so you can achieve pretty accurate electronic control. Digital pots do not provide particularly accurate overall resistance values (typically $\pm 20\%$), but they do guarantee accurate and stable control of divider ratio (1% or

---

[32] We'll see how you control these beasts with digital signals, in Chapter 10 and onward.

[33] They come in multiple units – singles, duals, quads, and even "sextets."

better); i.e., the resistors that make up the string are well matched. That is why they work well in this application, in which only the ratio is important.

Multiplying DAC Another way to effectively vary R_F in the state-variable filter is to use a multiplying DAC (digital-to-analog converter), rather than a programmable divider, to scale the op-amp's voltage output. The MDAC outputs a voltage (or a current, in some models) that is proportional to the product of an analog input voltage and a digital input quantity. Compared with the digital pot, the MDAC method provides higher resolution (finer step size), faster response, and (often) wider voltage range.

Analog switch If only a discrete set of filter parameters is desired, you can just use a set of MOSFET analog multiplexers to select among a preselected group of programming resistors. Don't forget to consider the effects of finite R_{ON}.

Integrated switchability There are a few active-filter ICs that provide for programmable cutoff frequency, by a digital code you apply to a set of programming pins. You don't get continuous control, but you sure save a lot of work (and a lot of parts). In this class are the LTC1564 (8-pole elliptic lowpass), which lets you select the cutoff frequency from 10 kHz to 150 kHz in 10 kHz steps, and the MAX270 (dual 2-pole lowpass), which lets you select the cutoff frequency among 128 steps going from 1 kHz to 25 kHz.

Electronic tuning alternatives: switched-capacitor filters and DSP

The above techniques achieve electronic tunability by presenting the continuous-time filter with an effectively variable set of programming resistors. When thinking about electronic tuning, it's wise to consider switched-capacitor filters and digital signal processing (DSP), in both of which electronic tunability is inherent. These are discussed later in this chapter (§6.3.6 and 6.3.7).

D. Multiple-feedback active filter

In addition to the VCVS (Sallen-and-Key) and state-variable (or biquad) active-filter circuit configurations, there's another active-filter circuit that's commonly used. It's called the "multiple-feedback" (MFB) active filter (also known as the "infinite-gain multiple-feedback"), and is shown in Figure 6.35. Here the op-amp is configured as an integrator, rather than as a voltage amplifier (or follower). Designing an MFB filter is no more difficult than designing a VCVS, and you can find nice filter software that supports both configurations, for example at the very fine website of

Uwe Beis (see §6.3.8). You can get nice MFB filter ICs, for example the LTC1563, an inexpensive ($2.30) linear-filter IC using the MFB configuration, convenient for making anti-alias filters, etc. The '1563-2 version makes 4- and 5-pole Butterworth filters, from 256 Hz to 360 kHz, and the -3 version makes Bessel filters. The ICs use internal 27 pF to 54 pF capacitors trimmed to 3%, combined with your external 7k to 10M 1% resistors. The datasheet is especially instructive.

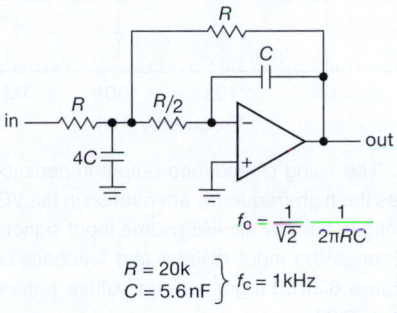

$$f_o = \frac{1}{\sqrt{2}} \frac{1}{2\pi RC}$$

$$\left. \begin{array}{l} R = 20\text{k} \\ C = 5.6\,\text{nF} \end{array} \right\} f_c = 1\,\text{kHz}$$

Figure 6.35. Multiple-feedback (MFB) active filter, here shown in a 2-pole lowpass configuration.

This configuration has an interesting advantage compared with the VCVS: as you go to high frequencies, approaching the bandwidth f_T of the op-amp, the degrading effects of rising op-amp output impedance are less severe. We ran SPICE simulations of VCVS and MFB 2-pole Butterworth lowpass filters (Figures 6.36 and 6.37), which show this effect nicely. We set the cutoff frequency at 4 kHz, well below the LF411's unity-gain frequency (f_T) of 4 MHz. In the VCVS configuration the op-amp's rising Z_{out} allows input signal to couple to the output through the first capacitor, a path that is absent in the MFB configuration.[34] In many applications, however, this is not a serious worry. And the effect is reduced as the resistor values of the filter are increased, as shown in Figure 6.36. The VCVS configuration is alive and well, and remains popular.[35]

---

[34] This effect is illustrated in the Texas Instruments documents "Analysis of the Sallen–Key Architecture" (SLOA024B) and "Active Low-Pass Filter Design" (SLOA049A), both by James Karki, and in Dave Van Ess' article in the EN-Genius analogZONE "What Sallen–Key Filter Articles Don't Tell You."

[35] A simple way to mitigate poor cutoff at high frequencies is to add a passive RC section at the active filter's output; for example, 200 Ω and 7.5 nF forms an additional 100 kHz cutoff stage.

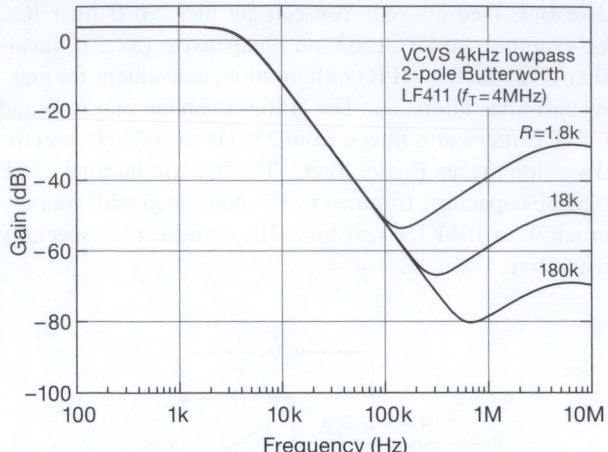

Figure 6.36. The rising closed-loop output impedance of the op-amp degrades the high-frequency attenuation in the VCVS (Sallen-and-Key) configuration, by allowing some input signal to couple to the output through the input resistor and feedback capacitor (R_1 and C_1 in Figure 6.28). Larger resistor values reduce the effect. See also Figure 6.37.

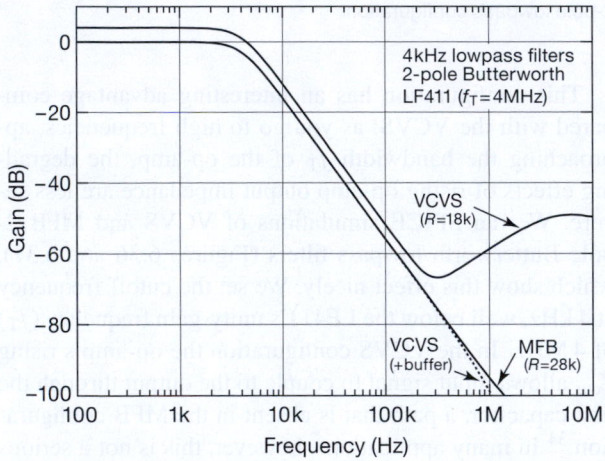

Figure 6.37. The stopband attenuation of the MFB configuration is not much affected by rising op-amp output impedance (e.g., as seen in Figure 4.53), compared with that of the VCVS. However, you can mitigate the effect in the VCVS by using a second op-amp to create a buffered output from the signal at the op-amp's noninverting input.

6.3.4 Twin-T notch filters

The passive *RC* network shown in Figure 6.38 has infinite attenuation at a frequency $f_c = 1/2\pi RC$. Infinite attenuation is uncharacteristic of *RC* filters in general; this one works by effectively adding two signals that have been shifted 180° out of phase at the cutoff frequency. It requires

good matching of components to obtain a good null at f_c. It is called a twin-T, and it can be used to remove an interfering signal, such as 60 Hz powerline pickup. The problem is that it has the same "soft" cutoff characteristics as all passive *RC* networks, except, of course, near f_c, where its response drops like a rock. For example, a twin-T driven by a perfect voltage source is down 10 dB at twice (or half) the notch frequency and 3 dB at four times (or one-fourth) the notch frequency. One trick to improve its notch characteristic is to "activate" it in the manner of a Sallen-and-Key filter (Figure 6.39). This technique looks good in principle, but it is generally disappointing in practice, owing to the impossibility of maintaining a good filter null. As the filter notch becomes sharper (more gain in the bootstrap), its null becomes less deep.

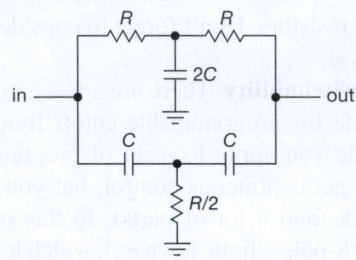

Figure 6.38. Passive twin-T notch filter.

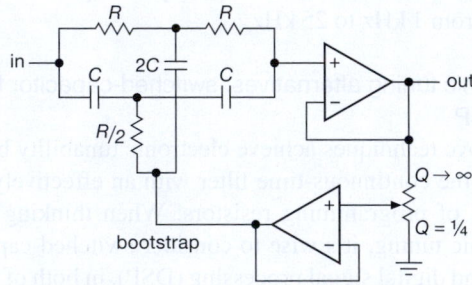

Figure 6.39. Bootstrapped twin-T.

Twin-T filters are available as prefab modules, going from 1 Hz to 50 kHz, with notch depths of about 60 dB (with some deterioration at high and low temperatures). They are easy to make from components, but resistors and capacitors of good stability and low temperature coefficient should be used to get a deep and stable notch. One of the components should be made trimmable.

The twin-T filter works fine as a fixed-frequency notch, but it is a horror to make tunable, because three resistors must be simultaneously adjusted while maintaining constant ratio. However, the remarkably simple *RC* circuit of Figure 6.40A, which behaves just like the twin-T, can be adjusted over a significant range of frequency (at least two octaves) with a single potentiometer. Like the twin-T (and most active filters) it requires some matching of components; in this case, the three capacitors must be identical, and the fixed resistor must be exactly six times the bottom (adjustable) resistor. The notch frequency is then given by

$$f_{notch} = 1/2\pi C\sqrt{3R_1 R_2}.$$

Figure 6.40B shows an implementation that is tunable from 25 Hz to 100 Hz. The 50k trimmer is adjusted (once) for maximum depth of notch.

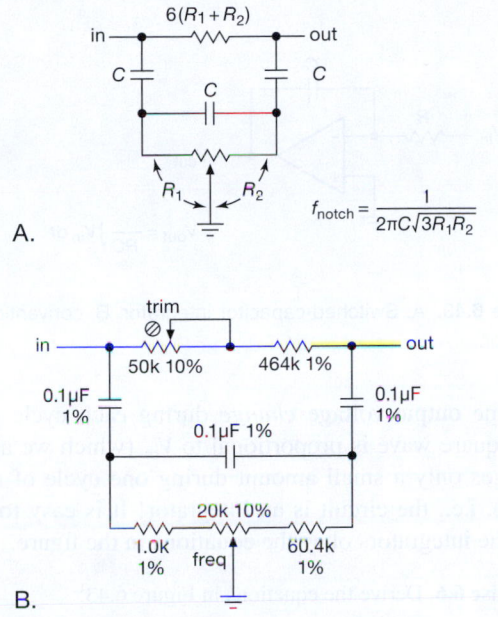

A.

B.

Figure 6.40. Bridged differentiator tunable-notch filter. The implementation in (B) tunes from 25 Hz to 100 Hz.

As with the passive twin-T, this filter (known as a *bridged differentiator*) has a gently sloping attenuation away from the notch and infinite attenuation (assuming perfect matching of component values) at the notch frequency. It, too, can be "activated" by bootstrapping the wiper of the pot with a voltage gain somewhat less than unity (as in Figure 6.39). Increasing the bootstrap gain toward unity narrows the notch, but also leads to an undesirable response peak on the high-frequency side of the notch, along with a reduction in ultimate attenuation.

6.3.5 Allpass filters

Allpass? *All*pass?! Whatever can that be? And why would you want such a thing, when a piece of wire does as well (and probably better)?

Allpass filters, also known as *delay equalizers* or *phase equalizers*, are filters with flat *amplitude* response, but with a phase shift that varies with frequency. They are used to compensate for phase shifts (or time delays) in some signal path.

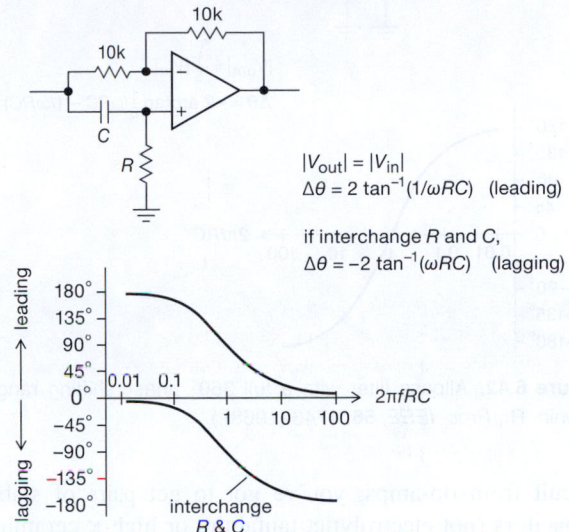

Figure 6.41. Allpass filter, also known as a *delay equalizer*, or *phase equalizer*.

Figure 6.41 shows the basic circuit configuration. Intuitively, it's easy to see that the circuit behaves like an inverter at low frequencies (where no signal is coupled to the noninverting input) and a follower at high frequencies (recall the optional inverter of §4.3.1A). By writing a few equations you can convince yourself that the circuit behaves as described in the figure. Interchanging R and C produces a similar characteristic, but with lagging (rather than leading) phase shifts between the extremes of inverting and following. The phase shift can be tuned by making R variable; but note that a small value of R makes the circuit's input impedance small at high frequencies (where the reactance of C goes to zero).

Figure 6.42 shows a variant that extends the range of phase shift to a full 360°. The downside is that you have to adjust two components simultaneously (e.g., the pair of equal-value resistors) to change its tuning. This can be done nicely, though, by using a digital dual potentiometer (an "EEpot") of the sort described in §3.4.3E.

6.3.6 Switched-capacitor filters

One drawback to these state-variable or biquad filters is the need for accurately matched capacitors. If you build the

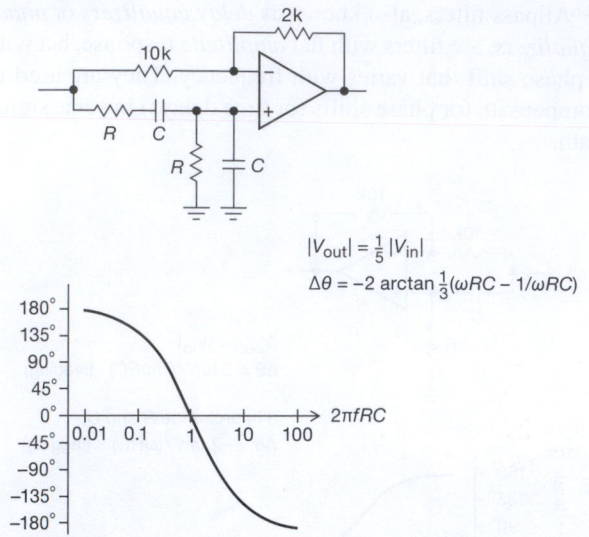

$$|V_{out}| = \tfrac{1}{5}|V_{in}|$$

$$\Delta\theta = -2\arctan\tfrac{1}{3}(\omega RC - 1/\omega RC)$$

Figure 6.42. Allpass filter with a full 360° phase shifting range. (Genin, R., *Proc. IEEE*, **56**, 1746 (1968).)

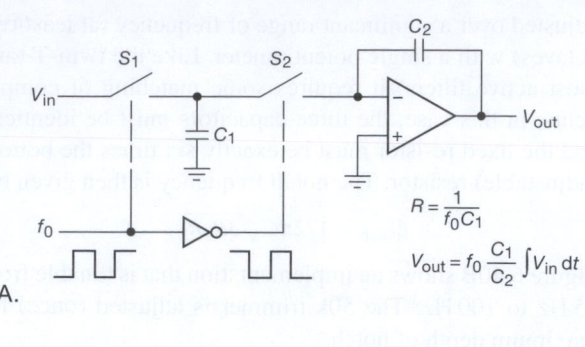

$$R = \frac{1}{f_0 C_1}$$

$$V_{out} = f_0 \frac{C_1}{C_2}\int V_{in}\,dt$$

A.

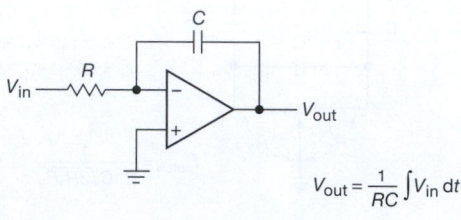

$$V_{out} = \frac{1}{RC}\int V_{in}\,dt$$

B.

Figure 6.43. A. Switched-capacitor integrator. B. conventional integrator.

circuit from op-amps, you've got to get pairs of stable capacitors (not electrolytic, tantalum, or high-κ ceramic), perhaps matched better than 1% for optimum performance. You also have to make a lot of connections, since the circuits use at least three op-amps and six resistors for each 2-pole section. Alternatively, you can buy a filter IC, letting the manufacturer figure out how to integrate matched 1000 pF ($\pm0.5\%$) capacitors into an IC. Semiconductor manufacturers have solved those problems, but at a price: the UAF42 and MAX274 "Universal Active Filter" ICs (mentioned earlier), implemented with hybrid or laser-trim technology, cost about $8–$16 apiece. These "continuous-time" filters also do not lend themselves to easy tunability.

A. Switched-capacitor integrator
There's another way to implement the integrators that are needed in the state-variable or biquad filter configurations. The basic idea is to use MOSFET analog switches, clocked from an externally applied square wave at some high frequency (typically 100 times faster than the analog signals of interest), as shown in Figure 6.43. In the figure, the funny triangular object is a digital *inverter*, which turns the square wave upside down so that the two MOS switches are closed on opposite halves of the square wave.

The circuit is easy to analyze: when S_1 is closed, C_1 charges to V_{in}, i.e., holding charge $C_1 V_{in}$. On the alternate half of the cycle, C_1 discharges into the virtual ground, transferring its charge to C_2. The voltage across C_2 therefore changes by an amount $\Delta V = \Delta Q/C_2 = V_{in}C_1/C_2$. Note

that the output-voltage *change* during each cycle of the fast square wave is proportional to V_{in} (which we assume changes only a small amount during one cycle of square wave), i.e., the circuit is an integrator! It is easy to show that the integrators obey the equations in the figure.

Exercise 6.6. Derive the equations in Figure 6.43.

Exercise 6.7. Here's another way to understand the switched-capacitor integrator: calculate the average current that flows through S_2 into the virtual ground. You should find that it is proportional to V_{in}. Therefore, the combination of S_1, C_1, and S_2 behaves like a resistor, forming a classic integrator. What is the value of that equivalent resistance, in terms of f_0 and C_1? Use that to arrive at the equation in the figure, $V_{out} = f_0(C_1/C_2)\int V_{in}dt$.

B. Advantages of switched-capacitor filters
There are two important advantages to using switched capacitors instead of conventional integrators. First, as hinted earlier, it can be less expensive to implement on silicon: the integrator gain depends only on the *ratio* of two capacitors, not on their individual values. In general, it is easy to make a matched pair of anything on silicon, but very hard to make a similar component (resistor or capacitor) of precise value and high stability. As a result, monolithic switched-capacitor filter ICs are inexpensive – TI's universal switched-capacitor filter (the MF10) costs $3.50

(compared with $16 for the conventional UAF42), and furthermore it gives you *two* filters in one package.

The second advantage of switched-capacitor filters is the ability to tune the filter's characteristic frequency (e.g., the center frequency of a bandpass filter, or the -3 dB point of a lowpass filter) by merely changing the frequency of the square-wave ("clock") input.[36] This is because the characteristic frequency of a state-variable or biquad filter is proportional to (and depends only on) the integrator gain.

C. Switched-capacitor filter configurations

Switched-capacitor filters are available in both dedicated and "universal" configurations. The former are prewired with on-chip components to form lowpass filters of the desired type (Butterworth, Bessel, Elliptic), whereas the latter have various intermediate inputs and outputs brought out so you can connect external components to make anything you want. The price you pay for universality is a larger IC package and the need for external resistors. For example, LTC's self-contained LTC1069-6 8-pole elliptic lowpass filter comes in an 8-pin package (about $9), compared with their LTC1164 quad 2-pole universal filter which requires 12 external resistors to implement a comparable filter and comes in a 24-pin package (about $15). Figure 6.44 shows just how easy it is to use the dedicated type. Look ahead to §7.1.5A to see an elegant and simple sinewave generator that uses a tracking switched-capacitor filter acting on a square wave at a fraction of the clock frequency (Figures 7.18 and 7.19).

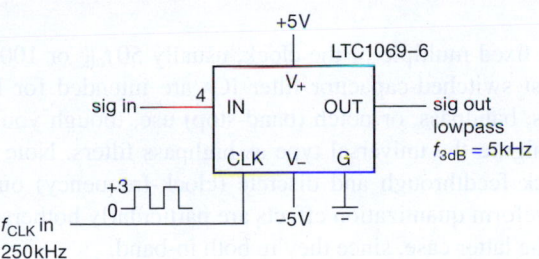

Figure 6.44. Switched capacitor dedicated lowpass filter, requiring no external components. The 8th-order elliptic response has ±0.1 dB passband ripple, and is more than 40 dB down at $1.3 f_{3dB}$.

Both dedicated and universal switched-capacitor filters use as the basic building block the 2-pole state-variable configuration, with switched-capacitor integrators replacing the resistor-fed op-amp integrators of the classic continuous-time state-variable active filter; see Fig-

ure 6.45. The universal filter ICs come with one to four such sections, which can be cascaded to form a higher-order filter (with each section implementing a quadratic term in the factored filter equation), or they can be used independently for multiple simultaneous channels (which must, however, share the common clocking input). The manufacturer's datasheets (or software or both) make filter design easy with these universal filter ICs.[37] And no design is needed at all for the dedicated filter – you just connect it up, and off you go.

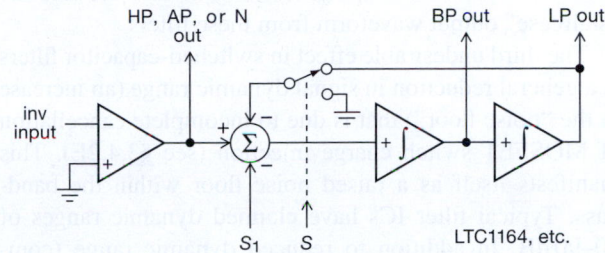

Figure 6.45. "Universal" second-order switched-capacitor building block. It can provide lowpass, highpass, bandpass, allpass, and notch, as determined by external connections. With its on-chip capacitors, the only external components required are a few resistors.

D. Drawbacks of switched-capacitor filters

Now for the bad news: switched-capacitor filters have three annoying characteristics, all related and caused by the presence of the periodic clocking signal. First, there is *clock feedthrough*, the presence of some output signal (typically about 10 mV to 25 mV) at the clock frequency, independent of the input signal. Usually this doesn't matter, because it is far removed from the signal band of interest. If clock feedthrough is a problem, a simple *RC* filter at the output usually gets rid of it.

The second problem is more subtle: if the input signal has any frequency components near the clock frequency, they will be "aliased" down into the passband. To state it precisely, any input signal energy at a frequency that differs from the clock frequency by an amount corresponding to a frequency in the passband will appear (unattenuated!) in the passband. For example, if you use a MAX7400 (dedicated 8-pole elliptic lowpass) as a 1 kHz lowpass filter (i.e., set $f_{clock} = 100$ kHz), any input signal energy in the

---

[36] Most switched-capacitor filters are configured with their characteristic frequency equal to 1/50 or 1/100 of the clock frequency.

[37] You simply specify characteristic frequencies, stopband attenuation, passband ripple, and gain; it tells you how many poles and gives you resistor values. The software packages give you plots of attenuation, phase, and delay versus frequency (on your choice of axes), and numeric tables as well.

range of 99–101 kHz will appear in the output band of dc–1 kHz. No filter at the output can remove it! You must make sure the input signal doesn't have energy near the clock frequency. If this isn't naturally the case, you can usually use a simple *RC* filter, because the clock frequency is typically quite far removed from the passband. The use of filter ICs with a high clock-to-corner frequency ratio (e.g., 100:1 instead of 25:1 or 50:1) simplifies input anti-alias filter design. You can get some nice filter ICs with 1000:1 clock ration from Mixed Signal Integration, for example their MSHN series.[38] A high clock ratio also reduces the "staircase" output waveform from these filters.

The third undesirable effect in switched-capacitor filters is a general reduction in signal dynamic range (an increase in the "noise floor") that is due to incomplete cancellation of MOSFET switch charge injection (see §3.4.2E). This manifests itself as a raised noise floor within the bandpass. Typical filter ICs have claimed dynamic ranges of 80–90 dB. In addition to reduced dynamic range (compared with continuous-time filters), switched-capacitor filters tend to have more distortion than you would expect, especially for output signals near the supply rails.

Like any linear circuit, switched-capacitor filters (and their op-amp analogs) suffer from amplifier errors such as input offset voltage and $1/f$ low-frequency noise. These can be a problem if, for example, you wish to lowpass filter some low-level signal without introducing errors or fluctuations in its average dc value. A nice solution is provided by the clever folks at Linear Technology, who dreamed up the LTC1062 "DC accurate lowpass filter" (or the MAX280, with improved offset voltage). Figure 6.46 shows how you use it. The basic idea is to put the filter outside the dc path, letting the low-frequency signal components couple passively to the output; the filter grabs onto the signal line only at higher frequencies, where it rolls off the response by shunting the signal to ground. The result is zero dc error, and switched-capacitor-type noise only in the vicinity of the rolloff[39] (Figure 6.47). You can cascade a pair of these filters to make higher-order filters, or a tunable sharp bandpass filter. The datasheet also shows you how to make a tunable notch filter.

Switched-capacitor filter ICs are widely available from manufacturers such as Linear Technology, TI, and Maxim. Typically you can put the cutoff (or band center) anywhere in the range of dc to a few tens of kilohertz, as set by the clock frequency. The characteristic frequency

[38] See the MSHN5 in action in John Ambrose's "Notch filter autotunes for audio applications," *EDN* Design Ideas, June 24, 2010.

[39] The same trick is used in multistage GIC-implemented FDNR filters.

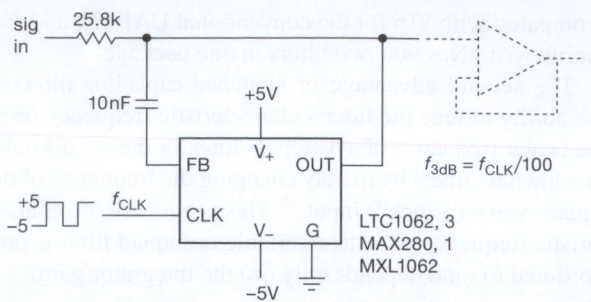

Figure 6.46. LTC1062 "dc-accurate" lowpass filter. The external clock input must swing rail-to-rail (add a small series resistor to protect the input); alternatively you can enable the internal oscillator by connecting a capacitor from CLK to ground.

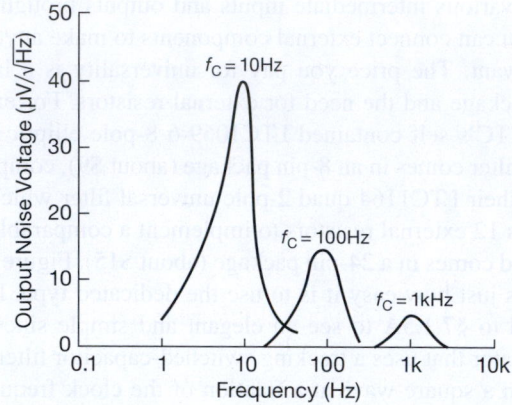

Figure 6.47. LTC1062 output noise spectra (see the datasheet).

is a fixed multiple of the clock, usually $50f_{clk}$ or $100f_{clk}$. Most switched-capacitor filter ICs are intended for lowpass, bandpass, or notch (band-stop) use, though you can configure the universal type as highpass filters. Note that clock feedthrough and discrete (clock-frequency) output waveform quantization effects are particularly bothersome in the latter case, since they're both in-band.

6.3.7 Digital signal processing

Our discussion of electronic filters in this chapter would be seriously incomplete without an introduction to the widespread technique of *digital signal processing* (DSP), also known as *discrete-time signal processing*. Contemporary systems that include microprocessors favor digital-filtering methods for their flexibility and performance. Digital signal processing is the manipulation of signals in the digital domain, in which a signal (for example, a speech waveform) has been converted to a sequence of numbers representing its sampled amplitude values at equally spaced intervals of time. The "manipulations" can be any of the things we've seen in the purely analog domain – filtering, combining, attenuation or amplification, nonlinear compression and clipping, and so on; but they can include also additional sophisticated operations made possible by the power of computation, such as coding, error correction, encryption, spectral analysis, speech synthesis and analysis, image processing, adaptive filtering, and lossless compression and storage.

We'll have much to say about digitizing and processing in Chapters 13 ("Digital meets analog") and 15 ("Microcontrollers"), when we'll have, respectively, the electronic tools for conversion between analog voltages and their digital representation, and the means to process those digital quantities. Here we would like simply to introduce the application of DSP to filtering and give a glimpse of its capabilities, particularly when compared with the analog filters we've seen. We'll stick to one-dimensional filters; that is, the filtering of "one-dimensional" signals such as speech (as contrasted with two-dimensional images), which are characterized by some voltage waveform $V(t)$ evolving in time.

A. Sampling

We mentioned earlier that a digitized representation of a continuous waveform involves sampling at a discrete set of (nearly always) uniformly spaced times, with a discrete set of (usually) uniformly spaced quantized amplitudes. These determine the fidelity of the quantization — in frequency (from the sampling rate, obeying the Nyquist sampling criterion, see Figure 13.60) and in dynamic range and noise (from the quantization precision); see §13.5.1. There's a lot to say about these; but at the most basic level you've got to sample at least twice the rate of the highest-frequency component that's in the input, and you've got to have enough precision in the n-bit amplitude quantization to preserve the dynamic range you want. Put compactly,

$$f_{\text{samp}} \geq 2f_{\text{sig}}(\text{max}),$$
$$\text{dynamic range} = 6n \text{ dB}.$$

Assuming you're beginning with an analog waveform, the sampling is done with an analog-to-digital converter (ADC), preceded by an anti-alias lowpass filter (LPF), if needed, to ensure that the waveform being digitized contains no signals of significance above the Nyquist frequency $f_{\text{samp}}/2$.

B. Filtering

The sequence of adequately sampled amplitudes (call it x_n, for the nth sample) represents the input signal. We want to do a filtering operation on the sequence, for example a lowpass filter. There are two broad classes of DSP filters: finite-impulse-response (FIR) and infinite-impulse-response (IIR). The FIR is easiest to understand – each output sample is simply a weighted sum of some number of input samples (see Figure 6.48):

$$y_i = \sum_{k=-\infty}^{\infty} a_k x_{i-k},$$

where the x_i are the input signal amplitudes, the a_k are the weights, and the y_i are the output of the filter. In real life there will be only a finite number of weights, and so the sum will run only over a finite set of input values, as in the figure. Speaking crudely, the set of coefficients is an approximation to the inverse Fourier transform of the desired filter function.

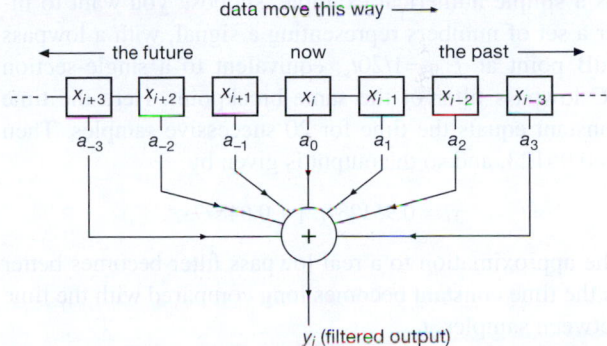

Figure 6.48. Finite impulse response (nonrecursive) digital filter.

Note an interesting – and important – feature of such a filter: its output is formed of samples both past and *future*. That is, it can generate an output that would appear to violate causality (effect must follow cause), but which is permitted here because the output signal has an overall delay with respect to the input. This ability to see into the future (a boast that no analog filter can make) allows digital filters to implement frequency- and phase-response characteristics that cannot be achieved with the (causal) analog filters we've seen up to this point.

The IIR filter differs in allowing the output to be included, with some weighting factor, along with the inputs in the weighted sum; this is sometimes called a *recursive* filter. The simplest example might be

$$y_i = by_{i-1} + (1-b)x_i,$$

which happens to be the discrete approximation to a continuous-time *RC* lowpass filter, in which the weighting factor b is given by $b = e^{-t_s/RC}$, where t_s is the sampling interval. Of course, the situation is not *identical* to an analog lowpass filter operating on an analog waveform because of the discrete nature of the sampled waveform.

Both FIR and IIR implementations have their pros and cons. FIR filters are generally preferred because they are simple to understand, easy to implement, unconditionally stable (no feedback), and they can be (and usually are) designed as linear-phase filters (i.e., the time delay is constant, independent of frequency). IIR filters are more economical, however, requiring fewer coefficients and therefore less memory and calculation. They also are easily derived from the corresponding classic analog filter; and they are particularly well suited to applications requiring high selectivity, for example, notch filters. However, they require more bits of arithmetic precision to prevent instabilities and "idle tones," and they are more difficult to code.

C. An example: IIR lowpass

As a simple numerical example, suppose you want to filter a set of numbers representing a signal, with a lowpass 3 dB point at $f_{3dB}=1/20t_s$, equivalent to a single-section *RC* lowpass filter of the same breakpoint. Here the time constant equals the time for 20 successive samples. Then $A=0.95123$, and so the output is given by

$$y_i = 0.95123y_{i-1} + 0.04877x_i \; .$$

The approximation to a real lowpass filter becomes better as the time constant becomes long compared with the time between samples, t_s.

You would probably use a filter like this to process data that are already in the form of discrete samples, e.g., an array of data in a computer. In that case the recursive filter becomes a trivial arithmetic pass once through the data.

D. An example: FIR lowpass

An ideal lowpass filter has unit response up to its cutoff frequency f_c and zero response for higher frequencies. That is, the response curve is rectangular, a "brick-wall" filter. To first order, the FIR coefficients a_k are the rectangle's Fourier transform, namely a $(\sin x)/x$ function (or *sinc* function), in which the scaling of the argument depends on the ratio of the cutoff frequency to the sampling frequency, namely,

$$a_k \propto \frac{\sin(2\pi k f_n)}{2\pi k f_n} \tag{6.18}$$

where the integers k go from $-\infty$ to ∞, and f_n is the normalized cutoff frequency, defined as $f_n = f_c/f_s$.

In a real-world implementation, of course, you get only a finite number of k's, say N of them. So the question: what set of truncated filter coefficients a_k, where k goes only from $-N/2$ to $N/2$, best approximates the ideal lowpass filter? This turns out to be more complicated than you

might at first imagine. Among other things, it depends on what you mean by "best."

If you simply truncate the a_k series, discarding coefficients beyond the length of your FIR sample string, the resulting filter's frequency response will exhibit large bumps in the stopband attenuation; that is, degraded rejection around those frequencies. This is exactly analogous to the problem of "spectral leakage" in digital spectrum analysis, or of diffraction sidelobes in optics, and the fix is the same: here you taper the coefficients a_k by multiplying them by a "window function" that goes smoothly toward zero at the ends (in spectrum analysis you multiply the incoming digitized signal amplitudes by an analogous windowing function, and in optics you "apodize" the aperture with a mask whose opacity increases toward the edges). The effect is to reduce greatly the stopband ripple, at the expense of a more gradual transition from passband to stopband (in spectral analysis the effect is greatly reduced spectral leakage into adjacent frequency bins, at the expense of broader bin width; in optics the sidelobes are attenuated, at the expense of decreased resolution in the form of a broader "point-spread function"). Typical window functions have names like Hamming, Hanning, and Blackman–Harris. There's no "best" window – it's always a tradeoff between the steepness of transition to the stopband versus the worst-case attenuation in the stopband. But most of the time it doesn't really matter a whole lot which of the standard windows you use.[40]

A second aspect of "best" is to choose a cutoff frequency f_n for which at least some of the coefficients are exactly zero; that way you can omit the multiply and add operations corresponding to those taps. This occurs, for example, with the choice $f_n=0.25$ (a sampling rate four times the cutoff frequency), for which the coefficients in eq'n 6.18 become

$$a_k \propto \frac{\sin(\pi k/2)}{\pi k/2}, \tag{6.19}$$

and therefore all the coefficients with even k (except a_0) are zero. You get a small bonus, also, by using a filter length N that is a multiple of 4; that makes the end coefficients (at $k = \pm N/2$) vanish, because their index k is then even.

Because a cutoff frequency of half the sampling frequency (i.e., $f_n=0.5$) is the maximum allowed by the Nyquist sampling theorem, a filter whose cutoff is at $f_n=0.25$ is known as a "half-band" filter. Figures 6.49 and 6.50 show the response of half-band filters with $N=8$, 16, 32, and 64, where the coefficients have been calculated

---

[40] To learn more than you ever wanted to know about window functions, see F. J. Harris, *Proc. IEEE*, **66**, 51–83 (1978).

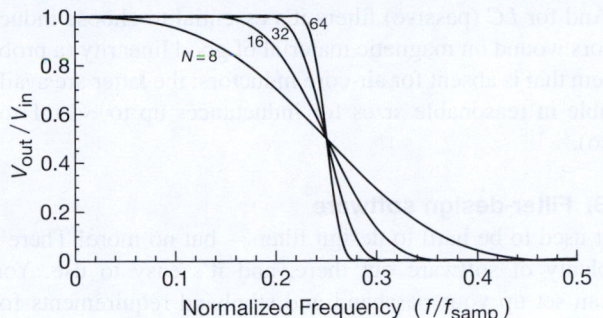

Figure 6.49. Half-band FIR digital filter response, plotted on a linear scale. A filter of order N requires $N/2+1$ coefficients.

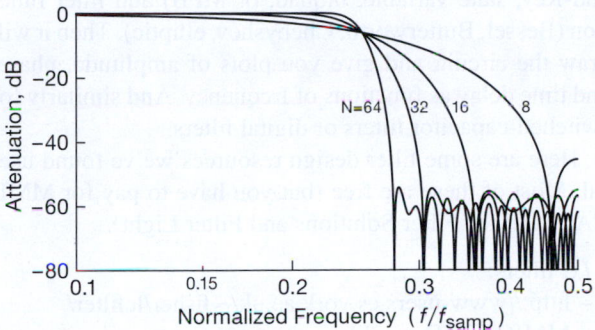

Figure 6.50. The half-band filters of Figure 6.49, plotted on a log–log scale to reveal the stopband response.

according to eq'n 6.19, weighted by a Hamming window. The latter is a raised cosine, given approximately by

$$w(k) = 0.54 + 0.46\cos(2\pi k/N). \qquad (6.20)$$

A final step is to normalize the coefficients (by multiplying each by the same factor) so that their sum is 1, giving the filter unit gain at dc. The recipe, then, is to choose an N (preferably a multiple of 4); then, for each positive odd k up to $N/2$, calculate the sinc function of equation (6.19) and multiply it by the Hamming coefficient of eq'n 6.20 to get the (not yet normalized) a_k. Note that the coefficients are symmetric ($a_{-k} = a_k$), and that the a_0 term will be 1.0 (because both sinc(0) and $w(0)$ have value unity). The final step is to normalize these coefficients by dividing each by their sum.

Because the even coefficients are zero, the resultant filters, though requiring N stages of memory, need approximately half that number of coefficients (those with odd subscripts, plus a_0), namely 5, 9, 17, and 33, respectively. You can check our recipe by calculating the coefficients for

the lowest-order filter in the figures ($N=8$); you should get

$$a_0 = +0.497374$$
$$a_1 = a_{-1} = +0.273977$$
$$a_2 = a_{-2} = 0$$
$$a_3 = a_{-3} = -0.022664$$
$$a_4 = a_{-4} = 0$$

An aside: window tradeoffs

We used a Hamming window as the coefficient multiplier for the filters of Figures 6.49 and 6.50, partly out of laziness (it's easy to calculate), and partly because it's a reasonably good window in terms of stopband attenuation ($\sim$60 dB). But, as we remarked above, you can produce better stopband attenuation at the expense of transition-region steepness. This is nicely illustrated in Figure 6.51, where we've redone the $N=32$ half-band lowpass FIR filter using three different window functions. The Blackman–Harris windows are a sum of two or three sinusoidal terms, weighted to produce the minimum sidelobe level. The exact form is

$$w(k) = a_0 + a_1 \cos(2\pi k/N) + a_2 \cos(4\pi k/N) + a_3 \cos(6\pi k/N),$$

where the a's are given by $[a_0, a_1, a_2, a_3]=[0.42323, 0.49755, 0.07922, 0]$ (3-term) and $[0.35875, 0.48829, 0.14128, 0.01168]$ (4-term). These windows produce impressive stopband attenuation ($\sim$85 dB and $\sim$105 dB), as compared with the Hamming's $\sim$55 dB, but with correspondingly softer transition regions. It's worth noting that these are *calculated* responses, and will be realized in practice only if the FIR multiply and add operations are done with adequate arithmetic precision, and if the upstream ADC has correspondingly accurate linearity.

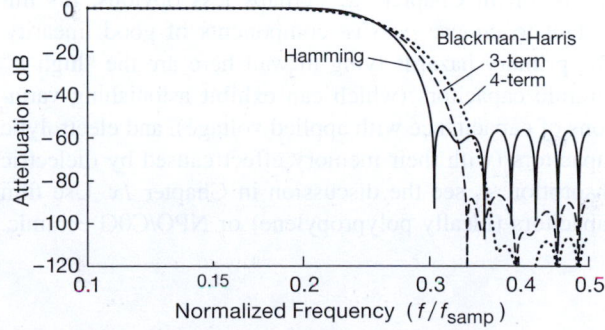

Figure 6.51. Half-band FIR lowpass filters of order $N = 32$, with three choices of coefficient window function. Note the change of vertical scale compared with those of Figures 6.49 and 6.50.

E. Implementation

You *could* set up a DSP filter with discrete hardware – shift registers, multipliers, accumulators, and the like – the stuff of Chapters 10 and 11. But any such attempt would seem quaint by current standards, where general-purpose processors (microprocessors and microcontrollers) let you do the same tasks, and with greater flexibility. Even better, there is a class of digital signal processor chips, optimized for the sort of multiply–accumulate operations you need to do, and generally arranged for efficient flow of lots of data in and out. An example is the TMS320 series from TI, which includes (at time of writing) chips like the TMS320C64xx, which can do a 1k-point fast Fourier transform (FFT) in about $1\,\mu s$ (!), or a 32-coefficient FIR on a 10,000-point data set in $108\,\mu s$. At the other end of the performance scale, the tiny QF1D512 from Quickfilter Technology is an inexpensive stand-alone chip that performs up to a 512-tap FIR filter on 12- to 24-bit serial data (at audio rates), with 32-bit programmable coefficients. It costs less than $2 in small quantities and comes with free design software; you can also get a variety of evaluation kits.

6.3.8 Filter miscellany

A. Linearity

In some filtering applications it is essential to maintain a high degree of amplitude linearity, even as the filter attenuates some frequencies more than others. This is necessary, for example, in high-quality audio reproduction. For such applications you should use op-amps designed for low distortion (which will be prominently featured on the datasheet), with adequate bandwidth, slew rate, and loop gain; some examples are the LT1115, OPA627, and AD8599; see Table 5.4 on page 310 and the discussion of high-speed op-amps and design issues in §5.8, and further discussion in Chapter *4x*. Perhaps less obvious, it's important to choose *passive* components of good linearity. The primary hazards lying in wait here are the "high-κ" ceramic capacitors (which can exhibit astonishing variations of capacitance with applied voltage), and electrolytic capacitors (with their memory effect caused by dielectric absorption – see the discussion in Chapter *1x*. Use film capacitors (ideally polypropylene) or NPO/C0G ceramic.

And for *LC* (passive) filters it's essential to choose inductors wound on magnetic material of good linearity (a problem that is absent for air-core inductors; the latter are available in reasonable sizes for inductances up to ~1 mH or so).

B. Filter-design software

It used to be hard to design filters – but no more! There's plenty of software out there, and it's easy to use. You can set up your passband and stopband requirements for a continuous-time filter (cutoff frequency, stopband frequency, ripple and attenuation, etc.), and the software obliges with a recitation of how many sections will be required, according to the circuit configuration (Sallen-and-Key, state variable, biquad, or MFB) and filter function (Bessel, Butterworth, Chebyshev, elliptic). Then it will draw the circuit, and give you plots of amplitude, phase, and time delay as functions of frequency. And similarly for switched-capacitor filters or digital filters.

Here are some filter design resources we've found useful. Most of these are free (but you have to pay for MMI-CAD, and for Filter Solutions and Filter Light).

- *LC* filters:
  - http://www-users.cs.york.ac.uk/~fisher/lcfilter/
  - MMICAD (Optotek)
- Analog active filters
  - FilterPro (TI)
  - FilterCAD (LTC)
  - ADI Analog Filter Wizard
  - http://www.beis.de/Elektronik/Filter/Filter.html
- Digital filters:
  - http://www-users.cs.york.ac.uk/~fisher/mkfilter/
- All types
  - Filter Solutions, Filter Light, and Filter Free (http://www.nuhertz.com/filter/)

Additional Exercises for Chapter 6

Exercise 6.8. Design a VCVS 6-pole highpass Bessel filter with cutoff frequency of 1 kHz.

Exercise 6.9. Design a 60 Hz twin-T notch filter with op-amp input and output buffers.

Review of Chapter 6

An A-to-J summary of what we have learned in Chapter 6. This summary reviews basic principles and facts in Chapter 6, but it does not cover application circuit diagrams and practical engineering advice presented there.

¶A. Filter Overview.

This chapter deals with signals in the frequency domain: by *filter* we mean a circuit with some deliberate passband and attenuation characteristics (both amplitude and phase) versus frequency. For some applications the filter's time-domain behavior is important also, i.e., the filter's transient response (overshoot and settling time) to a voltage-step input, and its in-band fidelity to an input waveform.

¶B. Filter Characteristics.

There are the basic shapes – *lowpass*, *highpass*, *bandpass*, and *band-stop* (notch). There's also the *all-pass* (or *delay equalizer*), which has a flat amplitude response, but a varying phase; and there are *comb* filters that pass (or block) an array of equally spaced frequencies. Important frequency-domain parameters include flatness of response in the passband, depth of attenuation in the stopband, and steepness of falling response in the transition region between. In the time-domain you care about overshoot, settling time, and linearity of phase across the passband (i.e., constancy of time delay).

¶C. Filter Implementations.

Filters can be built (a) entirely with passive components (R, L, and C); (b) with R's and C's, assisted with op-amps; (c) with C's alone, combined with periodically-clocked analog switches; or (d) with digital processing of the ADC-sampled input waveform. These are called *passive* filters, *active* filters, *switched-capacitor* filters, and *digital* filters, respectively. The term *continuous-time filter* is sometimes applied to filters of type (a) and (b), and *discrete-time filter* to (c) and (d). The *order* of a filter is equal to the number of C's plus the number of L's (or equivalent, if implemented digitally). An nth order lowpass filter has an ultimate rolloff of $6n$ dB/octave ($20n$ dB/decade).

¶D. Passive *RC* Filters.

Passive *RC* filters (§6.2.1) are the simplest, and good enough for applications such as blocking dc, suppressing high-frequency power-supply noise, or removing signals far from the band of interest. But *RC* filters, regardless of order, have a soft transition region (Figure 6.2) and are unsuited to separating signals that are nearby in frequency.

Their transfer function is far from the ideal "brick-wall" filter response (see for example Figure 6.11).

¶E. Passive *LC* Filters.

Perhaps surprisingly, combining inductors with capacitors allows you to make all manner of very sharp filters (§6.2.2; see for example Figure 6.5). The classic filter shapes, all implementable with *LC* filters, are the Butterworth (maximally flat passband), the Chebyshev (sharpest transition region, at the cost of amplitude ripple in the passband), and the Bessel (maximally flat time-delay in the passband). The performance characteristics of these filter types are compared in Table 6.1 and Figures 6.20, 6.21, 6.25–6.27, and 6.30.

¶F. Active Filters.

Inductors are non-ideal in several respects (size, linearity, electrical losses; see Chapter *1x*; but the combination of a capacitor and an op-amp (plus several resistors), in a configuration known as a *gyrator* (§6.2.4C), creates an electrical equivalent of an inductor. So you can make an inductor-less filter that mimics any *LC* filter, by simply substituting gyrators for the inductors. More generally, you can make such *active filters* (§6.3) with various configurations of op-amps, capacitors, and resistors that need not incorporate explicit gyrators.

¶G. Active Filter Circuits.

In §6.3.1 we showed how to design the simple and popular VCVS active filter, with tabulated data (Table 6.2) for 2nd-order to 8th-order lowpass or highpass filters, with Butterworth, Chebyshev, or Bessel response. Better performance and tunability is gotten with the *state-variable* and *biquad* active filters (§6.3.3), which require three op-amps for each 2nd-order section. These latter filter topologies are well suited for bandpass filters; see §6.3.3A, where explicit design equations are given. You can get nice state-variable ICs that include the op-amps and capacitors, and can be configured as lowpass, highpass, band-pass, or band-reject, requiring only a few external resistors to set the characteristic cutoff frequencies; examples are the UAF42 and the MAX274. The world is awash with active-filter implementations; some others seen in Chapter 6 include the Sallen-and-Key filter (Figure 6.16), a state-variable variant with independently settable gain and Q-factor (Figure 6.32), and the multiple-feedback filter (Figure 6.35).

¶H. Notch Filters.

In contrast to their generally "soft" frequency characteristics, the *RC* filter known as a *twin-T* (§6.3.4) produces a deep notch (limited only by component imperfection and mismatching). The twin-T is hard to tune (it requires three tracking adjustable resistors). But a similar *RC* notch filter (the *bridged-differentiator*, Figure 6.40) allows a modest range of tunability (5:1 or so) with a single pot.

¶I. Switched-capacitor Filters.

An op-amp combined with a pair of capacitors and a pair of analog switches forms a discrete approximation to a continuous-time integrator (Figure 6.43). This is the building block of the switched-capacitor filter (§6.3.6), easily incorporated into an IC, and conveniently tuned by varying the externally-applied switching frequency. The downside is the production of artifacts related to the switching operation: clock feedthrough, aliasing, and limited dynamic range.

¶J. Digital Filters and DSP.

With ubiquitous embedded microcontrollers and accompanying ADCs, it's natural to implement filtering operations with the machinery of digital signal processing (DSP, §6.3.7). If the signals to be filtered are in the form of ana-

log voltages, they first must be *digitized* (sampled at regular intervals and converted to a stream of numbers), taking care to sample at a high enough rate (at least twice the highest frequency present in the signal, $f_{samp} \geq 2f_{sig(max)}$) and with sufficient amplitude accuracy (*n*-bit quantization yields $6n$ dB dynamic range) to retain adequate fidelity. (If the input signal is already digitized, no sampling is needed, and digital filtering is particularly convenient.)

The stream of numbers representing the successive sampled and digitized signal voltages (call them x_i) is then subjected to a digital filtering operation. Easiest to understand is the *finite-impulse-response* (FIR) filter, where each output signal amplitude y_i is formed from a weighted sum of a finite number N of input samples; i.e., $y_i = \sum a_k x_{i-k}$, with k going from $-N/2$ to $N/2$. If the sum is permitted to include output samples, you've got a *recursive* filter, also known as an *infinite-impulse-response* filter. See Figure 6.49 for an example of an FIR lowpass filter. There's plenty of complexity, and plenty of math, in the business of digital filtering; this specialty appeals to EEs who are really applied mathematicians in disguise (you know who you are!).

OSCILLATORS AND TIMERS

In this chapter we meet oscillators and timers, the circuits that provide the essential "heartbeats" and timing of electronics. As we'll see, many of the important devices and techniques involve a blend of analog and digital electronics. We'll introduce the digital know-how you need for a first look, and point to later sections of the book where you can get a deeper knowledge of digital techniques.

7.1 Oscillators

7.1.1 Introduction to oscillators

Within nearly every electronic instrument it is essential to have an oscillator or waveform generator of some sort. Apart from the obvious case of signal generators, function generators, and pulse generators themselves, a source of regular oscillations is necessary in any cyclical measuring instrument, in any instrument that initiates measurements or processes, and in any instrument whose function involves periodic states or periodic waveforms. That includes just about everything. For example, oscillators or waveform generators are used in digital multimeters, oscilloscopes, radiofrequency receivers, computers, every computer peripheral (tape, disk, printer, terminal), nearly every digital instrument (counters, timers, calculators, and anything with a "multiplexed display"), every consumer electronic device (cell phone, camera, music or video player or recorder), and a host of other devices too numerous to mention. A device without an oscillator either doesn't do anything or expects to be driven by something else (which probably contains an oscillator). It is not an exaggeration to say that an oscillator of some sort is as essential an ingredient in electronics as is a regulated supply of dc power.

Depending on the application, an oscillator may be used simply as a source of regularly spaced pulses (e.g., a "clock" for a digital system), or demands may be made on its stability and accuracy (e.g., the time base for a frequency counter), its adjustability (e.g., the local oscillator in a transmitter or receiver), or its ability to produce accurate waveforms (e.g., the ramp generator in a dual-slope analog-to-digital converter).

In the following sections we treat briefly the most popular oscillators, from the simple *RC* relaxation oscillators to the stable quartz-crystal oscillators. Our aim is not to survey everything in exhaustive detail, but simply to make you acquainted with what is available and what sorts of oscillators are suitable in various situations.

7.1.2 Relaxation oscillators

A very simple kind of oscillator can be made by charging a capacitor through a resistor (or a current source), then discharging it rapidly when the voltage reaches some threshold, beginning the cycle anew. Alternatively, the external circuit may be arranged to reverse the polarity of the charging current when the threshold is reached, thus generating a triangle wave rather than a sawtooth. Oscillators based on this principle are known as *relaxation oscillators*. They are inexpensive and simple, and with careful design they can be made reasonably stable (better than 1%) in frequency.

A. Basic op-amp – comparator relaxation oscillator

In the past, negative-resistance devices such as unijunction transistors and neon bulbs were used to make relaxation oscillators, but current practice favors op-amps, comparators, or special timer ICs. Figure 7.1A shows a classic *RC* relaxation oscillator. The operation is simple: assume that when power is first applied, the comparator output goes to positive saturation (it's actually a toss-up which way it will go, but it doesn't matter). The capacitor begins charging up toward +5 V, with time constant *RC*. When it reaches one-half the supply voltage, the op-amp switches into negative saturation (it's a Schmitt trigger), and the capacitor begins discharging toward −5 V with the same time constant. The cycle repeats indefinitely, with period 2.2*RC*, independent of supply voltage.

Exercise 7.1. Show that the period is as stated.

Rail-to-rail CMOS output-stage comparators[1] (see §§4.3.2A. 12.1.7 and Table 12.2) were chosen because

---

[1] In this case the TLC3702 or LMC6762.

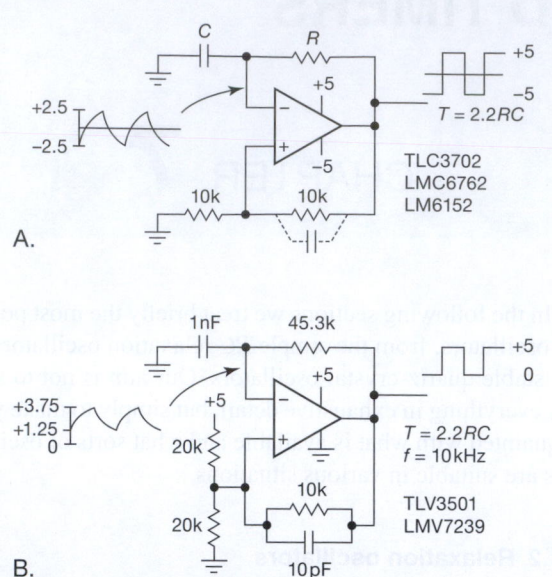

A.

B.

Figure 7.1. Classic op-amp (or comparator) relaxation oscillator, using comparators with rail-to-rail output stages. A. Dual supply with symmetrical bipolarity square-wave output. B. Single-supply variant, with parts values (for 10 kHz), including speed-up capacitor. See also Figure 4.39 in §4.3.3.

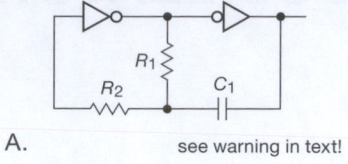

A.   see warning in text!

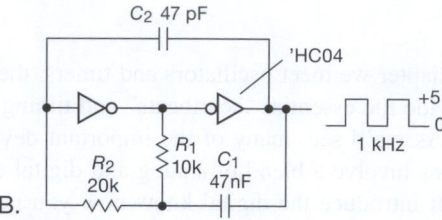

B.

Figure 7.2. Relaxation oscillator with CMOS digital logic inverters. A. Beware conventional wisdom! B. Taming parasitic instabilities with a small speed-up capacitor C_2.

their outputs saturate cleanly at the supply voltages. Comparators like the TLC3702 are much faster (its propagation time is ~5 ns) than op-amps of similar technology, because they do not need compensation for stable operation with negative feedback (see §4.9), so they are a good choice if you want to go above a few kilohertz. The bipolar LM6132–54 series of op-amps also swings rail-to-rail and, unlike their CMOS cousins, allows operation at a full ±15 V. However, if op-amps (rather than comparators) are used, this circuit makes considerable demands on the op-amp speed, because the output square wave goes rail-to-rail; even something like the LM6152, with its f_T of 75 MHz and large-signal slew rate of 45 V/μs, has enough speed for operation only up to 100 kHz or so. Note that this circuit does not operate the op-amp in the linear region, with the customary negative feedback;[2] so you could use an uncompensated op-amp (see §4.9) for better speed.

You can run this kind of circuit from a single supply voltage, as shown in Figure 7.1B, if you add one resistor. Here we've taken advantage of the faster comparators that are available for lower supply voltages only. The TLV3501 (supply voltage range of 2.7 V to 5.5 V only) has a propagation time of just 3 ns in this circuit, allowing operation

up to tens of megahertz; here it's loafing along at a mere 10 kHz.[3]

By using current sources (instead of resistors) to charge the capacitor, a good triangle wave can be generated. A clever circuit using that principle was shown in §4.3.3.

B. CMOS logic relaxation oscillators

You can build simpler *RC* relaxation oscillators by using CMOS digital logic inverters (Chapters 10–12) instead of an op-amp or comparator. Figure 7.2A shows a circuit often seen in the literature.[4] The good news is that it's simple; the bad news is that it doesn't work! Specifically, its output waveform has ragged edges, plagued with fast (~100 MHz) parasitic oscillations at each transition; this is due to the relatively slow rise time at the input to the first inverter (caused by capacitive loading). The final (good) news is that there's a simple fix, namely the inclusion of a small speed-up capacitor (C_2 in Figure 7.2B). Figure 7.3 shows measured waveforms, both for the conventional circuit (Figure 7.2A), and for the improved circuit.

One slightly worrisome aspect of this circuit is the fact that the input protection diodes are forced into conduction each cycle by the charged capacitor C_1; it's not really a

---

[2] The fast feedback is positive, and the output swing alternates between positive and negative saturation.

[3] If, instead, you're interested in very low operating current, you could use the impressive LPV7215 comparator: it runs at less than a microamp supply current, with propagation time ~10 μs. To exploit its very low power, of course, you must use high resistor values, say ~10 MΩ, which are OK, given the comparator's very low input current (<1 pA).

[4] For example, in the datasheet for the 74HC4060, and in our previous book edition.

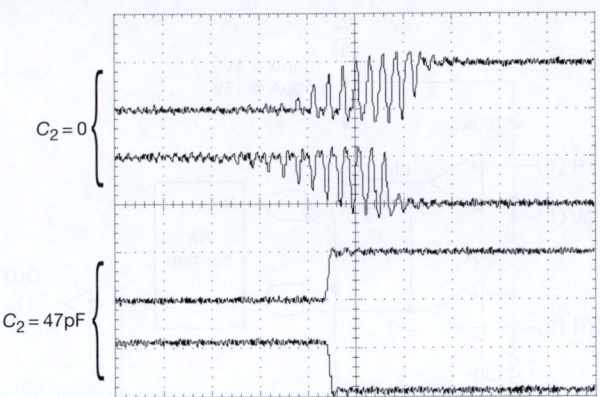

Figure 7.3. Parasitic oscillations plague the simple oscillator circuit of Figure 7.2A. The upper pair of waveforms show an ~90 MHz instability on the rising and falling edges of a 1 kHz oscillator (made with commonplace 74HC04 logic running at 5 V). Adding a 47 pF speed-up capacitor (Figure 7.2B) cleans it up nicely. Horizontal: 40 ns/div; vertical: 5 V/div

problem, however, because the current is safely limited by R_2. But if it bothers you, you'll like the circuit of Figure 7.4A, where a 2:1 voltage divider reduces the swing applied to the capacitor, preventing input clamping. Figure 7.4B is another variation on the logic oscillator theme, taking aim at the parasitic oscillation problem. When we breadboarded these two oscillators on the test bench, however, we found it still necessary to include the 47 pF speed-up capacitors to eliminate parasitics.

You can make an even simpler CMOS oscillator by just hooking RC feedback around a CMOS logic inverter with Schmitt trigger input (Figure 7.5). It is guaranteed to oscillate, with clean output transitions and a full logic swing. However, its frequency is not particularly well determined, because hysteresis is not a well-controlled parameter in logic ICs – it's provided to clean up slow inputs, not to do anything precise (you're lucky to get it at all!). The 74HC14, for example, specifies only that the hysteresis amplitude (that is, the difference between rising and falling thresholds) is somewhere between 0.5 V and 1.5 V![5] That means you can expect a frequency spread of 50% or more among oscillators with the same R and C values. The frequency will also vary with supply voltage; we found frequency approximately proportional to supply voltage for the oscillator of Figure 7.5. Finally, this oscillator generates a somewhat unstable output, with as much as a few percent "jitter" in the timing of successive edges (enough

---

[5] They specify the individual thresholds, also, with comparable imprecision: the rising threshold is somewhere between 1.8 V and 3.5 V, and the falling threshold is somewhere between 1.0 V and 2.5 V.

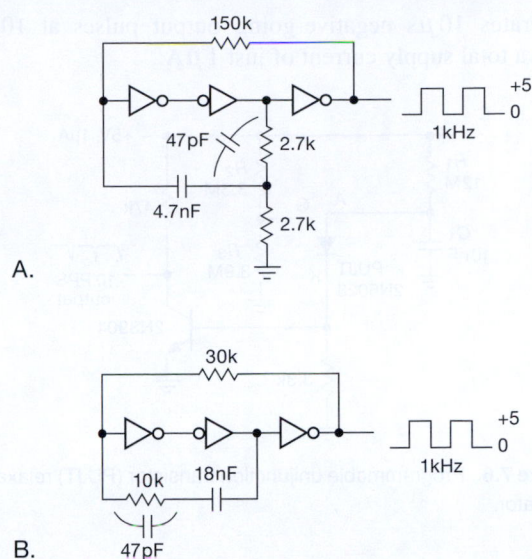

Figure 7.4. CMOS relaxation oscillator variants. A. Half-scale swing prevents input diode clamping (J. Thompson's design). B. Improved stability with one inverter and one buffer (E. Wielandt's design). In our test setup the 47 pF capacitor was needed to prevent parasitics.

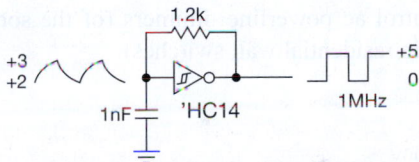

Figure 7.5. Simplest CMOS oscillator.

to see on an oscilloscope), with sensitivity to digital noise on the power supply.

C. Unijunction transistor relaxation oscillator

There are several ways to make a relaxation oscillator that exploit the "negative-resistance" characteristic of devices such as tunnel diodes, gas-filled discharge tubes, diacs, and unijunction transistors. In the circuit of Figure 7.6, for example, the programmable unijunction transistor (PUJT) is a 3-terminal, 4-layer (*pnpn*) device; it looks like an open circuit to the charging RC until the capacitor voltage reaches a diode drop above the gate (G) voltage (set by the R_2R_3 divider), at which point the PUJT conducts heavily from anode (A) to cathode (K), discharging the capacitor and beginning a new cycle. The discharge current turns on the output transistor switch, generating a saturated output pulse to ground. With the values shown, the oscillator generates $10\,\mu s$ negative-going output pulses at $10\,Hz$, with a total supply current of just $1\,\mu A$.

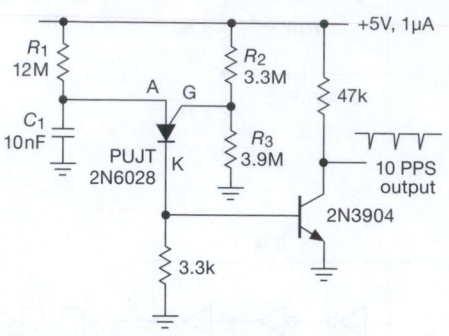

Figure 7.6. Programmable unijunction transistor (PUJT) relaxation oscillator.

Continuing this theme, Figure 7.7 shows a couple of funky oscillators that we couldn't resist hooking up, just to remind us of the old days of electronics. They also exploit "snapback" negative resistance, at somewhat higher voltages than the PUJT, in this case of a neon lamp and a 4-layer diac; the latter is widely used as a triac trigger in phase-control ac powerline dimmers (of the sort that are common in residential wall switches).

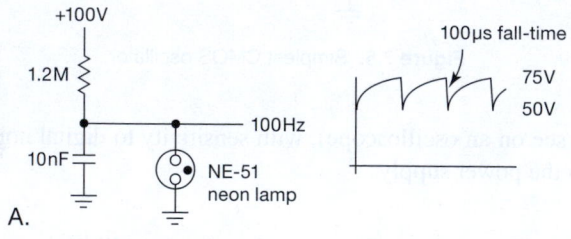

A.

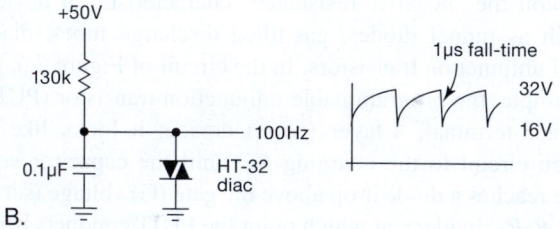

B.

Figure 7.7. Two unusual relaxation oscillators that exploit devices with a negative-resistance "snapback" VI characteristic. The Littelfuse HT32 and ST32 diacs were discontinued in 2009, but you can get the comparable DB3 from at least three manufacturers.

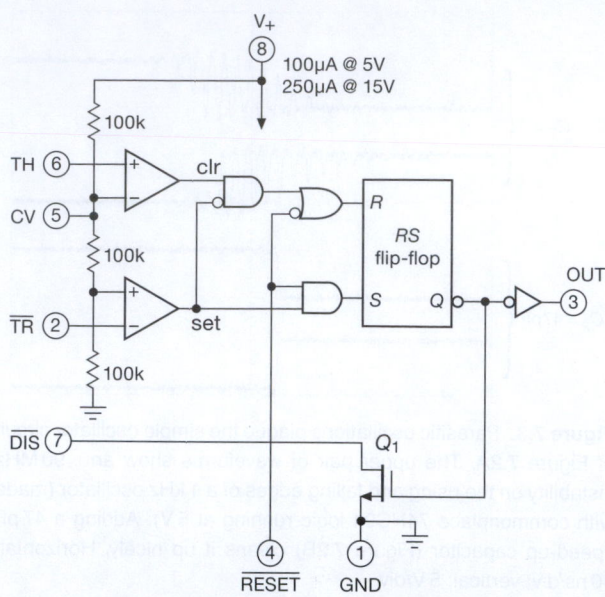

Figure 7.8. Block diagram of the legendary 555 in its contemporary CMOS implementation.

7.1.3 The classic oscillator–timer chip: the 555

The next level of sophistication involves the use of timer or waveform-generator ICs as relaxation oscillators. The most popular chip around is the legendary 555 (and its many successors), designed originally in 1970 by Hans Camenzind at Signetics. It is also a misunderstood chip, and we intend to set the record straight with the equivalent circuit shown in Figure 7.8. Some of the symbols belong to the digital world (Chapter 10 and following), so you won't become a 555 expert for a while yet. But the operation is simple enough: the output goes HIGH (near V_{CC}) when the 555 receives a $\overline{TRIGGER}$ input, and it stays there until the THRESHOLD input is driven, at which time the output goes LOW (near ground) and the DISCHARGE transistor is turned on. The $\overline{TRIGGER}$ input is activated by an input level below $\frac{1}{3}V_{CC}$, and the THRESHOLD is activated by an input level above $\frac{2}{3}V_{CC}$.

The easiest way to understand the workings of the 555 is to look at an example (Figure 7.9). Before power is applied, the capacitor is discharged; so when power is applied the 555 is triggered, causing the output to go HIGH, the discharge transistor Q_1 to turn off, and the capacitor to begin charging toward 15 V through R_A+R_B. When it has reached $\frac{2}{3}V_{CC}$ (+10 V), the THRESHOLD input is triggered, causing the output to go LOW and Q_1 to turn on, discharging C toward ground through R_B. Operation is now cyclic, with C's voltage going between $\frac{1}{3}V_{CC}$ and $\frac{2}{3}V_{CC}$, with period

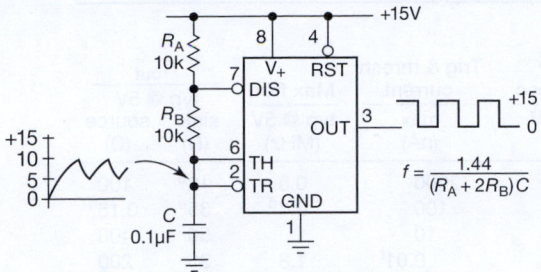

Figure 7.9. The 555 connected as an oscillator.

$T = 0.693(R_A + 2R_B)C$. The output you generally use is the square wave[6] at the output.

Exercise 7.2. Show that the period is as advertised, independent of supply voltage.

The original (bipolar-transistor version) 555 makes a respectable oscillator, with stability approaching 1%. It can run from a single positive supply of 4.5 to 16 V,[7] maintaining good frequency stability with supply-voltage variations because the thresholds track the supply fluctuations. The 555 can also be used as a timer to generate single pulses of arbitrary width (see §7.2.1E), as well as a bunch of other things. It is really a small kit, containing comparators, gates, and flip-flops. It has become a game in the electronics industry to try to think of new uses for the 555 – you too can become an alpha-nerd!

A caution about the bipolar 555: many versions of this IC (in common with some other timer chips) generate a big supply-current glitch (as much as 150 mA) during each output transition.[8] Be sure to use a hefty bypass capacitor near the chip. Even so, the 555 may have a tendency to generate double output transitions; the CMOS versions (discussed next) are better in this regard, but still not cured of all bad behavior.

A. CMOS 555s

Some of the less-desirable properties of the original bipolar 555 (high supply current, high trigger current, double

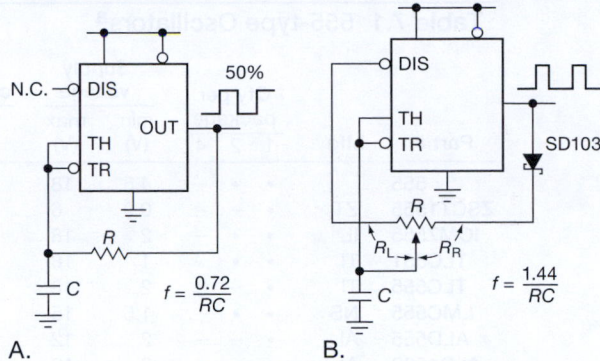

Figure 7.10. More CMOS 555 oscillator circuits. A. 50% duty cycle (square wave). B. Constant frequency, with fully variable duty cycle.

output transitions, and inability to run with very low supply voltage) have been remedied in a collection of CMOS successors. You can recognize these by the tell-tale "555" somewhere in the part number. Table 7.1 on the next page lists most of these that we could find, along with their important specifications. Note particularly the ability to operate at very low supply voltage (down to 1 V) and the generally low supply current. These chips also can run at higher frequency than the original 555. The CMOS output stages give rail-to-rail swing, at least at low load currents (but note that these chips don't have the output-current muscle of the standard 555). All chips listed are CMOS except for the original 555 and the ZSCT1555.

B. 50% duty cycle

The 555 oscillator of Figure 7.9 generates a rectangular-wave output whose duty cycle (fraction of time the output is HIGH) is always greater than 50%. That is because the timing capacitor is charged through the series pair R_A+R_B, but discharged (more rapidly) through R_B alone. But you can make a *CMOS* 555 (with its rail-to-rail output swing) give you exactly 50% duty cycle (a true square wave) with the circuit of Figure 7.10A. The trick is to use a single charge–discharge resistor, connected to the output; so the capacitor is either charging toward $+V_{CC}$ (with $\frac{2}{3}V_{CC}$ threshold) or discharging toward ground (with $\frac{1}{3}V_{CC}$ threshold). It's like two people throwing a ball back and forth, with a dog tirelessly zigzagging back and forth, trying to catch it.[9] You should be able to show that $f_{osc} = 0.72/RC$.

Exercise 7.3. Show that this result is correct.

---

[6] A *rectangular wave*, to be precise, because it spends 2/3 of the time HIGH and 1/3 of the time LOW. But it's conventional to use the term "square wave," as we do here, to distinguish a 2-level waveform (whatever its symmetry) from a continuous waveform like the exponentials at the capacitor.

[7] Variants like the bipolar ZSCT1555, as well as the CMOS versions, can run on lower voltages, down to 0.9 V for some types; see Table 7.1.

[8] Better-behaved versions will usually brag about it on the datasheet, for example "No output cross-conduction current spikes" (Micrel MC1555 IttyBitty™ RC Timer/Oscillator).

---

[9] Hope springs eternal in a charging capacitor.

Table 7.1 555-type Oscillators[a]

Part #	Mfg	Qty per package 1	2	4	Supply voltage min (V)	max (V)	Supply current/osc typ @ 5V (µA)	Trig & thresh current max (nA)	Max freq typ @ 5V (MHz)	R_{out} typ @ 5V sink (Ω)	source (Ω)
555		•	•	–	4.5	18	3000	2000	0.5	12[b]	100[c]
ZSCT1555	ZT	•	–	–	0.9	6	150	100	0.3[d]	35[e]	0.15[e]
ICM7555	IL	•	•	–	2	18	60	10	1	50	400
TLC551	TI	•	•	–	1	18	15[f]	0.01[t]	1.8	25	200
TLC555	TI	•	–	–	2	18	170	0.01[t]	2.1	25	200
LMC555	NS	•	•	–	1.5	15	100	0.01[t]	3	40	150
ALD555	AL	•	–	–	2	12	100	0.2	2	20	250
ALD1502	AL	•	•	•	2	12	50	0.4	2.5	20	200
MIC1555	MI	•	–	–	2.7	18	240	50	5[g]	25	100

Notes: (a) all are CMOS except first two (bipolar) entries. (b) I_O (mA) at V_O=0.3V. (c) I_O (mA) at V_{sat}=1.7V. (d) min, @ V_S=1.5V. (e) I_O (mA) at V_{sat}=±0.35V and V_S=1.5V. (f) at V_S=1V. (g) at V_S=8V. (t) typical.

C. Full duty cycle control

Figure 7.10B shows how to make a fixed-frequency output whose duty cycle can be varied from near 0% to near 100%. The frequency would be completely constant, and independent of the duty-cycle setting, except for the effect of the diode drop during charging; that's why we used a low-drop ($V_F = 0.3$ V at 10 mA) SD103C Schottky diode.

Exercise 7.4. Show that $f_{osc} = 1.44/RC$.

D. Sawtooth oscillator

By using a current source to charge the timing capacitor, you can make a ramp (or "sawtooth-wave") generator. Figure 7.11A shows how, using a simple *pnp* current source. The ramp charges to $\frac{2}{3}V_{CC}$, then discharges rapidly (through the 555's *npn* discharge transistor, pin 7) to $\frac{1}{3}V_{CC}$, beginning the ramp cycle anew. Note that the ramp waveform appears on the capacitor terminal and must be buffered with an op-amp since it is at high impedance. In practice, this circuit exhibits a subtle flaw: when a small value capacitor is used, the discharge is so rapid that the bottom of the sawtooth drops below $\frac{1}{3}V_{CC}$ before the discharge transistor shuts off. This is remedied, as shown, by the inclusion of a small resistor in series with the $\overline{\text{DIS}}$ pin, chosen for a discharge time constant of ~5 μs.[10]

Figure 7.11B shows an alternative, namely to delay the falling-edge signal to TR$'$ so that the discharge interval is extended long enough to ensure complete discharge; 1 μs was adequate, for the circuit values shown.

In Figure 7.11B we drew a current source symbol, be-

---

[10] We found, by measurement, that the minimum discharge times required to prevent undershoot were approximately 1 μs, 5 μs, and 10 μs, for samples of the LMC555, ICL7555, and bipolar 555, respectively.

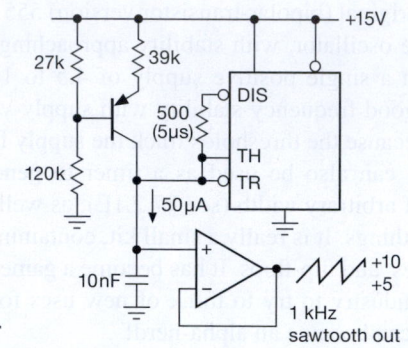

A.

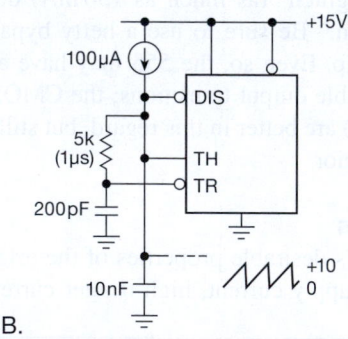

B.

Figure 7.11. Sawtooth oscillators with the CMOS 555. A. Discrete *pnp* current source charges C, whose discharge is slowed to prevent $V_+/3$ undershoot. B. Delayed TR$'$ on falling edge causes full discharge to 0 V.

cause there are several alternatives to the discrete *pnp* current source. Figure 7.12 shows some simple 2-terminal favorites, namely a JFET "current regulator diode," and two integrated circuits. As discussed in §3.2.2, a JFET with

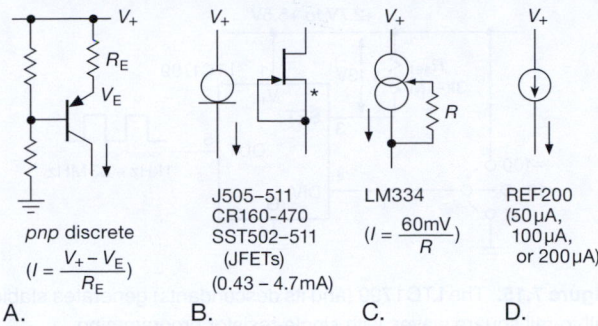

Figure 7.12. Current-source choices for the sawtooth oscillator.

gate tied to source runs at constant current when biased with $\gtrsim 1$ V; these are available as packaged 2-terminal devices, in a limited range of currents (0.43 mA–4.7 mA), and with operation up to 100 V. The LM334 is also a 2-terminal (floating) current source, with a third pin that lets you program the current by connecting a resistor as shown; the current is approximately[11] 60 mV/R, and, like the JFET, works down to ~1 V (its maximum is 40 V). The REF200 is a temperature compensated 2-terminal (floating) 100 μA current source (with several selectable multiples), operable from 2–40 V.

Ratiometric sawtooth oscillator

And now for an interesting variation. In many situations you *want* a stable current source like the REF200, designed to deliver a current that does not depend on the voltage across it (the REF200 is quite good in this regard, its current varying less than 0.1% for voltages from 2 V to 30 V; see Figure 9.37). And that would be fine here, delivering a constant frequency sawtooth in the circuit of Figure 7.11B – providing, of course, that the 555's thresholds stay constant. But if the supply voltage (here V_+=+15 V) were to change, the thresholds would follow proportionally (at $V_+/3$ and $2V_+/3$), and the frequency of oscillation would change. And this would happen if you were to use something like a 9 V battery for your V_+ supply.

There's an elegant way to finesse this problem, namely to make the current source's output proportional to supply voltage, which compensates exactly for the frequency change that would otherwise occur. This is the technique of *ratiometric* design, elegant and powerful. The simple *pnp* current source of Figure 7.12A is almost what you want: it would be exactly right except for the ~0.6 V base–emitter drop. And you can fix that by rigging up a transistor current source with V_{BE} cancellation.

<hr>

[11] It's actually "PTAT" — proportional to absolute temperature.

Look at Figure 7.13: in the first circuit a diode in the base divider adds a voltage drop that approximately matches the transistor's base–emitter drop. That's pretty good, but not ideal because (a) the V_{BE} match is imperfect, and (b) the diode's drop means that the voltage drop across R_1 is not quite proportional to V_+ (figure out why).

This is fixed in the second circuit (Figure 7.13B), where Q_1's base voltage tracks V_+ exactly, and its V_{BE} downward drop cancels Q_2's drop going back up. It's imperfect only insofar as the V_{BE}'s aren't equal, both because of transistor mismatch and because of I_C mismatch (à la Ebers–Moll). The third circuit fixes the V_{BE} mismatch (by using matched transistors operating at the same current), but it's got the not-quite-proportional current problem of Figure 7.12A; that is, the current through "programming resistor" R_p is proportional to $V_+ - V_{BE}$. Note, by the way, that this circuit is the classic current mirror that we saw back in §2.3.7.

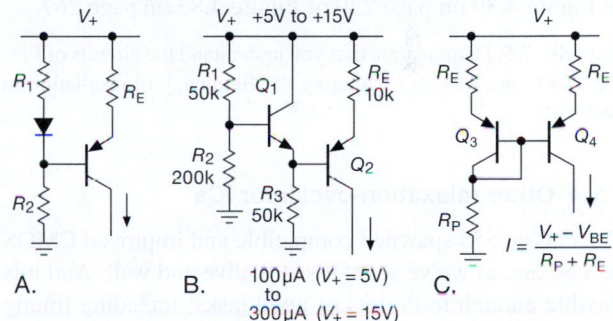

Figure 7.13. Current sources with output approximately proportional to the supply voltage ($I_{out} \propto V_+$), to make f_{out} in Figure 7.11 independent of supply voltage.

E. Triangle oscillator

Figure 7.14 shows a simple way to generate a *triangle* wave with a CMOS 555. The rail-to-rail output square wave is used to generate a current source–sink (of alternating polarity), producing a triangle waveform (going between the usual $\frac{1}{3}V_{CC}$ and $\frac{2}{3}V_{CC}$) at the capacitor. The configuration of diodes is the familiar *bridge rectifier* (§1.6.2), here used to trick the unipolar 2-terminal current-source device into thinking that the current is always flowing in the same (normal) direction, whereas to the outside world it's a bidirectional current source. (You can think of the bridge as presenting to the current source a rectified version of the alternating-polarity external current.) We used Schottky diodes here, to minimize the two-diode forward drop of the bridge. As with the sawtooth oscillator, you have to buffer the high-impedance waveform with an op-amp. This

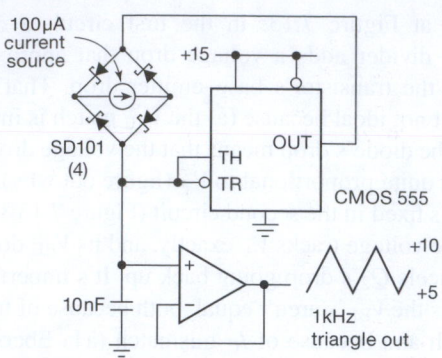

Figure 7.14. Triangle-wave oscillator with the 555. This circuit requires a floating (2-terminal) current source, as in Figure 7.12B–D.

circuit is simple (taking advantage of the oscillator-oriented innards of the 555), but its performance is not nearly as good as the more elaborate op-amp-based circuit of Figure 4.39 on page 239 or Figure 4.83 on page 267.

Exercise 7.5. Demonstrate that you understand the circuits of Figures 7.11 and 7.14 by calculating the frequency of oscillation in each case.

7.1.4 Other relaxation-oscillator ICs

The classic 555 spawned compatible and improved CMOS successors, as we've seen; it's still alive and well. And it is flexible enough to do lots of good tasks, including timing and pulse generation, which we treat later in the chapter. But there's been a lot of progress in semiconductor electronics since the 555's introduction in 1971, and you can get some very nice contemporary oscillator–timer chips that may well be your better choice.[12] Here are a few of our favorites.

A. LTC1799 and LTC6900 series

These elegant ICs come from the wizards at Linear Technology, who call them "silicon oscillators." The LTC1799 runs from a single positive 2.7 V to 5.5 V supply (drawing about a milliamp), it generates a 50% duty-cycle rail-to-rail square-wave output, and its output frequency is set by a single external resistor (or current source). It operates from 1 kHz to 33 MHz (it has an internal ÷1, ÷10, or ÷100 frequency divider, N, selected by tying the DIV input LOW, open, or HIGH, to produce its 33,000:1

---

[12] Being of recent design, they are invariably available in small surface-mount (SMT) packages. This can be a disadvantage, though, if you are looking for easy-to-prototype traditional *through-hole* packaging, because new designs increasingly come *only* in SMT.

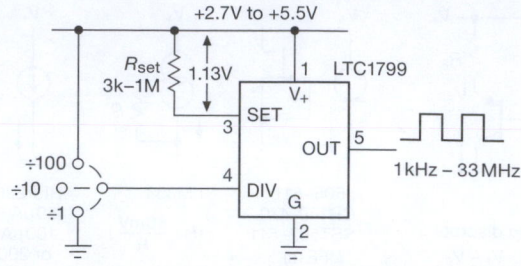

Figure 7.15. The LTC1799 (and its descendants) generates stable rail-to-rail square waves with single-resistor programming.

range). That's it – one external component, and off you go! It has very good accuracy ($\pm 0.5\%$, typ), temperature stability ($\pm 0.004\%/°C$, typ), and voltage coefficient ($\pm 0.05\%/V$) Figure 7.15 shows how to use it. The similar LTC6900 series includes the LTC6900 (somewhat lower power consumption), the LTC6905 (17 MHz–170 MHz), and the LTC6903/4 (1 kHz–68 MHz, programmed via a 3-wire serial digital connection). The latter part would be particularly well suited to a system with a resident microcontroller (Chapter 15), which can effortlessly send digital commands hither and yon.

Figure 7.16 shows the dependence of output frequency on resistor value. If you like equations, you may be happier with this:

$$f_{osc} = \frac{1}{N} \cdot \frac{100}{R(k\Omega)} \quad \text{MHz}.$$

Because the frequency is set by an input current at the SET input terminal, you can use instead an externally generated current to adjust the frequency (we'll call it an "ICO," for "current-controlled oscillator"), keeping in mind that the SET input sits approximately 1.13 V below the positive rail. For current programming the datasheet suggests currents in the range of 5 μA to 200 μA, with ×10 range switching as before (which could be done electrically, for example with 3-state logic or with a pair of transistor switches). The V_+-referenced voltage level at the SET input makes *voltage* programming a bit awkward: the datasheet suggests one method, namely to apply a control voltage through a second resistor to the SET input, to add or subtract a variable current from that supplied by R_{SET}; but this method has its problems, and you're probably best off just generating a current externally to drive SET.

B. LTC699x "TimerBlox"

A few years after their LTC1799/6900, Linear Technology introduced the "TimerBlox®" series of timing functions – oscillators, pulse-width modulators, one-shots

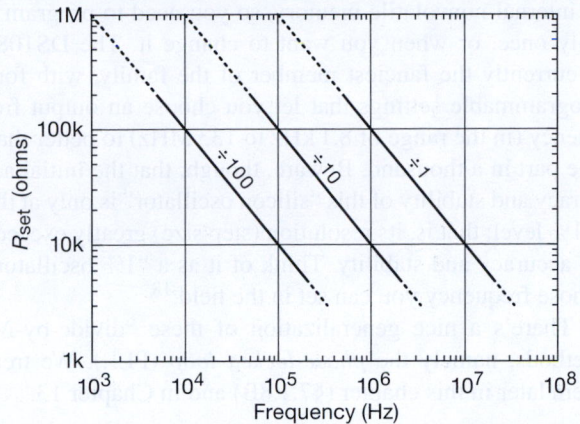

Figure 7.16. LTC1799 output-frequency programming. The manufacturer advises you to stay on the solid lines for best accuracy.

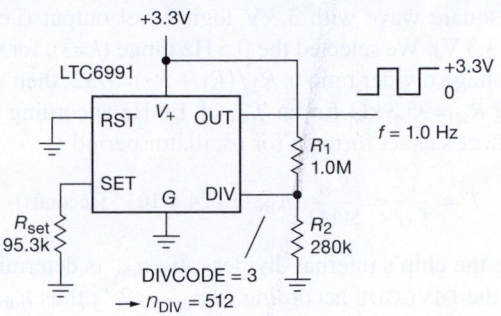

Figure 7.17. Resistor-programmed 1 Hz oscillator.

(monostables), and delay/debounce. Rather than trying to make a "one size fits all" chip (with lots of pins), they instead tailored separate chips, in the small 6-pin SOT23-6 package, or the even-smaller 2 mm × 3 mm DFN package. These chips share many properties, for example a single-supply (2.25–5.5 V), good out-of-the-box accuracy (~2% worst-case), single resistor tuning, and a single analog pin for user programming of 16 choices of range and mode (compare with the LTC1799/6900, where one pin sets one of three ranges). LTC has pulled out the stops to create a set of simple, function-specific, and small parts[13] to handle our oscillator and timing problems.

Here is a listing of the TimerBlox parts available at the time of writing:

Part #	Function	Total Range	Comments
LTC6990	VCO	488 Hz–2 MHz	16:1 tuning range, as eight octave (×2) ranges
LTC6991	LF osc; timer	29 µHz–977 Hz	1 ms–9.5 hr, as eight ×8 ranges
LTC6992-x	PWM	3.8 Hz–1 MHz	0%–100%, 5%–100%, 5%–95%, or 0%–95%
LTC6993-x	One-shot	1 µs–34 s	normal or retrig; rising or falling edge
LTC6994-x	delay/ debounce	1 µs–34 s	delay single or all edges; reject narrow pulse

The LTC6991 takes aim at the problem of very low-frequency square-wave oscillators, with a total output frequency range of approximately 30 *micro*hertz to 1 kHz;

---

[13] The SOT23 is the experimenter's ideal surface-mount part, large enough for ordinary soldering-iron assembly; LTC could have chosen a smaller machine-assembly part for the "larger" package.

thinking of this as a timer, that's a range of periods from 1 millisecond to 9 hours. You select one of eight ranges, whose "center" frequencies (corresponding to $R_{set}=200$ kΩ) are spaced apart by factors of eight: 0.00012 Hz, 0.001 Hz, 0.008 Hz, 0.064 Hz, 0.5 Hz, 4 Hz, 32 Hz, 250 Hz. Within a chosen range, the external resistor R_{set} (50–800 kΩ) tunes the oscillation frequency over a 16:1 range. For example, in the 4 Hz range you can tune continuously from 1 Hz ($R_{set}=800$k) to 16 Hz ($R_{set}=50$k).

Now for the mystery: how can you select one of 16 operating modes (eight ranges, two polarities) with a single pin, without the need for any kind of serial programming? Easy: these chips contain an internal 16-level (4-bit) ADC, with the power supply voltage V_+ as full-scale reference, to determine the DIVCODE value (an integer from 0 to 15). So you simply apply a dc voltage in the range 0 to V_+ to the DIV programming pin, most easily done via a pair of resistors from V_+ to ground.[14] That voltage divider's output (call it V_k) should aim for the midpoint of one of the sixteen subdivisions from 0 V to V_+. Stated precisely, you would use a pair of 1% resistors that produces an output voltage $V_k = V_+(2k+1)/32$ in order to select[15] DIVCODE $=k$. That DIVCODE then determines the frequency range and (in this case) the polarity of output and reset.

As an example, Figure 7.17 shows how to generate a

---

[14] You could instead use the voltage output of a DAC, if you want digital control.

[15] This is an, uh, *interesting* scheme. Why not use two more pins, for a total pincount of eight, with programming like the LTC1799 (each one tied to GND, V_+, or open)? Evidently, in their zeal to minimize pincount and package size, and simultaneously to avoid the need for any kind of serial programming, the designers chose this analog scheme, with internal ADC. We'd like to see them offer an 8-pin SOIC (and DFN) option, with 3-level selection on three pins; that way you'd get up to 27 mode choices, without the need for any external resistors. And, hey, a DFN-8 takes up less space than a DFN-6 plus two resistors! OK, you clever folks at LTC, what do you think?

1 Hz square wave with 3.3 V logic level output (i.e., 0 V and +3.3 V). We selected the 0.5 Hz range ($k=3$), for which the voltage divider ratio is $R_2/(R_1+R_2)$=7/32; then we selected R_{set}=95.3 kΩ for an f_{out} of 1.0 Hz according to the datasheet's exact formula for oscillator period

$$T = \frac{1}{f_{out}} = \frac{R_{set}}{50k\Omega} \cdot n_{div} \cdot 1.024 \times 10^{-3} \text{ (seconds)}$$

where the chip's internal divider ratio n_{div} is determined[16] from the DIVCODE according to $n_{div} = 2^{3k}$, thus n_{div}=512 here. These chips run on reasonably low power; this circuit draws about 0.1 mA when unloaded.

Later in the chapter, in connection with timers, we'll see another example in which the LTC6991 is used to generate a switched "hour of power" (Figure 7.65). And we'll see also how to use other members of this family, in VCO and timer applications (§7.2.4).

C. Oscillator + divider

Another class of oscillator–timers uses an oscillator (relaxation or otherwise) followed by a digital counter, to generate long delay times without resorting to large resistor and capacitor values. Examples of this are the 74HC4060 and the Maxim ICM7240/50/60. These CMOS parts generate an output pulse for some number N of oscillator cycles,[17] and they run on a fraction of a milliamp. These timers (and their near relatives) are great for generating delays from a few seconds to a few minutes.

Some more recent additions to this class of oscillator include the LTC6903/4 (described earlier), and the Maxim DS1070/80 series. The LTC ICs operate from the same supply voltages as the LTC1799/6900 and generate rail-to-rail square-wave outputs over the larger range of 1 kHz to 68 MHz; but they require *no* external components! The trick is that the frequency is set by sending a pair of numbers (a 4-bit scaling factor and a 10-bit frequency coefficient) via a bit-serial digital-input line. That might sound complicated; but in fact it is remarkably easy to do in any system where you have an embedded microcontroller (a small computer-on-a-chip), which in contemporary electronics includes just about any electronic circuit. We'll see much more of that in Chapter 15 ("Microcontrollers").

The DS1070/80 series "EconOscillators" is similar, with serial programming to set the frequency; that is, held in internal nonvolatile memory, so you need to program it only once, or when you want to change it. The DS1085 is currently the fanciest member of the family, with four programmable settings that let you choose an output frequency (in the range of 8.1 kHz to 133 MHz) to better than one part in a thousand. Beware, though, that the initial accuracy and stability of this "silicon oscillator" is only at the ±1% level; that is, its resolution (step size) greatly exceeds its accuracy and stability. Think of it as a "1% oscillator" whose frequency you can set in the field.[18]

There's a nice generalization of these "divide-by-N" methods, namely the *phase-locked loop* (PLL). We treat them later in this chapter (§7.1.8B) and in Chapter 13.

D. Voltage-controlled oscillators

Other IC oscillators are available as voltage-controlled oscillators (VCOs), with the output rate variable over some range according to an input control voltage. We saw the germ of this idea when we used a current source to charge the capacitor in the 555; with little additional effort we could have made the current proportional to a controlling input voltage. There are many uses for a VCO, and so there are many offerings from the chip manufacturers. Some of these have frequency ranges exceeding 1000:1. Examples are the original NE566 and later designs like the ICL8038, MAX038, XR2206/7, and 74LS624–9 series.

The 74LS624 series, for example, generates digital-logic-level outputs up to 20 MHz and uses external RC's to set the nominal frequency. Faster VCOs like the 1648 can produce outputs to 200 MHz, and there are much higher-frequency VCO techniques (like Gunn-diode oscillators and YIG oscillators) that operate in the many *giga*hertz range.

Where linearity is important, a precision voltage-to-frequency (V/F) converter like the AD537, LM331, or AD650 really does the job, with worst-case linearity of 0.15%, 0.01%, or 0.005%, respectively. Most VCOs use internal current sources to charge and discharge a capacitor, and many therefore provide triangle-wave outputs. The classic Exar XR2206 goes further – it includes a set of "soft" clamps to convert the triangle wave to a not-too-great sine wave; they call this a *sine shaper*, and it produces a sine-looking wave with <1% distortion. Depending on external timing components, it goes from fractional hertz

---

[16] Valid for k between 0 and 7; see the datasheet for more detail.

[17] Specifically, 10 choices of $N = 2^k$ (going from k=4 to k=14) for the 74HC4060; any of N=1 to 255 for the ICM7240; N=128 for the ICM7242; any of N=1 to 99 for the ICM7250; and N=60 for the ICM7260.

[18] For better accuracy, you can use a crystal oscillator (§7.1.6) upstream of the divider. A nice example, sadly discontinued, was the Epson SPG series of "selectable-output crystal oscillators," in which you chose the output frequency with six programming pins that you connect to ground or to the +5 V supply.

(at the low end) to 1 MHz (at the high end),[19] with a sweep range of 1000:1 and a temperature stability of frequency of 0.002%/°C. You can also use it as a triangle-wave generator, in which mode it lets you adjust the duty cycle from 1% to 99%.

In Chapter 13 we'll see some additional VCO methods, there in the context of "voltage-to-frequency conversion." Those methods are *synchronous* and require a fixed-frequency clocking input; those clocking pulses are passed through to the output, or not, such that the average output frequency is accurately proportional to the input voltage; see §13.8.1 and §13.9.

VCO chips sometimes have an awkward reference for the control voltage (e.g., the positive supply) and complicated symmetrizing schemes for sinewave output. It is our opinion that the ideal VCO has yet to be developed. Many of these chips can be used with an external quartz crystal, as we discuss shortly, for much higher accuracy and stability; in such cases the crystal simply replaces the capacitor.

You can make VCOs with techniques other than *RC* (or current-driven) relaxation oscillators. For example, the frequency of an *LC* oscillator (§7.1.5D) can be electrically tuned with a voltage-variable capacitor (a *varactor*), though the tuning range is much less (typically 1–10%) than a tunable relaxation oscillator. Likewise it is possible to "pull" the frequency of a quartz crystal over a narrow range of perhaps 0.01%. Other oscillator technologies (Gunn oscillators, dielectric resonator oscillators, YIG-tuned oscillators, "current-starved" inverter chains, and others) permit electrical tuning by various means, an essential element of phase-locked-loop frequency synthesis (§13.13).

7.1.5 Sinewave oscillators

For many applications you need a real *sine* wave, rather than the rectangular, triangular, or other waveforms you get from relaxation oscillators. Examples are in audio-frequency test and measurement, in radio and video communications, and in medical and scientific research and applications. It is common in those applications to talk about *spectral purity* or *harmonic distortion*, which are measures of departure from an ideal sinewave.

The *RC* relaxation oscillators we've been talking about don't generate sinewaves – their native waveforms are ramps (linear or *RC* exponential) and rectangular waves. The XR2206 in §7.1.4D illustrates one method of making an approximate sinewave, namely by clipping a triangle with a succession of soft clamps; this method is used in some analog function generators.[20] However, there are oscillator configurations that generate sinewaves directly, and there are other tricks for making sinewaves from square waves.

Examples of native sinewave oscillators are the *Wien bridge* (which uses humble *R*'s and *C*'s), the *resonator oscillators* (which use resonators such as an *LC* circuit, quartz crystal, coaxial or cavity resonator, or even an atomic–molecular resonator), and the method of *direct digital synthesis* (DDS).

In this section we look at ways to make sinewaves, beginning with a few tricks for making sinewaves from square waves (or other non-sinusoidal waveforms), then moving to techniques that generate sinewaves directly. In a later section (§7.1.9) we discuss *quadrature oscillators*, which generate a pair of signals with a 90° phase relationship.

A. Sine from square

The simple trick here is to lowpass filter the square (or whatever) wave, to remove all but the fundamental frequency. An easy way to think about it is to remember that any[21] periodic waveform can be deconstructed into a set of sinusoidal components (the waveform's "Fourier series"), each with some fixed amplitude and phase; when you add those components together, you reconstruct the original waveform. The lowest frequency component (at the periodicity of the original waveform) is the *fundamental*, and all higher components ("harmonics") are integer multiples of the fundamental frequency (that is, at $2f_0, 3f_0, 4f_0, \ldots$). So you can create a sinusoidal wave from an arbitrary periodic wave by lowpass filtering at a frequency higher than the

---

[19] The MAX038 was similar, but faster – 20 MHz. We are saddened by its passing. Its obituary, posted on the Maxim website, reads "This product was manufactured for Maxim by an outside wafer foundry using a process that is no longer available."

[20] Such techniques of "corrupting" a triangle wave with diode clamps do not produce a high-quality sinewave: the resulting distortion can rarely be reduced below 1%. By comparison, most audiophiles insist on distortion levels well below 0.1% for their amplifiers. To test such low-distortion audio components, pure sinewave signal sources with residual distortion less than 0.01% or so are required.

[21] Well, *nearly* any: mathematicians are adept at finding pathological functions to rebut any loosely-worded proposition. We should probably say "any *well-behaved* periodic waveform"; but we can say, confidently, "any periodic waveform you can actually create with electronics."

fundamental but lower than the second harmonic,[22] thus extracting only the fundamental sinusoidal component.

As long as the filter doesn't pass the harmonic spectrum significantly, you'll get a good sinewave. As we discussed in Chapter 6, you can make lowpass filters in several ways: as "continuous-time" analog filters (with a network of discrete L's and C's; or, at lower frequencies, with active filters; or with discrete-time switched-capacitor filters; or with the numerical methods of digital signal processing).

To demonstrate this technique, we hooked up a 555 as a 1 kHz oscillator (Figure 7.10A, with R=75 k and C=10 nF), and ran the square-wave output through a lowpass active filter (a continuous-time 8-pole Butterworth) with breakpoint at 1.5 kHz. The output looked like a pretty good sinewave, and in fact it measured just 0.6% distortion[23] This technique requires a somewhat complex analog circuit, and it lacks frequency agility (i.e., once you've chosen the filter's breakpoint frequency, you can vary the oscillator frequency by only a small amount, say ±25%).

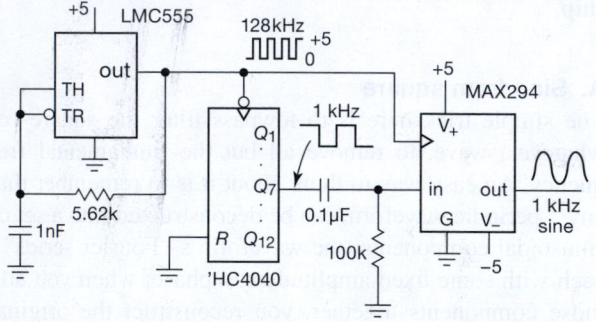

Figure 7.18. Sinewave generation with tracking lowpass filter. The MAX294 (or similar MAX293 or LTC1069-1) is an 8-pole elliptic lowpass switched-capacitor filter requiring no external components.

Switched-capacitor filters are simpler to use, being available as inexpensive clocked ICs; instead of requiring R's and C's to set the passband, the clock frequency tunes the filter – it determines the breakpoint frequency. Figure 7.18 shows a simple circuit in which a MAX294 switched-capacitor lowpass filter is used to convert an input square wave to an output sinewave. The MAX294's breakpoint is at $f_{CLK}/100$, so we've clocked it at 128 times the input square wave frequency. This puts the breakpoint at

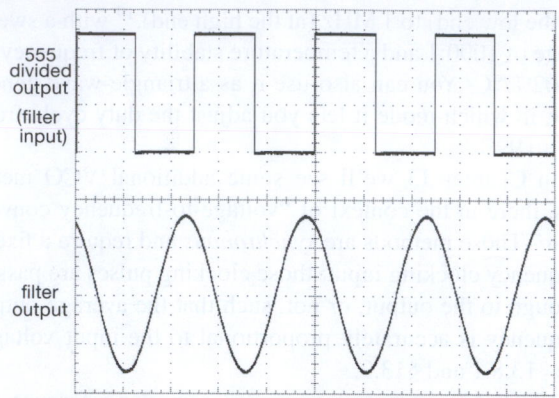

Figure 7.19. Waveforms from the circuit of Figure 7.18: the 1 kHz square-wave input ($f_{OSC}/128$), and the resulting sinewave after lowpass filtering. The output amplitude is equal to the fundamental component of the square wave input, which is a factor of $4/\pi$ larger. Horizontal: 400 μs/div. Vertical: 2 V/div.

$1.28 f_{IN}$, and results in the handsome sinewave shown in Figure 7.19; it measured a very clean 0.03% distortion.

Not only is this circuit simpler than the analog lowpass scheme, it provides highly predictable output amplitude[24], it provides tunability – that's because the filter tracks the input frequency, keeping its lowpass breakpoint at $1.28 f_{IN}$ as the input frequency is varied.[25] In this circuit we varied the 555's frequency by factors of ten, and measured a maximum of 0.1% distortion for output sinewave frequencies over the range of 100 Hz to 10 kHz.

B. Wien bridge oscillator

At low to moderate frequencies the Wien bridge oscillator (Figure 7.20) is a useful source of low-distortion sinusoidal signals. The idea is to make a feedback amplifier with 0° phase shift at the desired output frequency, then adjust the loop gain so that a self-sustaining oscillation just barely takes place. For equal-value R's and C's as shown, the voltage gain from the noninverting input to op-amp output should be exactly +3.00. With less gain the oscillation will cease, and with more gain the output will saturate. The distortion is low if the amplitude of oscillation remains within the linear region of the amplifier, i.e., it must not be allowed to go into a full-swing oscillation. Without some trick to control the gain, that is exactly what

---

[22] Depending on the waveform's symmetry, the Fourier series may consist of odd harmonics only, in which case the first higher harmonic is at $3f_0$; this is the case for a square wave of 50% symmetry.

[23] Putting the breakpoint at 1.2 kHz lowered the distortion to a mere 0.1%, or −60 dBc (dB relative to the carrier).

[24] Given by the first term of a square wave's Fourier series, $A_{PP}=(4/\pi)V_{cc}$, thus 2.25 Vrms.

[25] An elegant variation is illustrated in the LTC1799 datasheet, in which the tracking switched-capacitor lowpass filter is further configured with a stopband *notch* at the input signal's $3f_0$, thus providing additional attenuation at the square wave's strongest harmonic.

will happen, with the amplifier's output increasing until the effective gain is reduced to 3.0 because of saturation. The tricks involve some sort of long-time-constant gain-setting feedback, as we will see.

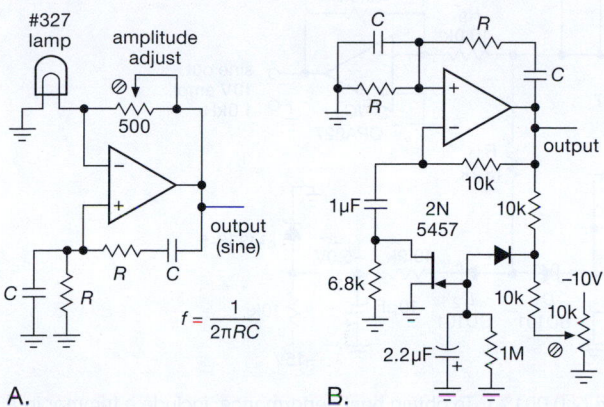

A.

B.

Figure 7.20. Wien-bridge low-distortion oscillators. A. Amplitude control with incandescent lamp. B. Amplitude control by a JFET's variable resistance.

In the first circuit, an incandescent lamp is used as a variable-resistance feedback element. As the output level rises, the lamp heats slightly, increasing its resistance and therefore reducing the noninverting gain. The circuit shown has less than 0.003% harmonic distortion for audio frequencies above 1 kHz; see LTC App. Note 5 (12/84) and App. Note 43 (6/90) for more details.[26]

In the second circuit, an amplitude discriminator consisting of the biased divider and diode charges a long-time-constant RC; this voltage adjusts the ac gain by varying the resistance of the FET, which behaves like a voltage-variable resistance for small applied voltages (see §3.2.7). Note the long time constant used (2 seconds); this is essential to avoid distortion, because fast feedback will distort the wave by attempting to control the amplitude within the time of one cycle.

Another interesting technique for amplitude control is shown in Figure 7.21, where a photoresistive optocoupler is used for gain feedback. These devices consist of an LED illuminating a resistive element, with the output terminals providing a resistance of good linearity (<0.1% distortion

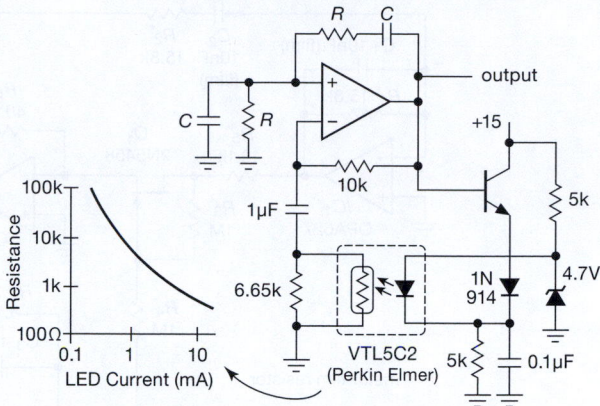

Figure 7.21. A photoconductive opto-isolator provides another method of amplitude control in the Wien-bridge oscillator. (Courtesy Steve Cerwin)

for applied voltages <1 Vrms), varied over several decades of resistance according to the LED current. Unlike silicon devices, they are intrinsically slow (tens of milliseconds for the device shown), which is helpful in this application.

It has been claimed that, with careful design, Wien-bridge oscillators can be built with distortion "well into the parts-per-billion range." Tricks to do this, which include the use of cascaded op-amps (for high loop gain, therefore low distortion), and cancellation of remaining harmonic distortion, are described in *Linear Technology Magazine*, Feb. 1994, pp. 26–28.

An ultralow-distortion design

We are skeptical of this claim.[27] However, with some attention to detail the distortion can be brought down to levels of a few parts per million (0.0001%) using quite conventional components and techniques. Figure 7.22 shows such a circuit, which we designed and tested to find out just how difficult this might be.

We began with a two-op-amp circuit variant, which has the advantage of using the inverting configuration; this reduces distortion by eliminating the common-mode signals present in the simpler non-inverting amplifier. The OPA627 is a fast (f_T=16 MHz), low-noise (e_n=4.5 nV/$\sqrt{\text{Hz}}$), low input-current ($I_B \approx 1$ pA) op-amp, with the particular advantages here of intrinsically low distortion (0.00003% as a unity-gain follower with a 10 V signal at 1 kHz) and the ability to operate with ±15 V supplies. (We want to operate with a large signal swing to minimize the effect of noise on the sinewave purity.)

---

[26] The use of a lamp to stabilize the Wien bridge oscillator was invented and patented (#2,268,872) by William Hewlett in 1942. The resulting model 200A audio oscillator was the first commercial product sold by Hewlett, with his partner David Packard; it cost $54.40 (but not because of cost calculations: they liked the slogan "54-40 or Fight!", presidential candidate Polk's aspirations for the northwestern border of the United States). The rest is history.

[27] We counsel a cautious skepticism when confronting any claims of parts-per-billion linearity in analog circuits.

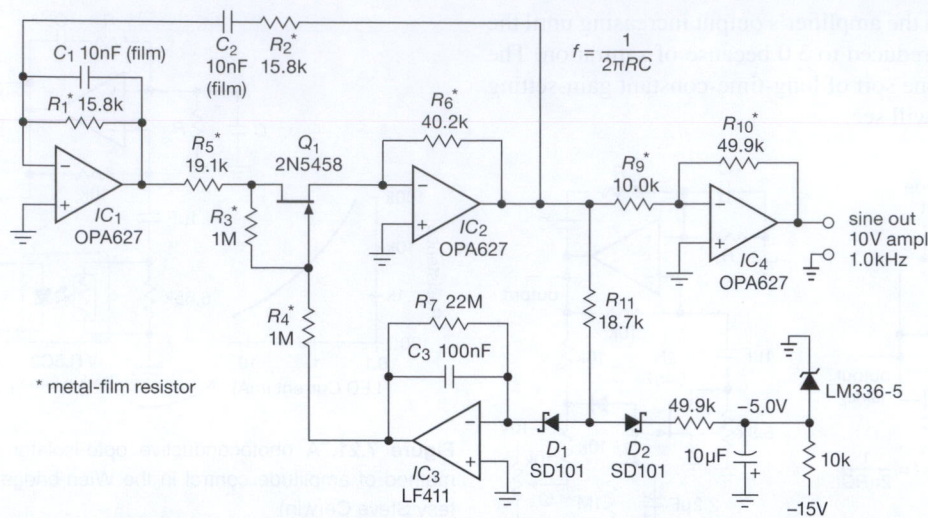

Figure 7.22. Wien-bridge (1 kHz) oscillator with unusually low distortion (<0.001%). To obtain best performance, include a trimmer in R_5, adjusted to optimize the controlled value of the JFET's resistance.

For this configuration, sustained oscillation occurs when IC_2's voltage gain is -2.00. We chose R_5 to be 5% smaller than critical, with the series JFET providing the adjustable 1k (nominal) additional resistance. That puts a 100 mV (peak-to-peak) sinewave across the JFET, which we judged small enough for good linearity, especially with the linearizing divider R_3R_4 (see §3.2.7A). Amplitude control is provided by integrator IC_3, which receives pulses of input current (via the divider to a stable -5 V reference) when the sinewave output of IC_2 reaches 2 V amplitude: its negative-going output back-biases the JFET's gate, relative to the source at virtual ground, raising the JFET's resistance and thus lowering the gain of IC_2 to maintain this output amplitude.[28] For the values shown, the JFET's minimum R_{ON} (i.e., at $V_{GS}=0$) must be less than 1k, which requires a minimum g_m of 1 mS (see §3.2.7); the 2N5458 specifies a minimum g_m of 1.5 mS, so the circuit is guaranteed to start up. We tacked on an inverting gain-of-5 stage to produce a healthy 10 V amplitude output.

The circuit worked "out of the box" – correct frequency and amplitude (1 kHz, 10 V) and a good-looking sinewave. The measured total harmonic distortion (THD) was an admirable 0.002%.[29] Before celebrating, though, we tried some variations: (a) Replacing the film capacitors with ceramic ("X7R" type) raised the distortion[30] a hundred-

fold, to 0.22%! (b) Dropping the swing across the JFET to 50 mVpp (by raising R_5 to 19.6k) halved the distortion, to 0.001%; from here on we stuck with this smaller JFET swing. (c) Next we trimmed the ratio of R_3/R_4 slightly, to minimize the (dominantly second harmonic) distortion, achieving a final THD figure of 0.0002%; that's -114 dB down from the signal, a mere 2 parts per million! (d) Finally, to see the effect of the linearizing gate divider, we omitted R_4, which raised the distortion 50-fold, to 0.01%.

Some important lessons from this exercise, if you want the lowest distortion, are (a) avoid inexpensive ceramic capacitors, (b) use the gate-linearizing trick (subtracting $V_{DS}/2$ from V_{GS}), and (c) keep the swing small across JFETs being used as resistors, preferably less than 100 mV (which, however, causes a rather lengthy amplitude settling time). Because JFET nonlinearity dominated the distortion, even when trimmed, we could have reduced the distortion still further by running the oscillator at lower amplitude, say 0.5 V, at the expense of added broadband noise produced by the fixed noise contribution of the op-amps.[31]

C. *RC* phase-shift oscillator

Unlike the relaxation oscillator (where an *RC* time constant is combined with voltage thresholds to make an

---

[28] We chose the integrator gain to put the unity-gain frequency of the control loop roughly at 50 Hz.

[29] Almost entirely second harmonic.

[30] Now dominated by third harmonic.

[31] Better yet, use a photoresistive gain control, as in Figure 7.21. Jim Williams did this, and he also added a lowpass filter between IC_3 and R_4 to attenuate the integrator's small cycle-by-cycle error-correcting waveform, achieving a measured distortion below 3 ppm; see LTC app-note AN132.

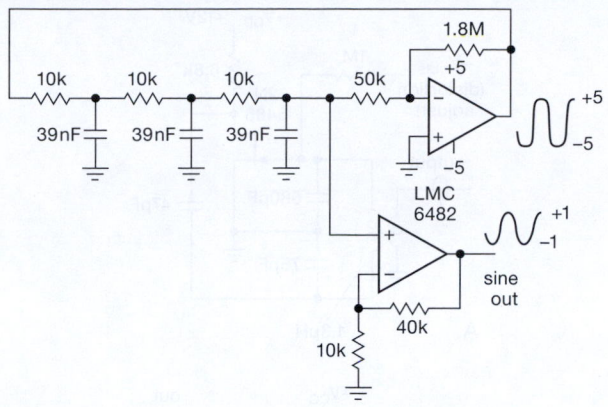

Figure 7.23. Phase-shift oscillator. Three *RC* sections produce a 180° phase shift, converted to positive feedback by the inverting amplifier.

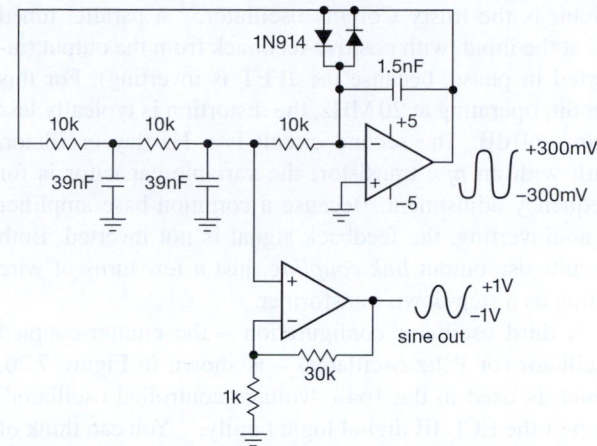

Figure 7.24. A variation on the phase-shift oscillator. An integrator adds 90° lagging phase shift (and inversion) to a 2-section *RC*. (Courtesy Tony Williams)

oscillation), the preceding Wien-bridge oscillator exploits the phase-shifting characteristics of an *RC* network, in a circuit with positive feedback, to select its operating frequency. This same idea is used in *phase-shift* oscillators: gain and feedback are applied around a network of several *R*'s and *C*'s, arranged so that the resulting loop oscillates at a frequency set by the network. Figure 7.23 shows a classic example.

The three-section *RC* produces a lagging phase shift that increases with frequency, reaching 180° at approximately $\omega = 2.4/RC$, where the loss through the network is a factor of 26.[32] The inverting amplifier provides the remaining 180° phase shift, and also the needed voltage gain (here a conservative $G_V = -36$). The circuit oscillates at 1 kHz, with a (rather distorted) clipped sinewave that swings rail-to-rail (i.e., ±5 V). However, the waveform at the last *RC* section is nicely sinusoidal, and, after an ×5 gain stage, emerges as a 1 V amplitude sinewave with just 0.9% distortion.

For phase-shift oscillator devotees, there is a world of possible variations: discrete transistor configurations, amplitude-limiting feedback schemes, and so on. Although we tried mightily, we are unable to resist the temptation to show another phase-shift oscillator (Figure 7.24). Here an (inverting) integrator provides 270° of lagging phase shift, so only two *RC* sections are needed to close the loop in-phase. This circuit illustrates also the use of back-to-back diode clamps for amplitude limiting. As with Figure 7.23, we've derived the output from the last *RC* section, where its distortion is minimum; the circuit's output is a 1 kHz sinewave of 1 V amplitude and 1% distortion. If

you were instead to put the diode limiters across the first 39 nF phase-shifting capacitor, the integrator output would be another sinusoidal wave of low distortion – in fact, it would be a 90° lagging wave (cosine, inverted), thus creating a "quadrature pair" (§7.1.9, though here of unequal amplitudes).

D. *LC* oscillators

At high frequencies (say above a megahertz) a favorite method of sinewave generation is to use a *resonator* of some sort to establish the frequency of oscillation. The resonator itself may be electrical (e.g., an *LC* circuit), or electromechanical (e.g., a piezoelectric quartz crystal), or even atomic or molecular (e.g., a hydrogen maser). Some resonators are easily tuned (e.g., *LC*), whereas others are quite stably fixed (e.g., a quartz crystal). Resonator-based oscillators are fundamentally different from the preceding *RC*-based oscillators, because they exploit a system that has an intrinsic resonant frequency (like a crystal resonator), compared with an *RC* circuit's nonresonant time constant (or phase shift). Because these resonances can be both narrow in frequency and stable over time, they are well suited for conscription into the noble service of oscillation.

We begin with *LC*-controlled oscillators, which play an important role in communications, and in which a tuned *LC* circuit is connected in an amplifier-like circuit to provide gain at its resonant frequency. Overall positive feedback is then used to cause a sustained oscillation to build up at the *LC*'s resonant frequency; such circuits are self-starting.

Figure 7.25 shows two popular configurations. The first

---

[32] The effects of loading cause those values to deviate from the ideal (fully isolated stages) values of $\omega = \sqrt{3}/RC$ and loss factor of 8.

circuit is the trusty Colpitts oscillator,[33] a parallel tuned *LC* at the input, with positive feedback from the output (inverted in phase, because the JFET is inverting). For this circuit, operating at 20 MHz, the distortion is typically less than −60 dB. The second circuit is a Hartley oscillator, built with an *npn* transistor; the variable capacitor is for frequency adjustment. Because a common-base amplifier is noninverting, the feedback signal is not inverted. Both circuits use output *link coupling*, just a few turns of wire acting as a step-down transformer.

A third oscillator configuration – the emitter-coupled oscillator (or Peltz oscillator) – is shown in Figure 7.26, which is used in the 1648 "voltage-controlled oscillator" chip of the ECL III digital logic family.[34] You can think of this as a feedback noninverting differential amplifier, with a parallel *LC* to set the frequency of oscillation. The 1648 will operate to 200 MHz, with the operating frequency set, as usual, by the resonant frequency of the parallel *LC*: $f_0 = 1/2\pi\sqrt{LC}$. Although the datasheet claims "high spectral purity," we found it to be mediocre at best when compared with a Clapp oscillator using a single JFET (see Figure 7.30 later in this section).

Electrical tunability

LC oscillators can be made *electrically* tunable over a modest range of frequency. The trick is to use a voltage-variable capacitor ("varactor") in the frequency-determining *LC* circuit. The physics of diode junctions provides the solution, in the form of a simple reverse-biased diode: the capacitance of a *pn* junction decreases with increasing reverse voltage (§1.9.5B). Although any diode acts as a varactor, you can get special varactor diodes designed for the purpose; Figure 7.27 shows the tuning characteristics of some representative types. And Figure 7.28 shows how to use a varactor to achieve ±1% electrical tunability, in this case with simple Armstrong-type JFET oscillator (with transformer-coupled feedback from the source). In this circuit the tuning range has been made deliberately small to achieve good stability, by using a relatively large fixed capacitor (100 pF) shunted by a small tunable capacitor (maximum value of 15 pF). Note the large biasing resistor (so the diode bias circuit doesn't load the oscillation) and the dc blocking capacitor.

Varactors typically provide a maximum capacitance of a few picofarads to a few hundred picofarads, with a tun-

[33] Edwin H. Colpitts, US patent 1624537, filed in 1918, but not granted until 1927.

[34] It is an *oscillator*; but to get voltage control you've got to use a tuning varactor, as explained below.

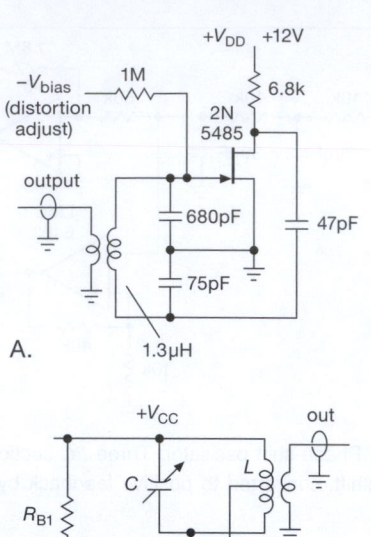

A.

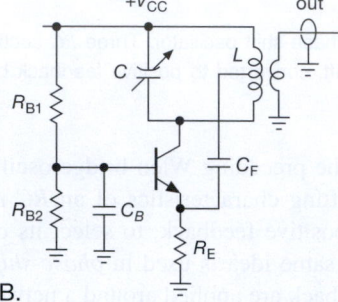

B.

Figure 7.25. Popular *LC* oscillator configurations. A. Colpitts with center-tapped resonating capacitor. B. Hartley with a center-tapped resonating inductor.

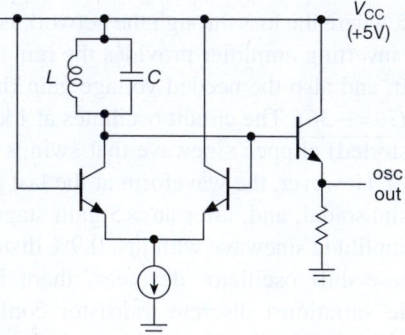

Figure 7.26. Emitter-coupled oscillator, a simplified version of that used in the MC1648 ECL-family IC.

ing range of about 3:1 (although there are wide range varactors with ratios as high as 15:1). Because the resonant frequency of an *LC* circuit is inversely proportional to the square root of capacitance, it is possible to achieve tuning ranges of up to 4:1 in frequency, though more typically you're talking about a tuning range of ±25% or so.

In varactor-tuned circuits, the oscillation itself (as well as the externally applied dc tuning bias) appears across the varactor, causing its capacitance to vary at the signal

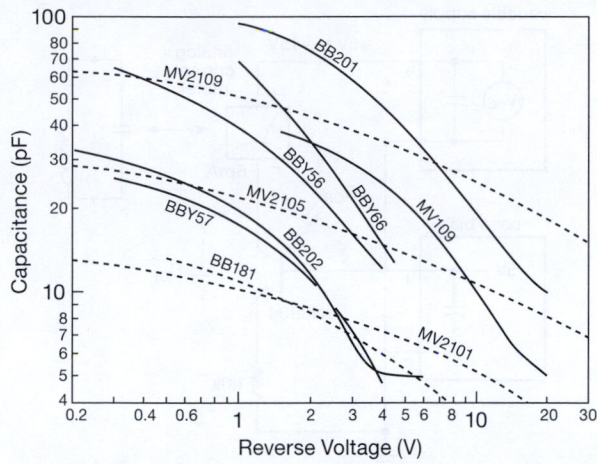

Figure 7.27. A reverse-biased diode exhibits capacitance that varies with applied voltage, shown here for several typical "varactor" tuning diodes. Those with steeper solid curves have "hyperabrupt" diode junctions.

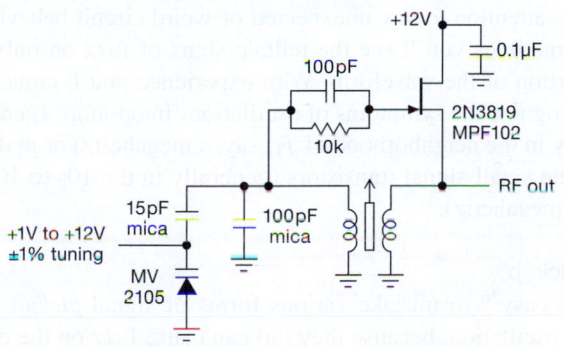

Figure 7.28. Varactor-tuned LC oscillator.

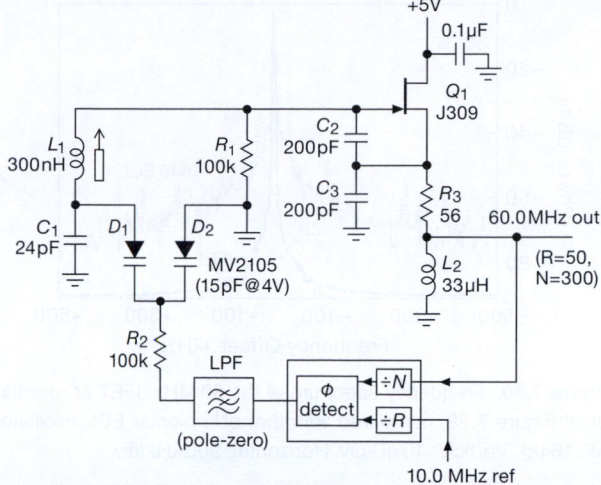

Figure 7.29. Low-noise JFET LC oscillator, used within a phase-locked loop (see §13.13). This design has unusually low sideband noise, as shown in the spectra of Figure 7.30.

frequency. This produces oscillator waveform distortion, and, more important, it causes the oscillator frequency to depend somewhat on the amplitude of oscillation. To minimize these effects, you should limit the amplitude of the oscillation (amplify in following stages, if you need more output); also, it's best to keep the dc varactor bias voltage above a volt or so, in order to make the oscillating voltage small by comparison.

An additional technique that helps mitigate this signal-biasing effect is to use a series pair of back-to-back varactors, so that the oscillating voltage seen by the two varactors acts to change their capacitances in opposite directions. This is illustrated in the low-noise oscillator of Figure 7.29 (see also Figure 6.8), used within a phase-locked loop (§13.13) to produce a clean 60 MHz "local oscillator" for a radio-astronomy receiver. This particular configuration is known as a Clapp oscillator, for which the fre-

quency is ordinarily set by the series resonant L_1C_1. Here we've added a parallel capacitance across C_1, consisting of the series pair of varactors. The tuning voltage is applied via R_2, which puts them both in equal back-bias (relative to their anodes at zero volts). The varactors, of equal capacitance, each take half the oscillation voltage, producing capacitance changes of opposite sign and (if the signal is not too large) of approximately equal magnitude. The net effect is a greatly reduced change in the capacitance of the series pair, and therefore lower distortion and frequency pulling. The measured signal purity was approximately 10 dB better than a good commercial frequency synthesizer (HP 3325A). A frequency spectrum of its output is shown in Figure 7.30, where its signal purity is compared with that of an MC1648 emitter-coupled oscillator using similar LC components and operating at approximately the same frequency.[35]

Electrically tunable oscillators are used extensively to generate frequency modulation, as well as in RF phase-locked loops like this one. PLLs are treated in Chapter 13.

For historical reasons we should mention a close cousin of the LC oscillator, namely the tuning-fork oscillator. It used the high-Q oscillations of a metallic tuning fork as the frequency-determining element of an oscillator, and it found use in low-frequency standards (stability of a few parts per million, if run in a constant-temperature oven) as well as in wristwatches. These objects have been super-

---

[35] To preserve the oscillator's free-running spectral purity when embedded in the PLL, we included an LC trap to suppress spurs at the phase-detector's 200 kHz reference frequency, see Figure 6.8.

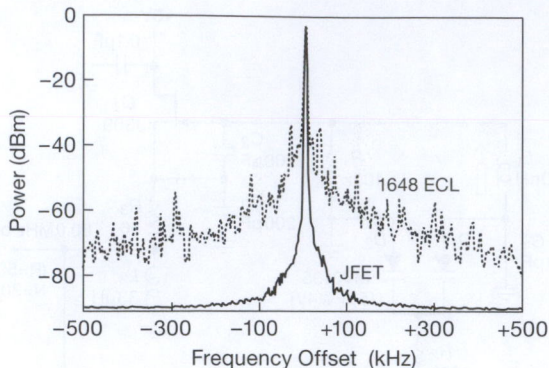

Figure 7.30. Frequency spectrum of the 60 MHz JFET *LC* oscillator of Figure 7.29, compared with that of a bipolar ECL oscillator (MC1648). Vertical: 10 dB/div. Horizontal: 200 kHz/div.

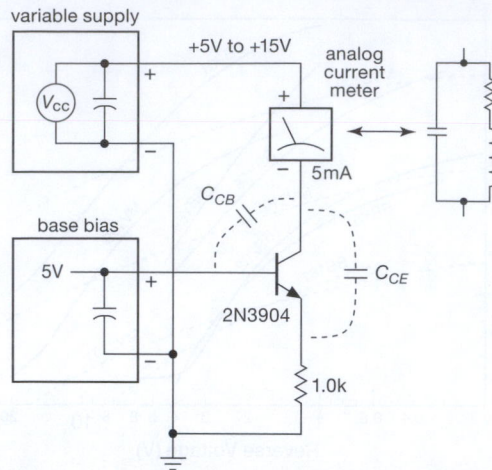

Figure 7.31. Unintended Hartley oscillator causes current-source mischief.

seded by quartz ("crystal") oscillators, which are discussed in §7.1.6. Interestingly, however, quartz crystals made for *low-frequency* operation (e.g., the 32.768 kHz used in wristwatches) oscillate in a mechanical tuning-fork mode.

E. Parasitic oscillations

Suppose you have just made a nice amplifier and are testing it out with a sinewave input. You switch the input function generator to a square wave, but the output remains a sinewave! You don't have an amplifier; you've got trouble.

Parasitic oscillations aren't ordinarily as blatant as this. They are usually observed as fuzziness on part of a waveform, erratic current-source operation, unexplained op-amp offsets, or circuits that behave normally with the oscilloscope probe applied, but go wild when the 'scope isn't looking. These are bizarre manifestations of untamed high-frequency parasitic oscillations caused by unintended Hartley or Colpitts oscillators exploiting lead inductances and interelectrode capacitances.

The circuit in Figure 7.31 shows an oscillating current source, born in an electronics lab course where an analog volt-ohm-milliammeter (VOM) was used to measure the output compliance of a standard transistor current source. The current seemed to vary excessively (5% to 10%) with load voltage variations within its expected compliance range, a symptom that could be "cured" by sticking a finger on the collector lead! The combined meter capacitance and the collector–base capacitance of the transistor resonated with the meter inductance in a classic Hartley oscillator circuit, with feedback provided by collector–emitter capacitance. Adding a small resistor in series with the base suppressed the oscillation by reducing the high-frequency common-base gain. This is one trick that often helps.

There are opportunities for parasitic oscillations in any active circuit that has gain. You've just got to be alert, and pay attention to any unexpected or weird circuit behavior. Sometimes you'll see the telltale signs of fuzz on only a portion of the waveform. With experience you'll come to recognize the symptoms of oscillations in op-amps (generally in the neighborhood of f_T, say, a megahertz) or in discrete small-signal transistors (generally in the 10s to 100s of megahertz).

"Pickup"

It's easy[36] to mistake various forms of signal *pickup* for an oscillation, because they too can cause fuzz on the displayed signal. If you suspect pickup, check to see if you've got 60 Hz[37] (or perhaps 120 Hz), a clear indication of powerline coupling. This can originate in capacitive coupling to a high-impedance point in the circuit. Or it can come from inductive coupling into a portion of your circuit that has some geometrical area linked by the alternating magnetic field. A third possibility is through a *ground loop* (portions of the circuit referenced to local grounds that are not at the same potential). Even in a well-layed-out circuit this latter problem can be severe, for example when it is connected to some external instrument that is plugged into a different ac power outlet. Higher-frequency pickup is common, too: coupling from switching power supplies (treated in Chapter 9), generally in the range of 20 kHz to 1 MHz; or

---

[36] Easier than you might think, because of (among other things) confusion caused by "aliasing" in digital oscilloscopes.

[37] 50 Hz if you live almost anywhere other than the Americas (or one of the other globally-scattered 60 Hz enclaves).

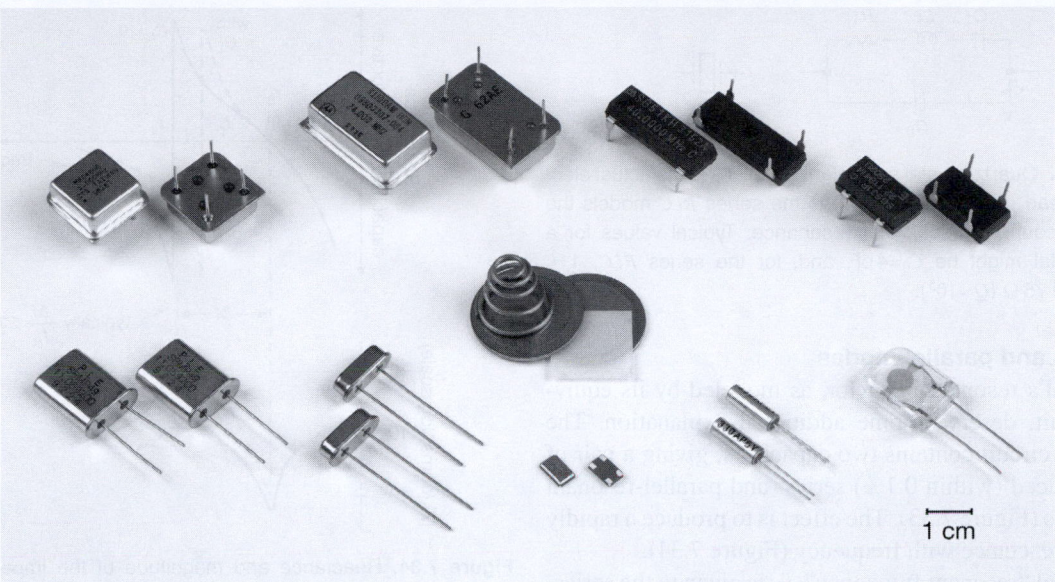

Figure 7.32. Quartz crystal packages. Across the top row are complete oscillator modules in DIP-8 and DIP-14 sizes; a much smaller alternative is the tiny 7mm×5mm surface-mount oscillator module at bottom center. The strange object in the middle is a bare crystal, shown with its spring-loaded electrode plates. You don't see those anymore; instead, crystals come in the popular sealed packages known as (left to right, bottom row) HC49/U, HC49/US, and 3mm tubular. We were lucky to find that oddball glass case at the right: inside you can see the gorgeous quartz disk with its plated electrodes.

modulated RF pickup from broadcast stations (US allocations are 0.5 MHz–1.7 MHz for AM, 88 MHz–108 MHz for FM, and anywhere from 55 MHz–700 MHz for television).

7.1.6 Quartz-crystal oscillators

RC (or capacitor plus current source) relaxation oscillators can easily attain stabilities approaching 0.1%, with initial predictability of 5% to 10%. That's good enough for many applications, such as a vacuum fluorescent display (VFD) in which the individual characters of a multicharacter display are driven sequentially in rapid succession (a 100 Hz overall rate is typical); this is called a *multiplexed display* (see the example circuit in §10.6.2). Only one character is lit at any time, but, if the entire display is "refreshed" rapidly enough, your eye sees the whole display without obvious flicker. In such an application the precise rate is quite irrelevant – you just want something in the ballpark. As stable sources of frequency, *LC* oscillators can do a bit better, with stabilities of 0.01% over reasonable periods of time. That's good enough for nondemanding applications like an inexpensive radio. Both kinds of oscillators are easily tunable – with a variable *R* or current source (for the

relaxation oscillator) and with a mechanically or electrically tunable capacitor, or a *slug-tuned* inductor (for the *LC* oscillator).

But for real stability there's no substitute for a crystal oscillator. This uses a piece of quartz (silicon dioxide, the primary ingredient of glass) that is cut and polished to vibrate mechanically at a certain frequency. Quartz is *piezoelectric* (a strain generates a voltage, and vice versa), so acoustic waves in the crystal can be driven by an applied electric field and in turn can generate a voltage at the surface of the crystal. By plating some contacts on the surface, you wind up with an honest electrical circuit element that can be modeled by a sharply resonant *RLC* circuit, pretuned to some frequency (which is the mechanical resonant frequency of the little slab of single-crystal quartz). Quartz crystals come packaged as bare components or as complete oscillator modules; some examples are shown in Figure 7.32.

The quartz crystal's high Q (typically around 10^4–10^5) and good stability make it a natural for oscillator control as well as for high-performance filters. As with *LC* oscillators, the crystal's equivalent circuit provides positive feedback and gain at the resonant frequency, leading to sustained oscillations.

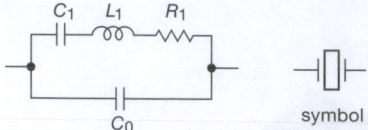

Figure 7.33. Quartz-crystal equivalent circuit. C_0 is the actual electrode and lead capacitance, whereas the series *RLC* models the electrically coupled *mechanical* resonance. Typical values for a 1 MHz crystal might be C_0=4 pF, and, for the series *RLC*, 1 H, 0.02 pF, and 75 Ω ($Q \sim 10^5$).

A. Series and parallel modes

The crystal's resonant behavior, as modeled by its equivalent circuit, deserves some additional explanation. The equivalent circuit contains two capacitors, giving a pair of closely spaced (within 0.1%) series- and parallel-resonant frequencies (Figure 7.33). The effect is to produce a rapidly changing reactance with frequency (Figure 7.34).

The label "resonant frequency" f_R is given to the series-resonant frequency of L_1 and C_1 (it's also called the series-resonant frequency f_S), so the net series reactance goes from being capacitive (below f_R) to inductive (above f_R). At f_R the net reactance of the series pair (L_1 and C_1) is zero, and the magnitude of the impedance is a minimum (and equals R_1).[38] Slightly above this in frequency (typically $\sim$0.1% higher) is the "antiresonant frequency" f_a, where the series combination of C_0 and C_1 (which is slightly less than C_1 alone) resonates with L_1. (Alternatively you can think of this as the parallel resonance of C_0 with the net series reactance of L_1 and C_1, which becomes increasingly inductive above f_R.) This is also called the parallel resonant frequency f_p, although that term should properly be reserved for the actual circuit situation in which an external load capacitance C_L is intentionally added in parallel (more on this soon). At this frequency (f_a or f_p) the net reactance again goes through zero, but this time with a peak in the magnitude of the impedance. When a crystal is operated in parallel resonance, the additional parallel capacitance added by the external circuit adds to the crystal's C_0 and lowers the resonant frequency somewhat. Crystals intended for operation in parallel resonant mode will specify a value of external shunt capacitance (typically in the range of 10–35 pF) for oscillation at the specified frequency stamped on the crystal.

This business of parallel and series is important, and any device that uses an external crystal will specify which mode it's using, along with some guidance on crystal pa-

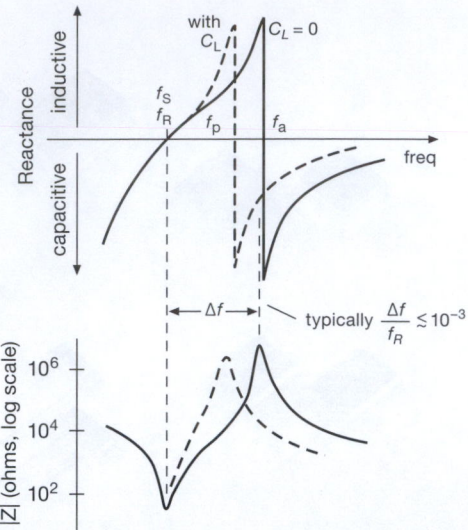

Figure 7.34. Reactance and magnitude of the impedance of a quartz crystal near its resonances, on a greatly expanded frequency scale. f_S and f_a are the series- and parallel-resonant frequencies [more precisely, the *resonant* (f_R) and *antiresonant* (f_a) frequencies, respectively]. An additional external capacitance C_L lowers the parallel-resonant frequency to f_p, when operated in that mode.

rameters (maximum allowable R_S, value of parallel capacitance). Better yet, you will see a listing of crystal manufacturers and part numbers that are known to work properly.

B. Exploring a quartz crystal

You can find plenty of sketches like Figure 7.34. But do you believe they accurately represent what real crystals do?

We weren't sure, so to find out we took a sample crystal (a CTS type MP100, specified as 10.0 MHz±45ppm, in series-resonant mode), and measured its impedance with a high-resolution vector impedance test instrument (an HP4192A). The latter can measure from 0.01 Ω to 200 kΩ, with a resolution of 1 Hz, and at frequencies to 13 MHz – perfect for this job. Our particular crystal had a measured series-resonant frequency of f_S=10.000086 MHz (that's a frequency error of +8.6ppm), an impedance (resistive) at resonance of R_1=4.736 Ω, and a parallel capacitance of C_0=5.5 pF. We know the product $L_1 C_1$ (from the resonant frequency), but not their individual values.[39] But we can get at them indirectly, by measuring the (unspecified) parallel-(anti)resonant frequency, and its variation

---

[38] Recall that the impedance of an ideal parallel *LC* goes to infinity at resonance, whereas that of a series *LC* goes to zero; §1.7.14.

[39] Recall there's no actual inductor or capacitor in there. These represent the electrical equivalent of the sharply resonant mechanical crystal, as piezoelectrically coupled by the attached electrodes. They're sometimes called a "motional" inductor and capacitor.

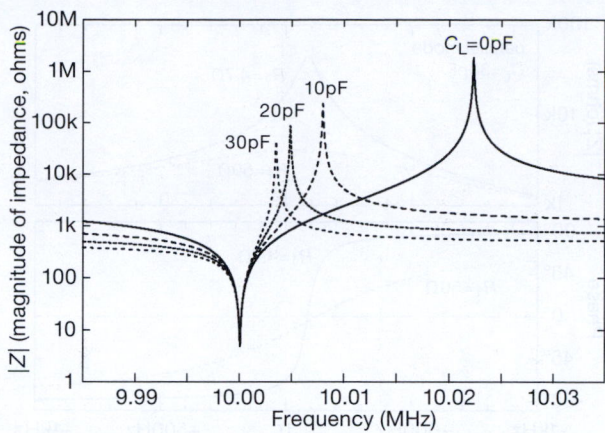

Figure 7.35. Impedance versus frequency for a sample 10.0 MHz series-resonant crystal, as modeled by SPICE from measured values of its *RLC* electrical model. Curves for four values of external parallel capacitance C_L are plotted.

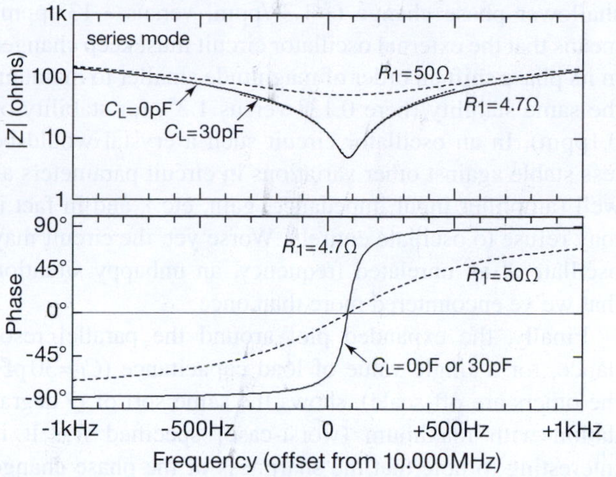

with added external parallel capacitance C_L. These we measured as f_a=10.02245 MHz (no C_L: C_0 only), and f_P=10.00355 MHz (with C_L=30 pF external parallel capacitance).

From these we can back out the values of L_1 and C_1, namely 10.3324 mH and 0.024515 pF.[40] And with those values we can enjoy many happy hours running the SPICE simulator, to learn how graphs of the crystal's impedance, phase shifts, and Q-value really look. Figures 7.35, 7.36, and 7.37 show such results.

These show the expected minimum at the series resonant frequency f_R (where $|Z|$=4.7 Ω), and that it varies very little with external capacitance (not visible, in fact, even in the expanded plot of Figure 7.36; that is, much less than 1 ppm going from C_L=0 pF to 30 pF). In contrast, the *parallel*-resonant frequency (impedance maxima) depends relatively strongly on external capacitance, which effectively "pulls" its resonance downward by ∼2000 ppm when 30 pF is added.

The fact that the parallel resonance is higher than the 10.0 MHz stamped on the case doesn't mean there's anything "wrong" with this crystal. Its frequency is simply specified for series-resonant circuit operation. If it were instead specified for parallel-resonant operation, this particular sample would be stamped "10.00355 MHz," and would specify "C_L=30 pF."[41] (Of course, you'd buy instead

---

[40] Formulas, for those who want them: $C_1 = 2(1 - f_B/f_A)/(1/C_A - 1/C_B)$, then $L_1 = 1/C_1(2\pi f_S)^2$.

[41] More formulas: the capacitive loading produces a parallel resonant frequency $f_P = f_S\{1 + C_1/2(C_0 + C_L)\}$. And, knowing the crystal's pa-

---

Figure 7.36. Impedance and phase in the neighborhood of the series resonance, for the 10.0 MHz crystal of Figure 7.35. Note that the graphs of impedance and phase are unaffected by external capacitance. The high-Q resonance of our sample, with its R_1=4.7 Ω, is considerably degraded for a crystal with worst-case specified R_1=50 Ω.

a 10.0 MHz standard frequency crystal specified for parallel mode; for this manufacturer that would have the part number MP101.)

From the relatively large frequency shifts with load capacitance, it's clear that you have to be careful to use the specified load capacitance (taking into account wiring and chip capacitances) when operating a crystal in parallel-resonant mode. Looking at the positive side, that means you can use an external variable capacitor to trim the operating frequency (or lock it, over a narrow range, using an electrically tunable varactor capacitor). On the negative side, it means that even small drifts in the external circuit's capacitance will cause frequency shifts. For example, to achieve 0.1 ppm frequency stability (assuming the crystal is that good over temperature or time) the external capacitance must not change by more than 0.002 pF; this is likely a difficult constraint for the external amplifier that closes the oscillation loop.

Our crystal measured an impressively low value of R_1=4.7 Ω, compared with the manufacturer's worst-case maximum of 50 Ω. To see how that changes things, we included that worst-case value in the expanded plots of the series and parallel resonances (Figures 7.36 and 7.37). For the series resonance the impedance minimum is shallower, and the phase shift versus frequency is more gentle. The

---

rameters, the load capacitance required for getting a parallel-resonant frequency f_P is $C_L = \{f_S C_1/2(f_P - f_S)\} - C_0$.

shallower phase change (~1.3°/ppm, versus ~13°/ppm) means that the external oscillator circuit must keep changes in its phase shift an order of magnitude smaller to maintain the same stability (here 0.13° versus 1.3°, for stability of 0.1ppm). In an oscillator circuit such a crystal would be less stable against other variations in circuit parameters as well (amplifier input impedance, gain, etc.), and in fact it may refuse to oscillate entirely. Worse yet, the circuit may oscillate at an unrelated frequency, an unhappy situation that we've encountered more than once.

Finally, the expanded plot around the parallel resonance, for a single value of load capacitance ($C_L=30$ pF; the others are off-scale), shows the same sort of Q degradation with maximum (worst-case) specified R_1. It is interesting to note that the sharpness of the phase change (and width of the impedance maximum) is similar to that of the series-resonant case, contrary to statements that you'll sometimes hear.

To complete the oscillator circuit, the crystal is connected within a positive feedback loop. Some common oscillator circuits are shown in §7.1.6D, where you'll see external loading capacitances, a large-value resistor to complete the bias path, and (sometimes) a smaller series resistor. You can simulate the full circuit configuration, if you like, by using the crystal model above.[42] We've done so, but we will show restraint and conclude this discussion, declaring victory in revealing the nature of those often misunderstood crystal resonances.

C. Standard crystal frequencies

Quartz crystals are available from about 10 kHz to about 30 MHz, with overtone-mode crystals going to about 250 MHz. Although crystals have to be ordered for a given frequency, most of the commonly used frequencies are available off the shelf. Frequencies such as 100 kHz, 1.0 MHz, 2.0 MHz, 4.0 MHz, 5.0 MHz, and 10.0 MHz are always easy to get. A 3.579545 MHz crystal was used in analog TV color-burst oscillators. Digital wristwatches use 32.768 kHz (divided by 2^{15} to get a useful 1 Hz), and other powers of 2 are also common. A crystal oscillator can be adjusted slightly by varying a series or parallel capacitor, for example one of the 27 pF or 32 pF capacitors in Figures 7.38D and E. Given the low cost of crystals (much less than a dollar), it is worth considering a crystal oscillator in any application where you would have to strain the capabilities of *RC* relaxation oscillators.[43]

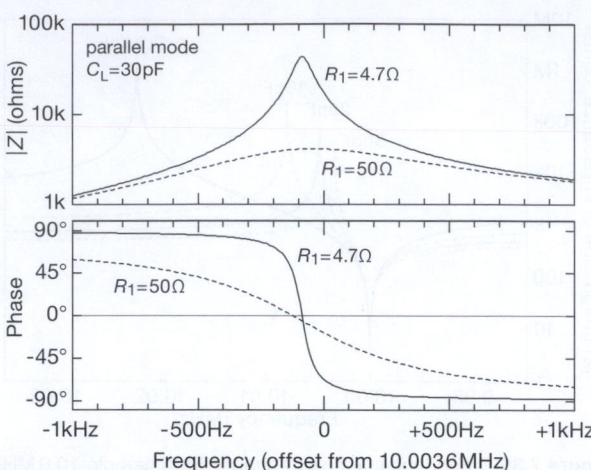

Figure 7.37. Impedance and phase in the neighborhood of the parallel resonance for the 10.0 MHz crystal of Figure 7.35. Here only one value of load capacitance ($C_L=30$ pF) is shown, because the other values are completely off-scale. Once again, the high-Q resonance of our sample is considerably degraded for a crystal with worst-case specified $R_1=50\,\Omega$.

Although crystal oscillators are not tunable, in the manner of relaxation oscillators or *LC*-based oscillators, you can use a varactor to vary some added external capacitance, thereby "pulling" the natural frequency of a parallel-mode quartz-crystal oscillator. The resulting circuit is called a "VCXO" (voltage-controlled crystal oscillator), and it augments the good-to-excellent stability of crystal oscillators with a small degree of tunability. The best approach is probably to buy a commercial VCXO, rather than attempt to design your own. Typically they produce maximum deviations of ±10 ppm to ±100 ppm from center frequency, though wide-deviation units (up to ±1000 ppm) are also available.

An alternative, and a popular one at that, is to synthesize (with either a phase-locked loop, §13.13, or by direct digital synthesis, §7.1.8) any desired output frequency, using a crystal oscillator's fixed-frequency output as a "reference." The synthesized frequency can be changed easily under digital control, and it is as stable as the crystal oscillator itself. As a consequence, most contemporary communications equipment (radios, televisions, transmitters, cell-phones, etc.) uses DDS or PLL synthesis to generate the needed internal frequencies.

---

[42] Readers unfamiliar with SPICE may benefit from Appendix J.

[43] A recent challenger in the low-cost resonator-based oscillator category is MEMS (microelectromechanical systems), in which the resonator is

made as a tiny etched silicon structure. Although these oscillators are not as stable as quartz, they can be *very* small (SiTime makes a series that measures 2 mm×2.5 mm×0.8 mm), and they can incorporate temperature compensation and frequency synthesis circuitry naturally within the same silicon technology.

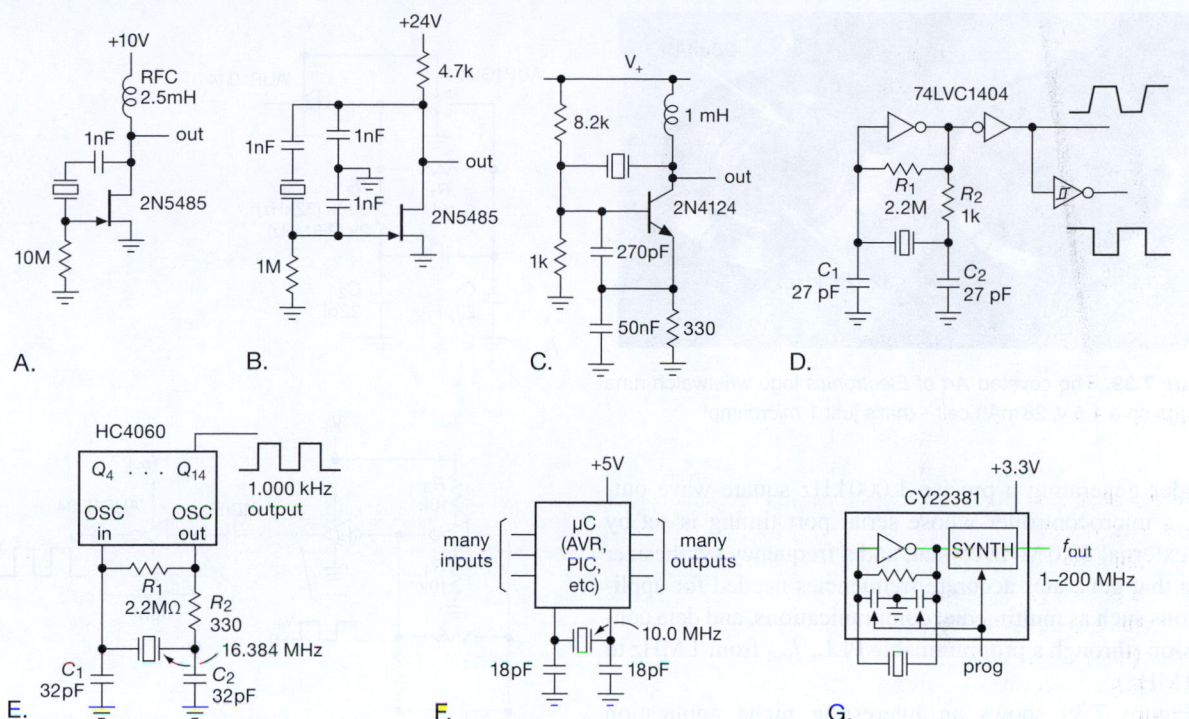

Figure 7.38. Various crystal oscillators. Circuits D–G exploit portions of digital logic circuits: an inverter, a 14-stage binary counter, a microcontroller, and a frequency synthesizer, respectively.

D. Crystal oscillator circuits

Figure 7.38 shows some crystal oscillator circuits. In Circuit A the classic Pierce oscillator implemented with the versatile FET (see Chapter 3). The Colpitts oscillator, with a crystal instead of an LC, is shown in circuit B. An *npn* bipolar transistor with the crystal as feedback element is used for circuit C. The remaining circuits generate logic-level outputs by using digital logic functions (circuits D through G). It's common to see an unbuffered logic inverter (i.e., a single CMOS transistor pair, as in Figure 3.90) used as a crystal oscillator (Figure 7.38D); in that application the inverter is biased into the linear region with a high-value feedback resistor, with the crystal providing (parallel-mode) resonant feedback. The LVC1404 is designed particularly for this application, with its pair of unbuffered inverters plus an optional Schmitt-trigger inverter (for generating abrupt output transitions); for low voltages (down to 0.8 V) the 'AUP1GU04[44] works well. In this circuit (and in Figure 7.38E) the series resistor R_2 should be

chosen comparable to the reactance of C_2 at the oscillator frequency.[45]

Let us pause for a moment to ask how these latter parallel-mode circuits can oscillate, given that the crystal has a 0° phase shift at its resonance (whether series or parallel mode; see Figures 7.36 and 7.37). What happens is that the load capacitance C_L is actually wired as a pair of capacitors (C_1 and C_2) in series, with the midpoint grounded. So when there's an oscillating voltage across the crystal, the two ends are seesawing 180° out of phase with each other; it's just like a center-tapped transformer winding. The inverting amplifier completes the 360° phase shift needed for a sustained oscillation.

Turning back to the remaining circuits in Figure 7.38, it's quite common to see a pair of "XTAL" terminals on more complex digital ICs (microprocessors, waveform synthesizers, serial communications chips, etc.), an invitation to use the chip's internal oscillator circuit (usually a prebiased unbuffered inverter). In Figures 7.38E–G we've shown three such examples – a 14-stage binary frequency

---

[44] Our practice is to strip off unimportant prefixes (hence the apostrophe) when talking about standard digital logic, as we'll explain in Chapter 10.

[45] For example, ∼330 kΩ for a 32 kHz oscillator, and ∼1 kΩ for a 5 MHz oscillator.

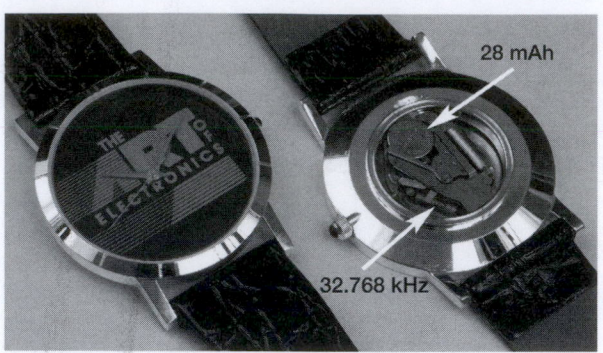

Figure 7.39. The coveted *Art of Electronics* logo wristwatch runs 3 years on a 1.5 V, 28 mAh cell – that's just 1 *micro*amp!

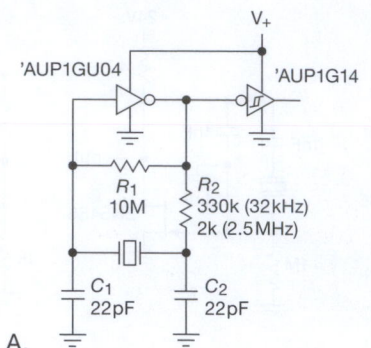

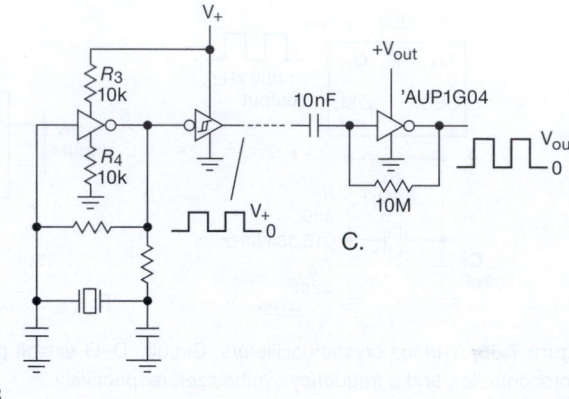

Figure 7.40. Micropower crystal oscillator. A. Unbuffered inverter oscillator with Schmitt second stage. B. Reducing linear "shoot-through" current. C. Low-voltage oscillator drives full-voltage output inverter.

divider generating a precise 1.000 kHz square-wave output, a microcontroller whose serial port timing is set by the external 10.0 MHz crystal, and a frequency synthesizer chip that generates accurate frequencies needed for applications such as multimedia, communications, and data conversion (through a programmable PLL, f_{out} from 1 MHz to 200 MHz).

Figure 7.39 shows an interesting niche application for crystal oscillators, the "quartz wristwatch." You need quartz stability here (there are 86,400 seconds in a day, so stability of "only" 1 part in 10^4 would cause a drift of a minute per week); and, you need *really* low power. These inexpensive mass-produced products run their oscillators, frequency divider electronics, and drive power for a tiny stepping motor on a power budget of about a microwatt.

A micropower design

Challenged by the astonishingly low power of highly customized quartz-oscillator wristwatch circuits, we looked further into what can be done using only standard components. We chose the low-voltage 74AUP logic family (specified for 0.8–3.3 V operation), and we tested the standard parallel-resonant Pierce configuration, using an unbuffered inverter for the oscillator followed by a Schmitt inverter second stage to generate a clean switching waveform (Figure 7.40A).

The measured total supply current versus supply voltage is plotted in Figure 7.41 for both a 32.768 kHz (wristwatch) crystal and a 2.5 MHz crystal; these curves (marked "$R_3=R_4=0$") show a rapid increase in supply current with increasing supply voltage, caused by the "class-A" oscillator current (the overlapping conduction of the nMOS and pMOS inverter pair for mid-supply input voltages; see Figure 10.101) during the transitions of the input waveform.

A nice trick to reduce greatly this effect is to add a pair of resistors in the supply leads (Figure 7.40B); this produced the supply-current curve marked "$R_3=R_4=10k$," a factor of 20–50 reduction for the 32.768 kHz oscillator. So we have a sub-microamp 32 kHz oscillator – but only for output voltages of less than a volt, which is lower than just about any logic device you might want the oscillator to drive. The final trick is to use that low-voltage oscillator to drive a full-voltage output stage (Figure 7.40C), with a blocking capacitor and large feedback resistor to bias the output stage in its linear region. With the oscillator and second stage running at 1.0 V, feeding a biased output inverter third stage running at 1.8 V, we measured 2.4 μA total supply current, compared with 12.8 μA for the 2-stage oscillator alone (Figure 7.40B) running at 1.8 V; this is a five-fold improvement.[46]

These modest experiments lead to the obvious question:

---

[46] The analogous figure for a 2.5 V output stage was 13.8 μA, a factor of 2.4 better than the two-stage circuit alone.

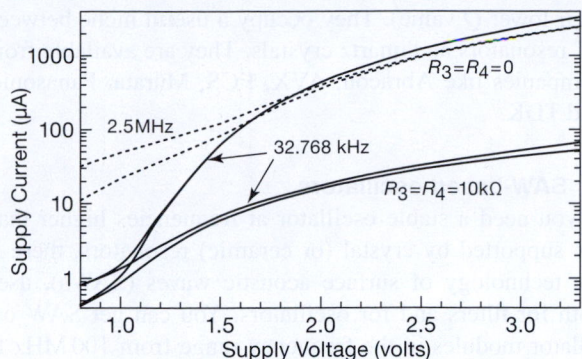

Figure 7.41. Measured power-supply current for the oscillators of Figure 7.40. Each curve pair plots first-stage (lower) and two-stage (upper) supply current.

how do the watchmakers do it? If you poke around a bit, you can find some very interesting datasheets. For example, a company called "EM Microelectronic" offers a tiny IC, the EM7604, which it describes as a "Low Power Crystal Oscillator Circuit 32.768 kHz." And they mean *low*: this puppy can run from 1.2 V to 5.5 V, and has a typical operating current of $0.3\,\mu$A or less at supply voltages between 3 V and 5 V. Perhaps a revealing hint as to their customer base is provided by the notation "Swatch Group Electronic Systems."

E. A caution
The proper design of crystal oscillators is not at all trivial – it is essential to ensure that the product of circuit gain A and crystal loss B (i.e., the loop gain AB) is greater than unity, and that the overall phase shift going around the loop is an integral multiple of $360°$, at the desired frequency of oscillation.[47] The crystal loss (caused by R_S in the equivalent circuit) may prevent proper oscillation, with the shunt capacitance C_P alone providing an adequate path for oscillation at a frequency unrelated to that stamped on the crystal. You must be careful, also, to select a crystal intended for operation at its series or parallel resonance, as required by the external oscillator circuit.[48] Oscillator chips and other ICs that accept an external crystal for their internal clocking (such as microcontrollers, see for example Figure 15.4) will state it clearly on the datasheet; here's an example, from the datasheet for the MPC9230 PLL synthesizer:

> ...As the oscillator is somewhat sensitive to loading on its inputs, the user is advised to mount the crystal as close to the MCP9230 as possible to

avoid any board level parasitics. ...Because the series resonant design is affected by capacitive loading on the XTAL terminals, loading variation introduced by crystals from different vendors could be a potential issue. For crystals with a higher shunt capacitance, it may be required to place a resistance across the terminals to suppress the third harmonic. ...The oscillator circuit is a series resonant circuit and thus for optimum performance a series resonant crystal should be used. Unfortunately, most crystals are characterized in a parallel resonant mode.

And here's an example from the 174-page (!) datasheet for the PIC16F7x microcontroller, for which you connect the crystal between two pins, and a capacitor from each to ground:

> The PIC16F7x oscillator design requires the use of a parallel cut crystal.

⋮

Capacitor selection for crystal oscillator (for design guidance only):
These capacitors were tested with the crystals listed below for basic start-up and operation. These values were not optimized. Different capacitor values may be required to produce acceptable oscillator operation. The user should test the performance of the oscillator over the expected V_{DD} and temperature range for the application. See the notes following this table for additional information.
[*a table of selected manufacturers' crystals follows, then more warnings.*]

These are not idle warnings (though the severity of their disclaimers may have been encouraged by the company's legal department). In various situations we've found that one company's crystals work fine, and another's, with similar specifications, do not. This is most likely due to poorly specified properties such as effective series resistance and mount capacitance. Our experience with discrete crystal-oscillator circuits has been, well, checkered.

F. Crystal oscillator modules
For reasons like these we favor the use of complete oscillator *modules*, to get bulletproof reliability. These cost more than bare crystals,[49] but they include oscillator circuitry

---
[47] These are the so-called *Barkhausen criteria* for oscillation.
[48] See, for example, RCA App. Note ICAN-6539.

[49] Typically $1.50 versus $0.30, in 100-piece quantities; twice that in single-piece quantities.

guaranteed to work, and they provide logic-level square-wave output. You can use them as the oscillator for any IC to which you could attach a bare crystal, because all such ICs will also accept a clocking square-wave input.

Oscillator modules come in IC-style packages, such as standard DIP and smaller 4-pin surface-mountable styles. They come off-the-shelf in lots of standard frequencies (e.g., 1, 2, 4, 5, 6, 8, 10, 16, and 20 MHz, up to 100 MHz or so), as well as some weird frequencies used in microprocessor systems (e.g., 14.31818 MHz, used for video boards; or 22.118 MHz, a favorite for 8051-type microcontrollers because of serial port timing). These "crystal clock modules" typically provide accuracies (over temperature, power-supply voltage, and time) of a modest 0.01% (100 ppm), but you get that performance inexpensively, reliably, and you don't have to wire up any circuitry.

If you need a non-standard frequency, you can get "programmable oscillator modules" that let you select the frequency, typically in the range of 1 MHz to 125 MHz or more; these cost roughly double that of standard modules (about $5 each in small quantities), and are "one-time" programmable (you select the frequency when you order; or you can buy a programmer for ~$500 and program the "blank" oscillator modules in the field). These use PLL techniques to synthesize the desired output frequency from the standard-frequency internal oscillator (§13.13).[50]

Some manufacturers of crystal (and ceramic) oscillator modules are Cardinal Components, Citizen, Connor Winfield, Crystek, CTS, Ecliptek, ECS, Epson, Fox, Seiko, and Vishay.

G. Ceramic resonators

Before going on to discuss oscillators of *higher* stability (§7.1.7), we must mention *ceramic resonators*. Like quartz crystals, these are piezoelectric mechanical resonators, with electrical properties similar to those of quartz. They come in a limited selection of frequencies from about 200 kHz to 50 MHz. However, they are less precise (typically ±0.3%), with correspondingly poorer stability (typically 0.2%–1% over temperature and time). The good news is that they are small, dirt cheap ($0.15–$0.25 in small quantities), available with built-in capacitors (for about $0.25–$0.50), generally interchangeable in any quartz crystal oscillator circuit, and can be "pulled" over a frequency range of a few parts per thousand (owing to

their lower Q value). They occupy a useful niche between *LC* resonators and quartz crystals. They are available from companies like Abracon, AVX, ECS, Murata, Panasonic, and TDK.

H. SAW-based oscillators

If you need a stable oscillator at frequencies higher than are supported by crystal (or ceramic) resonators, there is the technology of surface acoustic waves (SAWs), used both for filters and for oscillators. You can get SAW oscillator modules in the frequency range from 100 MHz to 1 GHz. They are tiny (just like crystal-oscillator modules) and they have comparable stabilities (50 ppm over temperature). The down side is that only a sparse set of standard frequencies is available, and these things tend to be somewhat pricey (upward of $50, in small quantities).

On the other hand, if you want to roll your own, you can get bare SAW *resonators* in the frequencies that are popular for garage-door openers, key fobs, and the like (433 MHz is widely used for these things) for about $1; add an inexpensive bipolar transistor and a few passive components and you've got an oscillator. Attach a short piece of wire, and you've got a transmitter!

7.1.7 Higher stability: TCXO, OCXO, and beyond

Without great care you can obtain frequency stabilities of a few parts per million over normal temperature ranges with crystal oscillators. By using temperature-compensation schemes you can make a TCXO (temperature-compensated crystal oscillator) with somewhat better performance. Both TCXOs and uncompensated oscillators are available as complete modules from many manufacturers, e.g., Bliley, Cardinal Components, CTS Knights, Motorola, Reeves Hoffman, Statek, and Vectron. They come in various sizes, ranging from modules down to SMT and DIP packages. TCXOs deliver stabilities of 1 ppm over the range 0°C to 50°C (inexpensive) down to 0.1 ppm over the same range (expensive).

A. Temperature-stabilized oscillators

For the utmost in stability, you may need a crystal oscillator in a constant-temperature oven ("OCXO"). A crystal with a zero temperature coefficient at some elevated temperature (80°C to 90°C) is used, with the thermostat set to maintain that temperature. Such oscillators are available as small modules for inclusion into an instrument or as complete frequency standards ready for rack mounting. The 1000B from Symmetricom is typical of high-performance

---

[50] Some manufacturers provide modules that offer a pin-selectable choice of several frequencies. An example is the ECS-300C series from ECS, which come in 8-pin packages, with 3 pins setting the binary division ratio (from 1/2 to 1/256 of the base frequency).

modular oscillators, delivering 10 MHz with stabilities of a few parts in 10^{11} over periods of seconds to hours.

When thermal instabilities have been reduced to this level, the dominant effects become crystal "aging" (the frequency tends to increase continuously with time), power-supply variations, and environmental influences such as shock and vibration (the latter are the most serious problems in quartz wristwatch design). To give an idea of the aging problem, the 1000B oscillator has a specified aging rate (after a month of break-in) of 1 part in 10^{10} per day, maximum. Aging effects are due in part to the gradual relief of strains, and they tend to settle down after a few months, particularly in a well-manufactured crystal.[51]

Oven-controlled crystal oscillators can be miniaturized, if need be, for portable applications that demand excellent oscillator stability. Valpey Fisher, for example, puts a resistive plating directly on the quartz crystal, creating a small OCXO ($\sim$1.3 cm^3) requiring just 0.15 W of heater power.

B. Atomic standards

Atomic frequency standards are used where the stability of ovenized-crystal standards is insufficient. These use a microwave absorption line in a rubidium (Rb) gas cell, or atomic transitions in an atomic cesium (Cs) beam, as the reference to which a quartz crystal is stabilized. Commercially available Rb and Cs frequency standards achieve accuracy and stability of a few parts in 10^{11} and 10^{13}, respectively. Cesium-beam standards are the official timekeepers in the United States, with timing transmissions from the National Institute of Standards and Technology (NIST) and the Naval Observatory.

Atomic hydrogen masers are yet another highly stable standard. Unlike Rb and Cs standards, the hydrogen maser is an actual oscillator (rather than a passive reference), with claimed stabilities approaching a few parts in 10^{14}. Recent research in stable clocks has centered on techniques using "cooled" trapped ions or atoms, or "atomic fountains," to achieve even better stability. These schemes are being used to create precise *optical* standards, which are then bridged to a radiofrequency reference via an "optical comb." Many physicists believe that ultimate stabilities of parts in 10^{17}–10^{18} are achievable.

Finally, you don't have to spend vast amounts of money to get an accurate frequency standard. Instead you can derive a precise 10 MHz clocking signal, along with 1 pps pulses, by receiving navigation signals from the Global Positioning System (GPS).[52] This is a constellation of 24 satellites in 12-hour orbits, covering all but the arctic latitudes, designed for precise navigation and timing. The satellites carry stable atomic clocks, and they transmit "navigation messages" using sophisticated dual-frequency spread-spectrum methods, at 1.575 GHz and 1.228 GHz. A GPS receiver on the ground, collecting signals from four satellites, can triangulate its position, and time as well. The inexpensive handheld GPS receivers you can buy (from companies like Garmin and Magellan, or included in cell phones) are intended for navigation, and they do not recover or regenerate a reference frequency. However, you can spend more and get a carefully engineered laboratory standard. An example is the Symmetricom 58503B "GPS Time and Frequency Reference Receiver," which delivers a 10 MHz output (stable to 1 part in 10^{12}, averaged over a day), and also 1 pps timing pulses (accurate to 20 nanoseconds); it costs \$4,500. All you need, to get this kind of precision, is a place to put the doorknob-size antenna; your tax dollars do the rest.

In Table 7.2 we've gathered together various oscillator technologies, and their characteristics. We hope you like it.

7.1.8 Frequency synthesis: DDS and PLL

A stable reference is *stable*, but not tunable. But, as we hinted earlier, there are two nice techniques that let you create an output frequency of your choosing, with the stability of the reference: direct digital synthesis (DDS), and phase-locked loop (PLL) synthesis. These are digital "mixed-signal" techniques, which we treat in detail in Chapters 12 and 13. But they are closely related to oscillators and frequency generation, and so we will describe them here at a basic level.

A. Direct digital synthesis

The idea here is to program a digital memory with the numerical values of sine and cosine for a large set of equally spaced angle arguments (say for every 1°). You then make sinewaves by rapidly generating the sequential addresses, reading the memory values for each address (i.e., each sequential angle), and applying the digital values to a digital-to-analog converter. (DAC)

Figure 7.42 shows the scheme, in both its simplest notional form (a counter increments addresses to a sine-lookup ROM), and in the (far better) method used in

---

[51] The Symmetricom 1000B datasheet suggests (but does not guarantee) that a well-aged specimen will typically mellow out at "parts in 10^{11} per day."

[52] There are two analogous navigation systems: the Russian GLONASS, and the European Galileo. The latter is slated to be fully operational by 2020.

Table 7.2 Oscillator Types[a]

Class	Type	Stability	Tunability	Agility	Freq Range	Cost	Notes
Relaxation and delay	RC relaxation	10^{-2} to 10^{-3}	wide (>10:1)	high	Hz to 10's MHz	low (<$10)	555, 1799, etc
	delay	10^{-2} to 10^{-3}	modest (~5:1)[b]	high	10 MHz to 100's MHz	low (<$10)	within an IC
	Wien bridge	10^{-3}	modest (<10:1)	slow-slewing	Hz to MHz	low (<$10)	sinewave output
Resonator	LC	10^{-3} to 10^{-5}	modest	high	kHz to 100's MHz	low (<$10)	
	ceramic	10^{-2} to 10^{-3}	small (<10^{-3})	high	100's kHz to 10's MHz	low (<$10)	
	crystal ("xtal")	10^{-5}	very small (10^{-4})	high	10's kHz to 100 MHz	low (<$10)	ubiquitous
	xtal - TCXO	10^{-6} to 10^{-7}	very small (10^{-4})	high	10's kHz to 100 MHz	medium ($10-100)	
	xtal - OCXO	10^{-8} to 10^{-9}	very small (10^{-4})	high	10's kHz to 100 MHz	high ($100-1000)	
	SAW	10^{-4}	very small (10^{-4})		100's MHz	low (<$10)	small, low jitter
	cavity[c]	10^{-5}	modest[d]	low to high[d]	10's MHz to 10's GHz	low to medium	
Atomic	Rb vapor	10^{-10}	N.A.	N.A.	10 MHz derived ref[e]	high ($1k)	
	Cs beam	$10^{-13(f)}$	N.A.	N.A.	10 MHz derived ref[g]	very high ($10k)	the def'n of the sec!
	H maser	10^{-14}	N.A.	N.A.	10 MHz derived ref[h]	even higher (>$100k)	
	GPS	$10^{-13(k)}$	N.A.	N.A.	10 MHz derived ref	high ($1k)	ref'd to NIST Cs stds
Reference-derived	PLL synthesis	equal to ref	wide	t_s= 0-100 ms	Hz to GHz	low-med ($10-100)	
	"direct" synth	equal to ref	wide	t_s= 5-10 ms	Hz to GHz	high (>$1k)	
	DDS	equal to ref	wide	immediate	Hz to GHz	low-high ($10-1000)	

Notes: (a) somewhat subjective evaluations. (b) tuning via operating current. (c) cylindrical cavity, waveguide, or dielectric "pill" resonator. (d) tuning via mechanical plunger, varactor, or YIG. (e) from 6.8346826128 GHz resonance. (f) long-term. (g) from 9.192631770 GHz resonance. (h) from 1.420405751767 GHz oscillation. (k) long-term.

practice. In the latter a ROM with an n-bit address space (thus values of the sine at 2^n phases within a single 360° cycle) is driven by a phase register that accumulates steps of phase according to the value of the "frequency tuning word" (FTW). At each clock the phase advances by $\Delta\phi=(360°/2^n)\cdot$FTW, generating an output frequency f_{out}=FTW$\cdot f_{clk}/2^n$, with any extra phase being carried forward from cycle to cycle.

This method has some drawbacks. The output is actually a staircase wave, since it is constructed from a set of discrete voltages, one for each table entry. You can, of course, use a lowpass filter to smooth the output; but having done so, you cannot span a wide range of frequencies, because the lowpass filter must be chosen to pass the sinewave itself while blocking the (higher) angle step frequency (the same problem applies to the switched-capacitor resonator; §7.1.9B). Decreasing the angular step size helps, but reduces the maximum output frequency.

Contemporary DDS chips include the sine lookup table, DACs, and everything else you need except for the stable fixed-frequency clocking input (usually provided by a simple crystal oscillator; §7.1.6). They are remarkably fast, and inexpensive. For example, the AD9850 series includes the AD9852, which goes to 150 MHz output frequency with 48-bit frequency resolution (that's a *micro*hertz!!); it costs $15 in quantity. If that isn't fast enough for you, perhaps the AD9912 ($37), with its 1 GHz clocking speed ($f_{out(max)}$ = 400 MHz) and 14-bit DAC will do the job.

DDS chips let you program frequency sweeps (a ramping frequency versus time); and also amplitude, frequency, and phase *modulation* (periodic variation over time). You can send frequency-changing commands at very high rates (100 million new frequencies per second, for the AD9852) to make an *agile oscillator*.

Various members of the family permit phase synchronization and precise phase offsetting, so you can make a precision programmable quadrature oscillator (i.e., simultaneous sine and cosine outputs) with stunning performance with just a few chips (and for a few dollars more).

B. Phase-locked loops

This mixed-signal technique synthesizes an output frequency f_{out} that is related to the reference oscillator frequency f_{osc} by a rational fraction; i.e., $f_{out}=(n/r)\times f_{osc}$, where n and r are integers. You can think of this as a generalization of the simple divide-by-N frequency divider (where $f_{out}=f_{osc}/N$). Because PLLs blend analog and digital techniques, we'll discuss them in detail only later, in §13.13.

What's important to know now is that this gives you wonderful flexibility in oscillator frequency generation. For example, if we use a fixed-frequency 16 MHz crystal oscillator to drive a PLL chip and set $r = 16$, then the output frequency will be exactly n MHz, thus creating a high-frequency single-chip oscillator with output frequency settable in 1 MHz steps (with a typical output-frequency range

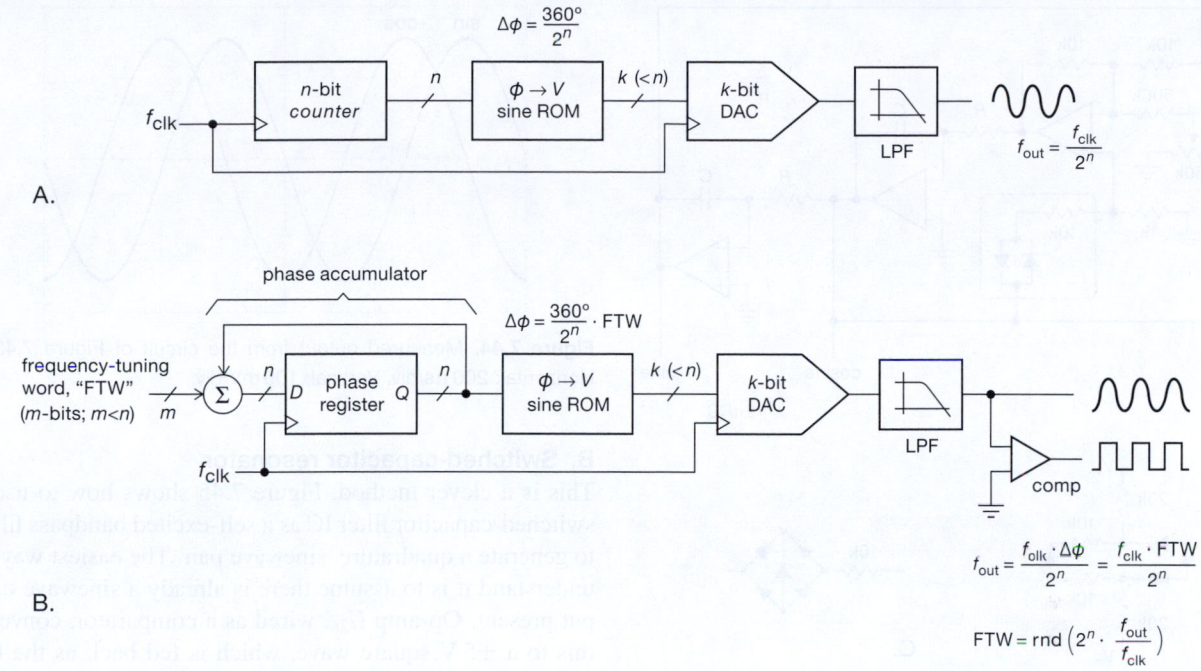

Figure 7.42. Direct digital synthesis creates its sinewave output from precomputed values of a sinusoid stored in ROM. The simplest configuration (A) increments a counter to address successive values from ROM. Far better is a configuration with a *phase accumulator* (B), which delivers *n* bits of output frequency resolution.

of 25–500 MHz or so). PLL synthesis is used in telecommunications devices (such as radios, televisions, and cellular telephones) to set the operating frequency for each channel.

As this example suggests, the somewhat complex PLL clock-synthesis technique is available in easy-to-use chips, with someone else having done all the hard work (specifically, making a phase detector and a voltage-controllable oscillator, and stably closing the loop). There are, in addition, the inexpensive "programmable oscillator modules" we mentioned in §7.1.6F that incorporate a crystal oscillator and a PLL, so you can stock a single module type, and program its frequency when you're ready to use it. You can get these from the companies that make fixed-frequency modules, for example Epson (SG8002 series), ECS (ECS-P series), Citizen (CSX-750P series), CTS (CP7 series), and Cardinal (CPP series). They come in the same DIP and SMT packages as conventional factory-set oscillators, with a slight price premium.

7.1.9 Quadrature oscillators

There are times when you need an oscillator that generates a simultaneous *pair* of equal-amplitude sine waves, *90° out*

of phase. You can think of the pair as sine and cosine (or *I* and *Q*, for in-phase and quadrature). This is referred to as a *quadrature pair* (the signals are "in quadrature"). One important application is in radio communications circuits (quadrature mixers, single-sideband generation). Of great utility, as we explain below, a quadrature pair is all you need to generate any arbitrary phase.

The first idea you might invent is to apply a sinewave signal to an integrator (or differentiator), thus generating a 90°-shifted cosine wave. The phase shift is right, but the amplitude is wrong (figure out why). Here are some methods that do work.

A. Paired integrators

Figure 7.43 is a variation on a circuit that has been bouncing around now for several decades. It uses a pair of cascaded integrators ($-90°$ phase shift each) inside a feedback loop, closed by a unity-gain inverting amplifier ($180°$ phase shift). Oscillation occurs at a frequency at which each integrator has unity voltage gain; that is, at a frequency at which the capacitive reactance $1/2\pi fC$ equals R. The diodes limit the amplitude to ~300 mV. We tested this circuit with LMC6482 op-amps running from ±5 V, and with $R=15.8$k and $C=10$ nF ($f_{osc}=1$ kHz), producing

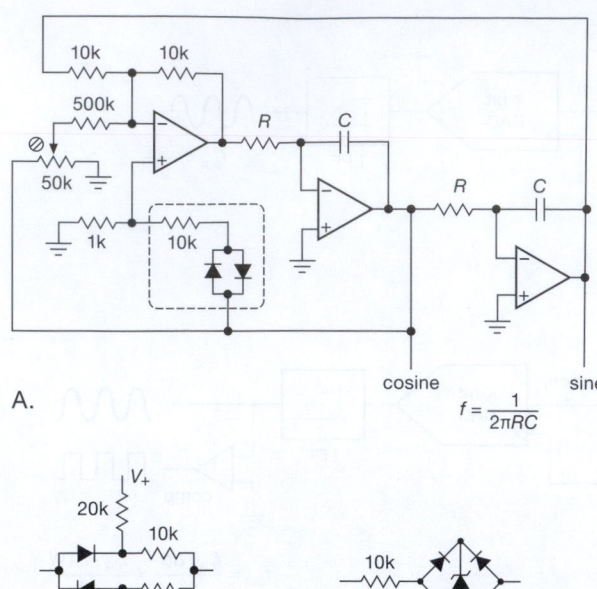

A.

$$f = \frac{1}{2\pi RC}$$

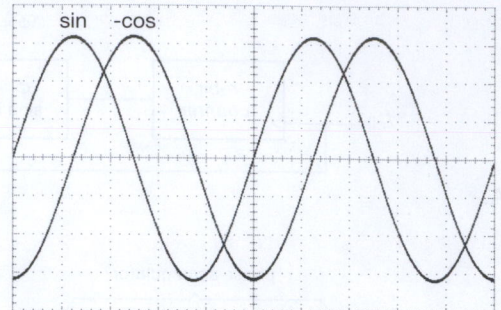

Figure 7.44. Measured output from the circuit of Figure 7.43A. Horizontal: 200 μs/div. Vertical: 100 mV/div.

B. Switched-capacitor resonator

This is a clever method: Figure 7.45 shows how to use a switched-capacitor filter IC as a self-excited bandpass filter to generate a quadrature sinewave pair. The easiest way to understand it is to assume there is already a sinewave output present. Op-amp U_{2a}, wired as a comparator, converts this to a ± 5 V square wave, which is fed back as the filter's input. The filter has a narrow bandpass ($Q=10$), so it converts the input square wave to a sinewave output, sustaining the oscillation. A square-wave clock input (CLK) determines the bandpass center frequency and hence the frequency of oscillation, in this case $f_{CLK}/100$. The circuit

Figure 7.43. Quadrature sinewave oscillator (adapted from a circuit by Tony Williams): A. Basic circuit, with diode limiter (dashed box). B. Biased limiter. C. Zener limiter.

B.

C.

the waveforms in Figure 7.44. The measured frequency was 997 Hz, with distortion of 0.006% and 0.02% for the sine and cosine outputs, respectively.

The simple diode limiter does not provide particularly good amplitude control, and it also limits the amplitude to ~300 mV. An improved limiter is shown in Figure 7.43B, in which the output is biased toward the supply rails by a pair of dividers, so that the diodes conduct only at a larger amplitude (set by the divider ratio). With the component values shown, the measured amplitude was 3.3 V. Figure 7.43C shows another limiter that some designers favor, using a zener wrapped in a diode bridge (therefore bidirectional) to set the amplitude. You can use regular zener diodes for larger amplitudes (5 V and above), but low-voltage zeners perform poorly, with their "soft knees" (see Figure 1.17). But you can use instead a low-voltage 2-terminal reference, which behaves like a nearly perfect zener; some examples are the LM385-1.2 and -2.5 (1.24 V and 2.50 V), the AD1580 (1.22 V), and the ADR510 (1.0 V). With any of these schemes it's a good idea to put a capacitor across the zener (or reference) to hold its voltage during zero crossings of the waveform.

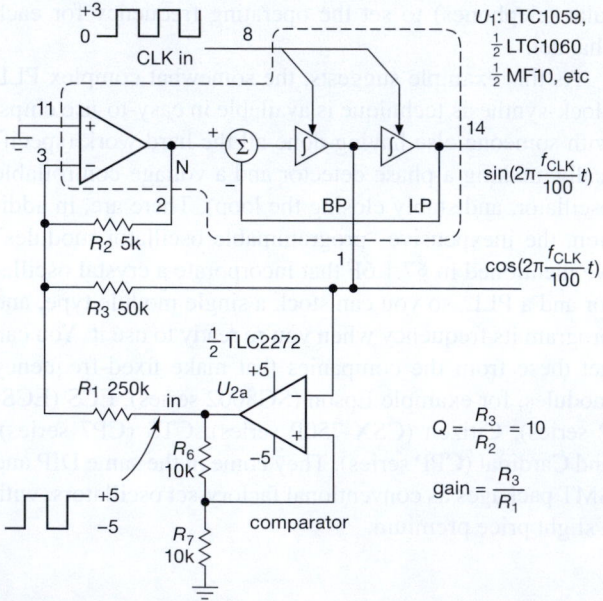

Figure 7.45. You can generate a pair of quadrature sinewaves by feeding back the squared-up output of a narrow bandpass filter, implemented in a switched-capacitor filter IC.

generates a quadrature pair of sinewaves of equal amplitude, and is usable over a frequency range of a few hertz to more than 10 kHz. Note that the output is actually a "staircase" approximation to the desired sinewave, owing to the quantized output steps of the switched filter – see Figure 7.46.

An interesting feature of this circuit is its ability to "calculate pi": the filter is configured with an accurate gain of $R_3/R_1=0.2$, and its input is an accurate ± 5 V square wave, so one might expect an output amplitude of ± 1 V. But no – the filter retains only the fundamental frequency component of the square wave, which has an amplitude equal to $4/\pi$ times that of the square wave. So, voilà, we get an output amplitude of $4/\pi$ volts (about 1.27 V), as seen in the 'scope traces.

C. Direct digital synthesis

We saw this popular method earlier (§7.1.8) as a general way to synthesize a sinewave (or any "arbitrary" wave, if you want) of precise frequency, relative to an accurate reference frequency input. Among its many virtues, this method lends itself well to the generation of a pair of signals in quadrature (or any other phase relationship, if you want). Figure 7.47 shows a nice chip from Analog Devices that is intended for quadrature sinewave generation; it permits frequency adjustment in steps of a *micro*hertz, along with a large bag of tricks that includes digital mixers (for quadrature amplitude modulation), and timers and accumulators (for frequency or phase modulation, nonlinear chirp, etc). The datasheet makes for some good reading!

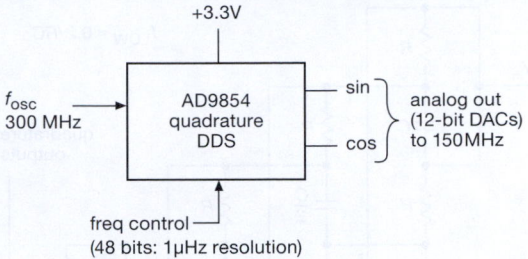

Figure 7.47. Quadrature generation by DDS. The AD9854 has many additional tricks up its sleeve (see text).

D. Phase-sequence filters

There are tricky *RC* filter circuits that have the property of accepting an input sinewave and producing as output a pair of sinewave outputs whose phase *difference* is approximately 90°. The radio hams know this as the "phasing" method of single-sideband generation (due to Weaver), in which the input signal consists of the speech waveform that you want to transmit. Unfortunately, this method works satisfactorily only over a rather limited range of frequencies and requires precision resistors and capacitors.

A better method for wideband quadrature generation uses "phase-sequence networks," consisting of a cyclic repetitive structure of equal resistors and geometrically decreasing capacitors, as in Figure 7.48. You drive the network with a signal and its 180°-shifted cousin (that's easy, since all you need is a unity-gain inverter). The output is a fourfold set of quadrature signals, with a 6-section network giving $\pm 0.7°$ error over a 40:1 frequency range; an 8-section network extends the range to 150:1. This is demonstrated in Figure 7.49, a SPICE simulation of an 8-section phase-sequence network, showing the behavior of a pair of quadrature outputs with good performance from 400 Hz to 50 kHz.

E. Quadrature square waves

For the special case of square waves, generating quadrature signals is a lead-pipe cinch. The basic idea is to generate four times the frequency you need, then divide by four with digital *flip-flops*. Figure 7.50 shows a simple circuit that does this, using type-D flip-flops (Chapter 10). This technique is essentially perfect from dc to at least 100 MHz.

F. Radiofrequency quadrature

At radio frequencies (upward of a few megahertz) the generation of quadrature sinewave pairs again becomes easy, using devices known as *quadrature hybrids* (or quadrature *splitter–combiners*). At the low-frequency end of the radio spectrum (from a few megahertz to perhaps 1 GHz) these

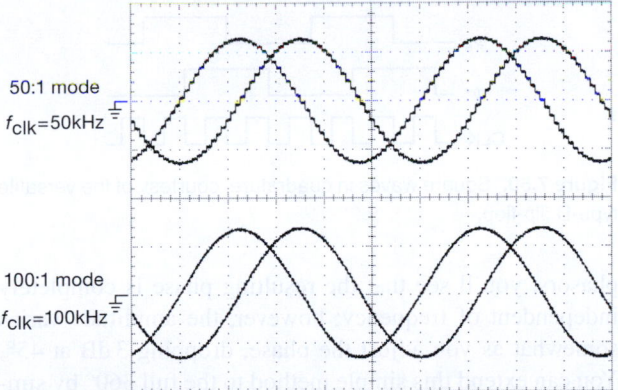

50:1 mode
f_{clk}=50kHz

100:1 mode
f_{clk}=100kHz

Figure 7.46. Observed waveforms from the circuit of Figure 7.45. The switched capacitor filter generates a staircase approximation to the ideal quadrature sinewaves, more evident when the filter is set to the coarser 50:1 f_{CLK}/f_{out} ratio. This filter knows the value of π: the output amplitude is $4/\pi$ volts! Horizontal: 200 μs/div; Vertical: 1 V/div.

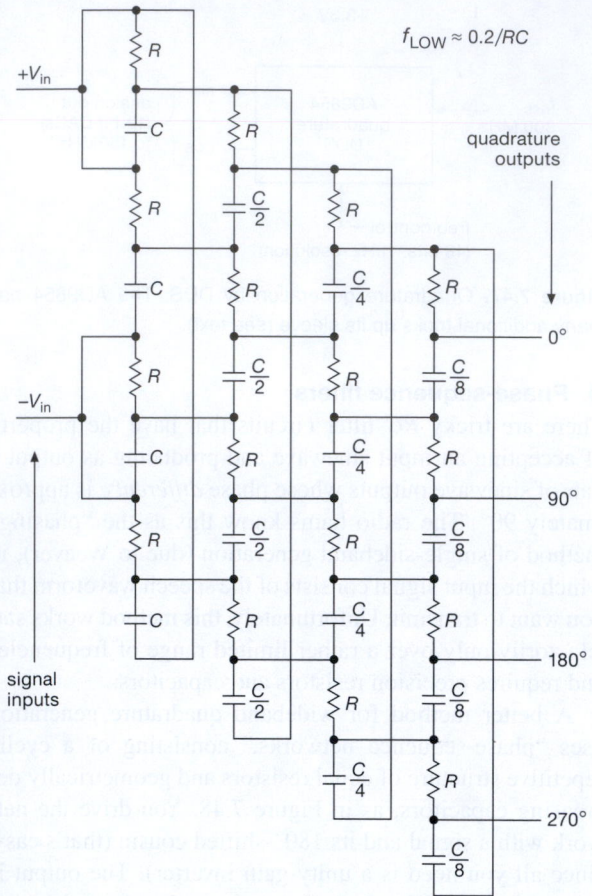

$f_{\text{LOW}} \approx 0.2/RC$

quadrature outputs

signal inputs

Figure 7.48. Four-section phase-sequence network. The network is effective for frequencies extending from a low frequency $f_{\text{LOW}} \approx 0.2/RC$ to a high end that depends on the ratio of capacitances of the first and last section, as $f_{\text{HIGH}}/f_{\text{LOW}} \approx C_{\text{first}}/C_{\text{last}}$.

take the form of small core-wound transformers, while at higher frequencies you find incarnations in the form of stripline (strips of foil insulated from an underlying ground plane; see Chapter *1x*) or waveguide (hollow rectangular tubing). These techniques tend to be fairly narrowband, with typical operating bandwidths of an octave (i.e., ratio of 2:1).

G. Generating a sinewave of arbitrary phase

Once you have a quadrature pair, it's easy to make a sinewave of *arbitrary* phase. You simply combine the in-phase (I) and quadrature (Q) signals in a resistive combiner, made most easily with a potentiometer going between the I and Q signals. As you rotate the pot, you combine the I and Q in different proportions, taking you smoothly from $0°$ to $90°$ phase. If you think in terms of

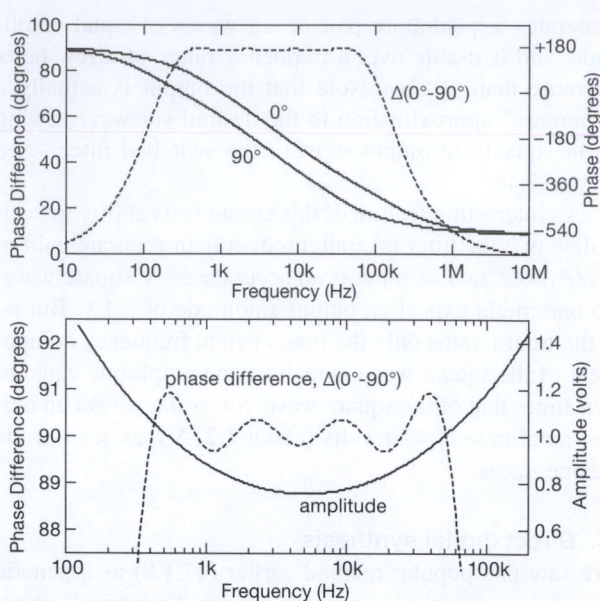

Figure 7.49. Phase and amplitude behavior of an 8-section phase-sequence network (with $R=10k$, $C=40\,$nF), as modeled in SPICE. The lower plot expands the region of accurate quadrature. Note that it is the *difference* of phases (dashed curves) that are in quadrature.

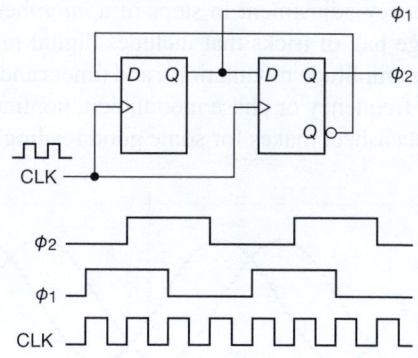

Figure 7.50. Square waves in quadrature, courtesy of the versatile type-D flip-flop.

phasors, you'll see that the resulting phase is completely independent of frequency; however, the amplitude varies somewhat as you adjust the phase, dropping 3 dB at $45°$. You can extend this simple method to the full $360°$ by simply generating the inverted ($180°$-shifted) signals, I' and Q', with an inverting amplifier of gain $G_V = -1$.

The generation of sinusoidal bursts of predetermined phase (and amplitude) is of great importance in digital communications. It is used in particular in the method of *quadrature amplitude modulation* (QAM, pronounced "quahm"), in which several bits are encoded in each QAM "symbol." For example, most digital cable television is encoded as 256-QAM, with each symbol (out of a constellation of 256) carrying 8-bits of information; you can think of the individual symbols as little bursts of several sinusoidal cycles, with a particular phase and amplitude characterizing each symbol.[53]

With DDS it is even easier to set the output phase (relative to a synchronizing pulse, or to a second DDS synthesizer that is synchronized to the first), because DDS chips let you add a user-specified offset to the internal accumulating phase that is used to generate the sinewave. For example, the AD9951 has a 14-bit *phase-offset word*, giving you a phase-settable step size of $360°/2^{14}$, or 0.02°. You can change the phase on-the-fly to produce phase modulation.

7.1.10 Oscillator "jitter"

In addition to the primary oscillator parameters – frequency, amplitude, and waveform – there is *stability*: if an oscillator drifts with time, temperature, or supply voltage, we talk about stability (or lack thereof), and we may assign corresponding coefficients. For example, a crystal oscillator might specify a tempco of 1 ppm/°C. That's important for applications such as timekeeping, communications, or spectroscopy.

But it's not the whole story. You could, for example, have an oscillator that maintains a precise 10.0 MHz average frequency, independent of time and temperature, but that exhibits variations in the timing of its zero crossings from cycle to cycle. (If you like, you can think of this as an instability on a short time scale, as compared with the long time scales of effects like drift with time and temperature.[54]) This undesirable property has several names, depending on the context: if it's a sinewave, and you're using it in a communications application, you call it *phase noise*, or *spectral purity*; if instead it's a square wave or digital pulse train, and you're using it for waveform sampling and reconstruction, or for clocking fast digital data links, you call it *jitter*.

A bit more on this business of jitter. Timing jitter arises whenever an output transition is created by a periodic signal crossing a decision threshold, as for example

---

[53] The constellation of phases and amplitudes are chosen to minimize error, given the characteristics of the cable.

[54] Those souls (you know who you are!) who derive pleasure from the misfortunes of other oscillators have more sophisticated measures of stability, notably the *Allan variance*, which is basically a plot of oscillator stability versus averaging time.

---

in an *RC* relaxation oscillator or a crystal oscillator. Figure 7.51 shows the situation. A signal of finite slew rate S (volts/second) is crossing a voltage threshold, and both the signal and the voltage threshold are imperfect, each with some additive noise voltage v_n; we'll call these $v_n(\text{sig})$ and $v_n(\text{th})$. So there's a timing uncertainty, such that the time at which the (noisy) signal crosses the (noisy) threshold can (and will) vary by

$$\Delta t = \frac{\sum v_n}{\text{slew rate}} = \frac{v_n(\text{sig}) + v_n(\text{th})}{S}$$

Jitter is greater with slowly slewing signals, for a given amount of signal or threshold noise. A sinewave of frequency f has a maximum slew rate of $S = 2\pi f V_0$, where V_0 is the peak signal amplitude (though less if the threshold is not at the amplitude midpoint); the slew rate of a pulse or step can be approximated by $S = V_{\text{step}}/t_r$, where t_r is the rise time (or, more generally, transition time).

Oscillator stability and jitter are important subjects, and we'll have more to say about them in Chapter 13, in connection with digital–analog conversion, serial digital communication, and phase-locked loops.

7.2 Timers

Oscillators generate a periodic signal, characterized by its waveform (sine, square, pulse), its frequency, and its amplitude. Closely related to oscillators are *timers*: circuits that generate a delayed pulse, or a pulse of a given width, following a triggering event. Here you talk about delay *times* and pulse *widths* rather than about *frequency*. The techniques, however, are quite similar, consisting mostly of *RC*-style analog waveforms (driving comparator–trigger circuits), or oscillator-driven digital counters or dividers.

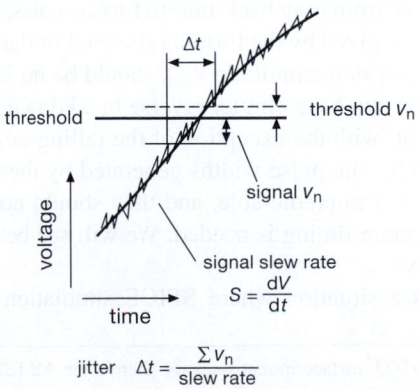

Figure 7.51. Timing jitter basics: a signal of finite slew rate passes through a decision threshold.

There's quite a range of techniques, depending on the time scale involved and the accuracy required. We visit most of the techniques in common use.

7.2.1 Step-triggered pulses

Sometimes you need to generate an output pulse, of some duration, from an input trigger. The input may be a short pulse, from which you want to generate a longer output pulse; alternatively, the input may be a "step" (usually a logic-level transition, longer than the desired output pulse), from which you want to generate a short pulse.

A. Short pulse from step: *RC* + discrete transistor

An *RC* differentiator turns a voltage step into a step with a trailing decay of time constant *RC* (Figure 7.52A). You can use the output directly, or you can run it through a transistor switch to generate something more like a rectangular pulse. Figure 7.52B shows the ubiquitous 2N7000 small MOSFET used to generate an output pulse of either polarity (swinging between a positive rail and ground); the pulse width is of order $\tau = RC$, but depends upon the ratio of gate threshold voltage to input step size. In Figure 7.52C we show the analogous circuits with a fast bipolar switching transistor (a PN2369: it uses gold-doping to reduce base charge storage time to 13 ns maximum, compared with 200 ns for a plain ol' jellybean like the 2N3904[55]). Note that a series resistor (R_1) is needed to limit the peak base current. The exact timing behavior with a BJT is more complicated than with a MOSFET, because of forward base-voltage clamping: for the circuit with rising-edge input, the output pulse width is determined primarily by R_1C (which sets the decay of forward base drive current); for the circuit with falling-edge input, on the other hand, the output pulse width is determined primarily by R_2C (which times the recovery from base back-bias to forward bias, assuming $R_2 \gg R_1$), as given by the formula shown. For the latter circuit the input step amplitude V_{step} should be no larger than ~6 V to prevent base-emitter reverse breakdown.

A caveat: with the exception of the falling-edge BJT in Figure 7.52C, the pulse widths generated by these circuits are somewhat unpredictable, and they should not be used where accurate timing is needed. We will see better methods shortly.

This is a situation where SPICE simulation provides

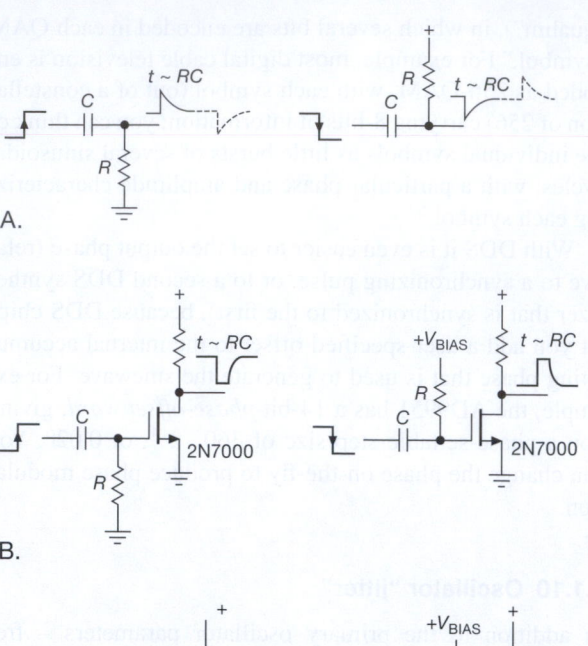

A.

B.

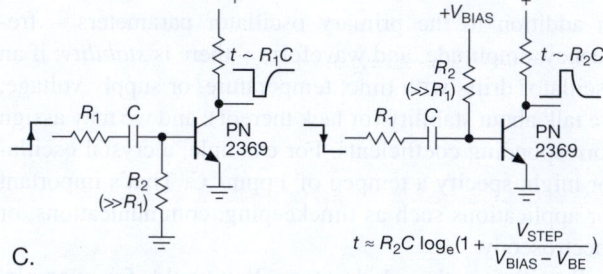

$$t \approx R_2 C \log_e \left(1 + \frac{V_{STEP}}{V_{BIAS} - V_{BE}}\right)$$

C.

Figure 7.52. Generating a pulse from a step with *RC*'s and transistors. A. Bare *RC* differentiator. B. *RC* drives MOSFET switch. C. *RC* drives BJT switch.

a useful tool for exploring detailed circuit behavior. Figure 7.53 shows a comparison of bench measurement and simulation for the falling-edge circuit of Figure 7.52C, running from a +5 V supply, when driven by a 5 V falling step of 5 ns fall time. The rather good agreement validates the use of the simulation models and tools. Note particularly the accurate simulation of delay time (the input step began at 20 ns), transistor saturation voltage, base voltage, and output waveform (which, though not visible in the printed figure, overshoots the +5 V rail by about 40 mV, owing to capacitive coupling of the rising base voltage).

We'll be seeing more accurate and predictable ways to generate output pulses from input edges. But this simple method is fine for noncritical jobs, for example driving a hefty latching power relay, as shown in Figure 7.54. This

---

[55] For a SOT-23 surface-mount package there's the MMBT2369. The 2N5771 (MMBT5771) and PN3640 (MMBT3640) are similar *pnp* types with 20 ns storage-time specs. These gold-doped transistors are all low-voltage types, rated at 12 to 20 V. Gold doping leads to low

beta and higher leakage current; these transistors are well suited for fast switching, but not much else.

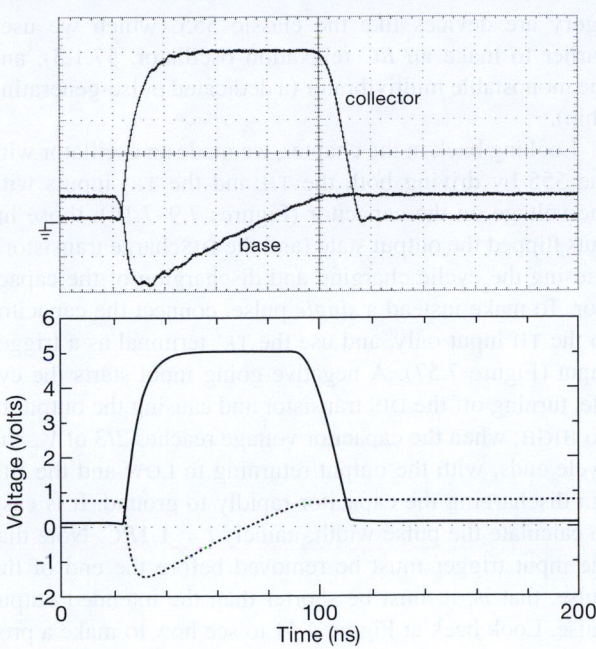

Figure 7.53. Simulated (lower) and measured (upper) waveforms of the right-hand circuit of Figure 7.52C, with R_1=220 Ω, C_1=47 pF, R_2=2.2k, R_C=470 Ω, and with 'scope probing capacitance to ground of 8 pF at base and collector. The SPICE plot and the 'scope trace have been scaled identically, at 20 ns/div and 1 V/div.

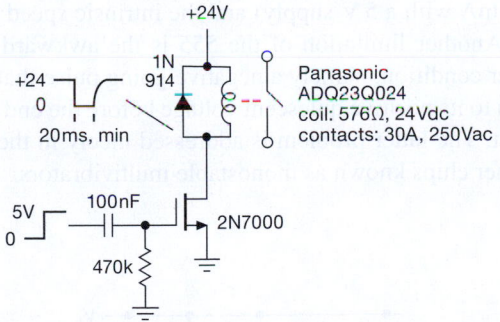

Figure 7.54. A 20 ms pulse actuates a latching power relay. This is a "2-coil" latching relay (for SET and RESET), only one of which is shown. This particular relay costs $23, and includes high-current PCB and FastOn (spade lug) contacts.

particular relay's contacts are rated at 30 A and 250 Vac; it has a pair of coils (set, reset), which you drive by applying 24 Vdc for a minimum of 20 ms. We tested the circuit in our lab – it made an impressive clatter that could be heard down the hallway.

B. Pulse from step or edge

With the addition of a couple more transistors you arrive, finally, at a pulse output circuit that is happy to trigger either on a step or on a short pulse; that is, it is *edge* triggered, and insensitive to whether the triggering edge belongs to a pulse that is shorter or longer than the desired output pulse length. We showed this classic BJT "one-shot" circuit earlier, in the transistor chapter – see Figure 2.12. You can wire up that configuration, if you want. But you're probably better off using an integrated *monostable multivibrator*: these devices take care of the details, and they provide flexible triggering options. We'll look at them next, after taking a brief look at the use of digital logic gates for generating short pulses.

C. Short pulse from step: logic gates

A closely related technique uses digital logic gates (Chapter 10) in place of discrete transistors. This is particularly useful if you want output signals to drive additional logic, because the output signals are of the correct logic voltages and speeds. In Figure 7.55 a logic inverter (§10.1.4D) with built-in Schmitt trigger (§4.3.2A) is used to create output pulses with abrupt transitions; although the edges are fast (owing to the Schmitt trigger), the timing is only approximate, because the Schmitt voltage thresholds are loosely specified (e.g., the specified hysteresis amplitude spans a range of 3:1).

In Figure 7.56A the short *propagation delay* of logic inverters (of order several nanoseconds), combined with the logic of *gates*, substitutes for RC delays to create the output pulse width; the three circuits shown respond to rising, falling, or both edges, respectively. The circuits in Figure 7.56B extend this gate idea to longer output pulse widths, by using an RC delay, sharpened up with a Schmitt inverter, to provide the delayed input to the gate.

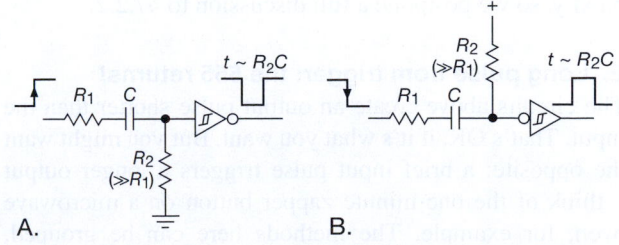

Figure 7.55. A logic inverter with Schmitt-trigger input creates a clean output pulse. R_1 limits gate input current when the input signal returns to its initial level, during which the gate input is driven below ground or above V_+, respectively.

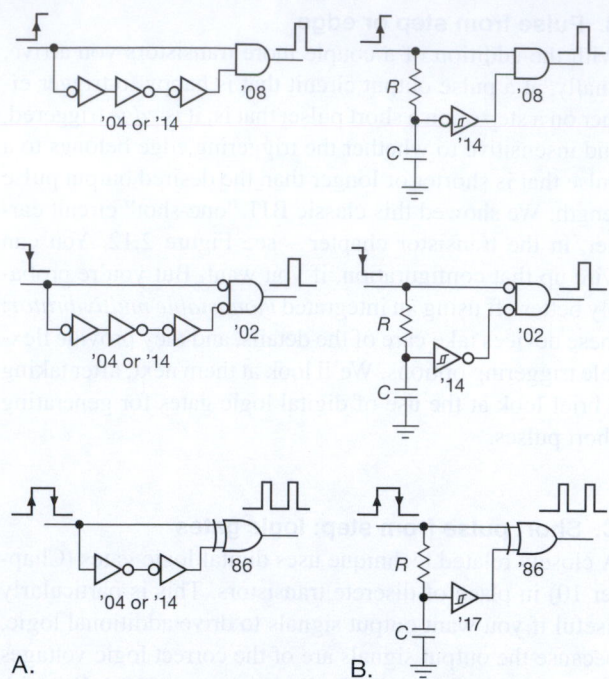

Figure 7.56. Generating a pulse from a step, with logic gates. A. Short pulses made with 2-input gates and cascaded inverter delays. B. Longer pulses made with 2-input gates and *RC* delays.

D. Short pulse from edge: monostable multivibrators

If these last circuits appeal to you, you're in luck: the semiconductor industry has created a class of integrated circuits that combines digital logic with an *RC* timing circuit, in the form of the *monostable multivibrator* (also known as a "one-shot," emphasis on the word "one"). These are edge triggered, with timing set with good accuracy by an external *RC* and with clean logic-level outputs. The output pulse can be short compared with the input duration (as with the circuits above); or it can be longer than the input (as with the circuits we discuss next). This topic has plenty of complexity, so we postpone a full discussion to §7.2.2.

E. Long pulse from trigger: the 555 returns!

The circuits above create an output pulse shorter than the input. That's OK, if it's what you want. But you might want the opposite: a brief input pulse triggers a longer output – think of the one-minute zapper button on a microwave oven, for example. The methods here can be grouped, roughly, into two categories: (a) *RC*-mediated analog timing circuits, and (b) oscillators followed either by dedicated digital counter–dividers, or by the full computational glory of microprocessors (computers-on-a-chip). In the first cat-

egory are devices like the classic 555 (which we used earlier to make an *RC* relaxation oscillator, §7.1.3), and the monostable multivibrator (a dedicated pulse-generating chip).

Looking back in the chapter, we made an oscillator with the 555 by driving both the TH and the TR′ inputs with the voltage on the capacitor (Figures 7.9–7.14); those inputs flipped the output state (and the DIScharge transistor), causing the cyclic charging and discharging of the capacitor. To make instead a *single* pulse, connect the capacitor to the TH input only, and use the TR′ terminal as a trigger input (Figure 7.57). A negative-going input starts the cycle, turning off the DIS transistor and causing the output to go HIGH; when the capacitor voltage reaches 2/3 of V_+ the cycle ends, with the output returning to LOW and the DIS pin discharging the capacitor rapidly to ground. It is easy to calculate the pulse width, namely $t = 1.1RC$. Note that the input trigger must be removed before the end of the pulse, that is, it must be shorter than the intended output pulse. Look back at Figure 4.41 to see how to make a programmable one-shot pulse generator (with "cold switching" control of the pulse width) with a 555.

Roughly speaking, a CMOS 555 (see Table 7.1 on page 430) can generate pulse widths from about $1\,\mu$s to 100 seconds. At the long end the limit is set by residual TH input current, which is less than 10 nA, so values of R as large as 10–$100\,M\Omega$ can be used. At the short end, the minimum resistor value is limited by the maximum DIS current ($\sim$15 mA with a 5 V supply) and the intrinsic speed of the 555. Another limitation of the 555 is the awkward input trigger condition, namely a negative-going pulse that must return to its positive quiescent voltage before the end of the output. The latter problem is addressed nicely in the class of timer chips known as monostable multivibrators.

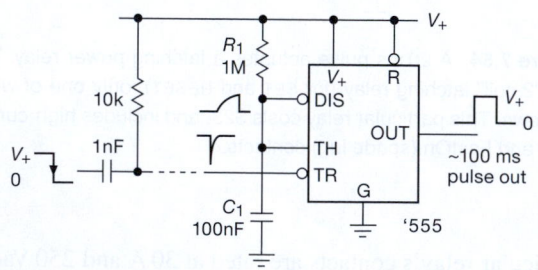

Figure 7.57. The venerable 555 generates a single positive output pulse when connected in *monostable* mode. The input *RC* differentiator converts a falling step into a trigger pulse; it can be omitted if a short trigger pulse is available to drive the TR′ input.

7.2.2 Monostable multivibrators

Within the various families of digital logic (§10.1.2B, §10.2.2) you can get *monostable multivibrators* (one-shots), which you can think of as logic-friendly versions of the 555 as it was used in Figure 7.57. They are edge triggered (by standard logic levels), and generate a logic-level output pulse Q (and its complement Q') whose width is determined by an external R and C. Monostables are very useful (some would say *too* useful!) for generating pulses of selectable width and polarity. Making one-shots with RC's combined with discrete transistors or gates (as we've just done) is tricky, and depends, for example, on the details of a gate's input circuit, since you wind up with voltage swings beyond the supply voltages. Rather than encouraging bad habits by illustrating more such circuits, we encourage you to adopt the one-shot as an available functional unit. In actual circuits it is best to use a packaged one-shot; you construct your own only if absolutely necessary, e.g., if you have a gate available and no room for an additional IC package (even then, maybe you shouldn't).

A. What's inside

Although you can happily use monostables without ever worrying about what's happening inside, it is an interesting place to visit. Figure 7.58 shows the internal circuit scheme used in most monostables. There are pins for an external C and R; the latter charges the capacitor toward V_+, which can range from +2 V to +15 V, depending on the particular logic family.[56] In the resting state the capacitor is fully charged, and the output flip-flop is reset, i.e., Q is LOW (ground) and Q' is HIGH (V_+).

Look at the waveforms in the figure, and assume for now that the three resistors R_1–R_3 are of equal resistance.[57] When the trigger condition is met (more on this below), for example by bringing input B HIGH while input A' is LOW, two things happen: (a) the output Q goes HIGH; and (b) the capacitor is rapidly discharged toward ground by the lower MOSFET.[58] When the capacitor's falling voltage reaches $\frac{1}{3}V_+$, the control circuitry removes gate drive from the lower MOSFET, allowing the capacitor to charge back up through R_{ext}; this begins the RC-mediated timing interval, which ends when the capacitor's voltage reaches $\frac{2}{3}V_+$,

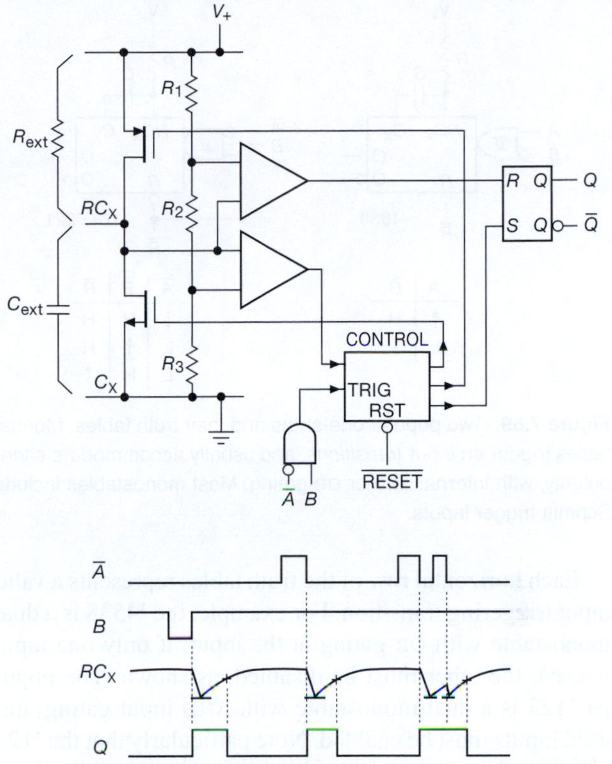

[56] Popular logic supply voltages are +5 V, +3.3 V, and +2.5 V.
[57] They usually aren't, though it's not important here. More on this later, in §7.2.2C.
[58] Typical discharge currents are 30–80 mA. So a 0.01 μF timing capacitor may take ~600 ns to discharge from 5 V down to the lower threshold at 1.6 V. This time is folded into the manufacturer's specified "K-factor" (see later discussion).

Figure 7.58. Internal circuit and waveforms for a monostable multivibrator. The waveforms show *retriggerability*.

at which time the output flip-flop is reset, ending the output pulse by bringing Q again LOW. It's a straightforward exercise in RC time constants to find that this takes a time

$$t = RC\log_e \frac{V_+ - V_L}{V_+ - V_H}, \tag{7.1}$$

where V_L and V_H are the lower and upper threshold voltages.

Exercise 7.6. Derive this formula.

B. One-shot characteristics

Inputs

One-shots are triggered by a rising or falling edge at the appropriate inputs. The only requirement on the triggering signal is that it have some minimum width, typically 25 ns to 100 ns; it can be shorter or longer than the output pulse. Two inputs are usually provided so that a signal can be connected to one or the other, to trigger the one-shot either on a rising edge or on a falling edge; alternatively, both inputs can be used, with a pair of separate triggering sources. The extra input can also be used to inhibit triggering. Figure 7.59 shows two examples.

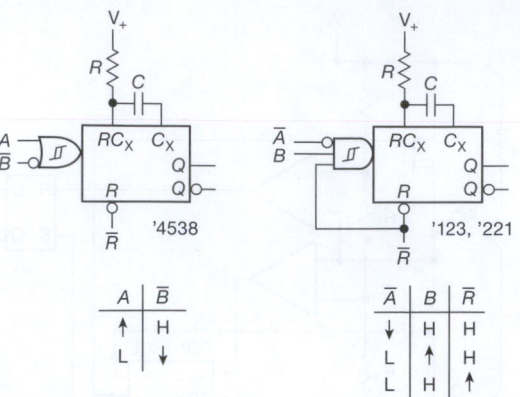

Figure 7.59. Two popular one-shots and their truth tables. Monostables trigger on input *transitions*, and usually accommodate either polarity, with internal AND or OR gating. Most monostables include Schmitt trigger inputs.

Each horizontal row of the truth tables represents a valid input triggering transition. For example, the '4538 is a dual monostable with OR gating at the input; if only one input is used, the other must be disabled, as shown. The popular '123 is a dual monostable with AND input gating; unused inputs must be enabled. Note particularly that the '123 triggers when RESET is disabled if both trigger inputs are already asserted. This is not a universal property of monostables, and it may or may not be desirable in a given application (it's usually not). The '423 is the same as the '123, but without this "feature."

When monostables are drawn in a circuit diagram, the input gating is usually omitted, saving space and creating a bit of confusion.

Retriggerability

Most monostables (e.g., the 4538, '123, and '423 mentioned earlier) will begin a new timing cycle if the input triggers again during the duration of the output pulse (as in Figure 7.58). They are known as *retriggerable* monostables. The output pulse will be longer than usual if they are retriggered during the pulse, finally terminating one pulse width after the last trigger. The '221, by contrast, is nonretriggerable: it ignores input transitions during the time that the output pulse is asserted.

Resettability

Most monostables have a jam RESET input that overrides all other functions. A momentary input to the RESET terminal terminates the output pulse. The RESET input can be used to prevent a pulse during power-up of the logic system; however, see the preceding comment about the '123.

Table 7.3 Monostable Multivibrators

Type	Retrig?	Trig	Families
'123[a]	●	!A & B & !R	AHC(T), HC(T), LS, LV, LVC, VHC
'221	-	!A & B	74C, HC(T), LS, LV, VHC
'423	●	!A & B	HC(T), LS
'4538	●	A or B	4000, HC(T)

Notes: (a) See Table 7.4.

Pulse width

Pulse widths from 40 ns up to milliseconds (or even seconds) are attainable with standard monostables, set by an external capacitor and resistor combination. A device like the 555 can be used to generate longer pulses, but its input properties are sometimes inconvenient. Very long delays are best generated digitally (see §7.2.4).

Table 7.3 lists the commonly available monostables. In addition to these traditional logic-family monostables, be sure to look at the LTC6993 "TimerBlox" one-shots (see table on page 433). It has only one trigger input, but four variants let you choose rising- or falling-edge triggering, and retriggerable or non-retriggerable. You set the pulse width with a range selection pin (via a voltage divider from V_+), and another external resistor that allows continuous tuning over a 16:1 span within the selected range. The eight digitally-divided ranges jump by successive factors of 8, thus a total range of 2^{21}:1. Output pulse widths range from 1 μs to 34 seconds, with worst-case timing accuracy of a few percent.

C. Cautionary notes about monostables

Monostables have some problems you don't see in other digital circuits. In addition, there are some general principles involved in their use. First, a rundown on monostable pathology.

Some problems with monostables

Timing One-shots involve a combination of linear and digital techniques. Because the linear circuits have the usual problems of V_{GS} (or V_{BE} and beta) variation with temperature, etc., one-shots tend to exhibit temperature and supply-voltage sensitivity of output pulse width. A typical unit like the '4538 will show pulse-width variations of a few percent over a 0–50°C temperature range and over a ±5% supply-voltage range. In addition, unit-to-unit variations give you a ±10% prediction accuracy for any given circuit. When looking at temperature and voltage sensitivity, it is important to remember that the chip may exhibit self-heating effects and that supply-voltage variations *during the pulse* (e.g., small glitches on the V_+ line) may affect the pulse width seriously.

Brand variation Monostables with the same generic part number, but made by different manufacturers, may have somewhat different specifications, particularly involving timing components. The usual way this is specified is by a "K-value," where the output pulse width (for all but small-value capacitors) is given approximately by $t_w = KRC$ (if the datasheet does not mention K, look for the pulse width with R=10k and C=100 nF). And here's how it goes: nearly all monostables fall into three clusters of K-values, either K=0.7 (all of the '4538 monostables), or K is either $\approx$0.45 or $\approx$1.0 (most other monostable part numbers).

The '4538 monostables are boring! They all have K=0.7 (hey, not a bad reason for choosing it...). But there's some excitement with the other part numbers, because the K-value for a given part number may be 1.0 or 0.45, depending on who makes it. For example, the 74HC123 is available from at least five manufacturers.[59] The FSC and Toshiba parts guarantee an output pulse width of 0.9–1.1 ms (with 1.0 ms typ) for 10k and 100 nF, but the NXP part instead specifies 0.45 ms (typ, with no minimum or maximum spec). The other two brands have chosen different RC combinations, at which they specify typical values only: ST uses 100 nF/100 kΩ (4.4 ms typ), and TI uses 10 nF/10 kΩ (45 μs typ). Evidently these parts are not fully interchangeable![60] This caution extends generally to analog and mixed-signal devices of all kinds (consider yourself warned).

To illustrate this point, we've collected in Table 7.4 a comprehensive listing of all available '123-style monostables. The data include the timing variable K that you can use to predict the pulse width according to $t_w = KRC + t_{min}$.[61] The effective value of K varies with supply voltage, so in Figure 7.60 we've made graphs showing this dependence for the monostables listed in Table 7.4.

Another parameter that varies with manufacturer, and that is not generally specified on datasheets, is the particular choice of threshold voltages – that is, the ratios of the internal resistors R_1–R_3 in Figure 7.58. That does matter, though, because the rapid discharging of a small-value tim-

Table 7.4 "Type 123" Monostable Timing

Mfg	Part Number[a]	T-variant[b]	Vsupply min (V)	Vsupply max (V)	K^c	t_{min}^d (ns)	K-graph[e]
multiple	74LS123[†]	●	4.5	5.5	0.37	-	-
Toshiba	TC74HC123	-	2	6	1.00	150	1
Renesas	HD74HC123[†]	-	2	6	1.00	390	2
Fairchild	MM74HC123	-	2	6	1.00	390	6
TI	CD74HC123	●	2	6	0.45	230	8
ST	M74HC123	-	2	6	0.44	230	9
NXP	74HC123	●	2	6	0.45	105	8
TI	SN74AHC123	●	2	5.5	1.00	110[f]	4
NXP	74AHC123	●	2	5.5	1.00	45	-
Fairchild	74VHC123	-	2	5.5	1.00	75	4
Toshiba	TC74VHC123	-	2	5.5	1.00	75	5
TI	SN74LV123	-	2	5.5	1.00	110	4
NXP	74LV123	-	1.2	5.5	0.43	70	7
TI	SN74LVC1G123[g]	-	1.65	5.5	0.95	95	3
Toshiba	TC7WH123[g]	-	2	5.5	1.00	75	-

Notes: (a) suffixes are omitted; all except † have Schmitt-trigger inputs, and all trigger on (!A) & B. (b) TTL thresholds, for example 74HCT123. (c) pulse width = $KRC + t_{min}$, spec'd at 5V for 10k and 0.1μF. (d) for RC=0, derived from the 2k and 28pF value by subtracting 56ns·K. (e) From the graph of K vs supply voltage. (f) 75ns for 'AHCT123. (g) single monostable in 8-pin package.

ing capacitor (say $\sim$1000 pF or less) overshoots the lower threshold (a similar effect bedevils the classic 555 sawtooth oscillator circuit, where the DIS pin pulls the timing capacitor rapidly toward ground). The result is an enlarged (and not terribly stable) pulse width. At least one manufacturer (TI, in their SN74HC4538) has addressed this problem by putting the lower threshold close to ground (at about 4.3% of V_+, or +0.2 V when running from a +5 V supply), so discharge overshoot has almost no effect – the capacitor charges back up from about the same voltage, overshoot or not.

Long pulses For the generation of long pulses, you can use large value timing resistors (up to 10M should be safe, even if the datasheet shows values only to 200k, say, because these are CMOS designs with low leakage currents). Even so, the capacitor value may be a few microfarads or more; in that case electrolytic capacitors are generally necessary.[62] You have to worry about leakage current (which is insignificant with the smaller capacitor types), and you should include a diode across R (Figure 7.61A); the latter is

---

[59] FSC, NXP, ST, TI, and Toshiba.

[60] Measurements made on samples of the FSC and TI parts, with R=10k and C=100 nF, yielded pulse widths of 1.05 ms and 0.42 ms, respectively; these are consistent with their individual datasheets.

[61] Datasheets routinely omit the t_{min} term, but it should be included for reasonable accuracy when designing for short pulse widths.

[62] If you insist on using a high-capacitance ceramic capacitor, your selection may be limited to types with a "high-κ" dielectric; if so, be aware of their characteristic large temperature and voltage coefficients, which can cause capacitance variations of 50% or more. See §*1x.3*.

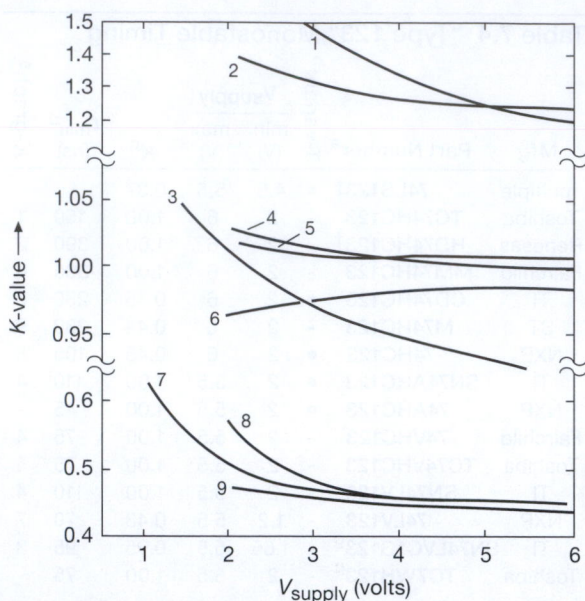

Figure 7.60. Datasheets plots of effective timing coefficient K versus supply voltage for the monostables in Table 7.4. Note the changes of scale, particularly the expanded set around $K=1.0$.

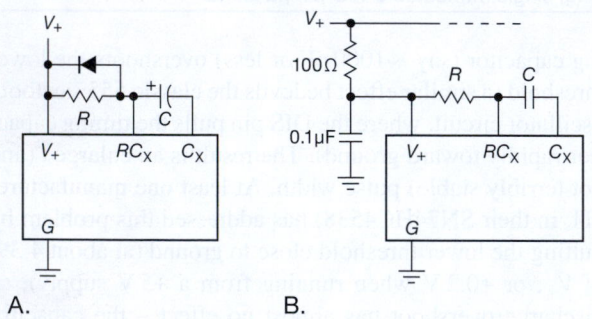

Figure 7.61. Monostable circuit variations. A. A diode prevents reverse conduction on power-down. B. A filtered "private V_+ supply" reduces supply-noise instabilities.

needed to prevent reverse conduction at the RC_X terminal by the charged timing capacitor if V_+ is turned off abruptly.

Duty cycle With some one-shots the pulse width is shortened at high duty cycle. For example, NXP's 74LV123, when powered from +3.3 V and using $R=10k$ and $C=100$ nF, has constant pulse width up to 95% duty cycle, decreasing about 1.5% near 100% duty cycle. The non-retriggerable '221 monostable is considerably worse in this respect, with erratic behavior at high duty cycles. By contrast, in our tests the Fairchild MM74HC123A maintained perfect timing to 99.98% duty cycle, along with a jitter-free

output pulse. (It, like the admirable TI SN74HC4538, uses a rather low V_L threshold, about 10% of V_+.)

Another thing to be aware of is the effect of capacitor size on retriggering recovery time. For example, TI's 74LVC1G123 datasheet has a plot of minimum retrigger time for various capacitances; it shows a minimum wait of 1 μs for a 10 nF timing cap, which is 1% of the 100 μs pulse width with a 10k resistor.

Triggering One-shots can produce substandard or jittery output pulses when triggered by too short an input pulse. There is a minimum trigger pulse width specified, e.g., 140 ns for the 4538 with +5 V supply, 60 ns with +15 V supply (4000-series "high-voltage" CMOS is faster and has more output drive capability when operated at higher supply voltages), 25 ns for the 'HCT423 at its specified +5 V supply, and 3 ns for the speedy 'LVC123 with +3.3 V supply.

Noise immunity Because of the linear circuits in a monostable, the noise immunity is generally poorer than in other digital circuits. One-shots are particularly susceptible to capacitive coupling near the external R and C used to set the pulse width. In addition, some one-shots are prone to false triggering from glitches on the V_+ line or ground. One way to avoid these problems is to create a "private" RC-filtered V_+ supply, as shown in Figure 7.61B; alternatively, you could supply the monostable from a separately regulated V_+, via a small linear voltage regulator, if you have available a higher-voltage supply rail.

Specsmanship Be aware that monostable performance (predictability of pulse width, temperature and voltage coefficients, etc.) may degrade considerably at the extremes of its pulse-width range. Specifications are usually given in the range of pulse widths where performance is good, which can be misleading. In addition, there can be a lot of difference from manufacturer to manufacturer in the performance of monostables of the same part number. Read the datasheets carefully!

Output isolation Finally, as with any digital device containing flip-flops, outputs should be buffered (by a gate, inverter, or perhaps an interface component like a line driver) before going through cables or to devices external to the instrument. If a device like a one-shot tries to drive a cable directly, the load capacitance and cable reflections may cause erratic operation to occur.

General considerations for using monostables

Be careful, when using one-shots, to generate a train of pulses, that an extra pulse doesn't get generated at the "ends." That is, make sure that the signals that enable the one-shot inputs don't themselves trigger a pulse. This is easy to do by looking carefully at the one-shot truth table, if you take the time.

Don't overuse one-shots. It is tempting to put them everywhere, with pulses running all over the place. Circuits with lots of one-shots are the mark of the neophyte designer. Besides the sort of problems just mentioned, you have the added complication that a circuit full of monostables doesn't allow much adjustment of the clock rate, since all the time delays are "tuned" to make things happen in the right order. In many cases there is a way to accomplish the same job without a one-shot, and that is to be preferred. Figure 7.62 shows an example.

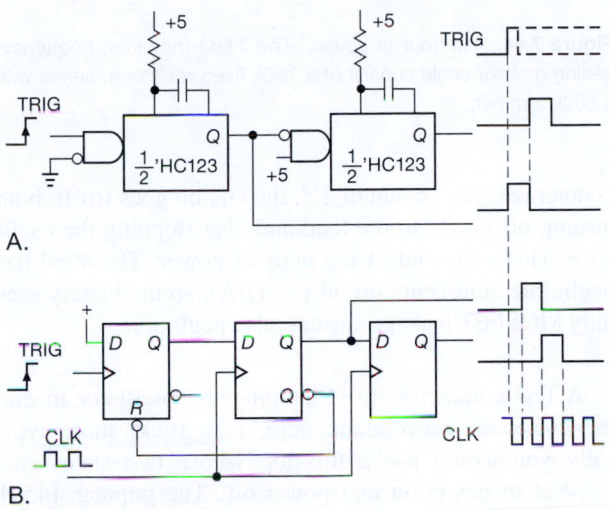

Figure 7.62. A digital delay can replace one-shot delays. Note that (unlike in circuit A, where the trigger initiates the output), in B the trigger input "arms" the circuit, whose digital output is synchronized to the next rising edge of CLK following the input trigger.

The idea is to generate a pulse and then a second delayed pulse following the rising edge of an input signal. These might be used to set up and initiate operations that require that some previous operation be completed, as signaled by the input rising edge. In the first circuit, the input triggers the first one-shot, which then triggers the second one-shot at the end of its pulse.

The second circuit does the analogous thing with type-D flip-flops, generating output pulses with width equal to one clock cycle. This is a synchronous circuit, as opposed to the asynchronous circuit using cascaded monostables. The use of synchronous methods is generally preferable from several standpoints, including noise immunity. If you wanted to generate shorter pulses, you could use the same

kind of circuit, with the system clock divided down (via several toggling flip-flops) from a master clock of higher frequency. The master clock would then be used to clock the D flip-flops in this circuit. The use of several subdivided system clocks is common in synchronous circuits. Note that there is up to one clock period of jitter in the digital delay circuit, unlike the "instantaneous" response of cascaded monostables.

§7.2.4 explores further this idea of "digital timers."

7.2.3 A monostable application: limiting pulse width and duty cycle

Here's a nice monostable application, simple and effective, which has saved the day (and much grief) on numerous occasions. It's useful whenever you're driving a device with high-current short pulses (for example solenoids or LEDs), and particularly in situations where those pulses are generated from software (in a microcontroller or FPGA). The danger, of course, is that a software bug or microcontroller crash can cause a destructively long pulse.

To set the stage with a specific example, researchers in one of our labs were photographing zebrafish with a microscope fitted with an illuminator ring of sixty LEDs. The camera shutter, operating at 120 frames/sec, generated 80 μs LED-driving pulses every 8 ms. Because the duty cycle (t_{ON}/T) was only 1%, it was OK to drive the LEDs at very high current (1 A, or ten times their continuous current rating of 100 mA), and to do so without heat sinking. Fine... but a screwup in camera programming generated a long pulse that destroyed the array of LEDs; the cost was $40... and a day's labor.

The solution is a circuit like Figure 7.63. Upon triggering, the first monostable generates a pulse T_1 that limits the maximum output pulse width T_{OUT}, by disabling the AND gate after time T_1. The second monostable inhibits retriggering until it has timed out, preventing further LED-driving pulses until a total time T_2 after the last trigger. Here we chose timing components RC to set $T_1=100\,\mu$s and $T_2=5$ ms. Fitted with this protection circuit, the LED illuminator has now captured many millions of portraits of zebrafish, with both fish and LED living happily.

7.2.4 Timing with digital counters

For timing jobs where you want a long delay (minutes to hours, or even days), or where you need real accuracy, stability, or predictability, these analog timing methods are inadequate. What you do instead is use digital counters in combination with a fixed oscillator (perhaps of high

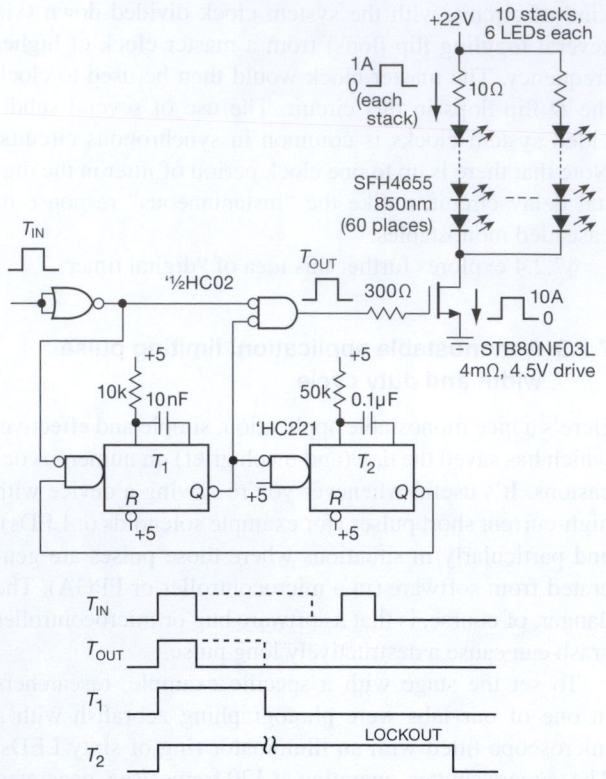

Figure 7.63. A simple protection circuit for high-current pulsed devices: limiting pulse width and duty cycle with a pair of monostables.

stability). Digital techniques are treated in detail in Chapters 10–13; but they form such an essential part of the business of timers that we must include them here.

A. An example: an "hour of power"

Suppose, to save battery power, you want a circuit to turn off a portable instrument after an hour. The circuit of Figure 7.64 will do the job. It exploits the CMOS 4060 binary counter chip (14 cascaded stages), which includes a pair of internal inverters intended for making an *RC* oscillator (the way we did in §7.1.2B). It belongs to the "4000B high-voltage CMOS" logic family, allowing operation from supply voltages of 3–18 V. That's handy here, because we can run it directly from the 9 V battery (whose terminal voltage begins life at about 9.4 V, and ends somewhere around 6 V).

At turn-on (or when the START button is pressed) the counter is reset to a count of zero (because the "R" reset input is brought momentarily HIGH), so all of the *Q*'s (the counter's binary outputs) are LOW. That turns on the *p*-channel MOSFET, supplying power to the load. When the

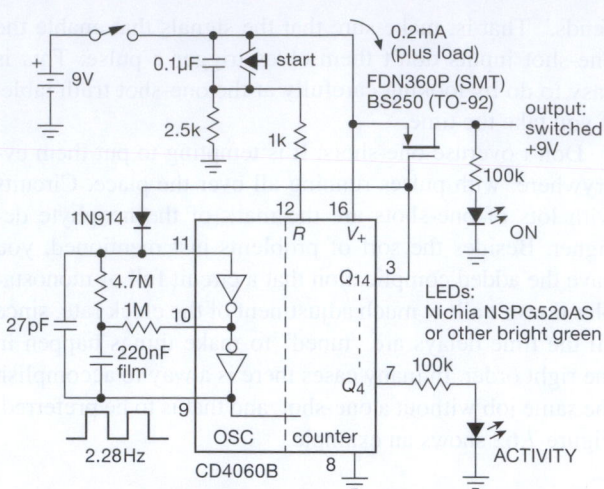

Figure 7.64. "An hour of power." The 1 MΩ (nominal) frequency-setting resistor could consist of a 750k fixed resistor in series with a 500k trimmer.

counter reaches a count of 2^{13}, the Q_{14} bit goes HIGH, both turning off power to the load and also stopping the oscillator. That's the end of the hour of power. The 4060 has negligible quiescent current ($\ll 1\,\mu$A), so the battery sees only MOSFET leakage current, also negligible.

A few comments. (a) Clamping the oscillator to end the cycle, as we've done here, is a trick; more typically you would use a flip-flop whose two states correspond to power-on and power-off. The popular 14541 "programmable-timer" chip offers such features internally, and is a better choice. (b) The fourth counter bit (Q_4) is used to signal activity, so you know the timer is running. We used a high-efficiency LED, running at $50\,\mu$A, to minimize current drain. (c) Tasks like this are easily accomplished with programmable *microcontrollers* (see Figure 7.69, and Chapter 15), which are flexibly reprogrammed, and which can do other tasks along the way. Our fondness for circuits like Figure 7.64 reveals, well, a certain nostalgia.

You can make an hour-of-power circuit with one of the elegant "TimerBlox" chips from LTC (introduced in §7.1.4B). The obvious choice would be the LTC6993 one-shot timer (see the table on page 433), but that part tops out at a pulse length of 34 seconds. Instead, you can press into service the LTC6991 low-frequency oscillator (maximum oscillator period of 9.5 hours!), rigged up as shown in Figure 7.65 so that it's held in the reset state after the

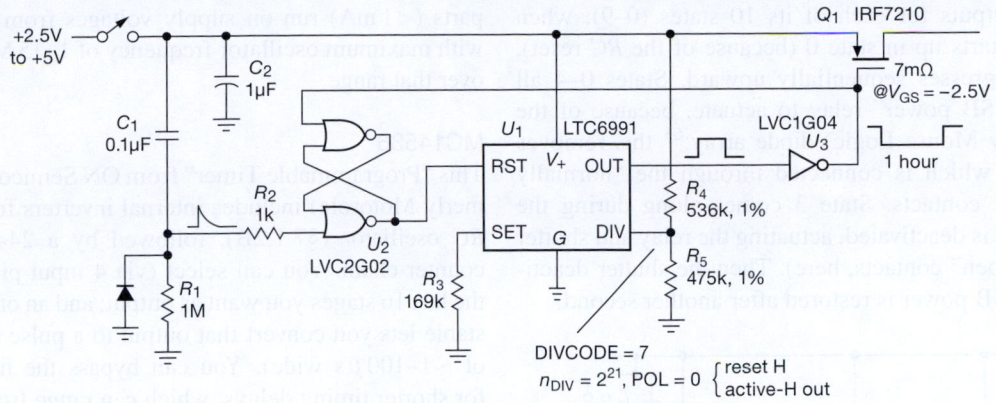

Figure 7.65. An hour of power, TimerBlox-style.

end of its first half cycle.[63] This circuit runs on the low voltages for which these chips are designed (single supply, +2.5 V to +5.5 V), which mandated a low-threshold pMOS switch for Q_1. A nice feature of this series of timers is their accuracy ($\pm 1.5\%$ worst case): your hour-of-power will terminate within one minute of its appointed time.

B. Another example: "one second per hour"

This circuit challenge appeared on the sci.electronics.design ("sed") newsgroup: generate a 1-second pulse once per hour, given a 1 Hz input clock; and do it with a minimum number of ICs. Figure 7.66 shows one way to do this,[64] once again using an integrated binary counter. This time we've used the 4040 12-stage counter, which lacks an oscillator but which provides outputs from every stage. The AND gates (§10.1.4) provide a HIGH output when the indicated Q's are all HIGH, i.e., the count $n = 2048 + 1024 + 512 + 8 + 4 + 2 + 1 = 3599$; that sets the output flip-flop (§10.4.1), which both generates one clock cycle of HIGH output and also resets the counter. Because the count is reset to zero, the full cycle is 3600 seconds, which, by great good fortune, happens to be the number of seconds in an hour.

---

[63] A fine point, but one that can bite: if the RST pin is found to be in the asserted state when the chip finishes its initialization (here ~1.7 ms), then (as casually stated in the datasheet) "the first pulse will be skipped." Sounds like no big deal – but that "first pulse" amounts to a full timing cycle, in other words 2 hours. That's a long time to be waiting for your hour of power! That's why we set a much longer ~100 ms power-on jam reset to *SR* flip-flop U_2, guaranteeing that U_1 won't see its RST pin improperly asserted until its OUT pin is happily into its hour-long HIGH state.

[64] As posted there by John Fields.

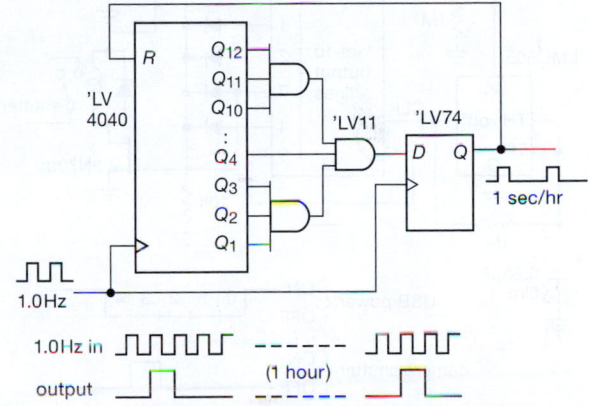

Figure 7.66. "One second per hour" timer, using digital logic.

C. A third example: remote camera control

We wanted to use a Panasonic DMC-LC1 digital camera to capture the faint light of star images as they transited an array of photomultipliers. The camera has a USB port for reading and deleting images from its memory card, but it does not provide a way to trip the shutter. However, it does have a separate connector for an electrical "shutter release," which can be activated with a switch or relay. So the idea is to use a relay to take a picture, then retrieve it via the USB connection.

Sounds easy – but there's a catch: you can't take a picture when the USB port is active. It turns out you have to deactivate the USB connection for at least 3 seconds before you take a picture, then you have to wait a second or two before reactivating it to get the image. Figure 7.67 shows a solution to this timing problem, relying again on digital counting methods. The circuit is timed by a CMOS 555, running at 1 Hz. It clocks a '4017 decimal counter, which has the interesting feature of providing individual outputs for each of its 10 states (0–9); when powered, it starts up in state 0 (because of the *RC* reset), and then progresses sequentially upward. States 0–4 all cause the "USB power" relay to actuate, because of the M²L (Mickey Mouse Logic) diode array;[65] this removes USB power, which is connected through the "normally closed" relay contacts. State 3 comes along during the time the USB is deactivated, actuating the relay and shutter ("normally open" contacts, here). Then the shutter deactivates, and USB power is restored after another second.

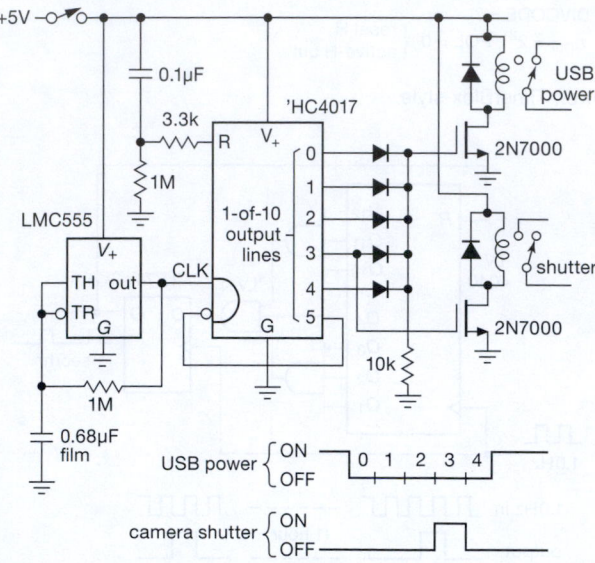

Figure 7.67. Camera control timer. The relays operate from 5 V (drawing 40 mA), and can switch up to 8 A; a suitable part is Omron G5C-2114P-US-DC5.

D. Other digital timing chips

There is a class of timer-oriented counter chips that are good for these kinds of tasks. Here are ones that we know and like (and see also the counter ICs in Table 10.5 on page 742).

ICM7240/42/50/60

This series of "Fixed and Programmable Timer/Counters" from Maxim includes an internal 555-like oscillator circuit, to which you attach an external frequency-setting R (to V_+) and C (to ground). The 8-pin 7242 has a fixed modulo-256 counter, with a pair of outputs at $f_{OSC}/2$ and $f_{OSC}/256$. The 16-pin 7240/50/60 parts allow you to set the divider modulus via programming pins: binary (1–255), decimal (1–99), and "real-time" (1–59), respectively. These low-power

---

[65] In the language of digital logic, this is a 5-input OR gate.

parts (<1 mA) run on supply voltages from 2 V to 16 V, with maximum oscillator frequency of 1–15 MHz (typical) over that range.

MC14536

This "Programmable Timer" from ON Semiconductor (formerly Motorola) includes internal inverters for making an *RC* oscillator (§7.1.2B), followed by a 24-stage binary counter chain. You can select (via 4 input pins) which of the last 16 stages you want as output; and an on-chip monostable lets you convert that output to a pulse (in the range of ∼1–100 µs wide). You can bypass the first 8 stages, for shorter timing delays, which can range from microseconds to days. This low-power part (∼1.5 µA/kHz, when driven from an external oscillator) runs on supply voltages from 3 V to 18 V, with maximum clocking frequency of 1–5 MHz (typical) over the range of 5–15 V.

MC14541

This "Programmable Timer" from ON Semiconductor is similar to the MC14536, but includes only 16 stages and limits the choices of divide ratios (2^8, 2^{10}, 2^{13}, and 2^{16}). In exchange, it gives you an internal power-up reset, a choice of output polarity, a choice of single-cycle or repeating-cycle modes, and somewhat lower operating current. It includes an output flip-flop, but unfortunately does not provide inputs to set or clear it.

LTC699x "TimerBlox"

This series includes oscillator and timer functions, with single-resistor frequency (or delay) programming, and voltage-divider (2 resistors) range and mode selection; see the table on page 433. They run on a single +2.5 V to +5.5 V supply, with excellent timing accuracy (<2%, worst-case, for the oscillators; 3.4% for the timers).

In addition to the oscillators and monostables in this family, there's the LTC6994-x "Delay Block & Debouncer." The -1 variant delays one edge only, while the -2 variant delays both edges (thus pulse-width preserving). The delay range τ is 1 µs to 34 seconds, selectable (along with polarity mode) via a voltage divider; a second resistor tunes the delay within the selected range, over a 16:1 span. Input pulses shorter than τ produce no output, which is handy for debouncing, or for "pulse qualification."

TPL5000/5100

These impressive nanopower timers from Texas Instruments run from 1.8 V to 5.5 V, draw a meager 40 nA (yes, *nano*amps!), and are 3-bit programmable (via three input pins) from 1 to 16 seconds or 16 to 1024 seconds

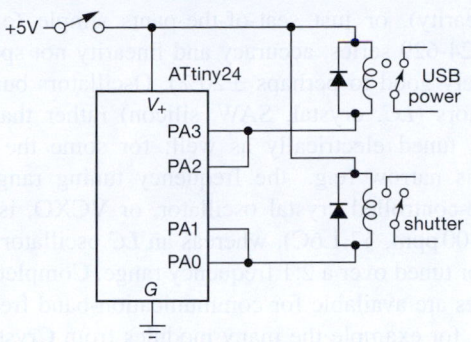

Figure 7.68. Microcontroller implementation of camera control timer.

(TPL5000 and 5100, respectively). The TPL5100 can drive a *p*-channel power MOSFET, to switch output loads.

Microcontrollers

A microcontroller is an inexpensive and flexible computer-on-a-chip, intended to be "embedded" into pretty much any kind of electronic device. We'll have much more to say about these wonderful devices in Chapter 15. But we cannot resist showing, in Figures 7.68 and 7.69, analogous microcontroller-based solutions to the camera shutter and "hour-of-power" examples whose implementation in discrete logic we showed in Figures 7.67 and 7.64, respectively.

In Figure 7.68 an Atmel "ATtiny24" microcontroller, with on-chip oscillator and timer functions (and *lots* more!) runs a program (which you have to write, of course) to do the shutter and USB power timing; its output pins can sink or source 20 mA, so we've used parallel pairs to handle the 40 mA relay current. This particular IC costs about a dollar, in small quantities. It, like all microcontrollers, is available in many variants, with additional memory, I/O, A/D converters, and so on. They are all remarkable bargains.

The microcontroller version of the "hour of power" (Figure 7.69) is a bit more complicated, because microcontrollers run only on supply voltages lower than the +9 V we

chose for the earlier example in Figure 7.64; typically they require a dc supply voltage in the range +1.8 V to +5 V. The approach we've taken here is to use a low-dropout linear regulator that has a shutdown mode (in which its standby current $I_{OFF} \sim 1\,\mu A$), and to use a microcontroller output pin to enable the regulator during operation. The circuit is rigged up for zero battery drain (other than leakage and I_{OFF}) except when timing.

A great virtue of programmable microcontrollers is their ability to give a fine performance in any of a variety of tasks. In this circuit you could program the controller to accept other inputs and produce other outputs (for example, sense temperature and humidity, and display values on an LCD display), and omit the output power switching entirely.[66] Or more simply, as a variation on the humble hour-of-power task, you could have several "mode" inputs that would set different timings, or patterns of power, or whatever.

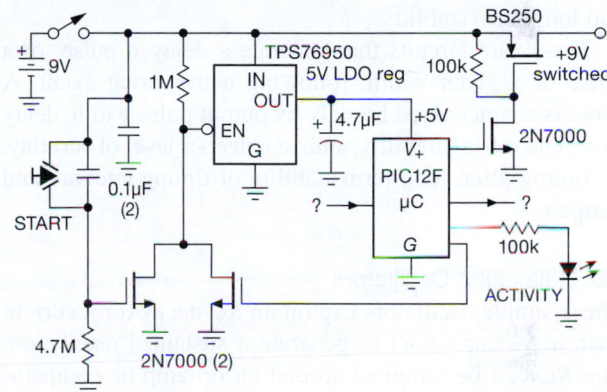

Figure 7.69. Microcontroller implementation of "an hour of power."

---

[66] For that kind of application you might prefer to run the whole circuit on +3 V, for example from a lithium cell, or a pair of alkaline cells.

Review of Chapter 7

An A-to-H summary of what we have learned in Chapter 7. This summary reviews basic principles and facts in Chapter 7, but it does not cover application circuit diagrams and practical engineering advice presented there.

¶A. Oscillator and Timer Overview.

Oscillators are circuits that create a periodic output waveform. The output may be as simple as a logic-level square wave (or train of pulses) for use as a "clock" in a digital system. Or it may be a highly accurate, stable, and perhaps programmable source of low-distortion sinewaves (or of other waveforms, for example a periodic ramp for use in an ADC or a PWM switchmode power converter). The signal from an oscillator is characterized broadly by its waveform, frequency, amplitude, and tunability, and, at a deeper level of scrutiny, by its phase noise, jitter, sideband suppression, distortion, temperature coefficient of frequency, and long-term stability.

Timers are circuits that generate a delayed pulse, or a pulse of a given width, following a triggering event. A timer is characterized broadly by output pulse width, delay time, and retriggerability, and, at a deeper level of scrutiny, by timing jitter, long-term stability of timing interval, and tempco.

¶B. Relaxation Oscillators.

These simple oscillators exploit an RC decay (or a current charging a capacitor) to generate a sustained oscillation. The RC can be wrapped around an op-amp or comparator, forming the classic pedagogical oscillator (§7.1.2A). More common is the use of logic inverters (§7.1.2B), or a timer IC like the ubiquitous 555 (§7.1.3 and Table 7.1) or its contemporary successors (e.g., the LTC6900-series, §7.1.4). Such timer or oscillator ICs deliver reasonable accuracy and predictability ($\sim$1%), over frequencies ranging from a few hertz to a megahertz or more. By adding a digital counter you can lengthen the period to minutes, hours, or just about forever; examples are the 74HC4060 (RC oscillator plus 14-stage binary counter; §7.1.4C), and the LTC6991 (resistor-programmed oscillator, periods from 1 ms to 9 hours; §7.1.4B). At the opposite end of the timescale, logic ICs extend the RC oscillator frequency range to tens or even hundreds of megahertz (§7.1.4D).

¶C. Voltage-controlled Oscillators.

Sometimes you want frequency tunability, via an input control voltage. That's a *VCO*. These can be highly linear and stable (e.g., the AD650 V-to-F converter, $\sim$0.01%

nonlinearity), or just seat-of-the-pants simple (e.g., the 74LS624-629 series, accuracy and linearity not specified, but likely good to perhaps $\pm$20%). Oscillators built with resonators (LC, crystal, SAW, silicon) rather than RC's can be tuned electrically as well; for some the tuning range is narrow (e.g., the frequency tuning range of a voltage-controlled crystal oscillator, or VCXO, is of order $\pm$100 ppm, §7.1.6C), whereas an LC oscillator can be varactor tuned over a 2:1 frequency range. Complete VCO modules are available for communication-band frequency ranges, for example the many modules from Crystek that together span the range from about 50 MHz to 5 GHz; the individual modules each tune over a modest range, from as little as $\pm$1% to as much as $\pm$25%.

¶D. Sinewave Oscillators.

The inherent waveforms of relaxation oscillators are segments of exponential charging and discharging (if built with an RC), or linear ramps (if the resistor is replaced by a current source); in either case these waveforms drive a comparator circuit to reverse the cycle. That is, the "natural" waveforms are square waves, triangle or sawtooth waves, or what might be called "shark's fin" waves. What you don't get are *sine* waves.

But sinewaves are essential for many tasks; and you *can* generate them, with some cleverness, with the frequency set by R's and C's only, analogous to the simple relaxation oscillator. Most famous is the *Wien-bridge oscillator* (§7.1.5B), an arrangement of two R's and two C's whose phase shifts cancel at a frequency $f=1/2\pi RC$, where the attenuation is exactly a factor of 1/3. The oscillator circuit wraps this network around a noninverting op-amp with gain equal to +3. The final trick is to sense the output amplitude, and maintain it at a preset (and nonsaturated) level by controlling the gain. A pair of fellows named Hewlett and Packard founded a business with this as their first product; the rest is history.

You can make sinewaves also by lowpass filtering a nonsinusoidal waveform. This is the basis of the *RC phase-shift oscillator* (§7.1.5C), and of the digital tracking-filter technique (§7.1.5A).

An oscillator that generates a pair of sinusoidal signals with a phase difference of 90° (i.e., sine and cosine) is called a *quadrature* oscillator (§7.1.9); the signals are "in quadrature." This can be done with analog techniques (integrator pair, switched-capacitor resonator, phase-sequence filter, RF quadrature hybrid), or with digital synthesis (see ¶G below).

¶E. Resonator-based Oscillators.
Relaxation oscillators, and *RC*-based oscillators in general, can deliver stabilities no better than 1% or so. That is because their accuracy depends on an exponential decay (in the time domain), or a gradual phase shift (in the frequency domain). For better stability you need to exploit a resonant system, whose natural physical frequency can be extremely well-defined. It can have a "quality factor" Q (which is its frequency selectivity $\Delta f/f$) of a million or more.

The simple *LC* resonant circuit is widely used in the frequency range from kilohertz to hundreds of megahertz (§7.1.5D), with stabilities of order 10–100 ppm, and with tunability via a mechanical-variable capacitor or inductor, or an electrically-tunable capacitor (a varactor). For higher frequencies the *LC* tuned circuit can be replaced by a coaxial or cavity resonator.

Better stability (but less tunability) is provided by *electro-mechanical* resonators: quartz-crystal (§7.1.6), ceramic (§7.1.6G), and surface-acoustic-wave (SAW; §7.1.6H). These are inexpensive and stable ($\sim$10ppm), and quartz-crystals and crystal-oscillator modules are available in a wide selection of standard frequencies in the range of 10 kHz to 100 MHz (and crystal manufacturers are happy to provide any frequency you want as a special-order product, at quite reasonable prices). Crystal oscillators can be "pulled" up to $\pm$100 ppm with a varactor; these are sometimes called VCXOs (voltage-controlled crystal oscillators). Ceramic-resonator oscillators can be pulled over a wider range ($\pm$1000 ppm), owing to their lower Q; the downside is poorer frequency stability. They are very inexpensive, but they come in only a limited set of standard frequencies.

¶F. High-stability Oscillators.
Crystal oscillators held at constant temperature are remarkably stable, as good as 1 part in 10^9 for a well-designed OCXO (oven-controlled crystal oscillator); the residual drift is due mostly to mechanical and diffusional "aging" effects. For still better stability you need an *atomic* standard (§7.1.7B), the most accessible being a rubidium-vapor-stabilized crystal oscillator. These deliver stabilities of order 1 part in 10^{10}, and their prices (around $1k, but you can get some real deals on eBay) allow inclusion in precision test and communication equipment. There are more precise (and exotic) frequency standards – cesium beam, hydrogen maser, atomic fountain, chilled ion – but, as they say, "if you have to ask the price, you can't afford them."

Happily, you can take advantage of your tax dollars at work, namely the precise timing provided by the GPS satellite constellation. For about $1k (and a place to put a little doorknob-size antenna) you can have your own 10 MHz reference, with long-term stability of order 10^{-12}.

¶G. Frequency Synthesis.
VCOs are *tunable*, but not particularly stable, nor capable of spanning many decades of frequency; by contrast, a crystal oscillator is *stable*, but tunable at most over parts-per-million. But you can have it both ways: from a stable reference oscillator at some standard fixed frequency (typically a highly stable 10.0 MHz crystal oscillator) you can generate an output frequency of your choosing, with one of several methods of *frequency synthesis* (§7.1.8). Simplest to understand is *direct digital synthesis* (DDS), in which successive sinusoidal amplitudes are read from a table at the appropriate rate and converted to an analog output signal voltage. Complete DDS chips that include all the hardware (counters, sine lookup table, ADC) are widely available, and inexpensive. You provide the reference frequency, and you talk to these things with a microcontroller (Chapter 15) to set the output frequency, phase offset, and so on. DDS chips let you synthesize output frequencies ranging from fractions of a hertz up to a gigahertz, with frequency *agility* (the ability to change frequency rapidly and precisely).

Phase-locked-loop (PLL) synthesis is the other common technique. In its simplest form, a VCO is controlled by a phase comparator whose inputs are the rth subdivision of the reference frequency and the nth subdivision of the VCO output frequency. Then $f_{out}=(n/r)f_{ref}$. There are many subtleties here, involving loop stability, sidebands and jitter, resolution, lockup and slewing times, and so on. See §13.13 for an extensive discussion.

¶H. Timers and One-shots.
You can make simple edge-triggered pulsers with as little as an *RC*, or (better) with an *RC* assisted by a BJT or MOSFET (§7.2.1A). But you get cleaner pulses with faster edge times if you use logic gates (§7.2.1C). For better predictability it's best to use an IC designed for timer use, either a monostable multivibrator (a "one-shot"), a 555-type timer, or one of a class of specialized timers like the LTC6991/3, ICM7240/50/60, or MC14536/41. Monostables, being mixed-signal components (logic parts with a linear function), have their "issues," notably their sensitivity to supply-rail noise; they also have a somewhat limited timing range (typically tens of nanoseconds to seconds). They come in both non-retriggerable and retriggerable types; the latter extend their pulse duration if another trigger arrives during the pulse. Pulse widths are set by an

Review of Chapter 7

external *RC*, with a multiplying constant *K* of order unity ($\tau=KRC+t_{min}$); see Table 7.4 and Figure 7.60.

You can do much more with timer chips that incorporate a digital counter (e.g., the LTC6991 extends the timing range to ten hours), or you can configure an external bi-nary counter. When thinking about extended timing intervals, or in fact pretty much any timing tasks, don't forget the versatile *microcontroller* (the subject of Chapter 15); there you'll find a timing example, in the form of a "suntan monitor" (§15.2).

LOW-NOISE TECHNIQUES

CHAPTER **8**

In many applications you're dealing with small signals, for which it is essential to minimize the degrading effects of amplifier "noise." Low-noise design is thus an important part of the art of electronics. The extensive detail (decorated with more than our usually paltry quota of equations!) in this chapter reflects the richness of the low-noise design. So does its length – it's the longest chapter in the book. Recognizing that many readers will have a less-than-passionate interest in the various subjects treated here, we offer the following guide.

A quick guide to this chapter The basics of noise are explained in §8.1 ("Noise"), which should be read first. Readers interested primarily in low-noise design with *op-amps* can then skip ahead to the discussion, tables, and graphs in §8.9 ("Noise in operational amplifier circuits"). Those interested in low-noise design with discrete transistors (or interested in gaining a fuller understanding of what's going on "under the hood" of op-amps) should read §8.5 ("Low-noise design with bipolar transistors") and §8.6 ("Low-noise design with JFETs"). Readers working with photodiode circuits and the like will want to read §8.11 ("Noise in transimpedance amplifiers").[1] For a discussion of noise *measurement*, go to §8.12 ("Noise measurements and noise sources") and §8.13 ("Bandwidth Limiting and RMS Voltage Measurement").

An even quicker guide to noise This chapter is lengthy, and it's filled with mathematical details and information about hundreds of transistors and op-amps. But noise doesn't have to be complicated. In a spirit of encapsulating the essence of noise, we offer a one-minute breathless end-of-class "takeaway":

> The random noise you care about is characterized by its *density* (rms noise amplitude in a 1 Hz band of frequency); voltage noise density is called e_n, and has units like $nV/\sqrt{Hz}$. Likewise, the symbol for noise *current* is i_n; a noise current at an amplifier's input flows through the signal's source resistance, creating its own noise voltage $e_n = i_n R_s$. If a noise source is uniform over frequency, it's called "white noise," and the rms noise voltage contained within a bandwidth B is just $v_n = e_n\sqrt{B}$. Knowing that, you can go to Tables 8.3a–8.3c on page 522, which lists e_n and i_n for a wide selection of op-amps, to figure out how much noise is added in an amplifier stage. Multiply by the amplifier's gain and, voilà, you've got the output noise.

> Amplifiers aren't the only source of noise. A resistor generates "Johnson noise", eq'n 8.4, and the discrete charges in a flow of current generate "shot noise", eq'n 8.6. Both of these are white noise.[2] Finally (with ten seconds remaining) – to figure the total noise in a circuit with multiple independent noise sources, you take the sum of the squares of each noise density, multiply by the bandwidth, then take the square root. Time's up. End of class.

8.1 "Noise"

In almost every area of measurement the ultimate limit of detectability of weak signals is set by noise – unwanted signals that obscure the desired signal. Even if the quantity being measured is not weak, the presence of noise degrades the accuracy of the measurement. Some forms of noise are unavoidable (e.g., real fluctuations in the quantity being measured), and they can be overcome only with the techniques of *signal averaging* and *bandwidth narrowing*.[3] Other forms of noise (e.g., radiofrequency interference and "ground loops") can be reduced or eliminated by a variety of tricks, including filtering and careful attention to wiring configuration and parts location. Finally, there is noise that arises in the amplification process itself, and it can be reduced through the techniques of low-noise amplifier design. Although the techniques of signal averaging

---

[1] And also the related discussion of stability and bandwidth in §4x.3.

[2] Things get more interesting when noise density varies with frequency, for example the notorious pink "flicker noise" that rises as $e_n \propto 1/\sqrt{f}$ at low frequencies. This fascinating (and infuriating) annoyance does not go unnoticed in this chapter!

[3] See §8.14, and also Chapter 15 of this book's second edition (1989).

can often be used to rescue a signal buried in noise, it always pays to begin with a system that is free of preventable interference and that possesses the lowest amplifier noise practicable.

We begin by talking about the origins and characteristics of the different kinds of noise that afflict electronic circuits. Then we launch into a discussion of bipolar-transistor (BJT) and field-effect transistor (FET) noise, including methods for low-noise design with a given signal source, and we present some design examples. After a short discussion of noise in differential and feedback amplifiers, we continue with low-noise design with op-amps, including transimpedance (current-to-voltage) amplifiers. Sections on noise measurements, bandwidth limiting, and lock-in detection follow, then a short discussion of power-supply noise. We conclude with a section on proper grounding and shielding and the elimination of interference and pickup.

Because the term *noise* can be applied to anything that obscures a desired signal,[4] noise can take the form of another signal ("interference"); most often, however, we use the term to describe "random" noise of a physical (often thermal) origin. Noise can be characterized by its frequency spectrum, its amplitude distribution, and the physical mechanism responsible for its generation. Let's next look at the chief offenders:

Johnson noise: Random-noise voltage created by thermal fluctuations in a resistor.

Shot noise: Random statistical fluctuations in a flowing current caused by the discrete nature of electrical charge.

Flicker noise: Additional random noise, rising typically as $1/f$ in power at low frequencies, with a multitude of causes.

Burst noise: low-frequency noise typically seen as random jumps between a pair of levels, caused by material device defects.

8.1.1 Johnson (Nyquist) noise

Any old resistor just sitting on the table generates a noise voltage across its terminals known as Johnson noise (or Nyquist noise).[5] It has a flat frequency spectrum, mean-

---

[4] As Lew Branscomb famously noted, "Nature does not 'know' what experiment a scientist is trying to do. God loves the noise as much as the signal." See L. Branscomb, "Integrity in Science," *Am. Sci.* **73**, 421–23 (1985).

[5] Experiment and formulas by J. B. Johnson, Letter to *Nature*, **119**, 50 (1927), *Phys. Rev.* **32**, "Thermal agitation of electricity in conductors,"

ing that there is the same noise power in each hertz of frequency (up to some limit, of course). Noise with a flat spectrum is also called "white noise." The actual open-circuit noise voltage generated by a resistance R at temperature T is given by

$$v_{\text{noise}}(\text{rms}) = v_{\text{n}} = (4kTRB)^{\frac{1}{2}} \quad \text{V(rms)}, \qquad (8.1)$$

where k is Boltzmann's constant, T is the absolute temperature in Kelvins (K=°C + 273.16), and B is the bandwidth in hertz. Thus $v_{\text{noise}}(\text{rms})$ is what you would measure at the output if you drove a perfect noiseless bandpass filter (of bandwidth B) with the voltage generated by a resistor at temperature T. At room temperature (68°F = 20°C = 293K),

$$
\begin{aligned}
4kT &= 1.62 \times 10^{-20} && \text{V}^2/\text{Hz} - \Omega, \\
(4kTR)^{\frac{1}{2}} &= 1.27 \times 10^{-10} R^{\frac{1}{2}} && \text{V}/\text{Hz}^{\frac{1}{2}} \qquad (8.2) \\
&= 1.27 \times 10^{-4} R^{\frac{1}{2}} && \mu\text{V}/\text{Hz}^{\frac{1}{2}}.
\end{aligned}
$$

For example, a 10k resistor at room temperature has an open-circuit rms voltage of $1.3\,\mu\text{V}$, measured with a bandwidth of 10 kHz (e.g., by placing it across the input of a good audio amplifier and measuring the output with a voltmeter). The source resistance of this noise voltage is just R. If you connect the terminals of the resistor together, you get a (short-circuit) current of

$$i_{\text{noise}}(\text{rms}) = v_{\text{noise}}(\text{rms})/R = v_{\text{nR}}/R = (4kTB/R)^{\frac{1}{2}}. \quad (8.3)$$

As we'll see in §8.2.1, it's convenient to express noise voltage (or current) as a *density* e_{n} (rms voltage per square root bandwidth). Johnson noise, with its flat (white) spectrum, has constant noise voltage-density

$$e_{\text{n}} = \sqrt{4kTR} \quad \text{V}/\text{Hz}^{\frac{1}{2}}, \qquad (8.4)$$

from which the rms noise voltage in some limited bandwidth B is then simply $v_{\text{n}} = e_{\text{n}}\sqrt{B}$. Likewise, the short-circuit noise-current density is

$$i_{\text{n}} = \sqrt{4kT/R} \quad \text{A}/\text{Hz}^{\frac{1}{2}}. \qquad (8.5)$$

Figure 8.1 plots the simple relationship between Johnson-noise voltage density and source resistance; also shown is the short-circuit noise current density. An easy number to remember, when choosing resistor values for low-noise amplifier designs, is that a 1kΩ resistor at room temperature generates an open-circuit noise voltage density of $4\,\text{nV}/\sqrt{\text{Hz}}$; scale by the square root of resistance for other values.[6]

97–109, (1928), subsequent theory by H. Nyquist, *Phys. Rev.*, "Thermal agitation of electric charge in conductors," **32**, 110–113, (1928).

[6] We find it handy to remember the values of q and of $4kT$ (which keeps

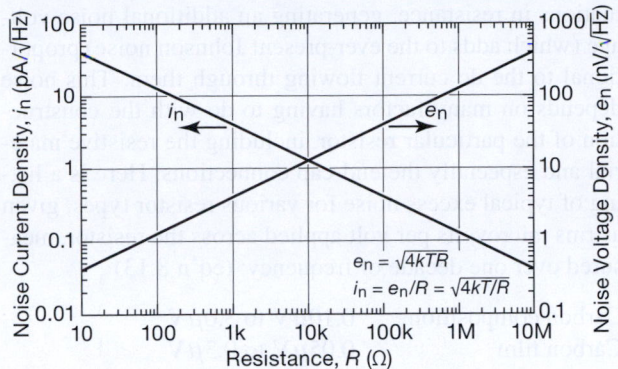

Figure 8.1. Open-circuit thermal noise-voltage and short-circuit thermal noise-current densities versus resistance at 25°C.

Here's a handy Johnson-noise mini-table, listing both voltage and current noise *densities* (units of $V/\sqrt{Hz}$ and $A/\sqrt{Hz}$), and noise within a 10 kHz band, for seven decade-related values of resistance:

		open circuit		short circuit
R	e_n (nV/$\sqrt{Hz}$)	$e_n\sqrt{B}$ $B=10\,kHz$ (μV)	i_n (pA/$\sqrt{Hz}$)	$i_n\sqrt{B}$ $B=10\,kHz$ (pA)
100 Ω	1.28	0.128	12.8	1280
1k	4.06	0.406	4.06	406
10k	12.8	1.28	1.28	128
100k	40.6	4.06	0.406	40.6
1M	128	12.8	0.128	12.8
10M	406	40.6	0.041	4.06
100M	1280	128	0.0128	1.28

Johnson noise, at $T=25°C$

The amplitude of the Johnson-noise voltage at any instant is, in general, unpredictable, but it obeys a Gaussian amplitude distribution (Figure 8.2), where $p(V)dV$ is the probability that the instantaneous voltage lies between V and $V+dV$, and v_n(rms) is the rms noise voltage, given earlier.[7]

The significance of Johnson noise is that it sets a lower limit on the noise voltage in any detector, signal source, or amplifier having resistance. The resistive part of a source impedance generates Johnson noise, as do the bias and load resistors of an amplifier. You will see how it all works out presently.

popping up) together, because in SI units they are 1.6×10^{-19} and 1.6×10^{-20}, respectively.

[7] See also Figure 8.115, which plots the odds (over 9 decades) that the instantaneous amplitude exceeds some multiple of the rms amplitude.

It is interesting to note that the physical analog of resistance (any mechanism of energy loss in a physical system, e.g., viscous friction acting on small particles in a liquid) has associated with it fluctuations in the associated physical quantity (in this case, the particles' velocity, manifest as the chaotic Brownian motion). Johnson noise is just a special case of this fluctuation–dissipation phenomenon.

Johnson noise should not be confused with the additional noise voltage created by the effect of resistance fluctuations when an externally applied current flows through a resistor. This "excess noise" has a $1/f$ spectrum (approximately) and is heavily dependent on the actual construction of the resistor. We will talk about it later.

8.1.2 Shot noise

An electric current is the flow of discrete electric charges, not a smooth fluidlike flow. The finiteness of the charge quantum results in statistical fluctuations of the current. If the charges act independently of each other, the fluctuating current's noise density is given by

$$i_n = \sqrt{2qI_{dc}} \quad A/Hz^{\frac{1}{2}}, \tag{8.6}$$

where q is the electron charge (1.60×10^{-19} coulomb). This noise, like resistor Johnson noise, is white and Gaussian. So its amplitude, taken over a measurement bandwidth B, is just

$$i_{noise}(rms) = i_{nR}(rms) = i_n\sqrt{B} = (2qI_{dc}B)^{\frac{1}{2}} \quad A(rms). \tag{8.7}$$

$$p(V, V+dV) = \frac{1}{V_n\sqrt{2\pi}} \exp(-V^2/2V_n^2)\, dV,$$

where V_n is the rms noise

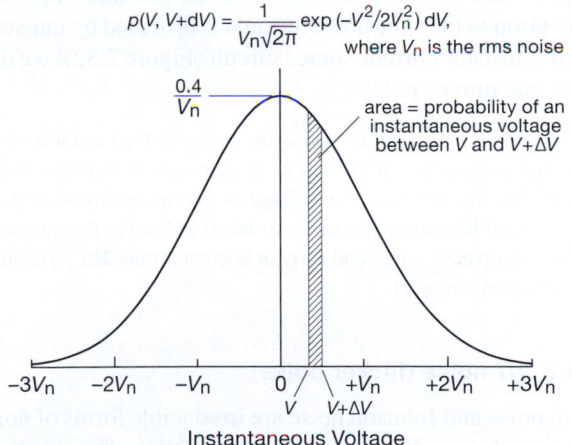

Figure 8.2. Johnson noise obeys a Gaussian distribution of amplitudes. The normalizing factor $0.4/V_n$ ensures dimensionless unit area under the bell curve (the "0.4" is actually $1/\sqrt{2\pi}$, about 0.3989).

For example, a "steady" current of 1 A actually has an rms fluctuation of 57 nA, measured in a 10 kHz bandwidth; i.e., it fluctuates by about 0.000006%. The relative fluctuations are larger for smaller currents: a "steady" current of 1μA actually has an rms current-noise fluctuation, measured over a 10 kHz bandwidth, of 0.006%, i.e., -85 dB. At 1 pA dc, the rms current fluctuation (same bandwidth) is 57 fA, i.e., a 5.7% variation! Shot noise is "rain on a tin roof."

Here's a handy minitable listing shot-noise current density and shot noise current in a 10 kHz band for decadal currents spanning 12 orders of magnitude:

	Shot noise current $B=10$ kHz		
I_{dc}	i_n	$i_n\sqrt{B}$ (10 kHz)	$\dfrac{i_n\sqrt{B}}{I_{dc}}$
1 fA	18 aA/$\sqrt{\text{Hz}}$	1.8 fA	+5 dB
1 pA	0.57 fA/$\sqrt{\text{Hz}}$	57 fA	-25 dB
1 nA	18 fA/$\sqrt{\text{Hz}}$	1.8 pA	-55 dB
1 μA	0.57 pA/$\sqrt{\text{Hz}}$	57 pA	-85 dB
1 mA	18 pA/$\sqrt{\text{Hz}}$	1.8 nA	-115 dB

An important point: the shot-noise formula given earlier assumes that the charge carriers making up the current act independently. That is indeed the case for charges crossing a barrier, for example the current in a junction diode, where the charges move by diffusion; but it is not true for the important case of metallic conductors, where there are long-range correlations between charge carriers. Thus the current in a simple resistive circuit has far less noise than is predicted by the shot-noise formula. Another important exception to the shot-noise formula is provided by our standard transistor current-source circuit (Figure 2.32); we discuss this further in §8.3.5.

Exercise 8.1. A resistor is used as the collector load in a low-noise amplifier; the collector current I_C is accompanied by shot noise. Show that the output noise voltage is dominated by shot noise (rather than Johnson noise in the resistor) as long as the quiescent voltage drop across the load resistor is greater than $2kT/q$ (50mV, at room temperature).

8.1.3 1/f noise (flicker noise)

Shot noise and Johnson noise are irreducible forms of noise generated according to physical principles. The most expensive and most carefully made resistor has exactly the same Johnson noise as the cheapest carbon resistor of the same resistance. Real devices have, in addition, various sources of "excess noise." Real resistors suffer from fluc-

tuations in resistance, generating an additional noise voltage (which adds to the ever-present Johnson noise) proportional to the dc current flowing through them. This noise depends on many factors having to do with the construction of the particular resistor, including the resistive material and especially the end-cap connections. Here is a listing of typical excess noise for various resistor types, given as rms microvolts per volt applied across the resistor, measured over one decade of frequency: (eq'n 8.13)

Carbon composition	0.10μV to 3.0μV
Carbon film	0.05μV to 0.3μV
Metal film	0.02μV to 0.2μV
Wire wound	0.01μV to 0.2μV

This noise has approximately a $1/f$ power spectrum (equal power per decade of frequency) and is sometimes called "pink noise." When plotted against voltage or current (rather than power) its *amplitude* falls as $1/\sqrt{f}$, as shown in Figure 8.3. Figure 8.4 shows how it looks in comparison with a sample of white noise and of what's sometimes called "red noise" ($1/f^2$ power spectrum); if you want to make your own, look ahead to Figure 8.93 to see how.

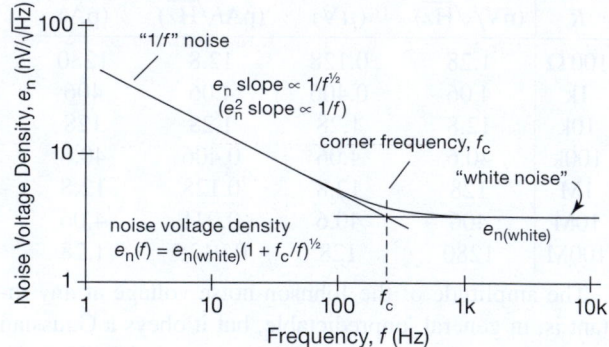

Figure 8.3. When plotted on log axes as noise voltage versus frequency, $1/f$ noise slopes downward with a slope of 1/2, i.e., as $1/f^{\frac{1}{2}}$ (it is noise *power* that goes as $1/f$).

You often see the notation f_c for the corner frequency at which the $1/f$ noise is the same as an underlying white-noise component.[8] The combined noise voltage density is

$$e_n(f)=e_{n(\text{white})}\sqrt{1+f_c/f}, \tag{8.8}$$

from which the rms integrated noise voltage in a band extending from f_1 to f_2 can be calculated; see eq'n 8.59 on page 565.

---

[8] You can estimate f_c with the help of eq'n 8.27 on page 491.

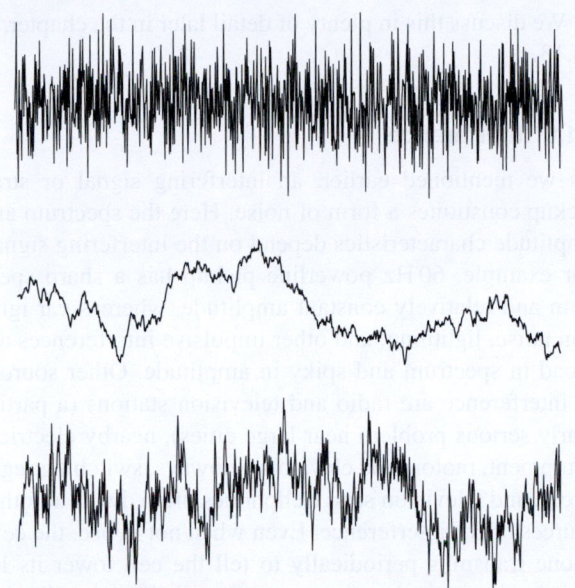

Figure 8.4. Three noises: top, "white noise" (uniform power per Hz); middle, "red noise" (power per Hz proportional to $1/f^2$); bottom, "pink noise" (or $1/f$ noise, power per Hz proportional to $1/f$).

Other noise-generating mechanisms often produce $1/f$ noise, examples being base-current noise in transistors and cathode current noise in vacuum tubes. Curiously enough, $1/f$ noise is present in nature in unexpected places, e.g., the speed of ocean currents, the flow of sand in an hourglass, the flow of traffic on Japanese expressways, and the yearly flow of the Nile measured over the last 2,000 years.[9] If you plot the loudness of a piece of classical music versus time, you get a $1/f$ spectrum! No unifying principle has been found for all the $1/f$ noise that seems to be swirling around us, although particular sources can often be identified in each instance.

8.1.4 Burst noise

Not all noise sources are characterized by a Gaussian (or even *smooth*) distribution of amplitudes. Most notorious among the exceptions is *burst noise* (also variously called *popcorn* noise, *bistable* noise, or *random telegraph signal* noise), seen occasionally in semiconductor devices (particularly in parts dating back to the 1970s and earlier). It consists of random jumps between two (usually) voltage levels, taking place on time scales of tens of milliseconds;

when played out on a loudspeaker it sounds like the birth throes of popcorn. Figure 8.5 shows a typical waveform, the output of a vintage[10] 741 op-amp wired as a noninverting amplifier with $G=100$.

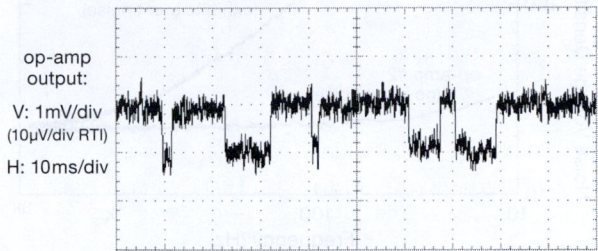

op-amp
output:

V: 1mV/div
(10µV/div RTI)

H: 10ms/div

Figure 8.5. Burst noise from a 1973-vintage 741 op-amp, configured as an ×100 noninverting amplifier with grounded input. The output was bandpass filtered to 0.1 Hz–3 kHz, with 6 dB/octave rolloffs.

Viewed in the frequency domain, the effect of burst noise is a raised low frequency portion, without any obvious spectral peaks. You can see this in Figure 8.6, where the voltage noise spectrum of the noisy and quiet op-amp specimens are plotted.[11]

8.1.5 Band-limited noise

All circuits operate within some limited frequency band. So, although it's nice to talk about (and calculate with) noise-*density* quantities, what you usually care about is the rms noise voltage contained within some signal band of interest (called B in eq'n 8.1). In many cases you are dealing with a white noise source (e.g., Johnson noise or shot noise). If that is passed through a perfectly sharp bandpass filter (a "brick-wall" filter), the band-limited rms amplitude is simply $v_{n(rms)}=e_n\sqrt{B}$. But analog brick-wall filters aren't practical – so what you want to know is the equivalent bandwidth of a real filter, say a simple RC lowpass. It turns out that the equivalent brick-wall bandwidth is given by

$$B = \frac{\pi}{2}f_{3dB} = 1.57f_{3dB} = \frac{1}{4RC} \quad \text{Hz,} \qquad (8.9)$$

---

[9] A delightful reference is W. H. Press, "Flicker noises in astronomy and elsewhere," *Comm. on Astrophys.* **7**, 103–119 (1978). Available at `http://www.nr.com/whp/Flicker_Noise_1978.pdf`.

[10] Semiconductor manufacturers have worked hard to alleviate this problem (believed to be caused by intermittent trapping of charge carriers at defects and interfaces), and popcorn noise is largely a thing of the past. We tested ten samples of 741s, from six different manufacturers, before we found this one. A second sample from the same manufacturer and with the same date code showed no evidence of popcorn noise, as seen in the pair of spectra of Figure 8.6.

[11] It is possible that a related mechanism is responsible for similarly shaped noise plateaus seen in some JFETs; see for example the LSK389 in Figure 8.47.

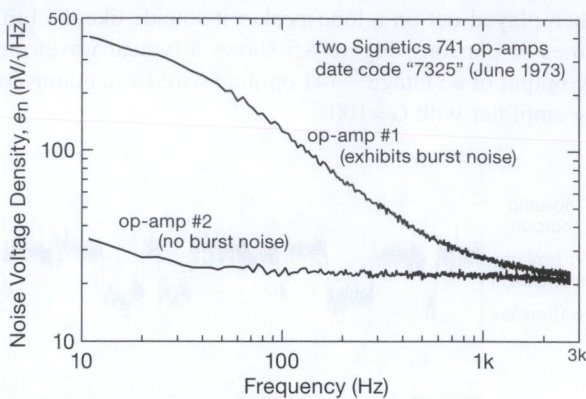

Figure 8.6. Spectrum of the burst noise produced by the same op-amp as used for Figure 8.5, along with that of a second op-amp from the same batch that exhibited no burst noise. The vertical scale shows RTI (referred to the input) rms noise densities.

where $f_{3dB}=1/2\pi RC$. You can use higher-order filters, of course, for example a 2-pole Butterworth lowpass filter; its equivalent brick-wall bandwidth is $B=1.11f_{3dB}$. For still higher-order filters (including bandpass) see the expressions in Table 8.4 on page 564. For slowly varying (or dc) signals you can instead do a simple averaging (as, for example, with an integrating ADC, see §13.8.3); in that case the equivalent noise bandwidth is $B=1/2T$, where T is the duration of the (uniform) averaging of the input signal. We'll have a bit more to say about this in §8.13.1.

Of course, the noise spectrum may be other than white (e.g., it may be $1/f$ noise or a combination of white noise with a $1/f$ rising tail at low frequencies). In such a case you cannot simply multiply the noise density by the square root of the bandwidth. Instead you have to integrate the (changing) noise density over the bandpass. For an ideal brick-wall bandpass that is just $v_n^2=\int e_n^2(f)df$ from the lower to upper frequency cutoff of the filter. For a realizable filter you have to integrate the noise density, multiplied by the filter's spectral response $H(f)$, over the bandpass: $v_n^2=\int |e_n(f)H(f)|^2df$. For an arbitrary noise spectrum this is what you would need to do. But life is simpler if you're dealing with classic noise spectra like $1/f$ flicker noise, in which case the noise integrals can be expressed analytically. We've gathered these together in Table 8.4 on page 564, which includes results for white-, pink-, and red-noise spectra, when band limited by brick-wall, single-pole, 2-pole Butterworth, and m-pole Butterworth bandpass filters (response from f_1 to f_2). Those tabulated formulas let you get the results for lowpass ($f_1=0$) or highpass filters ($f_2=\infty$), which are just special cases of the more general bandpass filter.

We discuss this in plenty of detail later in the chapter, at §8.13.

8.1.6 Interference

As we mentioned earlier, an interfering signal or stray pickup constitutes a form of noise. Here the spectrum and amplitude characteristics depend on the interfering signal. For example, 60 Hz powerline pickup has a sharp spectrum and relatively constant amplitude, whereas car ignition noise, lightning, and other impulsive interferences are broad in spectrum and spiky in amplitude. Other sources of interference are radio and television stations (a particularly serious problem near large cities), nearby electrical equipment, motors and elevators, subways, switching regulators, and television sets. Cellphones often dwarf all other sources of RF interference. Even when not in use, the cellphone transmits periodically to tell the cell tower its location, generating interference with a distinctive galloping rhythm.[12] The same goes for mobile computers that use the cellular network for Internet access.

In a slightly different guise you have the same sort of problem generated by anything that puts a signal into the parameter you are measuring. For example, an optical interferometer is susceptible to vibration, and a sensitive RF measurement (e.g., nuclear magnetic resonance, NMR or MRI) can be affected by ambient RF. Many circuits, as well as detectors and even cables, are sensitive to vibration and sound; they are *microphonic*, in the terminology of the trade.

Many of these noise sources can be controlled by careful shielding and filtering, as we discuss later in the chapter. At other times we are forced to take Draconian measures, involving massive stone tables (for vibration isolation), constant-temperature rooms, anechoic chambers, and electrically shielded ("Faraday cage") rooms.

8.2 Signal-to-noise ratio and noise figure

Before getting into the details of amplifier noise and low-noise design, we need to define a few terms that are often used to describe amplifier performance. These involve ratios of noise voltages, measured at the same place in the circuit. It is conventional to refer noise voltages to the input of an amplifier (although the measurements are usually made at the output), i.e., to describe source noise and amplifier noise in terms of microvolts *at the input* that would

---

[12] Which you can hear at www.covingtoninnovations.com/ michael/blog/0506/050622-cellnoise.mp3.

generate the observed output noise. This makes sense when you want to think of the relative noise added by the amplifier to a given signal, independent of amplifier gain; it's also realistic, because most of the amplifier noise is usually contributed by the input stage. Unless we state otherwise, noise voltages are referred to the input (RTI).

8.2.1 Noise power density and bandwidth

In the preceding examples of Johnson noise and shot noise, the noise voltage you measure depends both on the measurement bandwidth B (i.e., how much noise you see depends on how fast you look) and on the variables (R and I) of the noise source itself. So it's convenient to talk about an rms noise-voltage "density" e_n:

$$v_{n(rms)} = e_n B^{\frac{1}{2}} = (4kTR)^{\frac{1}{2}} B^{\frac{1}{2}} \quad \text{Vrms,} \quad (8.10)$$

where v_n is the rms noise voltage you would measure in a bandwidth B. White-noise sources have an e_n that doesn't depend on frequency, whereas pink noise, for instance, has a e_n that drops off at 3 dB/octave. You'll often see e_n^2, too, the mean squared noise density. Since e_n always refers to rms and e_n^2 always refers to mean square, you can just square e_n to get e_n^2! Sounds simple (and it is), but we want to make sure you don't get confused.

Note that B and the square root of B keep popping up. Thus, for example, for Johnson noise from a resistor R,

$$\begin{aligned}
e_{nR}(\text{rms}) &= (4kTR)^{\frac{1}{2}} & \text{V/Hz}^{\frac{1}{2}}, \\
e_{nR}^2 &= 4kTR & \text{V}^2/\text{Hz}, \\
v_{n(rms)} &= v_{nR}B^{\frac{1}{2}} = (4kTRB)^{\frac{1}{2}} & \text{V}, \\
v_n^2 &= v_{nR}^2 B = 4kTRB & \text{V}^2.
\end{aligned}$$

On datasheets you may see graphs of e_n or e_n^2, with units like "nanovolts per root Hz" or "volts squared per Hz." The quantities e_n and i_n that will soon appear work just the same way.

When you add two signals that are uncorrelated (two noise signals, or noise plus a real signal), you add their noise *power*; that is, their *squared* amplitudes add:

$$v = (v_s^2 + v_n^2)^{\frac{1}{2}},$$

where v is the rms signal obtained by adding together a signal of rms amplitude v_s and a noise signal of rms amplitude v_n. The rms amplitudes[13] *do not* add.

8.2.2 Signal-to-noise ratio

Signal-to-noise ratio (SNR) is simply defined as

$$\text{SNR} = 10\log_{10}(v_s^2/v_n^2) = 20\log_{10}(v_s/v_n) \quad \text{dB} \quad (8.11)$$

where the voltages are rms values, and some bandwidth and center frequency are specified; i.e., it is the ratio, in decibels, of the rms voltage of the desired signal to the rms voltage of the noise that is also present.[14] The "signal" itself may be sinusoidal, or a modulated information-carrying waveform, or even a noiselike signal itself. It is particularly important to specify the bandwidth if the signal has some sort of narrowband spectrum, because the SNR will decrease as the bandwidth is increased beyond that of the signal: the amplifier keeps adding noise power, while the signal power remains constant.

8.2.3 Noise figure

Any real signal source or measuring device generates noise because of Johnson noise in its source resistance (the real part of its complex source impedance). There may be additional noise, of course, from other causes. The *noise figure* (NF) of an amplifier is simply the ratio, in decibels, of the output of the real amplifier to the output of a "perfect" (noiseless) amplifier of the same gain, with a resistor of value R_s connected across the amplifier's input terminals in each case. That is, the Johnson noise of R_s is the "input signal":

$$\text{NF} = 10\log_{10}\left(\frac{4kTR_s + v_n^2}{4kTR_s}\right) \quad (8.12)$$

$$= 10\log_{10}\left(1 + \frac{v_n^2}{4kTR_s}\right) \quad \text{dB,} \quad (8.13)$$

where v_n^2 is the mean squared noise voltage per hertz contributed by the amplifier, with a noiseless (cold) resistor of value R_s connected across its input. This latter restriction is important, as you will see shortly, because the noise voltage contributed by an amplifier depends very much on the source impedance (Figure 8.7).

Noise figure is handy as a figure of merit for an amplifier when you have a signal source of a given source impedance and want to compare amplifiers (or transistors, for which the NF is often specified). The NF varies with frequency

---

[13] Which, we emphasize, are the convenient and familiar quantites found on datasheets, etc. For example, we're used to thinking of a 3 nV/$\sqrt{\text{Hz}}$ amplifier as quiet; it's hard to recognize a 0.9×10^{-17} V^2/Hz amplifier as the same thing.

[14] The expression in terms of squared amplitudes suggests a ratio in terms of *power*, which is the origin of the decibel ratio definition. But the "$20\log_{10}$" form is widely used, even when there is no actual power, for example with an open-circuit load (or, more confusingly, when the result is at odds with the actual power ratio, for example when expressing the ratio of amplitudes created by a signal transformer).

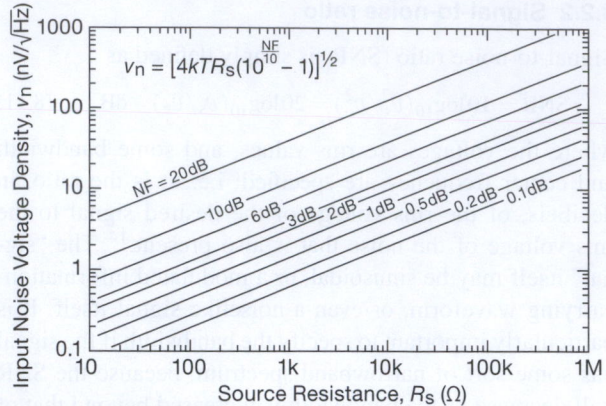

Figure 8.7. Effective input-noise-voltage density versus noise figure and source resistance.

and source impedance, and it is often given as a set of contours of constant NF versus frequency and R_s (we'll see examples later, in Figures 8.22 and 8.27). It may also be given as a set of graphs of NF versus frequency, one curve for each collector current, or a similar set of graphs of NF versus R_s, one for each collector current. *Note*: the foregoing expressions for NF assume that the amplifier's input impedance is much larger than the source impedance, i.e., $Z_{in} \gg R_s$. However, in the special case of RF amplifiers, you usually have $R_s = Z_{in} = 50\,\Omega$, with the NF defined accordingly. For this special case of matched impedances, simply remove the factor "4" from eq'ns 8.12 and 8.13.

Big fallacy: don't try to improve things by adding a resistor in series with a signal source to reach a region of minimum NF. All you're doing is making the source noisier to make the amplifier look better! Noise figure can be very deceptive for this reason. To add to the deception, the NF specification (e.g., NF = 2 dB) for a transistor or FET will always be for the optimum combination of R_s and I_C. It doesn't tell you much about actual performance, except that the manufacturer thinks the noise figure is worth bragging about.

In general, when evaluating the performance of some amplifier, you're probably least likely to get confused if you stick with a SNR calculated for that source voltage and impedance. Here's how to convert from NF to SNR:

$$\mathrm{SNR} = 10\log_{10}\left(\frac{v_s^2}{4kTR_s}\right) - \mathrm{NF(dB)(at}\ R_s)\quad \mathrm{dB}, \quad (8.14)$$

where v_s is the rms signal amplitude, R_s is the source impedance, and NF is the noise figure of the amplifier for source impedance R_s. See §§8.3.1 and 8.5.6 for some noise figure calculation examples.

8.2.4 Noise temperature

Rather than noise *figure*, you sometimes see noise *temperature* used to express the noise performance of an amplifier. Both methods give the same information, namely the excess noise contribution of the amplifier when driven by a signal source of impedance R_s; they are equivalent ways of expressing the same thing.

Look at Figure 8.8 to see how noise temperature works: we first imagine the actual (noisy) amplifier connected to a *noiseless* source of impedance R_s (Figure 8.8A). If you have trouble imagining a noiseless source, think of a resistor of value R_s cooled to absolute zero. There will be some noise at the output, even though the source is noiseless, because the amplifier has noise. Now imagine constructing Figure 8.8B, where we magically make the amplifier noiseless and bring the source R_s up to some temperature T_n such that *the output noise voltage is the same as in Figure 8.8A*. T_n is called the noise temperature of the amplifier, for source impedance R_s.

As we remarked earlier, noise figure and noise temperature are simply different ways of conveying the same information. In fact, you can show that they are related by the following expressions:

$$T_n = T\left(10^{\mathrm{NF(dB)}/10} - 1\right)\quad \mathrm{Kelvin}, \quad (8.15)$$

$$\mathrm{NF(dB)} = 10\log_{10}\left(\frac{T_n}{T} + 1\right), \quad (8.16)$$

where T is the ambient temperature, usually taken as $293°$K.

Generally speaking, good low-noise amplifiers have noise temperatures far below room temperature (or,

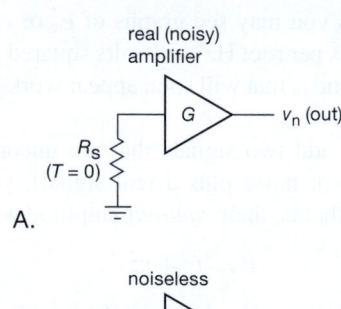

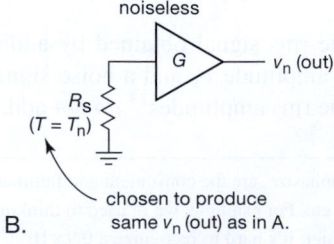

Figure 8.8. Noise temperature.

equivalently, they have noise figures far less than 3 dB). Later in the chapter we explain how you go about measuring the noise figure (or temperature) of an amplifier. First, however, we need to understand noise in transistors and the techniques of low-noise design. We hope the discussion that follows will clarify what is often a murky subject.

8.3 Bipolar transistor amplifier noise

The noise generated by an amplifier is easily described by a simple noise model that is accurate enough for most purposes. In Figure 8.9, e_n and i_n represent the transistor's internal noise, modeled as a noise voltage e_n in series with the input, combined with a noise current i_n injected at the input. The transistor symbol itself (or amplifier, in general) is assumed noiseless, and it simply amplifies the input-noise voltage it sees (caused by its own e_n, combined with its i_n flowing through the input signal's source impedance R_s). That is, the amplifier contributes a total noise voltage e_a, referred to the input, of

$$e_a(\text{rms}) = [e_n^2 + (R_s i_n)^2]^{\frac{1}{2}} \quad \text{V/Hz}^{\frac{1}{2}}. \quad (8.17)$$

The two terms are simply the squared values of the amplifier input-noise voltage, and the noise voltage generated by the amplifier's input noise current passing through the source resistance.[15] Because the two noise terms are usually uncorrelated, their squared amplitudes add to produce the effective noise voltage seen by the amplifier. For low source resistances the noise voltage e_n dominates, whereas for high source impedances the noise current i_n generally dominates.

Just to get an idea of what these look like, look at Figure 8.10, which shows a graph of e_n and i_n versus I_C and f for the excellent (but, sadly, discontinued[16]) low-noise *npn* 2SD786. We go into some detail now, describing these and showing how to design for minimum noise. It is worth noting that voltage noise and current noise for a bipolar transistor are in the range of nanovolts and picoamps per root hertz (i.e., nV/$\sqrt{\text{Hz}}$ and pA/$\sqrt{\text{Hz}}$); for FETs the current noise is lower, down in the fA/$\sqrt{\text{Hz}}$ range.

8.3.1 Voltage noise, e_n

The equivalent voltage noise looking in series with the base of a bipolar transistor arises from collector-current shot

---

[15] An additional term, important at higher frequencies or when i_n is small, is the noise current i_n generated by e_n in combination with the input capacitance: $i_n = e_n \omega C_{in}$. See §§8.11.3 and *4x.3.4*.

[16] The 2SD786 was for many years the standard to which low-noise transistors aspired. We feature it here because it is well characterized, with a wealth of information and excellent noise performance. Our nomination for its successor is the ZTX851 (with ZTX951 *pnp* complement), characterized and used in numerous examples later in the chapter.

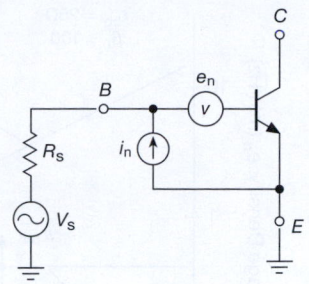

Figure 8.9. Noise model of a transistor.

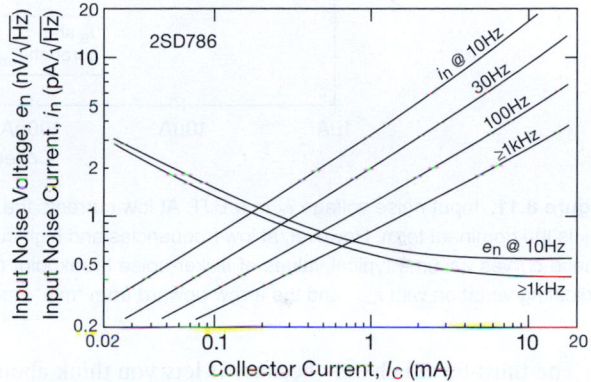

Figure 8.10. Equivalent rms input noise voltage (e_n) and noise current (i_n) versus collector current for a 2SD786 *npn* transistor, adapted from the datasheet.

noise generating a noise voltage across the intrinsic emitter resistance[17] r_e, Johnson noise generated in the base resistance $r_{bb'}$, and (as we'll see later) base current shot noise through that resistance. Deferring for the moment the base current term (which most often does not contribute significantly to a BJT's voltage noise), the input noise voltage density looks like this:

$$e_n^2 = 2qI_C r_e^2 + 4kT r_{bb'} \quad (8.18)$$

$$= 4kT \left(\frac{V_T}{2I_C} + r_{bb'} \right) \quad (8.19)$$

$$= 4kT \left(\frac{r_e}{2} + r_{bb'} \right) \quad \text{V}^2/\text{Hz}, \quad (8.20)$$

where we've eliminated r_e in the second form (eq'n 8.19) and I_C in the third form (eq'n 8.20) by remembering that $r_e = V_T/I_C = kT/qI_C$.

---

[17] Properly speaking, there is no "resistance" r_e; rather, it represents the inverse transconductance of the transistor: $r_e = v_b/i_c = 1/g_m$ (or the voltage change at the base corresponding to changes in collector current). It's really just a convenient way of speaking. Be careful, though, when thinking of it loosely as a resistor: it is a "noiseless" resistance; that is, it has no Johnson noise.

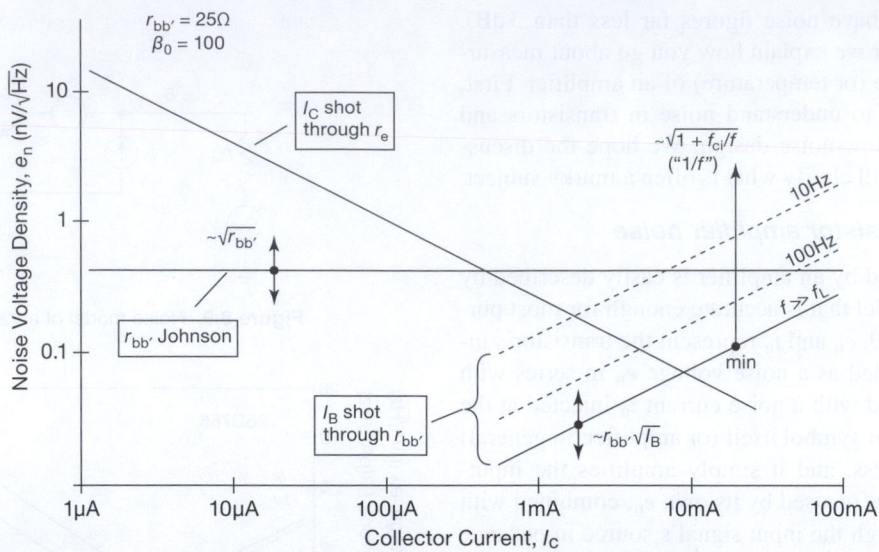

Figure 8.11. Input noise voltage e_n in a BJT. At low currents the shot noise in I_C through r_e dominates; otherwise the Johnson noise in $r_{bb'}$ is the dominant term. However, at low frequencies and high currents, the base-current shot noise through $r_{bb'}$ causes e_n to rise again. These curves assume typical values of flicker-noise breakpoint ($f_{ci} \sim 1\,\mathrm{kHz}$) and base resistance ($r_{bb'} \sim 25\,\Omega$), with double-sided arrows indicating variation with $r_{bb'}$, and the arrow upward from "min" indicating $1/f$ rise with decreasing frequency.

The third form is handy because it lets you think about the noise voltage as if arising from combining the separate Johnson noise from two resistors. To get a sense of the scale of BJT noise voltage, it's helpful to know that the base resistance $r_{bb'}$ for typical BJTs (see Table 8.1a on page 501) goes from a few ohms to a few hundred ohms; thus the second term's noise voltage contribution is typically in the range of $0.2\,\mathrm{nV}/\sqrt{\mathrm{Hz}}$ to $2\,\mathrm{nV}/\sqrt{\mathrm{Hz}}$ (we like to remember that a $100\,\Omega$ resistor has $e_n = 1.28\,\mathrm{nV}/\sqrt{\mathrm{Hz}}$, and that e_n scales as the square root of R).

As for the first term in eq'n 8.20, it's telling us that the *collector shot-noise current through r_e produces the same noise voltage as the Johnson noise of a fictional resistor of value $R = r_e/2$.* For example, in a BJT running at $100\,\mu\mathrm{A}$ (thus $r_e = 250\,\Omega$) this amounts to $e_n = \sqrt{4kT \cdot 125\,\Omega}$, or $1.4\,\mathrm{nV}/\sqrt{\mathrm{Hz}}$. This is a useful thing to know, for example when choosing resistor values so as not to compromise the noise performance of a low-noise amplifier. Lest we leave the wrong impression, we hasten to emphasize again that the intrinsic emitter resistance r_e is *not* a "real" resistor – it does not have Johnson noise; the noise-voltage term we're describing arises solely from the noise voltage generated by collector shot-noise current flowing through a noiseless r_e.[18]

Both voltage-noise terms in eq'ns 8.18 and 8.20 have a flat (white) spectrum, with a Gaussian distribution of

instantaneous amplitude. Because r_e falls inversely with collector current, a BJT's e_n falls as $1/\sqrt{I_C}$ with rising collector current, ultimately reaching a minimum that depends on $r_{bb'}$ (Figure 8.11). For this reason it's usually best to run at relatively high collector currents if the objective is to minimize noise voltage; the price you pay is increased base current and increased heating. Taking again the excellent 2SD786 as an example, at frequencies above $1\,\mathrm{kHz}$ it has an e_n of $1.5\,\mathrm{nV}/\sqrt{\mathrm{Hz}}$ at $I_C = 100\,\mu\mathrm{A}$ and $0.6\,\mathrm{nV}/\sqrt{\mathrm{Hz}}$ at $I_C = 1\,\mathrm{mA}$ (Figure 8.10). This transistor uses special geometry to achieve an unusually low $r_{bb'}$ of $4\,\Omega$, which is needed to realize the lowest values of e_n.

Of course, if you need to operate at a low collector current (where the effect of r_e dominates), it's of no particular benefit to have a low value of $r_{bb'}$. To make this point graphically, we plotted in Figure 8.12 the predicted and measured e_n versus collector current for six low-noise *npn* transistors, using a simple one-parameter model based upon $r_{bb'}$ alone. Selecting a transistor with low $r_{bb'}$ really matters at higher currents, though, which is where you have to operate if you want to achieve e_n values southward of $1\,\mathrm{nV}/\sqrt{\mathrm{Hz}}$.[19] We illustrate that point dramatically in §8.5.9, in an input stage with $e_n < 0.1\,\mathrm{nV}/\sqrt{\mathrm{Hz}}$.

The voltage-noise density e_n at $1\,\mathrm{kHz}$, as plotted in

---

[18] See §8.3.5 for some further discussion of this subtle point.

[19] The effective $r_{bb'}$ is said to increase slightly at low currents. For example, if the 2N5089 excess noise at $10\,\mu\mathrm{A}$ ($5.9\,\mathrm{nV}/\sqrt{\mathrm{Hz}}$ instead of $4.5\,\mathrm{nV}/\sqrt{\mathrm{Hz}}$) is due to $r_{bb'}$, we can calculate that $r_{bb'}$ increased from

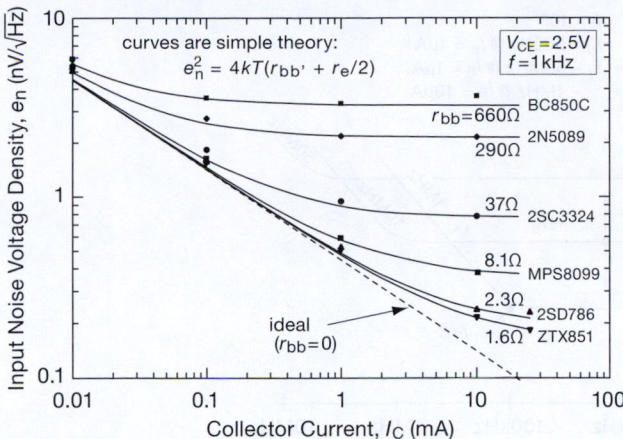

Figure 8.12. A simple model of BJT noise voltage (shot noise current through r_e, combined with Johnson noise of base resistance, solid curves) provides a good first approximation to measured noise values (datapoints at four currents), shown here for six selected low-noise BJTs. The dashed line shows the theoretical minimum noise voltage at current I_C, as given in eq'n 8.18. The measured noise is 10%–20% high at collector currents below $100\mu A$. See Figure 8.17 for the noise spectra of these six BJTs at $I_C = 10$ mA.

Figure 8.12, does not tell the whole story, of course. If you care about the noise at lower frequencies, you have to worry about a component of the transistor's noise that is frequency dependent, usually in the form of a rising $1/f$ flicker-noise tail (characterized by a $1/f$ corner frequency f_c). For BJTs, the $1/f$ noise in their e_n comes from one additional noise term that we've ignored thus far – the noise voltage produced by the base's noise current i_n flowing through the transistor's own $r_{bb'}$, i.e., $e_n = i_n r_{bb'}$. This latter term becomes a significant contributor to e_n only at low frequencies and at relatively high collector currents. But it is quite important for another reason: it flows through the input signal's source resistance, generating an additional noise voltage $v_n = i_n R_{sig}$. Let's look at the business of input noise current, after which we briefly revisit transistor e_n with this effect included.

8.3.2 Current noise i_n

The transistor's input noise current generates an additional noise voltage across the input signal source impedance.

The main source of base-current noise is shot-noise fluctuation in the steady base current,

$$i_n = \sqrt{2qI_B} = \sqrt{2qI_C/\beta_0} \quad \text{A/Hz}^{\frac{1}{2}}, \qquad (8.21)$$

which is Gaussian noise exhibiting a flat frequency spectrum (i.e., white noise); we've used the symbol β_0 here to emphasize that it's the beta *at dc*.

There is in addition a flicker-noise component that rears its ugly head at low frequencies. The latter exhibits the typical $1/f$ frequency dependence, with a corner frequency that we'll call f_{ci}; i.e., it contributes a term $2qI_B f_{ci}/f$ to the overall i_n^2. It rises somewhat more rapidly with I_C, owing to a rising $1/f$ breakpoint, typically with f_{ci} going as $I_B^{1/3}$ to $I_B^{1/4}$. See Figures 8.13 and 8.14. Typical current-noise corner frequencies are $f_{ci} \sim 50$–300 Hz at 1–10 μA, and 200 Hz–2 kHz at 1 mA.

Note that this simple base-current shot noise is *not* collector-current shot noise divided by beta. If it were, it would be $\sqrt{2qI_C}/\beta_0$ rather than the correct (and larger) $i_n = \sqrt{2qI_C}/\sqrt{\beta_0}$ (as given by eq'n 8.21). In fact, to first order the base-current shot noise is uncorrelated with collector-current shot noise.

However, at high frequencies this is no longer true, leading to one final current-noise term: at frequencies approaching the transistor's f_T (i.e., as the current gain approaches unity) the decreasing beta makes the collector-current's shot noise visible at the base.

Putting these terms together, we have, finally

$$i_n^2 = 2q\frac{I_C}{\beta_0}\left(1 + \frac{f_{ci}}{f}\right) + 2qI_C\left(\frac{f}{f_T}\right) \quad \text{A}^2/\text{Hz}, \qquad (8.22)$$

where the last term effectively represents a frequency-dependent beta.

Taking the example of the 2SD786 again (Figure 8.10), above 1 kHz i_n is about 0.25 pA/$\sqrt{\text{Hz}}$ at $I_C = 100\,\mu A$ and 0.8 pA/$\sqrt{\text{Hz}}$ at $I_C = 1$ mA. The noise current increases and the noise voltage decreases as I_C is increased. In the next subsection (§8.3.3) we see how this dictates operating current in low-noise design. Figure 8.15 shows graphs of i_n versus frequency and current for a pair of low-noise transistors.

Current-noise times input impedance   The input-noise current flowing through the input signal's source impedance generates a noise-voltage density of magnitude $v_n = i_n Z_s$, which combines (as sum of squares) with the transistor's input noise-voltage density.[20] Usually the source impedance is resistive, in which case you need to

---

290 Ω at 10 mA to about 900 Ω at 10 μA (threefold over 3 decades), or approximately as $(I_c)^{-1/6}$. (SPICE transistor noise models include a parameter for this effect.)

[20] The primary complicating issues are increases of i_n with temperature, and with decreasing frequency ($1/f$ behavior).

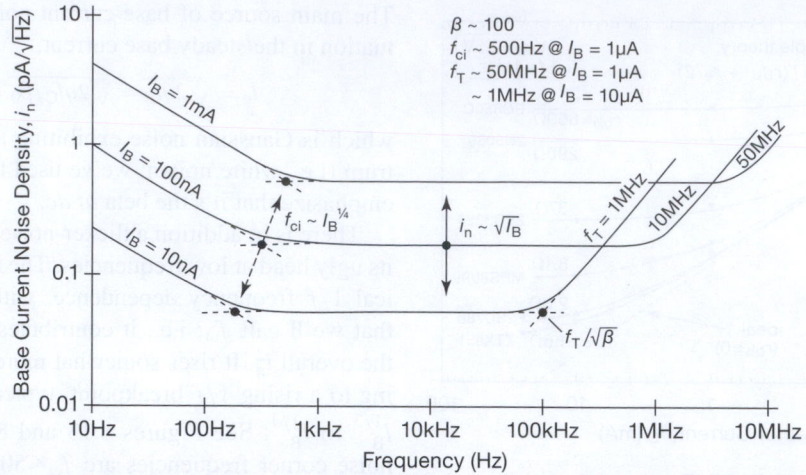

Figure 8.13. Input noise-current density i_n versus frequency in a BJT. At mid-frequencies it is entirely base-current shot noise, proportional to $\sqrt{I_B}$. At low frequencies i_n rises as $1/\sqrt{f}$ (it is the noise *power* that goes as $1/f$); however, the $1/f$ breakpoint (f_{ci}) rises with increasing current, so that at a given low frequency below f_{ci} the noise current rises faster than the square root of base current, as shown. At high frequencies the falling beta ($\beta \to 1$ at f_T) causes the noise density to rise $\propto f$. These curves assume a typical value of flicker noise breakpoint ($f_{ci} \sim 500\,$Hz at $I_B=1\,\mu$A), with the solid arrow indicating variation of i_n with base current, and the dashed arrow indicating variation of $1/f$ breakpoint with base current.

add its Johnson noise as well; in other words, the total input-referred squared noise voltage is

$$v_n^2 = e_n^2 + 4kTR_s + (i_n R_s)^2 \quad \text{V}^2/\text{Hz}. \tag{8.23}$$

If the source impedance is *reactive*, however, the current noise contribution (the last term in the equation) will be frequency dependent. A common situation in which this matters is an ac-coupled input signal. If you aren't thinking about noise, you'd normally choose the blocking capacitor value based on the downstream load impedance (transistor input impedance and bias resistor), to set the low-frequency rolloff somewhat below the lowest frequency of interest. That could be a rather small capacitance (if the amplifier has a high input impedance, or if the low-frequency cutoff isn't terribly low), through which the noise current would generate substantial noise voltage (and, of course, increasing proportional to the capacitor's reactance $X_C=1/2\pi fC$, i.e., proportional to $1/f$).[21]

So, you need to turn the process around: first, choose the capacitor value to keep its noise-voltage contribution $v_n=i_n X_C$ small enough at the lowest frequency of operation (and remember that base-current noise often exhibits a $1/f$ tail); then choose the transistor's input bias resistor value to get the desired low-frequency rolloff. We've been bitten by this issue, surprised by a $1/f$ noise voltage ($1/f$

---

[21] Note that an ideal capacitor (or inductor, for that matter) generates no Johnson noise.

in *amplitude*, not power) caused by the blocking capacitor's reactance versus frequency. To fix the problem we had to increase the blocking capacitor's value by a factor of 50!

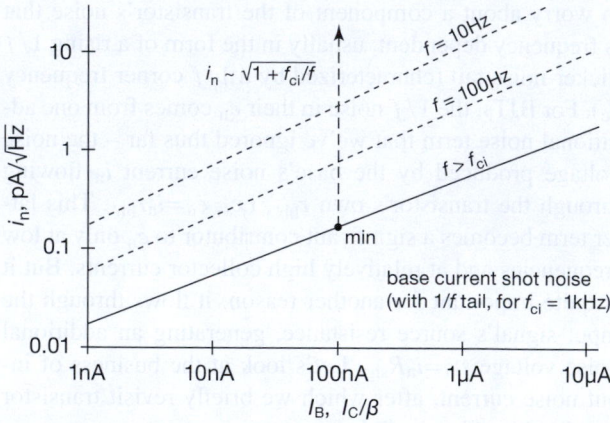

Figure 8.14. Input noise current in a BJT is shot noise, scaling as the square root of base current for frequencies above the $1/f$ corner frequency f_{ci}. At lower frequencies the curve is steeper, because f_{ci} itself rises with increasing current (see Figure 8.15).

8.3.3 BJT voltage noise, revisited

As suggested in §8.3.1, the transistor's input *current* noise i_n can become a significant contributor to the voltage noise e_n seen at its input terminals, owing to the noise voltage

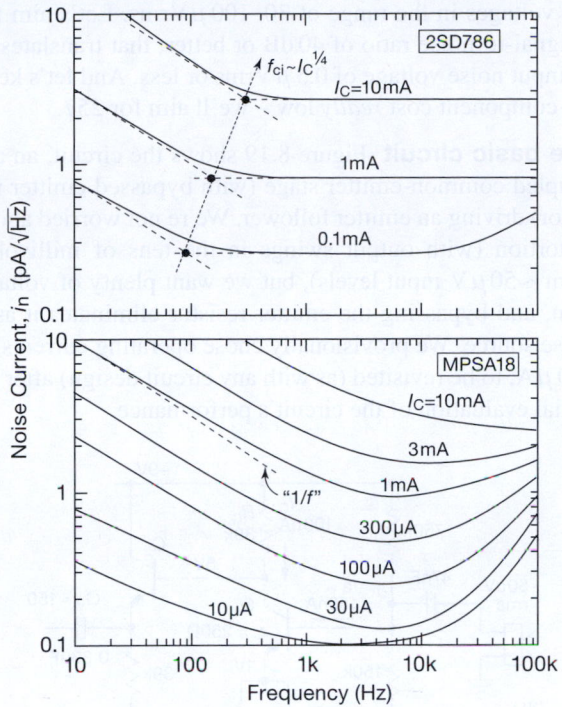

Figure 8.15. Noise current (i_n) versus frequency for two low-noise *npn* transistors. The low-frequency noise current exhibits a somewhat greater rise with collector current, owing to a rising $1/f$ corner frequency.

generated across its internal $r_{bb'}$ by its own input-noise current i_n. Because i_n is shot noise in the dc base current, it increases as the square root of collector current (for constant beta); and, because i_n of a BJT is seriously afflicted by a $1/f$ low-frequency tail, its contribution to the overall e_n is seen primarily (and often *only*) at low frequencies. A more detailed noise model of the bipolar transistor lets us incorporate this effect.

Adding this noise term to our original (incomplete) BJT noise voltage formula, eq'n 8.18, we have

$$e_n^2 = 2qI_Cr_e^2 + 4kTr_{bb'} + 2q\frac{I_C}{\beta_0}r_{bb'}^2\left(1+\frac{f_{ci}}{f}\right), \ \text{V}^2/\text{Hz}, (8.24)$$

where, as before, the first term can be replaced by $2kTr_e$, and, as in eq'n 8.22, the current-noise corner frequency f_{ci} rises approximately as the fourth root of I_C.

To see how this works out, look at Figure 8.16, where we've plotted separately (for three choices of collector current) the e_n from the first two terms (as dashed lines) and the e_n from the third term (as thin solid lines). The thick solid lines represent the total voltage noise, i.e., including all three terms of eq'n 8.24. For these calculated curves

we used typical BJT noise parameters ($\beta=200$, $r_{bb'}=50\,\Omega$, $f_{ci}=1$ kHz at $I_C=1$ mA).

This shows how the $i_nr_{bb'}$ noise-voltage term competes with the other two e_n noise terms, dominating at high collector current and low frequency. Although the noise-current $1/f$ breakpoint f_{ci} increases only mildly with I_C, its contribution acts to create a rapidly rising noise-*voltage* breakpoint f_c. Compare this with Figure 8.11, where the same effect can be seen: there, however, e_n is plotted versus collector current, with a family of several spot frequencies.

To maintain some perspective here, note that additional noise voltage is evident only at very low frequencies (Figure 8.16 goes down to 0.01 Hz!) and at relatively high current densities. It's not something over which to lose sleep.[22] This can be seen in the measured noise-voltage curves plotted in Figure 8.17, where the effect (at the substantial collector current of 10 mA) is serious above 10 Hz only for BC850 with its unusually high $r_{bb'}\approx750\,\Omega$. In fact, for transistors with attractively low $r_{bb'}$ the e_n improvement gained from running at high collector current more than compensates for the rise at low frequencies that comes from the base-current shot-noise contribution, as seen in Figure 8.18.

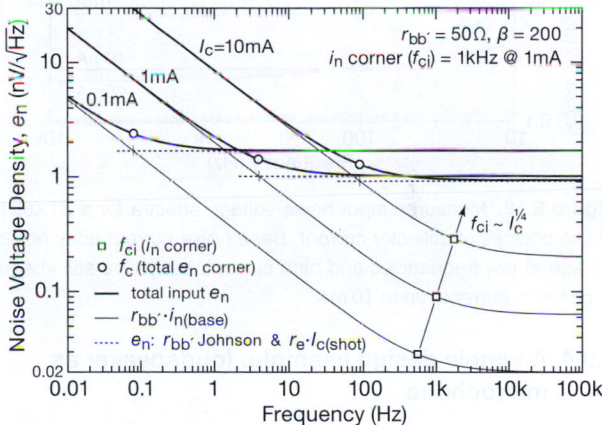

Figure 8.16. BJT noise-voltage density versus frequency (from eq'n 8.24) for a hypothetical transistor with the listed parameters, showing the effect of current noise i_n at low frequencies and high collector currents.

---

[22] But if you *are* losing sleep over $1/f$ problems in a high-I_C low-noise amplifier, there's a cute trick, namely to stack a pair of transistors so that their $1/f$ noise fluctuations cancel. See Broderson, Chenette, and Jaeger, "A superior low-noise amplifier," 1970 IEEE ISSCC, p 164. Suppression of low-frequency noise by a factor of five is possible. The penalty for using their two-transistor trick is a 3 dB increase in the higher-frequency (white noise) component of e_n.

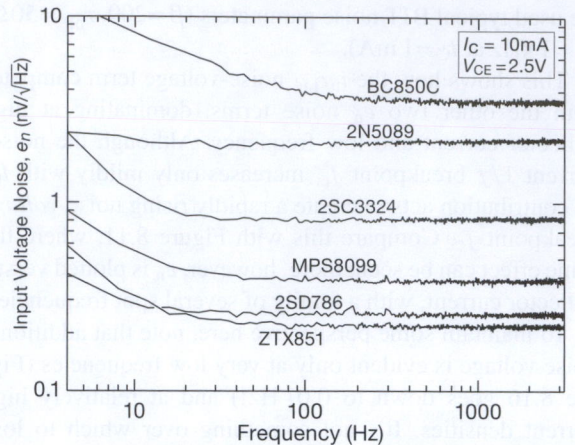

Figure 8.17. Measured input noise-voltage spectra (e_n versus frequency at $I_C=10$ mA) for the six BJTs of Figure 8.12.

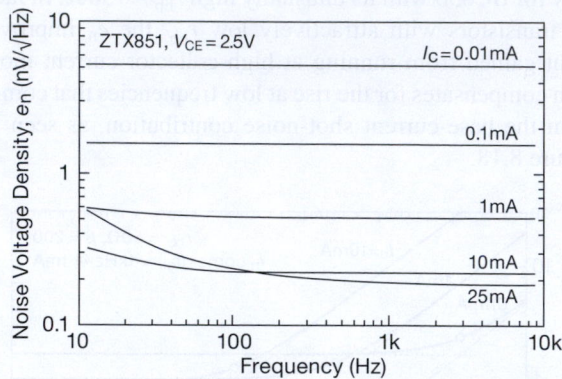

Figure 8.18. Measured input noise-voltage spectra for a ZTX851 at five choices of collector current. Base noise current adds noise voltage at low frequencies and high current, but you're still ahead at collector currents up to 10 mA.

8.3.4 A simple design example: loudspeaker as microphone

Let's put this noise theory into practice by designing a simple and inexpensive ac-coupled audio amplifier that runs on low current from a single +9 V supply (which could be a battery, or a "wall-wart" ac adapter), with input noise of just a few nV/$\sqrt{\text{Hz}}$. If you want to have an application in mind, think about a front-door intercom, where the small loudspeaker also serves reciprocally as a microphone.[23]

To get a handle on the noise specifications, we put a few small "8 Ω" loudspeaker samples on our lab bench, spoke to them in a normal voice, and measured output au-

dio voltages in the range of 30–100 μVrms. Let's aim for a signal-to-noise ratio of 40 dB or better; that translates to an input noise voltage of 0.5 μVrms or less. And let's keep the component cost *really* low – we'll aim for 25¢.

The basic circuit Figure 8.19 shows the circuit, an accoupled common-emitter stage (with bypassed emitter resistor) driving an emitter follower. We're not worried about distortion (with output swings in the tens of millivolts, from ~50 μV input levels), but we want plenty of voltage gain, and bypassing the emitter resistor eliminates it as a noise source. We provisionally chose operating currents of 100 μA, to be revisited (as with any circuit design) after an initial evaluation of the circuit's performance.

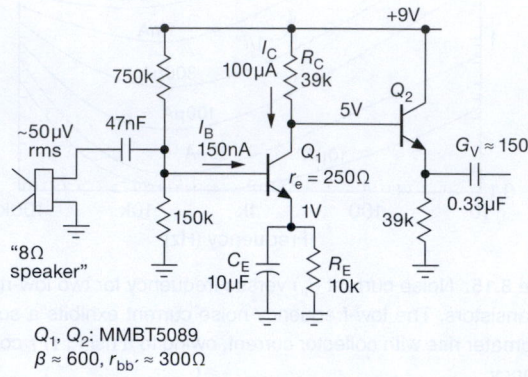

Figure 8.19. Low noise, and even lower cost: a simple audio preamp for running a loudspeaker "backward."

Noise calculation We chose the '5089 for this example because it's widely available (half a million in stock at Mouser and DigiKey, four different manufacturers), it's inexpensive ($0.026 in qty 100), and it promotes itself for low-noise amplifier applications.[24] It has lots of current gain (400 minimum, 600 typical at 100 μA), but it suffers from relatively high base resistance $r_{bb'}\approx300\,\Omega$.

From eq'n 8.24 it's easy to calculate the various noise voltage density (e_n) contributions (and the reader is encouraged to do so), referred to the input (RTI):

Shot noise of I_C through r_e	$r_e\sqrt{2qI_C}$	1.4 nV/$\sqrt{\text{Hz}}$
Johnson noise in $r_{bb'}$	$\sqrt{4kTr_{bb'}}$	2.2 nV/$\sqrt{\text{Hz}}$
Johnson noise in R_C	$\sqrt{4kTR_C}/G_V$	0.17 nV/$\sqrt{\text{Hz}}$
Shot noise of I_B through $r_{bb'}$	$r_{bb'}\sqrt{2qI_B}$	0.066 nV/$\sqrt{\text{Hz}}$

---

[23] To those who suspect that "the walls have ears," this may suggest interesting possibilities for surveillance.

[24] For example, ON Semiconductor explicitly calls it a "Low Noise Transistor"; Fairchild and NXP call it a "General Purpose Transistor," but promote it for "low noise, high gain, general purpose amplifier applications" and "low-noise input stages in audio equipment," respectively.

The dominant noise source is Johnson noise in Q_1's internal base resistance, followed by collector shot noise flowing through Q_1's intrinsic emitter resistance r_e of $250\,\Omega$ at its collector current of 0.1 mA. The last two terms are negligible by comparison: there's plenty of Johnson noise in Q_1's collector resistor, but owing to that stage's voltage gain of ~ 150 it is an insignificant contribution when referred to the input (RTI). Even smaller is the voltage noise produced by base-current shot noise flowing through $r_{bb'}$. There are other noise possibilities, but they are vanishingly small by comparison: as an example, the Johnson-noise current in the bias divider flowing through the speaker's $8\,\Omega$ source impedance generates a noise voltage density of just $0.003\,\text{nV}/\sqrt{\text{Hz}}$.

Performance   From these noise-voltage *densities* we arrive at the total rms noise voltage over the signal bandwidth: $v_n = e_n\sqrt{\Delta f} = 142\,\text{nVrms}$, where e_n is the combined noise-voltage density of $2.6\,\text{nV}/\sqrt{\text{Hz}}$, and Δf is the audio bandwidth (for which we've used the "telephone bandwidth" of 3 kHz). Compared with the nominal $50\,\mu\text{Vrms}$ audio input, that amounts to a signal-to-noise ratio of 51 dB – not bad for the price!

Speaking of which, component cost is a "performance" parameter, of sorts. A quick search of the usual distributors (DigiKey, Mouser, Newark, Future) turned up the following prices (SMT components – chip resistors and capacitors, SOT-23 transistors):

	qty 100	qty 1k
Q_1, Q_2	$0.026	$0.023
C_{in}	$0.032	$0.012
C_E	$0.068	$0.039
C_{out}	$0.034	$0.017
resistors	$0.012	$0.004
total parts	$0.246	$0.134

Variations   We exceeded our targets of SNR and cost, so we could leave it at that. But it's always good to review an initial design, looking to improve things. Here the noise budget is dominated by the relatively high base resistance of these "low-noise" transistors. We could improve the noise performance by choosing an input-stage transistor with lower $r_{bb'}$ – look ahead to Table 8.1a on page 501, and at Figure 8.12, to see some candidates. However, once you've driven down the $r_{bb'}$ Johnson noise, the collector shot noise through r_e becomes the dominant term, and you're forced to increase the collector current to reap further improvements in v_n.[25] In fact, as Figure 8.12 shows,

down at $I_C = 100\,\mu\text{A}$ the '5089 is hardly worse (not even a factor of 2) than transistors with the lowest $r_{bb'}$.

Looked at another way, we could instead stick with the inexpensive and widely available MMBT5089, and reduce somewhat the collector current (say, to $50\,\mu\text{A}$), with very little increase in overall noise: the noise density would increase from $2.6\,\text{nV}/\sqrt{\text{Hz}}$ to $3.0\,\text{nV}/\sqrt{\text{Hz}}$, which would reduce the audio–band SNR by just 1 dB (from 50.9 dB to 49.8 dB). Certainly worth it, if you're running from a battery.

8.3.5 Shot noise in current sources and emitter followers

We remarked earlier (in §8.1.2) that shot noise is suppressed in the classic BJT current source (redrawn in Figure 8.20A). You might at first think this is obvious, because the collector current is just the emitter current (apart from a small base current contribution), in which shot noise is absent (as it always is in metallic conductors). But at the collector you've got just the sort of barrier in which shot noise is unavoidable.

OK, you might say, perhaps the current source *does* exhibit shot noise, but surely at the emitter terminal you've got a quiet current (and thus an absence of noise voltage) – just look at that resistor!

Paradoxically, that intuition is wrong. It turns out that the current source is "quiet," but the emitter follower's output *does* indeed include a noise voltage just equal to a shot-noise current (calculated from I_C) flowing through r_e, i.e., just the $r_e\sqrt{2qI_C}$ we saw in eq'n 8.24.[26]

It's actually rather easy to test this. Look at Figure 8.20C, where a $\sim 10\,\mu\text{A}$ current source is loaded with a $1\,\text{M}\Omega$ collector resistor. Absent collector shot noise, we should see only a Johnson-noise voltage density of $181\,\text{nV}/\sqrt{\text{Hz}}$ at the collector,[27] to be compared with a noise voltage $10\times$ higher if the collector current has the statutory shot noise $i_n = \sqrt{2qI_C}$. Meanwhile, at the emitter, that shot-noise current, flowing through r_e, would produce a noise-voltage density of $4.8\,\text{nV}/\sqrt{\text{Hz}}$; in the absence of emitter shot noise the follower would be quiet (apart from the other sources of BJT noise in eq'n 8.24).

We built circuit C, and...(drumroll)...we measured $190\,\text{nV}/\sqrt{\text{Hz}}$ at the collector (consistent with Johnson

---

[25] To see how far you can take this, visit the "extreme challenge" in §8.5.9.

[26] To which you must of course add the additional terms, Johnson noise and base current shot noise in $r_{bb'}$.

[27] This is $\sqrt{2}$ times the $128\,\text{nV}/\sqrt{\text{Hz}}$ Johnson noise of a $1\,\text{M}\Omega$ resistor. The $\sqrt{2}$ arises because this is a unity-gain inverter, so R_E's Johnson noise contributes an equal uncorrelated noise voltage.

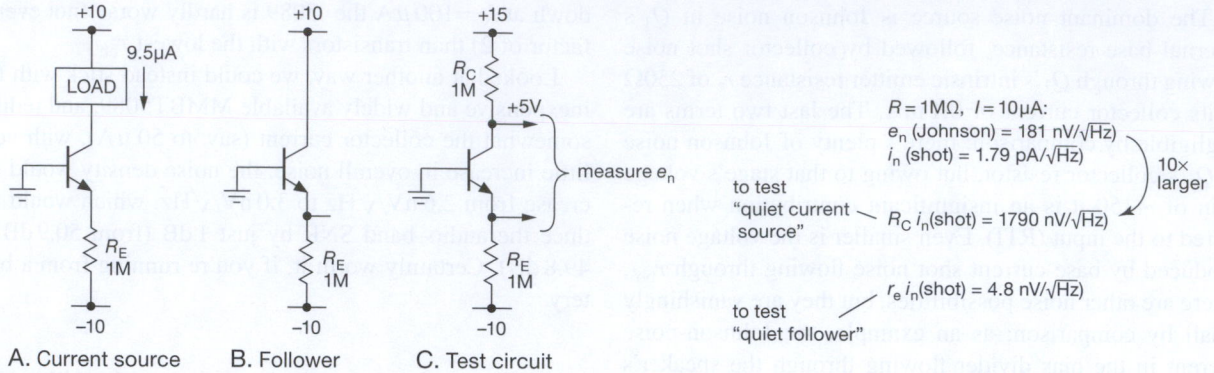

Figure 8.20. Shot noise is suppressed in the transistor current source, A, but not in the emitter follower, B. This is easily confirmed by measuring the noise-voltage density in the test circuit C.

noise alone, i.e., a quiet current source); and we measured $4.93\,\mathrm{nV}/\sqrt{\mathrm{Hz}}$ at the emitter (consistent with collector-current shot noise through r_e). Evidently shot noise is suppressed in the current source, but not in the follower. Interesting! And, to test the possibility that collector current *never* exhibits shot noise, we put a bypass capacitor between the base and emitter terminals. The noise voltage at the collector jumped up, to $1679\,\mathrm{nV}/\sqrt{\mathrm{Hz}}$, which agrees well with the shot-noise formula's predicted[28] $1664\,\mathrm{nV}/\sqrt{\mathrm{Hz}}$. How can we understand this?

To get an understanding of this puzzle, it's best to start with what *must* be true and see where it leads. Look at Figure 8.21, as we take it in steps.

1. We know for sure that the emitter current is quiet, because it's the current through a metallic conductor. And any worry we might have that this current could become noisy because of the fluctuations in the transistor's base-to-emitter *voltage* (i.e., input-voltage noise) is easily addressed: we can reduce any such effect (already negligible, because noise variations $v_{n(BE)}$ are down in the microvolts) by as much as we like by simply increasing both R_E and the negative supply voltage proportionally.

2. Somewhere within the transistor there must be a diffusive collector current that is subject to classic uncorrelated charge-flow shot noise, $i_n = \sqrt{2qI_{C(dc)}}$, and over which the transistor has no control.

3. But this is a three-terminal device with negligible base current, so we know that the current flowing into the collector terminal equals the current flowing out of the emitter terminal (in the limit of large beta); and the latter is "quiet."

4. So it must be the case that the controllable portion of the BJT's collector current is fluctuating in just such a way as to cancel the shot noise fluctuations.

For that to be so, the input voltage V_1 to the BJT transconductance model must be such that $g_m V_1 = I_E - i_n$, and so the input noise-voltage portion of V_1 is just

$$e_n = -i_n/g_m \qquad (8.25)$$

which (recall $r_e = 1/g_m$) is the familiar $i_n r_e$ form from eq'n 8.24.

This may seem like a circular argument; but in a real sense this is how this component of the base-to-emitter noise voltage is (and must be) created. This line of reasoning also explains nicely the presence of noise voltage e_n at the emitter of our test circuit (and, of course at the output of an emitter follower, or at the input of a common-emitter amplifier, etc.). And it explains also why bypassing the emitter liberates the full shot noise at the collector of our current source: such a connection invalidates the starting assumption (#1, above) because the capacitor diverts

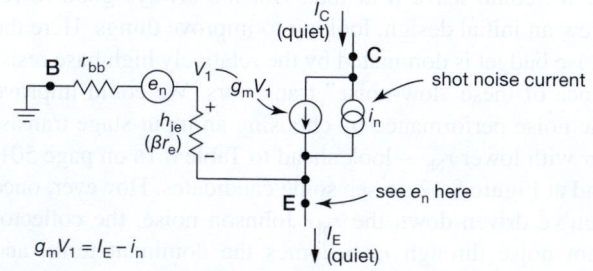

Figure 8.21. Three-terminal simplified "hybrid-π" model of the current-source–follower of Figure 8.20C. The noise voltage e_n reflects the unavoidable internal shot-noise current i_n. If the emitter current is quiet, so will be the resulting collector current I_C.

---

[28] For this test the collector current was $8.65\,\mu\mathrm{A}$, constrained by operation from batteries, thus the slightly smaller shot-noise prediction.

entirely the emitter current at signal frequencies. In fact, it creates a grounded-emitter amplifier (rather than a current source or follower), in which the full input-noise voltage as given above (eq'n 8.25) is amplified by the transconductance g_m to produce an output-noise current equal to $g_m e_n$. And that is exactly the full (unsuppressed) shot noise corresponding to the dc collector current I_C (rewind to step #2, above), which is what we measured in our test setup.

A final (and important) point: the shot-noise suppression in the current source requires a quiet emitter current sink, satisfied in this case by a large (compared with r_e) pull-down resistor R_E (here 1M versus 2.5k). But that's not the case for, say, an undegenerated current mirror. This raises the quantitative question, how large must R_E be, to be "large" in this sense?

This is easy enough to figure out: at the emitter we have a noise voltage e_n equivalent to the Johnson noise of a resistor of value $r_e/2$ and with source impedance equal to r_e; i.e., a noise current equivalent to the Johnson-noise current of a resistor of value $2r_e$. This is diluted by the quiet (i.e., shot-noise-free) pull-down resistor R_E, so that the reduction of current noise at the collector is in the ratio $\sqrt{2r_e/R_E}$. For example, at 1 mA collector current we have an impedance seen at the emitter of $r_e=25\,\Omega$, with a noise voltage equivalent to that of a $12.5\,\Omega$ resistor, or a noise current equivalent to that of a $50\,\Omega$ resistor. So a $50\,\Omega$ pull-down resistor reduces the noise by 3 dB, and a $4.95\,k\Omega$ pulldown reduces it by 20 dB, etc. Stated more generally, the reduction in collector noise current i_n is in the ratio $\sqrt{50mV/V_{R_E}}$, where V_{R_E} is the dc voltage across the emitter pull-down resistor R_E.

8.4 Finding e_n from noise-figure specifications

Transistor datasheets traditionally provided some tabulated values for e_n, and (often) graphs at selected collector currents of e_n and i_n versus frequency (or e_n and i_n versus collector current at selected frequencies), as we've seen for the 2SD786 and the MPSA18.

That was *then*! Now you see instead tabulations and graphs of noise figure (NF) – see for example the contours of constant NF plotted against I_C and R_s for the low-noise 2SC3324 *npn* bipolar transistor[29] from Toshiba (Figure 8.22).

There's a lot of information in these plots, even though they tell the story only at two frequencies (10 Hz and 1 kHz; it would sure be nice to see a plot of noise parame-

ters versus frequency). Let's see how much we can squeeze out of these two plots.

8.4.1 Step 1: NF versus I_C

We can make a plot of NF versus collector current, for each of a set of source resistances, by simply reading off values along a horizontal line in Figure 8.22. It's helpful to use a spreadsheet program with plotting capabilities such as Microsoft Excel, or (if you want to show off) a more sophisticated mathematical package such as MATLAB® or Mathematica®. Figure 8.23 shows what you get by reading across the 1 kHz plot of Figure 8.22 for six values of source resistance. Here we estimated NF values for half-decade steps of I_C (thus $10\,\mu A$, $30\,\mu A$, $100\,\mu A$, and so on), plotted them in Excel, then smoothed them with Bézier curves in Adobe Illustrator®. Noise-figure values below 1 dB should not be considered reliable, because Figure 8.22 (from which the data were derived) shows no contours below NF=1 dB.[30] The dashed line indicates NF=3 dB, where the noise contributed by the transistor is equal to the Johnson noise in the source resistor.

Before moving on, let's get some intuition about these curves. For low values of R_s the source has low Johnson-noise voltage (e.g., $R_s=100\,\Omega$ has $e_n=1.3\,nV/\sqrt{Hz}$), quite a bit less than the transistor's input e_n at small collector currents.[31] That's why the noise figure improves with increasing collector current. On the other hand, for large values of R_s the transistor's voltage noise is unimportant compared with the resistor's much larger Johnson noise. But now when we operate at large collector currents the transistor's input *current* noise (base current shot noise: $i_n \approx \sqrt{2qI_B}$) generates a substantial noise voltage across that larger R_s, such that the noise voltage at the input terminal is much larger than Johnson noise alone, hence a large (poor) noise figure.

8.4.2 Step 2: NF versus R_s

Figure 8.24 is an analogous plot, this time however showing NF versus *source resistance* (instead of I_C) for each of

---

[29] The *pnp* complement is the 2SA1312.

[30] The datasheet lists a value of 0.2 dB (typical) in the tabulated data, at 1 kHz, $100\,\mu A$ collector current, and $10\,k\Omega$ source resistance, a set of conditions that puts you nicely in the "eye" of the contour plot. But the datasheet also lists a worst-case (maximum) value of 3 dB! You'll have to do some cherry picking if you need NF=0.2 dB.

[31] Recall that the latter is approximately collector-current shot noise through r_e: $e_n \approx \sqrt{2qI_C} \cdot (kT/qI_C)$. That's equal to the Johnson noise of a fictitious resistor of value $R=r_e/2$, which is $100\,\Omega$ at $I_C=125\,\mu A$. So the transistor's e_n rises with falling collector current, as $e_n \propto 1/\sqrt{I_C}$.

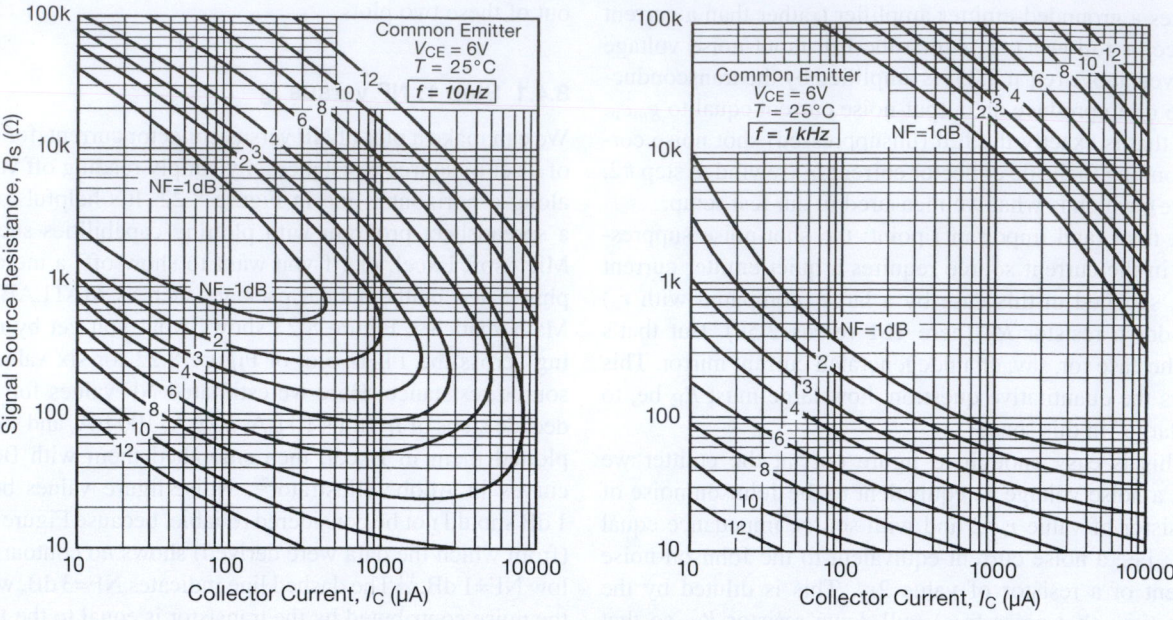

Figure 8.22. Toshiba's datasheet for its low-noise 2SC3324 *npn* transistor does not provide values or graphs for input voltage noise e_n. Instead you get these graphs of noise figure (NF) versus collector current and source resistance.

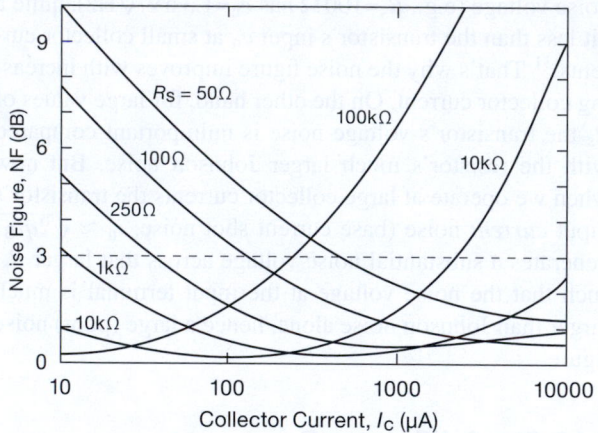

Figure 8.23. 2SC3324 noise figure versus collector current, from the datasheet contour plots shown in Figure 8.22.

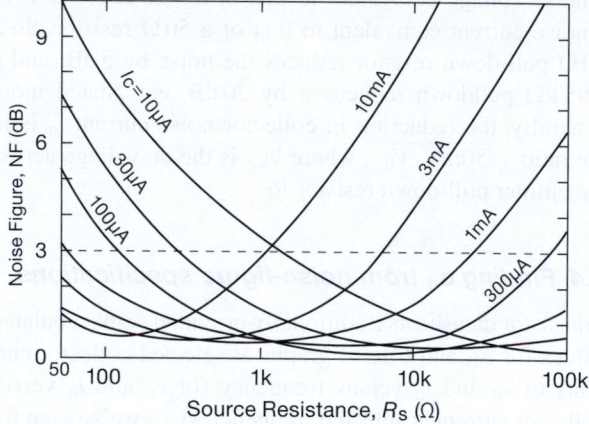

Figure 8.24. 2SC3324 noise figure versus source resistance, from the data of Figure 8.22.

a set of collector currents; it is taken from the same 1 kHz contour curves of Figure 8.22, this time reading off values along vertical lines of constant I_C. It's a useful graph for determining, approximately, the optimum operating current for a signal of a given source impedance.

8.4.3 Step 3: getting to e_n

The previous two graphs (NF versus I_C, NF versus R_s) are simply reorganizations of the manufacturer's 1 kHz noise

figure contour plot (Figure 8.22). We didn't need to calculate anything.

That's not the situation with noise voltage, e_n. Here we need to invert the defining equation (eq'n 8.13) for noise figure in order to find e_n:

$$e_n = \sqrt{4kTR_s}\ \sqrt{10^{NF/10} - 1}. \qquad (8.26)$$

The first term is the signal source resistor's Johnson-noise voltage density, and the second term is the multiplying factor attributable to the transistor's noise contribution, as given by its noise figure. The latter includes contributions from both e_n and i_n; so we must use the NF values for small source resistance so that the contribution from the transistor's i_n is negligible.[32]

Figure 8.25 shows what you get for e_n versus collector current, beginning with the noise-figure values corresponding to $R_s=50\,\Omega$ (from Figure 8.23 or 8.22). In common with most devices, this transistor exhibits the familiar "flicker-noise" excess below its "$1/f$ noise corner-frequency" (see §8.1.3, and also §5.10.6).

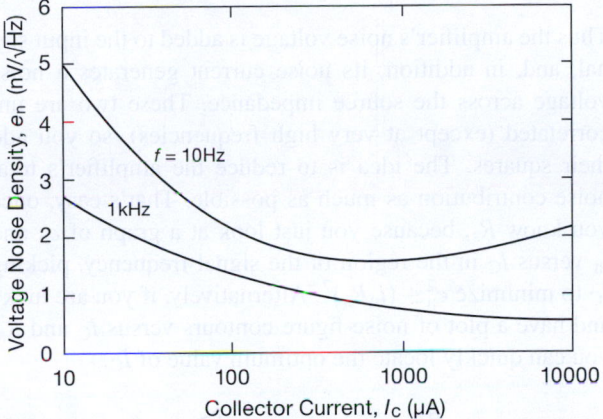

Figure 8.25. 2SC3324 noise voltage versus collector current, derived from the data of Figure 8.22.

8.4.4 Step 4: the spectrum of e_n

We complained earlier about the lack of noise data versus frequency – it's given only for 10 Hz and 1 kHz. But we can fill in the details, on the reasonable assumption that the noise power at low frequencies goes approximately as $1/f$ (that is, $e_n \propto 1/\sqrt{f}$), and that the two datapoints straddle the $1/f$ noise corner-frequency.

To do this, we first find the $1/f$ corner frequency f_c from the e_n values that straddle it. To a good approximation,

$$f_c = f_L \left(\frac{e_{nL}^2}{e_{nH}^2} - 1 \right), \qquad (8.27)$$

where e_{nL} is the noise density at a frequency f_L below the corner frequency, and e_{nH} is the noise density well above f_c (see the discussion on page 566). Once we've got f_c, we can find e_n versus frequency:

---

[32] Similarly, we could extract i_n from the NF values for large source resistance, where the transistor's e_n contribution is negligible.

$$e_n(f) = e_{nH}\sqrt{1 + f_c/f}. \qquad (8.28)$$

Because the corner frequency depends on collector current, we read off e_{nL} and e_{nH} from Figure 8.25 for each of four decade values of I_C, and then plot the corresponding e_n versus f. Figure 8.26 shows the resulting noise-voltage spectra.

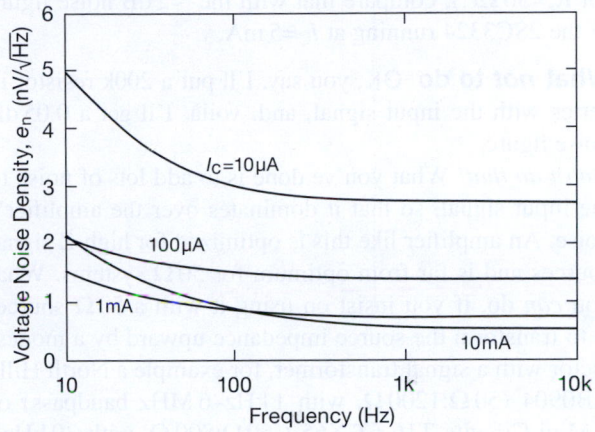

Figure 8.26. 2SC3324 noise voltage versus frequency, deduced from the curves of Figure 8.25.

8.4.5 The spectrum of i_n

By a similar procedure we could extract curves of current noise i_n versus frequency. Above the $1/f$ corner we would see base-current shot noise ($i_n=\sqrt{2qI_b}$), with the characteristic $1/f$ rise at low frequencies.

Once you've gotten plots of e_n and i_n versus frequency, you have all the basic information contained within a noise-figure contour plot. Look ahead to §8.9.1E, and especially Figure 8.58, to see how you can use e_n and i_n to predict the total effective input noise density versus source resistance. The graphic shown in the upper left of that figure shows how to find the breakpoints that separate the e_n-dominated (low-R_s) region, the source Johnson-noise-dominated (mid-R_s) region, and the i_n-dominated (high-R_s) region. We prefer that simpler approach, and we'll be using e_n and i_n extensively in the rest of the chapter.

8.4.6 When operating current is not your choice

Knowing how e_n and i_n vary with collector (or drain) current helps in setting the operating point for optimum noise performance, as we'll see in the next section.

Sometimes, though, the choice has already been made, by person or persons unknown. In such a case you live with what you're given. Figure 8.27 shows an example, with published noise-figure contours for the Stanford Research Systems model SR560 preamplifier (§8.6.4). It's clear from the high input noise resistance ($\sim$200 kΩ) that this is a JFET-input amplifier, and that its noise performance is far from optimum with a source of low impedance: NF=15 dB for R_s=50 Ω(!); compare that with the $\sim$2 dB noise figure of the 2SC3324 running at I_C=5 mA.

What *not* to do   OK, you say, I'll put a 200k resistor in series with the input signal, and, voilà, I'll get a 0.05 dB noise figure.

Don't do that! What you've done is to add lots of noise to the input signal, so that it dominates over the amplifier's noise. An amplifier like this is optimized for high-Z signal sources and is far from optimum for 50 Ω systems. What you *can* do, if you insist on using it with a 50 Ω source, is to transform the source impedance upward by a modest factor with a signal transformer, for example a North Hills HB0904 (50 Ω:1200 Ω, with 1 kHz–6 MHz bandpass) or a Mini-Circuits T16-6T-X65 ((50 Ω:800 Ω, with 30 kHz–75 MHz bandpass). This puts you in the $\sim$3 dB territory of noise figure, quite an improvement over the NF=15 dB you get with a bare 50 Ω source. For more discussion of signal transformers, see §8.10.

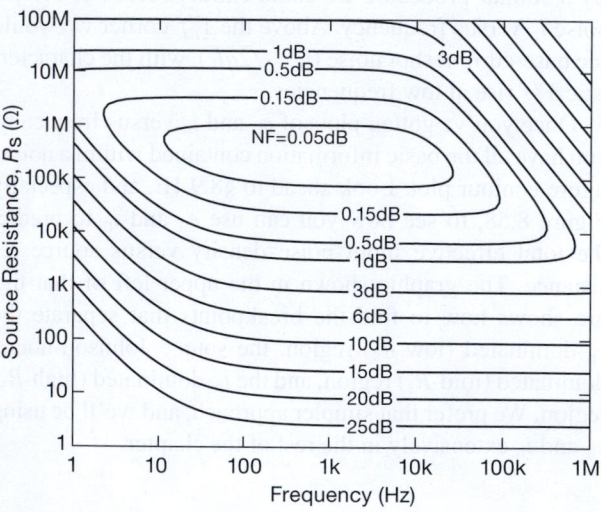

Figure 8.27. Noise figure versus frequency and source resistance for the SR560 low-noise preamp. (Courtesy of Stanford Research Systems.)

8.5 Low-noise design with bipolar transistors

The fact that e_n falls and i_n rises with increasing I_C provides a simple way to optimize transistor operating current to give lowest noise with a given source. Look at the model again (Figure 8.28). The noiseless signal source v_s has added to it an irreducible noise voltage from the Johnson noise of its source resistance:

$$e_R^2(\text{source}) = 4kTR_s \quad (\text{V}^2/\text{Hz}). \tag{8.29}$$

The amplifier adds noise of its own, namely,

$$e_a^2(\text{amplifier}) = e_n^2 + (i_n R_s)^2 \quad (\text{V}^2/\text{Hz}). \tag{8.30}$$

Thus the amplifier's noise voltage is added to the input signal, and, in addition, its noise current generates a noise voltage across the source impedance. These two are un-correlated (except at very high frequencies), so you add their squares. The idea is to reduce the amplifier's total noise contribution as much as possible. That's easy, once you know R_s, because you just look at a graph of e_n and i_n versus I_C in the region of the signal frequency, picking I_C to minimize $e_n^2 + (i_n R_s)^2$. Alternatively, if you are lucky and have a plot of noise-figure contours versus I_C and R_s, you can quickly locate the optimum value of I_C.

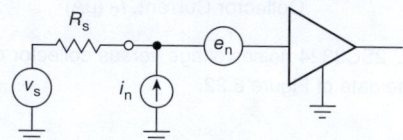

Figure 8.28. Amplifier noise model.

8.5.1 Noise-figure example

As an example, suppose we have a small signal in the region of 1 kHz with source resistance of 10k, and we wish to make a low-noise common-emitter amplifier with a 2N5087. From the datasheet graphs of e_n and i_n versus collector current (Figure 8.29), we see that the sum of voltage and current terms (with 10k source) is minimized for a collector current of about 20–40 μA. Because the current noise is falling faster than the voltage noise is rising as I_C is reduced, it might be a good idea to use slightly less collector current, especially if operation at a lower frequency is anticipated (i_n rises rapidly with decreasing frequency). We can estimate the noise figure using i_n and e_n at 1 kHz:

$$\text{NF} = 10\log_{10}\left(1 + \frac{e_n^2 + (i_n R_s)^2}{4kTR_s}\right) \text{ dB}. \tag{8.31}$$

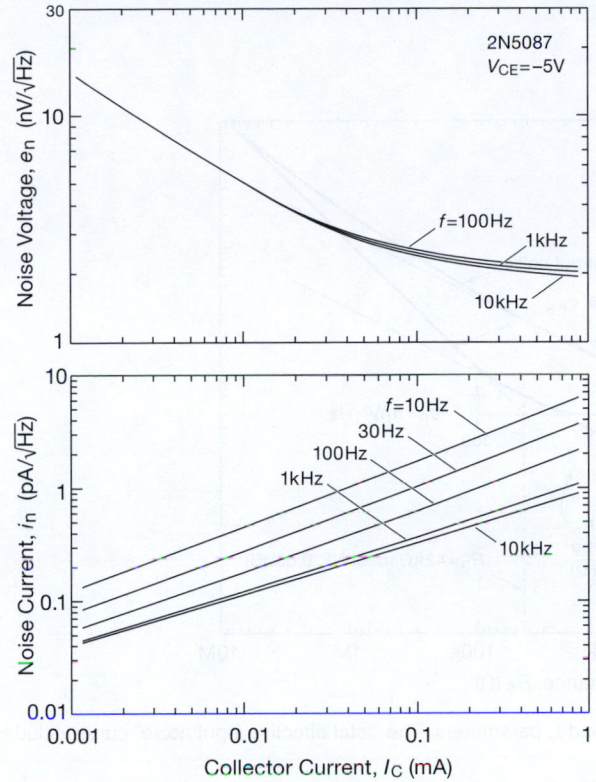

Figure 8.29. Voltage and current noise versus collector current for the 2N5087 *pnp* transistor.

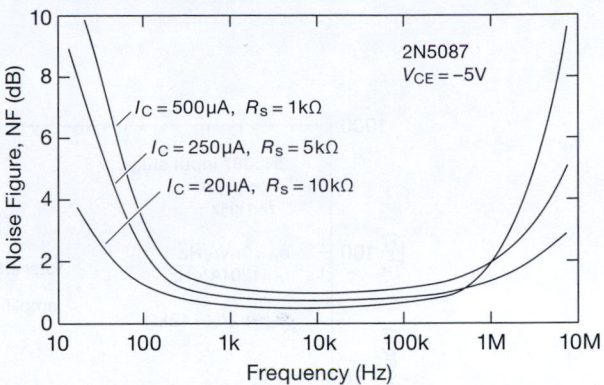

Figure 8.30. Noise figure (NF) versus frequency, for three choices of I_C and R_s, for the 2N5087.

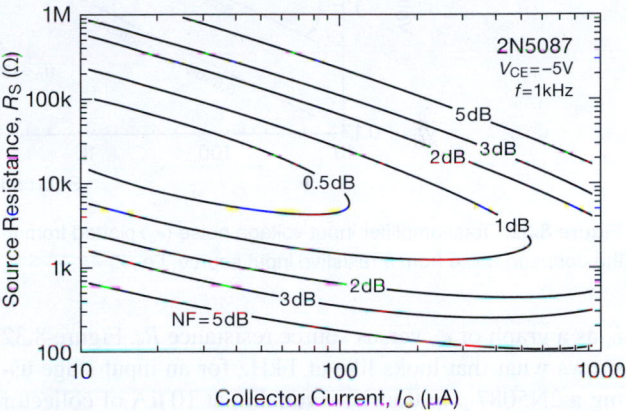

Figure 8.31. Contours of constant narrowband noise figure for the 2N5087 transistor. (From the ON Semiconductor datasheet.)

Reading from Figure 8.29, at $I_C = 20\,\mu A$, $e_n = 3.7\,nV/\sqrt{Hz}$ and $i_n = 0.17\,pA/\sqrt{Hz}$; and $4kTR_s = 1.65 \times 10^{-16}\,V^2/Hz$ for the 10k source resistance. The calculated noise figure is therefore 0.42 dB. This is consistent with the datasheet's graph (Figure 8.30) showing NF versus frequency, in which they have chosen this operating current for $R_s = 10k$. This choice of collector current is also roughly what you would get from the graph in Figure 8.31 of noise-figure contours at 1 kHz, although the actual noise figure can be estimated only approximately from that plot as being slightly less than 0.5 dB.

Exercise 8.2. Find the optimum I_C and corresponding noise figure for $R_s = 100k$ and $f = 1\,kHz$, using the graph in Figure 8.29 of e_n and i_n. Check your answer from the noise-figure contours (Figure 8.31).

For the other amplifier configurations (follower, grounded base) the noise figure is essentially the same, for given R_s and I_C, since e_n and i_n are unchanged. Of course, a stage with unity voltage gain (a follower) may just pass

the problem along to the next stage,[33] since the signal level hasn't been increased to the point that low-noise design can be ignored in subsequent stages.

8.5.2 Charting amplifier noise with e_n and i_n

The noise calculations just presented, although straightforward, make the whole subject of amplifier design appear somewhat formidable. If you misplace a factor of Boltzmann's constant, you suddenly get an amplifier with a 10,000 dB noise figure! In this section we present a simplified noise-estimation technique of great utility.

The method consists of first choosing some frequency of interest in order to get values for e_n and i_n versus I_C from the transistor datasheets. Then, for a given collector current, you can plot the total noise contributions from e_n and

---

[33] "Kicking the can down the road," in the political parlance of our time.

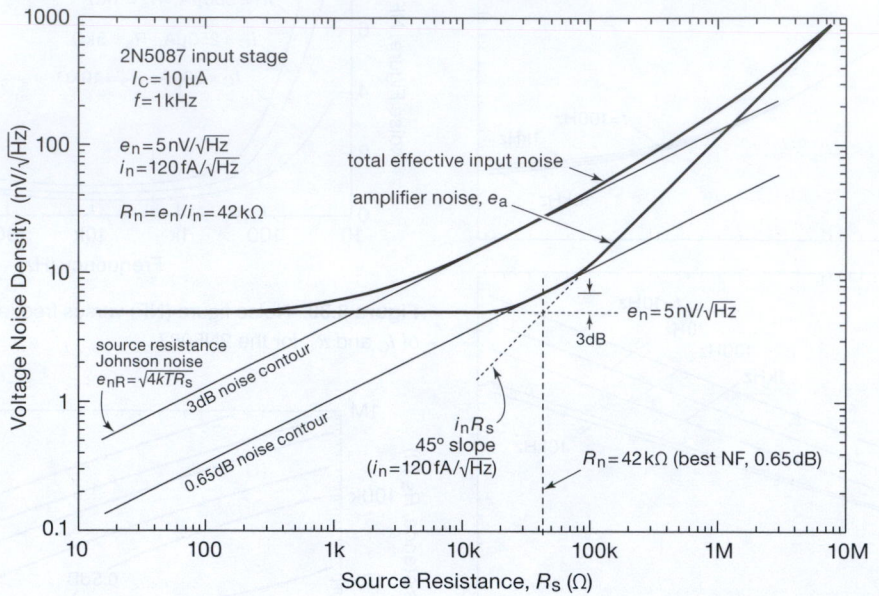

Figure 8.32. Total amplifier input voltage noise (e_a) plotted from the e_n and i_n parameters. The "total effective input noise" curve includes the Johnson noise from a resistive input source, i.e., $Z_s = R_s$.

i_n as a graph of e_a versus source resistance R_s. Figure 8.32 shows what that looks like at 1 kHz for an input stage using a 2N5087 *pnp* transistor running at 10 μA of collector current. The e_n noise voltage is constant, and the $i_n R_s$ voltage increases proportional to R_s, i.e., with a 45° slope. The amplifier noise curve is drawn as shown, with care being taken to ensure that it passes through a point 3 dB (voltage ratio of 1.4) above the crossing point of individual voltage- and current-noise contributions. Also plotted is the noise voltage of the source resistance, which is also the 3 dB NF contour. The other lines of constant noise figure are simply straight lines parallel to this line, as you will see in the examples that follow.

The best noise figure (0.65 dB) at this collector current and frequency occurs for a source resistance of 42 kΩ, and the noise figure is easily seen to be less than 3 dB for all source resistances between 2 kΩ and 1 MΩ, the points at which the 3 dB NF contour intersects the amplifier noise curve.

The next step is to draw a few of these noise curves on the same graph, using different collector currents or frequencies, or maybe a selection of transistor types, in order to evaluate amplifier performance. Before we go on to do that, let's show how we can talk about this same amplifier using a different pair of noise parameters, the noise resistance R_n and the noise figure NF(R_n), both of which pop right out of the graph.

We'll have some fun with this technique later (§8.7), after we've learned about noise in JFETs, by refereeing over a world championship low-noise shootout between a best-in-class BJT (2SD786) and a comparably excellent JFET (2SK170).

8.5.3 Noise resistance

The lowest noise figure in this example occurs for a source resistance R_s=42k, which equals the ratio of e_n to i_n. That defines the noise resistance:

$$R_n = \frac{e_n}{i_n} \quad \text{ohms.} \quad (8.32)$$

You can find the noise figure for a source of that resistance from our earlier expression, eq'n 8.31, for noise figure:

$$\text{NF(at } R_n) = 10\log_{10}\left(1 + 1.23\times10^{20}\frac{e_n^2}{R_n}\right) \text{ dB} \quad \approx \quad 0.31 \text{ dB}.$$

Noise resistance isn't actually a real resistance in the transistor, or anything like that. It is a tool to help you quickly find the value of source resistance for minimum noise figure, ideally so that you can vary the collector current to shift R_n close to the value of source resistance you're actually using. R_n corresponds to the point where the e_n and i_n lines cross.

The noise figure for a source resistance equal to R_n then follows simply from the preceding equation.

8.5.4 Charting comparative noise

It's easy to compare candidate transistors with this charting technique, plotting total amplifier noise for each of a selection of possible collector currents. We've done this in Figure 8.33, where we compare total amplifier noise (which includes source resistance Johnson noise) versus source resistance for the high-beta 2N5962 and the low-$r_{bb'}$ ZTX851 *npn* transistors, using measured values of beta and $r_{bb'}$ listed in Table 8.1a on page 501. You can see that the high-beta part operated at low collector current is the clear winner for high source resistances, where its relatively high $r_{bb'}$ (480 Ω) is harmless, being swamped by the Johnson noise of the signal source. By contrast, for low source resistances (say 1k and below) the ZTX851's admirably low $r_{bb'}$ ($\sim$1.7Ω) achieves the lowest noise, particularly when operated at a relatively high collector current to minimize the "r_e" noise term (recall that collector-current shot noise through $1/g_m$ produces a noise voltage density e_n equivalent to the Johnson noise of a resistor of value $r_e/2$, eq'n 8.20).

8.5.5 Low-noise design with BJTs: two examples

Let's put these ideas and equations to work, first by looking at a simple single-ended low-noise audio preamp that achieved popularity beginning in the 1980s, then by comparing with a classic differential design that addresses many of the shortcomings of the single-ended circuit.

A. The Naim preamp

Figure 8.34 shows the input stage used for many years in low-level preamps from the British manufacturer Naim Audio. It's an ac-coupled single-ended two-stage series feedback design (Figure 2.92), tailored for low input-noise voltage. The overall voltage gain is $G_V = 1 + R_f/R_E$ (here 30 dB), with R_E chosen rather small to keep its Johnson noise-voltage contribution down below a nanovolt per root hertz (15 Ω has e_n=0.5 nV/$\sqrt{\text{Hz}}$). The other significant noise term is the transistor's contribution from $r_{bb'}$, which is variously specified on datasheets either as an e_n value (at some specified collector current) or graph, or as a noise-figure value or graph, or (rarely) as a value for $r_{bb'}$ itself; or

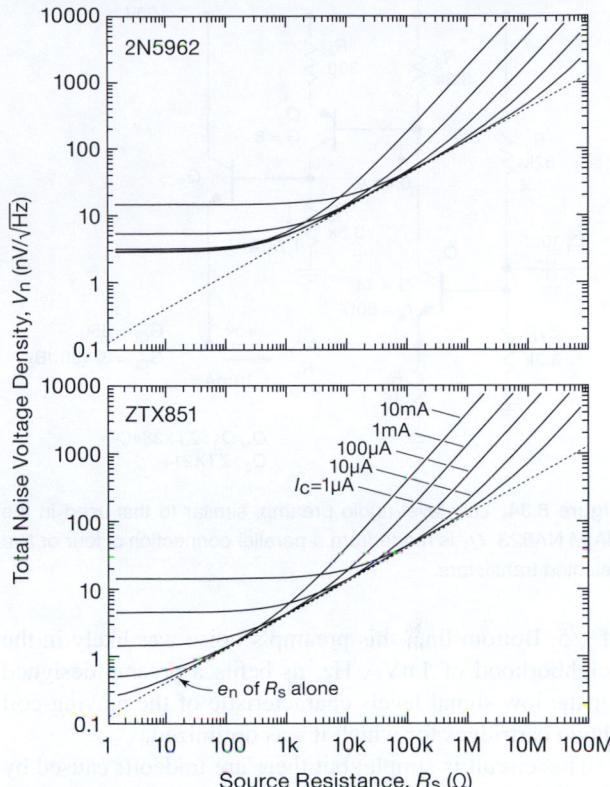

Figure 8.33. Comparing the total amplifier input-noise voltage versus source resistance of two candidate low-noise BJTs. The curves plot five decade values of collector current. For low source resistances the ZTX851's low $r_{bb'}$ yields low noise at high collector currents; by contrast, the 2N5962's higher $r_{bb'}$ limits the ultimate noise voltage, but its higher beta (thus lower base-current shot noise, $\sqrt{2qI_C/\beta}$) results in improved performance with high source resistances. These curves include the Johnson noise of the source resistance.

(worst of all) perhaps none of these. For the "Low Noise" ZTX384C used in this preamp, the datasheet is rather, uh, *taciturn*, revealing only that NF=4 dB (max) from 30 Hz–15 kHz for R_s=2k and I_C=0.2 mA. This is not terribly helpful, because that corresponds to a seriously large Johnson-noise voltage of 6.9 nV/$\sqrt{\text{Hz}}$.

The fact is that these amplifiers were pretty quiet. One reason is that worst-case specifications of difficult-to-measure parameters tend to be overly pessimistic. For example, the contemporary low-noise *npn* 2SC3324 specifies (for a particular test condition) NF=3 dB (max), but 0.2 dB (typ). Naim may have selected parts for low noise. The other reason is that Naim's circuit actually uses five transistors in parallel (each with a 15 Ω emitter ballasting resistor) for Q_1, which drives down its noise voltage by a factor

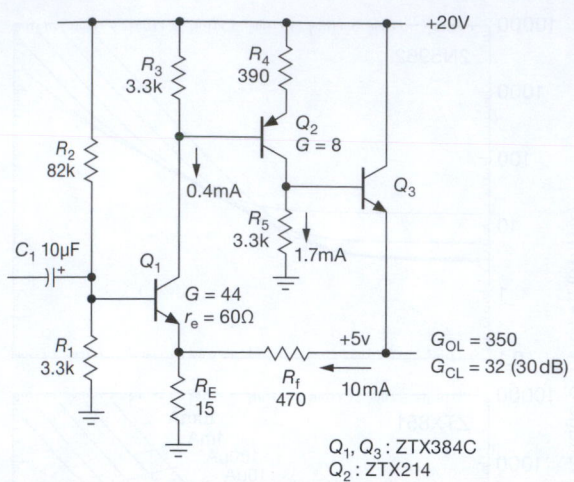

Figure 8.34. Low-level audio preamp, similar to that used in the NAIM NA323. Q_1 is made from a parallel connection of four or five selected transistors.

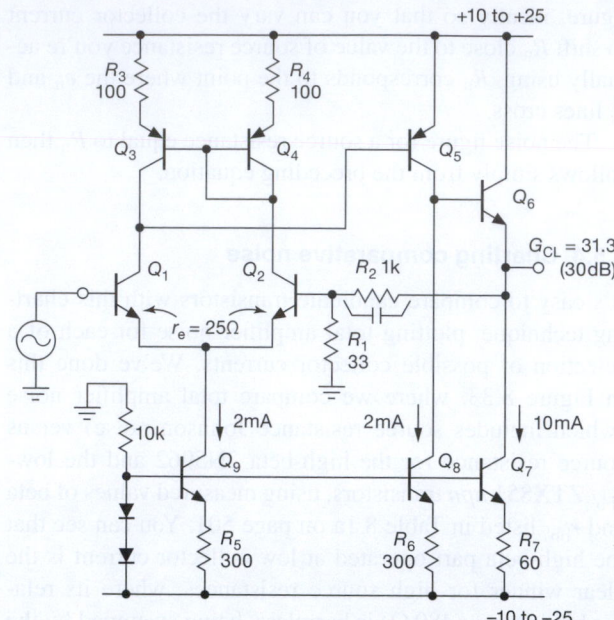

Figure 8.35. A low-noise dc-coupled audio preamplifier with predictable biasing.

of $\sqrt{5}$. Bottom line: this preamp's noise was likely in the neighborhood of $1\,\text{nV}/\sqrt{\text{Hz}}$, as befits a preamp designed for the low signal levels characteristic of the moving-coil phono cartridges for which it was optimized.

This circuit is simple; but there are tradeoffs caused by the small R_E that was chosen for low noise. It makes the bias scheme somewhat problematical, with the bias divider R_1R_2 acting more as a current source (compared with the classic voltage biasing, for example as in Figure 2.35), thus making the quiescent operating point rather dependent on Q_1's beta. The low value of R_E also results in a low input impedance, owing to the required stiff bias divider's impedance of $3.2\,\text{k}\Omega$.[34]

B. A better design: dc-coupled differential preamp

Many of the shortcomings of the Naim preamp are nicely addressed in the dc-coupled differential preamp shown in Figure 8.35, which retains the former's desirable low-noise performance. It's basically the classic 2-stage op-amp configuration (e.g., Figures 4.43 or 2.91), here simplified with a single-ended class-A output stage with current-sink active load (with the assumption that the next stage will not require substantial drive current).

This circuit eliminates electrolytic blocking capacitors, both at the input and in the gain-setting feedback divider R_1R_2; it minimizes Early effect nonlinearity in the input stage; and its use of active (current-source) loads both

improves its overall linearity (through improved single-stage linearity, plus higher loop gain) and provides stable and predictable biasing. The price you pay in exchange is small: increased circuit complexity, and the 3 dB increase in noise (owing to Q_2's contribution). At these collector currents the noise is dominated by the Johnson noise of the base resistance $r_{bb'}$. So, as with the Naim amplifier, the input transistors could be implemented as a parallel array of matched-V_{BE} transistors (even if single transistors are used, the pair should be matched to $\sim 10\,\text{mV}$ or so), or (better) implemented with larger die-size transistors with lower $r_{bb'}$.[35]

8.5.6 Minimizing noise: BJTs, FETs, and transformers

Bipolar transistor amplifiers can provide very good noise performance over the range of source impedances from about $200\,\Omega$ to $1\,\text{M}\Omega$, with corresponding optimum collector currents generally in the range of several milliamps down to a microamp. (With low source impedances you want to minimize e_n, whereas with high source impedances

---

[34] Which dominates the impedance seen looking into the base ($\beta R_E G_{\text{loop}}$), the latter about $50\,\text{k}\Omega$.

[35] For example the inexpensive 2SD2653, specified to 2A continuous collector current, but with high beta even at low currents: $\beta \approx 500$ at $I_C = 1\,\text{mA}$. See §8.5.9B.

you want to minimize i_n; as we've seen, that dictates high and low collector current, respectively).

If the source impedance is high, say greater than 100k or so, transistor *current* noise dominates, and the best device for low-noise amplification is a FET. Although their voltage noise is usually greater than that of bipolar transistors, the gate current (and its noise) can be exceedingly small, making them ideally suited for low-noise high-impedance amplifiers. We'll dive into FET noise shortly (§8.6), after an entertaining "challenge" interlude.

For very low source impedances (say $50\,\Omega$), transistor voltage noise will always dominate, and noise figures will be poor. One approach in such cases is to use a signal transformer to raise the signal level (and impedance), treating the signal on the secondary as before. Transformers have their drawbacks, of course: they're ac coupled; they operate over only a few decades of bandwidth (and never to dc); those intended for low-frequency operation are bulky and expensive and exhibit nonlinearities; and they are susceptible to magnetic pickup. Nevertheless, they can be the magic sauce when you're dealing with a low impedance signal (say less than $100\,\Omega$); see §8.10.

8.5.7 A design example: 40¢ "lightning detector" preamp

Here's an interesting design challenge, and a chance to exercise some of our noise theory. Imagine we want to manufacture an inexpensive photodiode amplifier that runs on low current from a 9V battery, with input noise of just a few $\mathrm{nV}/\sqrt{\mathrm{Hz}}$, and with response time of a few microseconds. You can think of this as a simple "lightning detector," because lightning is almost unique in delivering microsecond-scale pulses of light in the outdoor environment.[36] In a fully refined implementation this could serve as a useful early warning device to install in locations where you'd want to be indoors when the sky is becoming electric – golf courses, harbors, soccer fields, etc.

---

[36] Roughly 90% of lightning activity is cloud-to-cloud, stretching over large distances and exhibiting relatively long rise- and fall-times; their signals look not unlike typical background "cultural" noise. The remaining 10% are lightning strikes to ground (which matter to us!), with unique short rise-time signatures, typically less than $5\,\mu s$ in duration (even for distant ground strikes). This allows us to create filters to distinguish weak distant cloud-to-ground lightning strikes from nearby transient interference of other origins. The circuit described here is a simple low-noise preamp suitable for use with such filters and discriminators.

Here we explore the basics, a launching point to inspire the electronic hobbyist. It will turn out that we can do the job with standard parts, for a total component cost (in 1000-piece quantities) of just $0.40. It will be a "throwaway" device, manufactured in the millions, with commensurate profits and fame (dream on!). Let's take it in easy stages, as shown in Figures 8.36A–E (similarly labeled in the following paragraphs).

A. Block diagram We back-bias the PIN photodiode to lower its capacitance (thus lower noise and faster response), and we use a blocking capacitor into the summing junction of a transimpedance stage to eliminate the dc level from ambient light and leakage current. The feedback resistor R_f sets the gain ($G = -R_f$ volts/amp).

B. Discrete design, first iteration We need inverting feedback, so we begin with a grounded-emitter stage (Q_1), with an emitter follower (Q_3) to create a low-impedance output and feedback source.

C. Add cascode This circuit will be running at low currents (0.1 mA or so), thus relatively high impedances at which Miller effect causes a significant bandwidth reduction. So we add a cascode (Q_2) atop the gain stage's collector.

D. Inverted cascode A 9 V battery is down to 6 V at end-of-life, so now we're running out of headroom! We want to keep plenty of dynamic range, so we fix this problem by turning the cascode around, creating an "inverted cascode," in which collector current variations in Q_1 pass through *pnp* cascode transistor Q_2, while the latter continues to clamp Q_1's collector voltage. The output follower plays a dual role here, with its V_{BE} setting Q_2's collector current: $I_{C2} = V_{BE3}/R_3$.

E. Front-end follower and input current noise At the $\sim 50\,\mu A$ collector current in Q_1 that we'll need to get sufficient bandwidth, there would be too much current noise at the input. (Remember that the input signal is a *current*.) So we add a high-beta follower (Q_4) and choose a collector current of $1\,\mu A$ as an initial trial value. We also choose a large feedback resistor R_2 to minimize its noise current as seen at the input, with shunt capacitance C_c to limit the bandwidth to 100 kHz. The idea is to calculate the input noise to see where the dominant contribution comes from, then iterate the design toward an optimum. We've filled in component values for an initial design.

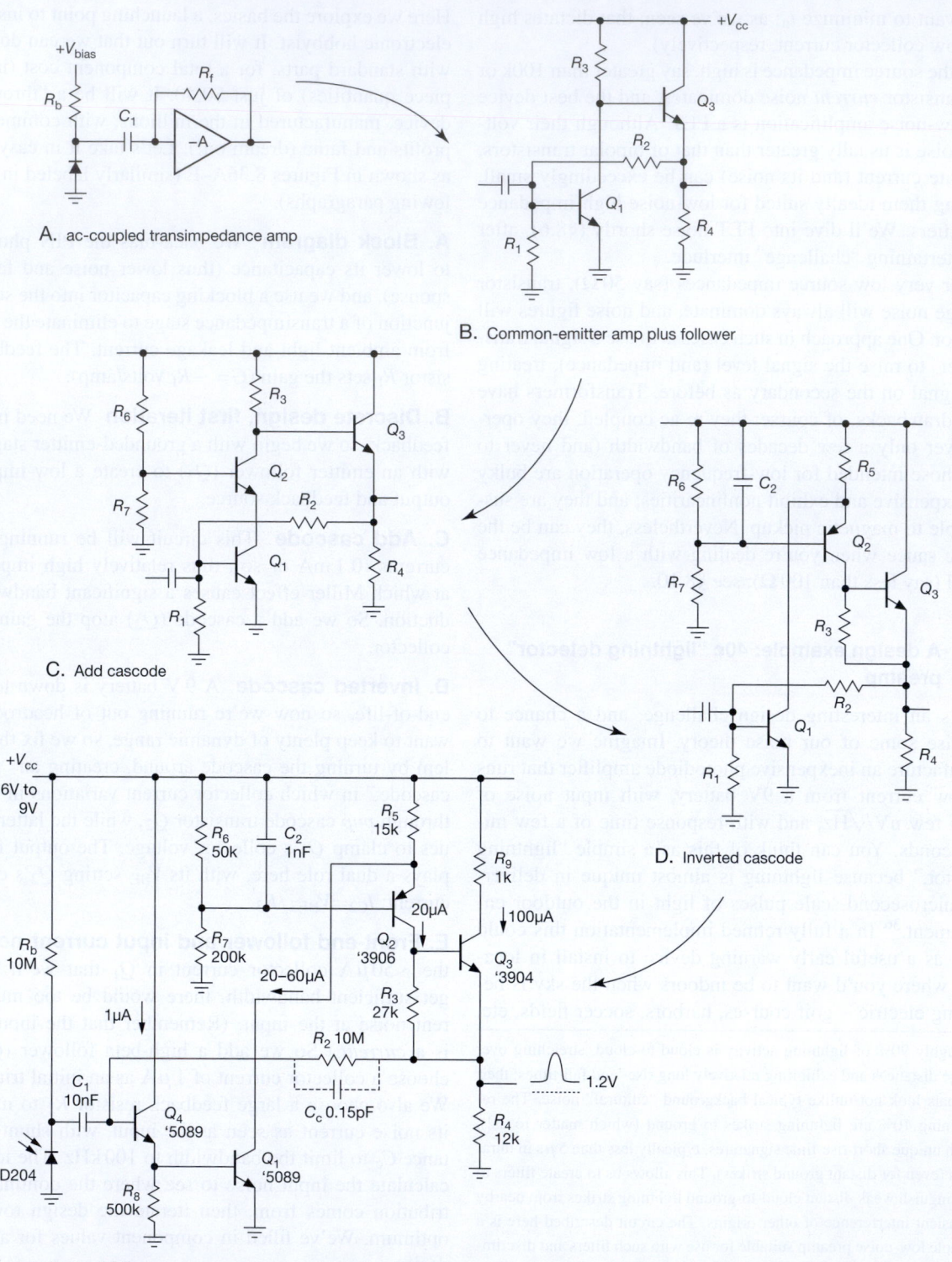

A. ac-coupled transimpedance amp

B. Common-emitter amp plus follower

C. Add cascode

D. Inverted cascode

E. Add frontend follower

Figure 8.36. Tutorial evolution of a low-noise and low-power photodiode amplifier built with low-cost discrete components. Apart from Q_1 and the photodiode, all parts are available in surface-mount packages (prepend "MMBT" to transistor part numbers, e.g., MMBT5089).

Calculating the noise performance Let's estimate the noise current at the input; that's what competes with the (current) signal from the photodiode. There's shot noise in Q_4's base current, and there's Johnson-noise current from the feedback resistor R_2 and the photodiode bias resistor R_b. In addition, we have to worry about noise *voltage* at the input: in combination with the input capacitance C_{in} it creates an input noise current $i_n = e_n \omega C_{in}$ (as we'll see in detail in §8.11). Input noise voltage is the combined e_n contributions of Q_1 and Q_4, each having Johnson noise in their $r_{bb'}$ and collector shot noise through their r_e. Let's take these in turn.

Noise current Estimating Q_4's $\beta \approx 350$ at its $1\,\mu A$ collector current, we find that its base current of 3 nA creates a shot-noise density $i_{n(shot)} = \sqrt{2qI_B} = 30\,fA/\sqrt{Hz}$. Feedback resistor R_2 and bias resistor R_b create a Johnson-noise current $i_{n(R)} = \sqrt{4kT/R} = 57\,fA/\sqrt{Hz}$. These combine (square root of sum-of-squares) to create $i_n(total) = 65\,fA/\sqrt{Hz}$.

Noise voltage Emitter follower Q_4, running at $1\,\mu A$, generates a shot noise voltage across its r_e of $e_{n(shot4)} = r_e\sqrt{2qI_C} = 14.3\,nV/\sqrt{Hz}$. Its internal base resistance of $\sim 300\,\Omega$ adds a relatively insignificant Johnson-noise voltage $e_{n(J4)} = \sqrt{4kTr_{bb'}} = 2.2\,nV/\sqrt{Hz}$, for a combined $e_{n4} = 14.5\,nV/\sqrt{Hz}$.

Grounded-emitter amplifier Q_1, running at nominal $40\,\mu A$ and with an $r_{bb'}$ of $\sim 300\,\Omega$, generates corresponding noise voltages $e_{n(shot1)} = r_e\sqrt{2qI_C} = 2.3\,nV/\sqrt{Hz}$ and $e_{n(J1)} = \sqrt{4kTr_{bb'}} = 2.2\,nV/\sqrt{Hz}$, for a combined $e_{n1} = 3.2\,nV/\sqrt{Hz}$.

Combining the noise voltages of Q_1 and Q_4, we find[37] $e_n(total) = 14.8\,nV/\sqrt{Hz}$. Evidently Q_4 is the big noise gorilla in the room here; but let's continue with the analysis.

In combination with the input capacitance of $\sim 10\,pF$ (5 pF for the photodiode, 2.5 pF for Q_4's C_{cb}, and 2.5 pF for wiring capacitance) that noise voltage creates an effective input-current noise (§8.11.3)of $i_n = e_n \omega C_{in} = 90\,fA/\sqrt{Hz}$, if we take for ω a characteristic frequency of 100 kHz.

Verifying feedback stability This is a feedback circuit, with the ever-present potential for oscillation. And, in a transresistance configuration like this, the shunt capacitance at the input combines with the large-value feedback resistor to introduce an additional lagging phase shift. We deal with this in detail in §§4x.3 and 8.11; the criterion for stability is that the open-loop amplifier's unity-gain bandwidth must satisfy

$$f_T(\text{open-loop}) > f_{R_2C_c}^2 / f_{R_2C_{in}}. \qquad (8.33)$$

(Stated differently, the feedback network's $-3\,dB$ rolloff frequency must be less than the geometric mean of (a) the amplifier's open-loop unity-gain frequency and (b) the $-3\,dB$ rolloff frequency of the feedback resistor in combination with the input capacitance.)

So, for the circuit of Figure 8.36E we require $f_T(\text{open-loop}) > 106\,kHz^2/1.6\,kHz$, or 7 MHz. We estimate the amplifier's open-loop unity gain as follows.

(a) The low-frequency voltage gain is $G = g_{m1}R_{load}$, where R_{load} is the impedance seen at cascode transistor Q_2's collector.

(b) The latter is $R_{load} = R_4 \cdot g_{m3}R_3 \approx 1.3\,M\Omega$, so the low-frequency open-loop gain G_{OL} ranges from 1000 to 3000 for Q_1 collector currents from $20\,\mu A$ to $60\,\mu A$, (battery voltage from 6 V to 9 V), respectively.

(c) This gain rolls off at 6 dB/octave, starting at a frequency set by R_{load} in combination with the capacitance seen at Q_2's collector; taking $C_{cb} \sim 2.5\,pF$ for each of Q_2 and Q_3, that's $f_{3dB} = 25\,kHz$.

(d) That brings the amplifier's open-loop gain down to unity at about 25 MHz (for Q_1's minimum collector current), a good margin of stability given our 7 MHz requirement.

The cascode was essential for getting this bandwidth at these low currents. Without it the gain stage (Q_1) would see its load capacitance multiplied by the voltage gain of 1000, owing to Miller effect.[38]

Optimization The contributions from amplifier noise currents ($65\,fA/\sqrt{Hz}$) and from noise voltage generating noise current through the input capacitance ($90\,fA/\sqrt{Hz}$) are in the same ballpark. The latter rises proportional to frequency, and dominates, slightly, at the high-frequency end of the bandpass; but when integrated over the amplifier's passband, the two noise terms contribute comparable integrated noise, totaling approximately $I_n = 30\,pArms$ (therefore 0.3 mVrms at the amplifier's output).

The lion's share of the noise current, then, comes from the 10M gain-setting and biasing resistors. These could be raised in value, though R_b must not be so large that the photodiode saturates in nighttime ambient light. The latter problem could be circumvented by sensing the voltage at the photodiode and using it to control a quiet biasing current source. If the resistor noise were brought down

---

[37] At these low collector currents we found actual measured noise voltages to be about 25% higher than those predicted by simple theory.

[38] See also the section "Analysis by Equivalent Circuits" in Chapter 2x for a discussion of classical feedback analysis in circuits like this.

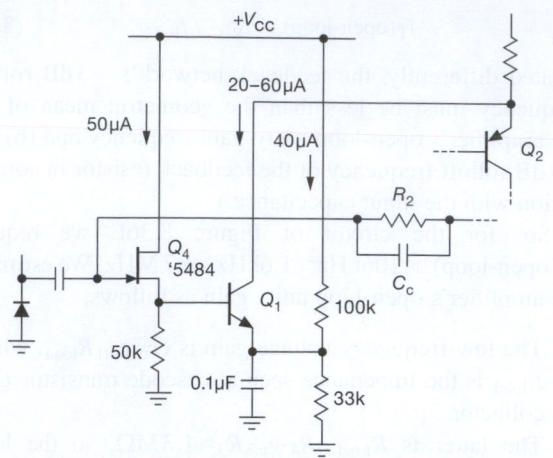

Figure 8.37. Replacing Figure 8.36's BJT input transistor with a low-capacitance JFET reduces input noise voltage and noise current by a factor of 3.

enough, the dominant remaining noise source would be Q_4, whose voltage noise (multiplied by ωC_{in}) and current noise are comparable. Raising its operating current reduces its voltage noise, but increases its current noise. We seem to be stuck.

But no, we're *not*! If we replace the BJT input follower with a low-capacitance JFET (Figure 8.37), we can score a significant improvement. The voltage noise of a 2N5484 ($C_{in} \approx 2.2\,\text{pF}$) running at $50\,\mu\text{A}$ is about $e_n = 5\,\text{nV}/\sqrt{\text{Hz}}$ (based on measurements at $100\,\mu\text{A}$), and its current noise is negligible; so we reduce the transistor noise contribution by a factor of 3 (for an additional cost of 5–10 cents). Of course, this makes sense only if the R_b and R_2 Johnson-noise contributions can be reduced by at least a factor of 4. With those improvements the amplifier's total integrated input noise becomes $I_n \approx 10\,\text{pArms}$. That's comparable to shot noise in the photodiode's leakage current (3 nA, typ, at $-10\,\text{V}$ and $25°\text{C}$) – no point in grooming the amplifier further.

We take up the subject of low-noise design with JFETs shortly, in §8.6.

8.5.8 Selecting a low-noise bipolar transistor

It's important to select the right transistors, and appropriate operating currents, in order to optimize a circuit's low-noise performance. This task has been made more difficult as some of the finest low-noise favorites have been recently discontinued by one manufacturer after another (most of these came in through-hole packages, not favored for machine assembly). Often this means you can't simply choose a part used in someone else's good design. The task is made difficult also by the lack of relevant information

in datasheets – noise-voltage curves, or even a mention of noise voltage is the exception in most contemporary BJT datasheets. We come to your rescue with Tables 8.1a,b, where we've listed a range of good candidate transistors,[39] along with measured noise parameters such as e_n, $r_{bb'}$, V_A (Early voltage), and plots of beta versus collector current. Herewith a guide to the table, and to the task of selecting and using a low-noise transistor.[40]

The **beta** or h_{FE} column shows the manufacturer's stated minimum values. While the *typical* beta is often much higher, you have to accept the reality that a minimum-beta part may end up in your circuit (unless you are willing to screen incoming parts). The manufacturer's beta specs are often at higher currents than you'll be using, and the beta may be severely degraded at lower currents, see the discussion of recombination and Gummel plots in §2x.13. To see which transistors are susceptible to this, we measured beta versus collector current, going from $1\,\mu\text{A}$ to 100 mA, with the results shown in Figure 8.39 (with corresponding labels in the **Beta plot** column of Tables 8.1a,b) . The falloff of beta at low current is summarized in the **Linearity** column in the table, rating the transistors on a scale of 1 (large beta falloff at microamp currents) to 5 (beta does not change with current). Note how some parts run out of steam at currents above 1mA; these are small-die parts running at high current density. Keep in mind our measurements were taken at $V_{CE} = 5\,\text{V}$, and that the high-current beta of some BJTs falls rapidly at lower collector voltages (a saturation effect).[41] Our measured plots go to 50 mA, but the datasheets of some larger-die parts (like the ZTX851) provide curves of beta going as high as 10 A.

Some BJT types are measured by the manufacturer and

---

[39] In many cases these transistors are not promoted by the manufacturer as "low noise." We made some guesses, then ordered and tested samples of some 100 different types.

[40] Some advice about finding parts. It's helpful to search using the inner (numeric) portion of the part name, because there may be versions with differing prefixes or suffixes. You'll have an easier time finding surface-mount parts than through-hole parts – see the notes in the SOT-23 column of Table 8.1a. Some parts are stocked only at specialty suppliers, such as B&D Enterprises, MCM, Donberg Electronics, Encompass, and Littlediode. Discontinued parts may also be available on eBay, in small quantities suitable for making specialized laboratory instruments. Some large Pacific rim distributors have good older-part inventory and sell from their websites, through Alibaba, or on eBay. The four Sanyo parts near the bottom of the table have been discontinued, but they make other parts that may offer similar performance.

[41] We've not provided measurements of beta at lower voltages, but we are confident that the low-noise winners (the *npn* ZTX851 and *pnp* ZTX951) do very well in this respect, having been designed as high-current transistors with low saturation voltage.

Table 8.1a Low-noise BJTs[a]

Part #	Manuf	npn	pnp	TO-92	SOT-23	Small SMT	SOT-223	TO-126	TO-220	Pinout[u]	V_CEO max (V)	f_T typ (MHz)	C_ob typ (pF)	beta min	@ I_C (mA)	Linearity[e]	Beta plot	Early V_A meas (V)	e_n @ (nV/√)	I_C (mA)	r_bb'[a] (Ω)	I_C (mA)	e_n (nV/√)	i_n (pA/√)	R_n (Ω)
BCX70J	m	•	-	-	•	-	-	-	-	D	45	250	1.7	250	2	4	37	215	3.54	10	760	0.015	5.1	0.14	37k
BC850C	I	•	-	b	•	a	-	-	-	D	45	250	2.5	420	2	4	37	220	3.32	1	650r	0.017	4.7	0.11	41k
BC860C	I	-	•	b	•	a	-	-	-	D	45	250	2.5	420	2	3	17	30	3.09	1	590r	0.02	4.5	0.38	12k
MPSA18	m	•	-	•	-	-	-	-	-	A	45	160	1.7	500	0.1	4	41	-	6.5d	0.1	-	-	-	-	-
2SC3624A	H	•	-	-	•	-	-	-	-	D	50	250	3	1000	1	-	-	60d	-	-	-	-	-	-	-
2N5962	m	•	-	•	m	-	-	-	-	A,D	45	100	1.5	500	0.1	2	42	60	2.77	10	480	0.02	4.0	0.38	10k
MMBT6429	O	•	-	-	m,p	-	-	-	-	D	45	400	1.3	500	0.1	5	39	140	2.27	10	310	0.04	3.2	0.15	21k
2N5089	m	•	-	•	m,p,c	-	-	-	-	A,D	25	50	1.3	450	1	5	38	200	2.15	10	290	0.04	3.1	0.17	19k
2N5088	m	•	-	•	m,p	-	-	-	-	A	30	50	1.3	300	0.1	5	38	180	1.97	10	240	0.05	2.9	0.22	13k
MPS8098	m	•	-	•	-	-	-	-	-	A	60	150	2.6	100	1	4	27	435	1.79	10	195	0.06	2.6	0.43	6.0k
2N5087	m	-	•	•	m	-	-	-	-	A,D	50	50	2.4	250	0.1	4	15	55	0.81	10	40	0.28	1.2	0.60	2.0k
2N5210	C	•	-	•	m	-	-	-	-	A,D	50	50	1.3	250	1	5	35	270	1.92	10	230	0.05	2.8	0.25	11k
2SC2412-R	R	•	-	a	•	a	-	-	-	D	50	180	2	180	0.1	5	32	360	1.66	10	165	0.07	2.4	0.35	6.8k
2SA1037-R	R	-	•	a	•	-	-	-	-	D	50	140	4	180	0.1	-	-	60d	-	-	-	-	-	-	-
2SC2712-GR	T	•	-	-	•	-	-	-	-	D	50	80	2	200	2	5	32	390	1.57	10	150	0.07	2.3	0.34	6.6k
2SA1162-GR	T	-	•	-	•	-	-	-	-	D	50	80	4	200	2	4	10	50	1.40	10	120y	0.09	2.0	0.38	5.2k
2SC3906K	R	•	-	a	•	a	-	-	-	D	120	140	2.5	180	2	5	35	280d	1.35	10	110	0.10	1.9	0.42	4.6k
2SA1514K	R	-	•	a	•	-	-	-	-	D	120	140	3.2	180	2	4	12	165	1.35	10	110	0.10	1.9	0.42	4.6k
2N5401		-	•	•	m	-	-	-	-	A,D	160	150	6	50	1	5	2	-	-	-	-	-	-	-	-
2N5551[s]		•	-	•	m	-	-	-	-	A,D	160	150	2.7	80	1	5	25	-	-	-	105	0.11	1.3	0.65	2.0k
2N3904	m	•	-	•	m	a	-	-	-	A,D	40	300	2.5	70	1	5	31	340	1.35	10	110	0.10	1.9	0.68	2.8k
2N3906	m	-	•	•	m	a	-	-	-	A,D	40	250	3	80	1	1	3	25	0.74	10	32	0.35	1.0	1.18	890
2SC3324[g]	T	•	-	-	•	-	-	-	-	D	120	100	3	200	2	5	35	560	0.78	10	35	0.32	1.1	0.71	1.5k
2SA1312	T	-	•	-	•	-	-	-	-	D	120	100	4	200	2	5	13	180	0.58	10	20	0.56	0.8	0.94	880
2SB1197K-Q[b]	R	-	•	-	•	-	-	-	-	D	32	200	12	120	100	5	6	110	0.60	10	20	0.56	0.8	1.22	680
2N4401	m	•	-	•	m	a	-	-	-	A,D	40	250	4	80	10	5	31	410	0.84	10	40	0.28	1.2	1.05	1.1k
2N4403	m	-	•	•	m	a	-	-	-	A,D	40	200	5.5	50	10	1	1	-	0.55	10	17	0.65	0.8	2.05	370
2SD2653	R	•	-	-	•	-	-	-	-	D	12	360	20	270	200	4	37	65	0.54	10	17	0.65	0.8	0.88	860
2SB1690K	R	-	•	-	•	-	-	-	-	D	12	360	15	270	200	5	19	10	0.52	10	15	0.74	0.7	0.94	760
2SB1424	R	-	•	a	-	•	-	-	-	F	20	240	35	120	100	4	5	30d	0.35	10	9.4	1.2	0.6	1.78	320
ZXTN19020D	Z	•	-	-	•	-	-	-	-	D	20	160	33	300	100	3	36	100	0.40	10	11	1.0	0.6	1.04	590
2SD1898	R	•	-	a	-	d	-	a	-	F	80	100	20	120	500	-	-	580	0.40	10	8.3	1.3	0.5	1.89	280
2SB1260	R	-	•	a	-	d	-	a	-	F	80	100	20	120	100	-	-		0.36	10	7	1.6	0.5	2.06	240
MPS8599	O	-	•	-	m	-	-	-	-	A,D	80	150	2.9	100	1	5	6	170	0.47	10	12	0.9	0.6	1.72	370
MPS8099	O	•	-	-	m	-	-	-	-	A,D	80	150	2.5	100	1	4	27	540	0.37	10	8	1.4	0.5	2.11	250
DSS20201L[b]	D	•	-	-	•	-	-	-	-	D	20	150	16	250	10	1	30	150	0.38	10	8.0	1.4	0.5	1.34	390
ZTX450	Z	•	-	•	-	-	-	-	-	A	45	150	15 m	100	150	3	26	730	0.39	10	8.5	1.3	0.5	2.05	260
ZTX550	Z	-	•	•	-	-	-	-	-	A	45	150	-	100	150	5	6	-	0.38	10	7.7	1.4	0.5	2.15	240
ZTX618	Z	•	-	•	-	f	-	-	-	A,D	20	140	23	200	10	3	36	90	0.41	10	9.3	1.2	0.6	1.38	410
ZTX718	Z	-	•	•	-	f	-	-	-	A,D	20	180	21	300	10	3	18	25	0.38	10	7.3	1.5	0.5	1.27	390
ZXTN19100C	Z	•	-	-	•	-	-	-	-	D	100	150	16	200	100	3	34	480	0.36	10	6.8	1.6	0.48	1.61	300
2SD1684[b,x]	S	•	-	-	-	-	•	-	•	E	100	120	11	100	100	2	30	380	0.29	10	4.0	2.8	0.37	2.98	125
2SB1243Q[b]	R	-	•	-	-	-	-	-	v	G2	50	70	50	120	500	3	7	180d	0.28	10	3.7	3.0	0.35	2.83	125
BD437	m	•	-	-	-	-	-	•	-	E	45	3	65	30	10	2	21	1100	0.27	25	3.9	2.8	0.36	5.51	65
BU406	O	-	•	-	-	-	-	•	-	G	200	10	80	45t	100	1	z	150d	0.26	10	3.1	3.5	0.33	5.02	65
2SC3955[v,x]	S	•	-	-	-	-	-	•	-	E	200	300	1.9	40	10	5	22	600d	0.24	10	2.3	4.8	0.28	6.20	45
2SD786-S[x]	R	•	-	-	-	•	-	-	-	B	40	100	13	270	10	4	33	320	0.24	10	2.3	4.8	0.28	2.39	120
2SB737[x]	R	-	•	-	-	•	-	-	-	B	40	100	25	270	10	4	7	140	0.21	10	1.7	6.5	0.24	2.78	85
ZXTN2018F	Z	•	-	-	•	-	-	-	-	D	60	130	28	100	10	2	28	4600	0.27	10	3.3	3.4	0.33	3.28	100
ZXTP2027F	Z	-	•	-	•	-	-	-	-	D	60	165	44	100	10	4	7	100	0.21	10	1.46	7.6	0.22	4.93	45
2SC6102[b,x]	S	•	-	-	-	-	•	-	-	E	30	290	40	200	500	5	32	280	0.23	10	2.0	5.5	0.26	2.98	88
2SC3601[v,x]	S	•	-	-	-	-	-	•	-	E	200	400	2	40	10	5	25	500d	0.22	10	1.61	6.9	0.23	7.43	31
ZTX851	Z	•	-	•	-	-	z	-	-	A,F	60	130	45	100	10	2	23	410	0.18	25f	1.67	6.7	0.24	4.61	52
ZTX951	Z	-	•	•	-	-	z	-	-	A,F	60	120	74	100	10	3	4	120	0.20	10	1.24	9.0	0.21	5.35	38

Manufacturers: **A** - Analog Devices, **C** - Central, **D** - Diodes, Inc., **H** - Renesas, **I** - Infineon, **In** - Intersil, **L** - Linear Integrated Systems, **m** - many, **N** - NSC, **O** - ON Semi, **R** - Rohm, **S** - Sanyo (ON Semi), **Th** - THAT, **T** - Toshiba, **Z** - Zetex (Diodes, Inc.).

Names & Packages: **a** - available, see datasheet for part numbers, most are pinout B. **b** - for TO-92 try BC550 and BC560, both pinout C. **c** - CMPTxxx, **d** - SOT-89, **f** - FMMTxxx, **m** - MMBTxxx, **p** - PMBTxxx or PMSTxxx, **v** - ATV, **z** - FZTxxx.

Notes: ◊=√Hz. (a) listed in order of decreasing $r_{bb'}$. (b) complement available, see datasheet. (c) complements grouped together. (d) from datasheet. (e) ranking of beta constancy over collector current, ranked on a scale from 1 to 5 (best). (f) lower 1/f noise and e_n=0.21nV/√Hz at I_C=10mA. (g) discussed extensively in this chapter's section "Finding e_n from noise figure." (m) maximum. (o) collector current I_C at which the r_e noise contribution raises the irreducible $r_{bb'}$ Johnson noise voltage by 50%; corresponding values of e_n, i_n, and $R_n=e_n/i_n$ are listed. (r) ON Semi BC550C datasheet has $r_{bb'}$=170Ω plot (drops 25% from 0.1 to 10mA). (s) 2N5550 is lower-beta version. (t) typical. (u) see figure. (v) video transistor, included for its low C_{cb}. (x) discontinued, included for comparison. (y) measured on a 2SA1162-Y. (z) measured beta≈25.

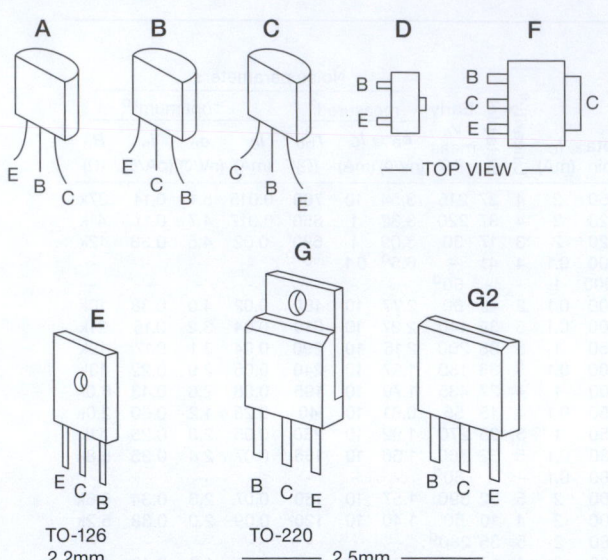

Figure 8.38. Pinouts for the BJTs in Table 8.1a.

Table 8.1b Dual Low-noise BJTs[a]

Dual part #	Manuf[b]	Single part #, etc.	npn/pnp	V_{CEO} (V)	Beta plot	Noise[x] e_n @ I_C (nV/◊) (mA)	$r_{bb'}$ (Ω)	Match[m] V_{OS} (mV)	h_{FE} (%)
BCM847	I	BC850B	N	45	37	3.3 1	650	2	10
BCM857	I	BC860B	P	50	17	3.1 1	590	2	10
LS301	L	-	N	18	-	- -	-	1	5
IT124	L	-	N	2	-	- -	-	5	10
CMKT5089M	C	2N5089	N	25	38	2.2 10	291	5	10
DMMT3904	D	2N3904	N	40	31	- -	-	1	2
DMMT3906	D	2N3906	P	40	3	- -	-	1	2
DMMT5551	D	2N5551	N	160	25	- -	-	1	2
DMMT5401	D	2N5401	P	160	2	- -	-	1	2
LM394C	N	obsolete	N	20	40	1.8 0.1	60	0.2	5
MAT12	A	-	N	40	-	0.85 1	28	0.2	5
HFA3134	In	f_T=8GHz	N	12	-	0.8 1	40	6	8
HFA3135	In	f_T=7GHz	P	12	-	1.3 1	105	6	8
SSM2212	A	-	N	40	-	0.85 1	28	0.2	5
SSM2220	A	-	P	36	-	0.7 1	25	0.2	6
THAT 300	Th	four npn	N	36	-	0.9 1	30	3	4
THAT 320	Th	four pnp	P	36	-	0.75 1	25	3	5
HN3C51F	T	2SC3324	N	120	-	0.78 10	35	-	-
HN3A51F	T	2SA1312	P	120	-	0.28 10	20	-	-

Notes: (◊) square-root Hz. (a) see Table 8.1a for single BJTs. (b) see footnotes in Table 8.1a. (m) maximum. (x) measured.

binned into beta categories. For example the BC850 has A, B, and C grades, with typical betas at 2 mA of 180, 290, and 520, respectively. It's tempting to go for the highest-grade part in your design, but often you'll find no availability at the distributors. The higher beta grades also suffer from degraded noise performance, lower V_{CEO} ratings and lower Early voltages (i.e., lower output impedances).[42]

Looking at the BC850 again, you see that it's a small-area transistor that you'd probably use at much lower currents than the 2 mA at which its beta is specified; so it's nice to see in Figure 8.39 that its beta drops by less than 10% at 1 μA, compared, say, with the 2N5962, whose beta drops by 50%. The popular *npn* 2N3904 holds steady, but the beta of its *pnp* sibling 2N3906 drops by a factor of 3 over four decades of current, a common situation with *pnp* parts. However, there's plenty of good news: the betas of the *pnp* BC860 and ZTX718 drop less than 20% at 1 μA and the beta of the lower-gain 2N5087 is flat clear down to 1 μA.

Before discussing the all-important noise measurements, let's look at the **Early voltage**, or V_A column. Look back to §§2.3.2 and *2x.8* for a discussion of the

---

[42] For example, the 2N5089 has a minimum beta of 450, compared with 300 for the '5088; its maximum voltage is reduced to 25 V from 30 V, and, most significantly, our samples showed somewhat (∼10%) higher measured e_n noise. Similarly, the D version of the ZXTN19020 has a beta of 300 (versus 200 for the C version), with 10% higher e_n. As another example, measuring the Q grade of the 2SB1197K (min beta of 120) gave us an Early voltage of V_A=110 V, whereas for the R grade (min beta of 180) the Early voltage dropped to 70 V.

Early effect. The Early voltage V_A provides an estimate of the BJT's output conductance (and output impedance): $g_{oe}=1/r_o=I_C/(V_A + V_{CE})$. We can also estimate a maximum possible single-stage gain (that is, with $R_L=\infty$), $G_{max}=g_m/g_{oe}$. By substituting $g_m=I_C/V_T$, we get (ignoring V_{CE} compared with the usually much larger V_A) $G_{max}=V_A/V_T$.

While it's possible to overcome the drawbacks of a low Early voltage, for example by adding a cascode stage (see §§2.4.5B and *3x.4*) or by adding emitter degeneration, it's frequently convenient not to have to do so. For such applications it's best to restrict your choices to BJTs with relatively high V_A. As the entries in Table 8.1a demonstrate, though, the *pnp* member of an *npn–pnp* complementary pair (they're grouped together in the table) usually suffers from dramatically lower Early voltages; e.g., the *pnp* BC860C has V_A=30 V compared with the *npn* BC850C's value of 220 V. This is unfortunate, because *pnp* transistors are useful in current mirrors for *npn* differential amplifiers, where their low Early voltage greatly reduces the stage gain. The problem can be solved in several ways, for example by adding emitter degeneration in the *pnp* mirror, as in Figure 8.35, or by using a Wilson mirror.

Looking at the table's **maximum collector voltage** (V_{CEO}) values, we see many low-noise BJTs with ratings

of 120 V or more, leading us to wonder why they have such a high voltage rating. Could it be to achieve a high Early voltage? Perhaps, but it's been observed that high-voltage "video-output" transistors often exhibit admirably low noise voltage. For example, Sanyo's 2SC3601 has a very low measured e_n of $0.22\,\text{nV}/\sqrt{\text{Hz}}$, with an $r_{bb'}$ of $1.7\,\Omega$.[43]

In general, high voltage transistors have lower betas; for example, the 300 V MPSA42 (#20 on our graph), has the lowest gain. A contender for the part with highest beta, the Linear Integrated Systems' IT124 (based on an old Intersil part), suffers from a V_{CEO} rating of just 2 V (!) in exchange for its superbeta gain. Happily, they've morphed the design into their LS301 with an 18 V rating, and it wins the beta contest (curve #44) with an astonishing $\beta=3000$ at $1\,\mu\text{A}$.[44]

A significant task is choosing the transistor's operating current. It's tempting to scan down the table's "Noise parameters" columns and pick a part near the bottom with an e_n under $0.5\,\text{nV}/\sqrt{\text{Hz}}$. But watch out for the high base-current noise that these transistors create, when operated at the relatively high collector currents needed for those low e_n values. For example, the winning ZTX851 has a very attractive $e_n=0.5\,\text{nV}/\sqrt{\text{Hz}}$ at $I_C=10\,\text{mA}$. But, with its β of 220 (see Figure 8.39), its base current of $45\,\mu\text{A}$ creates an input noise current i_n $(=\sqrt{2qI_C})$ of $3.8\,\text{pA}/\sqrt{\text{Hz}}$. That sounds small – but it creates an additional voltage noise of $0.19\,\text{nV}/\sqrt{\text{Hz}}$ across even a low $50\,\Omega$ source impedance, adding 3 dB to the transistor's input noise voltage. The problem only gets worse with higher source impedances, because the added $i_n Z_s$ noise voltage grows linearly with source impedance, compared with Johnson noise from a resistive source (which grows as the square root of R_s).[45] For example, this base current noise creates $2.3\,\text{nV}/\sqrt{\text{Hz}}$ with a $600\,\Omega$ source impedance.

We took the table's noise measurements at high-enough currents to eliminate the I_C term in the BJT noise equation $e_n=[4kT(r_{bb'}+0.5V_T/I_C)]^{\frac{1}{2}}$ (see Figure 8.10), because we wanted to expose their $r_{bb'}$ values. But usually you'll want to operate at lower currents. For example, reducing the ZTX851's current to 1 mA raises the noise voltage

by a factor of 2.3 (to $0.48\,\text{nV}/\sqrt{\text{Hz}}$), but the decrease in base current (to $5\,\mu\text{A}$) lowers the noise current by a factor of 3.2, to $1.2\,\text{pA}/\sqrt{\text{Hz}}$. At this lower current the $i_n R_s$ noise-voltage contribution drops to $0.72\,\text{nV}/\sqrt{\text{Hz}}$ for a $600\,\Omega$ source impedance, for a total amplifier noise of $0.86\,\text{nV}/\sqrt{\text{Hz}}$, considerably lower than the $2.3\,\text{nV}/\sqrt{\text{Hz}}$ at 10 mA collector current, and nicely lower than the $3.1\,\text{nV}/\sqrt{\text{Hz}}$ of $600\,\Omega$ resistive Johnson noise. Sometimes you may want to operate at a really low current, where the r_e $(=V_T/I_C)$ term dominates and the contribution of the $r_{bb'}$ term is insignificant.

The measured noise that is listed in Table 8.1a is based on a small sampling of contemporary parts. Because some of these parts are not specifically intended for low-noise applications, the manufacturer does not specify (or control) their noise properties. It is important to keep this in mind, because you may encounter batches of parts with degraded noise. Figure 8.40 shows an example: four parts from a single batch of power transistors of a type that have been favored by audio enthusiasts, but which exhibited disappointingly wide variation in low-frequency noise voltage.

These tradeoff considerations were discussed at length in §8.5, and they're illustrated in Figure 8.32, where the concept of a transistor's noise resistance $R_n=e_n/i_n$ is shown. Look back at Figure 8.33, where we compared the voltage noise of the small-die high-beta 2N5962 with that of the large-die low-$r_{bb'}$ ZTX851, at collector currents ranging from $1\,\mu\text{A}$ to 10 mA.[46] The '851 outperforms the '5962 for source impedances of less than $1\,\text{k}\Omega$ when operated at a collector current of $100\,\mu\text{A}$ or more. Conversely, the '5962 beats the '851 for source impedances above 100k for almost any value collector current.

Figure 8.41 compares the noise voltage versus source resistance for six exemplary low-noise transistors at operating currents of $10\,\mu\text{A}$ and 1 mA. The parts are ranked by increasing e_n at 1 mA collector current and with low source resistance, i.e., they are ranked by increasing $r_{bb'}$. While it's impressive to see $0.5\,\text{nV}/\sqrt{\text{Hz}}$ performance of the lowest-noise parts (like the ZTX851), it's sobering to realize that this holds only if the resistive component of the source impedance is below $10\,\Omega$. Note the generally inverse relationship of the ranking at high R_s relative to that at low R_s. Note also how at a collector current of $10\,\mu\text{A}$ all of the transistors have comparable noise voltages for source resistances less than 10k, although the high beta

---

[43] It also has a low $C_{ob}=2\,\text{pF}$, much lower than the other low-e_n BJTs listed near the bottom of Table 8.1a, demonstrating that it's possible to combine low capacitance with low $r_{bb'}$.

[44] We ordered some IT124s, but they sent us LS301 parts, tested to pass the easier IT124 specs!

[45] And, of course, you can have a high-impedance signal source that is not resistive and therefore has little or no Johnson noise.

[46] These curves were prepared with a spreadsheet, using as input our measured beta and $r_{bb'}$ data of Table 8.1a.

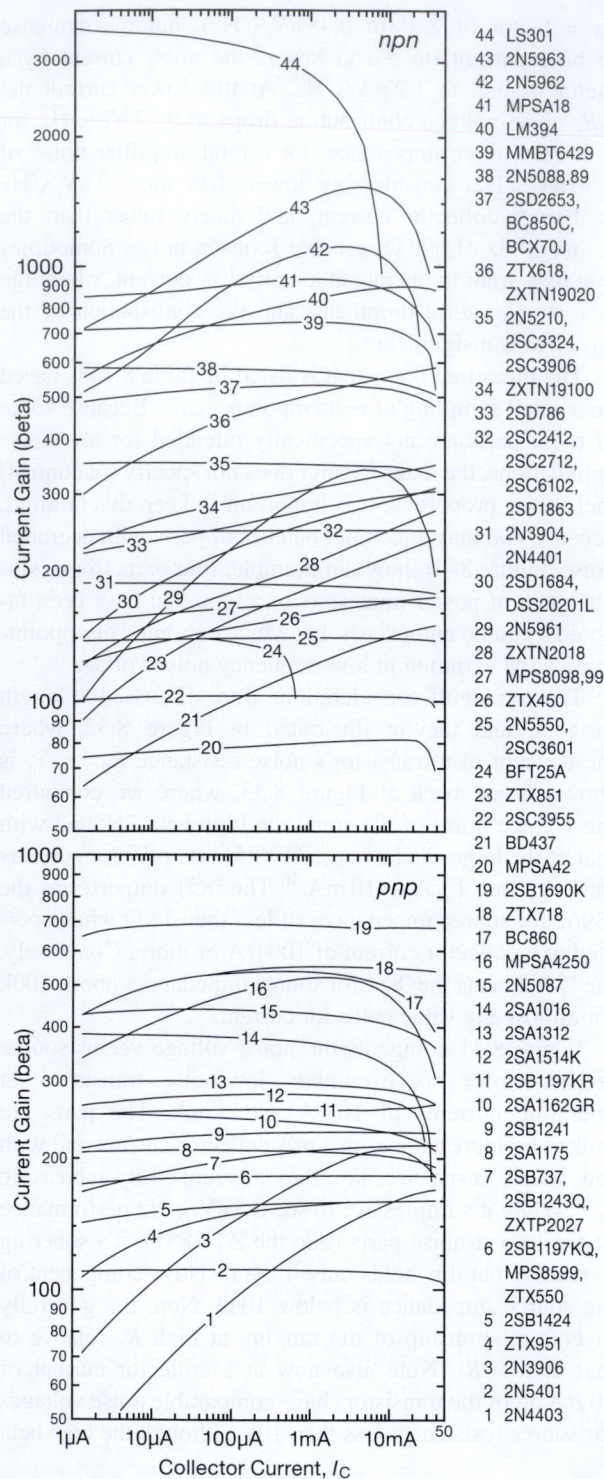

Figure 8.39. Measured beta versus collector current, at $V_{CE}=5$V, for the transistors in Table 8.1a. Use these data with $i_n=\sqrt{2qI_C/\beta}$ to find the base-current noise at the chosen collector current.

The list beside the npn plot:

44 LS301
43 2N5963
42 2N5962
41 MPSA18
40 LM394
39 MMBT6429
38 2N5088,89
37 2SD2653,
   BC850C,
   BCX70J
36 ZTX618,
   ZXTN19020
35 2N5210,
   2SC3324,
   2SC3906
34 ZXTN19100
33 2SD786
32 2SC2412,
   2SC2712,
   2SC6102,
   2SD1863
31 2N3904,
   2N4401
30 2SD1684,
   DSS20201L
29 2N5961
28 ZXTN2018
27 MPS8098,99
26 ZTX450
25 2N5550,
   2SC3601
24 BFT25A
23 ZTX851
22 2SC3955
21 BD437
20 MPSA42
19 2SB1690K
18 ZTX718
17 BC860
16 MPSA4250
15 2N5087
14 2SA1016
13 2SA1312
12 2SA1514K
11 2SB1197KR
10 2SA1162GR
9 2SB1241
8 2SA1175
7 2SB737,
  2SB1243Q,
  ZXTP2027
6 2SB1197KQ,
  MPS8599,
  ZTX550
5 2SB1424
4 ZTX951
3 2N3906
2 2N5401
1 2N4403

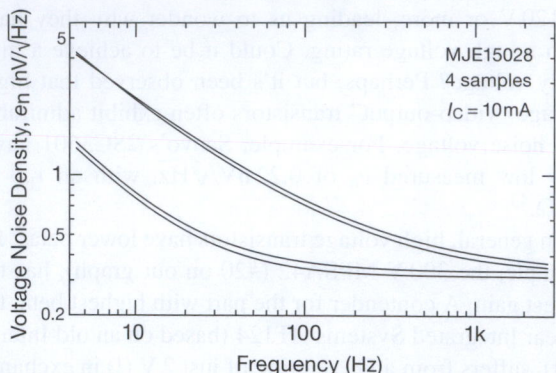

Figure 8.40. Because of their very low base resistance ($r_{bb'}$), the MJE15028-33 family of 8A bipolar power transistors have been used by audio experimenters to make low-noise audio amplifiers. That sounded good, to us... but we discovered >10 dB variations in their (unspecified) low-frequency noise, even within a single batch of parts.

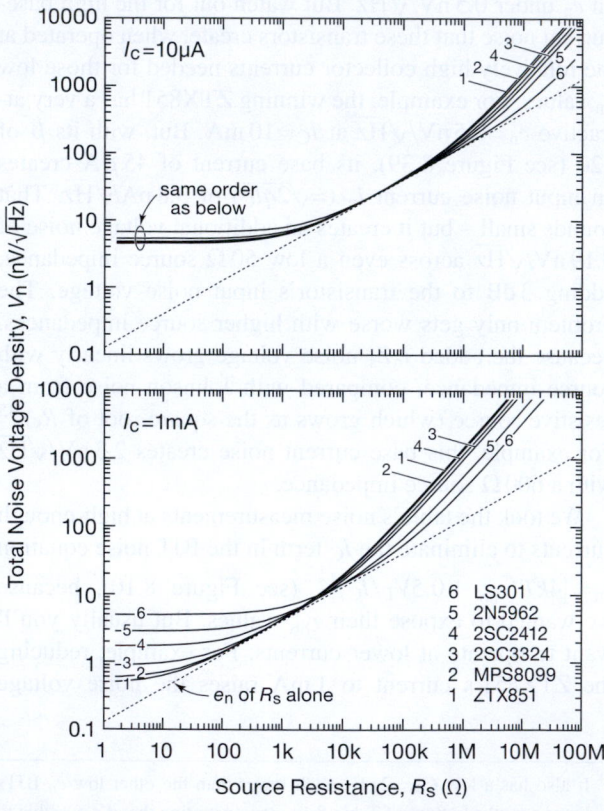

Figure 8.41. Comparing total input noise of six BJT candidates, plotted as in Figure 8.33. For low source resistance you want a transistor with low $r_{bb'}$, operated at high current; for high source resistance you want a high-beta transistor, operated at low current.

parts shine above $1\,M\Omega$.[47] These data and analysis illustrate how low-noise performance can be optimized for a given source impedance by selecting the right transistor and running it at the right current for the job.

The curves in Figures 8.33 and 8.41 include the Johnson noise from a resistive source, R_s, and would not be valid for reactive high-impedance sources such as capacitive sensors. In that case the e_n and the $i_n Z_s$ traces meet, similar to the 0.65 dB line in Figure 8.32, without being pushed up by an intervening R_s region.

The typical values of **capacitance** listed in Table 8.1a are from the datasheet specifications, or sometimes from plots, and generally at $V_{CE}=10\,V$. Most of the listed values are C_{ob}, although some are C_{cb}. These are related by $C_{ob}=C_{cb}+C_{ce}$, but the C_{ce} contribution is usually small, often well under 25%.

The four columns of optimum **noise parameters** deserve some explanation. They experiment with the concept of reducing I_C until the noise-voltage contribution from r_e (i.e., collector-current shot noise times $1/g_m$) raises the overall voltage noise by 50% over its high-current (i.e., $r_{bb'}$ Johnson noise alone) value. Put another way, the idea is to throttle down the collector current in order to reduce the transistor's noise *current*, but only to the extent that it increases the transistor's noise *voltage* by a modest amount. That collector current is listed, along with the corresponding e_n and i_n (conservatively calculated from the specified minimum beta), and the transistor's noise resistance $R_n=e_n/i_n$ at that current. If you buy into the idea of not running the BJT at an "excessive current" and are willing to compromise e_n to reduce i_n, these values give an upper limit for I_C and a lower limit for e_n.[48]

Looking at the bottom quarter of the table, now with

this "optimum" perspective, it's sobering to see the low R_n values, ranging from $30\,\Omega$ to $125\,\Omega$. And much of the next quartile has trouble getting above $400\,\Omega$. This means that for commonly encountered signal impedances above 100–$400\,\Omega$, the amplifier's noise contribution when operated at the "optimum" collector current would be dominated by the $i_n Z_s$ contribution (i.e., input-noise current flowing through the signal's source impedance). We'd be strongly motivated to drive down i_n, by finding parts with higher beta, and by lowering I_C. In this sense the "optimum" current is really a maximum practical current. Looking out of the corner into which we've painted ourselves, the alternative of a quiet JFET (instead of a BJT) looks seriously worth considering.

And that is the subject we take up in the next major section (§8.6), after taking a brief (but challenging) interlude.

8.5.9 An extreme low-noise design challenge: transformerless ribbon microphone preamp

We looked around for a challenging application, one that requires unusually low input noise voltage. A traditional application is a "moving-coil" phonograph cartridge, with its very low signal voltage ($\sim 1\,mV$ at maximum groove modulation, thus a mere $0.1\,\mu V$ for an 80 dB dynamic range). Problem is, it's *too* traditional![49]

A closely related technology is used in the *ribbon microphone*, in which an outrageously delicate strip of metal foil (typically $2\,\mu m$ thick – that's a mere 4 wavelengths of light!) is suspended in a magnet's gap, there wafted about by the ambient sound field. You can think of it as a single-turn generator, propelled by the sound's subtle vibrations. Ribbon mics were the first truly high fidelity microphones: they're the clunky big studio mics you see in old movies. These things go back to the 1930s, with models from RCA (the classic 44B) and BBC–Marconi (named, with British understatement, the "Type A"). They're clunky because they have a large magnet inside; they weigh typically 5–10 pounds. They continue to be built and used today,[50] favored by some for their "smooth sound," variously described (in the inimitable language of the audiophile) as "intimate, warm and detailed, yet never harsh."

The output signal from a ribbon mic is tiny by comparison even with the low-level moving-coil phono cartridge: at the standard sound pressure level of 1 Pa you

---

[47] You may be tempted to use a high-gain low-$r_{bb'}$ part at $10\,\mu A$, because its current noise would be no worse than that of a part with higher $r_{bb'}$; but notice the dramatically higher capacitances associated with low $r_{bb'}$, which can really degrade performance with input signals of high impedance.

[48] BJTs with high beta at low currents tend to have poor (i.e., high) $r_{bb'}$ values. Recognizing that high R_n and low $r_{bb'}$ are both desirable, we can define a figure of merit $FOM=R_n/r_{bb'}$. Going from the top to the bottom of Table 8.1a, the transistors have FOM values ranging from 20 to 70. The discontinued low-e_n favorites (2SD786 and 2SB737) stand out with FOM = 51. Others with FOMs greater than 50 are the ZTX718, DSS2020, ZXTN19020, 2SB1690, 2SD2653, 2N5088 and '5089. The MMBT6429 takes first place with FOM = 69, and the 2N5089 and BC850C take the second and third places. This would support the notion that the parts at the top of the table are the winners, rather than those at the bottom. This does reinforce the concept that you need to pick the right part for the job.

[49] Although vinyl records have a faithful base of devotees, who favor the "silky smoothness" and "warmer" sound, said to be best captured with a moving-coil cartridge's "clarity and transparency of tone."

[50] See, for example, Wes Dooley's website ribbonmics.com.

get 50–100 μV directly from the ribbon. That sounds like plenty of signal – until you realize that this reference level of one pascal corresponds to +94 dB SPL. That's *loud* – a jackhammer at 5 feet! A sensitive microphone needs to get down another 80 dB or so to capture the quietest sounds in a concert.[51] At those levels the signal directly from the ribbon is a mere 5–10 nVrms. To set the scale, an op-amp with the lowest voltage noise density (LT1028, e_n=0.85 nV/$\sqrt{\text{Hz}}$) has an input-noise voltage of about 100 nVrms, integrated over the audio band of 20 Hz–20 kHz; that's 20–25 dB larger than the quiet-level audio output directly from the ribbon.

So ribbon mics invariably include an audio step-up transformer, with a turns ratio typically of 1:30. That increases the signal amplitude by the same ratio, so that a well-designed low-noise audio amplifier does not compromise the low-level performance. But transformers can be problematical, both in terms of linearity and in maintaining a flat response over the 1000:1 frequency-range of high-quality audio. We could eliminate the transformer entirely by designing an amplifier whose input-noise voltage is at least 20 dB better than that of an LT1028, i.e., with $e_n \leq 0.1$ nV/$\sqrt{\text{Hz}}$.

A. A simple 70 *picovolt* per-root-hertz preamp test design

To get to this kind of noise level you've got to use a BJT (or, as we'll see, a bunch of them in parallel). You pay a price in terms of input current (i.e., relatively low input impedance); but we're aided here by the very low native source impedance of the ribbon, which is less than 1 Ω over the whole audio band. We seek a transistor with low base resistance ($r_{bb'}$), which we'll operate at relatively high collector current in order to drive down the r_e noise contribution (collector shot noise flowing through r_e); recall that the latter is numerically the same as the Johnson noise created by a "real" resistor of value $R = r_e/2$.

To set the scale, note that our target noise voltage of $e_n = 0.1$ nV/$\sqrt{\text{Hz}}$ corresponds to the Johnson-noise voltage created by a 0.6 Ω resistor! In other words, we need a transistor whose $r_{bb'}$ is significantly less than that, and we need to run it at a collector current of at least $\sim$50 mA (where the "r_e" noise contribution is equivalent to that of a 0.25 Ω resistor). And we need a circuit configuration that delivers on that low-noise promise.

There are several possible circuit configuration choices. We could try something like the Naim circuit of Figure 8.34

– but we'd have to reduce R_E to a fraction of an ohm, forcing Q_3 to drive a very stiff load. The same problem afflicts the "better" design of Figure 8.35. They have in common the need to keep the impedance of the feedback signal very low.

Two approaches we settled on are shown, in simplified form, in Figure 8.42. Both circuits dispense with feedback, arguing that a class-A preamp (of moderate gain, say $G \approx 100$) for microvolt-level signals is inherently linear. The single-ended common-emitter amplifier is simplest, and has a 3 dB noise advantage compared with a differential configuration. But it requires a *huge* input blocking capacitor ($\lesssim$150,000 μF to preserve the signal's very low source impedance down to a few hertz), which causes a long startup settling time; it's esthetically ugly, too. We

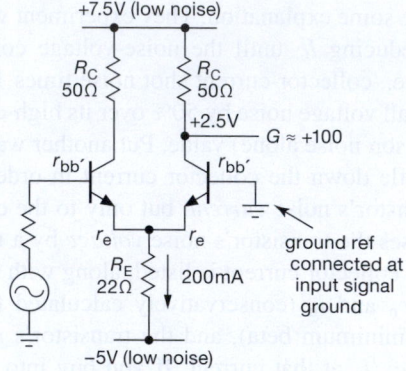

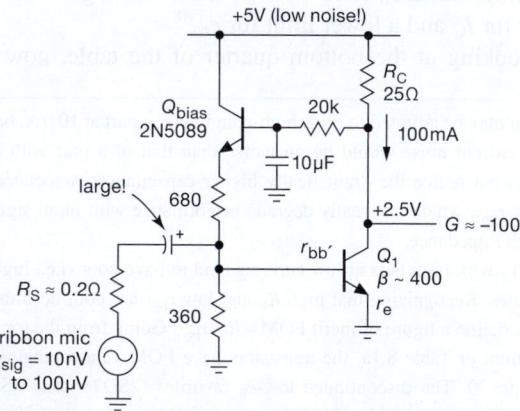

Figure 8.42. Ribbon microphone preamp configurations. Compared with the dc-coupled differential circuit, the single-ended circuit is quieter, but requires a very large blocking capacitor at the input. In either case it's necessary to use multiple paralleled transistors so that the effective base resistance $r_{bb'}$ is in the range of 0.1 Ω or less. See Figure 8.45.

---

[51] A *classical* music concert, that is. No problem of low levels in a rock concert.

did use this circuit to characterize a slew of candidate low-noise BJTs (look ahead to Figure 8.92 and associated discussion), but for our entry to the ribbon-mic preamp challenge we went with the dc-coupled differential open-loop configuration shown. No giant input capacitors here; but to deal with the doubled noise power we had to reduce $r_{bb'}$ and r_e by a factor of 2 (i.e., we doubled the number of paralleled transistors used for each side of the input pair).

B. Choosing a low-noise BJT

For a circuit like this it's essential to use a transistor (or several in parallel) that exhibits very low input-noise voltage. That requires a base resistance $r_{bb'}$ down in the range of just a few ohms. Sadly, $r_{bb'}$ is rarely specified; and the excellent low-$r_{bb'}$ transistors of yesteryear have largely gone extinct. A case in point is the fine Toyo–Rohm 2SD786, with a specified typical $r_{bb'}$ of $4\,\Omega$ (it's *npn*; the 2SB737 *pnp* complement is better still, with a specified typical $r_{bb'}$ of $2\,\Omega$). Actual samples of each performed even better, with measured $r_{bb'}$ values around $2.3\,\Omega$ and $1.2\,\Omega$, respectively.

You can't get these anymore (and we won't part with any of our precious and dwindling "lifetime" stock! Even if you beg us.). What about the "low-noise" small-signal BJTs you *can* get, such as the 2SC3324? The good news is that they *do* specify their noise performance, as we saw in §8.4. The bad news is that they don't promise a whole lot ("NF=0.2 dB typ, 3 dB max"); and these small-geometry transistors are generally optimized for operation at low current, where the r_e noise contribution is so large that there's no need to keep $r_{bb'}$ small. For example, the 2SC3324 achieves its minimum noise figure around $I_C = 30\,\mu$A; at that current r_e is $830\,\Omega$, and thus its $r_{bb'}$ noise contribution is insignificant as long as $r_{bb'}$ is kept below, say, $200\,\Omega$. Our samples measured $r_{bb'} \approx 40\,\Omega$ – plenty good enough for these transistors' intended low-current application, but not helpful here.

We embarked on a search for low-$r_{bb'}$ transistors and found that the situation is not bleak. It's just unspecified. In the circuit design community you'll find mention of good noise properties with large-geometry transistors (i.e., power transistors).[52] We measured the noise voltage (details later) of a few dozen promising candidates (most of which are listed in Table 8.1a on page 501), and found that, indeed, some of the large geometry ("power") transistors have $r_{bb'}$ values down below $10\,\Omega$, and deliver quite respectable input-referred e_n when operated at currents of

[52] Check out the website of the remarkable Uwe Beis (www.beis.de), who has used the MJE13007 (8 A 400 V); or Ovidiu's low-noise designs at www.synaesthesia.ca, with reported $r_{bb'}$ values below $2\,\Omega$ for the 2SC3601 and 2SC2547.

$10\,$mA or so (where the r_e contribution is equivalent to Johnson noise from a $1.25\,\Omega$ resistor). Figure 8.17 (back on page 486) shows that the ZTX851 (a 5 A, 60 V transistor, lacking any official low-noise endorsement) performs on a par with the legendary 2SD786; you can see, also, that the larger $r_{bb'}$ of the smaller geometry "low noise" parts like the 2N5089 and 2SC3324 makes them non-competitive in this high-I_C/low-e_n arena.

As we remarked earlier (§8.3.1), where we showed a plot (Figure 8.12) of e_n versus collector current for six candidate low-noise transistors, there's not much benefit in having a low $r_{bb'}$ if you are operating at low current (where the effect of r_e dominates); but it really matters at higher currents, which is where you have to operate if you want to achieve e_n values southward of $1\,$nV/$\sqrt{\text{Hz}}$.

To emphasize this latter point, we show in Figure 8.43 a plot of the voltage noise spectra of the ZTX851 over a large range of collector currents (seen earlier in Figure 8.18, and here extended down to 4 Hz). The lesson here is that you have to run plenty of current to get really low e_n. Don't overdo it, though: as seen in the earlier Figure 8.18, at still higher currents the low-frequency noise rises, owing to a rapidly increasing base-current flicker-noise contribution. For this transistor the "sweet spot" is at a collector current of $\sim$5 mA.

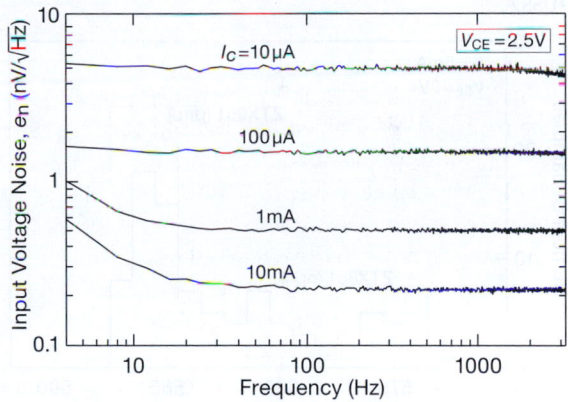

Figure 8.43. To reap the benefits of a transistor with desirable low base spreading resistance $r_{bb'}$ (thus low Johnson noise), you must operate at a high enough collector current to make the r_e noise contribution comparably small, as shown in these measured noise spectra of a ZTX851 *npn* transistor.

C. Beating the $0.1\,$nV/$\sqrt{\text{Hz}}$ goal

Among our candidates, we found that the quietest *pnp* transistors were somewhat better than their *npn* cousins. For example, at $I_C = 10\,$mA a typical (average of six samples) *pnp* ZTX951 measured $e_n = 0.20\,$nV/$\sqrt{\text{Hz}}$, compared with $0.21\,$nV/$\sqrt{\text{Hz}}$ for the ZTX851 *npn* complement (corresponding to $r_{bb'}$ values of $1.2\,\Omega$ and $1.5\,\Omega$, respectively). We're not sure why this is so, although as we remarked earlier the noise specifications of the legendary *npn* 2SD786 (with *pnp* 2SB737 complement) indicate the same trend.[53]

We used the differential circuit variant of Figure 8.42, with a parallel array of transistors for each of the input pair, followed with a $G=30$ quiet op-amp stage (made with the tried-and-tested LT1128, whose input noise voltage is $\lesssim 1\,\mathrm{nV}/\sqrt{\mathrm{Hz}}$ all the way down to 10 Hz). Ordinarily you'd include a small "emitter-ballasting" resistor in series with each emitter of the parallel array, chosen to drop $\sim 50\,\mathrm{mV}$ or so in order to equalize the currents and prevent current hogging; at 10 mA per transistor that would amount to $5\,\Omega$, adding an unacceptable amount of noise voltage. It's often said that transistors from a single production batch are inherently matched in their base-emitter voltage drops; out of curiosity we measured V_{BE} for a batch of 100 each of ZTX851 and ZTX951, with the results shown in the histogram of Figure 8.44. The matching is quite good, enough so that no emitter ballasting resistors are needed.[54] The same is *not* true for JFETs, by the way; see Figure 3.17's histogram of three hundred representative *n*-channel parts, painstakingly made[55] from three batches of 100 JFETs, in the popular 2N5457–59 series of *n*-channel parts (graded by I_{DSS}).

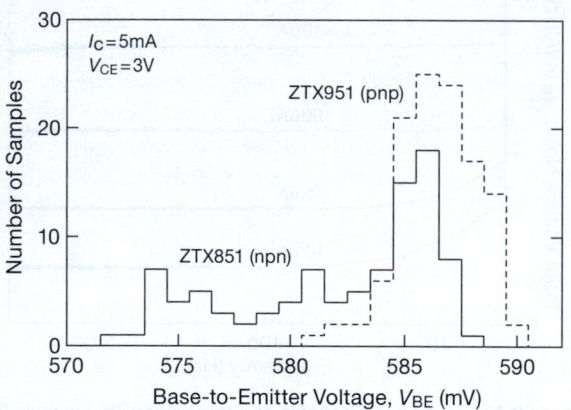

Figure 8.44. Distribution of measured V_{BE} in a batch of 100 each of *npn* and *pnp* parts.

For a circuit like this, with resistive collector loads and

[53] This may be related to the larger die size needed for a similarly-performing *pnp* transistor.

[54] We found similarly good matching for a batch of BD437s (an inexpensive and widely available BJT for "medium-power linear and switching applications").

[55] *Literally!* One's fingertips were rendered quite tender, after lead-forming and inserting three hundred transistors.

emitter pulldowns, the dc supply rails must be quiet, because noise on the rails appears unattenuated at the output. Here we used a so-called *capacitance multiplier* to eliminate supply noise; it's a useful subcircuit – see the description in §8.15.

The final results?...(drumroll)...here are measured noise voltages for several transistor configurations (with some graphs in Figure 8.45):

transistor	qty	I_C (mA)	e_n (nV/$\sqrt{\mathrm{Hz}}$) @1 kHz	@100 Hz
ZTX951	2×16	2×100	0.085	0.10
	2×32	2×200	0.070	0.09
BD437	2×24	2×100	0.095	0.17
	2×24	2×200	0.080	0.13
2SC3601E	2×12	2×200	0.093	0.18

This is impressively low input noise voltage – you rarely see figures below $0.5\,\mathrm{nV}/\sqrt{\mathrm{Hz}}$ – but it's only fair to point out the unusual nature of this challenge, with its ultralow source impedance of less than an ohm, thus tolerant of the low input impedance (and relatively high input current) of a grounded-emitter amplifier stage running at a collector current of 100 mA. Getting to much lower input currents, though, is best accomplished with JFETs, as we'll see presently.

D. Measuring BJT noise

How did we make these transistor noise measurements? There are fancy commercial instruments you can buy to measure noise parameters (e_n, i_n, NF) of discrete transistors (BJT or FET); these are generally aimed at RF and microwave characterization (although the discontinued

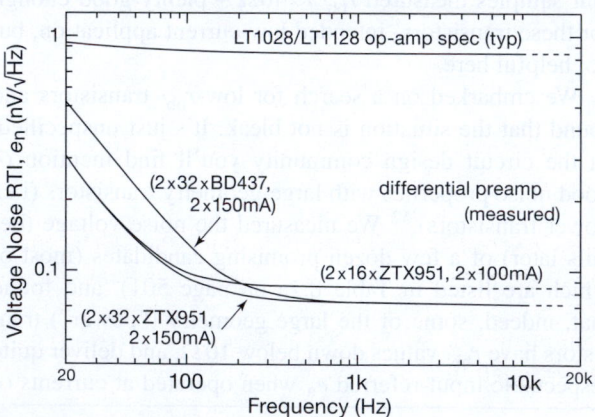

Figure 8.45. Measured noise voltage spectra for the amplifier of Figure 8.42, with three choices of input transistor and operating current. The preamp with 64 ZTX951 transistors is the winner.

HP/Agilent 4470A measured e_n, i_n, and NF at eleven spot frequencies, two per decade from 10 Hz to 1 MHz). We took a more modest route, building our own simple circuit. It's basically the device under test (DUT) configured as a grounded emitter amplifier stage, with settable collector current and collector-to-emitter voltage, and provision for determining its voltage gain. Knowing the latter, you measure the output noise voltage spectrum, first with the input bypassed (to get e_n), then with a series resistor at the input (to get i_n). It's described in detail in §8.12.2.

8.6 Low-noise design with JFETS

Picking up where we left off (at §8.5.6), for high source impedances it's transistor *current* noise that dominates, thus favoring FETs, usually in the form of a JFET. Compared with BJTs, JFETs generally have greater voltage noise, but very much lower gate current (and current noise), and so they're the universal choice for low-noise high-impedance amplifiers. In this context it is sometimes useful to think of Johnson-noise voltage, in combination with the signal's source resistance, as a *current* noise $i_n = v_n / R_s$ (as plotted earlier in Figure 8.1). This lets you compare source-noise contributions with amplifier current noise.

We can use the same amplifier noise model for FETs, namely a series noise-voltage source and a parallel noise-current source. You can analyze the noise performance with exactly the same methods used for bipolar transistors. For example, see the graphs in Figure 8.53 in §8.7 where we referee the "bipolar/FET shootout."

8.6.1 Voltage noise of JFETs

For JFETs the voltage noise e_n is essentially the Johnson noise of the channel resistance, given approximately by

$$e_n^2 = 4kT \left(\frac{2}{3} \frac{1}{g_m} \right) \quad \text{V}^2/\text{Hz}, \qquad (8.34)$$

where the inverse transconductance term takes the place of resistance in the Johnson-noise formula; in other words, the noise voltage is the same as the Johnson noise produced by a resistor of value $R = \frac{2}{3} \frac{1}{g_m}$ (analogous to the input noise voltage of a BJT, whose e_n is equivalent to the Johnson noise of a resistor of value $R = \frac{1}{2} \frac{1}{g_m}$). You can see from Figure 8.46 (measured noise voltage versus transconductance[56]) that this sets a reliable lower bound to the actual noise voltage, which can be somewhat greater in practice.

---

[56] More than 100 measurements on some 65 transistors chosen from 50 different types, and at various currents ranging from 75 µA to 50 mA, creating many different g_m test values. We followed the declining e_n

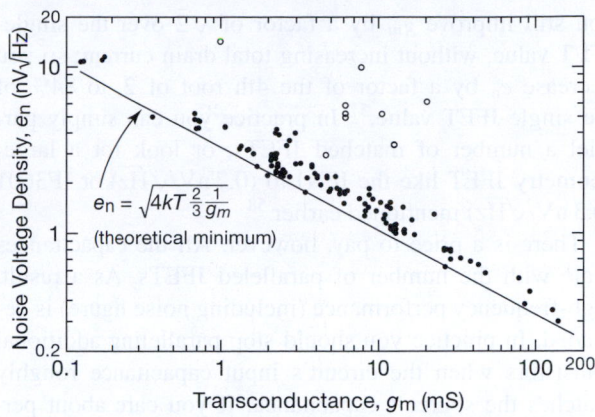

Figure 8.46. Testing the JFET noise formula: scatterplot of measured noise voltage density (well above the $1/f$ corner frequency) versus measured transconductance for a selection of 50 different JFET types, operated at various drain currents. Open circles indicate JFETs whose noise had not flattened to the white-noise floor at our maximum frequency.

In the figure: $e_n = \sqrt{4kT \frac{2}{3} \frac{1}{g_m}}$ (theoretical minimum)

Because the transconductance rises with increasing drain current (as $\sqrt{I_D}$, see §3.3.3, Figure 3.54, and Table 3.7), it is generally best to operate FETs at high drain current for lowest voltage noise. However, because the e_n is Johnson noise, which goes only as $1/\sqrt{g_m}$, and that in turn goes as $\sqrt{I_D}$, e_n is finally inversely proportional to the fourth root of I_D. With such a mild dependence of e_n on I_D it doesn't pay to run at a drain current so high that other properties of the amplifier are degraded. In particular, a FET running at high current gets hot, which (a) decreases g_m, (b) increases offset voltage drift and CMRR, and (c) raises gate leakage dramatically; the latter effect can actually *increase* voltage noise, because there is some contribution to e_n from flicker noise associated with the gate leakage current.

There is another way to increase g_m and therefore decrease JFET voltage noise: by paralleling a pair of JFETs, you get twice the g_m, but of course this is at twice the I_D. But now if you run the combination at the previous I_D,

---

with rising frequency until we found the high-frequency white-noise floor. This yielded a wide range of conditions under which to test the formula. No parts were better, but many were 10% to 30% worse. Others were much worse, up to ×2. For some the noise continued to fall with increasing frequency, so we couldn't find the noise floor; these are indicated with open circles on the graph. The takeaway lesson: get g_m as high as possible for low e_n. Remember, paralleling or using larger-die JFETs is a good way to increase g_m. But watch out for high capacitance and leakage. And keep in mind that $1/f$ noise and low-frequency noise plateaus can dominate your design. Examine our plots and the data for JFETs in Table 8.2 on page 516. For critical projects, take your own measurements.

you still improve g_m by a factor of $\sqrt{2}$ over the single-JFET value, without increasing total drain current; so you decrease e_n by a factor of the 4th root of 2, to 84% of the single-JFET value.[57] In practice you can simply parallel a number of matched JFETs, or look for a large-geometry JFET like the IFN146 (0.7 nV/$\sqrt{\text{Hz}}$) or IF3601 (0.3 nV/$\sqrt{\text{Hz}}$) mentioned earlier.[58]

There is a price to pay, however. All the capacitances scale with the number of paralleled JFETs. As a result, high-frequency performance (including noise figure) is degraded. In practice you should stop paralleling additional transistors when the circuit's input capacitance roughly matches the source's capacitance. If you care about performance at high frequencies, choose JFETs with high g_m and low C_{rss}; you might consider the ratios g_m/C_{rss} and g_m/C_{iss} to be high-frequency figures of merit (recall that $f_T = g_m/2\pi C$, where C is the input capacitance or Miller capacitance, depending on the circuit configuration). Note that circuit configurations can also play an important role; e.g., the cascode circuit can be used to eliminate the Miller effect (gain multiplication) on C_{rss}.

In Figure 8.52 we've plotted an extensive collection of measured noise voltage for many of the JFETs in Table 8.2. See also Table 3.7, with additional JFET characteristics (not related to noise), and the "Measured transconductance" graph in §3x.2 for plots of JFET transconductance versus drain current.

A. JFET 1/*f* voltage noise
Like their BJT brothers, most JFETs suffer also from rising noise voltage density at low frequencies, as seen in the measured data of Figure 8.47. In some cases (such as the 2SK170B) the $1/f$ noise is quite subdued and can be characterized by a single noise-corner frequency f_c, similar to BJTs. In such cases the noise corner can be estimated with eq'n 8.27, and, once f_c is known, the noise density can be calculated with eq'n 8.28. The plot shows 1 kHz noise voltage density ranging from 1 nV/$\sqrt{\text{Hz}}$ for the

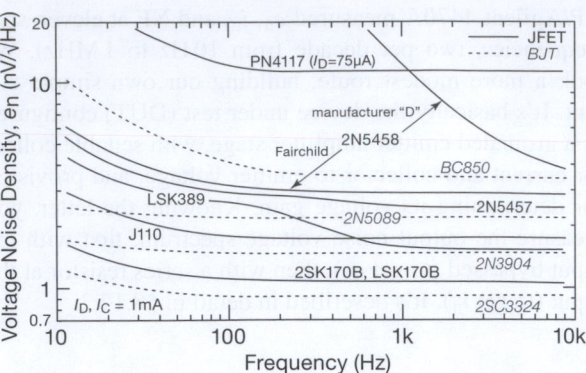

Figure 8.47. Measured voltage noise density versus frequency for several BJTs and JFETs, illustrating the $1/f$-like rise at low frequencies. See Figure 8.52 for an extensive set of JFET noise plots.

2SK170B, to 2.8 nV/$\sqrt{\text{Hz}}$ for the low-capacitance 2N5457, to 11 nV/$\sqrt{\text{Hz}}$ for the small-die PN4117 (for which the drain current is I_{DSS}, here 75 μA).

But some JFETs suffer from an elevated low-frequency noise plateau region,[59] like that of the LSK389. This part has an attractively low-noise density of 1.8 nV/$\sqrt{\text{Hz}}$ at 1 kHz and above, but it rises, in a snake-like curve, to twice that value in the 100 Hz region. This is not a terribly serious drawback for this otherwise excellent part, and it still enjoys low wideband noise levels of about 70 nVrms and 180 nVrms for 1 kHz and 10 kHz total bandwidths, respectively.

Sometimes this effect can get out of hand. For example, compare Fairchild's quiet version of the 2N5486, with $e_n = 3$ nV/$\sqrt{\text{Hz}}$, with the noisy sample we measured from "manufacturer D": the latter's 4.5 nV/$\sqrt{\text{Hz}}$ at 10 kHz climbs steeply at lower frequencies, going off-scale at 700 Hz. It climbs to 50 nV/$\sqrt{\text{Hz}}$ at 100 Hz, some 15$\times$ that of the quiet Fairchild JFET with the same part number. As a result the noisy part has about 1 μV of wideband noise (to 1 kHz). For a stunning comparison of noise variation

---

[57] The 4th root noise-to-current relationship holds for JFETs operated in their high current density region, say above $I_{DSS}/100$ (see the "Measured transconductance" plots of g_m versus I_D in §3x.2), but at low current densities (where $g_m \propto I_D$) the decrease in noise will approach a square-root relationship, just as with BJTs. For example, the measured noise in an IF3601 increased by a factor of 2.8 (nearly $\sqrt{10}$) when the drain current was reduced from 1 mA to 0.1 mA.

[58] Paralleling JFETs lowers the current density, pushing the g_m versus I_D plot up toward its maximum, where it would match that of a BJT, see Figure *3x.10*, which shows a given JFET's potential for improvement. Some parts (such as the IF3601) are already close to this limit, so that little further improvement can be expected.

[59] Called "roller-coaster" noise by Motchenbacher and Connelly. The mechanism for the excess $1/f$ noise was identified in the early 1960s as a bulk rather than surface effect, involving random capture or emission of electrons trapped in the depletion region. See, for example, P. O. Lauritzen, "Low-frequency generation noise in junction field effect transistors," *Solid-State Electron.*, **8**, 1, 41–58 (1965), or C. T. Sah, "Theory of low-frequency generation noise in junction-gate field-effect transistors," *Proc. IEEE*, **52**, 7, 795–814 (1994). In the time domain this noise can take the form either of discrete steps between two or more voltage levels ("popcorn" or "telegraph" noise) or random sudden voltage excursions (burst noise); see Figures 8.5 and 8.6 for examples of popcorn noise in a vintage 741 BJT op-amp.

in 2N5486s made by multiple manufacturers, see the measured spectra in Figure 8.51. The noisiest part in that group has a 210 nV/√Hz plateau, resulting in about 2.5 µVrms low frequency integrated voltage noise. In its defense, the 2N5486 "RF Amplifier" datasheet lists noise-figure specifications only at 100 MHz and 400 MHz. Note also that this low-frequency-noise excess does not affect operation at much higher frequencies (where "$e_n \cdot C_{in}$" noise, discussed in §8.11, rears its ugly head).

Although it's common to talk about $1/f$ noise and a corresponding corner frequency f_c, the reality is that many JFETs do not conform to such a neat model. In their datasheets, manufacturers deal with the issue of excess low-frequency noise in several ways. First, they may avoid a noise specification altogether. Second, they may give the e_n noise-density spec at relatively high frequencies – 10 kHz, 100 kHz, or even higher. Third, they may provide instead an rms noise voltage specification at some specified bandwidth. Fourth, they may give an intentionally high (conservative) specification, such as 115 nV/√Hz max, or perhaps a 3 dB noise figure for $R_s = 1$ MΩ (whose source noise-voltage density is a high 126 nV/√Hz). Fifth, they may give an e_n spec at say 100 Hz (where it's useful), but give a *typical* value only. Finally, for parts intended for high-frequency amplification, they may give their noise spec at *radio* frequencies – 100 MHz or higher.

8.6.2 Current noise of JFETs

At low frequencies the current noise i_n is extremely small, arising from the shot noise in the gate leakage current (Figure 8.48):[60]

$$i_n = \sqrt{2qI_G} = (3.2 \times 10^{-19} I_G)^{\frac{1}{2}} \quad \text{A/Hz}^{\frac{1}{2}}. \quad (8.35)$$

The noise current rises with increasing temperature, as the gate leakage current rises. Watch out for the rapidly increasing gate leakage in n-channel JFETs that occurs for operation at high V_{DG} and/or high I_D (see §3.2.8). A dose of reality: it's hard to estimate with any accuracy the level of current noise in JFETs, because the gate leakage is poorly specified. Too often you see unrealistically high worst-case specifications, for example $I_G = 1$ nA (max), whereas at normal ambient temperatures it's more typically down in the 1–10 pA range. With such low bias currents the input noise-current density is very small, e.g., 1.8 fA/√Hz for a gate leakage current of 10 pA. That noise current would generate an e_n just 1.8 nV/√Hz through a signal source

---

[60] In addition, there is a flicker ($1/f$) current-noise component in some FETs.

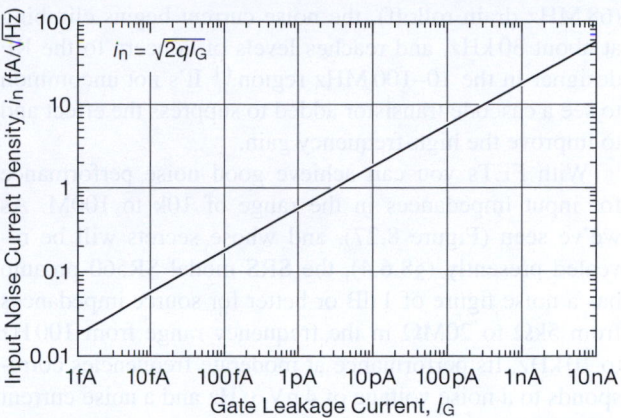

Figure 8.48. Input shot-noise current density versus FET gate leakage current.

impedance of 1 MΩ (which, by the way, would itself have a Johnson-noise voltage of 128 nV/√Hz; the source resistance would have to reach 5 GΩ before the noise voltage generated by the JFET's noise current would match the Johnson noise of the source).

At moderate to high frequencies there are additional noise and resistive terms.

(a) If there's a summing junction (for example in a transimpedance amplifier) then the input noise voltage e_n driving the input capacitance C_{in} generates a noise current of $i_n = e_n \omega C_{in}$, see §8.11 for details.

(b) In a common-source amplifier without cascode, there is an effective resistive part of the (normally capacitive) input impedance seen looking into the gate. This comes from the effect of feedback capacitance (Miller effect) when there is a phase shift at the output caused by load capacitance; i.e., the part of the output signal that is shifted 90° couples through the feedback capacitance C_{rss} to produce an effective resistance at the input, given by

$$R = \frac{1 + \omega C_L R_L}{\omega^2 g_m C_{rss} C_L R_L^2} \quad \text{ohms} \quad (8.36)$$

Both effects rise linearly with frequency above a breakpoint, and both have similar breakpoint frequencies, typically in the range of 2–100 kHz for low-leakage JFETs.

As an example, the 2N5486 n-channel JFET has a noise current of ~5 fA/√Hz and a noise voltage e_n of 2.5 nV/√Hz, both at I_{DSS} and 10 kHz. These figures are roughly 200 times better in i_n and 2 times worse in e_n than the corresponding figures for the 2N5087 BJT used in connection with the noise figure plots in §8.5.1 (Figure 8.29). If we assume a 470 Ω load with 5 pF of shunt capacitance

(68 MHz drain rolloff), the noise current begins climbing at about 30 kHz, and reaches levels of concern to the RF designer in the 10–100 MHz region.[61] It's not uncommon to see a cascode transistor added to suppress the effect and to improve the high-frequency gain.

With FETs you can achieve good noise performance for input impedances in the range of 10k to 100M. As we've seen (Figure 8.27), and whose secrets will be revealed presently (§8.6.4), the SRS model SR560 preamp has a noise figure of 1 dB or better for source impedances from 5kΩ to 20MΩ in the frequency range from 100 Hz to 10 kHz. Its performance at moderate frequencies corresponds to a noise voltage of $4\,\mathrm{nV}/\sqrt{\mathrm{Hz}}$ and a noise current of $0.013\,\mathrm{pA}/\sqrt{\mathrm{Hz}}$.

8.6.3 Design example: low-noise wideband JFET "hybrid" amplifiers

You can improve on the noise performance of commercial JFET preamps (and op-amps) by combining the best discrete JFETs with an op-amp in a "hybrid" design. Several nice circuit examples are shown in progressive examples in other chapters: (a) an ac-coupled low-noise wideband amplifier (to ~20 MHz) with $e_n \approx 1\,\mathrm{nV}/\sqrt{\mathrm{Hz}}$ in Figure 3.34; (b) an analogous dc-coupled amplifier with $e_n \approx 2\,\mathrm{nV}/\sqrt{\mathrm{Hz}}$ in Figure 3.37; and (c) a dc-coupled low-noise wideband differential amplifier with $e_n \approx 2\,\mathrm{nV}/\sqrt{\mathrm{Hz}}$ in §3x.4.

The reader is encouraged to spend a few minutes (or even a half hour!) reviewing those examples, which illustrate important techniques in low-noise amplifier design with discrete JFETs (and also when combined with an op-amp second stage). Here we complete the progression of such hybrid designs with a dc-coupled *instrumentation amplifier* (INA; see §5.15) of comparably low-noise and wide bandwidth.

The circuit in Figure 8.49 is an evolution of the single-ended dc-coupled hybrid amplifier of Figure 3.37, two of which here form the differential amplifier input of the classic three-amplifier INA. The voltage gain (difference input to single-ended output) is $G=100$, set by the ratio $1+2R_f/R_g$. The ±2% trim of R_{f2} is used to maximize the common-mode rejection ratio, a desirable feature that is responsible for much of the INA's fame. JFETs need help to achieve low offset voltage (Q_1 specifies 20 mV, worst case), so we've added an offset balance in $Q_{1a,b}$'s drain.

As for noise, the LSK389B[62] specifies a typical e_n of $0.9\,\mathrm{nV}/\sqrt{\mathrm{Hz}}$ at 1 kHz and $I_D=2\,\mathrm{mA}$. At our 5 mA drain current we can expect slightly better; but the noise of each differential pair (e.g., $Q_{1a,b}$) is larger by a factor of $\sqrt{2}$, with another factor of $\sqrt{2}$ to account for the combined noise of the upper and lower differential pairs. As with the single-ended amplifier of Figure 3.37, the op-amp's noise ($e_n \sim 12\,\mathrm{nV}/\sqrt{\mathrm{Hz}}$) is reduced by a factor of the JFET pair's gain (~20) when referred to the input. The overall noise voltage of this instrumentation amplifier is thus $\sim 2\,\mathrm{nV}/\sqrt{\mathrm{Hz}}$. Combined with its bandwidth (~20 MHz, for signal source impedance $\leq 1\,\mathrm{k\Omega}$), this is better performance than can be had in available instrumentation amplifiers.

Can we do better? The easiest way to beat down the JFET voltage noise is to add a second pair of JFETs at each input, paralleling the existing gate and drain terminals, but with each pair pulled down with its own 10 mA current sink (and with the drain resistors now 500 Ω). That reduces the input voltage noise by $\sqrt{2}$, and there is no need to worry about V_{GS} matching between the separate JFET matched pairs (figure out why). Finally, if you want to minimize input current, add a pair of cascode transistors to keep the V_{DS} of Q_1 below 5 V; see §3.2.8 and Figures 5.61 and 8.67.

8.6.4 Designs by the masters: SR560 low-noise preamplifier

It's always instructive to look at the innards of a good commercial product. Stanford Research Systems (SRS) has a nice line of scientific instruments, among which you'll find their SR560 low-noise preamplifier, in continuous production since 1989. It uses a differential JFET front-end for high input impedance (100 MΩ∥25 pF), and it has a nice set of panel controls that let you select things like the voltage gain (calibrated ×1 to ×50,000, in a 1–2–5 sequence or uncalibrated over the same range), frequency response (via switchable lowpass and highpass sections), ac or dc coupling, phase inversion, and so on. It's a good performer – noise-voltage density better than $4\,\mathrm{nV}/\sqrt{\mathrm{Hz}}$, response flat to ±0.5 dB to 1 MHz, 0.01% typical distortion, and output swing to 10 Vpp.

We showed this puppy's specified noise-figure contours in Figure 8.27. Let's look under the hood to see how they did it. Figure 8.50 shows the front-end amplifier circuit, in full detail (Figure 8.50A) and in simplified forms (circuit B, block diagram C).

---

[61] We made bench measurements that confirmed (within a factor of 2) the resistive input component predicted by eq'n 8.36. In a JFET common-source amplifier without cascode this effect could greatly lower the Q of an RF-tuned circuit.

[62] An available substitute for the excellent but discontinued 2SK389 from Toshiba.

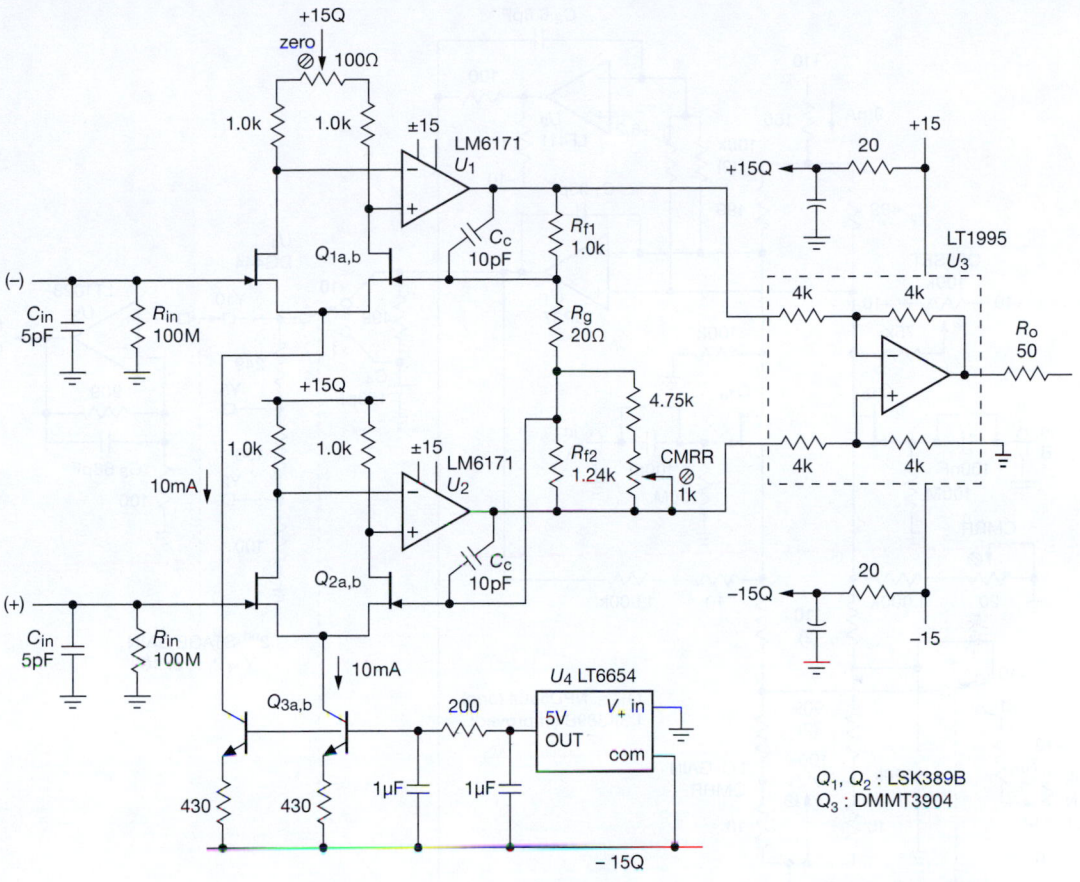

Figure 8.49. Hybrid JFET-input low-noise wideband instrumentation amplifier: $G=100$, $e_n \approx 2\,\mathrm{nV}/\sqrt{\mathrm{Hz}}$, BW$\approx 20\,\mathrm{MHz}$.

Overall topology   Matched low-noise JFET pair Q_{1ab}, running at 4.6 mA (close to its guaranteed minimum I_{DSS}, see below), forms the fully differential first stage, in a hybrid configuration in which the JFET differential output drives an op-amp (analogous to Figure 3.37), but here configured with feedback to Q_{1a}'s source terminal so that both JFET gates are available as external inputs. The overall configuration is therefore a differential "current-feedback amplifier" (CFB, see §4x.6), analogous to the single-ended hybrid circuit of Figure 3.34, and instrumentation amplifier configuration "C" in Figure 5.88. The first-stage native voltage gain is selectable as ×10 or ×2; for the latter a 499 Ω series resistor is switched in, and thus unity net gain is seen at the second stage's input. The second stage is an LT1028 low-noise op-amp ($e_n \sim 1\,\mathrm{nV}/\sqrt{\mathrm{Hz}}$ at 10 Hz), $G=10$, with $\lesssim 100\,\Omega$ source impedance at both inputs to preserve low noise voltage. With the input divider the overall second-stage gain is ×2, ×5, or ×10.

JFET bias   The designer nicely finessed the business of uncertain I_{DSS} by sensing the average drain voltage, comparing it with a +6.2 V reference, and closing the loop via integrator U_2. This is a reliable way to deal with loose I_{DSS} specs, for example those of the original NPD5564 dual JFET ($I_{DSS}=5\,\mathrm{mA}$ min, 30 mA max). That JFET is no longer available from the original manufacturer,[63] but fortunately there's a superior replacement (LSK389B), with pleasantly tighter I_{DSS} specs (6 mA min, 12 mA max), and, as a bonus, lower noise voltage (at $I_D=2\,\mathrm{mA}$ the NPD5564's typical noise-voltage densities at 10 Hz and at 1 kHz are $12\,\mathrm{nV}/\sqrt{\mathrm{Hz}}$ and $3.8\,\mathrm{nV}/\sqrt{\mathrm{Hz}}$, respectively; the LSK389B's corresponding figures are $2.5\,\mathrm{nV}/\sqrt{\mathrm{Hz}}$ and $0.9\,\mathrm{nV}/\sqrt{\mathrm{Hz}}$).

---

[63] Currently InterFET (direct sales only) is offering their similarly-spec'd IFN5564

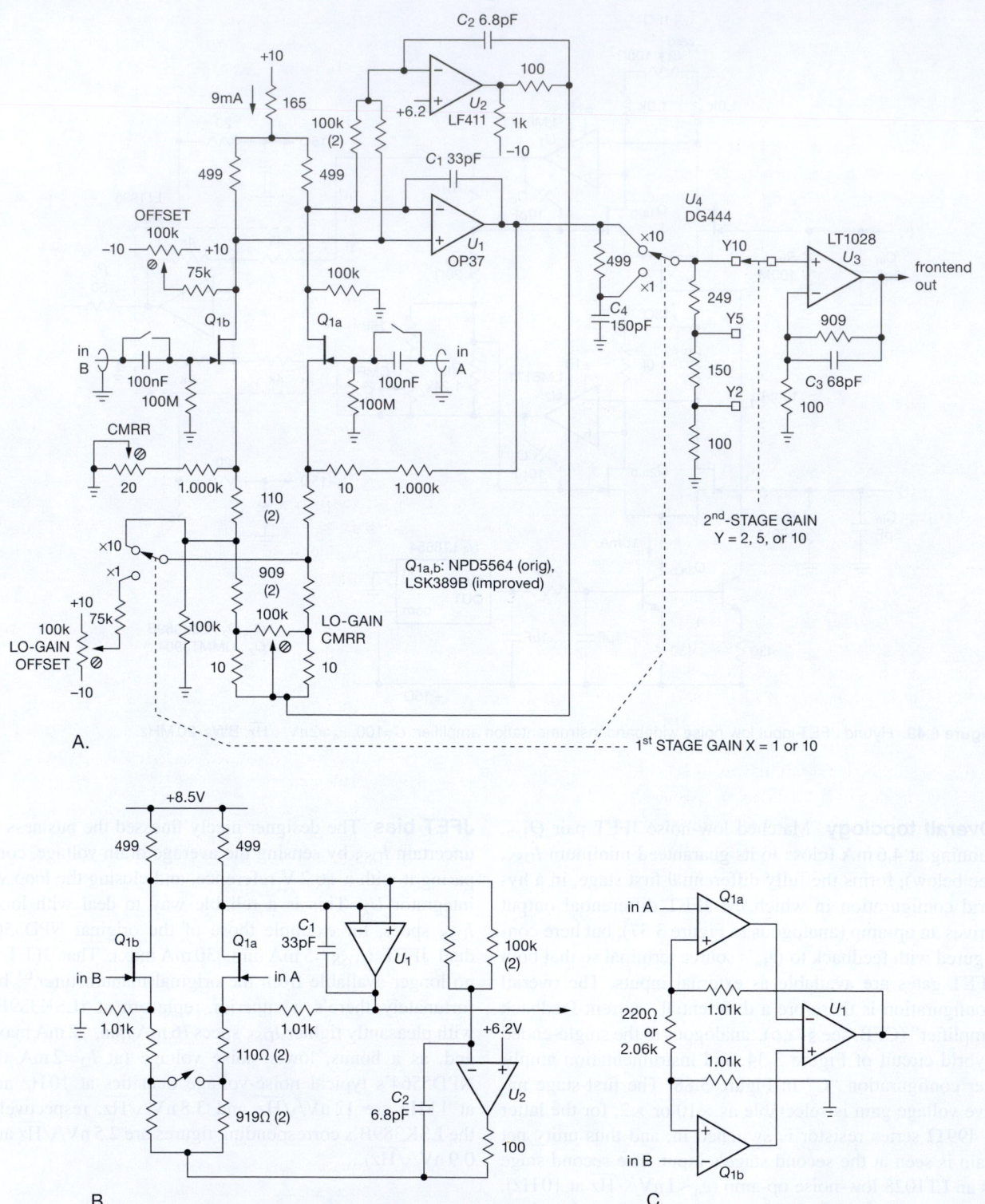

Figure 8.50. Front-end amplifier stages of the Stanford Research Systems SR560 Low-Noise Preamplifier (A: detailed; B and C: simplified). Some details of input signal switching have been omitted.

Bandwidth In the *signal path*, amplifier U_1 is a de-compensated low-noise ($e_n \sim 3.5\,\text{nV}/\sqrt{\text{Hz}}$ typ at 10 Hz) wideband ($f_T \sim 63\,\text{MHz}$) op-amp, here configured as an integrator with $f_T = 5\,\text{MHz}$ (33 pF and 1.0 kΩ). Its slew rate of 17 V/µs translates to a first-stage output of 5 Vpp at the instrument's rated 1 MHz bandwidth. In the second stage, capacitor C_3 limits the bandwidth to ∼2 MHz. In the first-stage *bias path* (which matters for common-mode rejection), integrator U_2 has an output slew rate of 10 V/µs (min), so it can follow a 3 Vpp common-mode input (the instrument's rated maximum) at the full 1 MHz rated bandwidth. The integrator's unity-gain bandwidth is 230 kHz (6.8 pF and 50 kΩ), thus a bias-setting loop gain of 230 at 1 kHz. That puts an increasing CMRR burden on U_1 at higher frequencies, evident in the instrument's CMRR specification: $> 90\,\text{dB}$ to 1 kHz, decreasing by 6 dB/octave (i.e., as $1/f$) above 1 kHz.[64]

Offset voltage These dual JFETs are not in the same league as precision-matched BJTs, and they don't claim to be. The NPD5564 (the best grade in its family) has a specified maximum offset voltage (i.e., gate-source mismatch, $|V_{GS1} - V_{GS2}|$) of 5 mV at $I_D = 2\,\text{mA}$; the LSK389B, though excelling in noise performance, is considerably poorer, with a maximum mismatch of 20 mV (specified at $I_D = 1\,\text{mA}$). The SR560 deals with this problem by throwing pots at it: there's a 10-turn OFFSET trimmer in Q_{1b}'s drain, with enough range ($\pm 0.13\,\text{mA}$) to balance a worst-case offset, and a second 10-turn trimmer down in the source resistor string to independently balance the offset when switched to low gain. There is a pair of analogous CMRR trimmers, 20-turns each. We've done the calibration routine (complete with some iteration of the not-very-orthogonal adjustments), and can attest to the fussiness of getting (and sustaining) deep offset and CMRR nulls.

Noise voltage To minimize noise voltage, the JFETs are operated at relatively high current, and the resistor values (in the source, drain, and signal path generally) are intentionally small. With the original noise budget of ∼4 nV/√Hz it was acceptable, for example, to use 110 Ω resistors in the JFET sources. But a series pair of that resistance produces a Johnson noise-voltage density of 2 nV/√Hz, a bit more than we'd like when paired with these quieter JFETs. An optimized design would reduce those values by perhaps a factor of 2.

Miscellany Low-level signal switching (e.g., ac/dc/gnd

input coupling, ×1/×10 gain selection, and A/B input interchange) is done with electromechanical relays. Not just any old relay, though – these guys use relays with gold-cladded twin contacts (intended for stable low resistance, even with low-level signals where there is no contact cleaning; see §1x.6.1). And these relays are the *latching* type, where a momentary pulse switches the state, thus eliminating noise-inducing coil currents in the steady state; this further eliminates local heating as a source of drift. In this instrument a microcontroller runs all relays and indicator lights, sensing a front-panel command (or external digital control), and appropriately pulses relays as needed to change their state.

8.6.5 Selecting low-noise JFETS

As we mentioned earlier, bipolar transistors offer the best noise performance with low source impedances, owing to their lower input-voltage noise. Voltage noise, e_n, is reduced by choosing a transistor with low base resistance, $r_{bb'}$, and operating at high collector current (as long as h_{FE} remains high). For higher source impedances the current noise can be minimized instead by operating at lower collector current.

At high values of source impedance, FETs are the best choices. Their voltage noise can be reduced by operating at higher drain currents, where the transconductance is highest. FETs intended for low-noise applications have high κ values (see §3.1.4), which generally means high input capacitance. For example, the low-noise 2SK170 has a typical C_{iss} of 30 pF, whereas the PN4117–9 series of low-current FETs has a *maximum* C_{iss} of just 3 pF.

Table 8.2 on the following page presents a selection of JFET candidates for low-noise circuits. It's meant to be used in conjunction with Table 3.7, which has more extensive tabular data, including single-sample measurements. Also refer to Figure *3x.10*'s transconductance plots, and Figure 8.52's noise plots. The entries near the bottom of Table 8.2 are dual JFETs, suitable for differential stages; but note that their V_{os} is not impressive – generally far inferior to even the poorest op-amps (the interdigitated LS840, whose layout is shown in Figure 3.18, is the best of the lot, with 2 mV typical, 5 mV worst-case offset voltage).

We favor JFETs with high transconductance and low gate-cutoff voltages, but sadly many of the old favorites with superior performance have been discontinued. NXP's BF862 has become a "go-to" low-noise part, both quiet and with attractively low capacitance. Toshiba's 2SK170 has been single-handedly kept in stock by large factory purchases by Mouser, and the second-source LSK170 is

---

[64] The high-frequency CMRR could likely be improved by configuring U_1's output to control a pair of current-sink transistors biasing Q_{1ab}.

Table 8.2 Low-noise JFETs[a]

Part number SOT-23	TO-92, etc	Polarity	V_{GS} max (V)	I_{DSS} min (mA)	I_{DSS} max (mA)	$V_{GS(off)}$ (10nA) min (V)	$V_{GS(off)}$ max (V)	g_m min (mS)	g_m max (mS)	@ I_D (mA)	R_{on} typ (Ω)	C_{iss} typ (pF)	Noise curve	e_n f=100Hz I_D=1mA	e_n f=10kHz 100µA	e_n f=10kHz 1mA	e_n f=10kHz 5,10mA
MMBF4117	PN4117	N	40	0.03	0.09	-0.6	-1.8	0.07	0.21	m	-	1.2	S	12	11nV/◊ at 0.07mA		
MMBF4118	PN4118	N	40	0.08	0.24	-1	-3	0.08	0.25	m	-	1.2	U	48	11nV/◊ at 0.2mA		
MMBF4119	PN4119	N	40	0.2	0.6	-2	-6	0.10	0.33	m	-	1.2	V	45	11.5nV/◊ at 0.3mA		
BFT46	-	N	25	0.2	1.5	-	-1.2	1		m	-	3.5		-	-	-	-
BF511	-	N	20	2.5	7	-1.5		4		m	-	5[m]		20	-	3.8	
MMBF5457	2N5457	N	25	1	5	-0.5	-6	1	5	m	-	4.5	P	2.8[f]	-	2.4	2.2
MMBF5458	2N5458	N	25	2	9	-1	-7	1.5	5.5	m	-	4.5		11	-	3	
MMBF5459	2N5459	N	25	4	16	-2	-8	2	6	m	-	4.5		-	-	-	
MMBF5460	2N5460	P	40	1	5	0.75	6	1	4	m	-	5		-			
MMBF5461	2N5461	P	40	2	9	1	7.5	1.5	5	m	-	5		-			
MMBF4416	-	N	30	5	15	-	-6	4.5	7.5	m	-	2.2	R	3.9	4.7	2.8	2.6
MMBF5484	2N5484	N	25	1	5	-0.3	-3	3	6	m	-	2.2		5	4.3	2.5	2.2
MMBF5485	2N5485	N	25	4	10	-0.4	-4	3.5	7	m	-	2.2		2.8	4.3	2.5	2.3
MMBF5486	2N5486[n]	N	25	8	20	-2	-6	4	8	m	-	2.2	Q	3[n]	5	3[f]	2.5
MMBF4392[p]	PN4392	N	40	25	75	-2	-5	16[t]		10	60	12		-	-	-	-
MMBF4393[p]	PN4393[o]	N	40	5	30	-0.5	-3	13[t]		10	100	12	J	1.8[f]	-	1.8	1.4
PMBF4393	-	N	*	*	*	*	*	*		*	*	*		22	14	5.2	3.4
MMBF5103	PF5103	N	40	10	40	-1.2	-2.7	7.5	-	2	30	16[m]	L	2	3.2	1.9	1.5
-	J107	N	25	100	-	-0.5	-4.5	75[t]		5	8	35	F	1.4	-	1.3	0.75
PMBFJ109	J109	N	25	40	-	-2	-6	26[t]		5	12	15		7.8	-	2.2	-
MMBFJ110[p]	J110	N	25	10	-	-0.5	-4	20[t]		5	18	15		1.6	3.2	1.5	0.9
MMBFJ112[p]	J112	N	35	5	-	-1	-5	17[t]		25	50[m]	6		2	3.3	1.8	1.5
MMBFJ113[p]	J113	N	35	2	-	-0.5	-3	12[t]		5	100[m]	6	H	1.8	3.2	1.7	1.4
MMBFJ309[p]	J309	N	25	12	30	-1	-4	13		10	50	4	G	18	-	2.6	1.3
MMBFJ310[p]	J310	N	25	24	60	-2	-6.5	13		10	50	4	K	2	4.8	2	1.5
BF861A	-	N	25	2	6.5	-0.2	-1	12	20	m	-	7.5		9	-	2.7	1.8
BF861B	-	N	25	6	15	-0.5	-1.5	16	25	m	-	7.5		9	-	1.5	1.2
BF862[b]	-	N	20	10	25	-0.3	-1.2	35	45[t]	m	-	10	E	1.3	2.2	1.1	0.8
-	2SK147B[d]	N	40	8	16	-0.3	-1.2	40	-	5		75		1.9	2.5	1.3	0.85
-	IFN147	N	40	5	30	-0.3	-1.2	30	40[t]	m	-	75[m]		1.5	2.4	1.1	0.75
-	2SK170BL[c]	N	40	6	12	-0.2	-1.5	10	-	1		30	B	1.2	2.3	1.1	0.78
-	LSK170B	N	40	6	12	-0.2	-2	10	-	1		20		1.3	2.3	1.0	0.78
-	IF3601	N	20	30	-	-0.04	-3	750		m	-	300	A	0.65	1.8	0.6	0.38

dual JFETs dual p/n	single p/n	V_{os}[e]	Polarity	V_{GS} max (V)	I_{DSS} min (mA)	I_{DSS} max (mA)	$V_{GS(off)}$ min (V)	$V_{GS(off)}$ max (V)	g_m min (mS)	g_m max (mS)	@ I_D (mA)	R_{on} typ (Ω)	C_{iss} typ (pF)	Noise curve	100Hz 1mA	10kHz 100µA	10kHz 1mA	10kHz 5,10mA
PMBFJ620[k]	'J310[k]	-	N	25	24	60	-2	-6.5	13		10	50	3	Y	33	*noisy*		5.2
LS840[g]	-	5	N	60	0.5	5	-1	-4.5	2	-	0.2	-	4		3.2	-	2.7	-
LS5912	-	15	N	25	7	40	-	-	7	-	5	-	5		3.1	-	2.7	-
LSK389B	LSK170	20	N	40	6	12	-0.15	-2	20	-	3	-	25	M	3.2	3.2	1.7	1.3
2SK146GR[d]	2SK147	20	N	40	5	10	-	-	-	-	-	-	75		1.3	-	1.1	0.85
IFN146	IFN147	20	N	40	-	30	-0.3	-1.2	30	40[t]	m	-	75[m]	C	1.3	-	1.1	0.85
IFN860	2N6550	25	N	20	10	-	-0.3	-3	25	40[t]	10		30	D	1.3	2.6	1.1	0.83
IF3602	IF3601	100	N	20	30	-	-0.04	-3	750	-	m	-	300	-	0.65	1.8	0.63	0.38

PREFIXES: IF = interFET, LS = Linear Integrated Systems, PMB = NXP.

NOTES: ◊ = √Hz. * = same as row above. (a) sorted by decreasing voltage noise, and also by part number within a family. (b) preferred part. (c) complement 2SJ74. (d) discontinued, shown for comparison. (e) mV, max. (f) Fairchild part, others are much noisier. (g) LS840 has an interdigitated die. (k) dual unmatched J310. (m) maximum, or g_m@I_{DSS}. (n) RF JFET, high 1/f noise from several manufacturers. (o) also MPF4393 from ON Semi. (p) also PMBF- from NXP, very noisy. (t) typical.

A selection of currently available low-noise discrete JFETs. See also Table 3.7, which includes single-sample measurements of I_{DSS}, $V_{GS(off)}$, g_m, and g_{os}. The "Noise curve" column refers to corresponding measured noise spectra plotted in Figure 8.52. Some JFET types (e.g., the 2N5484–6 series) are intended for RF applications, and may exhibit high noise voltage density at low frequencies, as seen in the measured spectra of Figure 8.51. See also §3x.2 for plots of measured transconductance versus drain current. Bad news: the excellent BF862 has been discontinued. Good news: the ON semiconductor CPH3910 is a good replacement (with dual: CPH6904, in SOT23–6).

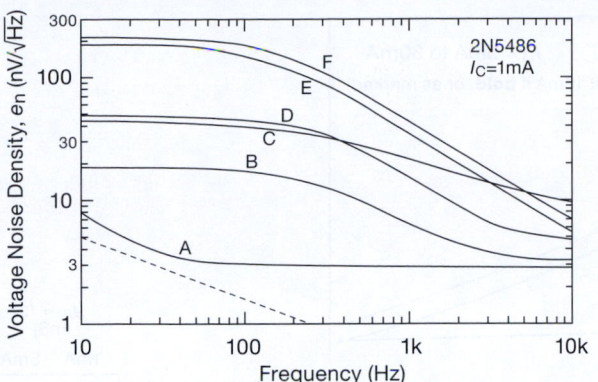

Figure 8.51. Beware unspecified noise parameters: measured voltage noise spectra of six 2N5486 JFET samples from five manufacturers (Fairchild's produced the winning "A" curve). Befitting its intended application as a "High-Frequency Amplifier," some datasheets specify noise performance only at high frequencies (e.g., ≥ 100 MHz). The slope of $1/f$ noise is indicated by the dashed line. JFETs "E" and "F" have 2.5 μVrms low-frequency noise (to 1 kHz).

available directly from Linear Integrated Systems, who also offer a matched dual version of the venerable '170, the LSK389. The excellent IFN146 and IFN147 are available directly from InterFET. Also consider the Fairchild J107, a low-cost switch that can moonlight as a low-noise JFET; with luck it will remain available.

Unspecified "low-noise" transistors *An important caution:* It is risky to rely on unspecified noise parameters, for instance the assumption (absent a manufacturer's e_n specification) that all transistors of a given part number will exhibit similar noise voltage to that measured in a sample batch. To see what can happen in practice, look at Figure 8.51, where we've plotted the measured noise-voltage spectra of six samples of a 2N5486 *n*-channel JFET. In fact, transistors even from the same manufacturer and production run may show large variations, as seen in Figure 8.40's measured noise spectra of four power transistors taken from a single batch. Figure 8.52 shows comparisons of the noise characteristics of a number of popular and useful JFETs.

Be aware of JFET weaknesses When we discussed the many considerations in choosing a low-noise *bipolar* transistor (§8.5.8), one factor of importance was the transistor's output resistance r_o (or, if you prefer, its output conductance, g_{oe}). For a BJT this is described by its Early voltage[65] V_A, whose measured values are listed in

Table 8.1a on page 501. Most *npn* BJTs have fairly high V_A values, which is to say they have a high output resistance, thus ordinarily not a serious concern.

But that's not the case for JFETs. If you tried measuring an analogous "Early voltage" for JFETs, you would be disappointed in their low values. The parameter commonly used to describe the variation of drain current with drain voltage is output conductance, g_{os}, which varies with both drain voltage and drain current. For many JFETs the output conductance (ideally zero) is high enough to have a serious affect on amplifier gain. Because the value of a JFET's g_{os} is roughly proportional to its transconductance (g_m), we've defined a parameter $G_{max}=g_m/g_{os}$, which represents the maximum voltage gain of a common-source amplifier with high-impedance drain load.[66] This parameter is not listed in Table 8.2, but you can find it in the JFET listing of Table 3.7. You need to take it into account when designing a JFET amplifier stage, perhaps circumventing its gain-killing effect by adding a cascode; see §3x.4.

Some JFETs suffer also from a severe drop in transconductance at drain voltages of less than 2 V, especially at high drain currents; see the plots of transconductance versus drain voltage in §3x.2. If you're thinking of doing anything unusual in a low-noise JFET design, you'd do well to study first the JFET-specific material in Chapters 3 and *3x*.

8.7 Charting the bipolar–FET shootout

Let's have some fun with the graphical noise-charting technique we introduced in §8.5.2. A perennial bone of contention among engineers is whether FETs or bipolar transistors are "better." We dispose of this issue with characteristic humility by matching two of the best contenders and letting them deliver their best punches.

In the bipolar corner we have the magnificent 2SD786, whose vital statistics are displayed in Figure 8.10 and listed in Table 8.1a on page 501. Its noise voltage is ~ 0.5 nV/$\sqrt{\text{Hz}}$ at 1 mA, with an admirably low $1/f$ noise corner well below 10 Hz; and its very low $r_{bb'}$ (we measured 2.3 Ω) allows us to get down to the 0.25 nV/$\sqrt{\text{Hz}}$ territory at higher collector currents. It has plenty of beta, too: above 200 at currents all the way down to 1 μA, which is helpful in achieving a low noise-*current*. It has been a perennial favorite among audio enthusiasts.

The FET entry is the 2SK170 *n*-channel JFET, known far and wide for its stunning low-noise performance, reputed to exceed that of bipolar transistors. Its very low

---

[65] Introduced in §2.3.2, and discussed in detail in §2x.8.

[66] We like G_{max} because it is largely independent of drain current, and varies in a predictable way (linearly) with drain voltage.

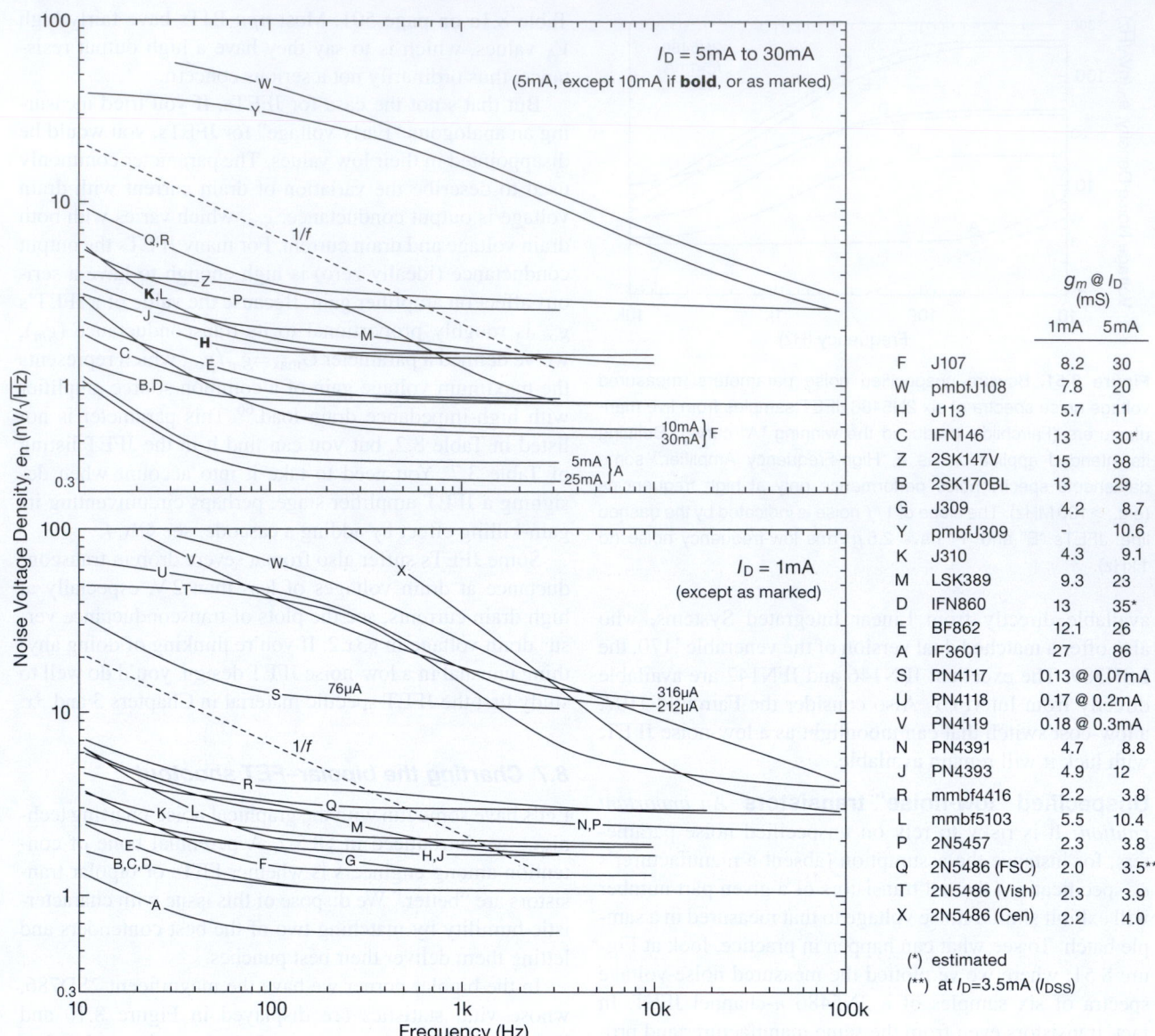

Figure 8.52. Measured noise voltage density versus frequency for a selection of JFETs listed in Table 8.2. All data taken at $V_{DS}=5$ V.

measured noise voltage can be seen in Figure 8.54, and its vital statistics are listed in Table 8.2 on page 516. According to its datasheet, it was trained only for drain currents ranging from $200\,\mu$A to 10 mA; but we've conservatively extrapolated its noise parameters down to $100\,\mu$A, so that we can evaluate it with drain currents ranging from $100\,\mu$A to 10 mA. We have high hopes for this fighter, given its very low input-current noise ($\sim1\,$fA$/\sqrt{\text{Hz}}$, the shot noise corresponding to its gate leakage current of $\sim3\,$pA). In Figure 8.53 we've charted their total amplifier noise voltage

versus source resistance at 1 kHz, just as we did for the 2N5087.[67]

---

[67] In our previous edition the contenders were the LM394 self-described "ultra well-matched monolithic NPN transistor pair" ($e_n=1\,$nV$/\sqrt{\text{Hz}}$ at $I_C=1\,$mA, versus $0.6\,$nV$/\sqrt{\text{Hz}}$ for this edition's 2SD786 championship contestant) and the 2N6483 matched monolithic n-channel JFET pair ($e_n=4\,$nV$/\sqrt{\text{Hz}}$ at $I_C=100\,\mu$A, versus $1.7\,$nV$/\sqrt{\text{Hz}}$ for this edition's 2SK170 entry). This edition's fighters are single transistors, rather than matched monolithic duals. Both contenders from the previous edition are long retired, having fought the good fight for several decades.

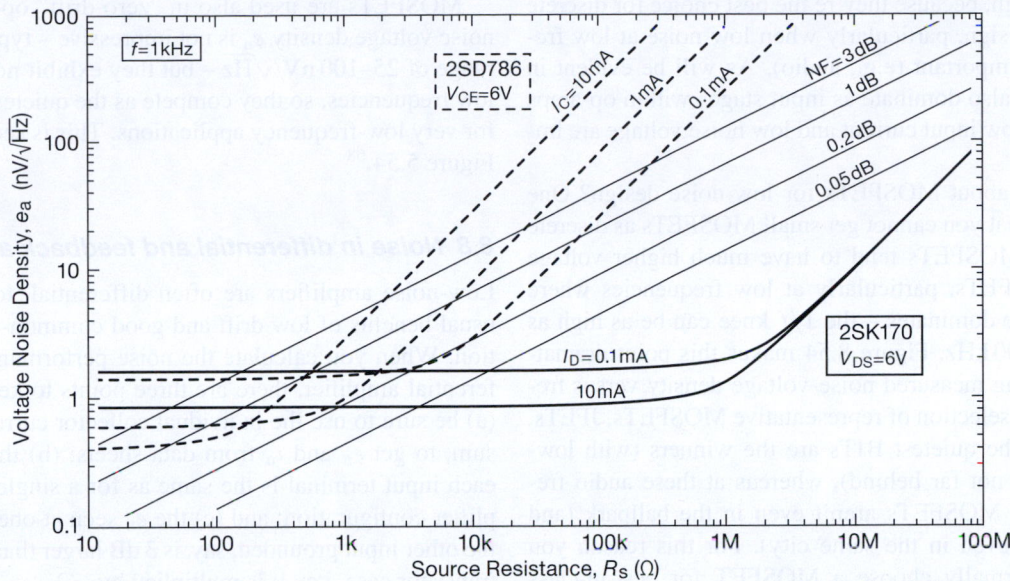

Figure 8.53. Comparison of total amplifier input voltage noise density (e_a) at 1 kHz, for the 2SD786 *npn* bipolar transistor (dashed lines) and the 2SK170 *n*-channel JFET (solid lines).

And the winner? Well, it's a split decision. The FET won points on lowest minimum noise figure, NF(R_n), reaching a phenomenal 0.0005 dB noise figure (that's a noise temperature of just 33 *milli*kelvin!), and dipping below 0.2 dB from 1k to 100M source impedance. For high source impedances, FETs remain unbeaten. The bipolar transistor is best at low source impedances, particularly below 5 kΩ, and it can reach a 0.2 dB noise figure for source resistances R_s from 1–10 kΩ, with a suitable choice of collector current. Although its victory in terms of noise alone is slim, it has other important virtues, notably its superior predictability of V_{BE} compared with a JFET's loosely characterized V_{GS}.

Just as in boxing, where the best fighters of yesteryear have retired from the rigors of competing, we must remind our readers, with some sadness, that the 2SD786 has been discontinued by Toyo–Rohm; and Toshiba's 2SK170 is sometimes available, sometimes not. However, also as in boxing, there are some younger contenders for the best low-noise transistor who haven't yet had a chance to compete in a world championship. The replacement LSK170 JFET from Linear Integrated Systems performs almost as well as its mentor; and the ZTX851 BJT from Zetex, though unspecified in noise performance, appears to outperform even the champion 2SD786 (as seen in the measured noise spectra of Figures 8.12 and 8.17).

And, if it's *really* low-noise voltage you want, consider the IF3601 from InterFET, with its 0.35 nV/√Hz typical noise voltage *even down at 30 Hz*! And this is a *JFET*, with low input current (100 pA typ, hence low i_n, about 6 fA/√Hz), and thus the noise resistance is about 60k. When used as an amplifier with a source impedance equal to its noise resistance (i.e., R_s=60k), its performance is unbeatable – the noise figure is 0.001 dB. InterFET and Linear Integrated Systems JFETs must be purchased directly from the manufacturer, but in our experience they're happy to accept small orders.

Before you go out and buy a bushel of these remarkable JFETs, consider the remarks of the critics, who claim it is muscle-bound – it has high input and feedback capacitance (650 pF and 80 pF, respectively), which limits its usefulness at high frequencies. Its relative, the IFN146, is better in this regard, at the expense of higher e_n (75 pF and 15 pF, respectively, with e_n=0.7 nV/√Hz). These same critics argue that a bipolar complementary pair like the ZTX851 and ZTX951, with e_n as low as 0.3 nV/√Hz, can offer even better performance at moderate source impedances and frequencies.

8.7.1 What about MOSFETs?

The dominant FET species on the planet, by a demographic ratio of at least 10^{12}:1, is the *MOS*FET, of which the JFET is the poor (and neglected) sister. *We* haven't neglected

JFETs, though, because they're the best choice for discrete low-noise design, particularly when low noise at low frequencies is important (e.g., audio). As will be evident in §8.9, JFETs also dominate as input stages within op-amps where both low input current and low noise voltage are important.

So, what about MOSFETs for low-noise design? One problem is that you cannot get small MOSFETs as discrete parts. And MOSFETs tend to have much higher voltage noise than JFETs, particularly at low frequencies where the $1/f$ noise dominates – the $1/f$ knee can be as high as 10 kHz to 100 kHz. Figure 8.54 makes this point dramatically, plotting measured noise-voltage density versus frequency for a selection of representative MOSFETs, JFETs, and BJTs. The quietest BJTs are the winners (with low-noise JFETs not far behind), whereas at these audio frequencies the MOSFETs aren't even in the ballpark (and perhaps not even in the same city). For this reason you wouldn't normally choose a MOSFET for best-in-class low-noise amplifiers below 1 MHz.

MOSFETs *are* used as low-noise linear amplifiers, however, in the form of integrated circuits intended for radiofrequency applications. At those frequencies their noise performance is good enough, and the CMOS process allows convenient integration and low cost.

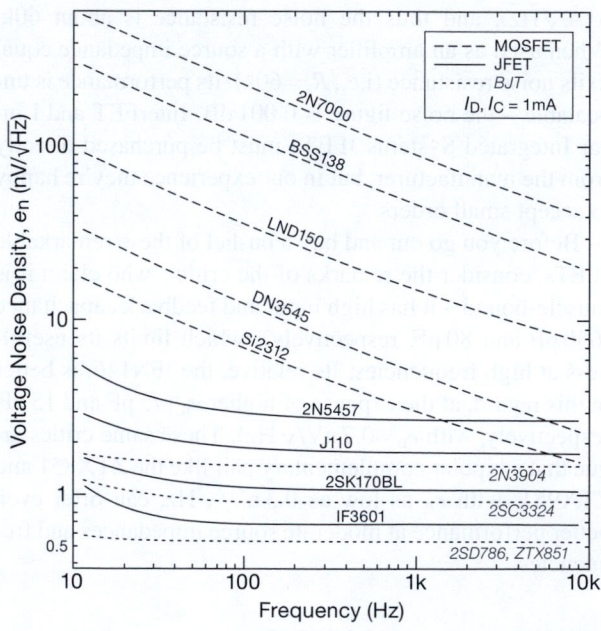

Figure 8.54. Comparison of measured noise-voltage spectra for some popular transistors: MOSFETs, JFETs, and BJTs. We included within each species both a "quietest" part and an inexpensive (<$0.05 in quantity) and popular "jellybean."

MOSFETs are used also in "zero-drift" op-amps: their noise voltage density e_n is not impressive – typically in the range of 25–100 nV/$\sqrt{Hz}$ – but they exhibit no $1/f$ rise at low frequencies, so they compete as the quietest amplifiers for very low-frequency applications. This is seen clearly in Figure 5.54.[68]

8.8 Noise in differential and feedback amplifiers

Low-noise amplifiers are often differential, to obtain the usual benefits of low drift and good common-mode rejection. When you calculate the noise performance of a differential amplifier, there are three points to keep in mind: (a) be sure to use the individual collector currents, not the sum, to get e_n and i_n from data sheets; (b) the i_n seen at each input terminal is the same as for a single-ended amplifier configuration; and (c) the e_n seen at one input, with the other input grounded, say, is 3 dB larger than the single-transistor case, i.e., it is multiplied by $\sqrt{2}$.

In amplifiers with feedback, you want to take the equivalent noise sources e_n and i_n out of the feedback loop, so you can use them as previously described when calculating noise performance with a given signal source. Let's call the noise terms brought out of the feedback loop e_A and i_A, for *amplifier* noise terms. Thus the amplifier's noise contribution to a signal with source resistance R_s is

$$e^2 = e_A^2 + (R_s i_A)^2 \quad V^2/Hz. \tag{8.37}$$

Let's take the two feedback configurations separately.

Noninverting   For the noninverting amplifier (Figure 8.55) the input noise sources become[69]

$$i_A^2 = i_n^2 \quad A^2 (\text{rms}), \tag{8.38}$$
$$e_A^2 = e_n^2 + 4kTR_\| + (i_n R_\|)^2 \quad V^2 (\text{rms}), \tag{8.39}$$

where e_n is the "adjusted" noise voltage for the differential configuration, i.e., 3 dB larger than for a single-transistor stage. The additional noise-voltage terms arise from Johnson noise and input-stage noise current in the feedback resistors. Note that the effective noise voltage and current are now not completely uncorrelated, so calculations in which their squares are added can be in error by a maximum factor of 1.4.

[68] The OPA188 and OPA2188 high-voltage CMOS autozero op-amps are among the best you may find, with an e_n of 9 nV/$\sqrt{Hz}$ and a noise corner at 0.4 Hz. We won't vouch for their input *current* noise.

[69] See C. D. Motchenbacher and J. A. Connelly, *Low-noise Electronic System Design*, Wiley (1993), who show the complete expression for a full differential amplifier.

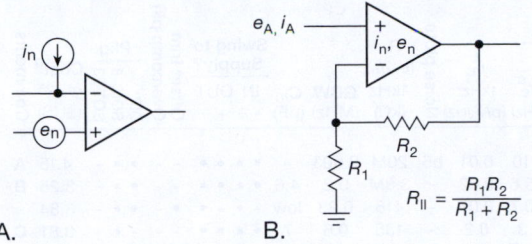

A.   B.

Figure 8.55. A. Noise model of the op-amp. B. Noise sources in the noninverting amplifier.

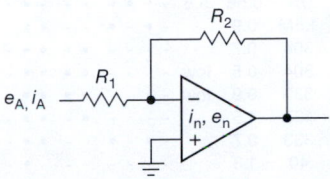

Figure 8.56. Noise sources in the inverting amplifier.

For a follower, R_2 is zero, and the effective noise sources are just those of the differential amplifier alone. Note that these drawings and formulas assume that the signal source resistance is zero (or at least small compared with $R_\parallel$, so as not to add Johnson noise).[70]

Inverting

For the inverting amplifier (Figure 8.56) the input noise sources become

$$i_A^2 = i_n^2 + 4kT\frac{1}{R_1} \quad \mathrm{A^2(rms)}, \tag{8.40}$$

$$e_A^2 = e_n^2\left(1 + \frac{R_1}{R_2}\right)^2 + 4kTR_1\left(1 + \frac{R_1}{R_2}\right)$$
$$+ (i_n R_1)^2 \quad \mathrm{V^2(rms)}. \tag{8.41}$$

For differential amplifiers and ordinary op-amps the two input terminals have comparable input noise voltages, and noise currents. This is not true, however, for current-feedback (CFB) op-amps (§4x.6), where the noise current at the inverting input is generally much larger than at the non-inverting input.

Operational amplifiers These have differential inputs, so they play by the same rules. Op-amps dominate most analog designs, and for good reason: they are highly

---

[70] If an additional series resistor is added at the noninverting input (to balance input-current offsets, see Figure 4.55), consider shunting it with a capacitor to quiet its Johnson noise (e.g., C_1 in Figure 8.78).

evolved and deliver excellent performance. There are thousands of op-amp choices, optimized for various combinations of parameters that include accuracy, speed, noise, power consumption, and supply voltage. If you can solve a design problem with op-amps, you probably should[71]. Following the theme of this chapter, low-noise design with op-amps is the subject of the next section.

8.9 Noise in operational amplifier circuits

Op-amp datasheets specify input noise in terms of e_n and i_n, just as with transistors and FETs. You'll see tabulated values, and usually plots of e_n and (sometimes) i_n versus frequency. Unlike design with discrete transistors, however, you don't get to adjust internal operating currents and component values – you only get to use them.

There are thousands of op-amps out there, with a good selection of parts aimed at low-noise applications. We've collected some 150 favorites in Tables 8.3a–8.3c, many of whose noise voltage and current density are plotted in Figures 8.60 and 8.61. How to choose among these? There's e_n and i_n, of course (and the corresponding $1/f$ low-frequency corner frequency f_c). But there are also all the usual performance tradeoffs: accuracy, speed, input current, power dissipation, and price. As an example, at $0.85\,\mathrm{nV}/\sqrt{\mathrm{Hz}}$ (and $f_c = 3.5\,\mathrm{Hz}$) the LT1028 is among the lowest e_n op-amps available (curve A in the upper plot of Figure 8.60). But it draws 8.5 mA of supply current, with ~ 1.8 mA devoted to the input stage alone. For all the reasons discussed in §8.3–§8.6, you can't match that kind of low-noise voltage in an op-amp whose total supply current is, say, 0.1 mA. And you may not want to: the LT1028's BJT inputs have plenty of bias current, with correspondingly high noise-current density ($\sim 1\,\mathrm{pA}/\sqrt{\mathrm{Hz}}$ – curve A again, this time in the lower plot of Figure 8.60). Put another way, the high-noise *current* of this low-e_n champ wipes out its noise advantage for an input signal of source impedance greater than $1\,\mathrm{k\Omega}$ (its noise resistance, $R_n \equiv e_n/i_n$).[72]

Let's look more closely at low-noise design with op-amps. We begin with a guide to the extensive listing in Table 8.3.

---

[71] Look at it this way: an entire op-amp in a SOT-23 package is the same size as a single transistor in a SOT-23 package, and it may not cost much more. Table 8.3 lists over 60 op-amps that are available in SOT-23 packages, with prices as low as $0.72 (qty 25).

[72] While we're complaining about the LT1028, we might add that it's a bit pricey, about $6 apiece in modest quantity.

Table 8.3a Low-noise BJT-input Op-amps[a]

□	Part #	# per pkg	Supply range (V)	Supply I_Q (mA[p])	Input I_B typ (nA)	Input I_B max (nA)	Bias cancel	V_{os} max (mV)	CMRR min (dB)	$V_{n(pp)}$ 10Hz[b] (µV)	e_n 1kHz (nV/√Hz)	f_C^c (Hz)	i_n 1kHz[t] (pA/√Hz)	Noise plots	R_n (e_n/i_n) 1kHz (kΩ)	GBW (MHz)	C_{in} (pF)	Cost qty 25 ($US)	Comments
	high-voltage BJT																		
	LT1495	1,2,4	2.2-36	0.001	0.3	1	-	0.38	90	4	185	10	0.01	b6	20M	0.003	-	4.15	A
□	LT1490A	2,4	2.5-44	0.04	1	8	-	0.5	84	1	50	5.6	0.02	-	3.3M	0.2	4.6	3.25	B
□	LT1077A	1	2.2-44	0.05	7	9	-	0.04	97	0.5	27	0.7	0.07	-	415	0.23	low	3.84	-
	ADA4096	2	3-36	0.06	10	15	-	0.3	73	0.7	27	3	0.2	-	135	0.6	7	3.81	C
□	LT6010A	1	2.7-40	0.14	0	0.11	•	0.04	107	0.4	14	3.6	0.1	34	140	0.35	4	2.22	D
	LT6011A	2,4	2.7-40	0.14	0	0.3	•	0.06	107	0.4	14	3.6	0.1	34	140	0.33	4	2.36	-
	LT6013A	1,2	2.7-40	0.15	0.1	0.25	•	0.06	115	0.2	9.5	2	0.15	-	63	1.6	4	2.36	E
	OP07C	1	8-44	2.7	1.8	7	•	0.15	97	0.38	9.8	0.7	0.13	-	75	0.6	-	0.62	F
	ISL28107	1,2	4.5-42	0.21	0.02	0.3	•	0.08	115	0.34	13	2	0.05	-	245	1	-	2.73	-
□	AD8622A	2,4	4-36	0.22	0.05	0.2	-	0.13	125	0.2	11	2	0.15	-	73	0.56	5.5	5.33	-
	LT1013A	2,4	3-44	0.35	12	20	-	0.15	100	0.55	22	1.3	0.02	19	1.5M	0.9	x	2.08	-
	LT1097	1	2.4-40	0.35	0	0.25	-	0.05	115	0.5	14	2.5	0.05	-	304	0.7	-	2.32	-
□	LT1012AC	1	2.4-40	0.37	0.03	0.1	•	0.05	114	0.5	14	2.5	0.05	M	304	0.5	low	5.11	-
□	OP97E	1,2,4	4.5-40	0.40	0	0.1	•	0.03	114	0.5	14	2.5	0.04	-	333	0.9	low	5.52	-
	LM741A	1,2,4	8-44	1.7	30	80	-	3	80	1.8	20	20	0.22	-	127	1.5	-	0.23	G
	LM358	1,2,4	4-32	0.7	45	250	-	7	70	1.8	40	10	0.12	24	333	0.7	-	0.14	H
	OP2177	1,2,4	5-36	0.5	2	5	•	0.08	120	0.4	8	5	0.2	-	40	1.3	-	3.51	-
□	OPA2188	1,2,4	4.0-40	0.42	0.2	0.85	•	0.03	120	0.25	8.8	0.4	0.75	J	12	2	9.5	3.43	I
□	OPA277P	1,2,4	4-36	0.79	0.5	1	•	0.02	130	0.22	8	20	0.2	13	40	1	low	3.17	J
	ISL28218	1,2	3-44	0.85	230	600	-	0.15	102	0.3	5.6	16	0.36	-	16	4	-	4.25	-
	ISL28127	1,2	4.5-42	0.85	1	10	•	0.07	115	0.085	2.5	6	0.4	-	6.3	10	-	2.50	-
	LT1468	1	7-36	3.9	3	10	•	0.08	96	0.3	5	27	0.6	9	8.3	90	4	4.26	K
	NE5534	1,2	10-44	4	0.5	1.5	-	4	70	0.4	3.5	50	4	G	8.8	10	-	0.85	L
□	LT1677	1	3-44	2.8	2e	20e	•	0.06	109	0.09	3.2	13	1.2	-	2.7	7.2	4.2	3.07	M
□	OP-27A[d1]	1	8-44	3	10	40	•	0.03	114	0.08	3	2.7	1	8	3	8	-	2.10	N
□	OPA227[d2]	1,2,4	5-36	3.7	2.5	10	•	0.08	120	0.09	3	4	0.4	F	7.5	8	12	1.67	-
	LME49710	1	5-36	4.8	7	72	•	0.7	110	0.35	2.5	70	1.6	-	1.6	55	-	1.97	O
	LM4562	2	5-36	5	10	72	•	0.7	110	0.4	2.7	75	1.6	b4	1.7	55	-	2.45	O
	ADA4075-2	2	9-40	1.8	30	100	-	1	110	0.06	2.8	3.6	1.2	b1	2.3	6.5	2.4	1.80	P
□	AD8675	1,2	9-36	2.5	0.5	2	•	0.08	114	0.1	2.8	6	0.3	-	9.3	10	-	2.22	Q
	AD8671	1,2,4	8-36	3	3	12	•	0.08	100	0.077	2.8	5	0.3	-	9.3	10	7.5	1.97	Q
	LT1124	2,4	8-44	2.75	8	30	•	0.1	108	0.07	2.7	2.3	0.3	-	9.0	12.5	low	4.91	R
□	LT1007[g,d4]	1	4-44	2.7	10	35	•	0.03	117	0.06	2.5	2		7,D	6.3	8	-	2.48	-
□	OPA209	1,2,4	4.5-40	2.2	1	4.5	•	0.15	120	0.13	2.2	16	0.5	-	4.4	18	4	2.27	S
□	ADA4004	1,2,4	9-36	2.2	40	90	•	0.13	110	0.15	1.8	2.5	1.2	b1	1.5	12	-	4.20	-
	OPA211	1,2	4.5-36	3.6	60	175	•	0.05	114	0.08	1.1	10	1.7	b2	0.65	45	8	12.87	T
□	AD8597	1	9-36	4.8	40	210	•	0.12	120	0.08	1.1	22	2.3	B	0.48	10	12	3.71	U
	MAX9632	1	4.5-40	3.9	30	180	•	0.13	120	0.065	0.94	22	3.8	-	0.25	55	-	5.76	V
	ADA4898-1	1,2	9-36	8.1	100	400	•	0.13	103	0.05[u]	0.9	14	2.4	4	0.38	65	3.2	5.37	-
	AD797	1	9-36	5.2	250	1500	-	0.08	114	0.05	0.9	30	2.0	-	0.45	110	-	8.03	-
	LT1115	1	9-44	8.5	50	380	-	0.2	104	0.04	0.9	3.5	1.2	3,A	0.75	70	5	3.73	W
	LME49990	1	10-38	9	30	500	•	1	118	0.03	0.88	10	2.8	-	0.31	110	-	3.98	X
□	LT1028A[h]	1	8-44	7.4	25	40	•	0.04	108	0.035	0.85	3.5	1.0	3,A	0.85	75	5	6.48	-
	low-voltage BJT																		
	TLV2401	1,2,4	2.7-17	880nA	0.1	0.3	-	1.2	63	35	400	30	0.01	30	50M	0.005	3	2.05	-
□	LT6003	1,2,4	1.5-18	850nA	0	0.14	-	0.5	73	3	325	-	0.01	b5	27M	0.003	6	1.60	-
	TLV2242	1,2,4	2.5-16	0.001	0.1	0.5	-	3	60	35	400	30	0.01	31	50M	0.005	3	2.42	-
	OP196	1,2,4	3-15	0.045	10	50	-	0.3	65	0.8	26	15	0.19	-	137	0.35	-	3.81	-
	EL8176	1	2.4-6	0.055	0.5	2	•	0.1	90	1.5	28	6	0.16	-	175	0.4	-	2.30	Y
	LMV358	1,2,4	2.7-6	0.11	15	250	-	7	65	-	39	8	0.21	28	186	1	-	0.74	Z
	LT1783	1	2.5-18	0.23	45	80	r	0.8	90	1	20	7.5	0.14	-	143	1.3	5	3.05	AA
	TLV2460	1,2,4	2.7-6	0.5	4.4	14	-	2	70	2.2	11	150	0.13	-	85	5.2	7	1.55	BB
	LT6220	1,2,4	2.2-13	0.9	15	150	•	0.35	85	0.5	10	30	0.8	-	12.5	60	2	1.75	-
□	LMP7731	1,2	1.8-6	2.2	1.5	30	•	0.5	101	0.08	2.9	1.4	1.1	b3	2.64	22	-	1.98	-
	MAX410	1,2,4	4.8-12	2.5	80	150	•	0.25	115	0.2	1.5	90	1.2	-	1.25	28	4	4.06	-
	LT6230	1,2,4	3-12.6	3.3	5µA	10µA	•	0.5	96	0.18	1.1	350	2.4	-	0.46	215	6.5	2.72	-

Notes: (a) see also op-amp tables in Chapters 4, 4x, and 5. (b) 0.01Hz or 0.1Hz to 10Hz. (c) calculated. (d) *italics* are lower-voltage JFET types. (d1) OP-37 decomp. (d2) OP228 decomp. (d3) OPA637 decomp. (d4) LT1037 decomp. (d5) OPA657 decomp. (d6) ADA4637 decomp. (d7) MAX4237 decomp. (d8) LMP7717 decomp. (g) LTC suggests LT1677, RRIO. (h) LTC suggests LT6200, 6230. (J) at 1kHz or 10kHz, except 10Hz for chopper types. (k) beyond the rail. (p) per amplifier. (q) *italics* are jellybeans. (s) SC70. (t) typical. (u) datasheet specs are 10x higher. (v) V_{os} vs V_{CM} stable. (w) to 5V above the rail. (x) SOIC is non-std pinout. (y) DIP discontinued, NRND. **A:** specs at 5V. **B:** over-the-top, low I_B for $V_{CM}<V_{CC}-0.7$. **C:** OVP to +32V. **D:** LT1012 replacement, w/RRO. **E:** decomp, $G>5$. **F:** original classic. **G:** noisy classic. **H:** noisy, cheap; LM321 single. **I:** zero-drift, CMOS(!). **J:** improved OP177. **K:** 0.7ppm dist. **L:** classic audio. **M:** "RRIO LT1007."

Table 8.3b Low-noise FET-input Op-amps[a]

Part #[d]	# per pkg	Supply range[d] (V)	I_Q[t] (mA[P])	I_B @25°C typ (pA)	I_B @25°C max (pA)	V_{os} max (mV)	TCV_{os} typ (µV/°C)	CMRR min (dB)	$V_{n(pp)}$ 10Hz[b] (µV)	e_n 1kHz (nV/√Hz)	f_c[c] (Hz)	$e_n \cdot C_{in}$ 1kHz[J] (nV·pF)	Noise plots	GBW (MHz)	C_{in} (pF)	Offset trim	Cost[q] qty 25 ($US)	Comments
JFET																		
OPA129B	1	10-36	1.2	30fA[z]	0.1	2	3	80	4	17	310	34	32,U	1	2	•	11.28	CC
AD549KH	1	10-36	0.6	75fA[z]	0.1	0.25	2	90	4	35	100	35	-	1	1	•	28.00	-
ISL28110	1,2	9-42	2.6	0.3	2	0.3	1	88	0.6	6	45	72	-	12.5	12		2.34	DD
OPA124PB	1	10-36	2.5	0.35	1	0.25	1	100	1.6	8	195	24	33,P	1.5	3	•	6.40	EE
OPA140	1,2,4	4.5-40	1.8	0.5	10	0.12	0.35	126	0.25	5.1	12	51	H	11	10		3.89	FF
AD8076	1	*5-24*	6.4	0.5	1	5	1	89	6	6.6	2000	17	-	350	2.5		5.32	-
AD795	1	8-36	1.3	1	2	0.5	3	90	1	8	50	18	-	1.6	2.2		7.97	GG
OPA141	1,2,4	4.5-40	1.8	2	20	3.5	2	120	0.25	6.5	13	52	-	10	8		2.36	FF
OPA1641	1,2,4	4.5-40	1.8	2	20	3.5	-	120	0.2	5.1	7	41	f1	20	8		2.76	FF
AD8620A	1,2	*10-27*	2.5	2	10	0.25	0.5	90	1.8	6	1200	90	-	25	15		11.86	-
OPA656[d5]	1	*9-13*	14	2	20	1.8	2	80	4.5	7[x]	1100	20	T	230	2.8		5.59	-
LF412C	1,2,4	10-44	3.6	50	200	3	7	80	2.5	25	45	75	21	3	3		*1.32*	HH
OPA171	1,2,4	2.7-40	0.48	8	15	1.8	0.3	104	3	14	170	42	18	3	3		0.91	ZZ
OPA604	1,2	9-50	5.3	50	-	5	8	80	1.3	11	60	110	16	20	10		2.57	JJ
OPA134	1,2,4	5-36	4	5	100	5	2	86	1.1	8	80	40	f3	8	5		*1.60*	KK
OPA2132	1,2,4	5-36	4	5	50	0.5	2	96	1.1	8	80	48	-	8	6		3.81	-
LME49880	2	10-36	14	5	-	10	3	90	1.9	7	250	-	-	25	-		2.43	O
ADA4627B[d6]	1	9-36	7	5	20	0.2	1	106	0.7	6	60	42	-	19	7		15.05	LL
LT1793	1	9-40	4.2	4	20	0.9	2	81	0.5[u]	5.8	30	9	10	4.3	1.5		3.15	-
OPA627A[d3]	1	10-36	7	5	20	0.28	0.5	106	0.6	4.5	90	32	L	16	7		20.82	-
LT1792	1	9-40	4.2	300	800	0.8	7	82	0.35[u]	4.0	30	56	f2	4.3	14		3.15	-
OPA827	1	8-40	4.8	15	50	0.15	1.5	104	0.25	3.8	25	34	-	22	9		9.22	MM
AD743	1	9.6-36	8.1	150	250	1	2	80	0.38	2.9	35	58	-	4.5	20	y y	7.83	-
CMOS																		
ISL28194	1	1.8-5.8	330nA	15	80	2	1.5	70	10	265	7	-	-	0.004	-		1.69	NN
LPV521	1	1.6-5.5	470nA	10fA	1	1	0.4	75	22	259	7	-	c3	0.006	-	s	1.66	-
LMC6442A	2	2.2-11	950nA	5fA	4	3	0.4	102	3	170	0.5	33	29	0.01	4.7		2.43	-
ISL28195	1	1.8-5.8	0.001	15	80	2	1.5	70	4	150	3	-	-	0.01	-		1.57	-
MAX9911	1,2	1.8-6	0.004	1	10	1	5	70	40	400	45	-	-	0.2	-		0.72	OO
ICL7612	1	2-16	0.006	0.5	30	5	15	70	40	100	500	-	c4	0.044	-		2.87	PP
LMP2232A	1,2,4	1.8-6	0.01	20fA	1	0.15	0.3	81	2.3	60	20	-	25	0.13	-		3.29	-
ISL28158	1,2	2.4-5.8	0.034	5	35	0.3	0.3	75	1.4	64	26	-	-	0.2	-		1.34	QQ
AD8603	1,2,4	1.8-6	0.04	0.2	1	0.05	1	85	2.3	25	50	18	20	0.4	2.5		1.36	-
LTC6078	2,4	2.7-6	0.055	0.2	1	0.03	0.2	95	1.0	18	25	126	-	0.75	18		3.54	RR
LMV358	1,2,4	2.7-6	0.1	<1nA	<1nA	7	6	50	55	33	7000	-	-	1.4	-	s	*0.45*	Z
TLV2221	1	2.7-12	0.11	1	150	3	1	70	3	19	100	42	-	0.51	6		1.20	-
MAX4236A[d7]	1	2.4-6	0.35	1	500	0.02	0.6	84	0.9	14	17	53	c2	1.7	7.5		1.78	-
LMC6001A	1	5-16	0.45	10fA	25fA	0.35	2.5	83	6	22	270	-	-	1.3	-		12.19	-
LMC6482A	2,4	3-16	0.5	20fA	4	0.75[v]	1	70	20	37	900	21	17	1.5	3		*1.88*	SS
LMV751	1	2.7-5.5	0.5	1.5	100	1		85	0.7	6.5	45	-	-	4.5	-		1.82	-
OPA376	1,2,4	2.2-7	0.76	0.2	10	0.03	0.26	76	0.8	7.5	50	91	c1	5.5	13		1.32	TT
OPA364	1,2	1.8-5.5	0.85	1	10	0.5	3	74	10	17	2200	21	-	7	3		2.18	UU
AD8692	1,2,4	2.7-6	0.85	0.2	1	2	0.3	68	1.6	8	215	35	-	10	5		1.36	Q
LMV791	1,2	1.8-6	0.95	0.05	1	1.35	1	80	1.3	6.2	170	105	-	14	15		1.42	-
TLC4501A	1,2	4-7	1	1	60	0.04	1	85	1.5	12	500	56	-	4.7	8		3.06	VV
OPA743	1,2,4	3.5-13	1.1	1	10	7	8	66	11	30	2400	28	26	7	4		*1.58*	-
LM6211	1	5-24	1.1	0.5	5	2.5	2	85	2.5	6	500	39	-	17	5.5		2.52	-
LMP7715[d8]	1,2	1.8-6	1.15	0.05	1	0.15	1	85	1	5.8	110	105	-	17	15		2.05	-
LMV710	1	2.5-5.5	1.2	4	-	3	-	50	14	20	1300	119	-	5	17		1.21	WW
LMP7721	1	1.8-6	1.3	3fA	20fA	0.15	1.5	83	2.3	7	360	-	-	17	-	x	11.89	-
OPA320	1,2	1.8-6	1.45	0.2	0.9	1.5	1.5	100	2.8	8.5	640	35	-	20	5		2.16	XX
AD8616	1,2,4	2.7-6	1.7	0.2	1	0.5	1.5	80	2.4	7	1300	49	11	24	7		*1.52*	-
TLC2272A	2,4	4.5-16	2.4	1	60	1.0	2	75	1.4	9	325	56	S	2.25	8		1.57	-
OPA365	1,2	2.2-5.5	4.6	0.2	10pA	0.2	1	100	5	4.5	5600	42	V	50	6		1.75	XX
OPA350	1,2,4	2.5-7	5.2	0.5	10	0.5	4	74	14	7	9000	46	12	38	6.5		1.67	-
LMP2021	1,2	2.5-5.8	1.1	25	100	0.005	0.004	105	0.26	11	flat	84	##	5	12		2.92	YY

N: classic industrial. **O:** 0.3ppm dist. **P:** cross-coupled. **Q:** 0.6ppm dist. **R:** replaces OP-27, OP270. **S:** 0.25ppm dist. **T:** precision. **U:** 1ppm dist. **V:** 0.4ppm dist. **W:** audio version of LT1028. **X:** 0.1ppm dist. **Y:** charge pump. **Z:** LMV321 single; BJT& CMOS versions (same p/n!). **AA:** above the rail to +18V. **BB:** 80mA output drive. **CC:** lowest I_B for an HV part. **DD:** bootstrapped JFETs. **EE:** substrate pin. **FF:** 0.5ppm dist. **GG:** replaces OPA111. **HH:** AoE jellybean. **JJ:** pro audio. **KK:** 0.8ppm dist. **LL:** replaces OPA627, 637. **MM:** costs less than OPA627. **NN:** <1µW. **OO:** 7.5µW. **PP:** programmable quiescent current. **QQ:** low V_{os} beyond rails. **RR:** V_{os} degrades near V+. **SS:** popular; LMC7101 (single) is similar. **TT:** e-trim. **UU:** charge pump, 20ppm dist. **VV:** 0.3sec self-cal at power-on. **WW:** BiCMOS; LMV711 has shdn. **XX:** charge-pump, RRI. **YY:** 2nd best chopper. **ZZ:** CMOS (among several, check out OPA170, 172, 192; also OPA2188).

Table 8.3c High-speed Low-noise Op-amps[a]

Part #	# per pkg	Supply range (V)	I_Q[t] (mA[p])	I_{BIAS} 25°C typ (µA)	CMRR min (dB)	e_n[t] 1kHz (nV/$\Diamond$)	f_C[c] (Hz)	i_n[t] 1kHz (pA/$\Diamond$)	C_{in} (pF)	$e_n C_{in}$ 0pF (pF·nV/$\Diamond$)	$e_n C_{in}$ +5pF	Min gain	GBW[t] (MHz)	Slew Rate (V/µs)	Distortion plot[d]	Iout min (mA)	Cost[q] qty 25 ($US)	Comments
HV bipolar VFB																		
LT1222	1	8-36	8	0.1	100	3	130	2	2	6.0	21	10	500	250	-	25	4.17	-
AD8021	1	4.5-26	7	7.5	86	2.1	2300	2.1	1	2.1	13	1	490	420	•	60	2.42	-
AD829	1	9-36	5	3.3	100	1.7	45	1.5	5	8.5	17	1	750[r]	150[g]	-	25	4.82	-
THS4031	1,2	9-33	8.5	3	85	1.6	390	1.2	2	3.2	11	1	100	100	-	60	5.60	-
THS4021	1,2	9-33	7.8	3	-	1.5	570	2	1.5	2.3	10	10	3500	470	-	80	8.20	-
MAX9632	1	4.6-40	3.9	0.03	120	0.94	16	3.8	-	-	-	1	55	30	-	53	5.76	-
AD797B	1	10-36	8.2	0.25	114	0.9	30	2	20	18	23	1	110	20	-	60	8.50	-
ADA4898	1,2	9-36	8.1	0.1	102	0.9	14	2.4	3.2	2.9	7.4	1	65	55	-	40	5.37	-
LME49990	1	10-38	9	0.03	110	0.88	10	2.8	-	-	-	1	110	22	•	50	4.16	A
LT1028A	1	8-44	7.4	0.03	108	0.85[v]	3.5[v]	1	5	4.3	8.5	1	75	15	-	25	6.48	B
HV bipolar CFB																		
THS3120	1	9-33	7	1	60	2.5	215	1	0.4	1.0	14	1	525	1500	-	425	5.24	C
AD844	1	9-36	6.5	0.2	-	2.0	-	10	2	4.0	14	2	330	2000	-	60	5.23	-
THS3091	1	10-33	9.5	4	62	2.0	800	14	1.2	2.4	12	1	950	7300	-	175	5.93	D
THS3001	1	9-33	6.6	2	65	1.6	250	13	7.5	12	20	1	1750	6500	-	85	6.93	-
HV JFET VFB																		
THS4631	1	10-33	11.5	50pA	95	7	340	20fA	3.9	27	62	1	210	900	-	80	8.81	-
AD8067	1	5-24	6.4	0.5	5	6.6	1650	0.6	2.5	17	50	8	540	640	-	25	4.63	-
ADA4637	1	10-36	7	1pA	100	6.1	80	1.6fA	8	49	79	5	80	170	•	45	10.12	E
LV bipolar VFB																		
AD8045	1	3-12.6	15	2	83	3	3800	3	1.3	3.9	19	1	800	1350	•	55	2.59	F
LT6202	1,2,4	3-12.6	3.5	1	83	2.8	940	1.1	1.8	5.0	19	1	95	22	-	30	1.90	-
LMH6628	2	5-13	9	0.7	57	2.0	4200	2	1.5	3.0	13	1	200	550	-	50	4.39	-
LMH6622	2	4-13	4.3	4.7	80	1.6	1000	1.5	1	1.6	10	1	180	85	-	100	3.20	-
LT6230	1,2,4	3-12.6	3.2	5	95	1.1	500	1	6.5	7.2	13	1	100	70	-	30	2.73	G
LT6230-10	1	3-12.6	3.2	5	95	1.1	500	1	6.5	7.2	13	10	1450	320	-	30	2.72	-
ISL28190	1,2	2.5-5.5	8.5	10	78	1.0	20	2.1	-	-	-	1	120	50	-	50	1.37	-
ADA4897	1,2	2.7-11	3	11	92	1.0	50	2.8	11	11.0	16	1	150	120	-	50	6.45	-
ADA4895	2	2.7-11	3	11	92	1.0	50	2.8	11	11.0	16	1	230	500	-	50	-	-
LT6200	1,2	3-12.6	16.5	10	75	0.95	4500	2.2	4	3.8	9	1	150	50	•	60	2.99	H
LT6200-10	1	3-12.6	16.5	10	75	0.95	4500	2.2	4	3.8	9	10	1600	450	•	60	2.99	H
AD8099	1	5-12.6	15	6	98	0.95	2200	5.2	2	1.9	6.7	2[x]	1200	1350	-	100	3.69	-
LMH6624	1,2	5-13	12	13	90	0.9	2200	2.3	2	1.8	6.3	10	1900	400	•	100	3.98	K
LMH6629	1,2	2.7-6	15.5	15	82	0.69	700	2.6	1.7	1.2	4.6	4	3200	530	-	200	4.00	-
LMH6629	1	2.7-6	15.5	15	82	0.69	700	2.6	1.7	1.2	4.6	10	9000	1600	-	200	4.00	-
LV bipolar CFB																		
AD8007	1,2	5-12	9	4	56	2.7	1300	2	1	2.7	16	1	800	1000	-	70	3.05	-
OPA694	1,2	7-13	5.8	5	60	2.1	600	22	1.2	2.5	13	1	2000	1700	-	80	3.38	-
AD8001	1,2	6-12.6	5	5	50	2.0	15	2	1.5	3.0	13	1	880	1200	-	85	2.72	-
MAX4224	1	6-12	6	2	55	2.0	-	3	0.8	1.6	12	2	1200	1400	-	60	6.34	-
AD8009	1	5-12.6	14	50	50	1.9	270	46	2.6	4.9	14	1	3500	5500	-	175	3.19	-
LMH6702	1	9-13.5	12.5	8	47	1.83	4600	18.5	1.6	2.9	12	1	1400	3100	-	50	3.11	-
OPA695	1,2,4	5-13	13	5	56	1.8	2100	18	1.2	2.2	11	1	5600	4300	-	90	3.42	-
OPA691	1,2,3	5-13	5.1	15	52	1.7	750	3.1	2	3.4	12	1	1050	2100	-	140	2.41	-
AD8000	1,3	4.5-13	13.5	5	52	1.6	3100	3.4	3.6	5.8	14	1	1300	4100	-	100	3.58	-
LV JFET																		
OPA656	1	9-13	14	2pA	80	7	1200	1.3fA	2.8	20	55	1	230	290	•	50	5.59	-
OPA657	1	9-13	14	2pA	80	4.8	670	1.3fA	4.5	22	46	7	1600	700	•	50	10.01	-
ADA4817	1,2	5-10.6	19	2pA	77	4	7000	2.5fA	1.5	6.0	26	1	410	870	-	70	4.93	-
LV CMOS																		
LTC6268	1,2	3-5.5	16	3fA	63	5.5[w]	150k	5.5fA	0.45[w]	1.8[w]	22	1	500	400	-	50	7.50	-
LMP7717	1,2	1.8-6	1.2	50fA	85	6.2	330	10fA	15	93	124	10	88	28	-	36	2.18	-
OPA380	1,2	2.7-7	7.5	3pA	100	5.5	600k	10fA	3	17	46	1	90	80	-	75	5.39	L
OPA365	1,2	2.2-5.5	4.6	0.2pA	100	4.5	8000	4fA	6	27	50	1	35	25	-	50	1.75	M

Notes: (a) listed within each category by decreasing e_n. (c) calculated. (d) see plots in Chapter 5. (g) for G=10 or G=20. (J) at 1kHz or 10kHz (above the 1/f corner), except 10Hz for auto-zero op-amps. (n) within 0.1V of V–. (p) per amplifier. (r) without C_{comp}. (s) SC70 available. (t) typical. (u) a pin sets G_{min}. (v) the LT1028 has a nasty ~10dB noise peak from 200kHz–600kHz. (w) at 100kHz. (x) G=1 with ext comp.

Comments: **A:** 0.1ppm distortion. **B:** lowest e_n; LTC suggests LT6200, LT6230. **C:** Hill favorite. **D:** use DDA pkg; Larkin favorite. **E:** OPA637 for DIP. **F:** improved SOIC pinout. **G:** LT6230-10 decomp. **H:** 1% dist at 50MHz. **K:** 0.03% dist at 10MHz. **L:** auto-zero. **M:** zero crossover.

8.9.1 Guide to Table 8.3: choosing low-noise op-amps

Table 8.3 lists a rich selection of op-amps suitable for low-noise applications. Here we provide a guide to the table entries, and advice for selecting op-amps. Because readers interested primarily in low-noise design with op-amps (rather than with discrete transistors) may have skipped over the earlier quantitative discussions of noise in connection with BJTs and JFETs, we restate below many of the basic ideas, with references to the relevant sections and figures. Readers whose route here (to what might be called "the promised land of op-amps") has taken them through the details of discrete design ("the valley of complexity") will find much of this discussion to be a welcome review.

A. Table sorting
In Table 8.3a we've listed BJT-input op-amps by increasing supply current I_Q, which correlates roughly with decreasing input noise voltage e_n. Once the latter falls below $5\,\mathrm{nV}/\sqrt{\mathrm{Hz}}$, the entries are roughly ordered by decreasing e_n. For JFET-input op-amps (Table 8.3b) the entries are listed by increasing input bias current (up to 1 pA), then by decreasing e_n. CMOS op-amps are sorted by increasing I_Q, which correlates approximately with decreasing e_n. Table 8.3c is devoted to high-speed op-amps of all types, listed within each category by decreasing e_n.

B. Supply voltage, high voltage, low voltage
The table presents a good collection of low-noise op-amps, with a number of representative popular op-amps (some of which might well be called "noisy") for comparison. The latter have other attractive features, such as low cost, high slew rate, low input bias current, low capacitance, low power, multiple manufacturers, and so on. The table separates "high-voltage" op-amps, capable of working with $\pm15\,\mathrm{V}$ supplies (or up to 36–44 V total supply), from op-amps that are limited to low voltages like 5–20 V maximum.

Note that many of the high-voltage op-amps (especially the bipolar types) can work also at very low supply voltages, as low as 2.2 V to 3 V total. So, for example, a precision "high-voltage" op-amp like the LT1677, which works fine with 3 V total supply, and which allows inputs and outputs to swing to either rail,[73] is also clearly an excellent low-voltage part. Op-amps whose inputs and outputs can swing to the negative rail qualify to be called "single-

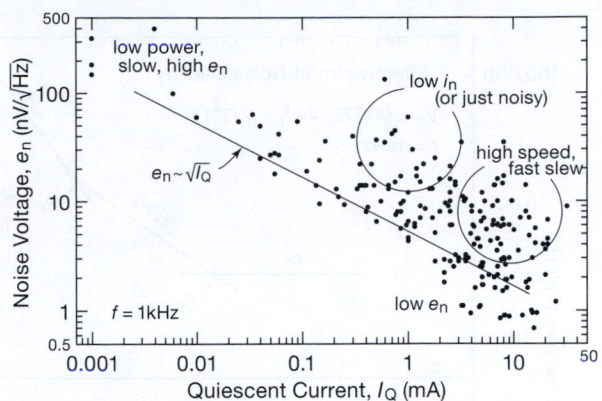

Figure 8.57. To minimize input noise voltage you have to run both BJTs and JFETs at relatively high quiescent currents. That trend extends to integrated op-amps, as seen here in a scatterplot of e_n versus I_Q for most of the op-amps in Table 8.3 and in Figures 8.60 and 8.61.

supply" op-amps, regardless of whether you run them from a single positive supply or from dual supplies.

C. Supply current I_Q, and voltage noise
High supply current is also a major consideration for low noise voltage, e_n. Table 8.3 includes op-amps with very low I_Q, even though these op-amps have much higher noise. For example, the ISL28194 uses only 330 nA but has a high $265\,\mathrm{nV}/\sqrt{\mathrm{Hz}}$ noise density. By comparison, the quiet LT1028 has $0.85\,\mathrm{nV}/\sqrt{\mathrm{Hz}}$ of noise density, but requires 7.4 mA, or ~20,000 times as much operating current.[74]

Figure 8.57 is a scatterplot of e_n versus I_Q for several hundred op-amps, showing the voltage-noise and supply-current tradeoff. You can see how the noise falls by approximately the square root of the operating current. Any part that's within a factor of two of the best-available part at a given supply current might be considered a low-noise op-amp. The correlation with supply current is far from perfect. That's because op-amp designers are balancing tradeoffs among many performance parameters, for example high slew rate, low input current, small die size (for small package size and low cost), and so on, at the expense of low noise voltage.

---

[73] The table has columns labeled "swing to supply" for inputs and outputs to both supply rails.

[74] Recall that collector [or drain] shot noise through r_e [or $1/g_m$] in BJTs [or JFETs] can be described equivalently as resistive Johnson noise created by a resistor of value $\frac{1}{2}r_e$ [or $\frac{2}{3}\frac{1}{g_m}$], according to eq'ns 8.20 or 8.34. Because higher collector [or drain] currents produce lower r_e [or higher g_m], high operating currents are needed to lower an op-amp's input-transistor noise voltage.

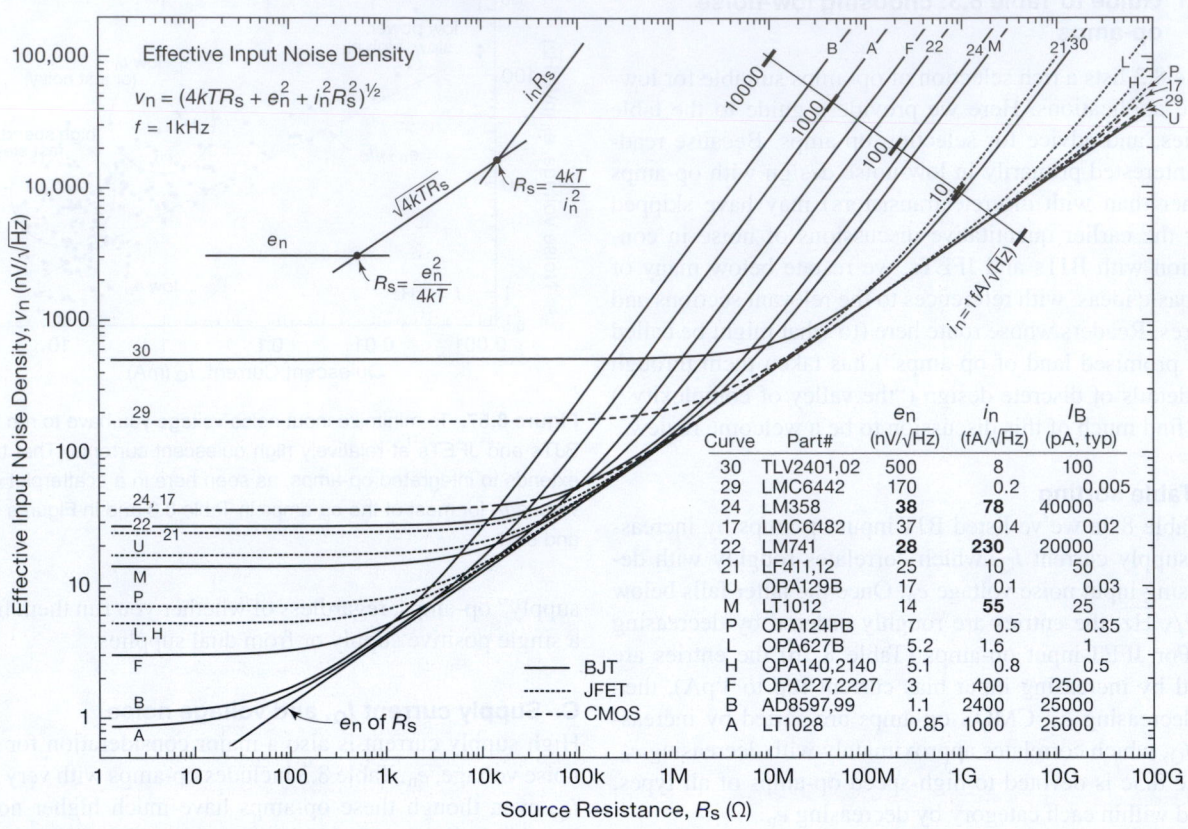

Figure 8.58. Total noise (source resistor plus amplifier, at 1 kHz) versus source resistance for a selection of low-noise op-amps (letter labels); we've included a few "ordinary" op-amps (number labels) for comparison. The curve labels correspond to those in Table 8.3 and in Figures 8.60 and 8.61. The curves are generated from datasheet values of e_n and i_n, except for measured values (shown in bold font) for the noise parameters of the classic LM741 and LM358 (unspecified) and the LT1012 (datasheet value in error); the current noise of the LMC6482 is an estimate, as the datasheet value cannot be correct.

D. Voltage-noise density, current-noise density: e_n, i_n

Just as with discrete transistors, op-amp noise performance is specified by *voltage* noise, e_n, and *current* noise, i_n, and their variation with frequency; see §§8.3.1 and 8.3.2. Figure 8.58 and its embedded formula show how to combine e_n and i_n with the Johnson noise of your circuit's effective source resistance, R_s, to find an effective total input noise density for your circuit. Stated simply, there are three sources of noise: the op-amp's input voltage noise, the op-amp's current noise flowing through the signal's source impedance, and Johnson noise in the source impedance. These are uncorrelated, so to find the total noise-voltage density you add their squares (units of V^2/Hz) and take the square root. Fourteen representative op-amps are shown as examples in Figure 8.58, with the BJT types best at low source resistances and the JFET and CMOS types best at

high resistances. Table 8.3 has datasheet specs for e_n and i_n at 1 kHz, from which you can determine what kind of trace a candidate part will make in such a plot.

E. Noise plots, e_n and i_n

It's one thing to look at reams of e_n and i_n values in datasheets, or to scan those values as organized in the three pages of Table 8.3; but it's another thing to compare noise-density plots, which include $1/f$ breakpoints and other unique information. Figures 8.60 and 8.61 show plots of datasheet e_n and i_n versus frequency for 60 of the op-amps in Table 8.3, where numbers and letters identify the corresponding traces in the plots.[75] Note how the voltage-noise plots range over a factor of 1000:1 (though most op-amps

---

[75] Figures 8.110 and 8.111 show *measured* noise spectra for a selection of these op-amps.

npn input pair   output power stage

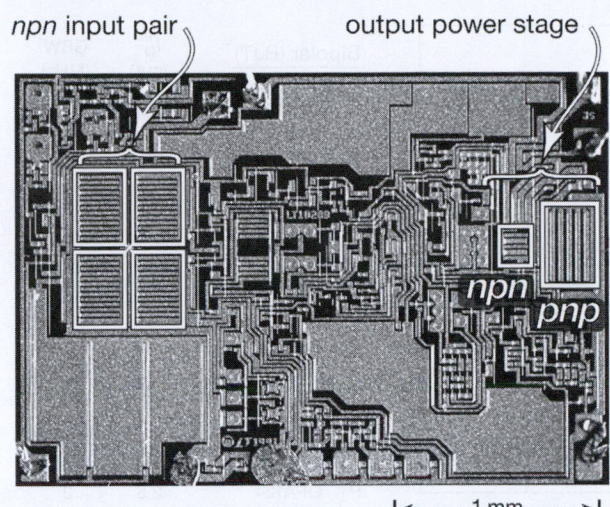

npn pnp

|◄——— 1 mm ———►|

Figure 8.59. The LT1028 op-amp, introduced in 1981, remains the low-noise winner, despite repeated attempts on its title. The input stage runs at high current (1.8 mA total) for low e_n; and the large-input transistors keep the current density low, to achieve an impressively low $1/f$ corner frequency of 3.5 Hz.

are in the range of 1–100 nV/$\sqrt{\text{Hz}}$), whereas the current-noise plots range over a factor of nearly 10^6:1. The latter reflects the very large range of dc input bias currents, from femtoamps (for some CMOS parts) to tens of microamps (for high-speed BJT parts), a ratio of 10^{10}:1. The *noise* current goes only as the square root of the *dc* current, but hey, the square root of 10^{10} is still pretty large.

An op-amp with very low voltage noise e_n, such as trace A or 3 for the exemplary LT1028 (whose portrait we present in Figure 8.59) in the top panels, generally correlates with a high current noise i_n, as seen in the lower panels. This is a common e_n versus i_n tradeoff, which you have to take into account when selecting parts from Table 8.3 and the corresponding noise plots. As an inducement to learn about low-noise design with discrete transistors, look at traces 1 and 2 of Figure 8.61, the noise voltage of a best-in-class BJT and JFET. With parts like these you can achieve lower noise voltage than with any op-amp, creating a "hybrid low-noise op-amp" in which a discrete front-end (for lowest e_n) combines with an op-amp second stage (to provide gain, and the output stage). We illustrate this in Figures 8.66 and 8.67 in §8.9.5. For a full understanding of discrete low-noise design, see §§8.5 and 8.6.

In general, BJT op-amps have low $1/f$ noise-voltage corner frequencies, typically in the range of 1–30 Hz; but beware their generally much higher noise-*current* corner frequencies, 30 Hz–1 kHz or more. This can seriously af-

fect designs with high-value feedback resistors or high signal source impedances.

If you are considering only op-amps with the lowest noise voltage (less than, say, 1.1 nV/$\sqrt{\text{Hz}}$), no others have a $1/f$ noise corner as low as the LT1028 and LT1128, at 3.5 Hz. These impressive parts also have a lower noise current than the other contenders. However, be warned that these op-amps have a nasty broad 15 dB noise peak (shown with admirable candor on the datasheet) starting at 150 kHz, and peaking at 400 kHz before fading away above 600 kHz. Other op-amps don't have this problem, but they have much higher corner frequencies (f_c), as high as 5 kHz for some of the high-speed op-amps.

F. I_B and i_n for BJT op-amps; I_B bias cancellation
Table 8.3 has typical and maximum input bias-current, or I_B, values.[76] In general, the listed current noise i_n comes from shot noise in the base current of a BJT op-amp or from gate leakage in a JFET or CMOS op-amp; so we'd expect the noise spec to be closely related to the bias current by the shot-noise equation $i_n = \sqrt{2qI_B}$. But many BJT op-amps use an input bias cancellation scheme (see §4x.10), to reduce greatly the dc bias current, I_B. However, the dc current cancellation in this scheme does not reduce the op-amp's i_n noise (in fact, it typically increases it, by $\sqrt{2}$). These op-amps are identified in the "bias cancel" column of Table 8.3a, and they typically have 10× to 40× more i_n noise than you'd expect from shot noise based on the specified bias current I_B.[77] This increased i_n noise is reflected in a lower R_n value, but ideally you'll not find the higher current noise and lowered feedback-resistor values to be a problem.

Bias cancellation near the rails
Be careful when relying on listed bias currents with bias-cancelled rail-to-rail input op-amps, because their input bias current rises dramatically as the inputs approach the supply rails (see §5.7.2). Datasheets show this in graphical form, but they generally do not reveal this undesirable behavior in the listed performance data.

G. Current-feedback op-amps
Unlike voltage-feedback (VFB) op-amps, current-feedback (CFB) op-amps (used in wideband applications) generally have much higher dc bias currents and current noise at their "− input" than at their corresponding

---

[76] For BJT op-amps (Table 8.3a) the I_B values are in nanoamps, whereas for the JFET-input op-amps (Table 8.3b) the units are *pico*amps.

[77] In a few cases the manufacturers have misstated the i_n noise; for those the table provides a measured value, indicated with italics.

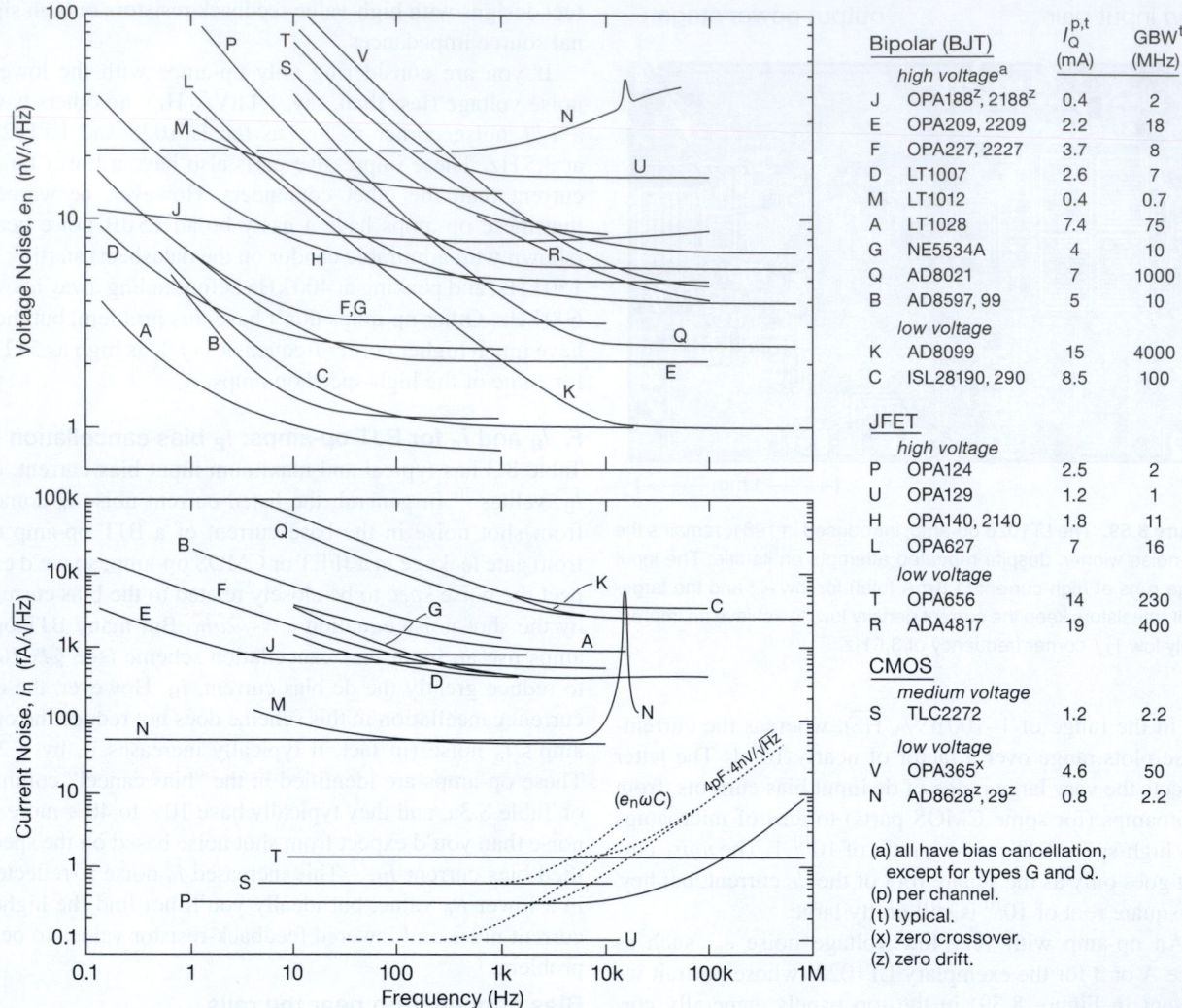

Bipolar (BJT)	$I_Q^{p,t}$ (mA)	GBW[t] (MHz)
high voltage[a]		
J OPA188[z], 2188[z]	0.4	2
E OPA209, 2209	2.2	18
F OPA227, 2227	3.7	8
D LT1007	2.6	7
M LT1012	0.4	0.7
A LT1028	7.4	75
G NE5534A	4	10
Q AD8021	7	1000
B AD8597, 99	5	10
low voltage		
K AD8099	15	4000
C ISL28190, 290	8.5	100

JFET		
high voltage		
P OPA124	2.5	2
U OPA129	1.2	1
H OPA140, 2140	1.8	11
L OPA627	7	16
low voltage		
T OPA656	14	230
R ADA4817	19	400

CMOS		
medium voltage		
S TLC2272	1.2	2.2
low voltage		
V OPA365[x]	4.6	50
N AD8628[z], 29[z]	0.8	2.2

(a) all have bias cancellation,
    except for types G and Q.
(p) per channel.
(t) typical.
(x) zero crossover.
(z) zero drift.

Figure 8.60. Voltage and current noise versus frequency for a selection of "low-noise" op-amps featured in Tables 8.3a–c on pages 522–524. All are adapted from datasheet plots, with the exception of measured i_n for op-amps G, J, and N. The labels are in order of ascending e_n at 10 Hz. See Figures 8.110 and 8.111 for measured noise spectra of selected op-amps.

"+input". For CFB op-amps, Table 8.3c gives the current-noise value for the input with the lowest current and noise. The difference can be as large as a factor of ten. As always, study the datasheet, *carefully!*

H. Noise resistance R_n

The noise-resistance parameter $R_n \equiv e_n/i_n$ has its own column. This is the value of source resistance that corresponds to the minimum possible amplifier noise figure; see for example §8.5.1 and Figure 8.31. But its real usefulness is to let you see at a glance the maximum feedback resistor values you can use: the impedance R_s seen by the op-amp at its inputs should be 5× to 30× smaller than the op-amp's

R_n value if you wish to ensure that e_n is the dominant noise source in your circuit; see Figure 8.58. The idea here is that you can't do anything about the e_n value you're stuck with, but you can lower your feedback resistor values to reduce the effect of the op-amp's i_n. But don't forget that you've got to keep the *signal's* source impedance small compared with R_n also. (Also, quite apart from the *noise* current, you may have to worry about the effects of the *dc* bias current flowing through the signal's source, which may be a sensor of relatively high dc resistance.) So in the end you may have to give up on your favorite low-e_n op-amp, and select one with lower i_n even if it has higher e_n.

There are no R_n values listed for the JFET and CMOS

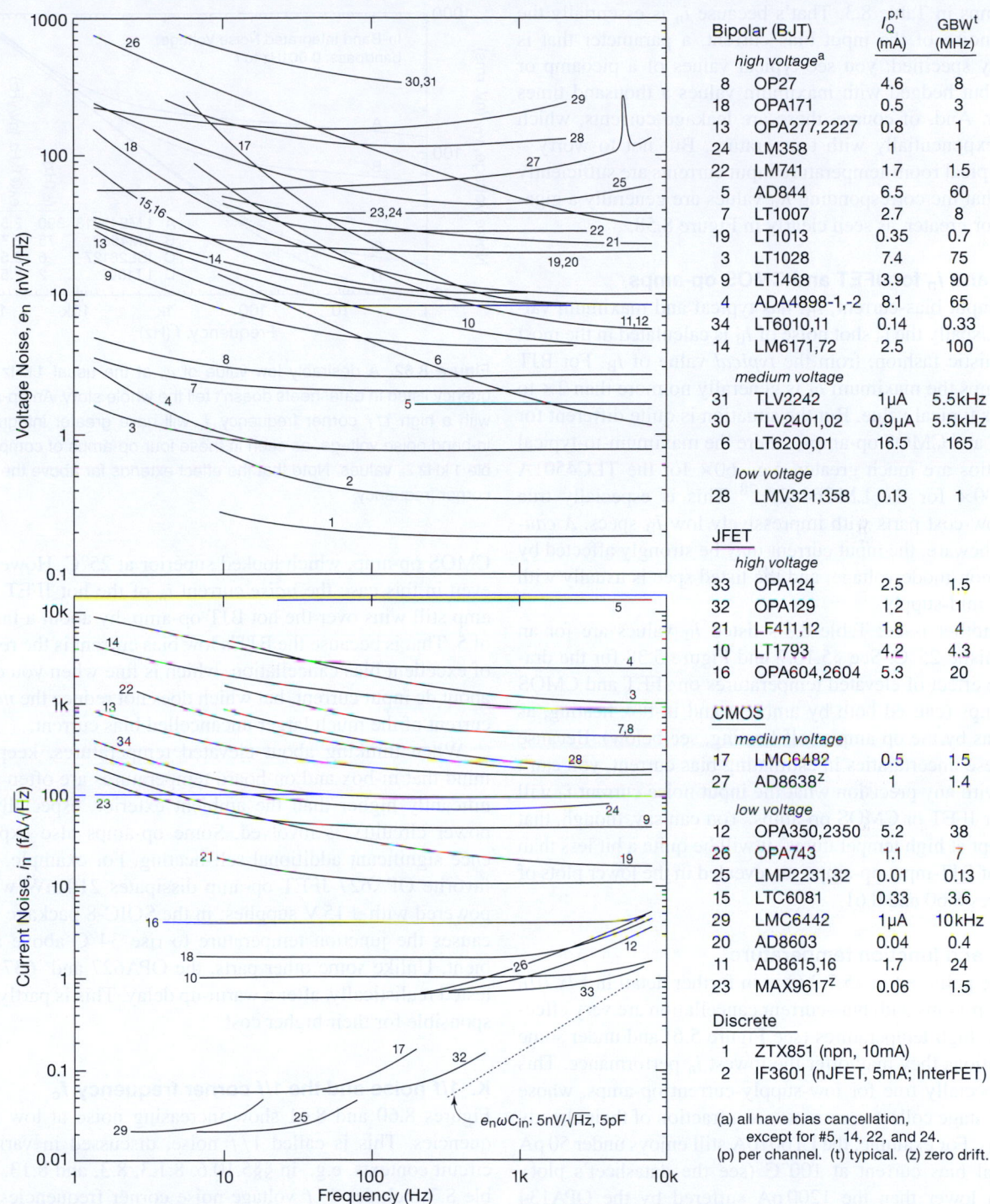

Figure 8.61. Voltage and current noise versus frequency, for a selection of "popular" op-amps featured in Table 8.3 on pages 522–524; a low-noise JFET and BJT are included for comparison. All are adapted from datasheet plots, with the exception of measured e_n for op-amps #22 and #24 and BJT #1, and measured i_n for op-amp #23. The labels are in order of ascending e_n at 10 Hz.

op-amps in Table 8.3. That's because i_n is essentially the shot noise of the input bias current, a parameter that is poorly specified: you see typical values of a picoamp or less, but hedged with maximum values a thousand times larger. And, of course, these are leakage currents, which rise exponentially with temperature. But not to worry – the typical room-temperature input currents are sufficiently low that the corresponding R_n values are generally a giga-ohm or greater, as seen clearly in Figure 8.58.

I. I_B and i_n for JFET and CMOS op-amps

The input bias-current, I_B, has typical and maximum values. Usually the i_n shot noise of I_B is calculated in the most optimistic fashion, from the *typical* value of I_B. For BJT op-amps the maximum I_B is generally no more than $2\times$ to $3\times$ its typical value. But the situation is quite different for JFET and CMOS op-amps, where the maximum-to-typical I_B ratios are much greater, e.g., $60\times$ for the TLC4501A or $800\times$ for the LMC6442A.[78] This is especially true for low-cost parts with impressively low I_B specs. *A caution*: beware, the input current may be strongly affected by common-mode voltage, and the listed spec is usually with V_{in} at mid-supply.

Another issue: Table 8.3's listed I_B values are for an optimistic 25°C. See §5.10.7 and Figure 5.38 for the dramatic effect of elevated temperatures on JFET and CMOS op-amps (caused both by ambient and in-box heating, as well as by the op-amp's self-heating, see below). Because of these uncertainties in estimating bias current, you can't say with any precision what the input noise current i_n will be for JFET or CMOS op-amps. You can say, though, that (except at high temperatures) it will be quite a bit less than that of BJT-input op-amps, as revealed in the lower plots of Figures 8.60 and 8.61.

J. I_B and junction temperature

As we point out in §5.7.2 (and in further detail in §4x.10), BJT op-amps with bias-current cancellation are very effective at high temperatures (see Figure 5.6) and under some conditions they may offer the lowest i_n performance. This is especially true for low-supply-current op-amps, whose input-stage collector currents are a fraction of their already low I_Q. For example, the LT6010A still enjoys under 50 pA typical bias current at 100°C (see the datasheet's plot), much lower than the 1200 pA suffered by the OPA134

---

[78] For very low input-current op-amps you'll often see an overly conservative (high) maximum I_B spec, dictated by the manufacturer's desire for fast and inexpensive automatic testing. Only high-priced parts get treated to slow and expensive low-current tests.

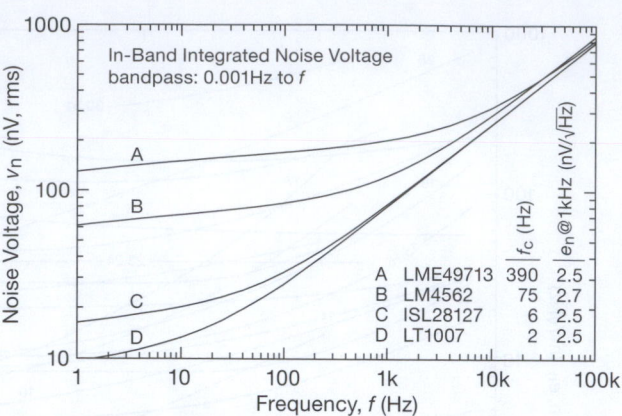

Figure 8.62. A desirably low value of e_n at the usual 1 kHz frequency listed in datasheets doesn't tell the whole story. An op-amp with a high $1/f$ corner frequency f_c will have greater integrated in-band noise voltage, as seen in these four op-amps of comparable 1 kHz e_n values. Note that the effect extends far above the $1/f$ corner frequency.

CMOS op-amp, which looked superior at 25°C. However, even in this case the *noise* current i_n of the hot JFET op-amp still wins over the hot BJT op-amp, by about a factor of 5. That is because the BJT's low bias current is the result of excellent bias cancellation, which is fine when you care about dc input current, but which does not reduce the *noise* current of the much larger uncancelled bias current.

When thinking about elevated temperatures, keep in mind that in-box and on-board temperatures are often significantly higher than the ambient exterior, especially if power circuitry is involved. Some op-amps also experience significant additional self-heating. For example, our favorite OPA627 JFET op-amp dissipates 210 mW when powered with ±15 V supplies; in the SOIC-8 package this causes the junction temperature to rise 34°C above ambient. Unlike some other parts, the OPA627 and '637 are tested realistically, after a warm-up delay. This is partly responsible for their higher cost.

K. $1/f$ noise and the $1/f$ corner frequency f_c

Figures 8.60 and 8.61 show increasing noise at low frequencies. This is called $1/f$ noise, discussed in various circuit contexts, e.g., in §§5.10.6, 8.1.3, 8.3, and 8.13. Table 8.3 includes $1/f$ voltage-noise corner frequencies f_c, calculated from datasheet noise parameters via eq'n 8.60. Watch out for op-amps that may have attractively low e_n (as specified at the usual 1 kHz), but that suffer from high $1/f$ corner frequencies. If your circuit's bandwidth is less than $\sim 10 f_c$ you should consider its effects.

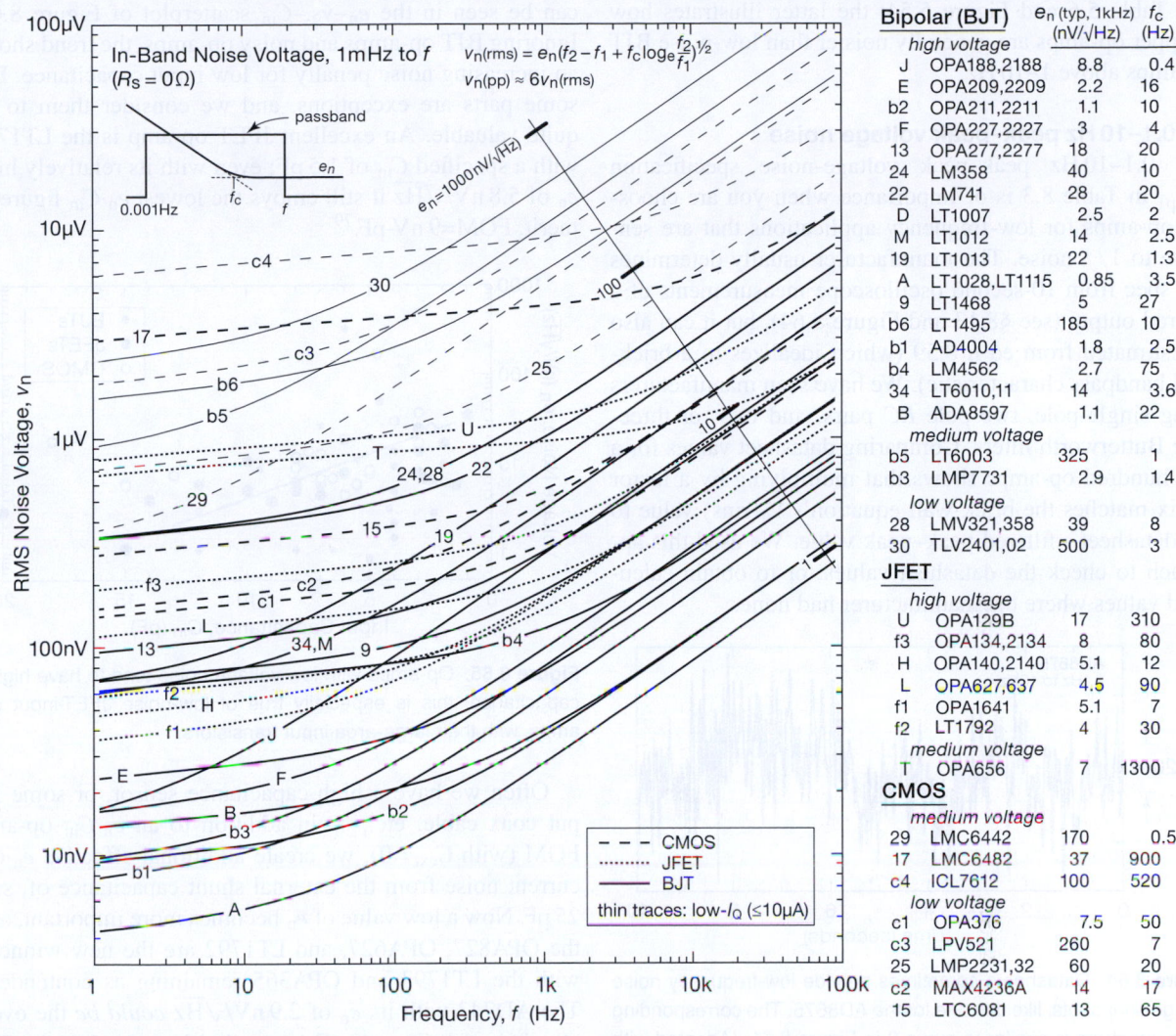

Bipolar (BJT)		e_n (typ, 1kHz) (nV/√Hz)	f_c (Hz)
high voltage			
J	OPA188,2188	8.8	0.4
E	OPA209,2209	2.2	16
b2	OPA211,2211	1.1	10
F	OPA227,2227	3	4
13	OPA277,2277	18	20
24	LM358	40	10
22	LM741	28	20
D	LT1007	2.5	2
M	LT1012	14	2.5
19	LT1013	22	1.3
A	LT1028,LT1115	0.85	3.5
9	LT1468	5	27
b6	LT1495	185	10
b1	AD4004	1.8	2.5
b4	LM4562	2.7	75
34	LT6010,11	14	3.6
B	ADA8597	1.1	22
medium voltage			
b5	LT6003	325	1
b3	LMP7731	2.9	1.4
low voltage			
28	LMV321,358	39	8
30	TLV2401,02	500	3
JFET			
high voltage			
U	OPA129B	17	310
f3	OPA134,2134	8	80
H	OPA140,2140	5.1	12
L	OPA627,637	4.5	90
f1	OPA1641	5.1	7
f2	LT1792	4	30
medium voltage			
T	OPA656	7	1300
CMOS			
medium voltage			
29	LMC6442	170	0.5
17	LMC6482	37	900
c4	ICL7612	100	520
low voltage			
c1	OPA376	7.5	50
c3	LPV521	260	7
25	LMP2231,32	60	20
c2	MAX4236A	14	17
15	LTC6081	13	65

Figure 8.63. Integrated in-band noise voltage for a selection of low-noise and popular op-amps, based on datasheet plots and tabulated values of noise-voltage density. The op-amp labels correspond to those in Figures 8.60 and 8.61 and the listings in Table 8.3. See also Figure 5.54, which includes auto-zero op-amps and extends down to 0.001 Hz.

L. Noise, "integrated noise"

Figure 8.62 shows the *integrated noise* for four low-noise op-amps, the result of adding up (i.e., integrating) the noise density (sometimes called "spot noise") over the operating bandwidth. It has units of voltage (e.g., μVrms), in contrast to noise density, which has units of voltage divided by square root of bandwidth (e.g., nV/$\sqrt{\text{Hz}}$). Sometimes integrated noise is loosely called the "noise voltage," or just the "noise." Integrated noise may be written as v_n, or V_n, but never e_n. In Figure 8.62 the four op-amps have comparable specified e_n *at 1 kHz*, but widely varying $1/f$ corner frequencies f_c, ranging from 2–400 Hz, producing dramat-

ically different integrated noise when used entirely below 1 kHz. In fact, the ghost of the low-frequency noise is seen lurking up to $\sim$10 kHz.

Figure 8.63 plots integrated noise for three dozen op-amps. In general JFET op-amps are worse than BJT op-amps at low frequencies, and CMOS op-amps are worse still. The plots were calculated from the table's e_n and f_c values, according to eq'n 8.59. At high frequencies, where white noise predominates, the plots simplify to $v_n = e_n\sqrt{f_2}$.

Chopper and auto-zero op-amps (not shown in the figure) do not suffer from $1/f$ noise, but they suffer from higher voltage noise, and generally excessive current noise.

See Table 5.6 and Figure 5.54; the latter illustrates how chopper op-amps are generally noisier than low-noise BJT op-amps above 1–10 Hz.

M. 0.1–10 Hz peak–peak voltage noise

The 0.1–10 Hz peak–peak voltage-noise specification $V_{n(pp)}$ in Table 8.3 is of importance when you are choosing op-amps for low-frequency applications that are sensitive to $1/f$ noise. The manufacturer usually determines this spec from 10-second oscilloscope measurements of a filtered output (see §8.13 and Figure 8.64), but it can also be estimated from eq'n 8.59 (which idealizes to a brick-wall bandpass characteristic). We have seen manufacturers using single-pole, two-pole RC pairs, and two- or three-pole Butterworth filters. Comparing datasheet values for a few hundred op-amps shows that multiplying by a factor of six matches the brick-wall equation's $V_n(rms)$ value to the datasheet's filtered peak–peak value. We used this approach to check the datasheet values, or to obtain calculated values where the manufacturer had none.

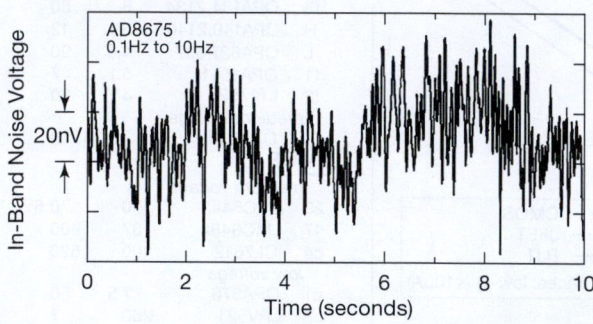

Figure 8.64. Datasheets sometimes provide low-frequency noise voltage snapshots, like this one for the AD8675. The corresponding noise spectrum is similar to curve 8 in Figure 8.61. (Adapted with permission of Analog Devices, Inc.)

N. Input capacitance C_{in}

Op-amp input capacitance is a serious issue when you are considering $e_n \cdot C_{in}$ noise in transimpedance amplifiers (TIAs); see §8.11. For some high-impedance sensors, input capacitance acts both as an additional high-frequency load, and it provides a way for high-frequency noise on the power-supply rails to be coupled to the input as noise current. Some op-amps provide both common-mode and differential capacitance values; for these we have taken the larger value for the table (as always, we suggest careful study of the datasheet before embarking on a design).

Quite often, op-amps with low e_n use large area input transistors, with correspondingly greater capacitance, as can be seen in the e_n–vs.–C_{in} scatterplot of Figure 8.65. Ignoring BJT op-amps and noisy op-amps, the trend shows an increasing noise penalty for low input capacitance. But some parts are exceptions, and we consider them to be quite valuable. An excellent JFET op-amp is the LT1793 with a specified C_{in} of 1.5 pF; even with its relatively high e_n of 5.8 nV/$\sqrt{Hz}$ it still enjoys the lowest $e_n \cdot C_{in}$ figure of merit, FOM=9 nV-pF.[79]

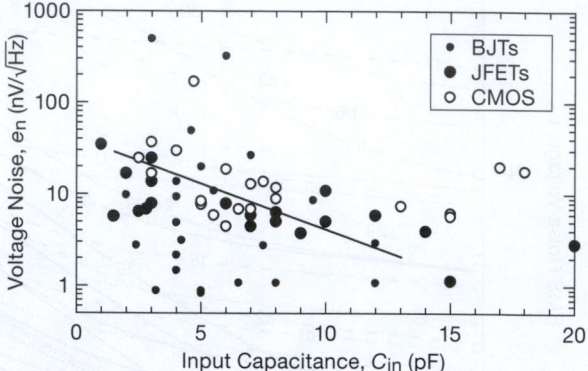

Figure 8.65. Op-amps with low voltage noise tend to have higher capacitance; this is especially true of low-noise JFET-input op-amps, with their large-area input transistors.

Often we have a high-capacitance sensor, or some input coax cable, etc., so in addition to an $e_n \cdot C_{in}$ op-amp FOM (with $C_{ext}=0$), we create additional effective $e_n \cdot C_{in}$ current noise from the external shunt capacitance of, say, 25 pF. Now a low value of e_n becomes more important, and the OPA827, OPA627, and LT1792 are the new winners, with the LT1793 and OPA365 remaining as contenders. The AD743 with its e_n of 2.9 nV/$\sqrt{Hz}$ *could be* the overall winner for $C_{ext}>25$ pF, but sadly it's entered the land of NRND ("Not Recommended for New Designs"). We recommend that you grab 'em if you can find 'em! Finally, for high-C_{ext} applications, consider discrete designs (see §8.3 and Figure 8.66) that can outperform the best IC op-amps.

Op-amps with small-geometry low-capacitance input transistors nonetheless suffer from the necessary added capacitance of their input protection devices. Truly low-capacitance op-amps may be more susceptible to static damage during handling. Bipolar transistors are less susceptible to static damage and tend to have much lower capacitance than low-noise JFET and CMOS op-amps, but their high I_B and i_n values will often preclude their use in

---

[79] Other parts to consider are the OPA124, OPA121, AD8067, and OPA656, the latter with an f_T of 230 MHz. For CMOS op-amps consider the OPA365, with a 0.2 pA I_B (typ) spec.

high-impedance circuits. An exception would be for transimpedance amplifiers above, say, 1–10 MHz, but for these applications you'll need to look at the table of high-speed op-amps, Table 8.3c; those parts aren't evaluated here.[80] If a bipolar op-amp's C_{in} is unstated, you can often assume that it's in the range of 2–5 pF.

8.9.2 Power-supply rejection ratio

Quite apart from noise sources *within* the op-amp, any noise (or interfering signals) on the power-supply rails will couple to the output, attenuated by the power-supply rejection ratio (PSRR). Typical op-amps have pretty good PSRR at low frequencies (roughly $1/G_{OL}$, so 80–140 dB), but the PSRR drops as $1/f$ at higher frequencies, allowing substantial supply-noise coupling. Often the high-frequency PSRR is particularly poor with respect to one or the other supply rail because of coupling by the internal compensation capacitor; see §4x.7. For example, at frequencies above 10 Hz, the LT1012 positive-rail PSRR is 25 dB worse than the negative-rail PSRR; and the micropower LT6003's negative-rail PSRR plummets to less than 10 dB at 1 kHz! Simple RC filtering for the sensitive stages (or a capacitance multiplier, see §8.15.1) can largely solve this problem. Be sure to read the datasheet, though, or you may not even know you *have* a problem.

8.9.3 Wrapup: choosing a low-noise op-amp

In summary, when choosing an op-amp for a low-noise application, begin by restricting your attention to op-amps that meet your other needs, such as accuracy, speed, power dissipation, supply voltage, input and output swing, and the like. Then choose among this subset, based upon their noise parameters. Generally speaking, *you want op-amps with low i_n for high signal impedances, and op-amps with low e_n for low signal impedances.* As we've seen, the total input-referred squared noise-voltage density is just

$$v_n^2 = 4kTR_{sig} + e_n^2 + (i_nR_{sig})^2 \quad V^2/Hz, \qquad (8.42)$$

where the first term is due to Johnson noise and the last two terms are due to op-amp noise voltage and current.[81] Clearly the Johnson noise sets a lower bound to the input-referred noise. Refer back to Figure 8.58 for an easy graphical look at v_n (at 1 kHz) as a function of R_{sig} for a selection

of low-noise op-amps from Table 8.3;[82] these span a representative range of e_n and i_n, from which you can interpolate for op-amps not shown explicitly.

This plot illustrates graphically the tradeoff between low noise *voltage* (where the BJT parts – solid lines – are king), and low noise *current* (where the CMOS parts – dashed lines – are the winners). A good low-noise JFET op-amp like the OPA140 combines the best (nearly) of both worlds. But note that the voltage noise of even a simple (and noisy) CMOS part like the LMC6482 is irrelevant when driven by a source resistance anywhere between 1 MΩ and 10 GΩ, where the op-amp's contribution to the total noise density is negligible.

See Figures 8.60 and 8.61 for op-amp spectral noise plots. For low-frequency amplifier applications see Figure 8.63, with its plots of total (integrated) rms noise v_n versus bandwidth.

Two cautions: (a) The total noise density plots (v_n versus R_s) of Figure 8.58 characterize the performance only at 1 kHz. So an op-amp with a high $1/f$ noise corner will look considerably worse at, say, 10 Hz. Taking the example of the LMC6482 (curve #17), its e_n at 10 Hz is ~170 nV/$\sqrt{Hz}$, aligning it with curve #29. (b) Likewise, at *high* frequencies you care about input capacitance (especially in combination with GΩ-scale input impedances), and you care about the noise current generated by the op-amp's noise voltage in combination with the input capacitance ($i_{nC} = e_n\omega C_{in}$). A quiet JFET op-amp like the OPA627 has nearly triple the input capacitance (because of its large-area input JFET) of its somewhat noisier (and less expensive) OPA656 cousin.

8.9.4 Low-noise instrumentation amplifiers and video amplifiers

In addition to the low-noise op-amps, there are some nice low-noise IC *instrumentation amplifiers* and low-noise *preamplifiers*. Unlike general-purpose op-amps, these generally have fixed voltage gain, or provision for attaching an external gain-setting resistor. Instrumentation amplifiers, intended for precision differential applications, are discussed extensively in Chapter 5 (§5.13). Those categorized as "video amplifiers" often have bandwidths into the tens of megahertz, though they can be used for low-frequency applications as well. Examples are the TI/Burr–Brown INA103 instrumentation amplifier, and the TI/National

---

[80] Table 8.3c features a new e_nC_{in} figure-of-merit, useful for TIA designs.

[81] In addition, as described in §8.11.3, in the inverting configuration an amplifier's input noise *voltage* creates a noise *current* through the input capacitance, to the tune of $i_n = e_n\omega C_{in}$.

[82] For comparison we included our jellybean JFET LF411, the legacy LM741 and LM358, and a few other popular op-amps that do not qualify as "low noise"; those have numeric labels.

LMH6517 or Linear Technology LTC6400-series video amplifiers. These amplifiers typically have e_n down in the $1 \, \mathrm{nV}/\sqrt{\mathrm{Hz}}$ territory, achieved (at the expense of high input noise *current*, i_n) by running the input transistor at relatively high collector current.

8.9.5 Low-noise hybrid op-amps

Back in §8.9.1E we hinted at the possibility of a hybrid low-noise design, in which a low-noise BJT or JFET differential front-end could combine with a conventional op-amp to provide superior noise performance. The simplified circuits in Figures 8.66 and 8.67 show how this can be done, with BJTs and JFETs, respectively. These amplifiers are also a form of composite amplifier, a subject discussed in §4x.5.

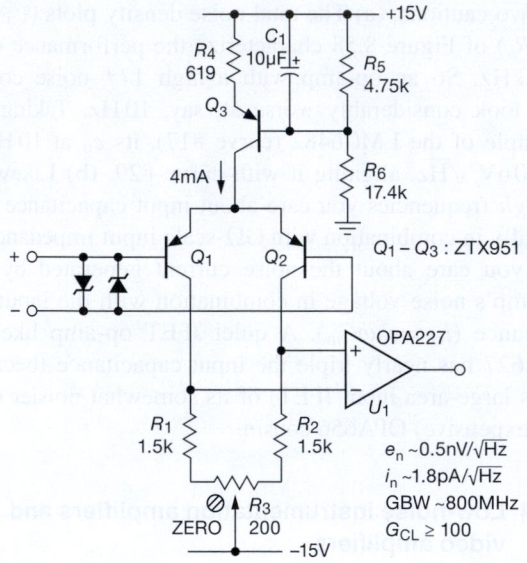

Figure 8.66. Combining the best of both worlds: a BJT hybrid wideband low-noise op-amp.

A. General design issues

Common-mode range In both cases we designed the circuit to provide a substantial common-mode input signal range (at least $\pm 10 \, \mathrm{V}$), intending this to serve as a general-purpose low-noise op-amp substitute. For that reason the current sources in the emitter (or source terminal) are biased for compliance to within 2.5 V of their respective rails, and the collector load resistors (R_1 and R_2) in the BJT design are sized to drop only 3 V (thus accommodating negative input swing to at least $-12 \, \mathrm{V}$). The JFET design is

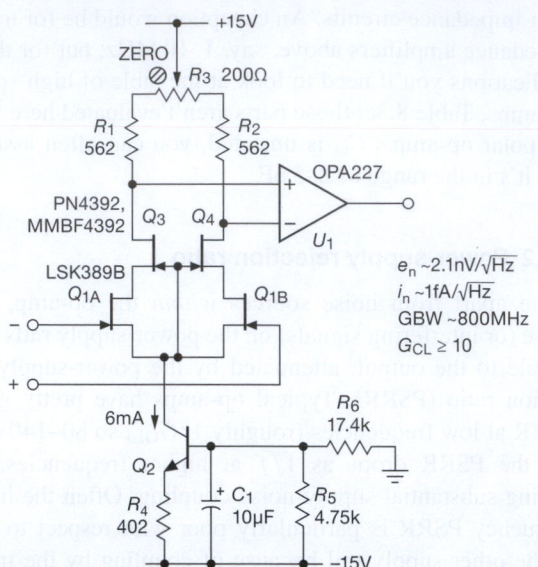

Figure 8.67. Another hybrid op-amp, this time with differential JFET cascode front-end.

similarly biased; however, a volt or two is lost in the cascode transistors (Q_3 and Q_4) and in the negative V_{GS} of the input stage, reducing the positive common-mode input range to approximately $+10 \, \mathrm{V}$.[83]

Offset trim In the BJT circuit we used one of the quietest available transistors we found in our tests (the Zetex ZTX951); it is not available as a matched pair, so we provided an offset trim R_3, with a relatively large adjustment range ($\pm 6\%$). For the JFET circuit we used the LSK389 "matched" dual, but its worst-case offset of $20 \, \mathrm{mV}$ mandates an even larger trim range, here $\pm 17\%$. Ironically, the discrete BJT pair's V_{BE} mismatch is likely to be less than that of the matched JFET dual; at least that's been our experience; see, for example, the histogram of measured V_{BE}'s in Figure 8.44.

It's important to realize that a simple offset-trim circuit that unbalances the load resistances, as shown here, greatly compromises the common-mode rejection ratio of the differential input. Additionally, the unbalanced load greatly compromises attenuation of noise in the current source (which would be at least $\times 50$ with 1% load resistors). A better method involves balancing the currents seen at the collector (or drain) load resistors; see for example Figure 8.80 later in the chapter.

Minimum closed-loop gain Because there's voltage

---

[83] If the cascode runs out of compliance it simply disappears, becoming a low-R_{on} switch.

gain in the transistor front-ends of these hybrid amplifiers (about $\times 120$ for the BJT, $\times 12$ for the JFET), the overall loop gain is substantially greater than that of the second-stage op-amp alone (which by itself is quite high, roughly 160 dB at low frequencies). To ensure stability (given that the OPA277 is unity-gain stable), these hybrid amplifiers should be configured with a closed-loop gain of at least $\times 100$ and $\times 10$, respectively.[84]

B. Some detailed design issues

Choice of transistors Although its noise voltage is unspecified, we found excellent noise performance (and good consistency) with some Zetex bipolar transistors, notably their ZTX851 (*npn*) and ZTX951 (*pnp*). The latter were slightly better, with a measured input noise voltage corresponding to an admirably low base resistance of $r_{bb'} \sim 1.2\,\Omega$ (thus, e.g., an e_n of $0.17\,nV/\sqrt{Hz}$ at 10 mA collector current), compared with $r_{bb'} \sim 1.4\,\Omega$ for the *npn* ZTX851.[85] For the JFET circuit we chose the low-noise LSK389B from Linear Integrated Systems; it specifies $e_n = 0.9\,nV/\sqrt{Hz}$ (typ) and $1.9\,nV/\sqrt{Hz}$ (max) at a drain current of 2 mA.[86] It's a large-geometry JFET (for low noise), with a correspondingly large capacitance of $C_{iss} = 25$ pF. It has pleasantly high transconductance, specified as 20 mS (min) at 3 mA drain current.

Input current The BJT has the expected high input current, of course, given the relatively high collector current needed to keep e_n low; with the specified $\beta = 200$ (typ) the input current is 10 µA. That's the price you pay for the lowest e_n. By comparison, the JFET circuit's input current is down in the 1 pA territory, judging from the graph of "excess gate current" shown for its predecessor 2SK389 (the LSK389 provides no analogous graphs or specification), and given the very low V_{DS} enforced by the cascode.

Overall noise voltage These are differential input stages, so you need to multiply the single-transistor e_n's by $\sqrt{2}$, which results in the values shown in the figures.

---

[84] The OPA277 has an excellent phase margin, about $60°$, to well past its 1 MHz GBW. If you need a lower closed-loop gain, you can add a series $R+C$ across the op-amp's inputs, effectively reducing the load resistance and open-loop gain at high frequencies.

[85] The *pnp* has lower noise than its *npn* sibling, but much poorer Early voltage, as Table 8.1a shows: $V_A = 120$ V versus 410 V. If the (relatively low) Early voltages of the *pnp* pair are not matched, the CMRR will be degraded. If this is an issue, the circuit could be recast with an *npn* pair, or with an added cascode.

[86] Our measurements landed right in the middle, at $1.34\,nV/\sqrt{Hz}$ ($1.9\,nV/\sqrt{Hz}$ total for the differential pair).

For the JFET version, with its lower front-end gain, there is a small contribution also from the op-amp's input noise ($3\,nV/\sqrt{Hz}$), but when referred to the input that amounts to just $0.3\,nV/\sqrt{Hz}$, negligible in the total when combined with the larger front-end squared noise-voltage density.

Note that these are quiet amplifiers – so you need to present their inputs with low signal-source impedances to retain their low e_n. For the BJT version, for example, that implies a resistor value down around $10\,\Omega$ in the lower leg of the feedback divider.

Differential input voltage If your application can expose the differential BJT hybrid op-amp to more than ± 5 V, add a pair of protective back-to-back diodes to prevent base-emitter breakdown and consequent transistor degradation. The differential JFET hybrid op-amp can be safely exposed to full-rail differential-input voltage. These op-amp input stages are running at rather high currents, so a substantial input voltage difference, maintained for more than a few milliseconds, will cause unbalanced heating and consequent input offset voltage. It's a good idea to thermally couple the input pair, and perhaps thermally isolate them from air currents as well.

8.10 Signal transformers

As mentioned earlier in §8.5.6, you're up against the wall when trying to minimize amplifier noise with a signal of very low source impedance Z_s, say less than $100\,\Omega$. The noise voltage of a $50\,\Omega$ resistor, for example, is just $0.9\,nV/\sqrt{Hz}$, which puts you down at the limit of the quietest op-amps. And there are signal transducers with considerably lower resistance, for example magnetic coil pickups. If you are interested only in ac signals (as you would be, for a magnetic coil), you can use a transformer to raise the signal level (by the turns ratio, $n{:}1$), simultaneously raising the source impedance seen at the amplifier's input by that ratio squared; i.e., the amplifier sees a signal of source impedance $n^2 Z_s$.

High-quality signal transformers are available from companies such as Jensen Transformers and Signal Recovery. For example, a signal in the audio band (say from 100 Hz to 10 kHz) with source impedance of $100\,\Omega$ would be a poor match for an amplifier like the SR560, whose lowest noise figure occurs for signals of source impedance around $500\,k\Omega$ (see Figure 8.27). The problem is that the amplifier's voltage noise is much larger than the signal source's Johnson noise; the resultant noise figure for that signal connected directly to the amplifier would be 12 dB. By using an external step-up transformer like the

Jensen JT-115K-E with its 1:10 turns ratio (150Ω:$15k\Omega$ impedances), the signal level is raised (along with its source impedance), thus overriding amplifier noise voltage. At this signal impedance the amplifier's noise figure is about 0.4 dB; however, the resistance of the transformer's windings[87] contributes to produce an overall noise figure of about 1.5 dB.

There's a tendency to associate transformers with mediocre performance; but well-engineered signal transformers like this one are pretty good: response ± 0.15 dB over the audio band (20 Hz–20 kHz), down 3 dB at 2.5 Hz and 90 kHz, distortion less than 0.1% even at 20 Hz,[88] and, as a bonus, common-mode rejection of 110 dB at 60 Hz. The Model 1900 signal transformer from Signal Recovery (originally from Princeton Applied Research) offers turns ratios of 1:100 and 1:1000. It has less bandwidth than the JT-115K-E, but it achieves a minimum noise figure of $\sim$0.5 dB for source resistances in the 0.8–10 Ω range owing to its low winding resistance of 0.04 Ω. This is somewhat better noise performance than the paralleled-BJT amplifier of §8.5.9; because of transformer limitations, however, it does not work over the full audio band.

At *radio* frequencies (e.g., beginning around 100 kHz) it is extremely easy to make good transformers, both for tuned (narrowband) and broadband signals. At these frequencies it is possible to make broadband "transmission-line transformers" of very good performance. It is at the very low frequencies (audio and below) that transformers become problematic.

Three comments:

(a) The voltage rises proportional to the turns ratio of the transformer, whereas the impedance rises proportional to the square of the ratio. Thus a 2:1 voltage step-up transformer has an output impedance four times the input impedance (this is mandated by conservation of energy).

(b) Transformers aren't perfect. They are susceptible to magnetic pickup, and they have trouble at low frequencies (magnetic saturation) and at high frequencies (winding inductance and capacitance), as well as losses from the magnetic properties of the core and from winding resistance. The latter is a source of Johnson noise, as well. Nevertheless, when dealing with a signal of very low source impedance, you may have

no choice, and transformer coupling can be very beneficial, as the preceding example demonstrates. Exotic techniques such as cooled transformers, superconducting transformers, and SQUIDs (superconducting quantum interference devices) can provide good noise performance at low impedance and voltage levels. With SQUIDs you can measure voltages of 10^{-15} V!

(c) Again, a warning: don't attempt to improve performance by adding a resistor in series with a low source impedance. If you do that, you're just another victim of the noise-figure fallacy.

8.10.1 A low-noise wideband amplifier with transformer feedback

Lest we leave the reader discouraged as to the merits of transformers in signal applications, we hasten to point out that some very impressive performance can be gotten with careful engineering. Figure 8.68 shows a nice example: a wideband low-noise amplifier with high input impedance,[89] with a transformer providing lossless (therefore noiseless) feedback to set the voltage gain via the turns ratio (here $\times 10$). The transformer nicely solves also the problem of distributing identical feedback to multiple paralleled input JFETs, each of which is biased to the same 10 mA drain current.

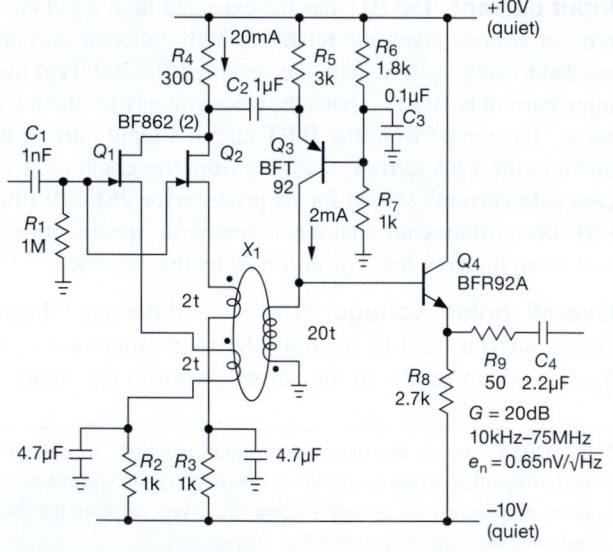

X_1: Vitrovac T60009 - E4006 - W650, $A_L = 13\mu H/t^2$

Figure 8.68. Transformer feedback to paralleled JFETs achieves 650 pV/$\sqrt{\text{Hz}}$ input noise voltage in this wideband amplifier.

---

[87] At low frequencies this is simply the dc resistance of the copper wire; but at higher frequencies the effective noise-generating resistance rises owing to the skin effect; see Appendix H (§H.1.4) and §*1x.1*.

[88] Low frequencies are a transformer's Achilles' heel; at 1 kHz this particular transformer's distortion is just 0.001%.

[89] From J. Belleman at CERN; see http://jeroen.home.cern.ch/jeroen/tfpu.

The transformer shown is a small (6.5 mm O.D.) high-mu tape-wound toroid, with just a few turns in the windings to set the high end of the operating band to 75 MHz. That choice limits the low frequency end to frequencies above $\sim$10 kHz, but the nearly 10,000:1 frequency range is mighty impressive. The source pull-down resistors R_2 and R_3 set the drain current of the BF862 input JFETs to 10 mA (their specified minimum I_{DSS}), where their voltage noise is $0.9\,\mathrm{nV}/\sqrt{\mathrm{Hz}}$; the parallel pair improves that by $\sqrt{2}$, thus $0.65\,\mathrm{nV}/\sqrt{\mathrm{Hz}}$. Bipolar *pnp* Q_3 forms an inverted cascode, with its r_e of 12 Ω diverting essentially all of the input transistors' drain signal. Both Q_3 and Q_4 are wideband transistors (5 GHz); Q_3's collector current is chosen to cancel the dc magnetization induced in the transformer by the input transistors to prevent core saturation. The supply rails need to be quiet, best achieved with a capacitance multiplier (§8.15.1). A caution: this feedback circuit has two low-frequency breakpoints (from C_2 and from the transformer's rather low magnetizing inductance), leading to potential instability and a low-frequency *motorboating*-type oscillation. This is prevented here by the large capacitance of C_2, which puts its breakpoint far below that of the transformer.

8.11 Noise in transimpedance amplifiers

Transresistance amplifiers (or *transimpedance* amplifiers, "TIA," or sometimes just "current amplifiers") produce a voltage output in response to a *current* input. Their gain is therefore V_{out}/I_{in}, with units of ohms, hence their name.[90] We introduced transresistance amplifiers in Chapter 4 (§4.3.1), we evolved a TIA design for an inexpensive lightning detector (§8.5.7, Figure 8.36); and we look at issues of their feedback *stability* in §4x.3. As a reminder, the basic circuit, which we've seen before, is shown in Figure 8.69. Assuming ideal components, a current I_{sig} produces an output $V_{out} = -I_{sig}R_f$, so the gain is simply $-R_f$. They are used extensively in circuits where the input is a current, for example from a photodiode or photomultiplier, charged particle detector, tunneling microscope, or patch-clamp amplifier.

The emphasis in this chapter is *noise*, which can arise in several ways in the TIA. The amplifier itself will have its input e_n and i_n; and the feedback resistor creates a Johnson noise voltage that equates to an input

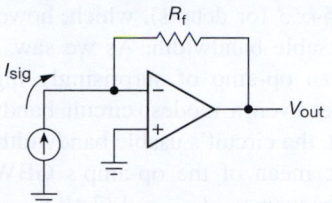

Figure 8.69. Transresistance amplifier: current in, voltage out.

noise current $i_{nR} = \sqrt{4kT/R_f}$ (thus favoring high values of R_f). Capacitance also plays an important role, not only in stability and bandwidth, but also in converting the amplifier's noise *voltage* into a noise current. There's also likely to be what we might call *signal noise*: shot noise in the signal current, Johnson noise in the signal source's resistance, and other forms of signal fluctuation.[91] Let's see how this all plays out.

8.11.1 Summary of the stability problem

As we saw back in §4.3.1C (and which we discuss in greater detail in §4x.3 in Chapter *4x*), capacitance to ground at the input (e.g., from a current-output sensor and its connecting cable), in combination with the (usually large) feedback resistance, produces a lagging phase shift in the feedback path. This is unstable when combined with the op-amp's 90° (or more) lagging phase shift. You cure this by putting a small capacitor C_f across the feedback

---

[90] This is the inverse of a device like a JFET, where an input *voltage* produces an output *current*. There the gain is I_{out}/V_{in}, with units of Ω^{-1} (siemens, previously called mhos), hence the name trans*conductance* amplifiers.

[91] Noise in the signal source itself may, in some situations, dominate over the noise contribution of a well-designed amplifier. Optical detectors, for example, are characterized by a "noise-equivalent power" (NEP), which is their electronic noise output stated in terms of optical input power (expressed as optical noise power density, generally in the range of $\mathrm{fW}/\sqrt{\mathrm{Hz}}$). The NEP arises from the detector's dark current (a perfect detector would have zero NEP), a "drift current" that can be thought of as a form of leakage. You get such current when photodiodes are operated with a deliberate reverse-bias voltage (the so-called *photo-conductive mode*), in which the capacitance is reduced, and the speed, linearity, and long-wavelength conversion efficiency are improved. If you want to detect very low light levels, and are willing to sacrifice speed, you should operate the photodiode with zero bias (the so-called *photovoltaic mode*), where the greatly reduced NEP arises mostly from a "diffusion current" related dark resistance, powered by the op-amp's offset voltage. For this purpose the photodiode's datasheet usually assumes a 10 mV offset (i.e., detector bias), but you can almost always do better than that. We've measured dark-adapted NEP levels four orders of magnitude better than specified, when driving an op-amp with $100\,\mu\mathrm{V}$ offset in a cool laboratory setting. Taking account of detector noise is an essential part of any system design whose ultimate job is to "convert photons into volts."

resistor (see §4x.3 for details), which, however, seriously reduces the usable bandwidth. As we saw, you therefore need to use an op-amp of surprisingly high bandwidth in order to get even a modest circuit bandwidth. To put numbers on it, the circuit's usable bandwidth f_c is roughly the geometric mean of the op-amp's GBW (or f_T) and the rolloff frequency, $f_{RC} = 1/2\pi R_f C_{in}$, of the input capacitance; that is

$$\text{GBW} = f_c^2 / f_{RCin} . \qquad (8.43)$$

8.11.2 Amplifier input noise

A transimpedance amplifier input stage, whether discrete or op-amp, will have some noise voltage and current, characterized by densities e_n and i_n; these will exhibit the, by-now, usual $1/f$ rise at low frequencies. And, in addition to the low-frequency noise tail, the current noise will depend on the dc input current (bias current, for BJTs; leakage current for JFETs or MOSFETs), because at mid-frequencies the input noise current is simply shot noise of the dc current.

How do e_n and i_n affect the TIA's overall input noise? We'll see presently (§8.11.3) that e_n generates a noise current through the capacitance seen at the input, and in fact this can easily become the dominant noise term, particularly at higher frequencies. Before worrying about that effect, note simply that the input e_n flows through the feedback resistor, generating a noise-current density $i_n = e_n / R_f$ (and if the signal source has a finite source resistance, replace R_f with $R_f \| R_s$). These terms are usually small compared with the corresponding Johnson-noise currents, but they can grow to significant levels at low frequencies where the $1/f$ behavior may raise e_n by factors of $10\times$ to $50\times$ over its mid-frequency ("white") value.[92]

The current noise i_n of the input stage needs no conversion – it contributes directly to the TIA's equivalent input noise current (as square root of sum of squares). Low-bias op-amps with FET inputs (as well as discrete JFETs) will generally have quite low input-noise currents, down in the $fA/\sqrt{Hz}$ territory. But watch out for rising $1/f$ noise: the datasheet for the low-e_n JFET AD743, for example, shows i_n rising from its mid-frequency 7 fA$\sqrt{Hz}$ value to

100 fA$\sqrt{Hz}$ at 1 Hz; that would correspond to the shot noise from a dc bias of 30 nA! (Its typical room temperature input current is specified as 0.15 nA.) Watch out, also, for rising i_n (from rising leakage current in FETs) at elevated temperatures: the AD743 specifies i_n only at 25°C (where it's 7 fA$\sqrt{Hz}$ value is consistent with shot noise in 0.15 nA dc input current); but you can use their plot of dc input current versus temperature to find that $i_n = 40$ fA$\sqrt{Hz}$ at 80°C (where the plot shows a dc input current of 5 nA) and rising further to 400 fA$\sqrt{Hz}$ at 125°C.

Temperature isn't the only contributor to JFET input current; there's the impact-ionization effect (§3.2.8), which can cause a devastating rise in input current (and noise) when a JFET is operated with drain-to-source voltages greater than a few volts. With a discrete JFET design you can prevent this by operating at low drain voltage (e.g., with a cascode); low-noise op-amps usually are designed with this effect in mind, but you will see increased input current (and current noise) as the input voltage approaches one of the rails (e.g., the positive rail for the LT1792 or ADA4627; the negative rail for the AD8610).

8.11.3 The $e_n C$ noise problem

In addition to the input stage's contributions of i_n (directly) and e_n (flowing through the feedback and input resistances), input *capacitance* (already an annoyance in terms of stability and bandwidth) creates interesting noise problems as well. For example, you might at first think that amplifier *voltage* noise is of little concern in an amplifier whose input is a current; because its feedback looks like a voltage follower, it would seem to produce at most an additive output voltage-noise contribution just equal to the input-voltage noise e_n (equivalent to an input-current-noise contribution of $i_n = e_n / R_f$). But you would be wrong! To see what happens, look at Figure 8.70, where the

---

[92] For example, the voltage noise of the AD743 (quietest JFET op-amp, with $e_n = 2.9$ nV/$\sqrt{Hz}$ at 10 kHz) grows to 23 nV/$\sqrt{Hz}$ at 1 Hz. The LT1792 grows to 30 nV/$\sqrt{Hz}$, the OPA627 to 33 nV/$\sqrt{Hz}$, and the ADA4627 to 42 nV/$\sqrt{Hz}$. And the speedy OPA656 grows to 75 nV/$\sqrt{Hz}$ at 10 Hz! Not to be outdone, the otherwise-admirable AD8610 and 8620 JFET op-amps grow from 6 nV/$\sqrt{Hz}$ to about 200 nV/$\sqrt{Hz}$. And when it comes to CMOS op-amps, whoa!

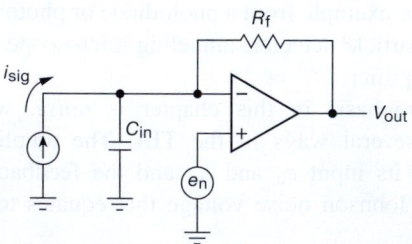

Figure 8.70. The capacitance at the input causes the amplifier's voltage noise e_n to create an input *current* noise $i_n = e_n \omega C_{in}$. Neither this "$e_n C$" current noise nor the amplifier's own input current noise is shown.

op-amp's internal differential-noise voltage e_n is modeled as a voltage in series with the noninverting terminal. Feedback forces the inverting terminal (with its capacitance C_{in} to ground) to follow, creating an actual input current $i_n(t) = C_{in}\, dv_n(t)/dt$ (where $v_n(t)$ is the op-amp's input noise voltage); expressed as a noise *density*, we get

$$i_n = e_n \omega C_{in} = 2\pi e_n C_{in} f. \tag{8.44}$$

That is, the amplifier's voltage noise creates a noise current proportional to the capacitance at the input, and rising proportional to frequency. We refer to this input-noise current, produced by the amplifier's internal noise voltage, as "e_nC noise."[93]

8.11.4 Noise in the transresistance amplifier

Let's apply this kind of thinking to figure out the noise performance of the transresistance amplifier we looked at in §4x.3. We've redrawn it in Figure 8.71, with the op-amp's input-noise voltage e_n, noise current i_n, and shunt feedback capacitor C_f shown explicitly. In a typical high-speed (small-area) photodiode application you might have $C_{in} \sim 10$–20 pF (but more if connected through a shielded cable), $R_f \sim 1$–10 MΩ, and, for a FET-input op-amp, $e_n \sim 3$–10 nV/$\sqrt{\text{Hz}}$ and $i_n \sim 1$–10 fA/$\sqrt{\text{Hz}}$.

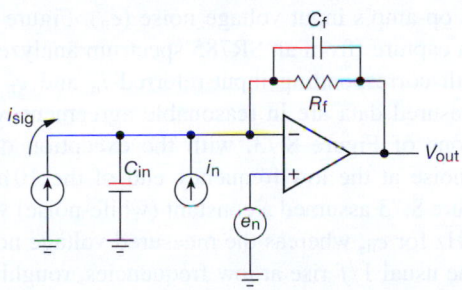

Figure 8.71. Noise in a photodiode amplifier. For the calculations we've used $R_f = 1$ MΩ. Here C_{in} is the total capacitance seen at the input (amplifier, wiring, and input device capacitances).

We'll calculate the noise contributions (from e_n, i_n, and from Johnson noise in the feedback resistor) as effective current noise (versus frequency) *at the input*; after all, that's where the input current signal that we care about enters. For the time being we ignore the capacitor C_f, and use typical circuit values of $R_f = 1$ M and C_{in}(circuit)=10 pF.

As we saw earlier, Johnson noise is flat with frequency, with a noise voltage $e_n = \sqrt{4kTR}$ volts per square root hertz; this translates to a short-circuit noise current of e_n/R; that is, $i_n = \sqrt{4kT/R}$ amps per square root hertz. So for a 1 MΩ feedback resistor at 25°C,

$$i_n = \left(\frac{4kT}{R_f}\right)^{\frac{1}{2}} = 1.28 \times 10^{-10} R_f^{-\frac{1}{2}} = 0.128\,\text{pA}/\sqrt{\text{Hz}}.$$

This is one to two orders of magnitude larger than the amplifier's input current noise, which we can therefore ignore. The last contribution is from the amplifier's e_n, which, as we remarked earlier, looks like an input current noise $i_n = 2\pi e_n C_{in} f$. It rises proportional to frequency, becoming dominant over resistor Johnson noise at some crossover frequency that we'll call f_X. By equating the Johnson-noise current with the e_nC current, you can find[94]

$$f_X = \frac{\sqrt{4kT/R_f}}{2\pi e_n C_{in}}. \tag{8.45}$$

It would continue to rise forever, except for the effect of the parallel capacitance C_f, which causes the noise current to flatten off at a frequency $f_c = 1/2\pi R_f C_f$ (caused by the $R_f C_f$ pole cancelling the +6 dB/octave rising e_nC noise). If we further choose C_f so that f_c equals the geometric-mean frequency[95]

$$f_{GM} = \sqrt{f_{RC_{in}} f_T}, \tag{8.46}$$

we produce a slight peaking at f_c, with the pair of poles causing the e_nC noise to fall at -6 dB/octave (i.e., $\propto 1/f$) at higher frequencies, as seen in Figure 8.72.

In Figure 8.72 we've plotted i_n(input) for the transresistance amplifier of Figure 8.71, with $R_f = 1$M, for the op-amp choices and corresponding data shown in the figure's table. The plot shows clearly the reduction in input-referred total noise current that you get by choosing an amplifier with low input capacitance and low noise voltage, assuming of course that you are planning to limit the output bandwidth with a later stage of lowpass filtering.

It's interesting to note that, if all other things were equal, an op-amp with greater amplifier bandwidth would not reduce the input-referred noise current; it would simply extend the bandwidth of the transresistance amplifier. However, because faster op-amps tend to have less input capacitance, there is some noise advantage, particularly when the external input capacitance is as low as we've assumed here.

---

[93] In the literature you'll see descriptions like "noise gain peaking at higher frequency" and "a complex response to the op-amp's input noise voltage." But we like "e-sub-n-C": it's easy to say, and easy to remember.

[94] For R_f=1M this becomes $f_X(\text{Hz})=2 \times 10^7/e_n C_{in}$, where e_n and C_{in} are in units of nV/$\sqrt{\text{Hz}}$ and pF respectively.

[95] See the stability criterion in §4x.3.

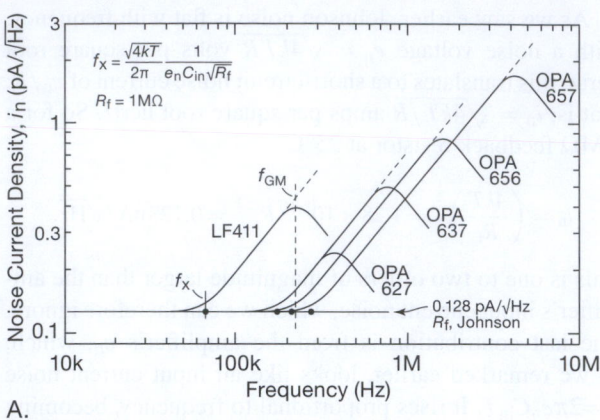

A.

$$f_X = \frac{\sqrt{4kT}}{2\pi} \frac{1}{e_n C_{in} \sqrt{R_f}}$$

$R_f = 1\text{M}\Omega$

0.128 pA/√Hz
R_f, Johnson

B.

Calculated f_C and f_X, with $C_{external} = 10\text{pF}$ and $R_f = 1\text{M}$

Op-amp	e_n (nV/√Hz)	C_{amp} (pF)	$C_{in(total)}$ (pF)	f_T (MHz)	f_C (kHz)	f_X (kHz)
LF411	25	2	12	4	230	67
OPA627	4.5	15	25	16	320	178
OPA637	4.5	15	25	80	715	178
OPA656	7	3.5	13.5	230	1650	212
OPA657	4.8	5.2	15.2	1600	4100	274

Figure 8.72. A. Input-referred noise-current spectrum for the photodiode amplifier in Figure 8.71. The crossover frequency f_X at which "$e_n C$" noise becomes dominant is marked with a dot for each configuration. For the LF411 example both f_X and the turnover frequency f_c are marked. In each case it is assumed that the compensation capacitor C_f is chosen so that f_c is equal to $f_{GM} = \sqrt{f_{RCin} f_T}$, thereby obtaining maximum amplifier bandwidth consistent with only modest peaking (damping ratio $\zeta = 0.7$); otherwise there will be a region of flat noise density, as in Figure 8.73. B. Input-referred current-noise parameters for the photodiode amplifiers of Figure 8.71, assuming an external input capacitance of 10 pF.

8.11.5 An example: wideband JFET photodiode amplifier

Continuing along this path, in Figure 8.73 we've plotted the transimpedance gain, noise gain, and effective input-noise current for a transimpedance amplifier made from an OPA656 ($f_T = 230$ MHz, $e_n = 7$ nV/√Hz), with a 1M feedback resistor and a conservative value of shunt capacitor (2 pF: $f_c = 76$ kHz) to ensure stability with input capacitances as high as 1000 pF.

Note that there are two "gains" here: the *transimpedance gain* (top plot) is the ratio of output signal voltage to input signal current, a plot that is nominally flat up to the rolloff at f_c, but with the additional constraint imposed by the finite (and falling) op-amp open-loop gain G_{OL}. The *noise gain* (middle plot) is the ratio of output signal voltage to input noise voltage, with its characteristic rising "$e_n C$" slope proportional to frequency. It flattens at f_c, but (for op-amps of modest open-loop gain[96]) is further limited by G_{OL}.

Finally, the effective input-referred noise current seen at the output (bottom plot) is the sum of the $e_n C$ noise and the feedback resistor's Johnson noise, as shaped by the amplifier's rolloff. Those noise-current terms (as seen at the input) are $e_n \omega C$ and $\sqrt{4kT/R_f}$, respectively. Here

---

[96] Open-loop gain limitations may be encountered with very wideband op-amps (e.g., parts like the OPA655/6/7, with GBW values in the GHz region), but such limitations are rare for lower-frequency op-amps such as the OPA637.

the Johnson noise dominates at low frequencies: a 1 MΩ resistor generates a (short-circuit) white-noise current of 0.13 pA/√Hz.

We measured this amplifier's output noise spectrum with the four values of input capacitance; we measured also the op-amp's input voltage noise (e_n). Figure 8.74 is a screen capture (from an SR785 spectrum analyzer), labeled with corresponding input-referred i_n and e_n scales. The measured data are in reasonable agreement with the predictions of Figure 8.73, with the exception of some excess noise at the low-frequency end of the 10 nF plot. But Figure 8.73 assumed a constant (white-noise) value of 6 nV/√Hz for e_n, whereas the measured voltage noise exhibits the usual $1/f$ rise at low frequencies, roughly triple its high-frequency value at 100 Hz.

8.11.6 Noise versus gain in the transimpedance amplifier

In the preceding discussion we casually took a round-number value of 1 MΩ for the feedback resistor R_f, without much consideration of the consequences for noise and bandwidth. From Figures 8.72, 8.73, and 8.74 it is easy to see that the low-frequency current noise floor is set by R_f's Johnson noise; so larger values of R_f would appear to be "better."

Not so fast! For a given input capacitance, larger values of R_f correspond to a lower input rolloff frequency ($\omega = 1/R_f C_{in}$), requiring more aggressive compensation

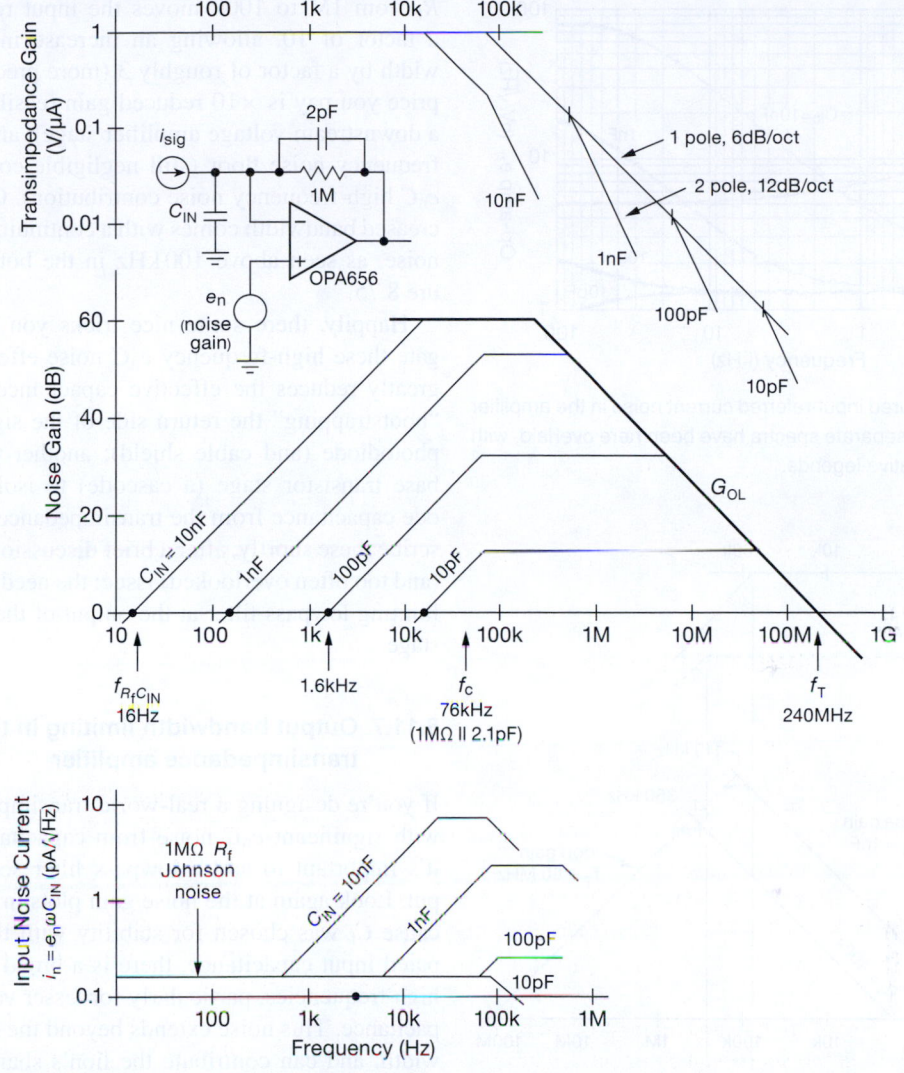

Figure 8.73. Charting the noise in a transimpedance amplifier. A 230 MHz op-amp is used to obtain ~75 kHz bandwidth for input capacitances from 10 pF to 1 nF. The transimpedance bandwidth is not much affected by input capacitance up to 1 nF, but a 10 nF capacitance reduces the bandwidth substantially. It does this without creating instability in this case, because of the OPA656's limited open-loop gain (65 dB).

(i.e., lower bandwidth f_c). If your noise is dominated by Johnson noise in R_f, and you don't care about bandwidth but want to minimize the low-frequency noise, then a larger R_f is good. But in a wideband photodiode amplifier in which the e_nC noise is dominant and in which you want to improve the bandwidth, it's better to reduce R_f.

But, *not too much!* Here's the reason: a goal in any transimpedance amplifier design should be to ensure that the amplifier adds insignificant noise to the inherent shot noise of the input signal. As we reduce R_f, however,

its Johnson-noise current $\sqrt{4kT/R_f}$ increases, ultimately dominating over the input signal's irreducible shot-noise current $\sqrt{2qI_{in}}$. By equating these noise currents we get the condition $I_{in}R_f=2kT/q=50$ mV. That is, to avoid adding amplifier noise, the feedback resistor should not be chosen so small that the voltage drop produced by the input (which could be a dc component of the input signal) is less than, say, 100 mV.

This noise–bandwidth tradeoff in the choice of feedback resistance is illustrated in Figure 8.75, where reducing

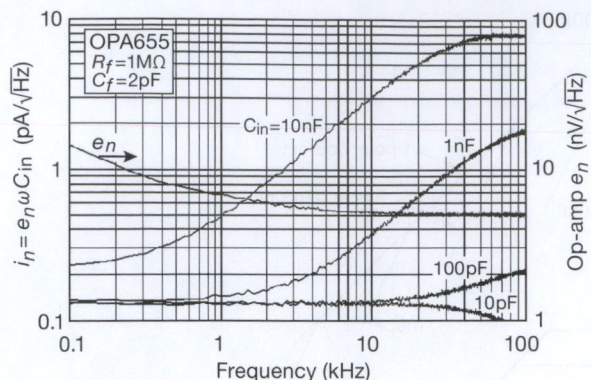

Figure 8.74. Measured input-referred current noise in the amplifier of Figure 8.73. The separate spectra have been here overlaid, with some helpful decorative legends.

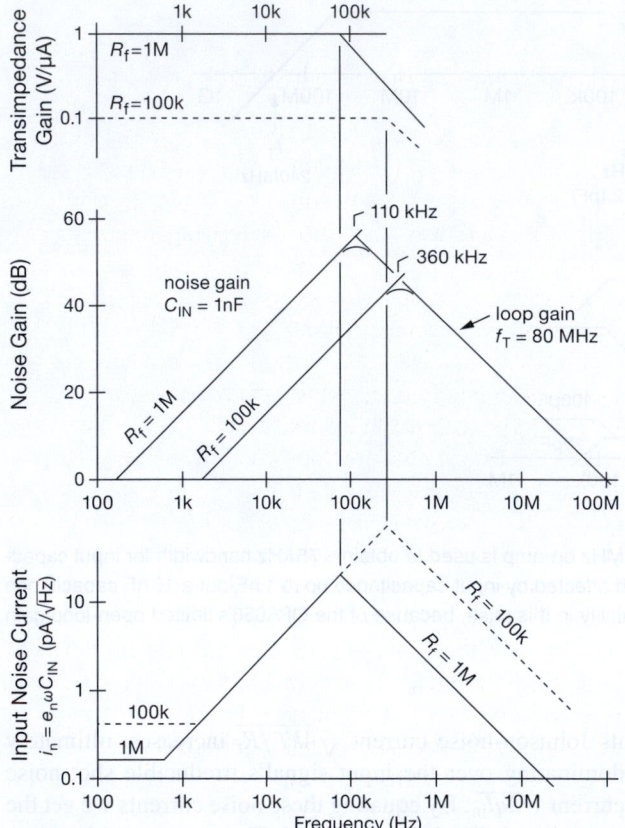

Figure 8.75. In a fast transimpedance amplifier, where e_nC noise dominates over R_f Johnson noise, you can increase the usable bandwidth by reducing R_f; recover the lost gain in a second-stage voltage amplifier.

R_f from 1M to 100k moves the input rolloff upward by a factor of 10, allowing an increase in amplifier bandwidth by a factor of roughly 3 (more precisely, $\sqrt{10}$). The price you pay is $\times 10$ reduced gain (easily recovered with a downstream voltage amplifier stage), and increased low-frequency noise floor (still negligible compared with the e_nC high-frequency noise contribution). Of course, the increased bandwidth comes with a continuing increase in e_nC noise, as seen above 100 kHz in the bottom plot of Figure 8.75.

Happily, there some nice tricks you can use to mitigate these high-frequency e_nC noise effects. One method greatly reduces the effective capacitance at the input by "bootstrapping" the return side of the signal source, e.g., photodiode (and cable shield); another uses a common-base transistor stage (a cascode) to isolate the photodiode capacitance from the transimpedance stage. We'll describe these shortly, after a brief discussion of an important (and too often overlooked) issue: the need for a bandwidth-limiting lowpass filter at the output of the transimpedance stage.

8.11.7 Output bandwidth limiting in the transimpedance amplifier

If you're designing a real-world transimpedance amplifier with significant e_nC noise from capacitance at the input, it's important to add a lowpass filter section at the output. Look again at the noise gain plots in Figure 8.73. Because C_f was chosen for stability with the largest anticipated input capacitance, there is a broad band of noise at high frequencies, particularly for lesser values of input capacitance. This noise extends beyond the amplifier's bandwidth, and can contribute the lion's share of total output noise (remember that logarithmic frequency plots tend to conceal the fact that most of the bandwidth is at the high end).

To see this more clearly, look at Figure 8.76, an analogous set of plots for the current-input amplifier of Figure 8.77.[97] Here we've used a decompensated low-noise JFET op-amp, with C_f chosen such that the amplifier is stable with input capacitances up to 1000 pF. Noise gain and effective input noise current are plotted, for both 1000 pF and 100 pF at the input. The solid lines show U_1's output, and the dashed lines show the signal after a simple RC lowpass. Because the amplifier is overcompensated when $C_{in}=100$ pF, there's a substantial region (the dotted area) of

---

[97] A simplified version of the exemplary photodiode amplifier discussed in §4x.3.

out-of-band noise that is eliminated by the R_1C_1 filter. (The effect is minimal when the amplifier is critically compensated, with f_c nearly equal to f_{GM}, as seen in the 1000 pF plots, because the op-amp's open-loop gain rolloff accomplishes the same thing.)

In this circuit we added an output ×10 gain stage, so that the overall gain is 1 V/μA. That provides another opportunity for an additional section of lowpass filtering, via C_2. This might seem extreme, but remember that a single-section RC has a gentle rolloff, such that equivalent white-noise bandwidth extends well beyond its characteristic frequency ($1.57f_{3dB}$, to be exact; see §8.13).

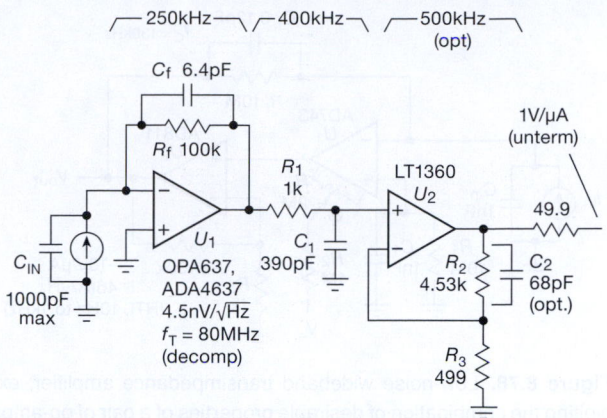

Figure 8.77. Current-input amplifier, BW=250 kHz, compensated for stability with input capacitances up to 1000 pF. The lowpass filters (R_1C_1 and R_2C_2) minimize out-of-band noise, particularly when C_{in} is less than the maximum.

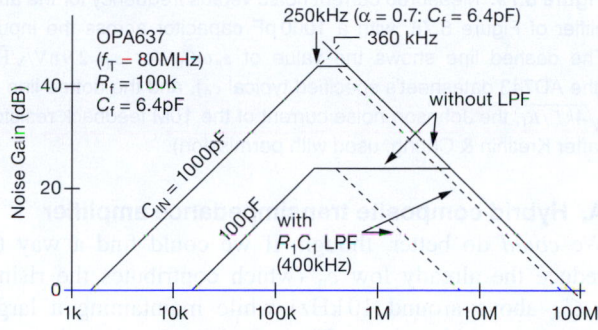

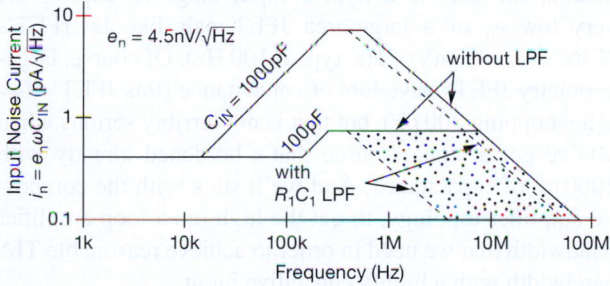

Figure 8.76. Noise plots for the current amplifier of Figure 8.77. Adding an output filter at or slightly above f_c greatly reduces the output noise.

8.11.8 Composite transimpedance amplifiers

When choosing an op-amp for use in a sensitive transimpedance amplifier, you want very low input noise current, thus a JFET or CMOS type. And if you want plenty of speed it's important to select an op-amp with low input noise voltage (to minimize the high-frequency e_nC_{in} noise current it produces), and especially so if the input capacitance is substantial. Finally, because the input noise current produced by the gain-setting feedback resistor R_f decreases

as $1/\sqrt{R_f}$, a low-noise TIA requires a large-value feedback resistor.

But a large R_f produces a low input rolloff frequency ($f_{RCin}=1/2\pi R_fC_{in}$), just what you *don't* need when you're aiming for lots of bandwidth. We'll see presently (in §8.11.9) techniques, such as bootstrapping and cascoding, that can be used in some situations for greatly reducing the effective input capacitance in a transimpedance amplifier. But another approach is simply to select an op-amp with high-enough bandwidth f_T to produce the required TIA bandwidth ($\sim \sqrt{f_{RCin}f_T}$), as we illustrated in §4x.3.3.

This is a reasonable approach – but the performance you get with available op-amps often falls short. Table 8.3 shows that most fast op-amps (say $f_T \geq 350$ MHz) have relatively high noise voltage (e_n of 6 nV/√Hz or greater), and quiet op-amps tend to be slow, like our low-noise winner, the AD743 (e_n=2.9 nV/√Hz and f_T=4.5 MHz). From the table you can see also that some of the fastest op-amps are low-voltage parts, for example the OPA657 with f_T=1500 MHz, whose total supply voltage range is restricted to 9–13 V. That matters, because the larger values of feedback resistor needed for reduced noise produce more gain, therefore larger output swings (and static dc levels, when there's a nonzero dc component to the input current), favoring op-amps that can run on ±15 V.

So, what can be done? A nice approach is to separate the input-stage performance from that of the output stage, to optimize overall noise and speed capability. This can be done with a "composite amplifier," a powerful technique seen, for example, in Figures 5.47 and 13.48, and treated in some detail in §4x.3.

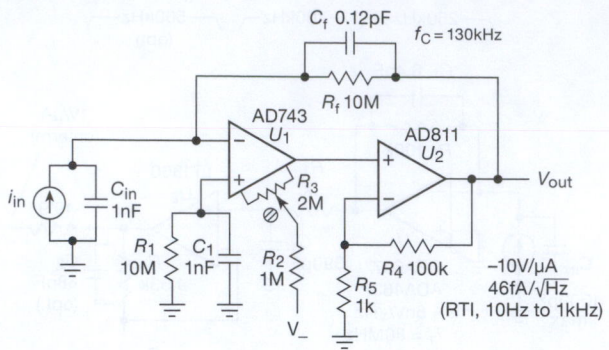

Figure 8.78. Low-noise wideband transimpedance amplifier, exploiting the combination of desirable properties of a pair of op-amps configured as a composite amplifier.

Figure 8.78 shows an example,[98] a TIA intended for low-level current signals burdened with a relatively high capacitance of 1000 pF. Here an AD743 low-noise input stage has been mated with a $\times 100$ wideband AD811 output stage, thus maintaining the low e_n while boosting the composite amplifier's f_T by a factor of 100, to 450 MHz. Although the output stage has considerable bandwidth (it's a current-feedback amplifier intended for video applications), the AD743's meager 4.5 MHz bandwidth (with additional poles around that frequency and above, as evidenced by the datasheet's open-loop gain and phase plots) causes the composite amplifier to exhibit a -12 dB/octave rolloff above 5 MHz, reaching a $180°$ phase shift around 20 MHz. This sounds dangerous, but it's OK because that's well above our f_c frequency (see §4x.3.2), and it's above the intercept of the flattened noise gain with the open-loop gain (see Figure 8.73). So the configuration is stable, and in fact the second-stage bandwidth is more than needed.

The performance is quite good, with $i_n \approx 50 \, \text{fA}/\sqrt{\text{Hz}}$ from 10 Hz to 2 kHz when driven with an input capacitance of 1000 pF, and with bandwidth out to 100 kHz. But the input-noise voltage ($2.9 \, \text{nV}/\sqrt{\text{Hz}}$) in combination with the large input capacitance causes a rapidly rising $e_n C_{in}$ noise current, reaching nearly $2000 \, \text{fA}/\sqrt{\text{Hz}}$ at 100 kHz (Figure 8.79). And the low-frequency noise floor is dominated by the Johnson-noise current of the feedback resistor R_f. Given that the AD743 is the quietest available JFET op-amp, we seem to have hit the limit of noise performance in a transimpedance amplifier driven with a highly capacitive source.

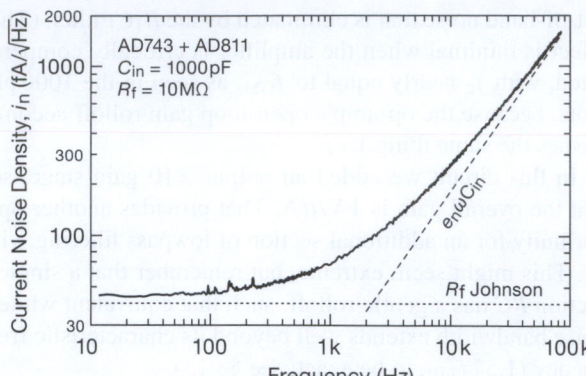

Figure 8.79. Measured current noise versus frequency for the amplifier of Figure 8.78, with a 1000 pF capacitor across the input. The dashed line shows the value of $e_n \omega C_{in}$ for $e_n = 2.9 \, \text{nV}/\sqrt{\text{Hz}}$ (the AD743 datasheet's specified typical e_n), and the dotted line is $\sqrt{4kT/R_f}$, the Johnson-noise current of the 10M feedback resistor (after Kretinin & Chung, used with permission).

A. Hybrid composite transimpedance amplifier

We *could* do better, though, if we could find a way to reduce the already low e_n (which contributes the rising $e_n C_{in}$ above around 10 kHz) while maintaining a large gain–bandwidth product. The technique that does the trick here is the use of a hybrid input stage to exploit the very low e_n of a large-area JFET pair like InterFET's IF3602 ($e_n = 0.3 \, \text{nV}/\sqrt{\text{Hz}}$, typ, at 100 Hz). Of course, large-geometry JFETs have lots of capacitance (this JFET's C_{iss} is a whopping 300 pF), but that is not terribly serious when you've got a signal source that's burdened already with 1000 pF of capacitance. And we'll stick with the composite amplifier topology, to get the high open-loop amplifier bandwidth that we need in order to achieve reasonable TIA bandwidth with a highly capacitive input.

Figure 8.80 shows such a design. The "op-amp" is a 3-stage composite amplifier whose input stage Q_{1ab} is a common-source differential JFET with current mirror drain load $Q_5 Q_6$, isolated with a cascode $Q_3 Q_4$. The overall GBW is 10 GHz, as seen in the Bode plot of Figure 8.80B; this composite op-amp is not stable (!) when used as a voltage amplifier (at least for closed-loop gains <45 dB), but when configured as a transimpedance amplifier (Figure 8.80C) with an effective compensation capacitance of 0.032 pF (set by R_2 in the circuit shown[99]) across the 20M

---

[98] A. Kretinin and Y. Chung, "Wide-band current preamplifier for conductance measurements with large input capacitance," arXiv:1204.2239v1 (2012).

[99] Alternatively you can make R_f from a series string of four or five resistors (spaced out and away from ground) to reduce the total capacitance of their individual ~ 0.15 pF parasitic shunt capacitances.

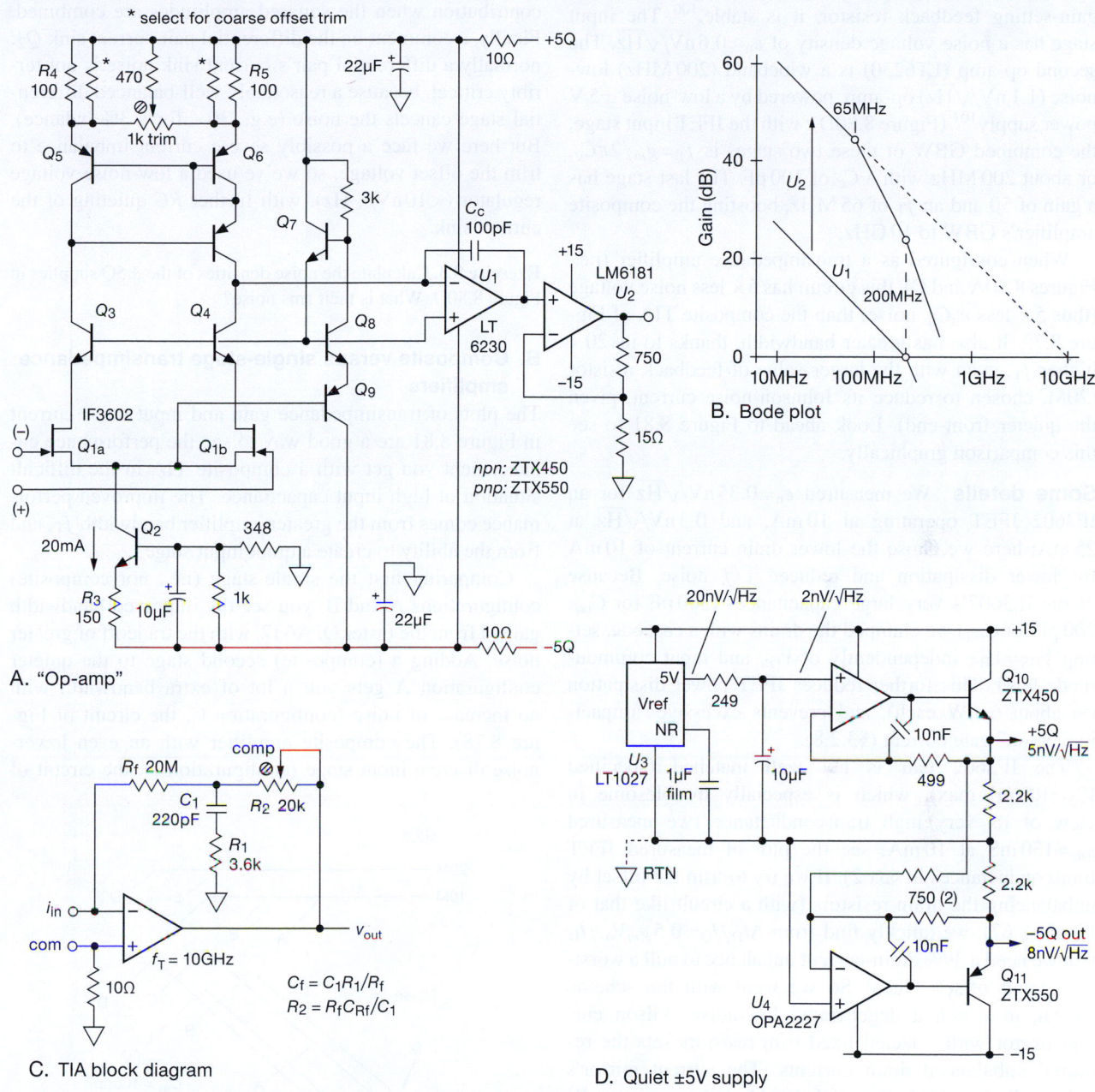

Figure 8.80. Hybrid transimpedance amplifier, optimized for lowest noise with capacitive inputs of order ~1000 pF by the use of large-area input-stage JFETs with extraordinarily low noise voltage ($e_n = 0.35\,\text{nV}/\sqrt{\text{Hz}}$). The composite amplifier (A) boosts the gain–bandwidth product f_T to 10 GHz, (B), to maintain an overall TIA bandwidth of ~250 kHz in spite of the large R_f and C_{in} (20 MΩ and 1000 pF). The transimpedance configuration (C) sets the gain at 20 V/μA, with an effective C_f of 32 fF (created by the pole-zero network $C_1 R_1 R_2$, needed to cancel R_f's excessive self-capacitance (which we'll call C_{Rf}); see §4x.3, setting a dominant pole at 200 kHz to ensure stability. A low-noise voltage reference (D) is used to create quiet ±5 V supply rails.

gain-setting feedback resistor, it is stable.[100] The input stage has a noise voltage density of $e_n \approx 0.6\,\mathrm{nV}/\sqrt{\mathrm{Hz}}$. The second op-amp (LT6230) is a wideband (200 MHz) low-noise (1.1 nV/$\sqrt{\mathrm{Hz}}$) op-amp, powered by a low-noise ± 5 V power supply[101] (Figure 8.80D); with the JFET input stage, the combined GBW of these two stages is $f_T = g_m/2\pi C_c$, or about 200 MHz with a C_c of 100 pF. The last stage has a gain of 50 and an f_T of 65 MHz, boosting the composite amplifier's GBW to 10 GHz.

When configured as a transimpedance amplifier (i.e., Figures 8.80A and C), this circuit has $5\times$ less noise voltage (thus $5\times$ less $e_n C_{in}$ noise) than the composite TIA of Figure 8.79. It also has greater bandwidth, thanks to its $20\times$ higher f_T, even with the larger value of feedback resistor (20M, chosen to reduce its Johnson-noise current, given the quieter front-end). Look ahead to Figure 8.81 to see this comparison graphically.

Some details   We measured $e_n = 0.35\,\mathrm{nV}/\sqrt{\mathrm{Hz}}$ for an IF3602 JFET operating at 10 mA, and $0.3\,\mathrm{nV}/\sqrt{\mathrm{Hz}}$ at 25 mA; here we chose the lower drain current of 10 mA for lower dissipation and reduced $1/f$ noise. Because of the IF3602's very large capacitances (300 pF for C_{iss}, 200 pF for C_{rss}) we clamped the drains with a cascode, setting $V_{DS} = V_{BE}$ independently of V_{GS} and input common-mode level. This further reduces JFET power dissipation (to about 6 mW each), and prevents excessive "impact-ionization" gate current (§3.2.8).

The IF3602 pair is not well matched (specified $V_{os} = 100$ mV max), which is especially troublesome in view of its very high transconductance (we measured $g_m = 130$ mS at 10 mA, see the plot of measured JFET transconductances in §3x.2). If we try to trim the offset by unbalancing the drain resistors (with a circuit like that of Figure 8.67), we quickly find from $\Delta I_D/I_D = 0.5\,g_m V_{os}/I_D$ that we need a 39% drain-current unbalance to null a worst-case input offset. Ouch! So we went with the scheme shown, in which a degenerated low-noise Wilson current mirror with selected fixed trim resistors sets the required unbalanced drain currents. The current mirror's noise-voltage contribution with 100 Ω emitter resistors R_4 and R_5, when referred to the JFET input terminals, is $e_n = \sqrt{4kT/R_4} \cdot 1/g_m$, or 0.1 nV/$\sqrt{\mathrm{Hz}}$, an insignificant 4%

[100] With $C_{in} = 1000$ pF, $f_{RCin} \approx 8$ Hz and $f_T = 10$ GHz, so $f_{GM} \approx 280$ kHz; a 0.032 pF effective capacitance across R_f makes $f_c = 200$ kHz.

[101] Where we've started with an LT1027 reference, which uses a buried zener for low $1/f$ noise, quieted further with a bypass capacitor at the noise-reduction pin. Adding a resistor to ground from the upper op-amp's inverting input lets you change the quiet split-supply voltages, e.g., to ± 6 V, or whatever.

contribution when the squared amplitudes are combined. Finally, a comment on the differential pair current sink Q_2: normally a differential pair's current sink noise is not terribly critical, because a reasonably well-balanced differential stage cancels the noise (e.g., $30\times$ for a 3% balance). But here we face a possibly severe current imbalance to trim the offset voltage, so we've used a low-noise voltage regulator ($<10\,\mathrm{nV}/\sqrt{\mathrm{Hz}}$), with further RC quieting of the current sink.

Exercise 8.3. Calculate the noise densities of the $\pm 5Q$ supplies in Figure 8.80D. What is their rms noise?

B. Composite versus single-stage transimpedance amplifiers

The plots of transimpedance gain and input noise current in Figure 8.81 are a good way to see the performance enhancement you get with a composite TIA in the difficult situation of high input capacitance. The improved performance comes from the greater amplifier bandwidth f_T, and from the ability to create a quiet input stage.

Comparing first the single-stage (i.e., not composite) configurations A and B, you see the improved bandwidth gained from the faster OPA637, with the tradeoff of greater noise. Adding a (composite) second stage to the quieter configuration A gets you a lot of extra bandwidth with no increase of noise (configuration C, the circuit of Figure 8.78). The composite amplifier with an even lower-noise discrete input stage (configuration D, the circuit of

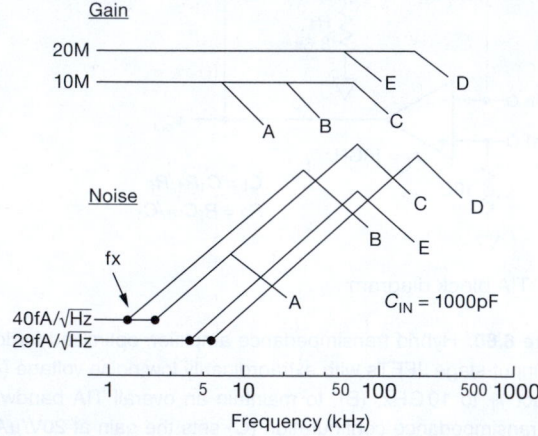

Figure 8.81. Comparing the performance of five transimpedance amplifiers when driven with a current signal of 1000 pF capacitance. A, AD743; B, OPA637; C, AD743 composite (Figure 8.78); D, IF3602 composite (Figure 8.80); E, OPA637 with BF862 bootstrap (similar to Figure 8.82); The feedback resistor R_f is 10 MΩ for circuits A–C, and 20 MΩ for D and E.

Figure 8.80) has yet more bandwidth, and can tolerate the larger R_f needed to fully exploit the amplifier's lower e_n. Finally, if you have the luxury of bootstrapping the input capacitance (see for example §8.11.9, Figure 8.82), the greatly lowered C_{in} lets you revert to the simple single-stage OPA637 configuration without a great sacrifice in performance.

This example illustrates once again a lesson we've learned over many decades of designing low-noise amplifiers, namely that circuit complexity increases rapidly as improvements are added to reach the limits of low-noise circuit performance.

8.11.9 Reducing input capacitance: bootstrapping the transimpedance amplifier

Input capacitance has not been our friend. As Figures 8.73, 8.74, and 8.76 make abundantly clear, input capacitance is the root of all evil, in terms of both noise and bandwidth. Large-area photodiodes are plagued with lots of capacitance (as much as 1000 pF or more); and, if the detector is at the end of a length of shielded cable, you can figure on an additional ~30 pF/ft of cable (this value is not arbitrary; see Appendix H).

As we hinted earlier, there are tricks by which you can greatly reduce the effective capacitance. Figure 8.82 shows an elegant solution,[102] namely an ac-coupled bootstrap of the return side of the capacitive input device (plus shielded cable, if any). In this circuit the JFET follower Q_1 drives the low side of the photodiode with a replica of any signal at the summing junction; Q_1's high transconductance (~25 mS) ensures a gain close to unity (output impedance ~40 Ω), thus a reduction of the photodiode's effective input capacitance (as seen at the summing junction) by at least a factor of 10.

But now we have to worry about noise introduced by Q_1 and associated circuitry. You might think first about Q_1's gate current, and also the capacitance C_{iss} that it adds to the input. For this particular JFET, things look pretty good: it has low gate input current (~1 pA, as long as you keep $V_{DS} < 5$ V), and low feedback capacitance (~2 pF). But the JFET's noise is another matter, because its e_n voltage noise generates an e_nC noise current in combination with the *un-bootstrapped* photodiode (and cable) capacitance. This is in contrast to the op-amp's e_n, which now sees the much lower bootstrapped capacitance; both of these noise currents must come through R_f, generating output noise.

[102] See Linear Technology's LTC6244 datasheet and their DN399 Design Note by Glen Brisebois.

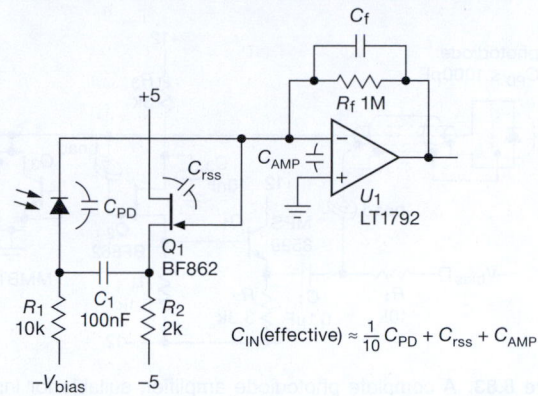

Figure 8.82. Follower Q_1 bootstraps the photodiode at signal frequencies, reducing its effective capacitance by a factor of 10 or more. The BF862 is especially well suited to this task, with its low capacitance and sub-nV/$\sqrt{\text{Hz}}$ noise. This lets us use a less-expensive op-amp (with less bandwidth and somewhat higher e_n) for U_1. We often add an emitter follower (not shown for simplicity). To find the break frequency of the e_nC_{in} noise, use eq'n 8.45 on page 539, with the "reduced" value of C_{in}; see text.

For this reason it's essential to choose a JFET with very low noise voltage, ideally much lower than the e_n of the op-amp. The BF862 is an excellent choice, with its impressively low[103] e_n~0.9 nV/$\sqrt{\text{Hz}}$. Even so, its noise contribution is larger than that of the op-amp (the latter benefitting, of course, from the JFET's successful reduction of effective photodiode capacitance), because the LTC1792 is a quiet (4.2 nV/$\sqrt{\text{Hz}}$) op-amp to begin with. For example, with a 1000 pF photodiode the op-amp's 4.2 nV/$\sqrt{\text{Hz}}$ acts on the (reduced) ~100 pF, while the JFET's 0.9 nV/$\sqrt{\text{Hz}}$ acts on the full 1000 pF of the photodiode. Making that quantitative, at 100 kHz the noise-current contributions are 0.26 pA/$\sqrt{\text{Hz}}$ and 0.57 pA/$\sqrt{\text{Hz}}$, respectively, for a combined noise current of 0.63 pA/$\sqrt{\text{Hz}}$. Although the JFET's noise dominates, we're still doing quite a bit better than the unbootstrapped configuration, with its noise current of 2.6 pA/$\sqrt{\text{Hz}}$.

We discuss transimpedance amplifiers in further detail (particularly issues of stability) in §4x.3. There we include a design for an input-capacitance-tolerant photodiode amplifier that exploits bootstrapping; the circuit is reproduced here in Figure 8.83. In this design we used the excellent (and not inexpensive) OPA637 (GBW=80 MHz,

[103] Inexpensive, too, about $0.50 in qty 25. And, according to Phil Hobbs, they can be paralleled without source ballasting (especially when taken from the same reel), because "they go from cutoff to I_{DSS} in about 400 mV. JFETs running near I_{DSS} have a low tempco as well. I use boatloads of them – they're Good Medicine."

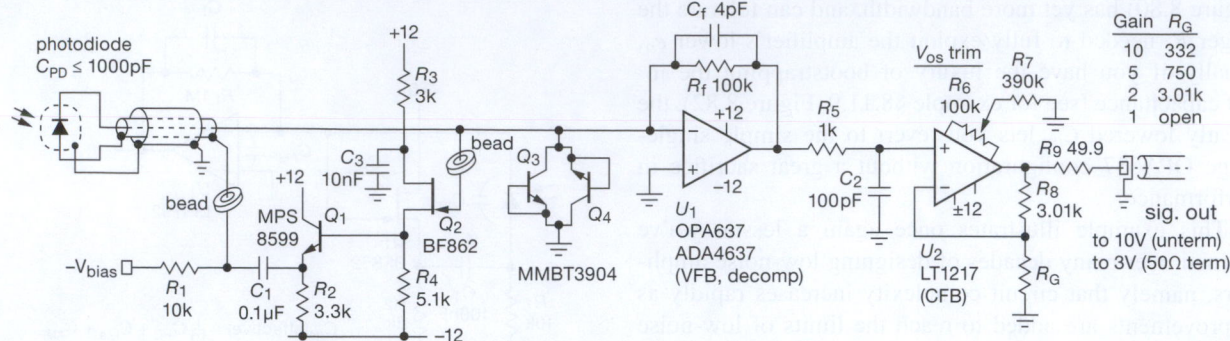

Figure 8.83. A complete photodiode amplifier, suitable for input capacitances up to 1000 pF. Input bootstrapping greatly reduces the effective photodiode and cable capacitance, for enhanced speed and reduced noise.

$e_n = 4.5$ nV/$\sqrt{\text{Hz}}$. In Figure 8.84 we plot the noise and bandwidth, under the conservative estimate of a tenfold reduction of input capacitance. Bootstrapping improves the bandwidth, reduces the noise, and leaves the transimpedance gain unchanged.[104] Not bad!

Exercise 8.4. Design a TIA with an OPA637 op-amp and a BF862 JFET bootstrap follower, for input signals with $C_{in} = 1$ nF. Use $R_f = 20$ MΩ, and assume the BF862 has a noise voltage of 0.85 nV/$\sqrt{\text{Hz}}$ and a voltage gain (when driving the bootstrap terminal) of $G_V = 0.95$. Evaluate your circuit's noise and gain performance, which should be not unlike that of curve E in Figure 8.81.

8.11.10 Isolating input capacitance: cascoding the transimpedance amplifier

Bootstrapping reduces the effective input capacitance (typically by an order of magnitude), allowing reduced noise and greater bandwidth from a capacitive current-output sensor such as a photodiode. But we can do even better: it's possible to isolate the input capacitance entirely, by interposing a common-base (cascode) stage between the input signal and the transimpedance amplifier.

There are some bumps along the way, so we'll take it in stages. We learned about this from Philip Hobbs, whose articles[105] make good reading for those interested in pursuing these techniques.

---

[104] Here the LT1792 costs $4.85 versus $18 for the OPA637 or $12 for the similar ADA4637, and the BF862 costs just $0.67. You often can do even better, with bootstrap-assisted op-amps of adequate performance down in the $2 territory.

[105] For a good first reading assignment, try his "Photodiode front ends – the REAL story," *Opt. Photon. News*, **12**, 42–45 (April 2001).

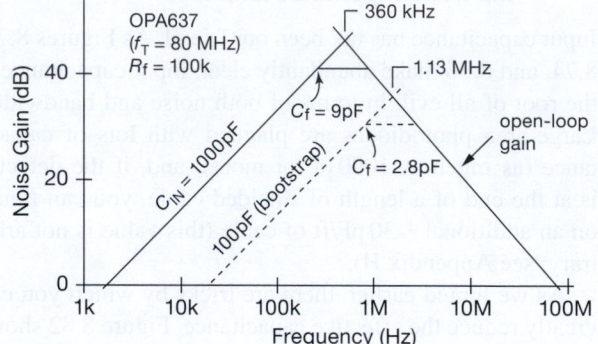

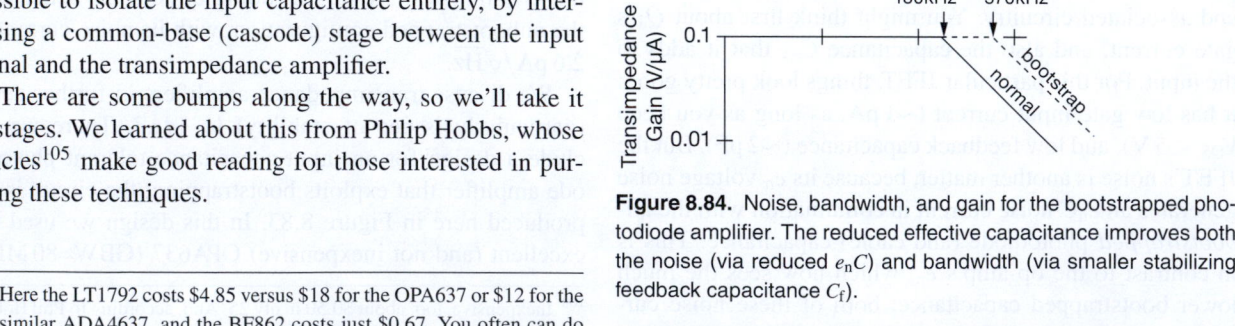

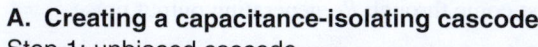

Figure 8.84. Noise, bandwidth, and gain for the bootstrapped photodiode amplifier. The reduced effective capacitance improves both the noise (via reduced e_nC) and bandwidth (via smaller stabilizing feedback capacitance C_f).

A. Creating a capacitance-isolating cascode
Step 1: unbiased cascode
Figure 8.86A shows the core idea, in which the current-sinking signal source drives the emitter of grounded-base

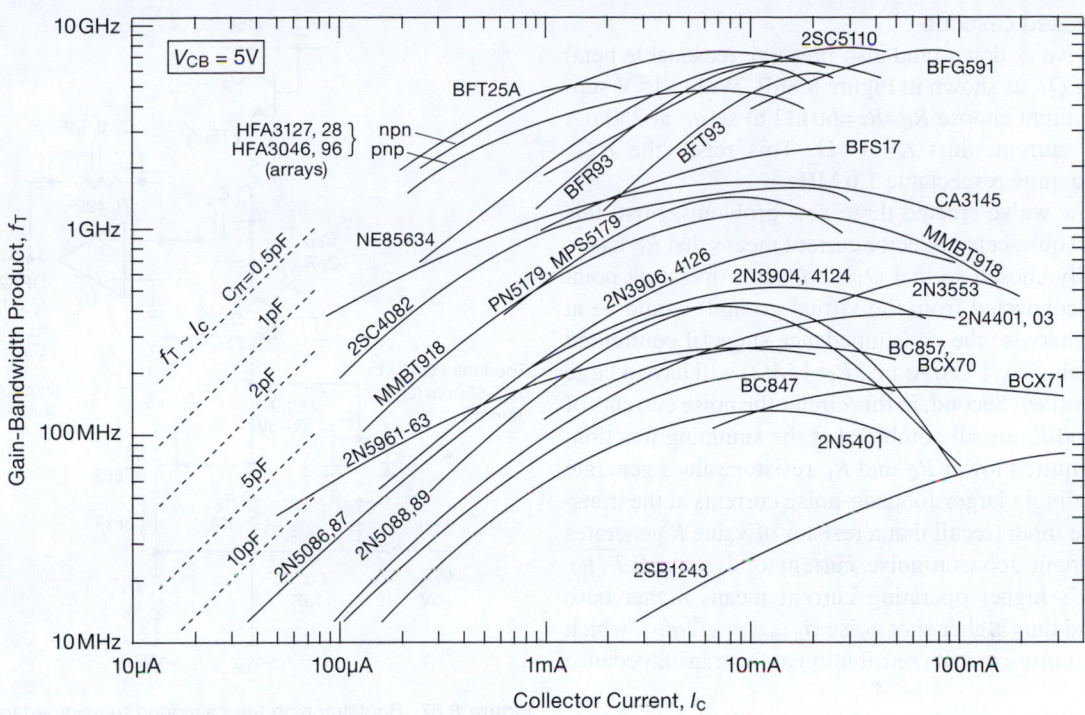

Figure 8.85. Gain-bandwidth product, f_T, versus collector current for selected bipolar transistors, as shown in the manufacturers' datasheets. Some wideband transistors are clear off the top.

npn transistor Q_1 (this is often called a "common-base amplifier"). Assuming reasonable transistor beta, most of that current appears at the collector – but with only the small collector output capacitance of Q_1, typically just a few picofarads.

A minor flaw here is that the input sits a diode drop below ground. But there are more serious snakes lurking in the Garden of Eden. We want good performance at very low signal currents, say fractions of a microamp. Down there we have to worry about falling transistor beta. Worse still, the input impedance r_e seen at the emitter is rising inversely with input current; at 1 μA it's 25 kΩ (recall $r_e = 25\,\Omega/I_C[\text{mA}]$). So the signal current is shunted to ground by the input capacitance at a frequency at which the reactance of $C_{in}=25$k or less; that is, the input rolls off at $f_{3dB} = 1/2\pi r_e C_{in}$, thus for example around 6.4 kHz for a 1 μA signal with $C_{in}=1000$ pF.[106]

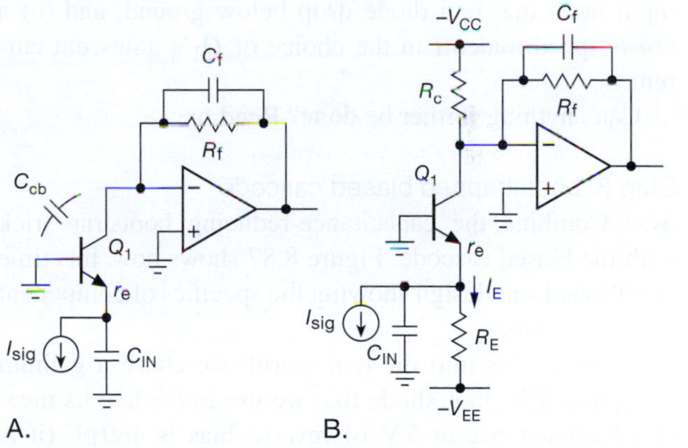

---

[106] You also have to worry about the transistor's falling f_T at low currents; the reduced beta diverts emitter current into the base at high frequencies. For discussion and measurements see the section "BJT Bandwidth and f_T" in Chapter *2x*, and see Figure 8.85. In the latter you can see that the 2N5089 has an (extrapolated) f_T down around

Figure 8.86. Isolating input capacitance with a common-base (cascode) input stage. A. Transistor Q_1 passes the signal current to the transimpedance amplifier, which sees only the small collector-to-base capacitance C_{cb}. B. Adding dc bias reduces r_e, greatly increasing the bandwidth.

---

2 MHz at 10μA, from which you can estimate that about 10% of the emitter current is diverted at 200 kHz.

Step 2: biased cascode

We can drive r_e down (and also preserve reasonable beta) by biasing Q_1, as shown in Figure 8.86B. With ± 15 V supplies, we might choose $R_C = R_E = 60$ kΩ to set I_C at 250 μA quiescent current; thus $r_e = 100\,\Omega$. This raises the $r_e C_{in}$ rolloff to a quite respectable 1.6 MHz.

But now we've created three new problems. First, this substantial quiescent collector current means that R_C has to be critically chosen so that Q_1's collector operating point (when disconnected from the virtual ground) would be at ground; otherwise the transimpedance stage (if configured for high gain, say 1 V/μA; i.e., $R_f = 1$ MΩ) will have a large output dc offset. Second, in this circuit the noise currents of R_C, R_E, and R_f are all combined at the summing junction; and the required lower R_C and R_E resistor values generate correspondingly larger Johnson-noise currents at the transimpedance input (recall that a resistor of value R generates a short-circuit Johnson-noise current of $i_n = \sqrt{4kT/R}$). Third, Q_1's higher operating current means higher base current and thus higher shot noise ($i_{n,base} = \sqrt{2qI_B}$), which is another noise-current contribution at the transimpedance input.[107]

So here's the situation so far: with the circuit of Figure 8.86B we have achieved better bandwidth with a given input capacitance, compared with the previous transimpedance designs, but at the cost of (a) dc offset, (b) an input node that is a diode drop below ground, and (c) a noise–speed tradeoff in the choice of Q_1's quiescent current.

Can anything further be done? Read on...

Step 3: bootstrapped biased cascode

Yes! Combine the capacitance-reducing bootstrap trick with the biased cascode. Figure 8.87 shows how, this time as a fleshed-out design showing the specifics of component values and types.

To bring this into the real world, we chose a gallium phosphide UV photodiode that we use in the lab; its measured capacitance at 5 V of reverse bias is 460 pF (it is not specified on the datasheet). The photodiode's low side is bootstrapped by JFET follower Q_2 (as in the earlier Figure 8.82), reducing its effective capacitance tenfold, to ~ 50 pF.

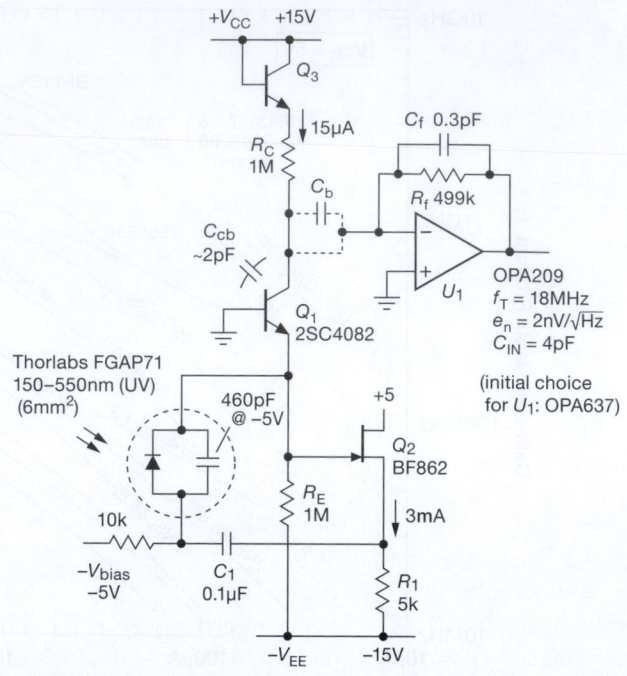

Figure 8.87. Bootstrapping the cascoded transimpedance amplifier to achieve wideband (1 MHz) low-noise performance with a photodiode of rather high capacitance. Optional blocking capacitor C_b can be used to eliminate dc offset in ac-coupled applications.

The next step is to choose the operating current for cascode transistor Q_1, such that its r_e is small enough so that the input $r_e C_{in}$ rolloff[108] does not compromise the bandwidth f_c set by the transimpedance stage U_1. We figure f_{GM} (eq'n 8.46) as usual (see §4x.3): the capacitance at U_1's input is the sum of Q_1's collector capacitance (2 pF) and U_1's own input capacitance (15 pF for our initial choice of an OPA637, whose low noise and wide bandwidth make it a favorite choice as an input stage for photodiode amplifiers); the 17 pF total puts U_1's input rolloff f_{RfCin} of 18.7 kHz, given our choice[109] of transimpedance gain (0.5 V/μA). The OPA637's decompensated f_T of 80 MHz gives us an $f_{GM} = \sqrt{f_{RfCin} f_T} = 1.22$ MHz; we choose a value of feedback capacitor C_f to set the critical bandwidth f_c to be 0.7 MHz (a conservative $\zeta = 1.2$, to ensure stability), hoping for a transimpedance bandwidth f_b of 1 MHz. That's 0.46 pF, of which the resistor itself provides ~ 0.1 pF. You can use a pair of PCB traces, an adjustable pot with $C_f = 0.5$ pF (see Figure 4x.18), or a "gimmick" (a few twists of a pair of insulated wires) to add the remaining 0.36 pF.

---

[107] There's a subtle point here: a current such as I_E that is generated by a voltage drop across a metallic conductor does not exhibit shot-noise current (the charge carriers do not act independently; see §8.1.2). But the base current *does* have the "full" shot noise given by the formula. That noise current manifests itself at the collector, because $I_C = I_E$(noiseless) $- I_B$(noisy).

[108] At high enough currents, transistor f_T is given by $f_T = 1/2\pi r_e C_{in}$.
[109] Arrived at through an analogous process of speed–noise tradeoff, the details of which we will spare the reader.

B. Iteration: better op-amp choice

Something's amiss here: the OPA637 is a fine op-amp (with a price to match!), but it uses large-area JFETs to achieve its admirably low e_n, resulting in a high input capacitance of 15 pF. That's not a worry when you have, say, an input signal accompanied by lots of capacitance (e.g., a photodiode and some shielded cable). But here it's at a low capacitance node (Q_1's collector) where it dominates the rolloff, forcing the high f_T requirement.

Let's redo the calculation, assuming we use an op-amp with lower capacitance, say $C_{in}=4$ pF. Now $f_{RfCin}=53$ kHz, and for an f_b of 1 MHz with $\zeta=1$ (so $f_c=0.7$ MHz) we need an op-amp with $f_T \geq 9$ MHz. Our effort is rewarded, for now there are many possibilities. We can choose the OPA209 from Table 8.3, with its f_T of 18 MHz and low cost of \$2.27.[110] That gives us $f_{GM}=0.98$ MHz, so we can set $f_b=f_{GM}=1$ MHz, and with $\zeta=1$ (critical damping) we have $f_c=0.7$ MHz and $C_f=3$ pF.

Knowing that the transimpedance stage has a usable bandwidth of 1 MHz, we next set the quiescent current of cascode transistor Q_1 high enough so that the rolloff of r_e, combined with the 50 pF (bootstrapped) effective photodiode capacitance, is at a somewhat higher frequency. Here an I_C of 15 µA gives us an r_e (the impedance the signal sees at the emitter) of 1.7 kΩ and thus an input rolloff frequency of 1.9 MHz. The reduced C_{in} produced by the bootstrap allows us to run Q_1 at this low current, with the benefit that the Johnson noise current contributed by Q_1's 1 MΩ collector and emitter resistors is no larger than that of R_f. The base current shot noise is similarly reduced; here it contributes an input-referred noise current of 0.1 pA/$\sqrt{\text{Hz}}$ (to be compared with the 0.26 pA/$\sqrt{\text{Hz}}$ combined contribution of R_C, R_E, and R_f).

Some details:[111]

(a) Bootstrap follower Q_2 must drive the low side of the photodiode with an output impedance that is much lower than Q_1's input impedance (i.e., r_e, here 1.7 kΩ); that's satisfied here, where Q_2's high transconductance ensures a low output impedance of $Z_o = 1/g_m \sim 40\,\Omega$.[112]

(b) We chose first a 2N5089 for the cascode transistor Q_1, because of its high beta at low currents ($\beta=400$ min at $I_C=100$ µA) and low output capacitance (2 pF), both of which promote low noise. But then we consulted Figure 8.85 and realized that it had too little GBW: f_T is just 2 MHz at 15 µA. So we looked next at an MMBT918 (2N918), with a healthy $f_T=13$ MHz at 15 µA. Its base would rob about f/f_T, about 8% at f_c and thus an insignificant contributor to the circuit's 3 dB rolloff at 1 MHz. But its beta is anemic, less than 40 at 15 µA. So we settled finally on the 2SC4082 shown in the figure: it has $f_T=20$ MHz at 10 µA, and a pleasantly flat beta curve with $\beta=90$ at 100 µA (and likely not much lower at 15 µA).

(c) There's an additional noise voltage contribution, in the form of the transistor's input noise voltage ($e_n=\sqrt{4kT[r_{bb'}+0.5r_e]}$) generating an "$e_nC$" noise current through the photodiode's effective input capacitance. These parameters don't look unfavorable for the 2SC4082.[113]

C. A final trick: "Regulated Cascode" (RGC)

Figure 8.88 shows a circuit configuration that's achieved considerable popularity in the world of photonics, where you're dealing with data rates of *giga*bits per second. Transistors Q_1 and Q_2 form a tight local feedback loop, with cascode transistor Q_1's base "regulated" by Q_2. (Q_1 and Q_2 can be replaced by MOSFETs, as indicated.) The regulated cascode improves on the simple cascode of Figure 8.86 in two important ways. (a) The circuit's input impedance (at Q_1's emitter) is reduced by a factor of Q_2's voltage gain, greatly increasing the $R_{in}C_{in}$–limited bandwidth; and (b) the input noise voltage is set by Q_2, rather than by Q_1, so the latter can be run at desirable low currents without the usual noise penalty.

To expand on this a bit, remember that input capacitance in the simplest transimpedance amplifier (Figure 8.70) is both a bandwidth killer (via R_fC_{in}) and a noise enhancer

---

[110] The OPA209 may even be *too good* for the job, with its e_n of 2.2 nV/$\sqrt{\text{Hz}}$. We require only that the op-amp's e_n be significantly less than that of Q_1 multiplied by the ratio of the reduced (bootstrapped) input capacitance to the capacitance at the op-amp's input node (here that works out to 7.5 nV/$\sqrt{\text{Hz}}$), and also that it be significantly less than the e_n of Q_2 multiplied by the ratio of the full input capacitance to the capacitance at the op-amp's input node (which usually works out to an even larger value of maximum e_n). In other words, we could have chosen an op-amp not intended for low-noise applications from the tables in Chapter *4x*.

[111] For a great read, see Phil Hobbs' book *Building Electro-Optical Systems, Making It All Work*, 2nd ed., Wiley (2009); a fine collection of tricks for designing cascode photodiode amplifiers, including series-peaking inductors and *T*-coils to extend the bandwidth, noise cancellers, and more.

[112] We usually add an emitter follower running at a few milliamps to further stiffen the guard signal, especially if it may be driving the external shield of a coax line where it is susceptible to pickup of AM radio stations, etc.; see Figure *4x.19*.

[113] The 2SC4082 has $r_{bb'}C_c=6$ ps, which (with $C_{ob}=0.9$ pF) implies an $r_{bb'}$ less than 10 Ω.

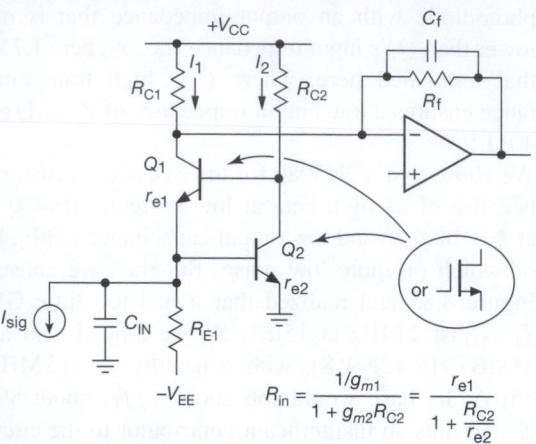

Figure 8.88. A "regulated cascode" input stage to transimpedance amplifier U_1 allows capacitance-isolating cascode Q_1 to operate at low current without a corresponding e_n penalty. It also reduces the impedance seen at the capacitive input (by a factor of G_{V2}), for enhanced bandwidth.

$$R_{in} = \frac{1/g_{m1}}{1 + g_{m2}R_{C2}} = \frac{r_{e1}}{1 + \dfrac{R_{C2}}{r_{e2}}}$$

(via $e_n C$). We struggled with this, and the stability issues it presents, first by adding enough feedback compensation to ensure stability, then piling on op-amp speed to recover some bandwidth. Then we fiddled with circuit configurations to address directly the problem of input capacitance: first we bootstrapped the low side of the photodetector to *reduce* the effective input capacitance; then we added a cascode input stage to *isolate* the input capacitance, but that raised the circuit's input impedance (sacrificing bandwidth); so we biased the cascode to reduce R_{in}; and finally we bootstrapped the biased cascode.

All well enough. But the result was an input stage that had to run at significant collector current, generating a difficult-to-tame dc offset that seriously limits the amount of gain in the transimpedance stage. This is not a good situation, particularly if you want to sense small input currents.

The regulated cascode elegantly addresses these problems, by allowing the cascode stage (Q_1) to run at low current (so the circuit works well with small-signal currents), while circumventing the bandwidth and noise penalties (by driving down R_{in}, and by allowing Q_2 to replace Q_1's voltage noise). To minimize input *current* noise you should operate Q_1 at low collector current, but make sure you don't reduce the bandwidth excessively;[114] see the discussion in the "BJT bandwidth and f_T" section of Chapter *2x*. A vari-

ation of this configuration, in integrated form, is used in most contemporary fiber-optic receivers.[115]

8.11.11 Transimpedance amplifiers with capacitive feedback

There is a way to eliminate completely the Johnson-noise contribution of the gain-setting feedback resistor R_f, namely to eliminate the resistor itself. Feedback is then provided by the capacitor C_f alone, forming an integrator. The output signal must be differentiated to recover an output proportional to the input signal current; and of course both the integrator and differentiator must be reset (interrupting its operation) often enough to prevent saturation. To preserve the low noise of the input stage, the voltage noise of the differentiator op-amp must be significantly less than that of the input (integrating) op-amp.

Although capacitance feedback may seem an unusual approach, in fact, for ordinary TIAs operating at modestly high frequencies with high-value feedback resistors (say 100 MΩ and above), the self-capacitance of the resistor conceptually turns the amplifier into an integrator, with the resistor playing something akin to a reset role. Once one starts thinking along these lines, the idea of deliberately raising the feedback capacitance to ∼1 pF is no longer so scary. See *§4x.8* for an example circuit and discussion.

This technique is often used in "patch-clamp" amplifiers[116] and other low-level current detectors, for example, chilled germanium or silicon x-ray detectors (called IGX and Si(Li) respectively), in which the integrator is reset by an optical pulse from an LED (thus avoiding switch leakage effects).

A more common application of such an integrating transimpedance amplifier is for the readout of an imaging detector, where the quantity of interest is the total charge delivered during the brief readout, rather than its current-versus-time waveform. For this application you need to know only the *change* in integrator output voltage caused

---

[114] The worry, of course, is that Q_1 will have poor f_T at low collector current. But these designers know how to make impressively good microwave transistors in their ICs.

[115] See, for example, E. Säckinger and W. Guggenbühl, "A high-swing, high-impedance MOS cascode circuit," *IEEE J. Solid-State Circuits* **25**, 1 (1990); S. M. Park, "1.25-Gb/s regulated cascode CMOS transimpedance amplifier for gigabit ethernet applications," *IEEE J. Solid-State Circuits* **39**, 1 (2004); or Z. Lu et al., "Broad-band design techniques for transimpedance amplifiers," *IEEE Trans. Circuits Sys.* **54**, 3 (2007).

[116] For example, the Axon 200B patch-clamp amplifier is claimed to achieve an open-input noise current of just $0.2\,\text{fA}/\sqrt{\text{Hz}}$ at 150 Hz when operated in capacitance-feedback mode and thermoelectrically cooled; that's equivalent to the shot noise of 0.1 pA of leakage current.

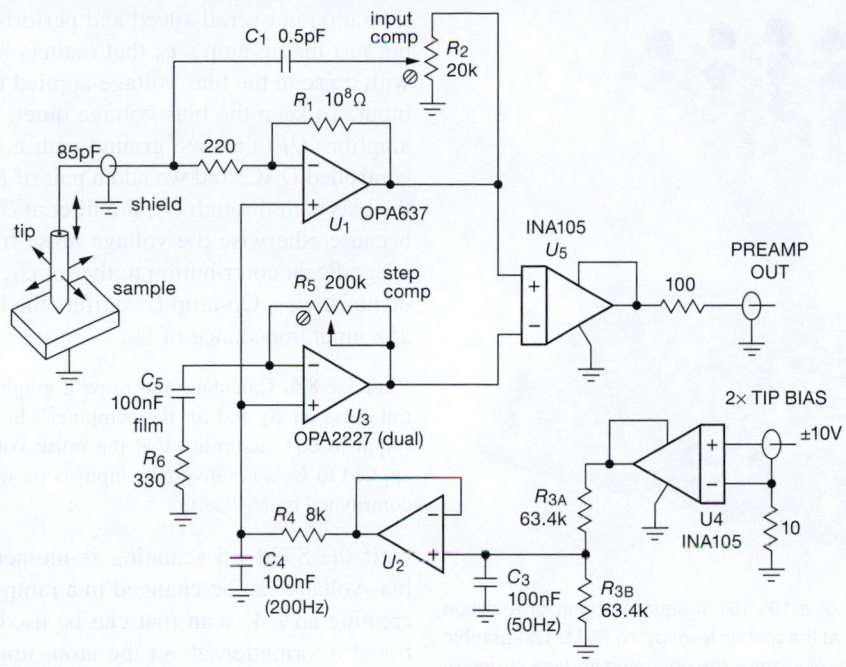

Figure 8.89. A low-noise transimpedance preamp for a scanning tunneling microscope (STM), where tip current signals are in the nanoamp range. The tip-to-sample voltage is set by the "tip-bias" input; the frequency response goes to ~10 kHz, and the gain is 0.1 V/nA. Trimmer R_2 adjusts the effective feedback capacitance, to compensate the 100 pF (or more) of capacitance seen at the shielded input line emerging from the cryogenic vacuum chamber.

by the charge delivery. This technique is known as "correlated double-sampling," and dates back to the 1950s; see the footnote on page 571.

8.11.12 Scanning tunneling microscope preamplifier

Scanning tunneling microscopy (STM), dating to Binnig and Rohrer's work in the early 1980s,[117] allows one to image the surface topography (and other properties) of a sample at the atomic level. Figure 8.89 shows a sharp metallic tip hovering above a sample in a vacuum chamber, connected to a current-measuring preamp. When the tip is close to the surface (~1 nm, about ten atom diameters) and biased to ~1 V, a current of order 1 nA flows via quantum-mechanical "tunneling" through the potential barrier. The current is very sensitive to tip-to-sample spacing, varying exponentially such that the current typically increases by an order of magnitude when the spacing is reduced by 0.1 nm.[118]

To form a topographic image, the tip is held at a fixed voltage and scanned across the sample, with its spacing controlled by a piezo actuator such that the measured tip current is held constant. The piezo drive voltage is then a measure of variations in the surface height, with a vertical resolution better than an atomic diameter. If the tip's point is made sufficiently small (no more than a few atoms wide), the horizontal resolution is likewise of atomic scale. Figure 8.90 shows an example, an STM image of a silicon surface.

The STM preamp is a transimpedance amplifier, usually mounted at a flange on the vacuum chamber, where it receives the tip's tunnel current signal through a length of shielded line. That adds input capacitance, typically in the range of 50–200 pF. Because the preamp is within the tip spacing servo loop, it needs plenty of bandwidth, say 20 kHz, to permit rapid scanning, thus requiring careful loop compensation (see §4x.3). The preamp also needs to apply the tip-bias voltage.

In Figure 8.89 the transimpedance stage U_1 is an OPA637, an 80 MHz (decompensated) JFET op-amp with 16 pF of input capacitance and 4.5 nV/$\sqrt{\text{Hz}}$ input voltage noise (see Table 8.2 on page 516). That's rather a lot of

---

[117] See for example their article "Scanning tunneling microscopy," *Helvetica Physica Acta* **55**, 726–735 (1982).

[118] See for example J. A. Golovchenko, "The tunneling microscope: a new look at the atomic world," *Science* **232**, 48–53 (1986).

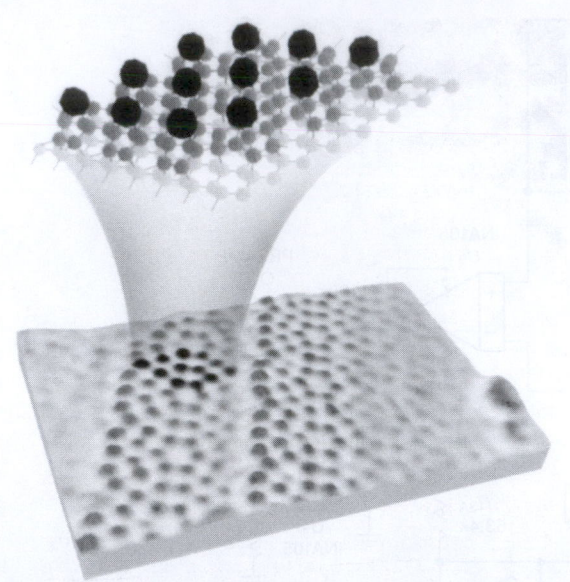

Figure 8.90. Structure of a 10×10 nm square region of a silicon crystal surface, imaged at the atomic level by an STM. The graphic model (expanded above) elucidates the observed surface structure (known as a *reconstruction*, because the surface atoms adopt a different arrangement compared with the bulk; here several atomic planes are seen, in a characteristic "7×7" reconstruction). This first image of an atomic step is adapted from the cover illustration accompanying the article referenced in Footnote 118 on page 553. (Courtesy J. Golovchenko)

input capacitance, but here it's insignificant compared with the cable capacitance, and in exchange we get a quiet wideband amplifier with low input current. Tip bias is set by the dc voltage at U_1's noninverting input; the op-amp's output is offset by the bias voltage, which we remove with difference amplifier U_5 (see §5.14 and Table 5.7).[119]

Exercise 8.5. Calculate the preamp's bandwidth for $C_1 = 0.1$ pF with $C_{in} = 100$ pF, and also calculate the maximum allowed input capacitance. Show your stability criteria. Make a graph of the total effective input-noise-current spectral density, with curves for several values of C_{in}, including the maximum value (don't forget R_1's Johnson noise). Calculate the corresponding f_x breakpoint frequencies.

This exercise demonstrates that $e_n C_{in}$ noise is a serious

---

[119] A higher-gain difference-amplifier could be used for U_5 (e.g., an INA106 with $G = 10$), as long as the tunneling currents are no more than a few nanoamps. STM preamps are often operated at very low current levels, with $R_1 = 10^9\,\Omega$. The summing-junction node connections (both sides of the $220\,\Omega$ resistor, along with the op-amp's inverting input) are often placed on Teflon standoffs to prevent PCB leakage currents.

limitation in overall speed and performance. Note that it's not just the op-amp's e_n that matters here – it's combined with noise in the bias voltage applied to U_1's noninverting input. To keep the bias voltage quiet, we use a difference amplifier U_4 to isolate ground path noise in its computer-controlled DAC, and we add a pair of RC lowpass filters in the bias path through U_2; the filter at U_2's output is needed because otherwise the voltage noise from U_2 would make a significant contribution to that of U_1, as the next exercise demonstrates. Op-amp U_3 buffers the bias voltage into the 25k input impedance of U_5.

Exercise 8.6. Calculate and draw a graph of the allowed spectral noise for U_2 and for the computer's bias-voltage DAC signal (input to U_4), assuming that the noise voltage of the final bias applied to U_1's noninverting input is no more than 30% of that contributed by U_1 itself.

If the STM tip scanning is momentarily stopped, the bias voltage can be changed in a ramp or a series of steps, creating an I–V scan that can be used to determine additional information about the atom immediately under the tip. One approach is to repeatedly stop and make I–V scans. In this way the STM can provide not only an elevation map of the surface, but also a map of elemental composition. But changing the tip-bias voltage causes a pulse of current $i = C_{in}\, dV/dt$ that is seen and amplified by the TIA op-amp U_1. A nice way to deal with this effect is to add R_5 and C_5 such that $R_5 C_5 = R_1 C_{in}$. This creates a cancelling pulse at the inverting input of U_5, allowing an accurate tunneling current to be measured before the voltage has completely settled. This greatly speeds up the full I–V scan.

This kind of circuitry is useful in other current-input applications, such as patch-clamp amplifiers in neurophysiology. It's basically a form of source-measure unit (SMU), a handy device we used to make the JFET transconductance measurements in §3x.2 and the BJT beta plots in Figure 8.39.

8.11.13 Test fixture for compensation and calibration

To adjust the input compensation (R_2 in Figures 8.80, 8.89 and *4x.17*) you'd like a clean source of nanoamp-scale square waves supplying a calibrated current at the input connector; with the input cable in place you'd then adjust R_2 for best step response. You might imagine rigging up a 1 V square wave in series with a 1 GΩ resistor. The problem is that the resistor's parasitic shunt capacitance of ∼0.1 pF causes input current spikes at each transition of the square wave; it takes only 0.01 V/µs of input slewing to produce 1 nA current pulses.

Figure 8.91 shows two solutions to this problem. In circuit A a tunable input series RC (lowpass, or "pole") compensates the current-scaling resistor R_2's parallel parasitic capacitance C_p (highpass, or "zero"). For this simple model of parasitic capacitance, cancellation requires $R_1C_1=R_2C_p$. The optional isolated input connector suppresses ground path current spikes between the signal generator and the preamp and oscilloscope. Circuit B takes a different approach, using a small series capacitor (into the summing junction at the output) as a differentiator; $i=C_2dV_{in}/dt$ predicts a ±1 nA output square wave for a 0.5 Vpp input triangle wave at 1 kHz. This circuit is the simpler, but its performance depends sensitively on the quality of the triangle wave at its turning points; we've gotten good results from Agilent 33120A (and later models) synthesized function generators. Your mileage may vary.

8.11.14 A final remark

We remind the reader that we've loosely referred to transimpedance amplifiers as "photodiode amplifiers"; that is, of course, just one application (though an important one) among many.

Our treatment in this chapter is concerned primarily with noise in transimpedance amplifiers, assuming familiarity with the basics. Be sure to visit §4x.3 for an introduction to transimpedance amplifiers, a treatment of stability

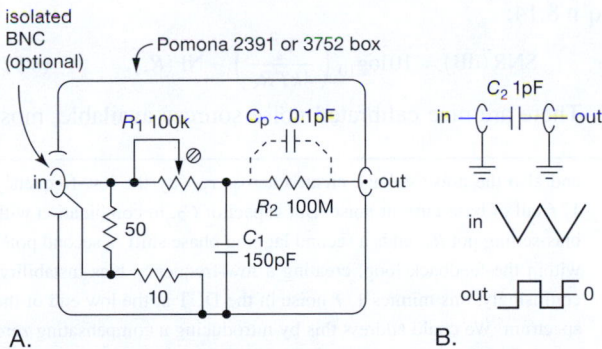

Figure 8.91. Test fixture for producing a nanoamp square-wave input for compensating and calibrating a transimpedance amplifier. A. Adjustable R_1C_1 (a "pole") cancels the peaking caused by R_2's shunt capacitance (a "zero"), so that a 0.1 V square wave creates a clean 1 nA output square wave of current into a summing junction. B. Capacitive differentiator converts a 500 mV (peak-to-peak) 1 kHz triangle voltage wave into a ±1 nA square current wave at the output summing junction.

and bandwidth, and a healthy collection of important and useful tricks.

8.12 Noise measurements and noise sources

It is a relatively straightforward process to determine the equivalent noise voltage and current of an amplifier, and from these the noise figure and signal-to-noise ratio for any given signal source. That's all you ever need to know about the noise performance of an amplifier. Basically the process consists of putting known noise signals across the input, then measuring the output-noise signal amplitudes within a certain bandwidth. In some cases (e.g., a matched input-impedance device such as a radiofrequency amplifier) an oscillator of accurately known and controllable amplitude is substituted as the input signal source.

Later we discuss the techniques you need to do the output voltage measurement and bandwidth limiting. For now, let's assume you can make rms measurements of the output signal with a measurement bandwidth of your choice.

8.12.1 Measurement without a noise source

For an amplifier stage made from an FET or transistor and intended for use at low to moderate frequencies, the input impedance is likely to be very high. You want to know e_n and i_n so that you can predict the SNR with a signal source of arbitrary source impedance and signal level, as we discussed earlier. The procedure is simple.

First, determine the amplifier's voltage gain G_V by actual measurement with a signal in the frequency range of interest. The amplitude should be large enough to override amplifier noise, but not so large as to cause amplifier saturation.

Second, short the input and measure the rms noise output voltage, e_s. From this you get the input noise voltage per root hertz from

$$e_n = \frac{e_s}{G_V B^{\frac{1}{2}}} \quad \text{V/Hz}^{\frac{1}{2}} \tag{8.47}$$

where B is the bandwidth of the measurement (see §8.13).

Third, put a resistor R across the input and measure the new rms noise output voltage, e_r. The resistor value should be large enough to add significant amounts of current noise, but not so large that the input impedance of the amplifier begins to dominate. (If this is impractical, you can leave the input open and use the amplifier's input impedance as R.) The output you measure is just

$$e_r^2 = [e_n^2 + 4kTR + (i_nR)^2]BG_V^2, \tag{8.48}$$

from which you can determine i_n to be

$$i_n = \frac{1}{R}\left[\frac{e_r^2}{BG_V^2} - (e_n^2 + 4kTR)\right]^{\frac{1}{2}}. \quad (8.49)$$

With some luck, only the first term in the square root will matter (i.e., if current noise dominates both amplifier voltage noise and source resistor Johnson noise).

Now you can determine the SNR for a signal V_s of source impedance R_s, namely,

$$\text{SNR(dB)} = 10\log_{10}\left(\frac{V_s^2}{v_n^2}\right)$$

$$= 10\log_{10}\left[\frac{V_s^2}{[e_n^2 + (i_nR_s)^2 + 4kTR_s]B}\right], \quad (8.50)$$

where the numerator is the signal voltage (presumed to lie within the bandwidth B) and the terms in the denominator are the amplifier noise voltage, amplifier noise current applied to R_s, and Johnson noise in R_s. Note that increasing the amplifier bandwidth beyond what is necessary to pass the signal V_s only decreases the final SNR. However, if V_s is broadband (e.g., a noise signal itself), the final SNR is independent of amplifier bandwidth. In many cases the noise will be dominated by one of the terms in the preceding equation.

8.12.2 An example: transistor-noise test circuit

For the measured noise data in Table 8.1a we used the circuit in Figure 8.92. It's basically an elaboration of the simple single-ended grounded-emitter amplifier of Figure 8.42, with provision for gain calibration, and sockets and test points for component substitution. The single-ended configuration requires a large input blocking capacitor. Follower Q_2 biases the device under test (DUT) to the nominal quiescent voltage point by way of R_8, with choice of R_C setting the quiescent current. Optional resistor R_B permits measurement of both β and i_n. The unusually long RC time constants throughout (capacitance multiplier, input and output blocking networks) are needed to suppress excess low frequency noise; the circuit takes many seconds to settle.

This circuit could be improved in several ways (but it was not, owing to operator exhaustion).[120]

---

[120] Some details, for the curious: the circuit sets both I_C and V_{CE} for the DUT by means of a collector-to-base feedback loop. Lowpass filter R_5C_3 attenuates negative feedback that would unduly lower the transistor's common-emitter gain. The job of the large-value capacitor C_2 is to ground the base at all signal frequencies (as low as 4 Hz for our measurements) so we can measure $r_{bb'}$ (from its Johnson noise),

8.12.3 Measurement with a noise source

The preceding technique of measuring the noise performance of an amplifier has the advantage that you don't need an accurate and adjustable noise source, but it requires an accurate voltmeter and filter, and it assumes that you know the gain versus frequency of the amplifier, with the actual source resistance applied. An alternative method of noise measurement involves applying broadband noise signals of known amplitude to the amplifier's input and observing the relative increase of output noise voltage. Although this technique requires an accurately calibrated noise source, it makes no assumptions about the properties of the amplifier, because it measures the noise properties right at the point of interest, at the input.

Again, it is relatively straightforward to make the requisite measurements. You connect the noise generator to the amplifier's input, making sure that its source impedance R_g equals the source impedance of the signal you ultimately plan to use with the amplifier. You first note the amplifier's output rms noise voltage, with the noise-source attenuated to zero output signal. Then you increase the noise source rms amplitude V_g until the amplifier's output rises 3 dB (a factor of 1.414 in rms output voltage). The amplifier's input noise voltage in the measurement bandwidth, for this source impedance, equals this value of added signal. The amplifier therefore has a noise figure

$$\text{NF(dB)} = 10\log_{10}\left(\frac{V_g^2}{4kTR_g}\right). \quad (8.51)$$

From this you can figure out the SNR for a signal of any amplitude with this same source impedance, using eq'n 8.14:

$$\text{SNR (dB)} = 10\log_{10}\left(\frac{V_s^2}{4kTR_s}\right) - \text{NF}(R_s). \quad (8.52)$$

There are nice calibrated noise sources available, most

---

and also the noise voltage created across $r_{bb'}$ by the low-frequency $1/f$ tail of base-current noise. But capacitor C_2, in combination with bias-setting pot R_8, adds a second lagging phase shift (a second pole) within the feedback loop, creating a low-frequency bias instability; confusingly, this mimics $1/f$ noise in the DUT at the low end of the spectrum. We could address this by introducing a compensating zero (e.g., a small resistor in series with C_3), but we were reluctant to tamper further with the circuit. The dominant pole is R_8C_2, and for low-frequency noise measurements down in the sub-nV/$\sqrt{\text{Hz}}$ territory we found ourselves increasing C_2, to 0.15 F and even 0.35 F (!), and waiting for the modest instabilities to settle. This defeated our original goal, which was to avoid the large emitter bypass capacitors normally used with other non-feedback bias configurations – but, whatever, we were able to get reliable noise data for the plots and tables. The reader is welcome to experiment further.

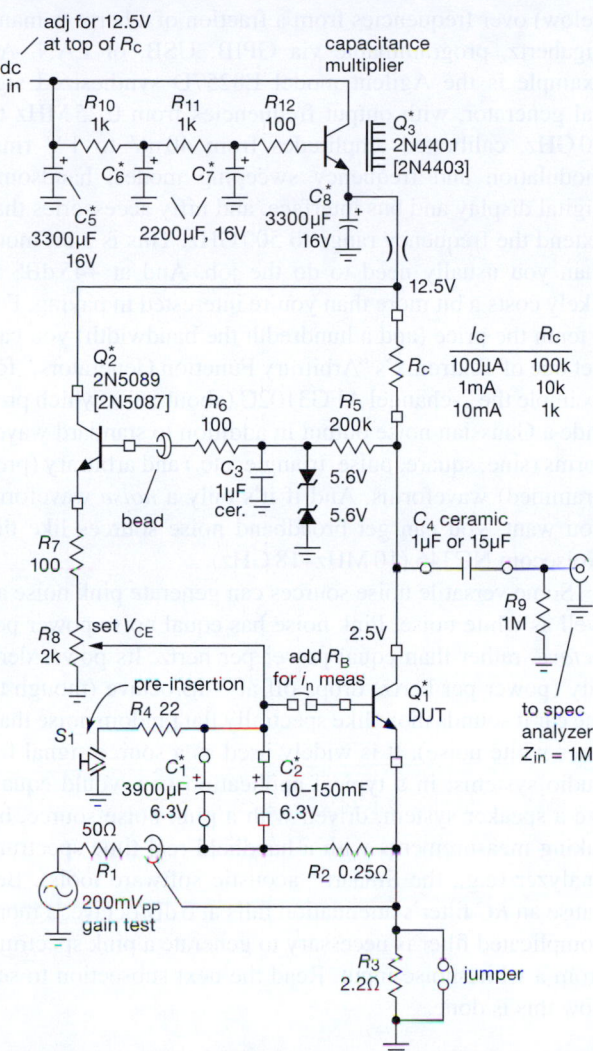

Figure 8.92. Transistor noise test circuit. Q_3 and associated components form a noise-eliminating capacitance multiplier. Polarities shown are for an *npn* device under test (DUT); for *pnp* reverse the polarity of components marked with asterisks. The gain-calibrating input signal is replaced by a $0\,\Omega$ shorting plug during noise measurement.

of which provide means for attenuation to precise levels in the microvolt range. Note: once again, the preceding formulas assume $R_{in} \gg R_s$. If, on the other hand, the noise-figure measurement is made with a *matched* signal source, i.e., if $R_s = Z_{in}$, then omit the factors "4" in the preceding expressions.

Note that this technique does not tell you e_n and i_n directly, just the appropriate combination for a source of impedance equal to the driving impedance you used in the

measurement. Of course, by making several such measurements with different noise source impedances, you could infer the values of e_n and i_n.

A nice variation on this technique is to use resistor Johnson noise as the "noise source." This is a favorite technique used by designers of very low noise radiofrequency amplifiers (in which, incidentally, the signal source impedance is usually $50\,\Omega$ and matches the amplifier's input impedance). It is usually done the following way: a dewar of liquid nitrogen holds a $50\,\Omega$ "termination" (a fancy name for a well-designed resistor that has negligible inductance or capacitance) at the temperature of boiling nitrogen, $77°K$; a second $50\,\Omega$ termination is kept at room temperature. The amplifier's input is connected alternately to the two resistors (usually with a high-quality coax relay), and the output noise power (at some center frequency, with some measurement bandwidth) is measured with an RF power meter. Call the results of the two measurements P_C and P_H, the output noise power corresponding to cold and hot source resistors, respectively. It is then easy to show that the amplifier's noise temperature, at the frequency of the measurement, is just

$$T_n = \frac{T_H - Y T_C}{Y - 1} \quad \text{Kelvin,} \qquad (8.53)$$

where $Y = P_H / P_C$, the ratio of noise powers. The noise figure is then given by eq'n 8.16, namely,

$$\text{NF(dB)} = 10\log_{10}\left(\frac{T_n}{290} + 1\right). \qquad (8.54)$$

Exercise 8.7. Derive the foregoing expression for noise temperature. *Hint:* begin by noting that $P_H = \alpha(T_n + T_H)$ and $P_C = \alpha(T_n + T_C)$, where α is a constant that will shortly disappear. Then note that the noise contribution of the amplifier, stated as a noise temperature, *adds* to the noise temperature of the source resistor. Take it from there.

Exercise 8.8. Amplifier noise temperature (or noise figure) depends on the value of signal source impedance, R_s. Show that an amplifier characterized by e_n and i_n (as in Figure 8.28) has minimum noise temperature for a source impedance $R_s = e_n/i_n$. Then show that the noise temperature, for that value of R_s, is given by $T_n = e_n i_n / 2k$.

If you don't want to bother with liquid nitrogen and are interested only in relatively low-frequency amplifiers, you can exploit the curious fact that a BJT's input noise voltage (at low currents, where $r_{bb'}$ effects are insignificant) is equal to the Johnson noise of a real resistor of value $r_e/2$. For example, if you ground the base, tie the collector to $+5\,V$, and pull down the emitter through $10k$ to $-5\,V$, you'll see at the emitter a noise signal with $50\,\Omega$ source impedance and $150K$ noise temperature. Add a blocking capacitor, and alternate this noise source with a real $50\,\Omega$ resistor (use a coax relay, not a CMOS switch!), and you've got a simple (and seriously inexpensive!) two-temperature noise calibrator.[121]

A. Amplifiers with matched input impedance

This last technique is ideal for noise measurements of amplifiers designed for a matched signal-source impedance. The most common examples are in radiofrequency amplifiers or receivers, usually meant to be driven with a signal source impedance of $50\,\Omega$, and that themselves have an input impedance of $50\,\Omega$. See Appendix H for an explanation of this departure from our usual criterion that a signal source should have a small source impedance compared with the load it drives. In this situation, e_n and i_n are irrelevant as separate quantities; what matters is the overall noise figure (with matched source) or some specification of SNR with a matched signal source of specified amplitude.

Sometimes the noise performance is explicitly stated in terms of the *narrowband* input signal amplitude required for obtaining a certain output SNR. A typical radiofrequency receiver might specify a 10 dB SNR with a $0.25\,\mu$V rms input signal and 2 kHz receiver bandwidth. In this case the procedure consists of measuring the rms receiver output with the input driven by a matched sinewave source initially attenuated to zero, then increasing the (sinewave) input signal until the rms output rises 10 dB, in both cases with the receiver bandwidth set to 2 kHz. It is important to use a meter that reads true rms voltages for a measurement where noise and signal are combined (more about this later). Note that radiofrequency noise measurements often involve output signals that are in the audiofrequency range.

8.12.4 Noise and signal sources

Broadband noise can be generated from the effects we discussed earlier, namely Johnson noise and shot noise. The shot noise in a vacuum diode is a classic source of broadband noise that is especially useful because the noise voltage can be predicted exactly; zener diode noise is also widely used in noise sources, as are gas discharge tubes. These extend from dc to very high frequencies, making them useful in audiofrequency and radiofrequency measurements. Look ahead to Figure 13.121 for an example of a "true random" noise generator.

Versatile signal sources are available with precisely controlled output amplitude (down to the microvolt range and

below) over frequencies from a fraction of a hertz to many gigahertz, programmable via GPIB, USB, or LAN. An example is the Agilent model E8257D synthesized signal generator, with output frequencies from 0.25 MHz to 20 GHz, calibrated amplitudes from 40 nV to 1 V rms, modulation and frequency sweeping modes, handsome digital display and bus interface, and nifty accessories that extend the frequency range to 500 GHz. This is a bit more than you usually need to do the job. And at +45 dB\$ it likely costs a bit more than you're interested in paying. For a tenth the price (and a hundredth the bandwidth) you can get one of Tektronix's "Arbitrary Function Generators," for example the 2-channel AFG3102C (about \$6k), which provide a Gaussian noise output in addition to standard waveforms (sine, square, pulse, triangle, etc.) and arbitrary (programmed) waveforms. And if it's only a *noise* waveform you want, you can get broadband noise sources like the Noisecom NC346 (10 MHz–18 GHz).

Some versatile noise sources can generate pink noise as well as white noise. Pink noise has equal noise power per *octave*, rather than equal power per hertz. Its power density (power per hertz) drops off at 3 dB/octave (though to the ear it sounds more like spectrally flat random noise than does white noise). It is widely used as a source signal for audio systems; in a typical application you would equalize a speaker system, driven with a pink noise source, by taking measurements with a handheld real-time spectrum analyzer (e.g., the Smaart® acoustic software tools). Because an *RC* filter's attenuation falls at 6 dB/octave, a more complicated filter is necessary to generate a pink spectrum from a white-noise input. Read the next subsection to see how this is done.

A. Pseudorandom noise source

We can make an interesting noise source by using digital techniques, in particular by connecting long shift registers with their input derived from a modulo-2 addition of several of the later bits (see §§11.3.1 and 13.14). The resultant output is a pseudorandom sequence of 1s and 0s that, after lowpass filtering, generates an analog signal of white spectrum up to the lowpass filter's breakpoint, which must be well below the frequency at which the register is shifted. These things can be run at very high frequencies, generating noise up to many megahertz or more. The "noise" has the interesting property that it repeats itself exactly after a time interval that depends on the register length (an n-bit maximal-length register goes through $2^n - 1$ states before repeating). Without much difficulty, that time can be made to be very long (years, or millennia), although most often a period of a second is long enough. For example, a 50-bit

---

[121] We learned this trick from Phil Hobbs, who suggests the use of a diode like an MBD301 (which doesn't suffer from high-level injection at $500\,\mu$A) instead of a BJT.

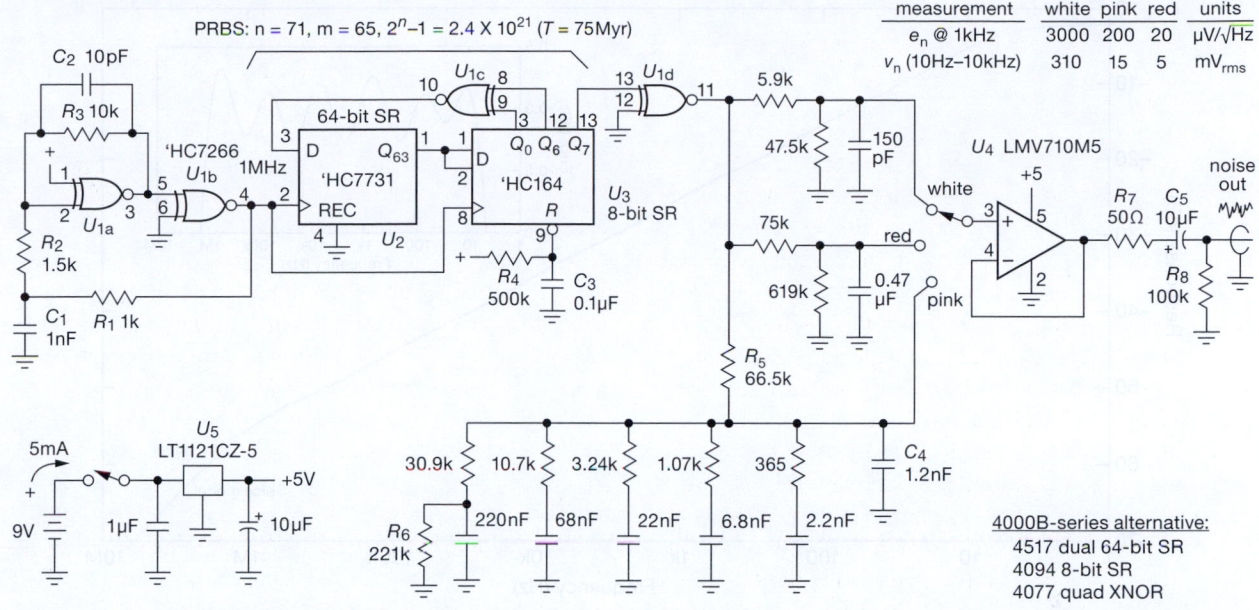

Figure 8.93. Pseudorandom noise source providing three colors of analog noise from 10 Hz to 100 kHz, with measured values of output noise density (at 1 kHz) and band-limited noise voltage (4-pole Butterworth, 10 Hz–10 kHz).

register shifted at 10 MHz will generate white noise up to 1 MHz or so, with a repeat time of 3.6 years.

The circuit shown in Figure 8.93 uses a 71-bit maximal-length shift register, clocked at 1 MHz, to generate a digital pseudorandom waveform that is spectrally flat (± 0.07 dB) out to 100 kHz. The pseudorandom bit sequence is ridiculously long – with a 1 MHz clock it repeats in about 75 million years. Creating either white or red noise is easy: the "white" selection simply filters the raw 2-level waveform through an *RC* lowpass at 200 kHz, above the band of interest, to suppress the (wideband) clocking edges. The "red" selection inserts instead a lowpass filter at 5 Hz, below the band of interest, so the output rolls off at the usual 6 dB/octave.

Pink noise is trickier; you need a filter that reduces the white-noise amplitude by a factor of $1/\sqrt{2}$ (rather than 1/2) for every doubling of frequency. The usual analog method is to use a parallel set of series *RC* sections (as in Figure 8.93), with the characteristic frequency of each successive section increasing by a fixed ratio (here $\times 10$, i.e., one decade) with an impedance that decreases by the square root of that same ratio (here $\sqrt{10}$). Even with such generous decade spacing you do remarkably well, as seen in the SPICE results of Figure 8.94, where the deviation from ideal -3 dB/octave behavior is just ± 0.25 dB over the 5-

decade frequency range of 10 Hz to 1 MHz.[122]

Simulations are fun, but the "rubber hits the road" when you actually build something and measure its performance. We did that, and Figure 8.95 shows the measured spectra of the circuit of Figure 8.93, which, to within the thickness of the (wiggly) lines, performs as advertised.

A few notes on the circuit

We cannot resist a few comments on the circuit design. As always, there are several choices at each step: a good design balances issues of performance, cost, complexity, availability of components, power, reliability, and (dare we say it?) *elegance*. For example, the ~ 1 MHz clocking signal could be supplied from a \$2 crystal-oscillator module; but for this application it needn't be accurate or stable, and so we chose to use two unused sections of the quad XNOR to make a simple *RC* relaxation oscillator. (We haven't seen this particular implementation used elsewhere, but it is completely straightforward: it is the topology of Figure 7.5, with U_{1a} configured as a noninverting Schmitt buffer, followed by the U_{1b} inverter. The hysteresis is set by $R_2 R_3$, to 0.65 V, with small speed-up capacitor C_2. The frequency is set by $R_1 C_1$.)

---

[122] A bit of candor here: to achieve this performance we did a bit of fine tuning to arrive at the circuit's choice of 1% resistor values.

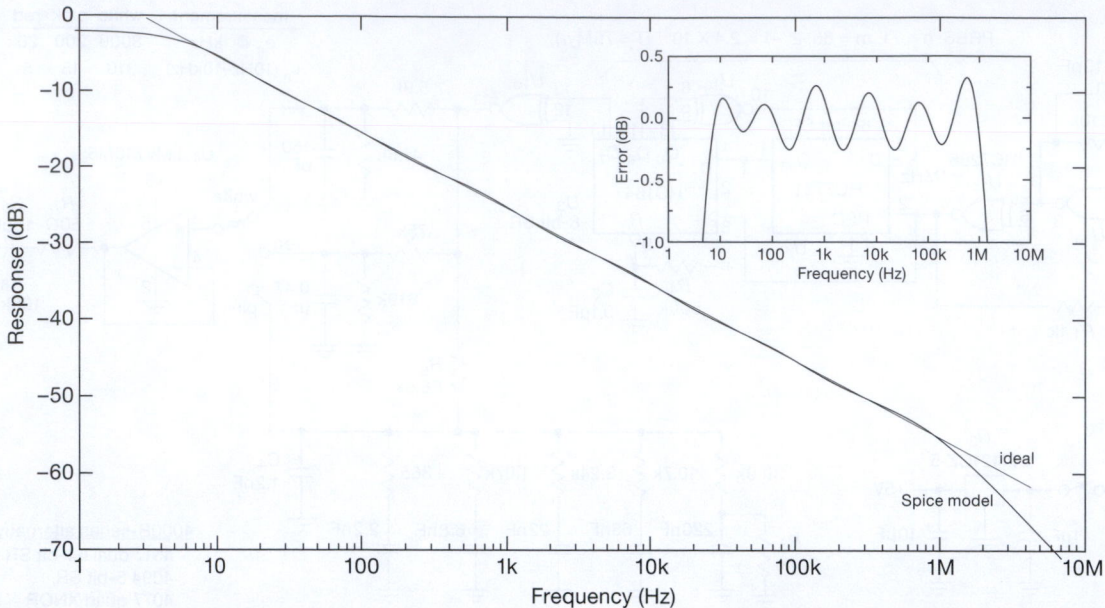

Figure 8.94. Simulation in SPICE of the pink-noise filter (R_5 and components below) in Figure 8.93.

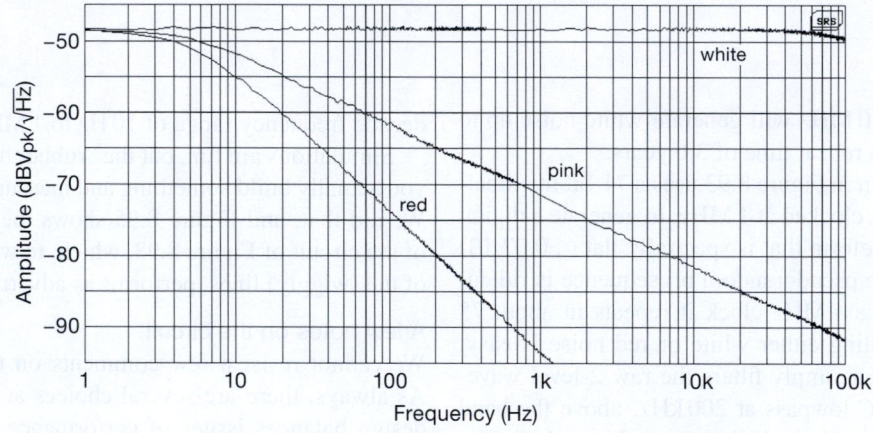

Figure 8.95. Measured spectra of the circuit of Figure 8.93. Each plotted spectrum is made from two 800-point FFT spectra to span the full 100,000:1 frequency range. The measured total noise voltage v_n and noise densities e_n are listed in Figure 8.93.

We chose the 74HC logic family because parts are readily available, easy to prototype (through-hole DIP), plenty fast, and of low power.[123] There are standard logic functions for what we needed, so we chose that path (rather than the alternatives: a cPLD, FPGA, or microcontroller). Available shift registers were either too small (8 stages), or outrageously large (the 'HC7731 we used is 256 stages, as four separate 64-stage shift-register banks). We went with the latter, using only one bank, and appending a 'HC164 8-stage parallel-out shift register so we could get at the needed taps (here $m=65$ and $n=71$; see Table 13.14). Note the use of exclusive-NOR (rather than exclusive-OR), to prevent the stuck state of all zeros and thus guarantee startup.

The output buffer is a nice BiCMOS rail-to-rail input/

---

[123] We could have used instead the "high-voltage" 4000B CMOS parts listed on the diagram, running directly from the 9 V battery. But there would be tradeoffs. (a) We would want to clamp the output to a stable amplitude, requiring some extra components; and (b) we would be clocking at a frequency uncomfortably close to the specified upper limit at low battery, foreclosing on any flexibility for going to a faster clock, say 10 MHz.

output op-amp, with 5 MHz GBW, low input current (4 pA, typ), plenty of output drive capability (± 20 mA), and modest power consumption (1.2 mA); it's noisy (20 nV/$\sqrt{\text{Hz}}$, with a high $1/f$ corner frequency), but that hardly matters when you're amplifying noise anyway. The series resistor R_7 ensures stability into capacitive loads and (if you care) provides source termination into $50\,\Omega$ cable.

For the 5 V regulator (U_5) we initially reached for something from the legacy LP2950 or LM2931 series of low-dropout linear regulators. But the LP2950 is intolerant of reversed input polarity (a feat easy enough to accomplish when replacing a 9 V battery); and both regulators are somewhat fussy when it comes to the output capacitor. In particular, they require some minimum amount of equivalent series resistance (ESR),[124] which is worrisome – how is one expected to deal with the low-ESR ceramic bypass capacitors that litter the rest of the circuit? Happily, there are nice LDOs that do not burden the designer with such worries: the LT1121-5 we chose is stable with output capacitors of zero ESR, and it's tolerant of reverse input (to -30 V); it also has the usual overcurrent and overtemperature protection.

There's more extensive discussion of pseudorandom noise, along with another pseudorandom noise source design (Figure 13.119), in §13.14.

8.13 Bandwidth limiting and rms voltage measurement

8.13.1 Limiting the bandwidth

All the measurements we have been talking about assume that you are looking at the noise output only in a limited frequency band. In a few cases the amplifier may have provision for this, making your job easier. If not, you have to hang some sort of filter on the amplifier output before measuring the output noise voltage.

RC filter   The easiest thing to use is a simple *RC* low-pass (or bandpass) filter, with 3 dB point(s) set at roughly

---

[124] From the LP2950 datasheet: "Ceramic capacitors whose value is greater than 1000 pF should not be connected directly from the LP2951 output to ground. Ceramic capacitors typically have ESR values in the range of 5 to 10 mΩ, a value below the lower limit for stable operation (see curve Output Capacitor ESR Range). The reason for the lower ESR limit is that the loop compensation of the part relies on the ESR of the output capacitor to provide the zero that gives added phase lead. The ESR of ceramic capacitors is so low that this phase lead does not occur, significantly reducing phase margin. A ceramic output capacitor can be used if a series resistance is added (recommended value of resistance about 0.1 Ω to 2 Ω)."

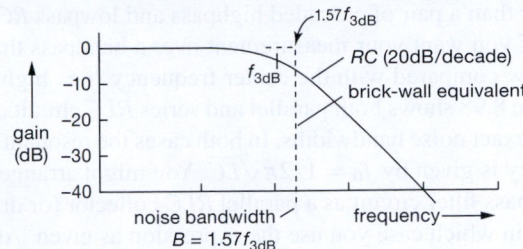

Figure 8.96. Equivalent brick-wall noise bandwidth for *RC* lowpass filter.

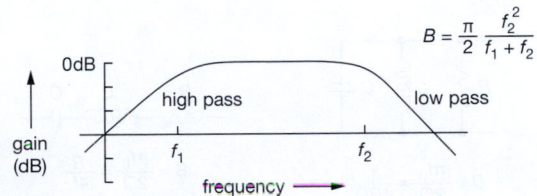

Figure 8.97. Equivalent brick-wall noise bandwidth for *RC* bandpass filter. For the case $f_1 = f_2$ the midband gain is -6 dB.

the bandwidth you want. For accurate noise measurements, you need to know the "equivalent noise bandwidth" (ENBW), i.e., the width of a perfect "brick-wall" filter that lets through the same noise voltage (Figure 8.96). This noise bandwidth is what should be used for B in all the preceding formulas. It is not terribly difficult to do the mathematics, and you find, for an *RC* lowpass filter,

$$B = \frac{\pi}{2} f_{3\text{dB}} = 1.57 f_{3\text{dB}}. \tag{8.55}$$

For a pair of cascaded lowpass *RC* sections (buffered so they don't load each other), the magic formula becomes $B = 1.22 f_{3\text{dB}}$. For the Butterworth lowpass filters discussed in §§6.2.6 and 6.3.2, the noise bandwidth is

$$
\begin{aligned}
B &= 1.57 f_{3\text{dB}} = \tfrac{1}{4RC} & \text{(1 pole)}, \\
B &= 1.11 f_{3\text{dB}} \approx \tfrac{1}{5.6RC} & \text{(2 poles)}, \\
B &= 1.05 f_{3\text{dB}} \approx \tfrac{1}{6RC} & \text{(3 poles)}, \\
B &= 1.025 f_{3\text{dB}} \approx \tfrac{1}{6.1RC} & \text{(4 poles)}.
\end{aligned}
$$

If you want to make band-limited measurements up at some center frequency, you can use a pair of *RC* filters (Figure 8.97), in which case the noise bandwidth is as indicated. You may wish to use higher-order Butterworth filters for more precise bandpass characteristics. In that case you'll need to know the corresponding equivalent noise bandwidths; don't panic – they're listed in Table 8.4 on page 564.

RLC filter   Another way to make a bandpass filter for noise measurements is to use an *RLC* circuit. This is better than a pair of cascaded highpass and lowpass RC filters if you want your measurement over a bandpass that is narrow compared with the center frequency (i.e., high Q). Figure 8.98 shows both parallel and series RLC circuits and their exact noise bandwidths. In both cases the resonant frequency is given by $f_0 = 1/2\pi\sqrt{LC}$. You might arrange the bandpass filter circuit as a parallel RLC collector (or drain) load, in which case you use the expression as given. Alternatively (recall Figure 1.107), you might interpose the filter as shown in Figure 8.99; for noise-bandwidth purposes the circuit is exactly equivalent to the parallel RLC, with $R = R_1 \| R_2$.

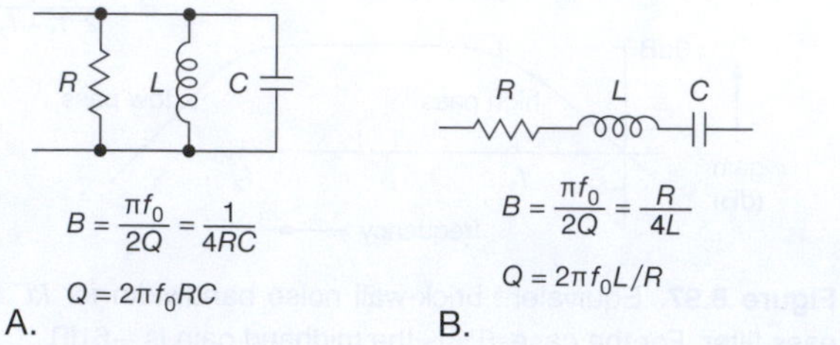

$$B = \frac{\pi f_0}{2Q} = \frac{1}{4RC}$$

$$Q = 2\pi f_0 RC$$

A.

$$B = \frac{\pi f_0}{2Q} = \frac{R}{4L}$$

$$Q = 2\pi f_0 L/R$$

B.

Figure 8.98. Equivalent *brick-wall* noise bandwidth for RLC bandpass filter. For the parallel circuit (A) the source signal is a current, and the output is the voltage across the terminals; for the series configuration (B) the input is a voltage applied across the circuit, and the output is the resulting current.

Averaging As we mentioned in §8.1.5, another way to accomplish lowpass filtering of a slow signal (a dc voltage, say, in the presence of additive white noise) is simply to average it over some time interval T; this is the same averaging operation as is done by an integrating ADC (e.g., in a digital voltmeter, see §13.8.3). In that case the equivalent noise bandwidth[125] is $B=1/2T$. Thus, for example, an averaging duration of 1 second admits a noise bandwidth of 0.5 Hz. This is a simple lowpass "filter" – but one that is less sharp even than a single-section RC (the latter falls off at 6 dB/octave, compared with 3 dB/octave for the time average).

---

[125] It is important to distinguish the operation of an averaging time T when measuring a static signal, as just described, from the use of a time "window" (again of duration T) to bound the measurement interval of an ac signal. In the latter case the use of such a *rectangular window* imposes a resolution bandwidth $B=1/T$ (i.e., twice that of the dc measurement). In digital signal processing the business of windowing plays an important role; see for example Harris, F. J., "On the use of windows for harmonic analysis with the discrete Fourier transform," *Proc. IEEE* **66** 51–83 (1978). We'll see this same theme, in connection with *synchronous signal averaging*, in §8.14, where synchronous (or "lock-in") detection of signals over some time duration T produces the same bandwidth reduction: $B=1/T$.

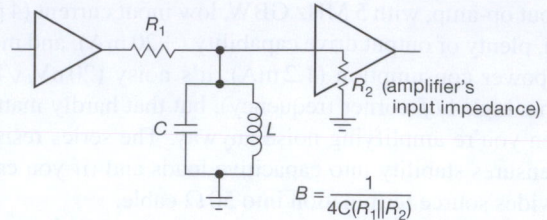

Figure 8.99. Interstage RLC bandpass filter.

Digital filter Digital signal processing (DSP) is an effective way to implement extremely well-defined filter functions, with their characteristics easily modified by changing the stored numerical coefficients. See §6.3.7 for more detail.

Frequency translation Imagine you want to make measurements of the "spot noise voltage" in some narrow bandwidth (say 10 Hz) centered at some relatively high frequency f_{in}, where the latter may be up at tens to hundreds of kilohertz, or perhaps even a megahertz or more. That is, the ratio $Q=f_{in}/\Delta f$ is very large, say greater than a thousand. Such a high-Q bandpass filter is pretty difficult to implement! But there's a good way to do the measurement without filter heroics.

The trick is to shift (translate) the frequency band of interest down to a much lower frequency, where it's easy to make a narrowband filter. The technique is called *heterodyning*, and it's a basic technique in most radiofrequency communications systems. It's easiest to understand if you think initially of a single input frequency. Figure 8.100 shows the basic scheme, in which an input signal voltage (at frequency f_{in}) is multiplied by a sinusoidal voltage (at frequency f_{LO}) from a *local oscillator* ("LO"), creating a pair of sinusoidal signals at frequencies $f_{in} \pm f_{LO}$.[126] The multiplier is called a *mixer*, and its output is filtered to eliminate one of the mixing products. Mixers can be in the form of an active circuit (a "4-quadrant multiplier"), or, for use at higher frequencies, a passive transformer-coupled diode arrangement called a *balanced mixer*.

In a communications application there may be several stages of frequency translation, passing through several intermediate-frequency ("IF") stages where amplification and filtering takes place.[127] For a simple noise-measuring

---

[126] From the identity $\cos x \cos y = \frac{1}{2}\cos(x+y) + \frac{1}{2}\cos(x-y)$, with $x=2\pi f_{in}$ and $y=2\pi f_{LO}$.

[127] For example, in an FM radio the LO is set 10.7 MHz below the desired station, with the 10.7 MHz IF signal amplified and demodulated. Radiofrequency and communications techniques are discussed in greater detail in Chapter 13 of this book's second edition.

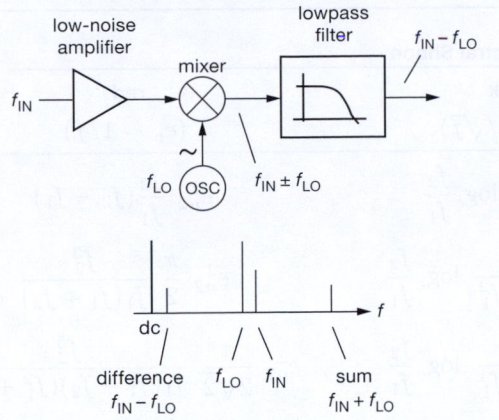

Figure 8.100. Heterodyne frequency shifting: the mixer produces sum and difference frequencies; only the latter survives the lowpass filter.

application it may be adequate to mix directly to "baseband," by tuning the LO to the frequency at which the spot noise voltage is to be measured. This is the same process used in *synchronous detection*, described later in §§8.14 and 13.13.6D. Commercial synchronous (lock-in) amplifiers let you make narrowband spot noise voltage measurements at any frequency in their range, extending to 100 kHz (for typical lock-in amplifiers) to as high as 200 MHz (e.g., for the SR844).

8.13.2 Calculating the integrated noise

Let's begin with the simple single-section RC lowpass band-limiting filter. From its equivalent brick-wall bandwidth (eq'n 8.55), the integrated noise-*voltage* output from filtering a white-noise input of noise *density* e_n is found by taking the square root of

$$v_n^2 = e_n^2 B = e_n^2 \frac{\pi}{2} f_{3dB} \quad \mathrm{V^2(rms)} \qquad (8.56)$$

and analogously for noise *current*.

Things get more complicated if the noise density e_n depends on frequency, as it does, for example, with $1/f$ ("pink") noise. In such a case you must integrate, over frequency, the squared noise density $e_n^2(f)$ times the filter's spectral power bandpass characteristic. And it's common to use a *bandpass* filter, with both lower and upper frequency limits (call them f_1 and f_2). Then an ideal "brick-wall" (or should it be "squash court"?!) bandpass filter has unit response between f_1 and f_2, and a noise bandwidth of $B = f_2 - f_1$.

Brick walls are hard to implement in analog technology, and a simple expedient, as suggested earlier, is simply

to use a cascaded pair of RC's, with 3 dB frequencies of f_1 and f_2. With a bit of mathematical finesse you could demonstrate proficiency by performing the appropriate integrals, thus figuring the output noise voltage for the "standard noises" (white, or $1/f$, or even $1/f^2$). A shortcut is to exploit the brilliance of Wolfram's impressive *Mathematica* program. A colleague[128] did this for us, and Table 8.4 on the next page summarizes the results, for those three colors of noise, and for four styles of bandpass filters (brick-wall, simple RC (single-pole; that is, a pair of cascaded RC sections), 2-pole Butterworth, and m-pole Butterworth.

The expressions listed give you the integrated squared noise voltage v_n^2, for given lower and upper -3 dB frequency limits f_1 and f_2, and noise voltage density e_n (Figure 8.101); the rms noise voltage is then gotten by taking the square root: $V_{n(rms)} = \sqrt{v_n^2}$. For pink noise ($1/f$ in power) and red noise ($1/f^2$ in power) the noise density e_n is a function of frequency; for these expressions we have used for the overall multiplying factor the noise density at the high band end; that is, $e_n^2(f_2)$ (abbreviated as e_{n2}^2 in the table), which is the squared noise density at $f = f_2$, in units of V^2/Hz.

Note that you can set $f_1 = 0$, in the expressions for white noise, to get the dc-to-f_2 (lowpass-limited) noise voltage. If you do so, you'll get the expressions shown on page 561. This does not work for pink or red noise, though, because the integral diverges at zero frequency, conveyed in these expressions by the embarrassment of a zero denominator in the log's argument (pink) or, worse, in the result itself (red). This is why $1/f$ noise is typically measured with a 0.1 Hz to 10 Hz limited bandwidth, etc.; see Figures 8.102 and 8.103.

Noise *current* is treated just the same: substitute everywhere the noise current density i_n in place of e_n, to find the integrated noise current $I_{n(rms)}$.

A. The high end matters most

To calculate the integrated noise voltage over some bandpass, you need to integrate the noise power density (e_n^2) over frequency, taking account of the filter's passband response (call that $H(f) \equiv V_{out}/V_{in}$, and don't worry about phase); that is,

$$v_n^2 = \int_{f_1}^{f_2} e_n^2 H^2(f)\, df \quad \mathrm{V^2(rms);} \qquad (8.57)$$

and then you take the square root: $V_{n(rms)} = \sqrt{v_n^2}$.

So, looking at log–log plots of filter passbands like

---

[128] Jason Gallicchio, again!

Table 8.4 Noise Integrals[a]

Filter Type	white (e_n = const)	pink ($e_n \sim 1/\sqrt{f}$)	red ($e_n \sim 1/f$)
		Noise Spectral Shape	
brickwall	$e_n^2\,(f_2 - f_1)$	$e_{n2}^2\, f_2 \log_e \dfrac{f_2}{f_1}$	$e_{n2}^2\, \dfrac{f_2}{f_1}(f_2 - f_1)$
1-pole (*RC*)	$e_n^2\, \dfrac{\pi}{2}\, \dfrac{f_2^2}{f_1 + f_2}$	$e_{n2}^2\, \dfrac{f_2^3}{f_2^2 - f_1^2}\, \log_e \dfrac{f_2}{f_1}$	$e_{n2}^2\, \dfrac{\pi}{2}\, \dfrac{f_2^3}{f_1(f_1 + f_2)}$
2-pole Butterworth	$e_n^2\, \dfrac{\pi}{2\sqrt{2}}\, \dfrac{f_2^4}{(f_1 + f_2)(f_1^2 + f_2^2)}$	$e_{n2}^2\, \dfrac{f_2^5}{f_2^4 - f_1^4}\, \log_e \dfrac{f_2}{f_1}$	$e_{n2}^2\, \dfrac{\pi}{2\sqrt{2}}\, \dfrac{f_2^5}{f_1(f_1 + f_2)(f_1^2 + f_2^2)}$
m-pole Butterworth	$e_n^2\, \dfrac{\pi/2m}{\sin(\pi/2m)}\, \dfrac{f_2 - f_1}{1 - (f_1/f_2)^{2m}}$	$e_{n2}^2\, \dfrac{f_2}{1 - (f_1/f_2)^{2m}}\, \log_e \dfrac{f_2}{f_1}$	$e_{n2}^2\, \dfrac{\pi/2m}{\sin(\pi/2m)}\, \dfrac{f_2 - f_1}{1 - (f_1/f_2)^{2m}}$

Notes: (a) Values of the mean-squared noise voltage, v_n^2, in a frequency band from f_1 to f_2, as bandpass limited by the indicated filter type. The noise voltage density e_{n2} is the value of e_n at $f = f_2$.

those in Figure 8.101, you might at first think that a bandpass filter with symmetrical and sharp rolloffs at both low and high ends is needed. But – surprise – the v_n^2 integral weights the high end disproportionately, as you can see from Figure 8.102, where we've plotted the value $e_n^2 H^2(f)$ for white noise (e_n=1) over a 100:1 bandpass, on *linear* frequency and amplitude scales (because that's what integrals do for a living). The integral is the area under the curve, lazily accumulating lots of unwanted spectrum for a single-section *RC* at the high end, but acquiring some discipline with higher orders; by contrast, the filter order at the low end matters hardly at all. And this general behavior persists, even with a narrower 10:1 passband (Figure 8.103).

We've taken the simple case of white noise, with its uniform spectral density. But the situation is not much changed even when the noise density is rising at low frequencies (e.g., pink noise, with $e_n^2 \propto 1/f$): the combination of a response $H^2 \propto f^2$, integrated over a small frequency span at low frequency, kills off the modestly increasing noise density.

8.13.3 Op-amp "low-frequency noise" with asymmetric filter

Because "the high end matters most," the *low-frequency noise-voltage* specifications (0.1–10 Hz) reported on many op-amp datasheets are measured with an asymmetrical filter, often a single-section highpass at 0.1 Hz cascaded with a 2-section (or sometimes higher-order) lowpass at 10 Hz. Here are the equivalent noise bandwidths for bandpass fil-

ters that are first-order (*RC*) highpass at f_1 and second-order Butterworth lowpass at f_2; as before, for pink or red noise the multiplying factor is the noise density at the upper frequency limit f_2, i.e., "e_{n2}"$\equiv e_n(f_2)$:

$$v_n^2 = e_n^2\, \frac{\pi}{4}\, \frac{\sqrt{2}f_1^2 f_2^3 - 2f_1 f_2^4 + \sqrt{2}f_2^5}{f_1^4 + f_2^4} \quad \text{(white)},$$

$$v_n^2 = e_{n2}^2\, \frac{\pi f_1^2 f_2^3 + 4f_2^5 \log_e(f_2/f_1)}{4(f_1^4 + f_2^4)} \quad \text{(pink)}, \quad (8.58)$$

$$v_n^2 = e_{n2}^2\, \frac{\pi}{4}\, \frac{\sqrt{2}f_1^3 f_2^3 - \sqrt{2}f_1 f_2^5 + 2f_2^6}{f_1^5 + f_1 f_2^4} \quad \text{(red)}.$$

Op-amp datasheets usually specify a 0.1 Hz to 10 Hz bandwidth for their listed low-frequency noise voltage; as often as not, it is defined with an asymmetric filter. But, curiously, they tend to list a peak-to-peak value (rather than rms), taken from a 10-second 'scope capture (similar to the bottom trace in Figure 8.4). It's common to estimate the rms noise voltage with this rule of thumb: $v_{n(rms)} \approx v_{n(pp)}/6$.

A. Op-amp low-frequency noise voltage

Operational amplifiers (with the exception of auto-zeroing op-amps) exhibit the, by now, familiar noise-density characteristic: flat at higher frequencies (call it e_{nH}), but rising approximately as $e_n \propto 1/\sqrt{f}$ (pink noise) for frequencies below the $1/f$ noise corner (call that frequency f_c). If you know f_c and e_{nH}, you can use the expressions in Table 8.4 to estimate the integrated noise voltage over any span of bandpass f_1 to f_2.

There are three possibilities. (a) The bandpass is entirely

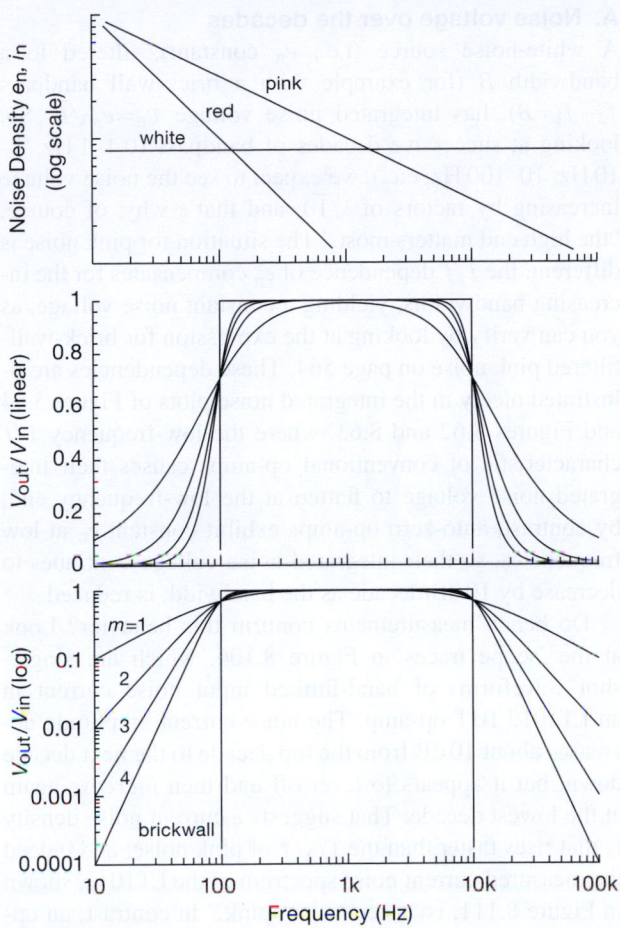

Figure 8.101. Spectral shapes of the three classic noises, and the bandpass filters used to evaluate their integrated noise voltage or current.

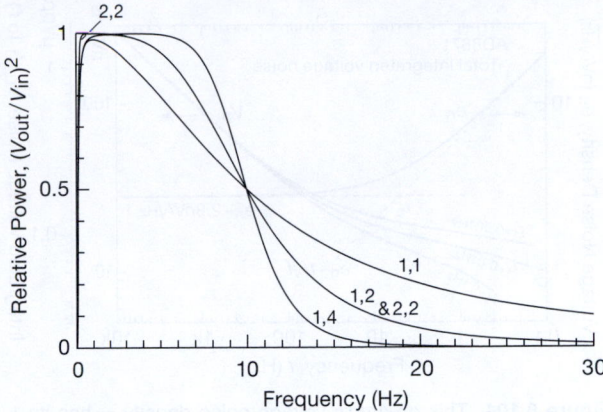

Figure 8.102. Power spectrum of filtered white noise, for 0.1–10 Hz Butterworth bandpass filters of the indicated orders (at their "low,high" ends; thus a "1,2" bandpass filter consists of a first-order highpass RC at f_1 cascaded with a second-order Butterworth low-pass at f_2).

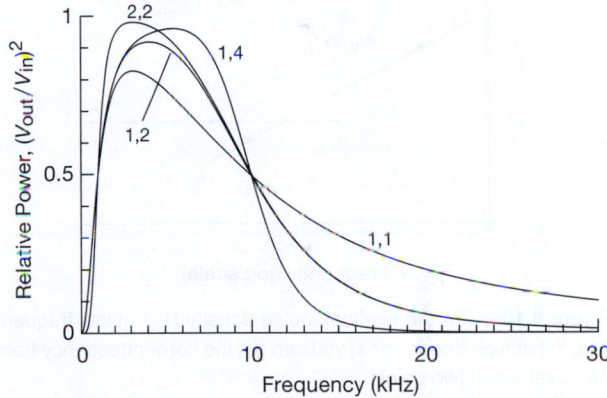

Figure 8.103. Power spectrum of filtered white noise, for 1 kHz–10 kHz Butterworth bandpass filters of the indicated orders ("low,high" ends).

in the white noise region, i.e., $f_1 > f_c$; (b) the bandpass is entirely in the pink noise region, i.e., $f_2 < f_c$; or (c) the bandpass straddles the $1/f$ corner frequency. For the first two cases, use the corresponding expression from Table 8.4 corresponding to the filter characteristic in use. For case (c), just calculate the separate white and pink v_n^2 contributions over the full passband (f_1 to f_2), and take their sum. In the idealized case of a brick-wall bandpass filter, this process gives an integrated noise voltage of

$$v_n^2 = e_{nH}^2 \left(f_2 - f_1 + f_c \log_e \frac{f_2}{f_1} \right) \quad \text{V}^2(\text{rms}). \quad (8.59)$$

This is what we've done to create the integrated noise curves in Figure 5.54 (in §5.11.1), based on the datasheet noise densities plotted in Figure 5.37 (in §5.10.6) and Fig-

ure 8.63 (from the data plotted in Figures 8.60 and 8.61).[129] Figure 8.104 shows an example, using the AD8671's datasheet e_n curve ($1/f$ corner at 5 Hz) to find the integrated-noise voltage v_n as a function of upper cut-off frequency (it's necessary to choose a nonzero low-frequency limit f_1 to prevent divergence).

---

[129] A spreadsheet program provides a handy way to make such calculations, and to plot the results.

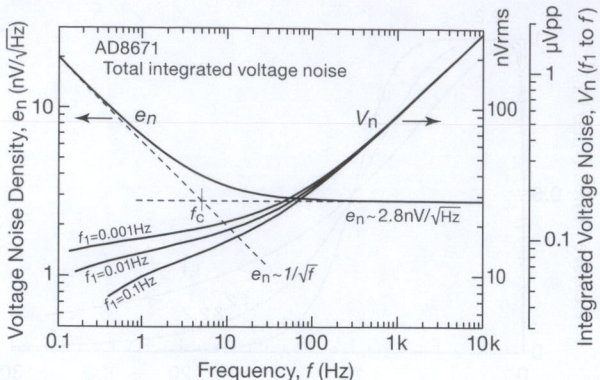

Figure 8.104. This op-amp's voltage noise density e_n has its $1/f$ corner frequency f_c at 5 Hz. Integrating the noise power ($\sim e_n^2$) from a low frequency f_1 up to a cutoff frequency $f_2=f$ gives the squared integrated noise voltage v_n^2, from which these curves of v_n result. If f_1 were set to zero, the v_n integral would diverge.

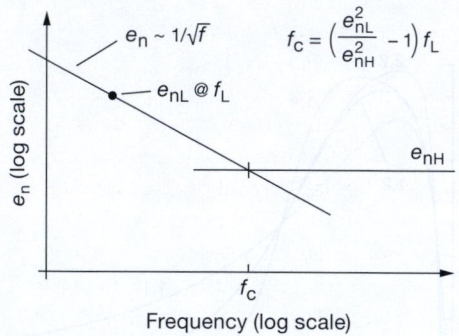

Figure 8.105. For the textbook noise density ($1/f$ at low frequencies, flat at high frequencies) you can get the corner frequency from the e_n values at two points.

8.13.4 Finding the $1/f$ corner frequency

If you're looking at a plot of noise density versus frequency, you can try to "eyeball" the $1/f$ corner frequency f_c. But it's nice to be able to find f_c from a pair of tabulated values when there's no graph available (often e_n is specified at 10 Hz and at 1 kHz in a datasheet's tabulated data); and anyway some folks are impressed by an equation or two in a not-inexpensive book like this one. You can figure it out, guided by Figure 8.105; the result is

$$f_c = \frac{e_{nL}^2 - e_{nH}^2}{e_{nH}^2} f_L = \left(\frac{e_{nL}^2}{e_{nH}^2} - 1 \right) f_L \quad \text{Hz,} \qquad (8.60)$$

where e_{nL} is the noise density at some frequency f_L that is below the corner frequency, and e_{nH} is the noise density well above f_c.

A. Noise voltage over the decades

A white-noise source (i.e., e_n constant), filtered to a bandwidth B (for example with a brick-wall bandpass $f_2-f_1=B$), has integrated noise voltage $v_n=e_n\sqrt{B}$. So, looking at successive decades of bandpass (0.1–1 Hz, 1–10 Hz, 10–100 Hz, etc.), we expect to see the noise voltage increasing by factors of $\sqrt{10}$; and that's why, of course, "the high end matters most." The situation for pink noise is different: the $1/f$ dependence of e_n^2 compensates for the increasing bandwidths, yielding a constant noise voltage, as you can verify by looking at the expression for brick-wall-filtered pink noise on page 564. These dependencies are illustrated nicely in the integrated noise plots of Figure 5.54 and Figures 8.62 and 8.63, where the low-frequency $1/f$ characteristic of conventional op-amps causes their integrated noise voltage to flatten at the low-frequency end; by contrast, auto-zero op-amps exhibit constant e_n at low frequencies, so their integrated noise voltage continues to decrease by 10 dB/decade as the bandwidth is reduced.

Do bench measurements confirm this behavior? Look at the 'scope traces in Figure 8.106, which are single-shot waveforms of band-limited input noise current in an LT1012 BJT op-amp. The noise-current amplitude decreases about 10 dB from the top decade to the next decade down, but it appears to level off and then *increase* again at the lowest decade. That suggests a current noise density i_n that rises faster than the $1/\sqrt{f}$ of pink noise; and indeed the measured current noise spectrum of the LT1012, shown in Figure 8.111, is "steeper than pink." In contrast, an op-amp whose low-frequency noise conforms to an ideal pink-noise spectrum would exhibit approximately constant noise amplitude per decade of bandwidth, once well below the $1/f$ corner frequency.

B. Forever $1/f$?

You often hear talk about low-frequency noise power conforming to a "$1/f$ law," as if there's some statutory requirement involved. You might at first think that this cannot possibly be true, because (you say to yourself) a $1/f$ power spectrum cannot continue forever, since it would imply unlimited noise amplitude. If you waited long enough, the input offset voltage (or input current, in this case) would become unbounded. In fact, the popular mythology of a low-frequency noise catastrophe (to which your thinking would have fallen victim) is quite without merit: even if the noise power *density* continues as $1/f$ all the way down to zero frequency, its total noise power (i.e., the *integral* of noise power density) diverges only logarithmically, given that $\int f^{-1} df = \log_e f$. To put some numbers to it, the total noise power in a pure $1/f$ spectrum between 1 *micro*hertz and

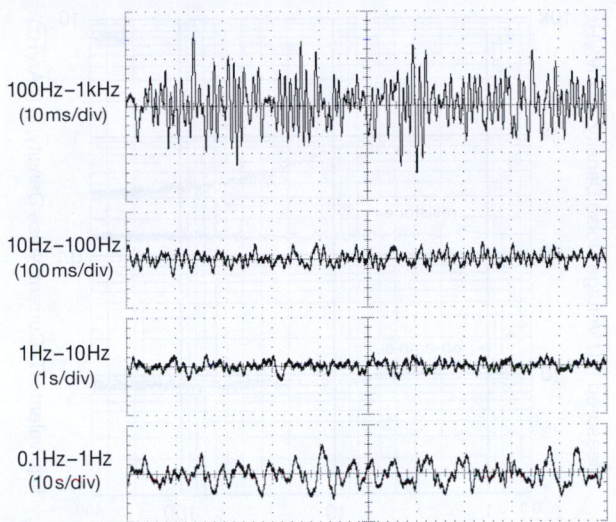

Figure 8.106. LT1012 noise current versus time, for successive decade bandpasses. Vertical: 5 pA/div. Horizontal: scaled to bandpass, as indicated.

10 Hz is only 3.5 times greater than that between 0.1 Hz and 10 Hz; going down another six decades (to 10^{-12} Hz), the corresponding ratio grows only to 6.5. Put another way, the $1/f$ total noise power, going all the way down to a frequency that is the reciprocal of 32,000 years (when Neanderthals still roamed the planet, and there were no op-amps), is just six times greater than that of the usual datasheet 0.1–10 Hz "low-frequency noise." So much for catastrophes.

To find out whether the low-frequency noise of real op-amps continues to conform to a $1/f$ spectrum, we measured the current noise spectrum of an LT1012 op-amp all the way down to 0.5 *milli*hertz,[130] with the result of Figure 8.107. As we remarked above, this op-amp is unusual in that its current noise density i_n rises faster than the usual $1/\sqrt{f}$ (pink noise) for a decade around 1 Hz; but even so it settles back to the canonical pink noise, and ultimately becomes something closer to "pale white" ($f^{-1/4}$ or slower).

You could conclude that this demonstrates the unphysical nature of $1/f$ behavior all the way down to zero. But there's another possible explanation, namely that this op-amp is afflicted with some mild burst noise. That would be consistent with the "faster-than-pink" slope around 1 Hz (recall the burst-noise spectrum in Figure 8.6), and it would also lead you to incorrectly attribute a "slower-than-pink"

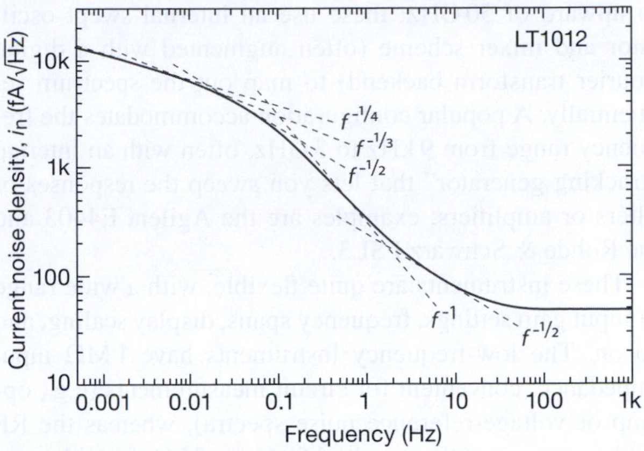

---

slope at the low frequency end of the spectrum in Figure 8.107.

If this latter explanation is correct, measurements at even lower frequencies (to 0.00001 Hz, say) would confirm a continuing $1/f$ (pink) slope. But it takes a full day to reach 10 μHz, which is why reliable long timescale data are hard to find. An interesting datapoint is provided by Daire's measurement of the "Spectral Noise Distribution" of a Keithley 6430 high sensitivity (0.05 fA resolution) source-measure instrument,[131] which displays a $1/f$ character all the way down to a microhertz (corresponding to a time scale of several weeks). With this in mind, it is quite possible that the low-frequency flattening seen in Figure 8.107 is indeed an artifact of a burst-noise plateau. Or maybe not – there's no statutory requirement that extremely-low-frequency noise must obey a $1/f$ (pink-noise) spectrum.

8.13.5 Measuring the noise voltage

There is a class of test instruments, called *spectrum analyzers* or *dynamic signal analyzers*, that measure and display the frequency spectrum of an input signal. One style is optimized for low-frequency and audio use, generally up to ~100 kHz or so, with spectral calculations done with a discrete Fourier transform; examples are the Stanford Research Systems SR780/5 and the Agilent U8903A. At the other end you find RF and microwave spectrum analyzers whose upper frequency limits range from ~3 GHz

---

Figure 8.107. Measured LT1012 noise current spectrum, extending down to 500 microhertz. Several slopes are shown, for your amusement.

---

[130] We averaged 100 power spectra, accumulating a 2000-second time series for each one: not a quick experiment!

[131] Adam Daire, "Counting electrons: how to measure currents in the attoampere range," Keithley Instruments, Inc., September 2005. Available as a pdf from www.keithley.com.

to upward of 50 GHz; these use an internal swept oscillator and mixer scheme (often augmented with a digital Fourier transform backend) to map out the spectrum sequentially. A popular configuration accommodates the frequency range from 9 kHz to 3 GHz, often with an internal "tracking generator" that lets you sweep the responses of filters or amplifiers; examples are the Agilent E4403 and the Rohde & Schwarz FSL3.

These instruments are quite flexible, with a wide range of input gain settings, frequency spans, display scaling, and so on. The low-frequency instruments have 1 MΩ input impedance, convenient for circuit measurements (e.g., opamp or voltage-reference noise spectra), whereas the RF analyzers present the standard 50 Ω (or 75 Ω, for video applications) input impedance. To measure an op-amp noise spectrum, for example, just use the circuit of Figure 8.108, with $R_s=0$; you'll get a spectrum like the lower curve in Figure 8.109. Voltage-noise spectra for some op-amps thus measured are shown in Figure 8.110.

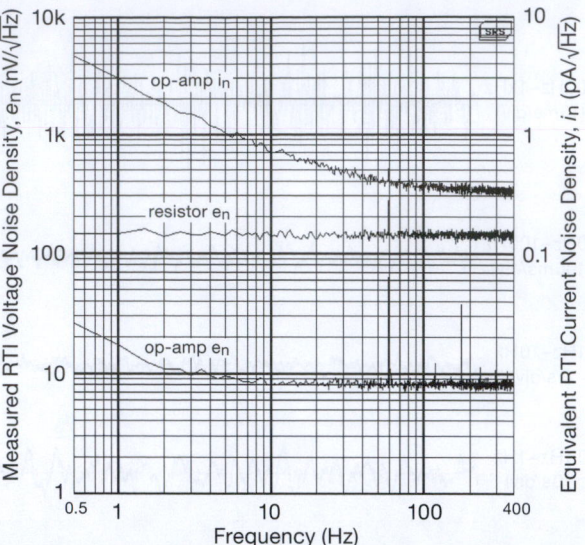

Figure 8.109. Voltage and current noise spectra of an OPA277 opamp, measured with the circuit of Figure 8.108. The Johnson noise of the 1 MΩ resistor R_s used for the i_n measurements sets a "noise floor," here seen to be well below the op-amp's measured i_n.

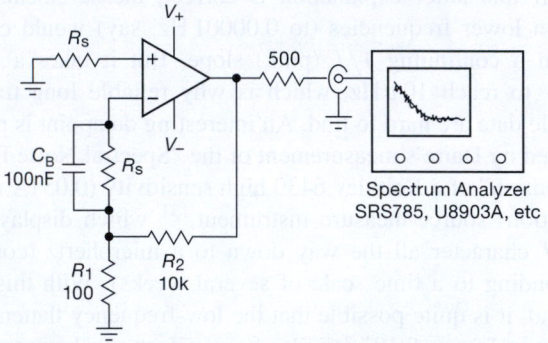

Figure 8.108. Measuring op-amp noise voltage and current spectra. For e_n, set $R_s=0$; for i_n, choose R_s substantially larger than the amplifier's noise resistance ($R_n=e_n/i_n$). Use a shielded box with filtered dc feedthroughs.

The most accurate way to make integrated output noise voltage measurements is to use a true rms voltmeter. These operate either by measuring the heating produced by the signal waveform (suitably amplified) or by using an analog squaring circuit followed by averaging. If you use a true rms meter, make sure it has response at the frequencies you are measuring; some of them go up to only a few kilohertz. True rms meters also specify a "crest factor," the ratio of peak voltage to rms that they can handle without great loss of accuracy. For Gaussian noise, a crest factor of 3 to 5 is adequate.

You can use a simple averaging-type ac voltmeter instead, if a true rms meter is unavailable. In that case, the values read off the scale must be corrected. As it turns out,

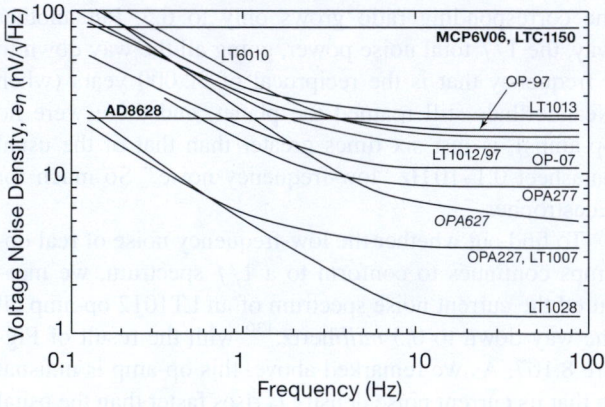

Figure 8.110. Measured voltage-noise-density spectra of a selection of op-amps. The **boldface** parts are auto-zero amplifiers, which circumvent the $1/f$ "flicker-noise" demon by repetitive offset correction; the *italic* OPA627 is a JFET-input op-amp. See also Figures 8.60 and 8.61.

all averaging meters (VOMs, DMMs, etc.) already have their scales adjusted, so what you read isn't actually the *average*, but rather the rms voltage *assuming a sine-wave signal*. For example, if you measure the powerline voltage in the United States, your meter will read something close to 117 V. That's fine, but if the signal you're reading is Gaussian noise, you have to apply an additional correction. The rule is as follows: to get the rms voltage of Gaussian noise, multiply the "rms" value you read on an averaging ac voltmeter by 1.13 (or add 1 dB). Warning: this works fine if the signal you are measuring is pure noise (e.g., the output of an amplifier with a resistor or noise source as input), but it won't give accurate results if the signal consists of a sinewave added to noise.

A third method, not exactly world famous for its accuracy, consists of looking at the noise waveform on an oscilloscope: the rms voltage is 1/6 to 1/8 of the peak-to-peak value (depending on your subjective reading of the pp amplitude). It isn't very accurate, but at least there's no problem getting enough measurement bandwidth.

8.13.6 Measuring the noise current

An easy way to measure input noise current in an op-amp is to use the circuit of Figure 8.108, with a large input resistor R_s. Its value has to be large enough so that the noise voltage generated across it by the op-amp's input noise current is at least comparable to (and preferably much larger than) the op-amp's noise voltage: $i_n R_s \gtrsim e_n$. Another way to put it is to say that $R_s \gtrsim R_n$, the op-amp's noise resistance.

We're not done yet. It's also necessary that $i_n R_s$ dominates over the resistor's Johnson-noise voltage density: $i_n R_s \gtrsim \sqrt{4kTR_s}$. That is, R_s looks like a current noise source of $i_n = \sqrt{4kT/R_s}$, so you have to choose a large-enough value so that the op-amp's input current noise dominates. We find it easiest to remember Johnson noise voltage and current values for a round-number resistance, and then scale it according to the square root of R. So, take note: the Johnson noise of a 1 MΩ resistor is $e_n = 127\,\text{nV}/\sqrt{\text{Hz}}$ (open-circuit), scaling as $\sqrt{R}$; and $i_n = 127\,\text{fA}/\sqrt{\text{Hz}}$ (short-circuit), scaling as $1/\sqrt{R}$.

Figure 8.111 shows input current noise spectra for a selection of op-amps (most of which have BJT inputs with bias-current cancellation), measured with the circuit of Figure 8.108 with $R_s = 100\,\text{M}\Omega$ (for which the resistor's Johnson-noise voltage is equivalent to a current-noise density of $12.7\,\text{fA}/\sqrt{\text{Hz}}$), and was so measured with an OPA627 JFET-input op-amp (whose voltage- and current-noise contributions are negligible by comparison). The two auto-zero (chopper-stabilized) parts exhibit a flat low-frequency noise spectrum, in contrast to the usual "pink" $1/f$ rising noise power density of conventional op-amps. But auto-zeros usually have nasty spectral peaks at higher frequencies, caused by the clocked switching at the input (those for the AD8628A are around 15 kHz, off the right-hand end of the graph; but Figure 5.52 shows them handsomely, for the MCP6V06).

Table 8.5 Auto-zero Noise Measurements

	Noise Voltage			Spectral pk	Noise Current		
	e_n		meas peak ampl		i_n		meas peak ampl
	spec[t] $\left(\frac{\text{nV}}{\sqrt{\text{Hz}}}\right)$	meas $\left(\frac{\text{nV}}{\sqrt{\text{Hz}}}\right)$	(μV)	(kHz)	spec[t] $\left(\frac{\text{fA}}{\sqrt{\text{Hz}}}\right)$	meas $\left(\frac{\text{fA}}{\sqrt{\text{Hz}}}\right)$	(pA)
AD8551	42	46	5	5.2	2	21	20
AD8572	51	55	a	a	2	16	a
AD8628	22	22	8	15	5	53	100
LMP2021	11	18	8	25	350	120	400
LTC1049	100	90	150	2.0	2	100	200
LTC1050	90	70	80	3.8	1.8	130	8000
LTC1150	90	92	2	0.6	1.8	70	15000[s]
MAX4239	30	28	20	18	-	24	b
MAX9617	42	42	20	60	100	74	50
MCP6V06	82	80	40	8.9	0.6	38	800
OPA335	55	52	5	11	20	12	50
OPA734	135	120	5	18	40	27	8000[s]
OPA2188	8.8	8.5	-	-	7	750	-
TLC4501A[c]	70/12[e]	60/11[e]	1/2/5[f]	-	0.6	0.6[g]	-
MAX4236A[d]	23/14[e]	24/16[e]	1/3/8[f]	-	-	0.45[g]	-

Notes: (a) broad spectrum, slight rise at 2.2 kHz and harmonics. (b) output noise dominated by V_n, unable to measure I_n separately. (c) auto-zero at power-up, no corrections thereafter. (d) precision conventional CMOS op-amp, for comparison. (e) at 10 Hz/1 kHz. (f) no spectral features; peak amplitudes for bandwidths to 1 kHz/10 kHz/100 kHz. (g) at 1 Hz. (s) spikey waveform, value listed is the peak amplitude. (t) typical.

A caution: datasheet values for input current noise are sometimes seriously in error, evidently because the manufacturer did not measure it, believing that it was accurately predicted by a shot noise calculation based on dc input current. We were curious, and measured the input noise of a dozen auto-zero op-amps, with the results in Table 8.5 on the current page. And how interesting! The datasheet values for *voltage* noise were correct; but for some of the parts, the specified *current* noise values were far too optimistic, sometimes by as much as a factor of fifty. Interestingly, some datasheets even admit what they've done; for example, the i_n entry for each of the three LTC parts is accompanied by a footnote saying "Current noise is calculated from the formula $i_n = \sqrt{2qI_B}$, where $q = 1.6 \cdot 10^{-19}$ coulombs." This same error afflicts some BJT-input op-amps, in particular those with input bias cancellation, where the (incorrect) current noise in the datasheet evidently was calculated from the shot noise corresponding to the *net* (i.e., cancelled) input current, rather than from the much larger uncancelled input current. See §§4x.10 and 5.10.8 for details.

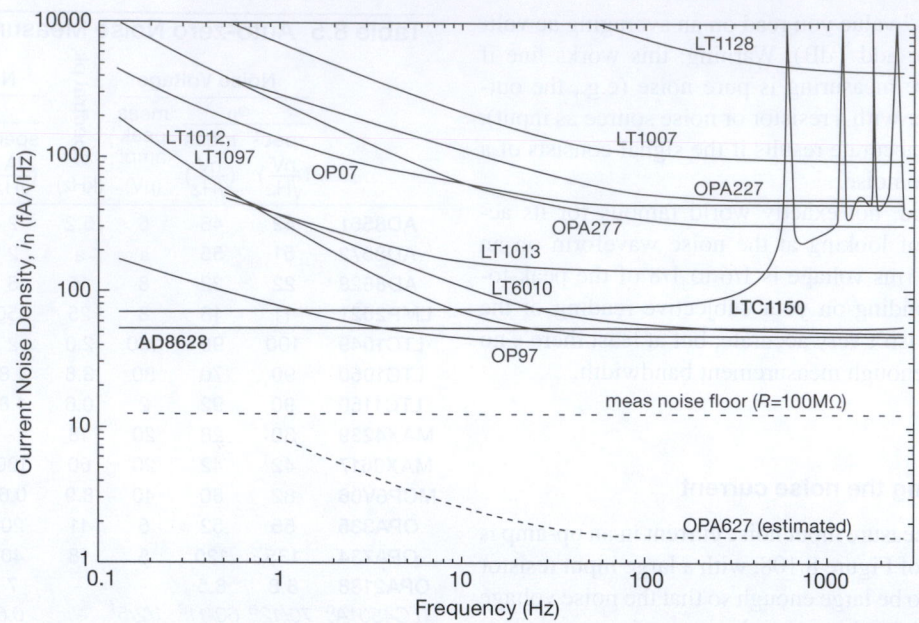

Figure 8.111. Measured current noise density spectra for most of the op-amps of Figure 8.110. The **boldface** parts are auto-zero amplifiers, which circumvent the $1/f$ "flicker-noise" demon by repetitive offset correction, but exhibit switching-induced clock noise at higher frequencies (see the expanded plot in Figure 5.52). See also Figures 8.60 and 8.61.

A. Some limitations: bandwidth, stability, dc offset

This simple i_n measurement scheme – letting the device under test amplify its own input current noise, as a voltage developed across a large input resistor – has some drawbacks, which seriously limit the ability to measure noise currents down in the low fA/$\sqrt{\text{Hz}}$ range. As just discussed, you've got to use sufficiently large values of R_s to overcome both the amplifier's e_n and the resistor's Johnson noise. For a noise current of 1 fA/$\sqrt{\text{Hz}}$, for example, that requires R_s at least 10 GΩ (equivalent i_n contribution is 1.3 fA/$\sqrt{\text{Hz}}$). But now you've got to worry about the dc voltage produced by the input bias current: a current of 10 pA, for example, causes 100 mV of dc input, thus saturation at the output after $G=100$. You've also got to worry about instability, because it doesn't take much feedback capacitance from the output pin to the noninverting input to make the amplifier into an oscillator. That can be tamed by a small shunt capacitor to ground, but the added capacitance reduces the (already small) bandwidth: with our $R_s=10$ GΩ, for example, a mere 1 pF of input capacitance limits the measurement bandwidth to 16 Hz! And in our measurements we needed additional shunt capacitance to prevent low-frequency oscillation in our (socketed) test jig.

B. Improved bandwidth with a current amplifier

The lesson here is that it's not easy to make measurements of low-level input current noise in good amplifiers. You can do better, though, by using a carefully designed *external* current amplifier (which presents a low impedance input, i.e., a virtual ground) connected directly to the noninverting input of the device under test, as in Figure 8.112. For example, the DL Instruments[132] Model 1211 "current preamplifier" has a specified input current noise of 1.5 fA/$\sqrt{\text{Hz}}$ (corresponding to a 10 GΩ feedback resistor to its input summing junction) and a bandwidth of 400 Hz; it is an "electrometer" configuration (dc negative feedback to an input summing junction), so it holds its input within 0.2 mV of ground at full-scale input current. Some other suppliers of low-noise-current amplifiers and current-measuring instruments are Laser Components (for example their Model DDPCA-300, with switchable gains from 10^4 V/A up to 10^{13} V/A, and with noise as low as 0.2 fA/$\sqrt{\text{Hz}}$ in the most sensitive ranges), and Keithley (for example their Model 428, with switchable gains from 10^3 V/A up to 10^{11} V/A).

---

[132] Formerly the instrumentation division of Ithaco, Inc.

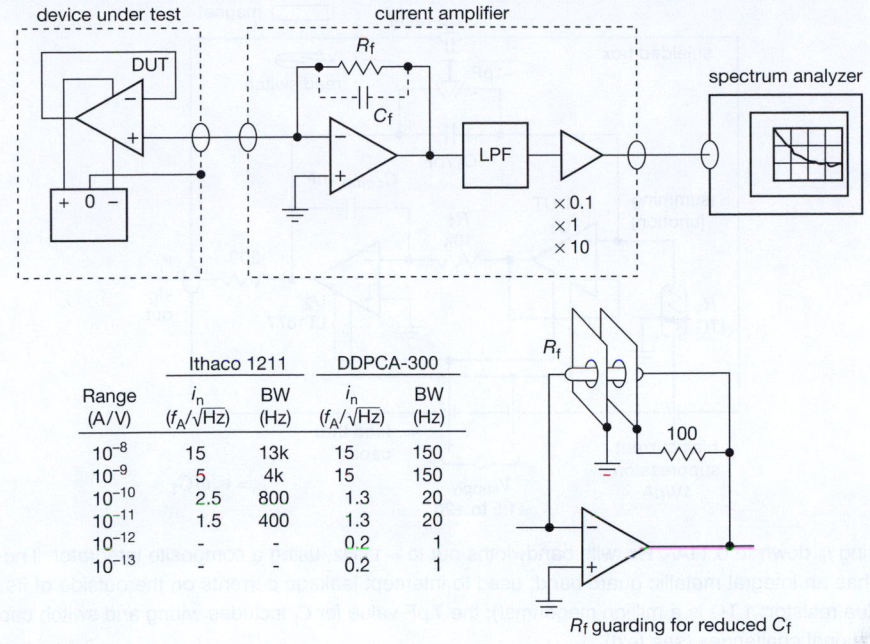

	Ithaco 1211		DDPCA-300	
Range (A/V)	i_n ($f_A/\sqrt{Hz}$)	BW (Hz)	i_n ($f_A/\sqrt{Hz}$)	BW (Hz)
10^{-8}	15	13k	15	150
10^{-9}	5	4k	15	150
10^{-10}	2.5	800	1.3	20
10^{-11}	1.5	400	1.3	20
10^{-12}	-	-	0.2	1
10^{-13}	-	-	0.2	1

R_f guarding for reduced C_f

Figure 8.112. Measuring op-amp current noise with a sensitive external current amplifier.

8.13.7 Another way: roll-your-own fA/√Hz instrument

Commercial high-performance current amplifiers can cost quite a bundle – you're talking numbers like 30–40 dB$. It's nice to have such a general-purpose bench instrument on hand, of course. But if all you want is to measure low-level op-amp input-current noise, and you're willing to make a special-purpose test jig, you can wire up something like the circuit in Figure 8.113. With the values shown you can measure noise currents down to 0.1 fA/√Hz (or less, with a larger R_s), with a component cost of just a few dollars.

This circuit takes an unusual approach, and is worth a bit of discussion. Our first idea was to eliminate entirely the noise-producing feedback resistor: recall that a resistor of value R has a Johnson noise-voltage density $e_n = \sqrt{4kTR}$, thus an equivalent noise-current density $i_n = \sqrt{4kT/R}$; so, for example, to keep that i_n contribution less than 1 fA/√Hz requires an R_f greater than 16,000 MΩ (with consequent problems of bandwidth, offset, and stability). So here we use instead a feedback *capacitor!*[133] This makes an integrator, here implemented as a composite amplifier in

which the device-under-test (DUT) (configured as a follower) drives an inverting op-amp stage whose unity-gain bandwidth is limited (by C_{comp}, in combination with R_1) to ~16 kHz. We don't need any more bandwidth, and this overcompensation ensures stability.

Ignoring for the moment the DUT's dc input leakage current, this integrator converts input current noise to output voltage noise according to $v_n = i_n/\omega C_f$. So, for example, a flat (white) i_n spectrum produces a v_n whose spectral amplitude falls as $1/f$, or, equivalently, its noise power density falls as $1/f^2$ (red noise). Of course, there's also voltage noise e_n in both op-amps, which gets added to v_n (in the usual square root of sum of squares manner of uncorrelated noise); by choosing a smaller value for C_f we can reduce the effect of the e_n's, because smaller values of C_f produce larger current-to-voltage "gain." (Another way to put it is that the effective current noise produced by the input noise voltage e_n is $i_n = e_n\omega C_f$. Equivalently, the current-noise measurements extend only up to a frequency $\omega = i_n/e_n C_f$,

acceptable. (An engineer would say that a capacitor has no Johnson noise; a physicist would say that a capacitor may have kT worth of thermal energy on average, but, being uncoupled from the thermal bath, that energy does not fluctuate.) This technique may have been originated by Garwin, who used it in the 1950s for high-energy physics instrumentation, and then in 1969 for image detector readout (where Garwin instructed "read twice to cancel kT").

---

[133] This isn't as nutty as it sounds. In fact, it is the method of choice ("correlated double-sampling") used in certain low-noise applications, for example photodiode or CCD imager readout amplifiers, where the thermal noise produced by a conventional feedback resistor is un-

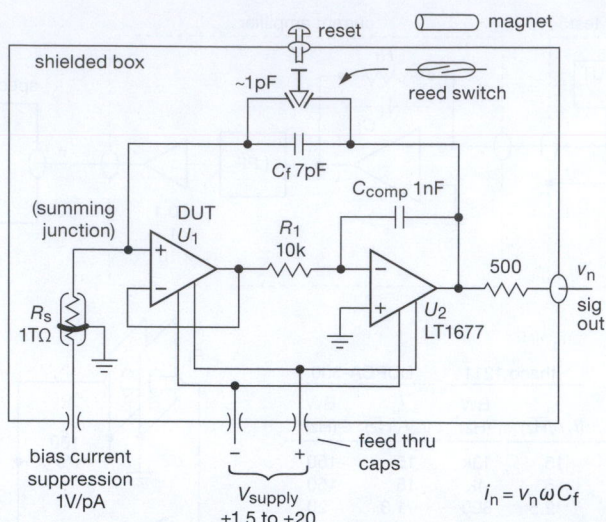

Figure 8.113. Measuring i_n down to 0.1 fA/$\sqrt{Hz}$, with bandwidths out to ~1 kHz, using a composite integrator. The suppression resistor R_s (made by Welwyn) has an integral metallic guard band, used to intercept leakage currents on the outside of its glass envelope (hey, this is a *really* high-value resistor: 1 TΩ is a million megohms!); the 7 pF value for C_f includes wiring and switch capacitances. The reset switch presented educational challenges (see text).

above which the amplifier's voltage noise dominates.) For this circuit we chose the LT1677 for U_2, because it combines pretty low voltage noise (3.2 nV/$\sqrt{Hz}$) with low bias current, wide operating voltage range (±1.5 V to ±20 V), and rail-to-rail output.

Figure 8.114 shows a screen capture of the voltage-noise spectrum of the circuit of Figure 8.113, when an LTC1049 auto-zero op-amp is plugged into the DUT

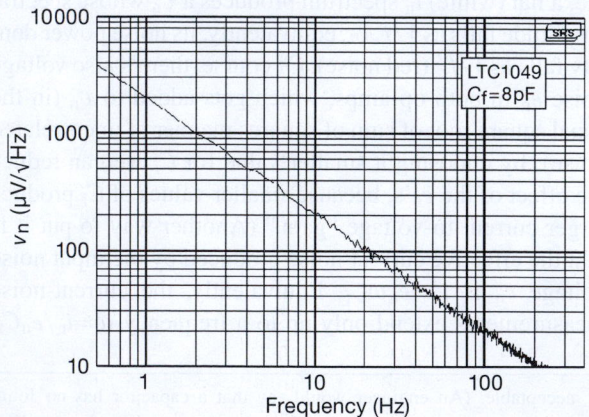

Figure 8.114. Measured voltage-noise density v_n from the circuit of Figure 8.113, for an LTC1049 auto-zero op-amp. The $1/f$ amplitude spectrum of the integrator's output (v_n) corresponds to a flat input current noise spectrum over this range of frequencies, with a value $i_n=100$ fA/$\sqrt{Hz}$.

socket. It conforms accurately to $1/f$, evaluating to $i_n=100$ fA/$\sqrt{Hz}$ over this frequency range, agreeing with the i_n measurement we made using the circuit of Figure 8.108 (and reported in Table 8.5 on page 569). There's no hint of a leveling off at the high frequency end, consistent with a predicted measurement bandwidth of ~10 kHz for the i_n and e_n of this op-amp.

Having calibrated the technique (and having confirmed the calibration by measuring also the AD8628 auto-zero op-amp), we used it to measure the tough cases – the conventional op-amps listed in Table 8.5, with current-noise values down in the ~1 fA/$\sqrt{Hz}$ territory – with the results listed. For those measurements the bandwidth extends only to ~1 kHz, above which the op-amp's voltage noise e_n, in combination with the input capacitances to ground (both within the op-amp, and externally as wiring, etc.), creates an "e_nC" equivalent input current noise (more precisely, $e_n\omega C_{in}$) that dominates the small current noise of the amplifier itself.

We also measured an LMC6081, which should have even lower current noise, given its specified typical input current I_b of 10 fA; for this op-amp the measured v_n corresponds to an i_n of 0.15 fA/$\sqrt{Hz}$. That sounds pretty good – but the expected current noise, calculated as I_B shot noise, should be quite a bit lower, about 0.06 fA/$\sqrt{Hz}$. So now we have to confess that the resistorless purity of this measurement technique has been compromised by the "current-suppression" resistor R_s, needed to cancel the dc input current. We used a very large value (1 TΩ, that's 10^6 megohms!) to reduce the added noise current, which here amounts to $i_n = \sqrt{4kT/R_s} = 0.13$ fA/$\sqrt{\text{Hz}}$. So for this op-amp what we've actually measured is the noise introduced by the bias-suppression resistor! We could do better, say with a 100 TΩ resistor (you can get these things, for example the 3810-series from Welwyn), which would push the added noise down by a factor of ten. We could do better, that is, if you really believe that you can keep all other insulation resistances (moisture, fingerprints, etc.) that high; a tall order.[134]

A. An amusing complication: the reset switch

Here's a little story about reality, and some considerable confusion on the way to ultimate enlightenment. Right "out of the box" this circuit worked nearly perfectly. But there was one curious complication: when the integrator RESET switch was pressed, the output went obediently to zero; but when the switch was released, the output took a large jump – typically to a value in the range of +1.5 V to +2 V. What could cause *that*? Most likely, some sort of electrostatic charging caused by the movement of the plastic parts, in a pushbutton switch series of which we've grown quite fond (Panasonic EVQ2130x–EVQ2150x series, which are compact, inexpensive, pleasantly tactile, and nearly free of annoying contact "bounce").

We pondered this, and decided to substitute a magnetically operated *reed switch* – a small sealed tubular glass envelope containing a pair of contacts, with a wire sticking out each end, and which is actuated by bringing a small permanent magnet in proximity (these things are widely used, most familiarly for door and window switches in home alarm systems). The reed switch has good properties: when OFF its capacitance is ~0.3 pF and its resistance is greater than $10^{12}\,\Omega$. And, with the switch inside the shielded aluminum box and the magnet outside, there would be no opportunity for accumulation of static charge. Better still, with no dc-actuated coil (as in a conventional relay, or a

reed relay) there would be no magnetic or electric coupling to the sensitive circuit.

This worked considerably better. But still not as expected: imagine our surprise when the output took a -50 mV step when the switch was opened. What could be causing *that*? We guessed at first that it might be due to a flux rearrangement when the (magnetic) contacts opened, injecting a pulse of charge equal to the change in flux that was linked by the circuit wiring. Easy to check: just move the switch in close proximity to the op-amp, reducing greatly the circuit's enclosed area. We did that, and, encouragingly, the output jump changed sign (now a positive jump, of about the same magnitude). OK, we thought, move the switch halfway back to where it was and the effect should go to zero. It didn't: still +50 mV.

This was really weird! On an inspiration we guessed that we might have reversed the ends when we made the first change, so we flipped the switch over and – aha! – the jump reversed sign (back to -50 mV).

So the reed switch seemed to have an asymmetry, and a memory of its own, such that it injected some charge (not much: $Q = C_f \Delta V = 0.35$ pC) when opened. What could be causing that? We spoke with some wise colleagues, who suggested we think about the difference of "work function" between the two contacts in the switch (which might plausibly be chosen of different metals, to prevent cold-weld sticking), and who mumbled terms like "Fermi levels," "contact potential difference," and the "Kelvin probe" technique for measuring the latter.

OK, so we tried a reed switch of different design, and also one from a different manufacturer (for the original Hamlin MDRR-4 we substituted first a Hamlin MDSR-10, then a Coto RI-01BAA), both of which exhibited the same curious effect, though with different amplitudes (15 mV and 100 mV, respectively).

Time to call for help! The Hamlin datasheet proclaims hospitably "For details on electrical specifications, contact Hamlin." So we did. We learned that the contact pairs are identical, in fact they just make a zillion of them, all the same, and take any two for each switch. There goes all that Fermi-level theory!

Finally we figured it out.[135] Glass is a darned good insulator, and it's easy to deposit charge onto the reed switch's glass envelope just by handling it (removing it from its plastic bag, etc.). There's a fancy name for this: it's called the *triboelectric effect*. A simpler name is static electricity. Glass is one of the classic materials, and in courses on

---

[134] Indeed! We tested this proposition by breathing heavily into the box, then remeasuring the bias suppression through R_s: it increased 20-fold, owing to the external leakage path evidently created by moisture condensation, returning to its dry value a minute later. That is why these "High-value Glass-sealed" resistors from Welwyn are outfitted with a conductive guard ring, which can be connected to ground (as shown) to shunt off such external leakage currents: the portion of the case between the guard ring and the summing junction then has no voltage drop across it (more precisely, a voltage equal to the op-amp's offset voltage, a millivolt or less); thus no leakage current flows. We confirmed this with an encore of the breathing test.

[135] Isn't it curious that it's always the *last* conjecture that proves to be correct?

electricity you see demonstrations of glass rods being rubbed with cat's fur.[136] With some charge stuck on the glass, the electric field that it produced caused the electrodes of the reed switch to acquire a small amount of opposing charge when separated.

Here's what we did to confirm this conjecture: we took an inch of woven metal cloth,[137] grounded one end with a clip lead, wrapped it around the reed switch, and slid it along the length several times, to provide an opportunity for any static charge on the glass surface to take a hike. And it did: after one treatment the output-voltage step was reduced to a negligible ∼2 mV.

The moral: sensitive measurements (here we're talking femtoamps and picofarads) can reveal effects that are so small that you've never thought about them. They can seriously disrupt your work… but there's a redeeming delight in discovering them for yourself. And then eliminating them.

8.13.8 Noise potpourri

Herewith a collection of interesting, and possibly useful, facts.

1. The averaging time required in an indicating device to reduce the fluctuations of a rectified noise signal to a desired level for a given noise bandwidth is

$$\tau \approx \frac{1600}{B\sigma^2} \text{ seconds,} \tag{8.61}$$

where τ is the required time constant of the indicating device to produce fluctuations of standard deviation σ percent at the output of a linear detector whose input is noise of bandwidth B.

2. For band-limited white noise, the expected number of maxima per second is

$$N = \sqrt{\frac{3(f_2^5 - f_1^5)}{5(f_2^3 - f_1^3)}}, \tag{8.62}$$

where f_1 and f_2 are the lower and upper band limits. For $f_1 = 0$, $N = 0.77 f_2$; for narrowband noise ($f_1 \approx f_2$), $N \approx (f_1 + f_2)/2$.

3. rms-to-average (i.e., average magnitude) ratios:

Gaussian noise: rms/avg $= \sqrt{\pi/2} = 1.25 = 1.96$ dB,
sinewave: rms/avg $= \pi/2^{\frac{3}{2}} = 1.11 = 0.91$ dB,

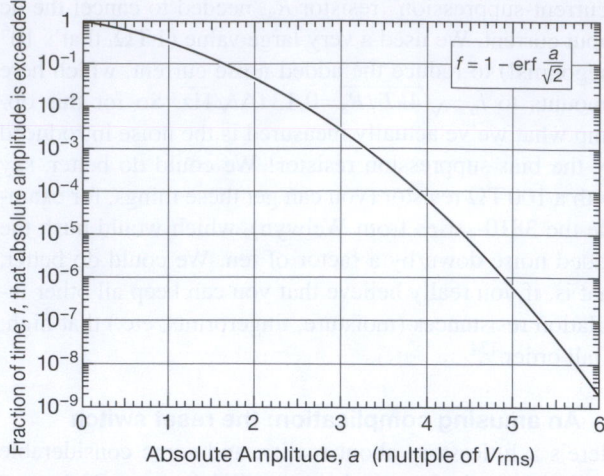

Figure 8.115. Relative occurrence of amplitudes in Gaussian noise. Potentially useful for estimating false trigger rates, required "crest factor" in rms measurements, and the like.

square wave: rms/avg $= 1 = 0$ dB.

4. Relative occurrence of amplitudes in Gaussian noise. Figure 8.115 plots the fractional time that a given amplitude level is exceeded by a Gaussian noise waveform of unit rms amplitude.

5. Positive-threshold crossing rate of lowpass-filtered Gaussian white noise of unit rms amplitude is

$$\text{TCR} = \frac{\text{BW}}{\sqrt{3}} \exp\left(-V_{th}^2/2\right) \text{ crossings/second,} \tag{8.63}$$

where V_{th} is the positive threshold voltage and BW is the brick-wall lowpass bandwidth.[138]

6. Standard deviation of noise resulting from quantization error is

$$\sigma_n = \frac{\text{LSB}}{\sqrt{12}} \approx 0.3 \text{ LSB.} \tag{8.64}$$

8.14 Signal-to-noise improvement by bandwidth narrowing

As luck would have it, the signals you often want to measure are buried in noise (where the "noise" may include other signals nearby in frequency, i.e., interference), frequently to the extent that you can't even see them on an

---

[136] Or is it a cat being rubbed with glass wool?

[137] Wrapped around some foam, this handy stuff is used to make flexible conductive shielding gaskets. Check out the self-stick "fabric-over-foam" gasket materials from Laird Technologies, e.g., their rectangular 4046 or D-shaped 4283.

[138] The crossing rate through a pair of symmetrical thresholds (i.e., the rate of crossings of *magnitude* greater than V_{th}) is double that obtained from the formula. We thank Phil Hobbs for this fact, found along with many others in his fine book referenced on page 551.

oscilloscope. Even when external noise isn't a problem, the statistics of the signal itself may make detection difficult, as, for example, when counting nuclear disintegrations from a weak source, with only a few counts detected per minute. Finally, even when the signal is detectable, you may wish to improve the detected signal strength in order to make a more accurate measurement. In all these cases some tricks are needed to improve the signal-to-noise ratio. They all amount to a narrowing of the detection bandwidth in order to preserve the desired signal while reducing the total amount of (broadband) noise accepted.

The first thing you might be tempted to try when thinking of reducing the bandwidth of a measurement is to hang a simple lowpass filter on the output, in order to average out the noise. There are cases where that therapy will work, but most of the time it will do very little good, for a couple of reasons. First, the signal itself may have some high frequencies in it, or it may be centered at some high frequency. Second, even if the signal is in fact slowly varying or static, you invariably have to contend with the reality that the density of noise power usually has a $1/f$ character, so as you squeeze the bandwidth down toward dc you gain very little. Electronic and physical systems are twitchy, so to speak.

In practice, there are a few basic techniques of bandwidth narrowing that are in widespread use. They go under names like signal averaging, transient averaging, boxcar integration, multichannel scaling, pulse-height analysis, lock-in detection, and phase-sensitive detection. All of these methods assume that you have a repetitive[139] signal; that's no real problem, since there is almost always a way to force the signal to be periodic, assuming it isn't already. Here we discuss an important one of these techniques, known as "lock-in" or "synchronous" detection.

8.14.1 Lock-in detection

This is a method of considerable subtlety. It consists of two steps. (1) Some parameter of the source signal is *modulated*; for example, an LED might be driven with a square wave at a fixed frequency. (2) The detected (and noisy) signal is *demodulated*, for example by multiplying it by a fixed-amplitude reference signal at the same modulating frequency. The modulation moves the source signal spectrum up to the modulating frequency, above noisy $1/f$ low-frequency backgrounds, and away from other noise sources (such as ambient-light fluctuations in the case of the LED

example). The demodulation step creates a dc output proportional to the signal, which can be lowpass filtered (a simple RC filter may be adequate) to narrow the detected bandwidth.

To understand the method, it is necessary to take a short detour into the phase detector, a subject we first take up in §13.13.2.

A. Phase detectors

In §13.13.2 we describe phase detectors that produce an output voltage proportional to the phase difference between two digital (logic-level) signals. For purposes of lock-in detection, you need to know about *linear* phase detectors, because you are nearly always dealing with analog voltage levels.

The simplest circuit[140] is shown in Figure 8.116. An analog signal passes through a linear amplifier whose gain is reversed by a square-wave "reference" signal controlling a FET switch (see Table 3.3 or 13.7 for candidates). The output signal passes through a lowpass filter, RC. That's all there is to it. Let's see what you can do with it.

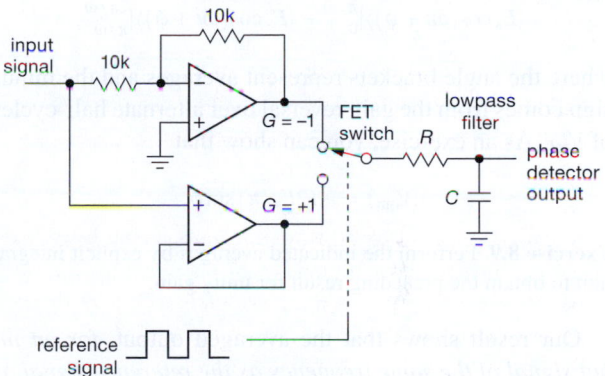

Figure 8.116. Phase detector for linear input signals. Most simply, you can implement this with a dual op-amp and a CMOS switch IC. This scheme is used in the monolithic AD630.

Phase-detector output

To analyze the phase-detector operation, let's assume we apply a signal

$$E_s \cos(\omega t + \phi)$$

to such a phase detector, whose reference signal is a square wave with transitions at the zeros of $\sin \omega t$, i.e., at

---

[139] Or, more generally, a *known* signal variation to which the measured signal can be correlated.

[140] But less than ideal: the square-wave modulation causes response at odd harmonics. The use of an analog multiplier like the AD633 or AD734, driven by a sinewave reference, eliminates this deficiency.

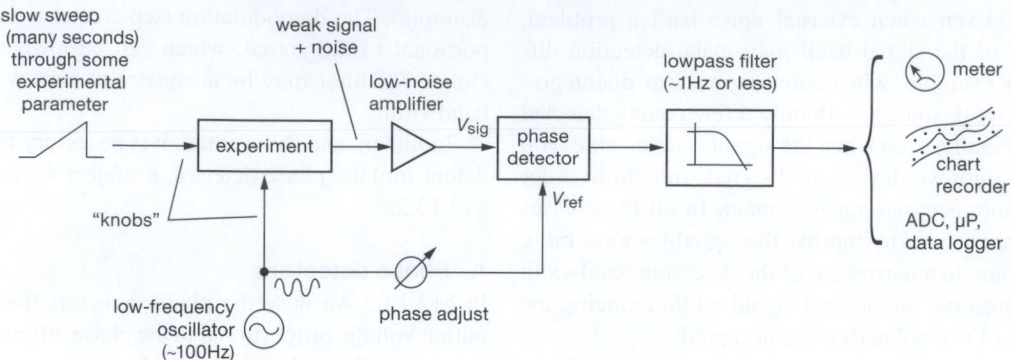

Figure 8.117. Lock-in amplifier detection.

$t = 0, \pi/\omega, 2\pi/\omega$, etc. Let us further assume that we average the output, V_{out}, by passing it through a lowpass filter whose time constant is much longer than one period:

$$\tau = RC \gg T = 2\pi/\omega.$$

Then the lowpass filter output is

$$\langle E_s \cos(\omega t + \phi)\rangle|_0^{\pi/\omega} - \langle E_s \cos(\omega t + \phi)\rangle|_{\pi/\omega}^{2\pi/\omega},$$

where the angle brackets represent averages and the minus sign comes from the gain reversal over alternate half-cycles of V_{ref}. As an exercise, you can show that

$$\langle V_{out}\rangle = -(2E_s/\pi)\sin\phi.$$

Exercise 8.9. Perform the indicated averages by explicit integration to obtain the preceding result for unity gain.

Our result shows that the averaged output, *for an input signal of the same frequency as the reference signal*, is proportional to the amplitude of E_s and sinusoidal in the relative phase.

We need one more result before going on: what is the output voltage for an input signal whose frequency is close to (but not equal to) the reference signal? This is easy, because in the preceding equations the quantity ϕ now varies slowly, at the difference frequency:

$$\cos(\omega + \Delta\omega)t = \cos(\omega t + \phi) \quad \text{with } \phi = t\Delta\omega,$$

giving an output signal that is a slow sinusoid:

$$V_{out} = (2E_s/\pi)\sin(t\Delta\omega),$$

which will pass through the lowpass filter relatively unscathed if $\Delta\omega < 1/\tau = 1/RC$ and will be heavily attenuated if $\Delta\omega > 1/\tau$.

B. The lock-in method

Now the so-called lock-in (or phase-sensitive) amplifier should make sense. First you make a weak signal periodic, as we've discussed, say at a frequency in the neighborhood of 100 Hz. The weak signal, contaminated by noise, is amplified and phase detected relative to the modulating signal. Look at Figure 8.117. In many cases you'll want to measure the weak signal as some experimental condition is varied – you'll have an experiment with two "knobs" on it, one for fast modulation to do phase detection, and one for a slow sweep through the interesting features of the signal (in NMR, for example, the fast modulation might be a small 100 Hz modulation of the magnetic field, and the slow modulation might be a frequency sweep 10 minutes in duration through the resonance). The phase shifter is adjusted to give maximum output signal, and the lowpass filter is set for a time constant long enough to give a good signal-to-noise ratio. The lowpass-filter rolloff sets the bandwidth, so a 1 Hz rolloff, for example, gives you sensitivity to spurious signals and noise only within 1 Hz of the desired signal. The bandwidth also determines how fast you can adjust the "slow modulation," because now you must not sweep through any features of the signal faster than the filter can respond; people use time constants of fractions of a second up to tens of seconds or more.[141]

Note that lock-in detection amounts to *bandwidth narrowing*, with the bandwidth set by the post-detection lowpass filter. Another way to reduce the detection bandwidth is with the technique of *signal averaging*, in which the results of repetitive measurements (e.g., frequency sweeps) are accumulated; this is a common option on instruments such as spectrum analyzers. In either case the effect of the modulation is to center the signal at the fast modulation

---

[141] In the old days the slow modulation was done with a geared-down clock motor turning an actual knob on something.

frequency, rather than at dc, in order to get away from $1/f$ noise (flicker noise, drifts, and the like).

C. Two methods of "fast modulation"

There are several ways to do the fast modulation: the modulation waveform can be either a very small sinewave or a very large square wave compared with the features of the sought-after signal (line shape versus magnetic field, for example, in NMR), as sketched in Figure 8.118. In the first case the output signal from the phase-sensitive detector is proportional to the *slope* of the line shape (i.e., its derivative), whereas in the second case it is proportional to the line shape itself (providing there aren't any other lines out at the other endpoint of the modulation waveform). This is the reason all those simple NMR resonance lines come out looking like dispersion curves (Figure 8.119).

For large-shift square-wave modulation there's a clever method for suppressing modulation feedthrough, in cases where that is a problem. Figure 8.120 shows the modulation waveform. The offsets above and below the central value kill the signal, causing an on–off modulation of the signal at *twice* the fundamental of the modulating waveform. This is a method for use in special cases only; don't get carried away by the beauty of it all!

Large-amplitude square-wave modulation is a favorite with those dealing in infrared astronomy, where the telescope's secondary mirrors are rocked to switch the image back and forth on an infrared source. It is also popular in radio astronomy, where it's called a Dicke switch.

Commercial lock-in amplifiers have a variable-frequency modulating source and tracking filter, a

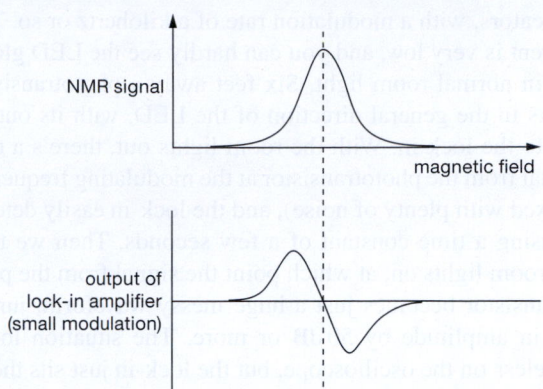

Figure 8.119. Line-shape differentiation resulting from lock-in detection.

switchable-time-constant post-detection filter, a good low-noise wide-dynamic-range amplifier (you wouldn't be using lock-in detection if you weren't having noise problems), and a nice linear phase detector. They also let you use an external source of modulation. The phase shift is adjustable, so you can maximize the detected signal. The whole item comes packed in a handsome cabinet, with a meter or digital display showing the output signal. Typically these things cost a few thousand dollars. Stanford Research Systems has a nice selection of lock-in amplifiers, including several that use digital signal-processing methods for enhanced linearity and dynamic range. In these the amplified input signal is accurately digitized (to 20 bits), the "oscillator" is a computed lookup table of (quadrature) sines and cosines, and the "mixer" is a numeric multiplier. Ordinarily lock-in amplifiers are of rather limited signal bandwidth, typically 100 kHz. But by using the radiofrequency "heterodyne" technique (input frequency band translation via linear mixing with a "local oscillator"), the lock-in method can be extended into high radio-frequencies. For example, the SR844 goes to 200 MHz; it uses a hybrid of analog techniques (input filtering and downconversion) and digital signal processing (baseband digitizing and synchronous detection).

To illustrate the power of lock-in detection, we set up a small demonstration for our students. We use a lock-in to modulate a small LED of the kind used for panel

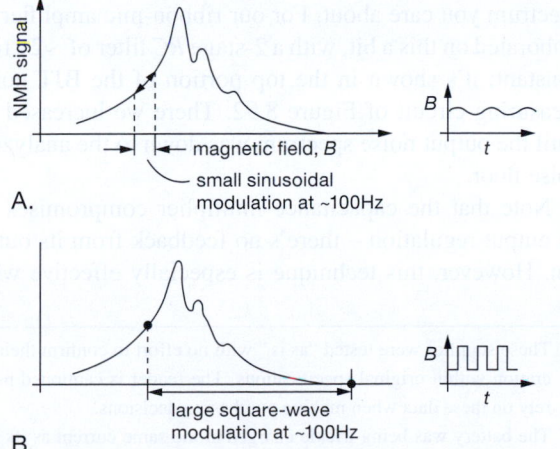

Figure 8.118. Lock-in modulation methods. A. Small sinusoid. B. Large square wave.

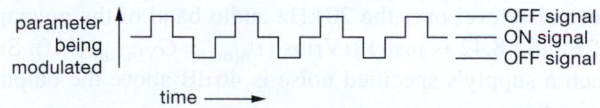

Figure 8.120. Modulation scheme for suppressing modulation feedthrough.

indicators, with a modulation rate of a kilohertz or so. The current is very low, and you can hardly see the LED glowing in normal room light. Six feet away a phototransistor looks in the general direction of the LED, with its output fed to the lock-in. With the room lights out, there's a tiny signal from the phototransistor at the modulating frequency (mixed with plenty of noise), and the lock-in easily detects it, using a time constant of a few seconds. Then we turn the room lights on, at which point the signal from the phototransistor becomes just a huge messy waveform, jumping in amplitude by 50 dB or more. The situation looks hopeless on the oscilloscope, but the lock-in just sits there, unperturbed, calmly detecting the same LED signal at the same level. You can check that it's really working by sticking your hand in between the LED and the detector. It's darned impressive.

At the other end of the cost spectrum, synchronous detection is used to accomplish the same rejection of ambient light in some inexpensive light-beam-detection components, for example the S6809/46 and S6986 from Hamamatsu. These ICs come in a clear plastic case (several available package styles) containing an integrated photodiode, preamp, and synchronous detector with logic-level output; included also is the internal oscillator and output driver for the external LED light source. They cost about $6 in small quantities.

8.15 Power-supply noise

Amplifier circuits that do not have a high degree of power-supply rejection are susceptible to noise (and signals) on the dc power supplies. If the dc rails are noisy, the output will be too, so you've got to keep them quiet. The problem is not as bad as it could be, though, because supply noise appears unamplified at the output, to be compared with the amplifier's signal gain of, say, $\times 100$. Still, dc power supplies are rarely quiet even to the level of $100\,\mathrm{nV}/\sqrt{\mathrm{Hz}}$ (100 times a reasonable input noise target of, say, $1\,\mathrm{nV}/\sqrt{\mathrm{Hz}}$). That is why the dc rails in many of the circuits in this chapter are marked "quiet."

How noisy are typical dc bench power supplies? You'll see specs like "0.2 mVrms, 2 mVpp," which sounds respectable enough until you realize that signal levels in a sensitive circuit may be far smaller. For example, the output noise level over the 20 kHz audio band of the preamp of Figure 8.42 is just $1\,\mu\mathrm{Vrms}$ ($v_{\mathrm{n(out)}}{=}G_{\mathrm{V}}e_{\mathrm{n(in)}}\sqrt{\Delta\mathrm{f}}$). So such a supply's specified noise is 46 dB above the output noise floor.

Specs are one thing, actual performance is another. To get a measure of the power supply noise scene we mea-

sured two dozen dc supplies from our lab's collection, with the resulting spectra of Figure 8.123.[142]

The spread in noise performance is stunning – the outstanding performer (#4) turned out to be a "precision power source" half a century old (we bought it in 1967), with an ovenized compartment for the zener reference and (discrete) error amplifier. By comparison, a contemporary bench supply with elegant digital readout like #12 is nearly 100 times (40 dB) as noisy. Not to be outdone, the real screamers turn out to be a simple switchmode cellphone charger (#22), and an unregulated wall-wart (#23) whose powerline-tracking output wanderings dominate the low-frequency spectrum. At the other end, nothing beats a lead-acid battery (#3, right down at our measurement noise floor) for the utmost in inherently quiet dc sources.[143]

What accounts for these large differences? Switching supplies are inherently noisy, of course. But even among the linear supplies there is a $100\times$ spread (40 dB) in noise voltage. A quiet dc regulated supply must have plenty of loop gain, implemented with low-noise amplifiers. And it's critically important to select a low-noise voltage reference (particularly down at low frequencies, where it can't be quieted with filters); see the discussion and plots in §9.10.

8.15.1 Capacitance multiplier

A nice trick for cleaning up a noisy supply is a "capacitance multiplier" circuit (Figure 8.121). We introduced this back in §8.5.9, where we explored the properties of a BJT input stage with $e_{\mathrm{n}}{\sim}0.07\,\mathrm{nV}/\sqrt{\mathrm{Hz}}$. Here we choose R small enough so that there is at most a volt or so drop at maximum load current, then choose C for a long-enough time constant RC to attenuate adequately the portion of the noise spectrum you care about. For our ribbon-mic amplifier we elaborated on this a bit, with a 2-stage RC filter of $\sim 2\,\mathrm{s}$ time constant; it's shown in the top portion of the BJT noise-measuring circuit of Figure 8.92. There we increased RC until the output noise spectrum was down to the analyzer's noise floor.

Note that the capacitance multiplier compromises the dc output regulation – there's no feedback from its output pin. However, this technique is especially effective when

---

[142] These supplies were tested "as is," with no effort to confirm their operation within original specifications. The reader is cautioned not to rely on these data when making purchasing decisions.

[143] The battery was being trickle charged at the same current as its load for curve #3. If the battery is being discharged (without replenishment) by the load, the slight downward voltage "tilt" appears as a low-frequency excess, seen here in curve #2.

added upstream of the regulating pass element (i.e., after the rectifier and storage capacitor). We have applied this simple modification to commercial power supplies and scientific instruments with great success. It's a far easier way to achieve $100\times$ lower output ripple than the alternative of raising the control-loop gain and bandwidth. If you do this, be sure there's enough drop across Q_1 to handle the full ripple amplitude – in this location the 120 Hz (or 100 Hz) ac ripple at full load current may be as much as a few volts. You can add a resistor from the base to ground to increase the dc drop across Q_1; or you can use a MOSFET, whose larger operating V_{GS} may provide adequate headroom.

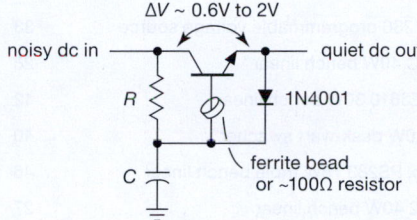

Figure 8.121. A "capacitance multiplier" for filtering incoming dc consists of an emitter follower biased with a smoothed (lowpass-filtered) replica of the noisy input. A bead or small resistor prevents oscillation.

Figure 8.122 shows the effect of Figure 8.92's capacitance multiplier, measured with two of the power supplies that appear in Figure 8.123's crowd. For each supply we measured (a) the spectrum directly from the output terminals (solid curves), (b) the spectrum downstream of the capacitance multiplier (dashed curves), and, for comparison, (c) the spectrum with $10,000\,\mu F$ across the power supply's terminals (dotted curve). In all cases the measured output was loaded with $100\,\Omega$. The capacitance multiplier is stunningly effective in eliminating power supply noise,[144] and it's a far better use of $10,000\,\mu F$ than the brute-force approach of putting the same capacitance directly across the dc terminals.

8.16 Interference, shielding, and grounding

"Noise" in the form of interfering signals, 60 Hz pickup, and signal coupling via power supplies and ground paths can turn out to be of far greater practical importance than the intrinsic noise sources we've just discussed. These interfering signals can all be reduced to an insignificant level (unlike thermal noise) with proper circuit design, layout,

---

[144] Except at very low frequencies, where there's no substitute for a power supply with a stable voltage reference.

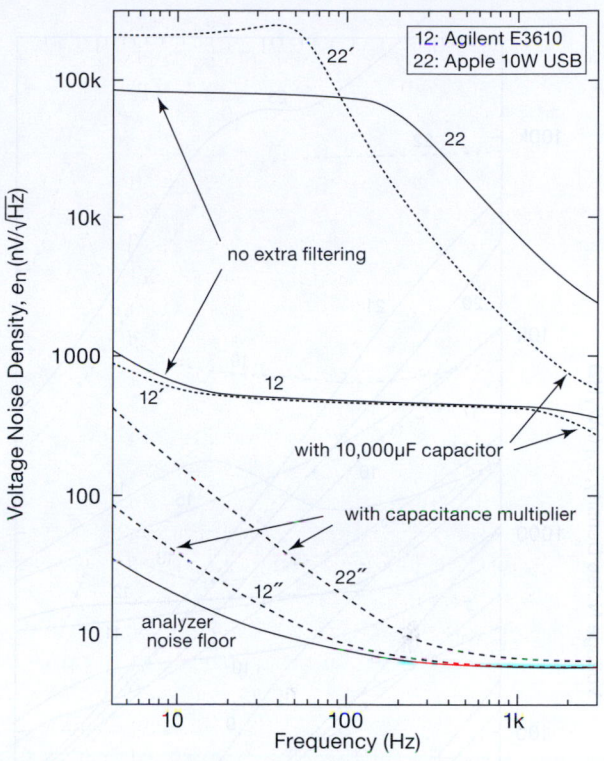

Figure 8.122. A capacitance multiplier is highly effective in eliminating power-supply noise, as shown in these measured noise spectra of two of the supplies whose spectra are included in Figure 8.123 (with same label numbers). Solid curves, power supply alone; dotted curves, (single-prime label), with added $10,000\,\mu F$ capacitor across the dc output terminals; dashed curves (double-prime label), with capacitance multiplier shown in Figure 8.92.

and construction. In stubborn cases the cure may involve a combination of filtration of input and output lines, careful layout and grounding, and extensive electrostatic and magnetic shielding. In these subsections we offer some suggestions that may help illuminate this dark area of the electronic art.[145]

8.16.1 Interfering signals

Interfering signals can enter an electronic instrument through the powerline inputs or through signal input and output lines. In addition, signals can be capacitively

---

[145] For more extensive advice see the popular classics: R. Morrison, *Grounding and Shielding: Circuits and Interference*, Wiley–IEEE Press (2007), and H. Ott, *Noise Reduction Techniques in Electronic Systems*, Wiley–Interscience (1988).

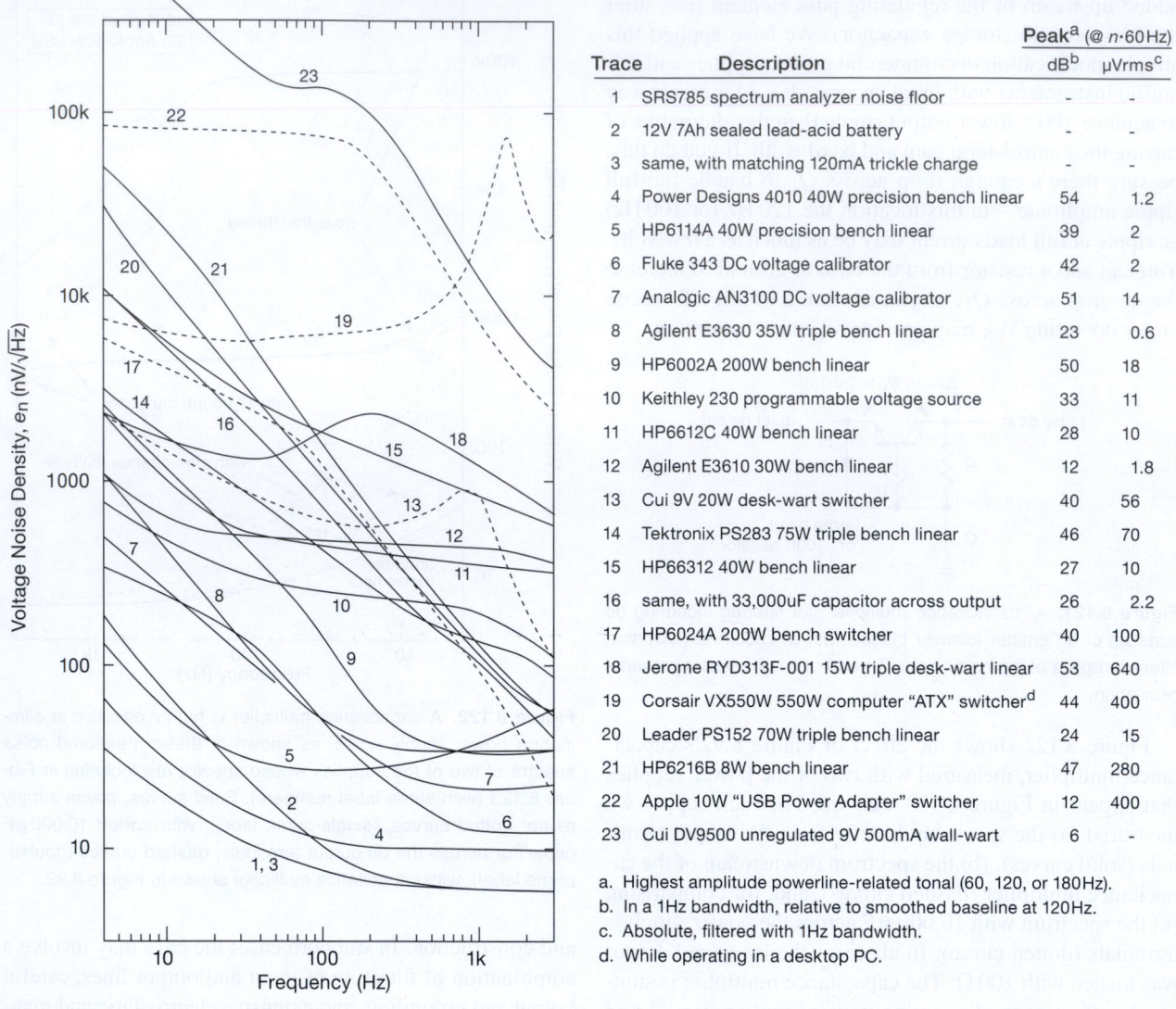

Trace	Description	Peak[a] (@ $n \cdot 60$Hz)	
		dB[b]	μVrms[c]
1	SRS785 spectrum analyzer noise floor	-	-
2	12V 7Ah sealed lead-acid battery	-	-
3	same, with matching 120mA trickle charge	-	-
4	Power Designs 4010 40W precision bench linear	54	1.2
5	HP6114A 40W precision bench linear	39	2
6	Fluke 343 DC voltage calibrator	42	2
7	Analogic AN3100 DC voltage calibrator	51	14
8	Agilent E3630 35W triple bench linear	23	0.6
9	HP6002A 200W bench linear	50	18
10	Keithley 230 programmable voltage source	33	11
11	HP6612C 40W bench linear	28	10
12	Agilent E3610 30W bench linear	12	1.8
13	Cui 9V 20W desk-wart switcher	40	56
14	Tektronix PS283 75W triple bench linear	46	70
15	HP66312 40W bench linear	27	10
16	same, with 33,000uF capacitor across output	26	2.2
17	HP6024A 200W bench switcher	40	100
18	Jerome RYD313F-001 15W triple desk-wart linear	53	640
19	Corsair VX550W 550W computer "ATX" switcher[d]	44	400
20	Leader PS152 70W triple bench linear	24	15
21	HP6216B 8W bench linear	47	280
22	Apple 10W "USB Power Adapter" switcher	12	400
23	Cui DV9500 unregulated 9V 500mA wall-wart	6	-

a. Highest amplitude powerline-related tonal (60, 120, or 180Hz).
b. In a 1Hz bandwidth, relative to smoothed baseline at 120Hz.
c. Absolute, filtered with 1Hz bandwidth.
d. While operating in a desktop PC.

Figure 8.123. All dc power supplies are not created equal! Measured noise voltage spectra of many varieties of supplies found in our lab. Dashed lines are switching supplies; the rest are linear. A 100 Ω load was used in most cases. Spectral peaks ("tonals") at powerline frequency and harmonics have been edited out for clarity; they range from 5 dB to more than 50 dB above the smoothed baseline, as listed. A drifting output voltage produces a rising low-frequency tail.

coupled (electrostatic coupling) onto wires in the circuit (the effect is more serious for high-impedance points within the circuit), magnetically coupled to closed loops in the circuit (independent of impedance level), or electromagnetically coupled to wires acting as small antennas for electromagnetic radiation. Any of these can become a mechanism for coupling of signals from one part of a circuit to another. Finally, signal currents from one part of the circuit can couple to other parts through voltage drops on ground lines or power-supply lines.

A. Eliminating interference
Numerous effective tricks have been evolved to handle most of these commonly occurring interference problems. Keep in mind the fact that these techniques are all aimed at reducing the interfering signal or signals to an acceptable level; they rarely eliminate them altogether. Consequently, it often pays to raise signal levels, just to improve the signal-to-interference ratio. Also, it is important to realize that some environments are much worse than others; an instrument that works just dandy on the bench may perform

miserably on location. Some environments worth avoiding are those (a) near a radio or television station (RF interference), (b) near a subway (impulsive interference and powerline garbage), (c) near high-voltage lines (radio interference, frying sounds), (d) near motors and elevators (powerline spikes), (e) in a building with triac lamp dimmers and heater controllers (powerline spikes), (f) near equipment with large transformers (magnetic pickup), and (g) near arc welders (unbelievable pickup of all sorts). Herewith a gathering of advice, techniques, and black magic.

B. Signals coupled through inputs, outputs, and powerline

The best bet for powerline noise is to use a combination of RF line filters and transient suppressors on the ac powerline. You can achieve 60 dB or better attenuation of interference above a few hundred kilohertz this way, as well as effective elimination of damaging spikes.

Inputs and outputs are more difficult, because of impedance levels and the need to couple desired signals that may lie in the frequency range of interference. In devices like audio amplifiers you can use lowpass filters on inputs *and outputs* (much interference from nearby radio stations enters via the speaker wires, acting as antennas). In other situations shielded lines are often necessary. Low-level signals, particularly at high impedance levels, should always be shielded. So should the instrument cabinet.

C. Capacitive coupling

Signals within an instrument can get around handsomely via electrostatic coupling: some point within the instrument has a 10 volt signal jumping around; a high-Z input nearby does some sympathetic jumping, too. The best things to do are to reduce the capacitance between the offending points (move them apart), add shielding (a complete metal enclosure, or even close-knit metal screening, eliminates this form of coupling altogether), move the wires close to a ground plane (which "swallows" the electrostatic fringing fields, reducing coupling enormously), and lower the impedance levels at susceptible points, if possible. Op-amp outputs don't pick up interference easily, whereas inputs do. More on this later.

D. Magnetic coupling

Unfortunately, low-frequency magnetic fields are not significantly reduced by metal enclosures. A dynamic microphone, audio recorder, low-level amplifier, or other sensitive circuit placed in close proximity to an instrument with a large power transformer will display astounding amounts of 60 Hz pickup. The best therapy here is to avoid large enclosed areas within circuit paths and try to keep the circuit

from closing around in a loop. Twisted pairs of wires are quite effective in reducing magnetic pickup, because the enclosed area is small, and the signals induced in successive twists cancel.

When dealing with very low level signals, or devices particularly susceptible to magnetic pickup (tape heads, inductors, wire-wound resistors), it may be desirable to use magnetic shielding. "Mu-metal shielding" is available in preformed pieces and flexible sheets. If the ambient magnetic field is large, it is best to use shielding of high permeability (high mu) on the inside, surrounded by an outer shield of lower permeability (which can be ordinary iron, or low-mu shielding material), to prevent magnetic saturation in the inner shield. Of course, moving the offending source of the magnetic field is often a simpler solution. It may be necessary to exile large power transformers to the hinterlands, so to speak. Shielded inductors (e.g., pot cores) are configured so that the magnetic material (usually ferrite) provides a closed magnetic path. Toroidal transformers have smaller fringing fields than the standard frame types, and a single reverse turn (or arranging the winding pattern to return back to where it begins) cancels the one-turn effective area of the toroid's winding.

E. Circuit boards and cables

Both capacitive and magnetic coupling occurs, also, between traces on circuit boards or between line pairs in cables and ribbons. Magnetic coupling is sometimes called *inductive* coupling, to distinguish it from capacitive coupling. This business of "crosstalk" is a rich subject, dealt with in good detail in references such as Johnson and Graham's *Handbook of Black Magic*.[146] Perhaps unintiutively, it turns out that the magnitudes of inductive and capacitive crosstalk are generally comparable, but they behave differently as seen at the two ends of a pair of closely spaced lines: forward-going crosstalk is proportional to the signal's slew rate, but the capacitive and inductive components are of opposite sign and tend to cancel. For reverse-going crosstalk the coupled signal looks like a flat-topped pulse (of width equal to the round-trip travel time), with capacitive and inductive components reinforcing. If the driving impedance does not match the line's characteristic impedance (see Appendix H), that "near-end crosstalk" (aka *NEXT*) signal will reflect and become forward going, contributing to (and often dominating) the "far-end crosstalk" (or *FEXT*).

---

[146] H. W. Johnson and M. Graham, *High-Speed Digital Design – A Handbook of Black Magic*, Prentice-Hall (1993). See also their follow-on, *High-Speed Signal Propagation – Advanced Black Magic*, Prentice-Hall (2003).

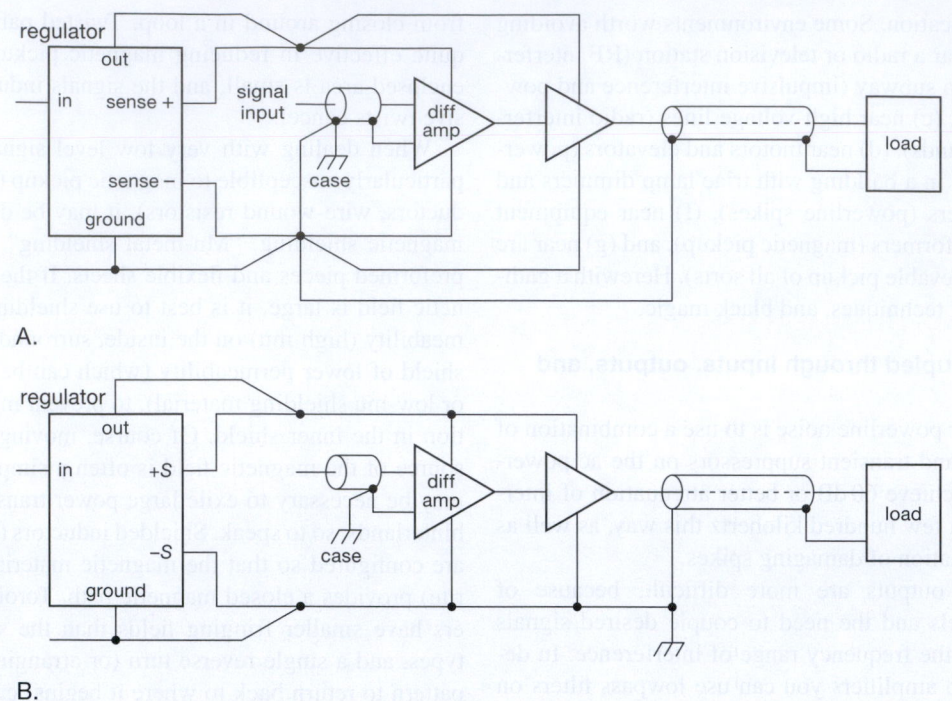

Figure 8.124. Ground paths for low-level signals. A. Right. B. Wrong.

See also additional discussion in Chapter 12 (§12.9, "Digital signals and long wires").

F. Radiofrequency coupling

RF pickup can be particularly insidious, because innocent-looking parts of the circuit can act as resonant circuits, displaying enormous effective cross-section for pickup. Aside from overall shielding, it is best to keep leads short and avoid loops that can resonate. Ferrite beads may help, if the problem involves very high frequencies. Doing good deeds can sometimes create havoc. For example, you may want to use several ceramic capacitors to improve power-supply bypassing; but in combination with the inductance of the connecting supply rails they can form a lovely parasitic tuned circuit somewhere in the HF to VHF region (tens to hundreds of megahertz), nudged by active circuits into ringing and even oscillation.[147]

8.16.2 Signal grounds

Ground leads and shields can cause plenty of trouble, and there is a lot of misunderstanding on this subject. The prob-

---

[147] One cure is to include some aluminum electrolytic capacitors – their lossy series resistance damps the resonance.

lem, in a nutshell, is that currents you forgot about that are flowing through a ground line can generate a signal seen by another part of the circuit sharing the same ground. The technique of a ground "mecca" (a common point in the circuit to which all ground connections are tied) is often seen, but it's a crutch; with a little understanding of the problem you can handle most situations intelligently.

A. Common grounding blunders

Figure 8.124 shows a common situation. Here a low-level amplifier and a high-current driver are in the same instrument. The first circuit is done correctly: both amplifiers tie to the supply voltages at the regulator (right at the sensing leads), so *IR* drops along the leads to the power stage don't appear on the low-level amplifier's supply voltages. In addition, the load current returning to ground does not appear at the low-level input; no current flows from the ground side of the low-level amplifier's input to the circuit mecca (which might be the connection to the case near the BNC input connector).

In the second circuit there are two blunders. Supply-voltage fluctuations caused by load currents at the high-level stage are impressed on the low-level supply voltages. Unless the input stage has very good supply rejection, this can lead to oscillations. Even worse, the load current

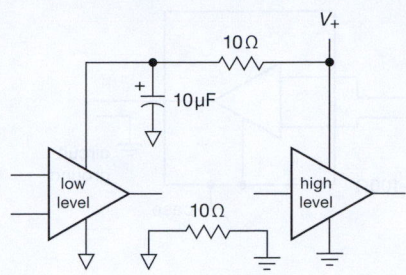

Figure 8.125. Decoupling the dc rail supplying low-level stages.

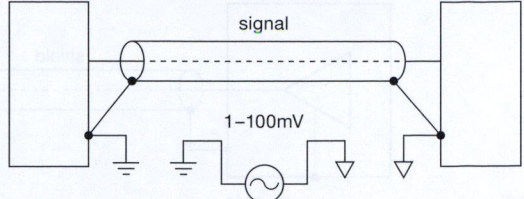

Figure 8.126. When connecting signals between instruments, you may see 100 mV (or more) differences (at powerline frequency) between their local grounds, even when they are bridged as shown.

returning to the supply makes the case "ground" fluctuate with respect to power-supply ground. The input stage ties to this fluctuating ground, a very bad idea. The general idea is to look at where the large signal currents are flowing and make sure their *IR* drops don't wind up at the input. In some cases it may be a good idea to decouple the supply voltages to the low-level stages with a small *RC* network (Figure 8.125). In stubborn cases of supply coupling it may pay to put a zener or 3-terminal regulator on the low-level-stage supply for additional decoupling.

8.16.3 Grounding between instruments

The idea of a controlled ground point within one instrument is fine, but what do you do when a signal has to go from one instrument to another, each with its own idea of "ground"? Some suggestions follow (and see also the extensive treatment of the transmission of *digital* signals in §12.9).

A. High-level signals
If the signals are several volts, or large logic swings, just tie things together and forget about it (Figure 8.126). The voltage source shown between the two grounds represents the variations in local grounds you'll find on different powerline outlets in the same room or (worse) in different rooms or buildings. It consists of some 60 Hz voltage, harmonics of the line frequency, some radiofrequency signals (the powerline makes a good antenna), and assorted spikes and other garbage. If your signals are large enough, you can live with this.

B. Small signals and long wires
For small signals this situation is intolerable, and you will have to go to some effort to remedy the situation. Figure 8.127 shows some ideas. In the first circuit, a coaxial shielded cable is tied to the case and circuit ground at the driving end, but it is kept isolated from the case

at the receiving end (use a Bendix 4890-1 or Amphenol 31-010 insulated BNC connector). A differential amplifier is used to buffer the input signal, thus ignoring the small amount of "ground signal" appearing on the shield. A small resistor and bypass capacitor to ground is a good idea to limit ground swing and prevent damage to the input stage. The alternative receiver circuit in Figure 8.127A shows the use of a "pseudodifferential" input connection for a single-ended amplifier stage (which might, for example, be a standard noninverting op-amp connection, as indicated). The 10 Ω resistor between amplifier common and circuit ground is large enough to let the signal source's reference ground set the potential at that point, because it is much larger than the impedance of the source's ground. Any noise present at that node, of course, appears also at the output. However, this becomes unimportant if the stage has sufficiently high voltage gain, G_V, because the ratio of desired signal-to-ground noise is reduced by G_V. Thus, although this circuit isn't truly differential (with infinite CMRR), it works well enough (with effective CMRR $= G_V$). This pseudo-differential ground-sensing trick can be used also for low-level signals *within* an instrument, when ground noise is a problem.

In the second circuit (Figure 8.127B), a shielded twisted pair is used, with the shield connected to the case at both ends. Because no signal travels on the shield, this is harmless. A differential amplifier is used as before on the receiving end. If logic signals are being transmitted, it is a good idea to send a differential signal (the signal and its inverted form), as indicated. Ordinary differential amplifiers can be used as input stages, or, if the ground interference is severe, special "isolated amplifiers" are available from manufacturers like Analog Devices, Inc. and TI/Burr–Brown. The latter permit kilovolts of common-mode signals. So do optoisolator modules (§12.7), a handy solution for digital signals in some situations.

At radio frequencies, transformer coupling offers a convenient way of removing common-mode signal at the

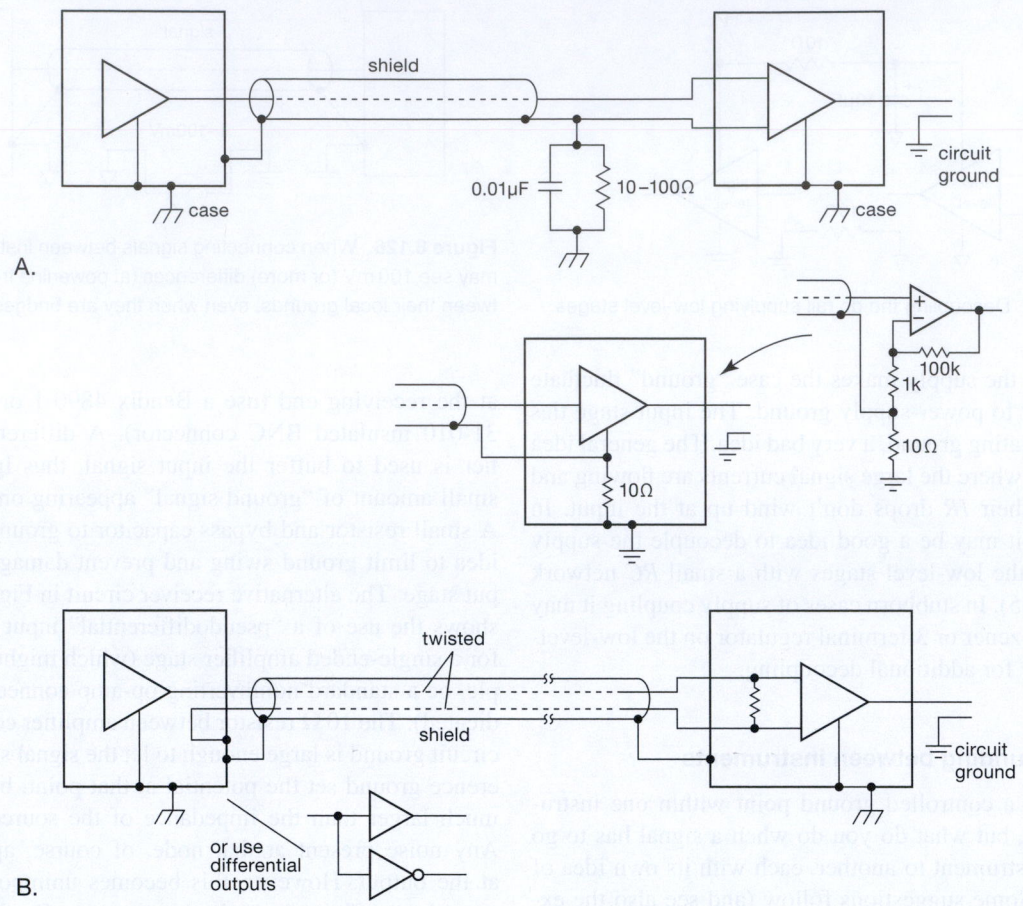

Figure 8.127. Ground connections for low-level signals through shielded cables.

receiving end; this also makes it easy to generate a differential bipolarity signal at the driving end. Transformers are popular in audio applications as well, although they tend to be bulky and lead to some signal degradation, as described in §8.10.

For very long cable runs (measured in miles) it is useful to prevent large ground currents flowing in the shield at radio frequencies. Figure 8.128 suggests a method. As before, a differential amplifier looks at the twisted pair, ignoring the voltage on the shield. By tying the shield to the case through a small inductor, the dc voltage is kept small while preventing large radiofrequency currents. This circuit also shows protection circuitry to prevent common-mode excursions beyond ±10 V.

Figure 8.129 shows a nice scheme to save wires in a multiwire cable in which the common-mode pickup has to be eliminated. Because the signals all suffer the same common-mode pickup, a single wire tied to ground at the

sending end serves to cancel the common-mode signals on each of the *n* signal lines. Just buffer its signal (with respect to ground at the receiving end) and use it as the comparison input for each of *n* differential amplifiers looking at the other signal lines.

The preceding schemes work well to eliminate common-mode interference at low to moderate frequencies, but they can be ineffective against radiofrequency interference, owing to poor common-mode rejection in the receiving differential amplifier.

One possibility here is to wrap the whole cable around a ferrite toroid (Figure 8.130). That increases the series inductance of the whole cable, raising the impedance to common-mode signals of high frequency and making it easy to bypass them at the far end with a pair of small bypass capacitors to ground. The equivalent circuit shows why this works without attenuating the differential signal: you have a series inductance inserted into both signal lines

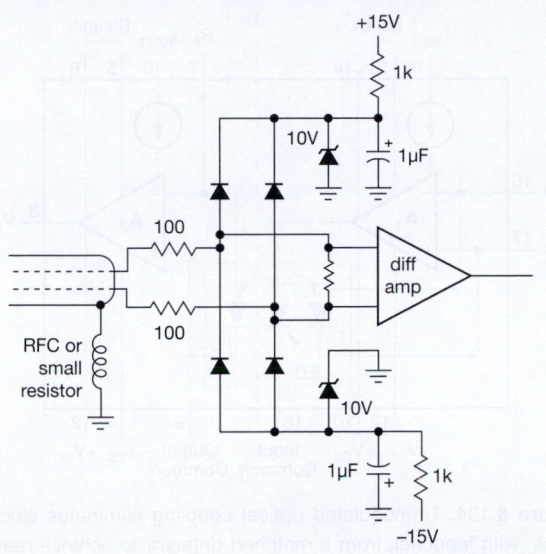

Figure 8.128. Input-protection circuits for use with very long lines.

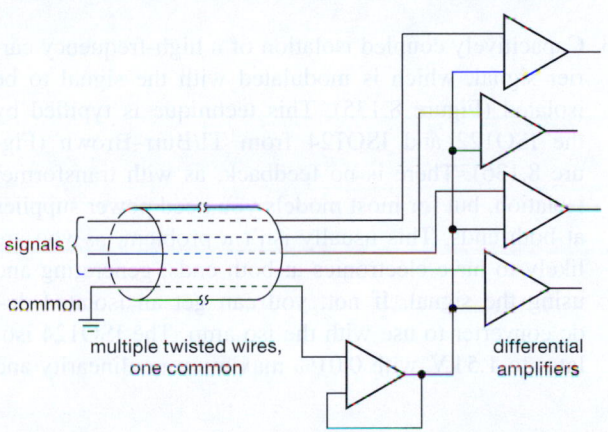

Figure 8.129. Common-mode interference rejection with long multiwire cables.

and the shield, but they form a tightly coupled transformer of unit turns ratio, so the differential signal is unaffected. This can be thought of as a "1:1 transmission-line transformer" (see §13.10 in the second edition of this book).

C. Floating signal sources

The same sort of disagreement about the voltage of "ground" at separated locations enters in an even more serious way at low-level inputs, just because the signals are so small. An example is a magnetic tape head or other signal transducer that requires a shielded signal line. If you ground the shield at both ends, differences in ground potential will appear as signal at the amplifier input. The best

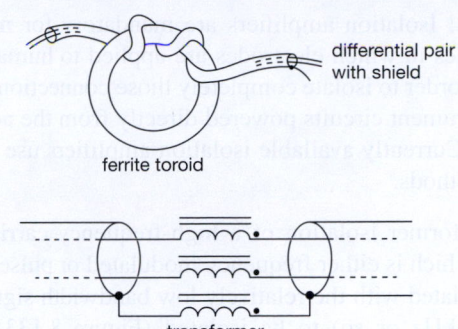

Figure 8.130. Wrapping a coaxial or multi-wire cable around a ferrite core for high-frequency common-mode suppression.

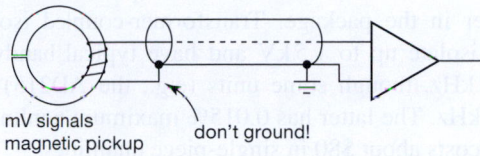

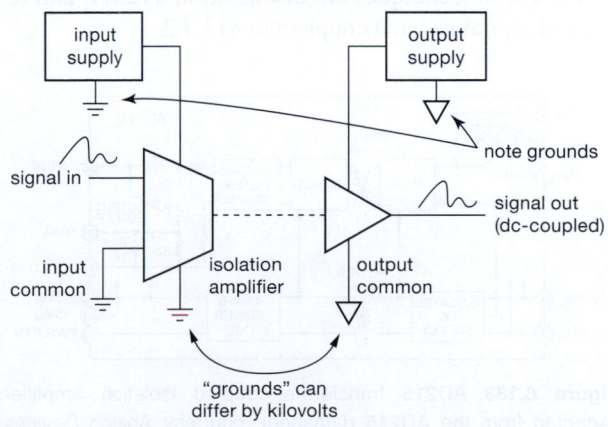

Figure 8.131. Preventing sneak ground currents: tie the shield to ground at the receiving end only.

approach is to lift the shield off ground *at the transducer* (Figure 8.131).

D. Isolation amplifiers

Another solution to serious ground-contention problems is the use of an "isolation amplifier." Isolation amplifiers (iso-amps) are commercial devices intended for coupling an analog signal (with bandwidth clear down to dc) from a circuit with one ground reference to another circuit with a completely different ground (Figure 8.132). In fact, in some bizarre situations the "grounds" can differ by many

Figure 8.132. Isolation amplifier concept.

kilovolts! Isolation amplifiers are mandatory for medical electronics in which electrodes are applied to human subjects, in order to isolate completely those connections from any instrument circuits powered directly from the ac powerlines. Currently available isolation amplifiers use one of three methods.

1. Transformer isolation of a high-frequency carrier signal, which is either frequency modulated or pulse-width modulated with the relatively low bandwidth signal (dc to 10 kHz or so) to be isolated (Figure 8.133). This method is used in isolation amplifiers from Analog Devices, Inc. Transformer-isolated iso-amps have the convenient feature of requiring dc power only on one side; they all include a transformer-coupled dc-to-dc converter in the package. Transformer-coupled iso-amps can isolate up to 1.5 kV and have typical bandwidths of 5 kHz, though some units (e.g., the AD215) go to 120 kHz. The latter has 0.015% maximum nonlinearity, and costs about $80 in single-piece quantities.

2. Optically coupled signal transmission via an LED at the sending end and photodiode at the receiving end. This technique is typified by the ACPL-C79 series from Avago. These devices use a delta–sigma modulation and demodulation scheme, and achieve a bandwidth from dc to 200 kHz. This series is inexpensive (about $5 in modest quantity), with maximum nonlinearity of 0.06%. An interesting alternative, which involves no clocking signal or carrier frequency, is to couple an LED at the sending end to a photodiode at the receiving end. To achieve good linearity you use feedback from a second matched photodiode at the transmitting side that receives light from the same LED, thus cancelling nonlinearities in both LED and photodiode (Figure 8.134; see also Figure 12.90 in Chapter 12). Some additional analog optoisolation techniques are discussed in §12.7.4, and related digital isolated couplers in §12.7.2.

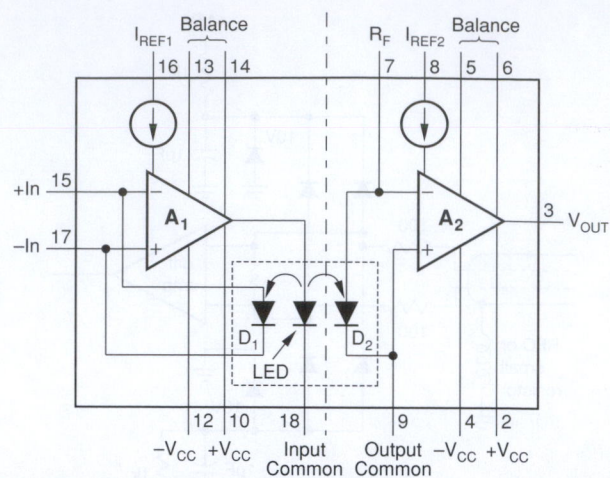

Figure 8.134. Unmodulated optical coupling eliminates clocking noise, with feedback from a matched detector to achieve reasonable linearity. (Adapted from the ISO100 datasheet, courtesy of Texas Instruments.)

3. Capacitively coupled isolation of a high-frequency carrier signal, which is modulated with the signal to be isolated (Figure 8.135). This technique is typified by the ISO122 and ISO124 from TI/Burr–Brown (Figure 8.136). There is no feedback, as with transformer isolation, but for most models you need power supplies at both ends. This usually isn't a problem, as you are likely to have electronics at both ends, generating and using the signal. If not, you can get an isolated dc–dc converter to use with the iso-amp. The ISO124 isolates to 1.5 kV, with 0.01% maximum nonlinearity and

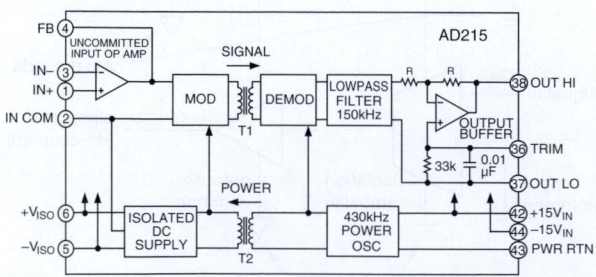

Figure 8.133. AD215 transformer-coupled isolation amplifier. (Adapted from the AD215 datasheet, courtesy Analog Devices, Inc.)

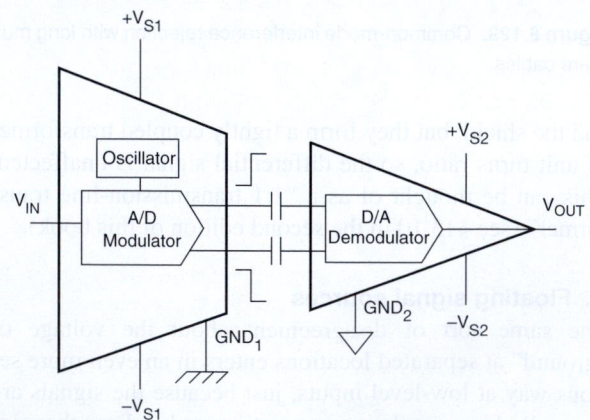

Figure 8.135. Capacitively-coupled isolation amplifier. (Adapted from the Burr–Brown Application Bulletin AB-047, courtesy of Texas Instruments.)

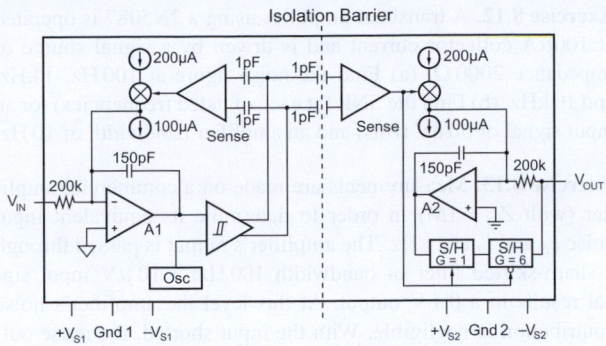

Figure 8.136. ISO124 capacitively-coupled isolation amplifier. (Adapted from the ISO124 datasheet, courtesy of Texas Instruments.)

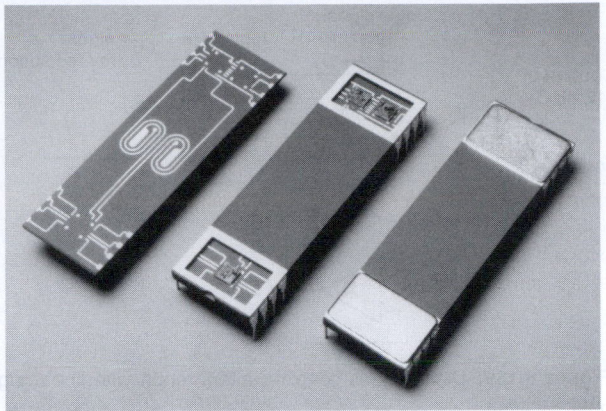

Figure 8.137. Burr–Brown (TI) ISO106 isolation amplifier, courtesy of Texas Instruments.

50 kHz bandwidth; it costs $18 in single-piece quantities. Figure 8.137 shows nicely the innards of one of these capacitively coupled devices.

These isolation amplifiers are all intended for *analog* signals. The same sorts of ground problems can arise in digital electronics, where the solution is simple and effective: optically coupled isolators ("optoisolators") are available, with plenty of bandwidth (10 MHz or more), isolation of several kilovolts, and low cost (a dollar or two). They are discussed extensively in Chapter 12.
A caution: isolation amplifiers can introduce noise of their own, particularly those that use some form of signal modulation (which is most of them!). For the latter you've got a residue of the modulating clock frequency; and all isolation amplifiers (whether clocked or not) introduce broadband noise of the usual sort. One remedy for clock noise is to add an analog lowpass filter at the output of the receiving side. For more detail see the helpful application document "Noise Sources in Applications Using Capacitive Coupled Isolated Amplifier" (Application Bulletin SBOA028, available at the Texas Instruments website, www.ti.com).

E. Signal guarding
A closely related issue is signal *guarding*, an elegant technique to reduce the effects of input capacitance and leakage for small signals at high impedance levels. You may be dealing with signals from a micro-electrode or a capacitive transducer, with source impedances of hundreds of megohms. Just a few picofarads input capacitance can form a lowpass filter, with rolloffs beginning at a few hertz! In addition, the effects of insulation resistance in the connecting cables can easily degrade the performance of an ultra-low input current amplifier (bias currents less than

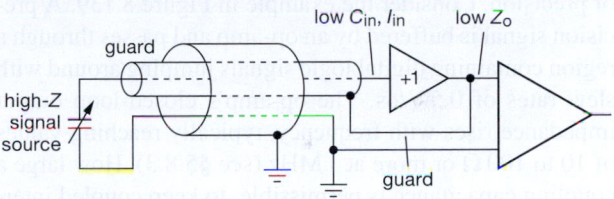

Figure 8.138. Using a guard to raise input impedance.

a picoamp) by orders of magnitude. The solution to both these problems is a *guard electrode* (Figure 8.138).[148]

A follower bootstraps the inner shield, effectively eliminating leakage current and capacitive attenuation by keeping zero voltage difference between the signal and its surroundings. An outer grounded shield is a good idea, to keep interference off the guard electrode; the follower has no trouble driving that capacitance and leakage, of course, given its low output impedance.

You shouldn't use these tricks more than you need to; it would be a good idea to put the follower as close to the signal source as possible, guarding only the short section of cable connecting them. Ordinary shielded cable can then carry the low-impedance output signal out to the remote amplifier.[149]

F. Coupling to outputs
Ordinarily the output impedance of an op-amp is low enough that you don't have to worry about capacitive signal coupling. In the case of high-frequency or fast-switching

---

[148] We saw this earlier, in §8.11.9.

[149] Discussed in more detail, in connection with high-impedance micro-electrodes, in §15.08 of the second edition of this book.

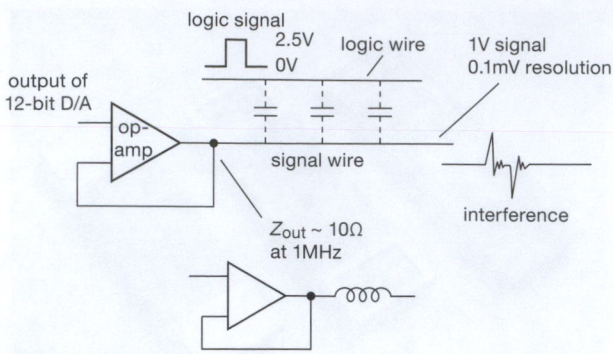

Figure 8.139. Digital cross-coupling interference with linear signals.

interference, however, you have just cause for alarm, particularly if the desired output signal involves some degree of precision. Consider the example in Figure 8.139. A precision signal is buffered by an op-amp and passes through a region containing digital logic signals jumping around with slew rates of 0.5 V/ns. The op-amp's closed-loop output impedance rises with frequency, typically reaching values of 10 to 100 Ω or more at 1 MHz (see §5.8.3). How large a coupling capacitance is permissible, to keep coupled interference less than the analog signal's resolution of 0.1 mV? The surprising answer is a maximum of 0.02 pF.

There are some solutions. The best thing is to keep your small analog waveforms out of the reach of fast-switching signals. A moderate bypass capacitor across the op-amp's output (with perhaps a small series resistor, to maintain op-amp stability) will help, although it degrades the slew rate. You can think of the action of this capacitor as lowering the frequency of the coupled charge bundles to the point where the op-amp's feedback can swallow them. A few hundred picofarads to ground will adequately stiffen the analog signal at high frequencies (think of it as a capacitive voltage divider). Another possibility is to use a low-impedance buffer such as the LT1010, or a power op-amp such as the LM675. Don't neglect the opportunity to use shielding, twisted pairs, and proximity to ground planes to reduce coupling.

Additional Exercises for Chapter 8

Exercise 8.10. Prove that SNR = $10\log_{10}(v_s^2/4kTR_s) -$ NF(dB) (at R_s).

Exercise 8.11. A 10 μV (rms) sinewave at 100 Hz is in series with a 1M resistor at room temperature. What is the SNR of the resultant signal (a) in a 10 Hz band centered at 100 Hz, and (b) in a 1 MHz band going from dc to 1 MHz?

Exercise 8.12. A transistor amplifier using a 2N5087 is operated at 100μA collector current and is driven by a signal source of impedance 2000 Ω. (a) Find the noise figure at 100 Hz, 1 kHz, and 10 kHz. (b) Find the SNR (at each of listed frequencies) for an input signal of 50 nV (rms) and an amplifier bandwidth of 10 Hz.

Exercise 8.13. Measurements are made on a commercial amplifier (with Z_{in} =1M) in order to determine its equivalent input noise e_n and i_n at 1 kHz. The amplifier's output is passed through a sharp-skirted filter of bandwidth 100 Hz: a 10 μV input signal results in a 0.1 V output. At this level the amplifier's noise contribution is negligible. With the input shorted, the noise output is 0.4 mV rms. With the input open, the noise output rises to 50 mV rms. (a) Find e_n and i_n for this amplifier at 1 kHz. (b) Find the noise figure of this amplifier at 1 kHz for source resistances of 100 Ω, 10k, and 100k.

Exercise 8.14. Noise measurements are made on an amplifier using a calibrated noise source whose output impedance is 50 Ω. The generator output must be raised to 2 nV/$\sqrt{\mathrm{Hz}}$ in order to double the output noise power of the amplifier. What is the amplifier's noise figure for a source impedance of 50 Ω?

Exercise 8.15. Your boss tells you she's working on a super-sensitive pressure-change instrument that uses a diaphragm-type pressure sensor with a 350 Ω strain-gage bridge. She tells you there's a 10 Hz measurement bandwidth, either from a filter or an integrating ADC, and she asks you to select an op-amp for the input stage. First consider auto-zero amplifiers (see §5.11 Table 5.6, and Figure 5.54). (a) Is a bipolar op-amp better than a JFET op-amp? Hint, don't forget i_n. (b) The bridge has a standard 2 mV/V full-scale output, and your boss is planning a 2.5 V excitation voltage. What will be your system's noise floor as a fraction of full scale? (c) Should you suggest to your boss that she consider running the ADC faster and averaging the results? (d) A fellow engineer (who suffers the indignities of the same boss) suggests that a silicon strain-gauge pressure sensor might be better, because they have higher output, 2.5 mV/V. You find sensors with R_s=1.4kΩ to 3kΩ. Choose a good op-amp for this case, and calculate the system performance.

Exercise 8.16. Digital oscilloscope input stage (*this is a difficult problem!*). Your assignment is to design an input stage for a low-cost "oscilloscope app," to be powered by a single +3.3 V supply from the mobile device's lithium-ion battery. The 'scope's architecture calls for attenuating its probe signals to 1 mV per division via a range switch, then amplifying and digitizing this signal, with a goal of peak-to-peak noise no more than 5% of a division. The screen shows ten vertical divisions, and you need to be able to shift it vertically through $\pm5\times$ over-range, plus there's a 40 mV budget for "software offset." (a) What is the total input range, and what is your rms input noise level goal? (b) If a 14-bit ADC is used, what is the LSB resolution?

An ADS7946 14-bit ADC has been selected; it draws 0.5 mA from +3.3V when converting at 100 kSPS and meets cost goals at

$6 in qty 1k. (c) What is a reasonable ADC input voltage range, and how much gain will you need to provide?

The ADC is capable of 2 Msps, for a 1 MHz bandwidth, but there are plans to use a faster ADC in a later revision; hence your design-spec bandwidth is 10 MHz. (d) What is your spec-tral noise-density goal?

Standard oscilloscopes are dc coupled, have a 1 MΩ∥15 pF input impedance, and do not exhibit visible dc shifts with source impedance. (e) Create a design that meets all your specs. Low power consumption is a big plus.

Review of Chapter 8

An A-to-Q summary of what we have learned in Chapter 8. This summary reviews basic principles and facts in Chapter 8, but it does not cover application circuit diagrams and practical engineering advice presented there.

¶A. Noise Basics.

See §8.1. The random noise you care about is characterized by its *density* (rms noise amplitude in a 1 Hz band of frequency), §8.2.1. Noise *voltage* density is called e_n, and has units like $nV/\sqrt{Hz}$. Likewise, the symbol for noise *current* density is i_n, with units like $fA/\sqrt{Hz}$. A noise current at an amplifier's input flows through the signal's source resistance, creating its own noise voltage density $e_n = i_n R_s$. Independent noise sources combine as the square root of the sum of their squares: $e_{n(total)} = \sqrt{e_{n1}^2 + e_{n2}^2 + \cdots}$. If a noise source is uniform over frequency, it's called "white noise," and the rms noise voltage (as contrasted with noise voltage *density*) contained within a bandwidth B is just $v_n = e_n \sqrt{B}$. Knowing that, you can go to Table 8.3a–c on page 522, which lists e_n and i_n for a wide selection of op-amps, to figure out how much noise is added in an op-amp amplifier stage. Compare this with the input signal's noise level; or multiply by the amplifier's voltage gain and, voilà, you've got the output noise voltage density.

¶B. Noise Spectra.

Whatever its source, noise density can vary with frequency, see §8.1. White noise (e_n constant over frequency, up to some cutoff) is common, for example a resistor's Johnson noise (¶E, below) or the shot noise fluctuations of a steady current (¶F, below). Also prevalent is "$1/f$ noise," sometimes called flicker noise, or pink noise; it is characterized by a $1/f$ *power* spectrum (equal power per octave, or per decade), thus a voltage noise density $e_n(f)$ proportional to $1/\sqrt{f}$. Most electronic circuits (and many other physical phenomena) exhibit $1/f$ noise, often characterized by the "$1/f$ corner frequency" at which the $1/f$ noise component is equal to the white noise component. Finally, the term *red noise* refers to a noise density e_n proportional to $1/f$ (thus a $1/f^2$ power spectrum); see for example Figure 8.95. The rms noise voltage v_n in some bandwidth B (extending from f_1 to f_2) is gotten by integrating $e_n^2(f)$ over frequency, then taking the square root: $v_n = \sqrt{v_n^2}$, where $v_n^2 = \int_{f_1}^{f_2} e_n(f)^2 \, df$. For a white spectrum this reduces to the simple $v_n = e_n \sqrt{B}$.

Actual circuit noise need not conform to these idealized spectra, which, however, are useful in characterizing the noise of real devices over chosen regions of frequency; see

for example Figure 8.107. In real life, spectra may show a noise "shelf" (e.g., curve Z in Figure 8.52), or a noise peak (e.g., Figures 5.52 or 8.72).

¶C. Noise Amplitude Distribution.

Quite apart from its spectrum, one can characterize the amplitude distribution of noise; that is, the distribution of instantaneous amplitudes sampled in time. Most noise sources obey a Gaussian distribution (Figure 8.2), a fact that is of lesser concern than the important properties of noise spectrum and amplitude. A notable exception is *burst* noise (also called *popcorn* noise, *bistable* noise, or *telegraph* noise), which jumps randomly between several voltage levels (Figure 8.5). Burst noise was prominent in the early decades of semiconductor technology, but it has been largely banished in contemporary products. One suspects some low-level remnants remain, though, as evidenced for example in a comparison of the measured burst-noise spectrum of Figure 8.6 with the measured JFET noise spectra of Figure 8.51.

¶D. Sources of Noise.

The dominant sources of noise in electronic circuits (detailed in paragraphs ¶E–H below) are:

Johnson noise. Thermal fluctuations generate a noise voltage in a resistor.
Shot Noise. The discrete nature of electric charge creates fluctuations in a "steady" current.
Excess Noise. Various semiconductor phenomena contribute additional noise (often $\sim 1/f$ in power density) at low frequencies.
Amplifier Noise. Transistors (both BJTs and FETs) add noise, traceable to the above sources (e.g., Johnson noise of the base resistance $r_{bb'}$, shot noise of the collector current, and excess noise terms).

¶E. Johnson Noise.

Thermal fluctuations cause a self-generated noise voltage across the terminals of an unloaded resistor; this is *Johnson noise*, of (white-) noise voltage density $e_n = \sqrt{4k_B T R}$, where k_B is Boltzmann's constant. Don't bother remembering the latter; just remember the value of e_n for a round-number resistance (we like $1.28\,nV/\sqrt{Hz}$ for $R=100\,\Omega$, or $4\,nV/\sqrt{Hz}$ for $R=1k\Omega$), and scale by the square root of R. If short-circuited, a resistor generates a Johnson noise current $i_n = e_n/R$; that is, $i_n = \sqrt{4k_B T/R}$. See the graph in Figure 8.1 and the mini-table in §8.1.1.

Johnson noise is a fundamental physical phenomenon, and does not depend on the particular construction of the resistor (or resistance). However, when a steady current flows through a resistance, you may get some additional noise current (which you can think of as due to resistance fluctuations), generally with something approximating a $1/f$ power spectrum. This *excess noise* varies with resistor construction, being worst in a granular "carbon composition" type, but insignificant in a wirewound resistor.

¶F. Shot Noise.
The finiteness of the charge quantum (electron charge) causes statistical fluctuations even in a steady current. If the charges act independently, the (white-) noise current density is $i_n = \sqrt{2qI_{dc}}$, where q is the electron charge (1.6×10^{-19} coulombs). As with Johnson noise, it's handy to simply remember the value of i_n for a round-number dc current (e.g., 18 pA/$\sqrt{Hz}$ for $I_{dc} = 1$ mA), and scale by the square root of current.[150] See the mini-table in §8.1.2. *Important*: the shot-noise formula assumes that the charge carriers act independently; the current noise is greatly reduced if there are long-range correlations, as for example in a metallic conductor.

¶G. BJT Noise.
See §8.3. The primary noise terms of a BJT are an input noise voltage e_n in series with the base, combined with an input noise current i_n injected at the base (Figure 8.9). The amplifier's input-referred noise (i.e., ignoring the Johnson noise of the signal source's R_s) is then $e_{a(rms)} = [e_n^2 + (R_s i_n)^2]^{\frac{1}{2}}$. As you increase the collector current, e_n decreases and i_n increases, so there's an operating-current tradeoff. The ratio e_n/i_n has units of resistance; it's called the *noise resistance*, and it's a useful quantity in circuit design; see ¶I below. Table 8.1a lists measured values of e_n for many low-noise BJT candidates.

Voltage noise, e_n. In the simplest model, the noise voltage term arises from two sources: collector shot-noise current flowing through the transistor's emitter resistance r_e, and Johnson noise in the transistor's internal base resistance $r_{bb'}$. Combining these independent noise terms (eq'n 8.20), we get the total input-referred squared noise voltage, $e_n^2 = 2qI_C r_e^2 + 4kT r_{bb'} = 4kT(r_e/2 + r_{bb'})$. Put another way, the input noise voltage is equal to the combined Johnson noise of the transistor's base resistance ($r_{bb'}$) and a fictional resistance equal to half its

intrinsic emitter resistance ($r_e/2$). The latter is inversely proportional to collector current, so a BJT's noise voltage decreases with increasing I_C, ultimately limited by its internal base resistance; see for example Figure 8.12. Thus to minimize BJT *voltage* noise, choose a part with a low $r_{bb'}$, and operate at relatively high collector current. A more refined model includes the effects of base current shot noise, important at low frequencies and high collector currents; see Figure 8.11 and eq'n 8.24.

Current noise, i_n. The primary noise current term is shot noise in the dc base current, $i_n = \sqrt{2qI_B}$ (eq'n 8.21). Taking this term alone, you minimize BJT *current* noise by operating at low collector current. A more refined model includes a rising $\propto 1/\sqrt{f}$ noise current term at low frequencies, and a rising $\propto f$ noise current term at high frequencies caused by falling beta; see eq'n 8.22 and Figure 8.13.

¶H. JFET Noise.
See §8.6. The low-noise choice for signals of high source impedance are FETs, owing to their very low input current noise; JFETs are quieter than MOSFETs, and, unlike the latter, they are available as discrete small-signal parts (see Table 8.2).

Voltage noise, e_n. The dominant noise voltage term is Johnson noise in the channel resistance (eq'n 8.34), $e_n^2 \approx 4kT(\frac{2}{3}\frac{1}{g_m})$. Put another way, the noise voltage is equivalent to Johnson noise in a resistor of value $R = \frac{2}{3}\frac{1}{g_m}$; see Figure 8.46. To minimize JFET voltage noise, choose a JFET of high transconductance, and operate it at a relatively high drain current (note, however, that e_n falls slowly, only as the fourth root of I_D). As with BJTs, JFETs exhibit a rising $1/f$-like noise tail at low frequencies (Figure 8.52), with enormous variation among types, and manufacturers.

Current noise, i_n. At low frequencies the current noise is low, just shot noise of the gate (leakage) current: $i_n = \sqrt{2qI_G}$, see Figure 8.48. To frame this in numbers, a typical gate leakage current of 10 pA has a noise current density i_n of just 1.8 fA/$\sqrt{Hz}$, generating only 1.8 nV/$\sqrt{Hz}$ of noise voltage through a 1MΩ source resistance. That's down at the low end of transistor noise voltage, and it's completely dwarfed by the 128 nV/$\sqrt{Hz}$ of Johnson noise produced by the source resistance itself. At rising frequencies there are some additional sources of input noise current. For example, in a transimpedance amplifier the gate input is a summing junction, at which the FET's *voltage* noise generates a noise current through

---

[150] Because both shot-noise current and Johnson-noise current go as $1/\sqrt{R}$, it's easy to derive this handy factoid: if the dc drop across a resistor is greater than 50 mV, shot noise dominates over the resistor's own Johnson noise.

the input capacitance, of magnitude $i_n = e_n \omega C_{in}$, see ¶N (below) and §8.11.

¶I. Noise Figure, Noise Temperature, and Noise Resistance.

See §8.2. Noise figure (NF) is a popular measure of added amplifier noise. It is the ratio (in dB) of the amplifier's output noise to the output of a noiseless amplifier of the same gain, each driven with a source resistance R_s: NF$=10\log_{10}(1 + v_n^2/4kTR_s)$, where v_n^2 is the mean squared noise voltage per hertz contributed by the amplifier when a noiseless (cold) resistor of value R_s connected across its input (see eq'n 8.13 and Figure 8.7). Another (and equivalent) way to quantify the excess noise contributed by an amplifier when driven by a signal of source resistance R_s is to state its noise *temperature* (T_n, see Figure 8.8 and eq'n 8.16).

An amplifier's noise figure (and its noise temperature) is minimum when driven by a signal of source resistance equal to its noise resistance, i.e., when $R_s = R_n = e_n/i_n$. Transistor datasheets sometimes provide contours of NF versus operating current and source resistance at a given frequency (e.g., Figure 8.22); for amplifiers where you have no control over operating current you'll find instead contours of NF versus frequency and source resistance (e.g., Figure 8.27). *Caution*: do not make the mistake of adding a series resistor to a signal of low source resistance to improve the noise figure (§8.4.6). Instead choose an amplifier that provides the noise figure you need with that input source resistance; in some situations (e.g., very low source resistance) you can use a signal transformer to make a lossless match to the amplifier's optimum source resistance.

¶J. Noise Sources and Measurements.

See §8.12. At low to moderate frequencies you can determine the noise properties of a transistor amplifier of known voltage gain by making two measurements of its rms output noise voltage in a known bandwidth, first with shorted input (to get e_n) and then with an appropriately chosen resistor across its input (to get i_n); see §8.12.1. Because you're measuring the integrated noise voltage, you need to know the equivalent noise bandwidth; see §8.13. A more general technique, applicable to amplifiers that require a matched source impedance (e.g., $50\,\Omega$ RF amplifiers) and insensitive to actual measurement bandwidth, is to drive the input with a calibrated noise source while observing the rms noise output; see §8.12.3. For frequencies up to a few tens of megahertz you can make your own "pseudorandom" noise source with a feedback shift register (§8.12.4A), or you can use an off-the-shelf noise source (a function gen-

erator, or noise diode; see §8.12.4) for frequencies well up into the gigahertz range. Resistor Johnson noise itself is the noise source for the "hot-load/cold-load" method, useful for low-noise microwave amplifiers; see eq'n 8.53 in §8.12.3.

¶K. Low-noise Design with Op-amps.

See §8.9. Just as with BJTs and FETs, op-amps exhibit input-referred voltage noise e_n and current noise i_n, whose magnitudes are characteristic of their input-stage transistor type. What's different, from a circuit designer's point of view, is that you don't have control over the stage's operating current. Instead you choose the op-amp type: FET-input for lowest i_n (for signals of high source impedance), BJT-input for lowest e_n (for signals of low source impedance). Among the FET-input types, the CMOS parts exhibit lower i_n but considerably higher e_n, compared with JFET types; the latter thus combine the best (nearly) of both worlds, see §8.9 the extensive Table 8.3 on page 522ff, and the noise plots in Figures 8.60, 8.61, and 8.63. A helpful graphical tool is the plot of *effective noise density* (v_n) versus source resistance, in which the op-amp's noise contributions (e_n, $i_n R_s$) are combined and plotted along with the source's Johnson noise, the latter setting a lower noise bound; i.e., $v_n^2 = 4kTR_s + e_n^2 + (i_n R_s)^2$, see Figure 8.58. Op-amp noise (with the exception of auto-zero op-amps) exhibits a typical rising $1/f$ low-frequency tail, characterized by a *corner frequency*, f_c. Auto-zero op-amps have no $1/f$ tail, but they have considerably higher broadband noise (both e_n and i_n), along with spectral noise peaks and chopping artifacts. Be sure to evaluate all noise sources in an op-amp design: a poor choice of component values may compromise the noise performance (e.g., Johnson noise from high-value resistors).

When choosing an op-amp for a low-noise application, begin by restricting your attention to op-amps that meet your other needs, such as accuracy, speed, power dissipation, supply voltage, input and output swing, and the like. Then choose among this subset, based upon their noise parameters.

¶L. Low-noise Design with BJTs.

See §8.5. Compared with op-amps, circuit design with discrete transistors gives you more control over noise parameters, but the price you pay is the additional labor associated with details of biasing and the like. A nice compromise is a *hybrid* approach, with a discrete front-end prepended to an op-amp, see §8.9.5. As with op-amps, a plot of total input noise versus source resistance is a useful graphical tool, see Figure 8.32 in §8.5.2, and Figure 8.41. The flexibility you get from control of operating current is illustrated nicely in Figure 8.33, where the curves shift downward with increasing collector current (lower e_n, higher i_n). Table 8.1a on page 501 lists datasheet and measured noise parameters for an extensive selection of low-noise BJTs; see also measured curves of current gain in Figure 8.39.

¶M. Low-noise Design with FETs.

See §8.6. You can't beat FETs for lowest input current noise, and for *discrete* amplifier circuits (or hybrid front-ends) you are limited to JFETs. In terms of device parameters, JFETs with large geometry have lowest e_n and highest g_m, but the large geometry means higher capacitance and leakage current (thus higher i_n), see Table 8.2 on page 516. If the JFET's input is a summing junction, the product of noise voltage and input capacitance creates a rising noise current $i_n = e_n \omega C_{in}$ (¶H, ¶N). As with BJTs, raising the operating current reduces a JFET's noise voltage, though not dramatically ($e_n \propto I_D^{-0.25}$). Watch out for some chronic problems with many JFETs: loose I_{DSS} and $V_{GS(th)}$ specs (a 5:1 range is typical), low transconductance, low output resistance, sharply rising gate current (thus i_n) at high temperatures and at elevated drain voltages (see §3.2.8), and poor low-frequency noise performance (see Figure 8.52).

¶N. Noise in Transimpedance Amplifiers.

See §8.11. Transimpedance amplifiers (TIA) convert an input current to an output voltage, with current feedback through a resistor R_f to the input summing junction, as introduced in §4.3.1, and elaborated in §4x.3. The noise sources are the amplifier's input i_n and e_n, and Johnson noise in R_f. Taking these in reverse order, the feedback resistor generates a Johnson noise *current* of $i_n = \sqrt{4kT/R_f}$; the amplifier's voltage noise e_n generates through the capacitance at the input a noise current of magnitude $e_n \omega C_{in}$, and through the feedback resistor a noise current of magnitude e_n / R_f; to these must be added the amplifier's own input noise current i_n. The choice of R_f is a compromise: the noise terms are minimized by choosing a high resistance, but, because of feedback stability requirements, the bandwidth suffers; see §§8.11.1, 8.11.4–8.11.6, and *4x.3*. These competing constraints can be mitigated with more complex configurations, used singly or in combination: a composite (2-stage) TIA (§8.11.8); or the use of a discrete first stage (i.e., a hybrid, §8.11.8A); or reduced capacitance by use of bootstrapping (§8.11.9) or a cascode connection (§8.11.10). The noise contribution of the feedback resistor can be eliminated entirely by replacing it with a *capacitor* (§8.11.11), forming an integrator; differentiating the output then recovers an output proportional to input current.

Capacitor feedback is used in low-level detectors, and in the technique known as *correlated double-sampling*.

¶O. Lock-in Amplifiers.

See §8.14. For the detection of slowly-varying signals it is desirable to minimize added noise by narrowing the measurement bandwidth; and because of typical $1/f$ rising noise at low frequencies, amplification of signals should take place at frequencies above the $1/f$ corner frequency. In the elegant technique of *lock-in detection*, the slowly-varying signal is modulated (typically at a few hundred hertz), amplified, then demodulated with a phase detector, and finally lowpass filtered with a bandwidth appropriate to the original signal; see Figure 8.117.

¶P. Power-supply Noise.

See §8.15. All power supplies are not created equal, as stunningly illustrated by the measured spectra of Figure 8.123. When designing a dc supply, be sure to use a quiet voltage reference, and adhere to good low-noise design practice. Switchmode power supplies are inherently noisy, but attention to current paths and the use of filters can reduce conducted and radiated noise levels. With power supplies of any design, the addition of an outboard *capacitance multiplier* can reduce noise by 40 dB or more (Figure 8.122). See also the extensive treatment of dc power regulation in Chapter 9.

¶Q. Shielding and Grounding.

See §8.16. To paraphrase a cliché,[151] "One man's signal is another man's noise." In the real world, unintentionally coupled signals can easily overwhelm the subtle forms of noise that dominate this chapter. Mitigation techniques include wiring and trace layout (to minimize both capacitive and inductive coupling), raised signal levels, careful attention to grounding (both within and between instruments), shielding, filtering, balanced lines, common-mode chokes, guarding, signal transformers, and isolation amplifiers.

---

[151] "One man's trash is another man's treasure," perhaps deriving from Lucretius' 1st-century BCE epic poem *De Rerum Natura*, where he writes "Ut quod ali cibus est aliis fuat acre venenum" (What is food to one, is to others bitter poison).

VOLTAGE REGULATION AND POWER CONVERSION

CHAPTER **9**

The control and conversion of power – power engineering – is a rich and exciting subfield of electrical engineering and electronic design. It encompasses applications ranging from high-voltage (kilovolts and upward) and high-current (kiloamperes and upward) dc transmission, transportation, and pulsing, all the way down to low-power fixed and portable (battery-operated) and micropower (energy-harvesting) applications. Perhaps of most interest to us in the context of circuit design; it includes the production of the voltages and currents needed in electronic circuit design.

Nearly all electronic circuits, from simple transistor and op-amp circuits up to elaborate digital and microprocessor systems, require one or more sources of stable dc voltage. The simple transformer–bridge–capacitor unregulated power supplies we discussed in Chapter 1 are not generally adequate because their output voltages change with load current and line voltage, and because they have significant amounts of powerline ripple (120 Hz or 100 Hz). Fortunately, it is easy to construct highly stable power supplies, by using negative feedback to compare the dc output voltage with a stable voltage reference. Such regulated supplies are in universal use and can be simply constructed with integrated circuit voltage-regulator chips, requiring only a source of unregulated dc input (from a transformer–rectifier–capacitor combination,[1] a battery, or some other source of dc input) and a few other components.

In this chapter we will see how to construct voltage regulators by using special-purpose integrated circuits. The same circuit techniques can be used to make regulators with discrete components (transistors, resistors, etc.), but because of the availability of inexpensive high-performance regulator chips, there is usually no advantage to using discrete components in new designs. Voltage regulators get us into the domain of high power dissipation,

so we will be talking about heatsinking and techniques like "foldback limiting" to limit transistor operating temperatures and prevent circuit damage. These techniques can be used for all sorts of power circuits, including power amplifiers. With the knowledge of regulators we will have at that point, we will be able to go back and discuss the design of the unregulated supply in some detail. In this chapter we will also look at voltage references and voltage-reference ICs, devices with plenty of uses outside of power-supply design (for example in analog–digital conversion).

We begin with the *linear* regulator, in which feedback controls conduction in a series voltage-dropping "pass transistor" to hold constant the output voltage. Later we treat the important topic of *switching* regulators, in which one or more transistors are switched rapidly to transfer energy, via an inductor (or capacitor) to the load, again with voltage-regulating feedback. In a nutshell, linear regulators are simpler and generate "cleaner" (i.e., noise-free) dc output; switchers (the nickname for switching regulators and converters) are more compact and efficient (Figure 9.1), but noisier and usually more complex.

It would be wrong to leave the impression that voltage regulators are used exclusively in ac-powered dc supplies. In addition to their use in creating stable dc voltages from the ac powerline, voltage regulators are used widely also to produce additional dc voltages from an existing *regulated dc* voltage within a circuit: it's common to see, for example, a regulator that accepts an existing +5 V input and generates a +2.5 V or +3.3 V output; this is easily done with a linear regulator, in which feedback controls the voltage drop to maintain constant (and reduced) output voltage. Perhaps more surprising, you can use a switching regulator to convert a given dc input to a *larger* output voltage, to an output voltage of opposite polarity, or to a constant current (for example, to drive a string of LEDs). These applications are particularly relevant to battery-powered devices. The more general term *power converter* is often used in such applications, which include also creating an ac output from a dc input.

---

[1] Sometimes the transformer can be omitted; this is most commonly done in *switchmode* power supplies (SMPSs), see §9.6.

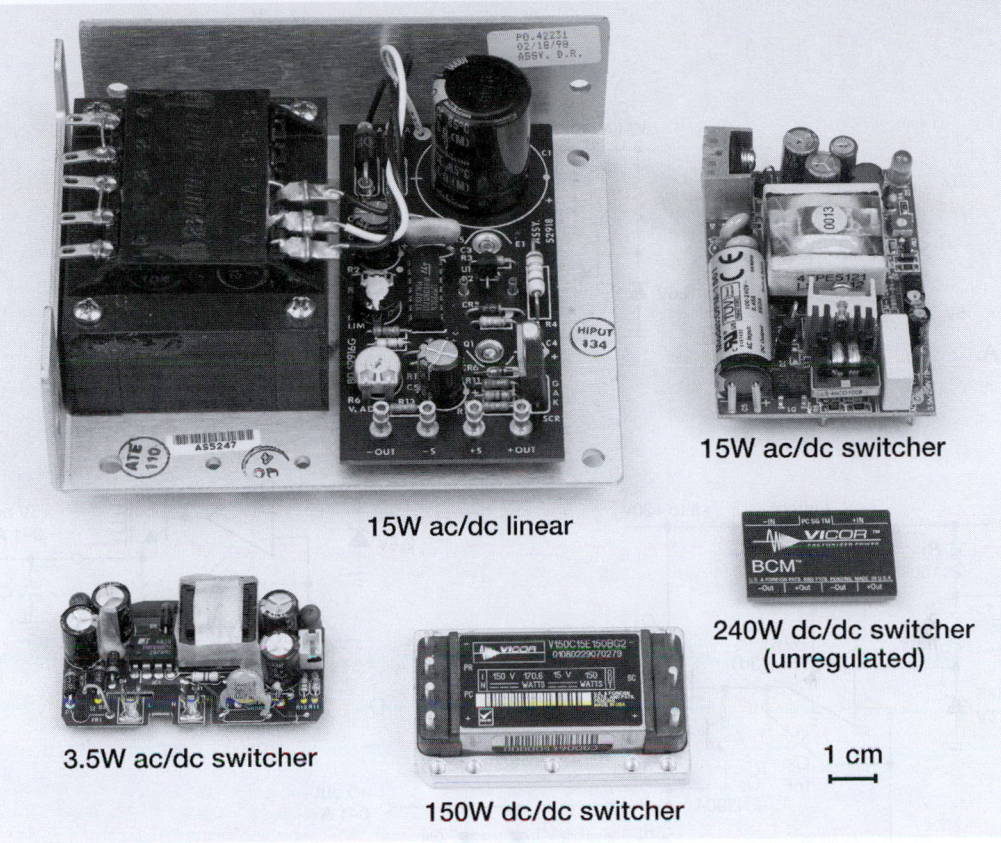

Figure 9.1. Switching power supplies ("switchers") are smaller and more efficient than traditional linear regulated power supplies, but the switching operation generates some unavoidable electrical noise.

9.1 Tutorial: from zener to series-pass linear regulator

To get started, let's look at the circuits in Figure 9.2. Recall that a zener diode is a voltage regulator, of sorts: it draws negligible current until the voltage across it is brought close to its zener voltage V_Z, at which point the current rises abruptly (look back at Figure 1.15 for a reminder). So a zener (or 2-terminal zener-like *reference* IC, see §9.10.2) biased through a resistor from a dc voltage greater than V_Z, as in Figure 9.2A, will have approximately V_Z across it, with the current set by the resistor:[2] $I_{zener} = (V_+ - V_Z)/R$. You can connect a load to this relatively stable output voltage; then, as long as the load draws less than I_{zener} (as just calculated), there will be some remaining zener current, and the output voltage will change little.

The simple resistor-plus-zener is occasionally useful as is, but it has numerous drawbacks: (a) you cannot easily change (or even choose precisely) the output voltage; (b) the zener voltage (which is also the output voltage) changes somewhat with zener current; so it will change with variations in V_+ and with variations in load current;[3] (c) you've got to set the zener current (by choice of R) large enough so that there's still some zener current at maximum load; this means that the V_+ dc supply is running at full current all the time, generating as much heat as the maximum anticipated load; (d) to accommodate large load currents[4] you would need a high-power zener; these are hard to find, and rarely used, precisely because there are much better ways to make a regulator, as we'll see.

Exercise 9.1. Try this out, to get a sense of the problems with

---

[2] With the exact I versus V curve of the zener in hand, you could determine the voltage and current exactly, using the method of *load lines*; see Appendix F, and §3.2.6B.

[3] These are called *line* and *load* variations, respectively.

[4] Or, more precisely, large *variations* in load current, and/or in V_+ dc input voltage.

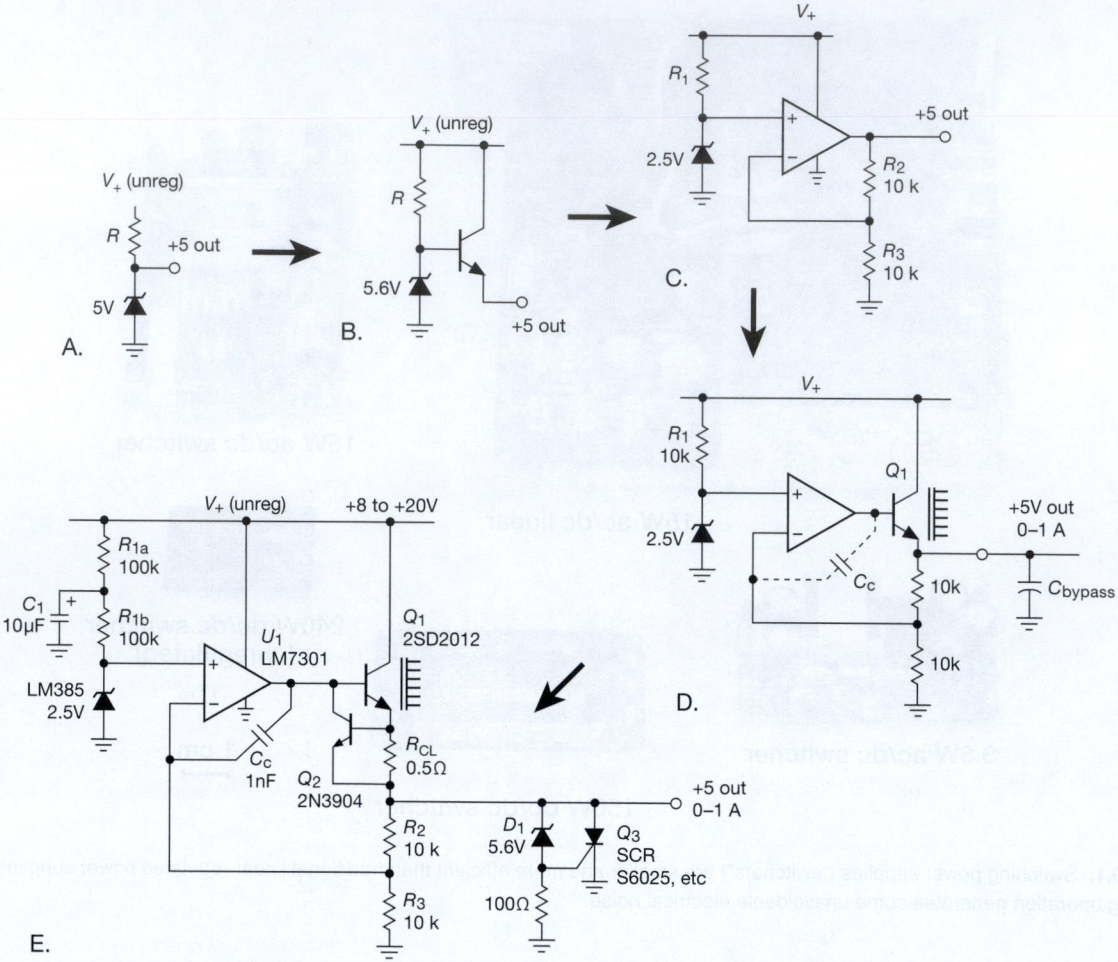

Figure 9.2. Evolving the (discrete-component) series-pass linear voltage regulator.

this simple regulator circuit: imagine we want a stable +5 V dc output, to power a load that can draw from zero to 1 A. We've built an unregulated dc supply (using a transformer, diode bridge, and capacitor) that puts out approximately +12 V when unloaded, dropping to +9 V at 1 A load. Those voltages are "nominal" and can vary ±10%.

(a) What is the correct resistor value, R, for the circuit of Figure 9.2A, such that the minimum zener current, under "worst-case" conditions, is 50 mA?

(b) What is worst-case (maximum) power dissipation in R and in the zener?

Contrasted with this approach – with its requirement for a 10 W zener at the desired output voltage, and nearly 10 W of power dissipation in each component, even at zero load – we'll see that it is a routine task to make a regulated power supply, with adjustable output voltage, without the need for

a power zener and with 75% or better efficiency over most of the load-current range.

9.1.1 Adding feedback

We could improve the situation somewhat by tacking an emitter follower onto a zener (Figure 9.2B); that lets you run at lower zener current, and low quiescent dissipation when unloaded. But the output regulation is still poor (because V_{BE} varies with output current), and the circuit still does not allow adjustment of output voltage.

The solution is to use a zener (or other voltage reference device; see §9.10.2) as a low-current voltage reference, against which we compare the output. Let's take it in a few easy steps.

A. Zener plus "amplifier"

First we solve the problem of *adjustability* by following the zener reference with a simple dc amplifier (Figure 9.2C). Now the zener current can be small, just enough to ensure a stable reference. For typical zeners this might be a few milliamps, whereas for an IC voltage reference, 0.1–1 mA will usually suffice. This circuit lets you adjust the output voltage: $V_{out} = V_Z(1 + R_2/R_3)$. But note that you are limited to having $V_{out} \geq V_Z$; note also that the output voltage comes from an op-amp, so it can at most reach V_+, with an output current limited by the op-amp's $I_{out}(max)$, typically 20 mA. We will overcome both these limits.

B. Adding outboard pass transistor

More output current is easy – just add an *npn* follower, to boost the output current by a factor of β. You might be tempted to just hang the follower on the op-amp's output, but that would be a mistake: the output voltage would be down by a V_{BE} drop, roughly 0.6 V. You could, of course, adjust the ratio R_2/R_3 to compensate. But the V_{BE} drop is imprecise, varying both with temperature and with load current, and so the output voltage would vary accordingly. The better way is to close the feedback loop around the pass transistor, as in Figure 9.2D; that way the error amplifier sees the actual output voltage, holding it stable via the circuit's loop gain. The inclusion of the output emitter follower boosts the op-amp's $I_{out}(max)$ by the beta of Q_1, giving us an available output current of an ampere or so. (We could use a Darlington, instead, for more current; another possibility is an *n*-channel MOSFET.) Q_1 will be dissipating 5–10 W at maximum output current, so you'll need a heatsink (more on this in §9.4.1). And, as we'll see next, you'll also need to add a compensation capacitor C_C to ensure stability.

C. Some important additions

Our voltage regulator circuit is nearly complete, but lacks a couple of essential features, related to loop stability and overcurrent protection.

Feedback loop stability

Regulated power supplies are used to power electronic circuits, typically festooned with many bypass capacitors between the dc rails and ground. (Those bypass capacitors, of course, are needed to maintain a pleasantly low impedance at all signal frequencies.) Thus the dc supply sees a large capacitive load, which, when combined with the finite output resistance of the pass transistor (and overcurrent sense resistor, if present), causes a lagging phase shift and possible oscillation. We've shown the load capacitance in Figure 9.2D as C_{bypass}, a portion of which might be included explicitly (as a real capacitor) in the power supply itself.

The solution here, as with the op-amp circuits we worried about earlier (§4.9), is to include some form of *frequency compensation*. That is most simply done (as it is within op-amps) with a Miller feedback capacitor C_C around the inverting gain stage, as shown. Typical values are 100–1000 pF, usually found experimentally ("cut-and-try") by increasing C_C until the output shows a well-damped response to a step change in load (and then doubling that, to provide a good margin of stability). The IC regulators we'll see later will either include internal compensation, or they'll give you suggested values for compensation components.

Overcurrent protection

The circuit as drawn in Figure 9.2D does not deal well with a short-circuit load condition.[5] With the output shorted to ground, feedback will act to force the op-amp's maximum output current into the pass transistor's base; so that I_B of 20–40 mA will be multiplied by Q_1's beta (which might range from 50 to 250, say), to produce an output current of 1 A to 10 A. Assuming the unregulated V_+ input can supply it, such a current will cause excessive heating in the pass transistor, as well as interesting forms of damage to the misbehaving load.

The solution is to include some form of overcurrent protection, most simply the classic current-limiting circuit consisting of Q_2 and R_{CL} in Figure 9.2E. Here R_{CL} is a low-value *sense resistor*, chosen to drop approximately 0.6 V (a V_{BE} diode drop) at a current somewhat larger than the maximum rated current; for example, we might choose $R_{CL} = 5\,\Omega$ in a 100 mA supply. The drop across R_{CL} is applied across Q_2's base–emitter, turning it on at the desired maximum output current; Q_2's conduction robs base current from Q_1, preventing further increase of output current. Note that the current-limit sense transistor Q_2 does not handle high voltage, high current, or high power; it sees at most two diode drops from collector to emitter, the op-amp's maximum output current, and the product of those two, respectively. During an overcurrent load condition, then, it typically would have to handle $V_{CE} \leq 1.5$ V at $I_C \leq 40$ mA, or 60 mW; that's peanuts for any general-purpose small-signal transistor.

Later we'll see variations on this simple overcurrent protection theme, including methods that limit to an

---

[5] Engineers like to refer to various bad situations such as this under the general rubric of *fault conditions*.

adjustable and stable current limit, and the technique known as *foldback current limiting* (§9.13.3).

Zener bias; overvoltage crowbar

We've shown two additional wrinkles in Figure 9.2E. First, we split the zener biasing resistor R_1 and bypassed the midpoint, to filter out ripple current. By choosing the time constant ($\tau = (R_{1a}\|R_{1b})C_1$) to be long compared with the ripple period of 8.3 ms, the zener sees ripple-free bias current. (You wouldn't bother with this if the dc supply V_+ were already free of ripple, for example a regulated dc supply of higher voltage.) Alternatively, you could use a current source to bias the zener.

Second, we've shown an "overvoltage crowbar" protection circuit consisting of D_1, Q_3, and the $100\,\Omega$ resistor. Its function is to short the output if some circuit fault causes the output voltage to exceed about 6.2 V (this can happen easily enough, for example if the pass transistor Q_1 fails by having a collector-to-emitter short, or if a humble component like resistor R_2 becomes open-circuited.). Q_3 is an SCR (silicon-controlled rectifier), a device that is normally nonconducting but that goes into saturation when the gate–cathode junction is forward-biased. Once turned on, it will not turn off again until anode current is removed externally. In this case, gate current flows when the output exceeds D_1's zener voltage plus a diode drop. When that happens, the regulator will go into a current-limiting condition, with the output held near ground by the SCR. If the failure that produces the abnormally high output also disables the current-limiting circuit (e.g., a collector-to-emitter short in Q_1), then the crowbar will sink a very large current. For this reason it is a good idea to include a fuse somewhere in the power supply, as shown for example in Figure 9.48. We will treat overvoltage crowbar circuits in more detail in §§9.13.1 and *9x.7*.

Exercise 9.2. Explain how an open circuit at R_2 causes the output to soar. What voltage, approximately, would then appear at the output?

9.2 Basic linear regulator circuits with the classic 723

In the preceding tutorial we evolved the basic form of the linear *series pass regulator*: voltage reference, pass transistor, error amplifier, and provisions for loop stability and overvoltage–overcurrent protection. In practice you seldom need to assemble these components from scratch – they are available as complete integrated circuits. One broad class of IC linear regulators might be thought of as flexible *kits*

– they contain all the pieces, but you've got to hook up a few external components (including the pass transistor) to make them work; an example is the classic 723 regulator. The other class of regulator ICs are complete, with built-in pass transistor and overload protection, and requiring at most one or two external parts; an example is the classic 78L05 "3-terminal" regulator – its three terminals are labeled *input*, *output*, and *ground* (and that's how easy it is to use!).

9.2.1 The 723 regulator

The μA723 voltage regulator is a classic. Designed by Bob Widlar and first introduced in 1967, it is a flexible, easy-to-use regulator with excellent performance.[6] Although you might not choose it for a new design nowadays, it is worth looking at in some detail, because more recent regulators work on the same principles. Its block diagram is shown in Figure 9.3. As you can see, it is really a power-supply *kit*, containing a temperature-compensated voltage reference (7.15 V$\pm$5%), differential amplifier, series pass transistor, and current-limiting protective circuit. As it comes, the 723 doesn't regulate anything. You have to hook up an external circuit to make it do what you want.

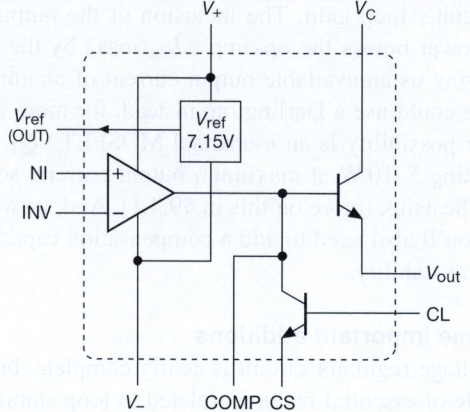

Figure 9.3. The classic μA723 voltage regulator.

The 723's internal *npn* pass transistor is limited to

---

[6] Building on the 723's success, other manufacturers introduced "improved" versions, such as the LAS1000, LAS1100, SG3532, and MC1469. However, while the 723 lives on, the improved versions are all gone! The 723 is "good enough," *very* inexpensive (about $0.15 in quantity), and is popular in many commercial linear power supplies, where the easily adjusted current limit is especially useful. It also has less noise than most modern replacements. And we like it for its pedagogic value.

150 mA, and it can dissipate about 0.5 W maximum. Unlike newer regulators, the 723 does not incorporate internal shutdown circuitry to protect against excessive load current or chip dissipation.

A. 723 regulator example: $V_{out} > V_{ref}$

Figure 9.4 shows how to make a positive voltage regulator with the 723 for output voltages greater than the reference voltage; it is the same circuit topology as the tutorial's Figure 9.2E. All the components except the three resistors and the two capacitors are contained in the 723. With this circuit, a regulated supply with output voltage ranging from V_{ref} up to the maximum allowable output voltage (37 V) can be made. Of course, the input voltage must stay a few volts more positive than the output at all times, including the effects of ripple on the unregulated supply. The "dropout voltage" (the amount by which the input voltage must exceed the regulated output voltage) is specified as 3 volts (minimum) for the 723. This is a bit large by contemporary standards, where the dropout voltage is typically 2 V, and much less for *low dropout* (LDO) regulators, as we'll see in §9.3.6. Note also that the 723's relatively high reference voltage means that you cannot use it in a power supply whose unregulated dc input is less than +9.5 V, its specified minimum V_+; this shortcoming is remedied in a large selection of regulators that use a lower-voltage *bandgap reference* (1.25 V or 2.5 V). And while we're complaining, we note that the reference is not exactly sterling in its initial accuracy – the production spread in V_{ref} is 6.8 to 7.5 volts – which means that you must provide for output-voltage trimming, by making R_1 or R_2 adjustable; we'll soon see regulators with excellent initial accuracy, for which no trim is needed.

It is usually a good idea to put a capacitor of a few microfarads across the output, as shown. This keeps the output impedance low even at high frequencies, where feedback becomes less effective. It is best to use the output capacitor value recommended on the specification sheet, to ensure stability against oscillations. In general, it is a good idea to bypass power-supply leads to ground liberally throughout a circuit, using a combination of ceramic types (0.01–0.1 μF) and electrolytic or tantalum types (1–10 μF).[7]

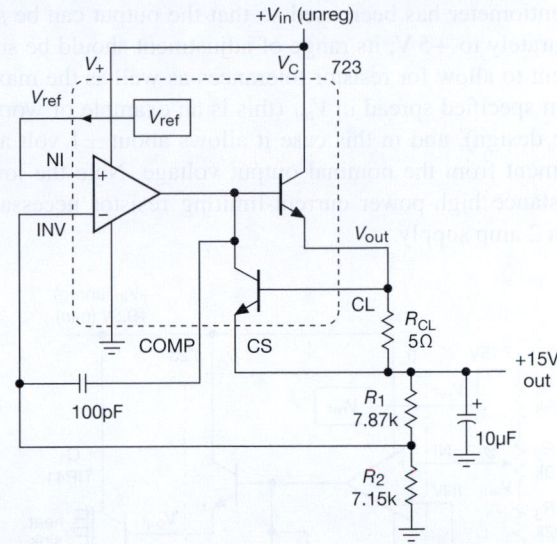

Figure 9.4. 723 regulator: configuration for $V_{out} > V_{ref}$, with 100 mA current limit.

B. 723 regulator example: $V_{out} < V_{ref}$

For output voltages less than V_{ref}, you just put the voltage divider on the reference (Figure 9.5). Now the full output voltage is compared with a fraction of the reference. The values shown are for a +5 V output. With this circuit configuration, output voltages from +2 V to V_{ref} can be produced. The output cannot be adjusted down to zero volts because the differential amplifier will not operate below 2 volts input, as specified on the datasheet. Note again that the unregulated input voltage must never drop below +9.5 V, the voltage necessary to power the reference.

For this example we've added an external pass transistor, in a Darlington configuration with the 723's small internal pass transistor, to get beyond the latter's 150 mA current limit. An external transistor is needed, also, because of power dissipation: the 723 is rated at 1 watt at 25°C (less at higher ambient temperatures; the 723 must be "derated" at 8.3 mW/°C above 25°C in order to keep the junction temperature within safe limits). Thus, for instance, a 5 volt regulator with +15 V input cannot deliver more than about 80 mA to the load. Here the external power transistor Q_1 will dissipate 14 W for $V_{in} = 12$ V and maximum load current (2 A); that requires a *heatsink*, most often a finned metal plate designed to carry off heat (alternatively, the transistor can be mounted to one wall of the metal chassis housing the power supply). We will deal with thermal problems like these later in the chapter.[8] A trimmer

---

[7] The ceramic capacitors provide low impedance at high frequencies, whereas the larger electrolytics provide energy storage, and also damping of oscillations (via their internal equivalent series resistance, or ESR).

[8] And for a table of bipolar power transistors see Table 2.2 on page 106.

potentiometer has been used so that the output can be set accurately to $+5\,V$; its range of adjustment should be sufficient to allow for resistor tolerances as well as the maximum specified spread in V_{ref} (this is an example of worst-case design), and in this case it allows about ± 1 volt adjustment from the nominal output voltage. Note the low-resistance high-power current-limiting resistor necessary for a 2 amp supply.

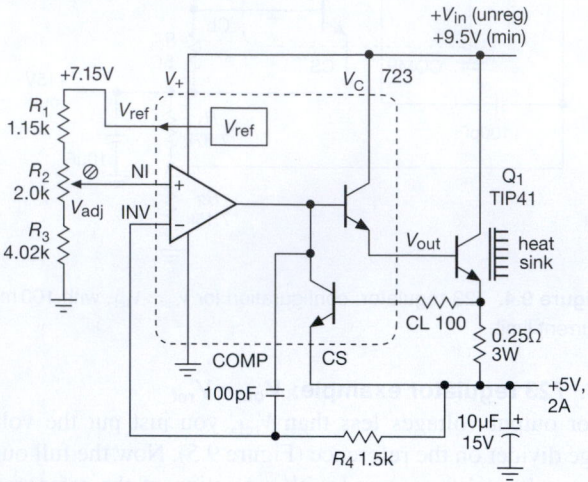

Figure 9.5. 723 regulator: configuration for $V_{\mathrm{out}} < V_{\mathrm{ref}}$, with 2 A current limit.

A third variation of this circuit is necessary if you want a regulator that is continuously adjustable through a range of output voltages around V_{ref}. In such cases, just compare a divided fraction of the output with a fraction of V_{ref} chosen to be less than the minimum output voltage desired.

Exercise 9.3. Design a regulator to deliver up to 50 mA load current over an output voltage range of $+5\,V$ to $+10\,V$ using a 723. *Hint*: compare a fraction of the output voltage with $0.5V_{\mathrm{ref}}$.

C. Pass-transistor dropout voltage

One problem with this circuit is the high power dissipation in the pass transistor (at least 10 W at full load current). This is unavoidable if the regulator chip is powered by the unregulated input, since it needs a few volts of "headroom" to operate (specified by the dropout voltage). With the use of a separate low-current supply for the 723 (e.g., $+12\,V$), the minimum unregulated input to the external pass transistor can be as little as 1.5 V or so above the regulated output voltage (i.e., two V_{BE}'s).[9]

9.2.2 In defense of the beleaguered 723

Lest we leave the wrong impression, we hasten to remark that rumors of the death of the vintage 723 regulator are greatly exaggerated.[10] We have been using dozens of linear regulated power supplies manufactured by Power One for more than three decades without a single failure. All of them use the humble 723 regulator chip, as do other OEMs ("original equipment manufacturers"). Here are some reasons not to overlook this remarkable design of the legendary Bob Widlar:

- very low cost, $0.17 (in qty 1000)
- many, many manufacturers
- fully settable current limit, including foldback
- good for pedagogy (that's why it's here!)
- the power dissipation is not in the control IC
- quiet voltage reference, plus can add filter
- works with *npn* or *pnp* pass transistors
- easily configured for negative outputs

9.3 Fully integrated linear regulators

The overall regulator circuit of Figure 9.5 has ten components, but only three terminals (IN, OUT, and GROUND), thus suggesting the possibility of an integrated solution, with on-chip voltage-setting resistors and with integrated components for current-limiting and loop compensation – a *3-terminal* regulator. The 723 is approaching half-century vintage (though still going strong!), during which the semiconductor industry has not been sleeping: contemporary linear regulator ICs generally integrate all regulator functions on-chip, including overcurrent and thermal protection, loop compensation, high-current pass transistor, and preset voltage divider for commonly used output voltages. Most of these regulators come also in adjustable versions, for which you provide only the voltage-setting resistor pair. And, with an additional terminal or two, you can get a "shutdown" control input and a "power-good" status output. Finally, a large and growing population of low-dropout regulators addresses low voltage applications, of increasing importance in low-power and portable electronic devices. Let's look at the choices favored for contemporary design.

---

[9] A trick you can use to reduce the minimum headroom to a single V_{BE} is to replace Q_1 with a *pnp* pass transistor (tie its emitter to V_{in} and

drive its base from the V_{c} pin of the 723), forming a Sziklai pair rather than a Darlington (see §2.4.2A and Figure 2.77). If the input comes from an unregulated dc supply, however, you will always have to allow at least a few volts of headroom, because worst-case design dictates proper operation even at 105 Vac line input.

[10] Paraphrasing Mark Twain's famous remark, upon opening the newspaper and reading his obituary.

9.3.1 Taxonomy of linear regulator ICs

As a guide to the following sections, we've organized the universe of integrated linear voltage regulators into a few distinct categories, here simply listed in outline form. For each category we've listed typical example part numbers of devices that we are fond of, and use often. Read on for explanations of when and how to use them, a description of their distinguishing features, and some important cautions.

3-terminal fixed

    pos: 78xx
    neg: 79xx

3-terminal adjustable

    pos: LM317
    neg: LM337

3-term "lower dropout" (adj & fixed)

    pos: LM1117, LT1083-85

3-term fixed & 4-term adj "true LDO"

    pos: LT1764A/LT1963 (BJT); TPS744xx (CMOS)
    neg: LT1175, LM2991 (BJT); TPS7A3xxx (CMOS)

3-term current reference

    pos: LT3080

9.3.2 Three-terminal fixed regulators

The original (and often good enough) 3-terminal regulator is the 78xx series (Figure 9.6), originated by Fairchild in the early 1970s.[11] It is factory trimmed to provide a fixed output, in which the voltage is specified by the last two digits of the part number, and can be any of the following: 05, 06, 08, 09, 10, 12, 15, 18, or 24. These regulators can supply up to 1 A of output current, and come in power packages (TO-220, DPAK, D^{2}PAK) that you attach to a heatsink or to an area of circuit-board copper. If you don't need much current, use the 78Lxx/LM340Lxx series, which come in small transistor packages, either surface-mount or TO-92 (through-hole). For negative output voltage use the 79xx/79Lxx (or LM320/320L) series. Figures 9.6 and 9.7 show, in simplified form, what's inside these inexpensive ($0.30) regulators.

Figure 9.8 shows how easy it is to make a +5 V regulator, for example, with one of these ICs. Here we've added also a 7905 negative regulator to create a −5 V regulated output from a more negative unregulated dc input. The bypass capacitors at the outputs ensure stability; they

---

[11] The LM340 series from National is essentially the same.

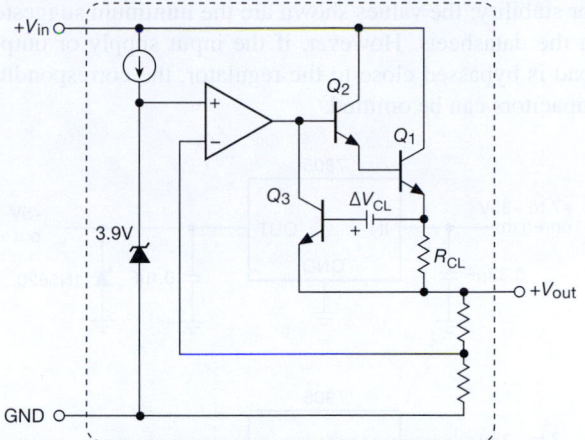

Figure 9.6. Simplified 78xx fixed 3-terminal positive voltage regulator. All components are internal, so only a pair of bypass capacitors is required (as in Figure 9.8). R_{CL}, the current-sensing resistor, is $0.2\,\Omega$, and develops somewhat less than a diode drop at full current; its drop is supplemented by an internal bias ΔV_{CL}, to turn on current-limiting transistor Q_3.

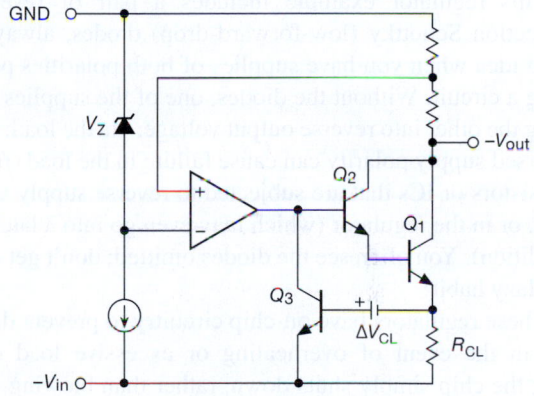

Figure 9.7. 79xx fixed 3-terminal negative voltage regulator.

also improve transient response, and maintain a low output impedance at high frequencies (where the regulator's loop gain is low).[12] The input bypass capacitors are also needed

---

[12] A regulator's datasheet will always specify minimum required capacitance. It may go into considerably more detail, in cases where stability is an important issue, for example with low-dropout regulators (see later discussion). Note the larger capacitance values in the negative regulator circuit: they are needed to ensure stability, because the 7905 regulator's output comes from the collector of a common-emitter amplifier output stage (whose gain depends on load impedance), rather than from the emitter follower output stage of the 7805 positive regulator (whose gain is near unity); the larger bypass capacitor kills the loop gain at high frequency, preventing oscillation.

for stability; the values shown are the minimum suggested in the datasheets. However, if the input supply or output load is bypassed close to the regulator, the corresponding capacitors can be omitted.

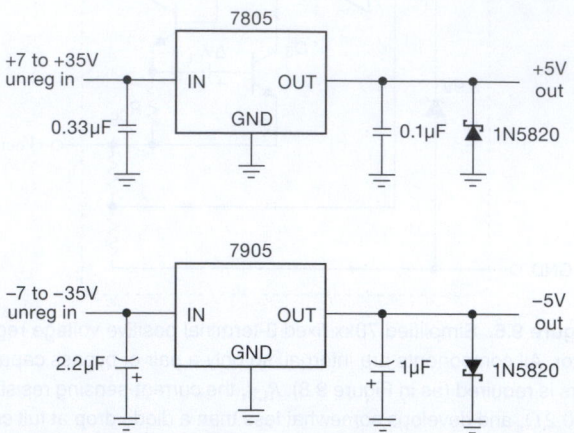

Figure 9.8. ±5 V regulated dc from a 7805/7905 regulator pair.

This regulator example includes a pair of reverse-protection Schottky (low-forward-drop) diodes, always a good idea when you have supplies of both polarities powering a circuit. Without the diodes, one of the supplies can bring the other into reverse output voltage, via the load; this reversed supply polarity can cause failure in the load (from transistors or ICs that are subjected to reverse supply voltage), or in the regulator (which may even go into a latchup condition). You often see the diodes omitted; don't get into this lazy habit!

These regulators have on-chip circuitry to prevent damage in the event of overheating or excessive load current; the chip simply shuts down, rather than blowing out. In addition, on-chip circuitry prevents operation outside the transistor safe operating area (see §9.4.2) by reducing available output current for large input–output voltage differentials. These regulators are inexpensive and easy to use, and they make it practical to design a system with many printed circuit boards (PCBs) in which the unregulated dc is brought to each board and regulation is done locally on each circuit card. Table 9.1 lists the characteristics of a representative selection of 3-terminal fixed regulators.

Three-terminal fixed regulators come in some highly useful variants. There are low-power and micropower versions (e.g., the LM2936 and LM2950, with quiescent current in the microampere range), and there are the very popular LDO regulators, which maintain regulation with only a few tenths of a volt input–output differential (e.g., the LT1764A, TPS755xx, and micropower LM2936, with typ-

ical dropout voltages ≈0.25 V). We'll discuss LDOs after taking a look at the very useful 3-terminal *adjustable* regulator.

Table 9.1 7800-Style Fixed Regulators[a]

Part #[c]	V_{in} max (V)	V_{out}[d] nom (V)	Tol (±%)	I_Q typ (mA)	I_{out} max (A)	Cost qty 25 ($US)
78L05	35	5	5	3	0.1	0.29
78L15	35	15	4	3	0.1	0.31
7805	35	5	4[b]	5[e]	1.0	0.47
7824	40	24	4[b]	5	1.0	0.49
79L05	−35	−5	5	2	0.1	0.30
79L15	−35	−15	4	2	0.1	0.30
7905	−35	−5	4[b]	3	1.0	0.47
7924	−40	−24	4[b]	4	1.0	0.56

Notes: (a) often called '7800 and '7900 series, e.g., "LM7800-series." L series available in TO-92, SO-8 and SOT-89 packages; regular series available in TO-220, DPAK, D2PAK, and TO-3. Some use buried-zener ref, some use bandgap. (b) A-suffix types are ±2% tol. (c) prefixes: uA, LM, MC, KA, NCP, L, NJM, etc. (d) L-series: 2.6 to 24V, regular: 5 to 24V. (e) some lower, 3.3mA to 4mA

9.3.3 Three-terminal adjustable regulators

Sometimes you want a nonstandard regulated voltage (say +9 V, to emulate a battery) and can't use a 78xx-type fixed regulator. Or perhaps you want a standard voltage, but set more accurately than the ±3% accuracy typical of fixed regulators. By now you're spoiled by the simplicity of 3-terminal fixed regulators, and therefore you can't imagine using a 723-type regulator circuit, with all its required external components. What to do? Get an "adjustable 3-terminal regulator"!

These convenient ICs are typified by the classic LM317 originally from National (Figure 9.9). This regulator has no ground terminal; instead, it adjusts V_{out} to maintain a constant 1.25 V (internal "bandgap" reference, §9.10.2) from the output terminal to the "adjustment" terminal. Figure 9.10 shows the easiest way to use it. The regulator puts 1.25 V across R_1, so 10 mA flows through it. The adjustment terminal draws very little current (50–100 μA), so the output voltage is just

$$V_{out} = 1.25\,(1 + R_2/R_1)\ \text{volts.}$$

In this case the output voltage is +3.3 V, with an untrimmed accuracy of ≈3% (from the ±2% internal 1.25 V reference and the 1% resistors). If you want accurate settability,

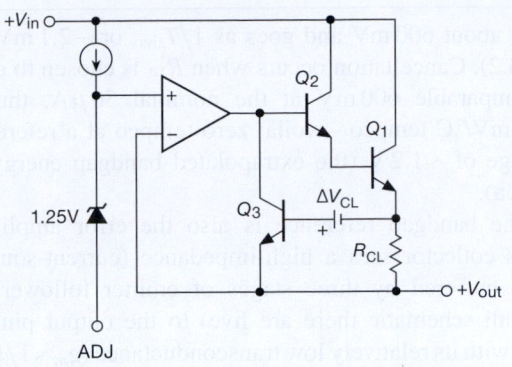

Figure 9.9. LM317 three-terminal adjustable positive voltage regulator.

replace the lower resistor with a 25 Ω trimmer in series with a 191 Ω fixed resistor, to narrow the trimmer's adjustment range to ±6%. If you want instead a wide adjustment range, you could replace the lower resistor with a 2.5k trimmer, for an output range of +1.25 V to +20 V. Whatever the output voltage, the input must be at least 2 V higher (the dropout voltage).

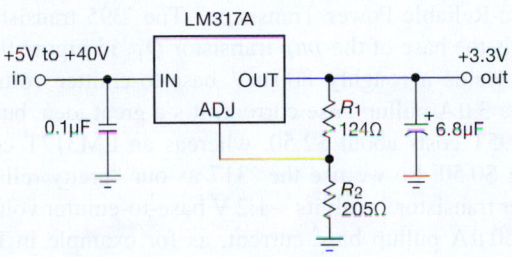

Figure 9.10. +3.3 V positive regulator circuit.

When using this type of regulator, choose your resistive divider values small enough to allow for a 5 μA change in adjustment pin current with temperature: many designers use 124 Ω for the upper resistor, as we've done, so that the divider alone sinks the chip's specified minimum load current of 10 mA. Note also that the current sourced out of the adjustment pin may be as large as 100 μA (the worst-case spec). The output capacitor, though not necessary for stability, greatly improves transient response. It's a good idea to use at least 1 μF, and ideally something more like 6.8 μF.

The LM317 is available in many package styles, including the plastic power package (TO-220), the surface-mount power package (DPAK and D²PAK), and many small transistor packages (both through-hole TO-92, and a half dozen tiny surface-mount styles). In the power packages it can deliver up to 1.5 amps, with proper heatsinking; the low-power variant (317L) is rated to 100 mA, again limited by

power dissipation. The popular LM1117 variant, also available from multiple manufacturers, improves on the dropout voltage of the classic 317 (1.2 V versus 2.5 V), but you pay a price (literally): in the TO-220 package it costs about $0.75 versus the 317's $0.20; it also has a more limited voltage range (see Table 9.2), and, in common with many low-dropout regulators, it requires a larger output capacitor (10 μF minimum).

Exercise 9.4. Design a +5 V regulator with the 317. Provide ±20% voltage adjustment range with a trimmer pot.

Three-terminal adjustable regulators are available with higher current ratings, e.g., the LM350 (3A), the LM338 (5A), and the LM396 (10A), and also with higher voltage ratings, e.g., the LM317HV (60V) and the TL783 (125V). We've listed their properties in Table 9.2 on page 605. Read the datasheets carefully before using these parts, noting bypass capacitor requirements and safety diode suggestions. Note also that the rated maximum output currents generally apply at lower values of $V_{in}-V_{out}$, and can drop to as little as 20% of their maximum values as $V_{in}-V_{out}$ approaches $V_{in(max)}$; the maximum output current drops also with increasing temperature.[13]

An alternative for high load currents is to add an outboard transistor (§9.13.4), though a high-current switching regulator (§9.6) is often a better choice. The LM317 family regulators are "conventional" (as opposed to low-dropout) linear regulators; typical dropout voltages are ≈2 V.

As with the fixed 3-terminal regulators, you can get lower-dropout versions (e.g., the popular LM1117, with 1.3 V maximum dropout at 0.8 A, or the heftier LT1083-85 series, with comparable dropout at currents to 7.5 A), and you can get micropower versions (e.g., the LP2951, the adjustable variant of the fixed 5V LP2950; both have $I_Q=75$ μA); see Figure 9.11. You can also get *negative* versions, though there's less variety: the LM337 (Figure 9.12) is the negative cousin of the LM317 (1.5 A), and the LM333 is a negative LM350 (3 A). There's more discussion ahead in §§9.3.6 and 9.3.9; see particularly Figure 9.24.

---

[13] As discussed later in §9.4.1, the junction temperature $T_J=P_{diss}(R_{\Theta JC}+R_{\Theta CS}+R_{\Theta SA})+T_A$, where the R_Θ are the thermal resistances from junction to case, case to heatsink, and heatsink to ambient. In situations with good heatsinking, you may choose to use a regulator of higher current rating and larger package style (e.g., the LM338K in its TO-3 metal can package) in order to take advantage of the much lower thermal resistance $R_{\Theta JC}$ (1°C/W versus 4°C/W for the LM317T in its TO-220 package). The larger parts also offer more relaxed safe-operating-area (SOA) constraints, e.g., at $V_{in}-V_{out}=20$ V the LM338 allows 3.5 A of output current, versus 1.4 A for the LM317.

Anatomy of a 317

The classic LM317, designed in 1976 by the legendary team of Widlar and Dobkin[14] has endured for more than four decades. Indeed, the generic 317 (along with the complementary LM337) has become the go-to part for linear regulators of modest current capability (to ~1 A) in situations where you've got a few volts of headroom. And it has spawned a host of imitators and look-alikes, spanning a range of voltages, currents, and package styles, with some variants of lower dropout-voltage; see Table 9.2.

Its design exhibits a nice elegance, for example by combining the functions of error amplifier and zero-tempco bandgap reference. It was also one of the first regulators to include thermal overload and safe-area protection. Figure 9.13 is a simplified circuit of its essential innards, with part designations following the schematic diagram in the National Semiconductor (TI) datasheet.

The transistor pair Q_{17} and Q_{19} forms the bandgap voltage reference, operating at equal currents from the $Q_{16}Q_{18}$ mirror. Because Q_{19} has 10× larger emitter area (or 10 emitters), it operates at 1/10 the current density of Q_{17}, thus a V_{BE} that is smaller by $(kT/q)\log_e 10$, about 60 mV (§2x.3.2). That sets its current (via R_{15}) to be $I_{Q19}=\Delta V_{BE}/R_{15}=25\,\mu A$, and thus the pair's total current is 50 μA.[15] Note that the current has a linear dependence on absolute temperature (because the drop across R_{15} is $\propto T_{abs}$) – it is "PTAT" (proportional to absolute temperature).

Now for the classic "bandgap reference" temperature compensation: the positive tempco of current is exploited to cancel Q_{17}'s negative tempco of V_{BE}, which is nomi-

nally about 600 mV and goes as $1/T_{abs}$, or −2.1 mV/°C (§2.3.2). Cancellation occurs when R_{14} is chosen to drop a comparable 600 mV at the nominal 50 μA, thus a +2.1 mV/°C tempco – voilà: zero tempco at a reference voltage of ~1.2 V (the extrapolated bandgap energy of silicon).

The bandgap reference is also the error amplifier: Q_{17}'s collector sees a high-impedance (current-source) load, buffered by three stages of emitter follower (in the full schematic there are five) to the output pin; so even with its relatively low transconductance ($g_m\sim 1/R_{14}$) there's plenty of loop gain in the error amplifier (whose input is the ADJ pin, offset by V_{ref}, relative to V_{out}).

Resistor R_{26} senses the output current, for current limiting via Q_{21}. A bias that depends on $V_{in}-V_{out}$ is added (the battery symbol), for safe-operating-area protection. Additional components add hysteretic overtemperature shutdown (Q_{21} is paired with a *pnp* to make a latch). One final note: the Widlar–Dobkin duo also created the LM395 and LP395 protected-transistor ICs; these include the current and thermal limiting from the 317, but without the bandgap reference. They call it, modestly, the "Ultra-Reliable Power Transistor." The '395 transistor's base is the base of the *pnp* transistor Q_{15} in Figure 9.13. This yields a roughly 800 mV base-to-emitter voltage, with a 3 μA pullup base current. It's a great idea, but an LM395T costs about $2.50, whereas an LM317T costs about $0.50. So we use the '317 as our "pretty reliable power transistor," with its −1.2 V base-to-emitter voltage and 50 μA pullup base current, as for example in Figures 9.16 and 9.18.

9.3.4 317-style regulator: application hints

The LM317-style adjustable 3-terminal regulators are delightfully easy to use, and there are some nice tricks you can use to make them do more than simply create a fixed dc output voltage. There are also some basic cautions to keep in mind. In Figure 9.14 we've sketched some helpful circuit ideas.

Herewith a quick tour (keyed to the figure parts), taking them in order.

A: The regulator requires some minimum load current, because the operating current for the internal circuitry returns through the load. So if you want it to work clear down to zero external load, you should choose the upper feedback resistor R_1 small enough, i.e., so that $V_{ref}/R_1 \geq I_{out(min)}$ for the worst-case (maximum) value of $I_{out(min)}$. For $V_{ref}=1.25$ V and the classic LM317's $I_{out(min)}=10$ mA, R_1 should be no larger than 125 Ω.[16] Of course, you could instead use a larger value of R_1 and add a load resistor to

---

[14] See Robert Widlar, "New developments in voltage regulators," JSSC, SC-6, pp 2–9, 1971, and US patent 3,617,859: "Electrical regulator apparatus including a zero temperature coefficient voltage reference circuit," filed 23 March 1970, issued 2 November 1971.

[15] The typical tolerance on resistor values in the planar silicon process (which is good for resistor *ratios*, but not absolute values) is ×0.5 to

×2, so the nominal 50 μA current sourced out of the ADJ pin can actually range from 25 μA to 100 μA.

[16] In contradiction to the LM117/317 datasheet's many circuit examples,

Table 9.2. 3-Terminal Adjustable Voltage Regulators ("LM317-style")[a]

Packages[z]						V_{in} max (V)	I_{out}[v] max (A)	V_{DO}[h] max (V)	I_{out} min[b] (mA)	C_{out} min (µF)	V_{ref} (V)	± (%)	I_{adj} typ (µA)	Temp stab[c] typ (%)	Ripple reject 120Hz typ (dB)	Line typ (%)	Load[e] typ (%)	Cost qty 25 ($US)	Comments
TO-92	SOIC	TO-220	TO-3, TO-3P	D-PAK	SOT-223														Part #
Positive																			
•	•	-	-	-	-	40	0.1	2.5[t]	5	0.1	1.25	4	50	0.5	80[g]	0.15	0.1	0.34	LM317L TO-92 lo-power '317
-	•	-	-	•	•	20	0.8	1.2	5	10	1.25	1	52	0.5	73	0.035	0.2	0.88	LM1117[n] low V_{DO} '317, popular
-	-	-	-	•	•	20	1.0	1.2	5	10	1.25	1	52	0.5	73	0.04	0.2	0.40	NCP1117 higher current '1117
-	-	-	-	d	•	20	1.0	1.2	5	10	1.25	1	60	0.5	75	0.035	0.2	0.92	LMS8117A higher current '1117
-	-	•	•	•	•	40	1.5[p]	2.5[t]	10	0.1	1.25	4	50	0.6	80[g]	0.01	0.1	0.15	LM317[k] orig, cheap, popular
-	-	•	-	•	-	30	1.5	1.5	10	22[u]	1.25	1	55	0.5	75	0.02	0.1	2.67	LT1086CP low dropout
-	-	•	•	-	-	35	3	2.5[t]	10	1	1.25	4	50	0.6	65	0.1	0.1	0.49	LM350T 3A monolithic
-	-	•	-	•	-	30	3	1.5	10	22[u]	1.25	1	55	0.5	75	0.02	0.1	4.50	LT1085CT 3A low dropout
-	-	•	-	•	-	30	5	1.5	10	22[u]	1.25	1	55	0.5	75	0.02	0.1	5.34	LT1084CP 5A low dropout
-	-	•	-	•	-	40	5	2.5[t]	5	1	1.24	4	45	0.6	80	0.1	0.1	1.62	LM338T 5A monolithic
-	-	-	•	-	-	30	7.5	1.5	10	22[u]	1.25	1	55	0.5	75	0.02	0.1	9.80	LT1083CP 7.5A low dropout
Positive, high-voltage																			
-	-	•	•	-	-	60	1.5	2.0[t]	12	0.1	1.25	4	50	0.6	80[g]	0.01	0.1	2.17	LM317HV high-voltage '317
•	•	-	-	•	-	100	0.05	12	0.5	0.1	1.20	5	10	1	60	0.003	1.4	1.39	LR12 Supertex
-	-	•	-	-	•	125	0.7	10	15	1	1.27	5	83	0.3	76	0.02	0.15	1.62	TL783C TI, MOSFET
•	-	-	-	•	•	450	0.01	12	0.5	1	1.20	5	10	1	60	0.003	1.4	0.72	LR8 Supertex
Negative																			
•	•	-	-	-	-	40	0.1	-	5	1	1.25	4	50	0.65	80[g]	0.02	0.3	0.65	LM337L low-power (neg 317L)
-	-	•	•	•	•	40	1.5[p]	2.0[t]	10	1	1.25	3	65	0.6	77[g]	0.02	0.3	0.28	LM337 negative 317

Notes: (a) all have V_{out} range from V_{ref} to $V_{in(max)}-V_{ref}$. (b) minimum current to operate the IC. (c) ΔV_{out} (%) for $\Delta T_J = 100°C$. (d) D²PAK. (e) for 10% to 50% I_{max}. (f) at 5V. (g) with V_{adj} bypass cap. (h) maximum dropout voltage at I_{max}. (k) JRC's NJM317F has isolated tab. (n) also with prefixes like TLV, LD, and REF. (p) for TO-220 and D-PAK packages. (t) typical. (u) 10µF min if low ESR tantalum; also requires 10µF input bypass. (v) maximum I_{out} at low $V_{in}-V_{out}$, e.g., $\Delta V<10V$; see text. (z) the metal case or tab (for TO-220, TO-3, D-PAK) is connected to V_{out} for positive regulators, and to V_{in} for negative regulators.
Beware differing pinouts: positive versus negative, and variants like LR8 and LR12.

make up the difference; but then you incur some additional uncertainty of output voltage, owing to the adjustment pin current of ~50 µA; see E below.

B: The standard 317-style regulator circuit (as in Figure 9.10) can adjust only down to V_{ref}. But you can trick a 317 into going down to zero by returning the lower leg of the output divider (R_2) to a negative reference. Be sure to sink enough current to bias that reference into conduction, as shown.

C: You can use a MOSFET switch (or a low R_{ON} analog switch) to shunt additional fixed resistors across the voltage-setting lower resistor, allowing selection of output voltage under logic-level control.

---

where the value of R_1 is 240 Ω. This design error most likely originated with illustrative circuit examples for the more tightly spec'd LM117 on the same datasheet, whose worst-case $I_{out(min)}$ is 5 mA (half that of the LM317). It's been 40 years, and no one at the factory seems to have noticed!

D: Alternatively, you can program the output voltage by applying a dc voltage via the ADJ pin; the output voltage will be V_{ref} greater. The programming voltage could be generated by a pot, as shown, or by a DAC. If programmed as in the circuit fragment shown, you would need to ensure that the external load satisfies the minimum load current specification (5 or 10 mA for most devices; see Table 9.2). You also need to take into consideration the effect of ADJ pin bias current through the larger than usual impedance, in this example rising to more than 1 kΩ at the pot's midposition; see E, next.

E: The ADJ pin sources ~50 µA (see the box titled "Anatomy of a 317"), which causes the output voltage to become

$$V_{OUT} = V_{ADJ}\left(1 + \frac{R_2}{R_1}\right) + I_{ADJ}R_2, \tag{9.1}$$

where the last "error" term is caused by the ADJ pin current. For the worst-case $I_{ADJ}=100$ µA and nominal $R_1=125$ Ω

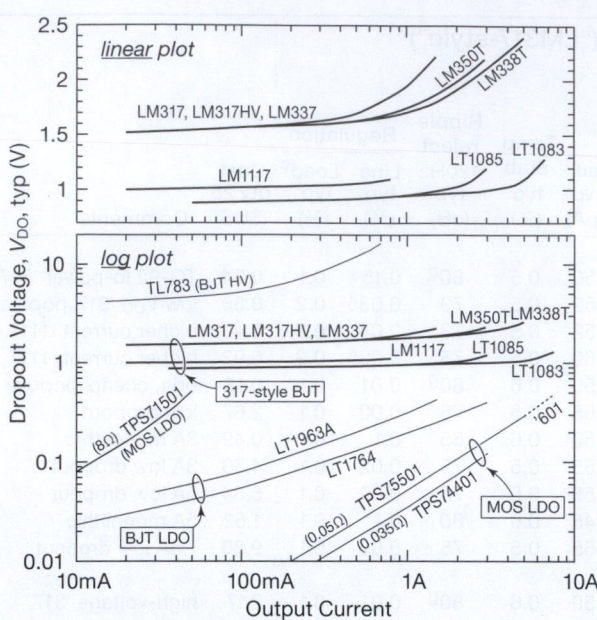

Figure 9.11. Typical dropout voltage ($V_{IN} - V_{OUT}$) versus load current for 317-style three-terminal regulators (bold curves). Representative low-dropout and high-voltage regulators are included for comparison. See also Figure 9.24.

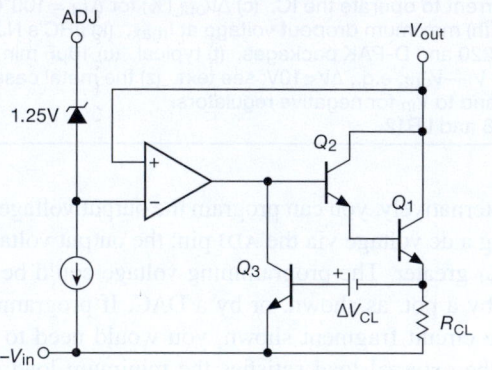

Figure 9.12. LM337 three-terminal adjustable negative voltage regulator. The common-emitter output stage requires at least 1 μF bypassing at input and output to ensure stability.

this amounts to a 1% increase in output voltage, above and beyond the initial V_{ref} uncertainty (usually 1% or 4%; see Table 9.2).[17] The current-induced error increases linearly with divider impedance, as indicated in the plot (which assumes a V_{ref} tolerance of 4%, a worst-case ADJ pin current

---

[17] If you care, you can calculate your resistor values from eq'n 9.1, using the datasheet's *typical* value for I_{ADJ}; that reduces the worst-case error by typically a factor of two (the ratio of maximum to typical I_{ADJ}).

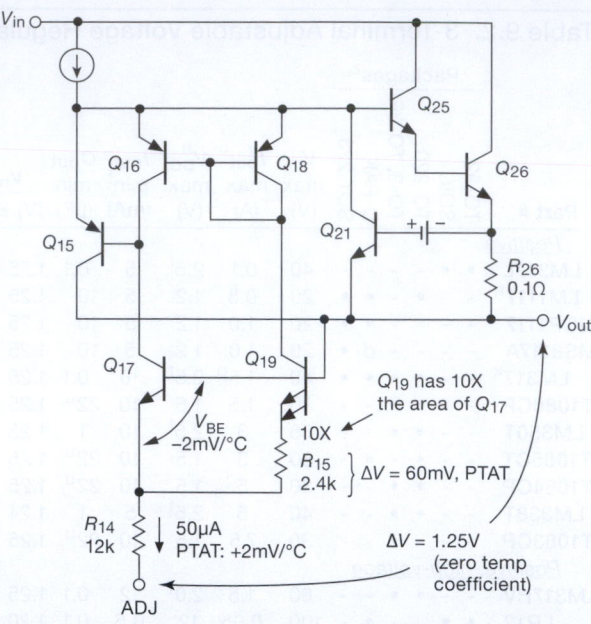

Figure 9.13. Simplified circuit of the 317-style linear regulator, illustrating its internal temperature-compensated bandgap reference; see the box "Anatomy of a 317."

of 100 μA, and no correction for the adjustment current, i.e., ignoring the last term of eq'n 9.1).

F: A linear regulator can be damaged by fault conditions in which bypass capacitors discharge suddenly through the regulator circuit, causing destructive peak currents. Diode D_2 prevents the output bypass capacitor from discharging through the regulator if the input is shorted; it never hurts to include such a diode, and it is definitely indicated for higher output voltages. Likewise, add diode D_1 if an optional noise reduction capacitor C_1 is used, to protect against input or output shorts.

G and **H:** You can extend the output-voltage ramp-up time[18] by bypassing the ADJ terminal with a large value

---

[18] Why would you do that? Perhaps this short story, from our research lab, will provide some motivation: we built a ±15 kV supply, using a pair of Spellman MP15 dc–dc HV converters (+24 V input, +15 kV and −15 kV maximum outputs, 10 W), powered from a commercial ac-powered +24 Vdc switching supply. We mounted the HV output connectors (type SHV; see Figure 1.125) a safe 2″ apart. Imagine our surprise, then, when we powered it up, and a huge spark jumped between the connectors; must have been at least 50 kV! Scary. And *worrisome* – can this thing (and its load) survive repeated startups? Our first attempted cure was to ensure that the MP15's 0–10 V control voltage was set to zero volts at startup. No joy. Finally we added an LT1085 three-terminal regulator with controlled ramp to the +24 V input, and, voilà, no more lightning.

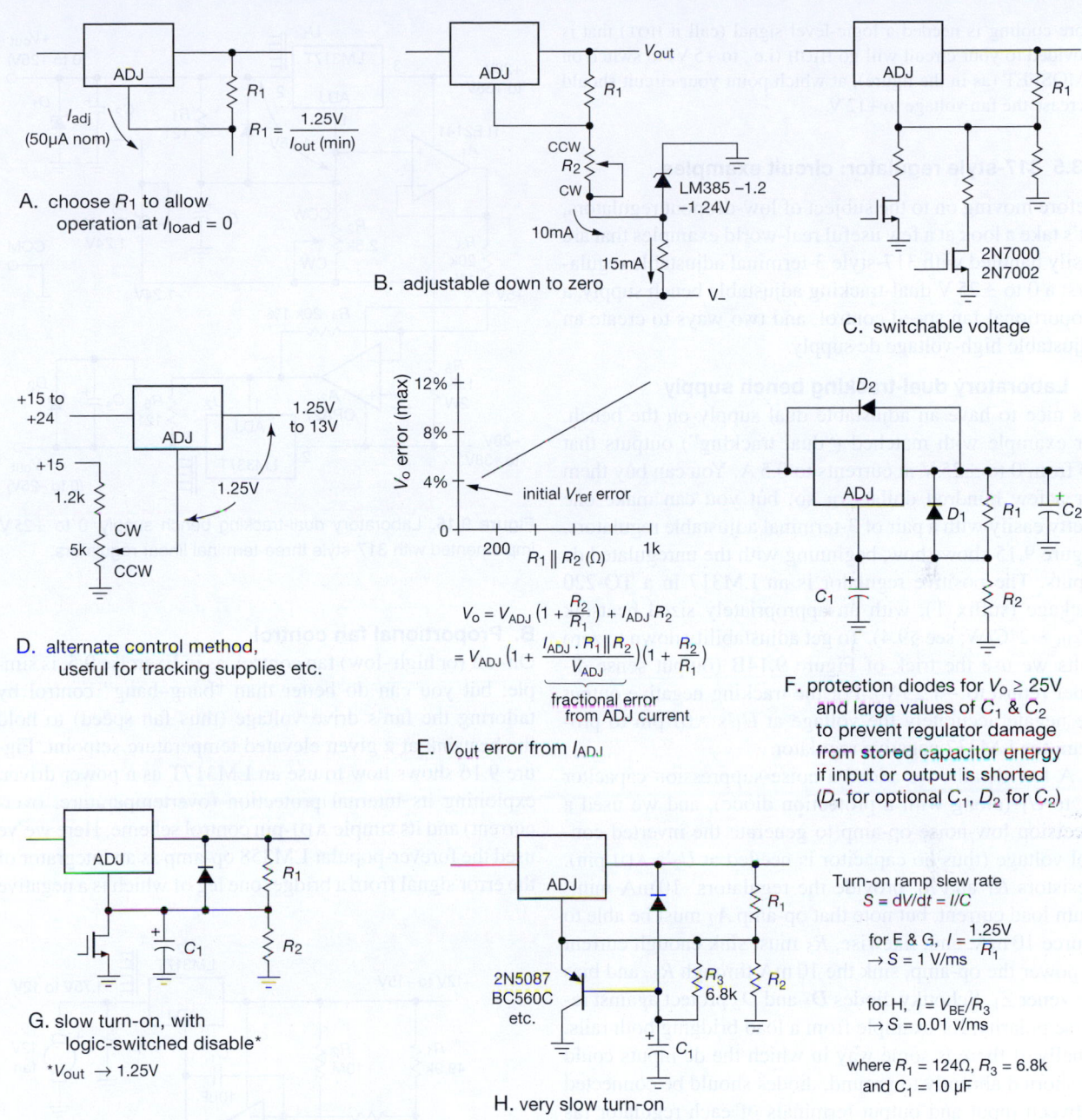

Figure 9.14. Application hints for the LM317-style three-terminal adjustable regulator, described in §9.3.4.

capacitor (be sure to add the protection diode; see **F** above). In both circuits the capacitor ramps up with a constant current, as indicated. Because R_1 is small, the capacitor value can become uncomfortably large (e.g., $100\,\mu F$ for a 10 ms/V ramp with $R_1 = 125\,\Omega$), so you may want to add a follower, as in **H**. Note that these circuits do not ramp from zero output voltage – in **G** it jumps to V_{ref} (1.25 V)

before ramping, and in **H** it jumps initially to $V_{ref} + V_{BE}$ (about 1.8 V). For the same reason the "disable" switch in **G** brings the output down only to V_{ref}.

Exercise 9.5. Draw a circuit (with component values), following the scheme of Figure 9.14C, to power a 12 V (nominal) dc cooling fan within an instrument: when little cooling is needed the circuit should provide +6 V (at which the fan runs, but quietly), but when more cooling is needed a logic-level signal (call it HOT) that is provided to your circuit will go HIGH (i.e., to +5 V) to switch on a MOSFET (as in the figure), at which point your circuit should increase the fan voltage to +12 V.

9.3.5 317-style regulator: circuit examples

Before moving on to the subject of low-dropout regulators, let's take a look at a few useful real-world examples that are easily handled with 317-style 3-terminal adjustable regulators: a 0 to $\pm$25 V dual-tracking adjustable bench supply, a proportional fan speed control, and two ways to create an adjustable high-voltage dc supply.

A. Laboratory dual-tracking bench supply

It's nice to have an adjustable dual supply on the bench, for example with matched ("dual tracking") outputs that go from 0 to $\pm$25 V at currents to 0.5 A. You can buy them for a few hundred dollars or so; but you can make one pretty easily with a pair of 3-terminal adjustable regulators. Figure 9.15 shows how, beginning with the unregulated dc inputs. The positive regulator is an LM317 in a TO-220 package (suffix T), with an appropriately sized heatsink ($R_{\Theta JC} \sim 2°C/W$; see §9.4). To get adjustability down to zero volts we use the trick of Figure 9.14B (output sense divider return to -1.25 V). For the tracking negative output we negate accurately the voltage at U_1's ADJ pin to program the LM337 negative regulator.

A few details: we added a noise-suppression capacitor C_1 to U_1 (along with a protection diode), and we used a precision low-noise op-amp to generate the inverted control voltage (thus no capacitor is needed at U_2's ADJ pin). Resistors R_1 and R_6 provide the regulators' 10 mA minimum load current, but note that op-amp A_2 must be able to source 10 mA, and, likewise, R_5 must sink enough current to power the op-amp, sink the 10 mA through R_2, and bias the zener Z_1. Schottky diodes D_1 and D_2 protect against reverse polarity, for example from a load bridging both rails. Finally, if there is some way in which the dc inputs could be shorted abruptly to ground, diodes should be connected between input and output terminals of each regulator (as in Figure 9.14F), to protect them from the fault current flowing back from the output bypass capacitors (including whatever you've got in the powered external circuit); a couple of 1N4004 rectifiers would be fine here.

Exercise 9.6. Zener reference Z_1 (actually it's a low-current shunt regulator) has a specified current range of 50 μA to 20 mA. Show that this circuit respects those limits, by calculating the zener current at both maximum and minimum negative input voltage (i.e., at -38 V and at -28 V). Assume the dual op-amp's supply current is in the range of 3 mA to 5.7 mA.

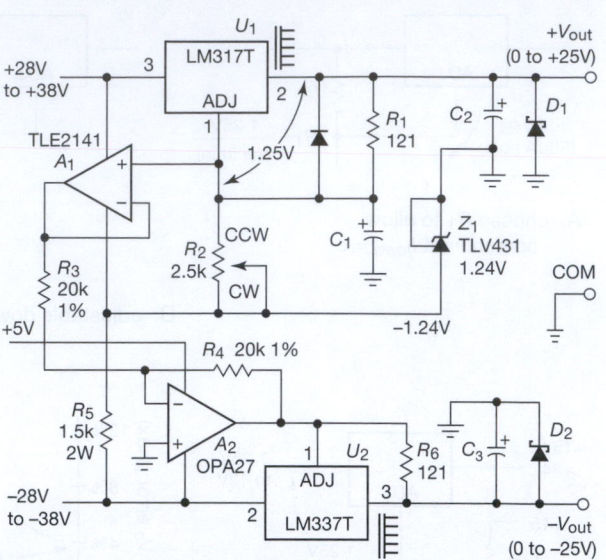

Figure 9.15. Laboratory dual-tracking bench supply, 0 to $\pm$25 V, implemented with 317-style three-terminal linear regulators.

B. Proportional fan control

On–off (or high–low) fan control, as in Exercise 9.5, is simple; but you can do better than "bang–bang" control by tailoring the fan's drive voltage (thus fan speed) to hold the heatsink at a given elevated temperature setpoint. Figure 9.16 shows how to use an LM317T as a power driver, exploiting its internal protection (overtemperature, overcurrent) and its simple ADJ-pin control scheme. Here we've used the forever-popular LM358 op-amp as an integrator of the error signal from a bridge, one leg of which is a negative

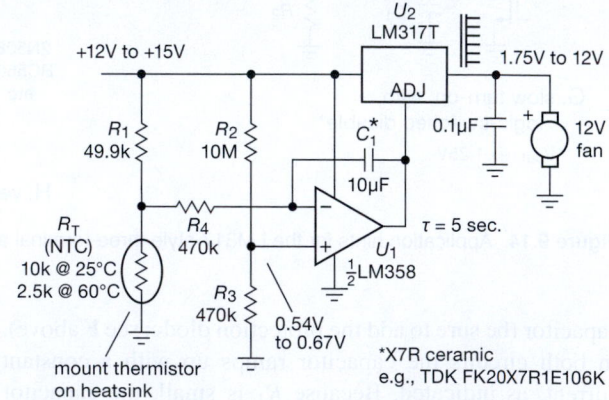

Figure 9.16. Controlling fan speed with analog feedback from a thermistor sensor, with 60°C setpoint. Fully analog control eliminates switching noise, and variable-speed operation minimizes acoustic noise.

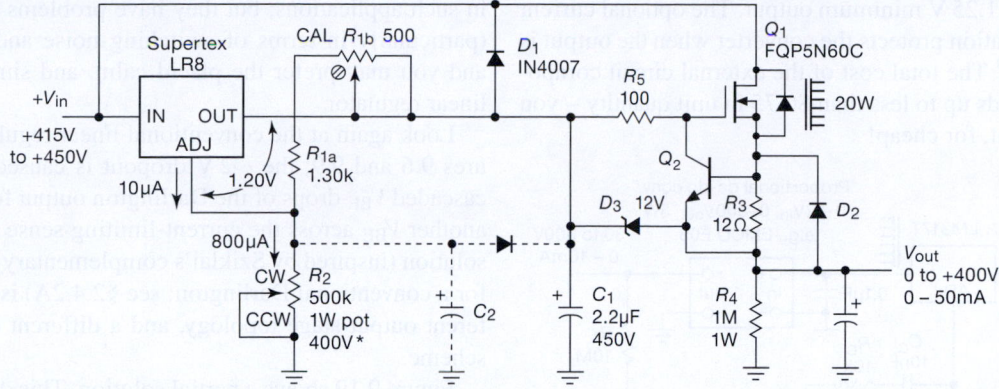

Figure 9.17. High-voltage adjustable supply I, 0 to 400 V, implemented with a low-current LR8 plus outboard MOSFET follower. * See text for R_2's voltage rating.

temperature coefficient (NTC) thermistor. The inputs are balanced at the 60°C setpoint, above which the integrator's output, and therefore the fan voltage, is driven positive. The integrator time constant R_4C_1 should be chosen somewhat longer than the thermal time constant from the heat source(s) to the thermistor sensor, to minimize feedback-loop "hunting."

We chose the LM358 for its low cost (as little as $0.10 in unit quantity), single-supply operation (input and output to the negative rail), and robust tolerance to supply voltages well beyond +15 V. But its 50 nA worst-case input offset current forced us to use a rather large integrating capacitor. Ideally one would like an inexpensive single-supply op-amp with bias-compensated or FET inputs. Happily, there's the unusual single-supply OPA171, whose offset current is down in the tens of picoamps even at elevated temperature, and which operates over the full supply voltage range of 3 V to 36 V; it costs about a dollar (but you come out ahead, with the less expensive 0.47 μF capacitor that goes with a 10 MΩ resistor R_4).

C. High-voltage supply I: linear regulator

Figure 9.17 shows a simple circuit that extends the output current of the LR8 high-voltage 3-terminal regulator from Supertex. This part operates to 450 V, but it's limited to 10 mA of output current, and further limited by power dissipation (~2 W in the D-PAK power package, depending on circuit-board foil pattern; see Figure 9.45) when operating at nearly its full rated voltage differential.

The regulator's minimum load current is specified as 0.5 mA, so we operate it at slightly less than 1 mA, with its output driving a high voltage power MOSFET follower. Feedback is local to the LR8, so the follower introduces an offset (slightly load dependent) that can be approxi-

mately trimmed to zero by R_{1b}. Q_2 provides current limiting, and diodes D_1–D_3 protect against the many injuries that are possible in a high voltage supply. Add capacitor C_2 (along with a protection diode) for noise reduction. The output-setting pot R_2 dissipates 320 mW at maximum output voltage, so it's wise to use a part rated at 1 W or more. Be sure to check the pot's specifications for a *voltage* rating, as well; for this application you could use a Bourns 95C1C-D24-A23 or a Honeywell 53C3500K.

D. High-voltage supply II: dc–dc switching converter

Another approach to generating regulated high-voltage dc is to use a switching dc–dc converter module. These come with a huge range of output voltages (to tens of kilovolts) and with selectable output polarity. You can get a wide variety of these with built-in regulator circuitry, intended to be powered by a low-voltage dc input (+5 V, or +15 V, etc.), and with the output voltage programmed via a variable resistor or by a low-voltage dc programming input. These are handy for generating the bias for a photomultiplier tube, avalanche photodiode detector, channel plate multiplier, or other devices that need a stable high-voltage low-current bias.

A less-expensive approach is to use a bare-bones dc–dc converter module that lacks internal regulation; for these the output voltage scales with input voltage – they are sometimes called "proportional" dc–dc converters. Typical units take a 0–12 V dc input, with output ranges going from as little as 100 V to as much as 25 kV, at power ratings from a fraction of a watt to 10 watts or so. Figure 9.18 shows how to do this, using one of EMCO's 3 W proportional converter modules. The output voltage is controlled by R_3, with the low end of the voltage range set by the LM317's 1.25 V minimum output. The optional current limit modification protects the converter when the output is overloaded.[19] The total cost of the external circuit components here adds up to less than $0.75 in unit quantity – you can't beat that, for cheap!

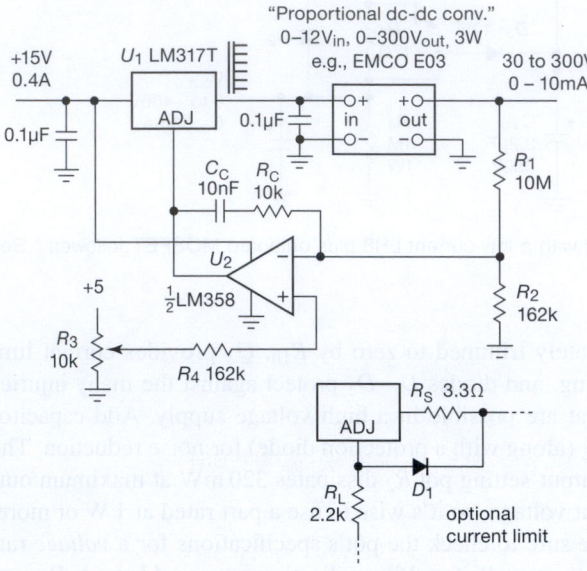

Figure 9.18. High-voltage adjustable supply II, implemented with a proportional dc–dc converter module powered by an LM317 with feedback control to the ADJ pin.

9.3.6 Lower-dropout regulators

There are applications where the $\sim$2 V dropout voltage (i.e., minimum input–output voltage differential) of these regulators is a serious limitation. For example, in a digital logic circuit you might need to create a +3.3 V supply from an existing +5 V; or (worse), a +2.5 V supply from an existing +3.3 V supply rail. Another application might be a portable device needing +5 V, and operating from a 9 V alkaline battery; the latter begins life at about 9.4 V, declining at end of life to 6 V or 5.4 V (depending on whether you subscribe to 1.0 V/cell or 0.9 V/cell as the definition of a fully exhausted battery). For these situations you need a regulator that can operate with a small input–output differential, ideally down to a few tenths of a volt.

One solution is to abandon linear regulators altogether and use instead a switching regulator (§9.6), which deals differently with voltage differential. Switchers (the nickname for switching, or switchmode, regulators) are popular

---

[19] We administered a torture test to an unprotected converter and measured an input current of 1.2 A.

in such applications; but they have problems of their own (particularly in terms of switching noise and transients), and you may prefer the placid calm, and simplicity, of a linear regulator.

Look again at the conventional linear regulators in Figures 9.6 and 9.9; the $\sim$2 V dropout is caused by the two cascaded V_{BE} drops of the Darlington output follower, plus another V_{BE} across the current-limiting sense resistor. The solution (inspired by Sziklai's complementary replacement for a conventional Darlington; see §2.4.2A) is to use a different output-stage topology, and a different current-limit scheme.

Figure 9.19 shows a partial solution. This design retains an *npn* output follower, but substitutes a *pnp* driver transistor; the latter can run close to saturation, eliminating one of the V_{BE} drops. Further, the current-limit sense resistor has been relocated in the collector ("high-side current sensing"), where it does not contribute to the dropout voltage (as long as its drop is less than a V_{BE} at current limit, which is easy to manage if a comparator is used to sense maximum current, as shown). With this circuit topology the LT1083-85 regulators (rated at 7.5 A, 5 A, and 3 A, respectively) achieve a typical dropout of 1 V at their maximum current.

We've used this series of regulators in many designs, with good success. Electrically they mimic the classic LM317 three-terminal adjustable regulator, with an internal 1.25 V referenced to the output pin. However, as with most low-dropout designs, these regulators are fussy about bypassing: the datasheet recommends 10 μF at the input, and at least 10 μF (tantalum) or 50 μF (aluminum electrolytic) at the output. If the ADJ pin is bypassed for noise reduction (see §9.3.13), the datasheet recommends tripling the output bypass capacitor value.

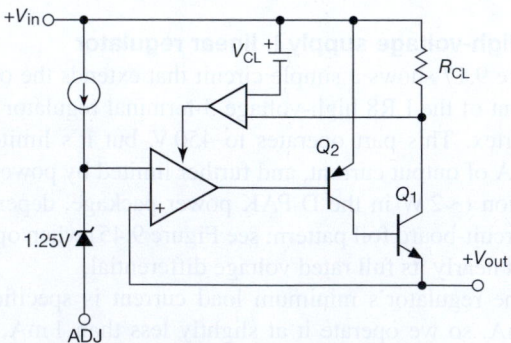

Figure 9.19. LT1083-85 series three-terminal positive regulators with reduced dropout voltage.

9.3.7 True low-dropout regulators

The dropout voltage can be reduced further, by replacing the *npn* (follower) output stage with a *pnp* (common emitter) stage (Figure 9.20A). That eliminates the V_{BE} drop, the dropout voltage now being set by transistor saturation. To keep the dropout voltage as low as possible, the current-limit circuit eliminates the series sense resistor, using instead a fractional output-current sample, derived from a second collector on Q_1. This is less accurate, but "good enough," given that its job is merely to limit destructive currents: the datasheet for the 3 A LT1764A, for example, specifies a current limit of 3.1 A (minimum) and 4 A (typical).[20]

Many contemporary low-voltage regulators use MOS-FETs, rather than bipolar transistors. The analogous LDO circuit is shown in Figure 9.20B. Like the bipolar LDO, these parts tend to be rather fussy about bypassing. For example, the TPS775xx regulators set requirements on both the capacitance and ESR of the output bypass: the capacitance must be at least $10\,\mu F$ (with an ESR no less than $50\,m\Omega$ *and* no greater than $1.5\,\Omega$), with the datasheet additionally showing regions of stability and instability in graphs plotting combinations of C_{bypass}, ESR, and I_{out}.

9.3.8 Current-reference 3-terminal regulator

All the regulators we've seen so far use an internal voltage reference (usually a 1.25 V "bandgap" reference), against which a voltage-divider fraction of the output is compared. The result is that you cannot have an output voltage less than that reference. In most cases this sets a lower bound of V_{out}=1.25 V (though a few can go down to 0.8 V, or even 0.6 V; see Table 9.3 on page 614).

Sometimes you want a lower voltage! Or you may want to have an adjustment range that goes clear down to zero volts. This has traditionally required an auxiliary negative supply, as for example in the book's previous edition's "laboratory bench supply" (its Figure 6.16).

A nice solution is the LT3080-style regulator, originated by Linear Technology (Figure 9.21). It is a 3-terminal adjustable positive regulator (with a fourth pin, in some package styles) in which the ADJ pin (called SET) sources an accurate *current* (I_{SET}=10 μA, $\pm2\%$); the error amplifier then forces the output to follow the SET pin. So, if you connect a resistor R from SET to ground, the output voltage will be simply $V_{out} = I_{SET}R$. The output-voltage range goes all the way to zero: when R=0, V_{out}=0.[21]

---

[20] The CMOS TPS755xx series of 5 A LDO regulators has a revealing current limit specification of 5.5 A (minimum), 10 A (typical), and 14 A (maximum).

[21] With one slight *gotcha*: the minimum load current is ~1 mA. So, for

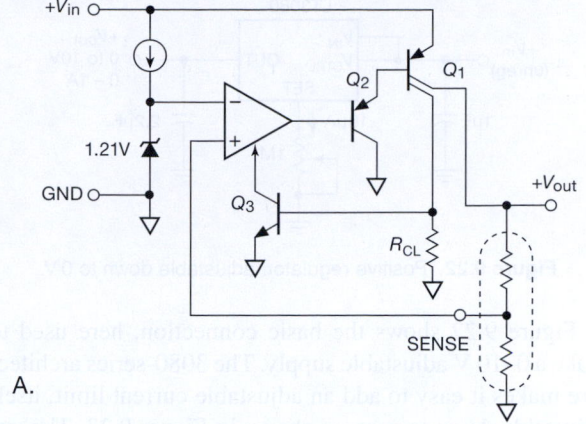

A.

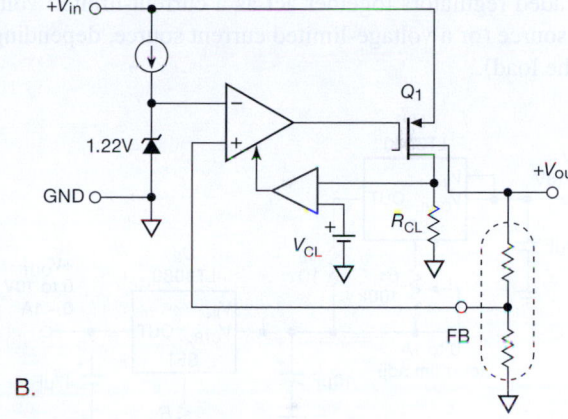

B.

Figure 9.20. A. LT1764 (bipolar) and B. TPS75xxx (CMOS) positive LDO regulators.

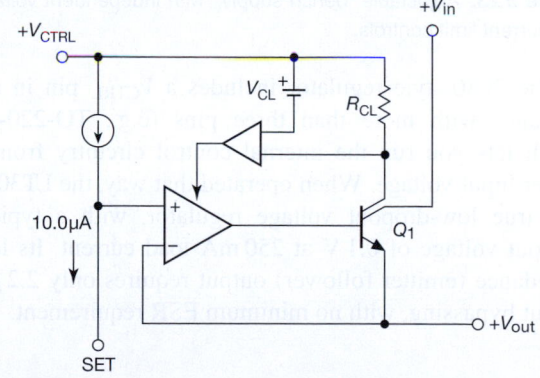

Figure 9.21. LT3080 "3-terminal" adjustable positive voltage regulator with precision current reference.

example, the output voltage into a $100\,\Omega$ load will not go below 0.1 V. To reach 0 V you need to sink a small current toward a negative supply.

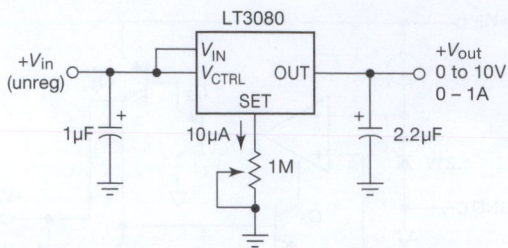

Figure 9.22. Positive regulator, adjustable down to 0 V.

Figure 9.22 shows the basic connection, here used to make a 0–10 V adjustable supply. The 3080-series architecture makes it easy to add an adjustable current limit, itself adjustable down to zero, as shown in Figure 9.23. The upstream regulator U_1 is by itself a 0–1 A current source; the cascaded regulators together act as a current-limited voltage source (or a voltage-limited current source, depending on the load).

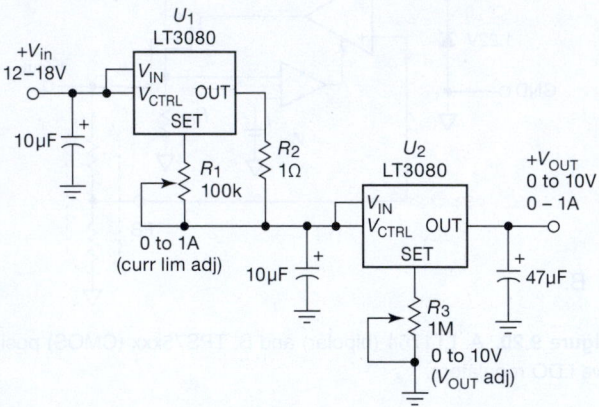

Figure 9.23. Adjustable "bench supply," with independent voltage and current limit controls.

The 3080-style regulator includes a V_{CTRL} pin in the packages with more than three pins (e.g., TO-220-5), which lets you run the internal control circuitry from a higher input voltage. When operated that way, the LT3080 is a true low-dropout voltage regulator, with a typical dropout voltage of 0.1 V at 250 mA load current. Its low impedance (emitter follower) output requires only 2.2 μF output bypassing, with no minimum ESR requirement.

9.3.9 Dropout voltages compared

To summarize the business of dropout voltage in these various regulator designs, we plotted in Figure 9.24 the dropout voltages of a representative regulator of each type. The curves are taken from "typical" dropout specifications in

the datasheets, all at 40°C, and are scaled to the rated maximum current of each device. Three categories are clearly seen: conventional regulators with Darlington *npn* pass transistor (top three curves); lower-dropout regulators with *pnp* driver and *npn* output follower (middle curve); and true low-dropout regulators with *pnp* or *p*MOS output stage (bottom four curves). Note particularly the resistive behavior of the CMOS regulators (bottom two curves), where the dropout voltage is linear in output current, and goes to zero at low current.

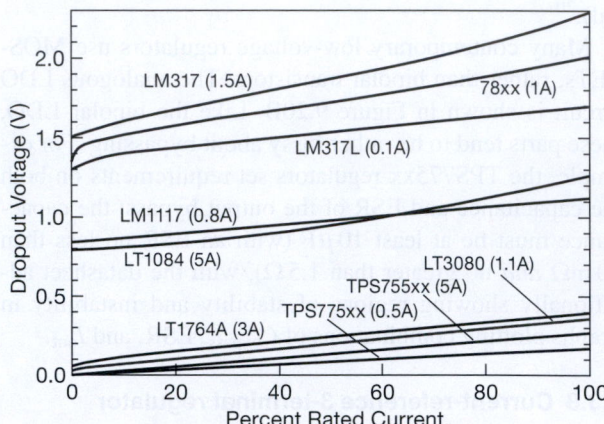

Figure 9.24. Linear regulator dropout voltage versus output current. The bottom pair of curves (TPS prefix) are CMOS; all others are bipolar. See also Figure 9.11.

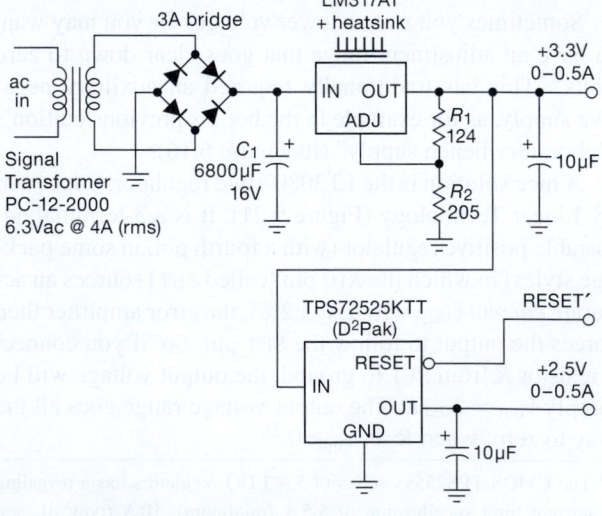

Figure 9.25. Dual low-voltage regulated supply.

9.3.10 Dual-voltage regulator circuit example

As an illustration, imagine we have a small digital circuit requiring +3.3 V and +2.5 V regulated supplies, each capable of supplying up to 500 mA. Figure 9.25 shows how to do this with a small PCB-mounted transformer driving an unregulated bridge rectifier, followed by a pair of linear regulators. The design is straightforward: (a) we began by choosing a transformer to deliver about +8 Vdc (unregulated) from Signal Transformer's nice selection; a unit with 6.3 Vrms is about right (its ac peak amplitude of $6.3\sqrt{2} \approx 9$ V is reduced by two diode drops); (b) we chose a conservative transformer current rating of 4 Arms, to allow for the extra heating caused by the relatively short current pulses in a bridge rectifier circuit ("small conduction angle," see §1.6.5); (c) the storage capacitor C_1 was then chosen (using $I = C\,dV/dt$) to allow ~1 Vpp ripple at maximum load current, with a voltage rating adequate to allow for the worst-case combination of high line voltage and zero output load; (d) for the +3.3 V output we used a 3-terminal adjustable regulator (LM317A, in a TO-220 power package), mounted on a small heatsink (10°C/W, adequate for the ~5 W maximum power dissipation; see §9.4; (e) finally, for the +2.5 V output we used a low-dropout CMOS fixed voltage regulator (the last two digits of the part number designate +2.5 V), taking its input from the regulated +3.3 V.

Several comments. (a) We have not shown details of the ac line input, including fuse, switch, and noise filter; (b) the bypass capacitor values shown are conservative (larger than the specified minimum), in order to improve transient response and provide robust stability; (c) the TPS72525 regulator includes an internal "supervisor" circuit providing a RESET′ output that goes LOW when the regulator falls out of regulation, often used to alert a microprocessor to save its state and shut down.

9.3.11 Linear regulator choices

Fixed or adjustable? 3-terminal or 4-terminal? Low dropout or conventional? How do you decide which type of integrated linear regulator to use? Here is some guidance.

- If you don't need low dropout, stick with a conventional regulator, either 3-terminal fixed (78xx/79xx style, Table 9.1) or 3-terminal adjustable (317/337 style, Table 9.2): they are less expensive, and they're stable with small-value bypass capacitors.
  - fixed: needs no external resistors; but limited voltage choices, and no trimmability.
  - adjustable: settable and trimmable, and fewer types to stock; but require a pair of external resistors.

- If you need adjustability down to zero volts, use a current-reference regulator (LT3080 style).
- If you need low dropout ($V_{DO} \leq 1$ V) you have many LDO choices (Table 9.3):
  - For input voltages ≥ 10 V, use bipolar types:
    * LT1083-85, LM1117, LM350, LM338 style (fixed or adj) for ~1 V dropout;
    * LT1764A style (fixed or adj) for ~0.3 V (but see §9.3.12);
  - For input voltages ≤ 10 V there are many MOSFET LDOs (fixed or adj).
- If you need high efficiency, high power density, voltage stepup, or voltage inversion, use a switching regulator/converter (§9.6).

For high-current low-voltage applications, consider the use of a regulator that has separate control and pass-element input pins, like that shown in Figure 9.21. These are indicated in the "boost, bias pin" column of Table 9.3.

9.3.12 Linear regulator idiosyncrasies

These integrated regulators are genuinely easy to use, and, with their built-in overcurrent and thermal protection circuits, there's not much to worry about. Circuit designers should, however, be aware of the following idiosyncrasies.

A. Pinout variations

Our students succumb, with distressing frequency, to this gotcha: complementary polarity regulators, such as our favorite LM317 (positive) and LM337 (negative) 3-terminal adjustables, often have *different pinouts* (Figure 9.26). In the case of the 78xx/79xx fixed regulators, for example, this creates real mischief: the mounting tab is ground for the 78xx positive regulator (so you can screw it to the chassis, or solder it to the circuit board's ground plane), but for the 79xx negative regulator the tab is connected electrically to the input voltage – ground that and you're in real trouble![22]

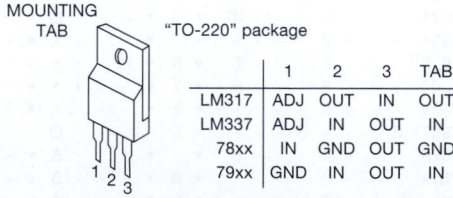

	1	2	3	TAB
LM317	ADJ	OUT	IN	OUT
LM337	ADJ	IN	OUT	IN
78xx	IN	GND	OUT	GND
79xx	GND	IN	OUT	IN

Figure 9.26. Never assume that negative regulators have the same pinouts as their positive twins. In fact, don't assume *anything* without consulting the datasheet.

---

[22] This comes about because the IC's substrate (normally at the most negative voltage) is soldered to a metal mounting frame, which is the best pathway for heat removal.

Table 9.3 Low-dropout Linear Regulators[a]

Package column group: TO-220 | DDPAK,DPAK | DIP | SOIC,MSOP,TSOP | SOT-223 | SOT-23 | TO-92 | smaller. Feature column group: adj version avail / #fixed[ti] | Enable | Filter pin | Soft start | UVLO | Reverse block | Sense pin (or func'n) | boost, bias pin | PG or /reset pin. Regulation column group: line max (ppm/V) | C_{out} min (µF) | ESR min,max (Ω).

Part #[n]	TO-220	DDPAK,DPAK	DIP	SOIC,MSOP,TSOP	SOT-223	SOT-23	TO-92	smaller	adj/#fixed	En	Filt	Soft	UVLO	Rev	Sense	boost	PG/rst	V_{in} min-max (V)	V_{out} (V)	I_{out} (A)	Acc'y 25C (%)	I_Q typ $I_{out}{=}0$ (mA)	V_n[b] @V_{out} typ (µVrms)	(V)	line max (ppm/V)	C_{out} min (µF)	ESR min,max (Ω)	Price[n] qty 25 ($US)
LM317-style																												
LT3080[A]	5	5	–	8	3	–	–	8	•	–	–	–	•	–	x	–	•	1.2-36	0-35	1.1	1	1[h]	40	all	50	2.2	0,0.5	2.71
LT1086	3	3	–	–	–	–	–	–	•5	–	–	–	–	–	Δ	–	–	3.5-25	1.25-25	1.5	1	10[h]	–	–	1000	22[f2]	?	2.67
LT3083A[A]	5	5	–	16	–	–	–	12	•	–	–	–	•	–	x	–	•	1.2-23	0-22	3	1	1[h]	40	all	10	10[c]	0,0.5	5.90
LT1083,84,85	3	3	–	–	–	–	–	–	•	–	–	–	•	–	•	–	•	2.7-25[g]	1.25-25	3,5,7.5	1	10[h]	150[e2]	5	300	22[f2]	0,–	5.20
LT1580	5,7	–	–	–	–	–	–	–	•1	–	–	–	–	–	•	•	–	3-5.5	1.25-6	7	0.6	6	–	–	1000	22	0.1	3.74
LT1581	7	–	–	–	–	–	–	–	•1	–	–	–	–	–	•	•	–	3.3-6	1.25-5	10	0.7	10[h]	–	–	300	22	0.1	8.93
High Voltage (≥10V)																												
TPS709xx	–	–	–	–	–	5	–	–	11	•	–	–	–	•	•	–	–	2.5-30	1.5-6.5	0.15	2[tt]	0.001	300	1.8	650	2.2	0,–	1.16
LT3008	–	–	–	–	–	8	–	6	•6	•	–	–	–	•	□	–	–	2-45[o]	0.6-45	0.02	1.6	0.003	92	0.6	250	2.2	0,3	2.66
TLV704xx	–	–	–	–	–	5	–	–	3	•	–	–	–	–	Δ	–	–	2.5-24	3-5	0.15	2	0.003	–	–	500	1	0,–	0.81
LT3014B[B]	–	–	–	–	–	5	–	8	•	–	•	–	–	–	□	–	–	3-80[o]	1.22-60	0.02	1.6	0.007	115	1.22	125	0.47	0,3	2.14
LT1521	–	–	8	3	–	–	–	–	•3	•	–	–	–	•	□	–	–	3.5-20	3-20	0.3	1.5	0.012	–	–	125	1.5	0,?	2.08
LM2936	–	3	–	8	3	–	3	–	3	q	–	–	–	–	□	–	–	4-26[o]	3.0-5	0.05	2	0.015	500	–	1000	22	0.3,6	1.03
TPS769,70xx	–	–	–	–	–	5	–	–	•9	•	–	–	–	–	Δ	v	–	2.7-10	1.2-5.5	0.05,0.1	3[tt]	0.017	190[e3]	all	400	4.7	0.2,5	0.77
TPS789,90xx	–	–	–	–	–	5	–	–	•5	•	•	–	–	–	Δ	–	–	2.7-10	1.5-3.0	0.1,0.05	3[tt]	0.017	56[f,e3]	3	1000	4.7	0.2,5	0.56
TPS788xx	–	–	–	–	–	5	–	–	2	•	•	s	–	–	Δ	–	–	2.7-13.5	2.5,3.3	0.15	3[tt]	0.017	56[f,e1]	all	1000	4.7	0.2,5	1.09
ADP667	–	–	8	8	–	–	–	–	•1	•	–	–	–	x	v	–	•	3.5-16.5	1.3-16	0.25	4[tt]	0.02	–	–	500	10	0,–	2.98
LT1761	–	–	–	–	–	5	–	–	•9	•	v2	–	–	–	□	–	–	1.8-20	1.22-20	0.1	1.3	0.02	20[f]	all	600	1	0.3	2.13
LM9076	–	5	–	8	–	–	–	–	2	q	–	–	–	–	□[y]	–	r	3.5-50[o]	3.3-5	0.15	1.5	0.025	–	–	1500	22[c]	0.1,3	1.31
LT1763	–	–	–	8	–	–	–	12	•6	•	•	•	–	–	□	–	–	1.8-20	1.22-20	0.5	1	0.03	20[f]	all	50	3.3	0,3	2.30
ADP3331	–	–	–	–	–	6	–	–	•	–	–	–	–	–	x	–	–	2.6-12	1.5-11.75	0.2	0.7	0.034	95	3	240	0.47	0,–	1.25
TPS765,66xx	–	–	–	–	–	8	–	–	•8	•	–	–	–	–	Δ	v	p	2.7-10	1.2-5.5	0.15,0.25	3[tt]	0.035	200[e3]	all	100[t]	4.7	0.3,20	1.50
TPS760,61xx	–	–	–	–	–	5	–	–	•5	•	–	–	–	–	Δ	–	–	3.5-16	3-5	0.05,0.1	1	0.09	190[e3]	all	300	2.2	0.1,20	0.90
TPS763,64xx	–	–	–	–	–	5	–	–	z9	•	z	–	–	–	Δ	z	–	3.3-10	1.5-6.5	0.15	2	0.09	140[e3]	all	700	4.7	0.3,10	0.90
LP2950A[C]	–	3	–	–	–	3	–	–	3	–	–	–	–	–	x	–	–	3.1-30	3.0-5.0	0.1	0.5	0.075	430	all	35	2.2	0,5	**0.37**
LP2951A[C]	–	–	8	8	–	–	–	8	2	•	v2	–	–	–	x	–	•	1.5-30	1.4-29	0.1	0.5	0.2 m	100[f]	all	40	2.2	0,5	**0.29**
LP2981A[C]	–	–	–	–	–	5	–	–	•9	•	–	–	–	–	Δ	–	–	2.1-16	2.5-5	0.1	0.75	0.095	160[e3]	all	140	3.3[c]	0.01,8	0.84
LP2985A	–	–	–	–	–	5	–	–	•9	•	–	–	–	–	Δ	–	–	2.2-16	1.8-10	0.15	1	0.065	30[f]	all	140	2.2	0.005-0.5	0.82
LM2931	3	3	–	8	–	3	3	6	•	–	–	–	–	–	□	v	–	3.3-26[o]	3.0-24	0.1	5	0.4	500	5	1500	47	0.03,0.4	**0.65**
LM2930	3	3	–	–	–	–	–	–	2	–	–	–	–	–	□[y]	–	–	5.3-26[o]	5.0-8.0	0.15	6	0.4	140	5	800	10	1	0.65
LP2986A	–	–	–	8	–	–	–	–	•3	•	v2	–	–	–	Δ	–	–	2.5-16	1.23-15	0.2	0.5[d]	0.1	160[e3]	all	140	4.7[c]	0.1,10	2.32
TPS7A4901	–	–	–	8	–	–	–	–	•	–	•	•	–	–	x	v	–	3-36	1.19-33	0.15	1.5	0.061	21[f]	5	32[t]	2.2[c]	0,0.2	2.14
TL750Lxx	3	3	8[y]	8	–	–	3	–	4	–	–	–	–	–	□[y]	–	–	6-26[o]	5,8,10,12	0.15	4	1.0	700	10	300	10	0,0.4	0.62
TPS72xx	–	–	–	8	–	–	–	–	•5	•	–	–	–	–	Δ	•	p	3-10	1.2-10	0.25	2[tt]	0.18	300	–	1400	10	0.1,1.3	2.00
REG101,102	–	–	–	–	5[y]	5	–	–	•6	•	d	–	–	x	–	•		1.8-10	2.5-5.5	0.1,0.25	1.5	0.4	18[f]	2.5	1000	0	–	2.22
REG103,04	–	5	–	8[y]	5	–	–	–	•5	•	d	–	–	x	–	•		2.1-15	1.3-5	0.5,1	2	0.5	25[f]	2.5	1000	0	–	3.78
REG113	–	–	–	8	–	–	–	–	•5	•	d	–	–	x	–	•		1.8-10	2.5-5	0.4	1.5	0.4	18[f]	2.5	1000	0	–	1.41
ADP3303	–	–	–	8	–	–	–	–	•5	•	•	–	–	x	–	•		3.2-12	2.7-5.0	0.2	0.8	0.25	100	5	100[t]	0.47	0,–	2.50
MIC5205,07	–	–	–	–	–	5	–	–	•13[y]	•	d	–	–	–	□	–	–	2.5-16	1.24-15	0.15	1	0.08	260[nV]	all	120	2.2	0,5	0.41
MIC5209	–	5	–	8	3	–	–	8	•7	q	d	–	–	–	□	–	–	2.5-16	1.24-15	0.5	1	0.08	300[nV]	all	500	2.2	1 rec	0.70
MIC5219	–	–	–	5	–	5	–	6	•11	•	d	–	–	–	□	–	–	2.5-12	1.24-12	0.5	1	0.08	300[nV]	all	500	2.2	0.1-1	1.50
LP38691,93	–	3[y]	–	4[y]	–	–	–	6	•4	z	–	–	–	–	Δ	q	–	2.7-10	1.8-5	0.5	2	0.055	70	3.3	1000	1[c]	0,100	0.90
LT3021	–	–	–	8	–	–	–	16	•3	–	–	–	–	–	□	–	–	0.9-10	0.2-9.5	0.5	2	0.11	300	1.2	650	3.3[c]	0,0.2	3.13
ADP3334	–	–	–	8	–	–	–	8	•	–	•	–	–	–	x	–	–	2.6-11	1.5-10	0.5	0.9	0.09	45	all	200	1	0,–	1.34
ADP3335,6[D]	–	–	–	8	–	–	–	8	•5	•	v2	–	–	–	x	–	–	2.6-12	1.5-10	0.5	0.9	0.08	47[f]	all	400[t]	1	0,–	2.13
ADP7102,04	–	–	–	8	–	–	–	8	•7	•	•	•	–	–	•	–	–	3.3-20	1.22-20	0.3,0.5	0.8	0.4	15	3.3	150	1	0,0.2	3.22
TPS73xx	–	–	8	8	–	–	–	–	•5	–	–	–	–	–	Δ	–	r	2.5-10	1.2-9.8	0.5	3	0.34	89	1.2	2000	10	0.1,4	2.25
LM2937	3	3	–	–	3	–	–	–	•5	–	–	–	–	–	□	–	–	7-26[o]	5-15	0.5	3	2.0	150	5	2500	10	–	0.76
LP3878	–	–	–	8P	–	–	–	8	•	–	•	•	–	–	Δ	•	–	2.5-16	1.0-5.5	0.8	1	0.18	18[f]	all	140	10[c]	0,0.4	1.44
TPS775xx-78xx	–	–	–	8	–	–	–	–	•6	•	–	–	–	–	Δ	v	r,p	2.7-10	1.2-5.5	0.5,0.75	2[tt]	0.085	53[e1]	all	100[t]	10	0.05,1.5	2.44
TPS767,68xx	–	–	–	8	–	–	–	–	•8	•	–	–	–	–	Δ	v	r,p	2.7-10	1.2-5.5	1.0	2[tt]	0.085	55[e1]	all	100[t]	10	0.05,1.5	2.77
LM2940	3	3	16	–	–	–	–	–	•6	–	–	–	–	–	□	–	–	7-26[o]	5-15	1.0	3	10	150	5	2000	22	0.1,1	1.71
TPS73801	–	–	–	5	–	–	–	–	•	–	–	–	–	–	–	v	–	2.2-20	1.2-20	1.0	1.5	1.0	45	1.2	230	10	0,0.3	2.34
LT1965	5	5	–	–	–	–	–	8	•4	•	–	–	–	–	□	–	–	1.8-20	1.2-19	1.1	1.5	0.5	40	all	600	10	0,3	2.77
UCC381	–	–	–	8	–	–	–	–	•2	•	–	v	•	•	•	–	–	3-9	1.2-8	1.0	1.5	0.4	–	–	150	0	–	4.61
TPS7A45xx	–	5	–	8	–	–	–	–	•4	•	•	–	–	–	•	–	–	2.2-20	1.2-20	1.5	1	1.0	35	all	140	10	0,3	3.12
LT1963A[E]	5	5	–	8	3	–	–	–	•4	•	–	–	–	–	□	–	–	1.9-20	1.21-20	1.5	1.5	1.0	40	all	240	10	0.02,3	3.57
LT1764A	3	3	–	16P	–	–	–	–	•4	•	–	–	–	–	□	–	–	2.7-20	1.2-20	3	1.5	1.0	40	all	600	10	0.005,3	4.97
UCC383	3,5	5	–	–	–	–	–	–	•2	•	•	v	•	•	v	–	–	3-9	1.2-8	3	1.5	0.4	–	–	400	0	–	4.32

Part #[n]	TO-220	DDPAK, DPAK	DIP	SOIC, MSOP, TSOP	SOT-223	SOT-23	TO-92	smaller	adj version avail / # fixed[ti]	Enable	Filter pin	Soft start	UVLO	Reverse block	Sense pin (or func'n)	boost, bias pin	PG or /reset pin	V_{in} min-max (V)	V_{out} (V)	I_{out} (A)	Acc'y 25C (%)	I_Q typ $I_{out}{=}0$ (mA)	V_n[b] typ (µVrms)	@V_{out} (V)	line max (ppm/V)	C_{out} min (µF)	ESR min,max (Ω)	Price[n] qty 25 ($US)
Low Voltage (≤8V)																												
ADP121	-	-	-	-	5	-	4		•8	•	•			x	v			2.3-5.5	1.2-3.3	0.15	1	0.01	52	2.5	300	0.7	0,1	0.87[u]
ADP150	-	-	-	-	5	-	4		-8	•	•			•	x			2.2-5.5	1.8-3.3	0.15	1	0.01	9	2.5	500	1	0,1	0.80
ADP130[F]	-	-	-	-	5	-	-		-5		•			x		•		1.2-3.6	0.8-2.5	0.35	1	0.025	61	2.5	1000	1	0,1	0.85
LTC1844	-	-	-	-	5	-	-		•5	•		•			x			1.6-6.5	1.25-6	0.15	1.5	0.035	65[f]	all	2000	1	0,0.3	1.45[k]
MAX8510	-	-	-	-	-	-	-	5,8	•10	•	d			x				2-6	1.45-6	0.12	1	0.04	11[f]	all	10[t]	1[c]	0,0.3	0.58[k]
ADP122,23	-	-	-	-	5	-	6		•10	•	•			x	v			2.3-5.5	0.8-5	0.3	1	0.045	45	2.5	500	0.7	0,1	0.88
ADP124,25	-	-	-	8	-	-	8		•8	•	•			x	z			2.3-5.5	0.8-5	0.5	1	0.045	45	2.5	500	0.7	0,1	1.14
MAX8867-68	-	-	-	-	5	-	8		•8	•	•			□				2.5-6.5	2.5-5	0.15	1.4	0.085	30[f]	all	1500	1[c]	0,2	0.88[k]
MIC5255	-	-	-	-	5	-	12		•12	•	•			Δ				2.7	2.5-3.5	0.15	1	0.09	30[f]	all	1000	1[c]	0,0.3	0.39[k]
TPS731xx	-	-	-	-	5	-	9		•9	•	d				v			1.7-5.5	1.2-5	0.15	0.5	0.4	21[f]	2.5	100[t]	0	-	1.08
TPS791,2,3xx	-	-	-	-	5,6	-			•5[y]	•				Δ	v			2.7-5.5	1.2-5.2	0.1, 0.2	2[tt]	0.17	16[f]	all	1200	1[c]	0.01,1	1.15
NCP700B	-	-	-	-	5	-	6		-4	•								2.6-5.5	1.8-3.3	0.2	2.5[tt]	0.07	10[f]	1.8	1000	1	0,1	0.33
TPS730xx	-	-	-	-	5,6	-			•7	•				Δ				2.7-5.5	1.2-5.5	0.2	2[tt]	0.17	33[f,e1]	all	500	2.2[c]	0.01,1	0.65
MIC5249	-	-	-	8	-	-			•6					x			•	2.7-6	1.8-3.3	0.3	1	0.085	-	-	3000	2.2[c]	0,0.3	0.98
TPS732, 36xx	-	-	-	5	5	-	8		•8	•	d		•	•	v			2.2-5.5	1.2-5.5	0.25, 0.4	0.5	0.4	21[f]	2.5	100[t]	1	0,-	1.41
TPS794, 95xx	-	-	8	5	-	-			•5	•	d,w			Δ	v			2.7-5.5	1.2-5.5	0.25, 0.5	3[tt]	0.17	33[f]	all	1200	1[c]	0.01,1	1.88
LP3879	-	-	8P	-	-	-	8		-2	•				Δ	•			2.5-6	1.0-1.2	0.8	1	0.2	18[f,e1]	all	140	10[c]	0,0.2	1.26
LP3961, 64	5	5	-	-	3	-			-4	•				Δ	z		z	2.5-7	1.2-5.5	0.8	1.5	3	150	all	30[t]	33[c]	0.2,5	1.83
TPS725xx	-	5	-	8	5	-			-4	•				Δ	v		r	1.8-6	1.2-5.5	1.0	2[tt]	0.075	-	-	1500	0[c]	-	2.77
TPS726xx	-	5	-	5	-	-			-5	•				Δ	v		p	1.8-6	1.2-2.5	1.0	2[tt]	0.075	-	-	1500	0[c]	-	3.37
TPS737xx[G]	-	-	-	5	-	-	8		•3	•	d			•	v			2.2-5.5	1.2-5.5	1.0	1	0.4	68	2.5	100	1	0,-	1.64
TPS796xx	-	5	-	5	-	-	8		•7	•	d			Δ	v			2.7-5.5	1.2-5.5	1.0	2[tt]	0.27	41[f,e1]	all	1200	1	0.01,1	2.44
LP3891, 92	5	5	-	8P	-	-			-3					x			•	1.3-5.5	1.2-1.8	0.8, 1.5	1.5	4	150	1.8	100[t]	10	0.01,4	2.25
ADP3338, 39	-	-	-	-	3	-			-7					x				2.7-8	1.5-5	1, 1.5	0.8	0.11	95	all	400[t]	1	0,-	2.30
LP3871 - 76	5	5	-	5	-	-			-4	•				Δ	z		z	2.5-7	1.22-5.5	0.8,1.5,3	1.5	5	150	2.5	40[t]	10[c]	0.1,4	2.23
LP3881,2,3	5	5	-	8P	-	-			-3	•				x			•	1.3-5.5	1.2-1.8	0.8,1.5,3	1.5	4	150	all	100[t]	4.7[e]	0.01,4	4.23
LP38851,2,3	7	7	-	8P	-	-			-3	•				•				3.0-5.5	0.8-1.8	0.8,1.5,3	1.5	10	150	0.7	400	10	0,-	1.77
LP3855	5	5	-	-	-	-			-3	•				Δ	v			2.5-7	1.22-5.5	1.5	1.5	3	150	2.5	35[t]	10[c]	0.1,5	3.86
TPS786xx	-	5	-	5	-	-			•5	•	d			Δ	v			2.7-6	1.2-5.5	1.5	2[tt]	0.26	49[f,e1]	all	1000[t]	1[c]	0.01,0.5	4.69
XRP6272	-	5	-	8P	-	-			•1	•							•	0.7-5	1.8-6	2	2	0.03	24[f]	all	3000	10[c]	0,-	1.19
LP38855 - 59	5	5	-	-	-	-			-3		z		z	•				3.0-5.5	0.8-1.2	1.5, 3	1	10	150	all	1000[t]	10[c]	0,-	1.53
LP3853, 56	5	5	-	-	-	-			-4	•				Δ	z		z	2.5-7	1.8-5	3	1.5	4	150	2.5	40[t]	10[c]	0.2,5	4.40
TPS744xx	-	7	-	-	-	-	25		-4	•				Δ	v			0.9-5.5	0.8-3.6	3	1[tt]	2	40[e1]	2.5	500	0	-	4.58
TPS755 -59xx	5	5	-	-	-	-			-4	•				Δ	v		d,z	2.8-6	1.2-5.5	3,5,7.5	2	0.125	35[e3]	1.5	1000	47[c]	0.2,10	6.27
TLE4275	5	5	-	14P	-	-			•4	•				□	v			6-42	1.2-5	5	2[tt]	0.275	-	-	600	22[c]	0.3,5	**1.14**
Negative LDOs																												
MAX664	-	-	8	8	-	-			•1	•				x		•		2-16	1.3-16	0.05	5[tt]	0.006	-	-	3500	0	-	2.78
LT1964	-	-	-	-	5	-	8		•1	•	q			□		•		2.8-20	1.22-20	0.2	1.5	0.03	30[f]	all	3000	1	0,3	2.85
TPS7A3001	-	-	-	8P	-	-			•1	•				x		•		3-35	1.18-33	0.2	1.5	0.055	18[f]	5	60	2.2[c]	0,-	3.89
MAX1735	-	-	-	-	5	-			•3	•				x		•		2.5-6.5	1.25-5.5	0.2	1	0.085	160[e4]	all	1500	1	0,0.1	2.33
TPS723xx	-	-	-	-	5	-			•1	•	d			x	d			2.7-10	1.2-10	0.2	1	0.13	60	all	400[t]	2.2	0,-	2.57
LT1175	5	5	8	8	3	-			•									4-20	3.8-18	0.5	1.5	0.045	-	-	150	0.1	0,10	3.78
UCC384	-	-	8	-	-	-			•2					x	d			3.2-15	1.25-15	0.5	1.5	0.2	200[e2]	5	200	4.7	0,-	4.82
TPS7A3301	7	-	-	-	-	-	20		•					x				3-35	1.2-33	1	1.5	0.21	16[f]	1.2	60	10[c]	0,-	4.95
LM2991	5	5	-	-	-	-			•	•				x				3-25	3-24	1	2	0.7	200	3	400	1	0.025,10	1.88
LM2990	3	3	16	-	-	-			-4					x				6-25	5-15	1	2	1	250	5	2000	10	0.025,10	1.88
LT3015	5	5	-	12	-	-	8		•6	•				□	v			1.8-30	1.22-29	1.5	1	1.1	60	all	800	10	0,0.5	4.81
LT1185	5	5	-	-	-	-			•					x				4-30	2.5-25	3	1	2.5	-	-	100	2	0.1,2	4.40

Notes: (a) "Low Dropout" (LDO) regulators typically drop <100mV, except at the highest currents; sorted approximately by ascending I_Q and I_{out} capability, with some family grouping; the ADJ versions of some parts have less accuracy than the fixed versions (which can be laser trimmed). (b) 10Hz-100kHz, with no noise reduction cap, unless indicated otherwise. (c) input and other capacitors also required. (d) fixed voltage versions only. (e1) 100Hz-100kHz or 200Hz-100kHz. (e2) 10Hz-10kHz. (e3) 300Hz-50kHz. (e4) 10Hz-1MHz. (f) with NR cap. (f2) if the ADJ pin is bypassed, larger output capacitors are required, see datasheet. (g) output-input differential. (h) minimum load current. (k) qty 1k or more. (m) min or max. (n) **bold italic** indicates an inexpensive "jellybean." (nV) nV/√Hz, with 10nF filter cap. (o) "load-dump": withstands input voltage transients (or continuous inputs) to 60V (or 40V for some parts), and does not pass the spike through to the output; often associated with "automotive" parts. (p) power good. (q) some pkgs only. (r) delayed reset output for µC. (rec) recommended. (s) slew control. (t) typical. (ti) some of TI's parts have an internal EEPROM, allowing quick custom factory voltage programming. (tt) over operating temperature range. (u) unit qty. (v) adjustable version only. (v2) filter cap across feedback resistor. (w) ADJ and filter avail in 8-pin pkg. (x) reversed input voltage with excess current flow may damage part, a Schottky diode is recommended; although not marked "Δ," these parts may have a reverse-diode function. (y) check datasheet. (z) choose between two versions.

Δ = input-output back diode conducts, limit current; also, reverse supply not allowed, see note x. □ = protected against reverse polarity input.

Comments: A: current-programmed; $V_{control}$ dropout 1.6V max; 1mA min load. **B:** 100V for -HV version. **C:** -C suffix for looser tolerance and tempco, etc. **D:** ADP7104 suggested alternative. **E:** see also TI's TL1963A, $2.65. **F:** control and pass-transistor inputs. **G:** internal charge pump.

B. Polarity and bypassing

As we mentioned earlier, the negative versions of common positive regulators have a different output topology (an *npn* common emitter stage), and require larger bypass capacitors to prevent oscillation. Always "go by the book" (the datasheet, that is) – don't assume you know better. Also, be careful to connect the bypass capacitors with correct polarity (and see next).

C. Reverse polarity protection

An additional caution with dual supplies (whether regulated or not): almost any electronic circuit will be damaged extensively if the supply voltages are reversed. The only way that can happen with a single supply is if you connect the wires backward; sometimes you see a high-current rectifier connected across the circuit in the reverse direction to protect against this error. With circuits that use several supply voltages (a split supply, for instance), extensive damage can result if there is a component failure that shorts the two supplies together; a common situation is a collector-to-emitter short in one transistor of a push–pull pair operating between the supplies. In that case the two supplies find themselves tied together, and one of the regulators will win out. The opposite supply voltage is then reversed in polarity, and the circuit starts to smoke. Even in the absence of a fault condition like this, asymmetrical loads can cause a polarity reversal when power is turned off. For these reasons it is wise to connect a power rectifier (preferably Schottky) in the reverse direction from each regulated output to ground, as we drew in Figure 9.8.

Some regulator ICs are designed to block any current flow if the input voltage is lower than the output; these are marked with a bullet dot symbol (●) in the "reverse block" column in Table 9.3. Other regulator ICs go further and also block current flow for reversed *input* polarity; these are marked with a square symbol (□).

D. Ground-pin current

A particular idiosyncrasy of bipolar low-dropout regulators with *pnp* output stages (Figure 9.20) is the sharp rise in ground-pin current when the regulator is close to dropout. At that point the output stage is near saturation, with greatly reduced beta, and therefore requires substantial sinking base current. This is particularly noticeable when the regulator is lightly loaded, or unloaded, when it would otherwise have only a small ground-pin or quiescent current. As an example, the bipolar LT1764A-3.3 (fixed 3.3 V LDO), driving a 100 mA load, has a normal ground-pin current of about 5 mA, rising to ~50 mA at dropout. The no-load quiescent current shows similar behavior, spiking to ~30 mA from its normal ~1 mA.[23] Manufacturers rarely advertise this "feature" on the front page of their datasheets, but you can find it inside, if you look. It is of particular importance in battery-operated devices. *A caution*: the quiescent ground-pin currents listed in the I_Q column of Table 9.3 are for a light load and with the input voltage above dropout.

E. Maximum input voltage

Table 9.3 lists the maximum specified input voltage for more than a hundred LDO linear regulators. CMOS regulators are good choices for low-voltage designs, and they come in a dazzling variety of fixed voltages (and, of course, adjustable versions): for example, the TPS7xxxx series from Texas Instruments includes a dozen types, each of which comes in a choice of 1.2, 1.5, 1.8, 2.5, 3.0, 3.3, or 5.0 volt outputs. But be careful, because many of these CMOS regulators have a specified maximum input voltage of just +5.5 V.[24] Some CMOS regulators, however, accept up to +10 V input; for higher input voltage you've got to use bipolar regulators, for example the LT1764A or LT3012, with input voltage ranges of +2.7–20 V and

---

[23] The load-induced additional ground current would normally be I_{load}/β driving the base of the *pnp* pass transistor, but during dropout the ever-zealous feedback loop provides a base drive appropriate for the LDO's maximum load-current rating. Some designs carefully limit this drive current, whereas others detect the saturation condition and limit the current accordingly. You've got to pay close attention to this behavior of candidate LDOs when designing battery operated devices, if you want to maximize remaining operating time after the battery voltage has dropped below the LDO's input criteria. Alternatively you may want to select an LDO that uses a *p*-channel MOSFET pass transistor and does not exhibit increased ground current at high loads or during dropout. For example, a 5 V regulator like the LT3008-5 runs at 3 μA, but this soars to 30 μA if the battery drops below 5 V. However, a TLV70450 (with internal *pMOS* pass transistor) suffers no increase at all, continuing to draw 3 μA under the same conditions.

[24] Contemporary regulator ICs, with their nanometer-scale features, are more susceptible to overvoltage transients and the like, as compared with robust legacy parts with their relatively large bipolar transistors. We've seen painful experiences, for example when a carefully designed and tested small PCB, filled to the brim with tiny parts, suffers unexplained failures in the field or at the customer's test sites. Sometimes this is due to uncontrolled (and arguably improper) user-supplied input transients. Adding a transient voltage suppressor (see Chapter 9x) at the input is a wise precaution for regulator ICs powered from an off-card dc source. Low-voltage ICs (such as those with 6 V or 7 V absolute-maximum ratings) are best powered from an on-card regulated 5 V, etc., rather than from external sources of power. (An exception might be made for a 3.7 V Li-ion cell.) Be very careful!

4–80 V, respectively. See §9.13.2 for an interesting way to extend the input-voltage range, to as much as +1 kV!

F. LDO Stability

It's worth repeating that low-dropout regulators can be quite fussy about bypassing (see the comments in §9.3.7) and that there are large differences among different types. For example, the LDO Selection Guide from Texas Instruments includes a C_{out} column, whose entries range from "No Cap" to "$100\,\mu$F tantalum." The symptoms of instability may manifest themselves as incorrect, or even zero, output voltage. The latter symptom flummoxed one of our students, who replaced an LP2950 (fixed 5 V LDO) several times before the real culprit was identified: he used a $0.1\,\mu$F ceramic bypass capacitor, which is less than the specified $1\,\mu$F minimum, and also whose equivalent series resistance (ESR; see §1x.3) is too low, a hazard discussed in the Application Hints section of the regulator's datasheet.[25] A more serious symptom of oscillation is output *overvoltage*: we had a circuit with an LM2940 LDO (+5 V, 1 A) that was bypassed in error with $0.22\,\mu$F (instead of $22\,\mu$F); its internal oscillation caused the measured dc output to go to 7.5 V!

Table 9.3 has two columns to assist in selecting a part, C_{out} (min) and ESR (min,max). But these numbers should be considered a rough guide, and they do not capture everything you need to know to ensure proper operation. You'll find more guidance (e.g., contours of stable operation versus capacitance, ESR, and load current; see for example Figure 9.27) in the plots and applications section of the datasheet – study it carefully!

G. Transient response

Because voltage regulators must be stable into any capacitive load (the sum of all downstream bypass capacitance, often many μF), their feedback bandwidth is limited (analogous to op-amp "compensation"), with typical loop bandwidths in the range of tens to hundreds of kilohertz. So you rely on the output capacitor(s) to maintain low impedance

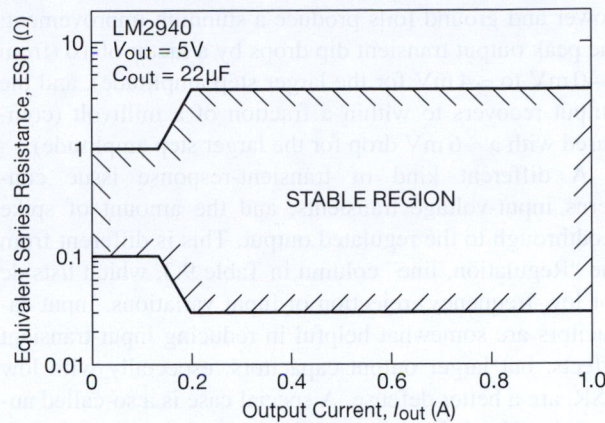

Figure 9.27. Low-dropout linear regulators can set rather fussy requirements for the ESR of the output capacitor, as seen for the LM2940; often you have to obey both minimum and maximum bounds – beware!

at higher frequencies. Or, to say it another way, the output capacitor(s) are responsible for holding the output voltage constant in the short term, in response to a step change in load current, until the regulator responds in the longer term. It's particularly important to include capacitors of low ESR (and equivalent series inductance, ESL) in the mix when you have low-voltage loads with abrupt current changes, as, for example, with microprocessors (which may generate steps of many amps).

We rigged up a 1 V 6 A LDO regulator, using the Micrel MIC5191 control chip, and measured the output response when we made abrupt load steps between 2 A and 4 A, and between 1 A and 5 A. We compared the transient response with two prototype configurations: (a) on a solderable protoboard, using mostly through-hole components; and (b) on a carefully laid-out[26] printed circuit board, using mostly surface-mount components, and with plenty of additional capacitance at both input and output.[27] Figures 9.28–9.31 show the measured step responses. The use of plentiful low-inductance surface-mount technology (SMT) capacitors and low resistance (and low inductance)

---

[25] In these words: "Ceramic capacitors whose value is greater than 1000 pF should not be connected directly from the LP2951 output to ground. Ceramic capacitors typically have ESR values in the range of 5 to 10 mΩ, a value below the lower limit for stable operation (see curve Output Capacitor ESR Range). The reason for the lower ESR limit is that the loop compensation of the part relies on the ESR of the output capacitor to provide the zero that gives the added phase lead. The ESR of ceramic capacitors is so low that this phase lead does not occur, significantly reducing phase margin. A ceramic output capacitor can be used if a series resistance is added (recommended value of resistance about $0.1\,\Omega$ to $2\,\Omega$)."

[26] Expertly done by our student Curtis Mead.

[27] Specifically, for the through-hole configuration we used a $10\,\mu$F radial tantalum and two $0.1\,\mu$F ceramic bypass capacitors at the input, and a $47\,\mu$F ceramic (X5R) SMT plus another $10\,\mu$F radial tantalum at the output. For the surface-mount configuration we used a $560\,\mu$F radial aluminum polymer capacitor, a $100\,\mu$F SMT tantalum polymer capacitor, and two $22\,\mu$F ceramic (X5R, 0805) SMT capacitors at both input and output, plus two more $10\,\mu$F ceramic (X5R, 0805) SMT capacitors at the output.

power and ground foils produce a stunning improvement: the peak output transient dip drops by a factor of 10 (from ~40 mV to ~4 mV for the larger step amplitude), and the output recovers to within a fraction of a millivolt (compared with a ~6 mV drop for the larger step amplitude).

A different kind of transient-response issue concerns input-voltage transients, and the amount of spike feedthrough to the regulated output. This is different from the "Regulation, line" column in Table 9.3, which lists dc (or low-frequency) rejection of input variations. Input capacitors are somewhat helpful in reducing input-transient effects, but larger output capacitors, especially with low ESR, are a better defense. A special case is a so-called automotive "load dump," a rapid input spike caused, for example, by accidental disconnection of the car battery (from a loose connection, or corrosion, or human error) while being charged by the alternator. This can cause the normal 13.8 V power rail to spike to amplitudes of 50 V or more, causing a spike at a regulator's output. Worse, it can destroy the IC by exceeding its maximum specified input voltage (the "V_{in} max" column of Table 9.3). Parts specifically designed to deal with load dumps are marked with an "o" note in the corresponding V_{in} entry.

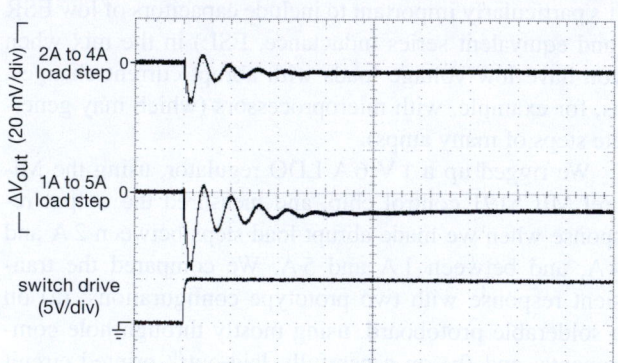

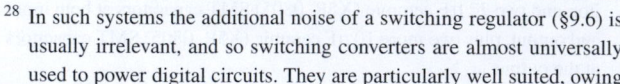

Figure 9.28. Output voltage response to a step increase in load current: 1 V 6 A LDO regulator, breadboarded primarily with through-hole components. Horizontal: 4 μs/div.

H. Noise

Linear regulators vary considerably in the level of output noise (that is, spectrum of output-voltage fluctuations). In many situations this may be unimportant, for example in a digital system, where the circuit itself is inherently noisy.[28] But for low-level or precision analog electronics where

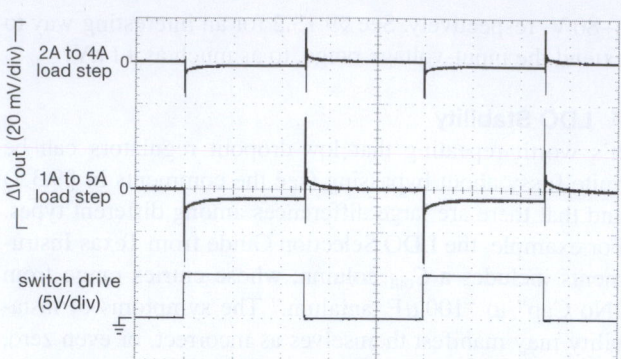

Figure 9.29. Same as Figure 9.28, with 'scope slowed to 400 μs/div to show full load cycle.

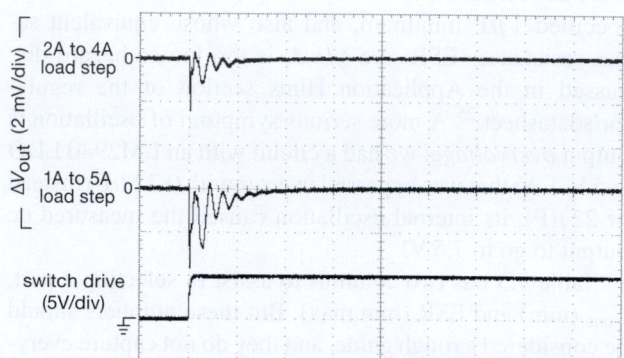

Figure 9.30. Same as Figure 9.28, but built on a printed circuit board using surface-mount capacitors. Note expanded vertical scale.

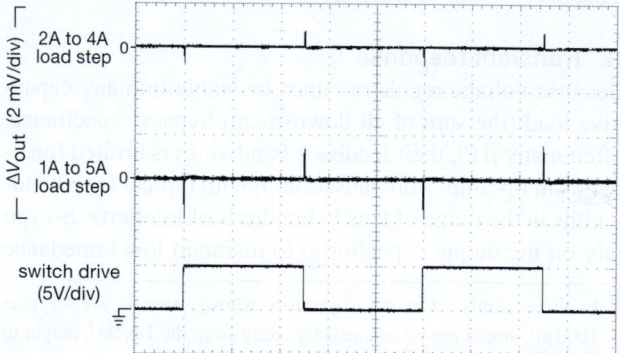

Figure 9.31. Same as Figure 9.30, with 'scope slowed to 400 μs/div to show full load cycle.

noise is important, there are regulators with superior noise specifications, for example the LT1764/1963 (40 μVrms,

---

[28] In such systems the additional noise of a switching regulator (§9.6) is usually irrelevant, and so switching converters are almost universally used to power digital circuits. They are particularly well suited, owing

to their small size and high efficiency, and especially so for the low dc supply voltages (~1.0–3.3 V) used in digital logic.

10 Hz–100 kHz) or the ADP7102/04 (15 μVrms). Additionally, some regulators provide access to the internal voltage reference, so that an external filter capacitor can be added to suppress all but the low-frequency end of the noise spectrum, for example the LT1964 negative regulator (30 μVrms with 10 nF capacitor); see §9.3.13.

Because manufacturers specify noise characteristics differently (bandwidth, rms versus peak-to-peak, etc.), it can be difficult to compare candidate parts. We've made an attempt in the V_n@V_{out} columns of Table 9.3, but be sure to consult the footnotes (and then the datasheets).

I. Shutdown protection

Some regulator types can be damaged if they see a large capacitance at their output, and the input voltage is brought abruptly to zero (e.g., by a crowbar, or an accidental short-circuit). In that situation the charged output capacitance can source a destructive current back into the regulator's output terminal. Figure 9.32 shows how to prevent such damage, in this case with the popular LM317. Although many engineers don't bother with this nicety, it is the mark of a careful circuit designer. A similar hazard exists when an external bypass capacitor is used to filter the voltage noise of the regulator's reference; see §9.3.13.[29]

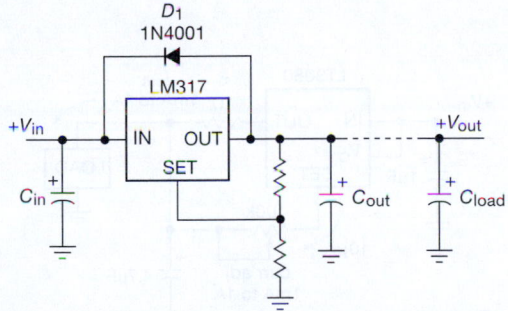

Figure 9.32. Diode D_1 protects the regulator if the input is suddenly grounded.

9.3.13 Noise and ripple filtering

The output noise from a linear regulator is caused by noise in the reference, multiplied by the ratio of V_{out}/V_{ref}, combined with noise in the error amplifier, and with noise and

ripple at the input terminal that is not completely suppressed by feedback.[30] Some regulators let you add an external capacitor for lowpass filtering of the internal voltage reference and thus the dc output. Figure 9.33 shows several examples. In Figure 9.33A the ADJ pin of the LM317-style 3-terminal adjustable regulator is bypassed to ground; this provides significant noise improvement by preventing multiplication of the reference noise voltage by the factor 1 + R_2/R_1 (the ratio of output voltage to 1.25 V reference voltage). It also improves the input ripple rejection ratio, from 65 dB to 80 dB (typical), according to the datasheet. Note the additional protection diode D_2, needed if the noise bypass capacitor C_1 is greater than 10 μF.

This scheme does not eliminate the reference noise, it just prevents it (an ac signal) from being "gained up" by the dc gain factor V_{out}/V_{ref}. The noise filtering in Figures 9.33B and C is more effective, because it filters the reference voltage directly. In Figure 9.33B the LT3080's SET pin, sourcing a stable 10 μA current, is converted to the output voltage by R_{SET}, filtered by C_1; the regulator's output is a unity-gain replica of this filtered voltage. With $C_1 = 0.1\,\mu$F, the reference noise is less than that of the error amplifier, producing an output noise of $\sim$40 μVrms (10 Hz–100 kHz). Note that the noise-filtering capacitor slows the regulator's startup: a 0.1 μF capacitor in a 10 V regulator circuit (R_{SET}=1 MΩ) has a startup time constant $R_{SET}C_1$ of 100 ms.

Finally, Figure 9.33C shows a CMOS low-dropout regulator with a dedicated noise-reduction (NR) pin, for filtering directly the reference voltage presented to the error amplifier. With the recommended 0.1 μF capacitor the output noise voltage is $\sim$40 μVrms (100 Hz–100 kHz). Regulators with this feature are marked in the "Filter pin" column of Table 9.3.

Prefiltering An effective way of reducing dramatically the output ripple at the powerline frequency (and its harmonics) is to prefilter the dc *input* to the regulator. This is also highly effective in attenuating broadband noise that may be present at the dc input; and it's easier than the alternative of increasing the regulator loop's gain and bandwidth. We discuss this in some detail in §8.15.1 ("Capacitance multiplier"), where we show the measured effects of prefiltering versus the brute-force approach of piling on lots of output capacitance (Figure 8.122).

See further discussion of noise in §9.10, in connection with voltage references.

---

[29] Many LDO regulator ICs include an internal diode sufficiently hardy to handle the reverse-discharge energy in modest (e.g., $\leq$10 μF) load capacitors. These are marked with a triangle symbol ($\triangle$) in the "reverse block" column in Table 9.3. Other parts do not discharge the output capacitor if the input voltage is taken below the output; these ICs are marked with either a bullet dot ($\bullet$) or a square ($\square$).

[30] Table 9.3 includes a "Line regulation" column, but note that this is at dc and low frequencies where the loop gain is high; it is not necessarily indicative of high-frequency supply-noise rejection.

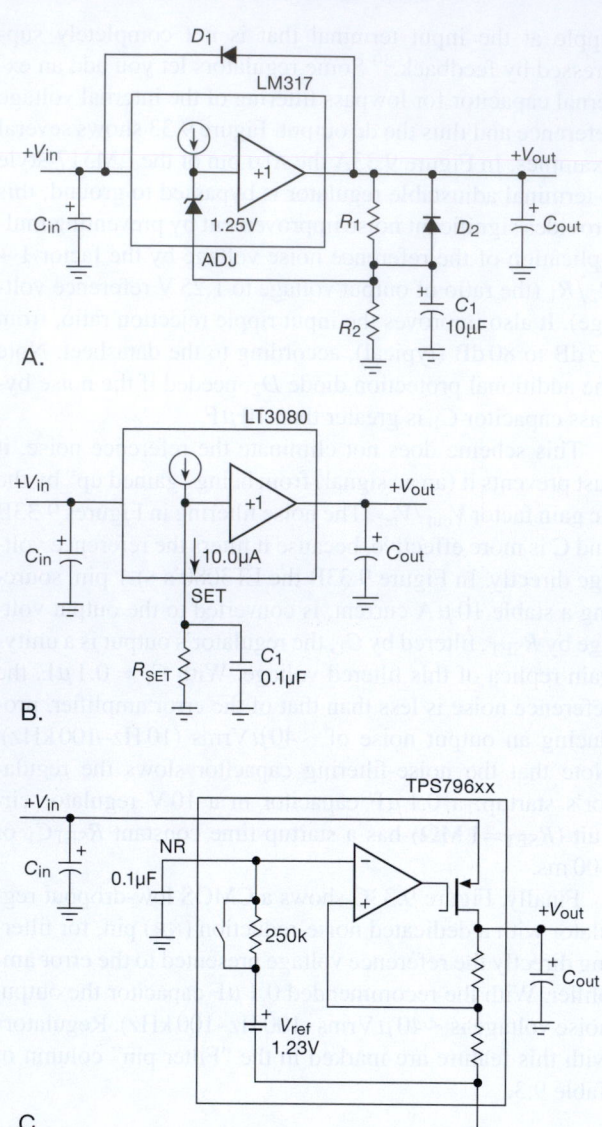

Figure 9.33. Reducing output-voltage noise (and improving transient line regulation) in linear regulators.

9.3.14 Current sources

A. Three-terminal regulators as current sources

A 3-terminal linear regulator can be used to make a simple current source by putting a resistor across the regulated output voltage (hence constant current $I_R = V_{reg}/R$), and floating the whole thing on top of a load returned to ground (Figure 9.34A). The current source is imperfect, however, because the regulator's operating current I_{reg} (which comes out of the ground pin) is combined with the well-controlled resistor current to produce a total output

current $I_{out} = V_{reg}/R + I_{reg}$. It's a reasonable current source, though, for output currents much larger than the regulator's operating current.

Originally this circuit had been implemented with a 7805, which has an operating current of ~3 mA and additionally has the disadvantage of squandering a rather large 5 V (the lowest-voltage part in the 78xx series) to define the

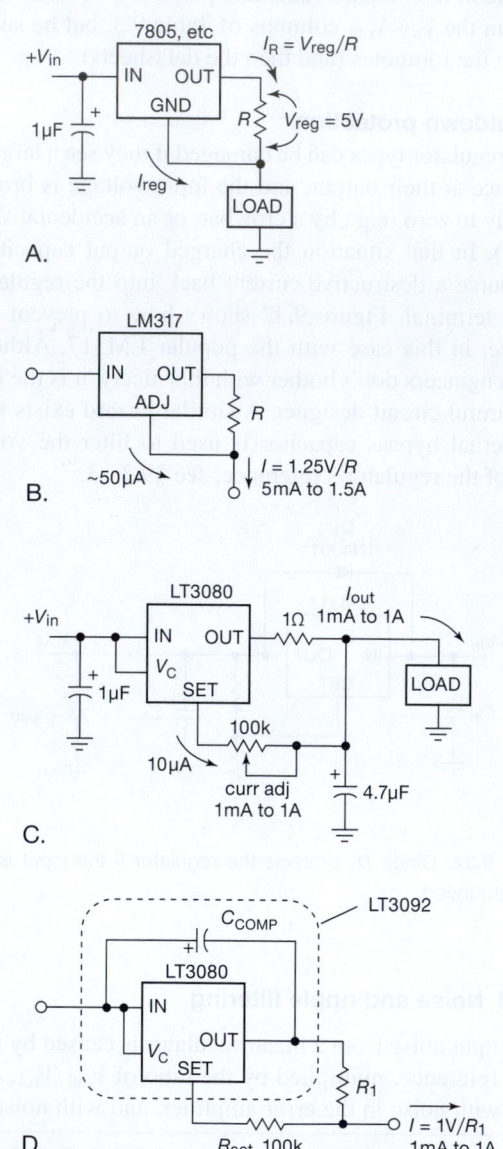

Figure 9.34. Three-terminal regulators used as current sources. The bypass and compensation capacitors can be eliminated in circuits C and D if the LT3080 is replaced by its internally compensated LT3092 variant (whose output current is limited to a maximum of 200 mA).

output current. Happily, with regulators like the LM317, this circuit (Figure 9.34B) becomes more attractive: only 1.25 V is used to set the current; and the regulator's operating current ($\sim$5 mA) emerges from the *output* pin, and thus is accurately accounted for in setting $I_{out} = V_{ref}/R$. The only error term is the ADJ pin current of $\sim$50 μA, which gets added to the current through R: $I_{out} = V_{ref}/R + I_{ADJ}$. Because the minimum output current of 5 mA is 100 times larger, that is a small error even at the minimum output current, and smaller still at currents up to the regulator's maximum of 1.5 A. For this circuit, then, the output-current range is 5 mA–1.5 A. It requires a minimum voltage drop of 1.25 V plus the regulator's dropout voltage, or about 3 V; the maximum voltage *across the two terminals* is limited either to 40 V or (at higher currents) to the maximum junction temperature of 125°C (as determined by power dissipation and heatsinking), whichever is less.[31]

With the admirable LT3080-style regulator you can do better still, because its 10 μA SET-pin current reference lets you set the voltage across the current-setting resistor to be much less than the 1.25 V of a 317-style voltage-reference regulator. Its operating current is smaller as well ($<$1 mA), and the SET pin current (which gets added to the output current) is a stable and accurate 10.0 μA. Figure 9.34C shows how to make a (1-terminal) current source to ground with an LT3080, and Figure 9.34D shows how to make a 2-terminal "floating" current source, analogous to the LM317 current-source circuit. As with the latter, the voltage drop is limited to a maximum of 40 V (less at higher currents) at the high end; its lower dropout and low SET-derived reference voltage allows operation down to $\sim$1.5 V drop. The LT3092-series is a nice variant of the LT3080, designed specifically for use as a 2-terminal current source. It uses the same 10 μA reference current, and operates from 1.2 V to 40 V drop; its internal compensation is configured to require *no* external bypass or compensation capacitors. Based on the LT3092's datasheet plot of output impedance, the device's effective parallel capacitance is approximately 100 pF at 1 mA, 800 pF at 10 mA, and 6 nF at 100 mA.

Figure 9.35 shows, on a greatly expanded scale, the measured output currents of an LT3092 and an LM317, configured as 10 mA current sources. In our measurements the latter does a better job of maintaining constant cur-

rent (versus voltage across it), but the LT3092 kicks in at a lower voltage.

Note that the current sources in Figures 9.34B and D are 2-terminal devices. Thus the load can be connected on either side. For example, you could use such a circuit to *sink* current from a load returned to ground, by connecting the load between ground and the input, and connecting the "output" to a negative voltage (of course, you could always use the negative-polarity 337, in a configuration analogous to Figure 9.34A.).

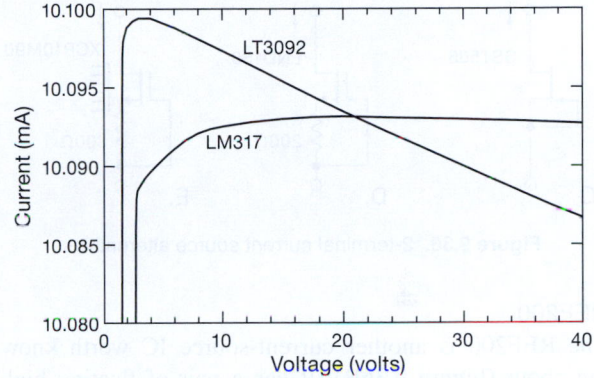

Figure 9.35. Measured current versus voltage drop for the current sources of Figures 9.34B and D, configured as 10 mA 2-terminal current sources. For the LM317, R_1=124 Ω; for the LT3092, R_1=20 Ω and R_{SET}=20k.

B. Lower currents
The above regulator-derived current sources are best suited for substantial output currents. For lower currents, or for higher voltages, there are some good alternatives.

LM334
It's worth knowing about the LM334 (originated by National Semiconductor), optimized for use as a low-power 2-terminal current source (Figure 9.36A). It comes in small-outline IC (SOIC) and TO-92 (transistor) packages and costs about $1 in small quantities. You can use it all the way down to 1 μA because the ADJ current is a small fraction of the total current; and it operates over a voltage range of 1–40 V. It has one peculiarity, however: the output current is temperature dependent – in fact, precisely proportional to absolute temperature (PTAT). So although it is not the world's most stable current source, you can use it as a temperature sensor! At room temperature (20°C, $\sim$293K) its tempco is about +0.34%/°C.[32]

---

[31] While we haven't experienced this ourselves, we've been told of possible issues with LM317-based current sources, such as long turn-on times, voltage retention, and poor voltage-compliance above a few kilohertz. It's always wise to test circuit performance fully (especially, uh, *creative* circuits).

[32] See also the discussion in §2x.3.

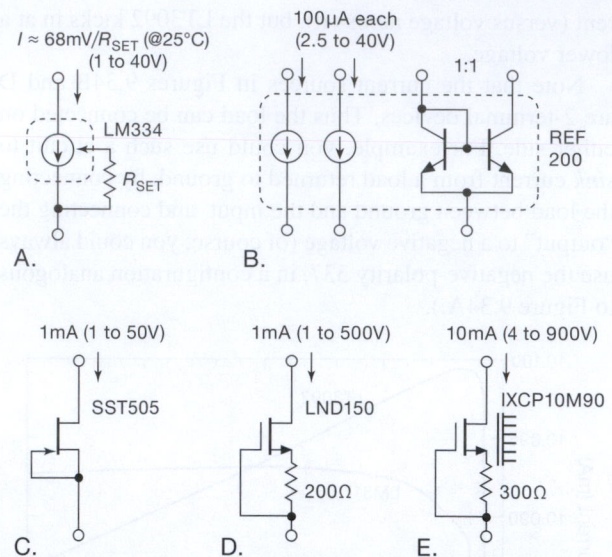

Figure 9.36. 2-terminal current source alternatives.

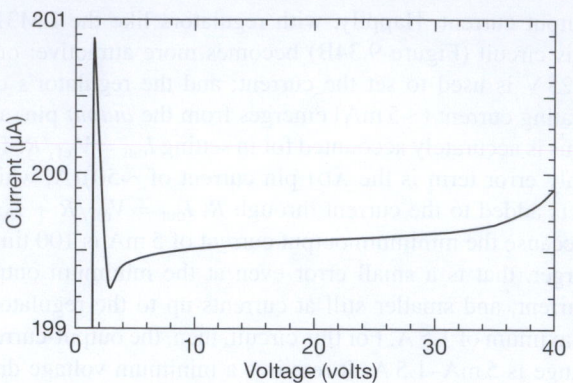

Figure 9.37. Measured current versus voltage for the REF200 two-terminal current source (parallel connection of the 100 μA pair).

REF200

The REF200 is another current-source IC worth knowing about (Figure 9.36B). It has a pair of floating high-quality 100 μA ($\pm0.5\%$) 2-terminal current sources (output impedance $>200\,\text{M}\Omega$ over a voltage range of 3.5 V–30 V). It comes in dual in-line packages (DIP) and SOIC packages and costs about $4 in small quantities. Unlike the LM334, the REF200's current sources are temperature stable ($\pm25\,\text{ppm}/^\circ$C, typ). It also has an on-chip unit-ratio current mirror, so you can make a 2-terminal current source with fixed currents of 50 μA, 100 μA, 150 μA, 200 μA, 300 μA, or 400 μA. The datasheet asserts that "applications for the REF200 are limitless," though we are skeptical. Figure 9.37 shows measured current versus voltage for the parallel connection of the 100 μA pair.

Discrete-component current sources

When thinking about current sources, don't forget about 2-terminal devices like

(a) the humble JFET "current-regulator diode" (§3.2.2), which makes a simple 2-terminal current source (Figure 9.36C) that operates nicely up to 100 V (we plotted measured current versus voltage in Figure 9.38);

(b) a discrete JFET (see Tables 3.1, 3.7, and 8.2), configured similarly as a 2-terminal current source;

(c) the analogous use of a depletion-mode MOSFET (see Table 3.6 on page 210) like the Supertex LND150 (Figure 9.36D), discussed on this page (§9.3.14C);

(d) the series of 2-terminal "Constant Current Regu-

lator and LED Driver" devices from ON Semiconductor. These are inexpensive ($0.10–$0.20 in qty 100), and are offered with selected currents (e.g., NSI50010YT1G: 10 mA, 50 V; NSIC2020BT3G: 20 mA, 120 V) and in adjustable versions (e.g., NSI45020JZ: 20–40 mA, 45 V). The datasheets don't say much about what's inside these things, but depletion-mode FETs are the likely culprits.

Op-amp current-source configurations

If the application does not require a floating current source, then consider also

(d) the simple BJT current source (§2.2.6), drawn schematically in Figure 9.39A;

(e) the op-amp assisted BJT current source (§4.2.5), Figure 9.39B;

(f) the Howland current source (§4.2.5B), Figure 9.39C.

In these last three figures the bias voltage that programs the current is drawn as a floating battery; in a circuit implementation it would be a voltage relative to ground or to a supply rail, derived from a voltage reference.

C. High-voltage discrete current source

As mentioned above in the paragraph on discrete-component current sources, a simple source-biased depletion-mode MOSFET (Figures 9.36D and E) forms a pretty good 2-terminal current source. These parts come in convenient packages (TO-92, SMT, TO-220, D^{2}PAK), with voltage ratings to 1.7 kV; familiar examples are the LND150 and DN3545 from Supertex and the IXCP10M45S and IXCP10M90S from IXYS – see Table 3.6 on page 210. Because of the uncertainty in I_D versus V_{GS}, this sort of current source is not particularly precise or

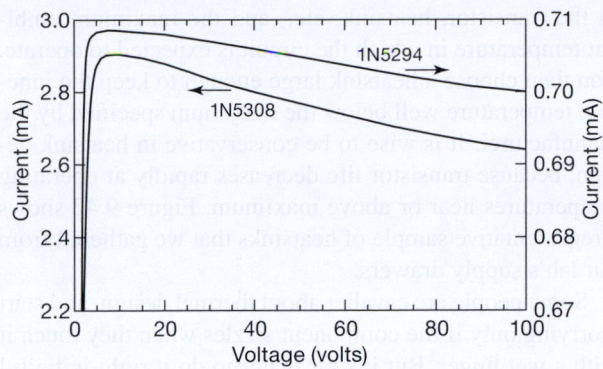

Figure 9.38. Measured current versus voltage for two members of the 1N5283 "current-regulator diode" (a JFET, actually) series.

predictable. But it's fine for noncritical applications, such as replacing a pullup resistor, and it has the advantage of operating to rather high voltages (500 V and 450 V for the Supertex parts; 450 V and 900 V for the IXYS). The IXYS datasheets call their product a "switchable current regulator." Figures 9.40 and 9.41 show some measured data for this simple circuit. The IXYS depletion-mode MOSFET line currently tops out at 1700 V (IXTH2N170). See §3x.6 for a discussion of yet-higher-voltage versions (to 3 kV or more).

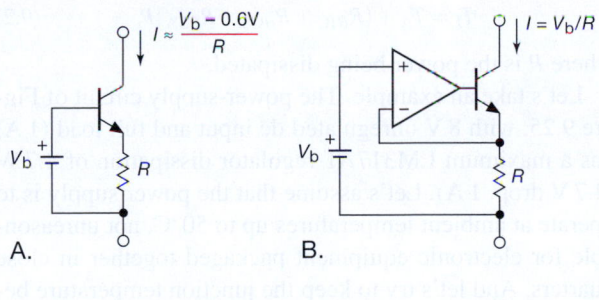

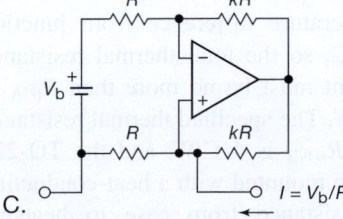

Figure 9.39. BJT and op-amp current sources, drawn in abbreviated form with a floating bias battery. For details see the relevant discussions in Chapters 2 (BJTs) and 4 (op-amps).

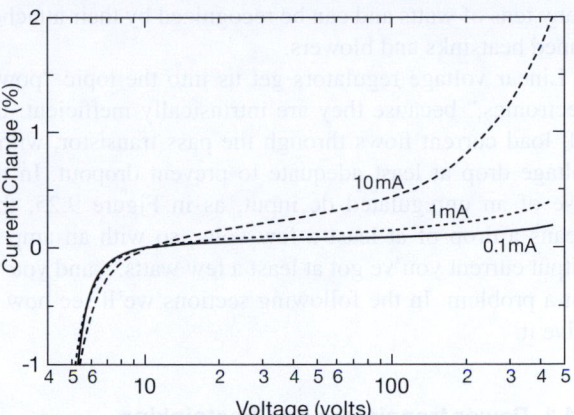

Figure 9.40. Measured current versus voltage for an IXCP10M45S depletion-mode power MOSFET, wired as a self-biased 2-terminal current source (as in Figure 9.36E).

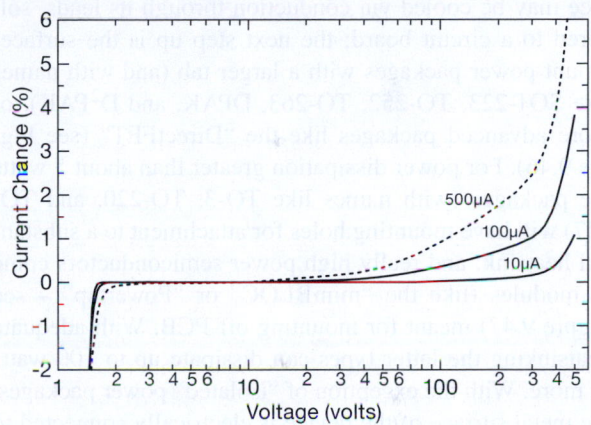

Figure 9.41. Measured current for the smaller LND150 depletion-mode MOSFET, a handy part for low-current applications (compare with Figure 9.40).

9.4 Heat and power design

Up to now we've been skirting the issue of *thermal management* – the business of dealing with the heat generated by transistors (and other power semiconductors) in which the power dissipation (the voltage drop times the current) is greater than a few tenths of a watt. The solution consists of some combination of passive cooling (conducting the heat to a heatsink or to the metal case of an instrument) and active (forced air or pumped liquid) cooling.

This problem is not unique to voltage regulators, of course – it affects linear power amplifiers, power-switching circuits, and other heat-generating components such as power resistors, rectifiers, and high-speed digital ICs. Contemporary computer processors, for example, dissipate many tens of watts and can be recognized by their attached finned heatsinks and blowers.

Linear voltage regulators get us into the topic "power electronics," because they are intrinsically inefficient: the full load current flows through the pass transistor, with a voltage drop at least adequate to prevent dropout. In the case of an unregulated dc input, as in Figure 9.25, that means a drop of at least a few volts; so with an amp of output current you've got at least a few watts... and you've got a problem. In the following sections we'll see how to solve it.

9.4.1 Power transistors and heatsinking

All power devices are packaged in cases that permit contact between a metal surface and an external heatsink. At the low-power end of the spectrum (up to a watt) the device may be cooled via conduction through its leads, soldered to a circuit board; the next step up is the surface-mount power packages with a larger tab (and with names like SOT-223, TO-252, TO-263, DPAK, and D^{2}PAK), or more advanced packages like the "DirectFET" (see Figure 9.46). For power dissipation greater than about 5 watts the packages (with names like TO-3, TO-220, and TO-247) will have mounting holes for attachment to a substantial heatsink; and really high power semiconductors come in modules (like the "miniBLOC" or "Powertap" – see Figure 9.47) meant for mounting off-PCB. With adequate heatsinking the latter types can dissipate up to 100 watts or more. With the exception of "isolated" power packages, the metal surface of the device is electrically connected to one terminal (e.g., for bipolar power transistors the case is connected to the collector, and for power MOSFETs to the drain).

The whole point of heatsinking is to keep the transistor junction (or the junction of some other device) below some maximum specified operating temperature. For silicon transistors in metal packages the maximum junction temperature is usually 200°C, whereas for transistors in plastic packages it is usually 150°C.[33] Heatsink design is then simple: knowing the maximum power the device will dissipate in a given circuit, you calculate the junction temperature, allowing for the effects of heat conductivity

---

[33] See Tables 2.2 and 3.4 for a selection of power transistors, including their maximum power dissipation assuming an (unrealistic) case temperature of 25°C. As we'll see, that's enough information to allow you to back out the thermal resistance $R_{\Theta JC}$, from which you can figure out realistic values of maximum power dissipation, and thus appropriate heatsinking.

in the transistor, heatsink, etc., and the maximum ambient temperature in which the circuit is expected to operate. You then choose a heatsink large enough to keep the junction temperature well below the maximum specified by the manufacturer. It is wise to be conservative in heatsink design, because transistor life decreases rapidly at operating temperatures near or above maximum. Figure 9.42 shows a representative sample of heatsinks that we gathered from our lab's supply drawers.

Some people are cavalier about thermal design, and start worrying only if the component sizzles when they touch it with a wet finger. But it's far better to do it right initially! Read on...

A. Thermal resistance

To carry out heatsink calculations, you use *thermal resistance*, R_θ, defined as heat rise (in °C) divided by power transferred. For power transferred entirely by heat conduction, the thermal resistance is a constant, independent of temperature, that depends only on the mechanical properties of the joint. For a succession of thermal joints in "series," the total thermal resistance is the sum of the thermal resistances of the individual joints. Thus, for a transistor mounted on a heatsink, the total thermal resistance from transistor junction to the outside (ambient) world is the sum of the thermal resistance from junction to case $R_{\theta JC}$, the thermal resistance from case to heatsink $R_{\theta CS}$, and the thermal resistance from heatsink to ambient $R_{\theta SA}$. The temperature of the junction is therefore

$$T_J = T_A + (R_{\theta JC} + R_{\theta CS} + R_{\theta SA})P, \qquad (9.2)$$

where P is the power being dissipated.

Let's take an example. The power-supply circuit of Figure 9.25, with 8 V unregulated dc input and full load (1 A) has a maximum LM317AT regulator dissipation of 4.7 W (4.7 V drop, 1 A). Let's assume that the power supply is to operate at ambient temperatures up to 50°C, not unreasonable for electronic equipment packaged together in close quarters. And let's try to keep the junction temperature below 100°C, well below its specified maximum of 125°C.

The allowable temperature difference from junction to ambient is thus 50°C, so the total thermal resistance from junction to ambient must be no more than $R_{\theta JA} = (T_J - T_A)/P = 10.6$°C/W. The specified thermal resistance from junction to case, $R_{\theta JC}$, is 4°C/W, and the TO-220 power transistor package mounted with a heat-conducting pad has a thermal resistance from case to heatsink of about 0.5°C/W. So we've used up $R_{\theta JC} + R_{\theta CS} = 4.5$°C/W of thermal resistance, leaving $R_{\theta SA} = 6.1$°C/W for the heatsink. A quick scan of the ever-helpful DigiKey

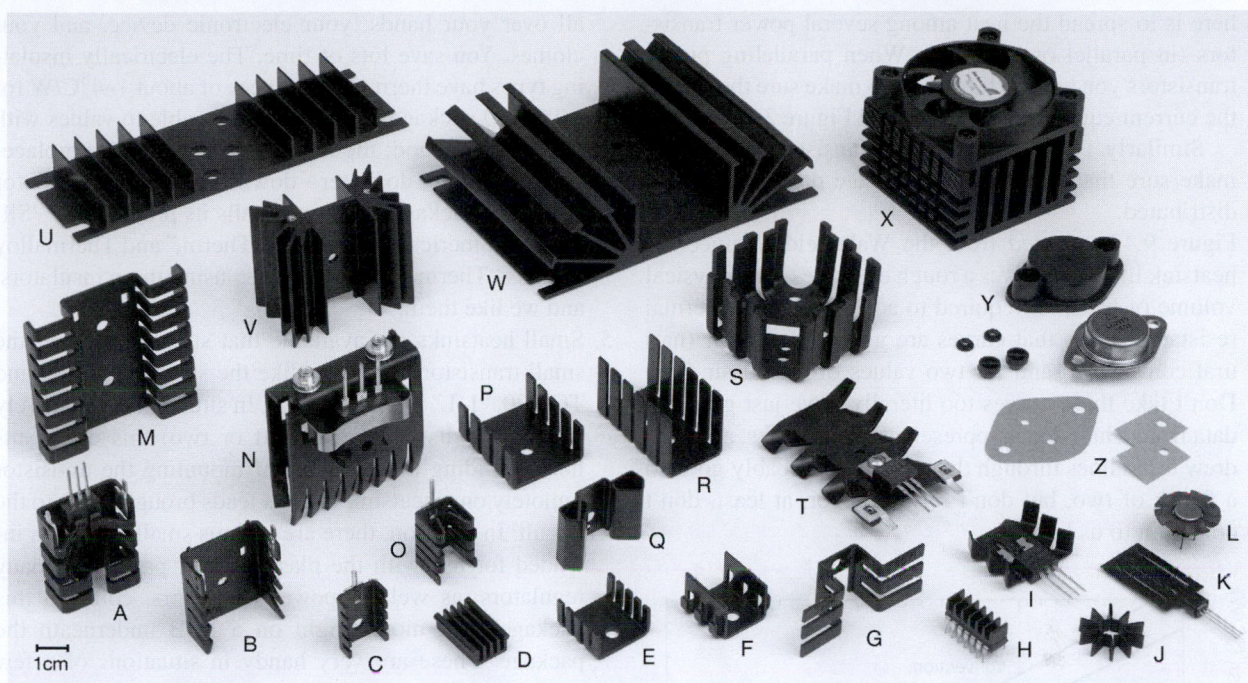

Figure 9.42. Heatsinks come in an impressive diversity, from little clip-on fins (I–L), to mid-sized PCB-mounting types (A–C, N, O, T), to large bolt-down units (U, W), to forced-air type used with microprocessors (X). The corresponding thermal resistance from sink to ambient, $R_{\Theta SA}$, ranges from about 50°C/W down to about 1.5°C/W. A TO-3 insulating cover is shown in (Y), along with shoulder washers and hole plugs; greaseless thermal insulating pads are shown in (Z). We've added alphabetic labels so readers can identify objects of interest when chatting on social media.

catalog finds many candidates, for example the Wakefield 647-15ABP "vertical board mounting" finned heatsink, with the requisite $R_{\Theta SA} = 6.1°C/W$ in still air ("natural convection"). They're priced at about $2, with 2000 pieces in stock. With a "forced convection" of 400 LFM (linear feet per minute) we could use instead the smaller (and cheaper, about $0.35) model 270-AB; DigiKey's got 4000 in stock today.

Here's a "sizzle-test" for checking for adequate heatsinking: touch the power transistor with a dampened finger – if it sizzles, it's too hot! (Be careful when using this "rule-of-finger" test to explore around high voltages.) More generally approved methods for checking component temperatures are (a) a contacting thermocouple or thermistor probe (these often come as standard equipment with handheld or benchtop digital multimeters); (b) special calibrated waxes that melt at designated temperatures (e.g., the Tempilstik® wax pencil kits from Tempil, Inc.); and (c) infrared non-contacting temperature probes,[34] for example

the Fluke 80T-IR, which generates 1 mV/°C or 1 mV/°F (switchable), operates from −18°C to +260°C, is accurate to 3% of reading (or ±3°C, if greater), and plugs into any handheld or benchtop DMM.

B. Comments on heatsinks

1. Where very high power dissipation (several hundred watts, say) is involved, forced air cooling is usually necessary. Large heatsinks designed to be used with a blower are available with thermal resistances (sink to ambient) as small as 0.05°C to 0.2°C per watt.

2. In cases of such high thermal conductivity (low thermal resistance, $R_{\Theta SA}$), you may find that the ultimate limit to power dissipation is in fact the transistor's own internal thermal resistance, combined with its attachment to the heatsink (i.e., $R_{\Theta JC} + R_{\Theta CS}$). This problem has been exacerbated in recent years by the evolution of smaller ("shrink") semiconductor chip sizes. The only solution

---

[34] Infrared ear thermometers use this method to measure core body temperature via infrared emission from the eardrum, evidently accurately

enough for clinical purposes; e.g., the Braun ThermoScan takes a measurement in one second, with an accuracy claimed to be significantly better than 1°C.

here is to spread the heat among several power transistors (in parallel or in series). When paralleling power transistors you have to be careful to make sure they share the current equally – see §2.4.4 and Figure 3.117.

Similarly, when connecting transistors in series, make sure that their off-state voltage drops are evenly distributed.

3. Figure 9.43, adapted from the Wakefield Engineering heatsink literature, gives a rough estimate of the physical volume of heatsink required to achieve a given thermal resistance. Note that curves are given for still air (natural convection) and for two values of forced air flow. Don't take these curves too literally – we just gathered data from a half dozen representative heatsinks, and then drew trend lines through them; they're probably good to a factor of two, but don't rely on it (or, at least, don't complain to us later!).

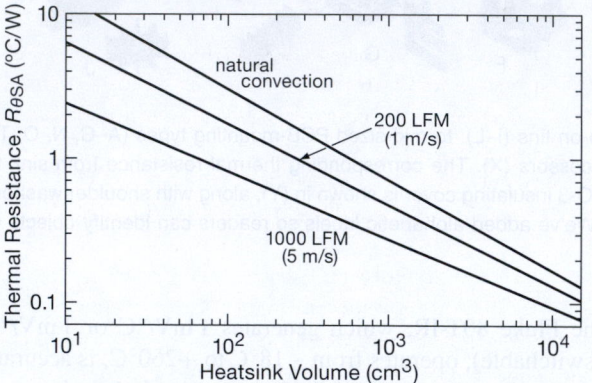

Figure 9.43. Rough guide to heatsink size needed for a given thermal resistance from sink to ambient ($R_{\theta SA}$).

4. When the transistor must be insulated from the heatsink, as is usually necessary (especially if several transistors are mounted on the same sink), a thin insulating washer is used between the transistor and sink, and insulating bushings are used around the mounting screws. Washers are available in standard transistor-shape cutouts made from mica, anodized (insulated) aluminum, beryllia (BeO), or polymer films such as Kapton®. Used with heat-conducting grease, these add from 0.14°C/W (beryllia) to about 0.5°C/W.

An attractive alternative to the classic mica-washer-plus-grease is provided by greaseless silicone-based insulators that are loaded with a dispersion of a thermally conductive compound, usually boron nitride or aluminum oxide ("Z" in Figure 9.42). They're clean and dry and easy to use; you don't get white slimy stuff

all over your hands, your electronic device, and your clothes. You save lots of time. The electrically insulating types have thermal resistances of about 1–4°C/W for a TO-220 package footprint, comparable to values with the messy method; the non-insulating ("grease replacement") varieties do better – down in the 0.1–0.5°C/W for a TO-220 package. Bergquist calls its product line "Sil-Pad," Chomerics calls its "Cho-Therm," and Thermalloy calls its "Thermasil." We've been using these insulators, and we like them.

5. Small heatsinks are available that simply clip over the small transistor packages (like the standard TO-92 and TO-220, "I–L" in Figure 9.42). In situations of relatively low power dissipation (a watt or two) this often suffices, avoiding the nuisance of mounting the transistor remotely on a heatsink with its leads brought back to the circuit. In addition, there are various small heatsinks intended for use with the plastic power packages (many regulators, as well as power transistors, come in this package) that mount right on a PCB underneath the package. These are very handy in situations of a few watts dissipation; a typical unit is illustrated in Figure 2.3. If you've got vertical space over the PCB, it's often preferable to use a PCB-mounting stand-up heatsink (like A–C, N, O, or T in Figure 9.42), because these types take up less area on the PCB.

6. Surface-mount power transistors (such as the SOT-223, DPAK, and D²PAK) carry their heat to the foil layer of a PCB via the soldered tab; here we're talking a few watts, not a hundred. You can see these packages in Figures 2.3 and 9.44. Figure 9.45 plots approximate values of thermal resistance versus foil area; these should be considered only a rough guide, because the actual heatsinking effectiveness depends on other factors such as the proximity of other heat-producing components, board stacking, and (for natural convection) board orientation.

7. Sometimes it may be convenient to mount power transistors directly to the chassis or case of the instrument. In such cases it is wise to use conservative design (keep it cool), especially because a hot case will subject the other circuit components to high temperatures and thus shorten their lives.

8. If a transistor is mounted to a heatsink without insulating hardware, the heatsink must be insulated from the chassis. The use of insulating washers (e.g., Wakefield model 103) is recommended (unless, of course, the transistor case happens to be at ground). When the transistor is insulated from the sink, the heatsink may be attached directly to the chassis. But if the transistor is accessible from outside the instrument (e.g., if the heatsink is

Figure 9.44. Power transistors come in convenient hand-solderable surface-mount packages that can dissipate up to several watts, through their mounting tab and leads, when soldered to a few square centimeters of foil area on a PCB (see Figure 9.45). The top three packages (SOT-223, DPAK, D^2PAK) are good for $\sim$3 W when mounted on 6 cm^2 of PCB foil area; the smaller packages in the row below can dissipate $\sim$0.5 W when similarly mounted. By way of comparison, the rectangular parts at lower right are chip resistors, tapering down from 2512 size to 0201 size (0603 metric).

mounted externally on the rear wall of the box), it is a good idea to use an insulating cover over the transistor (e.g., Thermalloy 8903N, "Y" in Figure 9.42) to prevent someone from accidentally coming in contact with it or shorting it to ground.

9. The thermal resistance from heatsink to ambient is usually specified for the sink mounted with the fins vertical and with unobstructed flow of air. If the sink is mounted differently, or if the air flow is obstructed, the efficiency

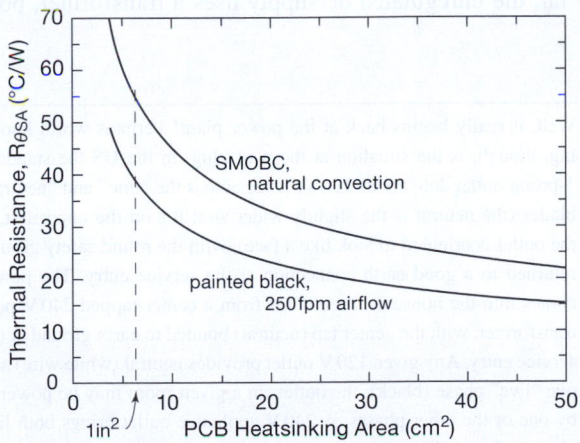

Figure 9.45. Approximate thermal resistance of isolated printed-circuit foil patterns. A soldermask layer (SMOBC means solder-mask over bare copper) reduces the effectiveness, particularly when compared with forced air over exposed copper.

will be reduced (higher thermal resistance[35]); usually it is best to mount it on the rear of the instrument with fins vertical.

10. In the second edition of this book there is additional information: see Chapter 6 (Figure 6.6, page 315) on heatsinks, and Chapter 12 (Table 12.2 and Figure 12.17, page 858) on cooling fans.

Exercise 9.7. An LM317T (TO-220 case), with a thermal resistance from junction to case of $R_{\Theta JC}$=4°C/W, is fitted with an Aavid Thermalloy 507222 bolt-on heatsink, whose thermal resistance is specified as $R_{\Theta SA}\approx$18°C/W in still air. The thermal pad (Bergquist SP400-0.007) specifies a thermal resistance of $R_{\Theta CS}\approx$5°C/W. The maximum permissible junction temperature is 125°C. How much power can you dissipate with this combination at 25°C ambient temperature? How much must the dissipation be decreased per degree rise in ambient temperature?

9.4.2 Safe operating area

The point of heatsinking is to keep junction temperatures within specified limits, given some ambient temperature and some maximum power dissipation, as just described. Of course, you must stay also within the specified voltage and current ratings of the power transistor. This is displayed graphically as a dc safe-operating-area (SOA) plot, on axes of transistor voltage and current, at some specified case temperature (usually an unrealistic T_C=25°C). For MOSFETs this plot (on logarithmic voltage and current axes) is bounded simply by straight lines representing maximum voltage, maximum current, and maximum power dissipation (at the specified T_C, as set by $R_{\theta JC}$ and $T_{J(max)}$) – see, for example, Figure 3.95.

There are two amendments to this basic picture.

A. Second breakdown

The bad news: in the case of *bipolar transistors* the SOA is further constrained by a phenomenon known as *second breakdown*, an important failure mechanism that you must keep in mind when designing power electronics with bipolar transistors. This is discussed in §3.6.4C, where the effect can be seen in the SOA plots in Figure 3.95 as a further reduction of allowable collector current at high voltages. Because MOSFET power transistors are largely immune to second breakdown, they are often favored over BJTs for power-regulator pass transistors (the exceptions are some newer small-geometry types, see IR App Note IN-1155).

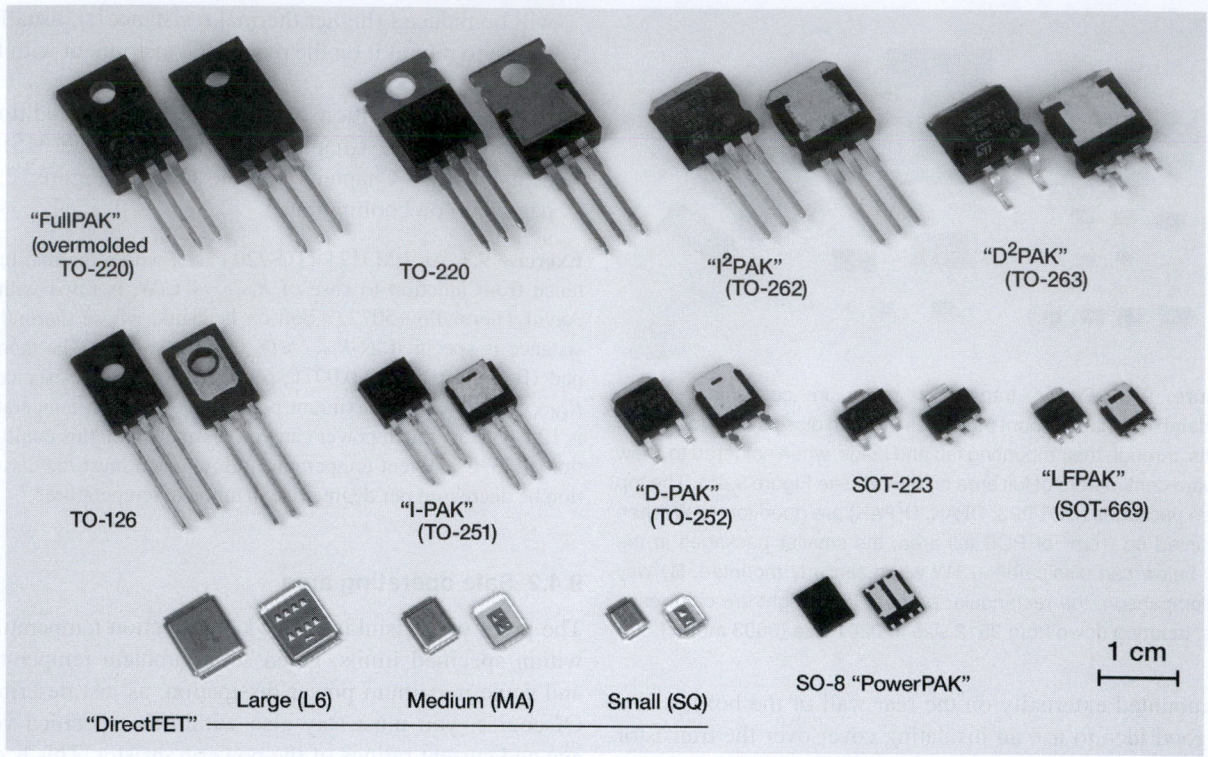

Figure 9.46. An extended selection of power packages, shown here and in Figure 9.47. The leadless packages in the bottom row require "reflow" solder techniques (trade in your soldering iron and wire solder for an oven and solder-paste dispenser!).

B. Transient thermal resistance

The good news: for pulses of short duration you can exceed the dc power-dissipation limit, sometimes by a large factor. That is because the mass of the semiconductor itself can absorb a short pulse of energy by heating locally ("heat capacity" or "specific heat"), limiting the temperature rise even if the instantaneous power dissipation is more than could be sustained continuously. This can be seen in the SOA plot (Figure 3.95), where the allowable power dissipation, for $100\,\mu$s pulses, is some 20 times greater than the dc value: an astonishing 3000 W versus 150 W. This is sometimes characterized on datasheets as a *transient thermal resistance* – a plot of R_θ versus pulse duration. The ability to dissipate very high peak power during short pulses extends to other electronic devices, for example, diodes, SCRs, and transient voltage suppressors. See §3.6.4C and the discussion in Chapter *9x*.

9.5 From ac line to unregulated supply

A regulated power supply that runs from ac line power

begins[36] by generating "unregulated" dc, a subject we introduced in §1.6.2 in connection with rectifiers and ripple calculations. For the linear voltage regulators we've seen so far, the unregulated dc supply uses a transformer, both

---

[36] Well, it really begins back at the power plant! Perhaps worth knowing, though, is the situation at the wall plug: in the US the standard 3-prong outlet delivers its 120 Vrms ac across the "line" and "neutral" blades (the neutral is the slightly wider slot; it's on the upper left, if the outlet is oriented to look like a face), with the round safety ground returned to a good earth connection at the service entry. The power comes into the house as three wires from a center-tapped 240 V pole transformer, with the center tap (neutral) bonded to earth ground at the service entry. Any given 120 V outlet provides neutral (white wire) and one "live" phase (black); the outlets in a given room may be powered by one or the other phases. A 240 V appliance outlet brings both live phases, along with safety ground, to a different style socket (this is in contrast to European 220 or 240 V outlets, which provide line, neutral, and safety ground). The in-wall wiring consists of "Romex"-style plastic-insulated oval cable with solid copper conductors: AWG14 for a 15 A residential circuit, and AWG12 for a 20 A circuit.

Figure 9.47. The larger cousins of the power packages in Figure 9.46, with three specimens from the latter shown for comparison. We're up in the tens to hundreds of watts with these when they are mounted on an appropriate heatsink.

to convert the incoming line voltage (120 Vrms in North America and a few other countries, 220 or 240 Vrms most everywhere else) to a (usually) lower voltage closer to the regulated output, and also to isolate the output from any direct connection to the hazardous line potentials ("galvanic isolation"); see Figure 9.48. Perhaps surprisingly, the switching power supplies we'll see shortly omit the transformer, generating line-derived dc at the powerline potential (∼160 Vdc or ∼320 Vdc), which powers the switching circuit directly. The essential galvanic isolation[37] is achieved instead with a transformer driven by the high-frequency switching signal.[38]

Transformer-isolated unregulated dc supplies are useful, also, for applications in which the stability and purity of regulated dc is unnecessary, for example high-power audio amplifiers. Let's look at this subject in more detail, beginning with the circuit shown in Figure 9.49. This is an unregulated ±50 volt (nominal) split supply, capable of 2 A output current, for a 100 watt linear audio amplifier. Let's

go through it from left to right, pointing out some of the things to keep in mind when you do this sort of design.

9.5.1 ac-line components

A. Three-wire connection
Always use a 3-wire line cord with ground (green or green/yellow) connected to the instrument case. Instruments with ungrounded cases can become lethal devices in the event of transformer insulation failure or accidental connection of one side of the powerline to the case. With a grounded case, such a failure simply blows a fuse. You often see instruments with the line cord attached to the chassis (permanently) using a plastic "strain relief," made by Heyco or Richco. A better way is to use an IEC (International Electrotechnical Commission) three-prong male chassis-mounted connector, to mate with those popular line cords that have the three-prong IEC female molded onto the end. That way the line cord is conveniently removable. Better yet, you can get a combined "power-entry module," containing IEC connector, fuse holder, line filter, and switch, as we've used here. Note that ac wiring uses a nonintuitive color convention: black = "hot" (or "line"), white = neutral, and green = ground (or "protective earth").[39]

---

[37] There are occasions where isolation is not needed; see Chapter *9x* for a discussion of some low-power off-line power supplies (including a step-down technique that uses a capacitor instead of a transformer or inductor.)

[38] The advantage of this peculiar arrangement is that the transformer, running at a high frequency (20 kHz–1 MHz) is much smaller and lighter.

[39] IEC cords use brown = line, blue = neutral, and green/yellow = ground.

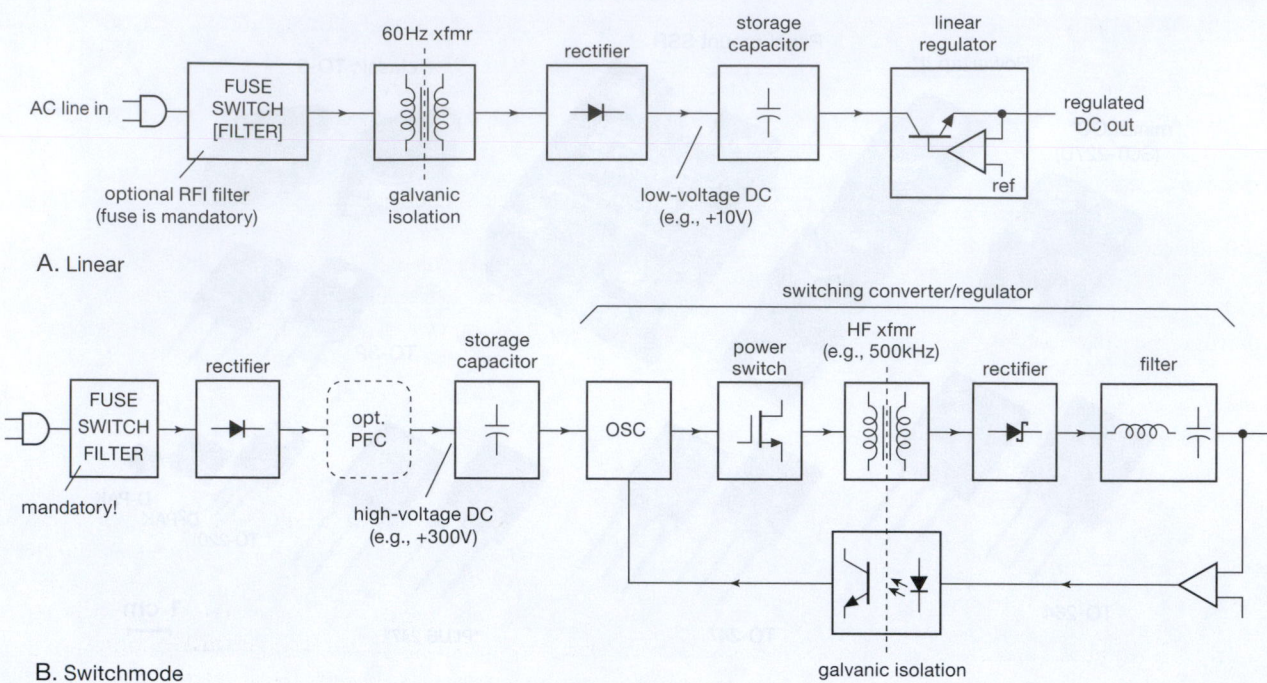

A. Linear

B. Switchmode

Figure 9.48. ac-line-powered ("offline") dc-output regulated supplies. A. In the linear supply the powerline transformer both isolates and transforms the input voltage. B. In the switching ("switchmode") converter the ac input is directly rectified to high-voltage dc, which powers the isolating switching converter. The PFC block performs power-factor correction, discussed later in §9.7.1C.

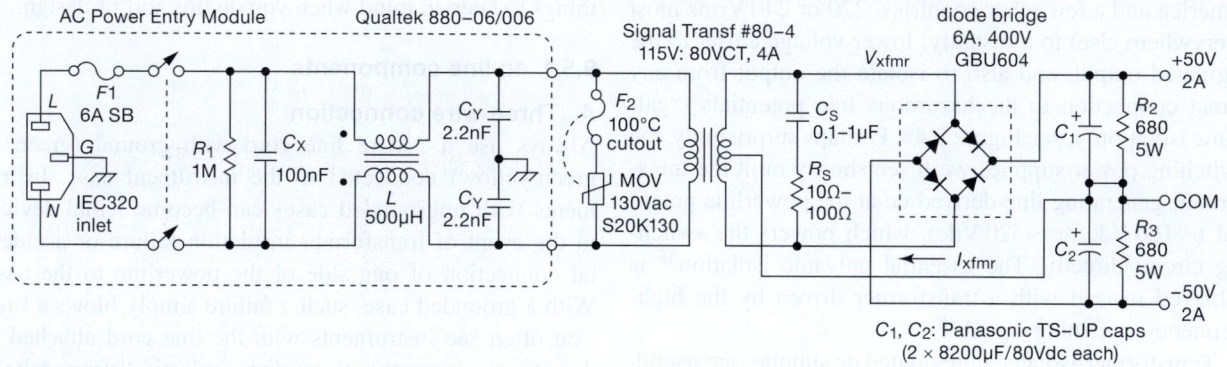

Figure 9.49. Unregulated ±50 V, 2 A power supply.

B. Fuse

A fuse, breaker, or equivalent function should be included in every piece of electronic equipment. A fuse holder, switch, and lowpass filter are often combined in the power entry module, but you can also wire them *à la carte*. The large wall fuses or circuit breakers (typically 15–20 A) in house or lab won't protect electronic equipment, because they are chosen to blow only when the current rating of the wiring in the wall is exceeded. For instance, a house circuit wired with 14-gauge wire will have a 15 A circuit breaker.

Now, if a storage capacitor in our unregulated supply becomes short-circuited someday (a possible failure mode), the transformer might then draw 10 A primary current (instead of its usual 2–3 A). The house breaker won't open, but your instrument becomes an incendiary device, with its transformer dissipating over a kilowatt.

Some notes on fuses. (a) It is best to use a "slow-blow" type in the power-line circuit, because there is invariably a large current transient ("inrush current") at turn-on, caused mostly by rapid charging of the power-supply filter capacitors. (b) You may think you know how to calculate the fuse current rating, but you're probably wrong. A dc power supply of this design[40] has a high ratio of rms current to average current, because of the small conduction angle (fraction of the cycle over which the diodes are conducting). The problem is worse if overly large filter capacitors are used. The result is an rms current considerably higher than you would estimate. The best procedure is to use a "true rms" ac current meter to measure the actual rms line current, then choose a fuse of at least 50% higher current rating (to allow for high line voltage, the effects of fuse "fatigue," etc.). (c) When wiring cartridge-type fuse holders (used with the popular 3AG/AGC/MDL type fuse, which is almost universal in electronic equipment), be sure to connect the leads so that anyone changing the fuse cannot come in contact with the powerline. This means connecting the "hot" lead to the rear terminal of the fuse holder (the authors have learned this the hard way!). Commercial power-entry modules with integral fuse holders are cleverly arranged so that the fuse cannot be reached without removing the line cord.

C. Switch

In Figure 9.49 the switch is integral with the power entry, which is fine, but it forces the user to reach around the back to turn the thing on. When using a front-panel power switch, it's a good idea to put a line-rated capacitor (called X1 or X2) across its terminals, to prevent arcing. For similar reasons the transformer primary should have some bridging capacitance, which in this case is taken care of by the lowpass filter in the entry module.

D. Lowpass filter

Although they are often omitted, such filters are a good idea, because they serve the purpose of preventing possible radiation of radiofrequency interference (RFI) from the instrument via the powerline, as well as filtering out incoming interference that may be present on the powerline. These filters typically use an *LC* "π-section" filter (as in the figure), with the coupled inductor pair acting as a common-mode impedance. Power-line filters with excellent performance characteristics are available from several manufacturers, e.g., Corcom, Cornell–Dubilier, Curtis, Delta, Qualtek, and Schurter.[41] Studies have shown that spikes as large as 1 kV to 5 kV are occasionally present on

the powerlines at most locations, and smaller spikes occur quite frequently. Line filters (in combination with transient suppressors, see below) are reasonably effective in reducing such interference (and thereby extending the life of a power supply and the equipment it powers).

E. Line-voltage capacitors

For reasons of fire and shock hazard, capacitors intended for line filtering and bypassing are given special ratings. Among other attributes, these capacitors are designed to be self-healing, i.e., to recover from internal breakdown.[42] There are two classes of line-rated capacitors: "X" capacitors (X1, X2, X3) are rated for use where failure would not create a shock hazard. They are used across the line (C_X in Figure 9.49; the common X2 type is rated for 250 Vac, with peak voltage of 1.2 kV). "Y" capacitors (Y1, Y2, Y3, Y4) are rated for use where failure would present a shock hazard. They are used for bypassing between the ac lines and ground (C_Y in Figure 9.49; the common Y2 type is rated for 250 Vac, with a peak voltage of 5 kV). Line-rated capacitors come in disc ceramic and in plastic-film flavors; the latter are usually a box geometry, with a flame retardant case. It's hard to miss these capacitors – they are usually festooned with markings proclaiming the various national certifications whose standards they meet[43] (Figure 9.50).

A *further* word about line-voltage capacitors: when the

1 cm

Figure 9.50. ac-line-rated capacitors flamboyantly display their safety ratings (right), compared with the minimalist decoration of a plain ol' film capacitor (left).

---

[40] Where the rectified input charges large storage capacitors at each voltage peak of the ac waveform. By contrast, switching power supplies with *power-factor correction* (PFC) nicely sidestep this problem; see §9.7.1C.

[41] Watch out, however, for misleading attenuation specifications: they are universally specified with 50 Ω source and load, because that's easy to

measure with standard RF instrumentation, and not because it has any resemblance to the real world.

[42] For example, line-rated plastic film capacitors are constructed so that a perforating breakdown causes the metal plating near the hole to burn away, clearing the short-circuit.

[43] Here are some of them: UL, CSA, SEV, VDE, ENEC, DEMKO, FIMKO, NEMKO, SEMKO, CCEE, CB, EI, and CQC.

instrument is unplugged, the X capacitor may be left holding the peak ac line voltage, up to 325 V, which appears across the exposed power plugs! This can cause electrical shocks, and spark discharges. That's why there's a parallel discharge resistor, sized for a safe time constant of less than a second.[44] Here the Qualtek RFI filter module uses 1 MΩ, and the Astrodyne switching supply (§9.8) uses 540 kΩ. The latter continuously dissipates 100 mW when powered with 220 Vac line input, which could constitute one of the largest standby power losses in an "Energy Star" design. Power Integrations offers their CAPZero™ IC to solve this problem. This clever part works by looking for an ac-line-voltage reversal every 20 ms or less, and if it fails to see one it turns on, connecting two discharge resistors across the X capacitor.

Some designs have substantial high-voltage dc storage capacitors that need to be discharged when the power is off. Because of their large capacitance, the ac-sensing CAPZero would not work. Here you could use normally-on relay contacts, and energize the relay when external ac power is present. Or, if you don't like moving parts, a high-voltage depletion-mode MOSFET (see Table 3.6 on page 210) and a photovoltaic stack (see §12.7.5, Figure 12.91A) can do the job.

F. Transient suppressor

In many situations it is desirable to use a "transient suppressor" (or "metal-oxide varistor," MOV) as shown in Figure 9.49. The transient suppressor is a device that conducts when its terminal voltage exceeds certain limits (it's like a bidirectional high-power zener). These are inexpensive and small and can shunt hundreds of amperes of potentially harmful current in the form of spikes. Note the thermal cutout fuse: that protects in a situation in which the MOV starts to conduct partially (for example if the line voltage becomes highly elevated, or if an aged MOV exhibits lowered breakdown voltage from having absorbed large transients). Transient suppressors are made by a number of companies, e.g., Epcos, Littelfuse, and Panasonic. Effective transient suppression is an interesting challenge, and we discuss it further in Chapter *9x*.

G. Shock hazard

It is a good idea to insulate all exposed line-voltage connections inside any instrument, for example by using polymer heat-shrink tubing (the use of "friction tape" or electrical tape inside electronic instruments is strictly bush-league).

Because most transistorized circuits operate on relatively low dc voltages (±15 V or less), from which it is not possible to receive a shock, the powerline wiring is the only place where any shock hazard exists in most electronic devices (there are exceptions, of course). The front-panel ON–OFF switch is particularly insidious in this respect, being close to other low-voltage wiring. Your test instruments (or, worse, your fingers) can easily come in contact with it when you go to pick up the instrument while testing it.

9.5.2 Transformer

Now for the transformer. Never build an instrument to run off the powerline without an isolating transformer! To do so is to flirt with disaster. Transformerless power supplies, which have been popular in some consumer electronics (radios and televisions, particularly) because they're inexpensive, put the circuit at high voltage with respect to external ground (water pipes, etc.).[45] This has no place in instruments intended to interconnect with any other equipment and should always be avoided. And use extreme caution when servicing any such equipment; just connecting your oscilloscope probe to the chassis can be a shocking experience.

The choice of transformer is more involved than you might at first expect. It may be hard to find a transformer with the voltage and current ratings you need. We have found the Signal Transformer Company unusual, with their nice selection of transformers and quick delivery. And don't overlook the possibility of having transformers custom-made if your application requires more than a few.

Even assuming that you can get the transformer you want, you still have to decide on the voltage and current rating. If the unregulated supply is powering a linear regulator, then you want to keep the unregulated dc voltage low, in order to minimize the power dissipation in the pass transistors. But you must be absolutely certain the input to the regulator will never drop below the minimum necessary for regulation (typically 2 V above the regulated output voltage, for conventional regulators like the LM317; or 0.5–1 V for low-dropout types) or you may encounter 120 Hz dips in the regulated output; in the design you need to allow for low line voltage (10% below nominal, say – 105 Vac in the US) or even brownout conditions (20% below nominal). The amount of ripple in the unregulated output is involved

---

[44] We have seen entirely too many designs that omit this discharge resistor; not good!

[45] Non-isolated off-line supplies are commonly found in some types of self-contained electronics, such as a screw-in LED light bulb, a wall clock, a smoke alarm, a Wi-Fi surveillance camera, a toaster or coffee-maker, and so on. We discuss some of these in Chapter *9x*.

here, because it is the *minimum* input to the regulator that must stay above some critical voltage (see Figure 1.61), but it is the *average* input to the regulator that determines the transistor dissipation.

As an example, for a +5 V regulator you might use an unregulated input of +10 V at the minimum of the ripple, which itself might be 1 to 2 volts peak-to-peak. From the secondary voltage rating you can make a pretty good guess of the dc output from the bridge, because the peak voltage (at the top of the ripple) is approximately 1.4 times the rms secondary voltage, less two diode drops. But it is essential to make actual measurements if you are designing a power supply with near-minimum drop across the regulator, because the actual output voltage of the unregulated supply depends on poorly specified parameters of the transformer, such as winding resistance and magnetic coupling (leakage inductance), both of which contribute to voltage drop under load. Be sure to make measurements under worst-case conditions: full load and low power-line voltage (105 V). Remember that large filter capacitors typically have loose tolerances: −30% to +100% about the nominal value is not unusual. It is a good idea to use transformers with multiple taps on the primary (the Triad F-90X series, for example), when available, for final adjustment of output voltage.

For the circuit shown in Figure 9.49, we wanted ±50 V output under full load. Allowing for two diode drops (from the bridge rectifier) we need a transformer with ∼52 V peak amplitude, or about 37 Vrms. Among the available transformer choices, the closest was the 40 Vrms unit shown, probably a good choice because of the effects of winding resistance and leakage inductance, which reduce slightly the loaded dc output voltage.

An important note: transformer current ratings are usually given as *rms* secondary current. However, because a rectifier circuit draws current only over a small part of the cycle (during the time the capacitor is actually charging), the secondary's rms current, and therefore the I^2R heating, will be significantly larger than the average rectified dc output current. So you have to choose a transformer whose rms current rating is somewhat larger (typically ∼ 2×) than the dc load current. Ironically, the situation gets worse as you increase capacitor size to reduce output ripple voltage. Full-wave rectification is better in this respect, because a greater portion of the transformer waveform is used. For the unregulated dc supply of Figure 9.49 we measured an rms current of 3.95 A at the transformer secondary when powering a 2 A dc load. The measured waveforms in Figure 9.51 show the pulsating nature of the current, as the rectified transformer output recharges the storage capacitors each half-cycle.

Naïvely you might expect that the conduction angle (fraction of the cycle during which the current flows) could be estimated simply by (a) calculating the capacitor's discharge between half-cycles, according to $I = C\,dV/dt$, then (b) calculating the time in the next half-cycle at which the rectified output exceeds the capacitor's voltage. However, this pretty scheme is complicated by the important effects of the transformer's winding resistance and leakage inductance, and the storage capacitor's ESR, all of which extend the conduction angle.[46] The best approach is to make measurements on the bench, perhaps informed by SPICE simulations using known or measured values of these parameters. In §9.5.4 we show results of such simulations.

9.5.3 dc components

A. Storage capacitor

The storage capacitors (sometimes called *filter* capacitors) are chosen large enough to provide acceptably low ripple voltage, with a voltage rating sufficient to handle the worst-case combination of no load and high line voltage (125–130Vrms).

At this point it may be helpful to look back at §1.7.16B, where we first discussed the subject of ripple. In general, you can calculate ripple voltage with sufficient accuracy by assuming a constant-current load equal to the average load current. (In the particular case that the unregulated supply drives a linear regulator, the load in fact is accurately a constant-current sink). This simplifies your arithmetic, since the capacitor discharges with a ramp, and you don't have to worry about time constants or exponentials (the measured waveforms in Figure 9.51, taken with a resistive load, illustrate the validity of this approximation).

For the circuit shown in Figure 9.49, we wanted approximately 1 Vpp output ripple at full 2 A load. From $I = C\,dV/dt$ we get (with $\Delta t = 8.33$ ms) $C = I\Delta t/\Delta V = 16{,}700\,\mu$F. The nearest capacitor voltage ratings are 63 V and 80 V; we chose the latter, in an abundance of caution. The available 16,000 μF/80 V capacitors are somewhat large physically (40 mm diameter × 80 mm long), so we decided to put a pair of 8200 μF capacitors (35 mm × 50 mm) in parallel (using smaller capacitors in parallel

---

[46] Although a transformer with a large leakage inductance might seem advantageous (because it increases the conduction angle losslessly), it has the undesirable effect of degrading voltage regulation under load; it also introduces a phase lag in the input current relative to the voltage, thus reducing the power factor. Furthermore, leakage inductance causes nasty voltage spikes to appear, owing to diode reverse recovery, as described in §9x.6.

also reduces the overall series inductance of the capacitor). Good design practice calls for the use of storage capacitors whose ripple current rating is conservatively larger than the value estimated from the dc output current and the conduction angle. In the above circuit, for example, we designed for a maximum dc load current of 2 A, from which we estimated an rms current of about 4 A in both the transformer secondary and the storage capacitor. The particular capacitors shown in the figure have a ripple current rating of 5.8 Arms at 85°C for each 8200 μF capacitor of the parallel pair, thus 11.6 Arms when combined to make either C_1 or C_2. This is definitely conservative! You can also calculate the heating, from the ESR specification of 0.038 Ω (maximum) per capacitor: each parallel pair has an ESR no greater than 19 mΩ, which produces a heating power of $P = I_{\mathrm{rms}}^2 R_{\mathrm{ESR}} \approx 0.15$ W in each capacitor.

When choosing filter capacitors, don't get carried away: an oversize capacitor not only wastes space but also increases transformer heating (by reducing the conduction angle, hence increasing the ratio of rms current to average current). It also increases stress on the rectifiers. But watch out for loose capacitance tolerance: although the capacitors we used here have a rated tolerance of $\pm 20\%$, electrolytic storage capacitors can be as loose as +100%/−30%.

The resistors R_2 and R_3 across the output in Figure 9.49 serve two purposes: they provide a minimum load (to keep the unloaded output from "soaring"); and they act as "bleeders" to discharge the capacitors when the unloaded supply is turned off. This is a good feature, because power supplies that stay charged after things have been shut off can easily lead you to damage some circuit components if you mistakenly think that no voltage is present.

B. Rectifier
The first point to be made is that the diodes used in power supplies (usually referred to as "rectifiers") are quite different from the small 1N914- or 1N4148-type signal diodes used in circuitry. Signal diodes are generally designed for high speed (a few nanoseconds), low leakage (a few nanoamps), and low capacitance (a few picofarads), and they can generally handle currents up to about 100 mA, with breakdown voltages rarely exceeding 100 volts. By contrast, rectifier diodes and bridges for use in power supplies are hefty objects with current ratings going from 1 A to 25 A or more, and breakdown voltage ratings going from 100 V to 1000 V or more. They have relatively high leakage currents (in the range of microamps to milliamps) and plenty of junction capacitance. General-purpose rectifiers of the sort used in Figure 9.49 are not intended for high speed, unnecessary for operation at the powerline

frequency of 60 Hz. By contrast, in *switching* power supplies it's necessary to use high-speed rectifiers because of the characteristic switching frequencies of 20 kHz–1 MHz; there the use of "fast recovery" or Schottky-barrier rectifiers (or MOSFETs used as "synchronous rectifiers"[47]) is universal.

Typical of general-purpose rectifiers are the popular 1N4001–1N4007 series, rated at 1 A, and the 1N5400–1N5408 series, rated at 3 A, with reverse-breakdown voltages ranging from 50 to 1000 volts. The 1N5817–1N5822 series of Schottky rectifiers come in axial lead packages, with current ratings of 1–3 A, and voltage ratings of 20–40 V. Rectifiers with higher current ratings require heatsinking, and come in packages similar to power transistors (TO-220, D^{2}PAK, stud-mount, etc). Examples are the MBR1545 and 30CTQ045 dual Schottky rectifiers (available in TO-220 or D^{2}PAK power packages), rated at 15 A and 30 A, respectively, at 45 V, and the MUR805 to MUR1100 6 A rectifiers (in TO-220 packages), with voltage ratings to 1 kV. Plastic-encapsulated bridge rectifiers are quite popular also, with lead-mounted 1 A to 6 A types, and heatsink mountable packages in ratings up to 35 A or more.[48]

C. Damping network
The series RC across the transformer secondary in Figure 9.49 is often omitted, but it shouldn't be. This simple linear unregulated dc supply has the surprising ability to generate substantial microsecond-scale voltage spikes, which can create strong 120 Hz interference and other forms of mischief. It turns out that a pair of non-ideal characteristics (transformer leakage inductance, combined with rectifier reverse recovery time) work together to create a train of periodic sharp spikes, whose amplitude can be tens of volts. This nasty effect is easily tamed with a series RC "snubber" network, as shown. There's some interesting stuff going on here; you can read more about it (and see a dramatic example) in §9x.6.

9.5.4 Unregulated split supply – on the bench!

We built the power supply of Figure 9.49, mostly out of curiosity to see how closely the actual device compared

---

[47] Sometimes called *active rectifiers*.

[48] An interesting option for implementing an efficient bridge rectifier is the use of four MOSFETs as synchronous switches; their gate control signals can be generated conveniently with an elegant device like the LT4320 "Ideal Diode Bridge Controller," which senses zero crossings and does the right thing on its gate-control output pins. Check out its datasheet.

with our predictions. Figure 9.51 shows the ac voltage and current at one end of the transformer secondary, and the positive dc output voltage, with the power supply driving ±2 A resistive loads. The waveforms are pretty much as expected: (a) the ripple voltage is about 0.8 Vpp, somewhat less than our 1 Vpp estimate; our calculation was conservative, though, because we assumed the storage capacitors had to supply output current for a complete half-period ($1/2f_{ac} \approx 8$ ms), whereas in reality recharging begins after ~6 ms; (b) the dc output voltage (54 V) is somewhat higher than expected, probably because the transformer voltage rating is for the full-rated load current of 4 A, and also because the powerline voltage in our lab was 3% above nominal; with no load the output rose to 60 V, typical of unregulated supplies; (c) the transformer current is confined to a fairly narrow conduction angle (about 60° of each 180° half-cycle), as expected; during conduction the ac waveform at the transformer secondary is flattened by the heavy load current because of the combined effects of leakage inductance and winding resistance.[49]

to rectify incoming powerline ac to dc while maintaining nearly unity power factor, by means of a "power-factor correction" (PFC) input circuit; we'll explore this cleverness, briefly, in §9.7.1C.

And on the *computer*! (SPICE)

To explore the effects of component imperfections (winding resistance and leakage inductance in the transformer, series resistance in the capacitors) we ran a SPICE simulation (see Appendix J) of this circuit, beginning with measured parameters where possible (e.g., transformer resistance and inductances), values found in the SPICE libraries (e.g., rectifier forward voltage versus current) and plausible guesses for series resistance in the storage capacitors. With just a small amount of adjustment we obtained the simulation shown in Figure 9.52 (presented at the same scale factors as in Figure 9.51). The agreement is impressive (although the simulation somewhat underestimates the conduction angle, thus higher-than-measured transformer current).[50]

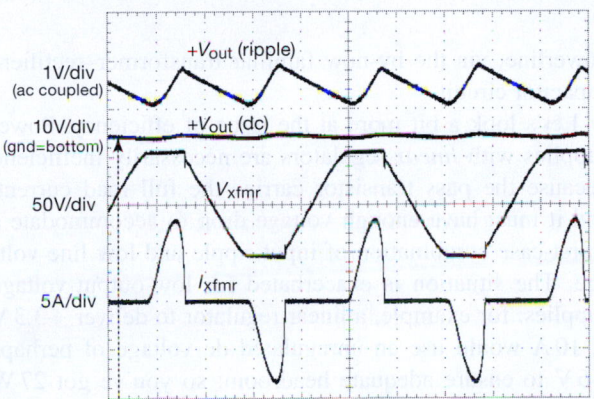

Figure 9.51. Measured waveforms for the unregulated dc power supply of Figure 9.49 driving ±2 A loads. Horizontal scale: 4 ms/div.

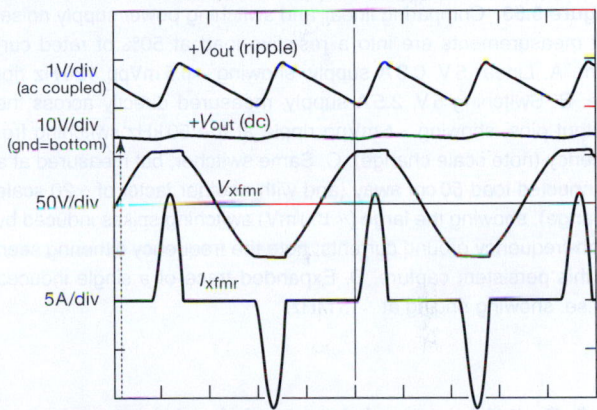

Figure 9.52. Waveforms from a SPICE simulation of the unregulated dc power supply of Figure 9.49, plotted on the same scales as those of Figure 9.51.

With a 2 A dc load on both outputs, the measured rms transformer current was 3.95 Arms. This doubling is caused by the shortened conduction angle: the *average* transformer current equals the dc output current, but the *rms* current is greater. This is sometimes described as a reduced *power factor* (the ratio of average input power to rms input power), an effect that is important in switching power supplies. With some cleverness it is possible

9.5.5 Linear versus switcher: ripple and noise

Coming next is the fascinating subject of *switching* regulators and power supplies. These have become dominant, owing to their combination of excellent efficiency, small

---

[49] For these waveform measurements we omitted Figure 9.49's R_SC_S damping network to reveal the spike (and jump) that is visible on the transformer ac voltage waveform, caused by the combination of transformer leakage inductance and diode recovery time; see §9.5.3C and §9x.6.

[50] The dominant circuit parameters used are: transformer primary $R=0.467\,\Omega$, $L_L=1.63\,\mu H$, $L_M=80\,mH$, turns ratio of 0.365, transformer secondary $R=0.217\,\Omega$, $L_{L(sec)}=20\,\mu H$, damping network $C_S=0.5\,\mu F$, $R_S=30\,\Omega$, rectifier "KBPC806" (Vishay 8A, 600V bridge), storage capacitor $C=14{,}000\,\mu F$, ESR=0.01$\,\Omega$, load resistors 27$\,\Omega$ (each side).

size and weight, and low cost. However, all is not roses: the rapid switching process generates transients at the switching frequency and its harmonics, and these can be extremely difficult to filter effectively. We'll discuss this soon enough... but it's worth taking a look now at Figure 9.53, where the bad stuff on the outputs of two 5 V power supplies are compared.

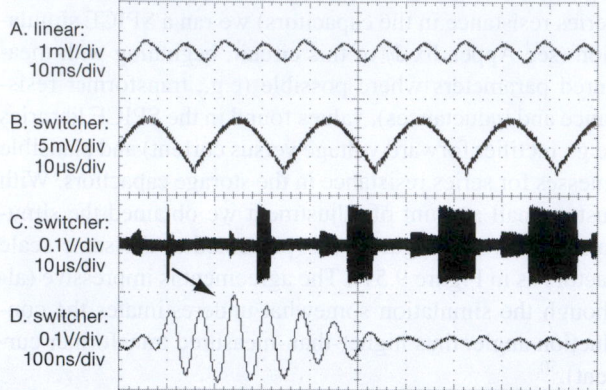

Figure 9.53. Comparing linear and switching power supply noise. All measurements are into a resistive load at 50% of rated current. A. Linear 5 V, 0.3 A supply, showing ~0.5 mVpp 120 Hz ripple. B. Switching 5 V, 2.5 A supply, measured directly across the output pins, showing ~6 mVpp ripple at the 50 kHz switching frequency (note scale change). C. Same switcher, but measured at a connected load 50 cm away (and with another factor of ×20 scale change), showing the large (~150 mV) switching spikes induced by high-frequency ground currents; note the frequency dithering seen in this persistent capture. D. Expanded trace of a single induced pulse, showing ringing at ~15 MHz.

9.6 Switching regulators and dc–dc converters

9.6.1 Linear versus switching

All the voltage-regulator circuits we have discussed so far work the same way: a linear control element (the "pass transistor") in series with the dc input is used, with feedback, to maintain constant output voltage (or perhaps constant current).[51] The output voltage is always lower in voltage than the input voltage, and significant power is dissipated in the control element, namely $P_{\mathrm{diss}} = I_{\mathrm{out}}(V_{\mathrm{in}} - V_{\mathrm{out}})$. As we've seen, the dc input to a linear regulator may be simply another (higher) regulated dc voltage within the system; or it may be unregulated dc that is derived from the

---

[51] A minor variation on this theme is the *shunt regulator*, in which the control element is tied from the output to ground, rather than in series with the load; the simple resistor-plus-zener is an example.

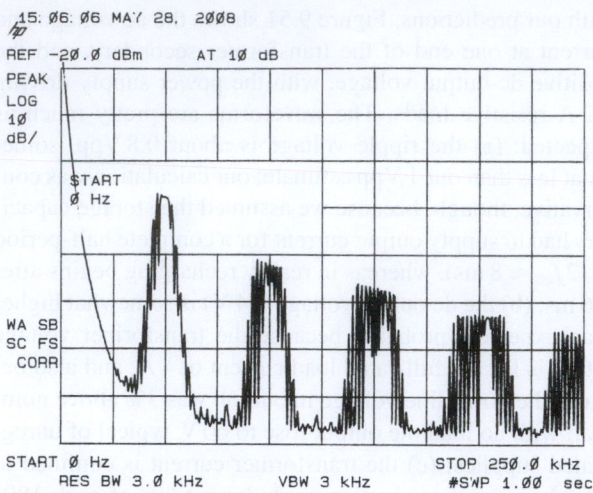

Figure 9.54. Averaged frequency spectrum of the switching power supply of Figure 9.53, showing the dithered ~50 kHz switching frequency and its harmonics.

powerline, via the by-now familiar transformer–rectifier–capacitor circuit.

Let's look a bit more at the issue of efficiency. Power supplies with *linear* regulators are necessarily inefficient, because the pass transistor carries the full load current, and it must have enough voltage drop to accommodate a worst-case combination of input ripple and low line voltage. The situation is exacerbated for low-output-voltage supplies: for example, a linear regulator to deliver +3.3 V at 10 A would use an unregulated dc voltage of perhaps +6 V to ensure adequate headroom; so you've got 27 W of pass-transistor dissipation while delivering 33 W to the load – that's 55% efficiency. You may not care that much about efficiency *per se*; but the wasted power has to be dissipated, which means a large heatsink area, blowers, etc. If you were to scale up this example to 100 A, say, you would have a serious problem removing the quarter of a kilowatt (!) of pass-transistor heat. You would have to use multiple pass transistors and forced-air cooling. The supply would be heavy, noisy, and hot.

There is another way to generate a regulated dc voltage (shown earlier in Figure 9.48B), which is fundamentally different from what we've seen so far – look at Figure 9.55. In this switching converter a transistor, operated as a saturated switch, periodically applies the full unregulated voltage across an inductor for short intervals. The inductor's current builds up during each pulse, storing $\frac{1}{2}LI^2$ of energy in its magnetic field. When the switch is turned off, some or all[52] of this stored energy is transferred to a filter capacitor at the output, which also smooths the output (to carry the output load between charging pulses). As with a linear regulator, feedback compares the output with a voltage reference – but in a switching regulator it controls the output by changing the oscillator's pulse width or switching frequency, rather than by linearly controlling the base or gate drive.[53]

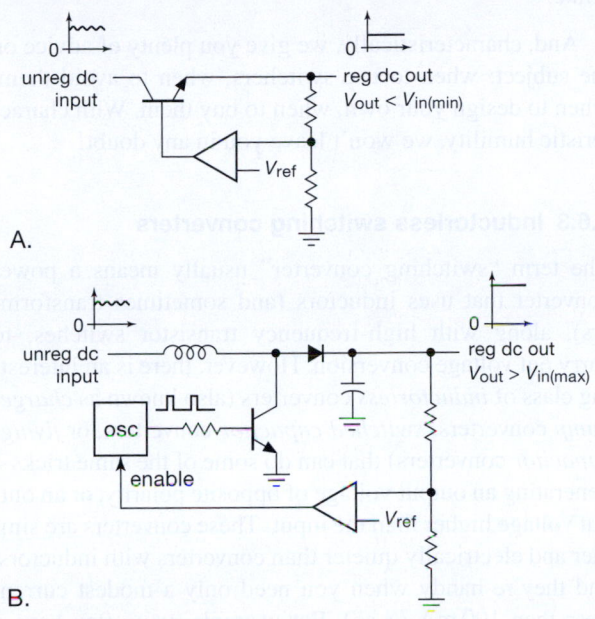

A.

B.

Figure 9.55. Two kinds of regulators: A. linear (series-pass); B. switcher (step-up, or "boost").

Advantages of switching converters

Switching regulators have unusual properties that have made them very popular:

(a) Because the control element is either off or saturated, there is very little power dissipation; switching supplies are thus very efficient, even when there is a large voltage difference between input and output. High efficiency translates to small size, because little heat needs to be dissipated.

(b) Switchers (slang for "switching power supplies") can generate output voltages *higher* than the unregulated input, as in Figure 9.55B; and they can just as easily generate outputs *opposite in polarity* to that of the input!

(c) The output storage capacitor can be small (in capacitance, and therefore in physical size), because the high operating frequency (typically 20 kHz–1 MHz) corresponds to a very short time interval (a few microseconds) between recharging.

(d) For a switching supply operated from the ac powerline input, the essential isolation is provided by a transformer operating at the switching frequency; it is *much* smaller than a low-frequency powerline transformer (see Figure 9.1).

The good news

The combination of small capacitor and transformer size, along with little power dissipation, permits compact, lightweight, and efficient ac-powered dc supplies, as well as dc-to-dc converters.[54] For these reasons, switching supplies (also known as *switchmode* power supplies, or SMPSs), are used almost universally in electronic devices such as computers, telecommunications, consumer electronics, battery-operated devices, and, well, just about everything electronic.

The bad news

Lest we leave too favorable an impression, we note that switching supplies do have their problems. The switching operation introduces "noise" into the dc output, and likewise onto the input powerline and as radiated electromagnetic interference (EMI); see Figures 9.53 and 9.54. Line-operated switchers (confusingly called "off-line") exhibit a rather large "inrush current" when initially powered on.[55]

---

[52] All of the stored energy goes forward if the inductor current is allowed to go to zero ("discontinuous-conduction mode," DCM); you get only a portion of the stored energy in "continuous-conduction mode" (CCM), in which the inductor's current does not go to zero before the next conduction cycle.

[53] One could object that we're unfairly comparing a *step-up* switching converter circuit with an inherently "step-down" linear pass regulator. Indeed, the switching topology that is analogous in function to the linear regulator is the *buck* regulator (shown presently, in Figure 9.61A). But we like the shock value of the boost switching converter, because it's unexpected that you can even do that if you've lived exclusively in the linear world.

[54] Examples of the former include the little power "bricks" that are used for laptop computers, cellphones, and the like, as well as the more substantial power supplies built into desktop computers. Examples of the latter are the "point of load" dc–dc converters that you find clustered around the processor on a computer motherboard: the processor might require 1.0 Vdc at 60 A (!); to generate that enormous current you use a set of 12 V to 1.0 V step-down converters, right at the point of load, supplied by a lower current 12 V "bus."

[55] For an example we opened to a random page in the power-supply section of the DigiKey catalog, and found a little ac-input 5 W switcher (5 Vdc, 1 A) with a specified powerline inrush current of…(drumroll)…40 A – that's a peak power of 4*kilo*watts!

And switchers have suffered from a bad reputation for reliability, with occasional spectacular pyrotechnic displays during episodes of catastrophic failure.

The bottom line

Fortunately, switching supplies have largely overcome the drawbacks of their earlier brethren (unreliability, electrical and audible noise, inrush current and component stress). Because they are small, lightweight, efficient, and inexpensive, switchers have largely replaced linear regulators over the full range of load power (from watts to kilowatts) in contemporary electronics, and particularly in large commercial production. Linear supplies and regulators are still alive and well, however, particularly for simple low-power regulation and for applications requiring clean dc power; and this last feature – the absence of pervasive switching noise – can be of major importance in applications that deal with small signals.

9.6.2 Switching converter topologies

In the following sections we tell you all about switching regulators and power supplies (collectively called "switching converters"), in several steps.

- First (§9.6.3) we look briefly at *inductorless* converters, in which the energy is carried from input to output by capacitors, whose connections are switched with MOSFETs. These are sometimes called "charge-pump converters," or "flying-capacitor converters." These simple devices can double or invert a dc input voltage, and they're useful for relatively low current loads (up to ~100 mA).
- Next (§9.6.4) we describe converter topologies that use inductors, beginning with the basic dc–dc non-isolated switching converter, of the sort you would use within a circuit, or with battery power. There are three basic circuit topologies, used for (a) step-down (output voltage less than input), (b) step-up (output voltage greater than input), and (c) inverting (output polarity opposite to input). All of these use an inductor for energy storage during the switching cycle.
- Next (§9.6.10) we look at dc–dc converters in which a transformer couples the input and output circuits. In addition to providing galvanic isolation (which may or may not be needed), the transformer is desirable when there's a large ratio between input and output voltages. That is because the transformer's turns ratio provides a helpful voltage conversion factor that is absent in the non-isolated (transformerless) designs. Transformer designs

also let you produce multiple outputs, and of either polarity.

- Finally (§9.7) we describe how the isolated converter permits power-supply designs that run straight from the rectified ac powerline. These "offline" supplies are, of course, the bread and butter of most line-powered electronics. And they have their special problems, related to safety, interference, inrush current, power factor, and the like.

And, characteristically, we give you plenty of advice on the subject: when to use switchers, when to avoid them; when to design your own, when to buy them. With characteristic humility, we won't leave you in any doubt!

9.6.3 Inductorless switching converters

The term "switching converter" usually means a power converter that uses inductors (and sometimes transformers), along with high-frequency transistor switches, to carry out voltage conversion. However, there is an interesting class of *inductorless* converters (also known as *charge-pump* converters, *switched capacitor* converters, or *flying-capacitor* converters) that can do some of the same tricks – generating an output voltage of opposite polarity, or an output voltage higher than the input. These converters are simpler and electrically quieter than converters with inductors, and they're handy when you need only a modest current (less than 100 mA or so). For example, you often have a source of +5 V (on a computer board, or a USB device), or perhaps +9 V from a battery, and you need a corresponding negative voltage because you want to run a dual-polarity op-amp. Just drop in a charge-pump inverter chip and two capacitors, and you're ready to go.[56]

Figure 9.56 shows how it goes: these devices have an internal oscillator and some CMOS switches, and they require a pair of external capacitors to do their job. When the input pair of switches is closed (conducting), C_1 charges to V_{in}; then, during the second half-cycle, C_1 is disconnected from the input and connected, upside-down, across the output. If $C_2 \ll C_1$, then the output voltage goes nearly to $-V_{IN}$ in one cycle of operation. In the more typical case of $C_2 \geq C_1$ it takes a number of cycles, from cold start, for the output voltage to equilibrate to $-V_{IN}$.

Similarly, you can create an output of $2V_{in}$, by arranging things so that C_1 charges as before, but then gets hooked in series with V_{in} during the second (transfer) half-cycle

---

[56] A good reference is M.D. Seeman & S.R. Sanders, "Analysis and optimization of switched-capacitor DC–DC converters," *IEEE Trans. Power Electron.* **23** (2) pp. 841–851 (2008).

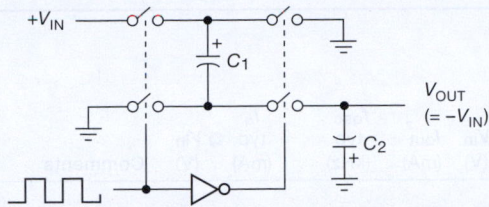

Figure 9.56. Charge-pump voltage inverter. An oscillator operates the switch pairs in alternation: the left-hand switches charge "flying capacitor" C_1 to a voltage of V_{IN}; the right-hand switches then apply that voltage, with reversed polarity, to the output storage capacitor C_2.

(Figure 9.57). The LT1026 and MAX680 conveniently integrate a positive doubler and an inverting doubler in one package: Figure 9.58 shows the simple circuitry required to generate an unregulated split supply from a single +5 V input.

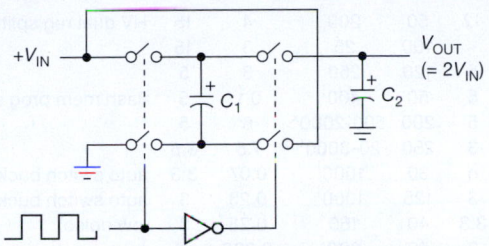

Figure 9.57. Charge-pump voltage doubler. Here the voltage on the flying capacitor, charged to V_{IN}, is added to the input voltage to generate an output voltage of twice V_{IN}.

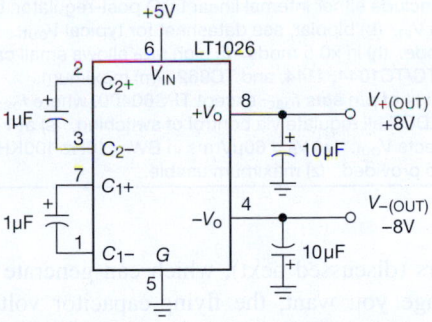

Figure 9.58. Generating a pair of unregulated ±8 V outputs from a single +5 V input.

A. Limitations of inductorless converters

This charge-pump technique is simple and efficient, and requires few parts and no inductors. However, the output is not regulated, and it drops significantly under load (Figure 9.59). Also, in common with other switching power

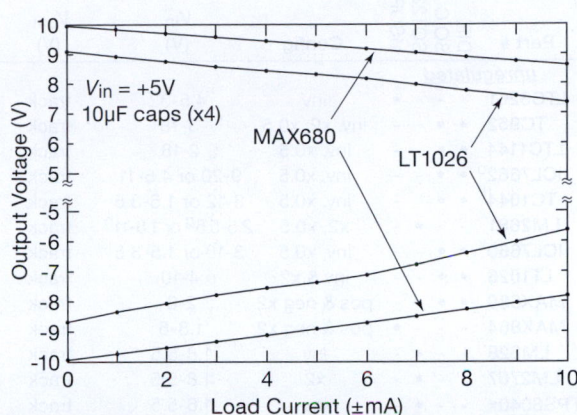

Figure 9.59. The output voltage of a charge-pump converter drops significantly under load, as seen here with measured data for the circuit of Figure 9.58, with either bipolar (LT1026) or CMOS (MAX680) devices. MOSFET switches have no voltage drop at zero current, where V_{out} is accurately equal to twice V_{in}.

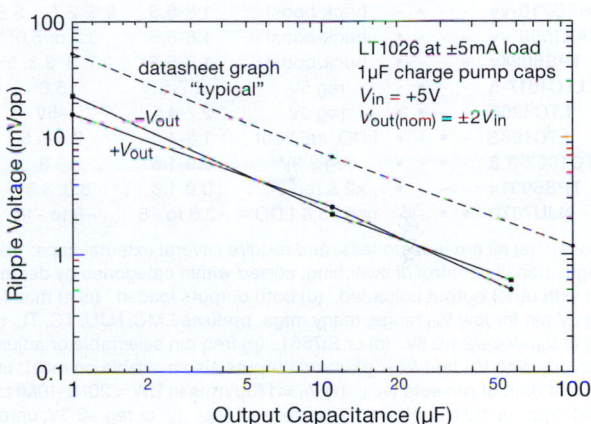

Figure 9.60. Reducing ripple with a larger output capacitor: measured peak-to-peak ripple voltage for the LT1026 doubler–inverter.

conversion techniques, the switching operation produces output ripple, which however can be reduced by using larger output capacitors (Figure 9.60), or by appending a low-dropout linear regulator (see below).[57] Furthermore, like most CMOS devices, charge pumps have a limited

---

[57] The ripple voltage is given approximately by $V_{ripple}(pp) = I_{out}/2f_{osc}C_{out} + 2I_{out} \cdot ESR$. The first term is just $I = C\,dV/dt$, and the second term adds the effect of the capacitor's finite equivalent series resistance.

Table 9.4 Selected Charge-pump Converters[a]

DIP	SOIC	SOT23	MSOP etc	Part #	Config	V_{in} (V)	V_{out} (V)	R_{out} typ (Ω)	@ V_{in} (V)	I_{out}[z] (mA)	f_{osc} typ (kHz)	I_q typ (mA)	@ V_{in} (V)	Comments
				unregulated										
-	-	-	•	LTC3261	inv	4.5-32	track	35	12	50[e]	50-500[p]	7	15	HV
•	•	-	-	TC962	inv, x2, x0.5	3-18	track	32	15	80	12 or 24[p]	0.5	15	improved 7660/2
•	•	-	-	LTC1144	inv, x0.5	2-18	track	56	15	50	10 or 100[p]	1.1[m]	15	HV ver of 1044/7660/62
•	•	-	-	ICL7662[o]	inv, x0.5	9-20 or 4.5-11	track	55	15	50	10	0.15	12.5	Maxim, orig Intersil
•	•	-	-	TC1044[k]	inv, x0.5	3-12 or 1.5-3.5	track	55	15	60	10 or 45[p]	0.15	12.5	improved 7660; see (k)
-	•	•	-	LM2681	x2, x0.5	2.5-5.5[g] or 1.8-11[h]	track	15	5	30	160	0.06	5	
•	•	-	-	ICL7660[k]	inv, x0.5	3-10 or 1.5-3.5	track	30	10	40	10 or 35[p]	0.08	5	*classic*, 5 manuf, see (k)
•	•	-	-	LT1026	inv & x2	4-10	track	b	-	20	-	15	15	
•	•	-	-	MAX680	pos & neg x2	2-6	track	100[c]	5	5	8	1	5	MAX864 for 200kHz
-	-	-	•	MAX864	pos & neg x2	1.8-6	track	40[c]	5	15[d]	7-185[p]	0.6-12[p]	5	
-	-	•	-	LM828	inv	1.8-5.5	track	20	5	25	12	0.04	5	
-	-	•	-	LM2767	x2	1.8-5.5	track	20	5	25	11	0.04	5	
-	-	•	-	TPS6040x	inv	1.6-5.5	track	10	3	60	20-250[i,q]	0.06-0.4[q]	5	f_{osc} variable ('60400)
•	•	-	-	MAX660	inv, x2	1.5-5.5	track	6.5	5	100	10 or 80[p]	0.12	5	
				regulated[r]										
-	-	-	•	LTC3260	dual LDO	4.5-32	1.2-32 & -1.2 to -32	0.03	12	50	200	4	15	HV dual reg split supply
•	•	-	-	LT1054	inv	3.5-15	$-V_{in}$, or adj reg	10	-	100	25	3	15	
•	-	•		ADP3605	inv	3-6	-3.0, or -3 to -6	0.3	5	120	250	3	5	
•	-	-	•	ST662	reg 12V	4.5-5.5	12	0.8	5	50	400[i]	0.1	5	flash mem prog supply
•	-	•		MAX889	reg adj -Vout	2.7-5.5	-2.5 to $-V_{in}$	0.05	5	200	500-2000[x]	6	5	
•	-	•		MAX682	reg 5V	2.7-5.5	+5V	<1	3	250	20-3000[i,p]	7.5	3.6	
-	-	•	•	REG710-vv	buck-boost	1.8-5.5	2.5, 2.7,...,5.5[vv]	2	n	30	1000[i]	0.07	3.3	auto switch buck/boost
-	-	-	•	MAX1595-vv	buck-boost	1.8-5.5	3.3 or 5.0[vv]	1	3	125	1000[i]	0.23	3	auto switch buck/boost
-	-	-	•	TPS6024x	buck-boost	1.8-5.5	2.7, 3, 3.3, 5[u]	0.7	3.3	40	160	0.25	3	low noise[v]
-	-	•	-	LTC1517-5	reg 5V	2.7-5	5.0	1	3	50	800	0.006	all	micropower reg 5V
-	•	•		LTC3200	reg 5V	2.7-4.5	+5V	0.4	3.6	100	2000[i]	3.5	3.6	
-	•	-	•	LTC1682	LDO, adj Vout	1.8-4.4	2.5-5.5	0.2	3	50	550	0.15	3	x2 to LDO; low noise[w]
-	-	•	•	LTC1502-3.3	reg 3.3V	0.9-1.8	3.3	<0.2	1	20[s]	500[i]	0.04	1	single-cell to reg +3.3V
-	-	•	•	TPS6031x	x2 & reg 3[y]	0.9-1.8	3.0, 3.3[u]	4[w], 0.3	1	50	700[i]	0.03	1.5	single-cell to reg +3V[y]
•	•	-	-	NJU7670	neg x3 & LDO	-2.6 to -6	-8 to -18	5	-5	20	2.5	0.08	-5	neg V_{in}, tripler plus LDO

Notes: (a) all are inductorless, and require several external caps; "regulated" types include either internal linear LDO post-regulator, or regulation via control of switching; sorted within categories by decreasing maximum V_{in}. (b) bipolar, see datasheet for typical V_{out}. (c) with other output unloaded. (d) both outputs loaded. (e) at max f_{osc}. (g) in x2 mode. (h) in x0.5 mode. (i) high f_{osc} allows small capacitors. (k) LV pin for low V_{in} range; many mfgs, prefixes LMC, NJU, TC, TL; see also MAX/LTC/TC1044, 1144, and TC962. (m) maximum. (n) at $V_{in}=V_{out}/2 + 0.8V$. (o) or Si7661. (p) freq pin selectable or adjustable. (q) last digit of p/n sets f_{osc}, except TPS60400, where f_{osc} varies cleverly with V_{in} and I_{out}. (r) unreg outputs also available on most; unless marked "LDO," all regulate via control of switching. (s) at $V_{in}=1.2V$. (u) last digit of p/n sets V_{out}. (v) $V_n = 170\mu Vrms$ in BW = 20Hz-10MHz. (vv) suffix selects V_{out}. (w) $V_n = 60\mu Vrms$ in BW = 10Hz-100KHz, $600\mu Vpp$ for 10Hz-2.5MHz. (x) suffix sets f_{osc}. (y) or reg +3.3V; unreg x2 output also provided. (z) maximum usable.

supply-voltage range: the original charge-pump IC (the Intersil ICL7660) allows V_{in} to range from +1.5 V to +12 V; and although some successor devices (e.g., the LTC1144) extend this range to as much as +18 V, the trend is toward lower-voltage devices with greater output current and with other features.[58] Finally, unlike *inductive* switching converters (discussed next), which can generate any output voltage you want, the flying-capacitor voltage converter can generate only small discrete multiples of the input voltage. In spite of these drawbacks, flying-capacitor voltage converters can be very useful in some circumstances, for example to power a split-supply op-amp or a serial-port chip (see Chapters 14 and 15) on a circuit board that has only +5 volts available. Table 9.4 lists a selection of charge-pump voltage converters, illustrating a range of capabilities (voltage, regulation, output current, and so on).

[58] For example, more than half of Maxim's offerings are limited to +5.5 V input; and of the 67 offerings from Texas Instruments, only 7 can operate above +5.5 V input (and 28 of them are limited to +3.6 V or less). The story is similar for Linear Technology's 62 charge-pump converter offerings.

B. Variations

There are interesting and useful flying-capacitor variations, many of which are listed in Table 9.4, which is organized into unregulated and regulated varieties (each sorted by maximum input voltage). The unregulated types represent variations on the original ICL7660, including its similarly named successors (from TI, NJR, Maxim, Microchip, etc.) and pin-compatible upgrades ('7662, '1044, '1144); such multiply-sourced jellybean parts are widely available and inexpensive. More recent parts, for example the low-voltage TPS6040x, offer flexibility in switching frequency, and generally lower output resistance. Operation at higher frequency reduces output ripple (e.g., 35 mV at 20 kHz, but 15 mV at 250 kHz for the TPS6040x series), but it increases quiescent current (which goes from 65 μA to 425 μA in this example).[59]

The regulated types, like the LT1054 from LTC (with a maximum output current of 100 mA), include an internal voltage reference and error amplifier, so you can connect feedback to regulate the output voltage; the internal circuitry accommodates this by adjusting the switching control. Other converters regulate the output by including an internal low-dropout linear regulator, for greatly reduced output ripple (at the expense of some additional voltage drop); examples are the LTC1550 and 1682 series, with less than 1 mV peak-to-peak output ripple. Note that most of the "regulated" types let you use them as unregulated converters, if you wish.

There are also converters that *reduce* the input voltage by a rational fraction, e.g., by a factor of 1/2 or 2/3 (see if you can figure out how that is done!). At the other end, there are converters that are voltage quadruplers, for example the LTC1502, which generates a regulated +3.3 V at 10 mA from an input of 0.9–1.8 V (e.g., to power digital logic from a single alkaline cell).[60] And there are convert-

ers that can provide up to 500 mA of output current. Some charge-pump converters include internal capacitors, if you want to be especially lazy; but the selection is limited, and the price is high.

Finally, there is the LTC1043 uncommitted flying-capacitor building block, with which you can do all kinds of magic. For example, you can use a flying capacitor to transfer a voltage drop measured at an inconvenient potential (e.g., a current-sensing resistor at the positive supply voltage) down to ground, where you can easily use it. The LTC1043 datasheet has eight pages of similarly clever applications.

Then there are integrated circuits that include charge pumps to power their primary functions:

(a) Many RS-232/485 driver–receiver chips are available with integral ±10 V charge-pump supplies, to run from a single +5 V or +3.3 V supply. An example of the latter is the MAX3232E from Maxim (the originator of the MAX232, now widely second sourced), which can run from a single supply between +3 V and +5.5 V.

(b) Some op-amps use integral charge pumps to generate a voltage beyond the supply rail, so their inputs can operate rail-to-rail while maintaining a conventional high-performance architecture (see §4.6.3B); examples are the OPA369, LTC1152, and MAX1462-4.

(c) Charge pumps are used in many MOSFET "high-side drivers" (like the HIP4080 series from Intersil) and in fully integrated power MOSFETs (like the PROFET series of "smart highside high-current power switches" from Infineon); these generate the necessary above-the-rail gate bias for an *n*-channel MOSFET operating as a follower up at the positive rail.[61]

(d) Some complex digital logic devices (processors, memory) require elevated voltages, which they generate on-chip with charge pumps. The manufacturers are modest, and you don't even hear about these things.

9.6.4 Converters with inductors: the basic non-isolated topologies

The term *switching converter* (or *switchmode converter*[62]) is generally understood to mean a converter that uses some arrangement of inductors and/or transformers, in combination with transistor switches (usually MOSFETs, but also

---

[59] You can reduce ripple by using much larger output capacitors (with low ESR, to minimize the effect of current spikes), or, perhaps better, an output filter stage.

[60] Sadly, there are no charge-pump converters that take a single-cell input (0.9 V at end of life) up to +5 V (that would require at least a $\times6$ voltage conversion), although you could accomplish that task in two steps by cascading, say, a TPS60310 (0.9–3.3 V) with a TPS60241 (3.3–5 V). That would require two ICs and seven capacitors. But, happily, such a task is easily done with a boost-mode *inductive* switching converter (§9.6.6). For example, TI's TPS61222 comes in a tiny 6-pin SC-70 package, requires only a single external 4.7 μH inductor (plus input and output bypass caps), and delivers +5 V at 50 mA with 0.9 V input (an alkaline cell's end-of-life voltage). It costs less than $2 in single-piece quantities. Another approach to powering from a single alkaline cell is to use a charge-pump converter to generate +3.3 V, which then

powers one or more inductive switching converters to generate the full set of voltages you need; the ENABLE input to the charge-pump converter can then be used to turn power on or off.

[61] For further detail see §3.5.3 and Figures 3.96 and 3.106.

[62] Switchmode power supplies are referred to as SMPSs, thus phrases like "SMPS technology."

IGBTs[63] for high voltages), to carry out efficient dc-to-dc conversion. A common characteristic of all such converters is this: in the first portion of each switching cycle the source of input power is used to increase the current (and therefore the energy) in an inductor; that energy then flows to the output during the second portion of the switching cycle. Switchmode power conversion is a major and vital area of electronics, and these converters are used in just about every electronic device.

There are literally hundreds of switchmode circuit variations, but they can be pared down to a few fundamental topologies. In this section we describe the three basic *non-isolated* designs – step-down, step-up, and invert – shown in Figure 9.61. After that we look at isolated converter designs; then we conclude with a look at the use of isolated converters fed from the ac powerline. Tables of selected switchmode converters (Tables 9.5a,b on pages 653 and 654) and controllers (Table 9.6 on page 658) appear later.

Along with the basic power-conversion *topologies* (which describe the circuitry that performs the voltage conversion itself), there is the important topic of *regulation*. Just as with linear voltage regulators, a sample of the output voltage is compared with a voltage reference in an *error amplifier*. Here, however, the error signal is used to adjust some parameter of the switching conversion, most often the pulse width; this is known as *pulse width modulation* (PWM).[64]

As we'll see, the pulse-width modulator circuits themselves fall into two categories (voltage mode and current mode), with important consequences in terms of response time, noise, stability, and other parameters. And, to introduce a bit of further complication, any of these switch-mode circuit combinations may operate in a mode with the inductor's current dropping fully to zero by the end of each switching cycle, or in a mode in which the inductor's current never drops to zero. These modes of operation are known as *discontinuous-conduction mode* (DCM) and *continuous-conduction mode* (CCM), respectively, and they have major effects on feedback stability, ripple, efficiency, and other operating parameters of a switchmode regulator. We describe the basics of PWM with a few examples; but we will touch only lightly on the more advanced topics of voltage- versus current-mode PWM, and on loop compensation.

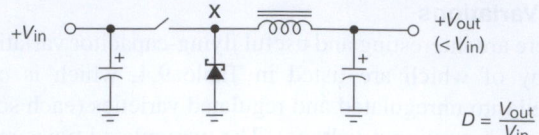

A. Buck (step-down)

$$D = \frac{V_{out}}{V_{in}}$$

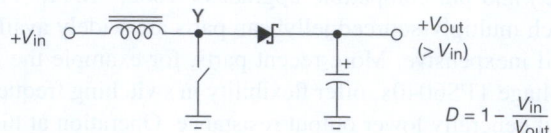

B. Boost (step-up)

$$D = 1 - \frac{V_{in}}{V_{out}}$$

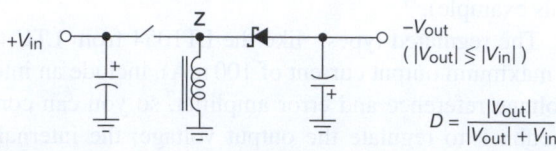

C. Invert (or inverting buck-boost)

$$D = \frac{|V_{out}|}{|V_{out}| + V_{in}}$$

Figure 9.61. The basic nonisolated switching converters. The switch is usually a MOSFET. Schottky diodes are commonly used for the rectifiers, as shown; however, a MOSFET can be used as an efficient synchronously switched "active rectifier."

9.6.5 Step-down (buck) converter

Figure 9.61A shows the basic step-down (or "buck") switching circuit, with feedback omitted for simplicity. When the switch is closed, $V_{out} - V_{in}$ is applied across the inductor, causing a linearly increasing current (recall $dI/dt = V/L$) to flow through the inductor. (This current flows to the load and capacitor, of course.) When the switch opens, inductor current continues to flow in the same direction (remember that inductors don't like to change their current suddenly, according to the last equation), with the "catch diode" (or "freewheeling diode") now conducting to complete the circuit. The inductor now finds a fixed voltage $V_{out} - V_{diode}$ across it, causing its current to decrease linearly. The output capacitor acts as an energy "flywheel," smoothing the inevitable sawtooth ripple (the larger the capacitor, the smaller the ripple voltage). Figure 9.62 shows the corresponding voltage and current waveforms, assuming ideal components. To complete the circuit as a *regulator*, you would of course add feedback, controlling either the pulse width (at constant pulse repetition rate) or the

---

[63] Insulated-gate bipolar transistors, §3.5.7A.

[64] In some switchmode converters the regulation is done instead by varying the pulse *frequency*.

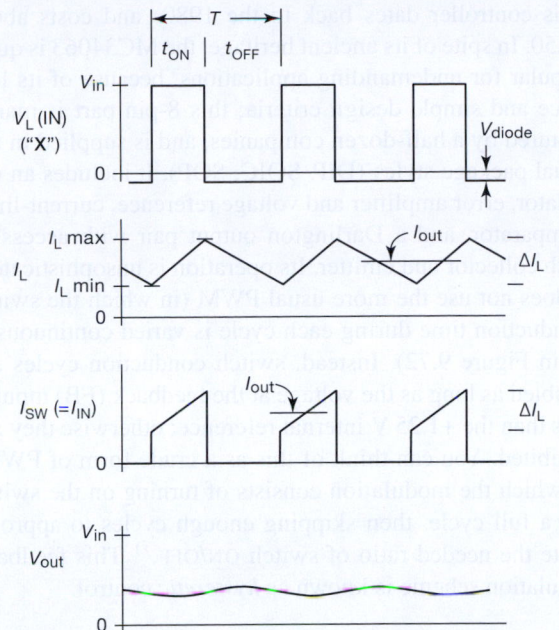

Figure 9.62. Buck converter operation. Inductor current ramps up during switch ON, and ramps down during switch OFF. The output voltage equals the input voltage times the duty cycle ($D \equiv t_{on}/T$). In the case of continuous inductor current (CCM; as shown here) the output current is equal to the average inductor current.

repetition rate (with constant pulse width) from an error amplifier that compares the output voltage with a reference.[65]

For all three circuits of Figure 9.61 the voltage drop across the catch diode wastes energy, reducing the conversion efficiency. Schottky diodes (as shown) are often used to mitigate this, but the best solution is to add a second switch across or in place of the diode. This is called *synchronous switching*; see the "synchronous" column in Tables 9.5a,b and 9.6.

Output voltage   What is the output voltage? In the steady state the average voltage across an inductor must be zero, because otherwise its current is continually growing (according to $V = L\,dI/dt$).[66] So, ignoring voltage drops in the diode and switch, this requires that $(V_{in} - V_{out})\,t_{on} = V_{out}t_{off}$, or

$$V_{out} = DV_{in}, \qquad (9.3)$$

where the "duty cycle" (or "duty ratio") D is the fraction of

the time the switch is ON, $D = t_{on}/T$, and T is the switching period ($T = t_{on} + t_{off}$).

You can think about this in another way: the LC output network is a lowpass filter, to which is applied a chopped dc input whose average voltage is just DV_{in}. So, after smoothing, you get that average voltage as the filtered output. Note that, assuming ideal components, the output voltage from a buck converter running at fixed duty cycle D from a fixed input voltage is intrinsically regulated: a change in load current does not change the output voltage; it merely causes the inductor's triangular current waveform to shift up or down, such that the average inductor current equals the output current. (This assumes continuous inductor current, or CCM, as we discuss below.)

Input current   What is the input current? If we assume ideal components, the converter is lossless (100% efficiency), so the input power must equal the output power. Equating these, the average input current is $I_{in} = I_{out}(V_{out}/V_{in})$.[67]

Critical output current   We've been assuming continuous inductor conduction in the waveforms of Figure 9.62, and also in deducing that the output voltage is simply the input voltage times the switch duty cycle. Look again at the graph of inductor current: its average current must equal the output current, but its peak-to-peak variation (call it ΔI_L) is completely determined by other factors (namely V_{in}, V_{out}, T, and L); so there is a *minimum output current* for which the inductor stays in conduction, namely when $I_{out} = \frac{1}{2}\Delta I_L$.[68] For output currents less than this critical load current, the inductor current reaches zero before the end of each cycle; the converter is then operating in discontinuous conduction mode, for which the output voltage would no longer remain stable at fixed duty cycle, but would depend on load current. Of greater importance, operating in DCM has a major effect on loop stability and regulation. For this reason many switching regulators have a minimum output current, in order to operate in CCM.[69] As the following expressions show, the minimum load current for CCM is reduced by increasing the inductance, increasing the switching frequency, or both.

---

[65] There is also hysteretic control, in which both pulse width and switching frequency may vary.

[66] Engineers like to say that the *volt–time product* (or the *volt–second product*) must average to zero.

[67] In real converters the efficiency is reduced by losses in the inductors, capacitors, switches, and diodes. It's a complicated subject.

[68] Operation at this current is called *critical conduction mode*.

[69] At load currents less than the minimum current for CCM they may enter other modes of operation, including "burst mode."

A. Buck converter equations (continuous-conduction mode)

From the preceding discussion and waveforms it is not terribly difficult to figure out that the ideal buck converter (Figure 9.61A), operating in continuous conduction mode, obeys these equations:

$$\langle I_{\text{in}} \rangle = I_{\text{out}} \frac{V_{\text{out}}}{V_{\text{in}}} = D I_{\text{out}}, \tag{9.3a}$$

$$\Delta I_{\text{in}} = I_{\text{out}}, \tag{9.3b}$$

$$V_{\text{out}} = V_{\text{in}} \frac{t_{\text{on}}}{T} = D V_{\text{in}}, \tag{9.3c}$$

$$D = \frac{V_{\text{out}}}{V_{\text{in}}}, \tag{9.3d}$$

$$I_{\text{out(min)}} = \frac{T}{2L} V_{\text{out}} \left(1 - \frac{V_{\text{out}}}{V_{\text{in}}}\right)$$

$$= \frac{T}{2L} V_{\text{out}} (1 - D), \tag{9.3e}$$

$$\Delta I_{\text{C(out)}} = \frac{T}{L} V_{\text{out}} (1 - D), \tag{9.3f}$$

$$I_{\text{L(pk)}} = I_{\text{out}} + \frac{T}{2L} V_{\text{out}} (1 - D), \tag{9.3g}$$

$$L_{\text{min}} = \frac{T}{2} \frac{V_{\text{out}}}{I_{\text{out}}} (1 - D), \tag{9.3h}$$

where $\langle I_{\text{in}} \rangle$ represents the time-averaged value of input current, and ΔI_{in} and $\Delta I_{\text{C(out)}}$ are the approximate peak-to-peak ripple currents at input and output (important for capacitor selection[70]). The first equation holds regardless of mode (CCM or DCM). The expressions for minimum inductance and minimum output current represent the critical values to maintain CCM; for these expressions use the minimum output current and the maximum value of V_{in}, respectively.

Exercise 9.8. Take the challenge: derive these equations (and be sure to tell us if we got them wrong). *Hint*: for $I_{\text{out(min)}}$ and L_{min} use the fact that the output current I_{out} equals half the peak-to-peak inductor current variation ΔI_L, at the threshold of CCM, as easily seen from the I_L waveform in Figure 9.62.

B. Buck converter example – I

Let's do a buck regulator design, using a very simple (and inexpensive) controller chip, the MC34063 (Figure 9.63).

This controller dates back to the 1980s and costs about $0.50. In spite of its ancient heritage, the MC34063 is quite popular for undemanding applications, because of its low price and simple design criteria; this 8-pin part is manufactured by a half-dozen companies, and is supplied in the usual package styles (DIP, SOIC, SOP). It includes an oscillator, error amplifier and voltage reference, current-limit comparator, and a Darlington output pair with access to both collector and emitter. Its operation is unsophisticated: it does not use the more usual PWM (in which the switch conduction time during each cycle is varied continuously, as in Figure 9.72). Instead, switch conduction cycles are enabled as long as the voltage at the feedback (FB) input is less than the +1.25 V internal reference; otherwise they are inhibited. You can think of this as a crude form of PWM, in which the modulation consists of turning on the switch for a full cycle, then skipping enough cycles to approximate the needed ratio of switch ON/OFF.[71] This feedback regulation scheme is known as *hysteretic* control.

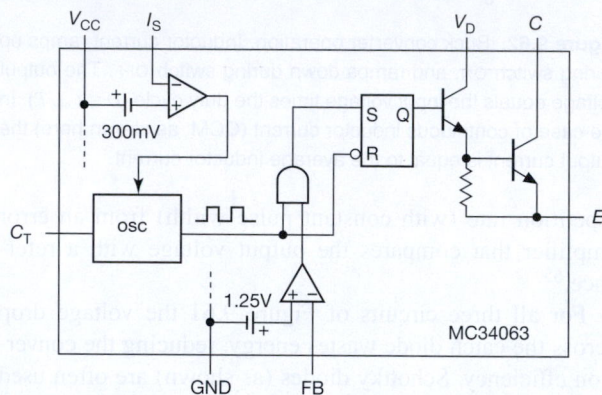

Figure 9.63. A popular $0.50 switching converter. The external connections to both collector and emitter of the 1.5 A switch make it easy to implement buck, boost, or inverting converters.

For our design let's assume a +15 V input, and produce a +5 V regulated output for load currents up to 500 mA. Figure 9.64 shows the circuit. The design is straightforward:
1. Choose an operating frequency: we picked 50 kHz, half the chip's recommended maximum. For that frequency the datasheet specifies $C_T = 470$ pF. The oscillator runs with a ratio $t_{\text{on}}/t_{\text{off}} = 6$, so the switch conduction time is $t_{\text{on}} = 17\,\mu\text{s}$.
2. Calculate the inductor value so the converter operates

---

[70] Note that capacitor datasheets specify maximum allowed *rms* ripple current, rather than peak-to-peak. Be sure to allow a large safety margin in this parameter when selecting input and output capacitors for power conversion.

[71] This is analogous to "bang–bang" feedback control, as contrasted with proportional (or PID) control in which the feedback signal operates in a continual manner.

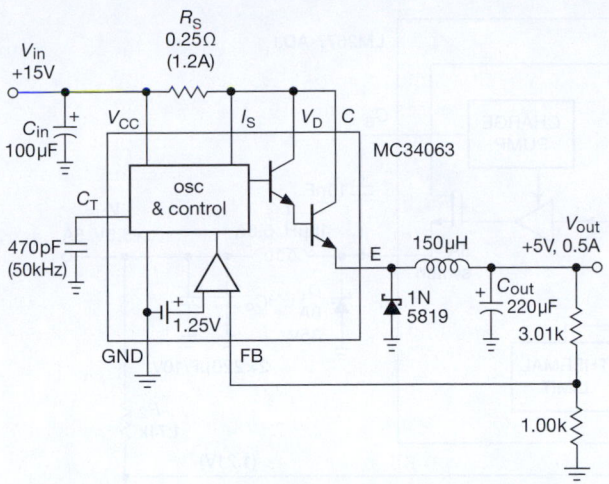

Figure 9.64. Step-down regulator using the MC34063. In contrast to proportional PWM, the chip's simple bang–bang control eliminates the need for feedback compensation components. But performance suffers.

in DCM,[72] assuming onset of CCM at minimum input voltage and maximum load current: at onset of CCM, the output current is half the peak inductor current, so, using $V = L dI/dt$ (and assuming a 1 V drop in the Darlington switch), we get $L = (V_{in} - V_{sw} - V_{out}) t_{on}/2 I_{out} = 153 \mu$H. We'll use a standard value of 150 μH.

3. Calculate the value of sense resistor R_S to limit the peak current I_{pk} to somewhat greater than the expected 1 A, but no greater than the chip's 1.5 A rating: $R_S = 300\,\mathrm{mV}/I_{lim} = 0.25\Omega$ (for a 1.2 A current limit).[73]

4. Choose an output capacitor value to keep the ripple voltage below some acceptable value. You can estimate the ripple by calculating the capacitor's voltage rise during one cycle of switch conduction (during which its current goes from 0 to I_{pk}), which gives a value $\Delta V = I_{pk} t_{on}/2 C_{out}$. So an output capacitor of 220 μF results in a peak-to-peak ripple voltage of ~40 mV.[74]

Several comments. (a) This simple design will work, but the performance will be far from ideal. In particular, the crude bang–bang control, combined with discontinuous-conduction operation, produces lots of output ripple, and even audible noise, caused by its intermittent pulsing.

---

[72] That is, the inductor current ramps completely to zero during each switch cycle.

[73] If you find that the expected peak current is greater than the chip's limit, you will have to append an outboard transistor, or (better) use a different chip.

[74] The actual ripple voltage will be higher because of the capacitor's ESR, an effect that can also be estimated.

(b) The Darlington output connection prevents saturation in the output stage, with some loss of efficiency; this could be remedied by connecting the driver collector line (V_D) to the input supply, through a current limiting resistor of the order of 200 Ω. (c) The internal switch is limited to 1.5 A peak current, which is inadequate for output currents greater than 0.75 A; this can be remedied with an external transistor switch, for example a *pnp* transistor or *p*-channel MOSFET (for this buck configuration). The main attractiveness here is the combination of very low cost, and lack of worries about feedback stability and compensation. You'll see this part used in relaxed applications such as cellphone chargers and the like.[75]

C. Buck converter example – II

Fortunately, there are very nice integrated switchers that implement proportional PWM and, furthermore, make it really easy to do a circuit design (many are listed in Tables 9.5a,b, discussed later). For example, National Semiconductor (part of Texas Instruments) has a series of "Simple Switcher™" ICs, individually configured for buck, boost, or invert topologies, that include all the necessary feedback loop compensation components on-chip.[76] They cover a voltage range up to 40 V or more, with currents to 5 A, and have built-in current limit, thermal limit, voltage reference, fixed-frequency oscillator, and (in some versions) features such as soft-start (see §9.6.8G), frequency synchronization, and shutdown. Best of all, they make it dead simple to design a converter either by following the step-by-step recipes in the datasheets or by using free web-based design tools: you get the component values (including recommended component manufacturers' part numbers) and performance data.

Figure 9.65 shows such a design, in this case converting a 14 V input (from an automobile battery) to a +3.3 V output that can supply up to 5 A (to power digital logic). We followed the datasheet's recipe to get the component values and part numbers shown. With these components the efficiency is 80% and the output ripple is less than 1% of V_{out} (~30 mV).

The LM2677 we used (and other "simple switcher" successors) follow on from the original LM2574,75,76 series

---

[75] Those who are struggling with an under-performing circuit based on an MC34063A should consider the NCP3063, a drop-in upgrade that operates to 150 kHz. This allows you to reduce the inductor size and deliver higher output currents.

[76] See for example the block diagram in the LM2677's datasheet, and associated patents for the active inductor (US patent 5,514,947) and active capacitor (US patent 5,382,918).

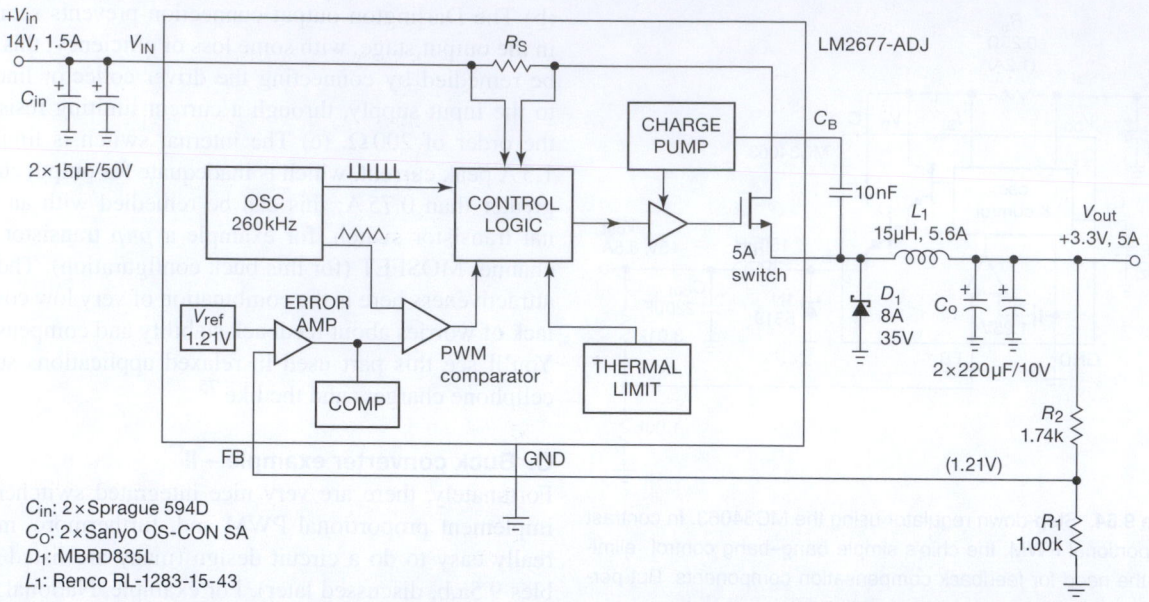

Figure 9.65. Step-down regulator using the LM2677 "Simple Switcher" (complete with elegant built-in compensation). We followed the datasheet's design recipe to get the component values and recommended part numbers shown.

(0.5 A, 1 A, and 3 A, respectively), which run at 52 kHz and which are widely popular "jellybean" parts – they are inexpensive and available from many manufacturers.[77] The LM2677 is a member of the improved LM2670 family, running at 260kHz, with output-current ratings to 5 A; it requires one additional capacitor (C_B in the figure) to drive the 5 A low-drop MOSFET.

Several comments:

(a) This converter provides ten times the output current of the previous design (Figure 9.64), and with significantly improved performance in terms of regulation, ripple, and transient response. That comes at a cost (literally), namely an IC that costs ten times as much (about $5, versus $0.50).[78]

(b) The good efficiency is due in part to the use of an *n*-channel MOSFET whose gate is driven from a voltage higher than V_{in}, thanks to an internal charge pump; that's the purpose of the boost capacitor C_B.

(c) Note the use of paralleled capacitors at the input and output. You see this often in switchmode converters, where it's important to keep ESR and ESL (equivalent series inductance) low: that reduces the voltage ripple caused by

---

[77] And ON semiconductor has introduced the compatible NCV2576 family, low-cost parts rated specifically for the automotive market.

[78] Power converter ICs vary over an enormous price range; the approximate prices listed in this chapter's tables can provide some guidance in their selection.

ripple current, and also keeps the capacitors within their ripple-current ratings.[79]

(d) For a standard output voltage like the +3.3 V here, you can save two resistors by selecting a fixed-voltage version (LM2677-3.3); but the adjustable version (LM2677-ADJ) lets you choose your output voltage, and you don't have to keep multiple versions in stock in your laboratory.

(e) Note that the input current is a lot less than the output current, representing a power-conversion efficiency of 80%; this is a major advantage over a linear regulator.

(f) Fixed efficiency means that if you increase the input voltage, the input current goes *down*: that's a negative resistance! This creates some amusing complications – for example you can get oscillation when the input is filtered with an *LC* network, a problem that applies to ac powerline input converters as well.

Exercise 9.9. What is the maximum theoretical efficiency of a linear (series pass) regulator, when used to generate regulated +3.3 V from a +14 V input?

Exercise 9.10. What does a step-down regulator's high efficiency imply about the ratio of output current to input current? What is the corresponding ratio of currents, for a linear regulator?

---

[79] It also assists in creating a desirably low physical profile.

9.6.6 Step-up (boost) converter

Unlike linear regulators, switching converters can produce output voltages higher than their input. The basic non-isolated step-up (or "boost") configuration was shown in Figure 9.61B (repeated here as 9.66, and seen earlier, in Figure 9.55, in comparison with the linear regulator). During switch conduction (point Y near ground) the inductor current ramps up; when the switch is turned off, the voltage at point Y rises rapidly as the inductor attempts to maintain constant current. The diode turns on, and the inductor dumps current into the capacitor. The output voltage can be much larger than the input voltage.

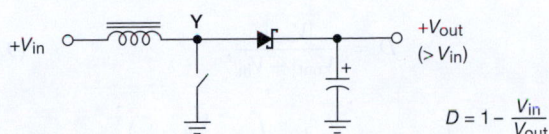

$$D = 1 - \frac{V_{in}}{V_{out}}$$

Figure 9.66. Basic boost (or "step-up") topology (non-isolated).

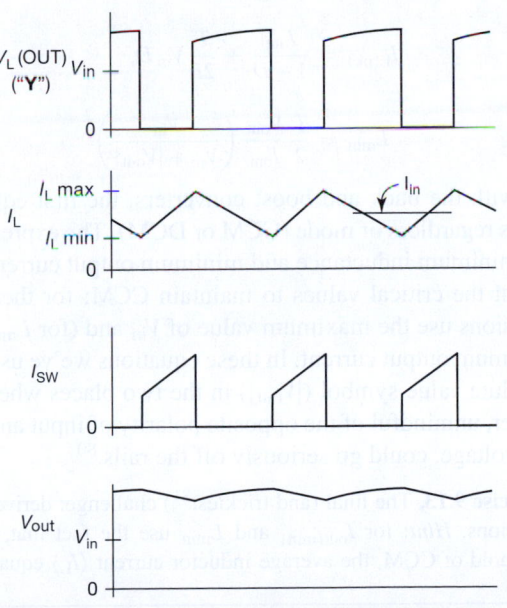

Figure 9.67. Boost converter operation. Inductor current ramps up during switch ON, and ramps down during switch OFF. The output voltage equals the input voltage divided by the fraction of the time the switch is OFF. In the case of continuous inductor current (CCM, as shown here) the input current is equal to the average inductor current.

A. Boost converter equations (continuous-conduction mode)

Figure 9.67 shows relevant voltage and current waveforms, assuming ideal components. As with the buck converter, it is not terribly difficult to figure out that the boost converter (Figure 9.61B), operating in continuous conduction mode, obeys these equations:

$$\langle I_{in} \rangle = I_{out} \frac{V_{out}}{V_{in}} = \frac{I_{out}}{1-D}, \tag{9.4a}$$

$$\Delta I_{in} = \frac{T}{L} V_{in} D, \tag{9.4b}$$

$$V_{out} = V_{in} \frac{T}{t_{off}} = \frac{V_{in}}{1-D}, \tag{9.4c}$$

$$D = 1 - \frac{V_{in}}{V_{out}}, \tag{9.4d}$$

$$I_{out(min)} = \frac{T}{2L}\left(\frac{V_{in}}{V_{out}}\right)^2 (V_{out} - V_{in}),$$

$$= \frac{T}{2L} V_{out} D (1-D)^2, \tag{9.4e}$$

$$\Delta I_{C(out)} = \frac{I_{out}}{1-D}, \tag{9.4f}$$

$$I_{L(pk)} = \frac{I_{out}}{1-D} + \frac{T}{2L} V_{in} D, \tag{9.4g}$$

$$L_{min} = \frac{T}{2I_{out}}\left(\frac{V_{in}}{V_{out}}\right)^2 (V_{out} - V_{in}). \tag{9.4h}$$

The first equation holds regardless of mode (CCM or DCM). The expressions for minimum inductance and minimum output current represent the critical values to maintain CCM; for these expressions use the maximum value of V_{in} and (for L_{min}) the minimum output current.

Exercise 9.11. Continuing the challenge: derive these equations. *Hint:* for $I_{out(min)}$ and L_{min} use the fact that, at the threshold of CCM, the input current I_{in} equals half the peak-to-peak inductor current variation ΔI_L, as easily seen from the I_L waveform in Figure 9.67.

Exercise 9.12. Why can't the step-up circuit be used as a step-down regulator?

The design procedures for step-up (and inverting) converters are analogous to those for the buck converter, and so we will resist the temptation to display actual circuit examples.

9.6.7 Inverting converter

The inverting circuit (also known as an "inverting buck–boost," or "negative buck–boost") was shown in Figure 9.61C (repeated here as 9.68). During switch conduction, a linearly increasing current flows from the input into the inductor (point Z) to ground. To maintain the current when the switch is open, the inductor pulls point Z negative, as much as needed to maintain continuous current flow. Now, however, that current is flowing into the inductor from the filter capacitor (and load). The output is thus negative, and its average value can be larger or smaller in magnitude than the input (as determined by feedback); in other words, the inverting regulator can be either step-up or step-down.

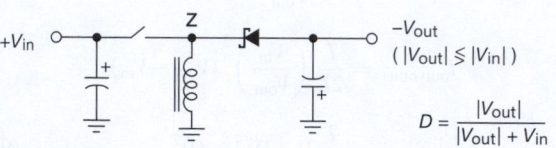

Figure 9.68. Basic inverting (or "inverting buck-boost") topology (non-isolated).

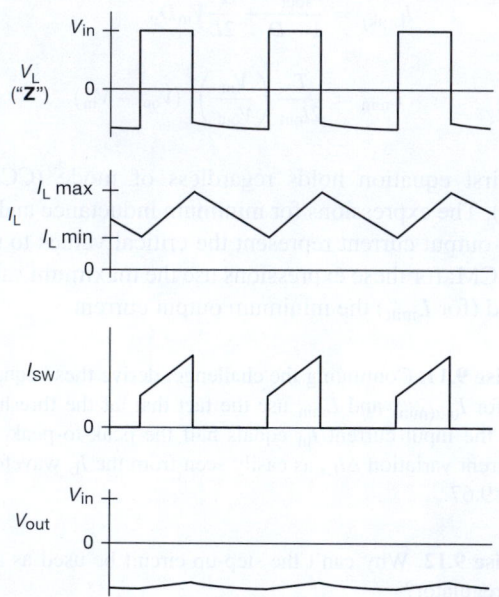

Figure 9.69. Inverting converter operation. Inductor current ramps up during switch ON, and ramps down during switch OFF. The output voltage is inverted in polarity, with a magnitude equal to the input voltage times the ratio of switch t_{on}/t_{off} (for CCM, as shown here).

A. Inverting converter equations (continuous-conduction mode)

Figure 9.69 shows the relevant voltage and current waveforms of the inverting regulator, once again assuming ideal components. With more than a bit of struggle you can figure out that the inverting converter (Figure 9.61C), operating in continuous-conduction mode, obeys these equations:

$$\langle I_{in}\rangle = I_{out}\frac{V_{out}}{V_{in}} = -I_{out}\frac{D}{1-D}, \quad (9.5a)$$

$$\Delta I_{in} = \frac{\langle I_{in}\rangle}{D}, \quad (9.5b)$$

$$V_{out} = -V_{in}\frac{t_{on}}{t_{off}} = -V_{in}\frac{D}{1-D}, \quad (9.5c)$$

$$D = \frac{|V_{out}|}{|V_{out}|+V_{in}}, \quad (9.5d)$$

$$I_{out(min)} = \frac{T}{2L}V_{out}\left(\frac{V_{in}}{V_{in}+|V_{out}|}\right)^2$$
$$= \frac{T}{2L}V_{out}(1-D)^2, \quad (9.5e)$$

$$\Delta I_{C(out)} = \frac{I_{out}}{1-D}, \quad (9.5f)$$

$$I_{L(pk)} = \frac{I_{out}}{1-D} + \frac{T}{2L}V_{in}D, \quad (9.5g)$$

$$L_{min} = \frac{T}{2}\frac{V_{out}}{I_{out}}\left(\frac{V_{in}}{V_{in}+|V_{out}|}\right)^2. \quad (9.5h)$$

As with the buck and boost converters, the first equation holds regardless of mode (CCM or DCM). The expressions for minimum inductance and minimum output current represent the critical values to maintain CCM; for these expressions use the maximum value of V_{in} and (for L_{min}) the minimum output current. In these equations we've used the absolute value symbol ($|V_{out}|$) in the two places where the reader, unmindful of the opposite polarity of input and output voltage, could go seriously off the rails.[80]

Exercise 9.13. The final (and trickiest[81]) challenge: derive these equations. *Hint*: for $I_{out(min)}$ and L_{min} use the fact that, at the threshold of CCM, the average inductor current $\langle I_L\rangle$ equals half

---

[80] Readers who feel insulted by such lack of trust should replace "$+|V_{out}|$" with "$-V_{out}$." They can argue, with some justification, that their signed equation correctly describes also an inverting converter that produces a positive output from a negative input rail.

[81] Dare we confess? It flummoxed more than a few of us before we got it right.

the peak-to-peak inductor current variation ΔI_L. Now figure out how $\langle I_L \rangle$ is related to I_{in} (or to I_{out}), and take it from there.

9.6.8 Comments on the non-isolated converters

This is a good place to pause, before moving on to the transformer-isolated switching converters, to discuss and review some issues common to these converters.

A. Large-voltage ratios
The ratio of output to input voltage in the basic non-isolated converters depends on the duty cycle ($D = t_{on}/T$), as given in the formulas above. For modest ratios that works fine. But to generate a large ratio, for example a buck converter converting a +48 V input to a +1.5 V output, you wind up with undesirably short pulse widths (hence greater transistor stress, in the form of high peak voltages and currents, and lower efficiency). A better solution is to take advantage of a transformer, whose turns ratio provides an additional voltage transformation. We'll see soon how this is done, in the analogous isolated converter topologies (buck converter → forward converter; inverting converter → flyback converter).

B. Current discontinuity and ripple
The three basic converters behave quite differently in terms of input- and output-current pulsation. In particular, assuming the preferred continuous-conduction mode, the buck converter has continuous current being supplied to the output storage capacitor, but pulsed input current from the +V_{in} supply; the boost converter has pulsed output current, but continuous input current; and the inverting converter has pulsed current at both input and output. Pulsed (discontinuous) currents are generally undesirable at high power levels because they require larger-value storage capacitors, with lower ESR/ESL, for comparable performance. There are some interesting converter topologies (discussed presently, §9.6.8H) that address these problems; in particular, the Ćuk converter (Figure 9.70) boasts continuity of current at both input and output.

C. Regulation: voltage mode and current mode
We've talked little about the details of feedback and voltage regulation in switchmode converters, though the examples above illustrate two approaches: the simple bang–bang pulse-skipping scheme of the MC34063-style regulator (Figure 9.64); and the more commonly used proportional PWM scheme implemented in Figure 9.65. In fact, PWM control can be done in two ways, known as *voltage mode* and *current mode*: in voltage-mode PWM, the error signal is compared with the internal oscillator's sawtooth (or triangular) waveform to set the switch-ON duration. By contrast, in current-mode PWM the switch's current, ramp-

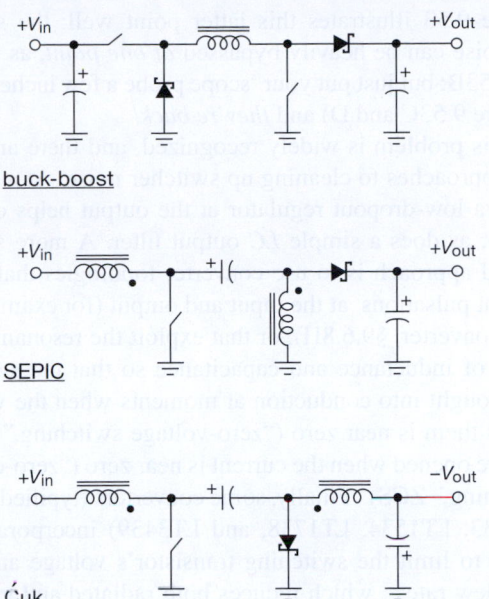

Figure 9.70. Converters allowing overlap of input and output voltage range. Both switches are operated together in the buck–boost (or "non-inverting buck–boost") configuration (A). The SEPIC (B) and Ćuk (C) configurations each use a single switch, but two (optionally coupled) inductors. The Ćuk "boost–buck" is inverting.

ing according to $V = L\,dI/dt$, replaces the sawtooth, and is compared with the error signal to terminate the switch's ON state, as shown below in Figure 9.71. We'll go into a bit more detail in §9.6.9.

D. Low-noise switchers
Switchers are noisy! Figure 9.53, which compared linear and switching 5 V power converters, shows several characteristics of this undesirable "feature": first, there is plenty of noise at the switching frequency, which typically falls in the 20 kHz–1 MHz range; second, the switching frequency may vary,[82] causing interference over a range of frequencies; and, third, (and most distressingly) the switching signals can be nearly impossible to eliminate, propagating both as radiated signals and through ground currents.

---

[82] This is often done intentionally, in order to meet regulatory standards on interference (EMI) by "spreading" the emitted switching signals over a range of frequencies (see Figures 9.53 and 9.54). Although there is some rationale for resorting to this measure when other options are exhausted, we're not wild about this practice, which paradoxically encourages sloppy design that emits *more* total radiated power. As NASA engineer Eric Berger remarked, "When I first heard about this practice, I was appalled. The radiated energy is not reduced, just the peaks in the frequency domain are. This is like getting rid of a cow pie by stomping on it."

Figure 9.53 illustrates this latter point well: the switching noise can be heavily bypassed *at one point*, as in Figure 9.53B; but just put your 'scope probe a few inches away (Figure 9.53C and D) and *they're back!*

This problem is widely recognized, and there are various approaches to cleaning up switcher noise. At a simple level, a low-dropout regulator at the output helps considerably, as does a simple *LC* output filter. A more sophisticated approach is to use converter topologies that avoid current pulsations at the input and output (for example the Ćuk converter, §9.6.8H), or that exploit the resonant properties of inductance and capacitance so that the switches are brought into conduction at moments when the voltage across them is near zero ("zero-voltage switching," ZVS), and are opened when the current is near zero ("zero-current switching," ZCS). Finally, some converters (typified by the LT1533, LT1534, LT1738, and LT3439) incorporate circuitry to limit the switching transistor's voltage and current slew rates, which reduces both radiated and ground-conducted switching noise.

When thinking about switching converter noise, keep in mind that it emerges in multiple ways, namely:
(a) ripple impressed *across* the dc output terminals, at the switching frequency, typically of the order of 10–100 mV peak-to-peak;
(b) *common-mode* ripple on the dc output (which you can think of as ground-line ripple current), which causes the kind of mischief seen in Figure 9.53C;
(c) ripple, again at the switching frequency, impressed onto the *input* supply;
(d) *radiated* noise, at the switching frequency and its harmonics, from switched currents in the inductors and leads.

You can get into plenty of trouble with switching supplies in a circuit that has low-level signals (say 100 μV or less). Although an aggressive job of shielding and filtering may solve such problems, you're probably better off with linear regulators from the outset.

E. Inductance tradeoffs
There's some flexibility in the choice of inductance. Usually you want to run PWM converters (but not bang–bang converters like the MC34063 in our first example) in continuous-conduction mode, which sets a minimum inductance for a given switching frequency and value of minimum load current. A larger inductor lowers the minimum load current, reduces the ripple current for a given load current, and improves the efficiency; but a larger inductor also reduces the maximum load current, degrades the transient

response,[83] and adds physical size to the converter. It's a tradeoff.

F. Feedback stability
Switching converters require considerably more care in the design of the frequency-compensation network than, say, an op-amp circuit. At least three factors contribute to this: the output *LC* network produces a "2-pole" lagging phase shift (ultimately reaching 180°), which requires a compensating "zero"; the load's characteristics (additional bypass capacitance, nonlinearities, etc.) affect the loop characteristics; and the converter's gain and phase versus frequency characteristics change abruptly if the converter enters discontinuous-conduction mode. And, to add a bit more complexity into an already-complex situation, there are important differences between voltage-mode and current-mode converters: for example, the latter, which are better behaved in terms of *LC*-network phase shifts, exhibit a "subharmonic instability" when operated at switch duty cycles greater than 50% (this is addressed by a technique called *slope compensation*).

The easiest approach for the casual user is to choose converters with built-in compensation (for example, the Simple Switcher series, as in Figure 9.65), or converters that provide complete recipes for reliable external compensation. Regardless, the circuit designer (you!) should be sure to *test* the stuff you've designed.[84]

G. Soft start
When input voltage is initially applied to any voltage-regulator circuit, feedback will attempt to bring the output to the target voltage. In the case of a switching converter, the effect is to command maximum duty cycle from the switch, cycle after cycle. This generates a large inrush current (from charging the output capacitor), but, worse, it can cause the output voltage to overshoot, with potentially damaging effects on the load. Worse still, the magnetic core of the inductor (or transformer) may saturate (reaching maximum flux density), whereupon the inductance drops precipitously, causing the switch current to spike. Core saturation is a major cause of component failure; you don't want it.

These problems are most severe in converters that run

---

[83] Transient speed is a major reason to use low inductance values in switching converters that power microprocessors, where you see the concept of *critical inductance*, i.e., an inductance small enough to handle the load step transients.

[84] When testing for stability, don't forget about the negative-resistance input characteristic of switching converters; be sure to test with whatever input filters you plan to use.

from the ac powerline, where the transformerless input stage (diode bridge and storage capacitor) causes additional inrush current, and where that input power source can deliver plenty of peak current. Many switching controller chips therefore incorporate "soft-start" circuitry, which constrains the switch duty cycle to ramp up gradually upon initial startup; these are indicated in the "soft start" column of Tables 9.5a,b and 9.6.

H. Buck–boost topologies

For the buck converter, V_{out} must be less than V_{in}, and for the boost converter V_{out} must be greater than V_{in}, required in both cases to reset the inductor current. Sometimes you'd like a converter that permits the input voltage to vary around both sides of the output voltage (for example in a battery-operated device with 2.5 V digital logic, powered by two AA cells, which begins life with 3 V input, and ends at about 1.8 V; or an automotive application, powered from a 12 V car battery, supplying 13.8 V running, but as little as 8 V starting and as much as 40 V in "load dump").

Although the inverting (buck–boost) converter (Figure 9.61C) allows the output voltage to be larger or smaller than the input, its polarity is reversed. Figure 9.70 shows three interesting configurations that allow overlap of the input- and output-voltage ranges. The first one is particularly easy to understand: both switches are operated simultaneously for a time t_{on}, applying V_{in} across the inductor; during t_{off} the inductor's current flows through the diode pair to the output. The output voltage, from the inductor's required volt–time equality (and ignoring voltage drops in the switch and diodes), is then simply $V_{out} = (t_{on}/t_{off})V_{in}$. Typical examples of buck–boost converter ICs are the LTC3534 (internal MOSFET switches) and the LTC3789 (external MOSFET switches); both use synchronous MOSFET switches in place of Schottky diodes, i.e., four MOSFETs in all. For other converters with synchronous switching see the "synchronous" column in Tables 9.5a,b on pages 653 and 654 and 9.6 on page 658.

The SEPIC (single-ended primary-inductance converter) and Ćuk[85] converters have the advantage of requiring only a single controllable switch. And the Ćuk converter has the remarkable property of producing *zero* output ripple current when the inductors are coupled (wound on the same core). This latter property was discovered accidentally, but is now part of the vocabulary of switchmode practitioners, who call it "the zero-ripple phenomenon." And while we're praising the Ćuk, it's worth noting that both input- and output-current waveforms are continuous, unlike the buck, boost, inverting, SEPIC, or buck–boost.

---

[85] Invented by Slobodan Ćuk (pronounced "chook") in 1976.

9.6.9 Voltage mode and current mode

There are two approaches to implementing pulse-width modulation, as we mentioned earlier in §9.6.8C; look at Figure 9.71.

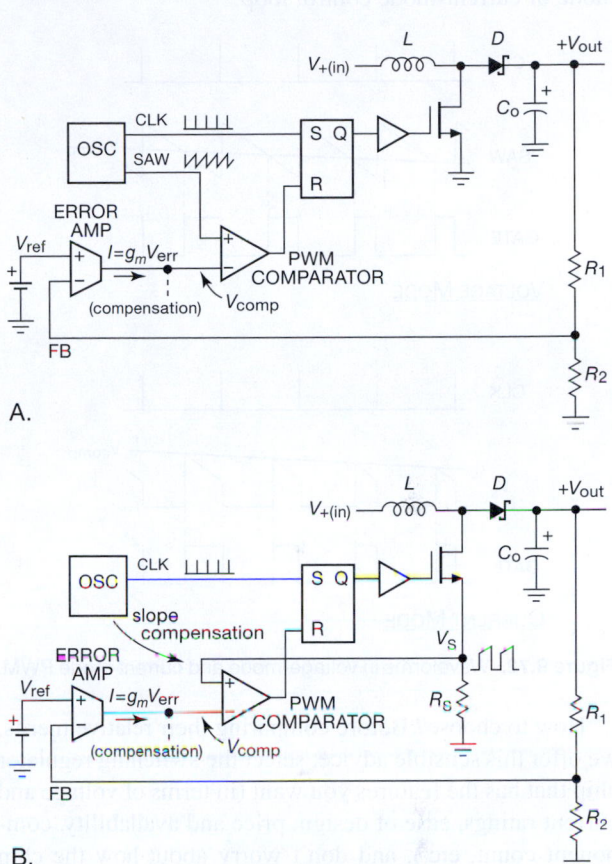

Figure 9.71. Pulse-width modulation in switchmode regulators. (A) Voltage-mode PWM compares the integrated error signal ($V_{err} = V_{ref} - FB$) with the oscillator's sawtooth whereas (B) current-mode PWM substitutes the switch's ramping current waveform.

At the top level, both methods compare the output voltage with an internal voltage reference to generate an error signal. That is, both methods are *voltage* regulators (don't confuse "current mode" with current *regulator*). The difference is in the way the error signal is used to adjust the pulse width: in *voltage-mode* PWM, the error signal is compared with the internal oscillator's sawtooth waveform to control the switch's ON duration.[86] In *current-mode* PWM, by

---

[86] Typically by using a *pulse* output from the oscillator to start the conduction cycle, and the output of the PWM comparator (which compares the error signal with the same oscillator's sawtooth) to end the conduction cycle, as shown in Figures 9.71A and 9.72.

contrast, the ramping current in the inductor replaces the sawtooth, with the internal oscillator used to *initiate* each conduction cycle[87] (Figures 9.71B and 9.72). Tables 9.5a,b and 9.6 indicate whether the SMPS IC employs a voltage-mode or current-mode control loop.

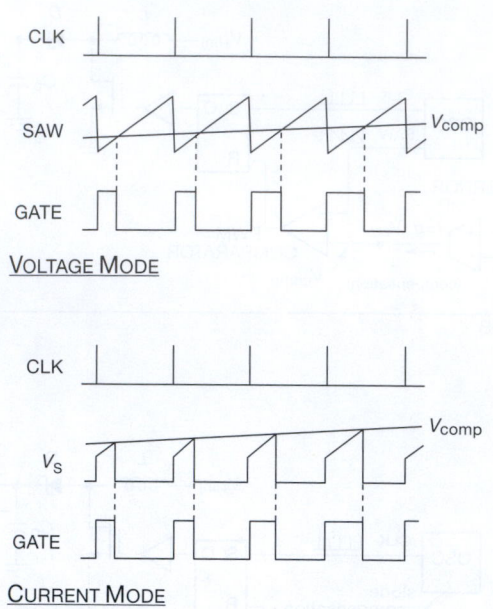

Figure 9.72. Waveforms in voltage-mode and current-mode PWM.

How to choose? Before comparing their relative merits, we offer this sensible advice: select the switching regulator chip that has the features you want (in terms of voltage and current ratings, ease of design, price and availability, component count, etc.), and don't worry about how the chip designers did their job.

Now for the comparison.

A. Voltage mode
This has been the traditional form of PWM. Its advantages include
(a) the simplicity of analyzing a single feedback path,
(b) low output impedance from the power stage, and
(c) good noise margins (because of the internally generated ramp).

Its disadvantages include
(a) the need for careful loop compensation (because of the 2-pole *LC* output filter),[88] and
(b) slow loop response (especially in response to input

changes), and
(c) the need for separate current-limiting circuitry for the switch transistor(s).

B. Current mode
Current-mode control became popular beginning in the 1980s when its benefits became apparent. They include
(a) rapid response to input changes,
(b) inherent pulse-by-pulse current limiting of the switch current,
(c) improved phase margin in the outer voltage-feedback loop (because the power stage's output, being current-like, effectively removes the inductor's phase shift; i.e., one pole instead of two in the feedback loop), and
(d) the ability to parallel the outputs of several identical converters.

The disadvantages of current-mode control include
(a) the greater difficulty of analyzing two nested feedback loops (mitigated by widely separating their characteristic frequencies),
(b) intrinsically higher output impedance of the power stage (the output is more affected by load changes because the fast loop tends toward a constant-current output),
(c) susceptibility to noise, particularly at low load, and to resonances (because the PWM depends on the current-derived ramp),
(d) premature termination of switch's ON-state caused by the leading-edge current spike (from parasitic capacitances and diode recovery effects), and
(e) instabilities and subharmonic resonances at high duty cycle.

Clever fixes Circuit designers are clever, and they've figured out some nice tricks to address the problems of each method. The slow response of voltage-mode controllers to input changes can be fixed by adding an input feedforward signal to the sawtooth ramp, and the slow loop response can be alleviated by running at a higher switching frequency. For current-mode control the bag of tricks includes leading-edge blanking (to ignore the switch-ON current spike), and "slope compensation" (to restore stability at high duty cycle).

Choice of control mode: both are viable In contemporary practice both modes are viable, and plenty of controller ICs are available using either technique. As a general statement, voltage-mode converters are favored
(a) in noisy applications, or in applications with light load

---

[87] And to generate the "slope-compensation" ramp signal.

[88] As the LT3435 datasheet succinctly puts it, "A voltage fed system will have low phase shift up to the resonant frequency of the inductor and output capacitor, then an abrupt 180° shift will occur. The current fed system will have 90° phase shift at a much lower frequency, but will

not have the additional 90° shift until well beyond the *LC* resonant frequency. This makes it much easier to frequency compensate the feedback loop and also gives much quicker transient response."

Table 9.5a Voltage-mode Integrated Switching Regulators[a]

Part #	TO220, DPAK	DIP	SOIC, MSOP	SOT23	smaller	Fixed-V versions[b]	Internal comp	Soft start	Burst mode, etc.	Shutdown	UVLO	Control mode[c]	Switch type[e]	Sync switching	V_{supply} min[o] (V)	V_{supply} max (V)	I_Q typ (mA)	V_{fb} typ (V)	V_{out} min (V)	V_{out} max (V)	f_{switch} (kHz)	I_{sw} max (A)	# external parts[g]	Comments
Buck																								
TPS62200	-	-	-	5	-	7	•	•	q	•	•	P	M	•	2.5	6	0.02	0.50	0.7	5.5	1000[t]	0.3	1,3	-
LT1934	-	-	-	6	•	-	•	-	•	•	•	H	B	-	3.2	34	0.012	1.25	1.25	28	~300	0.35	7	-
NCP1522B	-	-	5	-	•	-	•	•	q	•	•	P	M	•	2.7	5.5	0.05	0.90	0.9	3.3	3000[t]	1.2	4	-
CS51413	-	-	8	-	•	-	•	•	-	•	•	V2	B	-	4.5	40	4	1.27			520[t,p]	1.6	7	1
L4976	-	8	16	-	-	-	-	-	-	-	-	P	M	-	8	55	2.5	3.3	3.3	40	to 300	2	7	-
LM2574[h]	-	8	14	-	-	4	•	•	-	s		P	B	-	3.5	40,60	5	1.23	1.23	37,57	52[t]	1.0	2,4	-
LM2575[h]	5	-	-	-	-	4	•	•	-	s		P	B	-	3.4	40,60	5	1.23	1.23	37,57	52[t]	2.2	2,4	-
LM2576[h]	5	-	-	-	-	4	•	•	-	s		P	B	-	3.5	40,60	5	1.23	1.23	37,57	52[t]	5.8	2,4	2
NCP3125	-	-	8	-	-	-	-	•	•	•	•	P	M	•	4.5	13	4	0.80	0.8		350[t]	4 d	9	3
LT1074[h]	5	-	-	-	-	1	-	-	q	•	•	P	B	-	8	40,60	8.5	2.21	2.5	30[k]	100	5	6	4
LM2677	7	-	-	-	-	3	•	-	•	•	s	P	M	-	8	40	4.2	1.21	1.21	37	260	7	3,5	5
LMZ12010	11	-	-	-	-	-	•	•	•	•		P	M	•	4.3	20[f]	3	0.80	0.8	6	360[t]	10	3	-
Boost, Flyback, etc.																								
NCP1400A	-	-	-	5	-	9[n]	•	•	q	•	s	P	M	-	0.8[y]	5.5	0.03	-	1.9	5	180	0.1	2	6,8
NCP1423	-	-	-	-	10	-	•	•	•	•		P	M	•	0.8	6	0.01	0.50	1.8	3.3	to 600	0.4	4	7,8
L6920DC	-	-	8	-	-	2	•	-	-	•	•	T	M	•	0.8	5.5	0.01	1.23	1.8	5.5	1000	0.5	2,4	8
TPS61070	-	-	-	6	-	-	•	•	•	•		P	M	•	1.1[x]	5.5	0.02	0.50	1.8	5.5	1200	0.6	3	8
TPS61030	-	-	16	-	•	2	•	•	•	•		P	M	•	1.8	5.5	0.02	0.50	1.8	5.5	600	4.5	1,3	7,8

Notes: (a) all have integrated power switch(es), current-sensing, and (in some cases) loop compensation; listed in order of increasing switch current. (b) number of fixed voltages available; all except NCP1400A have adjustable versions. (c) H=hysteretic mode; P=PWM fixed frequency; T−min t_{off}, max t_{on}; V2=ONsemi "V²" control. (d) adjustable current limit. (e) B=BJT; M=MOSFET. (f) see LMZ23608 for V_{in} to 36V. (g) typical number of external parts (not counting bypass caps); two numbers indicate fixed/adjustable. (h) 60V for HV suffix. (m) adjustable current limit. (n) no adjustable version. (o) restart threshold. (p) CS51411 for 260kHz. (q) reduced freq or pulse skipping at low load. (s) parts with SHDN can have UVLO added with an ext circuit. (t) typical. (u) plus $I_{sw}/50$, etc., when the switch is ON (a power-dissipation issue if used with high V_{supply}). (v) plus BJT switch drive current, on BOOST pin, taken from low-voltage buck output. (x) runs down to 0.9 volts. (y) runs down to 0.3 volts. (z) runs down to 0.5 volts.

Comments: 1: pin compatible with LTC1375. **2:** many second sources. **3:** NCP3126 and 3127 for lower current. **4:** negative V_{out} to −35V (see datasheet); V_{in} comp; LT1076 for 2A. **5:** featured in text. **6:** NCP1402 for 200mA. **7:** 96% effy, low-batty comp. **8:** single-cell stepup.

conditions, or

(b) where multiple outputs are derived from a common power stage (that is, in converters that use a transformer with multiple secondary windings).

Current-mode controllers are favored

(a) where fast response to input transients and ripple is important,

(b) where it is desirable to parallel multiple power supplies (e.g., for redundancy),

(c) where you want to avoid the complexities in designing a proper pole-zero loop compensation network, and

(d) in applications where fast pulse-by-pulse current limiting is important for reliability.[89] Tables 9.5a and 9.5b list selected "integrated" switching regulators, i.e., with *internal* power switch(es). See also Table 9.6 on page 658 for switching regulators that drive external MOSFETS, Table 3.4 (MOSFETS, page 188), and Table 3.8 (drivers, page 218).

9.6.10 Converters with transformers: the basic designs

The non-isolated switching converters of the previous sections can be modified to incorporate a transformer within

---

[89] Evidently SMPS integrated circuit designers (and presumably their larger customers) prefer current-mode control over voltage-mode, as reflected in the shortened length of Table 9.5a compared with Table 9.5b, and by the paucity of voltage-mode controllers in the "control mode" column of Table 9.6.

Table 9.5b Selected Current-mode Integrated Switching Regulators[a]

Part #	TO220	DPAK, D2PAK	DIP	SOIC, TSSOP	SOT23, SC70	smaller	Fixed-voltage versions	Internal comp	Slope comp	Soft start	Burst mode etc	UVLO	OVP	Control mode[c]	DMOS switch	Synchronous	Vsupply min[o] (V)	Vsupply max (V)	IQ typ (mA)	VFB typ (V)	Vout max (V)	fswitch min-max (kHz)	Isw max (A)	# parts[g]	Comments
Buck																									
LTC1174[f]	-	-	8	8	-	-	2	•	-	-	•	•	-	P	•	-	4	13	0.45	1.25		200	0.3,0.6	2	1
LT1776	-	-	8	8	-	-	-	-	-	-	e	e	-	P	-	-	7.4	40	3.2	1.24		200	0.7	7	2
LT1933	-	-	-	6	6	-	•	•	•	•	p	•	-	P	-	-	3.6	36	1.6[v]	1.25		500[t]	1	6	-
ADP3050	-	-	-	8	-	-	2	•	-	-	•	•	-	P	-	-	3.6	30	0.7[v]	1.30		200	1.25	6/8	3
ADP2300	-	-	-	6	-	-	•	•	•	•	•	•	-	P	•	-	3	20	0.64	0.80		700[r]	1.2	5	4
ADP2108	-	-	-	5	5	11	•	-	-	•	•	•	-	P	•	•	2.3	5.5	0.02	fixed	3.3	3000[t]	1.3	1	5
LT1376	-	-	8	8	-	-	2	•	-	-	•	•	-	P	•	-	2.4	25	3.6[v]	2.42		500[t]	1.5	5/7	6
LMR12010	-	-	-	6	-	-	•	•	•	•	•	•	-	P	•	-	3	20	1.5	0.80	17	3000[t]	1.7	6	7
LT3500	-	-	-	-	12	-	•	•	•	•	•	•	-	P	•	-	3	36	2.5[v]	0.80		250-2000	2.8	12	8
NCP3170B	-	-	-	8	-	-	•	•	•	•	•	•	-	P	•	•	4.5	18	1.8	0.80	V_in	1000	3	5	9
A8498	-	-	-	8p	-	-	•	-	-	•	p	-	-	O	•	-	8	50	0.9	0.80	24	30-700	3.5	5	10
LT3435	-	-	-	16	-	-	•	•	•	•	•	•	-	P	•	-	3.3	60	3.3[v]	1.25		500[t,q]	4	12	11
LT1765	-	-	-	8	-	4	•	•	-	-	e	-	e	P	-	-	3	25	1.0[v]	1.20		1250	4	-	11
LT3690	-	-	-	-	16	-	•	•	•	•	•	•	-	P	w	•	3.9	36,60	0.1[v]	0.80	20	140-1500	5	8	12
Boost, Flyback, etc.																									
TPS61220	-	-	-	-	6	2	-	•	•	-	p	•	-	H	•	-	0.7	5.5	0.01	0.50	5.5	1000	0.2	1/3	13
TPS61040	-	-	-	-	5	2	-	•	-	-	-	•	-	M	•	-	1.8	6	0.03	1.233	30	1000	0.4	5	14
LT1613	-	-	-	-	5	-	•	•	•	•	p	-	-	P	-	-	1.1	10	3[u]	1.23	34[d]	1400[t]	0.8	4	15
LMR64010	-	-	-	-	5	-	•	•	•	•	•	•	-	P	•	-	2.7	14	2.1	1.23	40[d]	1600	1	5	16
LT1930A	-	-	-	-	5	-	•	•	•	•	•	•	-	P	•	-	2.6	16	5.5[u]	1.26	34[d]	2200[t]	1.2	5	-
ADP1612	-	-	-	8	-	-	•	-	•	•	•	-	-	P	•	-	1.7	5.5	4	1.235	20	650,1300	1.4	7	17
LTC3401	-	-	-	-	10	-	•	-	•	•	•	•	-	P	•	-	0.9	5.5	0.44	1.25	6	50-3000	1.6	7	18
LT1172[h]	5	5	8	16	-	-	-	-	-	-	e	-	-	P	-	-	3	40,60	6[u]	1.244	65,75[h]	100[t]	1.25	6	-
LT1171[h]	5	5	-	-	-	-	-	-	-	-	e	-	-	P	-	-	3	40,60	6[u]	1.244	65,75[h]	100[t]	2.5	6	-
LT1170[h]	5	5	-	-	-	-	-	-	-	-	e	-	-	P	-	-	3	40,60	6[u]	1.244	65,75[h]	100[t]	5	6	19
LT1534	-	-	-	16	-	-	•	•	•	•	-	•	-	P	-	-	2.7	23	12[u]	1.25	30[d]	20-250	2	12	20
LM2577[b]	5	5	16	-	-	-	2	•	-	•	•	•	-	P	-	-	3.5	40	7.5[u]	1.23	60[d]	52[t]	3	4/6	21
LM2586	7	7	-	-	-	-	3	•	-	•	•	•	-	P	-	-	4	40	11[u]	1.23	60[d]	200	3[k]	4/6	22
TPS61175	-	-	-	14p	-	-	•	•	•	•	•	•	-	P	•	-	2.7	18	3.5[m]	1.23	40[d]	200-2200	3.8	8	-
TPS55340	-	-	-	-	16	-	•	•	•	p	•	•	-	P	•	-	2.9	32	1.7	1.23	38[d]	100-1200	5	8	-
Push-pull																									
LT1533	-	-	-	16	-	-	-	-	-	-	-	•	-	-	-	-	2.7	23	12	1.25[n]	x	20-250[r]	1	13	23

Notes: (a) listed by increasing switch current; all have integrated switch, current-sensing, and in some cases loop compensation; all have shutdown capability except LM2577; all have thermal shutdown. (b) no power shutdown function; also UC2577. (c) H=hysteretic curr mode; M=Fixed peak current, with a minimum off time; O=var freq fixed off time; P=PWM fixed freq. (d) non-isolated boost higher voltages with a transformer. (e) with external parts. (f) suffix HV for 18V version. (g) typical number of external parts (not counting bypass caps); two numbers indicates fixed/adjustable. (h) suffix HV for 60V version. (k) 5A for LM2587. (m) maximum. (n) also negative, −2.5V. (o) restart threshold. (p) reduced freq or pulse skipping at low load. (r) reduced frequency during low V_{out}. (s) parts with SHDN can have UVLO added with an external circuit. (t) typical. (u) plus $I_{sw}/50$, etc., when the switch is ON (a power-dissipation issue if used with high V_{supply}). (v) plus BJT switch drive current, on BOOST pin, taken from low-voltage buck output. (w) low side. (x) transformer output.

Comments: 1: invert OK, especially +5V to −5V converter. **2:** 60V transients OK. **3:** 60V OK for 100ms; 3.3V, 5V, and adj versions. **4:** ADP2301 for 1.4MHz. **5:** just add external inductor; 11 fixed voltages, from 1.0V to 3.3V. **6:** 5V and ADJ, see LT1507 for 3.3V. **7:** "simple switcher" nano. **8:** buck plus LDO, ext sync to 2.5MHz. **9:** power-good output; 500kHz for "A" version. **10:** adj OFF time. **11:** 100µA no-load I_Q. **12:** 80µA no-load I_Q; transients OK to 60V. **13:** boost single-cell to 1.8V–5.5V out; 3.3V, 5V, and adj versions. **14:** good for LED constant current drive. **15:** single-cell boost or flyback. **16:** "simple switcher" nano. **17:** boost from single Li-ion cell. **18:** operates down to 0.5V input; 40µA in burst mode. **19:** can regulate output using transformer's primary voltage (no feedback resistors required). **20:** low-noise, slew-rate control. **21:** 12V, 15V, and ADJ versions. **22:** 3.3V, 5V, 12V, and ADJ versions. **23:** programmable slew rate, very quiet.

the switching circuitry. This serves three important purposes: (a) it provides galvanic isolation, which is essential for converters that are powered from the ac line; (b) even if isolation is not needed, the transformer's turns ratio gives you an intrinsic voltage conversion, so that you can produce large step-up or step-down ratios while staying in a favorable range of switching duty cycle; and (c) you can wind multiple secondaries, to produce multiple output voltages; that's how those ubiquitous power supplies in computers generate outputs of +3.3 V, +5 V, +12 V, and −12 V, all at the same time.

Note that these are not the heavy and ugly laminated-core transformers that you use for the 60 Hz ac powerline: because they run at switching frequencies of hundreds to thousands of kilohertz, they do not require a large magnetizing inductance (the inductance of a winding, with all other windings open-circuited), and so they can be wound on small ferrite (or iron powder) cores. Another way to understand the small physical size of the energy-storage devices in switchmode converters – that is, the inductors, transformers, and capacitors – is this: for a given power output, the amount of energy passing through these devices in each transfer can be much less if those transfers are taking place at a much higher rate. And less stored energy ($\frac{1}{2}LI^2$, $\frac{1}{2}CV^2$) means a smaller physical package.[90]

9.6.11 The flyback converter

The *flyback* converter (Figure 9.73A) is the analog of the inverting non-isolated converter. As with the previous non-isolated converters, the switch is cycled at some switching frequency f (period $T = 1/f$), with feedback (not shown) controlling the duty cycle $D = t_{on}/T$ to maintain regulated output voltage. As with the previous converters, the pulse-width modulation can be arranged as voltage mode or current mode; and the secondary current can be either discontinuous (DCM) or continuous (CCM) from each cycle to the next, depending on load current.

What is new is the transformer, which in the flyback converter topology acts simply as an inductor with a tightly coupled secondary winding. During the switch-ON portion of the cycle, the current in the primary winding ramps up according to $V_{in} = L_{pri} \, dI_{pri}/dt$, flowing into the "dotted" terminal; during that time the output diode is reverse biased because of the positive voltage on the dotted terminals of both windings.

---

[90] For the particular case of the flyback converter, discussed next, you can think of the transformer as formed by a second winding on the already-small inductor used for energy storage in the non-isolated inverting (buck–boost) converter.

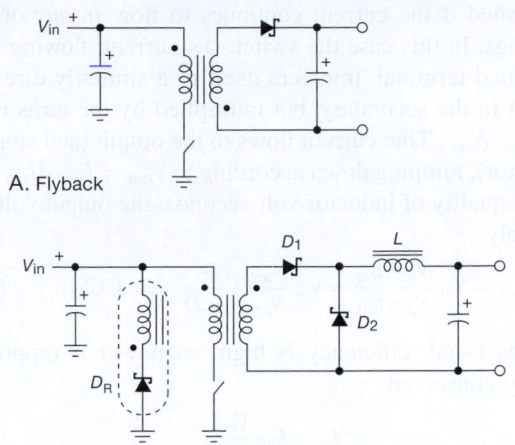

A. Flyback

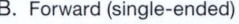

B. Forward (single-ended)

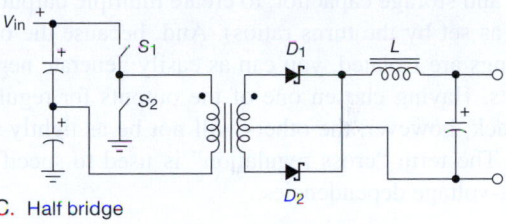

C. Half bridge

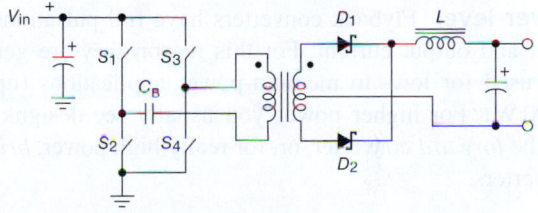

D. Full bridge ("H-bridge")

Figure 9.73. Isolated switching converters. The flyback converter (A) uses an energy-storage inductor with a secondary winding, whereas the forward and bridge converters (B–D) each use a true transformer with no energy storage (and thus require an output energy-storage inductor). The diode D_R and tertiary winding in the forward converter is one of several ways to reset the core in this single-ended design. The dc blocking capacitor C_B in the H-bridge prevents flux imbalance and consequent core saturation; for the half-bridge the series pair of capacitors serves the same function, while acting also as the input storage capacitor.

During this phase the input energy is going entirely into the magnetic field of the transformer's core. It gets its chance to go somewhere else when the switch turns OFF: unlike the situation with a single inductor, with *coupled* inductors the requirement of continuity of inductor current is satisfied if the current continues to flow in *any* of the windings. In this case the switch-ON current, flowing into the dotted terminal, transfers itself to a similarly directed current in the secondary, but multiplied by the turns ratio $N \equiv N_{pri}/N_{sec}$. That current flows to the output (and storage capacitor), ramping down according to $V_{out} = L_{sec}\, dI_{sec}/dt$. From equality of inductor volt-seconds, the output voltage is simply

$$V_{out} = V_{in}\frac{N_{sec}}{N_{pri}}\frac{t_{on}}{t_{off}} = V_{in}\frac{N_{sec}}{N_{pri}}\frac{D}{1-D} \quad \text{(in CCM)}. \quad (9.6)$$

And, as usual, efficiency is high, so power is (approximately) conserved:

$$I_{in} = I_{out}\frac{V_{out}}{V_{in}}. \quad (9.7)$$

You can wind additional secondaries, each with its diode and storage capacitor, to create multiple output voltages (as set by the turns ratios). And, because the output windings are isolated, you can as easily generate negative outputs. Having chosen one of the outputs for regulating feedback, however, the others will not be as tightly regulated. The term "cross regulation" is used to specify the output-voltage dependencies.

A. Comments on flyback converters

Power level Flyback converters have full pulsations of input and output current. For this reason they are generally used for low- to medium-power applications (up to ~200 W). For higher power you usually see designs using the *forward* converter, or, for really high power, *bridge* converters.

The transformer is an inductor The input energy each cycle is first stored in the transformer core (during switch-ON), then transferred to the output (during the switch-OFF). So the transformer design must provide the correct "magnetizing inductance" (acting as an inductor), as well as the correct turns ratio (acting as a transformer). This is quite different from the situation with the forward converter and the bridge converters, below, where the transformer is "just a transformer." We won't go into further detail about transformer design here, simply noting that the design of the "magnetics" is an important part of switching converter designs in general, and flybacks in particular. You have to worry about issues such as core cross-section, permeability, saturation, and deliberate "gapping" (in general, energy-storage inductors are gapped, whereas pure transformers are not). Extremely helpful resources for design are found in IC datasheets and design software (usually available at no charge from the manufacturer) that provide

specifics about the choice of magnetics. We explore this important topic further in §9x.4.

Snubbers With ideal components, the primary current would transfer completely to the secondary when the switch turns OFF, and you wouldn't have to worry about bad things happening on the dangling drain terminal of the switch. In reality the incomplete coupling between primary and secondary creates a series "leakage inductance," whose craving for current continuity generates a positive voltage spike at the switch, even though the secondary is clamped by the load. This is not good. The usual cure is to include a *snubber network*, consisting of an *RC* across the winding, or, better, a "DRC" network of a diode in series with a parallel *RC*.[91]

Regulation Flyback converters can be regulated with conventional PWM, either voltage mode or current mode, with a free-running oscillator calling the shots. Alternatively, you will see inexpensive designs in which the transformer itself becomes part of a *blocking oscillator*, thereby saving a few components. We cracked open some samples of low-power (5–15 W) "wall warts" and found, well, just about *nothing* inside! We reverse engineered them to look at the circuit tricks (Figure 9.74). They seem to work just fine.

Off-line converters This final circuit (Figure 9.74) is an example of a power converter that *requires* galvanic isolation. The transformer provides isolation for the power flow; in addition, the feedback signal from the dc output must be isolated as well on its way back to the primary side. This can be done with an optocoupler, as here, or with an additional small pulse transformer. We discuss these offline converters briefly in §9.7, and in Chapter *9x* we discuss high-efficiency ("green") power supplies, including a graph comparing the performance of this 5 W supply (whose standby power is 200 mW) with others.

9.6.12 Forward converters

The *single-ended forward converter* (Figure 9.73B) is the transformer-isolated version of the buck converter. It is helpful to refer back to the basic buck circuit (Figure 9.61A), to see how it goes. The transformer converts

---

[91] Leakage inductance values are typically ~1% of the magnetizing inductance. You can reduce leakage inductance greatly by splitting one of the windings (say primary) into two, with the other (secondary) sandwiched in between. And *bifilar* windings (wind primary and secondary as a pair of wires together) can reduce the leakage inductance to a low value. However these techniques increase inter-winding capacitance, and bifilar windings suffer from poor voltage insulation ratings.

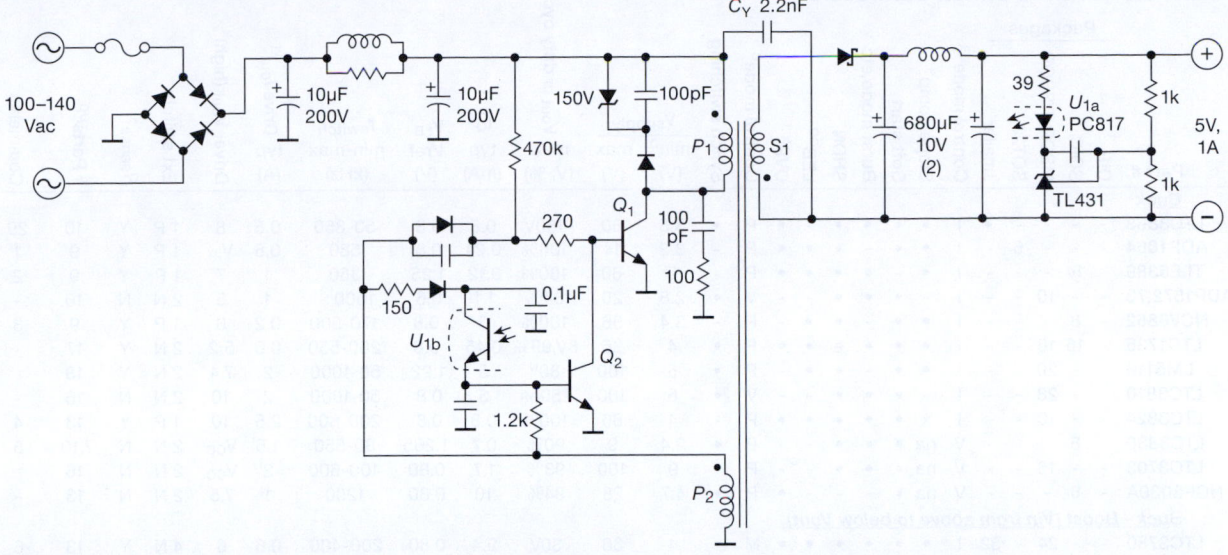

Figure 9.74. An inexpensive 5 W flyback converter, powered from 115 Vac line voltage, that uses a self-excited "blocking oscillator." Winding P2 provides positive feedback to sustain oscillation. The output voltage is sensed and compared with the TL431 shunt regulator, fed back via the optocoupler U_1 to adjust the conduction cycle.

input voltage V_{in}, during primary switch conduction, to a secondary voltage $(N_{sec}/N_{pri})V_{in}$. That transformed voltage pulse drives a buck converter circuit, consisting of catch diode D_2, inductor L, and output storage capacitor. The extra diode D_1 is needed to prevent reverse current into the secondary when the switch is OFF. Note that here, in contrast to the flyback converter, the transformer is "just a transformer": inductor L provides the energy storage, as with the basic buck circuit. The transformer does not need to store energy, because the secondary circuit conducts at the same time as the primary (energy goes "forward"), as you can see from the polarity marking.

Analogous to the buck converter, (eq'ns 9.3a–9.3h), the output voltage is simply

$$V_{out} = V_{in}\frac{N_{sec}}{N_{pri}}\frac{t_{on}}{T} = D\frac{N_{sec}}{N_{pri}}V_{in} \quad \text{(in CCM).} \quad (9.8)$$

Resetting the core In contrast to the flyback circuit, there's an additional winding in Figure 9.73B, which is needed to *reset* the transformer's core.[92] That is because the volt-second product[93] applied to the transformer must average zero (i.e., no average dc input) in order to prevent a continual buildup of magnetic field; but the input switch alone always applies voltage in one direction only. The tertiary winding fixes this by applying voltage in the opposite

direction during the switch-OFF portion of the cycle (when diode D_R conducts, from continuity of current in the winding as the magnetic field collapses).[94]

Additional comments (a) As with the flyback, and indeed with any transformer-coupled converter, the forward converter allows multiple independent secondaries, each with its inductor, storage capacitor, and pair of diodes. Regulating feedback then holds one output particularly stable. (b) The transformer isolates the output in a forward converter, if you happen to need isolation (as in a powerline-input converter); in that case you must galvanically isolate the feedback signal as well, typically with an optocoupler (as in the block diagram of Figure 9.48, or the detailed diagrams of Figures 9.74 and 9.83). On the other hand, if you do not need isolation you can have a common ground reference, and bring the error signal back to the PWM control circuit directly.

---

[92] Reset is inherent in the flyback, but not in the single-ended forward converter, as will become evident.

[93] Sometimes call "volt-time integral."

[94] There are clever circuits that reset the core without requiring a tertiary winding: one method uses a pair of primary switches, one at each end of the winding, in collaboration with a pair of diodes, to reverse the voltage across the single primary (see if you can invent the circuit!). Another method uses instead a second switch to connect a small capacitor across the primary during main switch-OFF; this clever method is known as "active clamp reset," and was devised independently by Carsten, Polykarpov, and Vinciarelli. It has the virtue of *reversing* the magnetic field in the transformer core, providing better performance by allowing double the normal flux excursion.

Table 9.6 External-switch Controllers[a]

Part #	DIP	SOIC	MSOP, T/SSOP	SOT23	smaller	Control mode[aa]	Slope comp[s]	Soft start	Burst mode etc	SHDN	LEB[oo]	OVP[x]	Control mode[c]	Synch switching	Vsupply min[o] (V)	Vsupply max (V)	Vout or duty cycle max (V, %)	IQ typ (mA)	VFB Vref (V)	fswitch min-max (kHz)	Drive Iout[d] typ (A)	Driver Vout(high)[k] typ	Ext switch	Rsense?	# Parts[pp]	Comments
Buck																										
LTC3863	-	-	-	-	-	I	•	•	•	•	•	•	P	•	3.5	60	−150V	0.8	0.8	50-850	0.5	8	1 P	Y	10	29
ADP1864	-	-	-	6	-	I	•	•	•	•	•	•	P	•	3.2	14	100%	0.24	0.8	580	0.6	Vin	1 P	Y	9	1
TLE6389	-	14	-	-	-	I	•	•	•	•	•	•	P^q	-	5	60	100%	0.12	1.25	360	1	7	1 P	Y	9	2
ADP1872,73	-	-	10	-	-	V	•	•	•	•	•	-	V	•	2.8	20	84%	1.1	0.6	1000^{f3}	1	5	2 N	N	10	-
NCV8852	-	8	-	-	-	I	•	•	•	•	•	•	P	•	3.4	36	100%	3	0.8	170-500	0.2	8	1 P	Y	9	3
LTC1735	-	16	16	-	-	I	•	•	e	•	•	•	P	•	4	36	6V,99%	0.45	0.8	200-550	0.6	5.2	2 N	Y	17	-
LM5116	-	-	20	-	-	I	•	•	•	•	•	•	P	•	5	100	80^u	5	1.22	50-1000	2	7.4	2 N	Y	18	-
LTC3810	-	-	28	-	-	I	•	•	•	•	•	•	V	•	6	100	250ns	3	0.8	50-1000	2	10	2 N	N	16	-
LTC3824	-	-	10	-	-	I	x	•	•	•	•	•	P	•	4	60	100%	0.8	0.8	200-600	2.5	10	1 P	Y	13	4
LTC3830	-	8	-	-	-	V	na	•	•	•	•	•	P	•	2.4	9	90%	0.7	1.265	80-550	1.5^i	Vcc	2 N	N	7,10	5
LTC3703	-	-	16	-	-	V	na	•	•	•	•	•	P	•	9	100	93%	1.7	0.80	100-600	2^i	Vcc	2 N	N	16	-
NCP3030A	-	8	-	-	-	V	na	•	•	-	-	-	P	•	4.7	28	84%	~10	0.80	1200	1^i	7.5	2 N	N	13	-
Buck - Boost (Vin from above to below Vout)																										
LTC3780	-	-	24	-	32	I	•	•	•	•	•	•	M	•	4	36	30V	2.4	0.80	200-400	0.6	6	4 N	Y	13	6
Boost, Flyback, etc.																										
UC384x	8	8	-	-	-	I	e	•	•	•	•	-	P	-	9,18^{ex}	30	50,100%	11	2.5	500 m	0.5	Vc	1 N	Y	20-30	7
MIC38HC4x	8	8	-	-	-	I	e	•	•	•	•	-	P	-	9,16^{ex}	20	50,100%	4	2.5	500 m	1	Vc	1 N	Y	20-30	8
ISL684x	8	8	8	-	8	I	e	•	•	•	•	-	P	-	9,15^{ex}	20	50,100%	4	2.5	2000 m	1	Vc	1 N	Y	20-30	8
UCC38C4x	8	8	8	-	8	I	e	•	•	•	•	-	P	-	9,16^{ex}	18	50,100%	2.3	2.5	1000 m	1	Vc	1 N	Y	20-30	9
UCC380x	8	8	8	-	8	I	e	•	•	•	•	-	P	-	5,14^{ex}	30	50,100%	0.5	2.5^h	1000 m	1	Vc	1 N	Y	20-30	9
TPS40210,11	-	-	10	-	10	I	•	•	•	•	•	-	P	-	4.5	52	80%	1.5	0.26^r	35-1000	0.4	8	1 N	Y	15	10
LTC3803	-	-	-	6	-	I	•	•	•	•	•	-	P	y	8.7	clamp	80%	0.24	0.80	200 t	0.7	Vcc	1 N	Y	8	11
MAX668, 69	-	-	10	-	-	I	•	•	•	•	•	-	P	-	1.8	28	90%	0.22	1.25	100-500	1	5.0	1 N	Y	10	12
LM3478	-	8	-	-	-	I	•	•	•	•	•	-	P	-	3	40	100%	3	1.25	100-1000	1	7	1 N	Y	9	13
LM5020	-	-	10	-	10	I	•	•	•	•	•	-	P	-	8	15	80%	2	1.25	50-1000	1	7.7	1 N	Y	many	14
LTC1872B	-	-	6	-	-	I	•	•	•	•	•	-	P	-	2.5	9.8	100%	0.27	0.80	550	1	Vin	1 N	Y	8	15
LM3481	-	8	-	-	-	I	•	•	•	•	•	-	P	-	3	48	85%	3.7	1.28	100-1000	1	5.8	1 N	Y	14	-
MAX15004	-	16	-	-	-	I	•	•	•	•	•	-	P	-	-	40	50,80%	2	1.23	15-1000	1	7.4	1 N	Y	20	-
ADP1621	-	-	10	-	-	I	•	•	•	•	•	-	P^q	-	2.9	5.5	95%	1.8	1.215	100-1500	2	Vin	1 N	N^w	10	16
LTC1871	-	-	10	-	-	I	•	•	•	•	•	-	P	-	2.5	36	92%	0.55	1.23	50-1000	2	5.2	1 N	N	13	-
NCP1450A	-	-	-	5	-	V	na	•	•	•	-	e	P	-	0.9	6	80%	0.14	f	180	0.05	Vin	1 N	N	3	17
Offline Flyback																										
FAN6300	8	8	-	-	-	I	•	•	•	•	•	-	Q	-	17°	25	70%	4.5	p	100^z	0.15	18	1 N	Y	20	18,28
NCP1252	8	8	-	-	-	I	•	•	•	•	•	-	P	-	8	28	80%	1.4	p	50-500	0.5	15	1 N	Y	20	19,20
NCP1237,38	-	7	-	-	-	I	•	•	•	•	•	w	P^q	-	13°	28	80%	2.5	p	65^{f3}	1	13.5	1 N	Y	20	20,28
L5991	16	16	-	-	-	I	e	•	•	•	•	j	P	-	9,16^o	20	50,100%	7	2.5	40-2000	1	Vc	1 N	Y	30	21
Push-Pull, Forward, Half-Bridge, etc.																										
MC34025	16	16	-	-	-	I	•	•	•	•	•	-	P	-	9.6	30	t, 45%	25	5.10	5-1000	0.33	Vc	2 N	Y	many	22
LM5041	-	16	-	-	16	I	-	•	•	•	•	-	P	-	9	15	t, 50%	3	0.75	1000	1.5	Vc	4 N	Y	many	23,14
TL594	16	16	-	-	-	V	na	e	•	-	-	e	P	-	7	40	t, 45%	9	5.0^w	1 - 300	0.2	b	2	-	9, 12	24
SG3525	16	16	-	-	-	V	na	•	•	-	-	-	P	-	8	40	t, 49%	14	5.10	0.1 - 400	0.2	Vc	2		many	24,25
LM5035	-	20	-	24	-	V	na	•	•	•	-	-	P	-	8	105	t, 50%	4	5.0	100-1000	1.25	Vcc	2,4 N	Y	many	26,14
NCP1395A	16	16	-	-	-	V	na	•	•	•	-	-	R	•	11	20	t, 50%	2.3	2.5	50-1000	ext	-	2 N	N	22	27

Notes: (a) all require external power switches (see listings in **Table 3.4**); all have undervoltage lockout (UVLO) and internal voltage references; listed within groups in approximate order of increasing drive current. (aa) I - current mode, V - voltage mode, P - fixed peak current, M - multiple modes. (b) uncommitted BJT output, sinks 200mA, 40V max. (c) P=PWM fixed freq; Q=quasi-res; R - resonant; V=var freq fixed width; (d) peak driver current, for controllers. (e) ext parts. (ex) lower voltage for x=3 or 5, higher voltage for x=2 or 4. (f) fixed only. (f3) three switching-frequency options. (g) unused footnote. (h) 2V for x=3 or 5. (i) adjustable current limit. (j) 25V zener clamp for Vcc. (k) to Vcc or voltage shown, whichever is less. (m) maximum. (n) nominal. (o) turn-on threshold. (oo) even with LEB (leading-edge blanking) an *RC* filter or at least a 100pF capacitor is often recommended. (p) ref pin is current-sourcing. (pp) [same note as integrated tables]. (q) reduced freq or pulse skipping at low load. (r) 0.7V for the '11. (s) helps stabilize the control loop against sub-harmonic oscillations. (t) transformer output. (u) a minimum off time (450ns) limits the duty cycle. (v) may not include dynamic gate-charge currents, etc. (w) for Vout below 30V, above 30V a current-sense resistor is required. (x) OVP = line over-voltage protection. (y) synchronous possible with low-voltage non-isolated flyback transfomer. (z) finds resonant frequency.

Comments: 1: LTC1772, LTC3801 second-source. **2:** fixed 5V version available. **3:** automotive. **4:** hi-side sense. **5:** LTC3832 goes down to 0.6V. **6:** single inductor, foldback current limit. **7:** jellybean. **8:** improved UC384x. **9:** UC384x with LEB, SS, low IQ. **10:** impressive 52V, LED drive. **11:** use with flyback xfmr. **12:** to 1.8V, slope-comp, soft-start, expensive. **13:** to 1MHz, advanced. **14:** HV pin, to 100V for startup. **15:** SOT23, low power, cute. **16:** can boost inputs as low as 1V. **17:** fixed voltage versions only, five choices 1.9V to 5.0V. **18:** quasi-resonant. **19:** inexpensive, ATX power supplies etc. **20:** freq dither. **21:** 25V zener clamp for Vcc. **22:** legacy, inexpensive, second sourced. **23:** programmable gap/overlap. **24:** legacy, inexpensive, flexible. **25:** also UC3525 etc. **26:** feed-forward ramp. **27:** resonant, use with FET driver IC. **28:** HV pin, to 500V for startup. **29:** optimized for inverting, Vout from −0.4V to −150V or more.

(c) As with all switchmode converters, snubber networks are needed to tame the voltage spikes caused by parasitic inductances (including particularly transformer leakage inductance).

(d) As with other converter types, PWM control can be either voltage mode or current mode. An alternative is to use pulse *frequency* modulation (PFM), with approximately constant pulse width, to take advantage of resonant behavior (thus avoiding "hard switching" by allowing the resonant ringing to charge and discharge parasitic capacitances, and thereby come closer to the ideal of zero-voltage/zero-current switching). (e) Single-ended forward converters are popular in the medium-power range (∼25–250 W).

9.6.13 Bridge converters

The last two transformer-isolated converters in Figure 9.73 are the *half-bridge* and *full-bridge* (H-bridge) converters. As with the single-ended forward converter, the transformer acts simply to effect voltage transformation and isolation; the secondary circuit's inductor does the energy storage, serving the same purpose as it does in the basic buck converter or single-ended forward converter. In fact, you can think of the bridge converters approximately as "double-ended forward converters." In both bridge circuits the capacitor(s) on the input side allow the voltage at the undotted end of the transformer primary to move up or down as needed to achieve zero average dc current, preventing transformer core saturation.

To understand the half-bridge converter, imagine first that switches S_1 and S_2 are operated alternately, with 50% duty cycle and with no gap or overlap. The voltage at the junction of input capacitors will float to half the dc input voltage, so what you've got is a center-tapped full-wave rectifier circuit, driven by a square wave. Power is transferred forward during both halves of each cycle, and the output voltage (ignoring diode drops) is just

$$V_{out} = V_{in}\frac{N_{sec}}{4N_{pri}}, \tag{9.9}$$

where the factor of 4 arises from the factor of $\frac{1}{2}$ for the applied input voltage and the same factor from the output center-tap. The operation of the full-bridge converter is similar, but its four switches enable it to apply the full dc input voltage across the primary during each half-cycle, so the 4 is replaced by 2 in the denominator.

Regulation With the switches operating in opposition, at 50% duty cycle, the output voltage is fixed by the turns ratio and the input voltage. To provide regulation you need to operate each switch for less than a half-cycle (Fig-

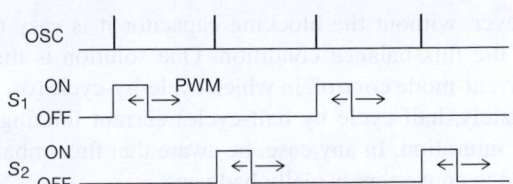

Figure 9.75. Pulse-width modulation in the half-bridge switching converter. The internal oscillator initiates switch conduction on alternate cycles, with feedback providing regulation by ending each switch's conduction according to the error signal.

ure 9.75), with a conduction gap ("dead time") whose length is adjusted according to the error signal. You can think of each half-cycle as a forward converter, of duty cycle $D = t_{on}/(t_{on} + t_{off})$, causing the converter to produce an output voltage (assuming CCM) of

$$V_{out} = DV_{in}\frac{N_{sec}}{4N_{pri}}. \tag{9.10}$$

Bridge converters are favored for high-power conversion (∼100 W and above), because they make efficient use of the magnetics by conducting during both halves of each cycle, and they cycle the magnetic flux symmetrically. They also subject the switches to half the voltage stress of a single-ended converter. By adding another pair of switches, you can convert it to a full-bridge (or H-bridge), in which the full dc input voltage is applied across the primary each half-cycle. (See the comments below, however, about flux balance.) The full-bridge configuration additionally allows another form of regulation, called "phase-shift control," in which a 50% duty cycle is maintained in each switch pair, but the relative phase of one pair is shifted relative to the other, to effectively produce a variable duty cycle.[95]

Additional comments (a) As with the single-ended forward converter, it is essential to maintain zero average voltage (or volt-time integral) across the transformer's primary. Otherwise the magnetic flux will grow, reaching destructive saturation. The H-bridge in Figure 9.73D includes a blocking capacitor C_B in series with the primary for this purpose; the pair of input capacitors serves the same function for the half-bridge (Figure 9.73C). That capacitor can be quite large, and it has to endure large ripple currents; so it would be nice to eliminate it, for example by connecting the bottom of the winding to a fixed voltage of $V_{in}/2$ (which is available automatically in an offline voltage-doubling input bridge). That configuration is known as "push-pull."

---

[95] Some phase-shift controller ICs we like are the UCC3895 from TI and the LTC3722 from Linear Technology.

However, without the blocking capacitor it is easy to violate the flux-balance condition. One solution is the use of current-mode control, in which cycle-by-cycle (or, more accurately, half-cycle by half-cycle) current limiting prevents saturation. In any case, be aware that flux imbalance in bridge converters is really bad news.

(b) In bridge converters the power switches are connected in series across the dc input supply. If there is conduction overlap, large currents can flow from rail to rail; this is known as "shoot-through" current. What you need to know is that you don't want it! In fact, turn-off delays in MOSFETs, and more seriously in BJTs, require that the control signals provide a short time gap to avoid shoot-through.

(c) Once again, snubbers are needed to tame inductive spikes.

(d) Full-bridge converters are favored for high-power converters, to 5 kW or more.

(e) At high load currents the output filter inductor has a continuous current flowing through it. During primary conduction cycles this is, of course, supplied either by D_1 or D_2, by normal transformer action. But what happens during primary *non*-conduction (the gaps in Figure 9.75)? Interestingly, the continuing inductor current flows through *both* D_1 and D_2, forcing the transformer secondary to act like a short-circuit (even though its primary is open), because equal diode currents flow in the same direction out of both ends of the center-tapped winding.

9.7 Ac-line-powered ("offline") switching converters

With the exception of Figures 9.48B and 9.74, all the switching converters and regulators we've seen so far are *dc-to-dc* converters. In many situations that's exactly what you want – for battery-operated equipment, or for creating additional voltages within an instrument that has existing dc power.[96]

Apart from battery-powered devices, however, you need to convert incoming powerline ac to the necessary regulated dc voltages. You could, of course, begin with an unregulated low-voltage dc supply of the sort in Figure 9.49,

followed by a switching regulator. But the better approach is to eliminate the bulky 60 Hz step-down transformer by running an isolated switching converter directly from the rectified (unregulated) and filtered ac power, as shown earlier in Figure 9.48.[97]

Two immediate comments. (a) The dc input voltage will be approximately 160 volts[98] (for 115 Vac power) – this is a dangerous circuit to tinker with! (b) The absence of a transformer means that the dc input is not isolated from the powerline, so it's essential to use a switching converter with an isolated power stage (forward, flyback, or bridge), and with isolated feedback (via an optocoupler or transformer).

9.7.1 The ac-to-dc input stage

A. Dual-voltage configurations

Figure 9.76 shows two common input-stage configurations. The simple bridge rectifier of Figure 9.76A is perfectly OK

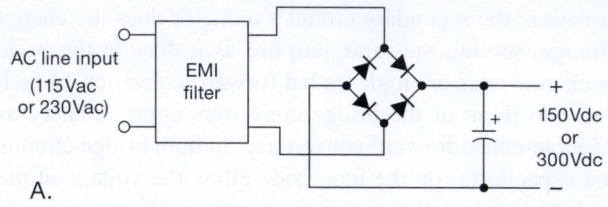

A.

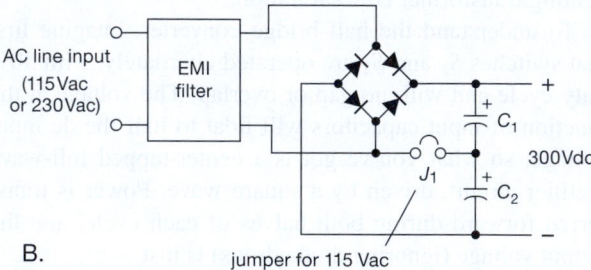

B.   jumper for 115 Vac

Figure 9.76. Switching power supplies run from the ac powerline (offline converters) use directly rectified dc to power an isolated converter. The jumper in the lower circuit selects bridge or voltage doubler configurations, so that either line voltage produces the same ~300 V dc output.

---

[96] A common application is within a computer, where the processor may require something like 1.0 V at 100 A (!). That's a lot of current to be running around a printed circuit board! What is done, instead, is to bring a higher "bus" voltage (usually +12 V) into the vicinity of the processor, where it powers a half-dozen or so 12 V-to-1.0 V buck converters that surround the power-hungry chip and that run in multiple phases to reduce ripple. This is called "point-of-load" power conversion. The benefit, of course, is the lower current in the bus, about 8 A in this example, combined with tight voltage regulation at the load itself.

[97] A story to prove us wrong: we routinely disassemble all sorts of commercial electronic gadgets, just to see how the other half lives. Imagine our surprise, then, when we cracked open a cellphone charger and found…a tiny ac power transformer, bridge rectifier, and low-voltage storage capacitor, followed by an MC34063 switching converter! Goes to show you.

[98] And, more commonly, 320 volts; see below.

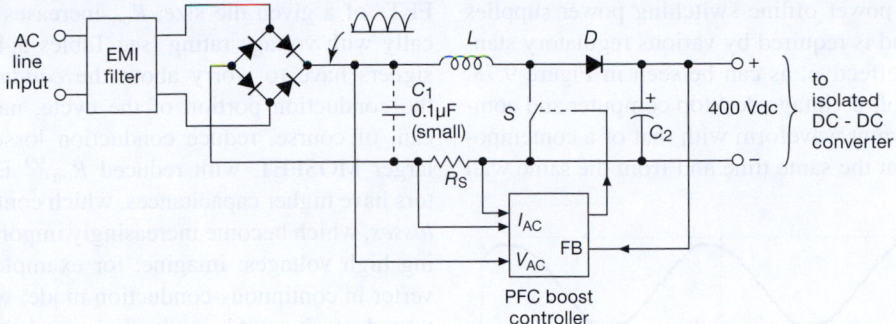

Figure 9.77. The direct rectifier circuits of Figure 9.76 create undesirable current pulses each half-cycle (low power factor). This is remedied with a power factor correction front end, consisting of a boost converter running from the (unfiltered) full-wave rectified line-voltage waveform, controlled by a special PFC chip that operates the switch to maintain the input current approximately proportional to the input voltage.

for devices intended for either 115 Vac or 230 Vac use, in which the switching converter that follows is designed for either ~150 Vdc or ~300 Vdc input, respectively. If you need a supply that can be switched to run on either input voltage, use the nice trick shown in Figure 9.76B: it's a simple full-wave bridge for 230 Vac input, but with the jumper connected it becomes a voltage doubler for 115 Vac input, thus generating ~300 Vdc on either continent. (The other popular approach is to design the switching converter to accommodate a wide dc input range; most low-power chargers for consumer devices like laptop computers and cameras work this way. Check the label, though, before you plug in to 230 Vac power. And don't expect more power-hungry electronic devices to work automatically on "universal" power; they usually have a recessed slide switch that is the jumper in Figure 9.76B.)

B. Inrush current

When you first turn on the power, the ac line sees a large discharged electrolytic filter capacitor across it (through a diode bridge, of course). The resulting "inrush" current can be enormous; even a tiny "wall-wart" can draw 25 A or more of instantaneous current when first plugged in. Commercial switchers use various soft-start tricks to keep the inrush current within civilized bounds. One method is to put a negative-tempco resistor (a low-resistance thermistor) in series with the input; another method is to actively switch out a small (10 Ω) series resistor a fraction of a second after the supply is turned on. The series inductance provided by an input noise filter helps somewhat, as well. But a very nice solution comes in the form of an input power-factor correction circuit, discussed next.

C. Power-factor correction

The pulsed current waveform of rectified ac, as seen for example in Figure 9.51, is undesirable because it produces larger resistive (I^2R) losses compared with the ideal of a sinusoidal current waveform that is in phase with the voltage. (This is why it's easy to make the mistake of choosing too small a fuse rating, as discussed earlier in §9.5.1B.) Another way to say it is that a pulsed current waveform has a low *power factor*, which is defined as the power delivered divided by the product $V_{rms} \times I_{rms}$. Power factor made its first appearance in Chapter 1 in connection with reactive circuits, in which the phase-shifted (but still sinusoidal) current created a power factor equal to the cosine of the phase difference between the ac voltage and current. Here the problem is not phase, it's the high rms/average ratio of the pulsed-current magnitudes.

The solution is to make the power supply's input look like a passive resistor, by devising a circuit that forces the input current waveform to be proportional to the input voltage over the ac cycle. That is known as a power-factor correction (PFC) circuit, and it is connected between the full-wave rectified ac input (but with the usual storage capacitor omitted) and the actual dc–dc converter, as shown in Figure 9.77. It consists of a non-isolated boost converter, operating at the usual high switching frequency, with the switching duty cycle continually adjusted to keep the sensed input current (I_{ac}) proportional to the instantaneous ac input voltage (V_{ac}) over the ac cycles. At the same time, it regulates its dc output to a voltage somewhat greater than the peak ac input, usually +400 V. This dc output then powers an isolated dc–dc converter to produce the final regulated voltages.

Power-factor correction is becoming standard in most

moderate-to-high-power offline switching power supplies (>100 W, say), and is required by various regulatory standards. It is quite effective, as can be seen in Figure 9.78, where we dusted off a vintage desktop computer and compared its input current waveform with that of a contemporary unit running at the same time and from the same wall outlet.

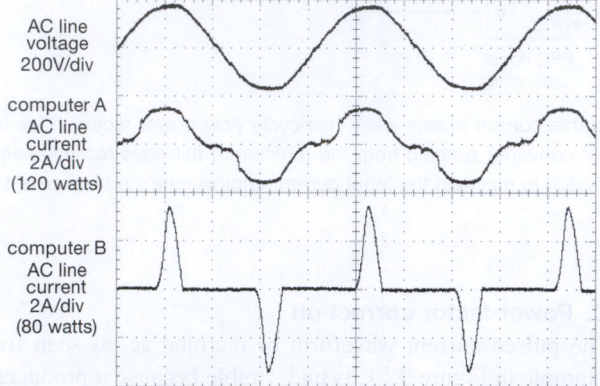

Figure 9.78. A tale of two computers. Computer A has a PFC-input power supply, causing its input current to track the input voltage. The power supply in computer B, built ten years earlier, lacks PFC; its input bridge rectifier charges the storage capacitor with short-duration current surges. Horizontal scale: 4 ms/div.

9.7.2 The dc-to-dc converter

There are some extra issues to contend with in the design of offline converters.

A. High voltage

Whether power-factor corrected or not, the dc supply to the converter–regulator will be at a substantial voltage, typically 150 V or 300 V, or somewhat higher if PFC is used. The converter itself provides the isolation, typically using one of the transformer configurations of Figure 9.73. The switch must withstand the peak voltages, which can be significantly larger than the dc supply. For example, in the forward converter with 1:1 tertiary reset winding (Figure 9.73B) the MOSFET drain swings to twice V_{in} during reset; and for the flyback the drain flies up to $V_{in} \cdot T / t_{off}$. Note also that these peak voltages assume ideal transformer behavior; leakage inductance and other non-ideal circuit realities further exacerbate the situation.

B. Switching losses

High-voltage MOSFETs do not have the extremely low R_{on} of their lower-voltage brethren. For high-voltage MOS-

FETs of a given die size, R_{on} increases at least quadratically with voltage rating (see Tables 3.4 and 3.5). So designers have to worry about the *conduction loss* during the conduction portion of the cycle, namely $I_D^2 R_{on}$. You can, of course, reduce conduction losses by choosing a larger MOSFET, with reduced R_{on}.[99] But larger transistors have higher capacitances, which contribute to *dynamic losses*, which become increasingly important when switching high voltages: imagine, for example, a forward converter in continuous-conduction mode; when the switch is turned ON, it must bring its drain (and attached load) from $+2V_{in}$ to ground. But there is energy stored in the switch's drain capacitance, as well as the parasitic capacitance of the transformer's winding, to the tune of $E = \frac{1}{2}CV^2$, which is squandered as heat each switching cycle. Multiply that by the switching frequency, and you get $P_{diss} = 2fCV_{in}^2$. It goes up quadratically with operating voltage, and it can be substantial: an offline forward converter, running from +300 V rectified line voltage, switching at 150 kHz, and using a 750 V MOSFET with drain (and load) capacitance of 100 pF would be dissipating 3 W from this dynamic switching loss alone.[100]

There are clever ways to circumvent some of these problems. For example, inductances can be exploited to cause the drain voltage to swing close to ground (ideally, zero-voltage switching) before the switch is activated; this is called "soft switching," and is desirable for reducing both $\frac{1}{2}CV^2$ switching losses and the component stress caused by hard switching. And the $V_D I_D$ switching loss during transitions can be minimized by driving the gate hard (to reduce switching time), and by exploiting reactances to bring about zero-current switching. These problems are not insurmountable; but they keep the designer busy, dealing with tradeoffs of switch size, transformer design, switching frequency, and techniques for soft switching. This kind of circuit design is not for the casual electronics tinkerer, nor for the faint of heart.

C. Secondary-side feedback

Because the output is deliberately isolated from the hazardous powerline input, the feedback signal has to cross

---

[99] Or, for high-enough voltages, use an IGBT instead; see §3.5.7.

[100] A second kind of dynamic switching loss occurs during the ramp-up and ramp-down of switch voltage, during which the instantaneous transistor power dissipation is the product of drain voltage and drain current. This is basically a dynamic conduction loss associated with switching *transitions*, to be distinguished both from the *static* conduction loss during the switch's ON state, and from the dynamic "hard-switching" losses associated with charging and discharging parasitic capacitances.

back over the same isolation barrier. The configuration in Figure 9.74 is typical: a voltage reference and error amplifier (here implemented with a simple shunt regulator) drives the LED of an optocoupler at the output, with the isolated phototransistor providing guidance to the switch control (usually PWM) on the drive side. A lesser-used alternative is a pulse transformer, driven from a "secondary-side controller" circuit. A third alternative, if a high degree of output regulation is not needed, is to regulate the output of an auxiliary winding that is not on the "output" side (for example, a winding like P2 in Figure 9.74); because it returns to the input-side common, no isolation of its feedback signal is needed. This is called *primary-side* regulation. Typically you'll get something like $\pm 5\%$ output regulation (over a load-current variation from 10% to 100% of rated current), compared with $\pm 0.5\%$ or better with secondary-side feedback.

D. The isolation barrier

Transformers and optocouplers provide galvanic isolation. Simple enough, it would seem. But, as with life itself, there's usually plenty of nuance lurking below the surface (and, as will become evident, *along* the surface as well).

There are two mechanisms by which an isolation barrier can be breached:

(a) High voltages can create a spark directly across an air gap (or through an insulating sheet); this kind of breakdown is called "arcing" (or "arc-over"), so you have to ensure a minimum *clearance* distance, defined as the shortest distance in air between a pair of conductors.

(b) A conductive path can develop on the surface of insulating material that separates a pair of conductors; this kind of breakdown is called "tracking,"[101] best prevented by ensuring a minimum *creepage* distance, defined as the shortest distance along the surface of insulating material between two conductors; see Figure 9.79. As will become evident, creepage is generally the greater worry (compared with clearance) in high-voltage circuit layouts.

It's bad news when there's breakdown of an isolation barrier; it will likely cause damage or destruction to downstream powered electronics. Worse yet, there's human safety – an electronic device whose isolation from the ac line power is lost can kill you. For these reasons there are guidelines and strict standards that govern the design of isolation barriers (codified by IEC, UL, DIN/VDE, etc.). Publications like IEC 60950 and IEC 60335 include extensive tables of minimum clearance and creepage, and web-

---

101 A colorful term that describes well the little carbonized tracks you tend to find in postmortem forensics of a high-voltage device that has failed.

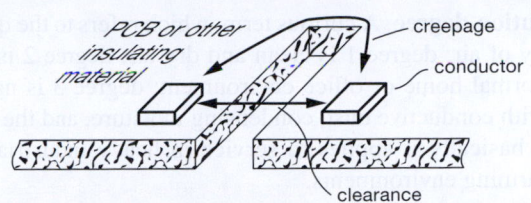

Figure 9.79. Two paths for breaching an isolation barrier: rapid arcing across the airgap (defined by the *clearance* distance), and conductive "tracking" along a path on the surface of the insulating material (defined by the *creepage* distance).

sites like www.creepage.com have delightful online calculators to keep your designs reliable and safe.

Generally speaking, clearances of 2 mm or so, and creepage distances of 4–8 mm or so, are appropriate for 120 Vac powered converters. However, there are additional variables that affect the required spacings. An example is "pollution degree" (referring to the presence of conductive dust, water, etc.); and there is the overall category of intended insulation (ranging from the merely "functional" to the strictest safety level of "reinforced"). Another factor is the intended application: for example, there are separate safety standards for products intended for household use (IEC 60335), and there are particularly strict standards for medical devices (IEC 60601). A full discussion of the subject is well beyond the scope of this book. The following treatment aims to alert the reader to the seriousness of high-voltage isolation, and some of the techniques that are used to deal with it.

The variables: insulation type, voltage, material group, pollution degree
These are the parameters you use with the tables or calculators.

Insulation type The overall level of required effectiveness, in five steps (functional, basic, supplementary, double, reinforced).

Voltage Arc-over in air or through an insulating sheet is rapid, so it's the *peak* voltage (or peak transient) that matters. By contrast, the deterioration or contamination that causes conductive creepage is slower, so you use rms or dc voltages when consulting the tables.

Material group This refers to the susceptibility of the particular insulating material to surface breakdown; the groups are called I, II, and III, going from least to most susceptible. Some standards prefer analogous parameters called "comparative tracking index" (CTI) and "performance level categories" (PLCs).

Pollution degree A curious term, which refers to the quality of air: degree 1 is clean and dry air; degree 2 is the normal home or office environment; degree 3 is nasty, with conductive dust, condensing moisture, and the like – basically it applies to service in heavy industrial or farming environments.

Increasing the creepage distance

If you've got a compact design, such that there's insufficient space to provide adequate creepage distances, you can use various measures. You'll frequently see gaps or slots cut through a printed circuit board, as in the offline switcher of Figure 9.80. You can also provide a protruding barrier to lengthen the surface-clinging path, a technique used in high-voltage optocouplers, transformer windings, and the like (see next paragraph). A conformal insulating coating applied over a populated circuit board is a particularly effective technique (but it must not delaminate, or it can be worse than no coating at all). Related techniques for individual components involve potting or molding.

Creepage considerations in component packaging and design

Components that bridge the isolation barrier, such as transformers and optocouplers, must be designed and packaged with appropriate clearances and creepage distances, both in the external leads and in the internal insulation. An example is the isolation-straddling Y-capacitor, with one foot on each side. As the photograph of Figure 9.81 shows, the leads of the disc-geometry Y-capacitor are oriented at right

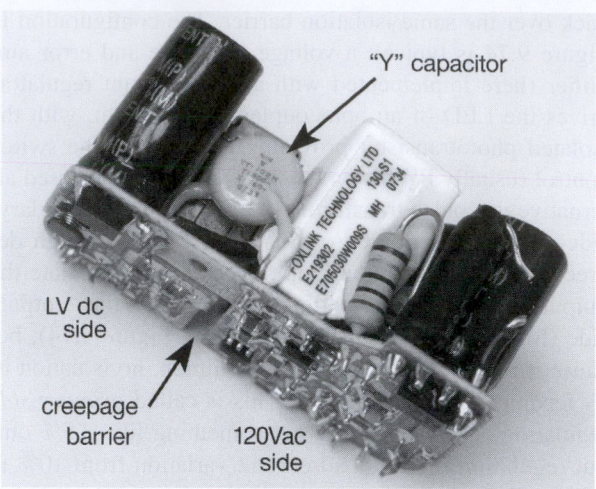

Figure 9.81. This edgewise view of the same converter reveals that the Y-capacitor's widely spaced leads preserve the 8 mm minimum creepage; by contrast, the converter's minimum *clearance* is just 1.5 mm.

angles and coated with a continuation of the same conformal insulation that covers the capacitor body. Components housed in DIP-style cases can achieve greater separation of input and output sections by omitting intermediate pins[102] (thus a "DIP-8" that's missing pins 2,3,6, and 7). An example of a fully specified high-voltage part comes from Avago, whose datasheet for an optocoupler (ACNV260E) includes an abundance of clearance and creepage specifications: both "external" and "internal" clearances (13 mm and 2 mm, respectively), and likewise for creepage distances (13 mm and 4.6 mm, described as "measured from input terminals to output terminals, shortest distance path along body" and "along internal cavity," respectively).

The leads of the switching transformer must similarly maintain adequate spacing and creepage distance. Equally important, the inter-winding insulation and winding geometry must create both appropriate insulation (by a sufficient number of layers of insulating tape, etc.) and also appropriate creepage standoff. To meet the creepage requirements, the windings may be arranged side-by-side (rather than coaxial), and separated with an insulating sheet that extends outward beyond the windings. This is good for creepage, but bad for the magnetic design, as it increases the leakage inductance. With a magnetically preferable coaxial geometry, the creepage distance can be extended by

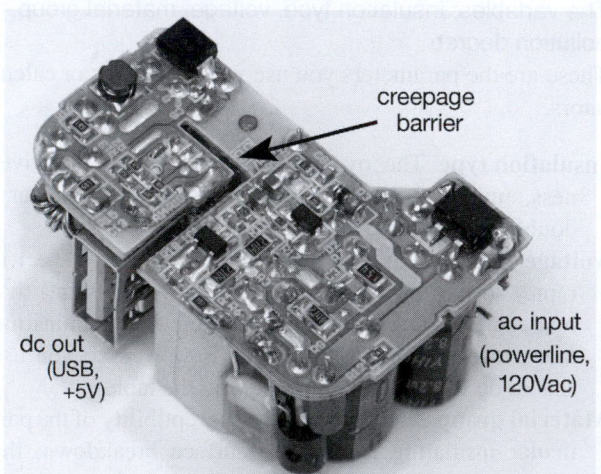

Figure 9.80. The designers of this switching converter included an L-shaped slot in the circuit board, greatly lengthening the creepage distance from the powerline circuitry to the isolated 5 V output.

---

[102] See for example the datasheets for the Vishay CNY64 coupler, the ON Semiconductor NCP1207 PWM controller, or the Power Integrations LNK-403 driver.

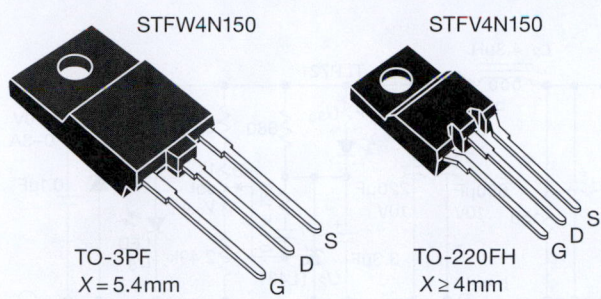

Figure 9.82. These 1500 V MOSFET packages employ shaped and grooved insulation to lengthen the creepage path length. (Adapted with permission of STMicroelectronics)

allowing the inter-winding tape to extend beyond the windings, or to wrap back around the outer winding.

Creepage effects are present whenever you deal with high voltage, whether or not an isolation barrier is involved. An example is shown in Figure 9.82, illustrating the pin configuration of two package styles of a 1500 V MOSFET. For the larger TO-3PF package (5.4 mm lead spacing) an extension of the plastic package material around the drain lead provides adequate creepage distance; for the smaller TO-220FH package (2.5 mm lead spacing) there's a grooved structure and offset lead geometry.

9.8 A real-world switcher example

To convey the additional complexity involved in a production-model line-powered switching power supply, we disassembled a commercial single-output regulated switching supply[103] (Astrodyne model OFM-1501: 85–265 Vac input, 5 Vdc @ 0–3 A output), another in our series of "Designs by the Masters," revealing the circuit of Figure 9.83.

9.8.1 Switchers: top-level view

Let's take a walk through the circuit to see how a line-powered switcher copes with real-world problems. The basic topology is precisely that of the switching converter in Figure 9.48, implemented with flyback power conversion (Figure 9.73A); there are, however, a few additional components! Let's take it first at the broad-brush level, circling back later to delight in the refinements.

At this very basic level it goes like this: the line-powered bridge rectifier D_1 charges the 47 μF storage capacitor[104] (rated at 400 Vdc, to accommodate the 265 Vac maximum

---

[103] Pictured in the northeast corner of Figure 9.1.
[104] The input storage capacitor is often called the *bulk capacitor*.

input), providing the unregulated high-voltage dc input (+160 Vdc or +320 Vdc, for 115 Vac or 230 Vac input, respectively) to the high side of the 70-turn primary winding of T_1. The low side of the winding is switched to input common (the $\perp$ symbol) at fixed frequency (but with variable pulse width) by the PWM switchmode controller chip U_1, according to feedback current at its FB terminal. On the secondary side the 3-turn paralleled secondaries are rectified by Schottky diode D_5, with "flyback" polarity configuration (i.e., nonconducting during the primary ON period). The rectified output is filtered by the four low-voltage storage capacitors (totaling 2260 μF), creating the isolated 5 Vdc output. This supply uses secondary-side regulation, comparing a fraction (50% nominal) of V_{out} with U_2's internal +2.50 V reference, turning on the LED emitter of optocoupler U_3 when the output reaches its nominal 5 Vdc. This couples to phototransistor U_{3b}, varying the feedback current into switchmode controller U_1, thus varying the ON pulse width to maintain regulated +5 Vdc output.

At this point we've accounted for perhaps a third of the components in Figure 9.83. The rest are needed to cope with issues such as (a) auxiliary power for the controller chip; (b) powerline filtering, mostly of *outgoing* switching noise; (c) protection (fusing, reverse polarity); (d) feedback loop compensation; and (e) switching transient snubbing and damping. And, although not obvious from the schematic, but most essential to the design – the choice of transformer parameters: core size and "gapping," turns ratios, and magnetizing inductance[105] L_M.

Before looking into those details, though, let's see how the basic converter works. We'll be able to figure out things like the voltage and current waveforms, peak voltages and currents, and the duty cycle as a function of input voltage and output current.

9.8.2 Switchers: basic operation

The control chip operates at a fixed frequency f_{osc} of 100 kHz, adjusting its primary switch conduction duty cycle ($D = t_{on}/T$) according to voltage feedback. We've drawn ideal waveforms for one cycle (duration $T = 1/f_{osc}$) in Figure 9.84. These are what you might expect in the

---

[105] The conventional symbols for magnetizing inductance and leakage inductance are L_m and L_l, respectively. But the lower-case L subscript can be hard on the eyes, especially in a footnote. In the interest of readability, therefore, we've adopted small upper-case subscripts: L_M and L_L throughout.

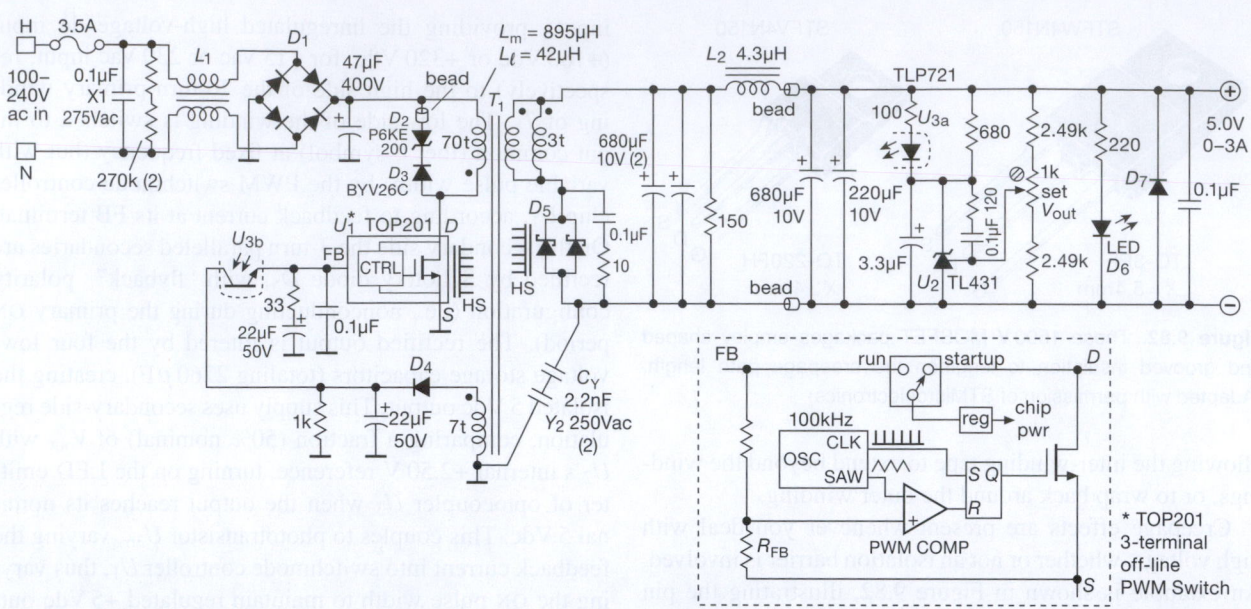

Figure 9.83. "Real-world" line-powered switching power supply. The circuit is relatively uncomplicated, thanks to its low power rating (15 W), and to the elegant 3-terminal switchmode controller U_1 from Power Integrations (with on-chip high-voltage power MOSFET). This is the open-frame "15W ac/dc switcher" shown in Figure 9.1.

absence of parasitic effects such as leakage inductance and switch capacitance.

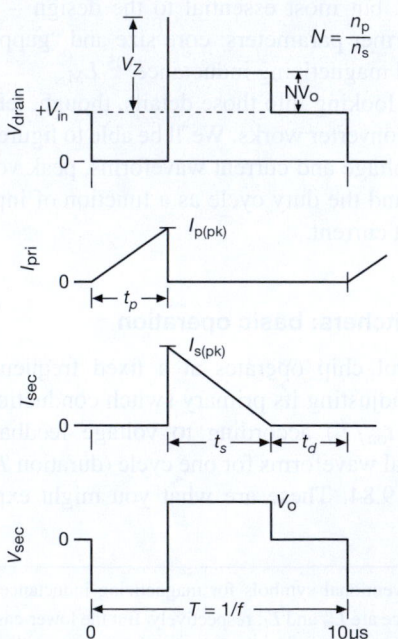

Figure 9.84. Ideal waveforms for an isolated flyback switching supply, operating in discontinuous-conduction mode.

A. The waveforms

We'll do the calculations shortly, but look first at the waveforms. (We've assumed the converter is operating in discontinuous-conduction mode, which will be borne out when we do the numbers.) During switch conduction the drain voltage is held at ground, putting $+V_{in}$ across the transformer primary and causing a ramp-up of primary current, according to $V_{in} = L_M \cdot dI_{pri}/dt$, where L_M is the primary "magnetizing inductance" (the inductance seen across the primary, with all other windings disconnected). That current ramps up to a peak value I_p, at which time there is a stored energy of $E = \frac{1}{2}L_M I_p^2$ in the transformer's core. When the switch turns off, the persistent inductive current transfers to the secondary winding, delivering that stored energy E to the output as the secondary current ramps down to zero, according to $V_{out} = L_{M(sec)} \cdot dI_{sec}/dt = (1/N^2)L_M \cdot dI_{sec}/dt$ (where $L_{M(sec)}$ is the magnetizing inductance seen at the secondary[106]). For the rest of the cycle there is no transformer current flowing.

The voltage waveforms are instructive. When the primary switch is turned off, at time t_p, the drain voltage rises

---

[106] Most of the time it's the magnetizing inductance seen at the *primary* that matters, for which we simply use L_M; in the few situations where we refer to the magnetizing inductance seen at the *secondary*, we add (sec) to the subscript: $L_{M(sec)}$.

well beyond the input supply voltage V_{in}: that is because the inductor tries to continue sourcing current into the drain terminal. The voltage would soar, but the secondary circuit goes into conduction instead (notice the polarity of "dotted" windings in Figure 9.83), clamping its output to V_{out}, which reflects back to the primary via the turns ratio N (shorthand for N_p/N_s). The brief spike shown in the figure is caused by some primary inductance[107] that is not coupled to the secondary, and therefore not clamped. This terrifying voltage spike is ultimately clamped by the zener clamp D_2 seen in the schematic (more on this later). When the secondary current has ramped down to zero, the voltage drop across both windings goes to zero; so the drain terminal sits at $+V_{in}$, and the voltage across the secondary winding goes to zero. Note that the latter is negative during primary switch conduction; it's a requirement that the "volt-time integral" (or "volt-second product") across any inductor average to zero, otherwise the current would rise without bound. That holds true for the primary also.

B. The calculations

Let us assume for simplicity that the converter is running at full load (5 V, 3 A) with nominal input voltage (115 Vrms or 160 Vdc).[108] We will calculate the switch duty cycle $D=t_p/T$, the secondary conduction duty cycle t_s/T, and the peak currents $I_{p(pk)}$ and $I_{s(pk)}$. It's easiest to take these in reverse order, doing the calculations from a simple energy standpoint.

The parameters   We measure the magnetizing inductance seen at the primary to be $L_M=895\,\mu H$, and the number of turns of primary and secondary to be $N_p=70t$ and $N_s=3t$. From this we get the turns ratio $N=N_p/N_s=23.3$, which sets the voltage and current transformation ratios. Finally, from the turns ratio we get the magnetizing inductance as seen at the secondary side: $L_{M(sec)}=L_M/N^2=1.65\,\mu H$ (impedances scale as N^2; see Chapter 1x). A final parameter that we will use later is the measured primary leakage inductance $L_L=42\,\mu H$.

Peak currents   The output circuit is delivering 15 W to the load; but, taking account of rectifier drop ($\sim$0.5 V) and

the combined resistive losses in the secondary winding and filter inductor L_2 (10 mΩ), the transformer secondary is delivering an average power of approximately 6 V$\times$3 A, or 18 W. So, at a switching frequency of f_s=100 kHz, the transformer must deliver an energy increment of $E=P/f_s=180\,\mu J$ during each switch cycle.

The rest is easy: we equate E to the magnetic energy in the core's magnetizing inductance, as seen at the secondary (because that's where it emerges). That is, $E=\frac{1}{2}L_{M(sec)}I^2_{s(pk)}$, from which we get $I_{s(pk)}=14.8$ A. Dividing by the turns ratio (N=23.3), we find that the peak primary current is $I_{p(pk)}=0.64$ A.

Conduction timing   The primary switch stays on for a duration that ramps its current up to this peak current. That is, $t_p=L_M I_{p(pk)}/V_{in(dc)}=3.6\,\mu s$. The secondary conduction commences when the primary switch turns off, and continues for the time duration t_s needed to ramp its current down from $I_{s(pk)}$ to zero: $t_s=L_{M(sec)}I_{s(pk)}/V_{sec}=4.1\,\mu s$. Note that the successive conduction of primary and secondary totals 7.7 μs, which is less than the cycle time of 10 μs; that is, the converter is running in discontinuous conduction mode, as we assumed at the outset (and drew in Figure 9.84). There is a "dead time" of about 2.3 μs before the next switch conduction.

C. Comparison with reality

How well did we do with this basic model? To find out, we measured voltage and current waveforms of this converter, at nominal input voltage and full output load. They are shown in Figure 9.85. The good news is that the timing and peak currents are in very good agreement with our

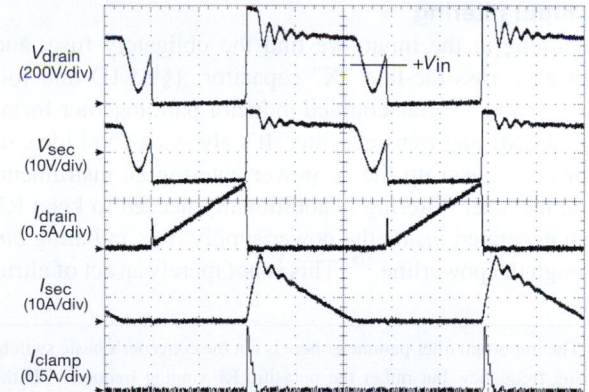

Figure 9.85. Measured waveforms for the switcher of Figure 9.83, running at full load (5 V, 3 A) and nominal input voltage (115 Vrms; V_{in}=160 Vdc). The arrows mark the location of zero voltage and current for each trace. Horizontal scale: 2 μs/div.

---

[107] This is in fact the infamous "leakage inductance" L_L. As with magnetizing inductance, we use the unadorned L_L to refer to leakage inductance seen at the primary winding; for secondary leakage inductance we add (sec) to the subscript: $L_{L(sec)}$.

[108] Of course, a full design analysis must consider operation at the extremes, in particular at minimum input with maximum load (hence maximum duty cycle), and for the full range of output current with maximum input.

calculations. The bad news is that there are some real-world "features" that are absent from our basic waveforms of Figure 9.84. Most prominent are

(a) a substantial drain voltage spike at turn-off, followed by
(b) some fast ringing on both windings during secondary conduction, and
(c) slower ringing during the dead time at the end of the cycle.

Visible also is

(d) a drain current spike at turn-on.

These are caused by non-ideal behavior of the MOSFET switch and the transformer, as we'll discuss soon; but, to put some names onto them, these effects are due to

(a) primary leakage inductance,
(b) resonance of drain (and other) capacitances with primary leakage inductance,
(c) resonance of drain (and other) capacitances with primary magnetizing inductance, and
(d) "hard switching" of the voltage across the drain and other capacitances.

9.8.3 Switchers: looking more closely

Let's go back and fill in the missing pieces. In the real world you cannot ignore important effects such as the voltage and current transients that we saw in Figure 9.85, and numerous other details that account for all the components you see in the circuit diagram.

A. Input filtering

Beginning at the input, we find the obligatory fuse, and then an across-the-line "X" capacitor (§9.5.1D and following) and a series-coupled inductor pair, together forming an EMI and transient filter. It's always a good idea, of course, to clean up the ac power entering an instrument; here, however, filtering is additionally needed to keep RF hash generated *inside* the power supply from radiating *out* through the powerline.[109] This is not merely an act of altru-

ism; there are regulatory standards governing permissible levels of radiated and conducted EMI.[110] The pair of 270k resistors discharges the X capacitor's residual voltage when the unit is unplugged.

B. Voltage range, inrush current, PFC

Note that this low-power (15 W) supply operates directly from a wide input voltage range (3:1), without a dual-voltage range switch in the manner of Figure 9.76B. Such wide-range operation is particularly convenient in chargers and power bricks for consumer electronics. It does, however, impose constraints on the design, because the converter must operate over a wide range of switch conduction duty cycle, and because the components must be sized for the wider range of peak voltages and currents. Absent, also, are any circuit elements to limit the inrush current during initial charging of the line-side storage capacitor. That's permissible in a small supply like this; but even with the relatively small 47 μF storage capacitor the specified typical inrush current is a hefty 20 A at 100 Vac input (and twice that for 200 Vac). Note also the absence of a PFC frontend; it's common practice to omit PFC in small supplies, but PFC is usually found in supplies of 50 W or more, at least in part from regulatory pressures. Note, by the way, that a PFC front-end reduces peak inrush current.

C. Auxiliary supply

Moving to the right, we see the interesting configuration of the "auxiliary supply," needed to power the internal circuits of the regulator–controller chip with low-voltage, low-power dc. An unattractive possibility would be to use a separate little linear supply, with its own line-powered transformer, etc. However, the temptation is overwhelming to hang another small winding (with half-wave rectifier D_4) on T_1, thus saving a separate transformer. That's what's been done here, with the 7-turn winding, which generates a nominal +12 V output.

Sharp-eyed readers will have noticed a flaw in this scheme: the circuit cannot start itself, because the auxiliary dc is present only if the supply is already running! This turns out to be an old problem,[111] solved with a "kick-

---

[109] The important filter parameter here is not the converter's basic switching frequency, but rather the parasitic RF ringing frequency. If the latter is 2.5 MHz, for example, a lowpass filter with 250 kHz cutoff will attenuate the RFI by approximately $(f_{RFI}/f_{LPF})^2$, or 100×. With the 100 nF "X1" capacitor shown, the series inductance of the common-mode choke (its transformer leakage inductance) need be only $L = 1/(2\pi f_{LPF})^2 C_X = 4\,\mu$H. Higher frequencies will be attenuated

[110] In the US, electronic equipment must meet FCC Class A (for industrial settings) and Class B (more stringent, for residential settings) limits; in Europe the analogous standards are set by VDE.

[111] For example, designers of traditional CRT-based television sets faced the same quandary, when they derived all their low-voltage dc supplies from auxiliary windings on the high-frequency horizontal drive transformer, the latter itself activated by those same supplies.

more, up to the frequency at which the PCB's wiring inductance and the choke's winding self-capacitance take over.

start" circuit that powers initially from the high-voltage un-regulated dc, switching over to its auxiliary dc power after things are running. We'd like to show you how this is implemented in detail, but we are frustrated in that worthy goal because in this supply those functions (and others) are cleverly integrated into the TOP201 controller chip (shown in simplified block diagram form in the dashed box).[112]

D. Controller chip: bias and compensation

Moving next to the controller chip itself, we see its internal high-voltage MOSFET (drawn explicitly, for clarity), which switches the low side of the primary to input common. The switch operates at fixed 100 kHz rate, varying the duty cycle according to the feedback, in a voltage-mode regulator. The chip is packaged in a 3-pin TO-220 plastic power package, and requires a small heatsink. Think about that – a *3-pin* switching regulator! Impossible, you say: it needs pins at least for common, drain, feedback, and chip power ("bias"). Surprisingly, this clever chip does it with just three, with the feedback terminal doing double duty as a bias pin. Feedback takes the form of a current into the FB pin, with an internal voltage divider to create the voltage-feedback signal that is presented to the PWM (duty-cycle) comparator, and a linear regulator to create the (higher) internal bias voltage. The remaining components on the primary side are for loop compensation (the series RC and C shunting the FB terminal), and for clamping and damping the inductive spike at the end of the conduction cycle (the 200 V zener transient suppressor and ferrite bead).

E. Input transient clamp (snubber)

At first you might reason that no clamp is needed, because the secondary circuit clamps the flyback voltage (as transformed to the secondary side by the turns ratio) to the output voltage. That is, after all, how a flyback works: the magnetic energy added to the core during switch conduction is stored in the transformer's magnetizing inductance ($E_M = \frac{1}{2}L_M I_p^2$), and released to the secondary circuit when the switch is turned OFF. But there is also "leakage inductance" (L_L, see Chapter *1x*), an effective series inductance caused by incomplete magnetic coupling between the windings.[113] The magnetic energy stored in L_L

---

[112] Look in our second edition, where we devote six pages (pp. 361–366) to a complex offline switcher, if you want to see the gory implementation details of these and other features.

[113] Referring all inductances to the primary side, the magnetizing inductance L_M is what you measure across the primary terminals with all other windings left open-circuited, and the leakage inductance L_L is what you measure with all other windings short-circuited.

($E_L = \frac{1}{2}L_L I_p^2$) is *not* transferred to, nor clamped at, the secondary, which is why you need the zener clamp on the primary side. (You can think of this unclamped energy as arising from the magnetic field of the primary that is not linked by the secondary.) This energy can be substantial – we'll see just how robust a zener is needed, even for this low-power switcher, when we do the clamp calculations in the next paragraph. It's worth noting that the effects of leakage inductance loom particularly large in a line-powered supply, because the required high-voltage insulation between primary and secondary mandates that the windings be physically well separated, causing incomplete flux coupling.

Let's take a moment to understand the drain voltage spike waveform in Figure 9.85. The primary-side leakage inductance, here measured to be 42 μH, though a smallish fraction ($\sim$5%) of the magnetizing inductance of 895 μH, stores that fraction of the total energy put into the transformer during primary switch conduction, and it is not transferred to the secondary; instead, it comes back out and is dissipated in the zener clamp D_2. That's about 0.84 W, which accounts for the robust zener that the designers chose. We can estimate the time duration of the primary current ramp to zero (call it t_{clamp}), mediated by the zener clamp. Look at Figure 9.86: the leakage inductance sees a clamp voltage equal to the zener voltage minus the reflected secondary voltage, which acts to ramp the primary

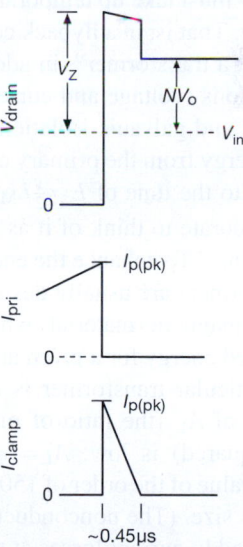

Figure 9.86. Drain-voltage spike caused by transformer leakage inductance. The zener clamp, whose voltage is higher than the reflected secondary output voltage, ramps the current to zero according to $V_Z - NV_{out} = L_L \, dI_D/dt$.

current down to zero from its starting value of $I_{p(pk)}$. So, from $V=L dI/dt$ we get $V_Z-NV_{out}=L_L I_{p(pk)}/t_{clamp}$, so $t_{clamp}=0.45\,\mu s$. This is in good agreement with the measured waveforms of Figure 9.85.

A final note on the clamp network: the zener D_2 is not a normal zener, but rather a "transient voltage suppressor" type (TVS; see discussion in Chapter 9x), designed and specified to absorb large pulses of energy. The series diode D_3 is needed to prevent conduction during the switch-ON cycle, when the zener would conduct as a normal diode. There's an interesting problem associated with D_3, namely the fact that ordinary diodes have a "reverse recovery time" after forward conduction, which is due to charge storage effects, before they become non-conducting (this is the origin of the curious microsecond-scale spikes seen in a simple 60 Hz unregulated power supply; see §9x.6). For this reason D_3 in this circuit is a "fast soft-recovery" rectifier: the "fast" means that it turns off quickly (<30 ns), and the "soft" means that it does so smoothly, not abruptly. That's useful, because an abrupt current transition to non-conduction produces large inductive spikes ($V = L dI/dt$). In addition, the designers added a ferrite bead to damp and suppress such effects.

F. The transformer

In a flyback converter the primary and secondary conduction cycles do not overlap (as they do in, say, a forward converter). So all the energy that is being moved from primary to secondary must take up temporary residence in the transformer's core. That is, in a flyback converter the transformer is not "just a transformer": in addition to the usual transformer functions (voltage and current transformation by the turns ratio, and galvanic isolation), it is also an *inductor*, storing energy from the primary cycle in its magnetizing inductance to the tune of $E=\frac{1}{2}L_M I_{p(pk)}^2$. In fact, it's probably more accurate to think of it as "an inductor with a secondary winding." To enhance the energy storage functions, such transformers are usually designed with a deliberate gap in the magnetic material, which has the effect of raising the stored energy for a given applied volt-second product. This particular transformer is evidently gapped, because its value of A_L (the ratio of magnetizing inductance to turns squared) is low: $A_L=L_M/N_p^2=183\,nH/t^2$, compared with a value of the order of 1500 for an ungapped ferrite core of this size. (The nonconducting ferrite core is used to eliminate eddy current losses at the high operating frequency.)

As we discovered above, this converter runs in discontinuous conduction mode at nominal input voltage and full load current. In fact, it stays in DCM even at minimum in-

put voltage (90 Vrms) and full load current, which is the combination that brings it closest to CCM. With a bit more transformer inductance it would enter CCM; presumably the design choice was based on the desire to keep it small, and also to avoid some issues associated with CCM.[114]

As we hinted earlier, the transformer's inductances are responsible for the ringing seen in the waveforms of Figure 9.85. Let's do a simple calculation of what frequencies we expect. During secondary conduction (immediately following primary switch turn-off) the primary circuit looks like a parallel *LC*, with leakage inductance L_L in parallel with the parasitic capacitances of the MOSFET and other components (clamp diode, primary winding). A reasonable estimate for the combined capacitances is something like 75 pF, due mostly to the transformer wiring and the clamp zener. So the parallel *LC* formed with the leakage inductance of $42\,\mu H$ resonates at about 2.8 MHz, in good agreement with the observed ringing ($\sim$2.5 MHz). At the completion of secondary conduction, the primary side no longer sees the leakage inductance (because the secondary is no longer clamped by the load); instead it sees the magnetizing inductance L_M of $895\,\mu H$ (because the secondary is now open-circuited).[115] The new calculated resonant frequency then drops to about 615 kHz. You can see the first half-cycle of this slower resonance in the measured waveforms, centered on the +160 V dc input voltage, and interrupted by the onset of the next conduction cycle. (We later ran the converter at 25% load, which allowed three cycles of ringing at $\sim$600 kHz, in excellent agreement with this estimate.)

While we're on the subject of parasitic capacitances, this is a good time to note the $\sim$0.3 A current spikes at switch turn-on. This occurs because the switch is abruptly shorting a charged capacitor (the parallel capacitance of

---

[114] Most notably some tendency for the output voltage to overshoot when there is an abrupt drop in load current (owing to a larger required inductance in a CCM design, perhaps influenced also by the nonzero magnetic field throughout the cycle), and also a change in the feedback loop behavior (because of the different functional dependence of output voltage versus duty cycle, and, more interestingly, the fact that in CCM the duty cycle is fixed, for a given output voltage, and is independent of load current). Given this last fact, it may seem paradoxical that regulation against changes in load current is even possible! What happens is that, once in CCM, a change in load current causes a *transient* change in duty cycle, such that the baseline (minimum) primary current moves up or down to accommodate the changed load current; having established that new baseline current, the duty cycle then returns to the fixed value appropriate to the regulated output voltage.

[115] Somewhat overdamped by the reflected impedance of about 5 kΩ in series with $\sim$200 pF from the $10\Omega+0.1\,\mu F$ secondary snubber network.

the switch itself, plus attached components). This is called "hard switching," and is responsible for significant power losses in converters running at high switching frequencies. Here, for instance, we can estimate the power dissipated in the switch by multiplying $\frac{1}{2}CV^2$ by the switching frequency, giving $P_{\mathrm{diss}} \approx 0.15\,\mathrm{W}$. That's not too serious at this modest switching frequency of 100 kHz, being just 1% of the output power; but its relative contribution is greater at low load current, and in any case it contributes to switch dissipation and stress. And it becomes increasingly important as you try to increase the switching frequency (in order to reduce size). The solution is to strive for "soft switching," in which the voltage across the switch is brought close to zero before switch activation (by exploiting reactive currents to discharge parasitic capacitances); this goal is called "zero-voltage switching" (ZVS).

G. Secondary power train

Moving to the secondary side, the rectifier is a Schottky type, which has both low forward voltage drop and zero recovery time (absence of charge storage).[116] Schottky rectifiers (also known as *hot carrier* rectifiers) are available at voltages up to $\sim$100 V; above that you'd use a "fast-recovery" (or "fast soft-recovery," like D_3) rectifier. Power rectifiers often come packaged as duals for applications that require two; here they've simply connected them in parallel. Note the heatsink: 3 A of average load current flowing through a 0.5 V (Schottky) forward drop dissipates 1.5 W, enough to merit a small heatsink. The series RC provides some damping and attenuation of switching transients, as do the ferrite beads. The series inductor L_2 filters ripple at the switching frequency: its reactance at the 100 kHz switching frequency is 2.7 Ω, compared with an impedance of $\sim$0.1 Ω (dominated by series resistance) for the downstream storage capacitors.

H. Secondary regulation

This power supply uses the popular TL431 "shunt regulator," which includes an internal voltage reference and error amplifier, and which goes into heavy conduction when the reference pin reaches 2.5 V above the ground pin. That turns on the optocoupler U_3's LED emitter, for currents above about 2 mA (the threshold set by the 680 Ω resistor). The resistive divider and trimmer allow $\pm$0.4 V output adjustment, and the series RC around the TL431 is a com-

pensation network to prevent oscillation. The large shunt capacitor limits the loop bandwidth, and also accomplishes "soft-start" at power-on: this it does by tricking the opto-emitter into thinking that the TL431 is conducting, when in fact the LED current is coming from the ramp-up of output voltage. It's easy to verify that an output ramp-up of 1.5 V/ms produces a sinking current of 5 mA at the cathode of the optocoupler LED, thus stretching the startup to about 3 ms, and therefore setting the secondary current needed to charge the four output storage capacitors to $\sim$3.4 A, roughly equal to the supply's maximum current rating.

I. Other design features

There are just a few additional goodies in this circuit. Capacitor C_Y is used to suppress conducted EMI. Because it bridges the isolation barrier, it must have appropriate "Y-capacitor" safety ratings (see §9.5.1). Rectifier D_7 protects against reverse polarity, in case some misbehaving load decides to create mayhem. The small output capacitor ensures low output impedance at high frequencies, where the large electrolytic capacitors become less effective (owing to internal inductance and ESR). And, finally, the switchmode controller itself (U_1) includes a host of nice features: internal oscillator requiring no external timing components, internal cycle-by-cycle current limit, overtemperature protection, automatic restart, internal regulator and dc source switching, and on-chip high-voltage power MOSFET, all integrated in an elegant 3-terminal configuration. Its high level of integration stole our thunder: it robbed us of the opportunity to show these important circuits explicitly!

9.8.4 The "reference design"

This is a nice power supply. We've bought a lot of them, and they work reliably and well. The circuit design might seem forbiddingly complicated, certainly for those inexperienced in offline switchmode supply design. In fact, we recommend strongly that the *user* of such supplies should not try to design and build them – *buy* them from the expert folks who do this for a living (see below).

But, how did those experts come up with this particular design? As it turns out, the manufacturers of interesting ICs have a great interest in making it painless to use their products. For this noble objective they provide what are known as *reference designs*, which basically consist of a complete circuit example (usually available from them as a "development board" or "evaluation board"). For the regulator chip used in this particular power supply, for example, Power Integrations (the TOP201 manufacturer) provides four example circuits, with increasing levels of regulation

---

[116] To deal with the high 15 A peak current in D_5, the designers selected a YG802C04 Schottky rectifier with its pair of 40 V 10 A sections paralleled (each of which specifies a forward drop of 0.53 V at 7 A), attached to its own heatsink.

stability (called "minimum parts count," "enhanced minimum parts count," "simple optocoupler feedback," and "accurate optocoupler feedback"). And in a pair of "Application Notes"[117] they provide a step-by-step recipe for these designs, complete with flowcharts, formulas, and graphs. You can hardly go wrong. The supply of Figure 9.83, in fact, closely follows the "accurate optocoupler feedback" design, differing primarily in the inclusion of soft start, ferrite beads, and reverse polarity protection. This is not to say that the design is a trivial exercise – the implementation of the transformer, the packaging and layout, and the process of testing and regulatory approval are all major challenges.

9.8.5 Wrapup: general comments on line-powered switching power supplies

- Line-powered switchers are ubiquitous, and for good reasons. Their high efficiency keeps them cool, and the absence of a low-frequency transformer makes them considerably lighter and smaller than the equivalent linear supply. As a result, they are used almost exclusively to power industrial and consumer electronics.
- Switchers are noisy! Their outputs have tens of millivolts of switching ripple; they put garbage onto the powerline; and they can even scream audibly! One cure for output ripple, if that's a problem, is to add an external high-current *LC* lowpass filter; alternatively, you can add a low-dropout linear post-regulator.[118] Some commercial converters include this feature, as well as complete shielding and extensive input filtering.
- Switchers with multiple outputs are available and are popular in computer systems. However, the separate outputs are generated from additional windings on a common transformer. Typically, feedback is taken from the highest current output (usually the +3.3 V or +5V output), which means that the other outputs are not particularly well regulated. There is usually a "cross-regulation" specification, which tells, for example, how much the +12 V output, say, changes when you vary the load on the +5 V output from 75% of full load to either 50% or 100% of full load; a typical cross-regulation specification is 5%. Some multiple-output switchers achieve excellent regulation by using linear post-regulators on the auxiliary outputs, but this is the exception. Check the specs!

- Line-powered switchers, like other switching converters, may have a minimum load-current requirement. If your load current could drop below the minimum, you'll have to add some resistive loading; otherwise the output may soar or oscillate.
- When working on a line-powered switcher, *watch out!* This is no empty warning – you can get yourself killed. Many components are at or above line potential and can be lethal. You can't clip the ground of your 'scope probe to the circuit without catastrophic consequences! (Use a 1:1 isolation transformer at the input, if you must go poking around.)
- Switchers usually include overvoltage "shutdown" circuitry, analogous to our SCR crowbar circuits, in case something goes wrong. However, this circuit often is simply a zener sensing circuit at the output that shuts off the oscillator if the dc output exceeds the trip point. There are imaginable failure modes in which such a "crowbar" wouldn't crowbar anything.[119] For maximum safety you may want to add an autonomous outboard SCR-type crowbar.
- Switchers used to have a bad reputation for reliability, but recent designs seem much better. However, when they decide to blow out, they sometimes do it with great panache! We had one blow its guts out in a "catastrophic deconstruction," spewing black crud all over its innards and innocent electronic bystanders as well.
- Line-powered switchers are definitely complex and tricky to design reliably. Our advice is to avoid the design phase entirely, by *buying* what you need! After all, why build what you can buy?
- A switching supply, operating at roughly constant efficiency, presents a load that looks like a negative resistance (averaged over the ac wave) to the powerline that drives it. It can cause some crazy effects, including (but not limited to) oscillations, when combined with the input reactance of noise filters.

9.8.6 When to use switchers

Luckily for you, we're not bashful about giving advice! Here it is.

---

[117] AN-14: "TOPSwitch Tips, Techniques, and Troubleshooting Guide"; AN-16: "TOPSwitch Flyback Design Methodology."

[118] You can get a switcher and LDO combined as a single regulator IC, for example in the Micrel "High Efficiency Low Dropout" (HELDO™) series.

[119] A personal anecdote: we smelled smoke, and found a dead 'scope in our lab one day. We opened it up, and found that the PFC's output storage capacitor (470 μF, 450 V) had failed, exuding lots of gooey stuff. No problem, we thought, we'll just replace it; especially since a new supply costs $800! Power on, looks good, go to lunch... come back, *smoke!* Turns out the PFC controller chip failed in a way that prevented both regulation and overvoltage shutdown, so the boost circuit just kept boosting, until the capacitor cried uncle.

- For *digital* systems, you usually need something like +2.5 V, +3.3 V, or +5 V, often at high current (10A or more). *Advice:* (a) Use a line-powered switcher. (b) Buy it (perhaps adding filtering, if needed).
- For analog circuits with low-level signals (small-signal amplifiers, signals less than $100\,\mu$V, etc.). *Advice:* Use a linear regulator; switchers are too noisy – they will ruin your life.[120] *Exception:* For some battery-operated circuits it may be better to use a low-power dc–dc switching converter.
- For high-power anything. *Advice:* Use a line-powered switcher. It's smaller, lighter, and cooler.
- For high-voltage, low-power applications (photomultiplier tubes, flash tubes, image intensifiers, plasma displays). *Advice:* Use a low-power step-up converter.

In general, low-power dc–dc converters are easy to design and require few components, thanks to handy chips like the Simple Switcher series we saw earlier. Don't hesitate to build your own. By contrast, high-power switchers (generally line-powered) are complex and tricky, and extremely trouble-prone. If you must design your own, be careful, and test your design very thoroughly. Better yet, swallow your pride and buy the best switcher you can find.

9.9 Inverters and switching amplifiers

The benefits of switchmode power conversion – high efficiency and small size – can be applied to the generation of a *time-varying* output voltage. You can think of this as "dc-to-ac" conversion, as contrasted with a dc-to-dc power converter. In essence, you can imagine substituting an input *signal* for the fixed dc voltage reference of a switchmode dc regulator; the output will follow the input signal as long as the input signal bandwidth is well below the switching frequency.

Switching converters of this sort are used widely, for example to provide multiphase ac power for motor driving, or to generate the individual winding currents for microstepping motors. A variable-frequency motor driver lets you control the motor speed. Powerline-frequency dc-to-ac converters are often called *inverters*, such as those used

in uninterruptible power supplies (UPSs) for computers. At higher power levels such inverters are used to generate powerline-frequency ac from the high-voltage dc that is shipped across the countryside (at dc voltages up to a *mega*volt, would you believe?). And, closer to home, switching audio amplifiers (known as "class-D" amplifiers; see §2.4.1C) are dominant in consumer electronics. In that application a passive *LC* lowpass filter smooths a rail-to-rail switching waveform (typically at frequencies of ~250 kHz or higher) whose duty cycle is modulated in accordance with the input signal. See Figure 2.73 for waveforms from a low-power class-D audio amplifier.

To get just a taste of this subfield of power electronics, take a look at Figure 9.87, where we've captured waveforms from two styles of uninterruptible 120 Vac power supplies, along with the raw 120 Vac wallplug power in our lab. You might guess that the clean middle waveform is the utility power, but in fact that waveform is the loaded output from a UPS that boasts "low-distortion sine wave." The top waveform is the wallplug power, showing rather typical levels of distortion. The 3-level waveform at the bottom is euphemistically called a "modified sinewave," and is typical of less-expensive inverters and UPSs. It isn't handsome, but it does the job: if switched to ± 170 V rails 25% of the time, and to zero (or unpowered) in between, it's easy to figure out that it has the same rms voltage (120 Vrms) and peak voltage (170 Vpk) as a 120 Vrms sinewave.[121] So it delivers the same power to resistive loads, etc., and it

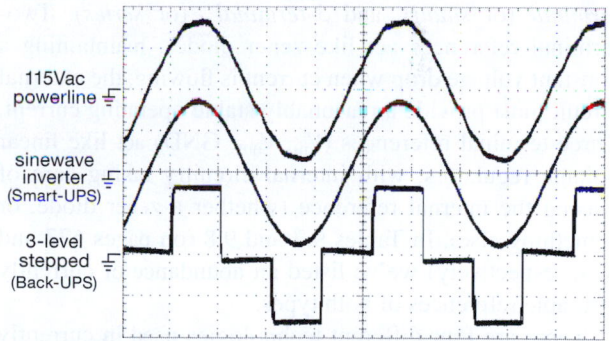

115Vac powerline

sinewave inverter (Smart-UPS)

3-level stepped (Back-UPS)

Figure 9.87. A true sinewave inverter generates a cleaner sinewave than wallplug ac. The 3-level waveform (sometimes called *modified sinewave*), though hardly a sinewave, has the same rms and peak voltages, and suffices for most loads. For these measured waveforms we loaded the APC UPSs with a 75 W incandescent lamp. Vertical: 100 V/div; horizontal: 4 ms/div.

---

[120] *Really!* Here's a pithy quote from James Bryant (from the Analog Devices "Rarely Asked Questions" series), in answer to the question "How can I prevent switching-mode power supply noise from devastating my circuit performance?" Answer: "With great difficulty – but it can be done." He continues: "Switching-mode power supplies are inherently the noisiest circuits imaginable. A large current from the supply is being turned on and off at high frequency with very fast dI/dt. There are inevitably large fast voltage and current transients."

[121] Just add up the squared voltages in equal time intervals, then take the square root of their average: $V_{\text{rms}} = \left[(V_1^2 + V_2^2 + \cdots + V_n^2)/n \right]^{1/2}$.

charges the input side of dc power supplies or converters to the same voltage as would the 120 Vrms utility line power.

There's more to think about than simply having the same rms and peak voltage, of course. There's *distortion*: the 3-level waveform has no even harmonics, but it has strong harmonics at all odd multiples of the fundamental frequency (there are various multilevel schemes to address this problem). Then there's the worry about systems that exploit zero crossings of the incoming ac power for timing, for which the 3-level waveform (or any stepped waveform with an odd number of levels) would wreak havoc.[122] There's plenty more to think about, even limiting ourselves to the subject of multilevel inverters.[123]

This area of power electronics is a rich subject; but, sadly, life is finite, and so (barely) is the size of this book.

9.10 Voltage references

Quite apart from their use in integrated voltage regulators, there is frequently the need for good voltage references within a circuit. For instance, you might wish to construct a precision regulated supply with characteristics better than those you can obtain using the best integrated regulators. Or you might want to construct a precision constant-current supply. Other applications requiring precision references (but not a precision power supply) include A/D and D/A converters, precision waveform generators, and accurate voltmeters, ohmmeters, or ammeters.

Integrated voltage references come in two styles: *2-terminal* (or *shunt*), and *3-terminal* (or *series*). Two-terminal references act like zener diodes, maintaining a constant voltage drop when current is flowing; the external circuit must provide a reasonably stable operating current. Three-terminal references (V_{in}, V_{out}, GND) act like linear voltage regulators, with internal circuitry taking care of biasing the internal reference (whether a zener diode, or something else). In Tables 9.7 and 9.8 (on pages 677 and 678, respectively) we've listed an abundance of currently available references of both types.

There are four different technologies used in currently available voltage references, all of which exploit some physical effect to maintain a well-defined and stable volt-age – *zener diodes*, *bandgap references*, *JFET pinchoff references*, and *floating gate references*. They are all available as stand-alone (2-terminal or 3-terminal) components; they are also commonly incorporated as an internal voltage reference within a larger IC such as an A/D converter. Let's take them in order.

9.10.1 Zener diode

The simplest form of voltage reference is the zener diode, a 2-terminal device we introduced in §1.2.6A. Basically, it is a diode operated in the reverse-bias region, where current begins to flow at some voltage and increases dramatically with further increases in voltage. To use it as a reference, you simply provide a roughly constant current; this is often done with a resistor from a higher supply voltage, forming the most primitive kind of regulated supply.

Zeners are available in selected voltages from 2 to 200 volts (they come in the same series of values as standard 5% resistors), with power ratings from a fraction of a watt to 50 watts, and tolerances of 1% to 20%. As attractive as they might seem for use as general-purpose voltage references (being simple, inexpensive, passive 2-terminal devices), zeners lose their luster when you look a bit more closely: it is necessary to stock a selection of values, the voltage tolerance is poor except in high-priced precision zeners, they are noisy (above 7 V), and the zener voltage depends on current and temperature. As an example of the last two effects, a 27 V zener in the popular 1N5221 series of 500 mW zeners has a temperature coefficient of +0.1%/°C, and it will change voltage by 1% when its current varies from 10% to 50% of maximum.

There is an exception to this generally poor performance of zeners. It turns out that in the neighborhood of 6 volts, zener diodes are quiet, become very stiff against changes in current, and simultaneously achieve a nearly zero temperature coefficient. The graphs in Figures 9.88 and 9.89 illustrate the effects.[124] If you need a zener for use as a stable voltage reference only, and you don't care what voltage it is, one possibility is to use one of the compensated zener references constructed from a 5.6 V zener (approximately) in series with a forward-biased diode – if you can find one! (Read on...) The zener voltage is chosen to give a positive coefficient to cancel the diode's temperature coefficient of −2.1 mV/°C. Temperature compensation can be

---

[122] One solution is a 6-interval, 4-level waveform: V_{pk}, $V_{pk}/2$, $-V_{pk}/2$, $-V_{pk}$, spending twice as much time at $V_{pk}/2$ as at V_{pk}, and never dwelling at zero. This eliminates the 3rd harmonic (as well as all the even harmonics), and produces 120 Vrms if V_{pk}=170 V.

[123] A nice review is found in J. Rodriguez *et al.*, "Multilevel inverters: a survey of topologies, controls, and applications," *IEEE Trans. Indus. Electronics.*, **49**, 724–738 (2002), complete with 78 references.

[124] This peculiar behavior comes about because there are two competing mechanisms going on in zener diodes: zener effect at low voltages, with negative tempco; and avalanche breakdown at high voltages, with positive tempco.

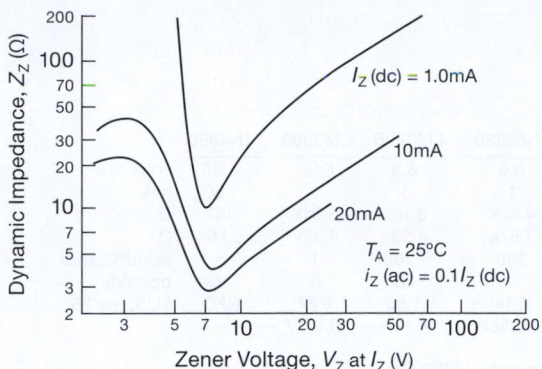

Figure 9.88. Zener diode dynamic impedance for zener diodes of various voltages. (Courtesy of Motorola, Inc.)

accomplished for other zener voltages also, for example in the 1N4057–85 series, which go from 12 V to 200 V, with tempcos of 20 ppm/°C.

Let's follow this thread – which, as we'll see, will lead us to a far better solution in the form of fully integrated voltage references (including those with a temperature-compensated zener on-chip) with superior characteristics. In fact, temperature-compensated zeners, as *discrete* devices, have become largely an extinct breed.

A. Providing operating current

A compensated zener could be used as stable voltage reference within a circuit, but it must be provided with constant

current.[125] The tightly specified[126] 1N4895, for example, is specified as 6.35 V±5% at 7.5 mA, with a tempco of 5 ppm/°C (max) and incremental resistance of 10 Ω (max). So a change in bias current of 1 mA can change the reference voltage by 10 mV, three times as much as a change in temperature from 0°C to +100°C. You can, of course, rig up a separate current source circuit to bias the zener; but you can do better – Figure 9.90 shows a clever way to use the zener voltage itself to provide a constant bias current. The op-amp is here wired as a noninverting amplifier in order to generate an output of +10.0 V. That stable output is itself used to provide a precision 7.5 mA bias current. This circuit is self-starting, but it can turn on with either polarity of output! For the "wrong" polarity, the zener operates as an ordinary forward-biased diode. Running the op-amp from a single supply, as shown, overcomes this nuisance problem.[127] Be sure to use an op-amp that has common-mode input range to the negative rail ("single-supply" op-amps).

There are compensated zeners available that characterize the stability of zener voltage with *time*, a specification that normally tends to get left out. The 1N4895, for example, specifies stability of better than 10 ppm/1000h. The ultimate example is probably the LTZ1000, a 7.15 V integrated zener whose datasheet specifies an astonishing long-term stability of 0.15 ppm/$\sqrt{\text{kHr}}$ (typ). This puppy includes an on-chip temperature stabilizing heater, and claims to deliver a tempco as low as 0.05 ppm/°C, if treated with respect.[128] Such zeners are not cheap: the LTZ1000 will set you back $50.

[125] Most small zener diodes are specified at an operating current of 20 mA (though you can run them at lower currents). But, happily for those seeking low-current zeners, there's the 1N4678 to 1N4713 family (MMSZ4678–4713 for surface-mount SOD-123 package), specified at 50 μA.

[126] And individually tested for 1000 hours! It's available only from the manufacturer, Microsemi.

[127] With a caveat: the circuit could get stuck at zero output if the op-amp's input offset voltage is greater than the ground-saturated output voltage. This could happen with a rail-to-rail CMOS output stage, which is why we chose a BJT op-amp (whose saturation voltage is at least a few millivolts from ground). If you select a CMOS op-amp (say a precision chopper), or if you're losing sleep over the remote possibility of a stuck circuit, you can force the circuit to start correctly with either of the dotted supplements to the circuit.

[128] An example of appropriate respect is the prevention of thermal gradients: a junction of two dissimilar metals (a "thermocouple") generates a thermal EMF, about 35 μV/°C for the connection of the LTZ1000's Kovar-alloy leads to a circuit board. That's about 7 ppm per °C temperature *difference* in the two leads, which is a hundred times larger than that of the chip's zener itself!

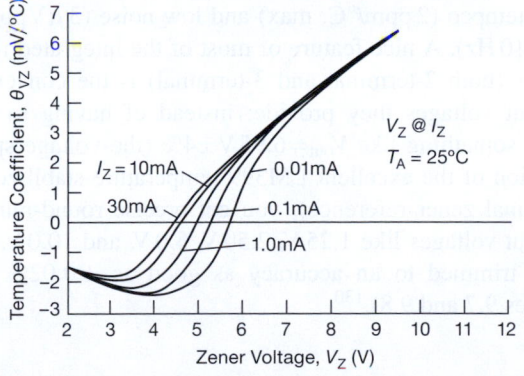

Figure 9.89. Temperature coefficient of zener diode breakdown voltage versus the voltage of the zener diode. (Courtesy of Motorola, Inc.)

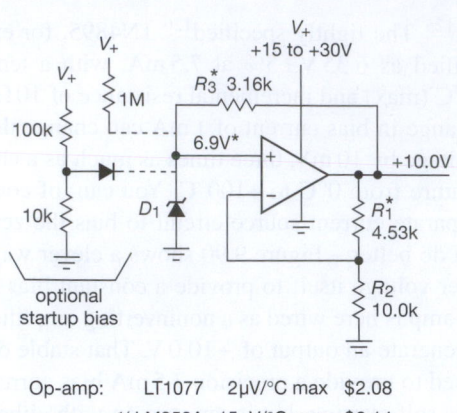

	1N5232B	LM329B	LM399A	1N4895	
V_Z	5.6	6.9	6.95	6.35	volts
I_Z	1	1	1	7.50	mA
R_3	4.42k	3.16k	3.01k	487	Ω
R_1	7.87k	4.53k	4.42k	5.76k	Ω
tempco	380	20	1	5	ppm/°C, max
drift	–	20	8	10	ppm/khr
price	0.14	1.80	9.37	RFQ	$US, qty 25
op-amp	LM358A	◄———	LT1077	———►	

Op-amp: LT1077   2μV/°C max $2.08
           ½LM358A   15μV/°C max $0.14

* values shown for LM329B

Figure 9.90. Stable output voltage provides a stable zener bias current over varying supply voltages V_+. The op-amp must operate to the negative rail. For a jellybean zener like the 1N5232 you can use an inexpensive op-amp like an LM358; but use a precision op-amp (e.g., the LT1077) to preserve the low tempco of a precision reference like the LM329 or LM399. Don't use a split supply here, because the output could happily settle to a negative output.

B. IC zeners

We hinted that precision-compensated zeners as *discrete* devices have largely disappeared; you can check this out for yourself by going to a site like Octopart.com, looking for once-popular parts like the 1N4895 or 1N821–29 series.

That's the bad news. The good news is that excellent compensated zeners now come in integrated form, as the internal reference within a variety of IC voltage references. Table 9.7 lists several of these, from the inexpensive (less than $1) LM329 to the spectacular LTZ1000. These include additional circuitry to yield improved performance (most notably, constancy of terminal voltage with applied current), in the form of an integrated circuit; it looks electrically just like a zener, with just two terminals, although internally it includes additional active devices. Being zener-based, these devices operate around the sweet spot of 7 V, although some (like the LT1236 in the table) include internal amplifier circuitry to create a round-number 10.0 V "zener."[129]

The jellybean LM329 is worth keeping in mind when you need only a "good-enough" zener reference; it has low noise, a zener voltage of 6.9 V, and in its best version it has a temperature coefficient of 10 ppm/°C (max), when provided with a constant current of 1 mA. Where better

performance is needed, consider the LT1236A or the thermally regulated (on-chip heater) LM399A, the latter with an admirable 1 ppm/°C worst-case tempco!

When thinking about 2-terminal zener references, don't overlook the other voltage reference technologies that are available as 2-terminal (shunt) devices (see Table 9.7 on the facing page). From the outside they behave just like zener diodes, but they use other tricks (e.g., a V_{BE} drop) internally to create their stable reference voltage. Among other benefits, such devices come in desirable low voltages (1.25 V and 2.5 V are common), and some can operate at currents as low as 1 μA. Read on!

And remember always not to limit yourself to *2-terminal* references – there are excellent 3-terminal references, both zener based and otherwise. A fine example is the LT1027B, a zener-based 5.0 V reference with excellent tempco (2 ppm/°C, max) and low noise (3 μVpp typ, 0.1–10 Hz). A nice feature of most of the integrated references (both 2-terminal and 3-terminal) is the convenient output voltages they provide: instead of having to deal with something like V_{out}=6.95 V±4% (the voltage specification of the excellent LM399 temperature stabilized 2-terminal zener reference), you get precise round-number output voltages like 1.25 V, 2.50 V, 5.0 V, and 10.0 V, factory trimmed to an accuracy as good as ±0.02% (see Tables 9.7 and 9.8).[130]

---

[129] Zener diodes can be very noisy, and some IC zeners suffer from the same disease. The noise is related to surface effects, however, and *buried* (or *subsurface*) zener diodes are considerably quieter; this is the technology that is used to achieve the very low noise of parts like the LT1236 and LTZ1000 references.

[130] Also powers-of-2 voltages (2.048 V, 4.096 V) to set round-number LSB steps in ADCs and DACs.

Table 9.7 Shunt (2-terminal) Voltage References[a]

Part #	TO-92	DIP	SOIC	SOT-23	SC70	other	1.235	2.048	2.50	3.0	4.096	5.00	other	Accy max (%)	trim pin	avail	V_{max}	min[q] (µA)	max (mA)	Noise 0.1-10Hz typ (µVpp)	noise density[x] (nV/√Hz)	Load cap C_L	Tempco typ (ppm/°C)	Tempco max (ppm/°C)	R_{out} typ[r] (Ω)	Price qty 25 ($US)	Comments
shunt / feedback references																											
TL431A	•	8	-	3,5	-	8	-	-	•	-	-	-	-	1	-	•	36	1000	100	10[o]	20	6uF[g]	6[f]	16[f]	0.22[b,u]	0.32	1,2
LMV431B	•	-	-	3,5	-	-	w	-	-	-	-	-	-	0.5	-	•	30	80	20	7[o]	195	2nF	4[f]	12[f]	0.25[b,u]	0.85	1,3
TLV431B[p]	-	-	8	3,5	6	-	w	-	-	-	-	-	-	0.5	-	•	16[v]	100	20	15[o]	220	20uF	6[f]	20[f]	0.25[b,u]	0.67	1,4
bandgap references																											
LM4431	-	-	-	3	-	-	-	-	•	-	-	-	-	2	-	-	-	100	15	-	170	N	30	-	1	0.75	
LM336B-2.5	•	-	8	-	-	-	-	-	w	-	-	-	-	2	•	-	-	400	10	-	95	-	1.8[f]	6[f]	0.27[u]	0.95	5
LM336B-5.0	•	-	8	-	-	-	-	-	-	-	-	•	-	2	•	-	-	600	10	-	95	-	4[f]	12[f]	0.6	0.95	5
LM336Z5	•	-	-	-	-	-	-	-	-	-	-	•	-	2	•	-	-	600	10	-	-	-	4[f]	12[f]	0.6	0.06	5,6
LM385B	•	-	8	-	-	20	w	-	-	-	-	-	-	1	-	•	5	11	20	-	400	N	-	150	0.4[b]	1.38	7
LM385B-1.2	•	-	8	-	-	8	•	-	-	-	-	-	-	1	-	-	-	15	20	-	490	N	20	-	0.4	0.54	-
LM385B-2.5	•	8	8	-	-	-	-	-	•	-	-	-	-	1.5	-	-	-	18	20	-	480	N	20	-	0.4	0.54	-
LT1034	•	•	-	-	-	-	w	-	•	-	-	-	•	1.2	-	-	-	30[g]	20	-	24	-	20	40	0.5	3.33	8
ADR510	-	-	-	3	-	-	1.00V	-	-	-	-	-	-	0.35	•	-	-	100	10	4	-	N	-	70	0.3	1.42	9
LT1004	•	-	8	-	-	-	-	-	•	-	-	-	-	0.8	-	-	-	20[g]	20	-	260	N[k]	20	-	0.2	1.74	10
LT1004 (TI)	-	-	8	-	-	-	-	-	•	-	-	-	-	0.3	-	-	-	10[h]	20	-	310	N[k]	20	-	0.2	0.94	10,11
MAX6006A	-	8	3	-	-	-	w	-	-	-	-	-	-	0.2	-	-	-	1.0[h]	2	30	high	R	30	-	1.5	1.56	12,13
LT1009	•	-	8	-	-	8	-	-	•	-	-	-	-	0.2	-	-	-	400	10	-	48	-	15	25	0.2	1.74	14
LT1029A	•	-	-	-	-	-	-	-	-	-	-	•	-	0.2	•	-	-	700	10	-	quiet	N[s]	8	20	0.2	3.34	-
LT1029	•	-	-	-	-	-	-	-	-	-	-	•	-	1.0	•	-	-	700	10	-	quiet	N[s]	12	34	0.2	2.07	15
LM4040A	-	-	3	5	-	-	-	-	-	-	-	-	2	0.1	-	-	-	75	15	-	165	N[s]	15g	-	0.3[g]	2.32	16
ADR5041B	-	-	3	3	-	-	-	•	•	•	•	•	-	0.1	-	-	-	50	15	3.2	600	N	10	75	0.2[u]	0.86	17
LM4041A	-	-	3	5	-	-	w	-	-	-	-	-	-	0.1	-	•	15	60	12	-	165	N[s]	15h	-	0.5[h]	1.55	16
LM4050A	-	-	3	-	-	-	-	•	•	-	•	•	2	0.1	-	-	-	60	15	-	180	N[s]	15g	50g	0.3[g]	2.43	-
AD1580B	-	-	3	3	-	-	w	•	•	-	-	-	-	0.1	-	-	-	50	20	5	160	N[k]	-	50	0.4	1.56	-
MAX6138	-	-	-	3	-	-	w	•	•	•	•	•	-	0.1	-	-	-	65	15	20	325	N[s]	4	25	0.3	2.08	-
LT1634A	•	8	8	-	-	-	w	-	•	-	•	•	-	0.05	-	-	-	8	20	15[g]	-	N[s]	4	10	0.15[g,u]	3.13	19
LT1389A	-	8	-	-	-	-	w	-	•	-	•	•	-	0.05	-	-	-	0.7	2	25	noisy	N[s]	4	10	0.25	7.43	20
buried zeners																											
LM329	•	-	-	-	-	-	-	-	-	-	-	6.9		5	-	-	-	600	15	-	11	-	50	100	1[u]	0.79	21
LT1236A-10	-	8	8	-	-	-	-	-	-	-	-	10.0		0.05	•	-	-	1700	20	6	13	-	2	5	0.5[b,u]	5.09	22
LM399AH	-	TO-46 metal										6.95		4	-	-	-	500	10	-	13	-	0.3	1	0.5	9.37	23
LTZ1000	-	TO-5 metal										7.15		4	-	-	-	1000 s	5	1.2	5.5	-	0.05	-	-	55.00	23,24
zener diodes																											
1N4370A	DO-35										2.4 V			5	-	-	-	20mA spec		-	-	-	−600	-	30	2.11	25
1N752A	DO-35										5.6 V			5	-	-	-	20mA spec		-	-	-	300	-	11	1.56	26
1N4895	DO-7										6.35 V			5	-	-	-	7.5mA spec		-	-	-	5e	-	10	na	27
1N3157	DO-7										8.4 V			5	-	-	-	10mA spec		-	-	-	10c	-	15[m]	na	28

Notes: (a) generally listing best accuracy grade.; sorted by type and increasing accuracy. (b) wired as a zener. (c) at 10mA. (C_L) load capacitor — R: >10nF required; N: not required but allowed, or recommended for transient-loads; µF = min required if more than a small cap is added, see datasheet; blank = no comment. The ac output impedance rises with frequency and will resonate with the load capacitor's reactance. A small resistor (22 to 100Ω, etc.) can isolate the capacitor and lower the resonance Q. (d) 5-10mA. (e) at I_z=7.5mA. (f) ΔV (mV) over temp. (g) for the 2.5V version (the 1.2V version is generally less). (h) for the 1.225 version, or V_{ref} for the adj version. (k) an RC is suggested, e.g. 22Ω. (m) min or max. (n) nominal. (na) not available. (o) of the 1.24V ref, gained up to V_{clamp}. (p) also TLVH431A. (q) minimum operating current (maximum, i.e., worst-case; often higher for higher fixed voltages. (r) usually at 1mA, but not current dependent. (s) see datasheet. (t) typical. (u) spec'd over operating range. (v) 6V for TI's TLV431, 16V for Onsemi TLV431 or TI TLVH431. (w) see datasheet for exact value, chosen for minimum tempco. (x) scaled to 1.0V output; multiply listed value by V_{out}.

Comments: 1: two resistors set V_{clamp}. **2:** I_{ref}=4µA max. **3:** I_{ref}=0.5µA max; complementary to LM385-adj. **4:** I_{ref}=0.5µA max; TLV432 is alternate pinout. **5:** LM336 has voltage-trim pin. **6:** multiple-source jellybean. **7:** -BX version is 30ppm/°C; I_{ref}=15nA. **8:** dual: bandgap and 7V zener (1.6%, 40ppm/°C typ, 90Ω), common neg terminal. **9:** lowest V_{ref} shunt ref. **10:** 1.235V is 0.3% tol, 2.45V is 0.8% tol. **11:** TI's -CDR suffix costs $0.25 (qty 25). **12:** nanopower, min I_z=1µA; 40ms turn-on settling time with 1.2µA bias and 10nF cap. **13:** MAX6007, 08, 09 for other voltages. **14:** LM336 upgrade. **15:** non -A version. **16:** -B, -C, -D suffix looser tolerance. **17:** -A suffix 0.2% tolerance. **19:** -C suffix for looser tolerance. **20:** nanopower. **21:** -A suffix is 5ppm/°C typ, 10ppm/°C max. **22:** series ref used in shunt mode. **23:** on-chip heater; lowest guaranteed tempco. **24:** factory purchase. **25:** low-voltage zeners are poor! **26:** optimum zener voltage. **27:** tested 1k hours; "reference zener," spec'd at 7.5mA only. **28:** temp comp zener reference.

Table 9.8 Series (3-terminal) Voltage References[a]

Part #	Ref[c]	Pkgs[b] (SOIC, SOT-23, TO-92, other)	Avail. V[d]	ADJ ver. other	Accy max (%)	Tempco typ (ppm/°C)	Tempco max	Supply min-max (V)	I_Q typ (µA)	I_{out} src (mA)	I_{out} sink (mA)	Trim	Filter	Shutdn	sense+	Noise 0.1Hz-10Hz typ (µVpp)	Noise 10Hz-10kHz typ (µVrms)	Reg max (ppm/V)	C_{in} min (µF)	C_{out} min (µF)	C_{out} max (µF)	Price qty 25 ($US)	Comments
LP2950A	B	TO-92•, P	5.0	-	0.5	20	100	5.4-30	100	100	-					-	430[r]	0.1%[q]	1	1	-	0.66	1
LP2951	B	8, 8, TO-92•, S	5.0	•	0.5	20	120	3.35-30	100	100	-			•		-	430[r,s]	0.1%[q]	1	1	-	0.56	2
LT6650	B	5	—	0.4 j	0.75	30	-	1.4-18	5.6	0.2	0.2	•				20	23[e]	900	0.1[w]	1	-	1.53[k]	3
LM4128A-2.5	B	8, 5	2.5	-	0.1	10	75	2.9-5.5	60	0.2	-					275	-	50 t	0.1[w]	-	10	1.80	4
AD680J	B	8, 8	2.5	-	0.4	10	25	4.5-40	200	10	-					8	7[e]	16	-	0.1	0.05	3.20	5
REF43G	B	8, 8	2.5	1.5	0.2	15	25	4.5-40	340	10	1.2					-	7[e]	2	-	-	10	6.66	6,7
ISL21010-25	B	3	2.5	-	0.2	15	50	2.6-5.5	46	25	-			•		67	37[e]	100	-	0.1	10	0.70[k]	8
LM4120A-2.5	B	5	2.5	-	0.2	14	50	2.7-12	160	5	5			•		20	36[p]	80	-	0.022	0.047	2.37	8
LM4125A-2.5	B	5	2.5	-	0.2	14	50	2.7-6	160	5	5					20	36[p]	80	-	0.022	0.1	1.75	8
LM4132A-2.5	B	5	2.5	1.2	0.05	-	10	2.9-5.5	60	20	-		•	•		240	-	50 t	-	-	10	2.60	4
ISL60002B-25	F	3	2.5	-	0.04	-	20	2.7-5.5	0.35	7	7					30	-	100	-	-	0.001[v]	2.07	4
REF3125	B	3	2.5	-	0.2	5	15	2.55-5.5	100	10	10			•		33	48	25	0.47	1	-	2.07	7
REF3225	B	6	2.5	-	0.2	4	7	2.55-6	100	10	5			•		33	48	65	0.47	1	10	3.60	7
AD584K	B	M	2.5/5.0/10.0	7.5	0.14	-	15	4.5-30	750	10	-	•				50	40	50	-	-	-	10.61	11
X60003B-50	F	3	5.0	-	0.02	-	10	5.1-9	0.5	5	10		•			30	80[e]	60	0.1	1	0.001[v]	2.68[h]	7
LT6656A-2.5	B	6, T	2.5	-	0.05	5	10	2.9-18	0.85	10	0.1			•		7.5	-	25	0.1	0.1	-	6.42	4,13
REF5025	B	8, S	2.5	-	0.05	-	6.5	5.2-18	800	10	10				•	18	42	1	1	1	50	2.97	6
ADR3425	B	6	2.5	-	0.1	2.5	8	2.7-5.5	85[m]	10	3			•		3	2.2[e]	50	0.1	0.1	-	2.34	-
LT1021B-5	Z	8, 8, S	5.0	7.0	0.1	2	5	7.2-36	800	10	-	•			•	1.5	2[e]	12	-	-	-	7.00	7,14
LT6654A-2.5	B	6, T	2.5	4.5	0.05	3	10	2.7-36	350	10	10			•		6.3	6.3[e]	5	0.1	1	-	5.66	4,7,15
LT1019A-2.5	B	8, 8	2.5	-	0.05	3	5	4-36	650	5	-	•				5	7.5[e]	3	-	-	y	7.50	6,14
LTC6652A-2.5	B	8, S	2.5	7.0	0.05	2	5	2.8-12	350	20	5			•	•	1.9	1.6[e]	50	-	-	y	4.80	4
ISL21090B-25	B	8, S	2.5	-	0.02	2	7	3.7-36	930	5	10			•		3	2.2[e]	18	-	0.1	10	7.56	7
LT1236A-5	Z	8, 8, S	5.0	7.0	0.05	1.5	5	7.2-36	800	10	-	•				16	12[e]	12	-	-	-	5.09	7
MAX6033A-25	B	6	2.5	-	0.04	-	7	2.7-12.6	75[m]	15	-			•		10	-	10	-	0.1	100	6.32	4
ADR02B	B	8, 5	5.0	-	0.06	1	3	7-36	650	10	-	•				2.2	-	30	-	-	-	5.00	4
LM4140A-2.5	J	8, S	2.5	-	0.1	1	3	2.7-5.5	230	8	0.01					1.2[g]	-	200	-	1[x]	4.7[x]	2.87	4,7
ADR441B	J	8, S	2.5	-	0.04	1	3	3-18	3000	10	5				•	4.5	2.2[e]	20	0.1	0.1	-	6.53	9,10
ISL21009B25	F	8	2.5	-	0.02	1	3	3.5-16.5	95	7	7					1.8	-	60	-	-	0.001[v]	6.11	4,7,12
ADR421B	J	8	2.5	-	0.04	1	3	5-15	390	10	-				•	3	2[e]	35	-	-	-	7.76	4,9
LT1027B	Z	8, 8, S	5.0	-	0.02	-	2	8-40	2200	15	-	•				0.6	1.7[e]	6	-	-	-	6.17	4,7
LTC6655B-2.5	Z	8, S	2.5	j	0.025	1	2	3-13	5000	5	5			•	•	0.25[g]	1.3[e]	25	0.1	2.7	100	8.16	7,9,16
ADR4525B	B	8	2.5	-	0.04	-	2	3-15	700	10	10					1.25[g]	-	10	1[u]	1[u]	100	5.85	9,17
MAX6325C	Z	8	2.5	-	0.04	0.5	1	8-36	1800	15	15	•			•	1.5	1.3[e]	5	0.1	0.1	100	13.40	4,7,18

Notes: (a) sorted approximately by tolerance, tempco and 0.1-10Hz noise; generally listing best accuracy grade. (b) other packages: M - TO-99 metal can; P - DPAK power pkg; S - small (micro8, MSOP); T - tiny (DFN, LCC). (c) B: bandgap; F: floating gate; J: JFET pinchoff; Z: zener. (d) tabulated data corresponds to the voltage choice indicated by a large bullet. (e) 10Hz-1kHz. (f) in LCC pkg. (g) for 2.5V version. (h) qty 3k. (i) adjustable via external resistors. (k) qty 1k. (m) min or max. (n) nominal. (o) $15nV/\sqrt{Hz}$ with $C_{NR}=1uF$. (p) peak-to-peak. (q) over V_{in} range. (r) 10Hz to 100kHz. (s) 100µV with 10nF filter cap. (t) typical. (u) 0.1µF for $V_{out} \geq 3V$. (v) up to 10uF with recommended pole-zero network. (w) a minimum of 0.1µF or C_{out}, whichever is larger. (x) ESR must fall in min-max range, see datasheet. (y) no min or max for all except the 2.5V version, which may oscillate with $400pF < C_{out} < 2µF$ when sinking between 1mA and 6mA. (z) a "," means no minimum or maximum; in all cases consult datasheet for details.

Comments: 1: inexpensive LDO reg/ref. 2: inexpensive LDO reg/ref with dropout flag; can add ext BJT. 3: lowest V_{ref}; op-amp FB input for V_{out} from 0.4V to rail. 4: other suffixes for relaxed tempco and accy. 5: -A grade for 0.2% accy. 6: temperature output. 7: load regulation 20ppm/mA or better. 8: no suffix for relaxed tempco and accy. 9: low noise, low tempco. 10: a favorite. 11: pin selectable V_{out}; ext resistors for variable V_{out}. 12: ISL21007 for $V_{in}=2.7$-$5.5V$ and $I_Q=75µA$. 13: 10mV no-load dropout. 14: can use in shunt mode. 15: low noise, wide supply. 16: very low noise, low tempco. 17: low noise, low tempco. 18: lowest noise and tempco.

Well, you say, I could do that with the circuit of Figure 9.90, which lets me set the dc output voltage via the ratio of R_1/R_2. Sure. But hold on – standard metal-film resistors come in 1% precision, with tempcos in the range of ± 50 ppm/°C. You *can* get fixed resistors and arrays with tempcos down in the ± 1 ppm/°C territory (see §5.6), but you'll pay a dear price, and the resistance selection is sparse. And don't forget that you've still got to trim the gain to reach the precise round-number output voltage. Use a trimpot? Not a good idea, because the tempco will suffer, and you will have to worry about resistance stability (wiper resistance, mechanical stability, etc.). You're likely to conclude that an on-chip factory-trimmed resistive divider (matched tempcos, thus very low tempco of gain) is the way to go. And it is.

9.10.2 Bandgap (V_{BE}) reference

This method exploits the $\sim$0.6 V base emitter voltage drop of a transistor operating at constant collector current (it should properly be called a V_{BE} reference), as given by the Ebers–Moll equation. Because that voltage has a negative temperature coefficient, the technique involves the generation of a voltage with a positive temperature coefficient the same as V_{BE}'s negative coefficient; when added to a V_{BE}, the resultant voltage has zero tempco.

Figure 9.91 shows how it works. We start with a current mirror with two transistors operating at different emitter current densities (typically a ratio of 10:1). Using the Ebers–Moll equation, it is easy to show that I_{Q2} has a positive temperature coefficient, because the difference in V_{BE}'s is just $(kT/q)\log_e r$, where r is the ratio of current densities (see the graph in Figure 2.62). You may wonder where we get the constant current to program the mirror. Don't worry – you'll see the clever method at the end. Now all you do is convert that current to a voltage with a resistor and add a normal V_{BE} (here Q_3's V_{BE}). R_2 sets the amount of positive-coefficient voltage you have added to V_{BE}, and by choosing it appropriately, you get zero overall temperature coefficient.[131] It turns out that zero temperature coefficient occurs when the total voltage equals the silicon bandgap voltage (extrapolated to absolute zero), about 1.22 V. The circuit in the box is the reference. Its own output is used (via R_3) to create the constant mirror programming current we initially assumed.

The classic bandgap reference requires three transistors, two for ΔV_{BE} and the third to add a V_{BE}. However, in 1974

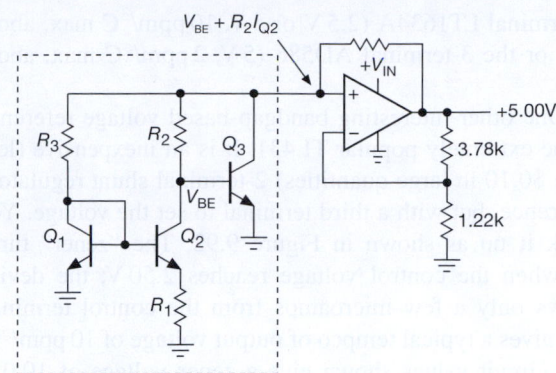

Figure 9.91. Classic V_{BE} bandgap voltage reference. The transistor pair $Q_1 Q_2$ is a ratio current mirror, typically $I_{Q1} = 10 I_{Q2}$; that ratio puts 60 mV across R_1, which sets the PTAT current I_{Q2}.

Paul Brokaw created an elegant 2-transistor version, first used in the LM317, see Figure 9.13.

IC bandgap references

An example of an IC bandgap reference is the inexpensive (about $0.50) 2-terminal LM385-1.2, with a nominal operating voltage of 1.235 V, $\pm 1\%$ (the companion LM385-2.5 uses internal circuitry to generate 2.50V), usable down to 10 μA. That's much less than you normally run zeners at, making these references excellent for micropower equipment.[132] The low reference voltage (1.235 V) is often more convenient than the approximately 5 V minimum usable voltage for zeners (you can get zeners rated at voltages as low as 1.8 V, but they are pretty awful, with very soft knees). The best grade of LM385 guarantees 30 ppm/°C maximum tempco and has a typical dynamic impedance of 1 Ω at 100 μA. Compare this with the equivalent figures for a 1N4370 2.4 V zener diode: tempco = 800 ppm/°C (typ), dynamic impedance $\approx 3000\,\Omega$ at 100 μA, at which the "zener voltage" (specified as 2.4 V at 20 mA) is down to 1.1 V! When you need a precision stable voltage reference, these excellent bandgap ICs put conventional zener diodes to shame.

If you're willing to spend a bit more money, you can find bandgap references of excellent stability, for example, the

---

[131] The complete expression for V_{ref} is therefore $V_{ref} = V_{BE3} + (V_{BE1} - V_{BE2})R_2/R_1$.

[132] But note that low-current references tend to be noisy: the LM385-2.5 (20 μA min) running at 100 μA has a noise voltage density of 800 nV/$\sqrt{\text{Hz}}$, compared with 120 nV/$\sqrt{\text{Hz}}$ for the analogous LT1009 or LM336-2.5 references (400 μA min) running at 1 mA. And, if you're willing to squander more operating current, the LTC6655 low-noise 3-terminal bandgap reference (with a quiescent current of 5 mA) has an output noise density of just 50 nV/$\sqrt{\text{Hz}}$, and, impressively, with a $1/f$ noise corner below 10 Hz.

2-terminal LT1634A (2.5 V or 5 V, 10 ppm/°C max, about $6), or the 3-terminal AD586 (5 V, 2 ppm/°C max, about $9).

One other interesting bandgap-based voltage reference is the extremely popular TL431. It is an inexpensive (less than $0.10 in large quantities) 2-terminal shunt regulator–reference, but with a third terminal to set the voltage. You hook it up as shown in Figure 9.92. The "zener" turns on when the control voltage reaches 2.50 V; the device draws only a few microamps from the control terminal, and gives a typical tempco of output voltage of 10 ppm/°C. The circuit values shown give a zener voltage of 10.0 V, for example. This versatile device comes in TO-92, mini-DIP, and a half-dozen surface-mount packages, and can handle currents to 100 mA and voltages to 36 V. Its low-voltage and low-power (80 μA min) cousins, the TLV431 and TLVH431, work the same way, but with a 1.25 V internal bandgap reference and limited output voltage and current.[133] Both types come in accuracy grades of ±2%, ±1%, and ±0.5%.

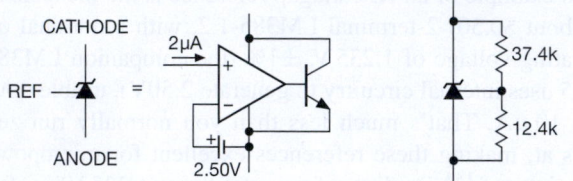

Figure 9.92. TL431 adjustable shunt regulator–reference. The resistive divider in the application circuit on the right sets the "zener" voltage to 10.0 V.

Bandgap temperature sensors

The predictable V_{BE} variation with temperature can be exploited to make a temperature-measuring IC. In Figure 9.91, for example, the difference in V_{BE}'s of $(kT/q)\log_e r$ implies that the current through Q_2 (and also Q_1) is proportional to absolute temperature (PTAT). The circuit can be rearranged (the Brokaw circuit) to produce simultaneously both an output voltage proportional to temperature, and a (fixed) bandgap voltage reference of 1.25 V. This is the case with a number of bandgap references, for example the AD680, a 2.50 V reference with an additional TEMP pin whose output voltage is 2.0 mV/K (thus 596 mV at 25°C). If you want only a temperature sensor, and don't need the bandgap reference, you can get nice stand-alone

temperature sensors, for example the LM35, a 3-terminal sensor with 10 mV/°C output (0 V at 0°C), or the LM61, whose output is offset by +600 mV so it can measure from −30°C to +100°C. The LM61 costs a half dollar, compared with $3 for the 8-pin multifunction AD680.

9.10.3 JFET pinch-off (V_P) reference

This recent technique is analogous to the V_{BE}-based bandgap reference, but uses instead the gate-source voltages of a pair of JFETs. A single JFET operating at fixed drain current has a wicked tempco of V_{GS}, but this can be circumvented by cleverly using a JFET pair. Figure 9.93 shows the configuration used in the ADR400-series of "XFET" voltage references from Analog Devices. The JFET pair Q_1Q_2 have identical geometry and run at equal drain currents; but their different channel doping produces a gate-voltage *difference* of ∼0.5 V that is quite stable, with a relatively small tempco of −120 ppm/°C. That's much smaller than the tempco of a V_{BE} drop (roughly −3000 ppm/°C), and requires only a small dose of positive tempco correction, here applied by the voltage drop across R_1.

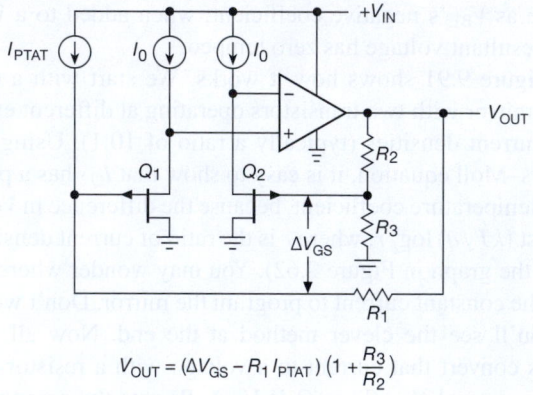

$$V_{OUT} = (\Delta V_{GS} - R_1 I_{PTAT})\left(1 + \frac{R_3}{R_2}\right)$$

Figure 9.93. JFET voltage reference. The asymmetrically doped JFET pair, running at the same drain current, generates a difference voltage ΔV_{GS} between the gates. The relatively small tempco is compensated by a current derived from a bandgap-type reference (not shown).

The result is a voltage reference with excellent tempco (e.g., 3 ppm/°C or 10 ppm/°C for the two grades in the ADR400-series). An important benefit of this technique is its unusually low noise (1.2 μVpp for the 2.5 V part[134]).

---

[133] 6 V and 15 mA for the TLV431; 18 V and 80 mA for the TLVH431. The adjustable LM385 works similarly, but with an operating current range of 10 μA–20 mA, and voltages to 5.3 V.

[134] Better specified as 0.5 ppm(pp), because the noise voltage scales linearly with output voltage.

Bandgap references cannot match this sort of noise performance, because the process of compensating for their large intrinsic temperature coefficient introduces the lion's share of their output noise.

9.10.4 Floating-gate reference

This most recent entry into the voltage reference sweepstakes is, well, bizarre. If you were challenged to come up with an idea that would most likely fail, you might invent the "floating-gate array" (FGA) reference. Intersil did that, but they made it succeed! The idea is to put some charge onto the buried and well-insulated gate of a MOSFET, during manufacturing, which puts it at some voltage (thinking of it as a capacitor); the MOSFET then acts as a voltage follower (or op-amp input) to create a stable output voltage.

The stability over time depends, of course, on the tiny capacitor not losing or gaining any charge. That's a tall order – you'd like it to remain stable to perhaps 100 ppm over several years, over the full operating temperature range. A gate capacitance of 100 pF charged to 1 V, for example, would require that gate leakage be no more than 10^{-22} A; that's about two electrons per hour!

Somehow the folks at Intersil have made it work. They also dealt with stability over temperature, with several tricks: one method uses capacitors of different construction to cancel the already-small tempco of approximately 20 ppm/°C; another method uses capacitors of one type only, cancelling the small residual tempco by adding a voltage of known tempco (as in the bandgap and JFET references).

The results are impressive: the ISL21009 series claim long-term stabilities of order 10 ppm per square-root-kilohour; tempcos of 3 ppm/°C, 5 ppm/°C, and 10 ppm/°C (max) for the three grades;[135] noise of 4.5 μVpp; and very low supply current of 0.1 mA (typ). They come in preset voltages of 1.250 V, 2.500 V, 4.096 V, and 5.000 V, each available in several grades of accuracy and tempco.

9.10.5 Three-terminal precision references

As we remarked earlier, these clever techniques make possible voltage references of remarkable temperature stability (down to 1 ppm/°C or less). This is particularly impressive when you consider that the venerable Weston cell, the traditional voltage reference through the ages, has a tempera-

ture coefficient of 40 ppm/°C. There are two methods used to make references of the highest stability.

A. Temperature-stabilized references

A good way of achieving excellent temperature stability in a voltage reference circuit (or any other circuit, for that matter) is to hold the reference, and perhaps its associated electronics, at a constant elevated temperature. In this way the circuit can deliver equivalent performance with a greatly relaxed temperature coefficient, because the actual circuit components are isolated from external temperature fluctuations. Of greater interest for precision circuitry is the ability to deliver significantly improved performance by putting an already well-compensated reference circuit into a constant-temperature environment.

This technique of temperature-stabilized or "ovenized" circuits has been used for many years, particularly for ultrastable oscillator circuits. There are commercially available power supplies and precision voltage references that use ovenized reference circuits. This method works well, but it has the drawbacks of bulkiness, relatively large heater power consumption, and sluggish warm-up (typically 10 min or more). These problems are greatly reduced if the thermal stabilization is done at the chip level by integrating a heater circuit (with sensor) onto the IC itself. This approach was pioneered in the 1960s by Fairchild with the μA726 and μA727 temperature-stabilized differential pair and preamp, respectively.

This technique is used in the LM399 and LTZ1000 references, which claim stabilized tempcos below 1 ppm/°C (max). Users should be aware that the subsequent op-amp circuitry, including gain-setting resistors, may degrade performance considerably unless extreme care is used in design. In particular, low-drift precision op-amps and matched-tempco resistor arrays are essential. These aspects of precision circuit design are discussed in Chapter 5.

B. Precision unheated references

Clever chip design has made possible unheated references of nearly comparable stability. For example, the MAX6325 series from Maxim have tempcos of 1 ppm/°C (max), with no heater power or warm-up delays. Furthermore, they exhibit low noise and long-term drift. Their chief drawback is the difficulty in getting them! These high-stability references (LTZ1000, LM399, and MAX6325) all use buried zeners.

---

[135] But see the paragraph and footnote about ionizing radiation on page 684.

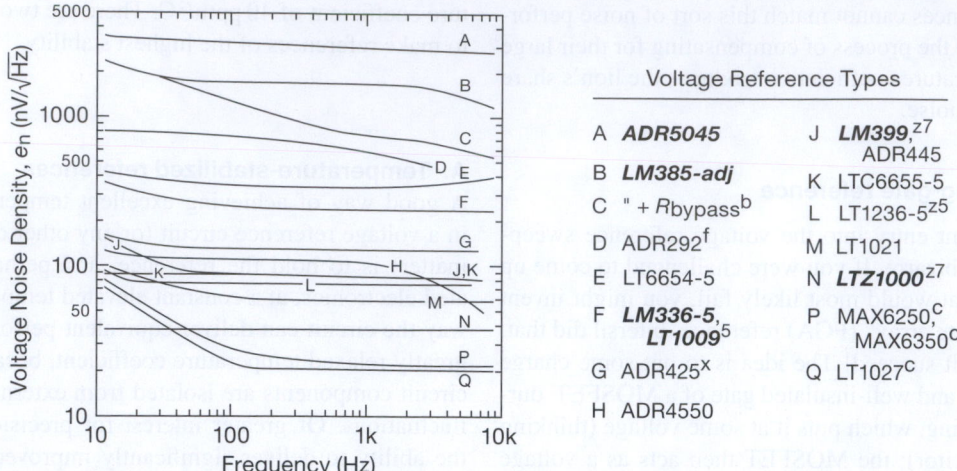

Figure 9.94. Noise density (e_n) versus frequency for a selection of voltage references. All are for 5 V output, except as noted; **boldface** part numbers are shunt type (2-terminal), the rest are series type (3-terminal). The LTZ1000 is operated at 4 mA. *Notes:* (5) 2.5 V reference, curve shown is 2× datasheet's e_n plot; (b) upper resistor bypassed; (c) with 1 μF noise reduction cap; (f) 4.096 V part; (x) "XFET" reference; (z5) buried zener, 5 V buffered output; (z7) 7 V buried zener.

9.10.6 Voltage reference noise

We mentioned briefly the business of *noise*, in connection with low-power references on page 679. You can always add filtering to suppress power supply or reference noise at higher frequencies (see the discussion of the *capacitance multiplier* in §8.15.1), but there's no substitute for a quiet reference at low frequencies, where the noise properties of the reference set a lower limit on output noise. The listings in Table 9.7 on page 677 and 9.8 on page 678 include datasheet values for integrated low-frequency noise (0.1–10 Hz, in units of μVpp), as well as rms noise voltage at somewhat higher frequencies (usually 10 Hz–10 kHz). In Figure 9.94 we've plotted noise-density curves (e_n, in units of nV/$\sqrt{\text{Hz}}$) for those references whose datasheets are considerate enough to provide such information. It's often useful to normalize the specified noise voltage values by the reference's voltage, to get a fair comparison among competing devices.

You can add lowpass filtering to reduce the noise from a voltage reference. Some references bring out an internal node on a "filter" pin (or "bypass," or "noise-reduction" pin) that you can bypass to ground; Table 9.3 indicates those in the "filter pin" column. Often the datasheets for such parts include numeric or graphical information to guide you in the choice of filter capacitor.[136]

Another technique you can use is the addition of an external lowpass filter, with an op-amp follower. Figure 9.95 shows this simple scheme, with an interesting twist: the basic lowpass filter is R_2C_2, with a time constant of 2.2 seconds (3 dB rolloff at 0.07 Hz). But why on earth is it perched atop C_1?! That's done in order to eliminate C_2's leakage current (which would produce an accuracy-damaging voltage drop across R_2) by bootstrapping the low side of C_2 – cute! The inclusion of R_1C_1 affects the rolloff and settling waveform, putting the 3 dB rolloff at 0.24 Hz and producing an 8% overshoot with a lengthened settling time (30 seconds) to 0.1%.[137] For this application you need a pretty good op-amp: input bias current low enough to prevent error from $\Delta V = I_B R_2$, and noise voltage low enough to add only insignificantly to the filtered reference's output. The OP-97E or LT1012A are similar and do the job well (or you could use a smaller R and larger C, permitting larger op-amp input current).

Generally speaking, a reference that runs at very low current will exhibit more noise, a trend evident in

---

[136] One of the more interesting pieces of such guidance appears in the LTC1844 datasheet, which cautions "Additionally, some ceramic capacitors have a piezoelectric response. A piezoelectric device gener-

ates voltage across its terminals due to mechanical stress, similar to the way a piezoelectric accelerometer or microphone works. For a ceramic capacitor the stress can be induced by vibrations in the system or thermal transients. The resulting voltages produced can cause appreciable amounts of noise, especially when a ceramic capacitor is used for noise bypassing." See the discussion in §1x.3.

[137] That is, with $R_1 = R_2$ and $C_1 = C_2$ the 3 dB rolloff frequency becomes $f_{-3dB} \approx 3.3/2\pi RC$, and the settling time to 0.1% (if you care) is approximately $\tau \approx 14RC$. Put a diode across R_2 to shorten the turn-on time.

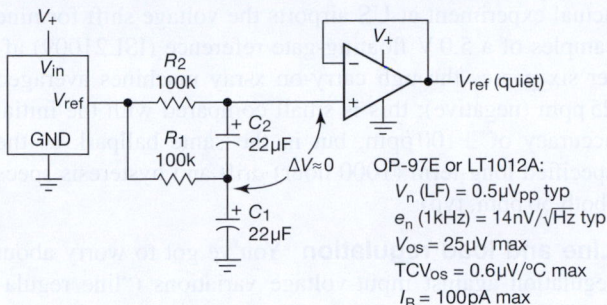

Figure 9.95. External lowpass filter with dc bootstrap quiets any voltage reference, while suppressing error from capacitor leakage currents. Use a quiet op-amp follower with low input current.

Table 9.8. This is easy to understand in the case of a band-gap (V_{BE}) reference, because BJT voltage noise decreases as the square root of collector current (see §8.3). You might conclude from this that a given shunt (2-terminal) IC reference would be quieter when biased at higher currents; but you would be wrong – a shunt reference runs its internal bandgap circuit at a current close to the part's "minimum operating current" (with corresponding noise voltage), and so running the reference at a higher current doesn't help.

9.10.7 Voltage references: additional comments

As should be evident from Tables 9.7 and 9.8, there are lots of things to think about when selecting a voltage reference. Here are some bits of advice, to help the bewildered circuit designer (that's *you*).

Accuracy and drift There's *initial* accuracy, of course, often with a choice of grades designed by a suffix (-A, -B, etc.), with prices to match. But parts *age*, and a well-specified part will include a "long-term drift" figure (usually parts-per-million per thousand hours, or, perhaps more properly,[138] per $\sqrt{\text{kHr}}$), and sometimes a "thermal hysteresis" specification (the voltage offset after thermal cycling

---

[138] From the LTC6655 datasheet: "Long-term stability typically has a logarithmic characteristic and therefore, changes after 1000 hours tend to be much smaller than before that time. Total drift in the second thousand hours is normally less than one-third of the first thousand hours with a continuing trend toward reduced drift with time. Long-term stability is also affected by differential stresses between the IC and the board material created during board assembly." Another bit of advice from LTC: "Significant improvement in long-term drift can be realized by preconditioning the IC with a 100–200 hour, 125°C burn in." A typical long-term drift spec is 60 ppm/$\sqrt{\text{kHr}}$. The datasheet for the REF5025 is instructive: it shows lower drift for the MSOP-8 package than an SO-8, 50 versus 90 ppm/$\sqrt{\text{kHr}}$, and it further shows 50 ppm/$\sqrt{\text{kHr}}$ for the first 1000 hours and 5 ppm/$\sqrt{\text{kHr}}$ for the

over the part's operating temperature range). Taking the LTC6655B (best grade) as an example, the initial accuracy is ±0.025%, the tempco is 1 ppm/°C (typ) and 2 ppm/°C (max), the long-term drift is 60 ppm/$\sqrt{\text{kHr}}$ (typ), and the thermal hysteresis is 35 ppm (typ) for thermal cycling between −40°C and +85°C. From these figures it's clear that initial accuracy is only a part of the story.

A caution about "temperature coefficient": most often we've used the description in terms of *slope*, i.e., ppm/°C (or µV/°C, etc.), giving perhaps a typical and a maximum (worst-case) value. But you'll sometimes see the description in terms of maximum deviation over the part's temperature range; an example is the LM385, where the specified worst-case *average* tempco of 150 ppm/°C is described in a footnote: "The average temperature coefficient is defined as the maximum deviation of reference voltage at all measured temperatures from T_{MIN} to T_{MAX}, divided by $T_{MAX} - T_{MIN}$." A maximum tempco so defined is guaranteed to be smaller than the maximum value of "slope" tempco (i.e., the maximum value of $\Delta V/\Delta T$ over the same temperature range), as you can convince yourself by drawing some wiggly curves. Figure 9.96 shows an example, adapted from the datasheet for the ADR4520–50 series of 3-terminal precision references.

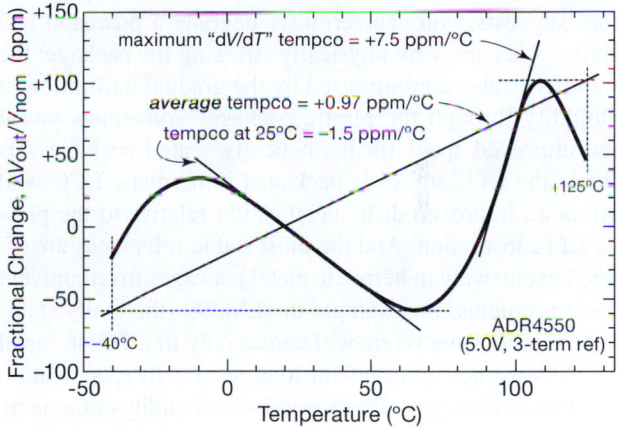

Figure 9.96. Three ways of defining the temperature coefficient of a voltage reference, illustrated with the ADR4550's serpentine datasheet curve. The datasheet's specified tempco is stated as 2 ppm/°C (max) over the full temperature range.

Some references include a "trim" terminal, which sounds like a great idea. But, as with op-amp offset trim configurations, it often provides *too much* trim range! You can try to remedy this by reconfiguring the trim network

---

1000 to 2000 hours. Better improvement than one gets from aging wine!

to supply less current. But a caution: some parts require that the external trim circuitry exhibit a specified temperature coefficient. Perhaps you're better off, as with op-amps, simply by choosing a reference with a tighter spec; a reference of better initial precision usually provides better tempco as well.

Self-heating   Voltage references are happiest when they are only lightly loaded. If a reference IC is used to power a load, on-chip heating produces thermal gradients that can seriously degrade the part's accuracy and drift. For such applications it's best to buffer the output with an op-amp. Most good op-amps have lower noise and offset voltages than the voltage reference itself (you can do the calculation!), so it does not degrade the reference voltage. Quite the opposite, in fact, considering the degrading effect of substantial load current with an unbuffered reference – that's the whole point. And even middle-of-the-road op-amps have far lower tempcos of offset voltage than most voltage references (but use a precision op-amp for a precision reference, as we did in Figure 9.90).

An op-amp buffer also provides an ideal opportunity to add an *RC* noise filter; see §9.10.6, with its unusual filter configuration (Figure 9.95).

External influences   As the footnote on the page before suggests, you can seriously degrade a precision reference's accuracy by physically stressing the package; the stability is also compromised by the gradual infiltration of humidity through the plastic package. Sometimes you'll see improved specs for hermetically sealed package versions: the LT1236LS8 is packaged in hermetic LCC, and offers an improved drift specification relative to the plastic LT1236 version. And the most stable references are offered exclusively in hermetic metal packages to circumvent these problems: for example the LM399 (thermally stabilized buried zener reference) comes only in a TO-46 metal can; it specifies an excellent long-term drift specification of 8 ppm/$\sqrt{\text{kHr}}$ (typ). Pretty good – but handily outdone by the LTZ1000 "Ultra Precision" shunt reference (also in a metal-hermetic package), with a spectacular 0.3 ppm/$\sqrt{\text{kHr}}$ (typ)!

A recent contribution to the gallery of badnesses is the exposure of floating-gate references (§9.10.4) to ionizing radiation, most seriously in the form of airport luggage inspection x-ray machines or PCB post-assembly x-ray inspection. According to Intersil's Application Note,[139] by

actual experiment at US airports the voltage shift for nine samples of a 5.0 V floating-gate reference (ISL21009) after six passes through carry-on x-ray machines averaged 25 ppm (negative); this is small compared with the initial accuracy of ±100 ppm, but in the same ballpark as the specified long-term (1000 hour) drift and hysteresis specs (both 50 ppm, typ).

Line and load regulation   You've got to worry about regulation against input-voltage variations ("line regulation") for a voltage reference that is powered from unregulated dc, for example in a battery-powered application. For such use, a reference isn't suitable that boasts a worst-case tempco of 3 ppm/°C, but whose output varies 200 ppm per volt of input change (these are the specs of an entry in Table 9.8), though that reference would be fine if powered from a reasonably regulated dc rail. For this reason we've listed the worst-case line regulation for the references in Table 9.8 – these vary over nearly a 1000:1 range!

Load regulation matters also if you are using the reference as a voltage regulator, i.e., to power a load that draws a few milliamps, perhaps with load-current variation. But we discourage such use of a precision reference, because it produces on-chip heating and drifts; for that reason (and lack of space) we've not listed load regulation specifications in the tables.

9.11 Commercial power-supply modules

Throughout the chapter we have described how to design your own regulated power supply, implicitly assuming that is the best thing to do. Only in the discussion of line-operated switching power supplies did we suggest that the better part of valor is to swallow your pride and buy a commercial power supply.[140]

As the economic realities of life would have it, however, the best approach is often to use one of the many commercial power supplies sold by companies such as Artesyn, Astec, Astrodyne, Acopian, Ault, Condor, CUI, Elpac, Globtek, Lambda, Omron, Panasonic, Phihong, Power-One, V-Infinity, and literally hundreds more. They offer

---

[139] Intersil App Note #1533, 23Feb2010, *X-Ray Effects on Intersil FGA References*, which explains "The floating gate capacitor is susceptible to radiation degradation from various particles and photons in exces-

sive doses, as the electrons generated in the silicon dioxide are collected in the storage cell. Normal radiation from cosmic rays or radon which exist in small amounts on earth will not cause the FGA reference voltage to drift appreciably for over 100 years. Artificial sources of radiation such as X-ray machines are capable of high enough doses to cause output voltage shift. Note that Flash memory devices are also susceptible to X-ray radiation degradation, although to a lesser degree as they are not precision analog devices."

[140] "Dude, this is a league game."

Figure 9.97. Commercial power supplies come in a variety of shapes and sizes, including potted modules, open-frame units, and fully enclosed boxes. (Courtesy of Computer Products, Inc.)

both switching and linear supplies, and they come in several basic packages (Figure 9.97).

- "Board-mount" supplies: these are relatively small packages, no more than a few inches on a side, with stiff wire leads on the bottom so you can mount them directly on a circuit board. Both ac–dc supplies and dc–dc converters come in this style, and they may be of "potted" or open construction. You can get linear or switchers, and they come with single or multiple output voltages. A typical PC-mounting open-case ac–dc triple-output switching supply provides $+5$ V at 2 A and ±12 V at 0.2 A and costs about \$30 in small quantities. Linear board-mount supplies fall in the 1 W to 10 W range, switchers in the

15 W to 50 W range. In the dc–dc category (which are always switching converters) you can get isolated or non-isolated converters.[141] These are commonly used to generate additional needed voltages (e.g., ±15 V from $+5$ V), as we've seen in this chapter. But another important use is point-of-load (POL) conversion, for example to create $+1.0$ V at 75 A right at the chip's pins, to power a high-performance microprocessor. POL converters come

---

[141] Common package styles include the "full-brick," "half-brick," and "quarter-brick" configurations ($4.6'' \times 2.2''$, $2.3'' \times 2.2''$, and $1.45'' \times 2.3''$, respectively), originated by Vicor; these span the range of 50–500 W, and may include an aluminum baseplate for heatsinking.

in regulated and unregulated versions, the latter with a fixed reduction ratio from a regulated dc input.[142]

- "Chassis-mount" supplies: these are larger power supplies, intended to be fastened to the inside of a larger instrument. They come in both "open frame" and "enclosed-case" styles; the former have all the components on display, whereas the latter (for example, the "ATX" power supplies you find in a desktop or server computer) are wrapped in a perforated metal box. They are available in an enormous variety of voltages, both single and multiple outputs. Chassis-mount linear supplies fall in the 10–200 W range, switchers in the 20–1500 W range.

- "External adapters": these are the familiar black "wall-wart" and desktop ("desk-wart"?) that come with small consumer electronic gadgets, and which are widely available from dozens of manufacturers. They actually come in three varieties, namely (a) step-down ac transformer only, (b) unregulated dc supply, and (c) complete regulated dc supply; the latter can be either linear or switcher. Some of the switching units allow a full 95 to 252 Vac input range, useful for traveling instruments.

- "DIN-rail mount" supplies: a popular way of mounting some kinds of industrial electronics (relays, circuit breakers, surge protectors, connectors, terminal blocks, and the like) is the European-originated DIN rail, which consists of a length of formed metal rail, 35 mm in width. Rail mounting makes it easy to assemble electrical equipment in industrial settings, and you can get a variety of switching supplies in this style.

9.12 Energy storage: batteries and capacitors

No chapter dealing with regulators and power conversion would be complete without a discussion of portable power. That usually means batteries (replaceable or rechargeable), sometimes assisted by energy-storage capacitors. Contemporary life is awash in portable electronic devices, which have driven the development and availability of improved batteries and capacitors. In this section we provide an introduction to battery choices and properties, and to the use of capacitors for energy storage. Because this chapter is already, well, *huge*, we defer further discussion of the care and feeding of batteries to Chapter 9x in the advanced volume.[143]

As we remarked in the previous edition, the (now out-of-print) Duracell "Comprehensive Battery Guide" listed 133 off-the-shelf batteries, with descriptions like zinc-carbon, alkaline manganese, lithium, mercury, silver, zinc-air, and nickel-cadmium. There are even subclasses, for example Li/FeS_2, Li/MnO_2, Li/SO_2, $Li/SOCl_2$, and "lithium solid state." And from other manufacturers you can get sealed lead-acid and gel-type batteries. For the truly exotic application you might even want to consider fuel cells or radioactive thermal generators. What are all these batteries? How do you choose what's best for your portable widget?

The foregoing list divides into so-called *primary* and *secondary* batteries. Primary batteries are designed for a single discharge cycle only, i.e., they're nonrechargeable. Secondary cells (lithium-ion, nickel metal-hydride, and sealed lead-acid gel-type), by comparison, are designed to be recharged, typically from 200 to 1000 times. You usually make your choice among battery types based on trade-offs among price, energy density, shelf life, constancy of voltage during discharge, peak current capability, temperature range, and availability. Once you've picked the right battery chemistry, you figure out which battery (or series combination of batteries) has enough energy content for the job.

Fortunately, it's pretty easy to eliminate most of the batteries in the catalogs, if you follow our first suggestion: *Avoid hard-to-get batteries.* Besides being hard to find, they're usually not fresh. So it's usually better to stick with the varieties available at the drugstore, or perhaps photography store, even if it results in somewhat less than optimum design. We particularly recommend the use of commonly available batteries in the design of any consumer electronic device; as consumers ourselves, we shun those inexpensive marvels that use exotic and expensive batteries. (Remember those early smoke detectors that used an 11.2 V mercury battery? They're better forgotten...)

---

[142] Although *unregulated* POLs might be expected to deliver poor "regulation," in fact they can surprise you: for example, Vicor's VTM series of high-efficiency fixed-ratio "current multipliers" (i.e., voltage step-down) include a unit rated at 130 A and 40:1 stepdown (thus 1.0 Vdc output from 40 Vdc input) with a worst-case output impedance of $0.00094\,\Omega$ at 100°C (thus < 0.1 V output change for a 100 A current step). Not bad. And you can always add feedback to the regulated supply that delivers the +40 Vdc input, to further stiffen the output for load change variations within the loop bandwidth. Moreover, these converters operate at 1.2 MHz switching frequency, so the residual ripple is at 2.4 MHz, conveniently bypassed with relatively small filter capacitors. In fact, the datasheet shows pretty good output-voltage waveforms with 0–130 A load steps, exhibiting essentially no overshoot even when the output is not filtered or bypassed with any external capacitor at all.

[143] See also the expansive discussion in Chapter 14 of the previous edition of this book.

9.12.1 Battery characteristics

If you want a primary (non-rechargeable) battery, your choices are essentially alkaline ("Zn/MnO$_2$") or one of the lithium chemistries ("Li/MnO$_2$," "Li/FeS$_2$," or "Li/SOCl$_2$"). Lithium batteries have a higher single-cell terminal voltage ($\sim$3 V), higher energy density, flatter discharge curves (i.e., constancy of voltage as their life ebbs, see Figure 9.98), better performance at low temperatures (where alkaline batteries just fade away) and higher price. By contrast, the alkaline types (the basic grocery-store battery) are cheap and plentiful, and you can buy them inexpensively at "big-box" stores (if you don't mind getting them in packages of several dozen); they're fine for undemanding applications.

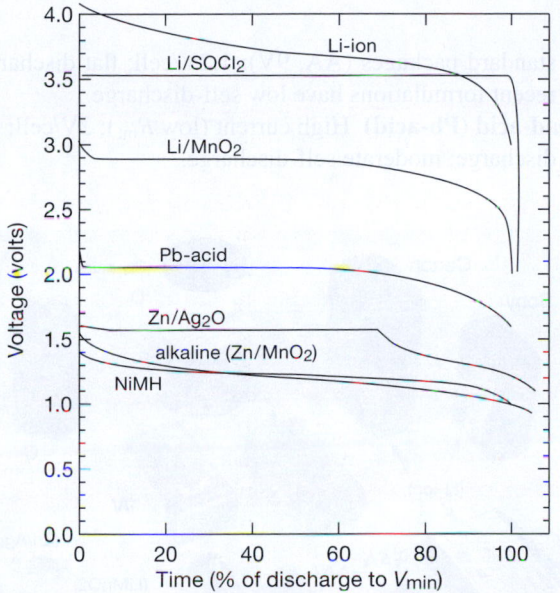

Figure 9.98. Battery discharge curves, as shown on the respective datasheets. In each case 100% discharge corresponds to the voltages as listed in the notes to Table 9.9.

Your choices for a secondary (rechargeable) battery are lithium-ion ("Li-ion"), nickel metal-hydride ("NiMH"), or lead-acid ("Pb-acid"). Li-ion batteries are lightweight, and provide the highest energy density and charge retention, but there are safety issues with lithium chemistries, and these aren't the kind of thing you buy off-the-shelf; they are the darlings of the manufacturers of smartphones, tablets, and laptop computers. Nickel metal-hydride batteries are the more common "consumer" rechargeables, and they come in standard form factors (AA, 9V sizes); early formulations had discouraging memory effects and self-discharge rates ($\sim$30%/month!), but recent versions

("LSD" – low self-discharge) are greatly improved. Lead-acid batteries are the heavy lifters, with their very low internal resistance; they are dominant in uninterruptible power supplies (UPSs) and other power-hungry devices (like boats and automobiles!); they don't come in tiny packages, but you can get them in sizes as small as D-cells.

Recharging secondary batteries can be a complicated business, particularly for fussy chemistries like Li-ion. Taking first the example of the humble lead-acid battery, a good charging method is the so-called two-step technique: after a preliminary "trickle" charge, you begin with a high-current "bulk-charge" phase, applying a fixed high current I_{max} until the battery reaches the "overcharge voltage," V_{OC}. You then hold the voltage constant at V_{OC}, monitoring the (dropping) current until it reaches the "overcharge transition current," I_{OCT}. You then hold a constant "float voltage," V_F, which is less than V_{OC}, across the battery. For a 12 V 2.5 Ah lead-acid battery, typical values are I_{max}=0.5 A, V_{OC}=14.8 V, I_{OCT}=0.05 A, and V_F=14.0 V.

Although this all sounds rather complicated, it results in rapid recharge of the battery without damage. TI makes some nice ICs, for example the UC3906 and BQ24450, that have just about everything you need to do the job. They include internal voltage references that track the temperature characteristics of lead-acid cells, and they require only an external *pnp* pass transistor and four parameter-setting resistors.

Charging Li-ion batteries requires a bit more care, but once again the semiconductor industry has responded with easy-to-use single-chip solutions. Figure 9.99 shows an example of the kind of thing that is commonly seen. Here the power from a USB port (+5 V nominal, able to supply 100 mA or 500 mA) is used to charge a single-cell 4.2 V Li-ion battery; the latter's output (which is down to $\sim$3.5 V when mostly discharged) is reduced to a stable +3.3 V logic-supply level with a linear LDO regulator. In this circuit the charger IC (U_1) takes care of the charging current and voltage profiles, and it includes safety features to sense over-voltage, short-circuit, and both chip and battery over-temperature (the latter uses the optional thermistor, which is found in many battery packs, or can be added externally). The cell temperature is also used to adjust the charging current or voltage when the temperature is outside the normal 10–45°C range, according to what's called JEITA standards. The ISET2 pin sets the input current limit as indicated; the USB protocol allows 100 mA drain initially, which can be increased to 500 mA by a negotiation through the USB data pins D− and D+ (this requires a microcontroller or other smart chip, not shown here). The LEDs indicate status (charging, input power good). The

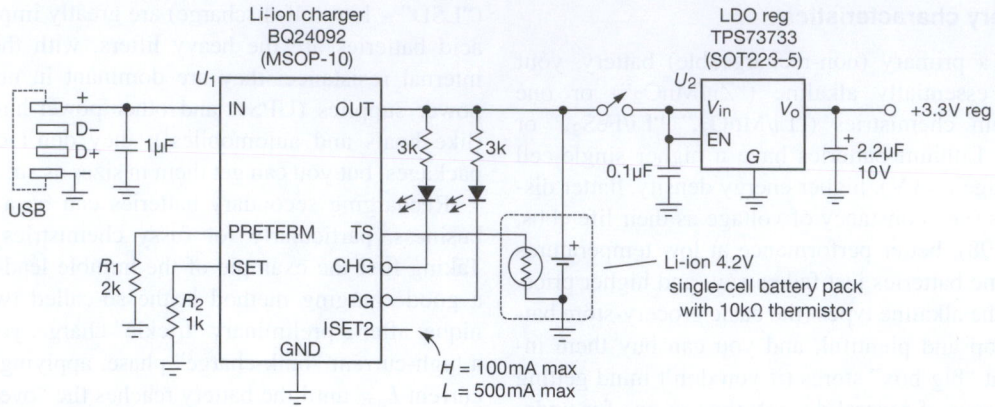

Figure 9.99. The +5 V provided by a USB port is ideal for charging a single-cell Li-ion battery; LDO regulator U_2 converts the battery's 3.5–4.2 V output to stable +3.3 V. See text for discussion of what to do with the USB's D+ and D− pins.

care and feeding of Li-ion batteries is discussed in more detail in §9x.2.

9.12.2 Choosing a battery

Table 9.9 lists characteristics of most of the batteries you might consider, and Figure 9.100 shows an assortment of common battery types. Here is a capsule summary of the most distinguishing characteristics of available batteries intended for use in electronic devices.

Primary Batteries (non-rechargeable)

Alkaline (Zn/MnO$_2$) Inexpensive; widely available (1.5V/cell AA, C, & D, and 9V pkgs); excellent shelf life; good low-temperature performance; sloping discharge.

Lithium (Li/MnO$_2$) High energy-density; good high-drain performance; 3V AA, C, & D, and 9V pkgs; excellent shelf life; excellent low-temperature performance; flat discharge.

Lithium (Li/FeS$_2$) Extraordinary shelf life (90% after 15 yrs); excellent low-temperature performance; flat discharge.

Lithium (LiSOCl$_2$) Extraordinary low-temperature performance (to −55°C); excellent shelf life; very flat discharge (but varies with I_{load}).

Silver (Zn/Ag$_2$O) Button cells; very flat discharge.

Zinc-air (ZnO$_2$) High energy-density (it breathes); flat discharge; short life after seal is removed.

Secondary Batteries (rechargeable)

Lithium-ion (Li-ion) High energy-density; popular; 3.6V/cell; flat discharge; very low self-discharge; safety issues.

Nickel metal-hydride (NiMH) Inexpensive & popular;

standard packages (AA, 9V); 1.2V/cell; flat discharge; recent formulations have low self-discharge.

Lead-acid (Pb-acid) High current (low R_{int}); 2V/cell; flat discharge; moderate self-discharge.

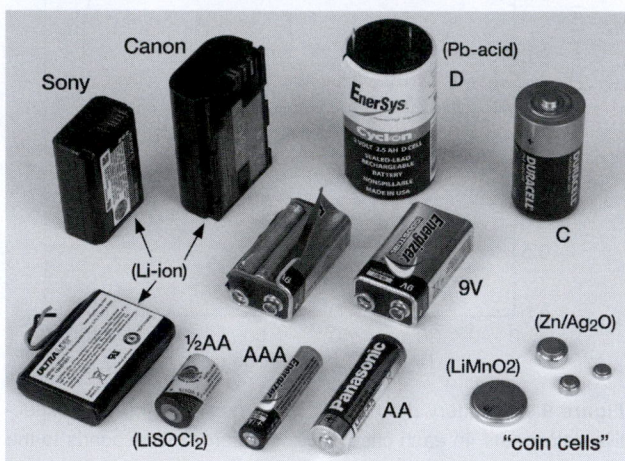

Figure 9.100. A collection of battery specimens. The Li-ion and Pb-acid types are rechargeable ("secondary"); the rest are non-rechargeable ("primary"). Those with unlabeled chemistry are alkaline. Some can opener activity revealed the 9 V innards – six itty-bitty alkaline cells.

9.12.3 Energy storage in capacitors

Batteries store energy *chemically*, either with reversible reactions (rechargeable batteries) or irreversible reactions (non-rechargeable). But batteries aren't the only way to store electrical energy: a charged capacitor stores $CV^2/2$ joules in its electric field, and a current-carrying inductor

Table 9.9 Battery Choices[a]

Chemistry	Part #	V_{nom} (V)	Discharge Capacity				Size (mm)	Weight (gm)	Comments
			(mAh)	at mA	(mAh)	at mA			
Primary (non-rechargeable)									
9V "1604"									
carbon-zinc	122	9	320[b]	5	150[b]	25	17.5 x 12.9 x 46	37	
alkaline MnO$_2$	MN1604	9	550[b]	10	320[b]	100	17.5 x 12.9 x 46	45	popular
lithium MnO$_2$	DL1604	9	1200[b]	20	850[b]	100	17.5 x 12.9 x 46	34	
zinc-air	146X	8.4	1300[b]	10	-	-	17.5 x 12.9 x 46	34	pull tab & it breathes!
cylindrical									
D alkaline	MN1300	1.5	11500[f]	250	3700[f]	1000	34D x 61L	139	D size
D LiMnO2	U10013	3	11100[c]	250	10400[c]	2000	34D x 61L	115	D size
C alkaline	MN1400	1.5	5100[f]	250	1300[f]	1000	26D x 50L	69	C size
AA alkaline	MN1500	1.5	2800[f]	10	2400[f]	100	14.5D x 50.5L	24	AA size; popular
AA LiFeS2	L91	1.5	3200[f]	25	3000[f]	500	14.5D x 50.5L	15	90% after 15y at 20°C
AA LiSOCl2	ER14505	3.6	2100[c]	1	1600[c]	16	14.5D x 50.5L	18	very flat discharge
AAA alkaline	MN2400	1.5	1200[f]	10	1000[f]	100	10D x 44L	11	AAA size
CR2 LiMnO2	CR2	3	850[c]	20	-	-	15.5D x 27L	11	popular
2/3A LiMnO2	CR123A	3	1550[c]	20	-	-	17D x 34.2L	17	popular
2/3A Li poly	BR-2/3A	3	1200[c]	2.5	-	-	17D x 33.5L	13.5	
button									
silver	357	1.55	195[d]	0.2	-	-	11.5D x 4.8H	2.3	
zinc-air	675	1.45	600[e]	2	-	-	11.6D x 5.4H	1.9	4yrs unactivated
LiMnO2	CR2032	3	225[c]	0.2	175[c]	2	20D x 3.2H	2.9	2032 size; popular
Li poly	BR2032	3	190[c]	0.2	90[c]	2	20D x 3.2H	2.5	2032 size
Secondary (rechargeable)									
cylindrical									
NiMH	HHR210AA/B	1.2	2000[f]	2000	-	-	14.5D x 50.5L	29	AA size; R_S=25mΩ; 80%/6mo
Li-ion	NCR18650	3.6	2900[h]	500	2500[h]	5000	18D x 65.2L	45	popular
Pb-acid	0810-0004	2	2500[g]	250	1900[g]	2000	34D x 61L	178	D size; R_S=5mΩ
button									
LiMn	ML2020	3	45[c]	0.12	40[c]	1	20D x 2.0H	2.2	memory backup
LiMnTi	MT621	1.5	2.5[f]	0.05	-	-	6.8D x 2.1H	0.25	memory backup
LiNb	NBL414	2	1[f]	0.004	-	-	4.8D x 1.5H	0.08	memory backup
LiV2O5	VL3032	3	100[c]	0.2	95[c]	1	30D x 3.2H	6.2	memory backup
rectangular									
Pb-acid	LC-R061	6	1200[k]	100	800[k]	1000	96 x 24 x 50	300	80% after 6mo; R_S=50mΩ
Pb-acid	LC-R127	12	7200[m]	500	5000[m]	5000	151 x 65 x 94	2470	80% after 6mo; R_S=40mΩ

Notes: (a) listed part numbers are representative (there are many manufacturers). (b) to 6V. [c] to 2V. (d) to 1.2V. (e) to 1.1V. (f) to 1.0V. (g) to 1.7V. (h) to 2.5V. (k) to 4.8V. (m) to 9.6V. (n) also Tadiran TL-2100.

stores $LI^2/2$ joules in its magnetic field. In quantitative terms these stored energies are dwarfed by those stored in batteries; but for some applications capacitors are just what you want. Among their other virtues, they have long lives, infinite endurance (charge/discharge cycles), the ability to be fully charged and discharged in seconds (or fractions of a second), and very high peak current capability (i.e., very low internal resistance, ESR). A storage capacitor, teamed with a conventional battery, can provide the best of both worlds: extraordinary peak power along

with substantial energy storage. Furthermore, the energy density of recent "supercapacitors" is creeping up on the tail end of batteries. These points are seen nicely in a *Ragone plot* (Figure 9.101). To put numbers on it, we've collected data on some real-world representative capacitors and batteries; these are listed in Table 9.10. Capacitors excel in low ESR and high peak current (and therefore in high *power* density: W/gm or W/m^3), but batteries beat the pants off capacitors in *energy* density (Wh/gm or Wh/m^3).

Table 9.10 Energy Storage: Capacitor versus Battery[a]

Parameter	Conditions	Ultracap Maxwell K2 3000F, 2.5V	Aluminum Electrolytic Panasonic T-UP 180,000µF, 25V	Lead-acid gel Yuasa NP7-12 12V, 7Ah	Lithium-ion Saft VL34570 3.7V, 5.4Ah	Alkaline[p] Duracell MN1500 AA: 1.5V, 2Ah
generic[a]						
Wh/kg	1hr discharge	6	0.05	16	150	40
Wh/m^3	1hr discharge	7800	73	44000	360000	120000
W/kg	maximum	1000	1400	170	500	65
kW/m^3	maximum	1300	2000	500	1100	180
charge time	fast charge	30s	0.25s	1hr	1hr	-
charge cycles		10^6	(note b)	500	500	0
self discharge	25°C	7×10^{5}hr	0.2hr	10^{4}hr	10^{5}hr[e]	10^{5}hr
life	float, 25°C	10yr	25yr	5yr	?	10yr
specific to exemplar						
ESR	maximum	0.29mΩ	9mΩ	25mΩ	30mΩ[e]	120mΩ
I_{max}	cont	210A	17A	40A	11A	1A
P_{max}	cont	525W	425W	440W	35W	1.2W
weight		510g	300g	2650g	125g	24g
energy (Wh)	1hr discharge	3.0	0.015	45	18.9	1
volume (cm^3)	with terminals	390	206	950	53	8.4

(a) from mfgrs' datasheets for listed parts.   (b) no wearout, limited by lifetime only.   (e) estimated.   (p) primary cell (non-rechargeable).

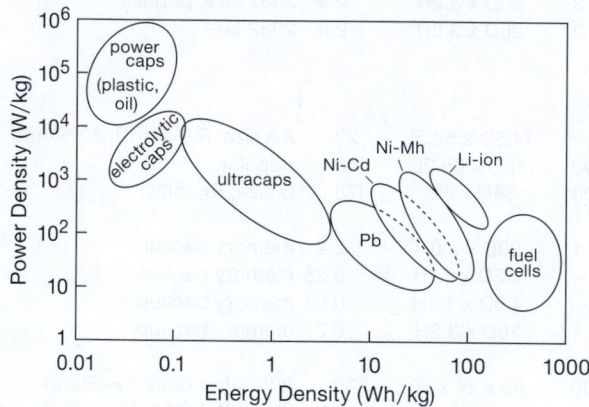

Figure 9.101. Energy-storage capacitors excel in delivering peak power, but batteries win out in energy storage, as seen in this "Ragone plot."

9.13 Additional topics in power regulation

9.13.1 Overvoltage crowbars

As we remarked in §9.1.1C, it is often a good idea to include some sort of overvoltage protection at the output of a regulated supply. Take, for instance, a +3.3 V high-current switching supply used to power a large digital system. Failure of a component in the regulation circuit (even something as simple as a resistor in the output voltage sensing divider) can cause the output voltage to soar, with devastating results.

Although a fuse probably will blow, what's involved is a race between the fuse and the "silicon fuse" that is constituted by the rest of the circuit; the rest of the circuit will probably respond first! This problem is most serious with low-voltage logic and VLSI, which operate from dc supply voltages as low as +1.0 V, and cannot tolerate an overvoltage of as much as 1 V without damage.[144] Another situation with considerable disaster potential arises when you operate something from a wide-range "bench" supply, where the unregulated input to the linear regulator may be 40 volts or more, regardless of the output voltage. We've encountered some aberrant bench supplies that soar to their full output voltage, briefly, when you switch them off. But "briefly" is all it takes to ruin your whole day!

A. Zener sensing

Figure 9.102 shows three classic crowbar circuits – (A) is simple and robust, but inflexible; (B) uses an IC trigger circuit that lets you set the trigger point more accurately; and

---

144 And sometimes much less: the datasheets for Xilinx's Virtex-5 and Virtex-6 FPGAs, for example, specify a core voltage of 1.0 V±5%, with an absolute maximum of 1.1 V! The Virtex-7 has the same 1.1 V limit and 1.0 V nominal core voltage, but it tightens the tolerance on the latter to ±3%; ouch!

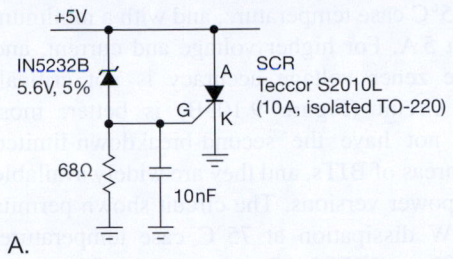

A.

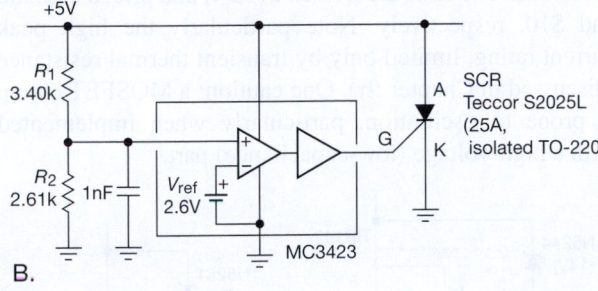

B.

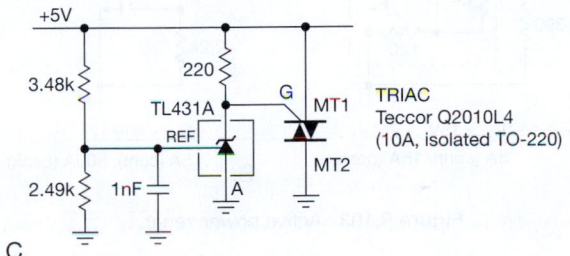

C.

Figure 9.102. Overvoltage crowbars.

(C) uses a popular and precise 3-terminal "shunt regulator," with 1% setpoint accuracy.

In each case you hook the circuit between the regulated output terminal and ground; no additional dc supply is needed – the circuits are "powered" by the dc line they protect. For the simple circuit (Figure 9.102A) the SCR is turned on if the dc voltage exceeds the zener voltage plus a diode drop (about 6.2 V for the zener shown), and it remains in a conducting state until its anode current drops below a few milliamps. An inexpensive SCR like the S2010L can sink 10 amps continuously and withstand 100 amp surge currents; its voltage drop in the conducting state is typically 1.1 V at 10 A. The particular unit here is electrically isolated, so you can attach it directly to the metal chassis (SCRs usually connect their anode to the attachment tab, so normally you'd have to use an insulating spacer, etc.). The 68 Ω resistor is provided to generate a reasonable zener current (10 mA) at SCR turn-on, and the

capacitor is added to prevent crowbar triggering on harmless short spikes.

There are several problems with this simple crowbar circuit, mostly involving the choice of zener voltage. Zeners are available in discrete values only, with generally poor tolerances and (often) soft knees in the VI characteristic. The desired crowbar trigger voltage may involve rather tight tolerances. Consider a 5V supply for digital logic, whose typical 5% or 10% tolerance demands a crowbar voltage at least 5.5V. But that minimum is raised because of transient overshoot of the regulated supply: the voltage can jump when there's an abrupt load current change, creating a spike followed by some ringing.

This problem is exacerbated by remote sensing via long (inductive) sense leads. The resultant ringing puts glitches on the supply that we don't want to trigger the crowbar. The result is that the crowbar voltage should be set not less than about 6.0 V, but it cannot exceed 7.0 V without risk of damage to the logic circuits. When you fold in zener tolerance, the discrete voltages actually available, and SCR trigger voltage tolerances, you've got a tricky problem. In the example shown earlier, the crowbar threshold could lie between 5.9 and 6.6 V, even using the relatively precise 5% zener indicated.

B. IC overvoltage sensing

The second circuit (Figure 9.102B) addresses these problems[145] by using a crowbar trigger IC, in this case the venerable MC3423, which has internal voltage reference (2.6 V±6%), comparators, and SCR drivers. Here we've set the external divider R_1R_2 to trigger at 6.0 V, and we've chosen a 25 A (continuous) SCR, also with an isolated mounting tab; it costs about a dollar. The MC3423 belongs to the family of so-called *power-supply supervisory* chips; the most sophisticated of these not only sense undervoltage and overvoltage, but can switch over to battery backup when ac power fails, generate a power-on reset signal on return of normal power, and continually check for lockup conditions in microprocessor circuitry.

The third circuit (Figure 9.102C) dispenses with a supervisor IC, using instead the wildly popular[146] TL431 shunt regulator to trigger a triac (a bidirectional SCR) when the voltage presented to the reference input exceeds the internal reference voltage of 2.495 V±1%; that causes heavy

---

[145] And others, for example the desirability of fast gate overdrive when crowbarring large capacitive loads; see ON Semiconductor MC3423 datasheet.

[146] A quick check of DigiKey's excellent website shows approximately a half million pieces in stock, in 134 variants, from five manufacturers. They cost as little as $0.09 in large quantities.

conduction from cathode (K) to ground, triggering the triac in what's known as "third quadrant" operation.[147] This circuit can be flexibly extended to higher supply voltages (the TL431 operates to 37 V); and, with the low-voltage TLV431 variant (whose internal reference is 1.240 V) to very low trigger voltages.

The preceding circuits, like all crowbars, put an unrelenting 1 V "short circuit" across the supply when triggered by an overvoltage condition, and it can be reset only by turning off the supply. Because the SCR maintains a low voltage while conducting, there isn't much problem with the crowbar itself failing from overheating. As a result, these are reliable crowbar circuits. It is essential that the regulated supply have some sort of current limiting, or at least fusing, to handle the short. There may be overheating problems with the supply after the crowbar fires. In particular, if the supply includes internal current limiting, the fuse won't blow, and the supply will sit in the "crowbarred" state with the output at low voltage, until someone notices. Foldback current limiting of the regulated supply would be a good solution here.

C. Clamps

Another possible solution to overvoltage protection is to put a power zener, or its equivalent, across the supply terminals. This avoids the problems of false triggering on spikes, because the zener will stop drawing current when the overvoltage condition disappears (unlike an SCR or triac, which have the memory of an elephant). However, a crowbar consisting of a simple power zener itself has its own problems. If the regulator fails, the crowbar has to contend with high power dissipation ($V_{zener}I_{limit}$) and may itself fail. We witnessed just such a failure in a commercial 15V 4A magnetic disk supply. When the pass transistor failed, the 16V 50W power zener found itself dissipating more than rated power, and it proceeded to fail too.

A better alternative, if you really want a power zener, is an "active zener" constructed from a small zener and a power transistor. Figure 9.103 shows two such circuits, in which a zener pulls the base or gate of a transistor into conduction, with a pull-down resistor to bring the zener current into the knee region at transistor turn-on. The TIP142 (in Figure 9.103A) is a popular Darlington bipolar power transistor, priced around $1, good for 75 W

dissipation at 75°C case temperature, and with a minimum beta of 1000 at 5 A. For higher voltage and current, and where effective zener voltage accuracy is not critical, the MOSFET circuit (Figure 9.103B) is better: most MOSFETs do not have the second-breakdown-limited safe-operating areas of BJTs, and they are widely available in robust high-power versions. The circuit shown permits 130 W or 300 W dissipation at 75°C case temperature, for the IRF1407 or IRFP2907, respectively. These are "automotive" MOSFETs, rated at 75 V, and priced at $2.50 and $10, respectively. Note particularly the high peak current rating, limited only by transient thermal resistance (discussed in Chapter *9x*). One caution: a MOSFET clamp is prone to oscillation, particularly when implemented with a high-voltage (low-capacitance) part.

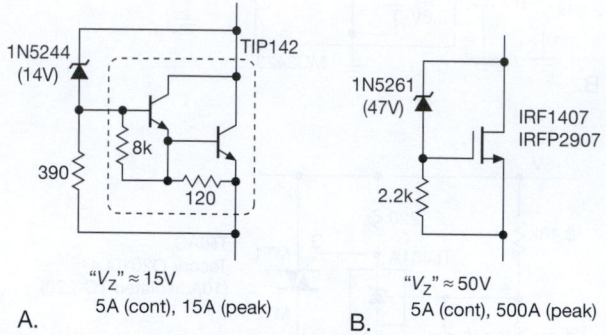

Figure 9.103. Active power zener.

D. Low-voltage clamp–crowbar

These techniques – zener crowbars, IC crowbars, and zener clamps – are generally inadequate for the low-voltage, high-current supplies used to power contemporary microprocessor systems; these may require +3.3 V (or lower) at 50–100 A: low-voltage zeners are imprecise and suffer from a soft knee, and crowbar trigger circuits like the MC3423 require too high a supply voltage (e.g., 4.5 V minimum for the 3423). Also, when an SCR triggers, it crowbars the supply until the power is cycled – not a good thing to do to a computer, particularly if the cause was a momentary (and harmless) transient.

We've wrestled with this problem, and, following the teachings of Billings,[148] have come up with a nice circuit for a low-voltage clamp–crowbar: it is adjustable, and it will operate down to 1.2 V. And (best of all) it operates in two steps – it *clamps* the transient, up to a peak current of 5–10 A; but if the transient persists, or rises above that current, it throws in the towel and fires a crowbar SCR that

---

[147] First and second quadrants have MT2 more positive than MT1, and trigger when the Gate is brought positive or negative with respect to MT1, respectively; third and fourth quadrants have MT2 more negative than MT1, and trigger when the Gate is brought negative or positive with respect to MT1, respectively. Quadrants two and four suffer from lower gate sensitivity.

[148] "Overvoltage clamping with SCR 'crowbar' backup," in K. Billings, *Switchmode Power Supply Handbook*, McGraw-Hill, 2nd ed. (1999).

can handle 70 A continuous (1000 A peak). Because you may be dealing with a high-power system, it also has provision to switch off the ac input. *That* sure is a belt-and-suspenders solution! Because this is a bit off the beaten path, we've grouped this material with other advanced topics in Chapter *9x*.

9.13.2 Extending input-voltage range

As mentioned in §§9.3.12 and 3.6.2, linear regulators have a limited range of input voltage, typically +20 V to +30 V for BJT types, or as little as +5.5 V for CMOS types. Figure 9.104 (a completion of the block diagram in Figure 3.114) shows a nice way to extend the permissible range of V_{IN}, to as much as 1000 V. Q_1 is a high-voltage depletion-mode MOSFET (see Table 3.6), here configured as an input follower to hold U_1's V_{IN} a few volts above its regulated output. For the parts shown, V_{GS} is at least −1.5 V, a comfortable margin for any LDO;[149] and their rated maximum V_{DS} of 400 V and 500 V provides plenty of input-voltage flexibility (substitute an IXTP08N100 if you want to go to 1 kV).

A few details. (a) In this circuit we've used a small gate resistor to suppress the oscillation tendency of high-voltage MOSFETs. (b) The source resistor R_S sets an output current limit of approximately V_{GS}/R_S, which is essential here because this regulator by itself is capable of output currents to 350 mA, which would cause more than 150 W dissipation in Q_1 for V_{IN}=500 V. Here we've chosen R_S for I_{lim}∼10 mA, thus 3.5 W maximum dissipation, handled easily with a modest heatsink attached to Q_1 in its TO-220 power package. (c) A power resistor could be added in Q_1's drain, to offload some of its power dissipation. (d) Low-dropout regulators specify (and require!) a minimum output capacitor C_{out} for stability (along with a specification of allowable range of its effective series resistance ESR); the value shown meets the specification for the TPS76301. (e) Some alternative choices for a fixed-voltage (+3.3 V; omit R_1 and R_2) low-power LDO are listed, with a few relevant parameters. All but the LM2936 are available also in adjustable versions, set with a resistive divider as we've done with the TPS76301 (but refer to their datasheets for V_{ref} and divider resistances).

---

[149] The dropout voltage available from the depletion-mode FET can easily be increased (by a factor of 2× or 3× if necessary) by connecting the gate to a resistive divider between the LDO's output and the FET's source terminal. Note that this raises the minimum output load current.

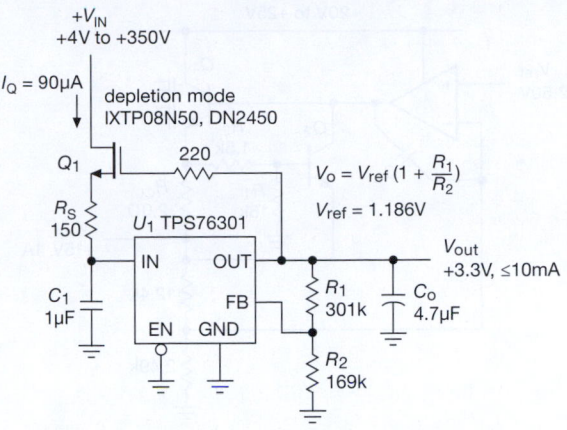

$$V_o = V_{ref}\left(1 + \frac{R_1}{R_2}\right)$$

$$V_{ref} = 1.186V$$

Fixed 3.3V LDO choices for U_1

Part #	I_Q typ (µA)	V_{IN} max (V)	C_O min (µF)	V_{DO} max (mV)	@I_{load} (mA)	Pkgs
TPS76333	85	10	4.7	450	150	SOT-23
LP2950/1–33	75	30	2.2	600	100	TO-92, DIP, SOIC
LM2936-3.3	15	40	22	400	50	TO-92, SOT-23, SOIC
TPS71533	3.2	24	0.47	740	50	SC-70

Figure 9.104. Extending LDO input-voltage range. The listing includes some other low-power 3.3 V fixed-voltage LDO choices.

9.13.3 Foldback current limiting

In §9.1 we showed the basic current-limit circuit, which is often adequate to prevent damage to the regulator or load during a fault condition. However, for a regulator with simple current limiting, transistor dissipation is maximum when the output is shorted to ground (either accidentally or through some circuit malfunction), and it exceeds the maximum value of dissipation that would otherwise occur under normal load conditions. Look, for instance, at the regulator circuit of Figure 9.105, designed to deliver +15 V at currents up to 1 A. If it were equipped with simple current limiting, the pass transistor would dissipate up to 25 watts with the output shorted (+25 V input, current limit at 1 A), whereas the worst-case dissipation under normal load conditions is 10 watts (10 V drop at 1 A). And the situation is even worse in circuits in which the voltage normally dropped by the pass transistor is a smaller fraction of the output voltage.

You get into a similar problem with push–pull power amplifiers. Under normal conditions you have maximum load current when the voltage across the transistors is minimum (near the extremes of output swing), and you have

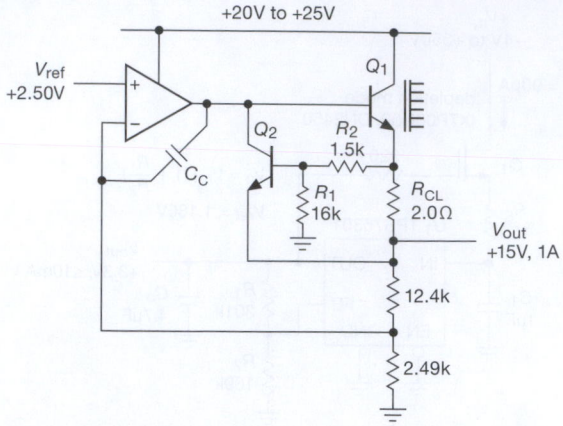

Figure 9.105. Linear regulator with foldback current limiting.

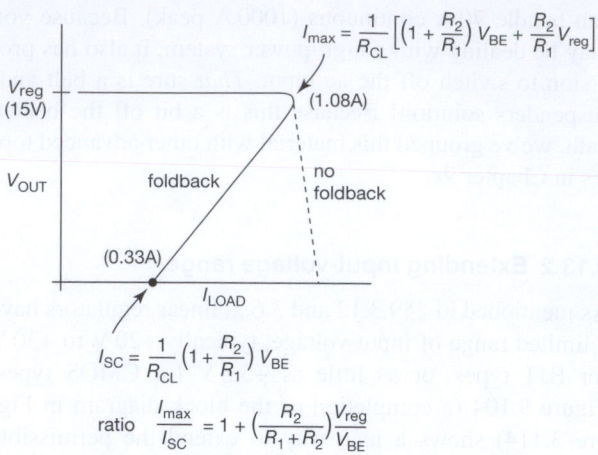

$$I_{max} = \frac{1}{R_{CL}}\left[\left(1 + \frac{R_2}{R_1}\right)V_{BE} + \frac{R_2}{R_1}V_{reg}\right]$$

$$I_{SC} = \frac{1}{R_{CL}}\left(1 + \frac{R_2}{R_1}\right)V_{BE}$$

$$\text{ratio}\quad \frac{I_{max}}{I_{SC}} = 1 + \left(\frac{R_2}{R_1 + R_2}\right)\frac{V_{reg}}{V_{BE}}$$

Figure 9.106. Foldback current limiting for the circuit of Figure 9.105.

maximum voltage across the transistors when the current is nearly zero (zero output voltage). With a short-circuit load, on the other hand, you have maximum load current at the worst possible time, namely, with full supply voltage across the transistor. This results in much higher transistor dissipation than normal.

The brute-force solution to this problem is to use massive heatsinks and transistors of higher power ratings (and safe-operating area; see §9.4.2) than necessary. Even so, it isn't a good idea to have large currents flowing into the powered circuit under fault conditions, because other components in the circuit may then be damaged. A better solution is to use *foldback* current limiting, a circuit technique that reduces the output current under short-circuit or overload conditions.

Look again at Figure 9.105. The divider at the base of the current-limiting transistor Q_2 provides the foldback. At +15 V output (the normal value) the circuit will limit at about 1 A, because Q_2's base is then at +15.55 V while its emitter is at +15 (V_{BE} is typically somewhat below the usual 0.6 V in the hot environment of power electronics). But the short-circuit current is less; with the output shorted to ground, the output current is about 0.3 A, holding Q_1's dissipation down to *less* (about 7.5 W) than in the full-load case (10 W). This is highly desirable, since excessive heatsinking is not now required, and the thermal design need only satisfy the full-load requirements. The choice of the three resistors in the current-limiting circuit sets the short-circuit current, for a given full-load current limit; see Figure 9.106.

An important caution: use care in choosing the short-circuit current, because it is possible to be overzealous and design a supply that will not "start up" into certain loads. Figure 9.107 shows the situation with two com-

mon nonlinear loads: an incandescent lamp (whose resistance rises with voltage), and a linear regulator's input (which begins as an open circuit, then looks like its load resistance while operating below dropout, and finally forms a constant-current load above dropout). As a rough guide, when designing a foldback circuit, the short-circuit current limit should be set no less than one-third to one-half the maximum load current at full output voltage.

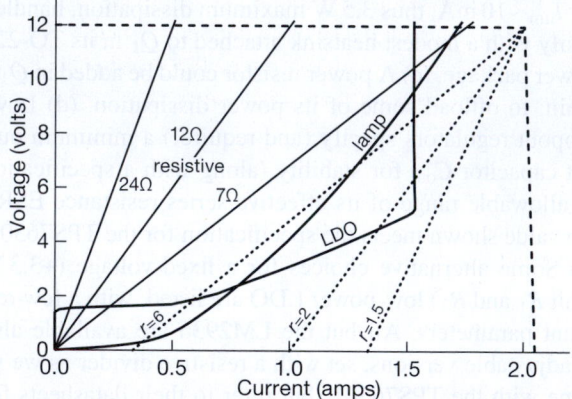

Figure 9.107. Excessive foldback may prevent startup into some loads. We measured VI curves for an automotive lamp (12 V, 21 W) and for a 5 V low-dropout regulator with 3.3 Ω load. The dashed line is normal 2 A current limiting, and the dotted lines show three values of foldback current limiting. A foldback current ratio of $r = I_{max}/I_{SC} = 6$ would fail to start up either load (it would get stuck at the lower intersection); for $r = 2$ the lamp is OK, but the LDO is not. A resistive load is never a problem.

9.13.4 Outboard pass transistor

Three-terminal linear regulators are available with 5 A or more of output current, for example the adjustable 10 A LM396. However, such high-current operation may be undesirable, because the maximum chip operating temperature for these regulators is lower than for power transistors, mandating oversized heatsinks. Also, they are expensive. An alternative solution is the use of external pass transistors, which can be added to integrated linear regulators like the 3-terminal fixed or adjustable regulators, whether of conventional or low-dropout configuration. Figure 9.108 shows the basic (but flawed!) circuit.

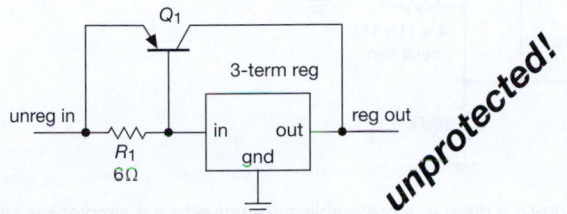

Figure 9.108. Basic three-terminal regulator with current-boosting outboard transistor. *Don't build this* – it has no current-limiting circuitry!

The circuit works normally for load currents less than 100 mA. For greater load currents, the drop across R_1 turns on Q_1, limiting the actual current through the 3-terminal regulator to about 100 mA. The 3-terminal regulator maintains the output at the correct voltage, as usual, by reducing input current and hence drive to Q_1 if the output voltage rises, and vice versa. It never even realizes the load is drawing more than 100 mA! With this circuit the input voltage must exceed the output voltage by the dropout voltage of the regulator (e.g., 2 V for an LM317) plus a V_{BE} drop.

In practice, the circuit must be modified to provide current limiting for Q_1, which could otherwise supply an output current equal to β times the regulator's internal current limit, i.e., 20 amps or more! That's enough to destroy Q_1, as well as the unfortunate load that happens to be connected at the time. Figure 9.109 shows two methods of current limiting.

In both circuits, Q_1 is the high-current pass transistor, and its emitter-to-base resistor R_1 has been chosen to turn it on at about 100 mA load current. In the first circuit, Q_2 senses the load current via the drop across R_{SC}, cutting off Q_1's drive when the drop exceeds a diode drop. There are a couple of drawbacks to this circuit: for load currents near the current limit, the input voltage must now exceed the regulated output voltage by the dropout voltage of the 3-

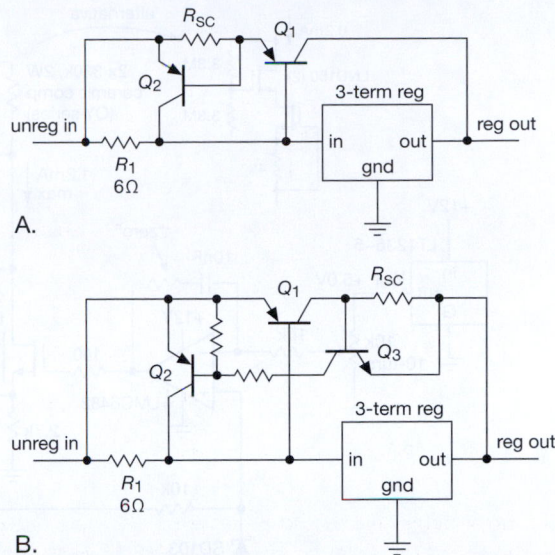

Figure 9.109. Outboard transistor booster with current limiting.

terminal regulator plus two diode drops. Also, the small resistor values required in Q_2's base makes it difficult to add foldback limiting.

The second circuit helps solve these problems, at the expense of some additional complexity. With high-current linear regulators, a low-dropout voltage is often important to reduce power dissipation to acceptable levels. To add foldback limiting to the latter circuit, just tie Q_3's base to a divider from Q_1's collector to ground, rather than directly to Q_1's collector. Note that in either circuit Q_2 must be capable of handling the full current limit of the regulator.

A caution: With an external pass transistor you do not get the overtemperature protection that is included in nearly all integrated regulators. So you have to be careful to provide adequate heatsinking for both normal and short-circuit load conditions.

9.13.5 High-voltage regulators

Some special problems arise when you design linear regulators to deliver high voltages, and you often need to resort to some clever circuit trickery. This section presents a few such techniques.

A. Brute force: high-voltage components

Power transistors, both bipolar and MOSFET, are available with breakdown voltages to 1200 volts and higher, and they're not even very expensive. And IGBTs are available at even higher voltage ratings, up to 6000 volts.

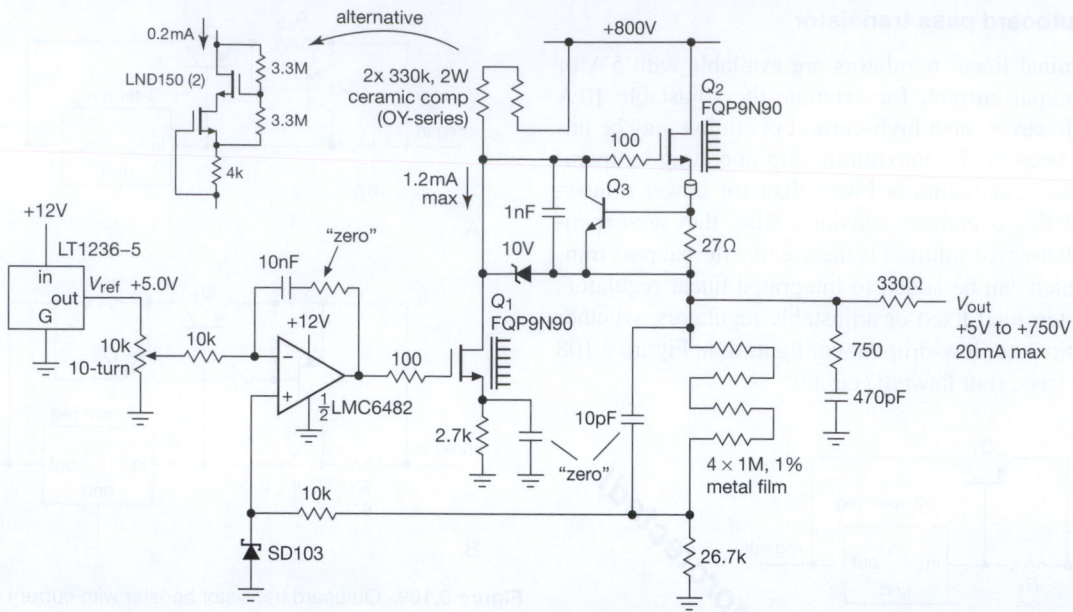

Figure 9.110. High-voltage regulated supply. The current-source pullup on Q_1's drain is a preferable alternative to the simpler resistive drain load; see §9.3.14C. See also Figure 3.111.

ON Semiconductor's MJE18004, for example, is a 5 A *npn* power transistor with conventional (V_{CEO}) collector-to-emitter breakdown of 450 V, and base back-biased breakdown (V_{CEX}) of 1000 volts; it costs less than a dollar in single quantities. And power MOSFETs are often excellent choices for high-voltage regulators owing to their excellent safe operating area (absence of thermally induced second breakdown); they are widely available with 800–1200 V ratings, and currents up to 8 A or more. For example, the FQP9N90 *n*-channel MOSFET (9 A, 900 V) from Fairchild costs about \$1.75. See listings in Tables 3.4b and 3.5.

By running the error amplifier near ground (the output-voltage-sensing divider gives a low-voltage sample of the output), you can build a high-voltage regulator with only the pass transistor and its driver seeing high voltage. Figure 9.110 shows the idea, in this case a +5 V to +750 V regulated supply using an NMOS pass transistor and driver. Q_2 is the series pass transistor, driven by inverting amplifier Q_1. The op-amp serves as error amplifier, comparing a fraction of the output with a precision +5 V reference. Q_3 provides current limiting by shutting off drive to Q_2 when the drop across the 27 Ω resistor equals a V_{BE} drop. The remaining components serve more subtle, but necessary, functions: the zener diode protects Q_2 from reverse gate breakdown if Q_1 decides to pull its drain down rapidly (while the output capacitor holds up Q_2's source); it also

protects against forward gate breakdown, for example if the output is abruptly shorted. The Schottky diode likewise protects the op-amp's input from a negative current spike, coupled through the 10 pF capacitor.

Note the use of several resistors in series, to withstand the large voltages; the OY-series of 1 W and 2 W ceramic composition resistors from Ohmite are non-inductive, as are the precision metal-film resistors in the output sensing divider. The various small capacitors in the circuit provide compensation, which is needed because Q_1 is operated as an inverting amplifier with voltage gain, thus making the op-amp loop unstable (especially considering the circuit's capacitive load). Likewise, the 330 Ω series output resistor promotes stability by decoupling capacitive loads (at the cost of degraded regulation). And Q_2's series gate resistor and source-lead ferrite bead suppress oscillations, to which high-voltage MOSFETs are particularly prone. *An important caution:* Power-supply circuits like this present a real electrical shock hazard – use care!

We can't resist an aside here: in slightly modified form (reference replaced by signal input) this circuit makes a very nice high-voltage amplifier, useful for driving crazy loads such as piezoelectric transducers; see Figure 3.111 for a simple 1 kV amplifier configured this way. For that particular application the circuit must be able both to sink and to source current into the capacitive load. Oddly enough, the circuit (called a "totem pole") acts like a "pseudo-push-pull" output, with Q_2 sourcing current and Q_1 sinking current (via the diode), as needed. See §3x.8 for a detailed discussion about capacitive loading of MOSFET source-followers.

If a high-voltage regulator is designed to provide a fixed output only, you can use a pass transistor whose breakdown voltage is less than the output voltage. For example, you could modify this circuit to produce a fixed +500 V output, using a 400 V transistor for Q_2. But with such a circuit you must ensure that the voltage across the regulator never exceeds its ratings, even during turn-on, turn-off, and output short-circuit conditions. A few strategically placed zeners can do the job, but be sure to think about unusual fault conditions such as an abrupt upstream short-circuit (a spark, or probing fault), as well as "normal" events like an output short. It's remarkably easy to have a pretty-good (and tested) high-voltage circuit fail, abruptly (and usually with a "snap" sound), leaving precious few clues as to the cause. Learn from our hard-earned experience here: use a MOSFET rated beyond the full supply voltage.

Exercise 9.14. Add foldback current limiting to Figure 9.110.

B. Transistors in series

Figure 9.111 shows a trick for connecting transistors in series to increase the breakdown voltage. In the circuit on the left-hand side, the equal gate resistors distribute the dc voltage drops across the series-connected MOSFETs, and the paralleled capacitors ensure that the divider action extends to high frequencies. (The capacitors should be chosen large enough to swamp differences in transistor input capacitance, which otherwise cause unequal division, reducing overall breakdown voltage.) The zener diodes protect against gate breakdown.[150] And the 100 Ω series gate resistors help suppress oscillations, common in high-voltage MOSFETs (use some ferrite beads on the source and gate leads, if there's an oscillation sighting).

For series-connected bipolar transistors you can distribute the voltage drops with resistors alone, as shown, because the rugged base–emitter junctions are not susceptible to damage analogous to the MOSFET's oxide punch-through (in the forward direction they simply conduct a small current; and the small reverse currents from the base divider are generally benign, and they can be prevented entirely with 1N4148-style small-signal diodes connected between base and emitter). Small-signal transistors

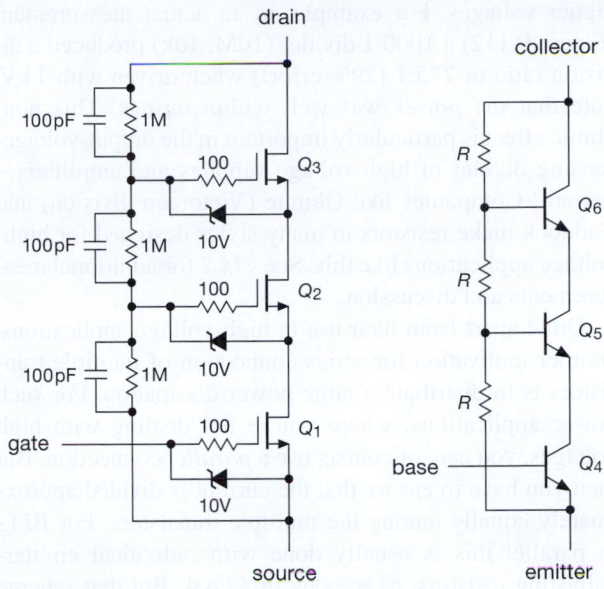

Figure 9.111. Connecting transistors in series to raise the breakdown voltage, to distribute the power dissipation, and (in power BJTs) to stay within the SOA.

like the 300 V MPSA42 and MPSA92 (*npn* and *pnp*, respectively) and 400 V MPSA44 (available in TO-92 and surface-mount packages) are usefully extended to higher voltages this way.

Note that the series-connected string has considerably poorer saturation voltage than that of an equivalent high-voltage transistor: for three MOSFETs (as shown) the ON-voltage is $V_{sat} \approx 3V_{DSon} + 3V_{GSon}$; for the BJT circuit it's $3V_{CEon} + 3V_{BE}$.

Series-connected transistors can, of course, be used in circuits other than power supplies. You'll sometimes see them in high-voltage amplifiers, although the availability of high-voltage MOSFETs often makes it unnecessary to resort to the series connection at all.

In high-voltage circuits like this, it's easy to overlook the fact that you may need to use 1 watt (or larger) resistors, rather than the standard 1/4-watt type. A more subtle trap awaits the unwary, namely the maximum *voltage* rating of a resistor, regardless of its power-dissipation rating. For example, standard 1/4-watt axial-lead resistors are limited to 250 V, and often less for surface-mount types.[151] Another underappreciated effect is the astonishing voltage coefficients of carbon composition resistors, when run at

---

150 Many designers use ordinary signal diodes, rather than zeners, assuming that there is no worry about *forward* gate breakdown, because the MOSFETs should turn on vigorously long before gate-channel breakdown. We're not so sure.

151 Specifically, 200 V, 150 V, 75 V, 50 V, and 30 V for Vishay's CRCW thick-film resistors in sizes 1206, 0805, 0603, 0402, and 0201, respectively.

higher voltages. For example, in an actual measurement (Figure 9.112) a 1000:1 divider (10M, 10k) produced a division ratio of 775:1 (29% error!) when driven with 1 kV; note that the *power* was well within ratings. This non-ohmic effect is particularly important in the output-voltage-sensing divider of high-voltage supplies and amplifiers – beware! Companies like Ohmite (Victoreen division) and Caddock make resistors in many styles designed for high-voltage applications like this. See §*1x.2* for additional measurements and discussion.

Quite apart from their use in high-voltage applications, another motivation for series connection of multiple transistors is to distribute a large power dissipation. For such power applications, where you're not dealing with high voltages, you can, of course, use a *parallel* connection. But then you have to ensure that the current is divided approximately equally among the multiple transistors. For BJTs in parallel this is usually done with individual emitter-ballasting resistors, as we saw in §2.4.4. But that scheme is problematic with MOSFETs, because they have a spread of gate–source voltages, forcing you to allow an uncomfortably large voltage drop across the source resistors. This can be addressed, though, with some cleverness; see Figure 3.117 for a nice solution.

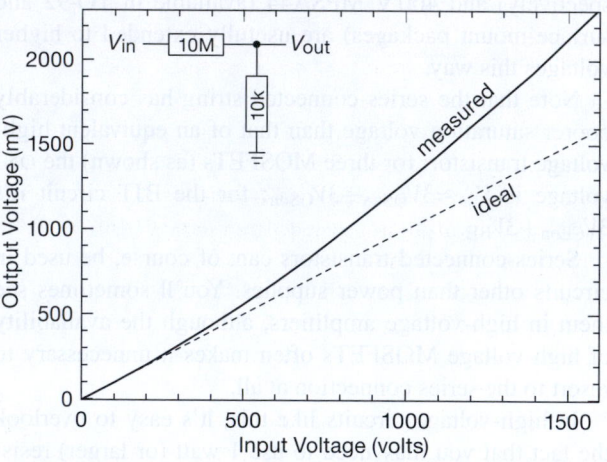

Figure 9.112. Carbon composition resistors exhibit a reduction in resistance as they approach their rated 250 V. Don't use resistors above their voltage rating!

C. Floating regulator

Another method sometimes used to extend the voltage range of integrated regulators, including the simple 3-terminal type, is to float the entire regulator above ground, for example as shown in Figure 9.113. Here the zener D_2 limits the drop across the 3-terminal regulator to just a

few volts (the zener voltage minus Q_1's gate–source voltage), with outboard MOSFET Q_1 taking up the rest of the voltage drop. The LT3080 is a nice choice here, with its simple 10 μA programming current setting the output voltage. We used the trick of a pair of resistors (a "current divider," which you can think of as a current multiplier, viewed from the 10 μA source side) to raise the effective programming current to 1 mA, so we could use a 500k pot to set the voltage (rather than the awkward – and unobtainable – 50 megohms that otherwise would be needed); the boosted programming current serendipitously provides the LT3080's 0.5 mA minimum load current as well. Zener current is provided by the handy LND150 depletion-mode MOSFET from Supertex, here throttled down to 0.2 mA with a source self-biasing resistor.

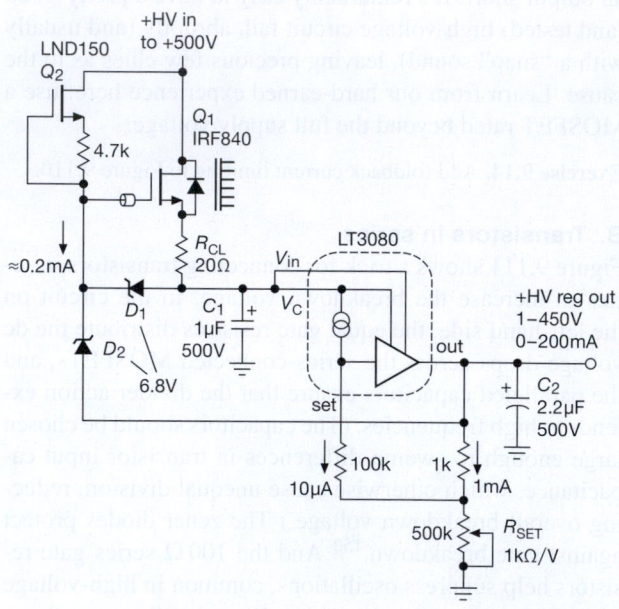

Figure 9.113. High-voltage floating three-terminal regulator.

The remaining components are easy to understand: D_1 protects Q_1's gate; the ferrite bead suppresses oscillations (you can use a 150 Ω gate resistor instead); and the LT3080 is outfitted with its minimum required input and output bypass capacitors. If there's an opportunity for the HV input to drop below the output voltage, add a diode across the LT3080 (heck, just do it anyway).

Exercise 9.15. Explore replacing Q_1 with a high-current depletion-mode MOSFET, like an IXTP3N50; see Table 3.6. Can you think of a way of using it to eliminate Q_2 and the zener diodes, despite the fact its $-V_{GS}$ may be less than the LT3080's required $V_{DO}(max)=1.6$ V at high currents? *Hint:* an LM385-2.5 may be useful.

Review of Chapter 9

An A-to-K summary of what we have learned in Chapter 9. This summary reviews basic principles and facts in Chapter 9, but it does not cover application circuit diagrams and practical engineering advice presented there.

¶A. Voltage Regulator Taxonomy.

Voltage regulators provide the stable dc voltages needed to power all manner of electronic circuits. The simplest (and least noisy) type is the *linear* regulator (Figure 9.2), in which the output error signal, suitably amplified and compensated, is used to control a linear "pass transistor" (BJT or MOSFET) that is in series with a higher (and perhaps unregulated) dc input voltage. Linear regulators are not power efficient, with dissipation $P_{diss} = I_{out}(V_{in} - V_{out})$, and they are not able to a produce dc output that is larger than the input, nor a dc output of reversed polarity.

The *switching* regulator (or switching *converter*, *switch-mode power supply*, SMPS, or just "switcher," §9.6) addresses these shortcomings, at the cost of some induced switching noise and greater complexity. Most switching power supplies use one or more inductors (or transformers), and one or more saturated switches (usually MOSFETs) operating at relatively high switching frequencies (50 kHz–5 MHz), to convert a dc input (which may be unregulated) to one or more stable and regulated output voltages; the latter can be lower or higher than the input voltage, or they can be of opposite polarity. The inductor(s) store and then transfer energy, in discrete switching cycles, from the input to the output, with the switch(es) controlling the conduction paths; with ideal components there would be no dissipation, and the conversion would be 100% efficient. The output error signal, suitably amplified and compensated, is used to vary either the pulse width ("PWM") or the pulse frequency ("PFM"). Switching converters can be *non-isolated* (i.e., input and output sharing a common ground, Figure 9.61), or *isolated* (e.g., when powered from the ac powerline, Figure 9.73); for each class there are dozens of topologies, see ¶D below.

A minor subclass of switching converter is the *inductorless converter* (or "charge-pump" converter; see §9.6.3), where a combination of several switches and one or more "flying" capacitors is used to create a dc output that can be a multiple of the dc input, or an output of opposite polarity (or a combination of both). For many of these, the output(s) track the dc input (i.e., unregulated), but there are also variants that regulate the output by controlling the switching cycle. See Table 9.4 and Figures 9.56 and 9.57.

¶B. The DC Input.

Regardless of the kind of converter or regulator circuit, you need to provide some form of dc input. This may be poorly regulated, as from a battery (portable equipment) or from rectified ac (line-powered equipment, Figures 9.25 and 9.48); or it may be an existing stable dc voltage already present within a circuit (e.g., Figure 9.64). For a line-powered instrument that uses a linear regulator, the "unregulated" dc input (with some ac ripple) consists of a transformer (for both galvanic isolation and voltage transformation) plus rectifier (for conversion to dc) plus bulk storage capacitor(s) (to smooth the ripple from the rectified ac). By contrast, in a line-powered (confusingly called "off-line") switcher the powerline transformer is omitted, because a transformer in the isolated switcher circuit provides galvanic isolation and it is far smaller and lighter since it operates at the much higher switching frequency; see Figure 9.48.

A diode bridge and storage capacitor converts an ac input to full-wave unregulated dc, whether from a powerline transformer or directly from the ac line. Ignoring winding resistance and inductances, the dc output voltage is approximately $V_{dc} = 1.4V_{rms} - 2V_{diode}$, and the peak-to-peak ripple voltage is approximately[152] $\Delta V_{ripple(pp)} \approx I_{load}/2fC$, where C is the capacitance of the output dc storage capacitor and f is the ac input frequency (60 Hz or 50 Hz, depending on geographic and political boundaries). The ac input current is confined to relatively short pulses during the part of the waveform leading up to the positive and negative peaks (see Figures 9.51 and 9.78). This low "power factor" waveform is undesirable, because it produces excessive I^2R heating and more stressful peak currents. For this reason all but the smallest switching converters use a power-factor correction (PFC) input stage (Figure 9.77) to spread out the current waveform and thus create an input current approximately proportional to the instantaneous ac input voltage.

Line-powered instruments need a few additional components, both for safety and convenience. These include a fuse, switch, and optional line filter and transient suppressor; these are often combined in an IEC "power entry module," see Figure 9.49.

¶C. Linear Voltage Regulators.

The basic linear voltage regulator compares a sample of the dc output voltage with an internal voltage reference (see ¶G, below) in an *error amplifier* that provides negative

---

[152] From $I = C\,dV/dt$, assuming approximately constant discharge current I.

feedback to a *pass transistor*; see Figure 9.2. The dc output voltage can be greater or less than the reference (Figures 9.4 and 9.5), but it is always less than the dc input. You can think of this as a feedback power amplifier, which is prone to instability with capacitive loads, thus the compensation capacitor C_c in Figure 9.2D,E. The final circuit in that figure shows a current limiting circuit (R_{CL} and Q_2), and also an *overvoltage crowbar* (D_1 and Q_3, see ¶J below) to protect the load in the event of a regulator fault.

All the components of the original 723-type linear voltage regulator can be integrated onto a single IC (Figure 9.6), forming a "3-terminal" fixed regulator, e.g., the classic 78xx-style (where "xx" is its output voltage). These require only external bypass capacitors (Figure 9.8) to make a complete regulator. There are only a few standard voltages available, however (e.g., +3.3 V, +5 V, +15 V); so a popular variant is the 3-terminal *adjustable* regulator (e.g., the classic 317-type, see Figure 9.9), which lets you adjust the output voltage with an external resistive divider (Figure 9.10). Figures 9.14, 9.16, and 9.18 show some application hints for this very flexible regulator. Both fixed and adjustable 3-terminal regulators are also available in negative polarities (79xx and LM337, respectively), as well as in low-current variants (78Lxx and LM317L, respectively).

One drawback of these classic linear regulators is their need for an input voltage that is at least ~2 V greater than the output (its *dropout voltage*); that is needed because their pass transistor operates as an emitter follower (thus at least a V_{BE} drop) and the current-limit circuit can drop up to another V_{BE}. Two volts may not sound like a lot, but it looms large in a low-voltage regulator circuit, e.g., one with a +2.5 V dc output. To circumvent this problem, you can use a *low-dropout* (LDO) regulator, in which the pass transistor (BJT or MOSFET) is configured as a common-emitter (or common-source) amplifier, see Figure 9.20; the resulting dropout voltages are down in the tenths of a volt (Figure 9.24). LDOs are nice, but they cost more, and they are more prone to instability because their high output impedance (collector or drain) causes a lagging phase shift into the substantial load capacitance. LDOs may require significant minimum output bypass capacitance (as much as $10\,\mu F$ or $47\,\mu F$), often constrained with both a minimum and maximum equivalent series resistance (ESR), e.g., $0.1\,\Omega$ min, $1\,\Omega$ max; see Table 9.3.

¶D. Switching Converter Topologies.

The basic *non-isolated* switcher topologies are the *buck* (or "step-down"), the *boost* (or "step-up"), and the *invert* (or "inverting buck–boost"); see §9.6.4 and Figure 9.61. The power train of these each uses one inductor, one switch, and one diode (or a second switch acting as an active rectifier), in addition to input and output storage capacitors. A complete converter requires additional components: an oscillator, comparator, error amplifier, drive circuits, and provisions for compensation and fault protection; see for example Figure 9.65. As with linear regulators, the semiconductor manufacturers have stepped in to provide most of the necessary components as packaged ICs, see Tables 9.5a,b and 9.6.

For the buck converter $V_{out} < V_{in}$, and for the boost converter $V_{out} > V_{in}$. The inverting converter produces an output of opposite polarity to the input, and whose voltage magnitude can be larger or smaller than the input voltage (this is true also of the remarkable Ćuk converter, §9.6.8H). The respective dc output voltages are $V_{out(buck)}=DV_{in}$, $V_{out(boost)}=V_{in}/(1-D)$, and $V_{out(invert)}=-V_{in}D/(1-D)$, where D is the switch-ON duty cycle $D=t_{on}/T$. There are also *non*-inverting buck-boost topologies that permit the output voltage range to bracket the input (i.e., able to go above or below the input). Examples are the 2-switch buck–boost (two switches, two diodes, one inductor), and the SEPIC (one switch, one diode, two inductors), see Figure 9.70. Of course, a switching converter with a transformer (whether isolated or not) provides flexibility in output polarity, as well as improved performance for large ratio voltage conversion.

Isolated switching converters use a transformer (for isolation), in addition to one or more inductors (for energy storage), see Figure 9.73. In the *flyback* converter (Figure 9.73A) the transformer acts also as the energy-storage inductor (thus no additional inductor), whereas in the *forward* converter and *bridge* converters (Figures 9.73B–D) the transformer is "just a transformer," and the diodes and inductor complete the energy storage and transfer. The respective dc output voltages are $V_{out(flyback)}=V_{in}[N_{sec}/N_{pri}][D/(1-D)]$ and $V_{out(forward)}=DV_{in}(N_{sec}/N_{pri})$. Speaking generally, flyback converters are used in low-power applications ($\lesssim 200\,W$), forward converters in medium-power applications (to ~500 W), and bridge converters for real power applications.

¶E. Switcher Regulation: Hysteretic, Voltage Mode, and Current Mode.

There are several ways to regulate a switching converter's dc output voltage. Simplest is *hysteretic* feedback, in which the error signal simply enables or disables successive switching cycles. It's a form of simple "bang–bang" control, with no stability issues that require a compensation network; see Figure 9.64 for a buck converter design with the popular MC34063. Proportional PWM control is better, and comes in two flavors: voltage mode and current mode. Both methods compare the output voltage with a fixed reference to regulate the output voltage, but they do it in different ways. In *voltage-mode* PWM, the output voltage error signal is compared with the internal oscillator's sawtooth waveform to control the primary switch's ON duration, whereas in *current-mode* PWM the comparison ramp is generated by the rising inductor current, with the internal oscillator used only to initiate each conduction cycle. See §9.6.9, and particularly Figures 9.71 and 9.72. In either case the controller terminates a conduction cycle if the switch exceeds a peak current, the input drops below an "undervoltage lockout" threshold, or the chip exceeds a maximum temperature. Figure 9.65 shows a simple voltage-mode PWM buck converter.

Voltage-mode and current-mode control loops both require compensation for stability, and each has its advantages and disadvantages. Current-mode controllers appear to be winning the popularity contest, owing to their better transient response, inherent switch protection (owing to pulse-by-pulse current limiting), improved outer-loop phase margin, and ability to be paralleled.

¶F. Switching Converter Miscellany.
Switching conversion is a rich subject, many details of which are well beyond the scope of this chapter (or this book). Some topics – ripple current and inductor design, core saturation and reset, magnetizing inductance and snubbing, soft start, diode recovery, CCM and DCM conduction modes, switching losses, loop compensation, burst mode, inrush current, isolation barriers, PFC, switching *amplifiers* – are treated lightly here and in Chapter *9x*. Consider this chapter's treatment of switching converters as a lengthy introduction to a specialty field that can easily consume a professional lifetime.

¶G. Voltage References.
A stable voltage reference is needed in any voltage regulator, as well as in accurate applications such as precision current sources, A/D and D/A conversion, and voltage- and current-measurement circuits. Often a good voltage reference is included in a regulator or converter IC (see for example Table 13.1), but you may want the improved performance you can get with a high-quality external reference. And, often, you need a stand-alone voltage reference for other uses in a circuit.

The simplest voltage reference is the discrete *zener diode* (§9.10.1), but most voltage references are multi-component integrated circuits that behave externally either like an extremely good zener ("2-terminal," or *shunt*; Table 9.7), or like an extremely good linear regulator ("3-terminal," or *series*; Table 9.8). Shunt references must be biased into conduction (just like a zener) by supplying current from a higher-voltage rail (use a resistor or a current source), while series references are powered by connecting their supply pin directly to the dc supply. References of either type are available in a small set of standard voltages, typically in the range of 1.25–10.0 V.

The discrete 2-terminal zener is fine for non-critical applications, but its typical accuracy of ±5% is inadequate for precision circuits. Integrated references of either kind are far better, with worst-case accuracies in the range of 0.02% to 1%, and tempcos ranging from 1 ppm/°C to 100 ppm/°C, as seen in the tables. Most integrated references are based on a circuit that temperature compensates the V_{BE} of a BJT (a so-called "bandgap reference"), generating a stable voltage of approximately 1.24 V; but others use a buried zener diode with $V_Z \approx 7$ V. The latter are generally quieter, but bandgap references can operate from low supply voltages and are widely available in voltages of 1.24 V, 2.50 V, etc. Two newer technologies with surprisingly good performance are the *JFET pinchoff reference* from ADI (the ADR400 "XFET" references), and the *floating-gate* reference from Intersil. Both exhibit very good tempco and low noise. Other important characteristics of voltage references are *regulation* (R_{out} for shunt types, PSRR for series types), minimum *load capacitance* and stability into capacitive loads, *trim* and *filter* pins, and *package* style.

¶H. Heat and Power Dissipation.
Along with power electronics comes... *heat!* You remove it with a combination of convection (air flow) and conduction (thermal contact with a heat-dissipating *heatsink*). Conductive heat flow is proportional to the temperature difference between the hot and cold sides (Newton's law of cooling), $\Delta T = P_{diss} R_\Theta$, where R_Θ is known as the "thermal resistance." For a succession of conductive joints the thermal resistances add; thus, for example, the junction temperature T_J of a power semiconductor dissipating P_{diss} watts is $T_J = T_A + P_{diss}(R_{\Theta JC} + R_{\Theta CS} + R_{\Theta SA})$, where T_A is the ambient temperature, and the successive R_Θ's represent the thermal resistances from junction to case, case to heatsink, and heatsink to ambient. Printed circuit foil patterns are often adequate for dissipation of a few watts or less (Figure 9.45); finned heatsinks or metallic chassis surfaces are used for greater heat removal, with forced airflow generally needed when the power dissipation reaches levels of 50 W or more (Figure 9.43). Semiconductor devices can

withstand considerably greater power dissipation during short pulses; this is sometimes specified as a graph of *transient thermal resistance* (i.e., R_Θ versus pulse duration and duty cycle), or as elevated contours on a plot of Safe Operating Area (see ¶I).

¶I. Safe Operating Area.
A power transistor (whether BJT or MOSFET) has specified maximum values of voltage and current, and also (because of maximum allowed junction temperature) a maximum product $V_{DS}I_D$ (i.e., power dissipation) for a given case temperature; the latter is just $V_{DS}I_D \leq (T_{J(max)} - T_C)/R_{\Theta JC}$. These limits define a *safe operating area* (SOA, §9.4.2), usually shown as contours on log–log axes of current versus voltage; see for example Figure 3.95. That plot shows two further features: (a) greater dissipation is allowed for short pulses; (b) the SOA of BJTs (but not MOSFETs) is further constrained by a phenomenon known as "second breakdown."

¶J. Overvoltage Crowbars.
Some failure modes of power converters cause output over-voltage, for example a shorted pass transistor in a linear regulator, or loss of feedback control in a switcher. This is likely to damage or destroy powered circuitry. An *overvoltage crowbar* (§9.13.1) senses overvoltage and triggers an SCR to short the output. A less brute-force technique shuts down conversion when an overvoltage is sensed; these are indicated in the "OVP" column in Table 9.5b.

¶K. Current Sources.
"Regulator" usually means a stable *voltage* source; but there are many uses for a controllable *current* source (§9.3.14). Three-terminal linear regulators are easily coaxed into current-source service (§9.3.14A). There are also dedicated current-source ICs like the LM334 and REF200. JFETs make convenient 2-terminal current sources, and depletion-mode MOSFETs make excellent current sources that can operate up to voltages as high as 1 kV, see §9.3.14C. And don't forget about the humble discrete BJT current source (§§2.2.6 and 2.3.7B), or the op-amp current-source circuits (§4.2.5: Howland; op-amp + transistor).

DIGITAL LOGIC

10.1 Basic logic concepts

10.1.1 Digital versus analog

Thus far we have been dealing mainly with circuits in which the input and output voltages have varied over a continuous range of values: *RC* circuits, amplifiers, integrators, rectifiers, op-amps, etc. This is natural when dealing with signals that are continuous (e.g., audio signals) or continuously varying voltages from measuring instruments (e.g., temperature-reading or light-detecting devices, or biological or chemical probes).

However, there are instances in which the input signal is naturally discrete in form, e.g., pulses from a particle detector, or "bits" of data from a switch, keyboard, or computer. In such cases the use of digital electronics (circuits that deal with data made of 1s and 0s) is natural and convenient. Furthermore, it is often desirable to convert continuous (analog) data to digital form, and vice versa, using analog-to-digital converters (ADCs) and digital-to-analog converters (DACs), in order to perform computations on the data (with a computer or signal processor) or to store large quantities of data as numbers. In a typical situation a microprocessor or computer might monitor signals from an experiment or industrial process, control the experimental parameters on the basis of the data obtained, and store for future use the results collected or computed while the experiment is running.

Another interesting example of the power of digital techniques is the transmission of analog signals without degradation by noise: an analog audio or video signal, for instance, picks up "noise" while being transmitted by cable or wireless that cannot be removed. If, instead, the signal is converted to a series of numbers representing its amplitude at successive instants of time, and these numbers are transmitted as digital signals, the analog signal reconstruction at the receiving end (done with DACs) will be without error, providing the noise level on the transmission channel isn't great enough to prevent accurate recognition of 1s and 0s. This technique, known as PCM (pulse-code modulation), is particularly attractive where a signal must pass through a series of "repeaters," since digital regeneration at each stage guarantees noiseless transmission. The information and stunning pictures sent back by planetary deep space probes, for example the Pioneer 10 mission to Jupiter in 1973, were stored and transmitted with PCM. Digital audio and video are now commonplace in the home, for example in the form of the humble 12 cm optical CD,[1] in which music is stored in the form of a stereo pair of 16-bit numbers every 23 microseconds, (1.4 megabits/sec), 6 billion bits (gigabits, Gb) of information in all. And by contemporary standards this is a storage medium of low speed and capacity: the corresponding figures for DVDs and Blu-rays are 10 and 48 megabits/sec (maximum), with a total storage of 38 and 200 gigabits per layer.

In fact, digital hardware has become so powerful that tasks that would have seemed well suited to analog techniques are usually better solved with digital methods. As an example, an analog temperature meter might incorporate a microprocessor and memory to improve accuracy by compensating for the instrument's departure from perfect linearity; likewise for something as commonplace as a digital bathroom scale. Because of the wide availability of inexpensive (less than $1) microcontrollers, such applications are ubiquitous. The average home is awash in devices with "embedded" processors: in every music player, television set, cellphone, dishwasher, washing machine and dryer, fax, copier, microwave oven, coffee maker,.... The list is long, and sometimes surprising.[2] Rather than attempt to enumerate what can be done with digital electronics, let's just start learning about it. Applications will emerge naturally as we go along.

---

[1] And their video counterparts: DVDs and Blu-ray (BD) discs.

[2] One might encapsulate the history of the "conquest of digital over analog" something like this: the 1970s – planetary photographs sent digitally from 800 million kilometers away, but domestic life still ruled by analog; the 1980s – digital audio (CDs) and the personal computer invade the home; the 1990s – digital video (DVDs), cellphones, and MP3s; the 2000s – HDTV, digital wireless connectivity, digital photography and videography, high-speed Internet, and *Google*; and the 2010s (OK, OK, we know the old saying "It's difficult to predict...especially the future") – universal digital dissemination and fusion of information and media.

10.1.2 Logic states

By "digital electronics" we mean circuits in which there are only two (usually) states possible at any point, e.g., a transistor that can either be in saturation or be nonconducting. We usually choose to talk about voltages rather than currents, calling a level HIGH or LOW. The two states can represent any of a variety of "bits" (binary digits) of information, such as the following:

one bit of a number;
whether a switch is opened or closed;
whether a signal is present or absent;
whether some analog level is above or below some preset limit;
whether or not some event has happened yet;
whether or not some action should be taken;
and so on.

A. HIGH and LOW

The HIGH and LOW voltage states represent the TRUE and FALSE states of Boolean logic, in some predefined way. If at some point a HIGH voltage represents TRUE, that signal line is called "active HIGH" (or "HIGH true") and vice versa. This can be confusing at first. Figure 10.1 shows an example. SWITCH CLOSED is true when the output is LOW; that's an "active-LOW" (or "LOW-true") signal, and you might label the lead as shown (a bar over a symbol means NOT; that line is HIGH when the switch is *not* closed). Just remember that the presence or absence of the negation bar over the label tells whether the wire is at a LOW or HIGH voltage state when the stated condition (SWITCH CLOSED) is true.[3] At first the idea of active-LOW may seem, well, *backward*; why not just keep it simple and outlaw this upside-down logic? As we'll see, though, there are good reasons to do things "backwards" sometimes. Be patient.

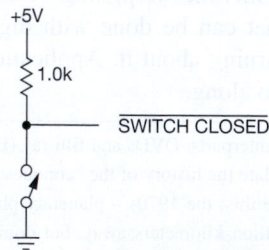

Figure 10.1. A LOW-true ("active-LOW") logic level.

A digital circuit "knows" what a signal represents by where it comes from, just as an analog circuit might know what the output of some op-amp represents. However, added flexibility is possible in digital circuits; sometimes the same signal lines are used to carry different kinds of information, or even to send it in different directions, at different times. To do this "multiplexing," additional information must also be sent (address bits, or status bits). You will see many examples of this very useful ability later. For now, imagine that any given circuit is wired up to perform a predetermined function and that it knows what that function is, where its inputs are coming from, and where the outputs are going.

To lend a bit of confusion to a basically simple situation, we introduce 1 and 0. These symbols are used in Boolean logic to mean TRUE and FALSE, respectively, and are sometimes used in electronics in exactly that way. Unfortunately, they are also used in another way, in which 1 = HIGH and 0 = LOW! In this book we try to avoid any ambiguity by using the word HIGH (or the symbol H) and the word LOW (or the symbol L) to represent logic states, a method that is in wide use in the electronics industry. We use 1 and 0 only in situations where there can be no ambiguity.

B. Voltage range of HIGH and LOW

In digital circuitry, the voltage levels corresponding to HIGH and LOW are allowed to fall in some range, according to the particular logic family.[4] For example, with high-speed CMOS ("HC" family) logic running from a +5 V supply, *input* voltages within about 1.5 volts of ground are interpreted as LOW, whereas voltages within 1.5 volts of the +5 V supply are interpreted as HIGH.[5] Those inputs are driven from the outputs of some other devices, for which typical LOW- and HIGH-state *output* voltages are usually within a tenth of a volt of 0 and +5 V, respectively (the output is a saturated transistor switch to one of the rails; see Figure 10.25). This allows for manufacturing spread,

---

[3] You'll sometimes see the terms "positive-true" and "negative-true" used for HIGH-true and LOW-true, respectively. Those terms are OK, but they can be confusing to the inexperienced, especially since no negative voltages are involved.

[4] A "family" is a particular hardware implementation of digital logic, characterized by operating voltage, logic voltage levels, and speed. For historical reasons, most logic families name their standard logic parts with a prefix of 74, followed by a few letters that name the family, and ending with some numbers that specify the logic function. The logical functions themselves are the same across families. For example, a 74LVC08 is a 2-input AND gate (actually, four of them in one package) in the low-voltage CMOS (LVC) electrical family; the "08" designates a quad 2-input AND, and the "74" designates a logic chip for operation at standard temperatures.

[5] See the box "Logic Levels" for additional examples.

variations of the circuits with temperature, loading, supply voltage, etc., and the presence of "noise," the miscellaneous garbage that gets added to the signal in its journey through the circuit (from capacitive or inductive coupling, external interference, etc.). The circuit receiving the signal decides if it is HIGH or LOW and acts accordingly.[6] As long as noise does not change 1s to 0s, or vice versa, all is well, and any noise is eliminated at each stage, where "clean" 0s and 1s are regenerated. In that sense, digital electronics is noiseless and perfect.

The term *noise immunity* is used to describe the maximum noise level that can be added to logic levels (in the worst case) while still maintaining error-free operation. As an example, the formerly popular family of logic known as "TTL" (transistor–transistor logic) struggled with just this issue, because it has just 0.4 V of noise immunity: a TTL *input* is guaranteed to interpret anything less than +0.8 V as LOW and anything greater than +2.0 V as HIGH, whereas the worst-case *output* levels are +0.4 V and +2.4 V, respectively (see the accompanying box on logic levels). In practice, noise immunity will be better than the worst-case 0.4 V margin, with typical LOW and HIGH voltages of +0.2 V and +3.4 V and an input decision threshold near +1.3 V. But always remember that if you are doing good circuit design, you use worst-case values. It is worth keeping in mind that different logic families have different amounts of noise immunity. CMOS has greater voltage-noise immunity than TTL, whereas the speedy ECL (emitter-coupled logic) family has less. Of course, susceptibility to noise in a digital system depends also on the amplitude of noise that is present, which in turn depends on factors such as output-stage stiffness, inductance in the ground leads, the existence of long "bus" lines, and output slew rates during logic transitions (which produce transient currents, and therefore voltage spikes on the ground line, because of capacitive loading). We will worry about some of these problems in Chapter 12 (Logic Interfacing).

10.1.3 Number codes

Most of the conditions we listed earlier that can be represented by a digital level are self-explanatory. How a digital

---

[6] Sometimes digital signals are sent as a differential voltage pair, rather than "single-ended." This is particularly popular with longer high-speed signals, or signals that go off-board over some distance, for example fast serial buses such as USB, Firewire, and SATA; it is also commonly used to distribute high-frequency clock signals. A popular format is "LVDS" (low-voltage differential signaling), in which the differential signal amplitude is ~0.3 V, centered at +1.25 V.

level can represent part of a number is a more involved, and very interesting, question. Put another way, we've seen *bits as indicators*; now we will see *groups of bits as a number*.

A decimal (base-10) number is simply a string of integers that are understood to multiply successive powers of 10, the individual products then being added together. For instance,

$$137.06 = 1 \times 10^2 + 3 \times 10^1 + 7 \times 10^0 + 0 \times 10^{-1} + 6 \times 10^{-2}.$$

Ten symbols (0 through 9) are needed, and the power of 10 each multiplies is determined by its position relative to the decimal point. If we want to represent a number using two symbols only (0 and 1), we use the *binary*, or base-2, number system. Each 1 or 0 then multiplies a successive power of 2. For instance,

$$1101_2 = 1 \times 2^3 + 1 \times 2^2 + 0 \times 2^1 + 1 \times 2^0$$
$$= 13_{10}.$$

The individual 1s and 0s are called "bits" (binary digits). The subscript (always given in base 10) tells what number system we are using, and often it is essential in order to avoid confusion, since the symbols all look the same.

We convert a number from binary to decimal by the method just described. To convert the other way, we keep dividing the number by 2 and write down the remainders. To convert 13_{10} to binary, therefore:

$$13/2 = 6 \quad \text{remainder} \quad 1,$$
$$6/2 = 3 \quad \text{remainder} \quad 0,$$
$$3/2 = 1 \quad \text{remainder} \quad 1,$$
$$1/2 = 0 \quad \text{remainder} \quad 1,$$

from which $13_{10} = 1101_2$. Note that the answer comes out in the order LSB (least significant bit) to MSB (most significant bit).

A. Hexadecimal ("hex") representation

The binary-number representation is the natural choice for two-state systems (although it is not the only way; we'll see some others soon). Because the numbers tend to get rather long, it is common to write them in hexadecimal (base-16) representation: each position represents successive powers of 16, with each hex symbol having a value from 0 to 15 (the symbols A–F are assigned to the values 10–15). To write a binary number in hexadecimal, just group it in 4-bit groups, beginning with the LSB, and write the hexadecimal equivalent of each group:

$$707_{10} = 1011000011_2 \, (= 10\ 1100\ 0011_2) = 2C3_{16}.$$

LOGIC LEVELS

The diagram in Figure 10.2 shows the ranges of voltages that correspond to the two logic states (HIGH and LOW) for popular families of digital logic. For each logic family it is necessary to specify legal values of both output and input voltages corresponding to the two states HIGH and LOW. The shaded areas above the line show the specified range of output voltages within which a logic LOW or HIGH is guaranteed to fall, with the pair of arrows indicating typical output values (LOW, HIGH) encountered in practice. The shaded areas below the line show the range of input voltages guaranteed to be interpreted as LOW or HIGH, with the arrow indicating the typical *logic threshold* voltage, i.e., the dividing line between LOW and HIGH. In all cases a logic HIGH is more positive than a logic LOW. Table 10.1 and Figure 10.26 provide additional information about these families, a subject we'll see in more detail in Chapter 12.

The meanings of "minimum," "typical,"and "maximum," in electronic specifications are worth a few words

of explanation. Most simply, the manufacturer guarantees that the components will fall in the range minimum-to-maximum, with many close to "typical." What this means is that typical specifications are an approximate guide to use when designing circuits; however, those circuits must work properly over the whole range of specifications from minimum to maximum (the extremes of manufacturing variability). In particular, a well-designed circuit must function under the worst possible combination of minimum and maximum values. This is known as *worst-case design*, and it is essential for any instrument produced from off-the-shelf (i.e., not specially selected) components.

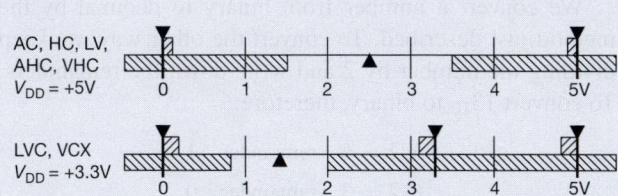

Figure 10.2. Logic levels of some popular logic families.

Table 10.1 Selected Logic Families

Family	Manufacturers	Supply voltage (V_{cc}) min (V)	max (V)	V_{in} max	t_{pd} @ Vcc (ns, typ)	(V)	Available packages DIP	SMT	1G, 2G
74HC00	9	2	6	V_{cc}	9	5	●	●	●
74AC00	5	3	6	V_{cc}	6	5	●	●	-
74AHC00	2	2	5.5	V_{cc}	3.7	5	●	●	●
74LV00	2	1.2[a]	5.5	5.5	3.6	5	●	●	●
74LVC00	4	1.7	3.6	5.5	3.5	3.3[b]	-	●	●
74ALVC00	3	1.7	3.6	3.6	2	3.3[c]	-	●	-
74AUC00	1	0.8	2.7	3.6	0.9	1.8	-	●	●

Notes: (a) NXP only (TI specs to 2V only). (b) 6ns @ 1.8V. (c) 2.7ns @ 1.8V.

Hexadecimal representation[7] is well suited to the popular "byte" (8-bit) organization of computers, which are most often organized as 16, 32, or 64-bit computer "words"; a word is then 2, 4, or 8 bytes. So in hexadecimal, each byte is 2 hex digits, a 16-bit word is 4 hex digits, etc. For example, the memory locations in a microcontroller with 65,536 ("64K") bytes of memory can be identified by a 2-byte address, because 2^{16}=65,536; the lowest address in hex is 0000h (the trailing "h" means hex), the highest address is FFFFh, the second half of memory begins at 8000h, and the fourth quarter of memory begins at C000h.

A byte sitting somewhere in computer memory can represent an integer number, or part of a number. But it can

also represent other things: for example, an alphanumeric character (letter, number, or symbol) is commonly represented as one byte. In the widely used ASCII representation (more in §14.7.8), lowercase "a" is represented as ASCII value 01100001 (61h), "b" is 62h, etc. Thus the word "nerd" could be stored in a pair of 16-bit words whose hex values are 6E65h and 7264h.

B. Binary-coded decimal

Another way to represent a number is to encode each decimal digit into binary. This is called BCD (binary-coded decimal), and it requires a 4-bit group for each digit. For instance,

$$137_{10} = 0001\,0011\,0111 \quad \text{(BCD)}.$$

---

[7] Alternative notations for a hexadecimal number (like $2C3_{16}$) are 2C3H, 2c3h, 2c3h, and 0x2C3.

Note that BCD representation is *not* the same as binary representation, which in this case would be $137_{10} = 10001001_2$. You can think of the bit positions (starting from the right) as representing 1, 2, 4, 8, 10, 20, 40, 80, 100, 200, 400, 800, etc. It is clear that BCD is wasteful of bits, because each 4-bit group *could* represent numbers 0 through 15, but BCD never represents numbers greater than 9. However, BCD is ideal if you want to display a number in decimal, since all you do is convert each BCD character to the appropriate decimal number and display it. For this reason, BCD is commonly used for input and output of numeric information. Unfortunately, the conversion between pure binary and BCD is complicated, because *each* decimal digit depends on the state of almost every binary bit, and vice versa. Nevertheless, binary arithmetic is so efficient that most computers convert all input data to binary, converting back only when data need to be output. Think how much effort and bother would have been saved if *Homo sapiens* had evolved with 8 (or 16) digits.[8]

Exercise 10.1. Convert to decimal: (a) 1110101.0110_2, (b) $11.01010101\ldots_2$, (c) $2A_H$. Convert to binary: (a) 1023_{10}, (b) 1023_H. Convert to hexadecimal: (a) 1023_{10}, (b) 101110101101_2, (c) 61453_{10}.

Table 10.2 4-Bit Signed Integers in Three Systems of Representation

Integer	Sign-magnitude	Offset binary	2's comp
+7	0111	1111	0111
+6	0110	1110	0110
+5	0101	1101	0101
+4	0100	1100	0100
+3	0011	1011	0011
+2	0010	1010	0010
+1	0001	1001	0001
0	0000	1000	0000
-1	1001	0111	1111
-2	1010	0110	1110
-3	1011	0101	1101
-4	1100	0100	1100
-5	1101	0011	1011
-6	1110	0010	1010
-7	1111	0001	1001
-8	-	0000	1000
(-0)	1000	-	-

C. Signed numbers
Sign–magnitude representation
Sooner or later it becomes necessary to represent negative numbers in binary, particularly in devices where some computation is done. The simplest method is to devote one bit (the MSB, say) to the sign, with the remaining bits representing the magnitude of the number. This is called "sign–magnitude representation," and it corresponds to the way signed numbers are ordinarily written (see Table 10.2). It is used when numbers are displayed, as well as in some ADC schemes. In general, it is not the best method for representing signed numbers (except for floating-point numbers), particularly where some computation is done, for several reasons: computation is awkward, and subtraction is different from addition (i.e., addition doesn't "work" for signed numbers). Also, there can be two zeros (+0 and −0), so you have to be careful to use only one of them.

Offset-binary representation
A second method for representing signed numbers is "offset binary," in which you subtract half the largest possible number to get the value represented (Table 10.2). This has the advantage that the number sequence from the most negative to the most positive is a simple binary progression, which makes it a natural for binary "counters." The MSB still carries the sign information, and zero appears only once. Offset binary is popular in A/D conversions, but it is still awkward for computation.

2's complement representation
The method most widely used for integer computation is called "2's complement." In this system, positive numbers are represented as simple unsigned binary. The system is rigged up so that a negative number is then simply represented as the binary number you add to a positive number of the same magnitude to get zero. To form a negative number, first complement each of the bits of the positive number (i.e., write 1 for 0, and vice versa; this is called the "1's complement"), then add 1 (that's the "2's complement").[9] As you can see from Table 10.2, 2's complement numbers are related to offset binary numbers by having the MSB complemented. As with the other signed-number representations, the MSB carries the sign information. There's only one zero, conveniently represented by all bits 0 ("clearing" a counter or register sets its value to zero). Because the 2's complement system is natural for computation, (i.e., it allows computers to treat negative and

---

[8] If the former, we'd be thumbless, thus spared the scourge of "texting."

[9] Alternatively, you can simply think of it as plain ol' (unsigned) binary, but with the MSB representing the negative of its usual value.

positive integers the same way), it is universally used for integer arithmetic in computers.[10]

D. Arithmetic in 2's complement

Arithmetic is simple in 2's complement. To add two numbers, just add bitwise (with carry), like this:

$5+(-2)=3$:

$$
\begin{array}{rr}
0101 & (+5) \\
+1110 & (-2) \\
\hline
0011 & (+3)
\end{array}
$$

To subtract B from A, take the 2's complement of B and add (i.e., add the negative):

$2-5=-3$:

$$
\begin{array}{rr}
0010 & (+2) \\
+1011 & (-5) \\
\hline
1101 & (-3)
\end{array}
$$

(Note in this last example that $+5=0101$, so its 1's comp is 1010 and its 2's comp is $-5=1011$.) Multiplication also "works right" in 2's complement representation. Try the following exercises.

Exercise 10.2. Multiply $+2$ by -3 in 3-bit 2's complement binary arithmetic. *Hint:* the answer is -6.

Exercise 10.3. Show that the 2's complement of of -5 is $+5$.

Because an n-bit integer can represent only 2^n numbers, you can get overflow or underflow when you add or subtract two numbers of fixed word size. To be precise, an n-bit unsigned integer can take on values from 0 to 2^n-1, and an n-bit 2's complement signed integer can take on values from -2^{n-1} to $+2^{n-1}-1$. For 8-bit integers those ranges are 0 to 256, and -128 to $+127$. To determine if an unsigned add has overflowed (and is therefore incorrect), simply note if it generated a carry out of the MSB. For signed 2's complement numbers the analogous criterion is slightly weird: if the sign bit (the MSB) is changed *by carries* (that is, if a carry in to the MSB is not balanced by a carry out, or vice versa), the result is incorrect.

Exercise 10.4. Check this unlikely criterion by doing a 2's complement addition of each of the following signed-number pairs, assuming a 4-bit word size: $7+(-6)$; $7+7$; $7+4$; $-7+(-8)$. Then repeat the same sums with a 5-bit word size, in which all the answers should fit.

E. Gray code

The following code is used for mechanical linear and shaft-angle encoders and for certain ADCs, among other things.

It is called a Gray code,[11] and it has the property that only one bit changes in going from one state to the next. This prevents incorrect codes at the transitions, which otherwise produce errors because there is no way of guaranteeing that all bits will change simultaneously at the boundary between two encoded values. If straight binary were used, it would be possible to generate an output of 7 in going from 3 to 4, for example. Here is a simple rule for generating Gray-code states: begin with a state of all zeros. To get to the next state, always change the single least significant bit that brings you to a new state.

State	Binary	Gray
0	000	000
1	001	001
2	010	011
3	011	010
4	100	110
5	101	111
6	110	101
7	111	100

Figure 10.3 shows how a Gray-coded angle encoder eliminates false codes at the transitions. Gray codes can be generated with any number of bits. They find use also in "parallel encoding" (also called *flash conversion*), a technique of high-speed A/D conversion that we will see later. We will talk about translation between Gray-code and binary-code representations in the next section (including a gate implementation in Figure 10.10).

10.1.4 Gates and truth tables

A. Combinational versus sequential logic

In digital electronics the name of the game is generating digital outputs from digital inputs. For instance, an *adder* might take two 16-bit numbers as inputs and generate a 16-bit (plus carry) sum. Or you might build a circuit to multiply two numbers. These are the kinds of operations a computer's processing unit should be able to do. Another task might be to compare two numbers to see which is larger, or to compare a set of inputs with the desired input to make sure that "all systems are go." Or you might want to compute a "parity bit" and attach it to a number to make the

---

[10] Note, however, that "floating-point" numbers are usually represented in a form of "sign magnitude," namely, sign-exponent-mantissa.

[11] Frank Gray, of the Bell Telephone Laboratories, received a patent for *Pulse Code Modulation* in 1953. It disclosed a "novel code" for A/D conversion, which he called a "reflected binary code," but known thereafter as a Gray code. These were evidently discovered earlier, however, by none other than Emile Baudot (from which "baud" derives), who used them in telegraphy in 1878.

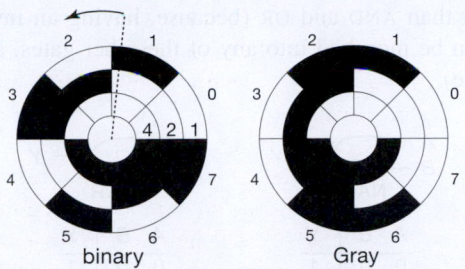

Figure 10.3. Two versions of a 3-bit angle encoder. Black or white sectors represent 1 or 0 for each of the three encoded bits (of value 4, 2, and 1 for the binary encoding). Binary encoding is prone to errors at transitions like 1→2 (shown), where more than one bit must change simultaneously; Gray code circumvents this problem.

total number of 1s even, say, before transmission over a data link; then the parity could be checked on receipt as a simple check of correct transmission. Another typical task is to take some numbers expressed in binary and display or print them as decimal characters. All of these are tasks in which the output or outputs are predetermined functions of the input or inputs. As a class, they are known as *combinational*[12] tasks. They can all be accomplished with devices called *gates*, which perform the operations of Boolean algebra applied to two-state (binary) systems.

There is a second class of problems that cannot be solved by forming a combinational function of the inputs alone, but require knowledge of past inputs as well. Their solution requires the use of *sequential* networks. Typical tasks of this type might be converting a string of bits in serial form (one after another) into a parallel set of bits, or keeping count of the number of 1s in a sequence, or recognizing a certain pattern in a sequence, or giving one output pulse for each four input pulses, or controlling the state of a system as time goes on. All these tasks require digital memory of some sort. The basic device here is the "flip-flop" (the fancy name is "bistable multivibrator").

We begin with gates and combinational logic, since they're basic to everything. Digital life will become more interesting when we get to sequential devices, but there will be no lack of fun and games with gates alone.

B. OR gate

The output of an OR gate is HIGH if either input (or both) is HIGH. This can be expressed in a *truth table*, as shown in Figure 10.4. The gate illustrated is a 2-input OR gate. In general, gates can have any number of inputs, but, when packaged in the form of "standard-logic" ICs, you get from

---

[12] Sometimes called *combinatorial*.

one to four gates in a single IC package.[13] For instance, a 4-input OR gate will have a HIGH output if any one input (or more) is HIGH.

The Boolean symbol for OR is +. "*A* OR *B*" is written as $A + B$ (in text), or as A | B (in the coding languages Verilog or C).

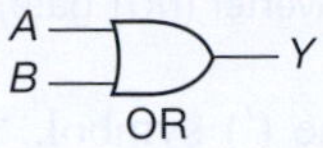

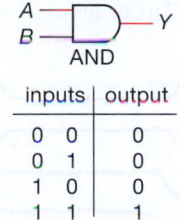

inputs		output
A	*B*	*Y*
0	0	0
0	1	1
1	0	1
1	1	1

Figure 10.4. 2-input OR gate and truth table.

C. AND gate

The output of an AND gate is HIGH only if both inputs are HIGH. The logic symbol and truth table are as shown in Figure 10.5. As with OR gates, AND gates are available with 3 or 4 (sometimes more) inputs. For instance, an 8-input AND gate will have a HIGH output only if *all* inputs are HIGH.

inputs		output
0	0	0
0	1	0
1	0	0
1	1	1

Figure 10.5. 2-input AND gate and truth table.

The Boolean symbol for AND is a dot ($\cdot$); this can be omitted, and usually is. "*A* AND *B*" is written $A \cdot B$, or simply AB (in text), or as A & B (in Verilog or C).

D. Inverter (the NOT function)

Frequently we need the complement of a logic level. That is the function of an inverter, a "gate" with only one input (Figure 10.6).

The Boolean symbol for NOT is a bar over the symbol,

---

[13] The alternative to standard logic is "programmable logic," in which you can connect up your own network of gates, up to hundreds of thousands within a single inexpensive IC. We'll have plenty to say about this later in the chapter, beginning at §10.5.4, and in detail in Chapter 11.

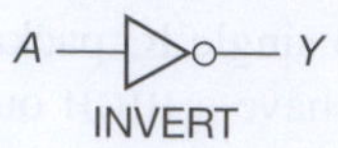

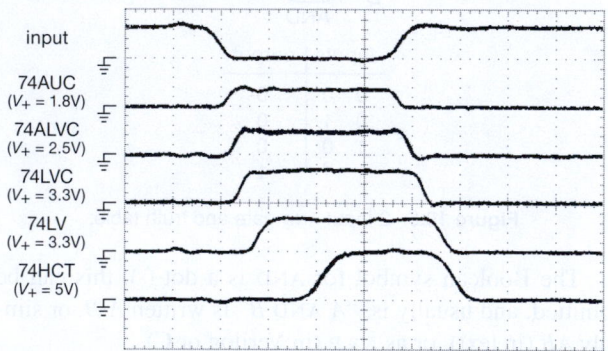

Figure 10.6. Inverter (NOT gate) and truth table.

or sometimes a prime (′) symbol. "NOT *A*" is written $\overline{A}$ or *A*′. For the convenience of typesetters, the symbols /, *, −, and ′ are often used in place of the overbar, to indicate NOT; thus, "NOT *A*" might be written as any of the following: *A*′, −*A*, ∗*A*, /*A*, *A*∗, *A*/. A given document will usually pick one of these alternatives and stick with it throughout. We have chosen the form *A*′ for this book. In coding languages NOT is written ! or ∼ .

An aside: propagation time

In the real world, logic devices like gates and inverters do not operate instantaneously when presented with a changed input level: it takes a *propagation time* (t_p) for the news to get from the input to the output. Figure 10.7 shows actual 'scope traces for inverter outputs from five logic families, when driven with a 15 ns active-LOW pulse. The newer low-voltage families (74AUC, 74AVLC) are fastest, with propagation times of 2 ns or less.

Figure 10.7. Real logic gates take a few nanoseconds to respond to a changing input. We drove inverters from five popular logic families with the active-LOW "input" pulse, and observed the outputs shown. The slowest family (74HCT, driven with a 5V swing) exhibits delays of 9 ns and 5 ns from leading and trailing edges, respectively, resulting in a shortened output pulse width. Vertical: 5 V/div; Horizontal: 4 ns/div.

E. NAND and NOR

The INVERT function can be combined with gates, forming NAND and NOR (Figure 10.8), which are somewhat more

popular than AND and OR (because, having an inversion, they can be morphed into any of the other gates, as we'll soon see).

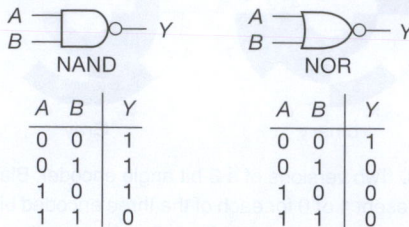

Figure 10.8. NAND and NOR, and truth tables.

F. Exclusive-OR

Exclusive-OR (XOR) is an interesting function, although less fundamental than AND and OR (Figure 10.9). The output of an XOR gate is HIGH if one or the other (but not both) input is HIGH (it never[14] has more than two inputs). Another way to say it is that the output is HIGH if the inputs are different. The XOR gate is identical with modulo-2 addition of two bits. The Boolean symbol for XOR is ⊕. "*A* XOR *B*" is written A ∧ B (in Verilog or C).[15]

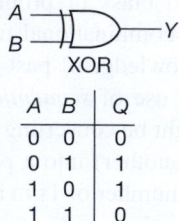

Figure 10.9. XOR and truth table.

Exercise 10.5. Show how to use the exclusive-OR gate as an "optional inverter," i.e., it inverts an input signal or buffers it without inversion, depending on the level at a control input.

Exercise 10.6. Verify that the circuits in Figure 10.10 convert binary code to Gray code, and vice versa.

---

[14] Well, hardly ever: the '1G386 claims to be a "3-input XOR gate," though we would call it a 3-input parity generator.

[15] Where, however, the ugly symbol ^ is used, probably because it happens to be included on standard keyboards, there intended as a foreign language accent, but evidently hijacked by the playful authors of programming languages. In the text we use the nicer-looking "wedge" symbol ∧, ignoring the derisive howls of C-language purists.

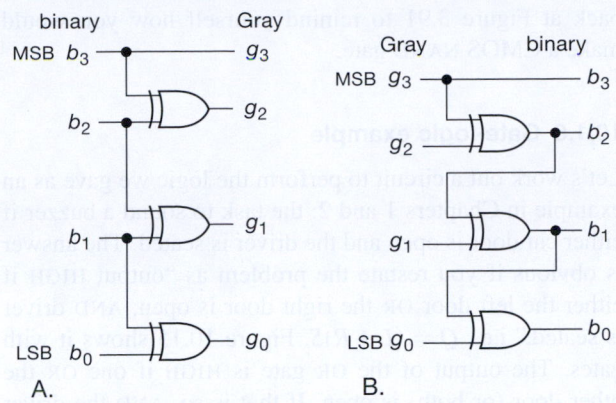

Figure 10.10. Parallel code converters: binary to Gray and Gray to binary.

	AND	OR	NOT	XOR
In text	A B	A + B	$\overline{A}$	A $\oplus$ B
ABEL	A & B	A # B	!A	A $ B
Verilog	A & B	A | B	~A	A $\wedge$ B
VHDL	A and B	A or B	not A	A xor B

Figure 10.11. Syntax for logical operations, as expressed in hardware description languages or in running text. For these operators CUPL shares the same symbols as ABEL. Alternative symbols for use in text are are the following: for AND A·B; for NOT any of the following: A', A*, *A, /A, A/, −A.

G. Basic logic: hardware description language

Thus far we've been using *schematic symbols* for the basic logic of gates. However, when you use *programmable logic devices* (PLDs: Chapter 11), rather than prefabricated standard logic, you have to enter, as text, the logic functions that you want to implement. This is done in a *hardware description language* (HDL), with names like Verilog or VHDL. Dedicated software then swallows and converts these expressions, creating a file that is used to program the actual part (or, for large volume applications, to create a "full custom" IC). The nomenclature shown in Figure 10.11 shows how these basic logical operations are expressed in the several programming languages used for programmable logic devices and custom ICs.

As a simple example, the Gray-to-binary logic shown schematically in Figure 10.10B could be written in the Ver-

ilog HDL (following some boring declarations) as:

```
assign b3 = g3;
assign b2 = g2 ∧ g3;
assign b1 = g1 ∧ (g2 ∧ g3);
assign b0 = g0 ∧ (g1 ∧ (g2 ∧ g3));
```

10.1.5 Discrete circuits for gates

Before going on to discuss gate applications, let's see how to make gates from discrete components. Figure 10.12 shows a diode AND gate. If either input is held LOW, the output is LOW. The output can go HIGH only when both inputs go HIGH. This circuit does work, but it has many disadvantages. In particular: (a) its LOW output is a diode drop above the signal holding the input LOW – obviously you couldn't use very many of these in a row! (b) there is no "fan-out" (the ability of one output to drive several inputs), since any load at the output is seen by the signal at the input; (c) it is slow, because of resistive pullup.

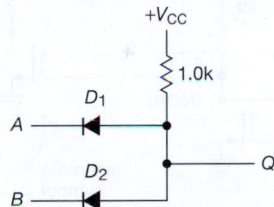

Figure 10.12. Diode AND gate.

Figure 10.13 shows how to fix some of these drawbacks, by using a pair of *npn* transistor switches to make a NOR gate, followed by an inverter to make it into an OR gate.[16] A HIGH at either input (or both) turns on at least one of the input transistors, pulling their common output LOW. Since this portion of the gate is intrinsically inverting (it's a NOR), you add an inverter, as shown, to make it into a (noninverting) 2-input OR gate.

Logic circuits using bipolar transistors have been almost entirely superseded by MOS circuits. Figure 10.14 shows the analogous NOR/OR gate circuit, with *n*-channel MOS transistor switches replacing the bipolar *npn* switches of Figure 10.13. The MOS implementation has the advantage of not requiring input current (though the input capacitance means that you do have to supply current during input transitions). But it still has some drawbacks, such as limited switching speed and significant power dissipation (due to

---

[16] This circuit was used in the family of logic known as RTL (resistor–transistor logic), which was popular in the 1960s because of its low price, but is now thoroughly obsolete.

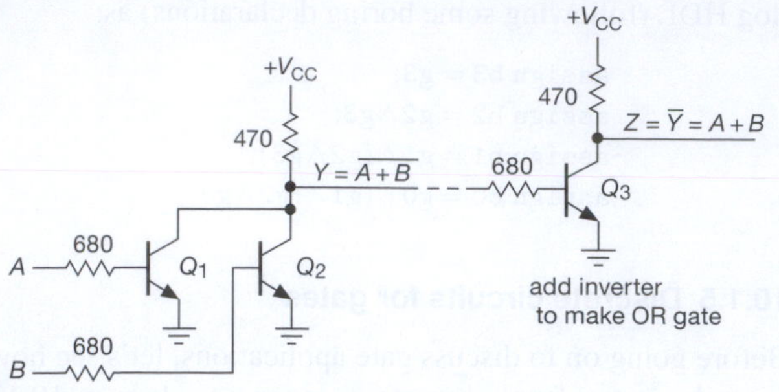

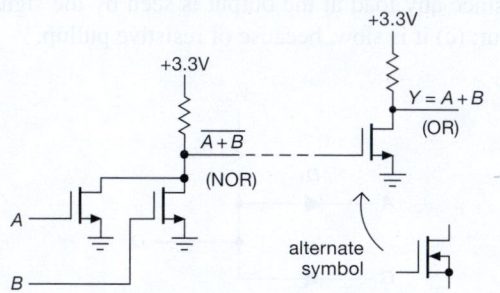

Figure 10.13. Resistor–transistor logic: NOR input stage, followed by inverter, makes a 2-input OR gate.

resistive pullup resistors). Both problems are elegantly finessed by the use of complementary MOS ("CMOS") transistors in a push–pull arrangement, as we discussed earlier in §3.4.4A.

Figure 10.14. An nMOS OR gate. Digital circuit designers simplify the MOSFET symbol, always omitting the substrate terminal; they usually center the gate attachment, too, and indicate polarity instead with a bubble on the gate, as shown in the bottom pair of Figure 3.6.

Although the discrete gate circuits just illustrated are simple to understand, you wouldn't use them in practice because of their disadvantages. In fact, except in rare circumstances[17] you would never construct gates (or any other logic) from discrete components, since a full range of excellent logic is available as inexpensive and compact ICs, as we will see shortly. Currently nearly all IC logic circuits are built with complementary MOSFETs (CMOS). Look

---

[17] In some industrial and consumer applications, where ruggedness and simplicity are important, you'll find circuits like Figure 10.13, constructed from discrete components. For this purpose you can get transistors with integrated base resistors (or base divider); these are called "bias resistor transistors" (BRTs), "pre-biased transistors," or, sometimes, "digital transistors." They're ridiculously inexpensive, as little as $0.02 apiece in quantity. They're made by companies like ON semiconductor, Diodes Inc., and Rohm.

back at Figure 3.91 to remind yourself how you would make a CMOS NAND gate.

10.1.6 Gate-logic example

Let's work out a circuit to perform the logic we gave as an example in Chapters 1 and 2: the task to sound a buzzer if either car door is open and the driver is seated. The answer is obvious if you restate the problem as "output HIGH if either the left door OR the right door is open, AND driver is seated," i.e., $Q = (L+R)S$. Figure 10.15 shows it with gates. The output of the OR gate is HIGH if one OR the other door (or both) is open. If that is so, AND the driver is seated, Q goes HIGH. With an additional transistor, this could be made to sound a buzzer or close a relay.[18]

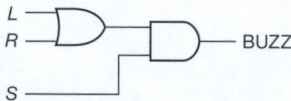

Figure 10.15. Car-door example: active-HIGH levels.

In practice, the switches generating the inputs will probably close a circuit to ground, to save extra wiring (among other reasons). This means, for example, that the inputs will go LOW when a door is opened. In other words, we have "active-LOW" inputs. Let's rework the example with this in mind, calling the inputs L', R', and S'.

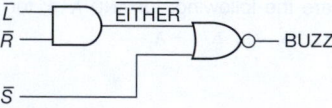

Figure 10.16. Car-door example: active-LOW levels cause confusion.

First, we need to know if either door input (L', R') is LOW; i.e., we must distinguish the state "both inputs HIGH" from all others. That's an AND gate. So we make L' and R' the inputs to an AND gate. The output will be LOW if either input is LOW; call that EITHER'. Now we need to know when EITHER' is LOW and S' is LOW; i.e., we must distinguish the state "both inputs LOW" from all others. That's an OR gate. Figure 10.16 shows the circuit. We have used a NOR gate, instead of an OR gate, to get the same output as

---

[18] This particular gate circuit just happens to be made, in a tiny minilogic package, as a 74LVC1G3208; they call it a "3-input positive OR–AND gate." But don't expect this level of service for all your gate circuit needs.

earlier, namely the Q output is HIGH when the desired condition is present. Something strange seems to be going on here, though. We have used AND instead of OR (and vice versa), as compared with the earlier circuit. Section 10.1.7 should clarify the matter. First, though, consider the following exercise.

Exercise 10.7. What do the circuits shown in Figure 10.17 do?

A. Gate interchangeability

When designing digital circuits, keep in mind that it is possible to form one kind of gate from another. For example, if you need an AND gate, and you have half of a 74LVC00 available (quad 2-input NAND), you can substitute as shown in Figure 10.18. The second NAND functions as an inverter, making AND. The following exercises should help you explore this idea.

Exercise 10.8. Using 2-input gates, show how to make (a) INVERT from NOR, (b) OR from NORs, and (c) OR from NANDs.

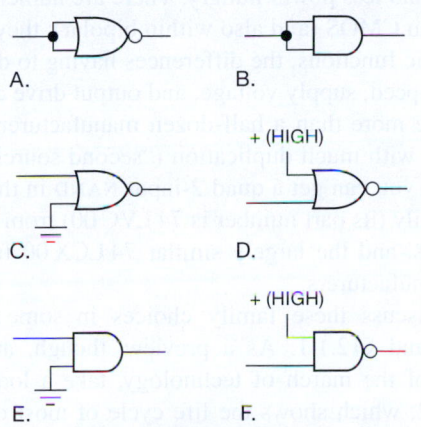

Figure 10.17. Gate configurations for Exercise 10.7.

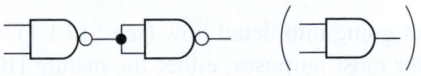

Figure 10.18. Making AND from NANDs.

Exercise 10.9. Show how to make (a) a 3-input AND from 2-input ANDs, (b) a 3-input OR from 2-input ORs, (c) a 3-input NOR from 2-input NORs, and (d) a 3-input AND from 2-input NANDs.

In general, multiple use of one kind of inverting gate (e.g., NAND) is enough to make any combinational function. However, this isn't true for a noninverting gate, because there's no way to make INVERT. This probably accounts for the greater popularity of NAND and NOR in logic design.

10.1.7 Assertion-level logic notation

An AND gate has a HIGH output if both inputs are HIGH. So if HIGH means "true," you get a true output only if all inputs are true. In other words, with active-HIGH logic, an AND gate performs the AND function. The same holds for OR.

What happens if LOW means "true," as in the last example? An AND gate gives a LOW if either input is true (LOW): it's an OR function! Similarly an OR gate gives a LOW only if both inputs are true (LOW). It's an AND function! Very confusing.

There are two ways to handle this problem. The first way is to think through any digital design problem as we did earlier, choosing the kind of gate that gives the needed output. For instance, if you need to know if any of three inputs is LOW, use a 3-input NAND gate. This method is still used by some misguided designers. When designing this way, you would draw a NAND gate, even though the gate is performing an OR function on the (active-LOW) inputs. You would probably label the inputs as in Figure 10.19. In this example, CLEAR′, MR′ (master reset), and RESET′ might be active-LOW levels coming from various places in a circuit. The output, CLR, is active-HIGH and will go to the devices that are to be cleared if *any* of the reset signals goes LOW (true).

Figure 10.19. Confusing notation for OR function of active-LOW signals.

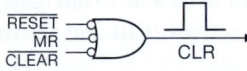

Figure 10.20. Use inverted input "bubbles" for active-LOW inputs.

The second way to handle the problem of active-LOW signals is to use "assertion-level logic." If a gate performs an OR function on active-LOW inputs, draw it that way, as in Figure 10.20. The 3-input OR gate with negated inputs is functionally identical with the preceding 3-input NAND. That equivalence turns out to be an important logical identity, as stated in DeMorgan's theorem, and we will spell out a number of such useful identities shortly. For now, it is enough to know that you can change AND to OR (and vice versa) if you negate the output and all inputs (see Table 10.3). Assertion-level logic looks forbidding at first, because of the proliferation of funny-looking gates. It is better, though, because the logical functions of the gates in the circuit stand out clearly. You'll find it friendly after you've used it for a while, and you won't want to use anything else.

Let's rework the car-door example again with assertion-level logic (Figure 10.21). The gate on the left determines if L or R is true, i.e., LOW, giving an active-LOW output. The second gate gives a HIGH output if both $(L + R)$ and S are true, i.e., LOW. From DeMorgan's theorem (after a while you won't even need that, you'll recognize these gates as equivalent) the first gate is AND and the second is NOR, just as in the circuit drawn earlier. Two important points.

1. Active-LOW (or LOW-true) is sometimes called "negative-true," but that doesn't mean that the logic levels are of negative *polarity*.[19] It means that the lower of the two states (LOW) stands for TRUE.
2. The symbol used to draw the gate itself assumes active-HIGH logic. A NAND gate used as an OR for active-LOW signals could be drawn as a NAND, or (better) using assertion-level logic, as an OR with negation symbols (bubbles) at the inputs. In the latter case you think of the bubbles as indicating inversion of the input signals, followed by an OR gate operating on active-HIGH input logic levels as originally defined.[20]

You might well ask why you shouldn't just keep things simple by doing all your designs with active-HIGH logic. In some cases you are constrained by the logic levels defined by the components themselves (for example, the common use of an active-LOW reset input on a microcontroller); and in other situations (like the car-door switches) it is better electrically to connect the common terminal to ground. In any case, you've got to be able to navigate in a digital world populated by both active-HIGH and active-LOW signals.

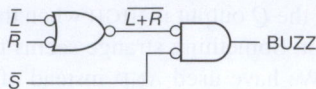

Figure 10.21. Car-door redux: assertion-level logic fixes confusing Figure 10.16.

10.2 Digital integrated circuits: CMOS and Bipolar (TTL)

Digital logic functions are implemented in hardware ICs, either as (small-scale) *standard logic* (e.g., the 74xx gates we've seen), or as *programmable logic* (e.g., an FPGA – field-programmable gate array; §11.2.3), or as a full-custom application-specific IC (ASIC, or ASSP[21]; e.g., a graphics processor). Because this book is intended primarily for the *circuit* designer (as opposed to the *chip* designer[22]), we will not discuss the design of ICs themselves.

CMOS dominates contemporary digital IC technology, having largely replaced the earlier bipolar ("TTL") logic. CMOS is faster, better adapted to operation at low supply voltage, and less power hungry. There are numerous families within CMOS (and also within bipolar); they offer the same logic functions, the differences having to do primarily with speed, supply voltage, and output drive capability. There are more than a half-dozen manufacturers of digital logic, with much duplication ("second sourcing"). For example, you can get a quad 2-input NAND in the popular LVC family (its part number is 74 LVC 00) from five manufacturers, and the largely similar 74 LCX 00 from three other manufacturers.

We discuss these family choices in some detail in §10.2.2 and §12.1.1. As a preview, though, and to get a sense of the march of technology, take a look at Figure 10.22, which shows the life cycle of most of the important digital logic families. Bipolar logic's days are numbered (except for BiCMOS – CMOS logic with bipolar output – and also some specialty families such as the speedy ECL).

Without going into detail now (see §12.1.1), we would suggest, for most purposes, either the mature HC(T) family, or the more recent (and faster) LVC/LVX family. The former is widely available, has an enormous variety of logic functions, and includes through-hole (dual in-line package or DIP) as well as surface-mount technology (SMT) packaging; the latter is faster, optimized for performance at lower supply voltages, but available in SMT only.

---

[19] A confusion that can lead to real damage: we loaned a digital frequency synthesizer to an inexperienced student, who read the manual and then applied -5 V to the programming inputs. He spent many days replacing burned-out circuitry.

[20] Logical AND and OR shouldn't be confused with the *legal* equivalents. The weighty legal tome known as *Words and Phrases* has over 40 pages of situations in which AND can be construed as OR. Among the more amusing is this gem: "OR will be construed AND, and AND will be construed OR, as the necessities of the case may require…." This isn't the same as DeMorgan's theorem!

[21] Application-Specific Standard Product.

[22] Sometimes you hear the terms *board-level* design and *chip-level* design.

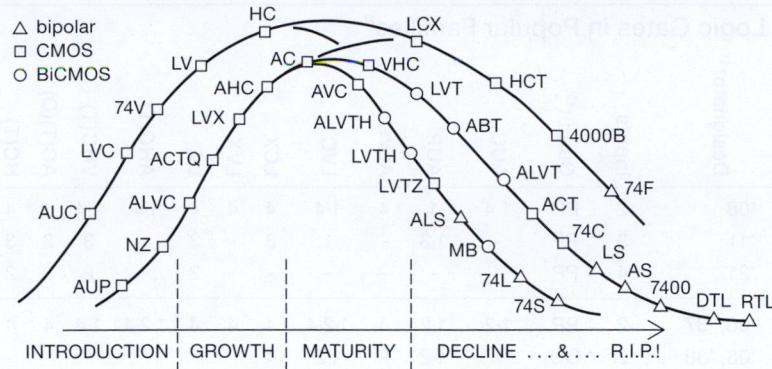

Figure 10.22. Life cycle of logic families, at a snapshot in time early in the third millennium. CMOS is in the ascendancy, bipolar in decline. We put this figure together by looking at analogous figures from NXP, TI, and others; by checking stock at distributors; and by factoring in our own biases.

10.2.1 Catalog of common gates

Table 10.3 shows the common gates you can get as standard digital logic. Each gate is drawn in its normal (active-HIGH inputs) incarnation, and also the way it looks for active-LOW inputs. These functions come in traditional 14- or 16-pin packages, with multiple gates per package (limited by total pin count); they are also available singly in tiny packages.[23] Figure 10.23 shows what these packages look like, going from the traditional through-hole DIP down to the tiny grain-of-sand CSP (chip-scale package) packages on the lower right; you could fit the latter parts nicely between any two leads of the former!

To fully specify one of these gates, you would form a part number beginning with 74, then add the several letters (like LVC, for low-voltage CMOS) to specify the family, then the numbers that designate the function (like 08, for 2-input AND). Additionally you would add some suffixes to specify package and temperature range, and perhaps a prefix like "SN" to specify the manufacturer: voilà, an SN74LVC08ADR, which is a quad 2-input AND in the LVC family, in reels of 2500 parts, packaged in a 14-pin "SOIC" (small-outline IC) temperature range of $-40°C$ to $+125°C$, manufactured by Texas Instruments. For simplicity (and prevention of insanity) we routinely omit most of this detail, indicating digital IC types with an apostrophe,

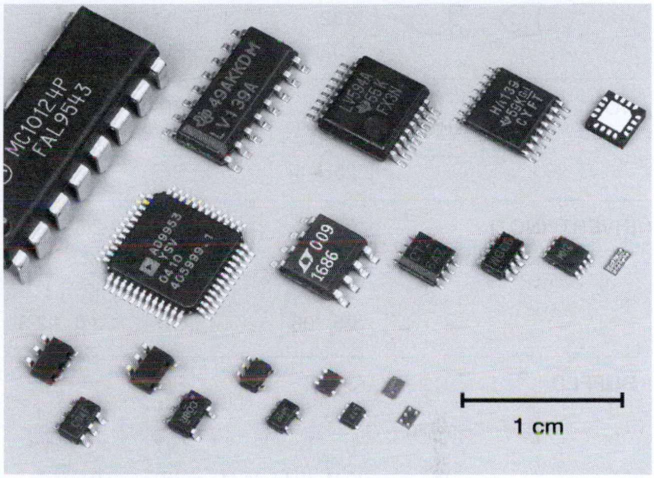

---

[23] In a curious sort of inversion, the part numbers are shortest for the large parts, and vice versa. Here's an example: a 2-input NAND gate, packaged as a quad (4 gates) in a 14-pin package, has part number SN74LVC00DR; the same function, packaged singly in a 5-pin package, is called an SN74LVC1G00MDBVREP. (The "1G" portion specifies what we'll call "minilogic" – the packaging of single logic units in small-pin-count packages.)

Figure 10.23. A selection of digital logic packages; all are surface mount, except for the DIP-16 cameo at upper left. Top row, left to right: DIP-16, SOIC-16, SSOP-16, TSSOP-16, QFN-16. Middle row: TQFP-48, SOIC-8, SSOP-8, SOT-23-8, US-8, WCSP-8 (DSBGA-8: die-size ball-grid array). Bottom row (two samples each): SOT-23-6, SOT-23-5, SC-70, SOT-533, WCSP-5 (DSBGA-5).

e.g., '08 for a 2-input AND function; in situations where the family type matters, we add the family designation, e.g., 'LVC08.

A. A "universal" gate?

Figure 10.24 shows an elegant trick, namely a tiny 6-pin part (called a '1G97) whose small block of logic lets you make any of nine logic functions, according to how you connect the inputs: inverter, noninverting buffer, 2-input multiplexer (MUX, see §10.3.3A), and six varieties of

Table 10.3 Standard Logic Gates in Popular Families[a]

Function	Symbol Active-H inputs	Symbol Active-L inputs	Designator[b]	Inputs	Output[c]	AUC	AUP	ALVC	LVC	LCX	LVX	LV	AHC(T)	VHC(T)	AC(T)(Q)	HC(T)	F	LS	7N, 7S, 7W	4000B	100E, EL, EP
AND			'08	2	PP	1·4	1	4	1·4	4	4	4	1·4	1·4	4	4	4	4	1·2	4	1·4·5
			'11	3	PP	-	1·3	-	1	3	-	3	-	3	3	3	3	3	1	3	-
			'21	4	PP	-	-	-	-	-	-	2	-	2	-	2	2	2	-	2	-
NAND			'00, '37	2	PP	1·2	1·2	4	1·2·4	4	4	4	1·2·4	1·4	4	4	4	4	1·2	4	1·4·5
			'03, '38	2	OD	-	1·2	-	1·2	4	-	-	-	-	-	-	-	-	1·2	-	-
			'10	3	PP	-	1	-	1	-	-	3	-	-	-	-	-	-	1	3	-
			'20	4	PP	-	-	-	-	-	-	2	-	-	2	2	2	-	-	2	-
			'30	8	PP	-	-	-	-	-	-	-	-	-	-	1	1	1	-	1	-
OR			'32	2	PP	1·4	1	4	1·2·3·4	4	4	4	1·4	1·4	4	4	4	4	1·2	4	-
			'332	3	PP	-	1	-	1	-	-	-	-	-	-	-	-	-	1	-	-
			'802	4	PP	-	-	-	-	-	-	-	-	-	2	-	-	-	-	2	1·4
NOR			'02	2	PP	1·2·4	1·2	-	1·2·4	4	4	4	1·4	1·4	4	4	4	4	1·2	4	-
			'27	3	PP	-	1	-	1	3	-	3	-	3	-	3	3	3	1	3	-
			'25	4	PP	-	-	-	-	-	-	-	-	2	-	-	-	-	-	2	1·4
INVERT (NOT)			'04		PP	1·2·6	1·2·3	6	1·2·3·6	6	6	6	1·6	1·6	6	6	6	6	1·2·3	6	-
			'14 (⎍)		PP	1·2	1·2	-	1·2·3·6	6	6	-	-	1·6	-	-	-	-	1·2·3	-	-
			'240		3S	1·2	1·2	-	1	6	-	-	-	6	-	-	-	-	2	-	-
			'05, '06		OD	1·2·6	1·2·3	-	1·2·3·6	6	-	6	6	1·6	6	6	-	6	1	-	-
BUFFER			'34		PP	2·6	1·2	-	1	-	-	-	-	-	-	-	-	-	1·2·3	6	-
			'125, '126		3S	1	1	4	1	4	4	-	1·4	1·4	-	4	4	4	1·2	4	-
			'07, '17		OD	1·2	1·2	-	1·2	6	-	-	1	1	-	-	-	-	1·2·3	6	-
			'241, '244, '541		3S	8	-	8	8	8	8	8	8	8	8	8	8	8	8	-	-
XOR			'86	2	PP	1·2	-	-	1·2·4	4	4	4	1·4	1·4	4	4	4	4	1·2	4	1·5
			'386	3	PP	-	-	-	-	-	-	-	-	-	-	-	-	-	1	-	-
UNIVERSAL			'57-8, '97-8	3	PP	-	1	-	1	-	-	-	-	-	-	-	-	-	-	1	-
			'99	4	3S	-	1	-	1	-	-	-	-	-	-	-	-	-	-	-	-

Notes: (a) available gates/pkg listed as *m·n* etc.; e.g., a 2-input OR ('32) in the AUC family is available with one or four gates in a package. (b) the digits signifying logic function follow the family designator (e.g., 2-input AND gates: *74LVC*08, *74LVC1G*08); common to all families except HV CMOS (4000B) and ECL (100E, EL, EP). (c) PP=push-pull (active pull-up & pull-down); OD=open-drain; 3S=3-state.

2-input gates (AND, OR, AND with one inverted input, OR with one inverted input, NAND with one inverted input, or NOR with one inverted input). The companion '1G98 (same logic, but with inverted output) also has nine disguises, three of which are different (namely NAND, NOR, and MUX with inverted output). The smallest package for these parts is a mere 0.9 × 1.4 mm, way too small a space to print its part number.[24] Going one step further, the '1G99 squanders two extra pins to give you (a) selectable output inversion (via an XOR) and (b) a "three-state" output

---

[24] Try it: a full part number is "SN74AUP1G97YZPR," and this ◇ is the package size (actually it's a bit generous, but we couldn't find a small enough symbol in the LATEX typesetting language).

(coming up in §10.2.4A). The similar '1G57 and '1G58 gates have Schmitt-trigger inputs.

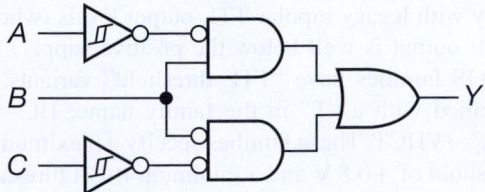

Figure 10.24. The '1G97 and '1G98 "configurable multiple-function gates" can perform any of nine logic functions. The '1G97 is shown; the '1G98 is identical, except with an inverted output.

Exercise 10.10. Show how to connect up a '1G97 to make each of the nine functions just listed.

10.2.2 IC gate circuits

Although a NAND gate, for instance, performs identical logic operations in the various family versions, the logic levels and other characteristics (speed, power, input current, etc.) are quite different. In general, you have to be careful when mixing logic-family types. To understand the differences, look at the schematics of a NAND gate in Figure 10.25.

The CMOS gate (by far the most common family type) is constructed from enhancement-mode MOSFETs of both polarities, connected as switches (rather than followers). An ON FET looks like a low resistance (R_{on}) to whichever supply rail it is connected. Both inputs must be HIGH to turn on the series pair Q_3Q_4 and to turn off both of the pullup transistors Q_1Q_2. That produces a LOW at the output (marked X), i.e., it is a NAND gate. Q_5 and Q_6 constitute the standard CMOS inverter, thus completing the AND gate. From this example it should be evident how to generalize to AND, NAND, OR, and NOR with any number of inputs.

Exercise 10.11. Draw the circuit of a 3-input CMOS OR gate.

Bipolar-transistor logic families are no longer preferred because they are outperformed by CMOS families;[25] but it is instructive to look at the legacy "TTL" family.[26] The once-popular LS (low-power Schottky) NAND gate (Figure 10.25A consists basically of the diode–resistor logic

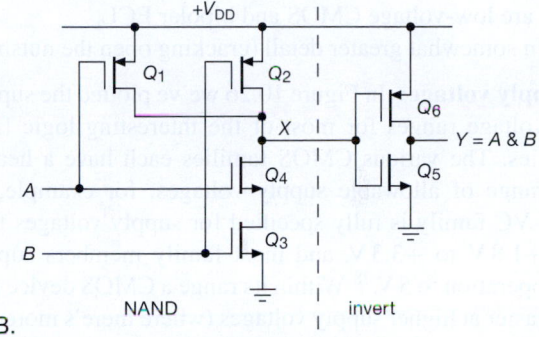

Figure 10.25. NAND/AND gates: two circuit implementations. A. Bipolar LS TTL NAND gate, with its "totem-pole" output stage; B. CMOS AND gate.

of Figure 10.12 driving a transistor inverter followed by a push–pull output. If both inputs are HIGH, the 20k resistor holds Q_1 on, thus producing a LOW output by saturating Q_4 and shutting off Darlington Q_2Q_3. If at least one input is LOW, Q_1 is held off, thus producing a HIGH output by follower action of Q_2Q_3 combined with Q_4 being held off. (Note that the HIGH output, coming from a Darlington follower, is at least two diode drops below the +5 V supply.) Schottky diodes and Schottky-clamped transistors are used throughout for enhanced speed.[27]

Note that both CMOS and bipolar TTL gates have an output circuit with "active pullup" to the positive supply rail, unlike our discrete gate examples (Figures 10.12–10.14).

---

[25] With the exception of the extremely fast ECL types and the hybrid "BiCMOS" hefty bus-driver families.

[26] Among other reasons, its ghost hovers over much contemporary logic, in the form of "TTL-input level" limits of ≤0.8 V (LOW) and ≥2.0 V (HIGH), seen for example in several logic families in Figure 10.2.

[27] Schottky diodes have no stored charge and therefore no reverse recovery delay (see, e.g., §9.5.3B or §9x.6); and Schottky clamping prevents transistor saturation, which otherwise causes a turn-off delay.

10.2.3 CMOS and bipolar ("TTL") characteristics

This section might be subtitled "Analog aspects of digital circuits." The same logic function (e.g., NAND) can be implemented in different ways; these can differ in their *electrical* characteristics, while performing the same *logic*. In a nutshell – **Supply voltage**: CMOS can operate over a range, TTL requires +5 V; **Input current**: CMOS inputs draw no steady current, TTL inputs require current; **Input voltage**: the various families have varying logic thresholds and allowable input voltage, therefore incompatibilities; **Output**: CMOS outputs are rail-to-rail, TTL cannot reach V_+; **Speed and power**: CMOS has only *dynamic* power consumption (proportional to frequency), whereas TTL has substantial quiescent power; and the fastest families are low-voltage CMOS and bipolar ECL.

In somewhat greater detail (cracking open the nutshell):

Supply voltage: In Figure 10.26 we've plotted the supply-voltage ranges for most of the interesting logic families. The various CMOS families each have a healthy range of allowable supply voltages; for example, the LVC family is fully specified for supply voltages from +1.8 V to +3.3 V, and most family members support operation to 5 V.[28] Within its range a CMOS device runs faster at higher supply voltages (where there's more gate drive voltage). The bipolar families run on a single voltage: +5 V ±5% for TTL, and either −5 V (sometimes −5.2 V) or +5 V for ECL (called NECL and PECL, for negative and positive ECL, respectively).

Input – Current: CMOS devices have no quiescent input current (apart from leakage); however, like all devices, their input capacitance (of order 4 pF) draws current during switching ($I = C\, dV/dt$; so, for example, a 2.5 V input transition in 2 ns would require ~5 mA of drive current). Bipolar logic does require quiescent input current: a TTL input held in the LOW state sources current into whatever drives it (e.g., 0.6 mA typ for the F family), so to hold it LOW you must sink current (in addition to the capacitive load current during switching).[29] In general, logic families have adequate output-current capability to drive additional logic; what matters more is compatibility of logic-level *voltages*.

– **Logic Levels:** CMOS families generally put their input threshold voltage at half the supply voltage (though with considerable spread, typically 1/3 to 2/3 of the supply voltage); this is a good choice, since CMOS outputs swing all the way to both rails. However, for compatibility with legacy bipolar TTL output levels (where the HIGH output is well below the positive supply), many CMOS families have "TTL threshold" variants, often specified with a "T" in the family name: HC→HCT, VHC→VHCT. These families specify a maximum LOW threshold of +0.8 V and a minimum HIGH threshold of +2.0 V (see Figure 10.2, and the more complete Figure 12.6). These duplicate the bipolar-TTL specification, where the input logic threshold is about two diode drops above ground (about 1.3V).[30]

– **Voltage Tolerance:** The world has not standardized on a single logic supply voltage (nor should it), so in a typical digital system you'll usually have several supply voltages (e.g., +5 V and +3.3 V). Hence the question: can the output of logic running at one supply voltage (call it X) drive the input of logic at a different supply voltage (call it Y)? The short answer (we'll have a longer answer in Chapter 12) is that two things are required: (a) X's output levels have to satisfy Y's input logic-level requirements; and (b) if Y's supply voltage is less than X's, Y's inputs must tolerate X's (larger) output voltages. This latter is called input-voltage tolerance, and you've got to respect it! For example, you can see in Figure 10.2 that the legacy HC(T) family will not tolerate inputs greater than its supply voltage,[31] whereas the newer LVC family accepts inputs to +5.5 V regardless of its own supply voltage (including when it is unpowered). Input voltage tolerance is essential when digital signals cross over supply-voltage boundaries.

CMOS inputs are susceptible to damage from static electricity during handling. Unused inputs should be tied HIGH or LOW, as necessary (more on this in §10.8.3B).

Output: CMOS outputs are driven by a pair of MOSFET switches, either to ground or to V_+; i.e., "rail-to-rail." The TTL output stage, by contrast, is a saturated transistor to ground in the LOW state and a (Darlington) follower in the HIGH state (two diode drops below V_+). The datasheet will usually give more details, specifying typical and worst-case output voltages for some typical load currents.[32]

---

[28] For most LVC logic devices the datasheets say the "recommended" operating range extends only to +3.6 V, but some LVC parts extend that to +5.5 V, with their operating characteristics specified at 5 V.

[29] The niche bipolar ECL is strange stuff – its outputs are "bare emitters," which are deliberately terminated with 50Ω resistors to −2 V (NECL) or +3 V (PECL).

[30] This bit of history has cast its shadow all over the digital world, with "TTL input levels" pretty much established as standard for nearly all devices with digital inputs.

[31] More precisely, the inputs may not go more than 0.5 V above V_+ or below ground.

[32] CMOS devices usually specify output voltage at symmetrical pairs of

In general, the faster families (ALVC, LVC, LCX; F, AS) have greater output drive capability than the slower families (CD4000, HC(T); LS).

Speed and power: All CMOS families of standard logic consume zero quiescent current.[33] However, their power consumption rises linearly with increasing frequency, because switching the capacitance of internal nodes and external capacitive loads requires current ($I = C\,dV/dt$). CMOS operated near its upper frequency limit may even dissipate more power than bipolar logic (see Figure 10.27). It's common to see this *dynamic current*[34] specified in terms of an effective "power-dissipation capacitance," C_{pd}, from which you can calculate the no-load dynamic power dissipation as $P_{diss} = C_{pd}V^2 f$ (there are two transitions per cycle, which cancels the usual factor of 1/2). For example, a 74LVC00 (quad NAND gate) specifies $C_{pd} = 19\,\text{pF}$ per gate, from which you get a power dissipation of 0.2 mW/MHz per gate (for 3.3 V supply); so one such IC with all four gates churning along at 100 MHz would dissipate 80 mW internally (and additional power from switching the external load capacitances). The speed range of CMOS standard-logic functions goes from about 2 MHz (for the high-voltage CD4000 series running at a low 5 V) to about 100 MHz (for AHCT/VHCT at 5 V) to about 150 MHz (for LVC/LCX at 3.3 V) to about 350 MHz (for AUC at 2.5 V). In contrast to CMOS's zero quiescent current, the bipolar TTL families consume considerable quiescent current, more for the faster families (AS, F, ABT); the corresponding speeds go from about 25MHz (for LS) to about 100 MHz (for AS and F).

In Figure 10.26 (seen again as Figure 12.3, with additional commentary) we've plotted worst-case gate propagation delays for commonly used standard logic families.

In general, the nice characteristics of the CMOS families (zero quiescent current, rail-to-rail output swings,

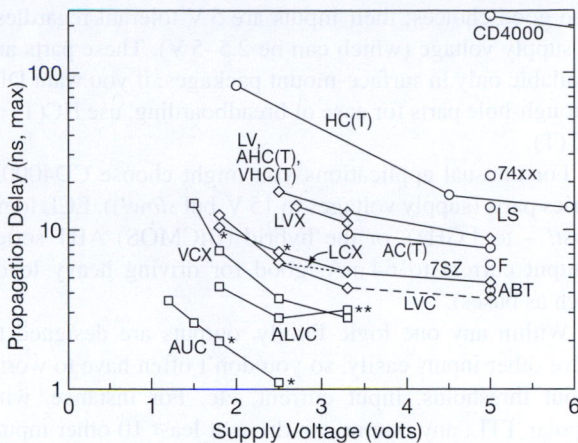

Figure 10.26. Gate speed versus supply voltage, for popular logic families. Maximum specified propagation delay ($t_{pd(max)}$) is shown for the standard supply voltages at which each family is specified. (As a rough guide, "typical" delays are in the range of 35–75% of $t_{pd(max)}$.) See the caption to Figure 12.3 for details.

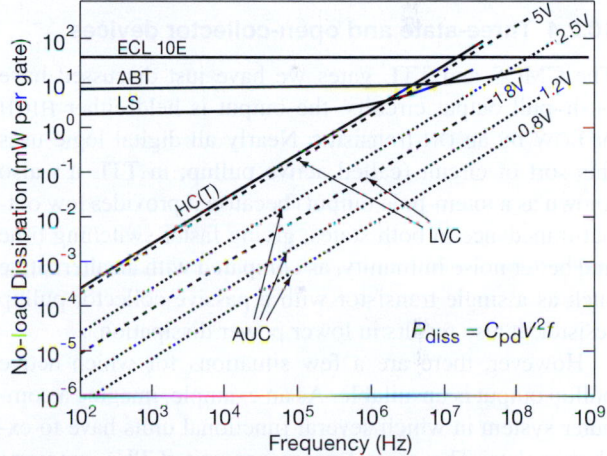

Figure 10.27. Gate internal power dissipation versus frequency for some CMOS and bipolar logic families. Note the dramatic (quadratic) dependence on power supply voltage, within any one family. See also Figure 12.2.

symmetrical sink–source output currents, and high speed) make them the logic of choice. Among them, the LVC and 7SZ families are nice,[35] with their 5 V tolerant inputs and good supply voltage ranges (1.8–3.3 V[36] and 1.8–5 V). For operation at 5 V the AHC(T), VHC(T), or LV families are

---

HIGH- and LOW-state load currents, for example at ±8 mA; TTL devices, with their unsymmetrical output, are usually specified at plenty of sink current, but very stingy source currents (e.g., at 8 mA and at −0.4 mA). This is important when a logic output is used to drive some external component, e.g., an LED indicator or solid-state relay: wire the component so the logic output *sinks* current (with the other end returned to the positive supply, through a current-limiting resistor if necessary).

[33] *Large-scale* CMOS circuits (for example gate arrays or microprocessors, as opposed to the basic "standard-logic" functions like gates and flip-flops) generally have nonzero (and sometimes quite substantial) quiescent current.

[34] When charging and discharging a capacitor from 0 to V at frequency f, the average current is $I = fCV$.

[35] The 7SZ and 17SZ families come only in minilogic packages.

[36] The maximum "recommended" supply voltage for some LVC devices is 5.5 V, whereas for others it is only 3.6 V; however, while the latter types give no specifications for operation above 3.6 V, *all* LVC devices grudgingly allow supply voltages to 5.5 V.

also good choices; their inputs are 5 V tolerant regardless of supply voltage (which can be 2.5–5 V). These parts are available only in surface-mount packages; if you want DIP through-hole parts for ease of breadboarding, use HC(T) or AC(T).

For unusual applications you might choose CD4000B series parts (supply voltages to 15 V, but *slow!*)), ECL logic (*fast!* – to 1 GHz), or the hybrid (BiCMOS) ABT series (output current to 64 mA, good for driving heavy loads such as buses).

Within any one logic family, outputs are designed to drive other inputs easily, so you don't often have to worry about thresholds, input current, etc. For instance, with bipolar TTL, any output can drive at least 10 other inputs (the official term for this is *fan-out*: TTL has a fan-out of 10), so you don't have to do anything special to ensure compatibility. In Chapter 12 we go into the issue of interfacing between logic families and between logic circuits and the outside world.

10.2.4 Three-state and open-collector devices

The CMOS and TTL gates we have just discussed have push–pull output circuits: the output is held either HIGH or LOW by an ON transistor. Nearly all digital logic uses this sort of circuit (called active pullup; in TTL it's also known as a totem-pole output) because it provides low output impedance in both states, giving faster switching time and better noise immunity, as compared with an alternative such as a single transistor with a passive collector-pullup resistor. It also results in lower power dissipation.

However, there are a few situations for which active pullup output is unsuitable. As an example, imagine a computer system in which several functional units have to exchange data. The central processor unit (CPU), memory, and various peripherals all need to be able to send and receive 16-bit words. It would be awkward (to put it mildly) to have separate 16-wire cables connecting each device to all others. The solution is the so-called *data bus*, a single set of 16 wires accessible to all devices. It's like the old-fashioned (now extinct?[37]) telephone "party line": only one device at a time may "talk" (assert data), but all may "listen" (receive data). With a bus system there must be agreement as to who may talk, and you'll see words like *bus arbitration* and *bus master*.

You can't use gates (or any other devices) with active pullup–pulldown outputs to drive a bus, since you couldn't disconnect your output from the shared data lines (you're holding it either HIGH or LOW at all times). What's needed is a gate whose output can be *open*. Such devices are available, and they come in two varieties, *three-state* devices and *open-collector* devices.

A. Three-state logic

Three-state logic, also called TRI-STATE® logic (a trademark of National Semiconductor Corporation, NSC), provides an elegant solution. The name is misleading; it is not digital logic with three voltage levels. It's just ordinary logic, with a third output state: open circuit (Figure 10.28). A separate *enable* input determines whether the output behaves like an ordinary active pullup output or goes into the "third" (open) state, regardless of the logic levels present at the other inputs. Three-state outputs are available on many digital chips, including counters, latches, registers, etc., as well as on gates and inverters. A device with three-state output behaves exactly like ordinary active pullup logic when enabled, always driving its output either HIGH or LOW; when disabled,[38] it effectively disconnects its output, so another logic device can drive the same line. Let's look at an example.

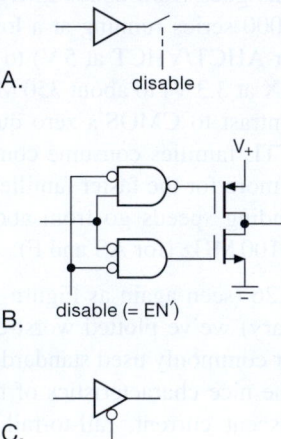

Figure 10.28. Three-state CMOS logic; A. Conceptual diagram. B. Realization with internal CMOS gates. C. Logic symbol.

B. A look ahead: data buses

Three-state drivers are widely used to drive computer data buses. Every device (memory, peripherals, etc.) that needs to put data on the (shared) bus ties onto it with three-state gates (or more complex functions such as "registers"). Things are cleverly arranged so that at most one device has its drivers enabled at any instant, all other devices being

---

[37] But playing a major role in the 1959 movie *Pillow Talk*.

[38] Mr. Lebowski is *disabled*, yes.

disabled into the open (third) state. In a typical situation the selected device "knows" to assert data onto the bus by recognizing its particular address on a set of address and control lines (Figure 10.29). In this simplified case the device is wired as port 6: it looks at address lines A_0–A_2 and asserts data onto data bus D_0–D_3 when it sees its particular address (i.e., 6) on the address lines *and* it sees a READ$'$ pulse. Such a bus protocol is adequate for many simple systems. Something like this is used in most microcomputers, as you will see in Chapter 14.

Note that there must be some external logic to make sure that three-state devices sharing the same output lines don't try to talk at the same time (that undesirable condition is officially called "bus contention"). In this case all is well as long as each device responds to a unique address.

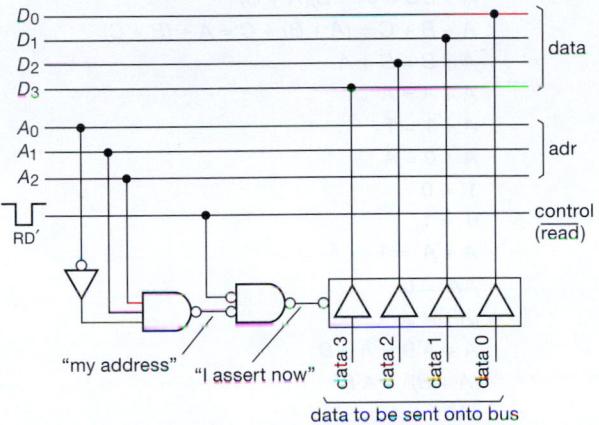

Figure 10.29. Data bus with address decode logic and three-state drivers.

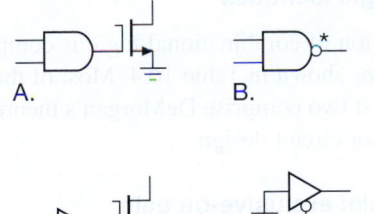

Figure 10.30. Open-drain logic: A. open-drain NAND; B. symbol; C. open-drain noninverting buffer; D. implementation with three-state buffer.

C. Open-collector and open-drain logic

The predecessor to three-state logic was "open-collector" logic, which allows you to share a single line among the

outputs of several drivers. An open-collector (or open-drain) output simply omits the active pullup transistor of the output stage (Figure 10.30). The name "open-collector" is a good one. When you use such gates, you must supply an external pullup resistor somewhere. Its value isn't critical; a small-value resistor gives increased speed and improved noise immunity, at the expense of increased power dissipation and loading of the driver. Values of a few hundred to a few thousand ohms are typical. If you wanted to drive a bus with open-collector gates (rather than with three-state drivers), you would substitute 2-input open-collector NANDs for the three-state drivers of Figure 10.29, bringing one input of each gate HIGH to enable the gates onto the bus; note that the data bits then asserted onto the bus are inverted. Each bus line would need a single resistive pullup to the positive supply.

The disadvantage of open-collector logic is that speed and noise immunity are degraded, when compared with logic constructed with active pullup devices, because of the resistive pullup circuit. That's why three-state drivers are nearly universally favored for computer bus applications. However, there are three situations in which you would choose open-collector (or open-drain) devices: driving external loads, "wired-OR," and external buses. Let's look at them briefly.

D. Driving external loads

Open-collector (O/C) logic is good for driving external loads that are returned to a higher-voltage positive supply. You might want to drive a low-current lamp or relay that requires 12 V, or perhaps just generate a 15 V logic swing by running a resistor from a gate's output to +15 V, as in Figure 10.31. A popular O/C device is the ULN2003/4, a seven-channel open collector Darlington array with internal clamp diodes (for inductive loads); it accepts direct logic drive, has a 50 V breakdown rating, and can switch up to 500 mA (the 75468/9 is similar, but with 100 V breakdown). More on these subjects in §12.4.

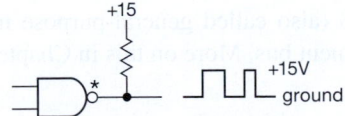

Figure 10.31. Open-collector logic as level translator.

E. Wired-OR

If you wire together some open-collector gates as shown in Figure 10.32, you get what's called "wired-OR" – the combination behaves in this case like a larger NOR gate, with the output going LOW if any input is HIGH. You can't do this with active pullup outputs, because there would be a contest of wills if all the gates didn't agree on what the output should be. You can combine NORs, NANDs, etc., with this kind of connection, and the output will be LOW if any gate asserts a LOW output. This connection is sometimes called "wired-AND," because the output is HIGH only if all gates have HIGH (open) outputs. Both names are describing the same thing: it's wired-AND for active-HIGH logic and wired-OR for active-LOW logic. This will make more sense to you after you've seen DeMorgan's theorem in the next section.

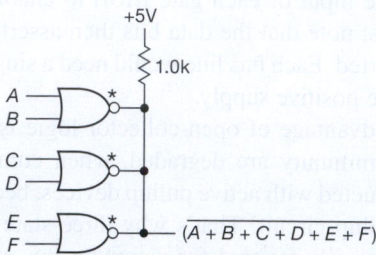

Figure 10.32. Wired-OR.

Wired-OR enjoyed some brief popularity in the early days of digital electronics, but it is not much used today, with two exceptions: (a) in the family of logic known as ECL (emitter-coupled logic) the outputs are what you might call "open-emitter," and can be wired-OR'ed painlessly; and (b) there are some shared lines in computer buses (most notably the line called *interrupt*) whose function is not to transfer data bits, but merely to indicate if *at least one* device is requesting attention; in that case you use wired-OR, because it does what you want and doesn't require external logic to prevent contention.

F. External buses

Where speed is not too important, you sometimes see open-collector drivers used to drive buses. Examples are the original SCSI bus used to connect disks and peripherals, and the IEEE-488 (also called general-purpose interface bus, GPIB) instrument bus. More on this in Chapter 14.

10.3 Combinational logic

As we discussed earlier in §10.1.4A, digital logic can be divided into *combinational* (sometimes called *combinatorial*) and *sequential*. Combinational circuits are those in which the output state depends on only the present input states in some predetermined fashion, whereas in sequential circuits the output state depends both on the input states

and the previous history. Combinational circuits can be constructed with gates alone, whereas sequential circuits require some form of memory (flip-flops). In these subsections we explore the possibilities of combinational logic before entering the turbulent world of sequential circuits.

Table 10.4 Logic Identities

$$ABC = (AB)C = A(BC)$$
$$AB = BA$$
$$AA = A$$
$$A1 = A$$
$$A0 = 0$$
$$A(B + C) = AB + AC$$
$$A + AB = A$$
$$A + BC = (A + B)(A + C)$$
$$A + B + C = (A + B) + C = A + (B + C)$$
$$A + B = B + A$$
$$A + A = A$$
$$A + 1 = 1$$
$$A + 0 = A$$
$$1' = 0$$
$$0' = 1$$
$$A + A' = 1$$
$$AA' = 0$$
$$(A')' = A$$
$$A + A'B = A + B$$
$$(A + B)' = A'B'$$
$$(AB)' = A' + B'$$

10.3.1 Logic identities

No discussion of combinational logic is complete without the identities shown in Table 10.4. Most of these are obvious. The last two comprise DeMorgan's theorem, the most important for circuit design.

A. Example: exclusive-OR gate

We illustrate the use of the identities with an example: making the exclusive-OR function from ordinary gates. Figure 10.33 shows the XOR truth table. From studying this and by realizing that the output is 1 only when $(A, B) = (0, 1)$ or $(1, 0)$, we can write

$$A \oplus B = \overline{A}B + A\overline{B},$$

from which we have the realization shown in Figure 10.34. However, this realization is not unique. Applying the identities, we find

$$A \oplus B = A\overline{A} + A\overline{B} + B\overline{A} + B\overline{B}$$
$$(A\overline{A} = B\overline{B} = 0)$$
$$= A(\overline{A} + \overline{B}) + B(\overline{A} + \overline{B})$$
$$= A(\overline{AB}) + B(\overline{AB})$$
$$= (A + B)(\overline{AB}).$$

(In the first step we used the trick of adding two quantities that equal zero; in the third step we used DeMorgan's theorem.) This has the realization shown in Figure 10.35. There are still other ways to construct XOR. Consider the following exercise.

A	B	$A \oplus B$
0	0	0
0	1	1
1	0	1
1	1	0

Figure 10.33. XOR truth table.

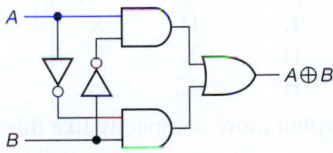

Figure 10.34. XOR realization.

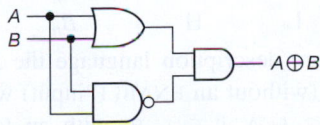

Figure 10.35. Another XOR realization.

Exercise 10.12. Show that

$$A \oplus B = \overline{AB + \overline{A}\,\overline{B}},$$
$$A \oplus B = (A + B)(\overline{A} + \overline{B}),$$

by logic manipulation. You should be able to convince yourself that these are true by inspection of the truth table, combined with suitable hand-waving.

Exercise 10.13. What are the following: (a) $0 \cdot 1$, (b) $0 + 1$, (c) $1 \cdot 1$, (d) $1 + 1$, (e) $A(A + B)$, (f) $A(A' + B)$, (g) A XOR A, (h) A XOR A'?

10.3.2 Minimization and Karnaugh maps

Because a realization of a logic function (even one as simple as exclusive-OR) isn't unique, it is often desirable to find the simplest, or perhaps most conveniently constructed, circuit for a given function. Many good minds have worked on this problem, and there are several methods available, including algebraic techniques widely available as software. For example, all of the "hardware description languages" (HDLs) used to enter circuitry going into programmable logic (see §11.2.6) include automatic logic minimization; you don't even see it happening.

Perhaps of historical interest only, you'll sometimes hear the term *Karnaugh map*, which is a simple tabular method for minimizing logic with four or fewer inputs; it also enables you to find a logic expression (if you don't know it) once you can write down the truth table.

We illustrate the method with an example (and then abandon it entirely!). Suppose you want to generate a logic circuit to count votes. Imagine that you have three active-HIGH inputs (each either 1 or 0) and an output (0 or 1). The output is to be 1 if at least two of the inputs are 1.

Step 1. Make a truth table:

A	B	C	Q
0	0	0	0
0	0	1	0
0	1	0	0
0	1	1	1
1	0	0	0
1	0	1	1
1	1	0	1
1	1	1	1

All possible permutations must be represented, with corresponding output(s). Write an X (= "don't care") if either output state is OK.

Step 2. Make a Karnaugh map. This is somewhat akin to a truth table, but the variables are represented along two axes. Furthermore, they are arranged in such a way that only one input bit changes in going from one square to an adjacent square (Figure 10.36).

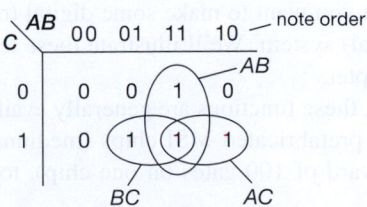

Figure 10.36. Karnaugh map for majority logic.

Step 3. Identify on the map groups of 1s (alternatively, you could use groups of 0s): the three blobs (called a "cover") enclose the logic expressions AB, AC, and BC. Finally, read off the required function, in this case

$$Q = AB + AC + BC$$

with the realization shown in Figure 10.37. The result seems obvious, in retrospect. We could have read off the pattern of 0s to get instead

$$Q' = A'B' + A'C' + B'C',$$

which might be useful if the complements A', B', and C' already exist somewhere in the circuit.

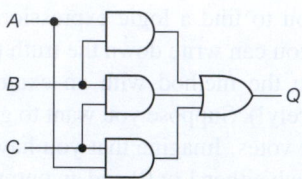

Figure 10.37. Majority (voting) logic.

Exercise 10.14. Just for fun, draw a Karnaugh map for logic to determine if a 3-bit integer (0 to 7) is prime (assume that 0, 1, and 2 are not primes). Show a realization with 2-input gates.

Exercise 10.15. Continuing Karnaugh-map fun: find logic to perform multiplication of two 2-bit unsigned numbers (i.e., each 0 to 3), producing a 4-bit result. *Hint*: use a separate Karnaugh map for each output bit.

10.3.3 Combinational functions available as ICs

With logic gates alone, you could construct logic to perform rather complicated functions such as binary addition or magnitude comparison, parity checking, multiplexing (selecting one of several inputs, as determined by a binary address), etc. In fact, that is just what is done when you implement complex logic in *gate arrays*[39] or in other forms of *programmable logic* (see §10.3.3F, §10.5.4, and Chapter 11.). Programmable logic (usually combined with a microcontroller) is quite often the method of choice when you want to make some digital (or combined analog/digital) system. We'll illustrate these techniques in the next chapter.

However, these functions are generally available also in the form of prefabricated MSI chips (medium-scale integration, upward of 100 gates on one chip), to be used as

standard-logic functions. Although many interesting MSI functions involve flip-flops (i.e., *sequential* circuits, which we will get to shortly), there are a number of them that are combinational functions involving gates only. Let's see what animals live in the MSI combinational zoo.

A. 2-input select (multiplexer)

The 2-input select (also called a 2-input *multiplexer*, or "MUX") is a very useful function. It is basically a two-position switch for logic signals. Figure 10.38 shows the basic idea, with both a discrete gate implementation and an IC that packages four 2-input multiplexers (a "quad MUX") in one IC. When SELECT is LOW, the A inputs are passed through to their respective Y outputs; when SELECT is HIGH, the B inputs appear at the output. Holding ENABLE' HIGH disables the device by forcing all outputs LOW. This is an important concept we'll see more of later. Here's the truth table, which illustrates the X (don't care) entry:

	Inputs			Outputs
E'	SEL	A_n	B_n	Y_n
H	X	X	X	L
L	L	L	X	L
L	L	H	X	H
L	H	X	L	L
L	H	X	H	H

It can be written more compactly like this:

	Inputs	Outputs
E'	SEL	Y_n
H	X	L
L	L	A_n
L	H	B_n

In a hardware description language the 2-input MUX logic function (without an ENABLE input) would be written as `Y = ~S & A | S & B`; with an ENABLE input it would become `Y = E & (~S & A | S & B)`.[40] Figure 10.38 and the preceding table correspond to the '157 quad 2-input select chip. The same function is also available with inverted output ('158) and with three-state outputs (true: '257; inverted: '258). It is also available as a tiny single-section MUX (without the EN' input), with part numbers '1G157 and '2G157. These chips work electrically like gates – they do the logic and regenerate a logic level at the output accordingly. Another way to implement this function is with a few *transmission gates*, in which the appropriate input signal is simply passed through to the

---

[39] A function so implemented is sometimes amusingly referred to as having been "Gatorade."

[40] For a somewhat obscure reason (having to do with a "logic-race" condition) it's a good idea to add the redundant term `A & B` to these expressions; thus `Y = E & (~S & A | S & B | A & B)`.

output (via MOS transistors) without regeneration; we'll see this next.

Exercise 10.16. Show how to make a 2-input select, using a pair of 3-state buffers and whatever other logic you need.

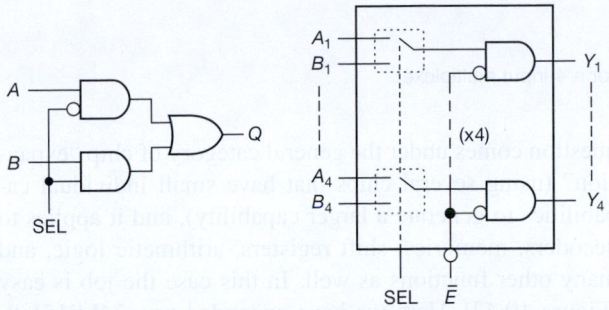

Figure 10.38. Two-input "select" gates: A. implemented with discrete gates; B. packaged together as a "quad," with a single SE-LECT line.

Although the function of a select gate can be performed by a mechanical switch in some cases, the gate is a far better solution, for several reasons:

(a) it is cheaper;
(b) all channels are switched simultaneously and rapidly;
(c) it can be switched, nearly instantaneously, by a logic level generated elsewhere in the circuit (most interestingly, from a microprocessor or other smart device);
(d) even if the select function is to be controlled by a front-panel switch, it is better not to run fast logic signals around through cables and switches, to avoid capacitive signal degradation and noise pickup.

With a select gate actuated by a dc level, you keep logic signals on the circuit board and get the bonus of simpler off-board wiring, namely a single line with pullup switched to ground by a single-pole single-throw (SPST) switch. Controlling circuit functions with externally generated dc levels in this manner is known as "cold switching," and it is a much better approach than controlling the signals themselves with switches, potentiometers, etc. Besides its other advantages, cold switching lets you bypass control lines with capacitors to eliminate interference, whereas signal lines cannot generally be bypassed. We'll see some examples of cold switching later.

B. Transmission gates

As we discussed in §3.4.1A, with CMOS it is possible to make "transmission gates," simply a pair of complementary MOSFET switches in parallel, so that an input (ana-

log) signal between ground and V_{DD} is either connected through to the output through a low resistance (less than a hundred ohms) or open-circuited (essentially infinite resistance). As you may remember, such a device is bidirectional and doesn't know (or care) which end is input and which end is output. Transmission gates work perfectly well with digital CMOS levels, and in fact they are used extensively in the internal circuitry of CMOS digital devices. You can get them also as standard logic ICs. Figure 10.39 shows the layout of the popular '4066 CMOS "quad bilateral switch." Each switch has a separate *control* input: HIGH closes the switch, and LOW opens it. They are available also in compact single-section and two-section packages ('1G66, '2G66). Note that transmission gates are merely switches, and therefore have no fan-out; i.e., they simply pass input logic levels through to the output without providing additional drive capability.[41]

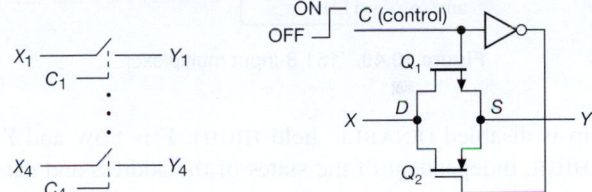

Figure 10.39. Quad transmission gate; a single gate, implemented with MOS transistors, is shown at right.

With transmission gates you can make 2-input (or more) select functions, usable with CMOS digital levels or analog signals. To select among a number of inputs, you can use a bunch of transmission gates (generating the control signals with a *decoder*, as will be explained later). This is such a useful logic function that it has been institutionalized as the multiplexer, discussed next.

Exercise 10.17. Show how to make a 2-input select with transmission gates. You will need an inverter.

C. Multiplexers with many inputs

Multiplexers are available with 4, 8, and 16 inputs. A binary address is used to select which of the input signals appears at the output. For instance, an 8-input MUX has a 3-bit address input to address the selected data input (Figure 10.40). The digital MUX illustrated is a '151. It has an active-LOW STROBE' (another name for ENABLE') input, and it provides true and complemented outputs. When the

---

41 Transmission gates are attractive to IC designers because their simple design requires little area on the silicon die, and they do not incur the switching delay of a conventional gate.

```
wire [1:0] A; // 2 address ("select") input lines
wire [3:0] D; // 4 input data lines
wire Y, YBAR, ENBAR // outputs (true and complemented), and enable
assign Y = ~ENBAR & ( D[0] & ~A[1] & ~A[0]
                    | D[1] & ~A[1] & A[0]
                    | D[2] & A[1] & ~A[0]
                    | D[3] & A[1] & A[0] )
assign YBAR = ~ Y;
```

Figure 10.41. Verilog code for a 4-input multiplexer.

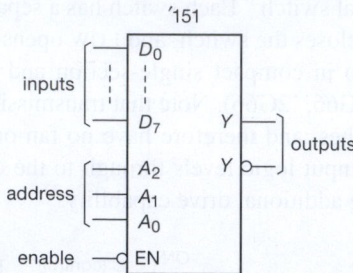

Figure 10.40. '151 8-input multiplexer.

chip is disabled (ENABLE' held HIGH), Y is LOW and Y' is HIGH, independent of the states of the address and data inputs.

It's a good idea to become somewhat familiar with description languages used for logic entry, if for no reason other than anxiety reduction. Figure 10.41 shows what Verilog code looks like, for the simpler example of a 4-input multiplexer.

Electrically, two varieties of multiplexers are available. One type is for digital levels only, with an input threshold and "clean" regeneration of output levels according to the input state: an example is the '153 logic MUX (available in both CMOS and bipolar logic families). The other kind of MUX is analog and bidirectional; it's really just an array of transmission gates. These come only in CMOS families, and can be used for both logic and analog signals. The '4051–'4053 CMOS multiplexers work this way. Remember that logic made from transmission gates has no fan-out. Since transmission gates are bidirectional, these multiplexers can be used as "demultiplexers," or decoders. We discuss them next.

Exercise 10.18. Show how to make a 4-input multiplexer using (a) ordinary gates, (b) gates with three-state outputs, and (c) transmission gates. Under what circumstances would (c) be preferable?

You might wonder what to do if you want to select among more inputs than are provided in a multiplexer. This

question comes under the general category of chip "expansion" (using several chips that have small individual capabilities to generate a larger capability), and it applies to decoders, memories, shift registers, arithmetic logic, and many other functions as well. In this case the job is easy (Figure 10.42). Here we have expanded two 74LS151 8-input multiplexers into a 16-input multiplexer (note that we've used lowercase designations for the input and output signals, to prevent confusion with the chip's similarly named pins).[42] There's an additional address bit, of course, and you use it to enable one chip or the other. The disabled chip holds its Y LOW, so an OR gate at the output completes the expansion. With three-state outputs the job is even simpler, because you can connect the outputs directly together.

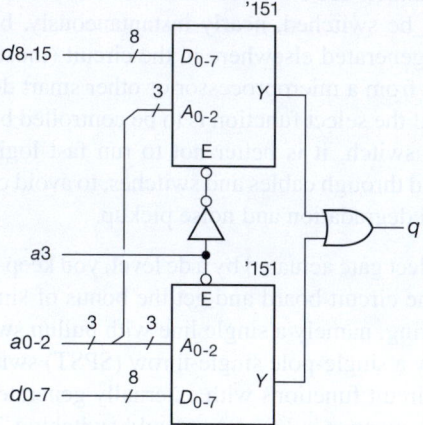

Figure 10.42. Multiplexer expansion.

D. Demultiplexers and decoders

A demultiplexer is the opposite of a multiplexer: it takes an input and routes it to one of several possible outputs,

---

[42] We've also introduced a common "bus" convention, namely the use of a single line with a diagonal slash to indicate a collection of similar signal lines; the number by the slash tells you how many signals are in the group, and the label tells you what they are.

```
wire [1:0] A; // 2-bit address: brackets show array range
wire [3:0] YBAR; // four active-LOW outputs
wire ENBAR; // active-LOW enable
assign YBAR[0] = ~(~ENBAR & ~A[1] & ~A[0]);
assign YBAR[1] = ~(~ENBAR & ~A[1] & A[0]);
assign YBAR[2] = ~(~ENBAR & A[1] & ~A[0]);
assign YBAR[3] = ~(~ENBAR & A[1] & A[0]);
```

Figure 10.44. Verilog code for one section of a '139 dual 1-of-4 decoder.

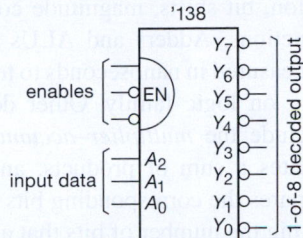

Figure 10.43. '138 1-of-8 decoder.

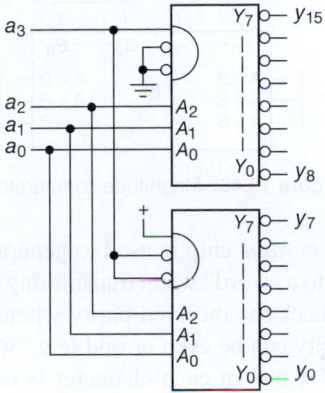

Figure 10.45. Decoder expansion.

according to an input binary address. The other outputs are either held in the inactive state or open-circuited, depending on the type of demultiplexer.

A decoder is similar, except that the address is the only input, and it is "decoded" to assert one of n possible outputs. Figure 10.43 shows an example. This is the '138 "1-of-8 decoder." The output corresponding to (addressed by) the 3-bit input data is LOW; all others are HIGH. This particular decoder has three ENABLE inputs, all of which must be asserted (two LOW, one HIGH); otherwise all outputs are HIGH.

Decoders are commonly used when interfacing to a data bus, to trigger different actions depending on the address; we'll treat this subject in detail in Chapter 13. Another common use of a decoder is to enable a sequence of actions in turn, according to an advancing address given by the output of a binary *counter* (§10.4.2E). A close cousin of the '138 is the '139, a dual 1-of-4 decoder with a single active-LOW ENABLE per section. Figure 10.44 illustrates what its Verilog code would look like.

Figure 10.45 shows how to use a pair of '138 1-of-8 decoders to generate a 1-of-16 decoder. No external gates are necessary, since the '138 has ENABLE inputs of both polarities.

Exercise 10.19. More expansion: make a 1-of-64 decoder from nine '138s. *Hint:* use one of them as an enabling switchyard for the others.

In CMOS logic, the multiplexers that use transmission gates are also demultiplexers, since transmission gates are bidirectional. When they are used that way, it is important to realize that the outputs that aren't selected are open-circuited. A pullup or pulldown resistor, or equivalent, must be used to assert a well-defined logic level on those outputs (the same requirement as with TTL open-collector gates).

Another kind of decoder is the '47 "BCD-to-7-segment decoder/driver." It takes a BCD input and generates outputs on 7 lines corresponding to the segments of a "7-segment display" that have to be lit to display the decimal character. This type of decoder is really an example of a *code converter,* but in common usage it is called a decoder.

Exercise 10.20. Design a BCD-to-decimal (1-of-10) decoder using gates.

Exercise 10.21. Design a "simple" encoder: a circuit that outputs the (2-bit) address telling which of 4 inputs is HIGH (all other inputs must be LOW).

Exercise 10.22. Figure out how to make a parity generator with XOR gates.

E. Other arithmetic chips

The *priority encoder* generates a binary code giving the address of the highest-numbered input that is asserted. It is particularly useful in "parallel-conversion" ADCs (see next chapter) and in microprocessor system design. An example is the '148 8-input (3 output bits) priority encoder. The '147 encodes 10 inputs.

Figure 10.46 shows a 4-bit *magnitude comparator*. It determines the relative sizes of the 4-bit input numbers A and B and tells you via outputs whether $A < B$, $A = B$, or $A > B$. Inputs are provided for expansion to numbers larger than 4 bits.

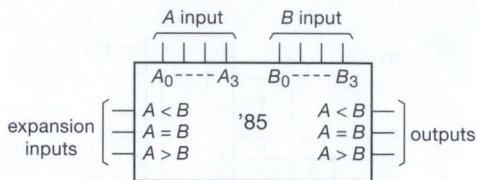

Figure 10.46. Magnitude comparator.

A *parity-generator* chip is used to generate a parity bit to be attached to a "word" when transmitting (or recording) data, and to check the received parity when such data are recovered. Parity can be even or odd (e.g., with odd parity the number of 1 bits in each character is odd). The '280 parity generator, for instance, accepts a 9-bit input word, giving an even- and an odd-parity bit output. The basic construction is an array of exclusive-OR gates.

Figure 10.47 shows a 4-bit *full adder*.[43] It adds the 4-bit number A_i to the 4-bit number B_i, generating a 4-bit sum S_i plus carry bit C_o. Adders can be "expanded" to add larger numbers: the "carry-in" input C_i is provided to accept the carry out of the next lower adder.

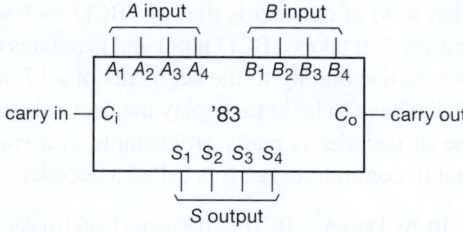

Figure 10.47. 4-bit full adder.

Figure 10.48 shows the Verilog code (here written out in somewhat longwinded form, for clarity) for a 4-bit full adder; see if you can understand how it figures out the sum.

---

[43] We show the '83 logic-family part number mostly from a sense of nostalgia. The part does not exist in the 'HC family, and the 7483 and 74LS83 parts are obsolete. In contemporary design such MSI (medium-scale integration) functions are most often implemented with some form of programmable logic (see §10.5.4 and Chapter 11). In fact, it is the very superiority of the latter that brought about the demise of many of the old MSI parts.

A nice thing about languages like Verilog or VHDL is that they understand higher levels of abstraction (and you can, too, once you get the hang of them). In this case all the stuff after the `wire` declarations can be written as a single line, namely `assign {COUT,S} = A + B + CIN;`.

A device known as an *arithmetic logic unit* (ALU) can be used as an adder, though it has the capability of performing a number of different functions. For instance, the '181 4-bit ALU (expandable to larger word lengths) can do addition, subtraction, bit shifts, magnitude comparison, and a few other functions. Adders and ALUs do their arithmetic in times measured in nanoseconds to tens of nanoseconds, depending on logic family. Other dedicated arithmetic chips include the *multiplier–accumulator* (MAC), which accumulates a sum of products, and the *correlator*, which compares the corresponding bits of a pair of bit strings, calculating the number of bits that agree.

However, owing to the development of large and fast microprocessors, contemporary digital design favors either general-purpose microprocessors or the more optimized digital signal processors (DSPs) for the kind of signal processing that involves extensive arithmetic functions. An attractive alternative is the FPGA, user-configurable to be just about anything; you can drop in a "soft" processor or other function, and you can get them with optimized "hard" functions already built in. We'll talk about configurable logic in the next chapter, and microprocessors in Chapter 15.

F. Programmable logic devices

You can configure your own custom combinational (and sequential) logic on a single chip by using ICs that contain an array of gates with programmable interconnections. These are known generically as *programmable logic devices* (PLDs). The popular varieties are cPLDs (complex PLDs) and field-programmable gate arrays (FPGAs). Both flavors are high achievers, flexible, and a delight to use. They are a part of every designer's toolbox of tricks. We'll meet them later in this chapter, and again in the next.

10.4 Sequential logic

10.4.1 Devices with memory: flip-flops

All our work with digital logic so far has been with combinational circuits (e.g., arrays of gates), for which the output is determined completely by the existing state of the inputs. There is no "memory," no history, in these circuits. Digital life gets really interesting when we add devices with memory. This makes it possible to construct counters, arithmetic

```
wire [3:0] A;
wire [3:0] B;
wire [3:0] S; // sum bits
wire CIN, COUT; // carrys
assign S[0] = A[0] ∧ B[0] ∧ CIN; // recall "∧" means XOR
assign C01 = A[0] & B[0] | A[0] & CIN | B[0] & CIN;
assign S[1] = A[1] ∧ B[1] ∧ C01;
assign C12 = A[1] & B[1] | A[1] & C01 | B[1] & C01;
assign S[2] = A[2] ∧ B[2] ∧ C12;
assign C23 = A[2] & B[2] | A[2] & C12 | B[2] & C12;
assign S[3] = A[3] ∧ B[3] ∧ C23;
assign COUT = A[3] & B[3] | A[3] & C23 | B[3] & C23;
```

Figure 10.48. Verilog code for a 4-bit full adder.

accumulators, and circuits that generally do one interesting thing after another. The basic unit is the flip-flop, a colorful name to describe a device that, in its simplest form, looks as shown in Figure 10.49.

Assume that both A and B are HIGH. What are X and Y? If X is HIGH, then both inputs of G_2 are HIGH, making Y LOW. This is consistent with X being HIGH, so we're finished. Right?

$$X = \text{HIGH},$$
$$Y = \text{LOW}.$$

Wrong! The circuit is symmetrical, so an equally good state is

$$X = \text{LOW},$$
$$Y = \text{HIGH}.$$

The states X, Y both LOW, and X, Y both HIGH, are not possible (remember, $A = B = $ HIGH). So the flip-flop has two stable states (it's sometimes called a "bistable"). Which state it is in depends on past history. It has memory! To write into the memory, just bring one of the inputs momentarily LOW. For instance, bringing A LOW momentarily guarantees that the flip-flop goes into the state

$$X = \text{HIGH},$$
$$Y = \text{LOW},$$

no matter what state it was in previously. You would describe it as an "SR flip-flop," which is Set or Reset with an active-LOW input pulse.

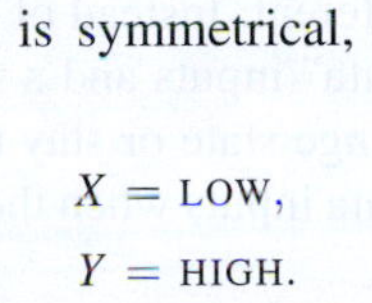

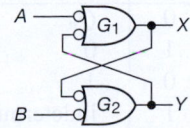

Figure 10.49. Flip-flop ("set-reset" type).

A. Switch debouncing

This kind of flip-flop (with a SET and RESET input, often labeled S and R) is quite useful in many applications. Figure 10.50 shows a typical example. This circuit is supposed to enable the gate and pass input pulses when the switch is opened.[44] The problem with this circuit is that switch contacts *bounce*. When the switch is closed, the two contacts actually separate and reconnect, typically 10 to 100 times over a period of about a millisecond. You would get the kind of waveforms shown; if there were a counter or shift register using the output, it would faithfully respond to all those extra "pulses" caused by the bounce.

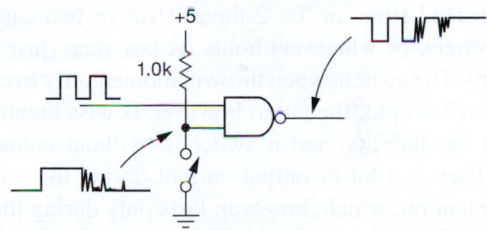

Figure 10.50. Mechanical-switch "bounce."

Figure 10.51 shows a cure. The flip-flop changes state when the contacts first close. Further bouncing against that contact makes no difference and SPDT (single-pole, double-throw) switches never bounce all the way back to the opposite position; the output is a *debounced* signal, as sketched. This debouncer circuit is widely used; the '279 "quad *SR* latch" lets you get four into one package, and the '1G74 (in a smaller package) can be used when only a single flip-flop is needed. Incidentally, the preceding circuit

---

[44] We generally like to tie a switch to ground (not +5V) for a couple of reasons: (1) ground is a convenient (and electrically "quiet") return path for switches and other controls; and (2) we got used to it, owing to a peculiarity of (current-sourcing) bipolar logic.

has a minor flaw: the first pulse after the gate is enabled may be shortened, depending on when the switch is closed relative to the input pulse train; the same holds for the final pulse of a sequence (of course, a switch that is not debounced has the same problem). A "synchronizer" circuit (see §10.4.4) can be used to prevent this from happening, for applications where it makes a difference.

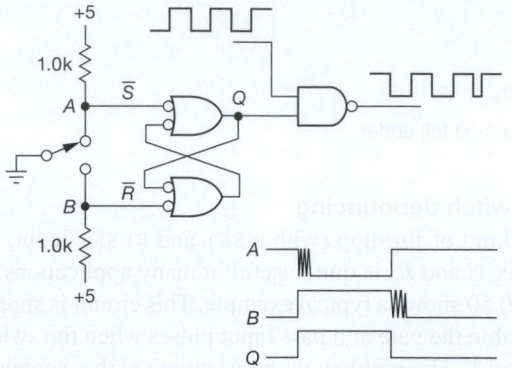

Figure 10.51. SR flip-flop switch debouncer. The signal at Q, though debounced, can corrupt the pulse train as indicated; use a synchronizer if this matters. The nodes labelled $\overline{S}$ and $\overline{R}$ are active-LOW SET and RESET inputs to the flip-flop.

Figure 10.52A shows a trick we use for making a simpler debouncer: the noninverting buffer (which could be made instead from an '08 2-input AND, or two cascaded '04 inverters, or whatever) holds its last state (just like a flip-flop). The switch, when thrown, momentarily overrides the buffer's output; the latter, however, is wise enough not to fight inevitability, and it switches (without bounce) to agree. There's a lot of output current during the momentary contention, which, however, lasts only during the gate transition time, of order a few nanoseconds. No harm done, and, despite one's feeling of discomfort at such a dirty trick,[45] it works just fine. With a pair of cascaded inverters (Figure 10.52B) you can avoid the connection to V_{DD}.

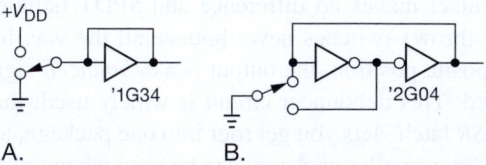

Figure 10.52. Simple switch debouncer.

---

[45] Timid engineers have been known to replace the feedback wire with a resistor, say 1 kΩ. This eliminates the momentary current spike on the power rail, simultaneously relieving your lingering anxiety.

B. Multiple-input flip-flop

Figure 10.53 shows another simple flip-flop. Here NOR gates are used; a HIGH input forces the corresponding output LOW. Multiple inputs allow various signals to set or clear the flip-flop. In this circuit fragment, no pullups are used, because logic signals generated elsewhere (by standard active pullup outputs) are used as inputs.

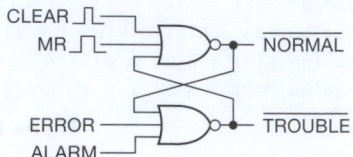

Figure 10.53. SR flip-flop with multiple inputs.

10.4.2 Clocked flip-flops

Flip-flops made with two gates, as in Figures 10.49 and 10.53, are known generically as *SR* (set–reset), or *jam-loaded*, flip-flops. You can force them into one state or the other whenever you want by just generating the right input signal. They're handy for switch debouncing and many other applications. But the most widely used form of flip-flop looks a little different. Instead of a pair of jam inputs, it has one or two "data" inputs and a single "clock" input. The outputs will change state or stay the same, depending on the levels at the data inputs when the clock pulse arrives.

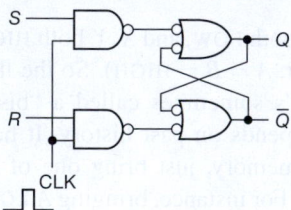

Figure 10.54. Clocked flip-flop: first approximation.

The simplest clocked flip-flop looks as shown in Figure 10.54. It's just our original flip-flop, with a pair of added gates (controlled by the clock) to enable the SET and RESET inputs. It is easy to verify that the truth table is

S	R	Q_{n+1}
0	0	Q_n
0	1	0
1	0	1
1	1	indeterminate

where Q_{n+1} is the Q output after the clock pulse and Q_n is the output before the clock pulse. The basic difference between this and the previous flip-flops is that R and S should now be thought of as *data* inputs (as opposed to flip-flop set–reset inputs). What is present on R and S when a short clock pulse comes along determines what happens to Q. The R and S inputs are otherwise ignored.

This flip-flop has one awkward property, however. The output can change in response to the inputs during the time the clock is HIGH. In that sense it is still like the jam-loaded *SR* flip-flop (it's also known as a *transparent latch*: the output "sees through" to the input when the clock is HIGH). The full utility of clocked flip-flops comes with the introduction of slightly different configurations: the master–slave flip-flop, and the edge-triggered flip-flop.

A. Edge-triggered: the type-D flip-flop

These are by far the most popular kind of flip-flop. The logic level present at the D (data) input just before[46] a clock transition, or "edge," determines the output state after the clock has changed.

Here's the truth table for the type-D flip-flop:

D	Q_{n+1}
0	0
1	1

You can think of the D-FF as copying the state of its input to its (latched) output. One of its many uses is, in fact, simply grabbing and holding a transient logic level,[47] as commanded by a separate clocking transition. These flip-flops are available as inexpensive packaged ICs and are always used in that form.

It is worth pausing for a moment to look at the innards of a D flip-flop, in order to understand what is going on. Figure 10.55 shows two circuit configurations, known as "master–slave" and "edge-triggered," respectively. The master–slave configuration is probably easier to understand. Here's how it works.

While the clock is HIGH, gates 1 and 2 are enabled, forcing the master flip-flop (gates 3 and 4) to the same state as the D input: $M=D$, $M'=D'$. Gates 5 and 6 are disabled, so the slave flip-flop (gates 7 and 8) retains its previous state. When the clock goes LOW, the inputs to the master are disconnected from the D input, while the inputs of the slave are simultaneously coupled to the outputs of the master. The master thus transfers its state to the slave. No further

changes can occur at the output, because the master is now stuck. At the next rising edge of the clock, the slave will be decoupled from the master and will retain its state, while the master will once again follow the input.

The edge-triggered circuit behaves the same externally, but the inner workings are different. It is not difficult to figure them out. The particular circuit shown happens to be the ever-popular '74 rising-edge-triggered type-D flip-flop. Flip-flops are available with either rising- or falling-edge triggering.[48] (The preceding master–slave circuit transfers data to the output on the *falling* edge.) In addition, most flip-flops also have SET and CLEAR jam-type inputs. They may be set and cleared on HIGH or on LOW, depending on the type of flip-flop. Figure 10.56 shows four flip-flop examples. The wedge means "edge-triggered," and the little circles mean "negation," or complement. Thus the '74 is a dual type D rising-edge-triggered flip-flop with active-LOW jam-type SET and CLEAR inputs; the '4013 is a dual type-D rising-edge-triggered flip-flop with active-HIGH jam-type SET and CLEAR inputs; the '1G79 is a single type-D rising-edge-triggered flip-flop with no SET or CLEAR inputs and without a Q' output; and the '112 is a dual JK master–slave flip-flop with data transfer on the falling edge and with active-LOW jam-type SET and CLEAR inputs.

JK and type-T flip-flops

The JK and T flip-flops work on principles similar to those of the type-D flip-flop. Here are the truth tables:

J	K	Q_{n+1}		T	Q_{n+1}
0	0	Q_n		0	Q_n
0	1	0		1	Q_n'
1	0	1			
1	1	Q_n'			

Thus, if J and K are complements, Q will go to the value of the J input at the next clock edge. If J and K are both LOW, the output won't change. If J and K are both HIGH, the output will "toggle" (reverse its state after each clock pulse).

The T (toggling-type) flip-flop toggles at each clock if T is HIGH; T LOW makes it stick.

B. Divide-by-2

In digital circuits you often want to create a signal that toggles at a subdivision of some higher-frequency "clock" signal that is already present. For example, digital wristwatches use as their timebase a 32,768 Hz crystal oscillator; that peculiar frequency is chosen because its 2^{15}

---

[46] This vague wording can, and will, be made precise, when we visit *setup* and *hold* times in §10.4.2C.

[47] Or perhaps a bunch of separate bits (e.g., an 8-bit data byte) in an array of D-flops (which is called a *register*).

[48] Sometimes loosely called *positive edge* and *negative edge*, respectively.

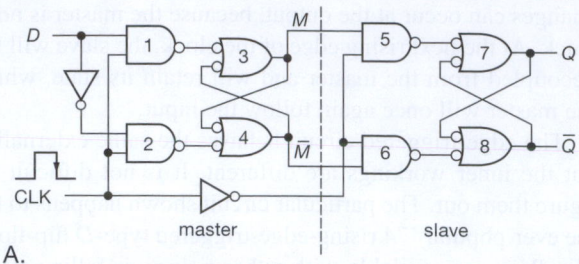

A.

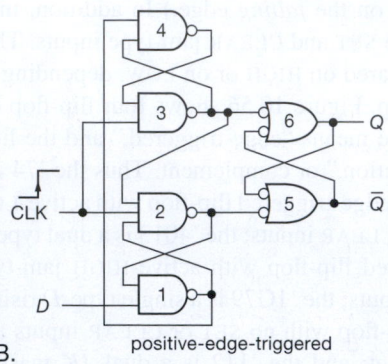

B.

Figure 10.55. True clocked flip-flops: master–slave and edge-triggered type-D flip-flops.

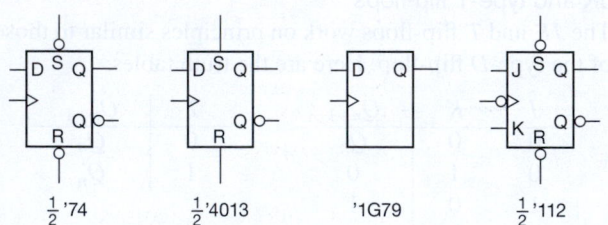

Figure 10.56. D-type and JK flip-flops.

subdivision is 1 Hz, just right for stepping the clock's second hand (or incrementing its displayed digital time). The basic trick here is to use the toggling capability of flip-flops: in Figure 10.57 the D flip-flop always sees, at its D input, the complement of its existing Q output. It therefore toggles at each clock pulse, generating an output at half the input clock frequency.

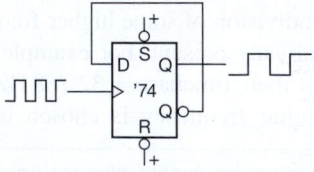

Figure 10.57. Toggling D flip-flop.

In Figure 10.58 you can see the essence of Verilog coding for the toggling D flip-flop.

```
wire CLKIN;
reg Q;
always @(posedge CLKIN)
    Q = ~Q;
```

Figure 10.58. Verilog code for the toggling D flip-flop.

C. Data and clock timing

This last circuit raises an interesting question: will the circuit fail to toggle, because the D input changes almost immediately after the clock pulse? In other words, will the circuit get confused, with such crazy things happening at its input? You could, instead, ask this question: exactly *when* does the D flip-flop (or any other flip-flop) look at its input, relative to the clock pulse? The answer is that there is a specified *setup time* t_s and *hold time* t_h for any clocked device. Input data must be present and stable from at least t_s before the clock transition until at least t_h after it, to guarantee proper operation. For the 74HC74, for instance, $t_s=20$ ns and $t_h=3$ ns (Figure 10.59). So, for the preceding toggling connection, the setup-time requirement is met if the output has been stable for at least 20 ns before the next clock rising edge. It may look as if the hold-time requirement is violated, but that's OK, also. The minimum *propagation time* from clock to output is 10 ns, so a D flip-flop connected to toggle as described is guaranteed to have its D input stable for at least 10 ns after the clock transition. Most devices nowadays have a zero hold-time requirement.

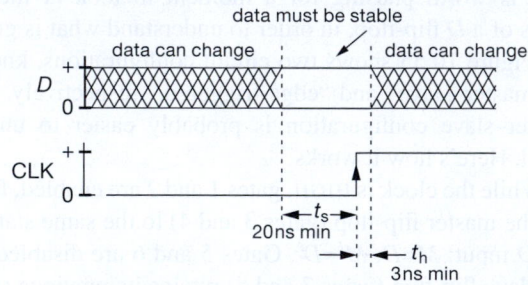

Figure 10.59. Data setup and hold times for the 74HC74 flip-flop.

D. Metastability

An interesting thing can happen if the level at the D input changes during the setup-time interval, namely a so-called *metastable* state in which the flip-flop can't make up its mind which state to go into. Figure 10.60 shows what happened when we deliberately violated the setup time for a 74HC74 type-*D* flip-flop (operating from a +3.3 V supply), bringing D_{in} HIGH at the last moment: the *Q* output took its good old time deciding which state to go to.[49] The 'scope trace accumulated about 2 seconds of data, during which there were some events in which the delay from clock to *Q* output stretched out to nearly 50 ns, compared with its normal value of about 16 ns; when we ran it for several minutes, its best cliffhanger clocked in at 75 ns (nearly at the right-hand edge of the screen). Faster logic families show a correspondingly shorter delay, and it is claimed that some are designed to be "metastable resistant."[50] We teased a 74LVC74 (at +3.3 V supply) into metastability (which occurred at $t_s\sim$0.4 ns!), and measured a leisurely (!) propagation delay of 2–4 ns, compared with its frenetic normal delay of just 1.4 ns.[51]

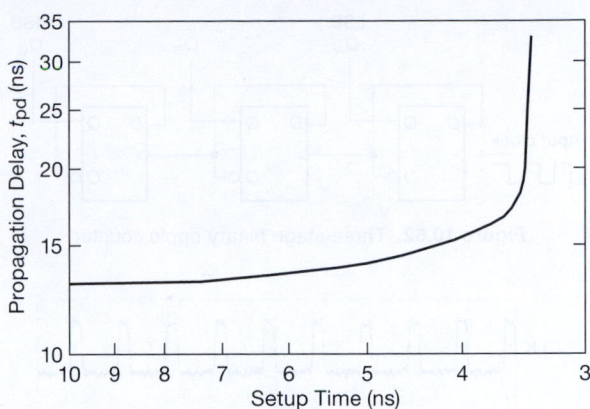

Figure 10.61. Twisting the dragon's tail: metastable propagation delay versus (violated) setup time for the 74HC74.

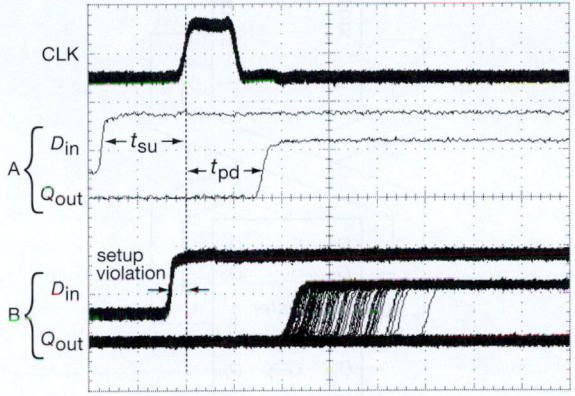

Figure 10.60. Timing violation produces metastability in clocked logic. Here a setup time violation at the *D* input of a 74HC74 (trace pair B) produces a much delayed output (Q_{out}) from CLK to *Q* compared with the normal delay of ~16 ns (trace pair A). Horizontal scale: 10 ns/div.

Just for fun we measured the lengthening of "decision time" for the 74HC74 flip-flop, as we tortured it with less and less setup time; Figure 10.61 shows the results.

E. Divide by more

By cascading several toggling flip-flops (connect each *Q* output to the next clock input), it is easy to make a divide-by-2^n, or binary, counter. Figure 10.62 shows a three-stage

ripple counter, which is a divide-by-8: the output waveform from the last flip-flop is a square wave whose frequency is 1/8 of the circuit's input clock frequency. Such a circuit is called a *counter* because the data present at the three *Q* outputs, considered as a single 3-bit binary number, goes through a binary sequence from 0 to 7, incrementing after each input pulse. (We've used the *Q'* outputs to clock successive stages, to make it count *up* rather than *down*.[52])

Figure 10.63 shows measured waveforms for this counter, zipping along at 50 MHz (the curved arrows were added to indicate what causes what, to aid in understanding): you can see the binary sequence of 3-bit numbers (0–7), where Q_A is the LSB, and Q_C is the MSB. You can also see that there is successive delay going from stage to stage (hence "ripple").[53]

In practice, the simple scheme of cascading counters by connecting each *Q* output to the next clock input creates certain problems related to the cascaded delays as the signal ripples down through the chain of flip-flops, and a *synchronous* scheme (in which all clock inputs see the same clocking signal) is usually better.[54] We'll see soon how to make a 3-bit synchronous counter.

The counter is a useful function, with many versions

---

[49] The abrupt *Q* output, buffered by the output stage, conceals the more interesting internal metastable behavior, in which the (unbuffered) voltage hovers between LOW and HIGH, balanced on a knife edge, trying to decide which way to fall.

[50] Metastable-*resistant* is not the same thing as metastable-*proof*; it's like water-resistant versus waterproof.

[51] The LVC family is a good overall choice for operation from 1.8–3.3 V, though by no means the fastest kid on the block (see Figure 10.26).

[52] There are two other configurations that produce an *up-counter*: (1) use the *Q* outputs to clock successive stages, but use the *Q'* signals as your binary outputs; or (2) use the *Q* outputs to clock successive falling-edge-triggered stages.

[53] In fact, the ripple delay is large enough here so that the count at any instant (i.e., a vertical slice through the traces) is never correct. This does not matter, however, if the counter is used only to generate an output of subdivided frequency, or if the counter is stopped prior to readout.

[54] However, if all you need is a 2^n subdivision of an input clock, with no

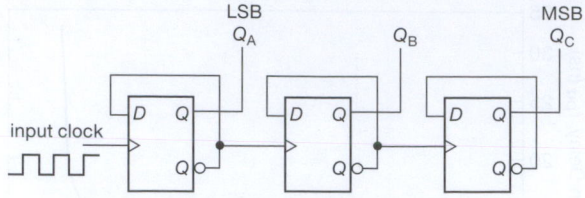

Figure 10.62. Three-stage binary ripple counter.

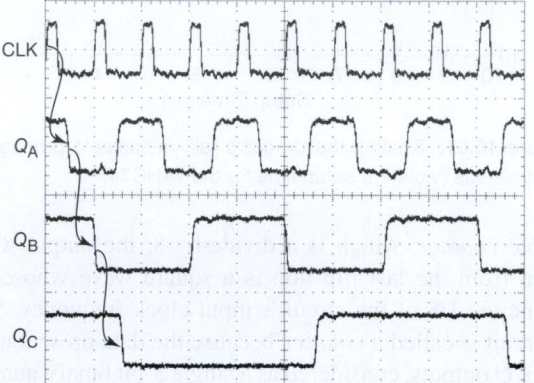

Figure 10.63. Ripple-counter waveforms. This 'scope trace (at 4 V/div vertical, 20 ns/div horizontal) shows a cascade of rising-edge triggered 74HC74 D flip-flops, clocked at 50 MHz, displaying successive ~10 ns delays per stage. The counter, beginning at maximum count $(1,1,1)$, is clocked to all zeros by the rising edge of CLK at the extreme left edge. See Figure 10.71 for analogous waveforms of a fully-synchronous counter.

available as standard logic, including 4-bit, BCD, and multidigit counting formats. By cascading several such counters and displaying the count on a numeric display device (e.g., an LED digital display), you can easily construct an event counter. If the input pulse train to such a counter is gated for exactly 1 second, you've got a frequency counter that displays frequency (cycles per second) by simply counting the number of cycles in a second.

10.4.3 Combining memory and gates: sequential logic

Having explored the properties of flip-flops, let's see what can be done when they are combined with the combinational (gate) logic we discussed earlier. Circuits made with gates and flip-flops constitute the most general form of digital logic.

---

particular phase relation to the input, a ripple counter is perfectly good; it is also simpler, and will operate to a higher maximum frequency.

A. Synchronously clocked systems

As we hinted in the previous subsection, sequential logic circuits in which there is a common source of clock pulses driving all flip-flops have some very desirable properties. In such a *synchronous system*, all action takes place just after each clocking pulse, based on the stable levels present just before each clock pulse. The system reaches its next stable state before the next clock pulse; it's a nice way of managing feedback; and, by clocking all flip-flops simultaneously, it takes advantage of the relative lack of digital noise just before each clock pulse (the calm before the storm).

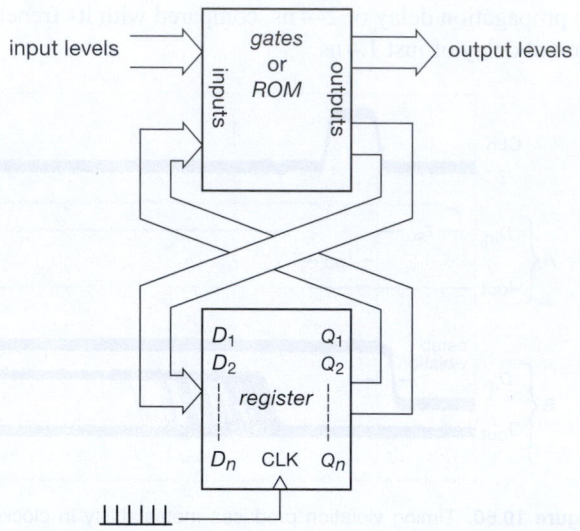

Figure 10.64. The classic sequential "state machine": a set of clocked flip-flops (a *register*) plus combinational logic. This scheme is easily implemented with single-chip programmable logic devices (PLDs or FPGAs).

Figure 10.64 shows the general scheme. The flip-flops have all been combined into a single *register*, which is nothing more than a set of type-D flip-flops with their clock inputs all tied together, and their individual D inputs and Q outputs brought out; i.e., each clock pulse causes the levels present at the D inputs to be transferred to the respective Q outputs. The box full of gates looks at both the Q outputs and whatever input levels are applied to the circuit and generates a new set of D inputs and logic outputs. This simple-looking scheme is extremely powerful; it is the basis for general digital processors. Let's look at an example.

B. Example: divide-by-3

Let's design a synchronous divide-by-3 circuit with two D flip-flops, both clocked from the input signal. In this case,

D_1 and D_2 are the register inputs, Q_1 and Q_2 are the outputs, and the common clock line is the master clocking input (Figure 10.65). The trick is to rig up gating so that the D inputs are presented with the desired next state.

1. Choose the three states. Let's use

Q_1	Q_2	
0	0	
0	1	
1	0	
0	0	(i.e., first state)

2. Find the combinational logic network outputs necessary to generate this sequence of states; i.e., figure out what the D inputs have to be to get those outputs:

Q_1	Q_2	D_1	D_2
0	0	0	1
0	1	1	0
1	0	0	0

3. Concoct suitable combinational logic (gates), using available outputs, to produce those D inputs. In general, you can use a "lookup table" (LUT) in ROM (read-only memory), programmed to hold the next-state D values (and addressed by the current Q value, along with any external inputs).[55] This example is simple, so you can do it with one logic gate: you can see by inspection that

$$D_1 = Q_2,$$
$$D_2 = (Q_1 + Q_2)',$$

from which the circuit of Figure 10.66 follows.

It is easy to verify that the circuit works as planned. Because it is a synchronous counter, all outputs change simultaneously (unlike the ripple counter). In general, synchronous (or "clocked") systems are desirable, since susceptibility to noise is improved: things have settled down by the time of the next clock pulse, so circuits that look at their inputs only at clock edges aren't troubled by capacitively coupled interference from other flip-flops, etc. A further advantage of clocked systems is that transient states (caused by delays, so that all outputs don't change simultaneously) don't produce false output, because the system is insensitive to what happens just *after* a clock pulse. We'll see some examples later.

---

[55] This example has no logic inputs, as you would have with, for example, an UP/DOWN counter, or a counter with a RESET input.

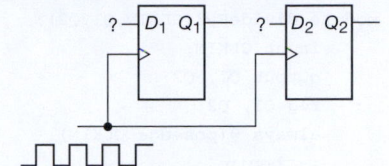

Figure 10.65. Divide-by-3: need logic!

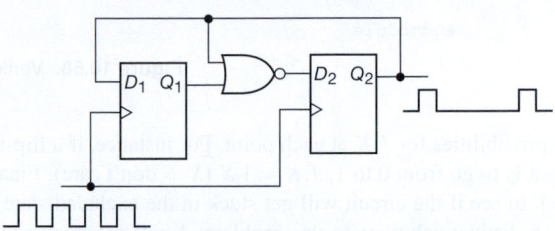

Figure 10.66. Synchronous divide-by-3.

C. Excluded states

What happens to the divide-by-3 circuit if the flip-flops somehow get into the state $(Q_1, Q_2) = (1,1)$? This can easily happen when the circuit is first turned on, since the initial state of a flip-flop is anyone's guess. From the diagram, it is clear that the first clock pulse will cause it to go to the state $(1,0)$, from which it will function as before. It is important to check the excluded states of a circuit like this, since it is possible to be unlucky and have it get stuck in one of those states. (Alternatively, and better, the initial design procedure can include a specification of all possible states.) A useful diagnostic tool is the *state diagram*, which for this example looks like Figure 10.67. Usually you write the conditions for each transition next to the arrows, if other variables of the system are involved. Arrows may go in both directions between states, or from one state to several others.

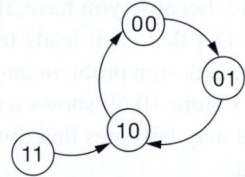

Figure 10.67. State diagram: divide-by-3.

The Verilog code for this circuit (this time a complete "module") is in Figure 10.68.

Exercise 10.23. Design a synchronous divide-by-3 circuit using two *JK* flip-flops. It can be done (in 16 different ways!) without any gates or inverters. *One hint:* when you construct the table of required J_1, K_1 and J_2, K_2 inputs, keep in mind that there are

```
module divideBy3(CLKIN,Q1,Q2);
    input CLKIN;
    output Q1, Q2;
    reg Q1, Q2;
    always @(posedge CLKIN)
        begin
          Q1 <= Q2;                   // the <= symbol is called a "nonblocking assignment"
          Q2 <= ~(Q1|Q2);    // it makes all steps happen at once, not sequentially
        end
endmodule
```

Figure 10.68. Verilog code for divide-by-3 circuit.

two possibilities for J, K at each point. For instance, if a flip-flop output is to go from 0 to 1, $J, K = 1, X$ (X = don't care). Finally, check to see if the circuit will get stuck in the excluded state (of the 16 distinct solutions to this problem, 4 will get stuck and 12 won't).

Exercise 10.24. Design a synchronous 2-bit UP/DOWN counter: it has a clock input, and a control input (U/D′); the outputs are the two flip-flop outputs Q_1 and Q_2. If U/D′ is HIGH, it goes through a normal binary counting sequence; if LOW, it counts backward – $Q_2 Q_1 = 00, 11, 10, 01, 00\ldots$.

D. State diagrams as design tools

The state diagram can be quite useful when designing sequential logic, particularly if the states are connected together by several paths. In this design approach, you begin by selecting a set of unique states of the system, giving each a name (i.e., a binary address). You will need a minimum of n flip-flops, or bits, where n is the smallest integer for which 2^n is equal to or greater than the number of distinct states in the system. Then you set down all the rules for moving between states, i.e., all possible conditions for entering and leaving each state. From there it is a straightforward (but perhaps tedious) job to generate the necessary combinational logic, because you have all possible sets of Q's and the set of D's that each leads to. Thus you have converted a sequential design problem into a combinational design problem.[56] Figure 10.69 shows a real-world example. Note that there may be states that don't lead to others, e.g., "with diploma."

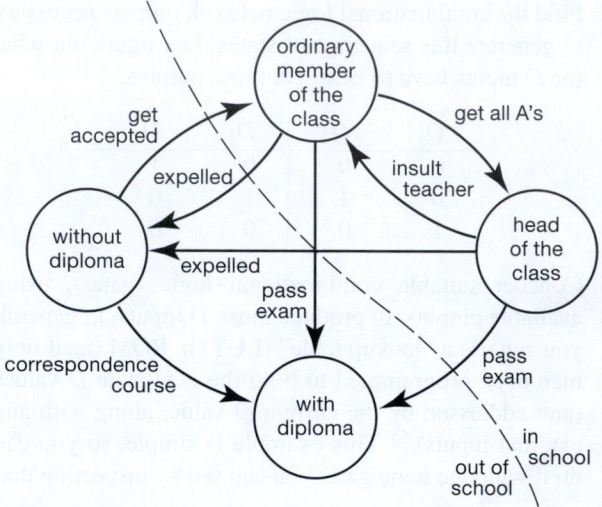

Figure 10.69. State diagram: going to school.

E. State machine design

Hardware description languages, used as algebraic entry tools for both programmable logic (cPLDs and FPGAs) and for full-custom integrated circuits (ASICs), provide a straightforward route to specify state machines. They include *if/elseif/else* statements that let you specify conditions for going to the next state, and for the outputs, given the current state and the inputs. The HDL software translates these specifications into a logic circuit, implemented with the usual gates and flip-flops.

One further point about state machines: the block diagram of Figure 10.64 allows for two possibilities, namely that (a) the outputs may depend on only the current state (defined by the Q's), and in fact may consist simply of the Q's themselves; or (b) for any given state of the Q's, the output may depend also upon the inputs, which are available to the combinational logic in the block labeled "*gates or ROM*." These are known as *Moore* and *Mealy* state

---

[56] You can always take the shortcut of using a lookup table, in the form of a ROM. Alternatively, there are design tools (e.g., hardware description languages (HDLs) used for programmable logic) that will minimize and instantiate the needed logic. Best of all, these HDLs always include an input syntax that lets you simply specify the states and their transition rules, so that you don't even have to figure out the necessary logic.

machines, respectively.[57] The Moore machine changes state only on clock edges, and, if the Q's themselves are considered the outputs, they are strictly synchronous. The Mealy machine responds asynchronously to input changes, regardless of the clock; it generally requires fewer flip-flops, because there are multiple possible outputs corresponding to a single state (defined by the Q's).

F. Synchronous binary counter

We promised we'd revisit our original 3-bit ripple counter (§10.4.2E) to make it fully synchronous. Figure 10.70 shows how it's done: the LSB always toggles, easily done by feeding back to its D an inverted Q. For higher-order bits the rule for counting in binary is to toggle only when all lower-order bits are 1s. Because an XOR is an "optional inverter," this is easily done by feeding each D input from an XOR, one of whose inputs comes from the corresponding Q, and the other from an AND of all lower-order Q's. It's worth seeing it also in a hardware description language:[58]

```
QA.d = !QA.q
QB.d = QB.q $ QA.q
QC.d = QC.q $ (QB.q & QA.q)
```

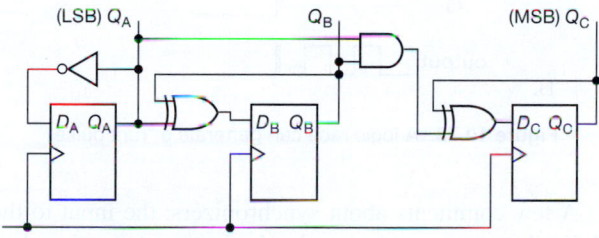

Figure 10.70. 3-bit synchronous binary counter.

As promised, the Q's all clock simultaneously,[59] as you can see in Figure 10.71, which shows measured waveforms for a 3-bit synchronous binary counter clocked at the same rate as the ripple counter of Figure 10.63. Here it is easy to see that the Q's of the successive clocked states are simply the numbers 0 to 7 in 3-bit binary.

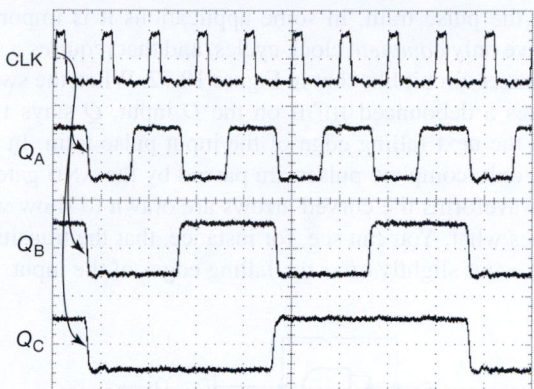

Figure 10.71. In contrast to a ripple counter (Figure 10.62), a *synchronous* counter (or, indeed, any synchronous system) has all flip-flops clocked from a common source. Here a 74HC161 4-bit synchronous counter (rising-edge triggered) displays a CLK-to-Q delay of about 14 ns, constant for all Q's, thus producing coincident changes. (Same conditions as Figure 10.63)

G. PLDs for state machine design

Programmable logic devices have just what you need for state machines – plenty of flip-flops and plenty of logic, all on one configurable (hence "programmable") chip. And the programming software includes tools that make state-machine design straightforward. We'll have much more to say about these remarkable devices in §10.5.4, and in the next chapter.

10.4.4 Synchronizer

An interesting application of flip-flops in sequential circuits is their use in a *synchronizer*. Suppose you have some external control signal coming into a synchronous system that has clocks, flip-flops, etc., and you want to use the state of that input signal to control some action. For example, a signal from an instrument or experiment might signify that data are ready to be sent to a computer. Because the experiment and the computer march to the beats of different drummers (in fancy language you would say they are *asynchronous* processes), you need a method to restore order between the two systems.

A. Example: pulse synchronizer

As an example, let's reconsider the circuit in which a debouncer flip-flop gated a pulse train (Figure 10.51 in §10.4.1A). That circuit enables the gate whenever the switch is closed, regardless of the phase of the pulse train being gated, so that the first or last pulse may be shortened. The problem is that the switch closure is asynchronous

---

[57] It's easy to get them mixed up; perhaps the mnemonic "Moore↔OutputsOnly" is helpful.

[58] We've chosen the primitive ABEL, rather than something sophisticated like Verilog, because it is "closer to the metal": it lets you specify D inputs and Q outputs, etc. In ABEL the XOR is written as $.

[59] Well, almost. At a fussier level of scrutiny you can worry about timing *skew*, the (relatively small) scatter of propagation times, among the various flip-flops, from the common clocking edge to the various Q outputs.

with the pulse train. In some applications it is important to have only *complete* clock cycles, and that requires a synchronizer circuit like that in Figure 10.72. When the switch creates a debounced HIGH on the *D* input, *Q* stays LOW until the next falling edge of the input pulse train. In that way, only complete pulses are passed by the AND gate. In the waveforms the curved arrows are drawn to show what causes what. You can see, for instance, that the transitions of *Q* occur slightly *after* the falling edges of the input.

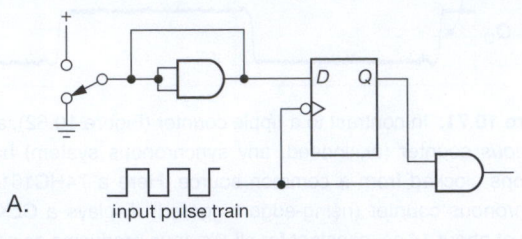

A.   input pulse train

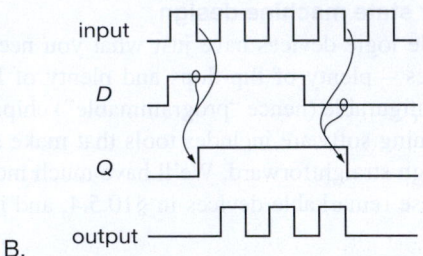

B.   output

Figure 10.72. Pulse train synchronizer: A. circuit; B. timing diagram.

B. Logic races and glitches

This example brings up a subtle but extremely important point: what would happen if a *rising*-edge-triggered flip-flop were used instead? If you analyze it carefully, you'll find that things work OK at the beginning of the pulse train, but that a bad thing happens at the end (Figure 10.73). A short spike (or "glitch" or "runt pulse") gets through because the final AND gate isn't disabled until the flip-flop output has a chance to go LOW, a delay of about 20 ns for HC family logic. This is a classic example of a *logic race*. With some care these situations can be avoided, as the example shows.[60] Glitches are terrible things to have running through your circuits. Among other things, they're hard to

---

[60] That example required a falling-edge-triggered flip-flop, which is a rare species – but a rising-edge flip-flop creates the glitch! What to do? With some thought you will realize that the glitch-free condition is this: *the clock level that enables the gate must precede the edge that*

see on an oscilloscope, and you may not know they are present. They can clock subsequent flip-flops erratically, and they may be widened – or narrowed to extinction – by passage through gates and inverters.

Exercise 10.25. Demonstrate that the preceding pulse-synchronizer circuit (Figure 10.72) does not generate glitches.

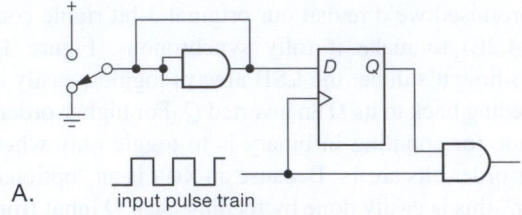

A.   input pulse train

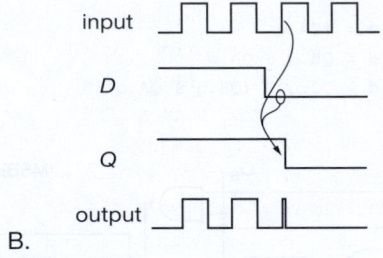

input

D

Q

output

B.

Figure 10.73. A logic race can generate a "runt pulse."

A few comments about synchronizers: the input to the *D* flip-flop can come from other logic circuitry, rather than from a debounced switch. There are applications in computer interfacing, etc., where an asynchronous signal must communicate with a clocked device; in such cases clocked flip-flops or synchronizers are ideal. In this circuit, as in all logic, unused inputs must be handled properly. For instance, SET and CLEAR must be connected so that they are not asserted (for a '74, tie them HIGH; for a 4013, they are grounded). Unused inputs that have no influence on the outputs can be tied to either level.

---

clocks the flip-flop. OK, you say, "I'll clock the rising-edge flip-flop with the input pulse train, and I'll pass that same input through an inverter on the way to the AND gate." Sounds fine, but there's a new problem, namely a logic hazard at the *beginning* of the synchronized output train. Challenge yourself by figuring it out, and drawing a circuit that works properly with a rising-edge flip-flop.

10.4.5 Monostable multivibrator

In §7.2.2 we introduced what is really a *mixed-signal* device (combined analog and digital), the monostable multivibrator (also known as a "one-shot," accent on the *one*); see Tables 7.3 and 7.4 on pages 462 and 463. Although we counseled some caution when considering monostables, there are times when they do just what you want. Let's interrupt our narrative here to look at a nice example, namely, a triggerable pulse generator.

A. Pulse-generator example

Figure 10.74 shows a square-wave generator with independently settable rate and duty cycle (ratio of HIGH to LOW), along with an input that permits an external signal to start and stop the pulse train synchronously. Current source U_1–Q_1 generates a ramp at C_1, with slew rate proportional the resistance of panel potentiometer R_2. When C_1's voltage reaches the upper comparator's threshold of 3.0 V, the one-shot is triggered and generates a 100 ns LOW pulse, putting n-channel MOSFET Q_2 into conduction and discharging the capacitor. C_1 therefore has a sawtooth waveform going from ground to +3 volts, with the rate set by potentiometer R_2. The lower comparator generates an output square wave from the sawtooth, with duty cycle adjustable linearly between 2% and 98% via R_7. Both comparators have a few millivolts of hysteresis (R_{10} and R_{11}) to prevent noise-induced multiple transitions. The TLV3502 is a fast (4.5 ns) CMOS dual comparator with input common-mode range to both rails (ground and +5 V), and rail-to-rail outputs (see Table 12.2 on page 813 for more comparator types).

A feature of this circuit is its ability to synchronize (start/stop) to an externally applied control level. The EN-ABLE input lets the driven circuit start the oscillator at a predictable phase (the time of an output pulse's falling edge), and command the oscillator to stop after the next complete pulse.

A few instructive details (which can be skipped over on a first reading).

- The additional input to the 3-input NAND from the comparator output ensures that the circuit won't get stuck with C_1 charged up.
- The one-shot pulse width has been chosen long enough to ensure that C_1 is fully discharged during the pulse: you can estimate the discharge time from looking at both the 2N7000's saturation drain current with 5 V gate drive (about 350 mA), and its R_{on} (5 Ω) when the discharge is nearly complete; they correspond to approximate discharge times of 50 ns and 17 ns, respectively, from which we conservatively set a fixed discharge time of 100 ns.

- The 50 Ω resistor at the output provides "source termination" for a 50 Ω cable (see Appendix H).
- We've arranged things so that the top of the sawtooth waveform corresponds to the short duty-cycle setting; that's because the waveform has a 100 ns flat bottom (during discharge), whereas the top is an accurate sawtooth.
- Note that the oscillation frequency is, to a good approximation, independent of supply-voltage variations, being set ratiometrically (both the charging current and the peak amplitude are set as a fraction of V_+, nominally 5 V).
- The frequency-setting voltage applied to U_1 is bypassed to the *positive* rail, because that is the reference for the current source; at the low-frequency end of the range, the voltage at R_{15} is just 5 mV below the positive rail, thus rather sensitive to supply-voltage noise.
- We've specified a log-taper potentiometer for the frequency control R_2; otherwise the low-frequency end of the range gets squished.

10.4.6 Single-pulse generation with flip-flops and counters

In §7.2.2 we discussed the generation of pulses and time delays with monostable multivibrators, and some reasons to use caution when considering whether to use these partly–analog devices. When a clocking signal is present (which is almost always), there are purely digital alternatives. Figure 10.75 shows how to make a single glitch-free pulse whose width equals one clock cycle. You'll have fun drawing the timing diagram for this circuit – take the challenge!

Exercise 10.26. Take the challenge!

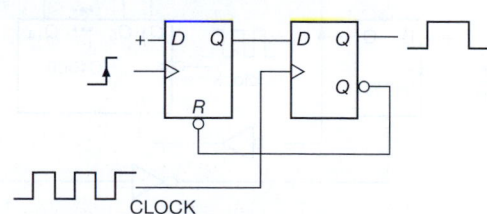

Figure 10.75. Single-pulse generator.

Figure 10.76 shows another case in which flip-flops and counters (cascaded toggling flip-flops) can be used in place of a monostable to generate a long output pulse. The '4060 is a 14-stage CMOS binary ripple counter (14 cascaded flip-flops). A rising edge at the input brings Q HIGH, enabling the counter. After 2^{n-1} clock pulses, Q_n goes HIGH,

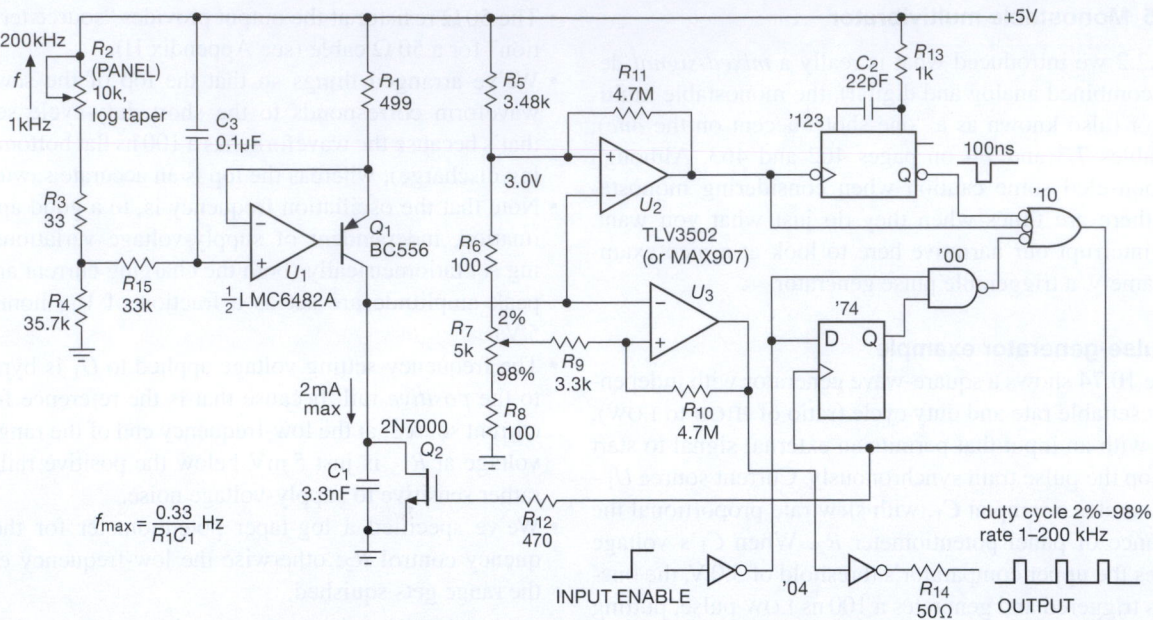

Figure 10.74. Autosynchronizing triggerable pulse generator. All digital logic is LVC family CMOS.

clearing the flip-flop and the counter. This circuit generates an accurate long pulse whose length may be varied by factors of 2. The '4060 also includes internal oscillator circuitry that can substitute for the external clock reference. Our experience is that the internal oscillator has poor frequency tolerance and (in some HC versions) may malfunction.

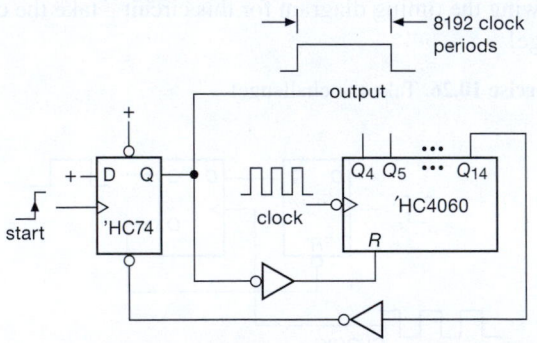

Figure 10.76. Digital generation of long pulses.

You can get complete ICs to implement timing with counters. The Maxim ICM7240/50/60 have 8-bit or 2-digit internal counters and the necessary logic to make delays equal to an integral number of counts (1–255 or 1–99 counts); you can set the number either with "hard-wired" connections or with external thumbwheel switches.

The ICM7242 is similar, but with prewired divide-by-128 counter.

10.5 Sequential functions available as integrated circuits

As with the combinational functions we described earlier, it is standard practice to integrate various combinations of flip-flops and gates onto a single chip to create sequential "standard logic." In the following subsections we present a brief survey of the most useful types. *As we feel obliged to mention often, you can also get* uncommitted *arrays of gates and flip-flops, the interconnections to be programmed by the user. These are the hugely popular FPGAs and cPLDs, which we discuss later in this chapter and in Chapter 11.*

10.5.1 Latches and registers

Latches and registers are used to "hold" a set of bits, even if the inputs change. A set of individual D flip-flops constitutes a register, but it has more inputs and outputs than necessary. Because you don't need separate clocks, or SET and CLEAR inputs, those lines can be tied together, requiring fewer pins and therefore allowing eight flip-flops to fit in a 20-pin package. The popular '574 is an octal D register with rising clock edge and three-state outputs; the '273 is similar, but has a reset instead of three-state outputs. For larger data widths you can get the '16374 (16-bit) and '32374 (32-bit) D registers. Figure 10.77 shows a quad D register with both true and complemented outputs. The Verilog code that describes it is in Figure 10.78.

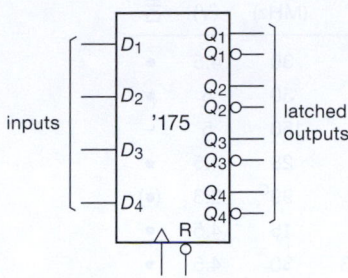

Figure 10.77. '175 4-bit *D* register.

```
// 4-bit D-register with true and complemented
// outputs, and with active-LOW asynchronous reset
reg [3:0] D;
reg [3:0] Q;
reg [3:0] QBAR;
wire CLKIN;
wire RESETBAR;
always @(posedge CLKIN) or negedge RESETBAR
    if (!RESETBAR)
        begin
            Q <= 4'b0000;
            QBAR <= 4'b1111;
        end
    else
        begin
            Q <= D;
            QBAR <= ~D;
        end
```

Figure 10.78. Verilog code for a quad *D*-register with true and complemented outputs.

The term "latch" is usually reserved for a special kind of register: one in which the outputs follow the inputs when enabled, and hold the last value when disabled. But the term "latch" has become ambiguous with use, so the terms "transparent latch" and "type-*D* register" are often used to distinguish these closely related devices. As an example, the '573 is the octal transparent latch analogous to the '574 *D* register, complete with three-state outputs. The 16-bit and 32-bit versions are called '16373 and '32373.

Some variations on the latch or register are as follows: (a) random-access memories (RAMs), which let you write to, and read from, a (usually large) set of registers, but only

one (or at most a few) at a time; RAM ICs come in sizes up to 1 Gbyte or more, and are used primarily for memory in microprocessor systems (see Chapter 14); (b) addressable latches, a multibit latch that lets you update individual bits while keeping the others unchanged; (c) a latch or register built into a larger chip, for example a digital-to-analog converter; such a device needs the input applied only momentarily (with appropriate clocking edge), since an internal register can hold the data.

When choosing a register or latch, look for important features such as input enable, reset, three-state outputs, and *broadside* pinout (inputs on one side of the chip, outputs on the other); the latter is convenient when you are laying out a printed-circuit board (PCB).

10.5.2 Counters

As we mentioned earlier, it is possible to make a *counter* by connecting flip-flops together. There is available an amazing variety of such devices as single chips, some of which are listed in Table 10.5. Here are some of the features to look for.

A. Size

You can get BCD (divide-by-10) and binary (or *hexadecimal*, divide-by-16) counters in the popular 4-bit category. There are larger counters, up to 24 bits (not all of them available as outputs); the 74LV8154 is a nice example – it has a pair of 16-bit synchronous counters with output registers and separate clocking inputs, and an 8-bit three-state output that forwards any selected byte. There are also modulo-*n* counters that divide by an integer *n*, with the modulus *n* specified as an input. For some applications (e.g., timing) you don't care about the intermediate bits, you just want a lot of internal stages, provided by chips like the ICM7240–60, MC14541, and MC14536; see §7.2.4D. You can always cascade counters (including synchronous types) to get more stages.

B. Clocking

An important distinction is whether the counter is a ripple counter or a synchronous counter. The latter clocks all flip-flops simultaneously, whereas in a ripple counter each stage is clocked by the output of the previous stage (Figures 10.62 and 10.63). Ripple counters generate transient states because the earlier stages toggle slightly before the later ones. For instance, a ripple counter going from a count of 7 (0111) to 8 (1000) goes through the states 6, 4, and 0 along the way. This doesn't cause trouble in well-designed circuits, but it would in a circuit that used gates to look for

Table 10.5 Selected Counter ICs[a]

Part number (74xxx)	Supply voltage min (V)	max (V)	Bits	Synchronous?	U/D	BCD	Sync clear	f_{max} min @ V_{CC} (MHz)	(V)	DIP avail?
HC4024	2.0	6.0	7	no	-	-	-	30	4.5	•
HC4040	2.0	6.0	12	no	-	-	-	30	5	•
VHC4040	2.0	6.0	12	no	-	-	-	150	5	-
HC4060	2.0	6.0	14[b]	no	-	-	-	28	4.5	•
LV4060	1.2	5.5	14[b]	no	-	-	-	99[c]	3.3	(•)
HC40103	2.0	6.0	8	•	D	'102	•	15	4.5	•
74HC161	2.0	6.0	4	•	-	'162	'163	30	4.5	•
74AC161	1.5	6.0	4	•	-	-	'163	90[c]	3.3	-
74LV161	2.0	5.5	4	•	-	-	'163	165[c]	3.3	-
74LVC161	1.2	3.6	4	•	-	-	'163	200[c]	3.3	-
74HC191	2.0	6.0	4	•	•	'192	'193	30	4.5	•
74AC191	1.5	6.0	4	•	•	-	-	133[c]	5	-
74HC590	2.0	6.0	8[d]	•	-	-	-	33	4.5	•

Notes: (a) all are binary, async reset, count up, unless otherwise marked. (b) no output pins for bits 0, 1, 2, and 10. (c) typical. (d) with 3-state outputs.

a particular state (this is a good place to use something like a D register, so that the counter output state is examined only at a safe clock edge). Ripple counters are slower than synchronous counters, because of the accumulated propagation delays. That is, it takes longer for all bits to "settle" to their next state; on the other hand, a ripple counter has a higher maximum counting rate (for the same speed of flip-flop). Ripple counters clock on falling edges for easy expandability (by connecting the Q output of one counter directly to the clock input of the next); synchronous counters clock on the rising edge.

We favor the '160–'163 family of 4-bit synchronous counters for most applications that don't require a special feature (Figure 10.79). The four family members include BCD and binary, each available with either synchronous or asynchronous reset.[61] They can be parallel loaded, and are easily cascaded via the carry output and enable inputs.

C. Up/down

Some counters can count in either direction, under control of some inputs. The two possibilities are (a) a U/D' input

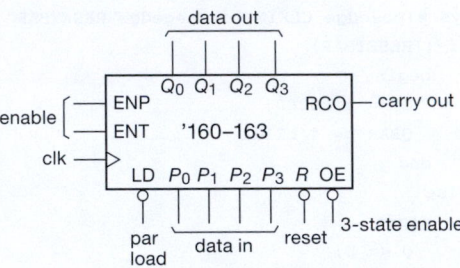

Figure 10.79. '160–'163 synchronous counters.

that sets the direction of count and (b) a pair of clocking inputs, one for UP, one for DOWN. Examples are the '191 and '193, respectively. The '579 and '779 are useful 8-bit up/down counters.

D. Load and clear

Most counters have data inputs so that they can be preset to a given count. This is handy if you want to make a modulo-n counter, for example. The load function can be either synchronous or asynchronous: the '160–'163 have synchronous load, which means that data on the input lines are transferred to the counter coincident with the next clock rising edge, if the LOAD' line is also asserted LOW; the '190–'193 are asynchronous, or *jam-load*, which means

---

[61] With the reset input R' asserted, a counter with synchronous reset waits until the next rising clock to reset, whereas a counter with asynchronous reset clears upon assertion of the R' input, regardless of the state of the clock.

that input data are transferred to the counter when LOAD$'$ is asserted, independent of the clock. The term "parallel load" is sometimes used, since all bits are loaded at the same time.

The CLEAR (or RESET) function is a form of presetting. The majority of counters have a jam-type (asynchronous) CLEAR function, though some have synchronous CLEAR; for example, the '160/'161 are jam CLEAR, whereas the '162/'163 are synchronous CLEAR.

Example: divide-by-3 (again)

Let's pause for a moment to look at this business of synchronous versus asynchronous control signals. Neither is "better" (after all, the '160–'163 family gives you both choices, and at the same price) – the choice depends on the application. Let's imagine we want to conscript a '161/'163-style 4-bit synchronous binary counter into service as a divide-by-3 (a challenge we took on earlier, with discrete flip-flops, in §10.4.3B). Because these are *up*-counters only, we'll use a NAND gate to detect the state count = 3, and use its active-LOW output to assert the RESET$'$ input of the counter. So it should count 0, 1, 2, and the next clock takes it to a count of 3, whereupon it is immediately reset. Because we want the reset to happen right away, we've chosen the '161, with its asynchronous RESET$'$ input (Figure 10.80A).

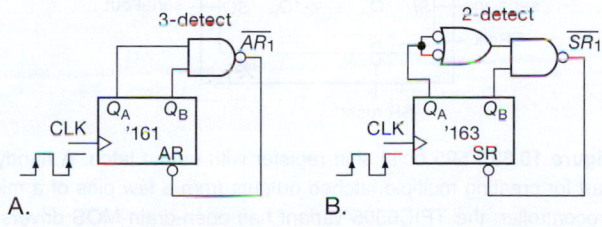

Figure 10.80. Two divide-by-3 circuits, from 4-bit synchronous counters: A. with asynchronous reset ('161); B. with synchronous reset ('163). We built both, with LV-A family logic running at 3.3 V; waveforms in Figures 10.81 and 10.82.

Sounds fine, and it does work, more or less. But there's a bit of a problem: look at Figure 10.81, which shows measured waveforms, as we clocked it at about 12 MHz. You can see it clock successively through states 0, 1, and 2; then, just past halfway across the trace you see it reach state 3 (Q_A and Q_B both HIGH), whereupon the NAND gate generates a LOW pulse (AR$'_1$), which resets the counter to 0. The problem is that the RESET$'$ pulse contains the seeds of its own destruction, so to speak: it brings the count to zero, which makes it promptly disappear. So it's possible for the pulse to be shorter than the minimum reset pulse

width, which could result in incomplete resetting of all the counter's flip-flops.[62]

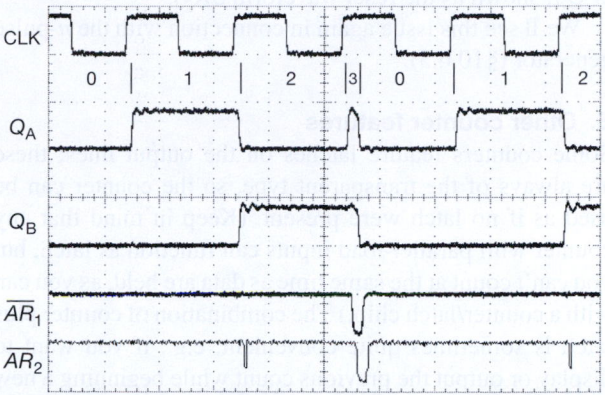

Figure 10.81. Measured waveforms from the divide-by-3 circuit of Figure 10.80A. The modified reset pulse AR$'_2$ resulted from 39 pF additional loading on Q_A. Horizontal: 40 ns/div. Vertical: 4 V/div.

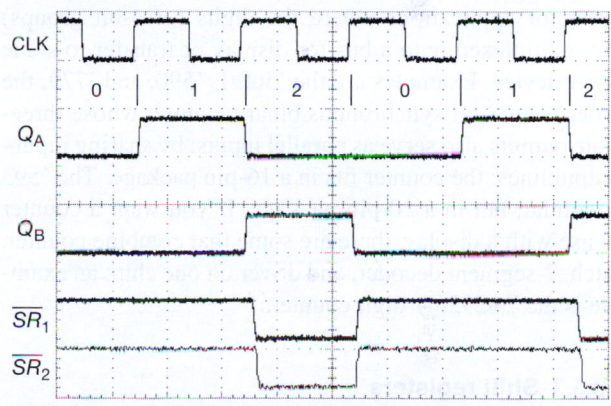

Figure 10.82. Measured waveforms from the divide-by-3 circuit of Figure 10.80B; same conditions as Figure 10.80A.

The scope trace shows another potential hazard of using asynchronous reset here: the trace labeled AR$'_2$ is the NAND gate output when the Q_A output sees more capacitive loading than Q_B (we put 39 pF to ground); that delayed Q_A enough to produce a transient false "3" state during the 1→2 transition. In this circuit it wasn't sufficient to prematurely reset the counter... but it sure looks bad!

The better solution is *synchronous* reset (Figure 10.80B), where the counter is reset at the rising clock *following* assertion of its RESET$'$ input. That means we need to detect count = 2 (rather than 3; i.e., $n-1$ rather

---

[62] Even if it works reliably in practice, you don't want little runt pulses running around your circuit, right?

than n).[63] Figure 10.82 shows the waveforms, free of glitches and runt pulses (even with extra capacitive loading of Q_A, shown as the reset waveform SR'_2).

We'll see this issue again in connection with the n-pulse generator (§10.6.3).

E. Other counter features

Some counters feature latches on the output lines; these are always of the transparent type, so the counter can be used as if no latch were present. (Keep in mind that any counter with parallel-load inputs can function as latch, but you can't count at the same time as data are held, as you can with a counter/latch chip.) The combination of counter plus latch is sometimes quite convenient, e.g., if you want to display or output the previous count while beginning a new counting cycle. In a frequency counter this would allow a stable display, with updating after each counting cycle, rather than a display that repeatedly gets reset to zero and then counts up.

There are counters with three-state outputs. These are great for applications where the digits (or 4-bit groups) are multiplexed onto a bus for display or transfer to some other device. Examples are the '560/1, '590, and '779; the latter is an 8-bit synchronous binary counter whose three-state outputs also serve as parallel inputs; by sharing input–output lines, the counter fits in a 16-pin package. The '593 is similar, but in a 20-pin package. If you want a counter to use with a display, there are some that combine counter, latch, 7-segment decoder, and driver on one chip; an example is the 74C926 4-digit counter.[64]

10.5.3 Shift registers

If you connect a series of flip-flops so that each Q output drives the next D input, and all clock inputs are driven simultaneously, you get what's called a *shift register*. At each clock pulse the pattern of 0s and 1s in the register shifts to

the right, with the data at the first D input entering from the left. As with flip-flops, the data present at the serial input just prior to the clock pulse are entered, and there is the usual propagation delay to the outputs. Thus they may be cascaded without fear of a logic race. Shift registers are very useful for conversion of parallel data (n bits present simultaneously, on n separate lines) to serial data (one bit after another, on a single data line), and vice versa. They're also handy as memories, particularly if the data are always read and written in order. As with counters and latches, shift registers come in a pleasant variety of prefab styles. The important things to look for are the following.

A. Characteristics
Size and Organization

The 4-bit and 8-bit registers are standard, with some larger sizes available (up to 64 bits or more).[65]

Shift registers are usually 1 bit wide, but there are also dual-, quad-, and hex-width registers. Most shift registers only shift right, but there are bidirectional registers like the '194 and '299.

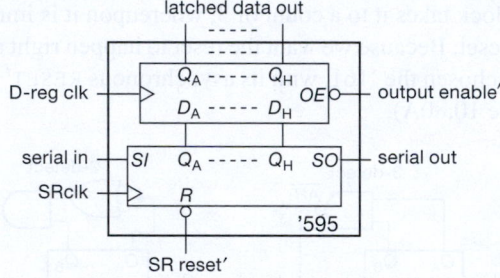

Figure 10.83. '595 octal shift register with output latch. A handy part for creating multiple latched outputs from a few pins of a microcontroller; the TPIC6595 variant has open-drain MOS drivers, good to 45 V and 250 mA, for driving heavy loads.

Inputs and outputs

Small shift registers can provide parallel inputs or outputs, and usually do; an example is the '395, a 4-bit parallel-in, parallel-out (PI/PO) shift register with three-state outputs. Larger registers may provide only *serial* input or output, i.e., only the input to the first flip-flop or the output from the last is accessible. In some cases a few selected intermediate taps are provided. One way to provide both parallel input and output in a small package is to share input and

---

[63] We could have ignored Q_A, and simply inverted Q_B (figure out why); however, for clarity we did full gating to detect $Q_A=0$ AND $Q_B=1$ for the synchronous reset signal SR'_1.

[64] We mourn the passing of the unusual TIL306/7, a counter *with display* on one chip: you just looked at the IC, which lit up with a digit telling the count! Another fine example of an integrated counter is Intersil's ICL7216, an 8-digit 10 MHz universal counter-on-a-chip (complete with 7-segment LED driver); it's shown in all its glory in our second edition (on p. 526). The bad news is that it was recently discontinued; but the good news is that you can do it yourself, in an FPGA or cPLD – see Chapter 11. (Maxim still makes the ICM7217 4-digit up-down counter with multiplexed LED drive, and the ICM7218 and 7228 8-digit-LED drivers are available from both Maxim and Intersil.)

[65] There are even variable-length registers (e.g., the 4557: 1 to 64 stages, set by a 6-bit input).

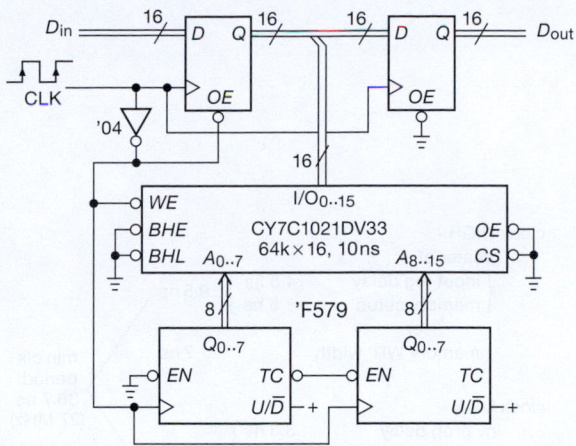

Figure 10.84. Large and wide shift register made from RAM plus counter; the slash indicates multiple lines, in this case 16-bit-wide data paths and a pair of 8-bit-wide addresses, for a total length of 65,536 16-bit words.

output (three-state) on the same pins, e.g., the '299, an 8-bit bidirectional PI/PO register in a 20-pin package. Some shift registers include a latch at the parallel input or output, so shifting can go on while data are being loaded or unloaded. A particularly nice example of the latter is the '595 8-bit shift register (Figure 10.83), nicely available in many logic families, including AHC(T), F, FCT, HC(T), LV, LVC, and VHC; it's great for creating a latched many-bit parallel output from a stingy bitstream coming from a microcontroller.[66] Similarly, the '597 is an 8-bit parallel-input shift register, convenient for getting data into a microcontroller via a single-bit serial input pin.

As with counters, parallel LOAD and CLEAR can be either synchronous or jam-load; for example, the '323 is the same as the '299, but with synchronous clear.

B. RAMs as shift registers

A random-access memory can always be used as a shift register (but not vice versa) by using an external counter to generate successive addresses. Figure 10.84 shows the idea. A pair of cascaded 8-bit synchronous counters generates successive addresses for a 64k-word×16-bit static RAM. The combination behaves like a 16-bit wide and 65,536-long shift register. By choosing a fast counter[67]

---

[66] The wonderful do-it-all microcontroller is the subject of Chapter 15, where we'll show an example.

[67] The '579 is available only in a 5 V logic family. The rest of the circuit is powered from +3.3 V; however, signal levels are compatible throughout (that is, the logic running on 3.3 V has "5V-tolerant" inputs; and the 5V logic accepts 3.3 V input levels; see §12.1.2A).

and memory, we were able to achieve a maximum clocking rate of 27 MHz (see timing diagram, Figure 10.85), which is comparable to that of an integrated (but much smaller) standard-logic shift register. (It's worth spending a few minutes studying that calculation – a good exercise in setup and hold times, propagation delays, and memory access timing.) This technique can be used to produce very large shift registers, if desired.

Exercise 10.27. In the circuit of Figure 10.84, input data seem to go into the same location that output data are read from. Nevertheless, the circuit behaves identically to a classic 64k-word shift register. Explain why.

10.5.4 Programmable logic devices

We've said it before (and we'll be saying it again) – contemporary digital design is moving relentlessly toward user-programmable ICs, typically containing from hundreds to hundreds of thousands(!) of gates and flip-flops,[68] in which the *connections* are programmable. Design entry is done in a hardware description language, processed by software to generate the connection *netlist*, then loaded into the chip via a serial interface (usually JTAG). We'll have more to say about these in the next chapter; but they do so many logic jobs so well that we cannot end this chapter without a short summary.

A. The bad news

To use these little puppies you usually need to learn an HDL like Verilog or VHDL, and you need a programming "pod" (or other link to the computer where the HDL is processed). And these devices are universally (well, almost) packaged only in surface-mount packages, which makes rough-and-ready prototyping more difficult.

B. The good news

You can preserve an ignorance of HDL programming by using instead schematic-entry alternatives that are available, both from the PLD manufacturers and from third-party suppliers (§11.3.3A). And for most digital jobs, PLDs really hit the spot. Here are the most important uses and advantages of PLDs.

State machines

The PLD is a natural for an arbitrary synchronous state machine. You would be foolish to use an array of D flip-flops and discrete combinational logic when a PLD does the job in one inexpensive and powerful package.

---

[68] Sometimes supplemented with dedicated functions such as RAM, interfaces, and processors.

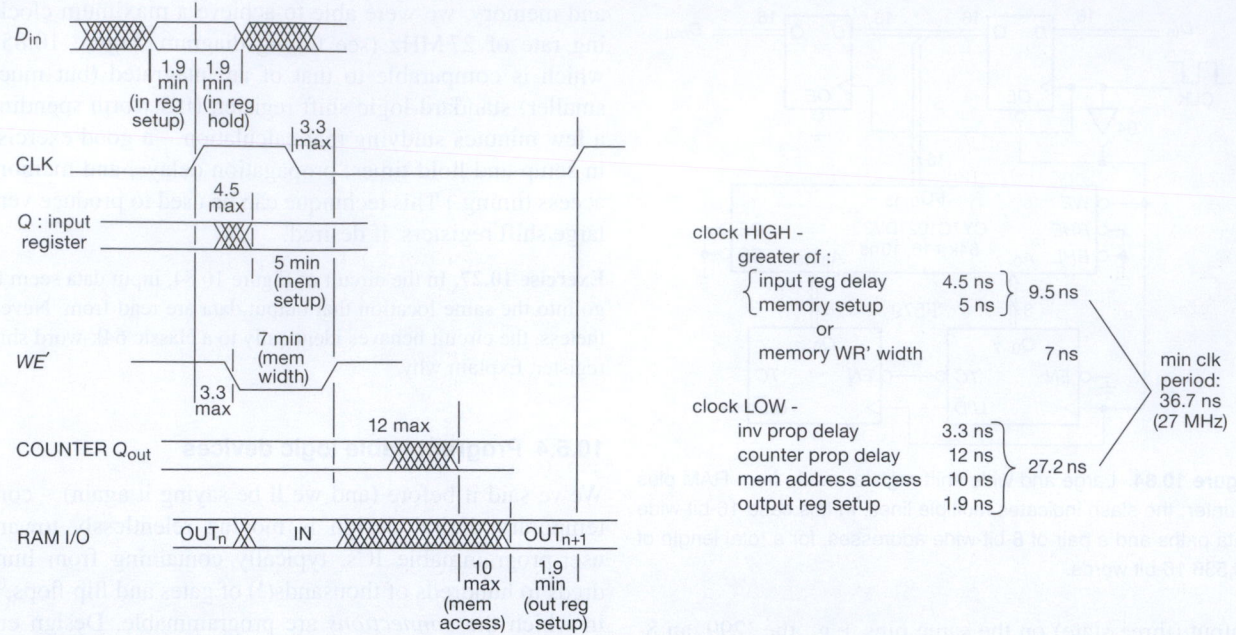

Figure 10.85. Timing diagram for the long shift-register implemented in RAM (the circuit of Figure 10.84), assuming worst-case timing specifications. This facilitates calculation of the maximum clock speed of 27 MHz.

Replacing "random" logic

Within many circuits you find little knots and tangles of gates, inverters, and flip-flops, scornfully referred to as *random logic*, or *glue logic*. A PLD will generally cut the package count by a factor of 10 or more.

Flexibility

Sometimes you're not quite sure how you ultimately want some circuit to work, yet you must finish the design so that you can play around with it. PLDs are great here, because you can reprogram them at some later stage, without the rewiring you'd have to do if you had used discrete logic. With PLDs, the circuit itself is a form of software.[69]

Multiple versions

PLDs make it possible to design a single circuit, then produce several different versions of the instrument by populating the board with differently programmed PLDs.

Speed and inventory

With PLDs you can generally get the design job done more quickly (once you've learned the ropes and set up your software tools). Furthermore, you need to stock only a few PLD types, rather than dozens of standard-function MSI logic types.

System-on-a-chip

The larger PLDs (in particular, FPGAs) have enough resources so that you may be able to do your whole design in one PLD. In particular, you can include functions such as an interface (Ethernet, USB, or whatever), memory, and even a microprocessor on the single FPGA. There are two possibilities: (a) if you include such functions in your HDL design and let the software implement them within the array of gates and flip-flops, you've got a "soft" implementation; that may include nicely implemented functions that you get from elsewhere, in which case it's called "intellectual property" (IP); (b) these functions may be hardwired into the FPGA (and not changeable) when you get it – that's a "hard implementation." More on this in the next chapter.

10.5.5 Miscellaneous sequential functions

With the relentless progress of the semiconductor industry, which routinely puts millions of transistors on a chip,[70] you

---

[69] More properly, *firmware*: in between unchangeable *hard*ware and easily changed *soft*ware.

[70] And *billions* in some of the larger microprocessors, graphic processors (GPUs), and FPGAs.

can get weird and wonderful gadgets all on one inexpensive chip. This brief subsection presents just a sampling.

First-in–first-out memory

A first-in–first-out (FIFO) memory is somewhat akin to a shift register in that data entered at the input appear at the output in the same order. The important difference is that with a shift register the data get pushed along as additional data are entered and clocked, but with a FIFO the data notionally *fall through* to the output queue. Input and output are controlled by separate clocks, and the FIFO keeps track of what data have been entered and what data have been removed.[71]

FIFOs are useful for buffering asynchronous data. The classic application is buffering a keyboard (or other input device, such as magnetic disk, or a fast external port such as Ethernet) to a computer or sluggish instrument. By this method, no data is lost if the computer isn't ready for each word as it is generated, provided the FIFO isn't allowed to fill up completely. The generic 7201–06 byte-wide series has been popular, and comes in several voltages and lengths; for example, 72V01–06 series of 3.3 V CMOS FIFOs has 0.5–16k words of 9 bits each, 40 MHz maximum speed, and zero *fall-through time* (an affliction suffered by early FIFOs, which were implemented as serial registers). You can get bidirectional FIFOs, synchronous FIFOs, and FIFOs up to 72 bits in width.

A FIFO is unnecessary if the device to which you are sending data can always get it before the next data arrive. In computer language, you must ensure that the maximum *latency* is less than the minimum time between data words. Note that a FIFO will not help if the data recipient is not able, *on the average*, to keep up with the incoming data.

Digital voltmeters

You can get complete digital voltmeters (DVMs) on a single chip (for some reason you'll find these things listed as "display drivers" when you go searching at distributors). They include voltage reference, high-impedance differential inputs, LCD drivers, and so on. An example is the MAX1495, a fully integrated $4\frac{1}{2}$-digit voltmeter in a 7 mm square package; it draws about a milliamp from a single 3–5 V supply. A popular generic part is the '7135 (TI, Intersil, and Maxim call it ICL7135; Microchip calls theirs TC7135; and TI also sells it as the TLC7135): it's a $4\frac{1}{2}$-digit voltmeter that runs from +5 V and drives 7-segment

LED displays. The '7136 is $3\frac{1}{2}$-digit and drives 7-segment LCDs.

Special-purpose circuits

There are nice collections of large-scale integration (LSI) chips for arcane jobs like radio communications (e.g., frequency synthesizers), digital signal processing (digital filters, correlators, Fourier transforms, arithmetic units), data communications (UARTs, modems, network interfaces, data encryption–decryption ICs, serial format converters, wireless protocols), and the like. Often these chips are used in conjunction with microprocessor-based devices, and many of them cannot stand alone.

Consumer chips

The semiconductor industry loves to develop ICs for use in large-market consumer products. You can get single chips to make digital (or "analog") watches, clocks, locks, calculators, smoke detectors, telephone dialers, music synthesizers, rhythm and accompaniment generators, etc. The guts of radios, TVs, music and video players, GPS navigators, and cellphones are nearly empty these days, thanks to large-scale integration. Speech synthesis and recognition is well developed: that's why devices like GPS navigators can speak to us and understand our analog-generated replies.[72] Automobiles are laden with dozens of processors, for tasks like engine control, braking, collision-avoidance systems, navigation, and so on. Even the lowly toothbrush has a processor chip, running on a few thousand lines of computer code.[73]

Microprocessors

The most stunning example of the wonders of very-large-scale integration (VLSI) is the microprocessor, a computer on a chip. At one extreme there are powerful number crunchers like the Intel eight-core Itanium, with more than 3 billion transistors; it has hundreds of internal registers, supports up to a petabyte (a million gigabytes) of RAM, and can be cobbled together into 512-processor architectures. At the other extreme are inexpensive single-chip processors with a rich load of input, output, and memory functions included on the same chip for stand-alone use.

---

[71] A FIFO can be implemented in software (as it usually is done in hardware) by creating a *ring buffer* in RAM, with a pair of pointers (write and read).

[72] One of the authors' children, enchanted with the talking Nissan Maxima we rented one summer, augmented its lexicon, which included messages such as "door is a-*jar*" and "tank is *low*," with his own: "You ran into a *di*nosaur."

[73] Check out the MC9RS08KA from Freescale Semiconductor, with internal 10 MHz oscillator and 2 KB of internal memory; it's "small enough to fit in the head of an electric toothbrush" and intended for "personal care appliances," among other uses. It costs a mere $0.40 in quantity.

An example (Figure 10.86) is the ARM7 LPC2458 from NXP (formerly Philips): 512kB/64kB memories, 72 MHz clock, 10/100 Ethernet, USB 2.0, 10-bit A/D and D/A, 2xPWM, 4 UARTs, 2xCAN, SPI, 2xSSP, 3xI^{2}C, I^{2}S, 136 bits of general-purpose I/O, and an external memory controller. This puppy costs just $10![74] This latter type is intended as a dedicated controller in an instrument rather than as a versatile computation device.

The microprocessor revolution hasn't begun to slow, and we have seen a doubling of computer power and memory size (now 8 Gbit per chip, compared with 1 Mbit/chip and 16 kbit/chip at the time the two previous editions of this book were written) every 18 months ("Moore's law"); at the same time, prices have dropped dramatically (Figure 10.87). Along with bigger and better processors and memory, recent activity in display devices, networking, and wireless data communication promises yet more excitement in coming years.

10.6 Some typical digital circuits

Thanks to the efforts of the semiconductor industry, digital design is wonderfully easy and pleasant. It's almost never necessary to "breadboard" a digital circuit, as so often is the case with analog design. Generally speaking, the only serious pitfalls involve timing and noise. We'll have more to say about the latter in Chapter 12.

This is a good place to illustrate timing with some sequential design examples. Some of these functions can be performed with LSI circuits or with programmable logic, but the implementations shown are reasonably efficient, and illustrate the kind of circuit design that's straightforward with widely available parts (and without mastery of any software languages or tools).

10.6.1 Modulo-*n* counter: a timing example

The circuit in Figure 10.88 produces one output pulse for every $n+1$ input clock pulses, where n is the 8-bit number you've set on the pair of hexadecimal thumbwheel switches. The '163s are 4-bit synchronous up-counters, with synchronous load (when LD$'$ is LOW) via the P_n inputs. The idea is to load the counters with the *complement*

---

[74] NXP's LPC1768 is a low-power part that's popular with the prototyping crowd. It has most of the features of the LPC2458, but lacks the external memory controller. It has only 70 GPIO pins (we're sure you'll not miss the rest), and it comes in an easy-to-solder 100-pin QFP package (0.5 mm pin spacing). Notably it has a 12-bit ADC, a 10-bit DAC, and six general-purpose PWM outputs.

of the desired count, then count up to FF_H, reloading at the next clock pulse. Because we've generated the preload levels with pullups to the positive supply (with the switch common grounded), those levels are LOW-true for the displayed switch settings; that makes the preload values, interpreted as active-HIGH, equal to the 1's complement of the switch settings.

Exercise 10.28. Show that the last statement is true by figuring out the active-HIGH value that will be loaded for the switch settings in Figure 10.88.

Circuit operation is entirely straightforward. To cascade synchronous counters, you tie all clocks together, then tie a "maximum-count" output of each counter to an enable of the successive counter. For an enabled '163, the RCO (ripple-clock output) goes HIGH at maximum count, enabling the second counter via the enabling inputs ENT and ENP. Thus IC$_1$ advances at each clock, and IC$_2$ advances at the clock after IC$_1$ reaches F_H. The pair thus counts in binary until the state FF_H, at which point the LD$'$ input is asserted. This causes synchronous preload at the next clock. In this example we've chosen counters with *synchronous* load to avoid the logic race (and runt-pulse RCO) that you would get with a *jam*-loaded counter. Unfortunately, this makes the counter divide by $n+1$ rather than by n.

Exercise 10.29. Explain what would happen if jam-load counters (e.g. '191s) were substituted for the synchronous-load '163s. In particular, show how a runt pulse would be created. Demonstrate also that the foregoing circuit divides by $n+1$, whereas the asynchronous-load version would divide by n (if it worked at all!).

A. Timing

How fast can our modulo-*n* counter count? Let's use LV-family logic at 3.3 V, at which the 74LV163A specifies a guaranteed f_{max} of 70 MHz.[75] However, in our circuit there are additional time delays associated with the cascading connection (IC$_2$ has to know that IC$_1$ has reached maximum count in time for the next clock pulse) and also the load-on-overflow connection. To figure the maximum frequency at which the circuit is guaranteed to work, we have to add up the worst-case delays and make sure there is enough setup time remaining. Look at Figure 10.89, where we've drawn a timing diagram showing the load sequence that occurs at maximum count.

---

[75] The '1G04 isn't available in LV, so we'll use a 74LVC1G04.

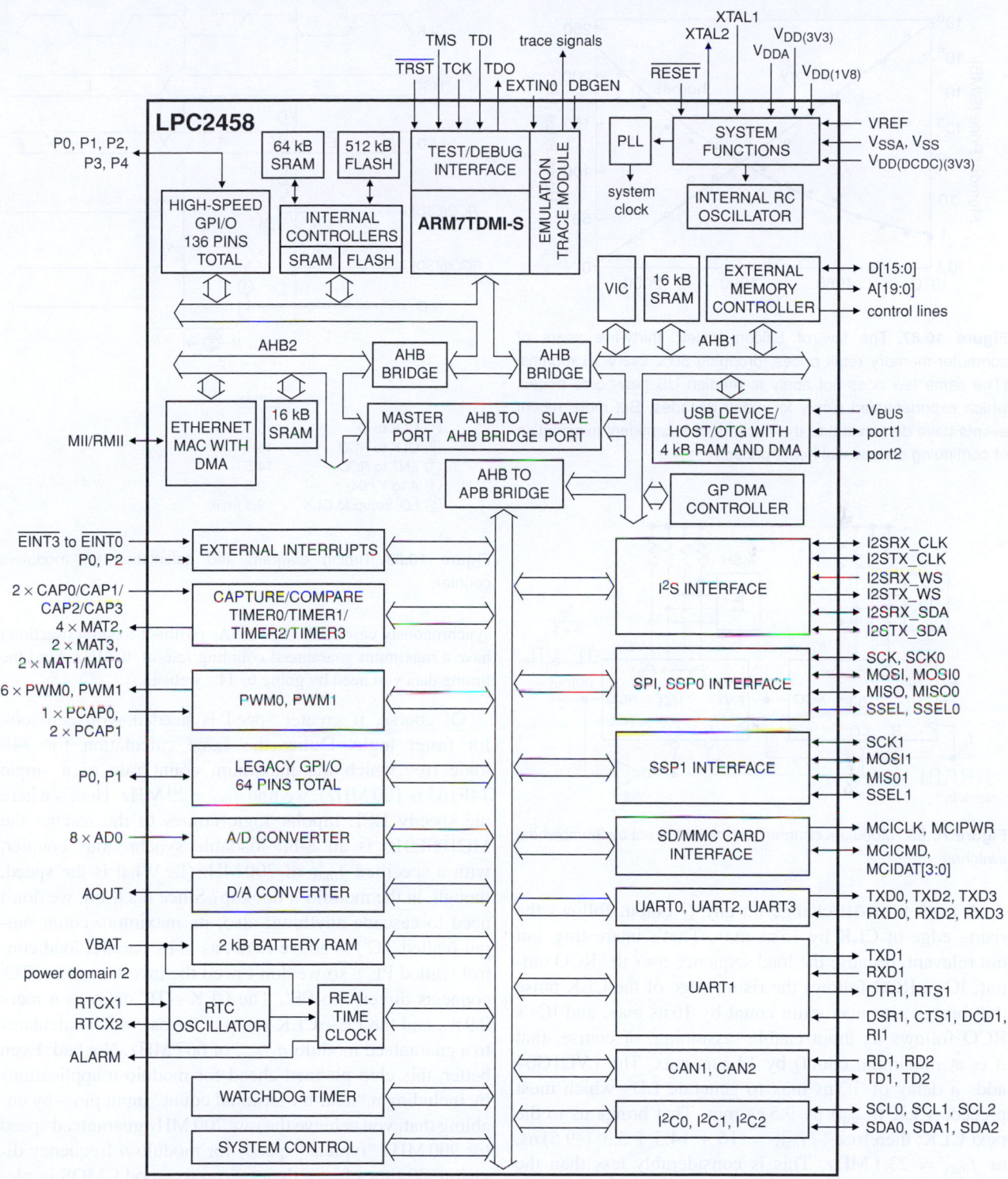

Figure 10.86. A preview of Chapter 13: an inexpensive microcontroller festooned with lots of cool stuff. Adapted from Document LPC2131_32_34 _36_38_4 ©NXP B.V. 2007)

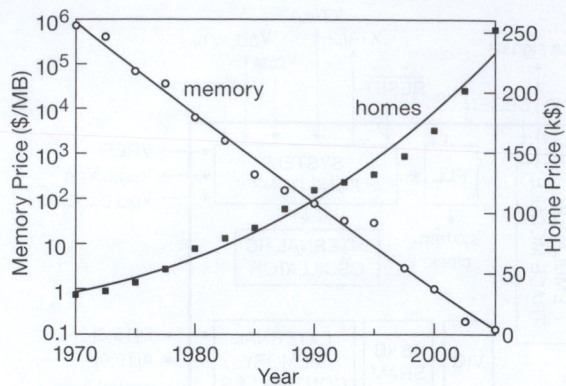

Figure 10.87. The law of Silicon Valley: thirty-five years of computer-memory retail prices, dropping 50% every 18 months. (The same law does not apply to median US new-home prices, which exponentiated nicely for many decades. But more recent events have demonstrated the hazards of a confident expectation of continuing exponential home prices.)

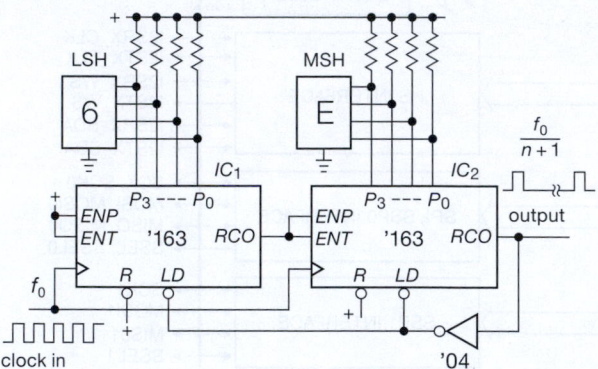

Figure 10.88. Modulo-n counter, with modulus set by thumbwheel switches.

A LOW-to-HIGH change on any Q output follows the rising edge of CLK by 15 ns max. That's interesting, but not relevant, because the load sequence uses the RCO output; IC_1's RCO follows the rising edge of the CLK pulse that brings it to maximum count by 16 ns max, and IC_2's RCO follows its input enable (assuming, of course, that it is at maximum count) by 14.5 ns max. The LVC1G04 adds a delay of 3.3 ns max to generate LD$'$, which must precede CLK (t_{setup}) by 9.5 ns min. That brings us to the next CLK; therefore $1/f_{max} = (16 + 14.5 + 3.3 + 9.5)$ ns, or $f_{max} = 23.1$ MHz. This is considerably less than the 70 MHz maximum guaranteed counting frequency of a single 74LV163A.

Exercise 10.30. Show by a similar calculation that a pair of

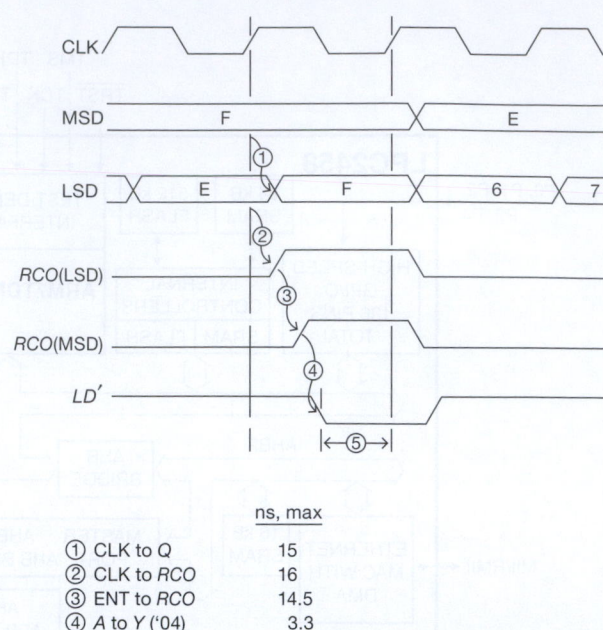

	ns, max
① CLK to Q	15
② CLK to RCO	16
③ ENT to RCO	14.5
④ A to Y ('04)	3.3
⑤ LD$'$ setup to CLK	9.5 (min)

Figure 10.89. Timing diagram and calculation for modulo-n counter.

synchronously cascaded 74LV163As (without load on overflow) have a maximum guaranteed counting rate of 40 MHz. Find the timing data you need by going to TI's website.

Of course, if greater speed is needed, you can look for faster logic. Doing the same calculation for 74F logic (for which the maximum count rate of a single 74F163 is 100 MHz), we find $f_{max}=29$ MHz. Here's where the speedy ECL bipolar logic comes to the rescue: the MC100E016 is an 8-bit loadable synchronous counter, with a specified f_{max} of 700 MHz (!). What is the speed, though, in the modulo-n hookup? Since it's 8-bit, we don't need to cascade anything; also, its maximum count output (called TC$'$) is active-LOW, as is its parallel load control (called PE$'$), so we don't need the inverter, either: TC$'$ connects directly to PE$'$. The CLK→TC delay is a mere 0.9 ns, and the PE→CLK setup is 0.6 ns, which calculates to a guaranteed modulo-n f_{max} of 667 MHz. Not bad. Even better, this chip planned ahead for modulo-n applications by including a "load-on-terminal count" input pin – by enabling that, you achieve the raw 700 MHz guaranteed speed (or 900 MHz "typical" speed) for modulo-n frequency division: 30 times faster than our pretty-good CMOS implementation!

Modulo-n counter devotees should take note of the 'HC40103, an 8-bit synchronous *down*-counter with

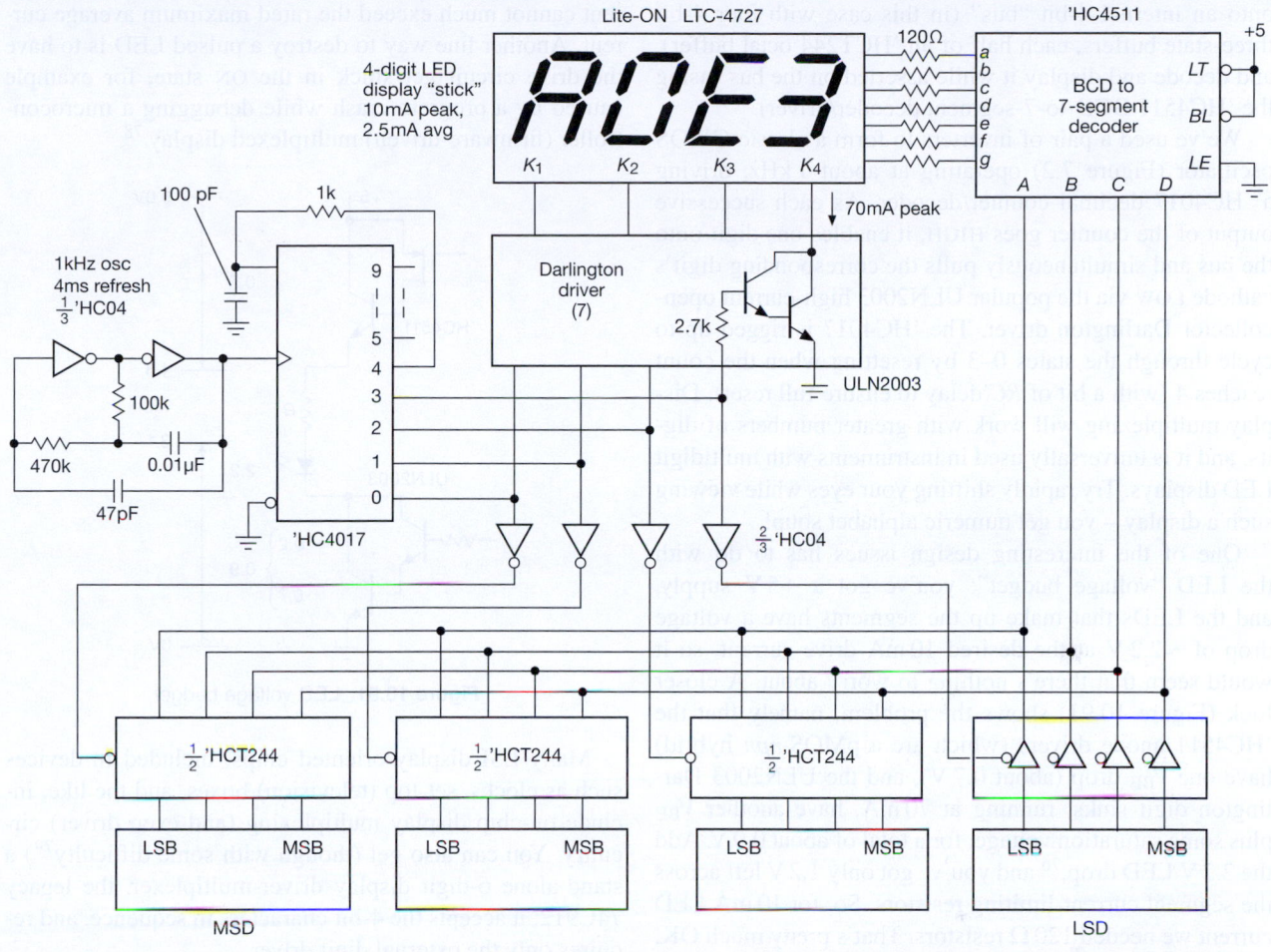

Figure 10.90. Four-digit multiplexed display. Whatever might that displayed quantity represent (and why is the 'HC4511 unhappy about it)?

parallel load (synchronous *or* jam-load!), decoded zero-state output, and reset-to-maximum input.

10.6.2 Multiplexed LED digital display

This example illustrates the technique of display multiplexing: displaying an *n*-digit number by displaying successive digits rapidly on successive 7-segment LED displays (of course, the characters need not be numbers, and the displays can have a different organization than the popular 7-segment arrangement). Display multiplexing is done for reasons of economy and simplicity: displaying each digit continuously requires separate decoders, drivers, and current-limiting resistors for each digit, as well as separate connections from each register to its corresponding decoder (4 lines) and from each driver to its corresponding display (7 wires); it's a mess!

With multiplexing, there's only one decoder/driver and one set of current-limiting resistors. Furthermore, because LED displays come in *n*-character "sticks" with the corresponding segments of all characters tied together, the number of interconnections is greatly reduced. An 8-digit display requires 15 connections when multiplexed (7-segment inputs, common to all digits, plus one cathode or anode return for each digit), rather than the 57 that would be required for continuous display (and, since most LED displays come in multiplexed varieties, you'll probably find them to be the best choice anyway).

Figure 10.90 shows the schematic diagram. The digits to be displayed are resident in registers at the bottom; they could be counters (e.g., if the device happened to be a frequency counter), or perhaps a set of latches receiving data from a computer, or possibly the output of an ADC, etc. In any case, the technique is to assert each digit successively onto an internal 4-bit "bus" (in this case with four 4-bit three-state buffers, each half of an 'HCT244 octal buffer), and decode and display it while asserted on the bus (using the 'HC4511 BCD-to-7-segment decoder/driver).

We've used a pair of inverters to form a classic CMOS oscillator (Figure 7.2) operating at about 1 kHz, driving a 'HC4017 decimal counter/decoder. As each successive output of the counter goes HIGH, it enables one digit onto the bus and simultaneously pulls the corresponding digit's cathode LOW via the popular ULN2003 high-current open-collector Darlington driver. The 'HC4017 is rigged up to cycle through the states 0–3 by resetting when the count reaches 4 (with a bit of *RC* delay to ensure full reset). Display multiplexing will work with greater numbers of digits, and it is universally used in instruments with multidigit LED displays. Try rapidly shifting your eyes while viewing such a display – you get numeric alphabet soup!

One of the interesting design issues has to do with the LED "voltage budget": you've got a +5 V supply, and the LEDs that make up the segments have a voltage drop of ~2.2 V at the desired 10 mA drive current, so it would seem that there's nothing to worry about. A closer look (Figure 10.91) shows the problem, namely that the 'HC4511 anode drivers (which are a pMOS-*npn* hybrid) have one V_{BE} drop (about 0.7 V); and the ULN2003 Darlington digit sinks, running at 70 mA, have another V_{BE} plus some saturation voltage, for a total of about 0.9 V. Add the 2.2 V LED drop,[76] and you've got only 1.2 V left across the segment current-limiting resistors. So, for 10 mA LED current we needed 120 Ω resistors. That's pretty much OK, although you'd be forgiven for worrying that some scatter in the LED forward drops, etc., would produce significant percentage changes in the small remaining voltage across the resistors; i.e., you might see unacceptable brightness variations across the digits. If you're still not worried, consider the effect of a 10% drop in the +5 V supply. Note, also, that this LED drive circuit would not work at all with a +3.3 V supply.[77]

In an LED the maximum allowable current is limited only by overheating. So it's OK to use rather high peak currents in a multiplexed display, as long as the average current stays within the rated value. You have to be a bit careful, though: the thermal time constant of the small LED semiconductor chip is somewhere down around a millisecond, so for significantly longer ON times the peak cur-

rent cannot much exceed the rated maximum average current. Another fine way to destroy a pulsed LED is to have the drive circuit get stuck in the ON state, for example caused by a program crash while debugging a microcontroller (firmware-driven) multiplexed display.[78]

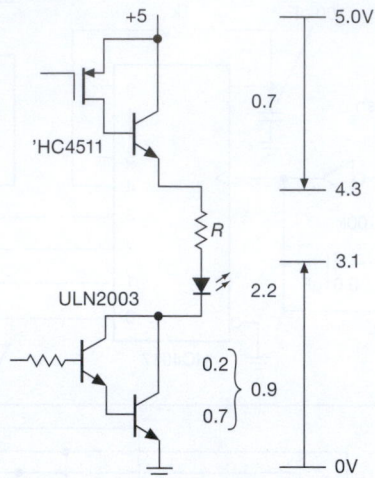

Figure 10.91. LED voltage budget.

Many LSI display-oriented chips, included in devices such as clocks, set-top (television) boxes, and the like, include on-chip display multiplexing (and even driver) circuitry. You can also get (though with some difficulty[79]) a stand-alone 6-digit display driver-multiplexer, the legacy 74C912; it accepts the 4-bit characters in sequence, and requires only the external digit driver.

10.6.3 An *n*-pulse generator

The *n*-pulse generator is a useful little test instrument. It generates a burst of *n* output pulses following an input trigger signal (or you can push a button), with a set of selectable pulse repetition rates. Figure 10.92 shows the circuit.[80] The 'HC190s are decade up/down counters (here wired to count down), clocked continuously by a selected power-of-10 subdivision of the fixed 10 MHz crystal oscillator, but disabled by having both the ALD' (asynchronous load) inputs asserted and the EN' (count enable) inputs disasserted. When a trigger pulse comes along, the first flip-flop enables the counter, and the second flip-flop synchronizes counting following the next rising edge of the clock.

---

[76] As specified for the yellow display stick shown in the figure; the forward drop of an LED depends on color; see Figure 2.8.

[77] Nor with a 5V supply and blue LED display, where the forward voltage $V_f \sim 3.5$ V.

[78] See §7.2.3 for a nice way to prevent this particular disaster.

[79] Jameco shows them in stock, $2 in unit quantity.

[80] Challenge for the reader: what is so "perfect" about the figure?

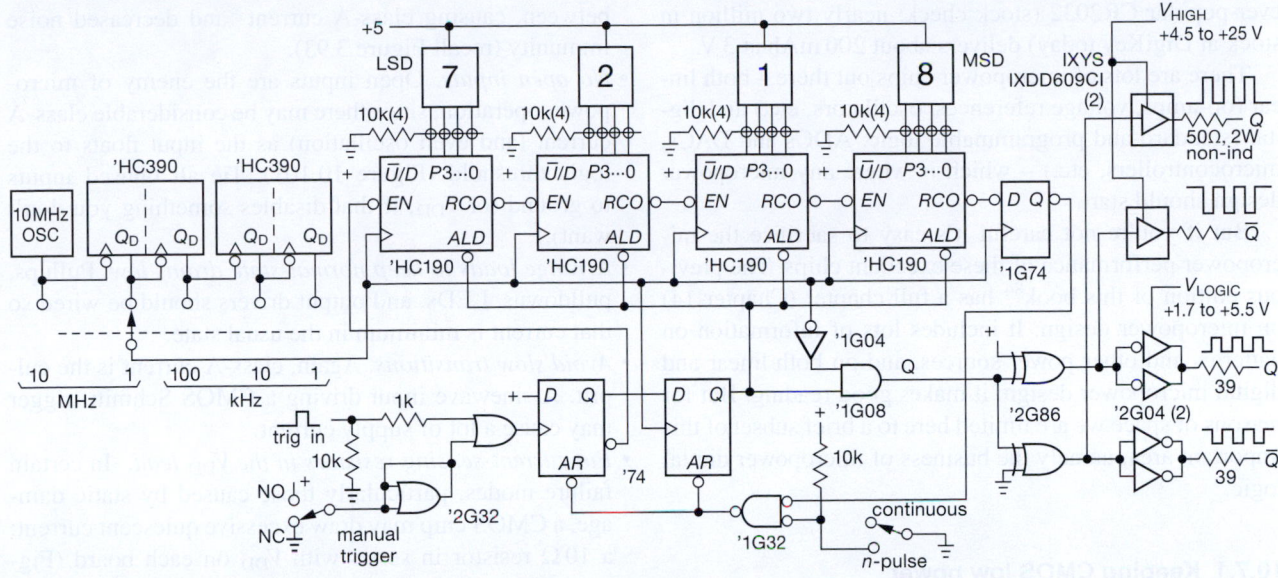

Figure 10.92. Perfect *n*-pulse generator. All logic is LVC family, running from +5 V, unless shown otherwise.

Pulses (during clock LOW) are passed by the AND gate until the counter reaches zero, at which state the RCO' output[81] is latched for one clock, and both flip-flops are reset; this parallel-loads the counter back to $n-1$ from the BCD switches, disables counting, and readies the circuit for another trigger.[82] Note that the use of pulldown resistors in this circuit means that true (rather than complemented) BCD switches must be used. Note also that the manual trigger input must be debounced, as shown, since it clocks a flip-flop. That is not necessary for the continuous/*n*-pulse switch, which simply enables a continuous stream of output pulses.

The output stage delivers two pairs of true/complement signals: the XORs generate complementary logic signals with equal delay; the paralleled 'LVC2G04 inverters give normal rail-to-rail logic swings, settable from +1.7–5.5 V via an external dc supply input. We used two inverters in parallel to increase drive capability (the paralleled inverter sections can sink or source 32 mA, staying within 0.5 V of the rails, for $V_+=3$ V); the 39 Ω resistors, in combination with the ~10 Ω inverter-stage output impedance, provide series termination for 50 Ω cable.

We added the boxed driver pair for serious driver tasks. It uses a hefty "MOSFET driver" chip, intended for rapid switching of highly capacitive MOSFET gate inputs. This particular specimen can sink or source up to 8 A, with a better-than-average switching time of ~10 ns; it is noninverting, and accepts standard 5 V logic swings. The outputs are series terminated, with non-inductive 50 Ω 2 W resistors.[83]

10.7 Micropower digital design

Small battery-powered gadgets of all sorts need to operate on very small currents, ideally down in the microamp range. To appreciate the scale, consider that a 9 V battery has a capacity of about 500 mAh, thus about 20 days at 1 mA current drain; and a small "coin cell" like the

---

[81] Indicating a ripple-through "terminal count," which for a down-counter (as here) is the 4-digit state 0000. A peculiarity of the '190 is that RCO' is enabled only during clock LOW.

[82] Because of the RCO' latch, the output pulse train has $n+1$ pulses (rather than n), so you have to set the switches to one fewer than you want. Omitting that latch, however, would trade this *numerical* embarrassment for a *logical* one, namely the generation of a terminal output runt pulse. Challenge yourself by figuring out why. However, further removing the 1G04 clock inversion would eliminate the output runt, producing instead a much-shortened reload pulse, while retaining the $n+1$ counting "feature." There's nothing wrong, really, with a reload pulse whose width is determined only by logic propagation delays – it's just a bit, well, *ugly*; by contrast, the latched RCO' for asynchronous reset–reload is neat and clean (and easy to see on a 'scope).

[83] Readers interested in constructing this circuit should consider, as an alternative to the '190 (or '192) counters, the elegant 'HC4059 4-decade loadable down-counter, which replaces four chips with one. If, however, *hexadecimal* counting is to your taste (perhaps you have 16 fingers?), use the 'HC191 (binary) alternative.

ever-popular CR2032 (stock check: nearly two million in stock at DigiKey today) delivers about 200 mAh at 3 V.

There are lots of micropower chips out there – both linear (op-amps, voltage references, oscillators, etc.) and digital (standard and programmable logic, ADCs and DACs, microcontrollers, etc.) – which is where any micropower design should start.

But if you're not careful it's easy to sacrifice the micropower performance of these excellent chips. The previous edition of this book[84] has a full chapter (Chapter 14) on micropower design. It includes lots of information on batteries and other power sources, and on both linear and digital micropower design; it makes good reading. But for reasons of space we are limited here to a brief subset of this important area, namely the business of micropower digital logic.

10.7.1 Keeping CMOS low power

There are several routine measures you should take to achieve low-current CMOS operation. In addition, it's worthwhile raising your CMOS pathology awareness.

A. Routine design considerations

* *Keep as few nodes as possible involved with high frequencies.* CMOS has no quiescent current (other than leakage), but current is required to charge internal (and load) capacitances during switching. Since the energy stored in a capacitor is $\frac{1}{2}CV^2$, and an equal amount of energy is dissipated by the resistive charging circuit, the power dissipated is

$$P = V_{\mathrm{DD}}^2 fC$$

for a switching frequency f. Thus CMOS devices consume power proportional to their switching frequency, as we saw in Figure 10.27. At their maximum operating frequency they may use more power than equivalent bipolar-TTL logic. The effective capacitance C is often given on datasheets as the "power-dissipation capacitance," C_{pd}, to which you must add the load capacitance C_{L} before applying the formula above.
* *Within a circuit, be careful whenever mixing supply voltages.* Otherwise you may have current flowing through input-protection diodes. Even worse, you may force a chip into SCR latchup (see §10.8.3B).
* *Make sure logic swings go all the way to the rails.* CMOS outputs swing rail-to-rail. Outputs from other devices – bipolar TTL, oscillators, NMOS chips – may hover in

---

[84] With continued (and perhaps perpetual) availability as an e-book.

between, causing class-A current and decreased noise immunity (recall Figure 3.93).
* *No open inputs.* Open inputs are the enemy of micropower operation, since there may be considerable class-A current (and even oscillation) as the input floats to the logic threshold (Figure 10.101). Tie all unused inputs to ground (or V_{DD}, if that disables something you don't want).
* *Arrange loads to keep normal-state drains low.* Pullups, pulldowns, LEDs, and output drivers should be wired so that current is minimum in the usual state.
* *Avoid slow transitions.* Again, class-A current is the culprit. A sinewave input driving a CMOS Schmitt trigger may cause a lot of supply current.
* *Put current-sensing resistors in the V_{DD} lead.* In certain failure modes, particularly those caused by static damage, a CMOS chip may draw excessive quiescent current; a $10\,\Omega$ resistor in series with V_{DD} on each board (Figure 10.93) makes it easy to see if that is happening (and you can use the stuck-node tracer in §4.8.2 to track down the offending component).

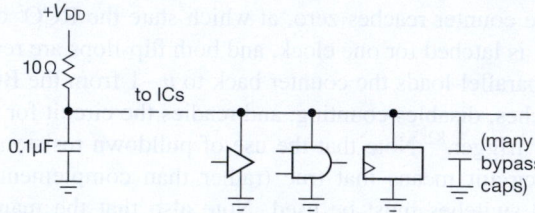

Figure 10.93. A "current spy" makes it easy to locate circuit subsections wherein lurks a current-hogging IC.

* *Quiescent current screening.* A typical CMOS logic chip (in any family – 4000B, HC, LVC, LCX, AUC, etc.) has a specified I_{Q} of 5–20µA *maximum*, but down around 0.04µA *typical* (if specified at all). Manufacturers evidently set a conservative spec on maximum leakage, probably because they don't want to bother testing to a much lower realistic value. Most of the time it is rare to have a quiescent current anywhere near the maximum, but it can happen. If you are operating at low switching frequencies (therefore low dynamic current), and require comparably low quiescent current, you may need to screen incoming chips. The use of small series resistors as recommended above makes the job much easier. We've noticed that in the case of CMOS LSI chips (such as large memories) the typical quiescent current may be close to the manufacturer's maximum leakage specifications – beware!
* *Time-out power switching.* You can save a lot of power by

making sure an instrument is turned off when no one is using it. Back in Chapter 7 we showed a simple time-out circuit, built with discrete logic, that turns off the switched +9 volt power an hour after the instrument has been turned on (§7.2.4A). Better yet, in any instrument with an embedded microcontroller (Chapter 15), use the controller's internal timer (or a programmed time-out loop) to command the power switching. In applications with limited battery energy, it's best to choose a micropower controller; alternatively, arrange for it to spend most of its time in a low-power mode (variously called "idle," "power-save," "power-down," "standby," "hibernate," or "sleep").

10.8 Logic pathology

There are interesting, and sometimes amusing, pitfalls awaiting the unsuspecting digital logician. Some of these, such as logic races and lockup conditions, can occur regardless of the logic family in use. Others (e.g., "SCR latchup" in CMOS chips) are genetic abnormalities of one logic family or another. In the following subsections we have collected our bad experiences in the hope that such anecdotes can help others avoid such problems.

10.8.1 dc problems

A. Lockup

It is easy to fall into the trap of designing a circuit with a lockup state. Suppose you have some gadget with a number of flip-flops, all going through their proper states. Everything seems to work fine. Then one day it just stops dead. The only way you can get it to work is to turn the power off and back on again. The problem is that there is a lockup state (an excluded state of the system that you can't escape from), and you got into it because of some powerline transient that sent the system into the forbidden state. It is important to look for such states when you design the circuit, and rig up logic so that the circuit recovers automatically. At a minimum, things should be arranged so that a RESET signal (generated manually, at startup, etc.) brings the system to a good state. This may not require any additional components (e.g., Exercise 10.23).

B. Startup clear

A related issue is the state of the system at startup. It is always a good idea to provide some sort of RESET signal at startup. Otherwise the system may do weird things when first turned on. One approach is to use an *RC* charging waveform, buffered by a Schmitt trigger (Figure 10.94).

In addition to requiring several discrete components, however, this circuit has the drawback of not reliably responding to a momentary voltage dip.

A better approach is provided by a power-supply *supervisory* circuit IC. These chips come in many varieties. The simplest are 3-pin parts that create a reset pulse on power-up, for example, the venerable MC34164, which comes in a convenient TO-92 leaded transistor package (in addition to the usual tiny surface-mount configurations), and holds its open-collector output LOW until the supply voltage rises above 4.3 V (2.7 V for the 34164-3 part); it includes an internal voltage reference, and some hysteresis. A more flexible part is typified by the Maxim MAX700, which comes in 8-pin packages (including DIP), and which provides both RESET and RESET′ active pullup outputs; it lets you set the threshold in the range of 1.2–4.7 V with external resistors (you can also set the hysteresis), and it has an input for a manual RESET switch (Figure 10.95). The MAX823 and ADM823 families are "jellybeans" that are widely second-sourced. Other supervisory chips include the so-called *watchdog* function: You have to pulse it at least once per second, or it does a reset; these are intended to sense a processor crash, and force a reboot (a function often integrated into contemporary microcontrollers, as well).

Many of these supervisory parts are made by multiple

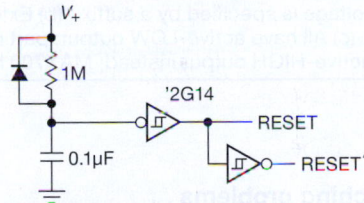

Figure 10.94. Simple power-on reset.

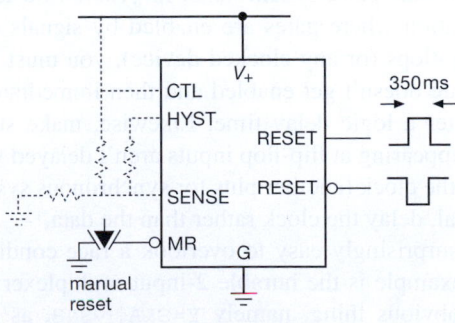

Figure 10.95. Supervisory IC provides power-on, voltage monitor, and manual reset functions. The optional (dotted) circuitry permits threshold and hysteresis adjustment.

manufacturers, but with different prefix part names. For example, the MAX809 is also available with APX, ADM, CAT, LM, STM, and TCM prefixes, and costs $0.14 in a roll of 3,000 pieces, thanks to all the competition. But be careful and read the datasheet before making a substitution! For example, NSC's MCP809 datasheet says it has a different pinout from the other '809 parts (it matches STM's jellybean STM1001). A drafting error? Maybe, but we warn you, be careful.

Advanced parts like the ADM690 series add features like supply switching to a backup battery, a second low-voltage warning comparator, and chip-enable gating. See Table 10.6 for characteristics of some favorite supervisory ICs.

Table 10.6 Selected Reset/Supervisors

Type	Pins	Package(s)	# Voltages[a]	Watchdog	Reset IN	Active High[c]	Supply current (µA)
MC34164	3	TO-92, SO-8	2	-	-	-	12
MAX809	3	SOT-23	7	-	-	'810	15
NCP303	5	SOT-23	7	-	-	'302	0.5
MAX700	8	SOIC-8	1[b]	-	●	both	100
ADM811	4	SOT-143	6	-	●	'812	5
ADM823	5	SOT-23, SC-70	7	●	●	'824	10

Notes: (a) Voltage is specified by a suffix. (b) External adjustment. (c) All have active-LOW output; part numbers listed have active-HIGH output instead; MAX700 has both.

10.8.2 Switching problems

A. Logic races

Lots of subtle traps lurk here. The classic race was illustrated with the pulse synchronizer in §10.4.4. Basically, in any situation where gates are enabled by signals coming from flip-flops (or any clocked device), you must be sure that a gate doesn't get enabled and then immediately disabled after a logic delay time. Likewise, make sure that signals appearing at flip-flop inputs aren't delayed with respect to the clock (another plus for synchronous systems!). In general, delay the clock rather than the data.

It is surprisingly easy to overlook a race condition. A classic example is the humble 2-input multiplexer: if you do the obvious thing, namely Y=S&A|∼S&B, as in Figure 10.96A, you are in trouble! The problem occurs when there is a HIGH level on both A and B inputs, and the SE-LECT input S transitions from HIGH to LOW, i.e., from se-

lecting A to selecting B. The inverter's delay causes the lower AND gate to be disabled before the upper AND is enabled, producing a transient LOW glitch at the output. The solution is to add the redundant term A&B, as drawn in Figure 10.96B. (This is done correctly in any commercial IC. The problem comes when you want to put a MUX into programmable logic,[85] for which you spell out the logic in a hardware description language. You have to instruct it, sternly, not to "optimize" away the redundant term. You have to tell it that you are wiser than it. You can tell it, if all else fails, that you've read this book.)

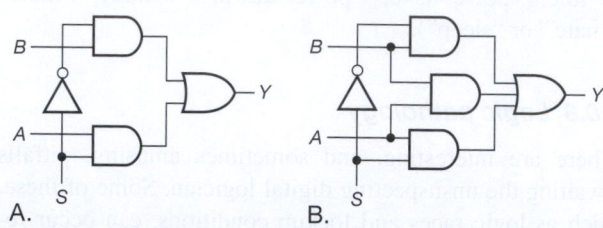

Figure 10.96. Adding a redundant term eliminates a logic-race glitch in the 2-input MUX.

B. Metastable states

As we discussed in detail earlier (§10.4.2D), a flip-flop (or any clocked device) can get confused if the data input changes during the setup-time interval preceding a clock pulse. As long as the flip-flop makes *some* decision promptly in this ambiguous case, all is well. However, there is a chance that the input may have changed at just the wrong time, at exactly the "moment of truth," such that the flip-flop can't make up its mind; its output can hover at the logic threshold for many times the usual propagation time (or it may sit in one logic state, then change its mind later, as in Figure 10.60).

This problem does not arise in properly designed synchronous systems in which setup times are always satisfied (by using logic fast enough so that inputs to flip-flops are stable by t_{setup} before the next clocking pulse). However, it can create problems in situations where asynchronous signals (e.g., going from device A, with its own clock, to device B, with a separate clock) must be synchronized. In these cases you cannot guarantee that input transitions do not occur during the setup interval; in fact, you can calculate how often they do![86] The metastability problem has

---

85 The subject of the next chapter.

86 The chance of landing in the "setup interval" Δt of the faster clock (from t_{su} before the clock to t_h after the clock) is $\Delta t / t_{clkF}$, where $t_{clkF} = 1/f_{clkF}$ is the period of the faster of the pair of asynchronous

been blamed for mysterious computer crashes, although we are skeptical. The cure generally involves a set of concatenated synchronizers, or a "metastable state detector" that resets the flip-flop. You may find claims of "metastable-resistant" logic, for example the AC(T) family of 5 V logic.

C. Clock skew

Clock skew problems arise when you have a clocking signal of slow rise time driving several interconnected devices (Figure 10.97). In this case two '595 octal shift registers have been cascaded to create a 16-bit parallel latched output; they are being clocked by a slowly rising edge, caused by capacitive loading of an anemic clocking signal (perhaps coming from a slow microcontroller's output pin). The problem is that the first register may have its threshold at a lower voltage than that of the second register (owing to process variations), and this causes it to shift earlier than the second register. The last bit of the first register is then lost. CMOS devices can display quite a spread of input threshold voltages, which compounds the problem (the threshold specification may be one-third to two-thirds of V_{DD}, and they mean it!). The best cure in such a situation is to drive the clock inputs from a nearby chip of adequate speed and without excessive capacitive loading. (Another way to fix the problem, if you've got clock skew, is to add some small delay in the data lines between successive clocked chips; but don't let that substitute for a clean clock.)

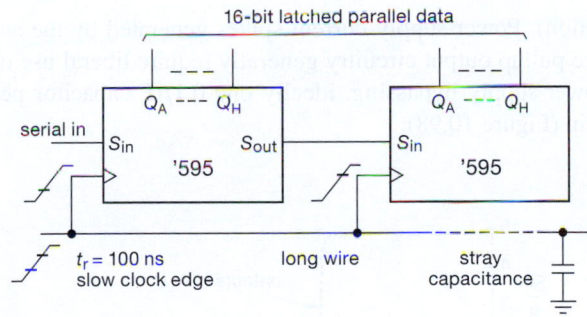

Figure 10.97. Slow risetimes can cause clocking skew, when the thresholds differ.

Speaking generally, edge-triggered clock inputs on any digital IC should always be treated with respect. For example, clock lines with noise or ringing should always be cleaned up with a gate (perhaps one with input hysteresis) before driving the clocked chip (but be careful not to vio-

late setup- and hold-time requirements). You're especially likely to have problems with clock lines that come from another board, or from a different family of logic. For example, slow 4000B or 74C logic driving the faster HC or AC families is likely to exhibit problems of clock skew or multiple transitions; ditto for HC driving LVC, and so on.

Clock-skew problems can occur, surprisingly, even within a single chip of programmable logic. An example we've encountered is the venerable 9500-series of cPLDs, in which individual flip-flops can be clocked either by one of the chip's distributed *global clock* signals, or, alternatively, by the output of internal logic (these are called *product terms*). Sounds fine. But if you use a product term to clock a set of flip-flops in a shift register, say, the chip may malfunction. That is because of routing delays in bringing that clocking signal to the several flip-flops. A synchronous circuit like this is only guaranteed to work if clocked by a global clock.[87] This "gotcha" is not prominently featured on the datasheets.

D. Runt pulses

In §10.6.1 (modulo-n counter) we used synchronous-load counters ('163), rather than a jam-load alternative (like the '191), because with the latter you would need to add some delay to prevent a pulse of substandard width (since the counter's output causes it to clear itself). The same comment goes for LOAD pulses when you are using counters or shift registers. Runt pulses will make your life miserable, because you may have marginal operation or intermittent failures. Use the worst-case propagation delay specifications when designing.

E. Unspecified rules

As the semiconductor industry was finding its way, beginning with the simplest RTL ICs of the 1960s (see §12.1.1 for a brief chronology), then the improved TTL and Schottky families, to the modern high-performance CMOS families, there was an understandable lack of standardization of pinouts, specifications, and functionality. As examples, the 7400 (NAND) had its gates pointing "down," but the 7401 (open-collector NOR) was built with the gates going the other way. This created so much confusion that it had to be mutated into the 7403, which is a 7401 with 7400-style pinouts; a similar disaster happened with the 7490 (BCD ripple counter), with power-supply pins in the middle instead of at the corners. (Ironically, mid-chip power-

---

clocks. Meanwhile, the slower clock is ticking along at f_{clkS}. So it lands in the metastable interval, on average, at a rate of $f_{clkS}f_{clkF}\Delta t$ per second.

[87] Which, unfortunately, you cannot drive from a product term, except by bringing the signal out on one pin and back in (to a global clock) on another.

supply pins have made a comeback in fast CMOS, because of their reduced inductance.)

An important legacy of this early anarchy is the hodge-podge of "unspecified rules" that we're stuck with. For example, the ever-popular '74 type-*D* flip-flop exists in every logic family; asserting both SET and CLEAR makes both outputs HIGH in every family except 74C, where it makes both outputs LOW! That's not exactly an unspecified rule, since if you look carefully in the fine print you'll find the inconsistency; the technical term for it is a *gotcha*. Another of our favorite gotchas is the '96, a 5-bit shift register with tricky jam-load inputs: they can SET, but they cannot CLEAR!

A genuine unspecified rule, and in fact a very important one, is *removal time*: that's the amount of time you must wait after disasserting a jam-type input before a clocked device is guaranteed to clock properly. Chip designers didn't bother specifying this until the logic families of the early 1980s (though circuit designers always wanted to know it), specifically the advanced Schottky and fast CMOS families. If you're designing with earlier logic, our advice is to be conservative; for example, assume that the removal time is the same as the data setup time.[88]

10.8.3 Congenital weaknesses of TTL and CMOS

We divide this section into nuisance problems and really bizarre behavior.

A. Nuisance problems
Bipolar (legacy) TTL
You have to remember that TTL inputs *source* current when held in the LOW state (e.g., 0.25 mA for LS, 0.5 mA for F). That makes it difficult to use *RC* delays, etc., because of the low impedances necessary, and in general you have to give some thought when interfacing linear levels to TTL inputs.

The TTL threshold (and that of its imitators, HCT and ACT) is too close to ground, making the whole logic family somewhat noise prone (more on this in Chapter 12). The high speed of these logic families makes them recognize short spikes on the ground line; those spikes, in turn, are generated by the fast output transition speeds, making the problem worse.

Bipolar TTL makes demands on the power supply (+5V, ±5%, with relatively high quiescent power dissi-

pation). Power-supply-current spikes generated by the active pullup output circuitry generally require liberal use of power-supply bypassing, ideally one $0.1\,\mu\mathrm{F}$ capacitor per chip (Figure 10.98).

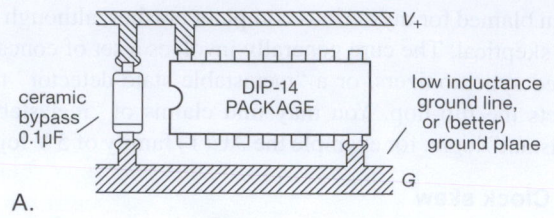

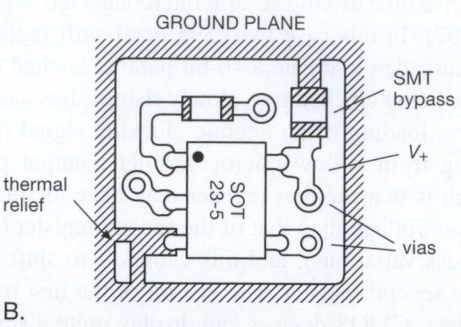

Figure 10.98. It's always a good idea to use robust low-inductance ground wiring, with liberal use of bypass capacitors. A. For an inexpensive two-sided board you should used gridded power and ground traces. B. Better yet, use a ground plane and surface-mount ceramic bypass caps. (The top layer is usually a signal layer, with power and ground planes stacked underneath, in a multilayer PCB. We've shown ground on top for clarity, though you might well do this in a two-layer PCB.)

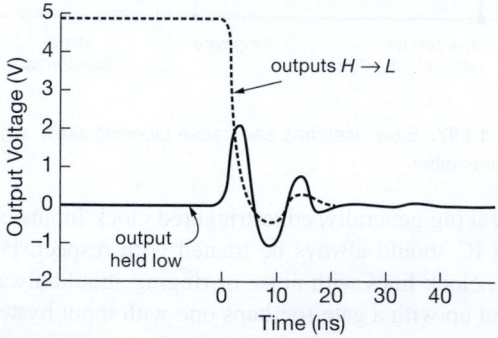

Figure 10.99. Ground bounce: 74AC244 octal buffer, driving seven 50 pF loads H→L, holding eighth output LOW. "Ground" is a copper foil plane (1 oz/ft²). (After Figure 1.1-4, *TI Advanced CMOS Logic Designer's Handbook.*)

---

[88] It's usually less; for example, the 74HC74 *D*-type flip-flop specifies a minimum removal time (preset or clear to clock) of 5 ns, whereas the minimum data setup time is 20 ns.

CMOS

CMOS inputs have traditionally been prone to damage from static electricity, with mortality rates really climbing in winter. Recent families with polysilicon gates and effective input protection networks are much more rugged than their metal-gate ancestors. CMOS inputs show a large spread in logic threshold, which can lead to problems of clock skew (§10.8.2C). Logic outputs can even exhibit double transitions when driven with slowly rising inputs. CMOS requires all unused inputs, even those of unused gate sections, to be connected to HIGH or LOW.

An interesting congenital problem with fast CMOS families is the presence of *ground bounce*: a fast CMOS chip driving its capacitive load generates enormous transient ground currents, causing the chip's ground line to jump momentarily, and thereby carrying with it LOW outputs that happen to be innocent bystanders on the same chip. Figure 10.99 shows the sort of thing you see. Notice particularly the magnitude of the effect: 1 to 2 volts is not uncommon! When you consider that a 3 ns, 5 V transition into 50 pF amounts to a transient current $I = C\,dV/dt = 83$ mA, and that an octal buffer might drive eight such loads simultaneously (total current of 2/3 A!), this behavior isn't surprising. When fast AC(T) logic first appeared, clothed in the traditional DIP corner-pin power and ground packaging, this problem turned out to be harder to solve than anticipated, leading to a new set of AC(T) circuits with "center-pin" power and grounds (for lower inductance). Additionally, logic IC manufacturers made design improvements to limit peak slew rates (sometimes called *edge rates*), and thus the consequent $C\,dV/dt$ capacitive-load currents: logic families such as AC(T)Q ("Q" for "quiet") helped considerably, with little speed compromise.

A better solution has evolved, namely the move to smaller surface-mount packages (with less lead inductance), and the widespread use of multilayer circuit boards (with dedicated power and ground layers), combined with surface-mount bypass capacitors. Recent logic ICs sometimes specify self-induced noise levels.[89] And VLSI chips with lots of pins usually dedicate multiple pins (sometimes dozens![90]) to ground. Ground-bounce problems have not been banished, though. Users should be aware of this serious problem and take measures to keep ground inductance as low as possible (Figure 10.98). It's best to use circuit boards with dedicated power and ground planes, and plenty of low-inductance bypass capacitors. Better still, if you don't need the speed, stay away from the fastest logic families entirely.

B. Bizarre behavior
Bipolar logic
TTL doesn't do many really weird things.[91] However, some TTL monostables will trigger on a glitch on the supply (or ground) line, and they generally behave somewhat fidgety. A circuit that works well with LS TTL may malfunction when replaced with AS TTL, because of faster edge times and consequently larger ground-line currents and ringing (74F TTL seems better in this regard). Most weird TTL operation can be traced to noise problems.

ECL involves very fast transition times, and interconnections longer than a few centimeters must be treated as terminated transmission lines (see Appendix H, and §*1x.1*).

CMOS
CMOS logic can drive you crazy! Open inputs on CMOS chips are bad news: you might have a circuit that intermittently misbehaves. You put your 'scope probe onto a point in the circuit, and it shows zero volts, as it should. Then the circuit works fine for a few minutes – before malfunctioning again! What happened was that the 'scope discharged the capacitance on the open input, and it took a long time to charge back up to the logic threshold.

Another amusing stunt is when a CMOS chip goes into "SCR latchup," caused by forcing an input (or output) beyond the supply rails momentarily. The resultant current (50 mA or so) through the input-protection diodes turns on a pair of parasitic cross-connected transistors that are a side effect of the "junction-isolated" CMOS process. This effectively shorts V_{DD} to ground; the chip gets hot, and you have to turn off the power supply before it will behave itself again. If you let this happen for more than a few seconds, you'll have to replace the chip. Some of the newer CMOS designs claim immunity to latchup, even with input swings 5 V *beyond* the rails, and to operate properly for input swings 1.5 volts beyond the rails.[92]

---

[89] Sometimes called *quiet-output* logic levels, or *ringback*. These specify the maximum dynamic swing on an output that should remain in one logic state, while all other outputs from the same chip make a logic transition.

[90] Or even *hundreds*: we counted 423 ground pins on Xilinx's Virtex-7 FPGA in a 1761-pin package.

[91] A cynic might add that it doesn't do many *good* things, either.

[92] SCR latchup happens when input *current* exceeds some threshold: manufacturers often guarantee no latchup if I_{in} is kept below some maximum "input clamp current," e.g., 20 mA. The datasheet will say something like "the input and output voltage ratings may be exceeded if the input and output current ratings are observed." This is good to know, because you can use a series input resistor to prevent latchup even with input-voltage overdrive.

CMOS has some strange and subtle failure modes. One of the output FETs can "fail open," giving pattern-sensitive failures that are difficult to detect. An input may begin to sink or source current. Or the whole chip may start drawing substantial supply current. Putting a 1 Ω resistor in series with each chip's V_{DD} lead (with downstream bypassing) makes it easy to locate faulty CMOS chips that are drawing quiescent supply current (for power drivers or chips driving many outputs, use 0.1 Ω sense resistors). Most of the time you don't bother with this precaution; but it can be a good idea if you're making a battery-operated device, where microamps matter.

Besides the input threshold variation between chips, a single chip can exhibit different thresholds for several internal chip functions driven from a single input. For example, the RESET input of a CD4013 can bring Q' HIGH before it brings Q LOW. This means that you should not terminate a reset pulse based on the output at Q', because the runt pulse that will be generated may actually fail to clear the flip-flop.

Open inputs would be bad news anyway, even on unused gate sections. That's because the input can float up toward mid-supply, putting both n-channel and p-channel MOSFETs into conduction. This "class-A current" causes unwanted quiescent current (hey, CMOS is zero power, right?), and it can cause oscillations and (in come cases) enough power dissipation to destroy the IC. You can see how this goes in Figure 10.100, where we measured separately the sinking and sourcing currents in one section of a 74LVC04 inverter running on +3.3 V. As long as the input is within about 0.7 V of either rail, the corresponding MOSFET is fully OFF; but in between there is some simultaneous conduction, or "shoot-through" current. In this case that current peaks at about 20 mA for $V_{in}{\sim}1.4$ V, causing 28 mW of dissipation in the inverter. If all six inverter sections floated, you could have nearly 200 mW of dissipation; and if the chip were running from a 5 V logic supply, the dissipation could reach destructive levels.

The maximum shoot-through current depends strongly on supply voltage, as shown in the measured data of Figure 10.101. At very low supply voltages, there is no input-voltage level that brings both MOSFETs into conduction simultaneously.

Here's the craziest of them all: you forgot to wire up the V_{DD} pin on a CMOS chip, but the circuit works just fine! That's because it is being powered by one of its logic inputs (via the chip's internal input-protection diodes from input to V_{DD}). You might get away with this for a long time, but suddenly the circuit reaches a state where all the logic inputs to the chip are simultaneously LOW; the chip loses its

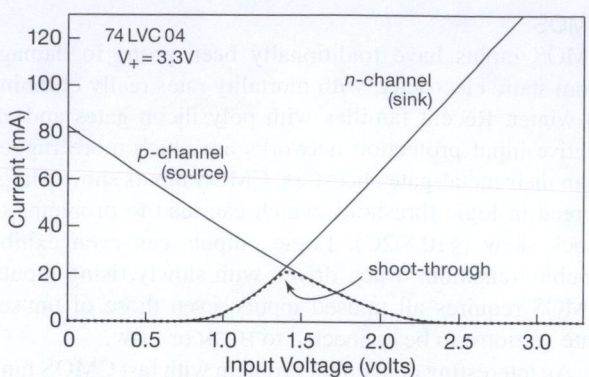

Figure 10.100. Measured sinking and sourcing currents in one section of a 'LVC04 inverter, as a function of logic input voltage. "Shoot-through current" is the colorful name given to simultaneous conduction, caused by a digital input voltage that is not near one of the rails.

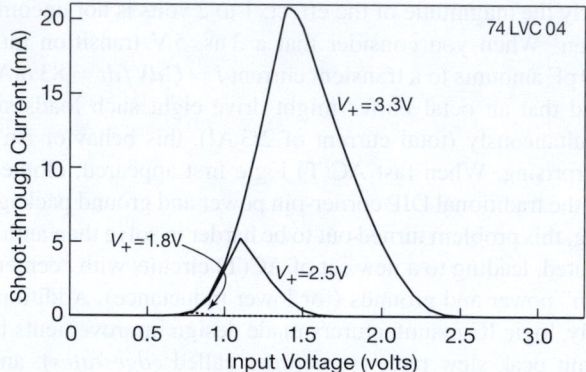

Figure 10.101. Measured shoot-through current versus digital input voltage level, for several supply voltages.

power and forgets its state. Of course, this is a bad situation anyway, since the output stage isn't adequately powered and can't source much current. The trouble is that this situation may produce symptoms only occasionally, and it can have you running around in circles until you figure out what's going on.

Additional Exercises for Chapter 10

Exercise 10.31. Show how to make a JK flip-flop by using a type-D flip-flop and a 4-input MUX. *Hint:* use the address inputs for J and K.

Exercise 10.32. Design a circuit that reads out, on 7-segment digits, how many milliseconds you've held a button down. The device should be smart enough to reset itself each time. Use a 1.0 MHz oscillator.

Exercise 10.33. Design a reaction timer. "A" pushes his button; an LED goes on, and a counter begins counting. When "B" pushes her button, the light goes out and an LED display reads the time, in milliseconds. Be sure to design the circuit so that it will function properly even if A's button is still held down when B's button is pushed.

Exercise 10.34. Design a period counter: a device to measure the number of microseconds in one period of an input waveform. Use a Schmitt comparator to generate logic levels; and use a 10 MHz clock frequency. Make it work so that pushing a button initiates the next measurement.

Exercise 10.35. Add latches to the period counter, if you haven't already.

Exercise 10.36. Now make it measure the time for 10 periods. Also, have it light an LED while it's counting.

Exercise 10.37. Design a true electronic stopwatch. Button *A* starts and stops the count. Button *B* resets the count. The output should be of the form *x.x* (seconds and tenths); assume that you have a 1.0 MHz square wave.

Exercise 10.38. Some stopwatches use a single button (start, stop, reset, start, etc., each time it is pushed). Design an electronic equivalent.

Exercise 10.39. Design a nice frequency counter to measure the number of cycles per second of an input waveform. Include lots of digits, latched count while counting the next interval, and a choice of 1 s, 0.1 s, or 0.01 s counting interval. You might add a good input circuit with several sensitivities, a Schmitt trigger with adjustable hysteresis and trigger point (use a fast comparator), and a logic signal input for TTL signals. How about a BCD output? Multiplex the digits on output, as well as parallel output. Spend some time on this one.[93]

Exercise 10.40. Design a circuit, using LVC logic at 3.3 V, to time a speeding bullet. The projectile breaks a thin wire stretched across its path; then, some measured distance farther along its path, it breaks a second wire. Beware of problems like "contact bounce." Assume that you have a 10 MHz logic square wave, and design your circuit to read out, in microseconds (four digits), the time interval between breaking the two wires. A pushbutton should reset the circuit for the next shot.

---

[93] H & H did – this was one of our first joint projects, some forty years ago.

Exercise 10.41. Invent a circuit to keep a running sum of successive 4-bit binary numbers that are input to it. Keep your result to only 4 bits (i.e., perform a sum modulo-16).[94]

Now add another feature to the circuit, namely, an output bit that is 1 if the *total number* of 1 bits in all the input numbers (since the last reset input) is odd, 0 if even. *Hint:* an XOR "parity tree" will tell you if the sum of 1s in each number is odd; figure it out from there.

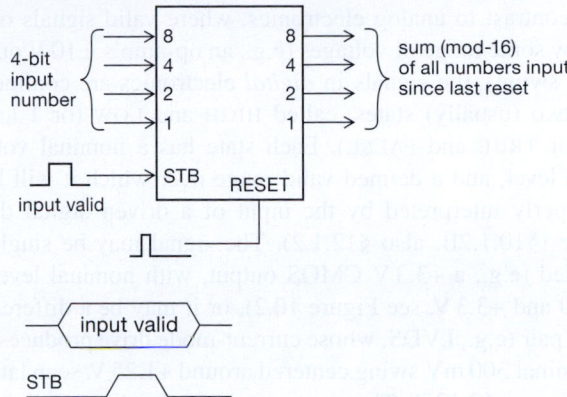

Figure 10.102. Checksum circuit block diagram.

Exercise 10.42. In Exercise 10.15 you designed a 2×2 multiplier by using a Karnaugh map for each output bit. In this problem you are to accomplish the same task by the process of "shift and add." Begin by writing out the product the way you would in elementary school (Figure 10.103). This process has a simple repeating pattern, requiring 2-input gates (what kind?) to generate the intermediate terms a_0b_0, etc., and 1-bit "half-adders" (adders that have a carry out but no carry in) to sum the intermediate terms.

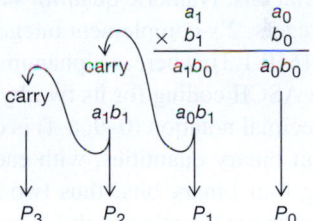

Figure 10.103. How to multiply.

---

[94] More sophisticated variants of such a sum, e.g., a CRC code (cyclic-redundancy checksum) are useful as a validity check against errors introduced into data files intended for storage or transmission.

Review of Chapter 10

An A-to-H summary of what we have learned in Chapter 10. This summary reviews basic principles and facts in Chapter 10, but it does not cover application circuit diagrams and practical engineering advice presented there.

¶A. Digital Voltage Levels.

In contrast to analog electronics, where valid signals occupy some range of voltages (e.g., an op-amp's ±10 V output swing), the signals in *digital* electronics are confined to two (usually) states, called HIGH and LOW (or 1 and 0, or TRUE and FALSE). Each state has a nominal voltage level, and a defined valid range over which it will be properly interpreted by the input of a driven digital device (§10.1.2B, also §12.1.2). The signal may be single-ended (e.g., a +3.3 V CMOS output, with nominal levels of 0 and +3.3 V, see Figure 10.2), or it may be a differential pair (e.g., LVDS, whose current-mode drive produces a nominal 300 mV swing centered around +1.25 V, seen later in Figure 12.135). There are numerous *families* of digital logic, distinguished by their supply voltage range, input thresholds, speed, power dissipation, and the like; see ¶G.

¶B. Meaning of Digital Bits.

Individual logic bits may represent part of a larger quantity (e.g., bits in a data byte), or they may represent a state (e.g., RESET, or RUN). A group of n bits can be sent simultaneously as a *parallel* quantity (on n wires) or sequentially in time (on a single wire or pair) as a *serial* string, comprising a parallel or serial *bus*. Numeric quantities are represented as unsigned integers, 2's-complement integers, or floating-point numbers (§10.1.3), whereas alphanumeric characters use single-byte ASCII coding (or its two-byte extensions). Base-16 hexadecimal notation (0–9, a–f) is ordinarily used to write multibit binary quantities, with each hex character representing four binary bits (thus two hex characters per byte). Whether representing multibit quantities or single state bits, digital voltage levels in a circuit can be assigned as active-HIGH or active-LOW. For example, one might have an active-LOW signal to reset a microprocessor; it would be called RESET′, and would ordinarily stay *disasserted* at the HIGH voltage state, going LOW only when it asserts a reset operation.

¶C. Combinational Logic: Gates.

Digital circuits whose state depends only on present inputs (i.e., not on past history) are called *combinational*. Their logical operations are performed with *gates*, whose basic forms (§10.1.4) are AND, OR, and NOT (or *invert*),

and whose standard graphic symbols and truth tables are shown in Figures 10.4, 10.5, and 10.6. These can be combined to form functions such as NAND and NOR (inverted-output AND and OR, Figure 10.8), or exclusive-OR (XOR, Figure 10.9). And they can be strung together to make any specified logical combinational output (or outputs) from a set of inputs.

¶D. Sequential Logic: Flip-flops.

The basic element of a sequential logic circuit is the *flip-flop* (§10.4), a device that maintains its state in the absence of external inputs; it is a "1-bit memory." The state of a circuit with flip-flops depends on both the present inputs and the previous state. For the ubiquitous *type-D* flip-flop ("D-flop"), the logic level present at the D (data) input at the time of[95] a clocking transition (edge) is captured and presented to the Q output. Once you have flip-flops, the world is your oyster – you can make counters, registers, arbitrary "finite-state machines" – and (drumroll) *computers!*

When a flip-flop is augmented with an RC, you get a *monostable*[96] *multivibrator* (or "one-shot") timing circuit, which generates an output logic pulse (of width set by the time constant RC) when triggered by an input edge. The one-shot is an example of a mixed-signal circuit, i.e., a combination of analog and digital techniques.

¶E. Sequential Logic: Registers and Counters.

A clocked D-flop captures and holds one bit; a collection of n D-flops with a common clock input is called a *register* (§10.5.1); it captures and holds n bits (a byte, for $n=8$). When combined with memory and gates, it can form a *finite state machine* (FSM, see Figure 10.64), just a step away from the microprocessor.

If the output bits Q_i of a D-register drive the subsequent inputs (i.e., Q_i is connected to D_{i+1}), you've got a *shift register*; these are handy for parallelizing a serial stream, or vice versa (if loaded in parallel and then shifted out the end). An interesting shift-register application is the generation of pseudo-random (noise-like) bit sequences, see §§8.12.4A and 13.14.

A collection of n D-flops can instead be connected to make an n-bit *counter*, such that successive input transitions cause the n-bit number represented by Q_0–Q_{n-1} to increment.[97] Simplest is the *ripple counter* (§10.4.2E), in

---

[95] More precisely, present and stable from a *setup time* t_s before the edge to a *hold time* t_h after the edge, to ensure proper operation.

[96] One-side stable, as contrasted with the ordinary *bistable* flip-flop (stable in either state), or the *astable* oscillator (stable in neither state).

[97] Or to decrement, for a *down counter*.

which the Q_i outputs drive the successive clock inputs; the alternative is a *synchronous* counter, in which all clock inputs are driven by the input signal, and gate logic is rigged up to present to the *D* inputs the levels corresponding to the next state. The latter is just a special case of a synchronous FSM. Synchronous counters (and synchronous systems in general) are favored, having the pleasant properties of simultaneous state changes (apart from *skew*), and of lack of digital noise in the setup interval before each system clock.

¶F. Standard Logic and Programmable Logic.

Digital logic is available as standard pre-wired functions (gates, flip-flops, counters, registers), in packages going from popular single-gate mini-logic "glue" in 5-pin or 6-pin packages, through the long-popular mid-sized (14-, 16-, or 20-pin) packages, to large multibyte-wide bus drivers in 48-pin or 96-pin packages; see, for example, the listings in Table 10.3 on page 716 (gates) and Table 10.5 on page 742 (counters).

These are good enough for many tasks; but an attractive alternative, particularly in a complex system, is the use of *programmable logic devices* (PLDs). These contain large numbers of uncommitted gates and flip-flops, whose interconnections are user-configurable (and re-configurable); see §10.5.4. The category known as *cPLDs* (complex PLDs) contain from a few dozen to a few hundred "macrocells" (a flip-flop and a collection of gates) and a few thousand gates, along with the non-volatile program memory, and they can replace much of the "random logic" (the tangle of gates and flip-flops) in a digital system; they are also ideal for implementing state machines. At a higher level of integration, the *field-programmable gate array* (FPGA) reigns supreme; the largest of these incorporate billions of transistors implementing millions of flip-flops, along with memory, serial transceivers, and all the resources needed to implement a microprocessor.

Most designers use a hardware description language (HDL), running on a PC platform, to specify the PLD's function, but there are available schematic-entry tools as well; these toolsets allow you to simulate the programmed functions, to verify that you've programmed what you intended. Contemporary PLDs are programmed in-circuit, usually via a JTAG serial port. We discuss PLDs in greater detail in Chapter 11.

Any discussion of digital logic must include *microcontrollers*, the increasingly capable and inexpensive computers-on-a-chip that can replace discrete and programmable logic devices in digital devices. Think of embedded microcontrollers not as exotic creatures, but simply as circuit components – they cost less than a precision op-amp, and they can do wonders. They are the subject of Chapter 15.

¶G. Logic Families.

Digital logic functions, whether standard logic or programmable logic, can be implemented in ICs with a variety of on-chip transistor technologies (§10.2). Earlier legacy families of standard logic (bipolar RTL, DTL, and TTL) have yielded their turf to the now-dominant CMOS logic families. The latter include the traditional 5 V HC[T] and AC[T] families (the T suffix indicates TTL thresholds, around 1.4 V), the high-voltage (but very slow) 4000B series, and a proliferation of CMOS families intended for lower-voltage operation; the latter include popular families like LVC (1.8–5 V) and AUC (1.2–2.5 V), among the dozens of choices. See Figures 10.22, 10.26, 10.27, and (in Chapter 12) §12.1 and Figure 12.2.

Programmable logic devices (cPLDs, FPGAs) and microcontrollers are invariably built with CMOS, with supply voltage ranging from as low as 1 V to as high as 5 V, and reflected in their speed and power dissipation. It is common to have separate "core" and I/O supply rails in many of these devices; for example a microcontroller or FPGA might have a 1.2 V core and a 1.6–3.6 V I/O (the latter for compatibility with external devices).

All these variations can be confusing. Chapter 12 deals in detail with logic family interfacing. At a simple level, though, what you need to know (summarized in §12.1.3, and encapsulated in Figure 12.9) is that (a) you can always make a direct connection between logic running at the same voltage; (b) logic running from a higher voltage (e.g., +5 V) can drive lower-voltage logic if the latter's input is "tolerant"; and (c) lower-voltage logic can drive higher-voltage logic if the latter has "TTL thresholds" and the former is powered from at least +2.5 V.

¶H. Digital Logic Cautions.

Digital design nicely sidesteps the complications inherent in linear circuits (biasing, thermal stability, etc.), and it's easy to become complacent, and lazy. But there are plenty of hazards in the digital playground, see §10.8. Among the *dc problems* are logical lockup states, SCR latchup, and undetermined startup states. More sinister are *switching problems* such as logic races and runt pulses, metastable states, clock and data skew, ground bounce, and power- and ground-conducted switching noise.

PROGRAMMABLE LOGIC DEVICES

In the previous chapter we introduced the basics of digital electronics – gates and combinational logic, flip-flops and sequential logic – and we illustrated their application with some examples: a modulo-*n* counter, a multiplexed LED display, and an *n*-pulse generator. In that chapter we used primarily *standard logic*; that is, small blocks of logic (gates, flip-flops, counters, registers) packaged as single integrated circuits.

However, as we remarked often, there is an alternative (and usually better) way to implement those kinds of circuits, namely, the use of *programmable logic devices* (PLDs).[1] A PLD consists of a chip with lots of logic (gates and registers, and sometimes much more), in which the *connections* are programmable. But it's not a computer program: in a computer the program tells the processor *what to do*; in a PLD the program tells the chip *how to connect* its component pieces.

11.1 A brief history

The first PLDs (1975) were the Signetics' "Integrated Fuse Logic" devices, with a handful of uncommitted gates and a matrix of fuses (literally!) that could be selectively blown to leave the desired interconnections. You "programmed" the things with a chart of fuse links, or (later) by designating the fuse locations manually on a computer screen. In subsequent developments, Signetics added flip-flops to the mix, enabling sequential circuits; and Monolithic Memories devised a simplified family known as PALs (programmable array logic), with a more usable text-entry design language called PALASM. By the middle of the 1980s we had Lattice's GALs (generic array logic), which used electrically reprogrammable memory and a chameleon-like output cell that could be either a flip-flop or a gate. There were improved programming languages, too, namely CUPL and ABEL; these, and their successors, are all generically called *HDLs* (hardware description languages).

Around the same time Xilinx introduced the FPGA (field-programmable gate array), an alternative programmable logic scheme with a "finer-grained" architecture – many more gates and registers, organized as a fabric of logic blocks, surrounded by I/O blocks, with more flexible interconnections – and whose configuration programming was held off-chip in a separate small non-volatile memory device (a serial "configuration ROM"), and loaded into on-chip (and volatile) SRAM at power-up.

FPGAs and complex PLDs, improving nicely with time, are both part of the essential toolkit of circuit designers. Contemporary cPLDs (the "c" stands for "complex") have from 32 to 2000 macrocells, with tens of thousands of gates, and operate at clock speeds up to a few hundred megahertz; they use on-chip nonvolatile program memory, and can be programmed and reprogrammed in-circuit, using simple serial protocols (e.g., JTAG). There are variants[2] with essentially zero quiescent current. They have highly predictable timing.

Contemporary FPGAs are denser, topping out at around a million flip-flops, in packages (see Figure 11.1) with as many as 1738 pins (!). They may include blocks with dedicated memory, arithmetic modules (ALUs, MACs, other DSP), interface (USB, Ethernet, PCI), and other specialized functions. With such extensive capabilities, it's now routine to program in standard designs for large modules such as a video processor, PCIe controller, Bluetooth, or even a complete microprocessor (and peripherals), with plenty of programmable logic left over to form a "system-on-a-chip." These standard programmed ("soft") designs are known as "IP" (intellectual property, some of which you may have to license); if you put in an IP microprocessor, it is called a *soft processor* core.[3] An efficient alternative, if you want a processor inside your FPGA, is the use of a hybrid FPGA, with the microprocessor and peripherals already hardwired.[4]

---

[1] Officially "PLDs," but generally meant to include cPLDs (complex PLDs) and FPGAs (field-programmable gate arrays).

[2] Notably the Xilinx (originally Philips) CoolRunner™ series, and the Lattice ispMACH™4000Z series.

[3] Examples are the Xilinx MicroBlaze, the Altera Nios-II, the Lattice Mico, and the Actel ARM.

[4] Examples are the PowerPC (in Xilinx FPGAs), the AVR (in Atmel's FPSLIC), and the ARM (in Altera's FPGAs).

Figure 11.1. A selection of programmable logic packages. At upper left are three PLCCs (84, 68, and 44 pins), to be compared with their denser QFP brethren at upper right (going clockwise: PQFP-208, TQFP-100, VQFP-44, TQFP-48, and TQFP-144). Down the middle are displayed the same sPLD type (22V10) in four increasingly dense packages: DIP-24, PLCC-28, SOIC-24, and TSSOP-24. Across the bottom are the densest packages (left to right): FTBGA-256, FBGA-100 (and the same IC in a TQFP-100), BGA-132, BGA-49 (CSP-49), and QFN-32; for the last three both top and underside are shown.

For such complex devices, the legacy programming tools (CUPL, ABEL) that evolved with small PLDs are hopeless.[5] Current practice favors either graphical schematic entry, or one of the two contemporary and powerful text-based HDLs (called *Verilog* and *VHDL*); presently we'll illustrate both. For most digital designs, programmable logic is generally a better choice than standard logic, because (a) one chip replaces many – less wiring, smaller finished product, less inventory, lower cost; and (b) it is easy to reprogram the part if the initial design is flawed, or if you want to add features.

11.2 The hardware

Contemporary PLDs are fiendishly complex... and they're not getting any simpler. High-end devices already have up-

wards of 5 billion transistors (and counting). To make it all understandable, we'll take it in easy stages: first the classic simple PAL architectures (e.g., 22V10) and the more flexible PLAs (e.g., Xilinx CoolRunner-II), and then the complex and register-rich FPGAs.

11.2.1 The basic PAL

A good starting point is the basic PAL, an example of a simple PLD. The classic device is the 22V10, the part number signifying 10 output *macrocells* (more on this soon), among a total of 22 inputs and outputs (originally offered in a 24-pin package). The PAL circuit scheme consists of a programmable array of connections, whereby both inputs and feedback outputs can be selectively connected to a set of many-input AND gates; the outputs from a fixed number of such AND gates then feed an OR gate. Its output can be used directly (a combinational output), or it can feed a type-D flip-flop (a "registered" output); the circuitry for these output options is called an "output macrocell." Figure

---

[5] You *can* still get (from Atmel or Lattice) the classic small "sPLDs" (*simple* PLDs): 16V8, 20V8, 22V10, and 26V12, for which CUPL and ABEL are completely adequate.

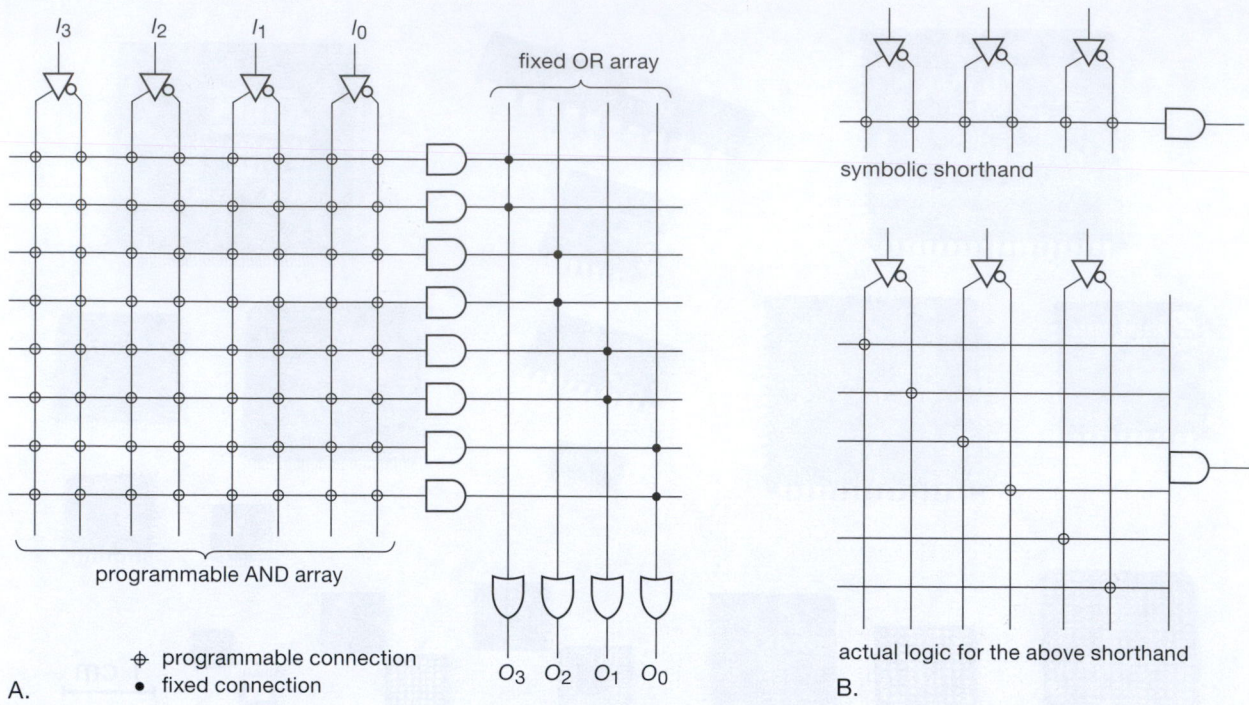

Figure 11.2. A. In a "PAL" (Programmed Array Logic) every available logic signal (or its complement) can be connected to many-input ANDs; the outputs of several such ANDs are OR-ed to form the logical outputs, each then passing through an output macrocell on the way to an output pin. B. Each AND gate has many inputs, shown in conventional shorthand and in fully expanded form.

11.2 shows the basic idea, in a simplified scheme with only four input signals; here each logic output comes from an OR gate fed from two of the 8-input ANDs, and that output would drive the output macrocell. (Note the shorthand notation used in Figure 11.2A for the many-input AND-gates and OR gates, shown expanded in Figure 11.2B.)

In real life a PAL has many more inputs and gates. Figure 11.3 shows the actual scheme for the 22V10, which, though impressive looking, is in fact tiny by contemporary standards. The 12 inputs (and their complements), along with feedback from the 10 outputs (and their complements) – thus 44 signals total – are all brought into the connection matrix, in which any set of them can be connected to any 44-input AND, whose output is then OR-ed with another 7 to 15 ANDs (more toward the center of the chip, fewer toward the ends) to create an output (one of ten). That output does not go directly off the chip, but rather gets massaged in an output macrocell (Figure 11.4), which consists of a bit of programmable logic that lets you latch the OR output in a flip-flop, or just pass it through without latching. You can see from Figure 11.4 that the output, whether "registered" or combinational, can be true or complemented, and further can be three-stated.

The 22V10 is adequate for many logic tasks: you can make a shift register, or a counter, or an address decoder, or just a collection of what's called "random logic." But by contemporary standards it is extremely limited, both by the small number of registers and I/O pins, and also by the constraint of a common CLK signal; you cannot, for example, make a ripple counter.

One evolutionary path has been the creation of what we might call a "super-PAL," in which the macrocells allow individual clocking from the logic array, and in which many more macrocells are packed into one IC. These often include schemes for expanding the logic by sharing product terms across macrocells. Such extensions of the basic PAL architecture are called "cPLDs" (complex PLDs), and are widely available.[6] They range in capacity up to 2000 or more macrocells (some of which may be "buried," i.e., available for internal interconnections but not connected to I/O pins), and they allow asynchronous clocking.

The "programmable-AND/fixed-OR" architecture

---

[6] For example, the Altera MAX7000 series, the Lattice Mach 4000 series, or the Xilinx CoolRunner-II series (the latter uses the more advanced PLA architecture; see §11.2.2).

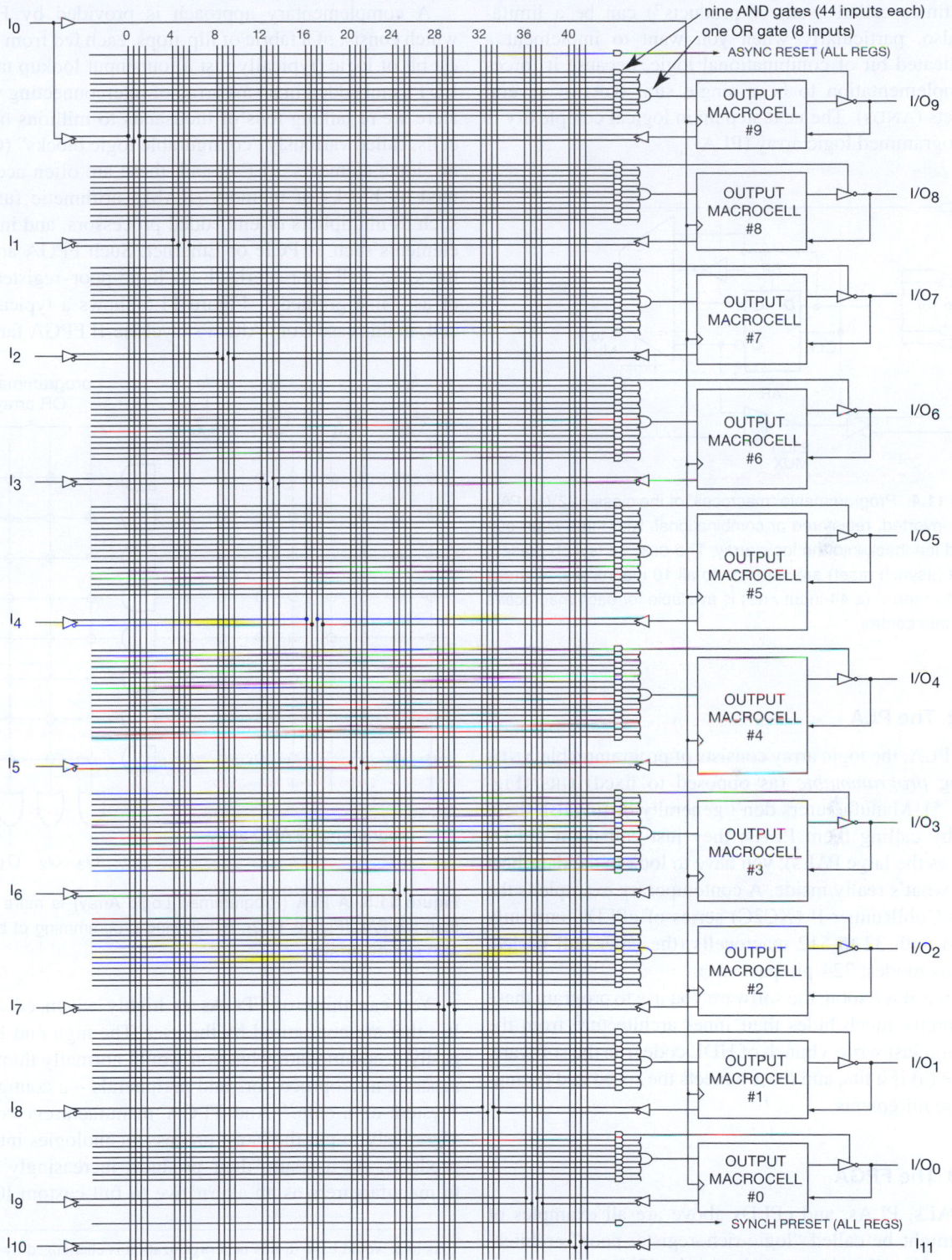

Figure 11.3. Logic array of the classic 22V10 PAL: programmable-AND/fixed-OR. The output macrocell can be programmed as a type-D flip-flop or a simple pass-through; see Figure 11.4. (Adapted from the Lattice ispGAL22V10 datasheet, ©Lattice Semiconductor Corporation, 2004.)

(sometimes called "sum of products") can be a limitation also, particularly when you want to implement a complicated bit of combinational logic, because it forces the implementation to be a single sum (OR) of several products (ANDs). The next step up in logical complexity is the programmed logic array (PLA).

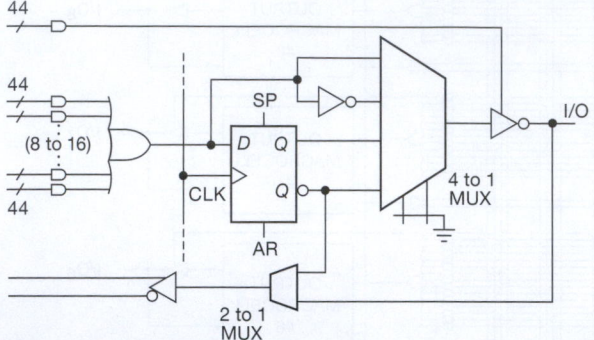

Figure 11.4. Programmable "macrocell" of the classic 22V10 PAL: true or inverted, registered or combinational, with three-state output and feedback into the logic array. The CLK, SP (synch preset), and AR (asynch reset) are common to all 10 macrocells, whereas a "product term" (a 44-input AND) is available for each macrocell's three-state control.

11.2.2 The PLA

In the PLA, the logic array consists of programmable ANDs feeding *programmable* (as opposed to fixed) ORs (Figure 11.5). Manufacturers don't generally distinguish these parts by calling them PLAs; they just call them cPLDs (same as the large PALs); you have to look in the datasheet to see what's really inside. A contemporary example is the Xilinx CoolRunner-II (XC2C) series of cPLDs, currently offered with 32 to 512 macrocells (the latter half buried, given its modest 324-pin package).

As we'll see soon, the software you use to program these parts pretty much hides their inner architecture from the user; you just write a bunch of HDL code, run the software, and see (a) if it fits, and (b) if it meets the speed and timing-delay requirements.

11.2.3 The FPGA

The PALs, PLAs, and cPLDs above are all examples of what might be called "logic-rich–register-poor" architectures. After all, with even the lowly 22V10 you have a dozen 44-input AND gates feeding each OR-gated macrocell – but only a handful (OK, *two* handfuls) of flip-flops.

A complementary approach is provided by FPGAs, which consist of a fabric of flip-flops, each fed from a modest bit of logic (typically just a four-input lookup table, or LUT), embedded in a vast array of interconnecting wiring. Here we're talking tens of thousands to millions of these cells, called variously "configurable logic blocks" (CLBs), or "logic elements" (LEs); and, these are often accompanied by blocks of memory (RAM), arithmetic functions such as multipliers or embedded processors, and interface elements such as PCIe or Ethernet. Such FPGA architectures are well characterized as "logic-poor–register-rich," or as "fine grained."[7] Figure 11.6 shows a typical logic cell, in this case from Altera's Cyclone-II FPGA family.

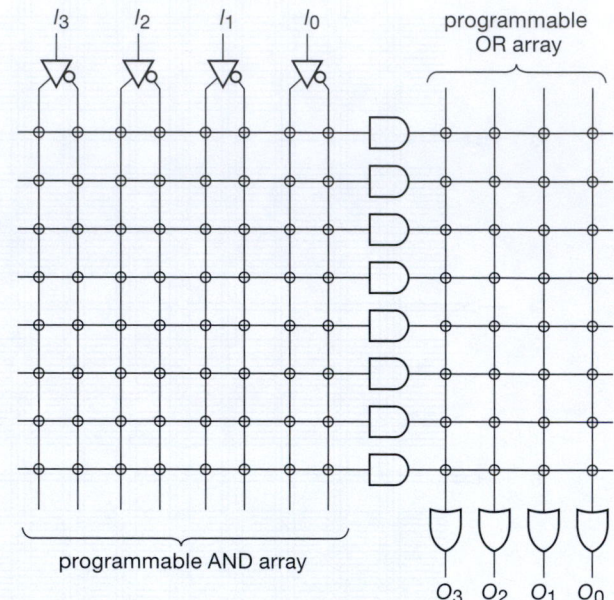

Figure 11.5. A PLA (Programmed Logic Array) is more flexible than the PAL (Figure 11.2), by allowing programming of both the AND and OR arrays.

You can think of FPGAs as highly advanced custom ICs that are configured by the user. The high-end FPGAs easily accommodate what you would normally think of as separate microprocessors and peripherals – a configurable "system-on-a-chip." And FPGA manufacturers compete vigorously to put the latest process technologies into their products. As a result, their products increasingly appeal to manufacturers as an alternative to full-custom ICs: the

---

[7] As luck would have it, this neat organization is challenged by some of the cPLDs that are actually chimeras, or FPGAs in disguise, for example the Altera Max-II cPLDs (which use 4-input LUTs), Cypress Delta39K cPLDs, or the Lattice MachXO "crossover cPLDs."

latter[8] involve a costly design cycle, provide less flexibility, and often delay the introduction of a new product into the marketplace. "Time-to-market" is critical in consumer electronics, making FPGAs cost effective, in spite of their higher per-unit costs. An additional benefit is the ability to upgrade their internal circuitry (for example to fix bugs, add signal processing options, or whatever).

Typically you'll pay a dollar or two for cPLDs and for small FPGAs, ten dollars or so for intermediate size FPGAs, and from hundreds of dollars up to a thousand dollars or more (!) for FPGAs of the largest size and highest performance. Traditionally FPGAs have not been low-power devices, with typical quiescent currents of the order of milliamps to tens of milliamps (and additional dynamic power dissipation, of course, when clocked rapidly). However, micropower FPGAs are becoming available, aimed squarely at portable applications: an example is the Actel IGLOO series, with quiescent power dissipation in the 5–50 μW range.

The big players in the high-end FPGA market are Altera and Xilinx (the originator), with smaller FPGAs from Actel, Atmel, Cypress, and Lattice.

11.2.4 The configuration memory

A programmable logic device has to have its program held *somewhere*. Traditionally, sPLDs and cPLDs have always kept their configuration on-chip, in non-volatile storage: the earliest PLDs (e.g., the original MMI PALs) used one-time on-chip fusible links, but they were quickly superseded by reprogrammable CMOS memory. Some early devices (e.g., the original Altera Max 7000) used UV-erasable (EPROM) memory (with a quartz window, so you could blast it with the appropriate amnesia-inducing dose – typically of 20 minutes duration – of ultraviolet light), but quickly enough the industry adopted *electrically*-erasable memory (EEPROM, or "flash" memory). That is how all sPLDs and cPLDs are now made; they can be erased in seconds, and reprogrammed typically in less than a minute. There are specifications for the minimum *data-retention* time of the on-chip memory (typically 20 years), and for its *endurance* (typically a minimum of 1000 to 10,000 program–erase cycles).

Early sPLDs had to be programmed in a *device programmer* (which you bought from a company such as BP Microsystems, Needhams Electronics, or DataIO), then placed in the target circuit. This worked OK for through-hole (DIP) packaged parts, for which you could provide

sockets in the final resting place; but it was awkward for soldered-in-place surface-mount parts, because reprogramming required removal of the IC. The solution is "in-system" programmability, which was retrofitted onto some sPLDs (e.g., Lattice's ispGAL22V10, an ISP version of the standard 22V10). All contemporary cPLDs are in-system programmable, using a simple serial data protocol such as JTAG (IEEE 1149) boundary scan.[9] You connect to the serial programming pins with a "pod," connected to the host computer that holds the programming data via a USB connection. Pods typically cost around $100; see the discussion of JTAG in §14.7.4.

For *FPGAs* the situation is somewhat different. Traditionally, FPGAs had no on-chip ROM; instead they inhaled their configuration data at power-up from an external serial "configuration ROM," then held the configuration on-chip in static (volatile) memory. Although this is still the protocol for many FPGAs, there are now available some FPGA families with on-chip nonvolatile flash memory, for example from Actel, Lattice, and Xilinx. As with contemporary cPLDs, programming is done in-circuit, via a comparably priced pod.

11.2.5 Other programmable logic devices

It's worth remembering that a simple digital memory is "programmable logic": it produces a stored n-bit output for each possible m-bit (address) input. That is how it is used in the general state machine of Figure 10.64, and, at a simpler level, a small (e.g., 16×1) lookup table is used in the logic elements of an FPGA.

There are recent developments in the form of *mixed-signal* programmable logic; an example is the Cypress PSoC™ mixed-signal programmable family, which incorporates amplifiers, filters, and other analog components on a programmable chip that includes a microprocessor.

11.2.6 The software

To use programmable logic devices, you've got to learn some software tools. You begin with design *entry*, either with a hardware description language (HDL) or with schematic entry. Then you run a *simulation* tool, to verify that the design does what you intend. Next is *synthesis*, which converts the design to a "netlist," which describes the logical connections. The netlist is then fitted to the target device, a process called *place and route*. Finally,

---

[8] Sometimes called an *ASIC*, for application-specific integrated circuit.

[9] The IEEE has developed an industry standard data format for using JTAG for in-system configuration, known as IEEE 1532.

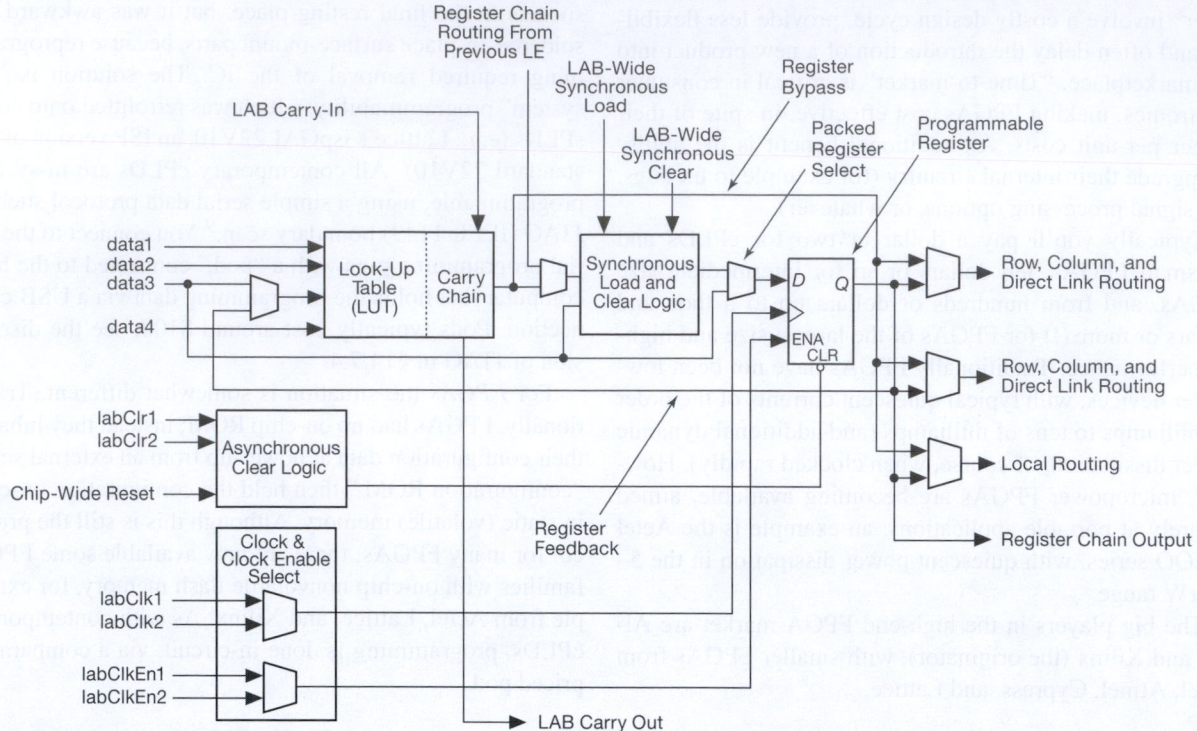

Figure 11.6. An FPGA logic element (LE), one of some tens of thousands in a typical FPGA, from the Altera "Cyclone II" family. The clocked flip-flop and lookup table (LUT) are decorated with extra logic and control signals from the larger logic array block (LAB), for flexible expansion. (Adapted from Altera Cyclone II Device Family datasheet, ©Altera Corporation, 2007.)

the fitted design is downloaded to the programmable logic chip, either to the chip itself (if it has on-chip non-volatile memory), or to a configuration ROM (from which it loads at power-up), or possibly hot-loaded from a processor directly to the powered-up chip.

FPGA manufacturers typically provide free software for their low-to-mid-range products,[10] but they may charge from several hundred dollars to several thousand dollars for software that supports their high-end offerings; some of the cost of the latter is attributable to the included software "cores" (code modules that creates blocks of useful hardware, sometimes called "IP," for intellectual property). There may be deep discounts offered to educational and nonprofit users.

These software tools are not terribly easy to learn, and they have a distressing tendency to change with time. They are often afflicted with bugs, requiring skillful workarounds. It can be a full-time job keeping up. It's

always helpful, therefore, to have some company (which misery loves); that is, to use the same software tools as those the vastly more knowledgeable people around you are using.

We will not attempt to teach these tools. However, to illustrate the general process we next show an example of a complete circuit design done five different ways (§11.3), one of which is with a cPLD (§11.3.3).

11.3 An example: pseudorandom byte generator

The quickest way to see how a circuit idea can be implemented in several ways is, well, simply to see an example. Ours is what might be described as the ultimate gibberish machine,[11] namely a circuit that sends out a succession of pseudorandom bytes, as standard RS-232 serial data. In this section we begin with a *block diagram*, then we show the circuit implemented first in (a) *standard logic*, then in (b) *programmable logic* (with entry in two popular methods, namely graphic *schematic entry*, and text-based

---

[10] An example being the Xilinx "WebPACK," which you get directly over the Internet (if you have the patience to download and install the >5 GByte set of files!).

[11] Or the proverbial monkey-at-a-typewriter.

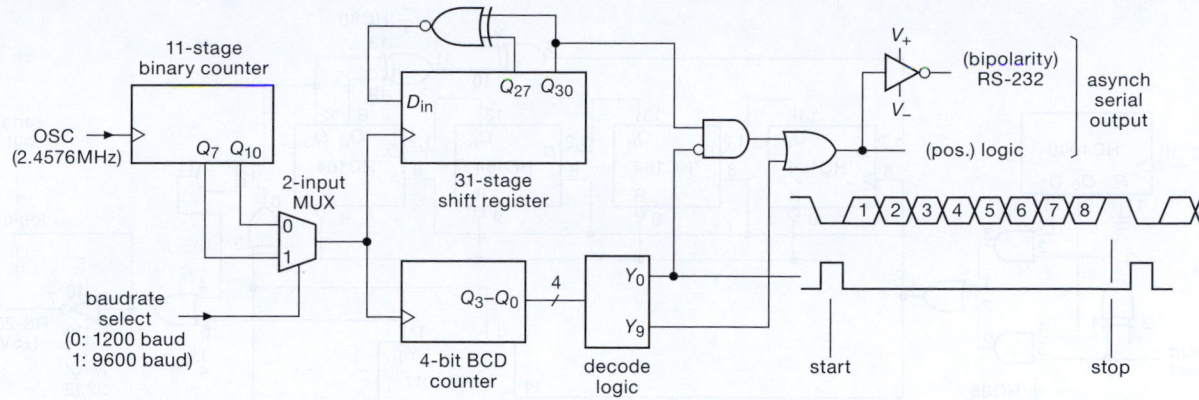

Figure 11.7. Block diagram of RS-232 pseudorandom byte generator. A 2-input MUX (multiplexer) selects the baudrate from a subdivided oscillator; the resulting clock shifts a maximal-length linear feedback shift register to produce pseudorandom bits, which are framed with START and STOP bits to produce 8n1-formatted serial bytes.

hardware description language.[12] Finally, we show how the same function is easily implemented with a *microcontroller*, the inexpensive (and indispensable) programmable computer-on-a-chip (complete with internal memory and I/O functions), intended for "embedded" use in any electronic device. Embedded microcontrollers are the subject of Chapter 15 – but we couldn't resist showing them now, simply because they are *the* attractive design alternative.

We wrap up this section with some advice – in particular, which hardware to use, paired with their appropriate design tools, and the pros and cons of each.

11.3.1 How to make pseudorandom bytes

Section 13.14 describes a simple method to produce a "pseudorandom" bit sequence (PRBS), using a shift register whose serial input comes from an XOR of two (or more) downstream bits.[13] The sequence so produced is *pseudo*random because, though it has many properties of randomness, it is not truly random; in fact, it is completely deterministic, with a repeat interval (with properly chosen XORed bits) of $2^n - 1$ for a register of n stages.[14]

Our example uses a register of length 31 (Figure 11.7), for which XOR feedback from bits 28 and 31 produces a maximal-length sequence. Its length ($2^{31} - 1$, or 2,147,483,647) corresponds to a repeat interval of 2.6 days

at the maximum serial output rate of 9600 baud.[15] The shift register is clocked at a selectable rate of either 1200 or 9600 shifts/second, with the output bits grouped into 8-bit bytes, each of which is framed by a "START" and "STOP" bit pair. This is known as *asynchronous serial* communication, typified by the familiar RS-232C serial computer port. This circuit allows selection of baud rate (1200 or 9600), and formats the output as 8-bit data, no parity, and one stop bit (this is usually abbreviated "8n1").

As explained in §14.7.8, the actual voltage levels for serial port communication, described in the RS-232C specification, are (annoyingly) of both polarities, with ± 5 V to ± 15 V levels corresponding to the two logic states; to add a bit of confusion, those outputs are logically inverted, with logic HIGH corresponding to the negative signaling level, and vice versa.[16] The good news is that there are plenty of RS-232C driver–receiver interface chips available, most of which include internal charge-pump voltage generators so that you can power them from a single positive logic supply (usually +3.3 V or +5 V); an example is the MAX3232 shown in Figure 11.8.

---

[12] In particular, *Verilog* and *VHDL*, the two popular choices.

[13] Sometimes called a *linear feedback shift register*, or LFSR, and introduced earlier in §8.12.4A.

[14] It should be noted that there are far better computational algorithms for generating pseudorandom bit sequences; this method was chosen for its abject simplicity.

[15] Readers unhappy with the concept that a random sequence eventually *repeats* might prefer something a bit longer, say a shift register of length 71 (with XOR tap at 65), whose repeat interval is approximately 8 billion years (about half the age of the universe). *Real* purists can elect a longer register, say of length 167 (XOR tap at 161), which won't repeat for 10^{38} years (about 10^{28} times the age of the universe).

[16] Sometimes called *MARK* and *SPACE*, respectively.

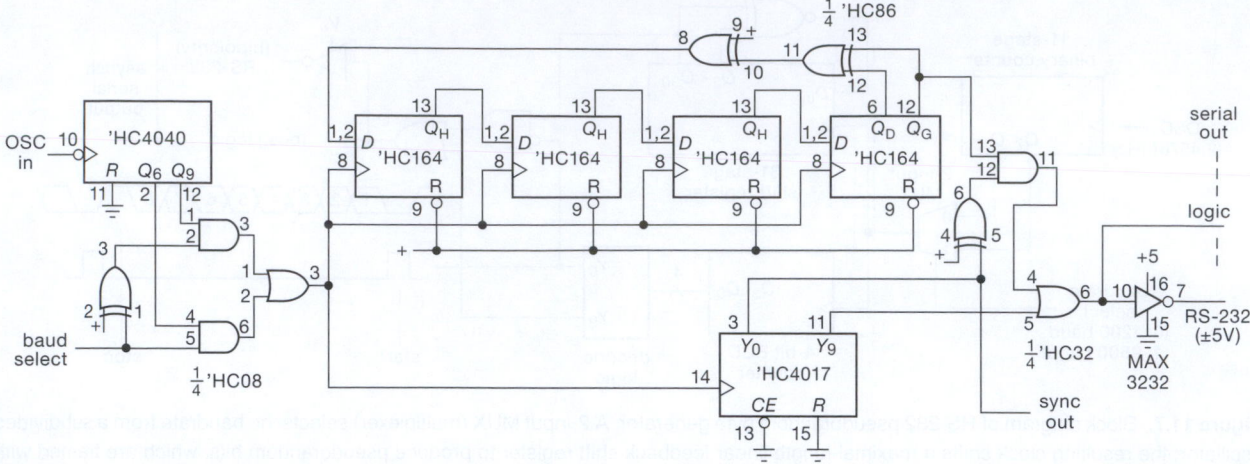

Figure 11.8. Implementation with standard logic (74HC family).

11.3.2 Implementation in standard logic

Gates, counters, shift registers – these are the stuff of "standard logic," most famously the legendary 74xxx logic family, with these discrete logic functions supplied in through-hole (DIP) or the many varieties of surface-mount (SOIC, SSOP, QFN, BGA) packages. In Figure 11.8 we've implemented the random generator in the 74HC subfamily (chosen for wide availability, adequate speed, low power, and DIP/SMT package options). We've indicated DIP pin numbers to emphasize that this is a complete and finished design.

Some comments.

- With standard logic you see tricks like unused XORs converted to inverters, to avoid extra components (such as a package of inverters). Another example is the use of discrete leftover gates (rather than an integrated 2-input MUX) for the baud-rate select.[17]
- We took advantage of some unusual packaged functions, for example the 'HC4017 fully decoded decimal counter, here used to frame the 8 data bits into 10-bit serial characters; the more conventional way would use a separate counter and decode logic. Another helpful packaged function is the 'HC4040 12-stage *ripple* counter (used for

the clock subdivision), rather than a cascade of smaller synchronous counters.
- Even with tricks like these, this design required many IC packages (nine, to be exact, not counting the RS-232 driver) – a hallmark of digital implementations done in standard logic. Those nine ICs have to be wired; and, once wired, any changes require at least a wiring change. Furthermore, there are six different IC types, so for this kind of implementation you have to keep a substantial inventory of different IC functions.

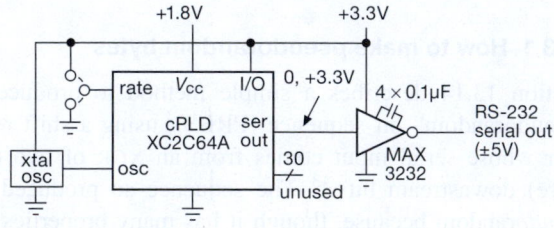

Figure 11.9. Programmable logic permits a far simpler circuit implementation. The connections within the cPLD can be specified, using software tools, either via graphical *schematic entry* or by a text-based HDL; we illustrate both. The resulting desired "fuse" configuration is programmed into the cPLD's internal non-volatile memory, with the cPLD in its actual circuit, via four dedicated pins (known as JTAG, not shown here).

11.3.3 Implementation with programmable logic

As we remarked, programmable logic devices usually provide a better hardware implementation of digital logic, and this example is no exception. Here programmable logic is

---

[17] Note, however, that you *can* get logic in convenient smaller bites, in the form of small pincount packages with 1–3 gates; these go by names like "TinyLogic™" (Fairchild's name), "PicoGate™" (NXP's name), "MiniGate™" (ON Semiconductor's name), "Little Logic™" (TI's name), Single Gate (ST's name), or LMOS™ (Toshiba's name). We'll call them all "mini-logic," a name evidently overlooked in the trademarking frenzy of the big guys. They come in handy when you need some glue between large digital chips.

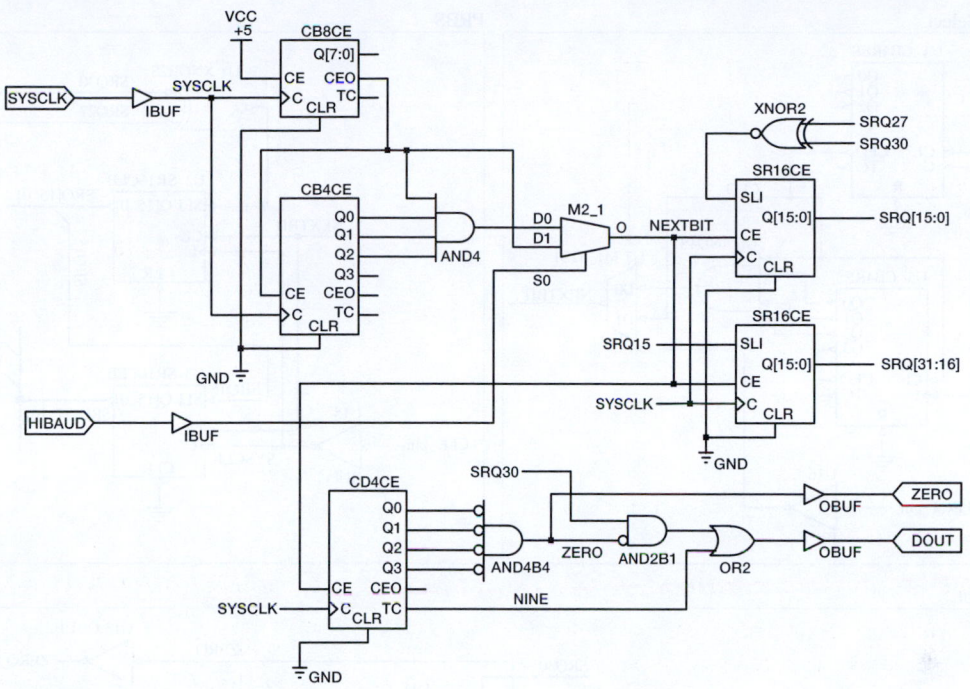

Figure 11.10. Graphical-entry schematic capture for programmable logic design, using Xilinx tools. The component blocks can be generic logic functions (e.g., CB4CE = counter, binary, 4 bits, with chip-enable), or specific standard logic part numbers (e.g., "163" = 74xx163 = 4-bit binary synchronous loadable counter with 3-state outputs and synchronous load and reset; not used in this design); or they can represent modules specified by a text-based hardware description language such as VHDL.

a *much* better choice than standard logic, with a single chip replacing nine; and, as with programmable logic generally, you can reprogram the circuit to fix errors, or to add features.

This design is small, about right for a cPLD, rather than the generally much larger FPGAs. Figure 11.9 shows the wiring, using a smallish cPLD (64 macrocells). The particular cPLD we chose runs its core at 1.8 V, but permits input and output operation from 1.5 V to 3.3 V. This task uses little of the chip's I/O capabilities; 30 pins (of 34 available) are unused. It does, however, use most of the 64 available flip-flops. The chip costs about $2.

A. Programmable logic – schematic entry

Many circuit designers favor a graphical "schematic-entry" technique for designing programmable logic. Figure 11.10 shows how it looks for our pseudorandom byte generator, entered using the tools supplied by Xilinx.

The individual symbols can be familiar gates (OR2: 2-input OR), or lopsided gates (AND2B1: 2-input AND with one bubbled input), or larger generic functions (CD4CE: counter, decimal, 4 bits, with chip enable). However, you

can draw a modular block to represent a larger thing, for example a number of interconnected blocks. Here, for example, the entire figure could be represented by a block with two inputs (SYSCLK and HIBAUD) and two outputs (ZERO and DOUT); that block could then be used as a component in a larger system diagram. Furthermore, you can use text-based HDL entry (see below) to define the guts of a block, which you then connect up graphically just like the blocks here. And you can even create the HDL (which defines such a block) with graphical tools. For example, you can implement a state machine by entering the state-transition diagram in a tool like Xilinx's StateCAD (or Altera's Quartus); it generates HDL output (Verilog, VHDL, or ABEL), which can then be represented as a graphical block, with input and output pins.

Graphical-entry software tools let you simulate the design, to make sure it does what you intended; then it performs a "fit" or "place and route" to create the connection pattern for a particular hardware PLD (cPLD or FPGA). Finally, as described earlier, you program the part itself, either in a "device programmer" (from a company such as BP Microsystems) or (more commonly) with the part already

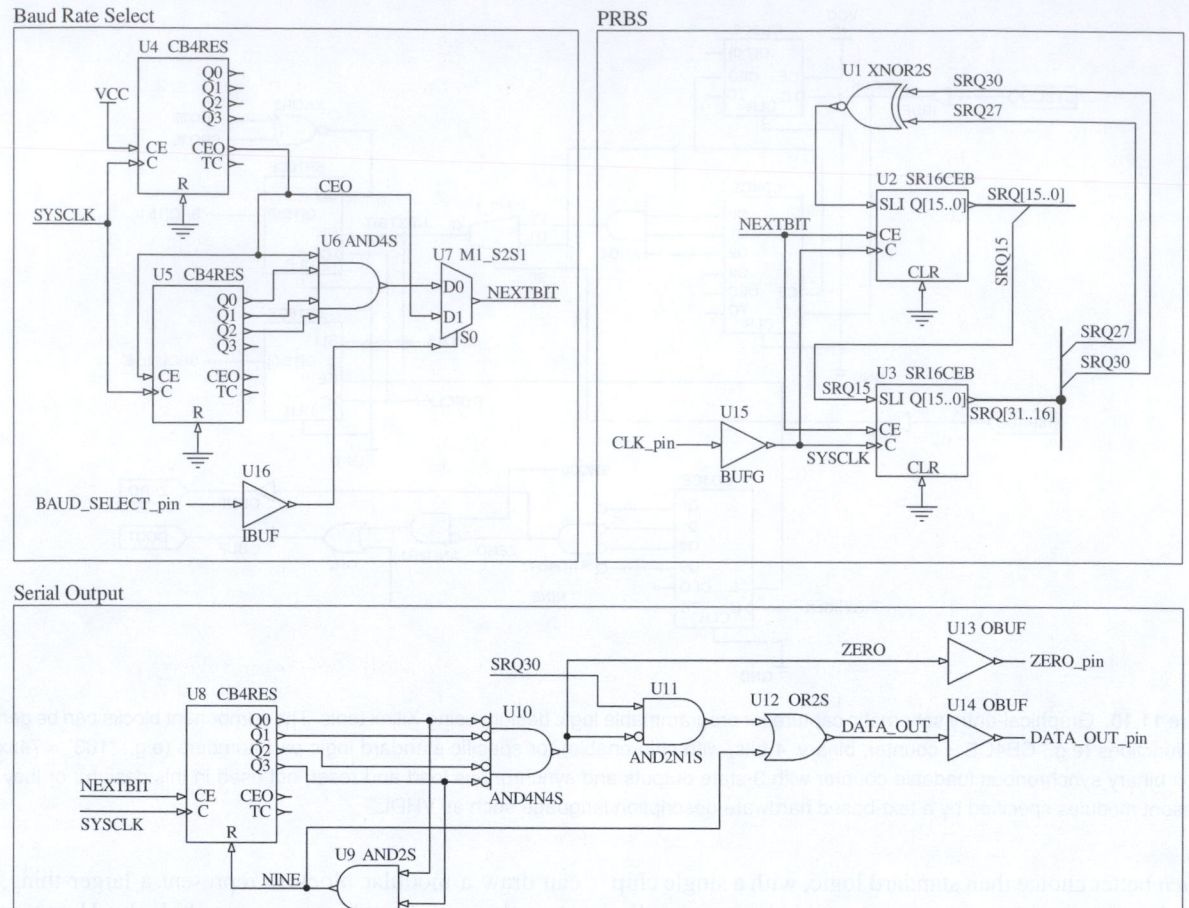

Figure 11.11. The same design as Figure 11.10, implemented with the Altium "Designer" software, and debugged using their NanoBoard development board.

mounted and powered in-circuit. The latter is done through a few dedicated programming pins, which you drive with a programming *pod* that connects to a computer running the design software (typically via USB).

In this example the schematic entry provides a reasonably readable circuit. The designer[18] chose a fully synchronous implementation (all clocked devices are driven from the same clock, namely the input oscillator), explaining that, for some PLDs, shift registers or counters are not guaranteed to work properly when driven from a derived clock (e.g., from the output of a ripple counter), even though this is normal practice when implemented with discrete standard logic.[19]

One traditional complaint with schematic entry has been that the software tools tend to be specific to one vendor's hardware (as in this example). That makes it hard to migrate a design to a different supplier's parts. But recently there have become available some reasonably priced schematic-entry and simulation suites that let you choose among several vendors' FPGAs. We've been trying out such a product – the "NanoBoard" series of hardware development boards supplied by Altium for use with their "Designer" software – that currently support FPGAs from Altera (Cyclone series), Lattice (ECP2 series), and Xilinx (Spartan and Virtex series). If you stick to the generic parts libraries, your design is portable. An example is shown in Figure 11.11, once again our friend the pseudorandom bit

---

[18] We thank Jim MacArthur for taking the challenge and producing both diagram and working chip in an hour or two.

[19] Evidently the signal delay as such a clock signal is routed in the PLD causes unacceptable skew (§10.8.2C); only the "global clock signals"

are guaranteed to work reliably, a fact somewhat buried in the fine print.

generator. The designer[20] chose to group the design into functional modules, to keep the schematic readable. With the exception of the input and output buffers (which were taken from the Xilinx-specific library), all the parts are generic, and so the design is portable with little effort.

At a higher level of abstraction (and cost), you can get synthesis software toolsets like the elegant "Synphony HLS" (High-Level Synthesis) from Synopsys. You start your design with MATLAB and Simulink (from The Math-Works™), which let you snap together Lego-like functional modules from a toolbox of blocks (digital filter, FFT, sampler, etc.). Synphony takes this description and targets (by way of an HDL) either an FPGA or a full ASIC implementation, along with C-language code for functional simulation, test vector generation, and timing. This is far less laborious than traditional HDL design entry. And, because the initial high-level description is verifiable (and its functional blocks are reliably correct), the result is far less prone to errors than a similarly complex chip that is created from an HDL in the conventional manner. Note, however, that the resulting implementation may not be terribly efficient, in terms of use of silicon resources, speed, and power.

Advantages of schematic entry

Those who favor graphical schematic entry generally explain their choice as (a) they come from a schematic background, so it is natural; (b) they find it easier to learn, to understand, and to explain to others; (c) they like having a graphical schematic as documentation; (d) because modules can be HDL-in-disguise, it has all the power of HDL for complex designs where that is needed; and (e) graphical entry is used in graphical programming languages (GPLs) like LabVIEW, MATLAB®, and Simulink®.

11.3.4 Programmable logic – HDL entry

The alternative to graphical design entry is one of the text-based Hardware Description Languages. The current favorites are Verilog and VHDL, with lots of activity also in applying the C programming language to high-level design.

We think it may be helpful to show first how this design looks in the legacy ABEL HDL, whose statements tend to be "close to the hardware": in ABEL you define *pins* (input and outputs) and *nodes* (internal flip-flops or gates), and then you write *equations* that connect up the signals, for example the gates that determine the D input and clock source

for a flip-flop. As with schematic entry, you can simulate (with *test vectors*), after which you run the *fitter* program that creates the connection netlist (called a *jedec file*) for your particular target PLD. Finally, you program the part itself, either in-circuit or (more rarely) in a separate stand-alone device programmer.

A. Programmable logic – HDL entry I (ABEL)

Figure 11.12 shows the ABEL source file for creating the pseudorandom byte generator, using a cPLD with 64 macrocells (each of which can implement a complicated logical function of available signals, delivering the result either combinationally or "registered" in a flip-flop).

A few explanatory comments (working from the top down): (a) lines end with a semicolon, and // signifies a comment; (b) the designations com and reg_D (or reg_T) following the directive istype specify outputs and nodes as combinational or registered (flip-flops); no designation means an input; (c) an array of signals (called a *set* in ABEL) is written as [sr30..sr0] (our 31-bit shift register); (d) NOT, AND, OR, and XOR are written as !, &, #, $; (e) variable names not declared as pins or nodes (like baudclk) are just *definitions*, to simplify later equations (like [sr30..sr0].clk = baudclk); thus, for example, sr$n$ is the name we have chosen for the nth shift register bit, and not a reserved name in the ABEL language; (f) the several pins for a named flip-flop are designated by "dot extensions," for example, sr0.d = !(sr30.q $ sr27.q) creates the XOR of the Q-outputs of shift register bits 30 and 27 and connects it to the D-input of bit 0 (that's the XOR feedback that generates the pseudorandom bits).

It's worth spending a bit of time puzzling through the code to get a feeling for how you do a circuit in a text-based language; we've commented the code liberally to make it (somewhat) understandable. It's not the most efficient implementation (see below for a truly inspired method!), but it works.

B. Programmable logic – HDL entry II (Verilog)

Simple languages like ABEL were OK for simple cPLD designs (we used it for designs with up to 128 macrocells, with some pain), but somewhere between frustrating and hopeless for the really complex designs for which contemporary FPGAs are so well suited. Although both ABEL and CUPL evolved from their roots in PALASM to incorporate higher-level constructions (like "If-Then-Else"), they were finally abandoned in favor of contemporary languages, principally Verilog and VHDL. These have similar capabilities, but, sadly, divide humanity into two camps,

---

[20] The ever-clever Curtis Mead, to whom we owe thanks.

```
// -------------------------------------------------------
module PRBYTES
title 'pseudo random byte generator - async serial
      Paul Horowitz, 10 Dec 07, rev 5'

// PRBYTES device 'XC2C64A';    xilinx zeropower cPLD


// Inputs
     osc       pin;   // 2.4576 MHz osc input (9600 baud x 256)
     rate      pin;   // LOW selects 1200 baud, HIGH selects 9600 baud

// Outputs
     serout    pin    istype 'reg_D';     // serial data
     syncout   pin    istype 'com';       // RS-232 START pulse (for testing)

// Nodes
     [sbc3..sbc0]   node   istype 'reg_D';   // serial bit counter, BCD (10 states, 0-9)
     [brd10..brd0]  node   istype 'reg_T';   // baud rate divider: divide by 256 or 2048
     [sr30..sr0]    node   istype 'reg_D';   // 31-bit shift register for pseudoran generator

// Constants & Declarations
     high = ^b1;              // high
     low = ^b0;               // low
     baudclk = rate & brd7.q # !rate & brd10.q;     // baud rate clock
     sbczero = !sbc3.q & !sbc2.q & !sbc1.q & !sbc0.q;  // serial bit counter is zero
     sbcnine = sbc3.q & !sbc2.q & !sbc1.q & sbc0.q;    // serial bit counter is nine


Equations
     // baud rate divider: 11 bit binary ripple ctr, use bits 7 or 10 for baudrate
     [brd10..brd0].t = [1,1,1,1,1,1,1,1,1,1,1];   // all flops set to toggle
     brd0.clk = osc;                              // clock lsb from external osc
     [brd10..brd1].clk = [brd9..brd0].q;          // all higher bits clock from previous Q output
                                                  // i.e., brd10.clk = brd9.q; brd9.clk = brd8.q; etc

     // serial bit counter: synch BCD up-counter (4-bit, 10 states: 0-9)
     [sbc3..sbc0].clk = baudclk;        // serial bits clocked at baud rate
     sbc0.d = !sbcnine & !sbc0.q # sbcnine & low;        // lsb toggle, except reset after state 9
     sbc1.d = !sbcnine & (sbc0.q $ sbc1.q);              // toggle if lower bits set
     sbc2.d = !sbcnine & ((sbc1.q & sbc0.q) $ sbc2.q);   // useless 'sbcnine & low' term of top line
     sbc3.d = !sbcnine & ((sbc2.q & sbc1.q & sbc0.q) $ sbc3.q);   // was included for clarity only

     // 31-bit xor-feedback shift register to generate pseudo-ran bits
     [sr30..sr0].clk = baudclk;       // prbs shift register clocked at baud rate
     [sr30..sr1].d = [sr29..sr0].q;   // compact form for sr30.d = sr29.q; sr29.d = sr28.q; etc
     sr0.d = !(sr30.q $ sr27.q);      // xor feedback to make prbs; negation ensures startup from reset

     // append start and stop bits to generate RS-232 style async serial output bytes; do syncout
     serout.clk = baudclk;  // serial output clocked at baud rate
     serout.d = (!sbczero & sr30.q # sbczero & low) # sbcnine & high;
     // stop bit (high) on nine, start bit (low) on zero, random on 1 through 8
     // RS-232 drivers invert; second term in parenthesis, and 'high', are both unnecessary

     syncout = sbczero;             // RS-232 start pulse


end PRBYTES
// -------------------------------------------------------
```

Figure 11.12. ABEL code for a pseudorandom byte generator.

with designers swearing passionate allegiance to one or the other.

To give a sense of these powerful text-based design languages, we've coded the pseudorandom byte generator design in both Verilog and VHDL.[21] Statements in these languages can be written in terms of the *structure* (e.g., what signals are connected to register inputs), or, at a somewhat higher level, in terms of *behavior* (e.g., what should happen next). These examples are primarily coded behaviorally, which is the mode generally favored by experienced users; but we show also a structural alternative for a small portion (the serial bit counter). Figure 11.13 shows the Verilog design, coded behaviorally; and Figure 11.14 shows alternative Verilog coding for the 4-bit BCD counter portion that is predominately structural.

Some explanatory comments. (a) The Verilog code is recognizably similar to the analogous ABEL code, particularly in the structural form. (b) The *wire* declaration creates an internal node, and *assign* makes a definition. (c) In common with computer programming languages, Verilog and VHDL make plentiful use of conditional structures like if, else if, else. (d) Designers tend to favor behavioral statements (being more compact and easily understood), at least through validation by simulation; however, structural coding often produces a more efficient implementation.[22]

A further detail: in this example there are two clocks – the fast ~2.5 MHz input clock, and a subdivided clock at the baud rate (1.2 kHz or 9.6 kHz, according to the selected baud rate) for the serial bit counter. So the system is not fully synchronous (although each counter is synchronous, by itself). This is OK for a design that will be implemented as a full-custom IC, but it may create difficulties if targeted for a cPLD, or some FPGAs.[23] A clean solution is to use a single fast global clock for both counters, and then to enable the (slower) serial bit counter only at the subdivided rate.

C. Programmable logic – HDL entry III (VHDL)

The *other* popular HDL these days is VHDL, a more strongly typed language with roots in the programming

language Ada (Verilog derives more closely from the C programming language). It tends to be more verbose. For the VHDL coding of the pseudorandom byte generator we've taken a fully synchronous approach; that is, the ~2.5 MHz input clock is used for both the baud-rate divider and the serial bit counter, with the latter enabled only for one clock cycle (when nextbit is true). This mirrors the scheme used in the schematic entry examples of Figures 11.10 and 11.11. The (rather lengthy) code is shown in Figure 11.15.

Advantages of HDL entry

Those who favor text-based HDL entry generally explain their choice thus: (a) design entry is rapid, particularly when portions of previous designs are used; (b) the design is more concise (thus easier to know that it is right), and self-documented (as a textual description); (c) the design cycle is streamlined, as a design is being iterated through simulation and prototype; (d) it is particularly easy to change parameters (such as number of bits in a register, ALU, etc) by merely resizing arrays (compared with rearranging graphical modules); (e) the languages are standardized and universal (compared with proprietary schematic-entry tools); (f) good simulators are available and free of charge; (g) HDL languages are best suited for high-level synthesis of complex designs (like a microprocessor), and prefab open-source IP cores are provided as HDL;[24] (h) HDL entry is well suited to implementation as a *full-custom* IC (ASIC), which is often desirable for a circuit design destined for large-quantity manufacture and a high degree of optimization in speed and power (but which start out as an FPGA during development and small-quantity production); and (i) for those with a programming background, HDL entry is more natural.

11.3.5 Implementation with a microcontroller

Micro*controllers* (the subject of Chapter 15) are inexpensive processors intended for use in things *other than* computers. You should think of them more as a circuit component and less as a computer. Compared with a micro*processor* intended for a computer, microcontrollers are intended to stand alone, and therefore always include on-chip program and data memory, along with a selection of "peripherals" such as communications (USB, firewire, Ethernet, UART, CAN, SPI, I^2C), ADCs, digital I/O, comparators, LCD display drivers, pulse-width modulators, timers, and the like. Many also include an internal oscillator (all they need is dc power!), in some cases accurate enough for

---

[21] We are indebted to our colleagues GuYeon Wei and Curtis Mead, who kindly coded this example.

[22] An example might be a 32-bit counter, where a behavioral statement like "count=count+1" might compile to a register and 32-bit full adder, the latter receiving a constant 00000001(hex) as one of its inputs.

[23] Some FPGAs let you drive internal clock distribution lines from logic outputs, but some don't, for example the Virtex-5 and Spartan-3, respectively; for the latter you would have to bring the clocking signal out to a pad.

[24] But IP cores that must be licensed are encrypted.

```verilog
// ------------------------------------------------------------------------
// Pseudo random byte generator - async serial
// Gu-Yeon Wei and Curtis Mead; "behavioral" coding

module PRBYTES(osc, rate, reset, serout);

// inputs
// osc: 2.4576 MHz oscillator input (9600 baud x 256)
// rate: 0 = 1200 baud, 1 = 9600 baud
// reset: active high reset to clear out counters
input osc, rate, reset;

// outputs
// serout: serial data
output serout;

// define registers
reg serout;      // a "reg" is a node that holds its value until overwritten
reg [3:0] sbc;   // RS-232 serial bit counter
reg [10:0] brd;  // baud rate divider
reg [30:0] sr;   // PRBS shift register

// define wires
wire baudclk, sbczero, sbcnine; // a "wire" is a combinational circuit node

// logic
assign baudclk = rate & brd[7] | ~rate & brd[10];        // baud rate clk, div 256 or 2k
assign sbczero = ~sbc[3] & ~sbc[2] & ~sbc[1] & ~sbc[0];  // serial bit counter zero flag
assign sbcnine = sbc[3] & ~sbc[2] & ~sbc[1] & sbc[0];    // serial bit counter nine flag

// baud rate divider: 11 bit binary ctr, use bits 7 or 10 for baudrate
always @(posedge osc)  // logic clocked by osc
  begin
    if (reset)
        brd <= 0;
    else
        brd <= brd+1;
  end

// serial bit counter: synch BCD up-counter (4-bit, 10 states:0-9)
// this is "behavioral"; see later program fragment for "structural"
always @(posedge baudclk)  // logic clocked by baudclk
  begin
    if (reset)
        sbc <= 0;
    else if (sbcnine)
        sbc <= 4'b0000;
    else
        sbc <= sbc + 1;
  end

// 31-bit xor-feedback shift register to generate pseudo-random bits
always @(posedge baudclk)
  begin
    if (reset)
        sr[0] <= 0;  // prevent stuck state of all ones
    else begin
        sr[30:1] <= sr[29:0];
        sr[0] <= ~(sr[30] ^ sr[27]);  // xnor makes all ones the stuck state
    end
  end

// serial output stream comes from last SR bit, but overridden by "0" start, and "1" stop
always @(posedge baudclk) begin
    serout <= (~sbczero & sr[30]) | sbcnine;
end

endmodule //PRBYTES
// ------------------------------------------------------------------------
```

Figure 11.13. Behaviorally-coded Verilog for pseudorandom byte generator.

```
// -------------------------------------------------------------------
// alternative structural coding of serial bit counter:
// synch BCD up-counter (4-bit, 10 stages:0-9)
always @(posedge baudclk) begin
   sbc[0] <= ~sbcnine & ~sbc[0] & ~reset;
   sbc[1] <= ~sbcnine & (sbc[0] ^ sbc[1]) & ~reset;
   sbc[2] <= ~sbcnine & ((sbc[0] & sbc[1]) ^ sbc[2]) & ~reset;
   sbc[3] <= ~sbcnine & ((sbc[0] & sbc[1] & sbc[2]) ^ sbc[3]) & ~reset;
end
// -------------------------------------------------------------------
```

Figure 11.14. Structurally-coded Verilog fragment for 4-bit BCD counter.

timing of on-chip serial ports (e.g., $\pm 2\%$). They are programmed (and reprogrammed) in-circuit, by a serial connection like the JTAG boundary-scan protocol.[25] Recent designs include internal circuitry that permits run-time debugging in-circuit (by the same JTAG port used for programming).

Microcontrollers (sometimes abbreviated μC) are inexpensive (typically ranging from less than a half dollar up to ten dollars), and are available in literally thousands of versions from dozens of manufacturers; popular offerings are the PIC series (from Microchip), the AVR series (from Atmel), the ARM series (from multiple vendors), and evolved incarnations of the legacy Intel 8051 series (multiple vendors). Low-end μCs are usually 8-bit processors, with perhaps 1 kB of program memory and 128 bytes of RAM, and clock speeds of 20 MHz; at the high end there are 32-bit processors with 512 kB program memory, 64 kB of RAM, and many integrated peripherals.[26]

Microcontrollers are wonderfully versatile and capable, and they are found in nearly everything electronic, from toasters and toothbrushes to trucks and traffic lights. They can do a pseudorandom byte generator with the greatest of ease. Figure 11.16 shows the "circuit" – basically just dc power to the chip, and bytes out! The rest of the task is the programming, which we'll illustrate two ways.

A. Microcontroller programming – assembly code
Microcontrollers, and indeed all processors, perform a sequence of operations, specified as instructions that are stored in some form of electronic memory (see Chapter 14); in the case of μCs the program is held on-chip in non-volatile memory (thus retained when power is off). The instruction set is particular to the processor type, and includes operations such as arithmetic (e.g. add), logical (e.g. shift), data transfer (e.g. move), and branching (e.g. brz, branch if zero). The instructions themselves reside in memory as a set of bytes, fetched by the processor during program execution.

As with computer programming generally, programmers rarely deal with this *object code* itself (which is both tedious and error-prone), programming and debugging instead at a more humanly understandable level: either in *assembly language*, or in a higher-level language (usually C or C++). The code so written is then *assembled* or *compiled* (using software) to produce the μC-readable binary object code that will reside in the processor's program memory. There are nice software tools to help you get the code right – for example, *simulators* that let you see how the candidate program runs, step-by-step; and *run-time debuggers* that let you watch what the μC is actually doing, in-circuit, when it runs your program. The term *Integrated Development Environment* ("IDE") is used to describe a software suite that includes compiler, assembler, loader, and run-time debugger. You want one of these.

Figure 11.17 is a listing of assembly code for the pseudorandom byte generator implemented in the microcontroller circuit of Figure 11.16; not to worry, we'll see it coded in C, finally, in Figure 11.18.

Just a few notes of explanation (hey, this isn't a book on software!). (a) All microcontrollers include a bunch of internal "special-function registers," whose contents you must initialize to set up timer modes, interrupts, location of initial stack pointer, and so on. That's what the first seven mov operations do, here. This can be fussy code, particularly involving timer modes and baud rates (best to copy it from a program known to work correctly). And,

---

[25] A 4-wire serial protocol, which can be chained through a succession of ICs, and which permits run-time loading of, examination of, and bypassing of some internal registers in complex ICs; see §14.7.4.

[26] A current example is the ARM7 LPC2458 from NXP (formerly Philips), displayed in Figure 10.86: 512kB/64kB memories, 72 MHz clock, 10/100 Ethernet, USB 2.0, 10-bit A/D and D/A, 2xPWM, 4 UARTs, 2xCAN, SPI, 2xSSP, 3xI^{2}C, I^{2}S, 136 bits of general-purpose I/O, and an external memory controller. This puppy costs just $10! For higher clock speeds you would choose instead the faster ARM9.

```
-------------------------------------------------------------------------
-- Pseudo random byte generator - async serial
-- Curtis Mead, modeled after Gu-Yeon Wei's Verilog
-- but implemented as a fully synchronous design
-- COMMENTS begin with two dashes (--)

library IEEE;
use IEEE.STD_LOGIC_1164.ALL;

entity PRBS is
    Port ( CLK           : in   STD_LOGIC;  -- 2.4576 MHz oscillator
           RESET         : in   STD_LOGIC;  -- Synchronous reset, active high
           BAUD_SELECT   : in   STD_LOGIC;  -- 0 = 1200 baud, 1 = 9600 baud
           ZERO_pin      : out  STD_LOGIC;  --
           DATA_OUT_pin  : out  STD_LOGIC); -- Serial data output
end PRBS;

architecture Behavioral of PRBS is
    signal sbc           : STD_LOGIC_VECTOR( 3 downto 0); -- serial bit counter
    signal brd           : STD_LOGIC_VECTOR(10 downto 0); -- baud rate divider
    signal prbs_bits     : STD_LOGIC_VECTOR(30 downto 0); -- prbs shift register
    signal baudtemp      : STD_LOGIC_VECTOR( 1 downto 0); -- brd temp register
    signal nextbit       : STD_LOGIC; -- pseudo clock signal for serial bit counter
    signal sbcnine, sbczero   : STD_LOGIC;

begin

-- PRBS with 32 bit shift register, bits 27 and 30 XNOR'ed into input
process (CLK)
begin
    if CLK'event and CLK='1' then
        if RESET ='1' then
            prbs_bits(0) <= '0';
        elsif nextbit = '1' then
            prbs_bits(30 downto 1) <= prbs_bits(29 downto 0);
            prbs_bits(0) <= prbs_bits(27) xnor prbs_bits(30);
        end if;
    end if;
end process;

-- baud rate divider, 11 bit counter, synchronous reset
-- use bit 7 for 9600 baud and bit 10 for 1200 baud
process (CLK)
begin
    if CLK='1' and CLK'event then
        if RESET='1' then
            brd <= (others => '0');
        else
            brd <= brd + 1;
        end if;
    end if;
end process;
```

Figure 11.15. Fully synchronous VHDL code for pseudorandom byte generator (continued below).

```
-- shift register used to set nextbit for one clock only,
--  triggered by baudrate divider bits
--  for clocking nextbit from baud rate divider output 'baudtemp'
process (CLK)
begin
    if CLK='1' and CLK'event then
        if RESET='1' then
            baudtemp <= "00";
        else
            baudtemp(1) <= baudtemp(0);
            baudtemp(0) <= (BAUD_SELECT and brd(7)) or (not BAUD_SELECT and brd(10));
        end if;
    end if;
end process;
nextbit <= baudtemp(0) and not baudtemp(1); -- nextbit high for one clock cycle


-- serial bit counter, 4-bit BCD up with synchronous clear
process (CLK)
begin
    if CLK='1' and CLK'event then
        if RESET = '1' then
            sbc <= (others => '0');
        elsif nextbit='1' then
            if sbcnine='1' then
                sbc <= (others => '0');
            else
                sbc <= sbc + 1;
            end if;
        end if;
    end if;
end process;
sbczero <= '1' when sbc = "0000" else '0'; -- serial bit counter zero flag
sbcnine <= sbc(0) and sbc(3); -- serial bit counter nine flag


-- outputs synchronous with clock signal CLK
process (CLK)
begin
    if CLK='1' and CLK'event then
        if RESET = '1' then
            DATA_OUT_pin <= '0';
            ZERO_pin <= '0';
        elsif nextbit = '1' then  -- nextbit is an "enable" (at the baudrate)
            DATA_OUT_pin <= (not sbczero and prbs_bits(30)) or sbcnine;
            ZERO_pin <= sbczero;
        end if;
    end if;
end process;

end Behavioral;
```

Figure 11.15 continued.

781

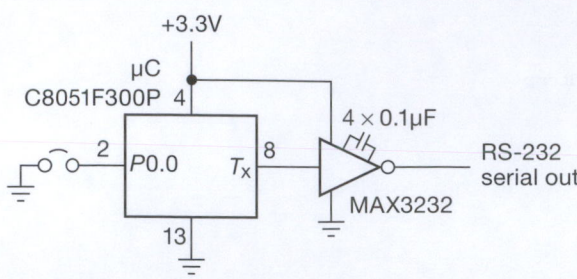

Figure 11.16. Microcontrollers represent a compact alternative to programmable logic, particularly when speed is not critical. This particular microcontroller choice (among literally thousands) includes an internal oscillator, so *no* external components are needed to generate the logic-level serial bitstream output.

surprise, there's absolutely no standardization across different microcontroller families. Fussy, fussy, fussy... (b) The program bookends `org 00h` and `end` tell the assembler where to put the first instruction, and where the code ends. That is, they are not instructions that the microcontroller executes, but rather directives to the assembler; they're sometimes called "pseudo-ops." (c) The rest of the code consists of arithmetic and logical operations, such as clear, copy (called `mov`), compare (`cjne`: compare and jump on not equal), loop counters (`djnz`: decrement and jump on nonzero), subroutine call (`acall`), and so on.

You really don't want to fall in love with this stuff – we show it mostly to warn you off!

B. Microcontroller programming – coding in C
Assembly language programming is fussy stuff, and it's easy to make trivial mistakes. Even when you get it right, it's tricky to modify. Large programs are laborious. The code must be rewritten if you want to use it with a different family of microcontroller. And, worst of all, you probably won't understand what you've written a few weeks later, when it needs to be changed.

For these reasons most programmers prefer a high-level language like C or C++ for coding.[27] It's easier to write and understand, and it's portable across microcontroller families (though with some changes required to accommodate architectural differences). Figure 11.18 list the PRBS generator, in the C programming language.

The programmer[28] noticed, in a burst of inspiration, that

the PRBS could be coded to perform a full byte at a time, rather than the ponderous bit-at-a-time coding we assumed necessary in the earlier examples. That's because the shift register has, at any moment, all the downstream bits you need to calculate all the fresh bits up to the first feedback tap (in this case 27 bits). Given the byte organization of 8-bit μCs and the fact that the built-in serial-port UART expects a full 8-bit quantity before transmission, the smart thing is to create a new upstream byte, using the bytewide XOR instruction (bitwise XOR, without carry). The resulting program runs wicked fast! See if you can figure out the coding.

11.4 Advice

So, what should you do? Here's a brief summary of recommendations, which we think fairly represents the ups and downs of the various choices you have for digital (and mixed-signal) implementations.[29] We'll do this two ways, first by *Technology*, then by *User Community*.

11.4.1 By *Technologies*
• **Standard logic**
  – `Upside`: OK if you want to put a simple circuit together quickly and you have the parts on hand; useful for bus drivers, buffers, and small logic "glue" (for sticking large chips together, or for fixing logic errors); available in through-hole, so you can use a pluggable breadboard, or a solder (or wirewrap) protoboard; no software design tools needed.
  – `Downside`: unwieldy for complex circuits; multiple functions to inventory; inflexible.
• **Programmable logic (cPLD, FPGA)**
  – `Upside`: generally preferred (over standard logic) for most digital designs – reduced package count, easily reprogrammed, flexible and inexpensive, small inventory needed; favor cPLDs for predictable timing and for small designs, FPGAs for large designs (though small FPGAs are emerging); best choice for fast parallel logic and state machines; good for timing-critical functions (such as communications protocols), bit-level operations (CRC, Viterbi), and functions with lots of I/O; designs can be migrated across PLD families, and to full-custom ICs.

---

[27] Perhaps with small blocks of assembly code inserted where speed is critical, for example an interrupt service routine (see Chapter 13), within a larger program.

[28] Jason Gallicchio, to whom we are indebted for this code and for other assistance with the microcontroller discussion.

[29] Note, however, that it's usually smoother sailing if you adopt the methods that people around you are using – their experience base and toolsets accelerate learning and debugging – even if some different technique potentially offers other advantages.

```
// --------------------------------------------------------
;; program to send pseudo-random bytes from serial UART   rev of 12/14/07

        org 00h;           starting address at power up or reset

        mov IE, #00h;      disable interrupts
        mov SP, #90h;      init stack pointer, used for subroutine
        mov PCON, #00h;
        mov TMOD, #20h;    timer 1: 8 bit, auto reload
        mov TCON, #40h;    timer 0 off, timer 1 on
        mov SCON, #40h;    8 bit UART, tx only, set by timer 1
        mov TH1, #0FDh;    9600 baud (reload value = 253 decimal)

        clr A;
        clr TI;
        jb P1.0, makebyte; logic HIGH input on P1.0 pin = 9600 baud
        mov TH1, #0E8h;    LOW: 1200 baud (reload value = 232 decimal)

makebyte: mov R3, #8;      setup loop counter - 8 shifts make a byte

          ; create feedback bit as XNOR of bits 27 and 30 (0..30)
makebit:  mov C, ACC.4;    that is, ACC.4 and ACC.1
          mov R2, #0;
          mov R2.1, C;     stick ACC.4 into R2.1, to XNOR with masked ACC
          anl A, #2;       mask ACC, preserving ACC.1 only
          clr C;
          cjne A, R2, loadreg;  XNOR=0, carry already cleared
          setb C;          XNOR=1, set carry

loadreg:  mov R0, #80h;    80h-83h will hold MSByte-LSByte of 32-bit SR

shift32:  mov A, @R0;      R0 points to one of 4 bytes at 80h..83h
          rrc A;           shift right through carry
          mov @R0, A;      stash it
          inc R0;
          cjne R0, #84h, shift32;  check for last byte of 4

          djnz R3, makebit; do it 8 times
          acall sendbyte;   send finished byte via serial UART
          sjmp makebyte;    and start next byte

          ; code for sending one byte
sendbyte: jnb TI, sendbyte; wait for last tx done
          mov SBUF, A;      load transmit buffer
          ret;

          end
// --------------------------------------------------------
```

Figure 11.17. Assembly code for pseudorandom byte generator.

```
// --------------------------------------------------------
#include <avr/io.h>

#define F_CPU  20000000   // 20 MHz
#define BAUD_RATE  9600      // or 115200

void USART_Init(void) {
  UBRR0 = F_CPU/BAUD_RATE/8 - 1; // Set baud rate
  UCSR0A = (1<<U2X0);   // Change baud divisor from 16 to 8
  UCSR0B = (1<<RXEN0) | (1<<TXEN0);    // enable Rx & Tx
  UCSR0C = (1<<UCSZ01) | (1<<UCSZ00); // 8N1, no parity
}


void putch(unsigned char ch)
{
  while ( !(UCSR0A & (1<<UDRE0)) );
  UDR0 = ch;
}



int main(void)
{
  unsigned char out;
  unsigned char d=0xff, c=0xff, b=0xff, a=0xff;

  USART_Init();

  while(1) {
    out =  ( (d<<1) | (c>>7) ) ^ c;  // 31 bit, tap 24
    d = c;
    c = b;
    b = a;
    a = out;

    putch(out);
  }
}
// --------------------------------------------------------
```

Figure 11.18. C code for pseudorandom byte generator.

– Downside: time to learn software tools (often proprietary), and their cost; surface-mount parts only (requires custom PCB or adapter); cost of programming
  pod and design tools; short life cycle (supply problems).
– PLD design--program tools
  * *Schematic entry*
    · Upside: easy to learn, understand, and explain
      to others (for circuit nerds especially); self-
      documenting; entry of parallel circuitry in a parallel
      way; analogous graphical programming languages

(GPLs) used in tools like LabVIEW, MATLAB®,
and Simulink®.
· Downside: proprietary design tools; cost; limited
  applicability.
* *HDL (Verilog/VHDL)*
  · Upside: natural migration for programmer nerds
    who can adapt to parallel thinking; concise and
    standardized; cost-free simulation and compilation
    tools; easy-to-iterate design modifications; can mi-
    grate across PLD families, and to full-custom ICs;

well suited for large complex designs; IP core availability.

- · Downside: time to learn, and lack of schematic documentation (especially for non-programmers).
- **Microcontrollers** (see Chapter 15!)
  - Upside: programmable, flexible embedded computer; ideal for complex state machines with lots of decisional branching; rapid iteration of code and (in-circuit) debugging; many on-chip peripherals (communication,[30] conversion, interface,[31] other[32]); familiar programming language (C, C++); best for embedded control, especially involving user communication (display and control); enjoyable!
  - Downside: slower than cPLD/FPGA; less suited to bit-level manipulation; less parallelism.
  - *μC program tools*
    * *Assembly language*
      · Upside: can handcraft optimized code, for example an interrupt service routine.
      · Downside: fussy and error prone; difficult to make major modifications; cannot port to a different processor family.
    * *C, C++*
      · Upside: standardized and portable (though some processor-specific changes needed); widespread programming expertise; structured language facilitates changes and upgrades; libraries of useful functions; well suited for complex tasks.
      · Downside: not much... compiled code may be less efficient in size and speed (can embed assembly code for critical loops and routines, and for access to special features); code can be hard to understand; requires compiler toolset (some are expensive, others are free).
    * *Other (Basic, Arduino Software; Java, Python)*
      · Upside: easy to understand, good for entry level.
      · Downside: limited range of microcontroller support, not generally portable across processor families, not for big projects.

11.4.2 By *User Communities*

All of the user communities are well served by *microcontrollers*, owing to low cost, ease of programming, and flex-

ibility. Among these communities, the following additional technologies are favored.

- **High-volume production**[33]
  Here unit cost and time-to-market are critical. In large volumes, unit cost is minimized with ASICs (unique Application Specific full-custom Integrated Circuits) and ASSPs (generally available Application-Specific Standard Products), whereas time-to-market is generally shortened by using FPGAs to eliminate the ASIC design–debug cycle. In large production it is common to use high-density BGA (ball-grid array) packages and other fine-pitch surface-mount (SMT) devices, along with the necessary multilayer circuit boards and advanced assembly techniques. And, of course, always use microcontrollers.
- **Complex prototype and small production**[34]
  Here you usually cannot afford the delays and costs associated with both ASICs and high-density BGA packages. So the favored technologies are cPLDs and FPGAs, along with assorted support chips, all in surface-mount packages, and assembled on multilayer boards.[35] And, of course, always use microcontrollers.
- **Laboratory and "1-Off"**[36]
  Here it's important to be able to put together an instrument relatively quickly, and to modify and evolve its circuitry as the need arises. We favor through-hole prototyping boards, populated with through-hole components, and (where essential) with a small number of SMT devices (on through-hole adapters[37]). This constrains you to standard logic and interface, a limited selection of microcontrollers, and perhaps some cPLDs/FPGAs, along with analog components (op-amps, etc). You can make greater use of FPGAs if you take advantage of third-party pre-fabricated daughterboards from companies like Digilent, DLP Design, or Opal Kelly: these range from bare-bones boards (FPGA, program ROM, and USB or JTAG interface) up to complex boards with displays, Ethernet, and so on.[38]

---

[30] For example, UART, SPI, I²C, CAN, USB, Ethernet, IrDA, WiFi, digital audio/video, Bluetooth, ZigBee.

[31] For example, keyboard, mouse, PCI, PCIe, PWM, SIM and Smart Card, GPS, PCMCIA/CF.

[32] For example, external DRAM/SDRAM, MMU, GPS, CCD/CMOS cameras, LCDs, graphic display.

[33] Cellphones, TV set-top boxes, etc.

[34] Oceanographic and scientific research instruments, MRI machines, experimental products, etc.

[35] If you need to use BGA or other high-pin-count parts, it's good to know about prototype assembly services such as Advanced Assembly™: they can handle fine-pitch BGA assembly (including X-ray inspection), and they will even take care of ordering the parts for you. Best of all, they don't mind orders of just a few boards.

[36] The *first* NMR, MRI, STM; atom traps, laser cooling, an astronomical telescope control, etc.

[37] For example from Aries Electronics, or Bellin Dynamic Systems.

[38] If you want to use FPGAs as building blocks in diverse projects, and

Alternatively, you can start with a fully populated microcontroller kit, including PLDs, that may include also unpopulated "prototyping" pads for a small number of additional devices.

- **Hobbyist**[39]
Here the idea is to have lots of inexpensive fun. The basic unit is the microcontroller, either as part of a prefabricated kit,[40] or used on a through-hole protoboard, or perhaps a custom printed-circuit board.[41] Through-hole components (see *Laboratory and "1-Off", above*) are always easy to use, as are PLDs included in microcontroller kits.

you can't find what you want from companies like these, you can follow the lead of one of our colleagues: he designed a small daughter-board to hold his chosen FPGA, along with a couple of EEPROMs, an oscillator, a JTAG header, and voltage regulators (so it runs from +5 V power). It has a pair of 2×20-pin headers (connected to lots of FPGA pins) aimed downward, to plug into a custom board.

[39] Robots, remote control, lighting systems, audio/radio/video, etc.

[40] Supplied by the μC manufacturer, or from an independent supplier like Parallax, Digilent, DLP Design, or the Arduino project. It's pleasant, also, to loiter at websites like sparkfun.com and adafruit.com.

[41] There are freeware versions of layout tools, such as Eagle (from Cad-Soft); there are open-source tools; and there are inexpensive PCB fabrication houses, for example Advanced Circuits (www.4pcb.com), who also offer their "PCB Artist" free PCB layout software (which, however, produces files in their proprietary format).

Review of Chapter 11

An A-to-G summary of what we have learned in Chapter 11. This summary reviews basic principles, facts, and application advice in Chapter 11.

¶A. Programmable Logic Devices (PLDs).

PLDs are digital integrated circuits, typically containing thousands to millions of gates (and, sometimes, additional functional blocks such as transceivers, RAM, DSP, or general-purpose microcontrollers), whose circuit interconnections are determined by user programming (§11.2). That is, in analogy with a computer program (which tells the processor *what to do*), the program for a PLD tells the chip *how to connect* its internal component pieces. Some PLDs include non-volatile (and reprogrammable) memory that holds the programmed configuration; others read their configuration from an external non-volatile ROM at power-up, and hold it in on-chip (and volatile) memory; and many allow programming via a serial configuration port (commonly a JTAG port).

¶B. PLD Applications.

Small PLDs (tens to hundreds of macrocells, thousands of gates) are commonly used as "glue logic," replacing discrete gates, flip-flops, registers, and the like. Larger PLDs, and particularly FPGAs (field-programmable gate arrays), can implement nearly a complete system-on-a-chip. The largest FPGAs contain millions of flip-flops, megabytes of RAM, hundreds of I/O ports (LVDS, PCIe, etc.), and dozens of serial transceivers – resources adequate to hold a "soft" processor (i.e., configured from gates and flip-flops) running embedded Linux. (And when thinking of PLDs for some application, be sure to consider the alternative of a general-purpose *microcontroller* – they're increasingly capable, inexpensive, and easy to program.)

¶C. Programming Overview.

PLD design (§11.2.6) starts with the functional specification, where the choice is either *schematic entry* or a *hardware description language* (HDL). Next you run a *simulation*, to verify that the design works as intended. After that a *synthesis* process converts the design to a *netlist* (a full set of logical connections). The netlist is then "fitted" to the target PLD, a process known as *place and route*. Finally, the fitted design is downloaded to the PLD's memory – either to the PLD itself (if it has on-chip non-volatile memory), or to an external configuration ROM (from which the PLD loads at power-up), or possibly hot-loaded from an external processor directly to the powered-up PLD.

PLD design tools are provided by the PLD manufacturers, and by some third parties. For low-end PLDs the tools may be free, but they can be quite expensive for the high-end parts. Considerable time may be needed to master the PLD design tools (which have a distressing tendency to change with time), and the software can be (and usually is) buggy.

¶D. Schematic Entry.

Most comfortable to non-coders is schematic entry, in which you wire up familiar logic symbols (e.g., Figure 11.10) in the same manner as you would do when entering a circuit for a PCB layout. There's additional flexibility, because you can define larger functional blocks that are specified with either schematic entry or HDL text entry.

Another graphical entry method, at a higher level of abstraction, exploits software synthesis tools that convert a MATLAB/Simulink Lego-like functional-block description to an HDL source-code output; the latter can target a PLD or a full-custom (ASIC) implementation, as well as providing simulation, timing, and test vectors.

Those who favor graphical schematic entry generally explain their preference as (a) coming from a schematic background, they find it natural; (b) they find it easier to learn, to understand, and to explain to others; (c) they like having a graphical schematic as documentation; (d) because modules can be HDL-in-disguise, it has all the power of HDL, for complex designs where that is needed; and (e) graphical entry is used in graphical programming languages (GPLs) like LabVIEW, MATLAB®, and Simulink®.

¶E. HDL Entry.

Hardware description languages like Verilog and VHDL dominate the PLD-entry scene. Statements in either language can be *structural* (e.g., what signals are connected from some output to another input) or *behavioral* (e.g., which state follows another, and under which conditions). See the illustrative examples in §§11.3.4B and 11.3.4C.

Those who favor text-based HDL entry generally explain their preference as (a) coming from a programming background, they find it natural; (b) design entry is rapid, particularly when re-using portions of previous designs; (c) the design is more concise (thus easier to know that it is right), and self-documented (as a textual description); (d) the design cycle is streamlined, as a design is being iterated through simulation and prototype; (e) it is particularly easy to change parameters (such as number of bits in a register, ALU, etc) by merely resizing arrays (compared with rearranging graphical modules); (f) the languages are standardized and universal (compared with proprietary schematic entry tools); (g) good simulators are available, and free of charge; (h) HDL languages are best suited for high-level synthesis of complex designs (like a microprocessor), and prefab open-source IP cores are provided as HDL; and (i) HDL entry is well-suited to subsequent implementation as a *full-custom* IC (ASIC), following a prototype stage implemented initially as an FPGA.

¶F. Advice, by *Technology*.
(Paraphrased shamelessly from §11.4.1.)

* **Standard logic**
  - Upside: OK for quick assembly of simple circuits, and useful for bus drivers, buffers, and small logic "glue"; available in through-hole (for quick prototyping); no software design tools needed.
  - Downside: unwieldy for complex circuits; multiple functions to inventory; inflexible.
* **Programmable logic (cPLD, FPGA)**
  - Upside: better for most digital designs – reduced package count, easily reprogrammed, flexible and inexpensive, small inventory needed; favor cPLDs for predictable timing and for small designs, FPGAs for large designs; best choice for fast parallel logic and state machines; good for timing-critical functions, and functions with lots of I/O; designs can be migrated across PLD families, and to full-custom ICs.
  - Downside: time to learn software tools (often proprietary), and their cost; surface mount parts only; cost of programming pod and design tools; short life cycle (supply problems).
  - PLD design/program tools
    * *Schematic entry*
      · Upside: easy to learn, understand, and explain to others; self-documenting; entry of parallel circuitry in a parallel way; analogous graphical programming languages (GPLs) used in tools like LabVIEW, MATLAB®, and Simulink®.
      · Downside: proprietary design tools – cost; limited applicability.
    * *HDL (Verilog/VHDL)*
      · Upside: natural migration for programmers; concise and standardized; cost-free simulation and compilation tools; easy to iterate design modifications; can migrate across PLD families, and to full-custom ICs; well-suited for large complex designs; IP core availability.
      · Downside: time to learn, and lack of schematic documentation.

* **Microcontrollers** (see Chapter 15!)
  - Upside: programmable, flexible embedded computer; ideal for complex state machines with lots of decisional branching; rapid iteration of code and (in-circuit) debugging; many on-chip peripherals (communication, conversion, interface, other); familiar programming language (C, C++); best for embedded control, especially involving user communication (display and control); enjoyable!
  - Downside: slower than cPLD/FPGA; less suited to bit-level manipulation; less parallelism.
  - µC program tools
    * *Assembly language*
      · Upside: can handcraft optimized code, for example an interrupt service routine.
      · Downside: fussy and error prone; difficult to make major modifications; cannot port to a different processor family.
    * *C, C++*
      · Upside: standardized and largely portable; widespread programming expertise; structured language facilitates changes and upgrades; libraries of useful functions; well suited for complex tasks.
      · Downside: not much... compiled code may be less efficient in size and speed (can embed assembly code where needed); code can be hard to understand; requires compiler toolset (some are expensive, others are free).
    * *Other (Basic, Arduino Software; Java, Python)*
      · Upside: easy to understand, good for entry level.
      · Downside: limited range of microcontroller support, not generally portable across processor families, not for big projects.

¶G. Advice, by *User Community*.
(Paraphrased shamelessly from §11.4.2.) *All* of the user communities are well served by *microcontrollers*, owing to low cost, ease of programming, and flexibility. Among these communities, the following additional technologies are favored.

* **High-volume production**
  Here unit cost and time-to-market are critical. In large volumes, unit cost is minimized with *ASICs* (unique Application Specific full-custom Integrated Circuits) and *ASSPs* (generally available Application-Specific Standard Products), whereas time-to-market is generally shortened by using *FPGAs* to eliminate the ASIC design–debug cycle. In large production it is common to use

high-density *BGA* (ball-grid array) packages and other fine-pitch surface-mount (SMT) devices, along with the necessary multilayer circuit boards and advanced assembly techniques. And, of course, always use *microcontrollers*.

- **Complex prototype and small production**
  Because of the delays and costs associated with both ASICs and high-density BGA packages, the favored technologies are *cPLDs* and *FPGAs*, along with assorted support chips, all in surface-mount packages, and assembled on multilayer boards. And, of course, always use *microcontrollers*.

- **Laboratory and "1-Off"**
  To meet the goals of quick assembly and easy modification, we favor through-hole prototyping boards, populated with through-hole components, and (where essential) with a small number of SMT devices (on through-

hole adapters). This constrains you to *standard logic* and *interface*, a limited selection of *microcontrollers*, and perhaps some *cPLDs/FPGAs*, along with *analog* components (op-amps, etc). You can make greater use of FPGAs if you take advantage of third-party pre-fabricated daughterboards. Alternatively, you can start with a fully populated *microcontroller kit*, including *PLDs*, and perhaps some unpopulated "prototyping" pads for additional devices.

- **Hobbyist**
  Here the idea is to have lots of inexpensive fun. The basic unit is the *microcontroller* (have we mentioned this before?), either as part of a prefabricated *kit*, or used on a through-hole protoboard, or perhaps a custom printed-circuit board. Through-hole components (see *Laboratory and "1-Off", above*) are always easy to use, as are PLDs included in microcontroller kits.

LOGIC INTERFACING

Although sheer "number crunching" is an important application of digital electronics, the real power of digital techniques is seen when digital methods are applied to analog (or "linear") signals and processes. In this chapter we begin with a brief chronology of the rise and fall of digital logic families and a review of the input and output properties of the surviving (mostly CMOS) families that you are likely to use in circuit design. This is essential for understanding how to interface logic families to each other and to digital-input devices (switches, rotary encoders, keypads, comparators, etc.) and output devices (indicator LEDs, relays, power MOSFETS, etc.). We continue with the important subject of bringing digital signals on and off circuit boards, in and out of instruments, and through cables. We discuss optoelectronic devices (fiber-optic drivers and receivers, optocouplers, LCD and LED displays, and solid-state relays). And in the next chapter we discuss the major subject of conversion between analog and digital signals. Finally, with an understanding of these techniques, we look at a number of applications in which combined analog and digital techniques provide powerful solutions to interesting problems. Much of this material is applicable not only to "discrete" digital logic, but also to the PLDs and FPGAs of Chapter 11 and to the microcomputers and microcontrollers of Chapters 14 and 15.

12.1 CMOS and TTL logic interfacing

12.1.1 Logic family chronology – a brief history

In the prehistoric early 1960s, adventurous people who didn't want to build their logic from discrete transistors struggled with RTL (resistor–transistor logic), a simple logic family introduced by Fairchild and characterized by poor fan-out and poor noise immunity. Figure 12.1 shows the problem, namely, a logic threshold at one V_{BE} above ground and miserable fan-out (in some cases one output could drive only one input!) caused by passive pullup and a low-impedance current-sinking load. Those were the days of small integration, and the most complicated function you could get was a dual flip-flop, which would operate

to 4 MHz (the MC790P, in case you want to look it up). We bravely built circuits with RTL; they sometimes malfunctioned when someone switched on a soldering iron in the same room.

The death knell for RTL came a few years later with the introduction by Signetics of DTL (diode–transistor logic) and, soon thereafter, Sylvania's SUHL – "Sylvania Universal High Speed Logic" – subsequently called TTL (transistor–transistor logic). Signetics had a popular mixture of the two, called 8000-series DCL Utilogic ("Designer's Choice Logic"). TTL caught on quickly, particularly in the "74xx" numbering system originated by Texas Instruments. (Although bipolar TTL is pretty much history now, the 74xx designations have remained healthy to this day, in contemporary CMOS incarnations.) These families used current-*sourcing* inputs with logic threshold at two V_{BE}'s and (usually) push–pull "totem-pole" outputs (Figure 12.1). DTL and TTL began the era of +5 V logic (RTL used +3.6 V), and offered speeds of 25 MHz and fan-outs of 10 (i.e., an output could drive 10 inputs). Designers rejoiced at the speed, reliability, and complex functions (divide-by-10 counters, for example) of these families. It seemed to us that you couldn't ask for more; TTL would live forever.

People are greedy, though. They wanted more speed. They wanted less power consumption. They soon got both, sort of. In the high-speed arena, a souped-up TTL (74H series; "High-speed" TTL) delivered roughly twice the speed, at twice the power! (It accomplished this underwhelming feat by cutting all resistor values in half.) Another family, ECL (emitter-coupled logic), delivered real speed (30 MHz in its original version), using a negative power supply and rather closely spaced logic levels (−0.9 V and −1.75 V); it consumed lots of power (30 mW/gate) and came only in small integration. For low power there was a souped-down TTL (74L series; "Low-power" TTL) with 1/4 the speed at 1/10 the power of the corresponding 7400 "standard" TTL.

Back at RCA, the first of the MOSFET logic families was developed, 4000-series CMOS. It had zero quiescent power consumption and a wide supply range (+3 V to

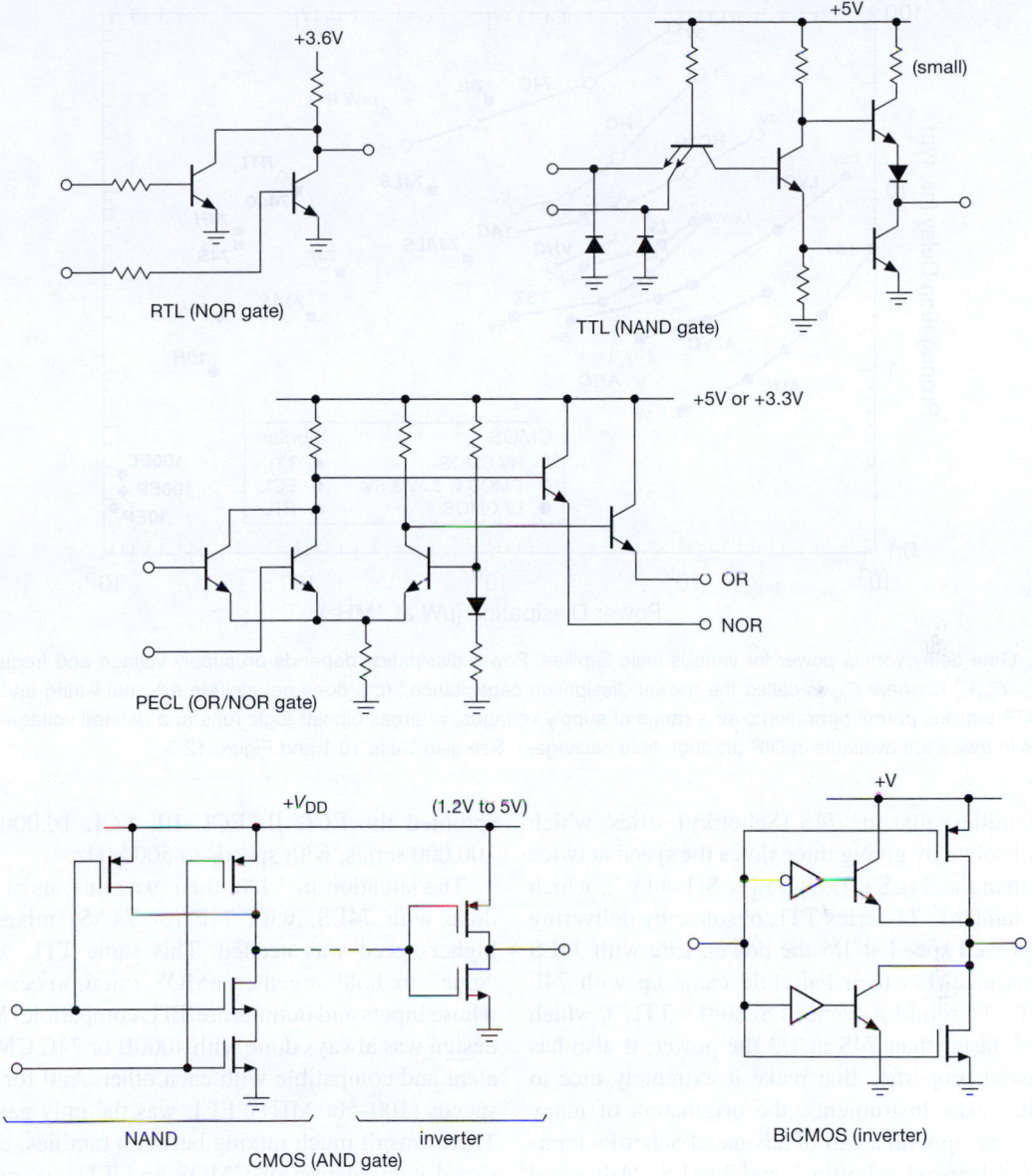

Figure 12.1. Simplified diagrams of several logic families.

+12V). The outputs swung rail-to-rail, and the inputs drew no current. That was the good news. The bad news was the speed (1 MHz at 5 V supply) and the price (about $20 for a package of four gates). In spite of the price, a whole generation of battery-powered instrument designers grew up on the micropower CMOS simply because there was no alternative. They learned the true meaning of static electricity as they worked with the easily damaged inputs.

This, then, was the situation at the beginning of the 1970s – two main lines of bipolar logic (TTL and ECL) and

the extraordinary CMOS. The TTL variants were essentially compatible, except that 74L TTL had feeble output drive (3.6 mA sink) and could drive only two standard (74-series) TTL loads (whose inputs sourced 1.6 mA when held LOW). There was almost no compatibility between the major families (though a pulled-up TTL could drive CMOS, and 5 V CMOS could just barely drive a single 74L TTL load).

During the 1970s there were steady improvements on all fronts. TTL sprouted the nonsaturating "Schottky-

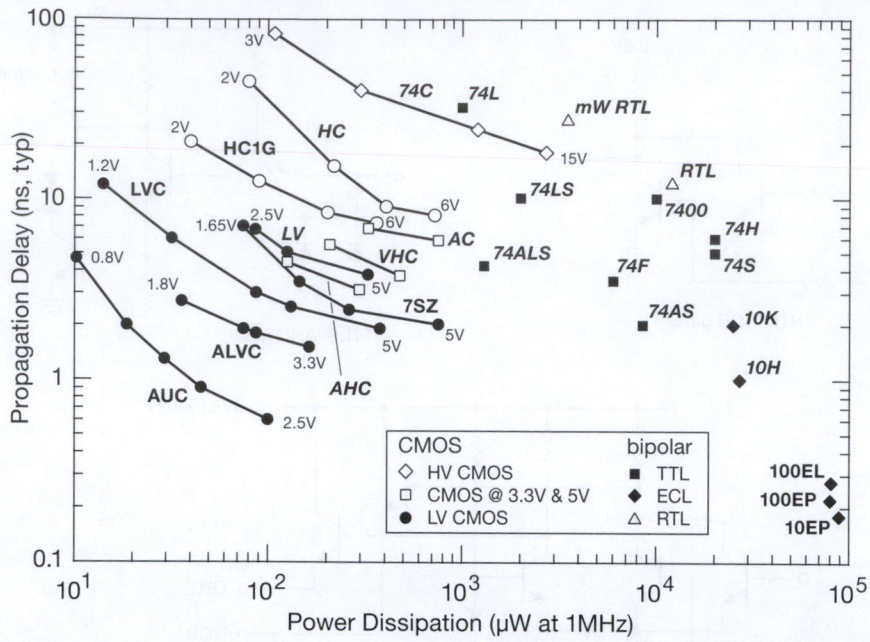

Figure 12.2. Gate delay versus power for various logic families. Power dissipation depends on supply voltage and frequency of logic switching: $P = C_{pd}V_{cc}^2 f$, where C_{pd} is called the "power dissipation capacitance." (C_{pd} does not include external wiring and load capacitances.) CMOS families permit operation over a range of supply voltages, whereas bipolar logic runs at a defined voltage (either 3 V or 5 V). Families in *italics* are available in DIP (through-hole packages). See also Table 10.1 and Figure 12.3.

clamped" families: first the 74S (Schottky) series, which made 74H obsolete by giving three times the speed at twice the power; then the 74LS ("Low-power Schottky"), which made the "standard" 74-series TTL obsolete by delivering slightly improved speed at 1/5 the power. Life with 74LS and 74S was nice, but then Fairchild came up with 74F (F for FAST: "Fairchild Advanced Schottky TTL"), which is up to 50% faster than 74S at 1/3 the power; it also has other improved properties that make it extremely nice to design with. Texas Instruments, the originators of many 74xx lines, came up with a pair of advanced Schottky families, 74AS ("Advanced Schottky") and 74ALS ("Advanced Low-power Schottky"); the former was intended to replace 74S, and the latter was intended to replace 74LS. All these TTL families have the same logic levels and plenty of output drive, and so they can be mixed within a circuit. Tables 10.1 and Figure 12.2 illustrate the speed and power of these families.

Meanwhile, the 4000-series CMOS evolved into the improved 4000B series, with wider supply range (3 V to 18 V), better input protection, and higher speed (3.5 MHz at 5 V). The 74C-series was essentially the same, with 74-family functions and pinouts to take advantage of the tremendous success of the 74-family bipolar logic. ECL

sprouted the ECL II, ECL III, ECL 10,000, and ECL 100,000 series, with speeds to 500 MHz.

The situation in 1980, then, was this: most design was done with 74LS, with 74F (or 74AS) mixed in where higher speed was needed. This same TTL was used as "glue" to hold together nMOS microprocessor circuits, whose inputs and outputs are TTL compatible. Micropower design was always done with 4000B or 74C CMOS, equivalent and compatible with each other. And for the highest speeds (100–500 MHz), ECL was the only game in town. There wasn't much mixing between families, except occasional combination of CMOS and TTL, or perhaps TTL interfaces to an ECL high-speed circuit.

During the 1980s came the remarkable development of CMOS logic with the speed and output drive of TTL: first 74HC ("High-speed CMOS"), with the same speed as 74LS, and, of course, zero quiescent current; then 74AC ("Advanced CMOS"), with the same speed as 74F or 74AS. With rail-to-rail output swings, and input threshold at half the supply voltage, this logic combined the best features of previous TTL and CMOS, and it essentially replaced bipolar TTL. However, there was an incompatibility when both kinds of logic were used within a circuit, because the +2.4 V (minimum guaranteed) logic HIGH output of traditional TTL standard logic (and also of more complex functions that were being done in nMOS) was insufficient to drive an HC or AC input (with their requirement of +3.5 V, min).

To address this problem, each CMOS family offered a variant with the lower input threshold. These are named 74HCT and 74ACT ("High-speed CMOS with TTL threshold"). During the 1980s, also, complex large-scale integrated (LSI) and very large-scaled integrated (VLSI) devices (microprocessors, memory, etc.) evolved from nMOS to CMOS (with consequent low power and, usually, full-swing output CMOS compatibility), at the same time increasing speed and complexity. And at the extreme high-speed end, there was some development of GaAs (gallium arsenide) logic devices (by companies like GigaBit Logic and Vitesse) to reach speeds of $\sim$3 GHz.

Predictably, things just got better and better in the following two decades. The most important development was the improvement in CMOS performance, brought about by reduced feature size on the silicon chip ("scaling"). First, as the length scales were shrunk, it became possible to put many more transistors on a reasonable-sized chip – this led to the development of massive processor, memory, and other complex function (e.g., video) chips, with transistor counts going from millions to billions. Second, and perhaps as important, the scaling increased the speed and also reduced the operating voltage and the power per gate.[1] The result is new families of low-voltage CMOS logic parts (74LVC, 74AUC, and so on; see Figure 12.3) with pin-to-pin speeds in the several-hundred megahertz range (delay times down to 1 ns or less).

These fast CMOS families have proliferated, with a head-spinning menagerie numbering into the dozens. Most of them operate over a range of supply voltages (e.g., 1.8–

5 V for 74LVC), and in most cases the input logic voltage can swing beyond the positive supply (e.g., 74LVC is "5 V tolerant," regardless of supply voltage; see Figure 12.3). With greater integration of most of the needed logic into the large VLSI components, there is little reliance on discrete logic, and these "standard-logic" parts are used mostly as occasional glue. For that purpose they come in small packages, holding one or two gates, or a single flip-flop, with names like TinyLogic, Little Logic, MiniGate, or PicoGate.

These packages share with other surface-mount ICs the advantage of low-inductance ground (and power) pins, which reduce "ground bounce" (§10.8.3 and Figure 10.99) and other transient symptoms caused by fast edge rates driving the combined capacitance of wiring and load. Ground bounce had become a serious headache in the decade of the 1990s, with the new 74AC and 74ACT families, which combined fast full-swing (between +5 V and ground) logic transitions in traditional DIPs with corner power and ground pins. Some manufacturers (notably TI) addressed this problem by adding extra power and ground pins, and moving them to the center (creating new part numbers like 74AC11004, a DIP-20 repinned version of the DIP-14 '04 hex inverter); others created "quiet" logic families with controlled edge rates (e.g., the 74ACTQ family from FSC/NSC). The situation improved considerably with the move to lower supply voltages, low-inductance SMT packaging, and good printed-circuit layout practices (especially the use of dedicated power and ground planes in multilayer PCBs). And the use of low-voltage differential signaling (LVDS) for fast signals and clocks pretty much eliminated the problem altogether, because of both the balanced current changes, and their relatively small swing ($\sim$0.4 V).

All CMOS families of standard logic (beginning with the original 4000A, and extending through HC, LVC, AUC, and a dozen others) have the pleasant characteristic of zero "static" (i.e., nothing happening) power dissipation, with typical quiescent currents less than a microamp. But CMOS does draw "dynamic" current when logic levels are switching, because of the combined effects of (a) transient rail-to-rail conduction of internal push–pull pairs during the middle of the logic swing and (b) dynamic current needed to charge and discharge internal and load capacitances. Dynamic supply currents are proportional to switching frequency and can rival bipolar logic as you reach maximum operating frequencies, as we discussed in Chapter 10 – see Figure 10.27. Note, however, that many VLSI functions implemented in CMOS (e.g., FPGAs and cPLDs, see Chapter 11) often have substantial quiescent current; this situation is changing, however, with a trend

---

[1] To see how this goes, consider the process of shrinking the linear feature size (channel length L and width W) on a CMOS IC by a factor of k ($k < 1$), while adjusting things to keep the electric field strengths constant; this is called "constant-field scaling." Then, with $L \propto k$ and $W \propto k$, constant-field scaling requires that $V_{DD} \propto k$; that makes the gate oxide insulation thickness scale as $t_{ox} \propto k$, which then makes the gate capacitance (which is proportional to LW/t_{ox}) go as $C_g \propto k$. A consequence of the geometry scaling is that the saturation drain current goes as $I_D \propto k$. Finally, with that drain current, a gate input can be driven through V_{DD} in a time $\tau \approx C_g V_{DD}/I_D$, which therefore scales as $\tau \propto k$. Voilà – the gate delay time scales as k; i.e., the speed increases proportionally to $1/k$. Even better, the power ($V_{DD}I_D$) decreases as $P \propto k^2$; and a reasonable figure-of-merit $1/P\tau$ (speed/power) goes as $1/k^3$. Perhaps this explains some of the semiconductor industry's relentless zeal always to move to the next-generation scaling "node" (shrink factor of $1/\sqrt{2}$).

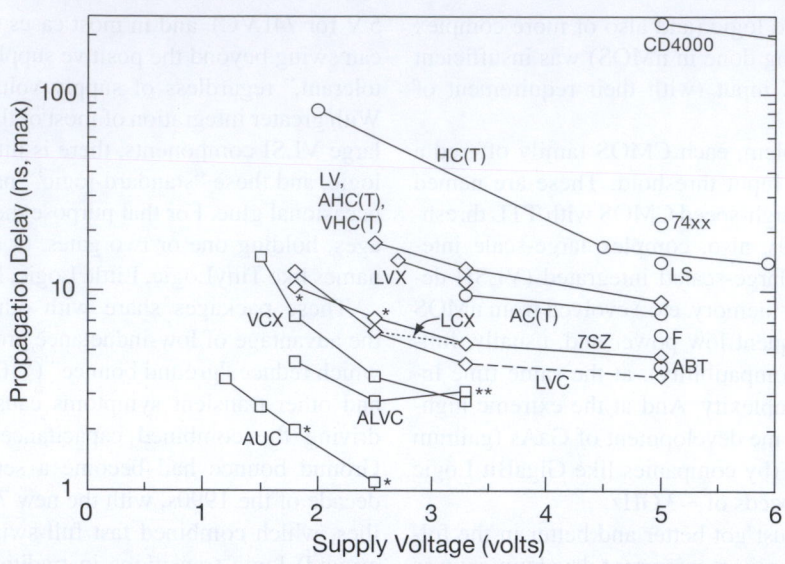

Figure 12.3. Gate speed versus supply voltage, for popular logic families. Maximum specified propagation delay ($t_{pd(max)}$) is shown for the standard supply voltages at which each family is specified. (As a rough guide, "typical" delays are in the range of 35–75% of $t_{pd(max)}$.) Open circles indicate families for which $V_{in(max)}$ is limited to V_{supply}, and data is for operation at 25°C; families with open diamonds have "5 V tolerant" inputs ($V_{in} \leq 5.5$ V, independent of supply voltage), and data is for operation over the "industrial" temperature range (−40°C to +85°C); families with open squares have "3.3 V tolerant" inputs ($V_{in} \leq 3.6$ V), data again over the industrial temperature range. Some families (e.g., LVC) have output-stage circuitry to ensure that the output does not load shared signal lines when unpowered. Plotted data is for load capacitance C_L=50 pF for 5 V and 3.3 V operation, 30 pF for 2.5 V and 1.8 V, and 15 pF for 1.5 V and below (except 50 pF for data points marked **, and 15 pF for *). The CMOS 4000-series logic runs to +15 V, at which $t_{pd(max)}$=70 ns. Only some LVC family members operate to +5V. Not shown are the speedy (and power-hungry) bipolar ECL families, operating at 5 V only: their maximum gate delays clock in at 0.6 ns (10E family), 0.44 ns (10EL), and 0.32 ns (10EP). Some families have evolved "enhanced" versions, for example LVC→LVCE, which operates down to 1.4 V and is 30% faster at low supply voltages. See also Figure 10.22 and §10.2.3.

toward micropower VLSI that is driven by battery-powered applications.

We wrap up our brief history with some recommendations.

- For simple logic circuits that are easy to assemble on a protoboard, and where blinding speed is not needed (nor desired), use 74HC or 74HCT logic (the latter for compatibility with either existing "TTL-level" signals, or with signals coming from low-voltage logic, e.g., powered from +3.3 V); you can substitute 74AC/74ACT/74ACTQ if you need the extra speed, but watch out for ground bounce.
- For low-voltage systems containing microcontrollers or other complex ICs, where some fast glue logic is needed, use a universal family like 74LVC, mindful that it is available only in surface-mount packages; these families are useful also for systems with several logic supply voltages.
- If you need to drive 5 V outputs (e.g., white LEDs, or solid-state relays) from a low-voltage system, use 74HCT

- for fast serial data and clock signals, use LVDS (or low-voltage PECL, "LVPECL") differential drivers, receivers, or SERDESs (serializer–deserializer).
- Choose the older 4000B/74C logic where the extended supply voltage range is needed and speed is unimportant (e.g., a portable device powered by an unregulated 9 V battery).
- Finally, use VLSI ICs (cPLDs, microcontrollers, ASSCs) rather than discrete logic – this reduces package count and wiring complexity and adds flexibility.

12.1.2 Input and output characteristics

Digital-logic families are designed so that the output from a chip can properly drive many inputs within the same logic family, powered from the same supply voltage. A typical fan-out capability is at least 10 loads, meaning that an output from a gate or flip-flop, for example, can be connected

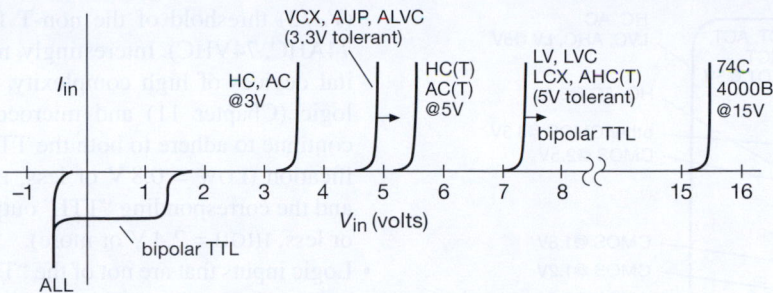

Figure 12.4. Logic gate input current versus input voltage. With the exception of bipolar "TTL," there is no static input current within the normal input-voltage range. All logic families include internal protective clamp diodes to ground. Some families (for example 74HC) clamp to the positive supply, and therefore draw input current when the input signal voltage exceeds a diode drop above V_+. More recent families (for example 74LVC or 74AUP) use internal zener-like protection, and permit inputs well above the supply voltage; these are called "5 V tolerant" or "3.3 V tolerant" (for example LVC and AUP, respectively), and permit such inputs even when unpowered (a situation in which a 74HC logic device would clamp at $+0.6V$ or, worse, cause the HIGH input to partially power the supply rail).

to 10 inputs and still perform within specifications.[2] In other words, in normal digital design practice you can get by without knowing anything about the electrical properties of the chips you're using, as long as your circuit consists only of digital logic driving more digital logic of the same type. In practice, this means that you don't often have to worry about what's actually going on at logic inputs and outputs.

However, as soon as you attempt to drive digital circuitry with externally generated signals, whether digital or analog, or whenever you use digital-logic outputs to drive other devices, you must face the realities of what it takes to drive a logic input and what a logic output can drive. Furthermore, when mixing logic families, or when you have logic running between different supply voltages, it is essential to understand the circuit properties of logic inputs and outputs. Interfacing between logic families is not an academic question. To take advantage of advanced VLSI chips, or special functions that are available in only one logic family, you must know how to mix logic types and voltages. In the next few subsections we will consider the circuit properties of logic inputs and outputs in detail, with examples of interfacing between logic families and between logic devices and the outside world.

A. Input characteristics

The graphs in Figures 12.4 and 12.5 show the important properties of digital-logic inputs: input current and output voltage (for an inverter) as functions of input voltage. We have extended the graphs to input voltages beyond the range normally encountered in digital circuits, because in interface situations the input signals may easily exceed the power-supply voltages. As the graphs imply, CMOS logic and bipolar TTL are normally operated with the negative supply pin connected to ground.

Input current (Figure 12.4)

Most logic these days is built with CMOS, where the inputs draw no current (except for leakage current, typically $10^{-5}\,\mu A$) for input voltages between ground and the supply voltage.[3] For voltages beyond the supply range, the input protection network looks like a clamping diode to ground, and either a diode to V_+ (e.g., the 74HCT family) or a zener-like clamp that permits input swings beyond the positive supply (e.g., the "5 V tolerant" 74LVC family); see Figures 12.3 and 12.4 for more details. Momentary currents greater than 20–50 mA through these diodes will damage or destroy the part, in some cases causing what's known as "SCR latchup" (§10.8.3B); you'll find such limits listed in the "Absolute Maximum Ratings" section of datasheets.

---

[2] For CMOS logic, which has zero dc input current, excessive loading merely slows down the transitions; in that sense the fan-out is "infinite." However, if you're driving bipolar logic, there's the additional issue of required dc input current (e.g., 1.6 mA for the 74LS family), which leads to a typical fan-out of 10.

[3] Note the peculiarity of the nearly obsolete bipolar TTL: an input sources a significant current (of order 0.1–1 mA) when held LOW, but draws only a small current when HIGH (typically a few microamperes, never more than $20\,\mu A$). To drive a TTL input, you must be able to sink a milliamp or so while holding the input below 0.4 V. Failure to understand this may lead to widespread circuit malfunction in interfacing situations!

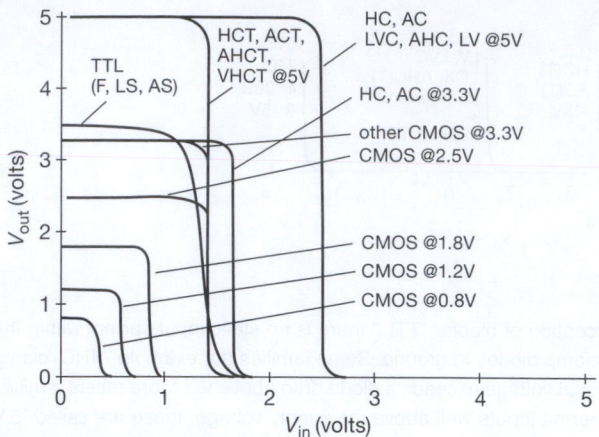

Figure 12.5. Output versus input ("transfer function") for logic inverters in commonly used families. In general, CMOS families that run at supply voltages of 2.5 V or less have their threshold at half the supply. At higher voltages most logic families (and many other more complex chips, such as microcontrollers and configurable logic) adhere to the "TTL" input specification, which guarantees that the threshold lies between 0.8 V and 2.0 V (and is typically 1.3–1.5 V). The HC/AC and 4000B families are exceptions, with their higher $V_+/2$ supply-tracking threshold.

Logic threshold (Figures 12.5 and 12.6)

The logic threshold voltage (the dividing line between logic LOW and logic HIGH inputs) depends on both the logic family, and on the supply voltage (for families where a range of supply voltages is permitted, as seen for example in Figure 12.3). The final authority is always the datasheet! But we can be of assistance here.

- Numerous digital devices adhere to what are known as "TTL thresholds," a legacy of the bipolar logic of the 1960s: $V < 0.8$ V is guaranteed to be interpreted as logic LOW, and $V > 2.0$ V is guaranteed to be interpreted as logic HIGH. (The voltages you send to such a logic device should stay outside those limits, to provide noise immunity; typically you'd provide a LOW level below 0.4 V, and a HIGH above 2.4 V.) The actual threshold voltage is typically around 1.35–1.5 V.[4] Common logic device families with TTL input thresholds (in addition to genuine bipolar TTL families such as 74F, 74LS, and 74AS) include the 74HCT, 74ACT, 74AHCT, and 74VHCT families; the "T" suffix signifies the family variant with (lowered) TTL thresholds, compared with the mid-supply

---

[4] We measured the threshold voltages for a sampling of "TTL threshold" inverters (by connecting the output back to the input), and here's what we found – 7404: 1.37 V; 74ACT04: 1.48 V; 74AS1004: 1.49 V; 74F04: 1.43 V; 74HCT04: 1.34 V; and 74LS04: 1.50 V.

($V_+/2$) threshold of the non-T families (74HC, 74AC, 74AHC, 74VHC). Interestingly, many contemporary digital devices of high complexity, such as programmable logic (Chapter 11) and microcontrollers (Chapter 15) continue to adhere to both the TTL input threshold specification (LOW = 0.8 V or less, HIGH = 2.0 V or more), and the corresponding "TTL" output levels (LOW = 0.4 V or less, HIGH = 2.4 V or more).

- Logic inputs that are not of the "TTL-compatible" variety have their thresholds typically at mid-supply; that is, at $V_+/2$. This is true of the older CMOS families, for example 74HC and 74AC (which can operate with supply voltages from 2 V to 6 V) and the high-voltage 4000B/74C families (which can operate with supply voltages from 3 V to 18 V). It also generally holds for most newer low-voltage families like 74LVC and 74AUC. Note, however, that the actual threshold voltage for a logic part with a "mid-supply threshold" can vary considerably: the datasheet specifications typically permit a range from about one-third to two-thirds of V_+ (V_+ is often called V_{CC} or V_{DD}).

B. Output characteristics

The output circuit of CMOS logic devices almost invariably uses a pair of complementary MOSFETs, one ON and the other OFF (Figure 12.1). The output looks like a MOSFET r_{ON} to ground or to V_+ when it is within a volt or so of the respective rail, becoming something like a current source when you draw so much current that the output is forced more than a volt or two away from the supply rails. Typical values of r_{ON} are 30 Ω for 74HC(T) operating at +5 V, 12 Ω 74AC(T) operating at +5 V, 10 Ω/15 Ω (sinking/sourcing) for 74LVC operating at 3.3 V, and 200 Ω for 4000B operating at 15 V.[5]

The output circuit of bipolar-TTL devices (a disappearing species) uses an *npn* transistor switch to ground and an *npn* follower (or Darlington) to V_+ with a current-limiting resistor in its collector. One transistor is saturated and the other is off. As a result, a TTL device can sink a large current (8 mA for 74LS, 24 mA for 74F) to ground with a small (saturation) voltage drop, but when sourcing current

---

[5] Watch out, though, for some "CMOS" digital functions (primarily older designs that operate from a +5 V supply, for example some cPLDs and microcontrollers) where the *p*-channel pullup switch is replaced with an *n*-channel source follower: for these devices the HIGH-state output does not reach the positive supply; instead it hovers around +3 V. You can recognize these parts from their datasheet's "DC Characteristics," which will say something like $V_{OH}(min)=2.4$ V, the telltale signs of an imitation legacy "TTL" output (where an *npn* follower forms the output pullup).

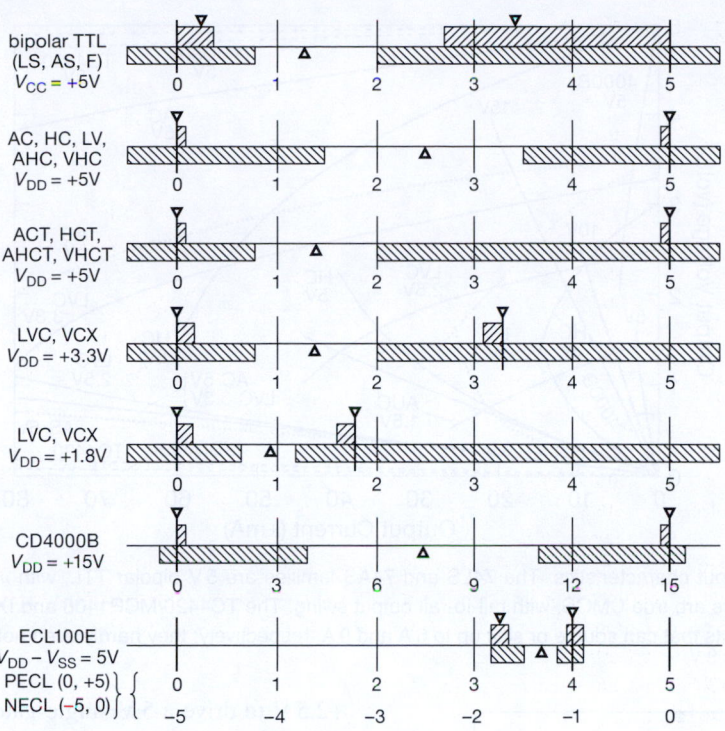

Figure 12.6. The range of voltages corresponding to the two logic states (HIGH and LOW) for popular families of digital logic. The shaded areas above the line show the specified range of output voltages within which a logic LOW or HIGH is guaranteed to fall, with the pair of arrows indicating typical output values (LOW, HIGH) encountered in practice. The shaded areas below the line show the range of input voltages guaranteed to be interpreted as LOW or HIGH, with the arrow indicating the typical *logic threshold* voltage, i.e., the dividing line between LOW and HIGH. In all cases a logic HIGH is more positive than a logic LOW. Table 10.1 and Figure 12.3 provide additional information about these families. See also Figure 12.135.

its output HIGH will be at least 1.5 V below the +5 V supply (see the 74AS and 74LS curves in Figure 12.7). The output circuit is designed to drive TTL inputs, or devices with "TTL input specifications" (<0.8 V guaranteed LOW, >2.0 V guaranteed HIGH) with a fan-out of 10.

In Figure 12.7 we have plotted the typical output voltage for both HIGH and LOW output states, against output current, for a selection of popular standard-logic families. To simplify the graph, output current is always drawn as positive. Note that true CMOS devices (that is, a complementary nMOS and pMOS push–pull switch pair) pull their outputs all the way to V_+ or ground, generating a full swing unless heavily loaded; so when driving only CMOS loads (zero dc current), the swing is fully rail-to-rail. Bipolar-TTL levels, by comparison, are typically 50–200 mV (LOW) or +3.5 volts (HIGH) when driving other TTL devices as loads. With a pullup resistor (discussed later), HIGH TTL outputs go all the way to +5 volts. We've also plotted two examples of MOSFET gate drivers, which

accept a TTL-compatible logic-level input, and use a high-current CMOS output stage to generate an output swing between ground and a chosen positive V_{DD} supply; for the TC4420 series that can range from +4.5–18 V, whereas for the IXDD509 series the range is +4.5–30 V.[6]

In contemporary digital electronics we are increasingly exploiting the nifty capabilities of programmable logic devices (Chapter 11) and of microcontrollers (Chapter 15). So you really need to know what their outputs can do, in terms of driving external loads. Figure 12.8 shows the output drive characteristics of several popular PLDs (Altera MAX7000A, Lattice Mach 4000, Xilinx Coolrunner-II) and microcontrollers (Atmel ATmega, Microchip PIC16F, TI MSP430). These all use true CMOS

---

[6] Note, however, that MOSFET drivers do not operate at full logic speeds: although their timing is usually specified into the large capacitive loading characteristic of power MOSFETs, you're generally looking at propagation delays in the neighborhood of 10–25 ns or more (to be compared with $t_{pd}{\sim}2$ ns for standard-logic families like 74LVC).

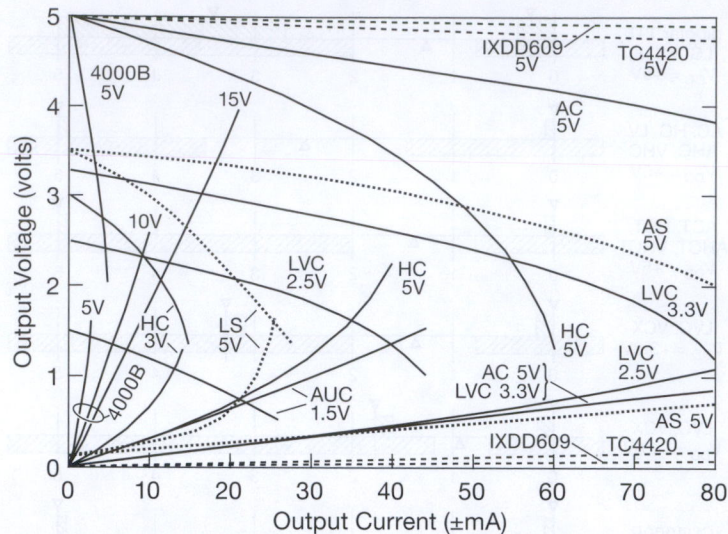

Figure 12.7. Logic gate output characteristics. The 74LS and 74AS families are 5 V bipolar TTL, with *npn* follower pullup, hence the ~3.5 V HIGH output. All others are true CMOS, with rail-to-rail output swing. The TC4420/MCP1406 and IXDD609 are "MOSFET driver" ICs, with robust CMOS outputs that can source or sink up to 6 A and 9 A, respectively; they hardly even notice an 80 mA load.

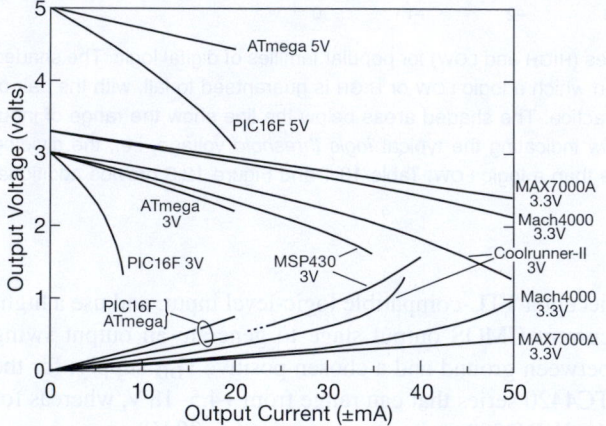

Figure 12.8. Output drive characteristics of selected PLDs and microcontrollers.

outputs with rail-to-rail voltage swing. As is evident from the graph, however, not all CMOS output stages are created equal.

12.1.3 Interfacing between logic families

It is important to know how to make different logic families talk to each other, because there are situations in which you must mix logic types, or portions of logic running from different supply voltages. In a typical situation you might want to use an output from a microcontroller running on

+2.5 V to drive a 5 V single-gate inverter, so that the final 5 V full-swing output can drive a 5 V mechanical relay, or a solid-state relay (SSR), or a white LED. Or you might want to go the other way: a full-swing 5 V logic output needs to get to a low-voltage part running on 1.8 V.

The three things that can keep you from connecting any pair of logic chips together are (a) input logic-level incompatibility, (b) output drive capability, and (c) supply voltages. Rather than bore you with pages of rules and of explanations of what works and what doesn't, we've boiled down the interface problem to a simple set of recommended methods (Figure 12.9). Let's take a quick tour.

A. CMOS at same voltage

You can always make a direct logic connection between CMOS logic devices running from the same supply voltage. The output is full-swing, and therefore happily drives another CMOS device, regardless of the latter's particular threshold voltage.

B. "5 V logic" driving lower-voltage CMOS

CMOS running from 5 V can be connected directly to what's called "5 V-tolerant" logic (for example the 74LVC family) running at a lower supply voltage. As shown here, either CMOS (with its full-swing output), or "TTL" output-level devices (including true bipolar TTL, or ICs with nMOS follower pullups, either of which have a HIGH-state output of ~3.5 V) satisfy the input voltage requirements

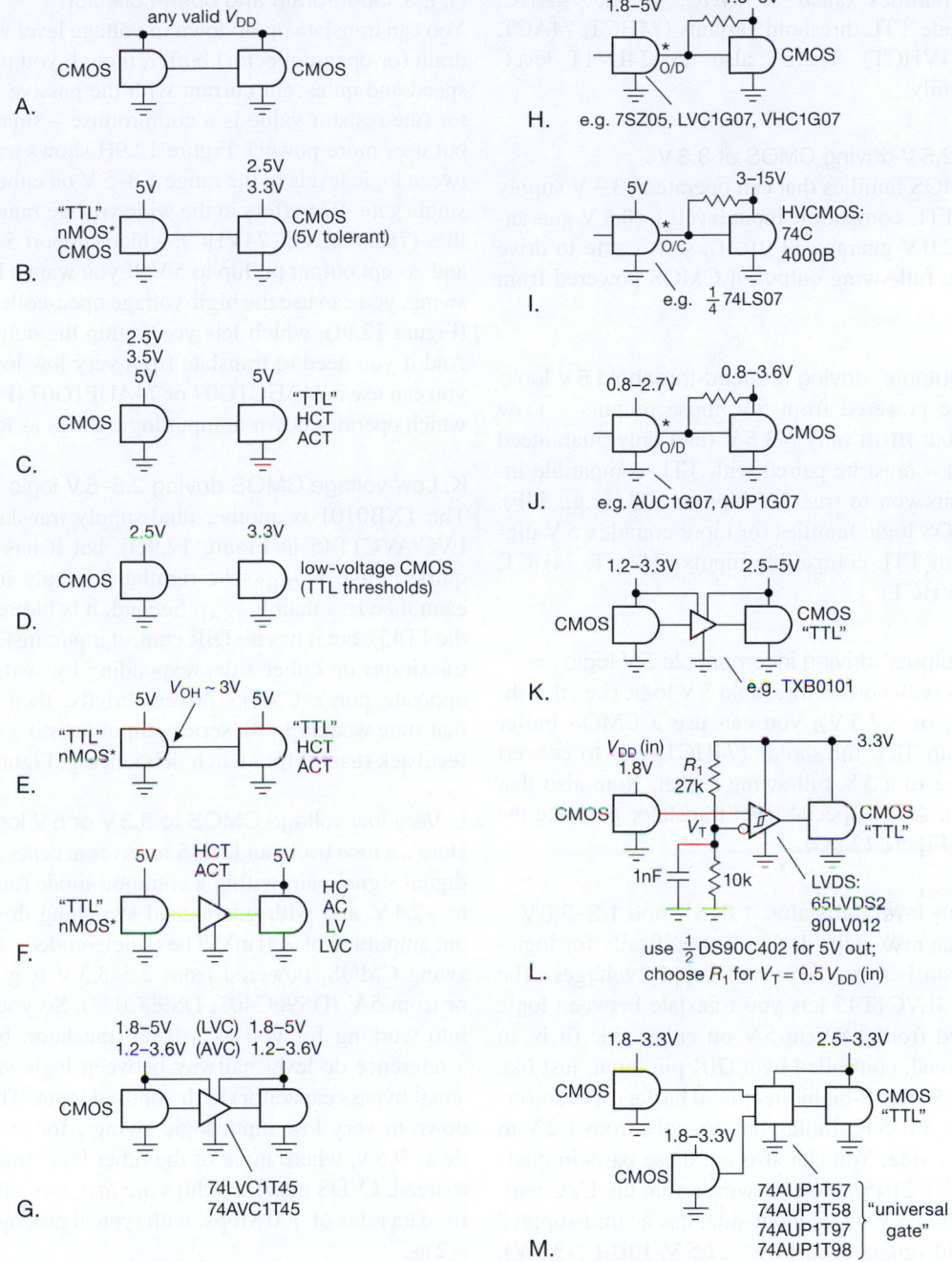

Figure 12.9. Logic family interconnections. See text for narrative.

of 5 V-tolerant CMOS running from supply voltages in the range of 2.5–3.3 V. There's also the 74LV1T level-translating family, which can do both "down translation" (as here) or "up translation" (as in C).

C. Lower-voltage CMOS driving 5 V logic

You can drive "TTL input" (reduced-threshold) 5 V logic directly from CMOS outputs that swing at least to 2.5 V, as shown. In addition to true bipolar-TTL parts, some

5 V CMOS families (such as 74HC, 74AC, 74AHC, 74VHC) include TTL threshold variants (74HCT, 74ACT, 74AHCT,[7] 74VHCT). There's also the 74LV1T level-translating family.

D. CMOS at 2.5 V driving CMOS at 3.3 V

Nearly all CMOS families that can operate at 3.3 V supply voltage have TTL-compatible input levels (<0.8 V guaranteed LOW, >2.0 V guaranteed HIGH), so it's safe to drive them from the full-swing output of CMOS powered from 2.5 V.

E. 5 V "TTL outputs" driving reduced-threshold 5 V logic

To drive logic powered from 5 V, those outputs – LOW close to 0 V, but HIGH only ∼3.5 V (and only guaranteed to be >2.4 V) – must be paired with TTL-compatible inputs; that limits you to true 5 V bipolar TTL (e.g., 74F), or to 5 V CMOS logic families (or more complex 5 V digital chips) with TTL-compatible inputs (74ACT, 74HCT, 74AHCT, 74VHCT).

F. 5 V "TTL outputs" driving incompatible 5 V logic

If you're stuck with normal threshold 5 V logic (i.e., threshold at $V_{DD}/2$, or ∼2.5 V), you can use a CMOS buffer or inverter with TTL thresholds (74HCT, etc) to convert the TTL swing to a 5 V full-swing signal. Note also that you can use instead a special level translator part like the 74LVC1T45 (Figure 12.9G).

G. Dual-supply level translator: 1.8–5 V and 1.2–3.6 V

There are some nice chips designed specifically for logic-level translation between a pair of supply voltages. The dual-supply 74LVC1T45 lets you translate between logic levels powered from 1.8 V to 5 V on either side (it is, in fact, bidirectional, controlled by a DIR pin input, just like the classic '245-style 8-bit bidirectional buffer). The lower-voltage 74AVC1T45 is similar, but operates from 1.2 V to 3.6 V on either side. You can also get these parts in duals (LVC2T45, AVC2T45). Note, however, that the LVC part, when operated at 5 V on its input side, has a "mid-supply" input threshold (guaranteed LOW <1.5 V, HIGH >3.5 V), and therefore cannot be driven from a TTL output level (with output HIGH only guaranteed to be ≥2.4 V). See also the TXB0101 dual-supply translator (Figure 12.9K).

H, I, J. Open-drain and open-collector

You can translate up or down in voltage level with an open-drain (or open-collector) buffer, though you pay a price in speed and quiescent current with the passive pullup resistor (the resistor value is a compromise – smaller is faster, but uses more power). Figure 12.9H shows translation between logic levels in the range 1.8–5 V on either side, using single gate '07 buffers in the wide voltage range logic families (7SZ, 74LVC, 74VHC), which support 5 V operation, and accept output pullup to 5 V. If you want a larger output swing, you can use the high-voltage open-collector 74LS07 (Figure 12.9I), which lets you pullup the output to +15 V. And if you need to translate from very low logic voltages, you can use a 74AUC1G07 or 74AUP1G07 (Figure 12.9J), which operates down to input logic levels as low as 0.8 V.

K. Low-voltage CMOS driving 2.5–5 V logic

The TXB0101 is another dual-supply translator (like the LVC/AVC1T45 in Figure 12.9G), but it has a couple of quirks. First, $V_{DD(B)}$ (the righthand supply in the figure) cannot be less than $V_{DD(A)}$. Second, it is bidirectional (like the 1T45), but it has no DIR control input; instead, it senses transitions on either side, responding by switching on the opposite port's CMOS drivers briefly, then maintaining that state weakly (∼4k series output resistor) with positive feedback (useful for switch debouncing, Figure 12.16).

L. Very low voltage CMOS to 3.3 V or 5 V logic

Here's a nice trick: an LVDS receiver accepts a differential digital signal pair, within a common-mode range from 0 V to +2.4 V, and with guaranteed switching down to an input amplitude of 200 mV. The (single-ended) output is full-swing CMOS, powered from 2.5–3.3 V (e.g., 65LVDS2) or from 5 V (DS90C402, DS90C032). So you can trick it into working for you as a level translator, by supplying a reference dc level halfway between logic states (with a small bypass capacitor) to the unused input. This will work down to very low input logic swings, for example as little as 0.5 V, where none of the other level translators dare to tread. LVDS interface chips are *fast*, too, often specified for data rates of 400 Mbps, with typical propagation delays <2 ns.

M. Low-voltage CMOS translator with configurable logic

The elegant "universal" gates that we mentioned in Chapter 10 (§10.2.1A) can be used to perform some logic while translating across logic voltage domains. When used this way, the gate runs from a single supply in common with the logic on the output side (which can range from 2.5–3.3 V), accepting inputs from logic that can be powered variously

---

[7] As an example, the 74AHCT1G125 is handy for converting a 3.3 V logic signal to a needed 5 V output swing; it comes in convenient SOT23-5 or SC70-5 packages, and costs less than $0.10.

from 1.8–3.3 V. These gates have Schmitt-trigger inputs, with about 0.4 V hysteresis centered about the threshold at ∼0.7 V.

A. Dynamic incompatibility: slow edge rates

Here's a problem that sometimes crops up when you drive an input of fast digital logic with a digital signal that does not slew through the threshold fast enough: the abrupt switching output of the driven device can couple back to the input (through the ground or power-supply connection, or internally on the chip itself, or just capacitively), causing multiple output transitions, as seen in the measured traces of Figure 12.10. To encourage such unseemly behavior we used a 74LS05 open-collector (OC) inverter with a lazy 5 kΩ pullup to +5 V, driving a frenetic 74AC04 gate.[8] You can clearly see the disruption of the leisurely input waveform, as the fast inverter furiously switches its output.

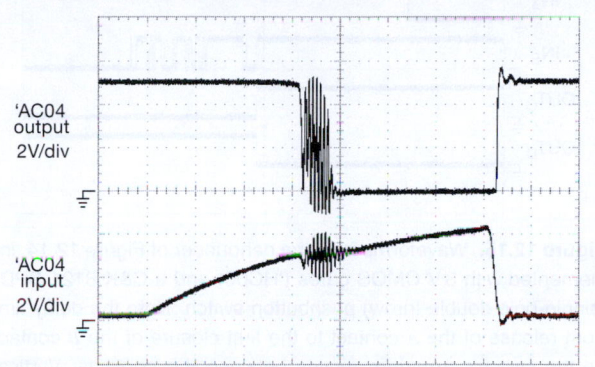

Figure 12.10. The relatively slow rising "edge" of a 'LS05 open-collector inverter with 5 kΩ pullup (lower trace) drives a fast 'AC04 inverter, causing multiple output transitions. The trailing edge exhibits no bad behavior, because the open-collector falling edge is fast. Horizontal: 40 ns/div.

Multiple output transitions may be merely ugly; but this problem becomes devastating when you are driving edge-sensitive inputs (e.g., the clock input of flip-flops or counters). A flip-flop may fail to toggle; or a counter or shift register may clock several times at each edge. To illustrate this effect, we drove the clock input of a 74AC74 toggling flip-flop from the square wave output of a CD4001B NOR gate, with both powered from +5 V (and with a reset pulse applied to the flip-flop before the next clock). Fig-

---

[8] We've found that 74AC and 74ACT logic is fussy stuff, particularly in corner-pin power and ground incarnations packaged in through-hole DIP. Stay away from this stuff unless you need the speed; and then be sure to use a ground plane, keep ground leads short, and bypass close to the chip.

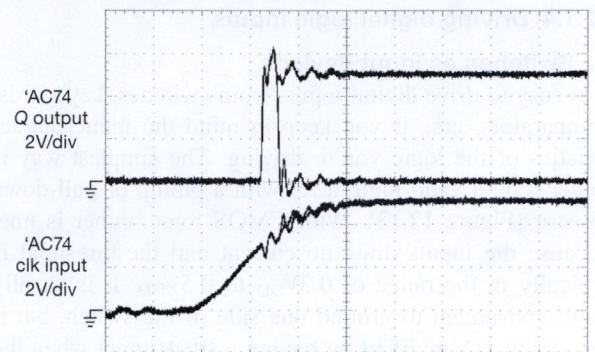

Figure 12.11. The slow rising edge of a CD4001B gate, loaded with 27 pF (a typical value of wiring and load capacitance), clocks a fast 'AC74 toggling flip-flop. The oscilloscope's persistence displays multiple clocking events, some more successful than others. Horizontal: 20 ns/div.

ure 12.11 shows the messy result: sometimes the flip-flop toggles correctly, but sometimes it toggles twice in rapid succession.

The lesson is clear: don't use slow edges to clock fast logic. It sometimes suffices to clean up the offending signal with a Schmitt-trigger inverter, available in most logic families as a '14 (e.g., 74LVC14). Figure 12.12 shows waveforms from the same setup as that of Figure 12.10, but with a 74AC14 (inverter with Schmitt) substituted for the 74AC04 (inverter without Schmitt). Better yet, stay away from slow edges completely.

You can get into the same kind of problems when you send digital signals between circuit boards, or between instruments, or over cables, an important set of topics we'll discuss presently (§12.9).

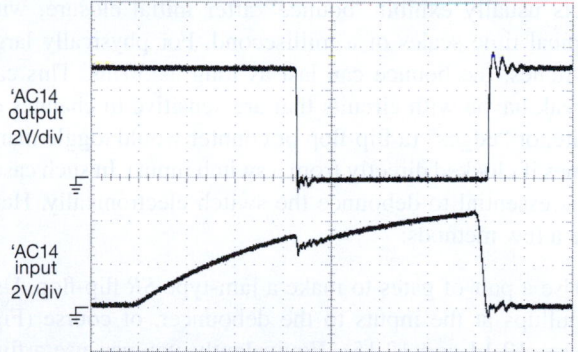

Figure 12.12. Same as Figure 12.10, but with the 'AC04 inverter replaced with an 'AC14 Schmitt-trigger inverter. The fast-switching output couples to the input, as before, but not enough to carry it back through the new (lower) threshold, thanks to the Schmitt's hysteresis. Horizontal: 40 ns/div.

12.1.4 Driving digital logic inputs

A. Switches as input devices

It is easy to drive digital inputs from switches, keyboards, comparators, etc., if you keep in mind the input characteristics of the logic you're driving. The simplest way is to generate a valid logic level with a pullup or pull-down resistor (Figure 12.13). With CMOS logic, either is fine, because the inputs draw no current and the threshold is typically in the range of $0.3V_{DD}$ to $0.5V_{DD}$. It is usually more convenient to ground one side of the switch, but if the circuit is simplified by having a HIGH input when the switch is closed, the method with pull-down resistor will be perfectly OK. Be careful, though, with bipolar TTL: their inputs *source* a substantial current (e.g., a 74F family input sources 0.6 mA when held LOW), so it's best to use the configuration with a pullup resistor and the switch returned to ground.

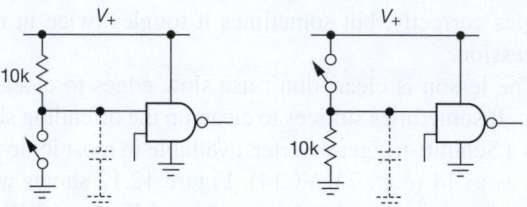

Figure 12.13. Mechanical switch to logic-level circuits (not debounced). If the switch is not close to the logic, it's common to see a small capacitor (~100–1000 pF) used to suppress capacitively coupled noise.

B. Switch bounce

As we remarked in Chapter 10, mechanical-switch contacts usually exhibit "bounce" after initial closure, with typical time scales of a millisecond. For physically large switches, the bounce can last as long as 50 ms. This can wreak havoc with circuits that are sensitive to changes of state, or "edges" (a flip-flop or counter would toggle many times if clocked directly from a switch input). In such cases it is essential to debounce the switch electronically. Here are a few methods:

- Use a pair of gates to make a jam-type *SR* flip-flop. Use pullups at the inputs to the debouncer, of course (Figures 12.14 and 12.15). Equivalently you can use a flip-flop with SET and CLEAR inputs (e.g., a '74); in that case, ground the clocking input.
- Use a noninverting buffer looped back into its input to make a latching "keeper" circuit, as in Figure 12.16. A buffer or noninverting gate (e.g., '1G34, '08, or '32) with

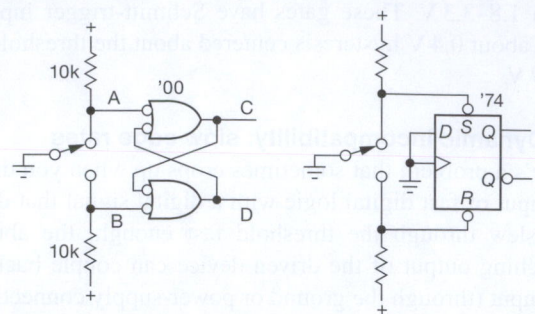

Figure 12.14. SR flip-flop switch debouncer, implemented either with cross-connected gates, or with a flip-flop with asynchronous SET and RESET inputs.

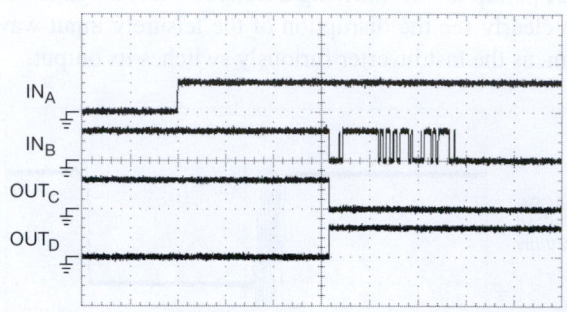

Figure 12.15. Waveforms from the debouncer of Figure 12.14, implemented with 3 V CMOS gates ('HC00) and a C&K 8121 SPDT (single-pole double-throw) pushbutton switch. Note the delay time from release of the *A* contact to the first closure of the *B* contact, as the switch's armature moves between the contacts. Vertical: 5 V/div; horizontal: 100 µs/div.

its output looped back works fine. Timid circuit designers include a resistor in the feedback path (as shown) to limit the momentary transient current when the switch changes state; but, trust us, you can happily omit it.[9] The TXB0101 (§12.1.3, Figure 12.9K) is one of many "automatic" bidirectional level-shifter chips that hold their state, but can be overdriven by a change of state on either side.

- Use an *RC* slowdown network to drive a CMOS Schmitt trigger (Figures 12.17 and 12.18). The lowpass filter smooths the bouncy waveform so that the Schmitt-trigger gate makes only one transition. A 1 ms to 10 ms *RC* time constant is generally long enough. This method isn't

---

[9] It's always OK to override a logic output by shorting it to V_+ or ground, *providing the duration is kept short*. In this circuit there is no problem, because the output is forced for only a gate propagation delay, after which it contentedly holds itself in the new state.

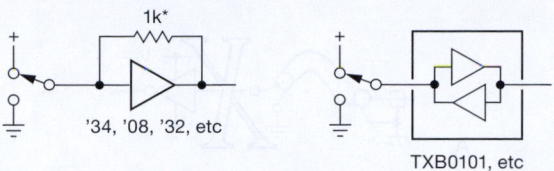

Figure 12.16. Switch debouncers using a "keeper" circuit that holds its logic state when the input is undriven.

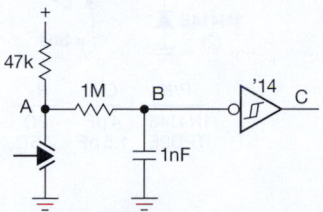

Figure 12.17. Debouncing an SPST (single-pole single-throw) switch with an *RC* smoothing circuit and a Schmitt trigger inverter.

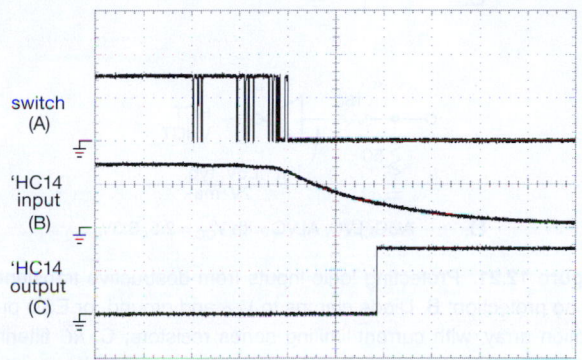

Figure 12.18. Waveforms from the debouncer of Figure 12.17, implemented with 3 V CMOS ('HC14) and a Microswitch 1PB13 pushbutton. Vertical: 2 V/div; Horizontal: 400 μs/div.

well suited to bipolar TTL because of the low driving impedance that TTL inputs require.

• Use a clocked circuit that samples the input level in a way that is not fooled by bounce. A simple way is to clock a *D*-type flip-flop with a period safely longer than the switch's bounce duration, e.g., with a clock frequency of 100 Hz (Figure 12.19). But there are dedicated debouncer chips, for example the MAX6816-8 (single, dual, and octal debouncers), which test for transition to a stable state for multiple clock periods (with internal oscillator, counters, and logic; effectively a digital lowpass filter) and generate a clean debounced output; they include internal pullup resistors, input protection to ± 25 V, and operation from 2.7 V to 5.5 V. You just tie your SPST switches

from input to ground – no external components needed (Figure 12.19). A similar part is the MC14490 hex (6-section) debouncer, which belongs to the CMOS 4000B-series and can run with supply voltages from 3 V to 18 V; it includes internal pullups, but it requires one external capacitor to set the clock rate. Another alternative is the so-called "power-supply supervisory" chip, which is ordinarily used to detect brownout (undervoltage) and to generate a clean reset pulse on power restoration (or initial power-up). Many of these chips include a manual reset input, to which you can connect a pushbutton, thus hijacking them into service as a debouncer.

• Use a microcontroller, with programming to carry out "software debounce" (Figure 12.20). Most microcontrollers include internal pullups; and simple code (either interrupt driven, or polled) can look for a stable change of state. This is a favorite method used by circuit designers for any gadget that needs a microcontroller anyway.

• Use a device with built-in debouncer. Keyboard encoders, for instance, are designed with mechanical switches in mind as input devices, and they usually include debouncing circuitry. Another example is shown in Figure 12.20, namely a pushbutton-controlled "digital potentiometer" (an internal resistor string, with tap selection via MOS-FET switches). Each press of a button increments or decrements the internal counter, so it's got to be bounce free. And it is.

A few general comments on switches as input devices: note that SPDT switches (sometimes called "form C") are necessary with the first two methods (SR flip-flop, "keeper" circuit), whereas the simpler SPST switches ("form A") can be used with the other methods. Keep in

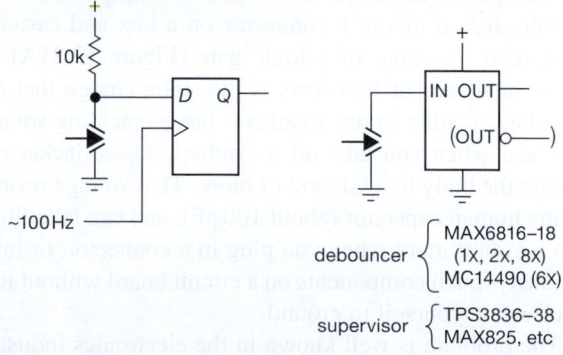

Figure 12.19. Clocked debouncers. Simplest is a slowly-clocked *D* flip-flop. There are better methods, embodied for example in special debouncer chips like the MAX6816-8 and MC14490, and in "supervisory" chips like the TPS3836-8.

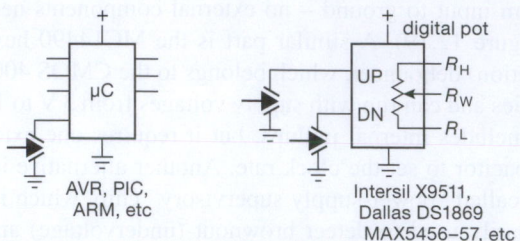

Figure 12.20. Debouncing with complex digital ICs: a microcontroller (μC) can use its user-written program (its *firmware*) to perform "software debounce"; and application-specific chips that accept pushbutton commands (like the digital potentiometers shown, with UP′ and DOWN′ inputs) often include internal pullup and debounce circuitry.

mind, also, that often it isn't necessary to debounce switch inputs, since they aren't always used to drive edge-sensitive circuitry. Another point: well-designed switches are usually "self-wiping" to maintain a clean contact surface (take one apart to see what that means), but it's a good idea to choose circuit values so that a current of at least a few milliamps flows through the switch contacts to clean them. With suitable choice of contact material (e.g., gold) and mechanical design, switches can be designed to avoid this "dry-switching" problem and will work properly even when switching zero current. More on this in Chapter *1x*.

12.1.5 Input protection

In these interfacing examples we've assumed that the signals being applied to the logic inputs are well behaved – that they do not have transient overvoltages or other destructive tendencies. That's not always the case. You can get into plenty of trouble with signals coming in from the outside, if you mount a connector on a box and casually hook it to the input of a logic gate (Figure 12.21A). A common source of transients is the static charge that accumulates readily in dry weather – those crackling sounds you hear when you take off a synthetic fleece jacket can elevate the body to a kilovolt or more. That voltage resides on the human capacitor (about 100 pF), and can take flight into a circuit input when you plug in a connector, or injudiciously touch components on a circuit board without first discharging yourself to ground.

The problem is well known in the electronics industry, and ICs are tested and rated for survival when subjected to electrostatic discharge. For this purpose the human body is modeled as shown in Figure 12.22. In the manner of the Physicist's "spherical cow," Electrical Engineers reduce all

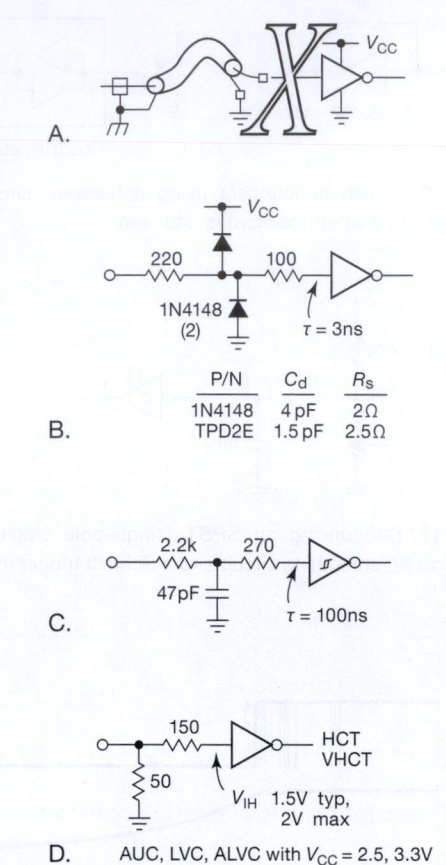

Figure 12.21. Protecting logic inputs from destructive transients: A. no protection; B. Diode clamps to V_{CC} and ground, or ESD protection array, with current-limiting series resistors; C. *RC* filtering plus Schmitt-trigger input; D. cable termination and series resistor, to dilute transient energy.

humans to a 100 pF capacitor in series with a 1.5 kΩ resistor,[10] and use that to test the robustness of their ICs. The usual charging voltages are 1–2.5 kV, but you see ICs rated up to 15 kV; for example, the MAX3232E RS-232 serial driver–receiver proclaims "ESD Protection for RS-232 Bus Pins – ±15 kV (HBM)." The figure shows the charged HBM (human body model) delivering a current pulse to the input of a logic-gate victim, which must sustain a transient current of an amp or more. That current is clamped to the supply rails, but the internal resistance of the clamp allows the input pin to go tens of volts beyond the rail (or below ground, for a negative transient), delivering a calculable energy to a very small area on the semiconductor die.

---

[10] JEDEC standard JESD22-A114D, dated 2006.

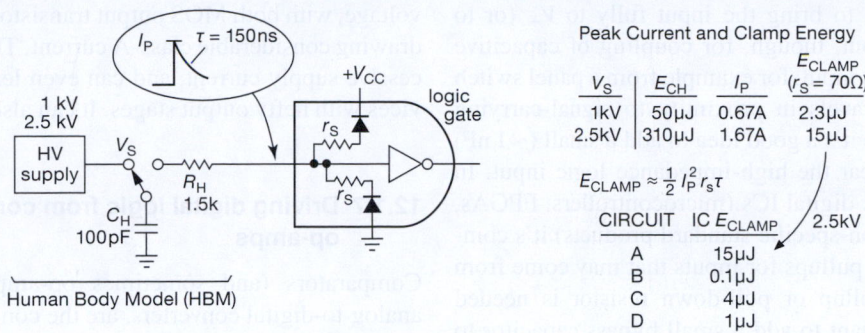

Figure 12.22. The Electrical Engineer's view of Biology: The "Human Body Model." ICs are tested and qualified at voltages of 1 kV, 2.5 kV, or more. The table lists estimated clamp energy values for the circuits in Figure 12.21. See also §3.5.4H.

ICs are rated to handle this. But plenty of damaged chips out there testify to the wisdom of adding external protection, especially if there's reason to expect exposure to high transient levels. Or inexperienced users. Look back at Figure 12.21. Circuit B is simple and effective. Ordinary diodes like the ubiquitous 1N4148 have more junction area than on-chip clamps, hence better clamping action; the upstream resistor limits the diode current, and the downstream resistor, fed from that clamped voltage, limits the chip's input current. You can get arrays of clamp diodes; these sometimes include a zener, as in Figure 12.23, for protecting differential logic like LVDS or RS-485.

Circuit C uses an *RC* transient filter to reduce the peak input current. To be effective, the time constant should be at least comparable to the HBM transient time scale, say 100 ns or so (and with an input resistance at least com-

parable to the HBM value of 1.5 k), and therefore should be followed by a Schmitt-trigger input. Circuit D takes a different approach: it combines a matched termination for 50 Ω coaxial cable (see Appendix H, and §12.10) with a larger series resistor, forming a "current divider." That dilutes the chip's transient input current by a factor of 4 (for the values shown), which is a 16× reduction of transient energy; and there's a further 2× to 5× reduction because the diluted energy is shared between the 150 Ω resistor and the chip's dynamic resistance during overvoltage. In Figure 12.22 we've estimated the delivered transient energy for each of these methods, compared with the 15 μJ estimated value for a 2.5 kV HBM with no protection.

Finally, for the ultimate in protection, use an optocoupler logic isolator, as in Figure 12.24. There's no galvanic connection at all (note the floating BNC connector), and these things can stand off potentials of many kilovolts (we'll see them later in the chapter, at §12.7). Note the current-limiting resistor and reverse protection diode; the latter is often omitted in error.

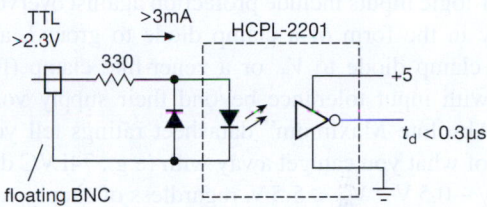

Figure 12.24. The ultimate in isolation: an optocoupler. This is more expensive than a diode clamp arrangement, but it's guaranteed to keep the bad stuff out.

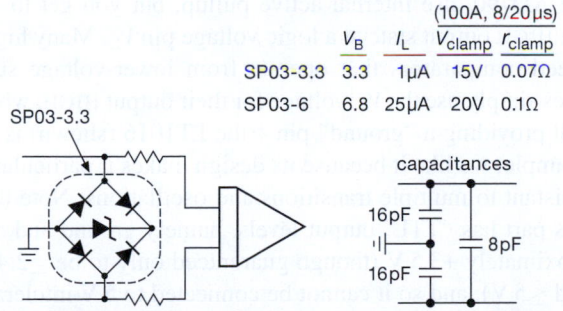

Figure 12.23. Diode protection arrays, like this one from Littelfuse, let you clamp single-ended or differential inputs, with very low clamp impedance. The internal zener sets the clamping voltage. You can get lower-current protection arrays with very low capacitance, for example the 5-volt CTLTVSS5-4 from Central Semiconductor: it has a maximum 0.8 pF to ground (and 0.4 pF line-to-line) at zero volts, a peak current rating of 2.5 A, and it includes two protective bridges in a single SMT package.

12.1.6 Some comments about logic inputs

A. Pullups and pull-downs

Most contemporary digital logic is CMOS, with essentially zero input current. So even a weak pullup (or pull-down) current is adequate to bring the input fully to V_+ (or to ground).[11] Watch out, though, for coupling of capacitive transients to such an input, for example from a panel switch with wiring that travels in proximity to signal-carrying lines. In such a case it's a good idea to add a small ($\sim$1 nF) bypass capacitor near the high-impedance logic input. In the case of complex digital ICs (microcontrollers, FPGAs, and other application-specific standard products) it's common to see internal pullups for inputs that may come from switches, so no pullup or pull-down resistor is needed (though you may want to add a small bypass capacitor to suppress coupled transients).

B. Input overdrive

Digital-logic inputs include protection against overvoltage, usually in the form of a clamp diode to ground, and either a clamp diode to V_+ or a zener-like clamp (for devices with input tolerance beyond their supply voltage). The "Absolute Maximum" datasheet ratings tell you the limits of what you can get away with (e.g., 74LVC devices specify $-0.5\,\text{V} < V_{\text{in}} < 5.5\,\text{V}$, regardless of supply voltage: they are "5 V tolerant."). However, it's often the input *current* that does the damage, which is duly noted in the same table: "The input and output negative voltage ratings may be exceeded if the input and output current ratings are observed." And the latter (input clamp current, output clamp current) are specified, in this case, as $-50\,\text{mA}$, maximum. Although it's a nice gesture to keep your input drive voltages within specified bounds, it's OK to go beyond if you have some series impedance to limit the current, as we've shown in Figure 12.21.

C. Unused inputs

Unused inputs that affect the logic state of a chip (e.g., a RESET input of a flip-flop) must of course be tied HIGH or LOW, as appropriate. Perhaps less obvious, even inputs that have no effect (e.g., inputs of unused gate sections in the same package) should be tied HIGH or LOW (your choice), because an open input to a CMOS device can float up to the logic threshold, causing the output to go to half the supply

---

[11] However, be aware that bipolar-TTL devices are not so friendly: their inputs source significant current (up to a milliamp) in the LOW state, and sink a small (but nonzero) current (up to a few tens of microamps) in the HIGH state. Because of this asymmetry, external digital signals used as inputs will therefore nearly always have a pullup resistor and will pull LOW (sinking current) when active, a convenient arrangement because switches, etc., can use a common ground return. It also leads to greater noise immunity, since a line held near the bipolar V_+ of $+5\,\text{V}$ has 3 V of noise immunity, as compared with the $\sim$0.8 V of noise immunity of a line held near ground.

voltage, with both MOS output transistors conducting, thus drawing considerable class-A current. This can result in excessive supply current, and can even lead to failure in devices with hefty output stages. It can also cause oscillation.

12.1.7 Driving digital logic from comparators or op-amps

Comparators (and sometimes op-amps), together with analog-to-digital converters, are the common input devices by which analog signals can interface into digital circuits (recall §4.3.2). If your circuit has a microcontroller ("μC," Chapter 15), then you can take advantage of built-in ADCs or comparators, which are common features of most μCs. But you'll sometimes want to go from a comparator (or op-amp) output directly into digital logic. This is not terribly difficult – but you've got to respect the allowable voltage input range of the driven logic. Let's look at some examples (for which it may be helpful to look ahead to Table 12.1 on page 812 and Table 12.2 on page 813, to keep score in the detailed discussion that follows).

A. Comparator driving logic

Figure 12.25 shows some common ways to connect comparator outputs to logic (and we'll have lots more to say about comparator interfacing in §12.3). The ever-popular and seriously inexpensive LM311 (and improved versions, like the LT1011) has a flexible open-collector output stage with a "ground" pin that sets the LOW state (which can be anywhere between V_+ and V_-); the pullup resistor, of course, sets the HIGH state, as shown. These are designated as output type "FL" in the tables. Some comparators, like the AD790, use internal active pullup, but you get to set the HIGH output state at a logic voltage pin V_L. Many high-speed comparators that operate from lower-voltage supplies simply use the V_+ voltage for their output HIGH, while still providing a "ground" pin – the LT1016 (shown) is an example; we like it because its design makes it particularly resistant to multiple transitions and oscillations. Note that this part has "TTL" output levels, namely ground and approximately $+3.5\,\text{V}$ (though guaranteed only to be $\geq 2.4\,\text{V}$ and $\leq 5\,\text{V}$), and so it cannot be connected to 5 V-intolerant logic. Next is a large class of low-voltage "single-supply" comparators (marked "CM to neg rail" in the tables; e.g., the TLC3702) that simply swing their outputs between ground and V_+. Finally, single-supply comparators that can run from higher voltages and that are not going for breakneck speed (e.g., the classic LM393) are usually configured with open-collector (or open-drain) outputs, with external

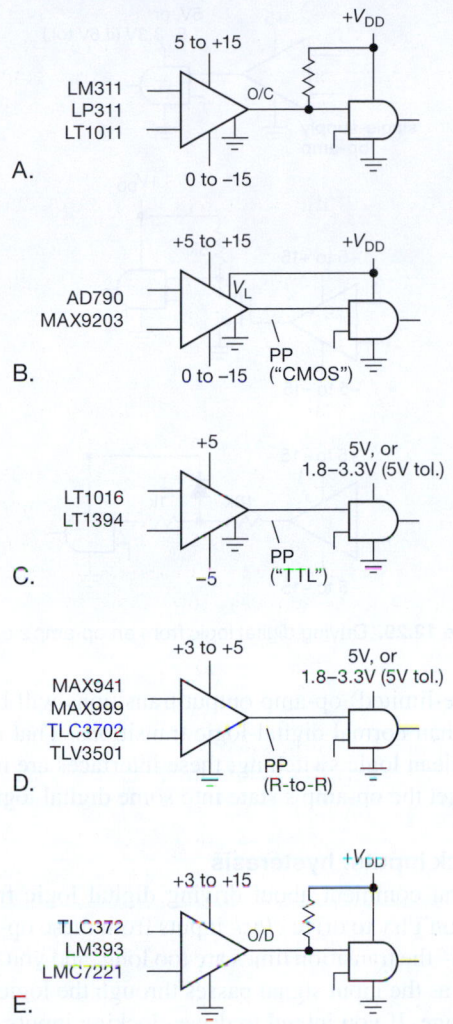

Figure 12.25. Driving digital logic from comparators.

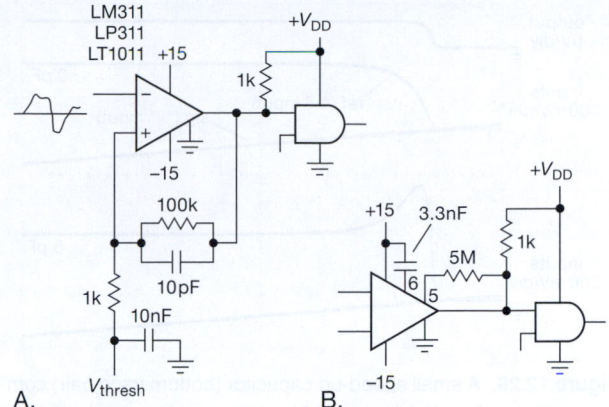

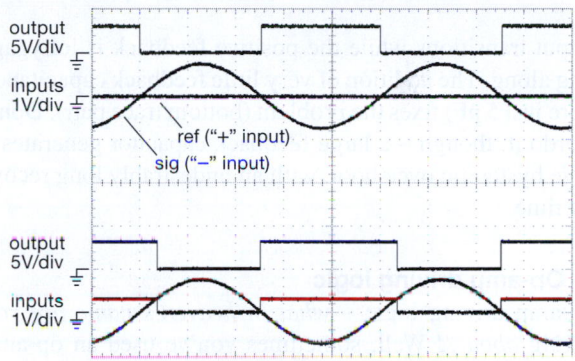

Figure 12.26. Threshold detector with hysteresis. A. Conventional circuit, for use with any comparator. B. Alternative method for the 311-type.

Figure 12.27. LM311 comparators driven with 1.5 V_{pp} 1 kHz sine wave, with output pullup to +3.3 V. Top: no hysteresis; bottom: 10% hysteresis (100k feedback, 11k to ground). Horizontal: 200 μs/div.

pullup resistor as shown; these are marked as output type "OC" or "OD" in the tables.

In Figure 12.25 we've ignored circuit configuration niceties like hysteresis, which we visited back in §4.3.2B. This is worth a reminder: Figure 12.26 shows the classic Schmitt-trigger threshold detector, configured with an amount of hysteresis equal to 1% of V_{DD}, and a small speed-up capacitor. In this circuit V_{thresh} must be provided from a low source impedance ($\ll$1k). That can be a serious disadvantage (although for slow or static V_{thresh} you can always use an op-amp buffer), in which case the alternative circuit shown may be just what you want. It takes advantage of the LM311's offset trim input terminals (pins 5 and 6) to produce both hysteresis (via the 5M resistor) and

speed-up (via the 3.3 nF capacitor).[12] Figure 12.27 shows measured waveforms for an LM311 configured without and with 10% hysteresis. Note, in the latter case, the asymmetrical trigger points (and output waveform timing). Although it's not visible in the captured waveforms, you are asking for trouble when you send slow input waveforms to a comparator without hysteresis (trust us, we've been there!).

Finally, Figure 12.28 shows in detail the effect of a small speed-up capacitor in the feedback path of the Schmitt trigger. Without it, the positive feedback to the noninverting input is slowed by input and wiring capacitance, so an input signal with some low-level fast noise may produce multiple

---

[12] This trick, and variations on it, are nicely described in the LT1011 datasheet; there's some discussion, also, in National's extensive LM311 datasheet.

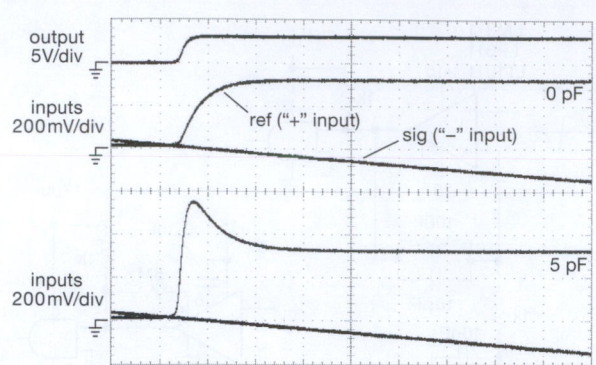

Figure 12.28. A small speed-up capacitor (bottom trace pair) compensates for the slowdown (middle trace pair) caused by input capacitance in the Schmitt trigger. Same circuit as Figure 12.27, but with expanded horizontal and vertical scales (to show detail of switching), and with 10 kHz sine wave input (so you can see the input signal's slope). Horizontal: 400 ns/div.

output transitions while the positive feedback is lollygagging along. The addition of very little feedback capacitance (here just 5 pF) fixes the problem (bottom trace pair). Don't overdo it, though – a large feedback capacitor generates a large hysteretic overshoot, with an undesirably long recovery time.

B. Op-amp driving logic

Op-amp driving logic – *what in heaven's name are you talking about?!* Well, sometimes you've used an op-amp section as a comparator, for example as a "battery-low" detector, or whatever. So its output swings between the rails, or nearly so (if it does not have a rail-to-rail output stage). All you want to do is get that output state into some digital logic. As with comparators, the only task is to make sure you respect the logic's input-voltage rules.

Figure 12.29 shows a few common situations. If the op-amp is running on a single low-voltage supply (in which case it likely has a rail-to-rail output stage) you can just connect it directly to logic that runs on that same V_+ voltage, or that is input tolerant to it. In the figure example, that could be 5 V logic, or 5 V-tolerant logic running on a lower supply (e.g., LV, LVC, or LVX: see Figure 12.3). If the op-amp has larger output swings, or swings both polarities, you need to limit it to logic bounds. One way is to interpose an nMOS inverter, as shown; alternatively you can use a passive clamp to the logic rail, combined with the logic's input protection negative clamp diode, as shown. We're not wild about this method, since it takes three components (and degrades the switching speed as well), but it does work. Whatever you do, be aware that the (analog and

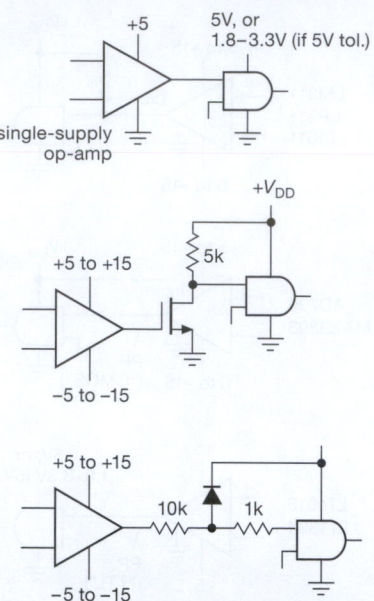

Figure 12.29. Driving digital logic from an op-amp's output.

slew-rate-limited) op-amp output transitions will be much slower than normal digital-logic transitions. That is, don't expect clean logic switching; these interfaces are merely a way to get the op-amp's state into some digital logic.

C. Clock inputs: hysteresis

A general comment about driving digital logic from op-amps: don't try to drive *clock* inputs from these op-amp interfaces – the transition times are too long, and you may get glitches as the input signal passes through the logic threshold voltage. If you intend to drive clocking inputs (of flip-flops, shift registers, counters, monostables, etc.), it is best to use a comparator with hysteresis (e.g., Figure 12.26), or to buffer the input with a gate (or other logic device) with Schmitt-trigger input (e.g., a '14 Schmitt-trigger inverter). The same comment goes for signals derived from transistor analog circuitry.

12.2 An aside: probing digital signals

We've had envious queries ("How do you guys get such nice traces? How do you get rid of all the wiggles?") about the clean digital 'scope waveforms in this book:[13] It's pretty easy, once you realize that you can't get away with the same lazy techniques that work OK at low frequencies.

---

[13] Take a look at some examples from this chapter, for example Figure 12.18 (slow) or 12.108 (fast).

In particular, the fast edges on digital signals produce probing artifacts (ringing, and soft edges) when you try to use a standard 10× passive 'scope probe with its 6″ grounding lead.[14] Figure 12.30 shows how the same logic waveform looks when probed three different ways.

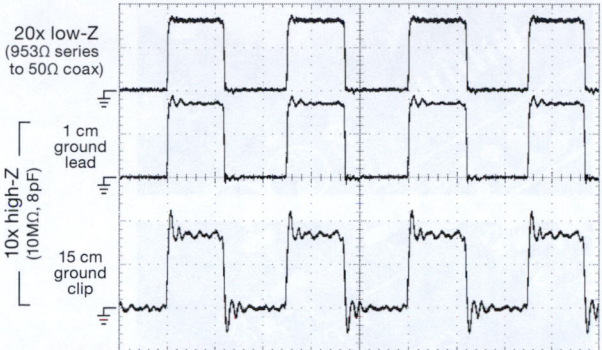

Figure 12.30. Logic waveform at 10 MHz from a 74AC14 inverter running at 3.3 V, as seen on a Tektronix TDS3044B 'scope when probed three different ways. Vertical: 2 V/div; horizontal: 40 ns/div.

This is a 10 MHz clock signal at the output of a 74AC14 inverter, wired up on a "solderless prototyping" board[15] (a dubious practice; but, hey, it works, most of the time anyway). We took the precaution of using an IC socket with integral SMT bypass capacitor, to put the best face on it. The bottom trace is business-as-usual, with a P6139A (500 MHz) 10× passive probe and 6″ ground lead. It's overshooting and ringing like crazy – can that be real, or is it an artifact of the probe's inductive ground path? You do considerably better by tossing the ground lead in the trash, removing the plastic sleeves, and using a little springy "ground tip contact." The middle trace shows the result: aha! Most of the ringing is gone. Better still, dump the 10× passive probe altogether (they don't come in speeds faster than 500 MHz anyway) and make your own by hooking a series resistor (we like 950 Ω) onto a length of skinny 50 Ω coax (we like RG-178); you temporarily solder the coax shield to a nearby ground, plug the other end into the 'scope (set for 50 Ω input) and voilà – a high-speed 20× probe![16]

---

[14] *Leads?* Yeah, sure. I'll just check with the boys down at the crime lab... they've got four more detectives working on the case... they got us working in *shifts!*

[15] For example, Global Specialties type UBS-100, or 3M type 923252.

[16] You can get these as a commercial product, the Keysight (Agilent) 54006A "6 GHz Passive Divider Probe Kit." It includes resistors for 10:1 and 20:1, and has a probe-tip capacitance of just 0.25 pF (as little as 0.1 pF, if you trim the tip). You can create greater ratios (thus higher loading resistance) by substituting other values of Caddock MG710-type resistors.

The top trace is about as clean as you can do, especially with through-hole (14-pin DIP) parts on a breadboard.

The homemade 50 Ω trick has the advantage of low cost, so you can do four traces easily; we used it for nearly all the digital 'scoping in this book. But because of its low input resistance, it is not useful for generalized circuit probing.

What happens with the more common configuration of surface-mount components on a printed-circuit board, and with much faster logic signals? Figure 12.31 shows what you see with four probing methods looking at the output of a 74AUC1G04 inverter, this time at 4 ns/div (and Figure 12.32 shows what the probes themselves look like).

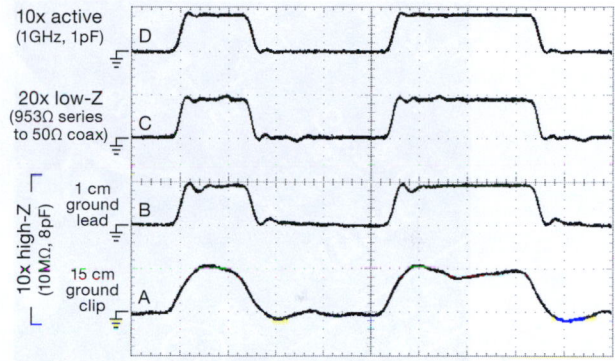

Figure 12.31. Logic waveform at a 6 ns clock rate from a 74AUC1G04 inverter running at 1.8 V, as seen on a Tektronix TDS3044B 'scope, when probed four different ways. The A–D labels match those in Figure 12.32. Vertical: 2 V/div; horizontal: 4 ns/div.

Here the passive-probe-with-ground-lead produces a mushy waveform (bottom trace) with moderate overshoot, considerably helped by using the shorter ground tip (next trace). The el-cheapo 20× trick looks better still. But best of all (if you've got the dough), is an "active probe" (an FET follower), with typical input capacitance less than 1 pF, and with speeds up into the high gigahertz (and prices to match). The top trace was made with a P6243 active probe (1 GHz bandwidth, less than – but not much less than – 1 kilodollar[17]).

12.3 Comparators

Comparators provide an important interface between analog (linear) input signals and the digital world, as we noted earlier in this chapter (§12.1.7A). In this section we would like to look at comparators in some detail, with emphasis on their output properties, their flexibility regarding power-supply voltages, and the care and feeding of input stages.

---

[17] But ~8 dB less expensive on used auctions.

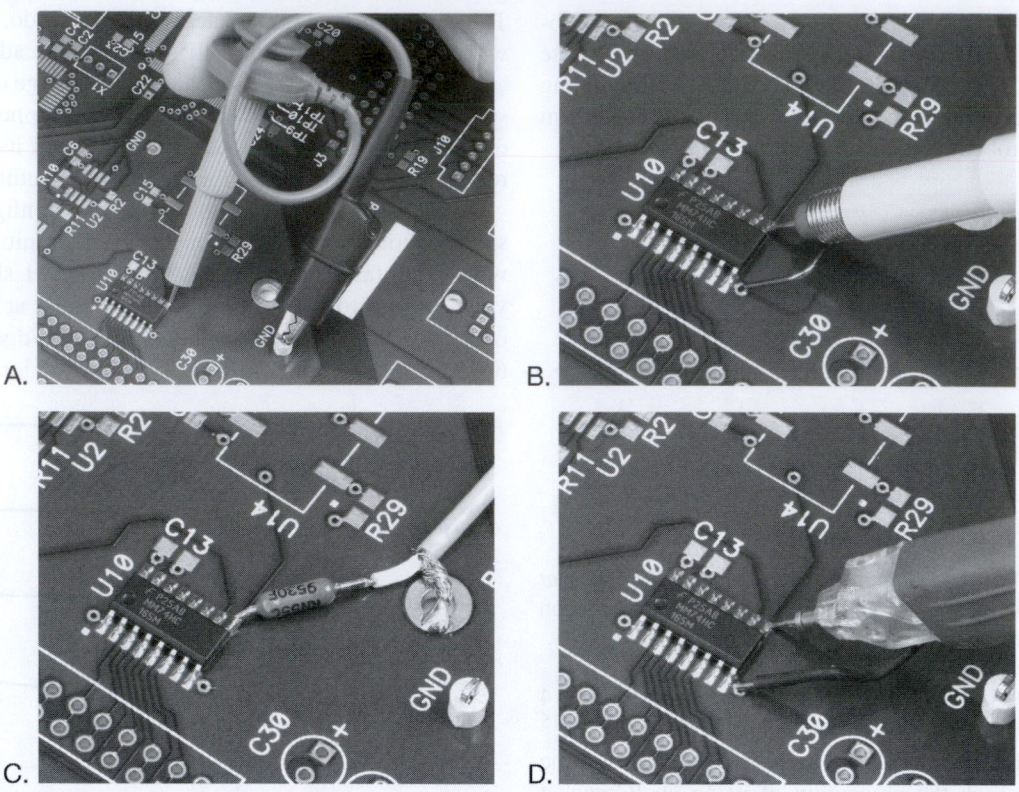

Figure 12.32. Probing digital signals: A. conventional 10× passive probe (Tektronix P6139A) with 6″ ground lead; B. short ground tip contact (Tek 016-1077-00) on 10× passive probe. C. Simple 20× passive "probe" for 50 Ω 'scope input: 953 Ω series resistor to coax; D. active probe (Tek P6243) with short ground tip. When probing fine-pitch ICs, it's a good idea to use a plastic guide (e.g., Tek "SureFoot" adapters) over the probe tip (not shown), so you don't short out adjacent contacts. The IC here is an SOIC-16, with 1.25 mm contact spacing.

Comparators were introduced briefly in §4.3.2A to illustrate the use of positive feedback (Schmitt trigger) and to show that special-purpose comparator ICs deliver considerably better performance than general-purpose op-amps used as comparators. These improvements (short delay times, high output slew rate, and relative immunity to large overdrive) come at the expense of the properties that make op-amps useful (in particular, careful control of phase shift versus frequency). Comparators are not frequency-compensated (§4.9) and cannot be used as linear amplifiers.

12.3.1 Outputs

We're used to op-amps, where the output can swing rail-to-rail (or nearly so), but where we usually stay in the linear region, deliberately avoiding saturation at the extremes of output swing. When the output is saturated, we're in trouble!

But comparators are different. Although the inputs are analog, the output is digital: it *lives* at the extremes. So what you care about is what the output does when LOW and when HIGH. As we've seen, the output may drive digital logic directly (Figure 12.25), in which case we need to bound its swing to that of the driven logic. Or we may want to drive an ON/OFF load, for example a relay (mechanical, or solid-state) or a bright LED, requiring plenty of output current, and perhaps powered from an external dc supply.

A. Output swing

Figure 12.33 shows the choices you've got to meet these various demands. In each case the comparator's analog circuitry is powered from a pair of supplies, V_+ and V_- (though for "single-supply" comparators, analogous to single-supply op-amps, the negative supply voltage V_- is ground). We'll have more to say soon about the input stage. What's interesting here are the output stages: one variety of comparators simply swing their output from rail to rail

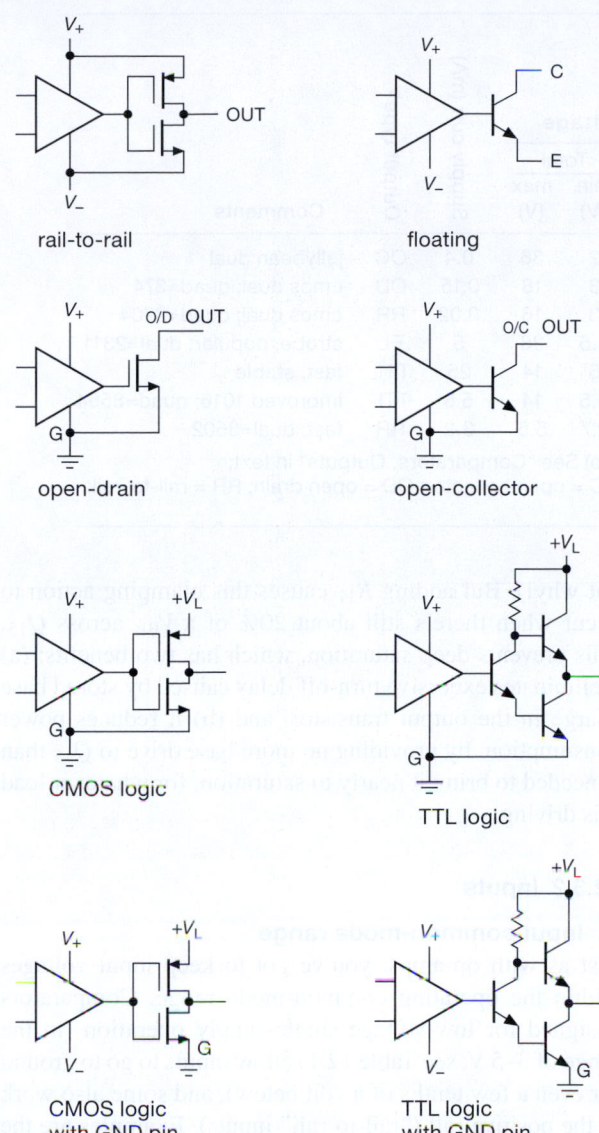

Figure 12.33. Don't let comparators confuse you: use this chart of simplified output stages, along with the listing in Tables 12.1 and 12.2.

("RR" in Tables 12.1 and 12.2), which is fine if that works for your application (e.g., if V_+ is +5 V, V_- is ground, and the output drives 5 V logic, or 5 V-tolerant logic; see Figure 12.25D). And the active pullup has the advantage of speed. But you may want to accommodate inputs that swing both sides of ground ("bipolarity") while driving digital logic (with its single positive supply), in which case you need to have V_- below ground; and so the right choice is either a comparator with floating output ("FL" in the tables) with resistive pullup (as in Figure 12.25A),

or a comparator with separate GND and V_L (logic voltage) pins ("RR-G" and "TTL" in the tables), as shown in Figure 12.25B.

Finally, if you're interested in single-supply operation, with input signals only between ground and a positive supply V_+, you can use (a) logic-output types with active pullup to a logic voltage V_L ("RR-G" or "TTL" in the tables); or (b) if V_+ is a low logic voltage already, you can use a rail-to-rail comparator with V_- connected to ground (as mentioned above); or (c) you can use a comparator with open-drain or open-collector output ("OD" or "OC" in the tables), with pullup to a logic supply (as shown in Figure 12.25E). The open-collector style is nice for driving power loads, or loads connected to a high output voltage (e.g., the classic LM311 can sink up to 100 mA, and its output can be pulled up to +40 V); but passive pullup is slow (compared with active pullup), so you're better off with a logic-output type when driving digital logic, unless you don't care about speed.

B. Output current

Comparators vary widely in their output-current capability. When driving digital logic this doesn't much matter, but it's important when driving high current loads such as relays or LEDs. Figure 12.34 shows how it goes, for most of the open-collector and open-drain comparators in Table 12.1. You can see some interesting trends here: (a) comparators with MOSFET outputs (e.g., the TLC393) behave resistively (R_{on}) at low voltages, trending downward

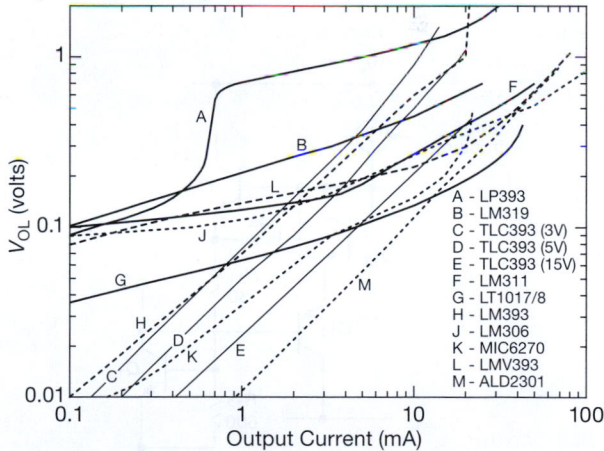

Figure 12.34. Saturated LOW output voltage versus sinking current, for a selection of open-collector and open-drain comparators. The popular TLC372-series is similar to the TLC393. Data are compiled from datasheet graphs, except for curve L and the low-current end of curves B and J, which were measured.

Table 12.1 Representative Comparators[a]

Type	t_d typ (ns)	V_{OS} max (mV)	Input current typ (nA)	CM to neg rail?	CM to pos rail?	Supply voltage V_+ max (V)	Supply voltage V_- max (V)	Total min (V)	Total max (V)	Supply curr (mA)	Output type[b]	Comments
LM393	600	5	25	●	-	36	-	2	36	0.4	OC	jellybean dual
TLC372	650	5	0.005	●	-	18	-	3	18	0.15	OD	cmos dual; quad=374
TLC3702	2500	5	0.005	●	-	16	-	3	16	0.02	RR	cmos dual; quad=3704
LM311	200	3	60	-	-	30	−30	4.5	36	5	FL	strobe, popular; dual=2311
LT1016	10	3	5000	-	-	7	−7	5	14	25	TTL	fast, stable
AD8561	7	7	3000	●	-	7	−7	3.5	14	5.6	TTL	improved 1016; quad=8564
TLV3501	4.5	6.5	0.002	●	●	5.5	-	2.7	5.5	3.2	RR	fast; dual=3502

Notes: (a) See also the more extensive listing in Table 12.2. (b) See "Comparators, Outputs" in text;
FL = floating *npn* output, collector and emitter output pins; OC = open collector; OD = open drain; RR = rail-to-rail;
TTL = logic swing, separate V_L pin.

to the left with unit slope; (b) comparators with bipolar *npn* output stages (e.g., the LT1017/8) tend toward a finite saturation voltage; and then there is (c) the LP393, which has a bipolar-output stage that behaves like a Darlington at higher currents (with saturation around a V_{BE}, or ~0.6 V), but becomes a simple grounded-emitter switch at low currents, which explains the weirdly gyrating curve.

The LM311's curve is worth an additional comment. The designers used a cute "antisaturation" circuit in the output stage (Figure 12.35): the series-connected V_{BE} drops of Q_{13} and Q_{14} are poised to rob base current from driver Q_{12} if Q_{15}'s collector gets too close to its emitter. If R_{11} were zero, this would happen just as Q_{15} saturates (figure

out why!). But adding R_{11} causes this clamping action to occur when there's still about 20% of a V_{BE} across Q_{15}. This prevents deep saturation, which has two benefits: (a) it eliminates excessive turn-off delay caused by stored base charge in the output transistor; and (b) it reduces power consumption, by providing no more base drive to Q_{15} than is needed to bring it nearly to saturation, for whatever load it is driving.

12.3.2 Inputs

A. Input common-mode range

Just as with op-amps, you've got to keep input voltages within the operating common-mode range. Comparators designed for low-voltage single-supply operation (in the range of 3–5 V; see Table 12.1) allow inputs to go to ground (or even a few tenths of a volt below), and some also work to the positive rail ("rail-to-rail" inputs). Examples are the TLC372/3702 and LMC7221/7211, respectively (with OD and RR versions listed in each pair). But these won't work with signals of both polarities (unless of course you tie the "ground" pin to a negative supply voltage, in which case you may be unhappy with an output that swings down to the negative supply). Instead you must use a dual-supply comparator (e.g., LM311, LT1016), many of which do not operate with inputs at or near the rails. Tables 12.1 and 12.2 include the common-mode operating range for the listed devices.

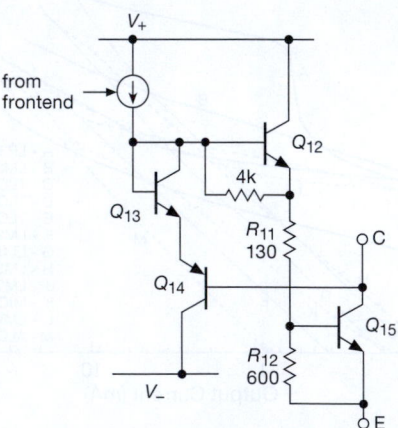

Figure 12.35. The output stage of the LM311 comparator incorporates a clever antisaturation circuit that limits base drive current when the output transistor is near saturation ($V_{CE} \approx 100$ mV). We've omitted here the output-stage current-limit circuit. Courtesy of Texas Instruments.

B. Offset voltage and trim

As with op-amps, garden-variety comparators have offset voltages in the millivolt range. You can do better with one of the "precision" comparators, from companies

Table 12.2 Comparators[a]

Type	Qty per pkg[h]	t_d typ (ns)	V_os max (mV)	Input current max (µA)	CM to neg rail?	CM to pos rail?	Cost qty25 ($US)	V_diff max (V)	Positive min (V)	Positive max (V)	Negative min (V)	Negative max (V)	Total min (V)	Total max (V)	Supply current max (mA)	Gain typ	Hyst (mV)	Bipolar sig	Offset trim	Q & Q* out	Shutdown Latch?	V_OH (max, V)	Output type[g]	DIP	SOIC	SOT-23	Smaller	Comments
TLV3701	1,2,4	37000	5	250pA	•	•	1.39	rails	2.5	16	-	-	2.5	16	0.6µA	1000k						40	RR	14	8	5	-	TLV3441 for OD
LT1017	2	20000	1	0.015	•		2.92	40	1.1	40	-	-	1.1	40	0.03	500k						40	OC	8	8	-	-	micropower
LP339	4	10000	5	0.025	•		0.45	36	2	36	-	-	2	36	0.06	500k						30	OC	14	14	-	14	low-power '339
TLV3491	1,2,4	6500	15	10pA	•	•	0.97	rails	1.8	5.5	-	-	1.8	5.5	0.9µA	-						-	RR	8	14	5	14	fastest nanopower
LT1018	2	6500	1	0.075	•		2.79	40	1.1	40	-	-	1.1	40	0.11	2M						40	RR	8	8	5	8	low power
LM393	2	600	5	0.25	•		0.36	36	2	36	-	-	2	36	0.4	200k						30	OC	8	8	-	8	jellybean dual
LM339	4	600	5	0.25	•		0.36	36	2	36	-	-	2	36	0.8	200k						30	OC	14	14	-	14	jellybean quad
MIC6270	1	600	5	0.25	•		0.69	36	2	36	-	-	2	36	0.3	200k						36	OC	-	8	5	-	low cost
LM311	1,2	200	3	0.1			0.77	30	5	36	0	-30	4.5	36	5	200k			•		•	40	FL	8	8	-	-	strobe, popular; dual=2311
LP311	1	2000	7.5	0.1			0.95	30	3	30	0	-30	3	30	0.15	200k			•		•	40	FL	8	8	-	-	strobe, low-power 311
LT1011	1	150	0.5	0.025			2.13	36	3	36	0	-36	3	36	3.0	500k			•		•	50	FL	8	8	-	-	strobe, improved '311
LM319	2	80	4	0.5			0.99	5[b]	5	36	0	-30	4.5	36	6	40k					•	40	FL-G	14	14	-	-	strobe, 0.1A current sink
LM306	1	28	6.5	5			1.04	5[b]	10	15	-3	-15	-	30	6.6	40k	0.4				•	24	TTL,OC	8	8	-	-	
AD790	1	45	0.25	5			6.61	16.5	4.5	15	-6	-15	4.5	30	10	3k		•		•	•	7	RR-G	8	8	5	-	strobe, aging, fastest 30V
LM361	1	14	5	30			2.21	5[b]	5	15	-6	-15	11	30	4.5	100k		•		•		7	TTL-G	14	14	5	-	low pwr; RR=7211, dual=6772
LMC7221	1,2	4000	5	40fA[t]	•	•	1.15	18	2.7	15	-	-	2.5	16	7µA	100k						16	OD	14	8	5	14	cmos 339; dual=393
TLC3702	4,2	2500	5	5pA[t]	•		1.02	18	3	16	-	-	3	16	0.04	-						18	RR	14	14	-	8	cmos; quad=3704
TLC372	2,4	2500	5	5pA[t]	•		0.72	18	3	16	-	-	3	16	0.20	200k						18	OD	8	8	-	8	cmos; quad=374
TLC352	2,4	650	5	5pA[t]	•		0.67	18	3	18	-	-	3	18	0.15	200k						18	OD	8	8	-	8	low power, low voltage
ALD2301A	1	650	2	10pA[t]			1.00	rails	1.4	18	-	-	1.4	18	0.07	150k						13	OD	8	8	-	-	will sink 50mA (600mV)
TS861A	1,2,4	300	7	1pA[t]			4.04	rails	3	12	-	-	3	13.2	0.11	-						-	RR	-	8	5	8	µPwr; dual=862, quad=64
MAX9203	1	500	7	1pA[t]			1.19	rails	2.7	10	-	-	2.7	12	7µA	-						-	RR-G	-	8	8	8	Vdd pin; dual='02, quad='01
LT1016	1	10	4	5			1.88	5	4.75	12	0	-6	4.75	12	1.3	3k				•		5.5	TTL	8	8	8	-	fast, stable single +5
TL3016	1	7.6	3	10			4.08	5	4.5	7	0	-7	5	14	25	-				•		-	TTL	8	8	-	-	replace LT1016
LT1394	1	7	2.5	4.5			1.76	12	4.5	7	0	-7	5	14	11	1600				•		-	TTL	8	8	-	8	improved LT1016
AD8611	1	4	7	4			3.42	4	4.4	8	0	-8	3[c]	12	6	-	3.3	•		•	•	-	TTL	8	8	5	8	really fast 1016; dual=8612
MCP6541	1	4000	7	1pA[t]	•		4.34	rails[d]	1.6	5.5	-	-	1.6	7	0.6µA	50k					•	-	RR	8	8	5	8	µPwr; dual=42, quad=44
LMV339	4,2,1	400	7	0.25	•		0.40	5.5	2.7	5.5	-	-	2.7	5.5	0.14	50k						5.5	OC	8,14	8,14	5	8,14	dual=393, single=331
TLV3501	1,2	4.5	6.5	2pA[t]	•		0.94	rails	2.7	5.5	-	-	2.7	5.5	3.2	-						-	RR	8	8	6	6	fast; dual=3502
ADCMP371	1	2000	9	0.05	•		2.97	22	2.25	6	-	-	2	6	4µA	-	6					-	RR	-	8	5	5	lopwr; ref=356, OC(22V)=370
MAX941	1,2,4	80	6	0.3			0.60	rails[d]	2.7	6	-	-	2.5	6.5	0.43	-	1				•	-	RR	8	8	5	5	dual=942, quad=944
TS3021	1	38	6	0.16	•	•	2.60	rails	1.8	5	-	-	1.8	5.5	0.07	-	•					-	RR	-	8	5	5	low voltage, low cost
LMV7219	1	7	6	0.95	•	•	0.92	1.5[f]	2.7	5.5	-	-	2.7	5.5	1.1	-	7.5					-	RR	-	8	5	5	low cost
MAX961	1	4.5	1.5	15			2.06	1.5[f]	2.7	5.5	-	-	2.7	6	7.2	-	3.5				•	-	RR	8	8	5	8	dual=963, quad=964
MAX999	1	4.5	1.5	15	•	•	5.00	1.5[f]	2.7	5.5	-	-	2.7	6	5	-	3.5				•	-	RR	-	8	5	8	shutdown=997
ADCMP600	1	3.5	2	10	•		5.64	rails	2.7	6	-	-	2.7	6	3.5	20k	2				•	-	RR	-	8	5	6,8	adj-hyst=601, shutdown=602
AD96685	1,2	2.5	2	10			3.09	5.5	4.8	5.2	-5	-	10	13	15	-		•		•	•	-	ECL	16	16	-	-	dual=AD96687
ADCMP604	1	1.6	5	5			4.77	5.5	2.5	5.5	-	-	2.5	5.5	15	-	•				•	-	RR	-	6	-	6	hyst+shutdown+latch=605
ADCMP565	1	0.3	6	40			3.92	7	4.8	5.2	-5	-	10	10.4	70	2k	adj			•	•	-	LVDS	-	-	20	20	5GHz input bandwidth
ADCMP553	1	0.5	10	5			7.97	3	3.2	5.5	-5	-5.4	3	6	35	1k	50			•	•	-	ECL	-	8	-	8	dual=551,552, hyst=552
65LVDS34	2	4	-	20			3.69	rails	3	3.6	-	-	3	3.6	16	1k						-	PECL	-	8	5	-	LVDS rcvr; int term=LVDT34
65LVDS2	1	1.8	1.8	4			1.84	rails	2.4	3.6	-	-	2.4	3.6	5.5	-	•					-	RR	-	8	5	8	LVDS receiver
FIN1018	1	1.5	1.5	20			0.97	rails	2.4	3.6	-	-	2.4	3.6	5	-	•					-	RR	-	8	5	8	LVDS receiver
HMC874LC	1	0.1	5[t]	15[t]			50.00	1.8	3.3[n]	-	-3.3[n]	-	6.6[n]	-	50	-	1					-	PECL	-	-	16	16	clocked, 10GHz BW

Notes: (a) ordered by supply voltage groups, then by speed. (b) note V_diff(max). (c) limited to 7V total analog supply voltage. (d) OK to rails, but 8.2kΩ for $V_{diff} > 1V$. (e) total, for listed part number. (f) back-to-back pairs of 2 clamp diodes, with 200Ω series resistor. (g) see "Comparators, Outputs" in text. (h) the first number corresponds to the listed part. (n) nominal. (t) typical.

like Analog Devices, Linear Technology, and Maxim; see Tables 12.1 and 12.2. A few comparators include external trim terminals; but, as with op-amps, an inexpensive trimmed comparator (e.g., an LM311) will have much larger tempco of V_{os} than an intrinsically accurate (i.e., precision) part. For example, the LT1011A "improved LM311" specifies a V_{os} tempco of $4\,\mu V/°C$, typ ($15\,\mu V/°C$ max), whereas for the generic LM311 none of the half-dozen manufacturers even bothers to specify it. In applications where you care about accurate input thresholds, it's always best to avoid heavy loads at the comparator's output.

Here's an interesting riff on offset voltage: thermal gradients set up on the chip from dissipation in the output stage can degrade input offset voltage specifications. In particular, it is possible to have "motorboating" (a slow oscillation of the output state) take place for input signals near zero volts (differential), because the state-dependent heat generated at the output can cause the input to switch.

C. Input current

Here, also, familiarity with op-amps can lead to trouble if you assume (incorrectly) that the inputs present essentially infinite impedance, and draw no current. An important feature of comparator inputs is the bias current at the input terminals and the way it changes with differential input voltage. For a wake-up call, look at Figure 12.36, a plot of measured current at the two inputs of the ever-popular LM311.[18] *It isn't zero!*

What's going on here? Many comparators use bipolar transistors for their input stages, with input bias currents ranging from tens of nanoamps to tens of microamps. Because the input stage is just a high-gain differential amplifier, the bias current changes as the input signal takes the comparator through its threshold. In addition, internal protection circuitry may cause a larger change in bias current a few volts away from threshold.

To see how this works in detail, look at the LM311's input circuit, which we've drawn in simplified form in Figure 12.37. The input stage consists of current-biased *pnp* followers driving an *npn* differential amplifier; the followers have an impressive $\beta\approx2000$, but even with that you've still got $-35\,nA$ of input current when the inputs are balanced. Of greater concern, the currents at the two inputs shift in opposite directions by roughly 10% when the inputs become unbalanced. This happens because the second-

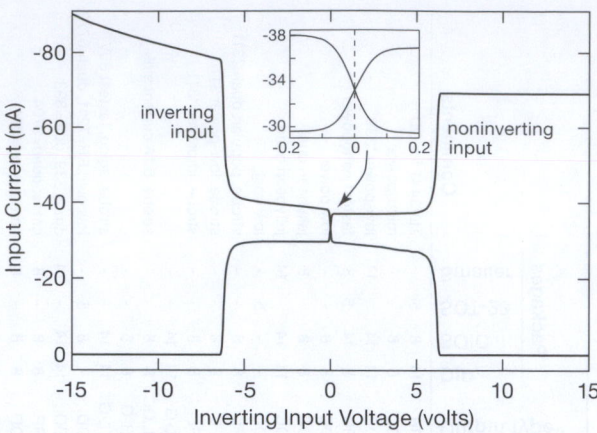

Figure 12.36. Measured input current of a specimen LM311N (NSC, with date code P134), with the noninverting input held at ground. The datasheet specifies an input current of 60 nA (typ), 100 nA (max).

stage differential amplifier transfers its operating current over to one side or the other, unbalancing its base current loading on the first stage. (The current "step" at zero volts differential is actually a smooth transition taking place over 100 mV or so, as seen in the expanded inset graph, and represents the voltage change necessary to switch the input differential amplifier stage fully from one state to the other.) So, unlike an op-amp (where feedback keeps the inputs balanced), a bipolar comparator's input current shifts at the input transition, which can cause trouble if the signal driving it is not of low source impedance.

For example, imagine you want to generate an output step when a slowly rising input signal (of finite source resistance) passes through zero volts. That's easy – you

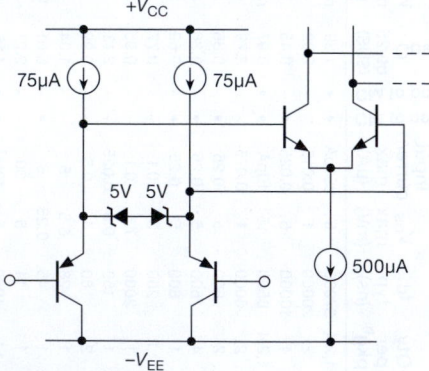

Figure 12.37. Simplified input stage of the classic LM311 bipolar analog comparator. Don't be alarmed by the 5 V zeners – the maximum allowed differential input is ±30 V.

---

[18] This is a true jellybean! A quick Internet search found at least a hundred 311 variants from a half-dozen manufacturers (Fairchild, NJR, National, ONsemi, ST, and TI), not even including "improved" 311s.

connect that signal to the inverting input, and you ground the noninverting input. Do this, and the output will likely exhibit fast multiple transitions as the input signal crosses through zero. The problem is that the drop in (negative) input current at zero volts causes the input voltage to reverse its rise, causing an extra transition; this continues several times, until the input signal has risen clear of the danger zone. Hysteresis (perhaps with a small speed-up capacitor) will generally cure this behavior, but it's helpful to understand its cause.

The graph in Figure 12.36 has some more surprises, namely the abrupt change in input currents when the differential input voltage reaches 6 V. This is caused by the symmetrical zener[19] clamp, included in the IC to prevent reverse base–emitter breakdown in the second-stage *npn* pair. Differential-input swings large enough to bring the clamp into conduction cause the *pnp* input transistor with the more negative input voltage to hog all the emitter current; so its base current doubles, and that of its twin drops to zero. The accurate graph of Figure 12.36 shows a feature you won't find in any of the official LM311 datasheets, namely the gradual increase in input current at large negative input voltages; this is evidently due to decreasing input-transistor beta at reduced V_{CE}. And a puzzle for the reader: why is this shape not mirrored in the noninverting input curve?

For comparator applications where extremely low input current is necessary, there are plenty of MOSFET-input comparators available, for example the TLC372, TLC3702, TLC393, and LMC7221. However, these are generally limited to a maximum total supply voltage of 16 V (compared with 36 V for "high-voltage" bipolar comparators); and, as with CMOS op-amps, they have poorer accuracy (V_{OS}) than precision bipolar comparators. In situations where the properties of a particular comparator are needed, but with lower input current, one solution is to add a matched-pair FET follower at the input.

D. Maximum differential-input voltage

Watch out for this one! Some comparators have a surprisingly limited differential input voltage range, as little as 5 V in some cases (e.g., the AD790, LM306, and LT1016), even though they can operate from a total supply voltage ($V_+ - V_-$) as high as 36 V. It may be necessary to use diode clamps to protect the inputs, because excessive differential input voltage will degrade beta, cause permanent input offset errors, and even destroy the base-emitter junctions of the input stage. General-purpose comparators that can op-

erate from total supply voltages up to 36 V are generally better in this respect, with typical differential-input voltage ranges of ± 30 V (e.g., the LM311, LM393, LT1011, etc.[20]).

E. Internal hysteresis

A little bit of hysteresis is usually a good thing. And some comparators (notably those intended for low-voltage single-supply operation) have a few millivolts of built-in hysteresis (see Tables 12.1 and 12.2). Some comparators (e.g., members of the ADCMP5xx and 6xx series from Analog Devices) permit adjustment of the amount of internal hysteresis.

12.3.3 Other parameters

A. Supply voltage

We've seen this already, because inputs have to stay in the "common-mode operating range," which at most extends slightly beyond the rails. Roughly speaking, there are three voltage ranges: (a) traditional bipolar comparators like the LM311 and LM393 can accept total supply voltages up to 36 V, and are now called "high-voltage" comparators; (b) a number of high-speed and CMOS comparators, for example the bipolar LT1016 and CMOS TLC/LMC parts, sit in a middle region, operating on total supply voltages up to 10–15 V; and (c) there has been a tremendous proliferation of "low-voltage" single-supply CMOS comparators like the LMV, TLV, and ADCMP600 series, which operate only to 6 V total supply. In the latter category there are some wickedly fast comparators (ADCMP572: 0.15 ns), and some micropower comparators (MCP6541: $0.6\,\mu$A typ; ISL28197: $0.8\,\mu$A typ). And there's pretty much everything in between.

B. Speed

It is convenient to think of a comparator as an ideal switching circuit for which any reversal in the differential-input voltage, however small, results in a sudden change at the output. In reality, a comparator behaves like an amplifier for small input signals, and the switching performance depends on the gain properties at high frequencies. As a result, a smaller input "overdrive" (i.e., more than enough signal to cause saturation at dc) causes a greater propagation delay and (often) a slower rise or fall time at the output. Comparator specifications usually include a graph of

---

[19] Implemented as a pair of back-to-back diode-connected transistors.

[20] These use integrated *pnp* input transistors, which tend to have high reverse emitter-base breakdown voltages, often more than 36 V (compared with typical *npn* breakdown around 6 V).

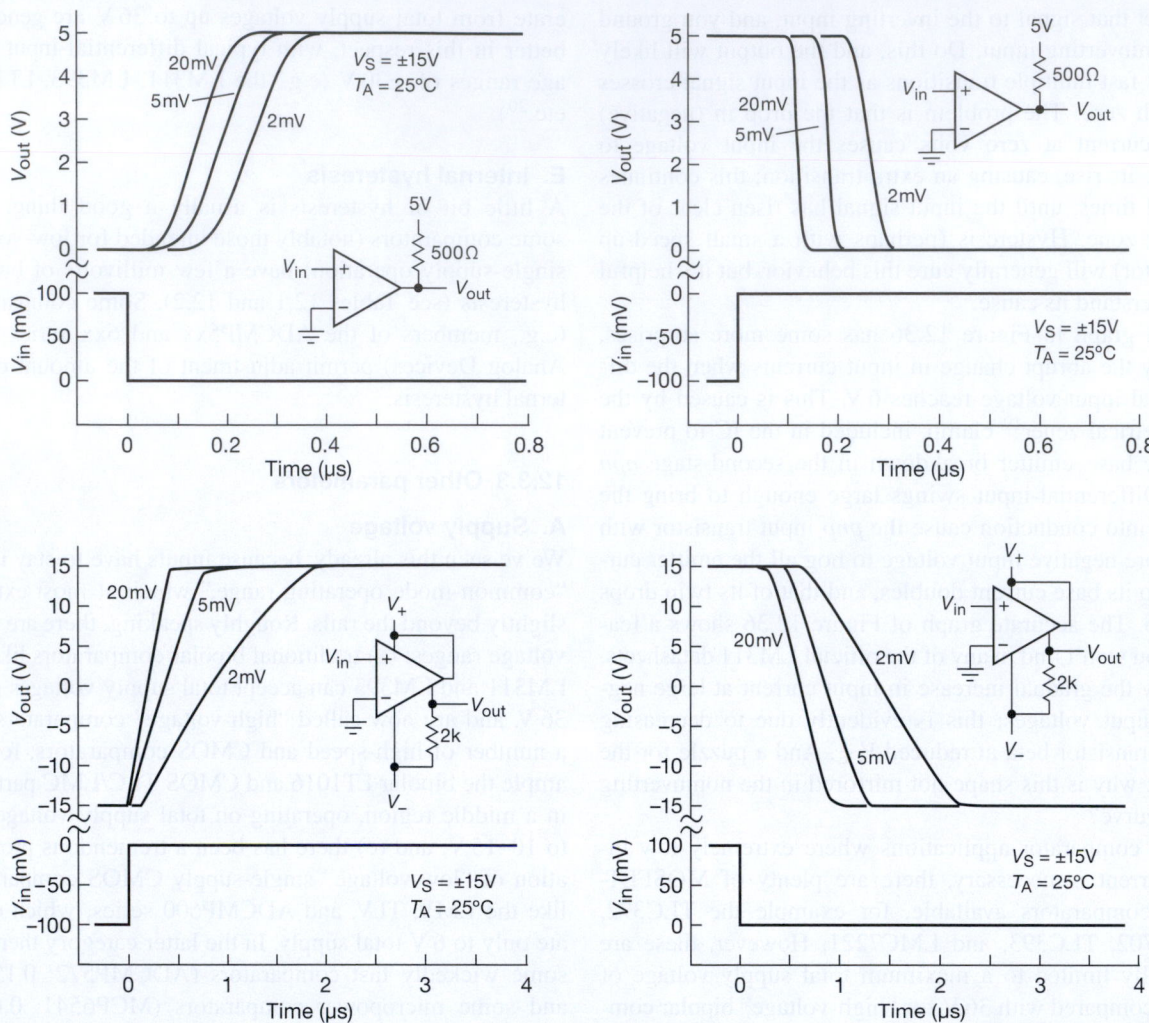

Figure 12.38. LM311 comparator response times for various input overdrives. Most comparators require considerable overdrive, 20 mV or more, for fast response. (Adapted from National Semiconductor Corp.)

"response times for various input overdrives." Figure 12.38 shows some for the LM311. Note particularly the reduced performance in the configuration in which the output transistor is used as a follower, i.e., with less gain. Increased input drive speeds things up because the amplifier's reduced gain at high frequencies is overcome by a larger signal. In addition, larger internal amplifier currents cause internal capacitances to charge faster.

The comparators in Tables 12.1 and 12.2 span a range of speeds (response time) from 0.3 ns to 300 μs – a ratio of a million to one! Haste makes waste – you pay the price in power dissipation (though low-voltage parts do much better), and in susceptibility to oscillations.

12.3.4 Other cautions

There are some general cautions concerning input circuits of comparators. Hysteresis (§4.3.2A) should be used whenever possible, because erratic switching is otherwise likely to result. To see why, imagine a comparator without hysteresis in which the differential-input voltage has just passed through zero volts, slewing relatively slowly, since it is an analog waveform. A mere 2 mV input differential causes the output to change state, with switching times of 50 ns or less. Suddenly you have 3000 mV fast digital-logic transitions in your system, with current pulses impressed on the power supplies, etc. It would be a miracle if some of these fast waveforms didn't couple into the input signal, at least to the extent of a few millivolts, overcoming the 2 mV input differential and thus causing multiple transitions and oscillations. This is why generous amounts of hysteresis (including a small capacitor across the feedback resistor), combined with careful layout and bypassing, are generally required for making sensitive comparator circuits function well. It is usually a good idea to avoid driving comparator inputs directly from high-impedance signals; use an op-amp output instead. It is also a good idea to avoid high-speed comparators, which only aggravate these problems, if speed is not needed. Then, too, some comparators are more troublesome in this regard than others; we have had plenty of headaches using the otherwise admirable LM311.

12.4 Driving external digital loads from logic levels

It isn't hard to use a logic-level output signal (coming from something as simple as a gate or a flip-flop, or from a fancier device like an FPGA or microcontroller) to control on/off devices like lamps (LEDs), relays, displays, and even ac loads. In some situations you can drive such loads directly from the logic-level signal; but more often you've got to add a few components to make it work. A classic example of the latter might be switching a load that returns to a *negative* supply voltage.

12.4.1 Positive loads: direct drive

Loads that don't require too much current and that return to a low-voltage positive supply can often be driven directly from a logic output. Figure 12.39 shows some methods. Circuit A shows the standard method of driving LED indicator lamps from logic running from a 3–5 V supply. You choose the current limiting resistor to set the LED current: LEDs behave like a diode with a forward drop of 1.5–3.5 V (depending on the semiconductor material and emitted color; see Figure 2.8). Contemporary high-efficiency LEDs look pretty bright at just a few milliamps, which is easy work for all logic-family outputs (as well as more complex digital chips such as FPGAs and microcontrollers); so the logic output stays valid even while driving the LED load (see the listed V_{OL} values). One caution here: the forward drop of $\sim$3.5 V of GaN-based LEDs (blue, white, and "bright green") requires 5 V logic, whereas lower-voltage LEDs can be driven by 3.3 V or 5 V logic.

For historical reasons connected with the highly asymmetrical output properties of earlier bipolar and nMOS logic families, designers tend to prefer the current *sinking* connection of Figure 12.39A; but for contemporary CMOS logic families it's OK to connect an LED in the sourcing configuration of Figure 12.39B. The output sourcing is somewhat less muscular than sinking (see the listed V_{OH}), but it's plenty good enough for the job.

Some panel-mount LEDs come with built-in current limiting resistors, intended for direct 5 V voltage drive. This saves a resistor, but the selection is limited, and you may not be happy with the manufacturer's choice of operating current (e.g., 10–12 mA for the CML 5100H-LC-series, or the Dialight 558-series).

You can drive small mechanical relays in a similar manner, providing their coil operating voltage is low (there are plenty of 5 V dc units, from Coto, Omron, Panasonic, Tyco-P&B, and other suppliers; and you can get relays that operate on as little as 1.5 V, for example the TXS2 series from Panasonic), and their operating current is low enough (i.e., high coil resistance). Figures 12.39C–F show several examples. "Signal" relays like the TXS2 series are intended for switching low voltage and current, and they have gold-plated contacts for "dry switching" (see §*1x.6*); their coil current of 10–20 mA can be sunk by the logic families shown (and others as well). You can also get logic-drivable relays that can handle power switching, for example the Omron G5- and G6-series 5 V-coil relays shown in the figure. These can handle up to 5 A when switching 115 Vac (or even 240 Vac) power.

For relays (and other loads) that require somewhat higher drive voltages or currents, you can use logic devices with open-collector outputs intended for that kind of job (Figures 12.39G,H). The venerable 74LS07 is an OC hex inverter good for output swings to +30 V and (LOW-state, sinking) load currents to 40 mA. The equally crusty (and wildly popular) ULN2003 is a 7-section ("heptal"?) grounded-emitter Darlington with input resistors (i.e., a logic inverter with open-collector) that can swing to +50 V and sink up to 350 mA; its equally muscular close cousin (the 75468) can swing to +100 V.[21] And if you want to drive these kinds of loads with a comparator, the vintage LM311 or LM306 can handle those kinds of currents, though the open-collector output swing is limited to 40 V above the negative supply, and +24 V above ground, respectively.

When driving relays and other power loads from a microcontroller (Chapter 15), it's worth knowing about a class of serial-input power registers. These are elaborations of the '595 8-bit serial-in parallel-out logic shift register, but

---

[21] On a more modest scale you can use the 75451–4 dual open-collector gates (AND, NAND, OR, and NOR, respectively) in 8-pin packages for loads to 30 V and 300 mA.

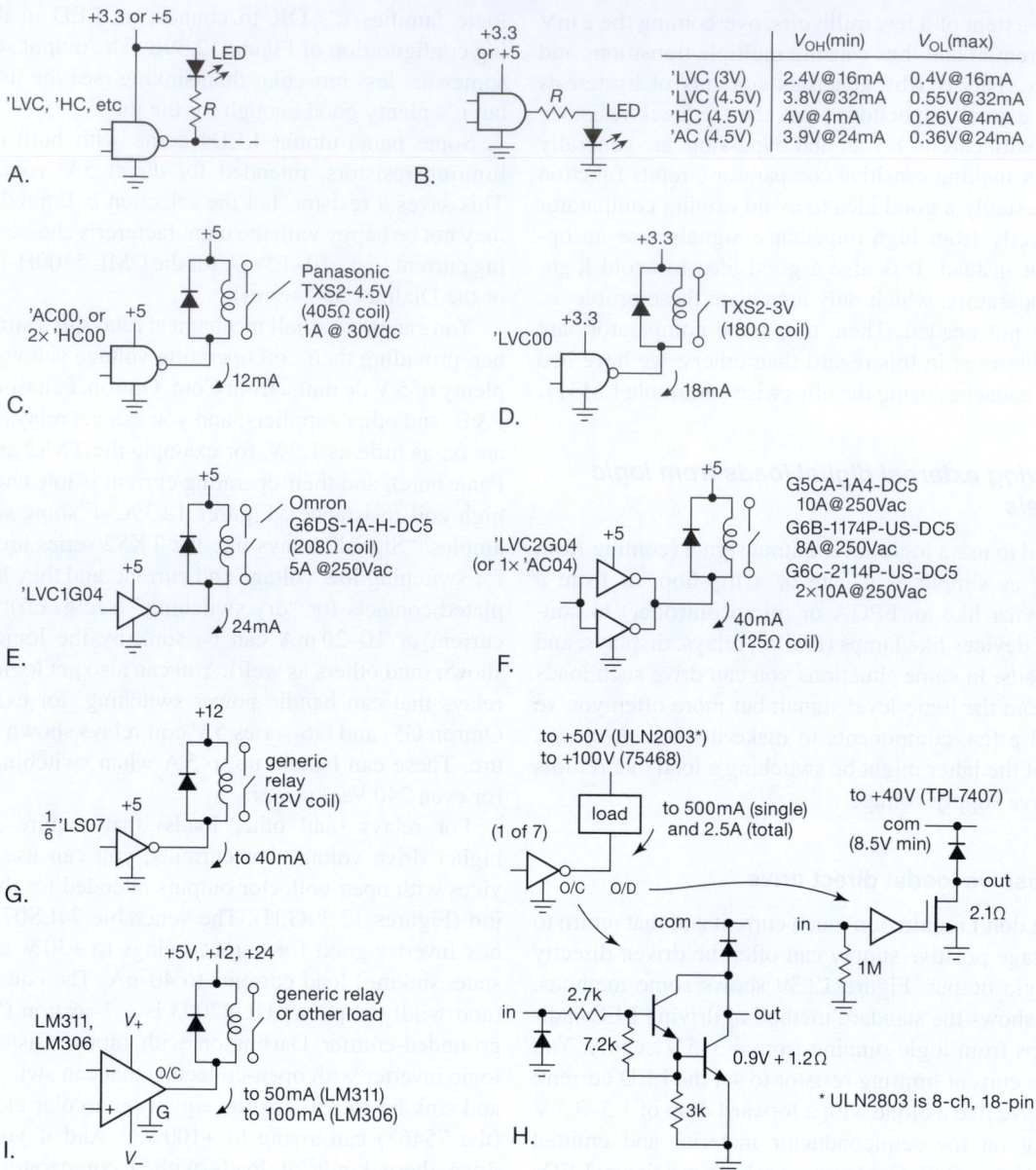

	V_{OH}(min)	V_{OL}(max)
'LVC (3V)	2.4V @ 16mA	0.4V @ 16mA
'LVC (4.5V)	3.8V @ 32mA	0.55V @ 32mA
'HC (4.5V)	4V @ 4mA	0.26V @ 4mA
'AC (4.5V)	3.9V @ 24mA	0.36V @ 24mA

Figure 12.39. Driving loads directly from logic outputs. Relay coils can demand a lot of drive current; be sure your choice of driver IC can deliver.

with open-drain outputs capable of sinking substantial currents, and with voltage ratings up to 50 V. Table 12.3 lists a nice selection of these, and Figure 12.40 shows what's inside these output devices, along with the companion '597 *input* register (i.e., parallel-in, serial-out). These are particularly helpful when you have a microcontroller with only a few I/O pins available because you can drive lots of power-hungry outputs (e.g., relays); and you can expand beyond

eight outputs by chaining the SDO (serial data out) from one register to the SDI (serial data in) of the next. Figure 12.41 shows the basic idea.

A few points about relays in general, and these small easy-to-drive PCB-mounting relays in particular.

• Note that you don't ordinarily use a series resistor, because the coil's resistance sets the operating current at the rated operating voltage; add a series resistor, though,

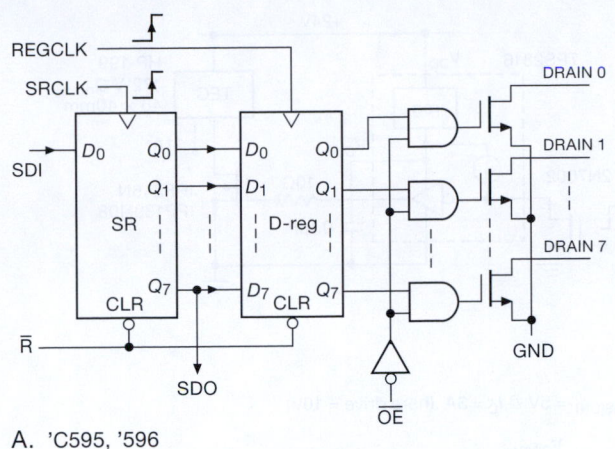

A. 'C595, '596

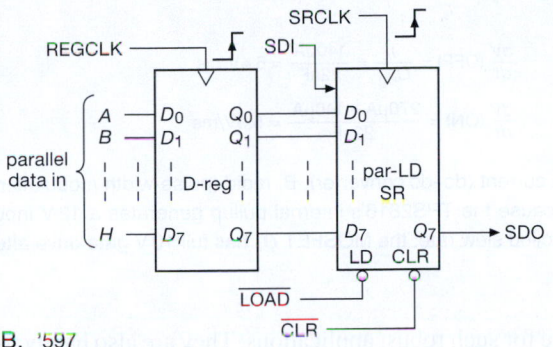

B. '597

Figure 12.40. The '595 and '596 serial-input power registers accept a logic-level bit-serial input, clocked into an internal shift register; the contents can be latched in an output *D*-type register with substantial drive capability. The '597 works the other way around, but accepts logic-level inputs only. See Table 12.3 for details.

Table 12.3 Power Logic Registers

Type[c]	Bits	Data comm[a]	V_O max (V)	I_O max (mA)	R_{DS} typ (Ω)	Cost qty25 ($US)	Reset	Prog Curr	Output Enable?	Output type[b]	Pkgs[d] DIP	SOIC
STP08CL596	8	SR	16	90	-	1.37	-	●	●	CS	(●)	●
STP08C596	16	SR	16	120	-	2.05	-	●	●	CS	(●)	●
TPIC6B259	8	AL	50	150	5	1.70	●	-	-	OD	●	●
74HC595	8	SR	V_{CC}	25	30	0.16	e	-	●	RR	●	●
TPIC6595	8	SR	45	250	1.3	2.50	e	-	●	OD	●	●
TPIC6B595	8	SR	50	150	5	1.41	e	-	●	OD	●	●
TPIC6C595	8	SR	33	100	7	1.09	●	-	●	OD	●	●
TPIC6C596	8	SR	33	100	7	0.98	●	-	●	OD	●	●
TPIC2810	8	I²C	40	210	5	2.15	f	-	●	OD	●	●
TPIC6B273	8	par	50	150	5	1.83	-	-	●	OD	(●)	●
TPIC6273	8	par	45	250	1.3	2.70	-	-	●	OD	●	●

Notes: (a) SR=shift register, AL=addressable latch. (b) CS=current sinking, adjust via ext resistor, range 15–90mA; OD=open drain; RR=rail-to-rail. (c) '596 types have registered data out. (d) "(●)" =no distributor stock. (e) resets SR, but not output latch. (f) power-on reset of SR and output latch.

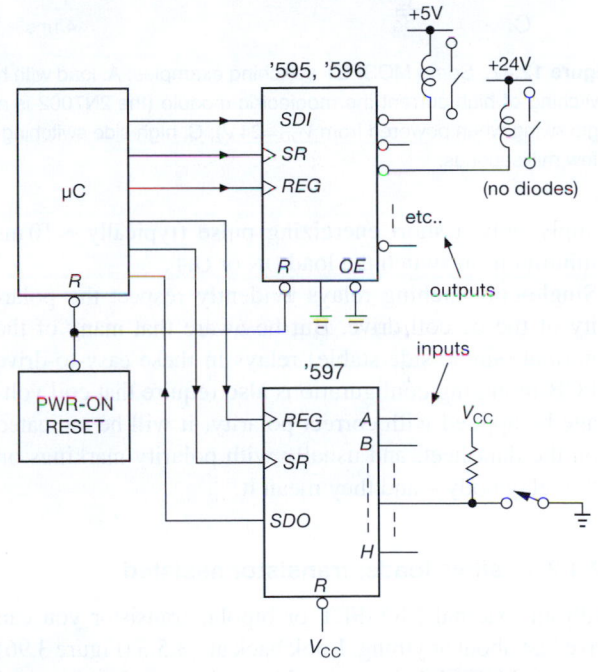

Figure 12.41. Driving relays and other power loads with a serial power register.

if you're running from a higher supply voltage (e.g., a 12 V relay operating from a 15 V supply). In any case, be sure to include the diode to clamp the inductive spike.[22]

- Relays are always available in the normal "single-side stable" (also known as *non-latching*) configuration, in which the contacts are held in the energized position only while coil power is being applied. But you can also get "latching" relays, which remain in whichever state they have been sent after the coil drive is removed. There are

two varieties of latching relays: "dual coil" (you drive one or the other, to set and reset the state), and "single coil" (you apply one polarity or the other, to set and reset). Latching relays are a good choice for something like a battery-powered lamp timer, because you need to

---

[22] Some driver parts include a diode, while other specifically give you an avalanche rating for the output MOSFET (with which to absorb the relay coil's flyback spike). For example, the '6B595 parts specify each output at 33 V and 30 mJ (some others allow 75 mJ), enough to handle the $E = LI^2/2$ inductive energy stored in a large coil. Always read the datasheet carefully.

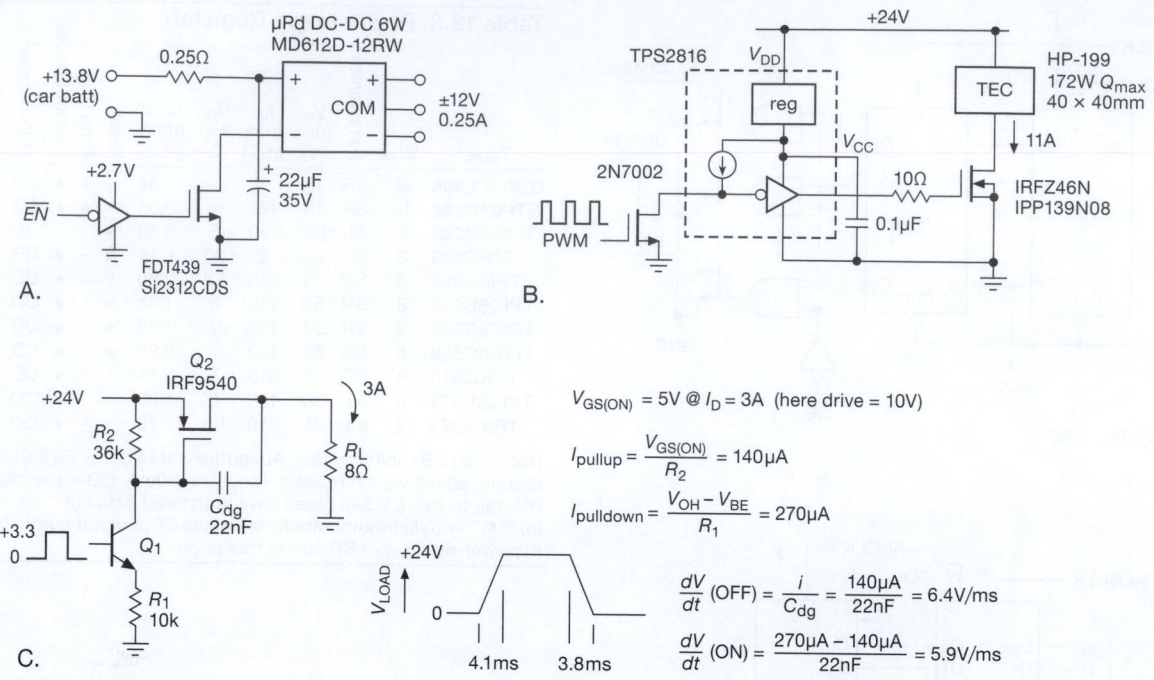

Figure 12.42. Some MOSFET switching examples: A. load with high inrush current (dc–dc converter); B. rapid (pulse-width modulation) switching of high-current thermoelectric module (the 2N7002 is needed because the TPS2816's internal pullup generates a 12 V input logic swing when powered from $V_{DD} = 24$ V); C. high-side switching with controlled slew rate; the MOSFET Q_2 has full 10 V gate drive after a few milliseconds.

apply only a short energizing pulse (typically ~ 10 ms minimum) to switch the load ON or OFF.[23]

- Single-coil latching relays evidently respect the polarity of the dc coil drive. But be aware that many of the normal (single-side stable) relays in these easy-to-drive PCB-mounting configurations also require that coil voltage be applied with correct polarity; it will be indicated on the datasheet, and usually with polarity markings on the relay body – and they mean it!

12.4.2 Positive loads: transistor assisted

With an external MOSFET or bipolar transistor you can drive just about anything. Look back at §3.5.3 (Figure 3.96) for some MOSFET drive circuits good for loads to hundred of volts and tens of amperes. MOSFETs (or insulated gate bipolar transistors, IGBTs; see §3.5.7) are the transistors of

choice for such robust applications. They are also handy for "high-side" switching, as we saw in §3.5.6 (Figure 3.106).

Figure 12.42 shows some additional configurations. The challenge in the first circuit is to handle the relatively large inrush current at startup of the dc–dc converter (5 A peak, compared with 0.8 A while running at full power). The mighty FDT439, in its small SOT-223 case, guarantees a maximum R_{ON} of $0.08\,\Omega$ at $V_{GS} = 2.5$ V (where its saturation drain current is around 20 A; see Table 3.4a). So it handles the inrush current with ease. The small input resistor and downstream bypass capacitor are included to filter switching noise and to isolate the turn-on transient to minimize disturbance of other electronic systems running from the car battery.

The challenge in circuit B is to do fast switching of a high power load (a thermoelectric cooling module), controlled by adjusting the duty cycle of its ON pulses (PWM, pulse-width modulation). Here we needed a hefty power MOSFET, and plenty of gate drive to switch the gate capacitance rapidly through its threshold (to minimize switching losses). The TPS2816 is a nice gate driver (see Table 3.8), with output HIGH voltage of $+10$ V (and peak current capability of ± 2 A), and with a built-in regulator for

---

[23] We installed several Intermatic ST01C "Digital In-Wall Timers" to control our exterior home lighting. These puppies include enough brainpower to compensate for seasonal variations in dusk and dawn, and they happily switch 15 A 120 Vac loads, using only a single CR2 lithium battery. We could hardly wait to crack one open, and voilà – a latching relay!

operation from supply voltages to +40 V. The power MOS-FETs shown are good to 30 A drain current, and come in power surface-mount packages (D^2Pak, TO-252, TO-262).

Finally, circuit C shows a high-side switch with controlled slew rate, to minimize production of transients. It's OK to switch power on time scales of milliseconds, as long as you don't try to do it at a high rate (as in PWM), and as long as the switch can handle the transient thermal pulse.[24] The slew rate is set by the gate drive current charging and discharging a "Miller capacitor" C_{dg}, here arranged for approximately equal net charging and discharging currents. The calculations in the figure say it all (five equations are worth a thousand words).

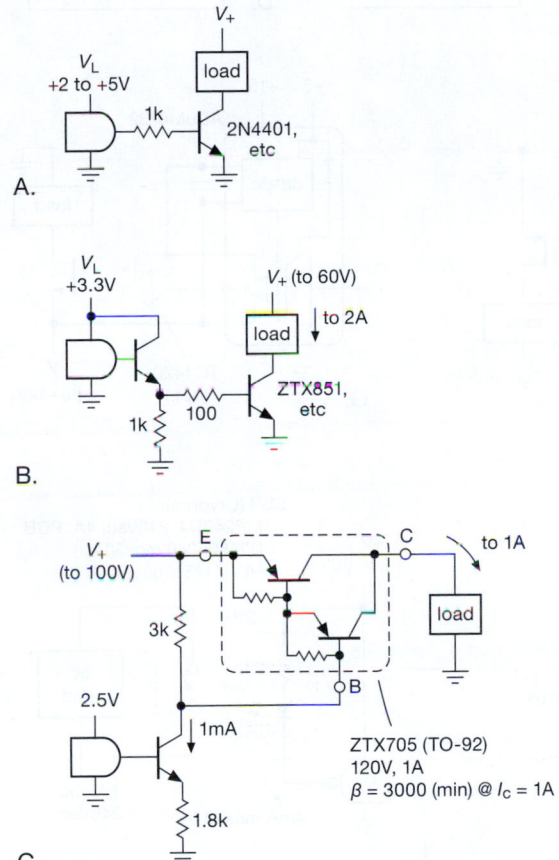

Figure 12.43. Bipolar transistors extend the voltage and current drive capability. Ditto for power MOSFETs – see Figures 3.96 and 12.42.

On a more modest scale you can use bipolar transistors (BJTs) for these same tasks, for example as shown in Figure 12.43. Something like a jellybean 2N4401[25] (about $0.06 in 100 qty) is good for up to +40 V and 500 mA; but beware – the minimum beta drops to 40 at that current, so you'll need to resize the series resistor to provide upward of 10 mA of base drive into the transistor's V_{BE}. And even then, the saturation voltage is not stunning: V_{CE}(sat)=0.75 V (max) at 500 mA with 15 mA of base drive (therefore 0.25 W transistor dissipation). A better choice for such currents is something like the ZTX851 from Zetex, which comes in a TO-92 variant that they call "E-line." It's good for 60 V and 5 A, minimum beta of 100 at 2 A, and V_{CE}(sat)=0.15 V (max) at 2 A with 50 mA of base drive. That's about the same dissipation as the 2N4401, but at four times the load current. To get the base drive you'd use an emitter follower driver, as in Figure 12.43B.

You can easily rig up a BJT as a high-side switch by using a logic-switched *npn* current sink to forward bias a high-side *pnp* switch (Figure 12.43C). Here we've used a Darlington from Zetex's excellent E-line series of BJTs, requiring only a milliamp of base drive to switch up to an amp of load current (where V_{CE}(sat)$\leq$0.75 V).

In these examples we've not worried about protecting the switches from fault conditions such as a shorted load. This should not be ignored; and we won't – see §12.4.4, coming up shortly.

12.4.3 Negative or ac loads

Figure 12.44 shows some methods by which a logic input can control loads that return to a negative supply rail; also shown is a common interface from logic to an ac power load. In circuits A and B a HIGH output state turns on the *pnp* transistor switch, pulling the collector into saturation at one diode drop above ground. In circuit A, the resistor (or gate output-current limit) sets the emitter current, and therefore the maximum collector (load) current, whereas in the more powerful circuit B, an *npn* follower is used as a buffer, and a diode in series with the output keeps the load from swinging above ground. In both cases the maximum load current is equal to the drive current to the *pnp* transistor's emitter. Circuit C requires a low-voltage negative bias supply voltage ($-V_{bias}$), but has the advantage of saturating cleanly to ground, and with a power MOSFET it can

---

[24] Datasheets specify a "transient thermal impedance" from junction to case, which for these transistors is about 0.5°C/W at millisecond time scales, fine for the 18 W (maximum) dissipation during switching. More on this in §3x.13.

[25] The 2N prefix is for the through-hole TO-92 package. For surface mount, the base part number becomes MMBT4401, with crazy suffixes for the different surface-mount packages (SOT-23, SOT-323/SC70, SOT-523, SOT-723).

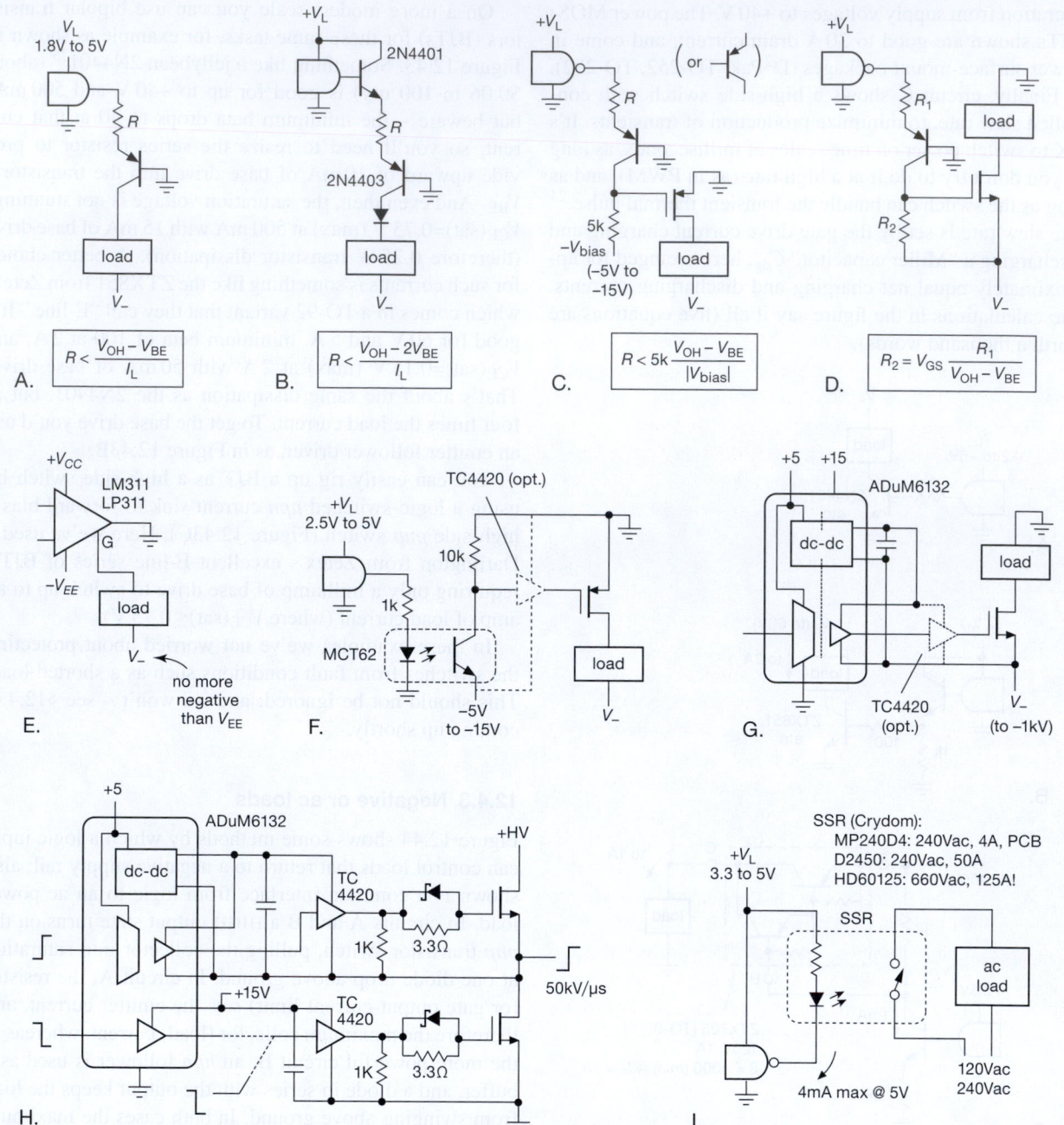

Figure 12.44. Driving negative and ac loads.

handle very large load voltages and currents, even when driven by low-voltage logic of minimal output drive current (but you may want to add a MOSFET gate driver IC, as in circuits F–H: its high current capability rail-to-rail output produces much faster switching into the MOSFETs large gate capacitance). Circuit D shows how to drive a ground-

returned load to a negative voltage; this has the pleasant property of not requiring a separate low-voltage negative supply. Circuits like this do not provide protection against load faults; be sure to read §12.4.4, which deals with this important issue.

Comparators with a floating output (LM311 and

cousins) can drive modest negative-referred loads, as in circuit E; but the negative return cannot be more negative than the comparator's negative rail, and the current is limited to 50 mA. Circuit F shows how to use an optocoupler (see §12.7) to translate a positive logic output to a negative-going level that can drive a *p*-channel power MOSFET gate directly (or through a MOSFET gate driver, for faster switching; see Figure 3.97). Because *n*-channel MOSFETs are better performers (lower R_{ON} and available up to 1000 V rating, compared with -300 V for pMOS), it's nice to be able to use them whenever you can. Circuit G does this by generating a gate drive signal down at the V_- rail, using the elegant ADuM6132 logic isolator from Analog Devices. The latter uses tiny on-chip transformers to generate both an isolated (floating) dc supply voltage, and also to couple the input logic signal to a similarly isolated output. Once again, you can interpose a MOSFET gate driver, powered by the same isolated power.

As an aside, circuit H shows how to use the same ADuM6132 to generate a *high-side* gate drive, so that you can use *n*-channel power MOSFETs for both switches in a high-voltage push–pull output stage. This isolator chip is notable for its ability to work properly even when the isolated output is flying around at slew rates up to 50 kV/μs. Note that the high-side MOSFET, though it looks like a follower, is actually a switch, because its gate is either at the same voltage as its source (when OFF), or it is 15 V above the source (when ON).

Finally, for driving ac loads the easiest method is to use a "solid-state relay," as in circuit I. These are optically coupled triacs, SCRs, or IGBTs with logic-compatible input and 1–50 A (or more!) load-current capability when switching a 115 Vac (or more!) load. The low-current varieties are available in SMT and DIP packaging (e.g., the photoMOS series from NAiS Aromat, the MOSFET relays from Omron, and the PV series of "photovoltaic switches" from International Rectifier; see §12.7.5 and 12.7.6 for more choices), while the heftier ones come as a chassis-mounting rectangular block roughly 2 inches square.[26]

Alternatively, you can switch ac loads with an ordinary relay, energized from logic. However, be sure to check the specifications, because most small logic-driven relays cannot drive heavy ac loads, and you may have to use a MOSFET or logic relay to drive a second larger relay. Most

solid-state relays use "zero-crossing" (or "zero-voltage") switching, which is actually a combination of zero-voltage turn-on and zero-current turn-off; it is a desirable feature that prevents spikes and noise from being impressed onto the powerline. Much of the "garbage" on the ac powerline comes from triac controllers that don't switch at zero crossings, e.g., phase-controlled dimmers used on lamps, thermostatic baths, motors, etc. As an alternative to the optical coupling used internally in circuit I, you sometimes see a pulse transformer used to couple triggering pulses to a triac or an SCR.

12.4.4 Protecting power switches

In these switch examples we've sidestepped an important topic: when driving power loads, you must worry about any of a variety of possible "fault conditions," for example a short-circuited load. This can happen more easily than you might expect, particularly with external loads wired through connectors and cables. Without some protective circuitry, the MOSFET is easily (and rapidly) destroyed, perhaps taking with it some additional circuitry. Let's take a closer look.

Figure 12.45 shows three versions of a pMOS high-

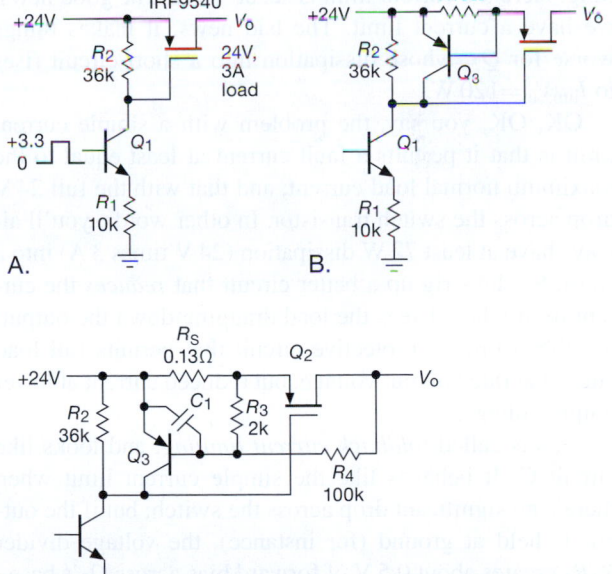

Figure 12.45. Current limiting the high-side switch: A. no protection; B. fixed 5 A current limit; C. foldback current limit.

$$I_{SC} = \frac{1}{R_S}\left(V_{BE} - \frac{R_3}{R_3 + R_4}(V_S - V_{BE} - V_o)\right)$$

---

[26] With high-current SSRs, the non-zero voltage drop can produce lots of power dissipation, requiring substantial heatsinks. For example, the 125 A HD60125 with its 1.7 V drop dissipates a bit over 200 W at full load! This is sobering: in spite of the no-moving-parts appeal of SSRs, you may decide that mechanical relays aren't so bad after all.

side power switch, intended for a ground-returned load that draws 3 A when powered by its rated +24 V. In all three circuits, *npn* transistor Q_1 converts the +3.3 V logic-level input into a 0.27 mA sinking current, which generates the ~10 V negative-going gate drive for the *p*-channel power MOSFET Q_2. The IRF9540 has a maximum R_{ON} of 0.2 Ω at that gate drive, so there's at most 0.6 V drop at full load, or 1.8 W dissipation. You hardly need a heatsink.

What happens if the output is shorted? Circuit A has no protection, so the current is limited either by the capability of the +24 V supply or Q_2's saturation drain current, whichever is less. For the latter, the datasheet shows ~50 A (at $V_{GS}=-10$ V and $V_{DS}=24$ V). For this 3 A application it's likely that the 24 V supply is less gutsy than that, perhaps good for 5–10 A. Taking the higher value, and using the datasheet's $R_{ON}=200$ mΩ at 25°C, we've got $I^2 R_{ON} \sim 20$ W dissipation in Q_2 (rising to 30 W at 100°C), not a happy situation for a transistor that's ordinarily dissipating less than 2 W.

OK, you say, let's add *current limiting* (Circuit B). This is the usual circuit, with a current-sensing resistor R_S sized to generate a V_{BE} drop at the limiting current, thus bringing Q_3 into conduction, robbing gate drive and preventing further rise of current. It's best to set the limit current enough above the normal maximum load so that variations in V_{BE} with temperature don't cause premature limiting. Here the current limit is set at ~5 A. The good news: we have a current limit. The bad news: it makes things worse for Q_2, whose dissipation into a short circuit rises to $I_{lim}V_{in}=120$ W.

OK, OK, you say, the problem with a simple current limit is that it permits a fault current at least equal to the maximum normal load current, and that with the full 24 V drop across the switch transistor. In other words, you'll always have at least 72 W dissipation (24 V times 3 A) into a short. So, let's rig up a better circuit that *reduces* the current limit when it sees the load dragging down the output; in other words, a protective circuit that permits full load current at rated output voltage, but reduced current at lower output voltage.

This is called *foldback current limiting*, and looks like circuit C. It behaves like the simple current limit when there's no significant drop across the switch; but if the output is held at ground (for instance), the voltage divider $R_3 R_4$ creates about 0.5 V of forward bias across Q_3's base–emitter. So it takes only a bit over 1 A to bring Q_3 into conduction, limiting the current to that lower value and the dissipation to about 30 W. It's common to set the short-circuit current limit somewhere in the range of 25–35% of the full load current. Capacitor C_1 provides some delay be-

fore triggering current limit or foldback; a time constant of ~1 ms guards against overzealous foldback while retaining protection.

Figure 12.46 shows the situation graphically. Among other things, you can see that the worst-case fault condition (in terms of transistor dissipation) occurs for a load that has a small resistance (rather than zero ohms). That's because the rising permitted load current more than compensates for the decreasing voltage drop across the switch. Even with this foldback scheme, we're once again faced with an enormous jump in switch dissipation: from 1.8 W (max) into a normal load, increasing to 34 W into a short, and 42 W into a 3 Ω fault-condition load. And foldback circuits have their disadvantages: if made sufficiently aggressive, they may prevent startup into a large capacitive load or other loads (like motors or dc–dc converters) that have a large inrush current.

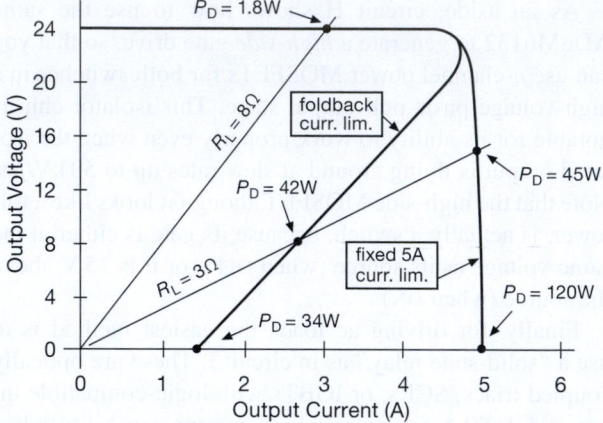

Figure 12.46. Foldback current limiting reduces the shorted-output switch dissipation by more than a factor of 3 compared with simple current limiting.

A. An easier way: protected switches

What to do? Foldback circuits can be made more precise,[27] for example by replacing Q_3 with a differential amplifier. And you can shape the foldback contour by including a zener in the feedback path. Then use a hefty heatsink, combined with temperature sensing to shut down the power when there's excessive heating, and you can make it all work reliably.

But there's a better way, in the form of "intelligent" pro-

---

[27] Our circuit suffers from uncertainties in V_{BE}, which is important: the short-circuit current is set by the *difference* between the actual V_{BE} and the 470 mV that the $R_3 R_4$ divider creates during shorted output.

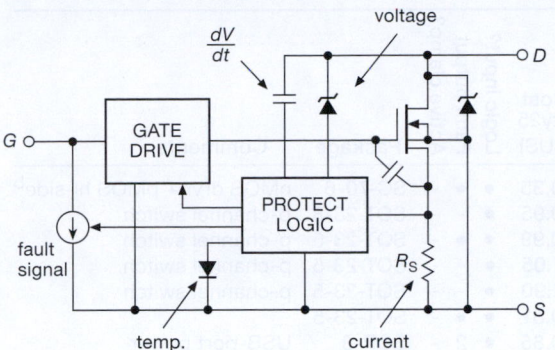

Figure 12.47. A protected MOSFET acts like an ordinary transistor, but it includes internal fault sensing and shutdown circuitry.

tected switches. One variety consists of *n*-channel MOS-FETs, decorated with internal circuitry that senses overvoltage, overcurrent, and overheating, shutting down the gate drive accordingly (and providing an indication in the form of reduced gate input resistance). These devices come in standard 3-terminal transistor packages, but internally they have an integrated circuit watching over the power MOSFET (Figure 12.47). Table 12.4 lists a sampling of typical parts.

These are suitable for any application in which you would use a relatively low-voltage *n*-channel power MOSFET. For example, they would be fine for switching the low side of a load that returns to a positive supply, as in Figures 12.42A,B; or you could use one for switching a

Table 12.4 A Few Protected MOSFETS[a]

Type	V_{DS} max (V)	I_D max (A)	R_{DS} max (mΩ)	Q_g typ (nC)	Cost qty25 ($US)	TO-220	TO-252	SOIC	SOT-223
BTS3207	42	0.6	500	-	0.60	-	-	-	●
VNN1VN04	40	1.7	250	5	0.62	-	3	●	●
VNN3VN04	40	3.5	120	8.5	1.08	-	3	●	●
IPS1041	36	4.5	100	-	1.92	-	3	d	●
BTS117	60	7	100	s	1.97	3	-	-	-
VNN7VN04	40	9	60	18	0.90	-	3	●	●
VNP10N07	70	10	100	30	1.36	3	-	-	-
VND14NV04	40	12	35	37	1.89	3	3	●	-
BTS133	60	21	50	s	1.27	3	-	-	-
BTS141	60	25	28	s	4.81	3	-	-	-
VNP35NV04	40	30	13	118	4.67	3	-	-	-

Notes: (a) All are *n*-channel types, with active current and overtemp limits; all require 5V min gate drive, and signal fault conditions with excess gate current; BTS types have slow *dV/dt*, while VN types are faster but have controlled *dI/dt*. (d) dual. (s) slew-rate limit 1V/μs.

negative voltage to a ground-returned load, in the manner of Figure 12.44D.

For a switching application like ours, though, you'd need a protected *p-channel* MOSFET, a breed that appears not to exist. For this application there's another variety of protected switches, intended specifically for *high-side* switching: these use an *n*-channel power MOSFET, suitably protected, with an internal charge pump and level-shifting circuitry to drive the gate ~10 V beyond the positive supply – see Figure 12.48. Once again there's sensing and protection against voltage, current, and temperature faults, sometimes responding also to undervoltage, reverse polarity, and loss of ground. So our problematic 3 A @ 24 V switching circuit becomes the thing shown in Figure 12.49: simple, inexpensive, and reliable.

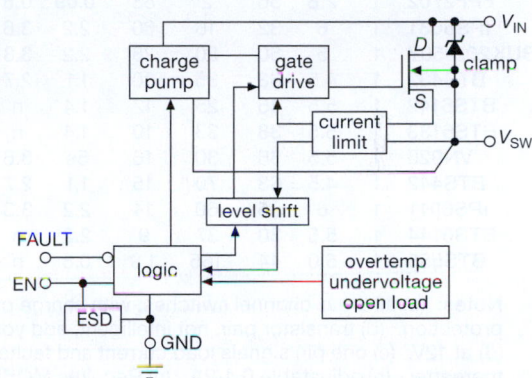

Figure 12.48. High-side intelligent switches have a charge pump and level-shifting circuits to drive the *n*-channel gate beyond the drain input supply; and they protectively monitor fault conditions such as overvoltage, overcurrent, and overtemp.

Table 12.5 lists the characteristics of selected high-side protected switches.[28] Within these there are two styles of logic input (Figure 12.50): one version accepts digital logic levels, relative to a ground pin; the other has no ground pin, instead requiring you to pull a current-sourcing pin to ground with a small external switch. This latter style provides a load-current indication, in the form of a "sense-output" pin that sources a current approximately proportional to load current.

[28] But a *caution*: the high-side intelligent-switch scene is strewn with discontinued parts. It's a healthy (but competitive) market sector (think automobiles), and new parts with improved performance (and pricing) are continually introduced. Table 12.5 gives a current list, and a sense of price and performance that's available; but you may have to look for comparable parts as these become consigned to the graveyard of obsoleted electronic components.

Table 12.5 Selected High-side Switches[a]

Type	Switches	V_{in} min (V)	V_{in} max (V)	I_o max (A)	R_{DS} typ (mΩ)	I_S typ (mA)	V_L min (V)	t_{ON} typ (ms)	Cost qty25 ($US)	Logic input?	Fault output?	Active clamp?	Package	Comments
FDG6323L	1	2.5	8	0.6	550	b	1.5	0.01	0.35	●	●	-	SC-70-6	nMOS drvr + pMOS hi-side[b]
TPS22960	2	1.8	6	0.5	435	0.00	1.6	0.08	0.95	●	-	-	SOT-23-8	p-channel switch
FPF2110	1	1.8	8	0.4	160	0.08	1.8	0.03	0.99	●	●	-	SOT-23-5	p-channel switch
FPF2123[g]	1	1.8	8	1.5[c]	160	0.08	1.8	0.03	1.05	●	-	-	SOT-23-5	p-channel switch
MIC2514	1	3	14	1.5	900[d]	0.08	2.3	0.01	1.90	●	-	-	SOT-23-5	p-channel switch
STMPS2151	1	2.7	6	0.5	90	0.04	2.2	1	0.81	●	●	-	SOT-23-5	
AP2156	2	2.7	5.5	0.8	100	0.09	2.2	0.6	0.85	●	2	-	SOP-8	USB-port power
TPS2041	1	2.7	5.5	0.7	80	0.08	2.2	2.5	2.59	●	●	-	DIP-8	
BTS452	1	6	62	1.8	150	0.8	2.5	0.08	1.97	●	●	●	TO-252-4	
BTS410	1	4.7	65	2.7	190	1.0	2.5	0.10	3.15	●	●	●	TO-220-5	see also BTS462T
BTS611	2	5	43	2.3	200	4	4	0.20	3.17	●	1	●	TO-220-7	
IPS511	1	6	32	5	135	0.7	3.3	0.05	1.76	●	●	●	TO-220-5	
FPF2702	1	2.8	36	2[h]	88	0.09	0.8	2.7	2.10	●	●	-	SO-8	adjustable current limit
IPS6031	1	6	32	16	60	2.2	3.6	0.04	2.52	●	●	●	TO-220-5	
BUK202-50Y	1	5	50	20	28	2.2	3.3	0.14	4.16	●	●	●	TO-220-5	
BTS432	1	4.5	63	35	30	1.1	2.7	0.16	4.85	●	●	●	TO-220-5	
BTS6142	1	5.5	45	25	12	1.4	n	0.25	2.65	n	e	●	TO-252-5	$I_S = I_L / 10k$ (±20% at 30A)
BTS6133	1	5.5	38	33	10	1.4	n	0.25	3.97	n	e	●	TO-252-5	$I_S = I_L / 9.7k$ (±10% at 30A)
VN920	1	5.5	36	30	16	5e	3.6	0.10	3.10	●	●	●	TO-220-5	
BTS442	1	4.5	63	70	15	1.1	2.7	0.35[f]	5.37	●	●	●	TO-220-5	
IPS6011	1	6	35	60	14	2.2	3.3	0.07	4.40	●	●	●	TO-220-5	
BTS6144	1	5.5	30	37	9	2.2	n	0.30	4.77	n	e	●	TO-220-7	$I_S = I_L / 12.5k$
BTS555	1	5.0	44	165	1.9	0.8	n	0.6[f]	5.80	n	e	●	TO-218-5	to 480A surge, $I_S = I_L / 30.2k$

Notes: (a) all are *n*-channel switches, with charge pumps, unless marked; all have overcurrent and overtemperature protection. (b) transistor pair, not intelligent, add your own source and gate resistors. (c) adjustable 0.15-1.5A. (d) at 12V. (e) one pin signals load current and faults. (f) max. (g) shuts down after 10ms, retries every 160ms thereafter. (h) adjustable 0.4-2A. (n) Requires MOSFET closure to GND, sinking I_S.

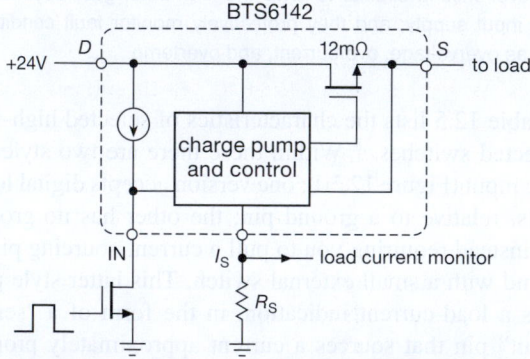

Figure 12.49. Problem solved! An intelligent high-side integrated switch provides a simple and reliable solution.

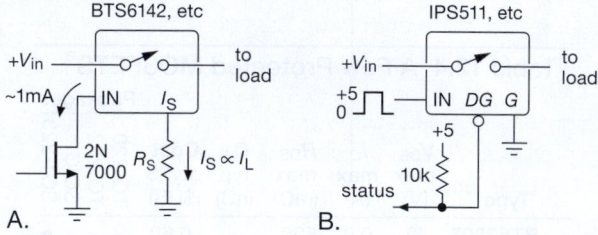

Figure 12.50. Two control schemes for high-side intelligent switches. A. No "ground" pin: enable by pulling input to ground, output current reported via proportional current-sourcing I_S pin. B. Logic-level voltage drive with respect to ground pin; active-LOW status output reports fault condition. This is the configuration shown in Figure 12.48.

12.4.5 nMOS LSI interfacing

Most LSI and VLSI circuits have true CMOS output drivers, with full rail-to-rail swings and with pretty much the same interfacing properties as the CMOS logic gates we just discussed. This is invariably true of ICs that op-erate from a supply voltage of +3.3 V or less. However, there are still useful ICs designed for 5 V supply operation that use a "totem-pole" *n*-channel MOSFET output stage (nMOS follower atop an nMOS switch; see Figure 12.51), thus producing a HIGH output voltage of only ~3.5 V, and

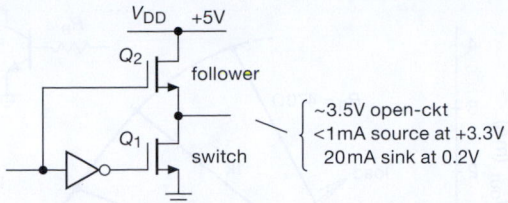

Figure 12.51. NMOS logic totem pole output circuit. The HIGH-state output voltage V_{OH} is around +3.5 V, with poor current-sourcing capability.

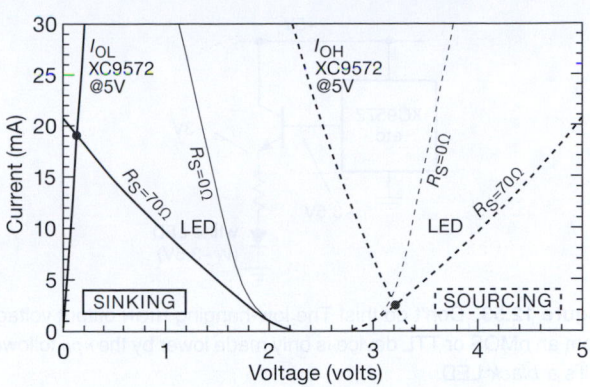

Figure 12.52. Driving a white LED with an nMOS output. The XC9500 5 V series of cPLDs provide good sinking all the way to ground, but weaker sourcing, and that only to ~3 V. Here we've used load lines (see Appendix F) to estimate the LED current for both the sinking and (undesirable) sourcing configurations, in each case with and without a series current-limiting resistor. (See Figures 2.8 and *1x.48* for LED *VI* curves.)

even that with hardly any sourcing current capability; you need to know about this quirky behavior. (The same goes for bipolar-TTL outputs, constructed with *npn* transistors in the same sort of totem pole.) While we're at it, we'll take a look also at a typical nMOS *input* stage, still widely used in ICs that can run over a range of supply voltages while maintaining an input logic threshold meeting the canonical TTL specifications (that is, anything less than +0.8 V is interpreted as LOW; anything above +2.0 V is HIGH).

A. nMOS and TTL outputs

The problem with an nMOS or TTL output is that its HIGH output, sitting anemically at ~3.5 V, is inadequate to drive loads like LEDs or relays; and it cannot even legally drive the input of a 5 V 'HC-logic device (with its mid-supply threshold). Taking the example of the XC95xx family of Xilinx cPLDs trying to drive a bright white LED, Figure 12.52 shows graphically the problem: the dashed curves show the output HIGH situation, where the cPLD's sourcing current versus voltage drops to zero around 3.4 V. We can figure out what will happen if we use that to drive a white LED whose cathode ("−" terminal) is grounded, by plotting on the same graph the "load line" of the LED; we've done that for a 70 Ω series resistor, and for no resistor at all. Either way we're lucky to get 4 mA of LED current, and without any confidence in the result. Contrast that with the *sinking* configuration (cPLD output connected to LED cathode, anode returned to +5 V via a series resistor), where a logic LOW output's clean saturation to ground produces a robust and predictable ~20 mA of drive with the series resistor (don't even *think* about omitting the resistor!). The results with a bipolar-TTL output structure are analogous (though with considerably poorer sourcing current).

The lesson is clear: when driving demanding loads directly from nMOS or TTL outputs, configure things so the output-LOW state does the heavy lifting.

If you insist on driving a load that returns to ground

from one of these wimpy outputs, you've got several choices. The *wrong* way is to say "hmmm, I'll just add a current-boosting emitter follower" (Figure 12.53). Nice try, but the additional V_{BE} drop only makes things worse. Figure 12.54 shows several ways that do work. In circuit A the nMOS output LOW sinks 2 mA, driving the *pnp* transistor into hard conduction; you can use a discrete resistor plus transistor, or a combination "digital transistor" (also known as a "prebiased transistor" or "bias–resistor transistor") like the DDTA123, good to 100 mA load current.[29] Circuit B substitutes a low-threshold *p*-channel MOSFET ($R_{ON} < 100$ mΩ at $V_{GS} = -2.5$ V), with a mild pullup resistor to ensure that output HIGH keeps the transistor OFF. Circuit C cheats a little, by interposing a CMOS inverter (or noninverting buffer) with TTL-compatible (therefore nMOS-compatible) input levels. The output can happily drive a 5 V load, sourcing (or sinking) tens of milliamps of current (see Figure 12.7). For more muscle, you can substitute a MOSFET gate driver like the ubiquitous TC4420-series (circuit D), which also lets you increase the output swing up to +18 V; these things hardly sweat at hundreds of milliamps source or sink (see Figure 12.7 again).

B. Weak CMOS outputs

Similar drive problems can crop up even with true CMOS rail-to-rail outputs. For example, look at the PIC16F sourcing (HIGH output) curve in Figure 12.8. We can use a

---

[29] Or the more than 300 alternatives in stock at DigiKey, from seven manufacturers.

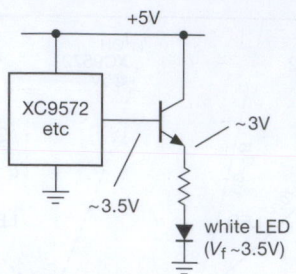

Figure 12.53. Don't do this! The low-hanging HIGH output voltage from an nMOS or TTL device is only made lower by the *npn* follower – it's a *black* LED.

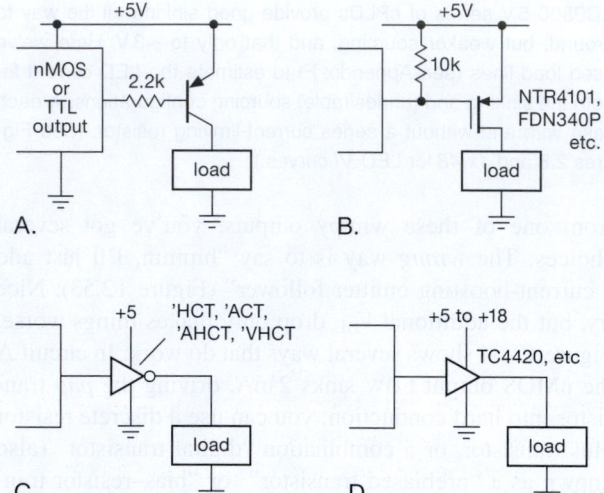

Figure 12.54. These nMOS logic outputs are driving loads that return to ground.

similar load-line graphical technique to see how such an output fares when driving a load like an *npn* switch with series base resistor (Figure 12.55), which you might use to switch on a 3 V relay coil (e.g., a Panasonic TXS2-3V, requiring 16.7 mA). The load lines are drawn for several choices of base resistor, allowing us to estimate the resulting base current from the intersection with the output-HIGH curve of the popular PIC10F microcontroller. Here a 2 kΩ series resistor produces about 1.7 mA of base drive, which (according to the transistor's datasheet) causes the collector to saturate at less than 50 mV with the relay coil as load. That's just fine – less than a milliwatt of transistor dissipation. But if you wanted to drive a heftier relay, for example the Omron G6RL-1A-3VDC, which has contacts rated at 8 A and 250 Vac, you'd have to contend with the 3 Vdc coil's resistance of 41 Ω, or 73 mA coil current. Now a 1 kΩ (or smaller) base resistor would be best: the

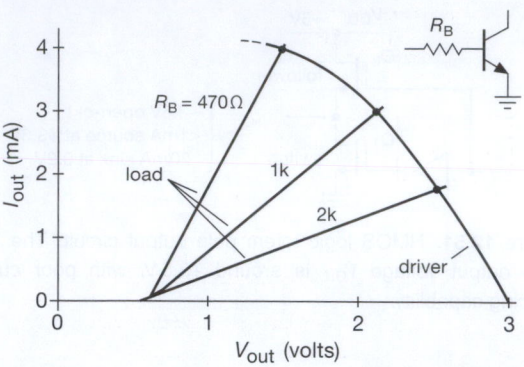

Figure 12.55. A weak CMOS output (here a PIC10F microcontroller, V_{DD}=3 V) can switch a "relay driver" transistor like the DRDNB16W (which includes both a 1k series base resistor and a clamp diode). Or you can use the resistorless DRDN005 with an external base resistor. Consider also an inexpensive 3-pin "digital transistor," which includes an integrated base resistor.

>3 mA base drive produces a collector saturation voltage of ∼100 mV, or 10 mW transistor dissipation.

C. nMOS inputs

You might think that nMOS is dead,[30] at the hands of the victorious CMOS. But you'd be wrong: many digital ICs that need to operate over a wide range of supply voltages use the simple input circuit of Figure 12.56. Q_1 is an inverter, and Q_2 is a small-geometry source follower providing pullup current (resistors take up too much space, so

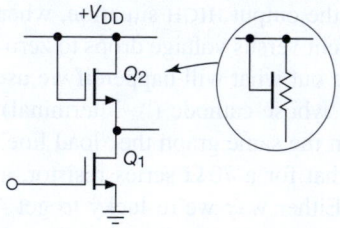

Figure 12.56. NMOS logic-input circuit.

MOSFETs are universally used as drain loads); the alternative symbol shown for Q_2 is widely used. The threshold voltage of the input transistor is in the range of 1–1.5 V, fully compatible with the long-standing "TTL input level" specification. A classic example of an IC that works this way is a MOSFET gate driver like the TC4420 (and its many cousins), which runs from a single +V_{DD} positive

---

[30] "She's not only merely dead, she's really most sincerely dead," as the Coroner proclaimed in *The Wizard of Oz.*

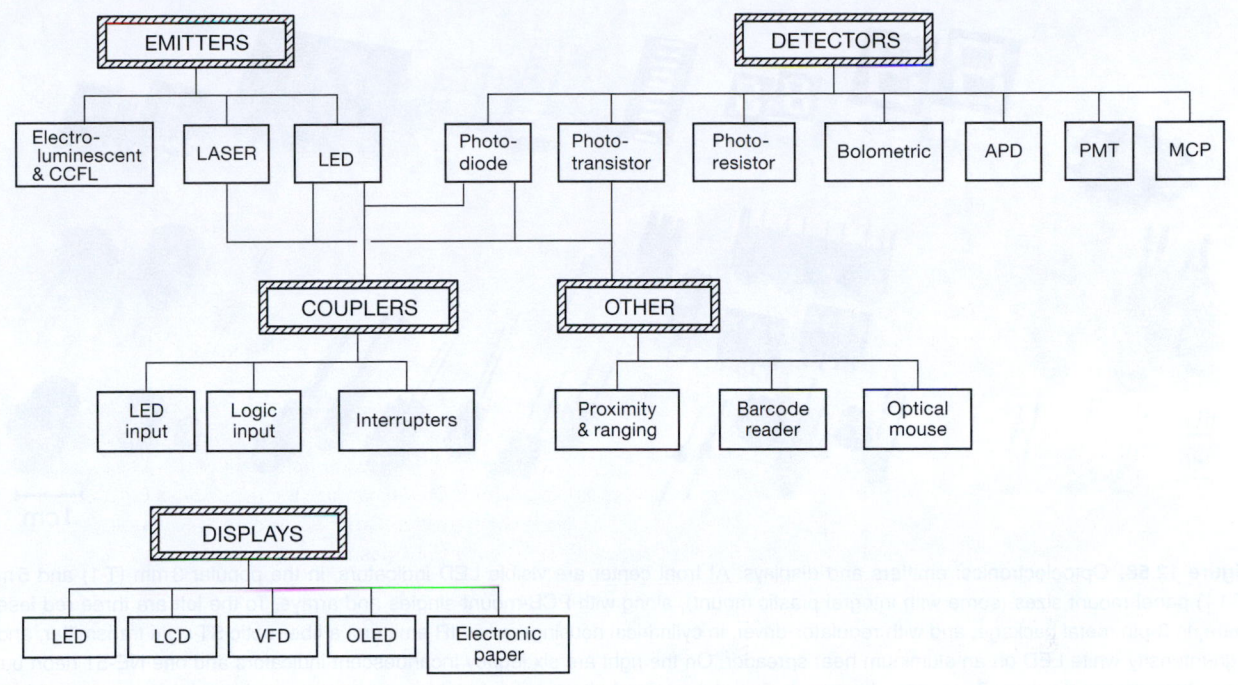

Figure 12.57. Optoelectronics family tree.

supply in the range of 4.5–18 V. Happily, such an input accepts pretty much any HIGH logic level you throw at it (including all the way to the elevated positive supply) while drawing essentially zero input current.

12.5 Optoelectronics: emitters

In the previous three chapters we have been using LED indicators and LED numeric display devices in various circuit contexts as we needed them. LEDs belong to the general area of *optoelectronics*, which includes displays based on other technologies as well, notably liquid crystals ("LCD") and gas discharge. It also includes optical electronics used for purposes other than indicators and displays: detectors (photodiodes and phototransistors), photomultipliers, array detectors such as "charge-coupled devices" (CCDs), light-coupled isolators ("opto-isolators"), solid-state relays, position and proximity sensors ("interrupters" and "reflective sensors"), diode lasers, image intensifiers, and a variety of components used in fiber optics.

Although we will continue to conjure up miscellaneous magic devices as we need them, this seems like a good place to pull together the area of optoelectronics, since it is related to the logic interface problems we have just been discussing.

To set the stage, we present a brief family tree of optoelectronics in Figure 12.57, and in the outline below. We've tried to make it inclusive, to provide perspective and orientation. And a family needs a *portrait*, so we gathered up a collection of optoelectronic goodies from our lab for Figures 12.58, 12.71, 12.80, 12.84, and 12.95. In the following sections we'll look at a subset of these devices, concentrating on the devices and techniques that are most important for day-to-day circuit and instrument design.

12.5.1 Indicators and LEDs

Electronic instruments look nicer, and are more fun to use, if they have pretty little colored lights on them. LEDs have replaced all earlier technologies (notably incandescent lamps) for this purpose. You can get red, yellow, green, blue, and white indicators, and you can get them in many packages, the most useful of which are (a) panel-mounting lights, and (b) PCB-mounting types. The catalogs present you with a bewildering variety of them, differing mostly in size, color, efficiency, and illumination angle. The latter deserves some explanation: a "flooded" (or "diffused") LED has some diffusing stuff mixed in, so the lamp looks uniformly bright over a range of viewing angles; that's usually best, but you pay a price in brightness.

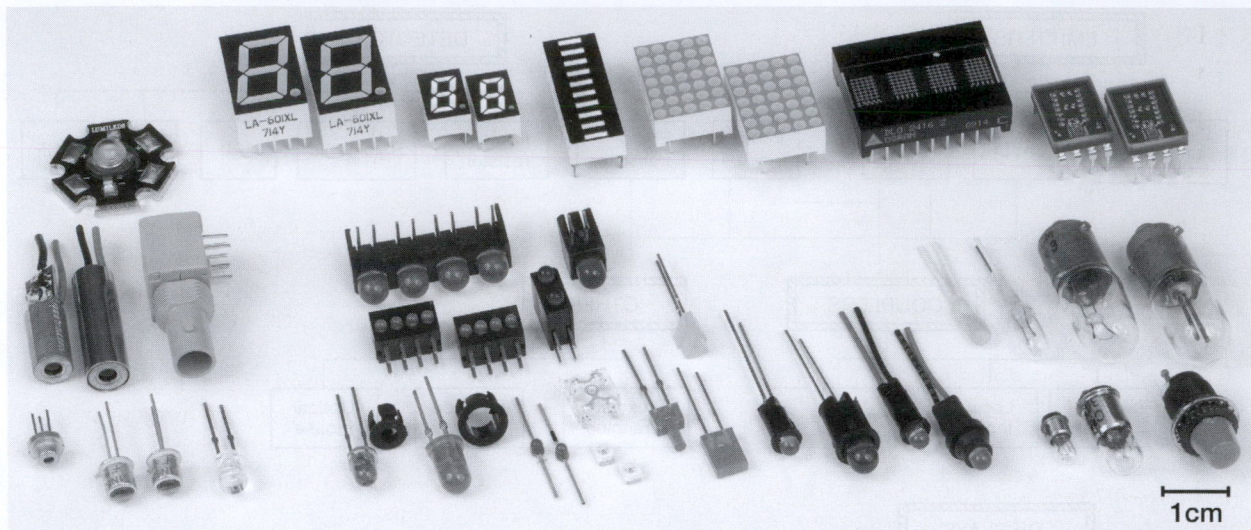

Figure 12.58. Optoelectronics: emitters and displays. At front center are visible LED indicators, in the popular 3 mm (T-1) and 5 mm (T-1$\frac{3}{4}$) panel-mount sizes (some with integral plastic mount), along with PCB-mount singles and arrays. To the left are three red lasers (bare, in 3-pin metal package; and with regulator-driver, in cylindrical housing), a few IR emitters, a fiber-optic ST-style transmitter, and a high-intensity white LED on an aluminum heat spreader. On the right are six legacy incandescent indicators and one NE-51 neon bulb. Along the rear are displays: 7-segment, bar graph, 5×7 dot matrix, 4-character dot matrix, and hexadecimal with latch–decoder–driver.

If the datasheet's "angle of half intensity" (or "viewing angle") specification is at least 90° (it may say "±45°"), and ideally 120° or more, it will look pretty good off-axis. Figure 12.59 shows a polar-plot comparison for LEDs with viewing angles of 30°, 60°, and 120°. These show relative intensity versus viewing angle, normalized to unit intensity on-axis (i.e., 0°). Without that normalization the plots for the 60° and 120° LEDs would shrink dramatically, as shown in the unnormalized plot of Figure 12.60.

An LED looks electrically like a diode, with a forward drop going from about 1.5 V (red) to 3.5 V (blue or white); they use semiconductors with a larger bandgap, and hence a larger forward drop, than silicon – see Figure 2.8 on page 76.[31] Panel-mount LEDs come mostly in 3 mm and 5 mm diameters (called T-1 and T-1$\frac{3}{4}$, respectively), sometimes with an integral mounting clip (see Figure 12.58). Typical flooded panel-type LED indicators look good at 4–10 mA forward current; on a board inside an instrument you can usually get away with 1 mA.

Figure 12.61 shows simple ways to drive small LED

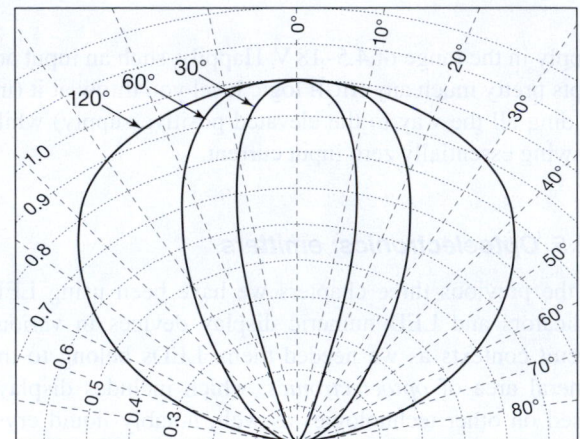

Figure 12.59. Intensity versus angle for three panel-mount LEDs with specified "viewing angles" (full-width to half-intensity) of 30°, 60°, and 120°, in each case "normalized" to unit intensity on-axis. *(After Vishay TLHx460, 520, and 640 series datasheets)*

indicators. Basically you just need to provide a few milliamps of operating current through their forward voltage drop (V_F). That's usually just a series current-limiting resistor of value $R = (V_+ - V_F)/I_{\text{LED}}$, typically a few hundred ohms to a few kΩ. Some LED indicators come with internal current-limiting resistors (or even internal constant-

---

[31] As an example, for the panel LEDs whose illumination patterns are plotted in Figure 12.59, the dependence of forward voltage V_F on wavelength is given approximately by $V_F(\text{volts}) \approx 1000/\lambda + 0.02 I_F$, where the wavelength λ is in nanometers, and the forward current I_F is in milliamps. The last term represents an effective series resistance of 20 Ω.

OPTOELECTRONICS FAMILY TREE

I. EMITTERS

LEDs (Light-Emitting Diodes) visible (red, yellow, green, blue, white) and infrared (IR); forward-biased diode, V_F ranges from ~1–3.5 V, depending on color; panel-mount and PCB-mount; many configurations; available as displays

Laser diodes IR, red, and blue; fiber-optic transmitters, laser pointers, CD/DVD/Blu-ray players, barcode readers

Electroluminescent Night lights, "Indiglo"™ low-power backlight

II. DISPLAYS

LED-based 7-segment (numeric), dot-matrix (character), and "smart" (decoded–latched); organized as individual characters or as arrays ("sticks")

LCD-based (Liquid-Crystal Display) Bare LCD (standard or custom), or "smart" (decoded with memory; parallel and/or serial interface); character only, character plus configurable graphics, or full graphics; backlit transmissive, or "transflective"; varying quality (view angle and contrast)

VFD-based (Vacuum-Fluorescent Display) Smart LCD emulation, with superior readability; custom configurations for high-volume users

OLED (Organic LED) Inexpensive alternative to semiconductor LEDs; graphics, cellphone screens etc; larger sizes for best flat-screen TV

Electronic paper For example, E-Ink® microcapsule technology used in e-book readers; image retention at zero power except during erase–rewrite

III. DETECTORS

Photodiode *pn* (or PIN, positive–intrinsic–negative) diode junction; self-generated photocurrent into short ckt ("photovoltaic" mode), or when back-biased ("photoconductive" mode); fiber-optic receivers (speeds to gigabits/s); solar cells are large-area photodiodes

    Arrays Linear (stripe); quad; proportional readout; full imaging array (CCD; CMOS)

Integrated Light → logic; light → voltage; light → current; light → frequency; synchronous detection pair

Phototransistor Transistor with base–emitter photodiode; higher current (factor of β), but slower; photo-Darlington even more so

Photoresistor Light-sensitive linear-resistive material (e.g., cadmium sulfide); slow response

Bolometric "Pyroelectric" material exhibiting large resistance change with temperature; motion detectors ("PIR", passive infrared)

APD (avalanche photodiode) High back-bias (~100 V) multiplies the collected charge per photon; can be linear, or saturating ("Geiger mode")

PMT (photomultiplier) Vacuum-tube device with a photocathode and array of electron-multiplying (gain~10^6) "dynodes"; operates at ~1 kV

HAPD (hybrid APD) Vacuum-tube device, combination of photocathode and APD target; operates at >5 kV

Pixelated PMT Multi-anode PMT for coarse imaging; 4×4, 8×8 array

Microchannel plate Vacuum-tube device, combination of photocathode and electron-multiplying capillary array; an "imaging PMT"

IV. COUPLERS

LED-input LED→photodiode; LED→phototransistor; LED→photo-Darlington; LED→FET (via PV stack, "PV" = photovoltaic); LED→photoresistor; LED→SCR/triac (via PV stack): a "solid-state relay"; LED→logic output (active pullup, or open-collector)

Logic input Logic input→logic output

Interrupters With gap, or reflective

V. OTHER

Proximity and ranging Emitter plus position- or intensity-sensing detector: faucets, paper towel dispensers, LCD monitor power-down... and your iPhone (it needs to know that your cheek isn't trying to control the touch-screen)

Barcode reader

Optical mouse LED or laser emitter, plus smart detector

---

current circuits) – with these you omit the external resistor. For higher currents (large LEDs) use a transistor switch, as in Figures 12.61C–F.

We've listed some of our favorite panel LED indicators in Table 12.6. You can see them in the front row of Figure 12.58.

With the trend toward lower-voltage logic, along with the higher forward voltage of gallium nitride type LEDs

Table 12.6 Selected Panel-mount LEDs[a]

| Color | LED only (bare wire leads) | | Integral mount, front entry[b] | | | |
| | | | short bare leads | | flying leads[c] | |
	3mm (T-1)	5mm (T-1 3/4)	3mm	5mm	3mm	5mm
RED	TLHR4605	LX5093SRD/D	5111F1	5101H1	5110F1	5100H1
YELLOW	TLHY4405	LX5093LYD	5111F7	5101H7	5110F7	5100H7
GREEN	TLHG4605	TLHG6405	5111F5	5101H5	5110F5	5100H5
BLUE	LX3044USBD	LX5093USBD				

Notes: (a) all are diffused, without internal resistor. manufacturers: 51xx = CML; TLHxxx = Vishay; LXxxx = Lumex, with prefix SSL-. (b) Front-mounting: T-1 size take 5/32" (4.0 mm) hole, T-1 3/4 size take 1/4" (6.4 mm) hole. (c) Insulated, 24 gauge stranded, 15 cm long.

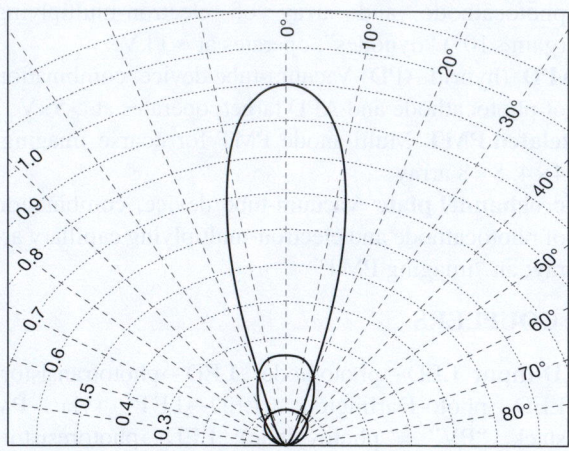

Figure 12.60. There's a price to pay in exchange for the pleasant appearance of a diffused ("flooded") LED, with its wide viewing angle. Here we've plotted intensity versus angle (relative to that of the 30° unit) for the same panel LEDs as in Figure 12.59, using datasheet values. It isn't quite as bad as it seems, though, owing to the eye's logarithmic-like sensitivity.

(used in blue, white, and bright-green LEDs – see Figures 2.8 and *1x.48*), you'll sometimes need a higher supply voltage. For example, +3.3 V is a popular supply voltage, but insufficient to drive a GaN-type LED. If you have a higher supply voltage (e.g., +5 V) available, use a transistor switch with pullup to that voltage (Figures 12.61E and F). If you don't, you'll need to generate the needed LED supply voltage. Figure 12.62 shows a few methods.

A favorite technique, particularly for driving several LEDs at once (e.g., a string of four or six white LEDs for backlighting) from a low supply voltage, is a non-isolated switching boost (step-up) converter, with feedback taken from a current-sensing resistor at the bottom of the

LED string (Figure 12.62A). There are dozens of choices from many manufacturers; we've listed a few in the figure. These things will run on voltages as low as +1 V (i.e., a single cell), putting out tens of milliamps into 20 V or more, as needed to provide the LED operating current $I_{LED} = V_{ref}/R_{CS}$; and they're pretty efficient, typically 80% or more. Some variations include high-side sensing, internal current sensing (Figure 12.62B), linear intensity control, internal Schottky diode, and internal zener clamp.[32] Note well the latter: a zener clamp (D_Z in the figure) is *mandatory* if there's a way that the LED load might be disconnected; otherwise the output voltage runs away and destroys the IC.

You can get tiny individual LEDs, and arrays of LEDs – sticks of 2, 4, or 10 LEDs in a row – designed for PCB mounting. The latter are actually intended for linear "bar-graph" readouts. They come in upright or right-angle mounting. You can also get panel-mounting indicators with multiple colored LEDs in one uncolored package: red/green is inexpensive and common; you can also get red/blue/green and yellow/blue/green.[33] These make an impressive panel, with lights changing color to indicate good or bad conditions. When you turn on more than one LED you get "additive color," so, for example, you can generate *yellow* as red plus green, or *white* as red plus blue plus green. You can do this by switching each

---

[32] In addition to these *boost* (step-up) converters, you can get *buck* (step-down) converters for LED driving from a higher voltage supply; and you can get *buck–boost* converters that work with supply voltages higher or lower than the output voltage driving the LED load. You'll find more than a thousand varieties on the DigiKey website alone.

[33] Multicolor LEDs that include blue cost significantly more, and require the ~3.5 V forward voltage characteristic of GaN-type LEDs.

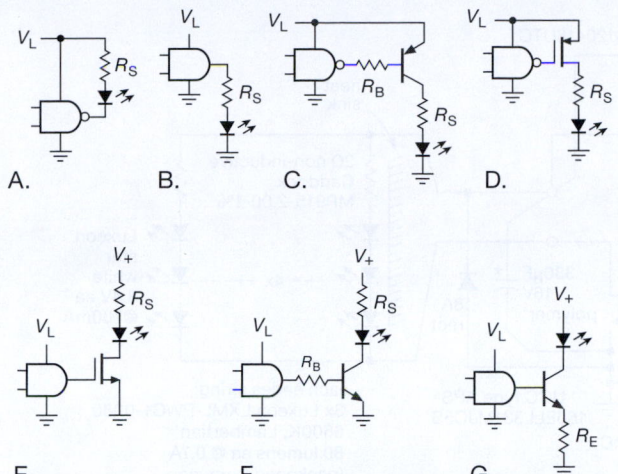

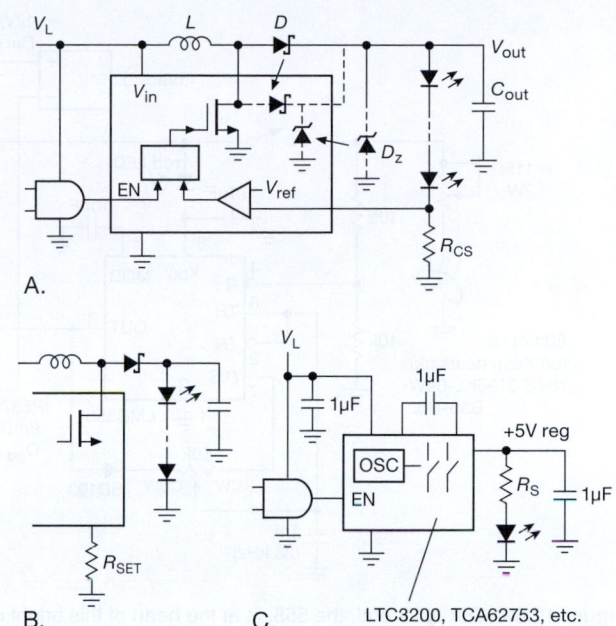

Figure 12.61. Driving LED indicators from logic – I: for LEDs with forward voltage V_F less than the logic supply voltage V_L, you can use the simple connections in circuits A and B to supply a few milliamps. For higher currents, use a transistor switch, as in circuits C–F (with V_+ tied to V_L). If you have a higher voltage available (e.g., +5 V) you can use circuits E–G to drive LEDs whose forward voltage is greater than the logic supply (e.g., a white LED from +2.5 V logic). Circuit G provides constant LED current, providing $V_+ \geq V_L + V_F$.

LED color fully ON or OFF; or you can produce intermediate levels of drive to each LED by varying the duty cycle (percentage of time ON) of a switching waveform (PWM). Multicolor LEDs come in both common-cathode (the − terminal) and common-anode (the + terminal) configurations; two-color LEDs also come as back-to-back diodes in a 2-pin package.

A. Example: "Superlamp"

Thanks to Shuji Nakamura's enabling breakthrough in GaN technology, white LEDs have reached impressive intensities and efficiencies, and they're becoming the preferred alternatives to incandescent and fluorescent lights for "area lighting" applications. We had a leftover Intel CPU heatsink, and we got to thinking, "'Hmmm, what hot thing can we stick onto *that*?" Nice bright white LEDs, of course!

Figure 12.63 shows the circuit: a CMOS 555 oscillator running at 28 kHz (chosen to be above audibility) and configured for a 0–100% duty cycle (see §7.1.3B) is used to switch a power MOSFET, generating a 12 V rectangular wave. This drives four strings of three white LEDs each (Philips Lumileds "Luxeon Star," mounted three-up on hexagonal metal heat spreaders) with $2\,\Omega$ current-limiting

part #	V_{in}^a (V)	I_{out}^b max (mA)	C_{IN}/C_{OUT} min (µF)	V_{ref} (V)	D_z max (V)	D	\$ea (25pc)	notes
FAN5331	2.7–5.5	35	$4.7^c/1.6$	1.22	20	ext	0.67	
TPS61041	1.8–6	30	$4.7^c/1$	1.22	24	ext	1.10	
LT1937	2.5–10	20	1/0.22	0.095	24	ext	2.70	
LT3465	2.7–16	20	1/0.22	0.20	30^d	int	3.00	lin.ctrl
LT3491	2.5–12	20	1/1	0.20^e	27^d	int	2.70	lin.ctrl
LT1932	1–10	30	2.2/1	int	32	ext	3.72	R_{set}

(a) V_{in} must be less than V_{out}. (b) at V_{in} (min). (c) typical. (d) internal. (e) R_S on high side, LED(s) return to GND.

Figure 12.62. Driving LED indicators from logic – II: there are dozens (maybe hundreds) of switching boost converters designed specifically for constant-current driving of strings of LEDs from low supply voltages, as in circuits A and B. For a single LED you can use an inductorless (charge-pump) regulated doubler (circuit C).

resistors, generating an ON current of ~700 mA per string. A couple of design details. (a) The oscillator frequency is high enough to generate a modest amount of "switching loss" during the intervals when the MOSFET is transitioning between fully ON and fully OFF; adding a MOSFET gate buffer (like the TC4420) reduces the switching time, hence the loss, if that were a serious issue (it isn't). (b) The use of a current-limiting series resistor (rather than a current source) in each LED string has an interesting side effect – it cancels (approximately) the reduction of light output as the LED heats up (if anyone cares); that's because the LED's lower forward voltage (at higher temperatures)

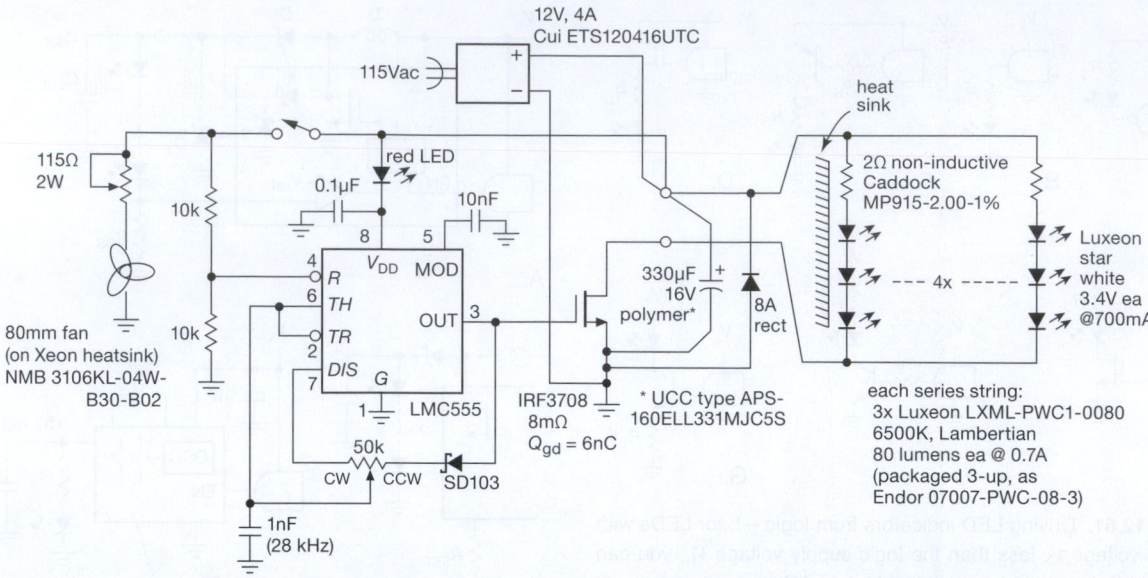

Figure 12.63. Our old friend, the 555, is at the heart of this bright desk lamp with 0–100% PWM dimming.

causes more drop across the resistors, hence a compensating rise in LED current. The compensation is nearly perfect for the values shown.

Visitors are amused by this off-the-beaten-path "desk lamp." One of them pointed out that our creation (Figure 12.64), though charming, is completely outclassed by the brilliant "light engines" from Enfis Limited; they make dense LED arrays (e.g., the "quattro mini") that put out as much as 5000 lumens of white light from 144 power LEDs packed into a 2 cm × 2 cm array (and their other colors put out as much as 24 *watts* of light from the same area[34]). We're loyal to our re-educated heatsink. And the ∼30 W of dissipated power is child's play for this Xeon heatsink, which was intended to handle four times as much in its regular day job; we substituted a much slower fan, and throttled it down to a bare whisper.

B. LED fast high-current driver
You can drive an LED with brief very high current pulses (far above its continuous rating), for example in a flash-lamp replacement application, exploiting the LED's thermal mass to absorb the energy. For this purpose you can

use a series resistor to set the current (Figures 12.62 and 12.63), but an active current regulator is better. An example is LTC's LT3743, showing off its prowess in Figure 12.65, where it is being instructed to make steps to 2 A and 20 A. Even though the PWM is running at 500 kHz, it's able to switch the LED current in just 2 μs. Amazing… check it out.

12.5.2 Laser diodes
We make a brief digression on semiconductor diode lasers. These are the heart of the ubiquitous handheld laser pointers, objects once exotic and expensive, but now commonplace and ridiculously inexpensive ($5 or less for a red pointer). They are also ubiquitous in optical storage devices such as CD, DVD, and Blu-ray players and recorders, as well as in fiber optic transmitters. Low-power red lasers come as little modules with circuitry to run on 3–5 V (e.g., Quarton VLM-650-03-LPA); or, for the adventurous, you can get the bare laser diode in a metal 3-lead transistor package (see Figure 12.58). These include an integral monitor photodiode, which is used as feedback to regulate the light output via the applied drive current. The threshold current at which laser action starts is somewhat variable, generally in the neighborhood of 20–40 mA; above that threshold the light output rises rapidly. This behavior is shown in Figure 12.66, a 'scope trace of emitted intensity versus linearly ramping laser-diode current. The output beam emerges directly from a tiny spot on the

---

[34] At a somewhat more mundane level, high intensity white LEDs are replacing residential incandescent bulbs, floodlights, and spotlights, delivering comparable illumination at 1/5 the ac wallplug power, and lifetimes of 25,000+ hours; many of these are selling for around $10. And the automotive industry has taken notice, using LEDs even for headlights (check out the Lumileds Altilon series).

Figure 12.64. Computer CPU heatsink hijacked to make cool desk lamp. We cranked down the current so you can see the four sets of LED triples, along with the surface-mount ballasting resistors. At full brightness (1000 lumens) you can't look at this thing! The switching driver circuitry of Figure 12.63 lives in the box at bottom.

light-emitting semiconductor (GaAlAs for infrared) chip, diverging with an angle of about 10–20°; it can be collimated with a lens to form a parallel beam or a very small focal spot.

It's easy to rig up a decent drive circuit, without even resorting to op-amps or special driver circuits. Just compare the monitor photodiode current with a threshold, and adjust the drive current accordingly. It's a good idea to include a current limit, so you don't blow out the diode; and you'll need to compensate the feedback loop to make it stable. Figure 12.67 shows a "good-enough" driver circuit: it converts the monitor current to a proportional voltage (switchable between two levels by Q_3), compared with a diode-compensated 1.2 V reference voltage. Q_5 drives the laser, with current limiting via Q_4. C_C and R_C compensate the loop, with final "cut-and-try" values as shown. This simple circuit runs fine from dc up to 1 Mb/s.

For such ON/OFF control you hardly need the whole

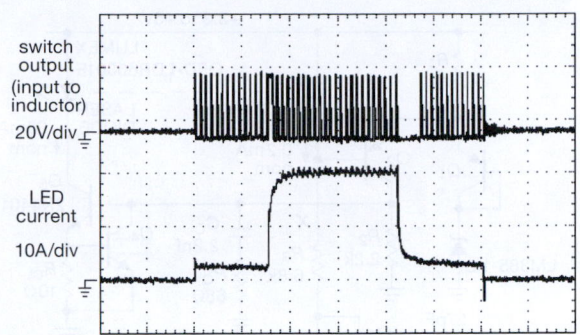

Figure 12.65. The LT3743 high-current LED driver uses cleverness to allow 2 μs switching over a 3000:1 dimming range. Horizontal: 20 μs/div. (adapted from LT3743 datasheet)

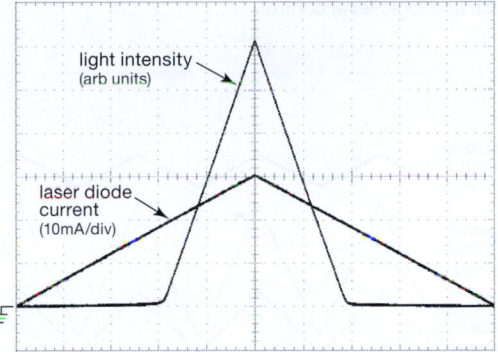

Figure 12.66. Diode laser output from a 0–30 mA triangular drive current. Horizontal: 100 μs/div.

loop business. But it's necessary if you want true *linear* control of laser intensity. Figure 12.68 shows waveforms for this same circuit, with an *npn* current sink attached at point X (replacing R_3, R_4, and Q_3). Here the programming current I_{prog} is a 200 kHz triangle wave with a 4:1 ratio of peak-to-valley current; the laser output is accurately proportional. Note that the laser drive current has a large offset (all three traces are positioned relative to the same ground).

This circuit is a simple example, and not intended for high performance. The real challenges come when you're trying to modulate a laser at hundreds of megabits per second or more. For this it's best to use an IC designed for fast laser modulation;[35] better yet, just buy the laser+driver module as a complete unit (see §12.8 on fiber optics for some examples).

---

[35] For example, the Maxim MAX3735-series, Micrel SY88722, or ADI ADN2870-series. The MAX3975-series and 3930-32-series go to 10 Gb/s; get their *Fiber Design Guide* at www.maxim-ic.com/design_guides/.

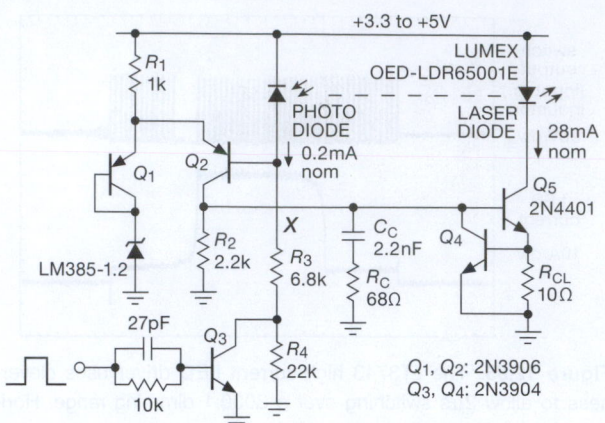

Figure 12.67. Simple laser drive circuit with monitor photodiode feedback and logic-level control.

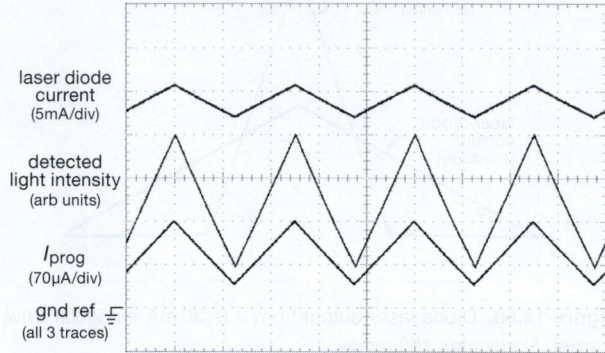

Figure 12.68. Linearizing the circuit of Figure 12.67 by replacing the components below "*X*" with an *npn* current sink: here a 200 kHz triangular programming current with 4:1 peak-to-valley ratio causes the offset laser current needed to produce an accurately proportional output intensity. Horizontal: 2 μs/div.

Figure 12.69 shows a simple alternative, if all you want is intensity modulation of a laser light source at some high frequency and you don't care much about linearity or response down to dc. The input signal, blocked by C_{block}, superposes a high-frequency variation of current at the laser diode, dc biased by the feedback circuit of Figure 12.67. Choose the blocking capacitor for the minimum modulating frequency (the value shown is good down to $\sim$100 kHz), and be sure to bypass the supply rail effectively. The inductor L isolates the transistor's shunt capacitance, and should have a reactance greater than 50 Ω at all modulating frequencies. For good performance the inductor should be constructed from multiple series sections, cleverly designed (in the manner of a "bias tee," Figure 12.70) so that self-resonances do not produce low-

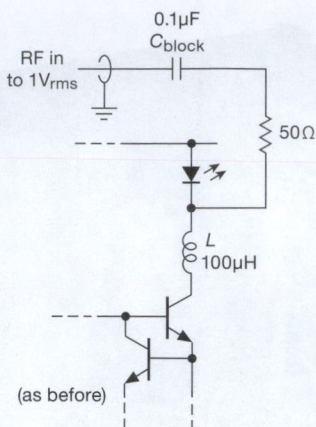

Figure 12.69. Adding a blocking capacitor, 50 Ω series resistor, and RF-isolating inductor to the laser drive circuit of Figure 12.67 provides a simple modulating input. We've used this circuit successfully with modulating frequencies up to $\sim$1 GHz.

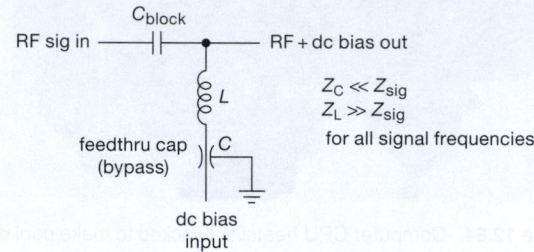

Figure 12.70. A "bias tee" lets you put a dc bias onto a signal (usually at radiofrequencies) that is coupled from input to output. Off-the-shelf bias tees are usually intended for line impedances of 50 Ω or 75 Ω, and they use careful inductor design for good performance over a wide frequency range (0.2 MHz to 12 GHz for the Minicircuits ZX85-12G, for example).

impedance minima across the frequency range of interest.[36]

12.5.3 Displays

A *display* is an optoelectronic device that can show one or more numbers ("numeric" display); or the set of hexadecimal digits (0–9 and a–f, "hexadecimal display"); or any combination of letters, numbers, and punctuation ("alphanumeric display"); or most generally any graphic that can be represented by a dot matrix ("graphic display").

---

[36] Instead of suppressing resonances by dividing such an inductor into multiple sections, you can wind it in a *tapered* geometry; see, for example, the GL series of "Ultra Broad Band Inductors" from AVX. These have excellent performance from 1 MHz to 40 GHz.

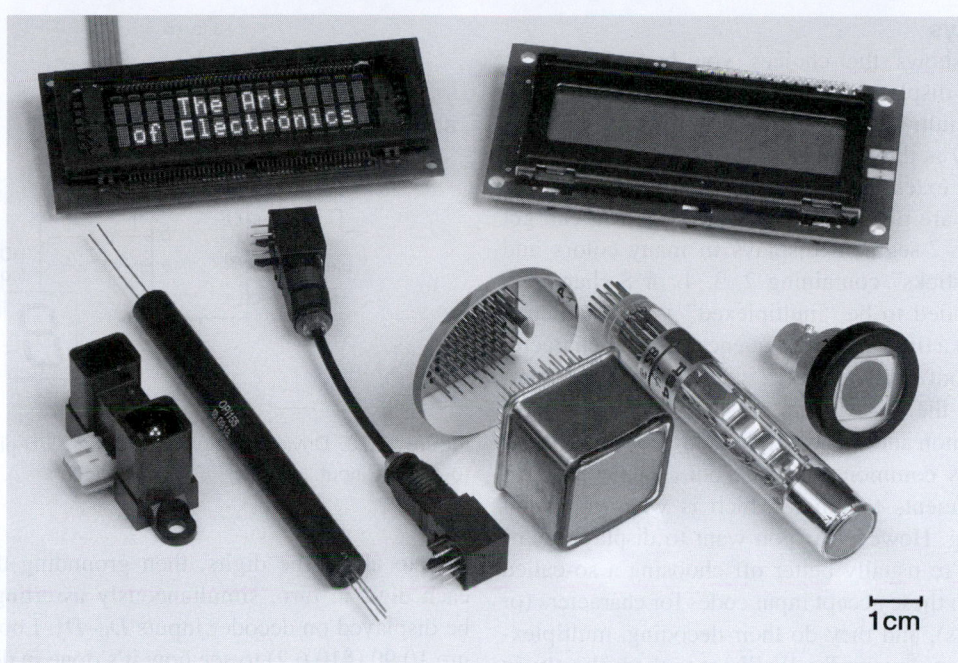

Figure 12.71. A collection of optoelectronic devices too large to fit in Figures 12.58, 12.80, and 12.84. At rear are VFD and LCD smart displays (Newhaven M0216SD-162SDARC-1 and Optrex DMC16207), the former with 3-wire serial interface. In front, left to right, are a ranging/proximity detector (Sharp GP2Y0D02YK0F), a 50 kV logic-output opto-isolator (Optek OPI155), a homemade TOSLINK isolator (Toshiba TOTX/TORX177PL), an 8×8 multianode photomultiplier (Hamamatsu R5900-00-M64), an end-on 13 mm PMT (Hamamatsu R647), and a 10 mm PIN-diode detector (OSI/UDT PIN-10D).

The dominant display technologies (in sizes relevant to electronic instruments) are LEDs, LCDs, VFDs (vacuum fluorescent displays), and OLEDs (organic LEDs). Figures 12.58 and 12.71 includes examples of all except OLED.

LED displays are bright, colorful, and available in large sizes; but they are power hungry, and not appropriate for graphics. LCD displays are quite popular – they are the monochrome yellowish or bluish backlit rectangular displays that commonly show one or two lines of 16 or 20 characters each. Depending on technology, backlit LCDs may be quite readable outdoors or in high ambient-lighting conditions ("transflective"), or they may look just terrible. LCDs intended for use without back-lighting (reflective) look pretty good. The clarity and off-axis readability depend considerably also on the liquid-crystal technology (active matrix versus passive matrix; "twisted nematic" versus "super-twisted nematic," and so on). Apart from the backlight, LCD displays can run on extremely low power (think of a digital wristwatch); and you can have them

made with custom shapes and symbols. VFD displays are similar to their LCD cousins (and emulate their interface), but they are self-luminous; we think they look terrific.[37] VFD display prices are coming down nicely – for example, the 2-row ×16 character alphanumeric "smart" display in Figure 12.71 costs $30 in single quantities; a bare 1-row ×20 character stick costs $7 in single quantities. OLEDs are making inroads, particularly in small sizes.[38] They are likely to be the large-screen display technology of the future (currently dominated by LCD and plasma), when manufacturing costs decline sufficiently to make them competitive.

---

[37] Some engineers have complained that they generate RF interference, and they avoid them in proximity to sensitive circuits.

[38] We are particularly impressed by the stunning clarity of the small (~1″) white OLED graphic modules (available from hobby suppliers like Adafruit, as well as from standard distributors like DigiKey, Mouser, and Newark); they can run from a single +3.3 V supply, drawing some 20 mA, and they communicate via SPI or I²C serial buses. They cost less than $20 in unit quantities.

A. LED displays

Figure 12.72 shows the choices you have in single-character LED displays (and the same layouts are used in integrated multi-character displays). The original 7-segment display is the simplest and can display the digits 0–9 and the hex extension (A–F), albeit somewhat crudely (the hex letters are displayed as "AbcdEF"). You can get single-character 7-segment displays in many colors and sizes, and in "sticks" containing 2, 3, 4, or 8 characters (generally intended to be "multiplexed" – the characters displayed one at a time in rapid sequence). Single-character displays bring out leads for the 7 segments and the common electrode; the two flavors are thus "common cathode" and "common anode." Multiple-character sticks bring each character's common electrode out, but tie the corresponding segments together, which is what you want for multiplexing. However, if you want to display lots of characters, you're usually better off choosing a so-called "smart" display: these accept input codes for characters (or graphic symbols), and they do their decoding, multiplexing, and displaying internally. We'll get to them shortly, in §12.5.3A.

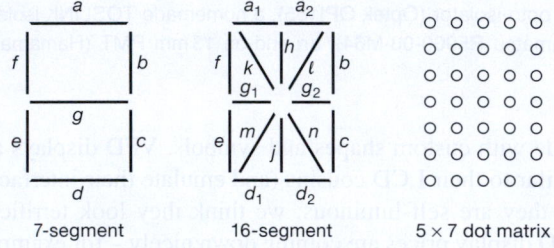

Figure 12.72. LED display layouts.

"Dumb" displays

Figure 12.73 shows how to drive a single-digit 7-segment common-cathode LED display. The 'HC4511 is a "BCD-to-7-segment latch–decoder–driver," able to source about 10 mA while holding its active outputs at +4.5 volts when powered from +5 V. The series resistors limit the segment drive current (to 2 mA or 4.7 mA, respectively, with 3 V or 5 V supply, assuming a typical LED drop of 1.5 V). Don't try to be clever and just stick a single resistor in the common cathode instead (why not?). You can get arrays of equal-value resistors in convenient SIPs (single in-line packages), in through-hole or surface-mount flavors.

You need only a single decoder–driver chip, even if you are displaying multiple digits, as long as you multiplex the display, i.e., illuminate only one displayed digit at a time. You do this by connecting the segment-driver out-

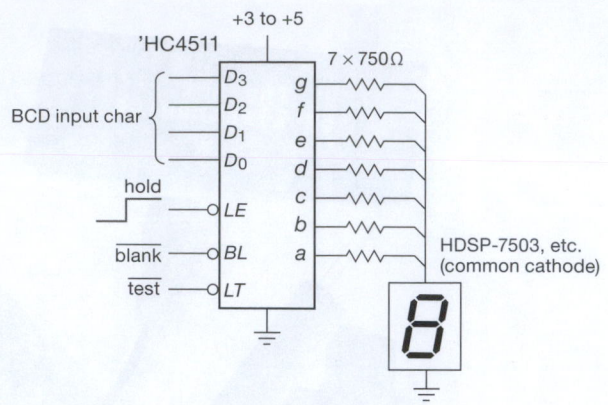

Figure 12.73. Driving a single 7-segment LED display from BCD (0–9) logic input.

puts to all of the digits, then grounding the cathode of each digit in turn, simultaneously asserting the value to be displayed on decoder inputs D_0–D_3. Look back at Figure 10.90 (§10.6.2) to see how it's done in detail.

Figures 12.74 and 12.75 show how to drive single-character LED displays in 16-segment or 5×7 configurations, respectively. Drivers for these displays generally assume you've got a microcontroller in your circuit somewhere, so they use serial input protocols such as SPI (serial peripheral interface; see §14.7.1). They include internal data registers, and current-source drivers programmed by a single external resistor. Note that the 16-segment driver in Figure 12.74 has no internal decoder – it just drives the segments you tell it to. So your microcontroller has to figure out which segments to drive; that's no big deal for these brainy parts, just be aware. The (smarter) 5×7 driver in

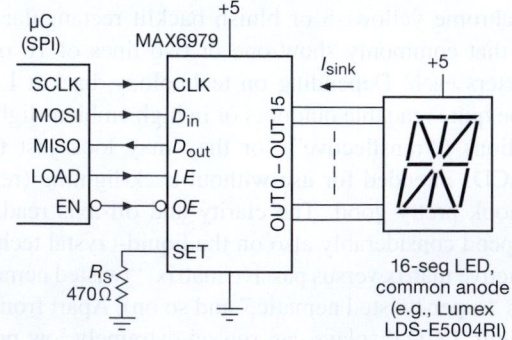

Figure 12.74. Driving a single 16-segment LED display from a serial (SPI) data input. There's no pre-stored character table in this particular driver chip; instead, you get to make up your own symbols, sending a HIGH or LOW bit for each segment.

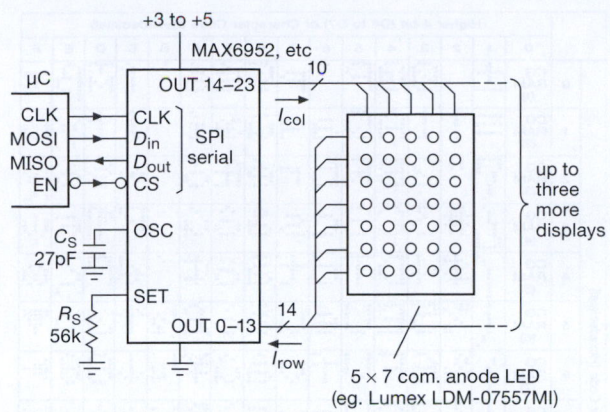

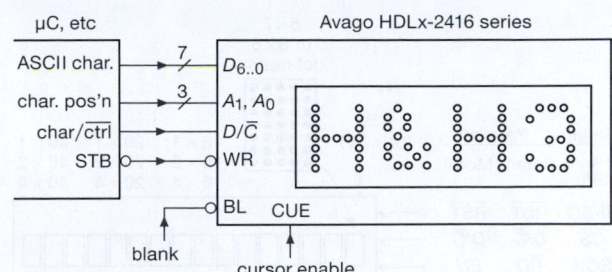

Figure 12.77. For half the price of Figure 12.76 you get a four-character smart LED display "stick." And you can display 128 ASCII characters with its built-in font ROM.

Figure 12.75. Driving up to four dot-matrix 5×7 LED displays from a serial data input (which can also control brightness and blinking). This driver has 104 predefined characters stored internally, and lets you create 24 more of your own.

Figure 12.75 *does* include an internal decoder with predefined character memory.

"Smart" displays

Except in high-volume applications that are cost-sensitive, it's usually a better idea to choose a display that integrates an array of characters (or graphics), with the decoders and drivers included. These come with LEDs, LCDs, VFDs, or OLEDs as the active display.

Single character LED The original smart display device is probably the single-character 5×7-dot LED, which delighted us back in the 1970s: you present it with a 4-bit code, and you can *see* (and latch) the result (Figure 12.76). These were made by HP and TI, and we liked them so much that we incorporated them in our electronic circuits lab course, confident that their ~$15 single-piece price would drop nicely over time. It didn't work out that way, and these puppies (still made by Avago Technologies, the

optoelectronics spinoff of HP) now cost double that price in single quantities.

Several character LED Usually you want to display a few characters, at least: so there are handsome four-character smart display sticks (in the usual LED colors) that are as easy to use as the single-character display. Figure 12.77 shows one of these, originated by HP (now Avago) and available for about $15. These display a full character set (7-bit ASCII: upper-case and lower-case alphabets, numbers, and symbols), latched with a WR′ pulse and sent to the character position given by a 2-bit address (a simplified form of "parallel bus interface" that we'll learn about in Chapters 14 and 15). There are additional control functions (dimming, cursor), written by holding LOW the input D/C′ (which the datasheet oddly calls CU′, for cursor "select").

Many character LCD/VFD/OLED As we remarked earlier, the best way to display one or several lines of text is to use an intelligent (smarter than smart) LCD, VFD, or OLED "character" module (Figure 12.78). There's lots of competition,[39] with prices beginning at $10 or less (for a 16×2-character backlit LCD). They come in standard configurations of 1, 2, or 4 rows, with 16, 20, or 40 characters per row. They include built-in character sets, and many of them let you supplement that with a few (usually eight) custom characters. Alternatively, you can choose a full graphic version, consisting of a dot matrix (e.g., 64×260), which you program as a bitmap. That can be a chore if you want to display mostly text, so there are combination text-plus-graphics units that include a pre-stored character set in

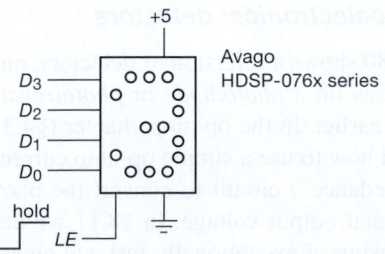

Figure 12.76. LED hexadecimal display with integral latch-decoder-driver. These are beautiful, and colorful, but expensive.

[39] Here's an alphabetical list from our favorite suppliers (DigiKey, Mouser, Newark): 4D Systems, Adafruit, AND Displays, Batron, Displaytech, Electronic Assembly, Everbouquet, Hantronix, Lumex, Matrix Orbital, Microtips, Newhaven Display, Optrex, Powertip, Trident, Varitronix, and Vishay/Dale.

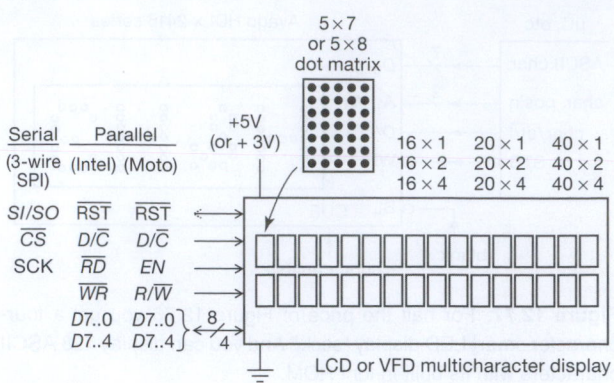

Figure 12.78. Better still, ride the popular wave with a smart LCD or VFD display stick. The character arrangements shown are standard. And some (as here) have both parallel and serial interfaces.

addition to full graphics capability. Examples of these, chosen from one company's offerings (Matrix Orbital, see Figure 12.79), are the LK162-12 (pre-stored text, plus eight user-defined symbols), the GLK24064-25 (full graphics, no built-in fonts), and GLK12232-25-SM (choice of pre-stored text or full set of user-defined symbols, plus full graphics).

To fully appreciate what these modules offer, it's important to understand that LCDs must be driven by an *ac* waveform; otherwise their liquid guts are ruined. So LCD driver chips usually have some way to generate a square-wave segment drive, synchronized to the LCD backplane waveform. An example is the 'HC4543, the LCD cousin of the 'HC4511 LED 7-segment latch–decoder–driver. Another complication, of course, is the need to continually drive the large array of dots with the pattern to be displayed. Add to that the need for an interface that lets you change individual displayed characters on command, or to scroll the display, or advance a cursor, and so on, and you can understand how complicated things can get. Fortunately, the manufacturers understand this, and therefore they provide complete displays that are more than intelligent; they're positively at the genius level.

These displays are invariably used in conjunction with a microcontroller, which communicates via a simple interface. The original LCD smart display sticks use a simple parallel interface, with 8 bits of data (either the character or control information, depending on a data/control′ line: D/C′ in the figure, but traditionally called "RS") and a few control lines.[40] Most now offer a 3-wire or 4-wire serial bus alternative, or provide both options (jumper selectable)

---

[40] They allow a "nybble" mode, in which 8-bit data is sent as two successive 4-bit quantities.

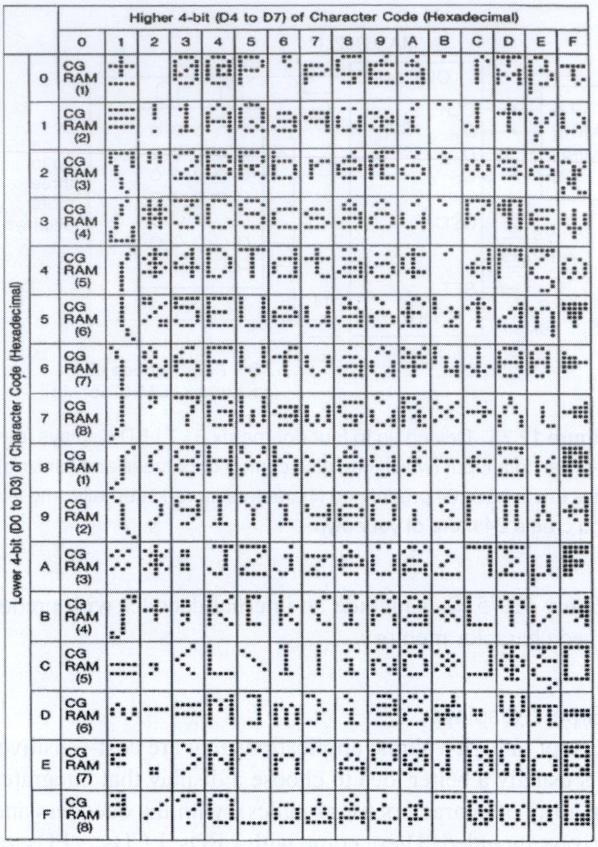

Figure 12.79. Display codes for an LCD alphanumeric "smart display" (Matrix Orbital LK-series) with a standard character set plus eight user-configurable characters (CG RAM). Reproduced with permission of Matrix Orbital.

on the same display, as in Figure 12.78 (which even offers a choice of "Intel" or "Motorola" parallel protocols, §14.3).

We like these displays a lot; you can see both LCD and VFD types in the photograph (Figure 12.71), the latter expressing fluorescent joy in the completion of this weighty tome.

12.6 Optoelectronics: detectors

Figure 12.80 shows a selection of detectors, most of which are variations on a *photodiode* or *phototransistor*. We've seen these earlier, in the op-amp chapter (§4.3.1C), where we showed how to use a simple op-amp current-to-voltage ("transimpedance") circuit to convert the photocurrent to a proportional output voltage. In §8.11 we dealt in detail with the design of exceptionally fast and quiet photodiode amplifiers. And in *§4x.3* there's a discussion of the thorny problem of stability in transimpedance amplifiers, taking

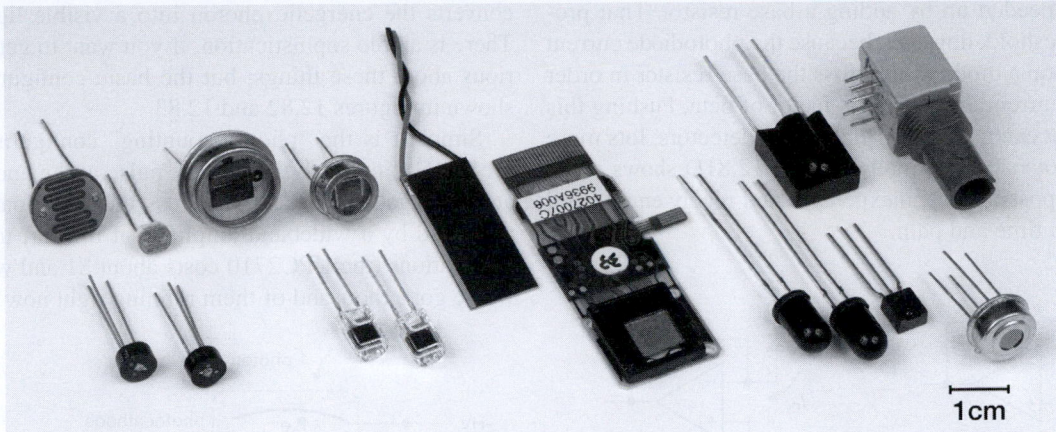

Figure 12.80. Optoelectronics: detectors. At left rear are a pair of cadmium sulfide photoresistive sensors, and a pair of photodiodes (GaAs, silicon) in hermetically sealed metal cases. Below them are plastic-case photodiodes (top- and side-looking), and to their right is a bare photodiode and a small CMOS image sensor (with attached flexible circuit connector) of the sort used in webcams and cellphones. The four opaque black objects toward the right are IR-detectors (filtered to eliminate interfering visible light); the small square one generates a logic-level output, and the large square one is used as a frequency selective (30–56 kHz) "clicker" receiver in consumer audio/video gear. At top right is an ST-style fiber-optic receiver, and below it is a pyroelectric thermal-IR sensor used in PIR motion detectors (featured in the geek movie classic "Sneakers").

into consideration the destabilizing effects of input capacitance.

Soon we'll see the ubiquitous use of photodetectors in the context of optocouplers (also called opto-isolators); and a bit later we'll see them again in connection with fiber optics. Here we simply summarize the earlier photodiode and phototransistor circuits (Figure 12.81), along with some nice parts you can get that integrate a photodiode with additional circuitry to produce either a digital-logic output, or a proportional output in the form of a voltage, a current, or a frequency.

This is a good place to say something also about *photomultipliers*: these sensitive and fast light detectors (two are shown in Figure 12.71) use a cascade of electron-multiplying electrodes (dynodes) to convert a single photoelectron (released, with about 20% probability, when a photon of light strikes the sensitive photocathode) into a fast (~ns) pulse of 10^5–10^6 electrons. The resultant current pulse has thus been amplified sufficiently before reaching any circuitry, so that amplifier noise is not a problem: a million electrons in a nanosecond is nearly 0.2 mA!

12.6.1 Photodiodes and phototransistors

Figure 12.81 reviews the standard ways that photodiode and phototransistors are used. In circuit A the photodiode operates in *photovoltaic mode*; that is, generating a photocurrent into a short circuit. The transimpedance amplifier generates a positive output voltage ($V_{out} = R_f I_P$), so you can operate from a single positive supply. You might think the op-amp is unnecessary – after all, Ohm's law gives you the same answer – but you'd be wrong! First, the diode's self-generated photocurrent drops to zero if allowed to develop a diode drop of forward voltage, so you'd have to use a small resistive load, such that the maximum output voltage (with maximum light input) is less than ~0.5 V; and second, the circuit would be slower, set by the time constant of the diode's capacitance and load resistance.

In circuit B the photodiode is back-biased (*photoconductive* mode), which increases the speed by reducing the diode's capacitance and by providing an electric field to sweep away charge. So this circuit is faster; but it's also noisier, and the diode's leakage current limits the low light-level performance. Capacitance can be a real problem in these circuits, especially when the photodiode sits at the far end of a connecting coax cable (§§8.11 and *4x.3*).

Circuit C uses a phototransistor, effectively boosting the photodiode current by the transistor's beta (the back-biased collector–base junction acts as a photodiode, whose photocurrent is multiplied by beta). So a small phototransistor under ordinary room illumination typically gives you currents in the ~100 μA range, compared with ~1 μA for photodiodes.[41] This configuration is the slowest of all, but

---

[41] Experimental data: we put an FPT110 phototransistor (lower left in

it can be speeded up by adding a base resistor. That produces a threshold, however, because the photodiode current must develop a diode drop across the base resistor in order to get the current amplification factor of beta. Pushing this further, you can get photo-*Darlington* detectors: lots more gain, but lots slower. Finally, Figure 12.81D shows some integrated possibilities, inexpensive and nicely engineered to save you time and pain.

converts the energetic photon into a visible light pulse). There is ample sophistication, if you want to get really serious about these things; but the basic configurations are shown in Figures 12.82 and 12.83.

Simplest is the "photon-counting" configuration (Figure 12.82), in which the current pulse at the anode generates a (negative) fast pulse across the $50\,\Omega$ load resistor, amplified by a wideband amplifier of the sort used in RF applications (the UPC2710 costs about $1 and works fine: we've got a thousand of them running right now).

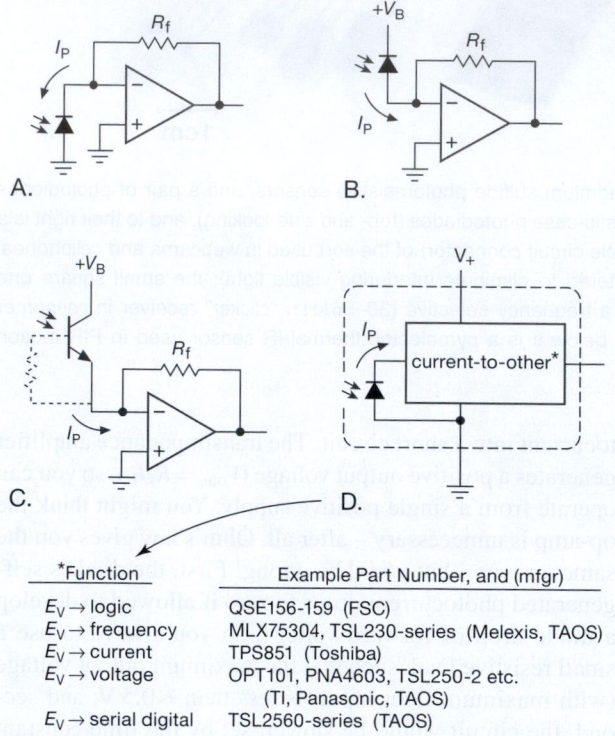

*Function	Example Part Number, and (mfgr)
$E_V \rightarrow$ logic	QSE156-159 (FSC)
$E_V \rightarrow$ frequency	MLX75304, TSL230-series (Melexis, TAOS)
$E_V \rightarrow$ current	TPS851 (Toshiba)
$E_V \rightarrow$ voltage	OPT101, PNA4603, TSL250-2 etc.
	(TI, Panasonic, TAOS)
$E_V \rightarrow$ serial digital	TSL2560-series (TAOS)

Figure 12.81. Photodiode and phototransistor circuits. A. Photovoltaic mode. B. Photoconductive mode. C. Phototransistor. D. Integrated light-to-other modules.

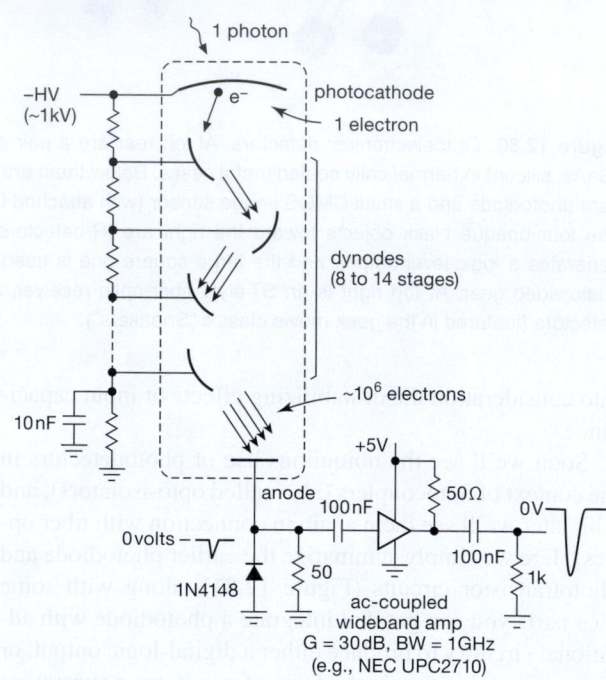

Figure 12.82. Photomultiplier "pulse counting" circuit. The cascade of electron-multiplying dynodes produces a ~ns pulse of ~10^6 electrons for each detected photon of light, which develops a ~mV negative pulse across $50\,\Omega$.

12.6.2 Photomultipliers

Photomultipliers are favorite detectors for photon-counting applications, low light-level detection and integration, and X-ray or gamma ray detection (via a scintillator, which

Figure 12.80) on our lab bench and measured its photodiode current as I_{CB}=0.4μA, compared with its phototransistor current (with 10 V bias) of I_{CE}=60μA. In hazy sunlight (an August day in Boston) the photodiode current was 250 times higher (100μA). Although the datasheet for this device doesn't tell you the active area, we measured it as 0.7 mm². That works out to an efficiency of about 5%, if you were to use the photodiode as a solar cell.

If you want instead to detect accurately very low light levels, and don't need nanosecond speed, you probably want a bandwidth-limited "integrating" amplifier like the one shown in Figure 12.83.[42] Here the excellent (and nearly unique) OPA656 wideband low-I_B (JFET) op-amp converts the anode current to a voltage waveform. Even though we're aiming for only a few megahertz of bandwidth, we need a fast op-amp, as explained in §§8.11 and *4x.3*. For reasons of bandwidth and noise, it's especially

---

[42] This is similar to the amplifier designed for our colleague Lene Hau, and was used in her pioneering "slow-light" experiment in 1999: 17 m/s (38 mph), slower than a speeding bicycle!

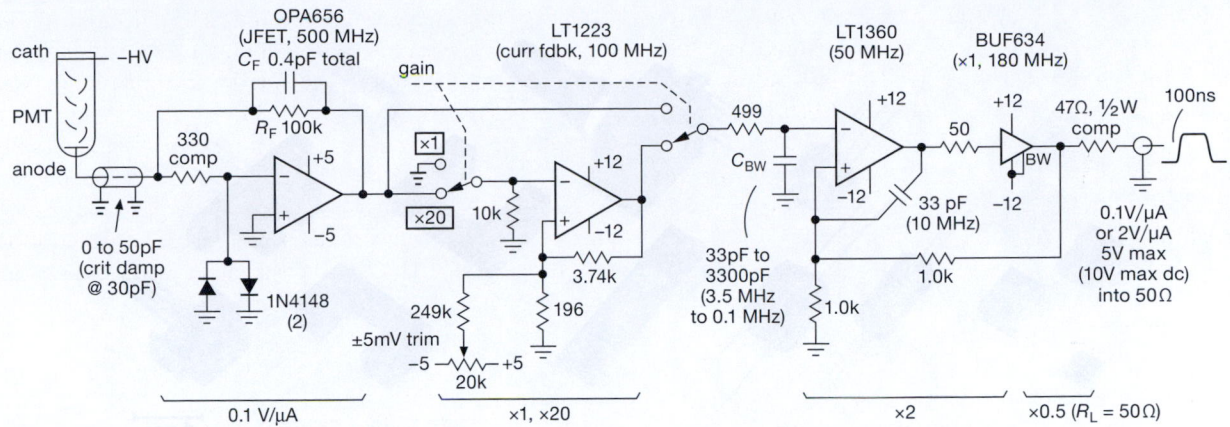

Figure 12.83. Low noise photomultiplier amplifier, with 3.5 MHz bandwidth and 2V/µA gain. See also the discussion in §8.11.

important to keep the input capacitance to a minimum, and to choose the stabilizing feedback capacitance C_F carefully.

The rest of the circuit is pretty straightforward: the series input resistor and clamp diodes protect the op-amp from input spikes, caused, for example, by PMT breakdown (it uses a kilovolt supply). The second stage is an optional ×20 gain boost, for which we used a current-mode op-amp. These have the interesting property that their closed-loop bandwidth is largely independent of closed-loop gain: the ratio of resistors sets the gain (as with the familiar "voltage-mode" op-amp), but the bandwidth is determined by the value of feedback resistor alone. For example, this particular op-amp has 100 MHz bandwidth with a 1k feedback resistor and $G_{CL}=1$, dropping modestly to 60 MHz for $G_{CL}=30$; by contrast, a voltage-feedback op-amp's gain would decrease inversely with closed-loop bandwidth. Here we chose a larger feedback resistor (3.74k), deliberately limiting the bandwidth to about 15 MHz, because that's all we need (and extra bandwidth just invites trouble).

This op-amp specifies ±3 mV (max) offset voltage, which, when combined with the ±2 mV (max) offset of the input stage, could produce as much as 100 mV of output offset; so we added an offset trim. There's a simple RC bandwidth-limiting lowpass on the way to the third stage; this could be elaborated into a switched set of 3 dB breakpoints, if desired. Note however that the maximum bandwidth of ~3.5 MHz is limited by the input-stage rolloff: $BW=(1/2\pi)R_F C_F$. Finally, the output stage consists of a wideband unity-gain power buffer (good to ±250 mA) enclosed within a gain-of-2 noninverting feedback stage. You need the op-amp, because the buffer alone does not have internal feedback to discipline its output: its offset is spec-

ified as ±100 mV (max). But you have to admire the brute – it's fast, and it's powerful.

12.7 Optocouplers and relays

An LED emitter, combined with a photodetector in close proximity, forms a very useful object known variously as an *optocoupler*, *opto-isolator*, or *photocoupler* (Figure 12.84). In a nutshell, optocouplers let you send digital (and sometimes analog) signals between circuits with separate grounds. This "galvanic isolation" is a good way to prevent ground loops in equipment that drives a remote load. It is essential in circuits that interact with the ac power mains. For example, you might want to turn a heater on and off from a digital signal provided by a microprocessor; in this case you would probably use a solid-state relay, which consists of an LED coupled to a high-current triac or SCR. Most ac-operated switching supplies (e.g., those used in computers, telecommunications, and instrumentation) use optocouplers for the isolated feedback path (see for example Figure 9.83 in §9.8). Similarly, designers of high-voltage power supplies sometimes use optocouplers to get a signal up to a circuit floating at high voltage.

Even in less exotic situations you can take advantage of opto-isolators. For example, an opto-FET lets you switch analog signals with essentially no charge injection (apart from effects of the fractional-pF isolation capacitance); the same goes for sample-and-hold circuits and integrators. Opto-isolators can keep you out of trouble when driving industrial current loops, hammer drivers, etc. Finally, the galvanic isolation of opto-isolators comes in handy in high-precision or low-level circuits: for example, it is difficult to take full advantage of a 16-bit analog-to-digital converter,

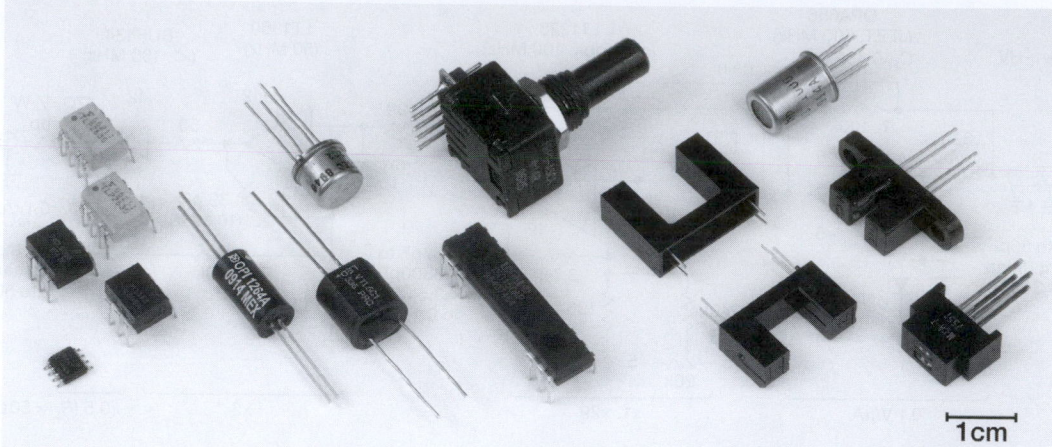

Figure 12.84. Optoelectronics: couplers and interrupters. The five ICs at left are optocouplers, with LED–photodiode pairs and (in some cases) logic inputs and outputs. They are good to several kilovolts or so of isolation, whereas the adjacent cylindrical object with whiskers is rated at 10 kV (and the stretch version in Figure 12.71 is good to 50 kV). The oval package (and metal can above it) are LED–photoresistor couplers, which use a CdS resistive sensor like the ones at top left in Figure 12.80. The stretched IC at center (ISO150) is also a digital isolator, but it uses a *capacitive* coupler; it doesn't belong at this party, but it's too elegant to omit! The three gapped objects to its right are opto-interrupters, and the boxy thing without a gap is a reflective sensor (as is the metal-can barcode reader at top). The panel control at top center is an optical incremental encoder, generating 120 quadrature cycles per turn.

because the digital output signals (and noise on the digital ground to which you connect the converter's output) get back into the analog front end. You can extricate yourself from "noise city" with optical isolation of the digital half. Optocouplers typically provide 2500 volts (rms) isolation, $10^{12}\,\Omega$ insulation resistance, and less than a picofarad coupling between input and output.

There are many varieties of optocouplers, the choice depending on the intended application – for example, the coupling of analog signals, or digital-logic signals, or ac power switching. In the sections below we've sorted these into seven categories, illustrated with examples of some of the more popular (or more interesting) parts currently on the market. The choice of categories is somewhat arbitrary, but it makes sense to us. They are (in order of appearance): I. Phototransistor output; II. Logic output; III. Gate driver output; IV. Analog-oriented; V. Solid-state relay with transistor output; VI. Solid-state relay with triac/SCR output; and VII. ac-input optocouplers.

12.7.1 I: Phototransistor output optocouplers

Figure 12.85 shows an assortment of optocouplers with bipolar transistor output. These are primarily intended for digital logic-level coupling (though it is possible to exploit the configuration in Figure 12.85C to make an approximately linear coupler; see Figures 12.88 and 12.89). The earliest (and simplest) is typified by the 4N35, an LED–

phototransistor pair with 40% (min) current transfer ratio (CTR) as a phototransistor, and sluggish $5\mu s$ turn-off time (t_{OFF}) into a $100\,\Omega$ load. Circuit A shows how to use it: a gate output and pullup resistor generate current-limited 8 mA drive, and a relatively large output-side collector resistor guarantees saturated switching between logic levels. Note the use of a Schmitt-trigger inverter, a good idea here because of the long switching times. You can get LED-phototransistor pairs with CTRs of 100% or more (e.g., the jellybean CNY17-4, with CTR=160% min), and you can get LED–photo-Darlingtons, as shown; they're even slower than phototransistors! To get improved speed, the manufacturers sometimes use separate photodiode and transistor, as shown in the 6N136 and 6N139 optotransistor and opto-Darlingtons. With optocouplers that provide access to the base, you can add a resistor from base to emitter to improve the speed (configurations B and F); however, this produces a threshold effect (as shown alongside circuit F), because the phototransistor doesn't begin to conduct until the photodiode current is large enough to produce a V_{BE} across the external base resistor. In digital applications the threshold can be useful, but in analog applications it is an undesirable nonlinearity.

12.7.2 II: Logic-output optocouplers

These preceding optocouplers are nice, but somewhat annoying to use because you have to supply discrete

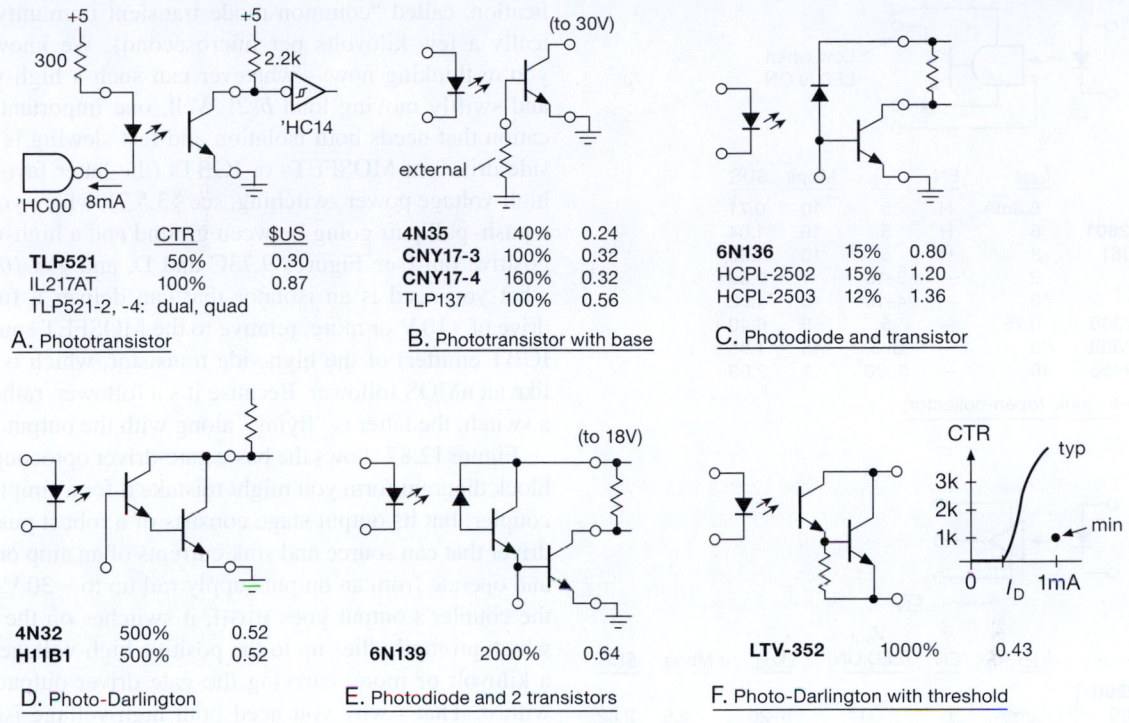

Figure 12.85. Optocouplers-I: Phototransistor output. Parts in **bold** are "jellybeans" – inexpensive and ubiquitous, though not necessarily the best performers.

components at both input and output. Furthermore, the current needed to drive the LED may exceed the drive capability of some logic families; and the passive pullup at the output side suffers from slow switching and mediocre noise immunity. To remedy these deficiencies the silicon wizards bring us "logic" optocouplers (Figure 12.86). The 6N137 and friends (configurations A and B) go halfway, with bare LED at the input, but buffered logic at the output. You still need plenty of input current (for the 6N137 it's specified as 6.3 mA, min, to guarantee output switching), but you get clean logic swings at the output (albeit open-collector), and speeds to 10 Mb/s. Note that you must supply +5 volts to the internal receive-side output-logic circuitry. Logic couplers with reduced drive are listed; among these are the classic (and curiously named) H11L1 and H11N1, descendants of the pioneering optoelectronic offerings from General Electric, happily enjoying great popularity well past their 30th anniversary. But note the cost penalty, particularly for high speed combined with low drive current.

As long as you have internal logic circuitry at the output, why not provide honest active pullup? Why not indeed.

That's configuration B, among which you'll find improved versions, some faster, some with impressive isolation slewing rates, and some with three-state outputs; but note the "improved" pricing!

The optocouplers in configuration C bring us to the promised land (where, however, the real estate values are somewhat elevated): these accept logic-level inputs, and produce logic-level outputs with active pullup. Because of the internal logic circuitry at both input and output, both sides of the chip require logic supply voltages. Some varieties (e.g., the ACPL-772L) will happily run on either 3 V or 5 V supplies at both ends, in any combination. These couplers are pretty fast, up to 50 Mb/s.

We've listed also three isolating logic couplers that perform similarly, but that use capacitive or transformer-based isolation techniques in place of light. They are faster, but note one complication: their isolating methods are all *ac-coupled*, using short pulses to transfer changes of state across the gap. That is, they are not intrinsically "dc correct," and they may exhibit artifacts such as timing-delay skew, or require an initializing signal to force the output to a known state.

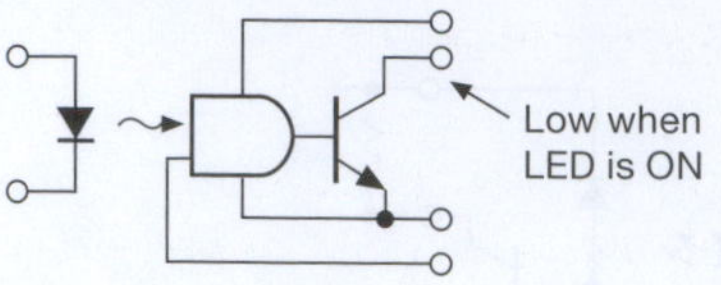

Low when
LED is ON

	I_{LED}	EN	V_{CC}	Mbps	$US
6N137	6.3mA	H	5	10	0.71
HCPL-2601	6	H	5	10	1.04
HCPL-061	3	H	5	10	1.90
H11L1	2	–	3–15	1	0.58
H11N1	3	–	4–16	5	4.02
HCPL-2300	0.75	–	5	8	6.40
ACPL-W60L	5	–	3–5	15	1.85
ACPL-P456	10	–	5–20	1	2.09

A. LED-to-logic (open-collector)

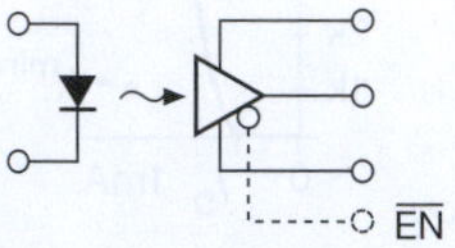

$\overline{EN}$

	I_{LED}	EN	V_{out} (LED ON)	V_{CC}	Mbps	$US
HCPL2200						
FOD2200	2.2mA	L	H	5–20	2.5	2.62
TLP2200						
HCPL-2201	2	–	H	5–20	5	2.00
HCPL-2400	4	L	L	5	40	3.92
ACPL-4800	6	–	H	5–20	3	1.85
(30kV/μs!)						

B. LED-to-logic (active pullup)

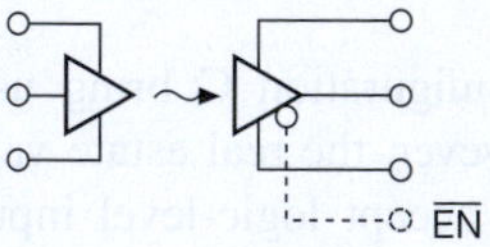

$\overline{EN}$

	EN	V_{CC}	Mbps	$US	coupling
HCPL-0721				4.07	
FOD0721	–	5	25	2.67	opto
PC412S	–	5	25	2.31	opto
ACPL-772L	–	3–5	25	2.68	opto
HCPL-7723	–	5	50	7.00	opto
HCPL-0900	L	3–5	100	5.30	xfmr
ISO721,722	–, L	3–5	150	2.80	cap
IL710	L	3–5	150	3.30	GMR

C. logic-to-logic (active pullup)

Figure 12.86. Optocouplers II: Logic output.

12.7.3 III: Gate driver optocouplers

Isolated optical coupling lets you float the output, relative to the input, at voltages up to the isolation rating of a few kilovolts (limited also by a maximum slewing speci-

fication, called "common-mode transient immunity," typically a few kilovolts per microsecond). We know what you're thinking now – whatever can such a high-voltage and swiftly moving load *be*?! Well, one important application that needs both isolation and fast slewing is "high-side drive" of MOSFETs or IGBTs (the latter favored for high-voltage power switching, see §3.5.7), where you have a push–pull pair going between ground and a high-voltage positive rail (see Figures 9.73C and D, and §9x.10). And what you need is an isolator that can deliver a full gate drive of +10 V or more, relative to the MOSFET source (or IGBT emitter) of the high-side transistor, which is acting like an nMOS follower. Because it's a follower, rather than a switch, the latter is "flying" along with the output.

Figure 12.87 shows the basic gate-driver optocoupler. In block diagram form you might mistake it for a simple logic coupler; but its output stage consists of a robust push–pull driver that can source and sink currents of an amp or more, and operate from an output supply rail up to ~30 V. When the coupler's output goes HIGH, it switches on the IGBT, which promptly flies up to the positive high-voltage rail at a kilovolt or more, carrying the gate driver output along with it. That's why you need both high-voltage isolation and common-mode transient tolerance.

Of course, the ~20 V supply for the isolator's output side has to fly also! Sounds like a serious problem – but there's a terribly clever solution, which exploits the fact that the output is switching between ground and +HV: if you hook a high-voltage diode to the coupler's V_{CC} from a plain ordinary +20 V supply (relative to circuit ground), as shown in Figure 12.87A, it will conduct during times when the output is LOW, charging up the bypass capacitor. Just make the latter large enough to keep the coupler's output powered during the periodic HIGH states, and you're done; that's easy enough, because the coupler output-stage quiescent current is only a few milliamps. This is basically a high-side "charge pump," sometimes called a "bootstrap supply." You can use one of these couplers to drive the low side also; in that case just power it directly from the +20 V, omitting the diode.

High-voltage switching is fraught with peril, and is not a recommended activity for the faint of heart: a momentary output short can annihilate things in a hurry. You need to protect against such "faults," for example with current-limiting circuitry. But even then you can destroy the IGBTs or MOSFETS, because a load short puts the full supply voltage across the high-side IGBT while it is running at that current limit. What you need to do is sense when the IGBT's gate is being driven, but its output has not gone promptly into voltage saturation. Happily, there

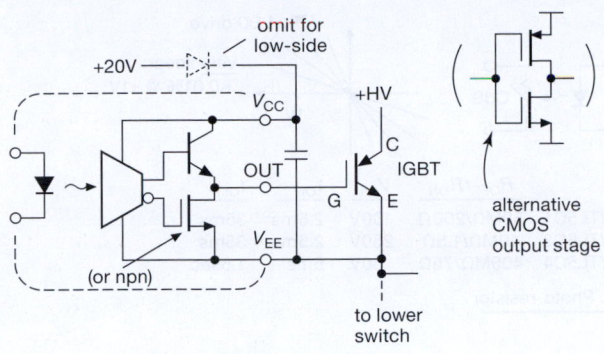

	I_O	max V_{CC}	\$US	
PC924	0.5A	35V	1.04	
ACPL-P302	0.4	30	1.26	
HCPL-314J	0.4	30	2.37	dual
FOD3184	3	30	1.14	
HCPL-3120	2	30	3.10	

A. Push-pull gate driver

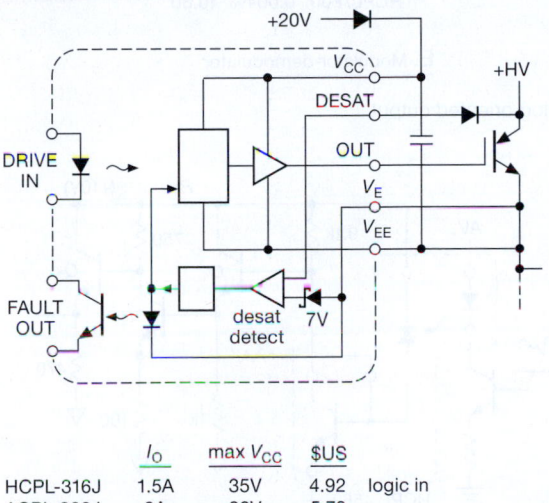

	I_O	max V_{CC}	\$US	
HCPL-316J	1.5A	35V	4.92	logic in
ACPL-332J	2A	33V	5.78	

B. Gate driver with fault shutdown

Figure 12.87. Optocouplers III: Gate driver output.

are improved gate-driver optocouplers with "desaturation" fault detection, as in Figure 12.87B: internal circuitry looks at the drop across the driven IGBT, and shuts off its gate drive during a non-saturating fault.[43] It also sends a fault indication back across the isolation gap, as shown.

---

[43] There's time to do this: an IGBT's thermal mass provides about $10\,\mu s$ at its maximum conduction current, with full rated voltage across it, before damage occurs.

12.7.4 IV: Analog-oriented optocouplers

So far we've seen only *switching* applications of optocouplers, where linearity is of little concern. But sometimes you need to isolate *analog* circuits. One method, of course, is to use a pair of converters, converting the analog quantity to a digital bitstream, coupling that via logic optocouplers, then converting back to analog. But there are analog-oriented optocouplers that do the job directly.

Figure 12.88 shows most of these. The classic H11F1 (another winner from the pioneers at GE, this one going back to 1979) is an opto-FET, in which the LED drive current affects the FET in the same way that a gate drive voltage would (we're not exactly sure what's inside this part). So increasing levels of LED drive increase the saturation current (that is, channel current at channel voltages greater than a few tenths of a volt or so), reaching about 1 mA for an LED drive of 25 mA. Notice that the output is completely symmetrical, and works up to ± 30 V across the output terminals. And, as with ordinary FETs, the output terminals look approximately resistive for small voltages across the channel; here, however, the R_{ON} value is set by the LED drive current, and varies from $>300\,M\Omega$ (with no LED drive) down to about $100\,\Omega$ (with 16 mA LED drive). Once again this property is symmetrical both sides of zero volts, but it's not particularly linear beyond about ± 50 mV, as seen in the graph of Figure 12.88A.

For really good linearity you can use an opto-photoresistor (configuration B), which is an LED illuminating a CdS photoresistor. The sensors are slow, and exhibit memory effects; but the output terminals behave like very linear resistors, with linearities of $\sim 0.01\%$ for voltage swings up to ± 1 V.[44] This would make an excellent amplitude limiter for a low-distortion Wien Bridge oscillator like the one in Figure 7.22.

There's a small class of "video optocouplers," which rely on the inherent linearity of LED intensity versus drive current (except at low currents), and the good linearity of photodiode current versus illumination. Figure 12.88C

---

[44] Sadly, the continued availability of these nice devices is threatened by legislated restrictions on hazardous substances (RoHS): cadmium is not a nice material, and worries about its toxicity put it on the no-no list. Ironically, while the milligrams of cadmium in a CdS photocell are a target of RoHS legislation, such potential exposure (from *eating* the things?) is dwarfed by the general population's exposure to cadmium via fossil-fuel burning and cigarette smoking. And, while we're whining, how about the vast quantities of lead that are permitted in automobile batteries? These weigh 50 pounds or more, most of it lead, and about 100 million are manufactured annually: that's more than a megaton of lead!

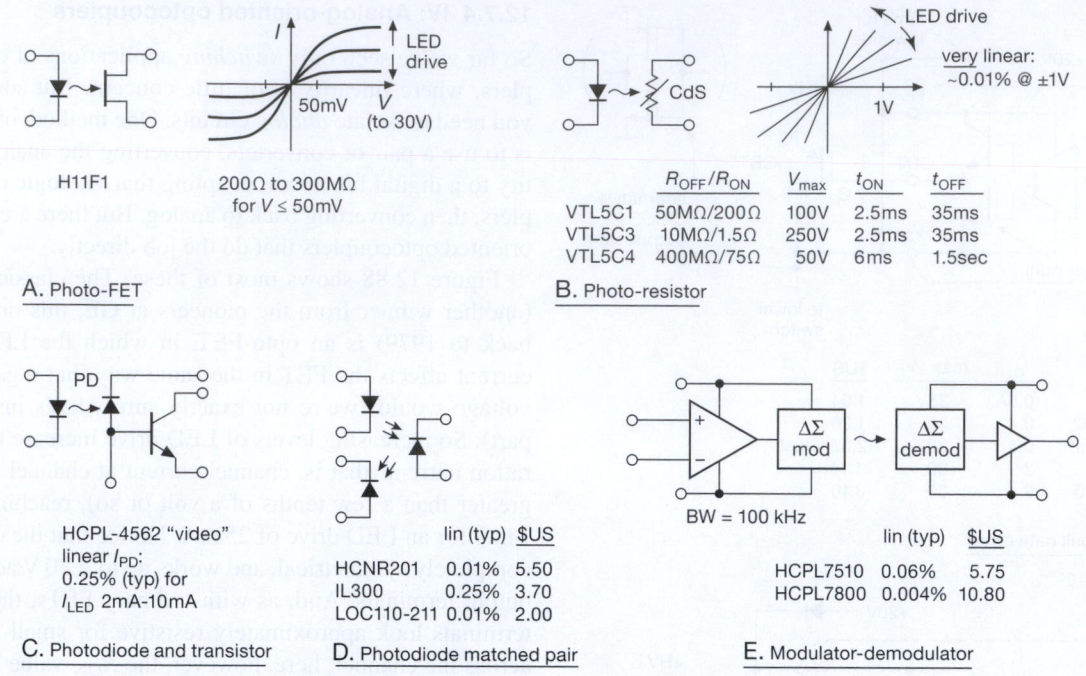

A. Photo-FET

H11F1   200 Ω to 300 MΩ
          for $V \leq 50\text{mV}$

B. Photo-resistor

	R_{OFF}/R_{ON}	V_{max}	t_{ON}	t_{OFF}
VTL5C1	50MΩ/200Ω	100V	2.5ms	35ms
VTL5C3	10MΩ/1.5Ω	250V	2.5ms	35ms
VTL5C4	400MΩ/75Ω	50V	6ms	1.5sec

C. Photodiode and transistor

HCPL-4562 "video"
linear I_{PD}:
0.25% (typ) for
I_{LED} 2mA-10mA

D. Photodiode matched pair

	lin (typ)	$US
HCNR201	0.01%	5.50
IL300	0.25%	3.70
LOC110-211	0.01%	2.00

E. Modulator-demodulator

BW = 100 kHz

	lin (typ)	$US
HCPL7510	0.06%	5.75
HCPL7800	0.004%	10.80

Figure 12.88. Optocouplers IV: Analog-oriented output.

shows on example, with a claimed bandwidth of 17 MHz when used in the circuit of Figure 12.89. Another way to achieve reasonable linearity is to package a matched pair of photodiodes with an LED (Figure 12.88D), then use one of them to provide feedback on the driving side (Figure 12.90); the far-side photodiode will then exhibit linearity limited only by the degree of matching. The devices listed in Figure 12.88D achieve linearities of better than 1% in such a configuration.

Finally, there's an interesting class of linear optocouplers that integrate an A-to-D converter and a D-to-A converter, with digital coupling. These use "delta–sigma" conversion (sometimes called "1-bit" conversion), which we discuss later in §13.9. Figure 12.88E shows the scheme. These are very linear, but the delta–sigma process leads to significant output noise (~30 mVrms, compared with 3 V full-scale output), and some signal delay (~5 μs). These parts have found widespread use in half-bridge three-phase variable-frequency-drive motor-power systems, where they measure the current in each leg. As a result they generally have full-scale ranges of only 300 mV.

12.7.5 V: Solid-state relays (transistor output)

Turning back again to couplers whose output is a switch, we have the class of "solid-state relays" (SSRs; Fig-

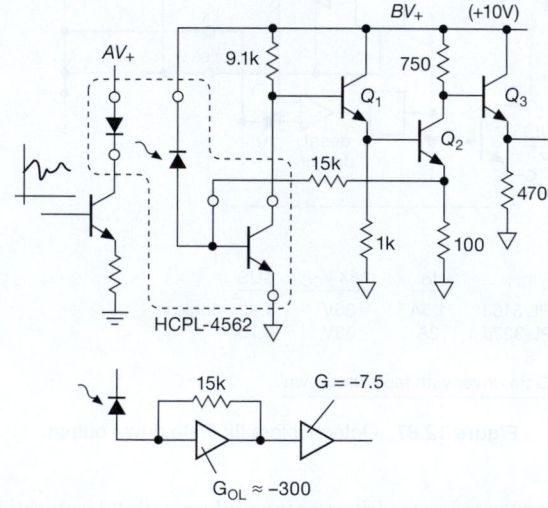

Figure 12.89. Using the HCPL-4562 "video optocoupler" in the datasheet's recommended transresistance configuration (~17 MHz bandwidth to −3 dB).

ures 12.91 and 12.92). These are characterized by isolated two-terminal outputs that are either open (nonconducting) or closed (conducting), depending on the state of input LED drive, and therefore can be thought of as substitutes for electromechanical relays (see §1x.6). They do not

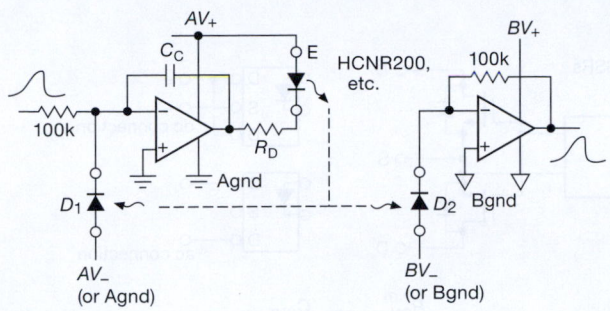

Figure 12.90. Linearizing an optocoupler with photodiode matched pair.

require any external source of supply voltage on the output side – they're "just a switch." One class of SSRs uses *thyristors* (SCRs and triacs) as the output switch; these devices, once triggered into conduction, remain conducting until the current is removed, and so they are suitable only for ac loads (where the current goes through zero twice per cycle). We treat these below, after looking at SSRs with MOSFETs as the output switches, which is the class of SSRs that is suitable for dc loads (and, as we'll see, these can also switch ac loads when configured as a series pair of MOSFETs).

To bring the output transistor(s) into conduction, MOSFET-output SSRs use a series string of a dozen or more photodiodes to generate the needed gate voltage. This "photovoltaic stack" is illuminated by the LED, generating 5–10 V of gate bias. It can supply only a few microamps to the gate, whose capacitance causes the turn-on and turn-off times to be typically in the range of 0.1–5 ms. You don't have to crack an SSR open to play with this portion of the relay, though, because you can buy it à la carte (Figure 12.91A). The datasheets for these devices don't tell you much about what's inside; but it's evident from the fast t_{OFF} specifications that most of them use an auxiliary bit of circuitry to discharge the gate for fast turn-off. This could be a *pnp* transistor that is pulled into conduction when the PV stack's output current ceases, perhaps aided by an SCR (indicated by dashed lines).

Most MOSFET SSRs use a series-connected pair of *n*-channel FETs, as shown in Figure 12.91D, driven by a photovoltaic stack.[45] For an ac load you use the top and bottom (drain) terminals. When the relay is OFF, one transistor or the other is acting as the open switch, depending on polarity; you need the series pair, because otherwise the reverse

(body) diode would conduct. When the relay is ON, both transistors act as switches, characterized by an R_{ON}.[46]

You can of course use this same connection for a dc load, but you do better by using the transistors in parallel (connecting the drains together as shown), which reduces R_{ON} by a factor of 4 (at the cost of increased output capacitance, if that matters to you). Note that enhancement-mode MOSFETs give you a "normally OFF" (form A) relay, whereas depletion-mode MOSFETs (with the gates connected to the negative end of the PV stack) make a "normally ON" (form B) relay.

Keep in mind that these "relays" come in an enormous range of current capabilities, and that the small ones can be used effectively for low-level switching: parts like the AQY221N3 or NEC/CEL PS7801-1A, for example, have an R_{ON} of 10 Ω or less while presenting a mere picofarad of output capacitance. Unlike CMOS analog switches (§3.4.1A), they have *zero* charge injection (apart from the isolation capacitance, about 1 pF for most parts, but a mere ~0.3 pF for the PS7801A) – you may want to use them even when you don't need the isolation. For example, you might use one of these in a sample-and-hold or integrator circuit.

There are a couple of oddball FET relays listed in Figures 12.91B and C: the H11F1 (featured earlier as "analog oriented") is both fast (~15 μs) and symmetrical (good to ±30 V when OFF, ±100 mA when ON). We're not sure how they do it. And the LH1514 has an interesting configuration, intended for balanced ac signals: it uses a "T-switch" arrangement, with a pair of normally-open series switches on each line, bridged across with a normally-closed switch (a depletion-mode MOSFET, driven from the same PV stack). This results in good signal attenuation when OFF, even for signals at radiofrequencies (65 dB at 1 MHz).

12.7.6 VI: Solid-state relays (triac/SCR output)

For powerline ac switching, it's common to use a solid-state relay with a triac or a pair of SCRs (collectively *thyristors*) as the switching device. The low-current SSRs shown in Figures 12.92A and B are used primarily as trigger devices to activate a high-current thyristor, as shown in Figure 12.93. But the big gorilla in the room is the integrated SSR, which includes the optocoupler and (usually)

---

[45] Some SSRs, intended for dc loads only, use only a single MOSFET; examples are the AQV100- and AQZ100-series from Panasonic.

[46] In §3.5.6B we suggested a general-purpose roll-your-own version of this SSR circuit (Figure 3.107). The advantage of providing your own MOSFETs is that you can select low-resistance, high-current parts (to over 1kA pulsed, 100 A continuous), or common low-cost parts like an IRF640 (200 V, 12 A, 0.25 Ω), or amazing high-voltage parts (to 4.5 kV); a handy table is provided there.

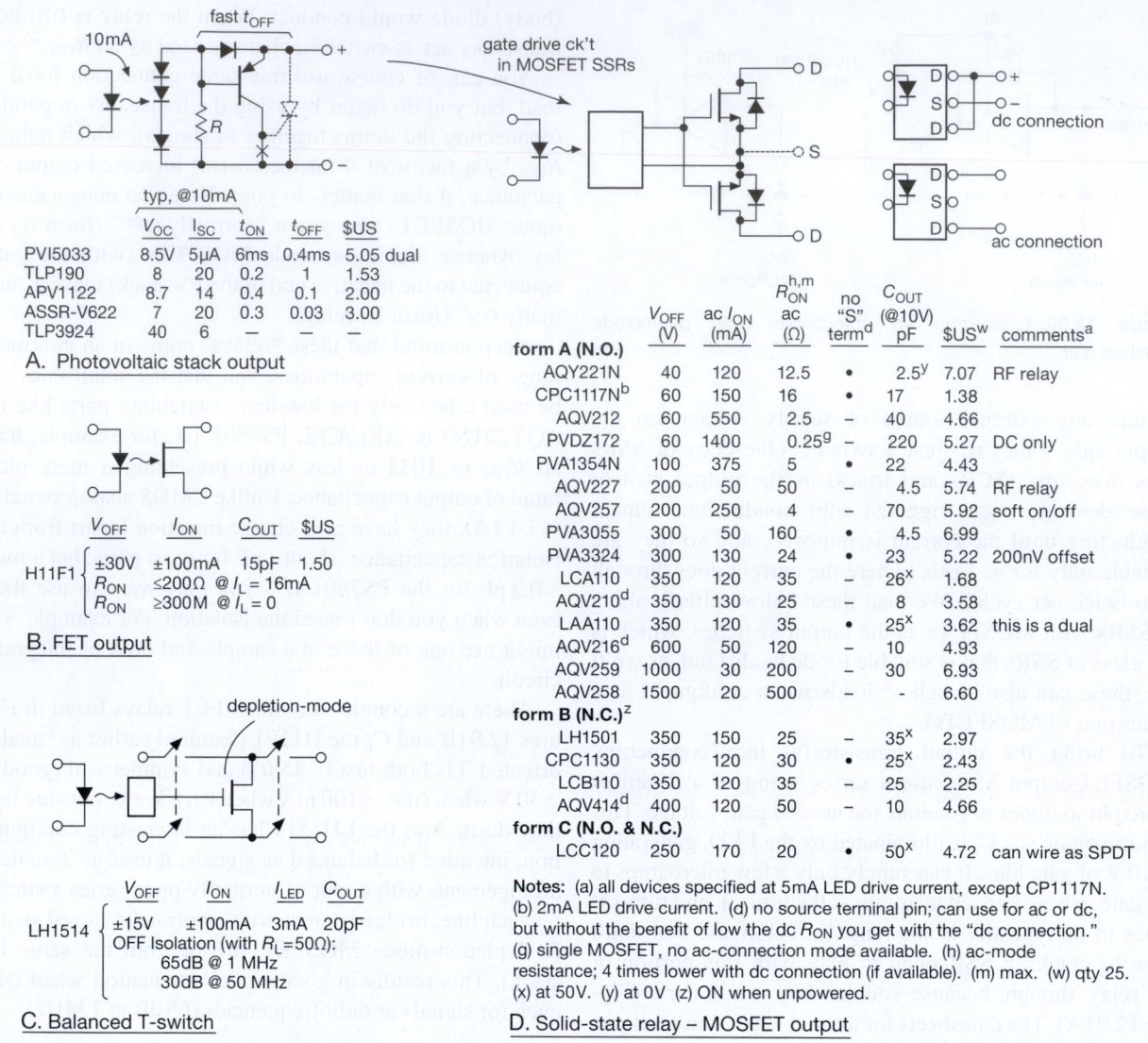

A. Photovoltaic stack output

	typ, @10mA				
	V_{OC}	I_{SC}	t_{ON}	t_{OFF}	$US
PVI5033	8.5V	5µA	6ms	0.4ms	5.05 dual
TLP190	8	20	0.2	1	1.53
APV1122	8.7	14	0.4	0.1	2.00
ASSR-V622	7	20	0.3	0.03	3.00
TLP3924	40	6	—	—	—

B. FET output

	V_{OFF}	I_{ON}	C_{OUT}	$US
H11F1	±30V	±100mA	15pF	1.50
	R_{ON} ≤200Ω @ I_L = 16mA			
	R_{ON} ≥300M @ I_L = 0			

C. Balanced T-switch

depletion-mode

	V_{OFF}	I_{ON}	I_{LED}	C_{OUT}
LH1514	±15V	±100mA	3mA	20pF
	OFF Isolation (with R_L=50Ω):			
	65dB @ 1 MHz			
	30dB @ 50 MHz			

D. Solid-state relay – MOSFET output

	V_{OFF} (V)	ac I_{ON} (mA)	$R_{ON}^{h,m}$ ac (Ω)	no "S" term[d]	C_{OUT} (@10V) pF	$US[w]	comments[a]
form A (N.O.)							
AQY221N	40	120	12.5	•	2.5[y]	7.07	RF relay
CPC1117N[b]	60	150	16	•	17	1.38	
AQV212	60	550	2.5	–	40	4.66	
PVDZ172	60	1400	0.25[g]	–	220	5.27	DC only
PVA1354N	100	375	5	•	22	4.43	
AQV227	200	50	50	–	4.5	5.74	RF relay
AQV257	200	250	4	–	70	5.92	soft on/off
PVA3055	300	50	160	•	4.5	6.99	
PVA3324	300	130	24	•	23	5.22	200nV offset
LCA110	350	120	35	–	26[x]	1.68	
AQV210[d]	350	130	25	–	8	3.58	
LAA110	350	120	35	•	25[x]	3.62	this is a dual
AQV216[d]	600	50	120	–	10	4.93	
AQV259	1000	30	200	–	30	6.93	
AQV258	1500	20	500	–	-	6.60	
form B (N.C.)[z]							
LH1501	350	150	25	–	35[x]	2.97	
CPC1130	350	120	30	•	25[x]	2.43	
LCB110	350	120	35	–	25	2.25	
AQV414[d]	400	120	50	–	10	4.66	
form C (N.O. & N.C.)							
LCC120	250	170	20	•	50[x]	4.72	can wire as SPDT

Notes: (a) all devices specified at 5mA LED drive current, except CP1117N. (b) 2mA LED drive current. (d) no source terminal pin; can use for ac or dc, but without the benefit of low the dc R_{ON} you get with the "dc connection." (g) single MOSFET, no ac-connection mode available. (h) ac-mode resistance; 4 times lower with dc connection (if available). (m) max. (w) qty 25. (x) at 50V. (y) at 0V (z) ON when unpowered.

Figure 12.91. Optocouplers V: "dc solid-state relay" (transistor output).

zero-voltage-switching (ZVS) trigger circuit, along with a substantial output triac or SCR pair. This is good: when driving ac loads, it's best to switch on the load during a zero crossing of the ac waveform, to avoid putting spikes onto the powerline; and the triac or SCR inherently switches off at zero current.[47] So we've got "ZVS/ZCS."

Figure 12.92C lists just a few of the hundreds of available types. High-current SSRs aren't cheap, but they're delightfully easy to use. The larger ones (10 A or more) come in a panel-mount "brick" package (about $1.75'' \times 2.25'' \times 1''$, intended for heatsinking), whereas the smaller ones come in several sizes of PCB-mounting "single in-line" packages (SIPs).[48]

---

47 An exception is when switching powerline ac across a transformer primary: there ZVS is the *worst* case, because the full unipolarity half-cycle applied voltage brings the core closest to (or into) saturation, with enormous peak currents. Ideally you'd want to switch the ac close to its *peak* voltage. We encountered this effect in our lab, when powering a 20 A Variac (§1.9.5D) – about half the time when you switched it on,

the 20 A circuit breaker on the wall would snap out (even though the Variac was unloaded); you had to be lucky and catch the ac waveform near a peak to successfully switch it on.

48 But see our warning footnote in §12.4.3 about heatsinks. Be sure to

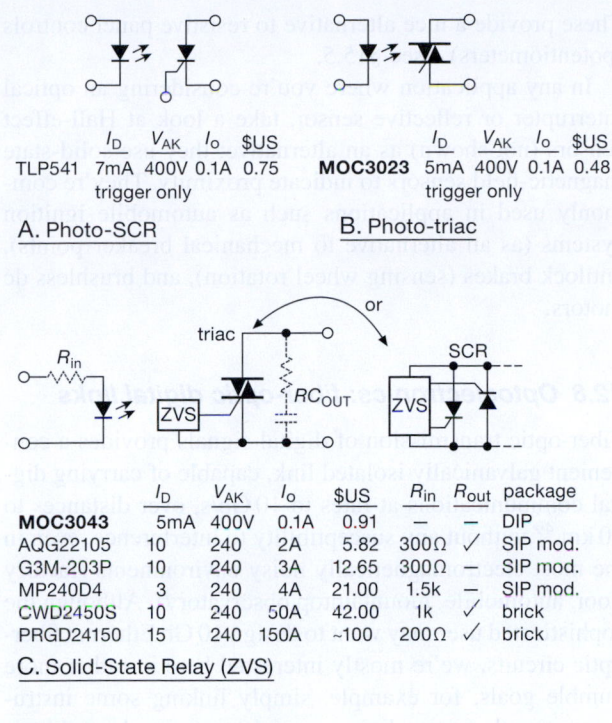

	I_D	V_{AK}	I_o	$US
TLP541	7mA	400V	0.1A	0.75
		trigger only		

A. Photo-SCR

	I_D	V_{AK}	I_o	$US
MOC3023	5mA	400V	0.1A	0.48
		trigger only		

B. Photo-triac

	I_D	V_{AK}	I_o	$US	R_{in}	R_{out}	package
MOC3043	5mA	400V	0.1A	0.91	—	—	DIP
AQG22105	10	240	2A	5.82	300Ω	✓	SIP mod.
G3M-203P	10	240	3A	12.65	300Ω	—	SIP mod.
MP240D4	3	240	4A	21.00	1.5k	—	SIP mod.
CWD2450S	10	240	50A	42.00	1k	✓	brick
PRGD24150	15	240	150A	~100	200Ω	✓	brick

C. Solid-State Relay (ZVS)

Figure 12.92. Optocouplers VI: "ac solid-state relay" (triac/SCR output).

12.7.7 VII: ac-input optocouplers

Finally, there is a class of optocouplers intended for ac input drive (Figure 12.94). Some use a pair of LEDs back-to-back, coupled to a phototransistor or photo-Darlington. The output phototransistor conducts according to the magnitude of LED current, with typical current-transfer ratios of 20–100%; this is useful for detecting zero crossings of the ac powerline.

Input–output modules

There is a category of opto-isolating "input–output modules" used in industrial settings: The *input* modules monitor an ac or dc signal (with a selection of voltage ratings, up to powerline voltage), producing an isolated low-voltage open-collector logic output to go to a computer or other industrial controller; the *output* modules use a low-voltage logic signal from a computer or controller to switch an ac or dc load, typically at powerline voltages. In other words, the output modules are SSRs with thyristor or transistor outputs (ac and dc, respectively), and the input modules are ac-input optocouplers with open-collector outputs. These

consult the datasheet for the ON voltage drop when designing an SSR into a system.

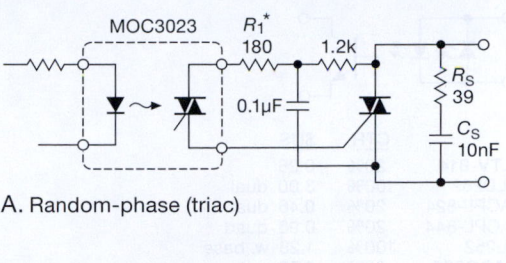

A. Random-phase (triac)

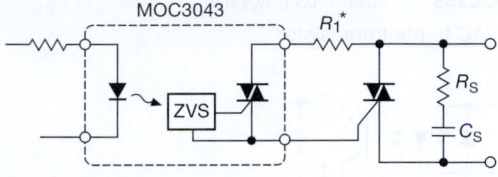

B. Zero-crossing (triac)

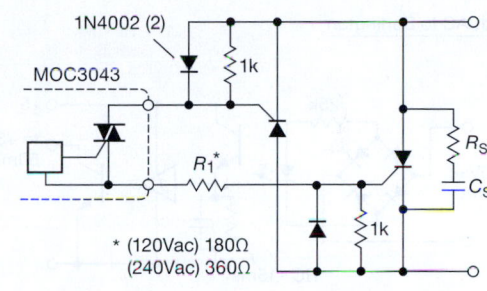

C. Zero-crossing (2 × SCR)

Figure 12.93. A small opto-triac triggers a large triac or SCR pair. Choose resistor R_1 according to the ac line voltage.

latter, therefore, are an easy way to use an ac powerline input to create an isolated logic-level swing (or higher, because the O/C output typically can go to +30 V), as shown in the figure. Note that these puppies include an internal *RC* filter, so the output indicates the presence or absence of an ac waveform, but does not capture the individual cycles. Input–output modules come in several standard upright module package sizes (e.g., $1.7'' \times 1.25'' \times 0.6''$ and $1.7'' \times 1'' \times 0.4''$), with a hold-down screw-fastener arrangement, and (often) an indicator LED on top.

12.7.8 Interrupters

You can use an LED–phototransistor pair to sense proximity or motion. An "optical interrupter" consists of an LED coupled to a phototransistor across an open slot. It can sense the presence of an opaque strip, for example, or the rotation of a slotted disk. An alternative form has the LED and photodetector looking in the same direction, and

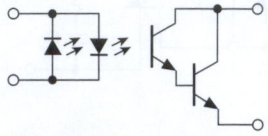

	CTR	$US	
LTV-814	20%	0.26	
ILD252	100%	3.00	dual
ACPL-824	20%	0.46	dual
ACPL-844	20%	0.93	quad
IL252	100%	1.28	w. base
MOC256	20%	0.77	w. base

A. AC to phototransistor

LTV-8141   600% $0.47

B. AC to Darlington

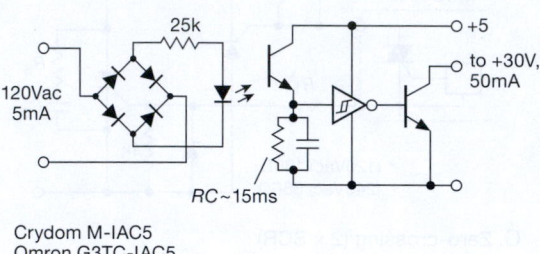

Crydom M-IAC5
Omron G3TC-IAC5
Tyco IAC-5

C. AC "input module"

Figure 12.94. Optocouplers VII: ac input.

it senses the presence of a reflective object nearby (most of the time, anyway!). As with optocouplers, you can get these with a simple phototransistor on the receive side, or you can get them with logic-level outputs (either open-collector, or active pullup). You can see some examples in Figure 12.84. Optical interrupters are used in mechanical devices such as printers, to sense the end of travel of the moving assembly. Optical interrupters can have problems when ambient light levels are high. A nice trick in such situations is to use *synchronous detection* (§8.14.1), making the detector selectively sensitive to the frequency at which the emitter is driven. Hamamatsu offers a nice selection (their S4282/89-, S6809/46/86-, and S7136-series) of detectors with built-in preamp and signal processing electronics. You can get optical "rotary encoders" that generate a quadrature pulse train (two outputs, 90° out of phase) as the shaft is rotated; there's an example of one in Figure 12.84.

These provide a nice alternative to resistive panel controls (potentiometers) – see §15.5.

In any application where you're considering an optical interrupter or reflective sensor, take a look at Hall-effect sensors (not shown) as an alternative; they use solid-state magnetic-field sensors to indicate proximity. They're commonly used in applications such as automobile ignition systems (as an alternative to mechanical breaker points), antilock brakes (sensing wheel rotation), and brushless dc motors.

12.8 Optoelectronics: fiber-optic digital links

Fiber-optic transmission of digital signals provides a convenient galvanically isolated link, capable of carrying digital communications at rates to 10 Gb/s, over distances to 10 km,[49] without any susceptibility to interference, even in the most electromagnetically noisy environments (factory floor, automobile, mountaintop observatory). Although the sophisticated user may want to design 10 Gb Ethernet fiber-optic circuits, we're mostly interested here in rather more humble goals, for example, simply linking some instruments together over distances of 10 m (or perhaps 1 km), with data rates of megabits per second (or perhaps with fast Ethernet, at 100 Mb/s). Let's see what off-the-shelf components will do the job.

12.8.1 TOSLINK

A very simple and inexpensive short-range digital-fiber link is provided by the TOSLINK™ family of transmitter–receiver pairs (known generically as "EIAJ optical," "JIS F05," "ADAT optical," or "Digital Audio Optical Cable"), see Figure 12.95. The TOSLINK standard was originated by Toshiba,[50] and is widely used for digital-audio interconnections, for example between audio and video components. It is one of the two "digital-audio" jacks you see on the rear of DVD and Blu-ray players, sometimes la-

---

[49] The speed record at time of publication is 1 petabit/s (10^{15} b/s) over a 50 km length of fiber optic cable, achieved by a collaboration of NTT (Japan), Fujikara Ltd., Hokkaido University, and the Technical University of Denmark. The "cable" is a 12-core bundle, each fiber of which carries 84.5 Tb/s by multiplexing upon 222 separate wavelengths a signal rate of 380 Gb/s per wavelength (the latter achieved by modulating eight separate carriers, each with polarization multiplexed 32QAM).

[50] **Caution**: we've found current availability largely through second-source suppliers such as Comoss, Sys Concept, and FiberFin; the parts used for the examples in this section appear to be unavailable from distributors, and may have been discontinued by Toshiba.

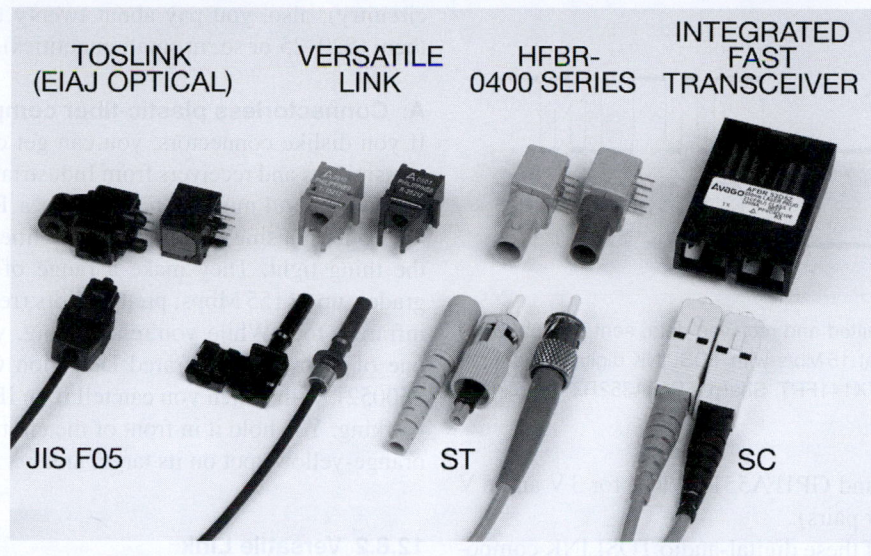

Figure 12.95. Some popular fiber-optic formats. The TOSLINK devices, widely used in consumer audio, are here shown in a shuttered PC-mount version (right), and a panel-mount unshuttered version (left). The other connector formats can be field terminated (bare connectors are shown for ST and Versatile Link), but it's easier to buy preterminated fiber cables.

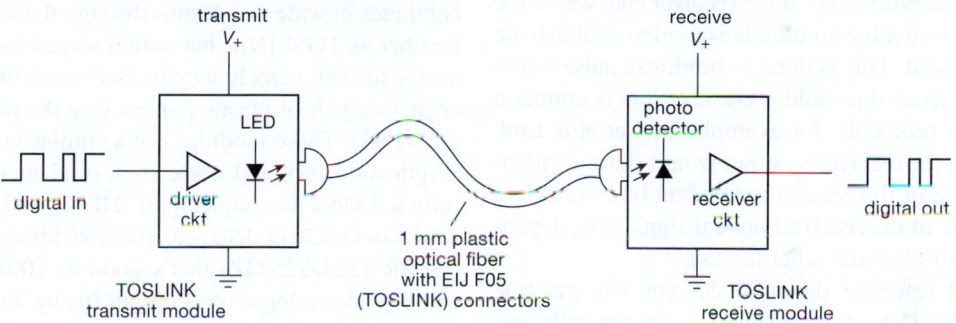

Figure 12.96. Inexpensive PC-mounting TOSLINK-style fiber-optic modules include all drive and receive electronics. They accept and reconstruct standard logic levels, at rates to 15 Mb/s over ranges to 5 m.

beled "digital optical" (the other being a coaxial electrical audio socket that is physically identical to the common "RCA" jack, usually placed alongside the TOSLINK connector, and that carries, electrically, the same digital stream as the optical port). TOSLINK uses visible red LEDs, operating at 650 nm; it's easy to see when it's working, and which end is the transmitter (unlike the case with invisible infrared fiber modules).

The charm of TOSLINK is its simplicity and low cost: the Toshiba TOTX147 and TORX147 are a typical transmitter–receiver pair, costing about $1 at each end (in small quantities). They run on a 2.7–3.6 V logic supply, with the transmitter accepting logic-level inputs, and the receiver replicating the logic levels at its output. That is,

all the logic interface circuitry is built in; all you do is connect it up (Figure 12.96). They are intended for short-range links, operating up to 15 Mb/s over a distance of 5 meters or less. You use an inexpensive 1 mm plastic fiber (APF, for all-plastic fiber, or POF, for plastic optical fiber), which you can get as a 2.2 mm (or thicker) sheathed cable with TOSLINK jacks at each end; we've found such "patch-cords" (e.g., at Amazon.com) for as little as a dollar or less, and are standard equipment at audio–video supply stores. Figure 12.97 shows logic input and output from a TOSLINK pair running at +3.3 V, driven with a 15 Mb/s data stream, and connected with an 8-foot fiber patch-cord. You can get 5 V versions as well, e.g., Toshiba part numbers TOTX/TORX177 (analogous Sharp part numbers are

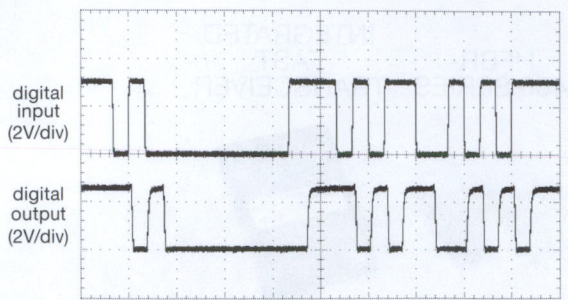

Figure 12.97. Transmitted and received data, sent through 8 feet of 1 mm plastic fiber at 15 Mb/s with TOSLINK digital-audio components (Toshiba TOTX141FPT, Sharp GP1FA352RZ). Horizontal scale: 200 ns/div.

GP1FA351TZ/RZ and GP1FA551TZ/RZ, for 3 V and 5 V transmitter–receiver pairs).

One drawback of these digital-audio TOSLINK components is that the receiver bandwidth does not extend down to dc, and is specified to operate properly only with a minimum data rate of 0.1 Mb/s.[51] Note that the transmitter end is dc-coupled; the problem is at the receiver end, where the optical signal's switching amplitude is used to establish the receiver's threshold. This is done to minimize pulse-width distortion: if a *fixed* threshold were used (as is common with other fiber protocols, for example the Versatile Link devices below), the receiver's output would show a widening or narrowing of the reconstructed data bits, according to the amplitude of the received optical signal (i.e., dependent on length of fiber and other losses).

If you need response down to dc, you can get true dc-coupled TOSLINK-style receivers, with generally improved performance: for example, the TOTX197 and TORX198 pair of "general-purpose" (as contrasted with "digital-audio") modules operate from dc to 6 Mb/s over a distance of 40 meters or less, using the same plastic-fiber cable and connector. The TOTX/TORX1350 series specify data rates of dc to 10 Mb/s over distances to 100 meters, again with all-plastic fiber (APF). And, by using a lower-loss plastic-clad glass fiber (PCF, with a fiber diameter of 0.2 mm) the range is extended to 1 km, for example, with the TOTX/TORX196 pair. As in life generally, there are tradeoffs: these devices exhibit greater pulse-width distortion (typically ±55 ns, compared with ±15 ns for the audio-style devices[52] with their adaptive threshold

circuitry); also, you pay about twenty times as much for them ($20–25 or so, in small quantities).

A. Connectorless plastic-fiber components

If you dislike connectors, you can get connectorless fiber transmitters and receivers from Industrial Fiber Optics, Inc. These accept 1 mm plastic fiber, in a PCB-mount cinch-nut collet housing; you just cut the fiber, insert, and twist the thing tight. They make a range of power and speed grades, up to 155 Mbps; pretty colors (red, green, blue, and infrared) too! While you're shopping, you might pick up one of their nifty "Infrared Detection Cards" (part # IF-850052), with which you can tell if an IR LED (or laser) is working. You hold it in front of the emitter, and you see an orange-yellow spot on its target area.[53]

12.8.2 Versatile Link

The Versatile Link (VL) series of optical-fiber transmit and receive modules was introduced by Hewlett–Packard (then spun off as Agilent, and finally Avago) around 1990, and continues in wide use. It runs through the same 1 mm plastic fiber as TOSLINK, but with a somewhat different connector format, namely a cylindrical snap-in connector that engages a pair of plastic pincers (see the photograph, Figure 12.95). These modules use a similar visible red wavelength (660 nm), and come in several models, most usefully a 5 Mb/s dc-coupled pair (HFBR-1521Z/2521Z) that works out to 20 m distance, and a 40 kb/s dc-coupled pair (HFBR-1523Z/2523Z) that's good to 100 m. The receive end provides a logic-level output (using an external or internal pullup resistor); but the transmit end is a bare LED, so you have to provide a current-limiting resistor and a saturated switch to ground, or the equivalent (Figure 12.98). This provides some flexibility in transmitted optical level, but does require extra components. Both transmit and receive modules cost less than $10 in small quantities. You can get the optical cable with connectors attached; or you can roll your own, which turns out to be simple and quick, using the "crimpless" (i.e., no tooling required) HFBR-4531Z series of connectors; they cost about $0.50 apiece, in small quantities.

For greater link speeds you can get the faster HFBR-

---

[51] It is not required that the data stream have balanced quantities of 1s and 0s, just that there be transitions at frequent-enough intervals.

[52] There has been some grumbling in the audiophile community about jitter in this interconnection technology. Although this can be addressed

with PLL clock regeneration or sample-rate conversion, it is an issue that does not arise with electrical (coaxial) interconnection of digital audio devices.

[53] Puzzle for the reader: converting a lower-energy photon into one of higher energy would appear to violate energy conservation. Figure it out!

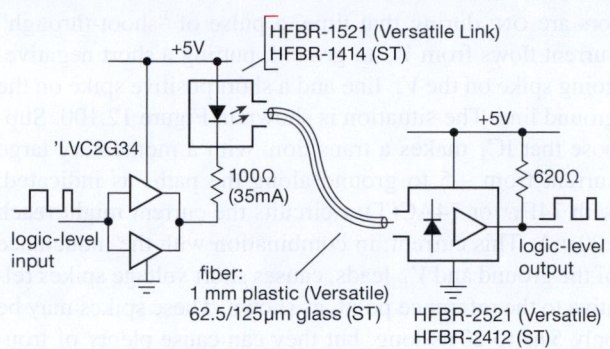

Figure 12.98. Typical drive and receive circuit for the Avago dc-coupled 5 Mb/s "Versatile Link" and ST-connected fiber-optic series. For 3.3 V logic levels use 50 Ω and +3.3 V on the driver side, and pullup to +3.3 V on the receiver side. Operating currents shown here are suitable for lengths to 10 m (Versatile Link: plastic fiber) or 1 km (ST: glass fiber); they can be altered, depending on link length and maximum data rate (see Avago datasheets and App Note 1035).

1527Z/2526Z modules, which use the same fiber and connector geometry, and which permit signal rates to 125 Mb/s (but not dc-coupled). These, and analogous devices (e.g., the TODX2402 or TOTX/RX1701 in the TOSLINK series; or the HFBR-1424/2426 in the 820 nm ST/SC-connector devices from Avago that use all-silica graded-index fiber, see below) are popular as fiber-based fast serial links, for example to carry Ethernet signaling, or serialized data between a SERDES (serializer–deserializer) pair.

12.8.3 ST/SC glass-fiber modules

For many years we've used the HFBR-14xx/24xx series of fiber transmitters and receivers from Avago. These are more expensive (about $15 in small quantity) than the plastic-fiber devices above, but by using graded-index silica-clad silica fiber (sometimes called "ASF," all-silica fiber, or "AGF," all-glass fiber) at 820 nm wavelength, they work happily even through a kilometer or more of fiber. An added bonus is the robustness of the fiber cables, which, with their very thin glass core/cladding (62.5 μm/125 μm, in the most common variety) and tough jacket, can withstand plenty of bending and pulling without damage.

We use the ST connector style, which is a locking bayonet (see Figure 12.95). You can buy preterminated fiber cables as simplex (one fiber, ST termination at each end), or duplex (also called "zip cord": a pair of fibers, each with its own ST connector at each end), from companies like Tyco/AMP, 3M, and Amphenol, for a few dollars per meter. If you're willing to spend a few hundred dollars on a

termination kit, and suffer a bit, you can even put the connectors on by yourself; look at connectors and tooling from companies like Tyco/AMP and 3M.

As with the Versatile Link components, you can get dc-coupled low-speed receivers, for example, the HFBR-2412Z, which goes to 5 Mb/s at distances to 2 km. Its open-collector output requires only a pullup resistor to generate the reconstructed digital logic levels. For high speeds use instead the HFBR-2416Z, which is good to 155 Mb/s at distances to 0.6 km. The latter provides a high-bandwidth "analog" output from its internal PIN-diode detector and preamp, which you ac-couple to external amplifier–comparator circuitry to generate the fast LVDS or ECL reconstructed digital stream. For either receiver you would use the HFBR-1414Z transmitter, which, like the analogous Versatile Link component, looks electrically like a bare LED; you have to provide the requisite drive, either with a transistor switch and resistor, or with a hefty logic gate like a 'LVC2G34 (see Avago App Note 1123 and App Brief 78).

12.8.4 Fully integrated high-speed fiber-transceiver modules

Why not integrate *all* the logic-level interfacing within the fiber-optic module itself? Indeed. With widespread use of duplex fiber to carry fast serial data – for example as optical fast Ethernet (125 Mb/s), or firewire (rates to 250 Mb/s), or simply between a pair of SERDES parallel–serial chips – there are now plenty of easy-to-use fiber-optic-transceiver ("FOT") modules. These devices have both transmitter and receiver optical modules and connectors (usually in duplex SC or ST format, for glass fiber; or as the duplex "SMI" connector for plastic fiber), along with driver and receiver circuitry using fast differential signaling (usually 5 V or 3.3 V PECL); see Figure 12.99 (and the portrait in Figure 12.95).

Contemporary examples are the Avago AFBR-5xxx-series, which come in duplex SC or ST connector styles, and which can handle 100 Mb/s data rates for fast Ethernet (100Base-FX) or ATM (asynchronous transfer mode); you connect the transceiver's PECL ports directly to the corresponding inputs or outputs on an Ethernet, firewire, or ATM "PHY" (physical layer IC). A typical member of this family, the AFBR-5803, good to 125 Mb/s, currently costs about $30; the faster AFBR-53D5 can handle gigabit Ethernet and costs about $60. You can use a fast fiber transceiver to connect to a SERDES like the Cypress CY7C924 or HDMP-1636, to link a pair of widely separated parallel ports, with data rates of 20–100 Mbytes/s.

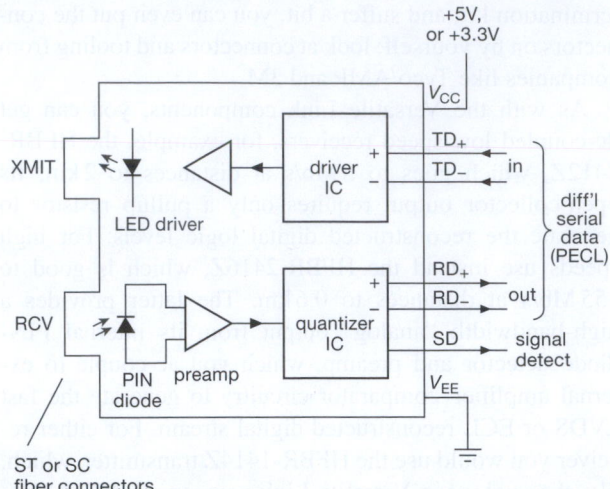

Figure 12.99. High-speed integrated fiber-optic transceivers like the Avago AFBR-5xxx-series, in the industry standard 1×9 SIP package, include all driver and receiver circuitry, to interface directly via 3 V or 5 V serial differential PECL.

For short links, fast transceivers that mate with plastic fiber are available, for example the Toshiba TODX2402 (PECL in–out, duplex SMI, rates to 250 Mb/s; about $25 in small quantity).

12.9 Digital signals and long wires

Special problems arise when you try to send digital signals through cables or between instruments. Effects such as capacitive loading of the fast signals, common-mode interference pickup, and "transmission-line" effects (reflections from impedance mismatching; see Appendix H) become important, and special techniques and interface ICs are often necessary to ensure reliable transmission of digital signals. Some of these problems arise even on a single circuit board, so a knowledge of digital transmission techniques is generally handy. We begin by considering on-card problems. Then we go on to consider the problems that arise when signals are sent between cards, on data buses, and finally between instruments via twisted-pair or coaxial cables.

12.9.1 On-board interconnections

A. Output-stage current transient

The push–pull output circuit for logic ICs consists of a pair of transistors going from V_+ to ground. As we remarked earlier (§3.4.4B and 10.8.3B), when the output changes state there is a brief interval during which both transis-

tors are ON; during that time, a pulse of "shoot-through" current flows from V_+ to ground, putting a short negative-going spike on the V_+ line and a short positive spike on the ground line. The situation is shown in Figure 12.100. Suppose that IC_1 makes a transition, with a momentary large current from +5 to ground along the paths as indicated; with 74F*xx* or 74AC(T)*xx* circuits the current might reach 100 mA. This current, in combination with the inductance of the ground and V_+ leads, causes short voltage spikes relative to the reference point, as shown. These spikes may be only 5 ns to 20 ns long, but they can cause plenty of trouble: suppose that IC_2, an innocent bystander located near the offending chip, has a steady LOW output that drives IC_3 located some distance away. The positive spike at IC_2's ground line appears also at its output, and, if it is large enough, it gets interpreted by IC_3 as a short HIGH spike. Thus, at IC_3, some distance away from the troublemaker IC_1, a full-size bona fide logic-output pulse appears, ready to mess up an otherwise well-behaved circuit. It doesn't take very much to toggle or reset a flip-flop, and this sort of ground-current spike can do the job nicely.

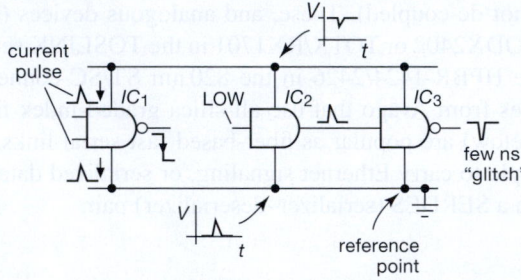

Figure 12.100. Ground-current noise, also known as *ground bounce*.

The best therapy for this situation consists of (a) using hefty ground lines throughout the circuit, or preferably a "ground plane" (one side of a double-sided PC board) or an inner layer of a multilayer PCB, and (b) using bypass capacitors liberally throughout the circuit. Large ground lines mean smaller current-induced spikes (lower inductance and resistance), and bypass capacitors from V_+ to ground sprinkled throughout the circuit mean that current spikes travel over only short paths, with the reduced inductance resulting in much smaller spikes (the capacitor acts as a local voltage source, since its voltage does not change appreciably during the brief current spikes). It is best to use a $0.1\,\mu\text{F}$ ceramic capacitor near each IC, although one capacitor for every two or three ICs may suffice. In addition, a few larger electrolytic capacitors ($\sim 10\,\mu\text{F}$ or so) scattered throughout the circuit for energy storage and resonance damping[54] is a good idea. We can hardly emphasize enough the importance of bypass capacitors from power-supply lines to ground in any circuit, digital or linear. They help make the supply lines low-impedance voltage sources at high frequencies, and they prevent signal coupling between circuits via the power supply. Unbypassed power-supply lines can cause peculiar circuit behavior, oscillations, and headaches. *Don't do it!*

B. Spikes caused by driving capacitive loads

Even with the supplies bypassed, your problems aren't over, as we mentioned earlier in §10.8.3A. Figure 12.101 shows why. A digital output sees the stray wiring capacitance and the input capacitance of the chip it drives (5–10 pF, typically) as part of its overall load. To make a fast transition between states it must sink or source a large current into such a load, according to $I = C(dV/dt)$. For instance, consider a 74LVCxx chip in a +3.3 V logic system, driving a total load capacitance of 25 pF (equivalent to three or four logic loads connected with short leads). With typical output rise and fall times of ∼2 ns, the current during the logic transition is 40 mA. This current returns through ground (HIGH-to-LOW transition) or the +3.3 V line (LOW-to-HIGH transition), producing those cute little spikes at the receiving end, as before. To get an idea of the effects of such current transients, consider the fact that wiring inductance is roughly 5 nH/cm. An inch of ground wire carrying this logic transition current would have a spike of $V = L(dI/dt) = 0.2$ V. And if the chip happened to be an octal buffer, with simultaneous transitions on a half dozen outputs, the ground spike would be over a volt; refer back to Figure 10.99. A similar (though generally smaller) ground spike is generated near the driven chip, where the drive-current spikes return to ground through the input capacitance of the driven device.

In a synchronous system, with a number of devices making output transitions simultaneously, the noise-spike situation can become so serious that the circuit will not function reliably. This is especially true in a large printed-circuit card, with long interconnections. The circuit may fail only occasionally, when a whole group of data lines unluckily happen to make a simultaneous HIGH-to-LOW transition, generating a momentary very large ground current.

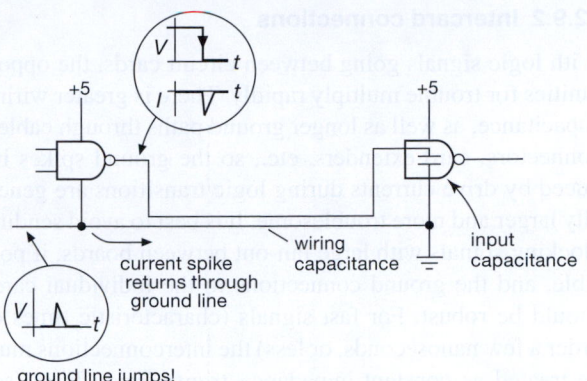

Figure 12.101. Capacitive-load ground current noise.

This kind of pattern sensitivity is characteristic of noise-induced error and is a good reason for running extensive memory tests on microprocessor systems (where you typically have 16 or 32 data lines and 32 address lines bouncing around in crazy patterns).

The best design approach is to use an internal ground-plane layer on a multilayer circuit board, or at least a perpendicular "gridded" arrangement of grounds on both sides of a simpler two-sided board. Copious use of bypass capacitors is mandatory. These problems have been mitigated considerably by measures such as (a) low-inductance surface-mount packages for discrete logic devices, (b) the use of multiple ground pins on complex logic devices,[55] (c) the nearly universal adoption of multilayer PC boards with dedicated power and ground planes, and with extravagant use of SMT chip bypass capacitors, and (d) controlled edge-rate chip designs (e.g., 74ACQ, 74ACTQ, or Gunning Transceiver Logic[56]) and redundant "center-pin" power–ground pinouts for situations in which fast logic chips will be used with less favorable layouts (i.e., through-hole packages, two-layer PCBs, etc.).

Because of these noise problems, it is generally best not to use a faster logic family than you need.[57] That's why we use 74HC(T) logic, rather than 74AC(T), for general-purpose use in our electronics course (where students put their circuits together using "solderless breadboards").

---

[54] A circuit with multiple 0.1 μF ceramic bypass capacitors, connected by inductive supply-bus wiring, is susceptible to ringing and even oscillation, owing to the multiple high-Q resonances; these are often in the 5–20 MHz range. They can be effectively damped by the equivalent series resistance (ESR) of one or more paralleled electrolytic capacitors.

[55] For example, the "Virtex-5" FPGA that we're using in our lab has 197 redundant ground pins; the larger "FF1760 package" version has 322!

[56] Read Fairchild's App Note AN-1072 to see how it works. GTL is a single-ended bus-driving logic family with reduced swing (<1 V) and controlled edge rates.

[57] This advice goes for *analog* circuits, too: don't use a 100 MHz op-amp, or a 2 ns comparator, when you don't need the speed.

12.9.2 Intercard connections

With logic signals going between circuit cards, the opportunities for trouble multiply rapidly. There is greater wiring capacitance, as well as longer ground paths through cables, connectors, card extenders, etc., so the ground spikes induced by drive currents during logic transitions are generally larger and more troublesome. It is best to avoid sending clocking signals with large fan-out between boards, if possible, and the ground connections to the individual cards should be robust. For fast signals (characteristic times of order a few nanoseconds, or less) the interconnections must be treated as constant-impedance *transmission lines* (see §12.10.1 and Appendix H), which may be single-ended (coaxial cable) or differential (twisted pair). We'll have much more to say about this soon (§12.10).

If clocking signals are sent between boards, it is important to use a gate (for single-ended signals) or a differential receiver (for differential signals such as LVDS) as an input buffer on each board. In some cases it may be best to use line driver and receiver chips, as we will discuss shortly. In any case, it is best to try to keep critical circuits together on one card, where you can control the inductance of the ground paths and keep wiring capacitance at a minimum. Fast signals (edge times of order 1 ns or less), and especially clocking signals, that go between the circuits on a single card will often be routed as "stripline" or "microstrip" transmission lines (§1x.1.3).[58] This can take the form of a single-ended trace above a ground plane (microstrip) or between ground planes (stripline); or it can be a differential pair, with two traces side-by-side or stacked vertically. Such constant-impedance routes will be terminated in their characteristic impedance, commonly $50\,\Omega$ (single ended) or $100\,\Omega$ (differential), either with "back termination" at the driving end, or with far-end termination, or both. The problems you'll encounter in sending fast signals around through several cards should not be underestimated; they can turn out to be the major headache of an entire project!

12.10 Driving Cables

You can't run digital signals from one instrument to another by just stringing a single conductor between them, because such an arrangement is prone to pickup of interference (as well as generating interference of its own), and also serious degradation of the digital signals themselves.

Instead, digital signals are generally piped through coaxial cables, twisted pairs, flat ribbon cables (sometimes with ground plane or shield), multiwire bundled cables, and, increasingly, fiber-optic cables.[59] Let's look at some of the methods used to send digital signals between boxes of electronics, since these methods constitute an important part of digital interfacing. In most cases there are special-purpose driver–receiver chips available to make your job easier.

12.10.1 Coaxial cable

If you've never dealt with fast signals going through cables, you're in for a surprise.

A. The wrong way

Here's a typical mistake that we see again and again: you've got some digital signals coming out of a "Digital I/O" interface card, for example the PCI-6509 from the popular series of data acquisition products made by National Instruments. This puppy plugs into a computer motherboard's PCI slot, and gives you 96 bits of bidirectional digital I/O, grouped in 12 bytes, each byte of which can be configured as input or output. As an output, each bit generates full-swing 5 V CMOS logic levels (i.e., 0 V and +5 V), with plenty of drive capability (24 mA sink or source), enough to drive easily loads like solid-state relays, small mechanical relays, bright LEDs, and the like.[60]

The mistake is to connect this digital output to a length of coax, and expect it to arrive safely at the far end, as in Figure 12.102A. The thinking goes like this: we've got at least ±24 mA of drive, which should be able to drive the ~200 pF of a 2-meter length of coax ($C{=}100$ pF/m) just fine; after all, $I = C\,dV/dt$ predicts a rise time of ~20 ns, assuming a typical switching output current of ~40 mA into the capacitive load. So, what's the problem?

The problem is that we have to treat a coax cable as a transmission line, rather than as the low-frequency approximation of a lumped capacitance, when dealing with signals that are changing on a time scale comparable with (or shorter than) the round-trip delay time of the cable. If you try to use the logic output to drive the cable directly, you'll get a messy waveform at the far end, with overshoots and polarity reversals, producing incorrect recovery of the waveform (and even destruction of the far-end gate). But you can magically fix the problem by simply adding a $50\,\Omega$

---

[58] This is mandatory, for example, with the fast logic families known as ECL-100K, ECL-100E, ECL-100EL, and ECL-100EP.

[59] Coaxial cables (affectionately called "coax") and twisted pairs are examples of *transmission lines*, discussed in detail in Appendix H.

[60] Look up the 74LVC4245A, if you want to see the specs on the voltage-translating (3 V ↔ 5 V) octal transceivers used in this product.

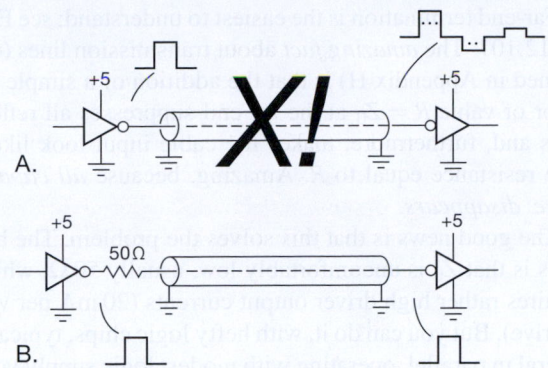

Figure 12.102. A. Driving a length of cable with a logic output produces a mangled waveform of overshoots and polarity reversals. B. Adding a ∼50 Ω series resistor effects a magic cure.

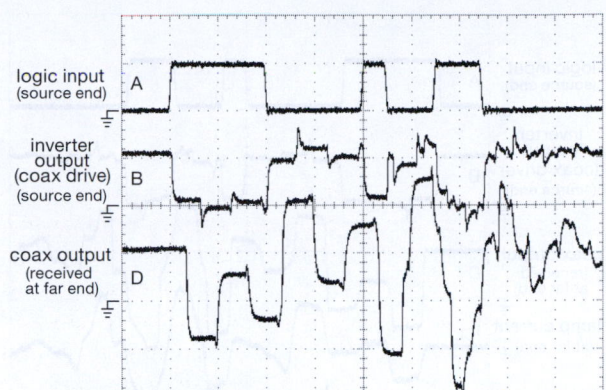

Figure 12.104. Waveforms seen in the circuit of Figure 12.103, when driven with a pulse pattern clocked at 20 ns, and with the far end of the 8 foot (2.4 m) RG-58 coax cable unconnected. The 3.3 V logic-level drive to the cable produces far-end swings of nearly 15 V peak-to-peak! Horizontal: 40 ns/div; Vertical: 3 V/div.

series resistor at the driving end (Figure 12.102B): the far-end waveform becomes a good replica of the driving waveform.

Let's look at this situation in more detail, beginning with the "wrong" way, then progressing through three configurations that solve the problem, each of which have advantages and disadvantages. We'll wind up with the magic fix of Figure 12.102B, known as "series termination," which is well suited to digital-logic signals.[61]

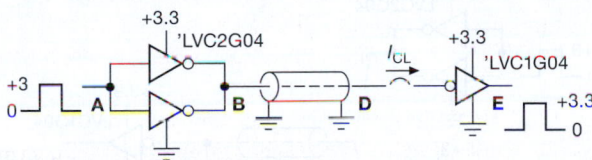

Figure 12.103. Test circuit that attempts to send digital logic signals directly through a length of cable whose far end is "unterminated". The disastrous results can be seen in Figures 12.104–12.106. *Don't do this!*

To illustrate the problem, we hooked up the circuit of Figure 12.103 and ran a pulse pattern through it. You can see the measured results in Figures 12.104–12.106. In Figure 12.104 the far end has been left unconnected: the first transition reaches the far end 12 ns later, where it bounces off the open end (with unchanged polarity), producing an output voltage nearly double the step size; things get messy as the signals bounce back and forth, reversed in polarity with each bounce off the source end, and decaying slowly with each bounce, but always adding the new source signal transitions. The signal at the far end looks terrible – and it

swings to +8 V and −6 V, even though we are driving the cable with 0 V and +3.3 V only. Imagine what will happen when we connect a logic inverter at the far end!

You needn't imagine. In Figure 12.105 we've attached the far end inverter, running also from +3.3 V. Its input protection diodes clamp the renegade waveform, limiting negative swings to roughly a diode drop, but allowing occasional positive swings to +8 V (the 'LVC1G04 has 5 V–compliant inputs, regardless of supply voltage). It's an ugly situation, and it's not surprising that the recovered logic output has some false transitions.

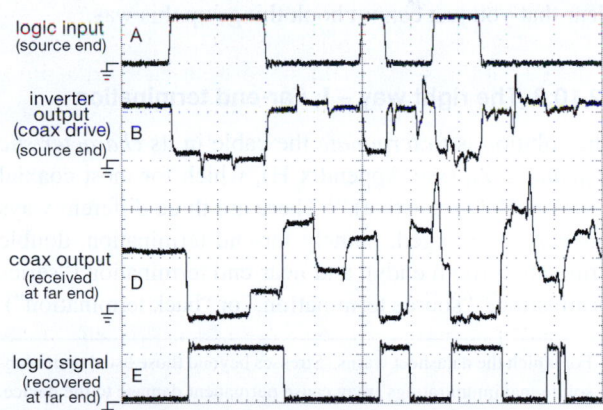

Figure 12.105. Same as Figure 12.104, but with the 'LVC1G04 inverter connected at the far end. The clamping effect of the inverter's input protection diodes reduces the swing at the coax output. This unhealthy situation produces some false transitions at the output; it can also destroy the output inverter.

---

[61] But not to *every* kind of signal: in the world of RF and video the method of "double-ended termination" is used universally.

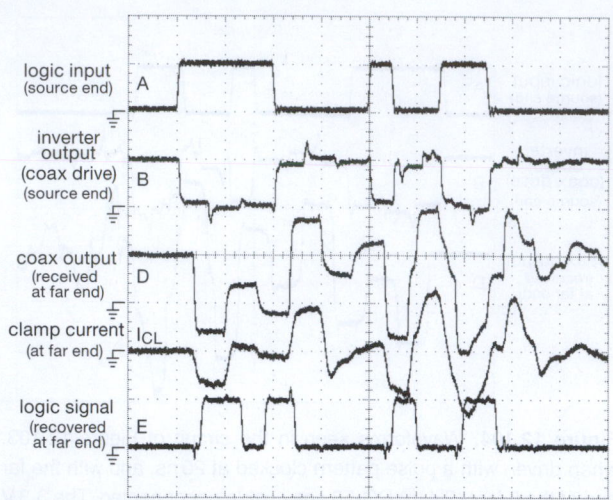

Figure 12.106. Same as Figure 12.105, but with 5 V logic ('LVC2G04 drivers, 'HCT04 far-end receiver). Note the substantial clamping currents, and the error-prone signal recovery. Horizontal: 40 ns/div; vertical: 5 V/div and 20 mA/div.

Turning now to the original scenario – a 2 m cable connecting a digital I/O card to some instrument – the driving signal comes from a hefty 5 V logic output, and the receiver (within some commercial instrument) is likely to be something like an 'HCT04 (5 V logic, with TTL threshold ~ 1.4 V). Figure 12.106 shows what happens. It's not a pretty picture: once again the flailing far-end signal is partially clamped, but reaches peaks of $+10$ V and -5 V, with corresponding clamp currents of ± 25 mA. This exceeds the "absolute maximum" input clamp current specification of ± 20 mA.[62] And the recovered output is a mess. It's pretty clear that you just cannot hook things up this way.

12.10.2 The right way – I: Far-end termination

The solution is to *terminate* the cable in its *characteristic impedance* $\mathbf{Z}_0$ (see Appendix H), which for most coaxial cables is 50 Ω (resistive).[63] There are three different ways this can be arranged, namely far-end termination, double termination (both ends), and near-end termination ("series termination," "source termination," or "back termination").

---

[62] For which the datasheet warns "Stresses beyond those listed under 'absolute maximum ratings' may cause permanent damage to the device. These are stress ratings only, and functional operation of the device at these or any other conditions beyond those indicated under 'recommended operating conditions' is not implied."

[63] A big exception is the video community, which has chosen 75 Ω (typified by RG-59 coax); and in pulse circuits you occasionally see 93 Ω (RG-62).

Far-end termination is the easiest to understand; see Figure 12.107. The *amazing fact* about transmission lines (explained in Appendix H) is that the addition of a simple resistor of value $R = Z_0$ at the far end suppresses all reflections and, furthermore, makes the cable input look like a pure resistance equal to R. Amazing, because *all capacitance disappears*.

The good news is that this solves the problem. The bad news is that Z_0 is uncomfortably low, usually 50 Ω, which requires rather high driver output currents (20 mA per volt of drive). But you can do it, with hefty logic chips, typically several in parallel, operating with modest logic supply voltages (for example, $+3.3$ V or less).[64]

Figure 12.107 shows one way to do this, using several sections of 'LVC04 or 'AC04 in parallel to drive the near end of a length of 50 Ω coax that is terminated with a 50 Ω resistor at the far end. The 'LVC2G04 is fully specified at ± 24 mA output current and 3 V supply; the corresponding figure for the 'AC04 is ± 12 mA. So we're pushing things just a bit here, demanding $+60$ mA of source current. (No worry, however, if we replaced the single 50 Ω termination with a 100 Ω–100 Ω divider at the far end. That would require ± 30 mA of drive, for which we could probably get away with a single 'LVC1G04.)

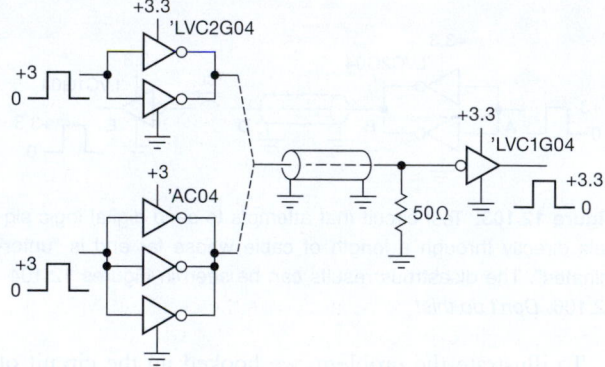

Figure 12.107. Digital logic levels driving end-terminated 50 Ω coax. Apart from losses, the received signals are full-swing replicas of the driving signal.

But it works: Figures 12.108 and 12.109 show the nice waveforms you get, in this case with a 10 m (33 foot) length, operating at the same 20 ns clocking rate. Note the clean waveforms throughout (with a small bit of ringing evident with the 'AC04 driver, probably caused by

---

[64] A nice trick that reduces the needed drive current by a factor of 2 is to use a pair of 100 Ω resistors at the far end, configured as a divider between the supply and ground.

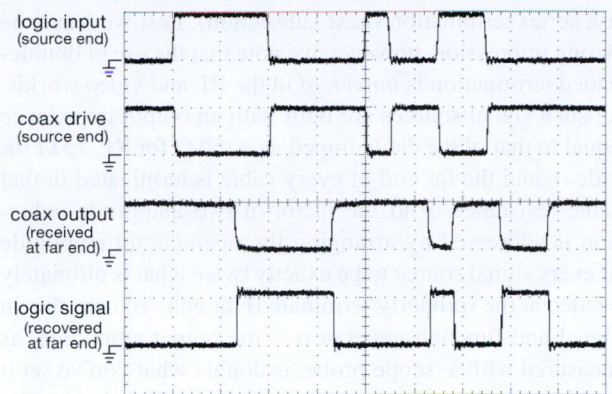

Figure 12.108. Waveforms from the circuit of Figure 12.107 ('LVC2G04 driver), with the same 20 ns clocked pulse pattern as in Figures 12.104–12.106, but with a 10 m length of coax. Horizontal: 40 ns/div; vertical: 3 V/div.

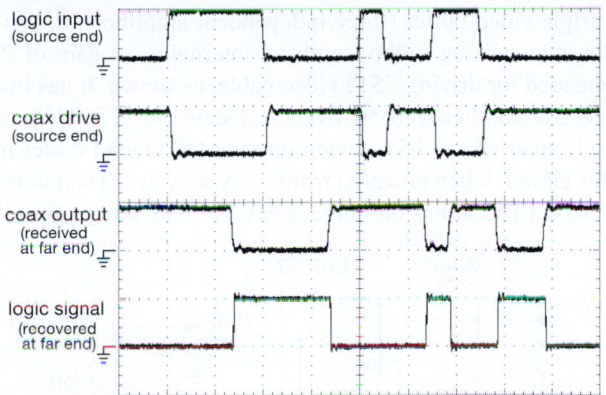

Figure 12.109. Same as Figure 12.108, but with 'AC04 driver.

the greater inductances in power and ground leads of its DIP-14 package, as compared with the compact SOT23-6 surface-mount package of the 'LVC2G04).

If you're worried about the substantial drive current needed to push full rail-to-rail logic-level swings into a 50 Ω load, and if you are not operating at maximum logic speeds, you can substitute a "MOSFET low-side driver" IC for the paralleled logic inverters we used. For example, the venerable TC4420-series of MOSFET drivers accept logic-level inputs, and generate a robust rail-to-rail output swing between ground and the supply voltage, which can range from +4.5 V to +18 V. The peak output current ranges from 1.5 A to 9 A in this series – no problem driving 50 Ω! These are dual drivers, available in many package styles (including DIP) and in inverting or non-inverting flavors; they cost about a dollar in small quantity (see Table 3.8 on page 218). There are literally hun-

dreds of MOSFET driver chips available, with some very nice choices from companies like Fairchild, IXYS, Microchip, ST, and TI, among others; most are happy to operate from 5 V supplies (where they have reduced output-drive capability, but they're still pretty muscular; the 9 A IXDD609, for example, can source or sink 2 A when powered from a 5 V supply). Don't be discouraged by the rather relaxed speed specifications on datasheets (e.g., rise and fall times of 20 ns or more), because these are usually specified into the horrendous capacitive loads (1000–10,000 pF, aided and abetted by the Miller effect) of switching power MOSFETs. You'll do much better when driving a 50 Ω far-end terminated coax, which presents a pure resistive load to the driver. For example, the IXYS IXDD509 9 A non-inverting driver specifies ~25 ns rise and fall, but that's into 10,000 pF! From the curves later in the datasheet, however, you'll discover rise and fall times of 4 ns or less for supply voltages anywhere between +5 V and +30 V.[65]

A. The right way – II: Double-ended termination

The high currents needed to drive end-terminated 50 Ω coax can be remedied somewhat by adding a series resistance at the cable's input, equal in value to its characteristic impedance (i.e., 50 Ω; see Figure 12.110). Then the driver sees a load of 100 Ω (the series resistor plus the 50 Ω seen at the cable's input). That's sometimes called "double-ended termination." It has an additional advantage, namely that any signals reflected from the far end are swallowed by the input resistor, which acts effectively as a termination for signals traveling backward.

It works, but now the output-signal amplitude is half that of the driver, because the cable's input resistance forms a voltage divider with the series resistor at the input. That's why we used 5 V logic at the input, combined with inverters at the output whose thresholds are in the range of 1.2–1.4 V: either the TTL threshold 'ACT04 (see Figure 12.111), or logic running from a lower +3 V supply (see Figure 12.112). The 'ACT04 specifies input logic levels of <0.8 V and >2.0 V; the corresponding figures for the 'LVC1G04 (powered from +3.0 V) are <0.8 V and >1.7 V.

---

[65] In spite of the good rise and fall times, there tend to be substantial delay times, regardless of load capacitance, at low supply voltages; for this IXYS part, the delay times are of order 30 ns with a +5 V supply, dropping to half that at V_S=10 V. From a practical point of view, this limits operation to speeds of a few megahertz. A more serious problem at high frequencies is the rising rail-to-rail supply current, and consequent power dissipation, at frequencies above a megahertz that are due to the high internal capacitance of the large output MOS devices. Some parts show plots of no-load supply current versus frequency in their datasheets.

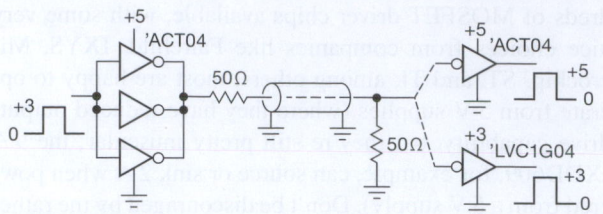

Figure 12.110. Digital logic driving double-end terminated 50 Ω coax. The received signal amplitude is half that of the driver logic output, therefore ∼2.5 V in this example.

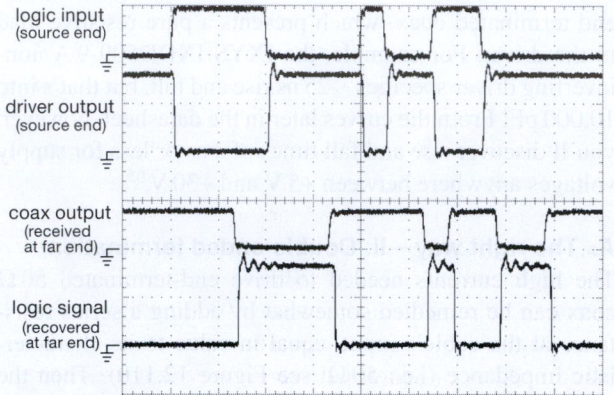

Figure 12.111. Waveforms from the circuit of Figure 12.110 ('ACT04 receiver), with the same 20 ns clocked pulse pattern as in Figures 12.104–12.109, and with a 10 m length of coax. Horizontal: 40 ns/div; vertical: 3 V/div.

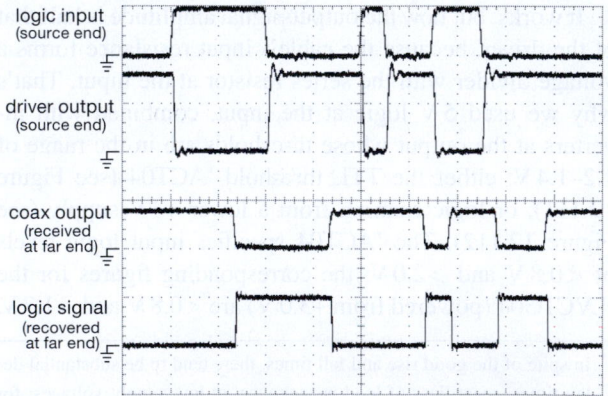

Figure 12.112. Same as Figure 12.111, but with 'LVC1G04 receiver.

The method of choice, for RF and video Because double-ended termination results in a received signal that is half the amplitude of the unloaded driver, it is not well suited for *digital logic* applications like this. It's better to

use series termination (next subsection). Lest we leave the wrong impression, however, we note that the use of double-ended termination is *universal* in the RF and video worlds: all such signal sources are built with an output impedance equal to that of the cable impedance (50 Ω for RF, 75 Ω for video), and the far end of every cable is terminated in that same resistance. And the factor-of-two amplitude reduction is addressed by arranging the open-circuit amplitude of every signal source to be exactly twice what is ultimately needed at the (properly terminated) far end. You see this in signal and function generators – the output amplitude, as measured with a 'scope probe, is double what you've set it for (because it assumes you've attached a 50 Ω load resistor, or a cable with 50 Ω far-end termination). And in the video world you'll find plenty of "video buffer amplifiers" that provide a gain of exactly ×2, to compensate for the corresponding loss when driving a terminated cable. Figure 12.113 shows an example from LTC: their LT6553 is a triple video buffer (three independent amplifiers, to handle color analog video), with an internally set gain of 2, intended for driving 75 Ω video cable, as shown. It has impressive bandwidth (650 MHz) and slew rate (2500 V/μs), and can drive a ±3.5 V swing into the 150 Ω load it sees in this circuit, when powered from ±5 V supplies. Their companion LT6554 has the same specs, but with unity gain.

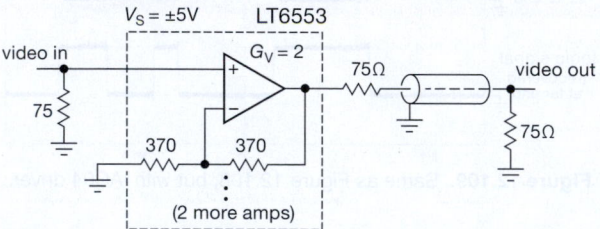

Figure 12.113. A video buffer amplifier drives a 75 Ω load through 75 Ω coax. The voltage gain of $G_V=2$ compensates for the ×2 signal attenuation caused by the series resistor at the amplifier's output.

B. The right way – III: Series termination

To summarize the previous methods: end termination with direct drive requires lots of drive current, but it delivers the full driver signal at the far end. Double-ended termination reduces the drive current, but attenuates the driver signal amplitude by a factor of 2.

There's a third way, which captures the best of each: use a series resistor at the source end and *no* termination at the far end. This is sometimes called "series termination" or "back termination" (Figure 12.114). It exploits a property of open-ended transmission lines, namely that there is a

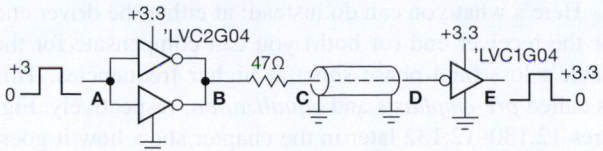

Figure 12.114. Series termination presents a load of twice the cable's characteristic impedance (therefore 100 Ω), while delivering the full driver amplitude at the far end. We recommend this method for logic outputs on instrument panels.

full-amplitude reflection from the far end, of the same sign as the incident signal, thus producing an output amplitude of twice the incident amplitude. But, because the incident amplitude is half the driver's output (owing to the voltage divider formed by the series resistor and the cable's input impedance), the net result is an output just equal to that of the driver. Voilà – full output swing, without the need to drive the cable's low impedance. And we don't get the ugly situation of signals bouncing back and forth (as in Figures 12.105 and 12.106), because the series resistor acts as a proper end termination for the backward-going signals. Instead you get one reflection off the far end, which gets swallowed at the source end, with the production of some steppy waveforms at the near end (Figure 12.115).

Series termination is the method of choice for piping logic signals through cable. This technique has the pleasant properties of (a) presenting a (less severe) load impedance

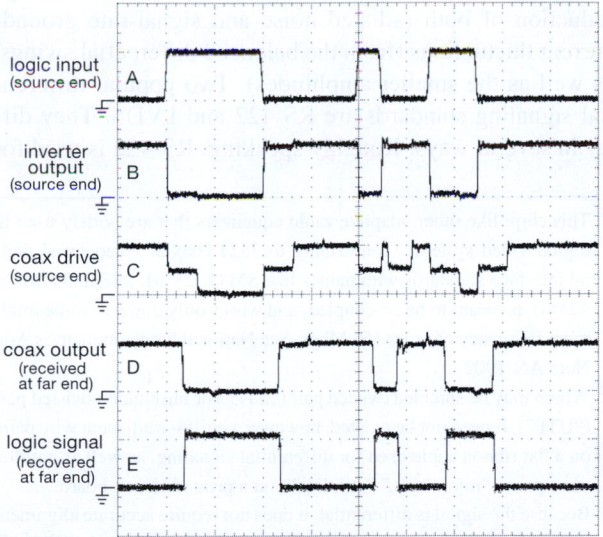

logic input (source end) — A

inverter output (source end) — B

coax drive (source end) — C

coax output (received at far end) — D

logic signal (recovered at far end) — E

Figure 12.115. Waveforms from the circuit of Figure 12.114, with the same 20 ns clocked pulse pattern as in Figures 12.104–12.112, and with an 8-foot (2.4 m) length of coax. Horizontal: 40 ns/div; vertical: 3 V/div.

equal to twice the cable's characteristic impedance, and (b) requiring *no* continuing drive current after a logic-level change has propagated down and back. To expand on this latter point: immediately after a step change, current flows from the driving gate, $I = V_{CC}/2Z_0$, or 50 mA for 5-volt logic driving a 50 Ω cable; but this ceases after a round-trip propagation delay. We have +50 mA for low-to-high steps, and −50 mA for high-to-low steps. Often we'll estimate the R_{OUT} of the driving gate and reduce the value of the added source resistor accordingly. But if the total source resistance is too low (less than 50 Ω), the returning echo will overshoot V_{CC} (for low-to-high steps) or undershoot ground (for high-to-low steps) for a time equal to a propagation delay. If the overshoot is high enough (i.e., more than 10% for a 5 V supply) it will cause current to flow in the gate's output clamp diode.

The logic family known as "AUC Little Logic" (one of TI's low-voltage logic families, one or two gates in 5-lead or 6-lead SMT packages) has an unusual output structure that is well suited for driving PCB traces or lengths of coax. Its output impedance is a reasonable approximation to a series termination, so you can drive a 50 Ω transmission line directly from the logic gate output, without any series resistor. The family is optimized for a +1.8 V supply voltage. As described in the Application Report[66] the output stage consists of three paralleled inverters, such that the driving impedance changes during a logic transition: it starts low (for high drive current), then becomes an approximate match to the transmission line, suppressing ringing or reflections. Also, there is no clamp diode from the output to the positive supply, so the output is "3.6 V tolerant" and is not damaged by back reflections from the open-ended transmission line.

Although they suggest cable or trace lengths only to 15 cm or so, we found that these devices work well for substantially longer runs of cable. Figure 12.116 shows the signals when driving a 30 cm length of 50 Ω cable (RG-316 "skinny" coax) with 100 MHz NRZ[67] data (twice the rate used in Figures 12.104–12.112 and 12.115). How far can you go with this driver, and at these speeds? Figure 12.117 shows the logic input pattern, and the (open-ended) coax output (points "A" and "D" in Figure 12.114), when driving lengths of 50 Ω coax[68] up to 5 m long. Looks pretty good, to us!

---

[66] *Application of the Texas Instruments AUC Sub-1-V Little Logic Devices*, SCEA027A (September 2002).

[67] "Non-return to zero," a fancy name for just sending each bit value, as a logic level, for one clock period.

[68] RG-141 for 500 cm, RG-316 for the rest.

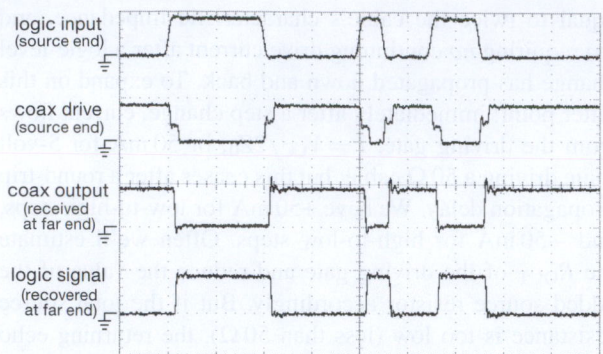

Figure 12.116. Waveforms from a 74AUC1G04 inverter operating at 1.8 V and directly driving a 30 cm length of coax, with the same pulse pattern as earlier, but at a 10 ns clock rate. The four traces (top to bottom) correspond to points A, C, D, and E in Figure 12.114, but with no resistor. The approximate half-step at the coax input confirms a drive impedance close to 50 Ω. Horizontal: 20 ns/div; vertical: 2 V/div.

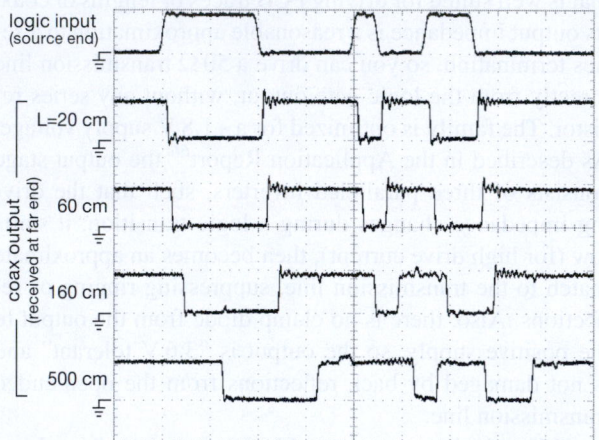

Figure 12.117. Same setup as Figure 12.116, showing received waveform at far end of coax cables with the indicated lengths. Horizontal: 20 ns/div; vertical: 2 V/div.

C. Driver pre-emphasis and receiver equalization

The measured waveforms above look pretty good. But we're not really pushing the limits, at these data rates of 50–100 Mbps NRZ. Things get pretty messy when you try to send upward of a few hundred megabits per second, because the coax cable itself becomes lossy and attenuates higher frequencies (see Appendix H). For example, the popular RG-58A (used for everybody's BNC patch cords) attenuates about 10 dB per 30 m at 500 MHz. You do better, of course, with cables of lower loss (e.g., RG-8 attenuates about 5 dB per 30 m). But these are bulkier – RG-8 cable is 10 mm in diameter, double that of RG-58.

Here's what you can do instead: at either the driver end or the receiver end (or both) you can compensate for the cable's loss (and phase shift) at higher frequencies. This is called *pre-emphasis* and *equalization*, respectively. Figures 12.130–12.132 later in the chapter show how it goes, in the case of a differential twisted-pair cable or printed-circuit differential stripline. For a (single-ended) coax line you could use a chip set like NSC's DS15BA101 and DS15EA101; the latter is a receiver with adaptive equalizer, which can apply up to 35 dB of boost at 750 MHz.[69] You can think of this trick as analogous to "turning up the treble," in an audio system, though here we need to worry about phase as well. It works quite well, owing to the very good signal-to-noise ratio of these large signal swings running through shielded (or twisted-pair) transmission lines. We'll see this next, in connection with LVDS signaling on differential pairs.

12.10.3 Differential-pair cable

There's another way to convey digital signals on cables, namely to use differential signaling, most often by way of a twisted-pair cable.[70] A common example of the latter is "Cat-5" (or Cat-6) Ethernet cable, which is an unshielded cable containing four independent twisted pairs, with characteristic impedances of 100 Ω. Some advantages of differential signaling are the suppression of common-mode interference and ground noise, the ability to use smaller signal swings (hence smaller drive currents),[71] and the large reduction of both radiated noise and signal-rate ground-current fluctuations (from the balanced differential swings, as well as the smaller amplitudes). Two popular differential signaling standards are RS-422 and LVDS. They differ in several ways. Roughly speaking, RS-422 is used for

---

[69] This chip, like other adaptive cable equalizers that are widely used in digital video systems (conforming to 75 Ω coaxial video serial digital interface standards with names like SMTE 259M, 292M, 344M, or 424M), is meant to be ac-coupled, and works only down to some minimum frequency of order 150 Mbps. See National Semiconductor's App Note AN-1909.

[70] Which may be shielded twisted pair ("STP") or unshielded twisted pair ("UTP"). It need not be twisted, however: you'll see adjacent wire pairs on a flat ribbon cable used for differential signaling, as well as parallel trace pairs ("microstrip," "stripline") on a printed circuit board.

[71] Because the signal is differential, it does not require accurate alignment of thresholds (as, for example, with the fussy single-ended 10K/100K ECL logic, whose ∼0.8 V swing mandates careful temperature compensation); the differential signal pair need stay only in the common-mode range of the receiver, e.g., 0 V to +2.4 V for LVDS, with its ∼0.35 V differential swing.

data rates up to a few megabits per second, and cable runs up to a kilometer. It uses differential *voltage* drive, and is common in industrial control applications. By contrast, LVDS is used for data rates to several gigabits per second, over distances to several meters. It uses differential *current* drive, and is common in short-range high-rate applications such as backplanes (e.g., PCIe[72]) and transmission of serial data (e.g., SATA, Firewire). Both are point-to-point connections, but provide multidrop variants (RS-422 → RS-485; LVDS → M-LVDS).

A. RS-422 and RS-485

These comprise a popular industrial data bus, improving on the venerable RS-232 data link standards (see discussion at §§12.10.4 and §14.7.8), and substantially extending the latter's capabilities, see Figure 12.134. The RS-422 and RS-485 standards[73] specify signal properties used in a differential voltage-drive arrangement like that shown in Figure 12.118. The differential outputs typically swing most of the way between ground and the +5 V rail, although the specification permits differential-output swings as small as ±2 V or as large as ±10 V. The receiver must respond to differential inputs as small as ±0.2 V, over a common-mode range of −7 V to +7 V (−7 V to +12 V for RS-485). The characteristic impedance of differential twisted pair cables is typically 100–120 Ω, so you've got to terminate the far end with that value resistor. You often see both ends terminated (as in Figure 12.118), which is not necessary (or desirable) with unidirectional RS-422, but required with the bidirectional (or multipoint) variant (RS-485).

Figure 12.119 shows measured signals traversing one of the four twisted pairs in a 140 m length of Cat-6 Ethernet cable, with a, by now, familiar bit pattern clocked at 10 MHz (i.e., an NRZ rate of 10 Mbps). Although this is ten times faster than allowed by the RS-422 specification (Figure 12.134), the differential nature of the received signal allows clean logic recovery nonetheless; note the propagation delay of ~700 ns (mostly caused by the 4.7 ns/m signal velocity in the cable).

Differential signaling provides impressive immunity to common-mode interference. The latter may arise from nearby signal-carrying wires or from radiated signals (from intentional emitters like radio/TV, wireless networks, etc.);

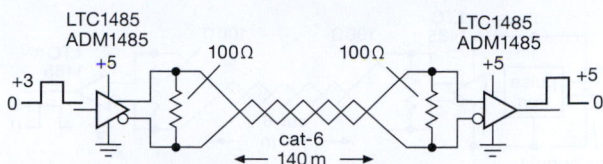

Figure 12.118. Differential voltage signaling with RS-422/485 over a length of unshielded twisted pair, using transceiver chips with 3-state outputs (each chip has a separately enabled Tx and Rx, sharing a common differential pair).

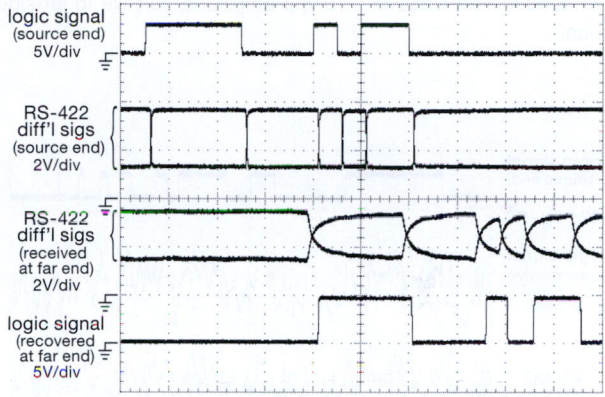

Figure 12.119. Waveforms from the circuit of Figure 12.118, with a 100 ns clocked pulse pattern, and a 140 m length of Cat-6 network cable. Horizontal: 200 ns/div.

it's also common to see a difference in ground potentials between equipment that is plugged into different power outlets. We have seen as much as a volt or two of 60 Hz ac between instruments in the same room!

We hooked up the circuit in Figure 12.120 to illustrate common-mode rejection in RS-422 signaling. The pseudo-random noise generator adds a ~15 Vpp noise voltage into the (floating) ground of the signal source, as seen from the destination end. The results are shown in Figure 12.121: both the 5 V logic signal and the RS-485 differential pair signals are hopelessly noised up, reaching peak levels of −7 V and +12 V (the specified receiver common-mode range) as seen from the destination end. But the magic of common-mode rejection recovers the original logic intact. Note that the noise is very much "in-band," with wiggles on the time scale of the digital data. Although the figure captures a single-shot data burst, the result is robust – we ran it many times, with never an error.

Slew rate limiting If you don't need the speed, it's wise to choose driver ICs of lower slew rate because you get less signal coupling from signals on adjacent pairs. RS-422 and

---

[72] Shorthand for "PCI Express," itself shorthand for "Peripheral Component Interconnect Express."

[73] Officially known as ANSI TIA/EIA-422 and TIA/EIA-485; however, most engineers continue to use the original designations "RS-422" and "RS-485," or, more loosely, they'll just say "422" or "485." In context it's clear what they mean.

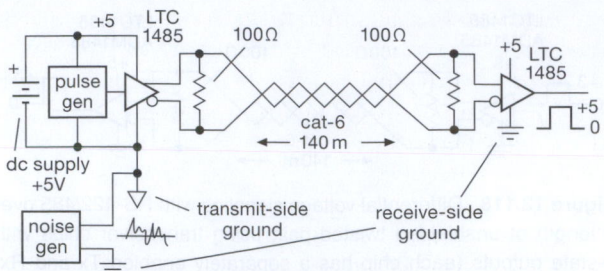

Figure 12.120. Torture test for RS-422 common-mode noise rejection. We floated the driver-side circuitry, and drove its "ground" with band-limited noise of ~15 Vpp amplitude, relative to far-side ground.

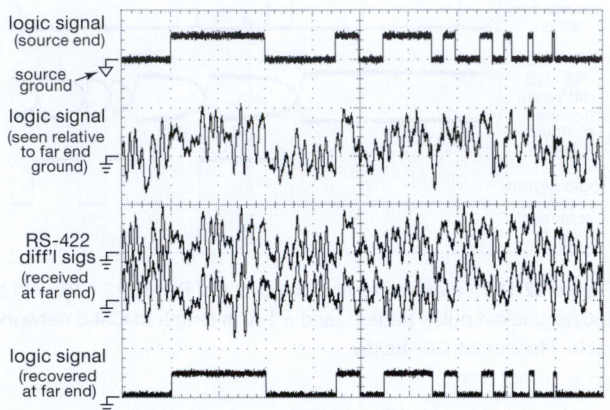

Figure 12.121. Waveforms from the circuit of Figure 12.120. As seen from the far end, the input logic signal is buried in the noise, extending from −7 V to +12 V. The individual differential signals are likewise a mess, but the recovered logic output is error free. The signal pattern includes pulses with widths from 200 µs down to 5 µs. Horizontal: 100 µs/div; vertical: 10 V/div.

RS-485 drivers are available with a selection of slew rates, for example the Maxim MAX3293-95 series, with specified rates of 250 kbps, 2.5 Mbps, or 20 Mbps. Other examples include the MAX481–489- and 1481-1487-series, the LTC2856–2858-series, and the 65ALS176 versus the 75ALS176B.

RS-422 versus RS-485 RS-485 is basically RS-422 with some additions that make it possible to have several drivers share a single signaling pair: this requires that the drivers have an ENABLE input, so they can be put into a high-impedance (noninterfering) state, analogous to the use of three-state drivers on a shared (single-ended)

data bus line.[74] RS-485 interface chips (e.g., the classic 75ALS176, or LTC1485) usually combine a transmit–receive pair, sharing the same differential signal lines ("half duplex"),[75] with complementary ENABLE pins (called DE and RE′); such a *transceiver* chip looks like Figure 12.122.

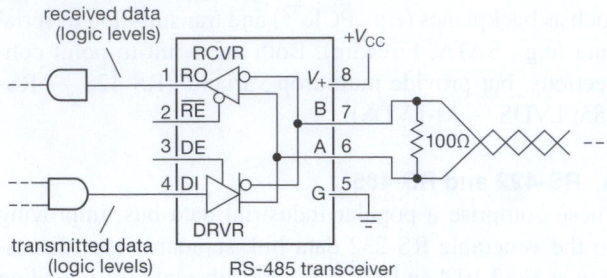

Figure 12.122. An RS-485 *transceiver* combines driver and receiver in one package, sharing a single twisted pair. Separate ENABLE pins (usually tied together) permit operation in either direction. The pin numbering shown is an industry standard.

Along with the ENABLE feature that is necessary for multiple drivers (called "multipoint"), the RS-485 specification beefs up some other RS-422 specifications: (a) it expands the receiver common-mode input range (−7 V to +12 V), thereby allowing symmetrical common-mode swings (up to ±7 V) around the traditional power-supply and signaling levels of 0 V and +5 V; (b) it lowers the minimum load resistance (to 54 Ω), which is necessary because a driver at either end requires end-termination at the other end (thus a 100–120 Ω at *both* ends). The RS-485 specification also (c) expands the allowable common-mode range that can be applied to a driver *output* (in the disabled state), from a diode drop beyond the rails (RS-422) to the full −7 V to +12 V range of an RS-485 input. This is necessary, of course, because the driver and receiver are both connected permanently to the signal lines in a typical bidirectional (or multipoint) arrangement; if the driver's output clamped to ground or the positive rail, it would defeat the receiver's wider common-mode range.

Most RS-485 transceivers satisfy also the narrower RS-422 specifications, so you might as well just use one of the hundreds of available RS-485 interface chips (from manufacturers like Analog Devices, Intersil, LTC, Maxim, or TI), even if you are sending data in only one direction; simply enable what you are using at each end. Some common choices are the 'ALS176 (and its many imitators –

---

[74] Good references: Analog Devices AN-960 and AN-727, National Semiconductor AN-759, Maxim AN-3776, and TI SLLA112.

[75] An interface chip that brings RS-485 driver and receiver pairs out separately is called "full duplex."

65LBC176, 65HVD1176, etc.), the 75176 and 75ALS180 classics, the LTC1480/5, and the ADM1485.

Isolated RS-422/485. Scenario: a "factory floor," with a host of automated machine tools and with products moving along on conveyor belts (think of a scene from the ever-informative "How It's Made" series on cable channels like Discovery Channel or Science Channel). Sensors send digital information through cables that snake along on overhead cable trays, and wind up at a central computer control; the same cables, or different ones, carry commands back to actuators. This centrally controlled network choreographs factory activities, ultimately directed toward the practical (and profitable) goal of... making stuff. It's common to see these signals piped around as differential RS-485 or a variant like the Process Field Bus (PROFIBUS®) that uses a similar "physical-layer" signal.[76]

These activities may be spread out over distances of hundreds of feet, and may go between buildings. Let's face it, with all that heavy machinery thumping along, you're likely to see common-mode transients that exceed even the "generous" specifications of RS-485.[77] The solution here is to use *isolated* interface chips, in which the RS-485 signals have their own independent ground, galvanically isolated from the ground of the logic signals. This requires a second dc power source, of course, floating with respect to the logic signal ground.

There are many isolated RS-485 interface chips to choose from. The clever designers use various tricks to make fast digital signals cross a gap that can withstand a kilovolt or so. For example, the LTC1535 (or MXL1535) "isolated RS485 Transceiver" uses small capacitors to couple the (modulated) digital signals (Figure 12.123). It also

helpfully includes a high-frequency ($\sim$420 kHz) oscillator, whose output you can use with a transformer to isolate and rectify to make the isolated dc for the RS-485 signal side. The ISO15/35 from TI is a less-expensive alternative, which omits the oscillator; you've got to provide isolated dc from scratch (either from an isolated dc–dc converter, or an ac-line-powered dc supply).[78]

Another approach is to couple with small (chip-scale) transformers. This technique is used in the ADM2485 series of isolated transceiver from Analog Devices, which also provides an oscillator output that can be used for generating the isolated dc (like the LTC1535). Or, you can add the popular MAX845 or MAX253 chips, which generate a pair of complementary square wave outputs at $\sim$0.75 MHz, suitable for directly driving a small isolation power transformer, good to a half-watt or so of isolated dc. The electronics industry makes it easy to use this part: major transformer companies even offer "MAX845 transformers."

An interesting technique is used in the IL3485 from NVE Corp, namely the "giant magnetoresistive effect" (GMR, widely used in hard-disk drives to sense the magnetically stored bits on the rotating platters). Instead of conventional transformer coupling, where a secondary winding senses flux *changes*, the NVE part uses GMR to sense the field directly.

Finally, isolated transceivers like the MAX1480/90 use optocouplers for the digital signals. And, to make your life really easy, they include an internal transformer to generate isolated dc power. These hybrid ICs include also the diodes and capacitor to complete the power circuit, so all you need to provide is a +5 V logic-side supply. That's the good news; the bad news is that these puppies will set you back twenty bucks (in small quantities), compared with $5–$8 for the capacitor- or transformer-coupled chips.

Ethernet PHY When thinking about moving digital data across isolation barriers, don't forget about the optical couplers and fibers earlier in this chapter, and also about isolation via pulse transformers, as used, for example, in local-area networks such as Ethernet. Figure 12.124 shows the physical layer ("PHY") of an Ethernet link, with pulse isolation transformers (which everyone just calls the "magnetics."). They provide excellent rejection of common-mode interference (and of course of differences in potential at the ends), using both a transformer and a common-mode

---

[76] PROFIBUS transceivers have higher maximum data rates, typically 30 Mbps or 40 Mbps, and generally meet the (less-rigorous) specifications of RS-422 and RS-485; check out parts like the 65ALS1176, ADM1486, ISO1176, or ISL4486. Some other industrial buses that use RS-485 signaling include BITBUS, Data Highway (DH-485), INTERBUS-S, Measurement Bus (DIN 66348), Optomux, P-NET, and Series 90 (SNP).

[77] As eloquently expressed in the MAX1480 datasheet, "The RS-422/485 standard is specified for cable lengths up to 4000 feet. When approaching or exceeding the specified maximum cable length, a ground-potential difference of several tens of volts can easily develop. This difference can be either DC, AC, at powerline frequency, or any imaginable noise or impulse waveform. It is typically very low impedance so that if a connection between the two grounds is attempted, very large currents may flow. These currents are by their nature unstable and unpredictable. In addition, they may cause noise to be injected into sensitive instrumentation and, in severe cases, might actually cause physical damage to such equipment." Amen.

[78] You sometimes see industrial RS-485 hookups with bused dc (+24 V or +48 V "telecom dc") bundled in the same cable as the RS-485 signal pair(s); hang an isolated dc–dc converter onto that, at each transceiver node, to power the RS-485 signal side of a transceiver like the ISO15/35.

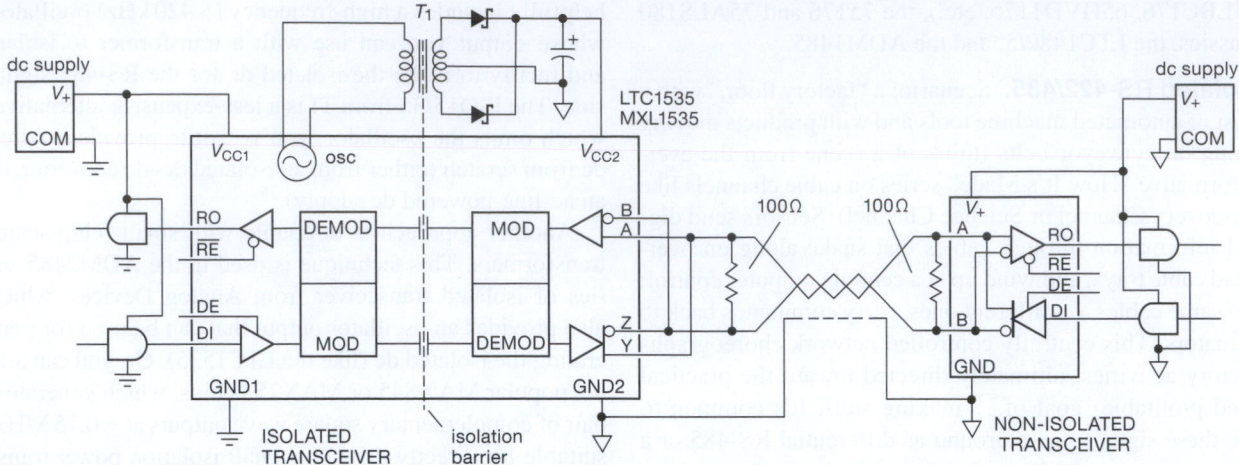

Figure 12.123. An "isolated" RS-485 transceiver galvanically separates the circuitry connected to the cable pair from the logic-level circuitry. The far-end ground ("GND2," triangular symbol) is brought back to the isolated transceiver, but is never connected to the local ground ("GND1," normal symbol). Such an arrangement prevents ground loops, and can accommodate common-mode and ground offset potentials of hundreds of volts. Additional isolated transceivers can be connected ("multidrop") along the span, referenced to the same GND2, each with its own isolated RS-485-side power supply; such midspan "stubs" must be kept short in length, and without $100\,\Omega$ terminating resistors.

choke, as shown.[79] If you've got a microprocessor in your system already, why invent your own isolation hardware, when Ethernet works just fine? And, if you need to go greater distances, you can use a media converter to carry Ethernet signals on fiber. Check out the offerings from manufacturers like Allied Telesis.

B. LVDS

In contrast to RS-422, which is intended for modest data rates (10 Mbps and below) over relatively long cables (to 1 km), the LVDS (low-voltage differential signaling, also known as RS-644) standard is aimed at much higher data rates (to 1 Gbps and above) over shorter cables (to ~10 m) or even shorter circuit-board trace runs. Rather than driving the wire pair with crisscrossing *voltages* of a few volts amplitude, LVDS switches *currents*: an LVDS driver sinks and sources 3.5 mA (nominal) into a wire pair that is terminated at the far end in its characteristic impedance (usually $100\,\Omega$). That produces a differential voltage at the far end of ±350 mV. The driver is obliged to maintain a common-mode voltage, nominally +1.2 V; so the receiver sees crisscrossing voltages of approximately +1.0 V and +1.4 V. The

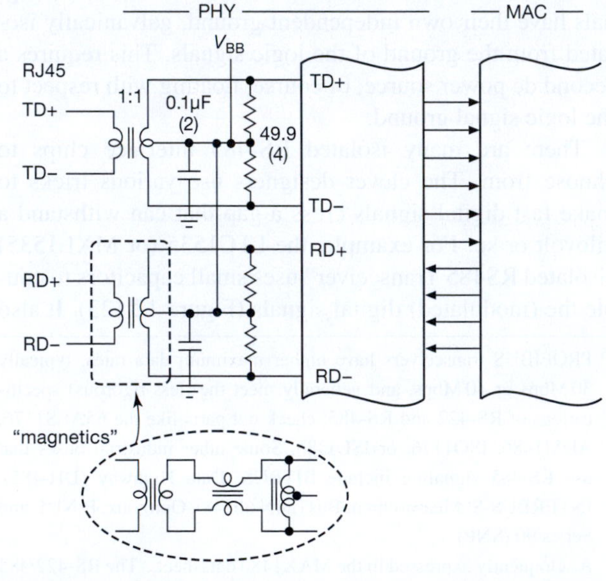

---

[79] You get additional robustness by adding surge-protecting components, notably the Transient Blocking Units (TBU™) from Bourns. These are little 2-terminal elements that go in series with the signal leads from the transformer: they act like a low resistance, up to a critical current, at which point they go into a high resistance state.

Figure 12.124. Ethernet uses transformer coupling and common-mode chokes for robust isolation of its differential signaling. Closest to the wiring is the physical layer ("PHY"), followed by the media-access control layer ("MAC," as in "MAC address").

relatively low common-mode voltage was chosen deliberately to accommodate both driver and receiver chips operating at low supply voltages. That's important because digital chips are moving steadily to lower voltages, and the

LVDS interface is often incorporated right into a complex chip that wants to run at low voltages.

We hooked up the circuit in Figure 12.125 (analogous to Figures 12.120 and 12.121) to illustrate LVDS's rejection of common-mode interference. We injected a trapezoidal waveform at ~30 MHz into unused pairs of a network cable, and cranked up the amplitude until the received signal reached the specified common-mode receiver limits of 0 V and +2.4 V. The results are shown in Figure 12.126: You can see the two received signal voltages weaving busily past each other, at the command of the input signal (delayed by that speed-of-light nuisance, of course), cleanly recovered by the LVDS receiver.

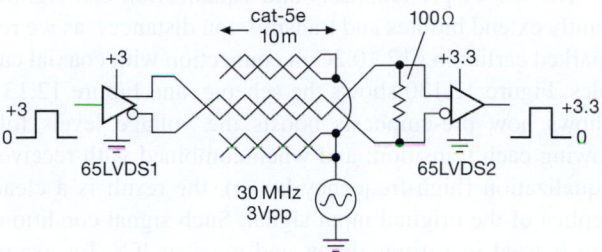

Figure 12.125. Torture test for LVDS common-mode interference rejection. We applied a ~30 MHz signal to two unused pairs at the far end of a 10 m length of Cat-5e network cable, while transmitting a pulse sequence (clocked at 20 ns) through a different pair.

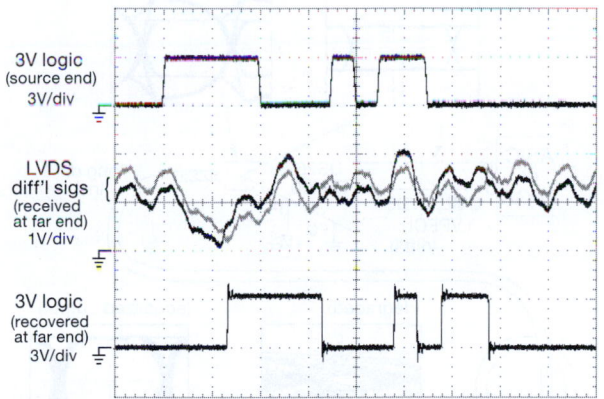

Figure 12.126. Waveforms from the circuit of Figure 12.125. The injected signal adds up to ±1 V of common-mode interference to the ~400 mV differential 50 Mbps received signal. Horizontal: 40 ns/div.

To illustrate the problem of conveying digital signals between instruments whose grounds differ by a small amount of 60 Hz ac, we rigged up the circuit of Figure 12.127. We "floated" the source device's ground, analogous to the arrangement earlier with RS-422 (Figure 12.120), then drove

it with a 2 Vpp sinewave. The results are shown in Figure 12.128, where we've clocked down the data rate so you can see the powerline frequency. Note that LVDS permits only 2 Vpp of added common-mode signal, compared with 14 Vpp for RS-422/485; but with LVDS you get lots of speed, and compatibility with low-voltage logic.[80]

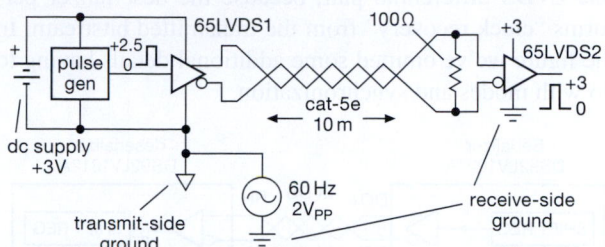

Figure 12.127. Powerline common-mode test setup, using a floating LVDS signal source (complete with pulse pattern generator and power supply).

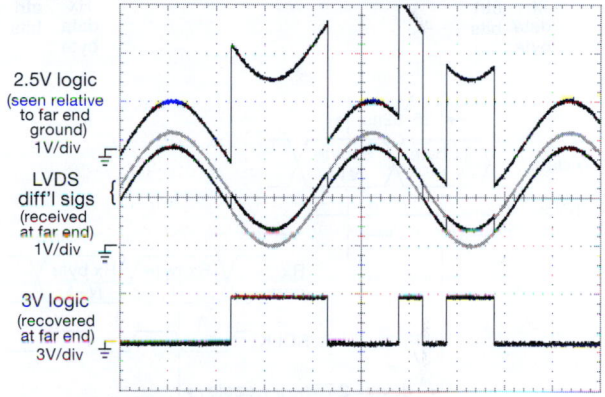

Figure 12.128. Waveforms from the circuit of Figure 12.127. The ~400 mV differential signal is superposed on the added 2 Vpp 60 Hz sinewave, reaching the specified receiver common-mode limits of 0 V and +2.4 V. Horizontal: 4 ms/div.

The LVDS protocol is widely used in serializer-deserializer links (see also §12.8.4, §12.10.5, and §14.7), in which a fast serial link connects a pair of separated parallel registers. At each end you think you're talking to a parallel port (and you are!), but in between the data travels as serial bits. A common data width is 10 bits, which lets you send a byte plus two extra bits to signify whatever you want (a new byte, or the beginning of a new "frame"

---

[80] You can get LVDS receivers with extended common-mode capability, for example the 65LVDS34, which specifies operating common-mode range of −4 V to +5 V (they do it with a resistive input divider, so R_{in}=250 kΩ).

of bytes). Figure 12.129 shows how it goes, in this case with a SERDES pair with a relatively relaxed speed specification: the transmit clock, which times the entry of 10-bit data symbols, can be in the range of 16–40 MHz. The serialized data bits that go over on the LVDS link are at ten times that rate, i.e., up to 400 Mbps.[81] You need only one LVDS differential pair, because the deserializer performs "clock recovery" from the transmitted bitstream. In the figure we've omitted some additional details having to do with modes and synchronization.

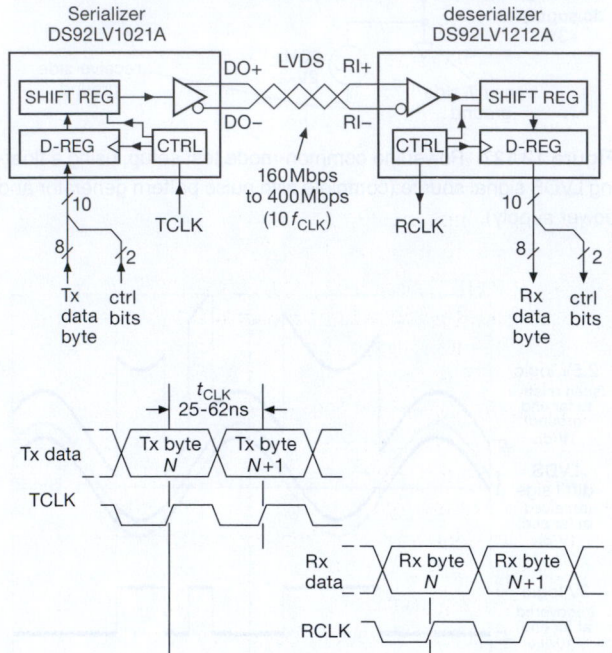

Figure 12.129. A serializer-deserializer pair lets you clock parallel data into a register at the transmit end; it appears, magically, as parallel data at the receive end. There are a few clock cycles of delay, but nothing to write home about.

LVDS signal conditioning As we remarked earlier (§12.10.2C), the use of transmit pre-emphasis and receive equalization can compensate for the frequency-dependent loss of the connecting cable. These *signal-conditioning* techniques reduced greatly the effects of data-dependent jitter, which is also known as *intersymbol interference*, or ISI.

This ISI effect is worth a few words of explanation: because of the lowpass filtering effect of the losses in ca-

bles or printed-circuit traces, particularly when operating at very high bitrates (gigabits per second), the initial received signal voltage at the beginning of each bit cell depends on the previous bit (or bits), and so the time to cross the threshold will vary somewhat, depending on the previous bit(s). Intersymbol interference plagues high-speed communication in all forms. You can see this effect in the 1.5 Gbps recovered LVDS waveforms in Figure 12.132, in which the 'scope traces, triggered by the clean clock signal, show the scatter of signal levels on the data line carrying a pseudo-random data pattern. This is called an "eye diagram;" it is a standard tool for visualizing noise and jitter in assessing signal quality in a clocked data stream.

The use of pre-emphasis and equalization can significantly extend bitrates and transmission distances, as we remarked earlier in §12.10.2C in connection with coaxial cables. Figure 12.130 shows the scheme, and Figure 12.131 shows how pre-emphasis boosts the voltage levels following each transition; and when combined with receiver equalization (high-frequency boost), the result is a clean replica of the original input signal. Such signal conditioning is used in various driver and receiver ICs, for example the DS25BR100/200/400 and DS25CP102 series from NSC; see their App Note AN-1957 for details and illustrative waveforms.

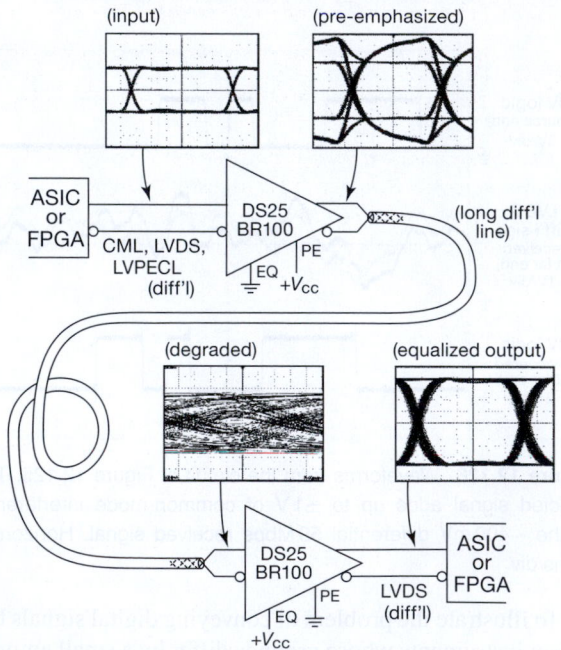

Figure 12.130. Driver pre-emphasis, receiver equalization, or both can compensate for cable losses when transmitting at high data rates over lossy media.

---

[81] The serial bitrate is actually 12 times the clock rate, because the serializer adds two clocking bits.

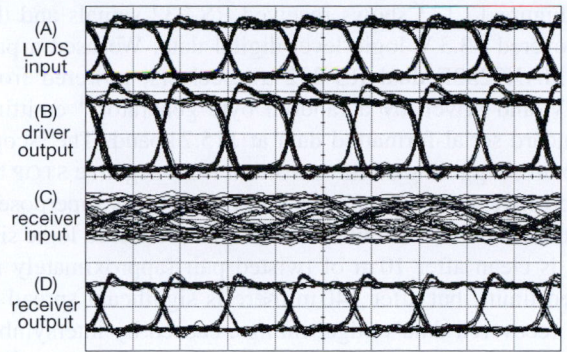

Figure 12.131. Waveforms from a 1.5 Gbps pseudorandom LVDS signal traversing a 2.5 m length of differential stripline on a printed circuit board. A. Input to DS25BR120 driver; B. Driver output, with pre-emphasis; C. Input to DS25BR110 receiver; D. Receiver output, with pre-emphasis and equalization. Vertical: 500 mV/div; horizontal: 500 ps/div.

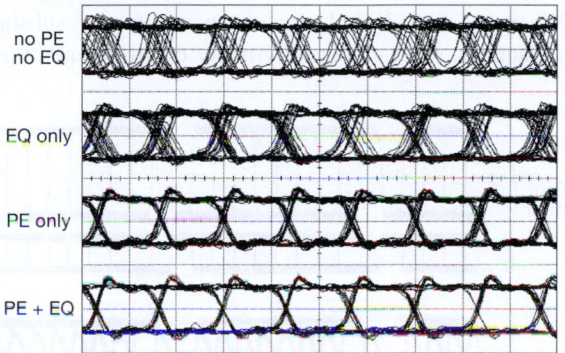

Figure 12.132. Receiver output waveforms ("eye diagrams") from the same setup as Figure 12.131, showing the effect of varying amounts of driver pre-emphasis and receiver equalization. The bottom trace represents PE=+9 dB and EQ=+8 dB at 1.5 GHz. Vertical: 500 mV/div; horizontal: 500 ps/div.

12.10.4 RS-232

This is a signaling format that goes back to the 1960s, intended originally for low-speed (<19.2 kbps) serial links between alphanumeric terminals (e.g., the legendary DEC VT-100) and computers. RS-232 has been revised several times and is now known officially as EIA232; although RS-232 ports are still seen on some computers and instruments, the standard is considered old-fashioned, and arguably headed for the trashcan. It's still with us, however (and perhaps always will be[82]). Although not part

of the standard, the data is usually sent as asynchronous 8-bit serial data bytes, with a synchronizing START bit, and one or two STOP bits (these extra bits allow the receiver to resynchronize after each byte; see §14.7.8). The usual bit rates (also not specified in the RS-232 standard itself) are power-of-2 multiples of 300 bps (thus 300, 600, 1200, 2400, 4800, 9600, and 19,200 bps, supplemented with an interleaved set consisting of 14.4 kbps, 28.8 kbps, 57.6 kbps, and 115.2 kbps). Thus, for example, when you set up a serial port for "9600 8N1," you are sending groups of 8 bits, with one START and one STOP bit, and no parity, at 9600 bits/s. Note that the synchronizing bits (START and STOP) are included in the overall bitrate measure; for this example, then, you're sending a payload of 960 bytes/s.

What the standard *does* specify are the signaling voltages, load resistance and capacitance, and slew rates, along with the connector pinouts. We'll see RS-232 again, in the context of computer communications, in Chapter 14. But here we stick to the use of RS-232, at the physical level (hence the term "PHY"), as a way to drive cables with digital data. The RS-232 voltage levels are bipolarity, with legal driver output voltages of either −5 V to −15 V (logical 1, also known as "MARK"), or +5 V to +15 V (logical 0, also known as "SPACE"). RS-232 drivers and receivers are inverting, so MARK corresponds to logic HIGH at the input of a driver or at the output of a receiver (see Figure 12.135). As indicated in the figure, RS-232 is single ended; and, because of the relatively low signaling rates, the cable is unterminated. That's a terrible thing to do, normally, because rapidly changing signals (i.e., on a time scale shorter than the round-trip signal travel time) reflect from the open end, as we've seen. RS-232 finesses that problem by specifying a maximum slew rate (30 V/μs), which is slow enough for typical cable runs of fifteen meters (the original specified maximum, later replaced with a maximum load capacitance) or less. To be in compliance with the standards, the load capacitance must be 2500 pF or less;[83] and the load resistance is a nominal 5kΩ (±2kΩ). RS-232 drivers must withstand a continuous short to ground or to any dc voltage in the range ±25 V.

---

[82] It's hard to completely dismiss RS-232, when so many test instruments are happily working well with it, as one of their several inter-

face choices. The move toward USB is being stolen by Ethernet instead (or other replacements), leaving an RS-232 connector as the easy-to-implement port. Unlike more advanced serial interfaces like USB and Firewire, RS-232 does not require any "smarts" – there's no initialization or negotiation needed. Partly for this reason, most microcontrollers include an easy-to-use serial port controller. And for connection to contemporary laptop computers and the like, you can get USB-to-RS-232 adapter pods, for example from FTDI.

[83] Which corresponds to 50 m of Cat-5 twisted pair (50 pF/m), or 25 m of 50 Ω coax (100 pF/m).

So, RS-232 rumbles along, slowly. The good news is that it can ride along in a multiwire unshielded cable, which need not act as a well-behaved transmission line; the limited slew rate minimizes both crosstalk and reflections. The bad news is that the driver needs both positive and negative supply voltages of at least ±5 V. The original driver–receiver chips (1488/1489, and their CMOS follow-ons DS14C88/89 and MC145406) required such dual supplies (±9 V, nominal), which is asking a lot in a digital system (like a computer motherboard) that runs on positive supply voltages only. Maxim was the first to introduce RS-232 drivers with an on-chip charge pump (flying capacitor) voltage doubler and voltage inverter, to generate ±10 V from a single +5 V; they named it (naturally) the MAX232. There are now dozens of such chips (some with the capacitors included in the package, e.g., the MAX203 or LT1039), spanning a range of maximum speed, power consumption, supply voltage, number of drivers and receivers in one package, and so on. As an example, the MAX3232E is a dual transceiver (two drivers, two receivers) that runs from a single +3 V to +5.5 V supply, requires four external 0.1 μF capacitors, and is guaranteed to meet RS-232 specifications to 120 kbps; it comes in five different packages (SMT and DIP). The -E suffix designates enhanced robustness to electrostatic discharge, namely ±15 kV ESD-protected, specified with the Human Body Model (HBM, recall §12.1.5) of a 100 pF charged capacitor in series with a 1.5kΩ resistor.

The RS-232 interface specification includes a number of additional control signals, intended for hardware "flow control" when a terminal is connected to a computer (see §14.7.8); these have names like Data Terminal Ready (DTR), Data Set Ready (DSR), Request to Send (RTS), and Clear to Send (CTS). You can ignore all of this if you simply want to use an RS-232 interface chip pair to send low-speed digital data over a wire connection. In fact, these are often ignored even in computer serial ports: you can simply use the transmit data (TD) and receive data (RD) lines, plus ground, and do the flow control with software (more on this in Chapter 14). Note, however, the curious (and confusing) official naming of the actual signal lines TD and RD: you would expect that TD is the output from a driver, which should be connected to RD of the (distant) receiver. Not so! In RS-232 nomenclature, a device is either Data Communications Equipment (DCE) or Data Terminal Equipment (DTE); the signal sent by the latter is called TD *at both ends of the link!* (and likewise for RD). Go figure. In practice, most engineers ignore this confusion; they call an outgoing signal TD and an incoming signal RD, at whichever end it's happening.

Figure 12.133 shows received RS-232 signals and the recovered +3.3 V logic-level digital data. We used a pair of MAX3232E dual RS-232 transceivers, powered from +3 V, and driven by a random byte generator[84] emitting standard serial-formatted data at 115.2 kbaud. The 'scope capture, triggering on the transition from negative STOP bit to positive START bit, shows many such bytes superposed, so the data bits attain both values. The recovered logic signal is clean after 10 m of twisted pair (approximately its rated limit), but after 140 m there is significant spread in the recovered data's edge timings, caused by intersymbol interference (§12.10.3B) in the slow-transitioning RS-232 signal. Here, the slow transition time is comparable to the time for one bit (1 "UI," or unit interval), causing the time to cross the 0 V threshold to vary somewhat, according to the preceding bit values. You can see this effect in the long cable data in Figure 12.133, where the received cable signal falls short of its asymptotic voltage levels (approximately ±4.5 V) by as much as a volt.[85]

When sending digital signals on wires over substantial distances, or in electrically noisy environments, there

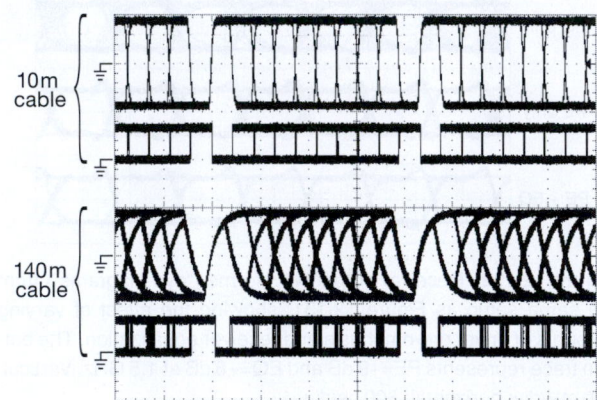

Figure 12.133. RS-232 signaling uses single-ended voltage drive, alternating between positive and negative voltage levels. These 'scope captures show received random bytes ("8N1") at 115 kbaud (top trace of each pair) after transmission through two lengths of Cat-5e/6 network cable; the drivers and receivers are inverting, as can be seen in the recovered logic signal (bottom trace of each pair). The marginal performance with the longer cable is consistent with Figure 12.134's limit of ~10 kbaud for this cable length. Driver/receiver: MAX3232E at +3 V. Horizontal: 20 μs/div; vertical: 5 V/div.

---

[84] The one used to generate the ~250 MB of random bytes on the *Numerical Recipes* CD-ROM.

[85] Users of RS-232 often use slower data rates, the most popular of which is 9600 baud, at which these effects are insignificant.

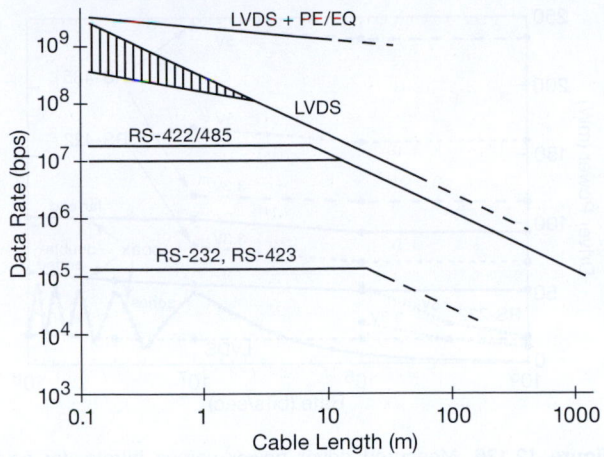

Figure 12.134. Approximate data rate limits versus cable length, for several signaling protocols. Your actual mileage may vary, depending on cable quality and interference environment. Note the speed improvement gained with pre-emphasis and equalization ("PE/EQ," see Figures 12.130–12.132).

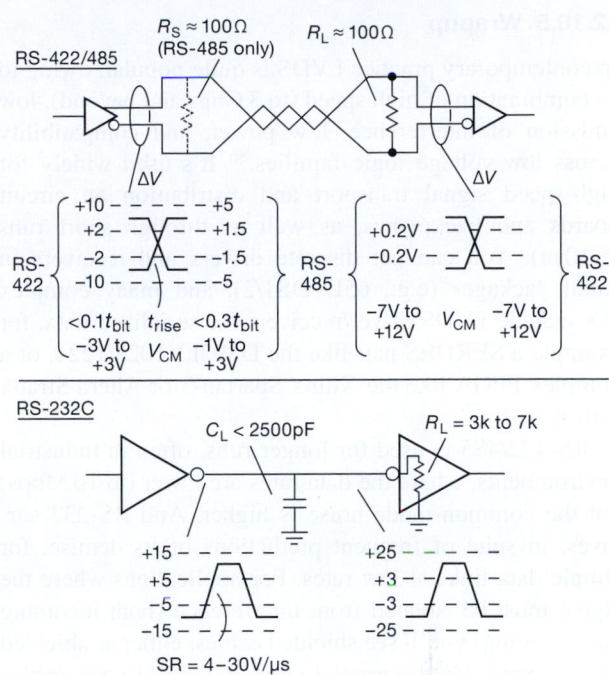

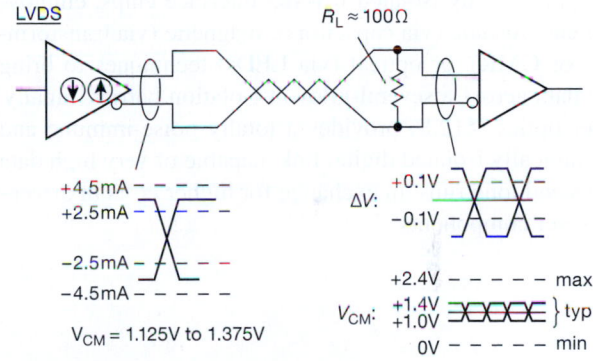

Figure 12.135. Allowable driver and receiver signal levels for LVDS, RS-422/485, and RS-232 signaling. Except for the bipolarity (and single-ended) RS-232, the output voltages from the drivers are of positive polarity only. LVDS outputs are *current-mode* drive (which are converted to a differential voltage signal at the cable's terminating resistance), at a common-mode voltage of +1.25 V; the others are *voltage-mode* outputs. All of the receivers respond to voltages, whether single ended (RS-232) or differential (LVDS, RS-422/485).

are always the problems of ground currents and of injected noise. Galvanic isolation is the best solution. Although RS-232 does not receive as much attention as RS-422/485, there is at least one good isolated driver, namely the ADM3251E from Analog Devices. It uses internal transformer coupling (with a modulator–demodulator pair) for the transmit and receive data, and an additional transformer (plus rectifier and regulator) to generate the isolated dc; the only external components are five $0.1\ \mu F$ capacitors for the (internal) charge pumps.

One final comment: we've been talking mostly about the RS-232 *physical layer* (the actual voltages on the cable), with its awkward bipolarity signaling, as a simple tool for direct transmission of digital data. In the real world, RS-232 is ordinarily used in conjunction with asynchronous serial data sources, for example between a host computer's "serial COM port" and a device such as a modem or programming pod. In that role, you need to use driver and receiver interface chips (like the MAX3232) at each end, to translate between the logic-level signals and the RS-232 signals. Although we're less than enthusiastic about RS-232 signals, we believe that the use of the simple asynchronous serial data protocol *without conversion to RS-232 voltage levels* will continue to be useful. That's because it is the last of the uncomplicated serial interface standards – later serial protocols such as USB, Firewire, and SATA require substantial brainpower to negotiate and operate the link. Most microcontrollers include one or more serial ports (called UARTs, or COM ports), which are easy to use,

and which can talk to any computer via a serial-to-USB converter such as the popular FTDI TTL-232R-3V3. This handy device plugs into a USB host at one end, and gives you a serial port with +3.3 V logic levels (for direct connection to a microcontroller, or whatever) at the other end.

12.10.5 Wrapup

In contemporary practice LVDS is quite popular, owing to its combination of high speed (to 3 Gbps and beyond), low emission of interference, low power, and compatibility across low-voltage logic families.[86] It's used widely for high-speed signal transport and distribution on circuit boards and backplanes, as well as through short runs ($\lesssim$10 m). You can get discrete drivers and receivers in small packages (e.g., 65LVDS1/2), and many complex ICs include LVDS driver/receivers for serialized data, for example a SERDES pair like the DS92LV1023/1224, or a complex FPGA like the Xilinx Spartan-3 or Altera Stratix series.

RS-422/485 is used for longer runs, often in industrial environments, where the data rates are lower (to 10 Mbps) but the common-mode noise is higher. And RS-232 survives, in spite of frequent predictions of its demise, for simple data links at low rates. For applications where the signal must be isolated from interference (both incoming and outgoing) you'll see shielded cables, either as shielded twisted pairs for differential signaling, or as coax cables for single-ended (or analog) signaling. It's also common to see galvanically isolated RS-485 interface chips, employing electrostatic (via capacitors), magnetic (via transformers, or GMR), or optical (via LEDs) techniques to bring the data across a several-kilovolt isolation barrier. Finally, fiber optics (§12.8) provides a totally noise-immune and galvanically isolated digital link, capable of very high data rates and long runs, in exchange for higher costs in driver–receiver components.

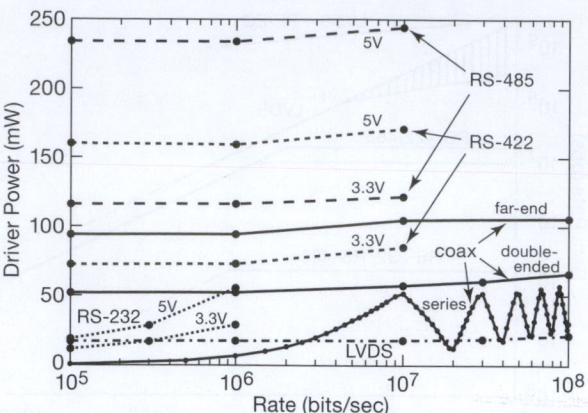

Figure 12.136. Measured driver power versus bitrate, for several single-ended (coax) and differential (twisted-pair) cable-driving configurations. All tests used a 10 m length of RG-58A coax or Cat-5e network cable, driven with alternating 1s and 0s (i.e., a square wave whose frequency is half the bit rate). The far-end termination resistance was 50 Ω for coax (except open circuit for "series" terminated), and 100 Ω for twisted pair (except 5 kΩ for RS-232) with another 100 Ω at the source end for RS-485. Note the effect of reflections from the mismatched end termination (5 kΩ) with RS-232, particularly evident here because of the single-frequency square-wave drive.

Figure 12.134 is a rough guide to speeds and lengths of LVDS, RS-422/485, and RS-232 signaling. Figure 12.135 summarizes the signal characteristics of these link standards. And Figure 12.136 compares their driver power requirements (i.e., the driver chip's supply voltage times current), along with that of 50 Ω coax.[87]

---

[86] There are several closely related protocols, namely PECL (positive emitter-coupled logic), LVPECL (low-voltage PECL), and CML (current-mode logic). The latter is used, for example, in digital-video links such as DVI and HDMI.

[87] The driver ICs we used for these power measurements were: coax – 74LVC2G04 (both sections in parallel); LVDS – 65LVDS1; RS-232 – MAX3232E; RS-422/485 – LTC1485 (5 V) or LTC1480 (3.3 V).

Review of Chapter 12

An A-to-S summary of what we have learned in Chapter 12. This summary reviews basic principles, facts, and application advice in Chapter 12.

¶A. Logic Interconnections.

The subject of this chapter is the interconnection of digital logic signals and logic devices to... *everything*, where "everything" includes (a) other logic devices, (b) input sources (switches, optoelectronics, cables), and (c) output devices (dc and ac power loads, optoelectonics, cables). So it's a long chapter, rich with multiple themes. Here we try to organize these diverse topics into manageable paragraphs.

¶B. Logic Families.

§12.1.1. Contemporary digital logic is owned by CMOS, with the minor exception of some emitter-coupled logic families (ECL, PECL, and LVPECL) and some BiCMOS (ABT, BCT). CMOS inputs draw no current, and their logic thresholds are usually close to mid-supply, except for the -T suffix types (HCT, ACT, AHCT, VHCT) where the threshold conforms to that of legacy bipolar TTL (~ 1.4 V); see Figure 12.5.

The 74HC[T] logic is good for easy prototyping and general-purpose use, and is available in DIP or SOIC; substitute 74AC[T] or 74LV for applications needing higher speed. For lower-voltage applications at moderate speeds, use 74LVC or 74VCX (available in SOIC and mini-logic packages only). For operation up to 15 V you can use 4000B logic, but this stuff is *slow*. Finally, for differential signaling use LVDS (or LVPECL, if specified) drivers and receivers. See Figures 12.2 and 12.3 for logic-family speed versus power, and speed versus supply voltage.

¶C. Input and Output Characteristics.

§§12.1.2A and 12.1.2B. CMOS *inputs* draw no current when $0V \leq V_{\text{in}} \leq V_{\text{DD}}$ (and even for inputs a few tenths of a volt beyond). For most logic families, the input protection diodes clamp to ground and V_{DD}, thus causing substantial input current for input voltages as much as a volt beyond V_{DD} or below GND; this can cause SCR latchup. However, some families use zener-like clamping, allowing input swings well beyond the rail (and even when unpowered); examples include 5V-tolerant 74LVC, and 3.3V-tolerant 74AUP. See Figure 12.4.

CMOS *outputs* swing to the rails, looking resistive for small load currents (Figure 12.7). The output current capability varies widely among families; within a family it increases with increasing supply voltage. See Figure 12.6

for a summary of thresholds, and valid input and output voltage ranges. Note that the outputs of (essentially obsolete) bipolar TTL logic (and also of digital ICs with nMOS outputs) do not saturate to the supply rail; see §12.4.5A.

¶D. Interfacing between Logic Families.

You can mix logic types as long as you respect logic-level input requirements. At the basic level this means that (a) you can always make a direct connection between logic running at the same voltage; (b) logic running from a higher voltage (e.g., +5 V) can drive lower-voltage logic if the latter's input is "tolerant"; and (c) lower-voltage logic can drive higher-voltage logic if the latter has "TTL thresholds" and the former is powered from at least +2.5 V. With dozens of logic families, most capable of operating over a range of supply voltages, you've got a lot of possibilities. Most are dealt with in §12.1.3, and presented pictorially in Figure 12.9.

¶E. Driving Digital Logic Inputs.

§12.1.4. A mechanical switch with pullup (§12.1.4A) generates the right logic levels, but with *bounce*. That may be OK for some applications; but for clean edges you need a debouncer, illustrated in a half-dozen variations in §12.1.4B. A logic input can come instead from a *comparator* (a subject summarized in ¶G below, and treated in detail in §12.3), whose output may have active pullup to the same V_{DD} as the driven logic, or it may come from an "open-drain" (or open-collector) terminal; in the latter case you need an external pullup to V_{DD}. Figure 12.25 shows representative circuit configurations.

When driving logic inputs from signals of any type, be careful not to overdrive the input (whether from a signal source of greater swing, or from transmission-line effects – see the egregious example in Figures 12.103 and 12.104); and do not leave unused logic inputs floating.

Contemporary logic devices are engineered to withstand substantial static-electricity insults (defined in terms of the human-body model, HBM, of 100 pF in series with 1.5kΩ, §12.1.5), but it's best not to tempt fate; be careful to discharge yourself, use antistatic materials, etc. See also the earlier discussion in §3.5.4H.

¶F. Driving External Loads from Logic Outputs.

§12.4. You can drive small loads (LEDs, SSRs, small mechanical relays) directly from logic outputs (§12.4.1 and Figure 12.39), taking care to respect the output drive capability (voltage swing, current), and, for a mechanical relay, adding a diode clamp across its coil. As shown in the figure, you can use an open-collector driver (ULN2003, 75468) to accommodate voltages and currents as high as 100 V and 350 mA respectively. There are analogous drivers for use with serial data (from a microcontroller), see Figures 12.40 and 12.41, and the listing in Table 12.3 on page 819.

For driving heftier loads you can append an external transistor or power-driving module. For driving *positive* loads see §12.4.2, and Figures 12.42 and 12.43. For driving *negative* or *ac* loads see §12.4.3 and Figure 12.44. Serious power switching requires *fault protection*, see §12.4.4 and Figures 12.45–12.48.

¶G. Comparators.

A comparator is a high-gain uncompensated differential amplifier, used to signal which of two analog voltage inputs is greater. It is an important interface between analog signals and the digital world. Comparators were introduced in §4.3.2, and treated in detail in §12.3 (with listings in Tables 12.1 on page 812 and 12.2 on page 813).

As with op-amps, comparators come in a range of speeds (<1 ns to tens of μs), supply voltages (total supply voltages from as little as 1.1 V to as much as 40 V), input offset voltages (0.25 mV to 10 mV), input bias currents (1 pA to $>10\,\mu$A), input common-mode voltage ranges (to both rails, one rail, or neither), and quiescent currents ($<1\,\mu$A to tens of mA). But, unlike an op-amp, which is normally operated in the linear region where the (analog) output voltage avoids saturation, a comparator's output is digital; it lives at the extremes. It may be used to drive logic, or it may drive an ON/OFF load like a relay or an LED. To accommodate various loads, comparator output stages come in a half-dozen variations (Figure 12.33), including (a) rail-to-rail (like an op-amp); (b) open-drain or open-collector; (c) logic-level with auxiliary logic-rail pin V_L; (d) logic-level with both V_L and GND pins; and (e) floating-transistor output stage.

Comparator miscellany: comparators are usually configured with some hysteresis (Schmitt trigger) to prevent multiple transitions and oscillation. Some comparators have quite limited differential input voltage ranges (as little as a few volts). For comparators with BJT input stages, the input currents may undergo an abrupt step at zero differential voltage (e.g., Figure 12.36). The delay and switching time of a comparator depends on the input overdrive (Figure 12.38).

¶H. Optoelectronics.

§12.5. Continuing the theme of "logic-to-everything," we must include among the latter *humans*, with senses that de-

light in visual presentations.[88] There's much richness here, including emitters, indicators, displays, detectors, and couplers, as illustrated in the photographs in Figures 12.58 (page 830), 12.71 (page 837), 12.80 (page 841), 12.84 (page 844), and 12.95 (page 853). To remind you of the breadth of optoelectronics, here is a condensed listing of devices (expanded in ¶¶I–M below) in the family tree:

Emitters: LED; Laser diode; Electroluminescent

Displays: LED-based; LCD-based; VFD-based; OLED; E-ink

Detectors: Photodiode; Phototransistor; Photoresistor; Bolometer; APD; PMT; HAPD; Microchannel plate

Couplers: LED-input or logic-input, with transistor-, thyristor-, or logic-output

Other: Opto-interrupter, proximity and ranging; Barcode reader; Optical mouse.

¶I. LED Indicators.

These have diode-like *I–V* curves, but with greater (and color-dependent) forward voltage drops (Figure 2.8). They come in many form factors (panel mount, PCB mount, arrays, 7-segment displays), sizes, viewing angles, and colors (§12.5.1). You can drive them from a logic signal of adequate swing, with a current-limiting series resistor (mindful that blue, bright-green, and white LEDs have forward drops of 3 V or more), or you can add an external transistor for higher voltage or current (see Figure 12.61). For serious illuminations purposes you're better off using a current-sensing dc-dc switching converter (Figure 12.62). By contrast, to drive *laser diodes* you use feedback from the integrated monitor photodiode to set the drive current (see Figure 12.67 and associated text).

¶J. Displays.

A popular device for numeric or hexadecimal data is the 7-segment (or 16-segment) *display* (§12.5.3A), available in multicharacter sticks of both "dumb" and "smart" varieties (see Figures 10.90 and 12.77, respectively). Once you've bought into the idea of a multicharacter display device, you should consider the multicharacter (and multiline) LCD smart display (Figure 12.78) and the nicer-looking (and compatible) VFD display, seen in Figures 12.71 and 15.25.

¶K. Detectors.

In the detector arena, *photodiodes* convert light to a proportional photocurrent, and can be operated in either of

---

[88] As a colleague once remarked, as we were admiring the sunset while walking along the beach, "nice graphics."

two modes – *photovoltaic* (self-generating, into a short circuit or virtual ground); or *photoconductive* (back-biased) – see Figure 12.81. The latter is faster, but suffers from leakage current and increased noise. A *phototransistor*, which operates only in photoconductive mode, combines a photodiode and transistor, for greater gain, but with reduced speed; similarly (and more so) for a *photo-Darlington*. The *avalanche photodiode* exploits the phenomenon of avalanche multiplication in a photodiode biased close to breakdown; each detected photon causes multiple electrons to be collected, thus providing higher gain (in linear operation), or (if biased still further) a full-sized "Geiger-mode" pulse for each detected photon. A different class of detector is the cadmium sulfide *photoresistive* sensor, which behaves like a linear resistance ($I \propto V$) that depends on illumination; CdS sensors are slow, and they are becoming a rarity owing to RoHS regulations (cadmium is not good for you), but they are useful in applications where you want a light-controlled linear resistance (see for example Figure 7.21). Finally, a still-popular old-school photodetector is the *photomultiplier* tube (PMT, §12.6.2), in which a photoelectron is accelerated and collides with successive electron-multiplying dynodes to produce a cascade of some 10^6 electrons at a collecting anode (Figures 12.82 and 12.83).

¶L. Optocouplers and Solid-state Relays.

Optocouplers (also known as opto-isolators, or photocouplers) consist of an opto-emitter combined with a detector in an opaque package (§12.7). They are used to convey digital (and sometimes analog) signals between circuits with separate grounds; such *galvanic isolation* prevents ground loops in sensitive circuits, it permits safe switching of ac powerline circuits, and it allows communication and control for circuits operating at a high voltage.

All optocouplers use an LED at the *input end*, in most cases simply providing the anode (+) and cathode (−) terminals (so you have to limit the driving current with an external resistor, see Figure 12.85A); some optocouplers include current limiting (e.g., SSRs, Figure 12.92C), while some others (e.g., high-speed logic-to-logic couplers, Figure 12.86C) accept a logic-level input signal. At the *output end* there are many configurations: transistor or Darlington (Figure 12.85), digital logic with open-collector or active pullup (Figure 12.86), active-pullup MOSFET-driver (Figure 12.87), ac-output SSR (Figure 12.92), and dc-output SSR in many flavors (Figure 12.91). There are also analog-oriented optocouplers (Figures 12.88, 12.89, and 12.90), ac-input optocouplers (Figure 12.94), and a whole range of industrial *input/output modules* (§12.7.7).

¶M. Opto-interrupters, Proximity Sensors, and Angle Encoders.

An *opto-interrupter* is an LED-detector pair with an open gap (Figure 12.84), widely used to sense end-of-travel in mechanical devices. An optical *proximity sensor* bounces a beam off an external object; the *distance measuring sensor* is a variant that incorporates a position-sensitive detector that uses parallax (triangulation) to measure object distance (to ∼1 m or so). Optical *shaft angle encoders* measure shaft position; they come in incremental (quadrature square-wave) and absolute versions, with resolutions of 32 to 128 counts per rotation (for panel controls) to 30,000 or more for high-resolution shaft encoders.

¶N. Fiber Optics.

Light travels happily, and unimpeded, through glass or plastic fiber. For long-haul high-datarate communications you use single-mode glass fiber (with rates to 100 Gbps or more per fiber); but for more modest data communications you can use multimode fibers, ranging from the simplest TOSLINK 1 mm plastic fiber to the very popular 62.5/125 µm graded-index glass fiber. Transmitter and receiver modules are widely available (§12.8.1–12.8.3), as well as high speed transceiver modules that include all necessary driver/receiver electronics (§12.8.4).

¶O. Digital Signals and Long Wires.

Fully 25% of the chapter (§12.9) is devoted to the problem of sending digital signals somewhere else, where the *somewhere* may be on the same PCB, or through a backplane, or through a cable to a remote electronic device. As simple as these may seem, numerous pitfalls await the unwary: for example, we were surprised (but should not have been) when encountering errors while sending parallel digital data of only modest speed (HC logic) through just 4″ of flat ribbon cable. Read the following paragraphs, to avoid membership in the club of the unwary.

¶P. Short Links.

For the *short links* that stay on a PCB, the distributed capacitive loading and consequent ground-current spikes cause logic glitches (§12.9.1A), requiring liberal power-supply bypassing (which is always salutary) and low-inductance ground plane layouts. The problem is less severe in synchronous systems; but it's more severe with fast logic, where interconnections may require constant-impedance transmission-line layouts (see §1x.1.4). These problems are more pronounced for signals moving between circuit boards or along backplanes, where you may need to use line driver and line receiver ICs, perhaps in combination with controlled impedance traces that are properly terminated; the latter is standard practice with LVDS (differential) signaling, as seen for example in Figure 12.127.

¶Q. Cable Runs.

Longer signal runs require *cable* (§12.10), usually *coaxial* (e.g., the ubiquitous RG58 BNC patch cords), multiwire *ribbon*, *twisted pair* (e.g., the comparably ubiquitous cat-5 or cat-6 network cable), or *fiber optic* (generally recognizable by their brightly colored svelte profile). With the exception of "slow" waveforms,[89] signal cables must be treated as *transmission lines* (Appendix H), with their *characteristic impedance*[90] Z_0. For coax, Z_0 is usually 50 Ω (but 75 Ω for video), while twisted-pair is usually 100 Ω. The significance of Z_0 is that *a signal applied to a cable, terminated with a load resistance equal to its characteristic impedance, is completely absorbed by the load*; it does not reflect. No other load impedance can make that statement.

¶R. Terminating Digital Coax.

So you can (a) put a 50 Ω resistor across the far end of a coax line (a "terminated line," or "matched line") and

---

[89] Those whose transition times are much longer than the round-trip travel time, $t = 2L\sqrt{\varepsilon}/c$, where the factor $\sqrt{\varepsilon}$ accounts for the slower-than-light "velocity factor" of wave propagation in the cable.

[90] Which should really be called characteristic *resistance*.

---

drive it with your digital signal (which sees a pure 50 Ω resistive load – the capacitance disappears!), as in Figure 12.107; or (b) drive such a terminated line with a signal of the same impedance, as in Figure 12.110, noting that the output signal amplitude is reduced by ×2; or (c) keep the matched source resistor but omit the load resistor, as in Figure 12.107, to preserve the full output amplitude. In the latter case there *is* a reflection of substantial amplitude, which however is hospitably absorbed by the source resistor – see the highly educational waveforms in Figure 12.115. See §12.10.1 for pretty pictures, and plenty of detail.

¶S. Differential Cable.

Differential signals are treated similarly, with a termination connected across the pair. For long hauls (to ~1 km) and modest speeds (10 kbps–10 Mbps) the long-lived RS-422/485 protocol is popular, whereas the LVDS protocol is used for higher datarates (10 Mbps–1 Gbps) over shorter distances; see the speed limit contours in Figure 12.134. To achieve the highest datarates over significant lengths of cable one commonly uses driver *pre-emphasis* and receiver *equalization*, see Figures 12.130–12.132 and associated discussion.

CHAPTER 13

DIGITAL MEETS ANALOG

Here we meet the major subject of conversion between analog and digital signals – analog-to-digital (A/D) and digital-to-analog (D/A) converters (ADCs and DACs) – as well as the important "mixed-signal" phase-locked loop (PLL). And we cannot resist a look at the fascinating topic of pseudorandom noise generation.

We live in a largely analog (continuous) world – of sounds, images, distances, times, voltages and currents, and so on – which would seem to call for analog circuits (oscillators, amplifiers, filters, combiners, etc.). But we live also in a partly digital (discrete) world – of numbers and arithmetic, of text and symbols, and the like – which would seem to call for digital circuits (arithmetic logic and storage, etc.). And that's how it was, for many years: analog amplifiers and filters for audio and video; analog oscillators, tuned circuits, and mixers for radio and television; and even *analog computers*, for solving differential equations[1] or for real-time control of flight or weaponry. Meanwhile digital techniques (initially with mechanisms and relays, then with vacuum tubes, followed by discrete transistors, small-scale ICs, and finally the large and fast microprocessors with a billion+ transistors that we take for granted) were used for computational tasks like keeping track of money, and of words.

But the almost miraculous improvements in the speeds and densities of purely digital electronics have produced a major paradigm shift, namely, the use of digital conditioning and processing for nearly every "analog" quantity. For example, audio engineers now digitize the individual microphone signals at the time of recording, and perform all subsequent mixing and conditioning (e.g., the addition of reverberation) as arithmetic on those numbers; the same goes for digital video. And at the everyday level, digital techniques have invaded our lives: the authors' bathroom scales indicate to 0.1 pound (sometimes to our regret) – that's a resolution[2] of a part per thousand; our porch light

is switched on and off by a digital wall switch that follows the seasonal variation of dusk and dawn; and our automobiles depend on a digital bus, to which are connected some 50 or more embedded digital controllers for functions like engine control and diagnostics, braking, air bags, entertainment, climate control, and so on.

The bottom line is that A/D and D/A conversion techniques have become central to every aspect of analog measurement and control. This is important stuff, and it is the major subject of this chapter. Let's go at it.

Our treatment of the various conversion techniques is not aimed at developing skill in converter design itself. Rather, we try to point out the advantages and disadvantages of each method, because in most cases the sensible thing is to buy commercially available chips or modules, rather than to build the converter from scratch. An understanding of conversion techniques and idiosyncrasies will guide you in choosing from among the thousands of available units.

13.1 Some preliminaries

13.1.1 The basic performance parameters

Before getting into lots of detail, we'd like to summarize the important performance parameters that you need to keep in mind when choosing ADCs and DACs. Knowing what you need makes it a lot easier to find what you want.

Digital-to-analog converters
  Resolution: number of bits
  Accuracy: monotonicity; linearity; dc stability
  Reference: internal or external; multiplying DAC (MDAC)?
  Output type: voltage output or current output
  Output scaling: unipolarity or bipolarity; V_{out} ranges; I_{out} compliance
  Speed: settling time; update rate

---

[1] There's a nice example of this in the section on Analog Function Circuits in Chapter *4x*: modeling the fascinating chaotic behavior of the system of nonlinear differential equations devised by Lorenz.

[2] To be distinguished from *accuracy* – recall the discussion in §5.1.1.

Although the bathroom scale reads with a *resolution* of 0.1 pound, its accuracy is likely poorer (perhaps to ± 1 pound), with some drift over time and temperature.

Quantity: single or multiple DACs/pkg

Digital input format: serial (I^2C, SPI, or a variant) or parallel

Package: module, through-hole, or various surface-mount packages

Other: glitch energy; power-on state; programmable internal digital scaling

Analog-to-digital converters

Resolution: number of bits

Accuracy: monotonicity; linearity; dc stability

Reference: internal or external

Input scaling: unipolarity or bipolarity; voltage range

Speed: conversion time and latency

Quantity: single or multiple ADCs/pkg

Digital-output format: serial (I^2C, SPI, or a variant) or parallel

Package: module, through-hole, or various surface-mount packages

Other: internal programmable gain amplifier (PGA); spur-free dynamic range (SFDR)

13.1.2 Codes

At this point you should review §10.1.3 on the various number codes used to represent signed numbers. Offset binary and 2s complement are commonly used in A/D conversion schemes, with sign-magnitude and Gray codes also popping up from time to time. Here is a reminder:

	Offset binary	2s Complement
+Full scale	11111111	01111111
+Full scale−1	11111110	01111110
↓	↓	↓
0+1 LSB	10000001	00000001
0	10000000	00000000
0−1 LSB	01111111	11111111
↓	↓	↓
−Full scale+1	00000001	10000001
−Full scale	00000000	10000000

13.1.3 Converter errors

The subject of ADC and DAC errors is a complicated one, about which whole volumes could be written. According to Bernie Gordon at Analogic, if you think a high-accuracy converter system lives up to its claimed specifications, you probably haven't looked closely enough. We won't go into the application scenarios necessary to sup-port Bernie's claim, but it's worth a first look at the four most common types of converter errors: offset error, scale error, nonlinearity, and nonmonotonicity, nicely illustrated in the self-explanatory Figure 13.1. Rather than boring you with a long-winded discussion, though, we'll move directly to a description of D/A converter techniques and capabilities. Then we'll revisit the business of converter errors (§13.4), which will make a lot more sense in context.

13.1.4 Stand-alone versus integrated

Sometimes an ADC or DAC (or both) is integrated into a fancier IC. The most common example is the microcontroller (Chapter 15), where you frequently see both ADCs and DACs integrated on the same chip as the processor and its other I/O peripherals. As far as we can tell, the least-expensive stand-alone ADC costs significantly more than the least-expensive microcontroller-*with*-ADC.[3] Microcontrollers are fond of integrating lots of useful peripherals, along with program and data memory, so that you've got essentially a "system-on-a-chip." Be aware, though, that these converters that come bundled with inexpensive general-purpose microcontrollers do not attain the excellent performance of a good stand-alone converter: you can get 8-bit or even 10-bit performance; but you won't get 16 bits, and nothing approaching the 24-bit performance of a high-quality audio ADC.[4]

For some classes of IC, though, an integrated converter delivers excellent performance. One example is a direct digital synthesis (DDS) chip (§7.1.8), where on-chip phase counters and a sine lookup table create digital values of the synthesized sinewave output; these things can clock at speeds of 1 GHz or more, with an on-chip 14-bit (for example) DAC generating the analog output signal. Another example comes from the video world, where it's common to see digital video processing and conversion functions combined on a single high-performance IC. And in the audio business you see parts like the Cirrus CS470xx-series (their "All-In-One Audio IC" system-on-a-chip), which includes multiple 24-bit ADCs and DACs with 105 dB dynamic range, integrated onto a chip that has a 32-bit DSP (with 32 kB RAM), audio codecs and sample rate

---

[3] To wit: National's ADC0831 8-bit ADC costs $1.85, whereas Microchip's PIC10F with its 8-bit ADC and 2-input multiplexer costs $0.48 (both in quantity 25).

[4] A shining exception is provided by Analog Devices' series of "Analog Microcontrollers," with honest performance to 16 or 24 bits. You might think of these as consisting of a high-quality converter core, with a quiet microcontroller tacked on.

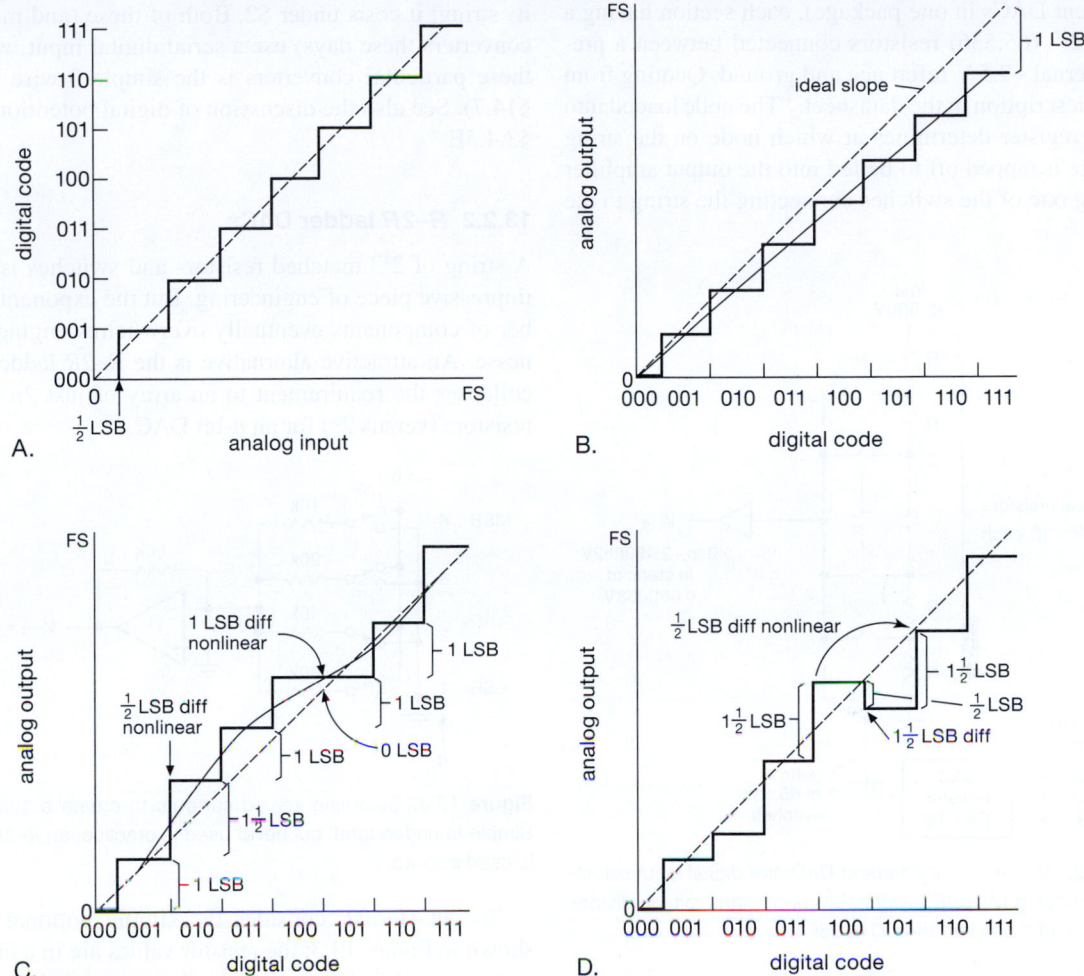

Figure 13.1. Graphs illustrating the definitions of four common digital conversion errors, for a 3-bit converter over its 8 levels from 0 to full scale (FS). A. ADC transfer curve, $\frac{1}{2}$LSB offset at zero. B. Linear, 1 LSB scale error. C. $\pm\frac{1}{2}$ LSB nonlinearity (implies 1 LSB possible error); 1 LSB differential nonlinearity (implies monotonicity). D. Nonmonotonic (must be $>\pm\frac{1}{2}$ LSB nonlinear). Used with permission of Texas Instruments Inc.

converters, digital audio ports (SPDIF), and an SPI/I²C control port.

Stand-alone converters are dominant, though, in high-accuracy and high-linearity applications (voltmeters; quality audio gear). They also provide a tremendous selection range, in terms of the many parameters just listed, as compared with the rather limited selection of on-chip converters you find in microcontrollers.

13.2 Digital-to-analog converters

The goal is to convert a quantity specified as a binary (or multidigit BCD, see §10.1.3B) number to a voltage, or to a current, proportional to the value of the digital input.

There are several popular methods: (a) resistor string with MOS switches; (b) R–2R ladder; (c) binary-scaled current sources; and (d) delta–sigma (and other pulse-averaging) converters. Let's take them in turns.

13.2.1 Resistor-string DACs

This method is about as straightforward as you can get. A string of 2^n equal-value resistors is connected between a stable voltage reference and ground, creating a very tall voltage divider; and a set of MOSFET analog switches is used to route the selected tap's voltage to an output voltage buffer (Figure 13.2). The figure shows the configuration of TI's impressive DAC8564, a quad 16-bit DAC (four independent DACs in one package), each section having a string of 2^{16} (65,536) resistors connected between a precision internal +2.5 V reference and ground. Quoting from the terse description in the datasheet, "The code loaded into the DAC register determines at which node on the string the voltage is tapped off to be fed into the output amplifier by closing one of the switches connecting the string to the amplifier."[5]

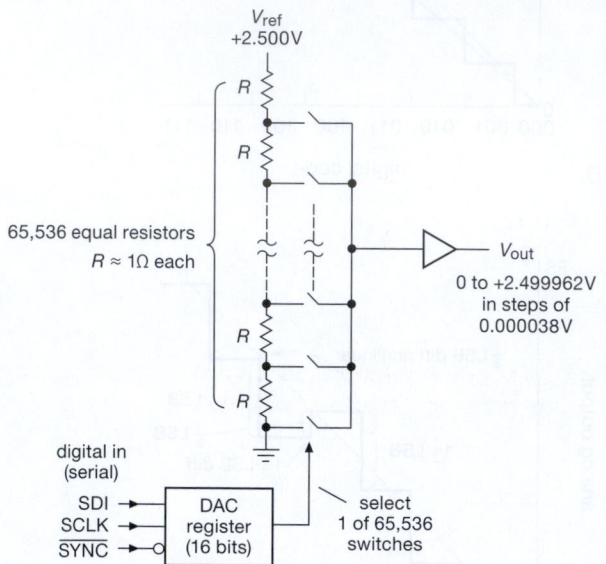

Figure 13.2. An easy-to-understand DAC: the digital input selects the corresponding MOSFET switch tap on a giant voltage divider. TI crams four of these into their DAC8564.

This method has the virtue of guaranteed monotonicity. As the datasheet puts it (even more tersely than before), "It is monotonic because it is a string of resistors." And this particular DAC exhibits other nice qualities, specifically low *glitch energy* (spikes that appear at the output during code transitions), excellent accuracy and stability (worst-case values of ±0.02% initial accuracy and 5 ppm/°C tempco), rail-to-rail output ("RRO") with a single positive supply (+2.7–5.5 V), and low power (1 mA, typ). This puppy costs about $12. The same method is used in DACs of more modest performance, for example National's DAC121: 12-bit, voltage output, micropower (150 μA) single-supply with rail-to-rail input and output (RRIO). The latter is a single DAC without internal reference (full scale is the positive supply), and of course it has "only" 4096 resistors in

its string; it costs under $2. Both of these (and most other converters these days) use a serial digital input, which for these particular converters is the simple 3-wire SPI (see §14.7). See also the discussion of digital potentiometers in §3.4.3E.

13.2.2 *R–2R* ladder DACs

A string of 2^{16} matched resistors and switches is a pretty impressive piece of engineering. But the exponential number of components eventually overwhelms engineering finesse. An attractive alternative is the *R–2R* ladder, which collapses the requirement to an array of just $2n$ matched resistors (versus 2^n) for an *n*-bit DAC.

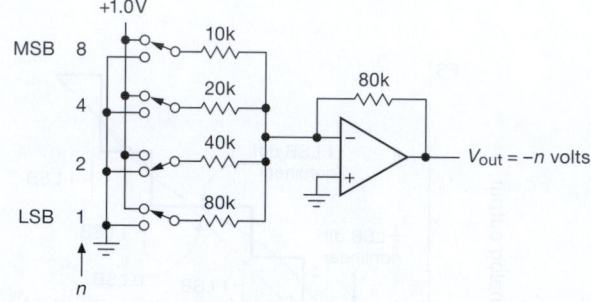

Figure 13.3. Summing scaled currents to create a simple DAC. Simple to understand, but never used in practice: an *R–2R* network is used instead.

To get started, consider the simple notional scheme shown in Figure 13.3: the resistor values are in a binary sequence, so their binary-weighted currents at the summing junction produce a binary-weighted voltage output. Simple, but not terribly practical with more than a few bits, because the resistor values must span a wide range, and with increasingly demanding accuracy for the lower resistance values; of greater worry still is the need for very low switch R_{on} corresponding to the low resistance values.[6]

Exercise 13.1. Design a 2-digit BCD DAC. Assume that the inputs are 0 or +1 volt; the output should go from 0 to 9.9 volts.

Instead, the scheme shown in Figure 13.4 is used. It's easy to convince yourself that this clever arrangement produces a binary-weighted current into the op-amp's summing junction, therefore a corresponding output voltage. And only two resistor values are needed (*R* and 2*R*, which, however, must both be accurately replicated and in a precise 2:1 ratio), regardless of the number of bits.

---

[5] These guys seem not to be fond of commas. We were tempted to add a few, to pace the breathless flow of the sentence. But, hey, a quote is a quote, right?

[6] This method does have the flexibility, however, of allowing arbitrary bit weights.

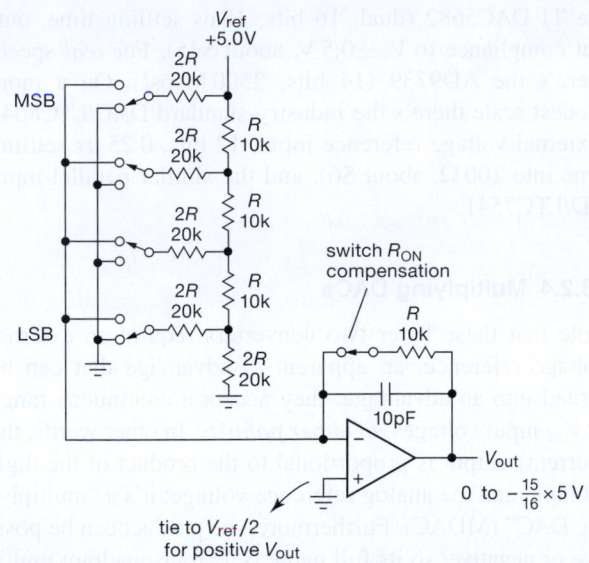

Figure 13.4. An *R–2R* ladder network generates a binary-scaled output current into the op-amp's summing junction, producing a voltage-output DAC.

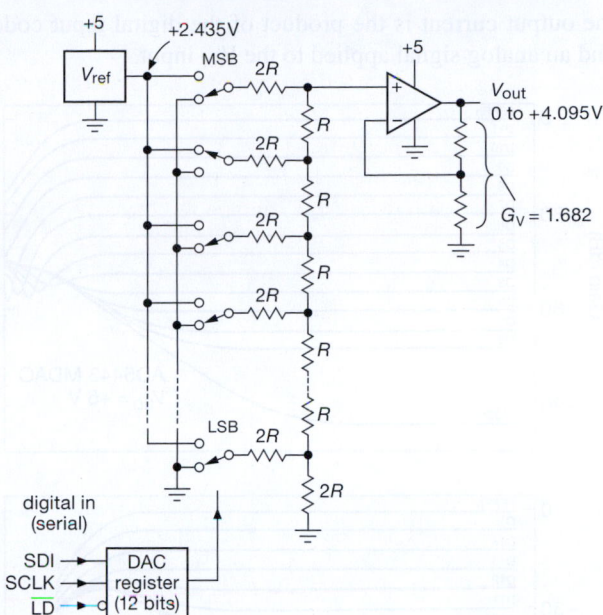

Figure 13.5. *R–2R* voltage-output DAC, in the more common voltage-combining configuration.

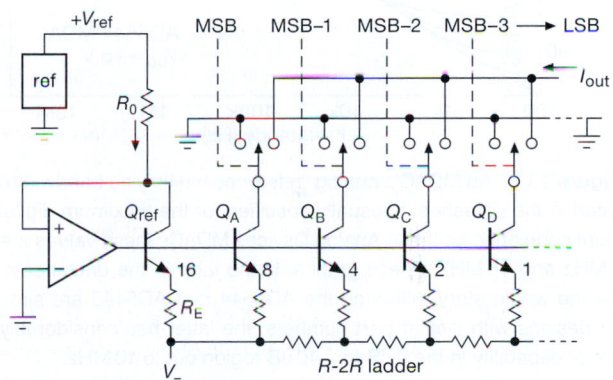

Figure 13.6. Classic current-switched DAC.

There are many excellent *R–2R* DACs out there. For example, TI's DAC9881 is an 18-bit voltage-output DAC with an SPI serial input; it is guaranteed monotonic,[7] with integral linearity to ± 2 LSB. It requires an external voltage reference (V_{ref}), which sets the full-scale voltage (the non-inverting input is biased at $V_{\text{ref}}/2$, for positive output polarity). It excels in accuracy and low noise, and costs about $30.

Exercise 13.2. Prove that the foregoing *R-2R* ladder works as advertised.

In practice, most *R–2R* DACs use the alternative configuration of Figure 13.5, in which the output of the *R–2R* network is itself a voltage. For example, the TI DAC7611 is a 12-bit voltage-output DAC (+4.095 V full-scale output, popular for DACs that run from +5 V) with an SPI serial input and on-chip voltage reference; it's linear and monotonic to its full 12 bits, comes in an 8-pin package, and costs about $4.

13.2.3 Current-steering DACs

The preceding converters generate *voltage* outputs. This is often most convenient, but the op-amp tends to be the slowest part of the converter circuit. In situations where you can use a converter with *current* output, you'll get better speeds, and usually at lower price. Some additional advantages of current-output DACs are: (a) flexible choice of current-to-voltage op-amp, for example to minimize noise, or to produce a larger output voltage swing; (b) the ability to combine several DAC outputs directly; and (c) the availability of *multiplying DACs* (see next subsection), in which

---

[7] Note that, unlike a resistor-string DAC, an *R–2R* DAC is not *guaranteed* monotonic when resistor tolerances are taken into account. The semiconductor industry does a good job, however, and most *R–2R* DACs are monotonic to 1 LSB.

the output current is the product of the digital input code and an analog signal applied to the V_{ref} input.

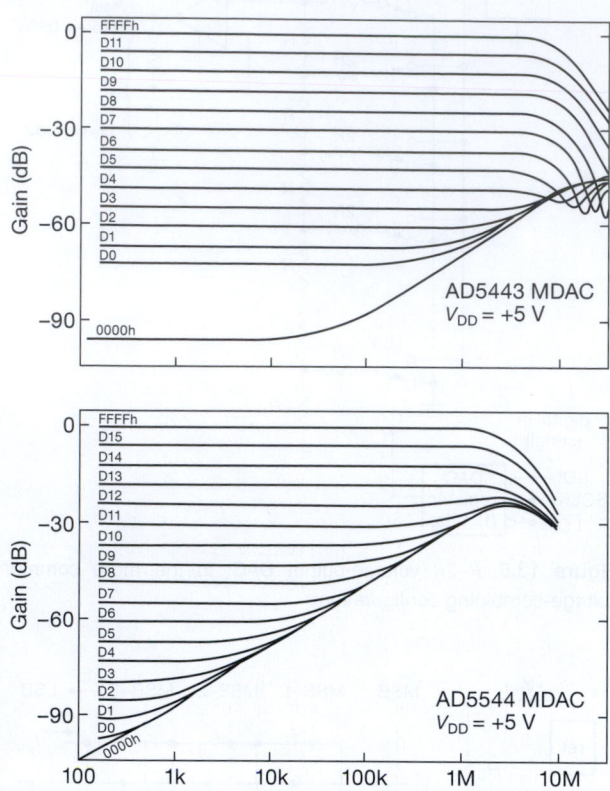

Figure 13.7. An MDAC's analog "reference multiplying bandwidth" listed in the datasheet is usually specified for the maximum digital input code only; for these Analog Devices MDACs those values are 2 MHz and 10 MHz. These graphs, found later in the datasheets, tell the whole story. Although the AD5544 and AD5443 are similar designs with similar part numbers, the latter has considerably better capability in the 0 dB to −40 dB region out to 10 MHz.

Figure 13.6 shows how these "current-steering" converters work. The currents can be generated by an array of transistor current sources with scaled emitter resistors, although IC designers usually use instead an R–$2R$ ladder of emitter resistors. In most converters of this type, the current sources are ON all the time, and their output current is switched to the output terminal or to ground, under control of the digital input code. Watch out for limited output compliance in current-output DACs; it can be as little as 0.5 V, though values of a few volts are typical.

Some examples of current-steering DACs (with serial digital inputs and internal voltage references) are the LTC1668 (16 bits, 20 ns settling time into a 50 Ω load as "voltage out," output compliance to ±1 V, about $20) and

the TI DAC5682 (dual, 16 bits, 10 ns settling time, output compliance to $V_+ \pm 0.5$ V, about $45). For *real* speed, there's the AD9739 (14 bits, 2500 Msps!). On a more modest scale there's the industry-standard DAC/LTC8043 (external voltage reference input, 12 bits, 0.25 μs settling time into 100 Ω, about $6), and the similar parallel-input AD/LTC7541.

13.2.4 Multiplying DACs

Note that these latter two converters require an external voltage reference, an apparent disadvantage that can be turned into an advantage: they accept a continuous range of V_{ref} input voltages, *of either polarity*. In other words, the (current) output is proportional to the product of the digital input and the analog reference voltage: it's a "multiplying DAC" (MDAC). Furthermore, the product can be positive or negative; so its full name is a "four-quadrant multiplying DAC." Examples of higher-resolution four-quadrant MDACs are the 16-bit DAC8814 (serial input) and closely similar DAC8820 (parallel input). Multiplying DACs specify both their conversion precision (linearity, monotonicity) and the bandwidth of the analog multiplying input (i.e., V_{ref}); for these two converters the "reference multiplying bandwidths" are 10 MHz and 8 MHz, respectively, with respective prices of $25 and $15.

Note that not all DACs are optimized for use in this way, so it is best to check the datasheets of the converters you're considering for details. A DAC with good multiplying properties (wide analog input range, high speed, etc.) will usually be called a "multiplying DAC" right at the top of the datasheet. Table 13.3 on page 894 lists selected Multiplying DACs.

A caution: the specified bandwidth can be seriously misleading, owing to the effects of capacitive feedthrough. Figure 13.7 shows this behavior, as illustrated in the respective datasheets of these two MDACs.

Multiplying DACs (and the ADC equivalent) open the possibility of *ratiometric* measurements and conversions. If a sensor of some sort (e.g., a variable-resistance transducer like a thermistor) is energized by the same voltage that also supplies the reference for the ADC, then variations in the reference voltage will not affect the measurement. This concept is very powerful, because it permits measurement and control with accuracy greater than the stability of voltage references or power supplies; conversely, it relaxes the requirements on supply stability and accuracy. The ratiometric principle is used in its simplest form in the classic *bridge* circuit, in which two ratios are adjusted to equality by nulling the differential signal taken

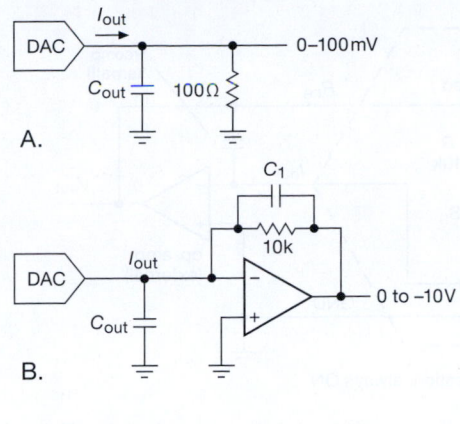

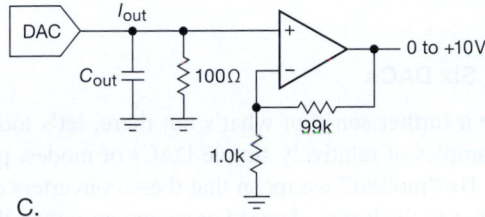

Figure 13.8. Generating voltages from current-output DACs.

between the two voltage-divider outputs. Devices like the 555 (see §7.1.3) achieve good stability of output frequency with large variations of supply voltage by using essentially a ratiometric scheme: the capacitor voltage, generated by an RC network from the supply, is compared with a fixed fraction of the supply voltage ($\frac{1}{3}V_{CC}$ and $\frac{2}{3}V_{CC}$), giving an output frequency that depends only on the RC time constant.[8] We will have more to say on this important subject in connection with ADCs later in this chapter.

13.2.5 Generating a voltage output

If you've chosen a voltage-output DAC, you're done! But with a current-output DAC you've got to use one of several schemes for generating a voltage output. Figure 13.8 shows some ideas. If the load capacitance is low and large voltage swings aren't needed, a simple resistor to ground will do nicely (but see warning, below). This is what's usually done with video DACs. For example, the THS8133 triple 10-bit video DAC generates full-scale output currents of 26.7 mA, which produce a standard 1.0 V analog video signal when driving doubly terminated 75 Ω coax. This method works fine also for general applications: with

the usual 1 mA full-scale output current, a 50 Ω load resistor will give 50 mV full-scale output with 50 Ω output impedance. If the capacitance of the DAC's output combined with the load capacitance doesn't exceed 100 pF, you will get 50 ns settling time in the preceding example, assuming the DAC is that fast. When worrying about the effect of RC time constants on DAC output response, don't forget that it takes quite a few RC time constants for the output to settle to within $\frac{1}{2}$ LSB of the final voltage. It takes 7.6 RC time constants, for instance, to settle to within 1 part in 2048, which is what you would want for a 10-bit converter output.

To generate large swings, or to buffer into small load resistances or large load capacitances, an op-amp can be used in the transresistance configuration (current-to-voltage amplifier), as shown. The capacitor across the feedback resistor may be necessary for stability, because the DAC's output capacitance in combination with the feedback resistance introduces a lagging phase shift; unfortunately, that compromises the speed of the amplifier. It is an irony of this circuit that a relatively expensive high-speed (fast-settling) op-amp may be necessary to maintain the high speed of even an inexpensive DAC.[9] In practice, the last circuit may give better high-speed performance, since no compensation capacitor is needed. Watch out for offset voltage error, because the op-amp's input offset voltage is amplified by the voltage gain (here a factor of 100).

An Important Warning: when using current-output DACs, note that both the initial accuracy (e.g., full-scale I_{out}) and the stability of the current output may be grotesquely poor, relative to the DACs resolution. It is not uncommon to see a spread of as much as 2:1 (!) in the full-scale current. What to do? Most current-output DACs include a built-in feedback resistor, closely matched to the R–$2R$ resistors, intended to be used with an external op-amp (Figure 13.9). If you don't use it, you may have gain errors of ±25% or more; and even if you trim away this gain error, there will be residual gain *drift* (not usually specified on datasheets) that typically will be 100 times larger than you get with the internal resistor.

To give an example, the LTC8043 is an improved version of the industry-standard DAC8043 12-bit multiplying

---

[8] We showed analogous ratiometric schemes in §§4.3.5, 4.6.4, 7.1.3D, and 10.4.5A.

[9] Some current-output DACs have surprisingly-high output capacitance, C_{out}, e.g., as much as 200 pF for the LTC7541 12-bit MDAC. So you need a stabilizing capacitor C_1, of value $C_1 > \sqrt{C_{out}/2\pi f_T R_f}$, as discussed before, see for example §§8.11.6 and *4x.3*. Choose an op-amp, then, with a high enough f_T so that C_1 is small enough for the desired speed. Some fast current-output DACs take pains to keep C_{out} small, as little as 5 pF.

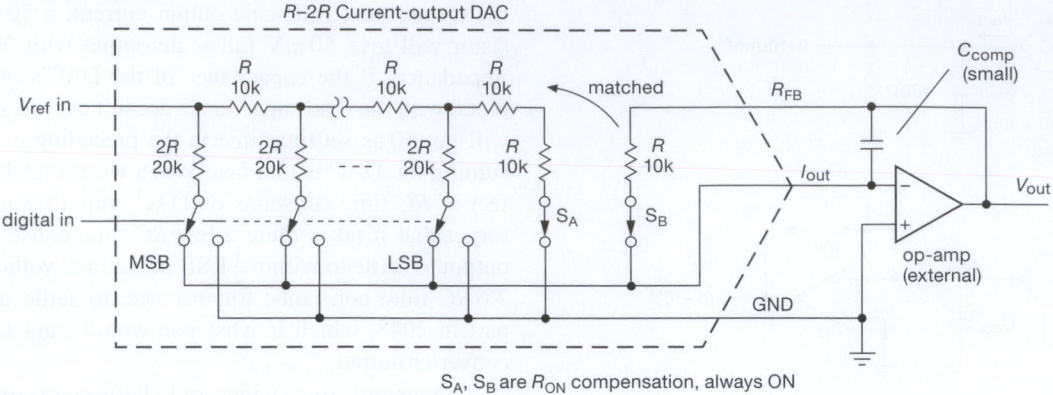

Figure 13.9. Current-output R–$2R$ DAC with internal feedback resistor matched to the precision network resistors both in initial resistance and in temperature coefficient. Ignore R_{FB} at your peril!

DAC. In the best grade (-E suffix) it specifies a gain error of $\pm 1\%$ (max), and a gain tempco of 5 ppm/°C (max). (It also guarantees maximum integral and differential nonlinearities of $\pm 0.5\%$; see the discussion in §13.4.) Impressive specs. Note, however, that the gain specifications state "Using internal feedback resistor." What if you don't? The datasheet doesn't say! But you can tease the answer from what it *does* say, namely that the input resistance of the R–$2R$ network (seen at the V_{ref} input) is 11 kΩ (nominal), with limits of 7 kΩ (min) and 15 kΩ (max). That is, this DAC, which guarantees an admirable gain accuracy of $\pm 0.024\%$ when the matched internal feedback resistor is used in conjunction with an external op-amp, would deliver a shockingly poor[10] $\pm 35\%$ absolute gain error as a current-output device (or, equivalently, with an external feedback resistor and op-amp, for voltage output).[11]

Bottom line: if a current-output DAC gives you an internal feedback resistor, you need to think long and hard before deciding not to use it.

Another bottom line: if you want a bipolarity output-voltage range, you might be tempted to sink a reference current (derived from V_{ref}) from the summing junction in Figure 13.9. *Don't do it!* Instead, append a difference amplifier to V_{out}, using an offset of $V_{ref}/2$ for the other input.

13.2.6 Six DACs

To give a further sense of what's out there, let's look at a few examples of relatively simple DACs of modest performance. By "modest" we mean that these converters do not push close to the limits of speed or accuracy; rather, they're inexpensive, compact, and easy to use. You can drop them onto a circuit board, and off you go. Later, in §13.3, we'll look at some applications that demand DACs of greater performance; in those situations you'll need carefully designed surrounding circuitry to exploit fully the advanced capabilities of the converter.

Take a look at Figure 13.10, with reference to Table 13.1. These were chosen somewhat arbitrarily from the many thousands (no kidding!) of available DACs, though we did make an effort to select devices from multiple manufacturers. They range from 8 to 14 bits, with settling times of 4–10 μs. Except for E, they are all voltage-output devices.

Taking it from the top of Figure 13.10: A–C are single-supply low-voltage parts with serial interfaces, so they can fit into tiny SOT23 or (even tinier) SC70 packages. The pair in A use the supply voltage as reference, with the full-scale output voltage V_{FS} equal to the supply V_{DD}; one uses an SPI interface (with the usual chip select), whereas the other uses an I²C interface (with the input pin A_0 selecting between factory preset bus addresses of decimal 76 or 77).[12] If you want a fixed and stable reference voltage, rather than using V_{DD}, a converter like the LTC2630 in B is a good choice. Its internal reference has adequate

---

[10] I'm shocked, shocked to find that gain inaccuracy and drift is going on in here!

[11] You might ask why an IC manufacturer whose chips deliver outstanding accuracy has a problem making resistors with even "pretty good" absolute accuracy. Good question. Turns out that the process is optimized for best *tracking*, with overall resistor scaling only of secondary importance.

[12] The DAC7512 from the same manufacturer substitutes SPI for I²C; and the AD5601/11/21 are less-expensive 8- to 12-bit versions of the 14-bit AD5641.

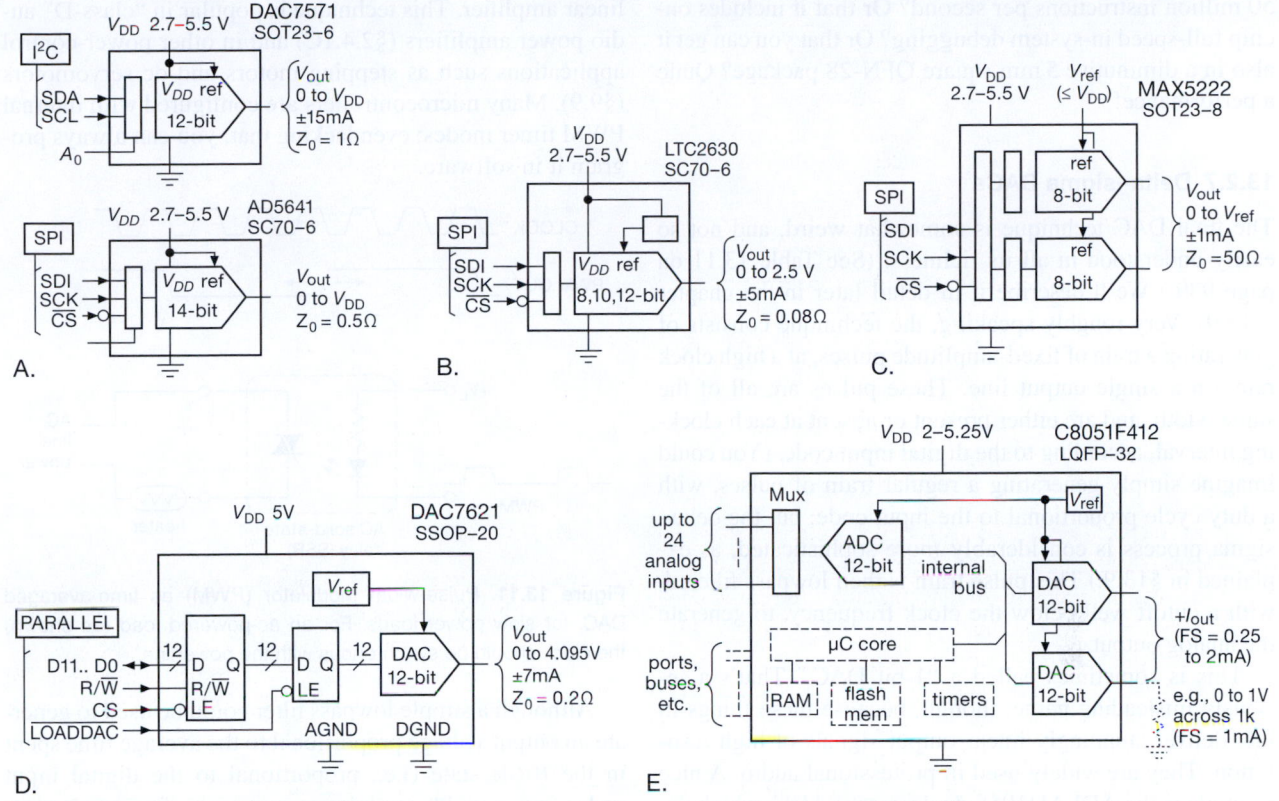

Figure 13.10. Six DACs, whose specifications are listed in Table 13.1. The DAC7571, LTC2630, and MAX5222 in the upper row are typical of inexpensive serial-input converters in small packages (see §14.7 for discussion of the SPI and I²C serial interfaces). The DAC7621 uses a traditional parallel data input; and the C8051 is a general-purpose microcontroller that includes a pair of current-output DACs among its many internal assets.

stability for a converter of this precision (± 10 ppm/°C, typ), though its absolute accuracy is only modest (full-scale error of $\pm 0.2\%$ typ, $\pm 0.8\%$ worst case). The LTC2630 family includes a "high-voltage" variant with a 4.096 V internal reference (thus 1.0 mV/LSB), for which the V_{DD} range is narrowed to 4.5–5.5 V (5 V nominal); it also includes 8- and 10-bit variants.

Although you *could* imagine squeezing two converters into a 6-pin package (e.g., with an I²C 2-wire interface and no choice of address), the converter in C takes the rational choice of adding two pins, which both accommodates the SPI port (three wires, including CS′) and allows an external reference.[13] By reducing the lead pitch (from 0.95 mm to 0.65 mm) this 8-pin device fits in the same overall package size (1.6×2.9 mm) as the 6-pin SOT23-6.

---

[13] One small fly in the ointment here is the lack of double buffering along with an 8-bit data payload; thus the two channels cannot be updated to different values simultaneously.

The bottom row illustrates a converter with a parallel input port, an interface that has largely gone out of popularity except for converters that operate at the highest speeds (e.g., the 2500 Msps AD9739 in §13.2.3). But there are applications for which this is useful, even at modest speeds, for example if you want to use the *n*-bit output of a counter directly: no microcontroller, no programming... just wires.

Finally, the last "converter" E is actually just some carry-on baggage belonging to a full-function microcontroller, complete with on-chip program memory (flash ROM), SRAM, timers, ports (parallel, SPI, UART), accurate oscillator, and even a 24-channel (multiplexed) 12-bit 200 ksps ADC. The dual DACs use an internal reference, and produce a current-sourcing output (with programmable full-scale ranges of 0.25–2 mA, by factors of 2) with compliance to 1.2 V below the supply rail (which runs over a wide range of 2.0–5.25 V). Oh, and did we mention that this thing includes a computational core that runs at

50 million instructions per second? Or that it includes on-chip full-speed in-system debugging? Or that you can get it also in a diminutive 5 mm square QFN-28 package? Quite a performance!

13.2.7 Delta–sigma DACs

The final DAC technique is somewhat weird, and not so easily understood in all its richness. (See Table 13.11 on page 939.) We'll describe it in detail later in the chapter (§13.9). Very roughly speaking, the technique consists of generating a train of fixed-amplitude pulses, at a high clock rate, on a single output line. These pulses are all of the same width, and are either present or absent at each clocking interval, according to the digital input code. (You could imagine simply generating a regular train of pulses, with a duty cycle proportional to the input code; but the delta–sigma process is considerably more sophisticated, as explained in §13.9.) This pulse train is then lowpass filtered, with a cutoff well below the clock frequency, to generate the analog output.

This is sometimes called a "1-bit DAC." That's a seriously misleading name, though, because these things in fact deliver stunningly linear output signals of high resolution. They are widely used in professional audio. A nice example is the ADI AD1955 dual (stereo) ADC, which delivers 20-bit analog audio output (120 dB dynamic range) when clocked at 12 MHz.[14]

13.2.8 PWM as digital-to-analog converter

One last method of driving an analog system from a digital quantity is via *pulse-width modulation* (PWM). This is qualitatively different from the true DACs just described, because it does not generate an analog output directly; but it is widely used for power loads such as a heater. The idea is to run a repeating cycle of N clock cycles, during which the load is switched ON for some smaller number of clock cycles k, and OFF for the rest, with the fraction (duty cycle) k/N proportional to the digital input (Figure 13.11). This is easily done with a counter, magnitude comparator, and high-frequency clock (see Exercise 13.3). The slowly responding load does the averaging over the full cycle. This is more efficient than driving the load with a properly smoothed analog signal, because the driver is a switch, with very little dissipation; a switch is also simpler than a

linear amplifier. This technique is popular in "class-D" audio power amplifiers (§2.4.1C) and in other power-control applications such as stepping motors and dc servomotors (§9.9). Many microcontrollers are configured with internal PWM timer modes; even lacking that, you can always program it in software.

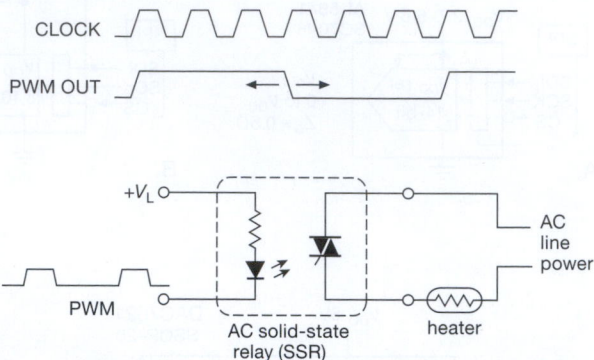

Figure 13.11. Pulse-width modulator (PWM) as time-averaged DAC, for slow power loads. For an ac-powered load (as shown) the clock should be synchronous with the powerline.

Although a simple lowpass filter could be used to generate an output voltage proportional to the average time spent in the HIGH state (i.e., proportional to the digital input code), pulse-width modulators are most often used when the load is itself a slowly responding system. The pulse-width modulator then generates precise parcels of energy, averaged by the system connected as a load. For example, the load might be capacitive (as in a switching regulator, see Chapter 9), thermal (a thermostated bath with heater), mechanical (a tape-speed servo, variable-speed motor, or stepping motor), or electromagnetic (a large electromagnet controller).

PWM outputs are attractive both for their simplicity and for their natural match to digital devices like counters and power-driving switches (MOSFETs); but there are some serious tradeoffs. For example, to get high PWM resolution for the k/N fraction, we need a large N. But the timer has a maximum clock rate f_{clk}, which sets a lower cycle rate $f_c = f_{clk}/N$. For a PWM that is within a feedback loop (as in the PWM discussion in §15.6) this implies reduced loop bandwidth and gain.

In practice you'll most likely encounter digital PWM hardware in a microcontroller. Sometimes you can choose a specific μC on account of its PWM capabilities, but more often not. For example, further along in §13.9.11B we choose TI's MSP430F2101 because it has an analog comparator. So, what kind of PWM does one get in the MSP430x2xx family? This information is not in the

---

[14] It does cheat, however, by generating several parallel 1-bit streams internally; it's a "multibit" delta–sigma ADC. More on this, and other delta–sigma amusements, in §13.9.

Table 13.1 Six Digital-to-Analog Converters

	Bits	Channels	Output	t_{settle} (µs)	Total V min (V)	Total V max (V)	I_s (µA)	bus/Mbps	Reference type	Reference error[m]	Output Z_{out} (Ω)	Output I_{out} (mA)	pkg	pins	Cost qty 1 $US
DAC7571	12	1	V	10	2.7	5.5	140	I2C / 4.8	V_{DD}	0.2%	1	±15	SOT23	6	4.65
AD5641	14	1	V	6	2.7	5.5	75	SPI / 30	V_{DD}	0.04%	0.5	±5	SC70	6	5.40
LTC2630	12[a]	1	V	4.4	2.7	5.5	180	SPI / 50	2.5[c]	0.8%	0.08	±15	SC70	6	3.70
MAX5222	8	2	V	10	2.7	5.5	380	SPI / 25	ext	10mV	50	±1	SOT23	8	3.00
DAC7621	12	1	V	7	4.7	5.3	500	parallel[d]	4.096	0.4%	0.2	±7	SSOP	20	7.00
C8051F412	12	2	I	10	2	5.3	I_{out}	µC internal	2mA[e]	2%	CS	NA	LQFP	32	7.80

Notes: (a) 10 bit and 8 bits avail, $2.63. (b) 5V logic OK. (c) choice of V_{DD} or V_{ref}, 2.5V or 4.096V avail. (d) double buffered. (e) software select 0.25mA to 2mA, by factors of 2. (f) compliance to $V_{CC}-1.2$V. (m) max.

52- or 88-page datasheets; we had to consult the 693-page "MSP430x2xx Family Users Guide," where 40 pages were devoted to Timers A and B.

Figure 13.12 shows the MSP430F2002 member of this microcontroller family (which has a 10-bit ADC instead of a comparator) driving a dc torque motor. We've arranged four MOSFETs with an H-bridge driver and the controller's PWM signal to set the current. Naively, you'd expect that a 50% duty cycle would correspond to zero motor current; but that's true only when the motor is stopped, because the motor's "back EMF" disturbs this simple concept. Here we've used two sense resistors to measure the forward or reverse currents, with a pair of $G=80$ instrumentation amplifiers (§5.15) to inform the µC's ADC so it can servo the PWM to set the desired motor current and torque.[15] The amplifiers can operate from a single supply if their output reference pin is at least +0.8 V (see Table 5.8 on page 363); here we used a 1.25 V zener-type IC reference.

The microcontroller has a pair of 16-bit timers with programmable input selectors and dividers. They can run as fast as 16 MHz, which results in a cycle frequency of $f_c=244$ Hz if we use the full 16-bit resolution. You can program the length N to less than 2^{16}, but remember that the timer's other users have to live with your choice. Timer A has two capture/compare registers (CCR1 and CCR2), which can be used in compare mode to generate two PWM outputs (CCR0 is already used to set N). For example, let's say we need a speedier cycle rate of $f_c=10$ kHz, and we stick with the maximum 16 MHz clock. We set the counter's modulus $N=f_{clk}/f_c=1600$ counts, and our PWM

resolution will be limited to... about 10 bits. My goodness, only 10 bits?[16]

This example illustrates that a favorable alternative to PWM may involve attaching a few external DACs to a microcontroller. Another possibility is to drive the MOSFET switch with the internal bitstream (if provided on an external pin) of a delta–sigma DAC, with its improved resolution–bandwidth tradeoff (compared with $f_c = f_{clk}/N$ for the simple PWM).[17]

Exercise 13.3. Design a circuit to generate a 10 kHz train of pulses of width proportional to an 8-bit binary input code. Use counters and magnitude comparators (suitably expanded).

A. An unusual PWM DAC
We close the PWM discussion by mentioning an unusual DAC from Linear Technology. Their LTC2644 (dual) and 2645 (quad) are PWM-to-DAC converters (a DAC with a digital PWM input). Each channel measures the duty cycle (fraction of time HIGH) during each incoming PWM cycle, and immediately presents *and holds* the corresponding correct output voltage. These particular devices are available in 8-, 10-, or 12-bit resolution versions, with internal 10 ppm/°C voltage reference and monotonic rail-to-rail voltage output. They are a great improvement over the classic technique of lowpass filtering a PWM input. The possibilities are mind-boggling!

---

[15] You could instead use a differential amplifier between the two sense resistors, with a single ADC channel digitizing the resulting bipolarity signal.

[16] In motor-drive applications it's often desirable to keep the drive frequency above audibility, so you don't drive people nuts. In this example we'd have to use a modulus $N=800$ or less, throwing out another bit of control resolution.

[17] There are families of ICs, intended for speaker-driving audio power output stages, that accept a digital input stream in a standard audio format (such as I2S), and create as output a $\Delta\Sigma$-like switching waveform to drive a MOSFET H-bridge. Some of these include the MOSFETs on-chip – a single-chip PCM-to-speaker solution.

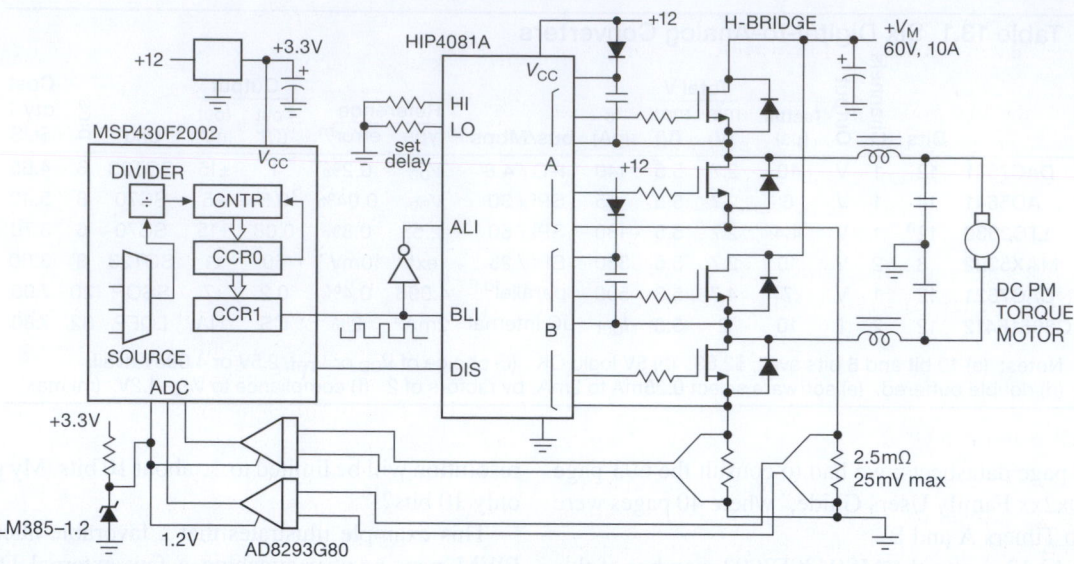

Figure 13.12. Controlling a torque motor with pulse-width modulation.

13.2.9 Frequency-to-voltage converters

In conversion applications the "digital" input may be a train of pulses or other waveform of some frequency; in that case, direct conversion to a voltage is sometimes more convenient than the alternative of counting for predetermined time, and then converting the binary count as in the preceding methods. In direct F/V conversion, a standard pulse is generated for each input cycle; it may be a voltage pulse or a pulse of current (i.e., a fixed amount of charge).

An *RC* lowpass filter or integrator then averages the pulse train, giving an output voltage proportional to the average input frequency. Of course, some output ripple results, and the lowpass filter necessary to keep this ripple less than the D/A precision (e.g., $\frac{1}{2}$ LSB) will, in general, cause a slow output response. To ensure less than $\frac{1}{2}$ LSB output ripple, the time constant τ of a simple *RC* lowpass filter must be at least $\tau \geq 0.69(n+1)T_0$, where T_0 is the output period of the *n*-bit F/V converter corresponding to maximum input frequency. The output of this *RC* network will settle to $\frac{1}{2}$ LSB, following a full scale change at the input, in $0.69(n+1)$ filter time constants. In other words, the output settling time to $\frac{1}{2}$ LSB will be approximately $t = 0.5(n+1)^2T_0$. A 10-bit F/V converter with 100 kHz maximum input frequency, smoothed with an *RC* filter, will have an output-voltage settling time of 0.6 ms. With more complicated lowpass filters (sharp cutoff) you can get improved performance. Before you get carried away with fancy filter design, however, you should remember that F/V

techniques are most often used when a voltage output is not what's needed. For some perspective, see the previous discussion about intrinsically slow loads in connection with pulse-width modulation (§13.2.8).

13.2.10 Rate multiplier

This is a somewhat rarified method, of occasional (*very* occasional!) utility. A "rate multiplier" is a bit of clocked synchronous logic that accepts a multibit digital input quantity (binary or BCD), and that passes (or blocks) clock pulses to its single output line with an *average* rate proportional to that digital quantity. You can get these as standard logic (CD4089, CD4527, or SN7497), or you can make your own. Then simple averaging, as in the preceding F/V converter, can be used to generate a dc output proportional to the digital input code, although in this case the resulting output time constant may be intolerably long, because the rate-multiplier output will have to be averaged for a time equal to the longest output period it can generate (i.e., $2^n/f_{clk}$ for a rate multiplier with an *n*-bit rate-setting input). As with PWM, rate multipliers are most useful when the output is intrinsically averaged by the slowly responding characteristics of the load itself.

An application for which this is suited is in digital temperature control, where complete cycles of ac power are switched across the heater for each rate-multiplier output pulse. In this application the rate multiplier is arranged so that its lowest output frequency is an integral submultiple of 120 Hz, and a solid-state relay (or triac) is used to switch the ac power (at zero crossings of its waveform) from logic signals.

Note that the last four conversion techniques involve some time averaging, whereas the resistor-ladder and current-source methods are "instantaneous," a distinction that also exists in the various methods of analog-to-digital conversion. Whether a converter averages the input signal or converts an instantaneous sample of it can make an important difference, as we will see shortly in some examples.

13.2.11 Choosing a DAC

To guide you in choosing a DAC for a particular application, we've assembled in Tables 13.2 and 13.3 a representative selection of DACs of various precisions and speeds. This listing is by no means exhaustive, but it does include many of the more popular converters and some more recent entries that were intended as improved replacements.

When looking for a DAC for some application, here are some issues to keep in mind:

1. resolution;
2. speed (settling time, update rate);
3. accuracy (linearity, monotonicity; external trimming required?);
4. input structure (parallel or serial? latched? CMOS/TTL/ECL-compatible?);
5. reference (internal, or externally supplied? if external, MDAC?);
6. output structure (current output? compliance? voltage output? range?);
7. required supply voltages and power dissipation;
8. single or multiple DACs per package;
9. package style;
10. price.

13.3 *Some DAC application examples*

It's always helpful to get a guided tour of a real-world application example, to get a sense of the details wherein the devil resides. It's remarkably easy to put together a circuit that delivers nowhere near the performance that its converter is capable of. The four examples in this section illustrate some of the things you've got to worry about when using a DAC.

13.3.1 General-purpose laboratory source

In our research laboratories it's common to control experimental parameters with low-noise analog voltages, which need to be highly stable over temperature and time. For example, electromagnetic traps for ions and molecules require precise voltages applied to pairs of electrostatic plates and precise currents through coils. Given the diversity of applications, the output range should be selectable in both polarity and span.

Figure 13.13 shows the business end of a popular product (the "BabyDAC"[18]) from our university's Electronic Instrument Design Laboratory. The core is the AD5544, a quad 16-bit current-output multiplying DAC that accepts an external reference voltage and outputs a set of four currents into an external node held at ground. You generate a voltage output with an external op-amp, using the matched internal feedback resistor. The internal structure for each channel is an R–$2R$ ladder driven by V_{ref}, and a set of switches that connect each $2R$ leg either to the I_o output or to ground. The external reference voltage can be of either polarity, in the range ±10 V. In fact, it can be a *signal* whose instantaneous voltage is multiplied by the digital input code to produce an output signal (hence a "multiplying DAC"); in such an application it has a signal bandwidth to audio frequencies and beyond.[19]

For this application the designer used a static V_{ref}, derived from the low-noise ADR440-series of voltage references (§9.10.3) based on a JFET analog of the standard BJT bandgap reference. The outboard circuitry is rigged up to provide jumper-selectable output-voltage ranges, both unipolarity and bipolarity; so J_1 selects a reference voltage of +5 V or +10 V. Jumper J_3 selects a gain of -2 for the output amplifier, doubling the span of U_{1a}'s output; and jumper J_2 offsets the output by the selected value of V_{ref}. With these three jumpers you can select any of the six V_{out} ranges listed in the figure.

That's the basic circuit. With ideal components the output would be precise, noiseless, and drift-free. We live in a real world, though, in which we have to choose among available components to provide the best balance of necessary compromises. For the kind of laboratory applications we have in mind, stability and low noise are paramount. In the noise department the voltage reference is usually the biggest troublemaker, thus the choice of the

---

18 One of hundreds of elegant circuits and instruments from the ever-prolific Jim Macarthur, EIDL's design guru.

19 Caution, here: the datasheet lists a "reference multiplying bandwidth" of 2 MHz, but that's measured at full-scale digital code. The bandwidth is more like 20 kHz if you want 0 to -50 dB digital control. See §13.2.4.

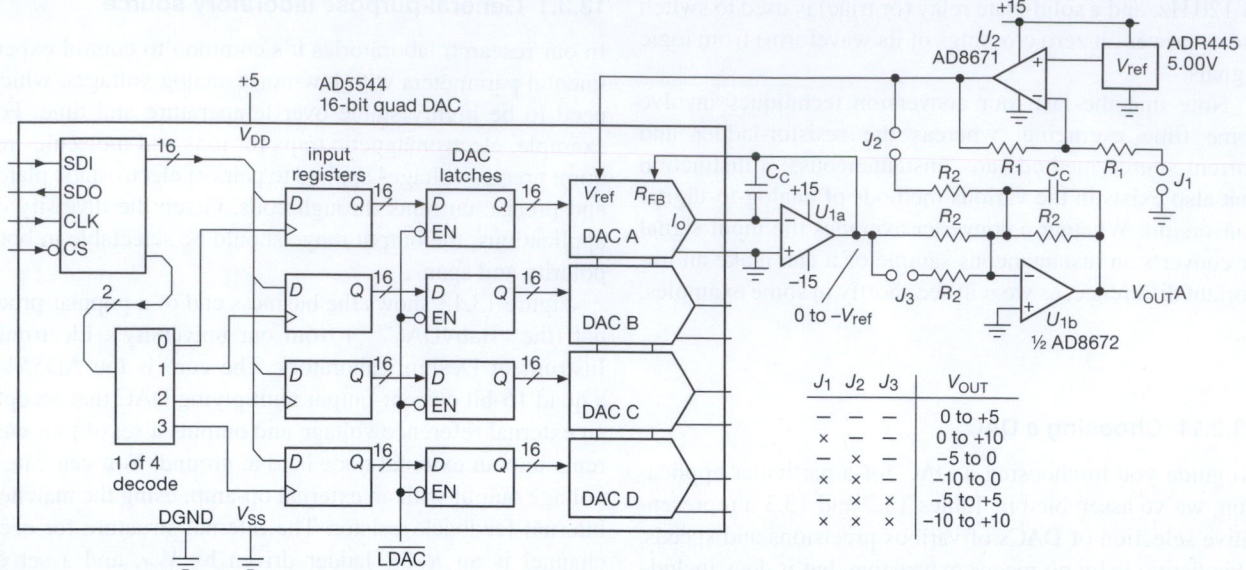

Figure 13.13. General-purpose programmable laboratory source, designed with the four-channel DAC5544 for low noise and good stability. Use an isolator for the SPI digital link signals to minimize coupled digital noise.

"ultra-low-noise" ADR445, among the quietest with its $\sim2\,\mu$Vpp typical low-frequency (0.1–10 Hz) noise. Its drift spec is also quite good (1 ppm/°C typ, 3 ppm/°C max).[20] Reference noise can be reduced by *RC* filtering (see Figure 13.19 on page 898 for an example), or, if you're desperate, by paralleling the outputs of several references (with small ballasting resistors to ensure current sharing). The op-amps are quiet by comparison: 0.1 μVpp (max) low-frequency noise; similarly, the noise contributed by the DAC is negligible.

In this circuit the op-amp gains are rolled off around 1 kHz to minimize high-frequency noise at the output (originating in the reference, and in the DAC's code-change glitches). This bandwidth was chosen somewhat arbitrarily, on the assumption that the outputs are not going to be changing rapidly. The output bandwidth could be extended by a decade if the DAC were being programmed near its maximum rate. Or, for an application that is basically quasi-static, you could limit the bandwidth still further.

For the output *stability* it's common to look for specifications of drift with temperature and with time. The typical voltage tempcos here are 1 ppm/°C for the reference and for the DAC (thus the gain: the corresponding full-scale tempco is 5 μV/°C or 10 μV/°C), and 0.3 μV/°C for the op-amps (which scales according to the gain jumper, either ×1 or ×2). With BJT-input op-amps you have to worry about input bias current and its tempco. For these op-amps the bias current is 3 nA (typ), with no tabulated tempco; however, there's an unsatisfying graph of I_B versus temperature that suggests a tempco in the range of 5 pA/°C. With the circuit's 5k (or less) impedance seen at the op-amp inputs, this amounts to a 25 nV/°C drift, entirely negligible in comparison with everything else that's going on.

Manufacturers tend to be shy about specifying drift over *time*. For the components here there's no long-term drift spec for the DAC or op-amps. The ADR445 voltage reference specifies 50 ppm typical drift over 1000 hours, but with an interesting footnote that reads "The long-term stability specification is noncumulative. The drift in the subsequent 1000-hour period is significantly lower than in the first 1000-hour period."[21]

---

[20] You'd get lower drift with the best-in-class MAX6350, at 1 ppm/°C max, but with somewhat greater noise: 3 μVpp.

[21] Curiously, some manufacturers prefer to specify the long-term drift per *square root* of time, suggestive either of a decreasing drift as the part ages, or perhaps a random walk. An example is the astonishing LTZ1000 ovenized zener, with a specified long-term drift of 2 μV/$\sqrt{\text{kHr}}$ (typ), and a claimed typical tempco of 0.05 ppm/°C.

Table 13.2 Selected D-to-A Converters[a]

Part #	Bits	settle[t] (µs)	Channels	Output[b]	Cost qty25 ($US)	Total Supply min (V)	max (V)	Single supply[c]	Supply Current (mA)	Interface[d]	Bipolar sig	Latch	Internal Ref[e]	External Ref[g]	Rail-rail out	Sleep Mode	Multiplying	5V inputs OK?	DIP	SOIC	TTSOP	SOT-23	Smaller	Monotonic	Comments
AD5300	8	6	1	V	2.28	2.7	5.5	●	0.14	S	-	-	-	V_S	●	●	-	-	-	-	8	6	-	●	A,J
MCP4706	8	6	1	V	0.60	2.7	5.5	●	0.21	I	-	-	●	-	●	●	-	-	-	-	-	6	6	●	M,Z6
AD8801	8	0.6	8	V	6.20	3	5.5	●	0.001	S	-	-	●	-	●	-	-	-	16	16	-	-	-	●	B,D
DAC0808	8	0.15	1	I	1.30	7	36	-	2.6	P	●	-	-	●	-	●	●	-	16	16	-	-	-	●	E
DAC08	8	0.085	1	I	1.61	9	36	-	3.6	P	●	-	-	●	-	-	●	-	16	16	-	-	-	●	E,F
LTC1663	10	30	1	V	3.70	2.7	5.5	●	0.06	2	-	●	1.25	●	●	●	-	-	-	8	-	5	-	●	A,G
LTC1669	10	30	1	V	2.39	2.7	5.5	●	0.06	I	-	●	1.25	●	●	●	-	●	-	8	-	5	-	●	A,G
AD5620	12	10	1	V	3.81	4.5	5.5	●	0.55	S	-	●	2.5	-	●	●	-	-	-	-	-	8	-	●	C,H
DAC121	12	8	1	V	2.07	2.7	5.5	●	0.20	S	-	●	-	V_S	●	●	-	-	-	-	8	6	-	●	A
DAC7611	12	7	1	V	6.05	4.75	5.25	●	0.50	S	-	●	4	-	-	-	-	●	8	8	-	-	-	●	K,L
MCP4725	12	6	1	V	0.80	2.7	5.5	●	0.20	I	-	●	-	V_S	●	●	-	-	-	-	-	6	-	●	M
AD7845	12	5	1	V	10.00	28	32	-	4.6	P	●	●	-	●	-	-	●	-	-	-	-	-	-	●	N
MCP4921	12	4.5	1,2	V	1.64	2.7	5.5	●	0.20	S	-	●	-	●	●	●	-	-	8	8	8	-	-	●	O,P
TLV5638	12	1	2	V	10.56	2.7	5.5	●	4.3	S	-	●	2	-	●	●	-	-	-	8	20	-	-	●	Q
TLV5630	12	1	8	V	14.87	2.7	5.5	●	16	S	-	●	2	-	●	●	-	-	-	20	20	-	-	●	A,G,Q,R
DAC5672	14	0.020	2	I	22.42	3.3[n]		●	100	P	-	●	1.2	●	-	●	-	-	-	-	48	-	-	-	S
AD9739	14	0.013	1	I	56.00	1.8 and 3.3[n]		-	400,70	L	-	●	1.2	●	-	●	-	-	-	-	-	-	160	-	T
AD5660	16	10	1	V	5.90	4.5	5.5	●	0.6	S	-	●	2.5	-	●	●	-	-	-	-	8	-	●	C,U
DAC8564	16	10	4	V	11.91	2.7	5.5	●	1.0	S	●	●	2.5	●	●	●	-	-	-	16	-	-	●	A,V
LTC2656	16	8.9	8	V	29.50	2.7	5.5	●	3.0	S	-	●	2.5	●	●	●	-	-	-	20	-	20	●	C,W
AD5686R	16	5	4	V	9.90	2.7	5.5	●	1.1	S	-	●	2.5	o	●	●	-	-	-	16	-	16	●	C,X
AD5541A	16	1	1	V	15.39	2.7	5.5	●	0.13	S	-	●	-	●	-	-	●	●	-	10	-	8	●	A,Y
LTC1668	16	0.020	1	I	18.81	+5 and -5[n]		-	3.33	P	●	●	2.5	●	-	-	-	-	-	28	-	-	-	Z
DAC5682	16	0.010	2	I	45.14	1.8 and 3.3[n]		-	500,133	L	-	●	1.2	●	-	●	-	-	-	-	-	64	-	Z2
AD5780	18	2.5	1	V	32.23	5	33	-	10,10	S	●	●	-	●	-	-	●	-	-	-	-	24	-	●	A,Z3
DAC9881	18	5	1	V	28.98	2.7	5.5	●	0.85	S	-	●	-	●	-	●	-	-	-	-	-	24	-	●	C,Z4
AD5791	20	1	1	V	53.53	10	33	-	4.2	S	●	●	-	●	-	-	-	-	-	20	-	-	●	Z5

Notes: (a) see also MDAC Table 13.3; listed by increasing resolution, then speed. (b) V-voltage; I-current. (c) operates from a single positive supply. (d) 2 - 2-wire serial; I - I²C; L - parallel LVDS; P - parallel; S - SPI. (e) 2 = 2.048; 4 = 4.096. (g) can use ext ref. (n) nominal. (o) non-R version. (t) typical.

Comments: **A:** power-on to 0V. **B:** power-on to midscale. **C:** power-on to 0V or midscale. **D:** TrimDAC, pot replacement, Z_{out}=5kΩ. **E:** multiplying, to ~1MHz. **F:** compliance to –10V and +18V. **G:** double buffered for simultaneous updating. **H:** 14-bit=AD5640, 16-bit=AD5660; 0.2%, 5ppm/°C ref. **J:** digital gain & offset adjustment. **K:** DAC8512 2nd source. **L:** power-on to 0V plus a CLR pin. **M:** power-on state from on-chip EEPROM. **N:** multiplying, to 600kHz. **O:** multiplying, to 450kHz; 14-pin dual=4922. **P:** power-on to hi-Z. **Q:** programmable settling time. **R:** 10-bit=TLV5631; 8-bit=TLV5632. **S:** 275Msps; 1.5ns to 90%; DAC2904, AD9767 2nd source. **T:** RF synthesis, 2.5Gsps. **U:** ext ref version =5662; 0.2%, 5ppm/°C ref. **V:** 14-bit=DAC8164; 12-bit=DAC7564; low glitch; 0.004%, 2ppm/°C ref. **W:** 0.2%, 2ppm/°C ref; also 4.096 Vref and 12-bit versions. **X:** 2ppm/°C ref. **Y:** multiplying, to ~1MHz; low glitch; 0.1ppm/°C; 12nV/√Hz; unbuffered output. **Z:** low glitch; 50Msps; 12-bit and 14-bit versions. **Z2:** 1Gsps; FIFO; clock PLL; on-chip digital filters. **Z3:** 8nV/√Hz; 0.02ppm/°C; 3 pwr supplies; "system ready." **Z4:** 0.3ppm/°C; 24nV/√Hz. **Z5:** 0.05ppm/°C; 7.5nV/rtHz. **Z6:** 10-bit=MCP4716; 12-bit=MCP4726.

As we remarked at the outset, for the intended application it is noise and drift that are most important. By contrast, the absolute accuracy is not particularly impressive: the worst-case specs are ±200 ppm for the voltage reference, ±75 µV for the op-amps, and ±3 mV full-scale gain error for the DAC. Translated to units of 16-bit LSB steps, and taking the output range to be ±10 V (for which an LSB is 0.3 mV), these correspond to ±13 LSB, ±0.5 LSB, and ±10 LSB.

13.3.2 Eight-channel source

If your application does not need the flexibility of output polarities and scale factors of the last example, and if you can tolerate a bit more noise and drift, you can do pretty well with a fully integrated multiple-section voltage-output DAC, for example the LTC2656 in Figure 13.14. This particular IC includes an internal voltage reference of good stability (±2 ppm/°C typ, ±10 ppm/°C max), with a 1 ppm/°C typical full-scale tempco for the DAC

Table 13.3 Multiplying D-to-A Converters[a]

Part #	Bits	t_{settle} typ (µs)	Channels	Output Type	Cost qty25 ($US)	Power Supplies Positive min (V)	Power Supplies Positive max (V)	Power Supplies Negative min (V)	Power Supplies Negative max (V)	V_{in} Range (V)	V_{out} Range (V)	I_{supply} typ (µA)	Interface[c]	Distortion (1kHz) typ (dB)	Distortion (1kHz) @V_{pp} (V)	Noise (nV/√Hz at 1kHz)	Bandwidth typ (MHz)	Bandwidth @V_{pp} (V)	Feedthrough[b] (MHz)	Packages, Pins DIP	Packages, Pins SOIC	Packages, Pins TTSOP	Packages, Pins Smaller	Comments
AD8842	8	2.9	8	V	13.71	4.75	5.25	4.75	5.25	±3	±3	10	S	-80	4	78	1.5	0.1	0.25	24	24	-		A
TLC7528	8	0.1	2	I	3.47	4.75	15	-	-	±25	±25	2000	P	-85	6	-	-	-	-	20	20	20	-	B
MCP4921	12	4.5	1	V	1.64	2.7	5.5	-	-	0 to V_S	0 to V_S	200	S	-73	0.4	-	0.45	0.4	-	8	8	8	-	C
AD/LTC7541	12	0.6	1	I	6.00	5	16	-	-	±10	±0.5	100	P	-	-	-	-	-	-	18	18	-	20	D
AD7943,5,8	12	0.6	1	I	7.78	3	5.5	-	-	±10	±0.3	0.005	S,P	-83	17	35	-	-	2	16	20	20	-	E
LTC8043	12	0.25	1	I	8.10	4.75	5.25	-	-	±10	±0.5	100[m]	S	-108	17	17[m]	-	-	1	8	8	8	-	F
AD5443	12	0.06	1	I	4.87	3	5.5	-	-	±10	±0.3	0.6[m]	S	-81	3.5	25	10	7	10[e]	-	-	10	-	G
DAC7821	12	0.05	1	I	6.38	2.5	5.5	-	-	±15	±0.3	0.8	P	-105	-	18	10	7	0.8	-	-	20	-	H
AD5544	16	0.9	4	I	27.43	2.7	5.5	0	5.5	±15	±0.3	5[m]	S	-98	5	7	5	5	-	-	-	28	32	I,J
DAC8814	16	0.5	4	I	29.06	2.7	5.5	-	-	±15	±0.3	2	S	-105	5	12	10	0.3	2	-	-	28	-	J
DAC8820	16	0.5	1	I	16.98	2.7	5.5	-	-	±15	±0.3	5[m]	P	-105	17	10	8	5	0.15	-	-	28	-	K
LTC2757	18	2.1	1	I	57.43	2.7	5.5	-	-	±15	±0.3	0.5	P	-110	5	13	-	-	1[e]	-	-	-	48	L

Notes: (a) listed by increasing accuracy and speed; all are monotonic, and all have latches except AD/LTC7541; see also DAC Table 13.2. (b) capacitive coupling causes a 6dB/octave rising output (from the desired digital value of attenuation) at high frequencies; the listed value is the frequency at which there is a +3dB increase relative to a −40dB programmed attenuation (i.e., an actual attenuation of −37dB). (c) 2 - 2-wire serial; I - I2C; P - parallel; S - SPI; L - parallel LVDS. (d) 65dB feedthrough at 100kHz. (e) 3dB loss, rather than feedthrough. (m) min or max. (n) nominal. (t) typical.

Comments: **A:** current conveyor, $V_{out}=V_{in}$ x (D/128 − 1). **B:** 2nd source for AD7528. **C:** MCP4922=dual. **D:** 80dB feedthru @ 10kHz and full-swing. **E:** improved AD7543, -45, and -48; AD7545 & -48 are parallel interface. **F:** also DAC8043. **G:** 10-bit=AD5432, 8-bit=AD5426. **H:** parallel readback. **I:** 14-bit=AD5554. **J:** reset to zero or midscale. **K:** reset to zero. **L:** double buffer, readback, reset to zero.

outputs. The DAC's low-frequency (0.1–10 Hz) output noise is 8 µVpp, typical. This is four times the noise voltage of the previous example; taking into account the more limited output span (0–2.5 V), it represents a larger relative noise contribution still. The good news is the relative simplicity of this part: no external reference or amplifiers, and operation from a single positive supply.

The SPI digital interface is simple and clean: every transfer is 24 bits, with 4 bits to specify the channel number (with an option to load all channels with the same value), 4 bits to specify the operation, and 16 bits to carry the digital value. Each channel is double buffered, so you can load the next value into each channel's input buffer, then transfer them simultaneously to the DAC's register so that all outputs change to their new values at the same time.

13.3.3 Nanoamp wide-compliance bipolarity current source

Here's an unusual application, and a circuit implementation of considerable subtlety, using a dual-supply current-output DAC: suppose you need a programmable current source that can operate over a wide voltage range (say ±10 V) while sourcing (or sinking) very small currents (in the nanoamp range, say). You might need this to measure the V–I characteristics of a semiconductor at the low

end of its current range, or perhaps for a research application such as the electrical properties of nanofibers. Another application might be to cancel the input leakage current of a high-impedance measuring device, for example an 8-digit digital multimeter with a JFET frontend (a discrete matched JFET pair or a precision JFET-input op-amp), in which the leakage current rises rapidly (but predictably) with temperature.[22] Such an instrument could store a table of leakage versus temperature, measured during initial calibration, to be used in conjunction with a temperature sensor to program the cancellation current during normal operation. Current-output DACs don't operate successfully down at these currents, and furthermore they don't generally provide outputs that can both source and sink current under control of a digital input code.

This circuit (Figure 13.15) is tricky, and seriously confusing at first. Work with us, here, and you'll get it (eventually). An essential component is the simple floating current source circuit (Figure 13.16), in which an op-amp follower with an added voltage V_0 at the output bootstraps a feedback resistor R to create a current V_0/R. That voltage can be created by biasing a zener-type reference, as shown; or

---

[22] It was an analogous circuit in Agilent's 34420A multimeter (see §5.12.5) that inspired this example. (Their circuit used a voltage reference in place of R_3, allowing the use of a single DAC output. They also used a smaller R_s, so that the output range was ±2 nA.)

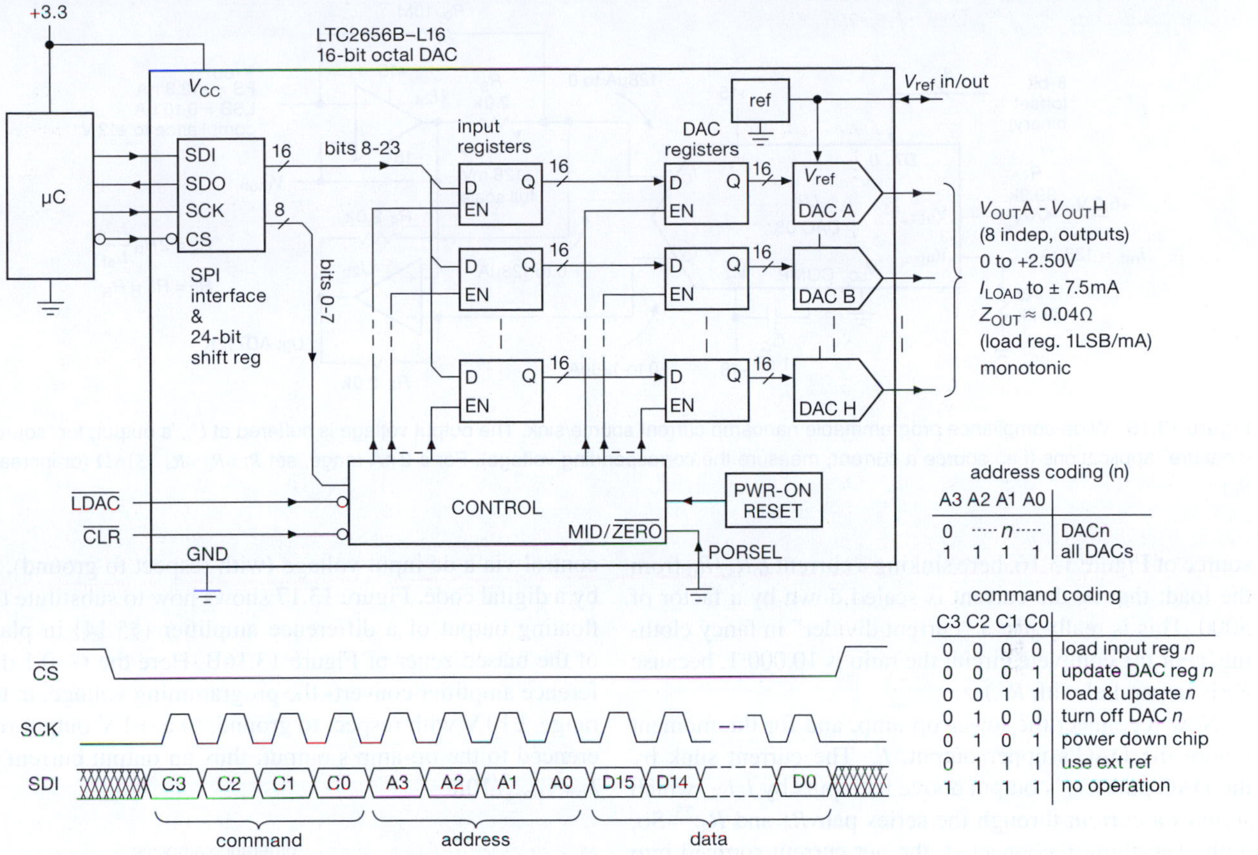

Figure 13.14. Eight-channel voltage-output DAC.

it can arise from a current flowing through a resistor that
returns to the op-amp's output.

Now for the complete circuit of Figure 13.15. The
DAC08 is an oldie (circa 1984), with a pair of opposite-
ramping current-sinking outputs that sum to a constant current I_{ref} (set by the current sourced through R_1, here equal
to 5 V/39.2 kΩ). An 8-bit offset-binary input code sets the
individual output currents. For example, the minimum code
(00h) causes I_o' to sink $128\,\mu A$ and I_o to sink 0; for a
quarter-scale code (20h, or 32 decimal) the corresponding currents are $96\,\mu A$ and $32\,\mu A$; and so on. With +18
and −15 V supplies, the output compliance extends from
−12 V to +18 V.

The external circuitry works the real magic, by (a) converting this unipolar current-sinking pair into a bipolarity
(sourcing or sinking) output current, and (b) simultaneously scaling down the current by a factor of 10,000, to
generate an output terminal that programmably sources or
sinks currents with a full-scale range of ±12.8 nA, in LSB
steps of 0.1 nA.

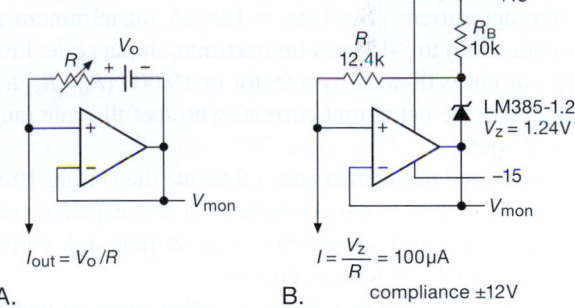

$I_{out} = V_o/R$

$I = \dfrac{V_z}{R} = 100\,\mu A$

compliance ±12V

A.   B.

Figure 13.16. A follower with a voltage applied in series with a
resistor creates a simple floating current source, and the op-amp's
output pin provides a "voltage monitor." A. The basic scheme. B.
Implementation with a bandgap voltage reference.

To understand first the current scaling, disconnect U_{2b}
and look at the upper op-amp alone: the current sunk by the
DAC generates a small voltage across R_s, approximately
$I_o R_s$ (for $R_o \gg R_s$). That's the V_o of the floating current

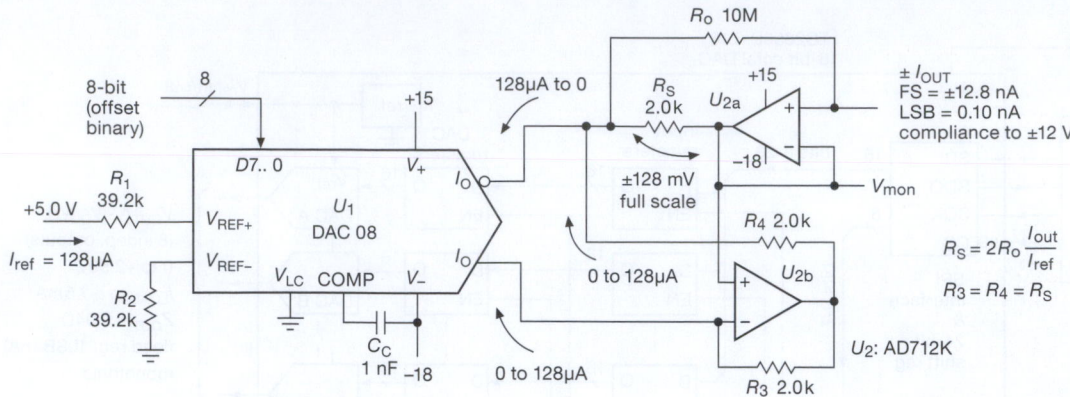

Figure 13.15. Wide-compliance programmable nanoamp current source/sink. The output voltage is buffered at U_{2a}'s output, for "source-measure" applications (i.e., source a current, measure the corresponding voltage). For a 2 nA range, set $R_s=R_3=R_4=316\,\Omega$ (or increase R_o).

source of Figure 13.16, here sinking a current I_oR_s/R_o from the load; that is, the current is scaled down by a factor of 5000. This is really just a "current divider" in fancy clothing. (For the complete circuit the ratio is 10,000:1, because R_4 is in parallel with R_s.)

Now reconnect the lower op-amp, and for the moment ignore the DAC's upper output, I_o'. The current sunk by the DAC puts U_{2b}'s output above its inputs by I_oR_3, which *sources* a current through the series pair R_4 and R_s.[23] So, with everything reconnected, the net current sourced into the left-hand side of R_s is the (positive) current equal to the magnitude of the DAC's I_o output, minus the (negative) current equal to the magnitude of the DAC's I_o' output. So the net current goes from $-128\,\mu A$ (at minimum input code, 00h) to $+128\,\mu A$ (at maximum input code, FFh). That current is divided by a factor of 10,000 ($R_o/[R_s\|R_4]$) to generate the net output current, with its full-scale range of ± 12.8 nA.

If you need more precision, substitute the similar 10-bit DAC10. It differs in having somewhat less negative compliance, and in providing a full-scale output sink current equal to *twice* the reference current.

And a final comment: there are other ways to make a wide-compliance nanoamp current source; see, for example, the relatively simple design of Figure 5.69A.

A. Variations on the floating current source
Going back to the simple floating current source circuit of Figure 13.16, there are several useful variations that permit

---

[23] This op-amp configuration is in fact a current mirror, sourcing into the node at its noninverting input a current proportional to the current sunk from its inverting input, in the ratio of the output resistor to R_3.

control via a dc input voltage (with respect to ground), or by a digital code. Figure 13.17 shows how to substitute the floating output of a difference amplifier (§5.14) in place of the biased zener of Figure 13.16B. Here the $G=0.1$ difference amplifier converts the programming voltage, in the range ± 10 V with respect to ground, to a ± 1 V output referenced to the op-amp's output, thus an output current of $I = V_{prog}/10R$.

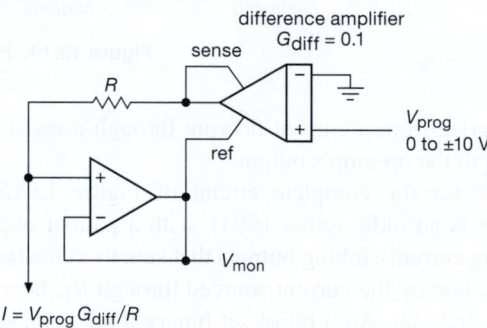

Figure 13.17. Floating current source analogous to Figure 13.16, programmed with a difference amplifier. See also Figure 5.69.

Some commercial instruments are capable of providing programmable currents over a very wide voltage range (say ± 200 V). This they do (Figure 13.18) by powering the op-amp with a floating dc supply (say ± 5 V), with a DAC (powered from the same supply) substituting for the difference amplifier in Figure 13.17. Both the op-amp and the DAC fly over the wide compliance range, with the DAC's digital input data fed via opto-isolators. In this scheme the op-amp's output is buffered and used to bootstrap the dc supply common, thus keeping the op-amp's inputs and outputs near mid-supply; the bootstrap acts also to prevent dynamic currents caused by the supply's capacitance to its primary power source. These programmable current sources with voltage-monitor outputs are examples of what's called a "source-measure unit" (SMU), typified by units from Keithley (their 2400 series) or Keysight (their B2900 series).

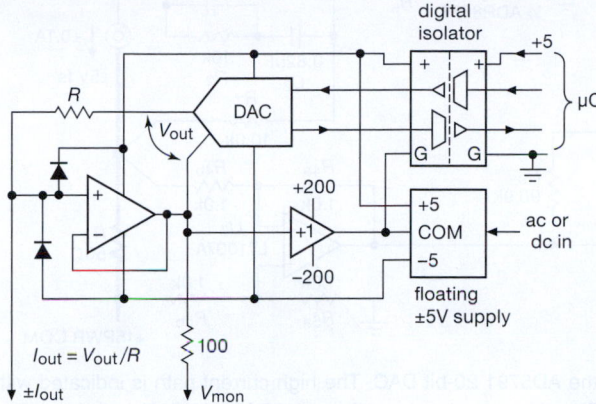

Figure 13.18. High-voltage floating current source scheme used in "source-measure" instruments. The unity-gain high-voltage buffer can be a simple push–pull follower, as its job is only to bootstrap the ± 5 V floating supply. Adding a divider and buffer to the V_{mon} output produces a low-voltage replica.

13.3.4 Precision coil driver

Here's an application that pushes the limits of DAC resolution and stability: a current-source driver to provide a settable and stable current (of either polarity) through a pair of coils that trim the magnetic field in a magnetic resonance apparatus. This kind of application can require parts-per-million (ppm) resolution and stability. It's an instructional example worthy of some detailed discussion.

A. DAC and voltage reference
Figure 13.19 shows an implementation, with the remarkable AD5791 20-bit DAC at its core. Looking first at the circuitry surrounding the DAC, we've again chosen the ADR445 reference (in the best-performing B-grade) for its low noise and excellent stability (an alternative is the MAX6350, with the recommended capacitor from its noise reduction pin to ground). We added a 10 Hz RC lowpass filter to suppress wideband noise: according to the datasheet the ADR445's noise in the band from 0.1–10 Hz is about 2.3 μVpp, increasing to 66 μV when that band is extended to 10 kHz.

When you're working at ppm error levels, you have to worry about everything! For example, just 1.5 nA of leakage current in the 10 μF filter capacitor (C_3) would cause nearly 1 ppm error in the +5.0 V reference driving the DAC (from the IR drop across R_{10}). The noise filter shown exploits a nice trick[24] to eliminate that error: the lower leg ($R_{11}C_4$) bootstraps the bottom of C_3 so that there's essentially no dc voltage across it, hence no leakage current; this is analogous to the technique of a *guard* electrode, which is used to eliminate leakage currents in sensitive low-current measurements (or to eliminate the effects of shunt capacitance where signals are present).

The AD5791 can operate with a single positive reference, or with both positive and negative references. For our application we want an output range of ± 5 V, but we can exploit the DAC's internal precision matched resistor pair (which returns to the positive reference) so that only a positive reference is needed. The DAC provides a "sense" output from both internal reference nodes, used with feedback as shown to eliminate IR errors. Note the matched input resistances at U_2's inputs, to take advantage of input current matching ($\Delta I_B < 1$ nA); the input-current tempco, estimated from graphs in the datasheet, is less than 10 pA/°C. The op-amp's offset voltage is 12 μV typ (50 μV max), and the offset voltage tempco is 0.6 μV/°C max (0.2 μV/°C typ).

On the scale of the +5.0 V reference that the op-amp is buffering, these figures translate into 0.7 ppm, 0.007 ppm/°C, 2.4 ppm, and 0.12 ppm/°C, respectively. In other words, the bias current and offset voltage errors add up to about 3 ppm, or 3 LSBs; but the drifts are just 0.13 ppm/°C (max), or an LSB for an 8°C change in temperature. We're happy with that stability, which is what we really care about for this application. The ~3 ppm scale error is unimportant, because in practice the current will be adjusted until the coil current is doing the right thing. Finally, we've got to consider the error and drift contributed by the DAC itself, which are comparable: its full-scale and zero-scale errors are each ± 2 ppm (2 LSB) max, and its output tempco (at zero, midscale, or full-scale) is ± 0.05 ppm/°C typ (± 0.5 ppm/°C max). Its typical low-frequency output noise is 0.6 μVpp at mid-scale, roughly half as much as that contributed by the reference.

B. Amplifier loop
The job here is to use the DAC's stable output voltage to control the coil current[25] over a range of ± 0.1 A while

---

[24] Originated by Walter Jung, as far as we know.

[25] We took as a reference design a pair of coils, each 30 cm in diameter with 500 turns, spaced axially by 15 cm in the so-called "Helmholtz"

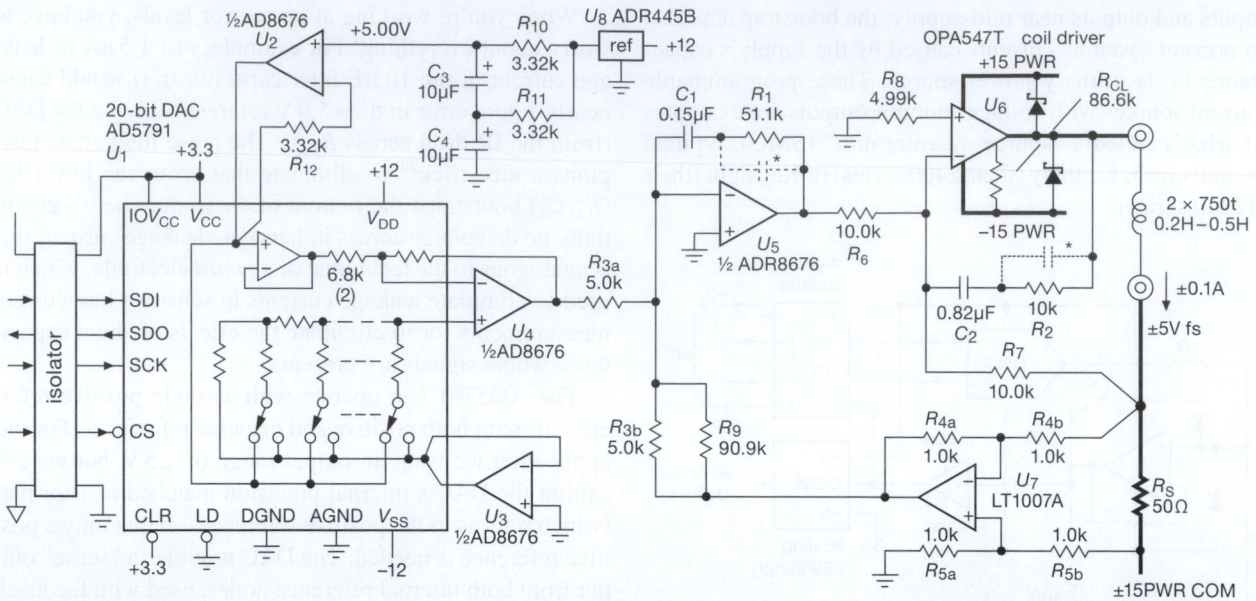

Figure 13.19. Precision programmable Helmholtz coil driver featuring the AD5791 20-bit DAC. The high current path is indicated with heavy lines. Resistors R_3–R_5 are Vishay MPM matched pairs (2 ppm/°C tracking), and R_s is a Vishay VPR221 (Y0926) bulk foil power resistor (2 ppm/°C tempco) with heat-sinking.

preserving part-per-million dc stability and noise. There's also the *other* kind of stability – freedom from oscillation.

Ignoring the latter for the moment (i.e., neglecting all of the capacitors in the circuit), at the top level the amplifier loop works like this: a temperature-stable current-sensing resistor $R_s = 50\,\Omega$ generates a full-scale voltage of ± 5 V, a replica of which is subtracted from the DAC's output by the matched resistor pair R_{3ab}. Error amplifier U_5 supplies the loop gain, applying the amplified error to the coil driver U_6, which operates as a unity-gain inverter. The phases are correct: too much current through R_s drives the output of U_7 down, U_5 up, and U_6 down.

At the next level there's the dc accuracy and drift to worry about. Once again we've chosen precision BJT op-amps, the now-familiar AD8676 for the error amplifier, and the vintage LT1007A for the unity-gain difference amplifier. The latter has the lower noise and offset voltage, at the expense of a higher input current; we circumvent the latter problem by reducing the resistance seen at the inputs, which is possible because of the low (50 Ω) source impedance. The power op-amp U_6 need not be accurate, because it is within an overall feedback loop whose gain rises at low frequencies as $1/f$.

configuration. With #20 gauge wire, the total dc resistance is 30Ω, the inductance is about 400 mH, and a current of 0.1 A produces a central field of 3 gauss (about six times Earth's field).

The sense resistor R_s is a Vishay "bulk metal foil" 4-wire (Kelvin connection) precision power resistor in a TO-220 package, rated at 8 W dissipation; the best grade is accurate to ±0.01%, with a typical tempco of 2 ppm/°C. The resistor pairs R_3–R_5 are each precision matched duals in convenient SOT23 packages, with 0.05% matching, and 2 ppm/°C (typ) tracking. The resistor shunting R_{3b} compensates for the reduction of R_s's effective resistance that is due to loading by R_7 and R_{4b}.

Assuming there's plenty of loop gain (and there is, see next paragraph), the ppm-scale dc stability of the DAC output is reasonably well preserved with these op-amps, assisted by resistors of this stability and tracking.

Finally, there's the serious business of stability against oscillations, which is complicated here by the inductive load. The latter alone causes a 6 dB/octave rolloff beginning at the frequency at which the inductive reactance equals the sense resistance, about 20 Hz for the nominal 400 mH coil pair. We've addressed it by making the error amplifier an integrator at lower frequencies (for plenty of dc gain), flattening to a gain of ×10 at 20 Hz.[26] That prevents the curve of overall loop gain versus frequency from dropping at more than 6 dB/octave, at least until well beyond the unity-gain frequency. The output amplifier U_6

---

[26] In fancy language, a pole at dc and a zero at 20 Hz.

gets the same treatment. An additional small capacitor across R_1 (and similarly across R_2) rolls off the local gain at higher frequencies still, to stabilize against high-frequency oscillations. Suitable values are 150 pF and 4.7 nF (rolling off at 20 kHz and 3 kHz, respectively).

Rather than prattle on about this, we present the self-explanatory Bode plots of Figure 13.20. Readers who are likely to encounter situations like this can benefit from the thinking presented there.

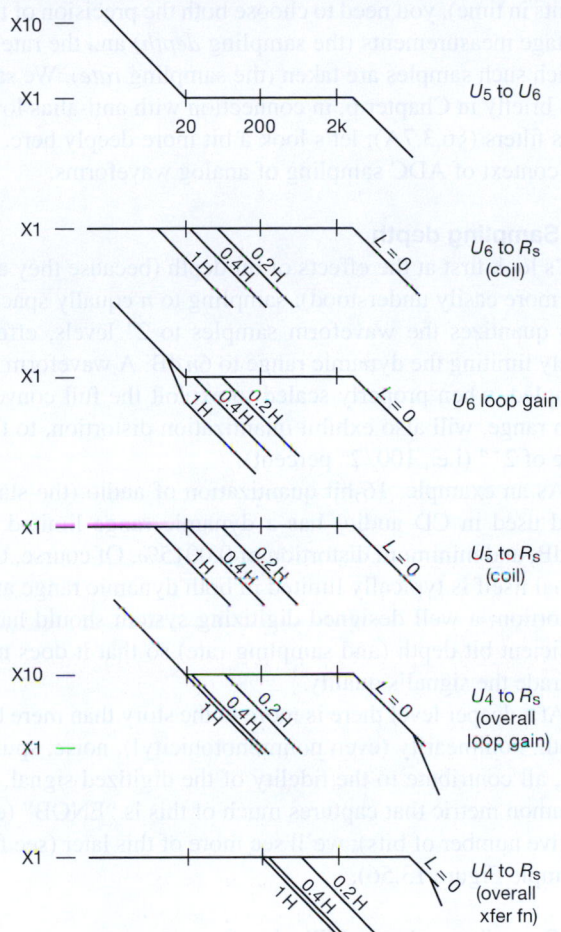

Figure 13.20. Bode plots for the coil driver amplifier in Figure 13.19, shown with several values of load inductance.

13.4 Converter linearity – a closer look

In §13.1.3 we mentioned, briefly, the sorts of errors that afflict DACs (and ADCs as well). The business of *linearity errors* deserves some more discussion.

Take a look at Figures 13.21 and 13.22. Both 3-bit DACs suffer from linearity errors. But there's some sub-

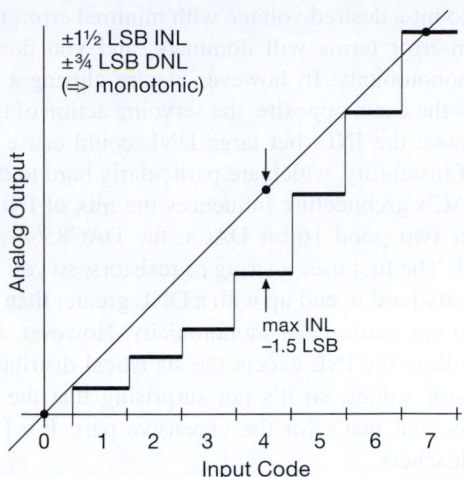

Figure 13.21. A DAC can exhibit both monotonicity (DNL<1 LSB) and relatively large integral nonlinearity (here INL=1.5 LSB).

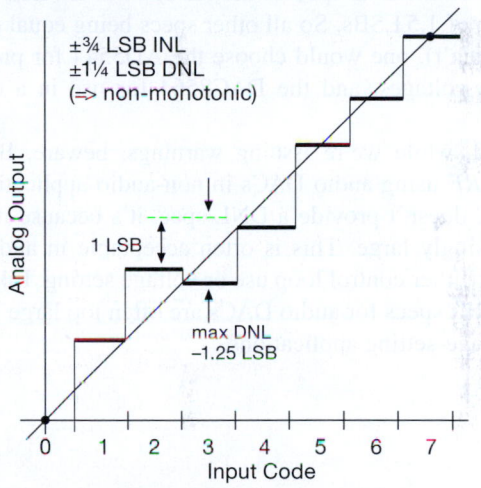

Figure 13.22. This converter has less INL (0.75 LSB) than in Figure 13.21, but its larger DNL (1.25 LSB) permits non-monotonicity. Which matters more depends on the application.

tlety here. We need a couple of definitions: *integral* nonlinearity (INL) is the maximum deviation from the ideal straight line of analog output versus digital input, over the full conversion range;[27] whereas *differential* nonlinearity (DNL) is the maximum error in any single digital step (i.e., from $n=2$ to $n=3$ in this example) from its proper step size of 1 LSB.

When do you care about INL versus DNL? If you need

---

[27] There's a bit of wiggle room here, because you can define the line as passing though the endpoints ("endpoint linearity," as used here), or you can make things look a bit better by using the best-fit straight line.

a DAC to hit a desired voltage with minimal error, the INL and gain-error terms will dominate, and you don't care about monotonicity. If, however, you're closing a control loop, it's the exact opposite: the servoing action of the loop will remove the INL, but large DNL could cause hidden zones of instability, which are particularly hard to debug.

A DAC's architecture influences the mix of INL/DNL. Consider two good 16-bit DACs, the DAC8564 and the AD5544. The first uses a string of resistors, so you have to work pretty hard to end up with a DNL greater than 1 LSB. And you are *guaranteed* monotonicity. However, nothing is controlling the INL except the statistical distribution of the resistor values, so it's not surprising that the INL is ± 8 LSBs, and that's for the expensive part. It's 12 LSBs in the bleachers.

In contrast, in the *R–2R* architecture a large INL will often translate to a large DNL; the same process that keeps the DNL under control also keeps the INL down, to some extent, so the INL spec of the AD5544 is ± 4 LSBs, with a DNL of 1.5 LSBs. So all other specs being equal (which they aren't), one would choose the AD5544 for precisely setting voltages, and the DAC8564 for use in a control loop.

And, while we're issuing warnings, beware, Beware, *BEWARE* using audio DACs in non-audio applications. If a DAC doesn't provide a DNL spec, it's because it's embarrassingly large. This is often acceptable in audio, but not for either control loop use or voltage setting. Likewise, gain-drift specs for audio DACs are often too large for use in voltage-setting applications.

13.5 *Analog-to-digital converters*

Look back at the earlier section on "Preliminaries" (§13.1) to remind yourself of some of the things to think about when choosing a converter (whether DAC or ADC). At the top level you're concerned much less with the details of how the thing actually does its conversion, and much more with the major questions of (a) performance (speed, accuracy, etc.), (b) digital interface (parallel or serial; single-ended or LVDS; etc) and (c) integration (single or multiple units; stand-alone, or integrated into a microcontroller or other complex function). In most cases you'll use a commercial ADC chip or module rather than building your own. But it's important to know about the inner workings of the various A/D conversion methods, in order not to be caught unaware by their idiosyncrasies.

13.5.1 Digitizing: aliasing, sampling rate, and sampling depth

We'll get into the nitty-gritty of analog-to-digital conversion soon enough, but first a short riff on the business of *sampling*, which will come up again and again as we visit various ADC methods.

When you convert an analog signal (e.g., an audio waveform) to a series of digital quantities (i.e., numbers corresponding to the instantaneous voltage at successive moments in time), you need to choose both the precision of the voltage measurements (the sampling *depth*) and the rate at which such samples are taken (the sampling *rate*). We saw this briefly in Chapter 6, in connection with anti-alias low-pass filters (§6.3.7A); let's look a bit more deeply here, in the context of ADC sampling of analog waveforms.

A. Sampling depth

Let's look first at the effects of bit depth (because they are the more easily understood): sampling to n equally spaced bits quantizes the waveform samples to 2^n levels, effectively limiting the dynamic range to $6n$ dB. A waveform so sampled, when properly scaled to exploit the full conversion range, will also exhibit quantization distortion, to the tune of 2^{-n} (i.e., $100/2^n$ percent).

As an example, 16-bit quantization of audio (the standard used in CD audio) has a dynamic range limited to 96 dB, and minimum distortion of 0.0015%. Of course, the signal itself is typically limited in both dynamic range and distortion; a well designed digitizing system should have sufficient bit depth (and sampling rate) so that it does not degrade the signal's quality.

At a deeper level there is more to the story than mere bit depth: nonlinearity (even nonmonotonicity!), noise, spurs, etc., all contribute to the fidelity of the digitized signal. A common metric that captures much of this is "ENOB" (effective number of bits); we'll see more of this later (see for example Figure 13.56).

B. Sampling rate and filtering

The story here is more subtle (and more interesting). Contrary to intuition, a waveform that is perfectly sampled at a rate at least twice that of the highest frequency component present suffers *no loss of information whatsoever*. Nothing is lost in the unsampled portion of the waveform in between the samples; this is the Nyquist/Shannon sampling theorem (which has its cadre of unbelievers, who swear that digital audio removes the very soul of music).[28]

---

[28] It can be shown mathematically that the original signal (excluding a set of pathological waveforms) is perfectly recovered:

Readers harboring contempt for authority may wonder what happens if one violates statute by *under*sampling. Easy enough to try: look at Figure 13.23, where we've sampled a 100 Hz sinewave (requiring $f_{samp} \geq 200$ sps) at 90 sps, in serious violation of Nyquist's prescription. The sampled points trace out a false signal, in this case 10 Hz. This is called an "alias," and most of the time it's something you don't want.[29] Put simply, for a given sampling rate f_{samp}, the analog input signal must be lowpass filtered (with an "anti-alias filter") such that no significant signal remains above $f_{samp}/2$. Conversely, for an analog signal that extends to some maximum frequency f_{max}, the minimum sampling rate is $2f_{max}$. (You can, of course, sample more rapidly than the Nyquist limit $f_{samp} > 2f_{max}$, and in fact it's wise to do a modest degree of "oversampling" because that permits more relaxed lowpass filtering of the analog signal, as we'll see presently.)

It's helpful to look at the business of aliasing in the *frequency* domain. In Figure 13.24A we've attached a wimpy 2-section RC filter to a broadband signal, putting the -3 dB point of each section at the Nyquist limit ($f_{samp}/2$). Frequency components in the forbidden region get falsely digitized as shown,[30] contaminating the intended signal band; they cannot be removed later by filtering – in the digitized output they are now "in-band."

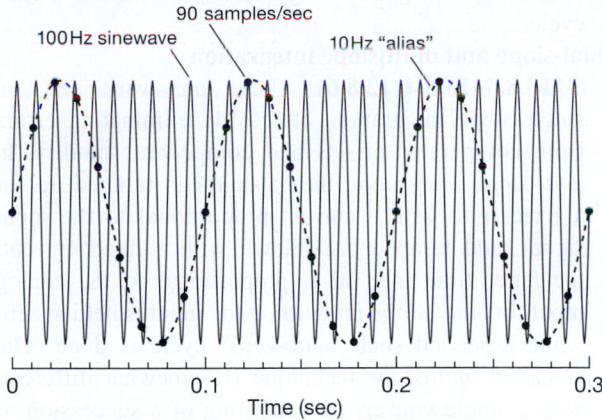

Figure 13.23. Digitizing at less than the Nyquist rate produces "aliases." A 100 Hz sinewave (solid line) sampled at 90 sps (far below the 200 sps Nyquist rate) produces a 10 Hz alias (dots, connected by a dashed line).

$v(t) = \sum v_i \operatorname{sinc} \pi(f_s t - i)$, where f_s is the sampling rate and v_i is the amplitude of the ith sample.

[29] See §13.6.3 for an important exception.

[30] To draw these "elevation sketches," just mirror the contours seen past the multiples of the Nyquist frequency.

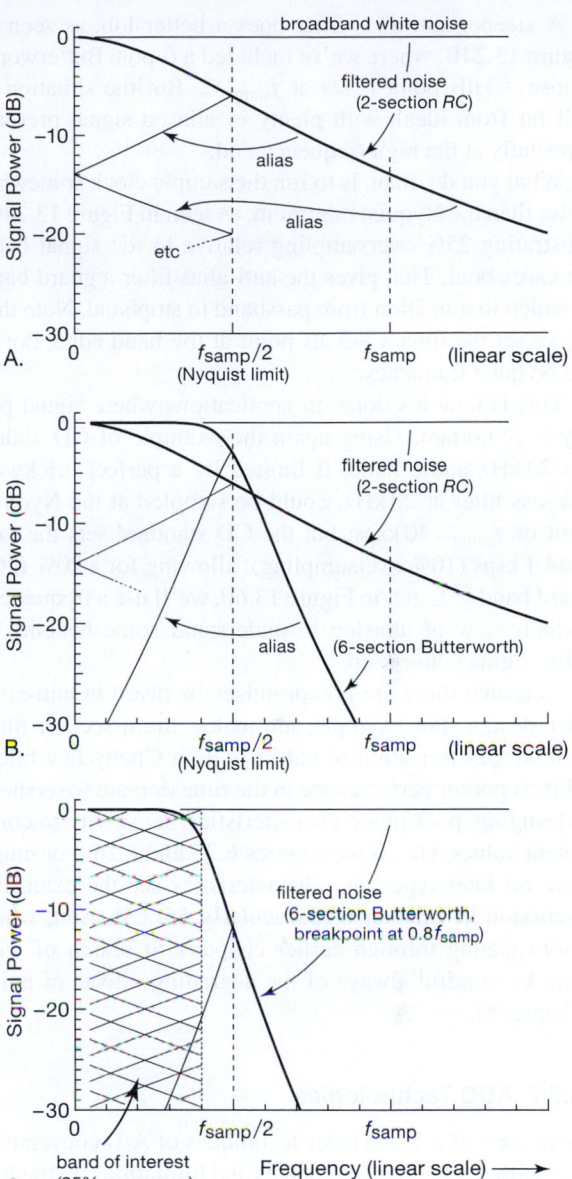

Figure 13.24. Undersampling, oversampling, and aliasing. A. Sampling a signal with frequency components above the Nyquist limit ($f_{samp}/2$) produces aliased digitized versions that fall within the properly-sampled signal; here a gentle 2-section RC rolloff allows significant aliased out-of-band signal energy to contaminate the intended signal. B. A steeper filter is more effective; but aliasing still contaminates the Nyquist band edge. C. Oversampling (setting the Nyquist frequency above the band of interest, here by 25%) gives the anti-alias filter a guard band in which to roll off, greatly reducing aliasing.

A steeper anti-alias filter does a better job, as seen in Figure 13.24B, where we've included a 6-pole Butterworth whose $-3\,\mathrm{dB}$ point is set at $f_{\mathrm{samp}}/2$. But the situation is still far from ideal, with plenty of aliased signal present, especially at the high frequency end.

What you do, then, is to run the sample clock somewhat faster than the Nyquist minimum, as seen in Figure 13.24C, illustrating 25% oversampling relative to the signal band we care about. That gives the anti-alias filter a guard band in which to transition from passband to stopband. Note that we've set the filter's $-3\,\mathrm{dB}$ point at the band edge, not at the Nyquist frequency.

This is how it's done, in applications where signal purity is important. Using again the example of CD audio, the 20 kHz audio band, if limited by a perfect brickwall lowpass filter at 20 kHz, could be sampled at the Nyquist limit of $f_{\mathrm{samp}}=40\,\mathrm{ksps}$; but the CD standard sets the rate at 44.1 ksps (10% oversampling), allowing for a 20% filter guard band.[31] Later, in Figure 13.60, we'll use a frequency-domain view of aliasing to understand some benefits of delta–sigma conversion.

Note that there are compromises involved in anti-alias filter design. For example, an analog multi-section filter with steeper transition to cutoff (e.g., a Chebyshev filter) exhibits poorer performance in the time domain (overshoot and ringing, poor phase characteristics, sensitivity to component values, etc.) – see Figures 6.25 and 6.26. For much more on filter types and characteristics see the extensive discussion in Chapter 6 (particularly §6.2.5). And, while you're paging through earlier chapters in search of wisdom, be mindful always of the degrading effects of noise (Chapter 8).

13.5.2 ADC Technologies

There are half a dozen basic techniques of A/D conversion, each with its peculiar advantages and limitations. In the following subsections we'll take each in turn, along with some application examples. Here, in outline form, is a compact summary of these techniques.

Flash, or "parallel" (§13.6) The analog input voltage is compared with a set of fixed reference voltages, most simply by driving an array of 2^n analog comparators to generate an n-bit result. Variations on this theme include pipelined or folded architectures, in which the conversion is done in several steps, each of which converts the "residue" of the previous low-resolution conversion.

Successive approximation (§13.7) Internal logic generates successive trial codes, which are converted to voltages by an internal DAC and compared with the analog input voltage. It requires just n such steps to do an n-bit conversion. The internal DAC can be implemented as a conventional n-stage R–$2R$ resistor ladder or, interestingly, as a set of 2^n binary-scaled capacitors; the latter method is known as a *charge-redistribution* DAC.

Voltage-to-frequency (§13.8.1) The output is a pulse train (or other waveform) whose frequency is accurately proportional to the analog input voltage. In an *asynchronous* V/F the oscillator is internal and free running. By contrast, a *synchronous* V/F requires an external source of clock pulses, gating a fraction of them through such that the *average* output frequency is proportional to the analog input.

Single-slope integration (§13.8.2) The time required for an internally generated analog ramp (capacitor charged by a current source) to go from zero volts to the analog input voltage is proportional to the value of the analog input. That time is converted to an output number by gating a fast fixed-frequency clock, and counting the number of clock pulses. Note that pulse-width modulation employs the same ramp-comparator scheme as single-slope integration to generate the ON time of each cycle.

Dual-slope and multislope integration (§§13.8.3–13.8.4, 13.8.6) These are variations on single-slope integration, effectively eliminating errors from comparator offsets and component stability. In *dual-slope integration* the capacitor is ramped up for a fixed time with a current proportional to the input signal, and ramped back down with a fixed current; the latter time interval is proportional to the analog input. In *quad-slope integration* the input is held at zero while a second such "auto-zero" cycle is done. The so-called *multislope* technique is somewhat different, with a single conversion consisting of a succession of fast dual-slope cycles (in which the input is integrated continuously, combined with subtractive fixed-current cycles), and with a correction based upon the partial-cycle residue at both ends. In some aspects it's a close cousin of the delta–sigma method.

Delta–sigma (§13.9) There are two parts: a *modulator* converts the analog input voltage to a serial *bitstream*; then a digital lowpass filter accepts that bitstream as input, producing the final n-bit digital output. Most simply (and it's never very simple!), the modulator consists of an integrator acting on the difference between the

---

[31] Much higher oversampling – at *many* times Nyquist rate – is exploited in the technique of delta–sigma conversion, see §13.9.

analog input voltage and the value of the 1-bit output serial bitstream to determine the next output bit. Variations include higher-order modulators (a succession of weighted integrators), or bitstreams that are several bits wide (a "wordstream"?), or both. Delta–sigma converters are both popular and confusing, and they deserve the extensive section later in the chapter.

13.6 ADCs I: Parallel ("flash") encoder

This is probably the simplest ADC concept; it's also the fastest (see Table 13.4). In this method the input signal voltage is fed simultaneously to one input of each of n comparators, the other inputs of which are connected to n equally spaced reference voltages. The output levels of the n comparators form a "thermometer code," which is converted (in a *priority encoder*) to a ($\log_2 n$)-bit binary output corresponding to the highest comparator activated by the input voltage. Figure 13.25 shows the idea *notionally* – here clumsily implemented with discrete comparators and standard logic. You wouldn't do that, of course; much better to let the silicon wizards do their integrated magic. In this simple (single-stage) scheme the delay time from input to output is the sum of comparator, encoder, and output latch (if provided) delays. An example of a commercial flash encoder using this scheme is the MAX1003: it does 6-bit conversions on each of two input channels, at sampling rates to 90 Msps, with the digitized and latched result available one clock cycle after sampling.

13.6.1 Modified flash encoders

In practice, the simple flash scheme has been largely supplanted by modified flash variants, with names like "half flash," "subranging flash," "folding/interpolating architecture," or "pipelined flash." These generally involve successive stages of partial conversion, so there is some amount of delay (or *latency*) from the moment of input sampling to valid digital output. This does not necessarily reduce the maximum sampling rate. In fact, quite the contrary: by subdividing the conversion into a succession of coarser quantizations, these converters can achieve very high sampling rates, with the partially quantized analog "residues" propagating along in a capacitor-based pipeline as newer samples begin their conversion. In such a converter an initial coarse conversion (say to 2-bit resolution) is followed by successive stages that operate on the residue (the difference between the analog input and that coarse estimate). An example is the ADC10D1500, a dual 10-bit 1.5 Gsps ADC

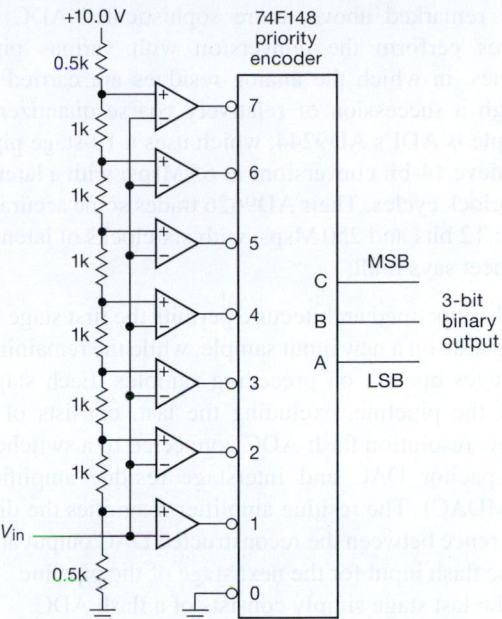

Figure 13.25. Parallel-encoded ("flash") ADC.

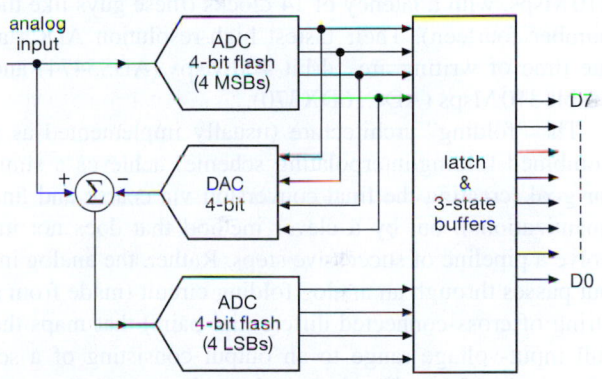

Figure 13.26. Half-flash ADC.

(the two sections can be interleaved, to achieve 3 Gsps); it has a latency of 35 clock cycles.

Perhaps the simplest of these converter architectures is the *half-flash*, a two-step process in which the input is flash-converted to half the final precision; an internal DAC converts this digital approximation back to analog, where the difference "error" between it and the input is flash-converted to obtain the least significant bits (Figure 13.26). This technique yields low-cost converters that operate at relatively low power. Examples are TI's TLC0820, ADI's AD7820, and TI's TLC5540, inexpensive 8-bit ADCs with latencies of two or three clock cycles, and modest conversion speeds (40 Msps for the latter).

As remarked above, more sophisticated ADC architectures perform the conversion with various pipeline schemes, in which the analog residues are carried along through a succession of relatively coarse quantizers. An example is ADI's AD9244, which uses a 10-stage pipeline to achieve 14-bit conversions at 65 Msps, with a latency of eight clock cycles. Their AD9626 trades some accuracy for speed: 12 bits and 250 Msps, with six clocks of latency. Its datasheet says it all:

> The pipelined architecture permits the first stage to operate on a new input sample, while the remaining stages operate on preceding samples. Each stage of the pipeline, excluding the last, consists of a low resolution flash ADC connected to a switched capacitor DAC and interstage residue amplifier (MDAC). The residue amplifier magnifies the difference between the reconstructed DAC output and the flash input for the next stage of the pipeline.... The last stage simply consists of a flash ADC.

Similar technology is offered by TI's ADS5547, a 14-stage pipelined ADC that achieves 14-bit conversions at 210 Msps, with a latency of 14 clocks (these guys like the number fourteen). Their fastest high-resolution ADCs at the time of writing are 14-bit 400 Msps (ADS5474) and 16-bit 370 Msps (ADC16DX370).

The "folding" architecture (usually implemented as a combined folding/interpolating scheme) achieves a similar goal (creating the final conversion via coarse and fine quantizations), but by a clever method that does not involve a pipeline of successive steps. Rather, the analog input passes through an analog folding circuit (made from a string of cross-connected differential pairs) that maps the full input-voltage range to an output consisting of a set of repeating folds. That output is flash-converted to produce the lower-order bits, while a coarse flash conversion of the full-range input signal simultaneously determines in which fold the signal lies (i.e., the higher-order bits); see Figure 13.27. The "Ultra High Speed ADC" family from National Semiconductor uses these techniques, with current offerings going to speeds of 3.6 Gsps at 12-bit resolution (ADC12D1800). There are lots of tricks involved in making this work well; as the saying goes, they are "well beyond the scope of this book."

Flash encoders are worth considering in waveform digitizing applications even when the conversion rate is relatively slow, because their high speed (or, more precisely, their short *aperture* sampling interval) ensures that the input signal is effectively not changing during the conversion. The alternative – the slower converters we describe

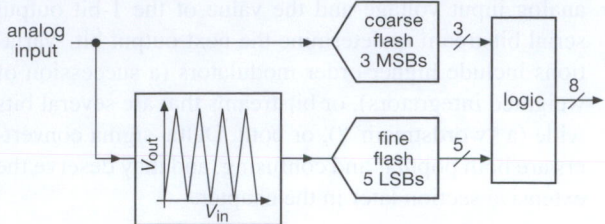

Figure 13.27. Flash ADC with "folding" architecture.

next – usually requires an analog sample-and-hold circuit to freeze the input waveform while conversion is going on. Note that a converter's latency may or may not matter, depending on the application: latency would not be of concern in an oscilloscope frontend, or in a "software radio"; but it would be a disaster in a fast digital control loop.

13.6.2 Driving flash, folding, and RF ADCs

Today's ADCs are not your father's simple converters. It's true they're much more capable than converters of previous generations, but they're cantankerous, they no longer observe the rule of speaking only when spoken to, and they may be very demanding of resources, especially power and digital assets. This is especially true for high-speed low-voltage converters with differential inputs. No longer can you count on simply taking an op-amp and connecting it to your ADC.

To illustrate with an example, Figure 13.28 shows a dual-channel 16-bit flash ADC capable of working up to low RF frequencies, ideal for digitizing I, Q signals for software radios. It employs a differential amplifier IC discussed in §5.17, performing the tasks shown in Figures 5.102 and 5.103. As we draw these circuits, at first glance they may look alike, but we find ourselves spending time on the details, looking up datasheet rules and suggestions, considering alternative parts with new rules, and honing the design.

The AD9269 converter[32] is a flash type, but later in the chapter we'll encounter the same issues with the other primary ADC types. For example, some successive-approximation register (SAR) converters (and many of the delta–sigma converters as well) will require the distinctive two-resistor plus capacitor configuration at their inputs. Many SAR types go further, being quite picky about

---

[32] The AD9269 (listed in Table 13.4) is a dual ADC with conventional CMOS logic outputs, but ADCs with slightly faster data rates generally upgrade to differential LVDS outputs, or 32 lines per 16-bit ADC channel.

Table 13.4 Selected Fast A-to-D Converters[a]

Part #	Mfg	ADC per pkg	Bits	Conv rate max (Msps)	Conv rate min (Msps)	Aperture jitter (ps, typ)	Analog BW (MHz, typ)	Latency (clks)	SFDR (dB, typ)	SNR (dB, typ)	@ MHz	ENOB, typ	@ MHz	DNL (LSBs, max)	INL (LSBs, max)	Input CM to rails?	Internal Vref?	External Vref OK?	Gain Error (%, max)	Drift (ppm/°C, typ)	Supply analog (+V)	Supply digital (+V)	Pdiss typ (mW)	Output	Pkg	Price qty 1 (US$)	Comments
ADC14L020	TI	1	14	20	5	0.7	150	7	93	74	10	12	10	1	3.8	-	●	●	3.3	2.5	3.3	2.5-V_{ana}	150	C	32LQFP	20	-
AD9225	ADI	1	12	25	5	1	105	3	83	70	10	11.3	10	1	2.5	B	●	●	1.7	0.4	5	3–5	280	C	28SO	25	-
ADC12040	NSC	1	12	40	0.1	1.2	100	6	-	69	10	11	10	1	1.8	N	-	-	2.1	-	5	2.5-V_{ana}	340	C	32LQFP	17	-
ADC14L040	TI	1	14	40	5	0.7	150	7	90	73	20	11.9	20	1	3.8	-	●	●	3.3	2.5	3.3	2.5-V_{ana}	235	C	32LQFP	20	-
TLC5540	TI	1	8	40	5	30	75	3	42	47	3	7.5	3	1	0.75	B	●	-	3.5	-	5	-	85	C	24SO	7	S
LTC2192	LTC	2	16	65	5	0.07	550	7	90	77	70	-	70	0.9	6	-	●	●	1.7	10	1.8	1.8	200	G	52QFN	140	-
AD9244	ADI	1	14	65	0.5	0.3	750	8	86	72	70	11.5	70	1	4	-	●	●	2	2.3	5	3–5	550	C	48LQFP	33	-
AD9269	ADI	2	16	80	3	0.1	700	9	90	76	70	12.3	70	1.7	6.5	-	●	●	2[t]	2	1.8	1.8–3.3	225	C,L	64LFCSP	77	A
ADS5562	TI	1	16	80	1	0.09	300	16	80	83	30	13.1	30	3	8.5	-	●	●	2.5	100	3.3	3.3	865	C,L	48VQFN	76	P
AD9283	ADI	1	8	100	1	5	475	4	-	47	40	7.3	40	1.25	1.25	-	-	●	6	80	3.3	3.3	90	C	20SOP	10	P
AD9650	ADI	2	16	105	10	0.075	500	12	90	82	30	13	30	1.3	6	-	●	●	1.3	15	1.8	1.8	328	C,L	64LFCSP	170	P
AD9268	ADI	2	16	125	10	0.07	650	12	88	78	70	12	70	1.2	5.5	-	●	●	2.5	15	1.8	1.8	750	C,L	64LFCSP	130	P
ADS6245	TI	2	14	125	5	0.25	500	12	86	72	100	11.7	100	2.5	5.5	-	●	●	1	50	1.8	3.3	1W	O	48VQFN	100	P
LTC2165	LTC	1	16	125	5	0.07	550	6	84	76	140	-	140	0.9	6	-	●	●	1.3	10	1.8	1.2–1.8	194	C,D	48QFN	125	A
ADC1610	NXP	1	16	125	100	-	650	13.5	84	75	170	11.4	170	0.95	4[t]	-	●	●	0.5[ty]	10	1.8	1.8–3.3	630	C,D	40QFN	40	R,S
ADS5485	TI	1	16	200	20	0.08	730	5	-	76	100	12.1	100	0.95	10	-	●	●	3	100	3	1.8–3.3	2.8W	D	64VQFN	130	P
ADC08200	NSC	1	8	200	20	2	500	6	54	44	400	7.3	400	2	1.9	B	●	-	3	-	3	2.5-V_{ana}	1/Msps	C	24SO	14	-
ADS5547	TI	1	14	210	1	0.15	800	14	70	68	300	11.8	300	2.5	5	-	●	●	2	100	3.3	3.3	1.2W	C,D	48QFN	120	P
ADS6149	TI	1	14	250	1	0.17	700	18	76	68	70	11.2	70	2	5	-	●	●	0.2[t]	10	3.3	1.8	687	C,D	48VQFN	140	P
AD9626	ADI	1	12	250	40	0.2	700	6	80	64	300	10.5	300	0.6	1.7	-	●	●	1.4[t]	210	1.8	1.8	364	C	56LFCSP	230	P
AD9467	ADI	1	16	250	50	0.06	900	16	90	74	600	12	600	1.3	10[t]	-	●	●	3.5[y]	360[y]	V2	1.8	1.3W	L	72LFCSP	210	P
ISLA216P25	ISL	1	16	250	40	0.075	700	10	67	69	170	10.7	170	0.99	12	-	●	●	6[y]	180[y]	1.8	1.8	770	C,D	72QFN	64	P
AD9211	ADI	1	10	300	40	0.2	700	7	80	59	230	9.6	230	0.5	0.7	-	●	●	4.3[y]	180[y]	1.8	1.8	420	L	56LFCSP	190	-
ADS5474	TI	1	14	400	20	0.1	1.4G	3.5	80	70	500	10.9	500	1.5	3	-	●	●	5[y]	200[y]	V1	3.3	2.5W	L	80TQFP	215	P
ISLA112P50	ISL	1	12	500	80	0.09	1.15G	17	71	65	850	10.3	850	0.8	3	-	●	●	2[ty]	325[y]	V1	1.8	475	C,L	72QFN	190	A
ADS5400	TI	1	12	1000	100	0.125	2.1G	7	66	58	750	9.3	750	0.8	4.5	-	●	●	5[y]	300[y]	V1	3.3	2.2W	L	100TQFP	930	A
ADC08D1520	NSC	2	8	1500	200	0.4	3.1G	13	55	47	750	7.4	750	0.6	0.9	-	●	●	3.3	-	1.9	1.8-V_{ana}	2W	L	128LQFP	830	I
ADC10D1500	NSC	2	10	1500	200	0.2	3.1G	13	55	55	750	8.8	750	0.55	1.4	-	●	●	3.3	-	1.9	1.8-V_{ana}	3.6W	L	292BGA	1k	I
ADC12D1800	NSC	2	12	1800	300	0.2	2.8G	34	66	57	500	9	500	0.4[t]	2.5[t]	-	●	●	3.3	-	1.9	1.8-V_{ana}	4.4W	L	292BGA	-	-
HMCAD5831	ADI	1	3	26G	-	-	20G	2	27	36	19G	2.9	12G	0.2[t]	0.2[t]	-	-	-	-	-	-5	-3.3	4.2W	L	64QFN	-	J
Fujitsu 'Robin'	FUJ	2	8	56G	-	0.1	15G	-	-	-	17G	-	-	0.5	1	-	-	-	-	-	±1.2	3.3	4W	F	flip-chip	-	K

Notes: (a) listed in order of increasing sample rate, for the fastest member of a family; unless noted otherwise, inputs are differential, with sampling capacitors at the inputs introducing switching transients that require low drive impedance and RC input networks; all guarantee no missing codes. (m) min or max. (t) typical. (y) includes int ref.

Comments: **A:** input buffer, isolates sampling transients. **B:** both. **C:** n-line parallel CMOS. **D:** $n/2$-line DDR LVDS/ch. **F:** 1024-lane LVDS. **G:** 1, 2, or 4 LVDS lanes/ch. **I:** interleave for 2x sampling rate. **J:** non-interleaved. **K:** fastest ADC; 320 interleaved 175Msps SARs. **L:** n-lane LVDS/ch. **N:** negative only. **O:** one LVDS lane. **P:** programmable operating parameters, usually via SPI. **R:** resistor ladder and comparators. **S:** single-ended input. **V1:** 3.3V and 5V. **V2:** 1.8V and 3.3V.

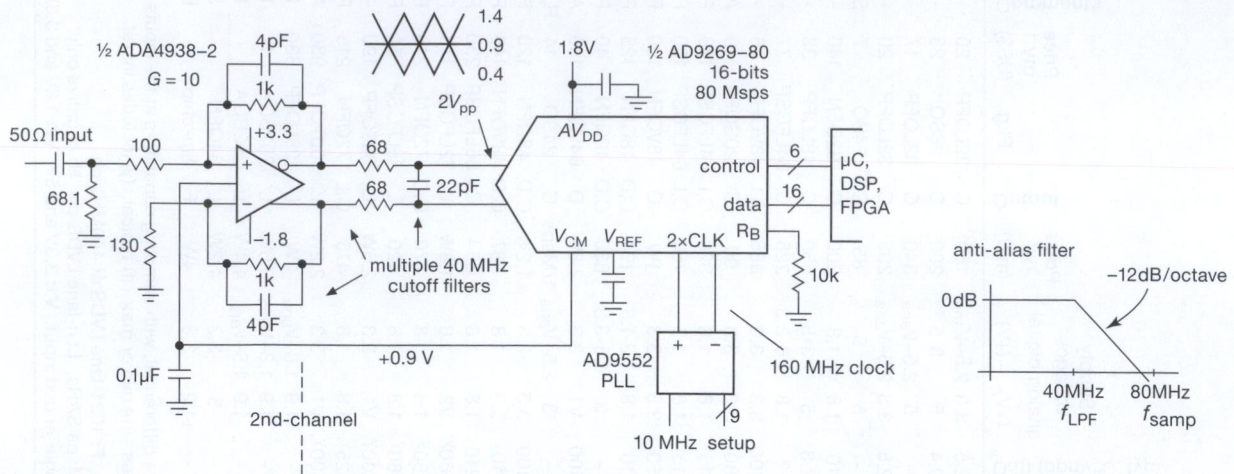

Figure 13.28. Fast ADCs are usually driven differentially, as in the two-channel 16-bit 80 Msps (40 MHz) RF digitizer. Radiofrequency applications like this require an accurate and stable clock source at some multiple of the conversion rate, here provided by a phase-locked-loop (§13.13).

what R and C values are allowed to achieve specified performance.

An 80 Msps converter has a 40 MHz Nyquist limit, set approximately with our $2R+C$ differential lowpass input filter. The filter R,C parts play two other roles: the ADC responds to noise (white and other) all the way up to its 700 MHz input bandwidth, so we need to quiet the output of the amplifier aggressively above 40 MHz; and the ADC's input switched-capacitor S/H needs to grab some charge from an input capacitor to work properly. The two R's also serve to isolate the capacitor from the op-amp, important because 1000 MHz op-amps don't kindly tolerate direct capacitive loads. So we have three motivations for these new parts, not seen in the good old days.

Why do we use a differential-*output* amplifier? Generally, when driving a device that presents a differential input, we have the option of grounding one side and feeding the other. But doing this with today's ADCs will cost us a substantial distortion penalty, and half the full-scale input range as well. But in choosing the differential amplifier IC we were boxed into a corner: looking in the 500–1500 MHz region of Table 5.10 on page 375, we couldn't find a part with high "$Z_{in(diff)}$." We wanted additional 40 MHz anti-alias filtering, which ruled out attractive parts with internal gain-setting resistors because the summing junctions aren't exposed. And we wanted a gain of at least 10, or 20 dB. So we chose, finally, an Analog Devices ADA4938 with a 1000 MHz rated bandwidth.[33] Looking at

its frequency-response plots, we see its GBW=800 MHz, thus an f_{-3dB}=GBW/G=80 MHz for G=10, so we have some loop gain left at 40 MHz.

Configuration-D amplifiers like this (Figure 5.96) have rather low input impedances, especially at high gains, because Z_{in}=$2R_g$, and $R_g = R_f/G$. Having thrown in the towel on high input impedances, we opt for matching the ubiquitous 50 Ω source impedance of wideband signals. If we choose R_g=100 Ω, the Johnson noise[34] of two of these will be 1.8 nV/$\sqrt{Hz}$, or nicely below the amplifier's rated e_n=2.6 nV/$\sqrt{Hz}$.

We have to provide a 50 Ω load for the input, and recognizing that Z_{in} isn't exactly R_g,[35] we turn to the formula provided in the datasheet, $Z_{in} = R_g/(1 - R_f/2[R_g + R_f])$, to determine that we need a 68 Ω load resistor; then we match the impedance seen driving the noninverting input by using a 130 Ω resistor to ground at the inverting input. This added resistance disturbs the usual $R_f = GR_g$ relationship, and we are forced to increase R_f by 11% to maintain $G = 10$, as detailed in §5.17.4. Finally we evaluate the single-ended to differential conversion by examining the datasheet's V_{ocm} specifications. The V_{ocm} −3dB spec is 230 MHz, which means that with our feedback attenuation the full-differential ADC drive will be down by 3 dB at 24 MHz.[36]

---

[33] We could also have chosen a TI LMH6552 or LMH6553.

[34] See §§8.1 and 8.2.

[35] A fraction of the differential-output voltage appears at the inputs as a common-mode signal, partially bootstrapping the voltage across the input resistor R_g.

[36] If this is not acceptable, we need to reduce our amplifier gain by half or

Our AD9269 ADC needs a sampling clock, for which we choose the capable AD9552 PLL upconverter (see §13.13.6H and Table 13.13). It seemed a good idea to take advantage of the ADC's internal divide-by-two option, to help ensure an internal 50% duty cycle, so we'll need a 160 MHz clock for 80 Msps sampling; thus, if we use a 10 MHz reference, we set the PLL multiplication to 16. If we want other sampling rates we can employ the AD9552's powerful delta–sigma modulator fractional-frequency synthesis capabilities, and we can also choose other ADC division ratios.

A final worry (if you haven't had enough already) is *clock jitter*. The AD9269 datasheet's graphs show that, to obtain the desirable best-available 75–78 dB signal-to-noise ratio (SNR), the clock jitter should be (gasp!) no more than 0.2 ps (about 15 ppm of the sampling period). Our AD9552 PLL datasheet specifies the jitter (for a 4–80 MHz reference input to be 0.11 ps, so we slip under the wire on that one (but without much to spare).

13.6.3 Undersampling flash-converter example

Figure 13.29 shows, in somewhat simplified form, an "undersampling converter" application, in which an input signal centered at some rather high frequency (say 500 MHz) is digitized at a rate far less (say 200 Msps) than would appear necessary from the Nyquist criterion. This works successfully if two conditions are satisfied: (a) the signal's *bandwidth* must satisfy the Nyquist sampling criterion, i.e., the sampling rate must be at least twice the bandwidth occupied by the signal; and (b) the signal's full spectrum (including the high carrier frequency) must fall within the ADC's analog input bandwidth.

The first condition requires that the input signal be strictly limited in bandwidth, usually with a bandpass filter. The second condition implies that the ADC has been designed for an undersampling application. The ADC08200 in the figure, for example, specifies a full-power bandwidth of 500 MHz, even though its maximum sampling rate is 200 Msps (which would normally be appropriate for signals only to 100 MHz). You can think of this as exploiting the aliased spectrum produced by undersampling; that's OK, as long as there are no other spectral components competing for that "baseband" piece of spectrum (see Figure 13.30).[37]

In the example circuit we've used a relatively slow

seek out a faster amplifier IC, such as an ADA4937, with a 440 MHz V_{ocm} spec.

[37] Sometimes called "super-Nyquist operation." See, for example, Application Note AN-939 from Analog Devices.

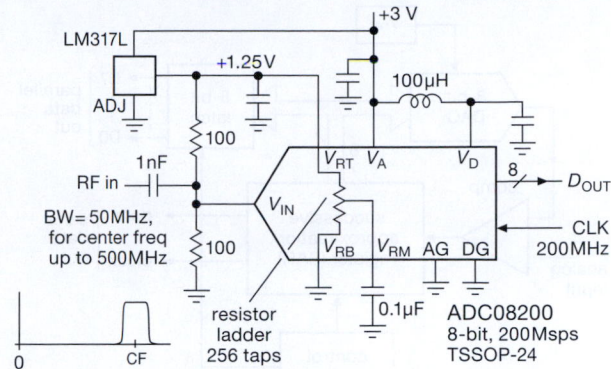

Figure 13.29. An inexpensive flash ADC digitizes a band-limited signal well above the Nyquist cutoff frequency, a job that traditionally calls for frequency downconversion with a local oscillator ("LO") and mixer.

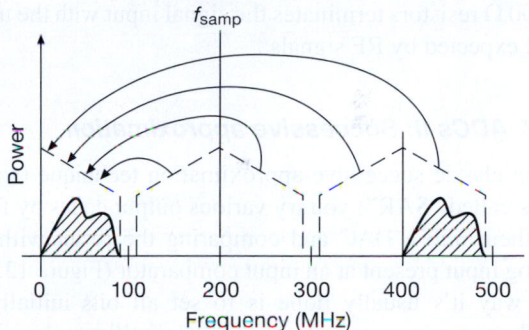

Figure 13.30. Putting the alias to work: a sampling rate of 200 Msps properly samples signals in the "baseband" extending to 100 MHz; but it creates aliases of successive 100 MHz bands. You can use this to your advantage by filtering out all input signals except those in the 400–500 MHz band (for example); that band is then properly digitized, and appears as a 0–100 MHz signal stream.

member from National Semiconductor's family of flash converters. This puppy runs from a single +3 V supply, converts at rates to 200 Msps with byte-wide output through a simple parallel output port, and costs under $15 in single quantities.[38] You get to supply the top and bottom of the conversion range (here ground and +1.25 V), and the datasheet sternly advises you to bypass the midpoint of the 256-tap resistor string. The 100 μH choke decouples the analog pin from the noisy V_D digital supply pin. With the positive-only conversion range it's necessary to bias the input signal to half the conversion range, as shown; the pair

[38] A related part, the ADC08B200, includes a 1024-byte output buffer, a handy thing if you want to sample in bursts and need to read the output stream at less than full speed.

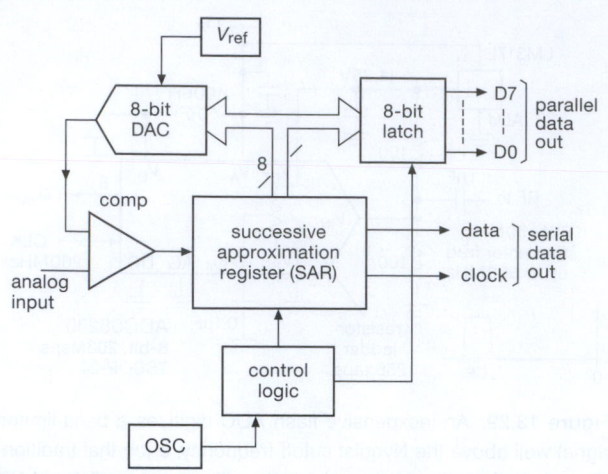

Figure 13.31. Successive-approximation ADC.

of $100\,\Omega$ resistors terminates the signal input with the usual $50\,\Omega$ expected by RF signals.

13.7 ADCs II: Successive approximation

In the classic successive-approximation technique (sometimes called "SAR") you try various output codes by feeding them into a DAC and comparing the result with the analog input present at an input comparator (Figure 13.31). The way it's usually done is to set all bits initially to 0. Then, beginning with the most significant bit, each bit in turn is set provisionally to 1. If the D/A output does not exceed the input signal voltage, the bit is left as a 1; otherwise it is set back to 0. For an n-bit ADC, n such steps are required. What you're doing is called a *binary search*, in the language of computer science.[39] A successive-approximation ADC has a BEGIN CONVERSION input and a CONVERSION DONE output. The digital output may be provided in parallel format (all bits at once, on n separate output lines), in serial format (n successive output bits, MSB first, on a single output line), or both.

In our electronics course the students construct a successive-approximation ADC, complete with DAC, comparator, and control logic. Figure 13.32 shows the successive outputs from the DAC, along with the eight clock pulses, as the trial analog output converges to the input voltage. And Figure 13.33 shows the full 8-bit "tree," a pretty picture you can generate by watching the DAC out-

---

[39] Historically this goes *way* back: in 1556 a mathematician by the name of Tartaglia proposed using a set of weights (1 lb, 2 lb, 4 lb...32 lb) in just such a manner to determine the weight of an object in the minimum number of trials on a balance.

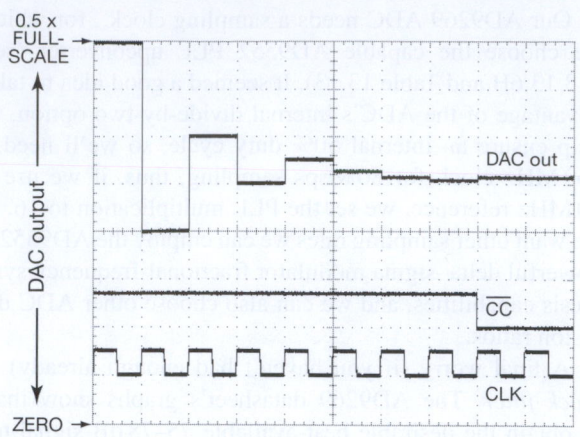

Figure 13.32. 'Scope trace of an 8-bit successive-approximation DAC's analog output converging to the final value. It's a binary search, with first guess equal to half of full scale. Note clock waveform and conversion-complete flag.

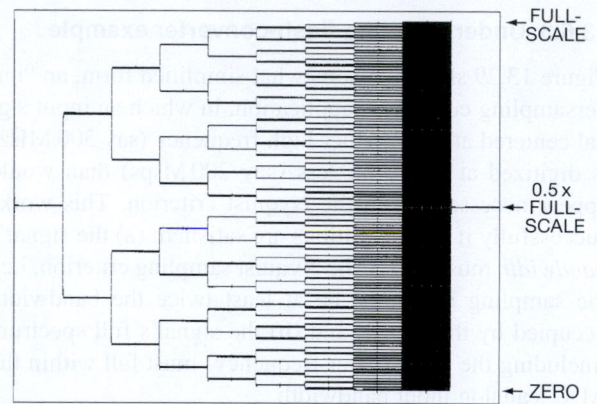

Figure 13.33. Accumulated 'scope trace of the 8-bit SAR's full "tree."

put while driving the input with a slow ramp that runs over the full analog input range.

Successive-approximation ADCs are intermediate in speed and accuracy (compared with the faster flash converters, or the more accurate but slower techniques used in "delta–sigma" converters, and multislope integrating converters); see Tables 13.5 and 13.6. They require n settling times of the DAC for n-bit precision. Typical conversion times are in the vicinity of $1\,\mu$s, with accuracies of 8 to 18 bits commonly available. This type of converter operates on a brief sample of the input voltage, and if the input is changing during the conversion, the error is no greater than the change during that time; however, spikes on the input are disastrous. Although generally quite accurate, these converters require accurately trimmed resistor networks, and they can have strange nonlinearities and "missing codes." One way to prevent missing codes is to use a chain of 2^n resistors and analog switches to generate the trial analog voltages, in the manner of the resistor-string DACs of §13.2.1; this technique was used in NSC's ADC0800-series 8-bit ADCs.

In contemporary successive-approximation ADCs, the conventional resistive DAC (R–$2R$ or resistor string, used internally to generate the analog voltages for the trial codes) is usually replaced with a *charge-redistribution* DAC architecture (Figure 13.34).[40] This scheme requires a set of binary-weighted *capacitors*, which nowadays is easy enough to fabricate and trim on-chip. (So an 18-bit converter like the AD7641 contains, remarkably enough, a set[41] of 18 binary-scaled capacitors going from some C_0, $2C_0$, ..., to a final $131{,}072C_0$; these capacitors are really small, with the capacitance of C_0 measured in the *femto*farads – f F, or 0.001 pF.)

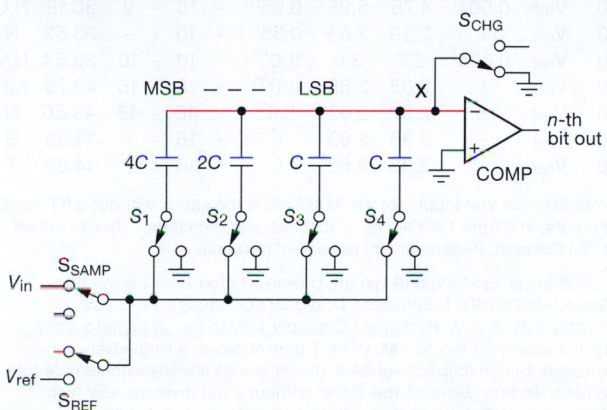

Figure 13.34. A capacitor-based "charge-redistribution" scheme replaces the R–$2R$ resistor ladder in many successive-approximation ADCs. The capacitor beyond the LSB is not used in the bit testing, but is needed to preserve the exact fractional ratios.

To understand how it works, look at the operation of the simplified 3-bit converter in the figure.

1. The switches are shown in the *sample* part of the cycle, during which the voltage across each of the capacitors follows (or *tracks*) the input signal.
2. Switch S_{SAMP} is opened, leaving the capacitors all holding the sampled input voltage.

3. Switch S_{CHG} is then opened, so the input to the comparator can move around as trial codes are applied to the bit switches S_1–S_3; for example, if the bit switches are all set to ground, the comparator input X would be at a voltage $-V_{in}$.
4. Now, to measure the held value of V_{in}, the bit switches are operated in turn: first the MSB switch S_1 is switched to $+V_{ref}$ (the full-scale range of the ADC), while S_2, S_3, and S_4 are switched to ground; this adds an offset of $V_{ref}/2$ to that $-V_{in}$ (it's a capacitive voltage divider: call it "charge redistribution" if you prefer).
5. The comparator's output now indicates the MSB: HIGH if $V_{in} > V_{ref}/2$, LOW otherwise.
6. As with the classic successive-approximation procedure, that switch is either returned to ground or left at V_{ref}, accordingly; the next-lower bit value is then tested similarly, with the process continuing in n steps (here $n=3$) to determine the full n-bit converted value.

13.7.1 A simple SAR example

Successive-approximation ADCs can be extremely easy to use, as illustrated by the circuit in Figure 13.35. The SPI serial interface is simplicity itself: assertion of CS' starts conversion, with successive bits clocked out by SCK (as each clock pulse triggers the SAR conversion of a new bit). The timing allows you to keep both serial lines quiet before assertion of CS', as shown, to minimize coupled digital noise. This family of relatively low-speed converters integrates on-chip track-and-hold, and includes three speed grades, three resolutions (8, 10, and 12 bits), and four packaging options (single, dual, quad, and octal): 36 choices! (The figure shows how to construct the part numbers.) The single units, like the 12-bit 1 Msps specimen in the figure, come in tiny SOT23-6 packages, with prices (in single quantities) ranging from about $2 (8 bit, 200 ksps) to $4.50 (12 bit, 1 Msps).

The input of an ADC is often less benign than something like an op-amp, where we've come to expect a high impedance (very low input current), and low capacitance. Figure 13.36 shows the equivalent input circuit of this converter, with its 26 pF sampling capacitor that the input signal must drive. This is not much of a burden at the relatively low frequencies here; but it's something to keep in mind, for example in the circuit of Figure 13.37 with its much higher resolution (18 bits) and somewhat higher speeds.

13.7.2 Variations on successive approximation

A variation known as a "tracking ADC" uses an up–down counter to generate successive trial codes; it is slow in responding to jumps in the input signal, but it follows

---

[40] There are also hybrid designs, in which a charge redistribution DAC is used to subdivide the steps of a coarse resistor-string DAC.

[41] Actually, *two* such sets, because its input is differential.

Table 13.5 Selected Successive-approximation A-to-D Converters[a]

Part #	ADCs per pkg	Bits	Conv Rate max (Msps)	Analog BW −3dB (MHz)	Channels/ADC	Single-ended	Differential	Internal Ref	Interface[k]	Power typ (mW)	Power at sps, V_S	V_{in} min (V)	V_{in} max (V)	I_{bias} max (µA)	V_S min (V)	V_S max (V)	I_{PD}[h] (µA)	SOIC	TTSOP	SOT-23	Smaller	Cost qty 25 ($US)	Comments
AD7927	1	12	0.2	8.2	8	•	-	-	S	3.6^m	200k, 3	0	V_{ref}^e	1	2.7	5.25	0.5^m	-	20	-	-	5.33	A
ADS7866	1	12	0.2	8	1	•	-	-	S	0.4	200k, 1.6	0	V_S	1	1.2	3.6	0.3^m	-	-	6	-	3.69	B
ADC121S	1	12	1	11	1	•	-	-	S	2	1M, 3	0	V_S	1	2.7	5.25	1^t	-	-	6	6	3.17	C
ADS7881	1	12	4	50	1	•	p	•	P	95	4M, 5	0	V_{ref}	0.5^t	4.75	5.25	2.5^m	-	48	-	48	14.52	D
MAX11131	1	12	3	50	16	•	•	-	S	15	3M	0	V_{ref}	1.5	2.4	3.6	6^m	-	28	-	28	12.24	E,Q
ADS7945[b]	1	14	2	15	2	-	•	-	S	20	2M, 5	0	V_{ref}	0.002	2.7	5.25	2.5^m	-	-	-	16	7.50	F
MAX1300[d]	1	16	0.11	0.7	8	•	•	•	S	17^m	100k, 5	−16	16	1250	4.75	5.25	0.5^t	-	24	-	-	11.00	M
LTC1609	1	16	0.2	1	1	•	•	•	S	65^c	200k, 5	−10	10	-	4.75	5.25	10^t	20	28	-	-	20.42	N
AD7685	1	16	0.25	2	1	-	•	•	S	2.7^c	200k, 2.5	0	V_{ref}	0.001^t	2.3	5.5	0.05^m	-	10	-	-	11.31	N
MAX11046	8	16	0.25	4	1	•	-	•	P	240	250k, 5	−5	5	1	4.75	5.25	10^m	-	64	-	56	19.48	O
MAX11166	1	16	0.5	6	1	-	•	•	S	26	500k, 5	−5	5	10	4.75	5.25	10^m	-	-	-	12	35.20	G,P
ADS8326	1	16	0.2	0.5	1	-	•	-	S	3.8^c	200k, 2.7	0	V_{ref}	0.05^t	2.7	3.6^g	0.1^t	-	8	-	-	9.88	N
AD7699	1	16	0.5	14	8	•	•	-	S	5.2	100k, 5	0	V_{ref}	0.001^t	4.5	5.5	0.05^t	-	-	-	20	12.00	H,N
ADS8319	1	16	0.5	15	1	-	•	-	S	18^c	500k, 5	0	V_{ref}	0.001^t	4.5	5.5	0.3^m	-	-	-	10	11.86	N,L
AD7985	1	16	2.5	19	1	-	•	•	S	28	2.5M, 5	0	V_{ref}	0.25^t	4.75	5.25	1^t	-	-	-	20	28.18	J,R
AD7690	1	18	0.4	9	1	-	•	•	S	4.3^c	100k, 5	0	V_{ref}	0.001^t	4.75	5.25	0.05^m	-	10	-	9	30.19	N,U
AD7982	1	18	1	9.0	1	-	•	-	S	7^c	1M, 2.7	0	V_{ref}	0.2^t	2.38	2.63	0.35^t	-	10	-	-	33.52	N
ADS8881	1	18	1	30	1	-	•	-	S	5.5^c	1M, 3	0	V_{ref}	0.005^t	2.7	3.6	100^m	-	10	-	10	36.54	N,V
LTC2379-18	1	18	1.6	34	1	-	•	-	S	18^c	1.6M	0	V_{ref}	1	2.38	2.63	100^m	-	16	-	16	42.79	K,N
AD7641	1	18	2	50	1	-	•	•	P,S	75^c	2M, 2.5	0	V_{ref}	18^t	2.37	2.62	0.6^t	-	48	-	48	43.56	N
AD7767-2	1	24	0.032	-	1	-	•	-	S	8.5	1M	0	V_{ref}	-	2.38	2.63	1^t	-	16	-	-	14.32	S
AD7767	1	24	0.128	-	1	-	•	-	S	15	1M	0	V_{ref}	-	2.38	2.63	1^t	-	16	-	-	14.32	T

Notes: (a) listed by accuracy and speed; all feature "no missing codes"; all permit external Vref input. (b) the ADS7946 is the same, without diff'l input. (c) power scales linearly with sample rate. (d) MAX1301 has half the number of inputs, in 20-pin TSSOP. (e) or to 2Vref, see datasheet. (f) with ext ref. (g) or 4.5-5.5V. (h) supply current during shutdown, power-down, or quiescent. (k) S=serial, P=parallel. (p) pseudo-differential.

Comments: A: sequencer; AD7928=1Msps. **B:** 10-bits='67, 8-bits='68, $1.80; 280ksps at V_S>1.6V; 8nA typ off, power-off after each conversion, 0.4µW at 100 per second, 44µW at 20ksps and V_S=1.2V. **C:** ADC121S051=500ksps, ADC121S021=200ksps. **D:** digital I/O supply 2.7V-5.25V. **E:** digital I/O supply 1.5V-3.6V. **F:** digital I/O supply 1.65V to V_S. **G:** digital I/O supply 2.3V-5.25V. **H:** digital I/O supply 1.8V to V_S. **J:** digital supply 2.4V-2.6V; digital I/O supply 1.8V-2.7V. **K:** digital I/O supply 1.7V-5.3V. **L:** digital I/O supply 2.4V-5.5V. **M:** PGA, 7 gain choices; 8 singled-ended or 4 diff'l inputs; V_{in} up to ±3Vref or 6Vref, or up to ±16V with V_S=5V. **N:** charge-redistribution (capacitive) SAR; power scales linearly with sample rate. **O:** 8 independent ADCs, simultaneous sampling; 6-ch=MAX11045, 4-ch=MAX11044. **P:** true "Beyond-the-Rails" without input dividers, ±5V with single V_S=+5; int V_{ref} 17ppm/°C max. **Q:** FIFO; averaging; 1, 2, 4, .. 32 channel sequencer. **R:** int ref 10ppm/°C,typ; digital supply 2.4-2.6V. **S:** 32x oversampling, on-chip FIR filter. **T:** 8x oversampling, on-chip FIR filter. **U:** 100dB min SNR, 125dB typ THD. **V:** digital I/O supply 1.65-3.6V; ADS886x are 16-bit diff'l and single-ended versions; family includes slower versions, to 100ksps.

smooth changes somewhat more rapidly than a successive-approximation converter. For large changes its slew rate is proportional to its internal clocking rate. The succession of up–down bits is itself serial, a simple form of *delta modulation*.

Another variation is *CVSD* (continuously variable-slope delta modulation), a simple scheme that is sometimes used for 1-bit serial encoding of speech, for example in wireless phones. With CVSD modulation the 1s and 0s represent steps (up or down) of the input waveform, but with the step size adaptively changing according to the past history of the wave. For example, the step size corresponding to a 1 increases if the last few bits have all been 1s, accord-

ing to a preset rule. The decoder knows the rule, so it can recreate an approximate replica of the (quantized) original analog input. In the past you could get CVSD chips, but in contemporary practice this is implemented in software in a microcontroller or DSP chip.

13.7.3 An A/D conversion example

Before continuing on to the important "integrating" conversion techniques (V-to-F, multislope, and *delta–sigma*), let's look at a demanding application example using a successive-approximation ADC: a low-noise, high-stability 18-bit converter with 2 Msps conversion rate.

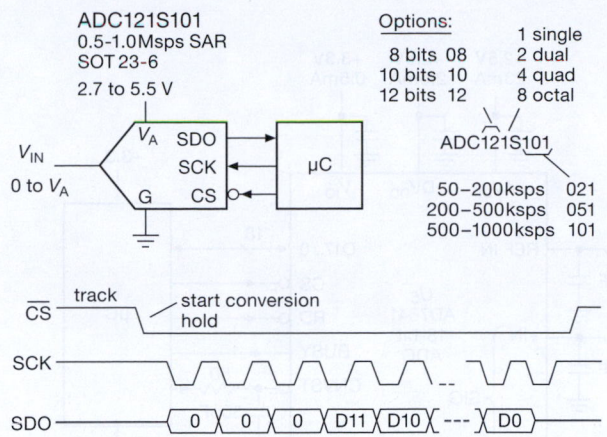

ADC121S101
0.5-1.0Msps SAR
SOT 23-6
2.7 to 5.5 V

Options:
		1 single
8 bits	08	2 dual
10 bits	10	4 quad
12 bits	12	8 octal

ADC121S101

50–200ksps	021
200–500ksps	051
500–1000ksps	101

Figure 13.35. National Semiconductor's ADC08/10/12S family of simple-to-use successive-approximation ADC with SPI serial output.

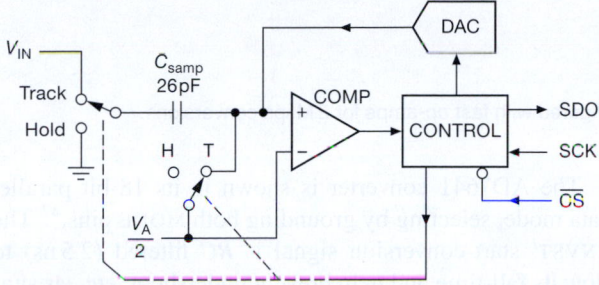

Figure 13.36. Block diagram of the ADC in Figure 13.35. The input signal drives the track-and-hold capacitor C_{samp} during acquisition.

Figure 13.37 shows a typical high-performance ADC, in this case Analog Devices' 18-bit 2 Msps PulSAR-series AD7641 converter. The AD7641 uses three positive power supplies,[42] with the nice characteristic that they may be turned on and off in any sequence.

The AD7641 has a full scale range of $\pm V_{ref}$, as is common with low-voltage ADCs. To maintain quiet conversions it's desirable to use a large voltage reference and signal voltage ranges. The AV_{DD} supply is 2.5 V, so the (differential) analog inputs are limited to 0 V to +2.5 V. If we use the maximum allowed +2.5 V reference, we get up to ± 2.5 V (differential) full scale: as +IN goes from 0 V to +2.5 V, −IN will have to go from +2.5 V to 0 V (otherwise we'd have only a 17-bit converter). For an 18-bit converter this corresponds to a differential LSB step of just 19 μV.

---

[42] Separate +2.5 V pins for the analog and digital sections, and a digital I/O pin that accepts +2.3 V to +3.6 V. Low-voltage ICs often need several supply voltages, requiring separate regulators.

You have to take great care with such small signals, especially when the converter's sample rate is 2 MHz, and the −3dB bandwidth corresponding to its aperture time is 50 MHz – there's plenty of analog noise in these bandwidths,[43] aided and abetted by coupled digital noise from the happenings at the output end.

Both the signal and voltage reference inputs experience charge-injection pulses from the charge-redistribution conversion process, so we use sizable capacitors (the datasheet's recommended 2.7 nF) on these pins to maintain a quiet voltage.[44] Op-amps don't like direct capacitive loads, because they cause ringing in combination with the inductive closed-loop output impedance (see §4.6.2 and the section on capacitive loads in Chapter *4x*), hence the 15 Ω series resistors. This *RC* also acts as a 4 MHz low-pass filter to reduce out-of-band noise; in this bandwidth an LSB corresponds to a more relaxed noise density of 9.6 nV/$\sqrt{\text{Hz}}$. Note that the series resistor in the V_{ref} path is larger (120 Ω) because we need to limit the peak current during power-supply startup, and the dc reference does not need the bandwidth of the signal paths.

The circuit shows an amplifier setup optimized for wideband operation with a single-ended input in the range 0 V to +1.25 V. The AD8021 is a wideband low-noise op-amp that is suggested in the ADC's datasheet. This may not be the best part to use,[45] but we'll continue our narrative with the manufacturer's suggested op-amp. The amplifier pair generates an accurate unipolarity differential output from the unipolarity single-ended input: the top stage has a noninverting voltage gain of +2, and the bottom inverting stage a gain of −2. Note the low resistor values, to maintain bandwidth and also reduce Johnson noise. To help ensure equal time delays, separate signal paths are used, instead of the alternative of cascaded amplifiers. Note how the inverting op-amp is biased at $V_{ref}/3$ to create the desired +2.5 V to 0 V signal. The two op-amp pathways have different noise gains, but the AD8021 lets us add a 10 pF capacitor to its compensation node to rolloff the response to achieve approximately equal bandwidths. To deal with the op-amp's high 7.5 μA input bias current, we've rigged

---

[43] In a 50 MHz bandwidth, 19 μV rms noise corresponds to a noise density e_n of just 2.7 nV/$\sqrt{\text{Hz}}$.

[44] Consider what's going on inside this successive-approximation ADC when operating at its full "warp-mode" speed of 2 Msps: its comparator has to make a new 19 μV decision every ~20 ns. Hectic!

[45] The choice is a bit curious, because this op-amp is not "precision" – its maximum offset voltage is an unimpressive 1 mV, and its input bias current is a rather high 7.5 μA – evidently design tradeoffs needed to achieve far more speed than is needed here (100 MHz bandwidth, 20 ns settling time).

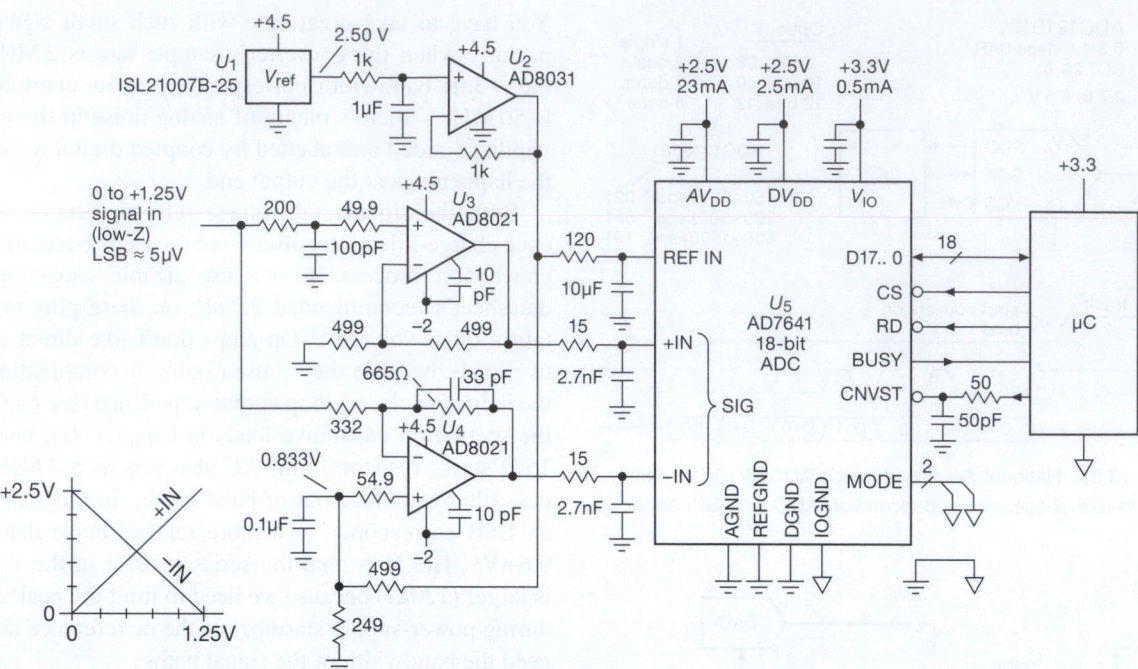

Figure 13.37. The AD7641 18-bit successive-approximation ADC, configured with fast op-amps for 2 Msps conversions.

up equal dc resistances seen at the inverting and noninverting inputs: this is effective here, because the typical offset current (0.1 μA) is 75 times smaller than the bias current itself.

The Intersil ISL21007/9BFB825 voltage reference (Table 9.8 on page 678) exploits floating-gate technology (i.e., a buried charged capacitor, §9.10.4) to achieve remarkably low drift over time (<10 ppm/√kHr). It has excellent initial accuracy (0.02%), and low tempco (3 ppm/°C). We've added a noise-quieting filter, and an op-amp is used to buffer the 3.3 mA load current to minimize power dissipation in the reference IC. The op-amps are powered from +4.5 V and −2.0 V, derived from the same ±5 V power that provides the regulated dc for the ADC (Figure 13.38), so that the op-amps are powered up at the same time as the ADC, thus minimizing clamp currents in the converter's input diodes at startup. Another way to prevent ADC input overdrive is to use a clamping op-amp like the AD8036, but this part has even larger dc errors than the AD8021. But there's a nice solution here, namely to clamp the C_{COMP} pin of the AD8021 op-amp with a pair of low-capacitance (2 pF) SD101 Schottky diodes, one to ground and the other to the ADC's +2.5 V supply, as shown in Figure 13.39.[46]

The AD7641 converter is shown in its 18-bit parallel data mode, selecting by grounding both MODE pins.[47] The CNVST' start-conversion signal is *RC* filtered (2.5 ns) to slow its fall-time and help prevent undershoot, etc., as suggested by Analog Devices. The CNVST' signal should not return to the HIGH state until conversion is finished, about 400 ns in its high-speed "warp" mode.

13.8 ADCs III: integrating

13.8.1 Voltage-to-frequency conversion

We continue our tour of A/D conversion techniques with the V-to-F (or V/F) converter. In this method an analog input voltage is converted to an output pulse train whose frequency is proportional to the input level. This can be done simply by charging a capacitor with a current proportional to the input level and discharging it when the ramp reaches a preset threshold. For greater accuracy, a feedback method is generally used. In one technique you compare

---

[46] The AD8021 datasheet does not tell you about this trick. But it does show a simplified schematic, from which you can see that the signal at

the C_{COMP} pin is the (high-impedance) output of the gain stages, on its way to the zero-offset complementary emitter followers that form the output stage (Figure 13.39); i.e., it is a clampable high-Z replica of the output signal.

[47] The other choices are 16-bit parallel (two READ cycles), 8-bit parallel (three READ cycles), or SPI (clocked out over 18 clock cycles).

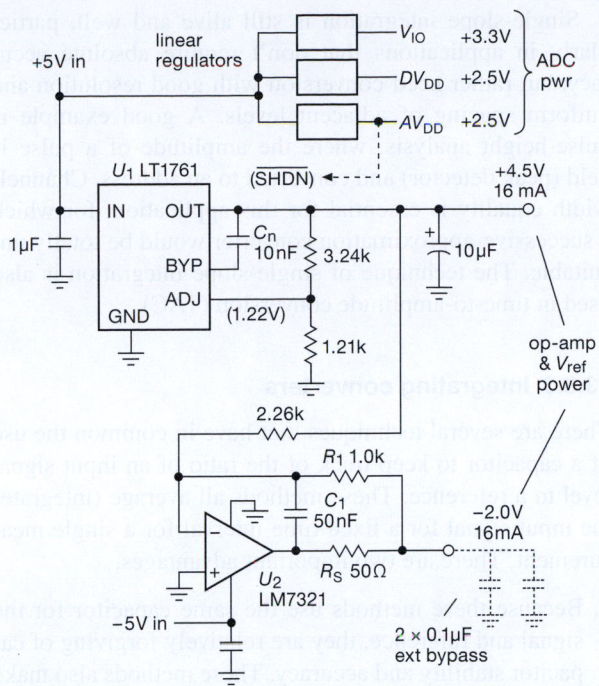

Figure 13.38. Linear regulators provide low-noise dc for the op-amps and ADC. The LM7321 is a high-current op-amp, good for 50 mA of output current. The split-feedback path (crossover at ~3 kHz) keeps it stable into the capacitive load of the op-amps' bypass capacitors while maintaining dc accuracy.

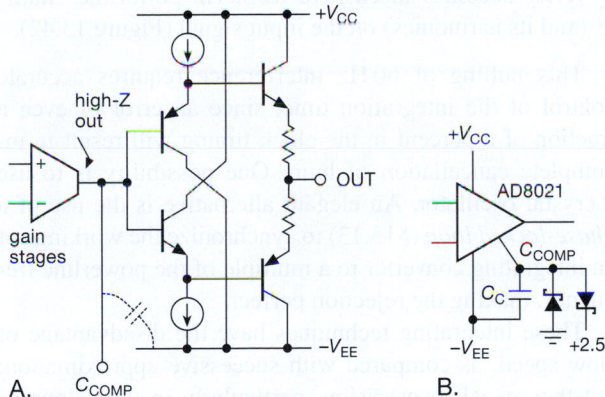

Figure 13.39. The compensation pin of some op-amps can be used to clamp the output signal. A. The AD8021's output stage is a push–pull arrangement of "zero-offset" complementary followers, with the C_{COMP} pin connected at the high-impedance output of the high-gain transconductance stages. B. Clamping that node with Schottky diodes limits the output swing to the reference voltages, here set to the ADC's conversion range.

the output of an F/V circuit with the analog input level and generate pulses at a rate sufficient to bring the comparator inputs to the same level. In the more popular methods, a "charge-balancing" technique is used, as will be described in greater detail later (in particular, the "capacitor-stored charge-dispensing" method).

Typical V/F output frequencies are in the range 10 kHz to 1 MHz for full-scale input voltage. Commercial V/F converters are available with the equivalent of 13-bit resolution (0.01% accuracy); they are examples of high-quality voltage-controlled oscillators (§7.1.4D). For example, the excellent AD650 from Analog Devices has a typical nonlinearity of 0.002% when operating from 0 to 10 kHz. They are inexpensive, and they are handy when the output is to be transmitted digitally over cables or when an output frequency (rather than digital code) is desired. If speed isn't important, you can get a digital count proportional to the average input level by counting the output frequency for a fixed time interval. This technique is popular in moderate-accuracy (3-digit) digital panel meters.

A VCO like the AD650 is an *asynchronous* V-to-F converter: Its oscillation is free running and internally generated, and it has no clocking input. But you can do things differently, namely by having a clock input and gating through clocking pulses such that the *average rate* coming out is proportional to an analog input voltage. For such a *synchronous* V/F converter, the output pulses, when present, occur coincident with the input clock; but the pulses are present or absent, as needed to keep their average rate proportional to V_{in}. In general the pulses are not equally spaced (though their spacings are exact multiples of the input clock period); that is, you don't get a single frequency out. The pulse train has "jitter." That's fine for some applications, particularly those that inherently average the output; an example is a resistive heater element, perhaps within a temperature controlled loop with an analog temperature sensor.

We rigged up an AD7741 synchronous V/F converter, clocked at 5 MHz, and measured its output frequency (averaged over several seconds) versus input voltage. It's pretty good (Figure 13.40).

The synchronous V/F converter is a simple example of a "1-bit" ADC. There are better ways to generate a bitstream whose average value represents the conversion of an analog input signal. In particular, the so-called *delta–sigma* converters do a far better job. Better job, but quite a bit harder to wrap your brain around. We'll get to these shortly, in §13.9, where we'll try valiantly (but perhaps unsuccessfully) to deconfuse the situation.

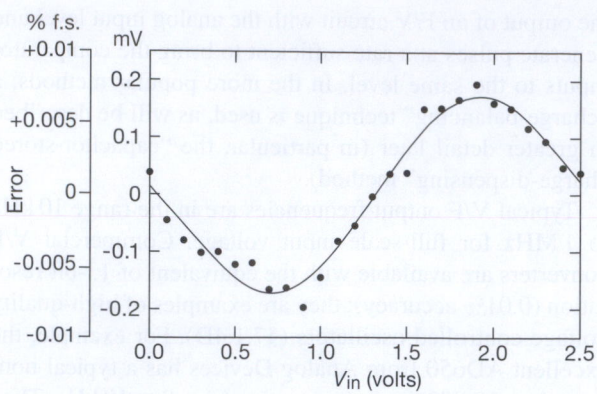

Figure 13.40. Measured nonlinearity of an AD7741 synchronous V/F converter as a function of input voltage. The specified linearity is ±0.015%.

13.8.2 Single-slope integration

In this technique an internal ramp generator (current source + capacitor) is started to begin conversion, and at the same time a counter is enabled to count pulses from a stable clock. When the ramp voltage equals the input level, a comparator stops the counter; the count is proportional to the input level, i.e., it's the digital output. Figure 13.41 shows the idea.

At the end of the conversion the circuit discharges the capacitor and resets the counter, and the converter is ready for another cycle. Single-slope integration is simple, but it is not used where high accuracy is required because it puts severe requirements on the stability and accuracy of the capacitor and comparator. The method of "dual-slope integration" eliminates that problem (and several others as well) and is now generally used where precision is required.

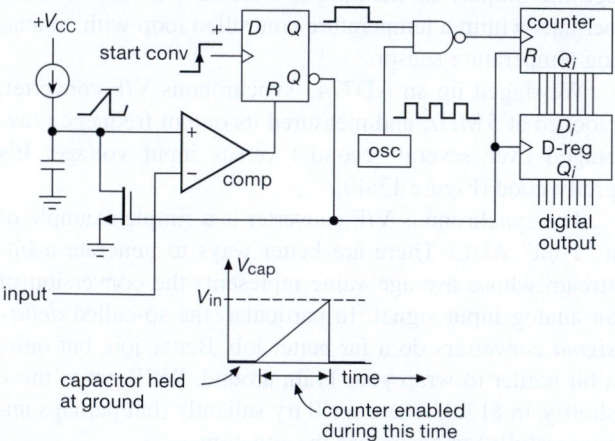

Figure 13.41. Single-slope ADC.

Single-slope integration is still alive and well, particularly in applications that don't require absolute accuracy, but rather need conversion with good resolution and uniform spacing of adjacent levels. A good example is pulse-height analysis, where the amplitude of a pulse is held (peak detector) and converted to an address. Channel-width equality is essential for this application, for which a successive-approximation converter would be totally unsuitable. The technique of single-slope integration is also used in time-to-amplitude conversion (TAC).

13.8.3 Integrating converters

There are several techniques that have in common the use of a capacitor to keep track of the ratio of an input signal level to a reference. These methods all average (integrate) the input signal for a fixed time interval for a single measurement. There are two important advantages.

1. Because these methods use the same capacitor for the signal and reference, they are relatively forgiving of capacitor stability and accuracy. These methods also make fewer demands on the comparator. The result is better accuracy for equivalent-quality components, or equivalent accuracy at reduced cost.
2. The output is proportional to the *average* input voltage over the (fixed) integration time. By choosing that time interval to be a multiple of the powerline period, the converter becomes insensitive to 60 Hz powerline "hum" (and its harmonics) on the input signal (Figure 13.42).

This nulling of 60 Hz interference requires accurate control of the integration time, since an error of even a fraction of a percent in the clock timing will result in incomplete cancellation of hum. One possibility is to use a crystal oscillator. An elegant alternative is the use of a *phase-locked loop* (§13.13) to synchronize the workings of an integrating converter to a multiple of the powerline frequency, making the rejection perfect.

These integrating techniques have the disadvantage of slow speed, as compared with successive approximation; but they excel in precision, particularly in dual-slope or multislope incarnations, or as sophisticated delta–sigma converters (§13.9).

13.8.4 Dual-slope integration

This elegant and very popular technique eliminates most of the capacitor and comparator problems inherent in single-slope integration. Figure 13.43 shows the idea. First, a current accurately proportional to the input level charges a

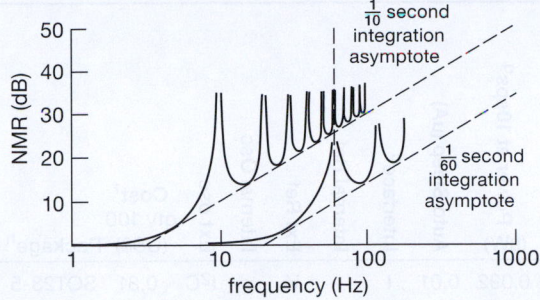

Figure 13.42. Normal-mode rejection with integrating A/D converters.

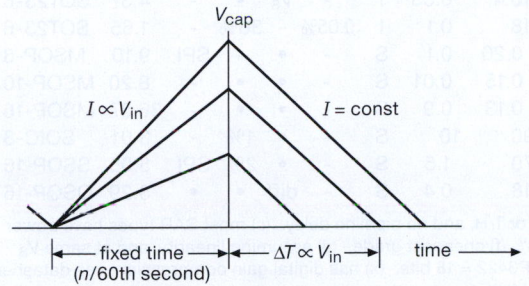

Figure 13.43. Dual-slope conversion cycle.

capacitor for a fixed time interval; then the capacitor is discharged by a constant current until the voltage again reaches zero. The time to discharge the capacitor is proportional to the input level and is used to enable a counter driven from a clock running at a fixed frequency. The final count is proportional to the input level, i.e., it's the digital output.

Dual-slope integration achieves very good accuracy without putting extreme requirements on component stability. In particular, the capacitor value doesn't have to be particularly stable, because the charge cycle and the discharge cycle both go at a rate inversely proportional to C. Likewise, drifts or scale errors in the comparator are cancelled out by beginning and ending each conversion cycle at the same voltage and, in some cases, at the same slope. In the most accurate converters, the conversion cycle is preceded by an auto-zeroing cycle in which the input is held at zero. Because the same integrator and comparator are used during this phase, subtracting the resulting "zero-error" output from the subsequent measurement results in effective cancellation of errors associated with measurements near zero; however, it does not correct for errors in overall scale.

Note that even the clock frequency does not have to be of high stability in dual-slope conversion, because the fixed

integration time during the first phase of the measurement is generated by subdivision from the same clock used to increment the counter. If the clock slows down by 10%, the initial ramp will go 10% higher than normal, requiring 10% longer ramp-down time. Since that's measured in clock ticks that are 10% slower than normal, the final count will be the same! Only the discharge current has to be of high stability in a dual-slope converter with internal auto-zeroing. Precision voltage and current references are relatively easy to produce, and the (adjustable) reference current sets the scale factor in this type of converter.

When choosing components for dual-slope conversion, be sure to use a high-quality capacitor with minimum dielectric absorption ("memory" effect; see §5.6.2 and the expanded discussion in §*1x.3*) – polypropylene, polystyrene, or Teflon capacitors work best. Although these capacitors are not polarized, you should connect the outside foil (indicated with a band) to the low-impedance point (the output of the integrator op-amp). To minimize errors, choose integrator R and C values to use nearly the full analog range of the integrator. A high clock frequency improves resolution, although you gain little once the clock period becomes shorter than the comparator response time.

When using precision dual-slope converters (and, indeed, any kind of precision converter) it is essential to keep digital noise out of the analog signal path. Converters usually provide separate "analog ground" and "digital ground" pins for this purpose. It is often wise to buffer the digital outputs (say with a '541 three-state octal driver, asserted only when reading the output) to decouple the converter from the digital roar of a microprocessor bus (see Chapters 14 and 15). In extreme cases you might use opto-couplers (§12.7) to quarantine the noise of a particularly dirty bus. Be sure to use liberal power-supply bypassing right at the converter chip. And be careful not to introduce noise during the critical endpoint of the integration, as the ramp reaches the comparator trip point. For example, some converters let you check for end-of-conversion by reading the output word: *don't do it!*[48] Instead, use the separate BUSY line, suitably isolated.

Dual-slope integration is used extensively in precision digital multimeters. It offers good accuracy and high stability at low cost, combined with excellent rejection of powerline (and other) interference, for applications where speed is not important. The digital-output codes are strictly monotonic with increasing input.

The alternative, for highest precision, is the delta–sigma

---

[48] OK, if you insist, go ahead and check it – but only after you're sure it's done.

Table 13.6 Selected Micropower A-to-D Converters[a]

Part #	Bits	Conv Rate max (ksps)	Channels	Delta-Sigma	Successive Approx[b]	Differential	PGA Gains	Supply Current,[c] I_S at V_S (V)	typ@speed (µA)	(sps)	@ max speed typ (µA)	Power at 10sps[g] (µW)	Auto Sleep (µA)	Interface[d]	Internal Ref	Ext Ref	Internal Osc	Ext osc	Cost[f] qty 100 (US$)	Package[h]	Comments
MCP3021	10	22	1	-	•	-	-	2.7	17	5k	175	0.092	0.01	I	-	V_S	-	I2C	0.81	SOT23-5	-
ADS7866	12	200	1	-	•	-	-	1.8	6.3	5k	275	0.023	0.01	S	-	V_S	-	SPI	2.75	SOT23-6	-
ADS7466	12	200	1	-	•	-	-	1.6	3.8	5k	150	0.012	0.01	S	-	V_S	-	SPI	4.80	SOT23-6	-
ADC121S	12	1000	1	-	•	-	-	2.7	15	5k	600	0.081	0.5	S	-	V_S	-	SPI	2.30	SOT23-6	-
AD7091R	12	1000	1	-	•	-	-	3.0	57	20k	350	0.086	0.26	S	1%	•	•	-	4.10	MSOP-10	-
ADS1100	16	0.128	1	•	-	•	1-8	3.0	70	128	70	16.4	0.05	I	-	V_S	•	-	4.37	SOT23-6	-
MCP3425	16	0.24	1	•	-	•	1-8	3.0	0.6	1[e]	145	18	0.1	I	0.05%	-	36%	-	1.65	SOT23-6	w
ADS8326	16	250	1	-	•	•	-	2.7	150	20k	1850	0.20	0.1	S	-	•	-	SPI	9.10	MSOP-8	-
AD7685	16	250	1	-	•	•	-	2.5	0.6	100	1350	0.15	0.01	S	-	•	•	-	8.20	MSOP-10	-
LTC2379-18	18	1600	1	-	•	•	-	2.5	25	5k	7200	0.13	0.9	S	-	•	•	-	36.37	MSOP-16	x
MCP3551	22	0.014	1	•	-	•	-	2.7	100	14	100	190	10	S	-	•	1%	-	3.01	SOIC-8	-
LTC2412	24	0.008	2	•	-	•	-	2.7	170	8	170	570	1.5	S	-	•	2%	SPI	5.05	SSOP-16	y
MAX11210	24	0.48	1	•	-	•	1-16	3.6	235	480	235	18	0.4	S	-	diff	•	•	3.29	QSOP-16	z

Notes: (a) sorted by resolution and maximum speed. (b) all SAR types have S/H or T/H, and no pipeline delay. (c) most SAR types have power proportional to sample rate. (d) I=I2C, S=SPI. (e) in 12-bit mode (10µA for 16-bits). (f) cheapest grade. (g) assuming linearity, and at same V_S as listed for I_S. (h) other packages may be available, consult datasheets. (w) MCP3422 = 18 bits. (x) has digital gain compression, see datasheet. (y) no latency, ping-pong channel selection. (z) four I/O bits, can use for external MUX.

converter. There's a lot of confusion surrounding this elegant technique. In a subsequent section (§13.9) we aim to blow away the smoke and provide some intuition into the workings of these things. First, though, a look at the ultimate in integrating converters – the so-called "multislope" technique devised by Hewlett–Packard (subsequently Agilent, now Keysight), and commercialized in their world-class $8\frac{1}{2}$-digit multimeters – preceded by a relevant detour into the use of analog switches in conversion applications.

13.8.5 Analog switches in conversion applications (a detour)

Analog switches, first seen in §3.4.1, figure importantly in conversion applications, both as components of the converter itself (see, e.g., Figures 13.2, 13.9,, 13.34, and 13.36), and as external helpers. In their former role they are an essential part of the precision multislope converter (§13.8.6), and of the delta–sigma converter (§13.9). Here we explore briefly some converter applications in which a discrete logic-family CMOS analog switch is particularly useful.

A. Logic-family analog switches

The widely available '4051 to '4053 family of CMOS switches is particularly useful for analog applications, because these parts have a negative V_{EE} supply line for the switches, along with internal logic-level shifters; so the switches work over an analog range of $-V_{EE}$ to $+V_{DD}$, and in fact an additional 0.25 V or so beyond these supply rails. Table 13.7 shows the logic families in which these switches are available. There are three parts in the family: the '4053 is especially attractive, with its three independently controlled SPDT switches; there's also the '4052 with a pair of 4-to-1-line switches, and the 4051 with a single 8-to-1 switch. Although these switches are attractive because they are inexpensive (less than $0.50) and available from a half-dozen companies, they're even more attractive to designers because they're very fast, and have low capacitance.

For example, the 74HC4053 typically has 40 Ω of ON resistance, switches in 20 ns, and has 8 pF of capacitance to ground. Compared with ICs officially intended for analog switching and multiplexing, a '4053 has a more limited voltage swing, and no electrostatic discharge (ESD) protection. When compared with CMOS power switches, it has higher ON resistance, but it doesn't suffer from their high capacitance. It's in a sweet spot that's ideally suited for switching signals between the circuits on the same circuit board.

Single SPDT versions are available in space-saving SOT23 and other SMT packages. These parts, for example the '1G3157 types (the 1G means single gate), do not

Table 13.7 4053-style SPDT Switches

	Supply[m]		TTL in?[a]	$t_{d(on)}$[f]	R_{ON}	ΔR_{ON}	@	Cap	leakage		Q_{inj}	avail in DIP?
	V_{cc} (V)	V_{tot}[c] (V)		typ (ns)	typ (Ω)	typ (Ω)	V_{tot}[c] (V)	typ[b] (pF)	typ (pA)	max[e] (nA)	(pC)	
16-pin triple												
CD4053	3-15	20	-	120	120	10	10	8	10	50		•
74HC4053	2-7.5	15	•	18	40	8	9	5[d]		100		•
74VHC4053	2-7.5	15	-	18	20	10	9	30[d]		500		•
74VHC4053[o]	2-7	14	-	49	120	10[m]	9	45[d]		100		•
DG4053A	2-7	14	•	31	66	3	10	4	20	1	0.25	-
MAX4053A	3-17	17	•	50	60	6[m]	10	2	2	0.1	2	•
ISL84053	2-15	15	•	75	60	6[m]	10	9	2	0.1	2	-
ADG633	2-6	13	•	70	52	0.8	9	7	5	0.2	2	-
MAX4583	2-12	13	•	90	80	4[m]	10	6		1	0.5	-
74LV4053	2-7	14	-	4.6	35	1.3	9	8[d]		100		-
74LVX4053	2-7	7	-	23[m]	26	10[m]	6	10		100	9	-
NLAS4053	1.7-7	7	-	23[m]	26	10[m]	6	10		0.1	9	-
MAX4693[g]	2-11	11	•	55	25	2.5	10	20		2	1.8	-
MAX4619	2-6	na	•	7	10	1	5	8.5	2	1	8	•
MAX4783	1.6-4.6	na	•	17	1	0.2	3	75	2	2	20	•
SMT, single												
ISL84544	2-15	na	•	35	30	0.8	5	8	10	0.1	1	•
74LVC2G53	1.7-6	na	-	2	13	3[m]	3	10		100		na
74AUC2G53	0.8-3.6	na	•	1.1	15	1[m]	2.3	4.5		100		na
74LVC1G3157	1.7-6	na	-	3.6	9	0.1	3	6[f]		2µA	4	na
NX3L1G3157	1.4-4.6	na	•	14	0.5	0.02	2.7	35		10	9	na

Notes: (a) TTL logic levels, or TTL available. (b) OFF, common node to gnd. (c) $V_{tot} = V_{CC} - V_{EE}$. (d) depends on manufacturer. (e) at 25°C if shown, reflects ATE capability. (f) delay time from logic input to switch ON, at V_{tot}; $t_{d(OFF)}$ is slightly less. (g) in 16-pin QFN pkg; quad also available. (m) maximum. (na) no V_{EE} pin. (o) ON Semi.

include the negative supply feature, so they do not use "4053" in their name.

Let's look at two examples in which '4053-style switches form a nice bridge between the analog and digital worlds. The second example (sawtooth generation with current-steering switches) will take us directly to the multislope and delta–sigma converters.

B. Programmable high-voltage pulse generator

It's nice to be able to generate a pulse that is gated by a logic signal, but whose amplitude is set separately. For the latter you might use a DAC under computer control, or maybe just a panel knob.[49] The simple circuit in Figure 13.44 does the job, in this case allowing output amplitudes up to +100 V.

The '4053-style analog switch applies the level selected by panel switch S_1 to the OPA454 high-voltage op-amp, here configured for a noninverting gain of 20. This op-amp

is not terribly fast ($\sim 10\,\mu s$ switching times), but it's quite inexpensive ($5), and it can supply 100 mA pulsed outputs to charge capacitive loads. You could substitute a faster op-amp to exploit the fast switching capability ($\sim 20\,ns$) of the analog switch.[50] Center-off panel toggle switch S_2 lets you turn pulses on or off, or enable continuous switch conduction for reading and setting the high-voltage (HV) level with a DMM, etc. We are fond of easy-to-use instruments that combine panel controls with CMOS signal switching.

C. Current-steering sawtooth generator

Here's a circuit (Figure 13.45) that exploits the nice switching characteristics of '4053-style analog switches, in the

---

[49] Back in Chapter 5 we showed a way to generate programmable high-voltage waveforms (Figure 5.47), however without the gating feature.

[50] For example, the Cirrus/Apex PA85 slews at 1000 V/μs (with closed-loop gain of 100), and can run from 450 V total supply. Here you could power it from +400 V and −15 V rails. Be prepared to shell out some serious cash, though: this puppy costs about $300. (See the selection of high-voltage op-amps in Table 4.2b.) A more economical solution is to build your own high-voltage amplifier, along the lines of Figure 3.111; see also the section on a precision high-voltage amplifier in Chapter 4x.

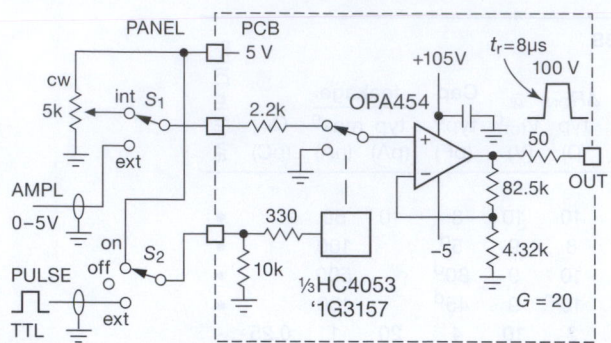

Figure 13.44. Simple high-voltage pulse generator, with programmable pulse amplitude and waveform. The '4053 switch includes a $-V_{EE}$ pin for bipolarity signal switching (to $\pm 5\,V$), whereas the single-section '3157 operates with positive polarities only.

same current-steering arrangement that's used in the remarkable multislope converter we'll see in the next subsection. Op-amp U_1 is an integrator, with its summing junction biased at half the supply voltage (for single $+5\,V$ supply operation). Switches S_1 and S_2 are sections of an 'HC4053, running at the same $+5\,V$; they individually program the rising and falling ramp rates, set by resistors R_1 and R_2. Closing S_1 sources a current $V_{cc}/2R_1$, causing the integrator to ramp down according to $dV_{ramp}/dt = I/C$; S_2 causes an analogous upward ramp. The comparator has thresholds at 1/3 and 2/3 of V_{cc}, turning the ramp around after it goes through $\Delta V = V_{cc}/3$. It's easy to show that the resulting ramp intervals are given by $t_{rise} = \frac{2}{3}R_2C$ and $t_{fall} = \frac{2}{3}R_1C$, and $f = 1.5/C(R_1+R_2)$.

Exercise 13.4. Go ahead, show it!

Both the switches and the comparator are fast, permitting operation to at least a few megahertz, for which suitable values might be resistors of a few thousand ohms, and C in the range of 100–500 pF. With such a small integrator capacitor, you have to worry about the effects of the switch capacitance C_{sw}, typically in the range of 5–10 pF. Consider for example switch S_1, in the position shown in the figure: its capacitance is charged to $+5\,V$, and so it transfers a slug of charge $\Delta Q = C_{sw}\Delta V$ (where $\Delta V = V_{cc}/2$) to the summing junction when the switch moves to the lower terminal. That charge transfer causes a step in the integrator's output, as in Figure 13.46. The cure here (and in the multislope ADC we'll see next) is to hold the other switch terminal at the same voltage as the summing junction (the dotted circuit).

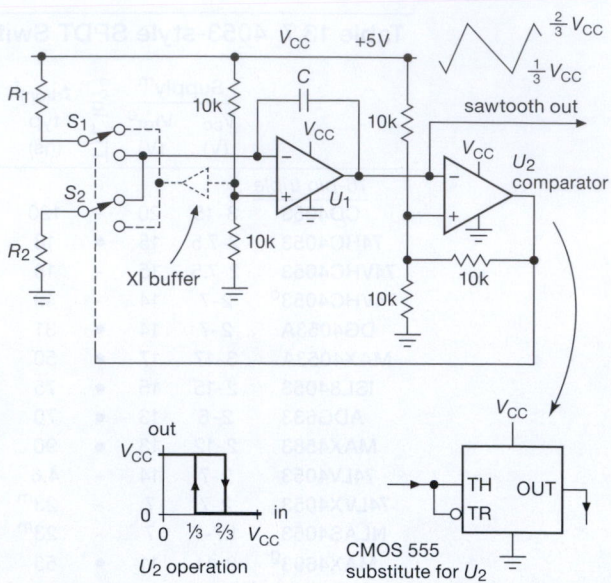

Figure 13.45. Sawtooth generation with current-steering switches. Comparator U_2 is configured as a Schmitt trigger with thresholds at $V_{cc}/3$ and $2V_{cc}/3$, for which a suitable part (with active rail-to-rail outputs) is the fast TLV3501 (t_p=4.5 ns); a CMOS 555 could be substituted, though it's not nearly as fast (t_p=100 ns).

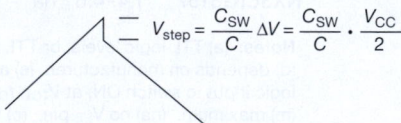

$$V_{step} = \frac{C_{sw}}{C}\Delta V = \frac{C_{sw}}{C}\cdot\frac{V_{cc}}{2}$$

Figure 13.46. Charge injection produces a step change in integrator output, when a circuit node (of capacitance C_{sw}) at a different voltage is switched on.

13.8.6 Designs by the masters: Agilent's world-class "multislope" converters

With these analog-switch applications in mind, we're in good shape to understand the "multislope" techniques used in instruments like the Keysight[51] 34420 $7\frac{1}{2}$-digit and 3458A $8\frac{1}{2}$-digit multimeters.[52] This has been Agilent's top-of-the-line instrument for more than twenty years, with a current price (year 2015) of $9.5k. A simplified variant ("Multislope III") is used in Keysight's contemporary high-performance DMM series of instruments (the 34420A 7.5-digit nanovolt and micro-ohm meter, the 34401A 6.5-digit

---

[51] Formerly Agilent, 1999–2014, and prior to that Hewlett–Packard or "HP," 1939–1999.

[52] The latter described on their website as "Recognized the world over as the standard in high performance DMMs."

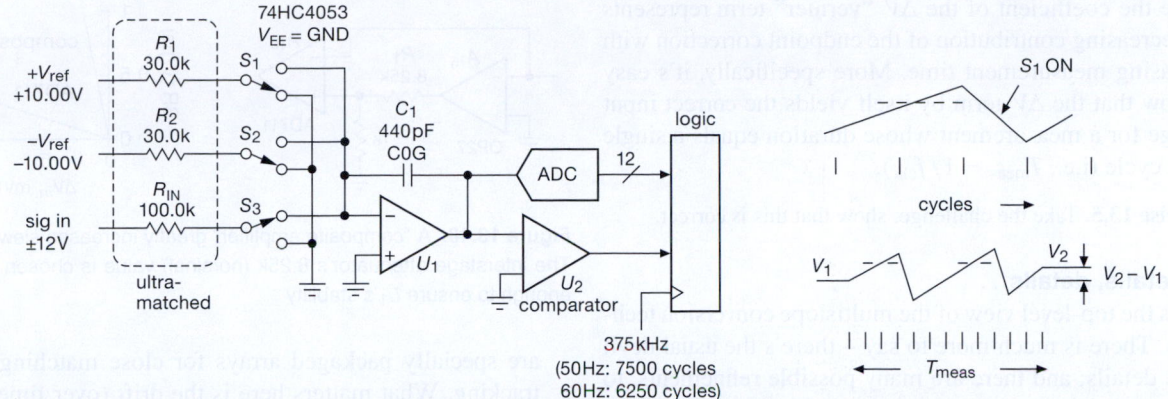

Figure 13.47. Keysight's "Multislope III" converter, a clocked charge-balancing integrator with endpoint correction via a low-precision ADC.

industry-standard bench DMM, and the 34970A 6.5-digit data-acquisition system). We'll see in detail how Multislope III works, and reveal a bit of the operation of next-generation Multislope IV (introduced in 2006).

A. The basic technique

To put it in perspective, the multislope technique is an evolution of the dual-slope integrator, exploiting a multicycle charge-balancing integration scheme that is both more forgiving of capacitor imperfections, and that takes account of the residue remaining after the final cycle of integration. It blends aspects of dual-slope and delta–sigma conversion, and it is a natural stepping-stone to the latter.

The basic circuit is extremely simple (Figure 13.47) and uses mostly low-cost parts (except for the voltage references and precision resistors). There's an integrator U_1, a "logic engine" that keeps track of the integrator output (via comparator U_2) at each 375 kHz clock tick, and a pair of switches (S_1 and S_2) that are operated synchronously by the logic to keep the integrator approximately balanced (by sinking or sourcing an accurate current into the integrator). There's also an ADC of only modest precision (12 bits) that is used to read the integrator's output voltage at the beginning and end of a multicycle measurement.

Here's how it works, at a basic level: to begin a measurement, switch S_3 is closed,[53] causing the integrator to ramp up or down (according to $dV/dt = -I_{in}/C_1 = -V_{in}/R_{in}C_1$). At each successive clock tick the logic engine causes either switch S_1 or S_2 to close (depending on the polarity of the integrator's output), thus adding or subtracting the corresponding reference current (± 10 V/30 kΩ) to force the in-

tegrator to come back toward ground. This goes on for many clock cycles (for maximum rejection of powerline pickup it's desirable to use a measurement time corresponding to an integral number of powerline cycles; e.g., 6250 clock ticks equals 1/60 of a second), after which the logic tallies up the number of positive (n_+) and negative (n_-) cycles. This gives a *first-order* estimate of the average input voltage during the measurement time:

$$V_{sig}(1) \approx V_{ref}\frac{n_- - n_+}{N_{cycles}}\frac{R_{in}}{R_1} = V_{ref}\frac{n_- - n_+}{6250}\frac{100k}{30k}.$$

This is not terribly accurate: a full-scale input of ± 12 V produces a net count ($n_- - n_+$) of ± 2250 (figure out why), so the resolution is approximately 12 bits. Now for a nice trick: because the measurement is timed by an integral number of clock cycles (rather than by a zero crossing, as in the dual-slope method), the residual integrator level contains additional information. It allows us effectively to perform a vernier-like subdivision of a clock cycle. That's the reason for the ADC in Figure 13.47, which is used to measure the integrator voltage at the beginning and end of the measurement cycle. For the 12-bit ADC in the figure, this provides some 512 levels of subdivision of the first-order LSB, adding 9 bits to the $\sim$12-bit first-order estimate, for a final result of $\sim$21 bit resolution.[54]

More precisely, the second-order (and final!) answer is given by

$$V_{sig}(2) = V_{sig}(1) + \frac{R_{in}C_1}{T_{meas}}(V_f - V_i) = V_{sig}(1) + 0.00264(V_f - V_i),$$

---

[53] We use the term "closed" to mean that a given switch is connected to the summing junction.

[54] The ADC's conversion range matches that of the integrator, but both are $\sim$8$\times$ larger than the integrator's ramp over a clock tick when the input signal is at $\pm$full scale. That is why the ADC effectively loses 3 bits of resolution when digitizing the residue ($V_f - V_i$).

where the coefficient of the ΔV "vernier" term represents the decreasing contribution of the endpoint correction with increasing measurement time. More specifically, it's easy to show that the ΔV term by itself yields the correct input voltage for a measurement whose duration equals a single clock cycle (i.e., $T_{\text{meas}} = 1/f_{\text{clk}}$).

Exercise 13.5. Take the challenge: show that this is correct.

B. Details, details…

That's the top-level view of the multislope conversion technique. There is much more to say – there's the usual devil in the details; and there are many possible refinements, to squeeze the most performance from this core idea. Here we restrain ourselves, restricting our commentary to a terse summary of the more interesting and instructive aspects.

Noncritical components For S_1–S_3, Agilent uses a standard 74HC4053 switch array (from NXP), and for C_1 a commodity ceramic chip capacitor of the stable NP0/C0G variety (from AVX). This type of capacitor, which is stunningly inexpensive,[55] exhibits a low temperature coefficient (± 30 ppm/°C) and negligible dielectric absorption ("memory," see Figure 5.4 in §5.6.2 and the more extensive figures in §*1x.3*), unmeasurable on the switching time scales here. Likewise, no great accuracy or stability is needed for the comparator or endpoint-quantizing ADC.

Critical components The voltage references set the scale of the measurement, and must be highly stable. In practice these instruments use a single 7.0 V zener-type reference and a pair of precision op-amps to produce the ± 10.0 V reference voltages.[56] The "10.0 V" voltage does not need to be precise to the ultimate accuracy of the instrument, which undergoes a factory calibration; it does need to be *stable*, of course, to maintain that calibrated accuracy.[57]

Two other critical components are the matched resistor arrays (R_1–R_3, and the gain-setting resistors in the voltage references) and the integrator op-amp. The latter is in fact a composite amplifier (an OP27+AD711), achieving high slew rate and high loop gain along with very low offset voltage (Figure 13.48). The resistors

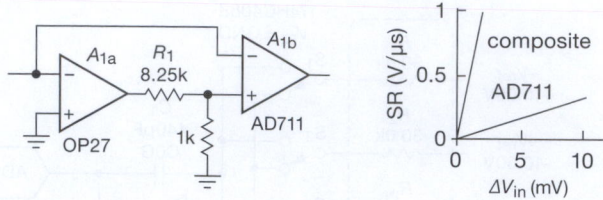

Figure 13.48. A "composite amplifier" greatly increases slew rate. The interstage attenuator's 8.25k (nominal) value is chosen large enough to ensure U_1's stability

are specially packaged arrays for close matching and tracking. What matters here is the drift (over time and temperature) of the resistor ratios, because minor mismatches in the initial resistor ratios are handled in the factory calibration.

Not shown, but just as critical, is the selectable-gain input amplifier. It must have precise and stable gains, and means to carry out calibrations and corrections; see §5.12.

The switches The 74HC4053 switch is used in a current-steering scheme like that in Figure 13.45, arranged so that the voltages on all of the '4053 switch pins remain always close to zero volts. The switches serve only to steer the currents, adding or subtracting a slew rate to the integrator output that is in a precise ratio to the slew rate produced by an input voltage. The on-resistance of the switches does, of course, figure into the value of these currents. But as long as the R_{on} of the switches is well matched, stable, and small compared with R_1–R_3, the effect is corrected by the calibration cycle that the instrument performs automatically before each measurement (see next paragraph). For the NXP 'HC4053 used in the Keysight instruments, for example, R_{on} is typically 85 Ω, matched to 8 Ω. It's important that these switches operate as "break-before-make," so that their output terminal pairs are not momentarily shorted (which would connect the integrator summing junction to ground, acting like a differential input signal equal to the op-amp's offset voltage). Some manufacturers include such a specification, others don't. For example, the same NXP datasheet specifies turn-ON and turn-OFF times, whose difference represents a break-before-make interval of 4 ns; but it does not specify that interval directly, whereas the DG4053 datasheet from Siliconix lists a "Break-Before-Make Time Delay" t_D of 6 ns typ (2 ns min).

Calibration The simple current-switched input topology is well suited to calibrating, and eliminating, the effects of resistor ratio mismatches, voltage reference

---

[55] $0.06 in 100-piece quantities, some 50,000 in stock at DigiKey this morning.

[56] We are told by reliable sources that Keysight uses the spectacular LTZ1000, see §9.10.1B.

[57] For the 34401A standard bench DMM, for example, the initial factory-calibrated dc accuracy is within ~2 ppm; it is specified to drift no more than $\pm 0.0015\%$ over 24 hours, but $\pm 0.0035\%$ after one year.

mismatches, op-amp offsets, switch delays, and the like. For example, when S_3 is turned off (i.e., no input signal) and S_1 and S_2 are alternated for some number of successive cycles, the endpoint ΔV is a measure of the mismatch of positive and negative reference currents. Likewise, by routing V_{ref} to the signal input and performing a voltage measurement, you get a measure of the mismatch of signal and reference currents. The Keysight instruments perform a suite of such calibrations before each measurement on its higher-resolution settings (where it makes a difference). Of course, there's no way it can determine the drift of its primary voltage reference; for that you need an external source of known voltage. This is the business model of calibration laboratories.

Measurement interval We used the example of a measurement time T_{meas} equal to the period of one power-line cycle (PLC), in this case 6250 cycles of 375 kHz, or 1/60 of a second. A measurement time that is an integral number of PLCs ("NPLCs") powerfully rejects coupled interference, and the longer measurement times improve the ultimate accuracy; see Table 13.8. But, as the table shows, you can make more rapid measurements, at the expense of powerline rejection and of accuracy.[58] It's also possible to make *continuous* measurements, in which signal switch S_3 is always on (in this mode the endpoint ADC must take accurately timed samples).

"*Multi*multislope" Circling back to the original game-changing HP3458A 8.5-digit DMM, it uses an amusing bit of trickery: it has four sets of input resistors and switches, such that it can reduce the integrator slew rate dramatically (by a factor of $\sim$600) as it nears the endpoint.[59] Oddly, it does not measure the endpoint residue; it ramps down to 0 V instead, missing out on a very powerful trick.

Miscellany There's a lot of detail in the final implementation of this conversion technique, as described in the service manuals and *HP Journal* articles, as well as the relevant patents.[60] For example, it's necessary only to prevent integrator saturation; so you can use more than one comparator, and turn on the reference current switches (S_1 and S_2) only as needed to keep the integrator in range. This minimizes the number of switching cycles

and the errors that go with it. There are also some curious circuit quirks; for example, there are lossy ferrite beads on the '4053 analog outputs, and there's a capacitor from the integrator summing junction to ground. Go figure....

Evolution of the technique In 2006 Agilent introduced faster "Multislope IV" versions,[61] the 34410A, 34411A, and the 34972A, all with USB and Ethernet data links; they cost more, and the classic 34401A and 34970A were reduced in price. They subsequently introduced the 34460A and 34461A models, with additional features such as display-panel signal processing and sensor interfaces (the 34461A has the same measurement speeds and capabilities as the 34401A, listed in Table 13.8). The 34420A remained the only 7.5-digit (20 ppm) meter in the lineup, and was not upgraded.

Table 13.8 Keysight's Multislope-III ADCs[a]

measurement duration			digits[d]	rdgs per sec[c]	50/60Hz reject[e] (dB)
(PLC[b])	(ms[c])	(clocks[c])			
0.024	0.4	150	4.5	1000	–
0.2	3	1500	5.5	300	–
1	16.7	6250	5.5	60	60
10	167	62.5k	6.5	6	95
100	1.67s	625k	6.5[f]	0.6	105
200[g]	3.33s	1.25M	7.5	0.3	110

Notes: (a) 34401A DMM, 34420A MicroVolt, 34970A DAQ.
(b) powerline cycles; selectable 50/60Hz via setup menu.
(c) when set for 60Hz; clock rate is 375kHz. (d) reported.
(e) normal-mode rejection, at selected powerline freq.
(f) 7.5 digits for the 34420A. (g) 34420A only.

C. From multislope to delta–sigma

The multislope converter takes us naturally to the wildly popular delta–sigma conversion technique, with which it has much in common. At the most basic level, both are

---

[58] Improving on the basic multislope technique, the recent Keysight 34411A bench DMMs do better: 1000 readings/s at 6.5-digit resolution, and 50,000 readings/s at 4.5-digit resolution.

[59] This suite of slopes may be the origin of the "multi" in multislope. They created the name "Multislope III" for the subsequent (and simpler) scheme outlined earlier.

[60] *HP Journal*, April 1989; US patents 4,357,600 and 4,559,521.

[61] Looking at Multislope IV waveforms with a scope, you see a much different beast. A_{1b} has been replaced with a faster AD829 op-amp (120 MHz, 230 V/μs), with the integrating capacitor C_1 reduced by 5$\times$. A hardware engine forces the error integrator to produce consistent 10 Vpp ramps with a 2 μs period, using coarse data from an 80 MHz AD9283 converter digitizing the 10 V ramp at 14 ns intervals, and fine data from an AD9200 10-bit converter with a limited 2 V range, clocked at 75 ns intervals near zero volts. Slope changes and the counter record are made at 75 ns intervals, and the AD9200 makes starting and ending readouts with 0.02% resolution. As a result, a Multislope IV converter can measure 4.5 digits in 20 μs (0.001 PLC), or 20$\times$ faster than Multislope III.

integrating methods in which a discrete offset is applied to the input at periodic intervals, based on the integrator's output level. As we'll see, however, the delta–sigma technique has several subtle tricks up its sleeve, enabling it to deliver astonishing performance.

13.9 ADCs IV: delta–sigma

Now, finally, an extended section on what has become a favorite A/D (and sometimes D/A) conversion technique: the "delta–sigma" converter. These things are confusing, but they're worth some serious effort to understand, because they deliver top-notch performance in resolution and accuracy (e.g., monotonic to 31 bits or more) from "voltmeter" through audio speeds and beyond. Their "oversampling" architecture greatly simplifies the input anti-alias lowpass filter, and performs some magic in shifting the noise spectrum out of the passband. And they provide this performance at surprisingly low cost. In the coming subsections we introduce the basic idea; then we take a serious look at how these converters deliver performance far better than would appear possible. We conclude with some application examples.

13.9.1 A simple delta–sigma for our suntan monitor

To get started, let's revisit the saga of our suntan monitor (see §§4.8.4; to be revisited yet once again, and finally set to rest, in §15.2), this time implemented with the simplest delta–sigma digital integrator. Figure 13.49 shows the implementation, using a clocked charge-dispensing integrator that operates in the same way as the multislope integrator. In fact, it is simpler, because it does not bother with the fractional-cycle endpoint correction: it simply accumulates the integrated sunlight dose, by counting the number of clock cycles during which it is obliged to inject a reference current (here V_{cc}/R) to balance the outgoing photodiode current I_{PD}. Not shown here is the circuitry to sound an alarm when the preset count is reached: the reader who has come this far knows well how to do that!

This is the simplest delta–sigma integrator: it accumulates (sigma) the difference (delta) between the analog input and the measured current that is combined at the summing junction. It *could* become a complete analog-to-digital *converter* (rather than a mere *integrator*), if some circuitry were added to (a) clear the counter to begin a conversion, and (b) read the counter's value after a fixed time interval that is much longer than the clock period. Indeed, that scheme would work as an ADC. But in practice you get far better performance by replacing the counter with a dig-

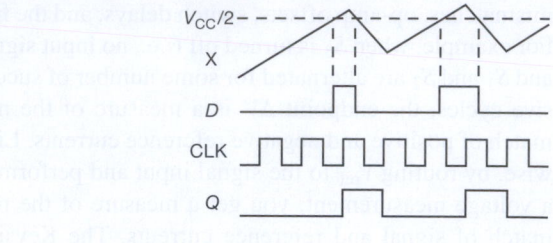

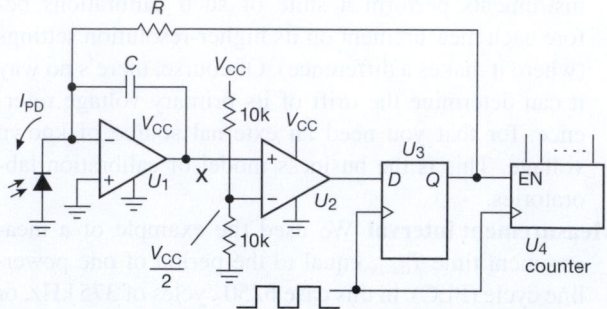

Figure 13.49. Discrete delta–sigma photocurrent-integrating suntan monitor.

ital filter; and you improve things still further by cascading several stages of difference amplifier plus integrator.

We'll get to all of that, soon enough. First, though, let's take the time to understand this simplest example.

In this circuit U_1 is a single-supply op-amp that operates with inputs to (and slightly beyond) the negative rail, and U_2 is a comparator with active pullup. For a low-speed application like this you could use a dual RRIO op-amp like our favorite LMC6482, running on the same +3.3 V or +5 V supply as the digital logic. The integrator ramps up (with slope proportional to the photodiode current I_{PD}) until the next clock rising edge at which its output is greater than $V_{CC}/2$, at which point it ramps down with slope proportional to the net current at the summing junction, $V_{CC}/R - I_{PD}$. The result is that the duty cycle D (fraction of time that the Q output of U_3 is HIGH), averaged over many cycles, is $D = I_{PD}R/V_{CC}$, thus $I_{PD} = DV_{CC}/R$. The duty cycle D is gotten from the count N in U_4 during time interval T by $D = N/f_{clk}T$; note that this result does not depend on the comparison voltage $V_{CC}/2$ (or the voltage at point X).

The design goes like this.

(a) Choose a clock period that is much shorter than the expected baking time, for example $f_{clk} = 10$ Hz; faster is OK, but then you need a larger counter.

(b) Choose R to source more current than the anticipated full-scale input current; for $I_{FS} = 1\,\mu A$ and $V_{CC} = 5$ V, R must be less than 5 MΩ.

(c) Choose C to keep the maximum integrator excursion safely less than $V_{CC}/2$ during one clock cycle.

Here we might choose f_{clk}=10 Hz, R=3.3 MΩ, and C=100 nF. The integrator ramps through at most 1.5 V per clock period (at minimum I_{PD}), so it cannot saturate. The peak count rate equals the clock frequency (and the average count rate is somewhat less, in this case $0.6f_{clk}$), so a 16-bit counter is conservatively adequate for bake settings up to 2 hours full-sunlight equivalent.

Some important points.

(a) The overall calibration depends on the supply voltage V_{CC}, which we've assumed is a stable +5 V; and we've taken advantage of CMOS logic's clean saturation to the rails.

(b) Note that the integrator waveform is not accurately periodic; its excursions above and below the threshold at $V_{CC}/2$ wander somewhat, the guarantee being only that it will be turned around at the next clock following a threshold crossing. This does not, however, degrade its overall accuracy, averaged over many cycles: the integrating nature of the delta–sigma system properly keeps track of the deficits and surpluses; the integrator gets credit for its extra mileage.

(c) The dynamic range of the converter is limited by the op-amp's offset voltage, which causes an equivalent input-current error of V_{os}/R; for this design that is about 0.2 nA (worst case) for the -A grade, thus a dynamic range of 5×10^3. The op-amp's bias current is negligible by comparison (4 pA, max, over temperature).

(d) The dynamic range would be greatly extended if R were replaced with a switched current source, assuming of course that the input signal remains in the form of a current.

(e) In this circuit the comparator U_2 need not be accurate; in fact, it could be omitted altogether, with the flip-flop's logic threshold taking its place. Similarly, the operation does not require an accurate comparison threshold voltage; we chose $V_{CC}/2$ for convenience.

We'll see some additional examples of delta–sigma conversion in §13.9.11. The impatient reader may wish to skip over the following subsections, in which we explore more deeply the operation and performance of the often-confusing delta–sigma technique.

13.9.2 Demystifying the delta–sigma converter

As we noted, the delta–sigma integrator becomes a *converter* of the average analog input voltage, if you cap-ture the count accumulated over a fixed measurement time T_{meas}. The measurement time must be much longer than the clock period, of course, to achieve decent resolution, because the maximum count is just T_{meas}/T_{clk}. So, for example, if you were to design an ADC to convert at 100 ksps, you might use a 10 MHz clock, cleared at the beginning of each conversion, and read out 10 μs later. The full-scale count would be 100, which you could describe as a (nearly) 7-bit conversion. To achieve 16 bits, you'd need to run the clock at $2^{16}\times100$ kHz, or 6.5536 GHz!

This does not sound promising. It just seems like a bad idea to design a "1-bit" converter – which is what you have in the stream of bits that is driving the counter in this design. So it will come as a surprise that the impossible *can* be done: there are plenty of 16-bit delta–sigma ADCs that convert at audio rates (e.g., 96 ksps), and in fact there are some that achieve 20 bits or more of resolution at that speed (see §13.10.1, and Tables 13.9 and 13.10). How can this be?! Read on....

Sometime in the 1990s the promotional materials for consumer-grade CD audio players began to trumpet their use of "1-bit digital-to-analog converters," as if there was something good about *reducing* the resolution from the previously trumpeted 16 bits.[62] This seemed puzzling to many of us; but we didn't complain because, well, the players sounded pretty good.

As Bob Pease would say, what's all this 1-bit converter stuff, anyhow?

13.9.3 ΔΣ ADC and DAC

As we'll see, ΔΣ (also known as ΣΔ) conversion can go either way – D/A, or A/D. In contemporary practice ΔΣ *DAC*s are used primarily for audio applications, where they excel in linearity, monotonicity, and low cost. A typical audio ΔΣ DAC might integrate six 24-bit 192 ksps converters, with 114 dB effective dynamic range, for about $10.[63] By contrast, ΔΣ *ADC*s cover a broad range of applications, ranging from precision (24-bit) dc-accurate slow converters, to audio-rate converters of high resolution (e.g., 24-bit, 96 ksps), and up to speedy ADCs with more accuracy than you would imagine (e.g., 16-bit, 20 Msps).

---

[62] Trumpet blasts are still echoing. Here's a contemporary example: "There is virtually no noise or sound degradation during the signal transmission and amplification process as 1-bit signals are digital." As if *n*-bit signals are *not* digital?!

[63] For audio applications the dc performance is irrelevant, and in general is not even specified. An exception is the excellent DAC1220 20-bit ΔΣ DAC from Texas Instruments.

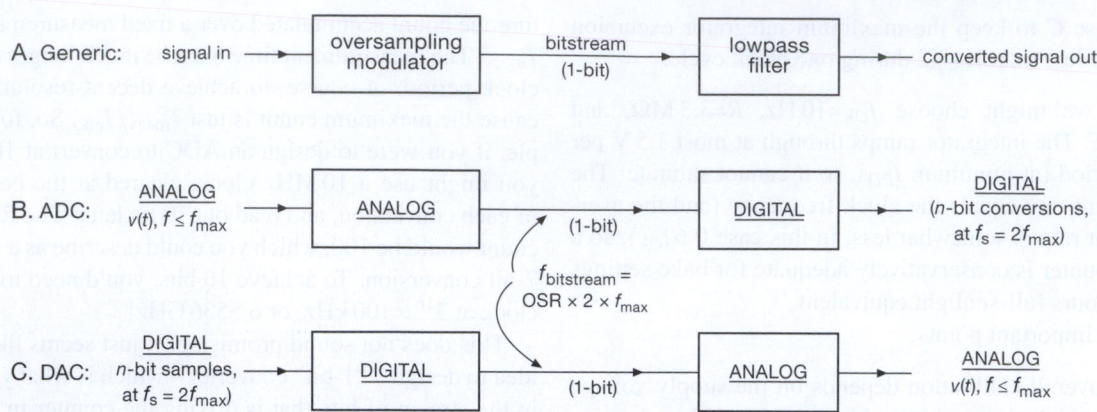

Figure 13.50. A delta–sigma converter, whether A/D or D/A, consists of two parts: an oversampling *modulator* that produces an intermediate *bitstream*, followed by a lowpass filter that recovers the converted output.

In the discussion that follows, we talk primarily about $\Delta\Sigma$ ADCs, both because of their importance and also because their architecture exploits the ideal characteristics of digital filtering.

Along the way, we will try to get to the bottom of what, to us, had seemed like a *big mystery*, namely: *How can it be that 1-bit conversions, at some modest "oversampling" rate (say 64 times the usual Nyquist rate of $2f_{max}$), can produce digitized output samples of great accuracy (say 16 bits)?* To rephrase the mystery: naively one might expect that 1-bit conversions at a 64-fold oversampling rate might allow us to recover a final digital output with 6-bit resolution (because $2^6=64$), but no better.[64] We'll see, though, that it's possible (and mandatory, for audio applications!) to do considerably better.[65]

13.9.4 The $\Delta\Sigma$ process

Figure 13.50 shows the basic $\Delta\Sigma$ process. An incoming signal, bandwidth limited to some maximum frequency f_{max}

(usually by an anti-alias filter[66]), is converted to a bitstream[67] by a *modulator*. The latter is clocked at some multiple of the minimum Nyquist sampling rate $2f_{max}$, generating an output bitstream of rate $f_{bit}=\text{OSR}\times 2f_{max}$, where OSR is called the *oversampling ratio*. This bitstream is the intermediate step in the overall converter: to get the converted output, the bitstream must be lowpass filtered.

Note that both the modulator and the lowpass filter may each be analog or digital, depending on the type of converter: a delta–sigma ADC consists of an analog modulator followed by a digital filter, whereas a delta–sigma DAC consists of a digital modulator followed by an analog filter.[68] In what follows we will deal primarily with the modulator portion of the overall converter.

A. The modulator

In either case, the lowpass filter is "just a filter," which simply bandwidth limits the incoming bitstream.[69] The

---

[64] That is, in fact, the case with filtered PWM, as used, for example, for motor control or LED dimming: there one might divide the Nyquist period into 64 time slots, setting the first few as 1s and the rest 0s. Delta–sigma is more subtle, and better, with 0s and 1s sprinkled through the Nyquist period in such a way as to produce a filtered output of high accuracy.

[65] Plot spoiler, for the impatient: thinking in the time domain, it's a willing conspiracy between an output lowpass filter that views a long stretch of the bitstream, and a very clever bitstream-building modulator whose filtered output represents an accurate conversion. A better (and quantitative) understanding comes in the frequency domain, where the oversampling modulator acts to reduce in-band quantization noise, "shaping" it to higher (out-of-band) frequencies.

[66] Which, as we'll see later, need not cut off sharply, thanks to the beneficial effects of oversampling; see Figure 13.60.

[67] Here shown as 1 bit wide, for simplicity, though in practice it may be several bits wide (i.e., more than 2 levels).

[68] An interesting third possibility (analog/analog) is exemplified by the Avago HCPL-7800A analog opto-isolator: an internal input modulator creates a bitstream that is optically coupled to an internal analog output demodulator, to deliver an accurate analog replica (0.004% nonlinearity, typical) of high stability (3 ppm/°C gain change, typ), kilovolt-level isolation, and 100 kHz bandwidth. Another example of delta–sigma "A-to-A" is the Super Audio CD (SACD), a CD-like audio storage format in which the 2.8 Mbps intermediate (encrypted) bitstream itself is recorded and distributed to the user, with lowpass filtering applied at playback.

[69] The bitstream can be thought of either as *digital* (1s and 0s), which is then filtered by a digital filter (if this is an ADC), or as an *analog wave-*

interesting action (and the *magic*) takes place in the modulator. Figure 13.51 shows a block diagram of a "first-order" oversampled modulator, which accepts analog input voltages between -1 V and $+1$ V, bandlimited to a maximum frequency of f_{max}, and produces a 1-bit output bitstream at a rate OSR time higher than the critical Nyquist sampling rate $2f_{max}$.

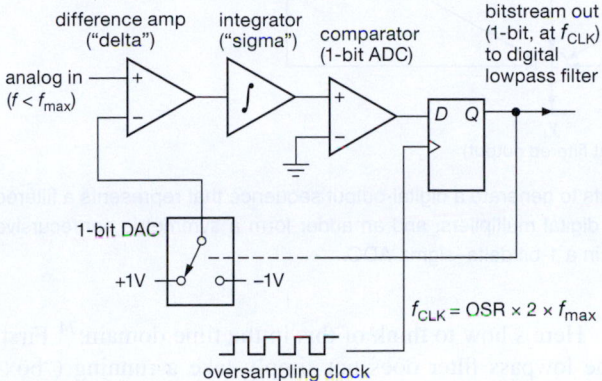

difference amp ("delta")　integrator ("sigma")　comparator (1-bit ADC)

analog in ($f < f_{max}$)

1-bit DAC

$+1$V　-1V

bitstream out (1-bit, at f_{CLK}) to digital lowpass filter

$f_{CLK} = $ OSR $\times 2 \times f_{max}$

oversampling clock

Figure 13.51. A first-order delta–sigma analog modulator.

At each clock cycle the current bitstream value, converted to an analog voltage (in this case ± 1 V), is subtracted from the analog input, the difference signal being integrated (in a standard op-amp analog integrator, here assumed to be noninverting) and presented to a latched comparator. The integrator gain is such that a full-scale analog input to the integrator (i.e., $+1$ V) produces a full-scale ($+1$ V) change in the integrator's output after one clock period. That is, you can think of the integrator as an "analog accumulator": for a (fixed) input voltage V, its output voltage increases by V during one clock period.

The result is a rapid stream of 1s and 0s (at, say, 64 times the usual $2f_{max}$ sampling rate), responding to the relatively slow (64 times slower, say) changes in the input signal. Thinking of these bits as ± 1 V, the modulator produces a stream whose *average value* matches the input signal. We can understand this by thinking of the modulator circuit as a negative-feedback loop that strives to minimize the averaged (i.e., integrated) error between the input signal and the output stream (which has been converted back to analog by a "1-bit DAC"). Looking more closely, though, we can see that it is doing a *terrible job*: sample by sample,

its output bitstream simply bounces between extremes. As Bob Adams has written,[70] pithily, "Oversampling converters achieve increased resolution not by decreasing the error between analog input and digital output, but by making the error occur more often."

B. The ADC's dynamic range (resolution)

The output lowpass digital filter (typically an FIR digital filter – see Figure 13.52 – in this case a 1-bit shift register with the marching 1- and 0-bit values turning on or off a set of fixed digital coefficients[71] that are added digitally to create the multibit output samples) creates the digital n-bit numbers that are the converter's output. Because they emerge from the filter at the full oversampling rate, they are subjected to a "decimation" operation, most simply by discarding superfluous outputs and outputting only one converted output for every OSR clock cycles.[72]

Naively, then, we achieve increased resolution by having plenty of 1-bit samples to average over, for each half-cycle of the highest frequency in the incoming waveform. Because the average value of the bitstream tracks the input signal, we understand Bob Adams' statement, and all is well.

Or is it? Consider an example: suppose we are digitizing audio, with an f_{max} of 20 kHz. A conventional ADC (say a successive approximation converter) might sample at 48 ksps, comfortably above the critical minimum of 40 kHz. Imagine instead that we rig up a ΔΣ ADC, with a typical oversampling ratio of 64; that is, we run the modulator at 3.072 Msps (64×48 kHz), creating the (1-bit) bitstream at that rate. Now we filter that bitstream, for example by taking a running average,[73] digitally, that captures 64 successive bits at a time. What does the output look like? Well, when you take the average value of 64 bits, there are only 64 possible values. So we've invented a paltry 6-bit ADC.

Following this logic, we'd need to oversample by 2^{16} (i.e., 64K) to achieve 16-bit conversion. That would require a sampling rate of roughly 3 *giga*hertz! Delta–sigma conversion is not looking good.

---

[70] "Design and implementation of an audio 18-bit analog-to-digital converter using oversampling techniques," *J. Audio Eng. Soc.* **34**, 153–166 (1986).

[71] To a first approximation the time-series coefficients are a (signed) sinc function, the Fourier transform of a "brickwall" lowpass function. See §6.3.7, on digital filters.

[72] In practical implementations the filtering and decimation are combined, using a "multirate decimation filter."

[73] Rather than the more complicated sinc-function weighted average that's required for implementing an ideal "brick-wall" lowpass; see §6.3.7.

---

form that switches between two fixed voltage levels (if this is a DAC). Note also that the phrase "just a filter" does not mean that the filter design is simple, or trivial. In particular, *digital* filter design is a sophisticated art, with issues of window functions, nulls in the response, and so on. See §6.3.7.

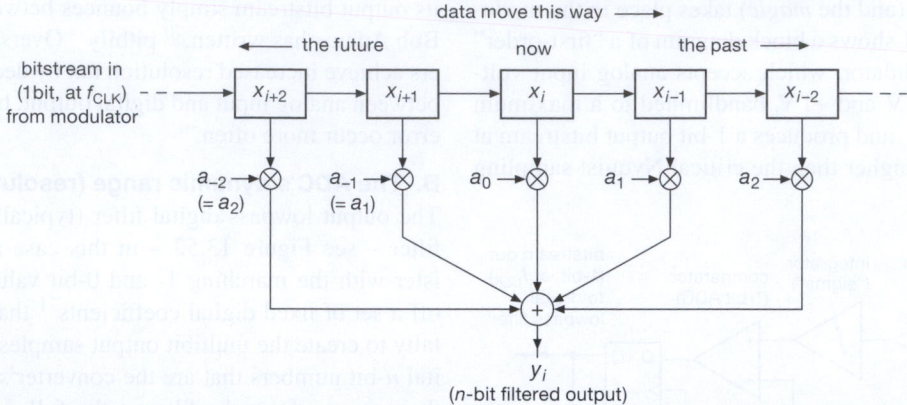

Figure 13.52. A digital filter uses digital memory and arithmetic elements to generate a digital-output sequence that represents a filtered version of a digital input sequence (see §6.3.7). Here a shift register, digital multipliers, and an adder form a symmetric nonrecursive (finite impulse response, FIR) filter, suitable as the output lowpass filter in a 1-bit delta–sigma ADC.

C. So, what's going on? (time-domain intuition)

The answer to this paradox can be stated in different ways. In the literature the usual approach is to say that the operation of the 1-bit digitizer consists of a perfect conversion, but with added broadband "noise" (consisting of the difference between the actual analog waveform and the 1-bit quantized samples). This "injected quantization noise" has a broad spectrum (because of the oversampled clocking frequency), extending to the clock frequency and beyond. Of greater importance, the resulting "output quantization noise" (what remains at the output) is lowest at low frequencies, being "shaped" by the modulation process such that most of the output quantization noise is well above f_{max}. Because of this so-called "noise shaping," the final lowpass filter acts to selectively eliminate most of the quantization noise from the bitstream, while preserving the converted signal. Voilà: resolution and dynamic range far better than our naive estimate above.

This is all quite correct, but, to us, unsatisfying (though we'll take a brief look at that approach in §13.9.5). We wanted to understand the secret of ADC dynamic range *in the time domain*, without recourse to the frequency domain. We struggled with this, reading exposés with titles like "Demystifying Sigma–Delta ADCs" and "Delta–Sigma ADCs in a Nutshell." Not much help – they all get to the critical step, then punt ("…sigma–delta converters overcome this limitation with the technique of noise shaping…" and "…you can see how the modulator shapes the noise to higher frequencies, facilitating the production of a higher resolution result," excerpts from those two articles, respectively).

Here's how to think of this in the time domain:[74] First, the lowpass filter does not simply take a running ("boxcar") average of the bitstream. Rather, it weights the individual 1-bit samples with carefully tailored coefficients to produce a better lowpass filter characteristic (see §6.3.7D). Because the individual bits are weighted differently, there are many more than 64 possible results (taking the example above). Furthermore, a typical FIR digital filter will weight and sum many more bits, and over a spread of time (samples) much longer than the oversampling rate (i.e., extending beyond what we might call a single "Nyquist interval," by which we mean the time interval equal to a half-period of f_{max}). For a 64× oversampling ADC, the digital lowpass filter might use of order a thousand "taps" (samples along the bitstream), each with its multiplying coefficient, and spanning perhaps ten to twenty Nyquist intervals, to generate each final (decimated) output number. So it is at least *plausible* that you could achieve resolution far greater than possible with simple averaging.

Continuing in this vein, it's worth noting that each bit in the bitstream contributes to many final n-bit output numbers (after decimation). So, in a conspiratorial vein, it's plausible that a cleverly contrived modulator could generate a bitstream that, after lowpass filtering, could produce an n-bit digitized output with quite considerably improved dynamic range. Thinking this way, we would not be wrong[75] in concluding that "the magic is in the modulator."

---

[74] We are indebted to Bob Adams of Analog Devices for helpful discussions, and the extermination of mental cobwebs. He is not responsible, however, for any errors (egregious or otherwise) here committed.

[75] "You're not *wrong*, Walter; you're just …."

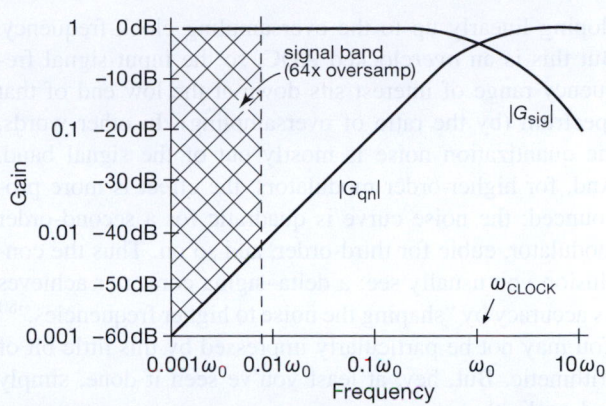

So the question becomes "How does such a simple device (Figure 13.51) behave so cleverly?"

13.9.5 An aside: "noise shaping"

As we remarked, the usual description of delta–sigma converters talks about *noise shaping* in the frequency domain: the flat-spectrum "quantization noise" introduced in the quantizer (the comparator of Figure 13.51) gets "pushed up" to high frequencies, mostly above the output sampling rate. And less in-band noise equates to higher accuracy – end of argument.

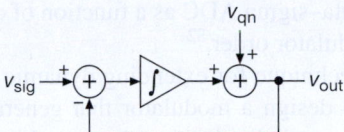

Figure 13.53. Noise shaping in a first-order delta–sigma ADC: an all-analog model, with quantizer replaced by an additive quantization noise source.

To many engineers this is a satisfactory explanation. But even if you aren't particularly convinced by this argument, it's worth understanding it. To see most simply how this works, look at Figure 13.53, in which the modulator is constructed with an *analog* integrator, and converts the (continuous) analog input to a 2-state analog (± 1 V) output. In this equivalent analog model we've replaced the 1-bit quantizer (comparator) with an additive quantization noise voltage v_{qn}, whose flat spectrum extends to the oversampling clock frequency (and beyond).[76] You can think of the integrator as being in the forward path of the loop for the signal input (thus lowpass), but in the feedback path for the noise input (thus highpass).[77] From this simple analog loop we can easily figure the frequency responses to both the input signal and to the quantization noise signal. There's only one gain parameter, namely that of the integrator, whose gain we write as $\mathbf{G} = \omega_0/j\omega$; that is, the integrator's gain (proportional to $1/\omega$) is such that its magnitude is unity at ω_0. From our earlier discussion, $\omega_0 \approx 1/2\pi f_{\text{overclock}}$, the frequency of the converter's oversampling input clock.

Let's figure the gains, for which we properly have to do

Figure 13.54. Signal gain and quantization noise gain versus frequency for a first-order delta–sigma ADC. The clock frequency $\omega_{\text{clk}} = \omega_0 = 2 \cdot \text{OSR} \cdot \omega_{\text{max}}$ here equals 128 times ω_{max}.

the full complex-number business. To get the input signal gain $\mathbf{G}_{\text{sig}}$ we set $v_{qn}=0$; then

$$v_{\text{out}} = \frac{\omega_0}{j\omega}(v_{\text{sig}} - v_{\text{out}}),$$

so

$$\frac{v_{\text{out}}}{v_{\text{sig}}} = \frac{\omega_0/j\omega}{1 + \omega_0/j\omega},$$

and

$$|\mathbf{G}_{\text{sig}}| \equiv \left| \frac{v_{\text{out}}}{v_{\text{sig}}} \right| = \frac{1}{\sqrt{1 + (\omega/\omega_0)^2}}.$$

That's a lowpass filter, with breakpoint at $\omega = 2\pi f = \omega_0$ (Figure 13.54).[78]

Similarly, for the quantization noise gain $\mathbf{G}_{\text{qn}}$ we set $v_{\text{sig}}=0$; then

$$v_{\text{out}} = v_{qn} - \frac{\omega_0}{j\omega} v_{\text{out}}$$

so

$$\frac{v_{\text{out}}}{v_{qn}} = \frac{1}{1 + \omega_0/j\omega}$$

and

$$|\mathbf{G}_{\text{qn}}| \equiv \left| \frac{v_{\text{out}}}{v_{qn}} \right| = \frac{\omega/\omega_0}{\sqrt{1 + (\omega/\omega_0)^2}}.$$

And that's a highpass filter, with the same breakpoint.[79]

So, in this lowest-order delta–sigma ADC the quantization noise is attenuated at low frequencies, its spectrum

---

[76] You usually think of additive "noise" as being small compared with the signal that it is guilty of degrading. Here the quantization noise (the difference between the analog signal and the 2-level output voltage v_{out}) is actually larger than the signal itself.

[77] Explained nicely by Ewe Beis, on his website at http://www.beis.de/Elektronik/Electronics.html.

[78] A bit of intuition may be helpful here: at *low* frequencies (well below ω_0) the integrator has lots of gain, so the loop is closed with plenty of loop gain, creating a unity-gain output (in spite of the integrator within). In fact, the combination of input adder and integrator is not unlike a standard op-amp with its $1/f$ (compensation) rolloff. But at high frequencies there's no loop gain, so you get the $\sim 1/f$ integrator rolloff.

[79] Intuition, again: this time the "signal" (i.e., the quantization noise v_{qn}) acts like an additive output disturbance to the closed-loop unity-gain amplifier. So it is *removed* by the loop gain, which is high at low frequencies (its magnitude is ω_0/ω), but ineffective above ω_0.

sloping linearly up to the oversampling clock frequency. But this is an *over*clocked ADC, so the input signal frequency range of interest sits down at the low end of that spectrum (by the ratio of oversampling). In other words, the quantization noise is mostly out of the signal band. And, for higher-order modulators, the effect is more pronounced: the noise curve is quadratic for a second-order modulator, cubic for third-order, and so on. Thus the conclusion you usually see: a delta–sigma converter achieves its accuracy by "shaping the noise to higher frequencies."[80] You may not be particularly impressed by this little bit of arithmetic. But, hey, at least you've seen it done, simply and explicitly.

13.9.6 The bottom line

Both from plausibility arguments in the time domain, and from explicit calculation in the frequency domain, it appears that the modulator circuit holds the key to the sigma–delta ADC's performance; that is, its ability to quantize an analog input signal with a resolution considerably greater than the oversampling ratio. Furthermore, that figure of merit ($N_{eff}/\log_2 OSR$, where N_{eff} is the effective number of bits in the quantized digital output) grows with modulator complexity: contemporary ADCs employ "higher-order" modulators, meaning that the single difference amplifier and integrator is replaced with several cascaded stages of difference amplifier plus integrator, each driven from the common bitstream (see Figure 13.55).[81] Higher-order modulators are widely used, because they extend the dynamic range without having to increase the oversampling ratio (see below); they also suppress to a large extent the *idle tones* (see §§13.9.9 and 13.9.10) that afflict first-order modulators.

Although our time-domain musings above may be helpful (if only to make plausible the excellent dynamic ranges that are claimed), any serious analysis must use the frequency-domain approach. The latter shows that a higher-order modulator (constructed with m integrators) modifies the noise shaping such that the in-band quantization noise (i.e., dc to f_{max}) is suppressed as $OSR^{m+0.5}$, where m is the modulator order ($m=1$ for Figure 13.51). Put another way, each doubling of the oversampling ratio suppresses quantization noise so as to increase the dynamic range by $m+\frac{1}{2}$ bits; or, to state it in terms of modulator order, the effective number of bits (ENOB) is the $\log_2$ of the oversampling ratio (e.g., 6 for OSR=64) multiplied by $m+\frac{1}{2}$ (thus, for example, ENOB≈15 for a second-order modulator with oversampling ratio of 64). The graph in Figure 13.56 shows the theoretical maximum dynamic range of a delta–sigma ADC as a function of oversampling ratio and modulator order.[82]

Another technique for extending dynamic range, speed or both, is to design a modulator that generates a modulated "wordstream" that is more than one bit wide. In Figure 13.51, for example, the 1-bit ADC, 1-bit DAC, and 1-bit register would be replaced with analogous 2-bit (4-level) components. There are lots of clever tricks to address imperfections in the multibit converters within the modulator (for example, cyclically exchanging bit positions to average out nonlinearities caused by offsets); they are well beyond the scope of this book.

13.9.7 A simulation

We wanted to see for ourselves just how the signals move through a delta–sigma ADC – particularly the bitstream produced by some random-looking analog input signal, and of course the resultant output numbers (plotted as discrete points alongside the analog input waveform). We also wanted to see what things look like in the frequency domain, where the noise shaping should be evident.

The simulation was coded[83] in Mathematica®, with the following recipe:

(a) a spectrally-flat pseudorandom waveform with Gaussian amplitude distribution was generated, evaluated at 8192 successive time steps;

(b) this waveform was filtered with an approximate ideal brick-wall lowpass, with cutoff at 1/8 of the maximum frequency; it was then normalized so that its amplitude was bounded by ±1, generating the "analog input

---

[80] Sometimes stated as "pushing the noise upward to higher frequencies." In our *linear* model here, nothing gets *pushed*; it just gets attenuated at the low-frequency end, and passes unattenuated at the high end. However, a fully accurate quantization noise model must properly take account of the fact that the 2-state bitstream has unit amplitude (always ±1), with the result that reducing the quantization noise power at the low-frequency end causes it to rise at the high end.

[81] A simple cascade of integrators can be used up to second order, but not beyond (because their accumulated phase shift produces instability); a weighted sum of cascaded integrator outputs is used instead, in a higher-order modulator. Contemporary audio ΔΣ ADCs typically use fifth-order modulators and 64× oversampling to achieve 20-bit effective dynamic range.

[82] Note, however, that practical modulators of order greater than 2 use a modified structure (weighted sum of integrator outputs), for which the formula is not strictly correct.

[83] By the ever-talented Jason Gallicchio, to whom we are indebted.

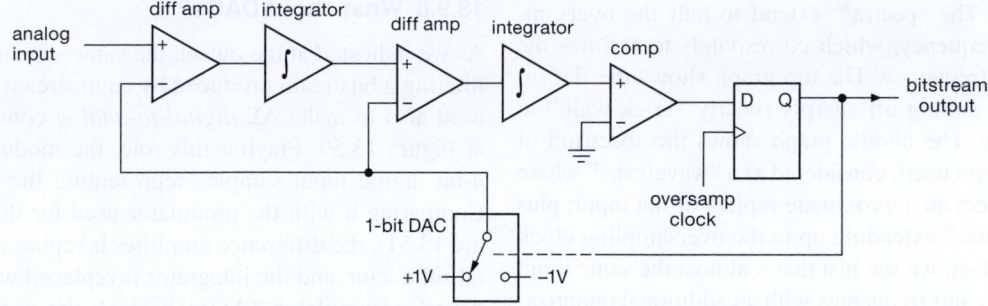

Figure 13.55. A second-order delta–sigma analog modulator. Lowpass filters can be substituted for one or more of the integrators.

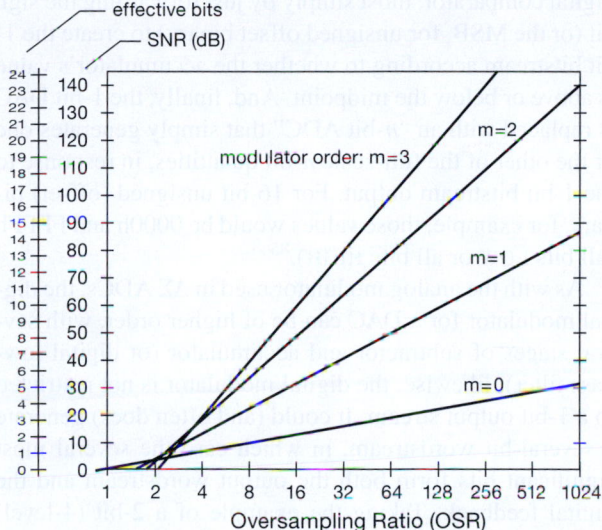

Figure 13.56. Dynamic range (SNR) and effective number of bits (ENOB), as functions of oversampling ratio (OSR) and modulator order (m), for a 1-bit oversampling ADC. (For a 2-bit modulator the slopes are doubled.)

signal"; the maximum frequency present in this signal was defined as the Nyquist frequency, f_{nyq};

(c) this signal is used to generate a bitstream with values of ± 1, by numerically simulating a first-order oversampling delta–sigma modulator (in which the integrator is realized as a discrete digital accumulator); the oversampling ratio is $8\times$, and thus the clock frequency $f_{clk}=16f_{nyq}$ (recall that critical "$1\times$" sampling requires $f_{clk}=2f_{nyq}$); finally,

(d) the bitstream, considered as an analog waveform itself, was lowpass filtered with the same filter function as in step (b), to produce the output samples; these emerge at the full $8\times$ oversampled rate, and would normally be decimated (e.g., by preserving only every eighth point) to produce the ADC's digitized output at the "$1\times$" rate

(twice the highest frequency present in the input waveform).

Figure 13.57 plots a typical portion of the (longer) simulation, showing what's happening in the time domain. The tick marks on the time axis correspond to $1\times$ (critical Nyquist) sampling, with individual dots plotted at the $8\times$ oversampling rate. The input signal is the wiggly solid line, closely approximated by the discrete dots that are the digitized output numbers (plotted at their full $8\times$ rate). You can see the bitstream waveform, on the same amplitude scale, as dots at ± 1. Finally, the error (i.e., digitized output minus analog input, at each oversampling point) is the dotted waveform of small amplitude. From these plots you can do an eyeball estimate of the converter's accuracy; to us it appears to exhibit something like $\pm 6\%$ peak-to-peak amplitude error, which translates to an amplitude SNR of 16:1 (24 dB), in good agreement with the graph of Figure 13.56.

From this same simulation we plotted also the frequency spectra of the input signal, the output digitized "waveform," and the difference (an "error signal"); see

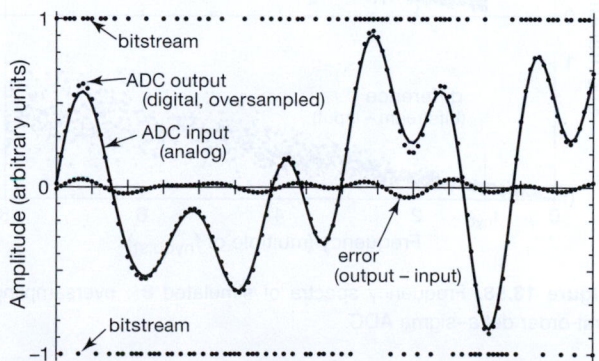

Figure 13.57. Numerical simulation of an $8\times$ oversampling first-order delta–sigma DAC.

Figure 13.58. The spectra[84] extend to half the oversampling clock frequency, which corresponds to 8 times the highest input frequency. The top graph shows the flat input spectrum, cutting off sharply (nearly "brick wall") at unit frequency. The middle graph shows the spectrum of the raw bitstream itself, considered as a "waveform," where we would expect an approximate replica of the input, plus additional "noise" extending up to the oversampling clock frequency. In fact, we see just that – almost the same input spectrum up to unit frequency, with an additional quantization noise that increases roughly proportional to frequency. By taking the difference of these two spectra you can extract approximately the spectrum of the added noise introduced by quantizing (bottom graph), showing an approximately linear growth from zero up to at least half the oversampling clock frequency. The quantization noise above unit frequency is, of course, removed by the digital lowpass filter that accepts the bitstream as input and that completes the converter circuit (Figure 13.50).

The linear shape of the quantization noise spectrum, for this first-order modulator, would be replaced with a quadratic shape for a second-order modulator, and so on for higher orders. That higher-order noise shaping corresponds to improved accuracy (or SNR, or effective number of bits), as shown graphically in Figure 13.56.

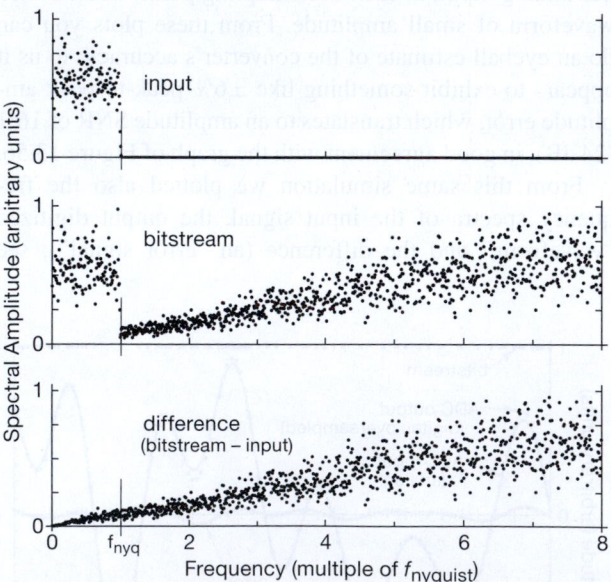

Figure 13.58. Frequency spectra of simulated 8× oversampling first-order delta–sigma ADC.

---

[84] The spectral amplitudes were binned in groups of four to generate a sparser, and therefore more easily plotted, figure.

13.9.8 What about DACs?

As we indicated at the outset, the same scheme of lowpass filtering a bitstream produced by an upstream modulator is used also to make $\Delta\Sigma$ *digital-to-analog* converters. Look at Figure 13.59. Playing this role, the modulator accepts *n*-bit digital input samples representing the input signal. Comparing it with the modulator used for the ADC (Figure 13.51), the difference amplifier is replaced with a digital subtractor, and the integrator is replaced with a clocked digital accumulator. (At each clock the accumulator replaces its current latched value by the sum of that value and the input value.) The analog comparator is replaced with a digital comparator, most simply by just forwarding the sign bit (or the MSB, for unsigned offset binary) to create the 1-bit bitstream according to whether the accumulator's value is above or below the midpoint. And, finally, the 1-bit DAC is replaced with an "*n*-bit ADC" that simply generates one or the other of the full-scale *n*-bit quantities, in response to the 1-bit bitstream output. For 16-bit unsigned (offset) binary, for example, those values would be 0000h and FFFFh (all bits LOW or all bits HIGH).[85]

As with the analog modulator used in $\Delta\Sigma$ ADCs, the digital modulator for a DAC can be of higher order, with several stages of subtractor and accumulator (or digital lowpass filter); likewise, the digital modulator is not restricted to a 1-bit output stream. It could (and often does) generate a several-bit wordstream, in which case the several most significant bits form both the output wordstream and the digital feedback. Taking the example of a 2-bit (4-level) modulator, the 2-bit output stream would be both (a) converted to a 4-level analog voltage with a resistor ladder, then analog lowpass filtered to form the (analog) output signal, and (b) simultaneously mapped to one of four *n*-bit codes spanning the full input range (e.g., 0000h, 5555h, AAAAh, and FFFFh), then used as input to the *n*-bit digital subtractor at the input.

The output stage of a $\Delta\Sigma$ DAC is, as with the ADC, a lowpass filter. Here, however, it is an *analog* filter, which denies you the benefits of sophisticated digital filtering. The result is some compromise in filter characteristics, including clock feedthrough, and (with an analog, or "continuous-time" filter) sensitivity to clock timing jitter.[86]

---

[85] For 2's complement 16-bit quantities, the corresponding values are 8000h and 7FFFh (corresponding to −32768 and +32767, the lowest and highest values).

[86] The lowpass filter can alternatively be implemented as a switched-capacitor (or "discrete-time") filter, which, sharing the same clock signal, effectively suppresses clock jitter.

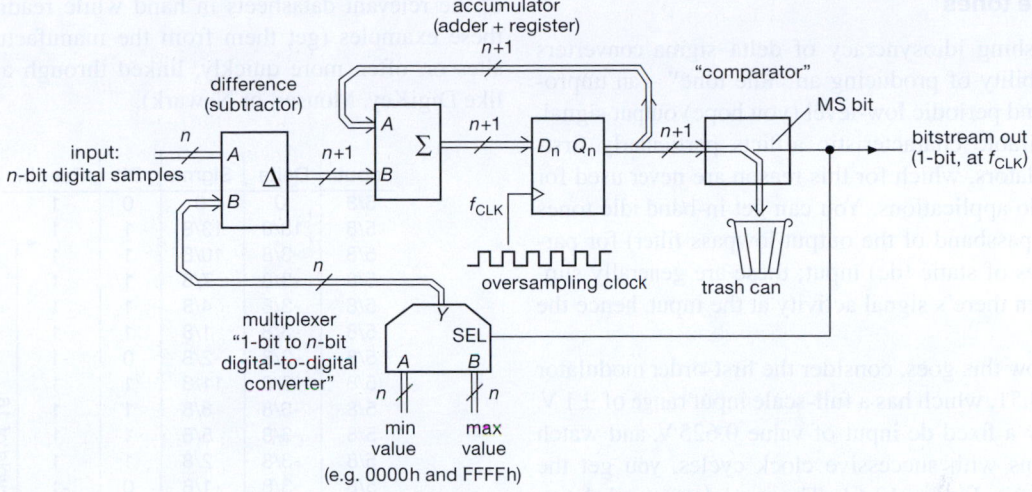

Figure 13.59. A first-order digital modulator in a delta–sigma DAC. The adder output is shifted one bit to prevent word growth.

13.9.9 Pros and Cons of ΔΣ oversampling converters

A. Advantages

Linear, monotonic, accurate Delta–sigma 1-bit converters are guaranteed monotonic; they are inherently linear, and can achieve 24-bit resolution at audio rates and below.

Inexpensive Delta–sigma ADCs use inexpensive (and accurate) *digital* lowpass filtering, and (because of oversampling) require only a low-order analog anti-alias filter at the input (see Figure 13.60)

B. Disadvantages

Limited bandwidth To ~10–100 Msps at most (limited by gigahertz-scale oversampling clock).

Time delay The built-in ADC post-conversion digital filter achieves nearly ideal "brick wall" cutoff by using many taps, resulting in significant delay (or "latency," typically tens of output sample times, thus ~millisecond for audio ADCs).[87]

DAC noise Delta–sigma DACs use an *analog* post-conversion lowpass filter, which permits some digital-switching feedthrough (by contrast, an R–2R DAC is completely "quiet").

Idle tones An ADC with first-order modulator can produce "idle tones" (see below) when presented with a

static input that causes the integrator output to cycle with a sufficiently long period to be in-band (and causing apoplexy among audio aficionados); higher-order modulators suppress this artifact, even if present in the quantizer, owing to the higher in-band loop gain.

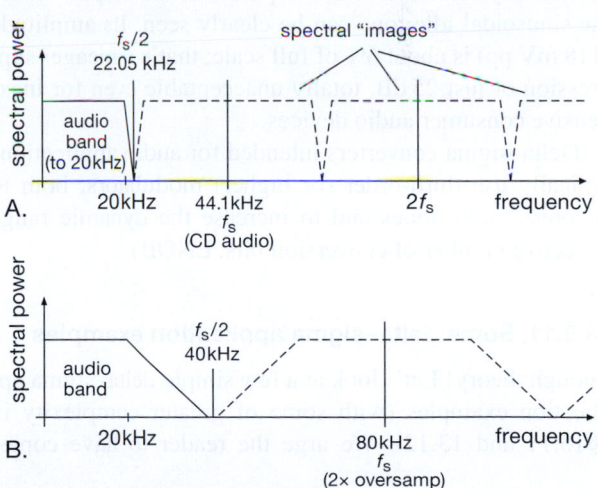

Figure 13.60. The spectrum of an analog signal that has been digitized ("sampled") periodically at a sampling frequency f_s includes mirrored copies ("images") centered at multiples of f_s. The analog signal must be lowpass filtered, before sampling, to eliminate components above $f_s/2$ (the "Nyquist frequency"); otherwise the mirrored bands create in-band "aliases" that cannot be subsequently removed. Oversampling relaxes the required steepness of such an "anti-alias LPF," as seen here for conventional CD-audio sampling (A, ≈10% oversampling) compared with 2× oversampling (B).

---

[87] However, note that the time delay is the same as would be produced by a sharp input anti-alias filter (of similar cutoff characteristics to that of the delta–sigma's post-conversion digital filter) followed by a conventional (zero-latency) ADC.

13.9.10 Idle tones

A distinguishing idiosyncracy of delta–sigma converters is the possibility of producing an "idle tone" – an unprovoked in-band periodic low-level (you hope) output signal. This undesirable characteristic afflicts particularly first-order modulators, which for this reason are never used for serious audio applications. You can get in-band idle tones (i.e., in the passband of the output lowpass filter) for particular values of static (dc) input; these are generally suppressed when there's signal activity at the input, hence the term "idle."

To see how this goes, consider the first-order modulator of Figure 13.51, which has a full-scale input range of ±1 V. If you apply a fixed dc input of value 0.625 V, and watch what happens with successive clock cycles, you get the states shown in Figure 13.61. The modulator (and therefore the output bitstream) goes periodically through a cycle of length 16 clocks. Such a repeating pattern can be of long enough period to produce a signal in the final lowpass-filtered output: if, in this example, the oversampling ratio were 4×, the idle tone would fall at the midpoint frequency of the output band.

This can be seen in Figure 13.62, where we've[88] calculated and plotted two complete cycles of this idle tone. Note particularly the brick-wall lowpass filtered output, where the sinusoidal idle tone can be clearly seen. Its amplitude (118 mV pp) is about 6% of full scale; that's a meager suppression of just 25 dB, totally unacceptable even for inexpensive consumer audio devices.

Delta–sigma converters intended for audio applications typically use third-order (or higher) modulators, both to suppress[89] idle tones and to increase the dynamic range (effective number of conversion bits, ENOB).

13.9.11 Some delta–sigma application examples

Enough theory! Let's look at a few simple delta–sigma application examples (with some of greater complexity in §§13.11 and 13.12). We urge the reader to have copies

---

[88] Jason, again.

[89] Higher-order modulators do not *eliminate* idle tones, they merely suppress them enough to render them almost harmless. As delta–sigma guru Bob Adams puts it, "Higher order modulators are less likely to fall into a simple repeating pattern, but it is still possible. The real benefit of higher-order modulators is that they have excellent suppression of quantization noise (because the loop gain at low frequencies is extremely high), so that, even if the quantizer decides to fall into an idle-tone-producing pattern, it will be suppressed by such a large amount that it will be buried in thermal noise."

---

of the relevant datasheets in hand while reading through these examples (get them from the manufacturers' websites or, often more quickly, linked through a distributor like DigiKey, Mouser, or Newark).

Input	Delta	Sigma	Bit	Feedbk
5/8	0	0	0	-1
5/8	13/8	13/8	1	1
5/8	-3/8	10/8	1	1
5/8	-3/8	7/8	1	1
5/8	-3/8	4/8	1	1
5/8	-3/8	1/8	1	1
5/8	-3/8	-2/8	0	-1
5/8	13/8	11/8	1	1
5/8	-3/8	8/8	1	1
5/8	-3/8	5/8	1	1
5/8	-3/8	2/8	1	1
5/8	-3/8	-1/8	0	-1
5/8	13/8	12/8	1	1
5/8	-3/8	9/8	1	1
5/8	-3/8	6/8	1	1
5/8	-3/8	3/8	1	1
5/8	-3/8	0/8	0	-1
5/8	13/8	13/8	1	1
5/8	-3/8	10/8	1	1
5/8	-3/8	7/8	1	1

(cycle of 16)

Figure 13.61. Sequence of states of the first-order delta–sigma modulator (Figure 13.51), with a fixed dc input of 625 mV. "Delta" and "Sigma" are the input and output, respectively, of the integrator, "Bit" is the bitstream output, and "Feedbk" is the feedback DAC's output voltage.

A. An even-simpler delta–sigma ADC

In §13.9.1 we presented "the simplest delta–sigma" in the form of the integrating charge-dispensing suntan monitor. With the sophistication of the preceding sections, we can now recast that example as a first-order delta–sigma modulator with simple accumulation of the 1-bit bitstream. If that's what you need, you can implement it even more simply, with a microcontroller.

These devices (Chapter 15) are the flexible putty of electronics, and one of them can easily replace the logic portion of the previous example, as in Figure 13.63. Here the pin labeled Q is an output-port bit that swings cleanly rail-to-rail, and A_{in} is an input that sets the threshold, analogous to U_2 in Figure 13.49. It could be an internal comparator, provided in many microcontrollers (see, e.g., Figure 13.64); or it could be a low-resolution ADC internal to the microcontroller (also quite common); or, at the crudest level, it could be a logic input, whose (inaccurate) threshold voltage replaces an honest comparator.

A microcontroller implementation like this lets you do

more: at the simplest level you could write software that implements the counter and other logic needed to complete the suntan monitor (including features such as a display of progress of baking, time remaining, clock time and date, your next appointment, and so on...). At a more sophisticated level you could implement a digital filter to create a sequence of converted values, in the manner of a fully integrated delta–sigma ADC. You could, but you probably shouldn't, because many highly skilled designers are turning out excellent delta–sigma ADCs, a sampling of which we'll visit soon enough. And there is a danger in hijacking a microcontroller to control the integrator charge cycles, namely that the accuracy of sigma-delta integration depends upon a stable switch ON-time; and the stability of a converter depends also upon a stable sampling clock frequency. You must ensure that the controller gives you

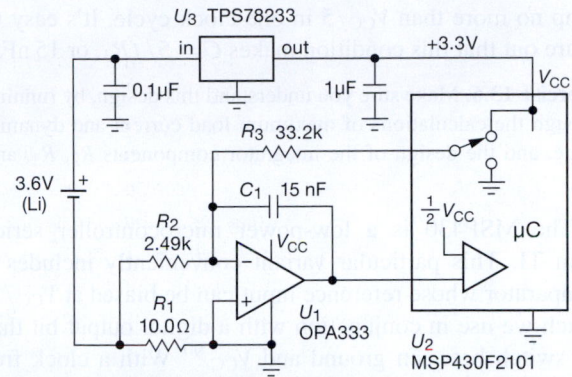

Figure 13.64. Keeping track of battery capacity with a discrete delta–sigma charge integrator.

control over that timing; and, even so, you may find that the firmware coding becomes inconveniently constrained. These examples are intended merely to illustrate the simplest kind of delta–sigma circuit. If what you want is a high-quality converter, you should choose one from the many hundreds that are available, that are easy to use, and that are remarkably inexpensive; several of these are illustrated below.

B. Coulomb counter

Here's an example of a microcontroller-assisted converter with low quiescent current and a wide dynamic range. In all candor, an admission: we designed the converter first, then we looked for a plausible application. Figure 13.64 shows what we came up with – a low-side sensing battery "gas gauge" to keep track of the state of discharge of a lithium cell that powers a portable instrument.

Here's how it works: we used a small ($10\,\Omega$) current-sensing resistor, to limit the voltage burden to 0.25 V at the maximum anticipated load current of 25 mA (caused, for example, by switched loads such as a $350\,\Omega$ strain gauge). Then we chose a single-supply chopper op-amp (maximum offset voltage of $10\,\mu V$) to minimize the error at low currents; here the offset voltage corresponds to a sensed current error of $1\,\mu A$ (max), or a dynamic range of 25,000:1. The voltage developed across the sense resistor drives the integrator through R_2, with a full-scale input current of $100\,\mu A$. (You can think of $R_1 R_2$ as a "current divider," if you like that concept.)

Next we chose R_3 to set the converter's full-scale input current ($I_{FS} = V_{CC}/R_3$), i.e., to source the full-scale current into the summing junction when the switch is on. Finally, taking a clock frequency of 10 kHz, we chose the value of the integrating capacitor C_1 such that the integrator would

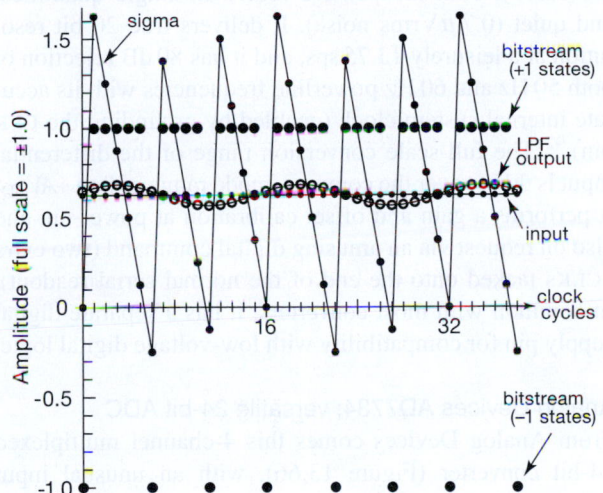

Figure 13.62. Modulator signals, along with time-aligned filtered-output waveform, for the idle-tone example of Figure 13.61. The tone is down ~25 dB, relative to full scale, and would fall at midband (half of Nyquist frequency) for 4× oversampling.

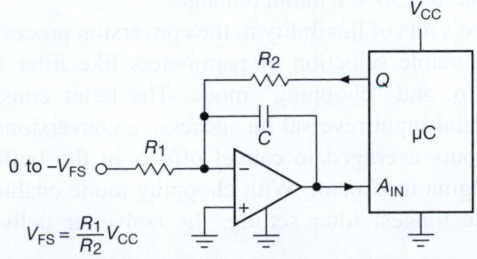

Figure 13.63. A microcontroller can replace the logic portion of a discrete delta–sigma ADC.

ramp no more than $V_{CC}/5$ in one clock cycle. It's easy to figure out that this condition makes $C = 5/fR_3$, or 15 nF.

Exercise 13.6. Make sure you understand this design, by running through the calculations of maximum load current and dynamic range, and the design of the integrator components R_2, R_3, and C_1.

The MSP430 is a low-power microcontroller series from TI. This particular variant conveniently includes a comparator whose reference input can be biased at $V_{CC}/2$, which we use in conjunction with a digital output bit that we switch between ground and V_{CC}.[90] With a clock frequency of 1 MHz the microcontroller draws 0.3 mA when active, and 25 µA when running in "low-power mode 2"; think of the latter as a "sleep" mode, during which an internal timer continues to run, and from which the processor can wake up to full alertness in one clock cycle. This is important, because it's common to put portable equipment into a low-power mode to conserve battery charge.

Note that this "Coulomb meter" keeps track of *all* loads, including the regulator's quiescent current, the microcontroller's operating current, additional loads powered from the regulated V_{CC} line, and even the current needed to run the delta–sigma's integrator op-amp itself. Exclusive of additional loads, the power budget is dominated by the processor, followed by the chopper op-amp (17 µA, typ) and the regulator (1.3 µA, max); that equates to a battery life of several months (with processor active) for a typical 1 Ah rechargeable Li-ion cell. The peripheral loads would likely reduce battery life considerably, and would typically be power switched by the processor.[91]

Note that the zero error of 1 µA (from the op-amp's worst-case offset) is completely negligible in comparison with the system current, even in sleep mode. The 25,000:1 dynamic range is extravagant for this application.

C. Three fully integrated delta–sigma ADCs

We conclude these delta–sigma examples with three elegant integrated ADCs from three different manufacturers. For more parts to consider, see Table 13.9.

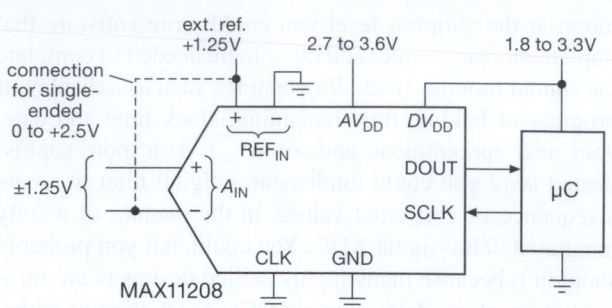

Figure 13.65. Maxim's MAX11208B 20-bit low-noise ADC with internal clock. The differential input can be reconfigured for single-ended 0 to $+2V_{ref}$ by connecting the A_{IN} input to the reference, as shown.

Maxim MAX11208B: inexpensive 20-bit ADC

From Maxim comes this compact (10-pin) converter (Figure 13.65), both inexpensive ($3.75 in single quantities) and quiet (0.7 µVrms noise). It delivers true 20-bit resolution at a leisurely 13.75 sps, and it has 80 dB rejection of both 50 Hz and 60 Hz powerline frequencies with its accurate internal system clock (enabled by grounding the CLK pin).[92] The full-scale conversion range of the differential input is $\pm V_{ref}$, over the common-mode range of 0 to $+AV_{DD}$. It performs a gain and offset calibration at power-on, and also on request via an amusing digital command (two extra SCLKs tacked onto the end of the normal serial readout). In common with most converters, it has a separate digital supply pin for compatibility with low-voltage digital logic.

Analog Devices AD7734: versatile 24-bit ADC

From Analog Devices comes this 4-channel multiplexed 24-bit converter (Figure 13.66), with an unusual input structure of trimmed resistors that provides a full ± 10 V input range (while operating from a single +5 V supply), and in fact with an over-range bit that extends the conversion range to 11.6 V, thus permitting accurate gain calibration at full-scale input. It is extremely tolerant of input overdrive: to ± 16.5 V without affecting the accuracy of the other inputs, and to ± 50 V without damage.

There's lots of flexibility in the conversion process, with programmable selection of parameters like filter length, auto-zero, and "chopping" mode. The latter consists of differential-input reversal on successive conversions, with the outputs averaged to cancel offsets in the buffer and delta–sigma modulator. With chopping mode enabled and with the longest filter setting, the converter delivers an

---

[90] This processor provides several "capture control registers" by which we can ensure that those pulses are of accurate and stable duration, a necessary condition for delta–sigma integration.

[91] With a much smaller battery, one could put the processor into sleep mode as well, waking up only often enough to ensure that the system current during sleep does not cause the integrator ramp to saturate between naps. For this system the ~ 45 µA total sleeping quiescent current would require measurements every 80 ms to limit the unobserved ramp amplitude to 1 V. This microcontroller can do that easily, because it has the nice property of waking up in a single 1 µs clock cycle.

[92] The $-A$ suffix version runs at 120 sps, with poor rejection of powerline frequencies, but deep rejection at 120 Hz and its harmonics.

Table 13.9 Selected Delta-sigma A-to-D Converters[a]

Part #	ADCs per pkg	Bits	Conv Rate max (ksps)	Channels per ADC	Single-ended	Interface[q]	Pwr typ mW	PGA Gains	Offset[b] typ (μV)	Offset[b] drift,typ (μV/°C)	Gain typ (ppm)	Gain drift, typ (ppm/°C)	Pwr Supply min (V)	Pwr Supply max (V)	Internal Ref	External Ref	Sequencer	Int osc	DIP	SOIC	TTSOP	SOT-23	Smaller	Cost qty25 ($US)	Comments
ADS1158	1	16	125	16	f	S	42	-	25	1[t]	10000	2	2.7	5.25[o1]	-	•	•	g,h	-	-	-	-	48	11.77	A
ADS1100	1	16	0.128	1	•	I	0.2	1-8	2500	1.5	1000	2	2.7	5.5	-	-	•		-	-	-	6	-	4.80	B
ADS1115	1	16	0.86	4	•	I	0.5[e]	1-16	50	0.4	100	5	2	5.5	-	•	-	•	-	-	10	-	10	5.53	C
AD73360	6	16	4	1	•	S	86[e]	1-80	20000	-	10000	-	2.7	5.5	•	-	-	-	-	-	28	44	-	9.24	-
MAX11208B	1	20	0.12	1	?	S	0.8	-	3	0.05	3[c]	0.05	2.7	3.6	-	•	-	•	-	-	-	10	-	2.18	-
CS5513	1	20	0.33	1	?	S	1.9	-	120	0.06	-	1	0	6[o3]	-	•	•	•	-	8	-	-	-	4.96	D
MCP3551	1	22	0.014	1	?	S	0.3[e]	-	3	0.1	2	0.028	2.7	5.5	-	•	•	•	-	8	8	-	-	3.31	E
LTC2412	1	24	0.008	2	?	S	0.2	-	0.5	0.01	2.5	0.03	2.7	5.5	-	•	•	•,h	-	16	-	-	-	6.34	F
AD7730	1	24	0.2	2	?	S	65[m]	four	2[c]	0.5	100[c]	2	4.75	5.25	-	•	-	g,h	24	24	24	-	-	16.10	G
AD7794	1	24	0.47	6	•	S	1[z]	INA	1[c,k]	0.01[k]	2[c]	1	2.7	5.25	•	•	-	•,h	-	-	24	-	-	10.80	H
MAX11210	1	24	0.48	1	•	S	0.25	1-16	30[m,c]	0.05	20[m,c]	0.05	2.7	3.6	•	•	-	•,h	-	-	16	-	-	3.32	J
ADS1246	1	24	2	1	?	S	1.4[e]	1-128	15[m,c]	0.05[b]	50	1	2.7	5.25[o1]	•	•	-	•,h	-	-	16	-	-	8.38	K
CS5532-BS	1	24	3.84	2	?	S	70	1-64	6	0.64[b]	16	2	0	5.5[o2]	-	•	•	g,h	-	-	20	-	-	12.80	L
AD7190	1	24	4.8	4	p	S	1[b]	1-128	0.5[k]	0.005	50[m]	1	4.75	5.25	•	•	•	•,g,h	-	24	-	-	-	10.89	M
ADS1259	1	24	14	1	d	S	13	-	40	0.05	500[y]	0.5	4.75	5.25[o1]	•	•	-	•,g,h	-	-	20	-	-	12.15	N
AD7734	1	24	15	4	f	S	85	-	13000[m,x]	2.5	4500[m,x]	3.2[m]	4.75	5.25	•	•	-	g,h	-	-	28	-	-	15.84	O
ADS1210	1	24	19.5	1	?	S	26	1-16	0.15[c]	1	0.06[c]	0.15	4.75	5.25	•	•	-	g,h	18	18	-	-	-	22.84	P
ADS1258	1	24	23.7	16	w	S	42	-	0.2	0.02	50	2	4.75	5.25[o1]	•	•	•	g,h	-	-	-	-	48	18.80	Q
ADS1298	8	24	32	1	d	S	10	1-12	500	2	2000	5	2.7	5.25	•	•	-	•	-	-	64	-	64	40.40	R
ADS1278	8	24	144	1	d	S	530	-	250	0.8	1000	1.3	v	v	-	•	-	-	-	-	-	-	64	39.57	S
AD7704	1	24	312	1	-	S	300	-	240	0.6	180	0.5	u	u	-	•	-	-	-	-	28	-	-	16.94	T
ADS1672	1	24	625	1	-	LC	350	-	2000[m]	2	10000	2	4.75	5.25	•	•	-	-	-	-	-	-	64	22.06	-
AD7760	1	24	2500	1	-	P	960	-	400	0.3	160	2	r	r	•	•	-	-	-	-	-	-	64	42.49	U
ADS1675	1	24	4000	1	-	LC	575	-	5000[m]	4	10000[m]	4	4.75	5.25	•	•	-	-	-	-	-	-	64	35.88	V
ADS1281	1	32	4	1	d	S	12	-	1[c]	0.06[t]	2[c]	0.4	4.75	5.25[o1]	•	•	-	-	-	-	24	-	-	49.68	W

Notes: (a) sorted by precision and speed; all except AD7734 have differential inputs; all have no missing codes except as noted. (b) at minimum PGA gain. (c) after calibration. (d) unused input biased at mid-supply. (e) at V_S=3.3V. (f) pseudo-diff'l. (g) external bare xtal. (h) opt ext osc input. (k) in chopper mode. (m) min or max. (n) 0.1 to V_-0.1 with buffers enabled. (o1) has neg supply, 0 to −2.6V, with 5.25V max total supply. (o2) has neg supply, 0 to −3.5V, with 5.5V max total supply. (o3) has neg supply, 0 to −6V, with 6V max total supply. (p) 2 differential, 4 pseudo-diff'l with single common return. (q) S=SPI, P=parallel, LC=LVDS or CMOS serial. (r) five supplies: +5 ±5%, +2.5 ±5%, +3.15 to +5.25 (2x), and +1.67 to +2.7. (s) at max decimation. (u) four supplies: +5 ±5% (3x), +2.5 ±5%. (v) 4.75-5.25V and 1.65-2.2V. (w) 8 diff'l, 16 single-ended (pseudo-diff'l). (x) before cal. (y) 2ppm after cal. (z) 0.4mW unbuffered.

Comments: **A:** 16 single-ended (pseudo-diff'l) or 8 diff'l; MUX output and ADC input pins. **B:** internal clk; self-cal. **C:** 4 single-ended (not pseudo-diff'l) or 2 diff'l; auto-shutdown in single-shot mode. **D:** CS5512 has ext osc; CS5510/11 = 16-bit. **E:** low-noise, 2.5μVrms; auto-shutdown in single-conv mode. **F:** no latency; 0.8μVrms noise; a favorite. **G:** bridge subsystem with offset DAC and ac excitation output. **H:** low-noise; opt input buffer for hi-Z; INA PGA; includes 2 curr srcs; chopper mode; auto-shutdown in single-conv mode; int ref 4ppm/°C typ; AD7793 has fewer channels; a favorite. **J:** opt input buffer for hi-Z. **K:** low noise; 20-pin ADS1247 has 2 channels, 28-pin ADS1248 has 4 channels, both have int 10ppm/°C ref. **L:** CS5534 4-ch diff'l; industrial workhorse; 6nV/√Hz. **M:** low-noise. **N:** off-scale detectors. **O:** has chopper mode; AD7732 has two diff inputs. **P:** ADS1211 has 4 ch diff'l MUX; sample-rate boost to 312kSps; no missing codes to 22 bits. **Q:** low noise; pins for MUX diff'l output and ADC input. **R:** biopotential measurements (EKG, EEG, etc); ADS1294=quad. **S:** ADS1274 = quad. **T:** diff'l input buffer; AD7765 to 156kSps. **U:** diff'l input buffer. **V:** pin-programmed (no registers); LVDS or CMOS serial output. **W:** no missing codes to 31 bits.

effective 21-bit conversion at 372 sps for input signals of ±10 V full scale; this figure comes from a consideration of the converter's noise level, which is 9.6 μVrms under these conditions (21 bits is the ratio of a ±10 V span to ~ 10 μV). The datasheet specifies resolution alternatively as "peak-to-peak resolution in bits," which for these same conditions is listed as 18.1 bits. This turns out to be the more conservative specification, which is explained as "representing [bit] values for which there will be no code flicker within a 6-sigma limit." In other words, you can rely on any single conversion to be accurate to 18 bits, taking into account the fact that the occasional peak excursions of a signal of given rms noise voltage are substantially larger than V_{rms}.[93] If you know the rms noise voltage V_{rms},

[93] The datasheet for Cirrus' CS5532 explains it this way: " 'Noise-free resolution' is not the same as 'effective resolution.' Effective resolution is based on the RMS noise value, while noise-free resolution is based on a peak-to-peak noise value specified as 6.6 times the RMS noise value."

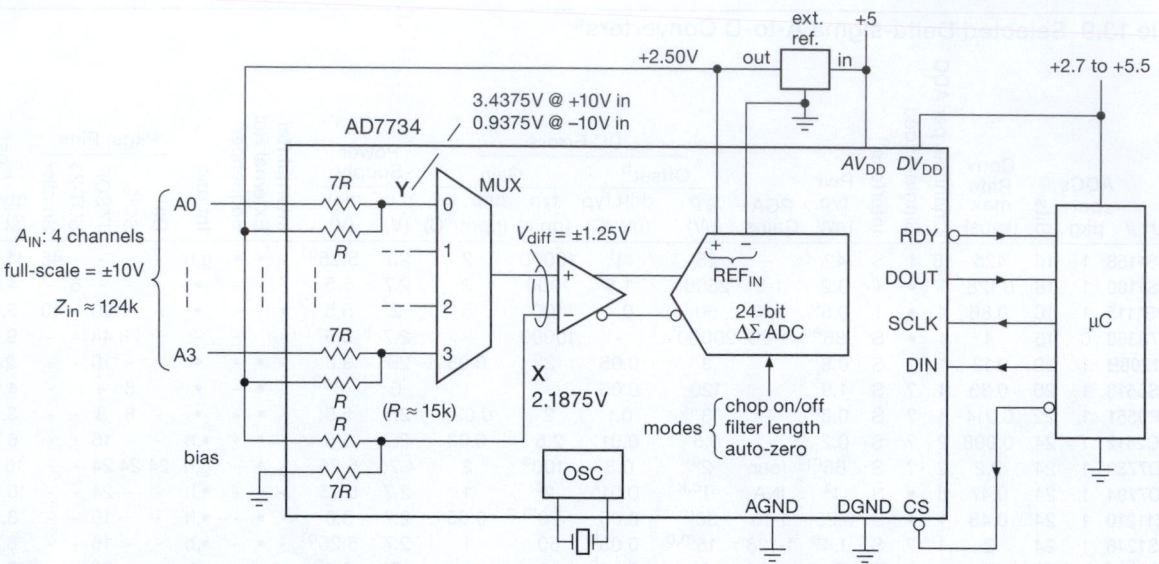

Figure 13.66. Analog Devices AD7734 multiplexed 4-channel 24-bit converter (21-bit ENOB) with wide input-voltage range and robust overdrive tolerance, courtesy of Analog Devices, Inc.

you can calculate the noise-limited effective resolution as $\mathrm{ENOB} = \log_2\left(V_{\mathrm{span}}/V_{\mathrm{rms}}\right) = 1.44\log_e\left(V_{\mathrm{span}}/V_{\mathrm{rms}}\right)$; from that you get the peak-to-peak resolution by subtracting 2.7 bits.

This converter can be operated at conversion rates to 12 ksps, with correspondingly degraded resolution. It has maximum offset and gain drifts of $\pm 2.5\,\mu\mathrm{V}/^{\circ}\mathrm{C}$ and $\pm 3.2\,\mathrm{ppm}/^{\circ}\mathrm{C}$, respectively. It comes in a 28-pin package and is currently priced at about \$15 in single quantities.

Cirrus CS5532 high-performance "industrial" ADC

From Cirrus Logic (formerly Crystal Semiconductor) comes the long-lived (circa 1999) CS5532-BS 24-bit delta–sigma converter (see Figure 13.67), with a chopper-stabilized PGA (gains of 2, 4, 8, 16, 32, and 64) and with particularly good noise, drift, and linearity characteristics: $e_{\mathrm{n}} = 6.4\,\mathrm{nV}/\sqrt{\mathrm{Hz}}$ (typ) at 0.1 Hz with $G = 64$,[94] $i_{\mathrm{n}} = 1\,\mathrm{pA}/\sqrt{\mathrm{Hz}}$ (typ), $\Delta V_{\mathrm{os}} = 15\,\mathrm{nV}/^{\circ}\mathrm{C}$ (typ) with $G = 64$, full-scale drift of 2 ppm/$^{\circ}$C (typ), and $\pm 0.0015\%$ (max) nonlinearity. It can convert at rates from 6.25 sps to 3.8 ksps. At the slowest rates it delivers noise-free resolution ranging from 20 bits (for $G = 64$) to 23 bits ($G \leq 8$); or, if you prefer, corresponding "effective resolutions" (ENOBs) of 23 and 24 bits, respectively.

With its low-noise, high-gain PGA and ability to run from split $\pm 3\,\mathrm{V}$ supplies,[95] this converter has what it

takes to process the low-level signals from a thermocouple ($\sim 40\,\mu\mathrm{V}/^{\circ}\mathrm{C}$) or a strain gauge (full scale $\Delta V \approx \pm 2\,\mathrm{mV}$ per volt of bridge excitation). With a PGA gain of 64, the full-scale span is $\pm 2.5\,\mathrm{V}/64$, or $\pm 40\,\mathrm{mV}$, and an LSB (at 20-bit resolution) corresponds to 80 nV which is $500\times$ less than the thermal-voltage change corresponding to a temperature change of 1°C. Likewise, at this gain a 20-bit LSB corresponds to 0.0008% of the strain gauge's full scale. At this gain, full-scale inputs from these sensors stay within the conversion range. Evidently there's no need for external front-end gain stages or the like. This converter costs about \$16 in unit quantities.

In our circuit we balanced the thermocouple signals about ground to minimize the effects of common-mode pickup in the lead wires, which are often unshielded. And for both sensors we added a simple *RC* filter (time constant 0.1 ms) to suppress spikes and protect the inputs. We chose the ADR441 because we needed a low-dropout voltage reference that would operate with 500 mV of headroom.

D. Pro-audio ADCs

Delta–sigma converters are adored by the professional audio community, owing to their combination of resolution,

---

[94] Analog Devices' AD7190 comes close, at 8.5 nV/$\sqrt{\mathrm{Hz}}$

[95] Other high-resolution delta–sigma converters that allow bipolar input

supplies and signals include (a) the 32-bit ADS1281, with ± 2.5 V analog supplies and input signals to those same limits (but sadly not to ± 3 V, with their total supply limit of 5.25 V), and (b) three 24-bit converters: the 16-channel ADS1258, the ADS1259 with its 2 ppm voltage reference, and the ADS1246 family with internal PGA.

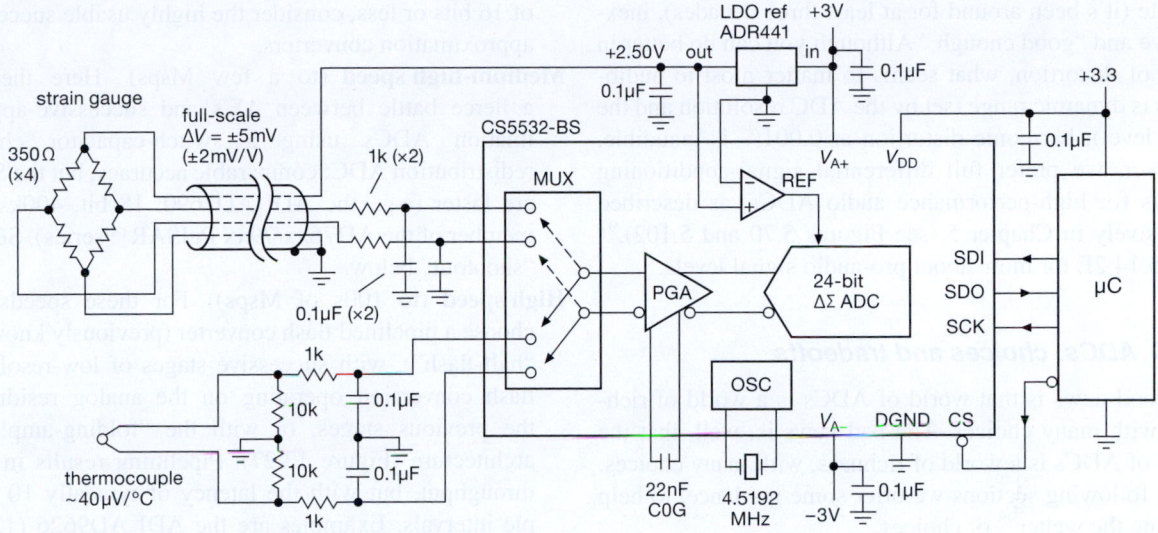

Figure 13.67. Cirrus' CS5532-BS precision low-noise 24-bit converter. The PGA permits gains of 1, 2, 4, ..., 64. This circuit omits the cold-junction compensation. The Maxim MAX31855 includes this important compensation for seven thermocouple types; see item 44 in §15.8.2 and accompanying figure.

Table 13.10 Selected Audio Delta-Sigma ADCs

	Channels	Bits	f_{samp} max (kHz)	SNR^k (dB^a)	$THD+N^k$ (dB^b)	Supply V_{cc} (V)	Supply P_{diss} (mW)	inputs, other
PCM1870A	2	16	50	90	-81	3	13	mic preamp
CS43432	4^c	24	96	105	-98	5	600	diff or SE; codec
PCM2906^h	2^d	16	48	89	-80	5	280	SE; codec, USB
AK5384	2	24	96	107^e	-94	5	275	diff
AK5388	4	24	192	120	-107	5	590	diff
AK5394A	2	24	192	123	-94^f	5	705	diff; popular
CS5381	2	24	192	120	-110	5	260	diff; preferred
AD1974	4	24	192	105	-96	3.3	430	diff, with PLL
PCM4204	4	24	216	117	-103	4	615	diff
PCM4222^g	2	24	216	123	-108	4	340	diff; configurable

Notes: (a) A-weighted filter. (b) at max sample rate. (c) codec: 4 in + 6 out. (d) codec: 2 in + 6 out. (e) at 48 kHz. (f) -110 dB at 96 kHz. (g) consider also the PCM4220. (h) stereo codec (ADC + DAC) with S/PDIF to USB in/out. (k) SNR and THD+N is with respect to full-scale; SNR is typically measured with a -60 dB input signal, processed with an A-weighted filter; THD is typically measured at 1 kHz, with a -1 dB signal.

inherent anti-aliasing, noise shaping, and monotonicity. (See Table 13.10.) If you open up pretty much anyone's high-quality audio gear, you'll probably find a circuit based around a 24-bit, 192 ksps 128× oversampling delta–sigma ADC; and chances are it will be a part from Cirrus (e.g., the CS5381) or AKM (e.g., the AK5394A). These parts seem to have "long legs" – they have been around for many

years and represent good value in terms of price and performance.

Audio ADCs universally use delta–sigma technology, but they differ greatly from their industrial ADC counterparts in the delta–sigma table. They generally have poor gain accuracy (5% to 10%) and dc offset (~25 mV), in part because these don't matter in the audio field. On the other hand, they do offer 0.1 dB or 1% stereo-channel gain *matching*. They're usually wired with ac-coupling, and they also have an internal digital highpass filter (typically ~1 Hz).[96] They are meant to operate at specific audio data rates, and they employ specialized audio PCM data-output interfaces (I²S, TDM, etc.). Compared with industrial ADCs they have high latencies (data-output delays) of 12 to 63 sample intervals, even though they may advertise "low latency" (by which they mean small compared with, say, one millisecond of time delay). They have unique audio specifications, like A-weighted SNR, and spectral-analysis-derived THD+N distortion specs.

Figure 13.68 shows a straightforward signal-conditioning circuit, adapted from AKM's evaluation board, not unlike the innards of many commercial audio digitizers. The 5534 op-amp seems to be the perennial

---

[96] You'll discover this if you bypass the coupling capacitors. However, many of these converters offer a dc mode, and some (like the CS5381 and AK5394A) provide a logic-triggered dc-auto-zero function, which is convenient for very slow or dc applications.

favorite (it's been around for at least three decades), inexpensive and "good enough." Although you can do better in terms of distortion, what seems to matter most to audiophiles is dynamic range (set by the ADC resolution and the noise level); harmonic distortion at 0.001% is inaudible. However, we prefer full differential signal-conditioning circuits for high-performance audio ADCs, as described expansively in Chapter 5 (see Figures 5.70 and 5.102).[97] See §5.14.2E for more about pro-audio signal levels.

13.10 ADCs: choices and tradeoffs

The good news is that world of ADCs is a world of richness, with many choices. The bad news is, well, that the world of ADCs is a world of richness, with many choices. In the following sections we offer some guidance, to help navigate the welter[98] of choices.

13.10.1 Delta–sigma and the competition

A. Analog-to-digital converters

Delta–sigma converters are one of several ADC conversion technologies, which (as we've seen) include also (a) dual-slope and quad-slope integrating converters, (b) successive-approximation converters, (c) flash converters, and (d) pipelined flash converters.

Low speed For "voltmeter-speed" conversion (e.g., 10/sec), multislope integrating converters have been the perennial favorite, but their dominance is being challenged by excellent $\Delta\Sigma$'s from LTC (e.g., the LTC2412: 24 bit) and ADI (e.g., the AD7732: 24 bit, ± 10 V range), among others.

Medium speed (to ~100s of ksps) Delta–sigma converter ICs dominate at resolutions above 16 bits, with nice products from companies like Cirrus and AKM (e.g., the AK5384: 24 bit, 96 ksps, 4 channels, or the converters in Figure 13.68). There are many good delta–sigma audio ADCs, but their dc specs tend to be poor (several percent) or nonexistent. For resolutions

of 16 bits or less, consider the highly usable successive approximation converters.

Medium-high speed (to a few Msps) Here there is a fierce battle between $\Delta\Sigma$'s and successive-approximation ADCs using a switch-capacitor charge-redistribution ADC: comparable accuracy, but the SARs are faster (e.g., the ADI AD7690: 18 bit, 400ksps; a member of the AD76xx/79xx PulSAR™series). See the "shootout" below.

High speed (to 100s of Msps) For these speeds you choose a pipelined flash converter (previously known as "half-flash"), with successive stages of low-resolution flash conversion operating on the analog residue of the previous stages, or with the "folding-amplifier" architecture (Figure 13.27). Pipelining results in high throughput, but with the latency of typically 10 sample intervals. Examples are the ADI AD9626 (12 bit, 250 Msps) and the TI ADS6149 (14 bit, 250 Msps).

Breakneck speed (>250 Msps) Variants on the basic flash (such as folding/interpolating) rule the roost here, but the tradeoff is modest resolution (6–10 bits). National has some nice ones, for example the ADC08D1520 (8 bits, 3000 Msps), ADC10D1500 (ditto, 10 bits), and ADC12D1800 (12 bits, 3600 Msps). These kinds of converters are used widely in 'scope front-ends,[99] and in digital radio. At the extreme (and we're not sure how they do it), Fujitsu is featuring a 56,000 Mbps (!) 8-bit converter.[100]

B. Digital-to-analog converters

Competing DAC technologies are (a) R–$2R$ ladder, (b) linear resistor ladder with switch array, and (c) current-steering switch array.

Highest linearity Delta–sigma DACs are best, with accuracy and linearity to 20 bits at audio speeds, and sometimes with excellent dc specs as well (e.g., the millisecond-speed 20-bit TI DAC1220); however, watch out for broadband and clock noise (the DAC1220 has ~ 1000 nV/$\sqrt{Hz}$ at 1 kHz compared with ~ 10 nV/$\sqrt{Hz}$ for resistor ladder converters).

Medium speed, high accuracy Many excellent R–$2R$ and linear ladder DACs compete, for example:

• TI DAC8552 (dual 16 bit, serial in, voltage out, ext

---

[97] Interestingly, both AKM and Cirrus use this simple scheme, and inexpensive '5534 amplifiers, in their ADC evaluation kits. However, the Cirrus "reference design" kit substitutes the lower-noise LT1128 opamp. By contrast, TI uses a true differential amplifier (the OPA1632) in the evaluation kit for their comparable PCM4222 audio delta–sigma converter.

[98] Merriam–Webster: "A large number of items in no order; a confused mass. *Synonyms:* confusion, jumble, tangle, mess, hodgepodge, mishmash, mass; *informal* rat's nest.

[99] Currently achieve digital 'scopes achieve analog bandwidths of 32 GHz with sampling rates of 80 Gsps (e.g., Agilent 90000X series), figures sure to increase with time.

[100] Evidently they're so excited about it that they have neglected, thus far, to assign a part number.

Table 13.11 Selected Audio D-to-A Converters[a]

Part #	Bits	# Channels	f_{samp} max (ksps)	THD+N typ (dB)	SNR typ (dB)	Technology[c]	Output[f]	Volume Control	Pwr Amp?	Ripple max (±dB)	Stop[n] min (dB)	Voltage analog (V)	digital (V)	Power[o] typ (mW)	Control Interface[b]	Pkg	Price qty 25 ($US)	Comments
AK4386	24	2	96	84[d]	100	mbDS	Vse	-	-	0.01	64	2.2-3.6	-	20	H	TSSOP-16	1.00	A,J
CS4334/5/8/9	24	2	96	88[d]	88[d]	DS	Vse	-	-	0.2	50	5	-	75	-	SOIC-8	1.63	B,J
PCM1753	24	2	192	94	106	mbDS	Vse	-	-	0.5	50	5	-	150	S	SSOP-16	1.65	C
TLV320DAC3100	24	2	192	82[e]	95	mbDS	SP/HP	•	p	-	-	3.3	1.8	-	I	QFN-32	3.03	D
PCM1789	24	2	192	94	113	C-seg	Vdiff	-	-	0.0018	75	5	3.3	168	H,I,S	TSSOP-24	4.38	E
LM49450	24	2	192	64g	88	DS	SP/HP	•	q	-	-	2.7-5.5	1.8-4.5	50[h]	I	WQFN-32	5.15[k]	F
AK4358	24	8	192	92[d]	112	mbDS	Vdiff	•	-	0.02	54	5	5	560	I,S	LQFP-48	5.22	G,J,N
PCM1690	24	8	192	94	113	DS	Vdiff	•	-	0.0018	75	5	3.3	620	H,I,S	HTSSOP-48	5.23	G
PCM4104	24	4	216	100[d]	118	DS	Vdiff	-	-	0.002	75	5	3.3	236	H,S	PQFP-48	7.74	H
CS4398	24	2	192	107	120	mbDS	Vdiff	•	-	0.01	102	5	1.8-5	192	H,I	TSSOP-28	8.86	N
PCM1794A	24	2	192	102[d]	127	AS	Idiff	-	-	0.00001	130	5	3.3	335	H	SSOP-28	10.83	L
ADAU1966	24	16	192	98	118	DS	Vdiff	•	-	0.01	68	5	2.5	521[m]	I,S	LQFP-80	11.71	J,M
AD1955	24	2	192	110	120	DS	Idiff	•	-	0.0002	110	5	5	210	S	SSOP-28	12.35	J,N
AK4399	32	2	192	102[d]	123	DS	Vdiff	•	-	0.005	95	5	5	530	S	LQFP-44	15.00	P
PCM1704K	24	1	96	102[d]	120	R-2R	Ise	-	-	-	-	±5	-	175[d]	-	SOIC-20	68.20	Q

Notes: (a) in order of increasing price. (b) H="hardware," I.e., pin-programmable; I=I²C; S=SPI. (c) AS=TI "advanced segment"; C-seg=current-segment; DS=delta-sigma; mbDS=multibit delta-sigma; R-2R=ladder. (d) at 96ksps. (e) 48ksps audio. (f) Idiff=differential current; Ise=singled-ended current; SP/HP=speaker or headphone; Vdiff=differential voltage; Vse=single-ended voltage. (g) 0.6W into 8Ω, with V_S=5V. (h) analog supply only; class-D amp has 80% effy. (k) qty 100. (m) at 48kHz. (n) stopband attenuation, in sharp rolloff mode, at f/f_S=0.546. (o) at 192ksps. (p) 2.5W mono. (q) 1.9W stereo, 1% THD into 4Ω with V_S=5V. (t) typical.

Comments: A: low power. **B:** 8-pin, "entry level"; choose p/n for data format. **C:** PCM1754=H/W interface. **D:** stereo headphone drivers, and class-D mono spkr amp, 2.7-5.5V. **E:** high performance consumer audio/video. **F:** portable consumer audio; 100dB SNR at headphone output. **G:** octal, multichannel consumer audio/video. **H:** quad, professional and high-end audio, high performance. **J:** low jitter. **L:** premium; DSD1794A=IIC/SPI. **M:** 16 channels; automotive, etc. **N:** supports SACD output. **P:** premium. **Q:** legendary, the best of the legacy R-2R audio DACs; needs external filter.

ref, very low glitch, 10 μs settle; DAC8560/4/5 similar, with int ref)

- ADI AD5544 or TI DAC8814 (quad 16 bit MDAC, serial in, current out, 0.5–2 μs settle with external I-to-V op-amp)
- LTC1668 (16 bit, parallel in, differential current out, 20 ns settle into 50 Ω as "voltage out")
- TI DAC9881 (18 bit serial in, rail-to-rail voltage out, ext ref, low-noise, 5 μs settle)

Highest speed Here you can't beat current-steering converters, for example the TI DAC5681/2 (16 bit, 1 Gsps) or the ADI AD9739 (14 bit, 2.5 Gsps).

C. Interlude: shootout at the ADC corral

To illustrate the important performance differences between delta–sigma and successive-approximation ADCs, we invited two able and well-matched contenders from the same training camp (Analog Devices) to slug it out. They came forward with their respective (and respectable) specifications, which look like this:

	SAR AD7641	ΔΣ AD7760	units
Introduced	2006	2006	year CE
Price	$47	$53	US dollars
Conv rate	2.0	2.5	Msps
Samp freq	2	40	MHz
Alias above	1	20	MHz
Resolution	18	24	bits
Zero error	60	200	ppm max
" tempco	0.5	0.1	ppm/°C typ
Gain error	0.25%	0.016%	max/typ
" tempco	1	2	ppm/°C typ
SNR	93	100	dB typ
THD	−101	−103	dB typ
INL	±7.6	±7.6	ppm typ
Data delay	0.5	12	μs
Reference	int	ext	
Supplies	1	3	
Power	75	960	mW

The delta–sigma contender put in a few good jabs, with its superior resolution and the ease with which the user could

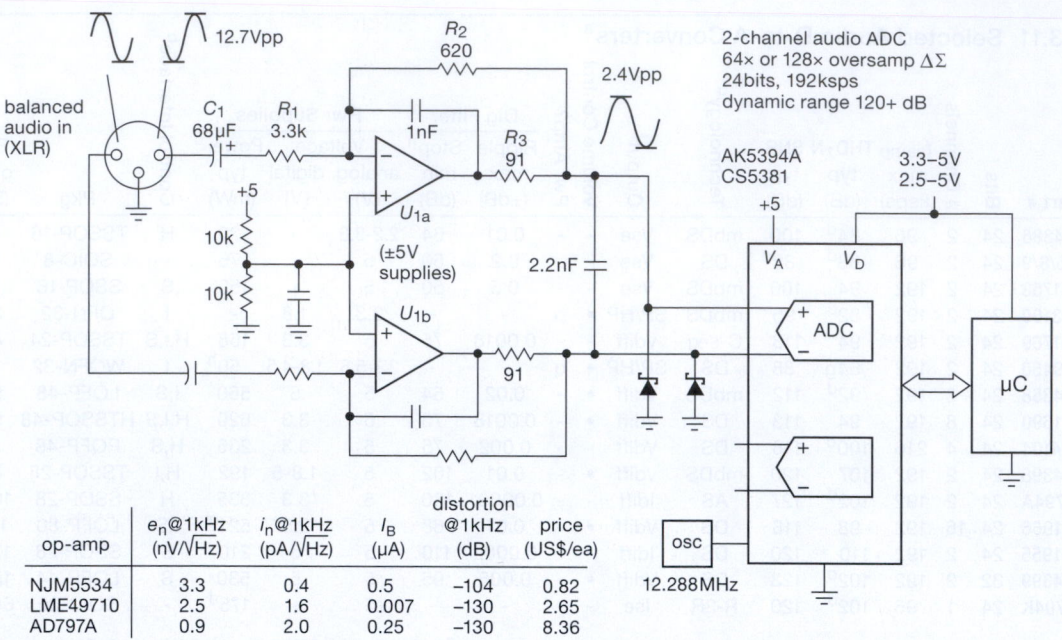

op-amp	e_n@1kHz (nV/√Hz)	i_n@1kHz (pA/√Hz)	I_B (µA)	distortion @1kHz (dB)	price (US$/ea)
NJM5534	3.3	0.4	0.5	−104	0.82
LME49710	2.5	1.6	0.007	−130	2.65
AD797A	0.9	2.0	0.25	−130	8.36

Figure 13.68. The AKM and Cirrus delta–sigma ADCs are ubiquitous in professional audio digitizers. They are often implemented with a simple analog frontend like this, though a better approach would exploit a true differential amplifier (§5.17). Reproduced with permission of Asahi Kasei Microdevices, Tokyo, Japan.

design an input bandwidth-limiting (anti-aliasing) lowpass filter (owing to its 8× oversampling ratio). The SAR responded that the number of bits isn't a big deal, it's really *linearity* (for which both are equal) that counts. And, by the way, the SAR produces output bits 25 times faster, with its superior low latency. The sigma–delta counterpunched with its claim to superior SNR, to which the SAR parried with disapproval at the ΔΣ's need for at least two power supplies and an external reference, and its profligate waste of 13× as much power. The sigma–delta, though somewhat chastened, bounced back with a claim of 15-fold smaller gain error, to which the SAR replied that the delta–sigma was cheating because it relied on a gain-correction register to do "smart" calibration. This set up the delta–sigma to deliver the ultimate insult, namely that the SAR wasn't even smart enough to cheat. Both contenders claimed victory (as they staggered back to their respective corners), but onlookers judged it a close match, with good punches on each side.

13.10.2 Sampling versus averaging ADCs: noise

Delta–sigma converters are inherently *integrating*; that is, a measurement takes account of the varying signal throughout the conversion time; you can think of this as simple *averaging*. With a SAR converter, by contrast, the instantaneous voltage of the input signal is captured in a track-and-hold (during the so-called *aperture* time) when the converter is triggered. This distinction has some important consequences, among them the ability of SAR converters to operate at exceptionally low average power consumption when sampling at a leisurely rate (see next section).

Another consequence of importance is the effective bandwidth at which the input signal is sampled. A short aperture corresponds to a wide bandwidth, and vice versa. Intuition serves well, here: high frequencies are washed out during a long averaging interval, whereas a quick sample can record the signal's amplitude as it gyrates rapidly. Put another way, averaging a signal over some time interval τ acts like a lowpass filter, whose bandwidth is very roughly of order $1/\tau$; mathematically speaking, they are related by the Fourier transform.[101]

To make these statements quantitative, look at Figure 13.69, which illustrates the lowpass filter function of an averaging window of time duration T. Low frequencies pass through, but higher frequencies suffer from averaging;

---

[101] And particularly by the convolution theorem, where the sampling interval is represented by a unit-amplitude rectangular window in time. Its Fourier transform is the sinc function, $(\sin t)/t$, with a first zero at $f = 1/\tau$.

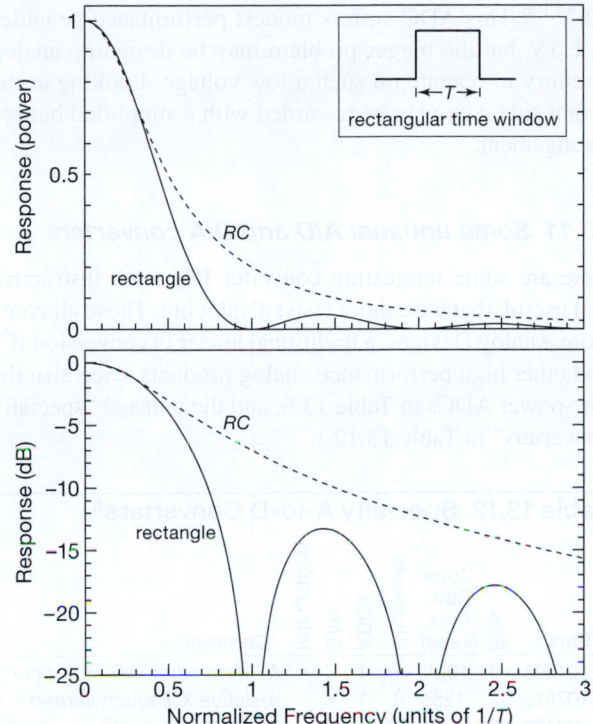

Figure 13.69. Spectrum of a rectangular pulse $\Pi(t)$ of length T. An input signal, gated (windowed, or multiplied) by $\Pi(t)$, is effectively lowpass filtered by the indicated power spectrum. An RC lowpass filter with 3 dB rolloff at $f=1/2T$ is shown for comparison.

a signal of frequency $f=1/T$ completes one full cycle during the window's time duration T, thus averages to zero. Additional zeros occur at multiples of $1/T$, where a signal of the corresponding frequency completes an integral number of cycles.[102]

So a short window admits wideband noise that may be present, degrading the accuracy of an intrinsically slow signal that would benefit from averaging. Keep this in mind when designing a conversion circuit, particularly one that operates on intermittent samples of a slowly varying signal (for example, a temperature sensor or strain gauge). It's OK to use a fast-sampling successive-approximation ADC, as long as you are willing to add a lowpass filter at the input. With an averaging converter (delta–sigma, dual-slope, or multislope) you get that benefit for free.

---

[102] It is this complete rejection of signal frequencies at $1/T$ and all its harmonics that lets you make benchtop DMM measurements without worrying about powerline pickup: the DMM's integrating interval is chosen to be an integral number of powerline cycles (PLCs), recall §13.8.3, and particularly Figure 13.42.

13.10.3 Micropower A/D converters

Small battery-powered devices often need some information about the real world, which they can get from a sensor signal and a low-power ADC. Frequently a simple 8- or 10-bit ADC included in the microcontroller IC will do, but for those who need better performance we offer a selection of micropower ADCs in Table 13.6 on page 916. The table lists both SAR and $\Delta\Sigma$ converters, with most of the entries appearing also in either Tables 13.5 or 13.9. Let's take a look at a comparison of these converter types for a typical micropower application.

SAR types are known for fast conversion speeds, but this comes at the expense of higher power consumption. SAR ADCs capture a sample of the signal when they are triggered, allowing you to turn the sensor off immediately; this saves power with something like a hungry strain-gauge bridge. Fast SAR converters draw more current, but they also finish faster and go to sleep. For example the AD7685 consumes 2.7 mW during continuous 16-bit conversions at 200 ksps (its maximum, when running on 3 V); but for a sensor application we can make measurements far less frequently, say at 100 samples per second where the average power dissipation is just 1.4 μW (2000× less). As noted in the table, for most SAR types the power dissipation is proportional to the sample rate, so there are plenty of other low-power candidates to be found in Table 13.5.

As remarked above, delta–sigma converters are integrating by nature, and they take longer to perform a conversion. Furthermore, a 16-bit conversion can take 16 times longer than a 12-bit conversion. But $\Delta\Sigma$ ADCs generally consume less operating power than comparable SARs. The delta–sigma MCP3425 dissipates 0.44 mW when operating continuously at its maximum 15 samples/second (for 16-bit conversions), six times less than the SAR above running at its maximum 200 ksps. At this stage you may conclude that the delta–sigma is the low-power winner. But the comparison is lopsided, because the sampling rates differ by a factor of more than 10,000. Note, by the way, that these power requirements are much larger than the promising figures you may find on the datasheet, where for example the MCP3425 claims the ability to operate at power levels as low as 1.8 μW average; but there's a gotcha, because that rosy figure applies to operation in 12-bit mode, and at just one sample per second.

Low-power ADC shootout. To compare these two converters fairly, then, let's assume we want to make ten measurements per second at 16-bit resolution, and we want to choose the converter that minimizes the power dissipation. At that rate the delta–sigma converter requires 290μW average power, to be compared with the SAR's required $0.14\,\mu\text{W}$. The SAR beats the pants off the $\Delta\Sigma$ – it's using just 1/2000 of the power! On this figure of merit alone, it's a slam-dunk.[103] But there's more to consider. The delta–sigma integrating conversion takes 66 ms, which yields a quieter measurement than a SAR ADC sampling the signal in a fraction of a microsecond, as described in §13.10.2.[104] By comparison, the AD7685 SAR mentioned above can perform 2000 measurements for the same total energy as one $\Delta\Sigma$ measurement – but you may need to take and average all of them to reduce the noise.

We always suggest a careful study of any candidate part's datasheet. When evaluating micropower ADCs, consider also whether they include an on-chip input amplifier, internal voltage reference, and conversion oscillator. If not, these functions may require additional external power. Some ADCs use the supply as the voltage reference, which is well suited for a ratiometric sensor like a thermistor or strain gauge. For other sensors you may have to operate the entire ADC from an external voltage reference. Some ADCs use the interface data shifting as the conversion clock, which may force the controller to waste time and power creating a slow data-clock rate. Note also that some external-clock converters require quite high frequencies, for example the AD7091R SAR ADC needs a 50 MHz clock to run at its 1 Msps maximum speed; requirements like this have a severe impact on power consumption.[105] When considering converters for applications where you are planning to power them intermittently, keep in mind also that some converters have a non-trivial startup delay time from sleep mode.

One last consideration is the supply voltage. Most of the parts in Table 13.6 require a moderately high minimum supply voltage, such as 2.7 V. But ADCs capable of operating at lower voltages can save considerable power. For example, the AD7466 consumes $620\,\mu\text{W}$ at 100 ksps when running with a 3.0 V supply, but it uses only $120\,\mu\text{W}$ at

1.6 V.[106] This ADC suffers modest performance penalties at 1.6 V, but the bigger problem may be designing analog circuitry to operate on such a low voltage. Looking at the bright side, you may be rewarded with a simplified battery arrangement.

13.11 Some unusual A/D and D/A converters

Here are some interesting converter ICs, both instructive and useful, that we cannot resist displaying. These all come from Analog Devices, a traditional leader in conversion ICs and other high performance analog products. (See also the low-power ADCs in Table 13.6, and the unusual "specialty converters" in Table 13.12.)

Table 13.12 Specialty A-to-D Converters[a]

Part #	Bits	Conv Rate max (ksps)	Channels	ADCs	SAR	Integrating	Comments
AMC1203	1	10M	1	1	•	-	$\Delta\Sigma$ modulator, AC motor curr
AD7873	12	125	6	1	•	-	resistive X,Y touch screen
AD7490	12	1000	16	1	•	-	flexible sequencer
AFE5401	12	25M	4	4	•	-	automotive-radar AFE[e]
AFE5804	12	40M	8	8	•	-	8-ch ultrasound, 0.9nV/√Hz
AD6620	12	67M	2	1	•	-	FIR filter + prog RAM, to FPGA
AD9869	12	80M	1	1	•	-	transceiver, 200Msps DAC
AD6655	14	150M	2	2	-	•	IF diversity rcvr, 32-bit NCO
LMP90080	16	0.21	8	1	•	-	sensor-AFE[e], many interfaces
ADE7753	16	14[c]	2	2	-	•	AC power mon, single phase
DDC316	16	100	16	16	-	•	16 current inputs, 3 to 12pC
AD7147	16	250	13	1	-	•	13 capacitance inputs, touch
AD7609	18	200[f]	8[f]	1	•	-	simultaneous-sampling, diff'l
DDC232	20	6	32	32	-	•	32 current inputs, 3 to 12pC
78M6631	22	2.5	6	1	-	•	3-phase power, w/ 8051 CPU
AD7746	24	0.09	2	1	-	•	precise capacitance, ±4pF FS
ISL26102	24	4	2	1	-	•	quiet, 7nV/√Hz, 2ppm linearity
LDC1000	24[b]	d	2	2	•	•	inductance, loss resistance
ADS1298	24	32	8	8	-	•	standard 12-lead ECG
TPA5050	24	192	2	2	-	•	audio, lip-sync delay to 120ms

Notes: (a) wherein we find inductance, capacitance, ultrasound, ECG, powerful sensor analog front ends, and RF communication, etc. (b) 24 bits for L(inductance), 16 bits for R_p (equiv parallel resistance). (c) 14kHz RMS bandwidth. (d) fastest response = 192 LC cycles, for example 10 ksps for 2MHz. (e) sensor analog front end. (f) simultaneous sampling, 200kHz all channels.

---

[103] And, of course, there's also the power saved if the external sensor is power-switched.

[104] Recall that bandwidth is related inversely to pulse width, as $\text{BW}\approx 1/\tau$, and that noise power grows with bandwidth ($P_n = e_n\cdot\text{BW}$ for white noise). So a slower measurement corresponds to reduced bandwidth; it averages out the high-frequency noise.

[105] Taking the AD7091R as an example, the SCLK driver consumes $P = CV^2f$; taking $C = 5\,\text{pF}$, that amounts to 2.25 mW for a 3 V logic swing at 50 MHz, which is significantly greater than the converter's own specified power dissipation of 1 mW. Happily, the 50 MHz clocking on the SCLK pin is needed only during the 12 or 13 shifts of output data, thus reducing the average SCLK power to 0.6 mW (i.e., a factor of 13/50); much better, but still a significant contributor to the total power.

---

[106] The 5.2× power ratio is more than the square of the 1.9× supply-voltage ratio, revealing that the operating current drops faster than linearly versus supply voltage. This should not be a surprise, considering the effect of class-A "shoot-through" current in CMOS logic, see Figure 10.101.

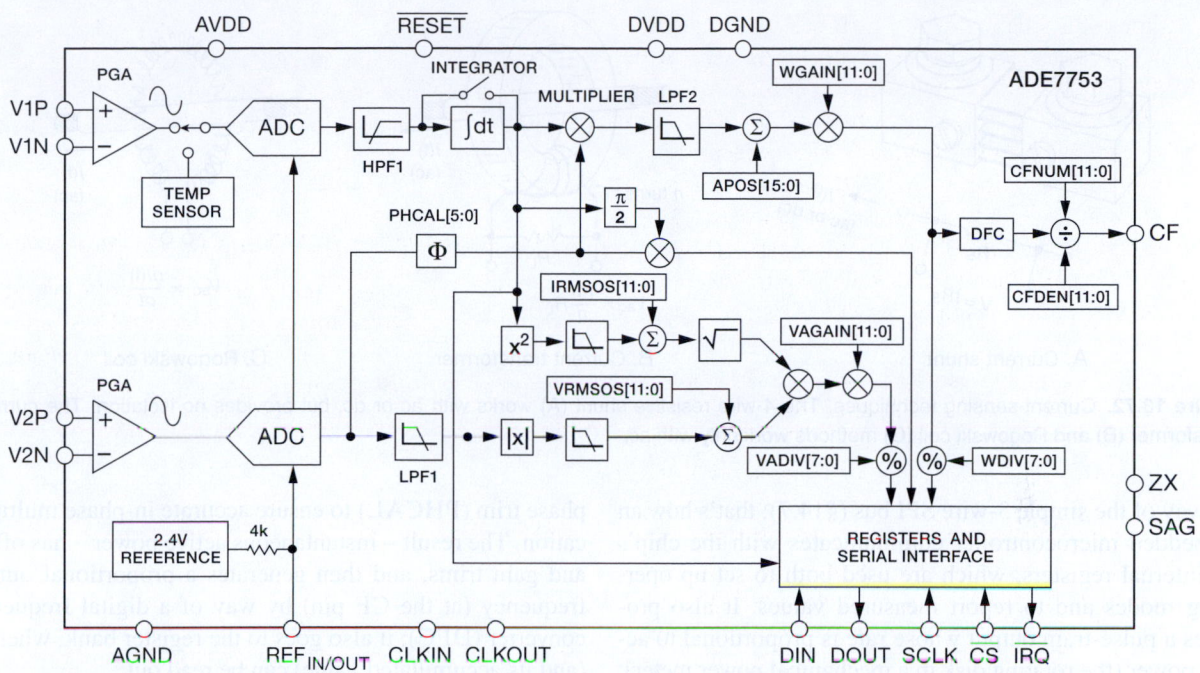

Figure 13.70. The Analog Devices ADE7753: an elegant ac power-monitor IC, courtesy of Analog Devices, Inc.

13.11.1 ADE7753 multifunction ac power metering IC

In industrial settings (and, increasingly, in the energy-aware residential context as well) it's important to keep track of electrical energy usage, in the manner of the traditional power meter with the rotating disk and the accumulated watt-hour dials. As important, or more so, is the need to monitor and minimize *reactive* power; that is, to compensate for reactive loads (such a motors) in order to keep the power factor (§1.7.6) close to unity. The power company cares about reactive power, and indeed conveys that care in the form of surcharges to industrial users, because it produces I^2R heating losses in their lines and transformers, even though it delivers no useful power to the load. It's nice also to be able to monitor the instantaneous power (both real and reactive), and, while we're at it, the presence of voltage dips (brownouts) or peaks (surges).

The elegant ADE7753 from Analog Devices[107] (Figure 13.70) is a fine example of A/D conversion, tailored specifically to this application. It normally would be paired with a microcontroller, as shown in Figure 13.71 (and in Figure 15.21). Here we'll simply admire its many thoughtful design features.

---

[107] We've chosen the single-phase part, for simplicity; the similar ADE7758 handles 3-phase power.

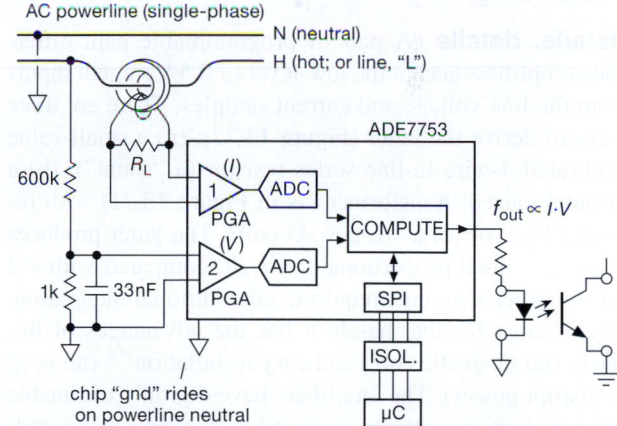

Figure 13.71. Basic powerline connection, with current transformer sensing of ac line current. The f_{out} pulse train provides a running count of energy usage, in the manner of the traditional rotating-disk power meters.

Overview From an analog input signal pair that provides a sample of line voltage and line current, this chip uses purely digital techniques (after the front-end amplifiers) to calculate continually the values of real ("active") power, reactive power, and the volt–amp product ("apparent power"). It also accumulates active and apparent power, and detects voltage sags and peaks. It is highly configurable

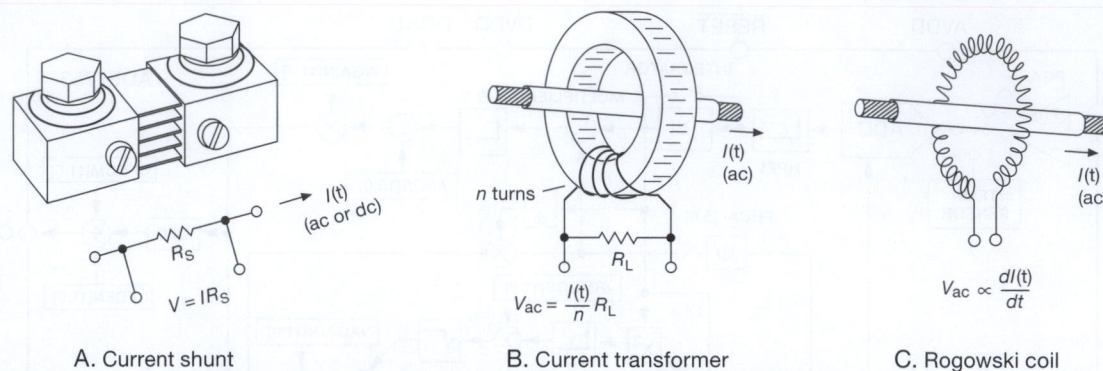

Figure 13.72. Current-sensing techniques. The 4-wire resistive shunt (A) works with ac or dc, but provides no isolation. The current-transformer (B) and Rogowski coil (C) methods work only with ac.

by way of the simple 3-wire SPI bus (§14.7); that's how an embedded microcontroller communicates with the chip's 64 internal registers, which are used both to set up operating modes and to report measured values. It also provides a pulse-train output whose rate is proportional to active power (the rotating disk in a mechanical power meter); so, once calibrated, it can be used in stand-alone mode (no microcontroller) if all you want is a running count of accumulated energy usage.

Details, details   A pair of programmable gain difference amplifiers accept the low-level (±0.5 V) signal inputs from the line voltage and current samples. There are three ways to derive the latter (Figure 13.72): (a) a small-value calibrated 4-wire in-line series resistor (a "shunt"); (b) a toroidal current transformer (as in Figure 13.71) with resistive load; or (c) a "Rogowski coil." The latter produces a voltage signal proportional to dI/dt, compared with $\propto I$ for the other two, thus requiring an additional integration. In exchange for that hassle it has the advantages of linearity (no magnetic core) and easy installation[108] (no need to disrupt power). The amplifiers have digitally trimmable offsets; their outputs are digitized with a pair of second-order 16-bit $\Delta\Sigma$ ADCs, producing the $\sim$28 ksps digitized stream of V and I samples.

Now comes the fun. The top path through the block diagram of Figure 13.70 is the active (real) power computation: channel 1 is the line current signal, with dc removed, feeding an optional integrator (for Rogowski); that is multiplied by the channel 2 voltage waveform, with a 0.05°

---

[108] It's even better than the figure suggests, because in practice one lead snakes back through the coil so that both leads come out at the same end. See datasheets for the RoCoil from DENT Instruments or the RopeCT from Magnelab.

phase trim (PHCAL) to ensure accurate in-phase multiplication. The result – instantaneous active power – has offset and gain trims, and then generates a proportional output frequency (at the CF pin) by way of a digital frequency converter (DFC); it also goes to the register bank, where it (and its accumulated value) can be read out.

The middle path is the reactive power computation: it's similar, but with a 90° phase shift in the current path. Finally, the bottom path is the volt–amp product (apparent power) computation, done as the product of the magnitude of voltage times the rms current, with the usual trimmable offsets and settable gain.

The block labeled "registers and serial interface" shyly conceals its considerable intelligence. It's really in charge here, with all those trims and gain settings, mode settings for things like the sag and peak detectors, the optional integrator, and the frequency scaling of the CF output. It also houses the 49-bit energy (power × time) accumulators (both real and apparent), and registers that hold data about sags and peaks. It can be configured to make an *interrupt* (§14.3.7) to the processor when bad things happen.

Overall, a pretty impressive performance for a part that costs about $4 in small quantities.

13.11.2 AD7873 touchscreen digitizer

A "touchscreen" is the familiar combination of a display device (usually a backlit color LCD), on top of which is an overlay that is sensitive to contact pressure (by fingertip or stylus). These things are used in smartphones, PDAs, tablet PCs, teller machines, point-of-sale terminals, and the like, to allow simple digital (that's a pun, get it?) manipulation of displayed objects. A simple and effective type, called a *resistive touchscreen*, consists of two thin sheets of transparent material, each with a conductive coating, which are pressed together by the finger's contact force.

How can you figure out where that happens? Easy: there is a metal-strip electrode along two opposing edges of each sheet; so if you apply a dc voltage across one such sheet, it acts like a resistive voltage divider, with a linear increase of voltage from one edge to the other. The touchscreen sandwich is a stacked pair, one oriented in the x direction, the other in the y direction. To read out the contact position, you first energize one sheet (say the x sheet) with a dc voltage, and read the voltage that the contacting point transfers to the other (y) undriven sheet; that gives you the x coordinate of the pressed location. Then you reverse the roles, energizing the y sheet and reading the voltage transferred to the x sheet.

The AD7873 (Figure 13.73) does everything you need, and more. It talks to the usual embedded processor (Chapter 15) via a 3-wire SPI serial port (§14.7 and §15.8.2), for both setup and readout. It includes internal MOSFET switches for energizing alternately the x and y sheets; an internal voltage reference; an internal temperature sensor; and a 12-bit SAR ADC with input multiplexer to select and digitize among (a) the undriven sheet, (b) the driving voltage, to make the measurement ratiometric, (c) the battery voltage, (d) the temperature, and (e) an uncommitted analog input of your choosing. This thing runs on a single supply ($+2.2$ V to $+5.25$ V), with power consumption of a few milliwatts and a price tag of about \$2 in modest manufacturing quantities (1000 pcs).

An alternative touchscreen technique substitutes *capacitance* for resistance, with various schemes for determining the point of proximity. You can get complete capacitance converters, with on-chip reference, excitation, delta–sigma converter, and serial interface, for example the AD7140/50 and AD7740-series from Analog Devices. These come in single- and multiple-channel varieties, with resolutions from 16 to 24 bits. They're not fast ($\sim$100 sps), but they're quite inexpensive (a dual 12-bit 200 sps or an 8-channel 16-bit 45 sps converter costs about \$2 in quantities of 25).

13.11.3 AD7927 ADC with sequencer

Many ADCs include an on-chip analog multiplexer, so you can sample and convert a succession of analog inputs. The AD7927 (Figure 13.74) lets you do this; but it adds a programmable sequencer mode, so that you can designate a subset (actually, two subsets) of input channels to be converted in sequence, over and over. Sampling and conversion are triggered by the chip select pin, without the pipeline delays characteristic of delta–sigma converters. It

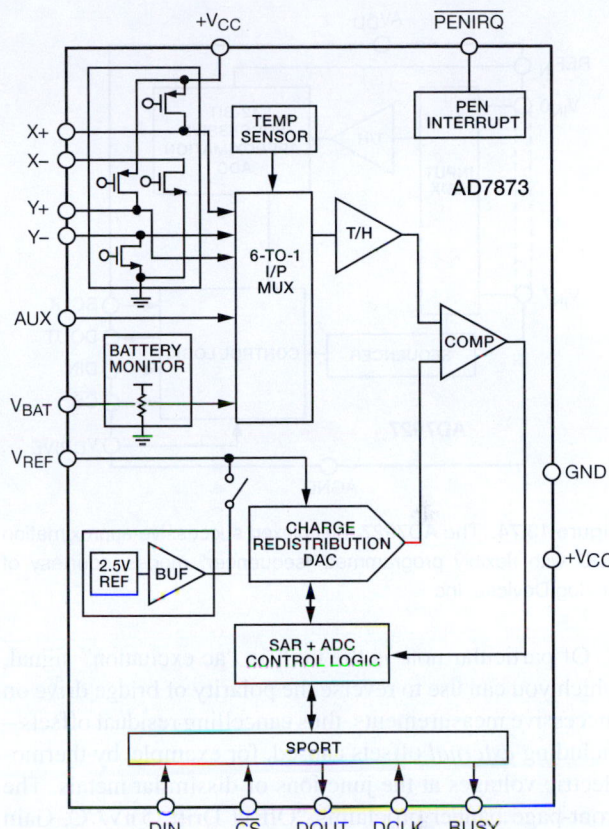

Figure 13.73. The AD7873 Resistive touchscreen digitizer. The fingertip's x and y position is determined in two phases, by energizing each plane in succession and reading the voltage-divided output from the other, courtesy of Analog Devices, Inc.

uses an SPI serial port for both control/programming and for data readout; it's described in this context in §14.7.1.

13.11.4 AD7730 precision bridge-measurement subsystem

Here's a nice chip (Figure 13.75), which takes aim at the weigh-scale market where resistive bridge transducers (strain gauges) are used. Its nicely organized datasheet makes it easy to navigate its host of clever features. It has programmable gain frontend differential amplifiers with adequate gain for 10 mV full-scale inputs, and a differential reference input for fully ratiometric measurements. It can be operated in a chopping mode, to minimize offset voltage errors and drift; and it has internal calibration modes to correct for scale errors. So you can connect the strain-gauge terminals directly to this chip, without any external preamps.

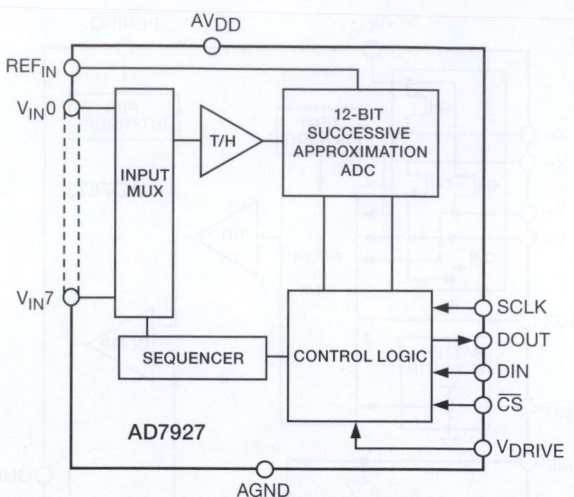

Figure 13.74. The AD7927 multiplexed successive-approximation ADC with flexibly programmed "sequencer" modes, courtesy of Analog Devices, Inc.

Of particular note is the built-in "ac excitation" signal, which you can use to reverse the polarity of bridge drive on successive measurements, thus cancelling residual offsets – including *external* offsets caused, for example, by thermo-electric voltages at the junctions of dissimilar metals. The front-page banner proclaims "Offset Drift: 5 nV/°C, Gain Drift: 2 ppm/°C." On the digital side of the fence, there's lots of programming flexibility in the control of its 24-bit delta–sigma converter and digital filter, by way of the SPI serial port (§14.7). This puppy runs on a single +5 V supply; it's available in both DIP and SMT package styles, and it costs about $15 in small quantities.

13.12 Some A/D conversion system examples

In this section we look at a few examples of complete conversion *systems*. These illustrate the kinds of design trade-offs and subcircuit interplay that you've got to worry about when you incorporate a converter within a larger system, one that includes front-end amplifiers, voltage references, and digital interfaces.

13.12.1 Multiplexed 16-channel data-acquisition system

This application example uses a successive-approximation ADC to create a 16-channel multiplexed A/D data-acquisition system (DAQ). Figure 13.76 shows the circuit configuration, which lets you digitize any combination of 16 differential analog inputs (or 32 single-ended inputs),

under flexible control of an embedded microcontroller (the latter is the subject of Chapter 15). For example, the various input channels can be configured on-the-fly as single ended or differential and with different front-end gains (set by the PGA programmable-gain amplifier), along with timestamps for each measurement; and the channels can be sampled (or skipped over) in any order, with programmable sampling intervals. A subsystem like this could form the "front end" of a microprocessor-controlled data-taking experiment, in which a quick scan of a dozen voltages is programmed to occur, with successive scans taken at intervals of 100 ms or so.

Although it looks pretty straightforward on its face, a finished circuit like this is usually the result of lots of juggling of features and compromises, as you struggle to find components with the right properties. This example was no exception. Let's take a "design walkthrough" to visit the various choices we made, and the resulting performance.

Input multiplexer Analog multiplexers are plentiful (DigiKey lists nearly a thousand); but less so in varieties that will handle a full ±10 V analog input range. And even less so in devices that permit swings beyond the ±15 V supply rails, or that do not clamp the input when unpowered. The long-lived MPC506[109] from TI (originally Burr–Brown, a leader in analog ICs) is outstanding in this regard, though at the expense of relatively high on-resistance (1.5 kΩ). Its dielectrically isolated CMOS process permits input swings 20 V beyond the rails, without "SCR latchup" or crosstalk between inputs; and its switches are "break-before-make," which means that the various input channels don't find themselves shorted together during address changes in the MUX. Watch out for considerations of this type when shopping for linear switches. They sometimes involve a compromise. For example, "break-before-make" results in a slower switching-time specification (here 0.3 µs typ) because the "make" must be delayed (by 80 ns here) to allow the switch to open.

A side note: what happens if an analog input swings more than 20 V beyond the MUX's supply rails? There will be some input current, beginning at about 15 V beyond the rails, and increasing to about 20 mA when you are 40 V beyond the rails. Beyond that you risk damaging the part. If you are expecting serious input over-voltages and want *real* protection, you can include an

---

[109] Or Intersil's original HI-506A, which is often listed with a break-the-search-engine part number like HI3-0506A-5Z. Ordinary ±15 V switches without beyond-the-rail capability have part numbers like DG506, HI-506, or ADG1206.

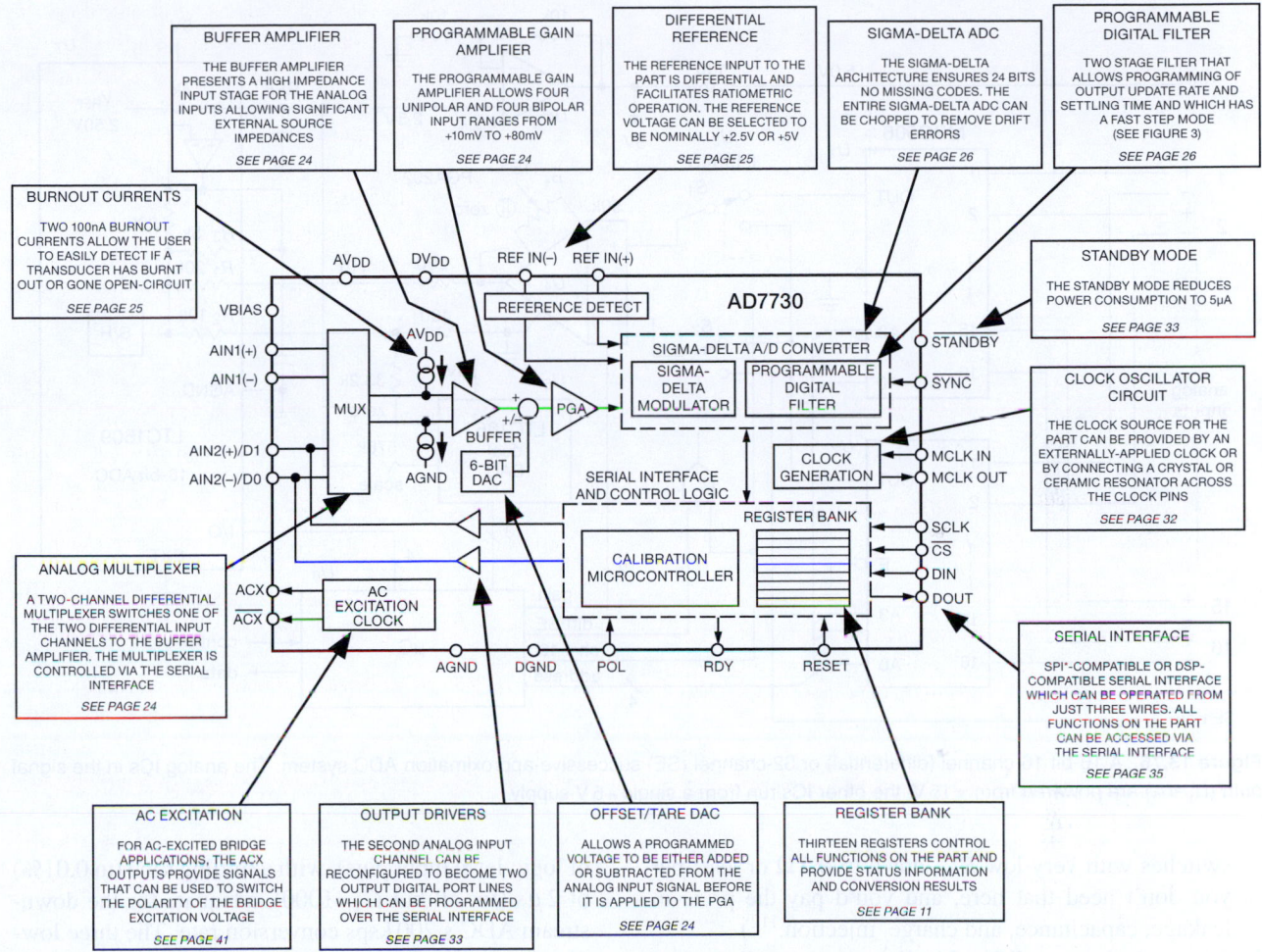

BUFFER AMPLIFIER

THE BUFFER AMPLIFIER PRESENTS A HIGH IMPEDANCE INPUT STAGE FOR THE ANALOG INPUTS ALLOWING SIGNIFICANT EXTERNAL SOURCE IMPEDANCES

SEE PAGE 24

PROGRAMMABLE GAIN AMPLIFIER

THE PROGRAMMABLE GAIN AMPLIFIER ALLOWS FOUR UNIPOLAR AND FOUR BIPOLAR INPUT RANGES FROM +10mV TO +80mV

SEE PAGE 24

DIFFERENTIAL REFERENCE

THE REFERENCE INPUT TO THE PART IS DIFFERENTIAL AND FACILITATES RATIOMETRIC OPERATION. THE REFERENCE VOLTAGE CAN BE SELECTED TO BE NOMINALLY +2.5V OR +5V

SEE PAGE 25

SIGMA-DELTA ADC

THE SIGMA-DELTA ARCHITECTURE ENSURES 24 BITS NO MISSING CODES. THE ENTIRE SIGMA-DELTA ADC CAN BE CHOPPED TO REMOVE DRIFT ERRORS

SEE PAGE 26

PROGRAMMABLE DIGITAL FILTER

TWO STAGE FILTER THAT ALLOWS PROGRAMMING OF OUTPUT UPDATE RATE AND SETTLING TIME AND WHICH HAS A FAST STEP MODE (SEE FIGURE 3)

SEE PAGE 26

BURNOUT CURRENTS

TWO 100nA BURNOUT CURRENTS ALLOW THE USER TO EASILY DETECT IF A TRANSDUCER HAS BURNT OUT OR GONE OPEN-CIRCUIT

SEE PAGE 25

STANDBY MODE

THE STANDBY MODE REDUCES POWER CONSUMPTION TO 5µA

SEE PAGE 33

CLOCK OSCILLATOR CIRCUIT

THE CLOCK SOURCE FOR THE PART CAN BE PROVIDED BY AN EXTERNALLY-APPLIED CLOCK OR BY CONNECTING A CRYSTAL OR CERAMIC RESONATOR ACROSS THE CLOCK PINS

SEE PAGE 32

ANALOG MULTIPLEXER

A TWO-CHANNEL DIFFERENTIAL MULTIPLEXER SWITCHES ONE OF THE TWO DIFFERENTIAL INPUT CHANNELS TO THE BUFFER AMPLIFIER. THE MULTIPLEXER IS CONTROLLED VIA THE SERIAL INTERFACE

SEE PAGE 24

SERIAL INTERFACE

SPI*-COMPATIBLE OR DSP-COMPATIBLE SERIAL INTERFACE WHICH CAN BE OPERATED FROM JUST THREE WIRES. ALL FUNCTIONS ON THE PART CAN BE ACCESSED VIA THE SERIAL INTERFACE

SEE PAGE 35

AC EXCITATION

FOR AC-EXCITED BRIDGE APPLICATIONS, THE ACX OUTPUTS PROVIDE SIGNALS THAT CAN BE USED TO SWITCH THE POLARITY OF THE BRIDGE EXCITATION VOLTAGE

SEE PAGE 41

OUTPUT DRIVERS

THE SECOND ANALOG INPUT CHANNEL CAN BE RECONFIGURED TO BECOME TWO OUTPUT DIGITAL PORT LINES WHICH CAN BE PROGRAMMED OVER THE SERIAL INTERFACE

SEE PAGE 33

OFFSET/TARE DAC

ALLOWS A PROGRAMMED VOLTAGE TO BE EITHER ADDED OR SUBTRACTED FROM THE ANALOG INPUT SIGNAL BEFORE IT IS APPLIED TO THE PGA

SEE PAGE 24

REGISTER BANK

THIRTEEN REGISTERS CONTROL ALL FUNCTIONS ON THE PART AND PROVIDE STATUS INFORMATION AND CONVERSION RESULTS

SEE PAGE 11

Figure 13.75. Frontend analog subsystem tailored for bridge-type transducers such as strain gauges, weigh scales, and pressure transducers. The AD7730 datasheet, from which this figure is taken, is a model of clarity, and a delight to read, courtesy of Analog Devices, Inc.

input current-limit circuit like the one in Figure 13.77: the back-to-back depletion-mode MOSFETs, in convenient TO-92 or SOT-23 packages, can hold off 500 V, and limit the current to $\sim$1 mA.[110] See §5.15.5 and Figure 5.81 and associated discussion for more detail.

Differential/single-ended selector switches Dual SPDT analog switches, under digital control, route the multiplexers' outputs: in 16-channel differential mode, S_1 and S_2 always look at U_1 and U_2; in 32-channel single-

ended (SE) mode S_2 looks at signal common, while S_1 looks at U_1 or U_2 for channels 1–16 or 17–32, respectively. (This is often called a *pseudo-differential* input, because the common terminal is shared by all the SE channels.) Mode and channel switching can be done on a channel-by-channel basis (explained below). Note the absence of an anti-alias filter, which would limit the multiplexing speed: we assume the input signals are appropriately band limited upstream of the multiplexers.

The IH5043[111] SPDT analog switches S_1 and S_2 were chosen for their low leakage, low capacitance, and low charge injection. These come at the expense of relatively high R_{on} (80 Ω max). (You can get plenty of analog

---

[110] Under normal conditions the series resistance is $2R_{on}+R_s$, or about 2.7kΩ. The added 1 kΩ of series resistance is not much of a compromise; you could omit the resistor, and the saturation $I_{DSS} \approx 2$ mA would still protect the MUX, but then the transistors' dissipation limit means that you'd need to limit sustained input overvoltages to 100 V or so.

[111] Alternative part numbers are HI5043 and DG403.

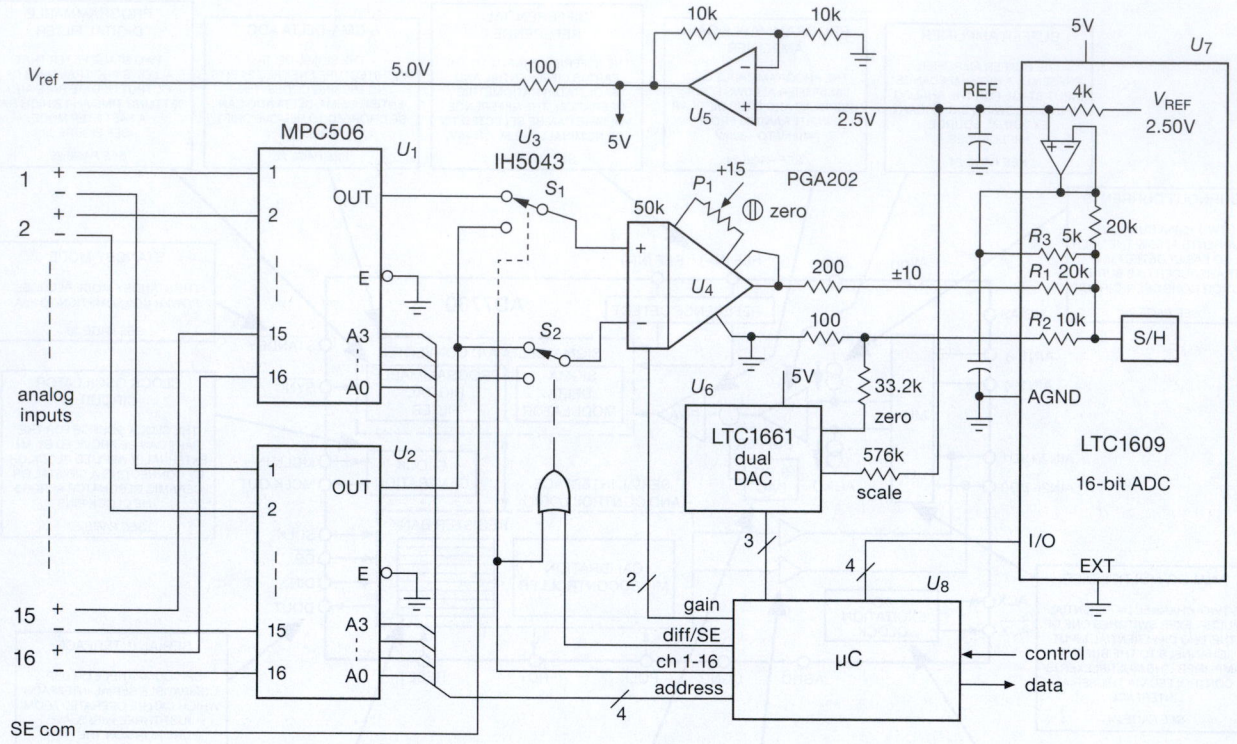

Figure 13.76. A 16-bit 16-channel (differential) or 32-channel (SE) successive-approximation ADC system. The analog ICs in the signal path (U_1–U_5) are powered from ±15 V; the other ICs run from a single +5 V supply.

switches with very low R_{on}, down to $0.5\,\Omega$ or less; but you don't need that here, and you'd pay the price in leakage, capacitance, and charge injection.[112])

Instrumentation amplifier Ideally we want an instrumentation amplifier (see §5.15 and Table 5.9) with digitally programmable gain (a PGA), that can accommodate the full ±10 V analog signal range, and that has fast settling time, stable gain, low offset voltage, low noise, and low bias current. It's fine to tick off a laundry list of desirable characteristics – but just how good does each of these need to be? What matters, ultimately, is that the amplifier's limitations do not degrade the overall system's performance.

Given the surrounding system, the PGA202 from TI (Burr–Brown) is a good choice here: it has programmable gains of 1, 10, 100, and 1000 (set by a pair

of logic-level input pins), with a settling time (to 0.01%) of $2\,\mu s$ (for all but $G_V=1000$), adequate for the downstream ADC's 200 ksps conversion rate. The three lowest gain settings correspond to full-scale input ranges of ±10 V, ±1 V, and ±0.1 V. Its high input impedance and low input current ($10\,G\Omega$, 50 pA) do not degrade the combined characteristics of the upstream multiplexer and switch (the MUX specifies a typical leakage current of 2 nA).

Finally, what about amplifier offset voltage and noise? We chose a JFET-input instrumentation amplifier for its high-Z input; but, looking at comparable amplifiers in Table 5.9, we see that we've paid a price, in terms of offset voltage and noise, compared with the bipolar-input PGA204. We need to compare these effects with the resolution (LSB step size) of the downstream ADC, which looks at the amplifier's output. But amplifier offset and noise is always specified at the *input* ("RTI," referred to the input); so we need to figure out the converter's RTI step size, which depends on the amplifier's gain. That's easy enough: the ±10 V converter input range, divided by its 2^{16} steps, amounts to an LSB

---

[112] For example, the ADG884 analog switch has a very low $R_{on} = 0.4\,\Omega$ (max); but its shunt capacitance $C_S(on)$ is 295 pF, compared with 22 pF for our chosen IH5043. It's also a low-voltage part, with a maximum analog swing of 5 Vpp. The ADG1413 operates over the full ±15 V range, with a low R_{on} of just $1.5\,\Omega$, but its charge injection (±300 pC) is five to ten times greater than that of the '5043/DG403.

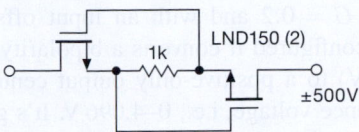

Figure 13.77. Input current-limiter circuit, good to ±500 V overdrive.

step size of 0.3 mV. So the step size, referred to the amplifier's input, is 300 μV, 30 μV, and 3 μV, for gains of 1, 10, and 100, respectively.

We're seriously challenged here, because the amplifier specifies a typical RTI offset voltage of $(0.5+5/G)$ mV; that is, 0.5 mV in its frontend amplifier, combined with 5 mV in its internal output stage. Without some additional tricks we're facing typical offset errors of 5.5 mV, 1 mV, and 0.55 mV at gains of 1, 10, and 100; that's 18, 33, and 180 times the converter's RTI step sizes, respectively. Clearly we need to manually trim its offset, and also include some electronic nulling circuitry;[113] and, even with that, we'll be limited ultimately by the amplifier's tempco and drift.

In our design we've got the recommended manual trim (set once, at maximum gain), and a 10-bit DAC to zero the offset; the latter has a 0–5 V output, for a trim range of ±7.5 mV at the amplifier's output ("RTO"). Its 10-bit resolution gives a 15 μV step size, far better than needed, given the converter's 300 μV LSB step size.

So, to use this thing for full-accuracy measurements, we zero the offset error (via the ADC) at the beginning of a suite of measurements, and we hope that short-term offset drifts are small. Are we on solid ground? Well, the amplifier's specified typical offset drift is $(3+50/G)$ μV/°C, 50 μV/month, and $(10+250/G)$ μV/V(supply). We use regulated supplies, and we are concerned with short-term drifts only; so we're generally OK, with thermal drifts a potential problem only at the highest gain (G=100), where a 1°C change causes a 1-LSB error. There's also drift of amplifier *gain* with temperature to worry about. Here that's 3 ppm/°C (typ) for G=1 or 10, and 40 ppm/°C for G=100. An LSB corresponds to 30 ppm at ±full scale, so once again we're in good shape except at the highest gain (G=100), where the amplifier's gain tempco causes a 1-LSB error for a 1°C temperature change.

Finally, what about the amplifier's voltage noise? The RTI specified values are 1.7 μVpp (typ) for the $1/f$-plagued low-end band of 0.1–10 Hz, and a density e_n=12 nV/$\sqrt{\text{Hz}}$ (typ) at 10 kHz (the $1/f$ corner is roughly 100 Hz). So a band extending from 0.1 Hz to ~10 kHz has an RTI noise voltage of about 3 μV, comparable to the RTI step size at G=100; it's negligible at lower gains.

Analog-to-digital converter We need a converter to handle the full ±10 V signal range, with enough speed for scanning at rates of 100 kHz or more. (There are scanning ADCs, with input multiplexers and sequencer logic, such as the 8-channel 16-bit 500 ksps AD7699, but they force you to live with their limited input range, e.g., 0–5 V or ±2.5 V, rather than the ±10 V as we have here.) The LTC1609 does the job, and runs on a single +5 V supply, which we can regulate and filter individually to keep it quiet. It converts up to 200 ksps, with 2 μs acquisition (input-settling) time, and a serial interface with multiple modes (internally or externally clocked serial data, etc.; see §15.9.2 for interfacing and programming details). Its worst-case zero-offset (±10 mV) and gain (±1.5%) errors have to be calibrated out (done here with the dual DAC, U_6), after which it has acceptably low offset and gain drifts of ±2 ppm/°C and ±7 ppm/°C, typ, respectively. (Recall that an LSB at full-scale is 30 ppm.)

Voltage reference Much of the gain error and drift is due to the internal +2.5 V reference, with its ±1% (worst case) accuracy and ±5 ppm/°C typical tempco. This is in fact a very good drift specification for an internal-reference ADC. But if you want better performance, use a precision external reference (see Tables 9.7 and 9.8), the best of which have ±0.02% (worst case) accuracy and ±1 ppm/°C or better typical tempcos. The LTC1609's REF pin invites an external reference, which simply overdrives the 4 kΩ internal source. With such an external voltage reference the untrimmed gain error is reduced to ±0.5% (which we can trim out with the DAC), and the gain drift is reduced to ±2 ppm/°C (typ). These gain drifts are comparable to those of the upstream amplifier, except at the latter's highest gain (G=100), where the amplifier's drift is an order of magnitude larger and becomes problematical.

Programming and operation With systems that include microcontrollers, there's still plenty of work to do! Here you'd run a calibration setup, storing the ADC trims in non-volatile microcontroller memory. Then at startup you would program the DAC as well as ADC ranges and communication modes. Look first at §15.9.2 and the

---

[113] Or software offset subtraction: we could devote one (shorted) channel to zero-error measurement. And we could use another to measure the voltage reference for full-scale calibration. Zero-offset error corrections are often stored as parameters that have been measured during calibration.

timing diagram in Figure 15.23 to see how a single ADC conversion is handled. But there's much more involved in controlling a full data-acquisition system like this: you need to set up in advance details such as the active input channels and their sequence, and for each channel whether it's single ended or differential, its gain, etc. This would typically be done with a lookup table, accessed at run time. You also need to specify scan rate, interrupt modes, what data are to be stored at run time (channel number, mode, gain, time stamp, and so on), along with header information that goes into the data file (such as machine ID, experiment type, date, operator, sensor configuration, and so on). We could ramble... but you get the picture: there's plenty of follow-on organization and programming to make it all fly. For these tasks we've found that it's nice to have dedicated graduate students, who have lots of time, lots of skills, and whose degree depends on making it all work.

13.12.2 Parallel multichannel successive-approximation data-acquisition system

The previous example is *multiplexed*, with a single ADC digitizing the input channels successively in some programmed sequence. That's OK in many applications; but sometimes it's important to capture *simultaneously* the input levels on multiple analog inputs. One way to do this is to capture each analog input on its own sample-and-hold (or track-and-hold), then multiplex these stable analog voltages into a single ADC. But ADCs are inexpensive, so it's often better (and always faster) to use a set of them to digitize the inputs simultaneously. In this example we illustrate this with successive approximation ADCs, and in §13.12.3 we'll do it with delta–sigma converters.

Figure 13.78 shows one implementation, using a set of chips from Analog Devices that play well together in this application. Let's walk through it.

Input amplifier Problem: you want to accept bipolarity input signals, say over a $\pm 10\,$V range – but you've got an ADC running from a single positive supply, which accepts positive signals only. Solution: use an op-amp input stage to offset the signal and reduce its swing. You can do this, but you'll need matched resistors of high precision in order not to compromise the accuracy of the ADCs.

Here we've taken advantage of the elegant AD8275 level-translating ADC driver, which does what you want: it's basically a single-supply difference ampli-

fier with $G = 0.2$ and with an input offset terminal. As here configured it converts a bipolarity input range ($\pm 10.24\,$V) to a positive-only output centered on half the reference voltage, i.e., 0–4.096 V. It's got plenty of bandwidth ($0.45\,\mu$s settling to 0.001%, which is less than an LSB), rail-to-rail output (so it can run from the same +5 V as the converter, protecting against ADC overdrive), accurate and stable gain ($G = 0.2\pm0.024\%$, tempco 1 ppm/°C max), and acceptably low offset and drift ($V_{os} < 0.5\,$mV max, tempco 7 μV/°C max).

Two important points. (a) The offset and drift specifications are *referred to the output* (RTO), not to the input. In other words, the input offset is five times larger, or $\pm 2.5\,$mV (max), and likewise for the drift. This amount of offset is not insignificant: the 16-bit converter's LSB, referred to the amplifier's input (RTI), corresponds to $2\times10.24\,$V$/2^{16}$, or 0.31 mV; so the amplifier's worst-case offset is about 8 LSBs. The drift, by comparison, *is* insignificant: it would take a 45°C change of temperature to move by 1 LSB. So the channels should be calibrated at zero input (and also at full-scale input). (b) The unround reference voltage (4.096 V) is popular because it produces a round-number conversion gain, namely precise 10.0 mV steps at the 11-bit level (which you can think of as further subdivided by 2^5 at the full 16-bit resolution). Also, because the full-scale range extends to $\pm 10.24\,$V, you can calibrate the system using a 10.0 V reference without driving it over-range.

ADC The AD7685 is a single-supply 16-bit successive-approximation converter with an SPI serial output fast enough to handle its maximum 250 ksps conversion rate. The SPI interface permits daisy chaining with additional converters, as shown; and the serial interface allows packaging in a small 10-pin package. It has good linearity and accuracy properties: no missing codes, ± 3 LSB (max) integral nonlinearity, ± 0.3 ppm/°C (typ) gain drift.[114]

Reference The ADC requires an external reference, which sets the positive full-scale range. The ADR440 series of "XFET" voltage references (§9.10.3) exploit the pinch-off voltage of a pair of JFETs, in a clever configuration that achieves low noise and low drift: 1.8 μVpp (typ) and 3 ppm/°C (max).

Serial port isolator Quiet systems get noisy if you allow digital hash and ground currents into the analog

---

[114] This is a charge-redistribution ADC, for which it is sometimes advisable to put a capacitor across the analog input, isolated from the driving amplifier with a small resistor, as in Figure 13.37; a suitable choice here would be 2.7 nF and 33 Ω.

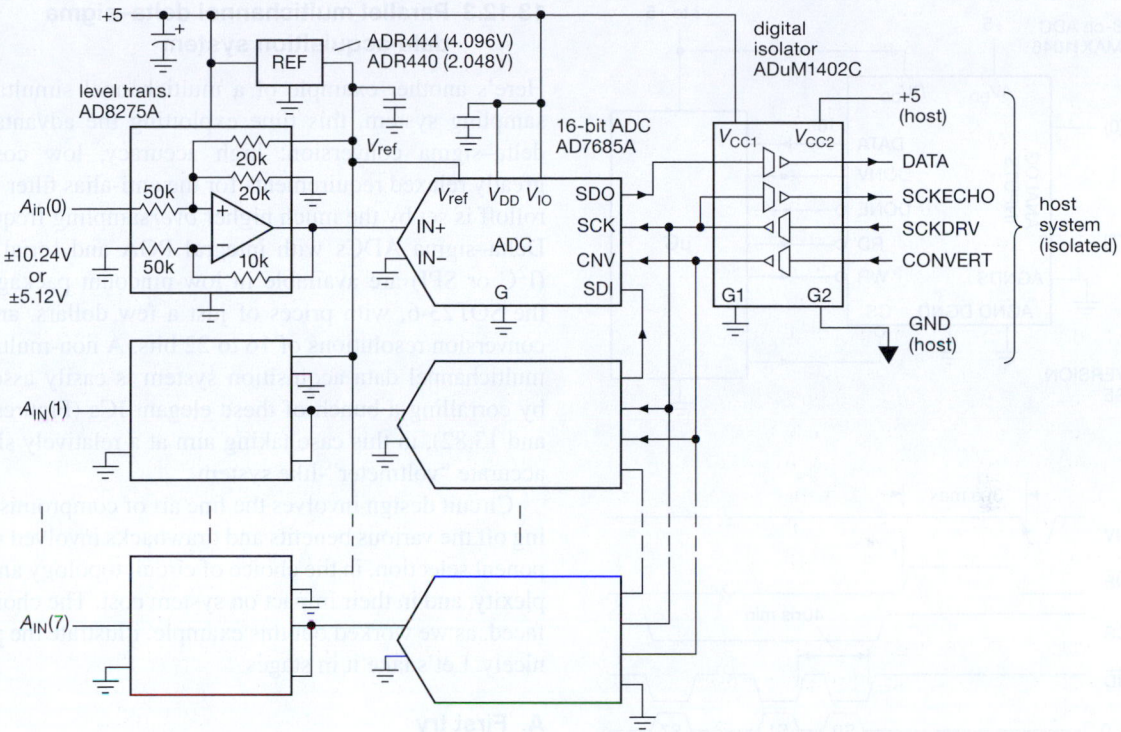

Figure 13.78. Multichannel parallel successive-approximation ADC with isolated SPI serial output port.

portion. With a 3-wire interface it's easy to add full galvanic isolation, in this case with the ADuM1402C 4-channel isolator. This puppy uses on-chip transformer coupling, and is good to 90 Mbps. Note the readback of the SPI clocking signal: that's needed because the signal delay through the isolator (27 ns typ) is comparable to the period of the maximum readout clock rate (~50 Mbps); by echoing back the SCK received by the ADC, the host system sees the output data accurately synchronized to that echoed clock. This is inaccurate only to the extent that the isolator delay time varies between channels (skew), which for this isolator is an impressive 2 ns (max).

Component cost The amplifier and ADC blocks, which are replicated for each channel, are not expensive: about $3 and $10, respectively (in quantities of 25). Adding $5 for the reference and $9 for the isolator, we have about $118 for an 8-channel system.

A. An integrated parallel multichannel SAR solution

Why build it, when you can buy it? As it happens, the clever folks at Maxim have integrated a simultaneous 8-channel 16-bit successive-approximation ADC system on a chip, with performance comparable to that in the previous

section: 250 ksps, single +5 V supply, bipolarity conversion range (±5 V), and good accuracy and linearity (±0.01% maximum offset, ±2.4 μV/°C typical offset tempco, no missing codes, ±2 LSB maximum integral nonlinearity). The MAX11046 uses a parallel data interface, with a separate digital supply pin for compatibility with a low-voltage microcontroller. Figure 13.79 shows the scheme.

This part is unusual in accommodating a bipolarity input voltage range while running from a single positive supply.[115] A single CONV pulse starts the conversion, simultaneous on all channels; the signals are sampled at the rising edge of CONV, with a typical timing skew of 0.1 ns (!). The conversions are complete 3 μs later, and are read out as 16-bit parallel words, one channel per RD′ pulse, as shown.

Figure 13.80 shows in more detail what's included on-chip. The analog inputs are clamped at about 0.3 V beyond the conversion range (i.e., ±5.3 V), but you have to include current-limiting series resistors to limit the clamp current to 20 mA. A set of fast (BW=4 MHz) track-and-hold

---

[115] The datasheet does not say how this is done, but it's most likely an on-chip charge-pump negative supply generator. The very high input impedance rules out something like the previous voltage-translation scheme.

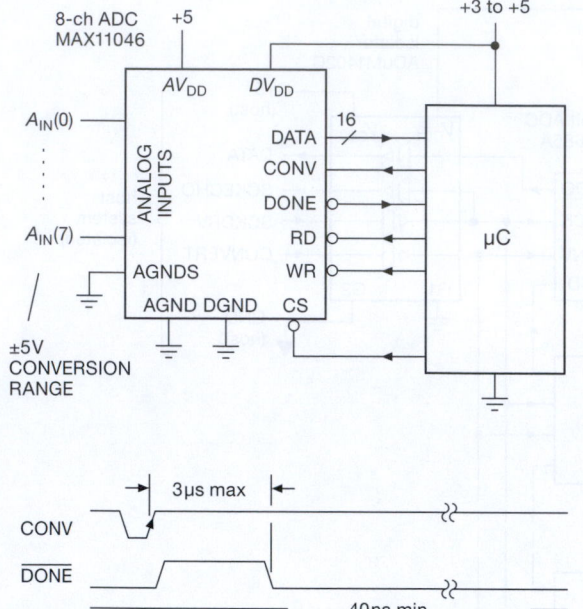

Figure 13.79. The MAX11046 integrates an 8-channel parallel successive-approximation ADC onto a single chip.

circuits capture the input signals, followed by the array of latched ADCs and the output multiplexer. The digital port allows some configuration, via the four low-order bidirectional data lines: internal or external reference, offset binary or 2's complement, and single or continuous conversion mode. You could think about adding galvanic isolation to the digital lines – but you'd have to use 21 channels of isolation, and you'd have to arrange for bidirectional signaling on the four low-order data lines D3..D0 (with the direction set by WR′). Ah, the elegance of serial communication!

Compared with our à la carte design of the previous section, this system is a bargain: about \$42 (single-piece price) for the complete 8-channel converter system. As one might expect in the competitive world of silicon, Maxim is not alone in integrating a simultaneous-sampling multichannel ADC. The AD7608 from Analog Devices provides eight channels of T/H (spanning a full ±10 V range) multiplexed into an 18-bit ADC capable of 200 ksps on all channels, with an on-chip digital filter and both parallel and serial output formats. Note the different approach: this latter part uses a single fast ADC to convert the levels captured on the eight T/Hs, whereas the Maxim part uses eight ADCs.

13.12.3 Parallel multichannel delta–sigma data-acquisition system

Here's another example of a multichannel simultaneous-sampling system, this time exploiting the advantages of delta–sigma conversion: high accuracy, low cost, and greatly relaxed requirements for the anti-alias filter (whose rolloff is set by the much higher *over*sampling frequency). Delta–sigma ADCs with integral PGA and serial output (I^2C or SPI) are available in low-pincount packages like the SOT23-6, with prices of just a few dollars, and with conversion resolutions of 16 to 22 bits. A non-multiplexed multichannel data acquisition system is easily assembled by corralling a bunch of these elegant ICs (Figures 13.81 and 13.82), in this case taking aim at a relatively slow but accurate "voltmeter"-like system.

Circuit design involves the fine art of compromise, trading off the various benefits and drawbacks involved in component selection, in the choice of circuit topology and complexity, and in their impact on system cost. The choices we faced, as we worked out this example, illustrate the process nicely. Let's take it in stages.

A. First try

We began with the notion of a bused I^2C array of ADCs (Figure 13.81), which minimizes the number of microcontroller pins needed (compared with an SPI interface), because there is no need for individual chip-select (CS′) lines. TI's ADS1100 converter (\$4.50 in unit quantity) looked good: it describes itself as a "self-calibrating, 16-bit ADC" that is a "complete data acquisition system in a tiny SOT23-6 package." It contains a differential input PGA (gains of 1, 2, 4, or 8), runs on a single supply from +2.7 V to +5.5 V (at 90 μA), includes an internal clock, and guarantees 16-bit conversions with no missing codes at its lowest conversion rate of 8 samples/s. It has a maximum integral nonlinearity (INL) of 0.013%, and a typical gain error of 0.01% (the positive supply V_{DD} is the reference, with full scale of $\pm V_{DD}/G$).

That's the good news. Here's the bad news: the I^2C protocol (§14.7.2) requires each bused device to have a unique address (of 128 possible); that's the price you pay for a 2-wire bus with no individual chip-select lines. This is usually accomplished by dedicating a few device pins to set the address (e.g., three pins to select 1 of 8 addresses within a subset of the full 128 possible addresses). But a part with only six pins cannot afford this luxury: count them – differential input (2 pins), power and ground (2 pins), I^2C bus (2 pins) – there are no pins left!

The ADS1100 solves this problem by pre-assigning the addresses, with a different part number for each of eight

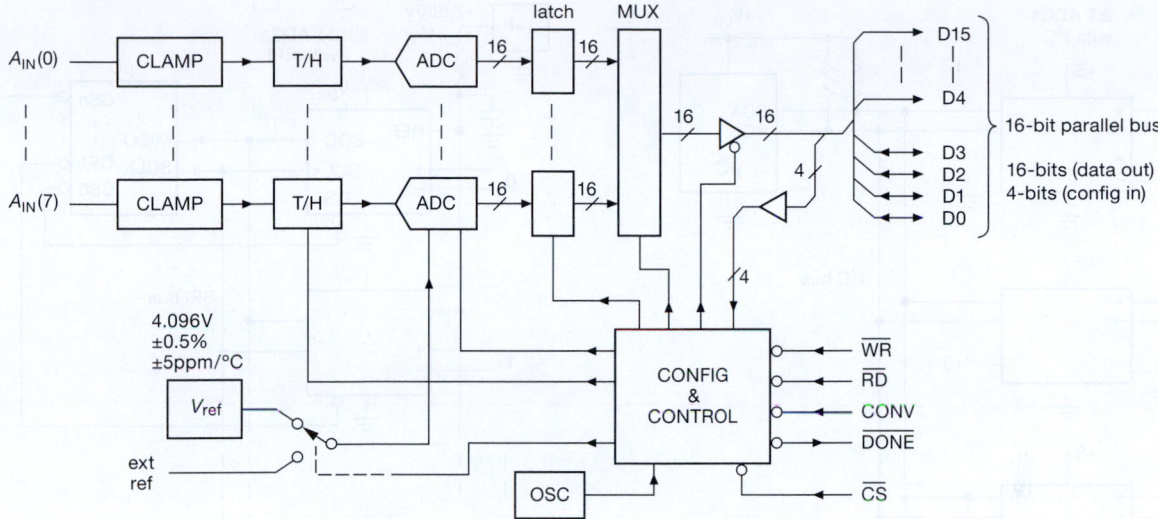

Figure 13.80. Block diagram of the MAX11046 innards.

possible addresses (decimal 72 through 79). For an eight-channel system, then, you've got to order eight different parts (if you can find them in stock: at the moment of writing, DigiKey has addresses 72 and 74; Mouser has addresses 75 through 79; and Newark has addresses 72 through 74), and with less opportunity for quantity pricing.

There's one more piece of bad news: the ADS1100 has an internal clock of coarse accuracy ($\pm20\%$), with no option for an external clock (no pins remaining!). So you cannot achieve the high rejection of powerline frequencies (60 Hz or 50 Hz) that comes with a sampling rate that is an integral number of powerline cycles (e.g., exactly 10 conversions/s). You wind up with a mediocre ~30 dB rejection of normal-mode signal at 50 Hz or 60 Hz.

B. Second try

The ADS1115 is a related part from the same manufacturer, priced around $5 (in quantities of 25), this time housed in a 10-pin package, with one of the additional pins allowing some degree of address selection. This is done cleverly: a single pin sets one of *four* addresses (decimal 72–75), according to whether it is tied HIGH, tied LOW, or connected to one of the two I^2C serial-interface lines (Figure 13.81). This is an improvement over the 6-pin ADS1100, but with only four possible I^2C addresses you'd have to use a second I^2C channel to get eight input channels.

There's a second input channel, which could be used for multiplexed conversion. But we want simultaneous conversions on all channels, so we've used the second channel

for zero calibration, as shown, which you'd do between active conversions. Some other nice features of this chip are its operation from +2.0–5.5 V, a wide range of PGA gains ($0.66\times$, and $1\times$ to $16\times$, by factors of two), a wide range of conversion rates (8–860 sps at full 16-bit resolution), and an internal clock and voltage reference.

So far, so good. But there's trouble still in paradise. As with the ADS1100, the internal clock is of only coarse accuracy ($\pm10\%$), so you get only some 30 dB of powerline rejection.[116] Likewise, the internal voltage reference is not of great precision, and there is no option for an external reference: this is specified as *gain error*, which for this part is 0.01% (typ), 0.15% (max). To give it perspective, an LSB (at 16-bit resolution) is 0.0015% (15ppm); so the worst-case full-scale error corresponds to 100 LSB steps. Furthermore the gain drift (tempco) is specified as 40 ppm/°C (max), which corresponds to 3 LSBs/°C.

C. Third try

Not satisfied with these choices, we next explored converters with an SPI digital interface. Instead of I^2C's bus addresses, you need to provide a separate chip-select line to each converter (Figure 13.82). That's the bad news. The good news is that there are some terrific converters out there.

We looked at many candidates; the first to pass muster

---

[116] You need to read the datasheet carefully, here: it lists 105 dB typical *common-mode* rejection at 50 Hz and 60 Hz, but no listed value for *normal-mode* (i.e., differential) powerline signals. Later there's a graph that shows the ~30 dB value.

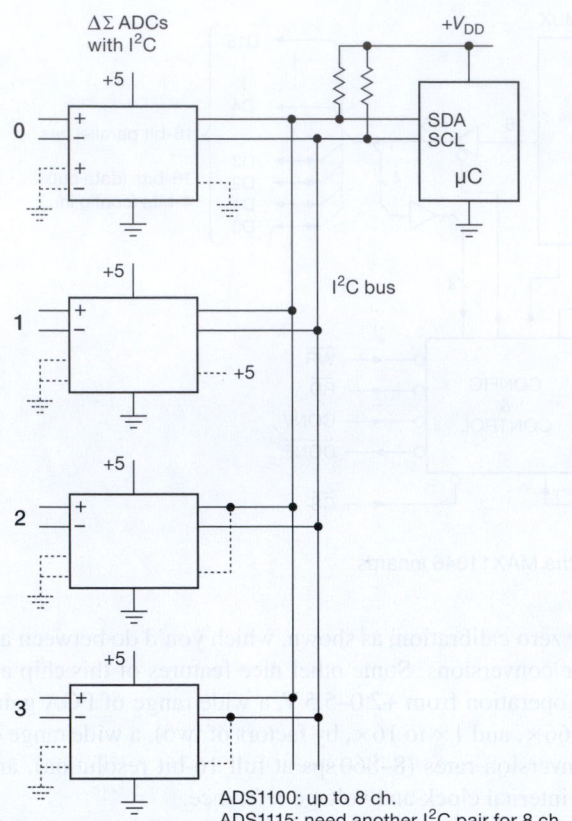

Figure 13.81. Multichannel parallel $\Delta\Sigma$ ADC with I^{2}C output port. The dashed lines show connections for the ADS1115 converter only.

was the CS5512 from Cirrus Logic. This is a single-supply (+5 V) 20-bit delta–sigma converter in an 8-pin SOIC, priced around $4.25 (in quantities of 25). An external clock source (32.768 kHz nominal) times the conversions, and also the SPI clocking. The accurate clock frequency is exploited for true *normal-mode* powerline rejection: the chip's digital filter is configured with a broad 80 dB (minimum) notch extending from 47 Hz to 63 Hz, with ~90 dB simultaneous rejection at both 50 Hz and 60 Hz.

This chip also excels in linearity ($\pm0.0015\%$ of full scale, maximum), and delivers 20-bit resolution with no missing codes. And its typical offset voltage and gain drifts are $0.06\,\mu$V/°C and 1 ppm/°C, respectively. This is one classy converter chip!

Now for the thunderclap: this chip requires an external voltage reference, which is normally a good feature (because you can use a high-quality reference, whose voltage also sets the analog scale). Imagine our surprise, then, when we read the datasheet's specification of the full-scale

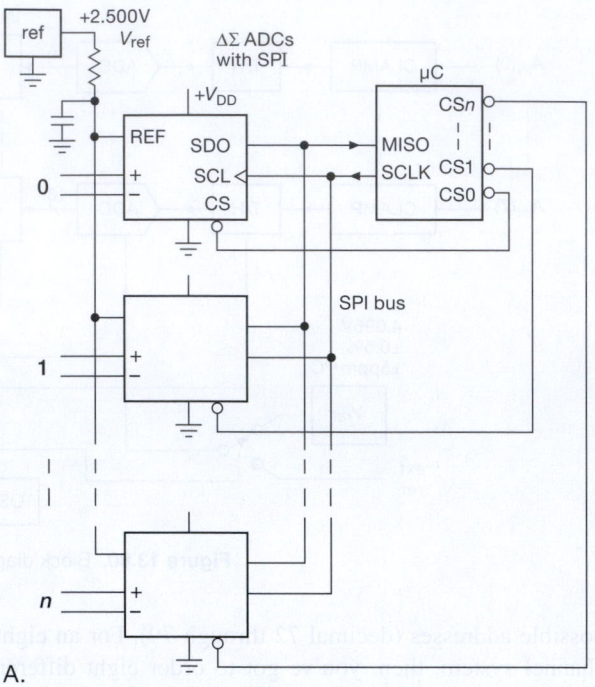

A.

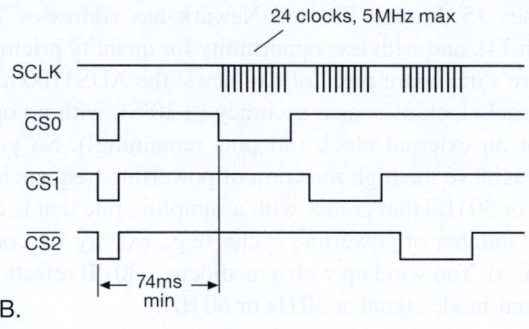

B.

Figure 13.82. Multichannel parallel $\Delta\Sigma$ ADC with SPI output port (e.g., a CS5512 or an MPC3551). The timing diagram applies to the MCP3551 family, for which the first assertion of CS' initiates conversion (simultaneous across all channels), with subsequent serial data readout (of the individual channels) clocked by SCLK during reassertion of CS'.

analog input range: $V_{\rm FS}=0.8V_{\rm REF}$, $\pm10\%$. In other words, this is a highly linear and stable converter, but with a conversion gain that is uncertain to $\pm10\%$.

D. Fourth try

From Microchip (a company traditionally known for its microcontrollers) comes the winning candidate. Their MCP3551 is a 22-bit delta–sigma converter in an 8-pin SOIC (or smaller MSOP), priced around $3.25 (in quantities of 25), with an accurate (±0.5%) internal clock that delivers excellent powerline rejection. It runs from a single +2.7 V to +5.5 V supply, drawing about 0.1 mA. It requires an external voltage reference, which (unlike the previous converter) accurately sets the analog scale factor; and it allows 12% over- and under-range inputs, signaled with two additional data bits. And it performs single-cycle conversions with no digital-filter settling time, during which it also carries out an offset and gain calibration.[117] Figure 13.82's timing diagram shows the scheme for simultaneous multichannel conversion, followed by sequential data readout.

The specifications are impressive: 22-bit resolution with no missing codes; V_{os} of ±12 μV (max), full-scale error of ±10 ppm (max), INL of 6 ppm (max), and normal-mode powerline rejection of 85 dB (typ) at both 50 Hz and 60 Hz (Figure 13.83).[118] The typical offset and gain drifts are 0.04 ppm/°C and 0.028 ppm/°C, respectively. What's not to like in this converter? The only thing we can really complain about is the absence of a PGA: lacking that, it takes the converter's full 22-bit capability to achieve 1.2 μV resolution, the same that you would get with a 16-bit converter having a built-in PGA with 64× gain.

Totaling up the cost for an 8-channel delta–sigma simultaneous-conversion system, we have $34 for the ADCs and about $4 for an ADR441A reference, or a very economical $38.

E. An integrated parallel multichannel ΔΣ solution

Never underestimate what can be done on a single piece of silicon. Here again the wizards of Silicon Valley have come through, with several interesting multichannel ADCs featuring simultaneous conversion on all channels.

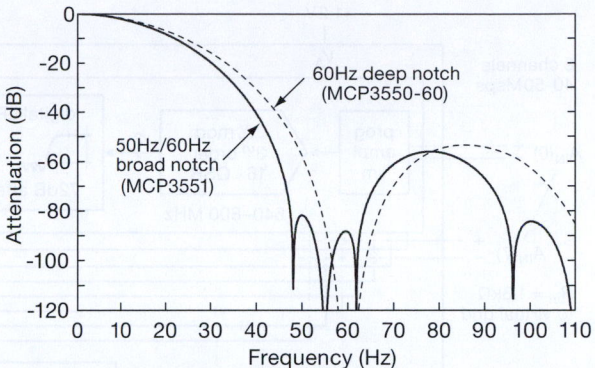

Figure 13.83. The MCP3551's digital filter is configured to create a broad notch covering 50 Hz and 60 Hz powerline frequencies, and a bit more. By contrast, the MCP3550-60 (or -50) is configured for a single deep notch.

For relatively low-speed conversion there's the AD73360, which houses six 16-bit 64 ksps delta–sigma converters, each with its own programmable-gain amplifier (0–38 dB). It comes in a convenient 28-pin SOIC and costs about $8 (in quantities of 25). It has a programmable sampling clock and a serial-output port optimized for transferring data automatically into a downstream DSP chip (from as many as eight cascaded converters). Its six channels are ideal for voltage and current measurements in a 3-phase motor drive, or for industrial power monitoring. It requires in-system calibration (±10% gain accuracy), and is usually used with ac coupling (worst-case dc offset ~10% of full scale).

What about *really* fast multichannel delta–sigma conversion? Our design runs at a leisurely conversion rate of just 15 samples per second (the AD73360 betters that by three orders of magnitude, but with lower resolution and degraded accuracy). The analogous task becomes extraordinarily difficult if you were to increase this by a factor of a million or so. Difficult, but evidently not impossible: the impressive ADC12EU050 from National Semiconductor (Figure 13.84) packs eight simultaneous 12-bit differential-input delta–sigma ADCs that run at 50 Msps (!). That creates a firehose of digital output, for which it devotes an LVDS pair to each output channel; it consumes about 0.4 W at full tilt, and costs about $100.

13.13 Phase-locked loops

13.13.1 Introduction to phase-locked loops

The phase-locked loop (PLL) is an interesting and useful building block, available both as a single stand-alone

---

[117] In the words of the datasheet, "A self-calibration of offset and gain occurs at the onset of every conversion. The conversion data available at the output of the device is always calibrated for offset and gain through this process. This offset and gain auto-calibration is performed internally and has no impact on the speed of the converter since the offset and gain errors are calibrated in real-time during the conversion. The real-time offset and gain calibration schemes do not affect the conversion process."

[118] How do they achieve simultaneous rejection at both frequencies? According to the datasheet, "The digital decimation SINC filter has been modified in order to offer staggered zeros in its transfer function. This modification is intended to widen the main notch in order to be less sensitive to oscillator deviation or line frequency drift. The MCP3551 filter has staggered zeros spread in order to reject both 50 Hz and 60 Hz line frequencies simultaneously." For maximum rejection at a *single* powerline frequency, use the MCP3550-50 or -60, which offer 120 dB (typ) normal-mode rejection at the corresponding frequency, once again with the internal clock.

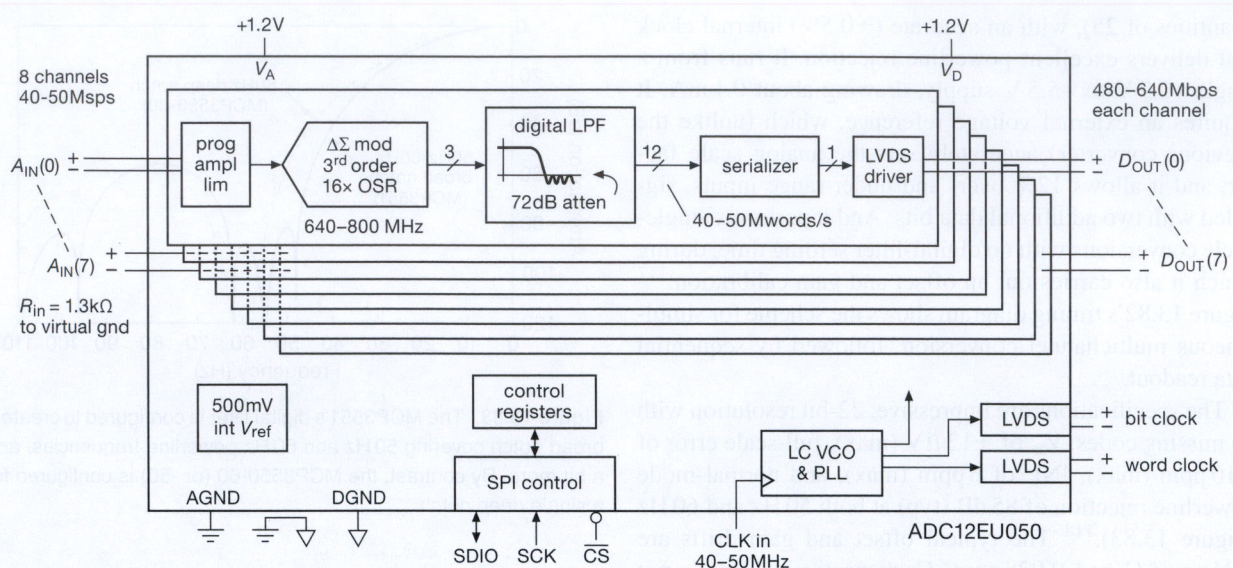

Figure 13.84. The ADC12EU050 is a fast 8-channel delta–sigma converter with 3-bit wordstreams generated by third-order modulators. Its PLL includes an on-chip *LC* VCO to generate the 16× oversampling clock from the 40–50 MHz sample-rate clock input.

integrated circuit, and also often incorporated within more complex ICs. A PLL contains a phase detector, amplifier, and voltage-controlled oscillator (VCO), and represents a blend (sometimes called "mixed signal") of digital and analog techniques. A few of its applications, which we will discuss shortly, are frequency multiplication and frequency synthesis, clock generation and clock recovery, tone decoding, and demodulation of AM, FM, and digitally modulated signals.

In the past there had been some reluctance to use PLLs, partly because of the complexity of discrete PLL circuits and partly because of a feeling that they cannot be counted on to work reliably. With inexpensive and easy-to-use PLLs now widely available, the first barrier to their acceptance has vanished. And with proper design and conservative application, the PLL is as reliable a circuit element as an op-amp or flip-flop.

Figure 13.85 shows the classic PLL configuration. The phase detector is a device that compares two input frequencies, generating an output that is a measure of their phase difference (if, for example, they differ in frequency, it gives a periodic output at the difference frequency). If f_{in} doesn't equal f_{VCO}, the phase-error signal, after being filtered and amplified, causes the VCO frequency to deviate in the direction of f_{in}. If conditions are right (lots more on that soon), the VCO will quickly "lock" to f_{in}, maintaining a fixed phase relationship with the input signal.

At that point the filtered output of the phase detector is

a dc signal, and the control input to the VCO is a measure of the input frequency, with obvious applications to tone decoding (sometimes used over telephone lines) and FM demodulation. The VCO output is a locally generated frequency equal to f_{in}, thus providing a clean replica of f_{in}, which may itself be noisy. Since the VCO output can be a triangle wave, sinewave, or whatever, this provides a nice method of generating a sinewave, say, locked to a train of input pulses.

In one of the most common applications of PLLs, a modulo-*n* counter is hooked between the VCO output and the phase detector, thus generating a multiple of the input reference frequency f_{in}. This is an ideal method for generating clocking pulses at a multiple of the power-line frequency for integrating ADCs (dual-slope, charge-balancing) in order to have infinite rejection of interference at the powerline frequency and its harmonics. It also provides the basic technique of frequency synthesizers.

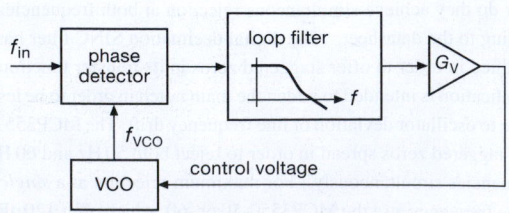

Figure 13.85. Phase-locked loop.

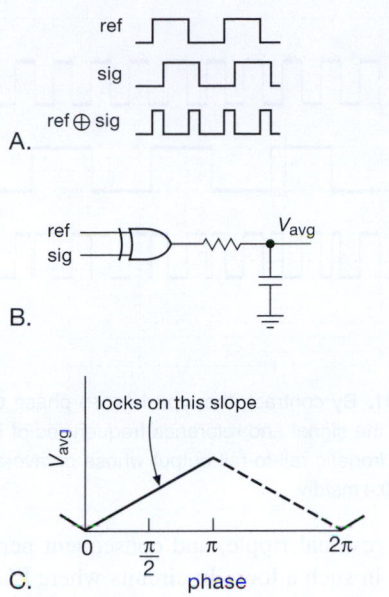

Figure 13.86. Exclusive-OR-gate phase detector (type I).

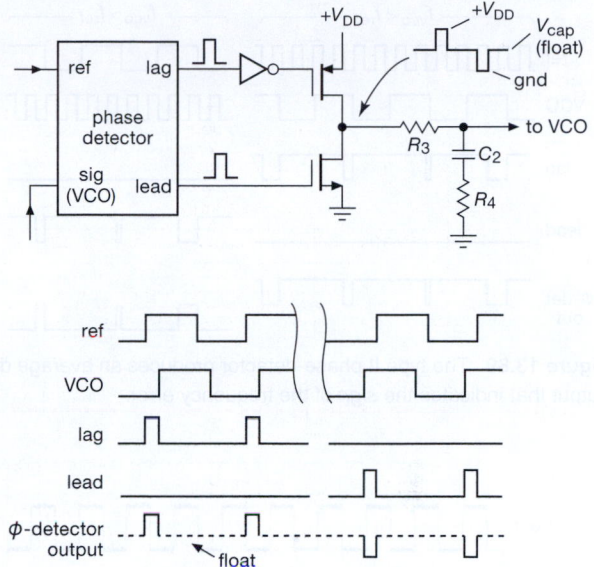

Figure 13.87. Edge-sensitive lead–lag phase detector (type II).

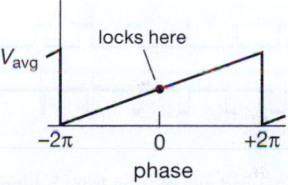

Figure 13.88. Type II phase detector output.

13.13.2 PLL components

A. The phase detector

Let's begin with a look at the phase detector (PD). There are two basic types, sometimes referred to as type I and type II.

The **type I phase detector** is applicable to either analog- or digital-input signals, and performs a simple multiplication of the inputs. For *digital* signals this is just an exclusive-OR gate (Figure 13.86). With lowpass filtering, the graph of the output voltage versus phase difference is a ramp, as shown, for input square waves of 50% duty cycle. For *analog* signals the type I "linear" phase detector is a true analog multiplier (called a "four-quadrant multiplier" or a "balanced mixer"), with similar output-voltage-versus-phase characteristics as with the digital XOR phase detector. Highly linear phase detectors of this type are essential for synchronous detection (also known as *lock-in detection*).

The **type II phase detector**, on the other hand, is a purely digital beast, driven by digital transitions (edges). It is sensitive only to the relative timing of *edges* between the signal and VCO input, as shown in Figure 13.87. The phase-comparator circuitry generates either *lead* or *lag* output pulses, depending on whether the transitions of the VCO output occur before or after the transitions of the reference signal, respectively. The width of these pulses is equal to the time between the respective edges, as shown. The output circuitry then either sinks or sources

current (respectively) during those pulses and is otherwise open-circuited, generating an average output-voltage-versus-phase difference like that in Figure 13.88. This is completely independent of the duty cycle of the input signals, unlike the situation with the type I phase detector.

Another nice feature of this phase detector is the fact that the output pulses disappear entirely when the two signals are in lock. This means that there is no "ripple" present at the output to generate periodic phase modulation in the loop, as there is with the type I phase detector. And while we're lavishing praise on the type II, let's point out that it has the pleasant property of producing an average dc output that is indicative of the *sign* of the frequency error (Figures 13.89–13.91). For this reason it's sometimes called a "phase-frequency detector" (PFD). We'll see how that ensures prompt lockup in a PLL.

The classic 74HC4046 PLL (which includes both oscillator and phase detector) gives you a choice (it contains both types of phase detector). Here is a comparison of the properties of the two basic types of phase detector:

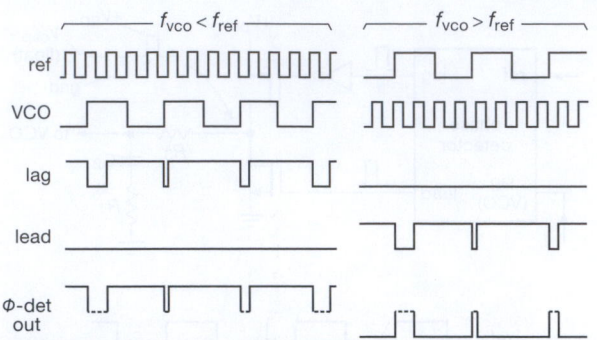

Figure 13.89. The type II phase-detector produces an average dc output that indicates the *sign* of the frequency error.

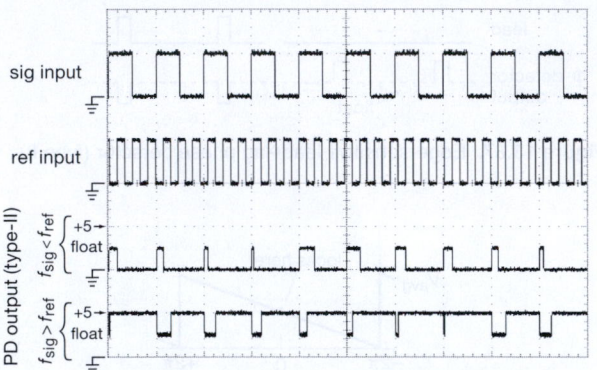

Figure 13.90. Measured waveforms from a type II phase detector driven with greatly mismatched frequencies. The 1 kHz signal and π-kHz reference inputs shown produce the phase-detector output shown in the third trace when driving a 10k–10k resistive divider that floats to +2.5 V. The bottom trace shows what happens when the inputs are interchanged. Horizontal: 1 ms/div.

Parameter	Type I exclusive-OR	Type II edge triggered ("charge pump")
Input duty cycle	50% optimum	Irrelevant
Lock on harmonic?	Yes	No
Rejection of noise	Good	Poor
Residual ripple at $2f_{IN}$	High	Low
Lock range (L)	Full VCO range	Full VCO range
Capture range	$fL(f<1)$	L
Output frequency when out of lock	f_{center}	f_{min}

There is one additional point of difference between the two kinds of phase detectors. The type I detector is always generating an output wave, which must then be filtered by the loop filter (much more on this later). Thus, in a PLL with type I phase detector, the loop filter acts as a lowpass filter, smoothing this full-swing logic-output signal. There will

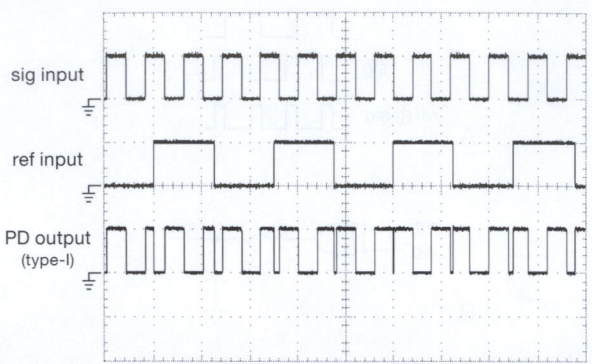

Figure 13.91. By contrast, the type I (XOR) phase detector, presented with the signal and reference frequencies of Figure 13.90, produces a frenetic rail-to-rail output whose dc average is $V_{DD}/2$. Horizontal: 0.4 ms/div.

always be residual ripple, and consequent periodic phase variations, in such a loop. In circuits where PLLs are used for frequency multiplication or synthesis, this adds "phase-modulation sidebands" to the output signal.

By contrast, the type II phase detector generates output pulses only when there is a phase error between the reference and the VCO signal. Since the phase-detector output otherwise looks like an open circuit, the loop filter capacitor then acts as a voltage-storage device, holding the voltage that gives the right VCO frequency. If the reference signal moves away in frequency, the phase detector generates a train of short pulses, charging (or discharging) the capacitor to the new voltage needed to put the VCO back into lock. It is a phase-error integrator.

"Dead zone" and "backlash"

A persistent problem with early PLLs using type II phase detectors was the presence of a *dead zone*: the phase pulses became vanishingly small at nearly zero phase error, so the loop tended to "hunt" (bounce back and forth), producing phase modulation and jitter. And this was exacerbated by the effects of capacitive loading at the phase-detector output. For applications that need a clean signal (for example, the synthesized oscillator in a cellphone, communications receiver, or RF frequency synthesizer) this was (and is) a serious problem. The solution, now nearly universally adopted, is to introduce some intentional overlap of the sourcing and sinking output pulses; to make this work you need to reconfigure the phase detector to produce *current* pulses (rather than voltage pulses).

Figure 13.92 shows how this is done: the current source or sink is activated by the first rising edge of the signals being compared (reference or signal, respectively), but it is not turned off until a short interval after the complementary

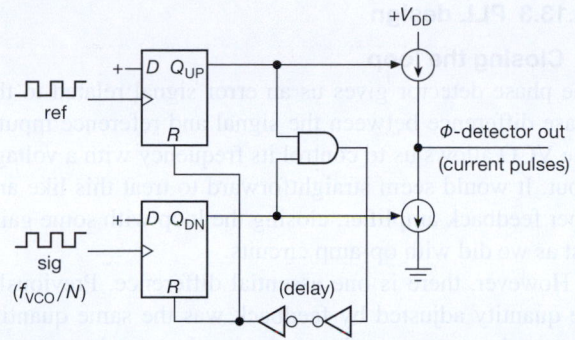

Figure 13.92. Improved type II phase detector (the '9046 version is shown here) replaces the switches with current sources and prevents a dead zone and backlash by creating intentional overlap of phase pulses.

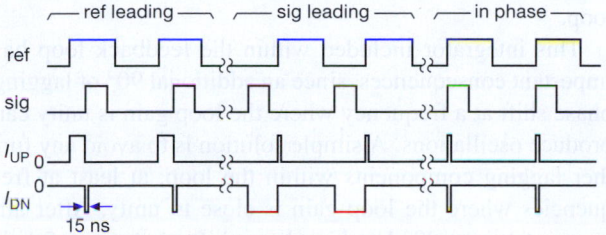

Figure 13.93. Current pulses (both sourcing and sinking) for the phase detector of Figure 13.92. The 15 ns pulses are created by the anti-backlash circuit.

current source is switched on. This "anti-backlash" circuit guarantees that the output pulses never vanish. When the two signals are exactly in phase (the loop is *locked*), the current pulses are short (~15 ns for the 74HCT9046, an improved version of the classic '4046) and of opposite sign, thus cancelling (Figure 13.93). Moving away from lock, a small phase difference produces an unbalanced pair of current pulses. This linear behavior around zero phase solves the problem; and capacitive loading does not cause trouble, because it behaves like a perfect integrator.

A cheap cure for backlash, if you need it, is to put a large resistor across the loop filter capacitor (C_2 in Figure 13.87), which biases the loop away from the dead zone. The trade-off is that you introduce a nonzero phase shift, which is not well defined; but at least you've gotten rid of the jitter.

B. The VCO

An essential component of a PLL is an oscillator whose frequency can be controlled by the phase-detector output. We discussed VCOs back in Chapter 7 (§§7.1.4D and 7.1.5D), and we'll see them again presently, in a PLL design example. For now, let's just look at the simple *RC* voltage-

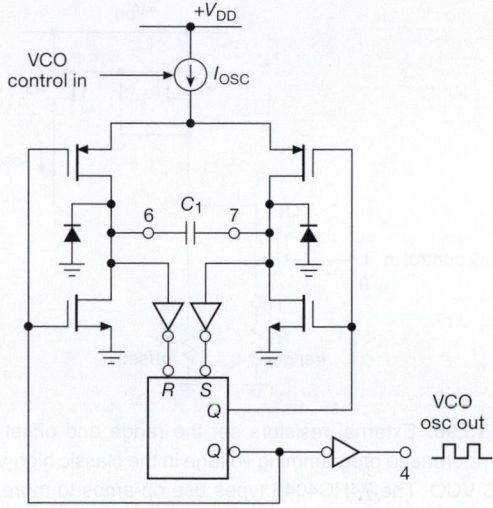

Figure 13.94. Voltage-controlled *RC* oscillator used in the classic '4046 PLL. Output frequency is approximately proportional to controlled current I_{osc}, which charges external capacitor C_1 alternately through the pMOS switches.

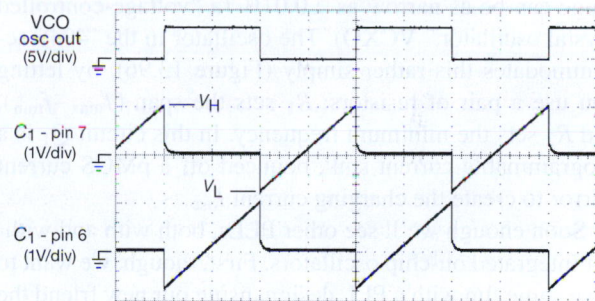

Figure 13.95. Observed waveforms of a 74HC4046 oscillator, with $V_{CC}=3.3$ V; with a 5 V supply the ramp starts at the same voltage but ends 0.2 V higher. Horizontal: 10 μs/div.

controlled oscillator used in the '4046 and its successors (Figure 13.94).

The operation is simple: the flip-flop's output holds one side of the external timing capacitor C_1 at ground (via an nMOS switch) while coupling charging current I_{osc} to the other side (via a pMOS switch). The cycle reverses when the rising voltage reaches the inverter's threshold, approximately +1.1 V. Figure 13.95 shows measured waveforms for an 'HC4046 powered from +3.3 V, with $C_1=10$ nF and $I_{osc}=0.85$ mA. Note that each cycle begins at approximately −0.7 V, clamped at a diode drop below ground when the high side is switched to ground.

In a PLL you often want to restrict the oscillator's tuning range, to span a modest range of frequency centered on the

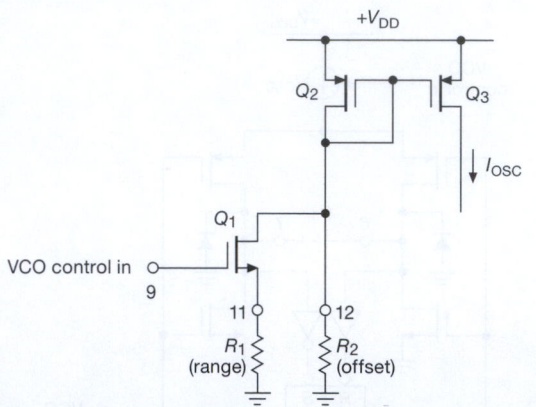

Figure 13.96. External resistors set the range and offset of the ground-referenced programming voltage in the classic high-voltage CD4046 VCO. The 74HC4046 types use op-amps to more tightly control the R_1 and R_2 currents.

desired output frequency. For example, the PLL in an FM radio needs to span ± 10 MHz about a central frequency of ~ 100 MHz; and we'll see examples later in which this range can be as narrow as $\pm 0.01\%$ (a "voltage-controlled crystal oscillator," VCXO). The oscillator in the '4046 accommodates this rather simply (Figure 13.96), by letting you use a pair of resistors: R_1 sets the span ($f_{\max}-f_{\min}$), and R_2 sets the minimum frequency. In this circuit Q_1 is a programmable current sink, bounced off a pMOS current mirror to create the charging current I_{osc}.

Soon enough we'll see other PLLs, both with and without integrated on-chip oscillators. First, though, we want to have some fun with a PLL design, using our new friend the '4046. Keep in mind, though, that PLLs (and their VCOs) don't have to be restricted to maximum speeds in the tens of megahertz. In fact, it's probably accurate to say that most PLLs in the world earn their livings at frequencies in the hundreds to thousands of megahertz. At those frequencies you don't use *RC* timing – instead you use *LC* circuits (tuned with a voltage-variable capacitor, known as a *varactor*), or a *ring oscillator* (a chain of inverters) tuned by adjusting the operating current (a "starved inverter chain"), or more exotic techniques such as a surface acoustic wave (SAW) delay-line oscillator or a resonator made from a silicon microelectromechanical system (MEMS). A VCO for use in a phase-locked loop doesn't have to be particularly linear in its frequency-versus-control-voltage characteristic, but if it is highly nonlinear, the loop gain (see below) will vary according to the signal frequency, compromising loop stability.

13.13.3 PLL design

A. Closing the loop

The phase detector gives us an error signal related to the phase difference between the signal and reference inputs. The VCO allows us to control its frequency with a voltage input. It would seem straightforward to treat this like any other feedback amplifier, closing the loop with some gain, just as we did with op-amp circuits.

However, there is one essential difference. Previously, the quantity adjusted by feedback was the same quantity measured to generate the error signal, or at least a proportional quantity. For example, in a voltage amplifier we measured output voltage and adjusted input voltage accordingly. But in a PLL there's an integration; we measure *phase*, but adjust *frequency*, and phase is the integral of frequency. This introduces a 90° lagging phase shift in the loop.

This integrator included within the feedback loop has important consequences, since an additional 90° of lagging phase shift at a frequency where the loop gain is unity can produce oscillations. A simple solution is to avoid any further lagging components within the loop, at least at frequencies where the loop gain is close to unity. After all, op-amps have a 90° lagging phase shift over most of their frequency range, and they work quite nicely. This is one approach, and it produces what is known as a "first-order loop." It looks just like the PLL block diagram shown earlier, with the lowpass filter omitted.

Although they are useful in many circumstances, first-order loops don't have the desirable property of acting as a "flywheel," allowing the VCO to smooth out noise or fluctuations in the input signal. Furthermore, a first-order loop will not maintain a fixed phase relationship between the reference and VCO signals, because the phase detector output drives the VCO directly. A "second-order loop" has additional lowpass filtering within the feedback loop (as drawn earlier), carefully designed to prevent instabilities. This provides flywheel action and also reduces the "capture range" and increases the capture time. Furthermore, with type II phase detectors, a second-order loop guarantees phase lock with zero phase difference between reference and VCO, as will be explained soon. Second-order loops are used almost universally, because the applications of PLLs usually demand an output frequency with low phase noise and some "memory," or flywheel action. Second-order loops permit high loop gain at low frequencies, resulting in high stability (in analogy with the virtues of high loop gain in feedback amplifiers). Let's get right down to business, illustrating the use of phase-locked loops with a design example.

13.13.4 Design example: frequency multiplier

Generating a fixed multiple of an input frequency is one of the most common applications of PLLs. This is done in frequency synthesizers, in which an integer multiple n of a stable low-frequency reference signal (1 Hz, say) is generated as an output; n is settable digitally, giving you a flexible signal source, easily controlled through a digital interface. In more mundane applications, you might use a PLL to generate a clock frequency locked to some other reference frequency already available in the instrument. For example, suppose we want to generate a 61.440 kHz clock signal for a dual-slope ADC. That particular choice of frequency permits 7.5 measurement cycles per second, allowing 4096 clock periods for the ramp-up (remember that dual-slope conversion uses a constant time interval) and 4096 counts full scale for the constant-current ramp-down. The unique virtue of a PLL scheme is that the 61.440 kHz clock can be locked to the 60 Hz powerline (61,440=60×1024), giving infinite rejection of 60 Hz pickup present on any signal input to the converter, as we discussed in §13.8.4.

We begin with the standard PLL scheme, with a divide-by-n counter added between the VCO output and the phase detector (Figure 13.97). In this diagram we indicate the units of gain for each function in the loop. That will be important in our stability calculations. Note particularly that the phase detector converts phase to voltage and that the VCO converts voltage to the time derivative of phase (i.e., frequency). This has the important consequence that the VCO is actually an integrator, with phase representing the variable in the lower part of the diagram; a fixed input-voltage error produces a linearly rising phase error at the VCO output. The lowpass filter and the divide-by-n counter both have unitless gain.

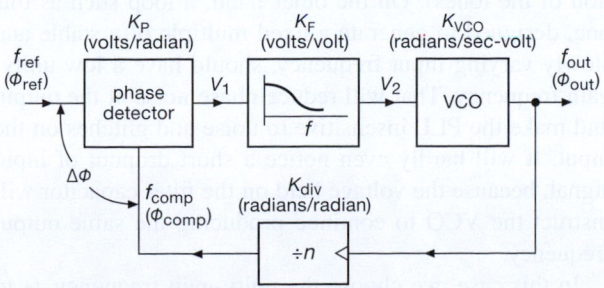

Figure 13.97. Frequency-multiplier block diagram.

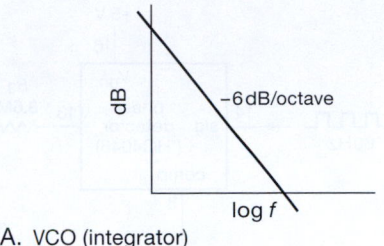

A. VCO (integrator)

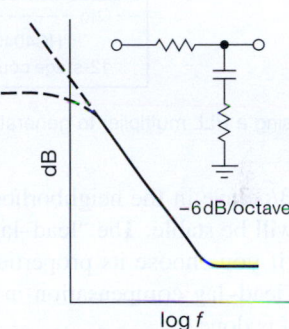

B. low-pass filter (lead-lag)

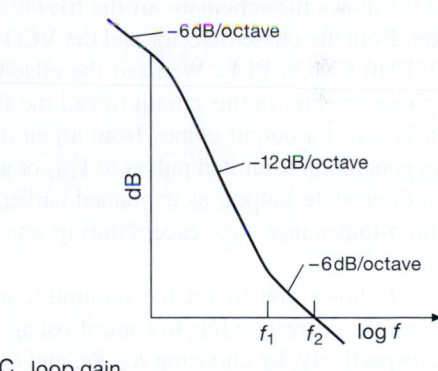

C. loop gain

Figure 13.98. PLL Bode plots.

A. Stability and phase shifts

The trick to a stable second-order PLL is shown in the Bode plots of loop gain in Figure 13.98. The VCO acts as an integrator, with $1/f$ response and 90° lagging phase shift (i.e., its response is proportional to $1/j\omega$, a current source driving a capacitor). In order to have a respectable phase margin (the difference between 180° and the phase shift around the loop at the frequency of unity loop gain), the lowpass filter has an additional resistor in series with the capacitor to stop the rolloff at some frequency (fancy name: a "zero"). The combination of these two responses produces the loop gain shown. As long as the loop gain

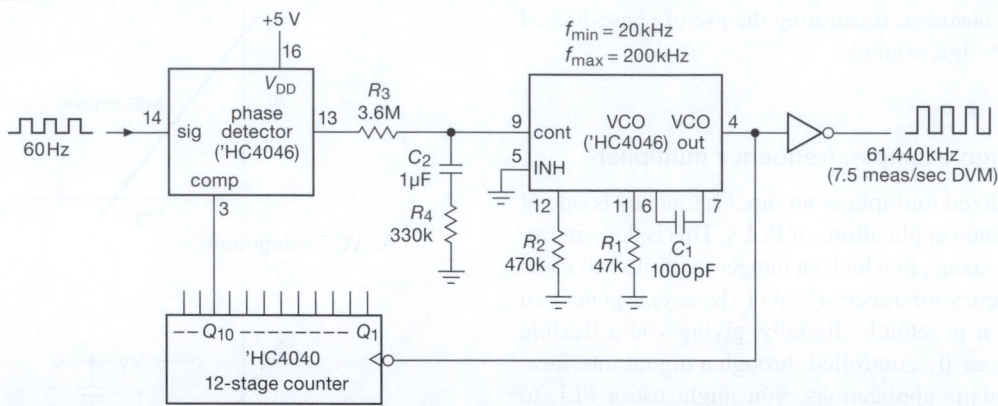

Figure 13.99. Using a PLL multiplier to generate a clock locked to the 60 Hz ac line. Parts values are for TI's CD74HC4046A.

rolls off at 6 dB/octave in the neighborhood of unity loop gain, the loop will be stable. The "lead–lag" lowpass filter does the trick, if you choose its properties correctly (this is the same as lead–lag compensation in op-amps). Next we'll see how it is done.

B. Loop-gain calculations

Figure 13.99 shows the schematic of the 61.440 kHz PLL synthesizer. Both the phase detector and the VCO are parts of an 'HC4046 CMOS PLL. We used the edge-triggered (type II) phase detector in this circuit (recall the 4046 contains both kinds). Its output comes from a pair of CMOS transistors generating saturated pulses to V_{DD} or ground. It is really a three-state output, as explained earlier, since it is in the high-impedance state except during actual phase-error pulses.

The VCO allows you to set the minimum and maximum frequencies corresponding to control voltages of zero and V_{DD}, respectively, by choosing R_1, R_2, and C_1 according to some design graphs. We have made the choices shown, based upon an initial design from the datasheet, validated (and adjusted!) by some bench measurements – see the "real-world" commentary that follows, in §13.13.4E. *Note*: the 4046 has chronic severe supply-sensitivity disease; check the graphs on the datasheet. The rest of the loop is standard PLL procedure.

Having rigged up the VCO range, the remaining task is the lowpass-filter design. This part is crucial. We begin by writing down the loop gain, as in the "PLL gain calculation" box, considering each component (refer to Figure 13.97). Take special pains here to keep your units consistent; don't switch from f to ω or (worse) from hertz to kilohertz. Having chosen the oscillator components, divider ratio, and supply voltages, we need to determine the

one remaining gain term (that of the loop filter), K_F. We do this by writing down the overall loop gain, remembering that the VCO is an integrator:

$$\phi_{out} = \int V_2 K_{VCO} dt.$$

The loop gain is therefore given by

$$\text{Loop gain} = K_P K_F \frac{K_{VCO}}{j\omega} K_{div}$$
$$= 0.40 \times \frac{1 + j\omega R_4 C_2}{1 + j\omega(R_3 C_2 + R_4 C_2)} \times \frac{3.77 \times 10^5}{j\omega} \times \frac{1}{1024}.$$

Now comes the choice of frequency at which the loop gain should pass through unity. The idea is to pick a unity-gain frequency high enough so that the loop can follow input-frequency variations you want to follow, but low enough to provide flywheel action to smooth over noise and jumps in the input frequency. For example, a PLL designed to demodulate an FM input signal, or to decode a rapid sequence of input tones, needs to have rapid response (for the FM input signal, the loop should have as much bandwidth as the input signal, i.e., response up to the maximum modulating frequency; while to decode input tones, its response time must be short compared with the time duration of the tones). On the other hand, a loop such as this one, designed to generate a fixed multiple of a stable and slowly varying input frequency, should have a low unity-gain frequency. That will reduce phase noise at the output and make the PLL insensitive to noise and glitches on the input. It will hardly even notice a short dropout of input signal, because the voltage held on the filter capacitor will instruct the VCO to continue producing the same output frequency.

In this case, we choose the unity-gain frequency f_2 to be 2 Hz, or 12.6 radians per second. This is well below

PLL GAIN CALCULATION

Component	Function	Gain	Gain calculation ($V_{DD} = +10V$)
Phase detector	$V_i = K_P \Delta\phi$	K_P	$K_P = \dfrac{V_{DD}}{4\pi}$ (0 to V_{DD} ↔ $-360°$ to $+360°$)
Lowpass filter	$V_2 = K_F V_1$	K_F	$K_F = \dfrac{1 + j\omega R_4 C_2}{1 + j\omega(R_3 C_2 + R_4 C_2)}$ volts/volt
VCO	$\dfrac{d\phi_{out}}{dt} = K_{VCO} V_2$	K_{VCO}	20kHz ($V_2 = 1V$) to 200kHz ($V_2 = 4V$) $\to K_{VCO} = 60$kHz/volt $= 3.77 \times 10^5$ radians/second-volt
Divide-by-n	$\phi_{comp} = \dfrac{1}{n}\phi_{out}$	K_{div}	$K_{div} = \dfrac{1}{n} = \dfrac{1}{1024}$

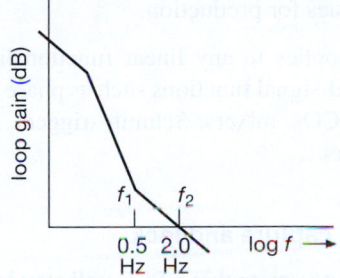

Figure 13.100. Stabilizing a second-order PLL: -6 dB/octave rolloff of loop gain around the frequency of unity gain.

the reference frequency, and you wouldn't expect genuine powerline frequency variations on a scale shorter than this (remember that the 60 Hz power is generated by enormous generators with lots of mechanical inertia). As a rule of thumb, the breakpoint of the lowpass filter (its "zero") should be lower by a factor of at least 3 to 5 for a comfortable phase margin. Remember that the phase shift of a simple RC goes from $0°$ to $90°$ over a frequency range of roughly 0.1 to 10 times the -3 dB frequency (its "pole"), with a $45°$ phase shift at the -3 dB frequency. In this case we put the frequency of the zero, f_1, at 0.5 Hz, or 3.1 radians per second (Figure 13.100). The breakpoint f_1 determines the $R_4 C_2$ time constant: $R_4 C_2 = 1/2\pi f_1$. Tentatively, take $C_2 = 1\mu F$ and $R_4 = 330k$. Now all we do is choose R_3 so that the magnitude of the loop gain equals 1 at f_2. In this case, that works out to $R_3 = 3.6\,M\Omega$.

Exercise 13.7. Show that these choices of filter components actually give a loop gain of magnitude 1.0 at $f_2 = 2.0$ Hz.

Sometimes the filter values are inconvenient, so you have to readjust them, or move the unity-gain frequency somewhat. With a CMOS phase-locked loop these values are acceptable (the VCO input terminal has a typical input impedance of $10^{12}\,\Omega$). For VCOs with a low input impedance you might want to use an external op-amp buffer.

We used an edge-triggered (type II) phase detector in this circuit example because of its simplified loop filter. In practice, that might not be the best choice for a PLL locked to the 60 Hz powerline because of the relatively high noise level present on the 60 Hz signal: many engineers have stumbled on this point, with a noisy reference signal causing false type-II triggering. With careful design of the analog input circuit (e.g., a lowpass filter followed by a Schmitt trigger) the type-II phase detector would likely perform satisfactorily; otherwise an exclusive-OR (type I) phase detector should be used.

C. "Cut and try"

For some people, the art of electronics consists of fiddling with filter component values until the loop "works." If you are one of those,[119] we will oblige you by looking the other way. We presented these loop calculations in detail because we suspect that much of the PLL's bad reputation is the result of too many people "looking the other way." Nevertheless, we can't resist supplying a hot tip for cut-and-try addicts: $R_3 C_2$ sets the smoothing (response) time of the loop,

---

[119] You can derive some comfort (and perhaps even hold your head high) from some remarks in the TI app note (referring to the loop-filter design) "Optimizing by trial and error should be considered in all cases."

and R_4/R_3 determines the damping, i.e., absence of overshoot for step changes in frequency. You might begin with a value of R_4 somewhere in the range of 10% to 20% of R_3.

D. Loop damping and jitter

A side effect of the nonzero "damping" resistor R_4 is the creation of some jitter in the PLL output. An easy way to see this is to realize that even at high frequencies the loop filter permits a fraction $R_4/(R_3 + R_4)$ of the raw phase-detector output to reach the VCO. For typical ratios, $R_3 \approx 10R_4$, this can add substantial jitter to the VCO output. The usual solution is to add a small capacitor ($\sim C_2/20$) from the VCO control input to ground, preferably close to the VCO pin to filter any other high-frequency noise as well.

E. PLL real-world design

We sailed through this design example, having faith that the information in the datasheets for the IC we chose (the popular 'HC4046) was reliable. That faith was, perhaps, a bit misplaced. To give this cautionary section some context, here's the story (the short version) of our oscillator component choices in Figure 13.99:

We wanted a $3\times$ safety margin for our 61 kHz center frequency, so we set f_{min}=20 kHz and f_{max}=200 kHz. We chose TI's CD74HC4046A because it was RCA's original time-proven design. Based on the datasheet's graphs, a good value for the timing capacitor C_1 was 1000 pF. For the timing resistors one of us started with the R_1 plots and came up with 30k and 300k for R_1 and R_2, and we edited the figure accordingly. The other author started with the R_2 plot (as suggested by two manufacturers, but it should not have mattered) and came up with 40 k and 410 k. Worried about these and other inconsistencies (e.g., one spec in the TI datasheet gives f_{osc}=400 kHz, typ, for C_1=1 nF and R_2=220k, while their Figure 27 suggests it should be more like 33 kHz), we went to the bench and found actual values of 45k and 482k for the desired frequency range. The values we chose initially would have worked, despite the factor of 1.5 deviation from reality, but they would have used up half of our $3\times$ safety margin.[120]

So, what's going on here? Each of the 'HC4046 manufacturers uses a different circuit for their VCO design.[121] Although intended to be predictable and linear, in practice the VCO control is nonlinear, and its parameters vary with control current, supply voltage, and operating frequency, especially above 10 MHz. Although you can find analytical expressions for the VCO's frequency (ON Semi app note AN1410), the recommended method is still to start with the timing component values (R_1, R_2, and C_1) from datasheet graphs; then the designer is sternly admonished to adjust and validate those values with careful bench measurements before committing to manufacturing.

This sort of variability and lack of confident predictability leads us to render this advice:

(1) choose one manufacturer for your production design, and do not allow substitutes;
(2) choose a wide safety margin for f_{min} and f_{max}, such as the $3\times$ factor in our Figure 13.99;
(3) replace your initial paper calculations with measured bench values for production.

Rule (1) applies to any linear functionality in a logic IC, e.g., mixed-signal functions such as phase comparators, oscillators, VCOs, mixers, Schmitt triggers, monostables, or comparators.

13.13.5 PLL capture and lock

Once locked, it is clear that a PLL will stay locked as long as the input frequency doesn't wander outside the range of the feedback signal and doesn't change more rapidly than the loop's bandwidth can track. An interesting question to ask is how PLLs get locked in the first place. After all, an initial frequency error results in a periodic output from the phase detector at the difference frequency. After filtering by the lowpass filter, it is reduced to small-amplitude wiggles rather than to a nice clean dc error signal.

A. Capture transient

The answer is a little complicated. First-order loops will always lock, because there is no lowpass attenuation of the error signal. Second-order loops may or may not lock, depending on the type of phase detector and the bandpass of the lowpass filter. In addition, the exclusive-OR (type I) phase detector has a limited *capture* range that depends on the filter time constant (this fact can be used to advantage

---

[120] We made additional bench measurements on both old and new 74HC4046A parts from TI, NXP, ON Semi, and Fairchild. We found good self-consistency within any single manufacturers parts; the variations in span and zero frequency within a batch and over a 15-year period were generally under 5%. Measured frequencies for the NXP, ON Semi, and Fairchild parts deviated by +5%, +160%, and −60%, respectively, from that of the TI parts, when configured with the timing component values shown in Figure 13.99.

[121] For example, TI biases R_2 from V_{DD}−0.7 V, whereas ON Semi biases it from $V_{DD}/3$. Their current mirrors have nominal gains of 7.5 and 25 respectively. We think NXP's 74HCT9046 is arguably the best mannered '4046 part available, and the NXP datasheets have better graphs.

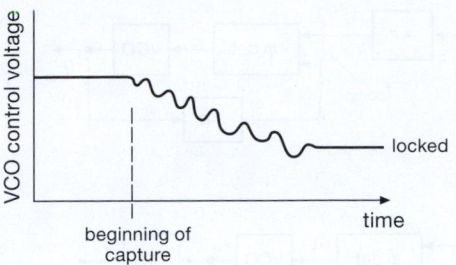

Figure 13.101. PLL capture transient.

if you want a PLL that will lock to signals only within a certain frequency range).

For a type I phase detector you might wonder how the loop can lock up at all, because, with the phase detector's output being periodic at the difference frequency, the VCO's frequency should just wiggle back and forth forever. But looking more closely, the capture transient goes like this: as the (phase) error signal brings the VCO frequency closer to the reference frequency, the error-signal waveform varies more slowly, and vice versa. So the error signal is asymmetric, varying more slowly over that part of the cycle during which f_{VCO} is closer to f_{ref}. The net result is a nonzero average, i.e., a dc component that brings the PLL into lock.[122] If you look carefully at the VCO control voltage during this *capture transient*, you'll see something like what is shown in Figure 13.101. That final overshoot has an interesting cause. Even when the VCO *frequency* reaches its correct value (as indicated by correct VCO control voltage), the loop isn't necessarily in lock, because the *phase* may be wrong. So it may overshoot. As with snowflakes, each capture transient is an individual – it looks a bit different each time.

For a PLL with a type II phase detector, the situation is considerably simpler: because this kind of detector produces a dc component indicating the direction of frequency error (recall it's a "phase-frequency detector"), the VCO's frequency is steered rapidly in the right direction.

B. Capture and lock range

For the exclusive-OR (type I) phase detector, the capture range is limited by the lowpass-filter time constant. This makes sense, because if you begin sufficiently far away in

frequency, the error signal will be attenuated so much by the filter that the loop will never lock. It should be evident that a longer filter time constant results in narrower capture range, as does reduced loop gain. The edge-triggered phase detector does not have this limitation, because it acts like a true integrator of the phase-error charge pulses. Both types have a lock range extending to the limits of the VCO, given the available control input voltage.

C. Capture time

PLLs with type II (integrating) phase detectors will always lock (assuming the VCO has sufficient tuning range, of course), with a time constant characteristic of the loop bandwidth.

A type I (multiplier or mixer) phase detector, if followed by an integrating loop filter, will also lock – but it can take a very long time if the loop bandwidth is narrow. It can be shown that the lockup time is roughly $(\Delta f)^2/\text{BW}^3$, where Δf is the initial frequency error and BW is the loop bandwidth. So a PLL with 100 Hz loop bandwidth and 100 kHz comparison frequency might take a minute to lock up if the VCO's initial frequency is 10% away from lock.

In such cases you sometimes see a cute trick: a slow full-range sawtooth is applied to the VCO control voltage of the unlocked loop, until lock takes place. For example, I have in my hand (I'm typing one-handed) an Efratom model FRS rubidium frequency standard, which uses the weak but extremely stable atomic resonance in an optically pumped vapor cell as a reference to which a high-quality crystal oscillator is phase locked. The 20 MHz ovenized crystal oscillator (XO) is voltage controlled (a VCXO), with a narrow tuning range (of order ± 1 kHz); it is the flywheel in a low-bandwidth PLL (the loop integrator has R=2M, C=1µF).

Without some help, this thing would take forever to lock. The helpful "Operation Manual" explains how they do it: "When no 'lock' signal is present... [there is] a slow down sweep of the crystal control voltage of about 250 mV per second. Sweeping continues until 'Lock' occurs. 'Lock' then disables the sweep circuit by disconnecting [analog switch] U3 pins 13 and 14, and connecting U3 pins 12 and 14. This switching places the integrator under fundamental loop control."

Such gimmicks aren't necessary with type II phase detectors, as we remarked above, thanks to their indication of both phase and (sign of) frequency difference. But passive mixer-type phase detectors (therefore type I) are prevalent in communications systems at very high radio frequencies, where the digital phase-frequency detector is impractical.

---

[122] Another way of looking at it is to realize that the error signal weakly modulates the VCO at the difference frequency $\Delta f = |f_{\text{ref}} - f_{\text{VCO}}|$. That frequency modulation puts weak symmetrical sidebands onto f_{VCO}, spaced apart by Δf. One of these is exactly at f_{VCO}, producing an average dc output component from the phase detector, which the loop filter integrates (type II) to bring the system into lock.

13.13.6 Some PLL applications

We have spoken already of the common use of phase-locked loops in frequency multiplication. The latter application, as in the preceding example, is so straightforward that there should be no hesitation about using these mysterious PLLs. In simple frequency-multiplication applications (e.g., the generation of higher clock frequencies in a digital system) there isn't even any problem of noise on the reference signal, and a first-order loop may suffice.

As will become evident, what you care about in a PLL depends on the application: there's a tradeoff among wide tuning range versus high quality (low phase noise, low jitter, low spurious frequency components) versus frequency step size versus loop bandwidth (and switching speed) versus low external component count. For example, for a microprocessor or memory clocking application you don't need high-quality waveforms, and you need only coarse tuning steps; for a cellphone synthesizer you want low phase noise and spurs, with tuning range and step size matched to the cell band and channelization; for a general-purpose sinewave synthesizer you want low phase noise and spurs, small tuning steps, and wide tuning range; for high-speed serial links you care about jitter, as you do when clocking high-quality ADCs (where jitter translates to distortion); and for a motherboard clock generator you'd like a single-chip solution that generates a suite of standard clocks (for the processor, memory, video, internal buses like PCIe and SATA, external serial ports like USB and Ethernet, and so on) without great concern about signal quality.

We would like now to describe two important variations (known as "n/m" and "fractional-n" synthesis) on this basic frequency-multiplication scheme; we'll continue with some other interesting applications of phase-locked techniques, to give an idea of the diversity of PLL uses. Finally, we'll conclude the discussion of phase-locked loops with examples of contemporary PLL ICs, which use a variety of clever tricks to create on-chip oscillators with admirable performance.

A. Fractional-*n* synthesis

The frequency multiplication scheme of Figure 13.97 generates an output frequency that is restricted to an integral multiple of the reference input: $f_{out}=nf_{ref}$. That's fine for an application like that of Figure 13.99, but it's not of much use for something like a general-purpose sinewave synthesizer, where you want to generate any old output frequency, perhaps settable down to 1 Hz, or even 0.001 Hz.

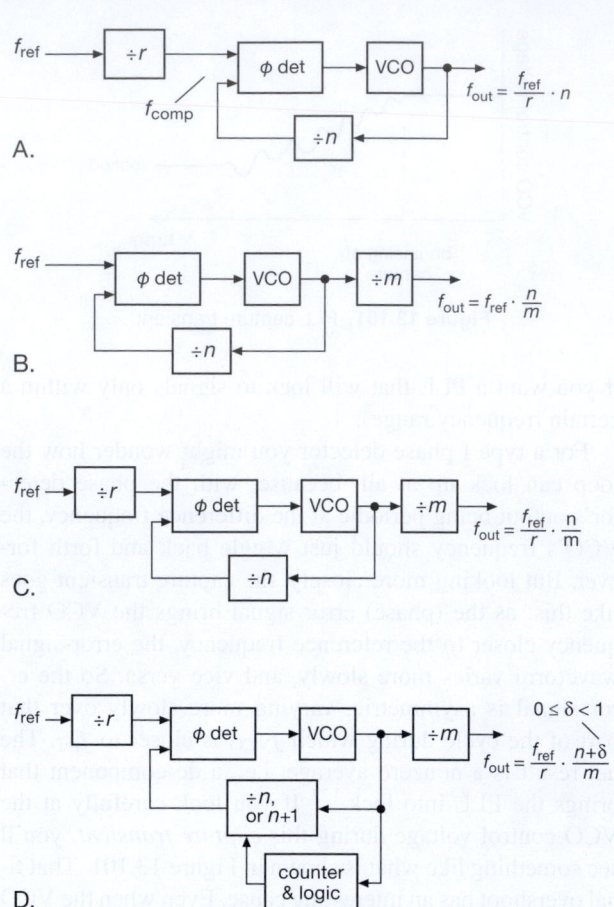

Figure 13.102. Getting to fractional-n. A–C. Integer-n with input prescaler, output scaler, and both. D. "Fractional-n" allows the feedback divider to effectively take on non-integer values. For simplicity the loop filter between phase detector and VCO has been omitted.

Input prescaler

Taking it in several steps (see Figure 13.102A), the first thing you might do is to reduce the reference frequency to the resolution step size, say 1 Hz. This can be done by "prescaling" the reference input frequency with a modulo-r counter: r is an integer, chosen so that $f_{comp}=f_{ref}/r$ equals the desired step size; for example, if we have a 10 MHz input reference (a common standard) and we want 1 Hz settability, we would choose $r=10^7$. The output frequency is then $f_{out}=n\cdot f_{ref}/r$.

OK, that would work. But the phase detector is now working on a pair of 1 Hz signals, which requires a very long loop time constant (many seconds). That's not good: it takes a long time to lock to a new frequency setting; and there's more phase noise, because the VCO's intrinsic instabilities are not corrected on short time scales (no loop gain at those frequencies). And (if you need more convincing), the phase detector's correction pulses to the VCO are at low frequency, producing spurious modulation sidebands ("spurs") that are close in frequency to the desired output frequency (to be precise, they are spaced apart by f_{comp}, going both up and down from f_{out}).

Output scaler

The next thing you might try is to keep a high reference frequency, but divide the *output* frequency instead (Figure 13.102B). The output frequency is now $f_{out}=f_{ref} \cdot n/m$. This looks pretty good: we can keep plenty of loop bandwidth (because the phase detector is operating at the high-frequency f_{ref}), and we get as small a step size as we want by choosing a large output divider modulus m.

And this works fine, as long as you're happy with low output frequencies. The problem is that the VCO now has to run at m times higher frequency to generate a given f_{out}. For example, with a 10 MHz reference input and with $m=10^7$ (for 1 Hz step size), the VCO would have to run at 1 GHz just to generate a 100 Hz output frequency ($n=100$). This is clearly a loser.

Input and output scalers

A compromise is in order: use integer frequency dividers at both input *and* output (Figure 13.102C). This way we can set the phase detector's comparison frequency somewhere between the (too small) output step size and the (larger than necessary) input reference frequency. The output frequency is now $f_{out}=(f_{ref}/r) \cdot (n/m)$. This is the standard configuration of an "integer-n" phase locked loop (because all three frequency dividers operate with an integer division ratio).

Taking our standard example with 10 MHz reference, we might choose $r=10^4$ (so $f_{comp}=1$ kHz) and $m=10^3$. The output step size is 1 Hz, the output frequency is n Hz, and we can generate output frequencies to 100 kHz (with 1 Hz resolution) with a VCO that goes to 100 MHz.

The whole enchilada: "fractional-n"

We've got a compromise among competing factors of step size, loop bandwidth, maximum output frequency, and maximum VCO frequency. In the example above we can get to a higher output frequency with the same 1 Hz step size and 10 MHz input frequency, (i.e., keeping the product $m \cdot r$ constant), but only at the expense of either smaller loop bandwidth (smaller m, larger r) or reduced maximum output frequency (larger m, smaller r).

Can we do better? The answer is yes, if we can somehow trick one of the dividers (say the n divider) into a non-integer "in-between" division ratio. We can do that *on the*

average with an integer-n divider, if we arrange things to change the modulus so that it spends some of its time as n, and the rest of its time as $n+1$.[123] This is *fractional-n* synthesis (Figure 13.102D). The output frequency is still $f_{out}=(f_{ref}/r) \cdot (n/m)$, but with n now permitted to take on a fractional value. With fractional-n synthesis you have (mostly) the best of both worlds: wide output-frequency range with high resolution (small step size), while retaining high f_{comp} (which permits plenty of loop bandwidth, and therefore fast lock and fast tracking, along with spurs widely offset from the synthesized frequency).

Fractional-n requires some additional counters and logic, to figure out how often to alternate between n and $n+1$. There's an analogous everyday situation: it's beneficial to keep the annual calendar (the kind you hang on a wall) synchronized to Earth's motion around the sun. Trouble is, there are not an integral number of days in a year. The Gregorian calendar's solution (leap years) is fractional-n: alternate the modulus between 365 and 366, with a 3:1 ratio, to achieve the (approximately) correct value of 365.25.[124,125]

Details, details...

Fractional-n synthesis[126] is a nice technique, but it's not without its own problems. For example, the phase detector is periodically presented with a phase discontinuity (i.e., each time the modulus is alternated), which creates periodic phase modulation at the output if not corrected or otherwise filtered. There are several tricks to fix this problem, involving either injecting compensating charge pulses at the phase detector's charge-pump output, or (probably better) a precomputed correction to the output waveform to create equally spaced output cycles (see later). Perhaps the best technique, though, is the use of delta–sigma

---

123 You sometimes hear the visually graphic term "pulse-swallowing" used for this subcircuit.

124 Astronomers, and other fussy readers, will complain that the number of Earth rotations in a year is actually one greater (366.25), and that Earth rotates once *not* in 24 hours, but in 23h 56m 4s (approximately). Of course they are right. But everyone hates a smarty-pants.

125 That gets the calendar close, but not perfect: there are 365.242374 solar days in a year. Hence the next-order fix: leap-years are *omitted* in years that are divisible by 100 (i.e., the turn of each century, which, being divisible by 4, would ordinarily be a leap year) unless they are also divisible by 400; we had a leap year in 2000, but we (more properly, our descendants; by then this book, and its authors, will be out of print) will not enjoy one in 2100, or 2200, or 2300. This will keep the calendar on track for about 8000 years.

126 Some PLLs do the fractional-n business on the input reference (r) divider; they're still called "fractional-n."

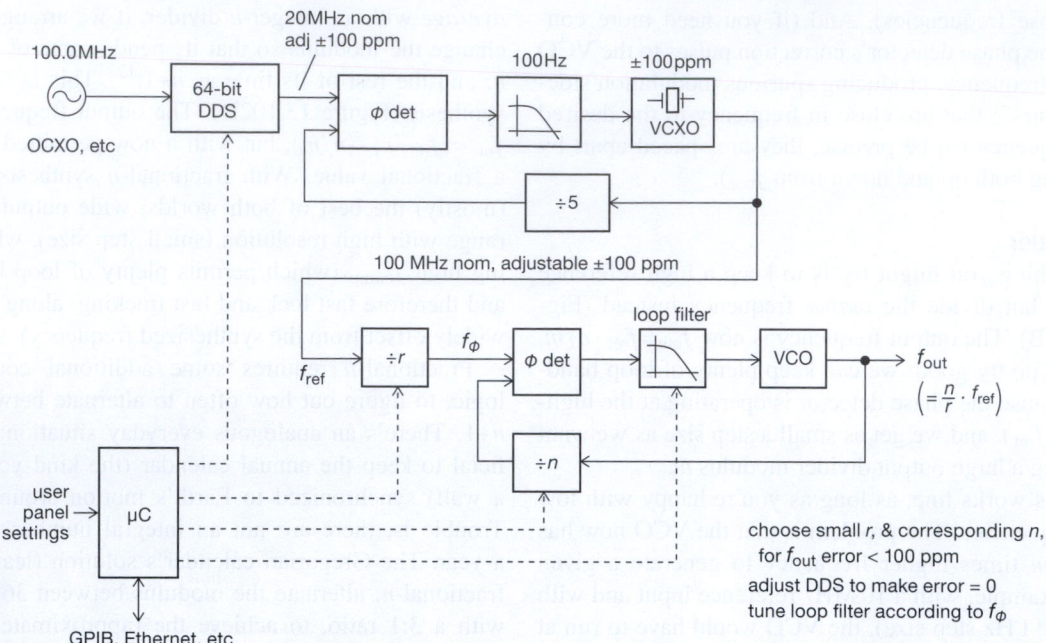

Figure 13.103. By making small frequency adjustments to its clean crystal-oscillator reference clock, the method of "rational approximation frequency synthesis" used in the SRS synthesizers achieves microhertz resolution while operating its PLL phase-detector at megahertz frequencies. The resulting output has excellent purity, with very low phase noise and absence of spectral "spurs."

modulation of the modulus: instead of simply alternating between the two moduli that surround the (fractional) desired modulus, the divider's modulus is distributed among a larger set, in such a way that the production of modulation sidebands is shaped to higher frequencies, and the production of discrete spurs is minimized. As with the delta–sigma modulators we saw earlier in the chapter, higher-order loops along with some randomization ("dithering") can be employed to reduce close-in spurs (analogous to its use there to suppress idle tones). This is a complicated business, and best left to the real professionals.[127] Bottom line: leave the converter design to others; but be aware of both the benefits and pitfalls, and examine the datasheets closely for the things you care about in your application.

B. Rational approximation synthesis

With an ingenious variation on integer-n synthesis, the ever-creative John Willison at Stanford Research Systems has devised a synthesizer that combines the best of both worlds: it operates with a small integer r value (thus a relatively high reference frequency input to the phase detector, for wide VCO loop bandwidth and thus low noise and jit-

ter sidebands), along with integer n (avoiding VCO phase modulation); but, with a bit of magic, it permits essentially infinite frequency resolution (*micro*hertz frequency setting) even though the phase detector's reference input is typically in the megahertz region.

How can this be? The trick is to choose a small integer r (and corresponding n) such that the synthesized frequency is *close to* (say within ±100 ppm of) the target frequency; you then adjust the master reference oscillator correspondingly, to bring the integer-synthesized output frequency on target. This technique, which they call "rational approximation frequency synthesis" (RAFS), was introduced in SRS's SG380 series of RF signal generators, which currently include models that provide outputs from dc to 6 GHz, settable with microhertz resolution. The use of integer PLL synthesis with a wide-bandwidth loop produces excellent output purity, seen for example in the phase noise specification of −116 dBc (relative to the "carrier," i.e., the signal amplitude) at an offset of 20 kHz from a 1 GHz output signal; and the use of a low-noise reference oscillator (the OCXO) keeps the close-in phase noise down to an impressive −80 dBc at an offset of just 10 Hz from a 1 GHz output.

Figure 13.103 shows the scheme, stripped down to its

---

[127] Take a look at National Semiconductor's App Note 1879, if you want to dip your toe into the turbulent waters.

basics. A microcontroller rides herd over the system, beginning with the choice[128] of r and n to get close to the desired f_{out}. The microcontroller also tunes the loop filter according to the resulting phase detector input frequency (here called f_ϕ). Finally, it fine-tunes the master clock, over the needed ±100 ppm range, by way of a 64-bit (thus effectively infinite resolution) direct digital synthesizer running from the clean fixed-frequency input clock. The DDS output is not as pure as its reference input (owing to irregular phase jumps inherent in the DDS process), so its output is cleaned up by phase locking a high quality crystal oscillator, whose frequency can be electrically "pulled" over a ±100 ppm range via a varactor (thus a "VCXO" – voltage-controlled crystal oscillator). With hindsight we can see that it's the VCXO's tuning range that constrains the initial selection of r and n.

To illustrate with an example, suppose we wish to synthesize an output at 1234.56789 MHz. You can bang away on a pocket calculator (does anyone under 50 still use those?), working your way up successive integer r values, until you find that $r=26$ gives you a "fractional-n" (320.9876514) that is within 100 ppm of an integer ($n=321$). So we go with the choice $[r,n]=[26,321]$, and we offset the master clock by -38.469 ppm (to 99.9961531 MHz) to get what we want. With this choice the phase detector's reference frequency f_ϕ is pleasantly high (~3.85 MHz), allowing plenty of loop bandwidth (thus low sideband noise, and absence of close-in spurs that would be caused by a low f_ϕ) in the synthesizing PLL.

In practice there are a host of details (as with any sophisticated and well engineered system) not seen in this simplified description. For example, (a) the output synthesizer tunes only over an octave (2:1 frequency range), driving a set of binary dividers and lowpass filters to generate the final output; (b) the actual production instruments use several staggered fine-tuned VCXOs, greatly relaxing the constraints on r and n (and resulting in values of f_ϕ typically greater than 10 MHz, with a worst case of 2.4 MHz); (c) there are additional DDSs and PLLs in the system, used among other things to create favorable clocking frequencies (which generally are chosen not to be the kind of "round-number" frequencies shown here, to prevent clock-collision artifacts); (d) the 64-bit DDS is dithered to reduce fixed-frequency spurious sidebands ("spurs"); and (e) there are additional subcircuits to provide for modulation, amplitude control, and the like. These are the sort of real-world

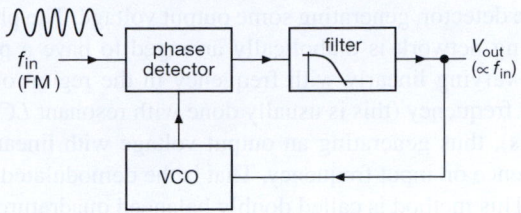

Figure 13.104. PLL FM discriminator.

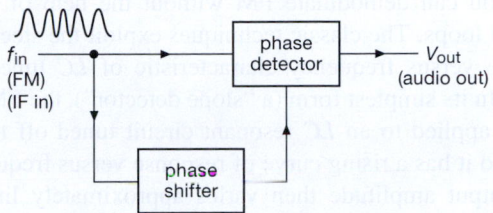

Figure 13.105. Quadrature FM detection.

issues that challenge the instrument designer, to whom great satisfaction can come when a nice solution is found.

C. FM detection

In frequency modulation, information is encoded onto a "carrier" signal by varying its frequency proportional to the information waveform. There are two methods of recovering the modulating information with phase detectors or PLLs. The word *detection* is used to mean a technique of demodulation.

In the simplest method, a PLL is locked to the incoming signal. The voltage controlling the VCO frequency is proportional to the input frequency and is therefore the desired modulating signal (Figure 13.104). In such a system you would choose the filter bandwidth to be wide enough to pass the modulating signal, i.e., the response time of the PLL must be short compared with the time scale of variations in the signal being recovered.[129] A high degree of linearity in the VCO is desirable in this method of FM detection, to minimize distortion in the audio output.

The second method of FM detection involves a phase detector, although not in a phase-locked loop. Figure 13.105 shows the idea. Both the input signal and a phase-shifted version of the signal are applied to a

---

[128] SRS uses an iterative approach for calculating r and n: the microcontroller chunks along, trying successive small-integer pairs until it's happy. This takes about a millisecond.

[129] The signal applied to the PLL does not have to be at the radiofrequency sent by the distant transmitter; it can be an "intermediate frequency" (IF) generated in the receiving system by the process of *mixing*. This *superheterodyne* technique was invented by Edwin H. Armstrong, who also invented FM. The big guys moved in, claimed his inventions, beat him up in court, and drove him to suicide.

phase detector, generating some output voltage. The phase-shifting network is diabolically arranged to have a phase shift varying linearly with frequency in the region of the input frequency (this is usually done with resonant *LC* networks), thus generating an output voltage with linear dependence on input frequency. That is the demodulated output. This method is called doubly balanced quadrature FM detection, and it is used in some IF amplifier/detector ICs.

Lest we leave the wrong impression, we hasten to add that you can demodulate FM without the help of phase locked loops. The classic techniques exploit the steep amplitude versus frequency characteristic of *LC* tuned circuits. In its simplest form (a "slope detector"), the FM signal is applied to an *LC* resonant circuit tuned off to one side, so it has a rising curve of response versus frequency; the output amplitude then varies approximately linearly with frequency, turning FM into FM+AM. An AM envelope detector completes the job of converting the AM to audio. In practice, a slightly more complicated arrangement (called a ratio detector, or Foster–Seeley detector) is used. Another (and simpler) technique uses averaging of a train of identical pulses at the intermediate frequency.

D. AM detection

Wanted: a technique to give an output signal proportional to the instantaneous *amplitude* of a high-frequency signal. The usual method involves rectification (Figure 13.106). Figure 13.107 shows a fancy method ("homodyne detection," or "synchronous detection") using PLLs. The PLL generates a square wave at the same frequency as the modulated carrier. Multiplying the input signal by this square wave generates a full-wave-rectified signal that only needs some lowpass filtering to remove the remnants of the carrier frequency, leaving the modulation *envelope*. If you use the exclusive-OR type of phase detector in the PLL, the output is shifted 90° relative to the reference signal, so a 90° phase shift would have to be inserted in the signal path to the multiplier.

E. Digital demodulation

A phase locked loop is an essential component in recovering ("demodulating") data from a carrier that has been

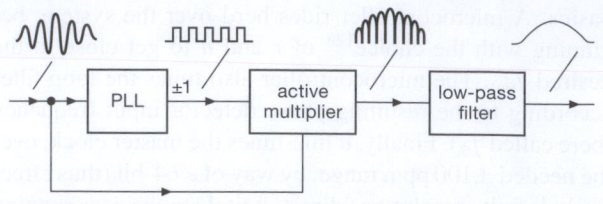

Figure 13.107. Homodyne detection.

modulated with a *digital* signal. In a simple form of digital modulation ("binary phase-shift keying," or BPSK), each bit to be transmitted either inverts, or not, the phase of a constant-amplitude carrier (Figure 13.108). These encoded bits are recovered at the receiving end by multiplying the received BPSK-modulated carrier by a signal at the same carrier frequency. Your first thought might be to use a PLL to recover a carrier replica. But that doesn't work, because the BPSK-modulated spectrum has no component at the carrier frequency.

One cute solution[130] is to notice that the *square* of the transmitted signal ignores the phase reversals, generating a signal at twice the carrier frequency. Pursuing that idea, you get the "squaring-loop" method of Figure 13.108. The first mixer M1 (a mixer is a multiplier) generates the doubled carrier frequency $2f_c$, which is cleaned up with a bandpass filter and used to lock a PLL, with the VCO acting as a flywheel (low loop bandwidth); a divide-by-2 then creates the carrier replica at f_c, with a phase trim to bring it into alignment with the (suppressed) underlying received carrier. Finally, multiplier M3 synchronously recovers the modulating bits, with a final lowpass filter to remove the $2f_c$ ripple.

If the phased bursts of cycles are thought of as *symbols*, BPSK encodes one bit per symbol. Commonly used digital-modulation schemes typically encode several bits per symbol. For example, you could encode symbols of 2 bits each by sending bursts of carrier cycles, each phased 0°, 90°, 180°, or 270°, according to the 2-bit symbol. This is called quadrature phase-shift keying (QPSK), also known as 4-QAM ("quadrature amplitude modulation," pronounced "quam"). More generally, you can create a "constellation" of symbols, each a (tapered) burst with some amplitude and phase. For example, cable television is commonly delivered as 256-QAM, each symbol carrying 8 bits of information. For all these modulation schemes you still need to recover a signal at the carrier frequency (or

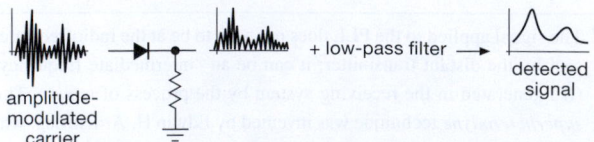

amplitude-
modulated
carrier

+ low-pass filter →

detected
signal

Figure 13.106. AM detection.

---

130 There's a more subtle method of BPSK demodulation, again using a PLL, known as a "Costas Loop." Its performance is comparable, but it's harder to understand what's going on. We like simplicity.

its frequency-shifted replica, an "intermediate frequency"), for which a PLL is essential. A trick that is sometimes used is to transmit a weak "pilot" signal at the carrier frequency, so that schemes like the squaring loop aren't needed. This is used, for example, in digital television broadcasting in the United States, where 3-bit symbols are encoded as amplitude modulation (four amplitude levels, at 0° or 180°), with a slight dc offset to create the pilot to which the receiver's PLL can lock.

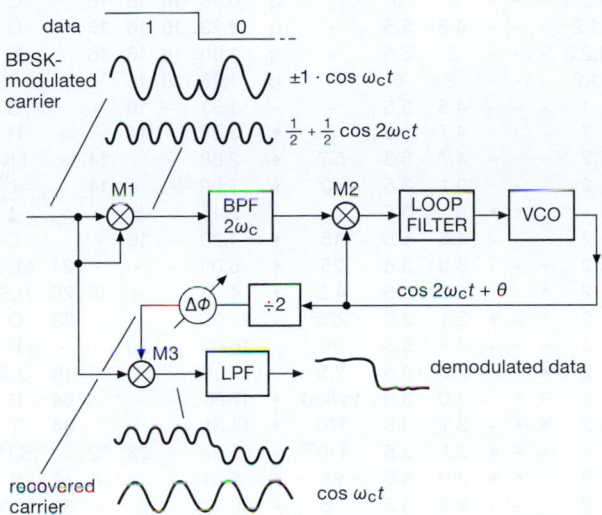

Figure 13.108. Squaring-loop demodulation of BPSK digital signal.

F. Other communications applications

As we suggested earlier, PLLs play an essential role in many aspects of communications. Channelized transmitters (think cellphones) must keep their signals at defined frequencies, with sufficient signal purity to prevent out-of-channel interference. And receivers (cellphones again; or FM radios, televisions, satellite receivers) use a *local oscillator* (LO) to determine their receiving frequency (that's Armstrong's superheterodyne technique, nearly a century old). Signal impurity (jitter, spurs) in the LO cause degradation of the received signal, in just the way that it would if it were the transmitter. For applications like these, signal quality is paramount and requires better VCOs than you get with simple capacitive charging circuits like those in the '4046.

For this kind of application you can get PLL chips that are intended for use with external VCOs and do not include an on-chip oscillator; examples are the NSC LMX2300-series, or the compatible ADI ADF4116–18. These families include members with phase detectors that can run to

6 GHz and beyond. With such PLL chips you can use any commercial VCO; or you can make your own (e.g., a JFET *LC* oscillator, electrically tuned with a varactor, §1.9.5B). An example of the latter is the PLL-disciplined JFET oscillator of Figure 7.29, with a noise spectrum as shown in Figure 7.30.

There's been considerable effort recently aimed at integrating high-quality VCOs directly onto the PLL chip, so you don't have to rig up a separate oscillator. Some of these need an external inductor (the hardest part to integrate with the required inductance, and with sufficient quality factor Q), for example the ADF4360-8. Others include all components on-chip, for example the LMX2531 or ADF4360-3; these latter are intended for cellphone use and have relatively narrow VCO oscillator tuning ranges, of order 5%. Other technologies used for on-chip oscillators include silicon micromachined (MEMS) resonators (e.g., the SiTime SiT3700-, 8100-, 9100-series), and surface acoustic wave (SAW) resonators (e.g., the IDT M680-series). These have very narrow VCO tuning ranges ($\sim$100 ppm), but you get very low phase noise and jitter, as you do with the competitive technology of crystal oscillators with a similar narrow range of voltage tuning (a VCXO; used, for example, within the IDT 810252 PLL).

G. Pulse synchronization and clean-signal regeneration

In digital signal transmission, a string of bits containing the information is sent over a communications channel. The information may be intrinsically digital or it may be digitized analog signals, as in "pulse-code modulation" (PCM). A closely related situation is the decoding of digital information from magnetic tape or disk, or optical disk storage. In such cases there may be noise or variations in pulse rate (e.g., from tape stretch), and it is desirable to have a clean clock signal at the same rate as the bits you are trying to read. PLLs work very nicely here. The PLL lowpass filter would be designed to follow rate variations inherent in the data stream (e.g., variations in tape or disk speed), while eliminating cycle-to-cycle jitter and noise that come from less-than-ideal received clock signal quality. This widespread application is often called "clock and data recovery" (CDR). An example in the audio world is the Burr–Brown (TI) DIR9001 Digital Audio Interface Receiver, which includes a low-jitter on-chip PLL/VCO clock-recovery subsystem, along with data demodulation. It is flexibly programmable to handle a wide range of data rates (28–108 ksps) and digital formats (with names like S/PDIF, AES3, IEC60958, and CPR-1205).

Table 13.13 Selected Phase-locked Loops[a]

Part #	VCO[b]	Output freq min (MHz)	Output freq max (MHz)	VCO freq min (MHz)	VCO freq max (MHz)	Ref input freq min (MHz)	Ref input freq max (MHz)	@V_S (V)	PD type	Fractional-n Synth	Clock Regen/Distrib	Ext bare XTAL ref	Supply Voltage min (V)	Supply Voltage max (V)	Supply Current typical (mA)	Power-down mode	Price qty25	DIP	SOIC	TSSOP	Smaller	Comments
LMC568	iRC	0	f	0	1	0	0.5	5	-	-	-	-	2	9	0.75	-	1.25	-	8	-	-	A
CD4046	iRC	0	0.6[m]	0	0.6[m]	0	0.6[m]	10	1,2	-	-	-	3	18	0.09[g]	q	0.48	16	16	16	-	B
74HC4046	iRC	0	12[m]	0	12[m]	0	12[m]	4.5	1,2,3	-	-	-	3	6	-	q	0.48	16	16	16	-	C
74HCT9046	iRC	0	11[m]	0	11[m]	0	11[m]	4.5	1,2	-	-	-	4.5	5.5	-	q	2.23	16	16	16	-	D
74LV4046	iRC	0	24	0	24	0	24	3.3	1,2,3	-	-	-	3	5.5	-	q	1.00	16	16	16	-	E
74HC7046	iRC	0	38	0	38	0	38	4.5	1,2	-	-	-	2	6	-	q	1.28	16	16	-	-	F
74ACT297	eV	0	55	-	-	0	55	5	1	-	-	-	4.5	5.5	-	-	1.53	-	16	-	-	G
TLC2932	iR	11	50	22	50	-	40	5	2	-	-	-	4.7	5.3	5	•	3.29	-	14	-	-	H
TLC2933	iR	64	96	64	96	-	50	5	2	-	-	-	4.7	5.3	5.7	•	2.88	-	-	14	-	I,K
TLC2934	iR	10	130	10	130	-	50	3.3	2	-	-	-	3.1	3.5	10	•	3.00	-	-	14	-	I
CY22800	int	1	200	-	-	0.5	100	3.3	-	-	•	•	3.1	3.5	-	•	2.45	-	8	-	-	J
ICS673	int	0.25	120	2	240	0.001	8	5	2	-	-	-	4.5	5.5	15	•	4.29	-	16	-	-	L
ADF4360-8	iL	65	400	65	400	10	250	3.3	2	-	-	-	3.0	3.6	25	•	6.02	-	-	-	24	M,S
ADF4110	eV	-	55[o]	50	550	5	104	3.3	2	•	-	-	2.7	5.5	4.5	•	4.72	-	-	16	20	N,S
IDS810252	iX	25[d]	312.5[d]	625	625	0.008	155	3.3	2	-	•	-	3.1	3.5	225[m]	-	29.00	-	-	-	32	O
SY89421	int/eV	20	1120	480	1120	30	560	5	2	-	-	-	4.7	5.3	28	•	16.00	-	-	20	-	P
LMX2316	eV	-	10[o]	100	1200	-	100	3	2	-	-	-	2.3	5.5	2.5	•	7.00	-	-	16	16	Q,S
CDCE72010	eVX	0	800	< 1	1500	-	500	3.3	2	-	•	•	3.0	3.6	to 880	•	17.96	-	-	-	64	R
AD9510	eVX	0	1200	0	1600	0	250	3.3	2	-	•	-	3.1	3.5	170	•	14.31	-	-	-	64	T
MPC9230	int	50	800	800	1800	10	20	3.3	-	-	•	-	3.1	3.5	110[m]	•	5.84	-	28[p]	32	-	S,U
CDCM61001	int	62	625	1750	2050	21.9	28.5	3.3	2	-	•	-	3.0	3.6	95	•	5.74	-	-	-	32	V
LMX2531	int	553	3132	1106	3132	5	80	3	2	•	-	-	2.8	3.2	38	•	10.44	-	-	-	36	S,W
AD9552	int	50	900	3350	4050	6.6	112	3.3	2	•	-	-	3.1	3.5	149	•	9.40	-	-	-	32	X
ADF4350	int	137	4400	2200	4400	10	105	3.3	2	•	-	-	3.0	3.6	120	•	9.03	-	-	-	36	S,Y
ADF4106	eV	-	100[o]	500	6000	20	300	3.3	2	•	-	-	2.7	3.3	10	•	4.59	-	-	18	20	Z
ADF41020	eV	-	100[o]	4000	18000	10	400	3	2	•	-	-	2.8	3.2	27	•	15.00	-	-	-	20	S

Notes: (a) sorted approximately by increasing VCO f_{max}. (b) eV - external VCO; eVX - external VCO or VCXO; int/eV - int or eV; iL - internal VCO with external inductor; int - internal VCO, no external components; iR - internal VCO with external resistor; iRC - internal VCO with external R and C; iX - internal VCO with external bare xtal. (d) 25, 125, 156.25, and 312.5MHz only. (f) FM and FSK demod, audio bandwidth. (g) at 10kHz and 10V. (m) min or max. (o) phase detector muxed to output. (p) PLCC. (q) no power-down mode, but quiescent current <1µA.

Comments: A: FM and FSK demod applications. **B:** classic 4000B ("HV") CMOS. **C:** classic HCMOS. **D:** improved 4046, no dead zone. **E:** LVCMOS. **F:** 74HCT also available. **G:** digital 1st-order PLL. **H:** can run at 3V, 14-21MHz. **I:** internal ring oscillator. **J:** f_{ref}=8-30MHz with bare xtal; ref can be VCXO; can generate spread spectrum. **K:** can run at 3V, 38-55MHz. **L:** can run at 3.3V, 0.25-100MHz; ICS663 (SOIC-8) lacks power-down and output enable. **M:** wireless local-osc with PLL synth; versions with other freq ranges. **N:** wireless local-osc with integer-n PLL synth; 1.2GHz, 3GHz, and 4GHz versions. **O:** 2-stage PLL (VCXO PLL drives PLL multiplier, with input and output dividers, for Gigabit and 10-Gig Ethernet. **P:** can run at 3.3V; complementary PECL outputs; ext osc to 2GHz. **Q:** wireless local-osc with PLL synth; 0.55GHz and 2.8GHz versions. **R:** eight individual dividers, output clock distribution, multiple refs, highly complex. **S:** SPI interface. **T:** eight LVDS outputs, 0.2ps jitter, adjustable delay. **U:** wireless local-osc with PLL synth; diff'l PECL; parallel and SPI interfaces; replaces MC12430. **V:** Ethernet clock generator, etc.; four outputs; <1ps jitter. **W:** wireless local-osc with PLL synth; stable low-noise; p/n selects ±5% freq band. **X:** delta-sigma fractional-n; 0.5ps jitter; includes xtal osc. **Y:** wireless local-osc with PLL synth; 0.5ps jitter. **Z:** ADF4107=7GHz, ADF4108=8GHz.

H. Clock generators

As we remarked earlier, there are plenty of applications crying out for a suite of standard clock frequencies, synthesized from a single oscillator input, in which the niceties of low phase noise and spurs, etc., are of less concern than minimum parts count and the ability to program among a few standard frequencies. See Table 13.13. An example is the IDT 8430S010i: it's a single chip PLL with multiple synthesized outputs intended for embedded processor ap-

plications. You connect a single 25 MHz crystal, and out comes (a) choice of two processor clock frequencies, (b) choice of four PCI or PCIe clock frequencies, (c) choice of four DDR DRAM clock frequencies, (d) gigabit Ethernet MAC clock and PHY clocks, and (e) choice of three SP14.2 link frequencies.

Parts like this may use a simple interchip SPI-like serial programming protocol, or they may be pin-strappable (as this one is). Or they may allow both, as for example

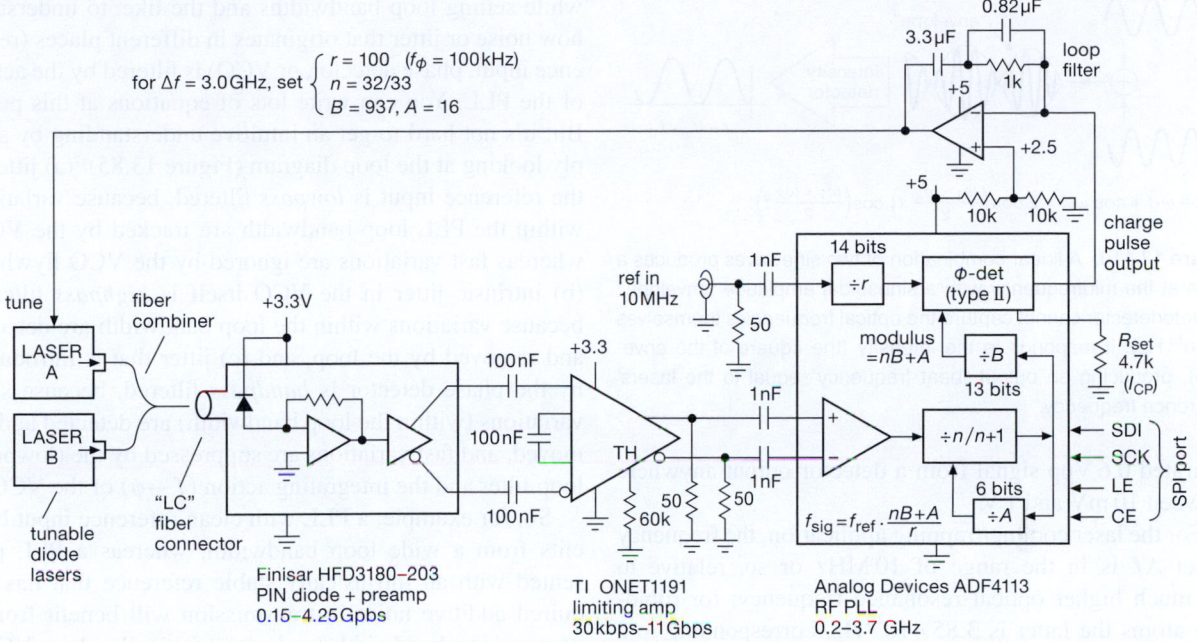

for $\Delta f = 3.0\,\text{GHz}$, set $\begin{cases} r = 100 \quad (f_\phi = 100\,\text{kHz}) \\ n = 32/33 \\ B = 937,\ A = 16 \end{cases}$

Figure 13.109. Controlling a diode laser to maintain a desired optical frequency difference relative to a reference laser. The components in this circuit cost less than \$40, exclusive of the lasers; the latter are in a completely different ballpark, roughly "40 dB\$."

the venerable NBC/MC12430 (or equivalent MPC9230), a simple integer-n PLL with a 9-bit n-counter and a 3-bit m-counter, programmable from 50–800 MHz. The on-chip VCO tunes from 400–800 MHz, and likely uses a starved inverter chain ring oscillator (the datasheet isn't talking). We used these as a clocking source in a quirky terasample/second data-acquisition instrument we built to detect intentional pulsed laser signals from possible extraterrestrial civilizations (no kidding).

I. Laser offset locking

In some scientific applications it's useful to be able to control a tunable laser, such that the frequency of its optical emission is offset by a specific frequency difference from that of a "reference" laser. As a specific example, a favorite technique in the business of "laser cooling" is to subject a beam of atoms to converging beams of laser light at a frequency slightly below that of a natural resonance of the atom. The Doppler effect causes an atom moving toward one of these lasers to see the light shifted upward in frequency, therefore more strongly absorbed by the atom, which is slowed by the transferred momentum.[131]

A phase-locked loop is perfect for this *offset locking*. Figure 13.109 shows how this goes, as implemented in a colleague's laboratory.[132] A portion of the light from the pair of tunable diode lasers is combined and sent to a wideband PIN-diode detector/amplifier module. What happens there is interesting; let's take it in two steps. (a) In a completely linear process, the two combined laser beams create a waveform consisting of a sinewave at the average laser frequency, modulated (multiplied) by a sinewave at half the difference frequency (Figure 13.110). (b) The detector cannot follow the *optical* waves, whose frequency is $\sim 10^{14}$ Hz. It responds only to the *intensity* of the light, proportional to the square of the "envelope" shown in Figure 13.110. And the square of a sinewave is just a sinewave at twice the frequency, plus a dc offset so that it sits atop the horizontal axis as shown.

In other words, at the output of the detector module you get a signal at the lasers' difference frequency (also called the *beat frequency*): $f_{\text{PDout}} = |f_2 - f_1| \equiv \Delta f$. The function of the rest of the circuit is simply to feed back a control signal to laser A so that the difference frequency Δf equals the desired offset. That's done with a fractional-n PLL, here preceded by a limiting amplifier that creates a clean

---

[131] This is known by the colorful term "optical molasses." Add some magnetic fields, and a few more tricks, and you've got a magneto-optical trap.

[132] The ever-capable Dr. Andrew Speck, to whom we are grateful for this and other excellent suggestions.

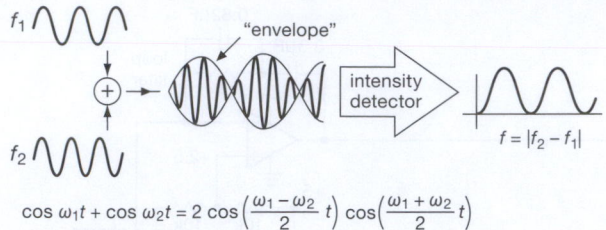

$$\cos \omega_1 t + \cos \omega_2 t = 2 \cos \left(\frac{\omega_1 - \omega_2}{2} t \right) \cos \left(\frac{\omega_1 + \omega_2}{2} t \right)$$

Figure 13.110. A linear combination of two sinewaves produces a wave at the midfrequency with a sinusoidal amplitude "envelope." A photodetector cannot capture the optical frequencies themselves ($\sim 10^{14}$ Hz); it responds to the intensity (the square of the envelope), producing an output "beat frequency" equal to the lasers' difference frequency.

saturated 0.6 Vpp signal from a detector output anywhere between 10 mV and 1 V.

For the laser cooling/trapping application, the frequency offset Δf is in the range of 10 MHz or so, relative to the much higher optical resonance frequency; for rubidium atoms the latter is 3.85×10^{14} Hz, corresponding to a wavelength of 780.24 nm.[133] As is often the case, there's much more to love (and hate) in this business: it turns out that there's a "hyperfine splitting" of the ground state of ^{85}Rb, so you need to kick out (the polite term is "optically pump") the atoms that happen to fall all the way to the bottom with some laser light that is offset by that energy difference. That's about 3 GHz,[134] which is why this circuit was designed for offsets in the gigahertz range, as indicated by the notations in the figure.

13.13.7 Wrapup: noise and jitter rejection in PLLs

We've seen applications where the reference is the higher-quality signal (e.g., multiple clocking signals derived from a single stable crystal reference), and applications where the opposite is true, namely that the PLL-generated signal is cleaner than the reference (e.g., clock recovery in a noisy channel, where the "flywheel" action of the VCO cleans up the output).

It's helpful, when thinking about these kinds of issues

while setting loop bandwidths and the like, to understand how noise or jitter that originates in different places (reference input, phase detector, or VCO) is filtered by the action of the PLL. You can write lots of equations at this point. But it's not hard to get an intuitive understanding by simply looking at the loop diagram (Figure 13.85): (a) jitter at the reference input is *lowpass* filtered, because variations within the PLL loop bandwidth are tracked by the VCO, whereas fast variations are ignored by the VCO flywheel; (b) intrinsic jitter in the VCO itself is *highpass* filtered, because variations within the loop bandwidth are detected and removed by the loop; and (c) jitter that is introduced by the phase detector is *bandpass* filtered, because slow variations (within the loop bandwidth) are detected and removed, and fast variations are suppressed by the (lowpass) loop filter and the integrating action ($f \rightarrow \phi$) of the VCO.

So, for example, a PLL with clean reference input benefits from a wide loop bandwidth, whereas a PLL presented with an intrinsically stable reference that has acquired additive noise in transmission will benefit from a narrow loop bandwidth (and an intrinsically clean VCO). And the "noise" can be more subtle: the divided VCO signal seen by the phase detector in a fractional-n PLL has jitter (introduced by the deliberate changes of modulus); a narrow loop bandwidth smooths this jitter source as well.

Of course, if the PLL output needs agility (as with tone decoding, or FM demodulation) then the loop bandwidth must be tailored accordingly, quite independent of noise and jitter tradeoffs.

13.14 Pseudorandom bit sequences and noise generation

13.14.1 Digital-noise generation

An interesting blend of digital and analog techniques is embodied in the subject of pseudorandom bit sequences (PRBSs).[135] It turns out to be remarkably easy to generate sequences of bits (or words) that have good randomness properties, i.e., a sequence that has the same sort of probability and correlation properties as an ideal coin-flipping machine. Because these sequences are generated by standard deterministic logic elements (shift registers, to be exact), the bit sequences generated are in fact predictable and repeatable, although any portion of such a sequence looks for all the world just like a random string of 0s and 1s.

---

[133] This happens to be the wavelength used in compact disc recorders; so the system of Figure 13.109 was implemented economically by using a pair of these ubiquitous laser diodes, which can put out 100 mW (caution, definitely *not* eye-safe!). To tune these things, you attach a piezo-tiltable external grating, adjusting the diode current in tandem to keep its wavelength tracking that of the external cavity. Less inventive souls can throw money at the problem: you can buy tunable diode lasers from companies like New Focus, ThorLabs, and Toptica.

[134] Actually, 3.035732439 GHz, if you really need to know.

[135] This is the example we used in Chapter 11 to illustrate programmable logic, where we contrasted PRBS implementations in discrete logic, in programmable logic, and in a microcontroller. See also its use as an analog noise generator in §8.12.4A.

With just a few chips you can generate sequences that literally go on for centuries without repeating, making this a very accessible and attractive technique for the generation of digital bit sequences or analog noise waveforms. When generating eye diagrams (e.g., Figures 12.132 and 14.33), or testing serial links for bit error rates (BERs), it is common to use a PRBS source. PRBSs are used also to "scramble" (deterministically) the serial data in gigabit Ethernet communications, in order to generate a lively bit pattern for the ac-coupled (transformer) physical link; the scrambling is reversed at the receive end, by XOR-ing with a synchronized PRBS running the same sequence.

A. Analog noise

Simple lowpass filtering of the output bit pattern of a PRBS generates bandlimited white Gaussian noise, i.e., a noise voltage with a flat power spectrum up to some cutoff frequency (see Chapter 8 for more on noise). Alternatively, a weighted sum of the shift-register contents (via a set of resistors) performs *digital filtering*, with the same result. Flat noise spectra out to several megahertz can easily be made this way. As you will see later, such digitally synthesized analog noise sources have many advantages over purely analog techniques such as noise diodes or resistors.

B. Other applications

Besides their obvious applications as analog or digital noise sources, pseudorandom bit sequences are useful in a number of applications that have nothing to do with noise. As just mentioned, they are used for pattern generation in serial link testing (eye diagrams, bit error rate), and for bit scrambling (as opposed to real encryption) in serial network protocols like Ethernet. They are used in "direct-sequence" spread-spectrum digital communications (in which each bit to be transmitted is sent as a predetermined sequence of shorter "chips"); such a technique is used, for example, in CDMA (code-division multiple-access) cellular phone systems, and in the airlink privacy cipher of the GSM cellular standard. They're also used in digital TV broadcasting. These sequences are used extensively in error-detecting and error-correcting codes, because they allow the transcription of blocks of data in such a way that valid messages are separated by the greatest "Hamming distance" (measured by the number of bit errors). Their good autocorrelation properties make them ideal for radar-ranging codes, in which the returned echo is compared (cross-correlated, to be exact) with the transmitted bit string. They can even be used as compact modulo-n dividers.

13.14.2 Feedback shift register sequences

The most popular (and the simplest) PRBS generator is the linear feedback shift register (LFSR, Figure 13.111). A shift register of length m bits is clocked at some fixed rate, f_0. An exclusive-OR gate generates the serial input signal from the exclusive-OR combination of the nth bit and the last (mth) bit of the shift register. Such a circuit goes through a set of states (defined by the set of bits in the register after each clock pulse), eventually repeating itself after K clock pulses; i.e., it is cyclic with period K.

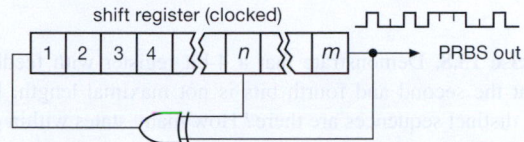

Figure 13.111. Pseudorandom bit sequence generator.

The maximum number of conceivable states of an m-bit register is $K = 2^m$, i.e., the number of binary combinations of m bits. However, the state of all 0s would get "stuck" in this circuit, because the exclusive-OR would regenerate a 0 at the input. Thus the maximum-length sequence you can possibly generate with this scheme is $2^m - 1$. It turns out that you can make such "maximal-length shift-register sequences" if m and n are chosen correctly, and the resultant bit sequence is pseudorandom.[136] As an example, consider the 4-bit feedback shift register in Figure 13.112. Beginning with the state 1111 (we could start anywhere except 0000), we can write down the states it goes through:

```
1111
0111
0011
0001
1000
0100
0010
1001
1100
0110
1011
0101
1010
1101
1110
```

We have written down the states as 4-bit numbers

---

[136] The criterion for maximal length is that the polynomial $1 + x^n + x^m$ be irreducible and prime over the Galois field GF(2).

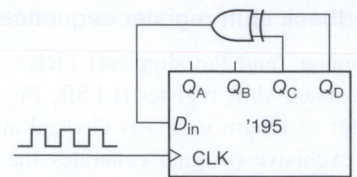

Figure 13.112. 4-bit linear feedback shift register ($m=4$, $n=3$; 15 states).

$Q_A Q_B Q_C Q_D$. There are 15 distinct states ($2^4 - 1$), after which it begins again; therefore it is a maximal-length register.

Exercise 13.8. Demonstrate that a 4-bit register with feedback taps at the second and fourth bits is not maximal length. How many distinct sequences are there? How many states within each sequence?

A. Feedback taps

Maximal-length shift registers can be made with exclusive-OR feedback from more than two taps (in these cases you use several exclusive-OR gates in the standard parity-tree configuration, i.e., modulo-2 addition of several bits). In fact, for some values of m, a maximal-length register can be made only with more than two taps. Table 13.14 is a listing of all values of m up to 167 for which maximal-length registers can be made with just two taps, i.e., feedback from the nth bit and the mth (last) bit, as previously. A value is given for n and for the cycle length K, in clock cycles. In some cases there is more than one possibility for n, and in every case the value $m-n$ can be used instead of n; thus the earlier 4-bit example could have used taps at $n=1$ and $m=4$. Since shift-register lengths in multiples of 8 are common, you may want to use one of those lengths. In that case, more than two taps are necessary. Table 13.15 gives the magic numbers.

It's rarely necessary to use a register much longer than 32 bits: when clocked at 1 MHz the repeat time is about an hour. Go to 64 bits and you can clock at 1 GHz for six centuries before it comes around again.

B. Properties of maximal-length shift-register sequences

We generate a string of pseudorandom bits from one of these registers by clocking it and looking at successive output bits. The output can be taken from any position of the register; it is conventional to use the last (mth) bit as the output. Maximal-length shift-register sequences have the following properties:

1. In one full cycle (K clock cycles), the number of 1s is

Table 13.14	Single-tap LFSRs							
m	n	*length*	m	n	*length*	m	n	*length*
3	2	7	49	40	5.6e14	108	77	3.2e32
4	3	15	52	49	4.5e15	111	101	2.6e33
5	3	31	55	31	3.6e16	113	104	1.0e34
6	5	63	57	50	1.4e17	118	85	3.3e35
7	6	127	58	39	2.9e17	119	111	6.6e35
9	5	511	60	59	1.2e18	121	103	2.7e36
10	7	1023	63	62	9.2e18	123	121	1.1e37
11	9	2047	65	47	3.7e19	124	87	2.1e37
15	14	32767	68	59	3.0e20	127	126	1.7e38
17	14	1.3e5	71	65	2.4e21	129	124	6.8e38
18	11	2.6e5	73	48	9.4e21	130	127	1.4e39
20	17	1.0e6	79	70	6.0e23	132	103	5.4e39
21	19	2.1e6	81	77	2.4e24	134	77	2.2e40
22	21	4.2e6	84	71	1.9e25	135	124	4.4e40
23	18	8.4e6	87	74	1.5e26	137	116	1.7e41
25	22	3.4e7	89	51	6.2e26	140	111	1.4e42
28	25	2.7e8	93	91	9.9e27	142	121	5.6e42
29	27	5.3e8	94	73	2.0e28	145	93	4.5e43
31	28	2.1e9	95	84	4.0e28	148	121	3.6e44
33	20	8.6e9	97	91	1.6e29	150	97	1.4e45
35	33	3.4e10	98	87	3.2e29	151	148	2.9e45
36	25	6.9e10	100	63	1.3e30	153	152	1.1e46
39	35	5.5e11	103	94	1.0e31	159	128	7.3e47
41	38	2.2e12	105	89	4.1e31	161	143	2.9e48
47	42	1.4e14	106	91	8.1e31	167	161	1.9e50

Table 13.15	Multiple-of-8 LFSRs				
m	*taps*	*length*	m	*taps*	*length*
8	4, 5, 6	255	96	47, 49, 94	7.9e28
16	4, 13, 15	64K	104	93, 94, 103	2.0e31
24	17, 22, 23	16M	112	67, 69, 110	5.2e33
32	1, 2, 22	4G	120	2, 9, 113	1.3e36
40	19, 21, 38	1.1e12	128	99, 101, 126	3.4e38
48	20, 21, 47	2.8e14	136	10, 11, 135	8.7e40
56	34, 35, 55	7.2e16	144	74, 75, 143	2.2e43
64	60, 61, 63	1.8e19	152	86, 87, 151	5.7e45
72	19, 25, 66	4.7e21	160	141, 142, 159	1.5e48
80	42, 43, 79	1.2e24	168	151, 153, 166	3.7e50
88	16, 17, 87	3.1e26			

one greater than the number of 0s. The extra 1 comes about because of the excluded state of all 0s. This says that heads and tails are equally likely (the extra 1 is totally insignificant for any reasonable-length register; a 17-bit register will produce 65,536 1s and 65,535 0s in one of its cycles).

2. In one full cycle (K clock cycles), half the runs of

consecutive 1s have length 1, one-fourth the runs have length 2, one-eighth have length 3, etc. There are the same numbers of runs of 0s as of 1s, again with the exception of a missing 0. This says that the probability of heads and tails does not depend on the outcome of past flips, and therefore the chance of terminating a run of successive 1s or 0s on the next flip is 1/2 (contrary to the person-in-the-street's understanding of the "law of averages").

3. If one full cycle (K clock cycles) of 1s and 0s is compared with the same sequence shifted cyclically by any number of bits n (where n is not 0 or a multiple of K), the number of disagreements will be one greater than the number of agreements. In fancy language, the autocorrelation function is a Kronecker delta at zero delay, and $-1/K$ everywhere else. This absence of "sidelobes" in the autocorrelation function is what makes PRBSs so useful for radar ranging.

Exercise 13.9. Show that the 4-bit shift-register sequence listed earlier (taps at $n = 3$, $m = 4$) satisfies these properties, considering the Q_A bit as the "output": 100010011010111.

13.14.3 Analog noise generation from maximal-length sequences

A. Advantages of digitally generated noise
As we remarked earlier, the digital output of a maximal-length feedback shift register can be converted to band-limited white noise with a lowpass filter whose cutoff frequency is well below the clock frequency of the register. Before getting into the details, we point out some of the advantages of digitally generated analog noise. Among other things, it allows you to generate noise of known spectrum and amplitude (see the circuit example in §8.12.4A), with adjustable bandwidth (via clock-frequency adjustment), using reliable and easily maintained digital circuitry. There is none of the variability of diode noise generators, nor are there interference and pickup problems that plague the sensitive low-level analog circuitry used with diode or resistor noise generators. Finally, it generates repeatable "noise" and, when filtered with a weighted digital filter (more about this later), repeatable noise waveforms independent of clocking rate (output noise bandwidth).

13.14.4 Power spectrum of shift-register sequences

The output spectrum generated by maximal-length shift registers consists of noise extending from the repeat fre-

quency of the entire sequence, f_{clock}/K, up to the clock frequency and beyond. It is flat within ± 0.1dB up to 12% of the clock frequency (f_{clock}), dropping rather rapidly beyond its -3 dB point of 44% f_{clock}. Thus a lowpass filter with a high-frequency cutoff of 5%–10% of the clock frequency will convert the unfiltered shift register output to a band-limited analog noise voltage. Even a simple RC filter will suffice, although it may be desirable to use active filters with sharp cutoff characteristics (see Chapter 6) if a precise frequency band of noise is needed.

To make these statements more precise, let's look at the shift-register output and its power spectrum. It is usually desirable to eliminate the dc offset characteristic of digital logic levels, generating a bipolarity output with 1 corresponding to $+a$ volts and 0 corresponding to $-a$ volts (Figure 13.113). This can be done in various ways, for example, referring to Figure 13.114, (A) with an 'HC4053 linear CMOS switch, operating from dual supplies and with a pair of inputs tied to $\pm a$ volts; (B) with a fast opamp with dc offset current into a summing junction; (C) with a fast rail-to-rail comparator running between $\pm a$-volt split supplies; (D) with a CMOS logic gate, powered from $\pm a$-volt supplies, driven by a properly shifted and scaled logic swing; or (E) with the same logic gate, driven with a diode-clamped (and current-limited) logic swing. This last method is a bit weird, and works only if the total supply voltage ($V_{total}=2a$) is in the range of a standard logic family (say 1–5 V); but it excels in speed. It's not dc-coupled, so it requires a logic input that's busy; that's OK here – a PRBS is peripatetic, and cannot rest for longer than m clock periods.[137]

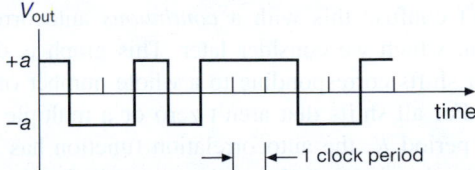

Figure 13.113. A symmetrical PRBS waveform eliminates the dc component.

As we remarked earlier, the string of output bits has a single peak in its autocorrelation. If the output states represent $+1$ and -1, the digital autocorrelation (the sum of the product of corresponding bits, when the bit string is compared with a shifted version of itself) is as shown in Figure 13.115.

---

[137] To paraphrase Woody Allen, "A clamped ac-coupled logic input is like a shark; it has to keep moving, or it dies."

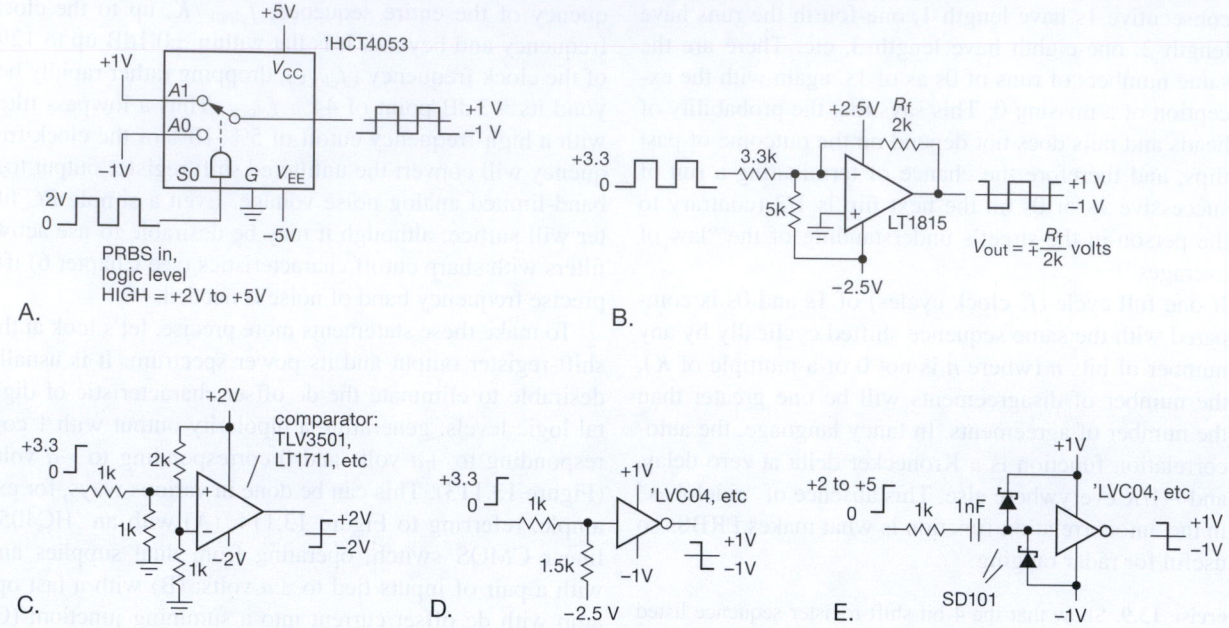

Figure 13.114. Converting a positive-only logic swing to a symmetrical voltage waveform.

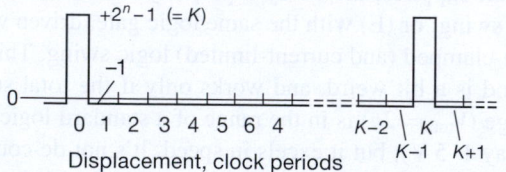

Figure 13.115. Full-cycle discrete autocorrelation for a maximal-length shift-register sequence.

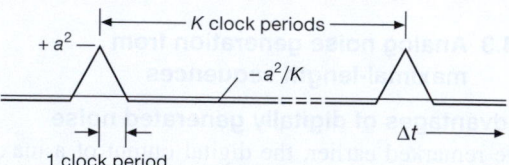

Figure 13.116. Full-cycle continuous autocorrelation for a maximal-length shift-register sequence.

Don't confuse this with a *continuous* autocorrelation function, which we consider later. This graph is defined only for shifts corresponding to a whole number of clock cycles. For all shifts that aren't zero or a multiple of the overall period K, the autocorrelation function has a constant -1 value (because there is an extra 1 in the sequence), negligible when compared with the zero-offset autocorrelation value of K. Likewise, if we consider the unfiltered shift-register output as an *analog* signal (whose waveform happens to take on values of $+a$ and $-a$ volts only), the normalized autocorrelation becomes a continuous function, as shown in Figure 13.116. In other words, the waveform is totally uncorrelated with itself when shifted more than one clock period forward or backward.

The power spectrum of the unfiltered digital output can be obtained from the autocorrelation by standard mathematical techniques. The result is a set of equally spaced series of spikes (delta functions), beginning at the frequency

at which the whole sequence repeats, f_{clock}/K, and going up in frequency by equal intervals f_{clock}/K. The fact that the spectrum consists of a set of discrete spectral lines reflects the fact that the shift-register sequence eventually (and periodically) repeats itself. Don't be alarmed by this funny spectrum; it will look continuous for any measurement or application that takes less time than the cycle time of the register. The envelope of the spectrum of the unfiltered output is shown in Figure 13.117. The envelope is proportional to the square of $(\sin x)/x$. Note the peculiar property that there is *no* noise power at the clock frequency or its harmonics.

A. Noise voltage

Of course, for analog noise generation you use only a portion of the low-frequency end of the spectrum. It turns out to be easy to calculate the noise power per hertz in terms of the half-amplitude (a volts) and the clock frequency

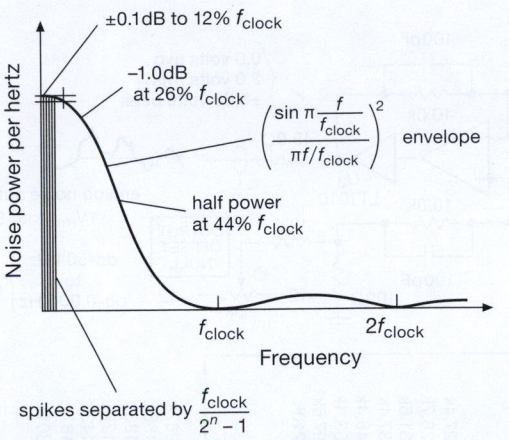

Figure 13.117. Power spectrum of unfiltered digital shift-register output signal.

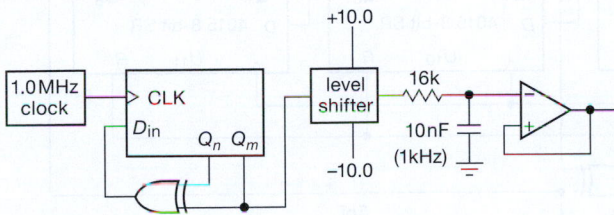

Figure 13.118. Simple pseudorandom noise source.

13.14.5 Low-pass filtering

A. Analog filtering

The spectrum of useful noise from a pseudorandom sequence generator extends from a low-frequency limit of the reciprocal repeat period (f_{clock}/K) up to a high-frequency limit of perhaps 20% of the clock frequency (at that frequency the noise power per hertz is down by 0.6 dB). Simple lowpass filtering with RC sections, as illustrated in the earlier example, is adequate provided that its 3 dB point is set far below the clock frequency (e.g., less than 1% of f_{clock}). To use the spectrum closer to the clock frequency, it is advisable to use a filter with sharper cutoff, e.g., a Butterworth or Chebyshev. In that case the flatness of the resultant spectrum depends on the filter characteristics, which should be measured, since component variations can produce ripples in the passband gain. Likewise, the filter's actual voltage gain should be measured if the precise value of noise voltage per root hertz is important.

B. Digital filtering

A disadvantage of analog filtering is the need to readjust the filter cutoff if the clock frequency is changed by large factors. In situations where you want to change the clocking frequency, an elegant solution is provided by discrete-time digital filtering, in this case performed by taking an analog-weighted sum of successive output bits (non-recursive digital filtering). In this way the effective filter cutoff frequency changes to match changes in the clock frequency. In addition, digital filtering lets you go to extremely low cutoff frequencies (fractions of a hertz) where analog filtering becomes awkward.

To perform a weighted sum of successive output bits simultaneously, you can simply look at the various parallel outputs of successive shift-register bits, using resistors of various values into an op-amp summing junction. For a lowpass filter the weights should be proportional to $(\sin x)/x$; note that some levels will have to be inverted since the weights are of both signs. Because no capacitors are used in this scheme, the output waveform consists of a set of discrete output voltages.

The approximation to Gaussian noise is improved by use of a weighting function over many bits of the sequence. In addition, the analog output then becomes essentially a continuous waveform. For this reason it is desirable to use as many shift-register stages as possible, adding additional shift-register stages outside the exclusive-OR feedback if necessary. As before, pullups or MOS switches should be used to set stable digital voltage levels (CMOS

(f_{clock}). Expressed as an rms noise *voltage*, the answer is

$$v_{rms} = a\left(\frac{2}{f_{clock}}\right)^{1/2} \text{ V}/\sqrt{\text{Hz}}, \qquad (f \leq 0.2 f_{clock}).$$

This is for the bottom end of the spectrum, the part you usually use (you can use the envelope function to find the power density elsewhere).

For example, suppose we run a maximal-length shift register at 1.0 MHz and arrange it so that the output voltage swings between +10.0 and −10.0 volts. The output is passed through a simple RC lowpass filter with 3 dB point at 1 kHz (Figure 13.118). We can calculate the rms noise voltage at the output exactly. We know from the preceding equation that the output from the level shifter has an rms noise voltage of 14.14 mV per root hertz. From §8.13 we know that the noise bandwidth of the lowpass filter is $(\pi/2)(1.0\,\text{kHz})$, or 1.57 kHz. So the output noise voltage is

$$V_{rms} = 0.01414 \cdot (1570)^{1/2} = 560\,\text{mV}$$

with the spectrum of a single-section RC lowpass filter.

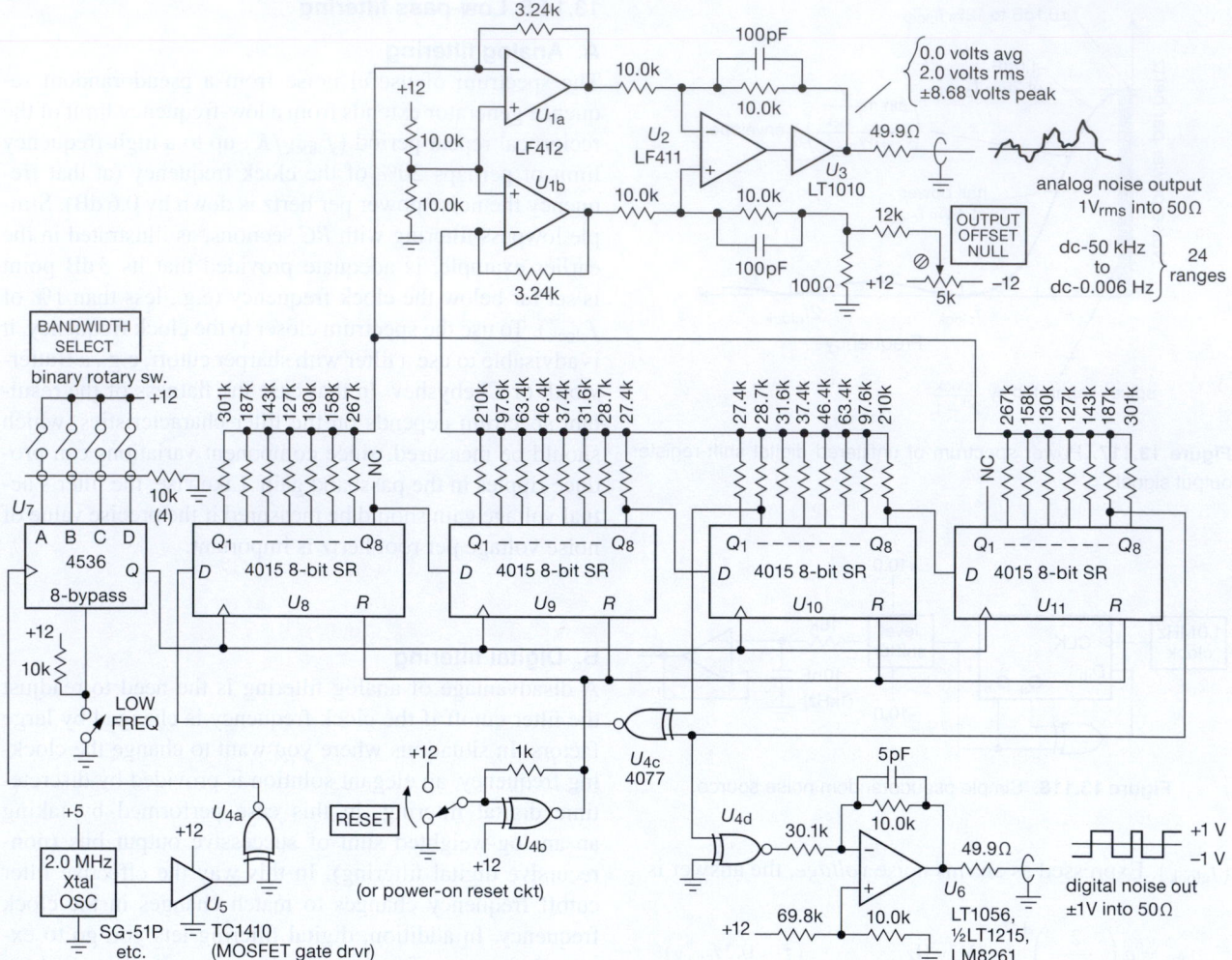

Figure 13.119. Wide-frequency-range laboratory pseudo-random noise source, inspired by Hewlett Packard's Model 3722A Noise Generator ("A precision digital instrument," in the parlance of its time, 1967). If you want to stick with discrete logic but need greater bandwidth, you could substitute 74LV164A 8-bit shift registers (with guaranteed clocking up to 125 MHz with a 5 V supply), with corresponding circuit changes throughout. Or you could use a fast cPLD or FPGA (§11.3.3) – but having gone that far, you might as well implement the digital lowpass filter in a DSP (digital signal processor) chip. Following this thread one step further, you could just code the whole thing in a fast microcontroller (§11.3.5 and Chapter 15), and use its on-chip D-to-A converter to generate the band-limited analog noise output.

logic is ideal for this application, because the outputs saturate cleanly at V_{DD} and ground).

The circuit in Figure 13.119 generates pseudorandom analog noise, with bandwidth selectable over an enormous range, using this technique. A 2.0 MHz crystal oscillator drives a 14536 24-stage programmable divider, generating selectable clock frequencies going from 1.0 MHz down to 0.12 Hz by factors of 2. A 32-bit shift register is connected with feedback from stages 31 and 18, generating a maximal-length sequence with 2 billion states (at the maximum clock frequency the register completes one cycle in

a half hour). In this case we have used a $(\sin x)/x$ weighted sum over 32 successive values of the sequence. Op-amps U_{1a} and U_{1b} amplify the inverted and noninverted terms, respectively, driving difference amplifier U_2. The gains are chosen to generate a 1.0 V rms output with no dc offset into a 50 Ω load impedance (2.0 V rms open circuit). Note that this noise amplitude is independent of the clock rate, i.e., the total bandwidth. This digital filter has a cutoff at about $0.05 f_{clock}$, giving a white-noise output spectrum extending from dc to 50 kHz (at maximum clock frequency) down to dc to 0.006 Hz (at minimum clock frequency), in 24 steps of bandwidth. The circuit also provides an unfiltered output waveform, going between $+1.0$ V and -1.0 V.

There are a few interesting points about this circuit. Note that an exclusive-NOR gate is used for feedback so that the register can be simply initialized by bringing it to the state of all zeros. This trick of inverting the serial input signal makes the excluded state the state of all 1s (rather than all 0s as with the usual exclusive-OR feedback), but it leaves all other properties unaffected.

A weighted sum of a finite number of bits cannot ever produce truly Gaussian noise, since the peak amplitude is limited. In this case it can be calculated that the peak output amplitude (into 50 Ω) is ±4.34 V, giving a "crest factor" of 4.34. That calculation is important, by the way, because you must keep the gains of U_1 and U_2 low enough to prevent clipping. Look carefully at the methods used to generate an output of zero dc offset from the CMOS levels of $+6.0$ V average value (LOW$=0$ V, HIGH$=12.0$ V).

13.14.6 Wrapup

A few comments about shift-register sequences as analog noise sources.[138] You might be tempted to conclude from the three properties of maximal-length shift registers listed earlier that the output is "too random," in the sense of having exactly the right number of runs of a given length, etc. A genuine random coin-flipping machine would not generate exactly one more head than tail, nor would the autocorrelation be absolutely flat for a finite sequence. To put it another way, if you used the 1s and 0s that emerge from the shift register to control a "random walk," moving forward one step for a 1 and back one step for a 0, you would wind up exactly one step away from your beginning point after the register had gone through its entire cycle, a result that is anything but "random".

However, the shift-register properties mentioned earlier are true only of the entire sequence of 2^n-1 bits, *taken as a whole*. If you use only a section of the entire bit sequence, the randomness properties closely approximate a random coin-flipper. To make an analogy, it is as if you were drawing red balls and blue balls at random from an urn initially containing K balls in all, half red and half blue. If you do this *without replacing them*, you would expect to find approximately random statistics at first. As the urn becomes depleted, the statistics are modified by the requirement that the total numbers of red and blue balls must come out the same.

You can get an idea how this goes by thinking again about the random walk. If we assume that the only nonrandom property of the shift sequence is the exact equality of 1s and 0s (ignoring the single excess 1), it can be shown that the random walk as described should reach an average distance from the starting point of

$$X = [r(K-r)/(K-1)]^{1/2}$$

after r draws from a total population of $K/2$ ones and $K/2$ zeros. Because in a completely random walk X equals the square root of r, the factor $(K-r)/(K-1)$ expresses the effect of finite urn contents. As long as $r \ll K$, the randomness of the walk is only slightly reduced from the completely random case (infinite urn contents), and the pseudorandom sequence generator is indistinguishable from the real thing. We tested this with a few thousand PRBS-mediated random walks, each a few thousand steps in length, and found that the randomness was essentially perfect, as measured by this simple criterion.

Of course, the fact that PRBS generators pass this simple test does not guarantee that they would satisfy some of the more sophisticated tests of randomness, e.g., as measured by higher-order correlations. Such correlations also affect the properties of analog noise generated from such a sequence by filtering. Although the noise amplitude distribution is Gaussian, there may be higher-order amplitude correlations uncharacteristic of true random noise. It is said that the use of many (preferably about $m/2$) feedback taps (using an exclusive-OR parity-tree operation to generate the serial input) generates "better" noise in this respect.

Noise-generator builders should be aware of the 4557 CMOS variable-length shift register (six input pins set its length, any number from 1 to 64 stages); you have to use it in combination with a parallel-output register (such as the '4015 or '164) in order to get at the n tap; another useful chip is the HC(T)7731, a quad 64-bit shift register (256 bits in all) that runs to 30 MHz (min). Noise-generator builders should be even more aware of the ease with which programmable logic devices (cPLDs or FPGAs) can be coaxed into PRBS duty, as we illustrate in Chapter 11. Microcontrollers will do the job, too, but a PLD solution will be faster.[139]

When going for absolute maximum speed, though, you may want to fall back on good ol' discrete logic: As an admirable example, the CG635 "Synthesized Clock Generator" from Stanford Research Systems[140] can deliver a

---

[138] We are indebted to our late colleague Ed Purcell for these insights.

[139] And a full-custom chip fastest of all… if you can afford the $50k buy-in.

[140] Who obligingly provide full schematic diagrams and parts lists for the instrument.

7-stage PRBS (i.e., $2^7 - 1$ states) at rates to 1.55 GHz.[141] This they do with some elegant tricks: (a) they use MC100EP emitter-coupled differential logic in an array of individually clocked flip-flops; this stuff is *fast* – the D-flops (MC100EP52D) have a setup time of 0.05 ns, zero hold time, and 0.33 ns (typ) propagation time; (b) they use differential clock and data lines (quiet, and fast); (c) the speed bottleneck is the XOR propagation time (0.3 ns), so they do a neat trick, namely (d) they arrange the FFs in a circle, with the data going around clockwise, and the clock going counterclockwise. This has the effect of delaying the clock to the first stage, relative to the clocking of the last two stages, by about 0.25 ns, thus allowing the XOR's (delayed) output to meet the first FF's setup requirement. And the cute geometry causes the clocks of the successive FFs to advance by about 0.05 ns each, thus effectively distributing the pain evenly around the array. Figure 13.120 shows how those clever folks laid out the printed circuit board to make it all happen.

13.14.7 "True" random noise generators

We remarked at the outset that algorithmically generated "noise" cannot be truly random – after all, you can *predict* exactly what's coming if you know the algorithm. Volumes have been written about this.[142] If you want *genuine* noise, you've got to look elsewhere.

A good place to start is some *physical* process (like radioactive decay) that is random *in principle*, and of course unpredictable. Circuit designers do not ordinarily include radioisotopes in their toolkits; but there are other physical processes that do just fine. For example, the noise voltage generated during avalanche conduction in a bipolar junction. We used this in the circuit responsible for the random byte collection that is shipped with the CD-ROM from *Numerical Recipes*, and available from Amazon. Figure 13.121 shows the circuit. Here is the description immortalized on the CD-ROM, where also are found the ~250 MB of what author William Press modestly refers to as "still the best 'random' bits anywhere:"

> Professor Paul Horowitz, of Harvard University, kindly constructed for us an electronic source of physical randomness. The analog noise source is a transistor junction that is biased to function as an

avalanche diode. Physically, the noise current from this junction is generated by the random creation of electron-hole pairs in the junction, due to thermal noise (and, ultimately, quantum mechanical processes). Experimentally, the output of the device, viewed with a spectrum analyzer, is flat from close to dc out to 50 MHz.

The "Hororan" circuit [Figure 13.121] samples the amplified analog voltage from the noise junction at a rate that is 8 times the desired baud rate for output (the latter typically 38.4 kbps or 115.2 kbps). The sample duration ("aperture") is very short, about 2 nanoseconds for the LT1016 latched comparator, so there is plenty of aliasing of the higher available frequencies into the sampled value. If the sample voltage is positive, a digital "1" bit is generated; if it is negative, a digital "0" bit is generated. The collected bits are continuously exclusive-OR'd (XOR) into a register. After every eight bits are collected, the state of the register is output as one bit in the raw file. After every eight output bits, the next two output bits are discarded during the formatting of the required RS-232 stop and start bits. The digital XOR and start/stop formatting functions are performed by a 26V12 PAL, whose programming is included here.

It is intentional that we do not do any more complicated digital processing (mixing, scrambling, encrypting, etc.) within the Hororan box, because we want to be able to measure, and characterize, the degree of randomness coming out of the box. The box indeed has a measurable nonrandomness in its 1-point statistic. That is, the number of output 1s and 0s are not exactly equal but differ by, typically, a few parts in 10^4. (Indeed, the box has a trim adjustment to minimize this nonrandomness.) Notice that it requires several times 10^8 collected bits even to measure this bias convincingly – but we have satisfied ourselves that it is present. The bias drifts slowly with time, presumably in response to thermal and other environmental changes.

We are unable to find, in numerical experiment, any trace of nonrandomness in the higher-point statistics of the raw collected bits. In particular, we have not been able to find (in several times 10^9 collected bits) any 2-point nonrandomness, either in the autocorrelation at small lags, or in the power spectrum at likely frequencies such as 60 Hz or its first few harmonics. On the basis of the physical construction of the Hororan box, we believe that M-point statistics with $M > 2$ (with the exception of high-point statistics that are

---

[141] Other sequence lengths commonly used for bit-error testing are $2^{23} - 1$ and $2^{31} - 1$.

[142] For example, Volume 2 of Donald Knuth's comprehensive classic, *The Art of Computer Programming*.

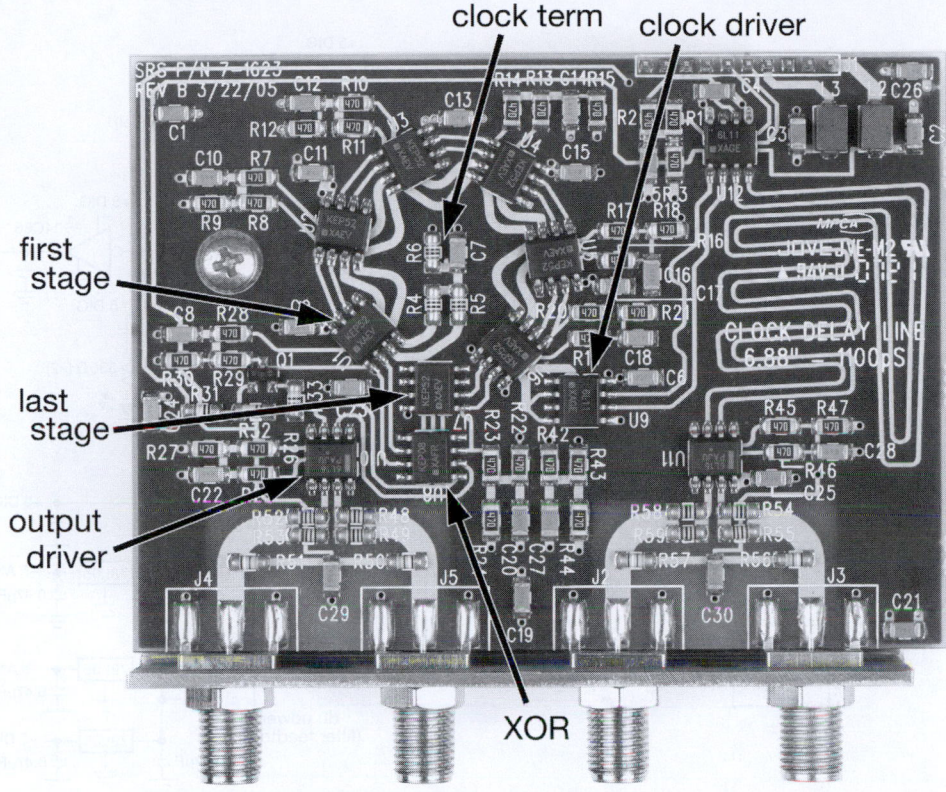

Figure 13.120. Racetrack layout of fast (to 1.55 Gbps) 7-bit PRBS, implemented with 100EP-series discrete LVPECL flip-flops and gates. Data goes clockwise, clock goes counterclockwise, and it all works!

the result of slow drifts in the 1-point bias) should be strongly decreasing with M. The reason is essentially that the transistor junction has no "memory" and has an internal timescale on the order of the reciprocal of its 50 MHz bandwidth.

Thus, while the output of the Hororan box is demonstrably non-random, we believe that its true "entropy per bit" is convincingly bounded as being close to 1 (the fully random value). If $1-e$ denotes the entropy per bit in a large file of N bits (a more detailed definition is given below), then we think that, with a high degree of experimental certainty (there is no way to "prove" this mathematically) we have $e < 0.01$. Incidentally, the raw output of the Hororan box easily passes all of the tests in Marsaglia's "Diehard battery of tests."

To return to that CD-ROM: the output of this hardware was merely the starting point (the "collection phase") for a complex sequence of operations on the way to the final published bytes: the "perfection phase" consists of mul-

tiple passes and shuffling through triple-DES encryption, XOR-ing with freshly collected bytes between each pass and again at the end. A full description can be found on the CD-ROM itself, in a folder labeled "Museum."

13.14.8 A "hybrid digital filter"

The example of Figure 13.119 revisits the interesting topic of *digital filtering*, discussed earlier in §6.3.7. What we have here is simply a 32-sample finite-impulse-response (FIR) lowpass filter, in this case implemented in hybrid fashion – the samples are *digital* (logic HIGH or LOW, in the 32 flip-flop outputs of the shift register U_8–U_{11}), but the weighted sum is performed as an *analog* summing of currents into a pair of op-amp (U_{1ab}) summing junctions. In the more general case, a digital filter would operate on data representing a sampled analog waveform, with each sample represented as a multi-bit quantity (e.g., 16-bits for each channel, in the case of stereo CD audio), sampled at a rate sufficient to retain the full input frequency range of interest (for CD audio $f_{samp} = 44.1$ kHz, about 10% above

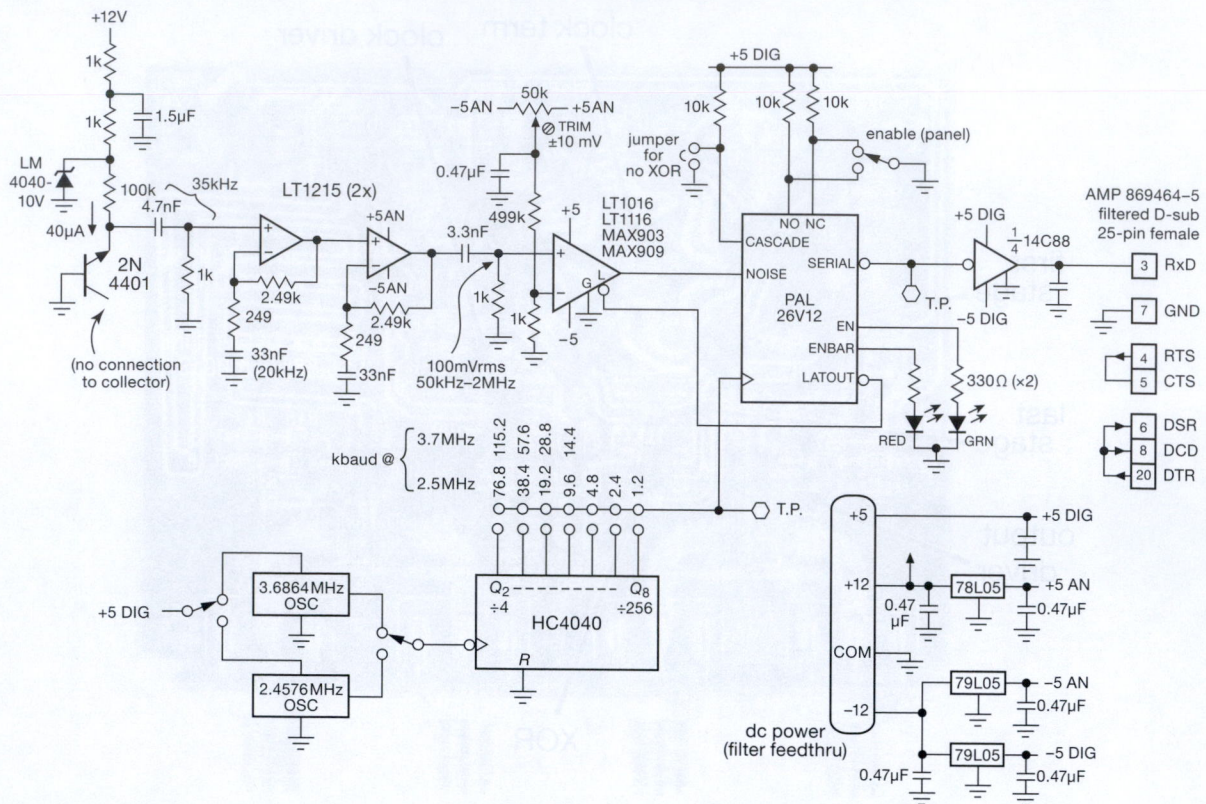

Figure 13.121. Using a physical process (base–emitter avalanche breakdown) to generate random bytes.

the minimum critical rate of $2 \times f_{max}$). And the weighting and summing would be done purely digitally (i.e., numerically), in multipliers and adders. Note however that high-quality digital processing does not *require* that the initial analog-to-digital sampling produce large word sizes: the discussion of delta–sigma conversion earlier in this chapter (§13.9) is a fine illustration of a "1-bit" digitized data stream, upon which digital filters can do their magic.

Additional Exercises for Chapter 13

Exercise 13.10. Why can't two n-bit DACs be used to make a $2n$-bit DAC by just summing their outputs proportionally (OUT$_1$ + OUT$_2/2^n$)?

Exercise 13.11. Verify that the peak output of the pseudorandom noise generator (Figure 9.94) is ± 8.68 V.

Review of Chapter 13

An A-to-I summary of what we have learned in Chapter 13. This summary reviews basic principles, facts, and application advice in Chapter 13.

¶A. DACs and ADCs – Performance Parameters.

The subject of conversion between analog and digital signals dominated the chapter, with extensive description of the internals of ADCs and DACs, their performance, and example applications. Essential ADC performance parameters include resolution (number of bits), accuracy (linearity, monotonicity, stability, ENOB,[143] and spur-free dynamic range), speed (conversion time and latency), input range, output format (serial I²C or SPI; or parallel), reference (internal or external), packaging, and additional features such as internal programmable-gain amplifier. The essential DAC performance parameters are similar, but include output type (voltage or current), glitch energy, and variants such as multiplying DAC (MDAC) or programmable internal digital scaling.

¶B. DAC Types.

The choice of converter for a particular application should be based mostly on its performance parameters (¶A), rather than on the method of conversion. But it's important to know about a converter's inner workings so that you won't be caught unaware by its idiosyncrasies.

Resistor-string DAC. This simplest technique (§13.2.1) consists of a series string of 2^n equal-value resistors, biased by a stable voltage reference, with digitally controlled MOS switches that pick off the dc voltage at a chosen tap as selected by the n-bit binary input code. These inexpensive DACs range from 8 to 16 bits of resolution; the output is strictly monotonic (a desirable feature for a digital control loop).

R–2R ladder DAC. This popular technique (§13.2.2) exploits the binary-weighting property of an R–$2R$ ladder (Figure 13.5), requiring only $2n$ (rather than 2^n) resistors and switches to create an n-bit DAC. These converters are inexpensive, but not inherently monotonic (although there are many that guarantee monotonicity). Their integral nonlinearity (INL) is generally superior to resistor-string

DACs, making them better for precise voltage-setting applications.

Current-steering DAC. In contrast to these voltage-output converters (resistor string and ladder types), the current-steering DAC (§13.2.3) excels in speed, and also permits easy combining of outputs. It uses an array of n switches to steer a set of binary-weighted currents (Figure 13.6). These DACs are wickedly fast, but their current output is limited in compliance; if you tack on a transresistance amplifier to create a voltage output, it will limit the dynamic performance (speed and settling time). Best to drive a load that wants a current input, or simply drive a $50\,\Omega$ load resistor ($75\,\Omega$ for video), appropriate for RF and high-speed applications. *A caution*: the stability and initial accuracy of current-output DACs can be quite poor if you are foolish enough to ignore the matched internal feedback resistor; scc the warning in §13.2.5 and the correct use of feedback shown in Figure 13.9.

Delta–sigma DAC. This technique (§§13.9 and 13.9.8), sometimes called a "1-bit DAC," is a two-step process in which a bitstream is first produced in an integrating digital modulator, then lowpass filtered to produce the analog output voltage; see the block diagram in Figure 13.50. It is popular in professional audio, where it can deliver stunningly linear high resolution output signals, for example dual (stereo) 20-bit resolution at 192 ksps. The delta–sigma technique is popular also for ADCs; in the pro-audio arena, for example, a popular part is the CS5381 dual 24-bit 192 ksps converter (§13.9.11D).

Multiplying DAC. An MDAC (§13.2.4) accepts a varying V_{ref} input, specifying both a voltage range (usually bipolarity, e.g., $-15\,V$ to $+15\,V$) and a reference-multiplying bandwidth (typically a megahertz or more). An MDAC permits digital amplitude control of signals within its bandwidth, and it lets you do ratiometric measurements with an imprecise reference voltage. A caution: capacitive feedthrough severely limits the bandwidth when operated with small-value digital codes; see Figure 13.7.

Pulse-width modulation and F-to-V conversion. When driving a load that is inherently slow (for example, a heater), a simple and effective form of D/A conversion is PWM (§13.2.8). Related techniques are F/V converters and rate multipliers (§§13.2.9 and 13.2.10).

¶C. DAC Application Examples.

In conversion applications the devil is very much in the details. In this chapter we explored four real-world applications, revealing their satanic complications.
(1) A four-channel general-purpose laboratory source (§13.3.1) illustrated precision and stable low-noise design,

---

[143] Where the "effective number of bits" ENOB=$1.44\log_e(V_{span}/V_{n(rms)})$. As explained in nice detail in Analog Devices' tutorial MT-003 ("Understand SINAD, ENOB, SNR, THD, THD+N, and SFDR so you don't get lost in the noise floor"), ENOB is related to SINAD (signal-to-noise+distortion) by the relation ENOB=(SINAD-1.76 dB)/6.02, where SINAD (in dB) is given, in terms of rms quantities for signal, noise, and distortion, by SINAD=$20\log_{10}S_{rms}/(N+D)$.

with an external current-to-voltage configuration providing flexible output ranges.

(2) A simpler eight-channel voltage source (§13.3.2) is easily designed with a fully integrated voltage-output DAC (LTC2656); it lacks the flexible output range capability of the previous example, along with some degradation in terms of noise and stability.

(3) A nanoamp-scale bipolarity current source of wide compliance (§13.3.3) is a circuit of considerable subtlety, offering low noise and low drift while sourcing or sinking nanoamps over a range of ± 10 V. There's additional teaching here, on the subject of floating current sources in general.

(4) A precision bipolarity coil driver (§13.3.4) illustrates a DAC application that pushes the limits of resolution (20 bit, i.e., parts-per-million), with corresponding control of noise and stability. This example includes plenty of discussion of requisite amplifier and reference choices, and of loop compensation and stability.

¶D. DAC Choices.

For highest linearity, delta–sigma DACs are best, with accuracy and linearity to 20 bits at audio speeds, and sometimes with excellent dc specs as well (e.g., TI's millisecond-speed 20-bit DAC1220); however, watch out for broadband and clock noise (the DAC1220 has ~ 1000 nV/$\sqrt{\text{Hz}}$ at 1 kHz, compared with ~ 10 nV/$\sqrt{\text{Hz}}$ for resistor ladder converters).

For high-accuracy applications at medium speed, there are many excellent R–$2R$ and linear ladder DACs, for example TI's DAC8552 (dual 16-bit, serial in, voltage out, ext ref, very low glitch, 10 μs settle; DAC8560/4/5 similar, with internal ref); or ADI's AD5544 or TI's DAC8814 (quad 16-bit MDAC, serial in, current out, 0.5–2 μs settle with external I-to-V op-amp); or LTC's LTC1668 (16-bit, parallel in, differential current out, 20 ns settle into 50 Ω as "voltage out"); or TI's DAC9881 (18-bit serial in, rail-to-rail voltage out, ext ref, low-noise, 5 μs settle).

For the highest speed you can't beat current-steering converters, for example the TI DAC5681/2 (16-bit, 1 Gsps) or the ADI AD9739 (14-bit, 2.5 Gsps).

¶E. ADC Types.

As with DACs, the choice of converter for a particular application should be based mostly on its performance parameters (¶A), rather than on the method of conversion. As with DACs, though, it's important to know about a converter's inner workings so that you won't be caught unaware by its idiosyncrasies.

Flash, or "parallel." In this simplest technique (§13.6)

the analog input voltage is compared with a set of fixed reference voltages, most simply by driving an array of 2^n analog comparators, to generate an n-bit result, with resolutions to $n=8$ bits and speeds to 20 Gsps (not in the same device). Variations on this theme include pipelined or folded architectures, in which the conversion is done in several steps, each of which converts the "residue" of the previous low-resolution conversion. These converters top out at 16-bit resolution and speeds to several Gsps (not in the same device); the pipelined architecture creates latency, which can be as much as 20 clock cycles.

Successive approximation. In this technique (§13.7) successive trial codes (generated by internal logic) are converted to voltages by an internal DAC and compared with the analog input voltage. It requires just n such steps to do an n-bit conversion. The internal DAC can be implemented as a conventional n-stage R–$2R$ resistor ladder, or, interestingly, as a set of 2^n binary-scaled capacitors; the latter method is known as a *charge-redistribution* DAC. SAR-type converters are intermediate in speed and accuracy (flash converters are faster; delta–sigma converters are more accurate). The input must be stable during the conversion (thus requiring some form of input track-and-hold); SAR converters can have missing codes.

Voltage-to-frequency. This technique (§13.8.1) produces an output pulse train (or other waveform) whose frequency is accurately proportional to the analog input voltage. In an *asynchronous* V/F the oscillator is internal and free-running; a *synchronous* V/F requires an external source of clock pulses, gating a fraction of them through such that the *average* output frequency is proportional to the analog input. V-to-F converters can be used for simple control of averaging loads (like a heater); but they barely deserve listing as an ADC, and are best thought of as a primitive approximation to the superior delta–sigma converter (see below).

Single-slope integration. In this technique (§13.8.2) an internally generated analog ramp (capacitor charged by a current source) goes from zero volts to the analog input voltage, timed by counting pulses from a fast fixed-frequency clock; the count is proportional to the value of the analog input. Single-slope integration is not particularly precise, yielding that turf to dual- or multi-slope integration (see below). But it is used in applications such as pulse-height analysis, where speed and uniformity of adjacent codes is required; it is also used for time-to-amplitude conversion (TAC).

Dual-slope and multi-slope integration. These techniques (§§13.8.3, 13.8.4, and 13.8.6) are variations on single-slope integration, effectively eliminating errors from comparator offsets and component stability. In *dual-slope integration* the capacitor is ramped up for a fixed time with a current proportional to the input signal, and ramped back down with a fixed current; the latter time interval is proportional to the analog input. In *quad-slope integration* the input is held at zero while a second such "auto-zero" cycle is done. The so-called *multi-slope* technique is somewhat different, with a single conversion consisting of a succession of fast dual-slope cycles (in which the input is integrated continuously, combined with subtractive fixed-current cycles), and with a correction based upon the partial-cycle residue at both ends. In some aspects it's a close cousin of the delta–sigma method (next). These integrating ADC methods are ideal for low-speed (milliseconds) high-resolution (20–28 bits) voltmeter-style measurements.

Delta–sigma ADCs. In this technique (§13.9) a *modulator* converts the oversampled analog input voltage to a serial *bitstream*; then a digital lowpass filter accepts that bitstream as input, producing the final *n*-bit digital output. In its simplest form the modulator consists of an integrator acting on the difference between the analog input voltage and the value of the 1-bit output serial bitstream, to determine the next output bit. Variations include higher-order modulators (a succession of weighted integrators), or bitstreams that are several bits wide, or both. Delta–sigma converters deliver excellent resolution (to 24 bits) and conversion rates to a few megasamples per second. They are hugely popular in professional audio.

¶F. ADC Application Examples.
In this chapter we illustrated several application examples. (1) A fast flash ADC (§13.6.2), driven differentially, illustrated input filtering and low-jitter clock generation. (2) A low-noise high-stability 18-bit successive-approximation converter with a 2 Msps conversion rate (§13.7.3) illustrated careful input conditioning to achieve low noise and low drift. (3) Three applications of delta–sigma converters (Figures 13.66, 13.67, and 13.68) illustrated a wide range (±10 V) multiplexed converter with good accuracy (18 bits or better); an industry-standard 24-bit ΔΣ converter of good noise, drift, and linearity specifications, and with chopper-stabilized programmable-gain amplifier for accurate measurements of low-level signals; and a pro-audio conversion application, where dc accuracy and latency are largely irrelevant, instead featuring excellent channel matching and with audio-type output formats.

In this chapter we illustrated further conversion examples, including some unusual ADCs and DACs (§13.11), and a few conversion *system* examples (§13.12). The latter included a multiplexed 16-channel data-acquisition system (§13.12.1), a parallel multichannel successive-approximation data-acquisition system (§13.12.2), and an analogous parallel multichannel delta–sigma data-acquisition system (§13.12.3). The last two examples included both external and integrated multichannel solutions, in tutorial narratives.

¶G. ADC Choices.
For low-speed (to kilosamples/sec) and high-accuracy (to 24 bits or more) applications, integrating converters (dual-slope, multislope) are best; there are also some excellent delta–sigma converters (which can be thought of as integrating converters).

For moderate-speed applications (to several Msps), both delta–sigma and successive-approximation converters are competitive, with resolution to 20 bits or more; delta–sigma converters have significantly longer latency.

For high-speed (100s of Msps) choose pipelined flash converters, mindful of their high latency. And for the highest conversion rates (>250 Msps) some variant on folding flash gets you to 8–12 bits and 3 Gsps. Simple flash is the fastest, but you pay the price in resolution (e.g., the Analog Devices HMCAD5831: 3-bit conversion at 20 Gsps).

¶H. Phase-locked Loops.
The chapter continued with the important mixed-signal subject of PLLs (§13.13), circuits in which feedback forces a signal derived from a voltage-controlled oscillator to match an input signal's frequency. PLL applications include frequency multiplication and frequency synthesis, clock generation and recovery, and demodulation of AM, FM, and digitally modulated signals. In addition to a voltage-controlled oscillator (VCO), the essential components of a PLL (Figure 13.85) are (a) the phase detector and (b) the loop filter.

The *phase detector* (PD) compares the phases of the input signal and the VCO-derived signal, generating an output signal representative of their relative phase. The simplest phase detector (type I) multiplies the input signals. It is applicable to either analog or digital input signals; for the latter it is just an exclusive-OR gate (Figure 13.86). The other common phase detector (type II) generates output pulses according to the relative timing of transitions at its two inputs (Figure 13.87). It works only with digital signals. Type II phase detectors have the benefit of locking with zero phase difference, and of introducing no ripple at the phase-detection frequency; however, they are more sensitive to input signal jitter than type I phase detectors.

The *loop filter* smooths the PD's output, with a time constant that sets the loop's response; the latter should be long compared with the comparison frequency, but fast enough to follow changes in the input frequency as required by the PLL's application. To clarify this last point, you'd want a long time constant if the PLL has a low-noise VCO and is used to generate a clean stable clock from a noisy input reference of fixed frequency; but if you want the PLL to follow a wobbling input frequency (e.g., clock recovery from a tape or disk drive) you would make the loop fast enough to respond to the changing input frequency. *A caution*: there's an inherent 90° lagging phase shift in the PLL: a measurement of phase is used to adjust frequency, but phase is the integral of frequency. So a simple lowpass loop filter (which introduces additional lagging phase shift, asymptotic to 90°) creates a marginally stable loop. The solution is to slow the rolloff in the region of unity loop-gain (a "zero," in the language of *s*-plane analysis; see the discussion in Chapter *1x*), as seen in Figure 13.98.

The procedure for designing the loop filter is illustrated in §13.13.3. And the use of PLLs for the important applications of frequency synthesis, analog and digital demodulation, pulse synchronization, and laser offset locking is discussed in some detail in §13.13.6.

¶I. Pseudorandom Digital Noise.

The chapter concluded with the fascinating topic of the generation of deterministic pseudorandom bit sequences (PRBSs) from linear feedback shift registers (LFSRs). These find application in communication channel testing (eye diagrams, see for example Figure 12.131) and in spectrum-spreading applications in digital communications (e.g., the transmitted navigational signals from GPS). And, they're just plain fun. The basic technique (described in some detail also in Chapter 11, in §11.3) is a clocked shift register whose input is derived from an XOR combination of two (or more) carefully chosen bits of the register (Figure 13.111).

As demonstrated in Chapter 11, PRBS generators are easily made with a tiny bit of PLD logic, or some microcontroller code, or even just a few logic chips (e.g., a 74HC7731, 74HC164, and 74HC86, clocked at 25 MHz, makes a PRBS whose cycle length is more than 10^{62} years; that's 10^{52} times the age of the universe). The output sequence can be lowpass filtered to generate an analog noise source of settable bandwidth; see the analog-filtered colored noise source in Chapter 8 (Figure 8.93) and the hybrid digitally-filtered white noise source of Figure 13.119.

Pseudorandom noise is not truly random, even though its statistical properties may mimic genuine randomness as measured by various tests (see §13.14.6), rendering it suitable for some applications. To generate honest random bit sequences you need to exploit the random properties of some physical process like β-decay in unstable isotopes. More convenient for the circuit designer is something like thermal noise in resistors, or avalanche noise in semiconductor junctions; the latter is illustrated in §13.14.7.

COMPUTERS, CONTROLLERS, AND DATA LINKS

In this chapter and the next we deal with the interesting subjects of computers and controllers (with the common alternative names of *microcomputers* and *microcontrollers*). These are huge subjects, and we do not attempt a complete treatment. Rather, our primary interests, in the context of electronic circuit design, are (a) how to interface external electronics to an existing computer, and (b) how to use a microcontroller as an "embedded" piece of electronics within a custom electronic device or instrument.

In this chapter, then, we look at computers and data *buses* (the mechanism for data interfacing), to understand how they work and how to create a data interface. Along the way we review, briefly, some basics of computers – their architecture, processor and memory hardware, bus timing, and instruction set – and we illustrate interfacing with hardware and code for simple input and output "ports" attached to a classic parallel data bus.[1] We end with a review of popular serial data buses: FireWire, USB, CAN-bus, Ethernet, and wireless (WiFi, Bluetooth, Zigbee).

In Chapter 15 we'll deal with embedded microcontrollers, the inexpensive and wonderful chips that let you put the intelligence of a computer into pretty much anything electronic.

Some terminology

Mainframes, CPUs, minicomputers, microcomputers, microprocessors, microcontrollers – what are all these names, anyway? History is partly to blame for this proliferation of confusing terms, because names like "microcomputer" were created (with understandable excitement) to describe a computer in which the central processor (the central processing unit, CPU) was integrated onto a single chip (the "microprocessor"), rather than constructed from a board full of smaller logic chips (as it was in a "minicomputer" or "mainframe" of the time).

This is no longer a big deal, and the important distinction is between *computers*, which are used primarily for computational processing of data, versus embedded *controllers*, which are incorporated within other electronic devices that are not themselves computers. Examples of the former are PCs, laptop and notebook computers, and the larger "mainframe" computers. Examples of controllers are the chips that run your electric toothbrush, bathroom scale, television and stereo sets, and the like.[2] In general, the microprocessors used in computers are optimized for computational performance, tend to be expensive, and require support complex "chipsets" in order to function. By contrast, the microcontrollers intended for embedded design trade off computational performance in favor of stand-alone capability, with as many peripheral functions as possible integrated onto the same chip; they tend to be inexpensive, low power, and completely self-sufficient. Figure 14.1 dramatically illustrates the progress in miniaturizing computer processors for embedded applications[3].

Generally speaking, with the computers we talk about in this chapter, the design of the computer itself (including the integration of memory, disks, and I/O, as well as system programming and utility program development) has been taken care of by the manufacturer (and suppliers of compatible hardware and software). The user need worry only about special-purpose interfaces and the job of user programming. By contrast, in an embedded controller application, the circuit design and choice of hardware, along with all programming, are done by the designer. Computer manufacturers are generally committed to providing system and utility software as part of a complete computing system (often including peripherals), whereas the manufacturers of microcontrollers (semiconductor companies) generally see the design and marketing of newer and better chips as their central tasks.

---

[1] We've chosen the PC104/ISA bus, both because of its simplicity, and its stability as an entrenched standard in the PC104 embedded PC world.

[2] Sometimes the line is a bit blurry: the processor inside a cable television "set-top" box or Blu-ray™ player, for example, is doing heavy processing of digital video (decryption, decompression, format conversion), yet the box is not a "computer."

[3] And a decade later the same $5 gets you a 100 MHz ARM microcontroller with 1 MByte of flash memory, 80 kBytes of SRAM, 47 parallel I/O ports, 11 channels of 10-bit ADC, a 10-bit DAC, six 16-bit timers, three I^2C and four SPI serial ports, four PWM controllers, four UARTs, and three USARTs. What's not to like about Moore's Law!

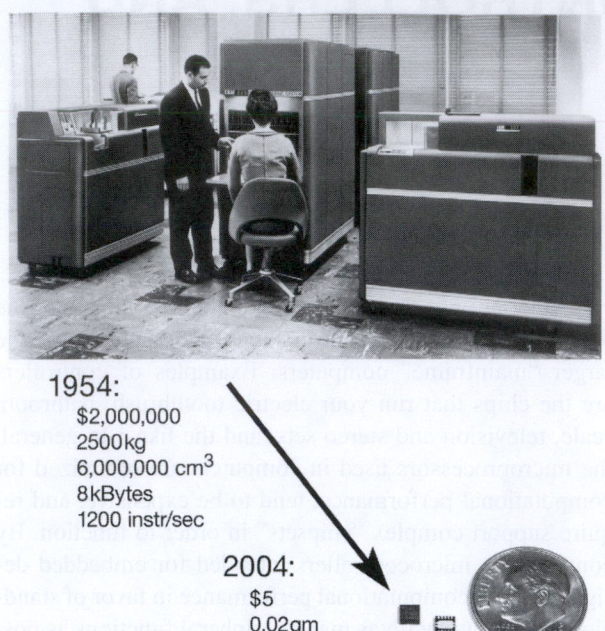

1954:
$2,000,000
2500 kg
6,000,000 cm^3
8 kBytes
1200 instr/sec

2004:
$5
0.02 gm
0.008 cm^3
8 kBytes
25,000,000 instr/sec

Figure 14.1. From mainframe to microcontroller: 50 years of computer progress (mainframe photograph courtesy IBM Corporation).

14.1 Computer architecture: CPU and data bus

In spite of rapid change in the field of computers and microprocessors, there are important basic ideas, many of which carry over to digital electronics generally. These are most easily understood from a look at a traditional bus-oriented architecture, shown in the block diagram of Figure 14.2.

The idea is to link all the connected devices via a shared set of lines – a data "bus" – rather than a rat's nest of direct connections. It uses far fewer wires; and, because traditionally the CPU was responsible for most transactions anyway, it was not necessary to have separate data paths to achieve good performance.[4] So, for example, if the CPU wants to store a byte at some address in random access memory (RAM), it would put the address and the data onto a set of bus lines (called ADR and DATA), then it would assert a WRITE signal; the RAM recognizes the address, and accepts the byte. If the CPU instead wants to fetch a byte, it would put the address only, then it would assert a READ signal; the RAM, recognizing a request for data from one of its addresses, responds by putting the corresponding byte on the DATA lines. Likewise for transfers to or from other bus-attached devices (each of which listens for its assigned range of addresses).

To understand what's going on, let's review the components in Figure 14.2, taking it from left to right.

14.1.1 CPU

The central processing unit, or CPU, is the heart of the machine. Computers do their computation in the CPU on chunks of data organized as computer *words*. Word size can range from 4 bits to 64 bits or more, with 32- and 64-bit word sizes being the most popular in contemporary computers. A *byte* is 8 bits (half a byte, or 4 bits, is sometimes called a "nybble"). A portion of the CPU called the *instruction decoder* interprets the successive instructions (fetched from memory), figuring out what should be done in each case. The CPU has an *arithmetic logic unit*, or ALU, that can perform the instructed operations, such as add, complement, compare, shift, move, etc., on quantities contained in *registers* (and sometimes in *memory*). The *program counter* keeps track of the current location in the executing program. It normally increments after each instruction, but it can take on a new value after a "jump" (branch) or a "call" instruction. The *bus control circuitry* handles communication with memory and I/O. Most computers also have a *stack pointer register* (more on that later) and a few *flags* (carry, zero, sign) that get tested for conditional branching. All high-performance processors also include *cache* memory, which holds values (both data and instructions) recently fetched from memory for quicker access.

Contemporary high-performance microcomputers exploit "parallel processing," in which several interconnected CPUs (each with multiple ALU paths) achieve greater computational power; high-density chips integrate this into a "multicore" architecture. However, to keep our discussion simple, and recognizing our interest in *interfacing* to a computer (rather than designing one), we consider only the

---

[4] This kind of single "multidrop" parallel bus was widely used throughout the minicomputer era (e.g., the DEC "Unibus") and well into the microcomputer era (in the form of the IBM "industry standard architecture" bus, or ISA, which is still used in the PC104 embedded-PC standard). But buses have evolved into more complex creatures, and contemporary PCs have multiple buses and connecting bridges, with names like "Front Side Bus," PCI and PCIe, "Northbridge," and "Southbridge," designed to optimize high-speed communication (such as memory transfers) without being slowed down by low-speed communication (such as peripherals). Also, as we'll see later, in spite of the elegance and econ-

omy of shared lines, a "point-to-point" (rat's nest) connection scheme is actually much better than a shared multidrop bus for high-speed communication.

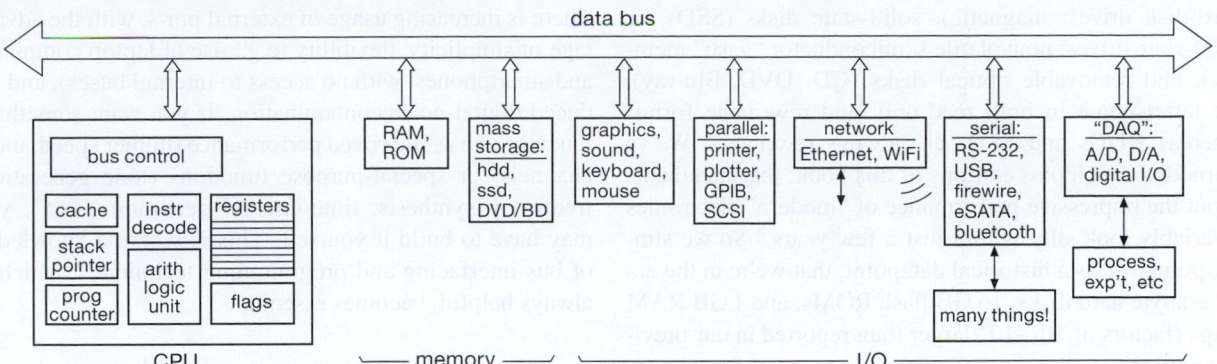

Figure 14.2. Traditional bus-oriented computer.

single-CPU machine, executing instructions serially: it's a world of relentless "read, do, write" operations.

14.1.2 Memory

All computers have some fast "random-access" memory,[5] called RAM (it used to be called "core," because tiny magnetic cores held the data, one bit per core). In an inexpensive computer this would typically amount to several gigabytes, whereas an embedded micro*controller* typically incorporates much less, as little as 4K–64K.[6] This memory can typically be read and written in about 100 ns. The high-density RAM modules usually used in computers (called DRAM[7]) are *volatile*, which means that its information evaporates when power is removed (maybe it should be called "forgettory"!). All computers therefore include some nonvolatile memory, usually "flash ROM" (read-only memory), to bootstrap the computer, i.e., get it started from a state of total amnesia when power is first turned on. (That nonvolatile memory additionally holds the entire program, in the case of an embedded controller.) Much more about memory in §14.4.

To get or store information in memory, the CPU "ad-

dresses" the desired word. Most computers address memory by bytes, beginning at byte 0 and going sequentially through to the last byte in memory. Since most computer words are several bytes long, you are usually storing or fetching a group of bytes at a time; this is usually expedited by having a data bus that is several bytes wide.

Both programs and data are kept in memory during program execution. The CPU fetches instructions from memory, figures out what they mean, and does the appropriate things, often involving data stored somewhere else in memory. General-purpose computers usually store programs and data in the same memory ("von Neumann architecture"), and in fact the computer doesn't even know one from the other. (Amusing things start to happen if a program goes awry and you "execute" data!) Embedded controllers, on the other hand, often use a "Harvard architecture," with separate data and program memory, the latter in non-volatile (usually flash or EEPROM) storage for obvious reasons.

Because computer programs spend most of their time looping through a relatively short sequence of instructions, you can enhance performance by providing a small, but fast, *cache* (pronounced "cash") memory, in which you routinely store copies of the most recently used memory locations. A cached CPU checks its local cache first, before fetching from (slower) main memory; when looping through familiar territory, you often achieve a cache "hit" rate of 95% or better, dramatically improving execution speed.

14.1.3 Mass memory

Computers intended for program development or computation, as opposed to embedded controllers, use nonvolatile mass storage devices such as hard disks (HDDs or

---

[5] So named because you can access data at any address in equal time, as distinguished from something like a shift register or FIFO.

[6] When used to describe memory sizes, K doesn't mean 1000, but rather 1024, or 2^{10}; thus, 16K bytes (16 KB) is actually 16,384 bytes. We employ the lowercase symbol k to mean 1000.

[7] For *dynamic* RAM: a form of memory in which data bits are stored, briefly, as a voltage state on tiny on-chip capacitors, whose charge must be periodically "refreshed." By contrast, *static* RAM ("SRAM") holds each bit as a flip-flop state, requiring no refresh. DRAM is more compact, requiring one transistor per bit ("1T"), versus 6T for SRAM. Both SRAM and DRAM are volatile, as contrasted with flash ROM or EEPROM ("E-squared").

hard-disk drives: magnetic), solid-state disks (SSDs, or solid-state drives: nonvolatile semiconductor "flash" memory), and removable optical disks (CD, DVD, Blu-ray); the latter come in both read-only and rewritable forms, whereas HDDs and SSDs are always rewritable. We've learned, through past editions of this book, that statements about the impressive performance of "modern" electronics invariably look silly within just a few years.[8] So we simply point out, as a historical datapoint, that we're in the era of terabyte hard disks, 16 GB flash ROMs, and 1 GB RAM chips (factors of 10^3–10^4 larger than reported in our previous edition).

14.1.4 Graphics, network, parallel, and serial ports

These are standard items in most computers, needed for communication with the user, with the network, and with external hardware. We presume the reader to be generally familiar with them. Worth noting, however, is the trend away from parallel interfaces (like internal IDE and SCSI, and external "LPT" printer ports), in favor of fast serial interfaces (SATA, USB, FireWire, Ethernet), the latter providing comparable (or better) performance, along with simplified wiring and reduced electrical noise. We'll talk about parallel-port and serial-port communications later, beginning in §14.5.

14.1.5 Real-time I/O

For experiment or process control and data logging, or for exotic applications such as speech or music synthesis, you need A/D and D/A devices that can communicate with the computer in "real time," i.e., while things are happening. The possibilities are almost endless here, although a general-purpose set of multiplexed A/D converters (ADCs), a few fast D/A converters (DACs), and some digital "ports" (serial or parallel) for exchange of digital data will permit many interesting applications.[9] Such general-purpose peripherals are commercially available for most popular computer buses, both internal (e.g., PCI and PCIe) and external (USB, FireWire, Ethernet), as exemplified by the extensive offerings from National Instruments.

There is increasing usage of external ports, with the advantage of simplicity, flexibility (e.g., use of laptop computers and smartphones, with no access to internal buses), and reduced digital-noise contamination. If you want something fancier, such as improved performance (higher speed, more channels) or special-purpose functions (tone generation, frequency synthesis, time-interval generation, etc.), you may have to build it yourself. This is where a knowledge of bus interfacing and programming techniques, which is always helpful, becomes essential.

14.1.6 Data bus

Communication between the CPU and memory or peripherals takes place on a *bus*, a set of shared lines for exchange of digital words. The use of a shared bus vastly simplifies interconnections, since otherwise you would need multi-wire cables connecting every pair of communicating devices.[10] With a little care in bus design and implementation, everything works fine.

Buses have evolved over time, reaching such levels of speed and complexity that interfacing now requires considerable expertise. For this reason we are going to adopt an older and simpler bus, called PC104/ISA, which makes interfacing a simple job. The ISA bus was originated in the IBM PC, and as the PC104 bus it is alive and well, and supported by more than 75 manufacturers. A colleague, when asked its likely lifetime, replied "PC104 will bury us all."[11]

In general, a data bus contains a set of DATA lines (generally the same number as bits in a word – 8 for small processors and controllers, 16 to 64 for more sophisticated microcomputers), some ADDRESS lines for determining who should "talk" or "listen" on the line, and a bunch of CONTROL lines that specify what action is going on (data going to or from the CPU, interrupt handling, direct memory access transfers, etc.). All the DATA lines, as well as a number of others, are *bidirectional* – they're driven by three-state devices, or in some cases by open-collector gates with resistor pullups somewhere (usually at the end of the bus, where they also serve as terminators to minimize reflections; see §12.10.1) and Appendix H); pullups may be necessary with three-state drivers also, if the bus is physically long.

---

[8] In our first edition (1980) we were excited about several-megabyte hard disks, 2 KB EPROMs, and 8 KB RAM; a decade later we were even more excited by several-hundred-megabyte hard disks, 128 KB EPROMs, and 512 KB RAM.

[9] And don't overlook the humble "sound card" (often integrated onto the motherboard), for lab use in noncritical audio-rate applications where dc coupling is not needed. The best of these can do 192 kb/s multichannel sampling at 16-bit resolution.

[10] This idea is repeated within the CPU itself, where internal data buses are used for communication between the arithmetic logic units (ALUs) and the registers (for example). And, in the case of a microcontroller, where "peripherals" like A/D converters and USB ports are integrated onto the same chip, the work of an external bus is carried out with on-chip buses.

[11] We think he meant that it will *outlive* us, not *kill* us. But we're not sure.

Three-state or open-collector devices are used so that devices connected to the bus can disable their bus drivers, because in normal operation only one device is asserting data onto the bus at any time. Each computer has a well-defined protocol for determining who asserts data, and when. If it didn't, total chaos would result, with everyone shouting at once. (Engineers cannot resist personalizing electronic circuits, in spite of advice like "Don't anthropomorphize computers – they don't like it.")

We'll return to the bus in detail, with 8-bit PC104 interface examples. First, though, we need to look at the CPU's instruction set.

14.2 A computer instruction set

14.2.1 Assembly language and machine language

To understand bus signals and computer interfacing, you've got to understand what the CPU does when it executes various instructions. At this point, therefore, we would like to introduce the instruction set that goes with the Intel x86-family processors used in PC104 computers. Unfortunately, the instruction sets of most real-world microprocessors tend to be rich with complexities and extra features (including the ghosts of processors past), and the Intel x86 series is no exception. However, since our purpose is only to illustrate bus signals and interfacing (not fancy programming), we'll take a shortcut by laying out a subset of x86 instructions. By leaving out the "extra" instructions we'll wind up with a compact set of instructions that is both understandable and complete enough to do most any programming task. We'll then use it to show some examples of interfacing and programming. These examples will help convey the idea of programming at the "machine-language" level, something quite different from programming in a high-level language like C or C++.

First, a word on "machine language" and "assembly language." As we mentioned earlier, the computer's CPU is designed to interpret certain words as instructions and carry out the appointed tasks. This "machine language" consists of a set of binary instructions, each of which may occupy one or more bytes. Incrementing (increasing by one) the contents of a CPU register would be a single-byte instruction, for example, whereas loading a register with the contents of a memory location would usually require at least two bytes, perhaps as many as five (the first would specify the operation and register destination, and four more would be necessary to specify an arbitrary memory location in a large machine). It is a sad fact of life that different computers have different machine languages, and there is no standard whatsoever.

Programming directly in machine language is extremely tedious, because you wind up dealing with columns of binary numbers, each bit of which has to be bit-perfect, so to speak. Instead, you can represent each instruction in its *assembly-language* readable-text form, using easily remembered mnemonics for the instructions, and symbolic names of your own choosing for memory locations and variables. This assembly-language program, consisting of a plain text file, is then massaged by a program called an *assembler* to produce as its output a finished program in machine-language *object code* that the computer can execute. Each line of assembly code corresponds to one machine-language instruction, which translates to a few machine-language bytes (1 to 15 bytes, for the x86). The computer cannot execute the assembly-language (text) instructions directly. To make these ideas concrete, let's look at our subset of the x86 assembly language and do a few examples.

14.2.2 Simplified "x86" instruction set

The x86-family processors (Intel, AMD, VIA) have a rich, and somewhat idiosyncratic, instruction set; part of its complexity stems from the designers' objective to maintain compatibility with the original 8080 8-bit processor. Later and more sophisticated processors in the x86 family (e.g., Pentium, Core 2, i3/i5/i7, Xeon) have many additional instructions but can still execute the original x86 instruction set. We've gone through those instructions with a machete, keeping just 10 arithmetic operations and 11 others (Table 14.1).

A. A quick tour

Some explanations: the first six arithmetic instructions in Table 14.1 operate on pairs of numbers ("2-operand" instructions), which we've abbreviated as b,a, and which can be any of the 5 pairs listed in the table footnotes; m means the contents of a memory location, r means the contents of a CPU register (there are 8 in the original 8086), and *imm* means an *immediate* argument, which is a number stored in the next 1 to 4 bytes of memory following the instruction. Thus, for example, the instructions

```
MOV    count,CX
ADD    small,02H
AND    AX,007FH
```

Table 14.1 Simplified x86 Instruction Set

Instruction		What you call it	What it does
arithmetic			
MOV	*b,a*	move	$a \rightarrow b$; *a* unchanged
ADD	*b,a*	add	$a + b \rightarrow b$; *a* unchanged
SUB	*b,a*	subtract	$b - a \rightarrow b$; *a* unchanged
AND	*b,a*	and	*a* AND *b* $\rightarrow$ *b* bitwise; *a* unchanged
OR	*b,a*	or	*a* OR *b* $\rightarrow$ *b* bitwise; *a* unchanged
CMP	*b,a*	compare	set flags as if $b - a$; *a,b* unchanged
INC	*rm*	increment	$rm + 1 \rightarrow rm$
DEC	*rm*	decrement	$rm - 1 \rightarrow rm$
NOT	*rm*	not	1's complement of $rm \rightarrow rm$
NEG	*rm*	negate	negative (2's comp) of $rm \rightarrow rm$
stack			
PUSH	*rm*	push	push *rm* onto stack (2 bytes)
POP	*rm*	pop	pop 2 bytes from stack to *rm*
control			
JMP	*label*	jump	jump to instr *label*
Jcc	*label*	jump conditional	jump to instr *label* if *cc* true
CALL	*label*	call	push next adr, jump to instr *label*
RET		return	pop stack, jump to that adr
IRET		return from int	pop stack, restore flags, return
STI		set interrupt	enable interrupts
CLI		clear interrupt	disable interrupts
input/output			
IN	AX,*port*	input	*port* $\rightarrow$ AX (or AL)
OUT	*port*,AX	output	AX (or AL) $\rightarrow$ *port*

Notes: *b,a*: any of *m,r r,m r,r m,imm r,imm*
rm: register or memory, via various addressing modes
cc: any of **Z NZ G GE LE L C NC S NS**
label: via various addressing modes
port: byte (via *imm*) or word (via DX)

have arguments of the form *m,r*, *m,imm*, and *r,imm*, respectively. The first copies the contents of register CX to a memory location that we've named "count"; the second adds 2 to the contents of another memory location called "small"; the third clears the top 9 bits of 16-bit register AX while preserving the bottom 7 bits unchanged (a so-called *masking* operation). Note Intel's argument convention: the first argument is replaced with or modified by the second argument.

The last four arithmetic operations in the table take only a single operand, which can be the contents of either a register or memory. Here are two examples:

```
INC   count
NEG   AL
```

The first adds 1 to the contents of memory-location "count," and the second changes the sign of the contents of register AL.

B. A detour: addressing

Before continuing, a word on registers and memory addressing. The original 8086 claimed to have eight "general-purpose" registers, but after reading the fine print you'll realize that most of them also have special uses (Figure 14.3). Four of them (A–D) can be used either as single 16-bit registers (e.g., AX; think of "X" as "extended") or as a pair of byte registers (AH, AL; "high" and "low" halves).[12] The BX and BP registers can hold addresses, as can the SI and DI registers, and tend to be used for addressing (see below). Special looping instructions (which we omitted from our short list) use register C, and multiply/divide and I/O instructions make analogous use of registers A and D.

Data used in instructions can be immediate constants, values held in a register, or values in memory. You

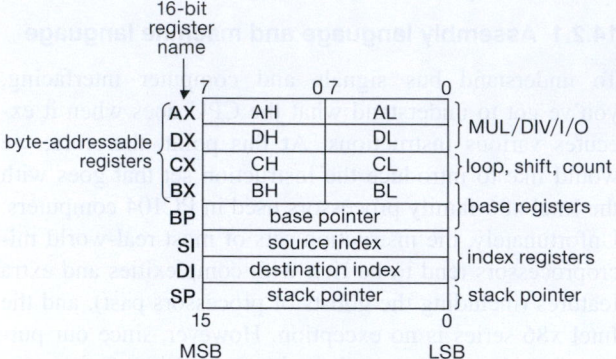

Figure 14.3. 8086 general-purpose registers.

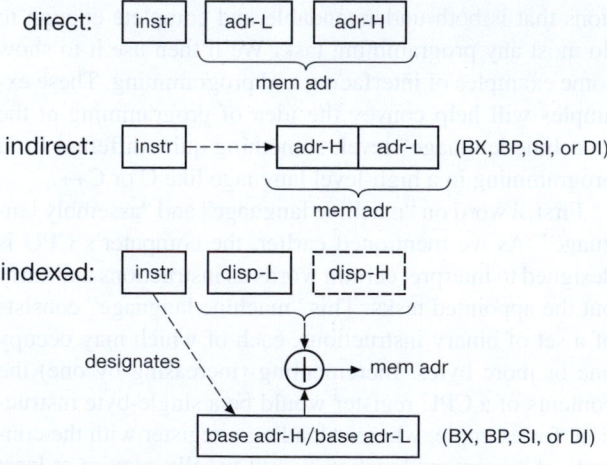

Figure 14.4. Some addressing modes.

---

[12] Later x86 CPUs extended these to 32 bits and to additional 64-bit registers.

specify immediates by value, and registers by name, as in the examples above. To address memory, the x86 provides six addressing modes, three of which are described by the diagrams in Figure 14.4. In assembly language you can just name the variable *directly*, in which case its address gets assembled as a pair of bytes immediately following the instruction; you can put the variable's address in an addressing register (BX, BP, SI, or DI), then use an instruction that specifies addressing *indirectly* through the register; or you can combine these actions, adding an immediate *displacement* to the value in a designated addressing register to get the variable's address. The indirect mode is faster (assuming the address has already been loaded into an addressing register) and much better if you want to do something to a whole set of numbers (a *string* or *array*). Here are a few addressing examples:

```
MOV    count,100H      (direct,immediate)
MOV    [BX],100H       (indirect,immediate)
MOV    [BX+1000H],AX   (indexed,register)
```

The last two assume you've already put an address into BX. The last instruction copies the contents of AX to a memory location 4K (1000 hex) higher than where BX points in memory; we'll give an example shortly, showing how you could use this to copy an array.

There's one other complexity of x86 memory addressing that we've swept under the rug: the "address" generated by any of these addressing modes is not actually the final address, as should be obvious from the fact that the address register BX has only 16 bits (which can address only 64K bytes of memory). In fact, it's called an *offset*; to get an actual address, you add to the offset a 20-bit *base* formed by shifting left 4 bits the contents of a 16-bit *segment register* (there are four such registers). In other words, the x86 lets you access groups of 64K bytes of memory at a time, with the location of those "segments" within a total memory size of 1MB set by the contents of the segment registers. The use of 16-bit addressing in the 8086 was basically a big mistake, inherited from earlier generations of microprocessors. Newer processors (80386 onward, and recent designs in other CPU families) are done right, with 32-bit or 64-bit addressing throughout.[13] Rather than complicate our examples, we'll simply ignore segments entirely; in real life you would, of course, have to worry about them.

---

[13] And with an additional level of indirection (virtual memory implemented with page tables) on the way to the actual physical memory. But you didn't really want to know about that, right?

C. Instruction-set tour (continued)

The *stack* instructions PUSH and POP come next. A stack is a portion of memory, organized in a special way: when you put data onto the stack (a *push*), it goes into the next available spot ("top" of the stack); and when you retrieve data (a *pop*), it is taken from the top, i.e., it is the item last pushed onto the stack. Thus a stack is a consecutive list of data, stored last-in, first-out (LIFO). It may help to think of a lunchroom tray dispenser.

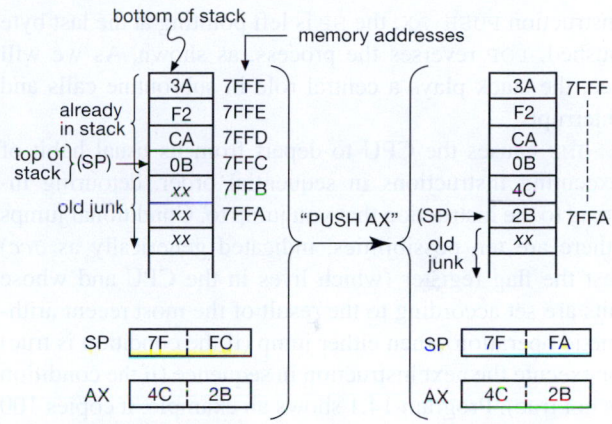

A. PUSH operation

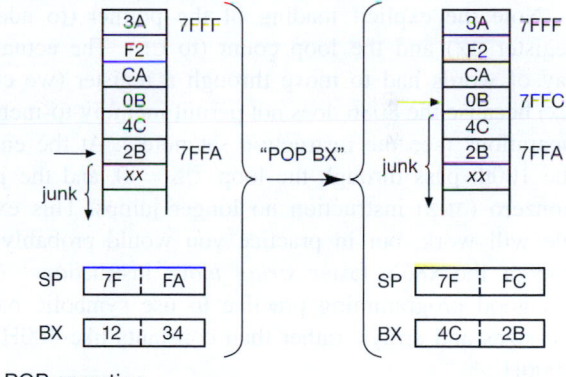

B. POP operation

Figure 14.5. Stack operation: A. effect of PUSH (illustrated with register AX); B. effect of POP (illustrated with register BX).

Figure 14.5 shows how it works. The stack lives in ordinary RAM, with the CPU's *stack pointer* (SP) keeping track of the location of the current "top" of the stack. The 8086 stack holds 16-bit words and grows *down* in memory as you push data onto it. The SP is automatically decremented by 2 before each PUSH, and incremented by 2 after each POP. Thus, in the example, the 16-bit data in register AX is copied onto the top of the stack by the

Program 14.1

```
          MOV   BX,1000H       ;put array address in BX
          MOV   CL,100         ;initialize loop counter
LOOP:     MOV   AX,[BX]        ;copy array element to AX
          MOV   [BX+400H],AX   ;then to new array
          ADD   BX,2           ;increment array pointer
          DEC   CL             ;decrement counter
          JNZ   LOOP           ;loop if count not zero
NEXT:     (next statement)     ;exit here when done
```

instruction PUSH AX; the SP is left pointing at the last byte pushed. POP reverses the process, as shown. As we will see, the stack plays a central role in subroutine calls and interrupts.

JMP causes the CPU to depart from its usual habit of executing instructions in sequential order, detouring instead to the instruction that you jump to. Conditional jumps (there are ten possibilities, indicated generically as J*cc*) test the flag register (which lives in the CPU and whose bits are set according to the result of the most recent arithmetic operation), then either jump (if the condition is true) or execute the next instruction in sequence (if the condition is not true). Program 14.1 shows an example. It copies 100 words from the array beginning at 1000 hex to a new array beginning 1kB (400H) higher.

Note the explicit loading of the pointer (to address register BX) and the loop count (to CL). The actual array of words had to move through a register (we chose AX) because the 8086 does not permit memory-to-memory operations (see the instruction set notes). At the end of the 100th pass through the loop, CL = 0, and the jump nonzero (JNZ) instruction no longer jumps. This example will work, but in practice you would probably use one of the x86's faster *string move* instructions. Also, it's good programming practice to use symbolic names for sizes and arrays, rather than constants like 400H and 1000H.

The CALL statement is a subroutine (or "procedure" or "function") call. It's like a jump, except that the return address (the address of the instruction that would have come next, except for the intervening CALL) is pushed onto the stack. At the end of the subroutine you execute a RET statement, which pops the stack so the program can find its way home (Figure 14.6). The three statements STI, CLI, and IRET have to do with interrupts, which we'll illustrate with a circuit example later in the chapter. Finally, the I/O instructions IN and OUT move a word or byte between the A register and the addressed port; more on this shortly.

14.2.3 A programming example

As the array-copy example above suggests, assembly language tends to verbosity, with a lot of little steps needed to do a basically simple thing. Here's another example. Suppose you want to increment a number, *N*, if it equals another number, *M*. This will typically be a tiny step in a larger program, and in higher level languages it will be a single instruction:

```
if (n==m) n++;      (C, C++, Java)
if n==m:            (Python)
  n+=1
IF (N.EQ.M) N=N+1   (FORTRAN)
if n=m then n:=n+1; (Pascal), etc
```

In x86 assembly language, it looks like Program 14.2. The assembler program will convert this set of mnemonics to machine language, generally translating each line of assembler *source code* to one *object code* instruction (occupying several machine-language bytes), and the resultant machine-language code will get loaded into successive locations in memory before being executed. Note that it is necessary to tell the assembler to assign some storage space for variables. This you do with the assembler *pseudo-op* "DW" (define word) (called a *pseudo*-op because it doesn't produce any executable code). Unique symbolic labels (e.g., NEXT) can be used to tag instructions; this is usually done only if there is a jump to that location (JNZ NEXT). Giving some locations understandable (to you!) names and adding comments (separated by a semicolon) make the job of programming easier; it also means that you have a chance of understanding what you've written a few weeks later.

Programming in assembly language can still be a nuisance, but it is sometimes preferable to write short routines in it, callable from a higher-level language, to code tight loops, or to handle unusual I/O. Assembly-language programs often run faster than programs compiled from a

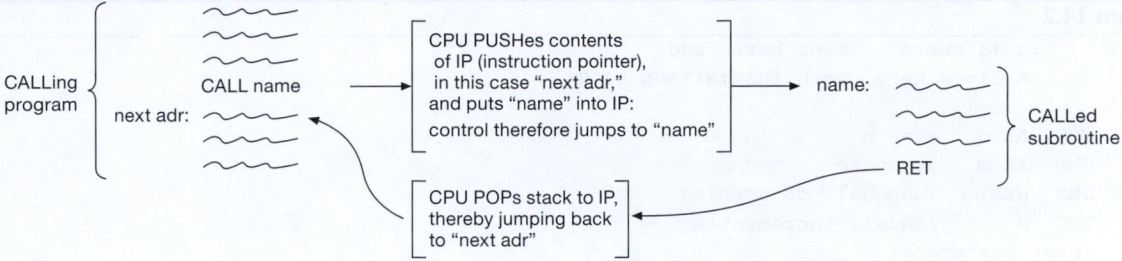

Figure 14.6. CALL operation.

higher language, so they are sometimes used where speed is crucial (e.g., the innermost loop of a long numerical calculation). To a great extent the development of the powerful C and C++ programming languages has minimized the occasions when you must use assembly code. In any case, you can't really understand computer interfacing without understanding the nature of assembly-language I/O. The correspondence between mnemonic assembly language and executable machine language is explored further in the student manual[14] in the context of microcontroller programming.

14.3 Bus signals and interfacing

A typical microcomputer data bus has 50–100 signal lines, devoted to the transfer of data, addresses, and control signals. The PC104/ISA bus is typical of a small machine, with 53 signal lines and 8 power and ground lines. Rather than throw them at you all at once, we will approach the subject by building up the bus, beginning with the signal lines necessary for the simplest kind of data interchange (programmed I/O) and adding additional signal lines as they become necessary. We will give some useful interface examples as we go along, to keep things comprehensible and interesting.

The PC104 bus signals are carried from board to board on a 64-pin (2 rows of 32 pins each) stack-through connector: it is a socket on the top side, and a plug on the bottom. Figure 14.7 shows a PC104 stacked pair, with CPU above a fast high-resolution ADC peripheral board.

14.3.1 Fundamental bus signals: data, address, strobe

To move data on a shared (party-line) bus, you have to be able to specify the data, the recipient, and the mo-

Figure 14.7. A PC104 CPU motherboard (top), with attached fast ADC (bottom). The Diamond Systems CPU includes serial and parallel I/O, USB, video, keyboard, mouse, hard-disk and floppy interfaces, and network ports; the Chase Scientific peripheral includes a pair of 14-bit 10 Msps ADCs, along with digital ports and memory. The PC104 stack-through bus can be seen along the southeast edge: 2×32 pins for the 8-bit bus, plus 2×20 pins for the 16-bit extension – voilà, 104 pins in all.

ment when data is valid. Thus a minimum bus must have DATA lines (for the data to be transferred), ADDRESS lines (to identify the I/O device or memory address), and some STROBE lines (which tell when data is being transferred). There are usually as many DATA lines as bits in the computer word, so a whole word can be transferred at once. In the 8-bit PC104,[15] however, there are only eight DATA lines (D0–D7); you can move a byte in one transfer, but to move a 16-bit word you have to do two transfers. The number of ADDRESS lines determines the number of addressable devices: if the bus is used for both I/O and memory (the usual situation) there will be 16 to 32 ADDRESS lines

---

[14] Hayes and Horowitz: *Learning the Art of Electronics – a Hands-On Course*, Cambridge University Press, 2015.

[15] The PC104/ISA specification allows both 8-bit and 16-bit buses; we will use the 8-bit version, for simplicity.

Program 14.2

```
n   DW 0        ;n (a "word") lives here, and
m   DW 0        ;m lives here, both initialized to 0


        MOV   AX,n    ;get n
        CMP   AX,m    ;compare
        JNZ   NEXT    ;unequal, do nothing
        INC   m       ;equal, increment m
NEXT:   (next statement)
          o
          o
          o
```

(corresponding to a 64 KB to 4 GB address space); a bus used for I/O only might have 8 to 16 ADDRESS bits (256 to 64K I/O devices). PC104/ISA talks to both memory and I/O on its bus, and has 20 ADDRESS lines (A0–A19), corresponding to a 1 MB address space.

Finally, data transfer itself is synchronized by pulses on additional "strobing" bus lines. There are two ways in which this can be done: by having separate READ and WRITE lines, with a pulse on one or the other synchronizing data transfer; or by having one DATA STROBE line (DS) and one READ/WRITE' line (R/W'), with a pulse on DS synchronizing data transfer in a direction specified by the level on the READ/WRITE' line. PC104/ISA uses the first scheme,[16] with (active-LOW) READ/WRITE lines called IOR', IOW', MEMR', and MEMW'; there are four because the PC distinguishes between memory and I/O addresses, with individual pairs of READ/WRITE strobes for each.

These bus signals – DATA, ADDRESS, and the four strobes – would normally be all you need to do the simplest kind of data transfers. However, on the PC104 bus you need one more, called ADDRESS ENABLE (AEN), to distinguish normal I/O transfers from what's called "direct memory access" (DMA). We'll get to DMA in §14.3.10; for now, all you need to know is that AEN is LOW for normal I/O, and HIGH for DMA. We now have 33 bus signals: D0–D7, A0–A19, IOR', IOW', MEMR', MEMW', and AEN. Let's see how they work.

14.3.2 Programmed I/O: data out

The simplest method of data exchange on a computer bus is known as "programmed I/O," meaning that data is transferred via an IN or OUT statement in the program (the direc-

[16] Because historically Intel processors used that scheme; Motorola chose R/W' and DS.

tions for IN and OUT are among the few things on which all computer manufacturers agree: IN always means *toward* the CPU, and OUT always means *from* the CPU). The whole process of data OUT (and memory write) is extremely simple and logical (Figure 14.8). The ADDRESS of the recipient and the DATA to be sent are put onto the respective bus lines by the CPU. A write strobe (IOW') is asserted (LOW) by the CPU to signal the recipient that data is good. On the PC104/ISA 8-bit bus, the address is guaranteed valid beginning ∼100 ns before IOW', and the data are guaranteed valid ∼500 ns before the end of IOW' (and for another 25 ns thereafter). To play the game, the peripheral looks at the ADDRESS and DATA lines. When it sees its own address, it latches the information on the DATA lines, using the trailing edge of the IOW' pulse as a clocking signal. That's all there is to it.

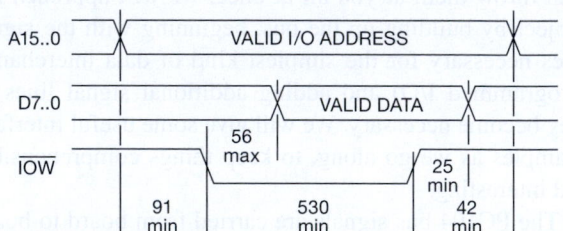

Figure 14.8. I/O WRITE cycle. Timing is in units of nanoseconds. Note that timing diagrams like these are rarely drawn to scale.

A. Example: byte-wide register

Figure 14.9 shows this simple logic: the box labeled "adr decode" produces a HIGH "adr match" output when the address lines A15..0 contain the peripheral's assigned address; this enables the gate to produce an output pulse when IOW' goes LOW (signaling a write to the peripheral), the trailing edge of which clocks the outgoing data into the byte-wide D-register. You can summarize this compactly by saying "the data is clocked into the register by the write pulse, qualified by the decoded address." Note that we've used the *trailing* edge of IOW' to clock the D-register: that is because the data is not yet valid on the leading edge, but is guaranteed valid by the trailing edge – in fact, the data has a generous setup time of ≥ 474 ns (and a hold time of ≥ 25 ns) relative to that edge. We've shown also the analogous gating for the alternative "Motorola" bus signal pair of direction and timing (R/W' and DS'), which replaces the "Intel" pair of IOR' and IOW'.

The assembly code for this interface is outrageously simple. If we want to send out the byte that's already in register AL, this is all it takes:

```
OUT 3F8h, AL ;(send to port adr = 3F8 hex)
```

The processor springs to action, with the prescribed bus hardware response of Figure 14.8: first it asserts the specific address 3F8 (hex)[17] on address lines A9..0, then it asserts IOW' and puts the byte in AL onto the data lines D7..0, and finally it disasserts IOW', then the data and address. Then it obediently fetches and executes the next instruction.

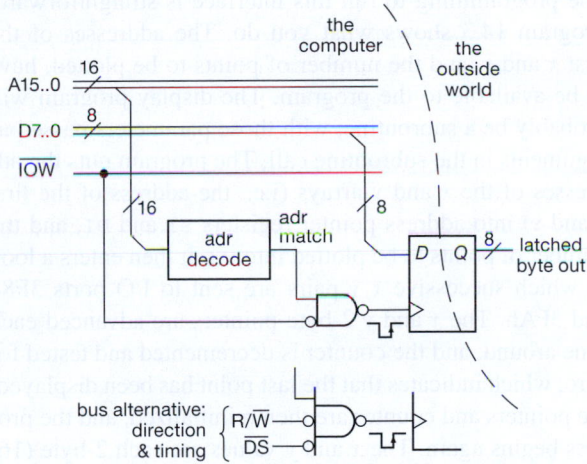

Figure 14.9. Programmed data out: a byte asserted by the CPU onto the data lines is clocked to an external device by the write pulse, if the address lines hold the peripheral's assigned address. For simplicity we've ignored the PC104's quirky AEN signal (see Figure 14.10).

B. Example: 16-bit XY vector graphic display

A more complete (and interesting) example is shown in Figure 14.10, where we've interfaced a pair of high-resolution (16-bit) D/A converters to the PC104 8-bit bus, for example to drive a high-resolution vector graphic dis-

play screen. This might be used in conjunction with a high-resolution (up to 64k×64k!) *XY* oscillographic display device that accepts analog *x* and *y* input voltages and that plots the corresponding point by unblanking the trace at each point when a *z*-input logic level is asserted.

The Analog Devices AD660 DACs have a pair of internal 8-bit registers that you load by asserting an enable (HBE' and LBE' for high and low byte) while clocking the WR' line; after that you transfer the pair to the internal 16-bit register that holds the value being converted to an output voltage (Figure 14.11). This scheme prevents false outputs: it lets you transfer the 16-bit values one byte after another, and then transfer them all, simultaneously, to the actual DAC output. It does, however, add a bit of complexity, when compared with the simple byte-wide register output port of Figure 14.9.

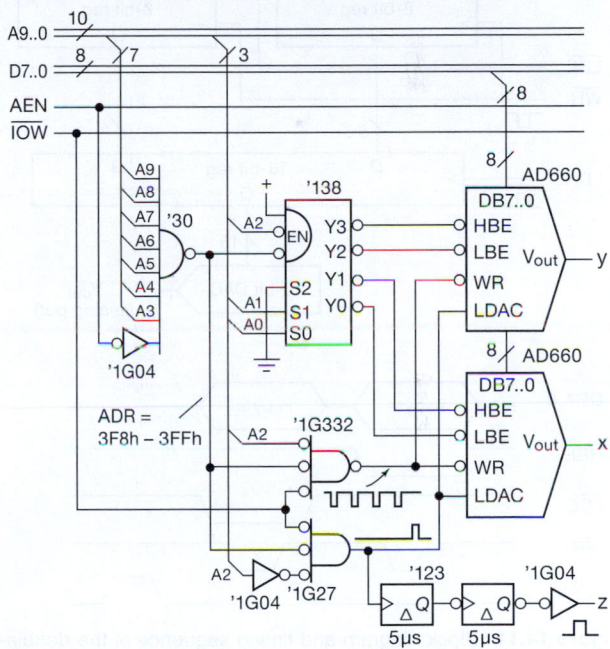

Figure 14.10. Dual-channel 16-bit DAC interface on the PC104 8-bit bus.

For this interface we use 8 successive bytes (addresses 3F8h–3FFh, or binary 11 1111 1xxx), choosing "base address" 3F8h (binary 11 1111 1000). The 8-input NAND detects this address range, qualified by the AEN bus signal asserted LOW.[18] Its output enables the '138 1-of-8 decoder (§10.3.3D), which responds to the two low-order address lines (A1 and A0) to enable the successive HBE' and LBE'

---

[17] We trust the reader will forgive our unlucky address choice: 3F8h is the x86's default serial port.

[18] AEN is asserted HIGH to signal a *DMA transfer* (§14.3.10); so for ordinary programmed I/O it must be LOW.

during byte-wide writes to addresses 3F8h, 3F9h, 3FAh, and 3FBh. After loading, these bytes are transferred to the business end of the DAC by a write to address 3FCh, which causes the lower gate to generate an LDAC (load DAC) pulse when it sees the IOW'. Note that this last "write" needs no data; the circuit ignores D7..0, and just uses the IOW', qualified by address 3FCh.[19] The LDAC pulse also generates a delayed "z-axis" unblanking pulse, providing time for the DACs and vector display screen to settle before displaying each x,y point.

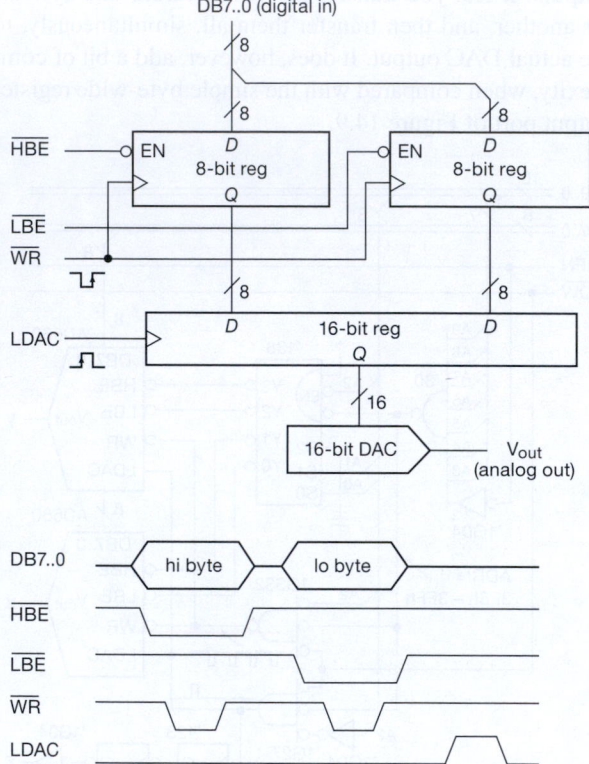

Figure 14.11. Block diagram and timing sequence of the double-buffered AD660 16-bit DAC with byte-wide parallel data input.

In practice you would probably combine all the logic, including address decoding, into a programmable logic device (PLD, Chapter 11), as in Figure 14.12; you might additionally include jumpers to set the base address.

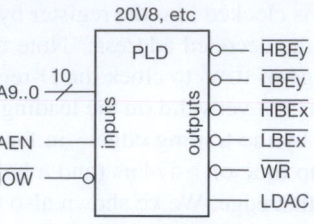

```
DACSEL = a9 & a8 & a7 & a6 & a5 & a4 & a3 & !aen;
wr_bar = DACSEL & iow_bar & !a2;
ldac = DACSEL & iow_bar & a2;
lbex_bar = DACSEL & !a2 & !a1 & !a0;
hbex_bar = DACSEL & !a2 & !a1 & a0;
lbey_bar = DACSEL & !a2 & a1 & !a0;
hbey_bar = DACSEL & !a2 & a1 & a0;
```

Figure 14.12. The logic of Figure 14.10 fits easily into a Lilliputian PLD like the 20V8. The code is written with logical assertion, assuming that active-LOW polarities are so defined in the HDL header file.

14.3.3 Programming the *XY* vector display

The programming to run this interface is straightforward. Program 14.3 shows what you do. The addresses of the first x and y, and the number of points to be plotted, have to be available to the program. The display program will probably be a subroutine, with those parameters passed as arguments in the subroutine call. The program puts the addresses of the x and y arrays (i.e., the address of the first x and y) into address pointer registers SI and DI, and the number of points to be plotted into CX. It then enters a loop in which successive x,y pairs are sent to I/O ports 3F8h and 3FAh. The x and y 2-byte pointers are advanced each time around, and the counter is decremented and tested for zero, which indicates that the last point has been displayed; the pointers and counter are then reinitialized, and the process begins again. The x and y values are each 2-byte (16-bit) integers; the code fetches them each with a single MOV (implicit in the use of the 16-bit internal AX register), and sends each out as two successive write-cycle transfers on the 8-bit bus. The x86 processors store multibyte quantities in successive memory locations in the byte order smallest-to-largest,[20] beginning always on an even-numbered byte. That is why we assigned the LBE and HBE pair addresses as shown.

A couple of important points: once started, this

---

[19] Actually, qualified by any address in the range 3FCh–3FFh, because the 3-input gate ignores A1 and A0, in what we call "lazy address decoding."

[20] Known as "little-endian," to be distinguished from the alternative scheme of "big-endian" whose byte order is largest-to-smallest in successive memory locations; see §14.8. (OK, it's a corny pun – but, hey, we didn't invent it, so don't shoot the messenger.)

program displays the x, y array forever. In real life the program would probably check the keyboard to see if the operator wants the plot terminated. Alternatively, the display could be terminated after a specified time had elapsed, or by an "interrupt," which we will discuss shortly. With this sort of *refreshed* display there usually isn't time to do much computing while displaying. A display device refreshed from its own memory takes that burden off the computer, and this is generally a better method. Nevertheless, if the objective is to make a precision plot for photographic hard copy, this program and interface would do the job.[21]

14.3.4 Programmed I/O: data in

The other direction (data *in*) of programmed I/O is similar. The interface looks at the ADDRESS lines as before. If it sees its own address (and AEN is LOW), it puts data onto the DATA lines coincident with the IOR′ pulse (Figure 14.13). Figure 14.14 shows the correspondingly simple hardware. This interface lets the processor read a byte latched in the '374 *D*-type register, which conveniently has three-state outputs.[22] Because the clock input and data inputs of the register are accessible to an external device, the register could hold just about any sort of digital information (the output of a digital instrument, A/D converter, etc.). The line labeled "ADR=200h" comes from an address decoder (e.g., a PLD, or discrete gates, that produce a HIGH when A9..0 contain binary 10 0000 0000); from now on we'll omit such minor implementation details.

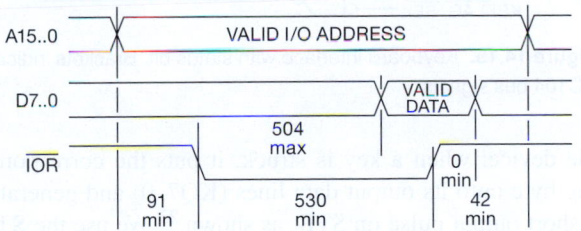

Figure 14.13. I/O READ cycle. Timing is in units of nanoseconds.

A15..0 ⟨ VALID I/O ADDRESS ⟩
D7..0 ⟨ VALID DATA ⟩
IOR‾
91 min   530 min   42 min
504 max   0 min

When an "IN AL,200H" instruction is executed, the CPU asserts 200_H (hexadecimal 200, sometimes written 0x200) on A9..0, waits a while, then asserts IOR′ for 530 ns. The CPU latches what it sees on the DATA lines (D7..0) at the trailing edge of IOR′, then disasserts A9..0. The peripheral's responsibility is to get the data onto D7..0 at least 26 ns before the end of IOR′; that's pretty relaxed timing, since it has known that data is being requested from it for at least 504 ns. With typical gate-logic propagation times of 10 ns, 500 ns looks like forever (an indication of its birthdate in 1981, an era of more relaxed timing).

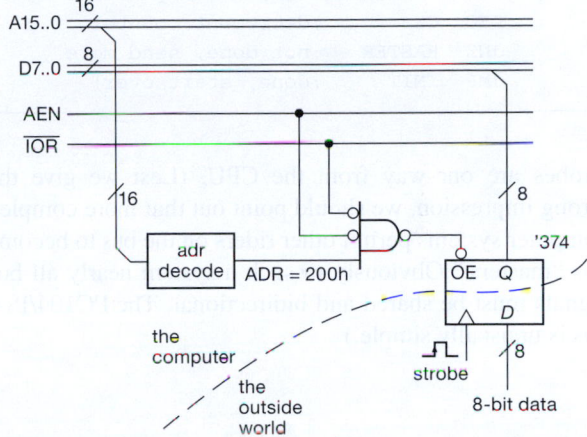

Figure 14.14. Programmed data in: an external device asserts its data on the data lines during the IOR′ pulse if the ADDRESS lines hold the peripheral's assigned address. The '374 is an octal *D*-register with three-state outputs.

A. Bus signals: bidirectional versus one-way

From the two examples we've done so far, you can see that some bus lines are *bidirectional*, for example the DATA lines: they are asserted by the CPU during write, but asserted by the peripheral during read. Both CPU and peripheral use three-state drivers for these lines. Others, like IOW′, IOR′, and the ADDRESS lines, are always driven by the CPU with standard two-state (active pullup) driver chips. It is typical of data buses to have both kinds of lines, using bidirectional lines for data that goes both ways and one-way lines for signals that are always generated by the CPU (or, more accurately, generated by the associated bus control logic). There is always a precise protocol, like our rules for asserting or reading according to IOW′, IOR′, and ADDRESS, to prevent "bus contention" on these shared lines.

Of the PC104 bus signals so far, only the DATA lines are bidirectional; the ADDRESS lines, AEN, and READ/WRITE

---

[21] In a more sophisticated implementation you might generate *software z*-axis pulses: you would assign a port address (say 3FDh) with a flip-flop to which you would write a HIGH bit followed, after a timed delay, by a LOW bit; the pulse width and delay would be created with CPU timers. Alternatively, you could decorate our circuit with a programmable hardware *z*-axis pulse generator, using a port address with a byte-wide register so that software could instruct the hardware to set the delay and width.

[22] It is for this very reason – this common use of shared (bused) data lines – that many chips include three-state outputs, controlled by an output enable (OE′) control pin.

Program 14.3

```
                ;routine to drive 16-bit xy DAC port
INIT:   MOV   SI,xpoint    ;initialize x pointer
        MOV   DI,ypoint    ;initialize y pointer
        MOV   CX,npoint    ;initialize counter

RASTER: MOV   AX,[SI]      ;get x word (2 bytes)
        OUT   3F8H,AX      ;send out (2 byte xfers)
        ADD   SI,2         ;advance x word pointer
        MOV   AX,[DI]      ;get y word (2 bytes)
        OUT   3FAH,AX      ;send out (2 byte xfers)
        ADD   DI,2         ;advance y word pointer
        OUT   3FCH,AL      ;load x and y to DAC
        DEC   CX           ;decrement counter
        JNZ   RASTER       ;not done, send more
        JMP   INIT         ;done, start over}
```

strobes are one-way from the CPU. (Lest we give the wrong impression, we should point out that more complex computer systems permit other riders on the bus to become bus "masters." Obviously in such a system nearly all bus signals must be shared and bidirectional. The PC104/ISA bus is unusually simple.)

14.3.5 Programmed I/O: status registers

In our last example, the computer can read a byte from the interface any time it wants to. That's nice, but how does it know when there's something worth reading? In some situations you may want the computer to read data at equally spaced intervals, as determined by its "real-time clock." Perhaps the computer instructs an ADC to begin conversions at regular intervals (via an OUT command), then reads the result a few microseconds later (via an IN command). That might suffice in a data-logging application. However, it is often the case that the external device has a mind of its own, and it would be nice if it could communicate what's happening to the computer without having to wait around.

A classic example is an alphanumeric input keyboard. You don't want characters to get lost; the computer has to get every character, and without much delay. With a fast device like a disk or high-speed serial interface the situation is even more serious; data must be moved at rates up to many megabytes per second without delay. There are actually three ways to handle this general problem: status registers, interrupts, and direct memory access. Let's begin with the simplest method – status registers – illustrated by the keyboard interface in Figure 14.15.

In this example, the raw "ASCII keyboard" is a sim-

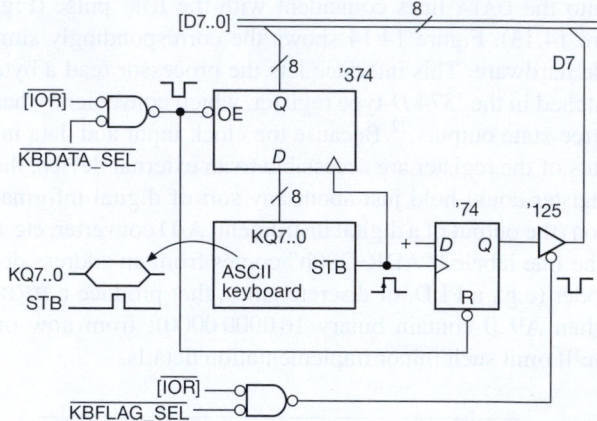

Figure 14.15. Keyboard interface with status bit. Brackets indicate PC104 bus signals.

ple device: when a key is struck, it puts the corresponding byte onto its output data lines (KQ7..0) and generates a short output pulse on STB, as shown.[23] We use the STB (for "strobe") to clock that byte-wide character code into an octal D-type register. We rig up the standard programmed data-in circuit, as shown, using the three-state outputs of the octal D-register to drive the DATA lines directly. The input labeled KBDATA_SEL′ comes from the usual address decoding circuit of the sort used in the previous examples, and it goes LOW when the particular address chosen for

---

[23] This keyboard lacks the intelligence of contemporary units, which generally include an onboard processor to convert the key codes to a bit-serial format, typically delivered via the USB hardware interface. Its very lack of intelligence, however, permits us to display our own.

this interface appears on the ADDRESS lines of the bus (in combination with AEN disasserted LOW).

What's new in this example is the flip-flop, which gets set when a character is struck, and cleared when a character is read by the computer. It's a 1-bit *status* register: HIGH if there's a new character available, LOW otherwise. The computer can query the status bit by doing a data IN from the other address of this device, decoded (with gates, decoders, or whatever) as KBFLAG_SEL'. You need only one bit to convey the status information, so the interface drives only the most-significant bit (MSB) on the bus (D7), in this case with a '125 three-state buffer. (*Never* drive a bidirectional line with an active-pullup (two-state) output! *Never!*) The line coming into the side of the buffer symbol is the three-state output enable, asserted when LOW, as indicated by the negation bubble.

This circuit could be implemented with standard logic, as indicated. Alternatively, it (and the address decoding logic) would fit easily in a small PLD (for example an XC2C32, XC9536, ATF2500, Mach4032, or even a lowly 22V10).

A. Program example: keyboard terminal

The computer now has a way to find out when new data is ready. Program 14.4 shows how. This is a routine to get characters from the keyboard, whose data port address is KBDATA (it's good programming style to define the actual numerical port addresses – which correspond to what the hardware decodes as KBDATA_SEL, etc. – in some statements near the beginning of the program, as shown); each character is "echoed" on the computer's display device (port address = OUTBYTE). When it has gotten a whole line, it transfers control to a line-handling routine, which might do just about anything, based on what the line says. When it's ready for another line, it types an asterisk "prompt." This sort of protocol should make sense to you if you've had some experience with computers.

The program begins by initializing the character buffer pointer (BP), by moving the *address* of the buffer that we just allocated to the address register BP. Note we can't just say "MOV BP,charbuf" because that would load the *contents* of charbuf, not its address; in x86 assembly language you use the word "offset" in front of a memory label to signify its address. The program then reads the keyboard status bit via an IN instruction, ANDs it with 80h to keep only the status bit (this is called "masking"), and tests for zero. Zero means the bit isn't set, so the program loops. When a nonzero status bit is detected, it reads the keyboard data port (which clears the status flag flip-flop), stores it consecutively in the line buffer, increments the

pointer (BP), and calls the routine that echoes the character to the screen. Finally, it checks to see if the line was terminated by a carriage return (CR): if it wasn't, it goes back and loops on the keyboard status flag again; if it was a CR, it transfers control to the line handler, after which it prints an asterisk and begins the entire process anew.

A subroutine has been used to display a character, since even that simple operation requires some flag checking and masking. The routine first saves the byte into AH, then reads and masks the screen's busy flag. A nonzero result means the screen is busy, so it keeps checking; otherwise it restores the character to AL, sends it to the screen's data port, and returns.

Some notes on the program. (a) We could have omitted the keyboard flag-masking step, because the MSB (where we put the flag bit in our hardware) is the sign bit; thus we could have used the instruction JNS KFCHK. However, this trick works only for testing the MSB and thus is somewhat specialized. (b) In keeping with good programming practice, the carriage return symbol (0Dh) and asterisk probably should be defined constants, similar to KBMASK. (c) The line handler probably should be a subroutine, also. (d) Characters will be lost if the line handler takes too long; this leads us to the more elegant approach of *interrupts*, which we'll take up shortly. (e) Keyboard and terminal handlers are used so often that microprocessor operating systems provide built-in handlers, accessed through "software interrupts" (we'll see them later); thus our program may not even be needed.

B. Status bits generalized

This keyboard example illustrates status bit protocol; but it's so simple that you may come away with the wrong idea. In an actual peripheral interface of some complexity, there will usually be several flags to signal various conditions. For example, in an Ethernet interface you will normally have individual status bits indicating a successful packet transmission, or (if you're not so lucky) any of various mishaps that may have occurred. As a specific example, Microchip's ENC28J60 Ethernet Controller IC has a 56-bit transmit status register; bit 20 indicates a "Transmit CRC Error," described as "The attached CRC in the packet did not match the internally generated CRC."[24]

For peripherals of modest complexity, the usual procedure is to put all the status bits into one byte or word, so that a data IN command from the status register gets all bits at once. Typically you would have a bit indicating any of a set

---

[24] These things can get humorously arcane: bit 27 ("Transmit Excessive Defer") signifies "Packet was deferred in excess of 24,287 bit times."

Program 14.4

```
                ;keyboard handler -- uses flags
KBDATA   equ ***H        ;put kbd data port adr here
KBFLAG   equ ***H        ;a different port for kbd flag
KBMASK   equ 80H         ;kbd flag mask
OUTBYTE  equ ***H        ;put disp port adr here
OUTFLAG  equ ***H        ;another for disp port flag
OUTMASK  equ ***H        ;disp port busy mask

charbuf DB 100 dup(0)    ;allocates buffer of 100 bytes

INIT:    MOV  BP,offset charbuf    ;init char buf pntr
KFCHK:   IN   AL,KBFLAG   ;read kbd flag
         AND  AL,KBMASK   ;mask unused bits
         JZ   KFCHK       ;flag not set -- no new data
         IN   AL,KBDATA   ;flag set -- get new kbd byte
         MOV  [BP],AL     ;store it in line buffer
         INC  BP          ;and advance pointer
         CALL TYPEIT      ;echo last char to display
         CMP  AL,0DH      ;was it carriage return (0Dh)?
         JNZ  KFCHK       ;if not, get next char
LINE:    o               ;if so, do something with line
         o               ;keep at it
         o               ;don't quit now
         o               ;done at last!
         MOV  AL,'*'
         CALL TYPEIT      ;type a "prompt" -- asterisk
         JMP  INIT        ;get another line

                         ;routine to type character
                         ;types and preserves AL
TYPEIT:  MOV  AH,AL       ;save the char in AH
PCHK:    IN   AL,OUTFLAG  ;check printer busy?
         AND  AL,OUTMASK  ;printer flag mask
         JNZ  PCHK        ;if busy check again
         MOV  AL,AH       ;restore char to AL
         OUT  OUTBYTE,AL  ;type it
         RET              ;and return
```

of error conditions as the MSB of the status word, so a simple check of sign tells if there are *any* errors; if there are, you test specific bits of the word (by ANDing with masks) to find out what's wrong. Furthermore, in a complex interface you probably wouldn't have the status bits reset "automatically," as we did with our single bit; instead, a data OUT statement might be used, each bit of which clears a specific flag.

Exercise 14.1. With our keyboard interface there is no way for the computer to know if it missed a character. Modify the circuit so there are two status bits: CHAR_READY (that's what we have already) and LOST_DATA. The LOST_DATA flag should be read-

able as D6 on the same status port as CHAR_READY; it is 1 if a key was struck before the previous character was fetched by the computer, 0 otherwise.

Exercise 14.2. Add a section of code to Program 14.4 to check for lost data. It should call a subroutine called LOST if it detects lost data; otherwise the program should work as before.

14.3.6 Programmed I/O: command registers

To summarize, a status bit (or a collection of bits: a status register) reports a condition to the CPU (when asked). Going in the opposite direction, the CPU can send out a bit (or collection of bits) to a peripheral, to tell it to do something. This is called a *command* bit (or command register). A simple example might be a bit that tells an *xy* positioning stage to move to the coordinates that have been deposited in a pair of data registers of the peripheral by a pair of previous programmed OUT statements. Or, taking the example of our Ethernet controller, the CPU deposits to a pair of registers in the IC the beginning and ending addresses of the data packet to be transmitted; then it sets a particular command bit (bit 3, to be exact) in the chip's "ECON1" register, which opens the floodgates by commanding the chip to send the packet out on the Ethernet port. The chip does as it has been told (in this case using DMA for highest speed; see §14.3.10), and reports back to the CPU (when asked; or more aggressively by using an *interrupt*; see next subsection) via the status registers.

14.3.7 Interrupts

The use of status flags just illustrated is one of three ways for a peripheral device to "tell" the computer when some action needs to be taken. Although it will suffice in many simple situations, it has the serious drawback that the peripheral cannot announce that some action needs to be taken – it has to wait to be asked by the CPU, via a data IN command from its status register. Devices that need quick action (such as disks or latency-sensitive real-time I/O) would have to have their status flags queried often, and with several such devices in a computer system the CPU would soon find itself spending most of its time checking status flags, as in the last example.

Furthermore, even with continual status flag checking you can still get in trouble: in the last example, for instance, the CPU will have no trouble keeping up with someone typing at the keyboard when it is in the main (flag-checking) loop. But what if it spends 1/10 second in the line-handling portion? Or what if the interfaced device is a slow one, making the program wait for its busy flag to clear?

What is needed is a mechanism for a peripheral to *interrupt* the normal action of the CPU when something needs to be done. The CPU can then check the status register to find out what the trouble is, take care of what needs to be done, and go back to its normal business.

To add interrupt capability to a computer, it is necessary to add a few new bus signals: at least one shared line for peripherals to signal an interrupt, and (usually) a pair of lines by which the CPU can determine who interrupted. As luck would have it, the PC104/ISA is not a very instructive example, because it does not implement a full interrupt capability. What it lacks in power, though, it more than makes up for in simplicity; implementing hardware interrupts in a PC104 peripheral interface is like falling off a log.

Here's how it works: the PC104 bus has a set of six *interrupt request* lines, called IRQ3–IRQ7 and IRQ9. They are active-HIGH inputs to the CPU's support circuitry. To make an interrupt, you simply bring one of the lines HIGH. If interrupts are enabled in general (along with the particular IRQ you assert), the CPU will break off after its next instruction, then (after saving its flags and current location onto the stack) jump to an "interrupt-handler" program somewhere in memory. You write the handler to do what you want (e.g., get keyboard data), and you can put the handler anywhere you wish, because the CPU figures out where to jump by looking for the handler's 4-byte address in a special ("vector") location in low memory. That location depends on which IRQ you've asserted; for the x86 it is given in hex by $20+4n$, where n is the interrupt level. For example, the CPU would respond to an interrupt on IRQ2 by jumping to the (4-byte) address stored in locations 28h through 2Bh (it's just like indirect addressing, except that the address is found in memory rather than in a register); of course, you would have cleverly arranged for the starting address of your handler to be there. At the end of your handler you execute an IRET instruction, which causes the CPU to restore the preexisting flag register and jump back to wherever it was when the interrupt happened.

A. Example: keyboard interface with interrupts

Let's illustrate by adding interrupts (Figure 14.16) to our keyboard interface circuit of Figure 14.15. We've kept the flag bit ("character ready") and programmed I/O circuitry essentially as before, except that we've ORed the flag clear with a new bus line, RESET, a bus signal that is momentarily asserted HIGH when the computer is turned on. This signal is generally used to force your flip-flops and other sequential logic into a known state at power-up. Obviously it should reset a flag that indicates a valid byte is ready to be claimed (and that, in our new interface, will further cause an interrupt).

The new interrupt circuitry consists of a three-state driver to assert IRQ3 when a character is ready. That's all the new hardware you need. Although not strictly necessary, we've added the capability to disable the interrupt driver buffer by sending a data byte with D0 LOW to the KBFLAG port address. This would be used if you wanted to plug in another peripheral with interrupts at the same IRQ level, allowing only one peripheral to use its interrupts at any given time (later we'll have further explanation on this awkward point).

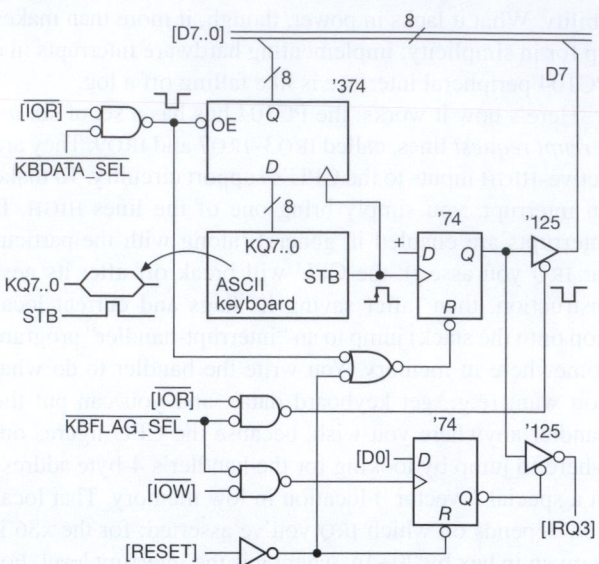

Figure 14.16. Keyboard interface with status bit and interrupt.

14.3.8 Interrupt handling

The PC104 bus signals, derived from the original IBM PC, make interrupt handling particularly easy, though limited in flexibility compared with the general (and more sophisticated) method described below (§14.3.9). The bus signals require, in addition to the CPU itself, some interrupt control circuitry on the motherboard.[25] This "support-chip" circuitry does most of the hard work, which consists of prioritizing, masking, and asserting vectors (we'll describe these after finishing the example). The CPU, for its part, recognizes the interrupt and responds by saving the instruction pointer and flag register, disabling further interrupts, then making a jump via the corresponding address stored in the low-memory vector area. Your handler program does the rest, namely: (a) saves (pushes) any registers you'll be using (remember that the interrupted program can't prepare for the interrupt, since it can happen anywhere in the running program – it's a bolt out of the blue); (b) figures out what needs to be done, by reading status registers if necessary; (c) does it; (d) restores the saved registers from the stack; (e) tells the interrupt control circuitry that you're done (by sending an "end-of-interrupt" byte 20h to its register at I/O address 20h); and, finally, (f) executes a return from interrupt instruction IRET; this causes the CPU to restore the old flag register that it saved on the stack and to jump (via the old instruction pointer it saved on the stack)

back to the program that was interrupted. Somewhere in the program, you must have (g) loaded the handler's address into the vector location corresponding to the IRQ level used by the hardware and told the interrupt control hardware to enable interrupts at that level.

Program 14.5 shows the code for the keyboard with interrupt. Here's the overall scheme: the main program sets things up, then loops on a flag (in memory, not hardware) that the interrupt handler sets when it recognizes a carriage return; when the main program sees that flag set, it goes off and does something with the line, then returns to the flag-checking loop. The handler, entered at each interrupt, puts a character into the line buffer, sets the flag if it was a carriage return, then returns.

Let's look at the program in some detail. After defining port addresses and the all-critical vector location for IRQ3, it allocates 100 bytes (initially filled with zeros) for the character buffer. The actual program execution begins by putting the buffer address in address register SI, zero in the end-of-line flag, and the address of the handler (which begins with KBINT) in location 2Ch. To enable level-3 interrupts in the interrupt controller, we clear bit 2 of its existing mask (IN, AND, OUT); then we enable CPU interrupts and send a 1 to KBFLAG, which enables the three-state driver. Now we're running. The program then loops, with interrupts secretly happening right under the main program's nose, until it mysteriously finds the "buflg" set. It resets the pointer and flag immediately (in case another interrupt occurs soon), then gobbles up the line. It would be well advised either to move quickly or to copy the line to another buffer, since another interrupt (with a new byte to go in the buffer) could come along in a few milliseconds; in that time you can execute a few hundred thousand instructions, however, more than enough to copy the line.

The interrupt handler is a separate little piece of code, with no entry from the main program. It gets entered upon a level-3 interrupt, via its address that we initially loaded into 2Ch. It knows exactly what it has to do, and it does it without complaining: it saves AX (because it plans to clobber it), reads the character from the keyboard data port, puts it in the buffer, increments the pointer, echoes the character to the screen, sets the flag if it was a carriage return, sends end-of-interrupt to the interrupt controller, restores AX, and returns.

If you look back at our list of handler tasks above, you'll see that we omitted just one step, namely, reading status flags to figure out which of several actions needed to be done. That's unnecessary here, though, because there's only one reason to interrupt, namely that a new keyboard character needs to be read. (The programmer obviously has

---

[25] Which was done by an 8259 interrupt-controller IC on the original PC motherboard.

Program 14.5

```
            ;keyboard handler -- uses interrupts
KBVECT equ word pntr 002CH    ;INT3 vector
KBDATA equ ***H            ;put kbd data port adr here
KBFLAG equ ***H            ;put kbd flag port adr here
buflg   DB   0             ;allocates "end-of-line" flag
bufpos  DW   0             ;allocates buffer pointer
charbuf DB 100 dup(0) ;allocates 100-byte char buf

        CLI ;disable interrupts
        MOV   bufpos,offset charbuf    ;initialize buf pntr
        MOV   buflg,0    ;clear end-of-line flag
        MOV   KBVECT,offset KBINT ;hndlr adr->vec area
        STI             ;enable interrupts
        MOV   AL,1
        OUT   KBFLAG,AL ;enable hardware 3-state drvr
PROMPT: MOV   AL,'*'
        CALL  TYPE        ;type prompt "*"
        IN    AL,21H       ;existing int ctrl mask
        AND   AL,0F7H      ;clear bit 3 to enable INT3
        OUT   21H,AL       ;and send back to intctrl OCW1
LNCHK:  MOV   AL,buflg
        OR    AL,AL        ;needed to set zero flag
        JZ    LNCHK        ;loop til end-of-line flag set
LINE:   MOV   bufpos,offset charbuf    ;reset pointer
        MOV   buflg,0      ;clear line flag
        o                  ;do something with line
        o
        o
        JMP   PROMPT       ;and wait for another line

; *** this ends the main program. ***
; *** the code below is completely independent ***
            ;keyboard interrupt handler
            ;an INT3 lands you here, via vect we loaded
KBINT:  STI               ;enable interrupts
        PUSH AX           ;save AX register, used here
        PUSH SI           ;save SI, possible other users
        MOV  SI,bufpos    ;and copy buffer pointer there
        IN   AL,KBDATA    ;get data byte from keyboard
        MOV  [SI],AL      ;put it in line buffer
        INC  SI           ;and advance pointer
        MOV  bufpos,SI    ;and copy to bufpos
        CALL TYPE         ;echo to screen
        CMP  AL,0DH       ;check for carriage return
        JNZ  HOME         ;not a CR -- return
        MOV  buflg,0FFH   ;CR -- set end-of-line flag
        IN   AL,21H       ;existing int ctrl mask
        OR   AL,08H       ;set bit 3 to disable INT3
        OUT  21H,AL       ;and send back to intctrl OCW1
HOME:   MOV  AL,20H
        OUT  20H,AL       ;end-of-int signal to int ctlr
        POP  SI           ;restore SI
        POP  AX           ;restore old AX
        IRET              ;and return
```

to understand under what conditions the hardware makes an interrupt, and what is required to service the interrupt.)

A few notes on this program: first, even though we're using interrupts, the program seems as dumb as before – it loops continually on the end-of-line flag. However, it could be doing other things, if there were things to do. In fact, it does just that beginning at statement LINE, where it processes the finished line; during that time, interrupts make sure that new characters are put into the buffer, whereas they would have been lost in our previous example without interrupts.

This brings up a second point, namely, even with interrupts we're still in trouble if the program is doing things with the previous line when the next line has been completely entered. Of course, *on the average* the program simply has to keep up with keyboard entry; but you could have a situation in which the line user occasionally spends a lot of time, and you need to buffer more than one line temporarily. One solution to this is to make a copy to a second buffer or to alternate between two buffers. An elegant alternative is to organize input as a queue, implemented as a "ring buffer" (or "circular buffer"), in which a pair of pointers keeps track of where the next input character goes, and from where the next character is removed. The interrupt handler advances the input pointer, and the line user advances the output pointer. Such a ring buffer might typically be 256 bytes long, permitting the line handler to get behind by a few lines.

A third point concerns the interrupt handler itself. It's usually best to keep it short and simple, perhaps setting flags to signal the need for complicated operations in the main program. If the handler does become long-winded, you risk losing data from other interrupting devices because interrupts are disabled (at that level and below) when the CPU jumps into the handler. The solution in this case is to re-enable interrupts *within* your handler with an STI instruction, after doing the critical things that have to be done first. Then, if an interrupt occurs, your interrupt handler will itself be interrupted. Because flags and return addresses are stored on the stack, the program will find its way back, first to your handler, finally to the main program.

14.3.9 Interrupts in general

Our keyboard example illustrates the essence of interrupts – a spontaneous hardware request for attention by a peripheral, producing a program jump to a dedicated handler routine (usually resulting in some programmed I/O), followed by a return to the code that was interrupted. Other examples of interrupting devices are real-time clocks, in which a

periodic interrupt (often 10 per second, but 18.2 per second in the original PC from which the PC104 bus derived) signals a timekeeping routine to advance the current time; another example is a printer port, which interrupts each time it is ready for a new character. By using interrupts, these peripherals let the computer interleave other tasks simultaneously; that's why you can be typing while your PC is printing a file (and, of course, keeping proper time throughout).

The PC104/ISA bus does not, however, illustrate the full generality of interrupts. As we saw, it has a set of six IRQ lines on the 8-bit bus (there are five more on the 16-bit extension), each one of which can be used only for a single interrupting device. The IRQ lines are numbered according to priority; in the event of multiple interrupts, the lowest-numbered interrupt is serviced first. And several of the IRQ lines are preassigned to essential peripherals, leaving precious few available. Furthermore, the interrupt is *edge-triggered*, which frustrates any reasonable possibility of using wired-OR to combine several peripherals on a single IRQ line. Evidently the original IBM PC designers, not anticipating the need for shared interrupts, designed a far-from-optimal interrupt scheme.

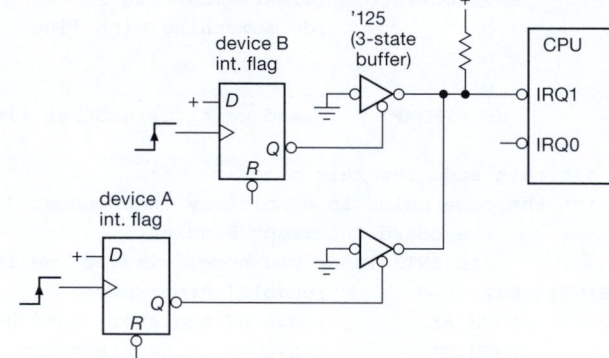

Figure 14.17. Sharing a level-sensitive interrupt line via wired-OR.

A. Shared interrupt lines

The usual interrupt protocol, as implemented on many microcomputers, circumvents these limitations. Look at Figure 14.17. There are several (prioritized) IRQ-type lines; these are active-LOW inputs to the CPU (or its immediate support circuitry). To request an interrupt, you pull one of the IRQ' lines LOW, using an open-collector (or three-state) gate, as shown (note the trick for using a three-state gate to mimic an open-collector gate). The IRQ' lines are shared, with a single resistive pullup, so you can put as many devices on each IRQ' line as you want; in our example, two ports share IRQ1. You would generally connect a latency-sensitive (impatient) device to a higher-priority IRQ′ line.

Because the IRQ′ lines are shared, there could always be another device interrupting on the same line at the same time. The CPU needs to know who interrupted so it can jump to the appropriate handler. There is a simple way, and a complicated way, to do this. The simple way is called *autovectored polling* and is used nearly universally (though not on the PC104 bus). Here's how it works.

Autovectored polling

Some circuitry on the CPU board instructs the microprocessor that it is to use autovectoring, which works just like the PC104 – each level of interrupt forces a jump through a corresponding vectoring location in low memory. You put the addresses of the handlers in those locations, just as in our previous example.

Once in the handler, you know which level of interrupt you're servicing; you just don't know which particular device caused the interrupt. To find out, you simply check the status registers of each of the devices connected to that level of interrupt (a device never requests an interrupt without also indicating its need by setting one or more readable status bits). If a bit is set indicating that something needs to be done, you do it, including whatever it takes to cause the device to disassert its IRQ′: some devices (like our keyboard) clear their interrupt when read, whereas others may need a particular byte sent to some I/O port address.

If the device you serviced was the only one interrupting at that level, that IRQ′ will now be HIGH on returning to the interrupted program, and execution will continue. However, if there had been a second interrupting device at the same level, that IRQ′ line would still be held LOW (by the wired-OR action of the shared IRQ′ line) on return from the service routine, so the CPU would immediately autovector back to the same handler. This time the polling would find the other interrupting device, do its thing, and return. Note that the order in which you poll status registers effectively sets up a "software priority," in addition to the hardware priority of the multiple IRQ′ levels.

Interrupt acknowledgment

We shouldn't leave the subject of interrupts without mentioning a more sophisticated procedure for identifying who interrupted – *interrupt acknowledgment*. In this method, the CPU doesn't need to poll the status registers of possible interrupters because the interrupting device *tells* the CPU its name, when asked. The interrupter does this by putting an "interrupt vector" (usually a unique 8-bit quantity) onto the DATA lines in response to an "interrupt-acknowledge"

signal that the CPU generates during the interrupt processing.

Nearly every microprocessor generates the needed signals. The sequence of events goes like this.

(a) The CPU notices a pending interrupt.
(b) The CPU finishes the current instruction, then asserts (i) bus signals that announce an interrupt, (ii) the interrupt level being serviced (on the low-order ADDRESS lines), and (iii) READ-like strobes that invite the interrupting device to identify itself.
(c) The interrupting device responds to this bus activity by asserting its identity (interrupt vector) onto the DATA lines.
(d) The CPU reads the vector and jumps into the corresponding unique handler for the interrupting device.
(e) The handler software, as in our last example, reads flags, gets and sends data, etc., as needed; among its other duties, it must make sure the interrupting device disasserts its interrupt.
(f) Finally, the interrupt handler software returns control to the program that was interrupted.

Sharp-eyed readers may have noticed a flaw in the procedure just outlined. In particular, there has to be a protocol to ensure that only one device asserts its vector, since there may be several simultaneous interrupting devices at the same IRQ level. The usual way to handle this is to have a bus signal (call it INTP, "interrupt priority") that is unusual in not being shared by devices on the bus, but rather is passed along *through* each device's interface circuit, beginning as a HIGH level at the device closest to the CPU and threading along through each interface. That's called a "daisy chain" in the colorful language of electronics. The rule for INTP hardware logic is as follows: if you have not requested an interrupt at the level being acknowledged, pass INTP through to the next device unchanged; if you *have* interrupted at that level, hold your INTP output LOW. Now the rule for asserting your vector goes like this: put your vector number onto the data bus when requested by the CPU only if (a) you have an interrupt pending at the level being acknowledged, and (b) your input INTP is HIGH. This guarantees that only one device asserts its vector; it also establishes a "serial-priority" chain within each IRQ level, with devices electrically closest to the CPU getting serviced first.

There is a nice alternative to the serial daisy-chain method of interrupt acknowledgment: instead of threading a line through each possible interrupter, you bring individual lines back from each one to a priority encoder (§10.3.3E), which in turn acknowledges the interrupt by asserting the identity of the highest-priority interrupting device. This scheme avoids the nuisance of daisy-chain jumpers.

In most microcomputer systems it isn't worth implementing the full-blown interrupt acknowledgment just described. After all, with 8-level autovectoring you can handle up to eight interrupting devices without polling, and several times that number with polling. Only in large computer systems, in which you demand fast response with dozens of interrupting devices present, might you succumb to the complexity of the interrupt-acknowledgment protocol. However, it is important to realize that even simple computers may be using vectored interrupt acknowledgment *internally*. For example, the simple 6-level autovector interrupt scheme of the PC104 is actually generated by an interrupt controller support chip that lives close to the CPU and generates the proper interrupt acknowledgment sequence just described (see below).

B. Interrupt masks

We put a flip-flop in our simple keyboard example so that its interrupts could be disabled, even though the interrupt controller chip itself lets you turn off ("mask") each level of interrupt individually. We did that so that some other device could then use IRQ3. For a bus with shared (*level*-sensitive) IRQ' lines, it is especially important to make each interrupt source maskable, using an I/O output port bit. For example, a printer port normally interrupts each time its output buffer is empty ("give me more data"); when you've finished printing, though, you don't care. The obvious solution is to turn off printer interrupts. Since there might be other devices hooked to the same interrupt level, you must not mask that whole level; instead, you just send a bit to the printer port to disable its interrupts.

C. How the PC104 ISA interrupts got the way they are

The ISA bus was created by the IBM PC's designers, and adopted without change (but with a different connector) by the PC104 consortium. The 8086/8 microprocessor used in the original IBM PC actually implemented the full vectored interrupt-acknowledgment protocol. To keep things simple, however, the PC designers used an 8259 interrupt controller IC on the motherboard. The way it was used in the PC, it had a set of IRQ inputs from the I/O bus card slots (that's where you make your interrupt requests), and it connected to the microprocessor's data bus and signal lines. When it received a request on an IRQ line from a peripheral, it figured out priority and went through the whole business of asserting the corresponding vector onto the data

bus. It had a mask register (accessible as I/O port 21h) so that you could disable any specified group of interrupts.

The 8259 let you select (through software) either *level*- or *edge*-triggered interrupts on its IRQ input lines, according to a byte sent to a control register (I/O port 20h). Unfortunately, the PC designers decided to use edge triggering, probably because that makes it a little easier to implement interrupts (for example, you can just connect the real-time clock's square-wave output directly to IRQ0). If they had selected level-sensitive interrupts instead, you would be able to hang multiple interrupting devices on each IRQ' line, with software polling as described above.

There is a partial solution to this problem. As long as there is an IRQ line available, you *can* combine several interrupting devices on a single PC board, with logic to generate edge-triggered interrupts on that single IRQ line. But, since the interrupting devices have to know about each other, you can't use this scheme for independent plug-in peripherals. Furthermore, you still use up an IRQ line per card, so in a complicated system there will not be enough to go around.

D. Software interrupts

The Intel x86-series of CPUs have an instruction ("INT n," where n is 0–255) that allows you to produce the same kind of vectored jump as an actual hardware interrupt. In fact, among its 256 possible jump vectors are duplicates of the eight levels of IRQ-requested hardware interrupts (INT 8 through INT 15, to be exact). Thus, you can make a "software interrupt" from a program statement.

Don't confuse this with the externally triggered hardware interrupts we've been talking about. Software interrupts turn out to be a handy way of implementing vectored jumps from user code into system software. But they are not real interrupts, in the sense of a hardware call for attention from an external autonomous device. On the contrary, you build these into your software, you know when they are coming (that's why you can pass arguments through registers), and they are merely the response (albeit identical with what follows a true interrupt) of the CPU to its own code. You might think of software interrupts as a clever way to extend the instruction set.

14.3.10 Direct memory access

There are situations in which data must be moved very rapidly to or from a device. The classic examples are fast mass-storage devices like magnetic and optical disk drives, and network connections. Interrupt-initiated programmed processing of each data transfer in these examples would be both awkward and, worse, too slow. For example, data comes from a typical magnetic disk drive at sustained rates up to 500MB per second. With the bookkeeping involved in handling an interrupt, the situation would be hopeless, even if the disk were the only interrupting device in the system. Devices like disks and tapes (not to mention real-time signals and data) can't stop in midstream, so a method must be provided for reliably fast response and high overall byte transfer rates. Even with peripherals with low average data transfer rates, there are sometimes requirements for short *latency*, the time from initial request to actual movement of data.

The solution to these problems is direct memory access (DMA), a method for direct communication from peripheral to memory. In some microcomputers the communication is actually handled by the CPU hardware, but that doesn't really matter. The important point is that no programming is involved in the actual transfer of data; bytes are moved between memory and peripheral via the bus, without program intervention. The only effect on the executing program is some slowing down of execution time, because DMA activity "steals" bus cycles that would otherwise be used to access memory for program execution. DMA usually involves more hardware complexity in the interface itself, and is not generally used unless necessary. However, it is useful to know what can be done, so we describe briefly what you need to make a DMA interface. As with interrupts, the PC104/ISA bus uses a streamlined DMA protocol, with a "DMA controller" chip on the motherboard doing the hard work for you, thus making a DMA interface relatively straightforward. We first explain the more usual "bus mastership" method of DMA, then the PC104's simplified DMA protocol.

A. A typical DMA protocol

In DMA transfers, the peripheral requests access to the bus via special "bus request" lines (prioritized like the IRQ lines) that are part of the bus. The CPU gives permission and releases control of address, data, and strobing lines. The peripheral then asserts memory addresses onto the bus and either sends or receives data, one byte at a time, according to the strobes it asserts; in other words, it takes over the bus (it becomes "bus master") and acts like a CPU, directing data transfers. The DMA bus master is responsible for generating addresses (usually a block of successive addresses, generated with a binary counter) and keeping track of the number of bytes moved. The usual way to do this is to have a byte counter and an address counter in the interface. These are initially loaded from the CPU, via programmed I/O, to set up the DMA transfer desired. On command from the CPU (via a command bit, written with programmed I/O), the interface makes its DMA request and begins to move its data. It may release the bus between each byte (allowing the CPU to sneak a few instructions in), or it may take the more antisocial approach of keeping the bus for a block of transfers. When all transfers are complete, it releases the bus for the last time and notifies the program that it is finished by setting a status bit and requesting an interrupt, whereupon the CPU can decide what to do next.

Getting data or programs from disk storage is a common example of DMA transfer: the executing program asks for some "file" by name; the operating system software translates this into a set of programmed data OUT commands to the disk interface's control (or "command") register, byte count register, and address register (specifying where to go on the disk, how many bytes to read, and where to put them in memory). Then the disk interface finds the right place on the disk, makes a DMA request, and begins moving blocks of data to the specified place in memory. When it's done, it sets bits in its status register to signify completion and then makes an interrupt. The CPU, which has meanwhile been executing other instructions (or possibly just waiting for data from the disk), responds to the interrupt, finds out from the status register of the disk interface that the data are now in memory, and then goes on to the next task. Thus, programmed I/O to the interface (the simplest kind of I/O) was used to set up the DMA transfer, DMA itself (stealing bus cycles from the CPU) was used for rapid transfer of data, and an interrupt was used to let the computer know the task was done. This sort of I/O hierarchy is extremely common, especially with mass-storage devices; you can expect maximum DMA transfer rates of hundreds of megabytes per second on a contemporary microcomputer bus like PCI express.

B. DMA on the PC104/ISA bus

The PC104/ISA bus, the descendant of an earlier (and simpler) time, has a simpler DMA protocol. A support chip that accompanies the microprocessor includes a DMA controller (an 8237 on the original IBM PC), with built-in address and byte counters, along with the logic to disable the CPU and take over the bus. A peripheral that wants to do DMA, therefore, doesn't have to generate addresses and drive the bus. Instead, it signals the controller (via one of the three DRQ1–DRQ3 "DMA request" lines), which in turn responds by returning the corresponding DACK0–3′ ("DMA acknowledge"). The controller then controls the transfers, asserting address and strobing lines, with the peripheral asserting (or receiving) data to (or from) memory. In this whole process the memory sees nothing unusual going on, since addresses and memory strobes (MEMW' or MEMR'), normally supplied by the CPU, are supplied by the DMA controller, and if it's DMA *to* memory, data are supplied by the peripheral. The peripheral, on the other hand, knows something special is happening because it requested DMA access (and received confirmation via DACK'); so when the DMA controller asserts IOR' (or IOW'), the peripheral supplies (or accepts) successive bytes. You might wonder why some innocent bystander peripheral doesn't get hurt in the DMA process, since both I/O strobes and addresses are being asserted, whereas the addresses are in fact the *memory* addresses that go with the memory strobes MEMW' or MEMR' asserted by the controller; they have nothing to do with I/O port addresses. The secret is our old friend AEN, specifically added to the bus just to solve this problem. AEN is asserted HIGH during DMA transfers, and all I/O port addressing must be qualified by ANDing with AEN LOW to prevent spurious responses to DMA memory addresses.

Even with the use of a separate controller chip, you still have to set up the starting address, byte count, and direction for the impending DMA transfer. These data go to the DMA controller, which is obliging, having a set of registers that you write (via programmed I/O) from the CPU. It's all pretty straightforward, except for some complexity created by a proliferation of "modes" (single transfer, block transfer, etc.). The PC104/ISA bus has a rather modest DMA capacity, about $2\,\mu s$ per byte transferred; and, as with interrupts, the PC104 bus is sparse on DMA channels.

14.3.11 Summary of PC104/ISA 8-bit bus signals

Through our examples – programmed I/O, interrupts, and DMA – we've seen most of the PC104 bus signals, which make a "multidrop" tour through the stacked peripherals (Figure 14.18). Table 14.2 lists the full bus, with pin connections. For completeness we summarize them all here, beginning with the ones we've already met.

A19–A0: Address lines   Three-state, out from bus master, active-HIGH. All 20 lines are used to address memory (with MEMR' and MEMW' as strobes, analogous to IOR' and IOW'), but only the 16 least significant lines are used during I/O access (64K port addresses); I/O devices should qualify address with AEN LOW.

D0–D7: Data lines   Three-state, bidirectional, active-HIGH. Asserted by CPU during memory or I/O write; asserted by memory during memory read or DMA from

memory; asserted by I/O port during I/O read or DMA to memory.

IOR', IOW', MEMR', MEMW': Data strobes   Three-state, output only, active-LOW. Asserted by bus master during read or write. On *writes*, data should be latched on trailing (rising) edge, qualified by address; on *reads*, data should be asserted during the strobe (and ready before the trailing edge), qualified by address.

AEN: Address enable   Two-state, output only, active-HIGH. Asserted by CPU during DMA cycles. I/O ports must not respond with normal address decoding to IOR' and IOW'; instead, I/O port that received DACK' looks at IOR' or IOW' to put or take DMA data bytes to or from DATA lines.

IRQ2–IRQ7: Interrupt request   Two-state, input only, rising-edge-triggered. Asserted by interrupting device. Prioritized, with IRQ2 highest, IRQ7 lowest. Maskable in the interrupt controller, via CPU write to port 21h. Each IRQ level can be used by only one device at a time.

RESET: Reset driver   Two-state, output only, active-HIGH. Asserted by CPU during power-on. Used to initialize I/O devices to known startup state.

DRQ1–DRQ3: DMA request   Two-state, input only, active-HIGH. Asserted by I/O device requesting DMA channel. Prioritized, with DRQ1 highest, DRQ3 lowest. Acknowledged by DACK1'–DACK3'.

DACK0'–DACK3': DMA acknowledge   Two-state, output only, active-LOW. Asserted by CPU (DMA controller) to indicate grant of corresponding DMA request.

ALE: Address latch enable   Two-state, output only, active-HIGH. The 8088 used a multiplexed data/address bus, and this signal corresponds to the 8088's strobing signal, used by latches on the motherboard to latch the address. Can be used to signal beginning of a CPU cycle; usually ignored in I/O design.

CLK: Clock   Two-state, output only. This is the CPU's clocking signal; it's asymmetrical, 1/3 HIGH and 2/3 LOW. The original PC used a 4.77 MHz clock, but higher speeds are common. CLK is used to synchronize wait-state requests (via IOCHRDY), in order to stretch an I/O cycle for slow devices.

OSC: Oscillator   Two-state, output only. This is a 14.31818 MHz square wave, which can be used (when divided by 4) as a color-burst oscillator for color display.

TC: Terminal count   Two-state, output only, active-

Table 14.2 PC104/ISA 8-bit Bus Signals

Signal	Qty	Active	Type[a]	Direction CPU[b] ↔ I/O	Pin #	Function
A[19..0]	20	H	3S	→	A12..21	address (A15..0 for I/O)
D[7..0]	8	H	3S	↔	A2..9	data
IOR#	1	L	3S	→	B14	I/O read strobe
IOW#	1	L	3S	→	B13	I/O write strobe
MEMR#	1	L	3S	→	B12	memory read strobe
MEMW#	1	L	3S	→	B11	memory write strobe
AEN	1	H	2S	→	A11	DMA address signal
IRQ[7..2]	6	↑	2S	←	B21..25, B4	interrupt request
RESET	1	H	2S	→	B2	power-on reset
DRQ[3..1]	3	H	2S	←	B16,6,18	DMA request
DACK[3..0]#	4	L	2S	→	B15,26,17,19	DMA acknowledge
ALE	1	H	2S	→	B28	"address latch enable"
CLK	1	–	2S	→	B20	CPU clock
IOCHCK#	1	L	OC	←	A1	I/O error — makes NMI
IOCHRDY	1	H	OC	←	A10	pull LOW for wait states
OSC	1	–	2S	→	B30	14.31818 MHz
TC	1	H	2S	→	B27	DMA terminal count
GND	4		PS		A32;B1,31,32	signal & power gnd
+5V	2		PS		B3,29	+5V supply
+12V	1		PS		B9	+12V supply
−5V	1		PS		B5	−5V supply
−12V	1		PS		B7	−12V supply

Notes: (a) OC = open-collector; PS = power supply; 2S = 2-state (active pullup); 3S = 3-state.
(b) address, data, and read/write lines can be driven by bus master (e.g., during DMA).

HIGH. Tells I/O port that a DMA block data transfer is complete. A DMA device must qualify it with DACK' for the channel in use, since TC is asserted when any of the DMA channels finishes a block transfer.

IOCHK': I/O channel check Open-collector, input only, active-LOW. Generates highest-priority interrupt (NMI, "nonmaskable interrupt"); used to signal error condition from some peripheral. CPU figures out who's in trouble by device polling (§14.3.9A); each peripheral that can assert IOCHCK' must therefore have a status bit that can be read by the CPU.

IOCHRDY: I/O channel ready Open-collector, input only, active-HIGH. The processor generates "wait states" if requested by a slow peripheral (that pulls it LOW) before the second CLK rising edge of a processor cycle (normally four CLKs). Used to extend bus cycle for slow I/O or memory.

GND, +5 Vdc, −5 Vdc, +12 Vdc, −12 Vdc: ground and dc supplies Regulated dc voltages that are bused for use by peripheral interface cards. Check the specifi-cations of the host processor for power limitations, which are machine dependent. Generally speaking, there should be enough power to run anything you can stack onto the PC104 bus.

14.3.12 The PC104 as an embedded single-board computer

The PC104 standardized bus has been implemented in numerous single-board computers (SBCs), with an impressive variety of compatible peripheral cards; these are created by more than 100 manufacturers. These little boards often wind up as *embedded systems*, that is, dropped into instruments as part of their intelligent design. Figure 14.19 shows a view looking down into the access hatch of a complex optical detector system on an astronomical telescope: there's a PC104 SBC, running embedded Linux from its piggyback flash-memory "disk." This particular SBC (from Diamond Systems) includes Ethernet and serial ports (and other stuff not used here). The box on the left is an Ethernet media converter that lets us use optical fiber back to the control room; that's a good idea when your observatory

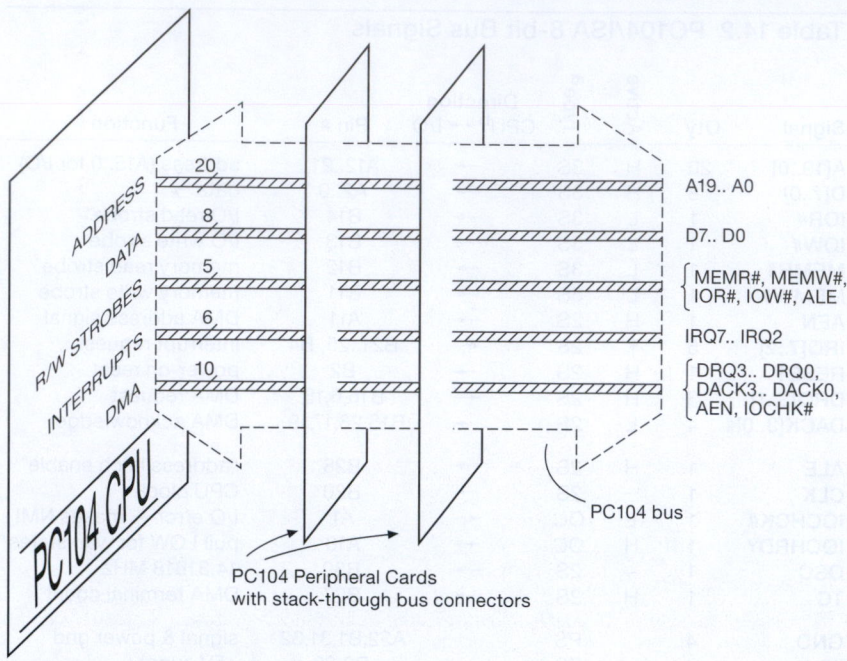

Figure 14.18. The PC104 parallel multidrop bus, making its grand tour.

14.4 Memory types

As we remarked back in §14.1.2, computers need fast "random-access" memory (i.e., memory in which you can get to any data directly, as compared with the sequential access of data stored on magnetic tape), which usually takes the form of *dynamic RAM modules*. These are the skinny circuit boards that plug into memory sockets on the computer motherboard, with typically 240 contacts along the mating edge. The most common variety currently is the "SODIMM" (small-outline dual-in-line memory module, 3×13.3 cm), on which is mounted an array of individual DRAM chips, with module capacities up to several gigabytes (organized as a 64-bit- or 72-bit-wide parallel data path). These are what you want if your task is simply to populate a commercial motherboard – just make sure you have chosen compatible memory modules from the bewildering variety[26] (check the computer motherboard specs

and/or the online product finder from manufacturers like Corsair, Crucial, or Kingston).

But there's lots more to know if you want to understand what's going on, and if you want to be able to design your own systems that need memory. In this chapter section we'll discuss the various kinds of memory: static RAM, dynamic RAM, and non-volatile memory types like "flash RAM."

14.4.1 Volatile and non-volatile memory

For many applications there's no need to retain memory contents when power is turned off. A computer, for example, freshly loads its operating system, application programs, and data into working memory during the boot process, so it's OK if that memory is volatile. Those

---

[26] Characterized by *too many* parameters: form factor (e.g., SIMM, RDIMM, SODIMM), pin count (e.g. 200, 204, or 240 pins), density

and data width, ECC or non-ECC, buffered or unbuffered, SDRAM generation (SDR, DDR, DDR2, DDR3, DDR4, DRDRAM), clock speed (e.g., PC3-10600) and data rate (e.g., 1333 MT/s), CAS latency (e.g., CL=4), voltage (e.g., 1.5 V), single or double rank, registered or unregistered, and parity or no parity. There are literally hundreds of choices, most of which will not work in a given motherboard. As a specific example, this footnote is being written into some portion of four banks of 1 GB density 64-bit wide DDR2 PC2-6400 CL=4 (5-5-5-15-2T) 1.9 V unregistered and unbuffered SDRAM with no parity and no ECC, on 240-pin SODIMMs.

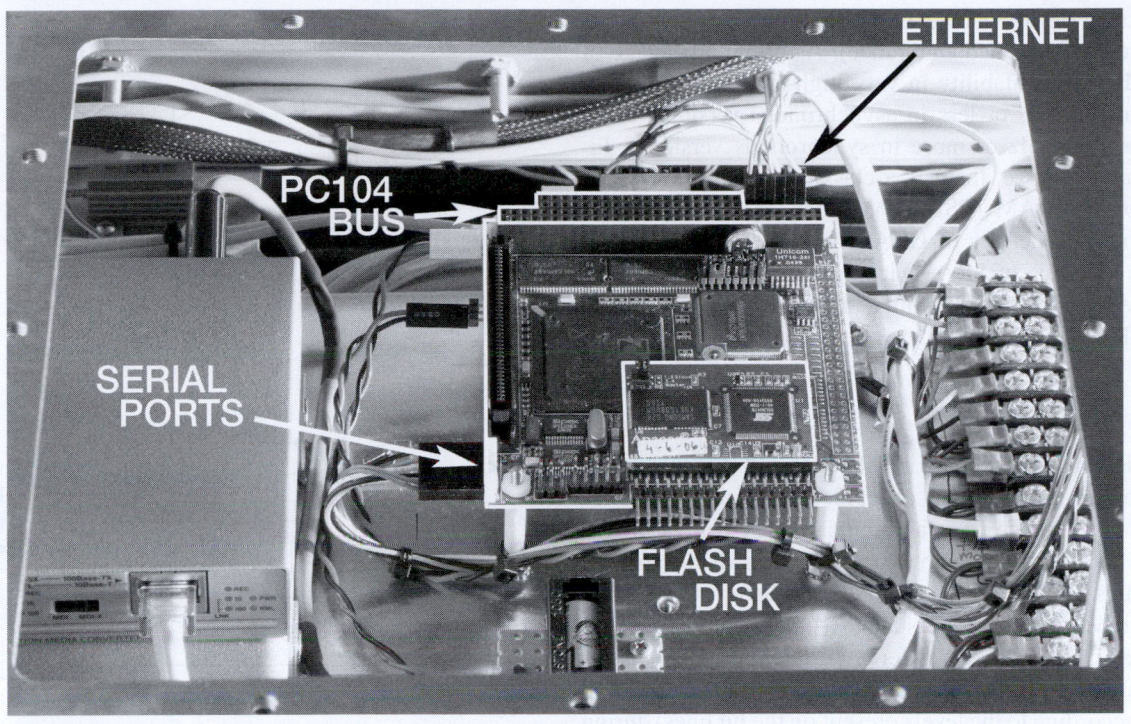

Figure 14.19. PC104 single-board computer embedded in an astronomical instrument.

programs and data must, of course, be retained somewhere when power is off, and that is the function of non-volatile mass memory, usually in the form of a hard disk (rotating magnetic storage) or a "solid-state disk" (a misnomer: it's an array of flash-memory chips, wrapped up in a disk-like enclosure for compatibility).[27] By accepting volatile memory you gain in speed, density, endurance (number of erase/write cycles before wearout), and price compared with those of currently available non-volatile technologies, as we'll see. Both static RAM (SRAM) and dynamic RAM (DRAM) are volatile,[28] whereas flash RAM and EEPROM (along with some interesting new technologies) are non-volatile.

14.4.2 Static versus dynamic RAM

Static RAM stores bits in an array of flip-flops, whereas dynamic RAM stores bits as charged capacitors. A bit once

written in a SRAM stays there until rewritten or until the power is turned off. In a DRAM the data will disappear in less than a second, typically, unless "refreshed." In other words, a DRAM is always busy forgetting data, and it is rescued only by periodic clocking through the "rows" of the two-dimensional pattern of bits in the chip. For example, you have to access each of 8192 row addresses in a 1 Gb DRAM every 64 ms (an average rate of one row every 7.8 μs).

You might wonder what would possess anyone to choose a DRAM: by not using flip-flops, the DRAM saves space, giving you more data on a chip and at something like a tenth the cost.

Now you might wonder why anyone would choose *static* RAM (fickle, aren't you?). The major virtue of SRAM is its simplicity, with no refresh clocks or timing complexity to worry about (the refresh cycle competes with normal memory access cycles and must be properly synchronized). Thus, for a small system with only a few memory chips, the natural choice is SRAM. Furthermore, SRAM's zero quiescent current (compared with significant idling current in standard DRAM) makes it desirable for battery-powered devices. In fact, CMOS static RAM, backed up with a battery when main power is off,

---

[27] In a small system, for example an embedded processor of the kind we'll see in the next chapter, the "mass memory" more often takes the form of a single flash-memory chip, or (better) some flash memory included on the microprocessor ("micro*controller*") chip itself.

[28] The volatile memories have no wearout mechanism – you can read and write to your heart's content.

constitutes an alternative nonvolatile memory (with advantages in speed and endurance). A further advantage of SRAM is its availability in very high-speed versions (access time of 8 ns or less in asynchronous versions; clock rates of 400 MHz or more in synchronous versions). Finally, as we'll see, there is a class of memory known as *pseudostatic* RAM (PSRAM), which combines the low cost and high density of DRAM with the low power and simple interface of SRAM; it might well be described as "a DRAM in SRAM's clothing." Let's take a closer look at both SRAM and DRAM.

14.4.3 Static RAM

Static RAM stores each bit in a flip-flop cell (Figure 14.20): the flip-flop itself consists of a pair of cross-connected inverters (each made from a complementary pair of pMOS and nMOS switches), with two additional nMOS transmission gates to couple it to the outside. The latter are switched on to convey the flip-flop's state to a pair of complementary *bit lines* (which drive a differential latching *sense amplifier*) during a READ, and to assert (overwrite) the flip-flop state (according to the levels present on the bit lines) during a WRITE. This is known as a "6T" (six-transistor) SRAM cell.

As with most varieties of memory, SRAM bit cells on a single chip are organized into multibit "words" that are read or written as a group. Common widths range from 8 bits to 32 bits, often accompanied by additional parity bits (thus 9-, 18- or 36-bit widths). These words are further organized logically into an array of rows and columns, as shown in Figure 14.21, such that the selected word's bit cells are coupled to the respective bit-line pairs (and sense amps) according to the word's multibit address. The figure represents a 4 Mb SRAM (512 k 8-bit words[29]), organized as eight planes (one for each bit) of 1024 rows and 512 columns.

A. Asynchronous SRAM timing

Traditional SRAM is asynchronous, meaning that there is no clocking input; instead, you apply address, data, and control signals with proper timing, and data comes out (READ cycle) or is written (WRITE cycle) accordingly. Using asynchronous SRAM is as easy as falling off a log: to

---

[29] We follow common usage here, with "Mb" (megabit) signifying 2^{20} bits, about 5% more than a *decimal* million (10^6). Purists prefer to use the prefix "Mi" (pronounced mebi, contraction of "mega binary") to mean 2^{20}, to keep pure the essence of mega as a good old-fashioned million. They would say that this memory is 4 Mib, or half a mebibyte; some would think they are speaking with food in their mouths.

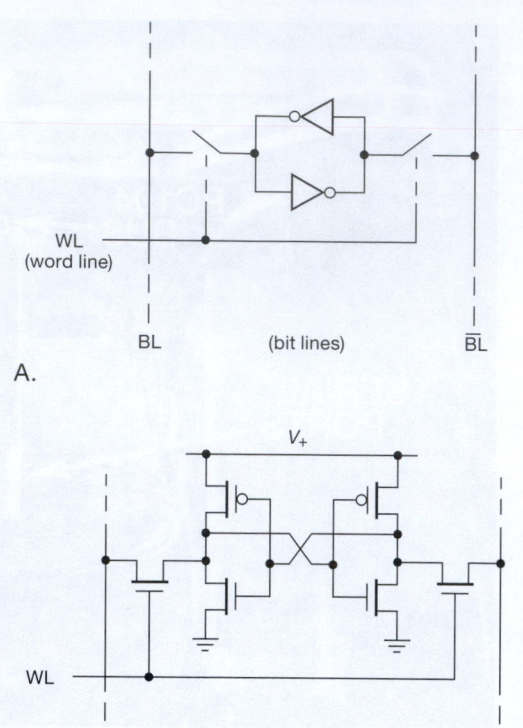

A.

B.

Figure 14.20. Static RAM holds each bit in a four-transistor CMOS flip-flop, read or written by a pair of one-transistor transmission gates. WL, word line; BL, bit line.

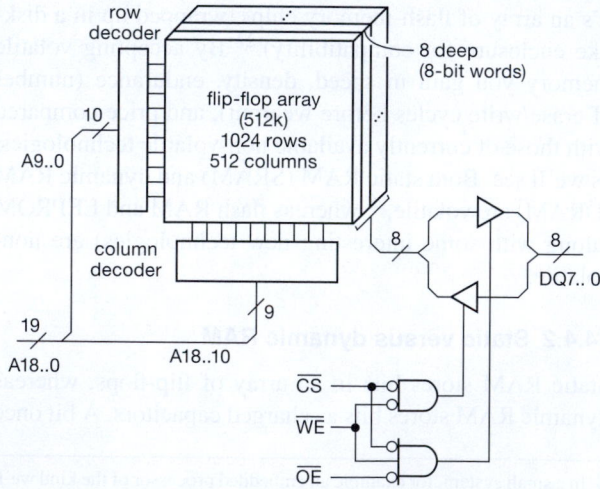

Figure 14.21. Asynchronous SRAM architecture: an array of 6T bit cells with an *n*-bit parallel address, and simple control by WE', CS', and OE'. Word widths of 8 bits and 16 bits are most common.

READ a word, you assert the address, chip enable (CE'), and output enable (OE'); the requested data obligingly appears on the three-state data lines a maximum of t_{AA} (address-access time) later. To WRITE a word, you assert address, data, and CE', then follow (after a minimum of an address setup time, t_{AS}) with a write enable (WE') pulse; valid data are latched into memory at the end of the WE' pulse. Figures 14.22 and 14.23 show the actual timing constraints for a fast static RAM ($t_{AA}=8$ ns, $t_{AS}=0$ ns), of the sort you might use in a switch or router, or as an external memory cache. The "speed" of the memory is set by the time from assertion of valid address to valid data (for READ) or to completion of the write cycle (for WRITE), assuming the other signals (CS', WE', and OE') are asserted when needed.

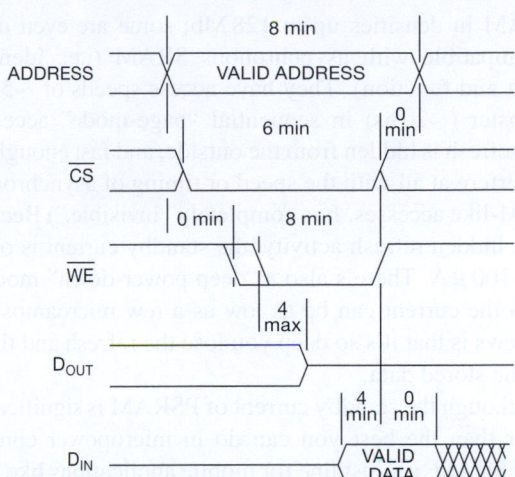

Figure 14.23. Write-cycle timing for the same memory as in Figure 14.22.

cell uses more area than the analogous one-transistor/one-capacitor DRAM cell (discussed below), so it costs more per bit and you can't put as much memory on a chip of a given size. However, the denser and less-expensive DRAM requires periodic refreshing, and, along with its multiplexed address scheme, it presents a more complicated interface.

Pseudo-static RAM combines their best features: it combines a DRAM core with hidden refresh logic, and wraps the combination in an interface that mimics the clean and simple asynchronous SRAM (Figure 14.24). It's a "DRAM in SRAM's clothing." Currently you can get

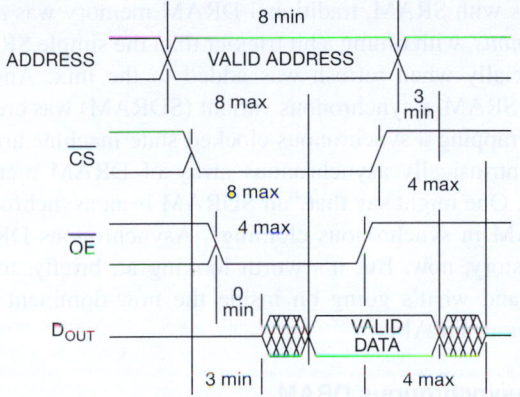

Figure 14.22. Asynchronous SRAM read cycle (during which WE' is held HIGH). The numbers shown are guaranteed worst-case timings, in nanoseconds, for this 512 kB fast (8 ns) SRAM (Samsung K6R4008V1D-08).

Static RAM is currently available in sizes up to 16 Mb, with word widths of 1 to 32 bits. Variations include memories with separate input/output pins, memories with very low current drain ($\sim 1\,\mu$A standby, ~ 1 mA when operating at 1 MHz), and memories with dual-port access (independent address lines, data lines, and control lines, sharing the same stored memory, with "semaphores" for software handshaking between ports).

For whatever it's worth, note that you don't have to hook up SRAM data lines to corresponding numbered data lines of a processor (or any other device needing memory) – you can scramble them any way you want, since they get unscrambled when you read back what you wrote.

B. Pseudo-static RAM

Static RAM has the advantages of a simple control interface and very low power consumption. But its six-transistor

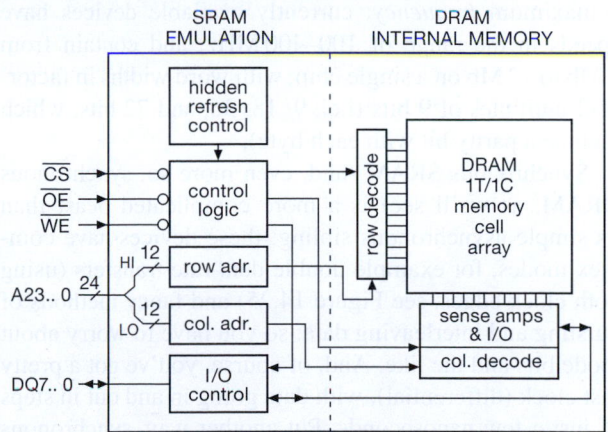

Figure 14.24. "Pseudostatic RAM" looks for all the world like a classic asynchronous SRAM. But its simple external interface conceals the truth: an efficient 1T/1C DRAM core, wrapped in an SRAM-emulating layer of logic (complete with hidden refresh).

PSRAM in densities up to 128 Mb; some are even drop-in compatible with asynchronous SRAM (i.e., identical pinout and function). They have access speeds of ∼50 ns, but faster (∼20 ns) in sequential "page-mode" accesses. (The refresh is hidden from the outside, and fast enough not to interfere at all with the speed or timing of asynchronous SRAM-like accesses. It is completely "invisible.") Because of the hidden refresh activity, the standby current is of order ∼100 μA. There's also a "deep power-down" mode in which the current can be as low as a few microamps; the bad news is that it's so deep you lose the refresh and therefore the stored data.

Although the standby current of PSRAM is significantly higher than the best you can do in micropower conventional SRAM, it's just fine for mobile applications like cellphones, with their typical battery capacity of ∼1000 mAh. The result is that pseudo-static RAM has largely replaced conventional asynchronous SRAM, except for applications like memory caches that require the blazing speed ($\leq$10 ns) of fast SRAM.

C. Synchronous SRAM

In previous chapters we sang the praises of synchronous clocked logic, with its benefits in terms of noise (things have settled down before each clock), predictable performance, pipelined architectures, lack of metastability, and so on. So, why is it that SRAM is *asynchronous*?

It doesn't have to be. You can wrap a synchronous clocked state machine, complete with its data registers, around the intrinsically asynchronous memory-array core. Then you've got *synchronous* SRAM.[30] Because it is clocked, the speed of synchronous SRAM is given as a maximum *frequency*; currently available devices have speeds in the range of 100–400 MHz, and contain from 1 Mb to 72 Mb on a single chip, with word widths in factor-of-2 multiples of 9 bits (i.e., 9, 18, 36, and 72 bits, which include a parity bit with each byte).

Synchronous SRAM (and, even more so, synchronous DRAM, as we'll see) is a more complicated beast than its simple asynchronous sibling: these devices have complex modes, for example double data-rate transfers (using both clock edges, see Figure 14.25) and fancy methods of bursting and interleaving data; so you have to worry about mode bits and the like. And, of course, you've got a pretty fast clock (differential), with data going in and out in steps of just a few nanoseconds. Put another way, synchronous

SRAM was devised for the highest speed and throughput, so the blessing of synchronous operation is mixed with the curse of fast clocks and tight timing margins.

14.4.4 Dynamic RAM

As we remarked earlier, you can save a lot of space on the chip by going to a one-transistor memory cell (with the state held on a small capacitor) if you're willing to carry out the periodic refresh of the capacitor's charge. That's dynamic RAM, with its "1T1C" memory cell (Figure 14.26). This is the workhorse of contemporary memory, with sizes of several gigabits per chip, typically supplied as plug-in memory modules with capacities currently up to 16 GB (128 Gb).[31]

As with SRAM, traditional DRAM memory was *asynchronous*, with timing a bit trickier than the simple SRAM, especially when refresh was added to the mix. And, as with SRAM, a synchronous variant (SDRAM) was created by wrapping a synchronous clocked state machine around the intrinsically asynchronous array of DRAM memory cells. One might say that "an SDRAM is an asynchronous DRAM in synchronous clothing." Asynchronous DRAM is history, now. But it's worth looking at, briefly, to understand what's going on inside the now-dominant synchronous DRAM.

A. Asynchronous DRAM

Figure 14.27 shows in simplified form a 16-word by 1-bit asynchronous DRAM, laughably small by any real-world standard, but just right for explaining how it works. The nMOS transistors each have a small capacitor (of order 30 femtofarads) from channel (call it the drain) to ground, and are arrayed as four rows and four columns. The row drivers use the latched row bits (high half of the address) to bring the corresponding nMOS gate terminals HIGH, turning on those transistors and therefore connecting their capacitors to the column lines, which in turn connect to a set of latching sense amplifiers (SA's). The output of one of the sense amps is selected by the latched column bits (low half of the address), passing through the bidirectional multiplexer to the output/input buffer. (Dynamic memory multiplexes the address bits, halving the number of address lines.)

Here's how it works. Let's take a READ cycle first, assuming that the various capacitors (one per bit) have been

---

[30] And the analogous thing is done with dynamic RAM, as we'll see, creating SDRAM in its many flavors: SDR (single data rate), DDR (double data rate), DDR3, DDR4, etc.

[31] *This* month, anyway; we've been riding Moore's law now for four decades, and we are certain that this will prove hopelessly quaint by the time the ink is dry.

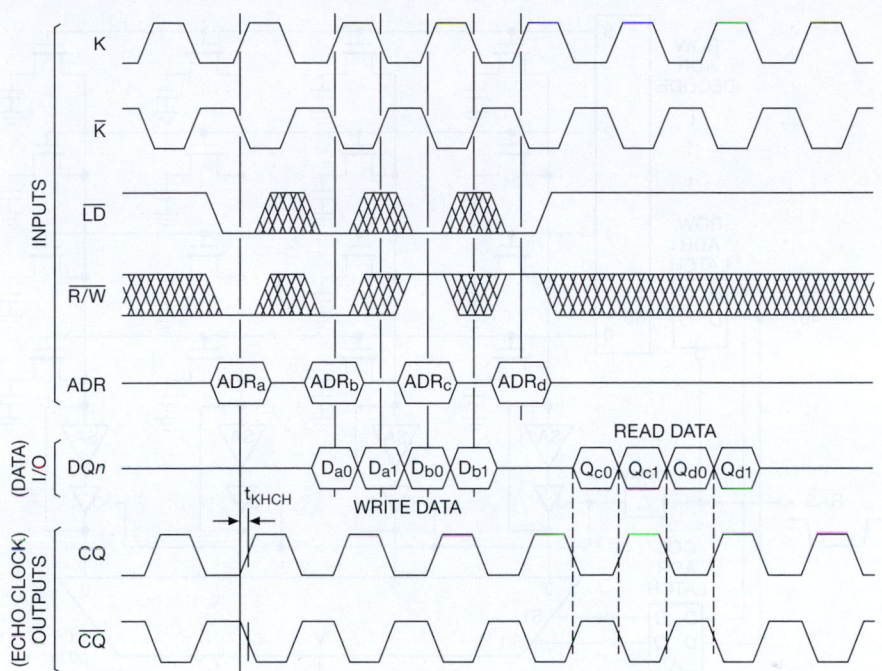

Figure 14.25. Synchronous SRAM timing, with double data-rate (DDR-2) clocking during both READ and WRITE. Data is clocked in two-word (successive addresses) bursts: for WRITE cycles the input data must be presented on both edges of the input clock $K/\overline{K}$, one cycle after the address is loaded; for READ cycles the data emerges 1.5 clock cycles after address loading, on both edges of the "echo clock" $CQ/\overline{CQ}$ (which regenerates the input clock $K/\overline{K}$, with a slight delay t_{KHCH}).

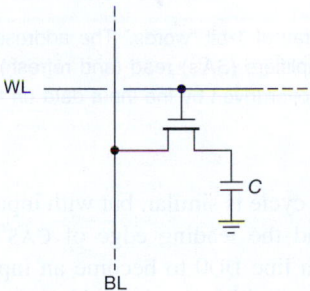

Figure 14.26. Dynamic-RAM "1T1C" bit cell. Each bit is held as a charged (~ 1 V) or discharged (0 V) capacitor, whose state is read, written, or refreshed by a bit line (BL) when asserted by a controlling word line (WL). Typically $C \approx 30$ femtofarads (1 fF=10^{-15} F).

previously written, therefore either positively charged (to ~ 1 V), or uncharged. Look at Figure 14.28, where we've shown the basic (single-address) READ and WRITE cycles for standard 70 ns asynchronous DRAM.[32] The address lines are multiplexed, with the two high-order address

bits (row address) asserted first, along with a row-address strobe (RAS'). Those bits are latched, causing the gates of the selected row to go HIGH, turning on the MOSFET transmission gates, and thus coupling the respective capacitors to the column sense amplifiers (SA). The sense amplifiers are latching devices, here drawn notionally as fed-back noninverting amplifiers. (In practice they are implemented as flip-flops that begin the cycle in a balanced state and become unbalanced by the bit-capacitor charge that is switched into them.[33]) During this part of the DRAM cycle the sense amplifiers in each column do two things: they latch into the state of the bits that were held in the cells of

---

[32] There are additional modes, with names like "page mode" and "extended data out," for more efficient access to data from several consecutive addresses.

[33] In a further dose of reality, things are a bit more complicated: the sense amplifiers are *differential*, and the DRAM array is usually built in a "folded-bit" arrangement so that any given row line activates only the even or odd cells; the inactive (neutral) bit line floats at the "precharge" level ($V_{\mathrm{DD}}/2$) and acts as a reference voltage by which the balanced sense amplifier compares the ΔV "bump" up or down from the capacitor's charge in the respective bit cell. The ΔV is quite a bit less than the ± 0.5 V you might expect, owing to the additional ~ 200 fF capacitive loading of the bit line and sense amplifier: in practice, memory designers aim for $\Delta V \geq 100$ mV, for a clean readout of 0 or 1 by the sense amplifier.

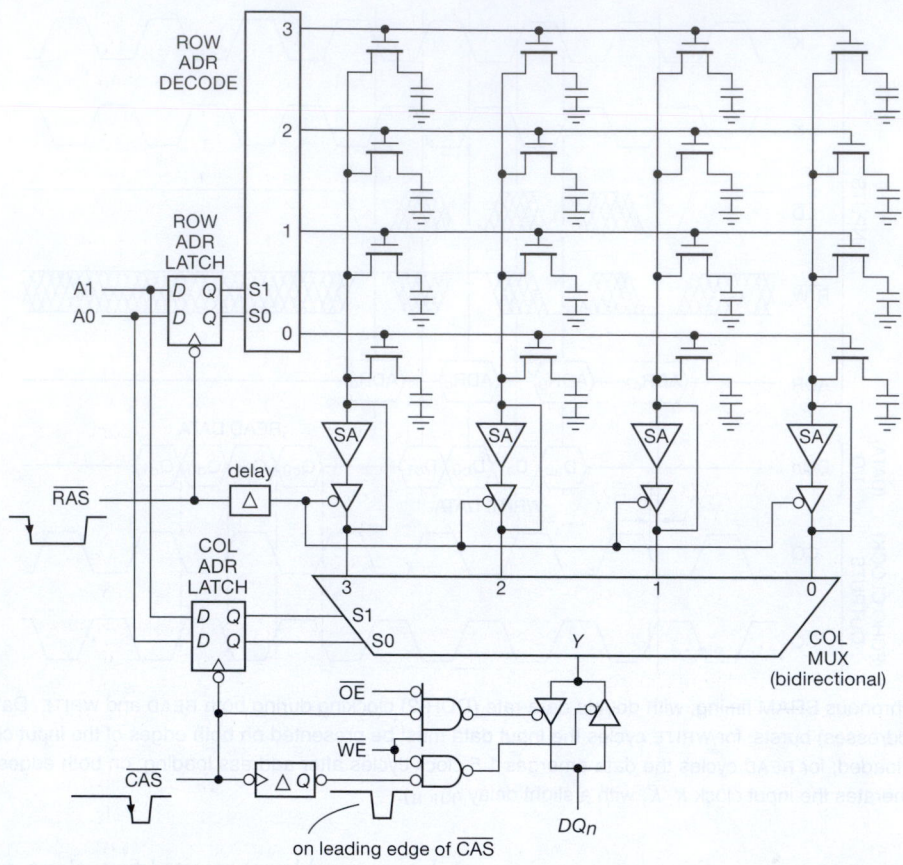

Figure 14.27. Asynchronous DRAM architecture, here illustrated with a tiny 4×4 array of 1-bit "words." The address lines are multiplexed, with row and column addresses registered internally. The latching sense amplifiers (SA's) read (and refresh) the state of the row-addressed bit lines during a READ cycle. During a WRITE cycle they are instead overdriven by the input data on the same shared input/output data line $DQ0$.

the selected row; and they "refresh" those cells by asserting their latched level back onto the respective cell capacitors.

In the second half of the DRAM cycle the address lines carry the two low-order address bits (column address), latched by the column-address strobe (CAS'). That latched address causes the column multiplexer to select the output from the selected sense amplifier, which (because WE' is disasserted) is asserted onto the bidirectional (input/output) data line (the single DQ0 here, in this 1-bit wide memory; most DRAMs are wider, with word sizes of 4, 8, or 16 bits). The memory is *asynchronous*, meaning (as with asynchronous SRAM) that the valid data appears on the data line(s) with some guaranteed maximum delay time following assertion of the various addresses and strobes. In contrast to the now-universal *synchronous* DRAM (next subsection), there's no master clock.

The WRITE cycle is similar, but with input data and WE' asserted around the leading edge of CAS'; the WE' line causes the data line DQ0 to become an input, whereupon the asserted data level forces (overrides) the selected sense amplifier into the state of the data input. That latched state then charges (or discharges) the respective bit cell capacitor.

Look at Figure 14.27 again. Once a row has been activated (in the RAS part of the cycle) and a column has been selected for reading (or writing), there's no reason to repeat the same row address if you want to read (or write) other columns within that same row. That's the idea behind "page-mode" and "extended-data-out" cycles; and it's of great practical utility, because computer memory accesses usually involve bursts of multiple successive addresses (for sequential instructions, data arrays, and so on).

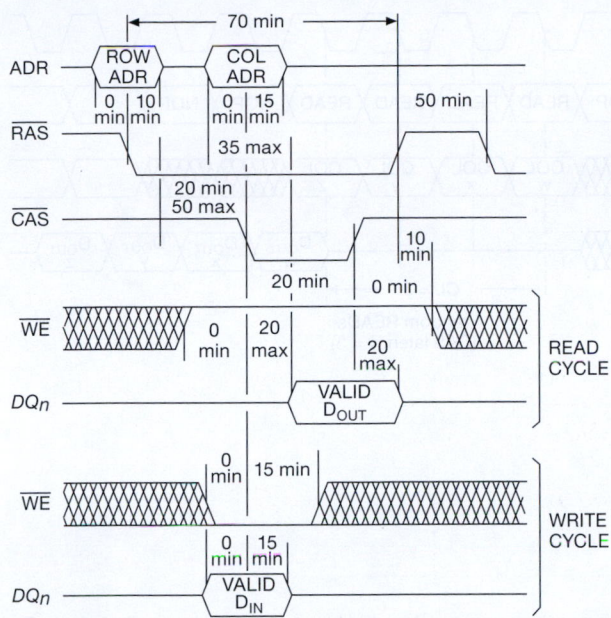

Figure 14.28. Asynchronous DRAM timing. RAS′ and CAS′ are the row and column address strobes for the multiplexed address. As with asynchronous SRAM, data in/out is not tied to a clocking edge; rather, it obeys worst-case timing specifications (here shown for standard 70 ns memory).

B. Synchronous DRAM

As we remarked above, all DRAM currently in common use is of the synchronous variety ("SDRAM"), in which an externally supplied clock signal[34] controls a synchronous state machine that is wrapped around the intrinsically asynchronous DRAM memory core. The original single-data-rate (SDR) SDRAM has evolved through several generations, with names like DDR (double data-rate, i.e., clocking data on both clock edges), DDR2, DDR3, and DDR4. The clock transitions are used for the obvious things: to load the row and column addresses and to clock data in (WRITE) or out (READ). Synchronous DRAM is usually operated in "burst mode," with multiple data words from consecutive column addresses clocked out in sequence (see the MODE description, below).

Unlike asynchronous DRAM, SDRAM is a complicated beast, with a set of "commands" (also clocked) that determine what is happening. These commands are defined by three bits, which rather curiously use the legacy signal names RAS′, CAS′, and WE′ that originated with asynchronous DRAM. Here, however, these three input signal

bits, set up before the next clocking edge, determine what happens on that clock edge. Look at Figure 14.29, where we've listed five basic commands. "Activate row" loads a row address, with subsequent "READ" and "WRITE" commands loading a column address and initiating the corresponding data transfer. During such transfers (which may be 4-, 8-, or 16-bits wide, according to the particular chip's architecture), the data moves out or in at the clock rate (SDR, as in Figure 14.29), or at double that rate (DDR), as timed by the clock.

However, note particularly the timing of a READ: although data moves out at the full clock rate, there is a delay (*latency*) of several clock cycles from assertion of the column address to the corresponding data. This is the "CAS latency," illustrated in Figure 14.29 for the case of CL=3. A given memory chip will specify (through its part number) a minimum CAS latency for a specified clock frequency (for example, an MT47H128M8HQ-25E is a 128 MB 8-bit-wide DDR2 SDRAM in a 60-ball FBGA package with CL=5 at t_{clk}=2.5 ns). Because the rest of the system needs to know (and agree on) the actual CAS latency, you tell the chip what CAS latency to use by sending it a "LOAD MODE" command (i.e., a clock cycle, during which RAS′, CAS′, and WE′ are all held LOW), with the mode defined by the bits asserted on the address lines. The mode includes not only the CAS latency, but also bits defining whether it's single-address access or burst mode (for the latter, a choice of two, four, or eight consecutive words), and some other control options that you really don't want to hear about. In a further bit of complication, some varieties of SDRAM (e.g., DDR2) have a clock latency during WRITE cycles; logically enough, that's called "write latency" (WL), which you respect by asserting the data delayed by WL clocks after the column address is clocked in.

Because it is clocked, the speed of synchronous DRAM is given in terms of *frequency*, with labels like "DDR3-1600." That would describe a double-data rate SDRAM, conforming to the DDR3 standard, with data clocked on both edges of an 800 MHz clock. Currently available devices have data rates in the range of 400–1600 MT/s (megatransfers per second), and contain from 1–4 Gb on a single chip, with word widths of 4, 8, or 16 bits. The next generation (DDR4) raises the transfer rate to 1600–3200 MT/s, and raises single-chip densities to 16 Gb or more.

14.4.5 Nonvolatile memory

Nonvolatile memory (NVM) – data storage that is retained in the absence of dc power – is essential in many everyday applications, for example (a) startup ("boot") code, and

---

[34] At these high speeds it's invariably configured as low-voltage differential: CK and $\overline{CK}$.

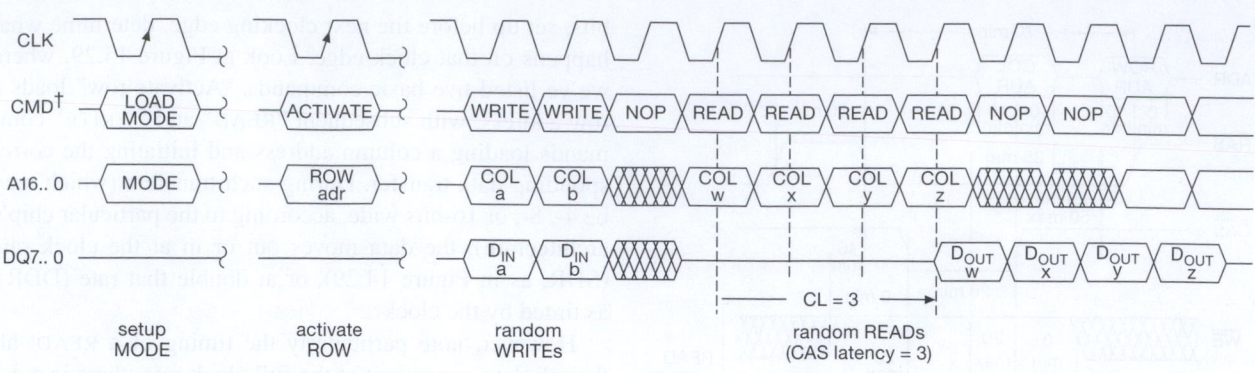

†CMD is given by $\overline{RAS}$, $\overline{CAS}$, and $\overline{WE}$:

	$\overline{RAS}$	$\overline{CAS}$	$\overline{WE}$	ADR lines
ACTIVATE ROW	L	H	H	row address
READ	H	L	H	column address
WRITE	H	L	L	column address
NOP	H	H	H	—
LOAD MODE	L	L	L	mode bits

Figure 14.29. Synchronous DRAM uses an external interface whose operations are synchronized by an externally applied clock. The simplest "single data rate" (SDR) SDRAM hijacks the RAS′, CAS′, and WE′ for use as a 3-bit command. Shown here are random (as opposed to *burst*) writes and reads for SDR SDRAM. Burst-mode operation is far more common, with data from several consecutive column addresses being clocked on successive clock edges (or on *both* edges, in the case of the popular DDR ("double-data rate") SDRAM.

various settings, in a general-purpose computer; (b) resident firmware for any device that uses an embedded processor; or (c) temporary storage on a USB "thumb drive" of documents, images, or other files. The semiconductor memories we've seen above are unsuitable, because they lose their stored data when dc power is removed.

One solution to the problem is to provide a battery, so power is never lost. That's "battery backed-up" memory (which is commonly used to store computer settings, where it's sometimes called, simply, "the CMOS"). The good news is that you can use any kind of memory with a micropower data-retaining standby mode (usually static or pseudostatic RAM), which means that you don't need to worry about limitations that afflict many forms of nonvolatile memory, for example the leisurely ($\sim$ms-scale) program or erase times, or the endurance (limited number of write cycles, typically $\sim 10^5$–10^6). The bad news is that you've got to provide a battery and make sure it is not discharged.[35]

The other solution is some form of truly non-volatile

storage. This is a very active area of semiconductor development, with most current devices using some form of charge storage on a MOSFET's "floating gate" (e.g., the ubiquitous "flash memory" in USB thumb sticks, in all manner of consumer electronics, and in computer solid-state drives known as SSDs). This is impressive technology, with single-chip storage capacities currently up to 1 Tb, data retention of at least 10 years, and retail prices as low as \$0.50 or less per gigabyte. But, as we'll see, floating-gate (FG) storage has some drawbacks, notably a finite number of erase/write cycles (endurance) and a relatively slow write (and erase) access time ($\sim$milliseconds); by contrast, standard SRAM and DRAM have unlimited endurance, and equally speedy read and write speeds ($\sim$10s of nanoseconds). Some NV technologies in development may remedy these shortcomings; they go by names like ferroelectric RAM (FRAM), magnetoresistive RAM (MRAM), and phase-change RAM (PRAM).

In the following subsections we describe briefly some of the precursors to current non-volatile memory technology; then we'll look at the 800-pound gorilla in the room, namely floating-gate flash memory.

---

[35] Not a completely trivial task: we bought four motherboards from a major manufacturer, all of which lose their CMOS settings if incoming ac power is lost for more than a few hours. This is quite an inconvenience (there are roughly 25 settings that must be restored), probably caused

by a batch of chips with out-of-spec leakage current or minimum retention voltage.

A. Legacy non-volatile memory

Mask ROM

First there was "mask ROM," which consists of gates with a hardwired connection pattern, built-in from birth, that cannot be altered. It's a fixed lookup table. And it's very much "read-only" (hence "ROM"). This is still a useful method for putting a conversion table, for example, into a custom IC. It's certainly non-volatile; but it's not really what we imagine as useful NV storage.

PROM

Next came programmable read-only memory (PROM), which you can program, once, in the field. It's also called "fuse ROM," because it uses an array of tiny on-chip fuses (metallic or semiconductor), which are selectively blown out to leave the desired fixed memory pattern. The subsequent development of reprogrammable (erasable) NV memory rendered PROM obsolete.

EPROM

Erasable programmable memory came next. Here the bit pattern is held on tiny capacitors, in the form of buried "floating gates" that alter the threshold voltage of the associated MOSFETs (Figure 14.30); this is the same technology as that used in contemporary flash memory. The "control gate" is then used to read out the stored bit: when the floating gate is charged (negatively) with electrons, the threshold voltage is several volts positive, whereas an uncharged floating gate produces a negative threshold voltage (therefore conduction with the control gate at 0 V). Because there's no connection to the floating gate, you use tricks for writing and erasing: to write, negative charge is put onto the gate by a process called "channel hot-electron injection" (CHE), which involves running drain current with elevated drain voltage (12–20 V); to erase you expose the chip to UV light for several minutes, which erases the entire chip at once. The erase process requires a transparent quartz window to let the UV light in (hence the alternative name, UV EPROM); the associated hermetically sealed ceramic packaging makes these things expensive, and large. And, they had to be removed from the circuit for erasure and reprogramming. These were made obsolete by in-circuit *electrically* erasable non-volatile memory.

OTP EPROM

What a magnificently self-contradictory name: "One-Time-Programmable Erasable Programmable Read-Only Memory! (Engineers *do* have a sense of humor, though perhaps a bit twisted.) OTP EPROM was the answer to the problem of EPROM packaging cost, namely, to put a UV-

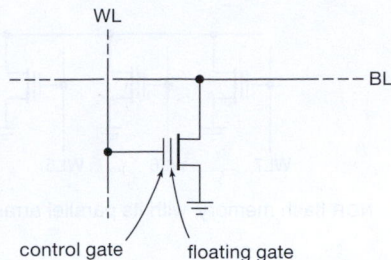

Figure 14.30. "Floating-gate" bit cell, used in non-volatile memories such as EEPROM and flash ROM. Data is written to the floating gate by tunneling or hot-electron injection; its charge alters the threshold voltage, as read out by the control gate and bit line. The leakage currents out of the floating gate are so low that data is retained for at least ten years, requiring no power or refresh.

erasable EPROM in an inexpensive plastic package. The only (!) drawback is that the package is opaque, so you can't erase the thing. What we did, in those days, was to buy one or two of the pricey UV EPROMs, use them to develop and test the program, then program the plastic parts with the fully debugged code. These devices, too, became obsolete with the development of electrically erasable non-volatile memory.

B. Electrically-erasable non-volatile memory

EEPROM

The modern era in non-volatile memory was ushered in with the electrically erasable programmable read-only memory[36] (EEPROM, or "E-squared"). No more UV sunbaths! No more removing and replacing the ICs – these puppies are *in-circuit* reprogrammable! The trick here is to use elevated voltages to discharge the floating gate, by a quantum-mechanical phenomenon known as *tunneling* (in which a particle, here an electron, with insufficient energy to overcome the potential barrier of the gate insulator can, under the right circumstances, just magically appear on the other side[37]). It's officially called Fowler–Nordheim tunneling, usually abbreviated F-N. EEPROMs include on-chip charge pumps to generate the elevated voltages needed to program like UV EPROMs and to erase via F-N

---

[36] Another name is EAROM – electrically alterable read-only memory.

[37] A macroscopic analogy: you need to get your oxcart over the Khyber Pass, but, because of poor planning, you lack an ox. Hoping for a miracle, however, you give the cart a shove. There is, according to quantum mechanics, a finite probability that the cart will vanish from your hands, and reappear on the other side of the pass. For oxcarts the probability is awfully small (and that's a real understatement!); but it works fine for electrons hoping to find their way to the buried gate.

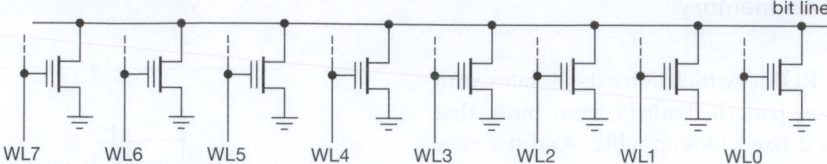

Figure 14.31. NOR flash memory, with its parallel arrangement of bit cells.

tunneling. Voilà – *in-circuit* programming (i.e., writing) and erasing.

EEPROMs use a memory-cell configuration (for example, a two-transistor bit cell) that permits bits or words to be erased and reprogrammed individually. That's a nice flexibility, but the circuitry that allows this flexibility takes up space that could be used for additional storage. The subsequent development of *flash* memory exploited a one-transistor bit cell, with its higher density (see flash memory, below); the price you pay is that erasure is carried out in much larger *blocks*, by way of their substrate contact.

EEPROMs, in common with flash memory, have a limited endurance, usually in the range of 10^5–10^6 erase/write cycles. Erasure and writing are slow ($\sim$10 ms), compared with reading ($\sim$100 ns). They have been largely superseded by flash memory. However, the flexibility of single-byte (or even single-bit) rewritability makes EEPROMs well suited for retaining small amounts of data, for example calibration parameters (think of a bathroom scale's strain gauge), settings (that scale, again: pounds or kilograms), or lookup tables (still thinking of that scale... which needs correction for ambient temperature). EEPROM for those kinds of applications is usually included on the same chip as the microcontroller itself (see Chapter 15). You can get stand-alone EEPROMs, also; these usually use a serial protocol (SPI, I^{2}C, UNI/O, Microwire), so they fit in a tiny package (SC-70, DFN, etc.). There's a good selection from companies like Atmel and Microchip, with densities from 128 bits to 1 Mb. They're inexpensive, too – a 1 kb I^{2}C EEPROM costs $0.17 in quantity, and twice that for a 64 kb part.

The EEPROM's intrinsic flexibility (erase/write individual bits or bytes) requires more silicon than the denser flash memory, with its *sector*-wide erase/write. Let's take a look.

C. Flash memory

Flash memory discards the bit or byte writability of EEPROM, instead performing block-sized erase. The good news is that it's faster (it erases "in a flash," hence the name), if you want to erase many bytes. The bad news is that you cannot modify smaller amounts of data (though

external logic, with some SRAM, can fake it so that you don't realize what's going on inside). But the big news is that the chip is denser, particularly in the form known as NAND flash, so you get a lot of memory on a single IC (up to 1 *tera*bit, currently; though the innards aren't an honest "chip," but rather a "stack" of thinned chips). Currently available flash memory comes in two flavors, called NOR and NAND.

NOR flash

Figure 14.31 is a simplified view of the bit-cell arrangement used in NOR flash memory, which was the original flash architecture. Because the storage transistors are in parallel, you read out by gating off all except the selected transistor. The floating gate cells are programmed via channel hot-electron injection (the same as with EPROM). However, as with all varieties of flash memory, erase must be done a block (or "sector") at a time, with the block's elevated substrate voltage causing F-N tunneling from the floating gates. Typical block sizes are 4–64 kB, often flexibly configured with several different block sizes on one chip. NOR flash uses a simple parallel SRAM-type interface (sometimes providing a choice of asynchronous or synchronous modes), and can be used directly to hold executable code. Note, however, that the limited endurance, as well as the sector-size erase, means that the executable code must treat the memory as read-only. Currently available NOR flash devices come in densities from about 1 Mb–1 Gb.

NAND flash

By contrast with NOR, the designers of NAND flash intended it as a dense mass-storage device, more akin to a magnetic hard drive. It's the memory that's used in USB thumb drives, compact flash (CF) cards, secure digital (SD) cards, solid-state drives (SSD), and the on-chip code memory used in microcontrollers (μC; see Chapter 15). The name NAND derives from the series connection of bit cells (Figure 14.32), which are read by holding HIGH all gates except the selected transistor; the selected transistor then determines the conduction of the series string, according to the charge on its floating gate. Both erasing and writing use tunneling; erasing is done in sectors. What's not obvious is the greater density you get with a series connection: no space-hogging source or drain contacts are needed, except at the ends of each string. To keep the pin count down, NAND uses a serial command interface; and, like NOR flash, it erases in sectors. Note, however, that a device like a USB thumb stick includes a memory controller, so that the electrical and timing details of erasing and data read/write are invisible to the external interface. Likewise, the specification for SD flash cards requires that they include not only the native SD interface protocol, but also the standard SPI (serial peripheral interface) protocol that is included in most contemporary microcontrollers.[38] Memory controllers used for flash storage devices do more than merely obey the rules: they use tricks like "wear leveling" (cycling through different blocks of storage to minimize reuse) to circumvent the endurance-limiting effects of insulator damage caused by tunneling; and they detect bad and degraded cells, whose addresses they then write to locations in sector 0 of the memory chip so that future memory accesses are diverted to spare cells.

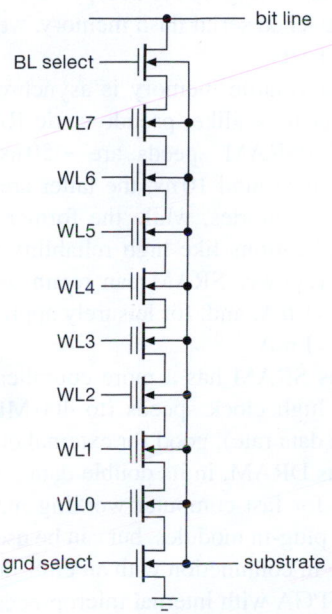

Figure 14.32. NAND flash memory, with its series arrangement of bit cells.

In a daring ploy to multiply memory density, manufacturers of NAND flash are using what's politely called "multilevel cell" (MLC) or "triple-level cell" (TLC) storage. Put bluntly, this means putting partial levels of charge on the floating gate, which are then read out by measuring the approximate threshold voltage. Current implementations use four (MLC) or eight (TLC) levels, therefore 2 or 3 bits per transistor. (Should we call this a "$\frac{1}{2}$T" or "$\frac{1}{3}$T" cell?!) This is scary stuff – your precious data is held on partially charged tiny floating gate capacitors, whose level of charge must not change too much during the next few years. To get a sense of the scale, the roughly 0.3 femtofarad capacitance must not leak (or acquire) more than ~3000 electrons in, say, 3×10^8 seconds (ten years). That's one electron per *day*! It's a miracle the stuff works at all.[39]

Currently available NAND flash comes in densities up to 1 Tb per IC; to get those densities, the manufacturers use 2-bit MLC storage, and multiple chip stacking within one IC package. To us, it is astonishing that a single IC can hold the equivalent of 16 bytes of information for every person on Earth.[40]

D. Future non-volatile memories

Flash memory is great stuff. But it has finite endurance, and the erase/write process is slow. An ideal non-volatile memory would be "SRAM without the volatility." That is, it would have random read and write access at full speed, unlimited endurance, and long retention.

There are several technologies being explored; the following appear to be the most promising.

Ferroelectric RAM (FRAM, FeRAM, or F-RAM)

A ferroelectric material is the electric analog of a ferromagnetic material; that is, it holds the state of electric polarization. The idea is to make a 1T1C-style DRAM bit cell, but with the capacitor replaced with a thin film of ferroelectric material (e.g., a few atom layers of strontium bismuth titanate). You write a bit as with DRAM, applying a field across the film. Readout is different, however: you read "destructively," by sensing whether a write changed the state (producing a pulse of current); then you restore the state.[41] FRAM potentially can deliver random

---

[38] The actual protocol that goes over the SD-card SPI interface isn't trivial, because it must support memory operations such as read, write, erase, check status, and so on. But it's straightforward, and you can get libraries for popular microcontrollers such as AVR and ARM (see Chapter 15).

[39] But it *does*! The several gigabytes of figures and text for this book edition made many journeys between office and home, in the form of partially charged femtofarad bits.

[40] We predict, with considerable confidence, that by the time you read this the bytes/IC will have increased far more than the people/Earth.

[41] This is analogous to the method used with the magnetic-core memory of yesteryear, in which the core's state was read by forcing it to a

ad and write speeds of <50 ns, many-year retention, and very high endurance. Companies like Fujitsu and Cypress are currently manufacturing FRAM, offering serial interface devices (SPI and I^2C) to 2 Mb with claimed endurance of 10^{14} cycles, and parallel (asynchronous SRAM-emulating) devices to 4 Mb. An example of the latter is the MB85RE4M2T, with specified random read and write cycle times of 150 ns, ten-year retention at 85°C, and (drumroll) endurance of 10^{13} read and write cycles (that's 300 years, at 1000 read/writes per second).

Magnetoresistive RAM (MRAM)

Magnetoresistance is the change of electrical conductivity of a material under an applied magnetic field. Variations of the effect ("giant magnetoresistance," "colossal magnetoresistance," "tunnel magnetoresistance") are used in magnetic disk readout heads, an improvement over the traditional coil-based readout (which generates a voltage proportional to *changes* in flux, rather than to flux itself) that has led to an explosion in hard-disk capacity. It's possible to fabricate tiny on-chip magnetoresistive cells (complete with some ferromagnetic material), written by pulsing a current, and read out by its magnetoresistance.

MRAM has been "in development" for many years, with some recent production parts. For example, Everspin (spinoff from Freescale) offers 8-bit and 16-bit word size MRAMs, with densities to 16 Mb. Their MR2A16A, for example, is a 4 Mb 16-bit-wide non-volatile MRAM, with a standard asynchronous SRAM parallel interface. It has specified read and write cycle times of 35 ns, 20-year retention, and (big drumroll) claimed "unlimited" endurance; it costs $20 in unit quantity. Other companies involved in MRAM include Hitachi, Hynix, IBM/TDK, Infineon, Samsung, and Toshiba/NEC (where a 1 Gb MRAM is rumored). MRAM is not cost competitive with flash, by a long shot; it's still at the starting gate – but it's worth keeping an eye on.

Phase-change RAM (PRAM or PCM)

Certain metallic alloys ("chalcogenide glasses") exhibit a large difference of electrical resistance between their crystalline and amorphous (glass-like) states. This can be exploited to make memory cells, in which the phase-change alloy is heated in controlled ways to convert between the two states. Heating sounds like a slow and diffuse process, but at the small scales here (tens of nanometers) things happen locally, and in tens of nanoseconds. Phase-

change memory is being pursued by companies like Samsung, Micron, IBM, and STMicroelectronics, and some prototype samples have been shipped (e.g., a 128 Mb 90 nm PRAM from Numonyx). No claims of unlimited endurance, yet; but PRAM potentially offers high-density non-volatile memory with fast read and write cycle times.

14.4.6 Memory wrapup

Encapsulating the long-winded narrative above, we can summarize the current state of memory and its usage this way.

- Much of the time the problem has been solved for you.
  - Microcontrollers (Chapter 15) include on-chip memory, both flash (for non-volatile program storage), SRAM (for working memory), and, often, EEPROM (for parameters, tables, etc.).
  - Computer motherboards are configured to use fast SDRAM, as plug-in SODIMM (or other form-factor) cards; just follow instructions and you'll be fine.
  - Field-programmable gate arrays (FPGAs) without on-chip flash are configured to accept a configuration code from an attached serial flash memory, well specified in their datasheets.
- The simplest volatile memory is asynchronous SRAM, or its popular look-alike, pseudo-static RAM. Standard SRAM and PSRAM speeds are ~50 ns, with "fast" SRAM speeds around 10 ns; the latter are good for external cache memories, while the former are still used in niche applications like high reliability medical electronics. Micropower SRAM can retain data at standby currents of ~1 μA, and, for leisurely applications, run at currents of ~1 mA.
- Synchronous SRAM has a more complicated interface, but runs at high clock speeds (to 400 MHz, combined with double data rate); good for external cache.
- Synchronous DRAM, in its double data rate varieties, is the favorite for fast computer working memory. Generally used as plug-in modules, but can be used as bare ICs, for example in conjunction with an embedded microprocessor (or FPGA with internal microprocessor core) in a stand-alone application like a wireless router, game console, display panel, or set-top box.
- Flash NAND is the current winner for dense non-volatile storage, and dominates in modules (USB thumbsticks, CF and SD cards) and as a hard-disk replacement (SSD), as well as for on-chip (microcontrollers, and application-specific ICs) and on-board (consumer electronics, video, network, etc.) applications. A particularly easy-to-use

---

known state, sensing if there was a change of state, then restoring (if needed) the original state.

form is provided by serial-interface flash, both as stand-alone ICs and, pleasantly, in the form of the SD card (which has an SPI-compatible interface); you can hook these serial memories directly to most microcontrollers, which include an integrated 3-wire SPI port.

- Flash NOR has the merit of a standard asynchronous SRAM interface, so you can substitute them for SRAM and execute read-only code directly.

- The byte rewritability of EEPROM is well suited to storage of parameters and tables, particularly where only a modest amount of data is involved.

- There is hope, in some recent technologies, for finding memory's holy grail: full-speed non-volatile memory with unlimited endurance. (We had that, a half-century ago, with magnetic-core memories. They're OK, if you're happy with cycle times measured in microseconds, densities measured in kilobytes, and prices in kilodollars. But, we've become jaded by the dazzling speed, density, and low cost of semiconductor memory; there's no going back... .)

14.5 Other buses and data links: overview

The PC104/ISA peripheral bus that we've seen in detail exemplifies a *parallel* multidrop bus architecture, with a group of shared DATA lines, ADDRESS lines, and some STROBE signals that signal direction and timing of data transfer (along with additional lines for interrupts and DMA). It originated more than 25 years ago, and has been superseded by faster and wider parallel buses, notably the PCI and PCIe[42] peripheral buses used in contemporary PCs.[43]

But the generality of a parallel data bus extends far beyond the innards of computer motherboards. Parallel data transfers, sharing the PC104's concepts of data–address–strobe, are used more generally in electronic devices for data transfer among devices such as liquid-crystal displays, video processing chips, and analog/digital converters, as we'll illustrate shortly. These connections between circuit devices often omit cumbersome addressing, and are better thought of as parallel data *links*. True parallel buses, however, can also extend to *external* data-hungry peripherals

such as tape and disk storage, or test and measurement instrumentation, in the form of the SCSI or general-purpose interface buses (GPIBs).

And then there are the increasingly popular *serial* buses and data links, going from the simple (and slow) legacy RS-232 ("COM" port), to the fast USB and FireWire protocols. Serial links send their data bits sequentially, rather than as a byte-wide (or more) wholesale multibit transfer; so you might expect them to suffer in speed. Remarkably, however, fast serial protocols recover most of that speed by using low-voltage differential signaling at very high bit rates. A fine example is the evolution of the 16-bit ATA (also called IDE, or PATA) parallel disk link (a total of 40 wires, in the once-familiar ribbon cable) into a serial "SATA" link with just two pairs in a skinny cable. The SATA link is actually the *faster*: it can transfer up to 6 gigabits/second, whereas the now-obsolete "PATA" tops out at 1 gigabit/second.

More speed in a *serial* link – how can that be? Differential low-voltage signaling certainly helps, but there are two additional factors favoring serial links like SATA: (a) such links are "point-to-point" (one driver at the transmit end and one receiver at the other end), rather than "multidrop" (one driver somewhere on the bus and multiple receivers along the bus); and (b) unlike a multiwire parallel bus, a single serial line has no timing *skew* (the scatter of propagation times between the clock lines and the several data lines). These factors really matter at contemporary data rates of gigabits per second: (a) At such speeds the bus lines are electrically "long" (electrical signals travel $\sim$20 cm in a nanosecond in cables and on circuit boards), and so you've got to think of bus wires as *transmission lines* (Appendix H), on which multiple taps (called "stubs") act as impedance mismatches and generate a sequence of reflected signals; by contrast, a point-to-point connection has but one recipient, permitting good impedance matching. And, (b), at these speeds the unavoidable timing skew sets an upper limit on data clocking rates (as revealed by an "eye diagram" – Figure 14.33) in parallel buses, nicely circumvented in serial links in which the clock is recovered from the data stream itself. As it turns out, the advantages of controlled impedance and absence of timing skew outweigh the advantages of parallelism.

This move from parallel (and multidrop) to serial (and point-to-point) is seen broadly: in computer buses (PCI $\rightarrow$ PCIe), in disk interfaces (PATA $\rightarrow$ SATA; SCSI $\rightarrow$ SAS), in Ethernet cabling (coaxial multidrop $\rightarrow$ twisted-pair point-to-point), and in external buses (e.g., GPIB $\rightarrow$ USB). In addition to performance benefits, these serial

---

[42] PCIe is a hybrid: serial communication on a set of parallel "lanes."

[43] The latter are considerably more complex; we chose the PC104/ISA for its simplicity, and also because it is well supported in contemporary PC104 offerings, which will likely ensure its survival well beyond the lives of its forgotten contemporaries. The latter had names like Unibus, STD bus, EISA bus, MicroChannel, Q-bus, Multibus, VAX BI, NuBus, Futurebus, and Fastbus.

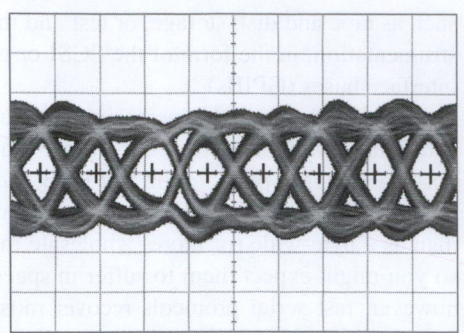
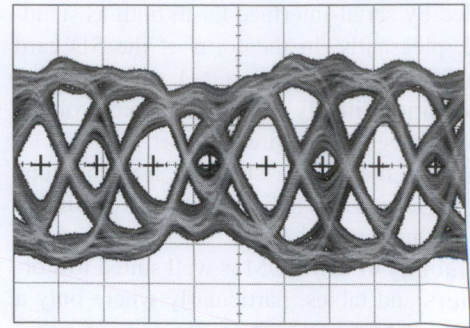

Figure 14.33. An *eye diagram* is formed from persistent 'scope captures of the voltage on a single data channel, triggered by the clock (whether recovered by the receiver or transmitted along with the data). This actual 'scope capture shows the signal at the far end of a 60 cm length of 0.085″ semi-rigid coax, driven by an 11.2 Gb/s (!) pseudorandom data stream. Left: with transmit equalization – the open "eyes" indicate adequate signal margin at the equally spaced clocking points (indicated with a "+" symbol). Right: without transmit equalization – a good example of a bad eye. Horizontal: 71.4 ps/div; vertical: 125 mV/div, differential. Oscilloscope bandwidth should extend to the 3rd (or even the 5th) harmonic of the clock frequency to display an accurate eye diagram. (Courtesy Hayun Chung.)

links are less expensive, with their small connectors and cabling.

Note that these serial links, though lacking address lines, can act logically like classical addressed buses, by streaming addressing bits along with the data. Examples are USB, FireWire, SAS (serial attached SCSI), and eSATA (external SATA).

Finally, serial protocols are commonly used for communication between ICs; these have names like SPI (Serial Peripheral Interface), I²C (Inter-IC), and JTAG (Joint Test Action Group).

Table 14.3 provides a helpful (we hope) organization and summary of the characteristics of commonly used data buses and links. In this section we'll look at these buses and data links, with examples to put some meat on the bones.

14.6 Parallel buses and data links

14.6.1 Parallel chip "bus" interface – an example

The familiar LCD "character displays" that you see on many devices originated with a simple parallel bus, which has become a standard (they now come also in serial bus variants). It is very simple: 8 DATA lines (D7..0), 1 (!) ADDRESS line ("RS"), a R/W′ line, and a data STROBE ("E"). The DATA lines are bidirectional; the other three are one-way, as shown in Figure 14.34. The 1-bit RS (address) selects either the display's internal *instruction* register (adr=0) or its *data* register (adr=1). We saw these earlier, in Chapter 12 – see, for example, Figure 12.78.

The operation is simplicity itself: a character is written to the display by setting its target address to the data register (RS=1=HIGH), asserting R/W′ LOW, then putting its *character code byte* on the DATA lines and pulsing the strobe line E. The display is then "busy" (which you can find out by doing a READ from address 0: the BUSY flag is returned as bit D7); you can send the next character when BUSY becomes false.

That's how you send characters. You can also do things like clearing the display, advancing the cursor, or determining where successive characters are displayed, by writing *command bytes* to the instruction register (adr=0). For example, writing 01h clears the display; writing 10h moves

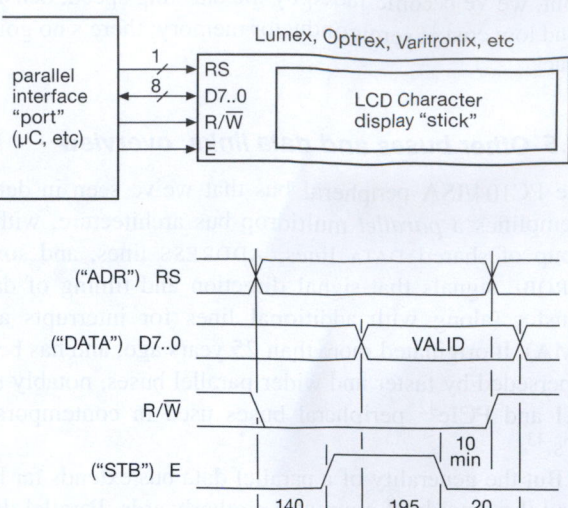

Figure 14.34. LCD module with parallel interface, with WRITE-cycle timings (in ns). The READ cycle is similar, but with R/W′ asserted HIGH (and D7..0 asserted by the LCD).

Table 14.3 Common Buses and Data Links[a]

	Par/Ser	PP/MD[b]	Mb/s (max)[d]	Devices per ch	Address bits	Length max (m)	Hot swap?	Used for	Comments
computer internal									
PC104/ISA	P8,16	MD	-	-	20, 24	-	N	peripheral cards	original IBM PC/XT (ISA), in PC104 format
PCI	P32,64	MD	1000, 2000	-	32	-	N	peripheral cards	obsolescent
PCIe	P1,2,4,8,16	MD	4000 per "lane"	1	-	-	N	graphic & peripheral cards	multiple serial unidirectional lanes [e]
IDE/PATA	P16	PP	1000	2	-	0.45	N	hard disk and optical drives	obsolescent
SATA	S	PP	1500, 3000, 6000	1	-	1	Y	hard disk, ssd, & optical drives	hot-swap; also external: eSATA
SCSI	P8/16	MD	320, 2500	8/16	-	12	note f	hard disk and tape storage	ultra2/ultra-320; also external
SAS	S	PP	3000, 6000	4	-	8	Y	hard disk drives	"serial attached SCSI"
computer peripheral (external)									
4-20mA	S	PP	110 baud	few	-	10k	Y	slow devices, "legacy"	"current loop," robust in noisy environments
RS-232C	S	PP	0.1	1	-	~30	Y	instruments, computer port	"COM" ports; USB-to-serial adapters
RS-485	S	MD	0.1, 10	32	-	1000, 10k	Y	process control	see graph, end Chapter 12
parallel (printer)	P8	PP	2.5	1	-	10	Y	printers; obsolete	original printer port; USB-to-parallel adapters
GPIB	P8	MD	8	15	-	20	Y	test/measurement insts	fat cable!
SCSI	P8,16	MD	320, 2500	8/16	-	12	Y	hard disk and tape storage	ultra2 / ultra-320; also internal
eSATA	S	PP	3000	15	-	2	Y	hard disk and optical drives	hot swap; external SATA
Firewire	S	PP	480,800,(3200)	63	64	10	Y	disk storage, live video	full duplex, repeater topology, IEEE-1394
USB 1.1	S	PP	12	127	11	5	Y	computer peripherals	half-duplex, tiered star topology
USB 2.0	S	PP	400	127	12	5	Y	...including hard drives, etc	
USB 3.0	S	PP	4800	127	-	3	Y	...including solid-state drives	full-duplex on added SuperSpeed pairs
Ethernet	S	PP	10, 100, 1000	1	48	100[h]	Y	computer network	original coax was multi-drop
WiFi	S	PP	6 to 54, to 600[i]	1	48	180	Y	wireless ethernet	IEEE 802.11a, g, n
Bluetooth	S	PP	1, 2, 3	8	48	10	Y	wireless peripherals	range: 1m at 1mW to 100m at 100mW
Zigbee	S	MD	0.045, 0.25	240[j]	64	30	Y	small nearby wireless devices	15-hop mesh network, IEEE 802.15.4
CAN	S	MD	1.0, 0.01	64	note k	1k, 40	Y	auto, factory floor, lab	8-byte payload per packet
LIN	S	MD	0.02	16	note l	40	Y	automobile	one wire, sub-network under CAN
inter-chip									
parallel[c]	P4,8,16	MD	to 3000	NA	-	-	-	general purpose, fast	
LVDS	S	PP	to 6000	NA	-	-	-	IC to IC, backplane, fast	
I2C / SMBus	S	MD	0.01,0.1,0.41,3.4	112	7	-	-	IC to IC, addressable	2-wire bidirectional, with handshake and adr
SPI	S	MD	>20[g]	1	-	-	-	simple IC to IC	4-wire, full-duplex, no handshake, no address
JTAG	S	PP	~100	many	7	-	Y	diagnostic, program loading	IEEE 1149.1
1-wire	S	MD	0.1	many	64	100	Y	sensors	Dallas; power and data on one line!

Notes: (a) there are more! these are commonly used, and likely to endure. (b) PP= point-to-point, MD = multidrop. (c) generic, no standards. (d) assumes no overhead.
(e) PCIe uses multiple parallel "lanes" of unidirectional serial links; the number of lanes is written "Xn": X1, X2, X4, X8, X16. Each lane accommodates 4 or 6 Gb/s.
(f) yes for SCA-2 connectors. (g) chip dependent, no established standard. (h) 100base-Tx. (i) four 64-QAM spatial streams, 40 MHz channel width.
(j) up to 65534 is possible. (k) with DeviceNet protocol layer. (l) communication initiated by master.

the cursor one character to the left; and writing 06h specifies that each successive character is written to the right of the previous character. The same protocols are used in a compatible display variant that substitutes a handsome "vacuum-fluorescent" technology. See §12.5.3; or take a look at §15.10.4 to see a frivolous example.

14.6.2 Parallel chip data links – two high-speed examples

The LCD hardly needs the extra speed of a parallel bus; in fact, its interface permits byte transfers to be carried out as two successive half-byte (4-bit "nybbles") transfers. But sometimes you really need the speed, for example with a really fast digital-to-analog converter like the one shown in Figure 14.35. The 8-bit AD9748 DAC converts at rates up to 210 Msps (megasamples/sec), so you've got to feed it new bytes every 5 ns. The chip obliges by having a byte-wide input port, combined with a differential clock input that accepts high-speed LVDS (low-voltage differential signaling). Unlike the LCD, this is not a *bus* in any sense, because there's no bidirectional communication and no addressing (not even a chip select or enable pin, as with the LCD). Note the speedy timing, with setup and hold times of 2.0 ns and 1.5 ns.

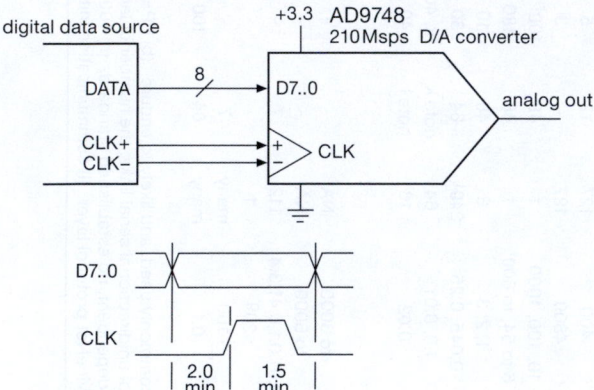

Figure 14.35. Fast parallel data link to a 210 Msps DAC.

Figure 14.36 shows another example with similar speed requirements – a video encoder chip that can handle the full bandwidth demands of high-definition video – but additionally with the need to specify literally hundreds of associated video parameters (such as video input and output formats, color and contrast corrections, test patterns, and so on). So, it has a 16-bit wide parallel input port, with frenetic nanosecond-scale setup and hold times. But it also has a leisurely serial port, which can accept either the I^2C

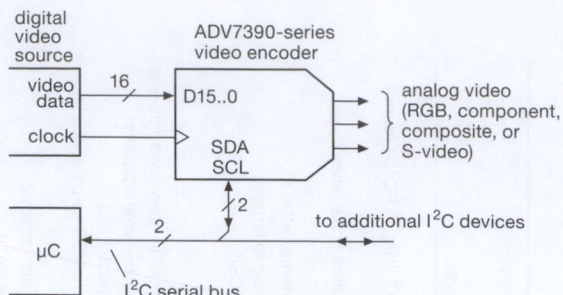

Figure 14.36. Video encoder with fast 16-bit parallel video-data input and 2-wire I^2C serial bus configuration port.

or SPI serial protocols and which provides access to the ~250 internal byte-wide configuration registers for setup. This hybrid approach takes advantage of the flexibility and compactness of a serial bus, while at the same time using a dedicated parallel input port to accept a firehose of digital video data.

We discuss serial buses and data links in §14.7, below.

14.6.3 Other parallel computer buses

As we remarked above, the buses within computers that are used to move data to peripherals and to memory have evolved, and continue to evolve. Contemporary buses like PCI and PCIe move data quickly (1 Gb/s to 32 Gb/s), with widths of 16 to 64 bits. PCIe ("PCI express") is actually a parallel/serial hybrid, and is emblematic of a shift toward *serial* data communication: the PCI that it replaces used a conventional bidirectional multidrop parallel data path (32 or 64 bits wide, with 32 address bits), whereas the newer PCIe uses 1 to 16 "lanes" of unidirectional point-to-point serial-link pairs, each of which runs at a speedy 2.5 Gb/s (PCIe v.1), 5 Gb/s (v.2), or 8 Gb/s (PCIe v.3). Timing is carried with the data, on each serial link, by coding the data so that the recipient can carry out "clock recovery" without separate clocking or strobing signals.[44] This is necessary for high bitrate communication; otherwise timing *skew* (the *difference* in propagation times of the clock and the several data lines) limits the maximum bit clocking rate.

Parallel data buses exist also within the silicon chips themselves: all processors contain internal data paths to send data among the registers and arithmetic units, to and from the on-chip memory caches, between the processor and external memory, and so on; you'll see names like

---

[44] PCIe uses "8b/10b" encoding, with each 8-bit byte sent as a 10-bit symbol, to generate a dc-balanced and "run-length-limited" bitstream in which there are never more than five consecutive 1s or 0s.

AMBA and Wishbone. As processors grow in speed, their external buses have followed; the current configuration is a fast and wide "Front Side" bus (FSB), which requires support chips ("Northbridge" and "Southbridge," likely to merge into a single unified support chip) to split off separate buses for memory, fast peripherals (PCIe bus), and slower peripherals (SATA, PCI, USB, Ethernet).

Another trend in contemporary computer designs is the integration of more functions, and more complexity, within the processor chip itself. Thus we have the delightful and inexpensive micro*controllers*, with their spectrum of on-chip "peripherals," ready to drop into a circuit or instrument with hardly a thought of buses or interfacing. Embedded microcontrollers are the subject of the next chapter.

Yet another approach dispenses with the dedicated processor altogether, exploiting the power of configurable logic (Chapter 11) in massive FPGA "system-on-a-chip" (SoC) devices. These permit flexible system design (including on-chip "soft" processor cores, internal buses, memory, and peripherals), with plentiful wide-bandwidth I/O pins. An example is the Xilinx Virtex series of FPGAs, topping out at several million flip-flops, tens of MB of on-chip RAM, and more than 500 LVDS differential ports. In additional to soft configurable SoCs, there are application-specific standard parts (ASSPs, see §11.4.2) for many consumer applications. For example, there's Broadcom's BCM7405 "Multiformat HD Digital Video/Audio SoC for Satellite, IP and Cable STBs with Picture-in-Picture" (the title says it all[45]).

14.6.4 Parallel peripheral buses and data links

Although the industry is moving from parallel to serial buses, there are still a few parallel buses in use.

A. PATA (IDE)

The 16-bit-wide PATA (parallel ATAPI, also called IDE) was for many years the standard connection to internal hard disk and optical (CD, DVD) drives. The wide 40-wire ribbon cables and connectors are somewhat cumbersome, and even after the introduction of 80-wire cables (adding interleaved ground lines to improve signal integrity) the data rates topped out at 133 MB/s ($\sim$1 Gb/s). PATA is obsolete, replaced with the serial SATA format.

B. SCSI

The 8/16-bit parallel small computer system interface dates from the 1980s, with enhancements in speed and signaling format (originally 5 V single-ended, with low-voltage differential added subsequently[46]), and was until recently the preferred interface for high-performance disks and tape storage devices. Internal SCSI buses use 68-wire ribbons with dual-row socket headers; external SCSI uses shielded multipair cable with high density 68- or 80-pin connectors.[47] Although SCSI bandwidth improved with each generation, it has been overtaken in speed and convenience by the serial formats: SAS (serial-attached SCSI), and SATA.

C. GPIB

The IEEE-488 general purpose interface bus was originated by HP for control and readout of test and measurement equipment; many instruments (e.g., from Agilent, Keithley, National Instruments, and Stanford Research Systems) continue to support GPIB control. It uses a robust stackable 24-pin connector and permits lengths to 20 meters. You can get GPIB adapter cards for the PCI and PCIe computer buses; or you can get adapters such as Ethernet-to-GPIB and USB-to-GPIB.

D. Printer ("Centronics") port

Last (and least) is the parallel "printer port" (LPT port), which was standard for two decades, beginning with the original IBM PC, where its I/O address was 378h (and still is). It was also used for external modems and for software "dongles." The original parallel interface was unidirectional (data OUT, with a few handshake lines), but extensions of the standard (IEEE 1284: ECP, EPP) made it bidirectional, faster, and more bus-like. It is quite thoroughly

---

[45] But in case you want to hear more, the product page has this to say: "The BCM7405 is a high-performance, high definition (HD) satellite, cable, and IP set-top box DVR System-on-Chip solution. It combines a fast 1100-DMIPS MIPS32/MIPS16e-class CPU, high-speed graphics processing, including video scaling and motion adaptive deinterlacing, a very flexible data transport processor, an MPEG-4/VC-1/MPEG-2/AVS-compliant video decoder, a programmable audio decoder, six video DACs, stereo high-fidelity audio DACs, dual Fast Ethernet ports, one with integrated PHY, triple USB 2.0, a PCI 2.3/Expansion Bus, a high-speed 400-MHz DDR2 memory controller, and a peripheral control unit that provides a variety of set-top box control functions. Integrated in 65-nm technology, it offers one of the highest levels of single-chip system performance available for STB applications."

[46] Differential signaling requires twice as many wires, but it is superior in terms of low crosstalk, excellent noise immunity, and far smaller ground and supply noise (because of balanced transitions); these improved characteristics let you use relatively small voltage swings, with correspondingly smaller drive currents.

[47] The miniature high-density connectors that were developed for advanced forms of parallel SCSI have come to be very useful in other fields, such as I/O connectors for computer DAQ boards for laboratory use by National Instruments and others. These are made by HRS Hirose, Honda Connectors, and others.

obsolete: printers use USB or Ethernet connections. Computers now rarely include a parallel port; if you need such a connection, you can get USB-to-parallel adapters.

That said, however, the parallel port (emulated with a USB adapter, say) is the simplest way to wiggle some bits from a PC. If you're running some version of Windows, you'll need a driver[48] that lets you get at the hardware (these operating systems don't let you get at hardware ports directly, as you were able in the good ol' days of DOS). Then you can write really simple code, like this example of Python[49] running in terminal mode:

```
>>> import parallel
>>> p = parallel.Parallel() # open LPT1
>>> p.setData(0x55)
```

You can use a compiled language, instead; some folks we know are fond of PowerBASIC, in which you can use in-line assembly language in a BASIC subroutine to send a value out to an addressed port. It looks like this:

```
Sub PortOut(ByVal PortNum as word, Byval Value
as byte)
    ASM MOV AL, Value
    ASM MOV DX, PortNum
    ASM OUT DX, AL
End Sub
```

14.7 Serial buses and data links

Serial buses and data links have several important advantages, two of which we saw earlier: (a) the convenience of fewer wires in the cable and connector (think of a skinny USB cable, compared with the elephant's trunk of GPIB or SCSI), as well as fewer pins on the driver and receiver chips; (b) high intrinsic bitrate, because of absence of timing skew (self-timed via clock recovery) and clean line termination (if point-to-point). In addition, (c) a one-wire serial link is easily conveyed by optical fiber or by wireless transmission. And, if you want parallel bits at either end, there are chips generically known as SERDES (serializer–deserializer, pronounced "ser'-deez"), which convert a serial stream to parallel, and vice versa (see also §12.8.4 and §12.10.3). As examples of the latter, FTDI Ltd. offers the popular FT245 and FT2232, which convert between a relatively low-speed USB and a simple byte-wide parallel port, with a built-in first-in, first-out (FIFO) buffer; high-speed examples include the DS92LV18 18-bit SERDES (with

speeds to 1.2 Gbps), or the generic SERDES used in the PHY (physical layer – i.e., the driver–receiver–switch ICs) of gigabit ("1G") and 10 gigabit ("10G") Ethernet links.

In the subsections that follow we describe most of the important themes in serial links, with examples from those in common use. As with the parallel buses, we look first at internal serial protocols (chip-to-chip, and within an instrument), then at the external serial buses. Within these categories we've ordered them roughly by increasing complexity, e.g., going from the simplest (and slow) 4-wire clocked link (SPI) to the complex (and fast) 1-wire 8b/10b-coded clock-recovery links used in SATA and PCIe.

See the next chapter for specific chip suggestions to implement these protocols with a microcontroller.

14.7.1 SPI

The serial peripheral interface (SPI) was introduced by Motorola, and is widely used for communication *between ICs* (for which the other popular standard is I^{2}C, discussed below).[50] It is organized as a master–slave protocol (like the PC104 bus), but uses only 4 wires (Figure 14.37): one clock, two data lines (one in each direction), and a chip select. They are named SCLK, MOSI (master out, slave in), MISO (master in, slave out), and SS' (slave select; active LOW).[51] There are several ways of hooking things up, but most commonly the scheme of Figure 14.38 is used: the clock and data lines are bused to all slave chips

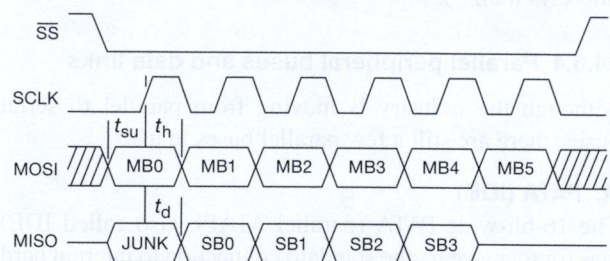

Figure 14.37. A typical SPI protocol, with bits in both directions clocked on the rising edge of SCLK. With SPI, the quantity and meaning of the bits are specific to the slave device, which here accepts 6 input bits from the master (MB5..0), and simultaneously outputs 4 other bits (SB3..0).

---

[48] Examples are DirectIO.exe, or InpOut32.dll.

[49] "setData(value)" is one of several bit-banging function calls in the py-Parallel API; the documentation describes it as "Apply the given byte to the data pins of the parallel port."

[50] You see SPI and I^{2}C on chips like sensors, converters, nonvolatile memory, analog switches, and digital potentiometers, to be controlled from a microcontroller, microprocessor, or other digital link.

[51] Sometimes with alternative naming: SDI, DI, or SI for data IN, and likewise for data OUT, with the signal names corresponding to the direction of data *at that IC*. For example, the MOSI pin on the master would connect to the DI pins on the slaves.

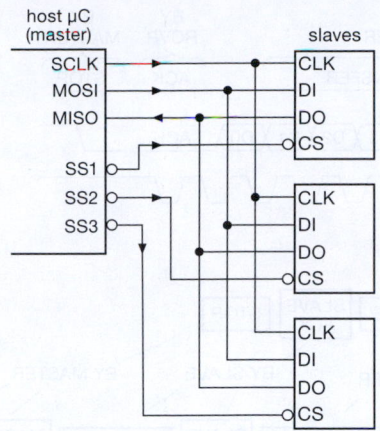

Figure 14.38. The common bused SPI configuration: clock and data lines are shared, with individual SS′ (slave-select) lines asserting the chip-select inputs of the corresponding slaves.

("multidrop"), with a separate dedicated select line to each slave.

The master controls all transfers, first asserting SS′ for the chosen slave IC (with the SCLK line in its resting state), then generating successive clock pulses, each of which enables a bitwise transfer of data on MOSI and MISO for that chip only. There is no fixed protocol for what the data represents, how many bits are to be sent, etc. What happens, instead, is that a particular chip specifies the meaning of the serial bits sent to it and of the bits it simultaneously sends back.

To give an example, the AD7927 is a 12-bit ADC of modest speed (200 ksps), with a built-in 8-input multiplexer, and with a SPI serial port (Figure 14.39); the latter both controls the conversions (e.g., selecting the input channel, voltage range, output coding, etc.) and also delivers the converted digital outputs. This particular chip loads the first 12 input bits (after SS′ assertion, which also initiates conversion) into its control register (ignoring later bits), and it simultaneously sends back the result of the previous conversion as a 16-bit string, as shown in Figure 14.39.[52] See Figure 15.21 for some examples of SPI peripheral chips that are well suited to microcontroller applications.

A. Some comments

The SPI protocol is free form in its content (how many bits are sent, and what they mean), as this example illus-

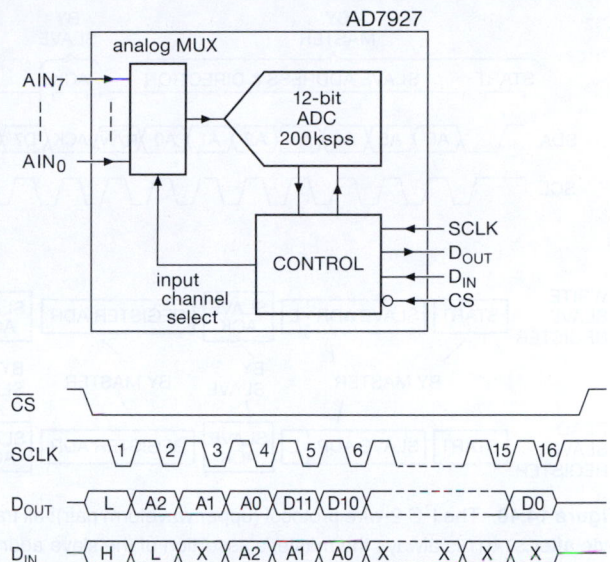

Figure 14.39. An 8-channel ADC with SPI control and readout. The protocol shown is the simplest of several allowed modes: the master's CS′ starts a conversion, whereupon the slave outputs both the channel address (3 bits) and the converted value (12 bits); the master simultaneously sends the address of the next input channel to be converted.

trates. SPI doesn't have any intrinsic addressing to designate where data is going within the target chip, so the usual scheme is to send a string of bits that gets shifted into sequential internal bit locations, with the datasheet defining how they sort out internally (the popular I^2C alternative, below, takes a different approach). Some chips may be write-only (e.g., an LCD with serial input), others read-only (e.g., the Maxim MAX6675 thermocouple-to-digital-converter chip, with SCLK, MISO, and SS′ only: you don't tell *it* what the temperature is, it tells *you*). Some chips may invert the clock polarity, and also which edge clocks the data (this produces four possibilities, known as SPI *modes*; the above illustration used mode-2). SPI has simple timing, and full-duplex data transfers (i.e., in both directions simultaneously and independently); it has no required "handshake": a master can send lots of data – to a nonexistent chip!

Because SPI (and SPI-like variants) do not conform to a well-defined standard, you have to read carefully the datasheet specifications for each interfaced chip. You'll discover, in addition to the polarity modes already mentioned, that the maximum (and minimum!) clock speeds can range from a few kilohertz to many megahertz. The AD7927, for example, specifies $f_{SCLK}=10$ kHz (min), 20 MHz (max). With several SPI chips in a system, you

---

[52] The AD7927 can do numerous additional tricks, such as cyclically converting a prescribed arbitrary sequence of input channels; you program that sequence by loading a "shadow register," which you access through two bits of the control register.

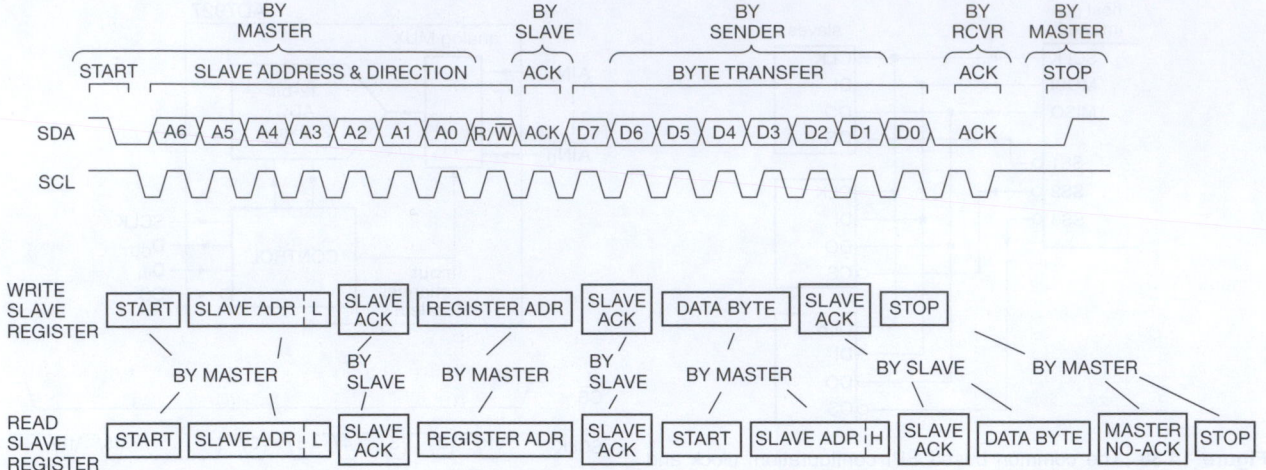

Figure 14.40. The I²C 2-wire protocol (upper waveform pair): all transfers are in 8-bit groups, with a 1-bit acknowledgment (ACK). The first byte after START is always the master's assertion of the slave address (A6..0) and direction (R/W′); subsequent bytes flow from sender to receiver (depending on that first R/W′), asserted by sender and acknowledged by receiver. The master can access registers within a slave device by sending their internal address as a data byte, as shown in the lower block diagrams; both a WRITE and a READ are illustrated, the latter requiring a "repeated START" to create a READ after writing the register's address in the initial transaction.

may have incompatibilities among them that force you to write code to assert and disassert port bits "by hand" (this is called *bit-banging*, rather than using the microcontroller's built-in SPI interface).

A widely used alternative to SPI is the I²C peripheral interface, discussed next.

14.7.2 I²C 2-wire interface ("TWI")

The Inter-Integrated-Circuit (IIC, I2C, or I²C) serial interface bus was originated by Philips (now NXP), for communication between chips.[53] It differs from SPI in several ways: (a) it uses only 2 wires, which are bused to all slave ICs (there is no separate chip select like the SS′ that is used in SPI); (b) addressing is sent (first) on the same line as data is sent or received; (c) the bus is "half-duplex" – that is, data can move in only one direction at a time (the direction is specified by a bit following the address); (d) although I²C is a master–slave architecture (like SPI), any device on the bus can become master when the current master relinquishes control (by sending the stop bit that terminates its session with a particular slave).

Figure 14.40 shows the protocol. The 2-wire I²C bus consists of a clock line (SCL) and a data line (SDA), both with resistive pullups to V_+. SCL is asserted by the master, whereas SDA is bidirectional: it is asserted by the master

to specify the slave's address (7 bits) and the direction of transfer (1 bit); the slave then sends an acknowledge (ACK) bit, following which one or more data bytes move from master to slave, or from slave to master (always clocked by the master), depending on the direction-of-transfer bit that was specified initially with the slave's address. The session ends when the master sends a stop bit following the last byte transferred.[54] START and STOP commands are created by violating the normal convention of "data can change only during clock LOW."

To give an example, the AD7294 "12-bit Monitor and Control System with Multichannel ADC, DACs, Temperature Sensor, and Current Sense" is a do-everything chip for applications such as automobiles, industrial controls, and cellular base stations (Figure 14.41). It isn't the fastest kid on the block – a mere 300,000 conversions per second on its ADC – but it will keep an eye on the whole shop, reporting back to the mother ship via an I²C port. The 44-page datasheet tells you how to communicate with its

---

[53] The closely related SMBus enforces tighter standards, both in protocol and in its electrical signaling.

[54] You can think of this whole process as the serial analog of a PC104/ISA data transfer: in the latter, the master asserts address on the A19..0 lines, and direction on the IOW′/IOR′ lines. If a WRITE, the master asserts the data on the bidirectional D7..0 lines; if a READ, the addressed slave asserts the data on those same line. In either case the transfer is clocked by the master's IOW′/IOR′ strobe. In I²C the same steps take place, but in serial sequence on the single bidirectional SDA data–address–direction line, clocked by the single unidirectional SCL clock line.

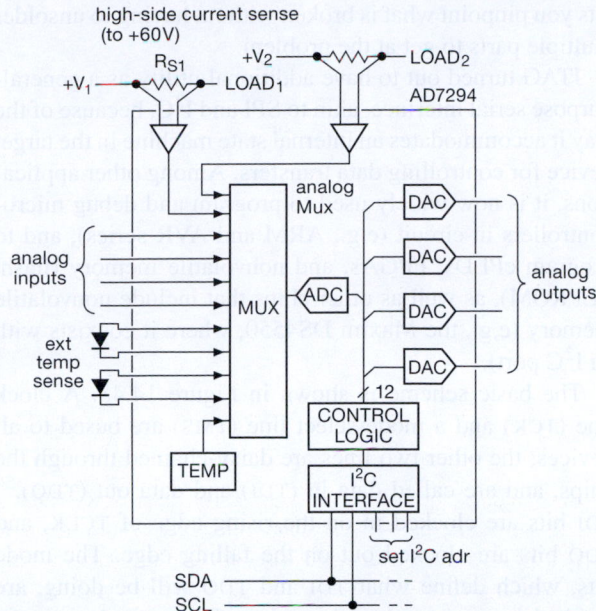

Figure 14.41. A multifunction monitor chip, rich with features but of modest speed only, with I²C control. The master can reach each of the 40 internal registers by sending the corresponding register address as the second byte of a transmission, as shown in Figure 14.40.

40 internal registers, with amusing names like "AlertRegisterA(R/W)" and "DATA$_{HIGH}$RegisterT$_{SENSE}$INT(R/W)." See Figure 15.22 for some examples of I²C peripheral chips that are well suited to microcontroller applications.

A. Some comments

The 2-wire I²C protocol is well defined and is economical of wiring, particularly when you need to include many chips on the bus, because the 2 wires carry all data, addressing, and clocking. In addition, it allows the slave to slow down the master by doing "clock-stretching" (by holding SCL LOW; this is called *flow control*), and it allows multiple bus masters. It is particularly well suited to tasks in which you want to take aim at a particular register in a chip that is endowed with many; with the AD7294, for example, you would have a 3- or 4-byte transaction: the first byte is the chip's bus address, the second is the address of the chip's internal register, and the last byte (or two) is a write (or read) to (or from) that internal register.

B. Comparison with SPI

When compared with SPI, however, I²C is a more complex protocol, and not as well suited to steady high-rate streaming of data. The flexibility of multiple bus master-

ship brings with it the problems of contention and arbitration. You have to give multiple devices unique addresses, which is commonly handled by including some dedicated pins to select among a built-in set (e.g., the AD7294 has 3 pins by which you can select any address from 61h to 7Bh), thereby defeating some of the advantage of low pin count. And the flexibility of addressing, bidirectional data, and bus mastership complicates debugging, as compared with the dead-simple SPI protocol.

Which to use? The choice is usually determined by the peripheral chip, which usually supports only one protocol or the other, whereas most microcontrollers include hardware support for both SPI and I²C (and if they don't, you can always do bit-banging in software).

14.7.3 Dallas–Maxim "1-wire" serial interface

The ultimate in reducing the number of wires is achieved in the 1-wire™ (plus ground) interface devised by Dallas (now merged with Maxim).[55] The single wire carries serial data and addresses, and also *power!* The way it does all this is by having data bits sent, bidirectionally, as brief pulses to ground, with each slave device having an on-chip capacitor to retain power. The goal is simplicity in interconnection to devices like temperature sensors, memory, converters, battery management, and so on (Figure 14.42). With only ground and data, the devices can be packaged in what Maxim calls iButton™, which looks just like a coin cell battery.

The protocol goes like this: multiple slave devices all bridge the common data line and ground, controlled by a master device (microcontroller or other digital interface). The line is pulled up to +5 V, which powers the slave devices and permits any device to assert a momentary LOW. The master initiates all transactions, asserting addresses and then either sending or receiving data. Data is encoded as pulse widths: a short pulse ($<15\,\mu$s) for 1, and a long pulse ($60\,\mu$s) for 0. If the master is sending, it simply asserts the bits with the corresponding pulse widths; if it is receiving, it sends the short pulse corresponding to a 1, and the addressed slave responds by either doing nothing (releasing the line to +5 V, therefore signaling a 1), or by holding the line low for $60\,\mu$s, creating a long pulse (therefore signaling a 0).

Each 1-wire device has a unique 64-bit address, given to it at the time of manufacture, which includes a byte indicating the type of device. The master resets the slaves

---

[55] Plenty of details in their Application Notes AN147, AN148, AN155, AN159, AN244, AN1796, and AN3358.

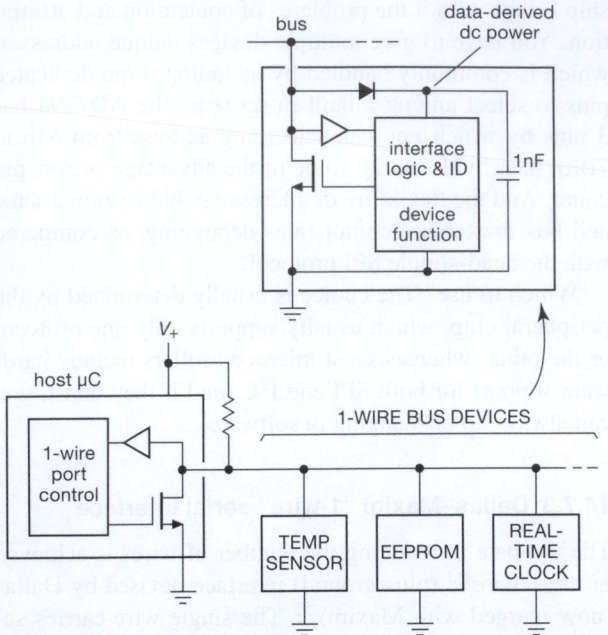

Figure 14.42. The 1-wire™data bus originated by Dallas–Maxim. An internal capacitor maintains slave power during the brief LOW data pulses.

with a long pulse (all "pulses" are LOW, i.e., to ground), then queries the bus to learn the attached addresses. The master can send broadcast messages to all attached devices, or it can carry out transactions with particular devices, according to their unique addresses. As with the I^2C bus, that address is sent (by the master) on the data line, followed by the data to be sent or received.

The Dallas–Maxim 1-wire bus can be used for connections to sensors, etc., outside of an instrument. It's normally limited to a maximum network distance of 30 meters, but it can work up to 500 meters with an appropriate driver (see AN244); and extensions by Ethernet are possible.

14.7.4 JTAG

A useful chip interface, with an interesting history, is the JTAG (Joint Test Action Group) standard, also known as "boundary scan," or IEEE 1149.1. It was devised in the 1980s to deal with the thorny problem of testing for component or connection failures in circuit boards using the then-new surface-mount and multilayer technologies; these made it increasingly difficult to get at the IC pin connections, not to mention the innards of the chips themselves. As such, it provides a way to look at the registers and data paths within the chips, using a simple 4-wire serial bus; that

lets you pinpoint what is broken without having to unsolder multiple parts to get at the problem.

JTAG turned out to have additional utility as a general-purpose serial interface, akin to SPI and I^2C, because of the way it accommodates an internal state machine in the target device for controlling data transfers. Among other applications, it is now widely used to program and debug microcontrollers in-circuit (e.g., ARM and AVR series), and to program cPLDs, FPGAs, and nonvolatile memory (flash, EEPROM), as well as other chips that include nonvolatile memory (e.g., the Maxim DS4550, where it coexists with an I^2C port).

The basic scheme is shown in Figure 14.43. A clock line (TCK) and a mode-select line (TMS) are bused to all devices; the other two lines are daisy-chained through the chips, and are called data in (TDI) and data out (TDO).[56] TDI bits are clocked in on the rising edge of TCLK, and TDO bits are clocked out on the falling edge. The mode bits, which define what TDI and TDO will be doing, are clocked in on TCLK's rising edge. (There is also an optional bused reset line, not surprisingly called TRST.) The bus runs at speeds in the range of 1–100 Mb/s, depending on the particular target chip(s).

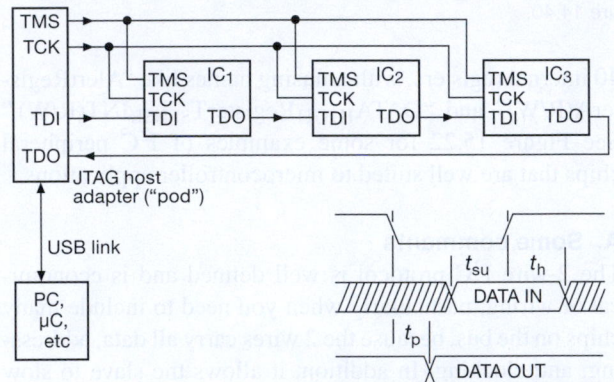

Figure 14.43. The JTAG "boundary scan" interface uses a pair of unidirectional bused lines (clock: TCK; mode select: TMS), and a daisy-chained data line passing through each slave (TDI, TDO).

Many PLDs and microcontrollers now use a JTAG port for programming; typically you get a programming "pod" that connects to a host computer via USB, and has a socket header to mate with a pin header on the target device's board. Sadly, the header configuration has not been standardized, although it's easy enough to make an adapter. However, the software that the chip manufacturers provide

---

[56] The letter "T" preceding each signal stands for "test," an indication of what the originators had in mind.

usually forces you to use their specific pod when talking to their chips. You use such vendor-supplied software, or open-source software, to load the compiled code into the target, and to carry out in-circuit real-time debugging, both via the same JTAG port.

14.7.5 Clock-be-gone: clock recovery

The SPI, I^{2}C, and JTAG serial links above, each use a separate clock line to time the data bits (SCLK, SCL and TCK, respectively). However, it is possible to do *clock recovery* from the data itself, if you arrange things correctly. This not only reduces the number of lines, it also permits higher ultimate data rates, because there is no issue of timing skew between clock and data. The remaining internal serial links (and, later, the external serial links) employ clock and data recovery (engineers' lingo: "CDR") from the transitions on the data line(s). The reader may wish to look ahead to §§14.7.9–14.7.12 for helpful information on how coding is used to make this work.

14.7.6 SATA, eSATA, and SAS

SATA is serial-ATA, and SAS is serial-attached SCSI. These are the fast serial buses for internal and external storage (disk, tape, and optical drives); they are replacements for the obsolete ATA (retroactively called PATA, or sometimes IDE) and SCSI, respectively. They share the same connector (though SAS offers additional connector types) and are hot-pluggable. Maximum data rate is currently 6 Gb/s, with upgrade paths to 12 Gb/s. SAS has some performance features not found in SATA, and is thus aimed at server (as opposed to consumer) applications; it basically continues the (parallel) SCSI protocol, but running on a serial connection interface.

These interfaces use low-voltage differential signaling, with 8b/10b encoding and clock recovery, and allow hot-swapping (although this feature requires operating system support to work correctly).

External SATA (eSATA) is an extension of the standard, for attaching external storage devices to a computer with SATA, using the same protocols. The physical connectors are different, however, being designed for greater ruggedness and signal integrity. External devices connected via eSATA currently require separate power, a nuisance likely to be remedied by the "Power Over eSATA" initiative.

14.7.7 PCI Express

PCI Express ("PCI-E," or "PCIe") was introduced in 2004, as a successor to the parallel PCI bus (and its brethren,

PCI-X and AGP), for interfacing to peripheral cards on a computer motherboard. It replaced the wide (32- or 64-bit) multidrop parallel architecture with a set of point-to-point serial "lanes," each lane consisting of two LVDS differential pairs (one pair for 1-bit serial communication in each direction, see §12.10.3). With multiple lanes there is parallelism (in that sense it is a hybrid), but the communication is basically serial, with 8b/10b coding and clock recovery, etc.

PCIe has blazing speed: currently 4 Gb/s *per lane* (PCIe v2.0, double that of the original version 1.1), with 1 to 16 lanes. Thus you could have up to 64 Gb/s to an x16 slot, if the hardware at both ends could handle it.

Let us take a brief aside, to wonder how we've come to this elaborate scheme, given the elegance of the traditional parallel bus with its economical sharing of data lines (as in the PC104/ISA bus). Roughly, it goes like this: in the beginning (say the 1970s) integrated circuits were not dense, and the DIP packaging limited the pin count to 16 pins (often) or 40 pins (occasionally). They also were not terribly fast, running at clock speeds of 10 MHz or so. These circumstances favored the shared parallel bus – less wiring, fewer driver–receiver pins, and still plenty of bus speed.

As chips became faster, and humans demanded more speed, the parallel buses were enhanced, with more width (the 8-bit ISA bus was broadened to 16 bits, then expanded to EISA: 32 bits), and, at the same time, with more speed. This was successful, up to a point. That point was near when the performance of the parallel PCI bus (the successor to the ISA family) had grown from its original 32 bits at 33 MHz (thus 133 MB/s) to 64 bits at 133 MHz ("PCI-X," 1064 Mb/s). A further enhancement ("PCI-X 2.0") quadrupled the speed, but was not widely adopted. Problems of timing skew and of reflections from the multidrop stubs prevented further gains from the bus. By this time (2003) computer designers saw a better way, namely the serial point-to-point PCIe architecture.

Of course, PCIe has its challenges: consider a motherboard with a couple of x16 slots (a common configuration, even in inexpensive computers). Each slot requires 32 differential pairs (64 wires), so you've got to connect to 128 wires, and deal with 4 Gb/s (actually, 5 Gb/s raw, because of the coding overhead) on each of the 64 pairs. That function (along with the memory bus) is carried out in the Northbridge chip, which has hundreds of pins and the necessary speed and complexity for the task: it even has its own heatsink. And one must not ignore the fabrication technology needed to connect all these pins: even inexpensive motherboards routinely have a half-dozen wiring layers, with traces 0.12 mm (5 mils) wide.

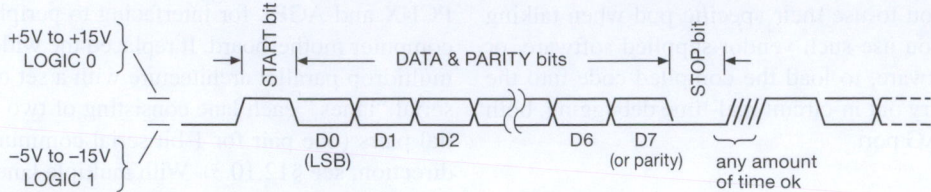

Figure 14.44. RS-232 serial data-byte protocol, which uses signal levels of both polarities. The two states are sometimes called *mark* (negative, logic 1) and *space* (positive, logic 0). You sometimes hear the descriptive phrases "LSB first" and "inverted levels."

In a sentence, then, the demand for greater bus performance coincided with the ability to proliferate point-to-point connections, to bring about the evolution from the elegant shared parallel bus... to the elegant coded point-to-point serial links!

This is clearly an area where casual circuit designers dare not tread – better just to buy some smarter person's finished design!

14.7.8 Asynchronous serial (RS-232, RS-485)

Turning now to *external* serial buses, we look first at the original (and long-lived) *asynchronous* serial link – used in the now-obsolete RS-232 serial port. We discussed RS-232 (and also RS-422 and RS-485) back in Chapter 12 (§12.10.3A), with emphasis on the physical layer (waveforms, noise immunity, and the like). Here we revisit RS-232, in the context of computer communication. These serial ports were popular for connection to external modems and to alphanumeric terminals like the legendary VT-100 (dismissively called *dumb* terminals). They have largely disappeared from computers, but you can get USB-to-RS-232 adapters; and you can get cards that plug into the internal bus and create a gaggle of RS-232 ports. The RS-232 designation ("Recommended Standard #232") refers to the *electrical* signaling scheme, which inconveniently uses voltages of both polarities to signal 1s and 0s (Figure 14.44; see also §12.10.4). However, the physical link need not be RS-232 – it can be the differential (and unipolarity) RS-422 or RS-485, or it can be an optical fiber (for galvanic isolation and immunity from environmental transients), or it can even be a 20 mA *current loop* (which is most often used for *analog* signaling).[57]

Whatever the physical transport medium, the signaling is straightforward: the transmitter rests in the *mark* state (logical 1), and springs to action with a START bit (logical 0, or *space*), followed (usually) by 8 data bits (which optionally may be 7 data bits plus a parity bit), followed by one or two trailing STOP bits (logical 1). The transmitter and receiver must agree on the bit rate and parity (if any); a common protocol, for example, is 9600 baud and "8N1," which means 8 bits, no parity, and one stop bit, transmitted at a rate of 9600 bits/s.[58] (Because of the overhead of the framing START and STOP bits, each byte during 8N1 transmission requires 10 bits, for a maximum net payload rate of one tenth of the baud rate, or 960 bytes/s.)

With this simple asynchronous serial coding, the receiver (whose clock runs at several times the baud rate) is triggered by the transition at the beginning of the START bit, waits for half a bit cell to be sure the START pulse is still present, then examines the bit value at the middle of each data cell (using the agreed upon baud rate interval); the STOP bit terminates the character, and is the resting state if no new characters are sent immediately. By resynchronizing on the START bit of each character, the receiver doesn't require a highly accurate clock; it need be only accurate and stable enough for the transmitter and receiver to stay synchronized to a fraction of a bit period over the duration of one character, i.e., an accuracy of a few percent.

There is one slight logical flaw in this pretty scheme, namely that a receiver may not be able to synchronize properly (that is, identify START/STOP) in an uninterrupted stream of data bytes. The best proof of this conjecture is a long sequence of the letter "U," which has the unfortunate distinction of being encoded as 01010101 (55h): plug that into Figure 14.44 (with the usual 8N1 setting) and you'll get... a *square wave*! A more serious flaw is the lack of standardization at the physical (electrical) level of RS-232: connector gender, hardware handshaking signals, and flavor of device ("DCE" and "DTE"; see Table 14.4 for official signal names and pin assignments). This is an eternal source of confusion because, too often, two RS-232 devices when connected together, won't work. We've all struggled with this, and readers have even complained to *us* (hey, give us a break, we didn't design RS-232; but we

---

[57] These alternatives permit longer cable runs (up to ~1 km) and, in the case of RS-422/485, a multidrop bus topology.

[58] The serial asynchronous standard permits a wider range: 5 to 8 data bits, with optional parity; so it would be legal to specify 8E1, for example. In practice you rarely see anything but 8N1.

Table 14.4 RS-232 Signals

Name	Pin number 25-pin	Pin number 9-pin	Direction (DTE ↔ DCE)	Function (as seen by DTE)	
TD	2	3	→	transmitted data	} data pair
RD	3	2	←	received data	
RTS	4	7	→	request to send (= DTE ready)	} handshake pair
CTS	5	8	←	clear to send (= DCE ready)	
DTR	20	4	→	data terminal ready	} handshake pair
DSR	6	6	←	data set ready	
DCD	8	1	←	data carrier detect	} enable DTE input
RI	22	9	←	ring indicator	
FG	1	-		frame ground (= chassis)	
SG	7	5		signal ground	

did at least provide a guide to "serial cables that actually work" – see Figure 10.17 in the previous edition of this book.). With finicky USB-to-RS232 emulators replacing the disappearing COM ports on contemporary computers, the problem has only gotten worse.

Figure 14.45 shows some 'scope captures of RS-232 waveforms: four single-byte captures from a stream of random[59] bytes, and a multibyte capture showing the invariant START and STOP bits.

The most common use of asynchronous serial links has been for alphanumeric data, in the standardized 7-bit ASCII (American Standard Code for Information Interchange) code of printable characters that is the norm for alphanumeric representation (Table 14.5).[60] However, any

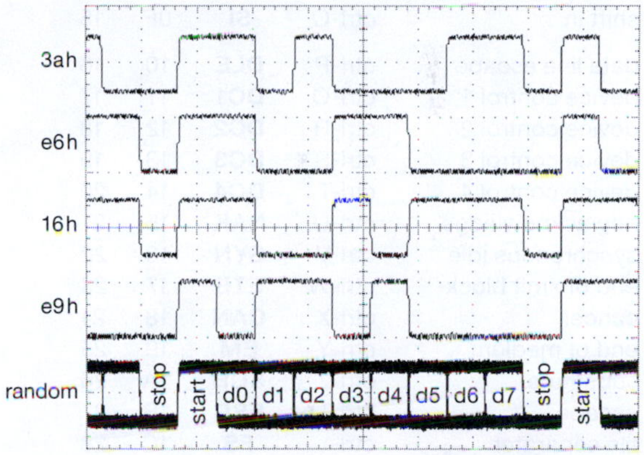

---

[59] *Truly* random: we fed analog noise to a comparator; see §13.14.7, and the "Museum" section on the *Numerical Recipes* Source Code CDROM (or www.nr.com).

[60] Standard ASCII has 128 characters, including the 32 "nonprinting" characters in the first column of Table 14.5. There are "extended ASCII" 8-bit character encodings that double the number of characters to include alphabetic characters from other languages, for example ë (decimal code 235), as well as symbols such as ° and ± (codes 176 and 177) and £ (code 163). A warning: there are multiple such extensions, for accommodating different language groups (some are gathered into Std. ISO-8859; others are proprietary); so they should be used with caution. Most software will accept these encodings, which you enter by holding the `alt` key while entering the 3-digit decimal code with a leading zero (thus `alt`-235 for e-umlaut). You can use this method also to enter (rather than act upon) any of the standard ASCII characters, including the nonprinting characters; thus, to enter a CR (carriage return) you would type `alt`-013. And good programming editors (such as Notepad++, UltraEdit, or Emacs) accept the insertion of control characters, in addition to displaying what's there. Try it!

Figure 14.45. Captured RS-232 waveforms from a random byte generator, 8N1 at 14.4 kbaud into a 2.2 nF load. The bottom waveform is an accumulation of multiple bytes, whereas the top four are single bytes (with indicated values). Horizontal: 100 μs/div; vertical: 10 V/div.

binary data can be so conveyed; you just won't be able to read it or print it directly. And serial communication via RS-232 is quite alive and well: many bench instruments include an RS-232 port for control and data transfer; and the serial format provides an easy way to communicate with microcontrollers (Chapter 15) with their (nearly universal) built-in serial ports (UARTs, universal asynchronous receiver–transmitters). See §12.10 for a discussion (with waveforms) of driver and receiver ICs.

14.7.9 Manchester coding

You don't need to frame data bytes with synchronizing START and STOP pulses, however, as long as you arrange

Table 14.5 ASCII Codes

	non-printing				printing			printing			printing		
Name	Control char	Char	Hex	Dec	Char	Hex	Dec	Char	Hex	Dec	Char	Hex	Dec
null	ctrl-@	**NUL**	00	00	**SP**	20	32	@	40	64	'	60	96
start of heading	ctrl-A	**SOH**	01	01	!	21	33	A	41	65	a	61	97
start of text	ctrl-B	**STX**	02	02	"	22	34	B	42	66	b	62	98
end of text	ctrl-C	**ETX**	03	03	#	23	35	C	43	67	c	63	99
end of xmit	ctrl-D	**EOT**	04	04	$	24	36	D	44	68	d	64	100
enquiry	ctrl-E	**ENQ**	05	05	%	25	37	E	45	69	e	65	101
acknowledge	ctrl-F	**ACK**	06	06	&	26	38	F	46	70	f	66	102
bell	ctrl-G	**BEL**	07	07	'	27	39	G	47	71	g	67	103
backspace	ctrl-H	**BS**	08	08	(	28	40	H	48	72	h	68	104
horizontal tab	ctrl-I	**HT**	09	09	)	29	41	I	49	73	i	69	105
line feed	ctrl-J	**LF**	0A	10	*	2A	42	J	4A	74	j	6A	106
vertical tab	ctrl-K	**VT**	0B	11	+	2B	43	K	4B	75	k	6B	107
form feed	ctrl-L	**FF**	0C	12	,	2C	44	L	4C	76	l	6C	108
carriage return	ctrl-M	**CR**	0D	13	-	2D	45	M	4D	77	m	6D	109
shift out	ctrl-N	**SO**	0E	14	.	2E	46	N	4E	78	n	6E	110
shift in	ctrl-O	**SI**	0F	15	/	2F	47	O	4F	79	o	6F	111
data line escape	ctrl-P	**DLE**	10	16	0	30	48	P	50	80	p	70	112
device control 1	ctrl-Q	**DC1**	11	17	1	31	49	Q	51	81	q	71	113
device control 2	ctrl-R	**DC2**	12	18	2	32	50	R	52	82	r	72	114
device control 3	ctrl-S	**DC3**	13	19	3	33	51	S	53	83	s	73	115
device control 4	ctrl-T	**DC4**	14	20	4	34	52	T	54	84	t	74	116
neg acknowledge	ctrl-U	**NAK**	15	21	5	35	53	U	55	85	u	75	117
synchronous idle	ctrl-V	**SYN**	16	22	6	36	54	V	56	86	v	76	118
end of xmit block	ctrl-W	**ETB**	17	23	7	37	55	W	57	87	w	77	119
cancel	ctrl-X	**CAN**	18	24	8	38	56	X	58	88	x	78	120
end of medium	ctrl-Y	**EM**	19	25	9	39	57	Y	59	89	y	79	121
substitute	ctrl-Z	**SUB**	1A	26	:	3A	58	Z	5A	90	z	7A	122
escape	ctrl-[	**ESC**	1B	27	;	3B	59	[	5B	91	{	7B	123
file separator	ctrl-\	**FS**	1C	28	<	3C	60	\	5C	92	|	7C	124
group separator	ctrl-]	**GS**	1D	29	=	3D	61	]	5D	93	}	7D	125
record separator	ctrl-^	**RS**	1E	30	>	3E	62	^	5E	94	~	7E	126
unit separator	ctrl-_	**US**	1F	31	?	3F	63	_	5F	95	**DEL**	7F	127

things so that there are enough transitions in a serial data stream so that the receiver can recover a clocking signal. A simple (though not terribly efficient) example is Manchester coding (Figure 14.46).

The successive bits are transmitted at fixed rate, with a required transition in the middle of each cell: a "1" is LOW-to-HIGH, and vice versa. There may or may not be a transition at the beginning of a cell, as required by the data. The guaranteed transitions at the bit rate facilitate clock recovery (with a phase-locked loop, or delay-locked loop), and the signal is dc balanced, so it can be transformer coupled.

Receiver synchronization with Manchester coding is not completely trivial, since, for example, a continuous sequence of 1s creates a simple square wave. If the receiver

chooses the wrong phase, it will interpret the stream as 0s. However, the presence of mixed 1s and 0s disambiguates the clock recovery, because the wrong phase creates violations of the required mid-cell transitions.

Manchester coding is used in low-speed Ethernet (10base-T), which carries 10 Mb/s in one direction on a cat-5 twisted pair (two pairs are used, for full duplex).[61]

---

[61] When faster Ethernet standards were created, the bandwidth-inefficient Manchester coding (100% overhead, compared with the stream of data bits themselves) was replaced with 4b/5b encoding (a 25% bandwidth overhead), conveyed with 3-level voltage signaling, still on a single twisted pair. The next step – to 1 Gb/s – required 5-level signaling and the simultaneous use of four pairs (with a "hybrid" coupler, permitting each pair to carry signals simultaneously in both directions).

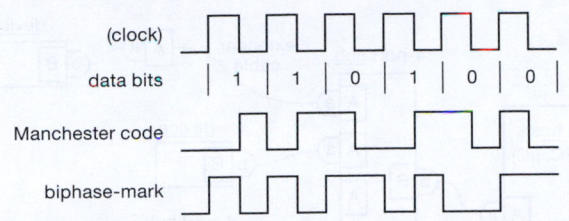

Figure 14.46. Biphase encodings. For Manchester ("biphase-level") coding, every bit cell has a transition in the middle, in a direction set by the bit's value; you can think of Manchester code as the exclusive-OR of a clock (not transmitted) and the data. For biphase-mark coding, the bit value is encoded as presence or absence of a mid-cell transition, following a mandatory transition at the beginning.

Because it is wasteful of bandwidth (by a factor of 2), the better "8b/10b" coding (see below) is used instead for high-performance self-clocking serial links such as SATA, HDMI (high-definition multimedia interface), PCIe, and gigabit Ethernet.

14.7.10 Biphase coding

Manchester is a special case of what is called *biphase coding*, and in fact is sometimes called *biphase-level*, to distinguish it from biphase-mark, biphase-space, and differential biphase codes. These all share the common feature of having a transition at every bit cell to facilitate receiving clock recovery. However, the latter three codes, which are closely similar to each other, have an important advantage over the vanilla Manchester code, namely that they are insensitive to an inversion of polarity (as can happen with transformer-coupled signals). Let's see how this works.

Look at the popular biphase-mark code in Figure 14.46: there is a required transition at the beginning of each cell, and a transition in mid-cell for a transmitted 1 (but no transition for a 0). It is the encoding of bits as the presence or absence of *transitions* (rather than as defined polarities, as in Manchester coding) that makes the biphase-mark code unambiguous, even with a polarity inversion of the transmitted levels.

Biphase-mark coding is used in digital audio links such as AES3, S/PDIF and Toslink, and in some magnetic stripe codes. It goes by the alternative names of *Aiken Biphase* or *F2F* coding.

14.7.11 RLL binary: bit stuffing

A simple serial scheme with reduced bandwidth is to transmit the data bits directly, adding occasional bits to ensure

transitions by which the receiver can synchronize its clock oscillator. "Sending the bits directly" has the acronym NRZ (non-return to zero), with numerous variants, notably NRZI (non-return to zero inverted) in which the encoded stream changes state for an input 0, but stays put for an input 1.[62]

The trouble with either scheme is that there can be long runs of data that cause no transitions in the encoded stream: a string of 1s (or of 0s) for NRZ, or a string of 1s for NRZI. This is bad not only for clock recovery, but also because it extends the transmission band to low frequencies, complicating transformer coupling. The way to fix this is to modify the data (either before or after coding) in order to limit the length of successive output states with no transitions. This is called *run-length limiting* (RLL).

There are various RLL coding schemes in use, for example 8b/10b encoding (coming next) in which each 8-bit string of data is encoded into a well-chosen 10-bit stream to be transmitted; or the 8-to-14 ETF code used for optical compact discs (in which the run length is bounded at both ends: $2 \leq RL \leq 10$). The simplest RLL scheme, though, is probably simple "bit stuffing," as used, for example, in the USB serial link. USB uses NRZI with bit stuffing to limit encoded runs to six consecutive 1s or 0s. This is done by inserting ("bit-stuffing") a 0 into the raw binary data stream after a run of six 1s, forcing a transition in the NRZI-encoded stream. The receiver, of course, knows to ignore a 0 that is preceded by six 1s in the decoded stream. Although the worst-case overhead could be 16% (for a data stream of all 1s), in fact the overhead is less than 1% with random input data.

14.7.12 RLL coding: 8b/10b and others

A more sophisticated way of generating run-length limited serial streams involves encoding the binary data stream in blocks, according to relatively complex algorithms that bound the encoded runs, while additionally controlling the spectral shape, and perhaps adding robustness against errors. For example, DVD optical discs use a scheme called EFMPlus, in which 8-bit groups of data are encoded into 16 serial bits, the latter run-length limited to $2 \leq RL \leq 10$, while simultaneously shaping the spectrum.

The 8b/10b scheme is a popular code in the serial bus and data link world, being used in FireWire, SATA/SAS,

---

[62] These history-encrusted terms are somewhat confusing (suggesting, for example, that NRZI is related to NRZ by inversion of the code or the data). We prefer the terminology used by Sklar: NRZ-L ("NRZ-level," for NRZ), and NRZ-M ("NRZ-mark," for NRZI).

gigabit Ethernet, DVI and HDMI, and in the multiple "lanes" of the internal PCIe (PCI Express) versions 1 and 2.. It encodes 8-bit input data groups into 10-bit serial stream bits, using the flexibility (of multiple possible 10-bit codes for each 8-bit input group) to balance the number of 1s and 0s: it keeps a running tally of bit inequality, and chooses accordingly. The resulting serial stream is guaranteed to have no more than five sequential 1s or 0s; and the number of 1s and 0s in any string of 20 or more serial bits is guaranteed to differ by no more than 2. An analogous 4b/5b code is used in 100 Mbps Ethernet.

8b/10b coding is used also in some serializer-deserializer chipsets, for example the CY7C924. However, the simpler asynchronous NRZ scheme, framed by START and STOP pulses, is used in other high-performance SERDES chips, for example the DS92LV18 series from TI. They claim that it's better, because the receiver will lock to random data without interrupting traffic with PLL training patterns and without a loss-of-lock feedback path from receiver to transmitter. The legacy 9600-baud asynchronous serial communication is thus reborn as a *gigabaud* serial link.

14.7.13 USB

The Universal Serial Bus (USB) was introduced in 1995, with the goal of simplifying connections between computers and peripherals by having a unified serial connection. It initially supported "low-speed" and "full-speed" devices (1.5 Mb/s and 12 Mb/s, respectively) in USB version 1, adequate for devices such as keyboard and mouse, and relatively modest speed transfers to external memory devices such as flash "thumb drives." Version 2 added a "hi-speed" category (480 Mb/s), suitable for serious data transfer to external hard drives, optical storage, and the like. Version 3 adds two "SuperSpeed" shielded twisted pairs (and a piggyback pair of connector pins), delivering both full-duplex communication and another factor of ten in speed (up to 4.8 Gb/s). And the latest revision – USB 3.1 – doubles the speed (to 10 Gb/s), increases the dc power options (to 5 V at 2 A, or optionally as much as 5 A at 12 V or 20 V), and promotes a new and greatly improved "type-C" connector.

USB versions 1 and 2 are organized as half-duplex master–slave, with point-to-point electrical connections via a 4-wire cable that carries power and ground, and one differential pair of data. The cable is unsymmetrical, with an *A* end (host, or master) and a *B* end (slave); each type (*A*, *B*) comes in a full-size and a mini-size version, the latter for connection to small equipment such as cameras and PDAs. A USB network is a star topology, with multiple devices

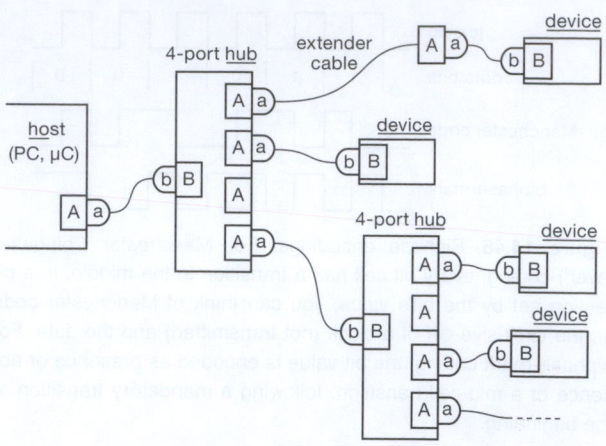

Figure 14.47. USB is a master–slave serial interface, arranged in a star topology (as opposed to a serial repeater topology like FireWire), with up to five repeater "hubs." The individual links are asymmetrical cables, with *A* at the master end and *B* at the slave end (a full-size pair is shown in Figure 1.123); here *a* and *b* represent plugs, *A* and *B* sockets. Individual links are limited to 5 m length (including passive extender cables), but an active hub resets the "ruler." An interesting variant (not shown) is the OTG (on-the-go) USB port, a chameleon that can masquerade as either an A or a B port.

connected via USB hubs, which replicate the host (*A*) sockets, allowing multiple slaves (up to a total of 127 devices from a single host controller port); see Figure 14.47. The supplied power is rather modest – a maximum of +5 V at either 100 mA ("low-power") or 500 mA ("high-power")[63] – and a single link can be no longer than 5 m (but can be extended with hubs, to a total of 20 m). USB version 3.0 introduced full-duplex (and higher speeds), along with somewhat greater (but not enough, in our opinion) dc power – up to 900 mA; happily, version 3.1 addresses that shortcoming, permitting as much as 10 W at 5 V, and an astounding 100 W at 20 V. USB devices are "hot-pluggable," with ground and power connecting before the signal lines.

14.7.14 FireWire

FireWire, known officially as IEEE 1394, was introduced also in 1995, and was designed as a high-speed serial bus for audio, video (including high-definition), and disk storage. It emerged from the starting gate with a 400 Mb/s full-duplex transfer rate (compared with USB's 12 Mb/s half-duplex), and with significant dc powering through its cable

---

[63] A USB device is allowed only low power, when first connected, and it must negotiate for high-power status. If successful, the controller turns on a 500 mA-capable power source. Table 12.4 lists a few protected power-switch ICs popular for this purpose.

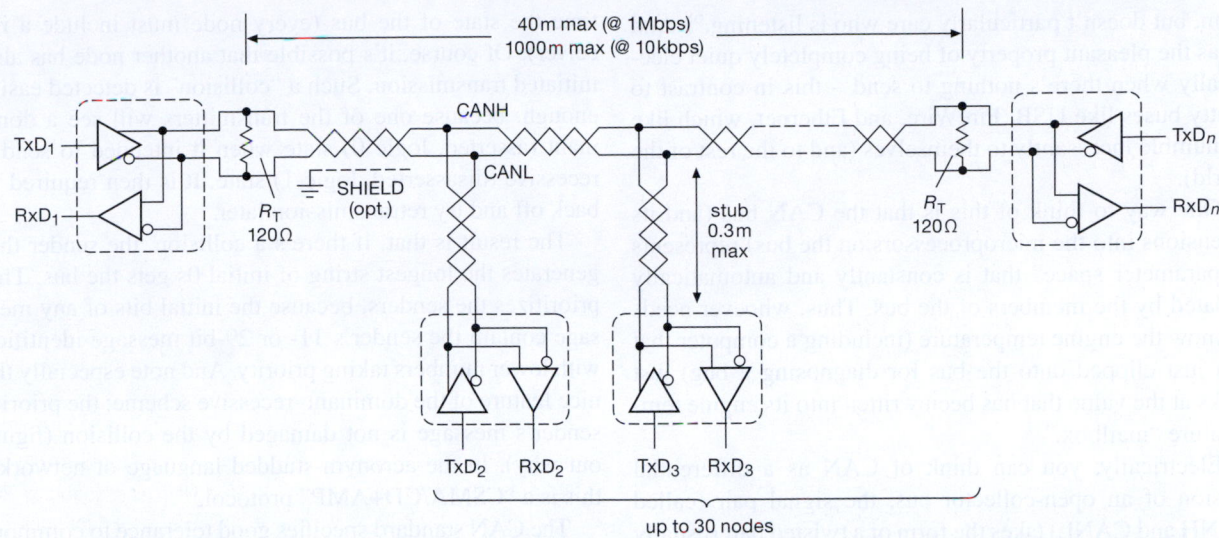

Figure 14.48. The Controller Area Network (CAN) is a multimaster, multidrop architecture that is used widely in automotive and factory-floor applications. It is optimized for short "broadcast" messages, and operates well in noisy environments.

(up to 45 W and 30 V, compared with USB's paltry 2.5 W). Subsequent revisions increased the data rate to full-duplex 800 Mb/s (FireWire 800), with a factor of four (to 3.2 Gb/s) promised in the next revision.

FireWire allows multiple hosts and peer-to-peer communication; its cables are symmetrical, and connect devices in a bused (repeater) configuration. Individual links can be no longer than 4.5 m (for FireWire 400), but can chain through repeaters to a total of 72 m. FireWire 800 permits longer links through copper, Cat-5e network cable, or optical fiber. Like USB, FireWire connections are hot-pluggable; sadly, however, FireWire has been largely overtaken by recent revisions of USB.

FireWire provides plenty of dc power, and works well with streaming media such as high-definition video. We have found FireWire both faster and more stable than USB. It has a nice robust connector design[64] that can be plugged in "blind," owing to its asymmetric shape. In spite of FireWire's technical advantages, the USB standard is becoming dominant, probably because of the complexity, hardware cost, and royalty burdens of FireWire.[65]

14.7.15 Controller Area Network (CAN)

Everyone knows about USB and FireWire; but who has heard about the *CAN* bus? This bus, sometimes simply

called "CAN," was originated by Bosch in the 1980s for automotive use and is now standardized as ISO 11898. It's all over the place. Several protocols have been layered on top of the CAN physical layer, notably DeviceNet™, which is in common use on factory floors.[66]

Unlike the other buses in this chapter, CAN can operate over distances to a kilometer (helpful for the factory application). It's not a point-to-point electrical link (like PCIe, Ethernet, or USB). Rather it's an egalitarian "multi-master" bus, accommodating up to 30 transceiver nodes along its length; that is, it is a "multidrop" bus (see Figure 14.48). The maximum bitrate is 1 Mbps for distances up to 40 m, dropping gradually to 10 kbps at the maximum specified bus length of 1000 m.

CAN is optimized for short "broadcast"-like transmissions, limited to 8 user data bytes per packet (plus some signaling overhead), directed to no one in particular. As the helpful Kvaser website (www.kvaser.com/can/) explains, a CAN message conveys the following sentiment: *"Hello everyone, here's some data labeled X, hope you like it!"*[67] This is just what you want in certain data acquisition and control networks: for example, a sensor generates a stream of temperature values and wants to tell the world about

---

[64] Said to be inspired by the kid-tested connector used in the Nintendo GameBoy.

[65] Which, however, managed to ship its billionth node in 2008!

[66] You'll find CAN in places other than the car and factory; one of our colleagues bought a laser scanning microscope (Zeiss Duoscan), and found…CAN bus!

[67] That's the sentiment; but these messages are *short!* – you'd run out of bytes partway into the second word.

them, but doesn't particularly care who is listening.[68] And it has the pleasant property of being completely quiet electrically when there's nothing to send – this in contrast to chatty buses like USB, FireWire, and Ethernet, which like to mumble incessantly to themselves (and to the rest of the world).

One way to think of this is that the CAN bus (and its extensions into the microprocessors on the bus) represents a "parameter space" that is constantly and automatically updated by the members of the bus. Thus, whoever needs to know the engine temperature (including a computer that you just clipped onto the bus for diagnosing a bug) just looks at the value that has been written into its engine temperature "mailbox."

Electrically, you can think of CAN as a differential version of an open-collector bus: the signal pair (called CANH and CANL) takes the form of a twisted pair (usually shielded – "STP"), terminated at both ends in its characteristic impedance (usually $120\,\Omega$). In the quiescent state both lines rest at ~2.5 V, and that's where they stay unless some node is talking. When that happens, data is driven onto the line pair in a curious asymmetric signaling scheme (Figure 14.49): a logic "0" is asserted by taking CANH ~1 V higher (to ~3.5 V) and CANL lower (to ~1.5 V), generating a ~2 V differential signal. But to assert logic "1," the talker does not reverse the sense of drive current, but rather *releases* drive current, which lets both lines of the pair go to their ~2.5 V resting state. These two signaling states are called *dominant* (bus driven, logic 0) and *recessive* (bus released, logic 1 or inactive).

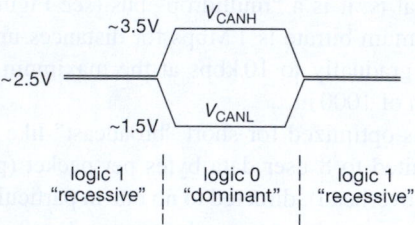

Figure 14.49. The CAN bus is differential, with an asymmetric signaling mode that simplifies arbitration.

This curious scheme was devised to simplify bus arbitration: there's no master controller in charge; instead any node can initiate a transmission, as long as it has seen the bus resting (recessive) for some minimum time. It then asserts its message bits sequentially, monitoring at the same

time the state of the bus (every node must include a receiver). Of course, it's possible that another node has also initiated transmission. Such a "collision" is detected easily enough, because one of the transmitters will see a dominant (asserted, logic 0) state when it intended to send a recessive (disasserted, logic 1) state. It is then required to back off and try retransmission later.

The result is that, if there's a collision, the sender that generates the longest string of initial 0s gets the bus. This prioritizes the senders, because the initial bits of any message contain the sender's 11- or 29-bit message identifier, with lower numbers taking priority. And note especially the nice feature of the dominant–recessive scheme: the priority sender's message is not damaged by the collision (figure out why). In the acronym-studded language of networks, this is a "CSMA/CD+AMP" protocol.[69]

The CAN standard specifies good tolerance to common-mode noise, requiring correct operation at least from -2 V to $+7$ V, but extended by many transceiver ICs to ranges of -7 V to $+12$ V, or -12 V to $+12$ V. And ISO 7637 prescribes a torture test consisting of a train of ±150 V nanosecond-scale pulses that a transceiver must survive. In addition, you can use an inexpensive diode and zener protective device like the NUP2105 or NUP2202 (see ON Semi's App. Note AND8169; see also §12.1.5). For long bus runs, however, it's best to provide true galvanic isolation (avoiding ground loops mandates that there be only one ground point).

In keeping with its original intended automotive use, the CAN bus includes robust error-detection mechanisms: at the *bit* level there's bit monitoring (with error flagging if a disagreement), and detection of a "bit-stuffing" error (an opposite bit must be inserted after 5 consecutive bits of the same level). And at the *message* level there is a CRC (cyclic redundancy checksum), along with checking of an ACK (acknowledge) field and of a set of certain designated message bits (a "form check") that must be recessive. By this mechanism a transmitter can know if its message was corrupted and needs to be resent.

You can get plug-in CAN interface cards for standard computer buses such as PCI/PCIe, PC/104, and PCMCIA, and adapters from USB. And many microcontrollers (e.g., Atmel's AT90CAN series) include an on-chip CAN protocol controller. At the component level there are many manufacturers of CAN transceiver chips (most are powered from a single +5 V, but some work at +3.3 V). You

---

[68] And, conversely, it's inefficient for transmitting large blocks of data (for example, digitized audio or video) between two nodes. You also would not use it to link two processors in the same box.

[69] Well, at least it's shorter than the tongue-twister "carrier-sense multiple-access with collision detection and arbitration on message priority."

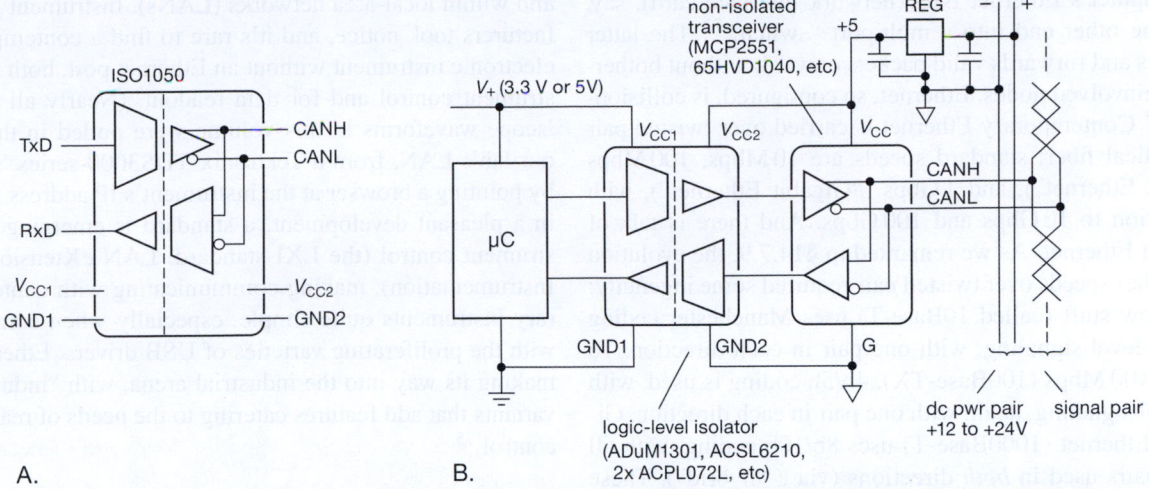

Figure 14.50. A. You can get single-chip galvanically isolated CAN transceivers, like TI's ISO1050, a good idea for long runs or noisy environments. B. Standard CAN cables include a second wire pair for isolated power, which you can use to power the bus side of the transceiver's isolation barrier. Here we've used a logic-level isolator between a microcontroller and the non-isolated CAN transceiver, the latter powered (along with the bus side of the isolator) from regulated +5 V derived from the bus dc power pair.

can also get galvanically isolated transceiver chips (e.g., the TI ISO1050, Figure 14.50), a good idea with long buses or noisy environments: you can power the bus side with an independent +5 V supply, or you can take advantage of the bused power pairs that often are included (for just such remote powering) along with the shielded and controlled-impedance signal pairs in CAN bus cables (e.g., Belden 3082/84 or Alpha 6451/52).

There are a couple of simplified CAN variants in use: single-wire CAN dispenses with CANL, and is limited to 40 kbps. And the closely related single-wire LIN (local interconnect network) bus dispenses with the regulated 5 V supply altogether, using a pullup to the +12 V battery and open-collector closure to ground; it's limited to 20 kbps. These inexpensive and simplified variants are sometimes used as a sub-bus in a CAN system; they both employ slew-rate limiting for reduced susceptibility to noise.

There is no defined CAN connector, but there are several common implementations, including a 9-pin D-subminiature, a 10-pin header, and a 4-pin open connector.[70]

14.7.16 Ethernet

Ethernet[71] is ubiquitous – those colorful telephone-like "RJ-45" network jacks – we all love it, we've mentioned it numerous times (for example in §12.10.3B, and in §14.7.9 in connection with Manchester coding), and we'll be mentioning it again in the next chapter. It was developed back in the 1970s at Xerox's famed Palo Alto Research Center ("Xerox PARC"), and at the physical layer (PHY) consisted originally of a shared coax cable (first "thick wire," officially called 10Base5; later "thin wire," or 10Base2), terminated at both ends, and tapped into by each node. Every node used transformer coupling, connecting right at the coax (no long stubs allowed), and (as with the CAN bus) there was a protocol for sensing collisions and backing off before retransmitting. To work properly in the face of inevitable (and data-destructive) collisions, there was established a minimum packet length (now standardized by IEEE Std. 802.3 as 74 bytes), and a maximum cable length (about 200 m for thin wire).

In contemporary practice the shared coaxial cable has been replaced with a point-to-point unshielded twisted-pair cable (UTP; Cat-5e or Cat-6), one end plugging into

---

[70] See   www.interfacebus.com/Design_Connector_CAN.html and   www.interfacebus.com/Can_Bus_Connector_Pinout. html. Additional information about the CAN bus can be found at www.kvaser.com/can/, www.can-cia.de, TI's App. Note SLOA101A, and Analog Devices' AN-770A.

[71] The whimsical name was chosen to convey the spirit of an all-present data-conveying medium, analogous to the "luminiferous ether" through which it was (incorrectly) thought that light traveled, before it was disproven by Michelson and Morley's famous experiment of 1887. With characteristic computer-nerd humor, the first two networked computers were named Michelson and Morley.

a computer's Ethernet NIC (network interface card), say, and the other end into a multiport "switch."[72] The latter buffers and forwards valid packets onward, without bothering uninvolved nodes. Ethernet, so configured, is collisionless.[73] Contemporary Ethernet is carried over twisted pair or optical fiber; standard speeds are 10 Mbps, 100 Mbps ("Fast Ethernet"), and 1 Gbps ("Gigabit Ethernet"), with evolution to 10 Gbps and 100 Gbps. And there is talk of *tera*bit Ethernet. As we remarked in §14.7.9, the evolution of higher speeds over twisted pair required some ingenuity: the slow stuff (called 10Base-T) uses Manchester coding and 2-level signaling, with one pair in each direction. To go to 100 Mbps (100Base-TX), 4b/5b coding is used, with 3-level signaling, again with one pair in each direction. Gigabit Ethernet (1000Base-T) uses 8b/10b coding, with all four pairs used in *both* directions (via a "hybrid"). These are descriptions of the *physical* layer; higher levels of the 7-layer OSI (open systems interconnection) network hierarchy don't know (and don't care) what's going on down there, so you can upgrade the hardware to your heart's content.

Twisted-pair Ethernet links are limited in length to ∼100 m owing to signal degradation and attenuation. Fiber does much better – to a kilometer or so with multimode fiber, and tens of kilometers with single-mode fiber.[74] You can get "media converters" to change between copper and fiber, or copper and wireless; some of these incorporate rate conversion as well. Look at the Ethernet offerings from Allied Telesis, TRENDnet, StarTech, or IMC Networks; or the broad range of converters and extenders (including products for USB and serial, in addition to Ethernet) from B&B Electronics. See also Chapter 15 for interface component suggestions, such as the Lantronix XPort or Silicon Labs 28-pin CP2201 chip (to which you can simply add an RJ-45 jack with an integrated transformer and indicator LEDs).

Ethernet is so widely supported that it has become the dominant communications medium between computers

and within local-area networks (LANs). Instrument manufacturers took notice, and it's rare to find a contemporary electronic instrument without an Ethernet port, both for instrument control and for data readout. (Nearly all of the 'scope waveforms in this volume were pulled in through our lab's LAN, from a Tektronix TDS3000-series 'scope, by pointing a browser at the instrument's IP address.) And, in a pleasant development, a standard is emerging in instrument control (the LXI standard: LAN eXtensions for Instrumentation), making communicating with contemporary instruments quite simple, especially when compared with the proliferating varieties of USB drivers. Ethernet is making its way into the industrial arena, with "industrial" variants that add features catering to the needs of real-time control.[75]

14.8 Number formats

We wind up this chapter with a short riff on number formats, meaning the way that numbers are represented internally during computation, or exchanged via digital media or communications ports. The scene is summarized in Figure 14.51, with some explanation in the following paragraphs.

14.8.1 Integers

Signed integers are always represented in 2's complement, using either 1, 2, or 4 bytes, or occasionally 8 bytes, as shown. The most-significant bit (MSB) tells the sign, even though 2's complement is not the same as sign–magnitude representation (e.g., −1 is 11111111, not 10000001; see §10.1.3). You can think of 2's complement as offset binary with inverted MSB; alternatively, you can think of it as an integer with the bit values as shown in Figure 14.51. Programming languages let you declare variables as *unsigned* integers, in addition to 2's complement signed integers. A 2-byte unsigned integer can have values from 0 to 65535. Quite apart from the number format itself, there's the hardware interface issue about how you pack integer data into a computer's larger word. For example, the binary integer

---

[72] This is a *star* topology, compared with the *bus* topology of coax Ethernet.

[73] Before switches were widespread and inexpensive, people used "hubs," which simply rebroadcast each packet to all connected nodes. With hubs you get collisions; with switches you don't.

[74] But twisted-pair Ethernet has a nice feature: it lets you send dc power ("Power over Ethernet," PoE), which is handy when connecting to remote devices like wireless access points, IP phones, or surveillance cameras. It uses the same signal pairs, applying the ∼48 V dc as common-mode "phantom power" between the two pairs, picked off between the transformer center taps at the remote end. Audio engineers have been using the same trick, for ages, to power their microphones.

[75] For example, implementations of the IEEE Std. 1588 "Precision Time Protocol" (PTP), which allows time synchronization to ∼100 ns via Ethernet (with dedicated hardware, for example the NSC DP83640 MAC/PHY chip, or included in embedded microcontrollers like TI's Luminary Stellaris ARM Cortex M3). PTP is currently being implemented by the LXI standard for interoperability between LAN-based instruments, as well as by several Ethernet-based Fieldbuses such as Profinet and CIP (Common Industrial Protocol).

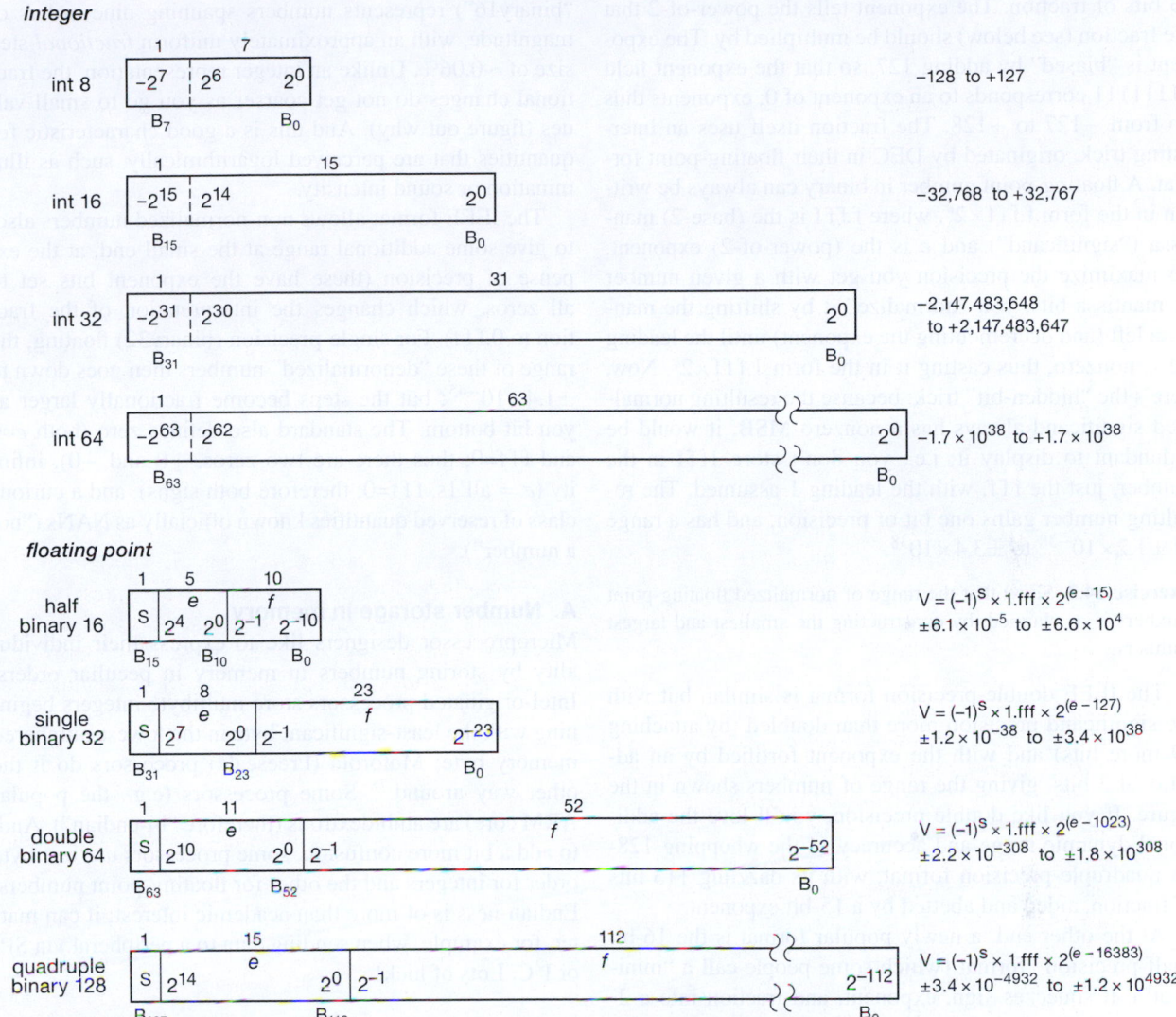

Figure 14.51. Commonly used number formats. The symbol "*e*" is the unsigned binary-integer value of the exponent field, used as shown to determine the value *V* of the various floating-point formats.

output from an ADC could be right-justified, so numbers go from zero up to the converter's full scale (0 to 4095, for a 12-bit ADC). But it's arguably better to left-justify the data, and think of the stored quantity as a *fraction*. This has a nice benefit, namely, if the ADC resolution is subsequently improved, it simply adds additional low-order fractional bits (rather than increasing the full-scale value).

14.8.2 Floating-point numbers

Floating-point numbers (sometimes called *real* numbers) are most commonly represented and stored as 32-bit

("single-precision") or 64-bit ("double-precision") quantities.[76] The bad news is that there are several incompatible representations in use. The good news is that the floating-point standard blessed by the IEEE (Std. 754-2008) has been implemented by nearly all processor families and is in nearly universal use.

Figure 14.51 shows the IEEE floating-point formats in detail. To see how it goes, look at the 32-bit single-precision format: it has 1 sign bit, 8 exponent bits, and

---

[76] Augmented by giant 128-bit ("quadruple precision"), and diminutive 16-bit (you guessed it – "half precision") formats.

23 bits of fraction. The exponent tells the power-of-2 that the fraction (see below) should be multiplied by. The exponent is "biased" by adding 127, so that the exponent field 01111111 corresponds to an exponent of 0; exponents thus go from -127 to $+128$. The fraction itself uses an interesting trick, originated by DEC in their floating-point format. A floating-point number in binary can always be written in the form f.fff$\times2^e$, where f.fff is the (base-2) mantissa ("significand") and e is the (power-of-2) exponent. To maximize the precision you get with a given number of mantissa bits, you "normalize" it by shifting the mantissa left (and decrementing the exponent) until the leading bit is nonzero, thus casting it in the form 1.fff$\times2^e$. Now, here's the "hidden-bit" trick: because the resulting normalized significand always has a nonzero MSB, it would be redundant to display it; i.e., you don't store 1fff in the number, just the fff, with the leading 1 assumed. The resulting number gains one bit of precision, and has a range of $\pm1.2\times10^{-38}$ to $\pm3.4\times10^{38}$.

Exercise 14.3. Show that the range of normalized floating-point numbers is as claimed, by constructing the smallest and largest numbers.

The IEEE double-precision format is similar, but with the significand precision more than doubled (by attaching 29 more bits) and with the exponent fortified by an additional 3 bits, giving the range of numbers shown in the figure. If you like double precision, you'll love the additional dynamic range and accuracy of the whopping 128-bit quadruple-precision format, with its dazzling 113 bits of fraction, aided and abetted by a 15-bit exponent.

At the other end, a newly popular format is the 16-bit "half-precision" format (which some people call a "minifloat"). It squeezes sign, exponent, and fraction into a 2-byte word. The largest values in this format are ±65504. This isn't a whole lot more than you can do with a 16-bit signed integer (±32767), so it might seem pointless. Not so: the smallest value of IEEE minifloat is $\pm6.1\times10^{-5}$. So, you get a large *dynamic range*, albeit without great precision.[77]

This last point is worth a bit more explanation. The IEEE half-precision floating-point format (officially called "binary16") represents numbers spanning nine orders of magnitude, with an approximately uniform *fractional* step size of $\sim0.06\%$. Unlike an integer representation, the fractional changes do not get coarser as you go to small values (figure out why). And this is a good characteristic for quantities that are perceived logarithmically, such as illumination or sound intensity.

The IEEE format allows non-normalized numbers also, to give some additional range at the small end, at the expense of precision (these have the exponent bits set to all zeros, which changes the interpretation of the fraction to 0.fff). For single-precision (binary32) floating, the range of these "denormalized" numbers then goes down to $\pm1.4\times10^{-45}$; but the steps become fractionally larger as you hit bottom. The standard also defines zero (both $e=0$ and fff=0; thus there are two zeros, $+0$ and -0), infinity ($e=$ all 1s, fff=0; therefore both signs), and a curious class of reserved quantities known officially as NANs ("not a number").

A. Number storage in memory

Microprocessor designers like to express their individuality by storing numbers in memory in peculiar orders. Intel-originated processors store multibyte integers beginning with the least-significant byte in the lowest-numbered memory byte; Motorola (Freescale) processors do it the other way around.[78] Some processors (e.g., the popular ARM core) are ambidextrous (therefore "bi-endian"). And, to add a bit more confusion, some processors use one byte order for integers and the other for floating-point numbers. Endian-ness is of more than academic interest; it can matter, for example, when sending data to a peripheral via SPI or I^{2}C. Lots of luck!

---

[77] Characteristics that are particularly useful in applications such as video imaging. In fact, it was devised by Industrial Light & Magic for just such uses. The half-float format is supported in graphics processor chips, which are becoming popular for large-scale computation. These chips, with hundreds of fast processing cores, are challenging traditional CPUs; throw a few dozen of them in a box, and you've got yourself a pretty impressive supercomputer.

[78] Engineers get their jollies by calling these "little-endian" and "big-endian," respectively.

Review of Chapter 14

An A-to-J summary of what we have learned in Chapter 14. This summary reviews basic principles, facts, and application advice in Chapter 14.

¶A. Processors and Data Buses.

This first of two processor-oriented chapters dealt with the subject of computer architecture and the interfaces (buses) over which data is exchanged, whereas the subsequent chapter is devoted to the use of microcontrollers as "embedded" components within a circuit or instrument. Given the omnipresence of computers in contemporary life, the reader is likely to be familiar with much of the material in this chapter.

¶B. Bus-oriented Computer Architecture.

In a classic computer architecture (§14.1) the processor unit (CPU or MPU) executes instructions fetched from memory, moving data on a set of lines called a *bus* (Figure 14.2). The instructions encode what is to be done, and the processor does it, as interpreted by its *instruction decoder*. Within the processor is an *arithmetic logic unit* (ALU), responsible for arithmetic and logical operations (e.g., ADD or COMPARE) performed on data held in *registers*; a *program counter*, which holds the memory address of the current instruction; a set of *flags* that are set according to the result of the last operation and which are tested for conditional branching; a *stack pointer* to address memory sequentially for temporary storage (e.g., of return address during interrupts and function calls); and, often, a *cache memory* to hold recent data and instructions for quicker access.

¶C. Peripherals.

A computation-oriented microprocessor (as would be used in a server or desktop computer) communicates with off-chip peripherals through one or more buses, whereas a processor aimed at embedded applications (a micro*controller*, Chapter 15) trades off some computational finesse by integrating instead a set of on-chip peripherals and memory. Typical peripherals include nonvolatile mass storage such as hard disk (hdd) and solid-state disk (ssd), random access memory (RAM), video graphics, network interface (Ethernet), data acquisition and control (ADC, DAC, digital I/O), and serial ports such as USB and SATA.

¶D. Instruction Set and "Machine Language."

A given processor is designed to interpret certain groups of bytes, taken together, as *instructions*, and to carry out the corresponding tasks (§14.2). Such *machine language* instructions can be of various lengths; for the Intel x86 32-bit processors, for example, instructions range from one to fifteen bytes in length. Programming directly in machine language is too tedious for humans; instead each instruction can be represented in a readable text-formatted *assembly language*, with understandable mnemonics (e.g., ADD AX,BX, which adds to the quantity held in register AX the quantity held in register BX). A program called an *assembler* then converts the assembly-language program (a text file created by the programmer, or by a *compiler* from a program written in a high-level language like C) into the processor's native machine language.

To illustrate processor instructions we chose the Intel x86, displaying a pared-down instruction set (§14.2.2 and Table 14.1). There are *arithmetic* instructions (for example ADD), *conditional branch* instructions (for example JNZ label, jump on non-zero), *stack* instructions (PUSH and POP), and I/O instructions (for example OUT port,AX). Along with the instructions, you've got to understand *addressing*, the various ways that locations in memory or registers can be designated. In the x86 language the basic possibilities are *direct* (the address itself), *indirect* (an address pointing to the place in memory where the address itself is held), *indexed* (a numerical offset from a base address, useful for "moving pointer" accesses to sequential addresses), and *immediate* (a numerical quantity contained within the multibyte instruction itself).

¶E. PC104/ISA: a Parallel Bus.

Bus interfacing can be a complicated business, with elaborate negotiation protocols and the like. To keep things simple we chose the legacy PC104/ISA 8-bit parallel bus (§14.3), an outgrowth of the original IBM PC bus that has been adapted to the industrial PC104 embedded system standard. It is in widespread use, and it illustrates well the basics of a multidrop bus (and parallel data transfer in general); for reference, the full set of bus signals is listed in §14.3.11.

WRITE cycle. To move some data out to a peripheral (a "write cycle"), the bus master (here the CPU) asserts the data onto a set of 3-state DATA lines (D0–D7) and the target address onto the ADDRESS lines (A0–A15), then it pulses a "write" line (IOW'), see Figure 14.8. The peripheral's corresponding hardware (Figure 14.9) responds by latching the data (on D0–D7) by clocking with IOW' if it sees its unique address (on A0–A15). This transaction takes place automatically when the CPU executes an OUT instruction to the peripheral's address.

READ cycle. To fetch data from a peripheral (a "read

cycle"), the bus master asserts the peripheral's address, but does not assert the (3-state) data lines; instead it pulses a "read" line (IOR′) and latches the data it finds on D0–D7 (asserted by the addresses peripheral) at the end of IOR′; see Figure 14.13. The peripheral's corresponding hardware (Figure 14.14) responds by asserting its data onto D0–D7 by enabling 3-state drivers with IOR′ if it sees its unique address (on A0–A15). This transaction takes place automatically when the CPU executes an IN instruction from the peripheral's address.

Command and status bits. The "data" that the CPU fetches during a read cycle can be ordinary data (e.g., a byte from an ADC); but it can instead indicate a status (e.g., new ADC data is available to be fetched). Such status data is essential in most cases; see, for example, the simple keyboard interface in Figure 14.15, where a flip-flop is set when a new character is struck and cleared when a character is read; its status is readable on bit D7 of its peripheral address (KBFLAG). The corresponding program code (§14.3.5A) repeatedly reads that *status byte*, looping until it finds bit 7 set, whereupon it fetches a data byte from the keyboard's data address (KBDATA) and appends it to data stored in a string buffer in memory. In analogous manner, bits that are sent *to* a peripheral can indicate an action to be initiated by that peripheral; these are *command* bits, for example to tell and ADC to begin conversion, or a serial port to send a data byte. Multiple status (or command) bits can be grouped into a status (or command) *register*, each bit of which can have its own function.

Interrupts. The use of status bits lets the CPU find out if action needs to be taken, but it has to ask (by reading the status register). Such *polling* is one way for a peripheral to signal the CPU; the alternative is the *interrupt*, in which the peripheral asserts one of several dedicated bus lines (called IRQ*n* on the PC104 bus). This signals the CPU's hardware, which (if interrupts are enabled) causes program execution to break off from what it's doing and jump to an "interrupt handler" routine; see §§14.3.7–14.3.9 for hardware and software examples.

Direct memory access. Most efficient of all for rapid transfer of multiple bytes (e.g., reading a whole file from disk) is DMA, in which the successive bus addresses and control pulses are generated by hardware, and thus do not require the processor cycles needed for the READ or WRITE cycles of *programmed I/O*; see §14.3.10.

¶F. Other Parallel Buses.
The PC104/ISA internal multidrop computer bus is slow (less than 10 MB/sec), and was superseded by the **PCI** (peripheral component interconnect) multidrop bus, which

reached 2 Gb/s in its 64-bit-wide incarnation. A PCI interface is considerably more complex than the simple PC104 described in ¶E, requiring some negotiation with the PCI controller, etc. PCI is now obsolete, having been replaced by the point-to-point serial PCIe, (see ¶H). For connection to disks, the legacy parallel interfaces are **IDE** (also called ATAPI or PATA) and **SCSI**; both are obsolete, having been replaced by serial buses: SATA and SAS. The original printer interface was parallel (a 25-pin D-connector or 36-pin "Centronics" micro-ribbon connector), replaced by serial buses (USB or Ethernet). And for connection to laboratory instruments the **GPIB** (general-purpose interface bus, originated by HP as the HPIB) is still alive and, well, surviving; it's a bit awkward, though, with its thick cables and hunky connectors. Most lab instruments now support Ethernet and USB. These and other buses are listed in Table 14.3.

¶G. Parallel Data Links.
At the much simpler level of data transfer between ICs on a circuit board (say between an embedded microcontroller and an ADC) the parallel bus's data–address–strobe protocol becomes very simple: for a point-to-point connection (i.e., a single peripheral chip) all you need are a byte-wide (or 4-bit-wide) data path plus direction and strobe; see for example Figure 14.34. This is hardly a *bus* – it's really just a data link.

Apart from some IC interfaces of this kind, and the wide and speedy CPU "Front-Side bus," nearly all parallel buses and data links have fallen by the wayside, yielding their dominance to *serial* buses and links (¶¶ H, I); ironically, a single serial link is usually faster than the byte-wide parallel link it replaces, with a lot less wiring and (for external interfaces) permitting much skinnier cabling.

¶H. Serial Buses.
There are many advantages to a *serial* data connection (§14.7): fewer wires, fewer pins on the driver and receiver chips, clean termination (in a point-to-point connection, i.e., a single driver and a single receiver), lack of timing skew (self-timed via clock recovery), and the flexibility of using an optical fiber or wireless channel. Counterintuitively, contemporary serial links deliver higher data throughput than their parallel forebears. For example, the last of the parallel disk interfaces topped out at 1–2 Gbps (PATA and SCSI, 16 bits wide), whereas their serial successors (SATA and SAS, a single differential pair in each direction) deliver 6 Gbps.

In the computer world the common *internal* serial buses are the **SATA** (serial ATA), **eSATA** (external SATA, a short extension of SATA), and **SAS** (serial-attached SCSI) disk buses, and the **PCIe** (PCI express) motherboard bus (see §§14.7.6 and 14.7.7). The latter is a hybrid of sorts – unlike the earlier parallel multidrop PCI bus, PCIe is point-to-point and serial (a differential pair in each direction), but it runs multiple such "lanes" to each motherboard slot, labeled with "×" (commonly ×1, ×4, and ×16, but also ×8 and ×32). Each lane can deliver up to 8 Gbps, with separate clock recovery at the receiver.

For communication *external* to the computer the common serial buses are USB, FireWire, Ethernet, and CANbus. In its current revision (3.0) **USB** (§14.7.13) is full duplex (a differential pair in each direction) with speeds to 3.2 Gbps; this improves on the previous version (2.0) which is half duplex with speeds 480 Mbps. FireWire (IEEE 1394, §14.7.14) is full duplex, with versions supporting 400 and 800 Mbps (and with rates to 3200 Mbps included in the standard).

Ethernet (§14.7.16) is the bus of choice for computer networks; the original coaxial multidrop 10 Mbps versions (10base2, thinnet, and 10base5, thicknet) have long since been replaced by full-duplex point-to-point twisted pair connections (10base-T, 100base-TX, 1000base-T), which both circumvent the problem of collisions and offer data transfer speeds to 1 Gbps.

CANbus (§14.7.15) is an industrial-strength multidrop and multimaster bus used in automotive and factory environments; it's robust, and permits links up to 1 km, though at greatly reduced data rate (10 kbps versus its 1 Mbps for links less than 40 m long).

¶I. Serial Data Links.

Largely superseding the parallel data transfer between ICs on a circuit board (¶G) are the popular serial data links: SPI, I^{2}C, and JTAG. **SPI** (§14.7.1) is drop-dead simple, with a master–slave architecture comprising three wires shared among multiple slave chips: one data wire in each direction (MOSI, MISO) and a clock (SCLK); there's a separate slave-select line (SS′) needed for each slave device; see Figure 14.38. A selected device receives data (MOSI) and sends data (MISO) on each clock, see Figure 14.37. There's no universal SPI standard, so you have to conform to each chip's protocols according to its datasheet. Take a look at Figure 15.21 in the next chapter to see some SPI peripheral chip examples.

By contrast, **I^{2}C** (§14.7.2) is a more sophisticated multidrop serial link, with two wires only (SDA, data; SCL, clock); there's a well-defined protocol by which the master (the microcontroller) communicates both target address and data direction to the various slave devices. The minimal wiring and communication flexibility of I^{2}C is offset by the need to give each device a unique address, and by the problems of contention and arbitration on the shared lines. Take a look at Figure 15.22 in the next chapter to see some I^{2}C peripheral chip examples.

The **JTAG** bus (§14.7.4), originally intended for in-circuit testing and debugging, has become a popular interface for loading and debugging microcontroller code, and for programming nonvolatile on-board memory. It buses a shared clock (TCK) and mode line (TMS), and daisy-chains a data line (TDI in to each device, TDO out); see Figure 14.43. You connect to the JTAG bus with a programming "pod" that attaches to a USB port on your host PC.

And don't forget the good ol' **RS-232** "serial COM port" link (§14.7.8), still widely supported by microcontrollers. You can talk to them with a PC running a terminal emulator, connected with a USB-to-RS232 adapter like FTDI's TTL-232R-3V3 or TTL-232R-5V (3.3 V and 5 V, respectively).

¶J. Memory.

Any talk of computers is incomplete without a discussion of *memory*, set forth in §14.4. The ideal memory would have fast random-access, persistent retention without power (i.e., nonvolatile), and infinite endurance (number of read–write cycles); ideally it should also be low power, inexpensive, and compact. Real-world memory has not yet achieved these goals, at least in a single memory type.

The current popular favorite in *nonvolatile* memory (§14.4.5) is **flash NAND** (the stuff of USB memory sticks, camera memory cards, and solid-state drives). It is limited in speed and endurance, and must be written or erased in sectors; but it's outrageously inexpensive (∼$0.50/GB in single quantities). For retention of small quantities of data there's **EEPROM**, whose 2-transistor cell permits single-bit or -byte erasability. Some new technologies that may deliver near-infinite endurance are **FRAM** (ferroelectric RAM), **MRAM** (magnetoresistive RAM), and **PCM** (phase-change memory, also called PRAM).

Contemporary memory types with fast access and infinite endurance are all *volatile*: SRAM (static RAM) and DRAM (dynamic RAM). The basic **SRAM** is an array of addressable 6-transistor ("6T") flip-flops (Figures 14.20 and 14.21), configured for asynchronous or synchronous (i.e., clocked) operation; see §14.4.3. Static RAM has a simple interface, fast access, and zero quiescent current, but its 6T cell takes up lots of space, so its density is considerably lower than that of dynamic RAM. For this reason a class of pseudo-static (**PSRAM**) memory has become popular: it exploits a hidden refresh mechanism to bring the density of DRAM to the simple interface of SRAM; it's a DRAM in SRAM's clothing. SRAM and PSRAM memories are useful in small or low-power systems where only a modest amount of memory is needed.

The other volatile memory type is **DRAM** (§14.4.4), an array of tiny ($\sim$30 femtofarads) capacitors whose charge state holds one bit, addressed with an array of MOS-FETs such that only one transistor is needed per bit, both for readout and for charge retention ("refresh"); this is a "1T1C cell" (Figures 14.26 and 14.27). DRAMs require periodic refreshing, and, having originated as asynchronous memory chips, they are now used in their synchronous (**SDRAM**) varieties: **SDR** (single data-rate), and evolving **DDR** (double data-rate) versions (**DDR2**, **DDR3**, and **DDR4**). DRAM has infinite endurance, fast access, high density, and low cost, and it is used for the main memory in PCs, laptops, and all variety of consumer electronics (e.g., cable and satellite boxes). It's what you get when you buy a DIMM module for your computer; these things are currently costing about \$5–\$10/GB.

MICROCONTROLLERS

15.1 Introduction

As we remarked in the previous chapter, microcontrollers are essentially stand-alone processors, intended to be *embedded* into some electronic device that is definitely not a "computer."[1] They are inexpensive and easy to use, and they let you put the intelligence of a computer into pretty much anything electronic. In so doing, they trade off some of the terrifying speed of computer-oriented processors in favor of built-in memory and peripherals.

To expand on this point, nearly all microcontrollers (including those that cost less than a dollar) include both data memory (static RAM) and non-volatile program memory ("flash") on-chip; many include additional non-volatile EEPROM for holding calibration data, system configuration settings, and so on.[2] Better still, you can choose among integrated "peripherals": communication links such as SPI, I^2C, USB, Ethernet, Bluetooth, and ZigBee; buses such as PCI and PCIe, SATA, PCMCIA, flash-memory cards, and external memory; analog interfaces such as comparators, multiplexed ADCs, DACs, and video and imaging sensors; and special-purpose interfaces such as pulse-width modulators, LCD drivers, GPS, digital audio, and WiFi.

To put it another way (Figure 15.1), a microcontroller includes within it the "CPU + memory + peripherals" that required an external data bus and separate components in the previous chapter. (You *can* implement an external data bus with a microcontroller, if you want to; but it's better to choose a microcontroller that has what you want on-chip.)

Of course, in an embedded system you don't boot up from a disk – you don't even *have* a disk! That's why the program memory is included on the controller, and also why it must be non-volatile: your dishwasher needs to know what it is supposed to do, even after a power failure. Because microcontrollers are built into the target device (and generally not removable from the circuit board) they must be programmable (and reprogrammable) *in-circuit*. You program them via an "in-circuit serial programming" interface ("ICSP," usually SPI or JTAG), with a hardware interface "pod," controlled by software running on a host computer.

Microcontrollers are fun, and easy. Think of them as a circuit "component," like an op-amp. The analogy is good: just as the op-amp is the universal analog component, the microcontroller might be thought of as the universal digital component.[3] And a microcontroller plus a few other components (e.g., a USB connector, the programming header, a few lights and buttons, and perhaps an alphanumeric LCD) put onto a small circuit board, is a "universal block" whose function is flexibly programmed to do whatever you want. The microcontroller manufacturers happily provide such objects (sometimes called "development kits," which often include programming hardware and software; see Figure 15.24 on page 1090), to encourage you to adopt their wares. There are also third-party and open-source products, such as "Ethernut" and "Arduino".[4] The former includes an Ethernet port; the latter offers serial, USB, SPI, and I^2C ports, with an optional stackable Ethernet adapter.

Let's launch into microcontrollers with a concrete example.

---

[1] A recent publication from Maxim Integrated Products (App Note 3967) begins with the sentence "The heart of today's advanced electronic products is a microcontroller (μC) that communicates with one or more peripheral devices." As a friendly amendment, we'd probably omit the word "advanced."

[2] *EEPROM* permits rewriting (or erasing) individual stored bytes, as contrasted with *flash* memory, in which erasure (required before rewriting) can be done only on a block of bytes at once. That is why the lower-density EEPROM is better for storing user data, whereas the higher-density flash is best for infrequently rewritten program code. Both memory types are non-volatile and in-circuit programmable, though flash typically has less rewriting "endurance," e.g., 10,000 erase–write cycles for flash, versus 100,000 for EEPROM.

[3] Although experienced users of FPGAs might well argue that *their* chips are "more universal": you can put a microcontroller "soft core" into an FPGA, but not vice versa.

[4] Here are a few more: BeagleBoard & BeagleBone, Odroid, Raspberry Pi, and Teensy.

MICROCONTROLLER ("µC")

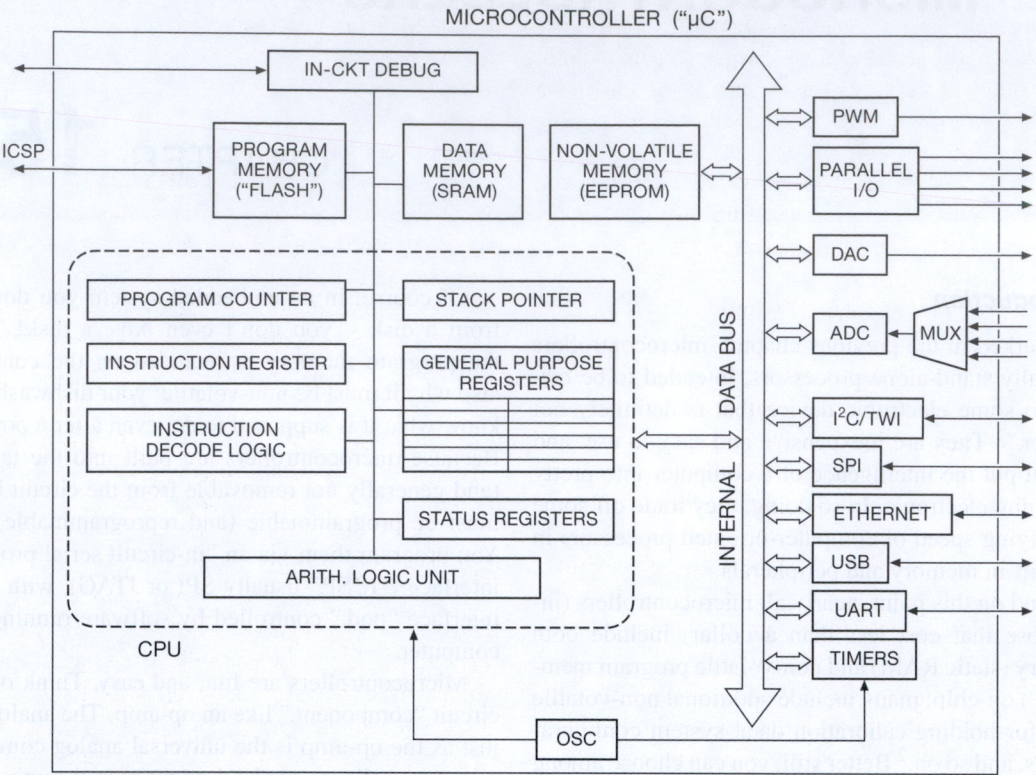

Figure 15.1. A microcontroller integrates memory and "peripherals" onto the same chip as the CPU. ICSP stands for "in-circuit serial programming."

15.2 Design example 1: suntan monitor (V)

We've visited this indispensable device for the beachgoer, first in Chapter 4 (where we explored three purely analog implementations), then again in Chapter 13 (with a photocurrent-integrating ADC). Those illustrated nicely the use of discrete analog and digital electronics. We conclude our pursuit of the perfect suntan integrator in this chapter, with (naturally) a microcontroller implementation. We take a first stab at it here, with a bit of refinement later in the chapter.

To review, the task is to inform the sunbather when to turn over (or go home), after having received the desired accumulated dose of sunlight, using as input a photodiode's current (which is proportional to suntanning intensity). The beachgoer sets the target FSE (full-sunlight-equivalent) dose with a pot, on a scale of 0–90 min FSE, pushes the START button, and begins the bake cycle. A piezo buzzer beeps when the job is complete.

15.2.1 Implementation with a microcontroller

An classic (but uninspired) way to approach this is to con-

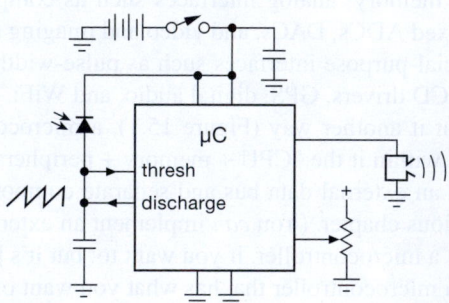

Figure 15.2. Suntan monitor, simply implemented with a microcontroller (µC).

vert the current to a voltage (with op-amp and feedback resistor), then use the microcontroller's on-chip ADC to generate digitized samples, which are then integrated numerically. But there's a simpler (and better) way, easily implemented even in a 50-cent microcontroller. Look at Figure 15.2. The idea is to use the photocurrent to ramp the voltage on a capacitor, then use the microcontroller's analog comparator to trigger a short discharge pulse on a three-state digital output port.[5] This generates a sawtooth whose frequency is proportional to light intensity; the microcontroller integrates this by simply counting cycles, sounding the horn at the desired terminal count, which it derives from a measurement of the "dose" pot's setting. A digital input-port bit senses the START button, and a digital output-port bit activates the piezo buzzer.

Not only is this simpler than the humdrum $I \to V \to ADC$ method, it is better: converting a current to a proportional voltage requires good dynamic range (and therefore very low offset voltages) to integrate accurately at low light intensities, whereas this $I \to f$ oscillator preserves accuracy by generating a fixed-amplitude sawtooth with excellent linearity at low currents. This departure from the usual modular engineering design approach illustrates nicely the sort of "holistic" creativity that microcontrollers stimulate in their devotees.

The circuit of Figure 15.2 is by far the simplest of the five iterations of a suntanning circuit, though, of course, there is the task of writing and downloading the embedded code (which we'll look at next). Before we do that, let's flesh out the circuit in all its glory – component values, choice of controller, PINs, etc. There's plenty to learn by doing this; in an important sense, dealing with the details and the compromises they compel is the essence of electronic circuit design.

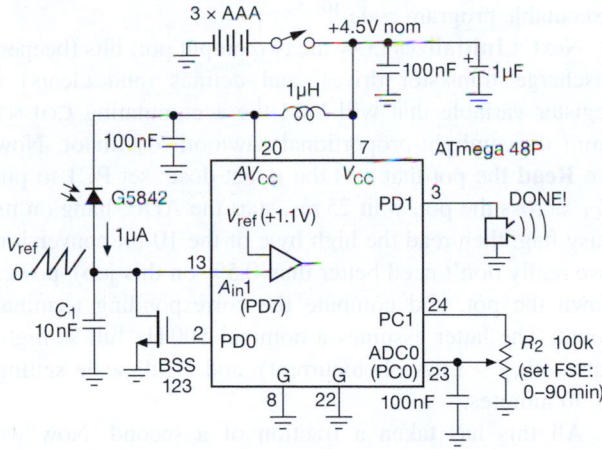

Figure 15.3. Detailed circuit of the suntan monitor. See Figure 4.93 for information about the G5842 GaAsP photosensor.

Figure 15.3 is a complete implementation, using parts available at the time this is being written. Let's look at some details.

A. Choice of microcontroller
This gadget doesn't demand much, in terms of speed or peripherals, and nearly all microcontrollers include a comparator and an ADC. However, we want low current consumption, and we'd like to run directly from a battery, without a voltage regulator. We chose this part because it is the smallest of the "picoPower" series of Atmel's AVR controllers, which have the virtue of low-power operation even when running from +5 V.[6] It operates from 1.8–5.5 V, and draws about 35–170 μA supply current (over that range of supply voltages) when running from its internal 128 kHz oscillator.

B. Discharge circuit
Although you could probably get away with using a three-state output to discharge the oscillator's capacitor directly (Figure 15.2), the worst-case specifications would deny you the satisfaction of a job well done: they specify a maximum leakage current of $\pm 1 \mu$A, comparable to the photocurrent in sunlight. You could choose a larger photocell, with correspondingly higher photocurrent; but an external n-channel MOSFET does the job nicely.[7] Here we've used a small BSS123 n-channel MOSFET as an external switch. It's a good choice: it has a satisfyingly low threshold, with ~ 0.5 A drain current at 2.5 V of gate drive, and a worst-case leakage of 10 nA at $V_{DS}=20$ V. It's stocked widely, priced at $0.05 in quantity.[8]

C. Dose-setting potentiometer
The simple diagram biased the pot from the positive supply rail. But good design requires its bias current to be large compared with the 1μA input leakage spec, which adds significant power drain. We can do better, though, by powering the pot only when we need to, i.e., when reading its value. That's easy – just tie its topside to a digital output-port bit; when the software sets that bit HIGH, the internal p-channel pullup brings it to $+V_{CC}$.

---

[5] You could arrange things to do the discharge on the same pin as the threshold comparison, by dynamically changing a bit in the register that sets the port-bit modes.

[6] Many controllers run at low power when operated at low supply voltage (say 1.8 V), but not at 3–5 V, as here.

[7] You could, instead, be reasonably confident that the typical leakage current is probably less than 10 nA, particularly since the specifications are given over the full temperature range of $-40°$C to $+85°$C; furthermore, semiconductor manufacturers are notoriously conservative when specifying leakage. You don't have to worry about the comparator's input pin, though: its specified leakage current is 50 nA (max).

[8] The BSS123 is similar to the 2N7000 and 2N7002 (see Table 3.4a).

D. Bypassing and decoupling

Figure 15.3 is decorated with decoupling capacitors, perhaps a bit conservatively. We decoupled the analog supply rail (AV_{CC}) from the noisy digital V_{CC} with the LC filter recommended in the datasheet. We also bypassed the pot readout, because it is biased from an equally noisy digital output. Finally, we used a parallel combination of supply-rail bypass capacitors to maintain low impedance over frequencies. (The series inductance and resistance of large-value capacitors reduce their effectiveness at high frequencies, remedied here by the smaller 100 nF ceramic bypass capacitor; see Chapter *1x*). Once again, we could probably get away with less, particularly since we need very little accuracy ($\sim$6 bits) from the part's too-good internal 10-bit ADCs, for which the bypass specifications were intended. But we like to play it safe.

E. Oscillator

There isn't one! That's because this controller, in common with many, includes internal oscillator options; for this chip there are two internal oscillators (8 MHz and 128 kHz), of modest accuracy ($\pm$10%).

15.2.2 Microcontroller code ("firmware")

The embedded software, called *firmware*, needs to do several tasks, namely,

(a) initializations at power-up (set up I/O port modes, ADC and comparator modes, and reset state of output port bits);

(b) assert V_{CC} to the pot, wait 25 ms, read its voltage, then disassert;

(c) use the measured voltage to compute the terminal count of oscillation cycles from the sawtooth oscillator, corresponding to the integrated sunlight exposure set by the pot;

(d) use the comparator output state to increment the cycle counter and to generate capacitor discharge pulses;

(e) when the counter reaches the computed terminal count, activate the piezo buzzer.

In addition, there are some operating characteristics (for example, oscillator frequency) that are not under program control and that are set instead at the time the program code is downloaded to the device. More, soon, about programming these so-called "fuses."

A. Pseudocode

We've written this in somewhat more detail in Pseudocode 15.1; pseudocode is a less formal substitute to a graphical *flowchart*, with more readable text.

A bit of explanation: the code begins with some **Setup** tasks. That is because contemporary microcontrollers are richly endowed, with built-in peripherals, operating modes, pin function options, and the like.[9] This flexibility and power are both a blessing and a curse. The blessings are obvious. The curse extracts its revenge, in requiring you to configure all those options. That is done by setting or clearing bits in an array of registers whose function is to configure and control the device's options. This particular processor has 256 such internal registers, with some 40 bytes worth of configuration and control bits. You generally don't have to worry about most of them, because they default to reasonable values; but you do have to set up the range and mode of the ADC and comparator, and the port bit directions.

The program code is executed on power-up, and is obliged to do first this fussy up-front housekeeping: microcontroller port bits can individually be configured as inputs or outputs (under program control), which you specify by setting bits in a port direction register (if an input, you can also enable or disable internal pullup). The setup code continues by initializing the buzzer output bit LOW and the discharge pin HIGH, and concludes with choice of ADC and comparator voltage references, scaling, and input sources. There are a few other configuration bits that have to be set, specifically the clock source (external or internal), the clock frequency, and the clock divider ratio. For this particular microcontroller, these "fuse" bits are set during hardware device programming, rather than in the executable program code.[10]

Next it **Initializes** LOW the two output port bits (beeper, discharge transistor drive), and defines (and clears) a register variable that will hold the accumulating COUNT from the sunlight-proportional sawtooth oscillator. Now we **Read** the pot that sets the target dose: set PC1 to put V_{CC} across the pot, wait 25 ms, start the ADC, hang on its busy flag, then read the high byte of the 10-bit conversion (we really don't need better than 0.5% on this job), power down the pot, and compute the corresponding terminal count. The latter assumes a nominal 100 Hz full-sunlight sawtooth, ($\sim$ 1 μA photocurrent), and a full-scale setting of 90 minutes.

All this has taken a fraction of a second. Now we

---

9 For example, this processor's pin 28 is bit 5 of byte-wide bidirectional port C; it can also be used as an analog input to the ADC (via the internal 8-input multiplexer), or as the SCL serial clock, or as a "pin-change interrupt" source.

10 For high-pincount microcontrollers it's common to see a few pins used to determine what happens at boot-up, with program control then in charge of those parameters.

Pseudocode 15.1 Suntan monitor pseudocode

Setup

> `Variables`: Define 32-bit integers `count` and `termcount`
> `Low Power`: Disable unused peripherals
> `Ports`: Set as outputs PC1 (pot bias), PD0 (cap discharge) and PD1 (buzzer);
>> set as analog inputs PC0 (pot readout) and PD7 (cap voltage);
>> initialize PD1 LOW (buzzer off), and PD0 HIGH (cap held at gnd)
> `Analog modes`: Set comparator mode & bandgap ref;
>> set ADC ref to Vcc, ADC mode to left-adjusted, and MUX to channel 0

Read "bake" setting

> Setbits PC1 (power to pot) and ADEN (ADC enable); wait 100 ms
> Start ADC conversion (setbit ADSC)
> Wait while ADSC=1 (busy), then read unsigned 8-bit result ("bake")
> Clearbits PC1 and ADEN, and disable ADC to save power
> Compute `termcount`=360000 x bake/256

Count Cycles

> Wait while comparator is HIGH ($V_{cap} < V_{ref}$), then:
>> setbit PD0, then clearbit PD0 (software discharge pulse)
>> increment `count`
>> if `count`<`termcount`, repeat **Count Cycles**
>> otherwise set bit PD1 (buzzer), clear 10 sec later

Count cycles of the sawtooth, whose ramp is generated by the photocurrent, abruptly discharged by a software pulse (HIGH, then LOW) to the MOSFET. Because there's nothing else to do anyway, we've used simple polling of the comparator's output bit, rather than the more elegant alternative of a counter-driven interrupt (our comparator mode setup disabled its interrupt). When the count reaches the computed target we turn on the beeper and hang "forever" (that is, until the bakee switches the thing off; restart at your peril – twice-baked human!).

B. Detailed C-code

Program 15.1 is a listing of the C-language source code. It should be readable by fluent C programmers, though there are a few quirks associated with microcontrollers. The `io.h` and `fuse.h` files contain the idiosyncratic register definitions, functions, etc., that deal with peculiarities such as fragmented address spaces and with bit variables. Bit twiddling in ports and registers is common in microcontroller applications; it's done here by OR and AND masking, getting the bit positions from the `defines`, and creating the OR mask with "1<<n" and the AND mask with "~(1<<n)". (For example, `PORTC1` has value 1, as defined in `io.h`; so the program statement "`#define POT (1<<PORTC1)`" sets `POT` equal to

0x02.) And a tricky *gotcha*: in an application like this, in which values are read from memory-mapped ports (such as a comparator or ADC), it is essential to declare those variables as `volatile` to prevent the compiler from "optimizing" away the reload from the associated port. This has been done for you in the `io.h` file, for the standard predefined memory locations; but you'd need to do it for any custom memory-mapped variables you create.

C. Some comments

(a) We used the power-on switch to initiate the timing cycle; another way to handle this is to use a pushbutton that brings an input pin to ground; you could poll that input bit, or you could use a level-sensitive interrupt. The latter is more elegant and has the benefit that the processor could put itself into a micropower ($<1\,\mu A$) "sleep" state when done; the processor would then wake up upon a pin LOW interrupt. In principle you could even omit the power switch! We'd advise against that, though, because the only recovery from a crashed processor would be to remove the battery. We've all experienced commercial electronic devices (answering machines, cameras, DVRs) that occasionally freeze up, requiring a "cold reboot" by removing the battery or unplugging the device.

Program 15.1

```c
#include <avr/io.h>
#include <avr/fuse.h>
#include <util/delay.h>

#define DISCHARGE    (1<<PORTD0)
#define BUZZER       (1<<PORTD1)
#define POT          (1<<PORTC1)

int main() {
    long termcount, count;  // Total timer counts and running counter

    // Power saving measures
    PRR = ~(1<<PRADC) & ~(1<<PRSPI); // Shut off peripherals except ADC & SPI
    DIDR0 = 0x3f;  // Disable digital input buffers on analog pins

    // Setup the pins
    DDRD   = DISCHARGE | BUZZER;  // Set two pins to output, rest to input
    DDRC   = POT;          // Set the POT pin to output, rest to input
    DIDR0 |= (1<<ADC0D);  // Use PC0 as ADC0 -- the ADC input
    DIDR1 |= (1<<AIN1D);  // Use PD7 as AIN1 -- the comparator input
    PORTD = BUZZER;  // Hold cap low, and start with buzzer off

    // Comparator Setup
    ACSR = (1<<ACBG); // Set the reference to the band gap (needs 70 us)

    // Read the desired exposure duration
    PORTC |= POT;  // Turn on the top of the resistor divider.
    _delay_ms(25);  // Delay 25 ms for 10RC settle
    ADMUX = (1<<REFS0) | (1<<ADLAR); // Use Vcc ref; left-adjusted result
    ADCSRA = (1<<ADEN) | (1<<ADSC);  // Enable, and start ADC

    /*** Wait until ADC conversion is done. ***/
    while ( ADCSRA & (1<<ADSC) ) { }

    termcount = (360000L * ADCH) >> 8; // Convert ADC result to timer count
    PORTC &= ~POT;  // Turn off the top of the resistor divider
    ADCSRA &= ~(1<<ADEN); // Disable ADC
    PRR |= (1<<PRADC);  // Enable power-reduction for ADC

    /*** Wait until desired sunlight exposure ***/
    for (count = 0; count < termcount; count++) {
        // Wait for cap to charge, then comparator output goes low
        while(ACSR & (1<<ACO)) { }
        PORTD |= DISCHARGE;  // Discharge the capacitor
        PORTD &= ~DISCHARGE; // And release it to recharge
    }

    // Buzz for 10 seconds
    PORTD |= BUZZER;     // Power the buzzer
    _delay_ms(10*1000);  // Delay 10 seconds
    PORTD = BUZZER;      // Turn off buzzer

    // Loop forever
    while (1) { }
}
```

(b) We polled the comparator to determine when the sawtooth reached the trigger voltage; you could, instead, use an interrupt. There's no benefit here, however, because there's nothing else to do anyway: with an interrupt you would discharge the capacitor and increment the counter in the interrupt service routine, while the main program would simply loop on testing the value in the counter.

(c) Microcontrollers are *smart*! With hardly any work we can add all kinds of cool features to this puppy: an LCD display with a "progress bar" to show the state of your tanning; a readout of sunlight intensity; an MP3 music player; a game of solitaire or scrabble… (but we digress: you get the point).

D. Downloading the code

The last step is to "program" the part (i.e., download the compiled code and set the fuses). Here's what you do.

(a) Connect the pod to the appropriate pins on the microcontroller (via a programming header), and to the PC (through serial or USB).

(b) Power up your board.

(c) Use the software on the PC to verify that the pod detects the correct device.

(d) Decide on the fuse-settable options,[11] and program them.

(e) Select the location of the compiled HEX file on the PC, and download it into the microcontroller flash memory; optionally select a HEX file with the EEPROM contents and download it.

(f) The pod should automatically reset the chip and begin running your program. At this point you can remove the pod. When you power cycle your board, your code will run, no strings attached.

E. Human-interface review

Our astute colleague Jim MacArthur offered the following commentary.

> If this book were about human interface (HI) design, design example #1 would be a textbook case of *spec drift*. The specification couldn't have been simpler: Duplicate the functionality of the analog solution. And yet, as your HI tester will doubtless tell you, this solution differs from the analog one in one important way: If you adjust the pot in the middle of the cooking cycle, the analog solution pays attention, whereas the digital one does not.[12] The

HI person would then go on, rather annoyingly, to prove that 47% of the users will operate the device in precisely that way – turning it on and then adjusting the pot – and that 13% of those customers will return the unit without trying to figure it out, costing your company millions of dollars… and you your bonus. The lessons are: (a) Never *ever* ship a consumer product without a human interface tester's blessing. (And that way, they take the blame.) (b) Analog designs are not as easy to simulate in the digital domain as they might first appear.

F. Design review

After the embarrassing HI experience, a full design review was scheduled. Someone remarked that anything sitting on the sun-baked sand would get very hot, to which someone else replied "Tell the user, who's sitting on a beach chair, to put the monitor in the shade underneath." This suggestion received a (wordless) withering stare. Moving along, the design temperature was set at 85°C. That's pretty hot, and (with a doubling every 10°C) the BSS123's leakage current would rise to an unacceptable $3.2\,\mu A$ ($64\times$ greater than the datasheet's 50 nA typ at 25°C). Oops! Interestingly, the μC's comparator is OK, because its 50 nA input spec extends to 85°C; similarly the photosensor's dark current (50 pA at room temperature) would increase to about 3 nA, also OK.

Exercise 15.1. Find an alternative to the BSS123 MOSFET reset switch: one possibility is to adapt the method of Figure 13.49 to a microcontroller. *Hint:* Replace U_2, U_3, and U_4 with the processor. Recall that the voltage at "X" doesn't matter. In that design we suggested an LMC6842 op-amp, which specifies 10 pA maximum input bias at 85°C; that's fine here, but the op-amp's supply current is 1 mA, dominating the battery drain. See if you can find a better choice.

15.3 Overview of popular microcontroller families

For the simple suntan monitor, we chose a controller from the popular Atmel 8-bit AVR series. We like them, because they exemplify a clean architecture, with plenty of general-purpose registers and a flat address space. They come in multiple package styles, including the through-hole dual in-line package (DIP) that we favor for quick prototyping. They are also well supported in terms of software tools, including open-source C-language compilers that are not too encumbered with controller-specific idiosyncrasies.

What other controllers are out there? We offer a listing of contemporary offerings below, with a bit of

---

[11] For the AVR these include brownout voltage level, on-chip debug on/off, clock source (internal, external square wave, external crystal, etc.), JTAG enable, SPI enable, and watchdog timer enable.

[12] The software could, of course, be modified to "pay attention."

commentary. This should be considered a snapshot in time, as microcontrollers constitute one of the faster-changing areas of electronic technology.[13] You might keep in mind the characteristics of an ideal microcontroller:

- plenty of on-chip flash (reprogrammable program memory) and RAM
- small versions available in easily prototyped DIP (or SMT on DIP adapter)
- internal oscillator
- minimum external parts count to get started
- fast, low power consumption, wide supply voltage range
- inexpensive programmer, serial or USB
- in-circuit programming
- free programming software
- free assemblers and compilers
- free Integrated Development Environment (IDE) to tie it all together
- in-circuit debug/emulate through the IDE
- open-source toolchain that runs in Windows, Mac, Linux
- active user community

AVR *(Atmel)* 8-bit RISC[14]; clean architecture, 32 general-purpose registers; in-circuit debugging; open-source (Linux, GCC, Arduino) tools supported by Atmel and integrated into their software; some chips include USB; many package styles (including DIP); available as hybrid/FPGA (Atmel); least expensive, less than $0.50; competes with Microchip's PIC. 32-bit series: AVR32. Programmers: simple serial ($12); Atmel AVR Dragon USB in-circuit debugger ($50). Arduino boards with USB power and programming are $20–$30. AVR Butterfly board with LCD and buzzer ($20). *Our recommendation for getting started.*

PIC *(Microchip)* 8, 16, and 32-bit; high-speed and low-power versions; some in-circuit debugging; many development kits; many package styles (including DIP); FPGA soft core; long time favorite of hobbyists; least expensive, less than $0.50.
8-bit: PIC10F, 12F, 16F, 18F. 16-bit: PIC24, dsPIC30, dsPIC33. 32-bit: PIC32MX.
Some chips include integrated USB or Ethernet MAC and PHY. Free chip samples from Microchip. The older 8-bit families, especially PIC16F84, used to be extremely popular (newer compatible versions have in-

ternal oscillators). Assembly-language coding can be tricky, owing to memory banking, a single accumulator, conditional skip (rather than jump) instructions, and an 8-level call stack in hardware (not RAM); but timing of instructions and interrupts is very predictable.
The 10F–16F devices are partially supported by C compilers by SourceBoost and Hi-Tech C, with prices ranging from free (limited versions) to $3000. Only 18F and above are well suited for C programming, and are supported by Microchip's C compilers (a free limited version is available to students). The 16-bit devices have the stack in RAM and no bank switching, and can be programmed with a variant of the GCC C-compiler. The 32-bit devices use an industry standard MIPS core (similar in features to the ARM), with a Microchip C compiler based on GCC.

ARM *(multivendor)* 32-bit RISC with optional 16-bit instructions; ARM Cortex M3, ARM7, and ARM9 series; least expensive, around $2; high-performance; in-circuit debugging; large range of included functions and I/O; open-source C/C++ with GCC and Eclipse debugging GUI (graphical user interface), also Arduino software for ARM-based *Due*; flat memory model that is easy to program; pincount ≥ 28 only (no DIP). Some can be programmed through USB, but need JTAG pod ($50) to debug. Available as prebuilt header boards from multiple vendors ($30), protoboards ($60); also Arduino Due (pronounced "doo-way," $50), UDOO (udoo.org, $99), Raspberry Pi ($30), Odroid ($65), and others.
Manufacturers: ADI, Atmel, Broadcom, Cypress, Freescale, Infineon, Microsemi, NXP, Renesas, Samsung, Silicon Labs, ST, TI, Toshiba; available as hybrid/FPGA (Altera SoC series, Xilinx Zynq series). From 8-bit replacements to iPhones. Pipelining and complex instructions provide lots of speed, but less predictable running time. *Would be our recommendation for getting started, if prototyping a design on your own board were easier.*

8051 *(multivendor)* 8-bit; evolution of early μC; mature development tools and development kits; some quirkiness in C-code, owing to complex address space; some in-circuit debugging (Silicon Labs); legacy instruction set limits memory, register, and I/O options; many package styles (including DIP); FPGA soft core. Good free development software (limited), e.g., "Ride" IDE (www.raisonance.com): C-compiler, assembler, simulator.

Rabbit *(Rabbit)* Emphasis on Ethernet and WiFi; available as modules, or as bare chips; C-language IDE; development kits and tools; popular.

---

[13] Stay abreast by looking at magazines like *Circuit Cellar*, *EDN*, and *Electronic Products*; distributors like DigiKey and Mouser; or do a search for "<μCname> tutorial" or "getting started with <μCname>."

[14] Reduced instruction-set computer; the opposite is a complex instruction set, CISC.

MSP430 *(TI)* Micropower 16-bit RISC; popular for low-power, RF, and LCD controllers; many package styles (including DIP). JTAG programming ($75–$100); open-source compiler MSPGCC and free code-size-limited versions of commercial compilers.

SH-4 *(Renesas, formerly Hitachi)* 32-bit RISC; smallest package is QFP-208 (LQFP-48 for the SH-2); popular in motor and engine control. M16C/R8C are 16-bit, available in DIP.

Coldfire *(Freescale, formerly Motorola)* Embedded 32-bit 68000-like architecture; no packages smaller than 64-LQFP; GNU C compiler; development kits (Tower System).

ST6/7 *(STMicroelectronics)* 8-bit, available in DIP; ST9/10 are 16-bit, smallest package LQFP-64. STR7/9/32 are 32-bit ARM7/9/Cortex, smallest package VFQFN-36 or LQFP-48.

PowerPC and MIPS *(multivendor)* Used in high-end embedded systems such as automobiles, networking, and video; by contrast, ARM dominates in cellphones and PDAs.

Blackfin *(Analog Devices)* 16/32-bit RISC plus DSP; fast and high-end, optimized for audio, video, and imaging. Easier than some other high-end controllers. Open-source GNU compilers, uClinux, FreeRTOS.

Propeller *(Parallax)* 8-core, 32-bit parallel processors; dc–80 MHz; small line of DIP-40/44 or SMD; philosophy of "why have lots of built-in hardware peripherals like UARTs and SPI when you can put eight real processors in there, give them all access to the I/O, and write good libraries to bitbang everything from mice and keyboard interface to analog video (VGA/NTSC)."

PSoC *(Cypress)* 8-bit microcontroller with configurable analog blocks: Op-amp, ADC/DAC, filter, modulator, correlator, peak detector, etc.

XMOS *(XMOS Ltd.)* Event-driven multithreaded 32-bit processors, with events and threading in hardware up to 400 MIPS; programming in XC, C/C++, assembly; free development tools.

15.3.1 On-chip peripherals

All microcontroller families include on-chip program and data storage, and timer/counters. They generally offer a selection of options such as comparator, ADC, and DAC; I^2C, SPI, and CAN buses; UARTs, USB and Ethernet; and support for external LCD, pulse-width modulation (PWM), video, and wireless devices. Here is a listing of some on-chip or low-level support found in contemporary microcontrollers, roughly in order of increasing level of complexity:

A. Low complexity

ADC, DAC, Analog Comparator
CANbus
Debug (JTAG or proprietary 1- or 2-wire interface)
I^2C / SMBus / TWI ("Two-Wire Interface")
Interrupts, sometimes with priority
Keypad matrix scan
Keyboard (serial), mouse
LCD (bare)
PWM
Real-Time Clock
SIM Card / Smart Card serial interface
Synchronous Serial Ports (SSPs): SSI, SPI, SSI Microwire
Timer, counters, watchdog
UART (RS232, RS485, IrDA), some with modem control

B. Mid-level complexity

AC97 (Intel Audio 20-bit 96 ksps stereo PCM)
Bluetooth wireless
Compact Flash (CF card)
Ethernet
External SRAM
GPS
I^2S (Inter-IC Sound: CD to DAC at 2.8 Mbps)
IrDA (high speed, to 4 Mbps)
Motor control, shaft encoder
PCMCIA[15] ("PC card")
S/PDIF (AES/EBU) audio (Dolby Digital or DTS surround)
SD/MMC flash cards
USB 1.1 (Host or Device, to 12 Mbps)
ZigBee wireless

C. High complexity

Camera and image sensors (CMOS, CCD)
External DRAM/SDRAM
Graphics display (bare color LCD)
MMU (Memory Management Unit) with OS protec
MPEG4 encode/decode
Operating System (Linux, Windows CE, Palm)
PCI, PCIe

---

[15] Sometimes translated as "Personal Computer Manufacturers Can't Invent Acronyms."

Storage Drives: ATA, IDE, SATA
USB 2.0 high speed (480 Mbps)
Video (NTSC, PAL, VGA, DVI, DV)
WiFi (802.11)

We predict, with considerable confidence, that the list will expand and that some items will be demoted.

We turn now to four more design examples to illustrate some common ways in which microcontrollers are used in embedded circuit design. These examples deal with
(a) ac power control (in which a microcontroller switches line power, under the direction of a serial link);
(b) a frequency synthesizer (in which a microcontroller runs errands between a user and a direct-synthesis chip that only understands cryptic serial commands);
(c) a precision temperature controller;
(d) a stabilized 2-wheel pseudo-Segway™contraption (which its inventor calls a "Psegué").

15.4 Design example 2: ac power control

At our astronomical observatory we have a dozen or more ac-powered devices that we need to control remotely – things like dome and telescope drives, mirror heaters, lights and cameras, power supplies for detectors and processors, and the like. Building a circuit to do this is a breeze with a microcontroller. And a benefit to having some intelligence in the control box is that it can remember a default configuration, and it can report status information (which outputs are powered, how much current they are using, and so on).

Let's do a fairly simple example here, namely a controller to switch 110 Vac power on two outlets, with readback of the commanded state, and with confirming readback that the actual ac output is present. In addition to remote control via serial RS-232 or USB, it will also permit "local" control via switches on its panel; some LEDs there will indicate which outputs are powered.[16]

15.4.1 Microcontroller implementation

Figure 15.4 shows the circuit. For variety we chose a modest member of the popular PIC family; the particular part (PIC16F627, DIP-18, $2) includes the needed serial port UART (universal asynchronous receiver–transmitter), an internal oscillator, nonvolatile data memory (EEPROM) that guarantees endurance of a million erase/write cycles, and plenty of digital port pins. We don't need much of anything else that the manufacturer has thrown in (analog multiplexer, comparator, PWM hardware, 20 MHz speed), but the microcontroller costs just $1.50, a small fraction of the total parts cost, so, hey, why complain.

The easiest way to switch ac power from a logic signal is to use a *solid-state relay* (SSR; see §§12.7.5 and 12.7.6), which consists of a sealed module containing an optocoupled SCR or triac. These devices are available from many manufacturers.[17] They accept a logic-level dc drive (often specified as 3–15 Vdc, or 3–32 Vdc), drawing somewhere in the range of 3–15 mA. The output is well isolated (typically specified as 3.5–5 kV), and most SSRs implement zero-voltage turn-on and zero-current turn-off. Typical load ratings are 10 A or 20 A at 280 Vac, but you can get standard units up to 100 A and 660 Vac.

In our example we've use a digital port bit to drive the SSR, configured to power on logic LOW; that choice of polarity is good here, because the port pins specify a typical saturation voltage of ∼0.35 V when sinking the required 15 mA, versus ∼1.3 V (below the positive rail) when *sourcing* that current. A panel LED rides along, indicating relay drive.

Solid-state relays have some leakage current when off, typically specified in the range of 0.1–10 mA; so we've bridged the switched ac with a 15k resistor.[18] The ∼8 mA current through it, when on, is used to light a panel LED (indicating delivery of ac to the load), and also to drive an opto-isolator to signal the microcontroller that all is well. This optocoupler is designed for ac operation, with its internal back-to-back LEDs; this particular unit has 5 kV isolation, 50% minimum current transfer ratio (ratio of phototransistor output current to LED drive current), and costs about $0.25 in moderate quantity.

Exercise 15.2. What is the purpose of the 1k resistor? Why is a diode needed across the indicator LED?

Moving back to the microcontroller circuitry, we've taken advantage of the internal "weak pullup" digital input mode (specified as 50–400 μA) for the input switches: a momentary contact pushbutton (for initiating local control), and a pair of spring-return momentary-contact lever

---

[16] If you want to save time (and cheat yourself out of an education), you can instead just throw money at the problem; for example from Pulizzi Engineering (Eaton Corporation) or APC (Schneider Electric). We opted for the educational opportunity, thanks in part to a grant from the Shulsky Foundation.

[17] For example, Crouzet, Crydom, Magnecraft, Omron, Opto22, P&B, and Teledyne.

[18] We measured the off-state leakage for two SSRs, and found ∼2 mA and ∼6 μA for two Crydom 240 Vac SSRs (a D2425 and EZ240D18, rated at 25 A and 18 A, respectively). They specify worst-case leakages of 10 mA and 100 μA, respectively.

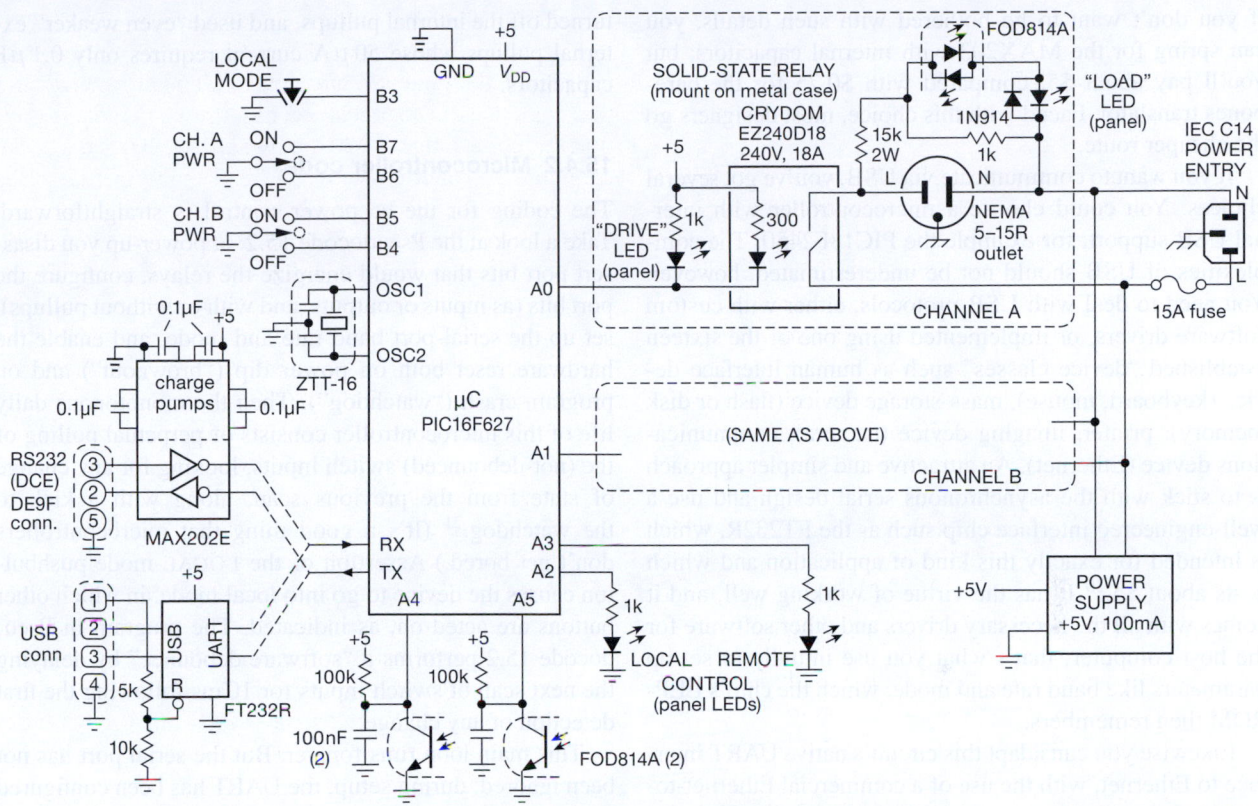

Figure 15.4. Dual ac-power control box, with remote (serial) or local controls. The SSR dissipates 10 W at a load current of 10 A, so it needs adequate heat-sinking. A mechanical relay could be used instead (e.g., Panasonic ALE12B05, contact rating of 277 Vac and 16 A, with faston tabs like the Crydom SSR). For the latter you'll need a transistor to drive its 80 mA coil; but the good news is its price: $1.60 in unit quantities, compared with $35 for the SSR.

switches for commanding the outputs on or off while in local control. A pair of panel LEDs indicate whether the unit is under local control (i.e., panel switches) or remote control (i.e., from a host computer, connected to this device via RS-232 or USB serial links).

We chose this particular controller partly on the basis of its internal oscillator...and then we discovered that it isn't accurate enough! The specified accuracy is ±7%, not good enough for 8N1 asynchronous serial communication: as explained in §14.7.8, the receiving device has to sample the 8 data bits in the middle of each bit cell, using its own asynchronous baud clock, having synchronized to the START bit at the beginning of each byte. So a 7% clock rate error between transmitter and receiver slides the last (8th) bit sample clear into the preceding or following bit cell.

The solution is either to use an external oscillator of adequate accuracy (for example, a crystal oscillator module, which costs about $2), or to put an external bare crystal or ceramic resonator across the OSC1 and OSC2 pins that

the microcontroller provides for this purpose. We chose the latter, using a convenient 3-pin resonator that includes the necessary capacitors; this puppy costs about $0.40, with a specified accuracy of ±0.5%.

This microcontroller has an internal UART that makes serial communication pretty simple.[19] The only external hardware needed to connect directly to a host computer RS-232 "COM" port is a level translator (RS-232 levels are bipolarity; see §14.7.8). We chose the popular MAX202, which runs from the single +5 V supply, and creates the needed ±10 V with an internal charge pump voltage   doubler–inverter; it requires four 0.1 μF capacitors.

---

[19] Although getting any microcontroller's UART settings right (timer mode and division ratios, serial data mode, interrupts, enables) is usually tricky. The best bet is to copy, shamelessly, the settings from the example that is invariably given in the applications section of the datasheet (more like a data*book*: typically 300–400 pages!), provided, no doubt, in the expectation that you stand only a small chance otherwise of getting it right.

If you don't want to be bothered with such details, you can spring for the MAX203 with internal capacitors; but you'll pay about $5, compared with $0.75 for the bare-bones translator. Faced with this choice, most designers go the cheaper route.

If you want to communicate via USB, you've got several choices. You could choose a microcontroller with internal USB support, for example the PIC18F2450. The complexities of USB should not be underestimated, however. You need to deal with USB protocols, either with custom software drivers, or implemented using one of the sixteen established "device classes" such as human interface device (keyboard, mouse), mass-storage device (flash or disk memory), printer, imaging device (camera), communications device (Ethernet). An attractive and simpler approach is to stick with the asynchronous serial design and use a well-engineered interface chip such as the FT232R, which is intended for exactly this kind of application and which costs about $4.[20] It has the virtue of working well, and it comes with all the necessary drivers and other software for the host computer; that's what you use initially to set up parameters like baud rate and mode, which the chip's EEPROM then remembers.

Likewise you can adapt this circuit's native UART interface to Ethernet, with the use of a commercial Ethernet-to-serial module like the Lantronix XPort. This elegant device looks like a somewhat elongated RJ45 (Ethernet jack), with all the electronics packed inside; it even has the usual green and yellow network activity indicator lights. It costs about $50.

We end the circuit tour with the microcontroller end of the opto-isolators. These have a floating phototransistor output, which we use to pull a pair of digital port input bits LOW when LED current flows at the ac-sensing end. (The LEDs shut off for a portion of each half-cycle, of order 1 ms, hence the 100 nF capacitors to hold the digital inputs LOW during those brief interruptions.) It's tempting to exploit the internal "weak" pullups, but they're not weak enough: according to the specifications, the pullup current can be as large as 0.4 mA, which would require 1 μF capacitors to keep the voltage at the input ports below 0.4 V during the $\sim$1 ms gaps. The solution here is to repudiate such bullying tactics: to regain control of the situation we turned off the internal pullups, and used "even weaker" external pullups whose 50 μA current requires only 0.1 μF capacitors.

15.4.2 Microcontroller code

The coding for the ac power control is straightforward. Take a look at the Pseudocode 15.2: at power-up you disassert port bits that would energize the relays, configure the port bits (as inputs or outputs, and with or without pullups), set up the serial-port baud rate and mode, and enable the hardware reset both on power dip ("brownout") and on program crash ("watchdog"). Then the monotonous daily life of this microcontroller consists of perpetual polling of the (not-debounced) switch inputs, looking for any change of state from the previous scan, along with a kick to the watchdog.[21] (It's a good thing that microcontrollers don't get bored.) Assertion of the LOCAL-mode pushbutton causes the device to go into local mode, in which other buttons are acted on, as indicated. The program in Pseudocode 15.2 performs a "software debounce" by delaying the next scan of switch inputs for 10 ms following the first detection of any change.

This main loop runs forever. But the serial port has not been ignored: during setup, the UART has been configured to generate an interrupt on receipt of an input character. So an incoming command on the serial port causes program diversion to the interrupt handler, where the seven possible commands (each consisting of a single byte printable character) are parsed and acted on. Those actions mostly consist of setting or clearing a single bit (either in a port, to control a relay; or a register, to set the local/remote mode). The "s" (request for status) command is more interesting, requiring the processor to collect the four bits indicating relay status (both the commanded output bit that drives the relay, and the optocoupled ac indication, for each of the two channels), and the mode bit; those are masked and shifted into a register, then sent back as a single binary byte.

A. Some comments

(a) Microcontrollers generally include some non-volatile data memory, which could be used to save the state of the outputs in the event of power failure or reboot; this controller has 128 bytes of EEPROM for this purpose.

---

[20] You can get it in several physical incarnations, for example, a cable with a USB connector on one end (in which the chip is concealed) and a 9-pin serial connector (with RS-232 voltage levels) on the other; alternatively, you can get it with flying leads at the serial end, configured for 3.3 V or 5 V logic-level operation, upgraded part. The upgraded part number is FT2232.

[21] Most engineers persist in calling this "kicking the dog." But our quirky colleague Frank Cunningham has pointed out that what you're really doing is *petting* the dog: when you kick a dog, it barks immediately; when you pet a dog, it barks only when you stop petting it.

Pseudocode 15.2 ac power control pseudocode

Setup

`Ports`: disassert (setbits) relay outputs (A0, A1)
  setports LEDs (A2, A3) and relays (A0, A1) as outputs
  setports switches (B3–B7) and opto (A4, A5) as inputs
  set switch input ports (B3–B7) as weak pullups
  set up UART: baudrate, 8N1, interrupts, enable
`Low Power`: disable unused peripherals
`Automatic Reboot`: enable "brownout reset" and watchdog

Switch Polling Loop

read switches (port B)
if (any switchbit has changed)
  if localmode switch (B3) asserted, set mode=local
  if mode=local and any switch went H→L
    turn ON or OFF that relay output
  delay 10 ms ("software debounce")
kick the watchdog
repeat **Switch Polling Loop**

Serial Interrupt Handler

get character
  if "r" or "R", set mode=remote
  if "l" or "L", set mode=local
  if "s" or "S", assemble 5-bit status byte and send
if mode=remote
  if "A", turn on A relay
  if "a", turn off A relay
  if "B", turn on B relay
  if "b", turn off B relay

(b) It is best to keep interrupt service routines short and simple; in particular, you would avoid tasks that tie up the processor in the service routine while waiting for a slow peripheral to complete (e.g., a serial port!), because that may cause the main loop to miss a switch-input event. Here it's OK, though, because every serial-port command and every response is a single byte, so the routine does not have to wait for a busy flag to clear.[22]

---

[22] There is a standard, known as SCPI (Standard Commands for Programmable Instruments), that specifies a set of commands, and a syntax, for communicating with programmable instruments. SCPI is a bit wordy, but it is clear, and it has been widely adopted by manufacturers of electronic instruments (e.g., Agilent, Fluke, Keithley, LeCroy, Rohde & Schwarz, and Tektronix). For example, to read a voltage you could send the command `MEASure:VOLTage:DC?` (the lowercase characters are optional). The measurement response is returned in a defined format.

15.5 Design example 3: frequency synthesizer

Our next example illustrates a common application of embedded microcontrollers, namely their use to manage communications between an instrument panel (or computer interface) and some circuitry that requires digital programming and control. Figure 15.5 shows a two-channel frequency synthesizer, using one of the elegant direct digital synthesis (DDS) chips from Analog Devices (§7.1.8). We've chosen the AD9954 (one of the three dozen current DDS offerings), which synthesizes a high-quality sinewave via internal table lookup and D/A conversion. It can generate output frequencies from dc to 160 MHz, with resolution set by the 32-bit "tuning word"; that corresponds to ∼0.1 Hz when a 400 MHz internal oscillator is used to clock the chip. Although you can get faster DDS chips, this one has excellent spectral purity, owing to its 14-bit output DAC; this translates to a clean output sinewave,

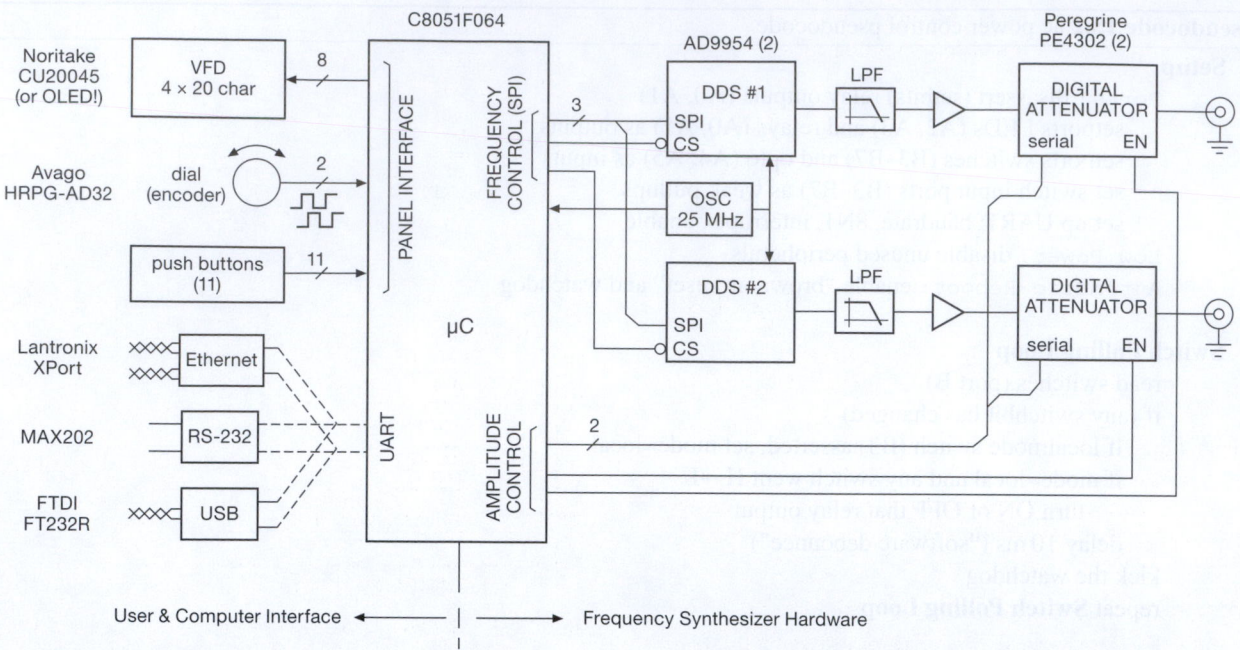

Figure 15.5. A microcontroller is ideal for interfacing between digitally programmed chips (here a pair of DDS frequency synthesizers, with programmable attenuators) and a set of human-operated panel controls and displays. And the μC's serial interface makes it easy to add remote computer control (here a pair of DDS frequency synthesizers, with programmable attenuators, see Figure 15.6).

with low phase noise and "spurs" (spurious frequency components).

This DDS chip, which costs about $25, won't do much of anything until you program it through its 4-wire SPI port. You could imagine interfacing the DDS to a laptop computer and running some software on it so you can program the chip for the frequency you want. But what you'd really like is a nice set of human-usable panel controls – a knob to dial up the frequency and amplitude and a numeric display to show those values (Figure 15.6). That's best done with an embedded microcontroller. And, of course, that controller can accept inputs via a USB or Ethernet port, so once you've built this instrument you can add remote digital control as well.

The block diagram (Figure 15.5) shows the microcontroller riding herd over the DDS chips (via their SPI serial ports) and the digital attenuators (via a similar 2-wire protocol, described in their datasheet). The synthesizer chips need an accurate 400 MHz clock reference. This is generated by its internal phase-locked loop (PLL) frequency multiplier from an external lower-frequency reference; we chose a 25 MHz crystal oscillator (with ×16 on-chip multiplication), which conveniently can clock the microcontroller as well.

On the user side there are buttons and a dial (an optical "incremental rotary encoder") for entering and changing parameters (frequency, amplitude, phase); and there's a vacuum fluorescent 4-line character display (VFD) that is pin compatible with the more usual LCD shown in the photograph, for telling the user what's going on. VFDs cost more than LCDs, but they are gorgeous. Remote control communicates through the on-chip serial UART, which requires only a voltage translation for a legacy RS-232 port; or you can use a smart serial format conversion device like the FT-232 from FTDI (for USB) or the XPort from Lantronix (for Ethernet).[23]

Although shown casually in the block diagram, there is additional detail in these user interface connections. The rotary encoder, for example, generates a pair of square waves, in 90° quadrature, from which you can figure out in which direction (and by how much) it has rotated. Inexpensive encoders use mechanical contacts, so you have to worry about the usual contact bounce. We've gone first class here, though, with an optical encoder (it lasts forever) that generates clean logic-level outputs (no bounce!). The

---

[23] Additional Ethernet interface devices are available from companies such as Digi International, Ipsil, Connect One, and WIZnet.

corresponding firmware simply notes the level on the other line when it senses a transition on either of the two inputs.

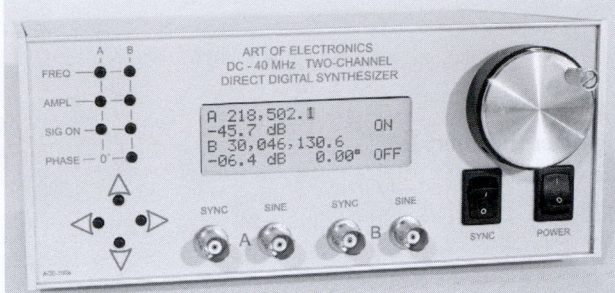

Figure 15.6. Dual DDS frequency synthesizer of Figure 15.5, with panel controls (eleven buttons and one rotary encoder dial) and readout. (Designed, built, coded, and documented with excellence by Jonathan Wolff; see www.artofelectronics.com/synthesizer)

The pushbuttons are mechanical, and prone to contact bounce; they must be debounced in software, as with the previous ac power control example. We've connected each button to its own input port pin (configured with weak internal pullup), which is the simplest configuration (though not the most economical). This particular microcontroller has 59 digital port pins scattered around its 100-pin package (Figure 15.7), so we're unlikely to run out. The alternative is a matrix-switch configuration, with row and column lines that are bridged by the individual switches, which here would require four fewer pins; we'll discuss that, along with other wrinkles, shortly.

15.5.1 Microcontroller code

The microcontroller's tasks are, briefly:

(a) set-up port bit modes, UART parameters and interrupt, SPI port, and watchdog;

(b) initialize DDS chips and attenuators to last stored state (held in nonvolatile memory), and send current operating parameters (frequency, amplitude, phase) to VFD display;

(c) execute forever a main loop that kicks the watchdog, checks for rotary dial movement, for button actuations, and for a serial command flagbit (set by the serial interrupt handler), and acts accordingly;

(d) service UART interrupts (independent of the main loop) by appending characters to a line buffer, and setting a software flag ("serial_cmd") upon receipt of a newline character (end of command).

The listing in Pseudocode 15.3, though somewhat abbreviated for this example, spells this out in greater detail.

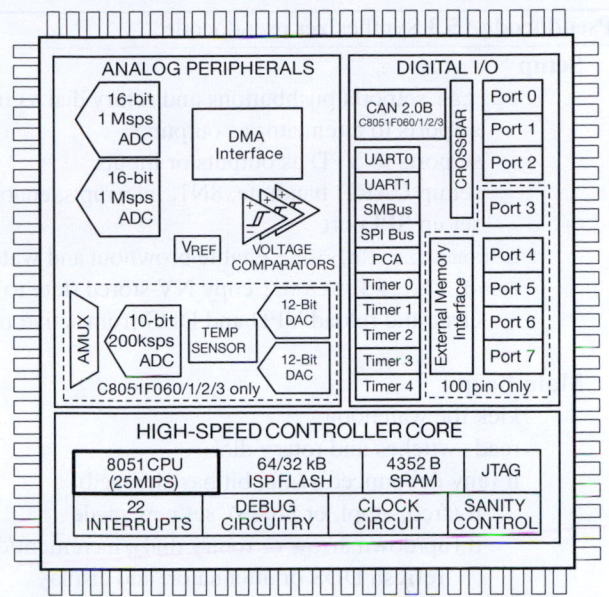

Figure 15.7. The venerable 8051 has come a long way since its 1980 debut, with many added functions, and one clock cycle per instruction (the original required 12) in contemporary variants like the C8051F060 series from Silicon Labs, from which this diagram is adapted, with their permission. (Curiously, the 328-page datasheet reveals nothing about the meaning or function of "sanity control.")

A. Some comments (Hardware)

(a) The block diagram shows relatively simple amplitude control, with a digital attenuator providing 0–30 dB attenuation in 0.5 dB steps. You can do much better, in range, step size, and calibration, by using the intelligence and memory of the microcontroller: the synthesizer in the photograph uses two additional layers of amplitude control, namely a fixed 30 dB attenuator that is switched in to reach 60 dB, and fine adjustment ($\pm6\%$) of the DDS chip's "reference current" (via an EEpot) to provide 0.1 dB resolution. One further trick is needed to make this all worthwhile: the user initially calibrates the instrument, by reporting measured signal levels at several frequency and amplitude settings over the full range, as prompted by the firmware; the microcontroller stores this calibration data in nonvolatile memory, retrieving it to produce a linear and accurate output-amplitude characteristic.

(b) The DDS output waveform updates at the internal reference clock frequency, in this case $f_{\rm ref}=400\,{\rm MHz}$, generating successive samples of the output "staircase" approximation to a sinewave. The resultant output signal, settable from 0 to 160 MHz, contains spurious out-of-band frequency components, the lowest of which is at

---

Pseudocode 15.3 Synthesizer pseudocode

Setup

    `Ports`: setports pushbuttons and rotary dial as input

        setports to attenuator as outputs

        setports to VFD as outputs or inputs

        set up UART: baudrate, 8N1, interrupts, enable

        set up SPI port

    `Automatic Reboot`: enable brownout and watchdog timer (timeout=1s)

    `Read Stored State`: copy NV-stored state to active registers

        clear and reload VFD and DDS from active registers

Main Loop

    kick the watchdog

    read switches and rotary dial

    if (any debounced switchbit has changed)

        if (freq, ampl, or phase), set new mode

        if (up/down arrow or rotary dial), increment/decrement that register

            refresh DDS or attenuator, and display

    if (serial_cmd bit) parse command, clear serial_cmd bit

        increment/decrement corresponding register

        refresh DDS or attenuator, and display

    repeat **Main Loop**

Serial Interrupt Handler

    append character to line buffer

    if newline, set flagbit serial_cmd

$f_{spur} = f_{ref} - f_{out}$. That's why lowpass filters are needed at the DDS's DAC outputs; typically you'd use an LC multisection elliptic filter with a sharp cutoff characteristic at $f_c \sim 180\,\text{MHz}$. The DAC outputs are differential *currents*, best filtered with a symmetrical differential filter, an example of which is shown in the 40-page datasheet.

(c) With dual DDS chips set to the same frequency, you can control the phase difference between the two outputs, providing the chips have a common reference clock. This particular DDS accepts a 14-bit phase offset word, equivalent to steps of 0.22° in relative phase.

(d) We used individual port input bits for each pushbutton, with a common ground return, a perfectly reasonable approach here with only 11 buttons, and a microcontroller with 100 pins. The alternative is to use a row and column *matrix*, with the switches bridging the intersections (Figure 15.8). Here the column lines are driven by output pins, and the rows are connected as inputs with internal pullups. To poll the switches the microcontroller pulls the column lines successively LOW, observing the returned input levels from the row lines; debouncing is done in software, as it

was with dedicated input lines. Matrix encoding is efficient when you have many switches, because the number of I/O pins used is the sum of the number of rows and columns, rather than their product. In some cases, moreover, you are forced to do it this way, because the switch array is built as a matrix; an example is a 4×4 hexadecimal keypad. One caution: a matrixed keypad generates false outputs if three (or more) keys are pressed simultaneously; figure this out for yourself.

B. Some comments (Firmware)

(a) Having a microcontroller in charge provides a nice opportunity to add features and functions that you wouldn't even think of otherwise. For example, the synthesizer in the photograph implements a nice algorithm (which we have not seen elsewhere) for determining how the digits are controlled by the dial and the buttons. The left- and right-arrow buttons do the obvious thing, namely choosing the ("active") digit being modified; and the up and down buttons increment and decrement that digit (with carry or

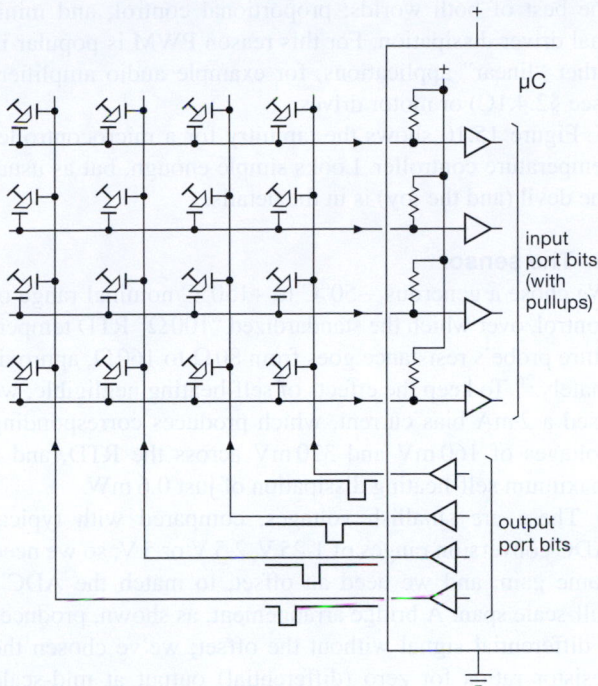

Figure 15.8. An array of switches can be polled in a row and column matrix arrangement, here illustrated with a 4×4 keypad. This reduces the number of port pins to $N_{\text{matrix}} = r + c$, compared with the $N_{\text{indiv}} = r \times c$ required for individual dedicated readouts.

borrow), assuming that you want to set an exact frequency. But the dial is treated differently: if you move the active digit leftward, the dial acts on that digit and zeros all digits to the right, on the assumption that you want to move the frequency faster, and without the baggage of leftover less-significant digits. On the other hand, moving the active digit rightward and spinning the dial preserves the more significant digits to the left, on the assumption that you are zeroing in on the target frequency. We wish commercial instruments would use this sensible algorithm!

(b) Just as with a computer's mouse, the panel dial's motion is translated by an acceleration algorithm, so that a fast spin of the dial takes you farther and faster.

(c) The firmware takes care of capturing initial calibration data and interpolating from that data during subsequent operation, as described above. A microcontroller lets you program frequency sweeps – linear ramps, cyclic ramps, or nonlinear sweeps (e.g., a logarithmic sweep, with equal time per octave).

(d) And, of course, a reprogrammable microcontroller lets you fix bugs, and dream up new features (with new bugs).

15.6 Design example 4: thermal controller

Here's an example of embedded control in the service of, well, *control*: imagine we want to hold steady the temperature of a well-stirred liquid bath, in which we've put both a heating element and an accurate temperature sensor. At first glance the solution seems easy: just apply plenty of feedback, using as the error signal the difference between the measured temperature (as reported by the sensor) and the desired target temperature. Raise the loop gain until the temperature error is small enough, just as you would with an op-amp voltage amplifier.

This is a classic "control" problem, encountered in industrial settings such as chemical plants (temperature and flow control), manufacturing and robotics (motion control), and the like (Figure 15.9). Simple feedback works poorly in such situations, because the controlled system has time delays (and therefore lagging phase shifts) that promote oscillation; and reducing the loop gain to prevent such oscillation leaves too little loop gain, so the controlled system both deviates from the desired condition and moves slowly to correct disturbances.

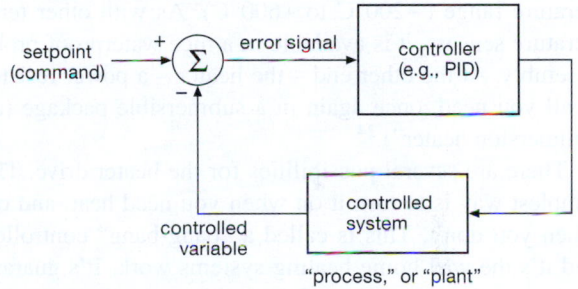

Figure 15.9. Classic control (or "servo-") system. For our example the setpoint is the desired temperature, the "plant" is the heater–waterbath–sensor, and the controller is implemented with a microcontroller. The μC's input compares the desired and actual temperature, and its PWM output powers the immersion heater.

There is a large literature on this subject, and on the feedback principles underlying the "tuning" of a controller to match the properties of the controlled system (called, curiously, the "plant," probably in a throwback to the industrial revolution). The general approach is to construct (with analog or digital methods) a so-called "PID" (proportional, integral, differential) controller, in which the error signal versus time is subjected to those three processes, and then combined in carefully chosen proportions to create the correction signal that is applied to the controlled system. The procedure for tuning the coefficients is basically straightforward (though it is considered by many to be a mysterious art), and is best understood with a Bode plot like those in Chapter 4. Although it's easy to make amplifiers, integrators, and differentiators with op-amps, microcontrollers are naturally suited to PID control; they have added benefits, too, such as the ability to report status, to accept and modify tuning parameters, and to carry out nonlinear control algorithms.

15.6.1 The hardware

Popular temperature sensors include a thermocouple (a junction of two dissimilar metals, which generates a voltage, typically $20–40\,\mu\text{V}/°\text{C}$), a thermistor (a resistive material of negative temperature coefficient, typically $-4\%/°\text{C}$), a silicon-based temperature sensor IC (that exploits the $-2.1\,\text{mV}/°\text{C}$ negative tempco of a diode's forward drop), or a platinum RTD ("Resistance Temperature Detector," a wirewound resistor, usually standardized at $100\,\Omega$ at $0°\text{C}$, whose resistance increases by $0.385\%/°\text{C}$). These sensors differ in temperature range, accuracy and stability, speed of response, size, and cost. For this example (and without excessively opening this substantial can of worms) we chose the RTD, primarily for its good stability; it also has excellent linearity, and a wide operating temperature range ($-200°\text{C}$ to $+600°\text{C}$). As with other temperature sensors, it is available in a nice waterproof probe assembly. At the other end – the heater – a power resistor is all you need, once again in a submersible package (an "immersion heater").[24]

There are several possibilities for the heater drive. The simplest way is to turn it on when you need heat, and off when you don't. This is called a "bang-bang" controller, and it's the way home heating systems work. It's guaranteed to produce a cycling of the temperature about the target, which is less than ideal. It is efficient, however, because the drive transistor operates as a switch, with very little power dissipation.

A better way, if the system permits, is proportional control. Here we could use a linear power output stage, varying the heater drive in a continuous manner, according to the controller's command. The drawback is that the linear driver stage has to dissipate lots of power. This is where pulse width modulation fits in. By operating the output driver as a switch, at, say, $10\,\text{kHz}$,[25] and varying the duty cycle (the ON fraction of each switch cycle), we get

the best of both worlds: proportional control, and minimal driver dissipation. For this reason PWM is popular in other "linear" applications, for example audio amplifiers (see §2.4.1C) or motor drives.

Figure 15.10 shows the circuitry for a microcontroller temperature controller. Looks simple enough, but as usual the devil (and the joy) is in the details.

A. The sensor

We chose a generous $-50°\text{C}$ to $+150°\text{C}$ nominal range of control, over which the standardized "$100\,\Omega$" RTD temperature probe's resistance goes from $80\,\Omega$ to $160\,\Omega$, approximately.[26] To keep the effects of self-heating negligible, we used a $2\,\text{mA}$ bias current, which produces corresponding voltages of $160\,\text{mV}$ and $320\,\text{mV}$ across the RTD, and a maximum self-heating dissipation of just $0.6\,\text{mW}$.

These are smallish voltages, compared with typical ADC conversion ranges of $1.25\,\text{V}$, $2.5\,\text{V}$ or $5\,\text{V}$; so we need some gain, and we need an offset, to match the ADC's full-scale span. A bridge arrangement, as shown, produces a differential signal without the offset; we've chosen the resistor ratios for zero (differential) output at mid-scale ($+50°\text{C}$), so the differential signal goes from $-80\,\text{mV}$ to $+80\,\text{mV}$. Note the 4-wire ("Kelvin") connection, to eliminate effects of cable and contact resistance. Now we need some gain.

B. The frontend amplifier

Microcontrollers come in such variety that you can usually find one with just the features you want. The ADuC800-series and ADuC7000-series of "analog microcontrollers" from Analog Devices is aimed at sensitive conversion applications, and this particular part has just the combination needed here: a true differential input stage, followed by a programmable gain low-noise amplifier and 16-bit ADC[27] that can be operated in a precision "chopping" mode (analogous to chopper op-amps) for low dc offset and drift. The datasheet includes tables of converter accuracy versus conversion rate, showing 15 bits of resolution for the $\pm80\,\text{mV}$ range at 50 conversions/sec in chopping mode; that corresponds to 6 millidegrees, plenty good enough for this job.

Without such a convenient microcontroller (which Analog Devices calls a "microconverter"), the alternative is to use a differential instrumentation amplifier, preferably one

---

[24] A good place to find these things in the impressive *Temperature Handbook* from Omega Engineering.

[25] Commercial temperature controllers intended for use in systems with large thermal mass and slow response typically perform their PWM at very low frequencies, in the Hz region or slower, allowing the use of a relay rather than an electronic switch output.

[26] Accurate tabulations of Pt-RTD resistance list nominal values of $80.31\,\Omega$ and $157.33\,\Omega$ at these temperatures, typically accurate to a few tenths of an ohm.

[27] The ADuC848 has a 16-bit ADC; the similar ADuC847 has a 24-bit ADC, and the ADuC845 has two 24-bit ADCs.

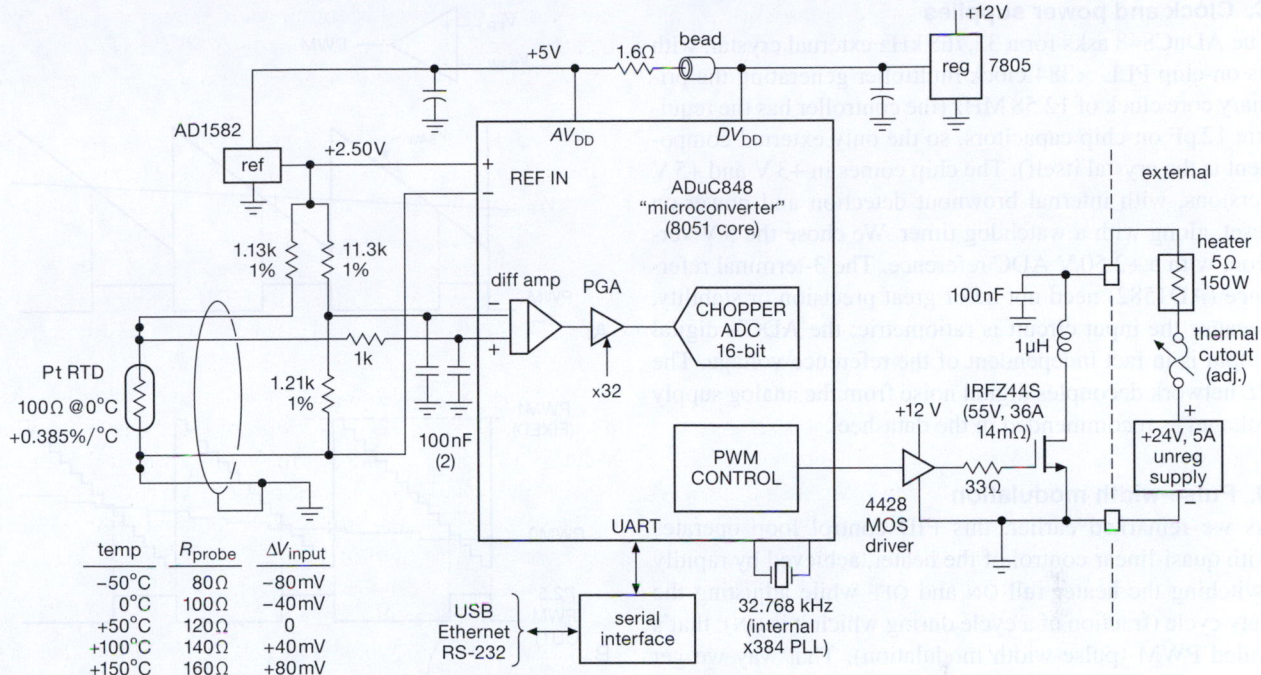

Figure 15.10. Temperature controller, with platinum RTD sensor and linear PWM heater. The ADuC848 "microconverter" includes a 16-bit ADC and programmable gain differential amplifier, aimed at analog applications requiring accuracy and low noise. The resettable thermal cutout is a good safety feature in the event of hardware or firmware failure.

with input common-mode range to its negative rail so it can be run from a single +5 V supply. Most instrumentation amplifiers (Table 5.8) don't have that feature; but a suitable candidate is the AD623 (Analog Devices strikes again!), which also has rail-to-rail output swing and can run from a single +3 V to +12 V supply.

Figure 15.11 shows how you would use it here, to drive the analog inputs of a more generic microcontroller: the output reference pin is tied to the midpoint (+1.25 V) of the desired conversion range, so that the differential input signal range of ±80 mV is mapped to the 0 to +2.5 V range of a generic single-ended ADC. (We could instead ground the reference pin, and replace the 1.21 kΩ resistor in Figure 15.10 with 1.0 kΩ, thereby balancing the bridge at the low temperature end.) For a situation like this, where you need accurate amplification of low-level signals, you can generally expect better performance with a dedicated instrumentation preamplifier, as compared with a bare microcontroller. However, the ADC in the particular microcontroller we chose is exceptional, and in fact has better accuracy and drift specifications than the outboard preamplifier: it specifies typical voltage offset and drift (in chopping mode) of 3 μV and 0.01 μV/°C, compared with the

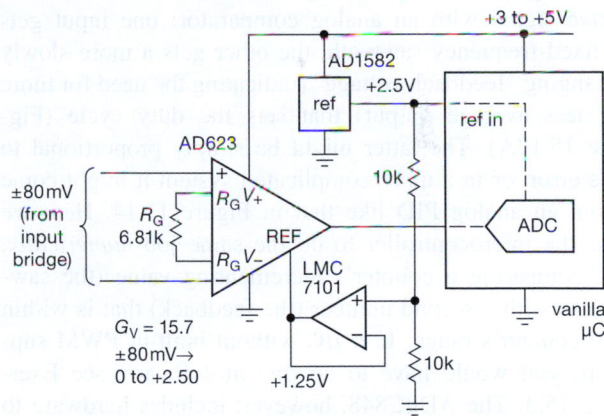

Figure 15.11. As an alternative, an outboard differential "instrumentation amplifier" could be used to map a small differential signal range (here ±80 mV) to the single-ended input range (here 0 to +2.5 V) of a generic microcontroller. In such an implementation, the gain-setting resistor sets the span, and the "reference" input pin sets the output to midscale when the input is balanced.

instrumentation amplifier's corresponding values of 25 μV and 0.1 μV/°C.[28]

---

[28] In fairness it should be noted that the chopping process, which is re-

C. Clock and power supplies

The ADuC848 asks for a 32.768 kHz external crystal, with its on-chip PLL ×384 clock multiplier generating the primary core clock of 12.58 MHz (the controller has the requisite 12 pF on-chip capacitors, so the only external component is the crystal itself). The chip comes in +3 V and +5 V versions, with internal brownout detection and power-on reset, along with a watchdog timer. We chose the 5 V version, with a +2.50 V ADC reference. The 3-terminal reference (AD1582) need not be of great precision or stability, because the input circuit is ratiometric: the ADC's digital output is in fact independent of the reference voltage. The *RL* network decouples digital noise from the analog supply voltage, as recommended in the datasheet.

D. Pulse-width modulation

As we remarked earlier, this PID control loop operates with quasi-linear control of the heater, achieved by rapidly switching the heater full ON and OFF while adjusting the duty cycle (fraction of a cycle during which it is ON); that's called PWM (pulse-width modulation). That way we get the benefits of linear control, along with the efficiency of saturated switching. For an application like this, with its long thermal time constants, the heater won't know the difference between PWM and honest linear voltage drive.

In a purely analog world, you'd create the PWM switch drive signal with an analog comparator: one input gets a fixed-frequency sawtooth, the other gets a more slowly changing "feedback voltage" (indicating the need for more or less average output) that sets the duty cycle (Figure 15.12A). The latter might be simply proportional to the error, or in a more complicated system it might come from an analog PID like that in Figure 15.14. Here we use the microcontroller to do the same job *numerically*, by comparing a counter's incrementing value (the sawtooth) with a second number (the feedback) that is within the counter's range. In a μC without built-in PWM support, you would have to do this in software; see Exercise 15.3. The ADuC848, however, includes hardware to make PWM a breeze: you initially set the counter's range (maximum count) by loading a 16-bit unsigned number into a terminal count register (called PWM1); the incrementing count is continually compared with the (slowly changing) 16-bit feedback value resident in another register (called PWM0), creating a HIGH PWM output on a

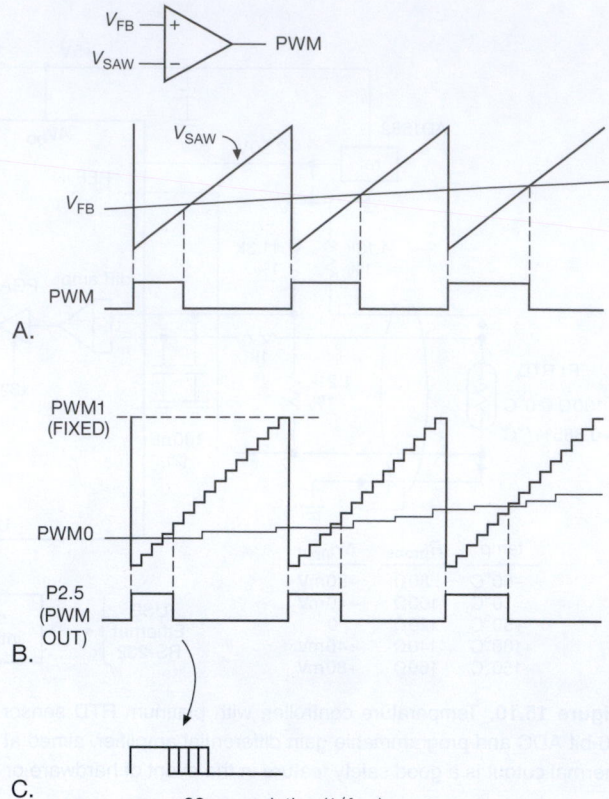

Figure 15.12. Pulse-width modulation (PWM). A. A slowly varying feedback signal V_{FB} is compared with a fixed-frequency sawtooth to produce the logic-level PWM output. B. Digital PWM substitutes a clocked counter and magnitude comparator. C. The digital PWM output has a granularity of $1/f_{CLK}$.

port pin (called P2.5) when the PWM count is less than the feedback (Figure 15.12B).[29]

For this application we chose a maximum count of 1258_{10} (by loading hex 04EAh into internal register PWM1), to produce a PWM frequency of 10 kHz (the 12.58 MHz clock, divided by 1258). The PWM output pulses thus come at a 10 kHz rate, with widths corresponding to an integer number of 12.58 MHz clock pulses; that is, they are quantized with a step size of 1/12.58 MHz, or 80 ns (Figure 15.12C).

Exercise 15.3. Problem for a rainy Sunday: imagine you've already used the ADuC848's two PWM pins, and you need to create two more 8-bit software PWMs. You're given the register values, and during software PWM operation you have full control of the

---

sponsible for the excellent accuracy and stability, reduces the bandwidth and conversion rate. For our thermal-control application, with time constants of order 250 ms, speed is not a problem.

[29] This "Single Variable Resolution PWM" mode is one of six modes offered by this elegant chip. Such flexibility is unusual in your average microcontroller.

μC. Devise workable code, and see how fast the PWMs can run. The "Optimized Single-Cycle 8051 Instruction Set" and timings are given in the datasheet (page 20 of the current revision). The ADuC848 has a 12.58 MHz clock.

E. Output circuit

From a microcontroller output port you can't switch a load that is powered from 24 volts. And you *really* can't switch 5 amps! So we need a transistor switch, here implemented with a modest-sized power MOSFET. And, to drive its gate fully (to +12 V) and fast, we added a gate driver chip.[30] The 4428 is an industry-standard MOSFET driver, good for converting a logic-level input swing to a full-voltage swing at the gate; this inexpensive part (about $1) has dual outputs (inverting and noninverting), good for 1.5 A of source or sink into the capacitive gate load. The particular transistor we chose (the IRFZ44, rated 55 V, 36 A, TO-220 power package) is inexpensive ($\sim$$2), with R_{ON} sufficiently low (14 mΩ max, at V_{GS}=10 V) that we don't need a heatsink. A full 12 V gate drive minimizes the on-resistance, and the 1.5 A gate drive capability ensures fast gate rise and fall times, reducing MOSFET "class-A" dissipation[31] during switching.

A switching MOSFET with healthy gate drive produces rapid (nanosecond-scale) drain transitions, potentially generating substantial electromagnetic interference. The small output inductor and shunt capacitor form a lowpass filter of characteristic frequency 0.5 MHz, limiting output slew rate. This eco-friendly filter has an interesting side effect, however; namely, the production of an inductive positive-going voltage transient at each switch-off (Figure 15.13). Unless clamped, the inductive spike causes avalanche breakdown in the MOSFET. That sounds bad, but in fact power MOSFETs are tolerant of such abuse, with a specified repetitive avalanche energy rating limited primarily by heating effects (see the section on Pulse Energy in Power MOSFETs in Chapter *3x*, and the section on Transient Thermal Response in Chapter *9x*). For this circuit the periodic delivery of the $\frac{1}{2}LI^2$ energy stored in the inductor merely adds some allowable increment of MOSFET average power dissipation.

In total, then, there are three contributions to MOSFET heating: (a) I^2R_{ON} dissipation during conduction; (b) I_DV_{DS} class-A dissipation during switching; and (c)

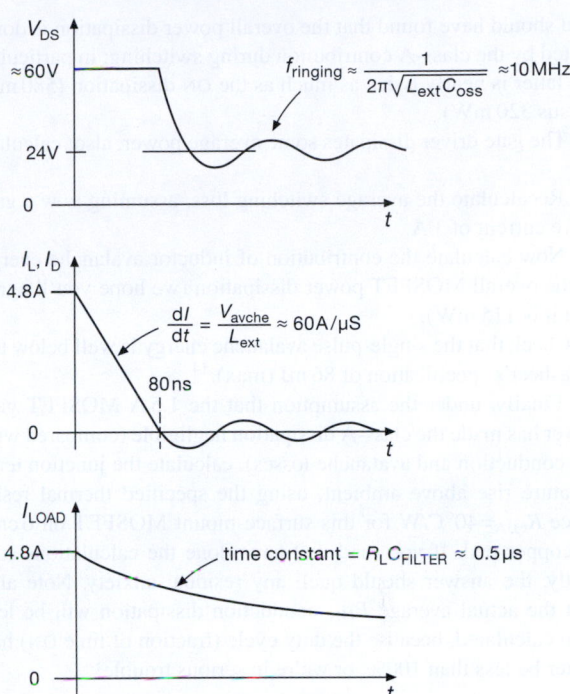

Figure 15.13. The output *LC* filter in the circuit of Figure 15.10 smooths voltage and current transients to the load (bottom graph); but the series inductor causes MOSFET avalanche breakdown at turn-off (top graph), during which the inductor current ramps to zero.

$\frac{1}{2}LI^2f_{osc}$ repetitive avalanche dissipation at switch turn-off. The following exercise explores these quantitatively for this circuit example.

Exercise 15.4. Let's flesh out this last statement, in several steps.[32]
(a) What is the MOSFET power dissipation when it is ON, assuming R_{ON}=14 mΩ?
(b) Now calculate the average power dissipation contributed by class-A conduction during transitions, assuming the PWM is running at 10 kHz. For this calculation use a gate-to-drain ("Miller") charge of Q_{GD}=15 nC, and assume for the moment that we have been foolish enough to drive the gate directly from a logic output that is able to source and sink just 10 mA (which is considerably better than this μC's specs: a feeble 1.6 mA at 0.4 V sinking, and even feebler 80 μA at +2.4 V sourcing). You can assume a linear ramp of voltage and current, but to do the calculation properly you'll have to integrate the product $V_{DS}I_D$ over the duration of the ramp (and don't forget that there are two ramps per cycle).[33]

---

[30] See §3.5.4 and Tables 3.4b and 3.8 on pages 189–191 and 218 respectively.

[31] That is, the power dissipated during the transition times when the transistor is neither fully ON or OFF, and thus has both a substantial drain current and a substantial drain-source voltage drop.

[32] Refer back to §9.7.2 if you are unsure how to go about these calculations.

[33] Spoiler (if you want to skip the calculus): $E = V_PI_PT_{ramp}/6$ per ramp.

You should have found that the overall power dissipation is dominated by the class-A contribution during switching; in particular, the latter is nearly twice as much as the ON dissipation (580 mW versus 320 mW).

(c) The gate driver dissipates some average power, also; calculate it.

(d) Recalculate the average switching loss, assuming now a gate drive current of 1 A.

(e) Now calculate the contribution of inductor avalanche energy to the overall MOSFET power dissipation (we hope you'll agree that it is 115 mW).

(f) Check that the single-pulse avalanche energy is well below the datasheet's specification of 86 mJ (max).[34]

(g) Finally, under the assumption that the 1.5 A MOSFET gate driver has made the class-A dissipation negligible (compared with the conduction and avalanche losses), calculate the junction temperature rise above ambient, using the specified thermal resistance $R_{\Theta JA}=40°C/W$ for this surface-mount MOSFET on 6 cm^2 of copper pad. If you (and we) have done the calculations correctly, the answer should quell any residual anxiety. Note also that the actual average R_{ON} conduction dissipation will be less than calculated, because the duty cycle (fraction of time ON) had better be less than 100%, or we're in serious trouble!

This exercise reveals some of the essential engineering process of tradeoff and iteration that goes into even a simple-looking output circuit like this.

(a) We chose a MOSFET with low enough R_{ON} to permit surface mounting without a heatsink, if desired (the IRFZ44S version: 40°C/W, good for ∼1 W dissipation).

(b) Budgeting about 1/3 W for conduction loss sets the maximum load current at about 5 A.

(c) Then we chose the heater supply voltage to produce 100+ W, to get things heated quickly.

(d) We added the *LC* filter to suppress RFI above ∼1 MHz.

(e) Then we confirmed that the avalanche power dissipated in the MOSFET from periodic dumping of the inductor's stored energy added only a modest additional dissipation, and stayed well within the ratings.

(f) Finally, we added a gate driver IC to minimize class-A conduction losses, which otherwise would dominate. During this process we made numerous adjustments to the heater supply voltage (+12 V, +24 V), heater current (2.5 A, 5 A), and inductor size (1μH, 5μH, 10μH), and toyed with the idea of using a MOSFET with lower R_{ON}, or even a TO-220 package MOSFET with a finned heatsink.

In the end we settled on the design shown in Figure 15.10 as a good overall compromise. However, it would be perfectly reasonable to choose a larger filter inductor, or run at higher current, or both, in which case you'd use a MOSFET with lower R_{ON}, perhaps in a larger package attached to a heatsink. The glib presentation of finished designs often conceals this kind of circuit design thinking.

15.6.2 The control loop

The microcontroller firmware's job is to implement a control loop, to maintain the bath at the desired "setpoint" temperature. Although this looks like plain ordinary feedback, of the sort used for an op-amp circuit, in reality it is bedeviled by thermal time delays; providing sufficient loop gain to stabilize the bath will invariably lead to thermal oscillations. This is a common problem in industrial control systems. As Jim Williams wrote,[35] "The unfortunate relationship between servo systems and oscillators is very apparent in thermal control systems."

The usual solution is what's known as a PID controller, in which ordinary proportional negative feedback (*P*) is augmented by negative feedback proportional to the rate of change of the error ("derivative," *D*), and also by a negative feedback term that grows with time according to the error ("integral," *I*). This can be done simply with op-amps, as shown in Figure 15.14.

The PID controller has to be "tuned" to the properties of the controlled system, to optimize the gains of the three feedback terms. An empirical procedure that works pretty well is the following: (a) with *I* and *D* terms turned off, increase the *P* gain until the system begins to oscillate, then back off a bit; the system will now exhibit overshoot and ringing, but not sustained oscillation, in response to a step change in setpoint; (b) now add *D* gain until the response to a step is critically damped;[36] (c) finally, while watching the error signal itself, add *I* gain to achieve minimum settling time.[37]

---

[34] As general guidance, if there's less than a factor-of-20 safety margin, you need to check the pulsed power dissipation using the MOSFET's specified "Transient Thermal Impedance." That's because datasheets use a particular set of operating conditions for their single-pulse avalanche spec; see §3x.13.

[35] In "Thermal techniques in measurement and control circuitry," Application Note 5, Linear Technology Corporation, December 1984.

[36] In the language of poles and zeros, you've brought in a zero to cancel the natural pole of lowest frequency in the physical system.

[37] Another way to tune a PID, which is due to Ziegler and Nichols (1942), uses the gain setting and frequency of oscillation in step (a) to determine, without further experimentation, the optimum *D* and *I* gains. We like the treatment in Tietze and Schenk's excellent *Electronic Circuits* (Springer, 2007).

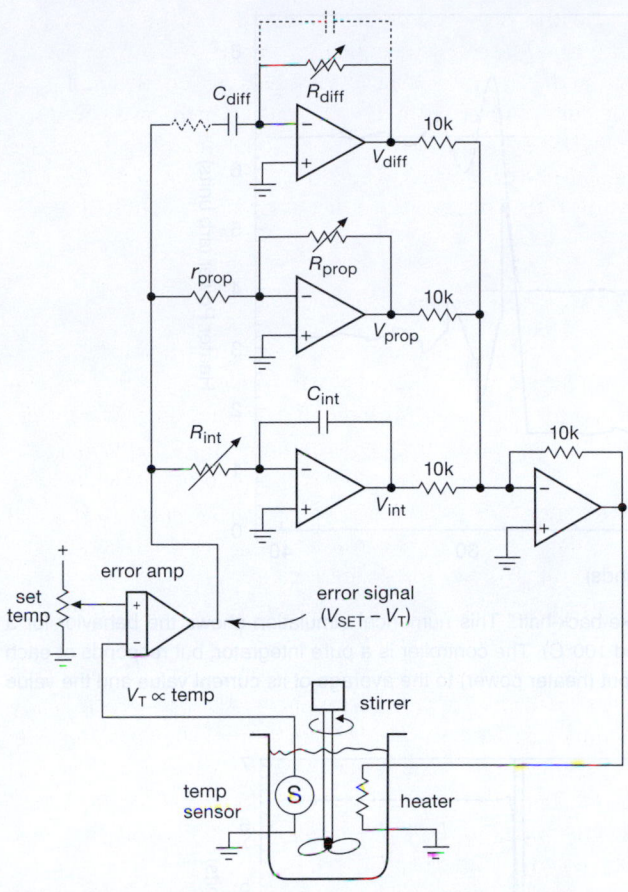

Figure 15.14. An analog PID control loop. The error signal (proportional to the difference between desired and actual temperature) drives an amplifier ("proportional"), an integrator, and a differentiator, with individual settable gains. Their combined outputs form the control signal to the heater. The dashed R and C are for differentiator stability. (You sometimes see the differentiator input taken separately from the upstream side of the error amplifier.)

15.6.3 Microcontroller code

The firmware to implement a PID in a microcontroller begins with the usual setups: ports, converters, timers, and communications. The interesting part is the main loop, where the measured temperature is digitized at equal time intervals (call the nth measurement T_n), and the results are used to compute numerically the P, I, and D outputs from the error "signal" $T_{\text{error}} = T_{\text{set}} - T_n$. The listing in Pseudocode 15.4 shows the simplest computation, assuming known tuning coefficients k_P, k_I, and k_D. In practice you would probably do some smoothing or filtering for the derivative term, which is prone to noise. You might also set up the ADC directly to perform periodic conversions, with

---

Pseudocode 15.4 PID main loop pseudocode

Initialize
    zero the integration accumulator: $I = 0$
Main Loop
    reset timer for 10 ms timeout
    read temp, compute nth $T_{\text{error}} = T_{\text{set}} - T_n$
    compute individual PID power output terms:
        $P = k_P T_{\text{error}}$
        $I = I + k_I T_{\text{error}}$
        $D = k_D (T_{\text{error}} - T_{\text{error}-1})$
        $T_{\text{error}-1} = T_{\text{error}}$
    combine and update PWM:
        $\text{PWM} = P + I + D$
    wait until timer times out
    repeat **Main Loop**

---

polling or interrupts to signal the main loop's PID computation.

A. Some comments (algorithm)
The PID control loop, though of great popularity, is not the only game in town. In particular, there are nonlinear algorithms that, though difficult to treat mathematically, seem to work just fine. An interesting example is Steve Woodward's "take-back-half" (TBH) algorithm,[38] which has the nice feature of "one knob" tuning: you don't have to know anything about the "plant."

The control algorithm has two parts. It begins with a pure I-loop (controller output proportional to the integral of the error), which has the advantages of simplicity (only one "tuning" knob: the integrator gain), and of zero averaged error. However, the bad news is that the controlled variable oscillates about the target setpoint forever. Woodward's fix is to make a step correction at each zero crossing of the error, replacing the controller's current output with the average of current value and the value at the last zero crossing. We were curious how this would look, so we dragooned a student skillful in Mathematica (that's him in Figure 15.25 on page 1094) into running a numerical simulation of a TBH temperature controller. You can see the results in Figure 15.15, where we plot the controller output (heater power) and the response of the controlled system (sensor temperature). We started it up at ambient temperature (20°C), teasing out its transient behavior by switching the setpoint between the two values shown. For this simulation we used heater and sensor time constants of 0.5 s, a delay time (from fluid flow) of 0.1 s, and a thermal

---

[38] "Temperature controller has "take-back-half" convergence algorithm," *EDN*, page 90, September 15, 2005.

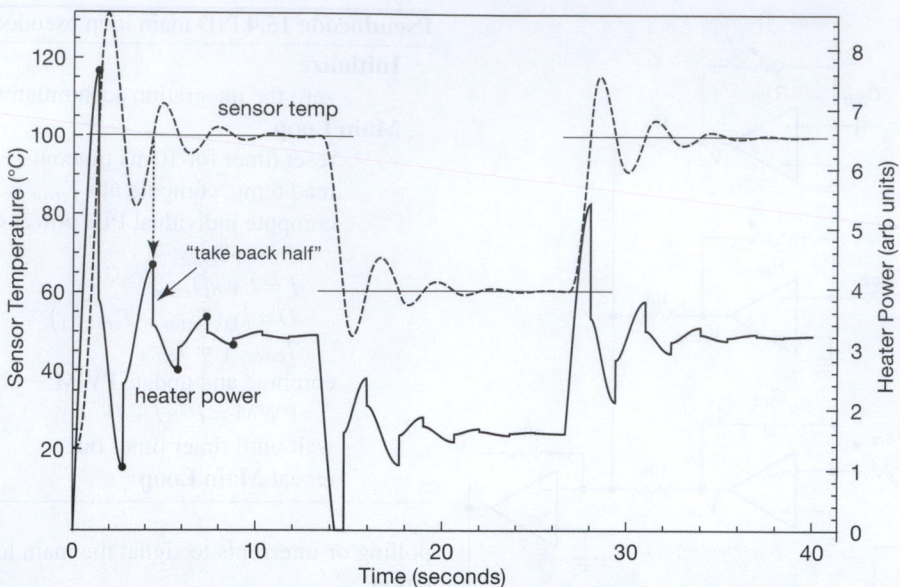

Figure 15.15. A nonlinear control algorithm: Steve Woodward's "take-back-half." This numerical simulation shows the behavior for a sequence of step changes in target temperature (alternating 60°C and 100°C). The controller is a pure integrator, but responds at each moment of zero temperature error (the black dots) by resetting its output (heater power) to the average of its current value and the value at the previous reset.

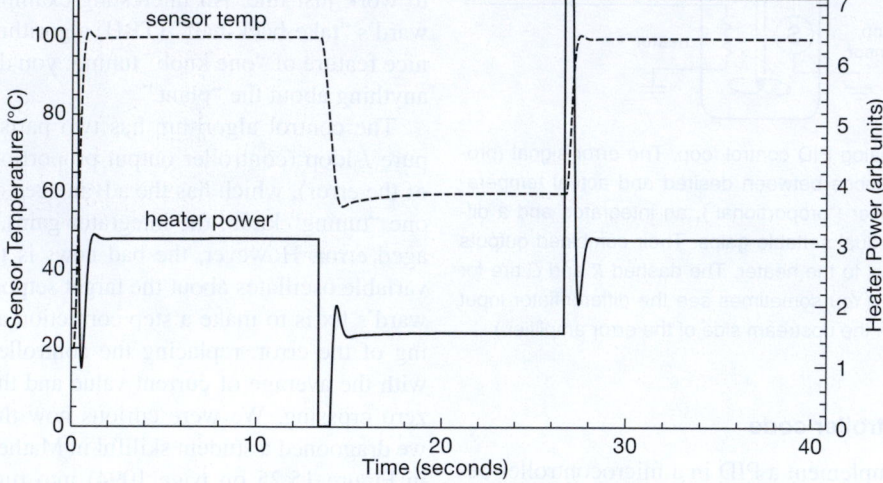

Figure 15.16. Numerical simulation of the system of Figure 15.15, this time with a classic tuned PID controller.

relaxation time to ambient temperature of 0.5 s. The one "knob" (integrator gain) was set to produce a reasonable convergence.

Having set up the TBH simulation, we couldn't resist running a classic PID. After fooling around with the tuning (*three* knobs, this time), we got the results shown in Figure 15.16. The PID does a considerably better job than TBH; but it requires some skill in tuning to find the *P*, *I*,

and *D* coefficients suited to the specific parameters of each controlled system.

B. Some comments (Hardware):

This example illustrates how a particularly good choice of microcontroller can greatly simplify the hardware (and coding): this microcontroller combines a good ADC (real 16-bit performance), with the low offset and drift you get with chopping, and an internal differential amplifier of programmable gain, so that the low-level RTD input can be used directly. The differential reference input makes it easy to rig up true ratiometric conversion, insensitive to variations of the reference voltage source. And internal PWM hardware dramatically eases the coding burden.

We've kept the block diagram lean, so as not to distract from the essentials of the analog input circuitry, and the PWM output drive circuitry. But it's easy enough to add some of the creature features shown in previous microcontroller examples: an LCD display of setpoint and actual temperatures; a keypad or thumbwheel entry for parameters such as a simple setpoint, or a more complex time/temperature profile; indicator LEDs and alarm buzzers for fault conditions, and the like.

15.7 Design example 5: stabilized mechanical platform

The final microcontroller example is fun – a motor-driven stabilized two-wheel contraption (which its inventor calls a *Psegué*), shown in action in Figure 15.17.

This thing is a tricycle without its front wheel, therefore unstable without active feedback. The prototype is, of course, Dean Kamen's "Segway™ Personal Transporter" that delighted the world at its debut in 2001. Weird and wonderful homemade variants are sprouting up, propelled by the enthusiasm of the lively hobbyist–nerd community. One such hobbyist is the young man who lives across the street, who asked if he could hang out in our lab to try out his ideas for making a stabilized motorized platform. We doubted he would succeed; we were wrong.

Figure 15.18 shows the system he finally arrived at (after some amusing mishaps). It is a digital PID control loop, implemented in an NXP ARM7 microcontroller. The latter comes nicely packaged on a 6 cm square board, which includes voltage regulators, connectors for analog, digital, and serial ports, and LCD display; it is called the MINI-MAX/ARM-C, sold by BiPOM Electronics for $100.

To sense the platform's instantaneous angle, Jesse used a two-axis solid-state accelerometer, rotated 45° to the vertical, feeding a difference amplifier. The output voltage is proportional to the sine of the platform angle, passing through zero when the platform is horizontal. This provides the analog input for the proportional and integral terms of the PID loop; for the derivative term he used a gyro, whose output is directly proportional to the rate of change of tilt. You steer this thing by pushing the vertical post sideways, which squeezes a pair of resistive force sensors under the post's base. The microcontroller has a Bluetooth link, for

playing around with the loop parameters while riding. Figure 15.19 shows the hardware.

Figure 15.17. Stabilized two-wheel contraption, demonstrated by its creator (Jesse Colman-McGill).

On the output side the microcontroller drives a pair of H-bridge dc motor drivers with PWM logic outputs, as commanded by the PID loop. The dc permanent-magnet gearmotors are substantial beasts, able to put out nearly a horsepower.

The PID loop is paced by one of the microcontroller's internal timers, operating at a 100 Hz heartbeat. The basic PID controller is augmented by some gimmicks that were arrived at by experiment, for example, an extra boost to get the thing moving from a dead stop, and some modifications to the PID gains according to the measured load. A fancy name for such unauthorized patches to the basic PID is *heuristics*; whatever you call them, they're needed, and with some luck they can work well. See Pseudocode 15.5.

Pseudocode 15.5 Psegué main loop pseudocode

Main Loop
    reset timer for 10 ms timeout
    read sensors
        tilt-sensing accelerometers (2)
        rotation-sensing gyro
        steering force sensors (2)
    compute PID with speed-dependent parameters
    apply heuristic rules
        dead-zone correction
        threshold boost
        load-dependent PID gain multiplier
    send updated torque command to motors
    if (logging) increment log_loop_count
        if (log_loop_count=10) log data & clear log_loop_count
    TimerCheck: if (timer not expired)
        if (command byte from wireless input FIFO buffer)
            append to line buffer
                if (newline) parse & execute
        write logging byte to wireless output FIFO buffer
    repeat **TimerCheck**
    repeat **Main Loop**

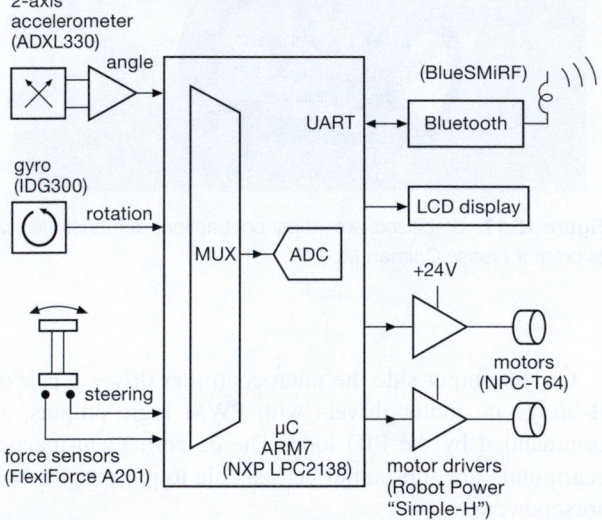

Figure 15.18. Block diagram of the stabilized scooter. The use of prefabricated hobbyist modules (gyro, accelerometer, wireless, microcontroller board, and motor drivers) simplified construction and wiring. The actual part numbers are in parentheses.

15.8 Peripheral ICs for microcontrollers

Our experience in electronics engineering comes from the real world, where an enormous variety of measurements and interactive control are needed. Having a programmable microcontroller is good, and having lots of built-in interfacing circuitry is great – but the world is bigger than that. When we talk about the rubber meeting the road, what we need is a tire, a wheel, and other specialized devices that aren't included inside a microcontroller.

In this section we've created three drawings that show sixty examples of specialized interface devices, complete with sample part numbers. It's our hope that you'll point your browser at Octopart, etc.,[39] and read the datasheets for some of these parts, as inspiration for searching out additional examples and alternative parts.

DACs and ADCs are a common peripheral for interact-

---

[39] Octopart is an excellent parts locator, showing availability and pricing of components at stocking distributors; the site also includes datasheets for most parts. An especially valuable feature is the ability to see the status of parts, and their activity at various distributors. If a part has been discontinued, you'll still find out what it was once possible to make (and what its specs were), and you can hope someone else will have created a similar part. *One caution*: as of this writing, Octopart only lists components that are carried by distributors, and not those sold directly from the factory. If Octopart doesn't show a part, that doesn't mean you can't get it. Sometimes the factory has to be your source, even for mainstream manufacturers who also sell through distributors. We have found that in most cases the manufacturers make it easy to buy directly, even in small quantities.

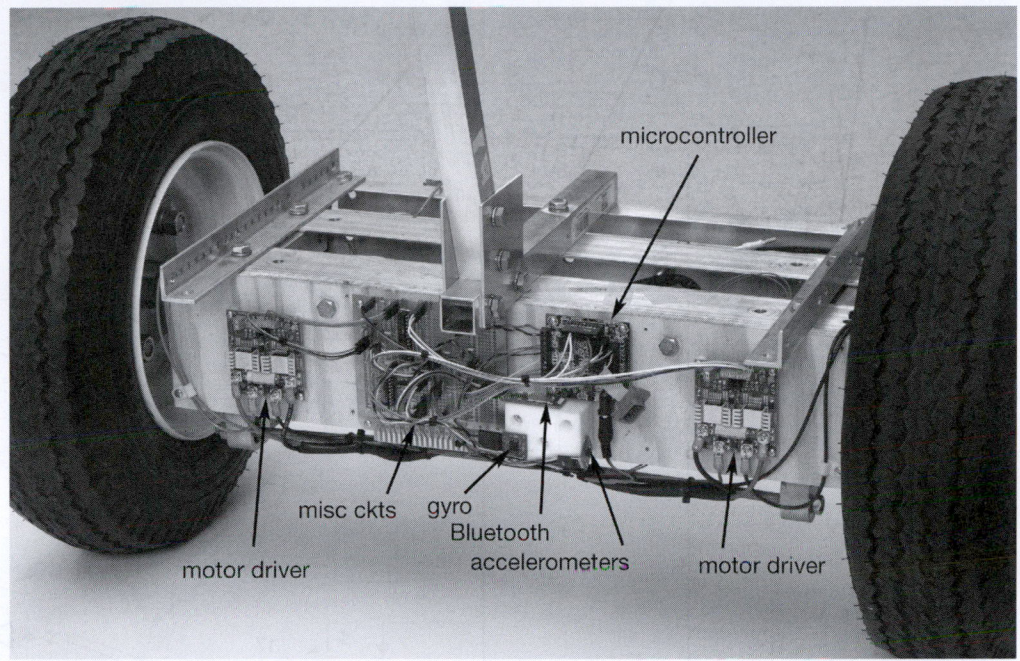

Figure 15.19. Closeup of the sensors and electronics. The difference signal from the accelerometer pair (each inclined 45° to the vertical) is a measure of fore–aft tilt angle, and the gyro provides a direct measure of the time derivative of tilt. The gel-cell batteries live in the space behind the electronics.

ing with real world stuff; if that's all you're interested in at the moment, use Tables 13.2, 13.3 and 13.11 (DACs), and Tables 13.5, 13.6 and 13.9 (ADCs). These tables include columns indicating SPI, I²C, or other interface methods.

When designing a gadget with an embedded microcontroller, there are lots of "devil in the details." We could prattle on for pages about how to connect to many wonderful peripheral devices. Instead we present a guided tour and navigation aid, in the form of Figures 15.20–15.22, tagged with numbers to match the brief outlines that follow. We've put plenty of cross references to relevant discussions elsewhere in the book, along with some selected part numbers. We've broken it down into several figures, because there's so much stuff that you can attach to a microcontroller: devices that can connect directly to the μC (Figure 15.20); devices that connect to an SPI bus or its near relatives (Figure 15.21); and devices that connect to an I²C bus (Figure 15.22). Your programming quality-of-life may be improved if your microcontroller includes a built-in interface for your chosen bus, like the one in Figure 10.86.

15.8.1 Peripherals with direct connection

Figure 15.20 shows a microcontroller, heavily encrusted

with devices that connect easily to standard "internal peripherals" such as digital I/O port pins, serial communication ports (UART, USB, Ethernet), ADCs and DACs, PWM outputs, and the like. It also shows important support chips such as power control, watchdog, and external oscillator. In outline form we've collected some explanatory comments, useful part numbers, and references to relevant discussions elsewhere in the book.

A. Outline for Figure 15.20

1. **Power-source selector.** If the system runs from battery power, an ICL7673 with two p-channel MOS-FETs makes a nice power selector, acting as an ideal diode-OR, automatically connecting to the higher of the two voltages. For rechargeable battery management see §9x.2.

2. **Power supervisor, power-ON-reset, watchdog.** Brownout protection, don't let the μC do anything important if its supply voltage is too low; reset it. See §10.8.1B, and Table 10.6 on page 756. There's a huge variety of choices available. Take a look at the ADM705 and TPS3306 supervisory ICs (which include a watchdog) and the ADM691 series (which also incorporate a battery-backup-switch feature).

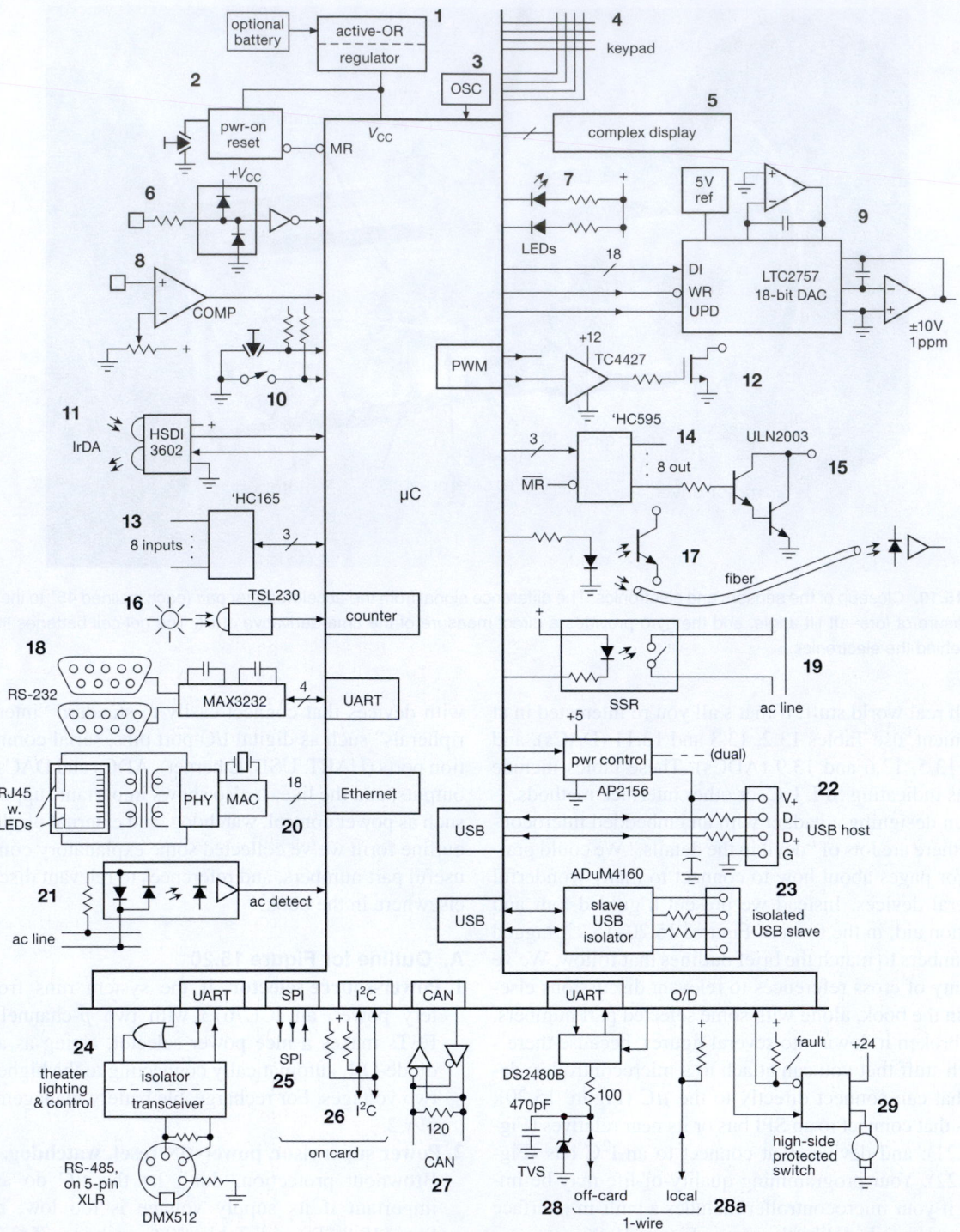

Figure 15.20. An assortment of peripheral devices that can be connected to a well-endowed microcontroller. Some of the devices require specialized interface circuits within the controller. The interface items marked with bold numbers **1** to **29** refer to descriptive paragraphs in §15.8.1.

3. **µP oscillator.** Internal, see §§15.2.1E and 15.9.3. External, see §7.1.6 for choices and cautions.

4. **Keypad scanning techniques.** See Figure 15.8.

5. **Panel-display choices.** LED, LCD, CFD, tricks. See §§10.6.2, 12.5.3, 14.6.

6. **Logic inputs, external.** Protect µC against ESD with separate user-repairable input gates, and the instrument's programmed µC survives; see §12.1.5.

7. **LED and indicator outputs.** See §§12.4, 12.4.5A, 12.5.1.

8. **Analog threshold detection, discriminators.** See §§12.1.7 and 12.3, and Table 12.2 on page 819.

9. **DACs.** See §§13.2 and 14.6.2, and Tables 13.2 on page 893 and 13.3 on page 894; fast 18-bit accurate parallel DAC LTC2757.

10. **Switch and pushbutton inputs.** See §12.1.4.

11. **IrDA infra-red transceiver.** HSDI-3602.

12. **Power MOSFET switches.** See §§3.5, 12.4, and 15.6, and Table 3.4b pages 189–191. MOSFET driver ICs, logic to 12 V MOSFET gate drive, see §3.5.3; classic TC442x family (see Figure 3.97 and Table 3.8 on page 218). For PWM see also §15.6.2 and §9x.5.

13. **8-bit parallel-in shift register.** See Figure 12.40, suited for onboard use. 74HC165 (and other '165 logic ICs), $0.35, compare with a simple 8-bit I^2C expander chip at ∼$2.30.

14. **8-bit parallel-out shift register.** See Figure 12.40, e.g., 74HC595 logic IC, $0.32; also check out the 74HC594 and '567, which have a double buffer. For SR chips with built-in power drivers (e.g., TI's TPIC6C595), see §12.4 and Table 12.3.

15. **Darlington and MOSFET driver ICs.** See §12.4. Drive loads with the venerable ULN2003 Darlington 7-unit array, $0.33, or with individual logic-level MOSFETs, see Table 3.4a, on page 188. Some other useful driver ICs include the ULN2803, SN75468, TPL7407, MC1413, ULN2068 and TD62783.

16. **Light-intensity sensor.** TAOS TSL230 provides a light-to-frequency signal, allowing for easy, accurate measurements over six decades of light level. See Figure 12.81.

17. **Optical isolation.** See §12.7 and the drawings and part numbers in Figures 12.85–12.88. Logic optocoupler, jellybean H11L1, HCPL-2201. For long-distance optical-fiber data links, with connectors, see §12.7 and Figure 12.98.

18. **Serial data-communication interfaces.** RS-232, with ±7 V levels; see §12.10.4. Use the classic DS14C88 and DS14C89 chips, or for simplified operation on +5 V, second-sourced MAX232, $0.90 from TI; or 15 kV ESD-protected MAX3232E runs from 3.3 V or 5 V, $3. These have 2 Tx and 2 Rx, thus one port plus two control lines, or two ports. For RS-485, see §§12.10.3, 14.7.8. A typical transceiver chip is the LTC1485.

19. **ac power switch.** Isolated SSR, zero-crossing triac or back-to-back SCRs: see §§12.7, 15.4, and Figures 12.91–12.93. Available as low-current ICs, e.g., Fairchild MOC3043, $0.80, or as powerful modules with screw terminals.

20. **Ethernet.** Protocol processing, transformer coupled, see §14.7.16 and Figure 12.124. Easy-to-use chip, Silicon Labs CP2201, about $4.50 in qty 100. You'll also need a transformer (e.g., Pulse PE-36023) and an RJ-45 connector. Most folks buy these combined into one compact part, complete with two indicator LEDs, e.g., Pulse J00-0065NL, $4.75.

21. **ac power detect.** To monitor relay outputs, blown fuse, respond to ac-line signals. Opto-isolated; see §12.7.7 and Figure 12.94A, e.g., MOC256.

22. **USB.** Supported by a controller in the µC, see Figure 10.86. USB hosts need a power supervisor with a 500 mA CL, etc., e.g., AP2156 dual (see Table 12.5). USB clients, see §§14.7.13 and 15.9.2.

23. **Isolated USB.** Analog Devices' ADuM4160.

24. **DMX512 theater lighting and control.** 5-pin XLR connectors, to 1200 meters. 250 kbaud (UART in µC), with OR gate for the packet-framing "MAB" symbol. Isolator and RS-485 transceiver, MAX1480 or MAX3480B, see §12.10.3. Near and far ends terminated. Theater electricians: with luck you'll be handed proper 120 Ω twisted cable. DMX512 is being superseded by DALI.

25. **SPI.** Serial microprocessor to chip interface. It's an SPI jungle out there, with many different interface rules grouped under the SPI label; read the IC datasheet very carefully. SPI interfaces are often implemented as individual bit-bang lines, and, as such, SPI may be more of an interface scheme than a multiple-device bus. See also §§14.7.1 and 15.8.2 (the next section).

26. **I^2C.** Chip-to-chip interface; multi-master. Two-wires, open-drain wired-AND operation, with pullup resistors; see §§14.7.2 and 15.8.3. Maximum bus capacitance 400 pF. Unlike SPI, the I^2C bus is well specified (and its use tightly controlled) for predictable results.[40]

27. **CAN-bus.** Two-wire bidirectional bus, differential signaling; see §14.7.15. Generally a CAN-bus

---

[40] The I2C-Bus Specification, Version 2.1, January 2000 (NXP Semiconductors).

transceiver is required, e.g., ON Semi's AMIS-42673 or AMIS-41683, and (if not built-in) an SPI-based controller like the Microchip MCP2515.

28. 1-wire bus. Power and bidirectional data signaling on one wire; see §14.7.3. Championed by Dallas, now Maxim. UART transceiver DS2480B. USB to 1-Wire, DS9490.

28a. Alternative simple 1-wire for μC to local chips, open-drain I/O and pullup resistor.

29. Power switches. High-side to 60 V, 550 A, protected, with fault feedback; see §12.4.4 and Table 12.5; e.g., BTS432, IPS6031.

15.8.2 Peripherals with SPI connection

Figure 15.21 shows a variety of available peripherals that communicate via a simple SPI serial 3-wire master–slave protocol. The SPI bus[41] (§14.7.1) normally consists of three lines: a clock (SCLK); data in (SDI), and data out (SDO'). But there needs to be an additional chip-select line from the processor to each slave device; this is usually called CS', but it's sometimes called other names like load (LD), or enable (EN), etc. This signal, by whatever name, generally transfers the contents of the device's serial-input shift register into an internal data register when CS' is disasserted (brought HIGH) after the data bits have been shifted in. You can dedicate one μC pin for the CS' of each device, or you can save a few pins by using a 74LVC138 (or other '138) decoder IC; see §10.3.3D. Furthermore, some peripheral chips on a bus may require additional individual connections to the microcontroller; for example, for interrupts, to reset the device, etc.

Signal-name usage in SPI may seem inconsistent; for example, the controller's data-in pin (SDI) is connected to the slave device's data-out pin (which might be called DO). Similarly, the master's SDO pin is connected to the slave's DI. A better scheme is to name your signals unambiguously as MISO (master-in, slave-out) and MOSI (master-out, slave-in); see §14.7.1. Sadly, you won't find these names on most device datasheets.

As we stated before (§14.7.1) the SPI "standard" is badly fragmented. For example, in some SPI slaves a single pin may be used for sending data in both directions. This requires the controller to reverse the direction of a pin, so it can receive data back on the same pin that it used for sending a command.[42]

A. Outline for Figure 15.21

30. Touchscreen. XY position digitizer, pressure sensitive; see §13.11.2 and Figure 13.73: AD7873.

31. Serial EEPROM. 8-pin package, §14.4.5B: 25LC080A, AT25080, M24C02 (256×8, \$0.09 in 2500-pc reel)

32. Accelerometer. Three-axis MEMS: ADXL345, Freescale MMA7455L.

33. SD card. "SecureDigital": SD,[43] miniSD, and some microSD memory cards; §14.4.5C. These cards communicate via standard SPI, so this interface consists simply of the socket, with no additional electronics needed! The miniSD and microSD cards have 10-pin connectors. Pin 1 is CS'; but it's also "card detect," with a 50k pullup resistor, to let you know that a card is plugged in. You pull pin 1 low to initiate SPI communication mode.

34. "SparkFun" modules. Wide selection of easy-to-use modules with through-hole connections; e.g., three-axis accelerometer (ADXL335), two-axis gyro rate sensor (LPY503A), or three-axis magnetometer (MAG3110 or HMC5883). Also see Adafruit modules (adafruit.com), e.g., OLED graphic displays available in both SPI and I²C interfaces.

35. Digital pot. DCP, EEPOT; accurate ratio ($\sim$1%), but overall tolerance typically $\sim$20%; low voltage; see §3.4.3E; e.g., 10kΩ, 10-bit, 1024-level MAX5481, \$2.60.

36. Digital resistor. Resistance tolerance 1%, operates to $\pm$16 V; e.g., AD5292. See the digital resistor section in Chapter *3x*.

37. ADCs. See §§13.11, 14.6.2, 14.7 and Tables 13.6–13.12; e.g., LTC2412, AD7927 (Figure 13.74), AD7734.

38. Digital capacitor. Give range and step size; see the digital resistor and capacitor sections of Chapter *3x*; quartz FLEcap: MAX1474.

39. ac power monitor. Including power factor, see §13.11.1: ADE7753, with 8052 μC, ADE7769; needs an *i*Coupler isolator, ADuM3260.

40. DDS RF frequency synthesizer. See §§7.1.8, 7.1.9C, 13.13.6, and 6.2 (on *LC* filters); e.g., AD9954.[44]

---

ers. It's harmless enough, because their chip-select pins are disasserted; but it's ugly.

[43] *SanDisk SD Card Product Family, OEM Product Manual*, Table 3-2, and *Physical Layer Simplified Specification*, Version 2.00, Chapter 7.

[44] Analog Devices is a giant in Direct Digital Synthesis ICs.

---

[41] Analog Devices' AN-877 is a useful application note, describing the use of the SPI bus in various of their high-speed conversion devices.

[42] And of course that data appears on the MOSI pins of the other bus rid-

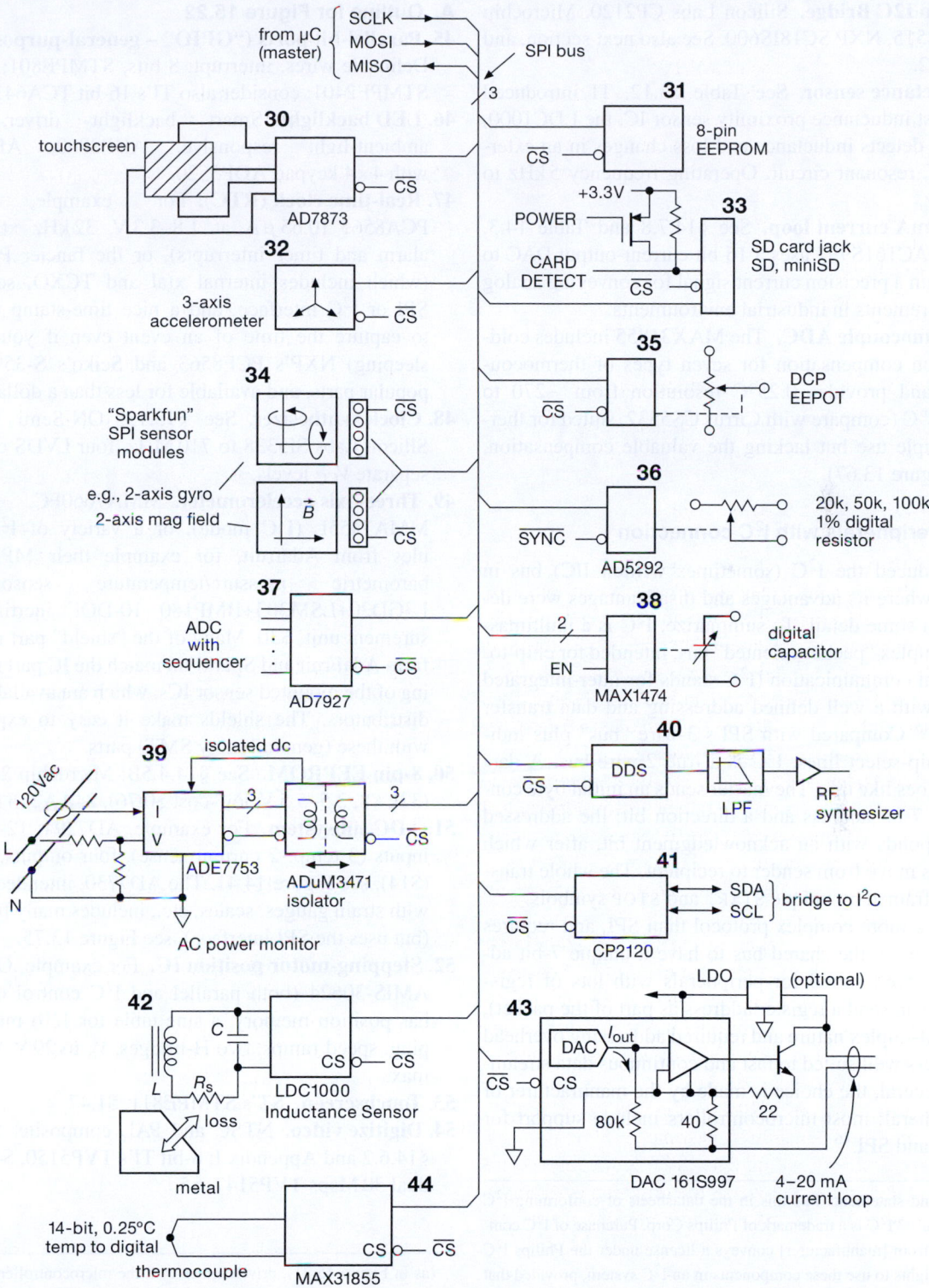

Figure 15.21. An assortment of SPI slave peripherals, well suited for microcontroller applications. The SPI interfaces marked with bold numbers **30** to **44** refer to descriptive paragraphs in §15.8.2.

41. **SPI to I2C Bridge.** Silicon Labs CP2120, Microchip MCP2515, NXP SC18IS600. See also next section, and §14.7.2.

42. **Inductance sensor.** See Table 13.12. TI introduced the first inductance proximity sensor IC, the LDC1000, which detects inductance and loss changes in an external *LC* resonant circuit. Operating frequency 5 kHz to 5 MHz.

43. **4–20 mA current loop.** See §14.7.8 and Table 14.3. The DAC161S997 uses a 16-bit current-output DAC to program a precision current signal for conveying analog measurements in industrial environments.

44. **Thermocouple ADC.** The MAX31855 includes cold-junction compensation for seven types of thermocouples, and provides 0.25°C resolution from −270 to +1372°C (compare with Cirrus CS5532, suited for thermocouple use but lacking the valuable compensation, see Figure 13.67).

15.8.3 Peripherals with I²C connection

We introduced the I²C (sometimes written IIC) bus in §14.7.2, where its advantages and disadvantages were described in some detail. To summarize, I²C is a multimaster half-duplex "packet-oriented" bus, intended for chip-to-chip serial communication (I²C stands for inter-integrated circuit), with a well-defined addressing and data transfer protocol.[45] Compared with SPI's 3-wire "bus" plus individual chip-select lines, I²C is a true 2-wire bus. A data transfer goes like this. The master sends an initial byte containing a 7-bit address and a direction bit; the addressed slave responds with an acknowledgment bit, after which data bytes move from sender to recipient. The whole transaction is framed by unique START and STOP symbols.

I²C is a more complex protocol than SPI, and requires each device on the shared bus to have a unique 7-bit address. It's well suited for peripherals with lots of registers (you can send a register address as part of the packet), but its half-duplex nature and required addressing overhead make it less well suited to fast and continuous data streaming. In general, the choice is made by the manufacturer of the peripheral; most microcontrollers include support for both I²C and SPI.[46]

---

[45] You'll find statements like this in the datasheets of conforming I²C peripherals: "I²C is a trademark of Philips Corp. Purchase of I²C components from [manufacturer] conveys a license under the Philips I²C Patent Rights to use these components in an I²C system, provided that the system conforms to the I²C Standard Specification as defined by Philips."

[46] For controllers without I²C support, you can use an SPI-to-I²C bridge

A. Outline for Figure 15.22

45. **Parallel-bit ports ("GPIO" – general-purpose I/O).** Definable wires, interrupt: 8 bits, STMPE801; 24 bits, STMPE2401; consider also TI's 16-bit TCA6416.

46. **LED backlight.** Smart backlight driver, fade, ambient-light responsive: ADP8860, ADP5501, with 4×4 keypad ADP5520.

47. **Real-time clock (RTC).** For example, NXP's PCA8565 (0.65 µA at 1.8–3.3 V, 32 kHz xtal, with alarm and timer interrupts), or the fancier PCF2129 (which includes internal xtal and TCXO, selectable SPI or I²C interface, and a nice time-stamp function to capture the time of an event even if your µC is sleeping) NXP's PCF8563 and Seiko's S-3590A are popular parts, and available for less than a dollar.

48. **Clock synthesizer.** See §13.13: ON-Semi FS714x, Silicon Labs Si5338 to 710 MHz, four LVDS out, with separate V_{CC} levels.

49. **Three-axis accelerometer.** MMA7660FC, $0.68, MMA7455L (I²C mode), or a variety of I²C modules from Adafruit, for example their MPL115A2 barometric pressure/temperature sensor, or L3GD20+LSM303+BMP180 10-DOF inertial measurement unit, $30. Many of the "shield" part numbers from AdaFruit and SparkFun match the IC part numbering of the mounted sensor ICs, which are available from distributors. The shields make it easy to experiment with these (generally tiny SMT) parts.

50. **8-pin EEPROM.** See §14.4.5B: Microchip 24LC256 (32k×8, 2.5–5.5 V, low-cost $0.76), 24AA256 (1.8 V).

51. **ADC subsystem.** For example, AD7294: 12-bits, six inputs (3 temp, 2 current sense), four outputs, alarms, ($14), see Figure 14.41. The AD7730, intended for use with strain gauges, scales, etc., includes many functions (but uses the SPI interface), see Figure 13.75.

52. **Stepping-motor position IC.** For example, ON-Semi AMIS-30624 (both parallel and I²C control options): has position memory, a sine table for 1/16 microstepping, speed ramps, two H-bridges, V_S to 29 V, 800 mA, max.

53. **Touchscreen.** ST's STMPE811, $1.47.

54. **Digitize video.** NTSC and PAL composite, S-video, §14.6.2 and Appendix I; 8-bit TI's TVP5150, $4; 10-bit dual 30 Msps TVP5147, $5.

(as in Figure 15.21), driven either with the microcontroller's internal SPI hardware (if present) or with bit-banging via individual port pins. The bridge takes care of the fussy I²C timing and half-duplex signal reversals.

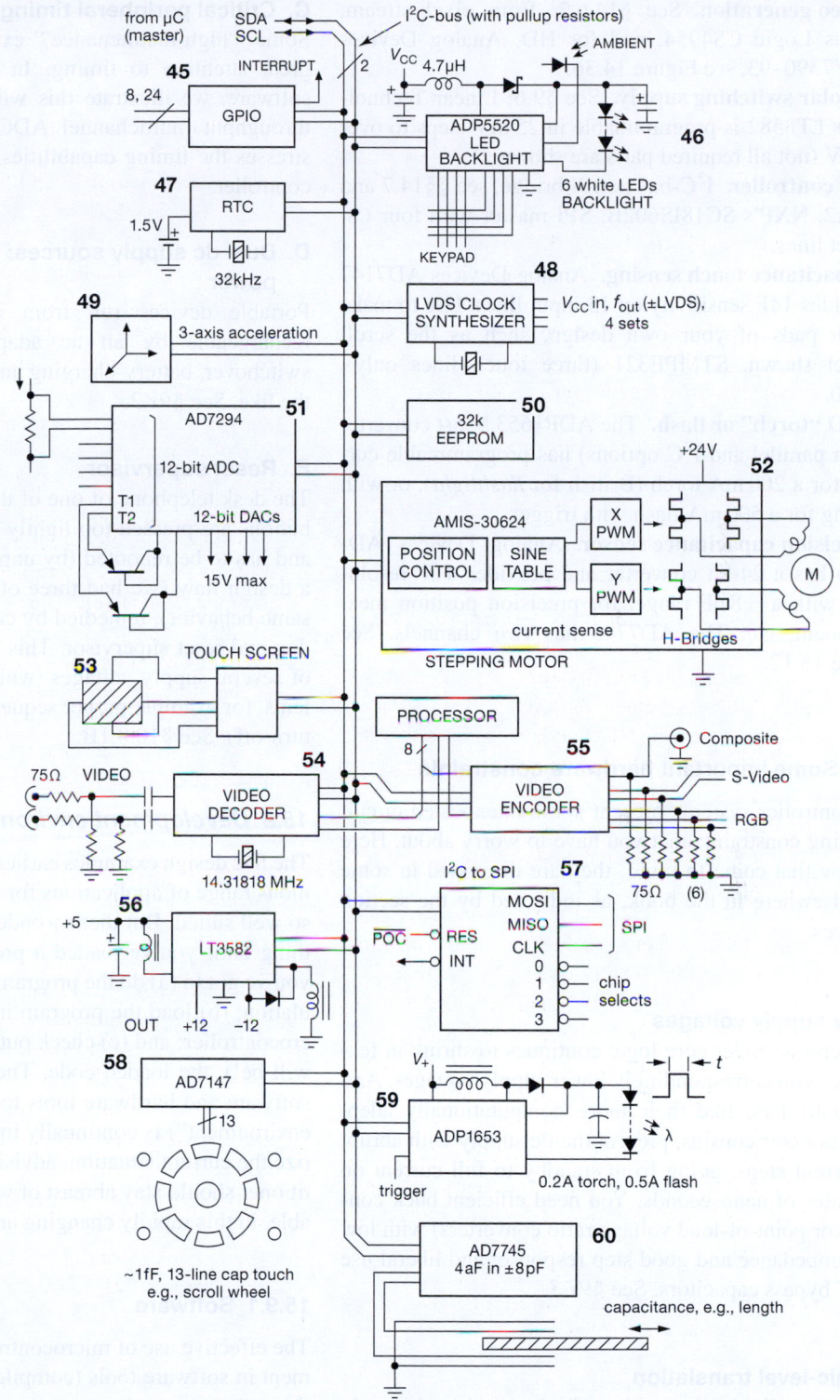

Figure 15.22. An assortment of I²C bus peripherals, well suited for microcontroller applications. The I²C interfaces marked with bold numbers **45** to **60** refer to descriptive paragraphs in §15.8.3.

55. **Video generation.** See §14.6.2: from pixel stream, Cirrus Logic CS4954; and for HD, Analog Devices ADV7390–93, see Figure 14.36.

56. **Bipolar switching supply.** See §9.6; Linear Technology's LT3582 is programmable in 25 mV steps to over ±12V (not all required parts are shown).

57. **SPI controller.** I^{2}C-bus to SPI bridge, see §§14.7 and 15.8.2. NXP's SC18IS602B, SPI master with four CS$'$ select lines.

58. **Capacitance touch sensing.** Analog Devices AD7147 provides 1 fF sensitivity on 13 input lines. Use to make touch pads of your own design, such as the scroll wheel shown. STMPE321 (three touch lines only), $1.30.

59. **LED "torch" or flash.** The ADP1653 boost converter (both parallel and I^{2}C options) has programmable current for a 200 mA torch (British for *flashlight*), or with timing for a 500 mA flash with trigger.

60. **Precision capacitance sensor.** Analog Devices AD-7745 has a 24-bit converter and provides 4 aF resolution with a ±8 pF range, for precision position measurement, etc. The AD7746 has two channels. See Table 13.12.

15.8.4 Some important hardware constraints

Microcontroller systems present some unexpected circuit and timing constraints that you have to worry about. Here are a few that come to mind; they are discussed in some detail elsewhere in the book, as indicated by the section references.

A. Low supply voltages

Fast microcontroller core logic continues to shrink in feature size, with correspondingly lower supply voltages. And microcontrollers, like their more computationally adept micro*processor* cousins, present the dc supply with abrupt load-current steps, going from standby to full current on time scales of nanoseconds. You need efficient buck converters (or point-of-load voltage ratio converters) with low output impedance and good step response; and liberal use of SMT bypass capacitors. See §9x.3.

B. Logic-level translation

You're generally dealing with multiple supply voltages in an embedded application, with some incompatible logic-level requirements. See §12.1.3 for lots of detail.

C. Critical peripheral timing

Some "high-maintenance" external peripherals require great attention to timing. In the upcoming section on software, we illustrate this with the example of a high-throughput multichannel ADC system (§15.9.2), which stresses the timing capabilities of a modest-speed microcontroller.

D. Dual dc supply sources: battery and ac line power

Portable devices run from internal batteries, usually rechargeable by an ac adaptor. You need seamless switchover, battery-charging and protection functions, and the like. See §9x.2.

E. Reset supervisor

The desk telephone of one of the authors crashes (!) if the buttons are pressed too lightly or too fast. It gets wedged and has to be rebooted (by unplugging the power). That's a design flaw (we had three of the same model, with the same behavior), remedied by careful design with a watchdog and reset supervisor. This task is complicated by use of several supply voltages (which present their own problems, for example proper sequencing during power-up and turn-off). See §10.8.1B.

15.9 Development environment

The five design examples earlier give a glimpse of the enormous range of applications for which microcontrollers are so well suited. But these wonderful gadgets won't do anything until you've loaded a program into them. For that, you've got to (a) do the program coding, and (usually) simulation; (b) load the program into the target in-circuit microcontroller; and (c) check out and debug, if necessary (it will be!), the loaded code. The detailed landscape of the software and hardware tools to do this (the "development environment") is continually improving. Here we summarize the current situation, advising as always that the practitioner should stay abreast of what's available, and affordable, in this rapidly changing area of electronics.

15.9.1 Software

The effective use of microcontrollers requires some investment in software tools (compilers, assemblers, debuggers) that run on some host-computer platform. Along with the cost of the tools comes a significant investment in time to learn how to use them. The great majority of contemporary programming is done in C/C++, although there is a place for programming at the assembly-language level.

A. C/C++

It's important to realize that there are vendor-specific variations in the C/C++ languages, when coding for a microcontroller: there are libraries that deal with the internal peripherals (SPI port, ADC, timers, and so on), and there are low-level issues such as the configuration of idiosyncratic memory spaces.[47] There are complications connected with timing (clock frequencies, baud rate, timer intervals), and also with processor word width (8/16/32 bits). The bottom line is that the code you are writing is not vanilla C, and consequently is not so easily ported from one microprocessor family to another.[48]

B. Assembly Code

Some microcontroller coding is done in the processor's native assembly code. This is fussy stuff, and it's generally not the right way to go when you are writing a program that is complicated, with complex branching and control, arithmetic functions, and the like. But it lets you get closer to the metal, with access to instructions (such as bit operations, or read–modify–write) that are inaccessible in a compiled language like C. And while it's perfectly OK to write in assembly code for simple microcontroller applications, you've got to remember that such code is difficult to port to a different processor family.

Assembly language coders can write highly optimized loops and timing-critical code (e.g., for a digital signal processor); and assembly-language routines can be invoked from C when needed.[49] Furthermore, a compiled program intended to run in stand-alone mode (i.e., without an operating system) will necessarily acquire some startup assembly code, put there by the development system, that performs tasks such as initializing memory and interrupt

vectors, copying executable code from nonvolatile memory to internal RAM (if so instructed), and the like. It's helpful to be able to understand this code, particularly when debugging.

C. BASIC

The BASIC "Stamp" series, made by Parallax, Inc., are tiny (stamp-sized) circuit boards based around PIC or Ubicom (formerly Scenix) microcontroller variants that include a BASIC interpreter in the μC's built-in ROM. They also include a voltage regulator, crystal, and nonvolatile EEPROM for user program and data storage. They come in 14-pin SIP, and in 24-pin- and 40-pin-wide (0.6") DIP, with prices beginning at $30.[50] These modules include things like digital I/O port pins, PWM, serial port, and I^2C, supported by Parallax's extensions[51] to the BASIC language (called PBASIC); the PBASIC source code is not compiled, but it is stored in a compressed ("tokenized") form, from which it is executed by the on-chip interpreter. Typical BASIC commands occupy 2 to 4 bytes, in their compressed form, when loaded into the Stamp's EEPROM. Because these devices run an embedded *interpreter* (rather than executing compiled code), they are not terribly fast – a few thousand instructions per second – but their simplicity has made them quite popular.

People tend to think of BASIC as an inefficient language (in both speed and memory requirements) – but in fact there are BASIC *compilers* that create good assembly code that runs fast. They don't require a preloaded run-time interpreter program, and they target code for many popular microcontrollers (e.g., AVR, PIC, and ARM).[52]

The BASIC language provides easy access to microcontrollers for the unsophisticated or beginning programmer. Other manufacturers have moved into this niche: you can get, for example, stamp-sized boards with a 32-bit ARM7 that runs a *compiled* BASIC for $50 (ARMexpress, from Coridium).[53]

[47] The 8051 architecture, for example, has internal and external code spaces, internal and external data spaces, bit-addressable areas, and special function registers. And, to complicate matters further, modern 8051 variants often have some *internal* "external" memory! The AVR and ARM processors are considerably simpler in this regard; the ARM, in particular, has a single "flat" address space.

[48] There are some pleasant exceptions to this complicated landscape, for example, the *Arduino* platform, which provides simple software and libraries that make programming as easy (or perhaps even *easier*) than the exemplary BASIC Stamp.

[49] But a *cautionary note* on inline assembly code within a C/C++ program: contemporary compilers will "optimize away" program code, assuming they know better than the human programmer; this can create serious havoc.

[50] Well, $29 actually; but we are weary of being *nined* to death, so we routinely round away the trailing 9.

[51] For example, "BUTTON" debounces an input button, performs an auto-repeat, and branches to an address if the button is in a target state.

[52] Examples are GCBASIC (open source), Swordfish, Proton PICBASIC, mikroBasic, and microEngineering Labs PicBasic compiler for the PIC; BASCOM-AVR and the GNU compilers for the AVR; OshonSoft for PIC, AVR and Z80; and several compilers for the ARM family.

[53] Some other BASIC-programmed microcontrollers are BasicX (NetMedia), PICAXE (Revolution Education), KicChip, C Stamp (A-WIT Technologies), CUBLOC (Comfile Technology), and ZBasic (ZBasic/Elba).

D. Java, Python

Some microcontrollers, notably those with ARM cores, include some hardware support for languages like Java. And some interpreted scripting languages have been ported to microcontrollers. These interpreted languages are convenient for high-level jobs that are not time-critical, for example a user interface; but be cautious when counting on them for timing dependent tasks like robotic control. It's often better to devote a separate microcontroller for the latter, with appropriate real-time programming.

15.9.2 Real-time programming constraints

Programming a microcontroller is similar to ordinary computer programming, but with a few important differences. We've already mentioned things like the need to initialize "internal peripherals" through special-function configuration registers, vendor-specific language augmentations, and the fact that run-time code is often just a stand-alone program, without anything like the usual operating system.

An important additional constraint in many applications is the need to conform to critical timing. Some examples are an ADC that must take accurately periodic samples, a serial port (USB, say) whose timing must meet a tight specification, or the generation of analog video.

When you're not dealing with breakneck speeds, the built-in timers offer a good solution: in the first case, you might use one of the μC's internal timers, brought to an output pin, to trigger the ADC at some leisurely rate, say 10 ksps; and the ADC's conversion complete could generate an interrupt, so you can fetch the result. But if you're pushing the speed limits, you may have to resort to carefully coded assembly language, taking account of the microcontroller's execution speed (clocks per instruction). This is fussy stuff, and you have to take care to equalize timings on branches and loops. And the code you generate is not portable, depending as it does on clock speed and processor type.

Timing example: 16-bit serial ADC

To illustrate this with a specific example, consider the LTC1609 200 ksps 16-bit successive-approximation ADC with serial data output, which we saw in Chapter 13 as the back end of a 16-channel data acquisition system (§13.12.1, Figure 13.76). Among other things in systems like this, you need to worry about digital noise contaminating the analog input. To deal with this problem, this converter happily offers several modes: for example, in "internal clock mode" it generates the conversion clocking pulses, and you grab the serialized bits flying your way at

exactly the right times; it's too fast for interrupts, so you have to code a highly constrained loop. In another mode, it does the conversion in a burst, while the interface is quiet; then you clock out the multibit data while the converter is idle.

Figure 15.23 shows how it works, in the internal clock mode. You initiate conversion with a CONV$'$ pulse, after which it gives you the serial data (MSB first, as with all successive-approximation converters) along with a clock. Although the timing is quite relaxed in terms of the speed of standard digital logic hardware (gates and flip-flops), it requires substantial minimum processor speed to handle properly in *software*.

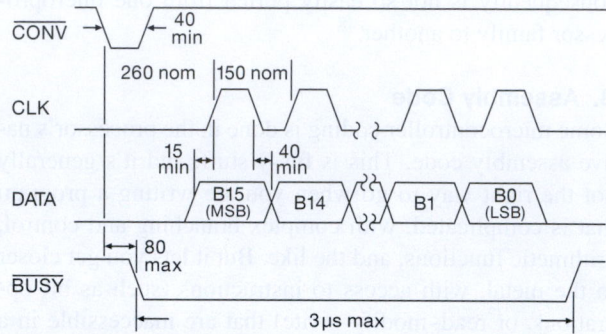

Figure 15.23. Serial data timing of the LTC1609 16-bit ADC when operating in "internal clock mode." Times are in nanoseconds.

Take a look at Pseudocode 15.6. The WAIT-FOR-DATA loop has to complete in 150 ns, which, as we'll see presently, requires at a minimum a processor with a ~30 MHz clock (with one instruction per clock, including a 16-bit shift instruction). Older PICs, etc., need not apply! Note that we command the channel switching and PGA (programmable gain amplifier) gain setting ("setup next channel") immediately after starting the current conversion, which may seem counterintuitive. But this is the right place to do that, because the PGA needs 2 μs to settle, and the ADC needs another 2 μs to acquire the PGA's output before the subsequent conversion is initiated. And we've got the time to send these setup commands, because there's a ~200 ns delay after CONV$'$ before the ADC starts gushing clocks and data.

After EXITing from this loop, you've got ~ 2 μs to do something with the data, and anything else that needs doing, before the converter is ready for its next conversion at its full speed of 200 ksps (of course, you're always free to run it slower).

Let's look more closely at the critical WAIT-FOR-DATA loop timing, taking it step by step. Our program takes one instruction to check if it's already gotten all the data bits,

Pseudocode 15.6 ADC serial loop pseudocode

Channel

wait until timer

set CONV' LOW

NOP

set CONV' HIGH

setup next channel

Wait-for-data

if BUSY' HIGH, Exit

if CLK LOW, repeat Wait-for-data

get & shift data bit into 16-bit word

repeat Wait-for-data

Exit

store 16-bit word

---

another to see if a new bit is ready (ADC output clock HIGH), one to get it, and one to go back for more. Assuming a 30 MHz single-cycle CPU clock, this loop takes 133 ns. Let's follow what happens as the μC grabs the ADC data clocked every 150 ns. It grabs the first bit someplace in the ADC's 75 ns clock HIGH time, and about 17 ns earlier on each successive grab. After a few bits it's too early, landing in an ADC clock LOW; so it has to loop back, costing it 66 ns and pushing it close to the end of the 75 ns ADC clock HIGH. The μC takes another 33 ns to grab the data bit, approaching the end of the 40 ns ADC data-valid hold time.

As the μC grabs the successive 16 bits, it's all over the place – not a pretty sight. But we're blind and can't see; short of having enough time to add two more instructions and output a marker pulse, there's no way from the outside to see exactly when the processor does what it does. Instead, it's necessary to do a careful thought-process analysis, a common situation with real-time programming and time-sensitive tasks. In this case we'd realize what needs to happen, and we'd calculate the minimum processor clock speed from the worst-case timing situation. That would happen when the ADC output clock is tested just before a rising edge; the loopback and read takes 100 ns to complete, at which time the data bit has only 5 ns of guaranteed life remaining.

So a 30 MHz CPU clock appears marginally fast enough – but it's a close call. And we're in trouble if the ADC's period is shorter than the "typical" value of 150 ns, or if its duty cycle is less than 50%, making its HIGH time 60 ns, say, instead of the 75 ns we assumed. To get a better safety margin we would need a faster processor. Probably better, though, would be to add an outboard D-FF (e.g., an

'LVC1G74) clocked on the rising edge of CLK, to capture each bit and make it available to the μC throughout the full clock period of 150 ns. This would give us an extra 35 ns (plus the flip-flop's delay time of $\sim$5 ns). If you don't need the processor speed for other reasons, adding a single flip-flop to relax the needed CPU clock frequency is a worthwhile tradeoff.

This example illustrates a common system design issue, namely the tradeoff between software (and processor cycles) and hardware. We could have used a pair of 'LVC595 or 'VHC595 shift registers (see §10.5.3) to clock in the 16 converted ADC bits, which the microcontroller then reads in byte-wise via a parallel port. This would reduce the software burden to fewer than ten instructions every 5 μs. But (apart from the coding task) program execution is free, whereas two chips take space and money; they also hog processor I/O pins. On the other hand, if you don't have much processor time available, or your processor is too slow, adding two chips makes good sense.

Standardized serial protocols

For standardized serial ports like UARTs, I²C, Ethernet, or USB, the best approach is to use a microcontroller with dedicated internal hardware. A second choice would be an external "bridge" that converts between USB, say, and a standard serial UART (for example the chips made by FTDI). Hardy souls have been known to implement serial ports in assembly language, but it can be a real tour-de-force. A stunning example is Paul Starkjohann's implementation[54] of 1.5 Mbps USB (version 1.1) in assembly language on an Atmel AVR running with a 12 MHz clock, which required precisely eight instructions per serial bit, during which he had to extract the bit stream from NRZI-encoded data, complete with bit stuffing and end-of-packet characters. This was used by Thomas Baier to control a DDS RF generator, part of a complete vector network analyzer.

Quite apart from the challenge of getting the timing right, serial communication via USB involves plenty of software complexity, in the form of drivers and the like. You should not feel bad about using the plain ol' UART, perhaps with a bridge to USB (with drivers provided by the manufacturer).

15.9.3 Hardware

Contemporary microcontrollers use internal nonvolatile (flash-memory) storage for program code, which you load

---

[54] See `http://www.obdev.at/products/vusb/index.html` and `http://www.obdev.at/articles/implementing-usb-1.1-in-firmware.html`.

Figure 15.24. Development kits make it easy to get started with a microcontroller family; you get them from the chip manufacturers, or from third-party suppliers. They often include software, cables, and power adaptors. At the top are kits for an Atmel ARM and a Microchip PIC24H; across the middle are PIC24F, Silicon Laboratories C8051F320, and Freescale ColdFire (M52259); across the bottom are Texas Instruments MSP430 (two views), Atmel AVR (ATmega168), and an FPGA devkit (Xilinx Spartan-3E, two views). Suppliers other than the manufacturers are Olimex (ARM), Arduino (AVR), and DLP Design (Spartan).

(while the μC is in-circuit) by one of several methods. You usually do the loading[55] with a commercial "pod" (officially called a "device programmer"), which you buy from the chip manufacturer or third party. If you buy a development kit (Figure 15.24), it will often include a programming pod, along with software (for compiling, simulating, assembling, and loading), and with a circuit board on which there's a microcontroller and other hardware (digital and analog ports, LEDs, a serial port, a programming header, and perhaps some display device). Here are the several loading protocols:[56]

---

[55] Which, confusingly, is called "programming" the device. We reluctantly use that term here (interchangeably with "loading," in this hardware context), to mean *physically* programming the microcontroller's flash memory. Do not confuse it with the software programming task of writing the code.

[56] See §§14.5, §14.6.4, and 14.7.4 for more detail on UART, SPI, and JTAG serial data links.

A. UART serial-port bootloader

Some microcontrollers include built-in serial port code (in ROM), so that they wake up listening for programming commands on the UART serial port (a "serial bootloader"). To activate this mode you have to assert one or more pins at reset to signal that you want to program via the UART, and thus put it into program mode. Examples include the Maxim–Dallas DS89C400 series, some Atmel AVRs, and some ARM7 controllers. The communication is via standard serial UART modes (usually 9600 8N1); however, because the microcontroller accepts (unipolarity) logic levels, not bipolarity RS-232, you have to use an interface chip (like a MAX232) between a PC's DE-9 serial-port connector and the target microcontroller.[57] You can, of course,

---

[57] PC serial ports are fast disappearing; you can substitute a USB-to-RS232 adapter, or better, a USB-to-UART chip like the FT232R; the latter matches the logic-level needs of the μC, so no MAX232 style chip is needed.

write your own bootloader code and load it into normal user code memory, so that a microcontroller that permits overwriting the flash code memory can be reprogrammed through the UART. With this method, however, you won't be able to load an unprogrammed chip initially through the UART (the method used by the Arduino project).

B. SPI serial-port bootloader

Some microcontrollers (e.g., the smaller Atmel AVR series) implement a bootloader through the chip's SPI port.[58] Contemporary programming pods connect to the host PC via a USB connection, replacing earlier versions that used the parallel (printer) port, or the serial (RS232) "COM" ports. As with the serial-port bootloaders, you have to assert a pin during boot-up to activate the SPI bootloader. For the AVR controllers, for example, you assert RESET'.

C. JTAG serial-port bootloader

The JTAG serial boundary-scan protocol (§14.7.4), originally intended for test and debugging, is used as a flash-memory bootloader by some microcontrollers. Examples include the higher-pincount AVRs, ARM7, Silicon Labs C8051F series, and Maxim–Dallas MAXQ series. The JTAG port may be one of several choices, as it is with the high-pincount AVRs, which allow bootloading via either JTAG or SPI (as well as parallel programming).

D. Proprietary serial-port bootloader

Some microcontroller manufacturers use their own simple serial protocols, which do not conform to standards like SPI, I^2C, or JTAG. It hardly matters, though, because you generally just use the appropriate pod, as supported by the vendor's development environment. Some of the PIC (Microchip) controllers are programmed this way; and for some of them you must apply a high voltage (+12 V) on a pin called V_{PP} ("high-voltage" programming), while others let you use normal logic-level supply voltages ("low-voltage" programming).

E. USB serial-port bootloader

Microcontrollers that include USB support often implement a USB bootloader option. Examples are some Cypress controllers, and the Atmel ARM and AVR32UC3 series, which can load program memory upon boot from either the USB port, the JTAG port, or the UART.

[58] Which may be one of several bootloader ports on the same μC: the larger Atmel AVR parts, for example, allow bootloader programming via either SPI or JTAG. Additionally, some μCs let you install a custom bootloader in the μC's flash memory, which can bootload from the USB, UART, or Ethernet ports.

F. Parallel loading

Finally, many microcontrollers provide a way to program the internal flash memory via a multiwire parallel connection. In some cases you must apply a higher voltage (e.g., +12 V) to one of the pins; this is sometimes called "high-voltage programming."

G. A "universal" pod?

The "Bus Pirate" project has what appears to be a great tool: for $30 you get a little USB-to-anything piece of hardware, with open-source support. You can open a terminal on your PC and pick from among 1-Wire, I^2C, SPI, JTAG, asynchronous serial (UART), MIDI, PC keyboard, HD44780 LCDs, generic 2- and 3-wire libraries for custom protocols, and scriptable binary bit-bang. You get software support for AVR, JTAG, and some flash-memory in-circuit serial programming. So it's useful for talking to many kinds of chips (not just microcontroller programming) – for example, when debugging the protocols for talking to some new "smart" chip that needs its operating modes set up (we spent plenty of time struggling with a smart alphanumeric LED display strip; this thing would have saved us hours). Bus Pirate version 4 is an upgrade that adds some features (e.g., OpenOCD JTAG mode); it currently costs $40 (http://dangerousprototypes.com).

H. Sawing off the branch...

You usually can choose which programming method to use when the μC supports several. However, in controllers where the bootloader can be overwritten, you can clobber the loader, and thus be forced to resort to parallel programming. There are other ways, also, to do what we call "sawing off the branch you're sitting on": we managed to do this with a small AVR chip, when we loaded a program that began by turning off most peripherals, *including the SPI*. That was a big mistake, because it disabled the only port available for serial bootloading. High-voltage parallel programming brought it back to the land of the living.

Another way to terminate your conversation with a microcontroller is to program it to use an external oscillator of the wrong type; that is, one that does not correspond to the actual hardware that is connected (bare crystal, ceramic resonator, or external oscillation signal from an outboard oscillator module). Some microcontrollers protect you from this trap: the Silicon Labs 8051-core processors and the ARM-core processors, for example, boot up in a known state, running from a slow internal oscillator; you have to specifically code any changes from this default. If you do it wrong, just fix the code, and reload from a cold boot.

I. In-circuit debugging

Microcontrollers of recent design include on-chip hardware that lets you debug code during in-circuit execution, by setting breakpoints, examining registers and memory, single-stepping, and so on. (This used to be a big deal, requiring special processor chips outfitted with extra leads, and requiring also considerable outlay of cash.) In-circuit debugging generally uses the same ports as you use for program loading; so you just keep the programming pod attached and iterate a debug–reprogram cycle until it all works right.

Some processor families that currently include these features are the Atmel AVR, processors based on the ARM core, some PIC processors, and some 8051 derivatives (notably those from Silicon Labs).

15.9.4 The Arduino Project

We've mentioned the Arduino Project, approvingly, several times in this chapter. What, exactly, is it?

In the words of the website (www.arduino.cc), "Arduino is an open-source electronics prototyping platform based on flexible, easy-to-use hardware and software. It's intended for artists, designers, hobbyists, and anyone interested in creating interactive objects or environments. ... The microcontroller on the board is programmed using the [C language with custom libraries] and the Arduino development environment. ... Arduino projects can be stand-alone or they can communicate with software running on a computer."

In *our* words, the Arduino *hardware* is a well-designed set of inexpensive boards, based on Atmel microcontrollers (the ATmega AVR series, and the SAM3X ARM Cortex-M series), with all the right external components: USB port (with the FTDI chip to convert to the μC's serial pins), 5 V regulator, SPI port, PWM output, ADC and digital I/O pins, LEDs, and a few additional components. You can buy the standard board (currently called "Uno") assembled, for about $30 from hobbyist suppliers like Adafruit and Spark-Fun; you can also buy it as a kit or even etch your own (they provide CAD files in Eagle format, or an image file in ∗.png format).

And the *software* development environment is open-source and free: it's a full-blown C, using the GNU C compiler targeted at the AVR and ARM processors. The Arduino software is a simple, user-friendly GUI wrapper around this compiler; it runs on Windows, Mac OS, or Linux, and allows easy code editing and project management. The Arduino firmware (written in C) that runs on the device includes a bootloader that listens for communica-tion over the USB serial link, and either downloads a new user program or times out after a second and runs the program already present in flash (code) memory.

There is a library of nice C functions that you can call from your own programs to do things like text I/O through the USB link, formatting numbers to be printed, setting up timers and interrupts, etc. For example, val=analogRead(3) puts the voltage read on analog pin 3 into the variable val.[59] Very little of this library code is tied to the Arduino hardware itself, and you can use it in your AVR-based projects. This works well because it's all open source and uses the open-source compiler.

Taken as a whole, the Arduino Project is simply all these components working together. An interesting metric of ease-of-use of any microcontroller hardware-plus-IDE might be the following: how long does it take, after you open the box, before it's doing something interesting? For the Arduino, the answer is "about 20 minutes": download the software, right-click to install the FTDI driver and mini-IDE; then open the Arduino program, type four lines, and off you go! And you don't have to know anything at all about hardware – the little board powers itself through the USB link – nor do you need any extra hardware (except for a standard USB cable).

There are Arduino look-alike products with special features. Many use the same microcontrollers, have the same shield (daughterboard) pinouts, and use the same or similar compilers. For example Linear Technology's "Linduino" board uses their LTM2884 USB isolation chip, providing an isolated USB 2.0 hub or USB peripheral (560 V peak or 2.5 kVrms for one second), complete with an isolated 5 Vdc (to 500 mA). The Linduino board also has LTC's standard DC590 interface board connector, and it's especially well suited for use with their high-resolution ADCs.

15.10 Wrapup

15.10.1 How expensive are the tools?

Both the microcontroller companies, and some third-party manufacturers, make it easy to get started. Many of them offer stripped-down development kits, typically in the range of $50–$200, consisting of (a) a programming pod, (b) a small board with the microcontroller and some trimmings (LEDs, serial port, USB port, analog I/O, and

---

[59] This ease-of-use comes with some loss of generality, for example in pin assignments.

programming header), and (c) software for compiling, assembling, loading, and debugging.[60]

The compiler that comes with such kits may be a free "evaluation" version of a more capable compiler that you can buy. For example, the inexpensive Silicon Labs kits come with a limited version of the commercial Keil compiler tools. It is full-featured, but limits you to 4 kB of compiled object code, and lacks the floating-point library. The unlimited Keil product lists for ~$2.5k, from which educational users receive large discounts. Similarly, the Raisonance software (see the 8051 description in §15.3) is available as a free download, again limited to 4 kB of object code, with the unlimited version priced comparably to the Keil product. Another purveyor of high-quality software tools is IAR Systems, with support for most of the popular microcontroller families: 8051, ARM, AVR, Coldfire, MAXQ, PIC, H8, and MSP430, among others. Other suppliers of compilers and debuggers for embedded systems include Green Hills Software, HI-TECH Software, Lauterbach Development Tools, and Rowley Associates.

In a pleasant development, the open-source community has ported the GNU C/C++ compiler to most operating systems, with the ability to target the AVR and ARM family microcontrollers. This free software, plus a programming pod, provides the least costly path to full-featured C-language microcontroller programming. Another encouraging trend is the growing acceptance by semiconductor manufacturers of the need for good software support. Traditionally the chip manufacturers have viewed software as a necessary evil, at best, which they would prefer to have someone else worry about; this seems, happily, to be changing. For example, Atmel integrated the GNU open-source tools into their Windows environment for compiling, simulating, and in-circuit debugging. And IBM's Eclipse open-source development environment has become popular, allowing both manufacturers and users to create plug-ins for specific families of processors. Two examples are Altera's NIOS soft-core μC Micrium, and the open-source ARM tools.

It's important to realize that the programming pods are not generally interchangeable across processor families. Pods aren't very expensive though, so this isn't a serious problem. In some cases the programming hardware allows more than one connection protocol. For example, the Atmel AVR Dragon, which connects to the host via USB

and costs about $50, lets you program in SPI, JTAG, parallel, and high-voltage serial modes; it also supports in-system debugging. The simpler AVR ISP pod supports SPI programming only. Programming pods are also sold by third-party and hobbyist suppliers such as Olimex Ltd and Sparkfun Electronics; you can find many more suppliers in the pages of enjoyable hobbyist publications such as *Circuit Cellar*, or *MAKE Magazine*.

15.10.2 When to use microcontrollers

Almost always!

Certainly for electronic systems that
(a) have character or graphic displays as part of their user interface;
(b) include chips requiring configuration of internal registers or operating modes;
(c) communicate with a host computer, stand-alone peripherals, network, or wireless devices;
(d) require some computation, storage, format conversion, signal processing, etc.;
(e) require calibration or linearization;
(f) involve sequencing events over time; or
(g) are subject to upgrades or feature revisions.
Microcontrollers should be considered even for traditional "analog" functions such as measurement and control, especially with the growing emphasis on analog-oriented internals in processors from companies such as Analog Devices and Cypress.

Programmable logic devices (PLDs, including FPGAs), by contrast, are generally preferred for tasks requiring critical timing or a high degree of parallelism. They are, however, considerably more difficult than microcontrollers to program and debug. Any system with PLDs will generally include a microcontroller as well; and the latter may take the form of a soft processor core (i.e., configured from the programmable resources in the FPGA) or a hard processor core (i.e., prewired within the "hybrid" FPGA). Examples of FPGA soft cores are the Actel ARM, Altera Nios-II, Lattice Mico, and Xilinx MicroBlaze; examples of prewired processors in hybrid FPGAs include and Altera's ARM, Atmel's FPSLIC with AVR, and Xilinx's PowerPC.

Another way to handle the dual demands of critical timing (e.g., in real-time video or wireless applications) along with the need for the versatility of an embedded microcontroller is to combine an application-specific standard product (ASSP; for example, an MPEG decoder, or cellphone RF subsystem) with a supervisory microcontroller.

---

[60] If you just want to dip your toe in, there are some demonstrator gadgets that let you program a few LEDs only; these usually plug into a USB port and cost around $10. You can't do much with them, though, and it's better to go with the real devkits.

15.10.3 How to select a microcontroller

There's a real benefit to using a processor family that the people around you are using, and therefore have available the necessary software and hardware tools, and the experience. Then look at factors like these (depending on your application):

(a) Ports (analog, digital, and communication);

(b) internal functions (e.g., converters, PWM, bare LCD drivers, etc.);

(c) compute speed;

(d) flash, EEPROM, and SRAM memory size;

(e) package configurations;

(f) power dissipation, low-power clock modes, and suspend modes;

(g) software programming, simulation, and in-circuit debugging tools.

The compendium in §15.3 may provide a good starting point.

When choosing a particular part within a microcontroller family, the choices can be overwhelming. It's usually easiest to start with the "premium" part in the series; that is, the one with the fastest clock and most data (RAM) and code memory (flash ROM). It is often the case that most of the family consists of reduced versions of a few premium parts, perhaps with specialized peripherals.

Early microcontrollers were organized around an 8-bit word size, but now there are many 32-bit controllers. Some advantages of the latter (a flat address space, more powerful instructions) are partially offset by considerations like the complexity of startup,[61] somewhat higher prices, and widespread familiarity with the 8-bit processor families.

Even if you code everything in C, most of your troubles will be microcontroller specific: initializing on-board I/Os and peripherals at startup, setting the "fuse bits" (independent of the program code) that control voltage levels and clocking, or dealing with multiple memory pages and idiosyncratic memory locations for peripherals and their configuration. As a result, a simple microcontroller architecture having a flat memory space and standard header files will be much easier to use.

---

[61] For example, an ARM peripheral needs a lot of code just to power it on, initialize it, connect it to the correct I/O pins, and so on.

The quality and quantity of included libraries can be quite important; for example, the included libraries for TCP/IP, SPI, I^{2}C, etc., makes it easier to program a Rabbit microcontroller. This is also an important advantage for the Arduino Project, with the large community providing code that has grown around it. Similar comments can be made on the importance of choosing a microcontroller that has a stable compiler–debugger. Similarly, there are a number of advantages with selecting a microcontroller board that is set up to run Linux (such as the Gumstix Overo, or the BeagleBoard). They come with a very stable OS and low-level drivers, they have native preemptive multitasking, they have some nice debugging advantages (one can just use SSH or a serial terminal and get console access), and finally they have impressive hardware (including fairly high-performance graphics controllers).

15.10.4 A parting shot

In case we haven't made it completely evident, microcontrollers are, well, *fun*! Figure 15.25 shows a wacky example, namely Jason Gallicchio's Halloween hat from 2004: it resembles (not coincidentally) an electrified colander, complete with flashing LEDs and a stack of batteries on top. Its notable characteristic, though, is the ability to read the mind of its wearer. The photo shows what happens when the wearer's interlocutor replies to the question "What college do you go to?"

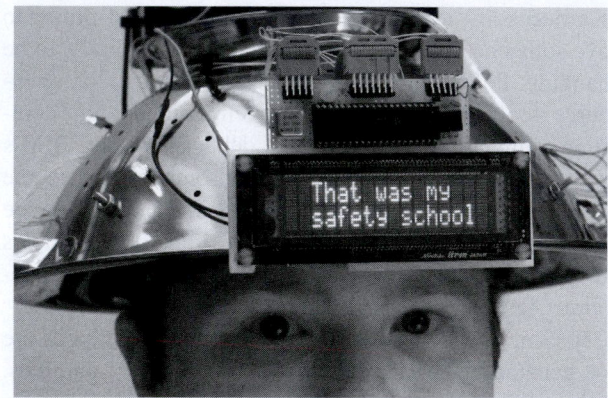

Figure 15.25. "This hat reads my mind."

Review of Chapter 15

The briefest of summaries, of Chapter 15's already brief treatment of microcontrollers.

¶A. Microcontrollers – the 10,000-foot View.

In 1960 the predictions of the future had folks flying around in jetpacks, traveling on supersonic passenger jets, and, most daring of all ... *a push-button phone in every home!* Put another way, they extrapolated the technologies they knew (transportation, wired telephones), but they missed the Big One (microelectronics, and particularly embedded microcontrollers). We don't have the personal jetpacks, but we've got portable and instant interactive access to news, information, and people. Hey, we don't *need* no stinkin' jetpack!

Microcontrollers (μCs) are standalone processors with a full suite of peripherals integrated on a single inexpensive chip (Figure 15.1). You get substantial computational performance (32-bit processor, 100 MIPs, no problem), plus an assortment of ADCs, DACs, Ethernet, USB, PWM, LCD controller, SPI, multiple UARTs and timers, and of course on-chip program memory and SRAM, all in one chip for a price well below $10; see, for example, the now-vintage example in Figure 10.86, or take a look at NXP's implementations of the ARM Cortex-M4 MCU (micro-controller unit), for example the LPC4088 series. Micro-controllers are useless without their stored *firmware* programs, whose creation can be the major stumbling block to a successful design. So, along with the creation of silicon with stunning integration and performance, the industry has been developing and streamlining the integrated development environment (IDE) – the process of initial code development, simulation, loading onto the target device, and in-circuit debugging.

To the circuit designer, the microcontroller should be thought of as a circuit *component*, like an op-amp and sometimes even less expensive. Even in their simplest varieties they are particularly useful as interfaces between the user and the other circuitry (see for example Figure 15.5); and in their complex varieties they can carry out most of the instrument's functions (as in Figure 15.18). In this chapter we introduced the vast subject of microcontrollers with some illustrative examples, to give a sense of what's possible; we included with each its corresponding *pseudocode*, and for the first example (suntan monitor) we listed the detailed C-language code.

¶B. Popular Microcontroller Families.

In §15.3 we provided an annotated listing of contemporary favorites. The dominant species are the simpler AVR (Atmel) and PIC (Microchip) devices, and the go-to choice for higher performance is the hugely popular ARM-derived processors (licensed by ARM Holdings to more than a dozen semiconductor manufacturers). The latter are used in most of the world's smartphones. The attractive Arduino platform (§15.9.4) includes both AVR- and ARM-based single-board computers (SBCs).

¶C. External Peripherals.

Microcontrollers love to pull the strings of other chips, easily done with a few direct connections (Figure 15.20, §15.8.1), or with simple inter-chip serial buses like SPI (Figure 15.21, §15.8.2) and I^2C (Figure 15.22, §15.8.3). The figures and notes list and describe more than 50 useful peripheral devices.

¶D. Design Hints (Hardware).

The five design examples described in this chapter (suntan monitor, ac power control, frequency synthesizer, thermal controller, and stabilized mechanical platform) included plenty of circuit designs, with corresponding lessons. Here are some of them:

(a) Most μCs include on-chip ADCs, which are attractive when dealing with analog inputs; but don't ignore the simple on-chip comparator, which can sometimes promote both simplicity and better performance (Figure 15.3).

(b) A digital output port bit can happily drive external MOSFETs or BJTs directly (Figure 15.3) or with the assistance of an external gate-driver IC (Figure 15.10); it can also drive a solid-state relay (Figure 15.4).

(c) Don't overlook the simplicity of simple RS-232 serial communication; it's supported in all μCs, and it's alive and well in laptops and PCs running a terminal emulator and connected via a USB or Ethernet adapter (Figure 15.5).

(d) Pushbutton switches can be connected in a matrix arrangement, to minimize wiring (Figure 15.8); they are read with digital port bit polling.

(e) Microcontrollers intended for sensing and control let you connect directly to low-level sensors (Figure 15.11); their PWM output provides an easy way to implement proportional control.

(f) Most μCs include ADCs, making it easy to attach analog-output sensors (e.g., a gyro or accelerometer, Figure 15.18). You can find a zillion little gadgets of this sort at hobbyist sites like sparkfun.com or adafruit.com.

(g) And an easy way to get started is with the Arduino boards and software (§15.9.4); open the box, and you're up and running in 20 minutes.

¶E. Programming Hints (Firmware).

You need microcontroller-specific software tools (compiler, assembler, simulator, debugger) that run on a host

PC; and you need a hardware pod (§15.9.3) to load the object code onto the target μC and to run debugging tools. You also need to be aware of ways in which microcontroller programming differs from ordinary computer programming. Here are some hints:

(a) There are vendor-specific variations in C/C++ having to do with idiosyncrasies of on-chip peripherals (ports, timers, converters, etc.); you're not writing in vanilla C.

(b) The code for a μC must do a significant amount of initialization of modes and internal peripherals; this is fussy stuff, requiring dozens of perfectly configured bytes.

(c) You many need to program in assembly language for time-critical tasks; if so, watch out for too-smart compilers that try to optimize away your code.

(d) Products intended for manufacture should be reviewed and blessed by someone skilled in human interfaces (see §15.2.2E).

(e) Enable the watchdog! Well coded microcontrollers should not crash, but they do.

(f) Keep interrupt routines short; in a simple system you may not need interrupts at all.

(g) Internal timers can generate signals at output pins, very useful when you want to trigger external devices (e.g., an ADC) with constant time intervals (and similarly for on-chip peripherals).

¶F. When to use Microcontrollers.
Almost always! Certainly for electronic systems that (a) have character or graphic displays as part of their user interface; (b) include chips requiring configuration of internal registers or operating modes; (c) communicate with a host computer, standalone peripherals, network, or wireless devices; (d) require some computation, storage, format conversion, signal processing, etc.; (e) require calibration or linearization; (f) involve sequencing events over time; or (g) are subject to upgrades or feature revisions. Microcon-

trollers should be considered even for traditional "analog" functions such as measurement and control, especially with the growing emphasis on analog-oriented internals in processors from companies such as Analog Devices and Cypress.

Programmable Logic Devices (including FPGAs, Chapter 11), by contrast, are generally preferred for tasks requiring critical timing, or a high degree of parallelism. They are, however, considerably more difficult than microcontrollers to program and debug.

¶G. How to select a Microcontroller.
Look first at a processor family that the people around you are using, and therefore have available the necessary software and hardware tools, and the experience. Then look at factors like these (depending on your application): (a) ports – analog, digital, and communication; (b) internal functions (e.g., converters, PWM, bare LCD drivers, etc.); (c) compute speed; (d) flash, EEPROM, and SRAM memory size; (e) package configurations; (f) power dissipation, low-power clock modes, and suspend modes; (g) software programming, simulation, and in-circuit debugging tools. The compendium in §15.3 may provide a good starting point.

The quality and quantity of included libraries can be quite important (e.g., the Arduino Project, with the large community providing code that has grown around it). Likewise, it's desirable to choose a microcontroller that has a stable compiler–debugger.

When choosing a particular part within a microcontroller family, the choices can be overwhelming. It's usually easiest to start with the "premium" part in the series, that is, the one with the fastest clock and most data (RAM) and code memory (flash ROM). It is often the case that most of the family consists of reduced versions of a few premium parts, perhaps with specialized peripherals.

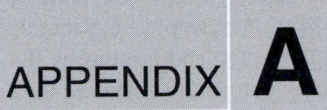

Some knowledge of algebra and trigonometry is essential for a full understanding of this book. In addition, a limited ability to deal with complex numbers and derivatives (a part of calculus) is helpful, although not entirely essential. This appendix is meant as the briefest of summaries of complex numbers and differentiation, preceded by a collection of useful formulas from trigonometry, exponentials, and logarithms. It is not meant as a textbook substitute. For a highly readable self-help book on calculus, we recommend *Quick Calculus*, by D. Kleppner and N. Ramsey, Wiley, 2nd ed., 1985.

A.1 Trigonometry, exponentials, and logarithms

Here is a collection of useful formulas:

$$x = \frac{-b \pm \sqrt{b^2 - 4ac}}{2a}$$

is the solution of the quadratic equation

$$ax^2 + bx + c = 0.$$

$$\sin(x \pm y) = \sin x \cos y \pm \cos x \sin y,$$

$$\cos(x \pm y) = \cos x \cos y \mp \sin x \sin y,$$

$$\sin 2x = 2 \sin x \cos x,$$

$$\cos x \cos y = \frac{1}{2}\left[\cos(x+y) + \cos(x-y)\right],$$

$$\cos x \sin y = \frac{1}{2}\left[\sin(x+y) - \sin(x-y)\right],$$

$$\sin x \sin y = \frac{1}{2}\left[\cos(x-y) - \cos(x+y)\right]$$

$$e^{x+y} = e^x e^y,$$

$$e^{x-y} = e^x / e^y,$$

$$x^{a/b} = \sqrt[b]{x^a},$$

$$e^{\log_e x} = x,$$

$$\log_e(xy) = \log_e x + \log_e y,$$

$$\log_e(x/y) = \log_e x - \log_e y,$$

$$\log_e x^n = n \log_e x,$$

$$\log_e e^x = x,$$

$$\log_e x = \log_e 10 \log_{10} x \approx 2.3 \log_{10} x,$$

$$a^x = e^{x \log_e a}.$$

A.2 Complex numbers

A complex number is an object of the form

$$\mathbf{N} = a + ib,$$

where a and b are real numbers and i is the square root of -1; a is called the real part, and b is called the imaginary part.[1] Boldface letters or squiggly underlines are sometimes used to denote complex numbers. At other times you're just supposed to *know*!

Complex numbers can be added, subtracted, multiplied, etc., just as real numbers:

$$(a+ib) + (c+id) = (a+c) + i(b+d),$$

$$(a+ib) - (c+id) = (a-c) + i(b-d),$$

$$(a+ib)(c+id) = (ac-bd) + i(bc+ad),$$

$$\frac{a+ib}{c+id} = \frac{(a+ib)(c-id)}{(c+id)(c-id)} = \frac{ac+bd}{c^2+d^2} + \frac{bc-ad}{c^2+d^2}i.$$

---

[1] Electrical engineers depart from the universal convention of $i \equiv \sqrt{-1}$, using instead the symbol j in order to avoid duplicating the use of the symbol i (which designates small-signal current). We follow the EEs in this book, but not in this Math appendix. Were we to do so, we would likely be disowned by our math colleagues.

All these operations are natural, in the sense that you just treat i as something that multiplies the imaginary part, and go ahead with ordinary arithmetic. Note that $i^2 = -1$ (used in the multiplication example) and that division is simplified by multiplying top and bottom by the *complex conjugate*, the number you get by changing the sign of the imaginary part. The complex conjugate is sometimes indicated with an asterisk. If

$$\mathbf{N} = a + ib,$$

then

$$\mathbf{N}^* = a - ib.$$

The magnitude (or *modulus*) of a complex number is a real number with no imaginary part:

$$|\mathbf{N}| = |a + ib| = \sqrt{(a+ib)(a-ib)} = \sqrt{a^2 + b^2},$$

i.e.,

$$|\mathbf{N}| = \sqrt{\mathbf{N}\mathbf{N}^*},$$

simply obtained by multiplying by the complex conjugate and taking the square root. The magnitude of the product (or quotient) of two complex numbers is simply the product (or quotient) of their magnitudes.

The real (or imaginary) part of a complex number is sometimes written as

real part of $\mathbf{N} = \mathcal{R}\rceil(\mathbf{N})$,
imaginary part of $\mathbf{N} = \mathcal{I}\Updownarrow(\mathbf{N})$.

You get them by writing out the number in the form $a + ib$, then taking either a or b. This may involve some multiplication or division, since the complex number may be a real mess.

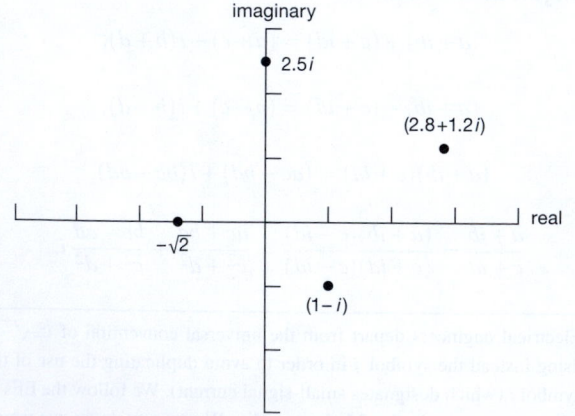

Figure A.1. Complex numbers in the "complex plane."

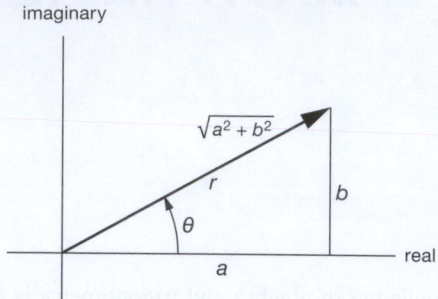

Figure A.2. Complex numbers, as magnitude and angle.

Complex numbers are sometimes represented on the complex plane. It looks just like an ordinary x, y graph, except that a complex number is represented by plotting its real part as x and its imaginary part as y, as shown in Figure A.1. In keeping with this analogy, you sometimes see complex numbers written just like x, y coordinates:

$$a + ib \leftrightarrow (a, b).$$

Just as with ordinary x, y pairs, complex numbers can be represented in polar coordinates; that's known as "magnitude, angle" representation. For example, the number $a + ib$ can also be written as (Figure A.2)

$$a + ib = r\angle\theta,$$

where[2] $r = \sqrt{a^2 + b^2}$ and $\theta = \tan^{-1}(b/a)$. This is usually written in a different way, using the astonishing fact that

$$e^{i\theta} = \cos\theta + i\sin\theta.$$

(You can derive the preceding result, known as Euler's[3] formula, by expanding the exponential in a Taylor series.) Thus we have the following equivalents:

$$\mathbf{N} = a + ib = re^{i\theta},$$
$$r = |\mathbf{N}| = \sqrt{\mathbf{N}\mathbf{N}^*} = \sqrt{a^2 + b^2},$$
$$\theta = \tan^{-1}(b/a),$$

i.e., the modulus r and angle θ are simply the polar coordinates of the point that represents the number in the complex plane. Polar form is handy when complex numbers have to be multiplied; you just multiply their magnitudes and add their angles (or, to divide, you divide their magnitudes and subtract their angles):

$$(r_1 e^{i\theta_1})(r_2 e^{i\theta_2}) = r_1 r_2 e^{i(\theta_1 + \theta_2)}.$$

---

[2] Caution: the formula for θ returns values only between $-\pi/2$ and $+\pi/2$; the signs of both a and b, and not merely their quotient, are required for a correct value of θ in all four quadrants.

[3] Leonhard Euler, pronounced like "oiler."

Finally, to convert from polar to rectangular form, just use Euler's formula:

$$re^{i\theta} = r\cos\theta + ir\sin\theta,$$

i.e.,

$$\mathscr{R}](re^{i\theta}) = r\cos\theta,$$

$$\mathscr{I}\Updownarrow(re^{i\theta}) = r\sin\theta.$$

(These can be used to easily derive the sum and difference of trigonometric functions, so you never have to remember those pesky formulas. Just work out $e^{i(x\pm y)}$.)

If you have a complex number multiplying a complex exponential, just do the necessary multiplications. If

$$\mathbf{N} = a + ib,$$

$$\mathbf{N}e^{i\theta} = (a+ib)(\cos\theta + i\sin\theta),$$

$$= (a\cos\theta - b\sin\theta),$$

$$+i(b\cos\theta + a\sin\theta).$$

When dealing with circuits and signals, the angular argument θ often takes the form of an evolving wave: $\theta = \omega t = 2\pi f t$; thus, for example, $V(t) = \mathscr{R}](V_0 e^{i\omega t}) = V_0\cos\omega t$, etc.

A.3 Differentiation (Calculus)

We start with the concept of a *function* $f(x)$, i.e., a formula that gives a value $y = f(x)$ for each x. The function $f(x)$ should be *single valued* i.e., it should give a single value of y for each x. You can think of $y = f(x)$ as a graph, as in Figure A.3. The derivative of y with respect to x, written dy/dx ("dee y dee x"), is the *slope* of the graph of y versus x. If you draw a tangent to the curve at some point, its slope is dy/dx *at that point*; i.e., the derivative is itself a function, since it has a value at each point. In Figure A.3 the slope at the point $(1,1)$ happens to be 2, whereas the slope at the origin is zero (we'll see shortly how to compute the derivative).

In mathematical terms, the derivative is the limiting value of the ratio of the change in y (Δy) to the change in x (Δx), as Δx goes to zero. To quote a song once sung in the hallowed halls of Harvard (by Tom Lehrer and Lewis Branscomb),

> You take a function of x, and you call it y
> Take any x-nought that you care to try
> Make a little change and call it delta x
> The corresponding change in y is what you find nex'
> And then you take the quotient, and now, carefully

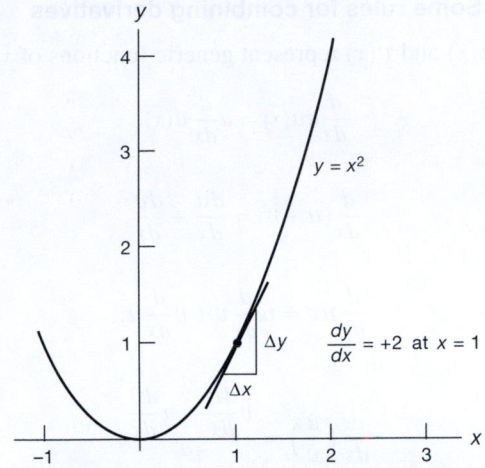

Figure A.3. A single-valued function: $f(x) = x^2$.

> Send delta x to zero, and I think you'll see
> That what the limit gives us (if our work all checks)
> Is what you call dy/dx...
> It's just dy/dx.

(*The Derivative Song*, sung to the tune of *There'll Be Some Changes Made*, W. Benton Overstreet).

Differentiation is a straightforward art, and the derivatives of many common functions are tabulated in standard tables and automatically calculated in programs like Mathematica®. Here are some rules (u and v are arbitrary functions of x, and a represents a constant).

A.3.1 Derivatives of some common functions

$$\frac{d}{dx}a = 0$$

$$\frac{d}{dx}ax = a$$

$$\frac{d}{dx}ax^n = anx^{n-1},$$

$$\frac{d}{dx}\sin ax = a\cos ax,$$

$$\frac{d}{dx}\cos ax = -a\sin ax,$$

$$\frac{d}{dx}e^{ax} = ae^{ax},$$

$$\frac{d}{dx}\log_e x = 1/x.$$

A.3.2 Some rules for combining derivatives

Here $u(x)$ and $v(x)$ represent generic functions of x:

$$\frac{d}{dx} au(x) = a\frac{d}{dx} u(x),$$

$$\frac{d}{dx}(u+v) = \frac{du}{dx} + \frac{dv}{dx},$$

$$\frac{d}{dx} uv = u\frac{d}{dx} v + v\frac{d}{dx} u,$$

$$\frac{d}{dx}\left(\frac{u}{v}\right) = \frac{v\dfrac{du}{dx} - u\dfrac{dv}{dx}}{v^2},$$

$$\frac{d}{dx}\log_e u = \frac{1}{u}\frac{du}{dx}$$

$$\frac{d}{dx}\{u[v(x)]\} = \frac{du}{dv}\frac{dv}{dx}.$$

The last one is very useful and is called the chain rule.

A.3.3 Some examples of differentiation

$$\frac{d}{dx}x^2 = 2x,$$

$$\frac{d}{dx}(1/x^{\frac{1}{2}}) = -\frac{1}{2}x^{-\frac{3}{2}},$$

$$\frac{d}{dx}xe^x = xe^x + e^x \qquad \text{(product rule)},$$

$$\frac{d}{dx}e^{-x^2} = -2xe^{-x^2} \qquad \text{(chain rule)},$$

$$\frac{d}{dx}a^x = \frac{d}{dx}(e^{x\log_e a}) = a^x\log_e a \quad \text{(chain rule)}.$$

Once you have differentiated a function, you often want to evaluate the value of the derivative at some point. Other times you may want to find a minimum or maximum of the function; that's the same thing as having a zero derivative, so you can just set the derivative equal to zero and solve for x. For example, you can easily determine that the slope of the function plotted in Figure A.3 equals 2 at $x=1$, and that its minimum occurs at $x=0$ (where its slope is zero).

HOW TO DRAW SCHEMATIC DIAGRAMS

A well-drawn schematic makes it easy to understand how a circuit works, and it aids greatly in troubleshooting. A poor schematic only creates confusion. By keeping a few rules and suggestions in mind, you can draw a good schematic in no more time than it takes to draw a poor one. In this appendix we dispense advice of three varieties: general principles, rules, and hints. We have also drawn some real knee-slappers to illustrate habits to avoid.

B.1 General principles

- Schematics should be unambiguous. Therefore pin numbers, parts values, reference designators, polarities, etc., should be clearly labeled to avoid confusion.
- A good schematic makes circuit functions clear. Therefore keep functional areas distinct; don't be afraid to leave blank areas on the page, and don't try to fill the page. There are conventional ways to draw functional subunits; for instance, don't draw a differential amplifier as in Figure B.1, because the function won't be easily recognized. Likewise, flip-flops are usually drawn with clock and inputs on the left, set and clear on top and bottom, and outputs on the right.

B.2 Rules

- Wires connecting are indicated by heavy black dots; wires crossing, but not connecting, have no dot (don't use a little half-circular "jog"; it went out in the 1950s).
- Four wires must not connect at a point; i.e., wires must not cross *and* connect. You sometimes see this rule violated, but it's poor practice (because a missing or undersized dot is a different circuit).
- Always use the same symbol for the same device; e.g., don't draw flip-flops in two different ways (exception: assertion-level logic symbols show each gate in two possible ways).
- Wires and components are aligned horizontally or vertically, unless there's a good reason to do otherwise.

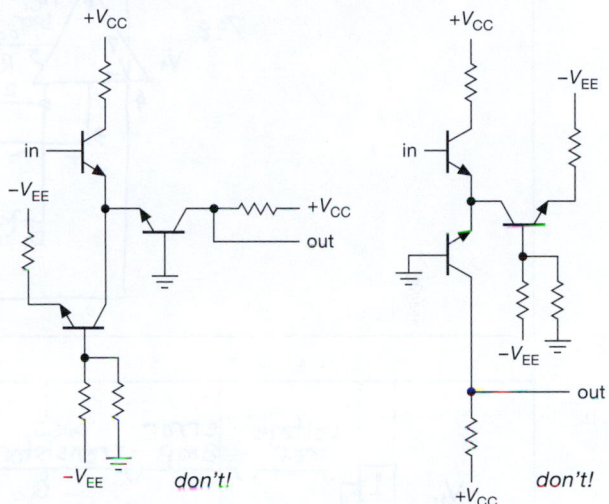

Figure B.1. Arrange components so that the function (here a differential amplifier) is clear. Don't corrupt the presentation to save space.

- Label pin numbers on the outside of a symbol, signal names on the inside.
- All parts should have values or types indicated; it's best to give all parts a label ("refdes"), too, e.g., R_7 or U_3.

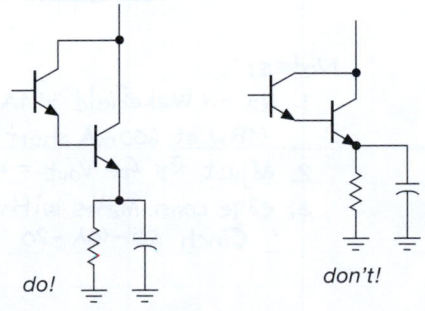

Figure B.2. Bring leads away from component symbols before connecting or jogging.

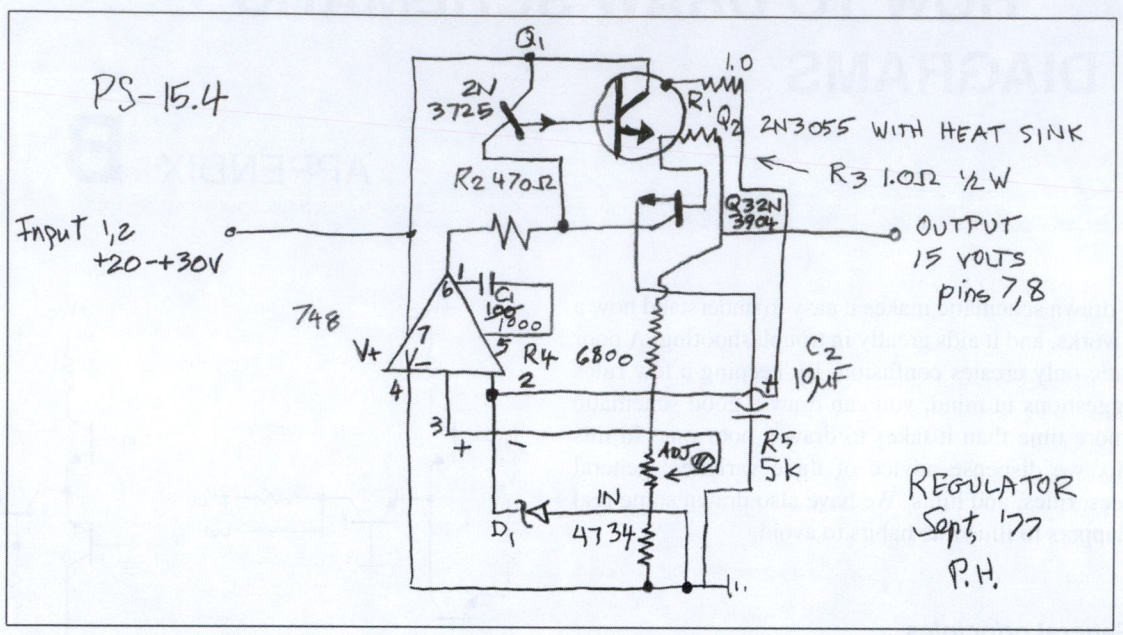

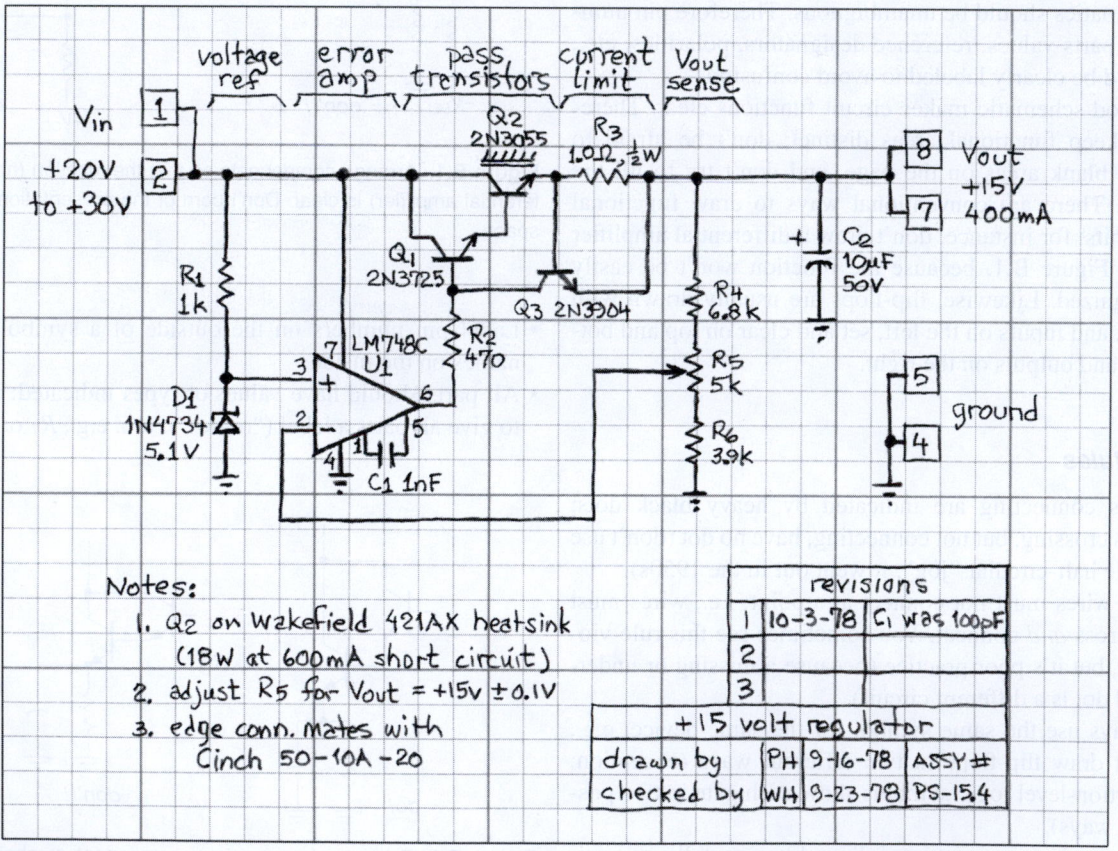

Figure B.3. Good drawing (perhaps too good) on "engineering paper," and *very* bad. Guess which is which.

B.3 Hints

- Identify parts immediately adjacent to the symbol, forming a distinct group giving symbol, label, and type or value.
- In general, signals go from left to right; don't be dogmatic about this, though, if clarity is sacrificed.
- Put positive supply voltages at the top of the page, negative at the bottom. Thus *npn* transistors will usually have their emitter at the bottom, whereas *pnp*s will have their emitter topmost.
- Don't attempt to bring all wires around to the supply rails, or to a common ground wire. Instead, use the ground symbol(s) and labels like $+V_{CC}$ to indicate those voltages where needed.
- It is helpful to label signals and functional blocks and show waveforms; in logic diagrams it is especially important to label signal lines, e.g., RESET' or CLK.
- It is helpful to bring leads[1] away from components a short distance before making connections or jogs. For example, draw transistors as in Figure B.2.
- Leave some space around circuit symbols; e.g., don't draw components or wires too close to an op-amp symbol. This keeps the drawing uncluttered and leaves room for labels, pin numbers, etc.
- Label all boxes that aren't obvious: comparator versus op-amp, shift register versus counter, etc. Don't be afraid to invent a new symbol.
- Use small rectangles, ovals, or circles to indicate card-edge connections, connector pins, etc. Be consistent.

---

[1] *Leads?* Yeah, sure... I'll just check with the boys down at the crime lab. They got four more detectives working on the case. They got us working in *shifts*. Hahahaha... *LEADS!*

- The signal path through switches should be clear. Don't force the reader to follow wires all over the page to find out how a signal is switched.
- Power-supply connections are normally assumed for op-amps and logic devices. However, show any unusual connections (e.g., an op-amp run from a single supply, where $V_- = $ ground), and the disposition of unused inputs.
- It is very helpful to include a small table of integrated circuit (IC) numbers, types, and power-supply connections (pin numbers for V_{CC} and ground, for instance).
- Include a title area near the bottom of the page, with name of circuit, name of instrument, by whom drawn, by whom designed or checked, date, and assembly number. Also include a revision area, with columns for revision number, date, and subject.
- We recommend drawing schematics freehand on coarse graph paper (pale gridlines, five per inch, for example National® Brand "Engineer's Computation Pad" in "Eye-Ease"® green), or on plain paper on top of graph paper. This is fast, and it gives very pleasing results. Use dark pencil (we like HB hardness, 0.5 mm diameter) or ink; avoid ballpoint or felt-tip pen.

B.4 A humble example

As an illustration, we've drawn a humble example (Figure B.3) showing "awful" and "good" schematics of the same circuit; the former violates nearly every rule and is almost impossible to understand. See how many bad habits you can find illustrated. We've seen all of them in professionally drawn schematics! (We drew the "bad" schematic in an airport while waiting for a flight. It was an occasion of great hilarity; we laughed ourselves silly.)

RESISTOR TYPES

C.1 Some history

For a half century people used "leaded" (pronounced lēd′-ed) resistors: if you look inside a very old radio (before ~1950), you'll see colored cylindrical objects with some colored dots painted on them and with a wire wrapped around each end that comes off perpendicularly to the axis ("radial leads"). These carbon-composition resistors evolved into the standard carbon-comp "axial-lead" resistors (still cylindrical, but with colored stripes all the way around, and with the wires now sticking straight out each end) that were dominant through the last half of the 20th century (and that we recommended for noncritical applications in our previous book editions). Axial-lead resistors are still popular for some uses, such as easy breadboarding in the lab. They're also used in applications that require very high resistance ($\geq 100\,\mathrm{M\Omega}$), or high voltage or power ratings, or for resistors of very high precision.

However, contemporary electronics has embraced surface-mount packaging, because of its high density (SMT devices are *small*, and you don't have to take up space with holes for the leads). Surface-mount resistors, in common with other two-terminal SMT components (capacitors, inductors), are available in a range of package sizes, characterized by a 4-digit code giving their length and width in units of 0.010″; for example, an "0603" package is $0.06'' \times 0.03''$ (1.5 mm × 0.75 mm). We favor that size, or the larger 0805 package, for general prototyping of surface-mount circuits. The smaller packages (0402, 0201, and even "01005") are a major pain – you basically have to work under a microscope (and don't sneeze).

C.2 Available resistance values

You can't get just any old resistance value. Available resistances fall into what's called an EIA Standard Decade, named by the number of values per decade (thus E24 – used for 5% tolerance resistors – has 24 values, spaced approximately 10% apart; see below). Resistors with 1% tolerance are quite inexpensive these days, costing hardly more than

an analogous 5% resistor,[1] so you might as well use 1% resistors by default. They come in the E96 set of standard values (96 values per decade, spaced approximately 2% apart; thus 481 values from $10\,\Omega$ through $1\,\mathrm{M\Omega}$, see below). Resistors of greater precision (e.g., 0.1%) are sometimes available in the **E192** superset,[2] and in convenient round-number values (e.g., 250, 300, 400, or 500) that are not included in the EIA sequences.

Here is the **E24** set of "5%" values (the **E12** subset, used for 10% components, is shown in **bold**):

10	16	**27**	43	**68**
11	**18**	30	**47**	75
12	20	**33**	51	**82**
13	**22**	36	**56**	91
15	24	**39**	62	**100**

And here is the **E96** set of "1%" values (the **E48** set, used for 2% components, or for a reduced set of 1% parts, is in **bold**):

100	137	**187**	255	**348**	475	**649**	887
102	**140**	191	**261**	357	**487**	665	**909**
105	143	**196**	267	**365**	499	**681**	931
107	**147**	200	**274**	374	**511**	698	**953**
110	150	**205**	280	**383**	523	**715**	976
113	**154**	210	**287**	392	**536**	732	
115	158	**215**	294	**402**	549	**750**	
118	**162**	221	**301**	412	**562**	768	
121	165	**226**	309	**422**	576	**787**	
124	**169**	232	**316**	432	**590**	806	
127	174	**237**	324	**442**	604	**825**	
130	**178**	243	**332**	453	**619**	845	
133	182	**249**	340	**464**	634	**866**	

---

[1] For example, the Digi-Key catalog shows a full selection of Vishay/Dale CRCW-series surface-mount resistors, in sizes from 1210 down to 0201. For the convenient 0603 size, the current prices for 1% and 5% resistors are $0.025 and $0.023 apiece, respectively, in quantity 200. (You'll pay about triple that, in quantity 10, and about a fifth as much, in a full reel of 5000 resistors.)

[2] The full E192 set, along with subsets, is nicely displayed at http://www.logwell.com/tech/components/resistor_values.html.

C.3 Resistance marking

Leaded resistors are marked in one of two ways: (a) with a set of four or five color bands, indicating resistance and tolerance; or (b) with a 4-digit resistance code, followed by a letter that indicates the tolerance. Surface-mount resistors use either (a) a 3- or 4-digit resistance code, or, for the smallest package sizes, (b) no marking at all!

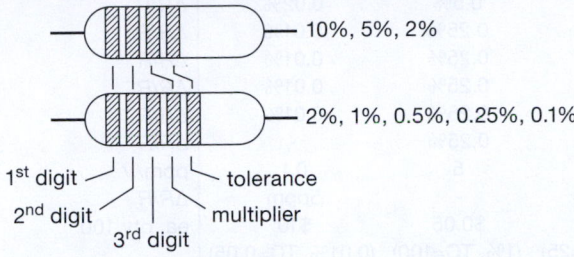

color	digit	multiplier	tolerance	(tol. suffix)
black	0	1	–	
brown	1	10	1%	F
red	2	100	2%	G
orange	3	1k	–	
yellow	4	10k	–	
green	5	100k	0.5%	D
blue	6	1M	0.25%	C
violet	7	10M	0.1%	B
gray	8	–	0.05%	A, W
white	9	–	–	
gold	–	0.1	5%	J
silver	–	0.01	10%	K
(none)	–	–	20%	M
			0.02%	N, Q, P
			0.01%	T, L
			0.005%	V
			0.0025%	X
			0.002%	U
			0.001%	S

Figure C.1. The resistor color code, used on some axial-lead resistors (notably carbon-film and carbon-composition types). The resistance is read as a 2- or 3-digit integer (depending on resistor precision) followed by a band indicating the power-of-10 multiplier. For example, yellow-violet-orange-gold is $47\,\mathrm{k\Omega}\pm5\%$, and yellow-white-white-black-brown is $499\,\Omega\pm1\%$. The alphabetic tolerance suffix is used on resistors with numerical printed resistance values.

Although it may seem diabolical to the beginner, the practice of color banding makes it easy to recognize resistor values in a circuit or parts bin, without having to search for a printed legend. Each color corresponds to a digit, in a sort of floating-point format (with the final digit indicating the power of ten); a last color band signifies the tolerance.

See Figure C.1. Resistors with numerical markings use the same system, but with the digits themselves printed along the body of the resistor (for leaded resistors), or on the top side of a surface-mount package; a final letter signifies the tolerance, as shown in the figure.

C.4 Resistor types

The usual choices for general-purpose use are metal-film (axial-lead) or thick-film (surface-mount) parts. Thin-film surface-mount resistors offer improved characteristics (accuracy, stability, and ability to operate in cryogenic environments). For power applications you usually use wire-wound resistors, either in an air-cooled ceramic package or a conduction-cooled ("Dale-type") metal package. High-value resistors ($>10\,\mathrm{M\Omega}$, say) are usually of metal-oxide construction (e.g., Ohmite "Mini-Mox" or "Super Mox," or Vishay RNX-series). Film resistors are not tolerant of high peak power; for such applications use something like ceramic or carbon composition, or other styles specified for peak-power use. For the utmost in stability and low temperature coefficient (tempco), you can't beat the excellent metal-foil types from Vishay. They exploit a clever design, in which the positive tempco of the resistive metal element (firmly attached to an insulating substrate) is cancelled by the negative strain-induced tempco caused by differential expansion of the substrate.[3] We've listed some comparative resistor properties in Table C.1; for much more detail see §*1x.2*.

General-purpose resistors are ridiculously inexpensive – thick-film surface-mount resistors cost a few cents apiece in small quantities, and just fractions of a cent apiece in full reel quantities (5000 pieces, for 0603 size). Distributors may be unwilling to sell fewer than 25 to 50 pieces of one value; thus an assortment box (e.g., from Yageo or Vishay/BC) may be a wise purchase. We particularly like the nice packaging and good pricing of the kits from SMT Zone (www.smtzone.com).

C.5 Confusion derby

Component markings should be clear and unambiguous. Sometimes it just ain't so! See Figure 1.130 for some real head-scratchers, both resistive and otherwise.

---

[3] Check it out: Felix Zandman's 1982 US patent #4,318,072, "Precision resistor with improved temperature characteristics."

Table C.1 Selected Resistor Types

	Resistor Type					
	carbon comp axial	thick film SMT-0603	thin film SMT-0603	metal film axial	metal foil SMT	
Parameter	(RC-07)	(Vishay CRCW)	(KOA Speer RN73)	(RN-55D)	(Vishay VSMP)	Units
Tolerances	5%, 10%	1%, 5%	0.05%-1%	0.1%-1%	0.01%-1%	$\Delta R/R$
Temp coef	~1000	100, 200	5, 10, 25, 50, 100	50, 100	0.05 (typ)	ppm/C
Load life	10%	2%	0.25%	0.5%	0.01%	$\Delta R/R$
Moisture	10%	2%	0.5%	0.5%	0.02%	$\Delta R/R$
Thermal cycle	2%	2%	0.25%	0.25%	0.01%	$\Delta R/R$
Low temp	3%	-	-	0.25%	0.01%	$\Delta R/R$
Overload	2%	0.5%	0.1%	0.25%	0.01%	$\Delta R/R$
Soldering	3%	0.5%	0.1%	0.25%	0.01%	$\Delta R/R$
Vibration	2%	-	-	0.25%	-	$\Delta R/R$
Voltage coef	-	-	-	5	0.1	ppm/V
Self-heating	-	-	-	-	5ppm	$\Delta R/R$
Price (approx)	$0.35	$0.025	$0.32	$0.05	$10	ea, qty 100
(for tol and TC)	(5%)	(1%, TC=200)	(0.1%, TC=25)	(1%, TC=100)	(0.01%, TC=0.05)	

Properties of selected resistor types. The legendary axial-lead "carbon-composition" resistors have been superseded by inexpensive metal-film (or carbon-film) types, with greatly improved properties (except for peak-power endurance, see Chapter *1x*). We like Vishay's CMF-55 metal-film resistors (industrial version of MIL RN-55D). For most surface-mount applications the "thick-film" (a metal-ceramic composite) types are fine, though thin-film and metal-film resistors have somewhat better properties. The extraordinary Vishay "Z-foil" ultraprecision hermetically sealed resistor is listed to show the best that is currently available (but if you have to ask the price, you probably can't afford it). It's useful to note that a parameter like a voltage coefficient of 5 ppm/V corresponds to a change of 0.1% over a full 200 V operating range.

THÉVENIN'S THEOREM

APPENDIX **D**

In Chapter 1 we stated (but did not "prove") Thévenin's Theorem, namely that any two-terminal network whose internal circuitry consists solely of resistors, batteries, and current sources, interconnected in any manner whatsoever, is equivalent (and indistinguishable) from the two-terminal network consisting of a single battery V_{TH} in series with a single resistor R_{TH};[1] see Figure D.1. We did not prove it, because, in the spirit of this book, we don't *prove* anything; we show you how to design circuits, instead. We make an exception here, because it's nice to see *something* proved, right?

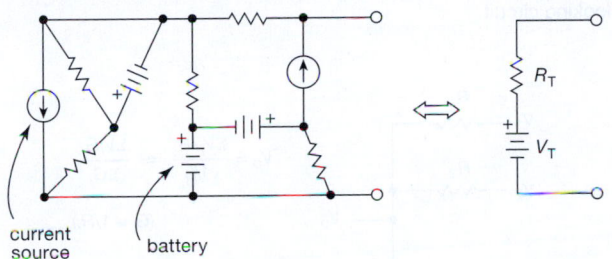

current source   battery

Figure D.1. Thévenin's Theorem: a single resistor in series with a single battery can mimic any mess of a two-terminal network made from resistors, batteries, and current sources.

D.1 The proof

For linear circuit elements (here resistors), the "nodal equations" (Kirchhoff's voltage law, KVL, and Kirchhoff's current law, KCL) are a set of linear equations. So we can find any circuit quantity (a voltage or a current), which depends on all the "independent sources" (batteries, current sources), by turning on each source in turn, and adding the partial contributions. (This is exactly analogous to using superposition to find, say, the electric field from a set of charges.) This technique is often useful in circuit analysis.

Here we wish to mimic the V versus I of the actual circuit with the (simpler) Thévenin equivalent of a single battery in series with a single resistor. Imagine we determine

---

[1] A related theorem is Norton's, where the equivalent circuit consists of a resistor R_N in parallel with a current source I_N.

that V versus I function by applying an external current I_{ext} that flows through the two-terminal circuit, and observing the resultant V across those same two terminals. V depends on I_{ext} *and* on all the internal batteries (V_{int}) and current sources (I_{int}).

1. Set all $V_{int} = 0$ and all $I_{int} = 0$; that is, replace all internal batteries with short circuits and all current sources with open circuits. Now, with a given applied I_{ext}, observe V_1.
2. Define $R_T = V_1/I_{ext}$. (They must be proportional, by linearity.)
3. Now set $I_{ext} = 0$, and turn on the internal batteries and current sources. Observe V_2, which we will call V_T.
4. Finally, by superposition it must be the case that

$$V(\text{actual}) = V_1 + V_2 = I_{ext}R_T + V_T.$$

This is true for all I_{ext}, and is exactly what you get with the Thévenin equivalent circuit, when connected to *any* load (which need not be linear); see Figure D.2.

To summarize: (a) you determine R_T and V_T by first finding the open-circuit voltage, which equals V_T; then (b) you find the short-circuit current, I_{SC}, which equals the ratio of V_T to R_T. In other words, $V_T = V_{OC}$ and $R_T = V_{OC}/I_{SC}$. You do this by analysis, if you know the "black-box" circuit; or by measurement, if you don't.

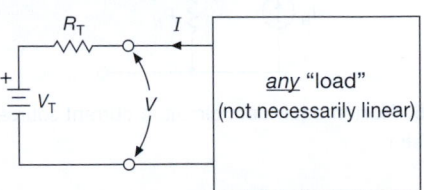

Figure D.2. The Thévenin equivalent circuit behaves exactly like the original network, regardless of the nature of the load.

D.1.1 Two examples – voltage dividers

Figures D.3 and D.4 show two simple examples, variations on the resistive divider. Interestingly, their Thévenin equivalent circuits are different, even though the resistor values and the open-circuit voltages are the same.

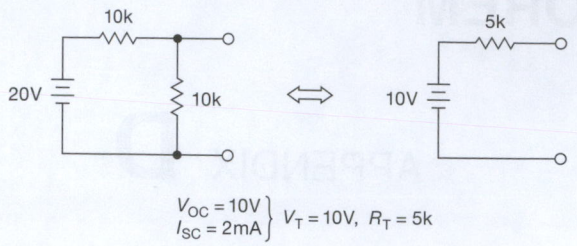

$$\left.\begin{array}{l} V_{OC} = 10\,V \\ I_{SC} = 2\,mA \end{array}\right\} V_T = 10\,V,\; R_T = 5k$$

Figure D.3. Thévenin equivalent of a simple resistive divider, Note that R_T is the parallel resistance of the divider (as if the voltage source were replaced with a short circuit).

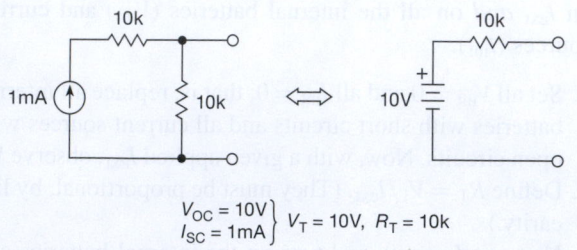

$$\left.\begin{array}{l} V_{OC} = 10\,V \\ I_{SC} = 1\,mA \end{array}\right\} V_T = 10\,V,\; R_T = 10k$$

Figure D.4. Note that the Thévenin equivalent resistance is here *not* equal to the parallel resistance of the divider components. Instead it equals the value of the resistor across the output alone (as if the current source were replaced with an open circuit).

D.2 Norton's theorem

You can replace a Thévenin circuit with a Norton circuit, which consists of a current source I_N in parallel with a resistor R_N (Figure D.5). It is easy to show that $I_N = I_{SC}$ and $R_N = R_T\,(= V_{OC}/I_{SC})$. So, for the two examples above, the Norton equivalents are as shown in Figure D.6.

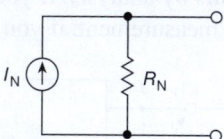

Figure D.5. Norton equivalent circuit: a current source in parallel with a resistor.

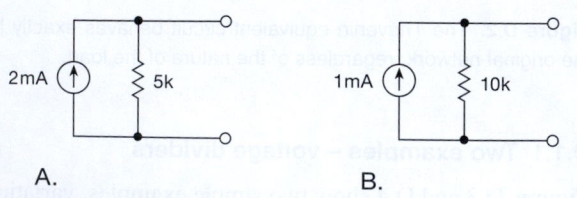

Figure D.6. Norton equivalents of the circuits of Figure D.3 (A) and Figure D.4 (B).

D.3 Another example

Figure D.7 shows a complicated-looking circuit, for which it is pretty easy to see that $V_{OC}=25\,V$ (the bottom of the 10k resistor sits at $+10\,V$, and 1.5 mA flows into the top) and that $I_{SC}=2.5\,mA$ (10 V across the 10k, plus the two current sources). From that you get the equivalent circuits shown.

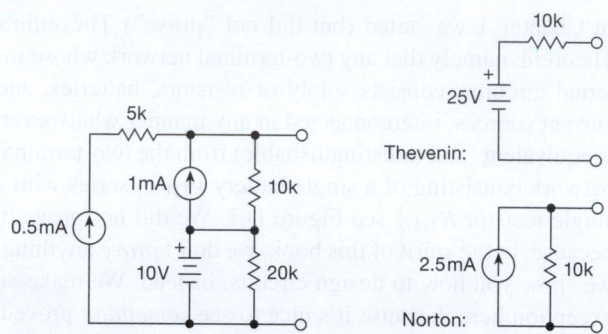

Figure D.7. Thévenin and Norton equivalents of a complicated-looking circuit.

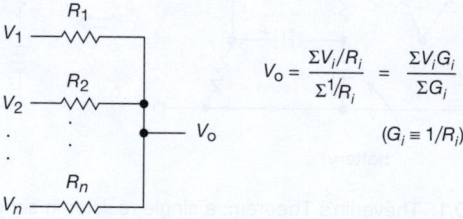

$$V_o = \frac{\Sigma V_i/R_i}{\Sigma 1/R_i} = \frac{\Sigma V_i G_i}{\Sigma G_i}$$

$$(G_i \equiv 1/R_i)$$

Figure D.8. Millman's theorem for parallel circuits.

D.4 Millman's theorem

A related – and useful – tool is *Millman's Theorem* (also known as the parallel generator theorem), which is helpful when dealing with circuits with several parallel branches. It's shown in Figure D.8, where a set of input voltages V_i are combined via resistors R_i, producing an output voltage V_o. The latter is just $V_o=(\Sigma V_i G_i)/\Sigma G_i$, where the G_i are the conductances $G_i\equiv1/R_i$. The input voltages V_i can of course include ground, forming a voltage divider. Millman's theorem, which comes from the more general class of network theorems, can be generalized to include input *currents* I_k, whose sum is added to the numerator (but whose series resistances, if any, do not appear in the denominator).

LC BUTTERWORTH FILTERS

Active filters (see Chapter 6) are convenient at low frequencies, but they are impractical at radio frequencies because of the slew-rate and bandwidth requirements they impose on the operational amplifiers. At frequencies of 100 kHz and above (and often at lower frequencies), the best approach is to design a passive filter with inductors and capacitors. (At UHF and microwave frequencies these "lumped-component" filters are replaced by stripline and cavity filters.)

As with active filters, there are many methods and filter characteristics possible with *LC* filters. For example, you can design the classic Butterworth, Chebyshev, and Bessel filters, each in lowpass, bandpass, highpass, and band-reject varieties. It turns out that the Butterworth filter is particularly easy to design, and we can present in just a page or two all the essential design information for lowpass and highpass Butterworth *LC* filters, and even a few examples.

E.1 Lowpass filter

Table E.1 gives the values of normalized inductances and capacitances for low-pass filters of various orders, from which actual circuit values are obtained by the frequency and impedance scaling rules

$$L_n(\text{actual}) = \frac{R_L L_n(\text{table})}{\omega},$$

$$C_n(\text{actual}) = \frac{C_n(\text{table})}{\omega R_L},$$

where R_L is the load impedance and ω is the angular frequency ($\omega = 2\pi f$).

Table E.1 gives normalized values for two-pole through eight-pole lowpass filters for the two most common cases, namely (a) equal source and load impedances and (b) either source or load impedance much larger than the other. To use the table, first decide how many poles you need, based on the Butterworth response (graphs are plotted in Figure 6.30). Then use the preceding equations to determine the filter configuration (T or π; see Figure E.1) and component values. For equal source and load impedances, either

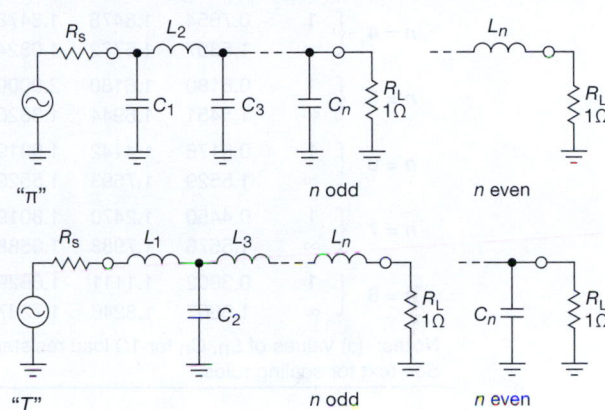

Figure E.1. π and T configurations. See Table E.1 and text.

configuration is OK; the π configuration may be preferable because it requires fewer inductors. For a load impedance much higher (lower) than the source impedance, use the T (π) configuration.

E.2 Highpass filter

To design a highpass filter, follow the same procedure to determine which filter configuration to use and how many poles are necessary. Then do the universal lowpass to highpass transformation shown in Figure E.2, which consists simply of replacing inductors by capacitors, and vice versa. The actual component values are determined from the normalized values in Table E.1 by the following frequency and impedance scaling rules:

$$L_n(\text{actual}) = \frac{R_L}{\omega C_n(\text{table})},$$

$$C_n(\text{actual}) = \frac{1}{R_L \omega L_n(\text{table})}.$$

E.3 Filter examples

Here are a few examples showing how to design both lowpass and highpass filters.

Example I. Design a five-pole lowpass filter for source

1109

Table E.1   Butterworth Lowpass Filters[a]

$\pi \longrightarrow$	R_S	C_1	L_2	C_3	L_4	C_5	L_6	C_7	L_8
$T \longrightarrow$	$1/R_S$	L_1	C_2	L_3	C_4	L_5	C_6	L_7	C_8
$n = 2$	1	1.4142	1.4142						
	∞	1.4142	0.7071						
$n = 3$	1	1.0000	2.0000	1.0000					
	∞	1.5000	1.3333	0.5000					
$n = 4$	1	0.7654	1.8478	1.8478	0.7654				
	∞	1.5307	1.5772	1.0824	0.3827				
$n = 5$	1	0.6180	1.6180	2.0000	1.6180	0.6180			
	∞	1.5451	1.6944	1.3820	0.8944	0.3090			
$n = 6$	1	0.5176	1.4142	1.9319	1.9319	1.4142	0.5176		
	∞	1.5529	1.7593	1.5529	1.2016	0.7579	0.2588		
$n = 7$	1	0.4450	1.2470	1.8019	2.0000	1.8019	1.2470	0.4450	
	∞	1.5576	1.7988	1.6588	1.3972	1.0550	0.6560	0.2225	
$n = 8$	1	0.3902	1.1111	1.6629	1.9616	1.9616	1.6629	1.1111	0.3902
	∞	1.5607	1.8246	1.7287	1.5283	1.2588	0.9371	0.5776	0.1951

Notes: (a) Values of L_n, C_n for 1Ω load resistance, and cutoff frequency (–3dB) of 1 rad/s. See text for scaling rules.

and load impedances of 75 Ω, with a cutoff frequency (−3 dB) of 1 MHz.

We use the π configuration to minimize the number of required inductors. The scaling rules give us

$$C_1 = C_5 = \frac{0.618}{2\pi \times 10^6 \times 75} = 1310\,\text{pF},$$

$$L_2 = L_4 = \frac{75 \times 1.618}{2\pi \times 10^6} = 19.3\,\mu\text{H},$$

$$C_3 = \frac{2}{2\pi \times 10^6 \times 75} = 4240\,\text{pF}.$$

The complete filter is shown in Figure E.3. Note that all filters with equal source and load impedances will be symmetrical.

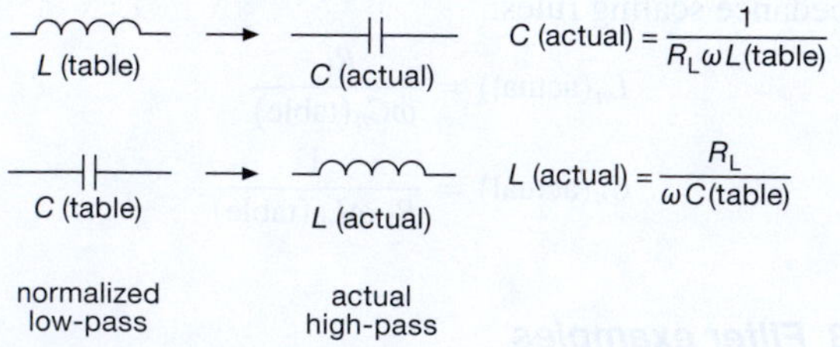

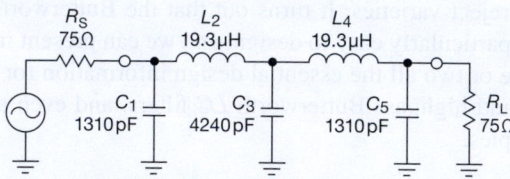

Figure E.3. Circuit for Example I. Five-pole 1 MHz lowpass with equal source and load impedances.

source impedance of 50 Ω and a load impedance of 10k, with a cutoff frequency of 100 kHz.

We use the T configuration, because $R_S \ll R_L$. For $R_L = 10\text{k}$, the scaling rules give

$$L_1 = \frac{10^4 \times 1.5}{2\pi \times 10^5} = 23.9\,\text{mH},$$

$$C_2 = \frac{1.3333}{2\pi \times 10^5 \times 10^4} = 212\,\text{pF},$$

$$L_3 = \frac{10^4 \times 0.5}{2\pi \times 10^5} = 7.96\,\text{mH}.$$

The complete filter is shown in Figure E.4.

Example III. Design a four-pole lowpass filter for a zero-impedance source (voltage source) and a 75 Ω load, with a cutoff frequency of 10MHz.

We use the T configuration, as in the previous example,

Figure E.2. Lowpass to highpass transformation.

normalized low-pass   →   actual high-pass

C (actual) $= \dfrac{1}{R_L \omega L\,(\text{table})}$

L (actual) $= \dfrac{R_L}{\omega C\,(\text{table})}$

Example II.   Design a three-pole lowpass filter for a

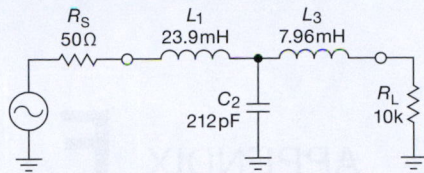

Figure E.4. Circuit for Example II. Three-pole 100 kHz lowpass with 50 Ω source and 10k load.

because $R_S \ll R_L$. The scaling rules give

$$L_1 = \frac{75 \times 1.5307}{2\pi \times 10^7} = 1.83\,\mu\text{H},$$

$$C_2 = \frac{1.5772}{2\pi \times 10^7 \times 75} = 335\,\text{pF},$$

$$L_3 = \frac{75 \times 1.0824}{2\pi \times 10^7} = 1.29\,\mu\text{H},$$

$$C_4 = \frac{0.3827}{2\pi \times 10^7 \times 75} = 81.2\,\text{pF}.$$

The complete filter is shown in Figure E.5.

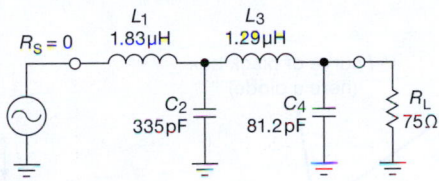

Figure E.5. Circuit for Example III. Four-pole 10 MHz lowpass with voltage source and 75 Ω load.

Example IV.
Design a two-pole lowpass filter for current-source drive and 1k load impedance, with a cutoff frequency of 10 kHz.

We use the π configuration because $R_S \gg R_L$. The scaling rules give

$$C_1 = \frac{1.4142}{2\pi \times 10^4 \times 10^3} = 0.0225\,\mu\text{F},$$

$$L_2 = \frac{10^3 \times 0.7071}{2\pi \times 10^4} = 11.3\,\text{mH}.$$

The complete filter is shown in Figure E.6.

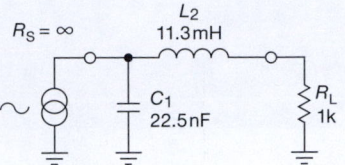

Figure E.6. Circuit for Example IV. Two-pole 10 kHz lowpass with current source drive and 1k load.

Example V.
Design a three-pole highpass filter for 52 Ω source and load impedances, with a cutoff frequency of 6 MHz.

We begin with the T configuration, then transform inductors to capacitors, and vice versa, giving

$$C_1 = C_3 = \frac{1}{52 \times 2\pi \times 6 \times 10^6 \times 1.0} = 510\,\text{pF},$$

$$L_2 = \frac{52}{2\pi \times 6 \times 10^6 \times 2.0} = 0.690\,\mu\text{H}.$$

The complete filter is shown in Figure E.7.

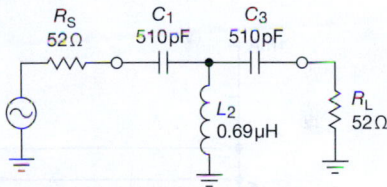

Figure E.7. Circuit for Example V. Three-pole 6 MHz highpass with equal source and load impedances.

We would like to emphasize that the field of passive filter design is rich and varied and that this simple table of Butterworth filters doesn't even begin to scratch the surface.

The graphic method of "load lines" usually makes an early appearance in electronics textbooks. We have avoided it because, well, it just isn't useful in transistor design the way it was in vacuum-tube circuit design. However, it is of use in dealing with some nonlinear devices (tunnel diodes, for example), and in any case it is a useful conceptual tool.

F.1 An example

Let's start with an example. Suppose you want to know the voltage across the diode in Figure F.1. Assume that you know the voltage-versus-current (V–I) curve of the particular diode (of course, it would have a manufacturing "spread," as well as a dependence on ambient temperature); it might look something like the curve drawn. How would you figure out the quiescent[1] point?

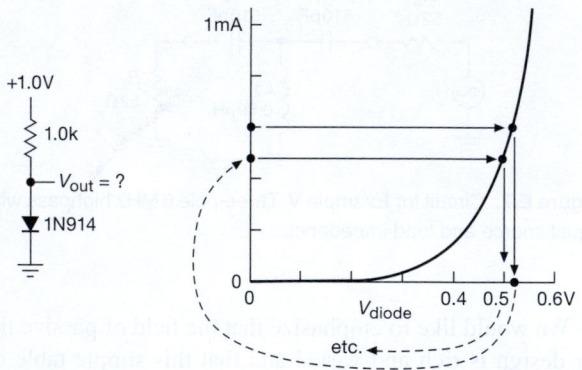

Figure F.1. Finding the operating point by iteration.

One method might be to guess a rough value of current, say 0.6 mA, then use the curve to get the drop across the resistor, from which you get a new guess for the current (in this case, 0.48 mA). This iterative method is suggested in

Figure F.1. After a few iterations, this method will get you an answer, but it leaves a lot to be desired.

The method of load lines gets you the answer to this sort of problem immediately. Imagine *any* device connected in place of the diode; the 1.0k resistor is still the load. Now plot, on a V–I graph, the curve of resistor current versus device voltage. This turns out to be easy: at zero volts the current is just V_+/R (full drop across the resistor); at V_+ volts the current is zero; points in between fall on a straight line between the two. Now, on the same graph, plot the V–I curve of the device. The operating point lies on both curves, i.e., at the intersection, as shown in Figure F.2.

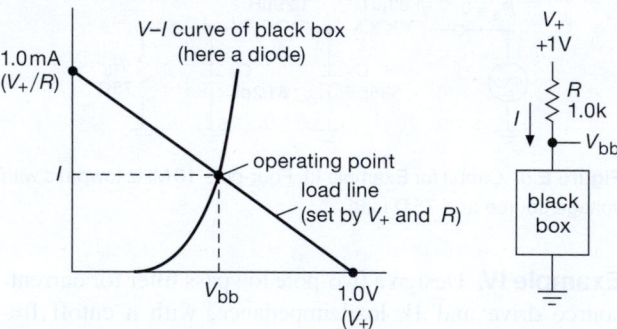

Figure F.2. A "load line" lets you find the operating point directly.

F.2 Three-terminal devices

Load lines can be used with a three-terminal device (tube or transistor, for example) by plotting a family of curves for the device. Figure F.3 shows what such a thing would look like for a depletion-mode field-effect transistor (FET), with the curve family parameterized by the gate-source voltage. You can read off the output for a given input by sliding along the load line between appropriate curves corresponding to the input you've got, then projecting onto the voltage axis. In this example we've done this, showing the drain

---

[1] The quiescent point, also known as the *operating point*, describes the various dc voltages and currents in a circuit with no ac signals applied.

voltage (output) for a gate swing (input) between ground and −2 V.

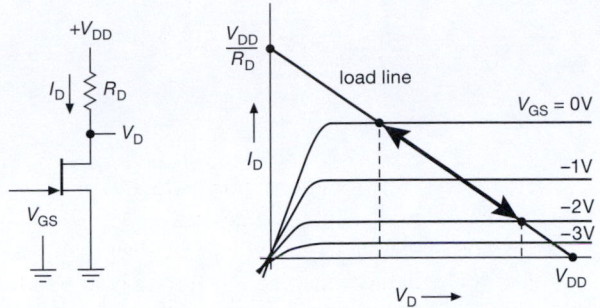

Figure F.3. Load-line solution for a three-terminal device.

As nice as this method seems, it has quite limited use for transistor or FET design, for a couple of reasons. For one thing, the curves published for semiconductor devices are "typical," with manufacturing spread that can be as large as a factor of five. Imagine what would happen to those nice load-line solutions if all the curves shrank to one-fourth their height! Another reason is that for an inherently logarithmic device like a diode junction, a linear load-line graph can be used to give accurate results over only a narrow region. Finally, the nongraphic methods we've used in this book are entirely adequate for handling solid-state design. In particular, these methods emphasize the parameters you can count on (r_e, I_C versus V_{BE} and T, etc.), rather than the ones that are highly variable (β, V_{th}, etc.). If anything, the use of load lines on published curves for transistors only gives you a false sense of security, since the device spread isn't also shown.

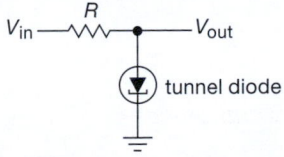

Figure F.4. The tunnel diode: a two-terminal nonlinear device with a region of negative resistance (see Figure F.5).

F.3 Nonlinear devices

Load lines turn out to be useful in understanding the circuit behavior of highly nonlinear devices. The example of tunnel diodes illustrates a couple of interesting points. Let's analyze the circuit in Figure F.4. Note that in this case, V_{in}

takes the place of the supply voltage in the previous examples. So a signal swing will generate a family of parallel load lines intersecting with a single device V–I curve (Figure F.5). The values shown are for a 100 Ω load resistor. As can be seen, the output varies most rapidly as the input swing takes the load line across the negative-resistance portion of the tunnel-diode curve. By reading off values of V_{out} (projection on the x axis) for various values of V_{in} (individual load lines), you get the "transfer" characteristics shown. This particular circuit has some voltage gain for input voltages near 0.2 V.

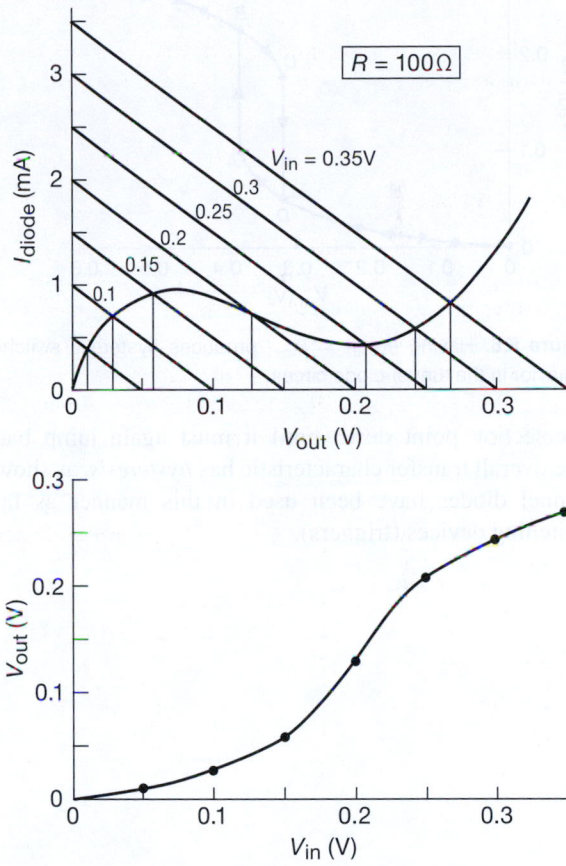

Figure F.5. Load lines and transfer characteristic for the tunnel-diode circuit.

An interesting thing happens if the load lines become flatter than the middle section of the diode curve. That happens when the load resistance exceeds the magnitude of the diode's negative resistance. It is then possible to have *two* intersection points, as in Figure F.6. A rising input signal carries the load lines up until the intersection point has nowhere to go and has to jump across to a higher V_{out} value. On returning, the load lines similarly carry the

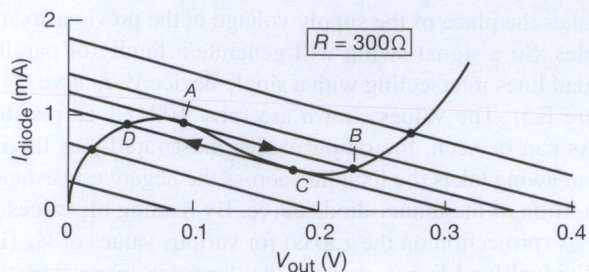

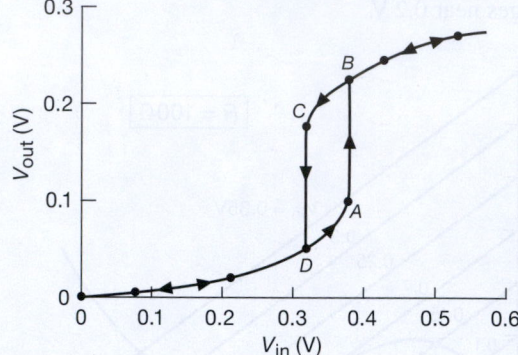

Figure F.6. Having $|R_{load}| > |R_{neg}|$ produces hysteretic switching behavior in the tunnel-diode circuit.

intersection point down until it must again jump back. The overall transfer characteristic has *hysteresis*, as shown. Tunnel diodes have been used in this manner as fast-switching devices (triggers).

THE CURVE TRACER

A handy instrument for exploring transistor behavior of both BJTs (Chapter 2) and MOSFETs (Chapter 3) is the *curve tracer*. Most simply, it plots collector current versus collector voltage for a family of equally spaced base currents (or, if you want to be an Ebers–Mollian, base *voltages*), and with a selectable current-limiting collector resistor.[1] Figure G.1 shows what you get from a random 2N3904, driven with seven successive base current steps of 5 μA each while sweeping the collector voltage from 0 to 50 V. You can see clearly the rise of beta with collector voltage, and the onset of breakdown somewhat below 50 V (maximum V_{CEO} is specified as 40 V). This particular curve tracer obligingly displays the scale factors, including "β per div," which is about 200 for this specimen (the datasheet specifies $100 \le \beta \le 300$ at $I_C = 10$ mA). A curve tracer makes it easy to select closely matched pairs.

Sadly, the traditional curve tracer has disappeared from the product lines of most T&M (test and measurement) manufacturers, including the venerable Tektronix. You can still find them used, for example on eBay, for a thousand dollars or so. Agilent offers some pretty fancy contemporary instruments that will do the job, though it's best to be sitting down when you ask the prices; they go by names like Semiconductor Parameter Analyzer (model 4155C), or Power Device Analyzer/Curve Tracer (model B1505A).

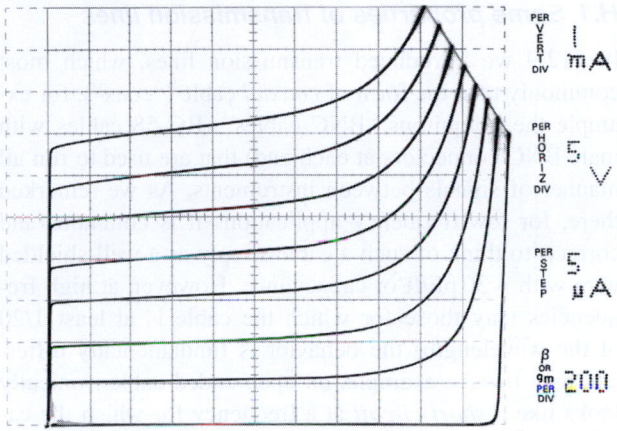

Figure G.1. Tektronix 576 Curve Tracer display of a 2N3904.

A less expensive alternative is to use a "source-measure unit" (SMU), a delightful instrument that lets you source voltages and currents to selected terminals of a device (or subcircuit), simultaneously measuring and logging other voltages or currents. You can program the excitation as dc, or ramps, steps, or pulses, and you can display the logged results via software running on an attached laptop computer; you can also save the logged data as a spreadsheet, to be manipulated to your heart's content. Take a look, for example, at Figure 8.39, or the figures in the "Power Transistors for Linear Amplifiers" section of Chapter *3x*, all of which plot data we collected with a SMU.

---

[1] You can run it as common base, if you like; and it has many amusing knobs to play with.

TRANSMISSION LINES AND IMPEDANCE MATCHING

H.1 Some properties of transmission lines

In §12.9 we introduced transmission lines, which most commonly take the form of *coaxial* cable ("coax"), for example the ubiquitous "BNC cables" (RG-58 cables with male BNC connectors at each end) that are used to run all manner of signals between instruments. As we remarked there, for low-frequency applications it is common (and correct) to think of such a cable simply as a well-shielded wire with ~30 pF/ft of capacitance. However, at high frequencies (say those for which the cable is at least 1/20 of the wavelength) the behavior is fundamentally different: as a bizarre example, an open-ended cable ironically looks like a *short circuit* at a frequency for which the cable's electrical length is $\lambda/4$. For a 5-foot length of coax, that happens at about 32 MHz. An important consequence is that you can't just hook such a BNC cable from a signal generator to some high-impedance circuit under test and assume that it will provide a nice signal source at the circuit's input; instead you will see huge dips and bumps as you tune the frequency, because the generator sees a load impedance that varies from a short circuit (at odd multiples of 32 MHz) to an open circuit (at multiples of 32 MHz). Perhaps surprisingly, if you were instead to connect a resistor of exactly 50 Ω across the circuit end of the cable, you would find that it now delivered a constant signal amplitude (equal to half the signal generator's open-circuit output amplitude) as you varied the frequency. This nonintuitive behavior is nicely illustrated in the measured data shown in Figure H.1. And even more nonintuitively, at the driving end of such a "terminated" cable the capacitance disappears entirely – you see instead a pure resistive load of 50 Ω!

H.1.1 Characteristic impedance

This simple example illustrates the importance of *termination*: coaxial cable is a form of *transmission line*, with a *characteristic impedance* Z_0 (which is always real: a resis-

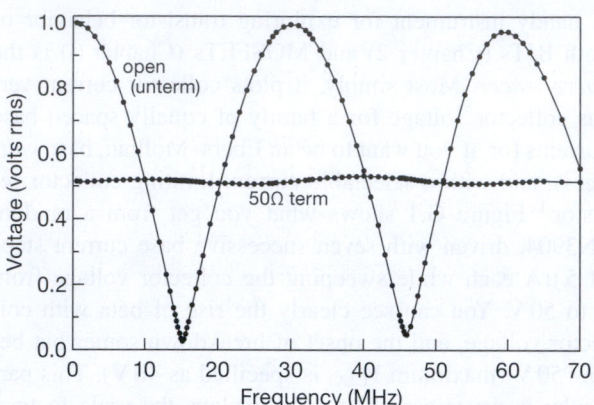

Figure H.1. Measured amplitude at the output connector of a sinewave oscillator of 1 V amplitude (open circuit), under two conditions: driving 10 feet of RG-58 (50 Ω) coaxial cable, open at the far end; and driving the same cable with a 50 Ω resistor connected across the far end.

tance) that depends on only its physical construction:

$$Z_0 = \sqrt{L/C} = \frac{138}{\sqrt{\varepsilon}} \log_{10} \frac{b}{a} \text{ ohms,}$$

where L and C are the inductance and capacitance per unit length, which as indicated depend on only the outer diameter, a, of the inner conductor, the inner diameter, b, of the outer conductor, and the dielectric constant, ε (relative to free space), of the insulating material that separates them. For a wave propagating along a transmission line, Z_0 is the ratio of signal voltage to signal current. The most popular coax line for general purposes is RG-58, with an impedance of 50 Ω (its dimensions are a=0.81 mm, b=2.95 mm, and ε=2.3, for which the above equation gives Z_0=51 Ω). This impedance has become the standard for radiofrequency use, except for video applications where the standard is Z_0=75 Ω; the corresponding popular cable type is called RG-59. In pulse electronics you sometimes see 93 Ω cable (RG-62).

The signal propagates along the cable at a speed

$$v_{\text{wave}} = \frac{c}{\sqrt{\varepsilon}} = \frac{1}{\sqrt{LC}},$$

which is a fraction $1/\sqrt{\varepsilon}$ times c (where c is the speed of light in vacuum, 3×10^8 m/s). The factor $1/\sqrt{\varepsilon}$ is called the *velocity factor*, and ranges from 0.66 (solid polyethylene) to 0.80 (polyethylene foam) for available flexible coaxial cables. In the absence of a dielectric the velocity factor is 1.0, i.e., waves travel at the speed of light in an air-spaced coaxial line. The "electrical length" seen by a propagating signal in a cable of physical length L is larger by the factor $\sqrt{\varepsilon}$, i.e., $L_{\text{elec}} = L_{\text{physical}} \sqrt{\varepsilon}$.

Note that the inductance and capacitance of the cable cannot take on any old values, because they are constrained such that their product LC is related to the speed of light. From this it is easy to show that if you know the characteristic impedance and the velocity factor then you can find the capacitance per unit length (or vice versa) by

$$C = \frac{\sqrt{\varepsilon}}{cZ_0} \text{ Farads/meter}.$$

For example, RG-8 has a specified impedance of $52\,\Omega$, and a velocity factor of 0.66; the above equation then gives $C = 97.1$ pF/m, or 29.6 pF/ft, in good agreement with the specified value of 29.5 pF/ft.

A. Twisted pair and PCB traces

Transmission lines are not *required* to be of coaxial geometry. An extremely popular form of contemporary transmission line is the *twisted pair*, which is just what it sounds like: a pair of insulated wires, gently twisted, and enclosed in an overall insulating jacket (often without any overall shield conductor). These may well be the dominant species of transmission line in our time, because they are the basic stuff of local area networks (LANs). You usually see four twisted pairs bundled into a single unshielded jacket; this is called "UTP" (unshielded twisted pair), and is the most common form of LAN cable. It is available also with a shield (shielded twisted pair, "STP"). Contemporary UTP and STP cables are $100\,\Omega$ nominal impedance, and are characterized (in terms of impedance and attenuation) for operation up to 10 Mbps (megabits per second) or 100 Mbps; these are called Category 3 and Category 5, respectively, and appear to differ primarily in the pitch of the twist. Those in the know refer to these casually as cat-3 and cat-5.[1] Ethernet LANs using these data rates

are called "10baseT" and "100baseTX," the "T" standing for "twisted"; the corresponding designation for "thinnet" coaxial Ethernet is "10base2."

In high-speed electronics on a printed circuit board it's often necessary to treat connection traces as transmission lines. See the discussion in §*1x.1*, where we described the *microstrip* transmission line, which consists of a thin conducting strip on an outer printed circuit layer, with an underlying ground-plane layer. A popular variant adds a pair of ground-trace shepherds on either side – that's called a *grounded coplanar waveguide* (GCPW). There's also the completely enclosed *stripline* geometry, where the trace(s) are sandwiched between ground-plane layers.

H.1.2 Termination: pulses

A transmission line at low frequencies (wavelength much longer than the cable length) looks simply like a capacitance, typically of order 30 pF/ft. However, at high frequencies, or, equivalently, when dealing with signals with fast rise times, the behavior is different. In order to understand the curious behavior illustrated in Figure H.1, it's helpful to look first at what happens when a simple *pulse* is applied to a length of transmission line. Suppose we connect a fast pulse generator with $50\,\Omega$ output impedance (the standard output impedance of signal generators, function generators, and pulse generators) to a length of $50\,\Omega$ coax, shorted at the far end. The pulse at first disappears into an impedance Z_0 (thus the signal amplitude is half that of the unloaded generator), but after a round-trip travel time a reflected pulse of opposite polarity returns (Figure H.2). If the signal applied is instead a fast step, the effect of the reflection is to convert the step into a pulse (Figure H.3). An open-ended line produces a reflection of the *same* polarity, with the effects shown in Figure H.4. For arbitrary load resistance R the ratio of reflected to incident wave amplitude (the reflection coefficient) is given by

$$\rho \equiv A_{\text{r}}/A_{\text{i}} = (R - Z_0)/(R + Z_0).$$

Note that a termination resistance of $R = Z_0$ produces no reflection. A signal applied to such a terminated line is absorbed by the terminating resistor (as heat) and disappears forever. The signal source sees a loading resistance equal to Z_0. (It is for this reason that we did not have to worry earlier about the reflected pulses reflecting again from the pulse

---

[1] Higher performance standards include Category 5e ("e" for *enhanced*), and Category 6, propelled by the development of Gigabit Ethernet –

literally 1 gigabit/sec over unshielded twisted pair – also known as 1000baseT. To achieve this data rate, all four pairs are used simultaneously with 5-level amplitude encoding.

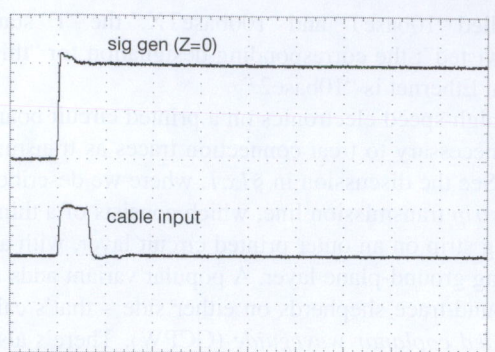

Figure H.3. 'Scope trace of a step waveform applied to an 8-foot length of RG-58A/U (solid polyethylene dielectric, velocity factor of 66%), shorted at the end. Vertical; 1 V/div; horizontal; 40 ns/div.

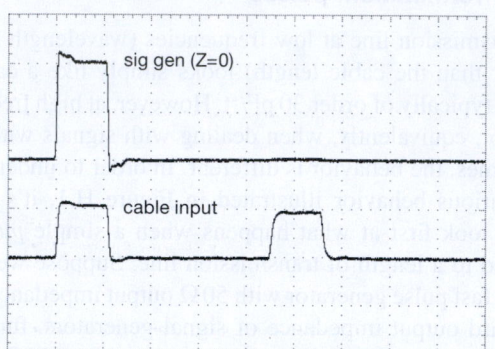

Figure H.4. 'Scope trace showing reflection from open-ended coax line. Same parameters as those in Figure H.2B.

generator – its 50 Ω source impedance swallows any signals returning from an improperly terminated cable, which is the reason most signal sources are standardized at 50 Ω impedance.)

A. Series termination

This last point – that returning (reflected) signals are completely absorbed if the signal source's impedance matches the line – leads to a nice technique called *series termination* (or *back termination*), frequently used for high-speed logic signals (and in other situations where the load has a high input impedance). Look at Figure H.5: a signal source in series with a resistor (equal to the line's impedance) drives a transmission line whose far end is unterminated (i.e., open). Now imagine a step input of amplitude V_{sig} at the signal source; it propagates down the line at half-amplitude, then reflects back from the far end at the full V_{sig} amplitude. Although any point along the line sees a two-step waveform, the surprising fact is that the waveform seen at the far end makes a single step from zero to the full

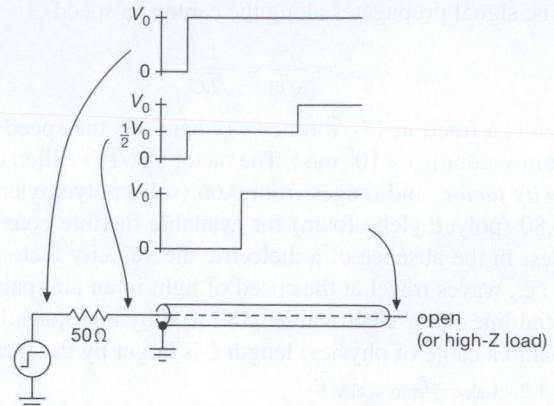

Figure H.5. Series termination of an open-ended line: the line is fed from a signal source of matched impedance; the half-sized step at the input propagates to the far end, from which it reflects in-phase to produce a returning step equal in amplitude to the *unloaded* amplitude of the generator. A high-impedance load at the far end sees only a single full-sized step.

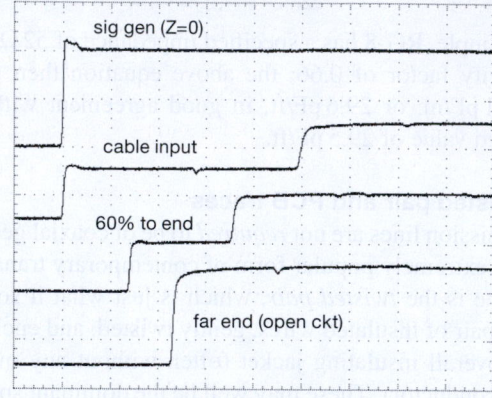

Figure H.6. 'Scope trace showing waveforms at cable input, midpoint, and far end, for series-terminated step input. The generator's zero-impedance signal is also shown; it was set to produce a 2 V step into an open circuit. The cable is 60 ft of RG-58/U (velocity factor of 66%), tapped at 36 ft with a high-impedance voltage probe. Vertical; 1 V/div; horizontal; 40 ns/div.

V_{sig}; at that place the half-sized incident wave arrives at the same time that the half-sized reflected wave departs. This interesting behavior is demonstrated in the 'scope traces in Figure H.6.

You can use this technique for sending CMOS logic signals through a length of coax: three paralleled 74HC buffers (for low source impedance, roughly 15 Ω) in series with a 33 Ω resistor nicely drives lengths of RG-174 or RG-316 (thin 50 Ω coax line), connected to the receiving gate at the far end *without termination*. The receiving gate sees

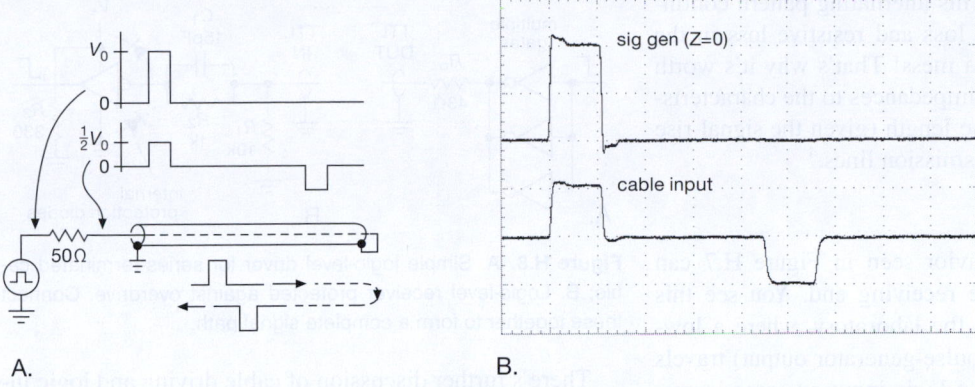

A.

B.

Figure H.2. A. A pulse driving a length of shorted transmission line reflects off the short and returns as a pulse of opposite polarity. B. Scope trace taken with 70 ft of RG-8/U with foam dielectric (velocity factor of 78%), shorted at end. Vertical; 1 V/div; horizontal; 40 ns/div. For this and following figures we used a high-impedance 'scope probe to avoid introducing additional transmission-line effects.

full-swing logic signals. This technique is often preferable to the matched-load alternative – directly driving a line that is terminated in $50\,\Omega$ – because with series termination the driver sees a load resistance twice as high ($100\,\Omega$ in this case), and that only for the round-trip duration of the signal (after which the load becomes an open-circuit).[2] For very fast logic signals (e.g., ECL 100K, or contemporary high-speed CMOS processors, memory, and peripherals), it may be necessary to treat a circuit trace of just a few inches as a transmission line. Typically printed circuit board (PCB) trace impedances are in the range of $50\,\Omega$ to $100\,\Omega$, but can be tailored to a specific impedance by proper choice of trace width and spacing above the ground plane; this specialty art is known as *microstrip* technique,[3] useful both for fast digital signals and for signals at frequencies above about 100 MHz (UHF and microwave).

This is all very nice in theory – but in practice you have to contend with sources of fast digital signals that are not matched to the line impedance. This happens often on digital PCBs where the digital output ports of speedy microprocessors and FPGAs are poorly matched to the PCB trace impedances. For example, a line driven by a signal of source impedance $Z_0/2$ produces lots of ringing at both ends of an unterminated line, including 33% overshoot at the far end; such ringing can produce false clocking.

To illustrate what this looks like, we drove a 2.4 m length of unterminated RG-58/U ($Z_0=50\,\Omega$) coax with a step input, probing both the input and output signals.

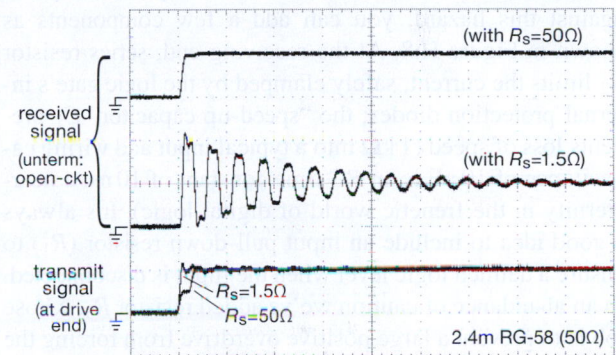

Figure H.7. Signals seen at the far end of an unterminated 8 ft length of $50\,\Omega$ line when driven with a unit step from a $50\,\Omega$ source (top trace) and from a low-impedance source (middle trace). The corresponding signals at the driving end are shown at bottom. Horizontal; 100 ns/div.

We did this under two conditions: (a) when driven with a $50\,\Omega$ series termination at the input (i.e., a "back-terminated" line); and (b) when driven with a low-impedance ($R_s=1.5\,\Omega$) voltage step.

Figure H.7 shows the results, where the drive signal's open-circuit amplitude (call it V_{OC}) is displayed as one vertical division. The matched series termination generates a clean received step to V_{OC} (and an input waveform with initial step first to $V_{OC}/2$, then to V_{OC} after a round-trip delay, just as seen in expanded scale in Figure H.6). By contrast, the low-Z drive imposes a full V_{OC} step at the input, which is first seen at the far end at $2V_{OC}$ (because the non-inverted reflected wave doubles the amplitude of the arriving wave), subsequently brought nearly back to zero (one round-trip time later) by the inverted wave reflected from

---

[2] See §12.10, where several methods of driving cables with logic levels are described and illustrated.

[3] If you sandwich the signal-carrying conductors between a pair of ground planes, you've got a *stripline*; see §1x.1.

the low-impedance driver. This alternating pattern continues, damped both by cable loss and resistive loss in the 1.5 Ω driver. This signal is a mess! That's why it's worth some effort to match driver impedances to the characteristic impedance of lines whose length (given the signal rise times) qualifies them as transmission lines.[4]

B. A robust logic link

The sort of overshoot behavior seen in Figure H.7 can destroy logic circuits at the receiving end. You see this vulnerability particularly in the laboratory, where a low-impedance logic signal (or pulse-generator output) travels through a length of coax to a logic input on some instrument. The latter is often unterminated, to keep the input impedance relatively high (to prevent heavy loading and attenuation for a signal source unable to drive 50 Ω).

If you want to make your own designs bulletproof against this hazard, you can add a few components as shown in Figure H.8. At the receiving end, series resistor R_2 limits the current, safely clamped by the logic gate's internal protection diodes; the "speed-up capacitor" C_1 prevents loss of speed (1 kΩ into a typical input and wiring capacitance of 10 pF is an RC time constant of 10 ns, a near-eternity in the frenetic world of digital logic). It's always a good idea to include an input pull-down resistor (R_1) to ensure a defined logic level when the input is disconnected. In an abundance of caution we've added resistor R_3, whose job is to prevent a large positive overdrive from forcing the V_+ rail to a positive voltage that can damage other ICs; this could normally be omitted, but it would be a good idea in an instrument with an inviting BNC connector on the front and that contains only a few ICs powered from a small regulator (like a 78L05 – see §9.3.2, Figure 9.6, and Table 9.1) whose dc output can be easily overdriven.

At the driver end the parallel connection of several logic gates generates a drive impedance down in the neighborhood of 5–10 Ω, which the added series resistor R_o brings up to a driving resistance close to the cable's 50 Ω characteristic impedance. That's what you want, of course, and that alone is enough to give you peace of mind as a respectable series-terminated driver. But it never hurts to protect the receiver end, as just discussed: you never know when someone will drive it with a pulse generator, inadvertantly set to deliver *negative* pulses, or 20 V positive pulses (as happened in our laboratory recently).

---

[4] An analogous effect in long-distance power transmission lines is known as the Ferranti effect; it is said that overvoltages caused by the Ferranti effect, if not properly compensated, can cause damage to power-line switch gear.

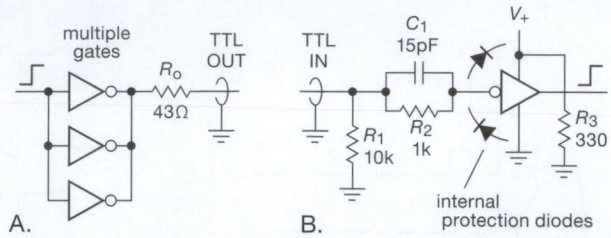

Figure H.8. A. Simple logic-level driver for series-terminated cable; B. Logic-level receiver protected against overdrive. Connect these together to form a complete signal path.

There's further discussion of cable driving and logic interfacing in Chapter 12, beginning at §12.10.

H.1.3 Termination: sinusoidal signals

We have been talking about signals propagating along transmission lines, for clarity using the particular case of pulses or voltage steps. Of course, a *sinewave* applied to a length of cable also produces reflections, unless of course the cable is properly terminated. The effect is to alter the input current, for an applied input voltage, in a way that depends on the (mismatched) load impedance Z_L, and also on the ratio of the signal's wavelength in the cable λ to the physical length of the cable l. The final effect is to produce an input impedance (complex in general) given by

$$Z_{in} = Z_0 \frac{Z_L + jZ_0 \tan(2\pi l/\lambda)}{Z_0 + jZ_L \tan(2\pi l/\lambda)}.$$

From this one can see that:
(a) a matched termination ($Z_L = Z_0 = $ [usually] 50 Ω) results in an input impedance equal to the characteristic impedance of the cable, independent of length or frequency;
(b) a quarter-wave line inverts the load impedance, i.e., $Z_{in} = Z_0^2/Z_L$;
(c) a half-wave line preserves the load impedance, i.e., $Z_{in} = Z_L$;
(d) a short length of open-circuited line $l \ll \lambda$ looks capacitive, viz., $Z_{in} \approx -j/\omega C'$, where C' (the effective capacitance) is the constant l/cZ_0;
(e) a short length of short-circuited line $l \ll \lambda$ looks inductive, viz., $Z_{in} \approx j\omega L'$, where L' (the effective inductance) is the constant $Z_0 l/c$.
The impedance-changing properties of transmission lines can be used to match impedances, though any such scheme will be frequency dependent; when you hear words like "stubs," you're dealing with transmission-line impedance matching. Virtuosos in this area make heavy use of network analyzers, and they will try to dazzle you with their

handsome "Smith Charts" (which are well beyond the humble scope of this book[5]).

When you have sinusoidal signals – with reflections – on a transmission line, you generate *standing waves*. That is, you can picture the net result of waves propagating in both directions (at the same frequency) as the sum of a nonpropagating (hence "standing") wave and some additional propagating wave. For example, an open-ended line produces a reflected wave of full amplitude; the result is a pure standing wave of the same frequency and twice the amplitude, with a maximum amplitude of oscillation at the open end (and repeating every half-wavelength), and complete nulls ("nodes" – places with no voltage) midway between. For a shorted-end line a similar thing happens, but the reflected wave is of opposite amplitude, producing a null at the far end (and repeating every half-wavelength), with maxima in between. (You get the same pattern if you tie a length of clothesline to a fence, then wiggle the end up and down at the right rate.) With a smaller termination mismatch you don't get complete cancellation anywhere.

Standing waves are not necessarily bad (though they are never[6] *good*!). But they do increase both the peak voltages and the resistive losses (see next section), relative to a matched line, for the same power transmitted. They are ordinarily seen as the symptom of a mismatched line. So in communications systems people try to minimize the *standing wave ratio* (abbreviated SWR, or sometimes VSWR – for *voltage* standing wave ratio – pronounced "VIZ-wahr"), which is defined as the ratio of the maximum amplitude to minimum amplitude:

$$\text{VSWR} = \frac{V_{\max}}{V_{\min}} = \frac{A_{\text{f}} + A_{\text{r}}}{A_{\text{f}} - A_{\text{r}}},$$

where V is the ac (signal) voltage, measured at points along the line; and A_{f} and A_{r} are the amplitudes of the forward and reflected waves, respectively. Voltage measurements along a cable with no standing waves will give a constant amplitude (hence VSWR=1.0).

The VSWR is a real number, between 1.0 (perfect match, no reflected wave) and ∞ ("perfect mismatch," reflected wave amplitude equal to incident wave amplitude). In terms of the reflection coefficient ρ, the VSWR is just

$$\text{VSWR} = \frac{1 + |\rho|}{1 - |\rho|}.$$

For a purely resistive mismatch we can use our earlier expression for ρ to find that

$$\text{VSWR} = \begin{cases} R/Z_0 & \text{if } R > Z_0 \\ Z_0/R & \text{if } Z_0 > R. \end{cases}$$

From the VSWR you can find the magnitude (but not the phase) of the reflection coefficient:

$$|\rho| = \frac{\text{VSWR} - 1}{\text{VSWR} + 1}.$$

The VSWR is measured with a directional power meter. Knowing the values of forward and reflected power, you know from the defining equation above that

$$\text{VSWR} = \frac{1 + \sqrt{P_{\text{r}}/P_{\text{f}}}}{1 - \sqrt{P_{\text{r}}/P_{\text{f}}}}.$$

H.1.4 Loss in transmission lines

In the real world of non-ideal transmission lines, things aren't quite as nice as we've led you to believe. Real transmission lines are *lossy*, meaning that signals are attenuated as they travel down the line; they are also slightly *dispersive*, meaning that signals of different frequency travel with slightly different speeds. The loss is *frequency dependent*: its value (often specified as attenuation in "dB per 100 ft") increases proportional to $\sqrt{f}$; i.e., a quadrupling of frequency doubles the loss (in dB) of a given length of line. This happens because the loss is dominated by the *skin effect*: when alternating current flows through a conductor, the current is not uniform throughout the bulk – it is confined to an outer layer (called the *skin depth*) of thickness $\delta = (\pi \sigma f)^{1/2}$, where σ is the conductivity and f is the frequency.[7] Because the skin depth decreases inversely with the square root of frequency, you have to quadruple the frequency to double the resistance (halve the skin depth),

---

[5] Smith charts are treated in the excellent reference *Fields and Waves in Communication Electronics* by Ramo, Whinnery, and Van Duzer (Wiley, 1994), as well as in the insightful and refreshing *Radio-Frequency Electronics* by Hagen (Cambridge University Press, 2009).

[6] Well, *hardly ever!* For operation over a narrow frequency range you sometimes exploit the impedance-changing properties of mismatched lines, which necessarily have standing waves. Examples are (a) the use of open or shorted lengths of line as high-Q capacitors or inductors, (b) a shorted half-wave or an open quarter-wave line used as an RF bypass capacitor, (c) an open half-wave or a shorted quarter-wave line used as an RF choke, (d) matching two different impedances (cables, signal sources, or loads) by interposing a quarter-wave section of transmission line whose characteristic impedance is the geometric mean of the two impedances being matched (this is analogous to a quarter-wave anti-reflection coating in optics), (e) the use of a slotted line and high-impedance probe for a direct measurement of wavelength, and (f) the use of transmission lines to make "ring hybrids" and "rat-race hybrids." We thank Jon Hagen and Darren Leigh for these applications of "good standing waves."

---

[7] To be precise, the current density decreases exponentially, falling to $1/e$ (37%) of its surface value at a depth equal to δ.

which is equivalent (in terms of loss) to doubling the length of the line. This explains the approximate slope of the attenuation curves for transmission lines (Figure H.9), where lines of larger diameter have lower losses. Dielectric losses contribute additional attenuation at the highest frequencies. The curves shown are for a matched line; if there are reflections (i.e., if the VSWR is greater than 1.0) then the loss, *for a given net power transmitted down the line*, will be greater.

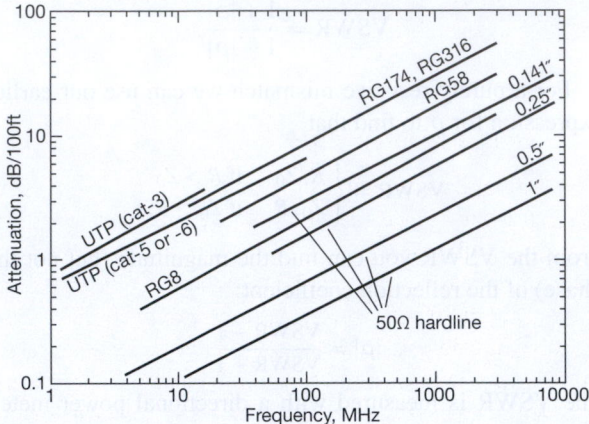

Figure H.9. Attenuation (dB/100 ft) as a function of frequency for several representative cable types.

It's useful to realize that skin-depth effects are significant even at low frequencies – current at the power-line frequency of 60 Hz is confined to a surface layer of roughly 1 cm in copper, for which the skin depth (in centimeters) at room temperature is given by $\delta(\text{Cu})=6.6/\sqrt{f}$; you don't reduce power transmission losses much by using wire thicker than that. At radio frequencies the skin depth is so shallow (e.g., $\delta=10\,\mu\text{m}$ at 40 MHz) that you can make low-loss connections, inductors, and so on, by silver plating a poor conductor. A common technique for shielding lightweight instruments and computers is to apply a thin metallic plating to a plastic enclosure. Figure H.10 plots the skin depth in copper conductors from 10 Hz to 10 GHz.

H.2 Impedance matching

Because you get reflections from unterminated (or incorrectly terminated) transmission lines, it's obviously a good idea to make sure you match impedances when you use coax lines whose electrical length is a significant fraction (at least 1/20, say) of the wavelength of the highest frequencies you're using. Stated in terms of time rather than frequency, you have to start worrying about termination

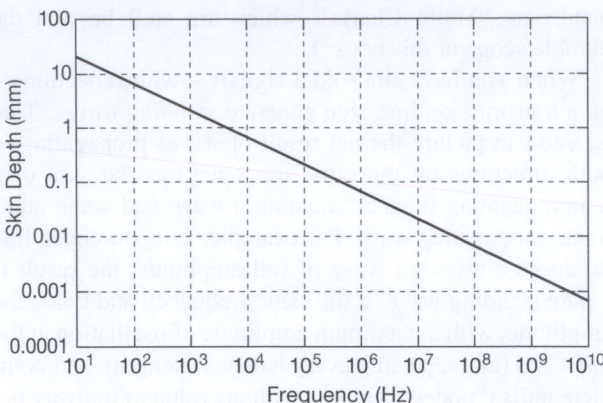

Figure H.10. Skin depth in copper as a function of frequency.

when the round-trip propagation time is about 20% of the signal rise time.

We've seen already that a simple way to do this is to terminate the line in its characteristic impedance (resistance), for example 50 Ω for most coax lines. Termination is not required at *both* ends, because a terminated end swallows any incident signals. Thus you can use a mismatched signal-source impedance to drive a line terminated at the far end; or, as we saw above, you can "series terminate" the driven end of a line whose far end is unterminated (i.e., fed to a load of much higher impedance than the line). The usual practice, however, is to terminate *both* ends in the line's characteristic impedance (a conservative instinct that ensures a minimum of reflections). For example, you usually use a 50 Ω cable to pipe the signal from a synthesizer or signal generator to a 50 Ω matched load at the far end; if your load is high impedance, you place a 50 Ω resistor across it (or use a 50 Ω coax feedthrough termination).

Sometimes you need to match a line to a load (or source) of a different impedance; for example, you might want to measure the properties of some 75 Ω video cable (that's the impedance that the video industry has chosen, much to the chagrin of the rest of the electronics community), and all you've got are 50 Ω test instruments. Or, you might want to match the output impedance of a high-frequency amplifier to a length of cable that goes to an antenna.

This brings us to the subject of matching networks. In the following subsections we treat (a) resistive (lossy) networks for broadband impedance matching and attenuation, (b) transformer (lossless) broadband matching, and (c) reactive (lossless) narrowband matching.

H.2.1 Resistive (lossy) broadband matching network

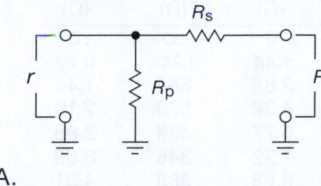

A.

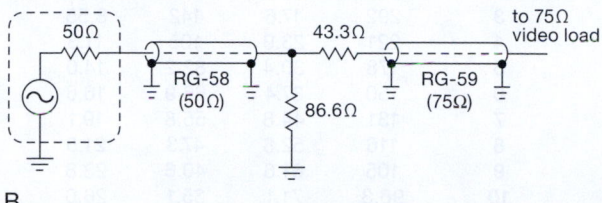

B.

Figure H.11. A. A resistive L-network can match any pair of real (resistive) impedances; the parallel resistor R_p goes across the port of lower impedance r. B. Example of matching a 50 Ω signal source and cable to a 75 Ω video cable and load (with a loss of 5.72 dB).

You can easily figure out that two resistors (in the form of an "L-network," Figure H.11A) is all it takes to match a pair of impedances r and R (assumed resistive, as all cables are); both sides are happy – each sees a matched load. The values of the matching resistors are

$$R_p = r\sqrt{\frac{X}{X-1}},$$
$$R_s = r\sqrt{X(X-1)},$$

where r is the smaller impedance and X is the ratio of impedances: $X=R/r$. Taking the earlier example, you can match a 50 Ω test instrument to a 75 Ω coax line (the common variety is called RG-59) by putting 86.6 Ω across the 50 Ω port and connecting it to the cable through a 43.3 Ω series resistor (Figure H.11B).

The good news is that such a resistive L-network is frequency independent; the bad news is that it's lossy. It's easy to show that the loss is

$$\text{loss} = 20\log_{10}\left(\frac{\sqrt{X}}{X+\sqrt{X(X-1)}}\right) \text{ dB}.$$

For example, the 50Ω-to-75Ω L-network above has a transmission loss of 5.72 dB for signals going in either direction. With a resistive match you have to accept this attenuation (this is sometimes called a *minimum loss pad*). We'll see below how to make lossless matching networks with trans-

formers or with networks of L's and C's ("reactive matching networks").

As you might imagine, you can do even *worse*, in terms of loss, with a network containing more resistors! In particular, you can add another resistor, making either a "T" network or a "Pi" network, that matches two resistive impedances to each other, with loss greater than the minimum loss we found above. Although this isn't something you ordinarily want to do, there is a variation on that theme that is often useful; namely, a resistive attenuation network between a pair of already-matched impedances.

H.2.2 Resistive attenuator

In radiofrequency circuits you sometimes need to attenuate a signal level – for example, to avoid overdriving a stage of gain. In other situations you need to use a resistive attenuator to provide some isolation between an impedance-sensitive component like an amplifier, mixer, or cable, say, and a component that is not impedance matched; an example of the latter is a filter, which typically is impedance matched in its passband, but reflective (a severe mismatch) in its stopband. Some amplifiers will oscillate if their output directly drives a sharp filter.

The solution to these problems is a resistive impedance-matched attenuator. The two topologies are T and Pi, named for their appearance on a schematic diagram (Figure H.12). It is not difficult to derive the resistor values:

$$R_p = \frac{1+x}{1-x}R_0,$$
$$R_s = \frac{1-x^2}{2x}R_0, \qquad (\text{Pi}-\text{network}),$$

and

$$R_s = \frac{1-x}{1+x}R_0,$$
$$R_p = \frac{2x}{1-x^2}R_0, \qquad (\text{T}-\text{network}),$$

where x is given by the attenuation: $\text{atten}(\text{dB})=-20\log_{10}x$ (or, equivalently, $x=10^{-\text{atten}(\text{dB})/20}$), and Z_0 (assumed resistive) is the impedance at both input and output. Tabulated values for 50 Ω source and load impedances are given in Table H.1.

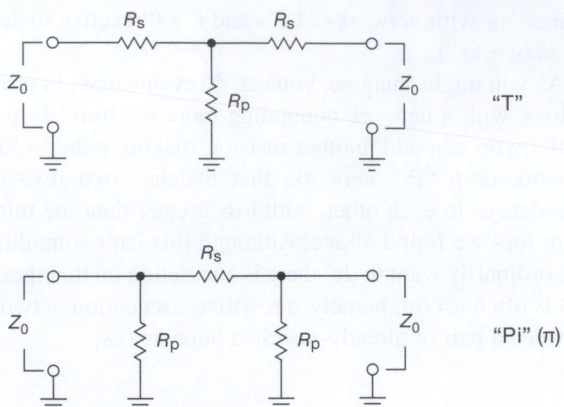

Figure H.12. Resistive T and Pi attenuators for equal input and output impedances.

H.2.3 Transformer (lossless) broadband matching network

If the unavoidable loss of a resistive matching network is unimportant in your application, that certainly is the simplest method. However, in many applications it is essential to minimize loss – for example, in a communications transmitter or in low-level circuits whose performance is limited by amplifier or thermal noise.

In that case you can use either a transformer or a reactive matching network; neither method provides coupling all the way down to dc, however. Transformer coupling is relatively broadband, but it is limited in impedance ratios; reactive matching, by contrast, flexibly matches impedances (including reactive impedances), but only around some chosen center frequency. We treat reactive matching in the next subsection.

Transformers for use at signal frequencies are similar in principle to ordinary ac power transformers, that is, they use a pair of windings that are coupled magnetically, and whose turns ratio is the desired voltage ratio. The impedance ratio is then the square of the turns ratio. (For example, a transformer with a 1:4 primary:secondary turns ratio, driven with a $50\,\Omega$ signal source, would present an output impedance of $800\,\Omega$, and should be loaded with that resistance.) However, because of the higher signal frequencies, it is necessary to use magnetic cores that do not have significant conductive paths for eddy currents. At audio frequencies the solution is to use the same sort of laminated metal stacks used in power transformers, but with much thinner laminations. At still higher frequencies, laminated cores are replaced either by powdered iron cores or by completely nonconducting magnetic "ferrite" materials. Because of the devastating effects of parasitic capac-

Table H.1 $50\,\Omega$ T and Pi Attenuators

Attenuation (dB)	Pi R_P (Ω)	Pi R_S (Ω)	T R_P (Ω)	T R_S (Ω)
0	∞	0	∞	0
0.25	3.47k	1.44	1.74k	0.72
0.50	1.74k	2.88	868	1.44
0.75	1.16k	4.32	578	2.16
1.00	870	5.77	433	2.88
1.25	696	7.22	346	3.59
1.50	581	8.68	288	4.31
1.75	498	10.1	247	5.02
2.0	436	11.6	215	5.73
2.5	350	14.6	171	7.15
3	292	17.6	142	8.55
4	221	23.9	105	11.3
5	178	30.4	82.2	14.0
6	150	37.4	66.9	16.6
7	131	44.8	55.8	19.1
8	116	52.8	47.3	21.5
9	105	61.6	40.6	23.8
10	96.3	71.1	35.1	26.0
15	71.6	136	18.4	34.9
20	61.1	248	10.1	40.9
25	56.0	443	5.64	44.7
30	53.3	790	3.17	46.9
35	51.8	1.41k	1.78	48.3
40	51.0	2.50k	1.00	49.0
45	50.6	4.45k	0.56	49.4
50	50.3	7.91k	0.32	49.7
55	50.2	14.1k	0.18	49.8
60	50.1	25.0k	0.10	49.9

Resistor values for T and Pi attenuators for use with $50\,\Omega$ source and load. The values shown can be scaled for use at some other impedance, assuming equal input and output impedances.

itance and inductance, transformers for use at high radio frequencies (say above 10 MHz) generally are constructed from transmission lines (coaxial or parallel) wound around a magnetic core.

Impedance-matching transformers are widely available commercially, though for special applications you may need to design and wind your own. At audio frequencies many manufacturers offer miniature impedance matching transformers with "telephone" bandwidths (200 Hz–4 kHz), or full audio bandwidth (20 Hz–20 kHz); impedances go from loudspeaker and microphone impedances (8–600 Ω) up to "hi-Z" values of 10k–50kΩ. There's further discussion in §8.10.

A nice series of radiofrequency transformers is made by North Hills, including models that transform $50\,\Omega$ or $75\,\Omega$ to impedances up to $1200\,\Omega$; these cover the frequency range between 20 Hz and 100 MHz, with a typical frequency range of 1000:1 or more for a given transformer. For higher frequencies you can get broadband matching transformers from Mini-Circuits, covering the range of 4 kHz to 2 GHz with impedance ratios from 1:1 to 16:1, and with frequency ranges of 1000:1 or more for a given transformer. These are constructed with transmission-line techniques.

It is worth noting that transformer coupling provides *galvanic isolation*: input and output circuits need not share the same ground. This is particularly useful when you need to send a signal (or distribute a "house clock") between instruments whose individual cases are each grounded through their power cords or enclosure rails. We have seen several instances in which instrument "ground," in the same laboratory, differed by as much as *several volts* of 60 Hz ac. Here an isolated $50\,\Omega{:}50\,\Omega$ broadband transformer is ideal, for example the Mini-Circuits FTB1-6 (10 kHz–125 MHz) or the North Hills 0016PA (20 Hz–20 MHz).

H.2.4 Reactive (lossless) narrowband matching networks

You can match *any* pair of impedances, real or complex, with just two reactive components. The resulting match is perfect only at a single frequency, but "good enough" over some modest band of frequencies. This can be considered an alternative to (broadband) transformer matching, with considerably greater flexibility in target impedances. It is worth noting that a lossless match between impedances that are not purely real (i.e., that have a reactive component) will always be narrowband.

The simplest reactive matching network is an L-network with one inductor and one capacitor (Figure H.13). You can choose either the inductor or the capacitor as the parallel element, but the network must have the parallel reactance located across the port with the larger impedance. The design procedure is straightforward, and is nicely motivated and explained in Hagen (see Appendix N). It amounts to choosing the parallel reactance to produce (in combination with the higher port's impedance R_{high}) the correct lower resistance R_{low} as its real part, then using the series reactance to cancel the resulting reactance.

The procedure goes like this.

1. Calculate the quantity

$$Q_{\text{EL}} = \sqrt{\frac{R_{\text{high}}}{R_{\text{low}}} - 1}$$

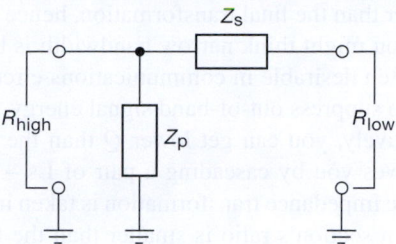

Figure H.13. Lossless reactive impedance-matching L-network.

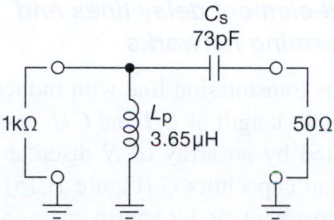

Figure H.14. Example of lossless network to match a 1 kΩ source impedance to a 50 Ω load, at a center frequency of 10 MHz.

(this will be twice the actual Q value, or frequency selectivity, of the matching network).

2. Now select the form of parallel reactance (i.e., inductor or capacitor), and set the magnitude of its reactance equal to $R_{\text{high}}/Q_{\text{EL}}$ at the center frequency. In other words, $L_{\text{parallel}} = R_{\text{high}}/2\pi f Q_{\text{EL}}$ or $C_{\text{parallel}} = Q_{\text{EL}}/2\pi f R_{\text{high}}$, respectively.

3. Finally, set the magnitude of the series reactance (i.e., capacitor or inductor, respectively) equal to $Q_{\text{EL}}R_{\text{low}}$ at the center frequency. In other words, $C_{\text{series}} = 1/2\pi f Q_{\text{EL}}R_{\text{low}}$ or $L_{\text{series}} = Q_{\text{EL}}R_{\text{low}}/2\pi f$, respectively.

As an example, let us match a 1000 Ω source (an amplifier output) to a 50 Ω load (an antenna) at a frequency of 10 MHz. We find $Q_{\text{EL}} = 4.36$, and, choosing a parallel inductor at the input, $L_{\text{parallel}} = 3.65\,\mu\text{H}$ and $C_{\text{series}} = 73$ pF (Figure H.14). The Q of the resultant coupled network is equal to $Q_{\text{EL}}/2$, roughly $Q \approx 2$; its bandwidth is thus about 50% between half-power points, though the match is perfect only at the center frequency. Note that the Q rises with increasing impedance ratio, and that you have no control over it. The network becomes a narrow bandpass filter for very large impedance-transformation ratios.

If you want a higher Q, you can get it by adding another reactive component, to form a Pi (or T) network. You can think of this as a pair of L-networks, going to an intermediate impedance that is much lower (or higher) than either port impedance. Each L-section then has an impedance ratio greater than the final transformation, hence the higher Q value. You might think narrow bandwidth is bad, but in fact it is often desirable in communications circuits where you want to suppress out-of-band signal energy.

Alternatively, you can get lower Q than the simple L-network gives you by cascading a pair of Ls – a "double L." Here the impedance transformation is taken in two half-steps – each section's ratio is smaller than the final ratio, hence a lower Q.

H.3 Lumped-element delay lines and pulse-forming networks

The continuous transmission line with inductance and capacitance per unit length of L/l and C/l, respectively, can be approximated by an array of N discrete series inductors L and shunt capacitors C (Figure H.15). It is easy to show that the resultant circuit approximates a transmission line whose propagation time per element is $\tau_i=\sqrt{LC}$, and whose impedance is $Z_0=\sqrt{L/C}$; the total propagation time is $t_p=N\tau_i=N\sqrt{LC}$.

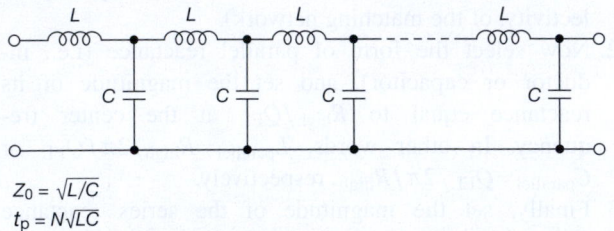

$Z_0 = \sqrt{L/C}$
$t_p = N\sqrt{LC}$

Figure H.15. Lumped-element delay line formed with an array of equal-value capacitors and inductors.

You can make a delay line this way, approximating a long transmission line. Roughly speaking, such a discrete approximation to a continuous transmission line will preserve details of the waveform only down to time scales of τ_i, or $1/N$th of the total propagation time. For example, a $1\,\mu s$ lumped delay line with 20 LC sections will swallow details shorter than about 50 ns. It is a lowpass filter that attenuates frequencies above $f=1/2\pi\sqrt{LC}$.

Lumped-element delay lines were used in early analog oscilloscopes to allow time for the sweep to begin before the (delayed) signal reached the deflection circuitry; this let you see the triggering event (and a bit before). Later analog scopes used lengths of a helical-conductor coaxial line for the same purpose. Take a look at §H.4.3 for a bit more on this fascinating application.

Lumped delay lines are useful as *pulse-forming networks*, as shown in Figure H.16. Here the parallel capaci-

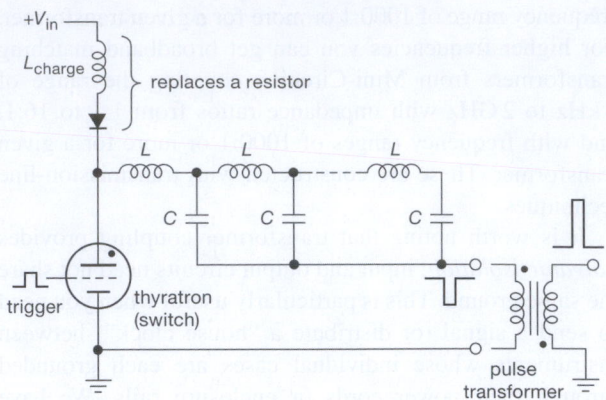

Figure H.16. Pulse-forming network for producing a high-voltage pulse of high energy. The thyratron is a special type of vacuum tube, containing a small amount of hydrogen or other gas, designed for switching really high voltages and high currents (10s of kV, 1000s of amperes: 10s of megawatts!). The inductor–diode fragment shown is a way to implement efficient "resonant charging" of a capacitor from a fixed dc voltage.

tance of a set of LC delay sections is charged to a high positive voltage; then the "center conductor" of the coax equivalent is switched to ground with a high-voltage switching element such as a thyratron. The common terminal (analogous to the coax "shield") then produces a negative voltage pulse, of duration equal to *twice* the delay-line propagation time; its source impedance is just that of the delay line. This can drive a load directly; often it is converted to a different amplitude (and perhaps opposite polarity) with a pulse transformer, as shown. Pulse-forming networks find use in radar and other applications in which the pulse voltage and/or duration are inconvenient to produce with the analogous transmission-line circuit.

Shown also in this diagram is the method of "resonant charging," in which an inductor L_{charge} plus diode replaces the conventional charging resistor for the purpose of charging a capacitor C_{total} (the N capacitors in parallel). This has several benefits: (a) the charging wastes no energy, whereas resistive charging wastes exactly 50%; (b) it is complete after a time equal to half the period of the resonant circuit formed by L_{charge} and C_{total}; and (c) it charges the capacitor(s) to twice the supply voltage. Resonant charging is a clever technique, and it is used also in switching converters and power supplies, flashlamp circuits, and elsewhere.

H.4 Epilogue: ladder derivation of characteristic impedance

In a formal course on "waves" you are usually subjected to an analysis that uses Maxwell's equations to derive the relationship between the traveling $\vec{E}$ field and $\vec{B}$ field, from which follows the relation between voltage and current, and hence impedance. As a bonus you get the capacitance and inductance per unit length, and the speed of propagation.

But there's a nice "circuit" way to convince yourself that a properly terminated transmission line looks like a pure resistance (equal in value to its "characteristic impedance," for example $50\,\Omega$), namely to model it as a discrete LC ladder (Figure H.17) consisting of little increments of length δx, each having a series inductance $\mathscr{L}\,\delta x$ and shunt capacitance $\mathscr{C}\,\delta x$, where $\mathscr{L}$ and $\mathscr{C}$ are the inductance and capacitance per unit length of the coaxial line. There are $l/\delta x$ of these in the whole length l of the line.

H.4.1 First method: terminated line

We start[8] by writing down the impedance looking into the last section of a terminated line ($\mathbf{Z}_1$ in Figure H.17), which will give us a condition on the ratio $\mathscr{L}/\mathscr{C}$ in order for $\mathbf{Z}_1$ to equal, approximately, R_0. Then we'll see that the impedance $\mathbf{Z}_{\text{in}}$ looking into the whole ladder converges exactly to R_0 as we convert the discrete ladder approximation to a continuous transmission line by taking the limit as δx goes to zero.

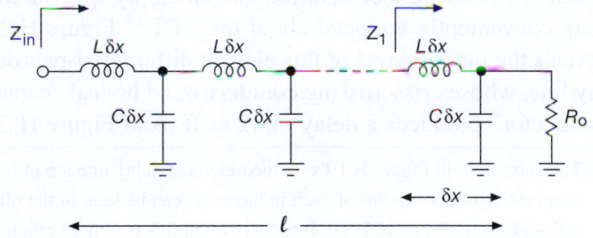

Figure H.17. *LC* ladder model of a transmission line of length l. We're ultimately interested in the limit $\delta x \to 0$, where the number of sections $N = l/\delta x \to \infty$.

Let's do it. $\mathbf{Z}_1$ is just the impedance of $\mathscr{L}\,\delta x$ in series with the parallel impedance of R_0 and $\mathscr{C}\,\delta x$:

$$\mathbf{Z}_1 = j\omega\mathscr{L}\delta x + \frac{R_0 \cdot (-j/\omega\mathscr{C}\delta x)}{R_0 - j/\omega\mathscr{C}\delta x}$$

$$= j\omega\mathscr{L}\delta x + \frac{R_0}{1 + j\omega\mathscr{C}R_0\delta x}$$

$$\approx R_0 + j\omega\delta x(\mathscr{L} - R_0^2\mathscr{C}),$$

where in the last step we've kept only the first term of the binomial expansion, i.e., $1/(1+\varepsilon) \approx 1-\varepsilon$.

The second term vanishes when $\sqrt{\mathscr{L}/\mathscr{C}} = R_0$, which is the formula for the characteristic impedance of a transmission line. But... not so fast – that term would have vanished anyway as $\delta x \to 0$; however, our transmission line would have vanished as well! What we need to do is to include the next-order binomial terms, and see what happens as we let δx go to zero, while holding the total line length l constant; the number of sections then increases, as $N = l/\delta x$.

You can do the math. You'll find that the next two terms add contributions[9] of order δx^2 and δx^3, making $\mathbf{Z}_1$ look like

$$\mathbf{Z}_1 \approx \left\{ R_0 + \mathscr{O}(\delta x^2) \right\} + j\left\{ \omega\delta x(\mathscr{L} - R_0^2\mathscr{C}) + \mathscr{O}(\delta x^3) \right\}$$

and so, cascading N such sections (where $N = l/\delta x$), with the condition $\sqrt{\mathscr{L}/\mathscr{C}} = R_0$, the higher order terms vanish as $\mathscr{O}(\delta x)$ and $\mathscr{O}(\delta x^2)$ (for the real and imaginary parts, respectively) as $\delta x \to 0$. Thus the input impedance of a transmission line of length l, terminated in its characteristic impedance (resistance) R_0, is pure resistive and equals R_0.

H.4.2 Second method: semi-infinite line

Here's a clever method[10] that doesn't require approximation or worries about convergence. The idea is to look into one end of a lumped-element transmission line that extends to infinity (Figure H.18), noticing that it looks just the same if we step one notch to the right. So, calling the (complex) input impedance $\mathbf{Z}_0$, we have, simply,

$$\mathbf{Z}_0 = j\omega L + \mathbf{Z}_0 \parallel \mathbf{Z}_C = j\omega L + \frac{\mathbf{Z}_0 \cdot (-j/\omega C)}{\mathbf{Z}_0 - j/\omega C}.$$

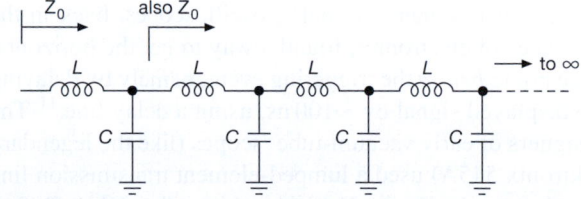

Figure H.18. Semi-infinite *LC* ladder model makes for an easy calculation.

---

[8] This treatment was inspired by Hagen's *Radio Frequency Electronics*; see Appendix N.

[9] They are $-R_0^3\,\omega^2C^2\delta x^2 + jR_0^4\,\omega^3C^3\delta x^3$, if you want to check your math (or ours!).

[10] Suggested to us by Jene Golovchenko.

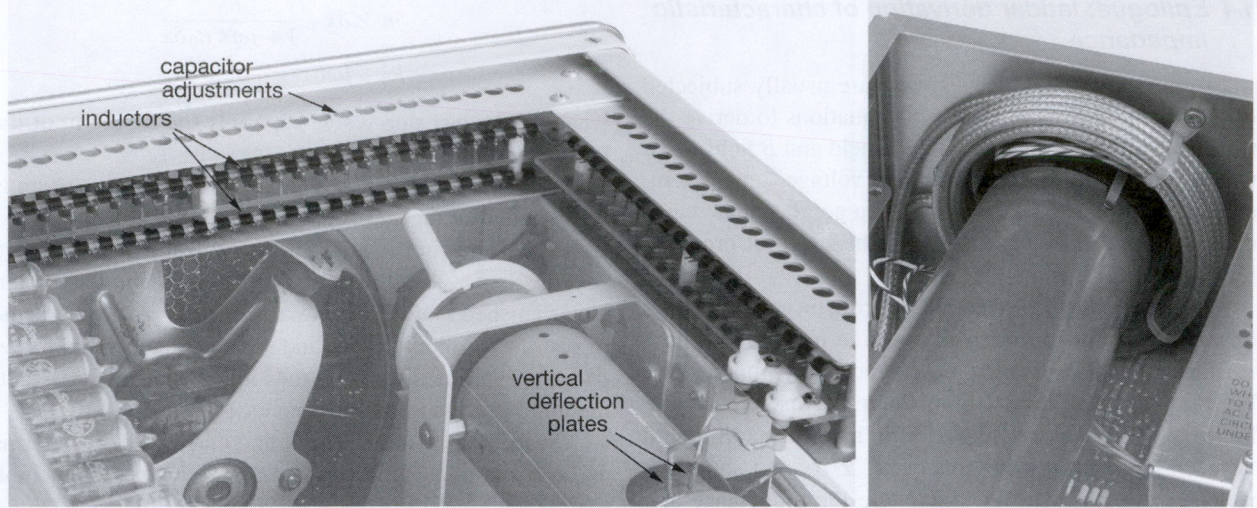

Figure H.19. Analog oscilloscopes used delay lines in the signal path, so you could see the triggering event. The 1959-vintage vacuum-tube Tektronix 545A (left) had a 30 MHz bandwidth, and used a two-channel lumped-element 200 ns delay line consisting of 50 inductor pairs and 50 (adjustable!) trimmer capacitors; (Figure H.20); their 1982 vintage solid-state 2213 'scope (right) had 60 MHz bandwidth, and used a 2.5 m length of spiral-conductor coaxial cable for its 100 ns delay line.

Multiplying through by the denominator of the last term and rearranging terms, we get a quadratic equation for $\mathbf{Z}_0$:

$$\mathbf{Z}_0^2 - j\omega L\mathbf{Z}_0 - L/C = 0,$$

with the solution

$$\mathbf{Z}_0 = j\omega L \pm \frac{\sqrt{4L/C - \omega^2 L^2}}{2}.$$

Now for the *coup de grâce*: we let the individual segments shrink toward zero while keeping the full line length. The individual L's and C's go to zero, but their ratio remains constant. Only the $4L/C$ term survives, giving us the (real) impedance $\mathbf{Z}_0 = \sqrt{L/C}$. No approximations!

H.4.3 Postscript: lumped-element delay lines

The clever designers of analog oscilloscopes, back in the dark ages of electronics, found a way to get the horizontal trace going *before* the triggering event, namely by delaying the displayed signal by ∼100 ns, using a delay line.[11] The designers of early vacuum-tube 'scopes (like the legendary Tektronix 545A) used a lumped-element transmission line like the one in Figure H.19 to achieve the delay (200 ns in this case) that otherwise would have required more than

100 feet of cable; presumably they also felt some satisfaction in exploiting the theory they learned in an electronics course they had taken years earlier.[12] Later analog 'scopes replaced the lumped-element delay line with a cleverly engineered coax delay line with a pair of spiral inner conductors; its larger inductance per unit length increased the signal delay,[13] and, happily, increased the characteristic impedance as well. The photograph shows an example, where the cable has been stuffed into an empty space at the rear, conveniently wrapped about the CRT.[14] Figure H.21 reveals the inner secrets of this elegant differential-pair delay line, whose crisscrossing counterwound helical "center conductor" produces a delay of 12 ns/ft. And Figure H.22

---

[11] Digital 'scopes finesse this problem by using digital memory to store some pretrigger digitized samples.

[12] The capacitors in Figure H.19's lumped-element delay line are in fact connected to the *midpoint* of each inductor, as can be seen in the official schematic of Figure H.20. It turns out that this is a more efficient implementation of a finite lumped line, as explained to us over a dinner of fine conversation and Persian cuisine by Larry Baxter ("Mr. Capacitive Sensors"; see the book by the same name).

[13] By a factor of $\pi n D$, approximately, where n is the number of turns per unit length and D is the spiral diameter. Because the coarse-pitch spiral is closely surrounded by the shield, you can think of the signal as propagating, corkscrew fashion, *along* the spiral; that approximation then gives you this simple expression, without the need to calculate the inductance and capacitance per unit length.

[14] That's *cathode ray tube*, for those born in this millennium, and thus deprived of the opportunity to admire one.

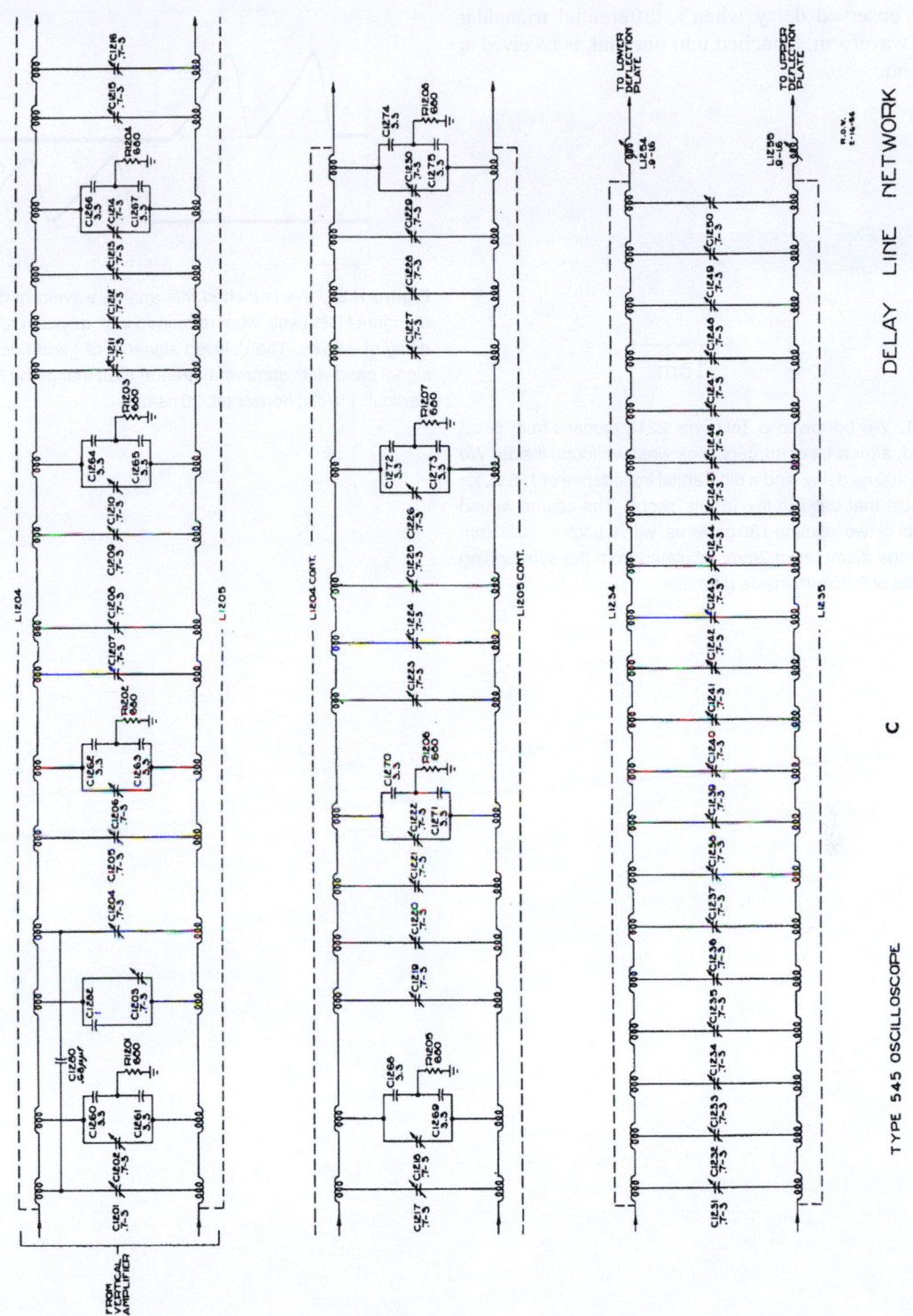

Figure H.20. Working hard to achieve perfection: Tektronix used fifty inductor pairs and fifty trimmers for the delay line in their type 545 'scope (pictured in Figure H.19) from the 1950s. Reproduced with permission of Tektronix Inc.

shows the observed delay when a differential triangular pulse-pair waveform, launched into one end, is received at the other end.

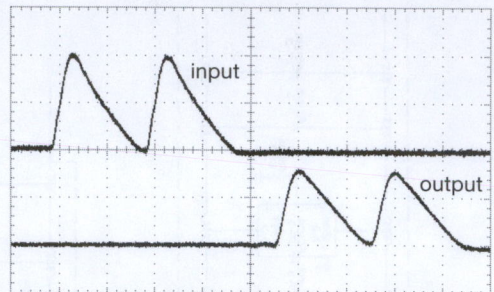

Figure H.22. We launched this analog waveform down the cable of Figure H.21 (well, what remained of it, anyway) to see the signal delay of ~95 ns. The delayed signal is of good fidelity, with a test signal bandwidth somewhat greater than that of the 60 MHz 'scope. Vertical: 1 V/div; horizontal: 20 ns/div.

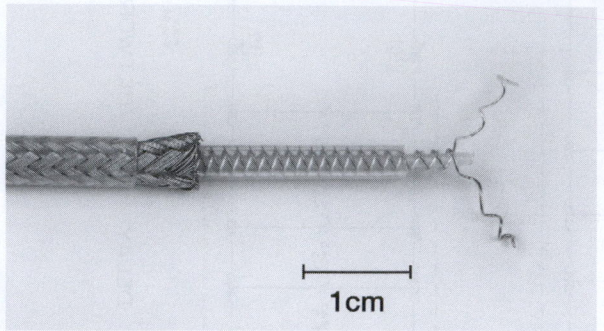

1 cm

Figure H.21. We borrowed a Tektronix 2213 carcass from Brian Shaban, and, after a bit of surgery, look what we found inside! We measured a 100 ns delay, and a differential impedance of 155 Ω, for the 8.5 ft cable that used to live in this 'scope. The counterwound helix consists of two insulated 30 ga wires, with a pitch of 1.125 mm and an average diameter of 2 mm, insulated from the surrounding braided shield of 3.25 mm inside diameter.

TELEVISION: A COMPACT TUTORIAL

This appendix evolved from a tutorial, originally written for a nontechnical audience.[1] It is organized as follows:

Television: video plus audio
Combining and sending the audio + video: Modulation
Recording analog-format broadcast or cable television
Digital Television: what Is it?
Digital Television: broadcast and cable delivery
Direct satellite television
Digital video streaming over internet
Digital cable: premium services and conditional access
Digital cable: video-on-demand
Digital cable: switched broadcast
Recording digital television
Display devices (CRT; LCD; plasma; OLED)
Video connections (analog; DVI/HDMI; DisplayPort)

I.1 Television: video plus audio

¶ **1.** Television involves the remote delivery of a moving picture plus sound. It is accurate to think of the *sound* as continuous; however, the *picture* is captured, and then delivered, as a succession of still images, at a rate fast enough that the viewer perceives a scene of continuous motion.[2]

¶ **2.** Television is distinguished further, of course, by the transmission of this movie-like content to the remote viewer. Originally this was carried out exclusively by terrestrial transmission, via radio waves, to the viewer's antenna and television set. Over time, other methods of transmission have been added – electrical cable,[3] optical fiber, direct satellite transmission via microwaves, and, of course, the Internet – along with recording methods such as magnetic videotape (Betamax, VHS, D-VHS), and optical discs (Laserdisc, VideoCD, DVD,[4] HD-DVD, Blu-Ray, and others).

I.1.1 The audio

¶ **3.** The *audio* portion of television is perhaps more easily understood, as it differs little from ordinary sound-recording techniques. A microphone converts the instantaneous sound pressure variations into an electrical signal; that is, it creates as its output an electrical voltage that at each moment is proportional to the pressure of the sound wave to which it is exposed. Contemporary audio recording and delivery usually employs two or more microphones, creating "stereo" sound (i.e., two channels), or multichannel sound (e.g., "5.1 channel sound").

¶ **4.** Traditionally these signals were processed, stored, and delivered by "analog" methods, which means simply that the signals were treated as smoothly varying voltages as they passed through the electronic innards of the amplifiers, recorders, modulators, and so on.[5] Contemporary "digital" technology does it differently: almost as soon as possible, the microphone's signal (the varying voltage that represents the sound) is converted to a succession of numbers (it is *digitized*), and everything that follows is some form of arithmetic on this torrent of numbers that come tumbling out. Only at the final stage – recreating the recorded sound for the listener – is the digital representation converted back to an analog voltage, and then, in the loudspeaker, to a reproduction of the original sound pressure wave.

¶ **5.** Just to give a sense of the quantity of numbers involved, in the standardized recording technology of the

---

[1] Following the stylistic conventions of that audience, the paragraphs are numbered sequentially.

[2] For conventional cinema-style movies, the rate is 24 frames/second; television in the United States uses a rate of approximately 30 frames/second.

[3] Known technically as *coaxial transmission line*.

[4] "Digital Versatile Disc."

[5] Common analog recording technologies, now nearly obsolete, include the vinyl record (where the audio signal waveform is embossed as small displacements of a fine groove), and the audio cassette tape (where the audio signal waveform is recorded as patterns of magnetization on a thin layer of magnetic oxide coating on a flexible plastic tape).

compact disc (CD), the instantaneous sound is *sampled* at a rate of 44,100 times per second (in both stereo channels simultaneously), and each such sample pair is converted ("digitized") to a 16-bit binary number.[6] So the bits are tumbling out at a rate of $2 \times 44{,}100 \times 16 = 1{,}411{,}200$ bits per second, or nearly 100 million bits per minute.[7]

¶ **6.** One might ask why any sane person would want to deal with such a quantity of numbers, when the original analog representation of the sound was so much simpler – just a pair of voltages that were varying at most 20,000 times per second.[8] The reasons are several, but they boil down to the contemporary ease and economy of digital processing, combined with the higher efficiency and quality of storage and transmission of audio (and video) that has been properly digitized. To get a sense of those advantages, one need only marvel at the gorgeous images transmitted daily from planetary probes visiting Mars and Saturn – images that are free of "snow" and other artifacts irreparably added to analog transmission by the effects of unavoidable electrical interference – to appreciate the benefits of error-free digital transmission. And, to get a sense of the density of digital storage, we note that a contemporary 5″ optical disc (dual-layer Blu-ray disc) holds 80 hours of CD-quality audio, or ten times that amount if modestly "compressed," compared with a mere hour's storage of analog audio on the 12″ vinyl LPs of yesteryear.[9]

I.1.2 The video

¶ **7.** The *video* is by far the more complicated part of television. The challenge is to reproduce a scene with motion, in color, while preserving adequate fidelity and absence of artifacts. And this must be done within the resources of the storage and delivery channels – that is, with finite disk storage and speed, and with finite transmission (via broadcast tower, cable, Internet, or satellite) *bandwidth*.[10]

---

[6] That is, a number ranging from −32768 to +32767, those bracketing the "full-scale" range of the recorded sound.

[7] The *recorded* bitrate is roughly triple this figure because of coding, error-correction redundancy, and the like.

[8] Or 20 kHz, the upper limit of human hearing; and that only for one of relative youth, such youth further possessed of sufficient wisdom to avoid deafening rock concerts.

[9] And a contemporary 3 terabyte 3″ magnetic hard drive that you can hold in your hand holds yet another factor of 60, or 50,000 hours of excellent quality (128 kbps AAC) compressed stereo audio; that's 15 years of 40-hour per week music!

[10] *Bandwidth* refers to the range of frequencies that can be carried on the cable or other transmission medium. It is technically accurate to think

¶ **8.** Video systems begin with a camera that has an electronic sensor (analogous to a digital camera) and that converts the two-dimensional color scene on that sensor into a succession of *frames*, each of which represents the image at those successive times (for US TV, the rate is approximately 30 frames per second). In traditional *analog* television, the two-dimensional image is converted into an electrical signal by the following method: imagine a single frame, that is, a still picture. To keep it simple, imagine further that it is monochrome; that is, "black and white."[11] We begin at the upper left, and move horizontally across the picture, generating an electrical voltage proportional to the brightness at each point as we pass by. When we reach the right-hand border, we jump back to the left edge, continuing with another horizontal path, slightly below the first. See Figure I.1. We continue in this way until we reach the bottom right-hand corner, at which time we have scanned the entire frame once, in what is known as a *raster* ("grid" in German) pattern.[12]

Figure I.1. A static two-dimensional image is "raster scanned" to create a video waveform (Figure I.2) representing intensity along the scan lines.

¶ **9.** What we have done, then, is to generate an elec-

---

of this, for example, as the range of stations on the radio dial that could be carried with fidelity by a single electrical cable (or other medium). The term is sometimes used loosely to refer to rate of data transfer.

[11] Or, more accurately, *grayscale*.

[12] Traditional standard definition television (SDTV, usually called "NTSC," for National Television System Committee, and going back to the 1940s) in the United States divides the whole picture into 480 horizontal lines, along each of which roughly 640 features (picture elements, or *pixels*) could be resolved; a computer user would not be terribly impressed – he or she would say that standard NTSC television has only "VGA" resolution (i.e., 640×480).

trical representation, in time (a varying voltage proportional to brightness at each point in the image) of a single two-dimensional image; that is, we've converted a two-dimensional image into a one-dimensional output voltage. See Figure I.2.

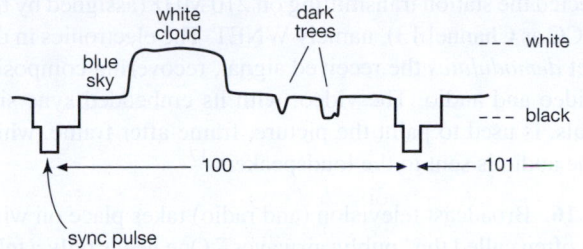

Figure I.2. A portion of the video waveform from Figure I.1, representing one of the 240 horizontal scan lines.

¶ **10.** This time-varying voltage is called the *video* signal, and it is the first step in creating a television image. In traditional NTSC analog television, this signal was transmitted by analog methods, after a process called *modulation* (more later), and was recovered and used by the television set to paint the picture on the screen, performing the same raster scan (left to right, top to bottom). Each frame follows in sequence, presenting a succession of 30 pictures per second on the television set's viewing screen.[13]

¶ **11.** To complete the video signal, it is necessary to add some synchronizing information, so the television set knows when to begin painting a frame, and also when each horizontal line begins. In traditional NTSC television this is done by adding a horizontal *sync pulse* at the beginning of each horizontal line, which is just a short[14] voltage pulse that, if it were in the middle of a picture, would be interpreted as "blacker than black." See Figure I.3 The television set detects these pulses and uses them to synchronize its scanning across each line. Likewise, a unique *vertical sync pulse* is transmitted for each field which informs the

---

[13] To complicate things, NTSC uses a method known as *interlacing*, in which a coarser raster – omitting every other horizontal line – is performed at twice the rate. Thus, in standard NTSC US television, the viewer sees 60 pictures ("fields") per second, each of which has only 240 horizontal lines; two such fields, with their interlaced lines, form one complete 480-line frame. This is sometimes called a "480i" format, to distinguish it from formats with higher resolution (e.g., high-definition TV, HDTV, with 720 lines or 1080 lines, or "4K Ultra HDTV," with 2160 lines), or from those without interlacing (which are known as *progressive*; e.g., *720p*).

[14] About 4.5 millionths of a second.

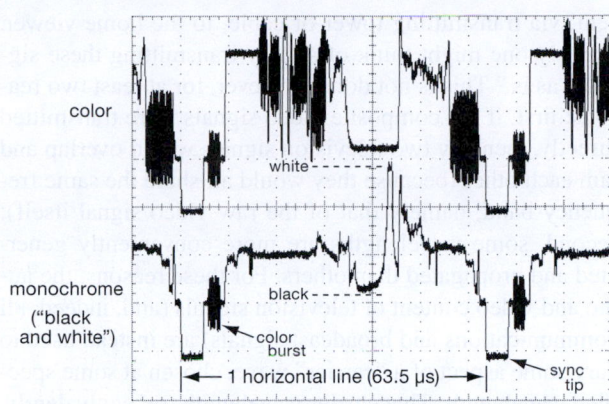

Figure I.3. Composite analog video signal of one horizontal line, framed by horizontal sync pulses. The brightness ("luminance") is represented by its amplitude. Color is accommodated by adding a modulated 3.58 MHz "chrominance" subcarrier, whose amplitude represents degree of saturation and whose phase encodes the colors.

television set when to return to the top to begin painting the next field or frame (see Figure I.4. The complete video picture signal, with its added sync pulses, is called *composite video*.

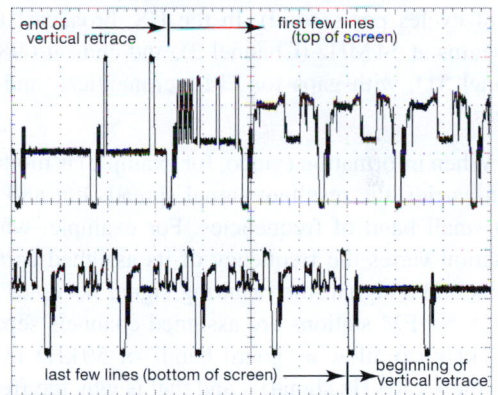

Figure I.4. The vertical retrace (beginning of a new field) is signaled by a set of tailored sync pulses, the first and last of which are shown here.

I.2 Combining and sending the audio + video: modulation

¶ **12.** Continuing for the moment with traditional NTSC television (as opposed to *digital* television, whose standards are known as *ATSC*, set by the Advanced Television Systems Committee, and which will be explained later), the composite video, along with the audio, must now be sent, via transmitting tower or cable, to the home viewer. Naively one might think of simply transmitting these signals "as is." This is not done, however, for at least two reasons: first, if the composite video signals were transmitted directly, then any two television signals would overlap and jam each other (because they would all share the same frequency band, namely that of the raw video signal itself); second, some wavelengths are more conveniently generated and propagated than others. For these reasons, the audio and video content of television signals (and, indeed, all communications and broadcast signals) are instead used to vary some aspect of a "carrier" wave, chosen at some specified wavelength. That carrier wavelength (or, equivalently, frequency) defines the "channel"; and the process of impressing the information (video and audio) onto the carrier wave is known as *modulation*.

¶ **13.** Radio stations use the same technique: AM stations vary the strength (amplitude) of the carrier (hence "amplitude modulation"), whereas FM stations vary the frequency ("frequency modulation"). The carrier frequency itself defines the channel: in the US, AM stations are assigned to carrier frequencies between 520 and 1710 kilohertz (kHz, thousands of cycles per second), while FM is assigned to the band of carriers from 88 to 108 megahertz (MHz, millions of cycles per second). In the US, broadcast television begins at 54 MHz (Channel 2), and ends at 698 MHz (Channel 51), with gaps for FM, aeronautical, and other services.[15]

¶ **14.** When information (video, for example) is modulated upon a carrier, the resultant signal spreads out and occupies a small band of frequencies. For example, when an FM station varies the frequency of its assigned carrier to carry its audio signal, the resulting signal occupies about 150 kHz. So FM stations are assigned channels separated by 200 kHz (to allow a "guard band" of 50 kHz in addition to their 150 kHz signal) – and that is why the frequencies of FM stations always end in an odd number after the decimal point (for example New York City's WNYC is at 93.9 MHz), ensuring a minimum spacing of 0.2 MHz (= 200 kHz).

¶ **15.** Traditional analog NTSC broadcast television used a variant of AM for the picture signal (composite video) and, separately, FM for the sound signal.[16] The assigned TV channels are spaced apart by 6 MHz, each station being

permitted to occupy nearly that amount, after allowing for a small guard band. Television sets "know" the frequencies allocated for each channel and tune to the correct frequency when the user chooses the channel number. For example, if (during television's analog era) one tuned to Channel 13 in the New York City area, the television set's electronics selected the station transmitting on 210 MHz (assigned by the FCC as Channel 13), namely WNET. The electronics in the set *demodulates* the received signal, recovering composite video and audio. The video, with its embedded sync signals, is used to paint the picture, frame after frame, while the audio is sent to the loudspeakers.[17]

¶ **16.** Broadcast television (and radio) takes place on what is often called the "public airwaves." One needs only a television set (or radio) and an antenna to receive these over-the-air (OTA) public transmissions. Although some countries require licensing of receiving devices (radios and television sets), in the United States the broadcast services are freely available to anyone within range of a transmitting tower.

¶ **17.** Depending on the distance and path from the broadcast station to the viewer, the "antenna" can be as simple as an indoor "bowtie" or pair of "rabbit ears," or as elaborate as a roof-mounted multi-element structure. Whatever its form, the antenna's function is to create an electrical signal on the feedline, induced by the speed-of-light broadcast signal passing by the antenna's site. Receiving antennas intended for broadcast television are designed to work over the range of frequencies used by broadcasters (see ¶13); thus the electrical signal delivered to the television set includes multiple stations, and it is the job of the TV tuner to select and process the channel to be viewed.

¶ **18.** *Cable* television sends traditional analog TV channel signals in almost exactly the same way as broadcast. An evident difference, however, is that the channelized signals are received at the viewer's end from a coaxial cable (rather than being received by the viewer's television antenna), and then connected to the television set directly (i.e., to its normal antenna connector on the rear). Alternatively, for additional cable services (such as premium channels) the incoming cable connects to a "set-top box" (STB) provided by the cable company, the output of which is connected to the viewer's television set (or flat-screen monitor,

---

[15] You can download a gorgeous multicolor wall-sized spectrum allocation figure from http://www.ntia.doc.gov/files/ntia/publications/spectrum_wall_chart_aug2011.pdf.

[16] That is, the picture and sound signals are carried simultaneously on a

pair of designated carrier frequencies within the single assigned television channel.

[17] In this primer we have ignored details associated with reproduction of *color* (versus black and white).

projector, etc.; see ¶56ff).[18] The channel *frequencies* are also somewhat different, with Channels 2–13 chosen the same as for terrestrial broadcast, but with the remaining channels reassigned to eliminate gaps; this can be done because the cable is a private world of its own, isolated so as not to interfere with, or be interfered by, other broadcast services.

¶ 19. A third difference is that some of the cable content is delivered as a subscription, for which the viewer pays additional fees; examples are premium services such as Home Box Office (HBO). These require some method for permitting or denying viewable delivery of selected channels or programs. Continuing for the moment with analog cable (whose days are numbered!), this can be done in several ways: the simplest is by installing filters (to block unsubscribed channel frequencies) at the utility pole, where the subscriber's cable splits off from the trunk running along the street;[19] A more sophisticated method involves scrambling the cable-borne analog signals of subscription programs,[20] and then instructing the set-top box (via digital communication from the cable provider to the individual STB) as to which programs may be unscrambled.

¶ 20. It is worth noting that cable companies have been required to carry the broadcast stations in their area, normally as analog cable channels.[21] Each such program occupies a cable channel (frequency). However, they may distribute additional services via digital methods ("digital cable") on additional channels (frequencies), which they much prefer; that is because, with digital methods, it is possible to carry up to *ten* NTSC-quality programs (i.e., SDTV, for Standard-definition TV) on a single channel. This is called *multicast*: the ability to carry multiple *programs* on a single channel (i.e., frequency). And, note that a cable can carry more than 100 such carriers – permitting more than 1000 simultaneous programs.

¶ 21. Previewing some additional characteristic of digital television: digital cable permits flexible subscriptions, with a program being authorized on-the-fly (e.g., pay-per-view, or video-on-demand). It also provides a natural mechanism for content protection via encryption. It allows for interactive participation, via a reverse channel back to the cable provider. It permits the delivery of high-definition content, with more than the 480 lines of NTSC (up to 1080 lines, at the highest quality currently supported). Finally, it provides a natural way to time-shift, pause, or replay live programs, via computer-type hard-disk storage.

¶ 22. Analog broadcast was sent into retirement in the United States in June of 2009, and all television broadcast delivery is now done by digital methods (more to follow). This conversion-to-digital process is taking place worldwide and will likely be complete by 2020 or earlier.

I.3 Recording analog-format broadcast or cable television

¶ 23. Video recording was complex and costly (and therefore confined to the broadcast studios) until 1975, when home video-recording devices were introduced in the US by Sony ("Betamax") and its competitors ("VHS," for video home system). These devices replicate the "front end" of a television set, to recover video and audio from the incoming signal (broadcast or cable), and use a clever spinning tape head arrangement[22] to capture on magnetic tape a reasonable replica of an NTSC analog television program. The technique is entirely analog (no digitization, no numbers) and records only onto special video-tape media (no computer media, no hard disks, etc.), as a magnetic recording (analogous to an analog audio tape recording; see the footnote at ¶4).

¶ 24. Videotape technology has been upstaged by digital alternatives such as optical disc recording (most famously in the form of DVDs and Blu-Ray discs – whether sold with prerecorded content, or recorded with a disk recorder), which creates a permanent copy of the video material; or by recording to a computer-type hard-disk drive (hdd), where the video copy is stored as a computer file. These digital methods require that the program material be converted from analog to digital form, if it is not already. (This is done internally and automatically in devices like TiVo®

---

[18] For better picture quality, the latter connection is usually made not to the set's antenna input (called "RF," for Radio frequency, meaning the modulated channels discussed above, in ¶¶12–15), but to special audio–video inputs, with names like *s-video, component video, composite video, DVI,* or *HDMI*; see ¶64ff. The latter pair are *digital* connections, discussed below in connection with digital TV.

[19] Vintage cable subscribers will remember calling the cable company to add a movie channel, whereupon a cable truck appeared, the cable guy went up the pole to fiddle with something (changing the filter), and, voilà, movies on your television!

[20] For example, by suppressing the horizontal sync pulses or inverting the video (interchanging black and white).

[21] Unless all subscribers are provided with STBs that can receive digital delivery.

[22] This is known as a "helical" tape head, which creates successive narrow slanted tracks across the slowly moving tape, each one holding one field of video. The use of a rapidly moving tape *head* eliminates the need for rapidly moving *tape*.

and other personal video recorders.) Digital television and digital video are discussed next.

I.4 Digital television: what is it?

¶ 25. Just as an *audio* signal can be digitized (i.e., its instantaneous amplitude is measured, at rapid intervals, and converted to a succession of numbers), and subsequently transmitted, stored, or processed (¶¶4–6, above), so it is possible to digitize the *video* signal that represents successive frames of picture. Although one could imagine simply sending the digitized version of traditional NTSC as "digital TV," in practice one takes advantage of the enormous processing finesse of contemporary digital electronics to economize by *compressing* the raw video signal to a small fraction of its native size before it is delivered. The use of compression, along with the fact that a digital signal is "just numbers," permits the delivery of multiple programs on what would otherwise carry just a single video signal (program), typically by a factor of five to ten.

¶ 26. There are several reasons for this improvement. One is the ability to detect and correct transmission errors by numerical techniques, allowing one to operate with received signal levels that are close to the "noise" (from interference, or signal loss owing to range or obstructions); with purely analog transmission a large received signal/noise ratio is necessary to reduce the visible effects of noise ("snow") to acceptable levels.

¶ 27. A second reason is the spectral efficiency of digital transmission – or, more accurately, its improvement compared with the *inefficiency* of analog signaling. This can be seen in Figures I.5 and I.6, a pair of spectra taken directly off the home antenna of one of the authors in March of 2009, a time during the switchover to digital when both analog and digital broadcasts were taking place (see also Figure I.7).

¶ 28. Compression aims to reduce by a large factor (tenfold or more) the quantity of numbers needed to describe the succession of picture frames, without significantly degrading the image quality. This seemingly impossible task takes advantage of redundancies in a moving picture, and of limitations in human visual perception.

¶ 29. Contemporary digital-video compression is a rich and mathematically complex subject, the result of enormous effort in the applied mathematics and electrical engineering communities over the last several decades. But the basic tricks are easy enough to understand. The pro-

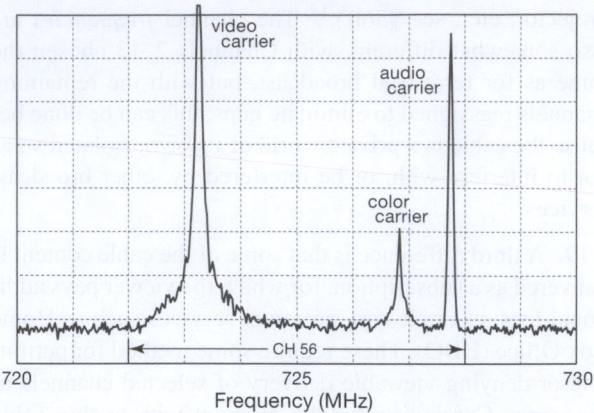

Figure I.5. The spectrum of 6 MHz-wide analog Channel 56 in Boston, as seen in May 2009. The video information resides in the sideband tails, while most of the transmitted power is wasted in the non-informational video carriers.

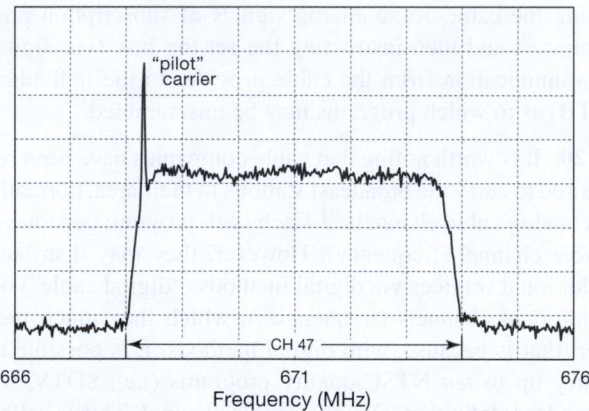

Figure I.6. Digital Channel 47, seen also in May 2009, fills its 6 MHz spectrum allocation with digitized video. It carries five times as many programs, with comparable (or better) picture quality.

cess begins by exploiting the fact that portions of an image near each other tend to be similar; so one can encode and send the (smaller) *differences* of brightness and color from a set of reference points, rather than the full description of brightness and color at each point. Likewise, successive frames tend to be similar, so one can define a sparse collection of index frames and send only the differences for intervening frames.[23] A further trick exploits the fact that the image often contains moving objects or a panning camera; so it is efficient to calculate "motion vectors" predicting

---

[23] More precisely, it is the corrections from an *interpolated guess* between index frames (or reference points within a frame) that are sent.

the approximate motions, and then send only the (smaller) corrections from the predicted values.

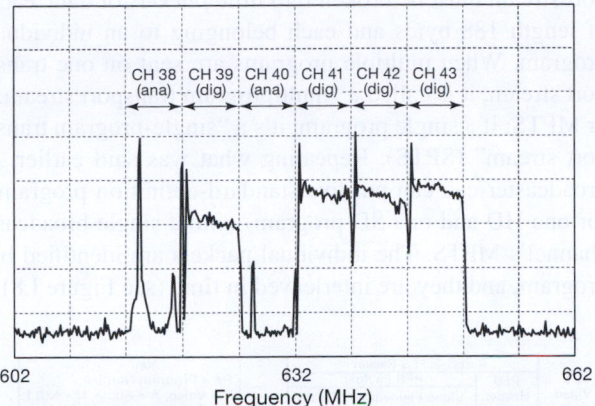

Figure I.7. A country in transition: RF spectrum of Channels 36–45 (each permitted 6 MHz of spectrum) as seen on our antenna feedline in Cambridge, Massachusetts, on May 6, 2009. Analog Channels 38 and 40 each carry one standard-definition (SDTV) program, whereas Digital Channels 39 and 41–43 can carry up to five SDTV programs each (though it's more common to see one HDTV and one SDTV program).

¶ **30.** These methods greatly reduce the needed bitrate (number of numbers per second), and they do so *without any loss of picture quality whatsoever* – they are "lossless." That is because the original digital image can be fully and exactly recovered by applying the numerical differences in the reverse order. However, further bitrate reduction is desirable (and often necessary), and this is accomplished by *lossy* compression. This consists essentially of discarding the less-important image information (from a psychovisual standpoint); the tradeoff is a somewhat degraded image (the degree of degradation depending on the degree of compression), which, however, can differ from the pristine original in ways that are hardly perceptible to the viewer.[24] The mathematics involves methods with names like discrete cosine transform, variable quantization, and Huffman coding; but the bottom line is that these methods permit a large reduction in bitrate with a relatively small reduction in perceived image quality.[25]

¶ **31.** The tradeoff of image quality with bitrate is gradual, and somewhere in the process a decision is made as to the desired final bitrate.[26] A major constraint is imposed by the fact that both digital broadcast and digital cable television in the US is sent on channels that conform to the same 6 MHz channel spacing that has been used for television since the 1940s. In practice (see ¶35, below) it is possible to send about 20 million bits per second (Mbps) on an over-the-air digital-broadcast channel, and nearly double that on a digital cable or satellite channel. A typical compressed bitrate for over-the-air NTSC-quality (SDTV) digital video is about 4 Mbps; thus digital-broadcast television stations are able to combine up to 5 or so NTSC-quality programs on a channel. (Recall that a frequency "channel" is no longer a single "program," because of multiplexing. More on this beginning at ¶33, below.) High-definition (HDTV) content requires nearly the full broadcast bitrate, so only one HDTV program can be broadcast on a channel. By contrast, cable or satellite systems, which are not constrained to MPEG-2 compression, are able to combine as many as eight HDTV programs onto one channel when using efficient H.264/MPEG-4 encoding. The second revision of the broadcast standard (ATSC 2.0) incorporates these more efficient codecs, as well as a host of transport and delivery enhancements that are aimed at mobile and interactive viewing, thus allowing over-the-air broadcasting to compete with services available on the internet.

¶ **32.** It is worth admiring the impressive bitrate reductions that these methods are achieving: a simple calculation[27] shows that digitizing an HDTV program without any compression would produce a bitrate of roughly 1000 Mbps, whereas contemporary compression methods reduce this to a modest (and deliverable) 20 Mbps, a 50-fold reduction! And comparable reductions are routinely achieved with SDTV.

---

[24] If such effects are noticeable, they are called *compression artifacts*; these are sometimes seen in over-compressed "jpeg" still photographs, as the patchy blocks or the "mosquito noise" around edges. Similar considerations apply to lossy audio compression, for example highly compressed "MP3" music files.

[25] The video-compression recipe currently used for all digital TV broad-

casting in the US is named "MPEG-2" and described in the Advanced Television Systems Committee documents A/53 and A/54 (see www.atsc.org). An improved set of compression methods is incorporated in the set of standards known as MPEG-4; these are widely used by the cable and direct broadcast satellite services, as well as for video streaming over the internet.

[26] Which is permitted to vary, as program content changes. This is known as *variable bitrate*, or VBR, as distinguished from *constant bitrate*, or CBR.

[27] Bitrate ≈ 1080 lines × 1920 pixels/line × 30 frames/second × 16 bits/pixel = 995,328,000 bits/second.

I.5 Digital television: broadcast and cable delivery

¶ **33.** Over-the-air digital television broadcast and digital cable television both use traditional frequency "channels," upon which they put a stream of numbers (the compressed video described in ¶¶25–32, plus the associated digitized audio[28]), instead of the continuous analog waveform that was used in traditional NTSC television. Because of the economical bitrates produced by compression, there is adequate capacity on a single cable (or broadcast) channel frequency to accommodate several simultaneous programs. This is called multicasting, and permits up to four or five SDTV programs (a "multiplex") to be carried on a single broadcast channel frequency.[29] One can think of these as subchannels.[30]

¶ **34.** For either OTA digital broadcast or digital cable, the set-top box (STB) or equivalent hardware within the television set receives the multiple channel frequencies, each with its multiplex of programs. The STB or television knows the program assignments within each channel and is able to pull out the subchannel that the viewer selects, which it identifies by assigning a "virtual channel number." That is what the viewer chooses – it is displayed on the STB and on the screen during selection. For example, the viewer might select HBO, which is assigned a virtual channel (e.g. 82) and which might actually be just one of ten subchannel programs carried on one digital cable channel frequency. The STB then captures the HBO stream, decrypts and decodes its MPEG-2 encoding, and converts it to displayable video for a television monitor (or flat screen, etc.).

¶ **35.** In more detail, and in the language of digital television engineering, the delivery channel (digital broadcast or digital cable) is called the "transport stream," which can be

thought of as a data pipe carrying some 20 Mbps (broadcast) or 38 Mbps (cable) in each frequency channel.[31] The ATSC specification dictates that the data put onto the transport stream must be broken into little *packets* of data, each of length 188 bytes and each belonging to an individual program. When multiple programs are sent on one transport stream, it is called a "multiprogram transport stream," or MPTS; if a single program, it's a "single-program transport stream" (SPTS). Repeating what was said earlier: a broadcaster can can put five standard-definition programs (or one HD and one SD program) onto a single broadcast channel's MPTS. The individual packets are identified by program, and they are interleaved in time (see Figure I.8).

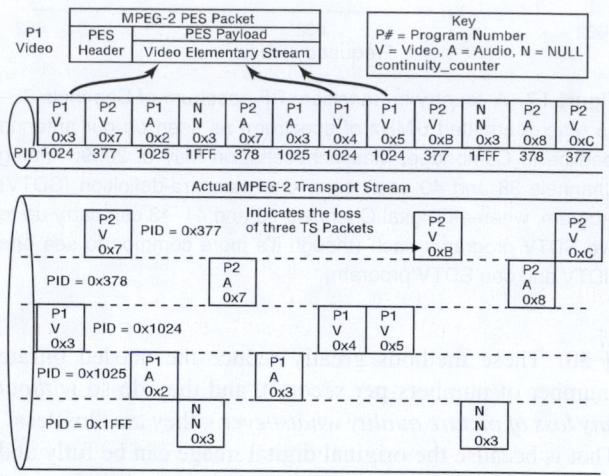

Figure I.8. Multiple programs can be interleaved into one digital transport stream, as a "multiprogram transport stream." Their individual video and audio packets are tagged with program identifiers (PIDs), by which they can be selected and reassembled (adapted from Figure 7.1 of ATSC Doc. A/54A, courtesy ATSC; readers are recommended to reference the current version of the standard or recommended practice available on the ATSC website).

¶ **36.** To put it another way, the several programs that will be put into a multiprogram transport stream (a "program multiplex") are cut into short pieces (packets, about 40 μs in length for digital cable), tagged with unique identifiers (called PIDs, for "program identifiers"), and then interleaved with the pieces (packets) from the other programs

---

[28] And also some "conditional access" (CA) information that enables legitimate subscribers to decrypt and view protected content; see ¶47.

[29] *Cable* delivery is more efficient, and permits as many as ten SDTV programs on a single channel.

[30] Because the cable (or broadcast channel) has a fixed total bitrate, the bitrates of the individual programs that are being multiplexed must be adjusted such that their total combined bitrate matches the channel capacity. This is called *bitrate grooming* and involves *null padding* (adding nulls, to increase a program's bitrate), on-the-fly compression (to reduce a program's bitrate), or even time shifting of program content (to prevent unlucky alignment of peak bitrates of the various programs). Digital television packets include "presentation time stamps," so it's OK to move things around a bit as they flow through the various digital pipes on their way to the television screen.

[31] The disparity has to do with the particular modulation schemes used: for broadcast it is called "8-VSB," whereas cable uses the more efficient "256-QAM" (pronounced "quahm"), thereby exploiting cable's better signal-carrying properties to carry roughly twice the information content.

that share the same transport stream. At the STB or television set the packets belonging to the selected program are identified ("filtered") by looking for their PIDs, and then reassembled into a single program transport stream to be decoded and displayed. In Figure I.8 there are two programs (P1 and P2), each with video (V) and audio (A), with their corresponding PIDs (1024, 1025, 377, 378); they are shown as the interleaved multiprogram stream at the top, and as filtered into their respective single-program streams at bottom.

I.6 Direct satellite television

¶ 37. Satellites provide an alternative to over-the-air or cable/fiber delivery of television programming, and satellite delivery is particularly welcome in areas not served by wired broadband connections. This is known variously as direct-to-home (DTH), direct broadcast satellite (DBS), or broadcast satellite services (BSS), and exploits (usually) satellites in the geostationary constellation, i.e., in the equatorial "Clarke" orbit[32] of radius 42,200 km, where a satellite's period matches Earth's rotational period of 23h56m4s.[33] As surprising as it may seem, a single satellite with a ∼100 W transmitter can deliver a half-dozen high-definition programs (or 30 "standard-definition" programs) simultaneously to small dish antennas on houses everywhere in the continental US. Typical direct-broadcast satellites are equipped with a dozen or more such "transponders," and the receiving dish antennas use multiple "feeds" (as many as four, for DirecTV or DISH Network) to point at several satellites, making available many hundreds of television programs.

¶ 38. Early DBS systems used a 4 GHz (C-band) downlink, and required large dishes (>3m diameter) and expensive RF electronics. Contemporary systems operate around 12 GHz (Ku-band), with mass-produced oval receiving dishes (typically 0.6m×0.8m) that incorporate several low-noise RF amplifiers, each with local oscillator, downconverting mixer, and IF amplifier in an integrated LNBF (low-noise block-downconverter plus feed) unit that sits at the focus of the parabolic dish. That's an impressive amount of hardware for $100.

¶ 39. One may be puzzled by the peculiar orientation of home dish antennas – why are they sometimes pointed so low that they seem to be aimed at or below the horizon? There are two parts to the answer: first, the constellation of geostationary satellites spans an arc across the southern sky, populated worldwide with some 200 satellites; over the longitudes of the US alone there are some 35 satellites parked[34] in geostationary orbit. The line of satellites across the southern sky dips down to the horizon at its eastern and western ends.[35] Second (and technologically more interesting), the geometrical arrangement of the receiving dish is what's called an "offset-feed paraboloid." That is, the little conical feeds are offset below so they do not block the incoming signal. This makes the dish appear to be pointing some 25° lower than it is, thus the explanation for the apparently "subterranean" satellites. This peculiar arrangement is used to eliminate blockage of incoming signal by the feedhorns, and also to reduce the encroachment of thermal radio noise that is emitted by the surrounding environment – another bit of thoughtful design that enables successful direct satellite broadcasting.

¶ 40. The individual transponder RF channels are 27 MHz wide, with as many as 32 such transponders on a satellite, which (with guard bands) adds up to a total downlink bandwidth of ∼1000 MHz. Satellite transponders typically use phase-shift keying (QPSK, 8PSK), with downlink bitrates of 40 Mbps per channel, adequate to deliver a half-dozen or more 1080i HD programs with efficient H.264/MPEG-4 encoding. The transponder channels are divided into two sets, one of each circular polarization at each transponder frequency, so that the total downlink spectrum occupied by a given satellite is about 500 MHz. This is downconverted in the LNBF to a pair of 500 MHz-wide IF bands, centered at 1.2 GHz and 1.9 GHz.

¶ 41. Rather more hardware resources are expended on the *up*link, with steerable dishes of ∼10 m-diameter class (Figure I.9) illuminating the uplink receiving dishes at the satellite, with transmitted power of the order of hundreds of watts per transponder; dishes of that size produce diffraction-limited beam diameters of about 0.15°

---

[32] See "Extra-terrestrial relays – can rocket stations give world-wide radio coverage?" [Sir] Arthur C. Clarke, *Wireless World*, October 1945.

[33] Ha! You thought it was 24 hours. Not so – we're in orbit around the sun, which adds an extra 4 minutes (1/365th part of a day) to Earth's true rotational period (the *sidereal* day) to arrive at the 24-hour *solar* day (the average time from solar noon to solar noon).

[34] Hardly "parked," of course – they are in equatorial orbits, whizzing around the Earth at nearly 7000 miles per hour, to keep up with Earth's rotation.

[35] You can get a good sense of the satellite arc with with mobile apps like DishPointer or Satellite Locator: when you point your smartphone into the sky it shows the satellites as bright red circles (with their locations) on a red arc, superimposed on the camera view (with obstructing trees, etc.) seen by the mobile device.

($\theta \approx \lambda / D$), small enough to prevent illumination of a satellite in an adjacent orbital slot, but requiring some active steering to maintain alignment on the desired satellite. Direct-broadcast providers such as DirecTV and DISH Network like to position their satellites close enough along the Clarke belt so that a single dish with several feeds can capture the downlinked signals from several of them; for example, both DirecTV and DISH Network offer an oval dish with three feeds, targeting satellites separated by 10° (currently at 110°W, 119°W, and 129°W longitudes).

Figure I.9. A few of the several dozen transmitting antennas at EchoStar's Network's uplink facility in Cheyenne, Wyoming.

¶ **42.** You hook the coax cables from the receiving dish to a set-top box (usually with DVR), similar to what's used for cable TV, but which is designed to power the LNBF electronics up there on the roof, to select from the dish's several feeds and polarizations, and to receive the intermediate-frequency (IF) bands coming down from the dish on standard $75\,\Omega$ video coax (quad-shielded RG-6, usually).[36] As with cable TV, the content providers control your available programming, via subscription-enabled decryption. Satellite systems are not bound by over-the-air terrestrial standards (e.g., MPEG-2 encoding), and they generally use more efficient schemes such as MPEG-4. When compared with cable or fiber television service, the inability to target millions of subscribers individually (along with the absence of a reverse uplink channel to the satellite) limits

the possibilities for interactive services such as video-on-demand.

I.7 Digital video streaming over internet

¶ **43.** With steady improvements in internet bandwidth (i.e., speed), it has become practical to deliver video (and associated audio) through the same internet infrastructure that serves personal computers and mobile devices (cellphones, tablets) with their email, web browsing, and so on. Some familiar examples of internet video streaming are news services such as CNN, government services such as NASA TV, movie streaming services such as Netflix and Hulu, and peer-to-peer services such as Skype. Just as with broadcast or cable delivery, the ultimate payload is a stream of numbers that constitute the video and audio content, in some efficient compressed format that takes advantage of sophisticated encoding schemes with names like "H.264" (also known as AVC, for advanced video coding), one of the current favorites. At the viewer's end the digital content is decoded[37] to recover the video and audio. For some services (e.g., Skype) a dedicated client program must be installed, whereas for others (e.g., NASA TV or CNN) a standard internet browser (such as Internet Explorer, Safari, or Firefox) suffices.

¶ **44.** Compared with delivery via broadcast or cable, however, internet delivery of the time-critical video data presents some unique challenges. That is because data sent over the internet (via "Internet Protocol," IP) travels as independent packets of data, each typically some thousand bytes in length, and each including headers that specify its destination (IP address). Packets make their way through the multiply-connected nodes of the internet, and (usually) reach their destination. But there is no reserved pathway for a stream of packets (as there is for the "circuit-switched" architecture of the telephone system), and no guarantee of speedy delivery, sequential delivery, error-free delivery, or, indeed, of delivery of any sort. Various schemes are used to circumvent these evident deficiencies, for example by requesting retransmission of missing packets (they are numbered) or of corrupted packets (they include error-revealing "checksums"). These work well, and a data file that is downloaded via "TCP/IP"[38] is effectively guaranteed to be bit perfect.

---

[36] There may be an intervening module or two, with names like "multi-switch" or "node," to deal with selection of polarizations and IF-band stacking.

[37] Hence "codec," a contraction for coder–decoder, usually appended to the name of the compression scheme, e.g., "the H.264 codec."

[38] Transmission Control Protocol/Internet Protocol, a universally used standard for transfer of data that needs to be transported and reconstituted without error.

¶ **45.** Because of this disorganized "packet switching" of internet data, streaming video may suffer interruptions or intervals during which the average delivery rate is reduced. For this reason it is common for the receiving end to store ("buffer") a few seconds of video beyond what is currently being displayed; and if the internet delivery speed is inadequate (as evidenced by buffer underruns), the sender will reduce the data rate (thus delivering lower-quality video). Contemporary broadcast-quality high-definition TV (known as 1080i, meaning that the picture consists of 1080×1920 pixels, delivered 30 frames/sec) requires upward of 2 megabits/sec of download speed, available now in most homes with a broadband internet connection (cable, optical fiber, or telephone-line DSL[39]). That speed is well within the capability of wireless (Wi-Fi) connections, so high-definition video can be delivered to mobile devices such as laptop computers and tablets.

I.8 Digital cable: premium services and conditional access

¶ **46.** Cable television providers offer premium services, such as Showtime, HBO, and pay-per-view, for which the subscriber pays additional monthly fees. A subscriber whose subscription includes Showtime, for example, is able to view (and record, if the STB includes a DVR) programming delivered on Showtime channels. The cable provider needs a method to control each subscriber's access to the full channel lineup. Although each customer has a cable coming into the home (and therefore one could imagine that different content is sent to each home), in fact the same signal is sent to all the homes within a neighborhood group, known as a "service group."[40]

¶ **47.** To limit access within the full suite of distributed programming, the cable provider includes *conditional access* information along with the video and audio. This is done by including "CA" packets, along with the usual V (video) and A (audio) packets that comprise one program within the multiprogram transport stream. The STB includes decryption hardware, and uses the CA packets to provide the key information to unlock the encryption that is imposed on the video and audio packets by the cable company.

I.8.1 Digital cable: video-on-demand

¶ **48.** How is it possible to provide program material specifically for an individual subscriber, for example "on-demand" delivery of a previously broadcast program, or of a movie, on a cable network that serves an entire city?[41] Such services go by names like "video-on-demand" (VOD), and are made possible by the fact that the cable provider is able to tailor the actual signals carried on its cables that go to different groups of cable subscribers in an area.

¶ **49.** A cable network in a metropolitan area is more complex than one might at first imagine: rather than a city-wide distribution of common program material, the network is organized into smaller groupings of *nodes* and *service groups*. A service group consists of segments of cable, typically running past no more than 500 homes, carrying identical material; those signals are fed into the cable at a "node" in the neighborhood, which in turn is fed via optical fiber from a more distant "hub" at which the provider inserts the group of channels that is to go to that particular service group.

¶ **50.** The trick to providing individual on-demand material, then, is first to ensure that there are at least a few extra channels (in addition to the standard lineup) available to carry such content; and, second, to divide up the city into many service groups, so that those extra channels can carry a different suite of on-demand content to the different service groups. For example, suppose there are five channels available for on-demand material and that each can carry ten programs (as described earlier); if a service group includes 500 houses, of which 200 are cable subscribers, then the cable provider can satisfy 25% of those subscribers with simultaneous on-demand programming (because its five extra channels, each sending ten custom programs, deliver 50 simultaneous programs).

¶ **51.** On-demand programming requires also a reverse channel for each subscriber, so that the subscriber can select programming, and also pause (or fast-forward or rewind) the material. However, these reverse channels need carry only a few simple commands up to the provider (as opposed to the high-bitrate video coming down in response), and are easily accommodated in an "interactive" uplink signaling band of the cable network.

---

[39] Digital subscriber line, technology for bidirectional digital transport over analog telephone lines.

[40] See ¶49, below, for more details.

[41] The term *unicast* is sometimes used to distinguish such individual delivery from *broadcast*.

I.8.2 Digital cable: switched broadcast

¶ **52.** Some cable providers use another service that exploits the flexibility inherent in the cable network's use of separate service groups, namely "switched broadcast." Switched broadcast delivers programs only when requested by viewers, on a service-group basis, as compared with delivering all programming to all customers. This allows the cable provider to offer more programming choices than could be carried simultaneously on the available number of channels.[42]

¶ **53.** Although switched broadcast and VOD use similar methods to deliver their content to a subscriber, it is worth noting a difference: switched broadcast content is delivered whenever that program material is normally scheduled, and not at the whim of the subscriber; what distinguishes it from normal program material is that it is not put onto the service group's cable at all, if no one in that group has tuned to the program. Once a subscriber has tuned that program (causing it to be sent to that service group, on a particular subchannel), that same program is present on the cable serving any additional subscribers in that service group, on the same subchannel. By contrast, VOD is delivered on-demand, at the time requested by the subscriber; likewise, it can be paused, or fast-forwarded, etc. (tasks that are performed on-the-fly by the cable provider, not by the set-top box). That is possible because VOD content is being sent on a particular subchannel on the subscriber's service group only, and is available for the requesting subscriber only (via encryption and user-specific "entitlement control," in the case of protected content). The terms *narrowcast* and *unicast* are sometimes used to refer to service-group-specific cable delivery of these two varieties of content, namely scheduled ("linear") material, and interactive user-specific on-demand material, respectively.

I.9 Recording digital television

¶ **54.** The conversion of analog material (audio and video of the real world) into digital form is hard work – but it makes the task of *recording* straightforward. That is because a single program received at the STB is, in essence, just a stream of numbers, which can be filtered from the multiprogram stream, assembled in a temporary memory "buffer," and written to a hard-disk file just like any computer file. In that sense, a set-top box with DVR is simply a special-purpose computer, with the usual processor, hard drive, etc., and having additional hardware to do the

special STB tasks – receive the cable signal, generate the displayable output, take control commands from the infrared remote "clicker," and so on. A typical contemporary set-top box with DVR contains a dual-processor chip, 64 MB program memory, a 160 GB (minimum) hard drive, and various video-related additional hardware (input tuner, video memory, display and audio drivers, etc.). Along with buffering and storing the video and audio content, the STB controls the access, decryption, and re-encryption of the (protected) video content.

¶ **55.** It is worth noting that any digital storage medium of adequate speed and capacity can be used to store digital video content; at the consumer level there are many "personal video recorders" (PVRs) that store programs on recordable DVDs or onto solid-state "flash" memory chips. There are also digital video tape recorders that can store both SDTV and HDTV onto a digital variant of VHS tape.

I.10 Display technology

¶ **56.** For over half a century television images were displayed with a cathode-ray tube (CRT), in which electrons emitted from a hot cathode and accelerated to potentials of kilovolts were deflected (usually magnetically) to paint a raster, at the video frame rate, on a phosphor surface coated on the interior of the viewing face of the evacuated tube. The intensity was modulated with a grid electrode near the cathode. Early CRTs were monochrome (black and white); color tubes used arrays (triads or stripes) of red, green, and blue phosphors, aligned with a metallic mask so that electrons from each of three electron-emitting cathodes (or, in Sony's Trinitron, steered from a single electron gun) struck only one color of phosphor.

¶ **57.** Cathode-ray tube displays worked; but they were heavy (over 100 lbs for a TV with 32″ display), bulky, and required fussy "convergence" adjustment to get the colors to track. Shortening the tube to reduce cabinet depth exacerbated the convergence and geometry problems. CRTs are now obsolete, replaced by several technologies, among them liquid-crystal displays (LCDs), plasma displays, and organic light-emitting diodes (OLEDs).

¶ **58.** In an LCD a liquid-crystal layer is sandwiched between a pair of crossed optical polarizers; an applied electric field alters the polarizing properties of the layer, thus varying the optical transmission. In the classic display there's a uniform white rear illumination (from white LEDs, or from one or more cold-cathode fluorescent lamps – CCFLs – combined with diffusers and light pipe

---

[42] This is sometimes called *narrowcast*, versus broadcast or unicast.

material); an array of electrodes applies local electric fields to the image pixels, which are overlaid with red, green, and blue color filters. The array of liquid-crystal pixels act as video-rate dimmable shutters; all the light originates in the rear illuminator.

¶ **59.** LCDs are dominant in computer displays and popular in televisions. They are thin (a centimeter or so) and bright. But they have somewhat limited dynamic range (or contrast ratio: ratio of maximum brightness to "maximum darkness") and speed, and their color balance and black level degrades when viewed off-axis. There have been great improvements in speed and in off-axis performance, owing to methods with names like in-plane switching (IPS), fringe-field switching (FFS), and the like. And the dynamic range can be improved by using LED-array backlighting, which can be dimmed locally and rapidly, adapting to the light and dark areas of the displayed image.[43]

¶ **60.** For the most realistic rendering of cinematic material, however, the plasma display is superior to the LCD. It consists of an array of tiny cells, in each of which a switchable gas discharge generates ultraviolet light that causes a spot of phosphor to glow. A high-definition display of 1080×1920 pixels has three such cells for each pixel (one each with red, green, and blue phosphors) to generate the pixel's overall emitted color. Unlike an LCD (which filters an underlying light source), the cells of a plasma display generate the emitted light directly. You can think of it as an array of 6 million little CRTs ($1080 \times 1920 \times 3$), each one time-switched to achieve the required light intensity.

¶ **61.** Plasma displays retain their color fidelity and contrast ratio regardless of viewing angle, and they have fast response. In larger sizes they are currently somewhat less expensive than LCDs. They are not as bright as LCDs, however, and a static pattern that is displayed for a long time can cause some image retention, or (in extreme cases) phosphor "burn." Their contrast ratio is very good, but not infinite, because the gas discharge in every pixel must be sustained at a low level (i.e., it cannot be switched off entirely) so that it can be rapidly modulated.

¶ **62.** LCD and Plasma have dominated display technology, but the future is likely to be ruled by OLED (organic LED), a direct-emitting array of tiny LEDs (either in three colors, or white LEDs with filters). These emerged first in small displays (e.g., cellphones and camera viewfinders), but by 2014 they had made the big time, with full 4K "Ultra HD" resolution (3840×2160), and screen sizes to 65″ and beyond. Unlike earlier display technologies, OLEDs can be made flexible, and the current fad is *curved* TV screens (capable of 3D, if you're interested in that). OLED is likely to be the ultimate winner, because of their very high dynamic range (1,000,000:1), wide viewing angle, low power consumption, elegant form factor (3 mm thick, lightweight, and nearly borderless), and potential manufacture by an inkjet-like process. These have been winning Best of Show awards, and they deserve it.[44]

¶ **63.** Two other technologies that looked good, but have fallen onto hard times, are field-emission display (FED) and surface-conduction electron-emitter display (SED). Both involve an array of phosphor cells (like the plasma display), but with electron (rather than ultraviolet) excitation of the phosphor. An SED prototype from Canon generated great enthusiasm[45] at the 2006 CES, but subsequent patent disputes and economic realities took their toll. FED and SED may rise again – but don't hold your breath.

I.11 Video connections: analog (composite, component) and digital (HDMI/DVI, DisplayPort)

¶ **64.** What are all those incompatible cables, anyway? To add to the analog–digital confusion, the *consumer* market (of large-screen TVs and flat panels, etc.) and the *computer* market (of LCD monitors) have gone their (mostly) separate ways. Here's a quick rundown of the most-used connections; their connectors are pictured in Figure I.10.

¶ **65.** In the *consumer television* world there are four types of connections (and connectors), the first two of which are nearly extinct.

Composite video. Low quality standard-definition (SDTV: 480i) analog video, recognizable by the yellow RCA-type connector ("phono jack"), usually bundled with an audio pair (red = right, white = left). The single yellow line carries bandwidth-limited luminance

---

[43] For marketing purposes these LCDs are sometimes called "LED TVs." Don't be fooled: it's an LCD, but with LED rear illumination replacing the CCFL. And it may or may not have local dimming – read the fine print.

[44] The reviewers are *gushing*: "The best picture I've seen on any TV, ever." (CNET), "The best-looking TV I've ever seen. Ever." (Digital Trends), and "The best direct-view display – of any size, at any price – we've ever laid eyes on." (HDTVtext, UK).

[45] A breathless review in SlashGear (19Oct2006) proclaimed "SED-TV is something that no amount of words can describe. It is something that must be SEEN to be believed; literally." and "SED-TV is the future of digital image displays; it's as simple as that."

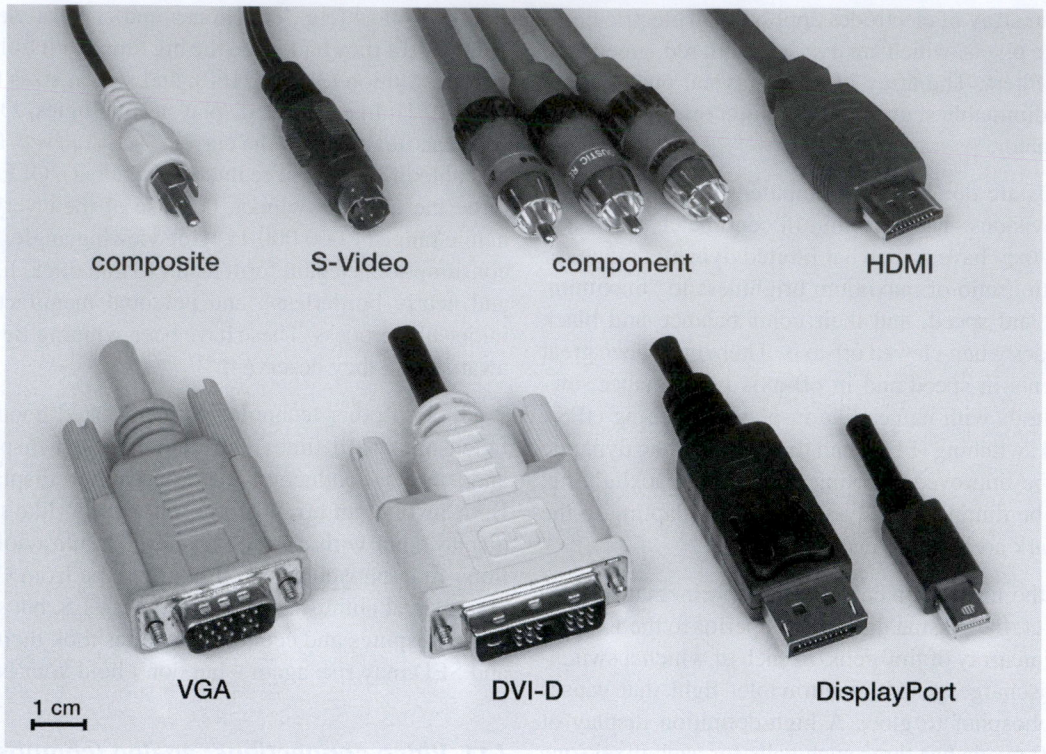

Figure I.10. Cable connectors for computer monitor and video signals. Top row: TV video connectors, from the legacy analog composite to contemporary HDMI high-definition digital video. Bottom row: computer monitor connectors, from the legacy analog VGA to the popular digital DVI and newer DisplayPort (standard and mini connectors).

("luma," abbreviated Y, essentially the grayscale image) and chrominance ("chroma," abbreviated C, the color difference signal pair that is modulated on a 3.58 MHz color subcarrier), along with line and frame synchronization pulses; it is sometimes called CVBS (composite video, blanking, and sync). Don't use such a connection unless there's nothing else available!

S-video. Somewhat better SDTV analog video, recognizable by the 4-pin miniDIN connector with fragile pins. It separates the luma (+ sync) and chroma signals, retaining more bandwidth. Avoid this one, too, unless you like fuzzy pictures.

Component video. Now we're talking! This analog format uses three 75 Ω coax lines, with (usually) RCA-type connectors (or, occasionally, BNCs) that are colored red, green, and blue, and that can carry full-bandwidth HDTV. The colors are misleading: the green line carries luminance (+ sync), while blue and red carry color *difference* signals. That is, the GBR-colored connectors denote "YPbPr," where Y is luminance (red+blue+green), Pb is blue-minus-Y, and Pr is red-minus-Y. As with composite

and S-video, component video carries only video; the audio needs it own cables. Component video does not know, or care, about things like content protection; for that reason it is not embraced by content providers, who may prevent full-resolution HDTV (1080×1920) output on the component jacks (e.g., on a Blu-Ray player).

HDMI. A purely digital format, whose initials stand for high-definition multimedia interface. HDMI is the digital alternative to component video. It is recognizable by the flat 19-pin USB-like connector (sadly without any required latching mechanism), and electrically it is equivalent to the DVI-I computer-monitor format (see below). It carries both audio (up to eight channels, digitized to 24 bits, up to 192 ksps) and video (digitized at 8–16 bits per component, at rates adequate for full HDTV ("1080p," 1080×1920 progressive at 60 Hz; HDMI versions 1.4 and later support full "4K," 4096×2160, with 60 fps progressive from version 2.0). The video data is a digitized version of analog video: uncompressed numerical data representing the amplitude of the color components, sent over four twisted pairs (R, G, B, clock). HDMI supports digital content protection (HDCP, high-bandwidth digital content protection), a protocol by which an HD video source authenticates a display device before sending (encrypted) data, so you are allowed to view the full-resolution video. HDCP seems to work, most of the time anyway (though you may get annoying messages and glitches).

¶ **66.** In the *computer monitor* world there are three types of connections (and connectors) in wide use.

Analog VGA. Legacy analog link, recognizable by the 15-pin D-type connector with locking screws. VGA (for video graphics array) carries separate RGB analog signals, plus Hsync and Vsync (thus "RGBHV"). In contemporary implementations it also has an I^2C channel for monitor identification and control. VGA will work up to resolutions of 1600×1200 or so (there's no specified limit, but you'll see smearing going rightward of sharp features when pushing the resolution), but many monitors have abandoned VGA altogether in favor of the digital formats: DVI and DisplayPort.

DVI. Currently the standard digital interface, recognizable by the 29-pin (maximum) connector with locking screws, looking somewhat like a "VGA on steroids." It is electrically similar to the more compact and inexpensive HDMI, above, that evolved from it for the consumer tele-

vision market, and that includes audio (thus a single connection from a cable box or Blu-Ray player to the television monitor). It comes in several variants, all using the same connector (in which some pins may not be loaded): DVI-D is video-only, and comes in single-link and dual-link varieties (the latter needed for resolutions greater than 1920×1200 at 60 Hz, for example the 2560×1600 at 60 Hz used in $30''$ displays); DVI-A is analog video only (for compatibility with analog monitors); and DVI-I ("integrated") has both digital (single- or dual-link) and analog video. DVI, like VGA, carries no audio.

DisplayPort. Newer standard, intended to supersede DVI; it uses a 20-pin USB-like connector with latching mechanism, and supports very high data rates (up to 4.3 Gbps on each of four differential pairs, thus 17.3 Gbps). It departs from the "digitized-raster" scheme of DVI/HDMI, using instead a packetized data protocol; but it's got enough bandwidth to handle the full video bandwidth of dual-link DVI (which can be converted to DisplayPort protocol). It supports up to 16 bits per color, and 8-channel audio at 24 bits and 192 ksps. It includes provision for fiber optics (instead of copper) for long cable runs (to 50 m or more), and it supports both existing 56-bit HDCP and its own DPCP (DisplayPort content protection, with stronger 128-bit AES encryption). The current revision of DisplayPort can handle 4K 60 fps progressive with ease.

SPICE PRIMER: GETTING STARTED WITH FREE ICAP/4 DEMO

Free SPICE is easy, and fun. It's available for several platforms (Macs, Linux, PCs). Here's how to get started in Windows.

J.1 Setting up ICAP SPICE

1. Download "ICAP/4Windows Demo" zip file (ICAP4Demo.zip) from intusoft.com.
2. Extract zip file to some temporary folder.
3. In that folder, double click (abbreviated hereafter as "**cc**") on setup.exe.
4. Launch ICAP/4 (you may have to be admin mode):
   (Programs → ICAP_4 Demo → Start ICAPS).

J.2 Entering a Diagram

- Add standard components with typing
  R resistor
  C capacitor
  L inductor
  o ground (the letter "oh," not the number)
  v signal or voltage source
  Y voltage test point.
  (These are case-insensitive: r = R, etc.)
- For a library component, Parts → Parts Browser,
  – browse by type, and subtype, or use Find,
  – then Place.
- Moving and orienting:
  – highlight component by click-hold, then can drag;
  – when highlighted, type + to rotate (90°CW), or − to flip 180°.
- Editing a component label:
  – highlight label by click-hold, drag to relocate;
  – when label highlighted, **cc** on label to edit values.
    * Note units multipliers: f, p, n, u, m, k, meg (not M).
    * Note that SPICE only needs the units *multiplier*, not the units themselves. For example, for a capacitor "1u" is OK for SPICE; but you can type "1.0uF" for clarity on the schematic, which SPICE will treat the same.

  * You can **cc** on the component, instead, to bring up the value dialog.
  – You can add parameters by entering the value and clicking Apply; this is necessary, e.g., for a voltage source (add dc or ac, give amplitude). Or you can click ≫Add≫, which also adds the parameter to the schematic drawing.
- Wires:
  – type w, then drag a wire between points;
  – you can drag multiple separate wires; ESC when done.
- Settings:
  – set "Rubberbanding" in Options, to maintain connections when dragging components.

J.3 Running a simulation

Can a passive circuit consisting only of resistors and capacitors have *voltage gain*? The surprising answer is yes. Let's use this simple example to illustrate an ICAP/4 SPICE entry and simulation.

J.3.1 Schematic entry

We launch ICAPS, and place two resistors (type the letter "r"), two capacitors (type the letter "c"), and a voltage source (type the letter "v") with the bottom terminal grounded (type the letter "o" right after typing "v"). If the pieces are too small to see on your screen, you can resize them by using Options→Zoom (or the function keys F6, F7, and F8). We drag them around to connect them as shown in Figure J.1, and add an output-voltage test point (type the letter "y" and change the label to Vout). We assign values by double clicking on each component, then entering the value in the value field (where "???" appears initially); note that pressing the Enter key after each value moves you along to another field in the dialog – you have to click OK to complete the operation.

For the capacitors we could have used microfarads (e.g., 0.01uF instead of 10nF); note that the "F" can be omitted, but including it makes the schematic marking clearer. For the voltage source we enter the value 1V (i.e., 1 volt) in the

"AC" field, and then we click on the ADD button so that the value shows on the schematic (SPICE uses the value, even if we do not ADD it; but it's nice to show it on the circuit diagram). As before, the unit symbol "V" can be omitted.[1]

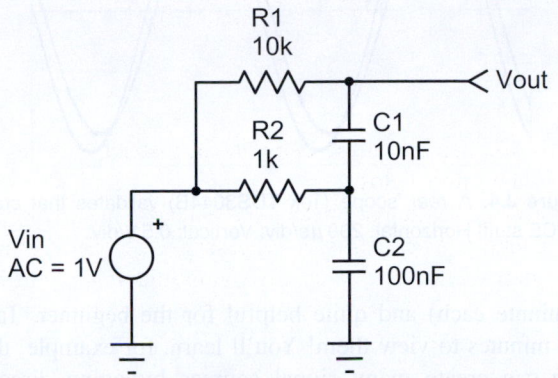

Figure J.1. An *RC* circuit with voltage gain! Schematic entry in Intusoft ICAP/4 is swift, taking no more than a couple of minutes. Printing the captured schematic to "Adobe PDF" produced this figure.

J.3.2 Simulation: frequency sweep

Now for the fun! We begin by saving the project to a folder of our choosing, using the File→Save As menu. Then we set up the simulation by Actions→Simulation setup→Edit (or by clicking on the button with a pencil over a wavy line). Choose and click "AC Analysis," then enter 20 points per octave, start at 100 Hz, end at 10 kHz (type 10k in the box), and click OK. Then click DONE. Now run the simulation by clicking on the little running person (or Actions→Simulate). This launches the Spice engine, with a small window ("IsSpice4") showing status and errors, and (if successful) a little graph of the result. Running the simulation also launches the "IntuScope" display program (which you can launch manually by clicking on the 'scope icon next to the running person, or by Actions→Scope). At this point there are multiple windows piled up, which you should resize and drag to convenient locations on the screen and then do Options→Save Preferences to stick them down. Now click the "Test Pts Only" box in the "Add Waveform" window that belongs to IntuScope, then highlight Vout[2] and click "Add." Voilà – this produces a nice

---

[1] For SPICE's ac analysis, the frequency-sweep gain computation is performed with infinitesimal small-signal amplitudes, which are then normalized to the signal amplitude you specified. Here, for example, SPICE uses a signal amplitude much smaller than 1 V.

[2] Or, if you have not renamed the output to Vout, select from among the signals with names like v2, etc., in the Y Axis output list.

plot of V_{out} versus frequency (see Figure J.2), with default axis labels, grids, and scaling. You can change the axis labels by double clicking on the label; and you can fiddle with the scaling in the "Scaling" window.

To add a plot of phase shift versus frequency, highlight "ph_Vout" in the Add Waveform window, click on the box labeled "With Like Traces" (unless you want a separate graph), and click "Add." We've done that to make Figure J.2, where we've also used the cursor tools on the IntuScope window to mark the point of maximum voltage gain.

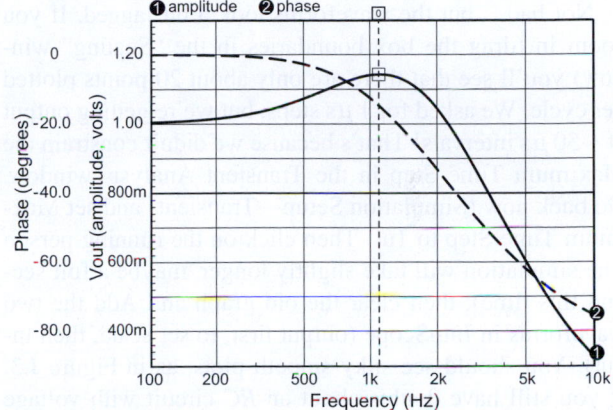

Figure J.2. "'Scope" output of the *RC* gain circuit, showing amplitude and phase ("AC Analysis") for the circuit of Figure J.1. The cursor indicates the peak of the voltage gain, V_{out}/V_{in}=1.142 at f=1.096 kHz. The original IntuScope output plotted phase in blue, which we've converted to a dashed black line; we enlarged the text, also, for readability.

J.3.3 Simulation: input and output waveforms

Do we really believe this gain-versus-frequency plot? Everyone knows that a simple *RC* network just mushes things out – shifted phase and reduced amplitude. We won't believe this thing has voltage gain unless SPICE shows us the actual input and output signal waveforms for a sinewave near the frequency of maximum voltage gain (and maybe not even then!).

Here's how you do it: first, add another voltage test point at the input ("y" again), and change the label to Vin. Then double click on the input voltage source and click on the field labeled "Tran Generators" (SPICE calls any old waveform, including a sinewave, a "transient"). Select SIN, set the peak amplitude to 1, the offset to 0, and the frequency to 1.096 kHz. Click OK, and OK. Then go to Simulation Setup (pencil over wavy line), deselect AC Analysis, and select Transient. In the Transient Analysis window that pops up, set Data Step Time to 1us, Total Analysis Time to 105ms, and Time to Start Recording Data at 100ms (this gives the circuit time to settle into its steady state). Leave the Maximum Time Step field blank (i.e., accept default value). Click the runner icon to run the simulation. At this point the IsSpice4 window should show a wiggly trace. Now go to the IntuScope window, clear the old graph by pressing the DEL key for each selected trace (you can save it, if you want, or start a new plot with File→New Graph), and Add "vin" and "vout" in the Add Waveform subwindow of IntuScope, with the "With Like Traces" box checked as before.

Not bad... but the waveforms look a bit jagged. If you zoom in (drag the box boundaries in the "Scaling" window) you'll see that there are only about 20 points plotted per cycle. We asked for 1 μs steps, but we're getting output at ~50 μs intervals! That's because we didn't constrain the Maximum Time Step in the Transient Analysis window. Go back now (Simulation Setup→Transient) and set Maximum Time Step to 1us. Then click on the running person (the simulation will take slightly longer, maybe a full second this time); then clear the old graph and Add the two waveforms in IntuScope (output first, to set scale, then input). You should see silky-smooth plots, as in Figure J.3. If you still have doubts about an *RC* circuit with voltage gain, you can hook up the circuit and check it out on a real-live oscilloscope, as we did (Figure J.4). Evidently this *RC* voltage-gain stuff is real!

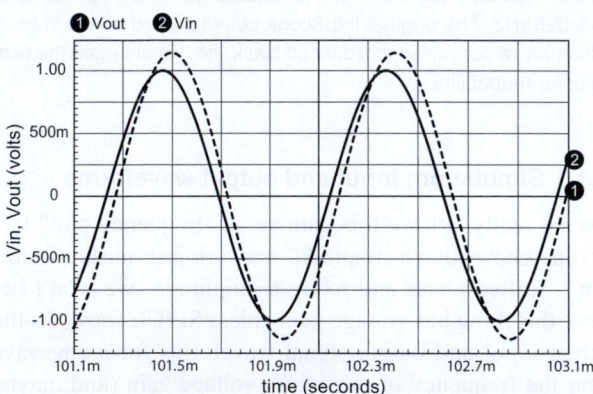

Figure J.3. "Scope" output of the *RC* gain circuit, showing sinewave response ("Transient Analysis") at the frequency of maximum gain for the circuit of Figure J.1. The original IntuScope output plotted V_{out} in blue, which we've converted to a dashed black line; we enlarged the text, also, for readability, and adjusted the axis scales to match the scope trace.

J.4 Some final points

The ICAP/4 demo program includes ten "tutorial movies" (in the Help menu), which are mercifully short (about

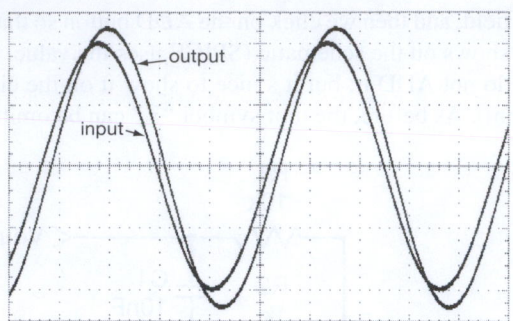

Figure J.4. A *real* 'scope (Tek TDS3044B) validates that crazy SPICE stuff! Horizontal; 200 μs/div; Vertical; 0.5 V/div.

a minute each) and quite helpful for the beginner. Take ten minutes to view them! You'll learn, for example, that you can create many signal sources by going directly to Parts→Parts Browser→!Generators. You'll also learn some tricks to simplify schematic entry (copying, zooming), and ways to make the parts labels more readable ("tall," "wide," and "split" options). And by experimenting a bit you can discover that both the schematic and the simulation graphs can be copied ("Print") to the clipboard and then pasted into Word or Wordpad, where they exist as scalable graphics. If you Print to Adobe PDF you'll get a scalable graphic that can be cleaned up in Adobe Illustrator.

We have been using the Intusoft "ICAP/4" full version of SPICE now for many years; it provides a convenient Windows-based schematic entry and simulation environment, which we use to explore circuit configurations. Typically we begin with a starting circuit design, making explanatory notations in a Word document as we go. We copy and paste the schematics and simulation graphics into the running text, making a "lab notebook" of sorts; see next.

J.5 A detailed example: exploring amplifier distortion

In Chapter *2x* ("BJT Amplifier Distortion: a SPICE exploration") we do some serious work with SPICE, in the form of an exploration of distortion in discrete BJT amplifier designs.[3] The amplifier stuff is lots of fun; and there are some nice SPICE tricks there revealed. Take a look!

---

[3] When we do this in real life, we like to paste circuit diagrams and screen shots of simulated performance into a Word document, annotating as we go. We've kept that spirit alive in the discussion in Chapter *2x*, albeit in the form of a LaTeX conversion of the original Word document.

J.6 Expanding the parts database

You can freely add SPICE models to the ICAP/4 parts library. An easy way to start the process is to download the folder AoE-PR from our website (www.artofelectronics.com), and place it under the default PR folder. Then do File→Update Part Database.

"WHERE DO I GO TO BUY ELECTRONIC GOODIES?"

APPENDIX K

Good question! Here are some hints, from our experiences.

I. Mail order and online

Digi-Key Corp (Thief River Falls, MN: 1–800–digikey). We used to say *"Get their catalog!!"* – but, sadly, they've abandoned paper, replacing it with an impressive search capability. You can get everything here, even in small quantities, with fast delivery. It's often worth designing with their webpage open in front of you. Online ordering and stock/price checking: www.digikey.com.

Mouser Electronics (www.mouser.com). Broad stocking distributor, with service comparable to Digi-Key's, and willingness to ship small quantities. Good selection of precision passives; and they are still printing a comprehensive paper catalog.

Newark Electronics + Farnell (1–800–2–newark; www.newark.com). Broadest stocking distributor, with service comparable to Digi-Key's and good selection of tools; paper catalog still in print.

"Stocking Distributors". These are the standard distribution channel for quantity buying; names like Allied (still publishing a paper catalog), Arrow, Avnet, FAI, Heilind, Insight, Pioneer, Wyle. Substantial minimum quantities – not generally useful for prototyping or small production.

Manufacturers' Direct. Many semiconductor manufacturers (Analog Devices, TI, Maxim,...) will not only send free samples with the slightest provocation, they will also sell in small quantities via credit card; check out Mini-Circuits for RF components, and Coilcraft for inductors, transformers, and RF filters.

Oddballs. Marlin P Jones, Jameco, B&D, Herbach & Rademan, Omnitron, ABRA, All Electronics; ephemeral collection of "surplus" stuff, some real bargains.

eBay (www.ebay.com). If you haven't been here, you've probably just arrived from Mars. LOTS of stuff, literally millions of items, an online auction. You can get plenty of electronic stuff, but CAVEAT BIGTIME EMPTOR. Feedback Forum helps.

Alibaba Small Pacific-rim companies that have stocks of obsolete components are easily found on Alibaba. You can place a quote request for a part number, and you'll get dozens of useful reasonably-priced suppliers.

II. Indexes and Locators

Octopart, FindChips, NetComponents (octopart.com, findchips.com, netcomponents.com). Give it a part number and it searches dozens of distributors, returning (sometimes unreliable) information on availability and pricing.

WhoMakesIt (www.whomakesit.com). A bit like the EEM catalog, helpful if you know the category of stuff you want, but not the manufacturer, etc.

Google (www.google.com). Our standard "portal," reads your mind and vectors you to the goodstuff. Helpful, sometimes, in finding parts and equipment manufacturers and vendors.

III. Local

Sometimes it's nice to shop in person; here are the sorts of places to go.

Radio Shack (www.radioshack.com). They call themselves "America's Electronic Supermarket"; we'd call them "America's Electronic Convenience Store." Their stores are everywhere, and they stock (pretty reliably) an idiosyncratic collection of parts and supplies, of uncertain quality or duration. However, in the changing marketplace of consumer electronics, their future path is unclear.

Electronics Flea Markets. Also known as "swap meets," perhaps somewhat in decline; two legendary meets are on opposite coasts: De Anza College (Cupertino, every second Saturday, March–October), and MIT (Cambridge, every third Sunday, April–October). What meets are three cultures (electronics, computers, hams); haggling is mandatory; caveat very emptor.

Electronics Surplus Supply Stores. These are incredibly cool! Several well-known haunts are Halted (www.halted.com; officially "HSC Electronics Supply"), in Santa Clara, Sacramento, and Santa Rosa; and Murphy's Surplus Warehouse (www.murphyjunk.net) in El Cajon.

IV. Miscellaneous

Obsolete ICs. Your best place to start is Rochester

Electronics (www.rocelec.com), a wonderful place that apparently buys up inventories of ICs being discontinued. Jameco also has a lot of obsolete parts. Freetradezone has broker lists for obsolete parts. Also try Interfet (www.interfet.com), a manufacturer of small-signal FETs, including ones that the big guys have abandoned.

PC Board Manufacture. We like a place called Advanced Circuits, www.4pcb.com; you can get online quotes, and they do a good job and deliver pronto. Another inexpensive and fast PC house is Alberta Printed Circuits in Canada (www.apcircuits.com).

WORKBENCH INSTRUMENTS AND TOOLS

Here are some electronic favorites, stuff that we really like using when we design and build electronic circuits. It's best to check current catalogs and websites – much of this sort of equipment becomes obsolete with terrifying speed.

Soldering iron
Metcal MX-500 (not cheap; it will change your life!)
Weller WSL (less expensive, variable temp)

Desoldering station
Pace MBT201-SD

Surface-mount prototype and rework
Zephyrtronics Airbath and Airpencil
Metcal offers SMT "tweezers" and other goodies

Bench DMM
Keithley 2100 ($6\frac{1}{2}$-digit, with USB control & readout)
Agilent 34410A ($6\frac{1}{2}$-digit, with LAN & USB control & readout)

Pocket DMM
Amprobe 37XR-A (cheap, good enough)
Fluke 289 (not cheap, very good)
Agilent U1252A/53A (everyone's favorite lately)
"Smart Tweezers" (auto-ranging SMT tweezer-style meter, neat!)

Triple bench power supply
HP E3630A (excellent performance, reasonable price)

High-voltage bench power supply
SRS PS300-series (single and split, to 20 kV)

Device programmer (if needed; JTAG in-ckt is taking over, pods from mfgrs)
BP Microsystems 1610 (universal and reliable; lifetime free algorithm updates)

LCR meter
HP 4263B (a cheaper one is the SRS model SR720)

Analog oscilloscope
B&K Precision and Hameg still offer some models, to 200 MHz bandwidth

Digital oscilloscope
Tek DPO2024B (cheap "lunchbox"); DPO/MSO 3k-, 4k-, 5k-series
Agilent DSO/MSO 5k-, 6k-, and 7k-series (Agilent's answer to the "lunchbox")
Lecroy WaveRunner, WaveSurfer, WaveJet series (lot of models, do your homework)

Arbitrary function generator
Tek AFG3000-series (single and dual channel, to 240 MHz and 2 Gs/s)

Low-distortion function generator
SRS DS360 (0.01 Hz–200 kHz, 0.001% distortion)

RF and microwave synthesizer
Agilent N9310A
SRS SG380-series (2, 4, and 6 GHz models, low phase noise, see §13.13.6B)

Low frequency spectrum analyzer
SRS model SR785

RF spectrum analyzer
Agilent ESA series (depends on frequency range, and $$)

Source measure unit
Agilent B2900-series (single and dual channel)
Keithley 2600-series (single and dual channel)

Precision time and frequency standard
Symmetricom 4411A (uses GPS constellation)

Engineering software
Altium System Designer, OrCad, or Eagle (for schematic capture and layout; includes simulation)
Xilinx WebPack (for PLD and FPGA design), and analogous tools from Altera, Lattice, Actel, etc.
ICAP/4, LTspice ($0!), MicroCap 9, MMICAD, PSpice (for simulation)
FilterCAD ($0!, from LTC), FilterPro ($0!, from TI) (simple analog filter design)
MathCAD, MATLAB, Mathematica (engineering/math worksheets)
LabVIEW™ (virtual instruments; control of real instruments)

CATALOGS, MAGAZINES, DATABOOKS

Here are some recommendations for data books, catalogs, and magazines; some have become on-line services only (e.g., IC Master). You will need some of these if you want to practice electronic design.

"Master" catalogs

EEM (Electronic Engineer's Master) *lists all categories of electronic stuff*
IC Master *ditto, for ICs*
Octopart *excellent on-line part (and datasheet) finder*

Parts and equipment catalogs

Digi-Key Corp (a distributor), *wide range, quick delivery*
Mouser Electronics (a distributor), *wide range, quick delivery*
Newark/Farnell Electronics (a distributor), *best broad-line catalog*
Allied Electronics (a distributor), *similar to Newark*
TechniTool (a distributor), *measurement, assembly, and test tools*
Stanley Supply & Services, (a distributor) *equipment and supplies for assembly*
Keysight Technologies (Agilent, HP), *broadest manufacturer of test and measurement equipment*
Tektronix, *test and measurement equipment, especially scopes*
Fluke, *test and measurement equipment*
Stanford Research Systems (SRS) *test and measurement equipment; great documentation and app notes*

Magazines and tabloids

EDN, *keeps you up to date on new products and methods*
Electronic Design, *ditto; a bit thin, lately*
Electronic Products, *ditto, emphasis on products only*
EE Times, *tabloid format*
Computer Applications Journal, *microcontroller project mania*

Circuit Cellar Ink, *emphasis on microcontrollers*
Nuts and Volts, *quirky do-it-yourself, tabloid format*
Make Magazine, *cool stuff you can build*

Data books; short-form/design guides
A starter selection, with product emphasis.
Altera: cPLD, FPGA
Analog Devices: all linear functions; converters; DSP
Atmel: PLDs and microcontrollers
Avago (←Agilent←HP←Avantek): opto, rf
Cirrus/Apex: power op-amps etc.
Cypress: memory, processors
Diodes Inc/Zetex: discretes, etc.
Fairchild: discretes, linear, digital
Freescale: processors, DSP, automotive
IDT, Micron, Samsung: memory
Infineon: discrete, power, processors, RF
Intel: microprocessors and microcontrollers
Linear Technology Corp: all linear functions
M/ACom: RF and microwave
Maxim/Dallas: linear, digital, μC
MiniCircuits: RF – broad line, inexpensive
NXP (Philips): logic, microcontroller
ON Semi/Sanyo: linear, logic, discrete
Renesas: discrete, memory, processors
TI/National/Burr–Brown: linear, logic, opto, power, processors, DSP
Xilinx: cPLD, FPGA

Datasheets (in *.pdf format) are available online from nearly every semiconductor manufacturer's website; try, e.g., www.analog.com, www.maxim-ic.com, www.linear.com, www.ti.com, etc.

Another way to find datasheets, prices, and availability is via a search website like octopart.com or findchips.com; these display price and inventory (and sometimes datasheets), and link you through to the stocking distributor.

FURTHER READING AND REFERENCES

APPENDIX N

General

Ashby, D., ed., *Circuit Design: Know It All*. Newnes (2008). A collection of fascinating electronics engineering stuff from 14 acclaimed authors.

Camenzind, H., *Much Ado About Almost Nothing, Man's Encounter with the Electron*. Booklocker.com (2007). Fascinating stories about electronics, by the famed designer of the 555.

Dobkin, B. and Williams, J., eds., *Analog Circuit Design: A Tutorial Guide to Applications and Solutions*. Newnes (2011). Excellent selection of informative and well-written application notes from Linear Technology. Lively and entertaining, too.

Dunn, P. C., *Gateways into Electronics*. Wiley (2000). Fascinating physics-based approach to electronics; deep coverage of critical areas.

Jones, R. V., *Instruments and Experiences: Papers on Measurement and Instrument Design*. Wiley (1988). Classic on instrument design, based on Jones' papers.

Lee, T. H., *The Design of CMOS Radio-Frequency Integrated Circuits*. Cambridge University Press (2nd ed., 2003). From the originator of gigahertz CMOS comes this delightful volume, covering much more than its humble title suggests. Terrific introductory chapter on the history of radio.

Pease, R. A., *Troubleshooting Analog Circuits*. Butterworth–Heinemann (1991). The curmudgeon-in-chief reveals his tricks.

Purcell, E. M., and Morin, D. J., *Electricity and Magnetism*. Cambridge University Press (2013). Excellent textbook on electromagnetic theory. Relevant sections on electrical conduction and analysis of ac circuits with complex numbers.

Scherz, P. and Monk, S., *Practical Electronics for Inventors*. McGraw-Hill (3rd ed., 2013). The title says it all.

Sedra, A. S. and Smith, K. C., *Microelectronic Circuits*. Oxford University Press (6th edition, 2009). Popular classic engineering text.

Senturia, S. D., and Wedlock, B. D., *Electronic Circuits and Applications*. Wiley (1975). Good introductory engineering textbook.

Sheingold, D. H., ed., *Nonlinear Circuits Handbook*. Analog Devices (1976). Highly recommended.

Sheingold, D. H., ed., *The Best of Analog Dialog, 1967 to 1991*. Analog Devices (1991). Outstanding collection of analog-engineering techniques.

Terman, F. E., *Radio Engineers' Handbook*. McGraw-Hill (1943). Three score and ten years later it continues to amaze, with excellent sections on passive circuit elements and other basic engineering.

Tietze, U., and Schenk, Ch., *Electronic circuits: Handbook for Design and Applications*. Springer-Verlag (2nd edition, 2008). Spectacular all-around reference.

Williams, J., ed., *Analog Circuit Design: Art, Science, and Personalities*. Butterworth–Heinemann (1991). Idiosyncratic collection of wisdom from 22 analog gurus.

Williams, J., ed., *The Art and Science of Analog Circuit Design*. Butterworth–Heinemann (1998). The sequel: 16 analog gurus dispense yet more wisdom.

Handbooks

Fink, D. G., and Beaty, H. W., eds., *Standard Handbook for Electrical Engineers*. New York: McGraw-Hill (16th ed., 2012). Tutorial articles on electrical engineering topics.

Jordan, E., ed., *Reference Data for Engineers: Radio, Electronics, Computer, and Communications*. Howard W. Sams & Co. (9th ed., 2001). General-purpose engineering data.

"Temperature Measurement Handbook." Stamford, CT: Omega Engineering Corp. (revised annually). Thermocouples, thermistors, pyrometers, resistance thermometers.

BJTs and FETS

Camenzind, H., *Designing Analog Chips*. Virtualbookworm.com and available online (2005). Inspiring book by a real world analog IC designer; includes the story of his design of the 555 at Signetics (now NXP).

Ebers, J. J., and Moll, J. L., "Large-signal behavior of junction transistors." *Proc. I.R.E.* **42**:1761–1772 (1954). The Ebers–Moll equation is born.

Gray, P. R., Hurst, P. J., Lewis, S. H., and Meyer, R. G., *Analysis and Design of Analog Integrated Circuits.* Wiley (5th ed., 2009). The classic go-to book for a real understanding of integrated linear circuit design.

Howe, R.T. and Sodini, C. G., *Microelectronics, an Integrated Approach.* Prentice-Hall (1996). Introductory IC design.

Mead, C. and Conway, L., *Introduction to VLSI Systems.* Addison-Wesley (1980). Device physics and circuit design; a classic.

Muller, R. S., and Kamins, T. I., *Device Electronics for Integrated Circuits.* Wiley (1986). Transistor properties in ICs.

Sze, S. M., *Physics of Semiconductor Devices.* Wiley (1981). The classic.

Tsividis, Y. P., and McAndrew, C., *Operation and Modeling of the MOS Transistor.* McGraw-Hill (3rd ed., 2010).

SPICE

Cheng, Y. and Hu, C., *MOSFET Modeling & BSIM3 User's Guide.* Springer (1999).

Kielkowski, R., *Inside SPICE.* McGraw-Hill (1998). A short book with hints about SPICE convergence, etc.

Liu, W., *MOSFET Models for SPICE Simulation: Including BSIM3v3 and BSIM4.* Wiley (2001). If you use SPICE to analyze MOSFET designs you need this book.

Massobrio, G. and Antognetti, P., *Semiconductor Device Modeling With SPICE.* McGraw-Hill (2nd ed., 1998). Modeling BJTs, JFETs and MOSFETs.

Ytterdal, T., Cheng, Y., and Fjeldy, T. A., *Device Modeling for Analog and RF CMOS Circuit Design.* Wiley (2003). MOSFET device physics, and SPICE modeling, noise in MOSFETs.

Amplifiers, Transducers, and Noise

Buckingham, M. J., *Noise in Electronic Devices and Systems.* Wiley (1983).

Hollister, A. L., *Wideband Amplifier Design.* Scitech (2007). Wideband amplifier design techniques using BJTs and FETs, with extensive SPICE analysis.

Morrison, R. *Grounding and Shielding Techniques in Instrumentation.* Wiley (1986).

Motchenbacher, C. D. and Connelly, J. A., *Low-noise Electronic System Design.* Wiley (1993). A serious in-depth treatment of low-noise amplifier design.

Netzer, Y., "The design of low-noise amplifiers." *Proc. IEEE* **69**:728–741 (1981). Excellent review.

Ott, H., *Noise Reduction Techniques in Electronic Systems.* Wiley (1988). Shielding and low-noise design.

Radeka, V., "Low-noise techniques in detectors." *Ann. Rev. Nucl. and Part. Physics*, **38**:217–277 (1988). Amplifier design, signal processing, and fundamental limits in charge measurement.

Op-amps

Applications Manual for Operational Amplifiers, for Modelling, Measuring, Manipulating, & Much Else. Philbrick/Nexus Research (1965). Charming compendium from the originators of the first commercial op-amp; these are collectors' items, long out of print.[1]

Carter, B., and Brown, T. R., *Handbook of Operational Amplifier Applications.* Rework of the classic Burr–Brown handbook, described by Carter as a "treasure...some of the finest works on op amp theory that I have ever seen."

Frederiksen, T. M., *Intuitive IC op-amps.* Santa Clara, CA: National Semiconductor Corp. (1984). Extremely good treatment at all levels.

Graeme, J. G., *Applications of Operational Amplifiers: Third Generation Techniques.* McGraw-Hill (1987). One of the Burr-Brown series.

Jung, W. G., ed., *Op Amp Applications Handbook.* Newnes (2004). Fascinating history section, excellent up-to-date detail.

Jung, W. G., *IC op-amp Cookbook.* Howard W. Sams & Co. (3rd ed., 1986). Lots of circuits, with explanations. See also Jung's *Audio IC Op-amp Applications.*

Soclof, S., *Analog Integrated Circuits.* Prentice-Hall (1985). Detailed IC-designer information, useful for IC users as well.

Zumbahlen, H., *Linear Circuit Design Handbook.* Newnes (2008). Things you need to know from Analog Devices engineers.

---

[1] But happily resurrected at `http://www.analog.com/library/analogdialogue/archives/philbrick/computing_amplifiers.html`.

Audio

Duncan, B., *High Performance Audio Power Amplifiers.* Newnes (1996). Excellent review of professional audio power amplifier design.

Hickman, I., *Analog Electronics.* Newnes (2nd ed., 1999). Interesting overview, one wishes he had written more.

Hood, J. L., *The Art of Linear Electronics.* Newnes (1998). Audio electronics, including FM.

Pohlman, K. C., *Principles of Digital Audio.* McGraw-Hill (3rd ed., 1995). All aspects of digital audio; non-mathematical, an easy read.

Self, D., *Small Signal Audio Design.* Focal Press (2010). Audio design basics and tricks, with a special view from an industry master.

Strawn, J., ed., *Digital Audio Signal Processing.* A-R Editions Inc. (Madison, WI; 1985). The first article (by Moore) is a magnificent introduction to the mathematics of digital signal processing. Sadly, this volume is out of print.

Watkinson, J., *The Art of Digital Audio.* Focal Press (3rd ed., 2000). Another nice book on digital audio.

Filters and Oscillators

Hilburn, J. L., and Johnson, D. E., *Manual of Active Filter Design.* McGraw-Hill (1982).

Lancaster, D., *Active Filter Cookbook.* Howard W. Sams & Co. (1979). Explicit design procedure; easy to read.

Matthys, R. J., *Crystal Oscillator Circuits.* Wiley (1983), Krieger Publishing (revised, 1992).

Parzen, B., *Design of Crystal and Other Harmonic Oscillators.* Wiley (1983). Discrete oscillator circuits.

Williams, A. and Taylor, F., *Electronic Filter Design Handbook.* McGraw-Hill (4th ed., 2006). Practical filter design, with formulas, tables, and many examples.

Zverev, A. I., *Handbook of Filter Synthesis.* Wiley (1967). Extensive tables for passive *LC* and crystal filter design.

See also Graeme, J. G., under op-amp listings.

Power, Regulation, and Control

Basso, C. P., *Switch-Mode Power Supplies: SPICE Simulations and Practical Designs.* McGraw-Hill (2008). The title says it all.

Billings, K. and Morey, T., *Switchmode Power Supply Handbook.* McGraw-Hill (3rd ed., 2010). Highly readable and comprehensive treatment of an often-confusing topic.

Erickson, R. W. and Maksimovic, D., *Fundamentals of Power Electronics.* Springer (2nd ed 2001). Learn how to compensate an SMPS feedback loop.

Grover, F. W., *Inductance Calculations.* Dover (2009 reprint of the 1946 classic). Formulas, tables, and graphs for the inductance of just about anything.

Hnatek, E. R., *Design of Solid-state Power Supplies.* Van Nostrand Reinhold (1989). Switching supplies.

MacFadyen, K. A., *Small Transformers and Inductors.* Chapman & Hall (1953). Learn how to calculate leakage inductance in your transformers.

Maniktala, S., *Switching Power Supplies: A to Z.* Newnes (2nd ed, 2012). Filled with useful unusual material, like magnetics design with the all-important AC-resistance loss analysis.

Pressman, A., Billings, K., and Morey, T., *Switching Power Supply Design.* McGraw-Hill (3rd ed., 2009). Standard comprehensive book for a two-week introductory course in SMPS design.

Rogers, G. and Mayhew, Y., *Engineering Thermodynamics: Work and Heat Transfer.* Prentice-Hall (4th ed, 1996). Develop a better understanding of thermal management in electronics.

Snelling, E. C., *Soft Ferrites.* Butterworth–Heinemann (2nd ed, 1988). The bible for inductor and transformer design.

Optics and Light

Friedman, E. and Miller, J. L., *Photonics Rules of Thumb: Optics, Electro-Optics, Fiber-Optics, and Lasers.* McGraw-Hill (2003). What's *this* doing here? Well, it's an amazing collection of coolstuff, both serious and quixotic (e.g., "crickets as thermometers.")

Graeme, J. G., *Photodiode Amplifiers: Op Amp Solutions.* McGraw-Hill (1995). The low-down on transimpedance amplifiers.

Hobbs, P. C. D., *Building Electro-Optical Systems: Making It All Work.* Wiley (2nd ed., 2009). Great balance of theory and practice.

Lenk, R. and Lenk, C. *Practical Lighting Design with LEDs.* Wiley (2011).

Schubert, E. F. *Light-Emitting Diodes.* Cambridge University Press (2nd ed., 2006). LED device physics, practical devices, color physics.

Yariv, A., *Introduction to Optical Electronics.* Rinehart & Winston (1976). Physics of opto-electronics, lasers, and detection.

High-speed Digital and RF

Hagen, J. B., *Radio-Frequency Electronics: Circuits and Applications*. Cambridge University Press (2nd ed., 2009). Refreshingly different, an insight per page.

Johnson, H. and Graham, M., *High Speed Digital Design: A Handbook of Black Magic*. Prentice-Hall (1993). Ringing, cross-talk, ground bounce, etc. – a must-have if you're doing fast digital design.

Johnson, H. and Graham, M., *High Speed Signal Propagation: Advanced Black Magic*. Prentice-Hall (2003). Techniques for pushing the limits of high-speed signal transmission.

Johnson, R. C., ed., *Antenna Engineering Handbook*. McGraw-Hill (3rd ed., 1992). Comprehensive, excellent tables and design information.

Krauss, J. D. and Marhefka, R. J., *Antennas for All Applications*. McGraw-Hill (3rd ed., 2001). Highly readable and usable text.

Ramo, R., Whinnery, J. R. and Van Duzer, T., *Fields and Waves in Communication Electronics*. Wiley (3rd ed., 1994). A classic electricity and magnetism text, with emphasis on communications.

Roy, K. and Prasad, S., *Low-Power CMOS VLSI Circuit Design*. Wiley (2000).

Sevick, J., *Transmission Line Transformers*. Noble (4th ed., 2001). Practical guide to understanding and building RF transformers.

Digital Signal Processing and Communication

Bracewell, R. N., *The Fourier transform and its applications*. McGraw-Hill (3rd ed., 1999). The classic in this field.

Brigham, E. O., *The Fast Fourier Transform and its Applications*. Prentice-Hall (1988). Highly readable.

Oppenheim, A. V. and Schafer, R. W., *Discrete-Time Signal Processing*. Prentice-Hall (3rd ed., 2009). Well-received classic on digital signal analysis.

Sklar, B., *Digital Communications: Fundamentals and Applications*. Prentice-Hall (2nd ed., 2001). Fine introduction to all aspects of digital communications.

Logic, Conversion, and Mixed-signal

Best, R. E., *Phase-locked Loops*. McGraw-Hill (6th ed., 2007). Advanced techniques.

Brennan, P. V., *Phase-Locked Loops: Principles and Practice*. McGraw-Hill (1966).

Gardner, F. M., *Phaselock Techniques*. Wiley (1979). The classic PLL book: emphasis on fundamentals.

Kester, E., ed., *The Data Conversion Handbook*. Newnes (2004). Includes an excellent history of data conversion, and extensive detail on the nuances of conversion, timing, bandwidth, etc.

Lancaster, D., *CMOS Cookbook*. Howard W. Sams & Co. (2nd ed., 1997). Good reading, down-to-earth applications. Includes widely used (but rarely mentioned) M^2L (Mickey Mouse logic) technique.

Rohde, U. L., *Digital PLL Frequency Synthesizers*. Prentice-Hall (1983). Theory and lots of circuit detail.

Sheingold, D. H., ed., *Transducer Interfacing Handbook*. Analog Devices (1980).

Computers and Programming

Hancock, L. and Krieger, M., *The C Primer*. McGraw-Hill (1982). Introduction for beginners.

Harbison, S. P. and Steele, G. L., Jr., *C: A Reference Manual*. Prentice-Hall (1987). Readable and definitive; has ANSI extensions.

Wescott, T., *Applied Control Theory for Embedded Systems*. Newnes (2006). From the author of *PID without a Ph.D.*

Miscellaneous

Grätzer, G., *More Math into LATEX*. Springer (4th ed., 2007). Best single reference on typesetting with LATEX (the software typesetting language in which this book was written).

Kleppner, D. and Ramsey, N., *Quick Calculus*. Wiley (2nd ed., 1985). The title is honest, it's the fastest way to learn calculus. Don't be put off by the book's vintage (hey, calculus itself goes back almost four centuries).

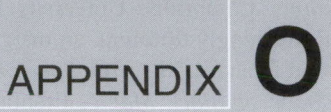
THE OSCILLOSCOPE

The oscilloscope ("scope" for short) is, by far, the most useful and versatile electronic circuit test instrument.[1] As usually used, it lets you "see" voltages in a circuit as a function of time, triggering on a particular point of the waveform so that a stationary display results. Contemporary scopes are almost invariably *digital* (input signals are digitized, processed, and displayed), and they do (and usually *better*) what their analog ancestors did. To understand how to use an oscilloscope, we think it best to start with the traditional (and nearly extinct) 2-channel *analog* scope, for which we've drawn a block diagram (Figure O.1) and typical front panel (Figure O.2). Digital scopes carry forward nearly all of its features, to which they add an impressive array of capabilities (and a few hazards).

O.1 The analog oscilloscope

O.1.1 Vertical

Beginning with the signal inputs, most analog scopes have two channels; that's very useful, because you often need to see the relationship between signals. Each channel has a calibrated gain switch, which sets the scale of **VOLTS/DIVISION** *on the screen*.[2] There's also a **VARIABLE** gain knob (concentric with the gain switch) in case you want to set a given signal to a certain number of divisions. Warning: be sure the variable gain knob is in the "calibrated" position when making voltage measurements! It's easy to forget. The better scopes have indicator lights to warn you if the variable gain knob is out of the calibrated position.

The scope is dc-coupled, an essential feature: what you see on the screen is the signal voltage, dc value and all. Sometimes you may want to see a small signal riding on a large dc voltage, though; in that case you can switch

the input to **AC COUPLING**, which capacitively couples the input with a time constant of about 0.1 second. Most scopes also have a grounded input position, which lets you see where zero volts is on the screen. (In **GND** position the signal isn't shorted to ground, just disconnected from the scope, whose input is grounded.) Scope inputs are usually high-impedance (1MΩ in parallel with about 20 pF), as any good voltage-measuring instrument should be.[3] The input resistance of 1 MΩ is an accurate and universal value, so that high-impedance attenuating probes can be used (as will be described later); unfortunately, the parallel capacitance is not standardized, which is a bit of a nuisance when changing probes.

The vertical amplifiers include a vertical **POSITION** control, an **INVERT** control on at least one of the channels, and an **INPUT MODE** switch. The latter lets you look at either channel, their sum (their difference, when one channel is inverted), or both. There are two ways to see both: **ALTERNATE**, in which alternate inputs are displayed on successive sweeps of the trace, and **CHOPPED**, in which the trace jumps back and forth rapidly (0.1–1 MHz) between the two signals. **ALTERNATE** mode is generally better, except for slow signals. It is often useful to view signals both ways, to make sure you're not being deceived.

O.1.2 Horizontal

The vertical signal is applied to the vertical deflection electronics, moving the dot up and down on the screen. The horizontal sweep signal is generated by an internal ramp generator, giving deflection proportional to time. As with the vertical amplifiers, there's a calibrated **TIME/DIVISION** switch and a **VARIABLE** concentric knob; the same warning stated earlier applies here. Most scopes have a **10× MAGNIFIER** and also allow you to use one of the input channels for horizontal deflection (this lets you generate those beloved but generally useless

---

[1] It is sometimes said that practitioners of other engineering disciplines are especially envious of EEs, because we are blessed with such a splendid instrument with which to visualize the happenings in our circuits.

[2] Note that the two channels can be set for different scale factors, offsets, and coupling. This goes also for digital scopes, which commonly have four channels.

---

[3] Scopes intended for high-frequency measurements, going beyond 100 MHz, say, offer also a 50Ω input impedance option.

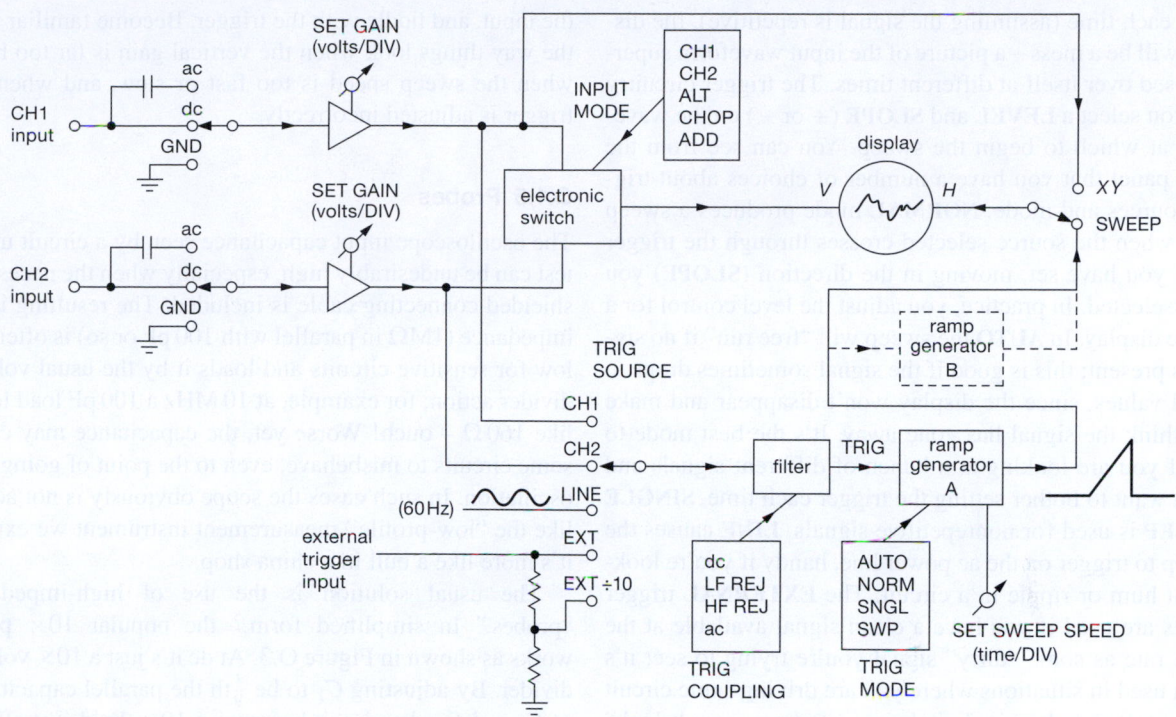

Figure O.1. Block diagram of a 2-channel analog oscilloscope.

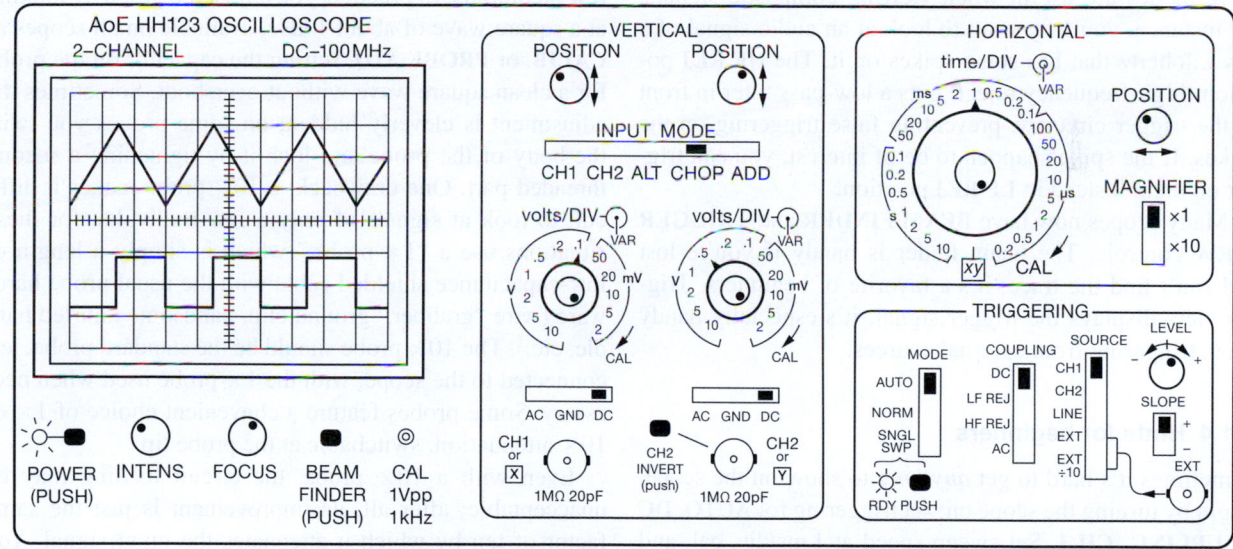

Figure O.2. Portrait of a 2-channel analog scope.

"Lissajous figures" featured in elementary books and science fiction movies).

O.1.3 Triggering

Now comes the trickiest part: *triggering*. We've got vertical signals and horizontal sweep; that's what's needed for a graph of voltage versus time. But if the horizontal sweep doesn't catch the input signal at the same point in its waveform each time (assuming the signal is repetitive), the display will be a mess – a picture of the input waveform superimposed over itself at different times. The trigger circuitry lets you select a **LEVEL** and **SLOPE** (+ or −) on the waveform at which to begin the sweep. You can see from the front panel that you have a number of choices about trigger sources and mode. **NORMAL** mode produces a sweep only when the source selected crosses through the trigger point you have set, moving in the direction (**SLOPE**) you have selected. In practice, you adjust the level control for a stable display. In **AUTO** the sweep will "free run" if no signal is present; this is good if the signal sometimes drops to small values, since the display won't disappear and make you think the signal has gone away. It's the best mode to use if you are looking at a bunch of different signals and don't want to bother setting the trigger each time. **SINGLE SWEEP** is used for nonrepetitive signals. **LINE** causes the sweep to trigger on the ac power line, handy if you're looking at hum or ripple in a circuit. The **EXTERNAL** trigger inputs are used if you have a clean signal available at the same rate as some "dirty" signal you're trying to see; it's often used in situations where you are driving some circuit with a test signal, or in digital circuits where some "clock" signal synchronizes circuit operations. The various coupling modes are useful when viewing composite signals; for instance, you may want to look at an audio signal of a few kilohertz that has some spikes on it. The **HF REJ** position (high-frequency reject) puts a low-pass filter in front of the trigger circuitry, preventing false triggering on the spikes. If the spikes happen to be of interest, you can trigger on them instead in **LF REJ** position.

Many scopes now have **BEAM FINDER** and **TRIGGER VIEW** controls. The beam finder is handy if you're lost and can't find the trace; it's a favorite of beginners. Trigger view displays the trigger signal; it's especially handy when triggering from external sources.

O.1.4 Hints for beginners

Sometimes it's hard to get *anything* to show on the scope. Begin by turning the scope on; set triggering for **AUTO, DC COUPLING, CH 1**. Set sweep speed at 1 ms/div, cal, and the magnifier off (×1). Ground the vertical inputs, turn up the intensity, and wiggle the vertical position control until a horizontal line appears (if you have trouble at this point, try the beam finder).[4] Now you can apply a signal, unground

the input, and fiddle with the trigger. Become familiar with the way things look when the vertical gain is far too high, when the sweep speed is too fast or slow, and when the trigger is adjusted incorrectly.

O.1.5 Probes

The oscilloscope input capacitance seen by a circuit under test can be undesirably high, especially when the necessary shielded connecting cable is included. The resulting input impedance (1 MΩ in parallel with 100 pF or so) is often too low for sensitive circuits and loads it by the usual voltage divider action; for example, at 10 MHz a 100 pF load looks like 160 Ω – ouch! Worse yet, the capacitance may cause some circuits to misbehave, even to the point of going into oscillation. In such cases the scope obviously is not acting like the "low-profile" measurement instrument we expect; it's more like a bull in a china shop.

The usual solution is the use of high-impedance "probes." In simplified form,[5] the popular 10× probe works as shown in Figure O.3. At dc it's just a 10× voltage divider. By adjusting C_1 to be $\frac{1}{9}$th the parallel capacitance of C_2 and C_3, the circuit becomes a 10× divider at all frequencies, with input impedance of 10 MΩ in parallel with a few picofarads. In practice, you adjust the probe by looking at a square wave of about 1 kHz, available on all scopes as **CALIB**, or **PROBE ADJ**, setting the capacitor on the probe for a clean square wave without overshoot. Sometimes the adjustment is cleverly hidden; on some probes you twist the body of the probe and lock it by tightening a second threaded part. One drawback: a 10× probe makes it difficult to look at signals of only a few millivolts; for these situations use a "1× probe," which is simply a length of low-capacitance shielded cable with the usual probe hardware (wire "grabber," ground clip, handsome knurled handle, etc.). The 10× probe should be the standard probe, left connected to the scope, with the 1× probe used when necessary. Some probes feature a convenient choice of 1× or 10× attenuation, switchable at the probe tip.

Even with a 10× probe, the circuit loading may be unacceptable; after all, its improvement is just the same factor of ten by which it attenuates the input signal. You *can* get 100× probes, with correspondingly higher input

---

[4] Curiously, some scopes (for example the once-popular Tektronix 400 series) don't sweep on **AUTO** unless the trigger level is adjusted correctly.

[5] In practice the cable itself is made from resistance wire, to damp transmission-line effects (frequency peaking and transient reflections, see Appendix H), an elegant 1959 invention by Kobbe and Polits (US Patent 2,883,619); you also see tricks such as a series *RC* across the scope terminals (e.g., 500 Ω and a trimmer capacitor), to provide a transmission-line match at high frequencies.

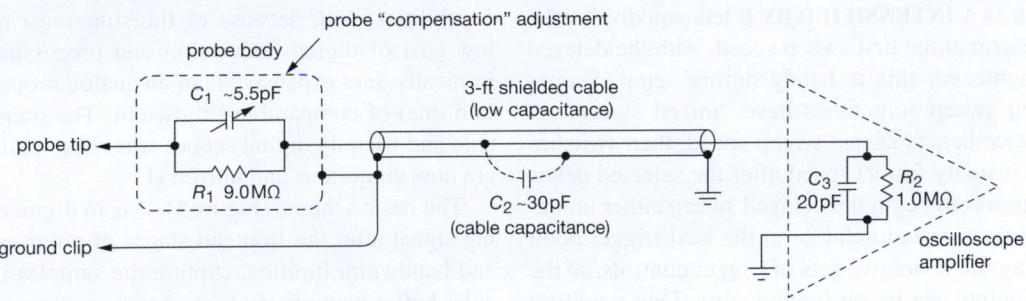

Figure O.3. A 10× passive scope probe attenuates signals by a factor of ten at all frequencies, conveniently raising the input impedance by the same factor. (In practice additional tricks are used to suppress transmission-line effects, see text.)

impedance (e.g., the Tektronix P5100 series), but these are intended primarily for viewing high-voltage signals (the scope itself is usually limited to a maximum of ±400 V at the input connector), and they do not excel in important features such as small physical size. What you do, instead, is to use an *active probe*, which uses a FET follower at the tip to achieve an input capacitance <1 pF.[6] Active probes, being intended for wideband use, are intended to drive a 50 Ω input (available on most high-speed scopes; if not, attach a 50 Ω pass-through terminator); they require a source of power, available at the scope's input connector (on digital scopes), or provided by a stand-alone box like the Tektronix 1103.

Any discussion of probes would be incomplete without a mention of *current* probes: these handy devices, when clipped around a wire in some circuit, convert the circuit's current to a voltage waveform that's displayed on the scope. The simplest current probes are inherently accoupled (they wrap a secondary winding around a magnetic split core that surrounds the one-turn wire "primary") and thus do not sense dc current; the fancier types use a combination of Hall effect and transformer coupling to achieve response down to dc. Examples of the latter are the Tektronix A622 (dc to 100 kHz) and the TCP312A (dc to 100 MHz); the latter requires the matching TCPA300 amplifier.

O.1.6 Grounds

As with most test instruments, the oscilloscope input is referred to the instrument ground (the outer connection of the input BNC connectors), which is usually tied electrically to the case. That, in turn, connects to the ground lead of the ac power line, via the 3-wire power cord. This means that you

cannot measure voltages between the two arbitrary points in a circuit, but are forced to measure signals relative to this universal ground.

An important caution is in order here: if you try to connect the ground clip of an oscilloscope probe to a point in the circuit that is at some voltage relative to ground, you will end up shorting it to ground. This can have disastrous consequences to the circuit under test; in addition, it can be downright dangerous with circuits that are "hot to ground" (for example line-powered switching power supplies). If it is imperative to look at the signal between two points, you can make a differential measurement by inverting one input channel and switching to ADD, or you can use an external differential preamp (e.g., the LeCroy DA1855A). In desperate situations we have been known to "float" the scope by lifting the ground lead at the power cord, but this is *not recommended*, unless you really know what you're doing (and agree to waive any liability on our part).

Another caution about grounds when you're measuring weak signals or high frequencies: be sure the oscilloscope ground is the same as the circuit ground where you're measuring. The best way to do this is by connecting the short ground wire on the probe body directly to the circuit ground,[7] then checking by measuring the voltage of "ground" with the probe, observing no signal. One problem with this scheme is that those short ground clips are usually missing, lost! Keep your probe accessories in a drawer somewhere.

O.1.7 Other analog scope features

Many scopes have a **DELAYED SWEEP** that lets you see a segment of a waveform occurring some time after the trigger point. You can dial the delay accurately with a multi-turn adjustment and a second sweep-speed switch. A delay

---

[6] One of our favorites is the Tektronix P6243, <1 pF and 1 GHz bandwidth.

[7] See the illustrations in Figure 12.32

mode known as **A INTENSIFIED BY B** lets you display the whole waveform at the first sweep speed, with the delayed segment brightened; this is handy during setup. Scopes with delayed sweep sometimes have "mixed sweep," in which the trace begins at one sweep speed, then switches to a second (usually faster) speed after the selected delay. Another option is to begin the delayed sweep either immediately after the selected delay or at the next trigger point after the delay; there are two sets of trigger controls, so the two trigger points can be set individually. (Don't confuse delayed sweep with "signal delay." All good analog scopes have a delay in the signal channel, so you can display the event that caused the trigger; it lets you look a little bit backward in time! See the photographs of the analog delay lines in Figures H.19 and H.21).

A common feature of analog scopes is a **TRIGGER HOLDOFF** control; it inhibits triggering for an adjustable interval after each sweep, and it is very useful when viewing complicated waveforms without the simple periodicity of, say, a sinewave. The usual case is a digital waveform with a complicated sequence of 1s and 0s, which won't generate a stable display otherwise (except by adjustment of the sweep-speed vernier, which means you don't get a calibrated sweep).

All scopes (analog and digital) include some **BAND-WIDTH LIMIT** vertical amplifier options (for simplicity, not shown in Figures O.1 and O.2), useful for reducing the amount of wideband "fuzz" on the displayed trace when you're working with relatively slow signals.

During the height of the analog scope era, you could get scopes with on-screen "storage" (for single-shot capture) and scopes with an impressive array of plug-in modules that let you do lots of interesting stuff, including display of eight traces, or spectrum analysis, or accurate (digital) measurements of voltage and time on waveforms, and so on. Happily, these functions and many others (e.g., looking far backward in time from the triggering event) are now embodied in the dominant oscilloscope species, the *digital* oscilloscope. Let's take a look.

O.2 The digital oscilloscope

Analog scopes are easy to use, but they are seriously limiting in what you can do. For example, (a) it's hard to see a "single-shot" event; (b) you can't store a trace, or compare a live trace with an earlier trace; (c) you can't extract a trace for measurement or illustration; and (d) you can't look back in time to see what happened before the triggering event.

Digital scopes effortlessly provide these and many other

capabilities; and, because of the stunning capability and low cost of digital conversion and processing, they are, ironically, less expensive than an analog scope (if you can find one) of comparable bandwidth. The transition to usable and friendly digital scopes was rocky at first, but they are now ubiquitous and universal.

The basic scheme (Figure O.4) is to digitize the incoming signal after the frontend stages of programmable gain and bandwidth limiting, capture the samples in a fast circular buffer memory, and then use a processor (or multiple processors) to do all the signal processing, measurements, conversion to a meaningful display, user interface, and I/O. We'll keep this section brief, and merely run through some capabilities of digital scopes.

O.2.1 What's different?

In no particular order:

Front-end: The signal emerging from the (variable-gain) input amplifiers is digitized at some sampling rate f_{samp} (typically 1 Gsps or more, but always above the minimum Nyquist rate of $2f_{\text{max}}$ when the scope is set at fast enough sweep rates to resolve the scope's bandwidth f_{max}. But – **important** – see "aliasing," below). The digital samples, typically of 8 bits resolution, are stored at full speed into a *sample memory* (or "capture memory"), often of length 1 Mpt or more per channel (and reaching 1 Gpt at the high end). Note that, even though digital scopes let you zoom in after a trace is captured, the resolution depends on the vertical scale factor, because of the fixed bit depth of conversion.

Simultaneous on all channels: Digital scopes digitize all channels simultaneously; there's no "alternate" or "chop." Most digital scopes come in 2- or 4-channel flavors, augmented in "mixed-signal" scopes by sixteen or more 1-bit (i.e., logic-level) channels.

Pre-trigger: Because digitized input signals are pouring into memory, you can set a trigger condition (most simply, level and slope; but see "Smart trigger," below) and, when it is satisfied, you've got substantial pre-history in the sample memory. From the user's point of view, you can simply set the displayed trigger pointer to the right portion of the screen to reveal what came before. And you can walk backward or forward to your heart's content through a saved single-shot capture (see "Single-shot capture," below).

Display: The time interval between points on the *displayed* waveform (the "waveform interval") is typically longer (often much longer) than the sampling interval

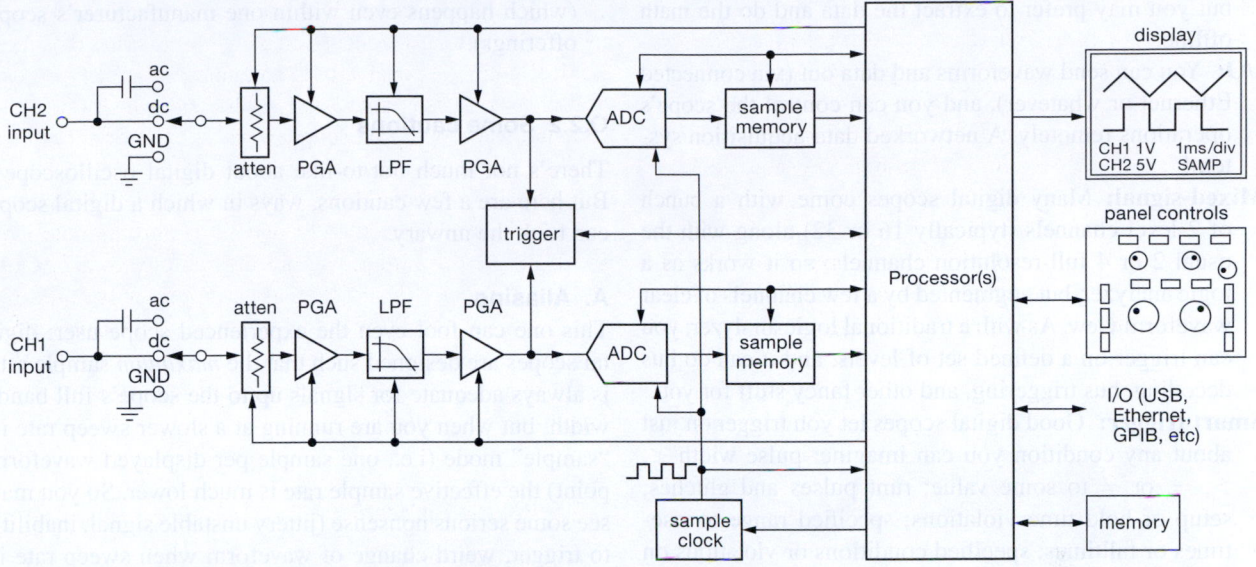

Figure O.4. Block diagram of a 2-channel digital oscilloscope.

$1/f_{\text{samp}}$. This allows for various modes of processing the sampled points to produce the displayed waveform, in particular:

sample one sample point per waveform interval is displayed; the rest are discarded. Simple, but susceptible to aliasing, see below.

peak detect the highest and lowest sampled points in two successive waveform intervals are displayed. Produces a thicker waveform, but no short spikes are lost.

envelope similar to peak detect, but combines min/max from multiple triggered acquisitions. Useful for seeing excursions from an ideal repetitive waveform.

average each point in the displayed waveform is the calculated average of single samples (as in sample mode) over many (a settable number, e.g, 2, 4, 8, …512) triggered acquisitions. Greatly reduces noise, without reducing bandwidth, but requires a repetitive signal.

high resolution each displayed point is calculated as the average value of the multiple samples captured within one waveform interval. Provides higher resolution, and does not require a repetitive signal, but reduces bandwidth.[8]

Persistence: Engineers with gray beards wax poetic about the beautiful gradations of intensity with which analog scopes display waveforms. Digital scopes took a while to catch up, but now they trumpet their ability to do the same, with terms like "digital phosphor," "persistence trace," and "digital persistence."

Single-shot capture: Digital scopes excel in single-shot capture. You can troll through the sampled data post-capture, either manually (by turning knobs for scrolling and magnifying) or with some helpful automated search tools (e.g., Tektronix's "Wave Inspector" – the name says it all).

Slow sweep: Analog scopes are hopeless when you want to view a waveform that takes many seconds; digital scopes couldn't care less. Use "rolling mode" at slow sweep rates.

Save/Recall: You can save one or more waveforms to memory, bring them back for comparison, etc. You can also save the scope's *state* (i.e., settings).

Measurements: The data's all there, so digital scopes have no problem measuring period, frequency, amplitude, time interval, duty cycle, etc. These measurements usually update continuously, and you can use settable horizontal and vertical cursors to define the measurement regions and intervals.

Math: Going further, digital scopes let you calculate products (e.g., to measure power from voltage and current), quotients (to normalize a waveform), jitter, histograms, frequency spectra, etc. Almost limitless possibilities,

---

[8] You can think of this as a "horizontal average" along one waveform capture, as compared with the "vertical averaging" of single sample points in successive stacked waveforms in *average* mode.

but you may prefer to extract the data and do the math offline.

I/O: You can send waveforms and data out (via connected Ethernet or whatever), and you can control the scope's operations remotely. A networked data-acquisition system!

Mixed-signal: Many digital scopes come with a bunch of 2-level channels (typically 16 or 32) along with the usual 2 or 4 full-resolution channels; so it works as a logic analyzer, but augmented by a few channels of clear waveform view. As with a traditional logic analyzer, you can trigger on a defined set of levels, and it can do bus decoding, bus triggering, and other fancy stuff for you.

Smart trigger: Good digital scopes let you trigger on just about any condition you can imagine: pulse width $<$, $>$, $=$, or $\neq$ to some value; runt pulses and glitches; setup or hold time violations; specified range of rise-times or falltimes; specified conditions or violations on serial buses; trigger after n events; and so on. (checkout the enjoyable reading in a datasheet from Tektronix, LeCroy, Keysight/Agilent, or Rohde & Schwarz)

Limit/Mask testing: You can set up a template and detect out-of-spec waveforms, for Go/No-Go testing on a production line; ditto for jitter and other measurable parameters.

Autoset: It's easy to get lost in this multi-dimensional wilderness; digital scopes provide a rescue button (autoset, autoscale, or some linguistic variant, depending on manufacturer), which will at least get something going on the screen (but see the Cautions, next).

Probe skew: When you're using several different probing systems (e.g., passive $10\times$ probe, active FET probe, current probe) the signal delays can vary by tens of nanoseconds or more, completely disrupting the fidelity of the multichannel display (hey, you may be tricked into thinking you've got a violation of causality – effect precedes cause![9]).

Probe readout, Probe power: Probes for use with contemporary digital scopes have extra connections by which they communicate their attenuation factor ($\times 1$, $\times 10$, $\times 100$) and other useful scale factor information (e.g., the amps/div of a current probe); they use such connections also to send power to the probe (needed, for example, with FET active voltage probes or Hall-effect current probes). This can be an annoyance, though, if your scope's input connectors are of the wrong format (which happens even within one manufacturer's scope offerings).

O.2.2 Some cautions

There's not much not-to-like about digital oscilloscopes. But here are a few cautions, ways in which a digital scope can trick the unwary.

A. Aliasing

This one can fool even the experienced scope user: digital scopes are designed such that the *maximum* sample rate is always adequate for signals up to the scope's full bandwidth; but when you are running at a slower sweep rate in "sample" mode (i.e., one sample per displayed waveform point) the effective sample rate is much lower. So you may see some serious nonsense (jittery unstable signal, inability to trigger, weird change of waveform when sweep rate is changed, etc.) if there's a high frequency signal present.[10]

If you suspect aliasing, try speeding up the sweep, or changing to **PEAK DETECT** mode. Aliasing can be really annoying when you're dealing with signals that combine timescales (the classic one was analog television, with a 3.59 MHz color carrier on a $\sim$15 kHz horizontal line frequency).

B. Dead time

For human visual perception it's necessary to update the scope display at only $\sim$100 times per second or so. If the scope captures input waveform data only at that rate, the fraction of time that it's sensitive to important signal events (like a glitch or timing violation) may be exceedingly low. For example, at a middling sweep rate of 1 μs/div (thus 10 μs per sweep) a scope that's updating 100 times/second is active only 0.1% of the time.

When scope users became aware of this ugly secret, scope manufacturers went at it, and they now provide some measure of true update rate (usually in the form of "waveforms per second," typically in the range of 100,000–1,000,000). Be careful when evaluating such metrics, because there's more than a little "specsmanship" going on.[11]

---

[9] As does the protagonist in Asimov's delightful short story from 1960: *Thiotimoline and the Space Age.*

[10] Dare we admit? One of the authors was testing a circuit that operates at frequencies in the kilohertz range, driving it with a digital function generator at 1.0 kHz. Go out to lunch, come back, look at the waveform – the scope is broken, won't trigger, jittery waveform sliding left and right. Weird. Tried everything. Then noticed that the generator had defaulted to its 1.0 *mega*hertz setting. Ha! Problem solved (and I won't ever tell anyone how dumb I was).

[11] Ask your scope salesperson, they love to flame the lying competition.

C. Lost in a multidimensional vector space

Analog scopes are simple, and you can see the full state of the instrument just by looking at the knobs and indicators. No such luck with the immense capabilities of digital scopes. Early digital scopes were particularly troublesome, lacking annunciators and (mostly) knobs. They've improved enormously,[12] but it's still awfully easy to be sitting in front of a scope that just isn't triggering, or showing significant vital signs. It takes some keen intuition to know which menu to pull down (horizontal? trigger? mode?) to get to the problem. It may be as simple as triggering on the wrong channel; or it might be that you've left the display in **AVERAGING** mode, and lack of a stable repetitive trigger produces a bunch of garbage. And you can even waste a minute not realizing that the thing is in **SINGLE-SWEEP** or **STOP** mode.

D. The scope is lying to you

When observing signals with a digital scope, you may be victimized by a blessing (a vast array of measurement capabilities and settings) that becomes a curse (the scope's settings are not what you think). It's easy to forget some obscure but important settings that falsify the measurements you think you're making. For example, it's easy to forget (we've done it, often) that you've left an earlier **PROBE SKEW** compensation in effect, or that some channels still have **BANDWIDTH LIMIT** set. Such oversights corrupt your measurements in less-than-obvious ways that you may not notice for quite a while; and when you do, you're sentenced to serve time repeating the measurements properly.

---

[12] We are particularly fond of the "QuickMenu" feature that was introduced by Tektronix in their original TDS3000-series "lunchbox" scopes. Inexplicably, this highly useful feature has been eliminated (despite our howls of protest) in Tek's successor scopes. We've been badgering them ever since.

ACRONYMS AND ABBREVIATIONS

APPENDIX P

Electrical engineers are fond (*too* fond, some would say) of acronyms and abbreviations, a familiarity that every educated circuit designer must necessarily acquire. To assist in that education, and for handy reference, we here provide a lightly annotated list of terms used in this book.

ac: literally "alternating current" (i.e., alternating *voltage*); more generally a varying signal

AC(T): advanced CMOS (logic family)

A/D: analog-to-digital

ADC: analog-to-digital converter

ADI: Analog Devices Inc.

AES: Audio Engineering Society

AFC: automatic frequency control

AGC: automatic gain control

AGF: all-glass fiber

AHC(T): advanced high-speed CMOS (logic family)

ALS: advanced low-power Schottky (logic family)

ALU: arithmetic logic unit (in a processor)

ALV: advanced low-voltage (logic family)

AM: amplitude modulation

ANSI: American National Standards Institute

APD: avalanche photodiode

APF: all-plastic fiber

ARM: a popular processor architecture from ARM Holdings

ASIC: application-specific full-custom integrated circuit

ASCII: American Standard Code for Information Interchange

ASF: all-silica fiber

ASSP: application-specific standard product

ATA: advanced technology attachment (a disk interface; see PATA, SATA)

ATAPI: ATA packet interface (a generalized ATA)

ATE: automated test equipment

ATM: asynchronous transfer mode

ATSC: Advanced Television Systems Committee (digital TV standards)

AUC: advanced ultra-low-voltage CMOS (logic family)

AVC: advanced (low)-voltage CMOS (logic family)

AVR: a microcontroller family from Atmel Corp.

AWG: American wire gauge

AZ: auto-zero

BBM: break-before-make (switch)

BCD: binary-coded decimal

BGA: ball-grid array (an IC package)

BJT: bipolar junction transistor

BNC: bayonet Neill–Concelman (connector)

BPSK: binary phase-shifting keying

BRT: bias resistor transistor

BSS: broadcast satellite services

BV: breakdown voltage

BW: bandwidth

C0G: low tempco (stable) ceramic dielectric

CA: conditional access

CAN: controller area network (bus)

CANH: controller area network high (a CAN signal)

CANL: controller area network low (a CAN signal)

CBR: constant bitrate (coding)

CCD: charge-coupled device

CCFL: cold-cathode fluorescent lamp

CCM: continuous-conduction mode (in a power converter)

CD: compact disc (optical storage)

CDMA: code-division multiple access

CDR: clock and data recovery

CES: Consumer Electronics Show

CF: compact flash (memory card)

CFB: current feedback (op-amp)

CHE: channel hot-electron

CLB: configurable logic block

CML: current-mode logic

CMOS: complementary metal-oxide semiconductor

CMRR: common-mode rejection ratio

codec: coder–decoder

cPLD: complex programmable-logic device

CPU: central processing unit

CR: carriage return

CRC: cyclic redundancy checksum

CRT: cathode-ray tube

CSMA: carrier-sense multiple-access

CSP: chip-scale package

CTR: current transfer ratio
CTS: clear to send (in a serial link)
CVBS: composite video, blanking, and sync
CVSD: continuously variable-slope delta-modulation

DA: dielectric absorption
D/A: digital-to-analog
DAC: digital-to-analog converter
DAQ: data-acquisition system
DBS: direct broadcast satellite
dc: direct current (i.e., a fixed voltage)
DCE: data communications equipment (in a serial link)
DCM: discontinuous-conduction mode (in a power converter)
DDR: double data rate (memory)
DDS: direct digital synthesis
DFC: digital frequency converter
DIN: Deutsches Institut für Normung (a German standards organization); a connector series
DIP: dual in-line package
DIR: direction (a control signal)
DMA: direct memory access
DMM: digital multimeter
DNL: differential nonlinearity
DPCP: display-port content protection
DPDT: double-pole double-throw (switch)
DRAM: dynamic random-access memory
DSBGA: die-size ball-grid array (an IC package)
DSL: digital subscriber line (for data over telephone line)
DSP: digital signal processing (or processor)
DSR: data set ready (in a serial link)
DTE: data terminal equipment (in a serial link)
DTH: direct-to-home (satellite TV)
DTL: diode–transistor logic
DTR: data terminal ready (in a serial link)
DUT: device under test
DVI: digital visual interface (for digital video)
DVM: digital voltmeter
DVR: digital video recorder

EAROM: electrically alterable read-only memory
EAS: avalanche energy specification
ECL: emitter-coupled logic
EEPROM: electrically erasable programmable read-only memory
EIA: Electronic Industries Alliance (standards and trade organization)
EMF: electromotive force (~voltage)
EMI: electromagnetic interference
ENOB: effective number of bits
EPROM: erasable programmable read-only memory

eSATA: external SATA interface
ESD: electrostatic discharge
ESL: equivalent series inductance
ESR: equivalent series resistance
ETF: 8-to-14 (digital coding)

FCC: Federal Communications Commission
FDNR: frequency-dependent negative resistor
FED: field-emission display
FET: field-effect transistor
FFS: fringe-field switching (an LCD display technology)
FFT: fast Fourier transform
FG: floating gate
FGA: floating-gate array
FIFO: first-in-first-out (memory)
FIR: finite-impulse-response (filter)
FM: frequency modulation
F-N Fowler–Nordheim tunnelling
FOT: fiber-optic transceiver
FPBW: full-power bandwidth
FPGA: field-programmable gate array
FR-4: "flame-retardant 4" (glass-epoxy PCB material)
FRAM (also FeRAM, F-RAM): ferroelectric random-access memory
FSB: front-side bus (of a computer processor)
FSE: full-sunlight equivalent (you saw it here first!)

GAL: generic array logic
GBP, GBW: gain-bandwidth product
GCC: GNU C-compiler
GCPW: grounded coplanar waveguide
GDT: gas-discharge tube
GIC: generalized impedance converter
GMR: giant magnetoresistance
GND: ground
GPIB: general-purpose interface bus
GPL: graphical programming language; general public license (in GNU)
GPS: global positioning system
GPU: graphics-processor unit
GUI: graphical user interface

HAPD: hybrid avalanche photodiode (detector)
HBM: human body model
HC(T): high-speed CMOS (logic family)
HDCP: high-bandwidth digital content protection
HDD: hard-disk drive
HDL: hardware description language
HDMI: high-definition multimedia interface (for digital display)
HDTV: high-definition television
HI: human interface

HP: Hewlett–Packard
HV: high voltage

IC: integrated circuit
ICSP: in-circuit serial programming
IDC: insulation displacement connector
IDE: integrated development environment (for coding)
IEC: International Electrotechnical Commission
IEEE: Institute of Electrical and Electronic Engineers
IF: intermediate frequency (in RF receiver)
IGBT: insulated-gate bipolar transistor
IGFET: insulated-gate field-effect transistor
IIC (I^2C): inter-integrated-circuit (a serial bus)
IIR: infinite-impulse-response (filter)
INA: instrumentation amplifier
INL: integral nonlinearity (in A/D conversion)
I/O: input-output
IP: Internet protocol; intellectual property
IPS: in-plane switching (an LCD display technology)
IR: infrared
ISA: International Society of Automation (a
    standard-setting organization)
ISI: intersymbol interference

JFET: junction field-effect transistor
JTAG: Joint Test Action Group (an IC interface)

KCL: Kirchhoff's current law
KVL: Kirchhoff's voltage law

LAB: logic array block (in programmable logic)
LAN: local area network
LCD: liquid- crystal display
LCX: low-voltage CMOS crossvolt (logic family)
LDO: low-dropout (linear voltage regulator)
LE: logic element (in programmable logic)
LED: light-emitting diode
LFSR: linear feedback shift register
LIFO: last-in first-out (memory)
Li-ion: lithium ion (battery)
LIN: local-interconnect network (bus)
LNA: low-noise amplifier
LNBF: low-noise block-downconverter plus feed (for
    satellite TV)
LO: local oscillator (in RF receiver)
LPF: lowpass filter
LPT: line printer (a parallel port)
LS: low-power Schottky (logic family)
LSB: least-significant bit
LSI: large-scale integrated circuit
LUT: lookup table
LV: low-voltage (logic family)
LVC: low-voltage CMOS (logic family)

LVDS: low-voltage differential signaling
LVPECL: low-voltage positive emitter-coupled logic
LVX: low-voltage crossvolt (logic family)
LXI: LAN eXtensions for Instrumentation

μC: microcontroller
MAC: multiplier-accumulator; media-access control
MBB: make-before-break (switch)
MCU: microcontroller unit
MDAC: multiplying digital-to-analog converter
MEMS: microelectromechanical system
MFB: multiple-feedback (active filter)
MIPS: mega-instructions per second
MLC: multilevel cell (in nonvolatile memory)
MMU: memory management unit
modem: modulator-demodulator
MOS: metal-oxide semiconductor
MOSFET: metal-oxide semiconductor field-effect
    transistor
MOV: metal-oxide varistor (surge protector)
MPTS: multiprogram transport stream (in digital TV)
MPU: microprocessor unit
MRAM: magnetoresistive random-access memory
MRI: magnetic resonance imaging
MSB: most-significant bit
MSI: medium-scale integration
MUX: multiplexer

NAN: not a number
NC: normally closed (switch)
NECL: negative emitter-coupled logic
NEMA: National Electrical Manufacturers Association (a
    standards-setting organization)
NIC: negative-impedance converter; network interface
    card
NiCd: nickel cadmium (battery)
NiMH: nickel metal-hydride (battery)
NIST: National Institute of Standards and Technology
NMI: nonmaskable interrupt
NMR: nuclear magnetic resonance
nMOS: n-type metal-oxide semiconductor
NO: normally open (switch)
NP0: low tempco (stable) ceramic dielectric
NRZ: nonreturn to zero (data code)
NRZI: nonreturn to zero inverted (data code)
NSC: National Semiconductor Corporation (now part of
    TI)
NTC: negative temperature coefficient
NTSC: National Television System Committee (analog
    TV standard)
NV: nonvolatile

NVM: nonvolatile memory

O/C: open-collector (logic output)

OCXO: oven-controlled crystal oscillator

O/D: open-drain (logic output)

OEM: original equipment manufacturer

OLED: organic light-emitting diode

op-amp: operational amplifier

OSI: open systems interconnection (network hierarchy)

OSR: oversampling ratio

OTA: over-the-air (broadcasting)

OTP: one-time-programmable NV memory

PAL: programmable array logic

PARC: Palo Alto Research Center

PATA: parallel ATA interface

PC: printed circuit; personal computer

PCB: printed circuit board

PCF: plastic-clad fiber

PCI: peripheral component interface (a computer bus)

PCIe (also PCI-E): extended peripheral component interface

PCM: pulse-code modulation

PCMCIA: Personal Computer Memory Card International Association (a card interface standard)[1]

PDA: personal digital assistant

PECL: positive emitter-coupled logic

PEN: polyethylene naphthalate (a capacitor dielectric)

PFC: power-factor correction (in ac-powered converters)

PFD: phase-frequency detector

PFM: pulse-frequency modulation

PGA: programmable gain amplifier

PID: proportional-integral-differential (in control systems); program identifier (in digital TV)

PIN: positive-intrinsic-negative (diode)

PI/PO: parallel-in-parallel-out

PIR: passive infrared (detector)

PIV: peak inverse voltage

PLA: programmable logic array

PLC: powerline cycles

PLD: programmable logic device

PLL: phase-locked loop

pMOS: p-type metal-oxide semiconductor

PMT: photomultiplier tube

POF: plastic optical fiber

POL: point-of-load

pp: peak-to-peak (voltage)

PPS: polyphenylene sulfide (a capacitor dielectric)

PRAM: phase-change random-access memory

PRBS: pseudorandom bit sequence

PROM: programmable read-only memory

PSRAM: pseudostatic random-access memory

PSRR: power-supply rejection ratio

PTAT: proportional to absolute temperature

PUJT: programmable unijunction transistor

PV: photovoltaic (light detector)

PVC: polyvinyl chloride (insulator)

PVR: personal video recorder

PWM: pulse-width modulation

QAM: quadrature amplitude modulation

QPSK: quadrature phase-shift keying

RAM: random-access memory

RCO: ripple-clock output

RD: receive data (in a serial link)

RF: radiofrequency

RFI: radiofrequency interference

RG-nn: "Radio Guide" (coax cable designators)

RGB: red–green–blue (video signals)

RISC: reduced instruction set computing

RLL: run-length limited (digital codes)

rms: root-mean-square

ROM: read-only memory

RRI: rail-to-rail input

RRIO: rail-to-rail input and output

RRO: rail-to-rail output

RTC: real-time clock

RTD: resistance temperature detector (or resistive temperature device)

RTI: referred to the input

RTL: resistor–transistor logic; register-transfer level (in an HDL)

RTO: referred to the output

RTS: request to send (in a serial link)

SA: sense amplifier

SACD: Super Audio compact disc

SAD: silicon avalanche device (i.e., a zener TVS)

SAR: successive approximation register

SAS: serial attached SCSI (interface)

SATA: serial ATA (interface)

SAW: surface acoustic-wave

SBC: single-board computer

SC: subscriber connector (a fiber-optic connector)

SCPI: Standard Commands for Programmable Instruments

SCR: silicon-controlled rectifier

SCSI: small computer system interface

SD: secure digital (memory card)

SDI: serial data in

---

[1] Whose awkwardness spawned jokes like "Personal Computer Manufacturers Can't Invent Acronyms."

SDO: serial data out
SDR: single data rate (memory)
SDRAM: synchronous dynamic random-access memory
SDTV: standard-definition television
SE: single-ended
SED: surface-conduction electron-emitter display
SEPIC: single-ended primary-inductance converter
SERDES: serializer–deserializer
S/H: sample-and-hold
SHV: "safe high voltage" (connector)
SI: serial input
SIP: single in-line package
SMA, SMB, SMC: subminiature RF coax connector series
SMI: small media interface (a fiber-optic connector)
SMPS: switch-mode power supply
SMT: surface-mount technology
SMU: source-measure unit
SNR: signal-to-noise ratio
SO: serial output; small-outline (IC package)
SOA: safe-operating area
SODIMM: small-outline dual-in-line memory module
SOIC: small-outline integrated circuit
SOT: small-outline transistor
SPDIF: Sony–Philips Digital Interconnect Format (for digital audio)
SPICE: "simulation program with integrated circuit emphasis" (analog circuit simulator software)
SPDT: single-pole double-throw (switch)
SPI: serial peripheral interface (a simple IC bus)
SPL: sound pressure level
sPLD: simple programmable logic device
SPST: single-pole single-throw (switch)
SPTS: single-program transport stream (in digital TV)
SR: slew rate
SRAM: static random-access memory
SSD: solid-state drive (an NV memory)
SSH: secure shell (a network protocol)
SSP: synchronous serial port
SSR: solid-state relay
ST: straight tip (a fiber-optic connector)
STB: set-top box (for cable or satellite TV)
STP: shielded twisted pair (cable)
SWR: standing-wave ratio (on a transmission line)

T&M: test and measurement
TAC: time-to-amplitude conversion
TBH: take-back-half (a control algorithm)
TCP: transmission control protocol (an Internet protocol)
TCXO: temperature-compensated crystal oscillator

TD: transmit data (in a serial link)
tempco: temperature coefficient
THD: total harmonic distortion
TI: Texas Instruments
TNC: threaded Neill-Concelman (connector)
TO: transistor outline (e.g., TO-92, TO-220)
TSSOP: thin-shrink small-outline package
TTL: transistor–transistor logic
TVS: transient voltage suppressor
TWI: two-wire interface (a serial bus)

UART: universal asynchronous receiver-transmitter
UDP: user datagram protocol (an Internet protocol)
UHF: ultrahigh frequency; also a legacy coaxial connector
UL: Underwriters Laboratories (a safety certification company)
UPS: uninterruptible power supply
USB: universal serial bus (a data interface)
UTP: unshielded twisted pair (cable)
UV: ultraviolet

VBR: variable bitrate (coding)
VCO: voltage-controlled oscillator
VCVS: voltage-controlled voltage-source (active filter)
VCXO: voltage-controlled crystal oscillator
VDE: Verband der Elektrotechnik, Elektronik und Informationstechnik (a German organization whose activities include safety standards)
V/F: voltage-to-frequency (converter)
VFB: voltage feedback
VFD: vacuum fluorescent display
VGA: video graphics array (640×480 analog video)
VHS: video home system (video recording)
VLSI: very-large-scale integration
VME: VERSAmodule Eurocard Bus (a card interface)
VOD: video-on-demand
VOM: volt-ohm-milliammeter
VSWR: voltage standing-wave ratio (on a transmission line)
VU: volume unit (an audio level)

WL: write latency (in computer memory)

X7R: a ceramic dielectric
XLR: a professional audio connector series
XO: crystal oscillator

Y5V: a ceramic dielectric
YIG: yttrium–iron garnet

Z5U: a ceramic dielectric
ZCS: zero-current switching
ZVS: zero-voltage switching

PARTS INDEX

SUBJECT INDEX